D1321462

TABLES OF ANTENNA CHARACTERISTICS

TABLES OF ANTENNA CHARACTERISTICS

Ronold W. P. King

Gordon McKay Professor of Applied Physics
Harvard University
Cambridge, Massachusetts

IFI/PLENUM • NEW YORK–WASHINGTON–LONDON • 1971

Library of Congress Catalog Card Number 74-157425

SBN 306-65154-8

A Subsidiary of Plenum Publishing Corporation
227 West 17th Street, New York, N. Y. 10011

United Kingdom edition published by Plenum Press, London
A Division of Plenum Publishing Company, Ltd.
Davis House (4th Floor), 8 Scrubs Lane, Harlesden, NW 10 6SE, England

Printed in the United States of America

Preface

Important practical properties of antennas are their driving-point admittances and far-field patterns. The accurate determination of these and other related characteristics requires the explicit or implicit solution of integral equations for the current distributions along the radiating structure. This can be accomplished with the help of analytical and numerical techniques; the validity of approximations can be checked experimentally.

In order to obtain specific data for practical applications, high-speed computers may be used to evaluate analytically derived formulas or, where these are unavailable, to obtain direct numerical solutions. Programs written for such a purpose are usually long and complicated, and they may require very fast machines with large storage capacity. Since these are not generally available outside of large organizations, a representative set of numerical tables to provide a variety of useful characteristics of cylindrical and loop antennas and dipole arrays should be of value.

Over a period of years extensive researches on antennas have been carried out at Harvard University with the support of the U.S. Navy, the U.S. Air Force, and the Signal Corps of the U.S. Army under Contracts N00014-67-A-0298-0005 and F19(628)-C-0030. A selection from the results of these investigations has been prepared, recomputed, and tabulated for this book. The researches include contributions by D. C. Chang, V. W. H. Chang, C. W. Harrison, Jr., S. S. Sandler, C. Y. Ting, and T. T. Wu. The programming was carried out primarily by Barbara Sandler and Georgia Efthymiopoulou, but important contributions were also made by E. A. Aronson at the Sandia Corporation, Margaret Owens, and Irma Rivera-Veve. The typing, checking, and proofreading were done by Margaret Owens. The figures were prepared with the assistance of Elmer Rising and his staff; photographic work was carried out by Armand Dionne.

Cambridge, Massachusetts — RONOLD W. P. KING
July 1970

Contents

1. The Complex Wave Number k and the Normalizing Factor Δ

The characteristics of antennas described and tabulated in this volume are obtained from solutions of Maxwell's equations in an infinite, homogeneous, isotropic medium characterized by the complex permittivity $\varepsilon = \varepsilon_0\varepsilon_r = \varepsilon_0(\varepsilon_r'' - j\varepsilon_r')$, the complex conductivity $\sigma = \sigma' - j\sigma''$, and the real permeability $\mu = \mu_0\mu_r$. In such a medium Maxwell's equations have the form

$$\nabla \times E = -j\omega B, \qquad \nabla \cdot E = 0 \tag{1.1}$$

$$\nabla \times B = \mu(\sigma + j\omega\varepsilon)E, \qquad \nabla \cdot B = 0 \tag{1.2}$$

In air $\sigma = 0$, $\varepsilon = \varepsilon_0$, $\mu = \mu_0$. The complex quantity $\sigma + j\omega\varepsilon$ can be separated into its real and imaginary parts such that

$$\sigma + j\omega\varepsilon = \sigma_e + j\omega\varepsilon_e \tag{1.3}$$

where the real effective conductivity σ_e and the real effective permittivity ε_e are given by

$$\sigma_e = \sigma' + \omega\varepsilon_0\varepsilon_r'', \qquad \varepsilon_e = \varepsilon_0\left(\varepsilon' - \frac{\sigma''}{\omega\varepsilon_0}\right) = \varepsilon_0\varepsilon_{er} \tag{1.4}$$

When the variables in Maxwell's equations are separated, these yield the second-order vector wave equation

$$\nabla^2 E + k^2 E = 0 \tag{1.5}$$

and a similar equation for B. The complex wave number or propagation constant k is related to $(\sigma_e + j\omega\varepsilon_e)$ by

$$k^2 = \omega^2\mu\varepsilon_e(1 - jp_e) = \omega^2\mu\varepsilon_e - j\omega\mu\sigma_e \tag{1.6}$$

where

$$p_e = \frac{\sigma_e}{\omega\varepsilon_e} = \frac{\sigma_e}{\omega\varepsilon_0\varepsilon_{er}} \tag{1.7}$$

is the loss tangent. The explicit formulas for the real and imaginary parts β and α of $k = \beta - j\alpha$ are obtained as follows:

$$\varepsilon_e > 0: \quad k = \omega\sqrt{\mu\varepsilon_e}\sqrt{1 - jp_e} = k_0\sqrt{\mu_r\varepsilon_{er}}[f(p_e) - jg(p_e)] \tag{1.8a}$$

$$\varepsilon_e = 0: \quad k = \sqrt{j\omega\mu\sigma_e} = \sqrt{\frac{\omega\mu\sigma_e}{2}}(1 - j) = k_0\sqrt{\frac{\mu_r\sigma_e}{2\omega\varepsilon_0}}(1 - j) \tag{1.8b}$$

$$\varepsilon_e < 0: \quad k = -j\omega\sqrt{\mu|\varepsilon_e|}\sqrt{1 + j|p_e|} = k_0\sqrt{\mu_r|\varepsilon_{er}|}[g(|p_e|) - jf(|p_e|)] \tag{1.8c}$$

where

$$k_0 = \omega\sqrt{\mu_0\varepsilon_0}, \qquad \mu_0 = 4\pi \times 10^{-7}\ \text{henries/m}, \qquad \varepsilon_0 = 8.854 \times 10^{-12}\ \text{farads/m} \tag{1.9}$$

The functions $f(p)$ and $g(p)$ are defined by

$$f(p) = \cosh(\tfrac{1}{2}\sinh^{-1} p) = \sqrt{\tfrac{1}{2}(\sqrt{1 + p^2} + 1)} \tag{1.10a}$$

$$g(p) = \sinh(\tfrac{1}{2}\sinh^{-1} p) = \sqrt{\tfrac{1}{2}(\sqrt{1 + p^2} - 1)} \tag{1.10b}$$

It follows that with $\zeta_0 = \sqrt{\mu_0/\varepsilon_0} \doteq 120\pi$ ohms,

$$\varepsilon_{er} > 0: \quad \beta = k_0\sqrt{\mu_r\varepsilon_{er}}\,f(p_e) = \sqrt{\frac{\sigma_e\zeta_0}{\varepsilon_{er}}}\,\frac{f(p_e)}{p_e} \tag{1.11a}$$

$$\alpha = k_0\sqrt{\mu_r\varepsilon_{er}}\,g(p_e) = \sqrt{\frac{\sigma_e\zeta_0}{\varepsilon_{er}}}\,\frac{g(p_e)}{p_e} < \beta \tag{1.11b}$$

$$\varepsilon_{er} = 0: \quad \beta = \alpha = \sqrt{\frac{\omega\mu\sigma_e}{2}} \tag{1.12}$$

$$\varepsilon_{er} < 0: \quad \beta = k_0\sqrt{\mu_r|\varepsilon_{er}|}\, g(|p_e|) = \frac{\sigma_e\zeta_0}{\sqrt{|\varepsilon_{er}|}} \frac{g(|p_e|)}{|p_e|} \tag{1.13a}$$

$$\alpha = k_0\sqrt{\mu_r|\varepsilon_{er}|}\, f(|p_e|) = \frac{\sigma_e\zeta_0}{\sqrt{|\varepsilon_{er}|}} \frac{f(|p_e|)}{|p_e|} > \beta \tag{1.13b}$$

In addition to the propagation constant k, the complex wave impedance

$$\zeta = \frac{\omega\mu}{k} = \frac{\omega\mu}{\beta(1 - j\alpha/\beta)} = \frac{\zeta_0}{(1 - j\alpha/\beta)} \frac{\omega\mu}{\beta\zeta_0} \tag{1.14a}$$

where

$$\zeta_0 = \sqrt{\mu_0/\varepsilon_0} = 376.7 \text{ ohms} \doteq 120\pi \text{ ohms} \tag{1.14b}$$

frequently occurs as a multiplier in amplitudes. It is convenient to use the quantity

$$\Delta = \frac{\beta\zeta_0}{\omega\mu} = \frac{\beta}{k_0\mu_r} \tag{1.15}$$

as a general normalizing factor. Then

$$\zeta = \frac{\zeta_0}{\Delta(1 - j\alpha/\beta)} \tag{1.16}$$

where

$$\varepsilon_e > 0: \quad \Delta = \sqrt{\varepsilon_{er}/\mu_r}\, f(p_e) \tag{1.17}$$

$$\varepsilon_e = 0: \quad \Delta = \sqrt{\sigma_e/2\omega\varepsilon_0\mu_r} \tag{1.18}$$

$$\varepsilon_e < 0: \quad \Delta = \sqrt{|\varepsilon_{er}|/\mu_r}\, g(|p_e|) \tag{1.19}$$

Note that as $\varepsilon_e \to 0$, $p_e \to \infty$, $f(|p_e|) \to g(|p_e|) \to \sqrt{|p_e|/2} = \sqrt{\sigma_e/2\omega\varepsilon_0|\varepsilon_{er}|}$. Thus, the value of Δ at $\varepsilon_e = 0$ is the limit as $|p_e| \to \infty$ of the values for $\varepsilon_e > 0$ and $\varepsilon_e < 0$. Clearly, when σ_e, ε_e, and μ are specified for any given medium, k and Δ can be determined from (1.11)–(1.13) and (1.17)–(1.19), respectively. For this purpose tables of $f(p)$ and $g(p)$ as defined in (1.10a,b) are convenient. The frequency dependence of α and β for any assigned set of values σ_e, ε_e, and μ is contained in $f(p)/p$ or $g(p)/p$ as seen from (1.11a,b) and (1.13a,b). The functions $f(p)$, $g(p)$, $f(p)/p$, and $g(p)/p$ are given in Table 1.1, in which p is the variable ranging from zero to large values. The following high- and low-frequency ranges and approximate formulas are useful:

$$0 \le p^2 \le 0.04: \quad f(p) \doteq 1, \quad g(p) \doteq p/2 \tag{1.20}$$

$$p^2 \ge 25: \quad f(p) \doteq g(p) \doteq \sqrt{p/2} \tag{1.21}$$

For convenience in visualizing the behavior of these functions, graphs of $f(p)$, $g(p)$, $f(p)/p$, and $g(p)/p$ as functions of p are given in Fig. 1.1. Low- and high-frequency ranges are indicated, and frequency scales for dry earth ($\sigma_e = 10^{-3}$ mho/m, $\varepsilon_{er} = 7$), moist earth ($\sigma_e = 1.2 \times 10^{-2}$ mho/m, $\varepsilon_{er} = 15$), and wet earth ($\sigma_e = 3 \times 10^{-2}$ mho/m, $\varepsilon_{er} = 30$) are given. In Fig. 1.2, $f(p)/p$ and $g(p)/p$ are shown as functions of the frequency for these three types of earth. In ordinary dielectrics $\varepsilon_{er} \ge 1$ so that (1.11a,b) apply.

The properties of certain types of plasma over limited ranges of the parameters can be approximated by introducing real effective permittivities and conductivities given by

$$\varepsilon_{er} = 1 - \frac{Ne^2}{\varepsilon_0 m(\nu^2 + \omega^2)} = 1 - \frac{\omega_p^2}{\nu^2 + \omega^2} \tag{1.22}$$

$$\sigma_e = \frac{Ne^2\nu}{m(\nu^2 + \omega^2)} \tag{1.23}$$

where N is the number of electrons per unit volume; e is the charge and m the mass of the electron; ν is the collision frequency; and $\omega_p = 2\pi f_p$, where f_p is the plasma frequency. Note that when $\omega_p^2 < (\nu^2 + \omega^2)$, $0 < \varepsilon_{er} \le 1$, so that (1.11a,b) apply; when $\omega_p^2 = \nu^2 + \omega^2$, $\varepsilon_{er} = 0$, so that (1.12) applies; and when $\omega_p^2 > (\nu^2 + \omega^2)$, $\varepsilon_{er} < 0$, so that (1.13a,b) apply.

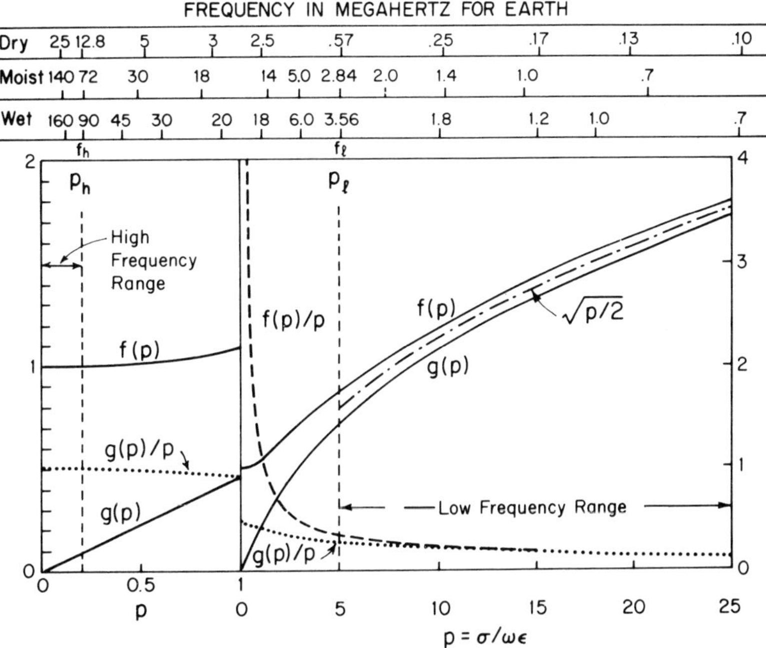

Fig. 1.1. The functions $f(p)$ and $g(p)$ and related quantities.

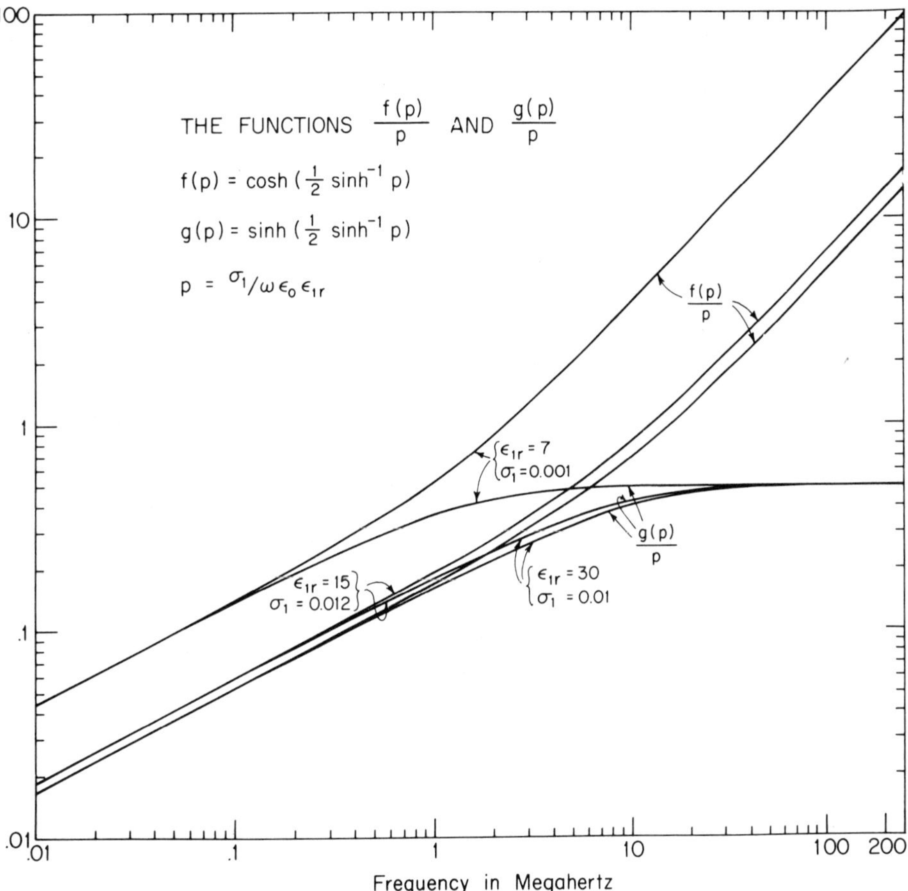

Fig. 1.2. The functions $f(p)/p$ and $g(p)/p$.

TABLE 1.1

TABLE OF F(P) AND G(P) FUNCTIONS

P	F(P)	G(P)	F(P)/P	G(P)/P	P	F(P)	G(P)	F(P)/P	G(P)/P	P	F(P)	G(P)	F(P)/P	G(P)/P
0.0	1.000	0.000	INF	0.500	10.0	2.351	2.127	0.235	0.213	30.0	3.938	3.809	0.131	0.127
0.1	1.001	0.050	10.012	0.499	10.2	2.372	2.150	0.233	0.211	30.2	3.951	3.822	0.131	0.127
0.2	1.005	0.100	5.025	0.498	10.4	2.392	2.173	0.230	0.209	30.4	3.963	3.835	0.130	0.126
0.3	1.011	0.148	3.370	0.495	10.6	2.413	2.195	0.228	0.207	30.6	3.976	3.848	0.130	0.125
0.4	1.019	0.196	2.548	0.491	10.8	2.434	2.219	0.225	0.205	30.8	3.988	3.861	0.129	0.125
0.5	1.029	0.243	2.058	0.486	11.0	2.454	2.241	0.223	0.204	31.0	4.001	3.874	0.129	0.125
0.6	1.041	0.288	1.735	0.480	11.2	2.474	2.263	0.221	0.202	31.2	4.013	3.887	0.129	0.125
0.7	1.054	0.332	1.505	0.475	11.4	2.494	2.285	0.219	0.200	31.4	4.026	3.900	0.128	0.124
0.8	1.068	0.375	1.335	0.468	11.6	2.514	2.307	0.217	0.199	31.6	4.038	3.913	0.128	0.124
0.9	1.083	0.416	1.203	0.462	11.8	2.534	2.328	0.215	0.197	31.8	4.051	3.925	0.127	0.123
1.0	1.099	0.455	1.099	0.455	12.0	2.554	2.350	0.213	0.196	32.0	4.063	3.938	0.127	0.123
1.1	1.115	0.493	1.014	0.448	12.2	2.573	2.371	0.211	0.194	32.2	4.075	3.951	0.127	0.123
1.2	1.132	0.530	0.943	0.442	12.4	2.592	2.392	0.209	0.193	32.4	4.088	3.963	0.126	0.122
1.3	1.149	0.566	0.884	0.435	12.6	2.611	2.412	0.207	0.191	32.6	4.100	3.976	0.126	0.122
1.4	1.166	0.600	0.833	0.429	12.8	2.630	2.433	0.206	0.190	32.8	4.112	3.988	0.125	0.122
1.5	1.184	0.634	0.789	0.422	13.0	2.649	2.453	0.204	0.189	33.0	4.124	4.001	0.125	0.121
1.6	1.201	0.666	0.751	0.416	13.2	2.668	2.474	0.202	0.187	33.2	4.136	4.013	0.125	0.121
1.7	1.219	0.697	0.717	0.410	13.4	2.687	2.494	0.201	0.185	33.4	4.148	4.026	0.124	0.121
1.8	1.237	0.728	0.687	0.404	13.6	2.705	2.514	0.199	0.185	33.6	4.160	4.038	0.124	0.120
1.9	1.254	0.757	0.660	0.399	13.8	2.724	2.533	0.197	0.184	33.8	4.172	4.051	0.123	0.120
2.0	1.272	0.786	0.636	0.393	14.0	2.742	2.553	0.195	0.182	34.0	4.184	4.063	0.123	0.119
2.1	1.290	0.814	0.614	0.388	14.2	2.760	2.572	0.194	0.181	34.2	4.196	4.075	0.123	0.119
2.2	1.307	0.842	0.594	0.383	14.4	2.778	2.592	0.193	0.180	34.4	4.208	4.087	0.122	0.119
2.3	1.324	0.868	0.576	0.378	14.6	2.796	2.611	0.192	0.179	34.6	4.220	4.100	0.122	0.118
2.4	1.342	0.894	0.559	0.373	14.8	2.814	2.630	0.190	0.178	34.8	4.232	4.112	0.122	0.118
2.5	1.359	0.920	0.544	0.368	15.0	2.831	2.649	0.189	0.177	35.0	4.243	4.124	0.121	0.118
2.6	1.376	0.945	0.529	0.363	15.2	2.849	2.668	0.187	0.176	35.2	4.255	4.136	0.121	0.118
2.7	1.393	0.969	0.516	0.359	15.4	2.865	2.686	0.186	0.174	35.4	4.267	4.148	0.121	0.117
2.8	1.409	0.993	0.503	0.355	15.6	2.884	2.705	0.185	0.173	35.6	4.279	4.160	0.120	0.117
2.9	1.426	1.017	0.492	0.351	15.8	2.901	2.723	0.184	0.172	35.8	4.290	4.172	0.120	0.117
3.0	1.443	1.040	0.481	0.347	16.0	2.918	2.741	0.182	0.171	36.0	4.302	4.184	0.119	0.116
3.1	1.459	1.062	0.471	0.343	16.2	2.935	2.760	0.181	0.170	36.2	4.314	4.196	0.119	0.116
3.2	1.475	1.085	0.461	0.339	16.4	2.952	2.778	0.180	0.169	36.4	4.325	4.208	0.119	0.116
3.3	1.491	1.106	0.452	0.335	16.6	2.969	2.795	0.179	0.168	36.6	4.337	4.220	0.118	0.115
3.4	1.507	1.128	0.443	0.332	16.8	2.985	2.813	0.178	0.167	36.8	4.348	4.232	0.118	0.115
3.5	1.523	1.149	0.435	0.328	17.0	3.002	2.831	0.177	0.167	37.0	4.360	4.243	0.118	0.115
3.6	1.539	1.170	0.427	0.325	17.2	3.019	2.849	0.176	0.166	37.2	4.371	4.255	0.118	0.114
3.7	1.554	1.190	0.420	0.322	17.4	3.036	2.855	0.174	0.165	37.4	4.383	4.267	0.117	0.114
3.8	1.570	1.210	0.413	0.318	17.6	3.052	2.883	0.173	0.164	37.6	4.394	4.279	0.117	0.114
3.9	1.585	1.230	0.406	0.315	17.8	3.068	2.901	0.172	0.163	37.8	4.405	4.290	0.117	0.113
4.0	1.600	1.250	0.400	0.312	18.0	3.084	2.918	0.171	0.162	38.0	4.417	4.302	0.116	0.113
4.1	1.616	1.269	0.394	0.309	18.2	3.101	2.935	0.170	0.161	38.2	4.428	4.314	0.116	0.113
4.2	1.631	1.288	0.388	0.307	18.4	3.117	2.952	0.169	0.160	38.4	4.439	4.325	0.116	0.113
4.3	1.645	1.307	0.383	0.304	18.6	3.133	2.969	0.168	0.160	38.6	4.450	4.337	0.115	0.112
4.4	1.660	1.325	0.377	0.301	18.8	3.149	2.986	0.167	0.159	38.8	4.462	4.348	0.115	0.112
4.5	1.675	1.343	0.372	0.299	19.0	[illegible]	[illegible]	[illegible]	[illegible]	[illegible]	[illegible]	[illegible]	0.115	0.112
[illegible]	[illegible]	[illegible]	[illegible]	0.298	19.2	3.180	3.019	0.156	0.157	39.2	4.484	4.371	0.114	0.112
4.7	1.704	1.379	0.362	0.293	19.4	3.196	3.035	0.155	0.156	39.4	4.495	4.383	0.114	0.111
4.8	1.718	1.397	0.358	0.291	19.6	3.211	3.052	0.154	0.156	39.6	4.506	4.394	0.114	0.111
4.9	1.732	1.414	0.354	0.289	19.8	3.227	3.058	0.153	0.155	39.8	4.517	4.405	0.114	0.111
5.0	1.746	1.432	0.349	0.286	20.0	3.242	3.084	0.152	0.154	40.0	4.528	4.417	0.113	0.110
5.1	1.760	1.449	0.345	0.284	20.2	3.258	3.100	0.151	0.153	40.2	4.539	4.428	0.113	0.110
5.2	1.774	1.465	0.341	0.282	20.4	3.273	3.116	0.150	0.153	40.4	4.550	4.439	0.113	0.110
5.3	1.788	1.482	0.337	0.280	20.6	3.288	3.132	0.150	0.152	40.6	4.561	4.450	0.112	0.110
5.4	1.802	1.499	0.334	0.278	20.8	3.303	3.148	0.159	0.151	40.8	4.572	4.462	0.112	0.109
5.5	1.815	1.515	0.330	0.275	21.0	3.318	3.154	0.158	0.151	41.0	4.583	4.473	0.112	0.109
5.6	1.829	1.531	0.327	0.273	21.2	3.333	3.180	0.157	0.150	41.2	4.594	4.484	0.112	0.109
5.7	1.842	1.547	0.323	0.271	21.4	3.348	3.196	0.155	0.149	41.4	4.605	4.495	0.111	0.109
5.8	1.855	1.563	0.320	0.269	21.6	3.363	3.211	0.156	0.149	41.6	4.616	4.506	0.111	0.108
5.9	1.869	1.579	0.317	0.268	21.8	3.378	3.227	0.155	0.148	41.8	4.627	4.517	0.111	0.108
6.0	1.882	1.594	0.314	0.266	22.0	3.393	3.242	0.154	0.147	42.0	4.637	4.528	0.110	0.108
6.1	1.895	1.610	0.311	0.264	22.2	3.408	3.257	0.153	0.147	42.2	4.648	4.539	0.110	0.108
6.2	1.908	1.625	0.308	0.262	22.4	3.422	3.273	0.153	0.146	42.4	4.659	4.550	0.110	0.107
6.3	1.921	1.640	0.305	0.260	22.6	3.437	3.288	0.152	0.145	42.6	4.670	4.561	0.110	0.107
6.4	1.934	1.655	0.302	0.259	22.8	3.451	3.303	0.151	0.145	42.8	4.680	4.572	0.109	0.107
6.5	1.946	1.670	0.299	0.257	23.0	3.466	3.318	0.151	0.144	43.0	4.691	4.583	0.109	0.107
6.6	1.959	1.685	0.297	0.255	23.2	3.480	3.333	0.150	0.144	43.2	4.702	4.594	0.109	0.106
6.7	1.972	1.699	0.294	0.254	23.4	3.494	3.348	0.149	0.143	43.4	4.712	4.605	0.109	0.105
6.8	1.984	1.714	0.292	0.252	23.6	3.509	3.363	0.149	0.143	43.6	4.723	4.616	0.108	0.106
6.9	1.997	1.728	0.289	0.250	23.8	3.523	3.378	0.148	0.142	43.8	4.733	4.627	0.108	0.106
7.0	2.009	1.742	0.287	0.249	24.0	3.537	3.393	0.147	0.141	44.0	4.744	4.637	0.108	0.105
7.1	2.021	1.756	0.285	0.247	24.2	3.551	3.407	0.147	0.141	44.2	4.755	4.648	0.108	0.105
7.2	2.033	1.770	0.282	0.246	24.4	3.565	3.422	0.146	0.140	44.4	4.765	4.659	0.107	0.105
7.3	2.046	1.784	0.280	0.244	24.6	3.579	3.437	0.145	0.140	44.6	4.776	4.670	0.107	0.105
7.4	2.058	1.798	0.278	0.243	24.8	3.593	3.451	0.145	0.139	44.8	4.786	4.680	0.107	0.104
7.5	2.070	1.812	0.276	0.242	25.0	3.607	3.465	0.144	0.139	45.0	4.796	4.691	0.107	0.104
7.6	2.082	1.826	0.274	0.240	25.2	3.621	3.480	0.144	0.138	45.2	4.807	4.702	0.106	0.104
7.7	2.093	1.839	0.272	0.239	25.4	3.635	3.494	0.143	0.138	45.4	4.817	4.712	0.106	0.104
7.8	2.105	1.853	0.270	0.238	25.6	3.648	3.509	0.143	0.137	45.6	4.828	4.723	0.106	0.104
7.9	2.117	1.866	0.268	0.236	25.8	3.662	3.523	0.142	0.137	45.8	4.838	4.733	0.106	0.103
8.0	2.129	1.879	0.266	0.235	26.0	3.676	3.537	0.141	0.136	46.0	4.848	4.744	0.105	0.103
8.1	2.140	1.892	0.264	0.234	26.2	3.689	3.551	0.141	0.136	46.2	4.859	4.755	0.105	0.103
8.2	2.152	1.905	0.262	0.232	26.4	3.703	3.565	0.140	0.135	46.4	4.869	4.765	0.105	0.103
8.3	2.163	1.918	0.261	0.231	26.6	3.716	3.579	0.140	0.135	46.6	4.879	4.775	0.105	0.102
8.4	2.175	1.931	0.259	0.230	26.8	3.730	3.593	0.139	0.134	46.8	4.889	4.786	0.104	0.102
8.5	2.186	1.944	0.257	0.229	27.0	3.743	3.607	0.139	0.134	47.0	4.900	4.796	0.104	0.102
8.6	2.197	1.957	0.256	0.228	27.2	3.756	3.621	0.138	0.133	47.2	4.910	4.807	0.104	0.102
8.7	2.209	1.969	0.254	0.226	27.4	3.769	3.634	0.138	0.133	47.4	4.920	4.817	0.104	0.102
8.8	2.220	1.982	0.252	0.225	27.6	3.783	3.648	0.137	0.132	47.6	4.930	4.828	0.104	0.101
8.9	2.231	1.994	0.251	0.224	27.8	3.796	3.662	0.137	0.132	47.8	4.940	4.838	0.103	0.101
9.0	2.242	2.007	0.249	0.223	28.0	3.809	3.675	0.136	0.131	48.0	4.950	4.848	0.103	0.101
9.1	2.253	2.019	0.248	0.222	28.2	3.822	3.689	0.135	0.131	48.2	4.960	4.859	0.103	0.101
9.2	2.264	2.032	0.246	0.221	28.4	3.835	3.703	0.135	0.130	48.4	4.970	4.869	0.103	0.101
9.3	2.275	2.044	0.245	0.220	28.6	3.848	3.715	0.135	0.130	48.6	4.980	4.879	0.102	0.100
9.4	2.286	2.056	0.243	0.219	28.8	3.861	3.729	0.134	0.129	48.8	4.991	4.889	0.102	0.100
9.5	2.297	2.068	0.242	0.218	29.0	3.874	3.743	0.134	0.129	49.0	5.001	4.900	0.102	0.100
9.6	2.308	2.080	0.240	0.217	29.2	3.887	3.756	0.133	0.129	49.2	5.010	4.910	0.102	0.100
9.7	2.319	2.092	0.239	0.216	29.4	3.900	3.759	0.133	0.128	49.4	5.020	4.920	0.102	0.100
9.8	2.329	2.104	0.238	0.215	29.6	3.913	3.783	0.132	0.128	49.6	5.030	4.930	0.101	0.099
9.9	2.340	2.115	0.236	0.214	29.8	3.925	3.796	0.132	0.127	49.8	5.040	4.940	0.101	0.099

TABLE 1.1

TABLE OF F(P) AND G(P) FUNCTIONS

P	F(P)	G(P)	F(P)/P	G(P)/P	P	F(P)	G(P)	F(P)/P	G(P)/P	P	F(P)	G(P)	F(P)/P	G(P)/P
50.0	5.050	4.950	0.101	0.099	100.0	7.107	7.036	0.071	0.070	200.0	10.025	9.975	0.050	0.050
50.5	5.075	4.975	0.100	0.099	101.0	7.142	7.071	0.071	0.070	201.0	10.050	10.000	0.050	0.050
51.0	5.100	5.000	0.100	0.098	102.0	7.177	7.107	0.070	0.070	202.0	10.075	10.025	0.050	0.050
51.5	5.124	5.025	0.099	0.098	103.0	7.211	7.142	0.070	0.069	203.0	10.100	10.050	0.050	0.050
52.0	5.148	5.050	0.099	0.097	104.0	7.246	7.177	0.070	0.069	204.0	10.124	10.075	0.050	0.049
52.5	5.173	5.075	0.099	0.097	105.0	7.280	7.211	0.069	0.069	205.0	10.149	10.100	0.050	0.049
53.0	5.197	5.099	0.098	0.096	106.0	7.315	7.246	0.069	0.068	206.0	10.174	10.124	0.049	0.049
53.5	5.221	5.124	0.098	0.096	107.0	7.349	7.280	0.069	0.068	207.0	10.198	10.149	0.049	0.049
54.0	5.244	5.148	0.097	0.095	108.0	7.383	7.315	0.068	0.068	208.0	10.223	10.174	0.049	0.049
54.5	5.268	5.172	0.097	0.095	109.0	7.416	7.349	0.068	0.067	209.0	10.247	10.198	0.049	0.049
55.0	5.292	5.197	0.096	0.094	110.0	7.450	7.383	0.068	0.067	210.0	10.271	10.223	0.049	0.049
55.5	5.315	5.221	0.096	0.094	111.0	7.483	7.416	0.067	0.067	211.0	10.296	10.247	0.049	0.049
56.0	5.339	5.244	0.095	0.094	112.0	7.517	7.450	0.067	0.067	212.0	10.320	10.271	0.049	0.048
56.5	5.362	5.268	0.095	0.093	113.0	7.550	7.483	0.067	0.066	213.0	10.344	10.296	0.049	0.048
57.0	5.386	5.292	0.094	0.093	114.0	7.583	7.517	0.067	0.066	214.0	10.368	10.320	0.048	0.048
57.5	5.409	5.315	0.094	0.092	115.0	7.616	7.550	0.066	0.066	215.0	10.392	10.344	0.048	0.048
58.0	5.432	5.339	0.094	0.092	116.0	7.649	7.583	0.066	0.065	216.0	10.416	10.368	0.048	0.048
58.5	5.455	5.362	0.093	0.092	117.0	7.681	7.616	0.066	0.065	217.0	10.440	10.392	0.048	0.048
59.0	5.478	5.386	0.093	0.091	118.0	7.714	7.649	0.065	0.065	218.0	10.464	10.416	0.048	0.048
59.5	5.500	5.409	0.092	0.091	119.0	7.746	7.681	0.065	0.065	219.0	10.488	10.440	0.048	0.048
60.0	5.523	5.432	0.092	0.091	120.0	7.778	7.714	0.065	0.064	220.0	10.512	10.464	0.048	0.048
60.5	5.546	5.455	0.092	0.090	121.0	7.810	7.746	0.065	0.064	221.0	10.536	10.488	0.048	0.047
61.0	5.568	5.478	0.091	0.090	122.0	7.842	7.778	0.064	0.064	222.0	10.559	10.512	0.048	0.047
61.5	5.591	5.500	0.091	0.089	123.0	7.874	7.810	0.064	0.063	223.0	10.583	10.536	0.047	0.047
62.0	5.613	5.523	0.091	0.089	124.0	7.906	7.842	0.064	0.063	224.0	10.607	10.559	0.047	0.047
62.5	5.635	5.546	0.090	0.089	125.0	7.937	7.874	0.063	0.063	225.0	10.630	10.583	0.047	0.047
63.0	5.657	5.568	0.090	0.088	126.0	7.969	7.906	0.063	0.063	226.0	10.654	10.607	0.047	0.047
63.5	5.679	5.591	0.089	0.088	127.0	8.000	7.937	0.063	0.062	227.0	10.677	10.630	0.047	0.047
64.0	5.701	5.613	0.089	0.088	128.0	8.031	7.969	0.063	0.062	228.0	10.701	10.654	0.047	0.047
64.5	5.723	5.635	0.089	0.087	129.0	8.062	8.000	0.062	0.062	229.0	10.724	10.677	0.047	0.047
65.0	5.745	5.657	0.088	0.087	130.0	8.093	8.031	0.062	0.062	230.0	10.747	10.701	0.047	0.047
65.5	5.767	5.679	0.088	0.087	131.0	8.124	8.062	0.062	0.062	231.0	10.770	10.724	0.047	0.046
66.0	5.788	5.701	0.088	0.086	132.0	8.155	8.093	0.062	0.061	232.0	10.794	10.747	0.047	0.046
66.5	5.810	5.723	0.087	0.086	133.0	8.185	8.124	0.062	0.061	233.0	10.817	10.770	0.046	0.045
67.0	5.831	5.745	0.087	0.086	134.0	8.216	8.155	0.061	0.061	234.0	10.840	10.794	0.046	0.045
67.5	5.853	5.767	0.087	0.085	135.0	8.246	8.185	0.061	0.061	235.0	10.863	10.817	0.046	0.045
68.0	5.874	5.788	0.086	0.085	136.0	8.277	8.216	0.061	0.060	236.0	10.886	10.840	0.046	0.045
68.5	5.895	5.810	0.086	0.085	137.0	8.307	8.246	0.061	0.060	237.0	10.909	10.863	0.046	0.045
69.0	5.916	5.831	0.086	0.085	138.0	8.337	8.277	0.060	0.060	238.0	10.932	10.886	0.046	0.046
69.5	5.937	5.853	0.085	0.084	139.0	8.367	8.307	0.060	0.060	239.0	10.954	10.909	0.046	0.046
70.0	5.958	5.874	0.085	0.084	140.0	8.397	8.337	0.060	0.060	240.0	10.977	10.932	0.046	0.045
70.5	5.979	5.895	0.085	0.084	141.0	8.426	8.367	0.060	0.059	241.0	11.000	10.954	0.046	0.045
71.0	6.000	5.916	0.085	0.083	142.0	8.456	8.397	0.060	0.059	242.0	11.023	10.977	0.046	0.045
71.5	6.021	5.937	0.084	0.083	143.0	8.485	8.426	0.059	0.059	243.0	11.045	11.000	0.045	0.045
72.0	6.042	5.958	0.084	0.083	144.0	8.515	8.456	0.059	0.059	244.0	11.068	11.023	0.045	0.045
72.5	6.062	5.979	0.084	0.082	145.0	8.544	8.485	0.059	0.059	245.0	11.091	11.045	0.045	0.045
73.0	6.083	6.000	0.083	0.082	146.0	8.573	8.515	0.059	0.058	246.0	11.113	11.068	0.045	0.045
73.5	6.104	6.021	0.083	0.082	147.0	8.602	8.544	0.059	0.058	247.0	11.136	11.091	0.045	0.045
74.0	6.124	6.042	0.083	0.082	148.0	8.631	8.573	0.058	0.058	248.0	11.158	11.113	0.045	0.045
74.5	6.144	6.062	0.082	0.081	149.0	8.660	8.602	0.058	0.058	249.0	11.180	11.136	0.045	0.045
75.0	6.165	6.083	0.082	0.081	150.0	8.689	8.631	0.058	0.058	250.0	11.203	11.158	0.045	0.045
75.5	6.185	6.104	0.082	0.081	151.0	8.718	8.660	0.058	0.057	251.0	11.225	11.180	0.045	0.045
76.0	6.205	6.124	0.082	0.081	152.0	8.747	8.689	0.058	0.057	252.0	11.247	11.203	0.045	0.044
76.5	6.225	6.144	0.081	0.080	153.0	8.775	8.718	0.057	0.057	253.0	11.269	11.225	0.045	0.044
77.0	6.245	6.165	0.081	0.080	154.0	8.804	8.747	0.057	0.057	254.0	11.292	11.247	0.044	0.044
77.5	6.265	6.185	0.081	0.080	155.0	8.832	8.775	0.057	0.057	255.0	11.314	11.269	0.044	0.044
78.0	6.285	6.205	0.081	0.080	156.0	8.860	8.803	0.057	0.056	256.0	11.336	11.292	0.044	0.044
78.5	6.305	6.225	0.080	0.079	157.0	8.888	8.832	0.057	0.056	257.0	11.358	11.314	0.044	0.044
79.0	6.325	6.245	0.080	0.079	158.0	8.916	8.860	0.056	0.056	258.0	11.380	11.336	0.044	0.044
79.5	6.345	6.265	0.080	0.079	159.0	8.944	8.888	0.056	0.056	259.0	11.402	11.358	0.044	0.044
80.0	6.364	6.285	0.080	0.079	160.0	8.972	8.916	0.056	0.056	260.0	11.424	11.380	0.044	0.044
80.5	6.384	6.305	0.079	0.078	161.0	9.000	8.944	0.056	0.056	261.0	11.446	11.402	0.044	0.044
81.0	6.403	6.325	0.079	0.078	162.0	9.028	8.972	0.056	0.055	262.0	11.467	11.424	0.044	0.044
81.5	6.423	6.345	0.079	0.078	163.0	9.055	9.000	0.056	0.055	263.0	11.489	11.446	0.044	0.044
82.0	6.442	6.364	0.079	0.078	164.0	9.083	9.028	0.055	0.055	264.0	11.511	11.467	0.044	0.043
82.5	6.462	6.384	0.078	0.077	165.0	9.111	9.055	0.055	0.055	265.0	11.533	11.489	0.044	0.043
83.0	6.481	6.403	0.078	0.077	166.0	9.138	9.083	0.055	0.055	266.0	11.554	11.511	0.043	0.043
83.5	6.500	6.423	0.078	0.077	167.0	9.165	9.111	0.055	0.055	267.0	11.576	11.533	0.043	0.043
84.0	6.519	6.442	0.078	0.077	168.0	9.192	9.138	0.055	0.054	268.0	11.597	11.554	0.043	0.043
84.5	6.539	6.462	0.077	0.076	169.0	9.220	9.165	0.055	0.054	269.0	11.619	11.576	0.043	0.043
85.0	6.558	6.481	0.077	0.076	170.0	9.247	9.192	0.054	0.054	270.0	11.640	11.597	0.043	0.043
85.5	6.577	6.500	0.077	0.076	171.0	9.274	9.220	0.054	0.054	271.0	11.662	11.619	0.043	0.043
86.0	6.596	6.519	0.077	0.076	172.0	9.301	9.247	0.054	0.054	272.0	11.683	11.640	0.043	0.043
86.5	6.615	6.539	0.076	0.076	173.0	9.327	9.274	0.054	0.054	273.0	11.705	11.662	0.043	0.043
87.0	6.633	6.558	0.076	0.075	174.0	9.354	9.301	0.054	0.053	274.0	11.726	11.683	0.043	0.043
87.5	6.652	6.577	0.076	0.075	175.0	9.381	9.327	0.054	0.053	275.0	11.747	11.705	0.043	0.043
88.0	6.671	6.596	0.076	0.075	176.0	9.408	9.354	0.053	0.053	276.0	11.769	11.726	0.043	0.042
88.5	6.690	6.615	0.076	0.075	177.0	9.434	9.381	0.053	0.053	277.0	11.790	11.747	0.043	0.042
89.0	6.708	6.633	0.075	0.075	178.0	9.461	9.408	0.053	0.053	278.0	11.811	11.769	0.042	0.042
89.5	6.727	6.652	0.075	0.074	179.0	9.487	9.434	0.053	0.053	279.0	11.832	11.790	0.042	0.042
90.0	6.746	6.671	0.075	0.074	180.0	9.513	9.461	0.053	0.053	280.0	11.853	11.811	0.042	0.042
90.5	6.764	6.690	0.075	0.074	181.0	9.539	9.487	0.053	0.052	281.0	11.874	11.832	0.042	0.042
91.0	6.783	6.708	0.075	0.074	182.0	9.566	9.513	0.053	0.052	282.0	11.895	11.853	0.042	0.042
91.5	6.801	6.727	0.074	0.074	183.0	9.592	9.539	0.052	0.052	283.0	11.916	11.874	0.042	0.042
92.0	6.819	6.746	0.074	0.073	184.0	9.618	9.566	0.052	0.052	284.0	11.937	11.895	0.042	0.042
92.5	6.838	6.764	0.074	0.073	185.0	9.644	9.592	0.052	0.052	285.0	11.958	11.916	0.042	0.042
93.0	6.856	6.783	0.074	0.073	186.0	9.670	9.618	0.052	0.052	286.0	11.979	11.937	0.042	0.042
93.5	6.874	6.801	0.074	0.073	187.0	9.695	9.644	0.052	0.052	287.0	12.000	11.958	0.042	0.042
94.0	6.892	6.819	0.073	0.073	188.0	9.721	9.670	0.052	0.051	288.0	12.021	11.979	0.042	0.042
94.5	6.910	6.838	0.073	0.072	189.0	9.747	9.695	0.052	0.051	289.0	12.042	12.000	0.042	0.042
95.0	6.928	6.856	0.073	0.072	190.0	9.772	9.721	0.051	0.051	290.0	12.062	12.021	0.042	0.041
95.5	6.946	6.874	0.073	0.072	191.0	9.798	9.747	0.051	0.051	291.0	12.083	12.042	0.042	0.041
96.0	6.964	6.892	0.073	0.072	192.0	9.824	9.772	0.051	0.051	292.0	12.104	12.062	0.041	0.041
96.5	6.982	6.910	0.072	0.072	193.0	9.849	9.798	0.051	0.051	293.0	12.124	12.083	0.041	0.041
97.0	7.000	6.928	0.072	0.071	194.0	9.874	9.824	0.051	0.051	294.0	12.145	12.104	0.041	0.041
97.5	7.018	6.946	0.072	0.071	195.0	9.900	9.849	0.051	0.051	295.0	12.166	12.124	0.041	0.041
98.0	7.036	6.964	0.072	0.071	196.0	9.925	9.874	0.051	0.050	296.0	12.186	12.145	0.041	0.041
98.5	7.054	6.982	0.072	0.071	197.0	9.950	9.900	0.051	0.050	297.0	12.207	12.166	0.041	0.041
99.0	7.071	7.000	0.071	0.071	198.0	9.975	9.925	0.050	0.050	298.0	12.227	12.186	0.041	0.041
99.5	7.089	7.018	0.071	0.071	199.0	10.000	9.950	0.050	0.050	299.0	12.247	12.207	0.041	0.041

2. Characteristics of Cylindrical Dipoles and Monopoles

a. THE APPARENT ADMITTANCE

The cylindrical dipole consists of a highly-conducting tube or rod with radius a and half-length h. In practice, it is center-driven from a balanced open-wire transmission line with a distance b between the axes of the identical conductors of the line as shown in Fig. 2.1a. The cylindrical monopole is essentially half a dipole. It consists of a highly-conducting tube or rod with radius a and length h, erected perpendicular to a sufficiently large (ideally infinite), highly-conducting ground plane in either of the arrangements shown in Figs. 2.1b and 2.1c. The axis of the single wire with radius a in Fig. 2.1b is at a distance $b/2$ from the ground plane; with its image in the highly-conducting plane it is equivalent to the open line in Fig. 2.1a. The inner radius of the outer conductor of the coaxial line in Fig. 2.1c is b; the radius of the inner conductor is a and its extension of length h above the ground plane is the monopole antenna. In order that radiation from a balanced open-wire line be negligible, the condition

$$k_0 b = \frac{2\pi b}{\lambda} \ll 1 \tag{2.1}$$

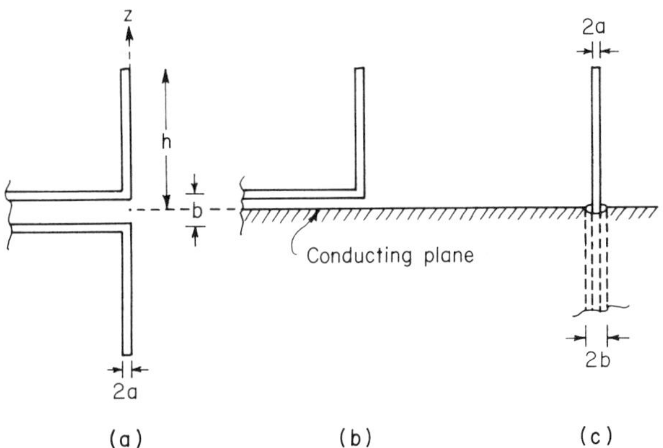

Fig. 2.1. Cylindrical antennas driven from open-wire and coaxial lines.

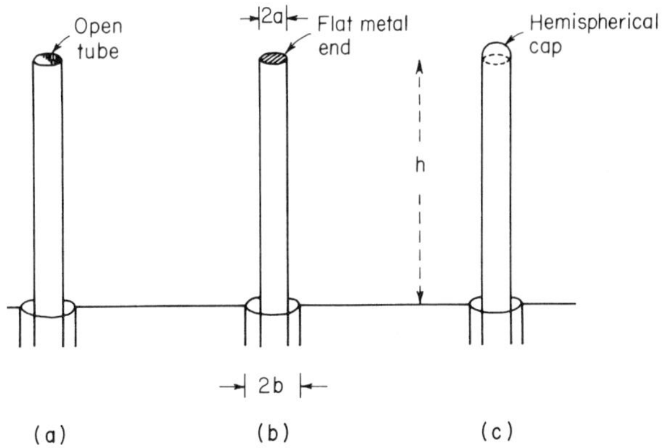

Fig. 2.2. Cylindrical monopoles with open, closed flat, and closed hemispherical ends.

must be satisfied. Similarly, in order to ensure that only the TEM mode can propagate in the coaxial line, the conditions

$$k_0 b < 1, \qquad k_0(b - a) < 1 \tag{2.2}$$

must be fulfilled. Of these, the first condition is required to exclude TE modes, the second, less severe condition, to exclude TM modes. If complete rotational symmetry is maintained in the entire generating–transmitting–radiating system, TE modes may be absent because they are nowhere generated. In this case, only the second condition in (2.2) is required.

The end (ends) of the monopole (dipole) may be open as in Fig. 2.2a, consist of a flat metal disk as in Fig. 2.2b, or be capped with a metal hemisphere as in Fig. 2.2c. In the case of the open tube or the rod with a flat metal end, the axial length (half-length) is h; with a hemispherical cap it is $h' = h + a$.

From the point of view of a generator supplying power at the input end of a transmission line, a cylindrical dipole or monopole terminating the other end of the line behaves like any other load in the sense that it is observed as an apparent admittance $Y_a = G_a + jB_a$. It can be determined experimentally by means of well-known techniques of measurement of the standing-wave pattern along the line.[1] Since conventional transmission-line theory, upon which most high-frequency measurements depend, is not accurate at and near the ends of the line,[2] the quantity actually measured is the admittance looking toward the load at a cross section of the line that is a half wavelength from the end. Owing to end effects on the transmission line and coupling between the antenna and the line over distances from their junction that are comparable with the line spacing (b for open wires, $b - a$ for coaxial lines), the apparent admittance Y_a depends on the physical properties of the junction region as well as on those of the antenna proper. It follows that the smaller b or $b - a$, the more nearly Y_a approaches a quantity characteristic exclusively of the antenna. Owing to the complicated nonrotationally symmetric properties of the antenna driven by an open-wire line, no mathematically useful limit is reached as $b \to 2a$. On the other hand, the limit $(b - a) \to 0$ for the coaxial line defines a physically unavailable but mathematically useful delta-function generator, which maintains at $z = 0$ a finite voltage across a "zero" gap that corresponds physically to an infinite knife-edge capacitance. The admittance seen by the emf of such a generator involves no transmission line but does include, in addition to the cylindrical surface of the antenna proper, the circular knife edges. Hence, the susceptance is infinite. However, for electrically thin antennas, the charging current associated with the infinite susceptance of these knife edges is highly localized in a narrow region very near the driving point and can be separated from the current associated with the cylindrical surface.[3,4] This process of "subtracting out" the current that charges the knife-edge capacitance is necessarily arbitrary and approximate. However, the various theoretical and experimental procedures that have been suggested are in good agreement with one another and the resulting finite susceptance is useful in determining the measurable apparent admittance of antennas driven from actual transmission lines in conjunction with lumped terminal-zone networks appropriate to the geometries of different lines.[5–7]. Only in certain special cases are terminal-zone effects sufficiently small to make the uncorrected ideal admittance a good approximation.[8]

b. THE MONOPOLE DRIVEN FROM A COAXIAL LINE

Owing to the complicated geometry of the junction region between the antenna and the transmission line, analytically accurate formulas for the apparent admittance of an antenna as a termination for the line are generally unavailable. An exception is the monopole driven from a coaxial line in the arrangement shown in Fig. 2.1c. For this rotationally symmetric configuration the apparent admittance has been determined for an infinitely long monopole subject to the following conditions:

$$(b - a)/a \ll 1, \qquad k_0(b - a) \ll 1 \tag{2.3}$$

[The severer condition ($k_0 b \ll 1$) need not be imposed if TE modes are nowhere generated.] The rigorous formula is[9]

$$Y_{a\infty}(k_0 a, b/a) = \frac{2\pi}{\zeta_0}\{F_{1\infty}(k_0 a) + jk_0 a F_{2\infty}(b/a)\} \tag{2.4}$$

where

$$F_{1\infty}(k_0 a) = \frac{2jka}{\pi}\left[1 - \ln\frac{4}{\pi} - \ln\frac{k_0 a}{2} - \gamma - C_0(k_0 a)\right] \tag{2.5a}$$

$$F_{2\infty}(b/a) = -\frac{2}{\pi}\ln(b/a - 1) \tag{2.5b}$$

and

$$C_0(k_0 a) = \int_0^\infty \left[\frac{H_1^{(1)}(\xi a)}{H_0^{(1)}(\xi a)} - i\right]\frac{dx}{\xi} + i\frac{\pi}{2} \tag{2.5c}$$

In these formulas $\xi = \sqrt{k_0^2 - x^2}$ and $\gamma = 0.577\ldots$. Note that $F_{1\infty}(k_0 a)$ is a complex function of $k_0 a$ and is independent of b/a, whereas $F_{2\infty}(b/a)$ is a real function of b/a alone.

The apparent admittance has also been evaluated[9] under the simplifying assumption that there are no end effects on the coaxial line, so that only the TEM mode exists at its end. The admittance $Y_{a\,\mathrm{TEM}}$ in this case is given by

$$\begin{aligned} Y_{a\infty} - Y_{a\mathrm{TEM}} = j(B_{a\infty} - B_{a\mathrm{TEM}}) &= -j\frac{4k_0 a}{\zeta_0}\ln\frac{4}{\pi} \\ &= -j16.1\frac{a}{\lambda}\ \text{millimhos} \end{aligned} \tag{2.6}$$

So far the discussion has been confined to infinitely long monopoles. However, since both transmission-line end effects and the coupling between the antenna and the coaxial line are limited to short distances from the junction of the monopole with the line, a formula like (2.4) must also be true for monopoles of finite length h provided that

$$(b - a)/h \ll 1 \tag{2.7}$$

That is,

$$Y_a(k_0h, k_0a, b/a) = \frac{2\pi}{\zeta_0}[F_1(k_0h, k_0a) + jk_0aF_2(k_0h, b/a)] \quad (2.8)$$

where $F_1(k_0h, k_0a)$ is a complex function of k_0h and k_0a that is independent of b/a, and $F_2(k_0h, b/a)$ is a real function of k_0h and b/a that is independent of k_0a. Since $F_2(k_0h, b/a)$ is real, the real and imaginary parts of $Y_a = G_a + jB_a$ are expressed as follows:

$$G_a(k_0h, k_0a) = \frac{2\pi}{\zeta_0} \operatorname{Re} F_1(k_0h, k_0a) \quad (2.9a)$$

$$B_a(k_0h, k_0a, b/a) = \frac{2\pi}{\zeta_0}[\operatorname{Im} F_1(k_0h, k_0a) + k_0aF_2(k_0h, b/a)] \quad (2.9b)$$

where Re and Im stand for the real and imaginary parts. Note that the apparent conductance $G_a(k_0h, k_0a)$ is independent of b/a, whereas the apparent susceptance $B_a(k_0h, k_0a, b/a)$ is not.

It is readily shown that the following relation is valid:[10]

$$B_a(k_0h, k_0a, b/a) = B_a[k_0h, k_0a, (b/a)_2] + [k_0a/(k_0a)_1]\{B_a[k_0h, (k_0a)_1, b/a] - B_a[k_0h, (k_0a)_1, (b/a)_2]\} \quad (2.10)$$

This formula is conveniently used to determine nm values of $B_a(k_0h, k_0a, b/a)$ from n known values of $B_a[k_0h, (k_0a)_i, (b/a)_2]$, $i = 1, 2, \ldots, n$, and m known values of $B_a[k_0h, (k_0a)_1, (b/a)_p]$, $p = 1, 2, \ldots, m$. Note that $(k_0a)_1$ and $(b/a)_2$ are arbitrarily selected values in the range $(k_0a)_i$, $i = 1, 2, \ldots, n$, and $(b/a)_p$, $p = 1, 2, \ldots, m$. The practical significance of (2.10) is that with it the tabulation of the $n + m$ values, $B_a[k_0h, (k_0a)_1, (b/a)_p]$, $p = 1, 2, \ldots, m$, and $B_a[k_0h, (k_0a)_i, (b/a)_2]$, $i = 1, 2, \ldots, n$, is sufficient to make available all of the nm values, $B_a[k_0h, (k_0a)_i, (b/a)_p]$, $i = 1, 2, \ldots, n$; $p = 1, 2, \ldots, m$.

The formulas (2.8)–(2.10) are valid for the exact apparent admittance $Y_a = G_a + jB_a$ and for the approximate values obtained if only the TEM mode is assumed to exist in the coaxial feed line at its junction with the antenna. When (2.3) is satisfied, a formula like (2.6) is valid for antennas of finite length. Specifically,

$$G_a(k_0h, k_0a) = G_{a\text{TEM}}(k_0h, k_0a) \quad (2.11a)$$

$$B_a(k_0h, k_0a, b/a) = B_{a\text{TEM}}(k_0h, k_0a, b/a) - j16.1\frac{a}{\lambda} \text{ millimhos} \quad (2.11b)$$

From these the exact values are readily obtained from the tabulated approximate TEM values in Table 2.1. A sample graph of the admittance in the complex plane is shown in Fig. 2.3. The theoretical points obtained from Table 2.1 are shown. From this graph the extreme values of G and B are readily obtained as well as values intermediate to those computed. Note that large parts of the curve are sections of circles so that only three or four points are needed to determine a wide range of values.

Distributions of current along tubular monopoles with selected lengths are given in Table 2.2. The admittances of hemispherically capped monopoles with axial lengths $h + a$ are given in Table 2.3. Owing to approximations in treating the hemispherical cap, these are somewhat less accurate than the corresponding values for the open tube in Table 2.1.

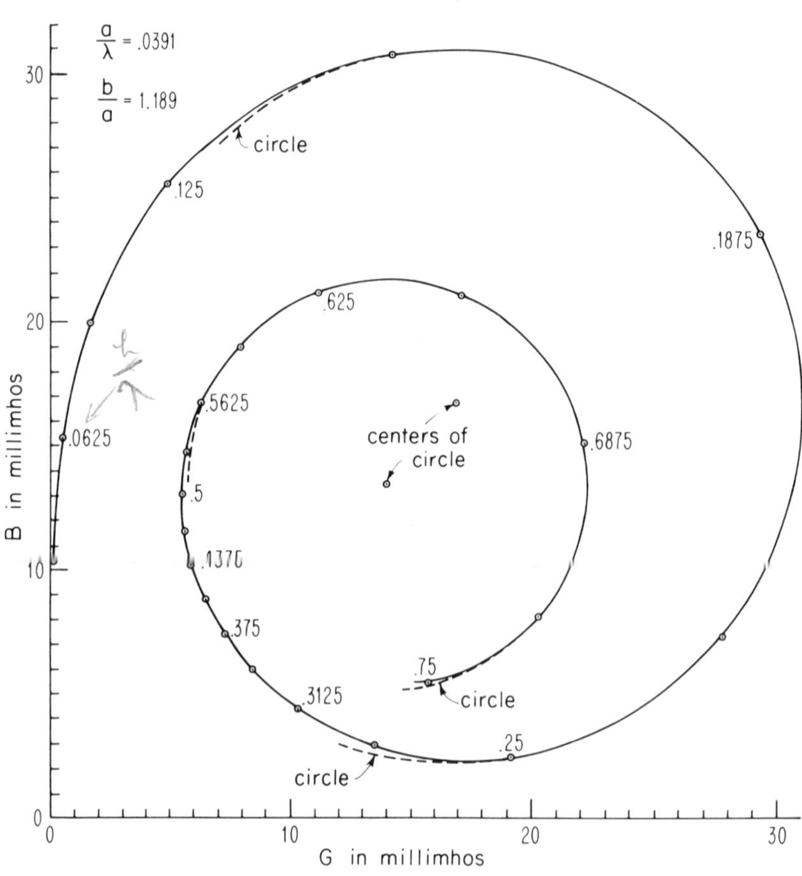

Fig. 2.3. Admittance of the tubular monopole.

TABLE 2.1

ADMITTANCE OF THE TUBULAR MONOPOLE

DIMENSIONS OF THE FEEDING COAXIAL LINE: b/a = 1.189

Y = G + jB (millimhos)

		a/λ (k_0a)				
h/λ	k_0h	.0064 (.0402)	.0127 (.0798)	.0190 (.1194)	.0254 (.1596)	.0318 (.1998)
.03125	0.196	0.00 + j3.14	0.01 + j4.87	0.02 + j6.43	0.04 + j7.89	0.06 + j9.22
.06250	0.393	0.03 + j4.86	0.07 + j7.23	0.14 + j9.36	0.22 + j11.38	0.33 + j13.28
.09375	0.589	0.14 + j6.65	0.30 + j9.63	0.51 + j12.32	0.79 + j14.87	1.13 + j17.28
.12500	0.785	0.49 + j9.00	1.02 + j12.81	1.70 + j16.22	2.56 + j19.42	3.57 + j22.39
.15625	0.982	1.76 + j12.68	3.59 + j17.68	5.82 + j21.84	8.40 + j25.35	11.13 + j28.18
.18750	1.178	7.47 + j18.88	13.92 + j23.29	19.60 + j24.81	23.95 + j24.79	26.97 + j24.18
.21875	1.374	27.60 + j12.61	29.30 + j7.66	28.85 + j6.00	28.30 + j5.81	27.95 + j6.29
.25000	1.571	17.24 - j7.12	17.21 - j4.85	17.49 - j2.86	17.92 - j1.03	18.43 + j0.64
.28125	1.767	8.24 - j5.62	9.65 - j3.70	10.72 - j1.94	11.67 - j0.29	12.55 + j1.24
.31250	1.963	5.10 - j3.21	6.47 - j1.61	7.47 - j0.06	8.51 + j1.43	9.40 + j2.84
.34375	2.160	3.70 - j1.46	4.89 + j0.08	5.86 + j1.59	6.75 + j3.05	7.57 + j4.43
.37500	2.356	2.95 + j0.16	3.99 + j1.45	4.87 + j2.99	5.68 + j4.47	6.43 + j5.87
.40625	2.552	2.49 + j0.89	3.43 + j2.60	4.23 + j4.21	4.98 + j5.75	5.69 + j7.21
.43750	2.749	2.20 + j1.80	3.07 + j3.64	3.81 + j5.35	4.52 + j6.96	5.19 + j8.49
.46875	2.945	2.01 + j2.65	2.82 + j4.64	3.54 + j6.45	4.22 + j8.17	4.89 + j9.78
.50000	3.142	1.89 + j3.50	2.69 + j5.65	3.40 + j7.60	4.09 + j9.44	4.76 + j11.15
.53125	3.338	1.84 + j4.41	2.66 + j6.76	3.40 + j8.88	4.13 + j10.85	4.86 + j12.70
.56250	3.534	1.89 + j5.45	2.79 + j8.06	3.63 + j10.38	4.47 + j12.52	5.33 + j14.52
.59375	3.731	2.12 + j6.76	3.22 + j9.70	4.29 + j12.25	5.40 + j14.59	6.53 + j16.71
.62500	3.927	2.76 + j8.55	4.37 + j11.87	5.98 + j14.63	7.64 + j17.04	9.30 + j19.10
.65625	4.123	4.58 + j11.10	7.45 + j14.55	10.17 + j16.98	12.69 + j18.73	14.92 + j19.98
.68750	4.320	10.24 + j13.64	14.98 + j14.85	18.00 + j14.93	19.90 + j14.82	21.13 + j14.84
.71875	4.516	19.60 + j6.41	20.07 + j4.95	19.98 + j5.06	19.94 + j5.78	20.00 + j6.76
.75000	4.712	13.74 - j3.63	13.86 - j1.56	14.18 + j0.31	14.61 + j2.05	15.10 + j3.68

TABLE 2.1

ADMITTANCE OF THE TUBULAR MONOPOLE

DIMENSIONS OF THE FEEDING COAXIAL LINE: b/a = 1.189

Y = G + jB (millimhos)

		a/λ (k_0a)				
h/λ	k_0h	.0391 (.2457)	.0426 (.2677)	.0509 (.3198)	.0565 (.3550)	.0635 (.3990)
.03125	0.196	0.10 + j10.62	0.13 + j11.80	0.20 + j13.42	0.26 + j14.48	0.36 + j15.78
.06250	0.393	0.48 + j15.36	0.57 + j16.30	0.81 + j18.47	1.00 + j19.85	1.30 + j21.80
.09375	0.589	1.61 + j19.92	1.87 + j21.13	2.56 + j23.90	3.09 + j25.67	3.81 + j27.78
.12500	0.785	4.91 + j25.53	5.60 + j26.93	7.37 + j30.00	8.63 + j31.87	10.25 + j33.98
.15625	0.982	14.27 + j30.67	15.72 + j31.60	18.96 + j33.32	20.92 + j34.15	23.11 + j34.94
.18750	1.178	29.22 + j23.45	29.99 + j23.17	31.32 + j22.79	31.96 + j22.74	32.59 + j22.86
.21875	1.374	27.79 + j7.22	27.78 + j7.74	27.91 + j9.06	28.09 + j9.99	28.39 + j11.16
.25000	1.571	19.06 + j2.43	19.38 + j3.23	20.15 + j5.06	20.69 + j6.22	21.37 + j7.61
.28125	1.767	13.50 + j2.89	13.93 + j3.64	14.95 + j5.35	15.61 + j6.44	16.43 + j7.75
.31250	1.963	10.35 + j4.38	10.79 + j5.08	11.81 + j6.69	12.47 + j7.72	13.30 + j8.97
.34375	2.160	8.46 + j5.94	8.88 + j6.62	9.84 + j8.21	10.48 + j9.22	11.26 + j10.43
.37500	2.356	7.26 + j7.40	7.65 + j8.10	8.56 + j9.70	9.16 + j10.73	9.90 + j11.97
.40625	2.552	6.47 + j8.79	6.84 + j9.51	7.70 + j11.16	8.27 + j12.22	8.99 + j13.50
.43750	2.749	5.94 + j10.14	6.30 + j10.89	7.13 + j12.62	7.69 + j13.73	8.40 + j15.07
.46875	2.945	5.63 + j11.52	5.98 + j12.32	6.82 + j14.14	7.38 + j15.31	8.10 + j16.72
.50000	3.142	5.52 + j13.01	5.89 + j13.85	6.77 + j15.79	7.36 + j17.02	8.12 + j18.51
.53125	3.338	5.70 + j14.68	6.10 + j15.59	7.08 + j17.64	7.74 + j18.95	8.59 + j20.51
.56250	3.534	6.33 + j16.64	6.81 + j17.60	7.99 + j19.76	8.79 + j21.12	9.81 + j22.73
.59375	3.731	7.85 + j18.93	8.49 + j19.90	10.02 + j22.06	11.06 + j23.38	12.34 + j24.90
.62500	3.927	11.16 + j21.12	12.03 + j21.96	14.02 + j23.73	15.29 + j24.75	16.77 + j25.87
.65625	4.123	17.09 + j21.02	18.00 + j21.41	19.86 + j22.22	20.92 + j22.70	22.07 + j23.28
.68750	4.320	22.07 + j15.09	22.42 + j15.29	23.12 + j15.93	23.53 + j16.45	24.01 + j17.19
.71875	4.516	20.21 + j8.01	20.35 + j8.63	20.74 + j10.11	21.05 + j11.10	21.48 + j12.32
.75000	4.712	15.69 + j5.42	15.99 + j6.21	16.71 + j8.13	17.22 + j9.16	17.85 + j10.54

TABLE 2.1

ADMITTANCE OF THE TUBULAR MONOPOLE

DIMENSIONS OF THE FEEDING COAXIAL LINE: b/a = 1.189

Y = G + jB (millimhos)

		a/λ (k_0a)				
h/λ	k_0h	.0726 (.4562)	.0847 (.5322)	.0924 (.5806)	.1129 (.7094)	.1270 (.7980)
.03125	0.196	0.51 + j17.44	0.78 + j19.59	1.00 + j20.94	1.76 + j24.47	2.47 + j26.82
.06250	0.393	1.65 + j23.53	2.43 + j26.82	2.94 + j28.55	4.55 + j32.90	5.87 + j35.67
.09375	0.589	4.83 + j30.34	6.30 + j33.42	7.27 + j35.18	10.51 + j39.98	12.68 + j42.60
.12500	0.785	12.37 + j36.36	15.11 + j38.97	16.76 + j40.35	20.66 + j43.19	23.67 + j45.00
.15625	0.982	25.54 + j35.71	28.14 + j36.47	29.50 + j36.90	32.28 + j38.07	33.68 + j38.93
.18750	1.178	33.27 + j23.29	34.05 + j24.15	34.51 + j24.83	35.70 + j26.89	36.50 + j28.44
.21875	1.374	28.88 + j12.68	29.64 + j14.65	30.17 + j15.88	31.70 + j19.03	32.80 + j21.10
.25000	1.571	22.27 + j9.33	23.49 + j11.48	24.26 + j12.78	26.34 + j16.04	27.76 + j18.14
.28125	1.767	17.48 + j9.37	18.86 + j11.41	19.72 + j12.65	22.00 + j15.74	23.55 + j17.74
.31250	1.963	14.34 + j10.52	15.72 + j12.47	16.58 + j13.65	18.86 + j16.63	20.41 + j18.54
.34375	2.160	12.26 + j11.98	13.59 + j13.90	14.42 + j15.08	16.63 + j18.01	18.16 + j19.89
.37500	2.356	10.86 + j13.52	12.13 + j15.47	12.93 + j16.66	15.08 + j19.62	16.57 + j21.52
.40625	2.552	9.91 + j15.10	11.14 + j17.11	11.93 + j18.33	14.04 + j21.38	15.52 + j23.31
.43750	2.749	9.31 + j16.74	10.54 + j18.82	11.32 + j20.09	13.45 + j23.25	14.94 + j25.25
.46875	2.945	9.03 + j18.46	10.29 + j20.65	11.10 + j21.97	13.30 + j25.25	14.84 + j27.31
.50000	3.142	9.11 + j20.35	10.46 + j22.64	11.33 + j24.01	13.70 + j27.39	15.35 + j29.49
.53125	3.338	9.71 + j22.43	11.23 + j24.80	12.21 + j26.20	14.85 + j29.60	16.66 + j31.66
.56250	3.534	11.15 + j24.67	12.95 + j27.01	14.08 + j28.37	17.06 + j31.58	19.04 + j33.48
.59375	3.731	13.98 + j26.67	16.08 + j28.74	17.36 + j29.91	20.55 + j32.59	22.54 + j34.16
.62500	3.927	18.55 + j27.14	20.66 + j28.59	21.87 + j29.42	24.70 + j31.42	26.39 + j32.70
.65625	4.123	23.34 + j24.04	24.78 + j25.08	25.59 + j25.77	27.54 + j27.70	28.78 + j29.06
.68750	4.320	24.62 + j18.23	25.44 + j19.71	25.98 + j20.67	27.48 + j23.24	28.56 + j24.98
.71875	4.516	22.10 + j13.88	22.99 + j15.87	23.59 + j17.09	25.29 + j20.18	26.50 + j22.18
.75000	4.712	18.70 + j12.25	19.84 + j14.39	20.58 + j15.68	22.57 + j18.90	23.96 + j20.95

TABLE 2.1

ADMITTANCE OF THE TUBULAR MONOPOLE

DIMENSIONS OF THE FEEDING COAXIAL LINE: b/a = 1.125

Y = G + jB (millimhos)

		a/λ (k_0a)				
h/λ	k_0h	.1016 (.6384)	.1180 (.7414)	.1290 (.8105)	.1410 (.8859)	.1530 (.9613)
.03125	0.196	1.29 + j25.07	1.97 + j28.29	2.54 + j30.41	3.27 + j32.66	4.11 + j34.90
.06250	0.393	3.60 + j33.19	4.98 + j37.02	6.03 + j39.46	7.28 + j41.98	8.63 + j44.40
.09375	0.589	8.81 + j40.30	11.26 + j44.11	12.96 + j46.41	14.81 + j48.68	16.66 + j50.78
.12500	0.785	18.58 + j44.34	21.48 + j46.85	23.15 + j48.26	24.74 + j49.64	26.06 + j50.97
.15625	0.982	30.87 + j40.02	32.81 + j41.48	33.86 + j42.52	34.81 + j43.69	35.61 + j45.02
.18750	1.178	35.05 + j28.32	36.00 + j30.56	36.63 + j32.12	37.29 + j33.86	37.94 + j35.71
.21875	1.374	30.86 + j19.92	32.11 + j22.90	32.98 + j24.84	33.92 + j26.93	34.86 + j29.08
.25000	1.571	25.21 + j16.88	26.88 + j19.91	27.99 + j21.88	29.19 + j23.96	30.39 + j26.10
.28125	1.767	20.76 + j16.67	22.58 + j19.57	23.79 + j21.45	25.10 + j23.45	26.41 + j25.49
.31250	1.963	17.62 + j17.62	19.44 + j20.43	20.65 + j22.24	21.98 + j24.16	23.31 + j26.13
.34375	2.160	15.43 + j19.02	17.20 + j21.80	18.30 + j23.50	19.70 + j25.40	21.01 + j27.40
.37500	2.356	13.91 + j20.62	15.64 + j23.42	16.80 + j25.22	18.09 + j27.11	19.38 + j29.05
.40625	2.552	12.89 + j22.33	14.59 + j25.18	15.75 + j27.02	17.02 + j28.94	18.31 + j30.89
.43750	2.749	12.28 + j24.14	14.00 + j27.08	15.17 + j28.96	16.46 + j30.92	17.77 + j32.90
.46875	2.945	12.09 + j26.08	13.87 + j29.11	15.08 + j31.03	16.42 + j33.03	17.78 + j35.03
.50000	3.142	12.39 + j28.17	14.31 + j31.27	15.60 + j33.21	17.03 + j35.22	18.46 + j37.22
.53125	3.338	13.40 + j30.38	15.52 + j33.47	16.93 + j35.38	18.47 + j37.34	20.00 + j39.27
.56250	3.534	15.44 + j32.48	17.80 + j35.39	19.33 + j37.17	20.96 + j38.98	22.53 + j40.76
.59375	3.731	18.84 + j33.77	21.30 + j36.27	22.83 + j37.81	24.41 + j39.38	25.90 + j40.97
.62500	3.927	23.21 + j32.94	25.34 + j34.99	26.63 + j36.31	27.95 + j37.73	29.20 + j39.23
.65625	4.123	26.51 + j29.22	28.01 + j31.29	28.97 + j32.70	29.99 + j34.24	31.01 + j35.88
.68750	4.320	26.65 + j24.42	27.88 + j26.97	28.73 + j28.67	29.69 + j30.48	30.67 + j32.36
.71875	4.516	24.35 + j21.10	25.74 + j24.01	26.70 + j25.89	27.77 + j27.87	28.86 + j29.89
.75000	4.712	21.48 + j19.76	23.09 + j22.75	24.17 + j24.68	25.37 + j26.69	26.58 + j28.74

TABLE 2.1

ADMITTANCE OF THE TUBULAR MONOPOLE

DIMENSIONS OF THE FEEDING COAXIAL LINE: b/a = 1.125

Y = G + jB (millimhos)

		a/λ (k_0a)				
h/λ	k_0h	.1640 (1.0304)	.1800 (1.1310)	.1920 (1.2064)	.2050 (1.2881)	.2150 (1.3509)
.03125	0.196	4.99 + j36.85	6.45 + j39.54	7.67 + j41.45	9.12 + j43.39	10.31 + j44.76
.06250	0.393	9.95 + j46.44	11.97 + j49.15	13.55 + j51.00	15.31 + j52.81	16.69 + j54.08
.09375	0.589	18.33 + j52.49	20.70 + j54.68	22.43 + j56.14	24.24 + j57.54	25.59 + j58.52
.12500	0.785	29.25 + j52.77	31.28 + j54.32	32.69 + j55.41	34.10 + j56.52	35.12 + j57.34
.15625	0.982	37.63 + j45.98	38.97 + j47.61	39.90 + j48.84	40.84 + j50.16	41.56 + j51.18
.18750	1.178	38.51 + j37.35	39.28 + j39.78	39.81 + j41.57	40.32 + j43.55	40.67 + j45.09
.21875	1.374	35.71 + j30.95	36.92 + j33.65	37.79 + j35.67	38.69 + j37.84	39.35 + j39.54
.25000	1.571	31.47 + j27.95	33.03 + j30.62	34.17 + j32.59	35.37 + j34.73	36.28 + j36.39
.28125	1.767	27.60 + j27.25	29.31 + j29.78	30.59 + j31.66	31.96 + j33.67	33.01 + j35.24
.31250	1.963	24.52 + j27.82	26.29 + j30.24	27.61 + j32.02	29.05 + j33.93	30.16 + j35.41
.34375	2.160	22.22 + j29.07	23.99 + j31.43	25.32 + j33.15	26.78 + j34.98	27.91 + j36.39
.37500	2.356	20.58 + j30.69	22.34 + j33.02	23.67 + j34.71	25.13 + j36.51	26.27 + j37.87
.40625	2.552	19.51 + j32.55	21.27 + j34.88	22.61 + j36.56	24.08 + j38.33	25.22 + j39.67
.43750	2.749	18.98 + j34.57	20.77 + j36.92	22.13 + j38.59	23.61 + j40.34	24.77 + j41.65
.46875	2.945	19.04 + j36.71	20.88 + j39.06	22.27 + j40.71	23.78 + j42.43	24.96 + j43.72
.50000	3.142	19.78 + j38.88	21.69 + j41.19	23.13 + j42.80	24.67 + j44.47	25.86 + j45.71
.53125	3.338	21.38 + j40.87	23.36 + j43.07	24.81 + j44.61	26.37 + j46.18	27.55 + j47.36
.56250	3.534	23.93 + j42.24	25.90 + j44.26	27.33 + j45.68	28.83 + j47.15	29.96 + j48.25
.59375	3.731	27.21 + j42.29	29.02 + j44.15	30.32 + j45.49	31.68 + j46.89	32.71 + j47.96
.62500	3.927	30.30 + j40.52	31.84 + j42.37	32.96 + j43.73	34.16 + j45.18	35.06 + j46.30
.65625	4.123	31.94 + j37.30	33.28 + j39.36	34.30 + j40.87	35.40 + j42.48	36.25 + j43.71
.68750	4.320	31.59 + j33.97	32.95 + j36.28	33.99 + j37.95	35.13 + j39.72	36.01 + j41.07
.71875	4.516	29.88 + j31.62	31.37 + j34.06	32.51 + j35.82	33.75 + j37.67	34.72 + j39.07
.75000	4.712	27.70 + j30.48	29.33 + j32.94	30.56 + j34.71	31.91 + j36.57	32.95 + j37.98

TABLE 2.2

DISTRIBUTION OF CURRENT ALONG TUBULAR MONOPOLES

$I = I'' + jI'$ (milliamperes per volt); $h/\lambda = 0.25$

	b/a = 1.189		b/a = 1.125	
	a/λ (k_0a)		a/λ (k_0a)	
z/h	.0254 (.1596)	.0509 (.3198)	.1016 (.6384)	.2050 (1.288)
0	17.92 - j1.03	20.15 + j5.06	25.21 + j16.88	35.37 + j34.73
.0625	17.85 - j5.02	20.08 - j0.65	25.13 + j7.63	35.26 + j23.28
.1250	17.66 - j6.59	19.88 - j3.38	24.89 + j2.76	34.92 + j15.37
.1875	17.33 - j7.73	19.54 - j5.29	24.49 - j0.58	34.37 + j9.77
.2500	16.87 - j8.44	19.06 - j6.59	23.94 - j2.97	33.59 + j5.42
.3125	16.29 - j8.92	18.46 - j7.56	23.24 - j4.81	32.59 + j1.93
.3750	15.59 - j9.18	17.73 - j8.24	22.38 - j6.33	31.37 - j0.90
.4375	14.77 - j9.25	16.87 - j8.69	21.37 - j7.39	29.94 - j3.21
.5000	13.84 - j9.14	15.89 - j8.91	20.21 - j8.15	28.30 - j5.05
.5625	12.80 - j8.86	14.00 - j8.99	18.90 - j8.60	26.44 - j8.48
.6250	11.65 - j8.42	13.59 - j8.75	17.44 - j8.92	24.36 - j7.44
.6875	10.40 - j7.82	12.26 - j8.38	15.81 - j8.88	22.04 - j8.17
.7500	9.04 - j7.05	10.78 - j7.78	13.98 - j8.52	19.44 - j8.34
.8125	7.58 - j6.12	9.18 - j6.96	11.94 - j7.85	16.56 - j8.05
.8750	5.84 - j4.87	7.18 - j5.70	9.39 - j6.59	12.96 - j7.00
.9375	4.06 - j3.51	5.06 - j4.21	6.62 - j4.97	9.09 - j5.45
1.0000	0.0 + j0.0	0.0 + j0.0	0.0 + j0.0	0.0 + j0.0

TABLE 2.2

DISTRIBUTION OF CURRENT ALONG TUBULAR MONOPOLES

I = I" + jI' (milliamperes per volt); h/λ = 0.375

	b/a = 1.189		b/a = 1.125	
	a/λ (k_0a)		a/λ (k_0a)	
z/h	.0254 (.1596)	.0509 (.3198)	.1016 (.6384)	.2050 (1.288)
0	5.68 + j4.47	8.56 + j9.70	13.91 + j20.62	25.13 + j36.51
.0417	5.66 + j0.44	8.54 + j3.96	13.88 + j11.34	25.08 + j25.02
.0833	5.62 - j1.24	8.48 + j1.11	13.78 + j6.36	24.90 + j16.98
.1250	5.56 - j2.59	8.39 - j0.98	13.63 + j2.82	24.61 + j11.18
.1667	5.47 - j3.57	8.26 - j2.55	13.42 + j0.17	24.21 + j6.56
.2083	5.36 - j4.41	8.09 - j3.85	13.15 - j2.00	23.70 + j2.70
.2500	5.22 - j5.09	7.89 - j4.94	12.83 - j3.93	23.08 - j0.58
.2917	5.07 - j5.66	7.66 - j5.87	12.45 - j5.49	22.36 - j3.42
.3333	4.89 - j6.13	7.40 - j6.64	12.02 - j6.81	21.54 - j5.89
.3750	4.69 - j6.49	7.10 - j7.28	11.55 - j7.99	20.64 - j8.04
.4167	4.47 - j6.77	6.78 - j7.79	11.03 - j8.97	19.66 - j9.82
.4583	4.24 - j6.93	6.44 - j8.17	10.47 - j9.77	18.60 - j11.51
.5000	3.99 - j7.02	6.07 - j8.44	9.88 - j10.40	17.47 - j12.81
.5417	3.73 - j7.01	5.69 - j8.58	9.25 - j10.85	16.28 - j13.83
.5833	3.45 - j6.92	5.28 - j8.61	8.59 - j11.15	15.05 - j14.57
.6250	3.17 - j6.74	4.86 - j8.53	7.91 - j11.27	13.77 - j15.06
.6667	2.88 - j6.48	4.43 - j8.33	7.21 - j11.22	12.46 - j15.28
.7083	2.58 - j6.13	3.99 - j8.02	6.50 - j11.01	11.12 - j15.22
.7500	2.27 - j5.70	3.54 - j7.60	5.77 - j10.61	9.77 - j14.87
.7917	1.97 - j5.20	3.09 - j7.06	5.02 - j10.02	8.41 - j14.21
.8333	1.65 - j4.61	2.62 - j6.39	4.26 - j9.21	7.03 - j13.19
.8750	1.34 - j3.94	2.15 - j5.59	3.48 - j8.17	5.65 - j11.80
.9167	1.00 - j3.09	1.62 - j4.48	2.62 - j6.64	4.17 - j9.65
.9583	0.67 - j2.19	1.10 - j3.25	1.75 - j4.86	2.72 - j7.10
1.0000	0.0 + j0.0	0.0 + j0.0	0.0 + j0.0	0.0 + j0.0

TABLE 2.2

DISTRIBUTION OF CURRENT ALONG TUBULAR MONOPOLES

$I = I'' + jI'$ (milliamperes per volt); $h/\lambda = 0.5$, $b/a = 1.189$

	a/λ (k_0a)			a/λ (k_0a)	
z/h	.0254 (.1596)	.0509 (.3198)	z/h	.0254 (.1596)	.0509 (.3198)
0	4.09 + j9.44	6.77 + j15.79	.53125	2.09 - j7.00	3.25 - j9.20
.03125	4.08 + j5.39	6.75 + j10.01	.56250	1.91 - j7.13	2.93 - j9.49
.06250	4.05 + j3.62	6.71 + j7.07	.59375	1.72 - j7.19	2.61 - j9.67
.09375	4.01 + j2.16	6.63 + j4.83	.62500	1.54 - j7.18	2.29 - j9.76
.12500	3.95 + j1.00	6.53 + j3.06	.65625	1.36 - j7.09	1.98 - j9.74
.15625	3.88 - j0.05	6.40 + j1.48	.68750	1.19 - j6.92	1.68 - j9.62
.18750	3.79 - j0.99	6.24 + j0.07	.71875	1.02 - j6.69	1.40 - j9.40
.21875	3.69 - j1.87	6.06 - j1.23	.75000	0.86 - j6.38	1.13 - j9.08
.25000	3.57 - j2.67	5.85 - j2.43	.78125	0.71 - j6.00	0.89 - j8.65
.28125	3.44 - j3.42	5.62 - j3.53	.81250	0.57 - j5.56	0.66 - j8.12
.31250	3.30 - j4.10	5.37 - j4.55	.84375	0.45 - j5.04	0.45 - j7.49
.34375	3.15 - j4.71	5.10 - j5.49	.87500	0.33 - j4.45	0.28 - j6.73
.37500	2.99 - j5.26	4.82 - j6.33	.90625	0.23 - j3.79	0.13 - j5.84
.40625	2.82 - j5.75	4.52 - j7.09	.93750	0.14 - j2.96	0.02 - j4.66
.43750	2.64 - j6.17	4.21 - j7.76	.96875	0.07 - j2.09	-0.05 - j3.36
.46875	2.46 - j6.52	3.90 - j8.34	1.00000	0.0 - j0.0	0.0 - j0.0
.50000	2.28 - j6.79	3.58 - j8.82			

TABLE 2.2

DISTRIBUTION OF CURRENT ALONG TUBULAR MONOPOLES

I = I" + jI' (milliamperes per volt); h/λ = 0.5, b/a = 1.125

	a/λ (k_0a)			a/λ (k_0a)	
z/h	.1016 (.6384)	.2050 (1.288)	z/h	.1016 (.6384)	.2050 (1.288)
0	12.39 + j28.17	24.67 + j44.47	.53125	5.27 - j12.69	9.43 - j18.05
.03125	12.36 + j18.86	24.61 + j32.94	.56250	4.63 - j13.26	8.05 - j18.92
.06250	12.27 + j13.76	24.41 + j24.78	.59375	3.98 - j13.68	6.69 - j19.59
.09375	12.12 + j10.04	24.09 + j18.78	.62500	3.36 - j13.95	5.37 - j20.02
.12500	11.91 + j7.13	23.64 + j13.86	.65625	2.75 - j14.08	4.09 - j20.23
.15625	11.64 + j4.62	23.06 + j9.64	.68750	2.17 - j14.05	2.88 - j20.21
.18750	11.32 + j2.29	22.37 + j5.91	.71875	1.62 - j13.87	1.75 - j19.96
.21875	10.95 + j0.27	21.57 + j2.55	.75000	1.11 - j13.53	0.72 - j19.47
.25000	10.52 - j1.58	20.67 - j0.51	.78125	0.65 - j13.04	-0.21 - j18.73
.28125	10.06 - j3.35	19.66 - j3.32	.81250	0.24 - j12.37	-1.01 - j17.74
.31250	9.55 - j4.97	18.58 - j5.84	.84375	-0.12 - j11.52	-1.66 - j16.48
.34375	9.01 - j6.46	17.41 - j8.32	.87500	-0.39 - j10.45	-2.12 - j14.91
.37500	8.43 - j7.83	16.18 - j10.47	.90625	-0.59 - j9.15	-2.39 - j13.01
.40625	7.83 - j9.06	14.89 - j12.40	.93750	-0.66 - j7.36	-2.32 - j10.42
.43750	7.21 - j10.17	13.56 - j14.12	.96875	-0.62 - j5.32	-1.98 - j7.48
.46875	6.57 - j11.15	12.20 - j15.64	1.00000	0.0 - j0.0	0.0 - j0.0
.50000	5.92 - j11.99	10.82 - j16.95			

TABLE 2.2

DISTRIBUTION OF CURRENT ALONG TUBULAR MONOPOLES

I = I" + jI' (milliamperes per volt); h/λ = 0.625, b/a = 1.189

	a/λ (k_0a)			a/λ (k_0a)	
z/h	.0254 (.1596)	.0509 (.3198)	z/h	.0254 (.1596)	.0509 (.3198)
0	7.64 + j17.04	14.02 + j23.73	.525	-1.11 - j10.65	-2.68 - j13.17
.025	7.61 + j12.95	13.97 + j17.92	.550	-1.61 - j11.18	-3.63 - j13.80
.050	7.52 + j11.08	13.80 + j14.88	.575	-2.06 - j11.60	-4.53 - j14.30
.075	7.38 + j9.44	13.53 + j12.46	.600	-2.48 - j11.92	-5.35 - j14.68
.100	7.18 + j8.04	13.16 + j10.45	.625	-2.86 - j12.13	-6.10 - j14.93
.125	6.93 + j6.69	12.69 + j8.57	.650	-3.19 - j12.23	-6.75 - j15.05
.150	6.63 + j5.37	12.11 + j6.79	.675	-3.47 - j12.22	-7.32 - j15.04
.175	6.28 + j4.06	11.45 + j5.06	.700	-3.69 - j12.10	-7.77 - j14.90
.200	5.89 + j2.77	10.71 + j3.38	.725	-3.85 - j11.86	-8.12 - j14.63
.225	5.45 + j1.49	9.88 + j1.73	.750	-3.95 - j11.51	-8.36 - j14.23
.250	4.98 + j0.23	8.99 + j0.13	.775	-3.99 - j11.05	-8.48 - j13.71
.275	4.48 - j1.02	8.04 - j1.43	.800	-3.97 - j10.49	-8.47 - j13.06
.300	3.95 - j2.23	7.03 - j2.95	.825	-3.88 - j9.81	-8.33 - j12.28
.325	3.40 - j3.41	5.98 - j4.41	.850	-3.71 - j9.04	-8.05 - j11.39
.350	2.83 - j4.55	4.90 - j5.80	.875	-3.48 - j8.16	-7.63 - j10.37
.375	2.25 - j5.63	3.80 - j7.13	.900	-3.17 - j7.16	-7.03 - j9.20
.400	1.67 - j6.66	2.69 - j8.38	.925	-2.78 - j6.07	-6.26 - j7.90
.425	1.09 - j7.62	1.57 - j9.54	.950	-2.23 - j4.73	-5.12 - j6.23
.450	0.51 - j8.50	0.47 - j10.60	.975	-1.62 - j3.32	-3.78 - j4.43
.475	-0.05 - j9.31	-0.62 - j11.57	1.000	0.0 + j0.0	0.0 + j0.0
.500	-0.59 - j10.03	-1.67 - j12.43			

TABLE 2.2

DISTRIBUTION OF CURRENT ALONG TUBULAR MONOPOLES

I = I" + jI' (milliamperes per volt); h/λ = 0.625, b/a = 1.125

	a/λ (k_0a)			a/λ (k_0a)	
z/h	.1016 (.6384)	.2050 (1.288)	z/h	.1016 (.6384)	.2050 (1.288)
0	23.21 + j32.94	34.16 + j45.18	.525	-3.87 - j15.34	-1.79 - j19.66
.025	23.12 + j23.61	34.04 + j33.66	.550	-5.47 - j15.82	-3.99 - j19.87
.050	22.86 + j18.47	33.70 + j25.51	.575	-6.99 - j16.16	-6.09 - j19.91
.075	22.43 + j14.66	33.15 + j19.53	.600	-8.40 - j16.36	-8.06 - j19.78
.100	21.84 + j11.64	32.37 + j14.64	.625	-9.69 - j16.43	-9.89 - j19.48
.125	21.08 + j8.99	31.38 + j10.45	.650	-10.85 - j16.37	-11.55 - j19.03
.150	20.17 + j6.48	30.19 + j6.76	.675	-11.86 - j16.18	-13.02 - j18.43
.175	19.11 + j4.26	28.81 + j3.44	.700	-12.72 - j15.85	-14.29 - j17.70
.200	17.92 + j2.18	27.24 + j0.43	.725	-13.40 - j15.41	-15.35 - j16.84
.225	16.60 + j0.15	25.52 - j2.34	.750	-13.90 - j14.84	-16.16 - j15.86
.250	15.16 - j1.75	23.63 - j4.81	.775	-14.21 - j14.16	-16.73 - j14.78
.275	13.63 - j3.55	21.62 - j7.25	.800	-14.31 - j13.36	-17.02 - j13.61
.300	12.01 - j5.25	19.48 - j9.38	.825	-14.20 - j12.46	-17.03 - j12.35
.325	10.32 - j6.84	17.25 - j11.28	.850	-13.84 - j11.44	-16.71 - j11.02
.350	8.57 - j8.33	14.93 - j12.99	.875	-13.22 - j10.32	-16.06 - j9.63
.375	6.78 - j9.70	12.56 - j14.51	.900	-12.28 - j9.07	-14.98 - j8.18
.400	4.97 - j10.96	10.14 - j15.84	.925	-11.01 - j7.69	-13.47 - j6.67
.425	3.15 - j12.09	7.71 - j16.98	.950	-9.04 - j5.99	-11.09 - j4.99
.450	1.34 - j13.10	5.28 - j17.93	.975	-6.69 - j4.18	-8.20 - j3.31
.475	-0.44 - j13.98	2.87 - j18.70	1.000	0.0 + j0.0	0.0 + j0.0
.500	-2.19 - j14.73	0.51 - j19.27			

TABLE 2.2

DISTRIBUTION OF CURRENT ALONG TUBULAR MONOPOLES

$I = I'' + jI'$ (milliamperes per volt); $h/\lambda = 0.75$, $b/a = 1.189$

	a/λ (k_0a)			a/λ (k_0a)	
z/h	.0254 (.1596)	.0509 (.3198)	z/h	.0254 (.1596)	.0509 (.3198)
0	14.61 + j2.05	16.71 + j8.01	.5208	-7.87 + j0.55	-7.42 - j1.57
.0208	14.55 - j1.96	16.65 + j2.28	.5417	-8.74 + j1.33	-8.40 - j0.72
.0417	14.38 - j3.60	16.47 - j0.53	.5625	-9.54 + j2.09	-9.30 + j0.13
.0625	14.10 - j4.86	16.17 - j2.56	.5833	-10.24 + j2.82	-10.11 + j0.96
.0833	13.70 - j5.74	15.75 - j4.03	.6042	-10.84 + j3.51	-10.81 + j1.77
.1042	13.20 - j6.44	15.22 - j5.23	.6250	-11.35 + j4.15	-11.42 + j2.55
.1250	12.59 - j6.97	14.58 - j6.19	.6458	-11.75 + j4.74	-11.92 + j3.29
.1458	11.89 - j7.36	13.84 - j6.97	.6667	-12.04 + j5.27	-12.30 + j3.97
.1667	11.10 - j7.62	12.99 - j7.58	.6875	-12.22 + j5.73	-12.57 + j4.60
.1875	10.22 - j7.76	12.06 - j8.04	.7083	-12.29 + j6.12	-12.73 + j5.16
.2083	9.26 - j7.79	11.05 - j8.36	.7292	-12.24 + j6.43	-12.77 + j5.64
.2292	8.23 - j7.71	9.96 - j8.55	.7500	-12.09 + j6.66	-12.70 + j6.05
.2500	7.15 - j7.53	8.81 - j8.62	.7708	-11.82 + j6.80	-12.50 + j6.37
.2708	6.01 - j7.27	7.60 - j8.57	.7917	-11.44 + j6.84	-12.19 + j6.60
.2917	4.84 - j6.91	6.34 - j8.41	.8125	-10.96 + j6.80	-11.77 + j6.73
.3125	3.63 - j6.48	5.05 - j8.14	.8333	-10.37 + j6.66	-11.24 + j6.75
.3333	2.40 - j5.97	3.74 - j7.78	.8542	-9.68 + j6.42	-10.59 + j6.67
.3542	1.16 - j5.39	2.41 - j7.33	.8750	-8.88 + j6.08	-9.84 + j6.48
.3750	-0.08 - j4.76	1.08 - j6.80	.8958	-7.99 + j5.63	-8.97 + j6.16
.3958	-1.30 - j4.08	-0.24 - j6.20	.9167	-7.00 + j5.08	-7.97 + j5.70
.4167	-2.51 - j3.36	-1.55 - j5.53	.9375	-5.91 + j4.41	-6.85 + j5.10
.4375	-3.68 - j2.61	-2.82 - j4.81	.9583	-4.59 + j3.52	-5.41 + j4.18
.4583	-4.81 - j1.83	-4.06 - j4.04	.9792	-3.21 + j2.54	-3.85 + j3.09
.4792	-5.89 - j1.04	-5.24 - j3.24	1.0000	0.0 + j0.0	0.0 + j0.0
.5000	-6.92 - j0.24	-6.37 - j2.41			

TABLE 2.2

DISTRIBUTION OF CURRENT ALONG TUBULAR MONOPOLES

$I = I'' + jI'$ (milliamperes per volt); $h/\lambda = 0.75$, $b/a = 1.125$

	a/λ (k_0a)			a/λ (k_0a)	
z/h	.1016 (.6384)	.2050 (1.288)	z/h	.1016 (.6384)	.2050 (1.288)
0	21.48 + j19.76	31.91 + j36.57	.5208	-7.22 - j5.19	-7.44 - j12.15
.0208	21.41 + j10.49	31.81 + j25.09	.5417	-8.43 - j4.24	-9.15 - j11.08
.0417	21.19 + j5.54	31.52 + j17.06	.5625	-9.56 - j3.27	-10.75 - j9.93
.0625	20.84 + j2.05	31.05 + j11.28	.5833	-10.58 - j2.28	-12.21 - j8.71
.0833	20.35 - j0.53	30.39 + j6.67	.6042	-11.49 - j1.29	-13.53 - j7.45
.1042	19.73 - j2.63	29.55 + j2.84	.6250	-12.30 - j0.30	-14.69 - j6.14
.1250	18.98 - j4.46	28.53 - j0.42	.6458	-12.97 + j0.66	-15.70 - j4.82
.1458	18.11 - j5.90	27.35 - j3.24	.6667	-13.53 + j1.59	-16.54 - j3.50
.1667	17.13 - j7.11	26.01 - j5.70	.6875	-13.95 + j2.48	-17.21 - j2.19
.1875	16.03 - j8.16	24.53 - j7.84	.7083	-14.25 + j3.32	-17.70 - j0.91
.2083	14.84 - j9.01	22.91 - j9.64	.7292	-14.41 + j4.09	-18.01 + j0.32
.2292	13.57 - j9.68	21.17 - j11.35	.7500	-14.43 + j4.78	-18.13 + j1.48
.2500	12.21 - j10.18	19.32 - j12.71	.7708	-14.32 + j5.39	-18.08 + j2.56
.2708	10.78 - j10.52	17.38 - j13.81	.7917	-14.08 + j5.91	-17.83 + j3.54
.2917	9.30 - j10.72	15.36 - j14.68	.8125	-13.69 + j6.32	-17.40 + j4.41
.3125	7.78 - j10.76	13.28 - j15.34	.8333	-13.17 + j6.61	-16.77 + j5.14
.3333	6.22 - j10.68	11.15 - j15.80	.8542	-12.52 + j6.78	-15.96 + j5.72
.3542	4.65 - j10.46	8.99 - j16.06	.8750	-11.72 + j6.80	-14.95 + j6.12
.3750	3.06 - j10.12	6.82 - j16.13	.8958	-10.77 + j6.67	-13.74 + j6.32
.3958	1.48 - j9.68	4.64 - j16.02	.9167	-9.64 + j6.34	-12.29 + j6.28
.4167	-0.08 - j9.13	2.49 - j15.74	.9375	-8.33 + j5.81	-10.60 + j5.96
.4375	-1.61 - j8.48	0.37 - j15.30	.9583	-6.62 + j4.86	-8.40 + j5.14
.4583	-3.10 - j7.76	-1.69 - j14.70	.9792	-4.72 + j3.67	-5.96 + j3.98
.4792	-4.54 - j6.96	-3.69 - j13.97	1.0000	0.0 + j0.0	0.0 + j0.0
.5000	-5.91 - j6.10	-5.61 - j13.12			

TABLE 2.3

ADMITTANCE OF A HEMISPHERICALLY CAPPED MONOPOLE

b/a = 1.189; AXIAL LENGTH IS h + a

Y = G + jB (millimhos)

		a/λ (k_0a)			
h/λ	k_0h	.0064 (.0402)	.0127 (.0798)	.0190 (.1194)	.0254 (.1596)
.03125	0.196	0.00 + j2.81	0.02 + j5.58	0.05 + j7.74	0.12 + j10.02
.06250	0.393	0.02 + j4.47	0.12 + j8.00	0.26 + j10.78	0.50 + j13.64
.09375	0.589	0.10 + j6.18	0.45 + j10.61	0.92 + j14.12	1.67 + j17.76
.12500	0.785	0.36 + j8.33	1.54 + j14.25	3.08 + j18.86	5.49 + j23.43
.15625	0.982	1.27 + j11.56	5.65 + j19.90	10.93 + j24.90	17.77 + j27.59
.18750	1.178	5.09 + j17.09	21.10 + j22.17	28.60 + j17.66	30.77 + j12.22
.21875	1.374	22.38 + j18.52	26.27 + j0.06	23.29 - j1.04	21.33 - j0.51
.25000	1.571	21.36 - j5.65	13.80 - j4.93	13.43 - j2.85	13.32 - j0.96
.28125	1.767	9.66 - j6.27	8.25 - j2.98	8.84 - j1.05	9.34 + j0.74
.31250	1.963	5.65 - j3.73	5.80 - j1.00	6.56 + j0.77	7.21 + j2.46
.34375	2.160	3.97 - j1.85	4.52 + j0.58	5.29 + j2.30	5.96 + j3.97
.37500	2.356	3.10 - j0.46	3.77 + j1.86	4.50 + j3.60	5.16 + j5.30
.40625	2.552	2.59 + j0.64	3.28 + j2.96	3.99 + j4.78	4.63 + j6.54
.43750	2.749	2.27 + j1.58	2.96 + j3.98	3.65 + j5.89	4.28 + j7.74
.46875	2.945	2.05 + j2.44	2.76 + j4.98	3.44 + j7.01	4.09 + j8.98
.50000	3.142	1.91 + j3.28	2.66 + j6.02	3.36 + j8.21	4.06 + j10.34
.53125	3.338	1.85 + j4.17	2.67 + j7.18	3.46 + j9.59	4.28 + j11.91
.56250	3.534	1.87 + j5.17	2.88 + j8.58	3.86 + j11.26	4.94 + j13.84
.59375	3.731	2.04 + j6.40	3.49 + j10.37	4.91 + j13.39	6.59 + j16.21
.62500	3.927	2.54 + j8.04	5.08 + j12.77	7.58 + j15.97	10.49 + j18.54
.65625	4.123	3.93 + j10.37	9.42 + j15.33	13.83 + j17.14	17.74 + j17.34
.68750	4.320	8.19 + j13.24	18.14 + j12.64	20.73 + j10.29	21.28 + j8.60
.71875	4.516	18.09 + j9.89	18.50 + j1.46	17.29 + j1.62	16.50 + j2.55
.75000	4.712	15.93 - j2.49	11.80 - j1.89	11.70 + j0.05	11.79 + j1.86

TABLE 2.3

ADMITTANCE OF A HEMISPHERICALLY CAPPED MONOPOLE

$b/a = 1.189$; AXIAL LENGTH IS $h + a$

$Y = G + jB$ (millimhos)

		a/λ (k_0a)			
h/λ	k_0h	.0318 (.1998)	.0391 (.2457)	.0426 (.2677)	.0509 (.3198)
.03125	0.196	0.23 + j12.40	0.45 + j15.32	0.60 + j16.80	1.13 + j20.59
.06250	0.393	0.87 + j16.60	1.53 + j20.15	1.97 + j21.91	3.42 + j26.27
.09375	0.589	2.81 + j21.50	4.75 + j25.86	5.98 + j27.92	9.80 + j32.58
.12500	0.785	8.96 + j27.63	14.23 + j31.40	17.12 + j32.59	24.27 + j33.42
.15625	0.982	24.56 + j27.25	30.23 + j24.09	31.88 + j22.14	33.59 + j17.90
.18750	1.178	30.15 + j8.67	28.56 + j6.73	27.81 + j6.34	26.27 + j6.22
.21875	1.374	20.10 + j5.36	19.22 + j1.93	18.95 + j2.62	18.52 + j4.24
.25000	1.571	13.35 + j0.76	13.48 + j2.58	13.58 + j3.40	13.84 + j5.25
.28125	1.767	9.80 + j2.40	10.29 + j4.18	10.52 + j4.99	11.06 + j6.83
.31250	1.963	7.79 + j4.06	8.40 + j5.80	8.68 + j6.60	9.32 + j8.43
.34375	2.160	6.56 + j5.56	7.20 + j7.30	7.49 + j8.11	8.17 + j9.97
.37500	2.356	5.76 + j6.92	6.41 + j8.71	6.71 + j9.54	7.41 + j11.46
.40625	2.552	5.23 + j8.22	5.88 + j10.08	6.19 + j10.95	6.92 + j12.96
.43750	2.749	4.89 + j9.51	5.56 + j11.48	5.89 + j12.39	6.67 + j14.53
.46875	2.945	4.72 + j10.87	5.45 + j12.98	5.81 + j13.95	6.89 + j16.25
.50000	3.142	4.77 + j12.38	5.61 + j14.66	6.04 + j15.72	7.11 + j18.20
.53125	3.338	5.14 + j14.16	6.22 + j16.64	6.78 + j17.79	8.25 + j20.46
.56250	3.534	6.15 + j16.30	7.72 + j18.97	8.55 + j20.18	10.74 + j22.83
.59375	3.731	8.51 + j18.77	11.02 + j20.28	12.31 + j22.29	15.50 + j24.14
.62500	3.927	13.61 + j20.32	17.10 + j21.35	18.62 + j21.49	21.63 + j21.21
.65625	4.123	20.56 + j16.54	22.41 + j15.30	22.89 + j14.78	23.40 + j13.93
.68750	4.320	21.02 + j7.88	20.51 + j7.88	20.28 + j8.09	19.85 + j8.91
.71875	4.516	16.05 + j3.75	15.80 + j5.23	15.76 + j5.95	15.78 + j7.64
.75000	4.712	11.98 + j3.55	12.28 + j5.37	12.45 + j6.19	12.88 + j8.07

TABLE 2.3

ADMITTANCE OF A HEMISPHERICALLY CAPPED MONOPOLE

$b/a = 1.189$; AXIAL LENGTH IS $h + a$

$Y = G + jB$ (millimhos)

		a/λ (k_0a)			
h/λ	k_0h	.0565 (.3550)	.0635 (.3990)	.0726 (.4562)	.0847 (.5322)
.03125	0.196	1.68 + j23.35	2.65 + j27.07	4.55 + j32.32	8.71 + j39.90
.06250	0.393	4.80 + j29.31	7.12 + j33.14	11.28 + j37.91	18.97 + j42.91
.09375	0.589	13.12 + j35.28	18.00 + j37.79	24.98 + j39.10	33.59 + j37.25
.12500	0.785	28.61 + j32.45	32.86 + j30.04	36.13 + j26.16	37.52 + j21.60
.15625	0.982	33.60 + j15.72	33.02 + j13.85	31.93 + j12.56	30.51 + j12.10
.18750	1.178	25.46 + j6.53	24.68 + j7.17	23.97 + j8.23	23.38 + j9.82
.21875	1.374	18.37 + j5.32	18.29 + j6.62	18.33 + j8.28	18.52 + j10.40
.25000	1.571	14.05 + j6.42	14.34 + j7.84	18.76 + j9.60	15.35 + j11.84
.28125	1.767	11.42 + j8.01	11.87 + j9.43	12.46 + j11.22	13.25 + j13.51
.31250	1.963	9.74 + j9.62	10.26 + j11.06	10.93 + j12.88	11.83 + j15.23
.34375	2.160	8.62 + j11.18	9.18 + j12.66	9.90 + j14.54	10.89 + j16.97
.37500	2.356	7.88 + j12.72	8.46 + j14.26	9.24 + j16.22	10.31 + j18.78
.40625	2.552	7.41 + j14.28	8.04 + j15.90	8.89 + j17.98	10.10 + j20.68
.43750	2.749	7.21 + j15.93	7.91 + j17.66	8.88 + j19.88	10.31 + j22.75
.46875	2.945	7.32 + j17.76	8.15 + j19.61	9.34 + j21.98	11.15 + j25.01
.50000	3.142	7.91 + j19.83	8.99 + j21.81	10.56 + j24.29	12.99 + j27.33
.53125	3.338	9.35 + j22.17	10.87 + j24.19	13.08 + j26.56	16.38 + j29.12
.56250	3.534	12.37 + j24.41	14.55 + j26.08	17.50 + j27.69	21.31 + j28.80
.59375	3.731	17.65 + j24.92	20.16 + j25.40	22.92 + j25.39	25.55 + j24.77
.62500	3.927	23.12 + j20.73	24.40 + j20.08	25.34 + j19.40	25.82 + j19.01
.65625	4.123	23.43 + j13.70	23.34 + j13.73	23.15 + j14.18	22.96 + j15.23
.68750	4.320	19.66 + j9.63	19.52 + j10.63	19.49 + j12.03	19.63 + j13.95
.71875	4.516	15.88 + j8.76	16.07 + j10.13	16.41 + j11.87	16.95 + j14.10
.75000	4.712	13.20 + j9.28	13.62 + j10.74	14.20 + j12.57	15.00 + j14.89

c. ELECTRICALLY THIN ANTENNAS

An electrically thin antenna is defined to be one with radius a sufficiently small to satisfy the inequality

$$k_0 a \leq 0.06 \qquad \text{or} \qquad a/\lambda \leq 0.01 \qquad (2.12)$$

Thin monopole antennas are commonly driven from coaxial lines with characteristic impedances Z_c with magnitudes between 25 and 100 ohms; that is, b/a ranges between about 1.5 and 5.3, so that

$$k_0 b \leq 0.32 \qquad \text{or} \qquad b/\lambda \leq 0.53 \qquad (2.13)$$

With a/λ as small as required by (2.12) the difference between the accurate apparent admittance and that calculated using the TEM approximation is usually smaller than experimental accuracy, so that the latter is quite adequate. The admittance of electrically thin tubular monopoles is given in Table 2.4 for four values of b/a corresponding to the characteristic impedances $Z_c = 25$, 50, 75, and 100. Distributions of current for selected lengths and thicknesses are given in Table 2.5.

When electrically thin antennas are driven from transmission lines with common characteristic impedances, the junction effects are highly localized so that account may be taken of them in terms of lumped reactive networks connected between an idealized antenna and an idealized transmission line.[2,6] The idealized antenna is one in which the admittance is determined entirely by currents on the cylindrical surface with the effects of feeding transmission lines or generators eliminated. The idealized transmission line is one with uniform parameters all the way to the end. In Fig. 2.4 are shown typical curves of the apparent susceptance of electrically thin tubular monopoles as a function of the ratio b/a of the feeding coaxial line. The length of the antennas is fixed at $h/\lambda = 0.5$. It is seen that the susceptance increases only slowly with decreasing b/a until this reaches approximately 1.5, when it begins to rise steeply to infinity at $b/a = 1$. The initial increase in the susceptance as b/a decreases from large values is the slower the thinner the antenna. With $a/\lambda = 0.001588$, the susceptance curve is practically linear with a very small slope as b/a decreases from large values to 1.5. It is quite easy to extrapolate the curve back to $b/a = 1$, and it is then reasonable to call the extrapolated value of B_a at $b/a = 1$ the independent susceptance of the isolated antenna. The difference between this extrapolated curve and the actual one is then assigned to the susceptance of end and junction effects. This ultimately becomes the infinite susceptance of adjacent knife edges in the actual limit $b/a \to 1$. Evidently this procedure is less and less satisfactory as a/λ increases.

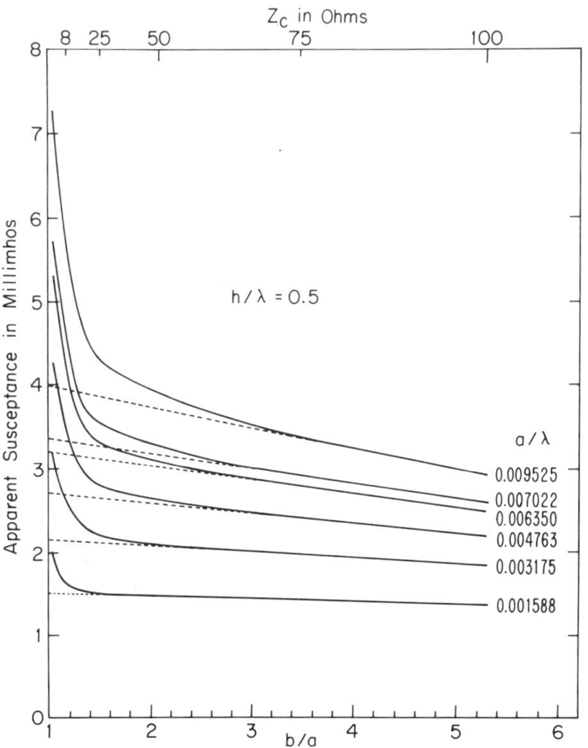

Fig. 2.4. Apparent susceptance of tubular monopole driven from coaxial line.

Although no rigorous analysis is available for antennas driven from open-wire lines, it is readily shown experimentally that for thin antennas the behavior of the apparent susceptance is similar to that shown in Fig. 2.4 when $b/2a$ is varied. When $b/2a$ is considerably greater than one, the susceptance varies only slowly and almost linearly, so that an extrapolation to $b/2a = 1$ can be made in order to define a hypothetical "ideal" susceptance. The value of susceptance so obtained is, of course, physically unavailable. Nevertheless, it is a convenient quantity to use in conjunction with a suitable lumped-constant network (which takes account of coupling and end effects) to determine the measurable apparent admittance of the antenna as a termination for the line. The theoretically determined admittances of electrically thin dipoles given in Table 2.6 have been normalized at one point (location of antiresonance) in order to provide quantities that conform to measured admittances extrapolated in the manner shown in Fig. 2.4. This table also provides the impedance $Z = Y^{-1}$, the electrical effective half-length $k_0 h_e = k_0 h_e(\pi/2)$, and the absolute directivity D. The effective half-length occurs in the formula

$$V = -2h_e E^{\text{inc}} \cos \psi \tag{2.14}$$

where V is the induced voltage in the equivalent circuit of a center-loaded receiving dipole in an electric field E^{inc} that is normally incident on the antenna. The polarization angle ψ is zero when E^{inc} is parallel to the axis of the antenna.

Note that the first part of Table 2.6 applies to antennas with assigned values of a/λ so that the thickness parameter $\Omega = 2\ln(2h/a)$ varies as k_0h is changed. These tables are useful when the length of an antenna with fixed radius is changed. In the latter part of Table 2.6 the constant parameter is Ω. It is convenient when the frequency is varied with both the length $2h$ and the radius a fixed.

The integral equation for the current in a thin antenna when center-driven by a delta-function generator, which maintains the electric field $E_z(\rho = a) = -V_0^e\,\delta(z)$ on the surface of the perfectly conducting cylinder, is[11]

$$\int_{-h}^{h} I(z')K(z, z')\,dz' = -j\frac{4\pi}{\zeta_0}[C_1 \cos k_0 z + \tfrac{1}{2}V_0^e \sin k_0|z|] \tag{2.15}$$

where C_1 is a constant to be determined from $I(h) = 0$ and

$$K(z, z') = \frac{1}{2\pi}\int_0^{2\pi} \frac{e^{-jk_0R(\theta)}}{R(\theta)}\,d\theta \doteq \frac{e^{-jk_0R}}{R} \tag{2.16}$$

In (2.16)

$$R(\theta) = \sqrt{(z - z')^2 + \left(2a \sin\frac{\theta}{2}\right)^2}$$
$$R = \sqrt{(z - z')^2 + a^2} \tag{2.17}$$

The exact kernel in (2.16) applies to a perfectly conducting tubular antenna. For electrically thin antennas the approximate kernel on the right is an excellent approximation.[5]

An exact solution of (2.15) is unavailable for finite h. Approximate solutions have been obtained by a variety of different methods. In each case the singularity in the current at $z = 0$ has been removed in one way or another so that the approximate current evaluated consists essentially of the total current in the tube minus the localized current charging the knife edges. The ideal admittance of the antenna is

$$Y = I(0)/V_0^e = Z^{-1} \tag{2.18}$$

where $I(0)$ is the current associated with the antenna at $z = 0$ and Z is the ideal impedance of the antenna. In the evaluation of Table 2.6, the equation (2.15) was solved by iteration using the King–Middleton procedure. Only two iterations and an empirical normalization were used. The actual formulas are given in the literature.[5]

TABLE 2.4

ADMITTANCE OF THIN TUBULAR MONOPOLES

$Y = G + jB$ (millimhos)

$a/\lambda = 0.001588$			B			
$k_0a = 0.00998$			b/a (Z_c in ohms)			
h/λ	k_0h	G	1.517 (25)	2.301 (50)	3.49 (75)	5.30 (100)
.03125	0.196	0.00	1.55	1.51	1.47	1.42
.06250	0.393	0.01	2.61	2.58	2.53	2.48
.09375	0.589	0.05	3.75	3.72	3.68	3.62
.12500	0.785	0.17	5.24	5.20	5.16	5.11
.15625	0.982	0.62	7.55	7.52	7.47	7.42
.18750	1.178	2.65	12.00	11.96	11.92	11.87
.21875	1.374	17.05	17.94	17.90	17.86	17.81
.25000	1.571	17.96	-9.34	-9.38	-9.42	-9.47
.28125	1.767	6.18	-7.10	-7.14	-7.18	-7.23
.31250	1.963	3.33	-4.26	-4.29	-4.33	-4.39
.34375	2.160	2.27	-2.54	-2.59	-2.62	-2.67
.37500	2.356	1.75	-1.39	-1.42	-1.46	-1.52
.40625	2.552	1.45	-0.51	-0.55	-0.59	-0.64
.43750	2.749	1.26	0.21	0.18	0.14	0.08
.46875	2.945	1.13	0.87	0.83	0.79	0.74
.50000	3.142	1.05	1.50	1.47	1.42	1.37
.53125	3.338	1.01	2.16	2.13	2.09	2.03
.56250	3.534	1.02	2.91	2.87	2.83	2.78
.59375	3.731	1.10	3.83	3.79	3.75	3.70
.62500	3.927	1.36	5.07	5.04	4.99	4.94
.65625	4.123	2.12	6.97	6.94	6.89	6.84
.68750	4.320	4.84	10.08	10.04	10.00	9.95
.71875	4.516	15.48	10.09	10.05	10.01	9.96
.75000	4.712	14.09	-5.49	-5.52	-5.57	-5.62

TABLE 2.4

ADMITTANCE OF THIN TUBULAR MONOPOLES

$Y = G + jB$ (millimhos)

$a/\lambda = 0.003175$			B			
$k_0a = 0.01995$			b/a (Z_c in ohms)			
h/λ	k_0h	G	1.517 (25)	2.301 (50)	3.49 (75)	5.30 (100)
.03125	0.196	0.00	2.09	1.99	1.89	1.77
.06250	0.393	0.02	3.41	3.31	3.20	3.08
.09375	0.589	0.07	4.80	4.71	4.60	4.47
.12500	0.785	0.27	6.63	6.53	6.42	6.29
.15625	0.982	0.98	9.48	9.38	9.27	9.15
.18750	1.178	4.23	14.80	14.71	14.60	14.47
.21875	1.374	22.60	16.64	16.54	16.43	16.30
.25000	1.571	17.58	-8.56	-8.66	-8.77	-8.90
.28125	1.767	7.11	-6.69	-6.79	-6.90	-7.02
.31250	1.963	4.08	-4.02	-4.12	-4.23	-4.36
.34375	2.160	2.86	-2.28	-2.37	-2.48	-2.61
.37500	2.356	2.23	-1.05	-1.14	-1.25	-1.38
.40625	2.552	1.87	-0.09	-0.19	-0.30	-0.42
.43750	2.749	1.63	0.72	0.62	0.51	0.39
.46875	2.945	1.48	1.46	1.36	1.25	1.13
.50000	3.142	1.38	2.19	2.09	1.98	1.85
.53125	3.338	1.33	2.95	2.85	2.74	2.62
.56250	3.534	1.35	3.82	3.72	3.61	3.49
.59375	3.731	1.48	4.90	4.80	4.69	4.56
.62500	3.927	1.87	6.37	6.27	6.16	6.03
.65625	4.123	3.00	8.58	8.48	8.37	8.24
.68750	4.320	6.89	11.74	11.64	11.53	11.40
.71875	4.516	17.91	8.49	8.39	8.28	8.15
.75000	4.712	13.89	-4.87	-4.96	-5.07	-5.20

TABLE 2.4

ADMITTANCE OF THIN TUBULAR MONOPOLES

$Y = G + jB$ (millimhos)

$a/\lambda = 0.004763$			B			
$k_0a = 0.02993$			b/a (Z_c in ohms)			
h/λ	k_0h	G	1.517 (25)	2.301 (50)	3.49 (75)	5.30 (100)
.03125	0.196	0.00	2.55	2.38	2.20	2.02
.06250	0.393	0.02	4.08	3.90	3.71	3.52
.09375	0.589	0.10	5.68	5.50	5.31	5.11
.12500	0.785	0.38	7.77	7.59	7.41	7.20
.15625	0.982	1.35	11.06	10.88	10.70	10.49
.18750	1.178	5.80	16.94	16.76	16.57	16.36
.21875	1.374	25.76	14.55	14.37	14.18	13.97
.25000	1.571	17.37	-7.90	-8.08	-8.27	-8.48
.28125	1.767	7.74	-6.24	-6.42	-6.61	-6.81
.31250	1.963	4.63	-3.71	-3.89	-4.08	-4.28
.34375	2.160	3.31	-1.96	-2.14	-2.33	-2.54
.37500	2.356	2.61	-0.69	-0.87	-1.06	-1.26
.40625	2.552	2.20	0.32	0.14	-0.05	-0.25
.43750	2.749	1.93	1.18	1.01	0.82	0.61
.46875	2.945	1.76	1.98	1.81	1.62	1.41
.50000	3.142	1.65	2.78	2.60	2.41	2.20
.53125	3.338	1.60	3.62	3.44	3.25	3.05
.56250	3.534	1.63	4.58	4.40	4.22	4.01
.59375	3.731	1.81	5.78	5.61	5.42	5.21
.62500	3.927	2.32	7.42	7.24	7.06	6.85
.65625	4.123	3.80	9.83	9.65	9.46	9.26
.68750	4.320	8.64	12.77	12.60	12.41	12.20
.71875	4.516	19.03	7.22	7.04	6.85	6.65
.75000	4.712	13.79	-4.32	-4.50	-4.68	-4.89

TABLE 2.4

ADMITTANCE OF THIN TUBULAR MONOPOLES

$Y = G + jB$ (millimhos)

$a/\lambda = 0.00635$			B			
$k_0a = 0.03990$			b/a (Z_c in ohms)			
h/λ	k_0h	G	1.517 (25)	2.301 (50)	3.49 (75)	5.30 (100)
.03125	0.196	0.00	2.97	2.71	2.46	2.21
.06250	0.393	0.03	4.68	4.41	4.14	3.88
.09375	0.589	0.14	6.46	6.19	5.92	5.65
.12500	0.785	0.49	8.80	8.53	8.26	7.98
.15625	0.982	1.75	12.47	12.20	11.93	11.65
.18750	1.178	7.42	18.66	18.39	18.12	17.83
.21875	1.374	27.56	12.50	12.23	11.96	11.66
.25000	1.571	17.25	-7.31	-7.58	-7.86	-8.15
.28125	1.767	8.23	-5.80	-6.07	-6.34	-6.63
.31250	1.963	5.08	-3.39	-3.66	-3.93	-4.21
.34375	2.160	3.69	-1.64	-1.91	-2.19	-2.47
.37500	2.356	2.94	-0.34	-0.61	-0.88	-1.17
.40625	2.552	2.48	0.71	0.44	0.16	-0.12
.43750	2.749	2.19	1.62	1.35	1.08	0.79
.46875	2.945	2.00	2.47	2.20	1.92	1.64
.50000	3.142	1.88	3.31	3.04	2.77	2.49
.53125	3.338	1.84	4.22	3.95	3.68	3.40
.56250	3.534	1.88	5.27	5.00	4.72	4.44
.59375	3.731	2.11	6.57	6.30	6.03	5.75
.62500	3.927	2.74	8.35	8.08	7.81	7.52
.65625	4.123	4.56	10.90	10.63	10.35	10.07
.68750	4.320	10.19	13.46	13.19	12.92	12.63
.71875	4.516	19.59	6.27	6.00	5.73	5.44
.75000	4.712	13.74	-3.81	-4.08	-4.35	-4.64

TABLE 2.4

ADMITTANCE OF THIN TUBULAR MONOPOLES

Y = G + jB (millimhos)

a/λ = 0.007022 k_0a = 0.04412			B			
			b/a (Z_c in ohms)			
h/λ	k_0h	G	1.517 (25)	2.301 (50)	3.49 (75)	5.30 (100)
.03125	0.196	0.00	3.13	2.84	2.55	2.28
.06250	0.393	0.04	4.92	4.61	4.31	4.01
.09375	0.589	0.15	6.77	6.46	6.16	5.85
.12500	0.785	0.54	9.21	8.90	8.59	8.28
.15625	0.982	1.92	13.03	12.72	12.41	12.10
.18750	1.178	8.11	19.29	18.98	18.67	18.35
.21875	1.374	28.06	11.72	11.40	11.09	10.75
.25000	1.571	17.21	-7.08	-7.39	-7.70	-8.03
.28125	1.767	8.41	-5.61	-5.93	-6.24	-6.56
.31250	1.963	5.26	-3.25	-3.56	-3.87	-4.18
.34375	2.160	3.83	-1.51	-1.82	-2.13	-2.44
.37500	2.356	3.06	-0.20	-0.51	-0.82	-1.13
.40625	2.552	2.60	0.87	0.55	0.25	-0.07
.43750	2.749	2.30	1.79	1.48	1.17	0.86
.46875	2.945	2.10	2.66	2.35	2.04	1.73
.50000	3.142	1.97	3.53	3.22	2.91	2.59
.53125	3.338	1.93	4.46	4.15	3.84	3.53
.56250	3.534	1.98	5.54	5.23	4.92	4.60
.59375	3.731	2.23	6.88	6.57	6.27	5.95
.62500	3.927	2.92	8.71	8.40	8.10	7.78
.65625	4.123	4.87	11.30	10.99	10.69	10.37
.68750	4.320	10.80	13.67	13.36	13.05	12.73
.71875	4.516	19.74	5.95	5.64	5.33	5.01
.75000	4.712	13.74	-3.61	-3.92	-4.23	-4.55

TABLE 2.4

ADMITTANCE OF THIN TUBULAR MONOPOLES

$Y = G + jB$ (millimhos)

$a/\lambda = 0.009525$			B			
$k_0a = 0.05985$			b/a (Z_c in ohms)			
h/λ	k_0h	G	1.517 (25)	2.301 (50)	3.49 (75)	5.30 (100)
.03125	0.196	0.01	3.72	3.27	2.87	2.52
.06250	0.393	0.05	5.76	5.29	4.87	4.47
.09375	0.589	0.21	7.87	7.40	6.96	6.55
.12500	0.785	0.74	10.65	10.17	9.74	9.32
.15625	0.982	2.62	14.97	14.50	14.06	13.63
.18750	1.178	10.70	21.21	20.73	20.29	19.84
.21875	1.374	29.05	9.27	8.79	8.34	7.86
.25000	1.571	17.16	-6.26	-6.74	-7.19	-7.64
.28125	1.767	9.00	-4.96	-5.44	-5.88	-6.33
.31250	1.963	5.83	-2.73	-3.21	-3.65	-4.09
.34375	2.160	4.33	-1.01	-1.48	-1.93	-2.37
.37500	2.356	3.50	0.33	-0.15	-0.59	-1.03
.40625	2.552	2.99	1.44	0.96	0.52	0.08
.43750	2.749	2.65	2.42	1.94	1.50	1.06
.46875	2.945	2.43	3.35	2.87	2.43	1.99
.50000	3.142	2.30	4.29	3.81	3.37	2.93
.53125	3.338	2.26	5.30	4.83	4.39	3.95
.56250	3.534	2.35	6.48	6.01	5.57	5.13
.59375	3.731	2.67	7.97	7.49	7.05	6.61
.62500	3.927	3.56	9.97	9.49	9.05	8.61
.65625	4.123	6.02	12.66	12.18	11.74	11.30
.68750	4.320	12.85	14.17	13.70	13.26	12.81
.71875	4.516	20.01	5.07	4.59	4.15	3.70
.75000	4.712	13.76	-2.89	-3.36	-3.81	-4.25

TABLE 2.5

DISTRIBUTION OF CURRENT ALONG THIN TUBULAR MONOPOLES

$I(z) = I''(z) + jI'(z)$ in milliamperes per volt; $b/a = 2.301$, $Z_c = 50$ ohms

	$h/\lambda = 0.125$	
z/h	$a/\lambda = 0.001588$	$a/\lambda = 0.007022$
0	0.17 + j5.20	0.54 + j8.90
.125	0.17 + j4.39	0.53 + j7.52
.250	0.16 + j3.83	0.51 + j6.54
.375	0.15 + j3.27	0.48 + j5.63
.500	0.13 + j2.70	0.43 + j4.73
.625	0.11 + j2.12	0.37 + j3.80
.750	0.08 + j1.51	0.28 + j2.79
.875	0.05 + j0.87	0.19 + j1.78
1.000	0.0 + j0.0	0.0 + j0.0

z/h	$h/\lambda = 0.250$	
0	17.96 - j9.38	17.21 - j7.39
.0625	17.88 - j10.11	17.14 - j8.67
.1250	17.65 - j10.42	16.93 - j9.35
.1875	17.26 - j10.56	16.59 - j9.77
.2500	16.72 - j10.54	16.10 - j9.98
.3125	16.04 - j10.37	15.49 - j10.01
.3750	15.22 - j10.06	14.74 - j9.89
.4375	14.26 - j9.62	13.88 - j9.61
.5000	13.18 - j9.06	12.89 - j9.20
.5625	11.97 - j8.37	11.80 - j8.64
.6250	10.65 - j7.57	10.59 - j7.95
.6875	9.23 - j6.66	9.29 - j7.13
.7500	7.70 - j5.64	7.88 - j6.18
.8125	6.08 - j4.51	6.38 - j5.11
.8750	4.33 - j3.26	4.69 - j3.83
.9375	2.50 - j1.90	3.01 - j2.50
1.0000	0.0 - j0.0	0.0 - j0.0

	$h/\lambda = 0.375$	
z/h	$a/\lambda = 0.001588$	$a/\lambda = 0.007022$
0	1.75 - j1.42	3.06 - j0.51
.0417	1.74 - j2.20	3.06 - j1.83
.0833	1.73 - j2.65	3.03 - j2.65
.1250	1.71 - j3.03	3.00 - j3.29
.1667	1.68 - j3.34	2.95 - j3.81
.2083	1.64 - j3.59	2.89 - j4.23
.2500	1.60 - j3.79	2.81 - j4.58
.2917	1.54 - j3.95	2.72 - j4.86
.3333	1.48 - j4.05	2.62 - j5.07
.3750	1.42 - j4.11	2.51 - j5.21
.4167	1.35 - j4.12	2.38 - j5.29
.4583	1.27 - j4.09	2.25 - j5.31
.5000	1.19 - j4.01	2.11 - j5.26
.5417	1.10 - j3.89	1.96 - j5.16
.5833	1.01 - j3.73	1.81 - j4.99
.6250	0.92 - j3.53	1.65 - j4.77
.6667	0.82 - j3.28	1.49 - j4.50
.7083	0.72 - j3.01	1.32 - j4.17
.7500	0.62 - j2.70	1.15 - j3.79
.7917	0.53 - j2.35	0.98 - j3.36
.8333	0.43 - j1.97	0.81 - j2.89
.8750	0.33 - j1.57	0.63 - j2.36
.9167	0.23 - j1.12	0.45 - j1.75
.9583	0.13 - j0.65	0.28 - j1.14
1.0000	0.0 - j0.0	0.0 - j0.0

TABLE 2.5

DISTRIBUTION OF CURRENT ALONG THIN TUBULAR MONOPOLES

$I(z) = I''(z) + jI'(z)$ in milliamperes per volt; $b/a = 2.301$, $Z_c = 50$ ohms

	h/λ = 0.5			h/λ = 0.5	
z/h	a/λ = 0.001588	a/λ = 0.007022	z/h	a/λ = 0.001588	a/λ = 0.007022
0	1.05 + j1.47	1.97 + j3.22	.53125	0.57 - j3.38	1.05 - j4.72
.03125	1.05 + j0.68	1.97 + j1.87	.56250	0.52 - j3.37	0.97 - j4.74
.06250	1.04 + j0.18	1.96 + j1.01	.59375	0.48 - j3.32	0.88 - j4.72
.09375	1.03 - j0.27	1.94 + j0.27	.62500	0.43 - j3.25	0.79 - j4.64
.12500	1.02 - j0.67	1.91 - j0.37	.65625	0.38 - j3.14	0.71 - j4.53
.15625	1.00 - j1.05	1.88 - j0.96	.68750	0.34 - j3.00	0.63 - j4.36
.18750	0.98 - j1.40	1.84 - j1.50	.71875	0.30 - j2.83	0.55 - j4.15
.21875	0.96 - j1.72	1.79 - j2.00	.75000	0.26 - j2.63	0.47 - j3.90
.25000	0.93 - j2.02	1.74 - j2.46	.78125	0.22 - j2.40	0.40 - j3.60
.28125	0.90 - j2.29	1.68 - j2.88	.81250	0.18 - j2.15	0.33 - j3.26
.31250	0.86 - j2.53	1.61 - j3.26	.84375	0.14 - j1.87	0.26 - j2.89
.34375	0.83 - j2.75	1.54 - j3.60	.87500	0.11 - j1.57	0.20 - j2.47
.37500	0.79 - j2.93	1.47 - j3.90	.90625	0.08 - j1.25	0.15 - j2.02
.40625	0.75 - j3.09	1.39 - j4.16	.93750	0.05 - j0.89	0.10 - j1.50
.43750	0.70 - j3.21	1.31 - j4.37	.96875	0.03 - j0.52	0.05 - j0.97
.46875	0.66 - j3.30	1.23 - j4.53	1.00000	0.0 - j0.0	0.0 - j0.0
.50000	0.61 - j3.36	1.14 - j4.65			

TABLE 2.5

DISTRIBUTION OF CURRENT ALONG THIN TUBULAR MONOPOLES

$I(z) = I''(z) + jI'(z)$ in milliamperes per volt; $b/a = 2.301$, $Z_c = 50$ ohms

	$h/\lambda = 0.625$			$h/\lambda = 0.625$	
z/h	$a/\lambda = 0.001588$	$a/\lambda = 0.007022$	z/h	$a/\lambda = 0.001588$	$a/\lambda = 0.007022$
0	1.36 + j5.04	2.92 + j8.40	.525	0.05 - j4.85	-0.10 - j7.02
.025	1.35 + j4.23	2.91 + j7.03	.550	-0.02 - j5.02	-0.27 - j7.31
.050	1.34 + j3.68	2.88 + j6.09	.575	-0.09 - j5.14	-0.43 - j7.53
.075	1.32 + j3.15	2.83 + j5.24	.600	-0.15 - j5.22	-0.57 - j7.69
.100	1.29 + j2.63	2.76 + j4.43	.625	-0.21 - j5.24	-0.70 - j7.77
.125	1.25 + j2.10	2.67 + j3.63	.650	-0.25 - j5.22	-0.81 - j7.78
.150	1.21 + j1.57	2.57 + j2.83	.675	-0.30 - j5.15	-0.90 - j7.71
.175	1.15 + j1.04	2.45 + j2.03	.700	-0.33 - j5.02	-0.98 - j7.57
.200	1.10 + j0.51	2.31 + j1.23	.725	-0.36 - j4.85	-1.04 - j7.36
.225	1.03 - j0.02	2.16 + j0.43	.750	-0.37 - j4.63	-1.07 - j7.08
.250	0.96 - j0.54	2.00 - j0.36	.775	-0.38 - j4.37	-1.08 - j6.73
.275	0.89 - j1.06	1.83 - j1.14	.800	-0.38 - j4.06	-1.08 - j6.31
.300	0.81 - j1.56	1.64 - j1.89	.825	-0.37 - j3.71	-1.05 - j5.82
.325	0.72 - j2.04	1.45 - j2.63	.850	-0.35 - j3.32	-0.99 - j5.27
.350	0.64 - j2.50	1.26 - j3.33	.875	-0.32 - j2.89	-0.92 - j4.66
.375	0.55 - j2.94	1.06 - j4.00	.900	-0.28 - j2.42	-0.82 - j3.98
.400	0.47 - j3.35	0.86 - j4.64	.925	-0.23 - j1.92	-0.69 - j3.25
.425	0.38 - j3.73	0.66 - j5.22	.950	-0.17 - j1.38	-0.53 - j2.41
.450	0.30 - j4.07	0.46 - j5.76	.975	-0.10 - j0.80	-0.36 - j1.55
.475	0.21 - j4.37	0.26 - j6.24	1.000	0.0 - j0.0	0.0 - j0.0
.500	0.13 - j4.63	0.08 - j6.66			

TABLE 2.5

DISTRIBUTION OF CURRENT ALONG THIN TUBULAR MONOPOLES

$I(z) = I''(z) + jI'(z)$ in milliamperes per volt; $b/a = 2.301$, $Z_c = 50$ ohms

	$h/\lambda = 0.75$			$h/\lambda = 0.75$	
z/h	a/λ = 0.001588	a/λ = 0.007022	z/h	a/λ = 0.001588	a/λ = 0.007022
0	14.09 - j5.52	13.74 - j3.92	.5208	-10.02 + j3.51	-8.96 + j2.56
.0208	14.02 - j6.28	13.68 - j5.23	.5417	-10.88 + j4.13	-9.79 + j3.25
.0417	13.83 - j6.67	13.50 - j5.98	.5625	-11.63 + j4.70	-10.54 + j3.90
.0625	13.52 - j6.94	13.21 - j6.53	.5833	-12.28 + j5.22	-11.18 + j4.50
.0833	13.08 - j7.09	12.80 - j6.91	.6042	-12.80 + j5.68	-11.72 + j5.06
.1042	12.52 - j7.15	12.28 - j7.16	.6250	-13.22 + j6.07	-12.15 + j5.55
.1250	11.85 - j7.11	11.66 - j7.31	.6458	-13.50 + j6.41	-12.47 + j5.98
.1458	11.07 - j7.00	10.93 - j7.35	.6667	-13.66 + j6.67	-12.68 + j6.34
.1667	10.20 - j6.80	10.12 - j7.30	.6875	-13.70 + j6.85	-12.77 + j6.63
.1875	9.23 - j6.52	9.21 - j7.15	.7083	-13.60 + j6.96	-12.74 + j6.84
.2083	8.17 - j6.18	8.23 - j6.92	.7292	-13.38 + j6.99	-12.59 + j6.96
.2292	7.05 - j5.76	7.18 - j6.61	.7500	-13.03 + j6.94	-12.32 + j7.01
[illegible]	[illegible]	[illegible]	[illegible]	[illegible]	[illegible]
.2708	4.62 - j4.75	4.91 - j5.76	.7917	-11.97 + j6.60	-11.44 + j6.84
.2917	3.34 - j4.17	3.71 - j5.24	.8125	-11.26 + j6.31	-10.84 + j6.63
.3125	2.03 - j3.54	2.49 - j4.67	.8333	-10.44 + j5.94	-10.13 + j6.33
.3333	0.70 - j2.88	1.24 - j4.04	.8542	-9.52 + j5.50	-9.32 + j5.94
.3542	-0.63 - j2.18	-0.01 - j3.37	.8750	-8.50 + j4.97	-8.42 + j5.47
.3750	-1.96 - j1.47	-1.26 - j2.66	.8958	-7.38 + j4.38	-7.42 + j4.91
.3958	-3.26 - j0.74	-2.49 - j1.93	.9167	-6.18 + j3.72	-6.32 + j4.26
.4167	-4.53 + j0.00	-3.70 - j1.18	.9375	-4.89 + j2.98	-5.14 + j3.53
.4375	-5.76 + j0.74	-4.86 - j0.42	.9583	-3.50 + j2.15	-3.80 + j2.65
.4583	-6.94 + j1.46	-5.98 + j0.34	.9792	-2.02 + j1.26	-2.44 + j1.74
.4792	-8.04 + j2.17	-7.05 + j1.10	1.0000	0.0 + j0.0	0.0 + j0.0
.5000	-9.08 + j2.86	-8.04 + j1.84			

TABLE 2.6

TRANSMITTING AND RECEIVING CHARACTERISTICS OF THIN CYLINDRICAL DIPOLES

k_0h	Y, millimhos	Z, ohms	k_0h_e	D	Ω
		a/λ = 0.001588			
0.5	0.013 + j1.615	4.8 - j619.1	0.237 - j0.001	1.393	9.21
0.6	0.026 + j1.911	7.2 - j523.1	0.290 - j0.001	1.413	9.58
0.7	0.051 + j2.254	10.1 - j443.4	0.347 - j0.002	1.433	9.89
0.8	0.097 + j2.666	13.6 - j374.6	0.406 - j0.004	1.454	10.15
0.9	0.183 + j3.185	18.0 - j312.9	0.470 - j0.006	1.476	10.39
1.0	0.350 + j3.872	23.2 - j256.1	0.538 - j0.010	1.499	10.60
1.1	0.705 + j4.836	29.5 - j202.5	0.612 - j0.015	1.523	10.79
1.2	1.543 + j6.256	37.2 - j150.7	0.692 - j0.022	1.548	10.97
1.3	3.843 + j8.241	46.5 - j99.7	0.780 - j0.033	1.575	11.13
1.4	10.141 + j8.514	57.8 - j48.6	0.878 - j0.047	1.603	11.27
1.5	13.898 - j0.676	71.8 + j3.5	0.986 - j0.066	1.633	11.41
π/2	9.611 - j4.752	83.6 + j41.3	1.071 - j0.084	1.656	11.50
1.6	7.943 - j5.109	89.1 + j57.3	1.108 - j0.092	1.666	11.54
1.7	4.394 - j4.506	110.9 + j113.7	1.246 - j0.130	1.699	11.66
1.8	2.808 - j3.506	139.2 + j173.7	1.405 - j0.183	1.732	11.78
1.9	2.009 - j2.712	176.3 + j238.1	1.589 - j0.259	1.764	11.88
2.0	1.552 - j2.110	226.2 + j307.5	1.803 - j0.371	1.797	11.99
2.1	1.267 - j1.642	294.6 + j381.8	2.051 - j0.535	1.830	12.08
2.2	1.075 - j1.265	390.1 + j459.0	2.334 - j0.783	1.864	12.18
2.3	0.939 - j0.951	525.7 + j532.3	2.641 - j1.161	1.899	12.27
2.4	0.839 - j0.682	717.7 + j583.3	2.927 - j1.736	1.936	12.35
2.5	0.763 - j0.445	977.9 + j570.7	3.072 - j2.583	1.977	12.43
2.6	0.703 - j0.232	1282.4 + j423.5	2.828 - j3.688	2.021	12.51
2.7	0.656 - j0.037	1520.8 + j84.8	1.918 - j4.750	2.070	12.59
2.8	0.617 + j0.147	1533.6 - j364.0	0.460 - j5.190	2.124	12.66
2.9	0.587 + j0.321	1312.2 - j717.6	-0.900 - j4.803	2.184	12.73
3.0	0.563 + j0.490	1011.4 - j880.4	-1.725 - j3.998	2.250	12.80
3.1	0.544 + j0.656	749.1 - j902.9	-2.071 - j3.194	2.321	12.86
π	0.538 + j0.725	660.1 - j888.9	-2.124 - j2.903	2.352	12.89
3.2	0.532 + j0.823	554.2 - j857.3	-2.145 - j2.543	2.397	12.93
3.3	0.525 + j0.993	416.2 - j786.9	-2.092 - j2.053	2.477	12.99
3.4	0.525 + j1.170	319.0 - j711.3	-1.991 - j1.687	2.561	13.05
3.5	0.532 + j1.360	249.6 - j637.8	-1.875 - j1.410	2.646	13.11
3.6	0.548 + j1.566	199.2 - j568.8	-1.760 - j1.196	2.729	13.16
3.7	0.578 + j1.797	162.1 - j504.3	-1.652 - j1.027	2.801	13.22
3.8	0.625 + j2.064	134.5 - j443.8	-1.551 - j0.888	2.852	13.27
3.9	0.703 + j2.381	114.0 - j386.4	-1.458 - j0.773	2.866	13.32
4.0	0.830 + j2.770	99.3 - j331.3	-1.371 - j0.674	2.823	13.37
		a/λ = 0.003175			
0.5	0.021 + j2.161	4.5 - j462.7	0.226 - j0.001	1.359	7.83
0.6	0.043 + j2.515	6.8 - j397.5	0.279 - j0.002	1.383	8.19
0.7	0.082 + j2.926	9.6 - j341.5	0.336 - j0.003	1.407	8.50
0.8	0.154 + j3.425	13.1 - j291.4	0.395 - j0.005	1.432	8.77
0.9	0.288 + j4.056	17.4 - j245.3	0.460 - j0.008	1.457	9.00

TABLE 2.6

TRANSMITTING AND RECEIVING CHARACTERISTICS OF THIN CYLINDRICAL DIPOLES

k_0h	Y, millimhos	Z, ohms	k_0h_e	D	Ω
		$a/\lambda = 0.003175$ (Continued)			
1.0	0.549 + j4.892	22.7 - j201.9	0.529 - j0.013	1.483	9.22
1.1	1.101 + j6.051	29.1 - j160.0	0.605 - j0.020	1.511	9.41
1.2	2.389 + j7.678	37.0 - j118.7	0.688 - j0.029	1.539	9.58
1.3	5.697 + j9.477	46.6 - j77.5	0.779 - j0.043	1.569	9.74
1.4	12.471 + j7.596	58.5 - j35.6	0.881 - j0.061	1.600	9.89
1.5	13.513 - j1.382	73.2 + j7.5	0.995 - j0.087	1.634	10.03
π/2	9.652 - j4.392	85.8 + j39.1	1.084 - j0.110	1.659	10.12
1.6	8.223 - j4.699	91.7 + j52.4	1.123 - j0.122	1.670	10.16
1.7	4.963 - j4.290	115.3 + j99.7	1.269 - j0.172	1.706	10.28
1.8	3.337 - j3.417	146.3 + j149.8	1.436 - j0.245	1.741	10.39
1.9	2.459 - j2.658	187.6 + j202.8	1.628 - j0.351	1.775	10.50
2.0	1.936 - j2.050	243.5 + j257.9	1.848 - j0.506	1.809	10.60
2.1	1.599 - j1.560	320.3 + j312.5	2.091 - j0.738	1.842	10.70
2.2	1.370 - j1.155	426.6 + j359.9	2.343 - j1.085	1.875	10.79
2.3	1.205 - j0.811	571.0 + j384.6	2.556 - j1.595	1.908	10.88
2.4	1.082 - j0.512	755.3 + j357.1	2.625 - j2.310	1.942	10.97
2.5	0.988 - j0.244	953.9 + j235.8	2.368 - j3.181	1.978	11.05
2.6	0.914 - j0.001	1094.2 + j1.2	1.631 - j3.967	2.018	11.13
2.7	0.855 + j0.224	1094.6 - j287.3	0.540 - j4.302	2.061	11.20
2.8	0.807 + j0.437	958.5 - j518.6	-0.496 - j4.074	2.109	11.27
2.9	0.769 + j0.640	768.5 - j639.4	-1.190 - j3.525	2.161	11.34
3.0	0.739 + j0.837	592.7 - j671.1	-1.544 - j2.928	2.218	11.41
3.1	0.718 + j1.032	454.4 - j653.1	-1.677 - j2.413	2.281	11.48
π	0.711 + j1.112	408.0 - j638.2	-1.693 - j2.231	2.307	11.50
3.2	0.704 + j1.227	352.0 - j613.2	-1.692 - j2.005	2.347	11.54
3.3	0.699 + j1.426	277.3 - j565.5	-1.651 - j1.691	2.417	11.60
3.4	0.703 + j1.633	222.5 - j516.8	-1.587 - j1.448	2.490	11.66
3.5	0.717 + j1.853	181.8 - j469.4	-1.517 - j1.257	2.563	11.72
3.6	0.746 + j2.093	151.1 - j424.0	-1.448 - j1.104	2.633	11.78
3.7	0.792 + j2.362	127.7 - j380.6	-1.382 - j0.978	2.693	11.83
3.8	0.866 + j2.672	109.8 - j338.7	-1.320 - j0.871	2.732	11.89
3.9	0.983 + j3.039	96.4 - j297.8	-1.262 - j0.778	2.736	11.94
4.0	1.173 + j3.489	86.6 - j257.5	-1.206 - j0.695	2.686	11.99
		$a/\lambda = 0.004763$			
0.5	0.030 + j2.684	4.2 - j372.5	0.217 - j0.001	1.335	7.02
0.6	0.060 + j3.081	6.4 - j324.5	0.269 - j0.002	1.362	7.38
0.7	0.114 + j3.545	9.1 - j281.8	0.324 - j0.004	1.388	7.69
0.8	0.212 + j4.111	12.5 - j242.6	0.384 - j0.006	1.415	7.96
0.9	0.394 + j4.829	16.8 - j205.7	0.449 - j0.010	1.443	8.19
1.0	0.748 + j5.780	22.0 - j170.2	0.519 - j0.016	1.472	8.40
1.1	1.491 + j7.078	28.5 - j135.3	0.597 - j0.024	1.501	8.59
1.2	3.197 + j8.794	36.5 - j100.4	0.682 - j0.036	1.532	8.77
1.3	7.261 + j10.179	46.5 - j65.1	0.776 - j0.052	1.564	8.93
1.4	13.707 + j6.717	58.8 - j28.8	0.882 - j0.075	1.598	9.08

TABLE 2.6

TRANSMITTING AND RECEIVING CHARACTERISTICS OF THIN CYLINDRICAL DIPOLES

k_0h	Y, millimhos	Z, ohms	k_0h_e	D	Ω
		$a/\lambda = 0.004763$ (Continued)			
1.5	13.264 - j1.575	74.3 + j8.8	1.000 - j0.107	1.634	9.22
π/2	9.717 - j4.047	87.7 + j36.5	1.093 - j0.137	1.661	9.31
1.6	8.424 - j4.327	93.9 + j48.2	1.134 - j0.152	1.673	9.34
1.7	5.352 - j4.025	119.4 + j89.8	1.287 - j0.216	1.711	9.47
1.8	3.716 - j3.237	153.0 + j133.3	1.461 - j0.309	1.748	9.58
1.9	2.792 - j2.508	198.2 + j178.1	1.658 - j0.446	1.784	9.69
2.0	2.226 - j1.902	259.7 + j221.9	1.875 - j0.649	1.818	9.79
2.1	1.855 - j1.400	343.5 + j259.2	2.097 - j0.949	1.851	9.89
2.2	1.598 - j0.977	455.5 + j278.4	2.286 - j1.386	1.883	9.98
2.3	1.413 - j0.612	595.9 + j258.1	2.356 - j1.990	1.915	10.07
2.4	1.274 - j0.290	746.3 + j169.9	2.166 - j2.723	1.946	10.16
2.5	1.167 + j0.000	857.2 - j0.1	1.594 - j3.405	1.979	10.24
2.6	1.082 + j0.266	871.5 - j214.6	0.715 - j3.757	2.014	10.32
2.7	1.014 + j0.515	784.0 - j398.1	-0.175 - j3.657	2.052	10.39
2.8	0.959 + j0.750	646.8 - j505.9	-0.826 - j3.256	2.094	10.46
2.9	0.916 + j0.976	511.2 - j544.9	-1.199 - j2.771	2.139	10.53
3.0	0.882 + j1.196	399.4 - j541.5	-1.371 - j2.326	2.189	10.60
3.1	0.858 + j1.413	314.0 - j516.9	-1.425 - j1.960	2.244	10.67
π	0.852 + j1.503	285.3 - j503.7	-1.427 - j1.832	2.267	10.69
3.2	0.845 + j1.630	250.5 - j483.5	-1.417 - j1.672	2.302	10.73
3.3	0.842 + j1.852	203.4 - j447.6	-1.382 - j1.448	2.363	10.79
3.4	0.851 + j2.081	168.2 - j411.7	-1.335 - j1.273	2.426	10.85
3.5	0.873 + j2.325	141.6 - j376.9	-1.286 - j1.133	2.490	10.91
3.6	0.914 + j2.590	121.1 - j343.3	-1.240 - j1.019	2.551	10.97
3.7	0.978 + j2.887	105.3 - j310.7	-1.197 - j0.923	2.603	11.02
3.8	1.077 + j3.228	93.0 - j278.7	-1.157 - j0.839	2.635	11.07
3.9	1.232 + j3.632	83.8 - j246.9	-1.120 - j0.764	2.634	11.13
4.0	1.480 + j4.121	77.2 - j214.9	-1.085 - j0.694	2.581	11.18
		$a/\lambda = 0.006350$			
0.5	0.041 + j3.270	3.8 - j305.7	0.204 - j0.001	1.318	6.44
0.6	0.079 + j3.702	5.8 - j270.0	0.255 - j0.002	1.345	6.81
0.7	0.149 + j4.211	8.4 - j237.2	0.310 - j0.004	1.374	7.12
0.8	0.274 + j4.836	11.7 - j206.1	0.369 - j0.007	1.402	7.38
0.9	0.505 + j5.629	15.8 - j176.2	0.434 - j0.012	1.432	7.62
1.0	0.955 + j6.675	21.0 - j146.8	0.505 - j0.019	1.462	7.83
1.1	1.892 + j8.077	27.5 - j117.4	0.584 - j0.029	1.493	8.02
1.2	3.994 + j9.805	35.6 - j87.5	0.672 - j0.044	1.526	8.19
1.3	8.620 + j10.655	45.9 - j56.7	0.770 - j0.064	1.560	8.35
1.4	14.433 + j6.075	58.9 - j24.8	0.880 - j0.093	1.596	8.50
1.5	13.098 - j1.508	75.4 + j8.7	1.004 - j0.133	1.634	8.64
π/2	9.791 - j3.641	89.7 + j33.4	1.102 - j0.171	1.663	8.73
1.6	8.594 - j3.903	96.5 + j43.8	1.145 - j0.190	1.675	8.77
1.7	5.662 - j3.674	124.3 + j80.7	1.305 - j0.272	1.715	8.89
1.8	4.025 - j2.952	161.6 + j118.5	1.485 - j0.394	1.754	9.00
1.9	3.069 - j2.249	212.0 + j155.3	1.684 - j0.574	1.791	9.11

TABLE 2.6

TRANSMITTING AND RECEIVING CHARACTERISTICS OF THIN CYLINDRICAL DIPOLES

k_0h	Y, millimhos	Z, ohms	k_0h_e	D	Ω
		<u>a/λ = 0.006350 (Continued)</u>			
2.0	2.471 - j1.646	280.4 + j186.7	1.887 - j0.840	1.826	9.22
2.1	2.073 - j1.136	371.0 + j203.2	2.061 - j1.226	1.859	9.31
2.2	1.795 - j0.699	483.7 + j188.3	2.131 - j1.755	1.891	9.41
2.3	1.593 - j0.317	603.8 + j120.2	1.981 - j2.396	1.921	9.49
2.4	1.441 + j0.023	694.0 - j11.2	1.508 - j3.000	1.950	9.58
2.5	1.323 + j0.333	711.1 - j179.0	0.765 - j3.338	1.979	9.66
2.6	1.229 + j0.619	649.0 - j326.9	-0.009 - j3.297	2.010	9.74
2.7	1.154 + j0.888	544.4 - j418.9	-0.602 - j2.984	2.043	9.82
2.8	1.093 + j1.143	436.8 - j456.9	-0.963 - j2.577	2.079	9.89
2.9	1.045 + j1.390	345.6 - j459.6	-1.145 - j2.189	2.119	9.96
3.0	1.008 + j1.630	274.5 - j443.7	-1.216 - j1.862	2.162	10.03
3.1	0.983 + j1.867	220.8 - j419.3	-1.227 - j1.600	2.209	10.09
π	0.976 + j1.965	202.7 - j408.1	-1.221 - j1.508	2.229	10.12
3.2	0.970 + j2.104	180.7 - j392.0	-1.207 - j1.394	2.259	10.16
3.3	0.970 + j2.345	150.5 - j364.1	-1.174 - j1.234	2.312	10.22
3.4	0.984 + j2.595	127.8 - j336.9	-1.137 - j1.108	2.367	10.28
3.5	1.015 + j2.859	110.3 - j310.6	-1.102 - j1.006	2.424	10.33
3.6	1.068 + j3.145	96.8 - j285.0	-1.071 - j0.923	2.478	10.39
3.7	1.151 + j3.465	86.3 - j260.0	-1.044 - j0.852	2.524	10.45
3.8	1.276 + j3.831	78.2 - j235.0	-1.020 - j0.789	2.552	10.50
3.9	1.468 + j4.263	72.2 - j209.7	-0.999 - j0.731	2.547	10.55
4.0	1.773 + j4.782	68.2 - j183.8	-0.980 - j0.676	2.493	10.60
4.2	3.151 + j6.128	66.4 - j129.0	-0.941 - j0.567	2.183	10.70
4.4	7.247 + j6.626	75.2 - j28.7	-0.897 - j0.459	1.620	10.79
4.6	10.021 + j0.186	99.8 - j1.9	-0.841 - j0.349	0.998	10.88
3π/2	7.365 - j2.274	124.0 + j38.3	-0.804 - j0.286	0.704	10.93
4.8	5.469 - j2.552	150.2 + j70.1	-0.770 - j0.235	0.518	10.97
5.0	3.125 - j1.746	243.9 + j136.3	-0.671 - j0.117	0.228	11.05
5.2	2.167 - j0.855	399.3 + j157.5	-0.523 - j0.004	0.082	11.13
5.4	1.708 - j0.154	580.7 + j52.2	-0.314 + j0.050	0.021	11.20
5.6	1.457 + j0.415	635.1 - j180.8	-0.119 - j0.010	0.003	11.27
5.8	1.306 + j0.905	517.3 - j358.6	-0.027 - j0.110	0.003	11.34
6.0	1.214 + j1.354	367.2 - j409.3	0.004 - j0.165	0.009	11.41
6.2	1.168 + j1.785	256.7 - j392.3	0.032 - j0.176	0.015	11.48
2π	1.162 + j1.964	223.2 - j377.1	0.048 - j0.173	0.017	11.50
6.4	1.168 + j2.220	185.6 - j352.8	0.075 - j0.166	0.021	11.54
6.6	1.225 + j2.686	140.6 - j308.2	0.135 - j0.147	0.034	11.60
6.8	1.369 + j3.222	111.7 - j262.9	0.213 - j0.128	0.066	11.66
7.0	1.681 + j3.899	93.2 - j216.3	0.315 - j0.110	0.143	11.72
		<u>a/λ = 0.007022</u>			
0.5	0.045 + j3.538	3.6 - j282.6	0.199 - j0.001	1.311	6.24
0.6	0.088 + j3.983	5.6 - j251.0	0.249 - j0.003	1.339	6.61
0.7	0.164 + j4.510	8.1 - j221.4	0.303 - j0.005	1.368	6.91
0.8	0.301 + j5.157	11.3 - j193.3	0.362 - j0.008	1.397	7.18
0.9	0.554 + j5.980	15.4 - j165.8	0.427 - j0.013	1.427	7.42

TABLE 2.6

TRANSMITTING AND RECEIVING CHARACTERISTICS OF THIN CYLINDRICAL DIPOLES

k_0h	Y, millimhos	Z, ohms	k_0h_e	D	Ω
		a/λ = 0.007022 (Continued)			
1.0	1.046 + j7.063	20.5 - j138.6	0.499 - j0.021	1.458	7.63
1.1	2.066 + j8.500	27.0 - j111.1	0.578 - j0.031	1.490	7.82
1.2	4.331 + j10.213	35.2 - j83.0	0.667 - j0.047	1.524	7.99
1.3	9.143 + j10.817	45.6 - j53.9	0.766 - j0.069	1.559	8.15
1.4	14.650 + j5.873	58.8 - j23.6	0.878 - j0.101	1.595	8.30
1.5	13.045 - j1.427	75.8 + j8.3	1.005 - j0.146	1.634	8.44
π/2	9.825 - j3.452	90.6 + j31.8	1.105 - j0.188	1.664	8.53
1.6	8.660 - j3.707	97.6 + j41.8	1.149 - j0.209	1.676	8.57
1.7	5.778 - j3.503	126.5 + j76.7	1.312 - j0.301	1.717	8.69
1.8	4.142 - j2.804	165.5 + j112.1	1.494 - j0.437	1.756	8.80
1.9	3.176 - j2.111	218.4 + j145.2	1.690 - j0.639	1.794	8.91
2.0	2.566 - j1.510	289.5 + j170.3	1.881 - j0.936	1.829	9.01
2.1	2.158 - j0.998	381.8 + j176.4	2.020 - j1.358	1.862	9.11
2.2	1.873 - j0.556	490.8 + j145.6	2.020 - j1.915	1.894	9.20
2.3	1.664 - j0.168	595.0 + j59.9	1.763 - j2.535	1.923	9.29
2.4	1.506 + j0.180	654.5 - j78.3	1.198 - j3.035	1.951	9.38
2.5	1.384 + j0.497	639.9 - j229.9	0.454 - j3.217	1.980	9.46
2.6	1.287 + j0.791	563.8 - j346.6	-0.228 - j3.064	2.009	9.54
2.7	1.209 + j1.068	464.5 - j410.3	-0.704 - j2.720	2.040	9.61
2.8	1.146 + j1.332	371.2 - j431.3	-0.978 - j2.336	2.073	9.69
2.9	1.096 + j1.587	294.8 - j426.6	-1.108 - j1.989	2.110	9.76
3.0	1.059 + j1.835	235.9 - j408.9	-1.154 - j1.702	2.150	9.83
3.1	1.033 + j2.080	191.5 - j385.7	-1.153 - j1.474	2.194	9.89
π	1.026 + j2.182	176.5 - j375.4	-1.145 - j1.394	2.213	9.92
3.2	1.020 + j2.325	158.2 - j360.7	-1.130 - j1.295	2.241	9.95
3.3	1.021 + j2.574	133.1 - j335.7	-1.099 - j1.155	2.291	10.02
3.4	1.038 + j2.832	114.1 - j311.3	-1.066 - j1.045	2.343	10.08
3.5	1.073 + j3.104	99.5 - j287.8	-1.035 - j0.957	2.397	10.13
3.6	1.132 + j3.398	88.2 - j264.9	-1.009 - j0.884	2.448	10.19
3.7	1.222 + j3.726	79.4 - j242.3	-0.987 - j0.822	2.492	10.24
3.8	1.358 + j4.102	72.7 - j219.7	-0.968 - j0.767	2.519	10.30
3.9	1.566 + j4.545	67.8 - j196.7	-0.953 - j0.715	2.514	10.35
4.0	1.896 + j5.074	64.6 - j172.9	-0.939 - j0.665	2.460	10.40
4.2	3.371 + j6.419	64.1 - j122.1	-0.912 - j0.565	2.154	10.50
4.4	7.591 + j6.712	73.9 - j65.4	-0.879 - j0.461	1.599	10.59
4.6	10.017 + j0.193	99.8 - j1.9	-0.835 - j0.351	0.986	10.68
3π/2	7.378 - j2.128	125.1 + j36.1	-0.802 - j0.286	0.696	10.73
4.8	5.525 - j2.387	152.5 + j65.9	-0.772 - j0.234	0.512	10.77
5.0	3.205 - j1.594	250.2 + j124.4	-0.677 - j0.108	0.225	10.85
5.2	2.240 - j0.704	406.2 + j127.7	-0.523 + j0.011	0.081	10.93
5.4	1.774 + j0.004	563.6 - j1.2	-0.304 + j0.059	0.020	11.00
5.6	1.517 + j0.582	574.5 - j220.5	-0.116 - j0.013	0.003	11.07
5.8	1.363 + j1.085	449.2 - j357.5	-0.033 - j0.111	0.004	11.14

TABLE 2.6

TRANSMITTING AND RECEIVING CHARACTERISTICS OF THIN CYLINDRICAL DIPOLES

k_0h	Y, millimhos	Z, ohms	k_0h_e	D	Ω
		a/λ = 0.007022 (Continued)			
6.0	1.270 + j1.546	317.3 - j386.3	-0.002 - j0.161	0.010	11.21
6.2	1.224 + j1.991	224.1 - j364.6	0.028 - j0.172	0.016	11.28
2π	1.218 + j2.176	196.0 - j349.9	0.044 - j0.170	0.019	11.30
6.4	1.226 + j2.440	164.5 - j327.2	0.071 - j0.163	0.023	11.34
6.6	1.291 + j2.919	126.7 - j286.5	0.131 - j0.147	0.037	11.40
6.8	1.449 + j3.471	102.4 - j245.4	0.209 - j0.129	0.070	11.46
7.0	1.787 + j4.162	87.1 - j202.9	0.309 - j0.113	0.149	11.52
		a/λ = 0.009525			
0.5	0.065 + j4.634	3.0 - j215.8	0.180 - j0.002	1.292	5.63
0.6	0.123 + j5.122	4.7 - j195.1	0.227 - j0.003	1.321	6.00
0.7	0.227 + j5.709	7.0 - j174.9	0.280 - j0.006	1.351	6.30
0.8	0.411 + j6.434	9.9 - j154.8	0.337 - j0.010	1.382	6.57
0.9	0.750 + j7.357	13.7 - j134.5	0.402 - j0.016	1.414	6.81
1.0	1.404 + j8.557	18.7 - j113.8	0.474 - j0.026	1.447	7.02
1.1	2.741 + j10.089	25.1 - j92.3	0.555 - j0.041	1.481	7.21
1.2	5.577 + j11.659	33.4 - j69.8	0.647 - j0.062	1.517	7.38
1.3	10.848 + j11.298	44.2 - j46.1	0.751 - j0.092	1.554	7.54
1.4	15.176 + j5.437	58.4 - j20.9	0.870 - j0.136	1.593	7.69
1.5	12.910 - j0.935	77.1 + j5.6	1.005 - j0.200	1.635	7.83
π/2	9.956 - j2.657	93.8 + j25.0	1.111 - j0.261	1.666	7.92
1.6	8.887 - j2.896	101.7 + j33.2	1.157 - j0.292	1.679	7.96
1.7	6.161 - j2.766	135.1 + j60.7	1.325 - j0.428	1.723	8.08
1.8	4.533 - j2.141	180.4 + j85.2	1.501 - j0.631	1.765	8.19
1.9	3.534 - j1.482	240.7 + j100.9	1.661 - j0.927	1.804	8.30
2.0	2.888 - j0.887	316.4 + j97.2	1.750 - j1.338	1.840	8.40
2.1	2.449 - j0.366	399.3 + j59.8	1.679 - j1.848	1.873	8.50
2.2	2.137 + j0.092	467.0 - j20.1	1.362 - j2.357	1.904	8.60
2.3	1.908 + j0.501	490.3 - j128.8	0.814 - j2.688	1.931	8.68
2.4	1.734 + j0.873	460.1 - j231.8	0.198 - j2.732	1.956	8.77
2.5	1.598 + j1.217	396.1 - j301.7	-0.308 - j2.538	1.980	8.85
2.6	1.490 + j1.539	324.8 - j335.5	-0.641 - j2.239	2.003	8.93
2.7	1.402 + j1.844	261.3 - j343.6	-0.826 - j1.932	2.027	9.00
2.8	1.332 + j2.137	210.1 - j337.1	-0.913 - j1.661	2.052	9.08
2.9	1.276 + j2.421	170.4 - j323.3	-0.942 - j1.437	2.079	9.15
3.0	1.234 + j2.699	140.1 - j306.5	-0.939 - j1.257	2.108	9.22
3.1	1.207 + j2.973	117.2 - j288.8	-0.920 - j1.115	2.141	9.28
π	1.200 + j3.087	109.4 - j281.4	-0.910 - j1.066	2.155	9.31
3.2	1.195 + j3.247	99.8 - j271.2	-0.895 - j1.004	2.176	9.34
3.3	1.202 + j3.525	86.7 - j254.2	-0.870 - j0.918	2.215	9.41
3.4	1.228 + j3.810	76.7 - j237.7	-0.847 - j0.851	2.257	9.47
3.5	1.279 + j4.110	69.1 - j221.8	-0.829 - j0.798	2.301	9.52
3.6	1.360 + j4.432	63.3 - j206.2	-0.816 - j0.756	2.345	9.58
3.7	1.481 + j4.790	58.9 - j190.6	-0.808 - j0.720	2.384	9.63
3.8	1.660 + j5.197	55.8 - j174.6	-0.805 - j0.687	2.407	9.69
3.9	1.931 + j5.670	53.8 - j158.0	-0.805 - j0.655	2.401	9.74

TABLE 2.6

TRANSMITTING AND RECEIVING CHARACTERISTICS OF THIN CYLINDRICAL DIPOLES

k_0h	Y, millimhos	Z, ohms	k_0h_e	D	Ω
		a/λ = 0.009525 (Continued)			
4.0	2.352 + j6.228	53.1 - j140.5	-0.808 - j0.621	2.351	9.79
4.2	4.176 + j7.528	56.4 - j101.6	-0.818 - j0.547	2.061	9.89
4.4	8.712 + j7.043	69.4 - j56.1	-0.823 - j0.458	1.534	9.98
4.6	10.000 + j0.423	99.8 - j4.2	-0.818 - j0.348	0.949	10.07
3π/2	7.434 - j1.489	129.3 + j25.9	-0.805 - j0.275	0.671	10.12
4.8	5.707 - j1.685	161.2 + j47.6	-0.787 - j0.212	0.494	10.16
5.0	3.465 - j0.921	269.5 + j71.7	-0.697 - j0.049	0.217	10.24
5.2	2.484 - j0.035	402.5 + j5.7	-0.502 + j0.087	0.077	10.32
5.4	1.996 + j0.697	446.6 - j155.9	-0.257 + j0.084	0.020	10.39
5.6	1.723 + j1.311	367.6 - j279.7	-0.107 - j0.020	0.004	10.46
5.8	1.558 + j1.855	265.5 - j316.0	-0.048 - j0.105	0.006	10.53
6.0	1.460 + j2.362	189.3 - j306.3	-0.018 - j0.145	0.014	10.60
6.2	1.415 + j2.855	139.4 - j281.2	0.016 - j0.156	0.021	10.67
2π	1.413 + j3.060	124.4 - j269.3	0.032 - j0.156	0.025	10.69
6.4	1.430 + j3.353	107.6 - j252.3	0.059 - j0.153	0.030	10.73
6.6	1.522 + j3.882	87.5 - j223.3	0.116 - j0.143	0.047	10.79
6.8	1.733 + j4.483	75.0 - j194.1	0.190 - j0.132	0.085	10.85
7.0	2.169 + j5.221	67.9 - j163.3	0.285 - j0.122	0.170	10.91

k_0h	Y, millimhos	Z, ohms	k_0h_e	D	a/λ
		Ω = 8.54			
0.5	0.016 + j1.847	4.7 - j541.4	0.232 - j0.001	1.377	0.002225
0.6	0.038 + j2.335	6.9 - j428.1	0.282 - j0.002	1.391	0.002670
0.7	0.081 + j2.903	9.6 - j344.2	0.336 - j0.003	1.408	0.003115
0.8	0.168 + j3.598	12.9 - j277.3	0.392 - j0.005	1.427	0.003560
0.9	0.343 + j4.471	17.0 - j222.4	0.454 - j0.009	1.449	0.004005
1.0	0.708 + j5.613	22.1 - j175.4	0.521 - j0.015	1.474	0.004451
1.1	1.524 + j7.165	28.4 - j133.5	0.595 - j0.024	1.500	0.004896
1.2	3.488 + j9.181	36.2 - j95.2	0.678 - j0.039	1.530	0.005341
1.3	8.158 + j10.513	46.1 - j59.4	0.772 - j0.060	1.562	0.005786
1.4	14.390 + j6.118	58.9 - j25.0	0.880 - j0.091	1.596	0.006231
1.5	13.071 - j1.460	75.6 + j8.4	1.005 - j0.140	1.634	0.006676
π/2	9.823 - j3.439	90.7 + j31.8	1.105 - j0.189	1.664	0.006991
1.6	8.669 - j3.654	98.0 + j41.3	1.150 - j0.214	1.676	0.007121
1.7	5.867 - j3.326	129.0 + j73.1	1.318 - j0.332	1.718	0.007566
1.8	4.304 - j2.531	172.6 + j101.5	1.505 - j0.517	1.760	0.008011
1.9	3.387 - j1.744	233.3 + j120.2	1.686 - j0.810	1.800	0.008456
2.0	2.812 - j1.040	312.8 + j115.7	1.798 - j1.248	1.838	0.008901
2.1	2.430 - j0.412	400.1 + j67.9	1.711 - j1.821	1.873	0.009346
2.2	2.164 + j0.157	459.7 - j33.3	1.292 - j2.372	1.905	0.009791

TABLE 2.6

TRANSMITTING AND RECEIVING CHARACTERISTICS OF THIN CYLINDRICAL DIPOLES

k_0h	Y, millimhos	Z, ohms	k_0h_e	D	a/λ
		$\Omega = 9.92$			
0.5	0.010 + j1.439	4.9 - j695.0	0.239 - j0.001	1.406	0.001116
0.6	0.024 + j1.810	7.2 - j552.4	0.291 - j0.001	1.419	0.001339
0.7	0.051 + j2.243	10.1 - j445.7	0.347 - j0.002	1.434	0.001563
0.8	0.104 + j2.772	13.5 - j360.2	0.405 - j0.004	1.451	0.001786
0.9	0.211 + j3.439	17.8 - j289.7	0.467 - j0.007	1.470	0.002009
1.0	0.432 + j4.319	22.9 - j229.2	0.534 - j0.011	1.492	0.002232
1.1	0.924 + j5.542	29.3 - j175.6	0.608 - j0.018	1.516	0.002456
1.2	2.132 + j7.284	37.0 - j126.5	0.689 - j0.027	1.542	0.002679
1.3	5.403 + j9.317	46.6 - j80.3	0.780 - j0.041	1.570	0.002902
1.4	12.419 + j7.626	58.5 - j35.9	0.881 - j0.061	1.600	0.003125
1.5	13.480 - j1.417	73.4 + j7.7	0.996 - j0.089	1.634	0.003348
π/2	9.664 - j4.317	86.3 + j38.5	1.086 - j0.116	1.660	0.003506
1.6	8.278 - j4.601	92.3 + j51.3	1.126 - j0.129	1.671	0.003572
1.7	5.128 - j4.186	117.0 + j95.5	1.277 - j0.189	1.708	0.003795
1.8	3.550 - j3.325	150.1 + j140.6	1.451 - j0.279	1.745	0.004018
1.9	2.690 - j2.561	195.0 + j185.6	1.650 - j0.415	1.781	0.004241
2.0	2.176 - j1.932	257.0 + j228.2	1.872 - j0.623	1.817	0.004465
2.1	1.844 - j1.408	342.6 + j261.6	2.098 - j0.940	1.851	0.004688
2.2	1.618 - j0.951	459.2 + j270.1	2.277 - j1.423	1.884	0.004911
2.3	1.457 - j0.544	602.4 + j224.8	2.281 - j2.102	1.916	0.005134
2.4	1.339 - j0.173	734.8 + j95.0	1.921 - j2.870	1.948	0.005357
2.5	1.249 + j0.171	785.8 - j107.8	1.131 - j3.418	1.979	0.005581
2.6	1.180 + j0.498	719.4 - j303.5	0.185 - j3.468	2.012	0.005804
2.7	1.127 + j0.812	584.1 - j421.1	-0.548 - j3.107	2.045	0.006027
2.8	1.085 + j1.119	446.7 - j460.5	-0.961 - j2.611	2.080	0.006250
2.9	1.055 + j1.430	334.0 - j452.9	-1.140 - j2.143	2.117	0.006474
3.0	1.034 + j1.747	250.9 - j423.8	-1.183 - j1.763	2.156	0.006697
3.1	1.025 + j2.066	192.7 - j388.3	-1.161 - j1.476	2.196	0.006920
π	1.025 + j2.201	173.9 - j373.4	-1.142 - j1.380	2.213	0.007013
3.2	1.029 + j2.391	151.8 - j352.9	-1.112 - j1.262	2.238	0.007143
3.3	1.047 + j2.725	122.9 - j319.8	-1.057 - j1.103	2.280	0.007367
3.4	1.082 + j3.071	102.0 - j289.6	-1.004 - j0.983	2.323	0.007590
3.5	1.139 + j3.438	86.9 - j262.1	-0.958 - j0.892	2.366	0.007813
3.6	1.225 + j3.832	75.7 - j236.8	-0.919 - j0.821	2.406	0.008036
3.7	1.351 + j4.265	67.5 - j213.1	-0.888 - j0.763	2.437	0.008259
3.8	1.535 + j4.752	61.5 - j190.5	-0.865 - j0.714	2.452	0.008483
3.9	1.812 + j5.312	57.5 - j168.6	-0.847 - j0.670	2.436	0.008706
4.0	2.243 + j5.962	55.3 - j146.9	-0.836 - j0.629	2.375	0.008929
4.2	4.129 + j7.464	56.7 - j102.6	-0.823 - j0.548	2.066	0.009376
4.4	8.831 + j7.071	69.0 - j55.3	-0.818 - j0.458	1.527	0.009822
		$\Omega = 12.5$			
0.5	0.005 + j1.022	5.0 - j978.2	0.244 - j0.000	1.443	0.000307
0.6	0.012 + j1.285	7.3 - j778.4	0.297 - j0.001	1.454	0.000369
0.7	0.026 + j1.589	10.2 - j629.1	0.353 - j0.002	1.466	0.000430
0.8	0.053 + j1.956	13.7 - j511.0	0.412 - j0.003	1.480	0.000492
0.9	0.105 + j2.416	18.0 - j413.1	0.474 - j0.005	1.496	0.000553

TABLE 2.6

TRANSMITTING AND RECEIVING CHARACTERISTICS OF THIN CYLINDRICAL DIPOLES

k_0h	Y, millimhos	Z, ohms	k_0h_e	D	a/λ
		<u>Ω = 12.5 (Continued)</u>			
1.0	0.213 + j3.024	23.2 - j329.0	0.541 - j0.008	1.514	0.000614
1.1	0.449 + j3.880	29.4 - j254.3	0.613 - j0.012	1.534	0.000676
1.2	1.027 + j5.170	37.0 - j186.1	0.692 - j0.018	1.556	0.000737
1.3	2.705 + j7.168	46.1 - j122.1	0.778 - j0.027	1.579	0.000799
1.4	8.223 + j8.722	57.2 - j60.7	0.874 - j0.040	1.605	0.000860
1.5	14.103 + j0.092	70.9 - j0.5	0.981 - j0.057	1.633	0.000922
π/2	9.612 - j4.909	82.5 + j42.1	1.064 - j0.073	1.655	0.000965
1.6	7.776 - j5.295	87.9 + j59.8	1.101 - j0.081	1.664	0.000983
1.7	4.096 - j4.548	109.3 + j121.4	1.238 - j0.115	1.695	0.001045
1.8	2.579 - j3.488	137.1 + j185.4	1.395 - j0.164	1.728	0.001106
1.9	1.846 - j2.688	173.6 + j252.8	1.578 - j0.236	1.760	0.001168
2.0	1.438 - j2.095	222.7 + j324.5	1.792 - j0.341	1.794	0.001229
2.1	1.186 - j1.637	290.3 + j400.6	2.043 - j0.499	1.827	0.001290
2.2	1.020 - j1.267	385.4 + j479.0	2.330 - j0.741	1.862	0.001352
2.3	0.903 - j0.958	521.3 + j552.8	2.646 - j1.115	1.898	0.001413
2.4	0.818 - j0.689	715.0 + j602.4	2.943 - j1.695	1.936	0.001475
2.5	0.754 - j0.450	978.1 + j583.7	3.092 - j2.560	1.977	0.001536
2.6	0.705 - j0.231	1281.7 + j419.9	2.820 - j3.693	2.021	0.001598
2.7	0.666 - j0.025	1499.4 + j55.3	1.828 - j4.745	2.070	0.001659
2.8	0.636 + j0.172	1465.9 - j396.2	0.320 - j5.082	2.123	0.001721
2.9	0.612 + j0.362	1210.2 - j715.3	-0.982 - j4.586	2.181	0.001782
3.0	0.595 + j0.549	907.9 - j837.6	-1.703 - j3.752	2.244	0.001843
3.1	0.583 + j0.736	661.7 - j834.4	-1.972 - j2.976	2.312	0.001905
π	0.580 + j0.814	580.6 - j814.4	-2.005 - j2.704	2.341	0.001930
3.2	0.578 + j0.925	485.5 - j777.5	-2.005 - j2.372	2.384	0.001966
3.3	0.579 + j1.121	363.6 - j704.5	-1.936 - j1.924	2.459	0.002028
3.4	0.587 + j1.326	279.0 - j630.7	-1.832 - j1.594	2.537	0.002089
3.5	0.604 + j1.546	219.1 - j561.3	-1.721 - j1.346	2.614	0.002151
3.6	0.632 + j1.787	176.0 - j497.5	-1.614 - j1.156	2.688	0.002212
3.7	0.677 + j2.057	144.3 - j438.7	-1.516 - j1.005	2.750	0.002274
3.8	0.745 + j2.368	120.9 - j384.2	-1.427 - j0.881	2.790	0.002335
3.9	0.853 + j2.737	103.8 - j333.0	-1.346 - j0.778	2.793	0.002396
4.0	1.026 + j3.187	91.5 - j284.3	-1.271 - j0.688	2.741	0.002458
4.2	1.846 + j4.466	79.0 - j191.2	-1.137 - j0.539	2.405	0.002581
4.4	4.875 + j6.027	81.1 - j100.3	-1.015 - j0.416	1.781	0.002704
4.6	9.994 + j0.786	99.4 - j7.8	-0.900 - j0.310	1.092	0.002827
3π/2	7.318 - j2.822	119.0 + j45.9	-0.835 - j0.255	0.769	0.002896
4.8	5.090 - j3.222	140.3 + j88.8	-0.785 - j0.214	0.566	0.002950
5.0	2.623 - j2.275	217.6 + j188.7	-0.664 - j0.124	0.252	0.003072
5.2	1.746 - j1.354	357.7 + j277.4	-0.525 - j0.039	0.093	0.003195
5.4	1.356 - j0.675	591.1 + j294.3	-0.351 + j0.026	0.025	0.003318
5.6	1.152 - j0.142	855.2 + j105.2	-0.147 + j0.018	0.003	0.003441
5.8	1.035 + j0.312	885.6 - j267.4	-0.005 - j0.084	0.001	0.003564

TABLE 2.6

TRANSMITTING AND RECEIVING CHARACTERISTICS OF THIN CYLINDRICAL DIPOLES

k_0h	Y, millimhos	Z, ohms	k_0h_e	D	a/λ
		Ω = 12.5 (Continued)			
6.0	0.967 + j0.727	660.6 - j496.8	0.032 - j0.168	0.005	0.003687
6.2	0.935 + j1.128	435.4 - j525.5	0.048 - j0.186	0.010	0.003810
2π	0.932 + j1.297	365.5 - j508.6	0.060 - j0.182	0.012	0.003861
6.4	0.938 + j1.538	288.9 - j473.9	0.084 - j0.170	0.015	0.003933
6.6	0.985 + j1.983	201.0 - j404.5	0.144 - j0.145	0.025	0.004056
6.8	1.101 + j2.501	147.4 - j335.0	0.226 - j0.120	0.053	0.004178
7.0	1.351 + j3.160	114.4 - j267.5	0.330 - j0.099	0.125	0.004301
		Ω = 15			
0.5	0.003 + j0.797	5.0 - j1255.2	0.247 - j0.000	1.464	0.000088
0.6	0.007 + j1.000	7.4 - j999.5	0.300 - j0.001	1.473	0.000106
0.7	0.016 + j1.237	10.3 - j808.5	0.356 - j0.001	1.484	0.000123
0.8	0.032 + j1.520	13.8 - j657.6	0.415 - j0.002	1.497	0.000141
0.9	0.064 + j1.875	18.1 - j532.9	0.477 - j0.004	1.511	0.000158
1.0	0.127 + j2.342	23.2 - j425.8	0.543 - j0.006	1.527	0.000176
1.1	0.266 + j2.998	29.3 - j330.9	0.614 - j0.009	1.544	0.000194
1.2	0.600 + j4.000	36.7 - j244.5	0.691 - j0.014	1.563	0.000211
1.3	1.579 + j5.672	45.6 - j163.6	0.775 - j0.021	1.584	0.000229
1.4	5.310 + j8.130	56.3 - j86.2	0.868 - j0.030	1.607	0.000246
1.5	14.086 + j2.121	69.4 - j10.5	0.971 - j0.043	1.632	0.000264
π/2	9.659 - j5.170	80.5 + j43.1	1.051 - j0.055	1.651	0.000277
1.6	7.387 - j5.637	85.6 + j65.3	1.086 - j0.060	1.660	0.000282
1.7	3.354 - j4.523	105.8 + j142.6	1.217 - j0.085	1.688	0.000299
1.8	1.959 - j3.325	131.5 + j223.2	1.367 - j0.120	1.718	0.000317
1.9	1.346 - j2.319	183.0 + j308.8	1.540 - j0.170	1.740	0.000334
2.0	1.022 - j1.958	209.5 + j401.4	1.747 - j0.243	1.781	0.000352
2.1	0.828 - j1.543	270.0 + j503.2	1.990 - j0.350	1.815	0.000370
2.2	0.702 - j1.219	354.7 + j615.9	2.282 - j0.512	1.850	0.000387
2.3	0.615 - j0.955	476.7 + j740.0	2.630 - j0.763	1.888	0.000405
2.4	0.552 - j0.731	658.1 + j870.9	3.037 - j1.164	1.929	0.000423
2.5	0.505 - j0.535	933.3 + j988.2	3.470 - j1.819	1.973	0.000440
2.6	0.469 - j0.359	1345.5 + j1029.3	3.793 - j2.879	2.022	0.000458
2.7	0.440 - j0.196	1893.6 + j844.9	3.614 - j4.439	2.077	0.000475
2.8	0.418 - j0.044	2366.4 + j249.2	2.313 - j6.065	2.136	0.000493
2.9	0.400 + j0.102	2345.7 - j598.5	0.036 - j6.561	2.202	0.000511
3.0	0.387 + j0.245	1846.2 - j1169.0	-1.822 - j5.627	2.274	0.000528
3.1	0.377 + j0.387	1292.9 - j1327.0	-2.636 - j4.289	2.352	0.000546
π	0.374 + j0.446	1104.0 - j1317.8	-2.753 - j3.792	2.386	0.000553
3.2	0.370 + j0.530	885.8 - j1268.0	-2.790 - j3.193	2.436	0.000563
3.3	0.368 + j0.678	618.3 - j1140.2	-2.676 - j2.418	2.525	0.000581
3.4	0.369 + j0.833	444.6 - j1003.8	-2.482 - j1.882	2.618	0.000599
3.5	0.375 + j0.999	329.4 - j877.3	-2.277 - j1.505	2.713	0.000616
3.6	0.388 + j1.181	250.9 - j764.3	-2.086 - j1.230	2.805	0.000634
3.7	0.409 + j1.386	196.0 - j663.8	-1.915 - j1.024	2.888	0.000651
3.8	0.444 + j1.622	156.9 - j573.7	-1.763 - j0.864	2.948	0.000669
3.9	0.499 + j1.902	129.0 - j491.8	-1.628 - j0.737	2.971	0.000687

TABLE 2.6

TRANSMITTING AND RECEIVING CHARACTERISTICS OF THIN CYLINDRICAL DIPOLES

k_0h	Y, millimhos	Z, ohms	k_0h_e	D	a/λ
		Ω = 15 (Continued)			
4.0	0.590 + j2.247	109.3 - j416.3	-1.508 - j0.632	2.934	0.000704
4.2	1.032 + j3.269	87.8 - j278.2	-1.299 - j0.470	2.608	0.000739
4.4	2.863 + j5.055	84.8 - j149.8	-1.122 - j0.348	1.952	0.000775
4.6	9.551 + j2.285	99.0 - j23.7	-0.965 - j0.251	1.205	0.000810
3π/2	7.349 - j3.084	115.7 + j48.6	-0.884 - j0.204	0.853	0.000830
4.8	4.574 - j3.632	134.1 + j106.5	-0.822 - j0.170	0.631	0.000845
5.0	1.986 - j2.440	200.6 + j246.5	-0.685 - j0.100	0.287	0.000880
5.2	1.226 - j1.516	322.5 + j398.7	-0.547 - j0.036	0.112	0.000915
5.4	0.915 - j0.911	549.1 + j546.2	-0.395 + j0.020	0.034	0.000951
5.6	0.758 - j0.468	954.7 + j589.8	-0.213 + j0.049	0.006	0.000986
5.8	0.668 - j0.111	1456.7 + j240.9	-0.020 - j0.005	0.000	0.001021
6.0	0.614 + j0.204	1465.4 - j487.7	0.070 - j0.133	0.002	0.001056
6.2	0.584 + j0.502	984.0 - j845.9	0.067 - j0.186	0.005	0.001092
2π	0.578 + j0.626	795.9 - j862.3	0.070 - j0.182	0.006	0.001106
6.4	0.575 + j0.804	588.7 - j822.9	0.087 - j0.167	0.007	0.001127
6.6	0.589 + j1.131	362.2 - j695.5	0.146 - j0.131	0.013	0.001162
6.8	0.638 + j1.516	235.7 - j560.2	0.234 - j0.099	0.033	0.001197
7.0	0.756 + j2.016	163.2 - j434.9	0.345 - j0.075	0.092	0.001232

d. ELECTRICALLY LONG DIPOLES IN DISSIPATIVE MEDIA AND IN AIR

Table 2.7 parallels Table 2.6 but provides the normalized admittances of dipoles in dissipative and dielectric media in addition to those in air. The numerical values have been computed from a formula derived by Wu[12,13] which is based on the Wiener–Hopf technique. It becomes increasingly accurate as the antenna is made longer but is somewhat inaccurate for antennas shorter than a wavelength. The values tabulated are idealized in the sense that a term which is responsible for the rapid rise of the current to infinity at the driving point of the assumed delta-function generator has been subtracted. Table 2.7 is in two parts: it begins with a/λ constant and concludes with Ω constant. The quantity tabulated is the normalized admittance Y/Δ, where the normalizing factor $\Delta = \beta\zeta_0/\omega\mu$ is defined in (1.15)–(1.19). It reduces to unity in air. The parameter is α/β in $k = \beta(1 - j\alpha/\beta)$.

Tables 2.8 and 2.9 extend Table 2.7 to much longer antennas, but only for those immersed in air. In Table 2.8 the constant parameter is a/λ; k_0h ranges from 25 to values slightly greater than 100. In Table 2.9 the constant parameter is $\Omega = 2\ln(2h/a)$; k_0h extends from 1.5 to the largest value permitted by the condition $\beta a \ll 1$. With $\Omega = 15$ this is about $k_0h = 100$; with $\Omega = 20$ it is about $k_0h = 1000$. Table 2.9 also provides the electrical effective half-length in radians, $k_0h_e = k_0h_e(\pi/2)$ as defined in (2.14).

The normalized radiation field of electrically long dipole antennas in air is listed in Table 2.10. Since the far-field pattern of a thin antenna is relatively insensitive to the thickness of the antenna, the field for $a/\lambda = 0.007022$ only is given. The numerical values in Table 2.10 are for the quantity

$$\lim_{R\to\infty} \frac{E_\theta(\theta)}{V_0} Re^{jk_0R} \tag{2.19}$$

in its real and imaginary parts and magnitude. R is the distance from the center of the dipole to the point where $E_\theta(\theta)$ is determined. The field defined in (2.19) differs from the conventionally defined far field,

$$\lim_{R\to\infty} \frac{-j2\pi E_\theta(\theta)}{\zeta_0 I(0)} Re^{jk_0R} \tag{2.20}$$

by the factor $-j2\pi/\zeta_0 Y$, where Y is the admittance of the antenna and $\zeta_0 \doteq 120\pi$ ohms. The electrical half-length of the dipole varies from $k_0h = 2\pi$ to $k_0h = 20.5\pi$ in steps of 0.5π. Note that the radian effective length $k_0h_e(\pi/2)$ is equal to the conventional far-field factor defined in (2.20) when $\theta = \pi/2$.

TABLE 2.7

NORMALIZED ADMITTANCES (Y/Δ) IN MILLIMHOS OF THIN DIPOLE ANTENNAS IN DISSIPATIVE MEDIA

$a/\lambda = 0.001190$

βh	α/β = 0.00	α/β = 0.01	α/β = 0.03	α/β = 0.05	α/β = 0.07	α/β = 0.10	α/β = 0.30	α/β = 0.50	α/β = 0.70	α/β = 1.00
1.5	12.265 -1.174	11.164 -0.736	9.445 -0.226	8.184 0.022	7.231 0.140	6.180 0.200	3.482 -0.066	2.815 -0.381	2.619 -0.680	2.619 -1.140
π/2	8.372 -4.287	8.069 -3.605	7.409 -2.575	6.770 -1.874	6.198 -1.395	5.481 -0.938	3.290 -0.320	2.710 -0.465	2.554 -0.709	2.586 -1.144
1.6	6.958 -4.526	6.844 -3.936	6.495 -2.981	6.084 -2.279	5.673 -1.767	5.115 -1.248	3.201 -0.402	2.665 -0.492	2.528 -0.718	2.574 -1.144
1.7	3.937 -3.960	4.032 -3.668	4.127 -3.121	4.125 -2.639	4.061 -2.230	3.902 -1.743	2.885 -0.591	2.519 -0.549	2.447 -0.731	2.538 -1.139
1.8	2.545 -3.092	2.648 -2.943	2.806 -2.644	2.908 -2.354	2.963 -2.086	2.981 -1.733	2.590 -0.665	2.387 -0.567	2.378 -0.727	2.511 -1.129
1.9	1.824 -2.397	1.911 -2.312	2.060 -2.136	2.177 -1.957	2.265 -1.783	2.350 -1.540	2.339 -0.663	2.275 -0.557	2.323 -0.712	2.491 -1.117
2.0	1.405 -1.864	1.477 -1.811	1.609 -1.699	1.720 -1.584	1.814 -1.467	1.922 -1.299	2.135 -0.614	2.183 -0.528	2.280 -0.690	2.477 -1.105
2.1	1.138 -1.446	1.201 -1.411	1.316 -1.337	1.420 -1.258	1.512 -1.178	1.627 -1.059	1.975 -0.540	2.111 -0.489	2.248 -0.665	2.469 -1.092
2.2	0.957 -1.109	1.013 -1.085	1.117 -1.033	1.214 -0.978	1.302 -0.921	1.419 -0.836	1.853 -0.453	2.057 -0.443	2.226 -0.638	2.464 -1.081
2.3	0.828 -0.828	0.879 -0.810	0.976 -0.773	1.067 -0.734	1.153 -0.693	1.269 -0.632	1.761 -0.360	2.018 -0.396	2.212 -0.613	2.461 -1.072
2.4	0.732 -0.587	0.780 -0.574	0.872 -0.547	0.960 -0.518	1.044 -0.489	1.160 -0.445	1.695 -0.268	1.992 -0.349	2.204 -0.589	2.461 -1.064
2.5	0.659 -0.375	0.705 -0.365	0.795 -0.345	0.881 -0.325	0.964 -0.304	1.081 -0.274	1.649 -0.177	1.977 -0.304	2.202 -0.568	2.462 -1.057
2.6	0.602 -0.184	0.647 -0.177	0.736 -0.162	0.822 -0.148	0.905 -0.134	1.024 -0.115	1.621 -0.090	1.971 -0.263	2.203 -0.549	2.464 -1.052
2.7	0.556 -0.008	0.602 -0.003	0.691 0.007	0.779 0.017	0.863 0.025	0.985 0.034	1.607 -0.008	1.972 -0.226	2.207 -0.534	2.466 -1.049
2.8	0.520 0.157	0.566 0.161	0.658 0.167	0.747 0.172	0.834 0.175	0.960 0.176	1.606 0.069	1.980 -0.193	2.213 -0.522	2.469 -1.046
2.9	0.490 0.315	0.538 0.317	0.633 0.321	0.726 0.321	0.817 0.319	0.947 0.312	1.615 0.140	1.991 -0.165	2.220 -0.512	2.471 -1.045
3.0	0.467 0.469	0.517 0.471	0.617 0.470	0.715 0.467	0.810 0.460	0.946 0.444	1.635 0.204	2.006 -0.143	2.228 -0.506	2.473 -1.044
3.1	0.449 0.623	0.502 0.623	0.608 0.619	0.712 0.611	0.812 0.599	0.956 0.573	1.662 0.262	2.023 -0.125	2.236 -0.501	2.475 -1.043
π	0.442 0.687	0.498 0.686	0.607 0.681	0.713 0.671	0.817 0.656	0.964 0.626	1.676 0.283	2.031 -0.119	2.239 -0.500	2.476 -1.043
3.2	0.435 0.778	0.493 0.777	0.607 0.769	0.718 0.756	0.826 0.737	0.978 0.701	1.697 0.311	2.042 -0.112	2.243 -0.499	2.477 -1.043
3.3	0.427 0.939	0.490 0.936	0.614 0.923	0.735 0.904	0.850 0.877	1.012 0.827	1.738 0.352	2.060 -0.104	2.250 -0.498	2.478 -1.044
3.4	0.423 1.108	0.494 1.103	0.631 1.084	0.763 1.056	0.888 1.020	1.061 0.953	1.783 0.385	2.078 -0.100	2.255 -0.498	2.479 -1.044
3.5	0.427 1.290	0.506 1.282	0.659 1.255	0.805 1.216	0.941 1.167	1.126 1.079	1.832 0.407	2.095 -0.099	2.260 -0.500	2.480 -1.044
3.6	0.439 1.491	0.529 1.478	0.703 1.439	0.865 1.384	1.014 1.318	1.211 1.202	1.882 0.420	2.110 -0.102	2.264 -0.502	2.480 -1.045
3.7	0.463 1.717	0.568 1.697	0.768 1.640	0.950 1.564	1.113 1.474	1.319 1.322	1.933 0.423	2.122 -0.107	2.267 -0.504	2.480 -1.045
3.8	0.505 1.978	0.630 1.948	0.863 1.864	1.069 1.755	1.245 1.632	1.456 1.434	1.981 0.416	2.133 -0.114	2.269 -0.506	2.480 -1.046
3.9	0.577 2.289	0.729 2.242	1.003 2.114	1.234 1.957	1.420 1.787	1.627 1.529	2.025 0.400	2.141 -0.122	2.271 -0.509	2.481 -1.046
4.0	0.697 2.671	0.887 2.593	1.212 2.393	1.465 2.163	1.651 1.928	1.833 1.596	2.064 0.376	2.147 -0.130	2.271 -0.511	2.481 -1.046
4.1	0.909 3.154	1.149 3.018	1.529 2.695	1.787 2.355	1.950 2.033	2.075 1.617	2.095 0.345	2.150 -0.139	2.271 -0.513	2.480 -1.047
4.2	1.299 3.779	1.601 3.528	2.014 2.990	2.230 2.489	2.322 2.065	2.342 1.572	2.117 0.310	2.152 -0.147	2.271 -0.515	2.480 -1.047
4.3	2.073 4.573	2.414 4.082	2.744 3.185	2.804 2.487	2.751 1.971	2.610 1.440	2.131 0.273	2.151 -0.155	2.271 -0.517	2.480 -1.047
4.4	3.701 5.371	3.868 4.417	3.747 3.059	3.457 2.228	3.175 1.701	2.840 1.214	2.135 0.236	2.150 -0.162	2.270 -0.518	2.480 -1.047
4.5	6.771 4.976	5.987 3.666	4.790 2.293	4.010 1.626	3.489 1.245	2.985 0.912	2.132 0.201	2.147 -0.167	2.269 -0.519	2.480 -1.047
4.6	8.998 1.102	7.215 1.047	5.255 0.900	4.217 0.778	3.587 0.685	3.015 0.578	2.122 0.170	2.144 -0.172	2.268 -0.520	2.480 -1.047
4.7	6.902 -2.446	6.084 -1.420	4.816 -0.429	3.992 -0.026	3.448 0.157	2.931 0.268	2.106 0.143	2.140 -0.175	2.268 -0.520	2.480 -1.047
3π/2	6.537 -2.638	5.858 -1.602	4.717 -0.554	3.941 -0.109	3.417 0.100	2.914 0.234	2.104 0.140	2.140 -0.175	2.268 -0.520	2.480 -1.047
4.8	4.365 -3.021	4.317 -2.220	3.931 -1.150	3.502 -0.560	3.147 -0.230	2.764 0.025	2.086 0.122	2.136 -0.177	2.267 -0.520	2.480 -1.047
4.9	2.895 -2.617	3.055 -2.140	3.095 -1.351	2.962 -0.807	2.789 -0.451	2.554 -0.137	2.065 0.107	2.132 -0.178	2.266 -0.520	2.480 -1.047
5.0	2.086 -2.108	2.271 -1.823	2.459 -1.288	2.490 -0.856	2.445 -0.537	2.338 -0.223	2.042 0.098	2.129 -0.179	2.266 -0.520	2.480 -1.047
5.1	1.612 -1.663	1.781 -1.485	2.007 -1.121	2.114 -0.796	2.149 -0.532	2.135 -0.249	2.020 0.094	2.126 -0.179	2.265 -0.520	2.480 -1.047
5.2	1.312 -1.296	1.460 -1.178	1.685 -0.926	1.827 -0.684	1.907 -0.473	1.958 -0.231	1.999 0.094	2.123 -0.178	2.265 -0.520	2.480 -1.047
5.3	1.111 -0.992	1.241 -0.911	1.455 -0.731	1.609 -0.550	1.714 -0.385	1.808 -0.185	1.980 0.098	2.120 -0.177	2.265 -0.519	2.480 -1.047
5.4	0.970 -0.735	1.086 -0.677	1.286 -0.546	1.443 -0.410	1.561 -0.282	1.684 -0.121	1.963 0.106	2.119 -0.175	2.264 -0.519	2.480 -1.047
5.5	0.866 -0.513	0.972 -0.470	1.160 -0.373	1.317 -0.271	1.443 -0.172	1.584 -0.045	1.949 0.115	2.117 -0.173	2.264 -0.519	2.480 -1.047
5.6	0.787 -0.318	0.886 -0.285	1.065 -0.212	1.220 -0.135	1.351 -0.061	1.505 0.036	1.938 0.127	2.116 -0.172	2.264 -0.519	2.480 -1.047
5.7	0.727 -0.141	0.820 -0.116	0.993 -0.062	1.147 -0.005	1.280 0.050	1.444 0.121	1.931 0.139	2.116 -0.170	2.264 -0.518	2.480 -1.047
5.8	0.679 0.022	0.769 0.041	0.938 0.081	1.092 0.122	1.227 0.160	1.399 0.208	1.926 0.152	2.116 -0.168	2.264 -0.518	2.480 -1.047
5.9	0.640 0.175	0.729 0.189	0.897 0.217	1.052 0.244	1.190 0.268	1.367 0.294	1.923 0.164	2.116 -0.167	2.264 -0.518	2.480 -1.047
6.0	0.610 0.321	0.659 0.331	0.868 0.350	1.024 0.364	1.166 0.374	1.348 0.380	1.923 0.176	2.116 -0.166	2.264 -0.518	2.480 -1.047
6.1	0.586 0.464	0.676 0.471	0.849 0.479	1.009 0.482	1.154 0.479	1.341 0.465	1.926 0.187	2.117 -0.165	2.264 -0.518	2.480 -1.047
6.2	0.567 0.606	0.660 0.609	0.839 0.608	1.004 0.599	1.153 0.583	1.345 0.548	1.930 0.197	2.117 -0.164	2.264 -0.518	2.480 -1.047
2π	0.556 0.725	0.652 0.725	0.837 0.716	1.008 0.696	1.161 0.668	1.357 0.615	1.934 0.204	2.118 -0.163	2.264 -0.518	2.480 -1.047
6.3	0.554 0.749	0.651 0.748	0.838 0.738	1.010 0.716	1.164 0.685	1.360 0.628	1.935 0.205	2.118 -0.163	2.265 -0.518	2.480 -1.047
6.4	0.546 0.896	0.649 0.892	0.847 0.870	1.027 0.834	1.187 0.787	1.386 0.706	1.942 0.212	2.119 -0.163	2.265 -0.518	2.480 -1.047
6.5	0.543 1.050	0.655 1.041	0.867 1.006	1.057 0.953	1.222 0.888	1.423 0.780	1.949 0.217	2.120 -0.163	2.265 -0.518	2.480 -1.047
6.6	0.547 1.215	0.671 1.200	0.901 1.148	1.102 1.073	1.272 0.986	1.471 0.849	1.957 0.220	2.120 -0.163	2.265 -0.518	2.480 -1.047
6.7	0.560 1.396	0.698 1.372	0.950 1.297	1.164 1.195	1.338 1.082	1.532 0.912	1.964 0.221	2.121 -0.163	2.265 -0.518	2.480 -1.047
6.8	0.584 1.597	0.741 1.562	1.021 1.454	1.247 1.318	1.421 1.172	1.604 0.965	1.972 0.221	2.122 -0.163	2.265 -0.518	2.480 -1.047
6.9	0.626 1.826	0.808 1.774	1.119 1.621	1.356 1.438	1.525 1.253	1.688 1.006	1.979 0.220	2.122 -0.163	2.265 -0.518	2.480 -1.047
7.0	0.694 2.095	0.908 2.015	1.256 1.796	1.496 1.551	1.652 1.320	1.783 1.031	1.985 0.217	2.122 -0.164	2.265 -0.518	2.480 -1.047
7.1	0.805 2.419	1.062 2.292	1.445 1.973	1.675 1.648	1.800 1.365	1.885 1.037	1.990 0.214	2.123 -0.164	2.265 -0.518	2.480 -1.047
7.2	0.991 2.817	1.301 2.611	1.706 2.141	1.896 1.715	1.970 1.378	1.990 1.020	1.994 0.209	2.123 -0.164	2.265 -0.518	2.480 -1.047
7.3	1.315 3.316	1.682 2.965	2.059 2.270	2.160 1.730	2.153 1.349	2.093 0.978	1.997 0.205	2.123 -0.165	2.265 -0.518	2.480 -1.047
7.4	1.914 3.927	2.300 3.309	2.519 2.305	2.454 1.665	2.336 1.266	2.185 0.910	1.999 0.200	2.123 -0.165	2.265 -0.518	2.480 -1.047
7.5	3.075 4.558	3.276 3.483	3.062 2.161	2.745 1.496	2.499 1.127	2.259 0.819	2.000 0.195	2.123 -0.165	2.265 -0.518	2.480 -1.047
7.6	5.214 4.597	4.599 3.081	3.584 1.753	2.981 1.214	2.617 0.938	2.307 0.711	1.999 0.190	2.123 -0.165	2.265 -0.518	2.480 -1.047
7.7	7.598 2.438	5.663 1.677	3.898 1.092	3.103 0.852	2.672 0.719	2.324 0.596	1.998 0.186	2.123 -0.166	2.265 -0.518	2.480 -1.047
7.8	7.154 -1.058	5.523 -0.160	3.862 0.356	3.082 0.473	2.655 0.497	2.311 0.483	1.996 0.182	2.122 -0.166	2.265 -0.518	2.480 -1.047
5π/2	5.961 -2.105	4.999 -0.887	3.708 0.011	3.015 0.286	2.618 0.386	2.292 0.426	1.995 0.181	2.122 -0.166	2.265 -0.518	2.480 -1.047
7.9	4.951 -2.477	4.465 -1.283	3.526 -0.229	2.934 0.146	2.574 0.300	2.270 0.381	1.994 0.179	2.122 -0.166	2.265 -0.518	2.480 -1.047
8.0	3.321 -2.441	3.381 -1.596	3.062 -0.568	2.708 -0.086	2.448 0.147	2.208 0.298	1.991 0.177	2.122 -0.166	2.265 -0.518	2.480 -1.047
8.1	2.372 -2.051	2.586 -1.515	2.609 -0.697	2.456 -0.220	2.299 0.044	2.132 0.236	1.988 0.175	2.122 -0.166	2.265 -0.518	2.480 -1.047
8.2	1.813 -1.646	2.048 -1.302	2.227 -0.693	2.213 -0.273	2.146 -0.012	2.050 0.196	1.985 0.175	2.122 -0.166	2.265 -0.518	2.480 -1.047
8.3	1.464 -1.293	1.683 -1.065	1.925 -0.617	1.998 -0.268	1.999 -0.029	1.968 0.177	1.982 0.174	2.122 -0.166	2.265 -0.518	2.480 -1.047
8.4	1.234 -0.994	1.431 -0.837	1.691 -0.507	1.816 -0.224	1.867 -0.016	1.889 0.176	1.980 0.175	2.122 -0.166	2.265 -0.518	2.480 -1.047
8.5	1.074 -0.741	1.250 -0.629	1.511 -0.382	1.666 -0.157	1.753 0.018	1.817 0.189	1.977 0.175	2.121 -0.166	2.265 -0.518	2.480 -1.047
8.6	0.958 -0.522	1.118 -0.439	1.373 -0.254	1.545 -0.076	1.656 0.068	1.754 0.214	1.975 0.176	2.121 -0.166	2.265 -0.518	2.480 -1.047
8.7	0.871 -0.329	1.019 -0.267	1.266 -0.126	1.447 0.012	1.575 0.128	1.700 0.248	1.973 0.178	2.121 -0.166	2.265 -0.518	2.480 -1.047
8.8	0.804 -0.155	0.943 -0.108	1.183 -0.001	1.370 0.105	1.510 0.195	1.655 0.288	1.972 0.180	2.121 -0.166	2.265 -0.518	2.480 -1.047
8.9	0.752 0.004	0.885 0.040	1.120 0.120	1.311 0.199	1.460 0.266	1.620 0.333	1.971 0.181	2.121 -0.166	2.265 -0.518	2.480 -1.047
9.0	0.711 0.153	0.840 0.180	1.072 0.238	1.266 0.294	1.422 0.339	1.594 0.380	1.970 0.183	2.121 -0.165	2.265 -0.518	2.480 -1.047
9.1	0.679 0.295	0.806 0.315	1.038 0.354	1.235 0.389	1.396 0.414	1.576 0.430	1.970 0.185	2.121 -0.165	2.265 -0.518	2.480 -1.047
9.2	0.653 0.432	0.780 0.446	1.015 0.469	1.216 0.484	1.381 0.489	1.567 0.480	1.971 0.187	2.121 -0.165	2.265 -0.518	2.480 -1.047
9.3	0.633 0.568	0.763 0.576	1.002 0.583	1.208 0.578	1.377 0.564	1.567 0.529	1.971 0.188	2.121 -0.165	2.265 -0.518	2.480 -1.047
9.4	0.619 0.705	0.753 0.707	1.000 0.697	1.210 0.672	1.382 0.638	1.574 0.577	1.972 0.190	2.121 -0.165	2.265 -0.518	2.480 -1.047
3π	0.616 0.739	0.751 0.740	1.001 0.725	1.213 0.695	1.386 0.656	1.577 0.589	1.972 0.190	2.121 -0.165	2.255 -0.518	2.480 -1.047
9.5	0.610 0.844	0.750 0.840	1.008 0.812	1.225 0.766	1.399 0.710	1.588 0.623	1.973 0.191	2.121 -0.165	2.265 -0.518	2.480 -1.047
9.6	0.607 0.990	0.756 0.978	1.027 0.930	1.251 0.859	1.425 0.780	1.611 0.665	1.974 0.191	2.121 -0.165	2.265 -0.518	2.480 -1.047
9.7	0.610 1.144	0.772 1.124	1.060 1.050	1.290 0.950	1.463 0.846	1.640 0.703	1.975 0.192	2.122 -0.165	2.265 -0.518	2.480 -1.047
9.8	0.621 1.311	0.799 1.279	1.108 1.173	1.343 1.040	1.512 0.907	1.675 0.736	1.976 0.192	2.122 -0.165	2.265 -0.518	2.480 -1.047
9.9	0.644 1.495	0.843 1.447	1.175 1.299	1.412 1.125	1.572 0.961	1.716 0.762	1.977 0.192	2.122 -0.165	2.265 -0.518	2.480 -1.047
10.0	0.681 1.703	0.906 1.632	1.265 1.427	1.500 1.204	1.643 1.006	1.762 0.780	1.978 0.192	2.122 -0.165	2.265 -0.518	2.480 -1.047
10.1	0.742 1.944	1.003 1.838	1.386 1.555	1.607 1.271	1.725 1.038					
10.2	0.838 2.228	1.144 2.067	1.544 1.675	1.735 1.322	1.817 1.055					
10.3	0.996 2.572	1.354 2.321	1.748 1.777	1.882 1.347	1.914 1.051					
10.4	1.261 2.995	1.671 2.590	2.006 1.840	2.046 1.338	2.012 1.026					
10.5	1.730 3.508	2.153 2.839	2.314 1.837	2.215 1.284	2.106 0.975					
10.6	2.597 4.064	2.864 2.964	2.654 1.727	2.375 1.180	2.187 0.901					
10.7	4.183 4.337	3.800 2.744	2.973 1.482	2.505 1.026	2.247 0.807					
10.8	6.399 3.251	4.682 1.915	3.196 1.106	2.585 0.835	2.281 0.700					
10.9	7.164 0.270	4.944 0.603	3.254 0.661	2.602 0.628	2.285 0.589					
7π/2	5.639 -1.815	4.457 -0.490	3.146 0.261	2.558 0.439	2.262 0.487					

βh	α/β = 0.00		α/β = 0.01		α/β = 0.03		α/β = 0.05		α/β = 0.07	
11.0	5.551	-1.865	4.424	-0.528	3.137	0.245	2.555	0.431	2.260	0.483
11.1	3.806	-2.281	3.591	-1.089	2.897	-0.069	2.458	0.266	2.211	0.391
11.2	2.692	-2.040	2.845	-1.216	2.606	-0.257	2.328	0.144	2.144	0.318
11.3	2.027	-1.675	2.286	-1.130	2.319	-0.339	2.184	0.068	2.067	0.266
11.4	1.617	-1.331	1.888	-0.964	2.065	-0.343	2.041	0.033	1.985	0.235
11.5	1.350	-1.032	1.604	-0.779	1.851	-0.300	1.908	0.030	1.904	0.224
11.6	1.166	-0.777	1.400	-0.596	1.677	-0.227	1.789	0.052	1.829	0.229
11.7	1.036	-0.556	1.249	-0.423	1.537	-0.139	1.686	0.092	1.760	0.247
11.8	0.939	-0.361	1.136	-0.262	1.426	-0.043	1.599	0.145	1.700	0.276
11.9	0.866	-0.187	1.049	-0.112	1.337	0.056	1.528	0.206	1.649	0.313
12.0	0.809	-0.029	0.983	0.029	1.269	0.157	1.471	0.272	1.607	0.355
12.1	0.764	0.119	0.932	0.163	1.216	0.258	1.426	0.343	1.574	0.402
12.2	0.729	0.259	0.893	0.291	1.176	0.359	1.394	0.415	1.550	0.452
12.3	0.701	0.394	0.864	0.417	1.149	0.459	1.372	0.489	1.535	0.503
12.4	0.680	0.526	0.844	0.541	1.133	0.559	1.361	0.564	1.528	0.555
12.5	0.664	0.658	0.832	0.665	1.128	0.660	1.360	0.638	1.530	0.606
4π	0.657	0.747	0.829	0.748	1.131	0.727	1.365	0.687	1.535	0.640
12.6	0.654	0.792	0.828	0.790	1.134	0.761	1.370	0.711	1.539	0.656
12.7	0.649	0.931	0.833	0.920	1.150	0.862	1.390	0.784	1.557	0.704
12.8	0.651	1.077	0.847	1.054	1.180	0.965	1.420	0.854	1.582	0.749
12.9	0.660	1.234	0.874	1.197	1.223	1.068	1.462	0.920	1.615	0.789
13.0	0.679	1.406	0.915	1.349	1.282	1.171	1.516	0.981	1.656	0.823
13.1	0.712	1.597	0.976	1.514	1.361	1.272	1.582	1.035	1.702	0.850
13.2	0.765	1.816	1.064	1.693	1.463	1.369	1.661	1.079	1.754	0.868
13.3	0.848	2.072	1.190	1.888	1.592	1.455	1.751	1.108	1.810	0.876
13.4	0.981	2.377	1.374	2.099	1.751	1.522	1.851	1.120	1.868	0.872
13.5	1.199	2.746	1.640	2.315	1.942	1.558	1.957	1.109	1.926	0.856
13.6	1.572	3.192	2.026	2.509	2.161	1.546	2.064	1.071	1.979	0.826
13.7	2.236	3.695	2.570	2.614	2.392	1.467	2.162	1.006	2.025	0.785
13.8	3.438	4.078	3.274	2.496	2.609	1.309	2.243	0.915	2.060	0.734
13.9	5.332	3.666	3.996	1.983	2.773	1.074	2.299	0.803	2.082	0.676
14.0	6.823	1.480	4.403	1.060	2.850	0.788	2.322	0.679	2.090	0.615
14.1	6.066	-1.079	4.246	0.067	2.823	0.495	2.310	0.556	2.083	0.555
9π/2	5.422	-1.624	4.065	-0.233	2.789	0.395	2.298	0.513	2.077	0.534
14.2	4.346	-2.050	3.683	-0.610	2.706	0.243	2.267	0.444	2.063	0.500
14.3	3.059	-2.025	3.040	-0.902	2.529	0.057	2.200	0.352	2.031	0.452
14.4	2.268	-1.721	2.494	-0.941	2.329	-0.057	2.117	0.283	1.991	0.414
14.5	1.781	-1.388	2.076	-0.855	2.131	-0.108	2.026	0.239	1.946	0.388
14.6	1.469	-1.088	1.768	-0.717	1.951	-0.110	1.935	0.218	1.899	0.372
14.7	1.258	-0.828	1.541	-0.563	1.794	-0.079	1.847	0.217	1.852	0.367
14.8	1.110	-0.602	1.372	-0.409	1.662	-0.026	1.767	0.231	1.806	0.371
14.9	1.002	-0.405	1.245	-0.260	1.552	0.041	1.696	0.259	1.765	0.383
15.0	0.920	-0.228	1.147	-0.119	1.463	0.117	1.635	0.297	1.728	0.402
15.1	0.858	-0.068	1.072	0.015	1.392	0.198	1.585	0.341	1.696	0.426
15.2	0.809	0.080	1.015	0.143	1.336	0.282	1.544	0.391	1.671	0.455
15.3	0.771	0.219	0.970	0.267	1.294	0.367	1.513	0.444	1.651	0.486
15.4	0.740	0.352	0.937	0.387	1.263	0.454	1.491	0.500	1.638	0.520
15.5	0.717	0.482	0.914	0.506	1.244	0.541	1.478	0.557	1.630	0.554
15.6	0.700	0.611	0.900	0.624	1.236	0.629	1.474	0.615	1.629	0.589
15.7	0.688	0.742	0.894	0.743	1.238	0.717	1.479	0.672	1.634	0.623
5π	0.687	0.752	0.893	0.753	1.238	0.724	1.480	0.676	1.634	0.626
15.8	0.681	0.875	0.896	0.865	1.251	0.805	1.493	0.728	1.644	0.655
15.9	0.681	1.015	0.909	0.991	1.275	0.893	1.515	0.781	1.660	0.686
16.0	0.688	1.164	0.933	1.123	1.312	0.981	1.546	0.831	1.681	0.712
16.1	0.704	1.325	0.971	1.262	1.363	1.066	1.586	0.876	1.707	0.735
16.2	0.732	1.504	1.027	1.411	1.429	1.148	1.635	0.915	1.737	0.752
16.3	0.778	1.706	1.107	1.570	1.514	1.223	1.692	0.945	1.769	0.764
16.4	0.849	1.939	1.220	1.740	1.618	1.288	1.755	0.965	1.804	0.769
[illegible]	[illegible]	[illegible]	[illegible]	[illegible]	1.763	[illegible]	1.824	0.971	1.840	0.767
16.6	1.141	2.542	1.605	2.100	1.888	1.359	1.896	0.963	[illegible]	[illegible]
16.7	1.441	2.937	1.921	2.260	2.048	1.348	1.967	0.939	1.907	0.741
16.8	1.960	3.393	2.352	2.354	2.215	1.293	2.032	0.897	1.935	0.717
16.9	2.883	3.817	2.901	2.295	2.371	1.186	2.086	0.841	1.957	0.688
17.0	4.430	3.804	3.494	1.969	2.497	1.031	2.126	0.771	1.972	0.655
17.1	6.188	2.435	3.931	1.324	2.572	0.838	2.148	0.694	1.979	0.620
17.2	6.371	-0.120	3.953	0.511	2.582	0.630	2.149	0.614	1.979	0.585
11π/2	5.262	-1.486	3.762	-0.053	2.546	0.475	2.136	0.554	1.973	0.558
17.3	4.912	-1.684	3.671	-0.177	2.530	0.435	2.131	0.538	1.971	0.551
17.4	3.482	-1.971	3.165	-0.582	2.427	0.274	2.096	0.471	1.955	0.521
17.5	2.545	-1.766	2.664	-0.733	2.293	0.157	2.048	0.416	1.935	0.495
17.6	1.966	-1.455	2.246	-0.728	2.147	0.086	1.992	0.376	1.909	0.475
17.7	1.598	-1.155	1.922	-0.643	2.003	0.056	1.931	0.350	1.882	0.461
17.8	1.354	-0.889	1.677	-0.523	1.870	0.057	1.869	0.339	1.853	0.454
17.9	1.185	-0.658	1.491	-0.390	1.752	0.082	1.810	0.341	1.824	0.452
18.0	1.063	-0.455	1.349	-0.256	1.650	0.125	1.755	0.353	1.797	0.457
18.1	0.972	-0.275	1.241	-0.124	1.565	0.179	1.706	0.374	1.771	0.466
18.2	0.903	-0.113	1.157	0.002	1.494	0.242	1.664	0.402	1.749	0.479
18.3	0.849	0.036	1.092	0.125	1.438	0.309	1.628	0.436	1.730	0.496
18.4	0.807	0.176	1.042	0.243	1.393	0.381	1.600	0.474	1.715	0.515
18.5	0.774	0.309	1.005	0.358	1.361	0.455	1.580	0.515	1.704	0.536
18.6	0.748	0.438	0.978	0.472	1.339	0.530	1.567	0.558	1.697	0.559
18.7	0.729	0.565	0.960	0.585	1.327	0.606	1.561	0.602	1.695	0.582
18.8	0.715	0.692	0.951	0.699	1.326	0.682	1.562	0.645	1.696	0.604
6π	0.711	0.756	0.950	0.756	1.329	0.720	1.565	0.666	1.699	0.615
18.9	0.707	0.822	0.952	0.814	1.334	0.758	1.570	0.688	1.702	0.626
19.0	0.705	0.957	0.962	0.933	1.353	0.834	1.586	0.729	1.711	0.646
19.1	0.710	1.099	0.983	1.056	1.384	0.908	1.608	0.766	1.724	0.664
19.2	0.723	1.252	1.017	1.185	1.426	0.980	1.637	0.800	1.741	0.679
19.3	0.746	1.420	1.068	1.320	1.481	1.047	1.672	0.829	1.759	0.691
19.4	0.785	1.607	1.140	1.464	1.550	1.108	1.712	0.851	1.780	0.699
19.5	0.846	1.822	1.241	1.615	1.634	1.159	1.758	0.865	1.802	0.703
19.6	0.941	2.072	1.381	1.771	1.733	1.196	1.806	0.871	1.824	0.703
19.7	1.091	2.368	1.573	1.925	1.844	1.213	1.855	0.866	1.845	0.698
19.8	1.334	2.722	1.836	2.062	1.966	1.205	1.904	0.850	1.866	0.688
19.9	1.746	3.137	2.186	2.149	2.090	1.167	1.949	0.824	1.883	0.675
20.0	2.464	3.565	2.627	2.127	2.208	1.093	1.987	0.788	1.898	0.658
20.1	3.697	3.772	3.116	1.918	2.307	0.986	2.017	0.743	1.908	0.639
20.2	5.408	3.068	3.534	1.465	2.373	0.852	2.035	0.693	1.914	0.618
20.3	6.363	0.912	3.714	0.826	2.399	0.703	2.041	0.641	1.916	0.597
20.4	5.445	-1.134	3.579	0.190	2.381	0.554	2.035	0.589	1.913	0.576
13π/2	5.136	-1.379	3.519	0.079	2.372	0.526	2.032	0.578	1.912	0.572
20.5	3.961	-1.842	3.220	-0.270	2.324	0.422	2.017	0.540	1.906	0.556
20.6	2.867	-1.793	2.790	-0.511	2.237	0.315	1.989	0.499	1.895	0.539
20.7	2.177	-1.524	2.393	-0.584	2.134	0.240	1.954	0.466	1.881	0.525
20.8	1.743	-1.228	2.065	-0.555	2.023	0.195	1.914	0.442	1.864	0.515
20.9	1.458	-0.958	1.806	-0.472	1.915	0.178	1.872	0.428	1.847	0.508

βh	α/β = 0.00		α/β = 0.01		α/β = 0.03		α/β = 0.05		α/β = 0.07	
21.0	1.263	-0.720	1.606	-0.364	1.814	0.183	1.830	0.423	1.829	0.505
21.1	1.125	-0.511	1.451	-0.246	1.723	0.206	1.789	0.426	1.811	0.505
21.2	1.023	-0.327	1.332	-0.126	1.644	0.242	1.751	0.437	1.794	0.509
21.3	0.946	-0.161	1.239	-0.008	1.577	0.288	1.717	0.454	1.779	0.516
21.4	0.887	-0.009	1.166	0.108	1.522	0.341	1.688	0.476	1.766	0.526
21.5	0.841	0.131	1.111	0.221	1.478	0.399	1.664	0.502	1.755	0.537
21.6	0.805	0.265	1.068	0.332	1.444	0.460	1.645	0.532	1.746	0.551
21.7	0.777	0.393	1.037	0.441	1.420	0.524	1.633	0.563	1.741	0.565
21.8	0.755	0.519	1.016	0.549	1.406	0.589	1.626	0.596	1.738	0.580
21.9	0.740	0.644	1.004	0.658	1.401	0.654	1.624	0.628	1.738	0.595
7π	0.730	0.759	1.002	0.758	1.404	0.714	1.628	0.658	1.740	0.608
22.0	0.730	0.770	1.002	0.768	1.405	0.720	1.629	0.661	1.741	0.609
22.1	0.725	0.901	1.009	0.880	1.419	0.785	1.639	0.692	1.746	0.622
22.2	0.728	1.037	1.027	0.995	1.443	0.848	1.654	0.720	1.754	0.635
22.3	0.738	1.183	1.057	1.115	1.478	0.909	1.674	0.746	1.763	0.645
22.4	0.757	1.342	1.103	1.240	1.523	0.966	1.699	0.768	1.775	0.653
22.5	0.790	1.518	1.168	1.370	1.579	1.016	1.728	0.785	1.788	0.659
22.6	0.842	1.717	1.257	1.506	1.646	1.057	1.760	0.796	1.801	0.662
22.7	0.922	1.947	1.380	1.644	1.724	1.088	1.794	0.801	1.815	0.662
22.8	1.048	2.216	1.545	1.780	1.812	1.102	1.829	0.799	1.829	0.660
22.9	1.248	2.536	1.767	1.901	1.906	1.098	1.863	0.789	1.842	0.655
23.0	1.578	2.914	2.056	1.982	2.001	1.071	1.895	0.772	1.854	0.647
23.1	2.143	3.327	2.417	1.984	2.092	1.020	1.923	0.749	1.863	0.637
23.2	3.117	3.644	2.825	1.849	2.171	0.945	1.945	0.720	1.870	0.626
23.3	4.624	3.405	3.206	1.530	2.229	0.849	1.960	0.687	1.875	0.614
23.4	6.029	1.858	3.441	1.039	2.259	0.740	1.967	0.651	1.877	0.600
23.5	5.846	-0.389	3.438	0.486	2.259	0.627	1.966	0.615	1.876	0.587
15π/2	5.033	-1.294	3.319	0.179	2.244	0.560	1.962	0.594	1.875	0.580
23.6	4.480	-1.595	3.213	0.019	2.230	0.521	1.958	0.581	1.873	0.575
23.7	3.242	-1.783	2.871	-0.283	2.176	0.429	1.942	0.550	1.867	0.564
23.8	2.423	-1.587	2.512	-0.424	2.104	0.358	1.920	0.524	1.859	0.554
23.9	1.907	-1.305	2.191	-0.451	2.022	0.308	1.895	0.504	1.850	0.547
24.0	1.573	-1.032	1.927	-0.409	1.937	0.281	1.866	0.490	1.840	0.541
24.1	1.348	-0.787	1.716	-0.328	1.853	0.273	1.836	0.483	1.828	0.538
24.2	1.190	-0.572	1.550	-0.230	1.775	0.281	1.806	0.482	1.817	0.537
24.3	1.075	-0.382	1.420	-0.123	1.705	0.302	1.778	0.486	1.806	0.539
24.4	0.990	-0.212	1.318	-0.014	1.643	0.334	1.752	0.496	1.796	0.542
24.5	0.924	-0.058	1.239	0.094	1.591	0.374	1.728	0.510	1.787	0.547
24.6	0.874	0.085	1.177	0.202	1.548	0.420	1.709	0.528	1.779	0.554
24.7	0.834	0.219	1.129	0.307	1.514	0.470	1.693	0.548	1.773	0.562
24.8	0.803	0.348	1.094	0.412	1.489	0.523	1.681	0.571	1.768	0.571
24.9	0.779	0.472	1.069	0.516	1.473	0.578	1.674	0.594	1.766	0.581
25.0	0.761	0.596	1.054	0.620	1.465	0.634	1.672	0.619	1.765	0.590

TABLE 2.7

NORMALIZED ADMITTANCES (Y/Δ) IN MILLIMHOS OF THIN DIPOLE ANTENNAS IN DISSIPATIVE MEDIA

a/λ = 0.001588

βh	α/β = 0.00	α/β = 0.01	α/β = 0.03	α/β = 0.05	α/β = 0.07	α/β = 0.10	α/β = 0.30	α/β = 0.50	α/β = 0.70	α/β = 1.00
1.5	11.940 -1.516	10.980 -1.038	9.431 -0.445	8.258 -0.129	7.352 0.040	6.335 0.153	3.652 -0.025	2.986 -0.342	2.800 -0.654	2.825 -1.138
π/2	8.243 -4.153	7.978 -3.525	7.389 -2.554	6.807 -1.876	6.275 -1.402	5.596 -0.938	3.450 -0.280	2.876 -0.427	2.732 -0.684	2.791 -1.142
1.6	6.936 -4.360	6.831 -3.814	6.515 -2.917	6.137 -2.246	5.754 -1.748	5.226 -1.232	3.357 -0.362	2.830 -0.454	2.705 -0.692	2.778 -1.142
1.7	4.067 -3.850	4.153 -3.568	4.237 -3.040	4.234 -2.575	4.172 -2.177	4.019 -1.700	3.032 -0.549	2.678 -0.511	2.621 -0.705	2.741 -1.137
1.8	2.686 -3.035	2.784 -2.885	2.934 -2.587	3.031 -2.301	3.084 -2.036	3.102 -1.687	2.729 -0.622	2.542 -0.528	2.551 -0.700	2.713 -1.127
1.9	1.948 -2.359	2.034 -2.272	2.180 -2.093	2.295 -1.913	2.382 -1.738	2.467 -1.495	2.472 -0.617	2.426 -0.516	2.494 -0.684	2.693 -1.114
2.0	1.511 -1.831	1.584 -1.776	1.716 -1.661	1.828 -1.543	1.922 -1.425	2.032 -1.254	2.263 -0.566	2.332 -0.486	2.450 -0.661	2.678 -1.101
2.1	1.230 -1.411	1.293 -1.375	1.411 -1.297	1.517 -1.216	1.610 -1.134	1.729 -1.012	2.100 -0.489	2.258 -0.445	2.417 -0.634	2.669 -1.088
2.2	1.037 -1.069	1.094 -1.043	1.202 -0.989	1.301 -0.931	1.392 -0.873	1.513 -0.785	1.974 -0.399	2.203 -0.398	2.395 -0.607	2.664 -1.077
2.3	0.898 -0.782	0.951 -0.763	1.052 -0.723	1.147 -0.682	1.236 -0.640	1.357 -0.577	1.879 -0.303	2.163 -0.348	2.380 -0.580	2.662 -1.067
2.4	0.795 -0.534	0.845 -0.520	0.942 -0.491	1.034 -0.461	1.121 -0.430	1.243 -0.385	1.811 -0.207	2.136 -0.300	2.372 -0.556	2.662 -1.058
2.5	0.716 -0.314	0.764 -0.304	0.859 -0.283	0.950 -0.261	1.037 -0.239	1.160 -0.208	1.764 -0.113	2.121 -0.253	2.370 -0.534	2.663 -1.052
2.6	0.654 -0.116	0.702 -0.108	0.796 -0.093	0.887 -0.077	0.975 -0.063	1.100 -0.043	1.735 -0.023	2.115 -0.210	2.371 -0.515	2.665 -1.047
2.7	0.604 0.067	0.653 0.073	0.748 0.084	0.840 0.094	0.930 0.102	1.058 0.112	1.721 0.063	2.117 -0.172	2.376 -0.499	2.668 -1.043
2.8	0.565 0.240	0.614 0.244	0.711 0.251	0.807 0.256	0.899 0.259	1.032 0.260	1.721 0.142	2.124 -0.138	2.382 -0.487	2.670 -1.041
2.9	0.533 0.406	0.584 0.408	0.685 0.411	0.784 0.412	0.880 0.410	1.019 0.402	1.731 0.216	2.137 -0.109	2.390 -0.477	2.673 -1.039
3.0	0.507 0.568	0.561 0.569	0.668 0.569	0.771 0.565	0.873 0.557	1.018 0.540	1.752 0.282	2.152 -0.086	2.398 -0.470	2.675 -1.038
3.1	0.488 0.729	0.545 0.729	0.658 0.725	0.769 0.716	0.876 0.703	1.029 0.675	1.781 0.342	2.170 -0.068	2.406 -0.465	2.677 -1.038
π	0.481 0.797	0.540 0.796	0.657 0.790	0.770 0.779	0.880 0.763	1.037 0.731	1.795 0.364	2.178 -0.062	2.409 -0.464	2.677 -1.038
3.2	0.473 0.893	0.535 0.892	0.657 0.883	0.776 0.869	0.890 0.848	1.052 0.809	1.817 0.393	2.189 -0.055	2.413 -0.463	2.678 -1.038
3.3	0.464 1.063	0.532 1.059	0.665 1.046	0.793 1.024	0.916 0.995	1.089 0.941	1.860 0.435	2.209 -0.047	2.420 -0.462	2.680 -1.038
3.4	0.462 1.243	0.537 1.236	0.683 1.216	0.824 1.185	0.957 1.145	1.141 1.073	1.908 0.468	2.227 -0.043	2.426 -0.463	2.681 -1.038
3.5	0.467 1.436	0.551 1.426	0.715 1.396	0.870 1.353	1.015 1.300	1.211 1.205	1.959 0.491	2.244 -0.042	2.431 -0.464	2.681 -1.039
3.6	0.481 1.649	0.578 1.634	0.763 1.590	0.936 1.531	1.094 1.458	1.302 1.334	2.011 0.504	2.260 -0.045	2.435 -0.466	2.682 -1.039
3.7	0.510 1.890	0.622 1.867	0.835 1.803	1.029 1.719	1.201 1.621	1.419 1.458	2.064 0.506	2.273 -0.051	2.438 -0.469	2.682 -1.040
3.8	0.560 2.168	0.693 2.134	0.941 2.039	1.158 1.920	1.344 1.785	1.565 1.572	2.114 0.498	2.284 -0.058	2.440 -0.471	2.682 -1.040
3.9	0.643 2.500	0.805 2.445	1.097 2.302	1.339 2.130	1.533 1.944	1.746 1.668	2.159 0.480	2.292 -0.066	2.441 -0.474	2.682 -1.041
4.0	0.784 2.907	0.986 2.816	1.328 2.593	1.590 2.340	1.781 2.086	1.965 1.732	2.198 0.455	2.298 -0.075	2.442 -0.476	2.682 -1.041
4.1	1.031 3.419	1.284 3.262	1.677 2.901	1.939 2.530	2.099 2.185	2.218 1.747	2.230 0.422	2.301 -0.084	2.442 -0.478	2.682 -1.041
4.2	1.486 4.072	1.796 3.785	2.207 3.190	2.411 2.652	2.490 2.203	2.495 1.690	2.252 0.386	2.303 -0.093	2.442 -0.480	2.682 -1.041
4.3	2.378 4.871	2.704 4.319	2.989 3.350	3.011 2.618	2.932 2.086	2.768 1.543	2.266 0.347	2.302 -0.100	2.442 -0.482	2.682 -1.041
4.4	4.194 5.554	4.265 4.538	4.022 3.145	3.671 2.311	3.356 1.786	2.995 1.301	2.270 0.309	2.301 -0.107	2.441 -0.483	2.682 -1.042
4.5	7.287 4.746	6.344 3.542	5.023 2.283	4.198 1.664	3.654 1.306	3.132 0.986	2.266 0.273	2.298 -0.113	2.440 -0.484	2.682 -1.042
4.6	8.954 0.762	7.283 0.871	5.387 0.867	4.360 0.802	3.728 0.734	3.152 0.645	2.255 0.241	2.294 -0.117	2.439 -0.485	2.682 -1.042
4.7	6.794 -2.373	6.067 -1.397	4.893 -0.409	4.103 0.012	3.570 0.210	3.058 0.334	2.238 0.214	2.290 -0.121	2.438 -0.485	2.682 -1.042
3π/2	6.453 -2.541	5.848 -1.561	4.793 -0.527	4.049 -0.068	3.538 0.154	3.040 0.300	2.236 0.211	2.290 -0.121	2.438 -0.485	2.682 -1.042
4.8	4.421 -2.882	4.374 -2.122	4.012 -1.088	3.603 -0.504	3.259 -0.170	2.883 0.092	2.218 0.193	2.286 -0.123	2.438 -0.485	2.682 -1.042
4.9	3.005 -2.514	3.153 -2.049	3.190 -1.278	3.063 -0.741	2.896 -0.385	2.668 -0.068	2.195 0.178	2.282 -0.124	2.437 -0.485	2.681 -1.042
5.0	2.200 -2.032	2.378 -1.748	2.560 -1.216	2.591 -0.787	2.550 -0.468	2.447 -0.152	2.172 0.169	2.279 -0.125	2.436 -0.485	2.681 -1.042
5.1	1.717 -1.601	1.884 -1.419	2.106 -1.053	2.214 -0.726	2.251 -0.462	2.242 -0.176	2.149 0.165	2.275 -0.124	2.436 -0.485	2.681 -1.042
5.2	1.407 -1.238	1.555 -1.116	1.781 -0.859	1.923 -0.614	2.006 -0.402	2.062 -0.158	2.127 0.165	2.272 -0.124	2.435 -0.485	2.681 -1.042
5.3	1.197 -0.933	1.329 -0.848	1.545 -0.663	1.701 -0.479	1.809 -0.312	1.909 -0.110	2.108 0.170	2.270 -0.122	2.435 -0.484	2.681 -1.042
5.4	1.048 -0.673	1.166 -0.612	1.371 -0.476	1.532 -0.337	1.654 -0.207	1.783 -0.043	2.091 0.178	2.268 -0.121	2.435 -0.484	2.681 -1.042
5.5	0.937 -0.447	1.046 -0.401	1.240 -0.300	1.402 -0.195	1.532 -0.094	1.680 0.034	2.077 0.188	2.267 -0.119	2.435 -0.484	2.681 -1.042
5.6	0.853 -0.246	0.955 -0.212	1.141 -0.136	1.302 -0.056	1.438 0.020	1.599 0.118	2.066 0.200	2.266 -0.117	2.435 -0.484	2.681 -1.042
5.7	0.788 -0.064	0.885 -0.038	1.066 0.019	1.226 0.078	1.365 0.134	1.537 0.206	2.058 0.212	2.265 -0.115	2.435 -0.483	2.681 -1.041
5.8	0.737 0.105	0.831 0.125	1.008 0.167	1.168 0.209	1.311 0.248	1.490 0.295	2.053 0.225	2.265 -0.114	2.435 -0.483	2.681 -1.041
5.9	0.696 0.264	0.789 0.279	0.965 0.308	1.127 0.336	1.272 0.360	1.458 0.385	2.050 0.238	2.265 -0.112	2.435 -0.483	2.681 -1.041
6.0	0.663 0.416	0.756 0.427	0.934 0.446	1.099 0.460	1.247 0.470	1.439 0.474	2.051 0.251	2.266 -0.111	2.435 -0.483	2.681 -1.041
6.1	0.637 0.565	0.732 0.572	0.914 0.581	1.082 0.583	1.235 0.579	1.432 0.561	2.053 0.262	2.266 -0.110	2.435 -0.483	2.681 -1.041
6.2	0.617 0.714	0.716 0.717	0.904 0.715	1.078 0.705	1.235 0.687	1.437 0.648	2.057 0.272	2.267 -0.109	2.435 -0.483	2.681 -1.041
2π	0.605 0.838	0.707 0.838	0.903 0.828	1.083 0.806	1.244 0.775	1.450 0.717	2.062 0.279	2.268 -0.109	2.435 -0.483	2.681 -1.041
6.3	0.603 0.864	0.706 0.863	0.904 0.851	1.085 0.827	1.247 0.793	1.453 0.731	2.063 0.280	2.268 -0.109	2.435 -0.483	2.681 -1.041
6.4	0.595 1.019	0.705 1.014	0.914 0.989	1.104 0.950	1.272 0.899	1.481 0.812	2.070 0.287	2.269 -0.108	2.435 -0.483	2.681 -1.041
6.5	0.593 1.182	0.712 1.171	0.936 1.132	1.137 1.074	1.310 1.003	1.521 0.888	2.078 0.292	2.269 -0.108	2.435 -0.483	2.681 -1.041
6.6	0.599 1.356	0.730 1.339	0.973 1.281	1.185 1.199	1.363 1.106	1.572 0.960	2.086 0.295	2.270 -0.108	2.435 -0.483	2.681 -1.041
6.7	0.614 1.547	0.760 1.520	1.027 1.437	1.252 1.326	1.434 1.204	1.636 1.023	2.094 0.297	2.271 -0.108	2.435 -0.483	2.681 -1.041
6.8	0.643 1.760	0.809 1.720	1.104 1.601	1.341 1.453	1.523 1.297	1.712 1.077	2.101 0.297	2.271 -0.108	2.435 -0.483	2.681 -1.041
6.9	0.692 2.003	0.884 1.943	1.212 1.775	1.458 1.577	1.633 1.379	1.801 1.118	2.108 0.295	2.272 -0.109	2.435 -0.483	2.681 -1.041
7.0	0.770 2.289	0.997 2.197	1.361 1.955	1.608 1.691	1.766 1.446	1.899 1.142	2.115 0.292	2.272 -0.109	2.435 -0.483	2.681 -1.041
7.1	0.899 2.631	1.170 2.487	1.566 2.136	1.798 1.788	1.923 1.489	2.005 1.146	2.120 0.288	2.272 -0.109	2.436 -0.483	2.681 -1.041
7.2	1.114 3.050	1.437 2.816	1.847 2.303	2.032 1.850	2.099 1.497	2.113 1.126	2.124 0.284	2.273 -0.110	2.436 -0.483	2.681 -1.041
7.3	1.488 3.567	1.860 3.172	2.223 2.422	2.307 1.855	2.288 1.460	2.218 1.079	2.127 0.279	2.273 -0.110	2.436 -0.483	2.681 -1.041
7.4	2.171 4.181	2.536 3.497	2.704 2.434	2.608 1.776	2.474 1.369	2.312 1.007	2.129 0.274	2.273 -0.110	2.436 -0.483	2.681 -1.041
7.5	3.462 4.741	3.570 3.599	3.256 2.254	2.900 1.588	2.636 1.221	2.385 0.911	2.129 0.269	2.273 -0.111	2.435 -0.483	2.681 -1.041
7.6	5.674 4.525	4.886 3.067	3.763 1.806	3.128 1.291	2.750 1.023	2.431 0.800	2.129 0.264	2.273 -0.111	2.435 -0.483	2.681 -1.041
7.7	7.737 2.122	5.812 1.583	4.041 1.124	3.237 0.918	2.799 0.798	2.446 0.681	2.128 0.260	2.272 -0.111	2.435 -0.483	2.681 -1.041
7.8	7.031 -1.096	5.556 -0.176	3.972 0.395	3.202 0.538	2.775 0.573	2.429 0.566	2.126 0.256	2.272 -0.111	2.435 -0.483	2.681 -1.041
5π/2	5.894 -2.013	5.027 -0.848	3.807 0.050	3.129 0.353	2.735 0.462	2.408 0.508	2.125 0.255	2.272 -0.111	2.435 -0.483	2.681 -1.041
7.9	4.950 -2.344	4.506 -1.212	3.621 -0.172	3.044 0.215	2.688 0.376	2.385 0.463	2.124 0.253	2.272 -0.111	2.435 -0.483	2.681 -1.041
8.0	3.407 -2.316	3.459 -1.503	3.155 -0.497	2.812 -0.014	2.558 0.224	2.320 0.379	2.121 0.251	2.272 -0.111	2.435 -0.483	2.681 -1.041
8.1	2.480 -1.955	2.682 -1.427	2.705 -0.621	2.557 -0.144	2.406 0.122	2.242 0.318	2.118 0.249	2.272 -0.111	2.435 -0.483	2.681 -1.041
8.2	1.919 -1.568	2.147 -1.223	2.324 -0.616	2.313 -0.195	2.249 0.068	2.158 0.279	2.115 0.248	2.272 -0.111	2.435 -0.483	2.681 -1.041
8.3	1.563 -1.223	1.779 -0.990	2.020 -0.540	2.096 -0.189	2.101 0.052	2.074 0.260	2.111 0.248	2.272 -0.111	2.435 -0.483	2.681 -1.041
8.4	1.324 -0.927	1.522 -0.764	1.783 -0.429	1.912 -0.144	1.967 0.066	1.994 0.260	2.109 0.248	2.271 -0.111	2.435 -0.483	2.681 -1.041
8.5	1.157 -0.672	1.338 -0.555	1.600 -0.303	1.759 -0.075	1.850 0.102	1.921 0.274	2.106 0.249	2.271 -0.111	2.435 -0.483	2.681 -1.041
8.6	1.034 -0.450	1.198 -0.364	1.458 -0.173	1.635 0.007	1.751 0.154	1.856 0.300	2.104 0.250	2.271 -0.111	2.435 -0.483	2.681 -1.041
8.7	0.942 -0.253	1.095 -0.188	1.348 -0.043	1.536 0.098	1.669 0.216	1.801 0.336	2.102 0.252	2.271 -0.111	2.435 -0.483	2.681 -1.041
8.8	0.872 -0.075	1.015 -0.026	1.263 0.085	1.457 0.193	1.603 0.284	1.755 0.377	2.101 0.254	2.271 -0.111	2.435 -0.483	2.681 -1.041
8.9	0.816 0.089	0.954 0.127	1.198 0.210	1.396 0.290	1.552 0.357	1.720 0.424	2.100 0.256	2.271 -0.111	2.435 -0.483	2.681 -1.041
9.0	0.772 0.243	0.906 0.272	1.148 0.332	1.351 0.388	1.513 0.433	1.693 0.473	2.099 0.257	2.271 -0.111	2.435 -0.483	2.681 -1.041
9.1	0.737 0.390	0.870 0.411	1.113 0.452	1.319 0.486	1.487 0.510	1.676 0.524	2.099 0.259	2.271 -0.111	2.435 -0.483	2.681 -1.041
9.2	0.710 0.534	0.843 0.548	1.089 0.571	1.299 0.584	1.472 0.588	1.667 0.575	2.099 0.261	2.271 -0.111	2.435 -0.483	2.681 -1.041
9.3	0.689 0.675	0.825 0.683	1.076 0.689	1.291 0.682	1.468 0.665	1.667 0.626	2.100 0.263	2.271 -0.111	2.435 -0.483	2.681 -1.041
9.4	0.674 0.818	0.815 0.820	1.074 0.808	1.295 0.780	1.475 0.742	1.675 0.676	2.101 0.264	2.271 -0.111	2.435 -0.483	2.681 -1.041
3π	0.671 0.854	0.814 0.854	1.075 0.838	1.298 0.804	1.478 0.760	1.678 0.688	2.101 0.264	2.271 -0.111	2.435 -0.483	2.681 -1.041
9.5	0.665 0.965	0.813 0.960	1.084 0.928	1.311 0.877	1.493 0.816	1.691 0.723	2.102 0.265	2.271 -0.111	2.435 -0.483	2.681 -1.041
9.6	0.662 1.117	0.820 1.105	1.105 1.050	1.339 0.973	1.521 0.888	1.714 0.766	2.103 0.266	2.271 -0.111	2.435 -0.483	2.681 -1.041
9.7	0.667 1.280	0.838 1.257	1.141 1.176	1.381 1.068	1.561 0.956	1.745 0.805	2.104 0.266	2.271 -0.111	2.435 -0.483	2.681 -1.041
9.8	0.681 1.456	0.869 1.420	1.193 1.304	1.438 1.160	1.613 1.019	1.782 0.838	2.105 0.267	2.271 -0.111	2.435 -0.483	2.681 -1.041
9.9	0.707 1.651	0.917 1.597	1.265 1.435	1.512 1.248	1.676 1.074	1.825 0.864	2.106 0.267	2.271 -0.111	2.435 -0.483	2.681 -1.041
10.0	0.751 1.871	0.990 1.791	1.363 1.567	1.604 1.328	1.751 1.119	1.873 0.882	2.107 0.266	2.271 -0.111	2.435 -0.483	2.681 -1.041
10.1	0.821 2.125	1.096 2.006	1.493 1.697	1.718 1.395	1.837 1.150					
10.2	0.932 2.425	1.252 2.244	1.662 1.818	1.852 1.444	1.932 1.165					
10.3	1.114 2.786	1.484 2.504	1.879 1.917	2.006 1.466	2.032 1.159					
10.4	1.418 3.224	1.833 2.773	2.149 1.973	2.174 1.451	2.133 1.130					
10.5	1.952 3.742	2.354 3.006	2.469 1.955	2.347 1.390	2.228 1.076					
10.6	2.918 4.257	3.104 3.085	2.813 1.826	2.507 1.277	2.308 0.997					
10.7	4.595 4.368	4.045 2.785	3.126 1.560	2.634 1.115	2.367 0.899					
10.8	6.668 3.015	4.859 1.884	3.334 1.169	2.709 0.918	2.399 0.789					
10.9	7.094 0.123	5.025 0.584	3.373 0.721	2.720 0.708	2.400 0.676					
7π/2	5.582 -1.725	4.507 -0.446	3.251 0.326	2.671 0.518	2.375 0.573					

βh	α/β = 0.00		α/β = 0.01		α/β = 0.03		α/β = 0.05		α/β = 0.07	
11.0	5.500	-1.770	4.473	-0.482	3.243	0.309	2.667	0.510	2.373	0.569
11.1	3.862	-2.146	3.659	-1.006	2.996	0.003	2.565	0.346	2.321	0.476
11.2	2.790	-1.928	2.931	-1.125	2.702	-0.179	2.432	0.226	2.252	0.403
11.3	2.133	-1.585	2.381	-1.043	2.415	-0.258	2.286	0.151	2.172	0.352
11.4	1.718	-1.252	1.983	-0.883	2.160	-0.261	2.141	0.117	2.089	0.322
11.5	1.444	-0.959	1.697	-0.700	1.945	-0.217	2.005	0.116	2.006	0.311
11.6	1.254	-0.704	1.488	-0.517	1.768	-0.143	1.885	0.139	1.929	0.317
11.7	1.117	-0.481	1.333	-0.344	1.626	-0.054	1.780	0.180	1.859	0.337
11.8	1.015	-0.284	1.216	-0.181	1.513	0.044	1.692	0.234	1.798	0.367
11.9	0.937	-0.106	1.126	-0.027	1.422	0.146	1.620	0.297	1.746	0.405
12.0	0.877	0.057	1.057	0.117	1.352	0.249	1.561	0.366	1.704	0.449
12.1	0.829	0.209	1.003	0.255	1.297	0.353	1.516	0.438	1.671	0.497
12.2	0.792	0.354	0.962	0.388	1.257	0.457	1.483	0.513	1.647	0.548
12.3	0.762	0.494	0.932	0.518	1.230	0.561	1.462	0.590	1.632	0.601
12.4	0.740	0.632	0.911	0.647	1.214	0.665	1.451	0.666	1.625	0.655
12.5	0.723	0.770	0.899	0.776	1.209	0.769	1.451	0.743	1.628	0.708
4π	0.715	0.863	0.896	0.863	1.212	0.838	1.457	0.794	1.634	0.742
12.6	0.713	0.910	0.896	0.907	1.215	0.874	1.462	0.819	1.638	0.759
12.7	0.708	1.056	0.902	1.042	1.234	0.979	1.483	0.894	1.657	0.808
12.8	0.711	1.210	0.918	1.183	1.266	1.085	1.516	0.966	1.684	0.854
12.9	0.723	1.375	0.947	1.332	1.312	1.192	1.560	1.034	1.719	0.895
13.0	0.745	1.556	0.993	1.452	1.376	1.298	1.617	1.096	1.761	0.929
13.1	0.783	1.758	1.060	1.664	1.461	1.402	1.687	1.150	1.809	0.956
13.2	0.844	1.989	1.158	1.850	1.569	1.500	1.769	1.194	1.863	0.974
13.3	0.940	2.258	1.297	2.052	1.706	1.585	1.863	1.222	1.921	0.980
13.4	1.093	2.577	1.498	2.267	1.874	1.650	1.967	1.231	1.981	0.975
13.5	1.342	2.961	1.786	2.482	2.073	1.680	2.076	1.216	2.039	0.957
13.6	1.766	3.414	2.203	2.665	2.298	1.659	2.184	1.175	2.093	0.925
13.7	2.509	3.894	2.776	2.739	2.532	1.569	2.283	1.105	2.139	0.882
13.8	3.803	4.175	3.490	2.567	2.747	1.399	2.363	1.009	2.174	0.829
13.9	5.670	3.535	4.179	1.996	2.904	1.153	2.416	0.893	2.195	0.769
14.0	6.862	1.269	4.519	1.058	2.970	0.861	2.436	0.768	2.202	0.707
14.1	5.984	-1.053	4.315	0.099	2.934	0.569	2.421	0.643	2.193	0.646
9π/2	5.372	-1.535	4.130	-0.184	2.896	0.469	2.408	0.599	2.187	0.624
14.2	4.363	-1.918	3.750	-0.539	2.809	0.319	2.376	0.531	2.172	0.590
14.3	3.142	-1.900	3.120	-0.815	2.628	0.137	2.306	0.438	2.139	0.542
14.4	2.370	-1.619	2.584	-0.852	2.426	0.026	2.220	0.370	2.097	0.504
14.5	1.884	-1.302	2.170	-0.768	2.227	-0.023	2.127	0.327	2.051	0.478
14.6	1.566	-1.009	1.861	-0.633	2.045	-0.024	2.034	0.307	2.002	0.463
14.7	1.349	-0.751	1.631	-0.480	1.887	0.008	1.946	0.306	1.954	0.458
14.8	1.194	-0.525	1.459	-0.326	1.753	0.062	1.864	0.322	1.908	0.463
14.9	1.081	-0.326	1.328	-0.176	1.642	0.131	1.792	0.351	1.866	0.476
15.0	0.995	-0.146	1.227	-0.032	1.552	0.208	1.731	0.390	1.828	0.496
15.1	0.929	0.017	1.150	0.105	1.479	0.291	1.679	0.436	1.796	0.521
15.2	0.877	0.170	1.089	0.236	1.422	0.378	1.638	0.488	1.770	0.550
15.3	0.836	0.313	1.044	0.364	1.379	0.466	1.606	0.543	1.751	0.583
15.4	0.804	0.452	1.009	0.488	1.348	0.555	1.584	0.600	1.737	0.617
15.5	0.779	0.587	0.985	0.611	1.329	0.646	1.572	0.659	1.730	0.653
15.6	0.761	0.721	0.970	0.734	1.321	0.736	1.569	0.718	1.730	0.689
15.7	0.749	0.857	0.965	0.858	1.324	0.827	1.574	0.777	1.735	0.724
5π	0.748	0.868	0.965	0.868	1.324	0.835	1.575	0.782	1.736	0.726
15.8	0.743	0.998	0.968	0.985	1.338	0.919	1.589	0.834	1.746	0.757
15.9	0.744	1.144	0.983	1.116	1.364	1.009	1.613	0.889	1.763	0.787
16.0	0.752	1.300	1.009	1.254	1.404	1.099	1.646	0.940	1.785	0.814
16.1	0.771	1.470	1.051	1.399	1.459	1.187	1.688	0.986	1.812	0.837
16.2	0.804	1.658	1.113	1.554	1.530	1.270	1.739	1.025	1.843	0.855
16.3	0.857	1.871	1.201	1.719	1.619	1.346	1.799	1.054	1.877	0.866
16.4	0.939	2.116	1.325	1.895	1.729	1.411	1.865	1.073	1.913	0.870
16.5	1.067	2.403	1.499	2.077	1.860	1.457	1.936	1.078	1.949	0.867
16.6	[illegible]	[illegible]	[illegible]	[illegible]	2.010	1.476	2.009	1.068	1.985	0.857
16.7	1.613	3.149	2.081	2.408	2.175	[illegible]	[illegible]	[illegible]	2.017	0.839
16.8	2.196	3.595	2.534	2.481	2.343	1.396	2.146	0.997	2.045	0.814
16.9	3.203	3.956	3.095	2.385	2.499	1.282	2.201	0.937	2.067	0.784
17.0	4.785	3.771	3.673	2.014	2.621	1.118	2.239	0.866	2.082	0.750
17.1	6.350	2.227	4.067	1.342	2.690	0.921	2.259	0.786	2.089	0.714
17.2	6.300	-0.104	4.083	0.541	2.694	0.711	2.259	0.705	2.087	0.678
11π/2	5.216	-1.397	3.837	0.001	2.653	0.556	2.244	0.644	2.081	0.651
17.3	4.889	-1.574	3.744	-0.117	2.636	0.517	2.239	0.629	2.078	0.644
17.4	3.543	-1.838	3.243	-0.500	2.529	0.357	2.202	0.561	2.062	0.613
17.5	2.641	-1.652	2.750	-0.645	2.392	0.242	2.152	0.507	2.041	0.588
17.6	2.068	-1.360	2.338	-0.639	2.245	0.173	2.094	0.467	2.015	0.568
17.7	1.698	-1.070	2.015	-0.556	2.099	0.144	2.032	0.442	1.986	0.554
17.8	1.448	-0.809	1.768	-0.437	1.965	0.146	1.969	0.431	1.957	0.547
17.9	1.272	-0.578	1.579	-0.304	1.846	0.173	1.909	0.433	1.927	0.546
18.0	1.145	-0.375	1.435	-0.169	1.743	0.217	1.853	0.447	1.899	0.551
18.1	1.049	-0.193	1.324	-0.036	1.656	0.272	1.804	0.469	1.874	0.560
18.2	0.976	-0.027	1.237	0.093	1.584	0.337	1.761	0.498	1.851	0.574
18.3	0.920	0.126	1.170	0.218	1.526	0.406	1.725	0.533	1.832	0.592
18.4	0.875	0.270	1.118	0.340	1.482	0.480	1.697	0.572	1.817	0.612
18.5	0.840	0.407	1.080	0.459	1.449	0.556	1.676	0.615	1.806	0.634
18.6	0.813	0.541	1.052	0.576	1.427	0.633	1.663	0.659	1.799	0.657
18.7	0.793	0.673	1.034	0.694	1.415	0.711	1.658	0.704	1.797	0.680
18.8	0.779	0.806	1.025	0.812	1.415	0.790	1.659	0.748	1.799	0.703
6π	0.774	0.873	1.025	0.871	1.418	0.829	1.663	0.770	1.802	0.714
18.9	0.771	0.941	1.026	0.932	1.424	0.869	1.669	0.792	1.805	0.725
19.0	0.769	1.082	1.038	1.055	1.445	0.947	1.685	0.834	1.815	0.746
19.1	0.775	1.232	1.062	1.183	1.478	1.023	1.709	0.872	1.829	0.764
19.2	0.791	1.392	1.100	1.317	1.523	1.096	1.739	0.906	1.846	0.779
19.3	0.819	1.569	1.155	1.458	1.582	1.164	1.776	0.935	1.865	0.791
19.4	0.863	1.766	1.234	1.607	1.654	1.226	1.818	0.957	1.886	0.799
19.5	0.933	1.991	1.344	1.762	1.742	1.276	1.865	0.971	1.909	0.803
19.6	1.042	2.253	1.496	1.921	1.845	1.312	1.914	0.975	1.931	0.802
19.7	1.213	2.562	1.703	2.075	1.961	1.326	1.965	0.969	1.954	0.796
19.8	1.490	2.925	1.983	2.206	2.085	1.314	2.014	0.952	1.974	0.786
19.9	1.953	3.338	2.351	2.277	2.212	1.271	2.060	0.924	1.992	0.772
20.0	2.744	3.729	2.803	2.229	2.329	1.192	2.098	0.886	2.006	0.755
20.1	4.040	3.815	3.288	1.984	2.426	1.079	2.127	0.840	2.016	0.735
20.2	5.664	2.916	3.679	1.504	2.489	0.941	2.144	0.788	2.022	0.713
20.3	6.357	0.769	3.822	0.862	2.510	0.789	2.149	0.735	2.023	0.691
20.4	5.387	-1.073	3.663	0.245	2.488	0.640	2.142	0.682	2.020	0.670
13π/2	5.095	-1.291	3.601	0.138	2.479	0.612	2.139	0.672	2.019	0.666
20.5	3.993	-1.709	3.299	-0.195	2.427	0.508	2.122	0.633	2.012	0.651
20.6	2.952	-1.669	2.874	-0.425	2.338	0.403	2.093	0.592	2.000	0.633
20.7	2.278	-1.420	2.482	-0.494	2.233	0.328	2.057	0.559	1.986	0.619
20.8	1.843	-1.138	2.156	-0.465	2.121	0.285	2.016	0.535	1.969	0.609
20.9	1.554	-0.874	1.897	-0.383	2.011	0.269	1.973	0.521	1.951	0.602

βh	α/β = 0.00		α/β = 0.01		α/β = 0.03		α/β = 0.05		α/β = 0.07	
21.0	1.354	-0.638	1.696	-0.276	1.909	0.275	1.930	0.517	1.933	0.600
21.1	1.210	-0.430	1.539	-0.158	1.817	0.299	1.888	0.521	1.915	0.600
21.2	1.103	-0.243	1.417	-0.037	1.737	0.336	1.850	0.533	1.898	0.605
21.3	1.023	-0.075	1.321	0.084	1.669	0.384	1.816	0.550	1.882	0.612
21.4	0.960	0.080	1.247	0.202	1.613	0.438	1.786	0.574	1.869	0.622
21.5	0.911	0.225	1.189	0.318	1.568	0.498	1.762	0.601	1.858	0.634
21.6	0.873	0.362	1.146	0.431	1.534	0.561	1.744	0.631	1.850	0.648
21.7	0.843	0.495	1.114	0.544	1.510	0.626	1.731	0.663	1.844	0.662
21.8	0.821	0.625	1.092	0.656	1.496	0.693	1.724	0.696	1.841	0.677
21.9	0.804	0.755	1.081	0.769	1.492	0.761	1.724	0.730	1.842	0.693
7π	0.795	0.876	1.079	0.873	1.497	0.823	1.728	0.760	1.844	0.706
22.0	0.794	0.887	1.079	0.863	1.498	0.829	1.729	0.763	1.845	0.707
22.1	0.791	1.024	1.087	0.999	1.513	0.895	1.739	0.795	1.850	0.721
22.2	0.794	1.167	1.107	1.119	1.539	0.960	1.756	0.824	1.858	0.733
22.3	0.806	1.320	1.141	1.244	1.575	1.022	1.777	0.850	1.869	0.744
22.4	0.829	1.487	1.191	1.373	1.623	1.079	1.803	0.872	1.881	0.752
22.5	0.868	1.672	1.261	1.508	1.682	1.130	1.833	0.888	1.894	0.758
22.6	0.927	1.881	1.359	1.647	1.752	1.171	1.866	0.899	1.908	0.761
22.7	1.019	2.121	1.491	1.788	1.834	1.200	1.901	0.904	1.922	0.761
22.8	1.162	2.402	1.669	1.924	1.924	1.213	1.937	0.900	1.936	0.758
22.9	1.389	2.732	1.904	2.040	2.020	1.206	1.971	0.890	1.949	0.752
23.0	1.761	3.112	2.208	2.110	2.117	1.176	2.004	0.872	1.961	0.744
23.1	2.388	3.505	2.580	2.092	2.209	1.120	2.031	0.848	1.970	0.734
23.2	3.432	3.743	2.985	1.931	2.296	1.041	2.053	0.818	1.978	0.722
23.3	4.931	3.332	3.354	1.587	2.342	0.942	2.068	0.783	1.982	0.710
23.4	6.123	1.681	3.561	1.085	2.370	0.831	2.074	0.747	1.984	0.696
23.5	5.781	-0.402	3.533	0.540	2.367	0.717	2.073	0.711	1.983	0.683
15π/2	4.996	-1.206	3.405	0.243	2.350	0.649	2.007	0.689	1.981	0.675
23.6	4.476	-1.476	3.297	0.088	2.335	0.610	2.063	0.676	1.979	0.670
23.7	3.309	-1.652	2.955	-0.200	2.278	0.519	2.046	0.645	1.973	0.659
23.8	2.519	-1.474	2.599	-0.336	2.204	0.448	2.024	0.619	1.965	0.649
23.9	2.008	-1.208	2.282	-0.361	2.121	0.400	1.997	0.599	1.955	0.642
24.0	1.671	-0.944	2.018	-0.319	2.034	0.373	1.968	0.585	1.944	0.637
24.1	1.441	-0.703	1.807	-0.239	1.950	0.366	1.937	0.578	1.933	0.634
24.2	1.278	-0.489	1.639	-0.140	1.871	0.375	1.907	0.578	1.922	0.633
24.3	1.158	-0.299	1.507	-0.032	1.799	0.398	1.878	0.583	1.910	0.635
24.4	1.068	-0.126	1.403	0.078	1.737	0.431	1.852	0.593	1.900	0.638
24.5	1.000	0.031	1.321	0.189	1.684	0.472	1.828	0.608	1.891	0.644
24.6	0.946	0.177	1.258	0.298	1.641	0.519	1.808	0.626	1.883	0.651
24.7	0.904	0.316	1.209	0.407	1.606	0.570	1.793	0.647	1.877	0.659
24.8	0.871	0.448	1.172	0.514	1.581	0.625	1.781	0.670	1.872	0.668
24.9	0.846	0.577	1.147	0.621	1.566	0.682	1.774	0.695	1.870	0.678
25.0	0.828	0.706	1.132	0.729	1.558	0.739	1.772	0.719	1.869	0.688

TABLE 2.7

NORMALIZED ADMITTANCES (Y/Δ) IN MILLIMHOS OF THIN DIPOLE ANTENNAS IN DISSIPATIVE MEDIA

$a/\lambda = 0.003175$

βh	α/β = 0.00		α/β = 0.01		α/β = 0.03		α/β = 0.05		α/β = 0.07		α/β = 0.10		α/β = 0.30		α/β = 0.50		α/β = 0.70		α/β = 1.00	
1.5	10.950	-2.042	10.318	-1.552	9.212	-0.864	8.300	-0.433	7.550	-0.162	6.663	0.071	4.120	0.143	3.483	-0.169	3.345	-0.514	3.461	-1.065
π/2	7.870	-3.679	7.687	-3.182	7.267	-2.375	6.828	-1.773	6.407	-1.327	5.843	-0.863	3.889	-0.108	3.360	-0.255	3.270	-0.543	3.424	-1.068
1.6	6.820	-3.814	6.741	-3.376	6.504	-2.634	6.214	-2.051	5.910	-1.599	5.473	-1.108	3.787	-0.187	3.309	-0.281	3.241	-0.551	3.411	-1.067
1.7	4.362	-3.431	4.426	-3.180	4.488	-2.710	4.483	-2.293	4.432	-1.932	4.304	-1.491	3.437	-0.364	3.144	-0.336	3.150	-0.562	3.371	-1.060
1.8	3.040	-2.753	3.124	-2.607	3.255	-2.321	3.341	-2.049	3.390	-1.797	3.412	-1.465	3.118	-0.427	2.998	-0.349	3.074	-0.554	3.340	-1.049
1.9	2.277	-2.143	2.357	-2.052	2.495	-1.870	2.605	-1.690	2.690	-1.517	2.777	-1.277	2.848	-0.414	2.874	-0.332	3.013	-0.535	3.319	-1.034
2.0	1.800	-1.637	1.873	-1.578	2.004	-1.456	2.118	-1.333	2.214	-1.211	2.329	-1.038	2.630	-0.355	2.774	-0.297	2.966	-0.508	3.304	-1.020
2.1	1.481	-1.220	1.548	-1.179	1.671	-1.093	1.782	-1.007	1.882	-0.920	2.009	-0.794	2.457	-0.270	2.696	-0.250	2.931	-0.479	3.295	-1.006
2.2	1.258	-0.869	1.319	-0.839	1.436	-0.778	1.544	-0.715	1.643	-0.652	1.777	-0.559	2.324	-0.172	2.637	-0.197	2.908	-0.449	3.289	-0.993
2.3	1.094	-0.567	1.152	-0.545	1.264	-0.500	1.370	-0.454	1.469	-0.408	1.606	-0.340	2.224	-0.067	2.595	-0.143	2.893	-0.419	3.287	-0.982
2.4	0.970	-0.302	1.026	-0.286	1.136	-0.252	1.240	-0.218	1.340	-0.184	1.479	-0.135	2.152	0.037	2.568	-0.089	2.885	-0.392	3.287	-0.973
2.5	0.874	-0.065	0.930	-0.052	1.038	-0.027	1.143	-0.002	1.243	0.022	1.386	0.056	2.103	0.140	2.553	-0.039	2.883	-0.369	3.289	-0.966
2.6	0.798	0.153	0.854	0.162	0.963	0.181	1.069	0.198	1.171	0.215	1.318	0.236	2.073	0.239	2.548	0.008	2.880	-0.348	3.291	-0.961
2.7	0.738	0.356	0.794	0.363	0.906	0.376	1.014	0.387	1.119	0.397	1.270	0.407	2.059	0.332	2.551	0.050	2.891	-0.331	3.294	-0.957
2.8	0.689	0.549	0.747	0.554	0.862	0.562	0.974	0.568	1.083	0.571	1.240	0.571	2.060	0.420	2.560	0.086	2.898	-0.318	3.297	-0.954
2.9	0.650	0.736	0.711	0.739	0.831	0.743	0.948	0.744	1.062	0.740	1.226	0.730	2.074	0.500	2.575	0.117	2.907	-0.308	3.300	-0.953
3.0	0.619	0.921	0.683	0.922	0.810	0.921	0.933	0.916	1.053	0.906	1.225	0.884	2.098	0.572	2.593	0.142	2.916	-0.301	3.302	-0.952
3.1	0.596	1.106	0.664	1.105	0.799	1.100	0.931	1.088	1.058	1.071	1.240	1.037	2.132	0.637	2.613	0.161	2.924	-0.296	3.304	-0.951
π	0.588	1.184	0.659	1.182	0.798	1.174	0.933	1.160	1.064	1.140	1.250	1.099	2.149	0.661	2.622	0.167	2.928	-0.295	3.305	-0.951
3.2	0.579	1.295	0.654	1.292	0.799	1.281	0.940	1.262	1.076	1.236	1.269	1.187	2.174	0.692	2.634	0.174	2.933	-0.294	3.306	-0.951
3.3	0.571	1.492	0.652	1.486	0.811	1.468	0.964	1.440	1.110	1.404	1.314	1.337	2.223	0.737	2.655	0.182	2.940	-0.293	3.307	-0.952
3.4	0.571	1.701	0.661	1.692	0.836	1.664	1.003	1.624	1.161	1.574	1.378	1.485	2.277	0.771	2.676	0.186	2.947	-0.294	3.308	-0.952
3.5	0.582	1.928	0.683	1.914	0.879	1.873	1.063	1.817	1.234	1.749	1.464	1.631	2.334	0.794	2.694	0.185	2.952	-0.295	3.309	-0.953
3.6	0.607	2.179	0.723	2.158	0.944	2.098	1.148	2.020	1.333	1.928	1.574	1.774	2.392	0.805	2.711	0.182	2.956	-0.298	3.309	-0.953
3.7	0.653	2.464	0.788	2.431	1.041	2.344	1.268	2.234	1.467	2.109	1.715	1.908	2.450	0.804	2.725	0.175	2.959	-0.301	3.310	-0.954
3.8	0.731	2.794	0.891	2.743	1.183	2.614	1.434	2.457	1.644	2.287	1.890	2.027	2.504	0.793	2.736	0.167	2.961	-0.304	3.310	-0.954
3.9	0.861	3.187	1.054	3.106	1.391	2.909	1.663	2.685	1.876	2.454	2.104	2.120	2.552	0.771	2.745	0.158	2.963	-0.306	3.310	-0.955
4.0	1.079	3.665	1.313	3.531	1.698	3.225	1.979	2.901	2.176	2.589	2.357	2.170	2.594	0.741	2.751	0.148	2.963	-0.309	3.310	-0.955
4.1	1.457	4.252	1.739	4.023	2.153	3.537	2.407	3.072	2.549	2.661	2.642	2.159	2.626	0.704	2.754	0.138	2.963	-0.311	3.310	-0.955
4.2	2.141	4.956	2.452	4.547	2.818	3.777	2.961	3.134	2.989	2.625	2.940	2.063	2.648	0.663	2.755	0.129	2.963	-0.313	3.310	-0.956
4.3	3.422	5.668	3.643	4.941	3.732	3.790	3.618	2.989	3.453	2.429	3.217	1.869	2.660	0.621	2.754	0.120	2.963	-0.315	3.309	-0.956
4.4	5.692	5.794	5.418	4.704	4.791	3.325	4.261	2.536	3.857	2.043	3.429	1.583	2.662	0.580	2.752	0.113	2.962	-0.316	3.309	-0.956
4.5	8.374	3.839	7.142	3.073	5.594	2.229	4.681	1.780	4.091	1.503	3.535	1.238	2.656	0.541	2.749	0.107	2.961	-0.317	3.309	-0.956
4.6	8.608	0.088	7.307	0.503	5.666	0.825	4.706	0.910	4.092	0.919	3.518	0.885	2.642	0.508	2.745	0.102	2.960	-0.318	3.309	-0.956
4.7	6.486	-2.063	5.972	-1.227	5.050	-0.276	4.367	0.182	3.880	0.416	3.394	0.577	2.623	0.430	2.741	0.099	2.959	-0.318	3.309	-0.956
3π/2	6.206	-2.176	5.775	-1.347	4.948	-0.374	4.309	0.110	3.843	0.363	3.373	0.543	2.620	0.477	2.740	0.098	2.959	-0.318	3.309	-0.956
4.8	4.522	-2.413	4.484	-1.763	4.195	-0.834	3.849	-0.273	3.544	0.066	3.199	0.343	2.600	0.458	2.736	0.097	2.958	-0.318	3.309	-0.956
4.9	3.264	-2.127	3.386	-1.700	3.420	-0.988	3.317	-0.477	3.174	-0.128	2.972	0.193	2.575	0.443	2.732	0.095	2.957	-0.318	3.309	-0.956
5.0	2.489	-1.717	2.648	-1.439	2.816	-0.927	2.853	-0.512	2.825	-0.199	2.743	0.117	2.550	0.435	2.728	0.095	2.957	-0.318	3.309	-0.956
5.1	1.995	-1.324	2.152	-1.138	2.367	-0.771	2.478	-0.449	2.524	-0.186	2.532	0.100	2.525	0.431	2.724	0.096	2.956	-0.318	3.309	-0.956
5.2	1.664	-0.977	1.810	-0.848	2.035	-0.583	2.183	-0.335	2.274	-0.122	2.346	0.124	2.502	0.433	2.721	0.097	2.956	-0.318	3.309	-0.956
5.3	1.431	-0.675	1.566	-0.582	1.789	-0.387	1.954	-0.197	2.072	-0.027	2.187	0.177	2.481	0.439	2.719	0.098	2.956	-0.318	3.309	-0.956
5.4	1.262	-0.411	1.387	-0.342	1.603	-0.196	1.775	-0.051	1.910	0.083	2.056	0.249	2.463	0.448	2.717	0.100	2.955	-0.317	3.309	-0.956
5.5	1.135	-0.176	1.252	-0.124	1.461	-0.014	1.637	0.098	1.781	0.202	1.950	0.333	2.449	0.459	2.715	0.102	2.955	-0.317	3.309	-0.956
5.6	1.037	0.037	1.149	0.076	1.353	0.160	1.530	0.245	1.681	0.324	1.865	0.424	2.438	0.473	2.714	0.104	2.955	-0.317	3.309	-0.956
5.7	0.960	0.233	1.068	0.263	1.269	0.326	1.447	0.389	1.604	0.447	1.799	0.519	2.430	0.487	2.714	0.106	2.955	-0.316	3.309	-0.956
5.8	0.899	0.416	1.005	0.439	1.205	0.486	1.385	0.530	1.546	0.570	1.751	0.615	2.425	0.501	2.714	0.108	2.955	-0.316	3.309	-0.956
5.9	0.851	0.591	0.956	0.608	1.156	0.640	1.340	0.669	1.505	0.692	1.718	0.713	2.423	0.515	2.714	0.109	2.955	-0.316	3.309	-0.956
6.0	0.812	0.760	0.919	0.772	1.122	0.792	1.310	0.806	1.480	0.813	1.699	0.810	2.424	0.528	2.715	0.110	2.955	-0.316	3.309	-0.956
6.1	0.781	0.927	0.891	0.934	1.100	0.942	1.294	0.941	1.468	0.932	1.694	0.906	2.427	0.540	2.715	0.111	2.955	-0.316	3.309	-0.956
6.2	0.759	1.095	0.873	1.097	1.090	1.093	1.291	1.077	1.471	1.051	1.702	1.000	2.432	0.550	2.716	0.112	2.956	-0.316	3.309	-0.956
2π	0.745	1.236	0.865	1.235	1.091	1.219	1.299	1.190	1.484	1.149	1.719	1.076	2.438	0.558	2.717	0.113	2.956	-0.316	3.309	-0.956
6.3	0.743	1.265	0.864	1.263	1.093	1.245	1.301	1.212	1.488	1.169	1.723	1.091	2.439	0.559	2.717	0.113	2.956	-0.316	3.309	-0.956
6.4	0.736	1.443	0.865	1.434	1.108	1.401	1.327	1.349	1.519	1.286	1.757	1.179	2.447	0.566	2.718	0.113	2.956	-0.316	3.309	-0.956
6.5	0.737	1.630	0.877	1.615	1.138	1.562	1.368	1.487	1.566	1.400	1.804	1.261	2.455	0.571	2.719	0.113	2.956	-0.316	3.309	-0.956
6.6	0.749	1.832	0.903	1.807	1.186	1.729	1.429	1.627	1.631	1.511	1.865	1.337	2.464	0.574	2.719	0.113	2.956	-0.316	3.309	-0.956
6.7	0.775	2.054	0.947	2.016	1.256	1.905	1.511	1.766	1.715	1.617	1.939	1.403	2.472	0.575	2.720	0.113	2.956	-0.316	3.309	-0.956
6.8	0.821	2.302	1.016	2.245	1.355	2.088	1.621	1.903	1.820	1.715	2.027	1.457	2.481	0.575	2.721	0.113	2.956	-0.316	3.309	-0.956
6.9	0.895	2.586	1.120	2.501	1.492	2.279	1.762	2.033	1.950	1.798	2.126	1.496	2.488	0.573	2.721	0.113	2.956	-0.316	3.309	-0.956
7.0	1.015	2.918	1.277	2.787	1.680	2.472	1.942	2.149	2.103	1.860	2.236	1.515	2.495	0.569	2.722	0.112	2.956	-0.316	3.309	-0.956
7.1	1.208	3.312	1.513	3.108	1.935	2.655	2.165	2.237	2.280	1.893	2.351	1.510	2.500	0.565	2.722	0.112	2.956	-0.316	3.309	-0.956
7.2	1.527	3.782	1.875	3.455	2.276	2.806	2.431	2.278	2.474	1.884	2.467	1.479	2.504	0.560	2.722	0.112	2.956	-0.316	3.309	-0.956
7.3	2.074	4.328	2.432	3.792	2.716	2.881	2.734	2.248	2.675	1.823	2.576	1.420	2.507	0.555	2.722	0.111	2.956	-0.316	3.309	-0.956
7.4	3.031	4.878	3.272	4.009	3.245	2.811	3.048	2.121	2.864	1.704	2.669	1.334	2.509	0.549	2.722	0.111	2.956	-0.316	3.309	-0.956
7.5	4.651	5.093	4.419	3.850	3.797	2.518	3.329	1.882	3.018	1.529	2.738	1.226	2.509	0.544	2.722	0.111	2.956	-0.316	3.309	-0.956
7.6	6.787	4.084	5.595	2.955	4.232	1.968	3.524	1.543	3.116	1.310	2.777	1.105	2.509	0.539	2.722	0.110	2.956	-0.316	3.309	-0.956
7.7	7.826	1.376	6.094	1.367	4.394	1.253	3.588	1.153	3.142	1.073	2.782	0.980	2.507	0.534	2.722	0.110	2.956	-0.316	3.309	-0.956
7.8	6.672	-1.040	5.581	-0.125	4.233	0.565	3.513	0.776	3.098	0.846	2.756	0.861	2.505	0.530	2.721	0.110	2.956	-0.316	3.309	-0.956
5π/2	5.695	-1.668	5.057	-0.649	4.043	0.264	3.424	0.600	3.048	0.736	2.731	0.803	2.503	0.529	2.721	0.110	2.956	-0.316	3.309	-0.956
7.9	4.915	-1.900	4.580	-0.931	3.848	0.060	3.329	0.470	2.994	0.653	2.704	0.757	2.502	0.527	2.721	0.110	2.956	-0.316	3.309	-0.956
8.0	3.598	-1.882	3.638	-1.159	3.387	-0.222	3.085	0.259	2.852	0.507	2.631	0.675	2.499	0.525	2.721	0.110	2.956	-0.316	3.309	-0.956
8.1	2.745	-1.589	2.918	-1.092	2.948	-0.327	2.825	0.141	2.692	0.412	2.547	0.615	2.496	0.523	2.721	0.110	2.956	-0.316	3.309	-0.956
8.2	2.194	-1.250	2.402	-0.909	2.574	-0.317	2.578	0.098	2.530	0.363	2.458	0.579	2.492	0.522	2.721	0.110	2.956	-0.316	3.309	-0.956
8.3	1.825	-0.930	2.033	-0.691	2.271	-0.240	2.358	0.110	2.377	0.352	2.369	0.564	2.489	0.522	2.721	0.110	2.956	-0.316	3.309	-0.956
8.4	1.569	-0.644	1.766	-0.472	2.031	-0.129	2.171	0.159	2.239	0.371	2.285	0.567	2.486	0.523	2.721	0.110	2.956	-0.316	3.309	-0.956
8.5	1.384	-0.390	1.568	-0.264	1.842	-0.001	2.014	0.232	2.119	0.411	2.209	0.585	2.483	0.524	2.721	0.110	2.956	-0.316	3.309	-0.956
8.6	1.246	-0.164	1.419	-0.068	1.694	0.134	1.886	0.319	2.017	0.468	2.142	0.615	2.481	0.525	2.720	0.110	2.956	-0.316	3.309	-0.956
8.7	1.141	0.042	1.304	0.114	1.577	0.270	1.782	0.416	1.932	0.535	2.085	0.654	2.479	0.527	2.720	0.110	2.956	-0.316	3.309	-0.956
8.8	1.059	0.230	1.216	0.286	1.486	0.405	1.700	0.517	1.863	0.609	2.039	0.699	2.478	0.529	2.720	0.110	2.956	-0.316	3.309	-0.956
8.9	0.995	0.407	1.146	0.450	1.416	0.539	1.636	0.622	1.810	0.688	2.002	0.750	2.477	0.531	2.720	0.110	2.956	-0.316	3.309	-0.956
9.0	0.943	0.574	1.093	0.607	1.362	0.671	1.588	0.728	1.771	0.770	1.976	0.804	2.476	0.533	2.720	0.110	2.956	-0.316	3.309	-0.956
9.1	0.903	0.736	1.052	0.760	1.324	0.802	1.555	0.834	1.745	0.854	1.959	0.859	2.476	0.535	2.720	0.110	2.956	-0.316	3.309	-0.956
9.2	0.871	0.895	1.022	0.911	1.300	0.932	1.536	0.941	1.731	0.938	1.952	0.914	2.477	0.537	2.720	0.110	2.956	-0.316	3.309	-0.956
9.3	0.847	1.054	1.003	1.062	1.288	1.063	1.531	1.048	1.730	1.022	1.954	0.969	2.477	0.538	2.720	0.110	2.956	-0.316	3.309	-0.956
9.4	0.831	1.215	0.993	1.215	1.288	1.194	1.538	1.155	1.740	1.105	1.965	1.022	2.478	0.540	2.720	0.111	2.956	-0.316	3.309	-0.956
3π	0.828	1.256	0.992	1.253	1.290	1.227	1.542	1.182	1.745	1.125	1.969	1.035	2.479	0.540	2.720	0.111	2.956	-0.316	3.309	-0.956
9.5	0.823	1.381	0.993	1.372	1.302	1.328	1.559	1.261	1.763	1.186	1.985	1.072	2.479	0.541	2.721	0.111	2.956	-0.316	3.309	-0.956
9.6	0.823	1.555	1.006	1.536	1.331	1.464	1.595	1.366	1.798	1.263	2.012	1.118	2.481	0.542	2.721	0.111	2.956	-0.316	3.309	-0.956
9.7	0.833	1.742	1.032	1.709	1.377	1.602	1.646	1.469	1.845	1.335	2.047	1.158	2.482	0.542	2.721	0.111	2.956	-0.316	3.309	-0.956
9.8	0.857	1.945	1.075	1.894	1.443	1.743	1.714	1.567	1.905	1.400	2.090	1.191	2.483	0.542	2.721	0.111	2.956	-0.316	3.309	-0.956
9.9	0.898	2.170	1.141	2.054	1.533	1.885	1.802	1.659	1.978	1.456	2.137	1.216	2.484	0.542	2.721	0.111	2.956	-0.316	3.309	-0.956
10.0	0.964	2.424	1.239	2.313	1.653	2.027	1.910	1.740	2.063	1.499	2.190	1.232	2.485	0.542	2.721	0.111	2.956	-0.316	3.309	-0.956
10.1	1.069	2.717	1.382	2.551	1.811	2.161	2.040	1.805	2.159	1.527										
10.2	1.235	3.059	1.590	2.809	2.013	2.278	2.192	1.846	2.262	1.535										
10.3	1.501	3.462	1.894	3.078	2.266	2.362	2.362	1.854	2.370	1.520										
10.4	1.940	3.926	2.337	3.327	2.571	2.387	2.542	1.819	2.475	1.479										
10.5	2.683	4.407	2.969	3.483	2.913	2.321	2.720	1.735	2.571	1.413										
10.6	3.920	4.706	3.799	3.396	3.257	2.131	2.876	1.597	2.649	1.323										
10.7	5.593	4.249	4.689	2.858	3.540	1.809	2.990	1.415	2.701	1.215										
10.8	7.130	2.351	5.261	1.811	3.692	1.387	3.046	1.204	2.725	1.097										
10.9	6.817	-0.091	5.180	0.607	3.676	0.941	3.036	0.989	2.717	0.980										
7π/2	5.411	-1.387	4.602	-0.244	3.518	0.569	2.971	0.802	2.683	0.876										

βh	α/β = 0.00		α/β = 0.01		α/β = 0.03		α/β = 0.05		α/β = 0.07	
11.0	5.342	-1.418	4.565	-0.272	3.508	0.554	2.967	0.794	2.681	0.872
11.1	3.973	-1.692	3.811	-0.693	3.246	0.275	2.852	0.636	2.621	0.780
11.2	3.024	-1.524	3.142	-0.788	2.951	0.112	2.710	0.523	2.545	0.709
11.3	2.400	-1.235	2.621	-0.714	2.666	0.045	2.559	0.454	2.461	0.661
11.4	1.984	-0.935	2.232	-0.565	2.412	0.047	2.410	0.425	2.373	0.634
11.5	1.696	-0.658	1.943	-0.390	2.195	0.095	2.271	0.428	2.287	0.627
11.6	1.491	-0.409	1.727	-0.211	2.016	0.171	2.148	0.456	2.207	0.636
11.7	1.339	-0.185	1.562	-0.036	1.870	0.264	2.041	0.502	2.135	0.659
11.8	1.225	0.018	1.436	0.131	1.752	0.366	1.950	0.560	2.072	0.693
11.9	1.136	0.204	1.338	0.291	1.658	0.474	1.875	0.628	2.019	0.735
12.0	1.067	0.378	1.262	0.445	1.583	0.584	1.815	0.702	1.976	0.783
12.1	1.012	0.542	1.203	0.593	1.526	0.696	1.769	0.780	1.943	0.835
12.2	0.968	0.700	1.158	0.737	1.485	0.808	1.736	0.861	1.919	0.891
12.3	0.934	0.854	1.125	0.880	1.457	0.921	1.715	0.944	1.905	0.948
12.4	0.909	1.007	1.102	1.022	1.442	1.034	1.706	1.027	1.901	1.005
12.5	0.891	1.161	1.091	1.165	1.439	1.148	1.709	1.110	1.906	1.062
4π	0.883	1.266	1.089	1.262	1.445	1.224	1.718	1.164	1.914	1.098
12.6	0.881	1.320	1.090	1.312	1.450	1.263	1.724	1.192	1.920	1.117
12.7	0.879	1.485	1.100	1.463	1.475	1.379	1.751	1.272	1.942	1.168
12.8	0.887	1.660	1.124	1.621	1.515	1.495	1.791	1.348	1.974	1.216
12.9	0.906	1.849	1.164	1.789	1.573	1.610	1.843	1.420	2.014	1.258
13.0	0.942	2.056	1.226	1.967	1.651	1.724	1.909	1.484	2.061	1.292
13.1	1.000	2.289	1.316	2.159	1.752	1.832	1.989	1.538	2.115	1.318
13.2	1.091	2.553	1.444	2.364	1.881	1.931	2.082	1.578	2.175	1.333
13.3	1.232	2.859	1.625	2.580	2.040	2.012	2.186	1.601	2.237	1.335
13.4	1.454	3.215	1.882	2.799	2.231	2.064	2.298	1.601	2.300	1.325
13.5	1.812	3.624	2.243	2.998	2.451	2.074	2.413	1.575	2.360	1.301
13.6	2.401	4.062	2.738	3.125	2.689	2.022	2.524	1.520	2.415	1.264
13.7	3.372	4.408	3.374	3.087	2.925	1.896	2.621	1.437	2.460	1.214
13.8	4.838	4.293	4.076	2.750	3.124	1.690	2.695	1.330	2.492	1.156
13.9	6.399	3.065	4.630	2.041	3.253	1.421	2.739	1.205	2.509	1.092
14.0	6.799	0.846	4.774	1.104	3.286	1.121	2.749	1.074	2.512	1.027
14.1	5.746	-0.865	4.462	0.264	3.221	0.835	2.724	0.947	2.499	0.964
9π/2	5.221	-1.201	4.268	0.028	3.175	0.740	2.706	0.904	2.490	0.942
14.2	4.381	-1.477	3.900	-0.264	3.078	0.599	2.669	0.836	2.472	0.907
14.3	3.332	-1.468	3.313	-0.491	2.888	0.430	2.591	0.747	2.435	0.860
14.4	2.624	-1.240	2.810	-0.519	2.683	0.331	2.500	0.682	2.390	0.824
14.5	2.149	-0.963	2.412	-0.441	2.482	0.289	2.403	0.643	2.341	0.799
14.6	1.824	-0.693	2.107	-0.312	2.299	0.293	2.307	0.626	2.290	0.785
14.7	1.594	-0.445	1.873	-0.163	2.139	0.328	2.215	0.629	2.239	0.783
14.8	1.426	-0.222	1.694	-0.010	2.003	0.386	2.132	0.649	2.191	0.790
14.9	1.300	-0.019	1.556	0.143	1.889	0.459	2.058	0.682	2.148	0.806
15.0	1.203	0.167	1.449	0.292	1.796	0.541	1.995	0.724	2.109	0.828
15.1	1.127	0.340	1.365	0.436	1.721	0.630	1.942	0.774	2.077	0.856
15.2	1.068	0.502	1.299	0.576	1.662	0.722	1.901	0.830	2.051	0.888
15.3	1.021	0.658	1.249	0.713	1.617	0.817	1.869	0.890	2.032	0.924
15.4	0.984	0.809	1.212	0.848	1.586	0.914	1.848	0.952	2.019	0.961
15.5	0.957	0.958	1.187	0.982	1.568	1.011	1.837	1.015	2.014	0.999
15.6	0.937	1.108	1.172	1.118	1.562	1.110	1.836	1.078	2.015	1.037
15.7	0.924	1.260	1.168	1.255	1.569	1.209	1.845	1.141	2.022	1.074
5π	0.924	1.272	1.169	1.266	1.570	1.217	1.846	1.146	2.023	1.076
15.8	0.920	1.418	1.176	1.397	1.589	1.307	1.864	1.202	2.036	1.108
15.9	0.925	1.584	1.197	1.543	1.622	1.405	1.893	1.259	2.056	1.140
16.0	0.941	1.762	1.234	1.697	1.671	1.502	1.932	1.312	2.081	1.167
16.1	0.971	1.956	1.290	1.858	1.736	1.594	1.980	1.359	2.111	1.190
16.2	1.021	2.171	1.371	2.029	1.820	1.680	2.038	1.397	2.145	1.206
16.3	1.099	2.414	1.485	2.209	1.924	1.756	2.103	1.424	2.182	1.216
16.4	1.219	2.691	1.643	2.395	2.050	1.816	2.175	1.439	2.220	1.218
16.5	1.405	3.012	1.862	2.580	2.196	1.853	2.252	1.439	2.259	1.212
16.6	[illegible]	[illegible]	2.161	2.747	2.360	1.858	2.329	1.422	2.295	1.199
16.7	2.174	3.782	2.561	[illegible]	[illegible]	[illegible]	[illegible]	[illegible]	2.328	1.178
16.8	2.948	4.149	3.065	2.852	2.704	1.734	2.467	[illegible]	[illegible]	[illegible]
16.9	4.151	4.241	3.633	2.638	2.853	1.597	2.518	1.268	2.376	1.117
17.0	5.658	3.537	4.137	2.149	2.959	1.415	2.553	1.190	2.389	1.080
17.1	6.577	1.724	4.390	1.434	3.008	1.208	2.567	1.107	2.394	1.042
17.2	6.064	-0.209	4.289	0.651	2.992	0.997	2.560	1.023	2.390	1.005
11π/2	5.080	-1.066	4.007	0.224	2.937	0.845	2.541	0.962	2.381	0.977
17.3	4.806	-1.192	3.912	0.124	2.917	0.808	2.534	0.946	2.378	0.970
17.4	3.674	-1.390	3.429	-0.197	2.799	0.656	2.491	0.879	2.360	0.939
17.5	2.871	-1.247	2.965	-0.317	2.657	0.550	2.437	0.826	2.336	0.913
17.6	2.329	-0.999	2.572	-0.308	2.505	0.488	2.375	0.789	2.307	0.894
17.7	1.959	-0.739	2.258	-0.228	2.358	0.464	2.310	0.767	2.277	0.882
17.8	1.699	-0.493	2.011	-0.112	2.222	0.470	2.245	0.759	2.246	0.876
17.9	1.511	-0.268	1.819	0.019	2.100	0.501	2.183	0.764	2.215	0.877
18.0	1.371	-0.064	1.669	0.155	1.995	0.548	2.125	0.780	2.186	0.883
18.1	1.265	0.123	1.552	0.292	1.906	0.608	2.074	0.805	2.160	0.894
18.2	1.183	0.296	1.460	0.426	1.833	0.677	2.031	0.837	2.137	0.910
18.3	1.118	0.458	1.388	0.558	1.774	0.752	1.995	0.876	2.118	0.929
18.4	1.067	0.613	1.332	0.688	1.729	0.830	1.967	0.918	2.103	0.951
18.5	1.027	0.762	1.291	0.817	1.696	0.912	1.947	0.964	2.093	0.975
18.6	0.997	0.908	1.261	0.945	1.675	0.995	1.935	1.011	2.087	0.999
18.7	0.974	1.054	1.243	1.073	1.665	1.080	1.931	1.059	2.086	1.024
18.8	0.960	1.202	1.237	1.203	1.668	1.165	1.936	1.106	2.090	1.049
6π	0.955	1.277	1.238	1.269	1.673	1.207	1.941	1.129	2.093	1.060
18.9	0.953	1.354	1.242	1.336	1.682	1.249	1.948	1.152	2.097	1.072
19.0	0.955	1.513	1.259	1.473	1.708	1.332	1.968	1.196	2.109	1.093
19.1	0.967	1.682	1.291	1.614	1.748	1.413	1.995	1.235	2.125	1.111
19.2	0.992	1.865	1.341	1.762	1.802	1.490	2.030	1.270	2.143	1.126
19.3	1.035	2.066	1.414	1.917	1.870	1.560	2.071	1.298	2.165	1.138
19.4	1.101	2.291	1.515	2.077	1.954	1.621	2.117	1.318	2.187	1.144
19.5	1.204	2.546	1.653	2.241	2.053	1.668	2.168	1.330	2.211	1.147
19.6	1.360	2.838	1.841	2.402	2.167	1.696	2.221	1.330	2.235	1.144
19.7	1.604	3.173	2.093	2.547	2.292	1.701	2.274	1.320	2.258	1.137
19.8	1.991	3.545	2.422	2.648	2.423	1.675	2.325	1.298	2.279	1.125
19.9	2.614	3.913	2.832	2.662	2.552	1.616	2.370	1.265	2.296	1.109
20.0	3.597	4.126	3.300	2.527	2.666	1.521	2.407	1.222	2.310	1.090
20.1	4.956	3.804	3.746	2.186	2.753	1.395	2.433	1.172	2.320	1.068
20.2	6.168	2.473	4.042	1.646	2.804	1.247	2.447	1.117	2.324	1.045
20.3	6.235	0.534	4.079	1.019	2.811	1.091	2.448	1.061	2.324	1.022
20.4	5.220	-0.808	3.861	0.465	2.776	0.943	2.436	1.008	2.319	1.001
13π/2	4.970	-0.962	3.794	0.372	2.765	0.915	2.433	0.997	2.317	0.996
20.5	4.048	-1.266	3.491	0.087	2.706	0.815	2.413	0.959	2.309	0.981
20.6	3.148	-1.241	3.084	-0.107	2.609	0.715	2.381	0.918	2.296	0.964
20.7	2.528	-1.037	2.710	-0.164	2.499	0.646	2.342	0.886	2.280	0.950
20.8	2.105	-0.790	2.395	-0.134	2.384	0.607	2.298	0.864	2.263	0.940
20.9	1.810	-0.547	2.140	-0.054	2.272	0.595	2.253	0.852	2.243	0.934
21.0	1.599	-0.321	1.937	0.052	2.168	0.605	2.208	0.850	2.224	0.932
21.1	1.443	-0.114	1.777	0.171	2.075	0.633	2.165	0.856	2.205	0.934
21.2	1.326	0.074	1.650	0.294	1.993	0.674	2.126	0.870	2.188	0.940
21.3	1.235	0.248	1.550	0.419	1.924	0.725	2.091	0.890	2.172	0.948
21.4	1.165	0.411	1.471	0.543	1.867	0.784	2.061	0.916	2.159	0.960
21.5	1.109	0.565	1.410	0.666	1.822	0.848	2.037	0.945	2.148	0.973
21.6	1.066	0.714	1.364	0.788	1.788	0.916	2.019	0.978	2.140	0.987
21.7	1.032	0.858	1.330	0.910	1.765	0.986	2.008	1.013	2.135	1.003
21.8	1.007	1.002	1.308	1.032	1.752	1.058	2.002	1.048	2.133	1.019
21.9	0.990	1.146	1.298	1.155	1.751	1.130	2.003	1.084	2.134	1.035
7π	0.981	1.280	1.299	1.269	1.759	1.196	2.010	1.115	2.137	1.049
22.0	0.981	1.293	1.300	1.280	1.760	1.202	2.011	1.118	2.138	1.050
22.1	0.980	1.447	1.313	1.408	1.780	1.273	2.024	1.151	2.145	1.064
22.2	0.988	1.608	1.341	1.540	1.811	1.342	2.043	1.181	2.154	1.077
22.3	1.009	1.782	1.385	1.677	1.855	1.406	2.067	1.207	2.165	1.087
22.4	1.045	1.971	1.449	1.818	1.910	1.464	2.096	1.229	2.178	1.095
22.5	1.101	2.181	1.538	1.963	1.977	1.514	2.129	1.244	2.193	1.100
22.6	1.188	2.417	1.660	2.110	2.056	1.553	2.164	1.253	2.208	1.102
22.7	1.321	2.686	1.822	2.253	2.145	1.577	2.202	1.255	2.223	1.101
22.8	1.525	2.993	2.036	2.381	2.242	1.582	2.239	1.249	2.237	1.097
22.9	1.843	3.338	2.311	2.475	2.343	1.566	2.275	1.236	2.250	1.091
23.0	2.349	3.697	2.652	2.503	2.443	1.524	2.307	1.215	2.262	1.082
23.1	3.151	3.977	3.042	2.420	2.532	1.458	2.334	1.187	2.271	1.070
23.2	4.327	3.913	3.433	2.182	2.605	1.368	2.354	1.154	2.278	1.058
23.3	5.635	3.043	3.737	1.776	2.653	1.261	2.367	1.118	2.282	1.044
23.4	6.214	1.307	3.858	1.262	2.671	1.145	2.371	1.080	2.283	1.030
23.5	5.581	-0.306	3.764	0.752	2.659	1.030	2.367	1.042	2.281	1.016
15π/2	4.881	-0.878	3.616	0.488	2.636	0.963	2.360	1.020	2.278	1.008
23.6	4.445	-1.072	3.502	0.353	2.618	0.924	2.354	1.007	2.276	1.003
23.7	3.458	-1.208	3.164	0.104	2.555	0.836	2.335	0.976	2.269	0.992
23.8	2.751	-1.069	2.823	-0.011	2.476	0.769	2.311	0.951	2.260	0.983
23.9	2.266	-0.843	2.517	-0.029	2.389	0.724	2.282	0.932	2.250	0.975
24.0	1.930	-0.605	2.259	0.014	2.299	0.702	2.251	0.919	2.238	0.970
24.1	1.691	-0.378	2.049	0.093	2.213	0.698	2.219	0.914	2.226	0.968
24.2	1.517	-0.169	1.880	0.192	2.132	0.710	2.187	0.915	2.214	0.968
24.3	1.387	0.022	1.744	0.302	2.060	0.736	2.158	0.921	2.203	0.970
24.4	1.287	0.198	1.636	0.415	1.996	0.773	2.131	0.933	2.192	0.975
24.5	1.210	0.362	1.551	0.531	1.943	0.817	2.107	0.950	2.183	0.981
24.6	1.150	0.517	1.484	0.646	1.899	0.868	2.087	0.970	2.175	0.989
24.7	1.102	0.665	1.432	0.762	1.865	0.924	2.072	0.993	2.169	0.998
24.8	1.065	0.808	1.395	0.878	1.841	0.982	2.061	1.018	2.165	1.008
24.9	1.037	0.950	1.369	0.994	1.826	1.043	2.055	1.044	2.163	1.018
25.0	1.017	1.091	1.355	1.111	1.821	1.104	2.054	1.070	2.163	1.028

TABLE 2.7

NORMALIZED ADMITTANCES (Y/Δ) IN MILLIMHOS OF THIN DIPOLE ANTENNAS IN DISSIPATIVE MEDIA

a/λ = 0.004763

βh	α/β = 0.00		α/β = 0.01		α/β = 0.03		α/β = 0.05		α/β = 0.07		α/β = 0.10		α/β = 0.30		α/β = 0.50		α/β = 0.70		α/β = 1.00	
1.5	10.226	-2.054	9.765	-1.610	8.919	-0.945	8.185	-0.496	7.557	-0.190	6.785	0.095	4.432	0.324	3.847	0.023	3.764	-0.341	3.970	-0.935
π/2	7.591	-3.224	7.451	-2.808	7.122	-2.112	6.769	-1.574	6.423	-1.160	5.944	-0.714	4.184	0.083	3.717	-0.061	3.685	-0.369	3.931	-0.937
1.6	6.696	-3.325	6.631	-2.954	6.441	-2.313	6.205	-1.796	5.953	-1.384	5.583	-0.923	4.077	0.008	3.663	-0.086	3.654	-0.376	3.916	-0.936
1.7	4.510	-3.017	4.562	-2.790	4.613	-2.364	4.609	-1.985	4.567	-1.654	4.461	-1.244	3.715	-0.156	3.491	-0.137	3.559	-0.384	3.875	-0.928
1.8	3.250	-2.432	3.326	-2.292	3.445	-2.019	3.525	-1.761	3.573	-1.523	3.600	-1.209	3.389	-0.209	3.339	-0.145	3.480	-0.374	3.844	-0.914
1.9	2.485	-1.874	2.562	-1.784	2.694	-1.603	2.801	-1.426	2.885	-1.258	2.976	-1.025	3.115	-0.189	3.211	-0.124	3.417	-0.352	3.822	-0.899
2.0	1.990	-1.393	2.062	-1.332	2.193	-1.207	2.308	-1.083	2.406	-0.962	2.526	-0.791	2.892	-0.124	3.108	-0.084	3.369	-0.323	3.807	-0.883
2.1	1.650	-0.984	1.718	-0.940	1.845	-0.852	1.960	-0.762	2.064	-0.674	2.199	-0.547	2.716	-0.033	3.028	-0.034	3.334	-0.291	3.797	-0.868
2.2	1.407	-0.632	1.472	-0.600	1.595	-0.535	1.709	-0.469	1.815	-0.404	1.958	-0.310	2.580	0.071	2.969	0.023	3.310	-0.259	3.792	-0.855
2.3	1.227	-0.325	1.289	-0.301	1.409	-0.252	1.523	-0.203	1.630	-0.155	1.779	-0.085	2.478	0.180	2.927	0.080	3.295	-0.228	3.791	-0.844
2.4	1.089	-0.051	1.150	-0.033	1.269	0.004	1.383	0.040	1.492	0.076	1.645	0.127	2.405	0.291	2.900	0.137	3.288	-0.200	3.791	-0.834
2.5	0.981	0.197	1.043	0.211	1.162	0.239	1.277	0.266	1.388	0.292	1.545	0.327	2.355	0.399	2.885	0.190	3.287	-0.175	3.793	-0.827
2.6	0.896	0.427	0.958	0.438	1.079	0.458	1.196	0.477	1.310	0.495	1.472	0.516	2.325	0.503	2.881	0.239	3.290	-0.153	3.795	-0.822
2.7	0.828	0.644	0.891	0.652	1.015	0.666	1.136	0.678	1.253	0.688	1.422	0.698	2.313	0.601	2.886	0.283	3.296	-0.136	3.798	-0.818
2.8	0.773	0.851	0.838	0.856	0.967	0.865	1.092	0.871	1.214	0.874	1.390	0.873	2.316	0.693	2.897	0.321	3.304	-0.122	3.801	-0.815
2.9	0.729	1.053	0.797	1.056	0.932	1.060	1.064	1.060	1.192	1.056	1.376	1.042	2.332	0.777	2.913	0.352	3.313	-0.112	3.804	-0.813
3.0	0.695	1.253	0.767	1.254	0.910	1.253	1.049	1.246	1.184	1.235	1.377	1.209	2.360	0.853	2.932	0.378	3.323	-0.105	3.807	-0.812
3.1	0.669	1.455	0.747	1.454	0.899	1.447	1.047	1.433	1.190	1.413	1.395	1.372	2.398	0.920	2.954	0.397	3.332	-0.100	3.809	-0.812
π	0.661	1.541	0.741	1.539	0.898	1.529	1.051	1.511	1.198	1.487	1.407	1.440	2.416	0.945	2.963	0.403	3.336	-0.099	3.810	-0.812
3.2	0.653	1.663	0.737	1.659	0.901	1.645	1.060	1.622	1.213	1.592	1.429	1.534	2.444	0.977	2.977	0.410	3.341	-0.098	3.811	-0.812
3.3	0.645	1.880	0.737	1.873	0.917	1.850	1.089	1.816	1.253	1.773	1.482	1.695	2.497	1.023	2.999	0.418	3.349	-0.098	3.812	-0.813
3.4	0.649	2.112	0.751	2.100	0.950	2.065	1.138	2.017	1.314	1.958	1.556	1.854	2.555	1.057	3.020	0.421	3.355	-0.099	3.813	-0.813
3.5	0.667	2.364	0.782	2.346	1.003	2.294	1.210	2.227	1.400	2.147	1.654	2.010	2.616	1.079	3.040	0.420	3.361	-0.101	3.814	-0.814
3.6	0.704	2.645	0.836	2.616	1.085	2.542	1.312	2.447	1.517	2.338	1.781	2.160	2.678	1.088	3.057	0.415	3.365	-0.103	3.814	-0.814
3.7	0.769	2.963	0.922	2.920	1.205	2.810	1.458	2.677	1.673	2.530	1.942	2.298	2.739	1.085	3.072	0.408	3.368	-0.106	3.815	-0.815
3.8	0.877	3.332	1.057	3.265	1.380	3.102	1.653	2.914	1.879	2.714	2.139	2.416	2.795	1.070	3.083	0.399	3.370	-0.109	3.815	-0.815
3.9	1.054	3.769	1.268	3.663	1.635	3.416	1.925	3.147	2.145	2.878	2.377	2.501	2.845	1.045	3.091	0.389	3.372	-0.112	3.815	-0.816
4.0	1.349	4.293	1.603	4.119	2.008	3.739	2.292	3.356	2.483	2.999	2.653	2.536	2.886	1.012	3.097	0.378	3.372	-0.115	3.815	-0.816
4.1	1.854	4.917	2.145	4.621	2.549	4.030	2.777	3.496	2.894	3.040	2.956	2.498	2.918	0.972	3.100	0.368	3.372	-0.118	3.815	-0.816
4.2	2.748	5.603	3.026	5.094	3.309	4.198	3.382	3.492	3.358	2.953	3.261	2.369	2.939	0.928	3.101	0.358	3.372	-0.120	3.815	-0.817
4.3	4.321	6.116	4.400	5.287	4.285	4.062	4.053	3.248	3.820	2.692	3.531	2.140	2.950	0.884	3.100	0.350	3.371	-0.121	3.814	-0.817
4.4	6.717	5.650	6.176	4.640	5.286	3.399	4.646	2.694	4.187	2.249	3.721	1.826	2.950	0.841	3.097	0.342	3.370	-0.123	3.814	-0.817
4.5	8.715	3.126	7.459	2.711	5.888	2.197	4.961	1.889	4.363	1.681	3.798	1.467	2.942	0.801	3.094	0.336	3.369	-0.124	3.814	-0.817
4.6	8.222	-0.114	7.184	0.397	5.770	0.870	4.888	1.044	4.307	1.105	3.753	1.114	2.926	0.767	3.089	0.331	3.368	-0.124	3.814	-0.817
4.7	6.257	-1.726	5.860	-0.989	5.102	-0.090	4.505	0.377	4.062	0.630	3.609	0.814	2.905	0.739	3.085	0.328	3.367	-0.125	3.814	-0.817
3π/2	6.015	-1.811	5.681	-1.084	5.001	-0.174	4.444	0.313	4.023	0.582	3.586	0.782	2.902	0.736	3.084	0.328	3.367	-0.125	3.814	-0.817
4.8	4.548	-1.991	4.518	-1.414	4.281	-0.564	3.984	-0.027	3.715	0.308	3.401	0.593	2.880	0.718	3.080	0.326	3.366	-0.125	3.814	-0.817
4.9	3.404	-1.750	3.511	-1.354	3.549	-0.691	3.466	-0.205	3.344	0.134	3.168	0.454	2.854	0.704	3.076	0.325	3.366	-0.125	2.814	-0.817
5.0	2.663	-1.387	2.809	-1.119	2.970	-0.629	3.015	-0.230	2.999	0.075	2.937	0.387	2.827	0.695	3.071	0.325	3.365	-0.125	3.814	-0.817
5.1	2.171	-1.025	2.321	-0.839	2.531	-0.479	2.646	-0.164	2.700	0.093	2.724	0.375	2.802	0.693	3.068	0.325	3.364	-0.124	3.814	-0.817
5.2	1.831	-0.693	1.975	-0.561	2.200	-0.295	2.352	-0.050	2.451	0.161	2.537	0.404	2.778	0.696	3.065	0.327	3.364	-0.124	3.814	-0.817
5.3	1.587	-0.398	1.723	-0.301	1.950	-0.102	2.121	0.089	2.247	0.258	2.377	0.461	2.757	0.703	3.062	0.328	3.364	-0.124	3.814	-0.817
5.4	1.407	-0.134	1.535	-0.061	1.759	0.090	1.939	0.237	2.083	0.372	2.244	0.536	2.739	0.713	3.060	0.330	3.364	-0.123	3.814	-0.817
5.5	1.269	0.105	1.392	0.160	1.611	0.275	1.797	0.389	1.952	0.494	2.136	0.624	2.724	0.725	3.059	0.332	3.363	-0.123	3.814	-0.817
5.6	1.162	0.324	1.280	0.366	1.497	0.454	1.686	0.541	1.849	0.621	2.050	0.719	2.713	0.739	3.058	0.334	3.363	-0.123	3.814	-0.817
5.7	1.078	0.528	1.193	0.560	1.409	0.627	1.601	0.691	1.770	0.749	1.984	0.819	2.705	0.754	3.057	0.336	3.363	-0.123	3.814	-0.817
5.8	1.011	0.720	1.125	0.745	1.341	0.794	1.536	0.839	1.711	0.878	1.935	0.921	2.701	0.769	3.057	0.338	3.363	-0.122	3.814	-0.817
5.9	0.957	0.905	1.072	0.924	1.290	0.957	1.490	0.986	1.670	1.007	1.903	1.023	2.700	0.784	3.058	0.340	3.364	-0.122	3.814	-0.817
6.0	0.914	1.086	1.031	1.099	1.254	1.119	1.459	1.130	1.645	1.134	1.886	1.126	2.701	0.797	3.058	0.341	3.364	-0.122	3.814	-0.817
6.1	0.881	1.266	1.002	1.273	1.232	1.279	1.444	1.275	1.635	1.261	1.883	1.227	2.705	0.809	3.059	0.342	3.364	-0.122	3.814	-0.817
6.2	0.857	1.447	0.984	1.448	1.224	1.440	1.443	1.419	1.641	1.387	1.894	1.326	2.711	0.820	3.060	0.343	3.364	-0.122	3.814	-0.817
2π	0.844	1.601	0.976	1.597	1.227	1.576	1.455	1.540	1.657	1.491	1.914	1.406	2.717	0.828	3.061	0.343	3.364	-0.122	3.814	-0.817
6.3	0.842	1.633	0.976	1.628	1.229	1.604	1.458	1.564	1.662	1.512	1.919	1.421	2.718	0.829	3.061	0.344	3.364	-0.122	3.814	-0.817
6.4	0.837	1.826	0.980	1.815	1.249	1.772	1.489	1.710	1.699	1.635	1.958	1.513	2.726	0.836	3.062	0.344	3.364	-0.122	3.814	-0.817
6.5	0.842	2.032	0.998	2.011	1.286	1.946	1.539	1.857	1.754	1.756	2.011	1.598	2.735	0.840	3.063	0.344	3.364	-0.122	3.814	-0.817
6.6	0.861	2.254	1.033	2.221	1.344	2.126	1.609	2.005	1.828	1.872	2.079	1.676	2.744	0.843	3.063	0.344	3.364	-0.122	3.814	-0.817
6.7	0.898	2.499	1.090	2.449	1.429	2.314	1.705	2.151	1.923	1.981	2.161	1.742	2.753	0.844	3.064	0.344	3.364	-0.122	3.814	-0.817
6.8	0.961	2.773	1.178	2.700	1.547	2.509	1.831	2.293	2.041	2.079	2.256	1.795	2.762	0.843	3.065	0.344	3.364	-0.122	3.814	-0.817
6.9	1.062	3.087	1.310	2.977	1.710	2.708	1.992	2.424	2.184	2.160	2.364	1.829	2.769	0.841	3.065	0.343	3.364	-0.122	3.814	-0.817
7.0	1.222	3.450	1.506	3.283	1.930	2.904	2.194	2.535	2.352	2.216	2.480	1.842	2.776	0.837	3.066	0.343	3.364	-0.122	3.814	-0.817
7.1	1.478	3.875	1.800	3.616	2.224	3.080	2.440	2.610	2.542	2.237	2.601	1.830	2.781	0.832	3.066	0.342	3.364	-0.122	3.814	-0.817
7.2	1.895	4.366	2.243	3.958	2.608	3.207	2.727	2.629	2.746	2.211	2.720	1.789	2.785	0.827	3.066	0.342	3.364	-0.122	3.814	-0.817
7.3	2.593	4.893	2.903	4.247	3.087	3.233	3.042	2.567	2.950	2.131	2.829	1.720	2.788	0.821	3.066	0.342	3.364	-0.122	3.814	-0.817
7.4	3.753	5.307	3.841	4.334	3.631	3.089	3.354	2.402	3.135	1.991	2.919	1.625	2.789	0.816	3.066	0.341	3.364	-0.122	3.814	-0.817
7.5	5.502	5.143	4.993	3.945	4.154	2.713	3.617	2.128	3.277	1.798	2.982	1.510	2.790	0.810	3.066	0.341	3.364	-0.122	3.814	-0.817
7.6	7.302	3.634	5.963	2.844	4.513	2.109	3.778	1.767	3.358	1.569	3.013	1.383	2.789	0.805	3.066	0.341	3.364	-0.122	3.814	-0.817
7.7	7.661	1.048	6.164	1.298	4.586	1.397	3.806	1.374	3.367	1.328	3.011	1.255	2.787	0.800	3.066	0.341	3.364	-0.122	3.814	-0.817
7.8	6.401	-0.847	5.538	0.022	4.369	0.756	3.702	1.010	3.306	1.104	2.978	1.136	2.785	0.796	3.065	0.340	3.364	-0.122	3.814	-0.817
5π/2	5.539	-1.322	5.035	-0.408	4.168	0.485	3.603	0.843	3.250	0.998	2.948	1.079	2.783	0.795	3.065	0.340	3.364	-0.122	3.814	-0.817
7.9	4.860	-1.501	4.593	-0.639	3.971	0.303	3.503	0.721	3.192	0.917	2.919	1.035	2.782	0.793	3.065	0.340	3.364	-0.122	3.814	-0.817
8.0	3.690	-1.483	3.727	-0.825	3.518	0.055	3.253	0.527	3.043	0.779	2.841	0.955	2.778	0.791	3.065	0.340	3.364	-0.122	3.814	-0.817
8.1	2.897	-1.230	3.055	-0.761	3.094	-0.035	2.993	0.421	2.879	0.691	2.753	0.898	2.775	0.789	3.065	0.340	3.364	-0.122	3.814	-0.817
8.2	2.363	-0.923	2.558	-0.591	2.730	-0.019	2.748	0.386	2.715	0.648	2.661	0.865	2.771	0.789	3.065	0.340	3.364	-0.122	3.814	-0.817
8.3	1.994	-0.624	2.195	-0.385	2.432	0.059	2.529	0.403	2.561	0.642	2.571	0.853	2.768	0.789	3.065	0.340	3.364	-0.122	3.814	-0.817
8.4	1.730	-0.349	1.925	-0.173	2.193	0.170	2.342	0.455	2.422	0.665	2.486	0.859	2.765	0.789	3.064	0.340	3.364	-0.122	3.814	-0.817
8.5	1.536	-0.099	1.722	0.033	2.003	0.299	2.185	0.531	2.301	0.709	2.409	0.880	2.762	0.791	3.064	0.340	3.364	-0.122	3.814	-0.817
8.6	1.390	0.129	1.567	0.229	1.852	0.435	2.055	0.621	2.198	0.769	2.341	0.913	2.760	0.792	3.064	0.340	3.364	-0.122	3.814	-0.817
8.7	1.277	0.338	1.446	0.415	1.732	0.575	1.950	0.721	2.113	0.839	2.284	0.955	2.758	0.794	3.064	0.341	3.364	-0.122	3.814	-0.817
8.8	1.188	0.532	1.353	0.592	1.639	0.715	1.867	0.827	2.044	0.917	2.237	1.003	2.757	0.796	3.064	0.341	3.364	-0.122	3.814	-0.817
8.9	1.118	0.716	1.279	0.762	1.566	0.854	1.802	0.936	1.991	1.000	2.201	1.056	2.756	0.798	3.064	0.341	3.364	-0.122	3.814	-0.817
9.0	1.062	0.892	1.222	0.927	1.511	0.992	1.754	1.047	1.952	1.086	2.176	1.113	2.756	0.800	3.064	0.341	3.364	-0.122	3.814	-0.817
9.1	1.017	1.064	1.179	1.086	1.473	1.130	1.722	1.159	1.927	1.174	2.160	1.171	2.756	0.802	3.064	0.341	3.364	-0.122	3.814	-0.817
9.2	0.983	1.234	1.148	1.249	1.448	1.268	1.704	1.272	1.915	1.262	2.155	1.229	2.756	0.804	3.064	0.341	3.364	-0.122	3.814	-0.817
9.3	0.958	1.404	1.128	1.410	1.438	1.407	1.701	1.385	1.916	1.350	2.159	1.286	2.757	0.806	3.064	0.341	3.364	-0.122	3.814	-0.817
9.4	0.942	1.578	1.120	1.575	1.442	1.547	1.712	1.497	1.930	1.437	2.173	1.340	2.758	0.807	3.064	0.341	3.364	-0.122	3.814	-0.817
3π	0.940	1.622	1.120	1.616	1.445	1.582	1.717	1.525	1.936	1.458	2.177	1.353	2.758	0.807	3.064	0.341	3.364	-0.122	3.814	-0.817
9.5	0.936	1.758	1.124	1.745	1.460	1.689	1.738	1.609	1.957	1.520	2.195	1.391	2.759	0.808	3.064	0.341	3.364	-0.122	3.814	-0.817
9.6	0.940	1.948	1.141	1.922	1.496	1.833	1.780	1.719	1.997	1.600	2.226	1.438	2.760	0.809	3.064	0.341	3.364	-0.122	3.814	-0.817
9.7	0.956	2.152	1.175	2.109	1.551	1.980	1.838	1.825	2.050	1.674	2.264	1.478	2.762	0.809	3.064	0.341	3.364	-0.122	3.814	-0.817
9.8	0.990	2.375	1.230	2.310	1.628	2.128	1.916	1.927	2.117	1.740	2.310	1.511	2.763	0.810	3.065	0.341	3.364	-0.122	3.814	-0.817
9.9	1.046	2.622	1.313	2.525	1.733	2.277	2.014	2.019	2.196	1.794	2.361	1.535	2.764	0.810	3.065	0.341	3.364	-0.122	3.814	-0.817
10.0	1.135	2.900	1.435	2.759	1.872	2.421	2.134	2.098	2.288	1.835	2.416	1.548	2.765	0.809	3.065	0.341	3.364	-0.122	3.814	-0.817
10.1	1.274	3.219	1.610	3.011	2.051	2.553	2.277	2.158	2.390	1.858										
10.2	1.491	3.586	1.863	3.275	2.278	2.662	2.440	2.190	2.499	1.859										
10.3	1.836	4.006	2.225	3.535	2.555	2.727	2.618	2.184	2.609	1.836										
10.4	2.393	4.459	2.740	3.746	2.879	2.721	2.803	2.133	2.715	1.787										
10.5	3.295	4.851	3.437	3.814	3.227	2.613	2.978	2.030	2.809	1.711										
10.6	4.671	4.892	4.281	3.580	3.557	2.380	3.126	1.876	2.882	1.613										
10.7	6.310	4.008	5.066	2.889	3.804	2.024	3.226	1.682	2.929	1.500										
10.8	7.194	1.976	5.440	1.807	3.910	1.592	3.265	1.466	2.944	1.379										
10.9	6.563	-0.057	5.218	0.715	3.854	1.159	3.240	1.252	2.929	1.261										
7π/2	5.276	-1.050	4.626	-0.008	3.674	0.812	3.163	1.070	2.888	1.159										

βh	α/β = 0.00		α/β = 0.01		α/β = 0.03		α/β = 0.05		α/β = 0.07	
11.0	5.216	-1.074	4.594	-0.033	3.664	0.798	3.158	1.062	2.886	1.154
11.1	4.015	-1.287	3.883	-0.384	3.395	0.543	3.036	0.912	2.821	1.065
11.2	3.152	-1.145	3.259	-0.461	3.103	0.397	2.890	0.806	2.742	0.998
11.3	2.560	-0.889	2.765	-0.390	2.822	0.340	2.736	0.744	2.654	0.953
11.4	2.151	-0.613	2.388	-0.249	2.572	0.348	2.587	0.721	2.565	0.929
11.5	1.861	-0.350	2.102	-0.081	2.358	0.398	2.448	0.728	2.478	0.925
11.6	1.648	-0.107	1.884	0.095	2.180	0.476	2.324	0.759	2.397	0.937
11.7	1.489	0.115	1.716	0.269	2.033	0.571	2.216	0.808	2.324	0.963
11.8	1.367	0.320	1.585	0.438	1.914	0.676	2.125	0.869	2.261	0.999
11.9	1.272	0.510	1.483	0.602	1.818	0.787	2.050	0.940	2.208	1.044
12.0	1.197	0.690	1.403	0.760	1.743	0.902	1.990	1.018	2.165	1.095
12.1	1.138	0.861	1.341	0.915	1.685	1.018	1.945	1.100	2.132	1.150
12.2	1.091	1.027	1.294	1.066	1.643	1.136	1.912	1.184	2.110	1.208
12.3	1.054	1.191	1.259	1.217	1.616	1.254	1.893	1.271	2.098	1.267
12.4	1.028	1.354	1.237	1.368	1.603	1.374	1.886	1.358	2.095	1.327
12.5	1.010	1.520	1.227	1.521	1.603	1.494	1.892	1.444	2.102	1.386
4π	1.003	1.633	1.227	1.625	1.612	1.574	1.903	1.501	2.113	1.423
12.6	1.001	1.691	1.229	1.678	1.618	1.615	1.911	1.530	2.119	1.442
12.7	1.002	1.870	1.244	1.841	1.649	1.737	1.943	1.612	2.145	1.495
12.8	1.016	2.061	1.276	2.011	1.696	1.858	1.988	1.691	2.181	1.543
12.9	1.045	2.267	1.326	2.191	1.763	1.979	2.047	1.763	2.224	1.584
13.0	1.093	2.494	1.402	2.381	1.853	2.095	2.120	1.827	2.275	1.618
13.1	1.171	2.747	1.512	2.584	1.968	2.205	2.206	1.879	2.333	1.641
13.2	1.290	3.034	1.666	2.798	2.111	2.300	2.306	1.915	2.395	1.653
13.3	1.474	3.362	1.883	3.018	2.286	2.374	2.416	1.931	2.460	1.653
13.4	1.760	3.734	2.185	3.229	2.492	2.413	2.533	1.924	2.524	1.638
13.5	2.212	4.139	2.595	3.399	2.723	2.402	2.650	1.888	2.585	1.609
13.6	2.933	4.520	3.143	3.465	2.964	2.325	2.759	1.824	2.638	1.567
13.7	4.041	4.690	3.796	3.326	3.192	2.172	2.852	1.732	2.681	1.514
13.8	5.496	4.229	4.445	2.873	3.373	1.944	2.920	1.616	2.710	1.453
13.9	6.671	2.728	4.870	2.103	3.476	1.663	2.955	1.487	2.724	1.386
14.0	6.630	0.731	4.884	1.209	3.483	1.365	2.956	1.355	2.723	1.320
14.1	5.565	-0.610	4.519	0.467	3.398	1.089	2.923	1.230	2.706	1.257
9π/2	5.100	-0.869	4.325	0.264	3.347	1.000	2.903	1.187	2.696	1.235
14.2	4.365	-1.084	3.971	0.015	3.243	0.868	2.861	1.122	2.676	1.201
14.3	3.429	-1.074	3.419	-0.177	3.050	0.713	2.780	1.036	2.636	1.155
14.4	2.771	-0.877	2.943	-0.197	2.844	0.623	2.686	0.976	2.589	1.121
14.5	2.314	-0.626	2.561	-0.121	2.646	0.589	2.587	0.940	2.538	1.098
14.6	1.990	-0.374	2.263	0.003	2.465	0.597	2.489	0.927	2.485	1.087
14.7	1.755	-0.136	2.031	0.149	2.305	0.635	2.397	0.934	2.434	1.086
14.8	1.580	0.084	1.851	0.301	2.169	0.696	2.313	0.956	2.386	1.095
14.9	1.448	0.287	1.709	0.454	2.055	0.771	2.239	0.992	2.342	1.113
15.0	1.345	0.476	1.598	0.605	1.962	0.857	2.176	1.037	2.304	1.137
15.1	1.264	0.653	1.511	0.753	1.886	0.948	2.124	1.090	2.272	1.167
15.2	1.200	0.821	1.444	0.898	1.827	1.045	2.083	1.148	2.246	1.201
15.3	1.149	0.984	1.392	1.041	1.783	1.144	2.052	1.210	2.228	1.238
15.4	1.110	1.144	1.354	1.183	1.753	1.245	2.033	1.275	2.217	1.277
15.5	1.081	1.302	1.328	1.326	1.737	1.347	2.023	1.341	2.212	1.316
15.6	1.061	1.462	1.315	1.469	1.733	1.450	2.025	1.407	2.215	1.355
15.7	1.050	1.626	1.314	1.616	1.744	1.554	2.037	1.472	2.224	1.393
5π	1.049	1.640	1.315	1.628	1.745	1.562	2.038	1.477	2.225	1.396
15.8	1.048	1.797	1.326	1.767	1.763	1.657	2.059	1.534	2.240	1.428
15.9	1.058	1.977	1.354	1.924	1.807	1.759	2.092	1.593	2.262	1.460
16.0	1.082	2.171	1.400	2.087	1.863	1.858	2.135	1.646	2.290	1.487
16.1	1.123	2.382	1.488	2.259	1.937	1.952	2.187	1.692	2.322	1.509
16.2	1.190	2.615	1.566	2.439	2.031	2.038	2.249	1.728	2.358	1.524
16.3	1.292	2.878	1.702	2.626	2.146	2.111	2.319	1.753	2.398	1.532
16.4	1.448	3.175	1.889	2.814	2.282	2.165	2.395	1.764	2.436	1.532
[illegible]	[illegible]	[illegible]	[illegible]	[illegible]	2.430	2.192	2.474	1.758	2.475	1.524
16.6	2.059	3.882	2.483	3.137	2.609	2.183	2.552	[illegible]	[illegible]	[illegible]
16.7	2.644	4.251	2.920	3.204	2.785	2.130	2.625	1.695	2.544	1.484
16.8	3.546	4.497	3.444	3.125	2.952	2.026	2.687	1.637	2.571	1.454
16.9	4.806	4.323	3.986	2.823	3.090	1.873	2.735	1.566	2.550	1.419
17.0	6.091	3.298	4.407	2.269	3.181	1.682	2.765	1.484	2.601	1.381
17.1	6.551	1.504	4.555	1.556	3.212	1.471	2.773	1.399	2.604	1.342
17.2	5.865	-0.074	4.384	0.874	3.180	1.264	2.761	1.316	2.598	1.304
11π/2	4.971	-0.737	4.088	0.464	3.117	1.118	2.739	1.255	2.587	1.276
17.3	4.730	-0.835	3.995	0.377	3.095	1.083	2.731	1.239	2.584	1.269
17.4	3.732	-0.991	3.530	0.100	2.971	0.940	2.685	1.175	2.563	1.239
17.5	3.001	-0.868	3.090	-0.001	2.825	0.842	2.628	1.124	2.538	1.215
17.6	2.487	-0.649	2.715	0.012	2.674	0.786	2.564	1.089	2.508	1.197
17.7	2.124	-0.409	2.410	0.091	2.526	0.767	2.497	1.069	2.477	1.185
17.8	1.863	-0.175	2.168	0.204	2.391	0.778	2.431	1.064	2.445	1.181
17.9	1.670	0.043	1.976	0.333	2.270	0.811	2.368	1.072	2.414	1.183
18.0	1.524	0.246	1.825	0.470	2.165	0.862	2.311	1.090	2.385	1.191
18.1	1.411	0.434	1.705	0.608	2.076	0.924	2.260	1.117	2.358	1.203
18.2	1.324	0.610	1.611	0.745	2.003	0.996	2.217	1.152	2.336	1.220
18.3	1.255	0.777	1.537	0.882	1.944	1.074	2.181	1.193	2.317	1.241
18.4	1.200	0.938	1.480	1.017	1.899	1.156	2.154	1.237	2.303	1.264
18.5	1.158	1.095	1.438	1.151	1.867	1.241	2.135	1.285	2.293	1.288
18.6	1.125	1.250	1.406	1.286	1.848	1.328	2.125	1.334	2.288	1.314
18.7	1.103	1.405	1.392	1.421	1.841	1.416	2.123	1.383	2.288	1.339
18.8	1.089	1.564	1.387	1.559	1.846	1.505	2.130	1.432	2.293	1.364
6π	1.085	1.644	1.390	1.629	1.854	1.548	2.136	1.455	2.297	1.376
18.9	1.084	1.728	1.396	1.700	1.865	1.592	2.144	1.479	2.302	1.388
19.0	1.090	1.899	1.419	1.845	1.896	1.678	2.167	1.523	2.315	1.409
19.1	1.109	2.083	1.459	1.996	1.941	1.761	2.197	1.562	2.332	1.427
19.2	1.144	2.281	1.520	2.152	2.002	1.838	2.235	1.596	2.352	1.441
19.3	1.201	2.499	1.607	2.314	2.077	1.908	2.279	1.623	2.374	1.452
19.4	1.288	2.742	1.727	2.480	2.169	1.966	2.328	1.642	2.398	1.458
19.5	1.421	3.015	1.889	2.646	2.276	2.009	2.381	1.650	2.423	1.459
19.6	1.622	3.323	2.105	2.802	2.398	2.030	2.435	1.648	2.447	1.455
19.7	1.931	3.665	2.389	2.929	2.526	2.025	2.489	1.634	2.470	1.446
19.8	2.410	4.018	2.748	2.997	2.659	1.988	2.539	1.608	2.491	1.432
19.9	3.149	4.306	3.178	2.958	2.785	1.916	2.583	1.571	2.508	1.415
20.0	4.222	4.323	3.635	2.755	2.893	1.811	2.618	1.526	2.521	1.395
20.1	5.490	3.693	4.030	2.354	2.971	1.677	2.642	1.473	2.529	1.372
20.2	6.316	2.218	4.246	1.795	3.011	1.525	2.653	1.417	2.533	1.349
20.3	6.070	0.535	4.213	1.198	3.007	1.369	2.650	1.360	2.531	1.326
20.4	5.091	-0.515	3.963	0.698	2.963	1.225	2.636	1.307	2.525	1.304
13π/2	4.869	-0.635	3.894	0.617	2.949	1.197	2.631	1.296	2.523	1.300
20.5	4.059	-0.874	3.595	0.367	2.885	1.102	2.610	1.259	2.514	1.284
20.6	3.253	-0.851	3.205	0.199	2.785	1.008	2.575	1.219	2.500	1.268
20.7	2.676	-0.674	2.848	0.152	2.672	0.944	2.534	1.189	2.483	1.255
20.8	2.268	-0.450	2.544	0.185	2.556	0.910	2.489	1.169	2.465	1.245
20.9	1.975	-0.222	2.295	0.265	2.444	0.902	2.443	1.159	2.445	1.240

βh	α/β = 0.00		α/β = 0.01		α/β = 0.03		α/β = 0.05		α/β = 0.07	
21.0	1.760	-0.004	2.095	0.370	2.341	0.916	2.397	1.159	2.425	1.239
21.1	1.599	0.198	1.934	0.488	2.247	0.946	2.354	1.167	2.406	1.242
21.2	1.476	0.387	1.806	0.612	2.166	0.990	2.315	1.182	2.389	1.249
21.3	1.381	0.563	1.705	0.739	2.097	1.044	2.280	1.204	2.373	1.258
21.4	1.306	0.730	1.624	0.867	2.040	1.105	2.251	1.231	2.360	1.270
21.5	1.246	0.890	1.562	0.994	1.996	1.172	2.228	1.263	2.349	1.284
21.6	1.200	1.045	1.515	1.121	1.963	1.243	2.211	1.297	2.342	1.299
21.7	1.165	1.197	1.481	1.248	1.942	1.316	2.200	1.333	2.337	1.316
21.8	1.139	1.350	1.461	1.377	1.931	1.391	2.196	1.369	2.336	1.332
21.9	1.122	1.504	1.453	1.507	1.932	1.466	2.199	1.406	2.338	1.348
7π	1.115	1.648	1.456	1.628	1.943	1.534	2.207	1.438	2.342	1.362
22.0	1.114	1.662	1.457	1.639	1.945	1.540	2.208	1.441	2.343	1.364
22.1	1.117	1.827	1.476	1.775	1.969	1.613	2.223	1.474	2.351	1.378
22.2	1.131	2.001	1.511	1.915	2.005	1.683	2.244	1.504	2.361	1.390
22.3	1.160	2.189	1.564	2.058	2.053	1.748	2.270	1.530	2.373	1.400
22.4	1.208	2.394	1.641	2.206	2.114	1.805	2.301	1.550	2.387	1.407
22.5	1.283	2.621	1.746	2.356	2.187	1.853	2.335	1.564	2.402	1.412
22.6	1.396	2.874	1.887	2.504	2.271	1.888	2.373	1.572	2.417	1.413
22.7	1.566	3.158	2.072	2.642	2.365	1.907	2.411	1.571	2.432	1.411
22.8	1.824	3.475	2.312	2.758	2.466	1.905	2.449	1.563	2.447	1.407
22.9	2.220	3.812	2.613	2.827	2.569	1.881	2.484	1.547	2.460	1.399
23.0	2.828	4.119	2.971	2.814	2.667	1.831	2.516	1.524	2.471	1.389
23.1	3.732	4.258	3.360	2.679	2.753	1.756	2.542	1.494	2.480	1.377
23.2	4.906	3.931	3.720	2.388	2.819	1.660	2.561	1.459	2.486	1.364
23.3	5.946	2.823	3.965	1.954	2.859	1.549	2.571	1.422	2.489	1.350
23.4	6.142	1.192	4.024	1.447	2.870	1.432	2.573	1.383	2.490	1.336
23.5	5.419	-0.108	3.888	0.975	2.850	1.318	2.567	1.345	2.487	1.322
15π/2	4.787	-0.552	3.731	0.737	2.824	1.253	2.559	1.323	2.484	1.314
23.6	4.404	-0.705	3.616	0.618	2.803	1.216	2.553	1.310	2.482	1.309
23.7	3.530	-0.812	3.285	0.398	2.735	1.131	2.532	1.280	2.474	1.298
23.8	2.884	-0.693	2.957	0.300	2.654	1.068	2.505	1.256	2.464	1.289
23.9	2.424	-0.492	2.663	0.288	2.565	1.028	2.475	1.238	2.453	1.282
24.0	2.095	-0.272	2.413	0.333	2.475	1.008	2.443	1.227	2.441	1.277
24.1	1.855	-0.056	2.206	0.412	2.388	1.008	2.411	1.223	2.429	1.276
24.2	1.677	0.147	2.038	0.511	2.308	1.023	2.379	1.225	2.417	1.276
24.3	1.541	0.337	1.902	0.621	2.235	1.052	2.349	1.233	2.405	1.279
24.4	1.436	0.514	1.794	0.736	2.172	1.091	2.322	1.247	2.395	1.284
24.5	1.354	0.681	1.707	0.854	2.119	1.138	2.299	1.264	2.385	1.291
24.6	1.290	0.840	1.639	0.974	2.076	1.191	2.280	1.286	2.378	1.299
24.7	1.239	0.994	1.587	1.094	2.043	1.249	2.265	1.310	2.372	1.309
24.8	1.200	1.145	1.549	1.214	2.020	1.310	2.255	1.336	2.369	1.319
24.9	1.171	1.294	1.524	1.336	2.008	1.373	2.250	1.362	2.367	1.330
25.0	1.152	1.445	1.512	1.459	2.005	1.436	2.250	1.389	2.367	1.340

TABLE 2.7

NORMALIZED ADMITTANCES (Y/Δ) IN MILLIMHOS OF THIN DIPOLE ANTENNAS IN DISSIPATIVE MEDIA

a/λ = 0.006350

βh	α/β = 0.00		α/β = 0.01		α/β = 0.03		α/β = 0.05		α/β = 0.07		α/β = 0.10		α/β = 0.30		α/β = 0.50		α/β = 0.70		α/β = 1.00	
1.5	9.637	-1.856	9.282	-1.464	8.611	-0.854	8.009	-0.420	7.479	-0.111	6.809	0.194	4.668	0.522	4.150	0.233	4.127	-0.142	4.425	-0.765
π/2	7.351	-2.756	7.237	-2.399	6.968	-1.792	6.677	-1.310	6.385	-0.931	5.976	-0.510	4.409	0.294	4.015	0.154	4.045	-0.168	4.385	-0.769
1.6	6.568	-2.839	6.513	-2.518	6.354	-1.957	6.158	-1.495	5.946	-1.120	5.629	-0.690	4.299	0.224	3.960	0.130	4.014	-0.174	4.370	-0.768
1.7	4.595	-2.592	4.638	-2.384	4.682	-1.996	4.681	-1.649	4.647	-1.345	4.560	-0.964	3.933	0.074	3.783	0.085	3.916	-0.180	4.328	-0.758
1.8	3.396	-2.082	3.465	-1.948	3.575	-1.689	3.651	-1.445	3.699	-1.220	3.733	-0.923	3.605	0.031	3.628	0.081	3.836	-0.167	4.296	-0.744
1.9	2.638	-1.571	2.712	-1.481	2.840	-1.304	2.944	-1.132	3.029	-0.969	3.123	-0.745	3.330	0.058	3.499	0.106	3.772	-0.143	4.274	-0.727
2.0	2.133	-1.114	2.205	-1.052	2.336	-0.927	2.451	-0.803	2.551	-0.684	2.676	-0.516	3.107	0.128	3.395	0.149	3.724	-0.112	4.259	-0.711
2.1	1.780	-0.716	1.849	-0.671	1.979	-0.580	2.098	-0.490	2.205	-0.401	2.347	-0.275	2.930	0.223	3.315	0.203	3.689	-0.078	4.250	-0.695
2.2	1.523	-0.368	1.590	-0.334	1.718	-0.266	1.838	-0.199	1.949	-0.132	2.101	-0.038	2.794	0.330	3.255	0.262	3.665	-0.044	4.245	-0.681
2.3	1.330	-0.058	1.396	-0.033	1.523	0.018	1.644	0.069	1.758	0.119	1.917	0.190	2.691	0.444	3.214	0.322	3.651	-0.012	4.244	-0.670
2.4	1.182	0.221	1.247	0.240	1.374	0.279	1.497	0.318	1.613	0.354	1.778	0.406	2.618	0.558	3.187	0.380	3.645	0.017	4.245	-0.660
2.5	1.065	0.476	1.131	0.492	1.260	0.521	1.384	0.549	1.504	0.576	1.675	0.612	2.568	0.670	3.174	0.436	3.644	0.042	4.247	-0.653
2.6	0.972	0.715	1.039	0.727	1.171	0.749	1.298	0.769	1.422	0.786	1.599	0.808	2.540	0.778	3.171	0.486	3.648	0.064	4.249	-0.647
2.7	0.898	0.941	0.967	0.950	1.102	0.965	1.234	0.978	1.362	0.988	1.547	0.997	2.529	0.879	3.177	0.531	3.654	0.082	4.253	-0.643
2.8	0.838	1.159	0.910	1.165	1.051	1.175	1.188	1.181	1.322	1.183	1.514	1.180	2.534	0.974	3.189	0.570	3.663	0.096	4.256	-0.641
2.9	0.791	1.373	0.866	1.377	1.014	1.380	1.158	1.380	1.299	1.374	1.500	1.358	2.553	1.061	3.207	0.602	3.673	0.106	4.259	-0.639
3.0	0.754	1.586	0.834	1.587	0.991	1.585	1.144	1.577	1.292	1.563	1.504	1.533	2.584	1.139	3.228	0.627	3.683	0.113	4.261	-0.638
3.1	0.728	1.803	0.814	1.801	0.981	1.792	1.144	1.775	1.301	1.752	1.525	1.705	2.625	1.207	3.250	0.646	3.693	0.117	4.264	-0.638
π	0.720	1.894	0.809	1.891	0.982	1.879	1.149	1.858	1.311	1.830	1.539	1.777	2.645	1.233	3.260	0.652	3.696	0.118	4.264	-0.638
3.2	0.712	2.026	0.805	2.021	0.986	2.003	1.161	1.976	1.329	1.941	1.565	1.876	2.674	1.265	3.274	0.659	3.701	0.119	4.265	-0.638
3.3	0.708	2.260	0.809	2.251	1.007	2.223	1.196	2.183	1.376	2.134	1.625	2.045	2.731	1.311	3.298	0.666	3.709	0.119	4.267	-0.639
3.4	0.717	2.511	0.829	2.496	1.047	2.453	1.253	2.397	1.446	2.329	1.709	2.212	2.793	1.344	3.320	0.669	3.716	0.118	4.268	-0.639
3.5	0.743	2.784	0.870	2.761	1.113	2.699	1.338	2.620	1.545	2.528	1.819	2.374	2.857	1.364	3.340	0.667	3.722	0.116	4.269	-0.640
3.6	0.794	3.089	0.939	3.054	1.211	2.963	1.458	2.853	1.678	2.728	1.961	2.527	2.921	1.371	3.357	0.661	3.726	0.113	4.269	-0.641
3.7	0.880	3.435	1.047	3.381	1.354	3.249	1.624	3.093	1.855	2.925	2.138	2.666	2.983	1.366	3.372	0.653	3.729	0.110	4.269	-0.641
3.8	1.020	3.835	1.215	3.752	1.563	3.555	1.851	3.336	2.086	3.110	2.354	2.780	3.041	1.348	3.383	0.644	3.731	0.107	4.269	-0.642
3.9	1.247	4.305	1.477	4.173	1.863	3.877	2.160	3.568	2.382	3.266	2.611	2.854	3.091	1.320	3.391	0.633	3.732	0.104	4.269	-0.642
4.0	1.622	4.858	1.887	4.642	2.296	4.193	2.571	3.759	2.750	3.367	2.903	2.869	3.132	1.284	3.397	0.622	3.733	0.101	4.269	-0.642
4.1	2.256	5.486	2.538	5.127	2.909	4.447	3.100	3.859	3.185	3.372	3.216	2.806	3.163	1.241	3.399	0.611	3.733	0.098	4.269	-0.643
4.2	3.339	6.098	3.558	5.509	3.735	4.527	3.732	3.788	3.659	3.236	3.520	2.647	3.183	1.196	3.400	0.602	3.732	0.096	4.269	-0.643
4.3	5.108	6.336	5.032	5.471	4.722	4.246	4.390	3.456	4.104	2.922	3.776	2.391	3.191	1.150	3.398	0.593	3.732	0.094	4.269	-0.643
4.4	7.392	5.343	6.675	4.489	5.622	3.438	4.917	2.831	4.427	2.440	3.942	2.059	3.190	1.106	3.395	0.585	3.731	0.093	4.269	-0.643
4.5	8.713	2.632	7.539	2.466	6.038	2.204	5.136	2.011	4.548	1.864	3.991	1.695	3.180	1.066	3.392	0.579	3.730	0.092	4.269	-0.643
4.6	7.852	-0.097	7.009	0.434	5.790	0.978	4.990	1.208	4.448	1.307	3.923	1.349	3.162	1.032	3.387	0.574	3.729	0.092	4.269	-0.643
4.7	6.062	-1.362	5.742	-0.706	5.105	0.136	4.581	0.600	4.181	0.862	3.763	1.062	3.140	1.005	3.382	0.571	3.728	0.091	4.269	-0.643
3π/2	5.848	-1.430	5.579	-0.783	5.008	0.064	4.520	0.542	4.141	0.818	3.739	1.031	3.137	1.002	3.382	0.571	3.727	0.091	4.269	-0.643
4.8	4.544	-1.574	4.521	-1.054	4.323	-0.272	4.068	0.239	3.830	0.568	3.548	0.854	3.114	0.984	3.377	0.569	3.727	0.091	4.269	-0.643
4.9	3.491	-1.367	3.588	-0.999	3.631	-0.378	3.566	0.083	3.464	0.411	3.313	0.726	3.087	0.970	3.373	0.568	3.726	0.091	4.269	-0.643
5.0	2.784	-1.042	2.920	-0.784	3.077	-0.315	3.128	0.067	3.125	0.361	3.083	0.666	3.059	0.963	3.368	0.568	3.725	0.091	4.268	-0.643
5.1	2.299	-0.705	2.444	-0.522	2.649	-0.171	2.768	0.135	2.831	0.385	2.871	0.660	3.033	0.962	3.365	0.569	3.725	0.092	4.268	-0.643
5.2	1.956	-0.389	2.099	-0.256	2.323	0.007	2.479	0.249	2.585	0.455	2.684	0.693	3.009	0.966	3.362	0.570	3.724	0.092	4.268	-0.643
5.3	1.706	-0.101	1.843	-0.002	2.072	0.198	2.248	0.387	2.382	0.554	2.525	0.753	2.987	0.973	3.359	0.572	3.724	0.092	4.268	-0.643
5.4	1.518	0.161	1.649	0.236	1.879	0.389	2.066	0.537	2.218	0.670	2.393	0.831	2.969	0.984	3.357	0.574	3.724	0.093	4.268	-0.643
5.5	1.373	0.401	1.500	0.458	1.728	0.576	1.923	0.691	2.087	0.795	2.285	0.922	2.955	0.997	3.356	0.576	3.724	0.093	4.268	-0.643
5.6	1.260	0.623	1.383	0.667	1.611	0.758	1.810	0.846	1.983	0.925	2.199	1.020	2.944	1.012	3.355	0.579	3.724	0.093	4.268	-0.643
5.7	1.170	0.832	1.292	0.866	1.519	0.935	1.723	1.000	1.903	1.057	2.133	1.123	2.937	1.027	3.355	0.581	3.724	0.093	4.268	-0.643
5.8	1.098	1.031	1.220	1.057	1.449	1.108	1.657	1.152	1.844	1.190	2.085	1.229	2.933	1.043	3.355	0.582	3.724	0.094	4.268	-0.643
5.9	1.041	1.224	1.164	1.243	1.397	1.277	1.611	1.304	1.803	1.323	2.054	1.335	2.932	1.058	3.355	0.584	3.724	0.094	4.268	-0.643
6.0	0.996	1.414	1.122	1.427	1.361	1.445	1.581	1.455	1.780	1.455	2.038	1.441	2.934	1.072	3.356	0.585	3.724	0.094	4.268	-0.643
6.1	0.961	1.603	1.092	1.610	1.339	1.614	1.567	1.606	1.772	1.587	2.038	1.545	2.939	1.084	3.357	0.586	3.724	0.094	4.268	-0.643
6.2	0.937	1.795	1.074	1.796	1.333	1.783	1.569	1.757	1.781	1.719	2.052	1.647	2.945	1.095	3.358	0.587	3.724	0.094	4.268	-0.643
2π	0.925	1.959	1.068	1.954	1.339	1.927	1.584	1.883	1.801	1.827	2.075	1.729	2.951	1.102	3.358	0.588	3.724	0.094	4.268	-0.643
6.3	0.923	1.993	1.068	1.986	1.341	1.956	1.588	1.908	1.806	1.848	2.081	1.746	2.953	1.103	3.358	0.588	3.724	0.094	4.268	-0.643
6.4	0.921	2.200	1.076	2.185	1.367	2.133	1.625	2.061	1.849	1.976	2.125	1.839	2.961	1.110	3.359	0.588	3.724	0.094	4.268	-0.643
6.5	0.931	2.421	1.100	2.395	1.412	2.316	1.682	2.214	1.911	2.100	2.183	1.926	2.971	1.114	3.360	0.588	3.725	0.094	4.268	-0.643
6.6	0.958	2.660	1.145	2.619	1.480	2.506	1.762	2.367	1.993	2.219	2.257	2.004	2.980	1.117	3.361	0.588	3.725	0.094	4.268	-0.643
6.7	1.008	2.924	1.216	2.862	1.579	2.703	1.870	2.517	2.098	2.329	2.345	2.069	2.989	1.117	3.362	0.588	3.725	0.094	4.268	-0.643
6.8	1.090	3.219	1.323	3.128	1.715	2.904	2.011	2.660	2.227	2.426	2.447	2.119	2.998	1.116	3.362	0.588	3.725	0.094	4.268	-0.643
6.9	1.218	3.555	1.483	3.420	1.901	3.107	2.189	2.789	2.381	2.502	2.560	2.149	3.005	1.113	3.363	0.587	3.725	0.094	4.268	-0.643
7.0	1.420	3.940	1.719	3.737	2.150	3.299	2.409	2.892	2.559	2.549	2.681	2.156	3.012	1.109	3.363	0.587	3.725	0.094	4.268	-0.643
7.1	1.739	4.383	2.067	4.071	2.477	3.461	2.672	2.952	2.758	2.558	2.804	2.136	3.017	1.104	3.363	0.586	3.725	0.094	4.268	-0.643
7.2	2.253	4.871	2.581	4.389	2.893	3.557	2.972	2.948	2.966	2.517	2.924	2.087	3.021	1.099	3.364	0.586	3.725	0.094	4.268	-0.643
7.3	3.084	5.340	3.321	4.609	3.394	3.533	3.290	2.856	3.169	2.419	3.030	2.010	3.024	1.093	3.364	0.586	3.725	0.094	4.268	-0.643
7.4	4.384	5.570	4.308	4.556	3.934	3.323	3.592	2.659	3.346	2.263	3.116	1.907	3.025	1.087	3.364	0.585	3.725	0.094	4.268	-0.643
7.5	6.112	5.047	5.397	3.976	4.412	2.885	3.831	2.359	3.475	2.058	3.172	1.787	3.025	1.081	3.363	0.585	3.725	0.094	4.268	-0.643
7.6	7.499	3.273	6.150	2.771	4.696	2.254	3.959	1.989	3.539	1.823	3.197	1.657	3.024	1.076	3.363	0.585	3.725	0.094	4.268	-0.643
7.7	7.421	0.941	6.138	1.326	4.698	1.565	3.956	1.601	3.532	1.583	3.187	1.529	3.022	1.071	3.363	0.585	3.725	0.094	4.268	-0.643
7.8	6.170	-0.583	5.465	0.228	4.444	0.972	3.831	1.253	3.459	1.365	3.148	1.412	3.019	1.068	3.363	0.584	3.725	0.094	4.268	-0.643
5π/2	5.401	-0.960	4.988	-0.134	4.238	0.728	3.726	1.096	3.398	1.264	3.116	1.356	3.018	1.066	3.363	0.584	3.725	0.094	4.268	-0.643
7.9	4.798	-1.105	4.579	-0.327	4.043	0.565	3.622	0.984	3.336	1.187	3.084	1.313	3.016	1.065	3.363	0.584	3.725	0.094	4.268	-0.643
8.0	3.741	-1.087	3.777	-0.482	3.603	0.345	3.371	0.805	3.183	1.058	3.003	1.237	3.013	1.062	3.363	0.584	3.725	0.094	4.268	-0.643
8.1	3.000	-0.864	3.145	-0.422	3.194	0.269	3.114	0.710	3.018	0.977	2.913	1.184	3.009	1.061	3.362	0.584	3.725	0.094	4.268	-0.643
8.2	2.484	-0.584	2.669	-0.263	2.842	0.289	2.872	0.682	2.854	0.939	2.819	1.154	3.005	1.060	3.362	0.584	3.725	0.094	4.268	-0.643
8.3	2.118	-0.303	2.313	-0.067	2.550	0.367	2.657	0.703	2.701	0.938	2.728	1.145	3.002	1.061	3.362	0.584	3.725	0.094	4.268	-0.643
8.4	1.852	-0.038	2.044	0.139	2.315	0.478	2.472	0.758	2.563	0.964	2.643	1.154	2.999	1.061	3.362	0.584	3.725	0.094	4.268	-0.643
8.5	1.653	0.207	1.839	0.341	2.125	0.607	2.316	0.836	2.442	1.011	2.566	1.178	2.996	1.063	3.362	0.584	3.725	0.094	4.268	-0.643
8.6	1.500	0.434	1.681	0.537	1.974	0.744	2.187	0.928	2.340	1.073	2.499	1.213	2.994	1.064	3.362	0.585	3.725	0.094	4.268	-0.643
8.7	1.382	0.644	1.557	0.724	1.853	0.886	2.082	1.031	2.255	1.146	2.442	1.257	2.992	1.066	3.362	0.585	3.725	0.094	4.268	-0.643
8.8	1.288	0.843	1.460	0.904	1.758	1.029	1.998	1.139	2.187	1.227	2.396	1.307	2.991	1.068	3.362	0.585	3.725	0.094	4.268	-0.643
8.9	1.214	1.031	1.384	1.079	1.685	1.171	1.934	1.252	2.135	1.312	2.361	1.363	2.990	1.071	3.362	0.585	3.725	0.094	4.268	-0.643
9.0	1.155	1.214	1.324	1.249	1.630	1.314	1.887	1.366	2.097	1.401	2.337	1.421	2.990	1.073	3.362	0.585	3.725	0.094	4.268	-0.643
9.1	1.109	1.393	1.280	1.417	1.591	1.457	1.856	1.482	2.074	1.492	2.323	1.480	2.990	1.075	3.362	0.585	3.725	0.094	4.268	-0.643
9.2	1.073	1.571	1.249	1.585	1.568	1.601	1.840	1.599	2.064	1.583	2.319	1.540	2.991	1.077	3.362	0.585	3.725	0.094	4.268	-0.643
9.3	1.048	1.750	1.230	1.755	1.560	1.745	1.840	1.716	2.068	1.673	2.326	1.598	2.991	1.078	3.362	0.585	3.725	0.094	4.268	-0.643
9.4	1.033	1.935	1.224	1.928	1.567	1.891	1.854	1.832	2.085	1.762	2.342	1.653	2.993	1.080	3.362	0.585	3.725	0.094	4.268	-0.643
3π	1.031	1.981	1.224	1.972	1.572	1.928	1.860	1.861	2.091	1.784	2.347	1.667	2.993	1.080	3.362	0.585	3.725	0.094	4.268	-0.643
9.5	1.029	2.127	1.231	2.108	1.591	2.040	1.885	1.947	2.116	1.848	2.367	1.705	2.994	1.081	3.362	0.585	3.725	0.094	4.268	-0.643
9.6	1.038	2.329	1.255	2.296	1.633	2.190	1.932	2.060	2.160	1.928	2.400	1.751	2.995	1.081	3.362	0.585	3.725	0.094	4.268	-0.643
9.7	1.062	2.547	1.297	2.494	1.696	2.343	1.998	2.169	2.218	2.002	2.441	1.791	2.996	1.082	3.362	0.585	3.725	0.094	4.268	-0.643
9.8	1.106	2.786	1.365	2.706	1.785	2.496	2.084	2.271	2.290	2.067	2.489	1.823	2.998	1.082	3.362	0.585	3.725	0.094	4.268	-0.643
9.9	1.179	3.050	1.464	2.933	1.903	2.646	2.191	2.362	2.375	2.120	2.543	1.844	2.999	1.082	3.362	0.585	3.725	0.094	4.268	-0.643
10.0	1.292	3.347	1.609	3.176	2.058	2.790	2.320	2.438	2.472	2.157	2.600	1.856	3.000	1.081	3.362	0.585	3.725	0.094	4.268	-0.643
10.1	1.465	3.683	1.815	3.433	2.256	2.917	2.471	2.491	2.578	2.175										
10.2	1.734	4.064	2.108	3.695	2.501	3.013	2.642	2.512	2.650	2.169										
10.3	2.154	4.483	2.522	3.934	2.796	3.057	2.825	2.494	2.801	2.139										
10.4	2.816	4.896	3.090	4.095	3.129	3.021	3.010	2.428	2.906	2.082										
10.5	3.836	5.162	3.822	4.070	3.473	2.877	3.180	2.310	2.996	2.000										
10.6	5.247	4.941	4.637	3.710	3.780	2.609	3.317	2.144	3.064	1.896										
10.7	6.645	3.766	5.298	2.924	3.990	2.234	3.402	1.943	3.103	1.779										
10.8	7.099	1.791	5.511	1.861	4.056	1.803	3.426	1.726	3.112	1.658										
10.9	6.327	0.096	5.202	0.880	3.970	1.390	3.388	1.516	3.090	1.541										
7π/2	5.155	-0.697	4.617	0.256	3.776	1.067	3.303	1.342	3.045	1.441										

βh	α/β = 0.00		α/β = 0.01		α/β = 0.03		α/β = 0.05		α/β = 0.07	
11.0	5.101	-0.717	4.587	0.235	3.765	1.054	3.298	1.334	3.042	1.437
11.1	4.027	-0.889	3.920	-0.064	3.495	0.822	3.170	1.192	2.973	1.351
11.2	3.234	-0.765	3.335	-0.128	3.208	0.691	3.022	1.094	2.891	1.287
11.3	2.673	-0.535	2.865	-0.059	2.934	0.643	2.869	1.039	2.802	1.246
11.4	2.273	-0.279	2.501	0.076	2.690	0.654	2.720	1.020	2.712	1.226
11.5	1.984	-0.029	2.220	0.238	2.480	0.707	2.582	1.030	2.625	1.224
11.6	1.768	0.207	2.002	0.410	2.304	0.787	2.459	1.064	2.545	1.239
11.7	1.604	0.426	1.833	0.582	2.158	0.883	2.352	1.116	2.472	1.267
11.8	1.478	0.631	1.700	0.752	2.039	0.990	2.262	1.180	2.410	1.306
11.9	1.378	0.823	1.596	0.918	1.943	1.103	2.188	1.253	2.357	1.352
12.0	1.300	1.007	1.514	1.079	1.868	1.221	2.129	1.333	2.315	1.405
12.1	1.237	1.183	1.450	1.238	1.811	1.340	2.084	1.417	2.284	1.462
12.2	1.188	1.356	1.402	1.395	1.770	1.462	2.053	1.505	2.263	1.522
12.3	1.150	1.526	1.368	1.552	1.744	1.584	2.036	1.593	2.252	1.583
12.4	1.123	1.698	1.346	1.710	1.733	1.708	2.032	1.683	2.252	1.644
12.5	1.106	1.873	1.338	1.870	1.736	1.833	2.041	1.772	2.261	1.704
4π	1.101	1.992	1.340	1.979	1.747	1.916	2.054	1.830	2.273	1.742
12.6	1.100	2.054	1.343	2.035	1.756	1.958	2.063	1.859	2.281	1.761
12.7	1.106	2.244	1.364	2.206	1.792	2.084	2.099	1.943	2.310	1.814
12.8	1.126	2.447	1.403	2.385	1.846	2.209	2.148	2.022	2.348	1.862
12.9	1.164	2.667	1.464	2.574	1.922	2.331	2.213	2.094	2.394	1.902
13.0	1.227	2.909	1.555	2.774	2.021	2.449	2.291	2.156	2.448	1.934
13.1	1.324	3.178	1.683	2.983	2.147	2.556	2.383	2.205	2.508	1.955
13.2	1.473	3.481	1.853	3.201	2.303	2.647	2.488	2.237	2.573	1.964
13.3	1.699	3.820	2.111	3.417	2.450	2.710	2.602	2.247	2.638	1.960
13.4	2.047	4.194	2.431	3.612	2.705	2.735	2.720	2.231	2.703	1.941
13.5	2.585	4.574	2.904	3.744	2.941	2.704	2.837	2.188	2.763	1.909
13.6	3.409	4.868	3.475	3.745	3.180	2.605	2.944	2.115	2.815	1.863
13.7	4.581	4.844	4.117	3.516	3.395	2.431	3.031	2.016	2.855	1.807
13.8	5.917	4.109	4.694	2.982	3.556	2.190	3.091	1.896	2.881	1.744
13.9	6.730	2.517	5.003	2.194	3.635	1.905	3.119	1.765	2.891	1.677
14.0	6.431	0.770	4.925	1.359	3.620	1.613	3.111	1.634	2.886	1.610
14.1	5.408	-0.314	4.533	0.701	3.520	1.350	3.072	1.512	2.866	1.548
9π/2	4.991	-0.522	4.343	0.525	3.466	1.266	3.050	1.470	2.856	1.527
14.2	4.335	-0.697	4.005	0.309	3.359	1.143	3.006	1.408	2.834	1.494
14.3	3.487	-0.686	3.484	0.145	3.165	1.001	2.921	1.327	2.792	1.450
14.4	2.872	-0.511	3.034	0.132	2.961	0.921	2.825	1.271	2.743	1.417
14.5	2.432	-0.281	2.668	0.206	2.766	0.892	2.726	1.239	2.691	1.396
14.6	2.113	-0.044	2.378	0.326	2.588	0.904	2.628	1.229	2.638	1.387
14.7	1.877	0.185	2.149	0.468	2.431	0.946	2.537	1.239	2.587	1.388
14.8	1.699	0.400	1.969	0.618	2.296	1.009	2.453	1.264	2.538	1.399
14.9	1.562	0.601	1.827	0.771	2.183	1.086	2.380	1.301	2.495	1.418
15.0	1.455	0.791	1.715	0.923	2.090	1.173	2.318	1.349	2.457	1.444
15.1	1.371	0.971	1.626	1.074	2.015	1.267	2.266	1.404	2.426	1.476
15.2	1.304	1.144	1.557	1.222	1.957	1.366	2.226	1.464	2.402	1.511
15.3	1.251	1.312	1.505	1.370	1.914	1.468	2.197	1.528	2.384	1.549
15.4	1.211	1.478	1.467	1.517	1.886	1.572	2.179	1.594	2.374	1.589
15.5	1.181	1.644	1.442	1.665	1.871	1.678	2.172	1.662	2.371	1.629
15.6	1.161	1.812	1.431	1.815	1.871	1.784	2.176	1.729	2.375	1.669
15.7	1.152	1.985	1.433	1.968	1.885	1.891	2.191	1.795	2.387	1.707
5π	1.152	1.999	1.434	1.981	1.887	1.899	2.192	1.800	2.388	1.709
15.8	1.154	2.165	1.450	2.126	1.914	1.997	2.216	1.858	2.404	1.742
15.9	1.170	2.357	1.485	2.290	1.959	2.101	2.252	1.917	2.428	1.773
16.0	1.202	2.563	1.539	2.461	2.021	2.201	2.298	1.969	2.457	1.799
16.1	1.255	2.787	1.620	2.640	2.103	2.295	2.355	2.013	2.491	1.820
16.2	1.339	3.035	1.734	2.825	2.205	2.378	2.420	2.048	2.528	1.834
16.3	1.467	3.311	1.890	3.014	2.328	2.447	2.493	2.069	2.568	1.840
16.4	1.658	3.620	2.102	3.199	2.472	2.494	2.570	2.076	2.608	1.838
[illegible]	[illegible]	[illegible]	[illegible]	[illegible]	[illegible]	[illegible]	[illegible]	[illegible]	[illegible]	[illegible]
16.6	2.392	4.316	2.756	3.481	2.808	2.489	2.728	2.038	2.683	1.809
16.7	3.066	4.625	3.216	3.500	2.982	2.421	2.799	1.992	2.715	1.784
16.8	4.046	4.729	3.738	3.355	3.142	2.303	2.859	1.931	2.740	1.752
16.9	5.271	4.318	4.238	2.986	3.268	2.140	2.903	1.856	2.758	1.716
17.0	6.292	3.115	4.578	2.400	3.344	1.943	2.927	1.773	2.767	1.677
17.1	6.433	1.446	4.643	1.710	3.360	1.734	2.931	1.687	2.767	1.638
17.2	5.684	0.140	4.427	1.088	3.316	1.533	2.915	1.605	2.759	1.600
11π/2	4.870	-0.394	4.127	0.724	3.246	1.394	2.889	1.546	2.748	1.573
17.3	4.655	-0.474	4.036	0.648	3.223	1.360	2.881	1.531	2.744	1.566
17.4	3.760	-0.600	3.592	0.407	3.095	1.227	2.832	1.468	2.722	1.537
17.5	3.086	-0.493	3.174	0.321	2.948	1.137	2.773	1.421	2.695	1.513
17.6	2.598	-0.294	2.816	0.338	2.798	1.087	2.708	1.389	2.665	1.497
17.7	2.245	-0.071	2.521	0.415	2.652	1.073	2.641	1.371	2.633	1.487
17.8	1.986	0.151	2.284	0.526	2.518	1.087	2.574	1.369	2.601	1.483
17.9	1.790	0.363	2.055	0.653	2.399	1.123	2.512	1.378	2.570	1.487
18.0	1.641	0.563	1.944	0.789	2.295	1.176	2.455	1.399	2.541	1.495
18.1	1.525	0.751	1.824	0.928	2.207	1.240	2.405	1.428	2.515	1.509
18.2	1.434	0.929	1.729	1.067	2.135	1.314	2.362	1.464	2.493	1.527
18.3	1.362	1.100	1.654	1.206	2.077	1.394	2.328	1.506	2.475	1.549
18.4	1.306	1.265	1.597	1.344	2.034	1.478	2.302	1.552	2.461	1.572
18.5	1.262	1.427	1.555	1.483	2.003	1.566	2.285	1.601	2.452	1.598
18.6	1.229	1.589	1.526	1.622	1.986	1.655	2.276	1.651	2.449	1.624
18.7	1.206	1.752	1.511	1.763	1.982	1.746	2.276	1.701	2.450	1.650
18.8	1.194	1.918	1.510	1.907	1.990	1.836	2.284	1.751	2.455	1.674
6π	1.192	2.003	1.514	1.979	1.999	1.881	2.292	1.775	2.460	1.686
18.9	1.192	2.091	1.523	2.054	2.012	1.926	2.301	1.798	2.466	1.698
19.0	1.203	2.273	1.552	2.206	2.048	2.013	2.326	1.842	2.480	1.719
19.1	1.229	2.468	1.600	2.362	2.098	2.096	2.359	1.881	2.498	1.736
19.2	1.275	2.678	1.671	2.524	2.164	2.173	2.399	1.914	2.519	1.750
19.3	1.347	2.909	1.771	2.690	2.246	2.241	2.445	1.939	2.542	1.759
19.4	1.455	3.165	1.908	2.858	2.343	2.296	2.496	1.955	2.566	1.764
19.5	1.618	3.449	2.090	3.020	2.456	2.333	2.550	1.961	2.591	1.764
19.6	1.863	3.763	2.331	3.166	2.580	2.347	2.605	1.955	2.615	1.759
19.7	2.232	4.097	2.639	3.272	2.712	2.332	2.659	1.938	2.638	1.749
19.8	2.789	4.410	3.017	3.305	2.844	2.286	2.708	1.909	2.658	1.734
19.9	3.608	4.595	3.450	3.215	2.966	2.204	2.750	1.870	2.674	1.716
20.0	4.699	4.419	3.882	2.958	3.067	2.091	2.783	1.822	2.687	1.695
20.1	5.801	3.575	4.222	2.520	3.135	1.952	2.803	1.768	2.694	1.672
20.2	6.310	2.098	4.369	1.962	3.164	1.799	2.811	1.711	2.696	1.648
20.3	5.897	0.653	4.286	1.402	3.151	1.646	2.806	1.655	2.694	1.625
20.4	4.975	-0.198	4.018	0.951	3.099	1.506	2.789	1.602	2.686	1.604
13π/2	4.777	-0.295	3.950	0.878	3.084	1.480	2.784	1.592	2.684	1.599
20.5	4.051	-0.490	3.658	0.658	3.017	1.390	2.761	1.556	2.675	1.584
20.6	3.318	-0.468	3.285	0.511	2.914	1.302	2.724	1.518	2.660	1.568
20.7	2.778	-0.309	2.943	0.474	2.801	1.244	2.682	1.490	2.642	1.556
20.8	2.386	-0.104	2.651	0.508	2.686	1.214	2.636	1.471	2.623	1.547
20.9	2.098	0.111	2.410	0.587	2.575	1.210	2.590	1.463	2.603	1.543
21.0	1.883	0.320	2.213	0.691	2.472	1.226	2.544	1.465	2.584	1.543
21.1	1.719	0.519	2.054	0.809	2.380	1.259	2.501	1.474	2.565	1.547
21.2	1.593	0.706	1.927	0.933	2.299	1.305	2.462	1.492	2.547	1.554
21.3	1.494	0.883	1.825	1.061	2.232	1.361	2.428	1.515	2.532	1.564
21.4	1.416	1.053	1.744	1.191	2.176	1.424	2.400	1.543	2.519	1.577
21.5	1.355	1.216	1.681	1.321	2.133	1.493	2.377	1.576	2.509	1.591
21.6	1.307	1.376	1.634	1.452	2.101	1.565	2.362	1.611	2.502	1.607
21.7	1.270	1.534	1.602	1.583	2.082	1.640	2.352	1.647	2.498	1.623
21.8	1.245	1.693	1.583	1.716	2.074	1.717	2.350	1.685	2.498	1.640
21.9	1.229	1.855	1.577	1.851	2.077	1.794	2.354	1.721	2.500	1.656
7π	1.224	2.006	1.585	1.977	2.091	1.863	2.363	1.754	2.505	1.670
22.0	1.224	2.021	1.586	1.989	2.093	1.870	2.365	1.757	2.506	1.672
22.1	1.231	2.195	1.610	2.130	2.120	1.943	2.381	1.790	2.514	1.686
22.2	1.251	2.380	1.651	2.275	2.160	2.013	2.404	1.819	2.525	1.697
22.3	1.289	2.579	1.714	2.423	2.213	2.077	2.432	1.844	2.538	1.707
22.4	1.350	2.796	1.802	2.574	2.278	2.133	2.464	1.863	2.552	1.714
22.5	1.443	3.034	1.921	2.725	2.355	2.178	2.500	1.876	2.567	1.718
22.6	1.582	3.299	2.079	2.870	2.444	2.209	2.538	1.881	2.583	1.718
22.7	1.788	3.591	2.284	3.001	2.541	2.222	2.576	1.879	2.598	1.716
22.8	2.098	3.904	2.544	3.100	2.643	2.214	2.614	1.869	2.613	1.710
22.9	2.562	4.215	2.861	3.142	2.745	2.183	2.649	1.851	2.625	1.702
23.0	3.248	4.449	3.225	3.091	2.841	2.126	2.680	1.826	2.636	1.692
23.1	4.199	4.437	3.600	2.911	2.922	2.045	2.704	1.794	2.645	1.679
23.2	5.292	3.899	3.923	2.585	2.982	1.945	2.721	1.759	2.650	1.666
23.3	6.064	2.688	4.114	2.139	3.015	1.833	2.730	1.720	2.653	1.651
23.4	6.015	1.215	4.124	1.651	3.018	1.716	2.730	1.681	2.652	1.637
23.5	5.274	0.145	3.962	1.216	2.992	1.604	2.721	1.644	2.649	1.623
15π/2	4.700	-0.215	3.799	1.001	2.963	1.541	2.712	1.622	2.646	1.615
23.6	4.357	-0.338	3.686	0.894	2.941	1.506	2.706	1.610	2.643	1.611
23.7	3.570	-0.426	3.365	0.700	2.870	1.426	2.683	1.581	2.635	1.600
23.8	2.973	-0.320	3.050	0.615	2.787	1.367	2.656	1.558	2.625	1.591
23.9	2.536	-0.138	2.767	0.608	2.698	1.331	2.625	1.541	2.613	1.585
24.0	2.216	0.068	2.525	0.655	2.608	1.315	2.593	1.531	2.601	1.581
24.1	1.978	0.273	2.323	0.734	2.522	1.317	2.560	1.529	2.589	1.579
24.2	1.798	0.471	2.158	0.833	2.442	1.335	2.528	1.532	2.576	1.580
24.3	1.660	0.658	2.023	0.943	2.371	1.365	2.499	1.542	2.565	1.584
24.4	1.552	0.835	1.915	1.059	2.309	1.406	2.472	1.556	2.555	1.589
24.5	1.468	1.004	1.828	1.179	2.257	1.455	2.449	1.575	2.546	1.597
24.6	1.401	1.166	1.761	1.300	2.215	1.510	2.431	1.597	2.538	1.605
24.7	1.348	1.324	1.709	1.423	2.183	1.570	2.417	1.622	2.533	1.615
24.8	1.308	1.480	1.672	1.547	2.162	1.632	2.408	1.648	2.530	1.626
24.9	1.279	1.636	1.649	1.673	2.151	1.696	2.404	1.676	2.529	1.636
25.0	1.260	1.793	1.639	1.800	2.151	1.761	2.406	1.703	2.530	1.647

TABLE 2.7

NORMALIZED ADMITTANCES (Y/Δ) IN MILLIMHOS OF THIN DIPOLE ANTENNAS IN DISSIPATIVE MEDIA

a/λ = 0.007022

βh	α/β = 0.00		α/β = 0.01		α/β = 0.03		α/β = 0.05		α/β = 0.07		α/β = 0.10		α/β = 0.30		α/β = 0.50		α/β = 0.70		α/β = 1.00	
1.5	9.414	-1.732	9.092	-1.362	8.480	-0.778	7.924	-0.356	7.430	-0.050	6.800	0.257	4.753	0.614	4.268	0.329	4.271	-0.050	4.609	-0.688
π/2	7.256	-2.550	7.150	-2.215	6.901	-1.641	6.631	-1.182	6.360	-0.817	5.976	-0.408	4.491	0.392	4.130	0.252	4.189	-0.075	4.569	-0.688
1.6	6.514	-2.629	6.461	-2.326	6.313	-1.794	6.130	-1.354	5.932	-0.994	5.636	-0.578	4.380	0.324	4.075	0.229	4.157	-0.081	4.555	-0.687
1.7	4.619	-2.406	4.660	-2.206	4.701	-1.832	4.700	-1.498	4.670	-1.204	4.590	-0.835	4.013	0.180	3.897	0.186	4.059	-0.085	4.512	-0.676
1.8	3.445	-1.924	3.512	-1.793	3.619	-1.539	3.693	-1.301	3.742	-1.082	3.778	-0.793	3.687	0.140	3.742	0.184	3.978	-0.072	4.481	-0.662
1.9	2.692	-1.432	2.764	-1.343	2.891	-1.167	2.995	-0.998	3.079	-0.837	3.175	-0.617	3.412	0.170	3.612	0.210	3.914	-0.046	4.458	-0.645
2.0	2.185	-0.986	2.256	-0.924	2.388	-0.799	2.503	-0.676	2.604	-0.557	2.731	-0.391	3.189	0.241	3.509	0.255	3.866	-0.015	4.444	-0.628
2.1	1.827	-0.593	1.897	-0.548	2.028	-0.456	2.148	-0.365	2.257	-0.277	2.402	-0.152	3.013	0.338	3.428	0.309	3.831	0.020	4.435	-0.612
2.2	1.566	-0.247	1.634	-0.213	1.764	-0.144	1.886	-0.076	2.000	-0.010	2.155	0.085	2.877	0.447	3.369	0.369	3.808	0.054	4.430	-0.598
2.3	1.368	0.063	1.436	0.089	1.566	0.141	1.689	0.192	1.806	0.242	1.969	0.313	2.775	0.562	3.328	0.430	3.794	0.086	4.429	-0.587
2.4	1.216	0.343	1.283	0.363	1.414	0.403	1.539	0.442	1.659	0.479	1.829	0.530	2.701	0.677	3.302	0.490	3.788	0.116	4.429	-0.577
2.5	1.096	0.601	1.164	0.617	1.296	0.647	1.425	0.676	1.548	0.702	1.724	0.738	2.652	0.790	3.289	0.545	3.788	0.141	4.432	-0.570
2.6	1.001	0.843	1.070	0.854	1.205	0.877	1.337	0.897	1.465	0.915	1.648	0.937	2.624	0.899	3.287	0.596	3.792	0.163	4.435	-0.564
2.7	0.924	1.072	0.995	1.081	1.135	1.097	1.272	1.110	1.404	1.119	1.595	1.128	2.615	1.002	3.293	0.641	3.799	0.181	4.438	-0.560
2.8	0.863	1.294	0.937	1.301	1.083	1.310	1.225	1.316	1.363	1.318	1.563	1.314	2.620	1.098	3.306	0.680	3.808	0.195	4.441	-0.557
2.9	0.814	1.513	0.892	1.516	1.045	1.520	1.195	1.518	1.340	1.512	1.549	1.495	2.640	1.185	3.324	0.712	3.818	0.205	4.444	-0.556
3.0	0.777	1.731	0.860	1.732	1.022	1.729	1.181	1.720	1.334	1.705	1.554	1.673	2.672	1.264	3.345	0.738	3.828	0.212	4.447	-0.555
3.1	0.751	1.952	0.840	1.950	1.014	1.940	1.182	1.922	1.345	1.897	1.576	1.849	2.714	1.332	3.369	0.757	3.838	0.216	4.449	-0.555
π	0.743	2.046	0.835	2.043	1.014	2.029	1.188	2.007	1.355	1.978	1.591	1.921	2.735	1.358	3.379	0.763	3.842	0.217	4.450	-0.555
3.2	0.736	2.181	0.832	2.175	1.020	2.156	1.201	2.128	1.374	2.091	1.619	2.022	2.765	1.390	3.393	0.769	3.847	0.218	4.451	-0.555
3.3	0.733	2.422	0.838	2.412	1.043	2.381	1.239	2.339	1.424	2.287	1.682	2.194	2.823	1.435	3.416	0.776	3.855	0.218	4.452	-0.556
3.4	0.744	2.680	0.861	2.664	1.087	2.618	1.300	2.558	1.499	2.487	1.770	2.363	2.886	1.468	3.439	0.778	3.861	0.217	4.453	-0.556
3.5	0.775	2.962	0.906	2.937	1.157	2.870	1.390	2.786	1.603	2.688	1.885	2.527	2.951	1.488	3.459	0.776	3.867	0.214	4.454	-0.557
3.6	0.832	3.276	0.982	3.237	1.263	3.140	1.517	3.022	1.743	2.891	2.032	2.681	3.016	1.494	3.476	0.770	3.871	0.212	4.454	-0.557
3.7	0.927	3.632	1.100	3.573	1.417	3.431	1.692	3.266	1.928	3.089	2.216	2.819	3.079	1.487	3.491	0.762	3.874	0.208	4.455	-0.558
3.8	1.082	4.043	1.283	3.953	1.639	3.742	1.932	3.510	2.170	3.273	2.439	2.930	3.137	1.468	3.502	0.752	3.876	0.205	4.455	-0.559
3.9	1.332	4.524	1.567	4.380	1.958	4.064	2.236	3.738	2.477	3.424	2.703	2.999	3.187	1.439	3.510	0.741	3.878	0.202	4.455	-0.559
4.0	1.743	5.083	2.009	4.850	2.415	4.373	2.684	3.920	2.855	3.515	3.000	3.006	3.228	1.402	3.515	0.730	3.878	0.199	4.455	-0.559
4.1	2.430	5.704	2.704	5.321	3.054	4.607	3.227	4.001	3.298	3.504	3.315	2.932	3.258	1.359	3.518	0.720	3.878	0.196	4.455	-0.560
4.2	3.585	6.269	3.773	5.654	3.901	4.646	3.865	3.900	3.771	3.348	3.616	2.762	3.277	1.313	3.518	0.710	3.877	0.194	4.454	-0.560
4.3	5.405	6.376	5.264	5.513	4.880	4.307	4.511	3.535	4.206	3.014	3.865	2.496	3.285	1.266	3.517	0.701	3.877	0.193	4.454	-0.560
4.4	7.589	5.195	6.823	4.416	5.728	3.452	5.006	2.888	4.509	2.521	4.020	2.159	3.283	1.222	3.514	0.693	3.876	0.191	4.454	-0.560
4.5	8.652	2.483	7.528	2.397	6.072	2.221	5.188	2.070	4.608	1.945	4.058	1.794	3.272	1.182	3.510	0.687	3.875	0.190	4.454	-0.560
4.6	7.701	-0.047	6.926	0.481	5.782	1.041	5.017	1.288	4.493	1.399	3.981	1.452	3.254	1.148	3.505	0.683	3.874	0.190	4.454	-0.560
4.7	5.985	-1.199	5.691	-0.573	5.097	0.245	4.601	0.704	4.219	0.968	3.817	1.171	3.231	1.121	3.500	0.679	3.873	0.189	4.454	-0.560
3π/2	5.782	-1.261	5.534	-0.645	5.002	0.176	4.541	0.649	4.178	0.924	3.792	1.142	3.227	1.118	3.500	0.679	3.872	0.189	4.454	-0.560
4.8	4.537	-1.395	4.516	-0.896	4.332	-0.140	4.093	0.359	3.858	0.684	3.600	0.969	3.204	1.101	3.495	0.677	3.872	0.189	4.454	-0.560
4.9	3.519	-1.200	3.612	-0.842	3.658	-0.238	3.598	0.212	3.505	0.535	3.365	0.846	3.177	1.088	3.491	0.677	3.871	0.189	4.454	-0.560
5.0	2.825	-0.889	2.558	-0.636	3.113	-0.176	3.168	0.199	3.170	0.489	3.135	0.790	3.149	1.081	3.486	0.677	3.870	0.190	4.454	-0.560
5.1	2.344	-0.563	2.487	-0.381	2.691	-0.035	2.812	0.267	2.878	0.514	2.924	0.785	3.123	1.080	3.483	0.678	3.870	0.190	4.454	-0.560
5.2	2.001	-0.253	2.143	-0.120	2.367	0.142	2.524	0.381	2.633	0.585	2.739	0.820	3.098	1.084	3.479	0.679	3.869	0.190	4.454	-0.560
5.3	1.749	0.031	1.886	0.131	2.117	0.331	2.295	0.519	2.432	0.684	2.580	0.881	3.077	1.092	3.477	0.681	3.869	0.191	4.454	-0.560
5.4	1.559	0.292	1.691	0.367	1.923	0.521	2.113	0.668	2.268	0.801	2.448	0.961	3.059	1.103	3.475	0.683	3.869	0.191	4.454	-0.560
5.5	1.412	0.532	1.540	0.590	1.771	0.709	1.969	0.823	2.137	0.926	2.340	1.052	3.045	1.116	3.474	0.685	3.869	0.191	4.454	-0.560
5.6	1.296	0.755	1.422	0.800	1.653	0.891	1.856	0.979	2.033	1.057	2.255	1.152	3.034	1.131	3.473	0.687	3.869	0.192	4.454	-0.560
5.7	1.204	0.966	1.329	1.001	1.561	1.070	1.769	1.134	1.953	1.191	2.189	1.256	3.027	1.147	3.473	0.689	3.869	0.192	4.454	-0.560
5.8	1.131	1.167	1.255	1.194	1.490	1.244	1.703	1.289	1.894	1.325	2.142	1.362	3.024	1.163	3.473	0.691	3.869	0.192	4.454	-0.560
5.9	1.073	1.363	1.199	1.382	1.438	1.416	1.656	1.442	1.854	1.460	2.111	1.469	3.023	1.177	3.473	0.693	3.869	0.192	4.454	-0.560
6.0	1.027	1.556	1.156	1.569	1.401	1.587	1.627	1.595	1.831	1.594	2.097	1.576	3.025	1.191	3.474	0.694	3.869	0.192	4.454	-0.560
6.1	0.992	1.749	1.126	1.755	1.380	1.758	1.614	1.748	1.825	1.728	2.097	1.682	3.030	1.204	3.475	0.695	3.869	0.192	4.454	-0.560
6.2	0.968	1.945	1.108	1.945	1.375	1.931	1.618	1.901	1.835	1.861	2.113	1.785	3.036	1.214	3.476	0.696	3.869	0.192	4.454	-0.560
2π	0.956	2.113	1.104	2.106	1.382	2.077	1.634	2.029	1.856	1.970	2.137	1.868	3.043	1.222	3.476	0.696	3.869	0.192	4.454	-0.560
6.3	0.955	2.148	1.104	2.140	1.385	2.106	1.638	2.055	1.862	1.992	2.144	1.884	3.044	1.223	3.477	0.696	3.869	0.192	4.454	-0.560
6.4	0.954	2.360	1.114	2.343	1.413	2.287	1.678	2.210	1.907	2.121	2.189	1.979	3.053	1.230	3.478	0.697	3.869	0.192	4.454	-0.560
6.5	0.967	2.586	1.141	2.557	1.461	2.473	1.738	2.366	1.971	2.246	2.250	2.065	3.062	1.234	3.478	0.697	3.870	0.192	4.454	-0.560
6.6	0.998	2.831	1.190	2.787	1.534	2.666	1.822	2.520	2.057	2.366	2.326	2.143	3.072	1.236	3.479	0.697	3.870	0.192	4.454	-0.560
6.7	1.053	3.102	1.267	3.036	1.639	2.866	1.935	2.671	2.166	2.475	2.416	2.207	3.061	1.236	3.480	0.697	3.870	0.192	4.454	-0.560
6.8	1.143	3.405	1.383	3.307	1.783	3.069	2.081	2.814	2.299	2.571	2.520	2.256	3.090	1.235	3.481	0.696	3.870	0.192	4.454	-0.560
6.9	1.284	3.748	1.555	3.603	1.978	3.271	2.266	2.940	2.457	2.644	2.635	2.284	3.097	1.232	3.481	0.696	3.870	0.192	4.454	-0.560
7.0	1.504	4.141	1.807	3.922	2.238	3.460	2.492	3.039	2.639	2.688	2.757	2.288	3.104	1.228	3.481	0.695	3.870	0.192	4.454	-0.560
7.1	1.851	4.586	2.178	4.253	2.577	3.615	2.761	3.092	2.840	2.691	2.881	2.265	3.109	1.223	3.482	0.695	3.870	0.192	4.454	-0.560
7.2	2.404	5.067	2.719	4.557	3.004	3.697	3.065	3.078	3.049	2.644	3.000	2.213	3.113	1.217	3.482	0.695	3.870	0.192	4.454	-0.560
7.3	3.286	5.501	3.485	4.742	3.510	3.651	3.383	2.974	3.250	2.539	3.105	2.132	3.115	1.211	3.482	0.694	3.870	0.192	4.454	-0.560
7.4	4.625	5.644	4.481	4.628	4.043	3.415	3.678	2.765	3.423	2.378	3.188	2.028	3.117	1.205	3.482	0.694	3.870	0.192	4.454	-0.560
7.5	6.308	4.984	5.528	3.981	4.500	2.956	3.906	2.458	3.546	2.168	3.243	1.906	3.116	1.200	3.482	0.694	3.870	0.192	4.454	-0.560
7.6	7.519	3.154	6.191	2.756	4.753	2.321	4.021	2.085	3.603	1.932	3.264	1.776	3.115	1.194	3.481	0.693	3.870	0.192	4.454	-0.560
7.7	7.309	0.943	6.109	1.362	4.730	1.644	4.006	1.701	3.589	1.694	3.251	1.648	3.113	1.190	3.481	0.693	3.870	0.192	4.454	-0.560
7.8	6.079	-0.454	5.429	0.330	4.464	1.073	3.874	1.361	3.512	1.480	3.209	1.532	3.111	1.186	3.481	0.693	3.870	0.192	4.454	-0.560
5π/2	5.345	-0.800	4.964	-0.007	4.257	0.838	3.767	1.209	3.449	1.380	3.177	1.477	3.109	1.184	3.481	0.693	3.870	0.192	4.454	-0.560
7.9	4.770	-0.933	4.567	-0.188	4.063	0.683	3.662	1.100	3.387	1.306	3.144	1.435	3.107	1.183	3.481	0.693	3.870	0.192	4.454	-0.560
8.0	3.755	-0.917	3.791	-0.332	3.630	0.474	3.411	0.928	3.233	1.180	3.062	1.360	3.104	1.181	3.481	0.693	3.870	0.192	4.454	-0.560
8.1	3.034	-0.705	3.175	-0.273	3.227	0.403	3.156	0.837	3.068	1.102	2.971	1.309	3.100	1.180	3.481	0.693	3.870	0.192	4.454	-0.560
8.2	2.526	-0.435	2.708	-0.118	2.880	0.424	2.916	0.812	2.905	1.067	2.878	1.280	3.097	1.179	3.480	0.693	3.870	0.192	4.454	-0.560
8.3	2.163	-0.161	2.355	0.074	2.593	0.502	2.703	0.834	2.752	1.067	2.787	1.273	3.093	1.179	3.480	0.693	3.870	0.192	4.454	-0.560
8.4	1.896	0.100	2.088	0.276	2.358	0.613	2.519	0.890	2.614	1.094	2.701	1.282	3.090	1.180	3.480	0.693	3.870	0.192	4.454	-0.560
8.5	1.695	0.343	1.882	0.477	2.170	0.742	2.364	0.969	2.495	1.142	2.625	1.307	3.087	1.181	3.480	0.693	3.870	0.192	4.454	-0.560
8.6	1.541	0.568	1.723	0.672	2.019	0.880	2.235	1.062	2.393	1.205	2.558	1.343	3.085	1.183	3.480	0.693	3.870	0.192	4.454	-0.560
8.7	1.421	0.779	1.598	0.860	1.898	1.022	2.131	1.166	2.308	1.279	2.502	1.387	3.083	1.185	3.480	0.693	3.870	0.192	4.454	-0.560
8.8	1.326	0.978	1.500	1.041	1.803	1.165	2.048	1.275	2.241	1.361	2.456	1.439	3.082	1.187	3.480	0.693	3.870	0.192	4.454	-0.560
8.9	1.250	1.169	1.423	1.217	1.730	1.309	1.984	1.388	2.189	1.447	2.422	1.495	3.081	1.190	3.480	0.693	3.870	0.192	4.454	-0.560
9.0	1.190	1.354	1.363	1.389	1.675	1.454	1.937	1.504	2.152	1.537	2.398	1.553	3.081	1.192	3.480	0.693	3.870	0.192	4.454	-0.560
9.1	1.143	1.535	1.318	1.560	1.637	1.598	1.907	1.621	2.129	1.629	2.385	1.614	3.081	1.194	3.480	0.694	3.870	0.192	4.454	-0.560
9.2	1.107	1.716	1.287	1.730	1.614	1.744	1.892	1.739	2.121	1.721	2.382	1.673	3.082	1.196	3.480	0.694	3.870	0.192	4.454	-0.560
9.3	1.082	1.899	1.269	1.903	1.607	1.890	1.893	1.857	2.125	1.812	2.389	1.732	3.083	1.197	3.480	0.694	3.870	0.192	4.454	-0.560
9.4	1.068	2.087	1.264	2.080	1.616	2.039	1.909	1.975	2.144	1.901	2.406	1.788	3.084	1.199	3.480	0.694	3.870	0.192	4.454	-0.560
3π	1.067	2.135	1.265	2.124	1.620	2.076	1.915	2.004	2.151	1.923	2.412	1.801	3.084	1.199	3.480	0.694	3.870	0.192	4.454	-0.560
9.5	1.066	2.283	1.273	2.262	1.642	2.189	1.941	2.091	2.176	1.987	2.432	1.839	3.085	1.200	3.480	0.694	3.870	0.192	4.454	-0.560
9.6	1.077	2.491	1.299	2.454	1.686	2.341	1.991	2.205	2.222	2.068	2.466	1.885	3.086	1.200	3.480	0.694	3.870	0.192	4.454	-0.560
9.7	1.104	2.714	1.346	2.656	1.753	2.496	2.059	2.314	2.282	2.142	2.509	1.925	3.088	1.201	3.480	0.694	3.870	0.192	4.454	-0.560
9.8	1.154	2.958	1.418	2.872	1.846	2.650	2.148	2.416	2.356	2.206	2.558	1.955	3.089	1.201	3.480	0.694	3.870	0.192	4.454	-0.560
9.9	1.233	3.229	1.525	3.103	1.970	2.801	2.259	2.506	2.443	2.257	2.612	1.977	3.090	1.201	3.480	0.694	3.870	0.192	4.454	-0.560
10.0	1.357	3.532	1.679	3.348	2.131	2.943	2.391	2.580	2.542	2.293	2.670	1.987	3.091	1.200	3.480	0.694	3.870	0.192	4.454	-0.560
10.1	1.545	3.874	1.898	3.606	2.336	3.067	2.546	2.629	2.649	2.308										
10.2	1.836	4.257	2.208	3.864	2.588	3.157	2.719	2.647	2.761	2.300										
10.3	2.287	4.670	2.641	4.092	2.888	3.191	2.903	2.623	2.873	2.267										
10.4	2.989	5.059	3.227	4.229	3.222	3.142	3.087	2.551	2.977	2.207										
10.5	4.046	5.264	3.965	4.164	3.563	2.984	3.254	2.428	3.065	2.122										
10.6	5.445	4.935	4.758	3.756	3.859	2.705	3.385	2.258	3.130	2.017										
10.7	6.725	3.677	5.365	2.944	4.053	2.325	3.465	2.054	3.167	1.899										
10.8	7.032	1.756	5.520	1.900	4.103	1.898	3.483	1.838	3.172	1.778										
10.9	6.232	0.186	5.186	0.965	4.006	1.493	3.440	1.631	3.148	1.662										
7π/2	5.106	-0.542	4.606	0.377	3.808	1.180	3.351	1.460	3.101	1.563										

βh	α/β = 0.00		α/β = 0.01		α/β = 0.03		α/β = 0.05		α/β = 0.07	
11.0	5.055	-0.559	4.577	0.358	3.798	1.168	3.346	1.453	3.098	1.559
11.1	4.027	-0.719	3.928	0.077	3.528	0.945	3.217	1.315	3.028	1.475
11.2	3.260	-0.601	3.359	0.018	3.244	0.820	3.069	1.220	2.946	1.413
11.3	2.711	-0.381	2.859	0.086	2.973	0.775	2.916	1.167	2.856	1.373
11.4	2.317	-0.133	2.540	0.218	2.732	0.789	2.768	1.150	2.766	1.354
11.5	2.028	0.112	2.262	0.378	2.523	0.842	2.631	1.162	2.679	1.354
11.6	1.812	0.345	2.045	0.548	2.348	0.922	2.509	1.197	2.599	1.370
11.7	1.647	0.563	1.876	0.720	2.203	1.019	2.403	1.249	2.527	1.399
11.8	1.519	0.767	1.742	0.889	2.085	1.127	2.313	1.314	2.465	1.438
11.9	1.418	0.960	1.637	1.055	1.990	1.241	2.239	1.388	2.413	1.486
12.0	1.338	1.145	1.555	1.218	1.915	1.359	2.181	1.469	2.372	1.539
12.1	1.274	1.323	1.491	1.378	1.858	1.480	2.137	1.554	2.341	1.597
12.2	1.225	1.498	1.443	1.537	1.817	1.602	2.106	1.642	2.320	1.657
12.3	1.187	1.671	1.405	1.696	1.792	1.726	2.090	1.732	2.311	1.718
12.4	1.160	1.846	1.388	1.856	1.782	1.851	2.087	1.822	2.311	1.780
12.5	1.144	2.024	1.391	2.019	1.787	1.977	2.097	1.912	2.321	1.840
4π	1.139	2.146	1.384	2.130	1.799	2.062	2.111	1.970	2.334	1.878
12.6	1.138	2.209	1.388	2.187	1.808	2.104	2.121	2.000	2.342	1.897
12.7	1.146	2.403	1.411	2.361	1.847	2.231	2.158	2.084	2.372	1.950
12.8	1.169	2.611	1.453	2.544	1.904	2.357	2.210	2.163	2.411	1.997
12.9	1.211	2.836	1.519	2.736	1.983	2.480	2.276	2.234	2.459	2.037
13.0	1.281	3.083	1.615	2.938	2.086	2.597	2.356	2.296	2.514	2.068
13.1	1.387	3.358	1.752	3.149	2.217	2.703	2.450	2.343	2.574	2.088
13.2	1.549	3.665	1.942	3.366	2.377	2.791	2.557	2.372	2.639	2.096
13.3	1.793	4.007	2.203	3.579	2.568	2.850	2.671	2.380	2.705	2.090
13.4	2.166	4.377	2.557	3.765	2.786	2.868	2.790	2.361	2.770	2.070
13.5	2.738	4.740	3.023	3.880	3.023	2.829	2.906	2.314	2.829	2.036
13.6	3.596	4.991	3.599	3.852	3.259	2.722	3.011	2.239	2.880	1.990
13.7	4.776	4.883	4.230	3.588	3.468	2.541	3.096	2.137	2.919	1.933
13.8	6.041	4.056	4.773	3.030	3.620	2.295	3.153	2.016	2.943	1.868
13.9	6.716	2.462	5.037	2.243	3.689	2.009	3.177	1.885	2.952	1.801
14.0	6.344	0.819	4.928	1.435	3.666	1.721	3.167	1.754	2.946	1.735
14.1	5.345	-0.177	4.531	0.810	3.561	1.465	3.125	1.634	2.925	1.673
9π/2	4.946	-0.369	4.344	0.644	3.506	1.383	3.102	1.593	2.914	1.652
14.2	4.319	-0.530	4.013	0.440	3.398	1.265	3.057	1.532	2.892	1.620
14.3	3.504	-0.520	3.504	0.286	3.204	1.128	2.971	1.453	2.849	1.577
14.4	2.906	-0.352	3.064	0.275	3.002	1.051	2.875	1.398	2.800	1.545
14.5	2.473	-0.131	2.705	0.349	2.808	1.025	2.776	1.368	2.747	1.525
14.6	2.157	0.101	2.418	0.467	2.632	1.038	2.679	1.360	2.694	1.517
14.7	1.921	0.326	2.192	0.608	2.476	1.081	2.588	1.371	2.643	1.519
14.8	1.742	0.538	2.012	0.757	2.343	1.144	2.505	1.397	2.595	1.530
14.9	1.604	0.739	1.870	0.910	2.230	1.222	2.432	1.435	2.552	1.550
15.0	1.496	0.929	1.758	1.062	2.138	1.310	2.370	1.483	2.515	1.577
15.1	1.410	1.110	1.669	1.213	2.064	1.405	2.320	1.539	2.484	1.608
15.2	1.343	1.284	1.600	1.363	2.006	1.505	2.280	1.600	2.460	1.644
15.3	1.289	1.454	1.547	1.511	1.963	1.608	2.252	1.664	2.443	1.683
15.4	1.249	1.622	1.510	1.660	1.936	1.713	2.235	1.731	2.433	1.723
15.5	1.219	1.790	1.486	1.810	1.922	1.819	2.228	1.799	2.431	1.763
15.6	1.200	1.961	1.475	1.962	1.923	1.927	2.233	1.867	2.436	1.803
15.7	1.192	2.138	1.479	2.118	1.939	2.034	2.249	1.933	2.448	1.841
5π	1.192	2.152	1.480	2.131	1.941	2.043	2.251	1.938	2.449	1.844
15.8	1.196	2.322	1.498	2.279	1.969	2.141	2.275	1.996	2.466	1.876
15.9	1.214	2.518	1.535	2.445	2.017	2.245	2.313	2.054	2.490	1.907
16.0	1.249	2.728	1.594	2.619	2.082	2.345	2.360	2.106	2.520	1.933
16.1	1.309	2.957	1.680	2.799	2.166	2.439	2.418	2.150	2.554	1.953
16.2	1.400	3.210	1.800	2.985	2.271	2.521	2.484	2.183	2.592	1.966
16.3	1.538	3.491	1.965	3.174	2.357	2.588	2.558	2.203	2.632	1.971
16.4	1.765	3.802	2.187	3.356	2.544	2.631	2.636	2.208	2.672	1.968
16.5	2.057	4.141	[illegible]	[illegible]	[illegible]	[illegible]	[illegible]	[illegible]	[illegible]	[illegible]
16.6	2.529	4.485	2.862	3.618	2.882	2.616	2.793	2.166	2.747	1.938
16.7	3.235	4.762	3.328	3.617	3.055	2.542	2.864	2.119	2.778	1.912
16.8	4.233	4.803	3.844	3.446	3.210	2.420	2.922	2.056	2.803	1.880
16.9	5.422	4.304	4.322	3.054	3.331	2.253	2.964	1.980	2.820	1.843
17.0	6.329	3.059	4.629	2.461	3.401	2.056	2.986	1.897	2.828	1.804
17.1	6.368	1.457	4.664	1.785	3.411	1.848	2.988	1.811	2.828	1.765
17.2	5.611	0.248	4.435	1.188	3.363	1.650	2.971	1.730	2.819	1.728
11π/2	4.829	-0.243	4.135	0.842	3.290	1.514	2.944	1.671	2.807	1.700
17.3	4.624	-0.316	4.046	0.770	3.267	1.481	2.935	1.656	2.803	1.694
17.4	3.765	-0.433	3.610	0.542	3.138	1.352	2.885	1.595	2.781	1.665
17.5	3.114	-0.331	3.202	0.461	2.992	1.265	2.826	1.549	2.753	1.642
17.6	2.637	-0.140	2.850	0.479	2.842	1.218	2.760	1.518	2.723	1.626
17.7	2.288	0.077	2.560	0.556	2.697	1.206	2.693	1.502	2.691	1.616
17.8	2.030	0.294	2.326	0.666	2.564	1.221	2.627	1.500	2.659	1.614
17.9	1.835	0.504	2.138	0.793	2.446	1.258	2.565	1.510	2.628	1.617
18.0	1.684	0.702	1.988	0.928	2.343	1.311	2.508	1.532	2.599	1.627
18.1	1.567	0.889	1.868	1.067	2.256	1.377	2.459	1.562	2.573	1.641
18.2	1.475	1.068	1.773	1.206	2.184	1.451	2.417	1.599	2.551	1.659
18.3	1.402	1.240	1.698	1.346	2.127	1.532	2.383	1.641	2.533	1.681
18.4	1.345	1.407	1.640	1.486	2.084	1.617	2.357	1.687	2.520	1.705
18.5	1.301	1.571	1.598	1.626	2.054	1.705	2.341	1.736	2.512	1.730
18.6	1.268	1.735	1.571	1.767	2.038	1.796	2.333	1.787	2.509	1.756
18.7	1.246	1.900	1.557	1.910	2.035	1.887	2.333	1.837	2.510	1.782
18.8	1.234	2.070	1.556	2.055	2.044	1.978	2.343	1.887	2.516	1.807
6π	1.233	2.156	1.562	2.129	2.054	2.023	2.350	1.911	2.521	1.819
18.9	1.234	2.246	1.571	2.205	2.068	2.068	2.360	1.934	2.527	1.830
19.0	1.247	2.432	1.603	2.358	2.105	2.155	2.386	1.978	2.542	1.851
19.1	1.277	2.630	1.655	2.517	2.158	2.233	2.420	2.016	2.560	1.868
19.2	1.327	2.845	1.730	2.680	2.226	2.315	2.460	2.049	2.581	1.882
19.3	1.405	3.080	1.836	2.847	2.310	2.381	2.507	2.073	2.604	1.891
19.4	1.523	3.340	1.979	3.014	2.409	2.434	2.559	2.088	2.629	1.895
19.5	1.699	3.628	2.170	3.174	2.523	2.469	2.613	2.093	2.654	1.894
19.6	1.962	3.941	2.419	3.314	2.649	2.480	2.668	2.086	2.678	1.889
19.7	2.355	4.267	2.735	3.411	2.781	2.462	2.722	2.068	2.701	1.878
19.8	2.941	4.559	3.119	3.427	2.912	2.411	2.770	2.038	2.720	1.863
19.9	3.784	4.696	3.549	3.318	3.032	2.326	2.812	1.998	2.736	1.845
20.0	4.864	4.442	3.968	3.041	3.129	2.210	2.843	1.949	2.748	1.823
20.1	5.885	3.533	4.284	2.594	3.193	2.070	2.863	1.895	2.755	1.800
20.2	6.280	2.080	4.405	2.040	3.218	1.917	2.869	1.838	2.757	1.777
20.3	5.823	0.728	4.305	1.497	3.201	1.766	2.863	1.782	2.754	1.754
20.4	4.929	-0.055	4.033	1.065	3.147	1.628	2.844	1.729	2.746	1.732
13π/2	4.738	-0.145	3.964	0.996	3.132	1.603	2.839	1.719	2.744	1.728
20.5	4.044	-0.325	3.677	0.786	3.063	1.514	2.816	1.684	2.734	1.713
20.6	3.338	-0.303	3.310	0.648	2.960	1.429	2.779	1.647	2.719	1.697
20.7	2.813	-0.152	2.976	0.614	2.847	1.374	2.736	1.619	2.701	1.685
20.8	2.427	0.047	2.689	0.649	2.732	1.346	2.690	1.602	2.682	1.677
20.9	2.142	0.257	2.450	0.728	2.622	1.343	2.644	1.594	2.662	1.674
21.0	1.927	0.462	2.256	0.831	2.519	1.360	2.598	1.596	2.642	1.674
21.1	1.763	0.659	2.098	0.948	2.428	1.394	2.555	1.607	2.623	1.678
21.2	1.636	0.845	1.971	1.073	2.348	1.441	2.517	1.625	2.606	1.685
21.3	1.536	1.023	1.869	1.201	2.281	1.497	2.483	1.648	2.591	1.695
21.4	1.457	1.193	1.786	1.331	2.226	1.561	2.455	1.677	2.578	1.708
21.5	1.395	1.357	1.726	1.462	2.184	1.630	2.433	1.710	2.568	1.723
21.6	1.347	1.519	1.679	1.594	2.153	1.704	2.418	1.745	2.562	1.739
21.7	1.310	1.679	1.647	1.727	2.134	1.779	2.409	1.782	2.558	1.755
21.8	1.285	1.840	1.629	1.861	2.127	1.856	2.407	1.819	2.558	1.772
21.9	1.270	2.005	1.625	1.998	2.132	1.934	2.412	1.856	2.561	1.788
7π	1.266	2.159	1.634	2.125	2.147	2.003	2.422	1.889	2.566	1.802
22.0	1.266	2.174	1.635	2.138	2.149	2.010	2.423	1.892	2.567	1.804
22.1	1.275	2.352	1.661	2.280	2.178	2.083	2.441	1.924	2.575	1.817
22.2	1.298	2.541	1.706	2.427	2.219	2.153	2.464	1.953	2.586	1.829
22.3	1.340	2.743	1.772	2.576	2.273	2.217	2.492	1.978	2.599	1.838
22.4	1.407	2.964	1.865	2.728	2.340	2.272	2.525	1.997	2.614	1.845
22.5	1.508	3.207	1.989	2.878	2.419	2.316	2.561	2.009	2.629	1.848
22.6	1.658	3.474	2.154	3.022	2.509	2.345	2.599	2.013	2.645	1.849
22.7	1.880	3.767	2.366	3.148	2.607	2.355	2.638	2.010	2.660	1.846
22.8	2.210	4.076	2.633	3.240	2.709	2.345	2.676	1.999	2.674	1.840
22.9	2.700	4.371	2.955	3.269	2.811	2.310	2.710	1.980	2.687	1.832
23.0	3.412	4.570	3.318	3.203	2.905	2.251	2.740	1.955	2.698	1.821
23.1	4.368	4.493	3.685	3.007	2.984	2.168	2.764	1.923	2.706	1.809
23.2	5.410	3.882	3.991	2.669	3.041	2.067	2.780	1.887	2.711	1.795
23.3	6.076	2.656	4.161	2.222	3.071	1.954	2.788	1.848	2.713	1.780
23.4	5.952	1.255	4.154	1.744	3.071	1.838	2.787	1.809	2.713	1.766
23.5	5.216	0.265	3.983	1.324	3.042	1.728	2.778	1.772	2.709	1.752
15π/2	4.664	-0.065	3.819	1.119	3.012	1.666	2.769	1.751	2.706	1.745
23.6	4.336	-0.180	3.706	1.017	2.990	1.631	2.762	1.739	2.703	1.740
23.7	3.581	-0.261	3.390	0.832	2.918	1.553	2.739	1.710	2.694	1.729
23.8	3.003	-0.160	3.081	0.752	2.835	1.497	2.711	1.687	2.684	1.721
23.9	2.575	0.016	2.803	0.748	2.745	1.462	2.680	1.671	2.673	1.715
24.0	2.259	0.216	2.564	0.795	2.656	1.447	2.648	1.662	2.660	1.711
24.1	2.023	0.417	2.365	0.874	2.570	1.450	2.615	1.660	2.648	1.710
24.2	1.843	0.612	2.201	0.972	2.491	1.469	2.584	1.664	2.636	1.711
24.3	1.704	0.798	2.068	1.082	2.420	1.501	2.554	1.674	2.624	1.715
24.4	1.595	0.974	1.960	1.199	2.359	1.542	2.528	1.689	2.614	1.721
24.5	1.510	1.144	1.873	1.319	2.307	1.592	2.505	1.708	2.605	1.728
24.6	1.442	1.307	1.806	1.441	2.266	1.647	2.487	1.731	2.598	1.737
24.7	1.389	1.467	1.755	1.565	2.235	1.707	2.474	1.756	2.593	1.747
24.8	1.349	1.624	1.718	1.690	2.215	1.770	2.465	1.782	2.590	1.757
24.9	1.320	1.782	1.696	1.817	2.205	1.835	2.462	1.810	2.589	1.768
25.0	1.302	1.942	1.687	1.946	2.206	1.900	2.464	1.837	2.590	1.778

TABLE 2.7

NORMALIZED ADMITTANCES (Y/Δ) IN MILLIMHOS OF THIN DIPOLE ANTENNAS IN DISSIPATIVE MEDIA

a/λ = 0.008496

βh	α/β = 0.00		α/β = 0.01		α/β = 0.03		α/β = 0.05		α/β = 0.07		α/β = 0.10		α/β = 0.30		α/β = 0.50		α/β = 0.70		α/β = 1.00	
1.5	8.962	-1.397	8.700	-1.073	8.192	-0.549	7.724	-0.159	7.300	0.132	6.752	0.434	4.916	0.831	4.510	0.556	4.577	0.169	5.008	-0.491
π/2	7.055	-2.080	6.964	-1.786	6.752	-1.279	6.521	-0.866	6.288	-0.533	5.956	-0.153	4.650	0.623	4.370	0.483	4.494	0.146	4.967	-0.490
1.6	6.392	-2.153	6.345	-1.885	6.216	-1.411	6.059	-1.015	5.889	-0.687	5.633	-0.303	4.539	0.561	4.314	0.462	4.462	0.141	4.953	-0.488
1.7	4.654	-1.978	4.689	-1.794	4.726	-1.450	4.728	-1.142	4.704	-0.871	4.640	-0.529	4.174	0.430	4.135	0.424	4.363	0.139	4.910	-0.477
1.8	3.535	-1.555	3.597	-1.430	3.697	-1.190	3.770	-0.965	3.819	-0.758	3.862	-0.485	3.851	0.398	3.980	0.426	4.282	0.156	4.879	-0.461
1.9	2.795	-1.103	2.865	-1.016	2.988	-0.845	3.091	-0.682	3.176	-0.527	3.277	-0.316	3.579	0.433	3.851	0.456	4.219	0.183	4.857	-0.443
2.0	2.286	-0.682	2.357	-0.619	2.488	-0.495	2.604	-0.374	2.707	-0.258	2.839	-0.097	3.359	0.508	3.748	0.503	4.171	0.216	4.842	-0.426
2.1	1.921	-0.302	1.992	-0.256	2.125	-0.164	2.248	-0.073	2.360	0.015	2.511	0.138	3.184	0.607	3.669	0.560	4.137	0.251	4.834	-0.410
2.2	1.650	0.038	1.721	0.073	1.856	0.143	1.982	0.212	2.100	0.279	2.264	0.373	3.049	0.718	3.610	0.621	4.114	0.286	4.830	-0.396
2.3	1.445	0.347	1.515	0.373	1.651	0.427	1.780	0.479	1.903	0.530	2.076	0.600	2.948	0.835	3.570	0.684	4.101	0.319	4.829	-0.384
2.4	1.285	0.629	1.356	0.650	1.494	0.691	1.626	0.730	1.753	0.768	1.933	0.820	2.876	0.952	3.545	0.744	4.096	0.349	4.830	-0.375
2.5	1.159	0.890	1.231	0.907	1.371	0.938	1.507	0.968	1.639	0.995	1.826	1.030	2.829	1.067	3.534	0.801	4.096	0.375	4.832	-0.367
2.6	1.059	1.137	1.132	1.150	1.276	1.173	1.417	1.194	1.553	1.212	1.749	1.233	2.802	1.178	3.532	0.852	4.101	0.397	4.835	-0.362
2.7	0.978	1.374	1.054	1.383	1.203	1.399	1.349	1.412	1.491	1.421	1.695	1.429	2.794	1.283	3.540	0.898	4.108	0.415	4.838	-0.358
2.8	0.913	1.604	0.993	1.610	1.149	1.619	1.301	1.625	1.450	1.626	1.663	1.620	2.802	1.380	3.554	0.936	4.118	0.428	4.842	-0.355
2.9	0.863	1.831	0.946	1.834	1.111	1.837	1.271	1.834	1.427	1.827	1.651	1.807	2.824	1.469	3.573	0.968	4.128	0.438	4.845	-0.354
3.0	0.825	2.058	0.914	2.059	1.088	2.054	1.258	2.043	1.423	2.026	1.658	1.990	2.859	1.548	3.595	0.993	4.138	0.445	4.847	-0.353
3.1	0.799	2.290	0.894	2.287	1.081	2.275	1.263	2.254	1.437	2.226	1.685	2.172	2.903	1.616	3.620	1.012	4.148	0.449	4.850	-0.353
π	0.792	2.389	0.890	2.385	1.083	2.368	1.270	2.343	1.449	2.309	1.702	2.246	2.925	1.642	3.630	1.018	4.152	0.450	4.850	-0.353
3.2	0.786	2.531	0.889	2.524	1.091	2.501	1.286	2.469	1.471	2.427	1.732	2.351	2.957	1.673	3.644	1.024	4.158	0.451	4.851	-0.353
3.3	0.787	2.785	0.900	2.773	1.120	2.737	1.330	2.689	1.528	2.630	1.802	2.528	3.017	1.718	3.668	1.030	4.166	0.450	4.853	-0.354
3.4	0.805	3.058	0.931	3.038	1.173	2.985	1.400	2.916	1.612	2.836	1.898	2.701	3.082	1.749	3.691	1.032	4.172	0.449	4.854	-0.354
3.5	0.846	3.357	0.987	3.326	1.255	3.248	1.502	3.152	1.727	3.044	2.024	2.867	3.149	1.767	3.711	1.029	4.178	0.446	4.855	-0.355
3.6	0.918	3.689	1.078	3.642	1.377	3.529	1.645	3.396	1.882	3.250	2.184	3.021	3.215	1.771	3.729	1.023	4.182	0.443	4.855	-0.356
3.7	1.036	4.065	1.220	3.995	1.554	3.830	1.842	3.643	2.086	3.448	2.381	3.155	3.279	1.762	3.743	1.014	4.185	0.440	4.855	-0.356
3.8	1.226	4.497	1.438	4.388	1.807	4.145	2.108	3.886	2.348	3.626	2.618	3.258	3.336	1.741	3.754	1.004	4.187	0.437	4.855	-0.357
3.9	1.529	4.994	1.771	4.824	2.167	4.463	2.462	4.103	2.677	3.763	2.894	3.313	3.386	1.709	3.762	0.992	4.188	0.433	4.855	-0.357
4.0	2.019	5.556	2.283	5.285	2.673	4.749	2.921	4.258	3.075	3.829	3.199	3.301	3.426	1.670	3.767	0.981	4.189	0.430	4.855	-0.357
4.1	2.821	6.140	3.068	5.706	3.362	4.930	3.489	4.293	3.527	3.784	3.514	3.204	3.455	1.625	3.769	0.970	4.188	0.428	4.855	-0.358
4.2	4.111	6.570	4.221	5.913	4.233	4.876	4.128	4.130	3.993	3.585	3.807	3.011	3.472	1.578	3.769	0.960	4.188	0.426	4.855	-0.358
4.3	5.973	6.378	5.701	5.552	5.172	4.419	4.737	3.701	4.399	3.215	4.036	2.728	3.478	1.531	3.767	0.951	4.187	0.424	4.855	-0.358
4.4	7.864	4.871	7.040	4.264	5.857	3.489	5.159	3.020	4.657	2.704	4.167	2.384	3.474	1.487	3.764	0.944	4.186	0.423	4.855	-0.358
4.5	8.440	2.270	7.438	2.315	6.099	2.290	5.265	2.218	4.711	2.137	4.182	2.023	3.462	1.448	3.760	0.938	4.185	0.422	4.855	-0.358
4.6	7.383	0.133	6.736	0.641	5.740	1.214	5.050	1.485	4.567	1.617	4.089	1.693	3.442	1.414	3.755	0.933	4.184	0.421	4.855	-0.358
4.7	5.823	-0.819	5.576	-0.254	5.066	0.507	4.627	0.952	4.283	1.216	3.915	1.426	3.418	1.388	3.750	0.930	4.183	0.421	4.855	-0.358
3π/2	5.641	-0.871	5.432	-0.316	4.975	0.447	4.568	0.902	4.242	1.176	3.890	1.399	3.415	1.385	3.749	0.930	4.183	0.421	4.855	-0.358
4.8	4.512	-0.988	4.495	-0.532	4.340	0.169	4.132	0.642	3.934	0.956	3.697	1.237	3.391	1.368	3.745	0.928	4.182	0.421	4.854	-0.358
4.9	3.566	-0.819	3.652	-0.484	3.699	0.084	3.656	0.511	3.580	0.821	3.463	1.124	3.363	1.356	3.740	0.928	4.181	0.421	4.854	-0.358
5.0	2.902	-0.537	3.028	-0.294	3.179	0.145	3.241	0.504	3.253	0.783	3.236	1.074	3.335	1.350	3.736	0.928	4.180	0.421	4.854	-0.358
5.1	2.431	-0.233	2.569	-0.055	2.770	0.281	2.894	0.573	2.968	0.811	3.029	1.075	3.309	1.350	3.732	0.929	4.180	0.422	4.854	-0.358
5.2	2.089	0.063	2.229	0.195	2.452	0.452	2.613	0.686	2.728	0.884	2.846	1.112	3.285	1.355	3.729	0.931	4.180	0.422	4.854	-0.358
5.3	1.834	0.339	1.971	0.439	2.204	0.638	2.387	0.822	2.529	0.984	2.690	1.176	3.264	1.363	3.727	0.933	4.179	0.422	4.854	-0.358
5.4	1.640	0.595	1.774	0.672	2.010	0.826	2.206	0.972	2.367	1.101	2.559	1.257	3.246	1.375	3.725	0.935	4.179	0.423	4.854	-0.358
5.5	1.488	0.834	1.620	0.894	1.857	1.013	2.062	1.127	2.237	1.228	2.453	1.350	3.232	1.389	3.724	0.937	4.179	0.423	4.854	-0.358
5.6	1.369	1.059	1.499	1.105	1.738	1.197	1.949	1.284	2.134	1.361	2.368	1.451	3.222	1.404	3.723	0.939	4.179	0.423	4.854	-0.358
5.7	1.274	1.273	1.403	1.308	1.644	1.378	1.862	1.442	2.055	1.496	2.304	1.557	3.215	1.420	3.723	0.941	4.179	0.424	4.854	-0.358
5.8	1.197	1.478	1.328	1.505	1.573	1.556	1.796	1.599	1.997	1.633	2.258	1.665	3.212	1.436	3.723	0.943	4.179	0.424	4.854	-0.358
5.9	1.137	1.679	1.269	1.699	1.520	1.732	1.750	1.756	1.958	1.770	2.229	1.775	3.212	1.451	3.723	0.945	4.179	0.424	4.854	-0.358
6.0	1.090	1.879	1.226	1.891	1.485	1.907	1.722	1.912	1.937	1.908	2.217	1.884	3.215	1.465	3.724	0.946	4.179	0.424	4.854	-0.358
6.1	1.055	2.079	1.197	2.084	1.465	2.083	1.711	2.069	1.933	2.044	2.220	1.991	3.220	1.477	3.725	0.947	4.179	0.424	4.854	-0.358
6.2	1.032	2.283	1.181	2.281	1.462	2.262	1.718	2.227	1.947	2.180	2.238	2.096	3.226	1.488	3.726	0.948	4.180	0.424	4.854	-0.358
2π	1.022	2.458	1.178	2.449	1.473	2.413	1.737	2.358	1.971	2.292	2.266	2.180	3.233	1.495	3.727	0.948	4.180	0.424	4.854	-0.358
6.3	1.021	2.494	1.179	2.484	1.476	2.444	1.743	2.385	1.978	2.314	2.273	2.196	3.235	1.496	3.727	0.948	4.180	0.424	4.854	-0.358
6.4	1.024	2.717	1.194	2.695	1.510	2.630	1.788	2.544	2.028	2.445	2.322	2.291	3.244	1.502	3.728	0.948	4.180	0.424	4.854	-0.358
6.5	1.043	2.954	1.228	2.920	1.565	2.823	1.855	2.702	2.098	2.572	2.387	2.377	3.253	1.506	3.729	0.948	4.180	0.424	4.854	-0.358
6.6	1.084	3.212	1.287	3.159	1.649	3.021	1.948	2.859	2.191	2.691	2.467	2.453	3.263	1.508	3.729	0.948	4.180	0.424	4.854	-0.358
6.7	1.153	3.496	1.378	3.418	1.766	3.224	2.072	3.010	2.307	2.800	2.562	2.515	3.272	1.508	3.730	0.948	4.180	0.424	4.854	-0.358
6.8	1.262	3.813	1.514	3.698	1.926	3.429	2.230	3.151	2.448	2.891	2.670	2.560	3.281	1.507	3.731	0.948	4.180	0.424	4.854	-0.358
6.9	1.432	4.170	1.712	4.000	2.141	3.629	2.427	3.271	2.614	2.958	2.788	2.583	3.288	1.503	3.731	0.947	4.180	0.424	4.854	-0.358
7.0	1.694	4.572	2.001	4.319	2.424	3.808	2.666	3.359	2.802	2.993	2.913	2.581	3.295	1.499	3.731	0.947	4.180	0.424	4.854	-0.358
7.1	2.102	5.015	2.419	4.635	2.786	3.942	2.944	3.395	3.007	2.984	3.037	2.552	3.300	1.494	3.732	0.947	4.180	0.424	4.854	-0.358
7.2	2.737	5.463	3.013	4.898	3.232	3.990	3.252	3.358	3.215	2.923	3.154	2.493	3.304	1.488	3.732	0.946	4.180	0.424	4.854	-0.358
7.3	3.713	5.802	3.823	5.000	3.740	3.896	3.564	3.229	3.410	2.805	3.255	2.407	3.306	1.482	3.732	0.946	4.180	0.424	4.854	-0.358
7.4	5.095	5.742	4.809	4.758	4.249	3.609	3.843	2.999	3.572	2.632	3.332	2.298	3.307	1.476	3.732	0.945	4.180	0.424	4.854	-0.358
7.5	6.622	4.829	5.743	3.990	4.654	3.116	4.044	2.678	3.681	2.417	3.380	2.173	3.306	1.470	3.732	0.945	4.180	0.424	4.854	-0.358
7.6	7.471	2.966	6.223	2.758	4.843	2.482	4.131	2.306	3.724	2.181	3.394	2.043	3.305	1.465	3.731	0.945	4.180	0.424	4.854	-0.358
7.7	7.055	1.024	6.022	1.486	4.772	1.839	4.093	1.934	3.698	1.949	3.377	1.918	3.303	1.461	3.731	0.945	4.180	0.424	4.854	-0.358
7.8	5.891	-0.141	5.341	0.585	4.489	1.313	3.949	1.613	3.612	1.742	3.330	1.805	3.300	1.457	3.731	0.945	4.180	0.424	4.854	-0.358
5π/2	5.227	-0.431	4.903	0.293	4.283	1.100	3.838	1.471	3.546	1.648	3.295	1.752	3.298	1.455	3.731	0.944	4.180	0.424	4.854	-0.358
7.9	4.707	-0.545	4.534	0.137	4.094	0.960	3.733	1.369	3.482	1.577	3.261	1.711	3.297	1.454	3.731	0.944	4.180	0.424	4.854	-0.358
8.0	3.775	-0.531	3.810	0.012	3.676	0.772	3.485	1.211	3.327	1.460	3.178	1.640	3.293	1.452	3.731	0.944	4.180	0.424	4.854	-0.358
8.1	3.096	-0.341	3.228	0.069	3.288	0.711	3.235	1.130	3.163	1.389	3.086	1.593	3.289	1.451	3.731	0.944	4.180	0.424	4.854	-0.358
8.2	2.606	-0.092	2.779	0.214	2.953	0.736	3.000	1.110	3.002	1.358	2.993	1.568	3.286	1.450	3.730	0.944	4.180	0.424	4.854	-0.358
8.3	2.248	0.167	2.436	0.397	2.673	0.814	2.791	1.136	2.851	1.362	2.902	1.563	3.282	1.451	3.730	0.944	4.180	0.424	4.854	-0.358
8.4	1.982	0.417	2.171	0.593	2.443	0.923	2.611	1.193	2.716	1.392	2.818	1.575	3.279	1.452	3.730	0.945	4.180	0.424	4.854	-0.358
8.5	1.779	0.655	1.966	0.789	2.257	1.051	2.458	1.273	2.598	1.442	2.742	1.601	3.277	1.453	3.730	0.945	4.180	0.424	4.854	-0.358
8.6	1.622	0.878	1.806	0.982	2.107	1.189	2.331	1.368	2.498	1.507	2.677	1.638	3.275	1.455	3.730	0.945	4.180	0.424	4.854	-0.358
8.7	1.499	1.088	1.680	1.170	1.987	1.331	2.228	1.472	2.415	1.582	2.621	1.685	3.273	1.457	3.730	0.945	4.180	0.424	4.854	-0.358
8.8	1.401	1.289	1.580	1.353	1.893	1.476	2.146	1.583	2.349	1.665	2.577	1.737	3.272	1.459	3.730	0.945	4.180	0.424	4.854	-0.358
8.9	1.323	1.483	1.501	1.531	1.819	1.623	2.084	1.698	2.298	1.753	2.544	1.794	3.271	1.461	3.730	0.945	4.180	0.424	4.854	-0.358
9.0	1.261	1.671	1.441	1.707	1.765	1.769	2.038	1.816	2.263	1.845	2.522	1.854	3.271	1.464	3.730	0.945	4.180	0.424	4.854	-0.358
9.1	1.213	1.858	1.396	1.882	1.728	1.917	2.010	1.936	2.242	1.938	2.510	1.915	3.271	1.466	3.730	0.945	4.180	0.424	4.854	-0.358
9.2	1.177	2.045	1.366	2.057	1.708	2.066	1.997	2.056	2.236	2.031	2.509	1.975	3.272	1.467	3.730	0.945	4.180	0.424	4.854	-0.358
9.3	1.153	2.235	1.349	2.236	1.703	2.216	2.001	2.176	2.243	2.123	2.518	2.034	3.273	1.469	3.730	0.945	4.180	0.424	4.854	-0.358
9.4	1.141	2.430	1.347	2.419	1.716	2.369	2.021	2.296	2.265	2.213	2.537	2.090	3.274	1.470	3.730	0.945	4.180	0.424	4.854	-0.358
3π	1.140	2.480	1.349	2.465	1.721	2.407	2.028	2.325	2.272	2.235	2.543	2.103	3.274	1.471	3.730	0.945	4.180	0.424	4.854	-0.358
9.5	1.142	2.635	1.361	2.608	1.746	2.523	2.057	2.414	2.300	2.300	2.565	2.141	3.275	1.471	3.730	0.945	4.180	0.424	4.854	-0.358
9.6	1.159	2.852	1.393	2.807	1.797	2.678	2.112	2.528	2.350	2.380	2.602	2.186	3.276	1.472	3.730	0.945	4.180	0.424	4.854	-0.358
9.7	1.195	3.085	1.449	3.017	1.872	2.835	2.186	2.637	2.414	2.453	2.646	2.224	3.278	1.472	3.730	0.945	4.180	0.424	4.854	-0.358
9.8	1.256	3.340	1.533	3.239	1.974	2.990	2.281	2.737	2.492	2.515	2.697	2.254	3.279	1.472	3.730	0.945	4.180	0.424	4.854	-0.358
9.9	1.353	3.622	1.656	3.475	2.110	3.141	2.398	2.825	2.582	2.563	2.752	2.273	3.280	1.472	3.730	0.945	4.180	0.424	4.854	-0.358
10.0	1.500	3.936	1.832	3.724	2.284	3.278	2.538	2.893	2.684	2.595	2.811	2.280	3.281	1.472	3.730	0.945	4.180	0.424	4.854	-0.358
10.1	1.723	4.285	2.078	3.979	2.502	3.393	2.698	2.935	2.793	2.605										
10.2	2.063	4.667	2.422	4.224	2.766	3.468	2.874	2.943	2.906	2.591										
10.3	2.579	5.056	2.891	4.420	3.074	3.481	3.059	2.907	3.016	2.552										
10.4	3.358	5.376	3.506	4.495	3.408	3.405	3.238	2.823	3.117	2.486										
10.5	4.465	5.435	4.242	4.348	3.734	3.219	3.397	2.689	3.200	2.397										
10.6	5.790	4.894	4.971	3.853	4.004	2.919	3.517	2.513	3.259	2.289										
10.7	6.803	3.523	5.460	3.009	4.164	2.534	3.583	2.308	3.290	2.170										
10.8	6.854	1.758	5.507	2.020	4.183	2.118	3.588	2.095	3.290	2.050										
10.9	6.031	0.427	5.134	1.180	4.067	1.736	3.536	1.895	3.261	1.937										
7π/2	5.001	-0.182	4.574	0.664	3.863	1.445	3.442	1.732	3.210	1.842										

βh	$a/\beta = 0.00$		$a/\beta = 0.01$		$a/\beta = 0.03$		$a/\beta = 0.05$		$a/\beta = 0.07$	
11.0	4.955	-0.197	4.547	0.647	3.852	1.434	3.437	1.725	3.207	1.838
11.1	4.018	-0.334	3.936	0.402	3.586	1.230	3.306	1.595	3.135	1.758
11.2	3.306	-0.230	3.400	0.353	3.308	1.117	3.158	1.507	3.051	1.699
11.3	2.783	-0.029	2.962	0.418	3.046	1.079	3.006	1.460	2.962	1.663
11.4	2.399	0.203	2.615	0.544	2.811	1.096	2.860	1.447	2.872	1.647
11.5	2.114	0.437	2.343	0.698	2.608	1.151	2.725	1.462	2.786	1.649
11.6	1.897	0.662	2.129	0.864	2.436	1.231	2.606	1.499	2.707	1.667
11.7	1.730	0.877	1.960	1.034	2.293	1.329	2.501	1.553	2.636	1.698
11.8	1.600	1.080	1.827	1.203	2.176	1.437	2.413	1.619	2.575	1.739
11.9	1.497	1.273	1.721	1.365	2.082	1.552	2.341	1.695	2.525	1.787
12.0	1.415	1.460	1.638	1.534	2.008	1.672	2.284	1.777	2.485	1.842
12.1	1.350	1.641	1.573	1.696	1.953	1.794	2.242	1.864	2.455	1.901
12.2	1.299	1.820	1.525	1.859	1.913	1.919	2.214	1.953	2.437	1.962
12.3	1.261	1.998	1.492	2.021	1.890	2.046	2.199	2.044	2.428	2.024
12.4	1.235	2.179	1.473	2.186	1.883	2.173	2.198	2.136	2.431	2.086
12.5	1.220	2.304	1.468	2.354	1.891	2.302	2.211	2.226	2.443	2.146
4π	1.218	2.490	1.474	2.469	1.906	2.387	2.228	2.285	2.457	2.184
12.6	1.218	2.556	1.480	2.528	1.916	2.431	2.238	2.314	2.466	2.203
12.7	1.231	2.758	1.508	2.708	1.960	2.560	2.279	2.398	2.498	2.255
12.8	1.261	2.975	1.558	2.895	2.023	2.686	2.335	2.477	2.539	2.301
12.9	1.314	3.209	1.634	3.092	2.109	2.809	2.405	2.547	2.589	2.340
13.0	1.398	3.466	1.744	3.298	2.220	2.924	2.489	2.605	2.646	2.368
13.1	1.525	3.750	1.898	3.511	2.359	3.026	2.587	2.649	2.708	2.386
13.2	1.715	4.064	2.109	3.725	2.528	3.106	2.695	2.673	2.773	2.391
13.3	2.000	4.406	2.396	3.927	2.726	3.155	2.812	2.674	2.839	2.382
13.4	2.427	4.760	2.777	4.088	2.949	3.158	2.930	2.649	2.903	2.359
13.5	3.064	5.074	3.265	4.161	3.184	3.103	3.044	2.596	2.961	2.322
13.6	3.977	5.216	3.843	4.072	3.412	2.979	3.144	2.514	3.009	2.273
13.7	5.135	4.933	4.438	3.742	3.606	2.784	3.223	2.409	3.045	2.215
13.8	6.217	3.954	4.902	3.144	3.738	2.532	3.272	2.286	3.066	2.150
13.9	6.629	2.411	5.077	2.375	3.787	2.250	3.289	2.155	3.072	2.083
14.0	6.150	0.982	4.916	1.630	3.748	1.972	3.273	2.027	3.063	2.017
14.1	5.213	0.146	4.513	1.071	3.634	1.730	3.226	1.911	3.040	1.957
9π/2	4.851	-0.016	4.333	0.924	3.577	1.653	3.202	1.872	3.028	1.937
14.2	4.281	-0.153	4.018	0.744	3.468	1.543	3.156	1.814	3.005	1.906
14.3	3.530	-0.144	3.537	0.609	3.276	1.417	3.069	1.739	2.961	1.865
14.4	2.967	0.007	3.119	0.603	3.078	1.348	2.973	1.689	2.911	1.835
14.5	2.551	0.212	2.774	0.675	2.889	1.327	2.874	1.663	2.859	1.817
14.6	2.241	0.432	2.496	0.790	2.716	1.344	2.778	1.657	2.806	1.810
14.7	2.007	0.648	2.274	0.926	2.564	1.388	2.688	1.670	2.755	1.814
14.8	1.828	0.856	2.097	1.074	2.433	1.453	2.607	1.698	2.707	1.827
14.9	1.688	1.054	1.956	1.225	2.323	1.532	2.535	1.738	2.665	1.848
15.0	1.578	1.243	1.843	1.377	2.232	1.621	2.475	1.788	2.629	1.876
15.1	1.490	1.425	1.754	1.529	2.159	1.717	2.425	1.844	2.599	1.909
15.2	1.421	1.602	1.685	1.681	2.103	1.818	2.388	1.906	2.576	1.945
15.3	1.367	1.775	1.633	1.832	2.062	1.923	2.361	1.972	2.560	1.984
15.4	1.326	1.948	1.596	1.984	2.037	2.029	2.346	2.040	2.552	2.024
15.5	1.297	2.121	1.574	2.138	2.026	2.138	2.341	2.108	2.551	2.065
15.6	1.279	2.298	1.566	2.294	2.029	2.247	2.348	2.177	2.557	2.105
15.7	1.274	2.481	1.573	2.454	2.048	2.355	2.366	2.243	2.571	2.142
5π	1.274	2.496	1.575	2.467	2.050	2.364	2.368	2.248	2.572	2.145
15.8	1.282	2.672	1.598	2.619	2.083	2.463	2.395	2.305	2.590	2.177
15.9	1.306	2.876	1.641	2.790	2.135	2.568	2.435	2.363	2.616	2.207
16.0	1.351	3.095	1.709	2.968	2.205	2.667	2.485	2.413	2.647	2.232
16.1	1.423	3.332	1.806	3.151	2.295	2.759	2.545	2.455	2.682	2.250
16.2	1.532	3.593	1.940	3.338	2.406	2.837	2.614	2.485	2.720	2.262
16.3	1.694	3.881	2.122	3.523	2.538	2.898	2.689	2.502	2.760	2.266
16.4	1.935	4.194	2.365	3.696	2.689	2.934	2.768	2.503	2.801	2.261
16.5	2.292	4.522	2.681	3.836	2.855	2.936	2.847	2.487	2.839	2.248
16.6	2.821	4.831	[illegible]	[illegible]	[illegible]	[illegible]	[illegible]	[illegible]	[illegible]	[illegible]
16.7	3.583	5.028	3.548	3.861	3.196	2.813	2.990	2.402	2.904	2.200
16.8	4.592	4.927	4.043	3.638	3.343	2.681	3.045	2.337	2.927	2.167
16.9	5.669	4.266	4.469	3.208	3.451	2.509	3.083	2.260	2.942	2.130
17.0	6.341	2.986	4.707	2.612	3.507	2.311	3.101	2.176	2.949	2.091
17.1	6.208	1.544	4.687	1.973	3.505	2.108	3.099	2.092	2.947	2.052
17.2	5.457	0.519	4.437	1.428	3.449	1.918	3.078	2.012	2.937	2.015
11π/2	4.741	0.106	4.141	1.118	3.372	1.789	3.049	1.955	2.924	1.989
17.3	4.555	0.044	4.055	1.054	3.348	1.758	3.040	1.941	2.919	1.982
17.4	3.770	-0.057	3.639	0.852	3.218	1.638	2.988	1.883	2.896	1.955
17.5	3.163	0.035	3.250	0.783	3.073	1.558	2.928	1.840	2.868	1.933
17.6	2.709	0.210	2.914	0.803	2.925	1.516	2.862	1.811	2.838	1.918
17.7	2.371	0.414	2.635	0.878	2.784	1.508	2.796	1.798	2.805	1.910
17.8	2.116	0.622	2.406	0.986	2.653	1.526	2.730	1.798	2.773	1.908
17.9	1.921	0.825	2.222	1.111	2.537	1.564	2.669	1.810	2.743	1.913
18.0	1.770	1.019	2.073	1.244	2.436	1.619	2.614	1.833	2.714	1.923
18.1	1.651	1.206	1.954	1.383	2.351	1.686	2.565	1.864	2.689	1.938
18.2	1.557	1.384	1.859	1.523	2.281	1.762	2.524	1.902	2.668	1.957
18.3	1.483	1.558	1.785	1.664	2.226	1.843	2.492	1.945	2.651	1.979
18.4	1.425	1.727	1.728	1.806	2.184	1.930	2.468	1.992	2.639	2.004
18.5	1.380	1.895	1.687	1.948	2.157	2.019	2.452	2.042	2.631	2.030
18.6	1.347	2.064	1.661	2.092	2.143	2.111	2.446	2.093	2.629	2.056
18.7	1.326	2.234	1.649	2.239	2.142	2.203	2.448	2.144	2.631	2.082
18.8	1.317	2.410	1.652	2.388	2.154	2.295	2.459	2.193	2.638	2.106
6π	1.318	2.499	1.660	2.463	2.166	2.340	2.468	2.217	2.644	2.118
18.9	1.321	2.593	1.672	2.541	2.181	2.385	2.479	2.240	2.650	2.129
19.0	1.340	2.785	1.709	2.698	2.222	2.473	2.507	2.283	2.665	2.149
19.1	1.377	2.991	1.768	2.860	2.279	2.555	2.542	2.320	2.684	2.166
19.2	1.438	3.214	1.853	3.025	2.351	2.629	2.584	2.351	2.706	2.178
19.3	1.531	3.457	1.971	3.192	2.439	2.693	2.632	2.373	2.730	2.186
19.4	1.670	3.723	2.128	3.357	2.543	2.741	2.684	2.386	2.754	2.190
19.5	1.875	4.013	2.335	3.509	2.660	2.770	2.739	2.388	2.779	2.188
19.6	2.176	4.321	2.601	3.635	2.787	2.774	2.794	2.379	2.803	2.181
19.7	2.619	4.622	2.932	3.707	2.918	2.748	2.846	2.358	2.825	2.170
19.8	3.259	4.855	3.321	3.689	3.046	2.690	2.893	2.326	2.844	2.154
19.9	4.132	4.880	3.740	3.539	3.160	2.598	2.932	2.284	2.859	2.135
20.0	5.156	4.473	4.124	3.227	3.249	2.477	2.961	2.234	2.870	2.113
20.1	5.990	3.470	4.387	2.766	3.304	2.336	2.978	2.179	2.876	2.090
20.2	6.177	2.104	4.458	2.230	3.320	2.185	2.982	2.123	2.877	2.067
20.3	5.662	0.936	4.328	1.725	3.296	2.038	2.973	2.068	2.873	2.044
20.4	4.829	0.276	4.051	1.333	3.236	1.907	2.953	2.017	2.864	2.023
13π/2	4.656	0.201	3.983	1.271	3.221	1.883	2.948	2.007	2.862	2.019
20.5	4.024	0.047	3.705	1.083	3.150	1.799	2.923	1.973	2.852	2.004
20.6	3.372	0.067	3.355	0.960	3.046	1.719	2.885	1.938	2.836	1.989
20.7	2.876	0.205	3.035	0.933	2.934	1.669	2.842	1.912	2.818	1.978
20.8	2.505	0.390	2.759	0.970	2.821	1.645	2.796	1.897	2.799	1.971
20.9	2.226	0.589	2.528	1.047	2.712	1.645	2.750	1.891	2.779	1.968
21.0	2.014	0.787	2.338	1.149	2.612	1.665	2.705	1.895	2.759	1.969
21.1	1.850	0.979	2.183	1.265	2.523	1.700	2.663	1.906	2.741	1.973
21.2	1.721	1.163	2.057	1.389	2.445	1.748	2.625	1.925	2.724	1.981
21.3	1.620	1.340	1.957	1.518	2.379	1.806	2.592	1.950	2.709	1.992
21.4	1.540	1.511	1.877	1.649	2.326	1.871	2.565	1.979	2.697	2.005
21.5	1.477	1.678	1.815	1.781	2.285	1.941	2.544	2.013	2.688	2.020
21.6	1.428	1.842	1.770	1.915	2.256	2.016	2.530	2.048	2.681	2.036
21.7	1.392	2.007	1.739	2.051	2.240	2.092	2.523	2.086	2.679	2.053
21.8	1.367	2.172	1.723	2.188	2.235	2.170	2.522	2.123	2.679	2.070
21.9	1.354	2.342	1.722	2.328	2.242	2.247	2.528	2.160	2.682	2.086
7π	1.353	2.502	1.734	2.457	2.259	2.317	2.539	2.192	2.688	2.100
22.0	1.354	2.517	1.736	2.470	2.261	2.324	2.541	2.195	2.689	2.101
22.1	1.367	2.701	1.767	2.616	2.293	2.397	2.559	2.227	2.698	2.114
22.2	1.397	2.897	1.818	2.765	2.338	2.466	2.584	2.255	2.709	2.125
22.3	1.448	3.107	1.893	2.916	2.395	2.528	2.613	2.278	2.723	2.134
22.4	1.528	3.335	1.995	3.068	2.465	2.580	2.647	2.296	2.737	2.140
22.5	1.647	3.584	2.132	3.216	2.547	2.621	2.683	2.306	2.753	2.143
22.6	1.822	3.855	2.310	3.353	2.638	2.645	2.722	2.309	2.768	2.143
22.7	2.077	4.144	2.536	3.468	2.737	2.651	2.760	2.304	2.783	2.139
22.8	2.451	4.437	2.815	3.540	2.839	2.635	2.797	2.292	2.797	2.133
22.9	2.991	4.689	3.143	3.543	2.939	2.595	2.831	2.271	2.810	2.124
23.0	3.743	4.801	3.500	3.444	3.029	2.530	2.859	2.244	2.820	2.113
23.1	4.683	4.590	3.844	3.217	3.103	2.444	2.882	2.212	2.827	2.100
23.2	5.593	3.854	4.111	2.861	3.153	2.341	2.896	2.175	2.832	2.086
23.3	6.050	2.640	4.238	2.419	3.177	2.228	2.902	2.137	2.834	2.072
23.4	5.805	1.396	4.197	1.967	3.171	2.115	2.900	2.098	2.833	2.058
23.5	5.092	0.557	4.013	1.580	3.138	2.008	2.889	2.062	2.828	2.044
15π/2	4.587	0.278	3.850	1.393	3.106	1.949	2.879	2.041	2.825	2.037
23.6	4.288	0.180	3.739	1.301	3.082	1.916	2.871	2.030	2.822	2.032
23.7	3.595	0.111	3.434	1.136	3.010	1.843	2.848	2.002	2.813	2.022
23.8	3.056	0.203	3.137	1.066	2.926	1.790	2.820	1.981	2.802	2.014
23.9	2.648	0.365	2.870	1.066	2.837	1.759	2.789	1.966	2.791	2.008
24.0	2.342	0.552	2.639	1.113	2.749	1.747	2.756	1.958	2.779	2.005
24.1	2.109	0.745	2.446	1.192	2.665	1.753	2.724	1.957	2.766	2.004
24.2	1.930	0.934	2.286	1.290	2.587	1.773	2.692	1.962	2.754	2.006
24.3	1.790	1.117	2.154	1.399	2.518	1.806	2.664	1.973	2.743	2.010
24.4	1.680	1.293	2.048	1.516	2.458	1.849	2.638	1.989	2.733	2.016
24.5	1.594	1.462	1.963	1.636	2.408	1.900	2.616	2.009	2.724	2.024
24.6	1.525	1.627	1.896	1.760	2.368	1.956	2.599	2.032	2.718	2.033
24.7	1.472	1.789	1.846	1.885	2.339	2.017	2.586	2.057	2.713	2.043
24.8	1.432	1.950	1.811	2.013	2.321	2.080	2.579	2.084	2.710	2.054
24.9	1.403	2.112	1.791	2.142	2.313	2.146	2.576	2.111	2.710	2.064
25.0	1.386	2.276	1.785	2.273	2.315	2.211	2.579	2.138	2.711	2.075

TABLE 2.7

NORMALIZED ADMITTANCES (Y/Δ) IN MILLIMHOS OF THIN DIPOLE ANTENNAS IN DISSIPATIVE MEDIA

$a/\lambda = 0.009525$

βh	α/β = 0.00	α/β = 0.01	α/β = 0.03	α/β = 0.05	α/β = 0.07	α/β = 0.10	α/β = 0.30	α/β = 0.50	α/β = 0.70	α/β = 1.00
1.5	8.671 -1.121	8.440 -0.826	7.993 -0.345	7.576 0.021	7.196 0.298	6.700 0.590	5.015 0.599	4.670 0.729	4.786 0.337	5.285 -0.336
π/2	6.920 -1.733	6.836 -1.466	6.645 -1.000	6.438 -0.618	6.229 -0.307	5.929 0.051	4.749 0.802	4.530 0.660	4.703 0.316	5.245 -0.335
1.6	6.304 -1.805	6.260 -1.559	6.143 -1.122	6.001 -0.754	5.849 -0.448	5.618 -0.087	4.635 0.743	4.474 0.640	4.671 0.312	5.230 -0.332
1.7	4.667 -1.662	4.699 -1.488	4.733 -1.164	4.737 -0.874	4.718 -0.618	4.664 -0.294	4.277 0.621	4.295 0.606	4.572 0.312	5.188 -0.320
1.8	3.586 -1.278	3.645 -1.157	3.742 -0.926	3.813 -0.710	3.864 -0.512	3.910 -0.250	3.957 0.595	4.140 0.611	4.492 0.330	5.157 -0.304
1.9	2.857 -0.853	2.926 -0.768	3.047 -0.601	3.149 -0.441	3.235 -0.291	3.338 -0.087	3.689 0.633	4.013 0.643	4.429 0.358	5.136 -0.286
2.0	2.348 -0.449	2.419 -0.387	2.550 -0.264	2.667 -0.145	2.772 -0.031	2.907 0.127	3.471 0.710	3.911 0.691	4.382 0.392	5.121 -0.268
2.1	1.979 -0.079	2.051 -0.033	2.187 0.059	2.311 0.150	2.426 0.236	2.582 0.358	3.298 0.810	3.832 0.749	4.348 0.428	5.113 -0.252
2.2	1.705 0.256	1.776 0.291	1.914 0.362	2.044 0.431	2.165 0.497	2.334 0.590	3.165 0.922	3.775 0.812	4.326 0.463	5.109 -0.238
2.3	1.494 0.562	1.567 0.590	1.706 0.644	1.840 0.697	1.967 0.747	2.145 0.817	3.065 1.040	3.736 0.875	4.314 0.497	5.108 -0.226
2.4	1.330 0.844	1.403 0.866	1.546 0.908	1.683 0.948	1.814 0.985	2.002 1.037	2.994 1.158	3.712 0.936	4.309 0.527	5.109 -0.216
2.5	1.200 1.108	1.275 1.125	1.421 1.157	1.562 1.186	1.699 1.214	1.894 1.248	2.948 1.274	3.701 0.993	4.310 0.553	5.112 -0.209
2.6	1.096 1.358	1.173 1.371	1.323 1.394	1.470 1.415	1.612 1.433	1.816 1.453	2.923 1.386	3.701 1.044	4.315 0.575	5.115 -0.203
2.7	1.013 1.598	1.093 1.608	1.249 1.624	1.401 1.637	1.549 1.646	1.763 1.652	2.916 1.491	3.709 1.090	4.323 0.592	5.119 -0.200
2.8	0.947 1.833	1.030 1.839	1.194 1.848	1.353 1.853	1.508 1.853	1.731 1.846	2.926 1.585	3.724 1.128	4.332 0.606	5.122 -0.197
2.9	0.896 2.065	0.983 2.068	1.155 2.070	1.323 2.067	1.486 2.058	1.720 2.035	2.949 1.678	3.744 1.160	4.343 0.616	5.125 -0.196
3.0	0.857 2.299	0.951 2.299	1.134 2.293	1.311 2.281	1.483 2.262	1.729 2.222	2.985 1.757	3.767 1.185	4.353 0.622	5.128 -0.195
3.1	0.832 2.538	0.932 2.534	1.128 2.519	1.318 2.496	1.500 2.465	1.759 2.407	3.032 1.825	3.792 1.203	4.363 0.626	5.130 -0.195
π	0.826 2.639	0.929 2.634	1.131 2.615	1.326 2.587	1.513 2.551	1.777 2.483	3.054 1.850	3.802 1.208	4.367 0.627	5.131 -0.195
3.2	0.821 2.786	0.930 2.777	1.141 2.752	1.344 2.716	1.538 2.671	1.810 2.589	3.087 1.882	3.817 1.214	4.373 0.627	5.132 -0.195
3.3	0.825 3.048	0.944 3.034	1.175 2.994	1.394 2.941	1.600 2.878	1.885 2.769	3.149 1.925	3.841 1.220	4.381 0.627	5.133 -0.196
3.4	0.849 3.331	0.981 3.308	1.233 3.248	1.470 3.174	1.690 3.088	1.987 2.943	3.215 1.955	3.864 1.221	4.387 0.625	5.134 -0.196
3.5	0.898 3.639	1.046 3.604	1.325 3.518	1.581 3.414	1.814 3.298	2.120 3.110	3.282 1.972	3.884 1.218	4.393 0.623	5.135 -0.197
3.6	0.982 3.983	1.149 3.930	1.459 3.805	1.736 3.661	1.979 3.505	2.287 3.263	3.349 1.974	3.901 1.211	4.397 0.619	5.135 -0.198
3.7	1.119 4.370	1.309 4.291	1.653 4.110	1.947 3.909	2.195 3.701	2.493 3.393	3.413 1.963	3.915 1.202	4.400 0.616	5.136 -0.198
3.8	1.334 4.812	1.552 4.691	1.928 4.425	2.230 4.147	2.470 3.872	2.739 3.489	3.470 1.940	3.926 1.191	4.402 0.613	5.136 -0.199
3.9	1.676 5.315	1.921 5.127	2.315 4.735	2.604 4.352	2.812 3.998	3.021 3.533	3.519 1.908	3.934 1.180	4.403 0.609	5.136 -0.199
4.0	2.223 5.869	2.481 5.572	2.852 4.999	3.082 4.486	3.220 4.045	3.329 3.508	3.558 1.867	3.938 1.169	4.403 0.606	5.136 -0.200
4.1	3.101 6.409	3.320 5.946	3.567 5.137	3.659 4.487	3.675 3.974	3.642 3.395	3.585 1.822	3.940 1.158	4.403 0.604	5.135 -0.200
4.2	4.461 6.724	4.510 6.052	4.441 5.015	4.290 4.281	4.130 3.748	3.925 3.188	3.601 1.774	3.940 1.148	4.402 0.602	5.135 -0.200
4.3	6.294 6.326	5.945 5.548	5.336 4.487	4.865 3.815	4.511 3.357	4.139 2.896	3.606 1.727	3.938 1.139	4.401 0.600	5.135 -0.200
4.4	7.944 4.670	7.114 4.176	5.970 3.528	5.235 3.121	4.737 2.841	4.253 2.550	3.601 1.683	3.935 1.131	4.400 0.599	5.135 -0.200
4.5	8.249 2.205	7.335 2.312	6.086 2.367	5.294 2.339	4.763 2.286	4.253 2.195	3.588 1.644	3.930 1.126	4.399 0.598	5.135 -0.200
4.6	7.170 0.308	6.597 0.794	5.695 1.362	5.056 1.644	4.602 1.787	4.150 1.874	3.568 1.611	3.925 1.121	4.398 0.597	5.135 -0.200
4.7	5.715 -0.535	5.495 -0.009	5.035 0.713	4.633 1.145	4.315 1.406	3.973 1.618	3.543 1.585	3.920 1.118	4.397 0.597	5.135 -0.200
3π/2	5.546 -0.582	5.359 -0.065	4.947 0.658	4.575 1.098	4.274 1.369	3.947 1.592	3.540 1.583	3.920 1.118	4.397 0.597	5.135 -0.200
4.8	4.450 -0.690	4.475 -0.262	4.337 0.402	4.150 0.856	3.970 1.161	3.754 1.439	3.516 1.567	3.915 1.117	4.396 0.597	5.135 -0.200
4.9	3.590 -0.538	3.671 -0.218	3.721 0.325	3.688 0.736	3.623 1.036	3.522 1.332	3.487 1.555	3.911 1.116	4.396 0.597	5.135 -0.200
5.0	2.947 -0.276	3.068 -0.040	3.217 0.385	3.283 0.732	3.303 1.003	3.299 1.287	3.459 1.550	3.906 1.116	4.395 0.597	5.135 -0.200
5.1	2.484 0.014	2.619 0.189	2.818 0.517	2.945 0.801	3.024 1.034	3.094 1.291	3.433 1.550	3.903 1.117	4.394 0.598	5.135 -0.200
5.2	2.144 0.299	2.282 0.430	2.505 0.684	2.668 0.913	2.787 1.107	2.914 1.330	3.409 1.555	3.899 1.119	4.394 0.598	5.135 -0.200
5.3	1.888 0.569	2.025 0.669	2.259 0.866	2.444 1.049	2.591 1.207	2.760 1.394	3.388 1.565	3.897 1.121	4.394 0.598	5.135 -0.200
5.4	1.691 0.822	1.826 0.900	2.065 1.053	2.265 1.197	2.431 1.325	2.631 1.476	3.371 1.577	3.895 1.123	4.394 0.599	5.135 -0.200
5.5	1.537 1.060	1.671 1.120	1.913 1.240	2.122 1.352	2.302 1.452	2.526 1.571	3.357 1.591	3.894 1.125	4.394 0.599	5.135 -0.200
5.6	1.416 1.285	1.548 1.332	1.793 1.424	2.010 1.510	2.200 1.585	2.442 1.673	3.348 1.606	3.893 1.128	4.393 0.599	5.135 -0.200
5.7	1.318 1.500	1.451 1.536	1.699 1.606	1.922 1.668	2.122 1.721	2.379 1.779	3.341 1.622	3.893 1.130	4.394 0.600	5.135 -0.200
5.8	1.241 1.708	1.375 1.735	1.627 1.785	1.857 1.827	2.065 1.859	2.334 1.888	3.338 1.638	3.893 1.131	4.394 0.600	5.135 -0.200
5.9	1.179 1.912	1.316 1.932	1.575 1.963	1.812 1.986	2.027 1.998	2.307 1.999	3.339 1.653	3.894 1.133	4.394 0.600	5.135 -0.200
6.0	1.132 2.115	1.273 2.127	1.540 2.142	1.785 2.144	2.007 2.137	2.296 2.108	3.342 1.667	3.895 1.134	4.394 0.600	5.135 -0.200
6.1	1.097 2.320	1.244 2.324	1.522 2.321	1.776 2.304	2.005 2.275	2.301 2.217	3.347 1.679	3.896 1.135	4.394 0.600	5.135 -0.200
6.2	1.075 2.529	1.230 2.525	1.521 2.503	1.785 2.463	2.021 2.412	2.322 2.322	3.354 1.690	3.897 1.136	4.394 0.600	5.135 -0.200
2π	1.067 2.709	1.229 2.698	1.534 2.656	1.807 2.597	2.048 2.525	2.351 2.406	3.361 1.697	3.897 1.137	4.394 0.600	5.135 -0.200
6.3	1.066 2.746	1.231 2.733	1.538 2.688	1.813 2.624	2.055 2.548	2.358 2.423	3.362 1.698	3.898 1.137	4.394 0.600	5.135 -0.200
6.4	1.072 2.973	1.249 2.950	1.575 2.878	1.862 2.785	2.109 2.680	2.410 2.517	3.372 1.704	3.899 1.137	4.394 0.600	5.135 -0.200
6.5	1.097 3.219	1.285 3.180	1.637 3.074	1.934 2.944	2.183 2.806	2.478 2.603	3.381 1.708	3.899 1.137	4.394 0.600	5.135 -0.200
6.6	1.145 3.485	1.355 3.425	1.727 3.275	2.033 3.102	2.280 2.925	2.561 2.677	3.391 1.709	3.900 1.137	4.394 0.600	5.135 -0.200
6.7	1.224 3.776	1.457 3.689	1.854 3.479	2.164 3.252	2.401 3.031	2.659 2.737	3.400 1.709	3.901 1.137	4.394 0.600	5.135 -0.200
6.8	1.349 4.101	1.607 3.973	2.025 3.684	2.329 3.389	2.547 3.119	2.769 2.779	3.409 1.707	3.901 1.136	4.394 0.600	5.135 -0.200
6.9	1.540 4.464	1.824 4.277	2.253 3.879	2.534 3.504	2.718 3.181	2.888 2.798	3.416 1.704	3.902 1.136	4.395 0.600	5.135 -0.200
7.0	1.833 4.868	2.138 4.551	2.550 4.049	2.780 3.583	2.909 3.209	3.013 2.792	3.423 1.699	3.902 1.135	4.395 0.600	5.135 -0.200
7.1	2.282 5.301	2.586 4.792	2.926 4.166	3.063 3.606	3.114 3.191	3.137 2.758	3.427 1.694	3.902 1.135	4.395 0.600	5.135 -0.200
7.2	2.971 5.716	3.212 5.119	3.379 4.188	3.371 3.554	3.320 3.121	3.252 2.695	3.431 1.688	3.903 1.135	4.395 0.600	5.135 -0.200
7.3	3.996 5.973	4.039 5.158	3.882 4.062	3.676 3.409	3.510 2.995	3.349 2.605	3.433 1.682	3.903 1.134	4.395 0.600	5.135 -0.200
7.4	5.371 5.772	4.995 4.831	4.369 3.745	3.941 3.166	3.663 2.817	3.423 2.494	3.434 1.676	3.902 1.134	4.395 0.600	5.135 -0.200
7.5	6.756 4.724	5.842 4.003	4.735 3.235	4.123 2.840	3.762 2.599	3.465 2.369	3.433 1.670	3.902 1.133	4.394 0.600	5.135 -0.200
7.6	7.382 2.896	6.207 2.793	4.883 2.610	4.192 2.472	3.795 2.365	3.475 2.240	3.432 1.665	3.902 1.133	4.394 0.600	5.135 -0.200
7.7	6.875 1.134	5.945 1.608	4.785 1.994	4.140 2.111	3.761 2.138	3.454 2.116	3.430 1.661	3.902 1.133	4.394 0.600	5.135 -0.200
7.8	5.766 0.103	5.274 0.788	4.454 1.499	3.988 1.803	3.670 1.938	3.405 2.006	3.427 1.657	3.902 1.133	4.394 0.600	5.135 -0.200
5π/2	5.146 -0.156	4.856 0.523	4.250 1.300	3.877 1.668	3.603 1.847	3.369 1.954	3.425 1.655	3.902 1.133	4.394 0.600	5.135 -0.200
7.9	4.660 -0.259	4.505 0.381	4.105 1.170	3.772 1.572	3.539 1.780	3.334 1.915	3.423 1.654	3.902 1.133	4.394 0.600	5.135 -0.200
8.0	3.781 -0.249	3.816 0.267	3.699 0.996	3.527 1.423	3.384 1.668	3.250 1.847	3.420 1.652	3.901 1.133	4.394 0.600	5.135 -0.200
8.1	3.130 -0.073	3.258 0.321	3.323 0.941	3.281 1.348	3.221 1.601	3.159 1.802	3.416 1.651	3.901 1.133	4.394 0.600	5.135 -0.200
8.2	2.653 0.162	2.822 0.460	2.996 0.967	3.051 1.332	3.062 1.574	3.066 1.780	3.412 1.651	3.901 1.133	4.394 0.600	5.135 -0.200
8.3	2.301 0.410	2.485 0.637	2.722 1.045	2.846 1.359	2.913 1.580	2.976 1.776	3.409 1.651	3.901 1.133	4.394 0.600	5.135 -0.200
8.4	2.036 0.654	2.223 0.828	2.496 1.153	2.668 1.418	2.780 1.612	2.892 1.790	3.406 1.652	3.901 1.133	4.394 0.600	5.135 -0.200
8.5	1.832 0.887	2.019 1.021	2.312 1.280	2.518 1.498	2.664 1.663	2.818 1.817	3.403 1.654	3.901 1.133	4.394 0.600	5.135 -0.200
8.6	1.674 1.107	1.859 1.213	2.164 1.417	2.393 1.593	2.565 1.729	2.753 1.856	3.401 1.656	3.901 1.133	4.394 0.600	5.135 -0.200
8.7	1.549 1.318	1.732 1.400	2.044 1.560	2.291 1.698	2.483 1.805	2.699 1.903	3.400 1.658	3.901 1.133	4.394 0.600	5.135 -0.200
8.8	1.449 1.519	1.631 1.583	1.950 1.705	2.210 1.810	2.418 1.889	2.656 1.956	3.398 1.660	3.901 1.133	4.394 0.600	5.135 -0.200
8.9	1.370 1.714	1.552 1.762	1.878 1.853	2.148 1.926	2.369 1.978	2.624 2.014	3.398 1.662	3.901 1.133	4.394 0.600	5.135 -0.200
9.0	1.307 1.905	1.492 1.940	1.824 2.001	2.104 2.045	2.335 2.070	2.603 2.074	3.398 1.664	3.901 1.133	4.394 0.600	5.135 -0.200
9.1	1.259 2.094	1.447 2.118	1.789 2.150	2.077 2.165	2.316 2.163	2.592 2.135	3.398 1.666	3.901 1.134	4.394 0.600	5.135 -0.200
9.2	1.224 2.285	1.418 2.296	1.769 2.301	2.067 2.286	2.311 2.257	2.592 2.196	3.399 1.668	3.901 1.134	4.394 0.600	5.135 -0.200
9.3	1.201 2.479	1.403 2.478	1.767 2.454	2.072 2.408	2.320 2.350	2.603 2.254	3.400 1.670	3.901 1.134	4.394 0.600	5.135 -0.200
9.4	1.191 2.679	1.403 2.665	1.782 2.608	2.094 2.528	2.344 2.440	2.623 2.310	3.401 1.671	3.901 1.134	4.394 0.600	5.135 -0.200
3π	1.190 2.730	1.406 2.712	1.788 2.646	2.102 2.558	2.352 2.462	2.629 2.323	3.401 1.671	3.901 1.134	4.394 0.600	5.135 -0.200
9.5	1.195 2.889	1.420 2.858	1.816 2.764	2.134 2.646	2.382 2.526	2.652 2.360	3.402 1.672	3.901 1.134	4.394 0.600	5.135 -0.200
9.6	1.216 3.112	1.458 3.061	1.871 2.921	2.192 2.761	2.434 2.606	2.690 2.405	3.403 1.673	3.901 1.134	4.394 0.500	5.135 -0.200
9.7	1.258 3.351	1.519 3.275	1.952 3.078	2.270 2.869	2.500 2.678	2.735 2.442	3.405 1.673	3.901 1.134	4.394 0.600	5.135 -0.200
9.8	1.328 3.613	1.613 3.501	2.061 3.233	2.369 2.968	2.580 2.738	2.787 2.470	3.406 1.673	3.901 1.134	4.394 0.500	5.135 -0.200
9.9	1.438 3.901	1.747 3.739	2.203 3.382	2.491 3.052	2.673 2.784	2.843 2.487	3.407 1.673	3.901 1.134	4.394 0.600	5.135 -0.200
10.0	1.603 4.219	1.938 3.988	2.386 3.515	2.634 3.116	2.776 2.812	2.902 2.493	3.409 1.672	3.901 1.134	4.394 0.600	5.135 -0.200
10.1	1.851 4.570	2.203 4.238	2.612 3.622	2.796 3.153	2.886 2.818					
10.2	2.224 4.944	2.568 4.469	2.883 3.685	2.974 3.153	2.998 2.801					
10.3	2.783 5.308	3.059 4.639	3.193 3.682	3.157 3.110	3.107 2.757					
10.4	3.604 5.569	3.685 4.674	3.523 3.588	3.333 3.018	3.205 2.688					
10.5	4.721 5.521	4.406 4.466	3.837 3.385	3.484 2.878	3.284 2.596					
10.6	5.963 4.856	5.082 3.923	4.086 3.076	3.595 2.699	3.339 2.487					
10.7	6.795 3.457	5.491 3.074	4.223 2.691	3.652 2.494	3.365 2.369					
10.8	6.712 1.816	5.476 2.132	4.223 2.285	3.650 2.285	3.361 2.250					
10.9	5.895 0.628	5.089 1.356	4.096 1.920	3.592 2.090	3.329 2.139					
7π/2	4.929 0.085	4.545 0.884	3.890 1.645	3.495 1.933	3.276 2.047					

βh	α/β = 0.00		α/β = 0.01		α/β = 0.03		α/β = 0.05		α/β = 0.07	
11.0	4.886	0.071	4.519	0.868	3.879	1.634	3.490	1.927	3.273	2.043
11.1	4.008	-0.053	3.935	0.643	3.617	1.442	3.358	1.803	3.200	1.966
11.2	3.330	0.042	3.421	0.600	3.345	1.338	3.211	1.720	3.116	1.910
11.3	2.825	0.231	2.998	0.663	3.088	1.304	3.061	1.677	3.027	1.877
11.4	2.449	0.452	2.660	0.785	2.858	1.323	2.917	1.666	2.938	1.863
11.5	2.167	0.678	2.393	0.936	2.659	1.379	2.785	1.683	2.853	1.867
11.6	1.951	C.898	2.181	1.099	2.490	1.460	2.667	1.722	2.775	1.886
11.7	1.783	1.109	2.013	1.266	2.350	1.557	2.564	1.777	2.706	1.917
11.8	1.652	1.311	1.880	1.434	2.234	1.666	2.477	1.844	2.646	1.959
11.9	1.547	1.504	1.774	1.601	2.141	1.781	2.406	1.920	2.596	2.009
12.0	1.464	1.692	1.691	1.766	2.068	1.901	2.351	2.003	2.557	2.064
12.1	1.399	1.875	1.627	1.930	2.014	2.025	2.310	2.090	2.529	2.123
12.2	1.348	2.056	1.579	2.094	1.976	2.151	2.283	2.180	2.511	2.184
12.3	1.310	2.237	1.547	2.259	1.954	2.278	2.270	2.272	2.505	2.246
12.4	1.285	2.421	1.529	2.426	1.949	2.407	2.271	2.363	2.508	2.308
12.5	1.272	2.610	1.527	2.597	1.960	2.537	2.286	2.454	2.522	2.368
4π	1.271	2.740	1.534	2.714	1.976	2.623	2.303	2.513	2.537	2.406
12.6	1.273	2.807	1.541	2.774	1.988	2.667	2.315	2.542	2.546	2.425
12.7	1.289	3.014	1.574	2.957	2.034	2.796	2.358	2.626	2.580	2.476
12.8	1.325	3.236	1.630	3.148	2.102	2.923	2.416	2.703	2.622	2.522
12.9	1.385	3.476	1.713	3.347	2.193	3.045	2.489	2.772	2.673	2.559
13.0	1.480	3.738	1.832	3.554	2.309	3.158	2.575	2.828	2.731	2.586
13.1	1.622	4.026	1.998	3.766	2.454	3.256	2.675	2.868	2.793	2.601
13.2	1.833	4.341	2.224	3.975	2.627	3.330	2.785	2.889	2.859	2.604
13.3	2.146	4.678	2.527	4.166	2.829	3.370	2.901	2.886	2.925	2.593
13.4	2.609	5.014	2.924	4.307	3.052	3.364	3.018	2.856	2.987	2.568
13.5	3.284	5.284	3.420	4.349	3.284	3.297	3.129	2.799	3.044	2.530
13.6	4.217	5.345	3.991	4.219	3.505	3.162	3.226	2.714	3.090	2.479
13.7	5.332	4.951	4.553	3.850	3.687	2.961	3.300	2.607	3.124	2.420
13.8	6.276	3.905	4.961	3.237	3.805	2.707	3.345	2.484	3.143	2.355
13.9	6.536	2.428	5.082	2.490	3.841	2.429	3.357	2.354	3.147	2.289
14.0	6.015	1.136	4.893	1.789	3.792	2.160	3.336	2.228	3.136	2.224
14.1	5.125	0.391	4.492	1.273	3.674	1.928	3.287	2.115	3.111	2.166
9π/2	4.786	0.246	4.317	1.137	3.616	1.855	3.263	2.078	3.099	2.146
14.2	4.252	0.122	4.014	0.972	3.508	1.750	3.215	2.022	3.075	2.116
14.3	3.542	0.131	3.553	0.849	3.318	1.632	3.128	1.951	3.031	2.076
14.4	3.002	0.271	3.150	0.845	3.123	1.568	3.032	1.904	2.981	2.048
14.5	2.598	0.466	2.815	0.916	2.937	1.550	2.934	1.880	2.928	2.031
14.6	2.293	0.676	2.543	1.028	2.768	1.569	2.839	1.876	2.876	2.026
14.7	2.061	0.887	2.325	1.162	2.619	1.614	2.751	1.890	2.825	2.031
14.8	1.881	1.091	2.150	1.307	2.490	1.680	2.670	1.919	2.779	2.045
14.9	1.741	1.287	2.009	1.457	2.381	1.759	2.600	1.960	2.737	2.067
15.0	1.630	1.476	1.897	1.609	2.292	1.849	2.541	2.011	2.702	2.095
15.1	1.541	1.658	1.809	1.761	2.220	1.946	2.493	2.068	2.673	2.128
15.2	1.472	1.836	1.740	1.914	2.165	2.047	2.456	2.131	2.651	2.165
15.3	1.417	2.011	1.689	2.067	2.126	2.152	2.431	2.197	2.636	2.204
15.4	1.376	2.186	1.653	2.220	2.102	2.260	2.417	2.265	2.629	2.244
15.5	1.348	2.362	1.632	2.376	2.093	2.369	2.414	2.333	2.629	2.285
15.6	1.332	2.543	1.626	2.535	2.098	2.479	2.423	2.401	2.636	2.324
15.7	1.329	2.729	1.636	2.697	2.120	2.588	2.442	2.467	2.650	2.362
5π	1.329	2.745	1.638	2.711	2.122	2.597	2.444	2.472	2.651	2.364
15.8	1.340	2.925	1.664	2.865	2.157	2.696	2.473	2.529	2.670	2.395
15.9	1.370	3.133	1.713	3.038	2.213	2.800	2.514	2.586	2.697	2.425
16.0	1.421	3.357	1.786	3.217	2.286	2.898	2.566	2.635	2.728	2.449
16.1	1.502	3.600	1.891	3.401	2.380	2.988	2.627	2.675	2.764	2.467
16.2	1.624	3.864	2.035	3.587	2.494	3.063	2.697	2.703	2.803	2.477
16.3	1.804	4.153	2.229	3.768	2.629	3.119	2.772	2.717	2.843	2.479
16.4	2.069	4.463	2.484	3.932	2.782	3.149	2.851	2.716	2.882	2.474
[illegible]	[illegible]	[illegible]	2.812	4.057	2.949	3.144	2.930	2.697	2.920	2.460
16.6	3.020	5.055	3.217	4.105	3.120	3.099	3.004	[illegible]	[illegible]	[illegible]
16.7	3.807	5.190	3.684	4.027	3.283	3.007	3.069	2.608	2.983	2.410
16.8	4.801	4.993	4.158	3.773	3.422	2.870	3.121	2.541	3.005	2.376
16.9	5.782	4.246	4.545	3.324	3.521	2.696	3.156	2.463	3.019	2.339
17.0	6.304	2.978	4.735	2.734	3.569	2.499	3.171	2.380	3.025	2.300
17.1	6.086	1.650	4.687	2.124	3.559	2.300	3.167	2.297	3.022	2.262
17.2	5.353	0.734	4.427	1.615	3.498	2.117	3.143	2.219	3.011	2.226
11π/2	4.681	0.365	4.137	1.328	3.419	1.993	3.113	2.164	2.997	2.200
17.3	4.506	0.309	4.053	1.269	3.395	1.964	3.104	2.150	2.992	2.194
17.4	3.768	0.218	3.651	1.082	3.265	1.849	3.052	2.094	2.969	2.167
17.5	3.189	0.303	3.277	1.020	3.121	1.774	2.991	2.053	2.941	2.146
17.6	2.751	0.468	2.951	1.042	2.976	1.736	2.926	2.027	2.910	2.132
17.7	2.421	0.662	2.679	1.116	2.836	1.730	2.859	2.015	2.878	2.125
17.8	2.169	0.863	2.456	1.222	2.708	1.750	2.795	2.016	2.846	2.124
17.9	1.975	1.062	2.274	1.345	2.594	1.790	2.734	2.030	2.816	2.129
18.0	1.824	1.254	2.127	1.477	2.495	1.845	2.680	2.053	2.788	2.140
18.1	1.704	1.439	2.009	1.615	2.411	1.913	2.632	2.085	2.763	2.156
18.2	1.610	1.618	1.915	1.755	2.343	1.989	2.592	2.124	2.742	2.175
18.3	1.535	1.792	1.841	1.897	2.289	2.071	2.561	2.167	2.726	2.198
18.4	1.476	1.963	1.785	2.040	2.249	2.158	2.538	2.215	2.714	2.222
18.5	1.432	2.133	1.745	2.184	2.223	2.248	2.524	2.265	2.708	2.248
18.6	1.400	2.303	1.720	2.329	2.210	2.340	2.519	2.316	2.706	2.274
18.7	1.380	2.477	1.710	2.477	2.211	2.433	2.522	2.366	2.709	2.300
18.8	1.373	2.656	1.716	2.629	2.226	2.525	2.534	2.415	2.717	2.324
6π	1.374	2.748	1.725	2.705	2.238	2.570	2.544	2.439	2.722	2.335
18.9	1.380	2.843	1.738	2.783	2.255	2.615	2.555	2.461	2.729	2.346
19.0	1.403	3.040	1.780	2.942	2.298	2.702	2.584	2.504	2.745	2.366
19.1	1.446	3.250	1.844	3.105	2.358	2.783	2.620	2.540	2.764	2.382
19.2	1.515	3.477	1.936	3.271	2.432	2.856	2.663	2.569	2.786	2.394
19.3	1.619	3.724	2.062	3.437	2.523	2.916	2.712	2.590	2.810	2.401
19.4	1.774	3.992	2.229	3.598	2.629	2.962	2.764	2.602	2.834	2.404
19.5	1.998	4.281	2.446	3.744	2.747	2.986	2.819	2.602	2.859	2.401
19.6	2.326	4.579	2.721	3.857	2.874	2.985	2.873	2.591	2.883	2.394
19.7	2.800	4.857	3.059	3.911	3.004	2.954	2.925	2.569	2.904	2.382
19.8	3.468	5.041	3.447	3.869	3.129	2.890	2.970	2.535	2.923	2.366
19.9	4.344	4.987	3.852	3.693	3.238	2.795	3.008	2.492	2.937	2.347
20.0	5.307	4.489	4.210	3.362	3.321	2.672	3.035	2.442	2.947	2.325
20.1	6.010	3.456	4.437	2.899	3.370	2.531	3.050	2.387	2.952	2.302
20.2	6.084	2.168	4.478	2.380	3.380	2.382	3.052	2.331	2.952	2.278
20.3	5.550	1.114	4.331	1.902	3.351	2.239	3.042	2.277	2.948	2.256
20.4	4.762	0.525	4.054	1.536	3.289	2.112	3.020	2.227	2.939	2.235
13π/2	4.599	0.457	3.988	1.478	3.273	2.089	3.015	2.218	2.936	2.231
20.5	4.006	0.319	3.718	1.305	3.202	2.009	2.989	2.185	2.926	2.217
20.6	3.388	0.337	3.379	1.192	3.098	1.933	2.951	2.151	2.910	2.203
20.7	2.913	0.466	3.069	1.168	2.987	1.886	2.908	2.127	2.892	2.192
20.8	2.553	0.642	2.801	1.206	2.875	1.865	2.862	2.113	2.872	2.185
20.9	2.278	0.833	2.575	1.283	2.768	1.867	2.816	2.108	2.853	2.183

βh	α/β = 0.00		α/β = 0.01		α/β = 0.03		α/β = 0.05		α/β = 0.07	
21.0	2.068	1.026	2.389	1.384	2.670	1.888	2.771	2.113	2.833	2.184
21.1	1.904	1.215	2.236	1.499	2.582	1.924	2.730	2.125	2.815	2.189
21.2	1.775	1.397	2.112	1.622	2.505	1.973	2.693	2.145	2.798	2.198
21.3	1.673	1.573	2.012	1.750	2.441	2.032	2.661	2.170	2.784	2.209
21.4	1.593	1.745	1.933	1.882	2.389	2.097	2.635	2.200	2.772	2.222
21.5	1.529	1.913	1.872	2.015	2.350	2.168	2.615	2.234	2.763	2.237
21.6	1.480	2.079	1.828	2.150	2.322	2.243	2.602	2.269	2.758	2.253
21.7	1.445	2.245	1.799	2.286	2.307	2.320	2.595	2.307	2.755	2.270
21.8	1.421	2.414	1.785	2.425	2.304	2.397	2.595	2.344	2.756	2.287
21.9	1.410	2.586	1.785	2.566	2.313	2.475	2.602	2.380	2.760	2.303
7π	1.411	2.749	1.800	2.698	2.332	2.544	2.614	2.412	2.766	2.316
22.0	1.412	2.765	1.802	2.711	2.334	2.551	2.616	2.415	2.767	2.317
22.1	1.429	2.953	1.837	2.858	2.368	2.624	2.635	2.446	2.776	2.330
22.2	1.464	3.153	1.893	3.007	2.415	2.692	2.660	2.474	2.788	2.341
22.3	1.523	3.367	1.973	3.159	2.474	2.752	2.690	2.496	2.801	2.350
22.4	1.612	3.599	2.082	3.309	2.546	2.803	2.724	2.513	2.816	2.355
22.5	1.744	3.850	2.227	3.455	2.629	2.840	2.761	2.522	2.831	2.357
22.6	1.936	4.120	2.413	3.586	2.721	2.862	2.799	2.524	2.847	2.357
22.7	2.215	4.404	2.648	3.691	2.820	2.864	2.837	2.518	2.862	2.353
22.8	2.616	4.679	2.932	3.749	2.921	2.844	2.874	2.504	2.876	2.346
22.9	3.185	4.895	3.261	3.733	3.018	2.800	2.907	2.483	2.888	2.337
23.0	3.950	4.942	3.610	3.613	3.105	2.733	2.934	2.455	2.897	2.326
23.1	4.858	4.645	3.935	3.369	3.175	2.645	2.955	2.422	2.904	2.313
23.2	5.667	3.851	4.174	3.006	3.221	2.541	2.968	2.386	2.909	2.299
23.3	5.997	2.671	4.273	2.572	3.240	2.430	2.973	2.347	2.910	2.285
23.4	5.697	1.532	4.214	2.138	3.231	2.318	2.970	2.310	2.908	2.271
23.5	5.007	0.782	4.023	1.774	3.195	2.214	2.958	2.274	2.904	2.257
15π/2	4.533	0.532	3.861	1.600	3.162	2.157	2.948	2.254	2.900	2.250
23.6	4.253	0.444	3.753	1.514	3.138	2.125	2.940	2.242	2.897	2.246
23.7	3.600	0.382	3.456	1.361	3.065	2.055	2.916	2.216	2.888	2.236
23.8	3.086	0.467	3.168	1.297	2.981	2.006	2.887	2.195	2.877	2.228
23.9	2.692	0.620	2.909	1.300	2.893	1.977	2.856	2.182	2.866	2.223
24.0	2.392	0.800	2.685	1.348	2.806	1.967	2.824	2.175	2.853	2.220
24.1	2.162	0.987	2.495	1.426	2.723	1.974	2.792	2.175	2.841	2.220
24.2	1.984	1.172	2.338	1.523	2.647	1.996	2.761	2.180	2.829	2.222
24.3	1.845	1.352	2.208	1.632	2.579	2.030	2.733	2.192	2.818	2.226
24.4	1.734	1.527	2.103	1.748	2.520	2.074	2.707	2.208	2.808	2.232
24.5	1.647	1.696	2.019	1.869	2.471	2.125	2.686	2.228	2.800	2.240
24.6	1.579	1.862	1.954	1.993	2.433	2.182	2.670	2.251	2.794	2.249
24.7	1.525	2.025	1.905	2.119	2.405	2.243	2.658	2.277	2.789	2.259
24.8	1.485	2.188	1.871	2.247	2.388	2.306	2.651	2.304	2.787	2.270
24.9	1.458	2.352	1.852	2.378	2.381	2.372	2.649	2.331	2.787	2.280
25.0	1.442	2.519	1.848	2.510	2.385	2.437	2.653	2.358	2.788	2.291

TABLE 2.7

NORMALIZED ADMITTANCES (Y/Δ) IN MILLIMHOS OF THIN DIPOLE ANTENNAS IN DISSIPATIVE MEDIA

Ω = 10.0

βh	α/β = 0.00		α/β = 0.01		α/β = 0.03		α/β = 0.05		α/β = 0.07		α/β = 0.10		α/β = 0.30		α/β = 0.50		α/β = 0.70		α/β = 1.00	
1.5	10.928	-2.047	10.302	-1.557	9.205	-0.869	8.298	-0.437	7.552	-0.164	6.668	0.071	4.129	0.148	3.493	-0.164	3.357	-0.509	3.476	-1.062
π/2	7.833	-3.623	7.657	-3.138	7.250	-2.347	6.823	-1.753	6.413	-1.310	5.861	-0.848	3.930	-0.086	3.408	-0.232	3.325	-0.523	3.491	-1.054
1.6	6.800	-3.733	6.724	-3.308	6.496	-2.585	6.217	-2.013	5.922	-1.568	5.497	-1.082	3.840	-0.157	3.372	-0.251	3.313	-0.525	3.498	-1.049
1.7	4.416	-3.309	4.475	-3.065	4.534	-2.609	4.529	-2.204	4.480	-1.853	4.359	-1.421	3.529	-0.305	3.255	-0.279	3.279	-0.512	3.528	-1.026
1.8	3.142	-2.618	3.222	-2.475	3.347	-2.194	3.430	-1.928	3.479	-1.682	3.503	-1.358	3.245	-0.336	3.154	-0.263	3.258	-0.480	3.567	-0.996
1.9	2.404	-1.995	2.482	-1.905	2.617	-1.723	2.725	-1.544	2.809	-1.373	2.898	-1.138	3.008	-0.289	3.073	-0.217	3.250	-0.434	3.613	-0.963
2.0	1.939	-1.470	2.012	-1.409	2.143	-1.285	2.257	-1.161	2.355	-1.039	2.474	-0.867	2.820	-0.195	3.015	-0.150	3.255	-0.381	3.664	-0.928
2.1	1.626	-1.025	1.694	-0.981	1.820	-0.893	1.935	-0.804	2.038	-0.716	2.172	-0.589	2.677	-0.073	2.978	-0.070	3.272	-0.323	3.720	-0.893
2.2	1.404	-0.639	1.468	-0.607	1.591	-0.542	1.705	-0.477	1.811	-0.411	1.954	-0.317	2.574	0.064	2.960	0.016	3.300	-0.264	3.779	-0.859
2.3	1.239	-0.298	1.302	-0.273	1.423	-0.224	1.537	-0.175	1.645	-0.126	1.795	-0.057	2.502	0.208	2.959	0.105	3.335	-0.206	3.841	-0.827
2.4	1.113	0.013	1.176	0.031	1.297	0.069	1.413	0.106	1.523	0.142	1.679	0.193	2.459	0.354	2.972	0.194	3.378	-0.149	3.904	-0.795
2.5	1.015	0.300	1.078	0.315	1.201	0.343	1.320	0.371	1.434	0.397	1.597	0.432	2.439	0.499	2.998	0.281	3.425	-0.096	3.968	-0.765
2.6	0.937	0.573	1.002	0.584	1.128	0.605	1.251	0.625	1.369	0.642	1.540	0.664	2.438	0.642	3.033	0.364	3.477	-0.045	4.032	-0.736
2.7	0.875	0.835	0.942	0.843	1.073	0.858	1.201	0.871	1.326	0.881	1.505	0.890	2.456	0.780	3.077	0.442	3.532	0.002	4.096	-0.708
2.8	0.825	1.091	0.895	1.097	1.034	1.107	1.169	1.113	1.300	1.115	1.489	1.112	2.489	0.912	3.128	0.514	3.588	0.046	4.160	-0.681
2.9	0.786	1.346	0.861	1.350	1.008	1.354	1.151	1.353	1.290	1.348	1.491	1.332	2.536	1.038	3.184	0.580	3.644	0.087	4.222	-0.655
3.0	0.757	1.604	0.837	1.605	0.995	1.603	1.148	1.595	1.297	1.581	1.510	1.550	2.595	1.155	3.243	0.641	3.701	0.125	4.285	-0.628
3.1	0.738	1.869	0.825	1.867	0.996	1.857	1.161	1.840	1.321	1.816	1.548	1.769	2.665	1.263	3.303	0.695	3.757	0.161	4.346	-0.602
π	0.734	1.982	0.824	1.979	1.001	1.965	1.172	1.944	1.336	1.915	1.570	1.860	2.697	1.305	3.329	0.715	3.781	0.175	4.372	-0.591
3.2	0.730	2.144	0.825	2.139	1.012	2.120	1.192	2.092	1.364	2.056	1.606	1.988	2.744	1.360	3.365	0.743	3.813	0.194	4.407	-0.575
3.3	0.735	2.436	0.840	2.425	1.046	2.395	1.242	2.352	1.428	2.300	1.687	2.207	2.831	1.446	3.426	0.786	3.867	0.227	4.467	-0.548
3.4	0.755	2.749	0.874	2.731	1.103	2.685	1.318	2.623	1.520	2.550	1.793	2.424	2.923	1.519	3.486	0.823	3.919	0.258	4.527	-0.521
3.5	0.798	3.091	0.932	3.064	1.189	2.993	1.427	2.906	1.644	2.805	1.931	2.638	3.017	1.578	3.543	0.857	3.970	0.288	4.586	-0.493
3.6	0.872	3.471	1.027	3.429	1.317	3.324	1.578	3.199	1.809	3.061	2.105	2.842	3.112	1.624	3.597	0.888	4.020	0.319	4.645	-0.465
3.7	0.994	3.900	1.174	3.834	1.501	3.678	1.785	3.500	2.026	3.311	2.319	3.027	3.204	1.656	3.648	0.916	4.068	0.349	4.703	-0.436
3.8	1.191	4.390	1.401	4.286	1.767	4.051	2.067	3.798	2.306	3.543	2.577	3.181	3.291	1.675	3.696	0.943	4.115	0.380	4.762	-0.407
3.9	1.510	4.953	1.752	4.785	2.148	4.428	2.444	4.070	2.659	3.733	2.877	3.285	3.369	1.685	3.740	0.969	4.161	0.412	4.820	-0.376
4.0	2.035	5.582	2.299	5.308	2.687	4.770	2.935	4.277	3.087	3.847	3.210	3.318	3.437	1.685	3.781	0.996	4.206	0.444	4.878	-0.345
4.1	2.902	6.220	3.141	5.778	3.422	4.991	3.539	4.350	3.571	3.839	3.552	3.259	3.493	1.681	3.819	1.023	4.250	0.477	4.936	-0.314
4.2	4.287	6.652	4.367	5.986	4.339	4.947	4.210	4.205	4.063	3.666	3.867	3.098	3.537	1.674	3.854	1.052	4.295	0.512	4.994	-0.281
4.3	6.206	6.345	5.878	5.551	5.291	4.468	4.830	3.781	4.479	3.315	4.110	2.846	3.569	1.668	3.888	1.082	4.338	0.547	5.052	-0.248
4.4	7.941	4.686	7.110	4.183	5.965	3.524	5.229	3.112	4.731	2.829	4.246	2.535	3.591	1.666	3.920	1.115	4.382	0.583	5.111	-0.214
4.5	8.224	2.201	7.321	2.315	6.083	2.378	5.296	2.355	4.769	2.305	4.261	2.216	3.603	1.669	3.951	1.149	4.426	0.621	5.170	-0.180
4.6	7.101	0.373	6.551	0.852	5.678	1.417	5.055	1.701	4.612	1.846	4.169	1.937	3.609	1.679	3.982	1.186	4.469	0.659	5.228	-0.145
4.7	5.658	-0.374	5.451	0.131	5.015	0.832	4.633	1.256	4.328	1.515	4.000	1.727	3.609	1.697	4.012	1.225	4.513	0.698	5.288	-0.109
3π/2	5.493	-0.411	5.317	0.085	4.928	0.785	4.575	1.217	4.288	1.484	3.976	1.707	3.609	1.699	4.016	1.230	4.519	0.703	5.295	-0.105
4.8	4.471	-0.459	4.458	-0.050	4.332	0.587	4.159	1.027	3.993	1.325	3.792	1.598	3.607	1.723	4.042	1.265	4.558	0.738	5.347	-0.073
4.9	3.607	-0.256	3.685	0.051	3.737	0.570	3.712	0.966	3.658	1.256	3.574	1.544	3.604	1.756	4.073	1.308	4.602	0.778	5.407	-0.035
5.0	2.992	0.049	3.108	0.276	3.255	0.684	3.326	1.017	3.355	1.278	3.365	1.552	3.601	1.797	4.104	1.352	4.647	0.820	5.467	0.003
5.1	2.549	0.379	2.679	0.549	2.876	0.866	3.006	1.139	3.093	1.362	3.176	1.609	3.600	1.845	4.136	1.397	4.692	0.862	5.527	0.041
5.2	2.222	0.706	2.358	0.835	2.580	1.082	2.747	1.302	2.873	1.488	3.013	1.701	3.602	1.898	4.169	1.443	4.737	0.905	5.588	0.081
5.3	1.975	1.021	2.113	1.121	2.348	1.314	2.539	1.490	2.694	1.642	2.876	1.820	3.607	1.956	4.202	1.491	4.783	0.949	5.649	0.121
5.4	1.786	1.323	1.923	1.400	2.168	1.551	2.375	1.691	2.550	1.813	2.765	1.956	3.616	2.017	4.237	1.539	4.829	0.993	5.711	0.162
5.5	1.638	1.612	1.776	1.673	2.027	1.791	2.246	1.900	2.436	1.994	2.677	2.104	3.628	2.081	4.272	1.589	4.876	1.038	5.773	0.204
5.6	1.522	1.892	1.661	1.940	1.917	2.031	2.147	2.113	2.350	2.183	2.612	2.261	3.645	2.147	4.308	1.638	4.923	1.084	5.836	0.246
5.7	1.430	2.166	1.571	2.202	1.835	2.270	2.074	2.329	2.288	2.376	2.567	2.424	3.666	2.213	4.344	1.689	4.970	1.131	5.899	0.290
5.8	1.359	2.437	1.503	2.463	1.775	2.510	2.023	2.547	2.248	2.573	2.542	2.590	3.690	2.281	4.381	1.740	5.018	1.178	5.963	0.334
5.9	1.305	2.707	1.454	2.725	1.736	2.751	1.994	2.767	2.228	2.771	2.534	2.758	3.717	2.348	4.419	1.791	5.066	1.227	6.028	0.379
6.0	1.267	2.980	1.423	2.988	1.716	2.995	1.985	2.988	2.228	2.970	2.545	2.926	3.747	2.414	4.457	1.844	5.114	1.276	6.093	0.425
6.1	1.245	3.258	1.409	3.257	1.717	3.242	1.997	3.212	2.248	3.170	2.572	3.092	3.780	2.479	4.496	1.896	5.163	1.326	6.158	0.473
6.2	1.240	3.545	1.414	3.533	1.738	3.493	2.029	3.436	2.288	3.368	2.617	3.255	3.814	2.543	4.534	1.950	5.213	1.376	6.225	0.521
2π	1.250	3.792	1.433	3.770	1.773	3.706	2.074	3.624	2.337	3.531	2.668	3.386	3.844	2.596	4.567	1.994	5.254	1.419	6.281	0.561
6.3	1.254	3.843	1.439	3.818	1.782	3.749	2.085	3.662	2.349	3.564	2.679	3.412	3.850	2.606	4.574	2.003	5.263	1.428	6.292	0.570
6.4	1.290	4.157	1.490	4.116	1.853	4.010	2.166	3.886	2.432	3.755	2.758	3.561	3.886	2.667	4.613	2.058	5.313	1.481	6.360	0.620
6.5	1.356	4.491	1.572	4.427	1.955	4.275	2.275	4.106	2.540	3.937	2.853	3.700	3.923	2.727	4.653	2.114	5.364	1.535	6.429	0.671
6.6	1.460	4.846	1.694	4.752	2.095	4.540	2.416	4.318	2.671	4.106	2.963	3.824	3.960	2.786	4.693	2.170	5.415	1.589	6.498	0.723
6.7	1.617	5.226	1.869	5.090	2.281	4.799	2.592	4.515	2.828	4.257	3.086	3.930	3.996	2.844	4.733	2.227	5.467	1.645	6.569	0.776
6.8	1.849	5.628	2.114	5.433	2.522	5.043	2.805	4.686	3.007	4.381	3.218	4.015	4.031	2.901	4.773	2.285	5.519	1.702	6.640	0.830
6.9	2.184	6.043	2.451	5.767	2.824	5.252	3.054	4.819	3.204	4.471	3.355	4.075	4.065	2.958	4.814	2.345	5.573	1.760	6.712	0.886
7.0	2.664	6.442	2.903	6.058	3.188	5.402	3.331	4.898	3.411	4.518	3.489	4.108	4.098	3.016	4.855	2.405	5.626	1.819	6.786	0.943
7.1	3.331	6.760	3.485	6.249	3.605	5.458	3.622	4.907	3.615	4.517	3.615	4.115	4.130	3.074	4.896	2.467	5.681	1.880	6.860	1.001
7.2	4.201	6.878	4.179	6.255	4.042	5.387	3.903	4.837	3.802	4.467	3.723	4.099	4.161	3.133	4.938	2.530	5.736	1.941	6.935	1.060
7.3	5.194	6.623	4.900	5.983	4.443	5.170	4.141	4.688	3.953	4.373	3.808	4.064	4.191	3.193	4.980	2.594	5.792	2.004	7.012	1.121
7.4	6.061	5.880	5.483	5.405	4.737	4.820	4.308	4.477	4.055	4.250	3.864	4.018	4.219	3.255	5.023	2.659	5.849	2.069	7.090	1.183
7.5	6.470	4.773	5.753	4.622	4.870	4.397	4.381	4.237	4.100	4.114	3.890	3.970	4.247	3.318	5.066	2.726	5.906	2.134	7.169	1.246
7.6	6.290	3.671	5.649	3.844	4.827	3.981	4.359	4.004	4.088	3.985	3.888	3.928	4.274	3.384	5.110	2.795	5.965	2.202	7.249	1.311
7.7	5.702	2.887	5.265	3.251	4.640	3.641	4.257	3.811	4.028	3.881	3.861	3.900	4.301	3.452	5.154	2.865	6.024	2.270	7.330	1.377
7.8	4.983	2.485	4.754	2.903	4.366	3.414	4.099	3.677	3.932	3.812	3.814	3.890	4.328	3.522	5.199	2.936	6.084	2.340	7.413	1.445
5π/2	4.606	2.399	4.470	2.809	4.201	3.340	3.999	3.632	3.870	3.791	3.783	3.894	4.343	3.561	5.224	2.976	6.117	2.379	7.458	1.482
7.9	4.306	2.378	4.236	2.771	4.058	3.303	3.910	3.609	3.814	3.782	3.754	3.903	4.355	3.595	5.245	3.010	6.146	2.412	7.497	1.514
8.0	3.736	2.455	3.769	2.793	3.755	3.290	3.713	3.603	3.685	3.793	3.687	3.938	4.382	3.670	5.291	3.084	6.208	2.485	7.583	1.586
8.1	3.276	2.636	3.374	2.916	3.476	3.354	3.523	3.651	3.557	3.841	3.619	3.994	4.410	3.748	5.338	3.161	6.271	2.560	7.670	1.658
8.2	2.913	2.871	3.049	3.099	3.232	3.473	3.349	3.741	3.436	3.921	3.552	4.070	4.438	3.828	5.386	3.239	6.336	2.637	7.759	1.733
8.3	2.625	3.134	2.785	3.318	3.025	3.632	3.195	3.865	3.327	4.027	3.492	4.164	4.468	3.910	5.435	3.319	6.401	2.715	7.850	1.809
8.4	2.398	3.411	2.572	3.559	2.851	3.817	3.063	4.015	3.232	4.154	3.439	4.273	4.498	3.995	5.485	3.401	6.468	2.796	7.942	1.887
8.5	2.218	3.693	2.402	3.812	2.710	4.022	2.954	4.184	3.152	4.299	3.395	4.395	4.529	4.081	5.535	3.484	6.536	2.878	8.036	1.968
8.6	2.075	3.979	2.267	4.074	2.597	4.239	2.866	4.368	3.089	4.457	3.362	4.527	4.561	4.170	5.586	3.570	6.605	2.962	8.132	2.050
8.7	1.963	4.267	2.161	4.340	2.509	4.467	2.799	4.562	3.042	4.625	3.340	4.667	4.594	4.261	5.639	3.658	6.675	3.048	8.230	2.134
8.8	1.876	4.558	2.081	4.612	2.445	4.702	2.752	4.765	3.011	4.801	3.328	4.812	4.629	4.355	5.692	3.748	6.747	3.137	8.330	2.220
8.9	1.813	4.851	2.024	4.888	2.403	4.942	2.724	4.973	2.996	4.982	3.328	4.962	4.664	4.450	5.746	3.840	6.821	3.227	8.432	2.309
9.0	1.770	5.149	1.989	5.168	2.382	5.188	2.715	5.186	2.996	5.167	3.338	5.115	4.700	4.547	5.802	3.934	6.896	3.320	8.536	2.400
9.1	1.748	5.452	1.976	5.453	2.381	5.436	2.724	5.401	3.011	5.353	3.359	5.269	4.737	4.647	5.858	4.030	6.972	3.415	8.642	2.493
9.2	1.748	5.761	1.984	5.743	2.402	5.688	2.751	5.616	3.041	5.539	3.390	5.421	4.776	4.748	5.916	4.129	7.050	3.513	8.751	2.589
9.3	1.770	6.079	2.016	6.039	2.445	5.940	2.797	5.830	3.085	5.722	3.430	5.572	4.814	4.852	5.975	4.231	7.130	3.613	8.862	2.687
9.4	1.819	6.405	2.074	6.339	2.511	6.191	2.861	6.040	3.143	5.901	3.478	5.720	4.854	4.958	6.035	4.335	7.211	3.716	8.976	2.788
3π	1.835	6.488	2.093	6.414	2.531	6.252	2.880	6.092	3.160	5.944	3.491	5.756	4.864	4.985	6.050	4.361	7.232	3.742	9.004	2.813
9.5	1.898	6.740	2.162	6.643	2.602	6.438	2.944	6.244	3.214	6.073	3.533	5.863	4.895	5.066	6.096	4.442	7.294	3.821	9.092	2.891
9.6	2.016	7.082	2.285	6.946	2.719	6.677	3.044	6.437	3.296	6.237	3.593	6.000	4.936	5.177	6.159	4.551	7.380	3.930	9.211	2.998
9.7	2.179	7.427	2.449	7.243	2.863	6.902	3.160	6.617	3.388	6.389	3.657	6.131	4.978	5.291	6.223	4.664	7.467	4.041	9.333	3.107
9.8	2.399	7.766	2.658	7.526	3.033	7.107	3.291	6.778	3.486	6.527	3.723	6.255	5.021	5.408	6.288	4.779	7.556	4.156	9.458	3.220
9.9	2.685	8.084	2.917	7.780	3.227	7.282	3.431	6.917	3.587	6.651	3.789	6.371	5.064	5.527	6.355	4.898	7.647	4.273	9.585	3.336
10.0	3.047	8.358	3.225	7.986	3.439	7.419	3.574	7.031	3.687	6.759	3.852	6.482	5.108	5.650	6.424	5.020	7.741	4.394	9.716	3.455

TABLE 2.7

NORMALIZED ADMITTANCES (Y/Δ) IN MILLIMHOS OF THIN DIPOLE ANTENNAS IN DISSIPATIVE MEDIA

Ω = 12.5

βh	α/β = 0.00		α/β = 0.01		α/β = 0.03		α/β = 0.05		α/β = 0.07		α/β = 0.10		α/β = 0.30		α/β = 0.50		α/β = 0.70		α/β = 1.00	
1.5	12.511	-0.828	11.287	-0.443	9.428	-0.025	8.099	0.156	7.113	0.227	6.040	0.242	3.343	-0.094	2.679	-0.406	2.476	-0.693	2.459	-1.133
π/2	8.457	-4.370	8.127	-3.649	7.416	-2.578	6.737	-1.863	6.137	-1.382	5.396	-0.931	3.182	-0.342	2.600	-0.485	2.437	-0.721	2.454	-1.138
1.6	6.967	-4.624	6.846	-4.005	6.477	-3.012	6.045	-2.291	5.617	-1.772	5.040	-1.251	3.104	-0.422	2.566	-0.510	2.422	-0.728	2.454	-1.139
1.7	3.877	-4.000	3.977	-3.704	4.076	-3.150	4.076	-2.662	4.012	-2.248	3.850	-1.757	2.823	-0.606	2.452	-0.562	2.374	-0.739	2.455	-1.136
1.8	2.510	-3.103	2.614	-2.954	2.775	-2.654	2.877	-2.365	2.933	-2.096	2.952	-1.742	2.557	-0.674	2.350	-0.575	2.338	-0.732	2.464	-1.128
1.9	1.816	-2.398	1.903	-2.314	2.053	-2.138	2.170	-1.959	2.257	-1.785	2.343	-1.542	2.330	-0.665	2.265	-0.559	2.312	-0.713	2.479	-1.117
2.0	1.416	-1.861	1.489	-1.808	1.620	-1.696	1.732	-1.580	1.826	-1.464	1.934	-1.295	2.149	-0.610	2.199	-0.524	2.298	-0.687	2.498	-1.105
2.1	1.163	-1.438	1.226	-1.403	1.342	-1.328	1.447	-1.248	1.539	-1.167	1.655	-1.048	2.009	-0.527	2.151	-0.478	2.294	-0.657	2.522	-1.092
2.2	0.992	-1.094	1.048	-1.069	1.154	-1.016	1.252	-0.960	1.341	-0.902	1.459	-0.816	1.905	-0.431	2.119	-0.425	2.298	-0.626	2.549	-1.081
2.3	0.869	-0.803	0.921	-0.784	1.020	-0.746	1.114	-0.705	1.201	-0.664	1.321	-0.601	1.830	-0.328	2.102	-0.370	2.310	-0.595	2.578	-1.070
2.4	0.778	-0.549	0.828	-0.535	0.923	-0.507	1.014	-0.477	1.101	-0.447	1.221	-0.402	1.780	-0.224	2.098	-0.314	2.327	-0.566	2.608	-1.061
2.5	0.709	-0.322	0.757	-0.312	0.851	-0.291	0.942	-0.269	1.028	-0.248	1.151	-0.217	1.750	-0.121	2.103	-0.260	2.349	-0.539	2.639	-1.053
2.6	0.655	-0.114	0.703	-0.106	0.797	-0.091	0.888	-0.076	0.976	-0.061	1.102	-0.041	1.738	-0.021	2.118	-0.209	2.375	-0.514	2.670	-1.047
2.7	0.612	0.081	0.661	0.086	0.757	0.097	0.850	0.107	0.940	0.116	1.070	0.126	1.740	0.075	2.140	-0.162	2.403	-0.492	2.701	-1.041
2.8	0.578	0.267	0.628	0.271	0.727	0.278	0.824	0.283	0.918	0.286	1.053	0.287	1.755	0.166	2.168	-0.119	2.433	-0.474	2.731	-1.037
2.9	0.551	0.448	0.603	0.451	0.707	0.454	0.809	0.455	0.907	0.452	1.050	0.444	1.781	0.252	2.200	-0.082	2.464	-0.458	2.761	-1.033
3.0	0.530	0.628	0.585	0.629	0.696	0.629	0.803	0.625	0.908	0.617	1.058	0.599	1.818	0.331	2.236	-0.050	2.495	-0.445	2.791	-1.030
3.1	0.514	0.809	0.574	0.809	0.692	0.804	0.807	0.795	0.919	0.781	1.079	0.752	1.863	0.403	2.273	-0.023	2.525	-0.434	2.820	-1.027
π	0.509	0.886	0.571	0.885	0.693	0.878	0.812	0.867	0.927	0.850	1.091	0.815	1.884	0.431	2.289	-0.013	2.537	-0.431	2.831	-1.026
3.2	0.504	0.995	0.569	0.993	0.697	0.984	0.822	0.968	0.942	0.947	1.113	0.905	1.916	0.467	2.311	-0.001	2.555	-0.426	2.848	-1.024
3.3	0.499	1.190	0.571	1.186	0.712	1.170	0.848	1.147	0.979	1.116	1.161	1.058	1.976	0.522	2.350	0.017	2.583	-0.419	2.875	-1.022
3.4	0.502	1.397	0.582	1.390	0.739	1.367	0.889	1.333	1.031	1.290	1.227	1.212	2.041	0.568	2.388	0.029	2.611	-0.413	2.901	-1.020
3.5	0.513	1.622	0.604	1.610	0.781	1.576	0.947	1.528	1.103	1.470	1.312	1.366	2.109	0.603	2.424	0.038	2.637	-0.408	2.927	-1.017
3.6	0.537	1.871	0.642	1.853	0.843	1.803	1.029	1.736	1.199	1.655	1.422	1.518	2.179	0.627	2.458	0.044	2.662	-0.404	2.953	-1.015
3.7	0.578	2.153	0.701	2.126	0.933	2.052	1.143	1.956	1.328	1.845	1.560	1.664	2.249	0.639	2.489	0.047	2.685	-0.400	2.978	-1.012
3.8	0.646	2.481	0.794	2.438	1.064	2.327	1.299	2.189	1.498	2.037	1.733	1.799	2.316	0.641	2.518	0.048	2.708	-0.396	3.002	-1.009
3.9	0.760	2.871	0.939	2.802	1.257	2.631	1.517	2.431	1.722	2.220	1.945	1.912	2.379	0.632	2.543	0.047	2.729	-0.393	3.026	-1.006
4.0	0.952	3.346	1.173	3.232	1.543	2.961	1.818	2.668	2.014	2.379	2.198	1.986	2.434	0.615	2.566	0.046	2.749	-0.389	3.050	-1.003
4.1	1.288	3.937	1.561	3.736	1.971	3.299	2.231	2.869	2.382	2.482	2.486	2.002	2.481	0.591	2.586	0.046	2.769	-0.384	3.073	-1.000
4.2	1.906	4.662	2.221	4.296	2.609	3.584	2.777	2.974	2.824	2.484	2.794	1.935	2.518	0.562	2.604	0.045	2.788	-0.380	3.096	-0.996
4.3	3.093	5.451	3.354	4.772	3.511	3.668	3.441	2.882	3.302	2.327	3.088	1.769	2.544	0.532	2.620	0.045	2.806	-0.375	3.119	-0.992
4.4	5.306	5.779	5.128	4.689	4.601	3.288	4.115	2.480	3.732	1.975	3.321	1.506	2.561	0.501	2.634	0.047	2.824	-0.369	3.142	-0.989
4.5	8.192	4.059	6.999	3.185	5.483	2.242	4.582	1.752	4.000	1.455	3.449	1.177	2.569	0.474	2.647	0.049	2.842	-0.363	3.164	-0.985
4.6	8.692	0.179	7.320	0.554	5.626	0.825	4.649	0.884	4.029	0.879	3.453	0.835	2.569	0.451	2.659	0.053	2.859	-0.357	3.186	-0.981
4.7	6.533	-2.121	5.990	-1.264	5.032	-0.305	4.332	0.149	3.837	0.378	3.346	0.534	2.564	0.433	2.670	0.059	2.876	-0.351	3.208	-0.976
3π/2	6.243	-2.239	5.790	-1.389	4.930	-0.405	4.275	0.077	3.802	0.326	3.326	0.501	2.563	0.431	2.671	0.059	2.879	-0.350	3.211	-0.976
4.8	4.514	-2.475	4.474	-1.812	4.177	-0.871	3.823	-0.306	3.513	0.032	3.164	0.308	2.554	0.421	2.681	0.065	2.894	-0.344	3.230	-0.972
4.9	3.245	-2.166	3.368	-1.735	3.403	-1.018	3.297	-0.505	3.152	-0.155	2.947	0.167	2.542	0.416	2.692	0.072	2.911	-0.337	3.252	-0.968
5.0	2.475	-1.738	2.635	-1.459	2.804	-0.946	2.841	-0.530	2.812	-0.216	2.728	0.100	2.529	0.418	2.703	0.081	2.928	-0.330	3.273	-0.963
5.1	1.989	-1.331	2.147	-1.145	2.362	-0.779	2.473	-0.456	2.518	-0.193	2.526	0.093	2.517	0.425	2.714	0.090	2.945	-0.323	3.295	-0.959
5.2	1.666	-0.974	1.813	-0.844	2.038	-0.579	2.186	-0.331	2.277	-0.118	2.348	0.128	2.506	0.436	2.726	0.100	2.962	-0.316	3.316	-0.954
5.3	1.441	-0.662	1.576	-0.568	1.798	-0.373	1.964	-0.183	2.082	-0.013	2.199	0.192	2.497	0.452	2.738	0.110	2.978	-0.308	3.337	-0.950
5.4	1.277	-0.386	1.402	-0.318	1.619	-0.171	1.792	-0.025	1.928	0.109	2.076	0.275	2.491	0.472	2.751	0.120	2.995	-0.301	3.358	-0.945
5.5	1.155	-0.140	1.273	-0.088	1.483	0.023	1.661	0.135	1.806	0.239	1.977	0.370	2.488	0.494	2.763	0.131	3.012	-0.293	3.378	-0.940
5.6	1.061	0.085	1.174	0.124	1.380	0.209	1.559	0.294	1.713	0.374	1.899	0.473	2.488	0.517	2.777	0.142	3.029	-0.286	3.399	-0.935
5.7	0.988	0.293	1.097	0.324	1.301	0.388	1.483	0.451	1.642	0.510	1.841	0.581	2.492	0.542	2.790	0.152	3.045	-0.278	3.419	-0.930
5.8	0.930	0.491	1.038	0.514	1.242	0.561	1.426	0.606	1.591	0.646	1.801	0.691	2.498	0.567	2.804	0.163	3.062	-0.271	3.440	-0.925
5.9	0.884	0.680	0.992	0.698	1.198	0.731	1.387	0.759	1.556	0.782	1.775	0.802	2.508	0.591	2.818	0.173	3.078	-0.263	3.460	-0.920
6.0	0.848	0.866	0.958	0.878	1.169	0.898	1.362	0.911	1.538	0.917	1.765	0.913	2.520	0.615	2.833	0.184	3.095	-0.256	3.480	-0.915
6.1	0.821	1.050	0.935	1.058	1.152	1.065	1.352	1.063	1.534	1.052	1.768	1.023	2.535	0.637	2.847	0.193	3.111	-0.249	3.500	-0.910
6.2	0.801	1.237	0.921	1.239	1.148	1.233	1.356	1.215	1.544	1.187	1.785	1.132	2.551	0.658	2.861	0.203	3.127	-0.241	3.520	-0.905
2π	0.791	1.396	0.916	1.393	1.154	1.375	1.371	1.343	1.564	1.299	1.809	1.220	2.566	0.674	2.873	0.211	3.140	-0.235	3.536	-0.901
[illegible]	[illegible]	[illegible]	[illegible]	[illegible]	[illegible]	[illegible]	1.375	1.369	1.569	1.321	1.815	1.238	2.569	0.678	2.876	0.213	3.143	-0.234	3.539	-0.900
6.4	0.787	1.628	0.923	1.618	1.179	1.580	[illegible]	[illegible]	[illegible]	[illegible]	[illegible]	[illegible]	[illegible]	[illegible]	2.890	0.222	3.159	-0.226	3.559	-0.894
6.5	0.794	1.840	0.942	1.822	1.218	1.763	1.461	1.681	1.668	1.586	1.917	1.437	2.607	0.710	2.904	[illegible]	[illegible]	[illegible]	[illegible]	[illegible]
6.6	0.814	2.069	0.978	2.040	1.278	1.953	1.534	1.839	1.745	1.714	1.990	1.527	2.627	0.723	2.918	0.240	3.191	-0.211	3.598	-0.884
6.7	0.851	2.322	1.035	2.277	1.363	2.151	1.631	1.998	1.843	1.836	2.077	1.607	2.646	0.735	2.932	0.249	3.206	-0.204	3.617	-0.878
6.8	0.912	2.604	1.121	2.537	1.480	2.358	1.758	2.153	1.965	1.948	2.178	1.673	2.665	0.744	2.945	0.257	3.222	-0.196	3.636	-0.872
6.9	1.009	2.927	1.250	2.825	1.642	2.571	1.921	2.299	2.113	2.044	2.291	1.722	2.683	0.752	2.959	0.266	3.237	-0.188	3.655	-0.867
7.0	1.163	3.301	1.442	3.145	1.861	2.783	2.126	2.427	2.285	2.116	2.415	1.750	2.700	0.759	2.972	0.275	3.252	-0.181	3.674	-0.861
7.1	1.411	3.740	1.731	3.495	2.156	2.978	2.376	2.520	2.481	2.153	2.544	1.751	2.716	0.765	2.985	0.283	3.267	-0.173	3.693	-0.855
7.2	1.818	4.250	2.167	3.858	2.542	3.127	2.670	2.558	2.693	2.144	2.671	1.725	2.730	0.770	2.998	0.292	3.283	-0.165	3.712	-0.850
7.3	2.505	4.804	2.825	4.175	3.027	3.176	2.994	2.514	2.907	2.079	2.789	1.669	2.743	0.775	3.010	0.301	3.298	-0.158	3.730	-0.844
7.4	3.661	5.259	3.770	4.297	3.584	3.055	3.317	2.366	3.102	1.954	2.888	1.586	2.755	0.779	3.023	0.309	3.312	-0.150	3.749	-0.838
7.5	5.431	5.146	4.945	3.939	4.125	2.695	3.593	2.105	3.255	1.773	2.961	1.483	2.765	0.784	3.035	0.318	3.327	-0.142	3.768	-0.832
7.6	7.282	3.658	5.947	2.850	4.500	2.101	3.766	1.754	3.346	1.554	3.001	1.367	2.774	0.789	3.047	0.327	3.342	-0.134	3.786	-0.826
7.7	7.666	1.052	6.163	1.299	4.583	1.394	3.802	1.369	3.363	1.323	3.007	1.250	2.782	0.795	3.059	0.336	3.357	-0.126	3.804	-0.820
7.8	6.397	-0.843	5.537	0.025	4.371	0.760	3.705	1.014	3.310	1.109	2.981	1.142	2.789	0.801	3.071	0.345	3.371	-0.118	3.823	-0.814
5π/2	5.533	-1.308	5.033	-0.398	4.171	0.494	3.609	0.853	3.257	1.008	2.956	1.090	2.793	0.805	3.078	0.350	3.379	-0.114	3.833	-0.810
7.9	4.857	-1.478	4.593	-0.622	3.976	0.318	3.511	0.736	3.201	0.933	2.929	1.051	2.796	0.809	3.083	0.354	3.386	-0.110	3.841	-0.808
8.0	3.697	-1.445	3.734	-0.792	3.528	0.082	3.266	0.553	3.058	0.806	2.858	0.981	2.802	0.817	3.095	0.363	3.401	-0.102	3.859	-0.801
8.1	2.913	-1.181	3.069	-0.716	3.109	0.006	3.011	0.459	2.900	0.729	2.777	0.936	2.808	0.826	3.107	0.372	3.415	-0.094	3.877	-0.795
8.2	2.387	-0.866	2.580	-0.535	2.752	0.034	2.772	0.436	2.742	0.698	2.691	0.915	2.814	0.835	3.119	0.382	3.429	-0.086	3.895	-0.789
8.3	2.023	-0.557	2.222	-0.319	2.460	0.123	2.559	0.465	2.593	0.704	2.607	0.915	2.820	0.846	3.131	0.391	3.444	-0.078	3.913	-0.783
8.4	1.763	-0.272	1.958	-0.096	2.226	0.246	2.377	0.530	2.460	0.739	2.528	0.932	2.827	0.857	3.142	0.400	3.458	-0.070	3.931	-0.776
8.5	1.573	-0.012	1.759	0.121	2.041	0.387	2.226	0.618	2.345	0.796	2.457	0.965	2.833	0.869	3.154	0.410	3.472	-0.062	3.949	-0.770
8.6	1.429	0.227	1.607	0.328	1.895	0.536	2.101	0.721	2.248	0.868	2.396	1.010	2.840	0.881	3.166	0.419	3.486	-0.054	3.967	-0.763
8.7	1.318	0.449	1.490	0.527	1.780	0.688	2.002	0.834	2.168	0.951	2.345	1.065	2.848	0.893	3.177	0.428	3.500	-0.045	3.985	-0.757
8.8	1.231	0.657	1.399	0.717	1.690	0.841	1.923	0.953	2.105	1.042	2.305	1.126	2.856	0.906	3.189	0.438	3.514	-0.037	4.002	-0.750
8.9	1.163	0.855	1.328	0.902	1.622	0.994	1.864	1.076	2.058	1.138	2.276	1.192	2.864	0.918	3.201	0.447	3.528	-0.029	4.020	-0.743
9.0	1.109	1.047	1.274	1.082	1.571	1.147	1.821	1.201	2.025	1.238	2.257	1.261	2.873	0.931	3.212	0.457	3.542	-0.021	4.038	-0.737
9.1	1.067	1.235	1.234	1.259	1.537	1.300	1.795	1.327	2.007	1.339	2.249	1.332	2.882	0.943	3.224	0.466	3.556	-0.012	4.055	-0.730
9.2	1.036	1.422	1.206	1.437	1.518	1.454	1.783	1.455	2.002	1.442	2.251	1.403	2.891	0.956	3.235	0.476	3.570	-0.004	4.073	-0.723
9.3	1.014	1.611	1.191	1.617	1.514	1.609	1.787	1.583	2.010	1.544	2.262	1.472	2.901	0.968	3.246	0.485	3.584	0.005	4.090	-0.716
9.4	1.002	1.805	1.188	1.800	1.524	1.767	1.805	1.711	2.032	1.644	2.284	1.540	2.911	0.980	3.258	0.495	3.598	0.013	4.108	-0.709
3π	1.001	1.855	1.189	1.847	1.529	1.806	1.812	1.742	2.039	1.669	2.290	1.556	2.914	0.983	3.261	0.497	3.601	0.015	4.112	-0.708
9.5	1.001	2.008	1.198	1.991	1.551	1.926	1.840	1.838	2.067	1.742	2.314	1.604	2.921	0.992	3.269	0.504	3.612	0.022	4.125	-0.703
9.6	1.011	2.221	1.224	2.190	1.596	2.089	1.891	1.963	2.116	1.835	2.353	1.662	2.931	1.003	3.281	0.514	3.625	0.030	4.142	-0.696
9.7	1.037	2.451	1.269	2.400	1.662	2.254	1.961	2.085	2.179	1.922	2.401	1.714	2.942	1.014	3.292	0.524	3.639	0.039	4.160	-0.689
9.8	1.083	2.701	1.338	2.624	1.754	2.420	2.051	2.200	2.256	2.000	2.455	1.758	2.952	1.025	3.303	0.533	3.653	0.047	4.177	-0.681
9.9	1.157	2.979	1.440	2.865	1.876	2.585	2.163	2.305	2.347	2.065	2.514	1.792	2.962	1.035	3.314	0.543	3.666	0.056	4.194	-0.674
10.0	1.272	3.290	1.587	3.123	2.035	2.743	2.297	2.394	2.450	2.115	2.578	1.816	2.971	1.046	3.325	0.553	3.680	0.065	4.211	-0.667
10.1	1.448	3.642	1.797	3.396	2.238	2.885	2.455	2.461	2.562	2.146										
10.2	1.721	4.040	2.096	3.674	2.490	2.995	2.632	2.496	2.680	2.153										
10.3	2.150	4.477	2.518	3.929	2.793	3.053	2.823	2.490	2.799	2.135										
10.4	2.827	4.906	3.099	4.104	3.135	3.028	3.015	2.435	2.910	2.089										
10.5	3.869	5.179	3.844	4.085	3.487	2.893	3.192	2.328	3.007	2.018										
10.6	5.297	4.940	4.668	3.721	3.800	2.633	3.334	2.172	3.080	1.926										
10.7	6.675	3.735	5.323	2.930	4.012	2.264	3.424	1.980	3.125	1.819										
10.8	7.072	1.773	5.516	1.876	4.077	1.843	3.451	1.773	3.138	1.709										
10.9	6.278	0.141	5.195	0.923	3.989	1.443	3.416	1.576	3.121	1.603										
7π/2	5.125	-0.604	4.611	0.329	3.796	1.135	3.333	1.413	3.079	1.515										

βh	α/β = 0.00		α/β = 0.01		α/β = 0.03		α/β = 0.05		α/β = 0.07	
11.0	5.073	-0.622	4.581	0.309	3.786	1.123	3.328	1.406	3.077	1.511
11.1	4.027	-0.770	3.926	0.034	3.519	0.908	3.204	1.278	3.012	1.438
11.2	3.255	-0.636	3.354	-0.012	3.237	0.793	3.060	1.194	2.935	1.387
11.3	2.707	-0.399	2.896	0.069	2.969	0.760	2.911	1.152	2.850	1.358
11.4	2.316	-0.137	2.539	0.214	2.731	0.785	2.767	1.146	2.765	1.351
11.5	2.031	0.122	2.264	0.387	2.526	0.851	2.634	1.171	2.683	1.363
11.6	1.818	0.367	2.052	0.570	2.355	0.944	2.516	1.218	2.607	1.391
11.7	1.657	0.597	1.886	0.754	2.214	1.053	2.414	1.283	2.540	1.432
11.8	1.532	0.814	1.756	0.937	2.100	1.174	2.329	1.361	2.483	1.484
11.9	1.434	1.021	1.655	1.116	2.009	1.301	2.260	1.448	2.436	1.544
12.0	1.357	1.218	1.575	1.292	1.938	1.432	2.206	1.541	2.400	1.610
12.1	1.296	1.411	1.515	1.466	1.885	1.566	2.167	1.640	2.374	1.681
12.2	1.249	1.600	1.470	1.639	1.849	1.703	2.142	1.741	2.359	1.754
12.3	1.215	1.788	1.440	1.813	1.829	1.841	2.131	1.844	2.355	1.828
12.4	1.191	1.979	1.423	1.988	1.824	1.980	2.133	1.948	2.361	1.902
12.5	1.178	2.174	1.421	2.168	1.835	2.121	2.149	2.051	2.377	1.975
4π	1.177	2.308	1.427	2.289	1.851	2.215	2.168	2.118	2.394	2.022
12.6	1.178	2.377	1.433	2.352	1.862	2.262	2.179	2.152	2.404	2.045
12.7	1.191	2.590	1.463	2.544	1.908	2.404	2.224	2.249	2.440	2.110
12.8	1.222	2.818	1.514	2.744	1.973	2.544	2.283	2.341	2.486	2.170
12.9	1.275	3.064	1.590	2.954	2.061	2.681	2.356	2.425	2.540	2.222
13.0	1.357	3.334	1.700	3.173	2.175	2.811	2.445	2.498	2.602	2.264
13.1	1.483	3.631	1.854	3.401	2.317	2.928	2.547	2.555	2.669	2.295
13.2	1.671	3.960	2.066	3.632	2.490	3.024	2.660	2.594	2.740	2.313
13.3	1.954	4.319	2.354	3.851	2.693	3.088	2.782	2.609	2.811	2.317
13.4	2.381	4.693	2.739	4.032	2.921	3.107	2.907	2.597	2.881	2.307
13.5	3.020	5.031	3.233	4.124	3.163	3.066	3.026	2.557	2.944	2.282
13.6	3.943	5.197	3.822	4.052	3.398	2.954	3.132	2.488	2.997	2.246
13.7	5.119	4.932	4.428	3.734	3.599	2.771	3.216	2.394	3.039	2.199
13.8	6.216	3.955	4.901	3.143	3.737	2.529	3.271	2.283	3.065	2.147
13.9	6.625	2.411	5.078	2.380	3.789	2.258	3.292	2.164	3.076	2.092
14.0	6.136	0.997	4.914	1.646	3.753	1.991	3.280	2.048	3.071	2.038
14.1	5.199	0.185	4.510	1.103	3.641	1.761	3.237	1.944	3.052	1.991
9π/2	4.839	0.032	4.330	0.962	3.585	1.690	3.214	1.910	3.041	1.975
14.2	4.275	-0.093	4.017	0.794	3.478	1.588	3.169	1.859	3.021	1.952
14.3	3.534	-0.067	3.542	0.676	3.289	1.476	3.086	1.798	2.981	1.923
14.4	2.980	0.097	3.130	0.685	3.094	1.422	2.994	1.762	2.936	1.907
14.5	2.570	0.313	2.791	0.770	2.909	1.416	2.899	1.749	2.887	1.902
14.6	2.266	0.543	2.518	0.898	2.741	1.446	2.807	1.757	2.839	1.909
14.7	2.036	0.771	2.301	1.048	2.593	1.505	2.721	1.784	2.792	1.926
14.8	1.859	0.991	2.128	1.208	2.467	1.583	2.644	1.826	2.750	1.953
14.9	1.722	1.202	1.991	1.373	2.361	1.676	2.577	1.879	2.712	1.987
15.0	1.614	1.405	1.882	1.539	2.274	1.780	2.522	1.943	2.680	2.028
15.1	1.529	1.601	1.796	1.705	2.206	1.890	2.477	2.014	2.655	2.075
15.2	1.463	1.793	1.730	1.871	2.154	2.005	2.444	2.090	2.638	2.125
15.3	1.411	1.982	1.682	2.038	2.119	2.124	2.423	2.169	2.627	2.177
15.4	1.373	2.171	1.649	2.206	2.098	2.246	2.413	2.251	2.624	2.231
15.5	1.348	2.362	1.632	2.376	2.093	2.369	2.414	2.333	2.629	2.285
15.6	1.335	2.557	1.629	2.549	2.102	2.493	2.427	2.415	2.640	2.338
15.7	1.335	2.760	1.643	2.727	2.128	2.616	2.451	2.494	2.659	2.388
5π	1.336	2.776	1.645	2.741	2.131	2.626	2.453	2.501	2.661	2.392
15.8	1.351	2.971	1.676	2.910	2.170	2.738	2.486	2.570	2.684	2.436
15.9	1.385	3.196	1.729	3.098	2.231	2.856	2.532	2.640	2.715	2.478
16.0	1.442	3.437	1.809	3.293	2.310	2.968	2.589	2.703	2.752	2.515
16.1	1.531	3.697	1.922	3.492	2.409	3.071	2.655	2.756	2.792	2.546
16.2	1.663	3.979	2.075	3.692	2.530	3.159	2.730	2.796	2.835	2.569
16.3	1.857	4.284	2.279	3.886	2.671	3.227	2.810	2.823	2.880	2.585
16.4	2.141	4.608	2.547	4.060	2.830	3.267	2.894	2.833	2.924	2.592
16.5	2.554	4.929	2.888	4.188	3.002	3.272	2.976	2.827	2.966	2.590
16.6	3.147	5.196	3.304	4.234	3.176	3.234	3.054	2.802	3.004	2.581
16.7	3.957	5.296	3.772	4.146	3.340	3.150	3.122	2.761	3.037	2.566
16.8	4.939	5.038	4.235	3.880	3.478	3.022	3.176	2.706	3.062	2.545
16.9	5.842	4.240	4.594	3.430	3.573	2.859	3.212	2.641	3.079	2.522
17.0	6.247	3.003	4.753	2.858	3.616	2.677	3.229	2.572	3.088	2.496
17.1	5.966	1.784	4.677	2.286	3.604	2.495	3.226	2.503	3.088	2.472
17.2	5.250	0.975	4.411	1.822	3.540	2.331	3.204	2.440	3.080	2.450
11π/2	4.617	0.657	4.127	1.568	3.462	2.223	3.175	2.398	3.069	2.436
17.3	4.454	0.610	4.045	1.517	3.438	2.198	3.166	2.388	3.065	2.432
17.4	3.762	0.544	3.660	1.361	3.310	2.104	3.117	2.348	3.045	2.420
17.5	3.215	0.639	3.302	1.321	3.171	2.048	3.060	2.323	3.021	2.415
17.6	2.798	0.809	2.991	1.359	3.032	2.029	2.998	2.313	2.994	2.416
17.7	2.479	1.009	2.731	1.448	2.898	2.040	2.936	2.317	2.966	2.424
17.8	2.235	1.217	2.516	1.566	2.776	2.077	2.876	2.335	2.938	2.438
17.9	2.046	1.425	2.342	1.703	2.668	2.133	2.821	2.364	2.913	2.458
18.0	1.898	1.628	2.201	1.849	2.575	2.205	2.771	2.403	2.889	2.484
18.1	1.781	1.827	2.087	2.001	2.497	2.288	2.729	2.450	2.870	2.515
18.2	1.689	2.020	1.998	2.156	2.435	2.379	2.694	2.504	2.854	2.549
18.3	1.617	2.210	1.929	2.313	2.387	2.477	2.668	2.563	2.843	2.586
18.4	1.561	2.399	1.877	2.473	2.353	2.580	2.651	2.626	2.836	2.625
18.5	1.520	2.587	1.843	2.634	2.333	2.686	2.643	2.691	2.835	2.666
18.6	1.493	2.778	1.824	2.797	2.327	2.793	2.644	2.756	2.838	2.706
18.7	1.480	2.972	1.822	2.964	2.335	2.900	2.653	2.821	2.846	2.746
18.8	1.481	3.173	1.836	3.133	2.358	3.007	2.672	2.884	2.859	2.784
6π	1.487	3.276	1.850	3.219	2.374	3.059	2.684	2.914	2.867	2.803
18.9	1.498	3.383	1.869	3.307	2.395	3.111	2.699	2.943	2.877	2.820
19.0	1.534	3.603	1.924	3.483	2.448	3.211	2.733	2.998	2.897	2.854
19.1	1.595	3.837	2.004	3.663	2.516	3.304	2.775	3.047	2.921	2.883
19.2	1.688	4.087	2.115	3.842	2.600	3.387	2.824	3.088	2.948	2.908
19.3	1.824	4.354	2.263	4.018	2.699	3.456	2.877	3.121	2.976	2.929
19.4	2.020	4.636	2.454	4.182	2.812	3.507	2.933	3.143	3.004	2.944
19.5	2.299	4.926	2.697	4.320	2.935	3.535	2.991	3.154	3.032	2.955
19.6	2.693	5.202	2.995	4.413	3.064	3.537	3.047	3.154	3.059	2.961
19.7	3.235	5.414	3.342	4.433	3.192	3.509	3.099	3.143	3.083	2.963
19.8	3.945	5.471	3.719	4.346	3.310	3.449	3.144	3.121	3.103	2.961
19.9	4.770	5.231	4.080	4.128	3.408	3.361	3.180	3.092	3.120	2.956
20.0	5.522	4.577	4.361	3.780	3.478	3.250	3.206	3.056	3.132	2.949
20.1	5.908	3.576	4.502	3.344	3.513	3.127	3.219	3.018	3.139	2.942
20.2	5.769	2.538	4.474	2.895	3.511	3.002	3.220	2.979	3.141	2.934
20.3	5.237	1.768	4.301	2.508	3.475	2.886	3.209	2.944	3.139	2.929
20.4	4.570	1.352	4.033	2.227	3.410	2.788	3.188	2.914	3.132	2.925
13π/2	4.435	1.305	3.972	2.184	3.394	2.771	3.183	2.908	3.130	2.925
20.5	3.942	1.214	3.727	2.059	3.324	2.715	3.159	2.891	3.122	2.924
20.6	3.417	1.246	3.425	1.988	3.225	2.667	3.123	2.878	3.109	2.927
20.7	3.001	1.369	3.148	1.992	3.121	2.646	3.084	2.874	3.095	2.934
20.8	2.675	1.537	2.907	2.048	3.018	2.649	3.043	2.879	3.080	2.945
20.9	2.421	1.723	2.702	2.139	2.921	2.673	3.002	2.894	3.064	2.960

βh	α/β = 0.00		α/β = 0.01		α/β = 0.03		α/β = 0.05		α/β = 0.07	
21.0	2.222	1.917	2.532	2.253	2.832	2.714	2.963	2.918	3.050	2.979
21.1	2.065	2.111	2.391	2.381	2.753	2.770	2.928	2.949	3.036	3.001
21.2	1.940	2.303	2.276	2.518	2.686	2.838	2.898	2.986	3.025	3.026
21.3	1.842	2.492	2.185	2.661	2.631	2.914	2.872	3.029	3.016	3.054
21.4	1.764	2.680	2.113	2.809	2.588	2.997	2.853	3.076	3.010	3.084
21.5	1.705	2.867	2.060	2.960	2.557	3.085	2.840	3.127	3.006	3.116
21.6	1.661	3.054	2.024	3.113	2.539	3.176	2.834	3.179	3.006	3.148
21.7	1.632	3.244	2.005	3.269	2.533	3.269	2.834	3.231	3.009	3.181
21.8	1.618	3.437	2.002	3.428	2.539	3.362	2.842	3.284	3.015	3.213
21.9	1.619	3.636	2.016	3.589	2.558	3.454	2.855	3.335	3.024	3.244
7π	1.635	3.824	2.045	3.738	2.586	3.536	2.873	3.379	3.035	3.272
22.0	1.638	3.842	2.049	3.753	2.590	3.544	2.875	3.384	3.036	3.274
22.1	1.676	4.058	2.103	3.917	2.634	3.628	2.901	3.429	3.050	3.302
22.2	1.740	4.285	2.182	4.082	2.691	3.706	2.931	3.469	3.066	3.328
22.3	1.836	4.524	2.288	4.244	2.760	3.775	2.966	3.504	3.084	3.351
22.4	1.974	4.775	2.427	4.398	2.840	3.832	3.004	3.532	3.102	3.372
22.5	2.168	5.035	2.603	4.537	2.928	3.873	3.043	3.554	3.121	3.389
22.6	2.438	5.294	2.819	4.650	3.024	3.897	3.083	3.568	3.139	3.404
22.7	2.808	5.529	3.076	4.719	3.121	3.901	3.122	3.574	3.156	3.415
22.8	3.300	5.698	3.365	4.725	3.216	3.883	3.157	3.574	3.172	3.425
22.9	3.918	5.725	3.669	4.647	3.302	3.844	3.188	3.567	3.186	3.432
23.0	4.614	5.508	3.955	4.471	3.374	3.786	3.214	3.555	3.197	3.438
23.1	5.243	4.970	4.179	4.201	3.426	3.713	3.232	3.539	3.206	3.442
23.2	5.599	4.161	4.301	3.866	3.455	3.630	3.242	3.521	3.212	3.446
23.3	5.554	3.287	4.302	3.515	3.458	3.546	3.245	3.503	3.215	3.450
23.4	5.171	2.582	4.190	3.198	3.438	3.466	3.241	3.487	3.215	3.455
23.5	4.628	2.148	3.996	2.951	3.396	3.398	3.229	3.474	3.213	3.461
15π/2	4.280	2.005	3.851	2.839	3.362	3.363	3.219	3.468	3.211	3.466
23.6	4.075	1.956	3.758	2.787	3.338	3.344	3.212	3.465	3.209	3.469
23.7	3.585	1.933	3.508	2.702	3.268	3.309	3.190	3.462	3.203	3.479
23.8	3.181	2.011	3.269	2.683	3.192	3.292	3.165	3.465	3.196	3.491
23.9	2.855	2.146	3.052	2.714	3.114	3.293	3.138	3.474	3.189	3.506
24.0	2.596	2.309	2.862	2.781	3.039	3.311	3.112	3.490	3.181	3.523
24.1	2.390	2.486	2.700	2.874	2.967	3.344	3.086	3.512	3.173	3.543
24.2	2.225	2.669	2.563	2.985	2.903	3.389	3.062	3.539	3.166	3.565
24.3	2.094	2.854	2.450	3.108	2.847	3.445	3.040	3.572	3.160	3.588
24.4	1.990	3.038	2.359	3.240	2.800	3.510	3.023	3.608	3.156	3.614
24.5	1.908	3.223	2.287	3.377	2.763	3.580	3.009	3.648	3.153	3.641
24.6	1.845	3.407	2.233	3.519	2.737	3.656	3.000	3.690	3.152	3.669
24.7	1.799	3.593	2.197	3.665	2.720	3.735	2.996	3.734	3.153	3.697
24.8	1.769	3.781	2.177	3.813	2.714	3.816	2.996	3.778	3.156	3.726
24.9	1.754	3.972	2.173	3.963	2.719	3.897	3.002	3.823	3.161	3.755
25.0	1.755	4.168	2.186	4.116	2.735	3.978	3.012	3.866	3.168	3.783

TABLE 2.7

NORMALIZED ADMITTANCES (Y/Δ) IN MILLIMHOS OF THIN DIPOLE ANTENNAS IN DISSIPATIVE MEDIA

Ω = 15.0

βh	α/β = 0.00		α/β = 0.01		α/β = 0.03		α/β = 0.05		α/β = 0.07		α/β = 0.10		α/β = 0.30		α/β = 0.50		α/β = 0.70		α/β = 1.00	
1.5	13.135	1.184	11.417	1.104	9.070	0.915	7.546	0.740	6.481	0.593	5.377	0.416	2.785	-0.165	2.158	-0.453	1.946	-0.692	1.881	-1.049
π/2	8.867	-4.711	8.365	-3.767	7.352	-2.487	6.467	-1.723	5.739	-1.253	4.897	-0.847	2.650	-0.401	2.086	-0.529	1.906	-0.720	1.870	-1.056
1.6	6.921	-5.101	6.773	-4.304	6.290	-3.092	5.741	-2.275	5.225	-1.725	4.567	-1.209	2.582	-0.479	2.055	-0.553	1.891	-0.728	1.866	-1.058
1.7	3.314	-4.191	3.454	-3.874	3.604	-3.272	3.623	-2.744	3.560	-2.301	3.388	-1.789	2.330	-0.662	1.950	-0.607	1.842	-0.743	1.858	-1.060
1.8	1.979	-3.126	2.099	-2.986	2.286	-2.695	2.407	-2.408	2.474	-2.138	2.499	-1.784	2.086	-0.737	1.854	-0.626	1.802	-0.743	1.855	-1.056
1.9	1.370	-2.382	1.460	-2.308	1.616	-2.150	1.739	-1.984	1.830	-1.819	1.918	-1.586	1.878	-0.739	1.773	-0.619	1.772	-0.732	1.858	-1.051
2.0	1.040	-1.853	1.111	-1.810	1.238	-1.714	1.347	-1.611	1.438	-1.506	1.542	-1.349	1.710	-0.697	1.708	-0.595	1.752	-0.714	1.866	-1.044
2.1	0.840	-1.458	0.898	-1.429	1.005	-1.367	1.102	-1.299	1.186	-1.229	1.292	-1.121	1.580	-0.630	1.660	-0.561	1.740	-0.693	1.877	-1.037
2.2	0.709	-1.146	0.758	-1.127	0.852	-1.084	0.939	-1.038	1.018	-0.988	1.122	-0.913	1.483	-0.552	1.625	-0.520	1.736	-0.671	1.891	-1.030
2.3	0.617	-0.891	0.661	-0.877	0.747	-0.847	0.827	-0.814	0.902	-0.779	1.003	-0.725	1.413	-0.468	1.604	-0.478	1.739	-0.649	1.906	-1.025
2.4	0.550	-0.673	0.591	-0.663	0.671	-0.641	0.747	-0.618	0.819	-0.593	0.919	-0.554	1.365	-0.383	1.594	-0.435	1.747	-0.628	1.923	-1.021
2.5	0.500	-0.483	0.539	-0.475	0.616	-0.459	0.690	-0.442	0.761	-0.425	0.861	-0.398	1.335	-0.299	1.593	-0.393	1.759	-0.609	1.941	-1.018
2.6	0.461	-0.311	0.499	-0.306	0.575	-0.294	0.648	-0.282	0.719	-0.270	0.820	-0.253	1.320	-0.219	1.599	-0.354	1.774	-0.592	1.960	-1.016
2.7	0.430	-0.153	0.469	-0.149	0.545	-0.140	0.619	-0.133	0.691	-0.126	0.794	-0.117	1.317	-0.142	1.612	-0.319	1.792	-0.578	1.978	-1.015
2.8	0.406	-0.003	0.445	-0.000	0.523	0.005	0.599	0.009	0.673	0.012	0.779	0.013	1.325	-0.069	1.630	-0.287	1.810	-0.566	1.996	-1.015
2.9	0.386	0.140	0.427	0.142	0.508	0.145	0.587	0.145	0.664	0.144	0.775	0.139	1.342	-0.002	1.652	-0.260	1.830	-0.557	2.014	-1.015
3.0	0.371	0.281	0.414	0.282	0.499	0.282	0.582	0.279	0.664	0.274	0.780	0.261	1.368	0.061	1.677	-0.236	1.850	-0.550	2.031	-1.016
3.1	0.359	0.422	0.405	0.422	0.496	0.419	0.585	0.413	0.671	0.402	0.795	0.381	1.402	0.117	1.703	-0.217	1.870	-0.545	2.048	-1.018
π	0.356	0.481	0.403	0.481	0.497	0.476	0.588	0.468	0.677	0.456	0.804	0.431	1.418	0.139	1.715	-0.211	1.878	-0.544	2.054	-1.018
3.2	0.351	0.565	0.401	0.564	0.499	0.558	0.595	0.547	0.688	0.532	0.819	0.501	1.442	0.167	1.731	-0.203	1.890	-0.542	2.064	-1.019
3.3	0.347	0.714	0.401	0.711	0.509	0.701	0.613	0.685	0.713	0.663	0.855	0.621	1.487	0.210	1.759	-0.192	1.908	-0.540	2.079	-1.021
3.4	0.346	0.871	0.407	0.866	0.526	0.852	0.641	0.828	0.750	0.798	0.902	0.742	1.538	0.245	1.787	-0.186	1.926	-0.540	2.095	-1.023
3.5	0.351	1.040	0.419	1.033	0.553	1.012	0.681	0.979	0.801	0.938	0.965	0.863	1.591	0.271	1.813	-0.182	1.943	-0.540	2.109	-1.025
3.6	0.362	1.226	0.441	1.216	0.593	1.185	0.737	1.139	0.869	1.083	1.045	0.984	1.647	0.289	1.837	-0.182	1.958	-0.541	2.123	-1.027
3.7	0.382	1.437	0.475	1.421	0.651	1.375	0.814	1.311	0.960	1.234	1.147	1.103	1.702	0.297	1.860	-0.184	1.973	-0.542	2.137	-1.029
3.8	0.417	1.680	0.528	1.656	0.735	1.587	0.920	1.496	1.081	1.390	1.275	1.216	1.757	0.296	1.880	-0.188	1.986	-0.543	2.151	-1.031
3.9	0.475	1.970	0.610	1.932	0.858	1.827	1.068	1.694	1.241	1.546	1.435	1.317	1.808	0.286	1.898	-0.193	1.999	-0.545	2.164	-1.032
4.0	0.572	2.326	0.743	2.263	1.041	2.098	1.276	1.900	1.453	1.693	1.630	1.395	1.853	0.269	1.914	-0.198	2.011	-0.546	2.177	-1.034
4.1	0.743	2.779	0.963	2.669	1.319	2.398	1.569	2.101	1.732	1.813	1.862	1.433	1.892	0.245	1.927	-0.204	2.022	-0.547	2.190	-1.035
4.2	1.060	3.373	1.345	3.168	1.751	2.708	1.978	2.260	2.085	1.871	2.123	1.411	1.923	0.216	1.939	-0.209	2.033	-0.548	2.202	-1.036
4.3	1.698	4.159	2.047	3.746	2.420	2.952	2.524	2.305	2.503	1.816	2.393	1.306	1.946	0.185	1.948	-0.214	2.043	-0.548	2.214	-1.037
4.4	3.096	5.073	3.362	4.216	3.384	2.930	3.173	2.114	2.934	1.591	2.632	1.107	1.960	0.153	1.956	-0.218	2.053	-0.548	2.226	-1.038
4.5	6.056	5.172	5.495	3.777	4.478	2.294	3.764	1.577	3.274	1.174	2.795	0.826	1.965	0.123	1.963	-0.221	2.063	-0.548	2.238	-1.039
4.6	8.978	1.571	7.087	1.280	5.075	0.945	4.034	0.756	3.409	0.632	2.847	0.507	1.965	0.097	1.969	-0.223	2.072	-0.548	2.250	-1.040
4.7	7.027	-2.510	6.093	-1.426	4.715	-0.438	3.856	-0.058	3.302	0.108	2.783	0.204	1.958	0.075	1.975	-0.224	2.081	-0.547	2.262	-1.041
3π/2	6.631	-2.734	5.859	-1.630	4.620	-0.571	3.809	-0.144	3.274	0.050	2.768	0.170	1.957	0.072	1.975	-0.224	2.083	-0.547	2.264	-1.041
4.8	4.291	-3.157	4.246	-2.311	3.834	-1.203	3.386	-0.609	3.021	-0.283	2.633	-0.036	1.948	0.058	1.980	-0.224	2.091	-0.546	2.274	-1.041
4.9	2.774	-2.703	2.946	-2.216	2.991	-1.411	2.854	-0.861	2.675	-0.506	2.437	-0.196	1.935	0.046	1.985	-0.223	2.100	-0.545	2.285	-1.042
5.0	1.972	-2.163	2.164	-1.879	2.359	-1.343	2.389	-0.910	2.342	-0.591	2.231	-0.280	1.920	0.040	1.990	-0.221	2.109	-0.544	2.297	-1.042
5.1	1.515	-1.705	1.686	-1.530	1.915	-1.170	2.022	-0.847	2.056	-0.584	2.039	-0.303	1.906	0.039	1.996	-0.218	2.118	-0.542	2.308	-1.043
5.2	1.232	-1.333	1.380	-1.219	1.604	-0.971	1.745	-0.732	1.823	-0.523	1.871	-0.284	1.893	0.043	2.001	-0.215	2.128	-0.541	2.319	-1.044
5.3	1.045	-1.028	1.173	-0.949	1.384	-0.774	1.536	-0.596	1.638	-0.433	1.729	-0.235	1.882	0.050	2.007	-0.211	2.137	-0.539	2.330	-1.044
5.4	0.914	-0.771	1.028	-0.715	1.225	-0.587	1.379	-0.454	1.495	-0.328	1.613	-0.169	1.873	0.060	2.014	-0.208	2.146	-0.538	2.341	-1.044
5.5	0.819	-0.550	0.922	-0.509	1.107	-0.414	1.260	-0.314	1.383	-0.217	1.521	-0.091	1.866	0.073	2.021	-0.203	2.155	-0.536	2.352	-1.045
5.6	0.748	-0.354	0.844	-0.323	1.019	-0.253	1.171	-0.177	1.298	-0.104	1.448	-0.008	1.863	0.088	2.028	-0.199	2.164	-0.535	2.362	-1.045
5.7	0.693	-0.178	0.784	-0.154	0.953	-0.101	1.103	-0.045	1.234	0.009	1.393	0.080	1.862	0.103	2.035	-0.195	2.173	-0.533	2.373	-1.046
5.8	0.650	-0.014	0.738	0.004	0.904	0.043	1.054	0.083	1.186	0.121	1.354	0.168	1.864	0.119	2.043	-0.191	2.182	-0.532	2.384	-1.046
5.9	0.616	0.140	0.703	0.154	0.868	0.181	1.018	0.208	1.154	0.232	1.327	0.258	1.868	0.135	2.051	-0.188	2.191	-0.530	2.394	-1.046
6.0	0.589	0.288	0.676	0.298	0.842	0.316	0.996	0.330	1.134	0.341	1.313	0.347	1.875	0.150	2.059	-0.184	2.200	-0.529	2.404	-1.046
6.1	0.569	0.433	0.657	0.439	0.827	0.448	0.984	0.451	1.127	0.449	1.311	0.435	1.883	0.164	2.068	-0.181	2.209	-0.527	2.414	-1.046
6.2	0.553	0.577	0.645	0.581	0.820	0.580	0.983	0.571	1.130	0.556	1.319	0.521	1.894	0.177	2.076	-0.177	2.218	-0.526	2.425	-1.047
2π	0.544	0.699	0.639	0.699	0.822	0.691	0.990	0.672	1.142	0.644	1.335	0.592	1.904	0.187	2.083	-0.175	2.225	-0.525	2.433	-1.047
[illegible]	[illegible]	[illegible]	[illegible]	[illegible]	[illegible]	[illegible]	[illegible]	[illegible]	[illegible]	[illegible]	[illegible]	[illegible]	[illegible]	[illegible]	[illegible]	[illegible]	2.226	-0.524	2.435	-1.047
6.4	0.537	0.875	0.640	0.871	0.835	0.850	1.014	0.814	1.171	0.768	1.369	0.688	1.919	0.199	2.093	-0.172	2.235	-0.523	2.445	-1.047
6.5	0.537	1.034	0.648	1.026	0.858	0.991	1.047	0.938	1.211	0.873	1.410	0.767	1.933	0.208	2.101	-0.169	2.243	-0.522	2.455	-1.047
6.6	0.543	1.205	0.666	1.190	0.895	1.138	1.095	1.064	1.265	0.977	1.464	0.841	1.947	0.214	2.109	-0.167	2.252	-0.520	2.464	-1.047
6.7	0.558	1.391	0.696	1.368	0.948	1.293	1.161	1.191	1.335	1.078	1.529	0.909	1.961	0.219	2.117	-0.164	2.260	-0.519	2.474	-1.047
6.8	0.586	1.600	0.743	1.565	1.022	1.457	1.249	1.320	1.423	1.174	1.607	0.967	1.974	0.223	2.124	-0.162	2.268	-0.517	2.484	-1.047
6.9	0.630	1.838	0.813	1.785	1.125	1.631	1.363	1.447	1.533	1.261	1.696	1.013	1.988	0.225	2.132	-0.160	2.276	-0.516	2.493	-1.047
7.0	0.703	2.117	0.918	2.035	1.268	1.814	1.509	1.567	1.665	1.334	1.796	1.044	2.000	0.225	2.140	-0.158	2.284	-0.514	2.503	-1.047
7.1	0.820	2.453	1.079	2.323	1.465	1.999	1.695	1.671	1.820	1.385	1.904	1.054	2.011	0.225	2.147	-0.156	2.292	-0.513	2.512	-1.047
7.2	1.016	2.865	1.329	2.653	1.735	2.174	1.925	1.743	1.997	1.403	2.016	1.042	2.021	0.224	2.154	-0.154	2.300	-0.512	2.522	-1.047
7.3	1.358	3.379	1.727	3.018	2.101	2.309	2.198	1.762	2.188	1.377	2.125	1.003	2.030	0.223	2.161	-0.152	2.308	-0.510	2.531	-1.047
7.4	1.990	4.004	2.370	3.367	2.575	2.344	2.501	1.698	2.378	1.297	2.224	0.939	2.038	0.221	2.168	-0.149	2.316	-0.509	2.540	-1.047
7.5	3.208	4.625	3.378	3.525	3.130	2.194	2.799	1.528	2.547	1.159	2.303	0.851	2.045	0.220	2.175	-0.147	2.324	-0.507	2.549	-1.046
7.6	5.399	4.574	4.714	3.078	3.656	1.774	3.039	1.244	2.670	0.971	2.356	0.745	2.050	0.218	2.181	-0.145	2.332	-0.506	2.558	-1.046
7.7	7.667	2.297	5.732	1.635	3.962	1.105	3.162	0.880	2.728	0.752	2.378	0.632	2.055	0.217	2.188	-0.143	2.339	-0.504	2.567	-1.046
7.8	7.095	-1.080	5.540	-0.170	3.916	0.374	3.141	0.503	2.713	0.533	2.368	0.522	2.059	0.217	2.194	-0.141	2.347	-0.502	2.576	-1.046
5π/2	5.927	-2.060	5.014	-0.869	3.759	0.035	3.074	0.319	2.678	0.423	2.351	0.467	2.061	0.217	2.198	-0.139	2.351	-0.502	2.581	-1.046
7.9	4.951	-2.407	4.488	-1.247	3.577	-0.200	2.993	0.181	2.635	0.339	2.331	0.423	2.062	0.217	2.201	-0.138	2.354	-0.501	2.585	-1.046
8.0	3.372	-2.371	3.427	-1.545	3.116	-0.529	2.768	-0.046	2.512	0.190	2.272	0.343	2.065	0.218	2.207	-0.136	2.362	-0.499	2.594	-1.045
8.1	2.439	-1.994	2.646	-1.463	2.669	-0.652	2.519	-0.175	2.365	0.091	2.200	0.285	2.068	0.220	2.214	-0.134	2.369	-0.498	2.603	-1.045
8.2	1.884	-1.596	2.114	-1.252	2.291	-0.644	2.279	-0.223	2.214	0.039	2.122	0.249	2.070	0.222	2.220	-0.131	2.377	-0.496	2.612	-1.045
8.3	1.534	-1.245	1.751	-1.014	1.992	-0.564	2.067	-0.214	2.071	0.026	2.042	0.234	2.073	0.225	2.226	-0.129	2.384	-0.495	2.620	-1.044
8.4	1.301	-0.945	1.499	-0.784	1.760	-0.450	1.887	-0.166	1.941	0.044	1.967	0.237	2.075	0.228	2.233	-0.126	2.391	-0.493	2.629	-1.044
8.5	1.139	-0.688	1.318	-0.572	1.581	-0.321	1.740	-0.094	1.829	0.083	1.899	0.255	2.078	0.232	2.239	-0.124	2.398	-0.491	2.638	-1.044
8.6	1.021	-0.464	1.185	-0.378	1.444	-0.188	1.620	-0.008	1.735	0.138	1.838	0.285	2.082	0.237	2.245	-0.121	2.406	-0.490	2.646	-1.043
8.7	0.933	-0.264	1.085	-0.199	1.338	-0.054	1.524	0.086	1.657	0.203	1.788	0.323	2.085	0.242	2.251	-0.119	2.413	-0.488	2.655	-1.043
8.8	0.866	-0.083	1.009	-0.034	1.256	0.077	1.449	0.185	1.595	0.276	1.746	0.369	2.089	0.247	2.258	-0.116	2.420	-0.487	2.663	-1.042
8.9	0.813	0.084	0.950	0.122	1.194	0.205	1.392	0.286	1.547	0.353	1.715	0.419	2.093	0.252	2.264	-0.114	2.427	-0.485	2.671	-1.042
9.0	0.772	0.242	0.906	0.271	1.148	0.331	1.350	0.387	1.512	0.432	1.692	0.472	2.098	0.257	2.270	-0.111	2.434	-0.483	2.680	-1.042
9.1	0.739	0.394	0.872	0.415	1.115	0.455	1.321	0.490	1.490	0.514	1.679	0.527	2.103	0.262	2.276	-0.109	2.441	-0.482	2.688	-1.041
9.2	0.714	0.541	0.848	0.556	1.094	0.579	1.305	0.592	1.479	0.596	1.674	0.583	2.109	0.267	2.282	-0.106	2.448	-0.480	2.696	-1.041
9.3	0.695	0.688	0.832	0.696	1.084	0.702	1.301	0.695	1.478	0.677	1.678	0.638	2.114	0.272	2.288	-0.104	2.455	-0.478	2.705	-1.040
9.4	0.682	0.836	0.824	0.838	1.085	0.826	1.308	0.797	1.489	0.758	1.690	0.692	2.120	0.276	2.294	-0.101	2.462	-0.477	2.713	-1.040
3π	0.680	0.874	0.824	0.874	1.088	0.857	1.312	0.823	1.493	0.778	1.695	0.705	2.122	0.277	2.296	-0.101	2.463	-0.476	2.715	-1.040
9.5	0.675	0.989	0.825	0.984	1.098	0.952	1.327	0.899	1.511	0.838	1.710	0.743	2.126	0.280	2.300	-0.099	2.469	-0.475	2.721	-1.039
9.6	0.675	1.149	0.835	1.135	1.123	1.080	1.359	1.001	1.544	0.915	1.738	0.791	2.133	0.284	2.306	-0.097	2.475	-0.473	2.729	-1.039
9.7	0.682	1.318	0.856	1.295	1.163	1.211	1.406	1.102	1.588	0.988	1.773	0.834	2.139	0.288	2.312	-0.094	2.482	-0.472	2.737	-1.038
9.8	0.699	1.503	0.891	1.466	1.219	1.346	1.467	1.199	1.644	1.055	1.815	0.872	2.145	0.291	2.318	-0.092	2.489	-0.470	2.745	-1.037
9.9	0.729	1.707	0.944	1.651	1.297	1.484	1.546	1.293	1.713	1.115	1.863	0.902	2.151	0.295	2.324	-0.089	2.495	-0.468	2.753	-1.037
10.0	0.778	1.938	1.022	1.855	1.401	1.623	1.645	1.378	1.793	1.165	1.915	0.923	2.157	0.298	2.330	-0.087	2.502	-0.467	2.761	-1.036
10.1	0.855	2.205	1.136	2.080	1.538	1.760	1.764	1.450	1.884	1.200										
10.2	0.977	2.519	1.303	2.328	1.717	1.886	1.906	1.503	1.985	1.219										
10.3	1.175	2.896	1.551	2.598	1.944	1.989	2.067	1.528	2.091	1.216										
10.4	1.507	3.350	1.921	2.873	2.226	2.046	2.243	1.514	2.197	1.189										
10.5	2.085	3.875	2.472	3.101	2.557	2.024	2.422	1.453	2.296	1.136										
10.6	3.120	4.367	3.251	3.156	2.908	1.887	2.586	1.339	2.381	1.059										
10.7	4.855	4.367	4.198	2.807	3.223	1.612	2.716	1.176	2.444	0.963										
10.8	6.818	2.852	4.969	1.863	3.423	1.215	2.791	0.979	2.478	0.854										
10.9	7.035	0.042	5.075	0.579	3.454	0.768	2.801	0.770	2.481	0.743										
7π/2	5.539	-1.649	4.539	-0.405	3.326	0.380	2.752	0.584	2.457	0.643										

βh	a/β = 0.00		a/β = 0.01		a/β = 0.03		a/β = 0.05		a/β = 0.07	
11.0	5.460	-1.690	4.505	-0.439	3.317	0.364	2.749	0.576	2.455	0.639
11.1	3.899	-2.032	3.706	-0.932	3.069	0.068	2.647	0.416	2.405	0.550
11.2	2.864	-1.826	2.997	-1.041	2.778	-0.107	2.515	0.301	2.338	0.481
11.3	2.218	-1.495	2.457	-0.958	2.494	-0.179	2.370	0.231	2.260	0.434
11.4	1.804	-1.169	2.064	-0.798	2.241	-0.178	2.227	0.202	2.179	0.408
11.5	1.528	-0.876	1.779	-0.614	2.028	-0.129	2.093	0.205	2.099	0.402
11.6	1.335	-0.620	1.570	-0.429	1.854	-0.051	1.975	0.233	2.024	0.413
11.7	1.195	-0.393	1.414	-0.251	1.713	0.043	1.872	0.280	1.956	0.437
11.8	1.091	-0.190	1.296	-0.083	1.600	0.146	1.786	0.339	1.898	0.472
11.9	1.012	-0.005	1.206	0.077	1.511	0.254	1.716	0.407	1.848	0.515
12.0	0.950	0.166	1.136	0.229	1.441	0.364	1.659	0.482	1.809	0.564
12.1	0.902	0.326	1.083	0.374	1.389	0.475	1.617	0.560	1.778	0.618
12.2	0.864	0.480	1.042	0.516	1.350	0.586	1.586	0.642	1.758	0.675
12.3	0.834	0.630	1.013	0.655	1.325	0.698	1.568	0.724	1.746	0.733
12.4	0.812	0.779	0.994	0.794	1.312	0.810	1.561	0.808	1.744	0.792
12.5	0.797	0.928	0.984	0.934	1.310	0.923	1.565	0.892	1.750	0.851
4π	0.791	1.029	0.982	1.028	1.316	0.998	1.574	0.947	1.759	0.889
12.6	0.788	1.081	0.983	1.076	1.321	1.036	1.580	0.975	1.765	0.908
12.7	0.787	1.240	0.993	1.223	1.345	1.151	1.607	1.056	1.789	0.963
12.8	0.794	1.409	1.015	1.377	1.383	1.267	1.646	1.135	1.822	1.013
12.9	0.811	1.590	1.052	1.540	1.438	1.382	1.697	1.209	1.862	1.059
13.0	0.841	1.790	1.107	1.714	1.512	1.497	1.762	1.277	1.910	1.098
13.1	0.891	2.013	1.188	1.902	1.608	1.609	1.840	1.335	1.965	1.128
13.2	0.968	2.267	1.303	2.104	1.730	1.712	1.931	1.382	2.025	1.148
13.3	1.089	2.563	1.467	2.320	1.882	1.801	2.034	1.412	2.089	1.157
13.4	1.280	2.910	1.700	2.545	2.066	1.865	2.146	1.420	2.154	1.153
13.5	1.589	3.317	2.033	2.760	2.281	1.889	2.263	1.404	2.217	1.136
13.6	2.108	3.776	2.499	2.921	2.519	1.856	2.376	1.359	2.275	1.105
13.7	2.990	4.202	3.118	2.941	2.759	1.749	2.478	1.286	2.324	1.062
13.8	4.415	4.273	3.840	2.677	2.971	1.562	2.559	1.186	2.361	1.009
13.9	6.149	3.265	4.461	2.020	3.116	1.304	2.611	1.068	2.384	0.950
14.0	6.851	0.979	4.687	1.074	3.167	1.009	2.628	0.942	2.391	0.889
14.1	5.838	-0.958	4.415	0.187	3.116	0.722	2.610	0.820	2.383	0.830
9π/2	5.280	-1.344	4.223	-0.067	3.074	0.626	2.596	0.777	2.377	0.810
14.2	4.380	-1.654	3.851	-0.381	2.982	0.483	2.562	0.711	2.362	0.777
14.3	3.271	-1.639	3.249	-0.622	2.798	0.312	2.490	0.623	2.329	0.733
14.4	2.541	-1.389	2.735	-0.651	2.596	0.211	2.404	0.560	2.289	0.700
14.5	2.063	-1.095	2.333	-0.567	2.398	0.170	2.311	0.523	2.243	0.678
14.6	1.742	-0.813	2.029	-0.433	2.218	0.175	2.219	0.508	2.196	0.667
14.7	1.518	-0.558	1.798	-0.279	2.061	0.213	2.131	0.514	2.149	0.667
14.8	1.356	-0.330	1.623	-0.121	1.927	0.273	2.050	0.536	2.104	0.677
14.9	1.235	-0.125	1.489	0.034	1.817	0.348	1.980	0.571	2.064	0.695
15.0	1.143	0.063	1.385	0.185	1.726	0.433	1.919	0.616	2.029	0.721
15.1	1.072	0.237	1.305	0.331	1.654	0.523	1.870	0.669	1.999	0.751
15.2	1.017	0.400	1.243	0.472	1.598	0.618	1.830	0.727	1.976	0.787
15.3	0.974	0.557	1.196	0.611	1.556	0.715	1.801	0.789	1.959	0.825
15.4	0.940	0.708	1.162	0.747	1.527	0.814	1.783	0.854	1.949	0.865
15.5	0.914	0.858	1.138	0.883	1.511	0.913	1.774	0.920	1.946	0.906
15.6	0.896	1.008	1.126	1.019	1.507	1.014	1.775	0.986	1.949	0.947
15.7	0.886	1.161	1.123	1.158	1.515	1.115	1.786	1.052	1.959	0.988
5π	0.885	1.174	1.124	1.169	1.516	1.123	1.787	1.057	1.960	0.991
15.8	0.883	1.320	1.132	1.301	1.536	1.217	1.806	1.116	1.975	1.026
15.9	0.888	1.486	1.154	1.448	1.570	1.317	1.837	1.177	1.997	1.061
16.0	0.904	1.665	1.190	1.603	1.620	1.417	1.877	1.233	2.025	1.092
16.1	0.934	1.859	1.246	1.767	1.686	1.513	1.927	1.284	2.057	1.119
16.2	0.983	2.075	1.325	1.940	1.770	1.603	1.986	1.326	2.093	1.139
16.3	1.058	2.318	1.437	2.122	1.874	1.683	2.054	1.358	2.133	1.153
16.4	1.174	2.596	1.593	2.313	2.001	1.749	2.128	1.377	2.174	1.159
16.5	1.353	2.918	1.808	2.503	2.148	1.791	2.207	1.382	2.215	1.157
16.6	1.638	3.290	2.103	2.678	2.313	1.802	2.287	1.369	2.254	1.148
16.7	2.099	3.702	2.500	2.801	2.490	1.772	2.363	1.339	2.290	1.130
16.8	2.856	4.088	3.004	2.809	2.664	1.692	2.431	1.292	2.321	1.107
16.9	4.051	4.220	3.579	2.612	2.817	1.560	2.486	1.229	2.345	1.077
17.0	5.587	3.567	4.096	2.135	2.928	1.384	2.523	1.155	2.361	1.044
17.1	6.570	1.760	4.366	1.423	2.982	1.180	2.541	1.076	2.368	1.010
17.2	6.084	-0.215	4.276	0.676	2.970	0.972	2.537	0.996	2.367	0.976
11π/2	5.090	-1.094	3.997	0.205	2.919	0.822	2.521	0.937	2.361	0.951
17.3	4.812	-1.222	3.903	0.104	2.899	0.785	2.514	0.922	2.358	0.944
17.4	3.668	-1.419	3.420	-0.218	2.784	0.636	2.475	0.858	2.343	0.917
17.5	2.861	-1.269	2.956	-0.336	2.644	0.532	2.424	0.808	2.321	0.895
17.6	2.320	-1.016	2.564	-0.323	2.496	0.473	2.364	0.774	2.296	0.879
17.7	1.951	-0.751	2.251	-0.240	2.350	0.452	2.302	0.755	2.268	0.870
17.8	1.694	-0.501	2.006	-0.121	2.216	0.462	2.239	0.751	2.240	0.868
17.9	1.508	-0.272	1.816	0.014	2.097	0.496	2.179	0.759	2.212	0.872
18.0	1.371	-0.065	1.668	0.154	1.994	0.547	2.125	0.779	2.186	0.882
18.1	1.266	0.125	1.553	0.294	1.908	0.611	2.076	0.807	2.162	0.897
18.2	1.186	0.302	1.463	0.432	1.837	0.683	2.035	0.843	2.141	0.916
18.3	1.123	0.468	1.393	0.568	1.780	0.761	2.001	0.885	2.125	0.939
18.4	1.073	0.626	1.339	0.702	1.736	0.844	1.976	0.931	2.113	0.964
18.5	1.035	0.779	1.299	0.835	1.706	0.929	1.958	0.981	2.105	0.991
18.6	1.006	0.930	1.272	0.967	1.687	1.017	1.949	1.032	2.102	1.020
18.7	0.985	1.081	1.256	1.100	1.680	1.105	1.948	1.083	2.103	1.048
18.8	0.972	1.234	1.251	1.234	1.685	1.194	1.954	1.135	2.109	1.076
6π	0.969	1.311	1.253	1.302	1.692	1.238	1.961	1.160	2.114	1.090
18.9	0.967	1.391	1.258	1.372	1.702	1.283	1.969	1.184	2.120	1.103
19.0	0.971	1.556	1.278	1.514	1.731	1.371	1.992	1.232	2.134	1.127
19.1	0.986	1.731	1.313	1.661	1.774	1.456	2.022	1.275	2.152	1.149
19.2	1.014	1.921	1.367	1.815	1.831	1.536	2.060	1.313	2.174	1.168
19.3	1.060	2.129	1.443	1.975	1.903	1.611	2.104	1.345	2.198	1.183
19.4	1.132	2.362	1.550	2.141	1.990	1.675	2.153	1.369	2.223	1.193
19.5	1.241	2.626	1.695	2.310	2.094	1.725	2.207	1.383	2.250	1.198
19.6	1.408	2.927	1.891	2.475	2.211	1.757	2.263	1.387	2.276	1.199
19.7	1.668	3.270	2.152	2.622	2.340	1.763	2.319	1.379	2.302	1.195
19.8	2.078	3.646	2.492	2.721	2.475	1.739	2.372	1.360	2.325	1.186
19.9	2.733	4.006	2.912	2.728	2.606	1.680	2.419	1.330	2.345	1.173
20.0	3.748	4.181	3.383	2.580	2.722	1.586	2.458	1.290	2.362	1.157
20.1	5.104	3.782	3.822	2.226	2.810	1.461	2.487	1.242	2.373	1.139
20.2	6.225	2.397	4.101	1.681	2.860	1.315	2.502	1.191	2.380	1.120
20.3	6.195	0.519	4.121	1.062	2.867	1.162	2.505	1.138	2.382	1.100
20.4	5.184	-0.733	3.894	0.525	2.832	1.018	2.495	1.087	2.379	1.081
13π/2	4.942	-0.876	3.826	0.436	2.820	0.990	2.491	1.077	2.378	1.078
20.5	4.054	-1.157	3.525	0.163	2.761	0.894	2.473	1.042	2.371	1.065
20.6	3.184	-1.128	3.124	-0.020	2.666	0.799	2.443	1.005	2.361	1.051
20.7	2.579	-0.929	2.757	-0.070	2.557	0.735	2.405	0.977	2.347	1.041
20.8	2.162	-0.686	2.447	-0.036	2.443	0.701	2.363	0.959	2.331	1.035
20.9	1.869	-0.444	2.196	0.048	2.333	0.693	2.320	0.951	2.314	1.033

βh	a/β = 0.00		a/β = 0.01		a/β = 0.03		a/β = 0.05		a/β = 0.07	
21.0	1.658	-0.217	1.995	0.157	2.231	0.708	2.277	0.953	2.297	1.035
21.1	1.502	-0.008	1.836	0.279	2.140	0.740	2.236	0.963	2.280	1.040
21.2	1.384	0.184	1.710	0.406	2.060	0.786	2.198	0.981	2.265	1.050
21.3	1.293	0.363	1.611	0.536	1.993	0.842	2.166	1.005	2.252	1.062
21.4	1.222	0.531	1.534	0.665	1.938	0.905	2.138	1.035	2.240	1.077
21.5	1.167	0.691	1.473	0.793	1.895	0.974	2.117	1.069	2.232	1.094
21.6	1.123	0.846	1.428	0.921	1.863	1.046	2.102	1.106	2.226	1.112
21.7	1.090	0.997	1.397	1.049	1.842	1.122	2.093	1.144	2.224	1.132
21.8	1.066	1.148	1.377	1.178	1.833	1.199	2.090	1.184	2.225	1.151
21.9	1.051	1.301	1.369	1.308	1.835	1.276	2.094	1.223	2.228	1.171
7π	1.044	1.444	1.373	1.429	1.846	1.346	2.103	1.259	2.234	1.188
22.0	1.044	1.458	1.374	1.441	1.847	1.353	2.104	1.262	2.235	1.190
22.1	1.046	1.621	1.392	1.576	1.871	1.429	2.120	1.299	2.244	1.207
22.2	1.058	1.793	1.424	1.716	1.907	1.502	2.142	1.332	2.256	1.223
22.3	1.084	1.978	1.475	1.861	1.955	1.570	2.170	1.362	2.271	1.237
22.4	1.128	2.180	1.547	2.010	2.015	1.632	2.202	1.386	2.286	1.243
22.5	1.195	2.403	1.646	2.162	2.087	1.685	2.238	1.405	2.303	1.256
22.6	1.297	2.654	1.780	2.314	2.172	1.726	2.277	1.417	2.321	1.261
22.7	1.452	2.937	1.958	2.460	2.266	1.751	2.317	1.421	2.338	1.263
22.8	1.687	3.256	2.189	2.586	2.368	1.756	2.357	1.417	2.355	1.263
22.9	2.052	3.604	2.483	2.671	2.473	1.738	2.395	1.406	2.371	1.259
23.0	2.622	3.943	2.837	2.681	2.574	1.696	2.429	1.387	2.384	1.253
23.1	3.494	4.150	3.232	2.570	2.664	1.628	2.458	1.361	2.396	1.244
23.2	4.687	3.933	3.610	2.303	2.736	1.537	2.480	1.331	2.404	1.235
23.3	5.847	2.905	3.882	1.880	2.781	1.431	2.493	1.297	2.410	1.224
23.4	6.181	1.218	3.967	1.370	2.797	1.317	2.498	1.261	2.413	1.213
23.5	5.480	-0.194	3.847	0.886	2.782	1.206	2.495	1.227	2.413	1.203
15π/2	4.822	-0.679	3.693	0.639	2.758	1.141	2.489	1.207	2.411	1.197
23.6	4.421	-0.844	3.579	0.515	2.739	1.105	2.483	1.195	2.410	1.193
23.7	3.508	-0.957	3.247	0.289	2.675	1.022	2.465	1.168	2.404	1.185
23.8	2.843	-0.827	2.915	0.188	2.596	0.961	2.442	1.147	2.397	1.179
23.9	2.375	-0.615	2.618	0.177	2.510	0.923	2.414	1.132	2.388	1.176
24.0	2.045	-0.385	2.367	0.225	2.422	0.905	2.384	1.124	2.379	1.175
24.1	1.807	-0.162	2.160	0.308	2.337	0.907	2.354	1.123	2.368	1.176
24.2	1.631	0.046	1.993	0.410	2.258	0.925	2.324	1.128	2.358	1.180
24.3	1.499	0.240	1.859	0.524	2.187	0.956	2.296	1.139	2.349	1.186
24.4	1.397	0.421	1.752	0.642	2.126	0.998	2.271	1.155	2.341	1.194
24.5	1.318	0.591	1.667	0.763	2.074	1.048	2.250	1.176	2.333	1.204
24.6	1.256	0.753	1.601	0.885	2.033	1.104	2.232	1.201	2.328	1.216
24.7	1.207	0.909	1.551	1.008	2.001	1.165	2.220	1.228	2.324	1.229
24.8	1.170	1.062	1.514	1.131	1.980	1.229	2.211	1.257	2.323	1.242
24.9	1.142	1.213	1.491	1.255	1.969	1.295	2.208	1.287	2.323	1.256
25.0	1.124	1.365	1.480	1.381	1.967	1.362	2.210	1.318	2.325	1.270

TABLE 2.7

NORMALIZED ADMITTANCES (Y/Δ) IN MILLIMHOS OF THIN DIPOLE ANTENNAS IN DISSIPATIVE MEDIA

Ω = 20.0

βh	α/β = 0.00	α/β = 0.01	α/β = 0.03	α/β = 0.05	α/β = 0.07	α/β = 0.10	α/β = 0.30	α/β = 0.50	α/β = 0.70	α/β = 1.00
1.5	11.916 5.007	10.144 3.640	7.784 2.168	6.312 1.420	5.316 0.978	4.314 0.581	2.077 -0.187	1.546 -0.421	1.353 -0.596	1.266 -0.850
π/2	9.355 -5.073	8.506 -3.711	6.974 -2.154	5.817 -1.388	4.963 -0.977	4.063 -0.664	1.975 -0.388	1.487 -0.485	1.318 -0.619	1.252 -0.856
1.6	6.527 -5.640	6.388 -4.541	5.774 -3.003	5.091 -2.082	4.491 -1.523	3.783 -1.045	1.920 -0.456	1.461 -0.505	1.304 -0.626	1.247 -0.857
1.7	2.376 -4.051	2.579 -3.747	2.817 -3.143	2.883 -2.602	2.843 -2.154	2.686 -1.649	1.713 -0.616	1.372 -0.552	1.260 -0.640	1.233 -0.859
1.8	1.259 -2.813	1.391 -2.702	1.602 -2.457	1.744 -2.202	1.827 -1.957	1.869 -1.631	1.509 -0.684	1.291 -0.570	1.222 -0.641	1.224 -0.857
1.9	0.821 -2.080	0.908 -2.028	1.061 -1.907	1.183 -1.774	1.275 -1.635	1.365 -1.435	1.334 -0.689	1.221 -0.567	1.193 -0.634	1.220 -0.853
2.0	0.603 -1.604	0.665 -1.575	0.779 -1.507	0.878 -1.429	0.961 -1.345	1.056 -1.218	1.194 -0.657	1.165 -0.551	1.171 -0.622	1.219 -0.848
2.1	0.476 -1.267	0.524 -1.249	0.615 -1.207	0.697 -1.158	0.769 -1.104	0.859 -1.019	1.087 -0.605	1.122 -0.526	1.157 -0.607	1.221 -0.843
2.2	0.396 -1.013	0.435 -1.001	0.511 -0.972	0.581 -0.940	0.645 -0.903	0.730 -0.846	1.007 -0.544	1.091 -0.496	1.148 -0.592	1.225 -0.839
2.3	0.341 -0.810	0.375 -0.801	0.441 -0.782	0.504 -0.759	0.562 -0.734	0.642 -0.694	0.949 -0.479	1.069 -0.465	1.145 -0.576	1.231 -0.836
2.4	0.302 -0.641	0.333 -0.635	0.393 -0.621	0.450 -0.606	0.505 -0.588	0.581 -0.560	0.908 -0.415	1.057 -0.433	1.146 -0.561	1.239 -0.833
2.5	0.272 -0.497	0.301 -0.492	0.358 -0.482	0.413 -0.471	0.465 -0.459	0.539 -0.440	0.882 -0.352	1.052 -0.403	1.150 -0.548	1.247 -0.831
2.6	0.250 -0.369	0.278 -0.365	0.332 -0.358	0.386 -0.351	0.437 -0.342	0.510 -0.330	0.867 -0.292	1.053 -0.375	1.157 -0.536	1.255 -0.831
2.7	0.232 -0.252	0.260 -0.250	0.314 -0.245	0.367 -0.240	0.418 -0.235	0.492 -0.229	0.862 -0.235	1.058 -0.349	1.166 -0.527	1.264 -0.831
2.8	0.218 -0.144	0.246 -0.143	0.301 -0.140	0.355 -0.137	0.407 -0.135	0.482 -0.134	0.865 -0.182	1.068 -0.327	1.176 -0.519	1.272 -0.831
2.9	0.207 -0.042	0.236 -0.040	0.292 -0.039	0.347 -0.039	0.401 -0.039	0.479 -0.043	0.875 -0.133	1.081 -0.307	1.187 -0.513	1.281 -0.833
3.0	0.198 0.058	0.228 0.059	0.287 0.059	0.345 0.057	0.401 0.053	0.483 0.045	0.892 -0.087	1.096 -0.291	1.198 -0.509	1.289 -0.834
3.1	0.192 0.157	0.223 0.157	0.286 0.155	0.347 0.151	0.406 0.145	0.492 0.131	0.914 -0.047	1.113 -0.278	1.209 -0.507	1.296 -0.836
π	0.189 0.198	0.222 0.198	0.286 0.195	0.349 0.190	0.410 0.182	0.498 0.166	0.925 -0.031	1.120 -0.274	1.214 -0.506	1.299 -0.837
3.2	0.186 0.256	0.220 0.256	0.288 0.252	0.353 0.245	0.417 0.236	0.508 0.216	0.941 -0.011	1.131 -0.269	1.220 -0.506	1.304 -0.839
3.3	0.183 0.358	0.220 0.357	0.293 0.351	0.365 0.342	0.434 0.328	0.532 0.301	0.972 0.020	1.149 -0.262	1.230 -0.506	1.311 -0.841
3.4	0.182 0.465	0.223 0.463	0.304 0.455	0.382 0.441	0.457 0.422	0.563 0.386	1.007 0.045	1.166 -0.259	1.240 -0.507	1.317 -0.843
3.5	0.182 0.580	0.229 0.577	0.319 0.565	0.407 0.545	0.490 0.520	0.604 0.471	1.044 0.064	1.183 -0.258	1.248 -0.509	1.323 -0.846
3.6	0.186 0.705	0.239 0.701	0.343 0.683	0.441 0.657	0.533 0.622	0.656 0.558	1.083 0.076	1.199 -0.259	1.256 -0.511	1.330 -0.848
3.7	0.194 0.846	0.256 0.839	0.375 0.814	0.488 0.776	0.590 0.729	0.723 0.644	1.123 0.082	1.213 -0.262	1.264 -0.513	1.335 -0.851
3.8	0.207 1.008	0.281 0.997	0.423 0.960	0.552 0.907	0.666 0.841	0.808 0.729	1.162 0.081	1.225 -0.267	1.270 -0.516	1.341 -0.853
3.9	0.230 1.199	0.321 1.182	0.491 1.127	0.641 1.049	0.768 0.958	0.915 0.808	1.199 0.072	1.236 -0.272	1.276 -0.518	1.346 -0.855
4.0	0.269 1.435	0.384 1.407	0.593 1.320	0.767 1.205	0.904 1.075	1.048 0.876	1.232 0.058	1.245 -0.278	1.282 -0.521	1.352 -0.857
4.1	0.338 1.736	0.488 1.687	0.749 1.545	0.948 1.368	1.088 1.183	1.212 0.921	1.261 0.038	1.253 -0.285	1.287 -0.523	1.357 -0.859
4.2	0.464 2.141	0.670 2.050	0.997 1.804	1.210 1.527	1.332 1.262	1.405 0.928	1.284 0.015	1.258 -0.291	1.291 -0.525	1.362 -0.861
4.3	0.721 2.718	1.013 2.529	1.404 2.077	1.587 1.640	1.643 1.277	1.618 0.877	1.300 -0.011	1.263 -0.296	1.295 -0.527	1.367 -0.863
4.4	1.321 3.582	1.722 3.135	2.075 2.271	2.097 1.621	2.004 1.173	1.827 0.750	1.310 -0.038	1.266 -0.302	1.299 -0.529	1.372 -0.865
4.5	2.999 4.761	3.278 3.605	3.069 2.106	2.677 1.329	2.346 0.899	1.992 0.544	1.313 -0.064	1.268 -0.306	1.303 -0.530	1.377 -0.866
4.6	7.485 4.038	5.844 2.372	4.020 1.140	3.094 0.687	2.551 0.463	2.072 0.283	1.311 -0.087	1.270 -0.309	1.307 -0.532	1.381 -0.868
4.7	7.660 -2.577	5.944 -1.216	4.038 -0.348	3.083 -0.106	2.530 -0.021	2.048 0.017	1.304 -0.108	1.271 -0.312	1.311 -0.532	1.386 -0.870
3π/2	7.067 -3.043	5.686 -1.549	3.962 -0.510	3.050 -0.196	2.512 -0.076	2.038 -0.014	1.303 -0.110	1.271 -0.312	1.311 -0.533	1.386 -0.870
4.8	3.647 -3.653	3.682 -2.571	3.198 -1.288	2.691 -0.698	2.313 -0.408	1.936 -0.206	1.294 -0.124	1.272 -0.314	1.314 -0.533	1.390 -0.871
4.9	1.952 -2.850	2.216 -2.358	2.325 -1.520	2.190 -0.970	2.008 -0.635	1.773 -0.359	1.282 -0.136	1.273 -0.315	1.318 -0.534	1.395 -0.873
5.0	1.247 -2.168	1.473 -1.926	1.710 -1.428	1.753 -1.018	1.708 -0.722	1.597 -0.442	1.268 -0.144	1.274 -0.316	1.321 -0.535	1.400 -0.874
5.1	0.899 -1.682	1.074 -1.548	1.312 -1.243	1.423 -0.955	1.454 -0.719	1.432 -0.468	1.255 -0.148	1.275 -0.316	1.325 -0.535	1.404 -0.876
5.2	0.703 -1.327	0.840 -1.245	1.052 -1.049	1.183 -0.846	1.253 -0.666	1.288 -0.456	1.242 -0.148	1.277 -0.315	1.329 -0.535	1.408 -0.877
5.3	0.580 -1.055	0.692 -1.002	0.878 -0.869	1.011 -0.725	1.098 -0.589	1.169 -0.419	1.230 -0.145	1.279 -0.314	1.332 -0.536	1.413 -0.878
5.4	0.499 -0.838	0.593 -0.801	0.757 -0.709	0.886 -0.604	0.980 -0.501	1.072 -0.367	1.220 -0.140	1.281 -0.313	1.336 -0.536	1.417 -0.880
5.5	0.441 -0.658	0.523 -0.632	0.671 -0.566	0.794 -0.489	0.891 -0.411	0.996 -0.307	1.213 -0.132	1.283 -0.312	1.340 -0.536	1.421 -0.881
5.6	0.399 -0.505	0.473 -0.485	0.609 -0.437	0.726 -0.380	0.824 -0.322	0.937 -0.242	1.207 -0.124	1.286 -0.311	1.343 -0.537	1.425 -0.883
5.7	0.367 -0.369	0.435 -0.355	0.563 -0.319	0.676 -0.277	0.774 -0.235	0.892 -0.176	1.203 -0.114	1.289 -0.309	1.347 -0.537	1.429 -0.884
5.8	0.342 -0.247	0.407 -0.236	0.529 -0.210	0.639 -0.180	0.737 -0.150	0.859 -0.109	1.202 -0.104	1.292 -0.308	1.351 -0.538	1.434 -0.885
5.9	0.323 -0.134	0.385 -0.125	0.504 -0.107	0.614 -0.087	0.712 -0.067	0.837 -0.042	1.203 -0.093	1.295 -0.307	1.354 -0.538	1.438 -0.886
6.0	0.307 -0.027	0.369 -0.021	0.487 -0.009	0.597 0.003	0.696 0.013	0.824 0.023	1.205 -0.084	1.299 -0.306	1.358 -0.538	1.442 -0.888
6.1	0.295 0.076	0.357 0.080	0.476 0.087	0.587 0.091	0.688 0.092	0.820 0.088	1.209 -0.074	1.302 -0.305	1.362 -0.539	1.445 -0.889
6.2	0.286 0.177	0.349 0.180	0.471 0.181	0.585 0.177	0.689 0.169	0.823 0.151	1.215 -0.066	1.306 -0.304	1.365 -0.539	1.449 -0.890
2π	0.280 0.262	0.345 0.263	0.471 0.259	0.588 0.249	0.695 0.233	0.832 0.202	1.221 -0.060	1.309 -0.304	1.368 -0.539	1.453 -0.891
6.3	[illegible]	[illegible]	[illegible]	0.590 0.263	0.697 0.246	0.834 0.213	1.222 -0.059	1.310 -0.304	1.369 -0.540	1.453 -0.891
6.4	0.274 0.382	0.344 0.381	[illegible]	[illegible]	[illegible]	[illegible]	[illegible]	[illegible]	[illegible]	1.457 -0.893
6.5	0.272 0.490	0.347 0.487	0.491 0.468	0.622 0.437	0.737 0.398	0.880 0.330	1.238 -0.048	1.317 -0.303	1.376 -0.540	1.461 -0.894
6.6	0.273 0.605	0.355 0.598	0.511 0.570	0.651 0.527	0.771 0.473	0.915 0.385	1.247 -0.044	1.321 -0.303	1.379 -0.541	1.464 -0.895
6.7	0.278 0.729	0.369 0.719	0.541 0.678	0.691 0.619	0.816 0.548	0.959 0.436	1.255 -0.042	1.325 -0.303	1.382 -0.541	1.468 -0.896
6.8	0.287 0.867	0.391 0.851	0.583 0.794	0.745 0.712	0.873 0.620	1.011 0.482	1.264 -0.041	1.328 -0.303	1.386 -0.542	1.472 -0.897
6.9	0.303 1.023	0.425 1.000	0.642 0.918	0.816 0.808	0.945 0.688	1.074 0.520	1.272 -0.042	1.331 -0.303	1.389 -0.542	1.475 -0.898
7.0	0.330 1.206	0.475 1.170	0.724 1.052	0.908 0.902	1.034 0.750	1.145 0.548	1.280 -0.043	1.335 -0.303	1.392 -0.543	1.479 -0.899
7.1	0.374 1.427	0.551 1.371	0.838 1.197	1.029 0.992	1.142 0.799	1.225 0.562	1.287 -0.045	1.338 -0.303	1.395 -0.543	1.483 -0.900
7.2	0.448 1.704	0.671 1.611	1.000 1.347	1.183 1.069	1.270 0.829	1.309 0.561	1.293 -0.047	1.341 -0.304	1.398 -0.543	1.486 -0.902
7.3	0.579 2.063	0.866 1.902	1.229 1.494	1.377 1.118	1.414 0.830	1.396 0.539	1.298 -0.050	1.344 -0.304	1.402 -0.544	1.490 -0.903
7.4	0.826 2.552	1.199 2.250	1.553 1.606	1.611 1.117	1.569 0.792	1.478 0.497	1.303 -0.054	1.347 -0.304	1.405 -0.544	1.493 -0.904
7.5	1.348 3.235	1.796 2.623	1.989 1.621	1.869 1.038	1.720 0.706	1.549 0.433	1.306 -0.057	1.349 -0.304	1.408 -0.544	1.496 -0.905
7.6	2.597 4.098	2.860 2.816	2.506 1.434	2.114 0.854	1.845 0.570	1.602 0.351	1.309 -0.060	1.352 -0.304	1.411 -0.545	1.500 -0.906
7.7	5.583 4.095	4.375 2.160	2.956 0.951	2.287 0.569	1.923 0.394	1.630 0.259	1.311 -0.063	1.355 -0.304	1.414 -0.545	1.503 -0.907
7.8	7.780 -0.239	5.079 0.166	3.103 0.251	2.332 0.231	1.938 0.203	1.632 0.164	1.312 -0.066	1.357 -0.304	1.417 -0.546	1.506 -0.908
5π/2	6.334 -2.439	4.669 -0.893	3.018 -0.121	2.298 0.052	1.920 0.103	1.622 0.115	1.312 -0.067	1.359 -0.304	1.418 -0.546	1.508 -0.908
7.9	4.791 -3.109	4.069 -1.483	2.872 -0.390	2.241 -0.086	1.891 0.024	1.608 0.076	1.313 -0.068	1.360 -0.304	1.420 -0.546	1.510 -0.909
8.0	2.620 -2.877	2.788 -1.889	2.436 -0.774	2.055 -0.321	1.795 -0.121	1.564 0.002	1.313 -0.069	1.363 -0.304	1.423 -0.546	1.513 -0.910
8.1	1.640 -2.272	1.936 -1.743	1.994 -0.914	1.833 -0.458	1.673 -0.220	1.506 -0.055	1.313 -0.070	1.365 -0.304	1.425 -0.547	1.516 -0.911
8.2	1.159 -1.778	1.427 -1.479	1.633 -0.906	1.616 -0.512	1.543 -0.275	1.441 -0.092	1.313 -0.071	1.368 -0.304	1.428 -0.547	1.519 -0.911
8.3	0.892 -1.403	1.115 -1.221	1.360 -0.826	1.427 -0.509	1.419 -0.294	1.374 -0.111	1.313 -0.071	1.370 -0.304	1.431 -0.547	1.523 -0.912
8.4	0.729 -1.114	0.914 -0.995	1.159 -0.719	1.270 -0.470	1.307 -0.285	1.310 -0.114	1.313 -0.070	1.373 -0.304	1.434 -0.547	1.526 -0.913
8.5	0.622 -0.883	0.779 -0.802	1.011 -0.603	1.144 -0.411	1.211 -0.256	1.252 -0.105	1.313 -0.069	1.375 -0.304	1.437 -0.548	1.529 -0.914
8.6	0.548 -0.692	0.684 -0.634	0.900 -0.489	1.044 -0.340	1.131 -0.215	1.201 -0.085	1.314 -0.068	1.378 -0.304	1.440 -0.548	1.532 -0.915
8.7	0.494 -0.529	0.615 -0.487	0.818 -0.379	0.965 -0.265	1.066 -0.165	1.157 -0.058	1.315 -0.066	1.380 -0.303	1.442 -0.548	1.535 -0.916
8.8	0.454 -0.387	0.564 -0.355	0.756 -0.275	0.905 -0.188	1.014 -0.110	1.121 -0.026	1.316 -0.064	1.383 -0.303	1.445 -0.549	1.538 -0.917
8.9	0.423 -0.259	0.526 -0.235	0.710 -0.175	0.859 -0.111	0.973 -0.052	1.093 0.011	1.317 -0.062	1.385 -0.303	1.448 -0.549	1.541 -0.918
9.0	0.399 -0.141	0.497 -0.123	0.675 -0.080	0.825 -0.034	0.944 0.007	1.072 0.050	1.319 -0.060	1.388 -0.303	1.450 -0.549	1.544 -0.919
9.1	0.380 -0.031	0.475 -0.017	0.651 0.013	0.801 0.043	0.923 0.068	1.059 0.090	1.321 -0.058	1.390 -0.303	1.453 -0.549	1.547 -0.919
9.2	0.365 0.075	0.459 0.084	0.635 0.103	0.786 0.118	0.911 0.128	1.051 0.131	1.323 -0.056	1.393 -0.303	1.456 -0.550	1.550 -0.920
9.3	0.353 0.178	0.448 0.184	0.626 0.192	0.780 0.193	0.908 0.188	1.051 0.171	1.326 -0.055	1.395 -0.303	1.458 -0.550	1.553 -0.921
9.4	0.345 0.281	0.442 0.284	0.624 0.280	0.782 0.268	0.912 0.248	1.057 0.211	1.329 -0.053	1.397 -0.302	1.461 -0.550	1.556 -0.922
3π	0.343 0.307	0.442 0.308	0.625 0.302	0.784 0.286	0.914 0.263	1.059 0.221	1.329 -0.053	1.398 -0.302	1.462 -0.550	1.557 -0.922
9.5	0.339 0.385	0.441 0.384	0.630 0.370	0.792 0.342	0.924 0.307	1.068 0.249	1.332 -0.052	1.400 -0.302	1.464 -0.550	1.559 -0.923
9.6	0.336 0.493	0.444 0.488	0.643 0.461	0.810 0.416	0.943 0.364	1.086 0.285	1.334 -0.051	1.402 -0.302	1.466 -0.551	1.562 -0.924
9.7	0.337 0.607	0.453 0.597	0.664 0.554	0.838 0.490	0.972 0.419	1.109 0.318	1.338 -0.050	1.405 -0.302	1.469 -0.551	1.565 -0.924
9.8	0.341 0.729	0.469 0.713	0.696 0.650	0.876 0.562	1.008 0.470	1.137 0.347	1.341 -0.049	1.407 -0.302	1.471 -0.551	1.568 -0.925
9.9	0.351 0.863	0.494 0.839	0.741 0.750	0.926 0.634	1.054 0.518	1.171 0.372	1.344 -0.048	1.409 -0.302	1.474 -0.551	1.570 -0.926
10.0	0.367 1.014	0.530 0.978	0.801 0.853	0.989 0.702	1.109 0.559	1.208 0.390	1.347 -0.048	1.412 -0.302	1.476 -0.552	1.573 -0.927
10.1	0.395 1.188	0.585 1.134	0.883 0.960	1.068 0.764	1.173 0.592					
10.2	0.439 1.395	0.665 1.313	0.991 1.067	1.165 0.816	1.246 0.614					
10.3	0.510 1.649	0.786 1.519	1.135 1.169	1.280 0.853	1.327 0.621					
10.4	0.632 1.972	0.973 1.756	1.324 1.253	1.413 0.865	1.411 0.611					
10.5	0.853 2.396	1.271 2.021	1.566 1.300	1.558 0.844	1.495 0.580					
10.6	1.287 2.964	1.757 2.276	1.858 1.274	1.706 0.780	1.571 0.528					
10.7	2.231 3.674	2.531 2.386	2.173 1.133	1.838 0.669	1.634 0.456					
10.8	4.356 4.009	3.568 2.009	2.444 0.850	1.934 0.515	1.675 0.368					
10.9	7.118 1.559	4.319 0.817	2.582 0.453	1.976 0.335	1.691 0.274					
7π/2	5.924 -2.113	4.076 -0.549	2.541 0.056	1.959 0.162	1.682 0.184					

βh	α/β = 0.00		α/β = 0.01		α/β = 0.03		α/β = 0.05		α/β = 0.07	
11.0	5.799	-2.205	4.044	-0.600	2.535	0.038	1.957	0.154	1.681	0.180
11.1	3.356	-2.809	3.140	-1.344	2.336	-0.292	1.885	-0.003	1.647	0.096
11.2	2.067	-2.366	2.316	-1.479	2.066	-0.493	1.777	-0.122	1.594	0.028
11.3	1.427	-1.881	1.746	-1.358	1.795	-0.579	1.652	-0.198	1.531	-0.021
11.4	1.078	-1.490	1.371	-1.168	1.558	-0.584	1.526	-0.234	1.462	-0.051
11.5	0.869	-1.183	1.123	-0.974	1.365	-0.542	1.409	-0.239	1.394	-0.064
11.6	0.735	-0.937	0.953	-0.794	1.211	-0.474	1.305	-0.221	1.329	-0.061
11.7	0.643	-0.734	0.833	-0.633	1.091	-0.393	1.216	-0.187	1.271	-0.047
11.8	0.577	-0.562	0.747	-0.489	0.997	-0.307	1.142	-0.142	1.220	-0.023
11.9	0.528	-0.413	0.682	-0.358	0.925	-0.221	1.082	-0.090	1.177	0.007
12.0	0.491	-0.279	0.634	-0.238	0.869	-0.134	1.034	-0.033	1.142	0.043
12.1	0.463	-0.157	0.597	-0.126	0.827	-0.049	0.997	0.026	1.114	0.083
12.2	0.440	-0.043	0.570	-0.020	0.797	0.035	0.970	0.087	1.094	0.124
12.3	0.423	0.065	0.550	0.082	0.775	0.118	0.952	0.149	1.081	0.168
12.4	0.409	0.170	0.536	0.181	0.763	0.200	0.943	0.211	1.075	0.211
12.5	0.399	0.274	0.527	0.280	0.758	0.282	0.941	0.272	1.076	0.255
4π	0.394	0.344	0.525	0.346	0.759	0.336	0.945	0.313	1.080	0.283
12.6	0.392	0.379	0.524	0.380	0.761	0.364	0.948	0.334	1.084	0.298
12.7	0.388	0.486	0.527	0.481	0.773	0.447	0.963	0.394	1.097	0.339
12.8	0.388	0.599	0.536	0.587	0.793	0.531	0.986	0.454	1.118	0.378
12.9	0.391	0.719	0.552	0.699	0.824	0.615	1.018	0.511	1.144	0.414
13.0	0.400	0.849	0.578	0.818	0.867	0.701	1.059	0.565	1.177	0.446
13.1	0.416	0.995	0.615	0.948	0.924	0.788	1.111	0.614	1.216	0.472
13.2	0.442	1.161	0.670	1.091	0.999	0.873	1.174	0.656	1.259	0.491
13.3	0.484	1.355	0.750	1.251	1.095	0.954	1.247	0.689	1.307	0.503
13.4	0.550	1.590	0.866	1.430	1.217	1.025	1.330	0.708	1.358	0.505
13.5	0.661	1.881	1.039	1.628	1.368	1.077	1.421	0.710	1.409	0.496
13.6	0.853	2.256	1.300	1.838	1.549	1.097	1.516	0.690	1.458	0.477
13.7	1.211	2.745	1.698	2.028	1.755	1.066	1.608	0.646	1.502	0.446
13.8	1.939	3.355	2.288	2.107	1.966	0.964	1.690	0.576	1.538	0.406
13.9	3.499	3.829	3.054	1.879	2.150	0.783	1.751	0.485	1.564	0.358
14.0	6.085	2.691	3.726	1.130	2.264	0.534	1.786	0.378	1.577	0.306
14.1	6.451	-0.939	3.818	0.038	2.278	0.258	1.788	0.267	1.577	0.253
9π/2	5.649	-1.897	3.675	-0.330	2.257	0.158	1.781	0.227	1.574	0.234
14.2	4.165	-2.565	3.295	-0.803	2.194	0.005	1.760	0.163	1.564	0.203
14.3	2.557	-2.425	2.602	-1.155	2.040	-0.186	1.707	0.075	1.541	0.160
14.4	1.725	-1.984	2.026	-1.188	1.856	-0.304	1.637	0.009	1.509	0.125
14.5	1.276	-1.583	1.612	-1.082	1.672	-0.357	1.558	-0.035	1.472	0.099
14.6	1.013	-1.259	1.324	-0.931	1.505	-0.362	1.477	-0.057	1.432	0.084
14.7	0.846	-0.998	1.123	-0.775	1.363	-0.334	1.400	-0.060	1.392	0.078
14.8	0.734	-0.783	0.979	-0.626	1.244	-0.285	1.330	-0.049	1.353	0.080
14.9	0.655	-0.602	0.874	-0.487	1.148	-0.224	1.268	-0.025	1.318	0.090
15.0	0.597	-0.445	0.795	-0.360	1.071	-0.157	1.215	0.007	1.286	0.106
15.1	0.553	-0.305	0.737	-0.241	1.010	-0.085	1.170	0.045	1.259	0.126
15.2	0.520	-0.179	0.692	-0.130	0.963	-0.012	1.135	0.088	1.237	0.151
15.3	0.494	-0.061	0.659	-0.025	0.927	0.062	1.108	0.134	1.220	0.178
15.4	0.473	0.050	0.634	0.076	0.902	0.136	1.089	0.182	1.208	0.207
15.5	0.457	0.157	0.616	0.175	0.885	0.210	1.078	0.232	1.202	0.238
15.6	0.445	0.262	0.605	0.273	0.878	0.284	1.073	0.281	1.201	0.268
15.7	0.437	0.367	0.600	0.371	0.878	0.358	1.077	0.331	1.205	0.298
5π	0.436	0.375	0.600	0.378	0.879	0.364	1.077	0.335	1.205	0.301
15.8	0.432	0.474	0.601	0.470	0.887	0.433	1.087	0.379	1.213	0.327
15.9	0.430	0.586	0.609	0.572	0.905	0.507	1.105	0.426	1.227	0.355
16.0	0.433	0.703	0.625	0.680	0.933	0.582	1.130	0.471	1.245	0.380
16.1	0.441	0.831	0.650	0.793	0.971	0.656	1.163	0.512	1.267	0.401
16.2	0.456	0.971	0.687	0.915	1.023	0.728	1.203	0.549	1.293	0.419
16.3	0.480	1.129	0.741	1.047	1.088	0.798	1.250	0.579	1.322	0.432
16.4	0.518	1.313	0.817	1.191	1.171	0.860	1.304	0.601	1.353	0.439
16.5	0.579	1.530	0.926	1.348	1.271	0.913	1.364	0.613	1.385	0.440
16.6	0.677	1.797	1.083	1.517	1.392	0.948	1.428	0.613	1.417	0.435
16.7	0.843	2.132	1.311	1.689	1.530	0.957	1.492	0.598	1.448	0.423
16.8	1.140	2.561	1.641	1.839	1.681	0.930	1.554	0.568	1.475	0.405
16.9	1.712	3.096	2.107	1.905	1.833	0.857	1.609	0.524	1.498	0.382
17.0	2.883	3.611	2.699	1.766	1.966	0.734	1.651	0.466	1.515	0.355
17.1	5.037	3.260	3.273	1.276	2.059	0.568	1.678	0.399	1.525	0.325
17.2	6.555	0.469	3.527	0.468	2.093	0.378	1.687	0.328	1.528	0.294
11π/2	5.447	-1.739	3.381	-0.176	2.075	0.228	1.681	0.273	1.526	0.270
17.3	4.984	-2.054	3.298	-0.324	2.064	0.189	1.678	0.259	1.524	0.264
17.4	3.126	-2.413	2.785	-0.805	1.980	0.029	1.651	0.196	1.514	0.236
17.5	2.068	-2.076	2.258	-0.975	1.861	-0.090	1.612	0.144	1.498	0.213
17.6	1.496	-1.681	1.832	-0.963	1.727	-0.162	1.564	0.105	1.478	0.194
17.7	1.167	-1.342	1.516	-0.867	1.594	-0.195	1.511	0.079	1.455	0.181
17.8	0.961	-1.066	1.288	-0.741	1.471	-0.196	1.457	0.067	1.431	0.174
17.9	0.826	-0.838	1.121	-0.609	1.363	-0.174	1.404	0.067	1.406	0.172
18.0	0.731	-0.647	0.998	-0.480	1.272	-0.135	1.356	0.077	1.383	0.175
18.1	0.663	-0.481	0.906	-0.358	1.195	-0.087	1.312	0.095	1.361	0.183
18.2	0.612	-0.336	0.836	-0.243	1.132	-0.031	1.275	0.119	1.342	0.194
18.3	0.573	-0.204	0.784	-0.134	1.083	0.029	1.243	0.149	1.325	0.209
18.4	0.543	-0.083	0.743	-0.030	1.044	0.092	1.218	0.182	1.312	0.226
18.5	0.520	0.030	0.713	0.070	1.015	0.157	1.200	0.218	1.303	0.245
18.6	0.501	0.139	0.692	0.167	0.996	0.222	1.188	0.256	1.297	0.265
18.7	0.487	0.246	0.678	0.263	0.985	0.289	1.182	0.295	1.295	0.286
18.8	0.477	0.351	0.670	0.359	0.983	0.355	1.183	0.334	1.296	0.307
6π	0.474	0.404	0.669	0.407	0.985	0.388	1.185	0.353	1.299	0.317
18.9	0.471	0.458	0.670	0.456	0.989	0.421	1.189	0.372	1.302	0.327
19.0	0.468	0.569	0.676	0.556	1.003	0.487	1.202	0.409	1.310	0.346
19.1	0.470	0.685	0.691	0.659	1.027	0.553	1.221	0.444	1.322	0.363
19.2	0.476	0.809	0.715	0.767	1.060	0.617	1.245	0.476	1.336	0.378
19.3	0.489	0.945	0.750	0.881	1.105	0.678	1.275	0.503	1.353	0.390
19.4	0.511	1.096	0.801	1.003	1.161	0.735	1.311	0.526	1.372	0.399
19.5	0.546	1.269	0.873	1.134	1.230	0.785	1.350	0.542	1.392	0.404
19.6	0.600	1.473	0.974	1.274	1.312	0.826	1.394	0.551	1.413	0.406
19.7	0.688	1.718	1.115	1.421	1.408	0.851	1.439	0.551	1.434	0.403
19.8	0.831	2.020	1.314	1.567	1.516	0.857	1.484	0.541	1.454	0.397
19.9	1.078	2.402	1.592	1.692	1.631	0.836	1.527	0.522	1.471	0.387
20.0	1.535	2.878	1.969	1.751	1.744	0.784	1.565	0.493	1.487	0.373
20.1	2.431	3.387	2.441	1.666	1.845	0.698	1.596	0.455	1.498	0.357
20.2	4.139	3.462	2.927	1.338	1.921	0.582	1.617	0.412	1.506	0.339
20.3	6.133	1.700	3.238	0.747	1.961	0.447	1.628	0.365	1.510	0.321
20.4	5.671	-1.231	3.205	0.065	1.958	0.307	1.626	0.318	1.509	0.302
13π/2	5.290	-1.615	3.156	-0.061	1.952	0.279	1.625	0.308	1.509	0.299
20.5	3.773	-2.276	2.873	-0.461	1.915	0.178	1.614	0.273	1.505	0.285
20.6	2.471	-2.142	2.432	-0.736	1.841	0.072	1.592	0.234	1.497	0.270
20.7	1.749	-1.777	2.023	-0.814	1.747	-0.004	1.563	0.203	1.486	0.257
20.8	1.336	-1.431	1.696	-0.779	1.646	-0.050	1.529	0.180	1.473	0.247
20.9	1.085	-1.140	1.447	-0.691	1.546	-0.069	1.492	0.165	1.459	0.241

βh	α/β = 0.00		α/β = 0.01		α/β = 0.03		α/β = 0.05		α/β = 0.07	
21.0	0.921	-0.899	1.260	-0.581	1.453	-0.066	1.454	0.159	1.444	0.238
21.1	0.809	-0.696	1.120	-0.466	1.370	-0.046	1.418	0.162	1.429	0.238
21.2	0.729	-0.523	1.015	-0.352	1.298	-0.014	1.384	0.171	1.415	0.241
21.3	0.670	-0.370	0.934	-0.242	1.237	0.028	1.354	0.185	1.402	0.248
21.4	0.625	-0.234	0.873	-0.136	1.187	0.075	1.328	0.205	1.390	0.256
21.5	0.590	-0.109	0.826	-0.035	1.147	0.127	1.306	0.228	1.381	0.267
21.6	0.563	0.008	0.791	0.063	1.117	0.182	1.289	0.255	1.374	0.279
21.7	0.542	0.119	0.765	0.159	1.095	0.239	1.278	0.283	1.369	0.292
21.8	0.526	0.227	0.747	0.253	1.082	0.297	1.271	0.312	1.367	0.306
21.9	0.514	0.333	0.737	0.347	1.076	0.355	1.270	0.342	1.367	0.320
7π	0.507	0.431	0.734	0.434	1.079	0.408	1.273	0.369	1.370	0.332
22.0	0.506	0.440	0.734	0.442	1.079	0.414	1.274	0.372	1.370	0.334
22.1	0.502	0.550	0.739	0.538	1.090	0.472	1.282	0.401	1.375	0.347
22.2	0.502	0.664	0.751	0.637	1.109	0.529	1.295	0.428	1.382	0.358
22.3	0.508	0.785	0.773	0.740	1.137	0.584	1.313	0.452	1.392	0.369
22.4	0.519	0.917	0.807	0.848	1.175	0.636	1.335	0.474	1.403	0.377
22.5	0.538	1.062	0.855	0.962	1.222	0.684	1.361	0.491	1.415	0.384
22.6	0.569	1.227	0.921	1.082	1.279	0.726	1.390	0.504	1.428	0.388
22.7	0.618	1.418	1.014	1.208	1.347	0.759	1.421	0.511	1.442	0.390
22.8	0.695	1.644	1.141	1.338	1.424	0.779	1.454	0.511	1.456	0.389
22.9	0.819	1.921	1.314	1.464	1.509	0.784	1.486	0.506	1.468	0.386
23.0	1.027	2.264	1.551	1.572	1.598	0.768	1.517	0.493	1.480	0.380
23.1	1.396	2.690	1.863	1.628	1.685	0.731	1.545	0.474	1.491	0.372
23.2	2.093	3.173	2.249	1.579	1.765	0.670	1.568	0.450	1.499	0.363
23.3	3.425	3.460	2.660	1.355	1.827	0.587	1.585	0.421	1.505	0.352
23.4	5.407	2.558	2.977	0.923	1.866	0.489	1.595	0.390	1.509	0.341
23.5	6.045	-0.160	3.062	0.363	1.876	0.384	1.597	0.357	1.510	0.330
15π/2	5.163	-1.514	2.978	0.030	1.867	0.320	1.595	0.338	1.509	0.323
23.6	4.465	-1.949	2.883	-0.149	1.857	0.282	1.592	0.326	1.508	0.319
23.7	2.943	-2.154	2.546	-0.487	1.812	0.193	1.580	0.297	1.504	0.309
23.8	2.042	-1.865	2.180	-0.642	1.749	0.122	1.563	0.273	1.499	0.300
23.9	1.529	-1.522	1.857	-0.670	1.674	0.072	1.541	0.253	1.492	0.294
24.0	1.220	-1.219	1.597	-0.624	1.596	0.043	1.516	0.240	1.483	0.289
24.1	1.023	-0.964	1.395	-0.541	1.519	0.034	1.489	0.232	1.474	0.286
24.2	0.890	-0.751	1.241	-0.443	1.447	0.040	1.463	0.231	1.465	0.285
24.3	0.796	-0.568	1.122	-0.340	1.382	0.059	1.438	0.234	1.456	0.287
24.4	0.727	-0.408	1.031	-0.237	1.325	0.087	1.414	0.243	1.447	0.290
24.5	0.675	-0.266	0.961	-0.136	1.277	0.123	1.393	0.255	1.440	0.295
24.6	0.636	-0.137	0.907	-0.038	1.237	0.165	1.376	0.271	1.433	0.301
24.7	0.605	-0.017	0.866	0.058	1.206	0.210	1.361	0.290	1.428	0.309
24.8	0.581	0.096	0.835	0.151	1.183	0.258	1.351	0.310	1.425	0.317
24.9	0.563	0.205	0.814	0.243	1.168	0.308	1.344	0.332	1.423	0.326
25.0	0.549	0.313	0.801	0.335	1.160	0.359	1.342	0.355	1.422	0.336

TABLE 2.8

ADMITTANCES IN MILLIMHOS OF LONG DIPOLE ANTENNAS IN AIR

k_0h	a/λ = 0.001588		a/λ = 0.003175		a/λ = 0.004763		a/λ = 0.007022		a/λ = 0.009525	
25.0	0.828	0.706	1.017	1.091	1.152	1.445	1.302	1.942	1.442	2.519
25.1	0.816	0.835	1.005	1.235	1.141	1.599	1.294	2.106	1.440	2.592
8π	0.813	0.878	1.003	1.283	1.139	1.650	1.294	2.161	1.442	2.750
25.2	0.810	0.968	1.001	1.383	1.140	1.758	1.298	2.277	1.451	2.873
25.3	0.811	1.106	1.007	1.539	1.150	1.925	1.316	2.457	1.479	3.063
25.4	0.819	1.253	1.022	1.704	1.173	2.104	1.350	2.650	1.529	3.266
25.5	0.838	1.411	1.052	1.884	1.213	2.298	1.407	2.858	1.605	3.485
25.6	0.870	1.585	1.100	2.081	1.277	2.511	1.493	3.086	1.718	3.722
25.7	0.921	1.780	1.174	2.301	1.373	2.747	1.621	3.337	1.883	3.977
25.8	0.999	2.003	1.287	2.550	1.517	3.012	1.810	3.611	2.120	4.247
25.9	1.119	2.261	1.458	2.834	1.734	3.307	2.089	3.904	2.462	4.517
26.0	1.307	2.562	1.722	3.154	2.064	3.626	2.500	4.197	2.947	4.751
26.1	1.609	2.912	2.137	3.499	2.568	3.938	3.101	4.432	3.612	4.869
26.2	2.111	3.293	2.792	3.810	3.325	4.151	3.935	4.480	4.445	4.727
26.3	2.947	3.606	3.785	3.912	4.369	4.045	4.935	4.115	5.293	4.155
26.4	4.241	3.525	5.052	3.425	5.487	3.290	5.774	3.153	5.826	3.137
26.5	5.652	2.431	5.994	2.033	6.060	1.848	5.976	1.806	5.774	1.979
26.6	5.581	0.403	5.833	0.304	5.679	0.387	5.462	0.652	5.229	1.086
26.7	4.959	-1.103	4.846	-0.785	4.753	-0.466	4.633	0.014	4.504	0.605
17π/2	4.512	-1.136	4.806	-0.808	4.719	-0.484	4.603	0.000	4.478	0.594
26.8	3.713	-1.576	3.799	-1.132	3.832	-0.740	3.841	-0.193	3.825	0.446
26.9	2.797	-1.511	3.002	-1.086	3.114	-0.700	3.209	-0.158	3.270	0.476
27.0	2.197	-1.278	2.446	-0.894	2.596	-0.532	2.736	-0.013	2.840	0.601
27.1	1.803	-1.017	2.062	-0.665	2.225	-0.323	2.385	0.173	2.512	0.766
27.2	1.537	-0.772	1.791	-0.439	1.956	-0.111	2.123	0.369	2.261	0.946
27.3	1.351	-0.553	1.596	-0.228	1.757	0.093	1.925	0.562	2.066	1.127
27.4	1.216	-0.356	1.450	-0.034	1.607	0.283	1.773	0.748	1.915	1.305
27.5	1.116	-0.180	1.340	0.145	1.492	0.462	1.654	0.924	1.795	1.478
27.6	1.039	-0.019	1.255	0.311	1.403	0.630	1.561	1.093	1.701	1.646
27.7	0.980	0.129	1.189	0.467	1.332	0.790	1.487	1.256	1.626	1.810
27.8	0.934	0.268	1.137	0.615	1.277	0.943	1.430	1.414	1.568	1.971
27.9	0.898	0.401	1.096	0.758	1.234	1.093	1.385	1.570	1.523	2.131
28.0	0.870	0.530	1.065	0.898	1.202	1.240	1.353	1.725	1.492	2.292
28.1	0.849	0.657	1.042	1.038	1.179	1.388	1.331	1.881	1.472	2.455
28.2	0.835	0.784	1.028	1.178	1.165	1.538	1.319	2.040	1.465	2.622
9π	0.828	0.880	1.022	1.285	1.161	1.652	1.318	2.163	1.468	2.751
28.3	0.827	0.914	1.021	1.322	1.160	1.692	1.319	2.206	1.471	2.796
28.4	0.825	1.048	1.023	1.473	1.166	1.854	1.332	2.379	1.493	2.979
28.5	0.831	1.189	1.035	1.632	1.184	2.025	1.359	2.563	1.533	3.173
28.6	0.846	1.340	1.059	1.802	1.218	2.209	1.407	2.761	1.598	3.381
28.7	0.873	1.505	1.099	1.989	1.271	2.410	1.480	2.976	1.695	3.605
28.8	0.916	1.688	1.162	2.195	1.353	2.632	1.590	3.212	1.836	3.847
28.9	0.982	1.896	1.258	2.428	1.476	2.879	1.751	3.470	2.040	4.103
29.0	1.083	2.134	1.402	2.691	1.660	3.155	1.987	3.748	2.331	4.366
29.1	1.240	2.411	1.622	2.989	1.936	3.456	2.335	4.034	2.746	4.610
29.2	1.488	2.733	1.964	3.316	2.355	3.765	2.843	4.290	3.321	4.776
29.3	1.892	3.094	2.502	3.638	2.987	4.020	3.561	4.421	4.070	4.751
[illegible]	[illegible]	[illegible]	3.330	3.840	3.893	4.066	4.482	4.248	4.908	4.374
29.5	3.639	3.562	4.479	3.840	4.990	3.814	[illegible]	[illegible]	[illegible]	[illegible]
29.6	5.049	2.956	5.612	2.643	5.829	2.450	5.883	2.348	5.769	2.437
29.7	5.927	1.247	5.929	0.985	5.832	0.953	5.643	1.099	5.410	1.438
29.8	5.382	-0.569	5.216	-0.387	5.086	-0.144	4.921	0.268	4.750	0.808
19π/2	4.841	-1.076	4.742	-0.749	4.660	-0.426	4.550	0.057	4.431	0.647
29.9	4.158	-1.412	4.168	-0.994	4.154	-0.619	4.116	-0.089	4.060	0.536
30.0	3.119	-1.518	3.284	-1.079	3.369	-0.686	3.434	-0.139	3.468	0.498
30.1	2.414	-1.340	2.650	-0.938	2.786	-0.565	2.911	-0.035	3.000	0.586
30.2	1.952	-1.092	2.209	-0.726	2.368	-0.374	2.521	0.132	2.640	0.733
30.3	1.643	-0.845	1.901	-0.502	2.065	-0.167	2.230	0.321	2.364	0.904
30.4	1.430	-0.620	1.680	-0.289	1.843	0.036	2.011	0.511	2.152	1.081
30.5	1.277	-0.418	1.517	-0.092	1.677	0.228	1.844	0.696	1.986	1.257
30.6	1.165	-0.236	1.395	0.090	1.550	0.408	1.714	0.872	1.856	1.428
30.7	1.080	-0.072	1.301	0.258	1.452	0.577	1.612	1.041	1.754	1.595
30.8	1.015	0.079	1.228	0.415	1.375	0.738	1.532	1.204	1.672	1.758
30.9	0.964	0.220	1.171	0.564	1.314	0.891	1.469	1.361	1.609	1.917
31.0	0.924	0.354	1.126	0.707	1.267	1.040	1.421	1.515	1.560	2.075
31.1	0.893	0.482	1.092	0.847	1.231	1.186	1.384	1.668	1.524	2.233
31.2	0.870	0.609	1.066	0.985	1.205	1.332	1.358	1.821	1.500	2.393
31.3	0.853	0.734	1.049	1.123	1.187	1.479	1.343	1.977	1.488	2.555
31.4	0.843	0.861	1.039	1.264	1.180	1.629	1.338	2.138	1.489	2.724
10π	0.842	0.882	1.038	1.287	1.179	1.654	1.339	2.164	1.491	2.751
31.5	0.839	0.992	1.038	1.410	1.181	1.785	1.346	2.305	1.505	2.900
31.6	0.842	1.128	1.046	1.563	1.195	1.950	1.368	2.481	1.538	3.086
31.7	0.853	1.273	1.065	1.726	1.222	2.126	1.407	2.670	1.592	3.284
31.8	0.875	1.431	1.099	1.903	1.267	2.317	1.470	2.875	1.675	3.497
31.9	0.911	1.604	1.152	2.098	1.337	2.526	1.563	3.097	1.797	3.727
32.0	0.967	1.798	1.233	2.316	1.441	2.758	1.701	3.341	1.971	3.972
32.1	1.053	2.019	1.355	2.561	1.597	3.017	1.902	3.605	2.221	4.226
32.2	1.184	2.275	1.540	2.839	1.829	3.301	2.196	3.882	2.576	4.473
32.3	1.389	2.572	1.824	3.148	2.180	3.602	2.626	4.146	3.073	4.670
32.4	1.718	2.911	2.267	3.468	2.708	3.878	3.242	4.332	3.736	4.730
32.5	2.257	3.262	2.953	3.727	3.483	4.023	4.067	4.301	4.532	4.514
32.6	3.137	3.500	3.955	3.725	4.498	3.807	4.996	3.843	5.282	3.884
32.7	4.419	3.262	5.130	3.097	5.481	2.954	5.682	2.845	5.679	2.883
32.8	5.624	2.013	5.841	1.673	5.849	1.552	5.734	1.589	5.533	1.829
32.9	5.666	0.114	5.512	0.123	5.371	0.268	5.181	0.587	4.980	1.059
21π/2	4.779	-1.024	4.686	-0.698	4.609	-0.376	4.504	0.105	4.390	0.693
33.0	4.618	-1.129	4.549	-0.774	4.488	-0.435	4.400	0.060	4.301	0.659
33.1	3.487	-1.478	3.598	-1.034	3.648	-0.641	3.677	-0.096	3.679	0.539
33.2	2.667	-1.389	2.879	-0.969	2.998	-0.586	3.102	-0.048	3.172	0.580
33.3	2.124	-1.166	2.374	-0.783	2.526	-0.422	2.669	0.094	2.778	0.704
33.4	1.764	-0.921	2.022	-0.566	2.185	-0.224	2.346	0.272	2.475	0.862
33.5	1.517	-0.690	1.771	-0.353	1.936	-0.023	2.103	0.458	2.242	1.034
33.6	1.344	-0.482	1.588	-0.153	1.750	0.170	1.918	0.642	2.061	1.207
33.7	1.217	-0.296	1.452	0.032	1.610	0.352	1.776	0.819	1.919	1.378
33.8	1.122	-0.127	1.348	0.203	1.502	0.523	1.665	0.988	1.807	1.544
33.9	1.050	0.027	1.268	0.363	1.417	0.685	1.578	1.151	1.719	1.705
34.0	0.994	0.171	1.205	0.513	1.351	0.839	1.509	1.308	1.649	1.864
34.1	0.950	0.306	1.156	0.657	1.299	0.988	1.455	1.461	1.595	2.020
34.2	0.916	0.435	1.118	0.796	1.259	1.133	1.414	1.612	1.555	2.175
34.3	0.890	0.561	1.090	0.933	1.230	1.277	1.384	1.763	1.527	2.332
34.4	0.871	0.685	1.069	1.069	1.209	1.421	1.365	1.916	1.511	2.491
34.5	0.858	0.810	1.056	1.207	1.198	1.568	1.357	2.072	1.507	2.654
11π	0.854	0.883	1.053	1.288	1.196	1.655	1.357	2.165	1.511	2.751
34.6	0.852	0.938	1.052	1.349	1.196	1.720	1.360	2.234	1.517	2.825
34.7	0.852	1.070	1.056	1.497	1.205	1.879	1.376	2.404	1.543	3.004
34.8	0.860	1.210	1.071	1.654	1.226	2.048	1.408	2.585	1.588	3.194
34.9	0.878	1.360	1.099	1.823	1.264	2.230	1.461	2.780	1.659	3.397
35.0	0.908	1.525	1.144	2.008	1.323	2.428	1.541	2.992	1.763	3.616
35.1	0.956	1.707	1.213	2.213	1.412	2.647	1.658	3.222	1.913	3.849
35.2	1.028	1.914	1.317	2.443	1.545	2.890	1.830	3.473	2.128	4.095
35.3	1.138	2.151	1.472	2.702	1.741	3.159	2.080	3.739	2.432	4.341
35.4	1.309	2.426	1.710	2.993	2.035	3.447	2.445	4.005	2.861	4.557
35.5	1.578	2.742	2.076	3.304	2.478	3.731	2.971	4.224	3.443	4.678
35.6	2.014	3.085	2.643	3.591	3.133	3.938	3.697	4.294	4.176	4.588
35.7	2.728	3.382	3.496	3.719	4.040	3.896	4.583	4.034	4.951	4.143
35.8	3.832	3.391	4.613	3.392	5.060	3.334	5.393	3.268	5.508	3.293
35.9	5.142	2.610	5.579	2.295	5.722	2.136	5.720	2.094	5.583	2.246
36.0	5.740	0.883	5.684	0.724	5.574	0.759	5.392	0.971	5.179	1.359
36.1	5.049	-0.700	4.918	-0.453	4.815	-0.175	4.682	0.265	4.542	0.823
23π/2	4.724	-0.979	4.637	-0.653	4.564	-0.332	4.464	0.148	4.355	0.734
36.2	3.896	-1.364	3.940	-0.935	3.949	-0.555	3.936	-0.021	3.901	0.604
36.3	2.959	-1.414	3.138	-0.978	3.233	-0.588	3.311	-0.045	3.357	0.588
36.4	2.322	-1.234	2.561	-0.834	2.701	-0.463	2.831	0.063	2.926	0.680
36.5	1.901	-0.997	2.157	-0.630	2.316	-0.279	2.472	0.225	2.594	0.823
36.6	1.616	-0.764	1.872	-0.418	2.036	-0.082	2.202	0.406	2.338	0.987
36.7	1.417	-0.550	1.666	-0.215	1.829	0.111	1.998	0.588	2.140	1.157
36.8	1.273	-0.357	1.513	-0.027	1.673	0.295	1.841	0.764	1.985	1.326
36.9	1.167	-0.184	1.398	0.147	1.554	0.468	1.719	0.934	1.863	1.491
37.0	1.087	-0.026	1.309	0.309	1.461	0.631	1.624	1.097	1.766	1.652
37.1	1.025	0.120	1.240	0.461	1.389	0.786	1.548	1.254	1.690	1.810
37.2	0.977	0.257	1.187	0.606	1.332	0.935	1.489	1.407	1.631	1.965
37.3	0.939	0.387	1.145	0.745	1.288	1.080	1.444	1.557	1.586	2.118
37.4	0.910	0.513	1.112	0.881	1.254	1.223	1.410	1.706	1.553	2.272
37.5	0.888	0.637	1.089	1.016	1.230	1.365	1.387	1.856	1.533	2.428
37.6	0.873	0.760	1.073	1.152	1.216	1.509	1.375	2.009	1.525	2.588
12π	0.865	0.884	1.066	1.289	1.210	1.656	1.374	2.165	1.529	2.751
37.7	0.864	0.885	1.066	1.291	1.210	1.657	1.374	2.167	1.529	2.753
37.8	0.862	1.015	1.067	1.434	1.215	1.811	1.385	2.331	1.549	2.926
37.9	0.867	1.150	1.077	1.586	1.231	1.974	1.410	2.505	1.586	3.108
38.0	0.881	1.294	1.100	1.748	1.262	2.148	1.454	2.692	1.645	3.303
38.1	0.906	1.451	1.137	1.924	1.312	2.337	1.522	2.893	1.735	3.512
38.2	0.946	1.623	1.196	2.118	1.388	2.544	1.623	3.112	1.864	3.736
38.3	1.007	1.817	1.284	2.334	1.501	2.773	1.770	3.350	2.048	3.972
38.4	1.101	2.039	1.416	2.577	1.668	3.027	1.983	3.606	2.311	4.214
38.5	1.243	2.294	1.615	2.850	1.916	3.303	2.294	3.868	2.680	4.440
38.6	1.466	2.588	1.920	3.148	2.287	3.586	2.742	4.106	3.188	4.602
38.7	1.821	2.916	2.390	3.445	2.840	3.828	3.372	4.245	3.850	4.609
38.8	2.399	3.237	3.106	3.652	3.630	3.906	4.188	4.140	4.609	4.326
38.9	3.320	3.397	4.112	3.550	4.613	3.590	5.046	3.600	5.271	3.648
39.0	4.578	3.006	5.190	2.797	5.466	2.656	5.597	2.581	5.552	2.668
39.1	5.576	1.639	5.694	1.365	5.662	1.304	5.528	1.408	5.331	1.704
39.2	5.384	-0.122	5.237	-0.024	5.112	0.170	4.947	0.533	4.774	1.033
25π/2	4.676	-0.939	4.593	-0.613	4.523	-0.292	4.428	0.185	4.323	0.769
39.3	4.331	-1.148	4.302	-0.765	4.266	-0.412	4.206	0.095	4.132	0.700
[illegible]	[illegible]	[illegible]	3.427	-0.955	3.492	-0.564	3.538	-0.020	3.555	0.612
39.5	2.553	-1.290	2.772	-0.874	2.897	-0.493	[illegible]	[illegible]	[illegible]	[illegible]
39.6	2.058	-1.072	2.309	-0.691	2.463	-0.332	2.609	0.182	2.722	0.787
39.7	1.726	-0.839	1.984	-0.484	2.147	-0.141	2.309	0.354	2.441	0.942
39.8	1.498	-0.620	1.750	-0.280	1.915	0.051	2.083	0.533	2.223	1.107
39.9	1.335	-0.422	1.579	-0.088	1.741	0.236	1.910	0.709	2.053	1.274
40.0	1.215	-0.243	1.451	0.089	1.609	0.411	1.776	0.879	1.920	1.439
40.1	1.126	-0.081	1.353	0.254	1.507	0.576	1.671	1.043	1.815	1.599
40.2	1.057	0.068	1.277	0.408	1.427	0.732	1.589	1.200	1.732	1.756
40.3	1.004	0.207	1.217	0.554	1.365	0.882	1.524	1.353	1.667	1.910
40.4	0.962	0.339	1.171	0.694	1.316	1.027	1.474	1.502	1.617	2.062
40.5	0.930	0.465	1.135	0.830	1.279	1.169	1.436	1.650	1.580	2.214
40.6	0.906	0.589	1.109	0.964	1.251	1.310	1.409	1.798	1.555	2.367
40.7	0.888	0.711	1.090	1.098	1.233	1.452	1.393	1.948	1.542	2.523
40.8	0.877	0.834	1.080	1.234	1.224	1.597	1.388	2.101	1.542	2.684
13π	0.874	0.885	1.078	1.290	1.223	1.657	1.389	2.166	1.545	2.751
40.9	0.872	0.961	1.077	1.374	1.225	1.746	1.394	2.261	1.555	2.851
41.0	0.874	1.092	1.084	1.521	1.237	1.903	1.413	2.429	1.585	3.028
41.1	0.885	1.231	1.101	1.677	1.262	2.071	1.449	2.608	1.635	3.215
41.2	0.905	1.381	1.133	1.845	1.303	2.251	1.507	2.800	1.711	3.415
41.3	0.938	1.545	1.182	2.029	1.368	2.448	1.593	3.009	1.822	3.629
41.4	0.991	1.728	1.257	2.233	1.464	2.665	1.719	3.236	1.981	3.857
41.5	1.069	1.935	1.369	2.461	1.606	2.904	1.901	3.480	2.207	4.093
41.6	1.189	2.172	1.537	2.718	1.816	3.167	2.166	3.736	2.525	4.323
41.7	1.374	2.446	1.791	3.002	2.129	3.444	2.549	3.983	2.967	4.512
41.8	1.665	2.756	2.182	3.298	2.594	3.704	3.092	4.167	3.556	4.590
41.9	2.135	3.081	2.780	3.549	3.273	3.862	3.823	4.178	4.274	4.441
42.0	2.892	3.327	3.655	3.602	4.175	3.736	4.673	3.838	4.987	3.936
42.1	4.014	3.219	4.731	3.156	5.116	3.075	5.378	3.016	5.440	3.071
42.2	5.208	2.281	5.535	1.982	5.617	1.861	5.572	1.876	5.420	2.084
42.3	5.552	0.571	5.461	0.504	5.347	0.597	5.176	0.864	4.982	1.292
42.4	4.758	-0.803	4.663	-0.505	4.584	-0.201	4.480	0.262	4.366	0.834
27π/2	4.632	-0.903	4.554	-0.578	4.487	-0.257	4.396	0.219	4.294	0.802
42.5	3.672	-1.322	3.746	-0.887	3.774	-0.502	3.781	0.034	3.766	0.659
42.6	2.820	-1.326	3.010	-0.895	3.115	-0.507	3.204	0.033	3.261	0.661
42.7	2.240	-1.143	2.481	-0.747	2.625	-0.378	2.760	0.145	2.861	0.757
42.8	1.853	-0.916	2.109	-0.549	2.269	-0.199	2.427	0.303	2.552	0.899
42.9	1.589	-0.693	1.844	-0.346	2.009	-0.009	2.175	0.478	2.312	1.058
43.0	1.403	-0.489	1.651	-0.152	1.815	0.177	1.984	0.653	2.126	1.222
43.1	1.268	-0.304	1.507	0.029	1.668	0.353	1.836	0.824	1.981	1.386
43.2	1.167	-0.137	1.399	0.197	1.555	0.520	1.721	0.988	1.866	1.546
43.3	1.091	0.015	1.315	0.354	1.468	0.678	1.631	1.146	1.775	1.702
43.4	1.032	0.157	1.249	0.501	1.399	0.829	1.560	1.299	1.704	1.855
43.5	0.986	0.290	1.198	0.642	1.345	0.974	1.505	1.448	1.648	2.007
43.6	0.951	0.418	1.159	0.779	1.304	1.116	1.463	1.595	1.607	2.157
43.7	0.923	0.541	1.129	0.912	1.273	1.256	1.432	1.741	1.577	2.308
43.8	0.903	0.663	1.107	1.045	1.251	1.396	1.411	1.888	1.560	2.461
43.9	0.890	0.785	1.094	1.179	1.239	1.538	1.402	2.039	1.554	2.618
14π	0.883	0.886	1.088	1.291	1.236	1.658	1.402	2.166	1.560	2.751

k_0h	a/λ = 0.001588		a/λ = 0.003175		a/λ = 0.004763		a/λ = 0.007022		a/λ = 0.009525	
44.0	0.882	0.908	1.088	1.316	1.236	1.684	1.403	2.194	1.562	2.780
44.1	0.882	1.037	1.091	1.458	1.243	1.836	1.417	2.357	1.585	2.951
44.2	0.889	1.171	1.104	1.609	1.262	1.997	1.446	2.529	1.627	3.131
44.3	0.905	1.315	1.130	1.770	1.297	2.170	1.495	2.713	1.691	3.323
44.4	0.933	1.472	1.171	1.945	1.352	2.358	1.568	2.912	1.787	3.528
44.5	0.977	1.645	1.235	2.138	1.434	2.564	1.676	3.128	1.924	3.747
44.6	1.044	1.839	1.330	2.354	1.555	2.791	1.832	3.362	2.119	3.977
44.7	1.145	2.060	1.472	2.595	1.733	3.040	2.059	3.610	2.393	4.207
44.8	1.299	2.315	1.686	2.864	1.997	3.308	2.385	3.859	2.776	4.413
44.9	1.539	2.606	2.011	3.153	2.390	3.575	2.852	4.071	3.296	4.542
45.0	1.923	2.924	2.510	3.426	2.967	3.782	3.496	4.164	3.956	4.497
45.1	2.541	3.213	3.256	3.579	3.770	3.794	4.299	3.989	4.678	4.152
45.2	3.499	3.291	4.257	3.377	4.715	3.382	5.087	3.376	5.256	3.435
45.3	4.717	2.750	5.231	2.514	5.441	2.384	5.513	2.346	5.435	2.480
45.4	5.505	1.298	5.547	1.095	5.487	1.090	5.344	1.254	5.153	1.598
45.5	5.124	-0.315	4.993	-0.144	4.885	0.089	4.745	0.487	4.596	1.011
29π/2	4.592	-0.871	4.518	-0.546	4.454	-0.226	4.366	0.250	4.268	0.831
45.6	4.082	-1.159	4.087	-0.756	4.074	-0.392	4.038	0.124	3.986	0.734
45.7	3.127	-1.330	3.278	-0.889	3.356	-0.498	3.416	0.044	3.447	0.673
45.8	2.451	-1.205	2.677	-0.792	2.807	-0.415	2.926	0.116	3.012	0.735
45.9	1.998	-0.992	2.250	-0.612	2.405	-0.254	2.555	0.257	2.672	0.859
46.0	1.691	-0.768	1.948	-0.412	2.112	-0.070	2.275	0.424	2.408	1.010
46.1	1.478	-0.559	1.730	-0.216	1.895	0.116	2.063	0.597	2.204	1.171
46.2	1.325	-0.368	1.569	-0.032	1.731	0.294	1.901	0.768	2.045	1.333
46.3	1.212	-0.196	1.448	0.139	1.607	0.463	1.774	0.933	1.919	1.492
46.4	1.127	-0.039	1.355	0.299	1.510	0.623	1.675	1.091	1.820	1.648
46.5	1.062	0.106	1.283	0.448	1.434	0.775	1.597	1.244	1.742	1.801
46.6	1.011	0.241	1.227	0.591	1.375	0.921	1.537	1.393	1.681	1.951
46.7	0.972	0.369	1.183	0.727	1.329	1.063	1.489	1.540	1.634	2.100
46.8	0.941	0.493	1.149	0.861	1.295	1.202	1.454	1.684	1.600	2.249
46.9	0.919	0.615	1.124	0.993	1.269	1.340	1.430	1.830	1.578	?.400
47.0	0.902	0.736	1.108	1.125	1.254	1.480	1.416	1.977	1.568	2.553
47.1	0.893	0.858	1.099	1.259	1.247	1.623	1.413	2.129	1.571	2.712
15π	0.891	0.887	1.098	1.292	1.247	1.658	1.414	2.166	1.573	2.751
47.2	0.889	0.983	1.099	1.398	1.250	1.772	1.422	2.287	1.587	2.877
47.3	0.893	1.114	1.108	1.544	1.264	1.928	1.445	2.454	1.621	3.051
47.4	0.906	1.252	1.128	1.699	1.293	2.094	1.485	2.631	1.676	3.236
47.5	0.929	1.402	1.163	1.867	1.338	2.273	1.548	2.821	1.757	3.434
47.6	0.966	1.566	1.217	2.050	1.408	2.469	1.640	3.028	1.876	3.644
47.7	1.022	1.750	1.298	2.254	1.512	2.684	1.774	3.251	2.044	3.866
47.8	1.108	1.957	1.418	2.481	1.663	2.921	1.968	3.491	2.281	4.094
47.9	1.237	2.195	1.598	2.736	1.887	3.179	2.247	3.737	2.613	4.309
48.0	1.437	2.468	1.871	3.014	2.219	3.445	2.648	3.965	3.068	4.472
48.1	1.751	2.772	2.286	3.294	2.708	3.680	3.209	4.114	3.664	4.508
48.2	2.256	3.078	2.915	3.508	3.407	3.788	3.942	4.066	4.363	4.301
48.3	3.055	3.269	3.807	3.484	4.301	3.579	4.752	3.651	5.016	3.744
48.4	4.185	3.039	4.833	2.924	5.158	2.831	5.356	2.784	5.372	2.873
48.5	5.247	1.963	5.476	1.695	5.509	1.616	5.433	1.686	5.272	1.943
48.6	5.362	0.301	5.252	0.318	5.141	0.461	4.983	0.775	4.809	1.235
31π/2	4.555	-0.842	4.485	-0.517	4.424	-0.197	4.340	0.278	4.244	0.857
48.7	4.499	-0.883	4.437	-0.546	4.382	-0.220	4.303	0.260	4.213	0.844
48.8	3.477	-1.282	3.576	-0.844	3.620	-0.457	3.646	0.080	3.647	0.706
48.9	2.697	-1.250	2.897	-0.823	3.010	-0.438	3.109	0.099	3.175	0.724
49.0	2.166	-1.064	2.410	-0.671	2.557	-0.304	2.696	0.215	2.802	0.824
49.1	1.808	-0.844	2.065	-0.478	2.226	-0.129	2.385	0.372	2.512	0.964
49.2	1.562	-0.631	1.817	-0.282	1.982	0.054	2.149	0.541	2.288	1.120
49.3	1.388	-0.435	1.636	-0.095	1.800	0.234	1.969	0.711	2.113	1.280
49.4	1.261	-0.257	1.501	0.080	1.661	0.405	1.830	0.877	1.975	1.438
49.5	1.166	-0.095	1.398	0.242	1.555	0.567	1.722	1.036	1.867	1.594
49.6	1.093	0.053	1.318	0.394	1.472	0.721	1.636	1.190	1.781	1.747
49.7	1.037	0.191	1.256	0.538	1.407	0.867	1.569	1.339	1.714	1.897
49.8	0.994	0.321	1.208	0.676	1.356	1.010	1.517	1.485	1.662	2.045
49.9	0.960	0.446	1.170	0.810	1.317	1.148	1.478	1.629	1.624	2.192
50.0	0.934	0.567	1.142	0.941	1.288	1.286	1.449	1.773	1.597	2.340
50.1	0.916	0.687	1.122	1.072	1.269	1.424	1.432	1.918	1.583	2.491
50.2	0.903	0.808	1.111	1.204	1.258	1.565	1.424	2.067	1.580	2.646
16π	0.899	0.888	1.107	1.293	1.257	1.659	1.426	2.167	1.586	2.751
50.3	0.898	0.931	1.107	1.340	1.258	1.709	1.429	2.220	1.591	2.807
50.4	0.899	1.058	1.112	1.482	1.267	1.861	1.446	2.382	1.618	2.975
50.5	0.908	1.192	1.128	1.632	1.290	2.021	1.478	2.552	1.663	3.154
50.6	0.926	1.336	1.156	1.792	1.328	2.193	1.531	2.735	1.733	3.344
50.7	0.957	1.493	1.202	1.967	1.387	2.380	1.610	2.933	1.835	3.546
50.8	1.005	1.666	1.271	2.160	1.476	2.585	1.725	3.147	1.980	3.761
50.9	1.077	1.862	1.373	2.375	1.605	2.810	1.891	3.377	2.185	3.984
51.0	1.187	2.084	1.526	2.616	1.795	3.056	2.131	3.618	2.472	4.203
51.1	1.353	2.339	1.754	2.881	2.076	3.317	2.474	3.853	2.869	4.389
51.2	1.613	2.628	2.102	3.160	2.490	3.566	2.959	4.039	3.400	4.484
51.3	2.025	2.934	2.629	3.407	3.092	3.737	3.615	4.086	4.056	4.390
51.4	2.684	3.188	3.402	3.504	3.904	3.682	4.401	3.841	4.739	3.988
51.5	3.674	3.177	4.392	3.202	4.804	3.178	5.117	3.165	5.236	3.240
51.6	4.834	2.490	5.254	2.244	5.404	2.132	5.427	2.134	5.324	2.314
51.7	5.413	0.986	5.399	0.857	5.321	0.905	5.175	1.122	4.994	1.506
51.8	4.881	-0.473	4.771	-0.243	4.682	0.024	4.564	0.450	4.438	0.993
33π/2	4.521	-0.815	4.455	-0.490	4.397	-0.171	4.315	0.303	4.223	0.881
51.9	3.861	-1.162	3.898	-0.745	3.904	-0.373	3.890	0.150	3.857	0.764
52.0	2.980	-1.268	3.146	-0.830	3.236	-0.441	3.308	0.100	3.351	0.726
52.1	2.360	-1.130	2.591	-0.721	2.727	-0.346	2.852	0.181	2.944	0.797
52.2	1.943	-0.920	2.196	-0.543	2.353	-0.186	2.505	0.322	2.625	0.921
52.3	1.658	-0.705	1.914	-0.349	2.078	-0.007	2.243	0.486	2.378	1.070
52.4	1.459	-0.503	1.710	-0.159	1.875	0.173	2.044	0.654	2.186	1.227
52.5	1.315	-0.320	1.558	0.019	1.721	0.346	1.890	0.820	2.035	1.385
52.6	1.208	-0.153	1.443	0.185	1.603	0.510	1.771	0.980	1.917	1.540
52.7	1.127	-0.001	1.355	0.340	1.511	0.665	1.677	1.135	1.823	1.693
52.8	1.065	0.140	1.287	0.485	1.439	0.814	1.604	1.285	1.749	1.842
52.9	1.017	0.272	1.234	0.624	1.383	0.956	1.546	1.430	1.692	1.989
53.0	0.979	0.398	1.192	0.759	1.340	1.095	1.502	1.574	1.648	2.135
53.1	0.951	0.520	1.161	0.890	1.308	1.232	1.469	1.716	1.617	2.282
53.2	0.929	0.639	1.138	1.020	1.285	1.369	1.448	1.860	1.598	2.430
53.3	0.915	0.759	1.123	1.150	1.271	1.507	1.436	2.006	1.591	2.582
53.4	0.906	0.880	1.116	1.284	1.266	1.649	1.436	2.156	1.596	2.739
17π	0.906	0.889	1.116	1.293	1.266	1.659	1.436	2.167	1.597	2.750
53.5	0.905	1.005	1.118	1.422	1.272	1.796	1.448	2.313	1.616	2.903
53.6	0.910	1.135	1.129	1.567	1.289	1.951	1.474	2.478	1.654	3.075
53.7	0.925	1.274	1.152	1.722	1.320	2.117	1.518	2.654	1.713	3.258
53.8	0.950	1.424	1.190	1.889	1.370	2.296	1.585	2.843	1.800	3.453
53.9	0.991	1.588	1.249	2.073	1.446	2.491	1.684	3.048	1.926	3.661
54.0	1.052	1.773	1.336	2.276	1.556	2.705	1.827	3.269	2.103	3.878
54.1	1.145	1.981	1.465	2.504	1.718	2.940	2.032	3.503	2.352	4.097
54.2	1.284	2.220	1.658	2.756	1.956	3.193	2.326	3.739	2.697	4.297
54.3	1.500	2.492	1.949	3.027	2.307	3.447	2.745	3.948	3.166	4.434
54.4	1.838	2.790	2.390	3.291	2.821	3.656	3.322	4.062	3.766	4.429
54.5	2.379	3.074	3.049	3.465	3.539	3.713	4.056	3.955	4.445	4.166
54.6	3.219	3.204	3.954	3.361	4.418	3.420	4.821	3.470	5.037	3.562
54.7	4.343	2.850	4.917	2.693	5.184	2.596	5.325	2.569	5.304	2.692
54.8	5.257	1.656	5.402	1.432	5.397	1.396	5.298	1.518	5.134	1.819
54.9	5.168	0.067	5.054	0.159	4.950	0.346	4.808	0.699	4.653	1.187
35π/2	4.490	-0.791	4.427	-0.465	4.371	-0.146	4.292	0.327	4.202	0.904
55.0	4.264	-0.943	4.235	-0.576	4.200	-0.234	4.144	0.260	4.075	0.853
55.1	3.303	-1.244	3.424	-0.804	3.483	-0.416	3.525	0.122	3.540	0.747
55.2	2.586	-1.181	2.796	-0.758	2.916	-0.376	3.023	0.158	3.098	0.779
55.3	2.098	-0.994	2.345	-0.603	2.495	-0.239	2.638	0.278	2.748	0.883
55.4	1.767	-0.780	2.023	-0.415	2.185	-0.067	2.346	0.432	2.476	1.022
55.5	1.538	-0.574	1.792	-0.225	1.957	0.112	2.125	0.598	2.265	1.175
55.6	1.374	-0.385	1.621	-0.044	1.785	0.286	1.955	0.763	2.099	1.331
55.7	1.254	-0.213	1.493	0.126	1.654	0.452	1.824	0.924	1.969	1.486
55.8	1.164	-0.057	1.395	0.283	1.553	0.610	1.721	1.080	1.867	1.638
55.9	1.095	0.087	1.320	0.432	1.474	0.759	1.640	1.230	1.786	1.787
56.0	1.041	0.222	1.261	0.572	1.412	0.903	1.576	1.376	1.722	1.934
56.1	1.000	0.350	1.215	0.707	1.364	1.042	1.527	1.519	1.674	2.080
56.2	0.968	0.472	1.180	0.839	1.328	1.179	1.490	1.661	1.638	2.225
56.3	0.944	0.592	1.154	0.968	1.301	1.315	1.465	1.803	1.614	2.371
56.4	0.926	0.711	1.136	1.098	1.284	1.451	1.449	1.946	1.602	2.520
56.5	0.915	0.830	1.126	1.229	1.276	1.590	1.444	2.093	1.603	2.673
18π	0.913	0.889	1.124	1.294	1.275	1.660	1.446	2.167	1.608	2.750
56.6	0.911	0.952	1.124	1.364	1.277	1.734	1.451	2.246	1.617	2.833
56.7	0.914	1.079	1.131	1.505	1.289	1.885	1.471	2.406	1.647	3.000
56.8	0.924	1.214	1.149	1.654	1.315	2.044	1.508	2.576	1.696	3.177
56.9	0.945	1.358	1.181	1.815	1.357	2.216	1.565	2.758	1.771	3.365
57.0	0.979	1.515	1.230	1.990	1.421	2.402	1.649	2.954	1.879	3.565
57.1	1.031	1.689	1.305	2.183	1.515	2.607	1.772	3.166	2.033	3.776
57.2	1.110	1.886	1.415	2.398	1.654	2.831	1.948	3.393	2.249	3.992
57.3	1.227	2.109	1.578	2.638	1.856	3.074	2.201	3.627	2.549	4.200
57.4	1.407	2.365	1.822	2.900	2.154	3.326	2.561	3.848	2.960	4.366
57.5	1.687	2.650	2.192	3.167	2.590	3.558	3.065	4.007	3.501	4.428
57.6	2.130	2.944	2.749	3.388	3.215	3.690	3.730	4.006	4.150	4.284
57.7	2.829	3.158	3.547	3.424	4.032	3.567	4.496	3.695	4.792	3.829
57.8	3.845	3.052	4.515	3.021	4.879	2.976	5.135	2.962	5.209	3.057
57.9	4.926	2.226	5.256	1.984	5.354	1.898	5.337	1.942	5.217	2.165
58.0	5.300	0.703	5.247	0.647	5.159	0.743	5.016	1.007	4.846	1.427
58.1	4.652	-0.603	4.567	-0.323	4.496	-0.030	4.401	0.420	4.296	0.979
37π/2	4.461	-0.768	4.401	-0.443	4.348	-0.124	4.271	0.348	4.184	0.924
58.2	3.663	-1.158	3.728	-0.731	3.752	-0.353	3.757	0.174	3.741	0.791
58.3	2.848	-1.211	3.028	-0.776	3.127	-0.389	3.211	0.150	3.264	0.774
58.4	2.278	-1.062	2.513	-0.657	2.653	-0.285	2.784	0.240	2.882	0.852
58.5	1.892	-0.856	2.146	-0.480	2.304	-0.125	2.460	0.381	2.583	0.978
58.6	1.627	-0.647	1.883	-0.291	2.047	0.050	2.213	0.542	2.350	1.124
58.7	1.440	-0.453	1.690	-0.107	1.855	0.225	2.025	0.706	2.168	1.278
58.8	1.304	-0.275	1.547	0.065	1.710	0.393	1.880	0.868	2.026	1.432
58.9	1.203	-0.114	1.439	0.226	1.598	0.553	1.767	1.024	1.913	1.584
59.0	1.126	0.034	1.355	0.377	1.511	0.705	1.678	1.175	1.824	1.733
59.1	1.067	0.171	1.290	0.520	1.443	0.850	1.608	1.322	1.755	1.880
59.2	1.021	0.301	1.239	0.656	1.390	0.990	1.554	1.465	1.700	2.024
59.3	0.986	0.425	1.200	0.788	1.349	1.126	1.512	1.606	1.660	2.168
59.4	0.958	0.545	1.170	0.917	1.319	1.261	1.482	1.747	1.632	2.313
59.5	0.939	0.663	1.149	1.046	1.298	1.396	1.463	1.888	1.615	2.459
59.6	0.925	0.781	1.136	1.175	1.286	1.533	1.454	2.033	1.611	2.610
19π	0.919	0.890	1.131	1.295	1.283	1.660	1.455	2.167	1.618	2.750
59.7	0.918	0.902	1.131	1.308	1.284	1.674	1.456	2.182	1.619	2.765
59.8	0.918	1.026	1.135	1.445	1.291	1.820	1.471	2.337	1.642	2.927
59.9	0.925	1.156	1.148	1.590	1.311	1.975	1.500	2.502	1.684	3.099
60.0	0.942	1.295	1.174	1.744	1.346	2.140	1.548	2.677	1.747	3.280
60.1	0.970	1.445	1.216	1.912	1.400	2.319	1.620	2.865	1.840	3.473
60.2	1.014	1.611	1.279	2.096	1.481	2.514	1.726	3.069	1.973	3.678
60.3	1.081	1.796	1.373	2.300	1.599	2.727	1.877	3.287	2.160	3.891
60.4	1.180	2.006	1.511	2.527	1.772	2.960	2.094	3.517	2.421	4.101
60.5	1.331	2.246	1.717	2.777	2.025	3.207	2.404	3.743	2.780	4.286
60.6	1.563	2.517	2.028	3.042	2.396	3.449	2.841	3.932	3.262	4.396
60.7	1.927	2.808	2.496	3.287	2.933	3.631	3.434	4.008	3.865	4.349
60.8	2.505	3.067	3.184	3.418	3.668	3.634	4.163	3.843	4.520	4.033
60.9	3.383	3.130	4.094	3.230	4.525	3.259	4.879	3.292	5.049	3.389
61.0	4.487	2.648	4.983	2.463	5.195	2.370	5.284	2.368	5.232	2.527
61.1	5.239	1.361	5.314	1.189	5.280	1.198	5.167	1.368	5.003	1.711
61.2	4.974	-0.133	4.865	0.025	4.772	0.249	4.646	0.635	4.510	1.147
39π/2	4.435	-0.747	4.377	-0.421	4.326	-0.103	4.252	0.369	4.167	0.944
61.3	4.049	-0.987	4.050	-0.597	4.036	-0.242	4.001	0.263	3.951	0.863
61.4	3.147	-1.205	3.287	-0.765	3.360	-0.377	3.415	0.161	3.444	0.785
61.5	2.487	-1.118	2.704	-0.700	2.830	-0.320	2.945	0.211	3.027	0.829
61.6	2.036	-0.930	2.285	-0.542	2.437	-0.180	2.584	0.334	2.698	0.937
61.7	1.729	-0.722	1.985	-0.357	2.148	-0.010	2.310	0.487	2.442	1.075
61.8	1.514	-0.522	1.767	-0.172	1.933	0.164	2.102	0.649	2.242	1.225
61.9	1.360	-0.340	1.606	0.003	1.770	0.334	1.941	0.811	2.086	1.378
62.0	1.246	-0.173	1.485	0.168	1.646	0.495	1.816	0.968	1.963	1.530
62.1	1.160	-0.021	1.392	0.322	1.550	0.649	1.719	1.120	1.866	1.679
62.2	1.095	0.120	1.321	0.467	1.475	0.796	1.642	1.267	1.789	1.825
62.3	1.044	0.252	1.265	0.604	1.417	0.937	1.582	1.411	1.729	1.969
62.4	1.005	0.377	1.221	0.737	1.372	1.074	1.536	1.551	1.683	2.113
62.5	0.974	0.498	1.188	0.866	1.337	1.208	1.501	1.691	1.650	2.256
62.6	0.951	0.616	1.163	0.994	1.313	1.342	1.478	1.831	1.629	2.400
62.7	0.936	0.733	1.147	1.122	1.297	1.477	1.465	1.973	1.620	2.548
62.8	0.926	0.852	1.139	1.253	1.291	1.615	1.462	2.119	1.623	2.700
20π	0.924	0.890	1.138	1.295	1.291	1.660	1.464	2.167	1.627	2.749
62.9	0.923	0.974	1.139	1.387	1.295	1.758	1.472	2.271	1.640	2.858
63.0	0.927	1.100	1.148	1.528	1.309	1.908	1.495	2.430	1.674	3.024
63.1	0.940	1.235	1.169	1.677	1.338	2.068	1.535	2.600	1.727	3.200
63.2	0.963	1.379	1.204	1.838	1.383	2.239	1.596	2.781	1.807	3.386
63.3	1.000	1.537	1.257	2.013	1.452	2.426	1.687	2.976	1.922	3.584
63.4	1.057	1.713	1.337	2.207	1.554	2.629	1.817	3.187	2.084	3.792
63.5	1.141	1.911	1.455	2.422	1.701	2.852	2.004	3.410	2.311	4.002
63.6	1.268	2.136	1.630	2.661	1.916	3.092	2.270	3.637	2.624	4.198
63.7	1.461	2.392	1.891	2.919	2.232	3.336	2.647	3.843	3.050	4.343
63.8	1.762	2.674	2.284	3.175	2.691	3.548	3.169	3.974	3.599	4.370
63.9	2.238	2.952	2.870	3.365	3.338	3.640	3.842	3.924	4.238	4.178

k_0h	a/λ = 0.001588		a/λ = 0.003175		a/λ = 0.004763		a/λ = 0.007022		a/λ = 0.009525	
64.0	2.978	3.121	3.689	3.337	4.154	3.446	4.581	3.547	4.836	3.673
64.1	4.009	2.913	4.625	2.835	4.940	2.776	5.142	2.768	5.176	2.886
64.2	4.993	1.959	5.239	1.736	5.293	1.680	5.244	1.767	5.111	2.031
64.3	5.170	0.447	5.093	0.462	5.002	0.602	4.865	0.908	4.709	1.359
64.4	4.435	-0.706	4.377	-0.387	4.325	-0.072	4.251	0.397	4.165	0.968
41π/2	4.409	-0.727	4.355	-0.402	4.305	-0.084	4.233	0.388	4.150	0.962
64.5	3.485	-1.148	3.574	-0.715	3.614	-0.332	3.636	0.198	3.635	0.816
64.6	2.728	-1.157	2.920	-0.726	3.028	-0.340	3.122	0.195	3.185	0.817
64.7	2.203	-0.999	2.442	-0.598	2.586	-0.228	2.722	0.293	2.825	0.902
64.8	1.845	-0.796	2.100	-0.423	2.259	-0.069	2.417	0.435	2.543	1.029
64.9	1.598	-0.594	1.853	-0.238	2.018	0.103	2.185	0.593	2.323	1.173
65.0	1.422	-0.406	1.672	-0.060	1.837	0.273	2.007	0.754	2.151	1.324
65.1	1.294	-0.234	1.536	0.108	1.699	0.437	1.870	0.912	2.016	1.475
65.2	1.198	-0.077	1.433	0.265	1.593	0.593	1.762	1.065	1.909	1.625
65.3	1.125	0.067	1.354	0.413	1.510	0.741	1.678	1.213	1.825	1.771
65.4	1.068	0.201	1.292	0.552	1.446	0.883	1.612	1.357	1.759	1.915
65.5	1.025	0.329	1.243	0.686	1.395	1.021	1.560	1.497	1.708	2.057
65.6	0.991	0.450	1.206	0.816	1.357	1.155	1.521	1.636	1.670	2.199
65.7	0.965	0.569	1.179	0.943	1.328	1.289	1.494	1.775	1.645	2.342
65.8	0.947	0.686	1.159	1.070	1.310	1.422	1.477	1.916	1.631	2.487
65.9	0.935	0.803	1.148	1.199	1.300	1.558	1.470	2.059	1.629	2.636
21π	0.930	0.891	1.145	1.295	1.298	1.661	1.472	2.167	1.636	2.749
66.0	0.929	0.923	1.144	1.331	1.299	1.698	1.474	2.207	1.640	2.791
66.1	0.930	1.047	1.150	1.468	1.309	1.844	1.492	2.362	1.667	2.952
66.2	0.939	1.177	1.166	1.612	1.332	1.998	1.524	2.526	1.711	3.122
66.3	0.958	1.316	1.194	1.767	1.370	2.163	1.576	2.700	1.780	3.303
66.4	0.989	1.467	1.240	1.935	1.429	2.342	1.654	2.888	1.879	3.494
66.5	1.037	1.634	1.308	2.120	1.515	2.537	1.766	3.090	2.019	3.696
66.6	1.108	1.821	1.408	2.324	1.641	2.750	1.927	3.306	2.216	3.904
66.7	1.216	2.033	1.556	2.551	1.825	2.981	2.156	3.531	2.489	4.106
66.8	1.378	2.274	1.777	2.799	2.093	3.223	2.481	3.747	2.862	4.275
66.9	1.627	2.543	2.107	3.057	2.485	3.451	2.936	3.915	3.356	4.358
67.0	2.019	2.826	2.603	3.282	3.046	3.604	3.543	3.952	3.960	4.267
67.1	2.635	3.056	3.319	3.366	3.794	3.549	4.265	3.727	4.588	3.901
67.2	3.546	3.044	4.226	3.091	4.620	3.094	4.926	3.116	5.053	3.223
67.3	4.612	2.434	5.030	2.234	5.191	2.152	5.235	2.180	5.158	2.375
67.4	5.194	1.081	5.214	0.968	5.159	1.020	5.038	1.236	4.879	1.615
67.5	4.779	-0.302	4.683	-0.088	4.603	0.167	4.495	0.581	4.378	1.113
43π/2	4.386	-0.709	4.333	-0.383	4.286	-0.065	4.216	0.405	4.135	0.978
67.6	3.852	-1.016	3.882	-0.610	3.885	-0.245	3.870	0.269	3.837	0.874
67.7	3.007	-1.166	3.163	-0.728	3.247	-0.340	3.316	0.197	3.356	0.820
67.8	2.396	-1.060	2.620	-0.645	2.752	-0.268	2.874	0.260	2.962	0.875
67.9	1.979	-0.870	2.229	-0.485	2.384	-0.125	2.534	0.386	2.652	0.986
68.0	1.693	-0.667	1.949	-0.304	2.112	0.042	2.277	0.538	2.410	1.123
68.1	1.492	-0.474	1.744	-0.124	1.910	0.212	2.079	0.697	2.221	1.271
68.2	1.346	-0.297	1.592	0.047	1.756	0.378	1.927	0.855	2.073	1.421
68.3	1.238	-0.135	1.477	0.207	1.639	0.536	1.809	1.009	1.956	1.570
68.4	1.157	0.013	1.389	0.358	1.547	0.686	1.716	1.158	1.864	1.717
68.5	1.094	0.150	1.321	0.499	1.476	0.830	1.643	1.302	1.791	1.861
68.6	1.046	0.280	1.267	0.635	1.420	0.968	1.586	1.443	1.734	2.003
68.7	1.008	0.403	1.226	0.765	1.378	1.103	1.543	1.582	1.692	2.144
68.8	0.980	0.522	1.195	0.893	1.345	1.236	1.511	1.720	1.661	2.285
68.9	0.958	0.639	1.172	1.019	1.323	1.369	1.490	1.859	1.643	2.428
69.0	0.944	0.755	1.158	1.146	1.309	1.503	1.479	2.000	1.636	2.574
69.1	0.936	0.873	1.151	1.276	1.305	1.640	1.479	2.145	1.642	2.725
22π	0.935	0.891	1.151	1.296	1.305	1.661	1.480	2.167	1.644	2.749
[illegible]	[illegible]	[illegible]	1.153	1.410	1.311	1.782	1.491	2.296	1.662	2.883
69.3	0.940	1.121	1.164	[illegible]	[illegible]	[illegible]	[illegible]	[illegible]	[illegible]	[illegible]
69.4	0.955	1.256	1.187	1.700	1.359	2.091	1.560	2.623	1.757	3.222
69.5	0.980	1.401	1.225	1.861	1.409	2.263	1.626	2.804	1.842	3.408
69.6	1.020	1.560	1.283	2.037	1.483	2.449	1.723	2.999	1.964	3.604
69.7	1.081	1.737	1.369	2.231	1.591	2.653	1.861	3.207	2.134	3.808
69.8	1.172	1.936	1.495	2.447	1.748	2.875	2.058	3.427	2.372	4.012
69.9	1.308	2.163	1.682	2.685	1.977	3.111	2.339	3.647	2.699	4.196
70.0	1.516	2.419	1.960	2.938	2.310	3.346	2.734	3.838	3.138	4.320
70.1	1.840	2.697	2.378	3.181	2.793	3.537	3.273	3.939	3.695	4.311
70.2	2.349	2.958	2.993	3.338	3.459	3.586	3.950	3.836	4.321	4.070
70.3	3.128	3.075	3.828	3.242	4.269	3.319	4.657	3.398	4.872	3.521
70.4	4.163	2.759	4.719	2.642	4.986	2.576	5.137	2.580	5.136	2.725
70.5	5.033	1.692	5.203	1.499	5.220	1.478	5.147	1.608	5.006	1.910
70.6	5.027	0.218	4.937	0.301	4.848	0.480	4.721	0.822	4.579	1.300
45π/2	4.363	-0.691	4.313	-0.366	4.267	-0.048	4.200	0.422	4.121	0.994
70.7	4.231	-0.788	4.200	-0.437	4.166	-0.104	4.112	0.379	4.045	0.962
70.8	3.322	-1.132	3.433	-0.695	3.488	-0.311	3.525	0.222	3.538	0.841
70.9	2.620	-1.106	2.822	-0.679	2.938	-0.295	3.041	0.238	3.112	0.857
71.0	2.134	-0.941	2.376	-0.544	2.524	-0.176	2.664	0.342	2.772	0.948
71.1	1.802	-0.742	2.056	-0.369	2.217	-0.018	2.377	0.484	2.506	1.076
71.2	1.570	-0.545	1.825	-0.189	1.990	0.151	2.158	0.640	2.298	1.219
71.3	1.405	-0.362	1.654	-0.015	1.819	0.318	1.990	0.798	2.135	1.368
71.4	1.283	-0.195	1.526	0.148	1.688	0.478	1.859	0.953	2.006	1.516
71.5	1.192	-0.043	1.428	0.302	1.587	0.630	1.757	1.103	1.905	1.662
71.6	1.123	0.098	1.352	0.446	1.509	0.775	1.677	1.248	1.825	1.806
71.7	1.069	0.230	1.293	0.583	1.447	0.915	1.614	1.389	1.762	1.948
71.8	1.027	0.355	1.247	0.714	1.400	1.051	1.566	1.528	1.714	2.089
71.9	0.995	0.475	1.212	0.842	1.363	1.183	1.529	1.665	1.679	2.229
72.0	0.971	0.592	1.186	0.969	1.337	1.315	1.504	1.803	1.656	2.370
72.1	0.954	0.708	1.168	1.095	1.320	1.448	1.489	1.942	1.645	2.514
72.2	0.943	0.825	1.158	1.222	1.312	1.583	1.484	2.084	1.646	2.662
23π	0.940	0.892	1.157	1.296	1.312	1.661	1.487	2.167	1.652	2.748
72.3	0.939	0.944	1.157	1.354	1.314	1.722	1.491	2.232	1.660	2.815
72.4	0.942	1.067	1.164	1.490	1.326	1.868	1.511	2.386	1.689	2.976
72.5	0.953	1.198	1.182	1.635	1.351	2.022	1.547	2.549	1.738	3.146
72.6	0.973	1.337	1.214	1.790	1.393	2.187	1.603	2.724	1.811	3.325
72.7	1.007	1.489	1.263	1.959	1.456	2.366	1.686	2.911	1.916	3.515
72.8	1.059	1.658	1.336	2.144	1.548	2.561	1.805	3.112	2.064	3.715
72.9	1.136	1.846	1.444	2.349	1.683	2.774	1.975	3.326	2.271	3.918
73.0	1.251	2.060	1.602	2.576	1.878	3.003	2.217	3.545	2.557	4.111
73.1	1.425	2.302	1.837	2.822	2.162	3.238	2.559	3.750	2.944	4.263
73.2	1.693	2.570	2.189	3.071	2.575	3.452	3.032	3.895	3.449	4.317
73.3	2.114	2.843	2.712	3.274	3.159	3.572	3.651	3.892	4.051	4.184
73.4	2.768	3.039	3.453	3.306	3.916	3.458	4.360	3.607	4.648	3.768
73.5	3.707	2.945	4.350	2.943	4.704	2.924	4.961	2.942	5.048	3.063
73.6	4.717	2.209	5.057	2.006	5.171	1.943	5.176	2.004	5.080	2.235
73.7	5.125	0.817	5.103	0.766	5.034	0.860	4.911	1.118	4.759	1.530
73.8	4.587	-0.445	4.509	-0.182	4.444	0.100	4.354	0.538	4.254	1.086
47π/2	4.342	-0.675	4.295	-0.350	4.250	-0.032	4.185	0.437	4.107	1.009
73.9	3.670	-1.034	3.726	-0.616	3.747	-0.244	3.749	0.277	3.722	0.886
74.0	2.879	-1.127	3.050	-0.691	3.144	-0.304	3.224	0.232	3.274	0.853
74.1	2.314	-1.005	2.543	-0.594	2.680	-0.220	2.808	0.306	2.903	0.917
74.2	1.927	-0.815	2.178	-0.433	2.335	-0.075	2.488	0.434	2.609	1.031
74.3	1.659	-0.617	1.915	-0.255	2.079	0.090	2.245	0.585	2.380	1.168
74.4	1.470	-0.430	1.722	-0.079	1.888	0.257	2.058	0.741	2.201	1.314
74.5	1.333	-0.257	1.578	0.088	1.743	0.419	1.914	0.896	2.060	1.462
74.6	1.231	-0.100	1.469	0.244	1.631	0.573	1.801	1.047	1.949	1.608
74.7	1.153	0.045	1.385	0.391	1.544	0.721	1.713	1.193	1.861	1.752
74.8	1.094	0.179	1.320	0.530	1.476	0.862	1.644	1.335	1.792	1.894
74.9	1.047	0.306	1.269	0.663	1.423	0.998	1.590	1.474	1.739	2.034
75.0	1.012	0.428	1.230	0.792	1.383	1.131	1.549	1.611	1.699	2.173
75.1	0.985	0.545	1.201	0.918	1.353	1.263	1.519	1.748	1.671	2.313
75.2	0.965	0.661	1.180	1.043	1.332	1.394	1.500	1.885	1.655	2.455
75.3	0.951	0.777	1.167	1.170	1.320	1.527	1.492	2.025	1.651	2.600
24π	0.945	0.892	1.162	1.296	1.318	1.661	1.494	2.167	1.659	2.748
75.4	0.944	0.894	1.162	1.299	1.318	1.664	1.494	2.170	1.659	2.750
75.5	0.944	1.015	1.166	1.432	1.326	1.806	1.508	2.320	1.682	2.907
75.6	0.952	1.142	1.179	1.573	1.346	1.955	1.537	2.478	1.723	3.071
75.7	0.968	1.277	1.205	1.722	1.380	2.115	1.585	2.647	1.785	3.245
75.8	0.996	1.423	1.246	1.884	1.433	2.286	1.655	2.828	1.876	3.430
75.9	1.040	1.583	1.308	2.061	1.513	2.473	1.758	3.022	2.004	3.624
76.0	1.106	1.761	1.400	2.256	1.628	2.677	1.905	3.229	2.184	3.825
76.1	1.203	1.963	1.535	2.473	1.795	2.898	2.113	3.445	2.433	4.022
76.2	1.349	2.191	1.734	2.710	2.037	3.130	2.408	3.656	2.774	4.193
76.3	1.573	2.447	2.031	2.958	2.390	3.356	2.821	3.831	3.226	4.294
76.4	1.921	2.720	2.474	3.186	2.896	3.523	3.376	3.900	3.788	4.249
76.5	2.465	2.960	3.117	3.306	3.580	3.525	4.054	3.747	4.398	3.959
76.6	3.281	3.020	3.962	3.136	4.377	3.185	4.724	3.246	4.898	3.371
76.7	4.307	2.590	4.797	2.444	5.016	2.378	5.121	2.400	5.090	2.574
76.8	5.045	1.428	5.149	1.275	5.136	1.290	5.046	1.463	4.903	1.802
76.9	4.872	0.017	4.781	0.160	4.699	0.374	4.584	0.749	4.457	1.250
49π/2	4.322	-0.660	4.277	-0.334	4.234	-0.017	4.170	0.452	4.095	1.023
77.0	4.038	-0.852	4.034	-0.475	4.019	-0.129	3.983	0.367	3.933	0.958
77.1	3.174	-1.112	3.304	-0.673	3.372	-0.287	3.423	0.247	3.449	0.866
77.2	2.522	-1.056	2.732	-0.633	2.855	-0.252	2.966	0.279	3.045	0.896
77.3	2.071	-0.886	2.316	-0.493	2.466	-0.127	2.611	0.388	2.723	0.992
77.4	1.761	-0.690	2.016	-0.320	2.178	0.031	2.340	0.531	2.471	1.120
77.5	1.544	-0.499	1.798	-0.143	1.964	0.196	2.133	0.684	2.274	1.262
77.6	1.388	-0.321	1.637	0.026	1.802	0.360	1.973	0.839	2.119	1.408
77.7	1.273	-0.159	1.515	0.186	1.678	0.516	1.849	0.991	1.997	1.554
77.8	1.187	-0.010	1.422	0.336	1.582	0.665	1.752	1.138	1.900	1.698
77.9	1.120	0.128	1.350	0.477	1.507	0.808	1.676	1.281	1.824	1.840
78.0	1.069	0.257	1.294	0.612	1.449	0.945	1.616	1.421	1.765	1.980
78.1	1.029	0.380	1.250	0.742	1.403	1.079	1.570	1.557	1.720	2.118
78.2	0.999	0.498	1.217	0.868	1.369	1.210	1.536	1.693	1.688	2.257
78.3	0.976	0.614	1.193	0.993	1.345	1.341	1.513	1.829	1.667	2.397
78.4	0.960	0.729	1.176	1.118	1.330	1.472	1.500	1.968	1.658	2.540
78.5	0.951	0.845	1.168	1.245	1.324	1.607	1.498	2.109	1.661	2.687
25π	0.949	0.892	1.167	1.297	1.324	1.662	1.500	2.167	1.666	2.748
78.6	0.948	0.964	1.168	1.376	1.327	1.746	1.507	2.256	1.678	2.840
78.7	0.952	1.088	1.178	1.513	1.342	1.891	1.530	2.410	1.711	3.000
78.8	0.965	1.219	1.198	1.658	1.370	2.045	1.569	2.573	1.764	3.169
78.9	0.988	1.359	1.233	1.813	1.415	2.211	1.630	2.748	1.841	3.348
79.0	1.025	1.512	1.286	1.983	1.483	2.390	1.718	2.935	1.953	3.536
79.1	1.080	1.682	1.364	2.169	1.581	2.585	1.844	3.135	2.109	3.733
79.2	1.163	1.872	1.479	2.375	1.724	2.797	2.024	3.346	2.326	3.932
79.3	[illegible]	[illegible]	[illegible]	[illegible]	1.932	3.024	2.279	3.560	2.024	4.115
79.4	1.474	2.331	1.898	2.845	[illegible]	[illegible]	[illegible]	[illegible]	[illegible]	[illegible]
79.5	1.762	2.597	2.272	3.084	2.667	3.451	3.128	3.873	3.541	4.274
79.6	2.213	2.857	2.824	3.261	3.273	3.536	3.757	3.827	4.138	4.096
79.7	2.905	3.014	3.587	3.237	4.034	3.359	4.447	3.483	4.700	3.634
79.8	3.864	2.831	4.462	2.785	4.774	2.751	4.984	2.770	5.034	2.909
79.9	4.800	1.975	5.064	1.783	5.136	1.743	5.110	1.840	4.999	2.106
80.0	5.033	0.573	4.982	0.584	4.907	0.717	4.786	1.015	4.643	1.456
80.1	4.399	-0.564	4.342	-0.260	4.293	0.044	4.221	0.502	4.138	1.064
51π/2	4.303	-0.645	4.260	-0.320	4.219	-0.003	4.156	0.466	4.082	1.037
80.2	3.503	-1.042	[illegible]	-0.615	3.618	-0.238	3.637	0.287	3.634	0.899
80.3	2.762	-1.087	2.946	-0.654	3.049	-0.268	3.139	0.266	3.199	0.885
80.4	2.238	-0.952	2.472	-0.546	2.613	-0.174	2.746	0.349	2.847	0.958
80.5	1.878	-0.763	2.131	-0.384	2.289	-0.027	2.445	0.479	2.569	1.074
80.6	1.628	-0.570	1.883	-0.208	2.048	0.136	2.215	0.629	2.352	1.211
80.7	1.450	-0.387	1.701	-0.036	1.867	0.300	2.038	0.783	2.182	1.355
80.8	1.320	-0.220	1.565	0.127	1.729	0.458	1.901	0.935	2.048	1.500
80.9	1.223	-0.056	1.462	0.280	1.623	0.609	1.794	1.083	1.942	1.644
81.0	1.149	0.075	1.382	0.424	1.540	0.754	1.710	1.227	1.859	1.786
81.1	1.093	0.207	1.319	0.560	1.476	0.892	1.644	1.367	1.793	1.926
81.2	1.048	0.332	1.271	0.691	1.425	1.027	1.593	1.504	1.743	2.064
81.3	1.014	0.451	1.234	0.818	1.387	1.158	1.554	1.639	1.706	2.202
81.4	0.989	0.568	1.206	0.943	1.359	1.288	1.527	1.774	1.680	2.341
81.5	0.970	0.682	1.187	1.067	1.340	1.419	1.510	1.911	1.667	2.482
81.6	0.958	0.798	1.176	1.193	1.330	1.551	1.504	2.050	1.665	2.626
26π	0.953	0.893	1.172	1.297	1.329	1.662	1.506	2.167	1.673	2.747
81.7	0.953	0.915	1.172	1.321	1.330	1.687	1.508	2.194	1.676	2.775
81.8	0.954	1.036	1.178	1.455	1.340	1.829	1.525	2.344	1.702	2.931
81.9	0.963	1.163	1.194	1.595	1.362	1.979	1.557	2.502	1.746	3.095
82.0	0.982	1.298	1.222	1.745	1.400	2.138	1.608	2.671	1.813	3.269
82.1	1.012	1.445	1.267	1.908	1.457	2.310	1.684	2.851	1.909	3.452
82.2	1.059	1.606	1.333	2.086	1.542	2.498	1.793	3.045	2.045	3.644
82.3	1.130	1.787	1.431	2.282	1.665	2.701	1.948	3.250	2.234	3.841
82.4	1.234	1.990	1.575	2.499	1.842	2.921	2.168	3.463	2.494	4.031
82.5	1.392	2.220	1.788	2.734	2.099	3.150	2.479	3.666	2.849	4.189
82.6	1.632	2.476	2.103	2.977	2.472	3.364	2.909	3.822	3.314	4.266
82.7	2.005	2.742	2.572	3.187	3.000	3.505	3.479	3.857	3.879	4.183
82.8	2.585	2.957	3.243	3.267	3.700	3.457	4.154	3.651	4.469	3.847
82.9	3.435	2.952	4.091	3.020	4.475	3.044	4.779	3.093	4.915	3.223
83.0	4.437	2.407	4.858	2.242	5.030	2.182	5.093	2.227	5.037	2.431
83.1	5.030	1.170	5.079	1.065	5.043	1.118	4.942	1.331	4.799	1.704
83.2	4.710	-0.159	4.626	0.038	4.553	0.284	4.452	0.686	4.340	1.207
53π/2	4.285	-0.631	4.244	-0.306	4.204	0.011	4.143	0.479	4.071	1.049
83.3	3.857	-0.899	3.879	-0.503	3.880	-0.146	3.863	0.360	3.829	0.958
83.4	3.038	-1.088	3.185	-0.650	3.264	-0.263	3.328	0.271	3.365	0.890
83.5	2.432	-1.008	2.650	-0.589	2.778	-0.211	2.896	0.318	2.982	0.932
83.6	2.013	-0.835	2.260	-0.444	2.413	-0.081	2.561	0.432	2.677	1.032
83.7	1.723	-0.642	1.979	-0.273	2.142	0.076	2.305	0.574	2.438	1.161
83.8	1.520	-0.455	1.773	-0.100	1.939	0.239	2.109	0.726	2.251	1.302
83.9	1.372	-0.283	1.620	0.066	1.786	0.399	1.957	0.879	2.104	1.446

k_0h	a/λ = 0.001588		a/λ = 0.003175		a/λ = 0.004763		a/λ = 0.007022		a/λ = 0.009525	
84.0	1.263	-0.124	1.505	0.222	1.668	0.553	1.839	1.028	1.988	1.590
84.1	1.181	0.021	1.416	0.369	1.576	0.699	1.747	1.172	1.896	1.732
84.2	1.118	0.156	1.347	0.508	1.505	0.839	1.674	1.313	1.824	1.872
84.3	1.069	0.283	1.294	0.640	1.449	0.974	1.618	1.450	1.767	2.010
84.4	1.031	0.404	1.252	0.768	1.406	1.106	1.574	1.586	1.725	2.147
84.5	1.002	0.521	1.221	0.893	1.374	1.236	1.543	1.720	1.695	2.285
84.6	0.981	0.636	1.199	1.017	1.352	1.366	1.521	1.855	1.677	2.424
84.7	0.966	0.750	1.184	1.141	1.339	1.496	1.511	1.993	1.670	2.566
84.8	0.958	0.866	1.177	1.268	1.334	1.630	1.511	2.134	1.676	2.712
27π	0.957	0.893	1.177	1.297	1.335	1.662	1.512	2.167	1.679	2.747
84.9	0.957	0.985	1.179	1.398	1.340	1.769	1.523	2.280	1.696	2.864
85.0	0.963	1.108	1.191	1.535	1.357	1.914	1.548	2.434	1.732	3.024
85.1	0.977	1.240	1.214	1.680	1.388	2.069	1.591	2.597	1.788	3.192
85.2	1.002	1.381	1.251	1.836	1.437	2.234	1.655	2.771	1.871	3.370
85.3	1.042	1.535	1.308	2.007	1.509	2.414	1.749	2.958	1.989	3.558
85.4	1.102	1.706	1.392	2.194	1.614	2.610	1.883	3.158	2.154	3.752
85.5	1.191	1.898	1.514	2.401	1.766	2.822	2.074	3.366	2.381	3.945
85.6	1.323	2.116	1.695	2.628	1.986	3.046	2.342	3.574	2.693	4.118
85.7	1.524	2.360	1.962	2.868	2.305	3.268	2.717	3.753	3.108	4.234
85.8	1.834	2.623	2.359	3.096	2.762	3.447	3.225	3.848	3.632	4.227
85.9	2.316	2.868	2.938	3.244	3.388	3.494	3.861	3.757	4.221	4.005
86.0	3.047	2.981	3.720	3.159	4.147	3.252	4.527	3.354	4.744	3.500
86.1	4.015	2.702	4.562	2.617	4.831	2.574	4.995	2.601	5.012	2.761
86.2	4.858	1.736	5.052	1.565	5.087	1.553	5.036	1.687	4.915	1.988
86.3	4.923	0.349	4.854	0.420	4.778	0.591	4.663	0.924	4.530	1.391
55π/2	4.268	-0.618	4.228	-0.293	4.190	0.024	4.131	0.492	4.060	1.061
86.4	4.215	-0.660	4.183	-0.324	4.149	-0.000	4.095	0.474	4.029	1.047
86.5	3.348	-1.042	3.449	-0.610	3.499	-0.229	3.533	0.300	3.543	0.914
86.6	2.655	-1.047	2.850	-0.618	2.961	-0.234	3.060	0.299	3.129	0.916
86.7	2.168	-0.902	2.407	-0.500	2.551	-0.130	2.689	0.390	2.795	0.996
86.8	1.833	-0.714	2.086	-0.337	2.246	0.018	2.405	0.522	2.532	1.115
86.9	1.599	-0.525	1.854	-0.164	2.019	0.179	2.187	0.670	2.326	1.251
87.0	1.431	-0.347	1.682	0.004	1.847	0.340	2.019	0.822	2.164	1.393
87.1	1.308	-0.184	1.552	0.163	1.716	0.495	1.889	0.972	2.036	1.536
87.2	1.216	-0.034	1.454	0.313	1.615	0.643	1.787	1.118	1.936	1.678
87.3	1.145	0.104	1.378	0.454	1.537	0.785	1.707	1.259	1.856	1.818
87.4	1.091	0.234	1.318	0.589	1.475	0.922	1.644	1.397	1.794	1.956
87.5	1.049	0.357	1.272	0.718	1.427	1.054	1.596	1.532	1.746	2.093
87.6	1.017	0.475	1.237	0.843	1.391	1.184	1.559	1.666	1.712	2.230
87.7	0.993	0.590	1.211	0.967	1.365	1.313	1.534	1.800	1.689	2.367
87.8	0.975	0.704	1.193	1.090	1.348	1.443	1.519	1.936	1.677	2.507
87.9	0.965	0.818	1.184	1.215	1.340	1.575	1.515	2.075	1.678	2.651
28π	0.961	0.893	1.182	1.298	1.340	1.662	1.518	2.167	1.685	2.747
88.0	0.960	0.935	1.182	1.343	1.341	1.710	1.522	2.218	1.692	2.800
88.1	0.963	1.056	1.190	1.477	1.354	1.852	1.541	2.368	1.721	2.955
88.2	0.974	1.183	1.207	1.618	1.379	2.002	1.577	2.526	1.768	3.119
88.3	0.994	1.319	1.238	1.768	1.419	2.162	1.632	2.695	1.840	3.292
88.4	1.028	1.467	1.287	1.932	1.481	2.335	1.712	2.875	1.942	3.474
88.5	1.079	1.630	1.358	2.110	1.571	2.522	1.828	3.068	2.085	3.665
88.6	1.154	1.812	1.463	2.308	1.702	2.726	1.992	3.272	2.283	3.858
88.7	1.266	2.018	1.616	2.525	1.890	2.945	2.224	3.480	2.556	4.040
88.8	1.435	2.250	1.843	2.759	2.163	3.168	2.550	3.674	2.924	4.184
88.9	1.693	2.505	2.179	2.995	2.556	3.370	2.998	3.810	3.401	4.235
89.0	2.093	2.762	2.674	3.186	3.106	3.483	3.581	3.809	3.968	4.113
89.1	2.709	2.948	3.371	3.219	3.817	3.382	4.248	3.549	4.533	3.731
89.2	3.585	2.870	4.213	2.893	4.563	2.896	4.824	2.938	4.923	3.078
89.3	4.549	2.209	4.901	2.037	5.029	1.990	5.055	2.062	4.978	2.298
89.4	4.989	0.921	4.995	0.870	4.943	0.960	4.836	1.213	4.697	1.617
89.5	4.544	-0.311	4.473	-0.065	4.411	0.207	4.326	0.633	4.229	1.172
57π/2	4.252	-0.606	4.214	-0.281	4.177	0.036	4.119	0.504	4.050	1.073
89.6	3.686	-0.932	3.734	-0.522	3.751	-0.156	3.750	0.358	3.731	0.960
89.7	2.913	-1.061	3.075	-0.624	3.165	-0.238	3.240	0.296	3.288	0.914
89.8	2.349	-0.962	2.573	-0.547	2.707	-0.171	2.831	0.356	2.923	0.967
89.9	1.959	-0.786	2.208	-0.398	2.363	-0.037	2.515	0.473	2.634	1.071
90.0	1.688	-0.596	1.944	-0.229	2.107	0.119	2.272	0.616	2.407	1.201
90.1	1.497	-0.414	1.750	-0.059	1.916	0.280	2.080	0.766	2.229	1.340
90.2	1.357	-0.246	1.605	0.103	1.770	0.437	1.942	0.915	2.089	1.483
90.3	1.254	-0.091	1.455	0.256	1.658	0.587	1.830	1.062	1.978	1.625
90.4	1.176	0.051	1.410	0.400	1.571	0.731	1.742	1.205	1.891	1.765
90.5	1.115	0.184	1.345	0.537	1.503	0.869	1.672	1.344	1.822	1.903
90.6	1.069	0.308	1.294	0.667	1.450	1.002	1.619	1.479	1.769	2.039
90.7	1.033	0.428	1.255	0.793	1.409	1.133	1.578	1.613	1.730	2.175
90.8	1.005	0.543	1.225	0.917	1.379	1.261	1.548	1.746	1.702	2.311
90.9	0.985	0.657	1.204	1.040	1.358	1.390	1.529	1.881	1.686	2.450
91.0	0.972	0.771	1.191	1.163	1.347	1.520	1.521	2.017	1.682	2.591
91.1	0.965	0.886	1.186	1.290	1.345	1.653	1.523	2.158	1.690	2.737
29π	0.965	0.894	1.186	1.298	1.345	1.662	1.523	2.167	1.691	2.746
91.2	0.965	1.005	1.190	1.420	1.352	1.792	1.537	2.304	1.713	2.888
91.3	0.972	1.129	1.203	1.557	1.372	1.937	1.566	2.458	1.752	3.048
91.4	0.989	1.261	1.228	1.703	1.406	2.092	1.612	2.621	1.813	3.216
91.5	1.016	1.403	1.269	1.860	1.458	2.259	1.681	2.795	1.901	3.393
91.6	1.059	1.558	1.330	2.031	1.535	2.439	1.781	2.982	2.025	3.579
91.7	1.123	1.731	1.420	2.220	1.647	2.635	1.923	3.181	2.199	3.771
91.8	1.219	1.926	1.551	2.427	1.809	2.846	2.123	3.387	2.437	3.958
91.9	1.361	2.145	1.743	2.654	2.042	3.068	2.406	3.587	2.762	4.121
92.0	1.576	2.390	2.027	2.890	2.379	3.281	2.798	3.752	3.190	4.216
92.1	1.909	2.649	2.448	3.105	2.858	3.440	3.322	3.818	3.721	4.176
92.2	2.424	2.876	3.055	3.220	3.503	3.445	3.962	3.680	4.298	3.910
92.3	3.191	2.938	3.850	3.070	4.254	3.137	4.598	3.221	4.779	3.366
92.4	4.158	2.556	4.647	2.442	4.874	2.395	4.994	2.436	4.981	2.619
92.5	4.890	1.494	5.021	1.355	5.026	1.373	4.956	1.545	4.829	1.880
92.6	4.797	0.147	4.722	0.275	4.648	0.480	4.542	0.844	4.422	1.335
59π/2	4.236	-0.594	4.200	-0.269	4.164	0.048	4.108	0.516	4.040	1.084
92.7	4.038	-0.738	4.030	-0.374	4.013	-0.036	3.977	0.452	3.926	1.035
92.8	3.204	-1.035	3.325	-0.600	3.388	-0.216	3.435	0.314	3.458	0.929
92.9	2.557	-1.007	2.761	-0.581	2.880	-0.199	2.987	0.332	3.063	0.947
93.0	2.104	-0.854	2.346	-0.456	2.494	-0.088	2.636	0.429	2.746	1.033
93.1	1.791	-0.668	2.045	-0.292	2.206	0.061	2.367	0.563	2.496	1.153
93.2	1.571	-0.483	1.826	-0.123	1.991	0.220	2.160	0.710	2.300	1.289
93.3	1.413	-0.309	1.663	0.042	1.829	0.378	2.001	0.860	2.146	1.430
93.4	1.296	-0.150	1.540	0.198	1.704	0.530	1.877	1.007	2.025	1.571
93.5	1.209	-0.004	1.446	0.345	1.608	0.676	1.780	1.150	1.929	1.711
93.6	1.142	0.132	1.374	0.484	1.533	0.816	1.703	1.290	1.853	1.849
93.7	1.090	0.260	1.317	0.616	1.474	0.950	1.644	1.426	1.794	1.986
93.8	1.050	0.381	1.273	0.743	1.429	1.081	1.598	1.560	1.749	2.121
93.9	1.019	0.497	1.240	0.867	1.395	1.210	1.564	1.693	1.717	2.256

k_0h	a/λ = 0.001588		a/λ = 0.003175		a/λ = 0.004763		a/λ = 0.007022		a/λ = 0.009525	
94.0	0.996	0.611	1.216	0.990	1.370	1.338	1.541	1.826	1.696	2.393
94.1	0.980	0.724	1.200	1.113	1.355	1.467	1.528	1.961	1.687	2.532
94.2	0.971	0.838	1.191	1.237	1.349	1.598	1.526	2.099	1.690	2.675
30π	0.968	0.894	1.190	1.298	1.349	1.662	1.529	2.167	1.697	2.746
94.3	0.968	0.955	1.191	1.365	1.352	1.733	1.535	2.242	1.707	2.824
94.4	0.972	1.076	1.201	1.499	1.367	1.875	1.557	2.392	1.739	2.979
94.5	0.984	1.204	1.221	1.640	1.394	2.025	1.596	2.550	1.790	3.142
94.6	1.007	1.341	1.255	1.791	1.439	2.186	1.655	2.719	1.866	3.315
94.7	1.043	1.490	1.306	1.956	1.505	2.359	1.740	2.899	1.974	3.497
94.8	1.098	1.655	1.383	2.136	1.601	2.547	1.863	3.092	2.125	3.685
94.9	1.179	1.839	1.495	2.334	1.740	2.751	2.037	3.294	2.334	3.874
95.0	1.299	2.046	1.658	2.552	1.940	2.968	2.281	3.498	2.619	4.049
95.1	1.480	2.279	1.900	2.784	2.228	3.187	2.623	3.680	3.001	4.176
95.2	1.757	2.533	2.257	3.012	2.642	3.374	3.089	3.795	3.488	4.201
95.3	2.185	2.780	2.779	3.180	3.214	3.456	3.683	3.755	4.053	4.039
95.4	2.838	2.932	3.498	3.163	3.932	3.298	4.337	3.441	4.590	3.612
95.5	3.740	2.774	4.327	2.755	4.640	2.742	4.858	2.782	4.922	2.936
95.6	4.642	2.000	4.924	1.832	5.013	1.803	5.007	1.906	4.914	2.173
95.7	4.925	0.686	4.898	0.690	4.837	0.817	4.729	1.106	4.596	1.539
95.8	4.377	-0.440	4.322	-0.153	4.274	0.142	4.204	0.589	4.123	1.142
61π/2	4.221	-0.583	4.186	-0.257	4.152	0.059	4.097	0.526	4.030	1.094
95.9	3.526	-0.954	3.598	-0.534	3.629	-0.161	3.644	0.359	3.639	0.965
96.0	2.797	-1.032	2.973	-0.597	3.072	-0.211	3.158	0.322	3.215	0.939
96.1	2.272	-0.916	2.502	-0.506	2.640	-0.132	2.770	0.392	2.868	1.001
96.2	1.909	-0.739	2.160	-0.354	2.317	0.004	2.471	0.513	2.594	1.108
96.3	1.655	-0.553	1.911	-0.186	2.075	0.160	2.242	0.655	2.378	1.238
96.4	1.475	-0.375	1.727	-0.020	1.893	0.318	2.065	0.803	2.209	1.377
96.5	1.343	-0.211	1.590	0.139	1.755	0.473	1.928	0.951	2.075	1.518
96.6	1.245	-0.060	1.485	0.289	1.648	0.620	1.821	1.096	1.970	1.658
96.7	1.170	0.080	1.405	0.430	1.565	0.762	1.737	1.236	1.886	1.796
96.8	1.113	0.210	1.342	0.564	1.500	0.898	1.671	1.373	1.821	1.932
96.9	1.068	0.333	1.294	0.693	1.450	1.029	1.620	1.507	1.771	2.067
97.0	1.034	0.451	1.256	0.818	1.412	1.158	1.581	1.640	1.734	2.202
97.1	1.008	0.565	1.229	0.940	1.383	1.286	1.554	1.772	1.709	2.337
97.2	0.989	0.678	1.209	1.062	1.365	1.414	1.537	1.905	1.695	2.475
97.3	0.977	0.791	1.198	1.186	1.355	1.543	1.530	2.041	1.693	2.616
31π	0.972	0.894	1.194	1.298	1.354	1.662	1.534	2.167	1.702	2.746
97.4	0.971	0.906	1.194	1.311	1.354	1.676	1.535	2.181	1.704	2.761
97.5	0.973	1.025	1.200	1.442	1.364	1.815	1.551	2.328	1.729	2.912
97.6	0.982	1.149	1.215	1.579	1.386	1.961	1.583	2.481	1.772	3.071
97.7	1.000	1.282	1.243	1.726	1.423	2.116	1.633	2.645	1.837	3.239
97.8	1.030	1.425	1.287	1.884	1.479	2.283	1.706	2.819	1.930	3.416
97.9	1.076	1.582	1.352	2.056	1.562	2.464	1.812	3.006	2.062	3.601
98.0	1.145	1.757	1.448	2.246	1.681	2.660	1.962	3.204	2.244	3.790
98.1	1.247	1.953	1.587	2.454	1.852	2.871	2.174	3.407	2.494	3.971
98.2	1.400	2.175	1.792	2.680	2.100	3.089	2.471	3.600	2.832	4.121
98.3	1.631	2.420	2.095	2.911	2.455	3.293	2.881	3.749	3.273	4.195
98.4	1.987	2.673	2.541	3.112	2.956	3.429	3.420	3.784	3.809	4.121
98.5	2.536	2.879	3.175	3.190	3.617	3.389	4.059	3.597	4.370	3.811
98.6	3.337	2.883	3.976	2.970	4.354	3.013	4.660	3.083	4.805	3.231
98.7	4.289	2.395	4.717	2.259	4.901	2.216	4.981	2.274	4.943	2.484
98.8	4.897	1.253	4.972	1.154	4.954	1.206	4.870	1.415	4.741	1.781
98.9	4.660	-0.033	4.586	0.148	4.519	0.383	4.424	0.775	4.316	1.286
63π/2	4.207	-0.572	4.174	-0.247	4.140	0.070	4.087	0.537	4.021	1.104
99.0	3.869	-0.798	3.885	-0.413	3.883	-0.062	3.864	0.436	3.828	1.027
99.1	3.071	-1.023	3.210	-0.586	3.284	-0.201	3.344	0.331	3.378	0.946
99.2	2.467	-0.967	2.679	-0.545	2.804	-0.165	2.918	0.364	3.002	0.977
99.3	2.044	-0.808	2.289	-0.413	2.440	-0.048	2.586	0.467	2.700	1.068
99.4	1.752	-0.624	2.006	-0.250	2.169	0.101	2.331	0.602	2.463	1.190
99.5	1.545	-0.442	1.799	-0.083	1.965	0.259	2.135	0.748	2.277	1.325
99.6	1.396	-0.273	1.645	0.079	1.811	0.414	1.983	0.895	2.130	1.464
99.7	1.285	-0.117	1.528	0.232	1.692	0.564	1.865	1.041	2.014	1.604
99.8	1.202	0.026	1.439	0.376	1.601	0.707	1.773	1.182	1.922	1.743
99.9	1.138	0.159	1.370	0.512	1.529	0.845	1.700	1.320	1.850	1.879
100.0	1.088	0.284	1.316	0.642	1.473	0.978	1.643	1.454	1.794	2.014
100.1	1.050	0.404	1.274	0.768	1.430	1.107	1.600	1.587	1.752	2.148
100.2	1.021	0.519	1.243	0.891	1.398	1.235	1.568	1.718	1.722	2.283
100.3	0.999	0.632	1.220	1.013	1.375	1.362	1.547	1.851	1.704	2.418
100.4	0.985	0.745	1.205	1.135	1.362	1.490	1.536	1.985	1.697	2.557
100.5	0.976	0.858	1.199	1.259	1.358	1.621	1.536	2.123	1.703	2.700
100.6	0.975	0.975	1.200	1.387	1.363	1.756	1.548	2.265	1.722	2.848
100.7	0.980	1.096	1.212	1.521	1.360	1.898	1.573	2.415	1.757	3.003
100.8	0.995	1.225	1.234	1.662	1.410	2.048	1.614	2.574	1.812	3.166
100.9	1.020	1.363	1.271	1.815	1.458	2.210	1.677	2.743	1.893	3.338

TABLE 2.9

ADMITTANCE AND EFFECTIVE LENGTH OF LONG DIPCLE ANTENNAS IN AIR

	OMEGA= 15.0					OMEGA= 20.0			
k_0h	ADMITTANCE		RAD.EFF.LENGTH		k_0h	ADMITTANCE		RAD.EFF.LENGTH	
1.5	13.125	1.183	1.011	-0.009	1.5	11.907	5.003	0.985	-0.000
π/2	8.860	-4.708	1.097	-0.024	π/2	9.348	-5.069	1.064	-0.010
1.6	6.916	-5.097	1.135	-0.031	1.6	6.522	-5.636	1.099	-0.014
1.7	3.311	-4.188	1.276	-0.062	1.7	2.374	-4.048	1.228	-0.033
1.8	1.977	-3.124	1.438	-0.104	1.8	1.258	-2.811	1.374	-0.058
1.9	1.368	-2.380	1.626	-0.163	1.9	0.821	-2.079	1.543	-0.092
2.0	1.039	-1.852	1.847	-0.248	2.0	0.602	-1.603	1.740	-0.140
2.1	0.840	-1.456	2.109	-0.372	2.1	0.476	-1.266	1.974	-0.208
2.2	0.708	-1.145	2.424	-0.558	2.2	0.396	-1.012	2.258	-0.307
2.3	0.616	-0.890	2.801	-0.846	2.3	0.341	-0.809	2.608	-0.457
2.4	0.550	-0.673	3.237	-1.308	2.4	0.301	-0.641	3.048	-0.694
2.5	0.499	-0.482	3.687	-2.063	2.5	0.272	-0.496	3.604	-1.086
2.6	0.460	-0.311	3.969	-3.282	2.6	0.250	-0.368	4.297	-1.777
2.7	0.430	-0.153	3.590	-5.009	2.7	0.232	-0.252	5.053	-3.061
2.8	0.405	-0.003	1.925	-6.576	2.8	0.218	-0.144	5.364	-5.423
2.9	0.386	0.140	-0.507	-6.705	2.9	0.207	-0.042	3.567	-8.653
3.0	0.371	0.281	-2.194	-5.513	3.0	0.198	0.058	-0.896	-9.493
3.1	0.359	0.422	-2.821	-4.137	3.1	0.191	0.157	-3.807	-7.015
π	0.355	0.481	-2.887	-3.654	π	0.189	0.198	-4.203	-5.885
3.2	0.351	0.565	-2.876	-3.080	3.2	0.186	0.256	-4.305	-4.571
3.3	0.346	0.713	-2.718	-2.346	3.3	0.183	0.358	-3.967	-3.052
3.4	0.346	0.870	-2.505	-1.840	3.4	0.181	0.465	-3.495	-2.153
3.5	0.350	1.039	-2.292	-1.483	3.5	0.182	0.580	-3.064	-1.598
3.6	0.361	1.225	-2.098	-1.223	3.6	0.186	0.705	-2.703	-1.235
3.7	0.382	1.436	-1.927	-1.027	3.7	0.193	0.845	-2.405	-0.985
3.8	0.416	1.679	-1.776	-0.876	3.8	0.207	1.007	-2.157	-0.804
3.9	0.474	1.968	-1.644	-0.755	3.9	0.230	1.198	-1.949	-0.669
4.0	0.572	2.324	-1.527	-0.657	4.0	0.269	1.434	-1.772	-0.564
4.1	0.742	2.776	-1.423	-0.574	4.1	0.337	1.735	-1.620	-0.481
4.2	1.059	3.370	-1.330	-0.504	4.2	0.463	2.140	-1.487	-0.412
4.3	1.697	4.156	-1.246	-0.442	4.3	0.720	2.716	-1.369	-0.355
4.4	3.093	5.069	-1.169	-0.387	4.4	1.320	3.579	-1.263	-0.306
4.5	6.051	5.168	-1.098	-0.338	4.5	2.997	4.758	-1.168	-0.263
4.6	8.971	1.570	-1.032	-0.292	4.6	7.480	4.035	-1.081	-0.225
4.7	7.021	-2.508	-0.969	-0.249	4.7	7.654	-2.575	-1.000	-0.191
3π/2	6.626	-2.732	-0.962	-0.244	3π/2	7.062	-3.041	-0.990	-0.187
4.8	4.288	-3.155	-0.910	-0.208	4.8	3.644	-3.650	-0.925	-0.160
4.9	2.771	-2.701	-0.853	-0.168	4.9	1.950	-2.848	-0.854	-0.131
5.0	1.971	-2.161	-0.796	-0.130	5.0	1.246	-2.166	-0.786	-0.104
5.1	1.514	-1.704	-0.740	-0.091	5.1	0.899	-1.681	-0.721	-0.079
5.2	1.231	-1.332	-0.683	-0.053	5.2	0.702	-1.326	-0.658	-0.055
5.3	1.044	-1.027	-0.623	-0.014	5.3	0.580	-1.054	-0.597	-0.031
5.4	0.913	-0.771	-0.558	0.025	5.4	0.498	-0.837	-0.535	-0.009
5.5	0.818	-0.550	-0.485	0.062	5.5	0.441	-0.658	-0.473	0.013
5.6	0.747	-0.354	-0.402	0.092	5.6	0.399	-0.504	-0.409	0.034
5.7	0.692	-0.177	-0.308	0.107	5.7	0.367	-0.369	-0.341	0.053
5.8	0.649	-0.014	-0.215	0.098	5.8	0.342	-0.247	-0.266	0.067
5.9	0.616	0.140	-0.138	0.068	5.9	0.322	-0.134	-0.186	0.069
6.0	0.589	0.288	-0.087	0.033	6.0	0.307	-0.027	-0.110	0.051
6.1	0.568	0.433	-0.053	0.009	6.1	0.295	0.076	-0.056	0.021
6.2	0.553	0.577	-0.025	-0.001	6.2	0.286	0.177	-0.024	0.002
2π	0.543	0.699	-0.000	-0.000	2π	0.280	0.262	-0.000	-0.000
[illegible]	[illegible]	[illegible]	[illegible]	[illegible]	[illegible]	[illegible]	[illegible]	[illegible]	[illegible]
6.4	0.536	0.875	0.042	0.008	6.4	0.274	0.382	0.042	0.009
6.5	0.536	1.034	0.084	0.018	6.5	0.272	0.490	0.087	0.020
6.6	0.543	1.204	0.132	0.029	6.6	0.273	0.604	0.138	0.031
6.7	0.558	1.390	0.186	0.040	6.7	0.277	0.728	0.193	0.040
6.8	0.585	1.599	0.245	0.049	6.8	0.287	0.866	0.253	0.047
6.9	0.630	1.837	0.309	0.056	6.9	0.303	1.022	0.316	0.052
7.0	0.702	2.116	0.379	0.060	7.0	0.330	1.205	0.384	0.055
7.1	0.819	2.451	0.454	0.061	7.1	0.374	1.426	0.454	0.055
7.2	1.015	2.863	0.536	0.059	7.2	0.448	1.702	0.530	0.053
7.3	1.357	3.377	0.625	0.052	7.3	0.578	2.062	0.610	0.048
7.4	1.988	4.001	0.721	0.039	7.4	0.826	2.550	0.696	0.040
7.5	3.205	4.621	0.826	0.020	7.5	1.347	3.232	0.788	0.028
7.6	5.394	4.571	0.942	-0.008	7.6	2.595	4.094	0.887	0.012
7.7	7.661	2.295	1.069	-0.047	7.7	5.578	4.091	0.996	-0.010
7.8	7.090	-1.079	1.209	-0.100	7.8	7.774	-0.239	1.115	-0.039
5π/2	5.923	-2.058	1.291	-0.137	5π/2	6.329	-2.437	1.185	-0.058
7.9	4.947	-2.406	1.364	-0.173	7.9	4.787	-3.107	1.247	-0.076
8.0	3.369	-2.369	1.535	-0.272	8.0	2.618	-2.874	1.394	-0.126
8.1	2.437	-1.992	1.723	-0.408	8.1	1.639	-2.270	1.560	-0.191
8.2	1.882	-1.595	1.926	-0.594	8.2	1.158	-1.776	1.748	-0.280
8.3	1.533	-1.244	2.135	-0.849	8.3	0.891	-1.402	1.962	-0.399
8.4	1.300	-0.945	2.334	-1.196	8.4	0.728	-1.113	2.206	-0.565
8.5	1.138	-0.688	2.484	-1.662	8.5	0.622	-0.882	2.483	-0.799
8.6	1.020	-0.463	2.518	-2.258	8.6	0.547	-0.691	2.790	-1.136
8.7	0.932	-0.264	2.341	-2.950	8.7	0.494	-0.529	3.105	-1.629
8.8	0.865	-0.083	1.866	-3.624	8.8	0.453	-0.386	3.367	-2.347
8.9	0.812	0.084	1.107	-4.092	8.9	0.423	-0.259	3.429	-3.357
9.0	0.771	0.242	0.225	-4.208	9.0	0.398	-0.141	3.014	-4.607
9.1	0.739	0.393	-0.564	-3.987	9.1	0.380	-0.031	1.856	-5.727
9.2	0.714	0.541	-1.135	-3.568	9.2	0.365	0.075	0.148	-6.098
9.3	0.695	0.688	-1.482	-3.092	9.3	0.353	0.178	-1.390	-5.547
9.4	0.682	0.836	-1.661	-2.645	9.4	0.345	0.281	-2.294	-4.549
3π	0.680	0.873	-1.686	-2.543	3π	0.343	0.306	-2.423	-4.293
9.5	0.675	0.988	-1.730	-2.259	9.5	0.339	0.385	-2.645	-3.572
9.6	0.674	1.148	-1.734	-1.936	9.6	0.336	0.493	-2.682	-2.788
9.7	0.682	1.317	-1.701	-1.671	9.7	0.337	0.606	-2.577	-2.200
9.8	0.699	1.502	-1.650	-1.452	9.8	0.341	0.728	-2.418	-1.763
9.9	0.728	1.706	-1.589	-1.271	9.9	0.351	0.862	-2.248	-1.437
10.0	0.777	1.937	-1.525	-1.119	10.0	0.367	1.013	-2.082	-1.188
10.1	0.854	2.203	-1.462	-0.990	10.1	0.394	1.187	-1.929	-0.995
10.2	0.976	2.517	-1.400	-0.878	10.2	0.438	1.394	-1.788	-0.841
10.3	1.174	2.894	-1.341	-0.781	10.3	0.510	1.648	-1.660	-0.716
10.4	1.506	3.347	-1.284	-0.694	10.4	0.632	1.970	-1.544	-0.613
10.5	2.084	3.872	-1.230	-0.616	10.5	0.852	2.394	-1.438	-0.527
10.6	3.118	4.363	-1.178	-0.545	10.6	1.286	2.962	-1.341	-0.452
10.7	4.852	4.364	-1.129	-0.478	10.7	2.229	3.671	-1.251	-0.388
10.8	6.813	2.850	-1.080	-0.415	10.8	4.352	4.006	-1.167	-0.331
10.9	7.030	0.042	-1.033	-0.355	10.9	7.112	1.558	-1.088	-0.280
7π/2	5.534	-1.648	-0.987	-0.299	7π/2	5.919	-2.111	-1.017	-0.236

	OMEGA = 15.0					OMEGA = 20.0			
k_0h	ADMITTANCE		RAD.EFF.LENGTH		k_0h	ADMITTANCE		RAD.EFF.LENGTH	
11.0	5.456	-1.688	-0.985	-0.296	11.0	5.794	-2.204	-1.014	-0.234
11.1	3.896	-2.031	-0.937	-0.238	11.1	3.354	-2.807	-0.943	-0.191
11.2	2.862	-1.824	-0.887	-0.180	11.2	2.065	-2.364	-0.875	-0.152
11.3	2.216	-1.494	-0.833	-0.122	11.3	1.426	-1.879	-0.808	-0.115
11.4	1.803	-1.168	-0.774	-0.065	11.4	1.077	-1.489	-0.743	-0.081
11.5	1.527	-0.876	-0.706	-0.008	11.5	0.869	-1.182	-0.678	-0.048
11.6	1.334	-0.619	-0.629	0.044	11.6	0.734	-0.936	-0.612	-0.016
11.7	1.195	-0.392	-0.539	0.088	11.7	0.642	-0.734	-0.544	0.013
11.8	1.090	-0.190	-0.440	0.115	11.8	0.577	-0.562	-0.473	0.040
11.9	1.011	-0.005	-0.338	0.120	11.9	0.528	-0.412	-0.397	0.062
12.0	0.949	0.166	-0.247	0.103	12.0	0.491	-0.279	-0.316	0.077
12.1	0.901	0.326	-0.174	0.072	12.1	0.462	-0.157	-0.232	0.078
12.2	0.863	0.480	-0.122	0.041	12.2	0.440	-0.043	-0.155	0.062
12.3	0.834	0.630	-0.084	0.017	12.3	0.422	0.065	-0.094	0.035
12.4	0.812	0.778	-0.052	0.003	12.4	0.409	0.170	-0.052	0.011
12.5	0.796	0.928	-0.022	-0.001	12.5	0.398	0.274	-0.021	-0.000
4π	0.790	1.028	-0.000	-0.000	4π	0.393	0.343	-0.000	-0.000
12.6	0.788	1.080	0.012	0.002	12.6	0.391	0.379	0.011	0.002
12.7	0.786	1.239	0.050	0.009	12.7	0.388	0.486	0.050	0.010
12.8	0.793	1.408	0.093	0.019	12.8	0.387	0.598	0.095	0.022
12.9	0.810	1.589	0.142	0.029	12.9	0.391	0.718	0.146	0.034
13.0	0.841	1.788	0.197	0.040	13.0	0.400	0.849	0.203	0.044
13.1	0.890	2.011	0.258	0.049	13.1	0.416	0.994	0.265	0.052
13.2	0.968	2.266	0.326	0.055	13.2	0.442	1.160	0.331	0.058
13.3	1.088	2.561	0.400	0.059	13.3	0.483	1.354	0.401	0.061
13.4	1.279	2.907	0.482	0.059	13.4	0.550	1.589	0.477	0.061
13.5	1.588	3.315	0.571	0.053	13.5	0.660	1.880	0.557	0.057
13.6	2.106	3.773	0.669	0.041	13.6	0.852	2.254	0.643	0.050
13.7	2.987	4.199	0.776	0.021	13.7	1.210	2.743	0.735	0.038
13.8	4.411	4.270	0.893	-0.010	13.8	1.938	3.353	0.834	0.022
13.9	6.145	3.263	1.022	-0.055	13.9	3.497	3.826	0.941	-0.002
14.0	6.846	0.978	1.163	-0.118	14.0	6.081	2.689	1.058	-0.033
14.1	5.834	-0.958	1.316	-0.204	14.1	6.446	-0.939	1.186	-0.074
9π/2	5.276	-1.343	1.376	-0.243	9π/2	5.645	-1.896	1.236	-0.092
14.2	4.376	-1.653	1.480	-0.321	14.2	4.162	-2.563	1.326	-0.128
14.3	3.269	-1.638	1.653	-0.480	14.3	2.555	-2.423	1.482	-0.199
14.4	2.539	-1.388	1.827	-0.693	14.4	1.723	-1.982	1.655	-0.293
14.5	2.062	-1.094	1.986	-0.973	14.5	1.275	-1.582	1.846	-0.418
14.6	1.741	-0.812	2.106	-1.335	14.6	1.012	-1.258	2.058	-0.587
14.7	1.516	-0.558	2.146	-1.780	14.7	0.846	-0.997	2.287	-0.817
14.8	1.354	-0.330	2.052	-2.286	14.8	0.733	-0.783	2.526	-1.132
14.9	1.234	-0.124	1.780	-2.796	14.9	0.654	-0.601	2.750	-1.567
15.0	1.143	0.063	1.323	-3.216	15.0	0.597	-0.444	2.911	-2.157
15.1	1.072	0.237	0.740	-3.455	15.1	0.553	-0.305	2.912	-2.922
15.2	1.016	0.400	0.136	-3.480	15.2	0.519	-0.179	2.607	-3.813
15.3	0.973	0.556	-0.393	-3.324	15.3	0.493	-0.061	1.871	-4.642
15.4	0.939	0.708	-0.798	-3.061	15.4	0.473	0.050	0.757	-5.105
15.5	0.914	0.857	-1.078	-2.755	15.5	0.457	0.157	-0.430	-5.020
15.6	0.896	1.007	-1.255	-2.452	15.6	0.445	0.262	-1.359	-4.500
5π	0.884	1.173	-1.362	-2.152	5π	0.436	0.375	-1.950	-3.753
15.7	0.885	1.160	-1.357	-2.173	15.7	0.436	0.367	-1.920	-3.809
15.8	0.882	1.319	-1.406	-1.926	15.8	0.431	0.474	-2.180	-3.140
15.9	0.888	1.485	-1.422	-1.712	15.9	0.430	0.585	-2.251	-2.572
16.0	0.904	1.664	-1.417	-1.527	16.0	0.433	0.703	-2.217	-2.115
16.1	0.933	1.858	-1.398	-1.366	16.1	0.441	0.830	-2.132	-1.753
16.2	0.982	2.073	-1.371	-1.227	16.2	0.455	0.970	-2.024	-1.467
16.3	1.057	2.316	-1.340	-1.105	16.3	0.479	1.129	-1.910	-1.239
16.4	[illegible]	[illegible]	[illegible]	[illegible]	[illegible]	[illegible]	[illegible]	[illegible]	[illegible]
16.5	1.352	2.916	-1.272	-0.899	16.5	0.578	1.529	-1.689	-0.902
16.6	1.637	3.288	-1.238	-0.811	16.6	0.677	1.795	-1.586	-0.775
16.7	2.097	3.699	-1.203	-0.729	16.7	0.843	2.130	-1.490	-0.668
16.8	2.854	4.085	-1.169	-0.652	16.8	1.139	2.559	-1.400	-0.576
16.9	4.048	4.216	-1.136	-0.580	16.9	1.711	3.094	-1.315	-0.496
17.0	5.583	3.564	-1.102	-0.510	17.0	2.881	3.608	-1.235	-0.425
17.1	6.565	1.758	-1.067	-0.441	17.1	5.033	3.258	-1.159	-0.362
17.2	6.079	-0.215	-1.031	-0.372	17.2	6.550	0.468	-1.087	-0.304
11π/2	5.086	-1.093	-1.001	-0.318	11π/2	5.443	-1.738	-1.031	-0.263
17.3	4.808	-1.221	-0.993	-0.304	17.3	4.980	-2.053	-1.017	-0.252
17.4	3.665	-1.418	-0.950	-0.234	17.4	3.123	-2.411	-0.949	-0.203
17.5	2.859	-1.268	-0.902	-0.163	17.5	2.066	-2.075	-0.882	-0.158
17.6	2.318	-1.016	-0.844	-0.092	17.6	1.495	-1.679	-0.816	-0.115
17.7	1.950	-0.750	-0.776	-0.023	17.7	1.166	-1.341	-0.750	-0.074
17.8	1.693	-0.500	-0.693	0.040	17.8	0.961	-1.065	-0.682	-0.036
17.9	1.507	-0.272	-0.597	0.090	17.9	0.825	-0.837	-0.612	-0.000
18.0	1.370	-0.065	-0.492	0.120	18.0	0.731	-0.646	-0.538	0.032
18.1	1.265	0.125	-0.387	0.127	18.1	0.663	-0.481	-0.460	0.059
18.2	1.185	0.302	-0.292	0.112	18.2	0.612	-0.336	-0.376	0.078
18.3	1.122	0.467	-0.216	0.084	18.3	0.573	-0.204	-0.290	0.084
18.4	1.073	0.625	-0.158	0.053	18.4	0.543	-0.083	-0.207	0.074
18.5	1.034	0.779	-0.115	0.027	18.5	0.519	0.030	-0.137	0.051
18.6	1.005	0.929	-0.080	0.010	18.6	0.501	0.139	-0.085	0.025
18.7	0.984	1.080	-0.049	0.001	18.7	0.487	0.246	-0.048	0.006
18.8	0.971	1.233	-0.017	-0.001	18.8	0.477	0.351	-0.016	-0.001
6π	0.968	1.310	-0.000	-0.000	6π	0.473	0.404	-0.000	-0.000
18.9	0.966	1.390	0.018	0.002	18.9	0.470	0.458	0.018	0.003
19.0	0.970	1.555	0.058	0.010	19.0	0.468	0.568	0.057	0.012
19.1	0.985	1.730	0.102	0.019	19.1	0.469	0.684	0.103	0.024
19.2	1.013	1.920	0.153	0.029	19.2	0.476	0.808	0.155	0.036
19.3	1.059	2.128	0.209	0.038	19.3	0.489	0.944	0.213	0.046
19.4	1.131	2.360	0.272	0.047	19.4	0.511	1.095	0.276	0.055
19.5	1.240	2.624	0.343	0.052	19.5	0.545	1.269	0.345	0.061
19.6	1.407	2.925	0.421	0.054	19.6	0.600	1.472	0.418	0.064
19.7	1.666	3.267	0.507	0.051	19.7	0.687	1.716	0.496	0.063
19.8	2.076	3.643	0.602	0.041	19.8	0.830	2.019	0.580	0.058
19.9	2.731	4.002	0.707	0.023	19.9	1.078	2.400	0.670	0.049
20.0	3.746	4.178	0.822	-0.006	20.0	1.534	2.876	0.767	0.034
20.1	5.100	3.779	0.948	-0.049	20.1	2.429	3.384	0.872	0.013
20.2	6.220	2.395	1.086	-0.112	20.2	4.136	3.460	0.985	-0.017
20.3	6.190	0.519	1.234	-0.199	20.3	6.129	1.699	1.107	-0.056
20.4	5.180	-0.733	1.392	-0.319	20.4	5.666	-1.230	1.242	-0.108
13π/2	4.939	-0.875	1.424	-0.348	13π/2	5.286	-1.614	1.270	-0.120
20.5	4.051	-1.156	1.553	-0.480	20.5	3.770	-2.274	1.388	-0.176
20.6	3.182	-1.127	1.709	-0.694	20.6	2.469	-2.140	1.549	-0.266
20.7	2.577	-0.928	1.842	-0.970	20.7	1.747	-1.776	1.724	-0.384
20.8	2.160	-0.685	1.926	-1.316	20.8	1.335	-1.430	1.914	-0.541
20.9	1.867	-0.443	1.926	-1.722	20.9	1.084	-1.139	2.116	-0.751

k_0h	ADMITTANCE (OMEGA = 15.0)		RAD.EFF.LENGTH		k_0h	ADMITTANCE (OMEGA = 20.0)		RAD.EFF.LENGTH	
21.0	1.657	-0.217	1.802	-2.161	21.0	0.920	-0.898	2.321	-1.031
21.1	1.501	-0.008	1.534	-2.577	21.1	0.808	-0.696	2.511	-1.405
21.2	1.383	0.184	1.132	-2.899	21.2	0.728	-0.522	2.648	-1.896
21.3	1.292	0.362	0.652	-3.076	21.3	0.669	-0.370	2.668	-2.516
21.4	1.221	0.530	0.169	-3.094	21.4	0.624	-0.234	2.478	-3.236
21.5	1.166	0.690	-0.257	-2.983	21.5	0.590	-0.109	1.986	-3.949
21.6	1.123	0.845	-0.596	-2.792	21.6	0.563	0.008	1.187	-4.469
21.7	1.090	0.997	-0.844	-2.564	21.7	0.542	0.119	0.227	-4.628
21.8	1.066	1.148	-1.016	-2.330	21.8	0.526	0.227	-0.665	-4.409
21.9	1.050	1.300	-1.128	-2.107	21.9	0.514	0.333	-1.329	-3.943
7π	1.043	1.443	-1.193	-1.920	7π	0.506	0.430	-1.710	-3.443
22.0	1.043	1.457	-1.197	-1.903	22.0	0.506	0.440	-1.737	-3.394
22.1	1.045	1.619	-1.236	-1.721	22.1	0.502	0.549	-1.944	-2.868
22.2	1.057	1.792	-1.253	-1.560	22.2	0.502	0.663	-2.016	-2.411
22.3	1.083	1.977	-1.257	-1.417	22.3	0.507	0.785	-2.008	-2.029
22.4	1.127	2.178	-1.251	-1.290	22.4	0.518	0.916	-1.955	-1.717
22.5	1.194	2.402	-1.239	-1.177	22.5	0.538	1.061	-1.879	-1.461
22.6	1.296	2.652	-1.224	-1.075	22.6	0.569	1.226	-1.794	-1.251
22.7	1.450	2.935	-1.206	-0.982	22.7	0.618	1.417	-1.705	-1.076
22.8	1.686	3.254	-1.187	-0.896	22.8	0.694	1.643	-1.616	-0.929
22.9	2.050	3.602	-1.168	-0.815	22.9	0.818	1.919	-1.530	-0.804
23.0	2.620	3.940	-1.148	-0.738	23.0	1.026	2.262	-1.448	-0.697
23.1	3.491	4.147	-1.127	-0.664	23.1	1.395	2.688	-1.369	-0.603
23.2	4.683	3.930	-1.106	-0.591	23.2	2.092	3.170	-1.293	-0.520
23.3	5.843	2.902	-1.084	-0.519	23.3	3.422	3.457	-1.220	-0.446
23.4	6.176	1.217	-1.059	-0.446	23.4	5.403	2.556	-1.150	-0.378
23.5	5.476	-0.194	-1.031	-0.370	23.5	6.040	-0.161	-1.082	-0.317
15π/2	4.819	-0.679	-1.011	-0.323	15π/2	5.159	-1.513	-1.041	-0.281
23.6	4.417	-0.844	-0.997	-0.293	23.6	4.461	-1.947	-1.016	-0.260
23.7	3.506	-0.956	-0.956	-0.213	23.7	2.941	-2.153	-0.950	-0.207
23.8	2.841	-0.827	-0.905	-0.132	23.8	2.040	-1.864	-0.884	-0.157
23.9	2.374	-0.614	-0.839	-0.052	23.9	1.527	-1.521	-0.818	-0.110
24.0	2.044	-0.384	-0.757	0.021	24.0	1.219	-1.218	-0.750	-0.065
24.1	1.806	-0.162	-0.660	0.080	24.1	1.022	-0.964	-0.680	-0.023
24.2	1.630	0.046	-0.552	0.118	24.2	0.889	-0.750	-0.605	0.015
24.3	1.498	0.240	-0.443	0.131	24.3	0.795	-0.567	-0.526	0.048
24.4	1.396	0.420	-0.344	0.121	24.4	0.726	-0.408	-0.442	0.073
24.5	1.317	0.590	-0.262	0.097	24.5	0.675	-0.266	-0.354	0.086
24.6	1.255	0.752	-0.198	0.067	24.6	0.635	-0.137	-0.267	0.084
24.7	1.206	0.908	-0.149	0.040	24.7	0.605	-0.017	-0.188	0.068
24.8	1.169	1.061	-0.110	0.019	24.8	0.581	0.096	-0.125	0.042
24.9	1.141	1.212	-0.077	0.006	24.9	0.562	0.205	-0.079	0.018
25.0	1.123	1.364	-0.045	-0.000	25.0	0.549	0.312	-0.044	0.003
25.1	1.113	1.520	-0.012	-0.001	25.1	0.539	0.420	-0.011	-0.001
8π	1.112	1.572	-0.000	-0.000	8π	0.537	0.455	-0.000	-0.000
25.2	1.113	1.681	0.025	0.003	25.2	0.533	0.528	0.024	0.004
25.3	1.123	1.850	0.066	0.010	25.3	0.532	0.641	0.065	0.013
25.4	1.147	2.031	0.112	0.019	25.4	0.536	0.760	0.112	0.025
25.5	1.187	2.227	0.163	0.028	25.5	0.545	0.887	0.165	0.037
25.6	1.249	2.442	0.222	0.037	25.6	0.562	1.027	0.224	0.048
25.7	1.344	2.680	0.287	0.043	25.7	0.590	1.184	0.288	0.057
25.8	1.486	2.948	0.360	0.047	25.8	0.633	1.364	0.358	0.063
25.9	1.699	3.247	0.441	0.047	25.9	0.701	1.575	0.434	0.065
26.0	2.025	3.571	0.531	0.040	26.0	0.807	1.829	0.515	0.064
26.1	2.523	3.893	0.631	0.026	26.1	0.983	2.140	0.602	0.057
26.2	3.276	4.120	0.742	0.001	26.2	1.285	2.523	0.696	0.046
26.3	4.324	4.034	0.864	-0.038	26.3	1.836	2.973	0.797	0.028
26.4	5.459	3.297	0.996	-0.096	26.4	2.872	3.353	0.906	0.002
26.5	6.055	1.854	1.139	-0.178	26.5	4.608	3.037	1.024	-0.033
26.6	5.682	0.380	1.290	-0.291	26.6	5.993	0.963	1.152	-0.080
17π/2	4.718	-0.496	1.449	-0.452	17π/2	5.053	-1.428	1.296	-0.145
26.7	4.753	-0.478	1.444	-0.445	26.7	5.113	-1.376	1.291	-0.142
26.8	3.828	-0.750	1.589	-0.649	26.8	3.484	-2.074	1.442	-0.224
26.9	3.110	-0.706	1.711	-0.911	26.9	2.385	-1.931	1.605	-0.332
27.0	2.593	-0.533	1.783	-1.233	27.0	1.749	-1.611	1.781	-0.473
27.1	2.224	-0.321	1.774	-1.606	27.1	1.370	-1.301	1.966	-0.659
27.2	1.957	-0.105	1.655	-1.999	27.2	1.132	-1.034	2.153	-0.904
27.3	1.760	0.101	1.411	-2.365	27.3	0.974	-0.809	2.328	-1.225
27.4	1.611	0.295	1.060	-2.647	27.4	0.864	-0.617	2.463	-1.641
27.5	1.498	0.477	0.646	-2.806	27.5	0.785	-0.450	2.514	-2.161
27.6	1.409	0.649	0.230	-2.835	27.6	0.726	-0.302	2.413	-2.771
27.7	1.340	0.812	-0.142	-2.758	27.7	0.681	-0.168	2.089	-3.410
27.8	1.286	0.969	-0.445	-2.611	27.8	0.646	-0.045	1.509	-3.955
27.9	1.245	1.123	-0.677	-2.429	27.9	0.619	0.071	0.733	-4.264
28.0	1.213	1.275	-0.845	-2.238	28.0	0.598	0.182	-0.087	-4.263
28.1	1.192	1.427	-0.962	-2.050	28.1	0.582	0.290	-0.793	-3.996
28.2	1.179	1.581	-1.041	-1.876	28.2	0.570	0.398	-1.305	-3.576
9π	1.176	1.699	-1.081	-1.756	9π	0.565	0.478	-1.559	-3.231
28.3	1.176	1.741	-1.092	-1.717	28.3	0.563	0.506	-1.625	-3.112
28.4	1.184	1.907	-1.124	-1.574	28.4	0.560	0.617	-1.795	-2.672
28.5	1.205	2.084	-1.141	-1.445	28.5	0.562	0.734	-1.863	-2.283
28.6	1.241	2.274	-1.149	-1.329	28.6	0.570	0.858	-1.866	-1.952
28.7	1.299	2.482	-1.150	-1.225	28.7	0.584	0.994	-1.831	-1.675
28.8	1.386	2.710	-1.147	-1.130	28.8	0.609	1.144	-1.774	-1.443
28.9	1.516	2.964	-1.142	-1.042	28.9	0.647	1.314	-1.707	-1.248
29.0	1.710	3.246	-1.134	-0.960	29.0	0.706	1.511	-1.634	-1.082
29.1	2.000	3.551	-1.126	-0.883	29.1	0.799	1.746	-1.559	-0.941
29.2	2.439	3.858	-1.117	-0.808	29.2	0.948	2.030	-1.485	-0.819
29.3	3.094	4.097	-1.107	-0.736	29.3	1.198	2.378	-1.413	-0.713
29.4	4.015	4.105	-1.097	-0.664	29.4	1.639	2.792	-1.342	-0.618
29.5	5.089	3.598	-1.085	-0.591	29.5	2.452	3.203	-1.274	-0.534
29.6	5.853	2.413	-1.071	-0.516	29.6	3.884	3.238	-1.207	-0.457
29.7	5.792	0.967	-1.053	-0.438	29.7	5.573	1.918	-1.141	-0.387
29.8	5.045	-0.062	-1.030	-0.356	29.8	5.592	-0.559	-1.077	-0.323
19π/2	4.632	-0.324	-1.017	-0.318	19π/2	4.963	-1.354	-1.048	-0.295
29.9	4.144	-0.502	-0.998	-0.270	29.9	4.082	-1.856	-1.013	-0.263
30.0	3.386	-0.560	-0.955	-0.181	30.0	2.791	-1.958	-0.948	-0.206
30.1	2.819	-0.440	-0.896	-0.093	30.1	2.007	-1.695	-0.883	-0.152
30.2	2.409	-0.252	-0.819	-0.010	30.2	1.542	-1.387	-0.816	-0.102
30.3	2.111	-0.046	-0.724	0.060	30.3	1.255	-1.110	-0.747	-0.054
30.4	1.891	0.157	-0.616	0.108	30.4	1.067	-0.872	-0.673	-0.010
30.5	1.726	0.351	-0.505	0.131	30.5	0.937	-0.670	-0.595	0.029
30.6	1.599	0.534	-0.400	0.129	30.6	0.845	-0.495	-0.511	0.061
30.7	1.501	0.707	-0.311	0.110	30.7	0.777	-0.341	-0.422	0.082
30.8	1.424	0.871	-0.240	0.082	30.8	0.726	-0.202	-0.332	0.090
30.9	1.364	1.030	-0.185	0.054	30.9	0.686	-0.075	-0.246	0.081

k_0h	ADMITTANCE (OMEGA = 15.0)		RAD.EFF.LENGTH		k_0h	ADMITTANCE (OMEGA = 20.0)		RAD.EFF.LENGTH	
31.0	1.317	1.184	-0.142	0.031	31.0	0.656	0.044	-0.173	0.060
31.1	1.282	1.336	-0.105	0.014	31.1	0.632	0.157	-0.116	0.035
31.2	1.257	1.488	-0.073	0.003	31.2	0.614	0.266	-0.073	0.013
31.3	1.242	1.642	-0.040	-0.001	31.3	0.600	0.374	-0.039	0.002
31.4	1.236	1.800	-0.006	-0.000	31.4	0.592	0.482	-0.006	-0.000
10π	1.236	1.825	-0.000	-0.000	10π	0.591	0.500	-0.000	-0.000
31.5	1.241	1.964	0.032	0.004	31.5	0.587	0.593	0.031	0.005
31.6	1.259	2.137	0.074	0.010	31.6	0.587	0.708	0.073	0.015
31.7	1.291	2.322	0.121	0.018	31.7	0.593	0.829	0.121	0.027
31.8	1.344	2.523	0.174	0.027	31.8	0.605	0.960	0.175	0.039
31.9	1.424	2.742	0.234	0.034	31.9	0.626	1.104	0.235	0.050
32.0	1.543	2.984	0.301	0.039	32.0	0.660	1.266	0.301	0.058
32.1	1.719	3.251	0.377	0.041	32.1	0.712	1.452	0.373	0.064
32.2	1.979	3.539	0.461	0.037	32.2	0.792	1.670	0.450	0.066
32.3	2.367	3.832	0.555	0.027	32.3	0.920	1.931	0.534	0.063
32.4	2.940	4.079	0.660	0.007	32.4	1.128	2.249	0.624	0.056
32.5	3.749	4.153	0.775	-0.027	32.5	1.486	2.629	0.721	0.042
32.6	4.743	3.825	0.902	-0.077	32.6	2.129	3.040	0.826	0.021
32.7	5.597	2.888	1.038	-0.151	32.7	3.281	3.265	0.939	-0.009
32.8	5.805	1.543	1.183	-0.254	32.8	4.945	2.583	1.061	-0.050
32.9	5.279	0.399	1.330	-0.395	32.9	5.771	0.408	1.193	-0.105
21π/2	4.556	-0.158	1.453	-0.555	21π/2	4.886	-1.289	1.317	-0.167
33.0	4.444	-0.209	1.471	-0.583	33.0	4.684	-1.446	1.336	-0.178
33.1	3.663	-0.385	1.588	-0.825	33.1	3.260	-1.918	1.490	-0.273
33.2	3.052	-0.329	1.661	-1.122	33.2	2.310	-1.763	1.655	-0.397
33.3	2.601	-0.173	1.659	-1.465	33.3	1.741	-1.472	1.828	-0.561
33.4	2.270	0.019	1.558	-1.827	33.4	1.392	-1.188	2.004	-0.774
33.5	2.025	0.216	1.345	-2.165	33.5	1.167	-0.940	2.172	-1.050
33.6	1.841	0.408	1.037	-2.430	33.6	1.016	-0.727	2.313	-1.405
33.7	1.701	0.591	0.671	-2.588	33.7	0.909	-0.544	2.392	-1.849
33.8	1.592	0.764	0.300	-2.634	33.8	0.831	-0.382	2.360	-2.377
33.9	1.507	0.930	-0.038	-2.585	33.9	0.772	-0.238	2.159	-2.954
34.0	1.440	1.090	-0.321	-2.471	34.0	0.727	-0.107	1.746	-3.500
34.1	1.389	1.245	-0.542	-2.323	34.1	0.693	0.015	1.132	-3.901
34.2	1.349	1.397	-0.708	-2.161	34.2	0.666	0.130	0.408	-4.064
34.3	1.320	1.549	-0.829	-1.999	34.3	0.645	0.241	-0.295	-3.973
34.4	1.302	1.702	-0.915	-1.846	34.4	0.630	0.350	-0.873	-3.689
34.5	1.293	1.858	-0.974	-1.704	34.5	0.619	0.458	-1.286	-3.306
11π	1.293	1.951	-0.999	-1.628	11π	0.615	0.521	-1.453	-3.072
34.6	1.295	2.020	-1.015	-1.575	34.6	0.613	0.568	-1.547	-2.901
34.7	1.309	2.190	-1.041	-1.458	34.7	0.611	0.681	-1.691	-2.518
34.8	1.339	2.371	-1.058	-1.352	34.8	0.615	0.800	-1.753	-2.178
34.9	1.387	2.565	-1.069	-1.256	34.9	0.625	0.927	-1.762	-1.884
35.0	1.460	2.777	-1.075	-1.167	35.0	0.643	1.066	-1.738	-1.633
35.1	1.568	3.008	-1.078	-1.085	35.1	0.672	1.221	-1.694	-1.420
35.2	1.728	3.262	-1.079	-1.009	35.2	0.718	1.396	-1.638	-1.237
35.3	1.961	3.535	-1.079	-0.936	35.3	0.788	1.600	-1.577	-1.081
35.4	2.305	3.816	-1.079	-0.865	35.4	0.897	1.842	-1.513	-0.945
35.5	2.808	4.064	-1.078	-0.796	35.5	1.073	2.133	-1.448	-0.826
35.6	3.521	4.185	-1.076	-0.726	35.6	1.366	2.482	-1.383	-0.721
35.7	4.427	3.995	-1.074	-0.655	35.7	1.880	2.877	-1.319	-0.627
35.8	5.312	3.280	-1.069	-0.581	35.8	2.798	3.195	-1.256	-0.541
35.9	5.733	2.086	-1.062	-0.503	35.9	4.273	2.962	-1.194	-0.463
36.0	5.444	0.891	-1.049	-0.420	36.0	5.603	1.353	-1.132	-0.391
36.1	4.717	0.129	-1.027	-0.331	36.1	5.198	-0.820	-1.071	-0.324
23π/2	4.489	0.003	-1.019	-0.305	23π/2	4.818	-1.231	-1.053	-0.306
36.2	3.938	-0.176	-0.994	-0.238	36.2	3.786	-1.770	-1.009	-0.261
36.3	3.291	-0.197	-0.945	-0.143	36.3	2.665	-1.800	-0.945	-0.202
36.4	2.799	-0.082	-0.876	-0.051	36.4	1.971	-1.553	-0.880	-0.145
36.5	2.434	0.090	-0.787	0.030	36.5	1.548	-1.270	-0.812	-0.092
36.6	2.163	0.278	-0.682	0.090	36.6	1.280	-1.012	-0.741	-0.042
36.7	1.959	0.467	-0.569	0.124	36.7	1.101	-0.789	-0.664	0.003
36.8	1.804	0.649	-0.460	0.132	36.8	0.976	-0.596	-0.581	0.042
36.9	1.683	0.823	-0.365	0.120	36.9	0.886	-0.427	-0.493	0.072
37.0	1.590	0.990	-0.286	0.097	37.0	0.820	-0.277	-0.401	0.089
37.1	1.516	1.150	-0.224	0.069	37.1	0.769	-0.141	-0.311	0.090
37.2	1.458	1.306	-0.175	0.044	37.2	0.730	-0.016	-0.228	0.076
37.3	1.414	1.458	-0.135	0.024	37.3	0.700	0.102	-0.160	0.053
37.4	1.382	1.610	-0.101	0.010	37.4	0.676	0.215	-0.108	0.028
37.5	1.360	1.763	-0.068	0.002	37.5	0.658	0.324	-0.068	0.010
37.6	1.348	1.918	-0.035	-0.001	37.6	0.646	0.433	-0.034	0.000
12π	1.347	2.076	-0.000	-0.000	12π	0.638	0.541	-0.000	-0.000
37.7	1.347	2.078	0.000	0.000	37.7	0.638	0.542	0.000	0.000
37.8	1.358	2.245	0.039	0.004	37.8	0.634	0.654	0.038	0.006
37.9	1.383	2.421	0.082	0.010	37.9	0.636	0.771	0.081	0.017
38.0	1.426	2.610	0.131	0.018	38.0	0.644	0.895	0.130	0.029
38.1	1.493	2.814	0.185	0.025	38.1	0.659	1.029	0.185	0.041
38.2	1.592	3.036	0.247	0.031	38.2	0.685	1.177	0.246	0.051
38.3	1.736	3.278	0.316	0.034	38.3	0.724	1.344	0.313	0.060
38.4	1.947	3.538	0.394	0.033	38.4	0.785	1.535	0.387	0.065
38.5	2.253	3.807	0.481	0.026	38.5	0.879	1.760	0.467	0.066
38.6	2.697	4.055	0.578	0.010	38.6	1.028	2.028	0.553	0.062
38.7	3.325	4.208	0.687	-0.017	38.7	1.271	2.349	0.645	0.053
38.8	4.144	4.121	0.806	-0.059	38.8	1.687	2.721	0.746	0.037
38.9	5.018	3.597	0.935	-0.123	38.9	2.419	3.078	0.854	0.013
39.0	5.590	2.579	1.073	-0.214	39.0	3.656	3.125	0.970	-0.022
39.1	5.534	1.396	1.215	-0.339	39.1	5.164	2.121	1.097	-0.069
39.2	4.952	0.511	1.352	-0.508	39.2	5.510	-0.012	1.232	-0.131
25π/2	4.429	0.161	1.438	-0.656	25π/2	4.757	-1.179	1.334	-0.187
39.3	4.206	0.071	1.470	-0.728	39.3	4.337	-1.473	1.379	-0.214
39.4	3.534	-0.041	1.548	-1.000	39.4	3.078	-1.785	1.535	-0.323
39.5	3.004	0.025	1.560	-1.316	39.5	2.241	-1.621	1.699	-0.465
39.6	2.604	0.170	1.483	-1.653	39.6	1.728	-1.351	1.868	-0.649
39.7	2.306	0.346	1.305	-1.972	39.7	1.406	-1.087	2.032	-0.887
39.8	2.081	0.529	1.036	-2.231	39.8	1.194	-0.854	2.178	-1.192
39.9	1.909	0.709	0.710	-2.397	39.9	1.049	-0.651	2.281	-1.573
40.0	1.776	0.883	0.371	-2.461	40.0	0.945	-0.475	2.302	-2.034
40.1	1.673	1.050	0.057	-2.438	40.1	0.869	-0.319	2.195	-2.554
40.2	1.591	1.211	-0.211	-2.352	40.2	0.812	-0.178	1.914	-3.083
40.3	1.528	1.367	-0.426	-2.230	40.3	0.768	-0.049	1.445	-3.535
40.4	1.479	1.521	-0.592	-2.091	40.4	0.733	0.072	0.830	-3.817
40.5	1.442	1.672	-0.716	-1.950	40.5	0.707	0.187	0.166	-3.879
40.6	1.416	1.824	-0.807	-1.813	40.6	0.687	0.297	-0.442	-3.733
40.7	1.401	1.978	-0.872	-1.685	40.7	0.672	0.406	-0.927	-3.447
40.8	1.397	2.136	-0.919	-1.567	40.8	0.662	0.515	-1.270	-3.095
13π	1.399	2.202	-0.934	-1.523	13π	0.659	0.560	-1.373	-2.947
40.9	1.405	2.300	-0.953	-1.460	40.9	0.657	0.626	-1.489	-2.733

k_0h	OMEGA = 15.0 ADMITTANCE		RAD.EFF.LENGTH		OMEGA = 20.0 ADMITTANCE		RAD.EFF.LENGTH	
41.0	1.426	2.473	-0.976	-1.363	0.657	0.741	-1.613	-2.393
41.1	1.465	2.657	-0.993	-1.273	0.662	0.863	-1.670	-2.089
41.2	1.525	2.854	-1.006	-1.192	0.675	0.993	-1.682	-1.823
41.3	1.615	3.068	-1.015	-1.115	0.697	1.135	-1.664	-1.593
41.4	1.746	3.299	-1.023	-1.044	0.731	1.294	-1.629	-1.395
41.5	1.935	3.548	-1.029	-0.976	0.784	1.475	-1.583	-1.224
41.6	2.209	3.806	-1.035	-0.910	0.866	1.685	-1.530	-1.074
41.7	2.602	4.051	-1.041	-0.845	0.992	1.933	-1.474	-0.944
41.8	3.158	4.225	-1.047	-0.779	1.195	2.229	-1.416	-0.828
41.9	3.894	4.216	-1.052	-0.711	1.535	2.576	-1.358	-0.724
42.0	4.730	3.848	-1.056	-0.641	2.122	2.941	-1.299	-0.630
42.1	5.396	3.013	-1.057	-0.565	3.133	3.147	-1.241	-0.545
42.2	5.548	1.896	-1.054	-0.483	4.587	2.643	-1.183	-0.466
42.3	5.139	0.927	-1.043	-0.394	5.537	0.867	-1.124	-0.392
42.4	4.456	0.358	-1.022	-0.299	4.851	-0.991	-1.064	-0.323
27π/2	4.374	0.317	-1.018	-0.288	4.703	-1.131	-1.058	-0.315
42.5	3.777	0.145	-0.984	-0.199	3.545	-1.688	-1.004	-0.258
42.6	3.215	0.149	-0.925	-0.100	2.557	-1.665	-0.941	-0.196
42.7	2.781	0.262	-0.845	-0.010	1.937	-1.428	-0.876	-0.137
42.8	2.453	0.421	-0.745	0.063	1.549	-1.164	-0.806	-0.081
42.9	2.206	0.596	-0.634	0.110	1.298	-0.923	-0.732	-0.030
43.0	2.016	0.773	-0.523	0.130	1.128	-0.711	-0.652	0.016
43.1	1.870	0.946	-0.421	0.127	1.008	-0.526	-0.566	0.054
43.2	1.756	1.113	-0.335	0.109	0.921	-0.363	-0.474	0.081
43.3	1.667	1.274	-0.266	0.084	0.856	-0.217	-0.381	0.092
43.4	1.597	1.431	-0.211	0.058	0.806	-0.083	-0.291	0.088
43.5	1.543	1.584	-0.167	0.036	0.768	0.040	-0.213	0.071
43.6	1.501	1.736	-0.130	0.019	0.738	0.157	-0.149	0.046
43.7	1.472	1.887	-0.096	0.007	0.715	0.270	-0.101	0.023
43.8	1.453	2.040	-0.063	0.001	0.698	0.379	-0.062	0.007
43.9	1.446	2.196	-0.029	-0.001	0.686	0.488	-0.028	-0.000
14π	1.448	2.329	-0.000	-0.000	0.680	0.579	-0.000	-0.000
44.0	1.450	2.358	0.007	0.000	0.679	0.599	0.006	0.001
44.1	1.468	2.527	0.046	0.005	0.677	0.712	0.045	0.008
44.2	1.502	2.706	0.090	0.010	0.681	0.831	0.089	0.018
44.3	1.556	2.898	0.140	0.017	0.691	0.957	0.139	0.030
44.4	1.638	3.104	0.196	0.022	0.709	1.095	0.195	0.042
44.5	1.756	3.326	0.259	0.026	0.739	1.247	0.258	0.053
44.6	1.927	3.564	0.330	0.027	0.785	1.418	0.326	0.061
44.7	2.172	3.812	0.411	0.023	0.855	1.615	0.401	0.065
44.8	2.522	4.052	0.500	0.012	0.964	1.846	0.483	0.065
44.9	3.014	4.241	0.601	-0.009	1.135	2.120	0.571	0.061
45.0	3.673	4.288	0.712	-0.044	1.414	2.443	0.667	0.050
45.1	4.457	4.047	0.834	-0.098	1.889	2.799	0.770	0.031
45.2	5.171	3.384	0.964	-0.175	2.707	3.087	0.881	0.003
45.3	5.493	2.374	1.100	-0.285	3.993	2.936	1.002	-0.036
45.4	5.269	1.368	1.233	-0.434	5.280	1.672	1.131	-0.089
45.5	4.682	0.683	1.352	-0.630	5.237	-0.326	1.270	-0.160
29π/2	4.324	0.473	1.403	-0.754	4.654	-1.087	1.348	-0.206
45.6	4.015	0.362	1.438	-0.875	4.048	-1.471	1.419	-0.253
45.7	3.429	0.295	1.467	-1.165	2.924	-1.668	1.576	-0.375
45.8	2.964	0.367	1.417	-1.479	2.180	-1.496	1.738	-0.534
45.9	2.607	0.505	1.273	-1.785	1.714	-1.242	1.900	-0.738
46.0	2.335	0.669	1.042	-2.042	1.415	-0.994	2.051	-1.000
46.1	2.127	0.840	0.752	-2.219	1.215	-0.774	2.171	-1.329
46.2	1.967	1.011	0.441	-2.303	1.076	-0.580	2.233	-1.731
[illegible]	[illegible]	[illegible]	[illegible]	[illegible]	[illegible]	[illegible]	[illegible]	[illegible]
46.4	1.744	1.339	-0.112	-2.244	0.902	-0.258	2.023	-2.697
46.5	1.667	1.496	-0.323	-2.144	0.846	-0.120	1.681	-3.169
46.6	1.606	1.649	-0.489	-2.025	0.803	0.007	1.181	-3.533
46.7	1.560	1.801	-0.616	-1.900	0.769	0.127	0.583	-3.720
46.8	1.527	1.952	-0.711	-1.777	0.744	0.241	-0.022	-3.708
46.9	1.504	2.104	-0.781	-1.660	0.724	0.351	-0.552	-3.531
47.0	1.493	2.259	-0.833	-1.553	0.710	0.461	-0.965	-3.249
47.1	1.494	2.418	-0.872	-1.454	0.701	0.571	-1.257	-2.923
15π	1.496	2.457	-0.879	-1.431	0.700	0.597	-1.310	-2.844
47.2	1.508	2.584	-0.900	-1.363	0.697	0.683	-1.444	-2.595
47.3	1.538	2.759	-0.922	-1.281	0.698	0.800	-1.552	-2.287
47.4	1.586	2.945	-0.940	-1.205	0.706	0.923	-1.604	-2.011
47.5	1.660	3.144	-0.954	-1.134	0.722	1.056	-1.617	-1.768
47.6	1.768	3.358	-0.966	-1.068	0.747	1.201	-1.604	-1.555
47.7	1.922	3.586	-0.977	-1.005	0.787	1.364	-1.575	-1.370
47.8	2.142	3.825	-0.988	-0.944	0.848	1.550	-1.536	-1.208
47.9	2.455	4.060	-0.999	-0.883	0.941	1.766	-1.490	-1.065
48.0	2.892	4.257	-1.011	-0.822	1.086	2.020	-1.440	-0.939
48.1	3.481	4.346	-1.022	-0.759	1.319	2.320	-1.388	-0.826
48.2	4.205	4.205	-1.033	-0.692	1.706	2.661	-1.335	-0.724
48.3	4.930	3.696	-1.042	-0.620	2.367	2.984	-1.281	-0.631
48.4	5.380	2.816	-1.047	-0.542	3.453	3.057	-1.227	-0.545
48.5	5.337	1.820	-1.047	-0.456	4.827	2.296	-1.172	-0.465
48.6	4.872	1.042	-1.036	-0.361	5.404	0.459	-1.116	-0.390
31π/2	4.277	0.628	-1.012	-0.265	4.610	-1.046	-1.061	-0.323
48.7	4.244	0.613	-1.011	-0.260	4.545	-1.100	-1.058	-0.319
48.8	3.646	0.466	-0.965	-0.156	3.343	-1.607	-0.998	-0.252
48.9	3.152	0.489	-0.896	-0.057	2.464	-1.545	-0.936	-0.188
49.0	2.766	0.599	-0.805	0.027	1.904	-1.317	-0.870	-0.127
49.1	2.470	0.749	-0.698	0.088	1.547	-1.068	-0.799	-0.069
49.2	2.242	0.913	-0.586	0.122	1.312	-0.840	-0.722	-0.017
49.3	2.066	1.080	-0.480	0.129	1.151	-0.638	-0.638	0.029
49.4	1.929	1.245	-0.388	0.119	1.036	-0.460	-0.548	0.065
49.5	1.822	1.406	-0.311	0.097	0.951	-0.301	-0.454	0.087
49.6	1.737	1.563	-0.250	0.073	0.888	-0.159	-0.360	0.094
49.7	1.671	1.717	-0.200	0.049	0.839	-0.028	-0.273	0.085
49.8	1.619	1.868	-0.160	0.030	0.802	0.095	-0.199	0.064
49.9	1.581	2.019	-0.124	0.015	0.773	0.211	-0.140	0.040
50.0	1.555	2.170	-0.091	0.006	0.750	0.323	-0.094	0.018
50.1	1.541	2.323	-0.058	0.001	0.734	0.433	-0.057	0.004
50.2	1.538	2.481	-0.024	-0.001	0.723	0.542	-0.023	-0.001
16π	1.543	2.587	-0.000	-0.000	0.718	0.615	-0.000	-0.000
50.3	1.548	2.644	0.013	0.001	0.717	0.654	0.013	0.002
50.4	1.573	2.815	0.054	0.005	0.716	0.768	0.052	0.009
50.5	1.616	2.996	0.099	0.010	0.722	0.889	0.097	0.020
50.6	1.683	3.189	0.150	0.015	0.734	1.018	0.148	0.032
50.7	1.780	3.395	0.207	0.019	0.756	1.158	0.206	0.044
50.8	1.920	3.614	0.272	0.021	0.791	1.314	0.269	0.054
50.9	2.118	3.844	0.345	0.019	0.843	1.490	0.340	0.061

k_0h	OMEGA = 15.0 ADMITTANCE		RAD.EFF.LENGTH		OMEGA = 20.0 ADMITTANCE		RAD.EFF.LENGTH	
51.0	2.398	4.073	0.427	0.011	0.924	1.692	0.416	0.065
51.1	2.787	4.275	0.519	-0.005	1.047	1.929	0.500	0.065
51.2	3.314	4.394	0.622	-0.032	1.242	2.208	0.590	0.059
51.3	3.977	4.330	0.735	-0.076	1.560	2.529	0.689	0.046
51.4	4.686	3.955	0.857	-0.142	2.096	2.863	0.795	0.024
51.5	5.223	3.214	0.986	-0.235	2.992	3.065	0.909	-0.007
51.6	5.344	2.268	1.115	-0.364	4.288	2.705	1.032	-0.051
51.7	5.021	1.430	1.235	-0.536	5.311	1.252	1.165	-0.110
51.8	4.455	0.899	1.328	-0.755	4.967	-0.560	1.307	-0.189
33π/2	4.234	0.784	1.351	-0.845	4.570	-1.007	1.361	-0.224
51.9	3.860	0.664	1.373	-1.017	3.802	-1.450	1.457	-0.293
52.0	3.344	0.630	1.349	-1.309	2.793	-1.562	1.614	-0.429
52.1	2.930	0.706	1.240	-1.601	2.125	-1.384	1.773	-0.604
52.2	2.609	0.837	1.047	-1.857	1.699	-1.143	1.927	-0.829
52.3	2.361	0.992	0.791	-2.046	1.421	-0.909	2.060	-1.113
52.4	2.169	1.154	0.507	-2.151	1.232	-0.698	2.152	-1.462
52.5	2.019	1.317	0.228	-2.176	1.100	-0.512	2.172	-1.877
52.6	1.901	1.477	-0.020	-2.138	1.003	-0.347	2.083	-2.340
52.7	1.808	1.634	-0.228	-2.060	0.931	-0.199	1.850	-2.811
52.8	1.735	1.787	-0.395	-1.958	0.877	-0.064	1.462	-3.223
52.9	1.679	1.938	-0.524	-1.848	0.835	0.061	0.949	-3.507
53.0	1.636	2.088	-0.623	-1.737	0.802	0.179	0.377	-3.617
53.1	1.606	2.239	-0.698	-1.631	0.777	0.293	-0.171	-3.553
53.2	1.588	2.391	-0.754	-1.532	0.758	0.404	-0.637	-3.357
53.3	1.581	2.546	-0.797	-1.440	0.745	0.513	-0.994	-3.082
17π	1.588	2.719	-0.832	-1.350	0.736	0.632	-1.259	-2.757
53.4	1.587	2.707	-0.830	-1.356	0.737	0.624	-1.245	-2.778
53.5	1.608	2.875	-0.855	-1.279	0.734	0.738	-1.407	-2.477
53.6	1.646	3.051	-0.876	-1.208	0.737	0.856	-1.503	-2.196
53.7	1.706	3.238	-0.894	-1.143	0.747	0.981	-1.550	-1.943
53.8	1.794	3.437	-0.910	-1.082	0.765	1.117	-1.564	-1.718
53.9	1.921	3.649	-0.924	-1.023	0.795	1.266	-1.554	-1.520
54.0	2.099	3.870	-0.939	-0.967	0.841	1.433	-1.530	-1.345
54.1	2.350	4.093	-0.954	-0.911	0.910	1.624	-1.496	-1.191
54.2	2.698	4.297	-0.969	-0.855	1.016	1.845	-1.456	-1.055
54.3	3.169	4.437	-0.985	-0.797	1.180	2.105	-1.411	-0.932
54.4	3.772	4.432	-1.001	-0.735	1.444	2.406	-1.364	-0.822
54.5	4.450	4.169	-1.017	-0.669	1.881	2.735	-1.316	-0.721
54.6	5.037	3.564	-1.030	-0.595	2.614	3.006	-1.266	-0.628
54.7	5.296	2.699	-1.038	-0.513	3.753	2.927	-1.214	-0.542
54.8	5.121	1.837	-1.038	-0.422	4.992	1.935	-1.162	-0.462
54.9	4.641	1.216	-1.025	-0.322	5.230	0.123	-1.108	-0.386
35π/2	4.194	0.941	-1.002	-0.240	4.532	-0.971	-1.064	-0.329
55.0	4.067	0.892	-0.993	-0.216	4.274	-1.166	-1.052	-0.313
55.1	3.538	0.792	-0.938	-0.111	3.170	-1.528	-0.993	-0.244
55.2	3.100	0.829	-0.858	-0.017	2.382	-1.437	-0.930	-0.178
55.3	2.754	0.936	-0.759	0.057	1.873	-1.216	-0.863	-0.115
55.4	2.484	1.078	-0.649	0.105	1.544	-0.979	-0.789	-0.057
55.5	2.275	1.234	-0.540	0.125	1.323	-0.762	-0.710	-0.004
55.6	2.111	1.393	-0.442	0.124	1.170	-0.568	-0.623	0.041
55.7	1.983	1.552	-0.359	0.108	1.059	-0.396	-0.530	0.074
55.8	1.881	1.708	-0.291	0.086	0.978	-0.242	-0.434	0.092
55.9	1.801	1.861	-0.236	0.063	0.916	-0.103	-0.341	0.094
56.0	1.739	2.012	-0.191	0.042	0.869	0.026	-0.257	0.081
56.1	1.691	2.161	-0.153	0.025	0.832	0.147	-0.186	0.058
56.2	1.657	2.311	-0.118	0.013	0.804	0.262	-0.131	0.034
[illegible]	[illegible]	[illegible]	[illegible]	[illegible]	0.783	0.376	-0.088	0.016
56.4	1.624	2.615	-0.052	0.000	0.767	0.484	-0.052	0.003
56.5	1.627	2.774	-0.018	-0.001	0.757	0.595	-0.017	-0.001
18π	1.633	2.853	-0.000	-0.000	0.754	0.649	-0.000	-0.000
56.6	1.643	2.938	0.020	0.001	0.752	0.707	0.019	0.002
56.7	1.676	3.110	0.061	0.004	0.753	0.823	0.060	0.011
56.8	1.730	3.292	0.107	0.008	0.760	0.946	0.106	0.022
56.9	1.809	3.485	0.159	0.012	0.775	1.077	0.158	0.034
57.0	1.924	3.689	0.218	0.015	0.801	1.221	0.216	0.045
57.1	2.085	3.903	0.284	0.014	0.840	1.380	0.281	0.055
57.2	2.311	4.120	0.358	0.009	0.900	1.560	0.353	0.062
57.3	2.622	4.323	0.442	-0.002	0.991	1.768	0.431	0.065
57.4	3.044	4.478	0.536	-0.024	1.130	2.010	0.517	0.064
57.5	3.590	4.518	0.640	-0.060	1.351	2.292	0.610	0.056
57.6	4.228	4.346	0.754	-0.113	1.710	2.608	0.710	0.041
57.7	4.835	3.867	0.875	-0.192	2.307	2.910	0.819	0.017
57.8	5.200	3.103	0.999	-0.302	3.272	3.010	0.937	-0.018
57.9	5.170	2.251	1.117	-0.450	4.533	2.440	1.063	-0.067
58.0	4.795	1.562	1.215	-0.642	5.273	0.871	1.198	-0.133
58.1	4.262	1.150	1.276	-0.877	4.708	-0.730	1.343	-0.221
37π/2	4.155	1.100	1.281	-0.927	4.498	-0.938	1.372	-0.241
58.2	3.731	0.979	1.276	-1.145	3.590	-1.417	1.494	-0.336
58.3	3.273	0.968	1.201	-1.422	2.680	-1.464	1.650	-0.485
58.4	2.903	1.048	1.045	-1.676	2.075	-1.282	1.804	-0.677
58.5	2.612	1.174	0.824	-1.874	1.684	-1.051	1.947	-0.921
58.6	2.385	1.320	0.566	-1.998	1.424	-0.828	2.061	-1.225
58.7	2.206	1.475	0.305	-2.047	1.246	-0.627	2.122	-1.591
58.8	2.066	1.631	0.065	-2.032	1.120	-0.448	2.100	-2.013
58.9	1.956	1.786	-0.140	-1.974	1.027	-0.288	1.959	-2.465
59.0	1.868	1.938	-0.307	-1.890	0.957	-0.143	1.677	-2.899
59.1	1.800	2.088	-0.440	-1.793	0.904	-0.010	1.258	-3.250
59.2	1.747	2.237	-0.542	-1.694	0.863	0.113	0.745	-3.463
59.3	1.708	2.386	-0.620	-1.597	0.832	0.231	0.206	-3.512
59.4	1.682	2.536	-0.681	-1.505	0.807	0.344	-0.291	-3.411
59.5	1.668	2.688	-0.727	-1.420	0.789	0.454	-0.703	-3.204
59.6	1.666	2.844	-0.763	-1.342	0.777	0.565	-1.015	-2.938
19π	1.677	2.990	-0.789	-1.277	0.771	0.665	-1.216	-2.682
59.7	1.678	3.006	-0.792	-1.270	0.770	0.676	-1.234	-2.654
59.8	1.707	3.174	-0.815	-1.204	0.769	0.791	-1.377	-2.375
59.9	1.754	3.351	-0.835	-1.143	0.774	0.911	-1.462	-2.116
60.0	1.826	3.539	-0.854	-1.086	0.786	1.039	-1.505	-1.881
60.1	1.929	3.736	-0.871	-1.032	0.807	1.177	-1.518	-1.672
60.2	2.075	3.943	-0.888	-0.981	0.841	1.329	-1.512	-1.486
60.3	2.278	4.153	-0.906	-0.930	0.893	1.501	-1.491	-1.321
60.4	2.558	4.355	-0.924	-0.879	0.971	1.696	-1.462	-1.174
60.5	2.936	4.519	-0.943	-0.826	1.090	1.922	-1.426	-1.043
60.6	3.430	4.593	-0.963	-0.770	1.275	2.186	-1.386	-0.924
60.7	4.022	4.494	-0.984	-0.709	1.573	2.487	-1.343	-0.816
60.8	4.627	4.127	-1.003	-0.641	2.061	2.798	-1.298	-0.717
60.9	5.068	3.472	-1.019	-0.565	2.862	3.002	-1.251	-0.625

	OMEGA = 15.0					OMEGA = 20.0			
k_0h	ADMITTANCE		RAD.EFF.LENGTH		k_0h	ADMITTANCE		RAD.EFF.LENGTH	
61.0	5.170	2.661	-1.028	-0.478	61.0	4.028	2.759	-1.203	-0.539
61.1	4.912	1.931	-1.027	-0.382	61.1	5.087	1.576	-1.153	-0.457
61.2	4.439	1.437	-1.008	-0.277	61.2	5.034	-0.147	-1.100	-0.380
39π/2	4.119	1.261	-0.986	-0.211	39π/2	4.466	-0.905	-1.067	-0.335
61.3	3.919	1.192	-0.967	-0.170	61.3	4.034	-1.199	-1.045	-0.306
61.4	3.447	1.128	-0.901	-0.068	61.4	3.022	-1.450	-0.986	-0.235
61.5	3.056	1.174	-0.812	0.018	61.5	2.309	-1.338	-0.923	-0.168
61.6	2.744	1.280	-0.708	0.080	61.6	1.845	-1.122	-0.854	-0.103
61.7	2.498	1.415	-0.600	0.114	61.7	1.539	-0.896	-0.779	-0.044
61.8	2.305	1.563	-0.498	0.123	61.8	1.332	-0.688	-0.696	0.009
61.9	2.153	1.716	-0.409	0.115	61.9	1.186	-0.502	-0.606	0.051
62.0	2.032	1.868	-0.334	0.098	62.0	1.080	-0.335	-0.511	0.081
62.1	1.937	2.020	-0.273	0.076	62.1	1.001	-0.185	-0.414	0.095
62.2	1.861	2.169	-0.224	0.054	62.2	0.942	-0.049	-0.322	0.092
62.3	1.803	2.318	-0.182	0.036	62.3	0.896	0.078	-0.241	0.076
62.4	1.760	2.466	-0.146	0.021	62.4	0.860	0.198	-0.175	0.052
62.5	1.729	2.615	-0.112	0.011	62.5	0.833	0.313	-0.123	0.029
62.6	1.711	2.765	-0.080	0.004	62.6	0.812	0.424	-0.082	0.011
62.7	1.706	2.919	-0.047	0.000	62.7	0.798	0.535	-0.046	0.001
62.8	1.714	3.078	-0.012	-0.000	62.8	0.789	0.646	-0.011	-0.001
20π	1.719	3.130	-0.000	-0.000	20π	0.787	0.681	-0.000	-0.000
62.9	1.737	3.243	0.026	0.001	62.9	0.785	0.759	0.026	0.004
63.0	1.779	3.416	0.068	0.004	63.0	0.787	0.877	0.067	0.012
63.1	1.843	3.598	0.115	0.007	63.1	0.797	1.001	0.114	0.023
63.2	1.937	3.789	0.168	0.009	63.2	0.814	1.135	0.168	0.035
63.3	2.069	3.989	0.228	0.010	63.3	0.844	1.281	0.227	0.047
63.4	2.252	4.194	0.295	0.006	63.4	0.888	1.445	0.294	0.056
63.5	2.504	4.392	0.371	-0.002	63.5	0.956	1.629	0.367	0.062
63.6	2.844	4.563	0.457	-0.019	63.6	1.058	1.842	0.447	0.065
63.7	3.289	4.662	0.552	-0.047	63.7	1.215	2.088	0.534	0.062
63.8	3.835	4.620	0.656	-0.091	63.8	1.462	2.372	0.629	0.053
63.9	4.422	4.350	0.769	-0.156	63.9	1.864	2.679	0.732	0.036
64.0	4.910	3.804	0.886	-0.248	64.0	2.524	2.940	0.844	0.009
64.1	5.124	3.058	1.001	-0.374	64.1	3.542	2.922	0.964	-0.030
64.2	4.987	2.313	1.102	-0.540	64.2	4.725	2.150	1.093	-0.085
64.3	4.592	1.752	1.173	-0.748	64.3	5.182	0.535	1.231	-0.158
41π/2	4.085	1.426	1.196	-0.997	41π/2	4.437	-0.875	1.381	-0.257
64.4	4.098	1.431	1.196	-0.990	64.4	4.465	-0.850	1.377	-0.254
64.5	3.622	1.309	1.151	-1.249	64.5	3.406	-1.373	1.529	-0.381
64.6	3.213	1.316	1.032	-1.497	64.6	2.580	-1.372	1.683	-0.544
64.7	2.881	1.397	0.846	-1.703	64.7	2.030	-1.188	1.830	-0.753
64.8	2.616	1.519	0.617	-1.844	64.8	1.669	-0.966	1.960	-1.015
64.9	2.407	1.659	0.374	-1.915	64.9	1.427	-0.752	2.053	-1.336
65.0	2.242	1.806	0.145	-1.923	65.0	1.258	-0.559	2.082	-1.715
65.1	2.111	1.957	-0.057	-1.885	65.1	1.138	-0.386	2.017	-2.138
65.2	2.007	2.106	-0.225	-1.817	65.2	1.048	-0.230	1.829	-2.571
65.3	1.925	2.255	-0.359	-1.734	65.3	0.981	-0.089	1.507	-2.964
65.4	1.861	2.403	-0.465	-1.646	65.4	0.930	0.042	1.069	-3.257
65.5	1.812	2.550	-0.547	-1.557	65.5	0.890	0.164	0.565	-3.408
65.6	1.777	2.697	-0.611	-1.473	65.6	0.859	0.281	0.060	-3.408
65.7	1.755	2.847	-0.660	-1.394	65.7	0.836	0.393	-0.390	-3.281
65.8	1.746	2.999	-0.699	-1.321	65.8	0.819	0.504	-0.756	-3.068
65.9	1.750	3.155	-0.730	-1.254	65.9	0.807	0.615	-1.031	-2.812
21π	1.762	3.273	-0.750	-1.208	21π	0.802	0.697	-1.180	-2.616
66.0	1.768	3.317	-0.756	-1.192	66.0	0.802	0.727	-1.224	-2.545
66.1	1.805	3.486	-0.779	-1.136	66.1	0.802	0.843	-1.350	-2.286
66.2	1.862	3.663	-0.799	-1.083	66.2	0.808	0.965	-1.427	-2.045
66.3	1.947	3.849	-0.818	-1.033	66.3	0.823	1.095	-1.466	-1.826
66.4	2.066	4.042	-0.837	-0.986	66.4	0.847	1.236	-1.480	-1.629
66.5	2.232	4.241	-0.856	-0.939	66.5	0.886	1.392	-1.475	-1.454
66.6	2.459	4.436	-0.877	-0.893	66.6	0.944	1.567	-1.458	-1.297
66.7	2.765	4.610	-0.898	-0.845	66.7	1.032	1.767	-1.432	-1.157
66.8	3.166	4.727	-0.921	-0.795	66.8	1.165	1.998	-1.399	-1.030
66.9	3.666	4.730	-0.945	-0.740	66.9	1.373	2.265	-1.363	-0.915
67.0	4.226	4.542	-0.968	-0.678	67.0	1.705	2.562	-1.324	-0.809
67.1	4.737	4.099	-0.990	-0.609	67.1	2.248	2.848	-1.282	-0.711
67.2	5.039	3.433	-1.007	-0.529	67.2	3.111	2.973	-1.238	-0.619
67.3	5.020	2.700	-1.016	-0.438	67.3	4.271	2.556	-1.192	-0.533
67.4	4.716	2.090	-1.010	-0.337	67.4	5.119	1.230	-1.144	-0.451
67.5	4.264	1.698	-0.984	-0.230	67.5	4.826	-0.360	-1.093	-0.373
43π/2	4.053	1.594	-0.964	-0.182	43π/2	4.409	-0.846	-1.069	-0.339
67.6	3.793	1.513	-0.933	-0.123	67.6	3.820	-1.209	-1.038	-0.298
67.7	3.372	1.476	-0.857	-0.028	67.7	2.892	-1.373	-0.980	-0.225
67.8	3.021	1.529	-0.762	0.046	67.8	2.245	-1.245	-0.916	-0.156
67.9	2.738	1.633	-0.657	0.094	67.9	1.818	-1.034	-0.845	-0.091
68.0	2.513	1.762	-0.554	0.116	68.0	1.534	-0.818	-0.767	-0.031
68.1	2.334	1.904	-0.460	0.118	68.1	1.339	-0.618	-0.681	0.021
68.2	2.192	2.051	-0.379	0.106	68.2	1.200	-0.438	-0.588	0.061
68.3	2.079	2.199	-0.313	0.087	68.3	1.099	-0.277	-0.491	0.087
68.4	1.989	2.346	-0.258	0.067	68.4	1.023	-0.130	-0.394	0.096
68.5	1.919	2.493	-0.213	0.047	68.5	0.965	0.004	-0.305	0.089
68.6	1.865	2.639	-0.174	0.031	68.6	0.921	0.129	-0.227	0.070
68.7	1.826	2.785	-0.139	0.018	68.7	0.886	0.248	-0.165	0.046
68.8	1.799	2.933	-0.106	0.009	68.8	0.860	0.362	-0.116	0.024
68.9	1.786	3.083	-0.074	0.003	68.9	0.840	0.473	-0.076	0.008
69.0	1.786	3.237	-0.041	0.000	69.0	0.826	0.584	-0.041	0.000
69.1	1.800	3.396	-0.006	-0.000	69.1	0.818	0.696	-0.005	-0.000
22π	1.804	3.421	-0.000	-0.000	22π	0.818	0.713	-0.000	-0.000
69.2	1.831	3.562	0.033	0.001	69.2	0.816	0.810	0.032	0.005
69.3	1.882	3.734	0.076	0.003	69.3	0.820	0.929	0.075	0.014
69.4	1.958	3.915	0.123	0.005	69.4	0.832	1.056	0.123	0.025
69.5	2.066	4.103	0.177	0.005	69.5	0.852	1.192	0.178	0.037
69.6	2.216	4.296	0.238	0.003	69.6	0.885	1.341	0.238	0.048
69.7	2.421	4.487	0.306	-0.003	69.7	0.936	1.508	0.306	0.057
69.8	2.697	4.662	0.383	-0.016	69.8	1.011	1.697	0.381	0.063
69.9	3.060	4.792	0.469	-0.038	69.9	1.125	1.914	0.463	0.064
70.0	3.516	4.831	0.564	-0.073	70.0	1.300	2.165	0.552	0.060
70.1	4.042	4.708	0.668	-0.126	70.1	1.577	2.449	0.649	0.049
70.2	4.558	4.360	0.777	-0.202	70.2	2.024	2.742	0.755	0.030
70.3	4.924	3.782	0.887	-0.309	70.3	2.744	2.951	0.869	-0.001
70.4	5.012	3.084	0.988	-0.451	70.4	3.798	2.801	0.992	-0.044
70.5	4.806	2.445	1.068	-0.631	70.5	4.863	1.847	1.124	-0.104
70.6	4.411	1.990	1.108	-0.847	70.6	5.054	0.246	1.264	-0.184
45π/2	4.022	1.766	1.097	-1.051	45π/2	4.383	-0.818	1.390	-0.273
70.7	3.958	1.742	1.091	-1.085	70.7	4.239	-0.932	1.411	-0.290
70.8	3.530	1.656	1.006	-1.323	70.8	3.244	-1.324	1.562	-0.428
70.9	3.163	1.676	0.855	-1.531	70.9	2.493	-1.285	1.713	-0.605

	OMEGA = 15.0					OMEGA = 20.0			
k_0h	ADMITTANCE		RAD.EFF.LENGTH		k_0h	ADMITTANCE		RAD.EFF.LENGTH	
71.0	2.863	1.759	0.656	-1.687	71.0	1.990	-1.100	1.852	-0.830
71.1	2.622	1.876	0.435	-1.778	71.1	1.656	-0.885	1.967	-1.109
71.2	2.429	2.010	0.217	-1.808	71.2	1.428	-0.680	2.036	-1.445
71.3	2.276	2.152	0.020	-1.790	71.3	1.269	-0.494	2.032	-1.833
71.4	2.153	2.297	-0.147	-1.740	71.4	1.153	-0.326	1.926	-2.251
71.5	2.056	2.443	-0.283	-1.670	71.5	1.068	-0.174	1.696	-2.661
71.6	1.979	2.588	-0.392	-1.593	71.6	1.003	-0.036	1.343	-3.010
71.7	1.920	2.733	-0.477	-1.513	71.7	0.953	0.093	0.895	-3.248
71.8	1.875	2.879	-0.544	-1.435	71.8	0.914	0.214	0.407	-3.346
71.9	1.845	3.025	-0.597	-1.362	71.9	0.885	0.330	-0.064	-3.307
72.0	1.827	3.174	-0.638	-1.294	72.0	0.862	0.442	-0.473	-3.161
72.1	1.823	3.325	-0.672	-1.232	72.1	0.846	0.553	-0.799	-2.947
72.2	1.833	3.482	-0.700	-1.174	72.2	0.836	0.664	-1.043	-2.701
23π	1.845	3.573	-0.714	-1.144	23π	0.832	0.728	-1.148	-2.557
72.3	1.859	3.644	-0.724	-1.122	72.3	0.831	0.777	-1.214	-2.448
72.4	1.904	3.812	-0.745	-1.073	72.4	0.833	0.894	-1.327	-2.206
72.5	1.972	3.988	-0.766	-1.027	72.5	0.841	1.018	-1.396	-1.980
72.6	2.070	4.170	-0.786	-0.983	72.6	0.858	1.150	-1.433	-1.775
72.7	2.205	4.358	-0.806	-0.941	72.7	0.887	1.294	-1.446	-1.590
72.8	2.391	4.545	-0.828	-0.899	72.8	0.930	1.453	-1.443	-1.424
72.9	2.640	4.720	-0.851	-0.856	72.9	0.995	1.632	-1.428	-1.275
73.0	2.968	4.860	-0.875	-0.811	73.0	1.093	1.837	-1.405	-1.140
73.1	3.383	4.925	-0.901	-0.761	73.1	1.241	2.072	-1.376	-1.017
73.2	3.873	4.856	-0.928	-0.707	73.2	1.473	2.341	-1.343	-0.904
73.3	4.380	4.590	-0.954	-0.644	73.3	1.843	2.631	-1.307	-0.800
73.4	4.788	4.101	-0.977	-0.572	73.4	2.439	2.886	-1.268	-0.704
73.5	4.966	3.457	-0.994	-0.488	73.5	3.356	2.916	-1.226	-0.613
73.6	4.860	2.811	-0.999	-0.393	73.6	4.477	2.325	-1.182	-0.526
73.7	4.535	2.306	-0.987	-0.289	73.7	5.097	0.908	-1.136	-0.444
73.8	4.112	1.996	-0.951	-0.181	73.8	4.618	-0.525	-1.086	-0.364
47π/2	3.992	1.943	-0.937	-0.152	47π/2	4.359	-0.791	-1.071	-0.343
73.9	3.686	1.858	-0.890	-0.080	73.9	3.630	-1.202	-1.032	-0.288
74.0	3.308	1.841	-0.807	0.005	74.0	2.778	-1.298	-0.973	-0.214
74.1	2.991	1.899	-0.709	0.066	74.1	2.187	-1.158	-0.907	-0.144
74.2	2.734	2.001	-0.607	0.102	74.2	1.794	-0.952	-0.835	-0.078
74.3	2.528	2.125	-0.511	0.114	74.3	1.529	-0.744	-0.754	-0.018
74.4	2.362	2.262	-0.426	0.110	74.4	1.345	-0.551	-0.665	0.032
74.5	2.230	2.403	-0.354	0.096	74.5	1.213	-0.377	-0.570	0.070
74.6	2.124	2.546	-0.294	0.078	74.6	1.116	-0.220	-0.471	0.092
74.7	2.041	2.690	-0.244	0.059	74.7	1.043	-0.077	-0.375	0.096
74.8	1.975	2.834	-0.203	0.042	74.8	0.987	0.055	-0.288	0.086
74.9	1.926	2.978	-0.166	0.027	74.9	0.944	0.179	-0.214	0.065
75.0	1.891	3.123	-0.132	0.016	75.0	0.910	0.297	-0.155	0.041
75.1	1.869	3.270	-0.100	0.008	75.1	0.885	0.410	-0.109	0.020
75.2	1.861	3.420	-0.068	0.003	75.2	0.866	0.521	-0.070	0.006
75.3	1.866	3.573	-0.035	0.001	75.3	0.854	0.632	-0.035	-0.000
24π	1.887	3.729	-0.000	-0.000	24π	0.847	0.743	-0.000	-0.000
75.4	1.887	3.732	0.001	0.000	75.4	0.847	0.745	0.001	0.000
75.5	1.926	3.897	0.040	0.000	75.5	0.846	0.860	0.039	0.006
75.6	1.987	4.068	0.083	0.001	75.6	0.852	0.981	0.083	0.016
75.7	2.076	4.246	0.131	0.002	75.7	0.865	1.109	0.132	0.027
75.8	2.199	4.428	0.186	0.000	75.8	0.889	1.248	0.188	0.039
75.9	2.366	4.611	0.247	-0.004	75.9	0.926	1.401	0.250	0.049
76.0	2.592	4.785	0.316	-0.014	76.0	0.983	1.571	0.319	0.058
76.1	2.888	4.931	0.393	-0.031	76.1	1.067	1.765	0.395	0.063
76.2	3.266	5.016	0.480	-0.060	76.2	1.193	1.986	0.479	0.063
76.3	3.719	4.991	0.574	-0.102	76.3	1.389	2.240	0.570	0.058
76.4	4.207	4.796	0.674	-0.165	76.4	1.696	2.521	0.669	0.045
76.5	4.639	4.392	0.778	-0.254	76.5	2.189	2.796	0.777	0.023
76.6	4.889	3.812	0.877	-0.373	76.6	2.966	2.941	0.894	-0.011
76.7	4.879	3.180	0.960	-0.528	76.7	4.034	2.648	1.020	-0.058
76.8	4.633	2.640	1.013	-0.717	76.8	4.947	1.542	1.154	-0.124
76.9	4.251	2.274	1.019	-0.933	76.9	4.900	0.004	1.296	-0.212
49π/2	3.964	2.125	0.989	-1.087	49π/2	4.336	-0.765	1.398	-0.288
77.0	3.836	2.082	0.965	-1.156	77.0	4.031	-0.984	1.444	-0.327
77.1	3.452	2.024	0.849	-1.362	77.1	3.101	-1.270	1.594	-0.477
77.2	3.122	2.054	0.681	-1.527	77.2	2.415	-1.202	1.739	-0.669
77.3	2.850	2.137	0.484	-1.635	77.3	1.953	-1.016	1.869	-0.909
77.4	2.629	2.251	0.281	-1.687	77.4	1.643	-0.809	1.967	-1.204
77.5	2.452	2.380	0.092	-1.689	77.5	1.429	-0.611	2.011	-1.553
77.6	2.309	2.517	-0.073	-1.656	77.6	1.278	-0.431	1.972	-1.945
77.7	2.195	2.657	-0.211	-1.601	77.7	1.168	-0.268	1.827	-2.353
77.8	2.104	2.799	-0.322	-1.534	77.8	1.085	-0.120	1.561	-2.734
77.9	2.033	2.941	-0.410	-1.463	77.9	1.023	0.016	1.185	-3.039
78.0	1.978	3.084	-0.480	-1.393	78.0	0.974	0.143	0.735	-3.226
78.1	1.938	3.227	-0.535	-1.326	78.1	0.937	0.263	0.267	-3.279
78.2	1.912	3.373	-0.579	-1.262	78.2	0.908	0.378	-0.170	-3.209
78.3	1.899	3.521	-0.615	-1.204	78.3	0.887	0.490	-0.542	-3.050
78.4	1.900	3.672	-0.644	-1.151	78.4	0.872	0.601	-0.835	-2.837
78.5	1.917	3.828	-0.670	-1.101	78.5	0.863	0.712	-1.052	-2.601
25π	1.928	3.891	-0.679	-1.083	25π	0.861	0.757	-1.120	-2.505
78.6	1.950	3.989	-0.693	-1.056	78.6	0.860	0.827	-1.206	-2.362
78.7	2.004	4.156	-0.714	-1.013	78.7	0.863	0.945	-1.307	-2.134
78.8	2.084	4.329	-0.735	-0.973	78.8	0.873	1.070	-1.370	-1.922
78.9	2.195	4.506	-0.756	-0.935	78.9	0.893	1.205	-1.404	-1.728
79.0	2.347	4.685	-0.778	-0.897	79.0	0.925	1.351	-1.417	-1.553
79.1	2.551	4.857	-0.802	-0.858	79.1	0.973	1.514	-1.415	-1.395
79.2	2.820	5.007	-0.827	-0.818	79.2	1.046	1.697	-1.402	-1.252
79.3	3.163	5.107	-0.854	-0.774	79.3	1.154	1.906	-1.381	-1.122
79.4	3.580	5.117	-0.883	-0.726	79.4	1.320	2.145	-1.355	-1.003
79.5	4.044	4.981	-0.912	-0.670	79.5	1.577	2.414	-1.325	-0.894
79.6	4.483	4.655	-0.940	-0.605	79.6	1.986	2.694	-1.291	-0.791
79.7	4.788	4.147	-0.963	-0.530	79.7	2.636	2.907	-1.255	-0.696
79.8	4.864	3.546	-0.977	-0.442	79.8	3.593	2.831	-1.215	-0.605
79.9	4.700	2.988	-0.977	-0.344	79.9	4.642	2.074	-1.173	-0.519
80.0	4.372	2.574	-0.956	-0.239	80.0	5.030	0.617	-1.128	-0.435
80.1	3.978	2.329	-0.911	-0.135	80.1	4.414	-0.649	-1.079	-0.355
51π/2	3.936	2.312	-0.905	-0.124	51π/2	4.315	-0.740	-1.073	-0.347
80.2	3.594	2.228	-0.841	-0.042	80.2	3.459	-1.182	-1.025	-0.277
80.3	3.254	2.226	-0.753	0.031	80.3	2.677	-1.225	-0.965	-0.202
80.4	2.968	2.288	-0.656	0.079	80.4	2.135	-1.075	-0.898	-0.131
80.5	2.734	2.388	-0.560	0.104	80.5	1.772	-0.874	-0.823	-0.064
80.6	2.544	2.508	-0.472	0.109	80.6	1.524	-0.673	-0.740	-0.005
80.7	2.391	2.639	-0.395	0.101	80.7	1.350	-0.487	-0.648	0.043
80.8	2.267	2.776	-0.331	0.087	80.8	1.224	-0.318	-0.551	0.078
80.9	2.169	2.915	-0.277	0.069	80.9	1.131	-0.165	-0.451	0.095

	OMEGA = 15.0		OMEGA = 20.0	
k_0h	ADMITTANCE	RAD.EFF.LENGTH	ADMITTANCE	RAD.EFF.LENGTH
81.0	2.091 3.056	-0.232 0.052	1.061 -0.025	-0.357 0.095
81.1	2.031 3.197	-0.193 0.037	1.007 0.105	-0.272 0.081
81.2	1.986 3.339	-0.158 0.024	0.965 0.228	-0.202 0.059
81.3	1.956 3.483	-0.126 0.015	0.933 0.344	-0.147 0.036
81.4	1.939 3.629	-0.094 0.008	0.909 0.457	-0.102 0.017
81.5	1.935 3.778	-0.062 0.003	0.891 0.569	-0.064 0.004
81.6	1.947 3.931	-0.029 0.001	0.879 0.680	-0.029 -0.001
26π	1.969 4.059	-0.000 -0.000	0.874 0.772	-0.000 -0.000
81.7	1.975 4.088	0.007 -0.000	0.874 0.793	0.007 0.001
81.8	2.023 4.252	0.046 -0.000	0.874 0.910	0.047 0.007
81.9	2.095 4.420	0.090 -0.001	0.882 1.032	0.091 0.017
82.0	2.195 4.593	0.139 -0.002	0.898 1.163	0.141 0.029
82.1	2.333 4.768	0.194 -0.006	0.925 1.304	0.198 0.040
82.2	2.518 4.937	0.255 -0.013	0.967 1.459	0.262 0.050
82.3	2.762 5.089	0.325 -0.027	1.029 1.634	0.332 0.058
82.4	3.073 5.201	0.402 -0.049	1.122 1.831	0.410 0.063
82.5	3.456 5.237	0.487 -0.084	1.263 2.057	0.495 0.062
82.6	3.892 5.151	0.579 -0.135	1.480 2.312	0.588 0.055
82.7	4.328 4.896	0.674 -0.208	1.820 2.589	0.690 0.040
82.8	4.669 4.460	0.769 -0.308	2.361 2.838	0.800 0.015
82.9	4.819 3.904	0.853 -0.438	3.189 2.909	0.919 -0.022
83.0	4.737 3.345	0.914 -0.602	4.245 2.466	1.047 -0.074
83.1	4.471 2.894	0.939 -0.793	4.980 1.243	1.184 -0.146
83.2	4.109 2.600	0.912 -0.998	4.732 -0.194	1.327 -0.242
53π/2	3.910 2.506	0.875 -1.104	4.295 -0.716	1.404 -0.303
83.3	3.732 2.452	0.828 -1.197	3.840 -1.012	1.476 -0.367
83.4	3.386 2.416	0.692 -1.366	2.974 -1.213	1.623 -0.529
83.5	3.088 2.454	0.521 -1.489	2.345 -1.123	1.763 -0.735
83.6	2.841 2.537	0.336 -1.559	1.919 -0.938	1.882 -0.991
83.7	2.639 2.647	0.156 -1.580	1.631 -0.737	1.961 -1.300
83.8	2.475 2.772	-0.005 -1.565	1.430 -0.545	1.977 -1.658
83.9	2.342 2.904	-0.142 -1.524	1.286 -0.371	1.905 -2.049
84.0	2.236 3.040	-0.255 -1.469	1.181 -0.212	1.723 -2.443
84.1	2.152 3.178	-0.345 -1.408	1.102 -0.067	1.426 -2.793
84.2	2.086 3.318	-0.418 -1.345	1.041 0.066	1.034 -3.054
84.3	2.036 3.458	-0.476 -1.283	0.994 0.192	0.589 -3.194
84.4	2.000 3.600	-0.522 -1.225	0.958 0.310	0.143 -3.209
84.5	1.979 3.744	-0.559 -1.171	0.931 0.425	-0.262 -3.116
84.6	1.972 3.891	-0.591 -1.121	0.911 0.537	-0.600 -2.948
84.7	1.979 4.042	-0.618 -1.075	0.897 0.648	-0.864 -2.737
84.8	2.002 4.196	-0.641 -1.033	0.889 0.760	-1.060 -2.510
27π	2.010 4.233	-0.647 -1.024	0.888 0.786	-1.096 -2.457
84.9	2.044 4.356	-0.664 -0.994	0.887 0.875	-1.197 -2.284
85.0	2.108 4.520	-0.685 -0.957	0.892 0.995	-1.289 -2.068
85.1	2.199 4.689	-0.707 -0.922	0.904 1.122	-1.347 -1.868
85.2	2.324 4.860	-0.729 -0.887	0.927 1.259	-1.378 -1.685
85.3	2.492 5.026	-0.753 -0.853	0.963 1.408	-1.391 -1.519
85.4	2.712 5.179	-0.778 -0.817	1.016 1.574	-1.389 -1.368
85.5	2.996 5.299	-0.806 -0.779	1.097 1.761	-1.378 -1.231
85.6	3.346 5.357	-0.835 -0.736	1.217 1.974	-1.360 -1.105
85.7	3.753 5.311	-0.866 -0.687	1.400 2.216	-1.336 -0.990
85.8	4.177 5.117	-0.896 -0.630	1.685 2.484	-1.308 -0.882
85.9	4.539 4.752	-0.924 -0.562	2.135 2.749	-1.277 -0.782
86.0	4.748 4.249	-0.945 -0.483	2.837 2.912	-1.243 -0.687
86.1	4.745 3.705	-0.955 -0.392	3.819 2.718	-1.205 -0.597
86.2	4.545 3.230	-0.948 -0.292	4.762 1.809	-1.165 -0.510
86.3	4.224 2.892	-0.917 -0.189	4.930 0.360	-1.120 -0.426
55π/2	3.885 2.707	-0.867 -0.098	4.275 -0.693	-1.075 -0.350
86.4	3.863 2.699	-0.863 -0.092	4.218 -0.740	-1.072 -0.345
86.5	3.516 2.626	-0.787 -0.010	3.306 -1.152	-1.017 -0.266
86.6	3.209 2.637	-0.698 0.050	2.586 -1.152	-0.957 -0.190
86.7	2.950 2.702	-0.605 0.086	2.089 -0.996	-0.888 -0.117
86.8	2.736 2.799	-0.516 0.102	1.751 -0.800	-0.811 -0.051
86.9	2.561 2.915	-0.436 0.102	1.519 -0.606	-0.725 0.007
87.0	2.419 3.042	-0.368 0.092	1.354 -0.425	-0.631 0.053
87.1	2.305 3.174	-0.310 0.078	1.235 -0.261	-0.531 0.084
87.2	2.214 3.310	-0.261 0.062	1.145 -0.111	-0.432 0.097
87.3	2.142 3.447	-0.220 0.047	1.078 0.026	-0.339 0.093
87.4	2.087 3.586	-0.183 0.033	1.026 0.154	-0.258 0.077
87.5	2.047 3.726	-0.150 0.022	0.985 0.275	-0.191 0.054
87.6	2.021 3.869	-0.119 0.014	0.954 0.391	-0.138 0.031
87.7	2.009 4.013	-0.088 0.008	0.931 0.504	-0.096 0.013
87.8	2.012 4.161	-0.056 0.004	0.914 0.615	-0.059 0.003
87.9	2.030 4.313	-0.023 0.001	0.904 0.727	-0.023 -0.001
28π	2.051 4.413	-0.000 0.000	0.901 0.800	-0.000 -0.000
88.0	2.066 4.469	0.013 -0.000	0.900 0.841	0.013 0.002
88.1	2.122 4.630	0.053 -0.002	0.902 0.959	0.054 0.009
88.2	2.205 4.795	0.097 -0.003	0.911 1.083	0.099 0.019
88.3	2.318 4.961	0.146 -0.007	0.930 1.215	0.151 0.030
88.4	2.471 5.125	0.201 -0.013	0.961 1.359	0.209 0.042
88.5	2.671 5.277	0.262 -0.024	1.007 1.518	0.273 0.051
88.6	2.929 5.403	0.331 -0.042	1.076 1.696	0.345 0.059
88.7	3.249 5.477	0.407 -0.070	1.179 1.897	0.425 0.062
88.8	3.627 5.464	0.490 -0.111	1.335 2.126	0.512 0.060
88.9	4.032 5.323	0.577 -0.171	1.574 2.383	0.607 0.052
89.0	4.404 5.023	0.666 -0.253	1.949 2.651	0.711 0.035
89.1	4.658 4.579	0.748 -0.362	2.538 2.868	0.824 0.007
89.2	4.724 4.063	0.814 -0.501	3.409 2.852	0.945 -0.034
89.3	4.594 3.578	0.852 -0.668	4.425 2.260	1.075 -0.091
89.4	4.322 3.204	0.847 -0.853	4.968 0.961	1.214 -0.169
89.5	3.984 2.968	0.792 -1.040	4.556 -0.354	1.358 -0.273
57π/2	3.861 2.916	0.760 -1.102	4.257 -0.670	1.411 -0.317
89.6	3.642 2.857	0.687 -1.208	3.666 -1.022	1.506 -0.409
89.7	3.330 2.838	0.543 -1.339	2.861 -1.154	1.651 -0.583
89.8	3.061 2.881	0.379 -1.425	2.283 -1.046	1.783 -0.803
89.9	2.836 2.964	0.212 -1.465	1.889 -0.862	1.889 -1.073
90.0	2.650 3.071	0.058 -1.466	1.619 -0.667	1.948 -1.395
90.1	2.499 3.191	-0.077 -1.441	1.430 -0.482	1.936 -1.760
90.2	2.376 3.319	-0.190 -1.398	1.294 -0.312	1.830 -2.147
90.3	2.278 3.451	-0.283 -1.347	1.193 -0.157	1.615 -2.521
90.4	2.199 3.586	-0.358 -1.291	1.117 -0.016	1.293 -2.837
90.5	2.139 3.723	-0.418 -1.236	1.058 0.116	0.892 -3.056
90.6	2.094 3.862	-0.466 -1.183	1.013 0.239	0.456 -3.155
90.7	2.064 4.002	-0.505 -1.133	0.979 0.357	0.039 -3.137
90.8	2.047 4.145	-0.538 -1.087	0.952 0.471	-0.341 -3.026
90.9	2.045 4.290	-0.566 -1.044	0.933 0.583	-0.649 -2.853

	OMEGA = 15.0		OMEGA = 20.0	
k_0h	ADMITTANCE	RAD.EFF.LENGTH	ADMITTANCE	RAD.EFF.LENGTH
91.0	2.059 4.439	-0.591 -1.005	0.920 0.694	-0.888 -2.646
29π	2.092 4.602	-0.615 -0.967	0.913 0.814	-1.074 -2.414
91.1	2.089 4.592	-0.614 -0.969	0.913 0.807	-1.065 -2.427
91.2	2.139 4.749	-0.636 -0.935	0.913 0.923	-1.189 -2.212
91.3	2.213 4.910	-0.657 -0.903	0.920 1.045	-1.273 -2.008
91.4	2.316 5.073	-0.680 -0.872	0.935 1.173	-1.326 -1.819
91.5	2.455 5.233	-0.703 -0.841	0.960 1.312	-1.355 -1.645
91.6	2.637 5.385	-0.729 -0.809	1.000 1.465	-1.367 -1.487
91.7	2.871 5.515	-0.756 -0.775	1.060 1.634	-1.367 -1.342
91.8	3.164 5.602	-0.785 -0.738	1.148 1.825	-1.357 -1.210
91.9	3.513 5.615	-0.816 -0.695	1.281 2.041	-1.341 -1.089
92.0	3.897 5.518	-0.848 -0.645	1.483 2.286	-1.319 -0.976
92.1	4.269 5.277	-0.879 -0.586	1.797 2.550	-1.294 -0.871
92.2	4.553 4.893	-0.905 -0.515	2.289 2.795	-1.264 -0.772
92.3	4.679 4.415	-0.923 -0.433	3.040 2.898	-1.232 -0.678
92.4	4.617 3.933	-0.927 -0.340	4.028 2.577	-1.196 -0.587
92.5	4.399 3.533	-0.910 -0.241	4.839 1.540	-1.157 -0.500
92.6	4.093 3.259	-0.871 -0.143	4.805 0.138	-1.113 -0.416
59π/2	3.838 3.133	-0.826 -0.074	4.240 -0.648	-1.076 -0.352
92.7	3.762 3.109	-0.810 -0.055	4.034 -0.802	-1.065 -0.334
92.8	3.449 3.058	-0.731 0.014	3.169 -1.115	-1.010 -0.254
92.9	3.172 3.078	-0.644 0.062	2.506 -1.082	-0.948 -0.177
93.0	2.937 3.145	-0.556 0.088	2.046 -0.921	-0.878 -0.104
93.1	2.741 3.240	-0.475 0.097	1.733 -0.729	-0.798 -0.037
93.2	2.581 3.352	-0.404 0.094	1.514 -0.540	-0.709 0.019
93.3	2.450 3.475	-0.343 0.084	1.358 -0.365	-0.612 0.062
93.4	2.344 3.604	-0.291 0.070	1.244 -0.206	-0.511 0.089
93.5	2.259 3.736	-0.247 0.056	1.159 -0.059	-0.412 0.098
93.6	2.193 3.870	-0.208 0.042	1.093 0.076	-0.322 0.090
93.7	2.143 4.007	-0.174 0.030	1.043 0.202	-0.244 0.072
93.8	2.108 4.145	-0.142 0.021	1.004 0.322	-0.181 0.049
93.9	2.087 4.286	-0.112 0.013	0.975 0.437	-0.131 0.027
94.0	2.081 4.429	-0.082 0.008	0.953 0.550	-0.090 0.011
94.1	2.089 4.576	-0.050 0.004	0.937 0.661	-0.053 0.001
94.2	2.114 4.726	-0.017 0.001	0.928 0.774	-0.018 -0.001
30π	2.133 4.799	-0.000 0.000	0.926 0.828	-0.000 -0.000
94.3	2.158 4.879	0.019 -0.001	0.925 0.888	0.020 0.002
94.4	2.224 5.036	0.059 -0.003	0.929 1.007	0.061 0.010
94.5	2.317 5.195	0.103 -0.007	0.940 1.133	0.108 0.020
94.6	2.443 5.353	0.152 -0.012	0.962 1.268	0.160 0.032
94.7	2.609 5.503	0.207 -0.021	0.996 1.414	0.220 0.043
94.8	2.822 5.635	0.268 -0.036	1.047 1.576	0.286 0.052
94.9	3.090 5.732	0.335 -0.058	1.123 1.757	0.359 0.059
95.0	3.411 5.766	0.409 -0.092	1.237 1.962	0.440 0.061
95.1	3.772 5.706	0.488 -0.140	1.409 2.194	0.529 0.058
95.2	4.136 5.518	0.569 -0.208	1.672 2.450	0.626 0.048
95.3	4.440 5.192	0.648 -0.298	2.083 2.707	0.732 0.029
95.4	4.614 4.759	0.715 -0.415	2.720 2.883	0.847 -0.002
95.5	4.615 4.292	0.761 -0.558	3.623 2.770	0.971 -0.047
95.6	4.454 3.878	0.773 -0.721	4.571 2.035	1.103 -0.109
95.7	4.186 3.570	0.743 -0.892	4.918 0.701	1.244 -0.194
95.8	3.874 3.383	0.667 -1.054	4.380 -0.478	1.389 -0.307
61π/2	3.815 3.360	0.648 -1.082	4.223 -0.627	1.416 -0.331
95.9	3.564 3.299	0.551 -1.190	3.507 -1.017	1.535 -0.453
96.0	3.283 3.294	0.409 -1.287	2.759 -1.094	1.676 -0.640
96.1	3.039 3.342	0.259 -1.343	2.227 -0.973	1.800 -0.873
96.2	2.834 3.424	0.114 -1.361	1.861 -0.790	1.891 -1.157
96.3	2.664 3.527	-0.017 -1.351	1.609 -0.600	1.928 -1.489
96.4	2.525 3.644	-0.129 -1.321	1.430 -0.421	1.888 -1.857
96.5	2.411 3.768	-0.223 -1.279	1.300 -0.255	1.749 -2.236
96.6	2.320 3.897	-0.299 -1.232	1.204 -0.104	1.503 -2.588
96.7	2.248 4.029	-0.361 -1.184	1.131 0.035	1.162 -2.869
96.8	2.193 4.163	-0.411 -1.136	1.075 0.164	0.758 -3.048
96.9	2.153 4.299	-0.453 -1.091	1.031 0.286	0.335 -3.110
97.0	2.128 4.438	-0.487 -1.048	0.998 0.403	-0.064 -3.066
97.1	2.117 4.579	-0.516 -1.009	0.973 0.517	-0.411 -2.940
97.2	2.121 4.723	-0.542 -0.973	0.955 0.628	-0.692 -2.764
97.3	2.141 4.870	-0.565 -0.939	0.943 0.740	-0.909 -2.561
31π	2.174 5.005	-0.585 -0.911	0.938 0.842	-1.054 -2.374
97.4	2.179 5.021	-0.587 -0.908	0.937 0.854	-1.069 -2.352
97.5	2.238 5.174	-0.609 -0.878	0.938 0.971	-1.182 -2.147
97.6	2.322 5.330	-0.631 -0.850	0.947 1.094	-1.258 -1.953
97.7	2.436 5.484	-0.654 -0.822	0.964 1.224	-1.307 -1.773
97.8	2.587 5.633	-0.679 -0.794	0.993 1.366	-1.334 -1.608
97.9	2.781 5.766	-0.706 -0.764	1.037 1.521	-1.346 -1.456
98.0	3.025 5.870	-0.734 -0.732	1.103 1.693	-1.347 -1.318
98.1	3.321 5.921	-0.765 -0.695	1.201 1.888	-1.338 -1.190
98.2	3.658 5.891	-0.797 -0.651	1.347 2.108	-1.324 -1.072
98.3	4.009 5.749	-0.829 -0.600	1.569 2.354	-1.304 -0.962
98.4	4.322 5.478	-0.859 -0.538	1.914 2.612	-1.280 -0.859
98.5	4.533 5.093	-0.882 -0.465	2.449 2.831	-1.253 -0.761
98.6	4.590 4.652	-0.895 -0.381	3.244 2.863	-1.222 -0.668
98.7	4.487 4.233	-0.891 -0.288	4.216 2.412	-1.187 -0.578
98.8	4.263 3.900	-0.866 -0.192	4.874 1.275	-1.149 -0.490
98.9	3.976 3.678	-0.819 -0.102	4.663 -0.048	-1.106 -0.405
63π/2	3.794 3.597	-0.781 -0.054	4.207 -0.606	-1.077 -0.355
99.0	3.675 3.562	-0.753 -0.026	3.860 -0.842	-1.057 -0.322
99.1	3.392 3.529	-0.674 0.032	3.046 -1.072	-1.002 -0.241
99.2	3.142 3.557	-0.591 0.068	2.433 -1.013	-0.939 -0.163
99.3	2.929 3.624	-0.511 0.087	2.008 -0.849	-0.866 -0.090
99.4	2.750 3.717	-0.439 0.091	1.715 -0.661	-0.784 -0.024
99.5	2.602 3.827	-0.375 0.086	1.510 -0.478	-0.692 0.031
99.6	2.481 3.945	-0.320 0.076	1.362 -0.307	-0.593 0.071
99.7	2.383 4.071	-0.273 0.063	1.253 -0.151	-0.491 0.093
99.8	2.305 4.200	-0.233 0.051	1.171 -0.008	-0.394 0.098
99.9	2.245 4.331	-0.197 0.039	1.108 0.125	-0.306 0.087
100.0	2.200 4.465	-0.165 0.028	1.060 0.250	-0.231 0.067
100.1	2.170 4.602	-0.135 0.020	1.023 0.368	-0.171 0.044
100.2	2.155 4.741	-0.105 0.013	0.994 0.483	-0.123 0.023
100.3	2.154 4.882	-0.075 0.008	0.973 0.595	-0.083 0.008
100.4	2.169 5.027	-0.044 0.004	0.959 0.707	-0.047 0.000
100.5	2.202 5.174	-0.011 0.001	0.951 0.820	-0.011 -0.001
32π	2.215 5.220	-0.000 0.000	0.949 0.855	-0.000 -0.000
100.6	2.254 5.324	0.025 -0.002	0.949 0.935	0.027 0.003
100.7	2.329 5.476	0.065 -0.006	0.955 1.056	0.069 0.012
100.8	2.433 5.628	0.108 -0.011	0.968 1.183	0.117 0.022
100.9	2.570 5.774	0.157 -0.018	0.993 1.319	0.170 0.033

OMEGA = 15.0					OMEGA = 20.0				
k_0h	ADMITTANCE		RAD.EFF.LENGTH		k_0h	ADMITTANCE		RAD.EFF.LENGTH	
101.0	2.747	5.908	0.211	-0.030	101.0	1.030	..469	0.231	0.044
101.1	2.970	6.017	0.271	-0.049	101.1	1.087	1.633	0.298	0.053
101.2	3.241	6.082	0.336	-0.076	101.2	1.171	1.818	0.373	0.059
101.3	3.555	6.077	0.407	-0.116	101.3	1.296	2.027	0.455	0.061
101.4	3.890	5.974	0.481	-0.171	101.4	1.485	2.261	0.546	0.056
101.5	4.204	5.753	0.553	-0.245	101.5	1.774	2.515	0.646	0.044
101.6	4.440	5.417	0.619	-0.342	101.6	2.224	2.756	0.754	0.022
101.7	4.545	5.009	0.670	-0.463	101.7	2.906	2.884	0.871	-0.012
101.8	4.498	4.596	0.694	-0.604	101.8	3.827	2.664	0.997	-0.061
101.9	4.321	4.246	0.684	-0.757	101.9	4.681	1.799	1.132	-0.129
102.0	4.065	3.995	0.633	-0.909	102.0	4.838	0.466	1.273	-0.221
$65\pi/2$	3.773	3.845	0.542	-1.045	$65\pi/2$	4.192	-0.586	1.421	-0.345
102.1	3.778	3.846	0.544	-1.043	102.1	4.207	-0.573	1.419	-0.343
102.2	3.498	3.786	0.426	-1.148	102.2	3.362	-1.001	1.563	-0.500
102.3	3.244	3.791	0.294	-1.216	102.3	2.668	-1.033	1.698	-0.699
102.4	3.024	3.842	0.161	-1.249	102.4	2.176	-0.902	1.812	-0.946
102.5	2.837	3.924	0.037	-1.254	102.5	1.836	-0.721	1.887	-1.242
102.6	2.681	4.024	-0.073	-1.237	102.6	1.599	-0.536	1.901	-1.582
102.7	2.552	4.137	-0.166	-1.206	102.7	1.430	-0.361	1.832	-1.951
102.8	2.448	4.258	-0.243	-1.168	102.8	1.307	-0.200	1.662	-2.317
102.9	2.364	4.384	-0.306	-1.126	102.9	1.215	-0.053	1.391	-2.644
103.0	2.298	4.513	-0.358	-1.084	103.0	1.144	0.084	1.035	-2.890
103.1	2.249	4.644	-0.401	-1.043	103.1	1.090	0.212	0.632	-3.030
103.2	2.214	4.779	-0.436	-1.005	103.2	1.048	0.333	0.224	-3.060
103.3	2.194	4.915	-0.467	-0.969	103.3	1.016	0.449	-0.151	-2.994
103.4	2.189	5.054	-0.493	-0.935	103.4	0.992	0.562	-0.471	-2.859
103.5	2.199	5.196	-0.517	-0.905	103.5	0.975	0.674	-0.729	-2.681
103.6	2.226	5.341	-0.539	-0.876	103.6	0.965	0.786	-0.926	-2.483
33π	2.257	5.447	-0.555	-0.856	33π	0.961	0.868	-1.036	-2.337
103.7	2.272	5.488	-0.561	-0.849	103.7	0.961	0.900	-1.072	-2.282
103.8	2.339	5.637	-0.583	-0.824	103.8	0.963	1.018	-1.175	-2.086
103.9	2.433	5.785	-0.605	-0.799	103.9	0.974	1.143	-1.245	-1.902
104.0	2.558	5.929	-0.630	-0.773	104.0	0.994	1.275	-1.290	-1.730
104.1	2.719	6.063	-0.655	-0.747	104.1	1.026	1.419	-1.316	-1.572
104.2	2.923	6.175	-0.683	-0.719	104.2	1.075	1.577	-1.327	-1.427
104.3	3.171	6.251	-0.713	-0.686	104.3	1.147	1.753	-1.328	-1.294
104.4	3.462	6.266	-0.745	-0.649	104.4	1.255	1.951	-1.321	-1.170
104.5	3.779	6.197	-0.777	-0.604	104.5	1.415	2.174	-1.308	-1.056
104.6	4.088	6.020	-0.808	-0.551	104.6	1.659	2.420	-1.290	-0.948
104.7	4.341	5.733	-0.835	-0.487	104.7	2.036	2.669	-1.268	-0.847
104.8	4.485	5.362	-0.854	-0.412	104.8	2.615	2.856	-1.242	-0.750
104.9	4.489	4.965	-0.860	-0.327	104.9	3.445	2.807	-1.213	-0.657
105.0	4.360	4.607	-0.847	-0.237	105.0	4.378	2.225	-1.179	-0.567
105.1	4.139	4.331	-0.814	-0.148	105.1	4.870	1.021	-1.142	-0.479
105.2	3.873	4.153	-0.762	-0.068	105.2	4.512	-0.203	-1.099	-0.393
$67\pi/2$	3.753	4.105	-0.734	-0.038	$67\pi/2$	4.178	-0.566	-1.079	-0.357
105.3	3.600	4.065	-0.694	-0.003	105.3	3.699	-0.863	-1.050	-0.309
105.4	3.345	4.047	-0.619	0.043	105.4	2.934	-1.025	-0.994	-0.227
105.5	3.119	4.080	-0.542	0.071	105.5	2.367	-0.945	-0.929	-0.149
105.6	2.925	4.148	-0.470	0.083	105.6	1.973	-0.779	-0.854	-0.076
105.7	2.761	4.239	-0.405	0.084	105.7	1.700	-0.595	-0.769	-0.011
105.8	2.626	4.345	-0.348	0.078	105.8	1.506	-0.417	-0.674	0.041
105.9	2.514	4.461	-0.299	0.069	105.9	1.366	-0.251	-0.573	0.078
106.0	2.424	4.582	-0.257	0.058	106.0	1.262	-0.099	-0.472	0.096
106.1	2.353	4.708	-0.220	0.046	106.1	1.183	0.042	-0.375	0.096
106.2	2.298	4.838	-0.186	0.036	106.2	1.123	0.173	-0.290	0.083
106.3	2.259	4.969	-0.156	0.027	106.3	1.076	0.296	-0.219	0.062
106.4	2.235	5.103	-0.126	0.019	106.4	1.040	0.414	-0.162	0.039
106.5	2.225	5.240	-0.098	0.013	106.5	1.013	0.528	-0.116	0.019
106.6	2.230	5.379	-0.068	0.008	106.6	0.993	0.640	-0.077	0.006
106.7	2.252	5.521	-0.038	0.004	106.7	0.980	0.752	-0.042	-0.000
106.8	2.291	5.665	-0.005	0.001	106.8	0.973	0.865	-0.005	-0.000
34π	2.299	5.686	-0.000	0.000	34π	0.972	0.882	-0.000	-0.000
106.9	2.352	5.811	0.031	-0.003	106.9	0.973	0.982	0.034	0.005
107.0	2.436	5.957	0.070	-0.008	107.0	0.980	1.104	0.077	0.013
107.1	2.550	6.099	0.113	-0.015	107.1	0.996	1.232	0.125	0.024
107.2	2.697	6.232	0.161	-0.026	107.2	1.023	1.371	0.180	0.035
107.3	2.883	6.347	0.213	-0.041	107.3	1.065	1.523	0.242	0.045
107.4	3.111	6.430	0.271	-0.063	107.4	1.128	1.691	0.311	0.054
107.5	3.379	6.463	0.334	-0.095	107.5	1.220	1.879	0.387	0.059
107.6	3.677	6.421	0.400	-0.140	107.6	1.357	2.091	0.471	0.059
107.7	3.977	6.284	0.466	-0.201	107.7	1.564	2.327	0.564	0.053
107.8	4.238	6.042	0.528	-0.281	107.8	1.881	2.576	0.665	0.039
107.9	4.412	5.711	0.580	-0.382	107.9	2.369	2.797	0.776	0.015
108.0	4.461	5.338	0.613	-0.502	108.0	3.094	2.867	0.895	-0.022
108.1	4.380	4.979	0.617	-0.636	108.1	4.016	2.533	1.023	-0.076
108.2	4.198	4.686	0.588	-0.774	108.2	4.753	1.557	1.160	-0.150
108.3	3.956	4.482	0.524	-0.902	108.3	4.733	0.260	1.302	-0.250
$69\pi/2$	3.734	4.378	0.446	-0.995	$69\pi/2$	4.164	-0.547	1.426	-0.358
108.4	3.695	4.366	0.430	-1.009	108.4	4.040	-0.643	1.447	-0.380
108.5	3.442	4.324	0.318	-1.087	108.5	3.230	-0.976	1.589	-0.548
108.6	3.213	4.338	0.199	-1.133	108.6	2.586	-0.972	1.718	-0.760
108.7	3.013	4.392	0.084	-1.151	108.7	2.130	-0.833	1.820	-1.020
108.8	2.843	4.472	-0.021	-1.146	108.8	1.813	-0.654	1.878	-1.327
108.9	2.700	4.570	-0.112	-1.127	108.9	1.591	-0.474	1.868	-1.673
109.0	2.582	4.680	-0.189	-1.098	109.0	1.431	-0.304	1.770	-2.038
109.1	2.486	4.797	-0.253	-1.064	109.1	1.313	-0.147	1.572	-2.390
109.2	2.409	4.920	-0.306	-1.028	109.2	1.225	-0.002	1.278	-2.689
109.3	2.350	5.046	-0.350	-0.992	109.3	1.157	0.133	0.913	-2.900
109.4	2.306	5.175	-0.387	-0.957	109.4	1.105	0.259	0.514	-3.006
109.5	2.277	5.307	-0.418	-0.925	109.5	1.065	0.378	0.124	-3.008
109.6	2.263	5.441	-0.445	-0.894	109.6	1.034	0.494	-0.228	-2.924
109.7	2.263	5.578	-0.470	-0.866	109.7	1.011	0.606	-0.524	-2.781
109.8	2.280	5.717	-0.492	-0.840	109.8	0.995	0.718	-0.760	-2.604
109.9	2.313	5.859	-0.514	-0.815	109.9	0.986	0.831	-0.941	-2.411
35π	2.340	5.938	-0.526	-0.802	35π	0.984	0.895	-1.020	-2.303
110.0	2.367	6.001	-0.535	-0.792	110.0	0.983	0.946	-1.073	-2.217
110.1	2.443	6.144	-0.558	-0.770	110.1	0.988	1.065	-1.168	-2.030
110.2	2.546	6.284	-0.581	-0.747	110.2	1.000	1.191	-1.233	-1.854
110.3	2.680	6.416	-0.605	-0.724	110.3	1.023	1.326	-1.274	-1.690
110.4	2.849	6.532	-0.632	-0.699	110.4	1.059	1.472	-1.299	-1.539
110.5	3.058	6.622	-0.661	-0.671	110.5	1.112	1.633	-1.310	-1.400
110.6	3.306	6.668	-0.691	-0.639	110.6	1.192	1.812	-1.311	-1.271
110.7	3.584	6.650	-0.723	-0.601	110.7	1.310	2.013	-1.305	-1.151
110.8	3.872	6.547	-0.754	-0.555	110.8	1.486	2.239	-1.294	-1.040
110.9	4.135	6.347	-0.783	-0.499	110.9	1.753	2.483	-1.277	-0.934

OMEGA = 15.0					OMEGA = 20.0				
k_0h	ADMITTANCE		RAD.EFF.LENGTH		k_0h	ADMITTANCE		RAD.EFF.LENGTH	
111.0	4.330	6.057	-0.806	-0.434	111.0	2.164	2.720	-1.257	-0.835
111.1	4.417	5.712	-0.819	-0.358	111.1	2.785	2.867	-1.232	-0.739
111.2	4.382	5.362	-0.817	-0.275	111.2	3.640	2.729	-1.204	-0.646
111.3	4.240	5.059	-0.797	-0.190	111.3	4.511	2.021	-1.172	-0.556
111.4	4.027	4.833	-0.758	-0.110	111.4	4.832	0.784	-1.135	-0.468
111.5	3.782	4.691	-0.703	-0.041	111.5	4.357	-0.327	-1.092	-0.381
$71\pi/2$	3.716	4.667	-0.686	-0.025	$71\pi/2$	4.151	-0.528	-1.080	-0.358
111.6	3.536	4.625	-0.637	0.013	111.6	3.550	-0.869	-1.043	-0.296
111.7	3.306	4.619	-0.566	0.049	111.7	2.833	-0.975	-0.985	-0.213
111.8	3.102	4.657	-0.496	0.070	111.8	2.308	-0.879	-0.918	-0.134
111.9	2.925	4.725	-0.432	0.078	111.9	1.942	-0.712	-0.841	-0.062
112.0	2.776	4.814	-0.374	0.077	112.0	1.685	-0.532	-0.753	0.002
112.1	2.652	4.917	-0.323	0.071	112.1	1.503	-0.358	-0.656	0.051
112.2	2.550	5.030	-0.279	0.063	112.2	1.369	-0.196	-0.554	0.084
112.3	2.467	5.148	-0.241	0.053	112.3	1.270	-0.047	-0.452	0.098
112.4	2.403	5.272	-0.207	0.043	112.4	1.194	0.091	-0.358	0.094
112.5	2.354	5.398	-0.175	0.034	112.5	1.136	0.220	-0.276	0.079
112.6	2.320	5.527	-0.146	0.026	112.6	1.091	0.342	-0.208	0.057
112.7	2.302	5.659	-0.118	0.019	112.7	1.057	0.459	-0.154	0.034
112.8	2.297	5.794	-0.090	0.013	112.8	1.031	0.572	-0.110	0.016
112.9	2.309	5.930	-0.061	0.009	112.9	1.012	0.685	-0.072	0.004
113.0	2.337	6.069	-0.031	0.004	113.0	1.000	0.797	-0.036	-0.001
36π	2.382	6.205	-0.000	0.000	36π	0.995	0.908	-0.000	-0.000
113.1	2.384	6.209	0.001	-0.000	113.1	0.995	0.911	0.001	0.000
113.2	2.452	6.348	0.036	-0.005	113.2	0.996	1.028	0.041	0.006
113.3	2.545	6.486	0.074	-0.012	113.3	1.005	1.151	0.085	0.015
113.4	2.667	6.616	0.116	-0.021	113.4	1.024	1.282	0.134	0.025
113.5	2.822	6.734	0.163	-0.033	113.5	1.054	1.423	0.191	0.036
113.6	3.013	6.828	0.214	-0.052	113.6	1.100	1.577	0.253	0.047
113.7	3.241	6.885	0.269	-0.078	113.7	1.169	1.748	0.324	0.054
113.8	3.500	6.887	0.327	-0.115	113.8	1.270	1.940	0.401	0.059
113.9	3.774	6.813	0.387	-0.165	113.9	1.420	2.155	0.487	0.058
114.0	4.035	6.651	0.444	-0.230	114.0	1.647	2.391	0.582	0.050
114.1	4.243	6.402	0.495	-0.313	114.1	1.993	2.634	0.685	0.034
114.2	4.361	6.088	0.532	-0.414	114.2	2.520	2.828	0.798	0.007

TABLE 2.9

ADMITTANCE AND EFFECTIVE LENGTH OF LONG DIPOLE ANTENNAS IN AIR

OMEGA= 20.0

k_0h	ADMITTANCE		RAD.EFF.LENGTH	
100.0	1.060	0.250	-0.231	0.067
100.1	1.023	0.368	-0.171	0.044
100.2	0.994	0.483	-0.123	0.023
100.3	0.973	0.595	-0.083	0.008
100.4	0.959	0.707	-0.047	0.000
100.5	0.951	0.820	-0.011	-0.001
32π	0.949	0.855	-0.000	-0.000
100.6	0.949	0.935	0.027	0.003
100.7	0.955	1.056	0.069	0.012
100.8	0.968	1.183	0.117	0.022
100.9	0.993	1.319	0.170	0.033
101.0	1.030	1.469	0.231	0.044
101.1	1.087	1.633	0.298	0.053
101.2	1.171	1.818	0.373	0.059
101.3	1.296	2.027	0.455	0.061
101.4	1.485	2.261	0.546	0.056
101.5	1.774	2.515	0.646	0.044
101.6	2.224	2.756	0.754	0.022
101.7	2.906	2.884	0.871	-0.012
101.8	3.827	2.664	0.997	-0.061
101.9	4.681	1.799	1.132	-0.129
102.0	4.838	0.466	1.273	-0.221
65π/2	4.192	-0.586	1.421	-0.345
102.1	4.207	-0.573	1.419	-0.343
102.2	3.362	-1.001	1.563	-0.500
102.3	2.668	-1.033	1.698	-0.699
102.4	2.176	-0.902	1.812	-0.946
102.5	1.836	-0.721	1.887	-1.242
102.6	1.599	-0.536	1.901	-1.582
102.7	1.430	-0.361	1.832	-1.951
102.8	1.307	-0.200	1.662	-2.317
102.9	1.215	-0.053	1.391	-2.644
103.0	1.144	0.084	1.035	-2.890
103.1	1.090	0.212	0.632	-3.030
103.2	1.048	0.333	0.224	-3.060
103.3	1.016	0.449	-0.151	-2.994
103.4	0.992	0.562	-0.471	-2.859
103.5	0.975	0.674	-0.729	-2.681
103.6	0.965	0.786	-0.926	-2.483
33π	0.961	0.868	-1.036	-2.337
103.7	0.961	0.900	-1.072	-2.282
103.8	0.963	1.018	-1.175	-2.086
103.9	0.974	1.143	-1.245	-1.902

k_0h	ADMITTANCE		RAD.EFF.LENGTH	
104.0	0.994	1.275	-1.290	-1.730
104.1	1.026	1.419	-1.316	-1.572
104.2	1.075	1.577	-1.327	-1.427
104.3	1.147	1.753	-1.328	-1.294
104.4	1.255	1.951	-1.321	-1.170
104.5	1.415	2.174	-1.308	-1.056
104.6	1.659	2.420	-1.290	-0.948
104.7	2.036	2.669	-1.268	-0.847
104.8	2.615	2.856	-1.242	-0.750
104.9	3.445	2.807	-1.213	-0.657
105.0	4.378	2.225	-1.179	-0.567
105.1	4.870	1.021	-1.142	-0.479
105.2	4.512	-0.203	-1.099	-0.393
67π/2	4.178	-0.566	-1.079	-0.357
105.3	3.699	-0.863	-1.050	-0.309
105.4	2.934	-1.025	-0.994	-0.227
105.5	2.367	-0.945	-0.929	-0.149
105.6	1.973	-0.779	-0.854	-0.076
105.7	1.700	-0.595	-0.769	-0.011
105.8	1.506	-0.417	-0.674	0.041
105.9	1.366	-0.251	-0.573	0.078
106.0	1.262	-0.099	-0.472	0.096
106.1	1.183	0.042	-0.375	0.096
106.2	1.123	0.173	-0.290	0.083
106.3	1.076	0.296	-0.219	0.062
106.4	1.040	0.414	-0.162	0.039
106.5	1.013	0.528	-0.116	0.019
106.6	0.993	0.640	-0.077	0.006
106.7	0.980	0.752	-0.042	-0.000
106.8	0.973	0.865	-0.005	-0.000
34π	0.972	0.882	-0.000	-0.000
106.9	0.973	0.982	0.034	0.005
107.0	0.980	1.104	0.077	0.013
107.1	0.996	1.232	0.125	0.024
107.2	1.023	1.371	0.180	0.035
107.3	1.065	1.523	0.242	0.045
107.4	1.128	1.691	0.311	0.054
107.5	1.220	1.879	0.387	0.059
107.6	1.357	2.091	0.471	0.059
107.7	1.564	2.327	0.564	0.053
107.8	1.881	2.576	0.665	0.039
107.9	2.369	2.797	0.776	0.015

k_0h	ADMITTANCE		RAD.EFF.LENGTH	
108.0	3.094	2.867	0.895	-0.022
108.1	4.016	2.533	1.023	-0.076
108.2	4.753	1.557	1.160	-0.150
108.3	4.733	0.260	1.302	-0.250
69π/2	4.164	-0.547	1.426	-0.358
108.4	4.040	-0.643	1.447	-0.380
108.5	3.230	-0.976	1.589	-0.548
108.6	2.586	-0.972	1.718	-0.760
108.7	2.130	-0.833	1.820	-1.020
108.8	1.813	-0.654	1.878	-1.327
108.9	1.591	-0.474	1.868	-1.673
109.0	1.431	-0.304	1.770	-2.038
109.1	1.313	-0.147	1.572	-2.390
109.2	1.225	-0.002	1.278	-2.689
109.3	1.157	0.133	0.913	-2.900
109.4	1.105	0.259	0.514	-3.006
109.5	1.065	0.378	0.124	-3.008
109.6	1.034	0.494	-0.228	-2.924
109.7	1.011	0.606	-0.524	-2.781
109.8	0.995	0.718	-0.760	-2.604
109.9	0.986	0.831	-0.941	-2.411
35π	0.984	0.895	-1.020	-2.303
110.0	0.983	0.946	-1.073	-2.217
110.1	0.988	1.065	-1.168	-2.030
110.2	1.000	1.191	-1.233	-1.854
110.3	1.023	1.326	-1.274	-1.690
110.4	1.059	1.472	-1.299	-1.539
110.5	1.112	1.633	-1.310	-1.400
110.6	1.192	1.812	-1.311	-1.271
110.7	1.310	2.013	-1.305	-1.151
110.8	1.486	2.239	-1.294	-1.040
110.9	1.753	2.483	-1.277	-0.934
111.0	2.164	2.720	-1.257	-0.835
111.1	2.785	2.867	-1.232	-0.739
111.2	3.640	2.729	-1.204	-0.646
111.3	4.511	2.021	-1.172	-0.556
111.4	4.832	0.784	-1.135	-0.468
111.5	4.357	-0.327	-1.092	-0.381
71π/2	4.151	-0.528	-1.080	-0.358
111.6	3.550	-0.869	-1.043	-0.296
111.7	2.833	-0.975	-0.985	-0.213
111.8	2.308	-0.879	-0.918	-0.134
111.9	1.942	-0.712	-0.841	-0.062
112.0	1.685	-0.532	-0.753	0.002
112.1	1.503	-0.358	-0.656	0.051
112.2	1.369	-0.196	-0.554	0.084
112.3	1.270	-0.047	-0.452	0.098
112.4	1.194	0.091	-0.358	0.094
112.5	1.136	0.220	-0.276	0.079
112.6	1.091	0.342	-0.208	0.057
112.7	1.057	0.459	-0.154	0.034
112.8	1.031	0.572	-0.110	0.016
112.9	1.012	0.685	-0.072	0.004

k_0h	ADMITTANCE		RAD.EFF.LENGTH	
113.0	1.000	0.797	-0.036	-0.001
36π	0.995	0.908	-0.000	-0.000
113.1	0.995	0.911	0.001	0.000
113.2	0.996	1.028	0.041	0.006
113.3	1.005	1.151	0.085	0.015
113.4	1.024	1.282	0.134	0.025
113.5	1.054	1.423	0.191	0.036
113.6	1.100	1.577	0.253	0.047
113.7	1.169	1.748	0.324	0.054
113.8	1.270	1.940	0.401	0.059
113.9	1.420	2.155	0.487	0.058
114.0	1.647	2.391	0.582	0.050
114.1	1.993	2.634	0.685	0.034
114.2	2.520	2.828	0.798	0.007
114.3	3.282	2.832	0.920	-0.034
114.4	4.187	2.381	1.050	-0.092
114.5	4.789	1.317	1.188	-0.173
114.6	4.612	0.083	1.331	-0.280
73π/2	4.138	-0.509	1.430	-0.371
114.7	3.881	-0.690	1.475	-0.420
114.8	3.110	-0.944	1.613	-0.599
114.9	2.511	-0.911	1.735	-0.823
115.0	2.088	-0.766	1.824	-1.095
115.1	1.792	-0.590	1.863	-1.412
115.2	1.583	-0.414	1.828	-1.762
115.3	1.431	-0.248	1.701	-2.120
115.4	1.319	-0.094	1.477	-2.454
115.5	1.234	0.048	1.165	-2.724
115.6	1.169	0.180	0.795	-2.902
115.7	1.119	0.305	0.405	-2.976
115.8	1.081	0.423	0.032	-2.954
115.9	1.051	0.538	-0.297	-2.856
116.0	1.029	0.651	-0.571	-2.707
116.1	1.015	0.763	-0.788	-2.531
116.2	1.007	0.876	-0.953	-2.344
37π	1.005	0.920	-1.005	-2.271
116.3	1.005	0.992	-1.075	-2.157
116.4	1.011	1.112	-1.161	-1.978
116.5	1.026	1.240	-1.221	-1.810
116.6	1.052	1.376	-1.260	-1.653
116.7	1.091	1.525	-1.283	-1.508
116.8	1.150	1.688	-1.294	-1.374
116.9	1.238	1.871	-1.296	-1.249
117.0	1.366	2.076	-1.291	-1.133
117.1	1.559	2.303	-1.281	-1.024
117.2	1.851	2.544	-1.265	-0.921
117.3	2.297	2.764	-1.246	-0.822
117.4	2.959	2.865	-1.223	-0.727
117.5	3.827	2.629	-1.196	-0.635
117.6	4.613	1.806	-1.165	-0.544
117.7	4.766	0.567	-1.128	-0.456
75π/2	4.126	-0.491	-1.081	-0.360
117.8	4.202	-0.424	-1.085	-0.368
117.9	3.412	-0.864	-1.035	-0.282

k_0h	ADMITTANCE		RAD.EFF.LENGTH	
118.0	2.741	-0.924	-0.976	-0.199
118.1	2.254	-0.814	-0.907	-0.120
118.2	1.913	-0.647	-0.827	-0.048
118.3	1.672	-0.470	-0.736	0.014
118.4	1.499	-0.301	-0.637	0.060
118.5	1.372	-0.143	-0.534	0.089
118.6	1.277	0.003	-0.433	0.098
118.7	1.205	0.139	-0.341	0.092
118.8	1.149	0.266	-0.262	0.074
118.9	1.106	0.387	-0.198	0.052
119.0	1.073	0.503	-0.146	0.030
119.1	1.048	0.616	-0.103	0.013
119.2	1.031	0.729	-0.066	0.003
119.3	1.020	0.841	-0.030	-0.001
38π	1.016	0.933	-0.000	-0.000
119.4	1.016	0.956	0.007	0.001
119.5	1.019	1.074	0.048	0.007
119.6	1.030	1.199	0.093	0.016
119.7	1.051	1.331	0.144	0.027
119.8	1.085	1.474	0.201	0.038
119.9	1.136	1.631	0.265	0.048
120.0	1.211	1.806	0.337	0.055
120.1	1.321	2.001	0.416	0.058
120.2	1.486	2.218	0.504	0.056
120.3	1.733	2.453	0.600	0.047
120.4	2.109	2.686	0.706	0.029
120.5	2.676	2.849	0.821	-0.001
120.6	3.469	2.778	0.944	-0.046
120.7	4.336	2.209	1.077	-0.110
120.8	4.791	1.085	1.216	-0.197
120.9	4.479	-0.066	1.359	-0.312
77π/2	4.115	-0.473	1.433	-0.384
121.0	3.731	-0.720	1.502	-0.462
121.1	3.000	-0.907	1.636	-0.653
121.2	2.443	-0.850	1.748	-0.889
121.3	2.049	-0.701	1.824	-1.172
121.4	1.773	-0.527	1.842	-1.497
121.5	1.575	-0.355	1.781	-1.847
121.6	1.431	-0.193	1.628	-2.196
121.7	1.324	-0.043	1.381	-2.509
121.8	1.243	0.097	1.055	-2.750
121.9	1.181	0.227	0.683	-2.896
122.0	1.133	0.350	0.304	-2.941
122.1	1.096	0.468	-0.051	-2.898
122.2	1.068	0.582	-0.359	-2.789
122.3	1.047	0.694	-0.612	-2.637
122.4	1.034	0.807	-0.811	-2.462
122.5	1.027	0.920	-0.963	-2.281
39π	1.026	0.946	-0.991	-2.241
122.6	1.027	1.037	-1.075	-2.101
122.7	1.035	1.159	-1.155	-1.929
122.8	1.052	1.288	-1.211	-1.768
122.9	1.080	1.426	-1.247	-1.618

k_0h	ADMITTANCE		RAD.EFF.LENGTH	
123.0	1.124	1.577	-1.269	-1.478
123.1	1.189	1.744	-1.280	-1.349
123.2	1.284	1.930	-1.282	-1.228
123.3	1.425	2.137	-1.278	-1.115
123.4	1.635	2.365	-1.269	-1.008
123.5	1.953	2.601	-1.255	-0.907
123.6	2.435	2.801	-1.237	-0.809
123.7	3.135	2.847	-1.215	-0.715
123.8	4.001	2.507	-1.189	-0.623
123.9	4.682	1.586	-1.158	-0.532
124.0	4.677	0.374	-1.121	-0.443
79π/2	4.103	-0.455	-1.082	-0.361
124.1	4.049	-0.499	-1.078	-0.355
124.2	3.285	-0.849	-1.027	-0.268
124.3	2.658	-0.870	-0.966	-0.184
124.4	2.205	-0.750	-0.895	-0.105
124.5	1.886	-0.584	-0.812	-0.034
124.6	1.660	-0.411	-0.718	0.025
124.7	1.497	-0.245	-0.617	0.068
124.8	1.376	-0.090	-0.514	0.093
124.9	1.285	0.053	-0.414	0.098
125.0	1.215	0.186	-0.325	0.089
125.1	1.161	0.312	-0.249	0.069
125.2	1.120	0.432	-0.188	0.047
125.3	1.088	0.547	-0.138	0.026
125.4	1.065	0.660	-0.097	0.011
125.5	1.049	0.772	-0.060	0.002
125.6	1.039	0.885	-0.024	-0.001
40π	1.037	0.958	-0.000	-0.000
125.7	1.036	1.000	0.014	0.002
125.8	1.041	1.120	0.055	0.008
125.9	1.054	1.246	0.101	0.018
126.0	1.078	1.380	0.153	0.029
126.1	1.115	1.526	0.212	0.039
126.2	1.171	1.686	0.277	0.049
126.3	1.253	1.863	0.350	0.055
126.4	1.374	2.061	0.431	0.058
126.5	1.553	2.280	0.521	0.054
126.6	1.823	2.513	0.619	0.043
126.7	2.231	2.733	0.727	0.022
126.8	2.837	2.858	0.844	-0.011
126.9	3.650	2.703	0.969	-0.060
127.0	4.459	2.022	1.103	-0.129
127.1	4.762	0.866	1.244	-0.222
127.2	4.341	-0.190	1.387	-0.346
81π/2	4.093	-0.438	1.436	-0.397
127.3	3.590	-0.735	1.527	-0.506
127.4	2.900	-0.865	1.656	-0.708
127.5	2.382	-0.790	1.758	-0.956
127.6	2.015	-0.638	1.818	-1.250
127.7	1.755	-0.467	1.815	-1.580
127.8	1.569	-0.298	1.729	-1.929
127.9	1.432	-0.140	1.551	-2.266

k_0h	ADMITTANCE		RAD.EFF.LENGTH	
128.0	1.330	0.008	[illegible]	[illegible]
128.1	1.252	0.145	0.947	-2.767
128.2	1.192	0.273	0.577	-2.883
128.3	1.146	0.395	0.210	-2.902
128.4	1.110	0.512	-0.126	-2.842
128.5	1.083	0.626	-0.414	-2.724
128.6	1.064	0.738	-0.648	-2.570
128.7	1.052	0.850	-0.832	-2.398
41π	1.047	0.971	-0.978	-2.212
128.8	1.047	0.964	-0.972	-2.222
128.9	1.048	1.082	-1.075	-2.049
129.0	1.058	1.205	-1.149	-1.884
129.1	1.078	1.336	-1.201	-1.729
129.2	1.109	1.477	-1.235	-1.584
129.3	1.157	1.630	-1.256	-1.450
129.4	1.228	1.800	-1.267	-1.325
129.5	1.332	1.988	-1.269	-1.208
129.6	1.485	2.198	-1.266	-1.097
129.7	1.714	2.426	-1.258	-0.993
129.8	2.060	2.654	-1.245	-0.893
129.9	2.578	2.829	-1.228	-0.797
130.0	3.310	2.812	-1.207	-0.703
130.1	4.159	2.366	-1.182	-0.611
130.2	4.719	1.366	-1.151	-0.520
130.3	4.572	0.204	-1.115	-0.430
83π/2	4.082	-0.421	-1.082	-0.362
130.4	3.902	-0.553	-1.071	-0.341
130.5	3.168	-0.826	-1.019	-0.253
130.6	2.582	-0.816	-0.956	-0.169
130.7	2.160	-0.688	-0.882	-0.090
130.8	1.862	-0.522	-0.796	-0.020
130.9	1.649	-0.353	-0.700	0.036
131.0	1.494	-0.190	-0.598	0.075
131.1	1.379	-0.039	-0.494	0.096
131.2	1.292	0.101	-0.396	0.097
131.3	1.225	0.233	-0.310	0.085
131.4	1.173	0.357	-0.237	0.065
131.5	1.134	0.476	-0.179	0.042
131.6	1.103	0.591	-0.131	0.022
131.7	1.081	0.703	-0.091	0.008
131.8	1.066	0.816	-0.054	0.001
131.9	1.058	0.929	-0.018	-0.001
42π	1.056	0.983	-0.000	-0.000
132.0	1.057	1.045	0.021	0.002
132.1	1.063	1.166	0.063	0.010
132.2	1.079	1.293	0.110	0.019
132.3	1.105	1.429	0.163	0.030
132.4	1.146	1.577	0.222	0.041
132.5	1.208	1.740	0.289	0.049
132.6	1.297	1.920	0.364	0.055
132.7	1.429	2.121	0.447	0.057
132.8	1.624	2.341	0.538	0.052
132.9	1.917	2.571	0.638	0.039

k_0h	ADMITTANCE		RAD.EFF.LENGTH	
[illegible]	[illegible]	1.776	0.748	0.016
133.1	3.000	2.853	0.867	-0.021
133.2	3.823	2.609	0.995	-0.074
133.3	4.554	1.825	1.130	-0.149
133.4	4.708	0.663	1.272	-0.249
133.5	4.200	-0.289	1.414	-0.382
85π/2	4.072	-0.404	1.439	-0.409
133.6	3.458	-0.738	1.551	-0.552
133.7	2.808	-0.821	1.673	-0.766
133.8	2.326	-0.731	1.765	-1.025
133.9	1.983	-0.576	1.808	-1.328
134.0	1.739	-0.408	1.783	-1.662
134.1	1.563	-0.243	1.672	-2.006
134.2	1.433	-0.088	1.469	-2.328
134.3	1.335	0.057	1.184	-2.594
134.4	1.260	0.192	0.841	-2.776
134.5	1.203	0.318	0.476	-2.864
134.6	1.158	0.439	0.123	-2.861
134.7	1.124	0.555	-0.194	-2.786
134.8	1.099	0.669	-0.463	-2.661
134.9	1.081	0.781	-0.680	-2.506
135.0	1.070	0.894	-0.850	-2.337
43π	1.066	0.995	-0.966	-2.186
135.1	1.066	1.009	-0.979	-2.166
135.2	1.070	1.127	-1.075	-1.999
135.3	1.081	1.252	-1.143	-1.841
135.4	1.103	1.384	-1.192	-1.692
135.5	1.138	1.527	-1.224	-1.552
135.6	1.191	1.683	-1.244	-1.423
135.7	1.268	1.855	-1.254	-1.302
135.8	1.381	2.047	-1.258	-1.188
135.9	1.548	2.259	-1.255	-1.080
136.0	1.797	2.485	-1.248	-0.978
136.1	2.172	2.703	-1.236	-0.879
136.2	2.727	2.846	-1.220	-0.784
136.3	3.484	2.759	-1.200	-0.690
136.4	4.297	2.207	-1.175	-0.598
136.5	4.726	1.151	-1.145	-0.507
136.6	4.454	0.059	-1.108	-0.416
87π/2	4.062	-0.388	-1.083	-0.363
136.7	3.761	-0.589	-1.064	-0.326
136.8	3.060	-0.796	-1.010	-0.238
136.9	2.513	-0.761	-0.945	-0.153
137.0	2.120	-0.627	-0.868	-0.075
137.1	1.840	-0.463	-0.780	-0.007
137.2	1.639	-0.296	-0.681	0.046
137.3	1.492	-0.137	-0.578	0.082
137.4	1.382	0.011	-0.474	0.098
137.5	1.299	0.149	-0.379	0.096
137.6	1.235	0.279	-0.295	0.081
137.7	1.185	0.402	-0.226	0.060
137.8	1.147	0.519	-0.170	0.038
137.9	1.118	0.634	-0.124	0.019

OMEGA = 20.0

k_0h	ADMITTANCE		RAD.EFF.LENGTH	
138.0	1.097	0.746	-0.084	0.006
138.1	1.083	0.859	-0.048	-0.000
138.2	1.077	0.972	-0.011	-0.001
44π	1.076	1.007	-0.000	-0.000
138.3	1.077	1.089	0.028	0.003
138.4	1.085	1.211	0.070	0.011
138.5	1.103	1.340	0.119	0.021
138.6	1.133	1.478	0.173	0.032
138.7	1.178	1.629	0.234	0.042
138.8	1.244	1.794	0.302	0.050
138.9	1.342	1.977	0.378	0.055
139.0	1.485	2.180	0.462	0.055
139.1	1.697	2.401	0.555	0.049
139.2	2.015	2.625	0.658	0.034
139.3	2.490	2.807	0.769	0.008
139.4	3.166	2.834	0.890	-0.032
139.5	3.985	2.495	1.020	-0.090
139.6	4.621	1.622	1.157	-0.170
139.7	4.632	0.480	1.299	-0.278
89π/2	4.053	-0.372	1.442	-0.421
139.8	4.059	-0.366	1.440	-0.420
139.9	3.335	-0.731	1.574	-0.601
140.0	2.725	-0.775	1.688	-0.825
140.1	2.274	-0.672	1.767	-1.095
140.2	1.954	-0.516	1.793	-1.406
140.3	1.724	-0.350	1.745	-1.742
140.4	1.557	-0.188	1.609	-2.078
140.5	1.434	-0.036	1.385	-2.384
140.6	1.340	0.105	1.085	-2.624
140.7	1.269	0.238	0.740	-2.779
140.8	1.213	0.363	0.382	-2.840
140.9	1.171	0.483	0.043	-2.817
141.0	1.138	0.598	-0.257	-2.730
141.1	1.114	0.711	-0.507	-2.600
141.2	1.097	0.824	-0.709	-2.445
141.3	1.088	0.937	-0.866	-2.280
45π	1.085	1.019	-0.955	-2.160
141.4	1.085	1.052	-0.985	-2.114
141.5	1.090	1.172	-1.074	-1.953
141.6	1.104	1.298	-1.138	-1.800
141.7	1.129	1.432	-1.183	-1.656
141.8	1.168	1.577	-1.213	-1.522
141.9	1.225	1.735	-1.233	-1.397
142.0	1.309	1.911	-1.243	-1.279
142.1	1.432	2.105	-1.247	-1.168
142.2	1.614	2.318	-1.245	-1.063
142.3	1.884	2.542	-1.238	-0.963
142.4	2.289	2.747	-1.227	-0.865
142.5	2.878	2.853	-1.212	-0.771
142.6	3.654	2.688	-1.193	-0.677
142.7	4.413	2.035	-1.169	-0.585
142.8	4.705	0.946	-1.139	-0.493
142.9	4.329	-0.063	-1.102	-0.402
91π/2	4.044	-0.356	-1.084	-0.363

k_0h	ADMITTANCE		RAD.EFF.LENGTH	
143.0	3.627	-0.611	-1.056	-0.311
143.1	2.961	-0.762	-1.001	-0.223
143.2	2.450	-0.706	-0.934	-0.138
143.3	2.082	-0.567	-0.854	-0.061
143.4	1.820	-0.404	-0.763	0.005
143.5	1.630	-0.241	-0.662	0.056
143.6	1.490	-0.085	-0.558	0.087
143.7	1.385	0.060	-0.455	0.099
143.8	1.305	0.196	-0.362	0.093
143.9	1.244	0.324	-0.281	0.077
144.0	1.196	0.446	-0.215	0.055
144.1	1.160	0.563	-0.161	0.034
144.2	1.132	0.676	-0.117	0.016
144.3	1.113	0.789	-0.078	0.005
144.4	1.100	0.902	-0.042	-0.001
144.5	1.095	1.016	-0.005	-0.000
46π	1.095	1.031	-0.000	-0.000
144.6	1.097	1.134	0.035	0.004
144.7	1.107	1.257	0.078	0.012
144.8	1.127	1.387	0.127	0.022
144.9	1.160	1.527	0.183	0.033
145.0	1.209	1.680	0.245	0.043
145.1	1.282	1.848	0.315	0.051
145.2	1.388	2.034	0.392	0.055
145.3	1.544	2.239	0.478	0.054
145.4	1.774	2.459	0.573	0.046
145.5	2.118	2.675	0.678	0.029
145.6	2.627	2.832	0.791	0.000
145.7	3.331	2.800	0.914	-0.044
145.8	4.132	2.363	1.046	-0.107
145.9	4.660	1.418	1.184	-0.193
146.0	4.539	0.317	1.326	-0.309
93π/2	4.035	-0.340	1.444	-0.433
146.1	3.922	-0.425	1.465	-0.460
146.2	3.221	-0.715	1.594	-0.651
146.3	2.648	-0.726	1.700	-0.887
146.4	2.228	-0.613	1.766	-1.167
146.5	1.927	-0.458	1.772	-1.483
146.6	1.711	-0.294	1.701	-1.819
146.7	1.553	-0.135	1.542	-2.146
146.8	1.435	0.014	1.299	-2.432
146.9	1.345	0.153	0.988	-2.647
147.0	1.277	0.284	0.642	-2.774
147.1	1.224	0.408	0.293	-2.812
147.2	1.183	0.526	-0.031	-2.772
147.3	1.152	0.641	-0.313	-2.675
147.4	1.129	0.754	-0.547	-2.541
147.5	1.114	0.866	-0.734	-2.387
147.6	1.105	0.980	-0.880	-2.225
47π	1.104	1.043	-0.944	-2.136
147.7	1.104	1.096	-0.990	-2.064
147.8	1.111	1.217	-1.073	-1.909
147.9	1.127	1.344	-1.132	-1.762

k_0h	ADMITTANCE		RAD.EFF.LENGTH	
148.0	1.155	1.480	-1.175	-1.623
148.1	1.197	1.627	-1.204	-1.494
148.2	1.260	1.788	-1.222	-1.372
148.3	1.351	1.966	-1.233	-1.258
148.4	1.485	2.163	-1.237	-1.150
148.5	1.682	2.377	-1.236	-1.047
148.6	1.975	2.597	-1.230	-0.948
148.7	2.410	2.785	-1.220	-0.852
148.8	3.033	2.847	-1.205	-0.757
148.9	3.816	2.599	-1.187	-0.664
149.0	4.504	1.852	-1.163	-0.572
149.1	4.660	0.755	-1.133	-0.480
149.2	4.200	-0.163	-1.095	-0.387
95π/2	4.027	-0.324	-1.084	-0.364
149.3	3.500	-0.620	-1.049	-0.296
149.4	2.870	-0.724	-0.991	-0.207
149.5	2.392	-0.651	-0.921	-0.122
149.6	2.048	-0.509	-0.839	-0.046
149.7	1.801	-0.347	-0.745	0.017
149.8	1.622	-0.187	-0.643	0.064
149.9	1.489	-0.034	-0.538	0.091
150.0	1.389	0.109	-0.437	0.099
150.1	1.312	0.242	-0.346	0.091
150.2	1.253	0.369	-0.268	0.072
150.3	1.207	0.489	-0.205	0.050
150.4	1.172	0.605	-0.153	0.030
150.5	1.146	0.719	-0.110	0.013
150.6	1.128	0.831	-0.072	0.003
150.7	1.117	0.944	-0.036	-0.001
48π	1.113	1.055	-0.000	-0.000
150.8	1.113	1.059	0.001	0.000
150.9	1.116	1.178	0.042	0.006
151.0	1.129	1.302	0.086	0.014
151.1	1.151	1.434	0.137	0.024
151.2	1.188	1.576	0.193	0.034
151.3	1.242	1.731	0.257	0.044
151.4	1.321	1.902	0.328	0.051
151.5	1.436	2.091	0.407	0.054
151.6	1.605	2.297	0.495	0.052
151.7	1.854	2.515	0.591	0.042
151.8	2.226	2.721	0.698	0.023
151.9	2.768	2.848	0.813	-0.009
152.0	3.495	2.749	0.938	-0.057
152.1	4.262	2.215	1.072	-0.125
152.2	4.671	1.217	1.211	-0.218
152.3	4.434	0.176	1.352	-0.341
97π/2	4.018	-0.309	1.446	-0.445
152.4	3.789	-0.467	1.489	-0.502
152.5	3.116	-0.692	1.613	-0.704
152.6	2.578	-0.676	1.709	-0.950
152.7	2.185	-0.556	1.759	-1.239
152.8	1.902	-0.400	1.746	-1.560
152.9	1.698	-0.239	1.653	-1.892

k_0h	ADMITTANCE		RAD.EFF.LENGTH	
153.0	1.548	-0.083	1.472	-2.208
153.1	1.436	0.063	1.211	-2.473
153.2	1.350	0.200	0.892	-2.663
153.3	1.284	0.329	0.548	-2.764
153.4	1.234	0.451	0.210	-2.780
153.5	1.194	0.569	-0.099	-2.726
153.6	1.165	0.683	-0.364	-2.621
153.7	1.143	0.796	-0.583	-2.484
153.8	1.129	0.909	-0.757	-2.331
153.9	1.122	1.023	-0.892	-2.173
49π	1.122	1.067	-0.935	-2.113
154.0	1.123	1.140	-0.995	-2.017
154.1	1.132	1.262	-1.071	-1.867
154.2	1.150	1.390	-1.127	-1.725
154.3	1.181	1.528	-1.167	-1.592
154.4	1.227	1.677	-1.195	-1.466
154.5	1.295	1.841	-1.213	-1.348
154.6	1.395	2.022	-1.223	-1.237
154.7	1.540	2.221	-1.228	-1.132
154.8	1.753	2.434	-1.227	-1.031
154.9	2.070	2.648	-1.222	-0.933
155.0	2.537	2.816	-1.213	-0.838
155.1	3.190	2.828	-1.199	-0.744
155.2	3.968	2.492	-1.181	-0.651
155.3	4.569	1.663	-1.157	-0.558
155.4	4.595	0.581	-1.127	-0.465
99π/2	4.010	-0.294	-1.085	-0.364
155.5	4.070	-0.243	-1.089	-0.372
155.6	3.382	-0.620	-1.040	-0.280
155.7	2.786	-0.683	-0.981	-0.191
155.8	2.339	-0.596	-0.909	-0.107
155.9	2.017	-0.451	-0.823	-0.032
156.0	1.784	-0.292	-0.727	0.029
156.1	1.614	-0.134	-0.623	0.071
156.2	1.488	0.016	-0.518	0.094
156.3	1.392	0.156	-0.418	0.098
156.4	1.319	0.288	-0.330	0.087
156.5	1.262	0.413	-0.256	0.068
156.6	1.218	0.532	-0.195	0.046
156.7	1.185	0.648	-0.146	0.026
156.8	1.160	0.761	-0.104	0.011
156.9	1.143	0.873	-0.066	0.002
157.0	1.133	0.987	-0.030	-0.001
50π	1.131	1.078	-0.000	-0.000
157.1	1.131	1.102	0.008	0.001
157.2	1.136	1.222	0.049	0.007
157.3	1.150	1.348	0.095	0.015
157.4	1.176	1.481	0.146	0.026
157.5	1.216	1.626	0.204	0.036
157.6	1.274	1.783	0.268	0.045
157.7	1.360	1.956	0.341	0.051
157.8	1.485	2.147	0.422	0.053
157.9	1.668	2.355	0.511	0.050

k_0h	ADMITTANCE		RAD.EFF.LENGTH	
158.0	1.938	2.570	0.610	0.038
158.1	2.338	2.762	0.718	0.017
158.2	2.913	2.853	0.836	-0.018
158.3	3.654	2.681	0.963	-0.071
158.4	4.371	2.054	1.097	-0.144
158.5	4.657	1.025	1.237	-0.244
158.6	4.321	0.056	1.378	-0.376
101π/2	4.003	-0.279	1.447	-0.456
158.7	3.661	-0.494	1.512	-0.546
158.8	3.018	-0.664	1.629	-0.758
158.9	2.513	-0.626	1.714	-1.015
159.0	2.145	-0.499	1.749	-1.311
159.1	1.880	-0.344	1.715	-1.635
159.2	1.687	-0.185	1.599	-1.963
159.3	1.544	-0.033	1.398	-2.264
159.4	1.437	0.111	1.122	-2.508
159.5	1.355	0.246	0.799	-2.672
159.6	1.292	0.373	0.459	-2.749
159.7	1.243	0.495	0.132	-2.746
159.8	1.206	0.612	-0.161	-2.679
159.9	1.178	0.726	-0.411	-2.568
160.0	1.158	0.838	-0.615	-2.429
160.1	1.145	0.951	-0.777	-2.278
160.2	1.140	1.065	-0.903	-2.124
51π	1.139	1.090	-0.925	-2.092
160.3	1.142	1.183	-0.998	-1.973
160.4	1.152	1.306	-1.070	-1.828
160.5	1.173	1.436	-1.122	-1.691
160.6	1.207	1.576	-1.160	-1.561
160.7	1.258	1.728	-1.187	-1.440
160.8	1.332	1.894	-1.204	-1.326
160.9	1.439	2.077	-1.215	-1.217
161.0	1.596	2.278	-1.219	-1.114
161.1	1.828	2.490	-1.219	-1.015
161.2	2.169	2.696	-1.215	-0.918
161.3	2.668	2.838	-1.206	-0.824
161.4	3.346	2.795	-1.193	-0.730
161.5	4.107	2.368	-1.175	-0.637
161.6	4.609	1.472	-1.152	-0.544
161.7	4.513	0.425	-1.121	-0.451
103π/2	3.995	-0.264	-1.085	-0.364
161.8	3.942	-0.305	-1.082	-0.357
161.9	3.271	-0.610	-1.032	-0.264
162.0	2.709	-0.640	-0.970	-0.175
162.1	2.290	-0.541	-0.895	-0.091
162.2	1.988	-0.395	-0.807	-0.018
162.3	1.768	-0.237	-0.708	0.039
162.4	1.607	-0.082	-0.603	0.078
162.5	1.487	0.065	-0.498	0.097
162.6	1.395	0.203	-0.401	0.097
162.7	1.325	0.333	-0.315	0.084
162.8	1.271	0.457	-0.244	0.063
162.9	1.229	0.575	-0.186	0.042

k_0h	ADMITTANCE		RAD.EFF.LENGTH	
163.0	1.197	0.690	-0.138	0.023
163.1	1.173	0.803	-0.098	0.009
163.2	1.158	0.915	-0.060	0.001
163.3	1.149	1.029	-0.024	-0.001
52π	1.148	1.102	-0.000	-0.000
163.4	1.148	1.145	0.015	0.002
163.5	1.156	1.266	0.056	0.008
163.6	1.172	1.393	0.103	0.017
163.7	1.200	1.528	0.155	0.027
163.8	1.244	1.675	0.214	0.037
163.9	1.308	1.835	0.280	0.046
164.0	1.401	2.010	0.354	0.051
164.1	1.536	2.203	0.437	0.052
164.2	1.734	2.411	0.528	0.047
164.3	2.026	2.622	0.629	0.034
164.4	2.455	2.797	0.739	0.010
164.5	3.060	2.846	0.859	-0.029
164.6	3.807	2.597	0.988	-0.085
164.7	4.459	1.884	1.123	-0.164
164.8	4.620	0.844	1.264	-0.271
164.9	4.202	-0.045	1.403	-0.412
105π/2	3.988	-0.249	1.448	-0.468
165.0	3.540	-0.509	1.533	-0.592
165.1	2.927	-0.632	1.643	-0.815
165.2	2.454	-0.575	1.716	-1.081
165.3	2.109	-0.443	1.734	-1.384
165.4	1.859	-0.289	1.680	-1.709
165.5	1.677	-0.132	1.542	-2.029
165.6	1.541	0.017	1.321	-2.314
165.7	1.439	0.159	1.034	-2.536
165.8	1.360	0.292	0.708	-2.675
165.9	1.300	0.417	0.374	-2.730
166.0	1.253	0.538	0.060	-2.710
166.1	1.217	0.654	-0.219	-2.632
166.2	1.190	0.767	-0.453	-2.515
166.3	1.172	0.880	-0.644	-2.377
166.4	1.160	0.993	-0.795	-2.228
53π	1.156	1.113	-0.917	-2.071
166.5	1.156	1.108	-0.912	-2.077
166.6	1.160	1.227	-1.001	-1.931
166.7	1.173	1.351	-1.068	-1.790
166.8	1.196	1.483	-1.118	-1.658
166.9	1.233	1.624	-1.154	-1.533
167.0	1.289	1.778	-1.179	-1.415
167.1	1.369	1.947	-1.196	-1.303
167.2	1.486	2.132	-1.206	-1.198
167.3	1.656	2.334	-1.211	-1.096
167.4	1.905	2.545	-1.212	-0.999
167.5	2.273	2.739	-1.208	-0.903
167.6	2.802	2.852	-1.200	-0.810
167.7	3.501	2.746	-1.187	-0.717
167.8	4.229	2.230	-1.170	-0.624
167.9	4.624	1.284	-1.146	-0.530

k_0h	ADMITTANCE		RAD.EFF.LENGTH	
168.0	4.418	0.287	-1.115	-0.436
107π/2	3.981	-0.234	-1.086	-0.365
168.1	3.816	-0.351	-1.075	-0.341
168.2	3.167	-0.593	-1.023	-0.248
168.3	2.638	-0.595	-0.959	-0.158
168.4	2.246	-0.487	-0.881	-0.076
168.5	1.961	-0.340	-0.790	-0.005
168.6	1.754	-0.184	-0.688	0.049
168.7	1.601	-0.031	-0.583	0.084
168.8	1.486	0.113	-0.479	0.098
168.9	1.399	0.249	-0.383	0.095
169.0	1.331	0.378	-0.301	0.080
169.1	1.279	0.500	-0.233	0.059
169.2	1.239	0.617	-0.177	0.037
169.3	1.209	0.732	-0.131	0.019
169.4	1.187	0.844	-0.091	0.007
169.5	1.172	0.957	-0.054	0.000
169.6	1.165	1.071	-0.018	-0.001
54π	1.165	1.125	-0.000	-0.000
169.7	1.166	1.188	0.021	0.002
169.8	1.175	1.310	0.064	0.009
169.9	1.194	1.438	0.112	0.018
170.0	1.225	1.576	0.165	0.028
170.1	1.273	1.724	0.225	0.038
170.2	1.342	1.886	0.293	0.046
170.3	1.443	2.065	0.368	0.051
170.4	1.589	2.259	0.452	0.051
170.5	1.803	2.467	0.545	0.045
170.6	2.118	2.671	0.648	0.029
170.7	2.577	2.826	0.760	0.002
170.8	3.209	2.827	0.882	-0.040
170.9	3.951	2.496	1.012	-0.101
171.0	4.523	1.707	1.149	-0.186
171.1	4.563	0.677	1.290	-0.301
171.2	4.081	-0.127	1.427	-0.450
109π/2	3.974	-0.220	1.449	-0.479
171.3	3.426	-0.514	1.553	-0.640
171.4	2.844	-0.596	1.654	-0.873
171.5	2.400	-0.523	1.714	-1.148
171.6	2.076	-0.388	1.714	-1.457
171.7	1.840	-0.235	1.639	-1.780
171.8	1.667	-0.081	1.480	-2.090
171.9	1.538	0.066	1.242	-2.358
172.0	1.440	0.206	0.946	-2.557
172.1	1.365	0.337	0.620	-2.673
172.2	1.307	0.461	0.294	-2.707
172.3	1.263	0.580	-0.008	-2.672
172.4	1.228	0.696	-0.272	-2.585
172.5	1.203	0.809	-0.492	-2.465
172.6	1.186	0.921	-0.670	-2.326
172.7	1.176	1.035	-0.811	-2.179
55π	1.173	1.136	-0.908	-2.051
172.8	1.173	1.150	-0.920	-2.033
172.9	1.179	1.270	-1.004	-1.891

k_0h	ADMITTANCE		RAD.EFF.LENGTH	
173.0	1.194	1.396	-1.066	-1.755
173.1	1.220	1.529	-1.113	-1.626
173.2	1.260	1.672	-1.147	-1.505
173.3	1.320	1.829	-1.172	-1.391
173.4	1.407	2.000	-1.188	-1.282
173.5	1.534	2.187	-1.199	-1.179
173.6	1.717	2.390	-1.204	-1.079
173.7	1.987	2.597	-1.205	-0.983
173.8	2.381	2.778	-1.202	-0.889
173.9	2.941	2.856	-1.194	-0.796
174.0	3.652	2.682	-1.182	-0.703
174.1	4.334	2.080	-1.165	-0.609
174.2	4.615	1.103	-1.141	-0.515
174.3	4.315	0.168	-1.109	-0.420
111π/2	3.967	-0.206	-1.086	-0.364
174.4	3.694	-0.384	-1.067	-0.325
174.5	3.071	-0.571	-1.014	-0.231
174.6	2.573	-0.549	-0.947	-0.142
174.7	2.204	-0.433	-0.866	-0.061
174.8	1.937	-0.285	-0.772	0.007
174.9	1.741	-0.131	-0.669	0.058
175.0	1.596	0.019	-0.562	0.088
175.1	1.486	0.161	-0.460	0.099
175.2	1.402	0.295	-0.367	0.092
175.3	1.338	0.422	-0.288	0.076
175.4	1.288	0.543	-0.222	0.054
175.5	1.249	0.659	-0.169	0.033
175.6	1.220	0.773	-0.124	0.017
175.7	1.200	0.886	-0.085	0.005
175.8	1.187	0.999	-0.048	-0.000
175.9	1.181	1.113	-0.011	-0.001
56π	1.181	1.147	-0.000	-0.000
176.0	1.184	1.231	0.028	0.003
176.1	1.195	1.354	0.072	0.011
176.2	1.216	1.484	0.120	0.020
176.3	1.250	1.623	0.175	0.030
176.4	1.302	1.774	0.236	0.039
176.5	1.378	1.938	0.305	0.047
176.6	1.487	2.119	0.382	0.051
176.7	1.645	2.315	0.468	0.050
176.8	1.876	2.521	0.563	0.041
176.9	2.214	2.717	0.668	0.024
177.0	2.702	2.847	0.782	-0.006
177.1	3.358	2.795	0.905	-0.052
177.2	4.082	2.380	1.037	-0.118
177.3	4.564	1.528	1.175	-0.210
177.4	4.491	0.526	1.315	-0.332
113π/2	3.961	-0.192	1.450	-0.490
177.5	3.961	-0.192	1.450	-0.490
177.6	3.318	-0.510	1.571	-0.690
177.7	2.767	-0.557	1.662	-0.933
177.8	2.350	-0.472	1.708	-1.216
177.9	2.045	-0.334	1.690	-1.528

OMEGA = 20.0

k_0h	ADMITTANCE		RAD.EFF.LENGTH	
178.0	1.823	-0.182	1.594	-1.848
178.1	1.659	-0.030	1.415	-2.147
178.2	1.536	0.115	1.162	-2.397
178.3	1.442	0.252	0.859	-2.573
178.4	1.370	0.381	0.535	-2.666
178.5	1.315	0.504	0.218	-2.681
178.6	1.272	0.622	-0.071	-2.632
178.7	1.239	0.737	-0.320	-2.538
178.8	1.215	0.850	-0.527	-2.415
178.9	1.199	0.963	-0.694	-2.277
179.0	1.191	1.077	-0.825	-2.133
57π	1.189	1.158	-0.900	-2.032
179.1	1.190	1.193	-0.927	-1.991
179.2	1.197	1.313	-1.005	-1.852
179.3	1.214	1.440	-1.065	-1.721
179.4	1.243	1.575	-1.109	-1.596
179.5	1.288	1.721	-1.141	-1.479
179.6	1.353	1.879	-1.165	-1.367
179.7	1.447	2.053	-1.181	-1.262
179.8	1.583	2.242	-1.192	-1.160
179.9	1.782	2.445	-1.198	-1.063
180.0	2.072	2.647	-1.199	-0.968
180.1	2.494	2.811	-1.196	-0.874
180.2	3.081	2.849	-1.189	-0.782
180.3	3.797	2.602	-1.177	-0.689
180.4	4.419	1.920	-1.160	-0.595
180.5	4.585	0.930	-1.136	-0.500
180.6	4.205	0.067	-1.103	-0.404
115π/2	3.954	-0.178	-1.086	-0.364
180.7	3.578	-0.404	-1.059	-0.308
180.8	2.981	-0.543	-1.004	-0.215
180.9	2.513	-0.501	-0.934	-0.126
181.0	2.167	-0.380	-0.850	-0.046
181.1	1.914	-0.232	-0.753	0.019
181.2	1.729	-0.080	-0.649	0.066
181.3	1.591	0.068	-0.542	0.092
181.4	1.486	0.208	-0.441	0.098
181.5	1.406	0.340	-0.351	0.089
181.6	1.344	0.465	-0.275	0.071
181.7	1.296	0.585	-0.212	0.050
181.8	1.259	0.701	-0.161	0.030
181.9	1.232	0.815	-0.118	0.014
182.0	1.213	0.927	-0.079	0.004
182.1	1.201	1.040	-0.042	-0.001
182.2	1.197	1.155	-0.005	-0.000
58π	1.197	1.170	-0.000	-0.000
182.3	1.201	1.274	0.035	0.004
182.4	1.214	1.398	0.080	0.012
182.5	1.238	1.529	0.129	0.021
182.6	1.276	1.670	0.185	0.031
182.7	1.332	1.823	0.248	0.040
182.8	1.414	1.990	0.318	0.047
182.9	1.532	2.172	0.397	0.050
183.0	1.702	2.370	0.484	0.048
183.1	1.952	2.573	0.581	0.038
183.2	2.315	2.758	0.688	0.018
183.3	2.832	2.860	0.804	-0.015
183.4	3.505	2.748	0.929	-0.065
183.5	4.200	2.251	1.062	-0.137
183.6	4.582	1.351	1.201	-0.234
183.7	4.405	0.392	1.340	-0.364
117π/2	3.948	-0.164	1.451	-0.501
183.8	3.841	-0.242	1.472	-0.532
183.9	3.216	-0.498	1.586	-0.742
184.0	2.695	-0.516	1.667	-0.994
184.1	2.303	-0.420	1.698	-1.285
184.2	2.017	-0.281	1.661	-1.599
184.3	1.807	-0.130	1.544	-1.913
184.4	1.651	0.020	1.347	-2.199
184.5	1.534	0.162	1.082	-2.429
184.6	1.445	0.297	0.774	-2.582
184.7	1.376	0.425	0.454	-2.654
184.8	1.322	0.547	0.147	-2.653
184.9	1.281	0.664	-0.129	-2.593
185.0	1.250	0.779	-0.365	-2.492
185.1	1.228	0.891	-0.559	-2.367
185.2	1.213	1.004	-0.715	-2.230
185.3	1.206	1.118	-0.838	-2.089
59π	1.205	1.181	-0.893	-2.014
185.4	1.207	1.235	-0.934	-1.950
185.5	1.216	1.357	-1.007	-1.816
185.6	1.235	1.485	-1.063	-1.688
185.7	1.267	1.622	-1.105	-1.567
185.8	1.315	1.769	-1.136	-1.453
185.9	1.386	1.930	-1.159	-1.345
186.0	1.488	2.106	-1.175	-1.242
186.1	1.635	2.297	-1.185	-1.142
186.2	1.849	2.498	-1.191	-1.046
186.3	2.161	2.694	-1.193	-0.953
186.4	2.611	2.838	-1.191	-0.860
186.5	3.223	2.831	-1.184	-0.767
186.6	3.934	2.507	-1.173	-0.674
186.7	4.483	1.754	-1.155	-0.580
186.8	4.536	0.770	-1.130	-0.485
186.9	4.093	-0.016	-1.096	-0.388
119π/2	3.942	-0.150	-1.086	-0.364
187.0	3.467	-0.413	-1.051	-0.292
187.1	2.898	-0.512	-0.993	-0.198
187.2	2.457	-0.453	-0.920	-0.109
187.3	2.131	-0.327	-0.833	-0.032
187.4	1.894	-0.179	-0.734	0.030
187.5	1.718	-0.029	-0.629	0.073
187.6	1.586	0.116	-0.523	0.095
187.7	1.486	0.254	-0.423	0.097
187.8	1.410	0.384	-0.336	0.086
187.9	1.350	0.508	-0.263	0.067
188.0	1.304	0.627	-0.202	0.046
188.1	1.269	0.742	-0.153	0.026
188.2	1.244	0.856	-0.111	0.012
188.3	1.226	0.968	-0.073	0.002
188.4	1.216	1.082	-0.036	-0.001
60π	1.213	1.192	-0.000	-0.000
188.5	1.213	1.197	0.002	0.000
188.6	1.219	1.317	0.043	0.005
188.7	1.234	1.442	0.088	0.013
188.8	1.261	1.575	0.138	0.023
188.9	1.302	1.718	0.195	0.032
189.0	1.363	1.873	0.259	0.041
189.1	1.451	2.042	0.331	0.047
189.2	1.578	2.226	0.411	0.049
189.3	1.763	2.424	0.501	0.046
189.4	2.031	2.624	0.599	0.034
189.5	2.419	2.795	0.708	0.011
189.6	2.964	2.863	0.826	-0.025
189.7	3.649	2.687	0.953	-0.079
189.8	4.300	2.109	1.088	-0.156
189.9	4.578	1.179	1.226	-0.261
190.0	4.310	0.275	1.364	-0.399
121π/2	3.936	-0.137	1.451	-0.512
190.1	3.725	-0.278	1.493	-0.576
190.2	3.121	-0.481	1.599	-0.796
190.3	2.629	-0.473	1.669	-1.057
190.4	2.261	-0.369	1.684	-1.353
190.5	1.991	-0.228	1.628	-1.667
190.6	1.792	-0.078	1.491	-1.975
190.7	1.644	0.069	1.276	-2.246
190.8	1.532	0.209	1.001	-2.455
190.9	1.447	0.342	0.691	-2.587
191.0	1.381	0.468	0.376	-2.639
191.1	1.330	0.589	0.080	-2.622
191.2	1.290	0.706	-0.183	-2.552
191.3	1.261	0.820	-0.406	-2.447
191.4	1.240	0.932	-0.588	-2.320
191.5	1.226	1.045	-0.735	-2.185
191.6	1.221	1.160	-0.850	-2.047
61π	1.221	1.203	-0.886	-1.996
191.7	1.223	1.278	-0.939	-1.911
191.8	1.234	1.400	-1.008	-1.781
191.9	1.256	1.530	-1.061	-1.657
192.0	1.291	1.668	-1.101	-1.540
192.1	1.344	1.818	-1.131	-1.429
192.2	1.420	1.981	-1.153	-1.323
192.3	1.530	2.159	-1.169	-1.222
192.4	1.689	2.351	-1.179	-1.125
192.5	1.920	2.551	-1.186	-1.030
192.6	2.255	2.738	-1.188	-0.937
192.7	2.732	2.858	-1.186	-0.845
192.8	3.366	2.799	-1.180	-0.753
192.9	4.059	2.397	-1.168	-0.660
193.0	4.525	1.585	-1.151	-0.565
193.1	4.471	0.624	-1.125	-0.469
123π/2	3.930	-0.123	-1.087	-0.364
193.2	3.979	-0.084	-1.090	-0.372
193.3	3.361	-0.414	-1.043	-0.275
193.4	2.821	-0.477	-0.982	-0.181
193.5	2.405	-0.404	-0.906	-0.093
193.6	2.099	-0.275	-0.816	-0.018
193.7	1.874	-0.128	-0.715	0.041
193.8	1.708	0.021	-0.608	0.079
193.9	1.583	0.164	-0.503	0.097
194.0	1.487	0.299	-0.406	0.096
194.1	1.414	0.428	-0.322	0.083
194.2	1.357	0.551	-0.251	0.063
194.3	1.313	0.669	-0.193	0.041
194.4	1.279	0.784	-0.146	0.023
194.5	1.255	0.897	-0.105	0.009
194.6	1.239	1.009	-0.067	0.001
194.7	1.230	1.123	-0.030	-0.001
62π	1.228	1.214	-0.000	-0.000
194.8	1.229	1.239	0.008	0.001
194.9	1.236	1.360	0.050	0.006
195.0	1.254	1.486	0.096	0.015
195.1	1.283	1.621	0.148	0.024
195.2	1.328	1.765	0.206	0.034
195.3	1.394	1.922	0.271	0.042
195.4	1.490	2.094	0.345	0.047
195.5	1.627	2.280	0.426	0.048
195.6	1.825	2.477	0.517	0.043
195.7	2.114	2.672	0.618	0.029
195.8	2.528	2.826	0.728	0.004
195.9	3.099	2.856	0.848	-0.036
196.0	3.787	2.611	0.977	-0.095
196.1	4.383	1.959	1.113	-0.177
196.2	4.554	1.014	1.252	-0.289
196.3	4.209	0.174	1.388	-0.435
125π/2	3.925	-0.110	1.451	-0.522
196.4	3.613	-0.302	1.511	-0.622
196.5	3.032	-0.457	1.610	-0.851
196.6	2.568	-0.429	1.668	-1.121
196.7	2.221	-0.318	1.666	-1.421
196.8	1.966	-0.176	1.590	-1.734
196.9	1.778	-0.028	1.434	-2.033
197.0	1.638	0.117	1.204	-2.288
197.1	1.531	0.255	0.920	-2.476
197.2	1.449	0.387	0.610	-2.586
197.3	1.386	0.511	0.302	-2.620
197.4	1.337	0.631	0.017	-2.589
197.5	1.299	0.747	-0.233	-2.511
197.6	1.271	0.861	-0.443	-2.402
197.7	1.252	0.973	-0.615	-2.275
197.8	1.240	1.086	-0.752	-2.141
197.9	1.236	1.201	-0.860	-2.006
63π	1.236	1.225	-0.879	-1.979
198.0	1.240	1.320	-0.944	-1.874
198.1	1.253	1.444	-1.009	-1.748
198.2	1.278	1.575	-1.059	-1.627
198.3	1.316	1.715	-1.097	-1.513
198.4	1.373	1.867	-1.126	-1.405
198.5	1.456	2.032	-1.147	-1.302
198.6	1.574	2.211	-1.163	-1.203
198.7	1.745	2.404	-1.174	-1.108
198.8	1.993	2.601	-1.181	-1.015
198.9	2.352	2.777	-1.183	-0.922
199.0	2.856	2.870	-1.182	-0.831
199.1	3.506	2.755	-1.176	-0.738
199.2	4.172	2.275	-1.164	-0.645
199.3	4.545	1.417	-1.146	-0.550
199.4	4.394	0.493	-1.120	-0.453
127π/2	3.919	-0.096	-1.087	-0.363
199.5	3.866	-0.136	-1.083	-0.355
199.6	3.262	-0.407	-1.034	-0.257
199.7	2.749	-0.440	-0.970	-0.164
199.8	2.358	-0.355	-0.891	-0.078
199.9	2.069	-0.223	-0.798	-0.005
200.0	1.857	-0.077	-0.695	0.050
200.1	1.699	0.070	-0.588	0.084
200.2	1.579	0.210	-0.484	0.098
200.3	1.488	0.344	-0.389	0.094
200.4	1.418	0.471	-0.308	0.079
200.5	1.363	0.593	-0.240	0.058
200.6	1.321	0.710	-0.185	0.038
200.7	1.289	0.825	-0.138	0.020
200.8	1.266	0.937	-0.098	0.007
200.9	1.251	1.050	-0.061	0.001
201.0	1.244	1.164	-0.024	-0.001
64π	1.243	1.236	-0.000	-0.000
201.1	1.245	1.281	0.015	0.001
201.2	1.254	1.403	0.057	0.008
201.3	1.274	1.530	0.105	0.016
201.4	1.306	1.666	0.157	0.025
201.5	1.355	1.813	0.217	0.035
201.6	1.427	1.972	0.283	0.042
201.7	1.530	2.146	0.358	0.047
201.8	1.677	2.333	0.442	0.047
201.9	1.891	2.529	0.535	0.040
202.0	2.200	2.717	0.637	0.024
202.1	2.641	2.852	0.749	-0.004
202.2	3.235	2.838	0.871	-0.047
202.3	3.917	2.522	1.001	-0.111
202.4	4.446	1.802	1.138	-0.199
202.5	4.512	0.860	1.276	-0.318
202.6	4.104	0.090	1.410	-0.473
129π/2	3.914	-0.083	1.451	-0.533
202.7	3.505	-0.316	1.529	-0.670
202.8	2.950	-0.430	1.618	-0.908
202.9	2.511	-0.384	1.662	-1.185
203.0	2.185	-0.267	1.643	-1.489
203.1	1.944	-0.125	1.548	-1.799
203.2	1.766	0.022	1.373	-2.086
203.3	1.632	0.165	1.130	-2.324
203.4	1.530	0.301	0.840	-2.492
203.5	1.452	0.431	0.532	-2.581
203.6	1.391	0.554	0.231	-2.598
203.7	1.344	0.673	-0.042	-2.555
203.8	1.308	0.788	-0.280	-2.470
203.9	1.282	0.901	-0.478	-2.358
204.0	1.264	1.014	-0.639	-2.231
204.1	1.253	1.127	-0.768	-2.099
65π	1.251	1.247	-0.872	-1.962
204.2	1.251	1.243	-0.869	-1.967
204.3	1.257	1.362	-0.948	-1.839
204.4	1.272	1.487	-1.010	-1.716
204.5	1.299	1.620	-1.057	-1.599
204.6	1.341	1.762	-1.093	-1.488
204.7	1.403	1.915	-1.121	-1.383
204.8	1.492	2.083	-1.142	-1.282
204.9	1.620	2.264	-1.158	-1.185
205.0	1.804	2.457	-1.169	-1.091
205.1	2.070	2.650	-1.176	-0.999
205.2	2.453	2.813	-1.179	-0.908
205.3	2.984	2.872	-1.177	-0.816
205.4	3.643	2.697	-1.172	-0.724
205.5	4.269	2.142	-1.160	-0.630
205.6	4.545	1.253	-1.142	-0.534
205.7	4.307	0.377	-1.114	-0.436
131π/2	3.909	-0.070	-1.087	-0.363
205.8	3.754	-0.176	-1.076	-0.337
205.9	3.169	-0.393	-1.024	-0.240
206.0	2.682	-0.400	-0.958	-0.147
206.1	2.314	-0.306	-0.876	-0.062
206.2	2.042	-0.172	-0.780	0.008
206.3	1.841	-0.027	-0.675	0.059
206.4	1.690	0.118	-0.568	0.089
206.5	1.576	0.257	-0.465	0.098
206.6	1.489	0.389	-0.373	0.092
206.7	1.422	0.514	-0.294	0.075
206.8	1.369	0.635	-0.229	0.054
206.9	1.329	0.751	-0.176	0.034
207.0	1.299	0.865	-0.131	0.017
207.1	1.277	0.978	-0.092	0.006
207.2	1.264	1.091	-0.055	-0.000
207.3	1.258	1.205	-0.017	-0.001
66π	1.258	1.258	-0.000	-0.000
207.4	1.261	1.323	0.022	0.002
207.5	1.272	1.445	0.065	0.009
207.6	1.294	1.575	0.113	0.017
207.7	1.330	1.712	0.167	0.027
207.8	1.383	1.861	0.228	0.036
207.9	1.460	2.022	0.296	0.043
208.0	1.571	2.197	0.372	0.047
208.1	1.730	2.386	0.457	0.045
208.2	1.960	2.580	0.552	0.037
208.3	2.291	2.759	0.656	0.019
208.4	2.758	2.871	0.771	-0.012
208.5	3.371	2.808	0.894	-0.059
208.6	4.037	2.418	1.026	-0.128
208.7	4.489	1.643	1.163	-0.223
208.8	4.455	0.718	1.301	-0.349
208.9	3.996	0.020	1.431	-0.513
133π/2	3.904	-0.057	1.451	-0.543
209.0	3.403	-0.321	1.544	-0.719
209.1	2.872	-0.399	1.624	-0.967
209.2	2.459	-0.338	1.653	-1.250
209.3	2.151	-0.217	1.616	-1.556
209.4	1.924	-0.075	1.502	-1.860
209.5	1.755	0.071	1.310	-2.136
209.6	1.627	0.212	1.055	-2.355
209.7	1.530	0.346	0.761	-2.502
209.8	1.455	0.474	0.456	-2.572
209.9	1.396	0.596	0.165	-2.574
210.0	1.352	0.714	-0.098	-2.520
210.1	1.317	0.829	-0.323	-2.429
210.2	1.292	0.942	-0.510	-2.315
210.3	1.276	1.054	-0.661	-2.189
210.4	1.267	1.168	-0.782	-2.059
67π	1.266	1.269	-0.866	-1.946
210.5	1.266	1.284	-0.877	-1.930
210.6	1.274	1.405	-0.952	-1.805
210.7	1.291	1.531	-1.010	-1.685
210.8	1.321	1.665	-1.055	-1.571
210.9	1.367	1.809	-1.090	-1.463
211.0	1.434	1.964	-1.117	-1.361
211.1	1.530	2.134	-1.138	-1.262
211.2	1.667	2.316	-1.153	-1.167
211.3	1.865	2.509	-1.164	-1.074
211.4	2.151	2.697	-1.171	-0.983
211.5	2.559	2.843	-1.175	-0.893
211.6	3.113	2.865	-1.174	-0.802
211.7	3.776	2.625	-1.168	-0.709
211.8	4.349	2.000	-1.156	-0.615
211.9	4.526	1.096	-1.137	-0.518
212.0	4.213	0.277	-1.108	-0.419
135π/2	3.899	-0.044	-1.087	-0.362
212.1	3.646	-0.204	-1.068	-0.320
212.2	3.081	-0.374	-1.014	-0.222
212.3	2.620	-0.359	-0.944	-0.130
212.4	2.273	-0.257	-0.859	-0.047
212.5	2.016	-0.122	-0.761	0.020
212.6	1.826	0.023	-0.655	0.066
212.7	1.683	0.165	-0.548	0.092
212.8	1.574	0.302	-0.447	0.098
212.9	1.490	0.433	-0.358	0.089
213.0	1.426	0.557	-0.282	0.071
213.1	1.376	0.677	-0.219	0.050
213.2	1.337	0.792	-0.168	0.030
213.3	1.309	0.906	-0.125	0.015
213.4	1.289	1.018	-0.086	0.004
213.5	1.277	1.131	-0.049	-0.001
213.6	1.272	1.246	-0.011	-0.001
68π	1.273	1.279	-0.000	-0.000
213.7	1.277	1.365	0.029	0.003
213.8	1.290	1.488	0.073	0.010
213.9	1.315	1.619	0.122	0.019
214.0	1.354	1.758	0.177	0.028
214.1	1.411	1.909	0.239	0.037
214.2	1.495	2.072	0.308	0.043
214.3	1.614	2.249	0.386	0.046
214.4	1.785	2.438	0.473	0.044
214.5	2.031	2.629	0.570	0.033
214.6	2.385	2.797	0.676	0.013
214.7	2.878	2.881	0.792	-0.021
214.8	3.506	2.765	0.917	-0.073
214.9	4.145	2.303	1.050	-0.146
215.0	4.512	1.483	1.188	-0.248
215.1	4.385	0.590	1.324	-0.382
137π/2	3.894	-0.031	1.450	-0.554
215.2	3.889	-0.035	1.451	-0.555
215.3	3.305	-0.318	1.557	-0.770
215.4	2.800	-0.365	1.626	-1.026
215.5	2.410	-0.291	1.641	-1.315
215.6	2.120	-0.167	1.585	-1.621
215.7	1.905	-0.025	1.452	-1.919
215.8	1.744	0.119	1.245	-2.180
215.9	1.622	0.258	0.980	-2.381
216.0	1.529	0.391	0.684	-2.508
216.1	1.457	0.517	0.384	-2.560
216.2	1.402	0.638	0.102	-2.547
216.3	1.359	0.755	-0.149	-2.485
216.4	1.326	0.870	-0.363	-2.389
216.5	1.303	0.982	-0.539	-2.273
216.6	1.288	1.095	-0.681	-2.147
216.7	1.280	1.209	-0.795	-2.020
69π	1.280	1.290	-0.860	-1.931
216.8	1.281	1.326	-0.885	-1.894
216.9	1.291	1.447	-0.955	-1.772
217.0	1.311	1.575	-1.010	-1.656
217.1	1.343	1.710	-1.053	-1.545
217.2	1.393	1.856	-1.087	-1.440
217.3	1.465	2.013	-1.113	-1.339
217.4	1.569	2.184	-1.133	-1.243
217.5	1.717	2.368	-1.149	-1.150
217.6	1.929	2.560	-1.160	-1.058
217.7	2.235	2.740	-1.167	-0.968
217.8	2.668	2.867	-1.171	-0.878
217.9	3.244	2.848	-1.170	-0.787

OMEGA = 20.0

k_0h	ADMITTANCE		RAD.EFF.LENGTH	
218.0	3.901	2.540	-1.164	-0.694
218.1	4.412	1.852	-1.152	-0.599
218.2	4.490	0.947	-1.132	-0.502
218.3	4.114	0.191	-1.103	-0.402
139π/2	3.889	-0.018	-1.087	-0.362
218.4	3.542	-0.222	-1.060	-0.302
218.5	2.998	-0.350	-1.003	-0.204
218.6	2.563	-0.317	-0.931	-0.113
218.7	2.235	-0.208	-0.842	-0.033
218.8	1.993	-0.072	-0.742	0.031
218.9	1.812	0.072	-0.635	0.073
219.0	1.676	0.212	-0.528	0.094
219.1	1.572	0.347	-0.429	0.097
219.2	1.492	0.476	-0.343	0.085
219.3	1.430	0.599	-0.270	0.067
219.4	1.382	0.718	-0.210	0.046
219.5	1.345	0.833	-0.160	0.027
219.6	1.318	0.946	-0.118	0.012
219.7	1.300	1.059	-0.080	0.003
219.8	1.289	1.172	-0.043	-0.001
219.9	1.287	1.287	-0.005	-0.000
70π	1.287	1.301	-0.000	-0.000
220.0	1.293	1.407	0.036	0.004
220.1	1.308	1.531	0.081	0.011
220.2	1.336	1.663	0.131	0.020
220.3	1.378	1.805	0.187	0.029
220.4	1.440	1.957	0.250	0.038
220.5	1.530	2.122	0.321	0.043
220.6	1.658	2.301	0.401	0.045
220.7	1.842	2.489	0.489	0.041
220.8	2.107	2.677	0.588	0.029
220.9	2.483	2.831	0.696	0.006
221.0	3.000	2.884	0.814	-0.031
221.1	3.637	2.710	0.941	-0.087
221.2	4.240	2.177	1.075	-0.166
221.3	4.515	1.326	1.212	-0.274
221.4	4.305	0.476	1.347	-0.417
141π/2	3.884	-0.006	1.450	-0.564
221.5	3.782	-0.078	1.470	-0.599
221.6	3.213	-0.308	1.568	-0.823
221.7	2.733	-0.329	1.625	-1.087
221.8	2.365	-0.245	1.624	-1.380
221.9	2.091	-0.117	1.550	-1.685
222.0	1.887	0.024	1.399	-1.974
222.1	1.734	0.166	1.177	-2.220
222.2	1.618	0.303	0.905	-2.402
222.3	1.529	0.435	0.609	-2.509
222.4	1.461	0.560	0.315	-2.544
222.5	1.407	0.680	0.042	-2.519
222.6	1.366	0.796	-0.197	-2.448
222.7	1.335	0.910	-0.399	-2.348
222.8	1.313	1.023	-0.566	-2.232
222.9	1.300	1.135	-0.700	-2.108
223.0	1.294	1.250	-0.807	-1.982
71π	1.294	1.312	-0.854	-1.916
223.1	1.296	1.367	-0.891	-1.859
223.2	1.308	1.490	-0.958	-1.740
223.3	1.330	1.618	-1.010	-1.627
223.4	1.366	1.755	-1.051	-1.520
223.5	1.420	1.903	-1.084	-1.417
223.6	1.498	2.063	-1.109	-1.319
223.7	1.609	2.235	-1.129	-1.224
223.8	1.768	2.420	-1.145	-1.132
223.9	1.996	2.609	-1.156	-1.042
224.0	2.323	2.781	-1.164	-0.953
224.1	2.780	2.885	-1.167	-0.863
224.2	3.375	2.819	-1.167	-0.772
224.3	4.016	2.443	-1.161	-0.679
224.4	4.456	1.700	-1.149	-0.583
224.5	4.439	0.809	-1.128	-0.485
224.6	4.013	0.120	-1.096	-0.385
143π/2	3.880	0.007	-1.087	-0.361
224.7	3.442	-0.230	-1.052	-0.284
224.8	2.921	-0.323	-0.992	-0.187
224.9	2.510	-0.273	-0.916	-0.096
225.0	2.200	-0.159	-0.825	-0.019
225.1	1.971	-0.022	-0.722	0.041
225.2	1.799	0.120	-0.615	0.079
225.3	1.670	0.259	-0.509	0.096
225.4	1.570	0.392	-0.412	0.095
225.5	1.494	0.519	-0.328	0.082
225.6	1.434	0.641	-0.258	0.062
225.7	1.389	0.759	-0.201	0.042
225.8	1.354	0.873	-0.153	0.024
225.9	1.328	0.986	-0.111	0.010
226.0	1.311	1.099	-0.073	0.002
226.1	1.302	1.213	-0.036	-0.001
72π	1.301	1.322	-0.000	-0.000
226.2	1.301	1.329	0.002	0.000
226.3	1.309	1.449	0.043	0.005
226.4	1.327	1.575	0.089	0.013
226.5	1.357	1.708	0.140	0.021
226.6	1.403	1.851	0.198	0.030
226.7	1.470	2.005	0.262	0.038
226.8	1.567	2.172	0.335	0.043
226.9	1.705	2.352	0.416	0.044
227.0	1.902	2.540	0.506	0.039
227.1	2.185	2.722	0.606	0.025
227.2	2.585	2.860	0.716	-0.001
227.3	3.125	2.877	0.836	-0.042
227.4	3.764	2.642	0.965	-0.102
227.5	4.319	2.043	1.099	-0.187
227.6	4.501	1.175	1.237	-0.302
227.7	4.217	0.376	1.369	-0.453
145π/2	3.875	0.019	1.449	-0.574
227.8	3.678	-0.109	1.487	-0.645
227.9	3.127	-0.292	1.577	-0.878
228.0	2.671	-0.290	1.621	-1.149
228.1	2.323	-0.197	1.603	-1.445
228.2	2.064	-0.068	1.511	-1.746
228.3	1.871	0.073	1.343	-2.025
228.4	1.726	0.213	1.108	-2.256
228.5	1.615	0.349	0.830	-2.418
228.6	1.530	0.479	0.536	-2.507
228.7	1.464	0.602	0.248	-2.526
228.8	1.413	0.721	-0.014	-2.489
228.9	1.374	0.837	-0.242	-2.412
229.0	1.344	0.950	-0.434	-2.309
229.1	1.324	1.063	-0.590	-2.192
229.2	1.312	1.176	-0.717	-2.069
229.3	1.307	1.291	-0.818	-1.946
73π	1.308	1.333	-0.849	-1.902
229.4	1.311	1.409	-0.897	-1.826
229.5	1.325	1.532	-0.960	-1.710
229.6	1.350	1.662	-1.010	-1.600
229.7	1.389	1.801	-1.050	-1.495
229.8	1.448	1.950	-1.081	-1.395
229.9	1.531	2.112	-1.106	-1.299
230.0	1.651	2.286	-1.126	-1.206
230.1	1.822	2.471	-1.141	-1.116
230.2	2.066	2.657	-1.152	-1.026
230.3	2.415	2.818	-1.160	-0.938
230.4	2.896	2.895	-1.164	-0.848
230.5	3.504	2.779	-1.164	-0.757
230.6	4.121	2.333	-1.158	-0.664
230.7	4.481	1.548	-1.145	-0.567
230.8	4.376	0.684	-1.123	-0.468
147π/2	3.871	0.032	-1.087	-0.360
230.9	3.910	0.063	-1.090	-0.367
231.0	3.347	-0.231	-1.043	-0.266
231.1	2.849	-0.292	-0.980	-0.169
231.2	2.460	-0.229	-0.901	-0.080
231.3	2.168	-0.111	-0.807	-0.006
231.4	1.951	0.026	-0.702	0.050
231.5	1.788	0.167	-0.594	0.084
231.6	1.664	0.304	-0.490	0.097
231.7	1.569	0.436	-0.396	0.093
231.8	1.496	0.562	-0.315	0.078
231.9	1.439	0.683	-0.247	0.058
232.0	1.395	0.800	-0.192	0.038
232.1	1.362	0.914	-0.146	0.021
232.2	1.338	1.026	-0.105	0.008
232.3	1.322	1.139	-0.067	0.001
232.4	1.315	1.253	-0.030	-0.001
74π	1.315	1.344	-0.000	-0.000
232.5	1.315	1.370	0.009	0.001
232.6	1.325	1.491	0.051	0.006
232.7	1.345	1.618	0.097	0.014
232.8	1.379	1.753	0.150	0.023
232.9	1.429	1.897	0.208	0.031
233.0	1.501	2.054	0.274	0.039
233.1	1.605	2.222	0.348	0.043
233.2	1.753	2.403	0.431	0.043
233.3	1.965	2.590	0.523	0.036
233.4	2.267	2.764	0.625	0.020
233.5	2.691	2.884	0.737	-0.009
233.6	3.251	2.861	0.858	-0.053
233.7	3.884	2.562	0.988	-0.118
233.8	4.381	1.902	1.124	-0.209
233.9	4.471	1.032	1.260	-0.332
234.0	4.125	0.290	1.390	-0.491
149π/2	3.867	0.044	1.448	-0.584
234.1	3.577	-0.130	1.503	-0.692
234.2	3.045	-0.272	1.582	-0.934
234.3	2.612	-0.250	1.613	-1.211
234.4	2.284	-0.150	1.578	-1.509
234.5	2.039	-0.019	1.468	-1.805
234.6	1.856	0.121	1.284	-2.073
234.7	1.718	0.259	1.039	-2.286
234.8	1.612	0.393	0.757	-2.430
234.9	1.530	0.521	0.465	-2.500
235.0	1.467	0.644	0.185	-2.505
235.1	1.418	0.762	-0.067	-2.458
235.2	1.381	0.877	-0.284	-2.375
235.3	1.353	0.990	-0.465	-2.270
235.4	1.334	1.103	-0.613	-2.153
235.5	1.324	1.216	-0.732	-2.032
235.6	1.321	1.331	-0.827	-1.911
75π	1.321	1.354	-0.843	-1.888
235.7	1.327	1.450	-0.903	-1.794
235.8	1.343	1.575	-0.963	-1.681
235.9	1.370	1.706	-1.010	-1.574
236.0	1.413	1.847	-1.048	-1.471
236.1	1.476	1.998	-1.078	-1.374
236.2	1.566	2.161	-1.103	-1.279
236.3	1.695	2.337	-1.122	-1.188
236.4	1.878	2.521	-1.138	-1.099
236.5	2.140	2.703	-1.149	-1.011
236.6	2.510	2.851	-1.157	-0.923
236.7	3.014	2.898	-1.161	-0.833
236.8	3.630	2.726	-1.161	-0.742
236.9	4.213	2.214	-1.155	-0.648
237.0	4.488	1.398	-1.141	-0.551
237.1	4.303	0.571	-1.118	-0.451
151π/2	3.863	0.057	-1.087	-0.359
237.2	3.808	0.017	-1.083	-0.349
237.3	3.256	-0.225	-1.033	-0.248
237.4	2.782	-0.258	-0.967	-0.152
237.5	2.414	-0.184	-0.885	-0.065
237.6	2.138	-0.063	-0.788	0.007
237.7	1.932	0.075	-0.682	0.059
237.8	1.777	0.214	-0.574	0.088
237.9	1.659	0.350	-0.472	0.097
238.0	1.568	0.480	-0.380	0.091
238.1	1.498	0.604	-0.302	0.074
238.2	1.443	0.724	-0.237	0.054
238.3	1.402	0.840	-0.183	0.034
238.4	1.370	0.954	-0.138	0.018
238.5	1.348	1.066	-0.099	0.007
238.6	1.334	1.179	-0.061	0.000
238.7	1.328	1.294	-0.024	-0.001
76π	1.328	1.365	-0.000	-0.000
238.8	1.330	1.411	0.016	0.001
238.9	1.342	1.533	0.058	0.007
239.0	1.364	1.661	0.106	0.015
239.1	1.401	1.798	0.159	0.024
239.2	1.455	1.944	0.219	0.032
239.3	1.533	2.102	0.286	0.039
239.4	1.645	2.272	0.362	0.043
239.5	1.803	2.454	0.446	0.041
239.6	2.030	2.638	0.540	0.033
239.7	2.353	2.803	0.644	0.014
239.8	2.800	2.901	0.758	-0.017
239.9	3.377	2.833	0.881	-0.066
240.0	3.996	2.469	1.012	-0.136
240.1	4.426	1.758	1.148	-0.233
240.2	4.426	0.898	1.284	-0.363
240.3	4.029	0.218	1.410	-0.531
153π/2	3.858	0.069	1.447	-0.594
240.4	3.479	-0.142	1.516	-0.741
240.5	2.968	-0.248	1.585	-0.991
240.6	2.558	-0.209	1.602	-1.273
240.7	2.248	-0.103	1.550	-1.571
240.8	2.016	0.029	1.422	-1.861
240.9	1.842	0.168	1.222	-2.116
241.0	1.710	0.305	0.968	-2.312
241.1	1.609	0.437	0.684	-2.437
241.2	1.531	0.564	0.396	-2.490
241.3	1.471	0.685	0.125	-2.482
241.4	1.424	0.803	-0.116	-2.426
241.5	1.388	0.917	-0.323	-2.338
241.6	1.362	1.030	-0.494	-2.231
241.7	1.345	1.143	-0.634	-2.115
241.8	1.336	1.256	-0.746	-1.995
77π	1.335	1.375	-0.838	-1.874
241.9	1.334	1.372	-0.836	-1.877
242.0	1.342	1.492	-0.907	-1.763
242.1	1.360	1.618	-0.964	-1.653
242.2	1.391	1.750	-1.010	-1.548
242.3	1.438	1.892	-1.046	-1.448
242.4	1.505	2.045	-1.076	-1.353
242.5	1.602	2.210	-1.100	-1.261
242.6	1.740	2.387	-1.119	-1.171
242.7	1.936	2.571	-1.134	-1.083
242.8	2.216	2.747	-1.146	-0.995
242.9	2.609	2.879	-1.155	-0.908
243.0	3.135	2.891	-1.159	-0.818
243.1	3.753	2.662	-1.158	-0.727
243.2	4.290	2.087	-1.152	-0.632
243.3	4.478	1.253	-1.137	-0.534
243.4	4.221	0.472	-1.113	-0.433
155π/2	3.854	0.081	-1.087	-0.358
243.5	3.708	-0.017	-1.075	-0.331
243.6	3.170	-0.213	-1.023	-0.230
243.7	2.719	-0.223	-0.954	-0.134
243.8	2.371	-0.139	-0.868	-0.049
243.9	2.110	-0.015	-0.769	0.019
244.0	1.914	0.122	-0.662	0.066
244.1	1.767	0.260	-0.554	0.091
244.2	1.654	0.394	-0.453	0.097
244.3	1.568	0.523	-0.364	0.088
244.4	1.500	0.646	-0.289	0.070
244.5	1.448	0.765	-0.227	0.050
244.6	1.408	0.880	-0.175	0.031
244.7	1.378	0.994	-0.132	0.015
244.8	1.358	1.106	-0.092	0.005
244.9	1.345	1.219	-0.055	-0.000
245.0	1.340	1.334	-0.017	-0.001
78π	1.341	1.386	-0.000	-0.000
245.1	1.345	1.452	0.023	0.002
245.2	1.358	1.575	0.066	0.008
245.3	1.384	1.705	0.115	0.016
245.4	1.423	1.843	0.169	0.025
245.5	1.482	1.991	0.230	0.033
245.6	1.566	2.151	0.299	0.040
245.7	1.686	2.322	0.376	0.042
245.8	1.856	2.504	0.462	0.040
245.9	2.099	2.684	0.558	0.029
246.0	2.442	2.839	0.663	0.008
246.1	2.912	2.911	0.779	-0.026
246.2	3.501	2.795	0.904	-0.079
246.3	4.098	2.366	1.036	-0.154
246.4	4.453	1.613	1.172	-0.258
246.5	4.369	0.775	1.306	-0.396
246.6	3.931	0.158	1.429	-0.573
157π/2	3.850	0.093	1.445	-0.603
246.7	3.386	-0.146	1.527	-0.791
246.8	2.896	-0.220	1.585	-1.049
246.9	2.508	-0.167	1.587	-1.335
247.0	2.214	-0.056	1.517	-1.632
247.1	1.995	0.077	1.372	-1.915
247.2	1.829	0.215	1.159	-2.155
247.3	1.704	0.350	0.898	-2.333
247.4	1.607	0.481	0.613	-2.440
247.5	1.532	0.606	0.331	-2.477
247.6	1.474	0.727	0.068	-2.457
247.7	1.430	0.843	-0.163	-2.394
247.8	1.396	0.957	-0.359	-2.302
247.9	1.371	1.070	-0.521	-2.194
248.0	1.356	1.183	-0.653	-2.078
248.1	1.348	1.296	-0.759	-1.960
79π	1.348	1.396	-0.833	-1.861
248.2	1.348	1.413	-0.844	-1.845
248.3	1.358	1.534	-0.912	-1.733
248.4	1.378	1.660	-0.966	-1.626
248.5	1.412	1.795	-1.010	-1.524
248.6	1.462	1.938	-1.045	-1.426
248.7	1.535	2.093	-1.074	-1.333
248.8	1.639	2.260	-1.097	-1.242
248.9	1.787	2.437	-1.116	-1.154
249.0	1.998	2.619	-1.132	-1.067
249.1	2.296	2.788	-1.144	-0.980
249.2	2.712	2.901	-1.152	-0.893
249.3	3.256	2.875	-1.156	-0.803
249.4	3.869	2.586	-1.156	-0.711
249.5	4.352	1.953	-1.149	-0.616
249.6	4.453	1.114	-1.133	-0.517
249.7	4.135	0.386	-1.107	-0.416
159π/2	3.847	0.105	-1.087	-0.357
249.8	3.610	-0.041	-1.068	-0.313
249.9	3.089	-0.196	-1.012	-0.212
250.0	2.660	-0.185	-0.940	-0.117
250.1	2.331	-0.093	-0.851	-0.035
250.2	2.084	0.033	-0.749	0.030
250.3	1.898	0.169	-0.641	0.073
250.4	1.758	0.306	-0.535	0.094
250.5	1.650	0.438	-0.436	0.096
250.6	1.567	0.566	-0.350	0.085
250.7	1.503	0.688	-0.277	0.066
250.8	1.453	0.806	-0.217	0.046
250.9	1.415	0.921	-0.168	0.028
251.0	1.387	1.034	-0.125	0.013
251.1	1.367	1.146	-0.086	0.004
251.2	1.356	1.259	-0.049	-0.001
251.3	1.353	1.374	-0.011	-0.001
80π	1.354	1.407	-0.000	-0.000
251.4	1.359	1.493	0.030	0.003
251.5	1.375	1.617	0.074	0.010
251.6	1.403	1.748	0.123	0.018
251.7	1.446	1.888	0.179	0.026
251.8	1.510	2.038	0.241	0.034
251.9	1.600	2.199	0.311	0.040
252.0	1.728	2.372	0.390	0.042
252.1	1.911	2.553	0.478	0.037
252.2	2.170	2.729	0.575	0.025
252.3	2.534	2.871	0.683	0.002
252.4	3.026	2.913	0.801	-0.036
252.5	3.623	2.745	0.927	-0.093
252.6	4.187	2.253	1.060	-0.174
252.7	4.463	1.469	1.196	-0.285
252.8	4.301	0.664	1.328	-0.431
161π/2	3.843	0.117	1.444	-0.613
252.9	3.833	0.110	1.446	-0.617
253.0	3.297	-0.143	1.536	-0.844
253.1	2.828	-0.189	1.582	-1.108
253.2	2.461	-0.124	1.568	-1.397
253.3	2.183	-0.009	1.481	-1.691
253.4	1.975	0.124	1.320	-1.964
253.5	1.818	0.261	1.095	-2.190
253.6	1.698	0.395	0.828	-2.350
253.7	1.605	0.524	0.544	-2.439
253.8	1.534	0.648	0.268	-2.462
253.9	1.478	0.768	0.015	-2.431
254.0	1.436	0.884	-0.206	-2.361
254.1	1.403	0.997	-0.393	-2.266
254.2	1.381	1.110	-0.546	-2.157
254.3	1.366	1.222	-0.671	-2.042
254.4	1.360	1.337	-0.771	-1.927
81π	1.361	1.417	-0.828	-1.848
254.5	1.362	1.454	-0.851	-1.813
254.6	1.374	1.575	-0.916	-1.704
254.7	1.397	1.703	-0.967	-1.600
254.8	1.434	1.839	-1.009	-1.500
254.9	1.488	1.984	-1.043	-1.405
255.0	1.567	2.141	-1.071	-1.313
255.1	1.678	2.309	-1.094	-1.224
255.2	1.837	2.487	-1.114	-1.137
255.3	2.062	2.666	-1.129	-1.051
255.4	2.379	2.826	-1.141	-0.965
255.5	2.817	2.918	-1.150	-0.878
255.6	3.377	2.850	-1.154	-0.788
255.7	3.977	2.498	-1.153	-0.696
255.8	4.397	1.816	-1.146	-0.600
255.9	4.413	0.984	-1.129	-0.500
256.0	4.044	0.312	-1.101	-0.397
163π/2	3.839	0.129	-1.086	-0.356
256.1	3.515	-0.056	-1.059	-0.294
256.2	3.013	-0.174	-1.001	-0.193
256.3	2.605	-0.146	-0.925	-0.100
256.4	2.294	-0.047	-0.833	-0.021
256.5	2.060	0.080	-0.730	0.040
256.6	1.884	0.216	-0.621	0.079
256.7	1.750	0.351	-0.515	0.096
256.8	1.647	0.482	-0.419	0.094
256.9	1.567	0.608	-0.335	0.081
257.0	1.506	0.729	-0.266	0.062
257.1	1.458	0.846	-0.208	0.042
257.2	1.422	0.961	-0.160	0.024
257.3	1.395	1.073	-0.118	0.011
257.4	1.377	1.186	-0.080	0.003
257.5	1.368	1.299	-0.043	-0.001
257.6	1.367	1.415	-0.004	-0.000
82π	1.367	1.427	-0.000	-0.000
257.7	1.374	1.535	0.037	0.004
257.8	1.393	1.660	0.082	0.011
257.9	1.423	1.792	0.132	0.019

OMEGA = 20.0

k_0h	ADMITTANCE		RAD.EFF.LENGTH	
258.0	1.470	1.933	0.189	0.027
258.1	1.538	2.085	0.253	0.035
258.2	1.635	2.248	0.324	0.040
258.3	1.773	2.422	0.405	0.040
258.4	1.968	2.601	0.494	0.035
258.5	2.245	2.772	0.594	0.021
258.6	2.631	2.898	0.703	-0.005
258.7	3.143	2.907	0.822	-0.047
258.8	3.741	2.684	0.950	-0.109
258.9	4.263	2.132	1.084	-0.195
259.0	4.457	1.329	1.219	-0.313
259.1	4.225	0.565	1.349	-0.467
165π/2	3.835	0.141	1.442	-0.622
259.2	3.736	0.074	1.461	-0.662
259.3	3.212	-0.134	1.543	-0.897
259.4	2.765	-0.156	1.576	-1.167
259.5	2.417	-0.080	1.545	-1.459
259.6	2.154	0.038	1.441	-1.748
259.7	1.956	0.171	1.265	-2.010
259.8	1.807	0.307	1.029	-2.220
259.9	1.692	0.439	0.758	-2.363
260.0	1.604	0.567	0.477	-2.435
260.1	1.535	0.690	0.207	-2.444
260.2	1.482	0.808	-0.036	-2.403
260.3	1.442	0.924	-0.247	-2.327
260.4	1.411	1.037	-0.424	-2.230
260.5	1.390	1.149	-0.569	-2.121
260.6	1.377	1.262	-0.687	-2.007
260.7	1.372	1.377	-0.782	-1.894
83π	1.373	1.438	-0.824	-1.835
260.8	1.376	1.495	-0.858	-1.783
260.9	1.390	1.617	-0.919	-1.676
261.0	1.416	1.746	-0.969	-1.575
261.1	1.456	1.884	-1.009	-1.477
261.2	1.514	2.031	-1.042	-1.384
261.3	1.599	2.189	-1.069	-1.294
261.4	1.718	2.358	-1.092	-1.207
261.5	1.888	2.536	-1.111	-1.121
261.6	2.128	2.712	-1.127	-1.036
261.7	2.466	2.861	-1.139	-0.950
261.8	2.926	2.928	-1.148	-0.863
261.9	3.498	2.813	-1.152	-0.773
262.0	4.075	2.400	-1.151	-0.680
262.1	4.427	1.676	-1.143	-0.583
262.2	4.361	0.863	-1.125	-0.482
262.3	3.951	0.251	-1.095	-0.379
167π/2	3.832	0.153	-1.086	-0.355
262.4	3.423	-0.063	-1.050	-0.275
262.5	2.941	-0.149	-0.989	-0.175
262.6	2.554	-0.106	-0.909	-0.084
262.7	2.259	-0.002	-0.815	-0.007
262.8	2.037	0.127	-0.709	0.049
262.9	1.870	0.262	-0.601	0.083

k_0h	ADMITTANCE		RAD.EFF.LENGTH	
263.0	1.742	0.396	-0.497	0.096
263.1	1.644	0.526	-0.403	0.093
263.2	1.568	0.650	-0.322	0.078
263.3	1.509	0.770	-0.255	0.058
263.4	1.463	0.887	-0.199	0.039
263.5	1.429	1.000	-0.153	0.022
263.6	1.404	1.113	-0.112	0.009
263.7	1.387	1.225	-0.074	0.002
263.8	1.379	1.339	-0.036	-0.001
84π	1.379	1.448	-0.000	-0.000
263.9	1.380	1.456	0.002	0.000
264.0	1.390	1.576	0.044	0.005
264.1	1.410	1.702	0.090	0.012
264.2	1.444	1.836	0.142	0.020
264.3	1.495	1.979	0.199	0.028
264.4	1.568	2.132	0.265	0.035
264.5	1.672	2.296	0.338	0.040
264.6	1.819	2.471	0.419	0.039
264.7	2.028	2.649	0.511	0.032
264.8	2.323	2.812	0.612	0.016
264.9	2.730	2.920	0.723	-0.013
265.0	3.260	2.892	0.844	-0.059
265.1	3.853	2.612	0.974	-0.125
265.2	4.325	2.004	1.108	-0.218
265.3	4.436	1.194	1.242	-0.343
265.4	4.144	0.479	1.369	-0.505
169π/2	3.828	0.165	1.441	-0.632
265.5	3.641	0.047	1.475	-0.708
265.6	3.132	-0.120	1.546	-0.952
265.7	2.706	-0.120	1.566	-1.227
265.8	2.376	-0.036	1.519	-1.519
265.9	2.127	0.084	1.398	-1.803
266.0	1.939	0.218	1.207	-2.053
266.1	1.797	0.352	0.963	-2.246
266.2	1.687	0.483	0.690	-2.371
266.3	1.603	0.610	0.412	-2.427
266.4	1.537	0.731	0.150	-2.424
266.5	1.486	0.849	-0.084	-2.374
266.6	1.448	0.964	-0.285	-2.294
266.7	1.419	1.077	-0.453	-2.195
266.8	1.399	1.189	-0.591	-2.086
266.9	1.388	1.302	-0.702	-1.973
267.0	1.385	1.417	-0.792	-1.862
85π	1.386	1.459	-0.819	-1.823
267.1	1.391	1.536	-0.864	-1.754
267.2	1.407	1.659	-0.922	-1.650
267.3	1.435	1.790	-0.970	-1.550
267.4	1.478	1.928	-1.009	-1.455
267.5	1.542	2.077	-1.041	-1.364
267.6	1.632	2.237	-1.068	-1.276
267.7	1.760	2.407	-1.090	-1.190
267.8	1.942	2.584	-1.109	-1.105
267.9	2.198	2.756	-1.125	-1.020

k_0h	ADMITTANCE		RAD.EFF.LENGTH	
268.0	2.556	2.892	-1.137	-0.935
268.1	3.036	2.930	-1.146	-0.848
268.2	3.615	2.766	-1.150	-0.758
268.3	4.163	2.292	-1.149	-0.664
268.4	4.440	1.538	-1.140	-0.566
268.5	4.299	0.754	-1.120	-0.465
171π/2	3.825	0.177	-1.086	-0.354
268.6	3.857	0.201	-1.088	-0.360
268.7	3.336	-0.063	-1.041	-0.256
268.8	2.873	-0.121	-0.976	-0.157
268.9	2.506	-0.065	-0.893	-0.068
269.0	2.227	0.044	-0.796	0.006
269.1	2.016	0.173	-0.689	0.058
269.2	1.857	0.308	-0.581	0.087
269.3	1.735	0.440	-0.478	0.097
269.4	1.641	0.568	-0.387	0.090
269.5	1.568	0.692	-0.309	0.074
269.6	1.512	0.811	-0.244	0.054
269.7	1.468	0.927	-0.191	0.035
269.8	1.436	1.040	-0.145	0.019
269.9	1.412	1.152	-0.105	0.007
270.0	1.397	1.265	-0.068	0.001
270.1	1.391	1.379	-0.030	-0.001
86π	1.392	1.469	-0.000	-0.000
270.2	1.393	1.496	0.009	0.001
270.3	1.405	1.617	0.052	0.006
270.4	1.428	1.745	0.099	0.013
270.5	1.465	1.880	0.151	0.021
270.6	1.520	2.024	0.210	0.029
270.7	1.598	2.179	0.276	0.036
270.8	1.710	2.345	0.351	0.039
270.9	1.868	2.520	0.435	0.038
271.0	2.091	2.695	0.528	0.029
271.1	2.404	2.849	0.631	0.010
271.2	2.833	2.936	0.744	-0.022
271.3	3.377	2.868	0.867	-0.071
271.4	3.958	2.528	0.997	-0.142
271.5	4.371	1.873	1.132	-0.241
271.6	4.401	1.068	1.265	-0.374
271.7	4.058	0.404	1.387	-0.545
173π/2	3.821	0.189	1.439	-0.641
271.8	3.549	0.029	1.487	-0.757
271.9	3.056	-0.101	1.548	-1.007
272.0	2.650	-0.084	1.553	-1.287
272.1	2.338	0.008	1.489	-1.578
272.2	2.102	0.131	1.352	-1.855
272.3	1.924	0.263	1.148	-2.092
272.4	1.788	0.397	0.897	-2.268
272.5	1.683	0.527	0.622	-2.376
272.6	1.602	0.652	0.349	-2.417
272.7	1.539	0.772	0.095	-2.402
272.8	1.491	0.889	-0.129	-2.345
272.9	1.454	1.003	-0.321	-2.261

k_0h	ADMITTANCE		RAD.EFF.LENGTH	
273.0	1.427	1.116	-0.480	-2.160
273.1	1.408	1.228	-0.610	-2.051
273.2	1.399	1.342	-0.716	-1.941
273.3	1.397	1.457	-0.801	-1.831
87π	1.398	1.479	-0.815	-1.811
273.4	1.405	1.577	-0.870	-1.725
273.5	1.423	1.701	-0.925	-1.624
273.6	1.454	1.833	-0.971	-1.527
273.7	1.501	1.973	-1.008	-1.434
273.8	1.570	2.123	-1.040	-1.344
273.9	1.667	2.285	-1.066	-1.258
274.0	1.804	2.456	-1.088	-1.173
274.1	1.998	2.632	-1.107	-1.089
274.2	2.271	2.798	-1.123	-1.005
274.3	2.650	2.918	-1.136	-0.920
274.4	3.149	2.925	-1.144	-0.833
274.5	3.729	2.708	-1.149	-0.742
274.6	4.238	2.177	-1.147	-0.648
274.7	4.437	1.403	-1.136	-0.549
274.8	4.229	0.656	-1.116	-0.446
175π/2	3.818	0.200	-1.086	-0.353
274.9	3.763	0.162	-1.081	-0.342
275.0	3.252	-0.057	-1.031	-0.238
275.1	2.810	-0.090	-0.962	-0.139
275.2	2.461	-0.023	-0.876	-0.052
275.3	2.197	0.090	-0.776	0.017
275.4	1.997	0.220	-0.669	0.065
275.5	1.845	0.353	-0.561	0.091
275.6	1.729	0.484	-0.460	0.096
275.7	1.639	0.611	-0.372	0.087
275.8	1.569	0.733	-0.296	0.070
275.9	1.515	0.851	-0.234	0.050
276.0	1.474	0.966	-0.183	0.032
276.1	1.443	1.080	-0.138	0.016
276.2	1.421	1.192	-0.099	0.006
276.3	1.408	1.305	-0.061	0.000
276.4	1.403	1.419	-0.024	-0.001
88π	1.404	1.489	-0.000	-0.000
276.5	1.407	1.537	0.016	0.001
276.6	1.421	1.659	0.059	0.007
276.7	1.446	1.788	0.107	0.014
276.8	1.486	1.924	0.161	0.022
276.9	1.545	2.070	0.221	0.030
277.0	1.630	2.226	0.289	0.036
277.1	1.749	2.393	0.365	0.039
277.2	1.918	2.568	0.450	0.036
277.3	2.156	2.740	0.545	0.025
277.4	2.489	2.883	0.650	0.004
277.5	2.938	2.946	0.765	-0.031
277.6	3.493	2.833	0.889	-0.084
277.7	4.054	2.435	1.020	-0.161
277.8	4.402	1.740	1.155	-0.267
277.9	4.355	0.950	1.286	-0.407

k_0h	ADMITTANCE		RAD.EFF.LENGTH	
278.0	3.969	0.342	1.405	-0.586
177π/2	3.815	0.212	1.437	-0.650
278.1	3.459	0.019	1.496	-0.807
278.2	2.984	-0.079	1.546	-1.064
278.3	2.598	-0.045	1.536	-1.347
278.4	2.302	0.052	1.456	-1.635
278.5	2.078	0.177	1.303	-1.903
278.6	1.909	0.309	1.087	-2.126
278.7	1.779	0.441	0.830	-2.286
278.8	1.679	0.570	0.556	-2.377
278.9	1.602	0.694	0.289	-2.404
279.0	1.542	0.813	0.043	-2.378
279.1	1.495	0.929	-0.172	-2.315
279.2	1.460	1.043	-0.354	-2.227
279.3	1.435	1.155	-0.505	-2.126
279.4	1.418	1.268	-0.629	-2.018
279.5	1.410	1.381	-0.729	-1.909
89π	1.410	1.500	-0.811	-1.800
279.6	1.410	1.498	-0.809	-1.802
279.7	1.420	1.618	-0.875	-1.698
279.8	1.440	1.744	-0.928	-1.599
279.9	1.474	1.876	-0.972	-1.504
280.0	1.525	2.018	-1.008	-1.413
280.1	1.599	2.170	-1.039	-1.325
280.2	1.703	2.333	-1.065	-1.240
280.3	1.850	2.504	-1.087	-1.156
280.4	2.057	2.678	-1.106	-1.073
280.5	2.347	2.837	-1.122	-0.990
280.6	2.747	2.940	-1.134	-0.905
280.7	3.262	2.911	-1.143	-0.817
280.8	3.838	2.639	-1.147	-0.726
280.9	4.299	2.056	-1.144	-0.631
281.0	4.420	1.273	-1.133	-0.531
281.1	4.152	0.569	-1.110	-0.428
179π/2	3.812	0.224	-1.085	-0.352
281.2	3.671	0.133	-1.074	-0.323
281.3	3.172	-0.046	-1.020	-0.219
281.4	2.750	-0.057	-0.948	-0.122
281.5	2.420	0.020	-0.859	-0.037
281.6	2.169	0.135	-0.757	0.028
281.7	1.979	0.265	-0.648	0.072
281.8	1.834	0.398	-0.541	0.093
281.9	1.723	0.528	-0.443	0.095
282.0	1.637	0.653	-0.357	0.084
282.1	1.571	0.774	-0.285	0.066
282.2	1.519	0.892	-0.225	0.047
282.3	1.479	1.006	-0.175	0.028
282.4	1.450	1.119	-0.132	0.014
282.5	1.430	1.231	-0.093	0.004
282.6	1.418	1.344	-0.055	-0.000
282.7	1.415	1.459	-0.017	-0.001
90π	1.416	1.510	-0.000	-0.000
282.8	1.421	1.577	0.023	0.002
282.9	1.437	1.701	0.067	0.008

k_0h	ADMITTANCE		RAD.EFF.LENGTH	
283.0	1.465	1.830	0.116	0.015
283.1	1.509	1.968	0.171	0.024
283.2	1.572	2.116	0.232	0.031
283.3	1.663	2.274	0.301	0.036
283.4	1.791	2.442	0.379	0.038
283.5	1.971	2.615	0.466	0.034
283.6	2.224	2.782	0.563	0.021
283.7	2.576	2.914	0.669	-0.002
283.8	3.045	2.949	0.786	-0.041
283.9	3.607	2.789	0.912	-0.099
284.0	4.140	2.333	1.044	-0.181
284.1	4.418	1.607	1.178	-0.293
284.2	4.298	0.842	1.307	-0.441
284.3	3.879	0.290	1.420	-0.629
181π/2	3.808	0.235	1.434	-0.659
284.4	3.373	0.016	1.504	-0.858
284.5	2.916	-0.053	1.541	-1.121
284.6	2.550	-0.006	1.516	-1.406
284.7	2.269	0.097	1.419	-1.691
284.8	2.057	0.222	1.251	-1.949
284.9	1.895	0.354	1.025	-2.157
285.0	1.772	0.485	0.764	-2.300
285.1	1.676	0.612	0.492	-2.375
285.2	1.602	0.735	0.231	-2.388
285.3	1.544	0.854	-0.006	-2.354
285.4	1.500	0.969	-0.212	-2.285
285.5	1.466	1.082	-0.386	-2.194
285.6	1.443	1.195	-0.529	-2.092
285.7	1.427	1.307	-0.646	-1.985
285.8	1.421	1.421	-0.740	-1.878
91π	1.422	1.520	-0.806	-1.788
285.9	1.423	1.538	-0.817	-1.773
286.0	1.435	1.659	-0.879	-1.671
286.1	1.458	1.786	-0.930	-1.574
286.2	1.495	1.920	-0.973	-1.482
286.3	1.550	2.063	-1.008	-1.393
286.4	1.629	2.217	-1.038	-1.306
286.5	1.740	2.381	-1.063	-1.223
286.6	1.897	2.552	-1.085	-1.140
286.7	2.118	2.723	-1.104	-1.058
286.8	2.427	2.873	-1.120	-0.975
286.9	2.846	2.956	-1.133	-0.890
287.0	3.376	2.888	-1.142	-0.802
287.1	3.940	2.560	-1.145	-0.710
287.2	4.346	1.930	-1.142	-0.614
287.3	4.390	1.150	-1.130	-0.514
287.4	4.071	0.494	-1.105	-0.409
183π/2	3.805	0.247	-1.085	-0.350
287.5	3.581	0.112	-1.065	-0.303
287.6	3.097	-0.029	-1.008	-0.200
287.7	2.694	-0.022	-0.933	-0.104
287.8	2.381	0.063	-0.841	-0.023
287.9	2.143	0.181	-0.737	0.039

k_0h	ADMITTANCE		RAD.EFF.LENGTH	
288.0	1.962	0.311	-0.628	0.078
288.1	1.824	0.442	-0.522	0.095
288.2	1.718	0.571	-0.426	0.094
288.3	1.636	0.695	-0.343	0.081
288.4	1.572	0.815	-0.273	0.062
288.5	1.522	0.932	-0.215	0.043
288.6	1.485	1.046	-0.167	0.025
288.7	1.457	1.158	-0.125	0.012
288.8	1.438	1.270	-0.086	0.003
288.9	1.428	1.384	-0.049	-0.001
289.0	1.427	1.499	-0.011	-0.001
92π	1.428	1.530	-0.000	-0.000
289.1	1.435	1.618	0.030	0.003
289.2	1.453	1.742	0.075	0.009
289.3	1.484	1.873	0.125	0.017
289.4	1.531	2.013	0.181	0.025
289.5	1.600	2.162	0.243	0.032
289.6	1.697	2.321	0.314	0.036
289.7	1.833	2.490	0.393	0.037
289.8	2.026	2.662	0.482	0.031
289.9	2.296	2.823	0.581	0.017
290.0	2.667	2.939	0.689	-0.010
290.1	3.154	2.944	0.808	-0.052
290.2	3.718	2.734	0.935	-0.114
290.3	4.214	2.223	1.067	-0.202
290.4	4.419	1.477	1.201	-0.321
290.5	4.232	0.745	1.327	-0.478
185π/2	3.802	0.258	1.432	-0.668
290.6	3.789	0.249	1.435	-0.674
290.7	3.291	0.019	1.509	-0.910
290.8	2.853	-0.024	1.533	-1.179
290.9	2.505	0.035	1.492	-1.464
291.0	2.238	0.142	1.379	-1.744
291.1	2.036	0.268	1.197	-1.991
291.2	1.883	0.399	0.963	-2.183
291.3	1.764	0.529	0.699	-2.310
291.4	1.673	0.655	0.430	-2.369
291.5	1.602	0.776	0.175	-2.371
291.6	1.547	0.894	-0.053	-2.328
291.7	1.505	1.009	-0.250	-2.254
291.8	1.473	1.122	-0.415	-2.162
291.9	1.451	1.234	-0.550	-2.059
292.0	1.437	1.347	-0.661	-1.953
292.1	1.432	1.461	-0.751	-1.848
93π	1.434	1.540	-0.802	-1.777
292.2	1.436	1.578	-0.824	-1.745
292.3	1.450	1.700	-0.884	-1.646
292.4	1.476	1.828	-0.933	-1.551
292.5	1.516	1.964	-0.973	-1.460
292.6	1.575	2.109	-1.008	-1.373
292.7	1.660	2.264	-1.037	-1.288
292.8	1.779	2.428	-1.062	-1.206
292.9	1.947	2.600	-1.084	-1.124

k_0h	ADMITTANCE		RAD.EFF.LENGTH	
293.0	2.182	2.767	-1.103	-1.043
293.1	2.509	2.906	-1.119	-0.960
293.2	2.948	2.966	-1.132	-0.875
293.3	3.488	2.855	-1.141	-0.787
293.4	4.034	2.472	-1.144	-0.694
293.5	4.379	1.802	-1.140	-0.597
293.6	4.348	1.034	-1.126	-0.496
293.7	3.986	0.431	-1.099	-0.390
187π/2	3.799	0.270	-1.085	-0.349
293.8	3.493	0.099	-1.057	-0.284
293.9	3.025	-0.009	-0.996	-0.181
294.0	2.642	0.015	-0.917	-0.088
294.1	2.344	0.106	-0.822	-0.009
294.2	2.118	0.226	-0.716	0.048
294.3	1.947	0.356	-0.608	0.082
294.4	1.815	0.486	-0.503	0.095
294.5	1.714	0.613	-0.410	0.092
294.6	1.635	0.737	-0.329	0.078
294.7	1.574	0.856	-0.262	0.059
294.8	1.526	0.972	-0.207	0.039
294.9	1.490	1.085	-0.160	0.023
295.0	1.464	1.198	-0.118	0.010
295.1	1.447	1.310	-0.080	0.002
295.2	1.439	1.423	-0.043	-0.001
94π	1.440	1.551	-0.000	-0.000
295.3	1.439	1.539	-0.004	-0.000
295.4	1.449	1.659	0.038	0.004
295.5	1.470	1.784	0.083	0.010
295.6	1.504	1.917	0.134	0.018
295.7	1.555	2.057	0.191	0.026
295.8	1.628	2.208	0.255	0.032
295.9	1.732	2.368	0.327	0.036
296.0	1.878	2.537	0.408	0.035
296.1	2.083	2.708	0.498	0.028
296.2	2.370	2.862	0.599	0.012
296.3	2.761	2.961	0.709	-0.017
296.4	3.264	2.931	0.829	-0.063
296.5	3.823	2.668	0.957	-0.130
296.6	4.275	2.107	1.090	-0.224
296.7	4.406	1.350	1.223	-0.351
296.8	4.160	0.658	1.346	-0.515
189π/2	3.796	0.281	1.430	-0.677
296.9	3.700	0.217	1.447	-0.720
297.0	3.212	0.028	1.511	-0.964
297.1	2.793	0.007	1.521	-1.237
297.2	2.462	0.076	1.464	-1.522
297.3	2.209	0.186	1.335	-1.794
297.4	2.017	0.313	1.141	-2.029
297.5	1.871	0.443	0.899	-2.206
297.6	1.758	0.572	0.634	-2.316
297.7	1.671	0.696	0.370	-2.361
297.8	1.603	0.817	0.122	-2.351
297.9	1.550	0.934	-0.097	-2.301

OMEGA = 20.0

k_0h	ADMITTANCE		RAD.EFF.LENGTH	
298.0	1.510	1.048	-0.285	-2.223
298.1	1.480	1.161	-0.442	-2.129
298.2	1.459	1.273	-0.571	-2.027
298.3	1.447	1.386	-0.676	-1.922
298.4	1.444	1.501	-0.761	-1.818
95π	1.445	1.561	-0.799	-1.766
298.5	1.449	1.619	-0.831	-1.718
298.6	1.465	1.742	-0.888	-1.621
298.7	1.494	1.871	-0.935	-1.528
298.8	1.538	2.008	-0.974	-1.439
298.9	1.601	2.154	-1.007	-1.354
299.0	1.692	2.310	-1.036	-1.271
299.1	1.819	2.476	-1.061	-1.189
299.2	1.998	2.646	-1.083	-1.109
299.3	2.249	2.809	-1.102	-1.028
299.4	2.595	2.936	-1.118	-0.945
299.5	3.052	2.969	-1.131	-0.860
299.6	3.599	2.813	-1.139	-0.771
299.7	4.118	2.374	-1.142	-0.678
299.8	4.397	1.675	-1.137	-0.580
299.9	4.296	0.928	-1.122	-0.477
300.0	3.900	0.377	-1.093	-0.371
191π/2	3.793	0.293	-1.084	-0.348
300.1	3.409	0.094	-1.047	-0.265
300.2	2.958	0.014	-0.983	-0.163
300.3	2.592	0.053	-0.901	-0.071
300.4	2.310	0.149	-0.803	0.003
300.5	2.096	0.271	-0.696	0.056
300.6	1.932	0.400	-0.588	0.086
300.7	1.807	0.530	-0.485	0.096
300.8	1.710	0.656	-0.394	0.090
300.9	1.634	0.778	-0.316	0.074
301.0	1.576	0.896	-0.252	0.055
301.1	1.530	1.011	-0.198	0.036
301.2	1.496	1.125	-0.152	0.020
301.3	1.472	1.237	-0.112	0.008
301.4	1.456	1.349	-0.074	0.001
301.5	1.450	1.463	-0.036	-0.001
96π	1.451	1.571	-0.000	-0.000
301.6	1.452	1.579	0.003	0.000
301.7	1.463	1.700	0.045	0.005
301.8	1.487	1.826	0.091	0.011
301.9	1.524	1.960	0.143	0.019
302.0	1.579	2.102	0.201	0.026
302.1	1.657	2.254	0.267	0.033
302.2	1.769	2.416	0.340	0.036
302.3	1.925	2.585	0.423	0.034
302.4	2.143	2.752	0.515	0.025
302.5	2.447	2.897	0.617	0.006
302.6	2.858	2.976	0.730	-0.026
302.7	3.374	2.909	0.851	-0.076
302.8	3.923	2.593	0.980	-0.148
302.9	4.323	1.987	1.114	-0.248

k_0h	ADMITTANCE		RAD.EFF.LENGTH	
303.0	4.379	1.230	1.244	-0.382
303.1	4.083	0.583	1.363	-0.555
193π/2	3.790	0.304	1.427	-0.686
303.2	3.612	0.194	1.457	-0.768
303.3	3.137	0.042	1.510	-1.018
303.4	2.736	0.040	1.506	-1.295
303.5	2.422	0.117	1.434	-1.578
303.6	2.182	0.231	1.290	-1.842
303.7	2.000	0.358	1.084	-2.064
303.8	1.860	0.487	0.836	-2.225
303.9	1.752	0.614	0.571	-2.319
304.0	1.669	0.738	0.311	-2.350
304.1	1.603	0.858	0.072	-2.331
304.2	1.553	0.974	-0.139	-2.274
304.3	1.515	1.088	-0.318	-2.192
304.4	1.485	1.200	-0.467	-2.097
304.5	1.467	1.312	-0.590	-1.995
304.6	1.457	1.425	-0.689	-1.892
304.7	1.455	1.541	-0.770	-1.790
97π	1.457	1.581	-0.795	-1.756
304.8	1.463	1.659	-0.837	-1.691
304.9	1.481	1.783	-0.891	-1.597
305.0	1.512	1.914	-0.937	-1.506
305.1	1.560	2.052	-0.975	-1.419
305.2	1.628	2.200	-1.007	-1.335
305.3	1.725	2.357	-1.036	-1.253
305.4	1.861	2.523	-1.060	-1.173
305.5	2.052	2.692	-1.082	-1.093
305.6	2.319	2.849	-1.101	-1.012
305.7	2.683	2.961	-1.117	-0.930
305.8	3.158	2.964	-1.130	-0.845
305.9	3.706	2.761	-1.139	-0.756
306.0	4.191	2.270	-1.141	-0.662
306.1	4.401	1.548	-1.135	-0.562
306.2	4.235	0.832	-1.117	-0.459
195π/2	3.788	0.316	-1.084	-0.346
306.3	3.813	0.334	-1.086	-0.352
306.4	3.327	0.095	-1.037	-0.245
306.5	2.894	0.041	-0.970	-0.144
306.6	2.546	0.092	-0.884	-0.056
306.7	2.279	0.193	-0.784	0.015
306.8	2.075	0.316	-0.676	0.064
306.9	1.919	0.444	-0.568	0.089
307.0	1.799	0.573	-0.467	0.095
307.1	1.706	0.698	-0.379	0.087
307.2	1.634	0.819	-0.304	0.070
307.3	1.578	0.936	-0.242	0.051
307.4	1.535	1.051	-0.190	0.032
307.5	1.502	1.164	-0.145	0.017
307.6	1.480	1.276	-0.105	0.007
307.7	1.466	1.388	-0.068	0.001
307.8	1.460	1.502	-0.030	-0.001
98π	1.463	1.591	-0.000	-0.000
307.9	1.464	1.620	0.010	0.001

k_0h	ADMITTANCE		RAD.EFF.LENGTH	
308.0	1.478	1.741	0.053	0.006
308.1	1.504	1.869	0.100	0.012
308.2	1.544	2.003	0.153	0.020
308.3	1.603	2.147	0.212	0.027
308.4	1.688	2.300	0.279	0.033
308.5	1.807	2.463	0.354	0.035
308.6	1.973	2.631	0.438	0.032
308.7	2.206	2.795	0.532	0.021
308.8	2.528	2.930	0.636	0.000
308.9	2.957	2.986	0.750	-0.035
309.0	3.483	2.878	0.873	-0.089
309.1	4.014	2.509	1.003	-0.166
309.2	4.357	1.864	1.136	-0.273
309.3	4.342	1.117	1.265	-0.415
309.4	4.003	0.518	1.379	-0.596
197π/2	3.785	0.327	1.424	-0.695
309.5	3.526	0.179	1.465	-0.817
309.6	3.066	0.059	1.507	-1.073
309.7	2.683	0.075	1.488	-1.352
309.8	2.385	0.159	1.400	-1.632
309.9	2.157	0.275	1.241	-1.887
310.0	1.983	0.402	1.025	-2.095
310.1	1.850	0.531	0.773	-2.240
310.2	1.747	0.657	0.510	-2.318
310.3	1.667	0.779	0.255	-2.337
310.4	1.605	0.898	0.023	-2.308
310.5	1.556	1.014	-0.178	-2.246
310.6	1.520	1.127	-0.349	-2.162
310.7	1.493	1.239	-0.491	-2.066
310.8	1.476	1.351	-0.607	-1.964
310.9	1.467	1.465	-0.702	-1.862
311.0	1.467	1.580	-0.779	-1.763
99π	1.468	1.601	-0.791	-1.745
311.1	1.477	1.700	-0.842	-1.666
311.2	1.498	1.825	-0.895	-1.573
311.3	1.532	1.956	-0.938	-1.485
311.4	1.583	2.096	-0.975	-1.399
311.5	1.656	2.245	-1.007	-1.317
311.6	1.760	2.404	-1.035	-1.236
311.7	1.905	2.570	-1.060	-1.157
311.8	2.109	2.737	-1.082	-1.078
311.9	2.391	2.887	-1.101	-0.998
312.0	2.775	2.982	-1.117	-0.915
312.1	3.265	2.952	-1.130	-0.830
312.2	3.809	2.699	-1.138	-0.740
312.3	4.252	2.159	-1.139	-0.645
312.4	4.391	1.426	-1.132	-0.545
312.5	4.168	0.746	-1.113	-0.440
199π/2	3.782	0.338	-1.083	-0.345
312.6	3.726	0.300	-1.079	-0.332
312.7	3.249	0.101	-1.027	-0.226
312.8	2.834	0.070	-0.955	-0.127
312.9	2.503	0.131	-0.866	-0.040

k_0h	ADMITTANCE		RAD.EFF.LENGTH	
313.0	2.249	0.237	-0.764	0.026
313.1	2.055	0.360	-0.655	0.070
313.2	1.906	0.488	-0.548	0.092
313.3	1.792	0.616	-0.450	0.094
313.4	1.703	0.739	-0.364	0.084
313.5	1.634	0.860	-0.292	0.067
313.6	1.580	0.976	-0.232	0.047
313.7	1.539	1.090	-0.182	0.029
313.8	1.509	1.203	-0.138	0.015
313.9	1.487	1.315	-0.099	0.005
314.0	1.475	1.427	-0.061	-0.000
314.1	1.472	1.542	-0.023	-0.001
100π	1.474	1.611	-0.000	-0.000
314.2	1.477	1.660	0.017	0.001
314.3	1.493	1.782	0.060	0.007
314.4	1.522	1.911	0.108	0.013
314.5	1.565	2.047	0.162	0.021
314.6	1.629	2.192	0.223	0.028
314.7	1.719	2.347	0.291	0.033
314.8	1.846	2.510	0.368	0.034
314.9	2.024	2.677	0.454	0.030
315.0	2.272	2.836	0.549	0.017
315.1	2.611	2.959	0.655	-0.006
315.2	3.059	2.989	0.771	-0.045
315.3	3.591	2.838	0.896	-0.103
315.4	4.096	2.416	1.026	-0.186
315.5	4.377	1.741	1.159	-0.300
315.6	4.294	1.013	1.285	-0.449
315.7	3.920	0.463	1.394	-0.639
201π/2	3.779	0.349	1.421	-0.703
315.8	3.443	0.171	1.471	-0.867
315.9	2.998	0.081	1.501	-1.129
316.0	2.634	0.111	1.467	-1.409
316.1	2.350	0.202	1.362	-1.684
316.2	2.134	0.319	1.190	-1.929
316.3	1.968	0.446	0.966	-2.122
316.4	1.841	0.574	0.711	-2.251
316.5	1.742	0.699	0.450	-2.315
316.6	1.666	0.821	0.201	-2.322
316.7	1.606	0.938	-0.022	-2.285
316.8	1.560	1.053	-0.216	-2.218
316.9	1.525	1.166	-0.378	-2.131
317.0	1.500	1.278	-0.513	-2.035
317.1	1.485	1.391	-0.623	-1.934
317.2	1.477	1.504	-0.713	-1.834
101π	1.479	1.621	-0.787	-1.735
317.3	1.479	1.620	-0.787	-1.736
317.4	1.491	1.741	-0.847	-1.641
317.5	1.514	1.867	-0.898	-1.551
317.6	1.551	1.999	-0.940	-1.464
317.7	1.606	2.141	-0.976	-1.380
317.8	1.685	2.291	-1.007	-1.299
317.9	1.796	2.451	-1.035	-1.220

k_0h	ADMITTANCE		RAD.EFF.LENGTH	
318.0	1.951	2.617	-1.059	-1.141
318.1	2.167	2.781	-1.081	-1.063
318.2	2.467	2.922	-1.100	-0.983
318.3	2.869	2.998	-1.117	-0.900
318.4	3.372	2.932	-1.129	-0.814
318.5	3.906	2.627	-1.137	-0.724
318.6	4.300	2.044	-1.138	-0.628
318.7	4.369	1.309	-1.129	-0.526
318.8	4.095	0.670	-1.108	-0.421
203π/2	3.777	0.361	-1.083	-0.343
318.9	3.641	0.275	-1.071	-0.313
319.0	3.175	0.112	-1.015	-0.207
319.1	2.777	0.101	-0.940	-0.109
319.2	2.463	0.172	-0.848	-0.026
319.3	2.221	0.280	-0.744	0.037
319.4	2.036	0.404	-0.635	0.076
319.5	1.895	0.532	-0.529	0.093
319.6	1.785	0.658	-0.433	0.093
319.7	1.700	0.781	-0.350	0.081
319.8	1.634	0.900	-0.281	0.063
319.9	1.583	1.016	-0.223	0.044
320.0	1.544	1.130	-0.174	0.026
320.1	1.515	1.242	-0.132	0.013
320.2	1.495	1.354	-0.093	0.004
320.3	1.485	1.467	-0.055	-0.000
320.4	1.483	1.582	-0.017	-0.001
102π	1.485	1.631	-0.000	-0.000
320.5	1.490	1.700	0.024	0.002
320.6	1.509	1.824	0.068	0.008
320.7	1.540	1.953	0.117	0.015
320.8	1.587	2.091	0.172	0.022
320.9	1.656	2.237	0.234	0.029
321.0	1.752	2.393	0.304	0.033
321.1	1.888	2.557	0.382	0.033
321.2	2.077	2.722	0.470	0.027
321.3	2.340	2.876	0.567	0.013
321.4	2.698	2.984	0.675	-0.013
321.5	3.161	2.986	0.792	-0.056
321.6	3.695	2.789	0.918	-0.118
321.7	4.169	2.316	1.049	-0.207
321.8	4.384	1.619	1.181	-0.328
321.9	4.238	0.917	1.304	-0.485
322.0	3.836	0.418	1.407	-0.683
205π/2	3.774	0.372	1.418	-0.712
322.1	3.363	0.169	1.475	-0.919
322.2	2.934	0.105	1.491	-1.185
322.3	2.587	0.148	1.442	-1.464
322.4	2.318	0.244	1.322	-1.734
322.5	2.112	0.363	1.138	-1.967
322.6	1.954	0.490	0.906	-2.146
322.7	1.832	0.617	0.649	-2.259
322.8	1.738	0.741	0.391	-2.309
322.9	1.665	0.861	0.150	-2.305

k_0h	ADMITTANCE		RAD.EFF.LENGTH	
323.0	1.608	0.978	-0.066	-2.261
323.1	1.564	1.093	-0.251	-2.189
323.2	1.531	1.205	-0.406	-2.101
323.3	1.508	1.317	-0.534	-2.004
323.4	1.493	1.430	-0.638	-1.905
323.5	1.488	1.544	-0.724	-1.806
103π	1.490	1.642	-0.784	-1.725
323.6	1.492	1.660	-0.794	-1.710
323.7	1.505	1.782	-0.852	-1.617
323.8	1.531	1.909	-0.901	-1.528
323.9	1.571	2.043	-0.941	-1.443
324.0	1.631	2.185	-0.977	-1.361
324.1	1.715	2.337	-1.007	-1.282
324.2	1.833	2.497	-1.035	-1.204
324.3	1.999	2.663	-1.059	-1.126
324.4	2.229	2.823	-1.081	-1.048
324.5	2.545	2.954	-1.100	-0.968
324.6	2.965	3.008	-1.116	-0.885
324.7	3.478	2.903	-1.129	-0.799
324.8	3.995	2.547	-1.136	-0.707
324.9	4.336	1.926	-1.136	-0.610
325.0	4.335	1.198	-1.126	-0.508
325.1	4.018	0.604	-1.102	-0.401
207π/2	3.771	0.383	-1.082	-0.342
325.2	3.557	0.258	-1.062	-0.293
325.3	3.104	0.128	-1.003	-0.188
325.4	2.724	0.134	-0.925	-0.092
325.5	2.425	0.213	-0.829	-0.012
325.6	2.195	0.324	-0.723	0.046
325.7	2.019	0.448	-0.615	0.081
325.8	1.884	0.575	-0.510	0.094
325.9	1.779	0.700	-0.417	0.091
326.0	1.698	0.822	-0.337	0.078
326.1	1.635	0.940	-0.270	0.059
326.2	1.586	1.056	-0.214	0.040
326.3	1.548	1.169	-0.167	0.024
326.4	1.521	1.281	-0.125	0.011
326.5	1.503	1.393	-0.086	0.003
326.6	1.494	1.506	-0.049	-0.001
326.7	1.494	1.621	-0.010	-0.001
104π	1.496	1.652	-0.000	-0.000
326.8	1.504	1.741	0.031	0.003
326.9	1.524	1.865	0.076	0.009
327.0	1.559	1.996	0.126	0.016
327.1	1.610	2.135	0.182	0.023
327.2	1.683	2.283	0.245	0.029
327.3	1.786	2.439	0.316	0.033
327.4	1.931	2.603	0.396	0.032
327.5	2.132	2.767	0.486	0.025
327.6	2.411	2.913	0.585	0.008
327.7	2.787	3.005	0.694	-0.021
327.8	3.265	2.975	0.813	-0.067
327.9	3.795	2.730	0.940	-0.135

k_0h	ADMITTANCE		RAD.EFF.LENGTH	
328.0	4.230	2.210	1.072	-0.230
328.1	4.377	1.501	1.202	-0.357
328.2	4.174	0.831	1.322	-0.523
209π/2	3.769	0.394	1.415	-0.720
328.3	3.752	0.383	1.418	-0.728
328.4	3.286	0.173	1.476	-0.971
328.5	2.874	0.133	1.478	-1.241
328.6	2.543	0.187	1.414	-1.519
328.7	2.287	0.287	1.279	-1.781
328.8	2.091	0.407	1.083	-2.002
328.9	1.941	0.533	0.845	-2.166
329.0	1.825	0.659	0.588	-2.263
329.1	1.734	0.782	0.335	-2.300
329.2	1.664	0.902	0.100	-2.286
329.3	1.609	1.018	-0.107	-2.236
329.4	1.568	1.132	-0.284	-2.160
329.5	1.536	1.244	-0.431	-2.071
329.6	1.515	1.356	-0.553	-1.974
329.7	1.502	1.469	-0.652	-1.876
329.8	1.499	1.583	-0.734	-1.779
105π	1.501	1.662	-0.781	-1.715
329.9	1.504	1.701	-0.801	-1.685
330.0	1.520	1.823	-0.857	-1.594
330.1	1.548	1.951	-0.903	-1.507
330.2	1.592	2.086	-0.943	-1.424
330.3	1.656	2.230	-0.977	-1.343
330.4	1.746	2.383	-1.008	-1.265
330.5	1.872	2.544	-1.034	-1.188
330.6	2.048	2.708	-1.059	-1.111
330.7	2.293	2.864	-1.081	-1.033
330.8	2.627	2.983	-1.100	-0.953
330.9	3.064	3.011	-1.116	-0.870
331.0	3.582	2.865	-1.129	-0.783
331.1	4.076	2.458	-1.135	-0.691
331.2	4.358	1.807	-1.134	-0.593
331.3	4.292	1.095	-1.122	-0.489
331.4	3.939	0.548	-1.096	-0.382
211π/2	3.766	0.405	-1.082	-0.340
331.5	3.476	0.248	-1.053	-0.273
331.6	3.037	0.147	-0.990	-0.169
331.7	2.674	0.169	-0.908	-0.075
331.8	2.389	0.254	-0.810	0.001
331.9	2.171	0.368	-0.703	0.054
332.0	2.003	0.492	-0.595	0.085
332.1	1.874	0.618	-0.492	0.095
332.2	1.774	0.742	-0.401	0.089
332.3	1.696	0.863	-0.324	0.074
332.4	1.635	0.981	-0.259	0.055
332.5	1.589	1.095	-0.205	0.037
332.6	1.553	1.208	-0.159	0.021
332.7	1.528	1.320	-0.118	0.009
332.8	1.512	1.432	-0.080	0.002
332.9	1.504	1.545	-0.043	-0.001

k_0h	ADMITTANCE		RAD.EFF.LENGTH	
106π	1.507	1.672	-0.000	-0.000
333.0	1.506	1.661	-0.004	-0.000
333.1	1.518	1.781	0.038	0.004
333.2	1.541	1.907	0.084	0.010
333.3	1.578	2.039	0.135	0.017
333.4	1.633	2.179	0.193	0.024
333.5	1.711	2.328	0.257	0.029
333.6	1.822	2.486	0.330	0.032
333.7	1.976	2.649	0.411	0.030
333.8	2.190	2.810	0.502	0.021
333.9	2.485	2.947	0.603	0.002
334.0	2.879	3.020	0.714	-0.030
334.1	3.369	2.956	0.835	-0.080
334.2	3.889	2.662	0.963	-0.152
334.3	4.279	2.100	1.094	-0.253
334.4	4.359	1.386	1.223	-0.388
334.5	4.105	0.755	1.338	-0.562
213π/2	3.764	0.417	1.412	-0.729
334.6	3.669	0.355	1.427	-0.775
334.7	3.212	0.182	1.474	-1.024
334.8	2.817	0.162	1.463	-1.297
334.9	2.502	0.226	1.383	-1.572
335.0	2.259	0.330	1.234	-1.826
335.1	2.072	0.451	1.028	-2.034
335.2	1.929	0.577	0.785	-2.182
335.3	1.817	0.702	0.529	-2.265
335.4	1.731	0.824	0.280	-2.289
335.5	1.664	0.942	0.053	-2.266
335.6	1.612	1.058	-0.146	-2.210
335.7	1.572	1.171	-0.315	-2.132
335.8	1.542	1.283	-0.455	-2.041
335.9	1.522	1.395	-0.571	-1.945
336.0	1.511	1.508	-0.666	-1.848
336.1	1.509	1.623	-0.743	-1.752
107π	1.512	1.682	-0.777	-1.705
336.2	1.517	1.741	-0.807	-1.660
336.3	1.535	1.864	-0.861	-1.571
336.4	1.566	1.993	-0.906	-1.486
336.5	1.614	2.129	-0.944	-1.404
336.6	1.682	2.275	-0.978	-1.325
336.7	1.778	2.429	-1.008	-1.248
336.8	1.913	2.590	-1.035	-1.172
336.9	2.100	2.753	-1.059	-1.096
337.0	2.359	2.902	-1.081	-1.018
337.1	2.711	3.007	-1.100	-0.938
337.2	3.164	3.009	-1.116	-0.855
337.3	3.683	2.818	-1.128	-0.767
337.4	4.148	2.363	-1.134	-0.674
337.5	4.367	1.689	-1.132	-0.575
337.6	4.240	1.001	-1.118	-0.470
337.7	3.858	0.502	-1.089	-0.362
215π/2	3.761	0.428	-1.081	-0.339
337.8	3.397	0.244	-1.043	-0.253
337.9	2.973	0.170	-0.976	-0.150

OMEGA = 20.0

k_0h	ADMITTANCE		RAD.EFF.LENGTH	
338.0	2.626	0.205	-0.891	-0.059
338.1	2.356	0.296	-0.791	0.013
338.2	2.148	0.411	-0.683	0.062
338.3	1.988	0.535	-0.575	0.088
338.4	1.865	0.661	-0.475	0.094
338.5	1.769	0.784	-0.386	0.086
338.6	1.695	0.904	-0.311	0.070
338.7	1.637	1.020	-0.249	0.052
338.8	1.592	1.135	-0.197	0.033
338.9	1.558	1.247	-0.152	0.018
339.0	1.535	1.359	-0.112	0.008
339.1	1.520	1.471	-0.074	0.001
339.2	1.514	1.584	-0.036	-0.001
108π	1.517	1.692	-0.000	-0.000
339.3	1.518	1.701	0.003	0.000
339.4	1.532	1.822	0.046	0.004
339.5	1.557	1.948	0.092	0.011
339.6	1.598	2.082	0.145	0.018
339.7	1.657	2.223	0.203	0.024
339.8	1.741	2.373	0.269	0.030
339.9	1.859	2.532	0.343	0.032
340.0	2.023	2.694	0.426	0.028
340.1	2.250	2.851	0.519	0.018
340.2	2.561	2.979	0.622	-0.003
340.3	2.973	3.030	0.735	-0.039
340.4	3.472	2.929	0.857	-0.093
340.5	3.977	2.586	0.985	-0.171
340.6	4.316	1.986	1.116	-0.278
340.7	4.329	1.278	1.243	-0.421
340.8	4.032	0.688	1.353	-0.602
217π/2	3.759	0.439	1.409	-0.737
340.9	3.587	0.336	1.434	-0.823
341.0	3.142	0.196	1.469	-1.078
341.1	2.763	0.194	1.444	-1.352
341.2	2.463	0.266	1.348	-1.623
341.3	2.232	0.373	1.186	-1.867
341.4	2.054	0.494	0.971	-2.062
341.5	1.917	0.619	0.725	-2.195
341.6	1.811	0.743	0.471	-2.263
341.7	1.728	0.865	0.228	-2.276
341.8	1.664	0.982	0.008	-2.245
341.9	1.614	1.097	-0.183	-2.184
342.0	1.576	1.210	-0.344	-2.103
342.1	1.548	1.322	-0.478	-2.012
342.2	1.530	1.434	-0.588	-1.916
342.3	1.521	1.547	-0.678	-1.821
342.4	1.520	1.662	-0.752	-1.727
109π	1.523	1.702	-0.774	-1.696
342.5	1.530	1.781	-0.813	-1.636
342.6	1.551	1.905	-0.865	-1.549
342.7	1.585	2.035	-0.908	-1.466
342.8	1.636	2.173	-0.946	-1.386
342.9	1.709	2.319	-0.979	-1.308

k_0h	ADMITTANCE		RAD.EFF.LENGTH	
343.0	1.812	2.475	-1.008	-1.232
343.1	1.955	2.636	-1.035	-1.156
343.2	2.155	2.796	-1.059	-1.081
343.3	2.429	2.939	-1.081	-1.004
343.4	2.798	3.028	-1.100	-0.924
343.5	3.265	2.998	-1.116	-0.840
343.6	3.781	2.762	-1.128	-0.752
343.7	4.209	2.262	-1.133	-0.657
343.8	4.364	1.574	-1.130	-0.557
343.9	4.180	0.915	-1.114	-0.451
219π/2	3.757	0.450	-1.081	-0.337
344.0	3.777	0.464	-1.082	-0.342
344.1	3.321	0.245	-1.032	-0.233
344.2	2.912	0.195	-0.962	-0.132
344.3	2.582	0.242	-0.873	-0.044
344.4	2.324	0.337	-0.771	0.024
344.5	2.127	0.454	-0.662	0.069
344.6	1.974	0.578	-0.555	0.090
344.7	1.857	0.703	-0.457	0.094
344.8	1.765	0.825	-0.372	0.084
344.9	1.693	0.944	-0.299	0.067
345.0	1.638	1.060	-0.239	0.048
345.1	1.595	1.174	-0.189	0.030
345.2	1.564	1.286	-0.145	0.016
345.3	1.542	1.397	-0.106	0.006
345.4	1.529	1.510	-0.068	0.000
345.5	1.525	1.624	-0.030	-0.001
110π	1.528	1.712	-0.000	-0.000
345.6	1.530	1.741	0.010	0.001
345.7	1.546	1.863	0.053	0.005
345.8	1.574	1.990	0.101	0.012
345.9	1.618	2.125	0.154	0.019
346.0	1.682	2.267	0.214	0.025
346.1	1.771	2.419	0.281	0.030
346.2	1.897	2.578	0.357	0.031
346.3	2.071	2.739	0.441	0.026
346.4	2.313	2.891	0.536	0.014
346.5	2.641	3.007	0.641	-0.010
346.6	3.068	3.034	0.755	-0.049
346.7	3.573	2.893	0.878	-0.107
346.8	4.057	2.501	1.008	-0.190
346.9	4.340	1.872	1.138	-0.305
347.0	4.289	1.177	1.262	-0.455
347.1	3.956	0.631	1.367	-0.645
221π/2	3.754	0.461	1.405	-0.745
347.2	3.507	0.323	1.439	-0.873
347.3	3.074	0.213	1.462	-1.132
347.4	2.712	0.227	1.421	-1.406
347.5	2.427	0.306	1.311	-1.673
347.6	2.207	0.416	1.137	-1.906
347.7	2.037	0.538	0.914	-2.087
347.8	1.907	0.662	0.665	-2.204
347.9	1.805	0.785	0.414	-2.259

k_0h	ADMITTANCE		RAD.EFF.LENGTH	
348.0	1.726	0.905	0.177	-2.261
348.1	1.664	1.022	-0.035	-2.223
348.2	1.616	1.137	-0.217	-2.157
348.3	1.580	1.249	-0.371	-2.074
348.4	1.555	1.361	-0.499	-1.983
348.5	1.538	1.473	-0.603	-1.888
348.6	1.530	1.586	-0.689	-1.794
348.7	1.532	1.702	-0.760	-1.702
111π	1.533	1.722	-0.771	-1.687
348.8	1.543	1.821	-0.819	-1.613
348.9	1.566	1.946	-0.868	-1.528
349.0	1.603	2.078	-0.910	-1.446
349.1	1.658	2.217	-0.947	-1.367
349.2	1.737	2.364	-0.979	-1.291
349.3	1.847	2.520	-1.009	-1.216
349.4	1.999	2.681	-1.035	-1.141
349.5	2.211	2.839	-1.059	-1.066
349.6	2.501	2.973	-1.081	-0.989
349.7	2.887	3.043	-1.101	-0.909
349.8	3.366	2.981	-1.117	-0.825
349.9	3.873	2.698	-1.128	-0.735
350.0	4.258	2.156	-1.132	-0.640
350.1	4.349	1.462	-1.127	-0.538
350.2	4.115	0.839	-1.109	-0.431
223π/2	3.752	0.472	-1.080	-0.335
350.3	3.696	0.435	-1.075	-0.322
350.4	3.248	0.252	-1.021	-0.214
350.5	2.855	0.223	-0.947	-0.114
350.6	2.540	0.280	-0.855	-0.030
350.7	2.295	0.380	-0.751	0.034
350.8	2.107	0.497	-0.642	0.074
350.9	1.961	0.621	-0.536	0.092
351.0	1.849	0.745	-0.441	0.092
351.1	1.761	0.866	-0.358	0.081
351.2	1.693	0.984	-0.288	0.063
351.3	1.640	1.100	-0.230	0.044
351.4	1.599	1.213	-0.181	0.027
351.5	1.569	1.325	-0.138	0.014
351.6	1.549	1.436	-0.099	0.005
351.7	1.537	1.549	-0.061	-0.000
351.8	1.535	1.663	-0.023	-0.001
112π	1.538	1.732	-0.000	-0.000
351.9	1.542	1.781	0.017	0.001
352.0	1.561	1.904	0.061	0.006
352.1	1.592	2.032	0.109	0.013
352.2	1.639	2.168	0.164	0.019
352.3	1.707	2.312	0.225	0.026
352.4	1.803	2.464	0.293	0.030
352.5	1.937	2.623	0.370	0.030
352.6	2.122	2.783	0.457	0.024
352.7	2.378	2.929	0.553	0.009
352.8	2.723	3.031	0.660	-0.017
352.9	3.166	3.032	0.776	-0.059

k_0h	ADMITTANCE		RAD.EFF.LENGTH	
353.0	3.672	2.848	0.900	-0.122
353.1	4.127	2.410	1.030	-0.211
353.2	4.351	1.758	1.160	-0.332
353.3	4.241	1.083	1.280	-0.490
353.4	3.878	0.584	1.379	-0.688
225π/2	3.750	0.483	1.402	-0.753
353.5	3.429	0.317	1.441	-0.923
353.6	3.011	0.234	1.451	-1.186
353.7	2.665	0.261	1.396	-1.460
353.8	2.393	0.347	1.271	-1.720
353.9	2.183	0.459	1.085	-1.942
354.0	2.022	0.581	0.856	-2.108
354.1	1.897	0.704	0.607	-2.210
354.2	1.800	0.827	0.359	-2.253
354.3	1.724	0.946	0.129	-2.245
354.4	1.665	1.062	-0.075	-2.200
354.5	1.619	1.176	-0.250	-2.130
354.6	1.585	1.288	-0.397	-2.046
354.7	1.561	1.400	-0.518	-1.955
354.8	1.546	1.512	-0.618	-1.861
354.9	1.540	1.625	-0.700	-1.768
113π	1.543	1.742	-0.768	-1.677
355.0	1.543	1.742	-0.768	-1.677
355.1	1.557	1.862	-0.824	-1.590
355.2	1.583	1.988	-0.872	-1.507
355.3	1.623	2.120	-0.913	-1.427
355.4	1.682	2.261	-0.948	-1.350
355.5	1.766	2.409	-0.980	-1.274
355.6	1.883	2.566	-1.009	-1.200
355.7	2.045	2.726	-1.035	-1.126
355.8	2.270	2.880	-1.060	-1.051
355.9	2.576	3.004	-1.082	-0.974
356.0	2.979	3.054	-1.101	-0.894
356.1	3.466	2.955	-1.117	-0.809
356.2	3.959	2.625	-1.128	-0.719
356.3	4.296	2.047	-1.131	-0.622
356.4	4.322	1.356	-1.124	-0.519
356.5	4.046	0.772	-1.104	-0.412
227π/2	3.747	0.494	-1.079	-0.334
356.6	3.616	0.413	-1.066	-0.302
356.7	3.178	0.264	-1.009	-0.194
356.8	2.801	0.253	-0.931	-0.097
356.9	2.501	0.319	-0.836	-0.016
357.0	2.268	0.422	-0.730	0.044
357.1	2.088	0.540	-0.622	0.079
357.2	1.949	0.664	-0.518	0.093
357.3	1.842	0.787	-0.424	0.091
357.4	1.758	0.907	-0.344	0.077
357.5	1.692	1.024	-0.277	0.060
357.6	1.641	1.139	-0.221	0.041
357.7	1.603	1.252	-0.173	0.025
357.8	1.575	1.363	-0.132	0.012
357.9	1.556	1.475	-0.093	0.004

k_0h	ADMITTANCE		RAD.EFF.LENGTH	
358.0	1.546	1.588	-0.055	-0.000
358.1	1.546	1.703	-0.017	-0.001
114π	1.548	1.751	-0.000	-0.000
358.2	1.555	1.821	0.024	0.002
358.3	1.576	1.945	0.069	0.007
358.4	1.610	2.074	0.118	0.014
358.5	1.661	2.211	0.174	0.020
358.6	1.734	2.356	0.236	0.026
358.7	1.836	2.510	0.306	0.029
358.8	1.979	2.669	0.385	0.028
358.9	2.176	2.826	0.473	0.021
359.0	2.446	2.965	0.571	0.004
359.1	2.808	3.052	0.679	-0.025
359.2	3.264	3.023	0.797	-0.071
359.3	3.767	2.795	0.922	-0.139
359.4	4.188	2.313	1.052	-0.234
359.5	4.351	1.646	1.180	-0.362
359.6	4.186	0.998	1.297	-0.528
359.7	3.800	0.544	1.389	-0.733
229π/2	3.745	0.505	1.398	-0.761
359.8	3.354	0.317	1.441	-0.974
359.9	2.950	0.258	1.438	-1.241
360.0	2.620	0.297	1.367	-1.512
360.1	2.361	0.388	1.229	-1.764
360.2	2.161	0.501	1.032	-1.974
360.3	2.007	0.623	0.798	-2.125
360.4	1.888	0.746	0.549	-2.213
360.5	1.795	0.868	0.306	-2.244
360.6	1.722	0.986	0.082	-2.227
360.7	1.666	1.102	-0.114	-2.176
360.8	1.622	1.215	-0.281	-2.103
360.9	1.590	1.327	-0.421	-2.018
361.0	1.568	1.438	-0.537	-1.927
361.1	1.554	1.551	-0.632	-1.834
361.2	1.550	1.664	-0.710	-1.742
115π	1.554	1.761	-0.764	-1.668
361.3	1.555	1.781	-0.775	-1.654
361.4	1.571	1.903	-0.829	-1.569
361.5	1.599	2.029	-0.875	-1.487
361.6	1.643	2.163	-0.915	-1.408
361.7	1.706	2.305	-0.950	-1.332
361.8	1.796	2.455	-0.981	-1.258
361.9	1.920	2.611	-1.010	-1.185
362.0	2.093	2.770	-1.036	-1.111
362.1	2.332	2.919	-1.060	-1.037
362.2	2.654	3.032	-1.082	-0.960
362.3	3.072	3.058	-1.102	-0.879
362.4	3.565	2.921	-1.117	-0.794
362.5	4.038	2.544	-1.128	-0.702
362.6	4.322	1.936	-1.130	-0.605
362.7	4.287	1.257	-1.121	-0.500
362.8	3.973	0.714	-1.098	-0.392
231π/2	3.743	0.516	-1.079	-0.332
362.9	3.537	0.399	-1.057	-0.281

k_0h	ADMITTANCE		RAD.EFF.LENGTH	
363.0	3.111	0.279	-0.996	-0.175
363.1	2.750	0.285	-0.915	-0.080
363.2	2.464	0.358	-0.817	-0.003
363.3	2.242	0.464	-0.710	0.052
363.4	2.071	0.583	-0.602	0.083
363.5	1.938	0.706	-0.500	0.094
363.6	1.835	0.828	-0.409	0.089
363.7	1.755	0.948	-0.331	0.074
363.8	1.692	1.064	-0.267	0.056
363.9	1.644	1.178	-0.212	0.038
364.0	1.607	1.291	-0.166	0.022
364.1	1.581	1.402	-0.125	0.010
364.2	1.564	1.514	-0.087	0.003
364.3	1.556	1.627	-0.049	-0.001
364.4	1.557	1.742	-0.010	-0.001
116π	1.559	1.771	-0.000	-0.000
364.5	1.568	1.861	0.031	0.003
364.6	1.591	1.986	0.077	0.008
364.7	1.628	2.116	0.127	0.015
364.8	1.683	2.255	0.184	0.021
364.9	1.761	2.401	0.247	0.026
365.0	1.870	2.555	0.319	0.029
365.1	2.022	2.713	0.399	0.027
365.2	2.231	2.868	0.489	0.018
365.3	2.517	2.999	0.589	-0.001
365.4	2.895	3.067	0.699	-0.033
365.5	3.362	3.007	0.818	-0.083
365.6	3.858	2.734	0.945	-0.156
365.7	4.238	2.211	1.074	-0.257
365.8	4.339	1.537	1.200	-0.392
365.9	4.124	0.922	1.313	-0.566
233π/2	3.741	0.527	1.394	-0.769
366.0	3.721	0.514	1.397	-0.779
366.1	3.282	0.322	1.438	-1.026
366.2	2.893	0.284	1.421	-1.294
366.3	2.577	0.334	1.336	-1.563
366.4	2.331	0.429	1.184	-1.806
366.5	2.141	0.544	0.978	-2.003
366.6	1.994	0.666	0.740	-2.140
366.7	1.879	0.788	0.493	-2.214
366.8	1.790	0.909	0.254	-2.233
366.9	1.721	1.026	0.038	-2.208
367.0	1.667	1.141	-0.151	-2.152
367.1	1.626	1.254	-0.311	-2.076
367.2	1.595	1.366	-0.444	-1.990
367.3	1.574	1.477	-0.554	-1.899
367.4	1.563	1.589	-0.644	-1.808
367.5	1.560	1.704	-0.719	-1.718
117π	1.564	1.781	-0.761	-1.660
367.6	1.567	1.821	-0.781	-1.631
367.7	1.585	1.943	-0.833	-1.547
367.8	1.616	2.071	-0.878	-1.467
367.9	1.664	2.206	-0.917	-1.390

k_0h	ADMITTANCE		RAD.EFF.LENGTH	
368.0	1.731	2.349	-0.951	-1.315
368.1	1.827	2.500	-0.982	-1.242
368.2	1.960	2.657	-1.010	-1.170
368.3	2.144	2.814	-1.037	-1.097
368.4	2.396	2.957	-1.061	-1.022
368.5	2.734	3.056	-1.083	-0.945
368.6	3.167	3.057	-1.103	-0.864
368.7	3.661	2.879	-1.118	-0.778
368.8	4.108	2.457	-1.127	-0.686
368.9	4.335	1.826	-1.128	-0.586
369.0	4.242	1.164	-1.118	-0.481
369.1	3.898	0.665	-1.092	-0.371
235π/2	3.739	0.538	-1.078	-0.330
369.2	3.461	0.391	-1.048	-0.261
369.3	3.047	0.298	-0.982	-0.156
369.4	2.702	0.318	-0.898	-0.064
369.5	2.429	0.398	-0.798	0.009
369.6	2.218	0.506	-0.689	0.060
369.7	2.054	0.626	-0.582	0.086
369.8	1.928	0.748	-0.482	0.093
369.9	1.829	0.869	-0.394	0.086
370.0	1.752	0.988	-0.319	0.071
370.1	1.692	1.104	-0.256	0.052
370.2	1.646	1.218	-0.204	0.035
370.3	1.611	1.329	-0.159	0.020
370.4	1.587	1.441	-0.118	0.009
370.5	1.571	1.553	-0.080	0.002
370.6	1.565	1.666	-0.042	-0.001
118π	1.569	1.791	-0.000	-0.000
370.7	1.568	1.782	-0.003	-0.000
370.8	1.582	1.902	0.039	0.003
370.9	1.607	2.027	0.085	0.009
371.0	1.647	2.159	0.136	0.015
371.1	1.706	2.298	0.194	0.022
371.2	1.790	2.446	0.259	0.027
371.3	1.906	2.600	0.332	0.028
371.4	2.067	2.758	0.414	0.025
371.5	2.289	2.908	0.505	0.014
371.6	2.590	3.030	0.607	-0.007
371.7	2.984	3.078	0.719	-0.042
371.8	3.460	2.983	0.839	-0.096
371.9	3.942	2.664	0.967	-0.174
372.0	4.277	2.106	1.096	-0.282
372.1	4.316	1.434	1.220	-0.425
372.2	4.058	0.854	1.327	-0.606
237π/2	3.736	0.549	1.390	-0.777
372.3	3.643	0.490	1.403	-0.827
372.4	3.213	0.331	1.433	-1.079
372.5	2.838	0.312	1.401	-1.348
372.6	2.537	0.372	1.301	-1.611
372.7	2.303	0.471	1.137	-1.845
372.8	2.121	0.587	0.924	-2.029
372.9	1.981	0.708	0.683	-2.151

k_0h	ADMITTANCE		RAD.EFF.LENGTH	
373.0	1.872	0.830	0.438	-2.212
373.1	1.786	0.949	0.205	-2.220
373.2	1.720	1.066	-0.004	-2.187
373.3	1.668	1.181	-0.185	-2.127
373.4	1.629	1.293	-0.338	-2.049
373.5	1.600	1.405	-0.465	-1.963
373.6	1.581	1.516	-0.570	-1.872
373.7	1.571	1.628	-0.656	-1.782
373.8	1.571	1.743	-0.728	-1.694
119π	1.574	1.801	-0.759	-1.651
373.9	1.580	1.861	-0.787	-1.608
374.0	1.600	1.984	-0.838	-1.526
374.1	1.634	2.113	-0.881	-1.448
374.2	1.685	2.249	-0.919	-1.372
374.3	1.758	2.393	-0.952	-1.299
374.4	1.859	2.545	-0.983	-1.226
374.5	2.001	2.701	-1.011	-1.155
374.6	2.196	2.856	-1.038	-1.082
374.7	2.462	2.992	-1.062	-1.008
374.8	2.817	3.076	-1.084	-0.931
374.9	3.262	3.049	-1.103	-0.849
375.0	3.754	2.829	-1.118	-0.762
375.1	4.168	2.364	-1.127	-0.668
375.2	4.338	1.717	-1.127	-0.568
375.3	4.190	1.080	-1.114	-0.461
375.4	3.822	0.624	-1.085	-0.351
239π/2	3.734	0.559	-1.077	-0.328
375.5	3.387	0.388	-1.037	-0.241
375.6	2.987	0.320	-0.968	-0.138
375.7	2.657	0.353	-0.880	-0.048
375.8	2.397	0.438	-0.778	0.021
375.9	2.195	0.548	-0.669	0.066
376.0	2.039	0.668	-0.563	0.089
376.1	1.918	0.790	-0.465	0.093
376.2	1.824	0.910	-0.379	0.083
376.3	1.750	1.028	-0.307	0.067
376.4	1.693	1.144	-0.247	0.049
376.5	1.649	1.257	-0.196	0.031
376.6	1.616	1.368	-0.152	0.017
376.7	1.593	1.479	-0.112	0.007
376.8	1.579	1.591	-0.074	0.001
376.9	1.575	1.705	-0.036	-0.001
120π	1.579	1.811	-0.000	-0.000
377.0	1.580	1.822	0.004	0.000
377.1	1.595	1.942	0.046	0.004
377.2	1.623	2.069	0.093	0.010
377.3	1.667	2.201	0.146	0.016
377.4	1.730	2.342	0.205	0.022
377.5	1.819	2.490	0.271	0.027
377.6	1.943	2.645	0.345	0.027
377.7	2.114	2.801	0.429	0.023
377.8	2.349	2.947	0.522	0.010
377.9	2.666	3.057	0.626	-0.013

OMEGA = 20.0

k_0h	ADMITTANCE		RAD.EFF.LENGTH	
378.0	3.075	3.083	0.739	-0.052
378.1	3.556	2.951	0.861	-0.110
378.2	4.019	2.588	0.989	-0.194
378.3	4.304	2.000	1.117	-0.308
378.4	4.283	1.336	1.238	-0.458
378.5	3.988	0.796	1.340	-0.648
241π/2	3.732	0.570	1.386	-0.785
378.6	3.566	0.474	1.407	-0.875
378.7	3.146	0.345	1.424	-1.131
378.8	2.787	0.343	1.378	-1.400
378.9	2.500	0.410	1.264	-1.658
379.0	2.276	0.512	1.088	-1.881
379.1	2.103	0.629	0.868	-2.051
379.2	1.969	0.750	0.626	-2.159
379.3	1.865	0.871	0.384	-2.207
379.4	1.783	0.990	0.157	-2.205
379.5	1.719	1.106	-0.045	-2.166
379.6	1.670	1.220	-0.218	-2.102
379.7	1.633	1.332	-0.364	-2.023
379.8	1.606	1.443	-0.485	-1.936
379.9	1.588	1.555	-0.585	-1.846
380.0	1.580	1.667	-0.667	-1.757
380.1	1.581	1.783	-0.736	-1.670
121π	1.584	1.821	-0.756	-1.642
380.2	1.592	1.901	-0.793	-1.586
380.3	1.615	2.025	-0.842	-1.506
380.4	1.652	2.155	-0.884	-1.429
380.5	1.707	2.292	-0.921	-1.355
380.6	1.785	2.437	-0.954	-1.282
380.7	1.893	2.589	-0.984	-1.211
380.8	2.043	2.746	-1.012	-1.140
380.9	2.250	2.897	-1.039	-1.068
381.0	2.531	3.025	-1.063	-0.994
381.1	2.902	3.092	-1.085	-0.916
381.2	3.358	3.033	-1.104	-0.834
381.3	3.843	2.771	-1.119	-0.746
381.4	4.219	2.266	-1.127	-0.651
381.5	4.328	1.611	-1.125	-0.549
381.6	4.132	1.004	-1.110	-0.442
243π/2	3.730	0.581	-1.076	-0.327
381.7	3.745	0.592	-1.078	-0.331
381.8	3.315	0.391	-1.026	-0.221
381.9	2.929	0.345	-0.953	-0.120
382.0	2.614	0.388	-0.862	-0.034
382.1	2.366	0.479	-0.758	0.031
382.2	2.174	0.591	-0.649	0.072
382.3	2.025	0.710	-0.544	0.091
382.4	1.909	0.832	-0.448	0.092
382.5	1.819	0.951	-0.365	0.081
382.6	1.748	1.068	-0.295	0.064
382.7	1.693	1.183	-0.237	0.045
382.8	1.651	1.295	-0.188	0.029
382.9	1.620	1.407	-0.145	0.015

k_0h	ADMITTANCE		RAD.EFF.LENGTH	
383.0	1.599	1.518	-0.106	0.006
383.1	1.587	1.630	-0.068	0.000
383.2	1.584	1.744	-0.030	-0.001
122π	1.589	1.831	-0.000	-0.000
383.3	1.591	1.861	0.011	0.001
383.4	1.609	1.983	0.054	0.005
383.5	1.640	2.110	0.102	0.011
383.6	1.687	2.244	0.155	0.017
383.7	1.755	2.386	0.215	0.023
383.8	1.850	2.535	0.283	0.027
383.9	1.982	2.590	0.359	0.026
384.0	2.164	2.844	0.444	0.020
384.1	2.412	2.985	0.539	0.006
384.2	2.744	3.081	0.644	-0.020
384.3	3.168	3.082	0.759	-0.063
384.4	3.650	2.911	0.882	-0.125
384.5	4.089	2.504	1.010	-0.215
384.6	4.320	1.893	1.138	-0.336
384.7	4.243	1.245	1.255	-0.494
384.8	3.916	0.745	1.351	-0.691
245π/2	3.728	0.592	1.382	-0.792
384.9	3.491	0.464	1.408	-0.924
385.0	3.083	0.362	1.413	-1.184
385.1	2.738	0.375	1.352	-1.451
385.2	2.465	0.449	1.224	-1.703
385.3	2.252	0.554	1.038	-1.914
385.4	2.086	0.671	0.813	-2.070
385.5	1.958	0.792	0.570	-2.163
385.6	1.858	0.912	0.332	-2.200
385.7	1.780	1.030	0.112	-2.189
385.8	1.719	1.146	-0.083	-2.144
385.9	1.672	1.259	-0.249	-2.077
386.0	1.637	1.371	-0.388	-1.996
386.1	1.612	1.482	-0.504	-1.909
386.2	1.596	1.593	-0.599	-1.820
386.3	1.589	1.706	-0.678	-1.732
386.4	1.592	1.822	-0.743	-1.647
123π	1.594	1.841	-0.753	-1.634
386.5	1.605	1.942	-0.798	-1.565
386.6	1.631	2.066	-0.845	-1.486
386.7	1.671	2.197	-0.886	-1.411
386.8	1.730	2.336	-0.922	-1.338
386.9	1.813	2.482	-0.955	-1.266
387.0	1.928	2.634	-0.985	-1.196
387.1	2.088	2.790	-1.013	-1.125
387.2	2.307	2.937	-1.040	-1.054
387.3	2.603	3.056	-1.064	-0.979
387.4	2.989	3.103	-1.086	-0.901
387.5	3.453	3.011	-1.105	-0.819
387.6	3.925	2.704	-1.119	-0.729
387.7	4.259	2.165	-1.126	-0.633
387.8	4.309	1.510	-1.122	-0.530
387.9	4.069	0.936	-1.105	-0.421
247π/2	3.726	0.603	-1.075	-0.325

k_0h	ADMITTANCE		RAD.EFF.LENGTH	
388.0	3.669	0.566	-1.070	-0.310
388.1	3.246	0.399	-1.014	-0.201
388.2	2.875	0.372	-0.937	-0.102
388.3	2.573	0.425	-0.843	-0.020
388.4	2.337	0.520	-0.737	0.041
388.5	2.154	0.633	-0.629	0.077
388.6	2.012	0.752	-0.525	0.092
388.7	1.901	0.873	-0.432	0.090
388.8	1.815	0.992	-0.352	0.078
388.9	1.747	1.108	-0.284	0.060
389.0	1.695	1.222	-0.228	0.042
389.1	1.655	1.334	-0.180	0.026
389.2	1.625	1.446	-0.138	0.013
389.3	1.606	1.557	-0.099	0.004
389.4	1.596	1.669	-0.061	-0.000
389.5	1.595	1.783	-0.023	-0.001
124π	1.598	1.851	-0.000	-0.000
389.6	1.603	1.901	0.018	0.001
389.7	1.624	2.024	0.062	0.006
389.8	1.658	2.152	0.110	0.012
389.9	1.708	2.287	0.165	0.018
390.0	1.781	2.430	0.226	0.023
390.1	1.882	2.580	0.295	0.026
390.2	2.022	2.734	0.373	0.025
390.3	2.215	2.886	0.460	0.018
390.4	2.477	3.020	0.557	0.001
390.5	2.825	3.101	0.663	-0.028
390.6	3.261	3.075	0.780	-0.074
390.7	3.741	2.863	0.904	-0.142
390.8	4.149	2.415	1.032	-0.237
390.9	4.325	1.787	1.158	-0.365
391.0	4.194	1.161	1.271	-0.530
391.1	3.843	0.703	1.360	-0.735
249π/2	3.724	0.614	1.378	-0.800
391.2	3.418	0.459	1.407	-0.975
391.3	3.022	0.382	1.398	-1.237
391.4	2.693	0.408	1.324	-1.501
391.5	2.431	0.489	1.182	-1.745
391.6	2.228	0.596	0.987	-1.944
391.7	2.071	0.713	0.757	-2.086
391.8	1.948	0.834	0.515	-2.166
391.9	1.852	0.953	0.281	-2.191
392.0	1.777	1.070	0.068	-2.172
392.1	1.719	1.185	-0.119	-2.122
392.2	1.674	1.298	-0.278	-2.051
392.3	1.641	1.409	-0.411	-1.970
392.4	1.617	1.520	-0.522	-1.883
392.5	1.603	1.632	-0.612	-1.795
392.6	1.599	1.745	-0.688	-1.708
125π	1.603	1.861	-0.750	-1.626
392.7	1.603	1.862	-0.751	-1.625
392.8	1.619	1.982	-0.804	-1.544
392.9	1.647	2.108	-0.849	-1.467

k_0h	ADMITTANCE		RAD.EFF.LENGTH	
393.0	1.690	2.240	-0.889	-1.393
393.1	1.753	2.379	-0.924	-1.321
393.2	1.842	2.526	-0.957	-1.251
393.3	1.965	2.679	-0.986	-1.181
393.4	2.134	2.833	-1.015	-1.111
393.5	2.366	2.976	-1.041	-1.039
393.6	2.677	3.083	-1.065	-0.965
393.7	3.078	3.108	-1.087	-0.887
393.8	3.548	2.981	-1.106	-0.803
393.9	4.002	2.631	-1.120	-0.713
394.0	4.287	2.062	-1.125	-0.615
394.1	4.280	1.413	-1.120	-0.511
394.2	4.003	0.876	-1.100	-0.401
251π/2	3.722	0.625	-1.075	-0.323
394.3	3.594	0.548	-1.061	-0.290
394.4	3.180	0.411	-1.001	-0.182
394.5	2.823	0.401	-0.921	-0.085
394.6	2.535	0.462	-0.824	-0.006
394.7	2.310	0.561	-0.717	0.049
394.8	2.135	0.675	-0.609	0.081
394.9	1.999	0.794	-0.507	0.092
395.0	1.894	0.914	-0.416	0.088
395.1	1.811	1.032	-0.339	0.074
395.2	1.746	1.148	-0.274	0.057
395.3	1.696	1.261	-0.219	0.039
395.4	1.658	1.373	-0.173	0.023
395.5	1.630	1.484	-0.131	0.011
395.6	1.613	1.595	-0.093	0.003
395.7	1.604	1.708	-0.055	-0.000
395.8	1.605	1.823	-0.016	-0.001
126π	1.608	1.871	-0.000	-0.000
395.9	1.616	1.941	0.025	0.002
396.0	1.639	2.065	0.070	0.007
396.1	1.675	2.194	0.119	0.013
396.2	1.730	2.330	0.175	0.019
396.3	1.807	2.474	0.237	0.024
396.4	1.915	2.624	0.308	0.026
396.5	2.064	2.778	0.387	0.024
396.6	2.268	2.927	0.476	0.014
396.7	2.545	3.052	0.574	-0.004
396.8	2.908	3.117	0.683	-0.036
396.9	3.354	3.061	0.801	-0.086
397.0	3.828	2.808	0.925	-0.159
397.1	4.200	2.321	1.053	-0.260
397.2	4.318	1.684	1.177	-0.395
397.3	4.140	1.084	1.286	-0.569
397.4	3.768	0.669	1.367	-0.781
253π/2	3.720	0.635	1.374	-0.807
397.5	3.347	0.461	1.404	-1.025
397.6	2.965	0.405	1.381	-1.289
397.7	2.649	0.443	1.292	-1.550
397.8	2.400	0.529	1.138	-1.784
397.9	2.206	0.637	0.935	-1.971

k_0h	ADMITTANCE		RAD.EFF.LENGTH	
398.0	2.056	0.755	0.702	-2.098
398.1	1.939	0.875	0.462	-2.165
398.2	1.847	0.994	0.233	-2.180
398.3	1.775	1.110	0.026	-2.153
398.4	1.719	1.224	-0.154	-2.098
398.5	1.677	1.337	-0.306	-2.026
398.6	1.645	1.448	-0.433	-1.943
398.7	1.623	1.559	-0.538	-1.857
398.8	1.611	1.671	-0.625	-1.770
398.9	1.608	1.785	-0.697	-1.685
127π	1.613	1.880	-0.747	-1.617
399.0	1.615	1.901	-0.757	-1.603
399.1	1.633	2.020	-0.808	-1.524
399.2	1.663	2.149	-0.853	-1.448
399.3	1.710	2.282	-0.891	-1.376
399.4	1.778	2.423	-0.926	-1.305
399.5	1.872	2.570	-0.958	-1.235
399.6	2.003	2.723	-0.988	-1.166
399.7	2.183	2.875	-1.016	-1.097
399.8	2.427	3.013	-1.042	-1.025
399.9	2.754	3.107	-1.067	-0.951
400.0	3.168	3.108	-1.089	-0.872
400.1	3.640	2.943	-1.107	-0.787
400.2	4.070	2.551	-1.120	-0.696
400.3	4.305	1.959	-1.124	-0.597
400.4	4.242	1.323	-1.117	-0.491
400.5	3.933	0.825	-1.094	-0.381
255π/2	3.718	0.646	-1.074	-0.321
400.6	3.520	0.536	-1.051	-0.269
400.7	3.117	0.426	-0.988	-0.163
400.8	2.774	0.431	-0.904	-0.069
400.9	2.499	0.501	-0.804	0.006
401.0	2.285	0.602	-0.696	0.057
401.1	2.118	0.717	-0.589	0.085
401.2	1.988	0.836	-0.489	0.092
401.3	1.887	0.955	-0.401	0.086
401.4	1.807	1.072	-0.326	0.071
401.5	1.745	1.187	-0.264	0.053
401.6	1.697	1.300	-0.211	0.036
401.7	1.661	1.412	-0.165	0.021
401.8	1.636	1.523	-0.125	0.010
401.9	1.620	1.634	-0.087	0.002
402.0	1.613	1.747	-0.049	-0.001
402.1	1.615	1.862	-0.010	-0.000
128π	1.618	1.890	-0.000	-0.000
402.2	1.629	1.981	0.032	0.003
402.3	1.654	2.106	0.078	0.007
402.4	1.694	2.236	0.128	0.013
402.5	1.752	2.373	0.185	0.019
402.6	1.835	2.518	0.249	0.024
402.7	1.950	2.669	0.321	0.025
402.8	2.108	2.822	0.401	0.022
402.9	2.324	2.967	0.492	0.011

k_0h	ADMITTANCE		RAD.EFF.LENGTH	
403.0	2.615	3.083	0.592	-0.010
403.1	2.993	3.129	0.702	-0.045
403.2	3.447	3.040	0.822	-0.099
403.3	3.909	2.745	0.947	-0.177
403.4	4.241	2.224	1.074	-0.284
403.5	4.302	1.585	1.196	-0.427
403.6	4.080	1.016	1.300	-0.608
257π/2	3.716	0.657	1.370	-0.814
403.7	3.694	0.643	1.372	-0.827
403.8	3.279	0.466	1.397	-1.076
403.9	2.910	0.431	1.361	-1.340
404.0	2.608	0.478	1.258	-1.596
404.1	2.371	0.569	1.092	-1.821
404.2	2.186	0.679	0.881	-1.994
404.3	2.042	0.797	0.647	-2.108
404.4	1.930	0.916	0.409	-2.162
404.5	1.842	1.034	0.186	-2.167
404.6	1.773	1.150	-0.014	-2.134
404.7	1.720	1.264	-0.187	-2.075
404.8	1.679	1.376	-0.332	-2.000
404.9	1.650	1.487	-0.453	-1.918
405.0	1.630	1.598	-0.554	-1.832
405.1	1.619	1.710	-0.637	-1.746
405.2	1.618	1.824	-0.706	-1.662
129π	1.622	1.900	-0.745	-1.609
405.3	1.627	1.941	-0.764	-1.582
405.4	1.647	2.063	-0.813	-1.504
405.5	1.680	2.191	-0.856	-1.430
405.6	1.731	2.325	-0.894	-1.359
405.7	1.803	2.466	-0.928	-1.289
405.8	1.903	2.615	-0.960	-1.220
405.9	2.042	2.767	-0.989	-1.152
406.0	2.233	2.917	-1.017	-1.083
406.1	2.491	3.047	-1.044	-1.011
406.2	2.833	3.127	-1.068	-0.936
406.3	3.259	3.102	-1.090	-0.857
406.4	3.728	2.898	-1.108	-0.771
406.5	4.130	2.466	-1.120	-0.679
406.6	4.312	1.856	-1.123	-0.578
406.7	4.197	1.240	-1.113	-0.472
406.8	3.862	0.782	-1.087	-0.360
259π/2	3.715	0.668	-1.073	-0.319
406.9	3.448	0.530	-1.041	-0.249
407.0	3.057	0.445	-0.973	-0.144
407.1	2.728	0.464	-0.886	-0.053
407.2	2.465	0.539	-0.784	0.017
407.3	2.261	0.643	-0.676	0.064
407.4	2.101	0.758	-0.570	0.087
407.5	1.977	0.877	-0.472	0.092
407.6	1.880	0.996	-0.387	0.083
407.7	1.804	1.112	-0.314	0.068
407.8	1.745	1.227	-0.254	0.050
407.9	1.699	1.339	-0.203	0.033

k_0h	ADMITTANCE		RAD.EFF.LENGTH	
408.0	1.665	1.451	-0.158	0.018
408.1	1.641	1.561	-0.118	0.008
408.2	1.627	1.673	-0.080	0.002
408.3	1.622	1.786	-0.042	-0.001
130π	1.627	1.910	-0.000	-0.000
408.4	1.626	1.902	-0.003	-0.000
408.5	1.642	2.022	0.039	0.003
408.6	1.670	2.147	0.086	0.008
408.7	1.713	2.278	0.138	0.014
408.8	1.775	2.416	0.195	0.020
408.9	1.863	2.562	0.261	0.024
409.0	1.986	2.713	0.334	0.024
409.1	2.153	2.864	0.416	0.020
409.2	2.382	3.005	0.508	0.007
409.3	2.688	3.110	0.610	-0.017
409.4	3.080	3.135	0.722	-0.055
409.5	3.539	3.012	0.843	-0.113
409.6	3.984	2.675	0.969	-0.196
409.7	4.271	2.125	1.095	-0.310
409.8	4.276	1.490	1.213	-0.460
409.9	4.016	0.956	1.312	-0.649
261π/2	3.713	0.679	1.365	-0.822
410.0	3.621	0.623	1.375	-0.875
410.1	3.213	0.476	1.388	-1.128
410.2	2.858	0.459	1.338	-1.391
410.3	2.570	0.515	1.221	-1.641
410.4	2.343	0.609	1.045	-1.855
410.5	2.167	0.721	0.828	-2.015
410.6	2.029	0.838	0.592	-2.114
410.7	1.922	0.957	0.358	-2.157
410.8	1.838	1.075	0.141	-2.153
410.9	1.772	1.190	-0.052	-2.113
411.0	1.721	1.303	-0.218	-2.051
411.1	1.682	1.414	-0.357	-1.975
411.2	1.654	1.525	-0.473	-1.892
411.3	1.636	1.636	-0.568	-1.807
411.4	1.627	1.749	-0.648	-1.722
411.5	1.628	1.863	-0.714	-1.640
131π	1.632	1.920	-0.742	-1.601
411.6	1.639	1.981	-0.769	-1.561
411.7	1.661	2.104	-0.817	-1.485
411.8	1.698	2.232	-0.859	-1.412
411.9	1.752	2.368	-0.896	-1.342
412.0	1.829	2.510	-0.930	-1.273
412.1	1.936	2.659	-0.961	-1.205
412.2	2.084	2.811	-0.991	-1.138
412.3	2.285	2.957	-1.019	-1.069
412.4	2.558	3.080	-1.045	-0.997
412.5	2.914	3.143	-1.070	-0.922
412.6	3.350	3.089	-1.092	-0.842
412.7	3.813	2.845	-1.109	-0.755
412.8	4.182	2.376	-1.120	-0.661
412.9	4.308	1.756	-1.121	-0.560

k_0h	ADMITTANCE		RAD.EFF.LENGTH	
413.0	4.146	1.164	-1.109	-0.451
413.1	3.790	0.746	-1.080	-0.339
263π/2	3.711	0.689	-1.072	-0.317
413.2	3.378	0.530	-1.030	-0.228
413.3	2.999	0.466	-0.958	-0.125
413.4	2.684	0.497	-0.868	-0.038
413.5	2.434	0.578	-0.764	0.028
413.6	2.238	0.684	-0.656	0.070
413.7	2.086	0.800	-0.551	0.089
413.8	1.967	0.919	-0.455	0.091
413.9	1.874	1.036	-0.373	0.080
414.0	1.802	1.152	-0.303	0.064
414.1	1.745	1.266	-0.244	0.046
414.2	1.701	1.378	-0.195	0.030
414.3	1.669	1.489	-0.151	0.016
414.4	1.647	1.600	-0.112	0.007
414.5	1.634	1.712	-0.074	0.001
414.6	1.631	1.825	-0.036	-0.001
132π	1.636	1.930	-0.000	-0.000
414.7	1.638	1.941	0.004	0.000
414.8	1.655	2.062	0.047	0.004
414.9	1.686	2.188	0.094	0.009
415.0	1.732	2.320	0.147	0.015
415.1	1.799	2.460	0.206	0.020
415.2	1.893	2.606	0.272	0.024
415.3	2.023	2.757	0.347	0.023
415.4	2.201	2.906	0.431	0.017
415.5	2.442	3.041	0.525	0.003
415.6	2.763	3.134	0.629	-0.024
415.7	3.168	3.135	0.742	-0.066
415.8	3.629	2.977	0.864	-0.128
415.9	4.052	2.598	0.990	-0.217
416.0	4.290	2.024	1.115	-0.338
416.1	4.242	1.401	1.230	-0.495
416.2	3.950	0.904	1.322	-0.691
265π/2	3.709	0.700	1.360	-0.829
416.3	3.548	0.609	1.376	-0.923
416.4	3.150	0.490	1.376	-1.179
416.5	2.809	0.488	1.311	-1.440
416.6	2.533	0.552	1.181	-1.683
416.7	2.317	0.650	0.996	-1.885
416.8	2.148	0.762	0.774	-2.032
416.9	2.017	0.880	0.539	-2.118
417.0	1.914	0.998	0.309	-2.150
417.1	1.834	1.115	0.097	-2.137
417.2	1.771	1.229	-0.089	-2.092
417.3	1.722	1.342	-0.247	-2.027
417.4	1.686	1.453	-0.380	-1.950
417.5	1.660	1.564	-0.491	-1.867
417.6	1.643	1.675	-0.582	-1.782
417.7	1.636	1.787	-0.658	-1.699
417.8	1.638	1.902	-0.721	-1.618
133π	1.641	1.940	-0.739	-1.593
417.9	1.651	2.021	-0.775	-1.541

OMEGA = 20.0

k_0h	ADMITTANCE		RAD.EFF.LENGTH	
418.0	1.676	2.145	-0.821	-1.466
418.1	1.716	2.274	-0.862	-1.395
418.2	1.774	2.411	-0.898	-1.326
418.3	1.856	2.554	-0.932	-1.258
418.4	1.970	2.703	-0.963	-1.191
418.5	2.127	2.854	-0.992	-1.124
418.6	2.340	2.996	-1.020	-1.055
418.7	2.626	3.110	-1.047	-0.983
418.8	2.997	3.155	-1.071	-0.907
418.9	3.441	3.070	-1.093	-0.826
419.0	3.893	2.785	-1.110	-0.739
419.1	4.223	2.282	-1.120	-0.643
419.2	4.295	1.659	-1.120	-0.540
419.3	4.089	1.096	-1.105	-0.431
267π/2	3.707	0.711	-1.071	-0.316
419.4	3.718	0.718	-1.072	-0.319
419.5	3.311	0.534	-1.018	-0.208
419.6	2.944	0.491	-0.943	-0.108
419.7	2.643	0.532	-0.849	-0.024
419.8	2.404	0.618	-0.744	0.038
419.9	2.217	0.725	-0.636	0.075
420.0	2.072	0.841	-0.532	0.090
420.1	1.958	0.960	-0.439	0.089
420.2	1.869	1.077	-0.359	0.078
420.3	1.799	1.192	-0.292	0.061
420.4	1.745	1.305	-0.235	0.043
420.5	1.704	1.417	-0.187	0.027
420.6	1.673	1.528	-0.145	0.014
420.7	1.653	1.638	-0.105	0.005
420.8	1.642	1.750	-0.068	0.000
420.9	1.641	1.864	-0.029	-0.001
134π	1.646	1.950	-0.000	-0.000
421.0	1.649	1.981	0.011	0.001
421.1	1.669	2.103	0.055	0.005
421.2	1.702	2.229	0.103	0.010
421.3	1.753	2.363	0.156	0.016
421.4	1.824	2.503	0.217	0.021
421.5	1.924	2.650	0.285	0.023
421.6	2.062	2.800	0.361	0.022
421.7	2.250	2.947	0.447	0.014
421.8	2.505	3.076	0.542	-0.002
421.9	2.840	3.154	0.647	-0.031
422.0	3.257	3.130	0.762	-0.077
422.1	3.716	2.934	0.885	-0.144
422.2	4.112	2.516	1.011	-0.239
422.3	4.299	1.925	1.135	-0.366
422.4	4.200	1.319	1.246	-0.531
422.5	3.881	0.860	1.331	-0.735
269π/2	3.705	0.722	1.356	-0.836
422.6	3.477	0.601	1.374	-0.972
422.7	3.090	0.507	1.361	-1.230
422.8	2.762	0.519	1.282	-1.488
422.9	2.499	0.590	1.140	-1.723

k_0h	ADMITTANCE		RAD.EFF.LENGTH	
423.0	2.292	0.690	0.946	-1.913
423.1	2.131	0.804	0.721	-2.046
423.2	2.006	0.921	0.486	-2.120
423.3	1.907	1.039	0.261	-2.141
423.4	1.830	1.155	0.056	-2.120
423.5	1.770	1.269	-0.123	-2.071
423.6	1.724	1.381	-0.275	-2.003
423.7	1.689	1.492	-0.402	-1.925
423.8	1.665	1.602	-0.508	-1.842
423.9	1.650	1.714	-0.595	-1.759
424.0	1.645	1.826	-0.658	-1.677
424.1	1.649	1.942	-0.728	-1.597
135π	1.650	1.960	-0.737	-1.586
424.2	1.664	2.061	-0.780	-1.521
424.3	1.692	2.186	-0.825	-1.448
424.4	1.735	2.316	-0.865	-1.378
424.5	1.797	2.453	-0.901	-1.310
424.6	1.885	2.598	-0.934	-1.243
424.7	2.006	2.747	-0.965	-1.176
424.8	2.172	2.896	-0.994	-1.110
424.9	2.397	3.034	-1.022	-1.041
425.0	2.697	3.137	-1.049	-0.969
425.1	3.082	3.162	-1.073	-0.893
425.2	3.530	3.044	-1.095	-0.811
425.3	3.967	2.718	-1.111	-0.722
425.4	4.254	2.186	-1.120	-0.625
425.5	4.272	1.566	-1.117	-0.521
425.6	4.029	1.036	-1.100	-0.410
271π/2	3.704	0.733	-1.070	-0.314
425.7	3.646	0.697	-1.064	-0.298
425.8	3.246	0.542	-1.006	-0.188
425.9	2.892	0.517	-0.926	-0.090
426.0	2.603	0.568	-0.830	-0.010
426.1	2.375	0.658	-0.723	0.046
426.2	2.197	0.767	-0.616	0.079
426.3	2.058	0.883	-0.514	0.091
426.4	1.949	1.000	-0.424	0.087
426.5	1.864	1.117	-0.346	0.074
426.6	1.797	1.232	-0.281	0.057
426.7	1.746	1.344	-0.226	0.040
426.8	1.706	1.456	-0.179	0.024
426.9	1.678	1.566	-0.138	0.012
427.0	1.659	1.677	-0.099	0.004
427.1	1.650	1.789	-0.061	-0.000
427.2	1.650	1.903	-0.023	-0.001
136π	1.655	1.969	-0.000	-0.000
427.3	1.661	2.021	0.018	0.001
427.4	1.683	2.143	0.062	0.005
427.5	1.720	2.271	0.111	0.011
427.6	1.774	2.405	0.166	0.016
427.7	1.850	2.546	0.228	0.021
427.8	1.956	2.694	0.297	0.023
427.9	2.103	2.843	0.375	0.020

k_0h	ADMITTANCE		RAD.EFF.LENGTH	
428.0	2.302	2.987	0.462	0.011
428.1	2.570	3.108	0.559	-0.007
428.2	2.919	3.170	0.666	-0.039
428.3	3.346	3.118	0.783	-0.089
428.4	3.799	2.883	0.906	-0.161
428.5	4.164	2.430	1.032	-0.262
428.6	4.298	1.827	1.153	-0.397
428.7	4.152	1.244	1.260	-0.569
428.8	3.811	0.823	1.337	-0.779
273π/2	3.702	0.743	1.351	-0.842
428.9	3.408	0.599	1.369	-1.021
429.0	3.033	0.528	1.343	-1.280
429.1	2.718	0.552	1.251	-1.534
429.2	2.466	0.629	1.097	-1.760
429.3	2.269	0.731	0.895	-1.938
429.4	2.116	0.845	0.667	-2.057
429.5	1.995	0.962	0.435	-2.119
429.6	1.901	1.079	0.214	-2.130
429.7	1.827	1.194	0.016	-2.102
429.8	1.770	1.308	-0.156	-2.049
429.9	1.726	1.420	-0.301	-1.979
430.0	1.693	1.530	-0.423	-1.900
430.1	1.670	1.641	-0.524	-1.818
430.2	1.657	1.752	-0.607	-1.735
430.3	1.653	1.865	-0.677	-1.654
137π	1.660	1.979	-0.734	-1.578
430.4	1.660	1.981	-0.735	-1.577
430.5	1.677	2.102	-0.785	-1.502
430.6	1.708	2.227	-0.829	-1.430
430.7	1.754	2.358	-0.868	-1.361
430.8	1.821	2.497	-0.903	-1.294
430.9	1.914	2.641	-0.936	-1.228
431.0	2.042	2.790	-0.966	-1.162
431.1	2.218	2.938	-0.996	-1.096
431.2	2.456	3.070	-1.024	-1.027
431.3	2.771	3.161	-1.051	-0.955
431.4	3.167	3.163	-1.075	-0.878
431.5	3.618	3.010	-1.096	-0.795
431.6	4.035	2.645	-1.112	-0.705
431.7	4.276	2.089	-1.119	-0.607
431.8	4.241	1.479	-1.115	-0.501
431.9	3.965	0.983	-1.094	-0.390
275π/2	3.700	0.754	-1.069	-0.312
432.0	3.575	0.682	-1.054	-0.277
432.1	3.183	0.554	-0.992	-0.169
432.2	2.843	0.545	-0.909	-0.074
432.3	2.566	0.604	-0.810	0.002
432.4	2.349	0.698	-0.703	0.054
432.5	2.179	0.808	-0.596	0.083
432.6	2.046	0.924	-0.497	0.091
432.7	1.942	1.041	-0.409	0.085
432.8	1.860	1.157	-0.334	0.071
432.9	1.796	1.271	-0.271	0.054

k_0h	ADMITTANCE		RAD.EFF.LENGTH	
433.0	1.747	1.383	-0.218	0.037
433.1	1.709	1.494	-0.172	0.022
433.2	1.683	1.605	-0.131	0.011
433.3	1.666	1.716	-0.093	0.003
433.4	1.658	1.828	-0.055	-0.000
433.5	1.660	1.943	-0.016	-0.001
138π	1.664	1.989	-0.000	-0.000
433.6	1.673	2.061	0.025	0.002
433.7	1.698	2.184	0.070	0.006
433.8	1.738	2.313	0.120	0.012
433.9	1.795	2.448	0.176	0.017
434.0	1.877	2.590	0.239	0.021
434.1	1.990	2.737	0.310	0.022
434.2	2.145	2.886	0.389	0.019
434.3	2.356	3.026	0.478	0.008
434.4	2.637	3.138	0.577	-0.013
434.5	3.000	3.182	0.686	-0.048
434.6	3.434	3.100	0.803	-0.102
434.7	3.878	2.826	0.927	-0.179
434.8	4.206	2.340	1.052	-0.286
434.9	4.287	1.732	1.171	-0.428
435.0	4.098	1.175	1.273	-0.608
435.1	3.741	0.794	1.341	-0.825
277π/2	3.699	0.765	1.346	-0.850
435.2	3.341	0.601	1.362	-1.071
435.3	2.978	0.550	1.322	-1.330
435.4	2.676	0.586	1.217	-1.579
435.5	2.436	0.668	1.052	-1.795
435.6	2.248	0.772	0.844	-1.960
435.7	2.101	0.886	0.614	-2.066
435.8	1.986	1.003	0.385	-2.115
435.9	1.895	1.119	0.170	-2.117
436.0	1.825	1.234	-0.022	-2.084
436.1	1.770	1.347	-0.187	-2.026
436.2	1.728	1.458	-0.326	-1.955
436.3	1.697	1.569	-0.442	-1.875
436.4	1.676	1.679	-0.539	-1.794
436.5	1.665	1.791	-0.618	-1.712
436.6	1.663	1.904	-0.685	-1.633
139π	1.669	1.999	-0.732	-1.570
436.7	1.671	2.021	-0.742	-1.556
436.8	1.691	2.142	-0.790	-1.483
436.9	1.724	2.268	-0.833	-1.413
437.0	1.774	2.401	-0.870	-1.345
437.1	1.845	2.540	-0.905	-1.279
437.2	1.944	2.685	-0.938	-1.213
437.3	2.081	2.834	-0.968	-1.148
437.4	2.267	2.978	-0.998	-1.082
437.5	2.518	3.104	-1.026	-1.013
437.6	2.846	3.181	-1.053	-0.941
437.7	3.254	3.158	-1.077	-0.863
437.8	3.704	2.969	-1.098	-0.779
437.9	4.095	2.567	-1.113	-0.688

k_0h	ADMITTANCE		RAD.EFF.LENGTH	
438.0	4.287	1.992	-1.119	-0.588
438.1	4.202	1.397	-1.112	-0.481
438.2	3.899	0.938	-1.088	-0.369
279π/2	3.697	0.776	-1.068	-0.310
438.3	3.505	0.672	-1.044	-0.256
438.4	3.123	0.570	-0.978	-0.150
438.5	2.796	0.575	-0.892	-0.058
438.6	2.532	0.641	-0.791	0.014
438.7	2.324	0.738	-0.683	0.061
438.8	2.161	0.849	-0.577	0.085
438.9	2.034	0.965	-0.479	0.091
439.0	1.934	1.082	-0.394	0.083
439.1	1.856	1.197	-0.322	0.068
439.2	1.795	1.310	-0.261	0.051
439.3	1.748	1.422	-0.210	0.034
439.4	1.712	1.533	-0.165	0.020
439.5	1.688	1.643	-0.125	0.009
439.6	1.672	1.754	-0.086	0.002
439.7	1.667	1.867	-0.049	-0.001
439.8	1.671	1.982	-0.009	-0.000
140π	1.673	2.009	-0.000	-0.000
439.9	1.686	2.101	0.033	0.002
440.0	1.713	2.225	0.078	0.007
440.1	1.756	2.355	0.129	0.012
440.2	1.818	2.491	0.186	0.017
440.3	1.905	2.633	0.250	0.021
440.4	2.025	2.781	0.323	0.021
440.5	2.189	2.928	0.404	0.016
440.6	2.411	3.064	0.494	0.004
440.7	2.707	3.164	0.595	-0.020
440.8	3.083	3.189	0.705	-0.058
440.9	3.522	3.076	0.824	-0.116
441.0	3.951	2.762	0.948	-0.198
441.1	4.238	2.247	1.072	-0.312
441.2	4.267	1.641	1.188	-0.461
441.3	4.040	1.114	1.284	-0.648
281π/2	3.695	0.786	1.341	-0.857
441.4	3.670	0.771	1.344	-0.872
441.5	3.277	0.608	1.352	-1.121
441.6	2.925	0.575	1.299	-1.379
441.7	2.636	0.620	1.180	-1.621
441.8	2.407	0.707	1.005	-1.827
441.9	2.227	0.813	0.792	-1.979
442.0	2.087	0.927	0.562	-2.072
442.1	1.977	1.044	0.336	-2.110
442.2	1.890	1.160	0.127	-2.103
442.3	1.822	1.274	-0.058	-2.064
442.4	1.770	1.386	-0.217	-2.004
442.5	1.730	1.497	-0.350	-1.930
442.6	[illegible]	1.607	-0.461	-1.851
442.7	[illegible]	1.718	-0.553	-1.770
442.8	[illegible]	1.830	-0.629	-1.690
442.9	1.672	1.944	-0.693	-1.612
141π	1.678	2.019	-0.729	-1.563

k_0h	ADMITTANCE		RAD.EFF.LENGTH	
443.0	1.683	2.061	-0.748	-1.537
443.1	1.705	2.183	-0.795	-1.465
443.2	1.741	2.310	-0.836	-1.396
443.3	1.795	2.443	-0.873	-1.329
443.4	1.871	2.583	-0.907	-1.264
443.5	1.976	2.729	-0.940	-1.199
443.6	2.121	2.876	-0.970	-1.134
443.7	2.318	3.018	-1.000	-1.068
443.8	2.581	3.136	-1.028	-0.999
443.9	2.924	3.197	-1.055	-0.927
444.0	3.341	3.148	-1.079	-0.848
444.1	3.785	2.922	-1.099	-0.763
444.2	4.146	2.484	-1.113	-0.670
444.3	4.288	1.897	-1.117	-0.569
444.4	4.157	1.322	-1.108	-0.461
444.5	3.831	0.900	-1.081	-0.348
283π/2	3.694	0.797	-1.067	-0.308
444.6	3.437	0.668	-1.033	-0.236
444.7	3.065	0.589	-0.963	-0.131
444.8	2.751	0.607	-0.874	-0.043
444.9	2.499	0.679	-0.770	0.024
445.0	2.300	0.779	-0.662	0.067
445.1	2.145	0.890	-0.558	0.087
445.2	2.023	1.006	-0.463	0.090
445.3	1.927	1.122	-0.380	0.080
445.4	1.853	1.237	-0.310	0.065
445.5	1.794	1.350	-0.251	0.047
445.6	1.749	1.461	-0.202	0.031
445.7	1.716	1.572	-0.158	0.017
445.8	1.693	1.682	-0.118	0.008
445.9	1.679	1.793	-0.080	0.002
446.0	1.675	1.906	-0.042	-0.001
142π	1.682	2.029	-0.000	-0.000
446.1	1.681	2.022	-0.003	-0.000
446.2	1.699	2.141	0.040	0.003
446.3	1.729	2.266	0.087	0.008
446.4	1.775	2.397	0.139	0.013
446.5	1.841	2.534	0.197	0.018
446.6	1.934	2.677	0.262	0.021
446.7	2.061	2.824	0.336	0.020
446.8	2.235	2.969	0.418	0.014
446.9	2.469	3.099	0.511	-0.000
447.0	2.779	3.188	0.613	-0.026
447.1	3.167	3.191	0.725	-0.068
447.2	3.608	3.044	0.845	-0.130
447.3	4.018	2.692	0.969	-0.219
447.4	4.261	2.153	1.092	-0.339
447.5	4.239	1.555	1.204	-0.495
447.6	3.979	1.061	1.293	-0.690
285π/2	3.692	0.808	1.336	-0.864
447.7	3.601	0.754	1.344	-0.919
447.8	3.214	0.619	1.340	-1.171
447.9	2.876	0.603	1.273	-1.426

k_0h	ADMITTANCE		RAD.EFF.LENGTH	
448.0	2.599	0.656	1.141	-1.662
448.1	2.380	0.746	0.957	-1.856
448.2	2.208	0.854	0.740	-1.994
448.3	2.074	0.968	0.511	-2.075
448.4	1.968	1.084	0.289	-2.102
448.5	1.885	1.200	0.085	-2.088
448.6	1.821	1.313	-0.093	-2.044
448.7	1.770	1.425	-0.245	-1.981
448.8	1.733	1.536	-0.372	-1.906
448.9	1.705	1.646	-0.478	-1.827
449.0	1.688	1.757	-0.566	-1.747
449.1	1.680	1.869	-0.639	-1.668
449.2	1.682	1.983	-0.701	-1.591
143π	1.687	2.039	-0.727	-1.556
449.3	1.695	2.101	-0.753	-1.517
449.4	1.719	2.223	-0.799	-1.447
449.5	1.758	2.351	-0.839	-1.379
449.6	1.816	2.486	-0.876	-1.313
449.7	1.897	2.626	-0.910	-1.249
449.8	2.009	2.772	-0.942	-1.185
449.9	2.162	2.919	-0.972	-1.121
450.0	2.370	3.057	-1.002	-1.055
450.1	2.647	3.166	-1.030	-0.986
450.2	3.003	3.210	-1.057	-0.912
450.3	3.428	3.132	-1.081	-0.833
450.4	3.863	2.867	-1.101	-0.747
450.5	4.189	2.397	-1.114	-0.653
450.6	4.279	1.804	-1.116	-0.550
450.7	4.106	1.254	-1.104	-0.440
450.8	3.762	0.869	-1.074	-0.327
287π/2	3.690	0.819	-1.066	-0.306
450.9	3.371	0.669	-1.022	-0.215
451.0	3.010	0.610	-0.948	-0.113
451.1	2.709	0.640	-0.855	-0.028
451.2	2.468	0.717	-0.750	0.034
451.3	2.278	0.819	-0.642	0.073
451.4	2.129	0.931	-0.539	0.089
451.5	2.013	1.047	-0.447	0.089
451.6	1.921	1.162	-0.367	0.078
451.7	1.850	1.276	-0.299	0.061
451.8	1.794	1.389	-0.242	0.044
451.9	1.751	1.500	-0.194	0.028
452.0	1.720	1.610	-0.151	0.015
452.1	1.698	1.720	-0.112	0.006
452.2	1.686	1.832	-0.074	0.001
452.3	1.684	1.945	-0.036	-0.001
144π	1.691	2.049	-0.000	-0.000
452.4	1.692	2.061	0.004	0.000
452.5	1.712	2.182	0.048	0.004
452.6	1.745	2.307	0.095	0.008
452.7	1.795	2.439	0.148	0.014
452.8	1.865	2.577	0.207	0.018
452.9	1.964	2.720	0.274	0.020

k_0h	ADMITTANCE		RAD.EFF.LENGTH	
453.0	2.099	2.867	0.349	0.019
453.1	2.283	3.010	0.433	0.011
453.2	2.530	3.133	0.527	-0.005
453.3	2.852	3.209	0.631	-0.034
453.4	3.252	3.188	0.745	-0.079
453.5	3.692	3.006	0.865	-0.146
453.6	4.077	2.617	0.990	-0.240
453.7	4.274	2.059	1.111	-0.367
453.8	4.203	1.474	1.219	-0.531
453.9	3.915	1.015	1.301	-0.733
289π/2	3.689	0.829	1.331	-0.870
454.0	3.532	0.743	1.341	-0.967
454.1	3.155	0.633	1.325	-1.220
454.2	2.828	0.632	1.244	-1.472
454.3	2.564	0.693	1.101	-1.700
454.4	2.354	0.786	0.909	-1.882
454.5	2.190	0.895	0.688	-2.007
454.6	2.061	1.009	0.460	-2.075
454.7	1.960	1.125	0.243	-2.093
454.8	1.881	1.239	0.046	-2.072
454.9	1.819	1.352	-0.126	-2.023
455.0	1.771	1.464	-0.271	-1.958
455.1	1.735	1.574	-0.393	-1.883
455.2	1.710	1.684	-0.494	-1.804
455.3	1.694	1.795	-0.579	-1.724
455.4	1.688	1.907	-0.649	-1.646
455.5	1.692	2.022	-0.708	-1.571
145π	1.695	2.059	-0.725	-1.548
455.6	1.707	2.141	-0.759	-1.499
455.7	1.734	2.264	-0.803	-1.429
455.8	1.777	2.393	-0.843	-1.363
455.9	1.838	2.528	-0.878	-1.298
456.0	1.925	2.669	-0.912	-1.234
456.1	2.044	2.815	-0.944	-1.171
456.2	2.206	2.960	-0.974	-1.107
456.3	2.425	3.094	-1.004	-1.041
456.4	2.715	3.193	-1.032	-0.972
456.5	3.084	3.217	-1.059	-0.898
456.6	3.514	3.108	-1.083	-0.818
456.7	3.935	2.806	-1.102	-0.730
456.8	4.223	2.307	-1.114	-0.634
456.9	4.262	1.715	-1.114	-0.530
457.0	4.051	1.193	-1.100	-0.419
291π/2	3.687	0.840	-1.065	-0.304
457.1	3.694	0.844	-1.065	-0.306
457.2	3.307	0.674	-1.009	-0.195
457.3	2.958	0.634	-0.931	-0.096
457.4	2.669	0.674	-0.836	-0.015
457.5	2.438	0.756	-0.730	0.043
457.6	2.257	0.860	-0.623	0.077
457.7	2.115	0.972	-0.521	0.090
457.8	2.003	1.087	-0.431	0.087
457.9	1.916	1.202	-0.354	0.075

OMEGA = 20.0

k_0h	ADMITTANCE		RAD.EFF.LENGTH	
458.0	1.847	1.316	-0.288	0.058
458.1	1.794	1.428	-0.233	0.041
458.2	1.753	1.539	-0.186	0.026
458.3	1.723	1.649	-0.144	0.014
458.4	1.704	1.759	-0.105	0.005
458.5	1.694	1.870	-0.068	0.000
458.6	1.694	1.984	-0.029	-0.001
146π	1.700	2.068	-0.000	-0.000
458.7	1.704	2.101	0.011	0.001
458.8	1.726	2.222	0.055	0.004
458.9	1.762	2.348	0.104	0.009
459.0	1.815	2.481	0.158	0.014
459.1	1.891	2.620	0.218	0.018
459.2	1.995	2.764	0.286	0.020
459.3	2.138	2.910	0.363	0.017
459.4	2.333	3.049	0.449	0.008
459.5	2.592	3.165	0.544	-0.010
459.6	2.928	3.225	0.650	-0.042
459.7	3.337	3.178	0.765	-0.091
459.8	3.772	2.961	0.886	-0.163
459.9	4.129	2.537	1.010	-0.263
460.0	4.277	1.966	1.129	-0.397
460.1	4.161	1.400	1.233	-0.568
460.2	3.850	0.976	1.307	-0.776
293π/2	3.685	0.851	1.326	-0.877
460.3	3.465	0.737	1.336	-1.015
460.4	3.097	0.650	1.306	-1.269
460.5	2.784	0.662	1.212	-1.517
460.6	2.530	0.730	1.058	-1.735
460.7	2.330	0.826	0.859	-1.905
460.8	2.173	0.936	0.637	-2.017
460.9	2.050	1.050	0.411	-2.074
461.0	1.953	1.165	0.199	-2.082
461.1	1.877	1.279	0.008	-2.054
461.2	1.818	1.392	-0.157	-2.002
461.3	1.772	1.503	-0.297	-1.935
461.4	1.738	1.613	-0.413	-1.859
461.5	1.715	1.723	-0.510	-1.780
461.6	1.701	1.834	-0.590	-1.702
461.7	1.697	1.946	-0.658	-1.625
461.8	1.703	2.062	-0.715	-1.551
147π	1.704	2.078	-0.722	-1.541
461.9	1.720	2.181	-0.764	-1.480
462.0	1.750	2.305	-0.807	-1.412
462.1	1.795	2.435	-0.846	-1.347
462.2	1.861	2.571	-0.881	-1.283
462.3	1.953	2.713	-0.914	-1.220
462.4	2.079	2.858	-0.946	-1.157
462.5	2.251	3.001	-0.977	-1.094
462.6	2.482	3.129	-1.006	-1.028
462.7	2.786	3.217	-1.035	-0.958
462.8	3.166	3.220	-1.061	-0.883
462.9	3.598	3.079	-1.085	-0.802

k_0h	ADMITTANCE		RAD.EFF.LENGTH	
463.0	4.001	2.740	-1.103	-0.713
463.1	4.247	2.217	-1.114	-0.616
463.2	4.237	1.630	-1.112	-0.510
463.3	3.992	1.139	-1.094	-0.399
295π/2	3.684	0.861	-1.063	-0.302
463.4	3.626	0.826	-1.056	-0.285
463.5	3.245	0.683	-0.996	-0.176
463.6	2.908	0.660	-0.914	-0.079
463.7	2.631	0.708	-0.816	-0.002
463.8	2.411	0.795	-0.710	0.051
463.9	2.237	0.900	-0.603	0.081
464.0	2.101	1.013	-0.504	0.090
464.1	1.994	1.128	-0.416	0.085
464.2	1.911	1.242	-0.341	0.072
464.3	1.845	1.355	-0.278	0.055
464.4	1.794	1.467	-0.225	0.038
464.5	1.755	1.577	-0.179	0.023
464.6	1.728	1.687	-0.138	0.012
464.7	1.710	1.798	-0.099	0.004
464.8	1.702	1.909	-0.061	0.000
464.9	1.703	2.023	-0.022	-0.001
148π	1.709	2.088	-0.000	-0.000
465.0	1.715	2.141	0.019	0.001
465.1	1.740	2.263	0.063	0.005
465.2	1.779	2.390	0.112	0.010
465.3	1.836	2.523	0.167	0.015
465.4	1.917	2.663	0.229	0.018
465.5	2.028	2.807	0.299	0.019
465.6	2.179	2.951	0.377	0.016
465.7	2.384	3.087	0.464	0.005
465.8	2.657	3.194	0.562	-0.016
465.9	3.006	3.238	0.669	-0.051
466.0	3.421	3.163	0.785	-0.104
466.1	3.848	2.909	0.907	-0.181
466.2	4.172	2.454	1.030	-0.287
466.3	4.272	1.876	1.146	-0.428
466.4	4.113	1.332	1.245	-0.606
466.5	3.783	0.944	1.311	-0.821
297π/2	3.682	0.872	1.321	-0.883
466.6	3.400	0.737	1.328	-1.064
466.7	3.042	0.671	1.285	-1.317
466.8	2.741	0.694	1.178	-1.559
466.9	2.499	0.767	1.014	-1.768
467.0	2.307	0.866	0.809	-1.925
467.1	2.157	0.976	0.586	-2.025
467.2	2.039	1.091	0.363	-2.070
467.3	1.947	1.205	0.156	-2.070
467.4	1.874	1.319	-0.028	-2.036
467.5	1.817	1.431	-0.187	-1.981
467.6	1.774	1.542	-0.320	-1.911
467.7	1.742	1.652	-0.432	-1.836
467.8	1.720	1.762	-0.524	-1.758
467.9	1.708	1.873	-0.601	-1.680

k_0h	ADMITTANCE		RAD.EFF.LENGTH	
468.0	1.706	1.985	-0.666	-1.605
149π	1.713	2.098	-0.720	-1.534
468.1	1.713	2.101	-0.721	-1.532
468.2	1.733	2.221	-0.769	-1.462
468.3	1.765	2.346	-0.811	-1.396
468.4	1.815	2.477	-0.849	-1.331
468.5	1.885	2.614	-0.884	-1.268
468.6	1.983	2.756	-0.917	-1.206
468.7	2.117	2.901	-0.948	-1.144
468.8	2.298	3.041	-0.979	-1.080
468.9	2.541	3.163	-1.009	-1.014
469.0	2.858	3.237	-1.037	-0.944
469.1	3.249	3.217	-1.064	-0.869
469.2	3.680	3.042	-1.087	-0.786
469.3	4.061	2.667	-1.105	-0.696
469.4	4.261	2.126	-1.113	-0.597
469.5	4.204	1.551	-1.109	-0.490
469.6	3.931	1.092	-1.088	-0.377
299π/2	3.681	0.883	-1.062	-0.300
469.7	3.558	0.814	-1.047	-0.264
469.8	3.186	0.696	-0.982	-0.156
469.9	2.861	0.688	-0.897	-0.063
470.0	2.595	0.744	-0.797	0.010
470.1	2.384	0.834	-0.689	0.058
470.2	2.219	0.941	-0.584	0.083
470.3	2.089	1.054	-0.487	0.090
470.4	1.986	1.168	-0.402	0.083
470.5	1.906	1.282	-0.329	0.068
470.6	1.843	1.395	-0.268	0.051
470.7	1.794	1.506	-0.216	0.035
470.8	1.758	1.616	-0.171	0.021
470.9	1.732	1.726	-0.131	0.010
471.0	1.716	1.836	-0.093	0.003
471.1	1.710	1.948	-0.055	-0.000
471.2	1.713	2.063	-0.016	-0.001
150π	1.717	2.108	-0.000	-0.000
471.3	1.728	2.181	0.026	0.002
471.4	1.755	2.303	0.071	0.006
471.5	1.797	2.431	0.121	0.010
471.6	1.858	2.566	0.177	0.015
471.7	1.944	2.706	0.240	0.018
471.8	2.062	2.850	0.311	0.018
471.9	2.222	2.993	0.391	0.013
472.0	2.438	3.124	0.480	0.001
472.1	2.723	3.221	0.579	-0.022
472.2	3.085	3.246	0.688	-0.060
472.3	3.505	3.142	0.805	-0.118
472.4	3.919	2.851	0.927	-0.200
472.5	4.207	2.367	1.049	-0.312
472.6	4.257	1.788	1.163	-0.460
472.7	4.061	1.270	1.256	-0.646
301π/2	3.679	0.894	1.315	-0.890
472.8	3.716	0.918	1.312	-0.867
472.9	3.336	0.740	1.318	-1.112

k_0h	ADMITTANCE		RAD.EFF.LENGTH	
473.0	2.990	0.693	1.262	-1.364
473.1	2.701	0.727	1.142	-1.600
473.2	2.469	0.805	0.969	-1.798
473.3	2.286	0.906	0.759	-1.942
473.4	2.142	1.017	0.536	-2.030
473.5	2.029	1.131	0.317	-2.064
473.6	1.941	1.245	0.115	-2.056
473.7	1.871	1.358	-0.063	-2.018
473.8	1.817	1.470	-0.215	-1.959
473.9	1.776	1.580	-0.343	-1.888
474.0	1.745	1.690	-0.450	-1.813
474.1	1.725	1.800	-0.538	-1.735
474.2	1.715	1.911	-0.612	-1.659
474.3	1.715	2.025	-0.674	-1.584
151π	1.722	2.118	-0.718	-1.527
474.4	1.724	2.141	-0.727	-1.513
474.5	1.746	2.262	-0.774	-1.445
474.6	1.782	2.387	-0.814	-1.379
474.7	1.835	2.519	-0.852	-1.316
474.8	1.910	2.656	-0.886	-1.254
474.9	2.014	2.799	-0.919	-1.192
475.0	2.155	2.943	-0.951	-1.130
475.1	2.347	3.080	-0.981	-1.067
475.2	2.602	3.194	-1.011	-1.001
475.3	2.932	3.254	-1.040	-0.930
475.4	3.332	3.209	-1.066	-0.854
475.5	3.759	3.000	-1.089	-0.770
475.6	4.112	2.591	-1.106	-0.679
475.7	4.267	2.035	-1.113	-0.578
475.8	4.165	1.477	-1.106	-0.470
475.9	3.867	1.052	-1.082	-0.356
303π/2	3.678	0.904	-1.061	-0.298
476.0	3.492	0.807	-1.036	-0.243
476.1	3.128	0.712	-0.967	-0.137
476.2	2.815	0.718	-0.879	-0.048
476.3	2.561	0.781	-0.777	0.021
476.4	2.360	0.874	-0.669	0.065
476.5	2.201	0.981	-0.565	0.086
476.6	2.077	1.094	-0.470	0.089
476.7	1.979	1.208	-0.388	0.080
476.8	1.902	1.322	-0.317	0.065
476.9	1.841	1.434	-0.258	0.048
477.0	1.795	1.545	-0.208	0.032
477.1	1.761	1.654	-0.164	0.019
477.2	1.737	1.764	-0.124	0.009
477.3	1.722	1.875	-0.086	0.002
477.4	1.718	1.987	-0.048	-0.001
477.5	1.723	2.102	-0.009	-0.000
152π	1.726	2.128	-0.000	-0.000
477.6	1.740	2.221	0.033	0.002
477.7	1.770	2.344	0.079	0.006
477.8	1.815	2.473	0.130	0.011
477.9	1.881	2.608	0.187	0.016

k_0h	ADMITTANCE		RAD.EFF.LENGTH	
478.0	1.972	2.749	0.252	0.018
478.1	2.097	2.892	0.324	0.017
478.2	2.266	3.033	0.405	0.011
478.3	2.494	3.159	0.496	-0.003
478.4	2.792	3.245	0.597	-0.029
478.5	3.165	3.250	0.707	-0.071
478.6	3.588	3.114	0.825	-0.132
478.7	3.985	2.787	0.948	-0.220
478.8	4.233	2.280	1.068	-0.339
478.9	4.234	1.705	1.178	-0.494
479.0	4.004	1.216	1.265	-0.687
305π/2	3.676	0.915	1.310	-0.896
479.1	3.649	0.899	1.312	-0.913
479.2	3.275	0.748	1.305	-1.161
479.3	2.940	0.718	1.236	-1.410
479.4	2.662	0.761	1.104	-1.639
479.5	2.441	0.844	0.922	-1.825
479.6	2.266	0.947	0.709	-1.957
479.7	2.128	1.058	0.486	-2.032
479.8	2.020	1.171	0.271	-2.057
479.9	1.935	1.285	0.075	-2.041
480.0	1.868	1.398	-0.096	-1.998
480.1	1.817	1.509	-0.242	-1.937
480.2	1.777	1.619	-0.364	-1.866
480.3	1.749	1.729	-0.466	-1.790
480.4	1.731	1.839	-0.551	-1.713
480.5	1.723	1.950	-0.622	-1.638
480.6	1.724	2.064	-0.682	-1.565
153π	1.730	2.138	-0.715	-1.520
480.7	1.736	2.181	-0.733	-1.495
480.8	1.760	2.302	-0.778	-1.428
480.9	1.799	2.429	-0.818	-1.363
481.0	1.856	2.561	-0.855	-1.301
481.1	1.935	2.699	-0.889	-1.239
481.2	2.046	2.842	-0.921	-1.178
481.3	2.196	2.985	-0.953	-1.117
481.4	2.398	3.118	-0.984	-1.054
481.5	2.666	3.224	-1.014	-0.987
481.6	3.008	3.267	-1.042	-0.916
481.7	3.415	3.195	-1.069	-0.839
481.8	3.834	2.950	-1.091	-0.754
481.9	4.156	2.510	-1.106	-0.661
482.0	4.263	1.947	-1.112	-0.559
482.1	4.119	1.409	-1.103	-0.449
482.2	3.803	1.019	-1.075	-0.335
307π/2	3.675	0.926	-1.060	-0.296
482.3	3.428	0.805	-1.024	-0.223
482.4	3.073	0.731	-0.952	-0.119
482.5	2.773	0.749	-0.860	-0.033
482.6	2.529	0.818	-0.756	0.031
482.7	2.337	0.914	-0.649	0.070
482.8	2.185	1.022	-0.547	0.087
482.9	2.066	1.135	-0.454	0.088

k_0h	ADMITTANCE		RAD.EFF.LENGTH	
483.0	1.972	1.248	-0.374	0.078
483.1	1.898	1.361	-0.306	0.062
483.2	1.840	1.473	-0.249	0.045
483.3	1.796	1.583	-0.200	0.029
483.4	1.764	1.693	-0.157	0.017
483.5	1.741	1.803	-0.118	0.007
483.6	1.729	1.914	-0.080	0.002
483.7	1.726	2.026	-0.042	-0.001
154π	1.734	2.148	-0.000	-0.000
483.8	1.734	2.142	-0.002	-0.000
483.9	1.753	2.261	0.041	0.003
484.0	1.785	2.385	0.087	0.007
484.1	1.834	2.515	0.139	0.012
484.2	1.904	2.651	0.198	0.016
484.3	2.001	2.792	0.263	0.018
484.4	2.133	2.935	0.337	0.016
484.5	2.313	3.073	0.420	0.008
484.6	2.552	3.193	0.513	-0.008
484.7	2.863	3.266	0.615	-0.037
484.8	3.246	3.248	0.727	-0.081
484.9	3.668	3.080	0.845	-0.148
485.0	4.044	2.718	0.968	-0.241
485.1	4.249	2.191	1.087	-0.367
485.2	4.204	1.626	1.193	-0.529
485.3	3.945	1.168	1.272	-0.729
309π/2	3.673	0.937	1.304	-0.902
485.4	3.583	0.885	1.309	-0.960
485.5	3.216	0.760	1.289	-1.209
485.6	2.892	0.745	1.207	-1.455
485.7	2.626	0.796	1.064	-1.675
485.8	2.414	0.883	0.875	-1.850
485.9	2.247	0.987	0.659	-1.969
486.0	2.115	1.098	0.438	-2.032
486.1	2.012	1.212	0.227	-2.047
486.2	1.930	1.325	0.038	-2.026
486.3	1.866	1.437	-0.127	-1.978
486.4	1.817	1.548	-0.267	-1.915
486.5	1.780	1.658	-0.385	-1.843
486.6	1.753	1.767	-0.482	-1.767
486.7	1.737	1.877	-0.563	-1.691
486.8	1.730	1.989	-0.631	-1.617
486.9	1.734	2.103	-0.689	-1.546
155π	1.739	2.158	-0.713	-1.513
487.0	1.748	2.221	-0.739	-1.477
487.1	1.774	2.343	-0.782	-1.411
487.2	1.816	2.470	-0.822	-1.347
487.3	1.877	2.603	-0.858	-1.286
487.4	1.962	2.742	-0.891	-1.225
487.5	2.079	2.884	-0.924	-1.165
487.6	2.238	3.026	-0.955	-1.104
487.7	2.451	3.155	-0.986	-1.040
487.8	2.731	3.251	-1.016	-0.974
487.9	3.085	3.276	-1.045	-0.902

k_0h	ADMITTANCE		RAD.EFF.LENGTH	
488.0	3.497	3.175	-1.071	-0.824
488.1	3.904	2.895	-1.093	-0.738
488.2	4.192	2.427	-1.107	-0.643
488.3	4.251	1.861	-1.110	-0.539
488.4	4.070	1.348	-1.098	-0.428
488.5	3.737	0.992	-1.066	-0.314
311π/2	3.672	0.947	-1.059	-0.294
488.6	3.365	0.807	-1.012	-0.202
488.7	3.021	0.753	-0.936	-0.102
488.8	2.732	0.781	-0.841	-0.019
488.9	2.499	0.855	-0.736	0.040
489.0	2.315	0.953	-0.629	0.074
489.1	2.170	1.062	-0.529	0.088
489.2	2.055	1.175	-0.438	0.086
489.3	1.965	1.288	-0.361	0.075
489.4	1.895	1.401	-0.295	0.059
489.5	1.840	1.512	-0.240	0.042
489.6	1.798	1.622	-0.193	0.027
489.7	1.767	1.732	-0.151	0.015
489.8	1.747	1.841	-0.112	0.006
489.9	1.736	1.952	-0.074	0.001
490.0	1.735	2.065	-0.035	-0.001
156π	1.743	2.168	-0.000	-0.000
490.1	1.745	2.181	0.005	0.000
490.2	1.766	2.301	0.048	0.003
490.3	1.801	2.426	0.096	0.008
490.4	1.854	2.557	0.149	0.012
490.5	1.929	2.693	0.208	0.016
490.6	2.031	2.834	0.275	0.017
490.7	2.172	2.977	0.351	0.014
490.8	2.361	3.112	0.435	0.005
490.9	2.612	3.224	0.529	-0.013
491.0	2.936	3.283	0.633	-0.045
491.1	3.327	3.241	0.746	-0.093
491.2	3.746	3.039	0.866	-0.164
491.3	4.096	2.644	0.988	-0.264
491.4	4.256	2.103	1.104	-0.396
491.5	4.167	1.553	1.206	-0.566
491.6	3.884	1.128	1.277	-0.772
313π/2	3.670	0.958	1.299	-0.908
491.7	3.518	0.876	1.303	-1.007
491.8	3.159	0.774	1.271	-1.256
491.9	2.847	0.774	1.176	-1.497
492.0	2.592	0.832	1.022	-1.708
492.1	2.389	0.922	0.827	-1.871
492.2	2.229	1.027	0.609	-1.978
492.3	2.103	1.139	0.390	-2.030
492.4	2.004	1.252	0.185	-2.037
492.5	1.926	1.365	0.001	-2.009
492.6	1.865	1.476	-0.157	-1.958
492.7	1.817	1.587	-0.292	-1.893
492.8	1.782	1.696	-0.404	-1.820
492.9	1.758	1.806	-0.497	-1.745

k_0h	ADMITTANCE		RAD.EFF.LENGTH	
493.0	1.743	1.916	-0.575	-1.670
493.1	1.738	2.028	-0.640	-1.597
493.2	1.743	2.143	-0.696	-1.527
157π	1.747	2.178	-0.711	-1.506
493.3	1.760	2.261	-0.744	-1.459
493.4	1.789	2.384	-0.786	-1.394
493.5	1.835	2.512	-0.825	-1.332
493.6	1.899	2.646	-0.860	-1.271
493.7	1.990	2.785	-0.894	-1.212
493.8	2.114	2.927	-0.926	-1.152
493.9	2.281	3.066	-0.958	-1.091
494.0	2.506	3.190	-0.989	-1.027
494.1	2.799	3.275	-1.019	-0.960
494.2	3.164	3.280	-1.048	-0.888
494.3	3.578	3.149	-1.074	-0.808
494.4	3.969	2.834	-1.094	-0.721
494.5	4.218	2.342	-1.107	-0.624
494.6	4.231	1.779	-1.109	-0.519
494.7	4.016	1.293	-1.093	-0.407
315π/2	3.669	0.969	-1.057	-0.292
494.8	3.672	0.971	-1.058	-0.292
494.9	3.304	0.813	-0.999	-0.182
495.0	2.971	0.777	-0.919	-0.085
495.1	2.693	0.814	-0.822	-0.006
495.2	2.470	0.893	-0.716	0.048
495.3	2.294	0.993	-0.610	0.078
495.4	2.155	1.103	-0.511	0.089
495.5	2.046	1.215	-0.423	0.085
495.6	1.959	1.328	-0.348	0.072
495.7	1.892	1.440	-0.285	0.056
495.8	1.839	1.551	-0.231	0.039
495.9	1.799	1.661	-0.185	0.024
496.0	1.771	1.770	-0.144	0.013
496.1	1.752	1.880	-0.105	0.005
496.2	1.743	1.991	-0.067	0.001
496.3	1.744	2.105	-0.029	-0.001
158π	1.751	2.188	-0.000	-0.000
496.4	1.756	2.221	0.012	0.001
496.5	1.780	2.342	0.056	0.004
496.6	1.818	2.468	0.104	0.008
496.7	1.875	2.599	0.159	0.013
496.8	1.954	2.736	0.219	0.016
496.9	2.063	2.877	0.288	0.017
497.0	2.211	3.018	0.364	0.013
497.1	2.411	3.150	0.450	0.002
497.2	2.674	3.254	0.546	-0.019
497.3	3.010	3.297	0.652	-0.053
497.4	3.405	3.228	0.766	-0.106
497.5	3.820	2.992	0.886	-0.182
497.6	4.140	2.566	1.007	-0.287
497.7	4.255	2.017	1.121	-0.427
497.8	4.125	1.486	1.217	-0.604
497.9	3.821	1.093	1.280	-0.816
317π/2	3.668	0.980	1.293	-0.914

OMEGA = 20.0

k_0h	ADMITTANCE		RAD.EFF.LENGTH	
498.0	3.454	0.873	1.295	-1.054
498.1	3.104	0.792	1.250	-1.303
498.2	2.803	0.804	1.142	-1.538
498.3	2.559	0.869	0.979	-1.740
498.4	2.365	0.961	0.778	-1.890
498.5	2.212	1.068	0.560	-1.984
498.6	2.091	1.179	0.344	-2.026
498.7	1.996	1.292	0.144	-2.025
498.8	1.922	1.404	-0.033	-1.991
498.9	1.863	1.515	-0.186	-1.937
499.0	1.818	1.625	-0.314	-1.871
499.1	1.785	1.735	-0.422	-1.798
499.2	1.762	1.844	-0.511	-1.723
499.3	1.750	1.955	-0.586	-1.649
499.4	1.746	2.067	-0.648	-1.577
499.5	1.754	2.182	-0.702	-1.508
159π	1.756	2.198	-0.709	-1.499
499.6	1.773	2.301	-0.749	-1.442
499.7	1.805	2.425	-0.790	-1.378
499.8	1.853	2.553	-0.828	-1.317
499.9	1.923	2.688	-0.863	-1.257
500.0	2.019	2.828	-0.897	-1.198
500.1	2.150	2.969	-0.929	-1.138
500.2	2.327	3.106	-0.961	-1.078
500.3	2.562	3.223	-0.992	-1.014
500.4	2.868	3.295	-1.022	-0.946
500.5	3.243	3.279	-1.051	-0.873
500.6	3.657	3.117	-1.076	-0.793
500.7	4.028	2.768	-1.096	-0.704
500.8	4.236	2.256	-1.108	-0.605
500.9	4.204	1.702	-1.106	-0.499
501.0	3.959	1.245	-1.088	-0.386
319π/2	3.666	0.990	-1.056	-0.289
501.1	3.607	0.956	-1.048	-0.271
501.2	3.245	0.824	-0.985	-0.163
501.3	2.923	0.802	-0.902	-0.068
501.4	2.657	0.849	-0.802	0.006
501.5	2.443	0.932	-0.696	0.055
501.6	2.275	1.033	-0.591	0.081
501.7	2.142	1.143	-0.494	0.089
501.8	2.037	1.256	-0.409	0.082
501.9	1.954	1.368	-0.336	0.069
502.0	1.889	1.480	-0.275	0.052
502.1	1.839	1.590	-0.223	0.036
502.2	1.801	1.699	-0.178	0.022
502.3	1.774	1.809	-0.137	0.011
502.4	1.758	1.919	-0.099	0.004
502.5	1.751	2.030	-0.061	0.000
502.6	1.754	2.144	-0.022	-0.001
160π	1.760	2.208	-0.000	-0.000
502.7	1.767	2.261	0.019	0.001
502.8	1.794	2.382	0.064	0.004
502.9	1.835	2.509	0.113	0.009

k_0h	ADMITTANCE		RAD.EFF.LENGTH	
503.0	1.896	2.641	0.168	0.013
503.1	1.980	2.779	0.230	0.016
503.2	2.096	2.920	0.300	0.016
503.3	2.253	3.059	0.378	0.011
503.4	2.463	3.186	0.466	-0.002
503.5	2.738	3.280	0.563	-0.025
503.6	3.086	3.306	0.670	-0.063
503.7	3.489	3.210	0.786	-0.119
503.8	3.889	2.939	0.906	-0.201
503.9	4.176	2.486	1.026	-0.312
504.0	4.245	1.933	1.137	-0.459
504.1	4.078	1.424	1.227	-0.642
504.2	3.758	1.065	1.281	-0.860
321π/2	3.665	1.001	1.287	-0.920
504.3	3.392	0.874	1.284	-1.101
504.4	3.051	0.813	1.226	-1.348
504.5	2.762	0.835	1.107	-1.577
504.6	2.529	0.906	0.935	-1.768
504.7	2.343	1.001	0.729	-1.906
504.8	2.196	1.108	0.512	-1.989
504.9	2.081	1.219	0.300	-2.020
505.0	1.990	1.332	0.105	-2.011
505.1	1.918	1.444	-0.067	-1.973
505.2	1.862	1.554	-0.213	-1.916
505.3	1.819	1.664	-0.336	-1.849
505.4	1.788	1.773	-0.439	-1.776
505.5	1.767	1.883	-0.524	-1.702
505.6	1.756	1.994	-0.596	-1.629
505.7	1.755	2.106	-0.656	-1.558
161π	1.764	2.218	-0.707	-1.493
505.8	1.764	2.222	-0.708	-1.490
505.9	1.786	2.341	-0.754	-1.425
506.0	1.821	2.466	-0.794	-1.363
506.1	1.873	2.595	-0.831	-1.302
506.2	1.947	2.731	-0.866	-1.243
506.3	2.049	2.870	-0.899	-1.185
506.4	2.187	3.011	-0.932	-1.125
506.5	2.374	3.144	-0.963	-1.065
506.6	2.621	3.255	-0.995	-1.001
506.7	2.939	3.313	-1.025	-0.933
506.8	3.323	3.273	-1.053	-0.858
506.9	3.733	3.079	-1.078	-0.777
507.0	4.080	2.697	-1.097	-0.686
507.1	4.245	2.171	-1.107	-0.586
507.2	4.169	1.629	-1.103	-0.478
507.3	3.900	1.203	-1.081	-0.364
323π/2	3.663	1.012	-1.055	-0.287
507.4	3.543	0.946	-1.038	-0.250
507.5	3.188	0.837	-0.971	-0.144
507.6	2.877	0.830	-0.884	-0.053
507.7	2.622	0.884	-0.782	0.017
507.8	2.418	0.971	-0.676	0.062
507.9	2.256	1.074	-0.572	0.084

k_0h	ADMITTANCE		RAD.EFF.LENGTH	
508.0	2.129	1.184	-0.477	0.088
508.1	2.028	1.296	-0.395	0.080
508.2	1.949	1.408	-0.325	0.066
508.3	1.887	1.519	-0.265	0.049
508.4	1.839	1.629	-0.215	0.033
508.5	1.804	1.738	-0.171	0.020
508.6	1.779	1.847	-0.131	0.010
508.7	1.764	1.957	-0.092	0.003
508.8	1.758	2.069	-0.055	-0.000
508.9	1.763	2.183	-0.015	-0.001
162π	1.768	2.228	-0.000	-0.000
509.0	1.779	2.301	0.026	0.001
509.1	1.809	2.423	0.072	0.005
509.2	1.853	2.551	0.122	0.009
509.3	1.918	2.683	0.178	0.013
509.4	2.008	2.821	0.242	0.016
509.5	2.130	2.962	0.313	0.015
509.6	2.296	3.099	0.393	0.008
509.7	2.517	3.221	0.482	-0.006
509.8	2.804	3.304	0.581	-0.032
509.9	3.163	3.311	0.689	-0.073
510.0	3.568	3.185	0.805	-0.134
510.1	3.953	2.881	0.926	-0.220
510.2	4.204	2.404	1.044	-0.338
510.3	4.227	1.853	1.152	-0.492
510.4	4.026	1.369	1.236	-0.682
325π/2	3.662	1.023	1.281	-0.926
510.5	3.694	1.043	1.280	-0.905
510.6	3.332	0.879	1.271	-1.148
510.7	3.001	0.835	1.200	-1.392
510.8	2.723	0.868	1.069	-1.614
510.9	2.499	0.943	0.889	-1.794
511.0	2.322	1.041	0.681	-1.920
511.1	2.181	1.148	0.464	-1.991
511.2	2.071	1.260	0.256	-2.013
511.3	1.983	1.372	0.067	-1.997
511.4	1.915	1.483	-0.098	-1.954
511.5	1.861	1.593	-0.239	-1.895
511.6	1.821	1.703	-0.357	-1.827
511.7	1.792	1.812	-0.455	-1.754
511.8	1.772	1.922	-0.537	-1.681
511.9	1.763	2.032	-0.606	-1.609
512.0	1.764	2.145	-0.664	-1.539
163π	1.772	2.238	-0.704	-1.486
512.1	1.775	2.262	-0.714	-1.473
512.2	1.799	2.382	-0.758	-1.409
512.3	1.837	2.507	-0.798	-1.347
512.4	1.893	2.637	-0.835	-1.288
512.5	1.972	2.773	-0.869	-1.229
512.6	2.080	2.913	-0.902	-1.171
512.7	2.227	3.052	-0.934	-1.112
512.8	2.423	3.182	-0.966	-1.052
512.9	2.682	3.284	-0.998	-0.988

k_0h	ADMITTANCE		RAD.EFF.LENGTH	
513.0	3.012	3.327	-1.028	-0.919
513.1	3.402	3.262	-1.056	-0.844
513.2	3.806	3.034	-1.081	-0.760
513.3	4.124	2.622	-1.099	-0.668
513.4	4.246	2.086	-1.107	-0.567
513.5	4.130	1.562	-1.100	-0.457
513.6	3.839	1.168	-1.075	-0.343
327π/2	3.661	1.034	-1.053	-0.285
513.7	3.480	0.941	-1.027	-0.230
513.8	3.133	0.854	-0.956	-0.125
513.9	2.834	0.859	-0.865	-0.038
514.0	2.589	0.920	-0.762	0.027
514.1	2.393	1.010	-0.656	0.067
514.2	2.239	1.114	-0.554	0.085
514.3	2.117	1.224	-0.461	0.087
514.4	2.021	1.336	-0.381	0.078
514.5	1.945	1.447	-0.313	0.063
514.6	1.885	1.558	-0.256	0.046
514.7	1.840	1.668	-0.207	0.031
514.8	1.806	1.777	-0.164	0.018
514.9	1.783	1.886	-0.124	0.008
515.0	1.770	1.996	-0.086	0.002
515.1	1.766	2.108	-0.048	-0.000
515.2	1.773	2.223	-0.009	-0.000
164π	1.776	2.248	-0.000	-0.000
515.3	1.792	2.341	0.034	0.002
515.4	1.824	2.464	0.080	0.006
515.5	1.872	2.592	0.131	0.010
515.6	1.941	2.726	0.188	0.014
515.7	2.036	2.864	0.253	0.015
515.8	2.166	3.004	0.326	0.013
515.9	2.340	3.138	0.407	0.005
516.0	2.572	3.254	0.498	-0.011
516.1	2.872	3.325	0.599	-0.039
516.2	3.240	3.311	0.708	-0.083
516.3	3.646	3.155	0.825	-0.149
516.4	4.012	2.817	0.945	-0.241
516.5	4.224	2.321	1.062	-0.366
516.6	4.202	1.776	1.166	-0.527
516.7	3.972	1.321	1.242	-0.724
329π/2	3.659	1.044	1.276	-0.932
516.8	3.630	1.027	1.276	-0.951
516.9	3.273	0.888	1.255	-1.195
517.0	2.953	0.860	1.172	-1.435
517.1	2.686	0.901	1.030	-1.649
517.2	2.472	0.981	0.843	-1.817
517.3	2.302	1.080	0.632	-1.931
517.4	2.167	1.188	0.418	-1.990
517.5	2.061	1.300	0.214	-2.003
517.6	1.978	1.411	0.031	-1.981
517.7	1.912	1.522	-0.128	-1.935
517.8	1.861	1.632	-0.263	-1.874
517.9	1.823	1.742	-0.376	-1.805

k_0h	ADMITTANCE		RAD.EFF.LENGTH	
518.0	1.795	1.851	-0.470	-1.733
518.1	1.778	1.960	-0.549	-1.660
518.2	1.770	2.071	-0.615	-1.589
518.3	1.773	2.185	-0.671	-1.521
165π	1.780	2.258	-0.702	-1.479
518.4	1.787	2.301	-0.720	-1.455
518.5	1.813	2.422	-0.763	-1.393
518.6	1.854	2.548	-0.802	-1.332
518.7	1.914	2.679	-0.838	-1.274
518.8	1.998	2.815	-0.872	-1.216
518.9	2.112	2.955	-0.905	-1.158
519.0	2.267	3.092	-0.937	-1.100
519.1	2.474	3.218	-0.969	-1.039
519.2	2.745	3.311	-1.001	-0.974
519.3	3.086	3.336	-1.031	-0.905
519.4	3.481	3.245	-1.059	-0.828
519.5	3.874	2.984	-1.083	-0.744
519.6	4.161	2.545	-1.100	-0.650
519.7	4.239	2.004	-1.106	-0.547
519.8	4.085	1.501	-1.096	-0.436
519.9	3.777	1.139	-1.067	-0.321
331π/2	3.658	1.055	-1.052	-0.283
520.0	3.419	0.941	-1.015	-0.209
520.1	3.081	0.873	-0.940	-0.107
520.2	2.792	0.890	-0.846	-0.024
520.3	2.558	0.956	-0.742	0.036
520.4	2.371	1.049	-0.636	0.072
520.5	2.223	1.154	-0.536	0.087
520.6	2.106	1.264	-0.446	0.086
520.7	2.013	1.376	-0.368	0.075
520.8	1.941	1.487	-0.303	0.060
520.9	1.884	1.597	-0.247	0.043
521.0	1.841	1.706	-0.199	0.028
521.1	1.809	1.815	-0.157	0.016
521.2	1.788	1.925	-0.118	0.007
521.3	1.776	2.035	-0.080	0.002
521.4	1.775	2.147	-0.042	-0.001
166π	1.784	2.268	-0.000	-0.000
521.5	1.784	2.262	-0.002	-0.000
521.6	1.805	2.381	0.041	0.002
521.7	1.839	2.505	0.088	0.006
521.8	1.891	2.634	0.140	0.010
521.9	1.965	2.768	0.199	0.014
522.0	2.066	2.906	0.265	0.015
522.1	2.203	3.045	0.339	0.012
522.2	2.387	3.177	0.422	0.002
522.3	2.630	3.285	0.514	-0.016
522.4	2.942	3.343	0.616	-0.047
522.5	3.318	3.306	0.727	-0.095
522.6	3.721	3.119	0.845	-0.165
522.7	4.064	2.749	0.965	-0.263
522.8	4.235	2.238	1.079	-0.395
522.9	4.171	1.704	1.178	-0.562

k_0h	ADMITTANCE		RAD.EFF.LENGTH	
523.0	3.914	1.279	1.247	-0.766
333π/2	3.657	1.066	1.270	-0.938
523.1	3.567	1.016	1.271	-0.997
523.2	3.217	0.900	1.237	-1.241
523.3	2.907	0.887	1.141	-1.476
523.4	2.651	0.936	0.989	-1.681
523.5	2.446	1.019	0.797	-1.837
523.6	2.283	1.120	0.584	-1.939
523.7	2.154	1.229	0.372	-1.988
523.8	2.053	1.340	0.173	-1.993
523.9	1.972	1.451	-0.004	-1.965
524.0	1.909	1.562	-0.157	-1.916
524.1	1.861	1.671	-0.286	-1.853
524.2	1.825	1.780	-0.395	-1.783
524.3	1.799	1.889	-0.485	-1.711
524.4	1.784	1.999	-0.560	-1.640
524.5	1.778	2.110	-0.624	-1.570
524.6	1.783	2.224	-0.678	-1.503
167π	1.788	2.278	-0.700	-1.473
524.7	1.799	2.341	-0.725	-1.439
524.8	1.827	2.463	-0.767	-1.377
524.9	1.872	2.590	-0.805	-1.317
525.0	1.936	2.721	-0.841	-1.260
525.1	2.025	2.858	-0.875	-1.203
525.2	2.146	2.997	-0.908	-1.145
525.3	2.310	3.132	-0.940	-1.087
525.4	2.527	3.252	-0.972	-1.026
525.5	2.810	3.335	-1.004	-0.961
525.6	3.161	3.342	-1.034	-0.891
525.7	3.559	3.222	-1.062	-0.813
525.8	3.938	2.928	-1.085	-0.727
525.9	4.190	2.465	-1.101	-0.632
526.0	4.223	1.925	-1.104	-0.527
526.1	4.036	1.446	-1.092	-0.415
526.2	3.715	1.116	-1.058	-0.300
335π/2	3.655	1.077	-1.050	-0.281
526.3	3.359	0.945	-1.002	-0.189
526.4	3.031	0.894	-0.923	-0.090
526.5	2.753	0.922	-0.827	-0.011
526.6	2.528	0.993	-0.722	0.044
526.7	2.349	1.088	-0.617	0.076
526.8	2.207	1.194	-0.518	0.087
526.9	2.095	1.304	-0.431	0.084
527.0	2.007	1.415	-0.356	0.072
527.1	1.937	1.526	-0.292	0.056
527.2	1.883	1.636	-0.238	0.040
527.3	1.842	1.745	-0.192	0.026
527.4	1.812	1.854	-0.150	0.014
527.5	1.793	1.963	-0.111	0.006
527.6	1.783	2.074	-0.073	0.001
527.7	1.783	2.187	-0.035	-0.001
168π	1.793	2.288	-0.000	-0.000
527.8	1.795	2.302	0.005	0.000
527.9	1.818	2.422	0.049	0.003

k_0h	ADMITTANCE		RAD.EFF.LENGTH	
528.0	1.856	2.546	0.097	0.007
528.1	1.911	2.676	0.150	0.011
528.2	1.989	2.810	0.209	0.014
528.3	2.096	2.948	0.277	0.014
528.4	2.241	3.086	0.352	0.010
528.5	2.435	3.214	0.437	-0.001
528.6	2.690	3.315	0.531	-0.022
528.7	3.013	3.357	0.635	-0.056
528.8	3.396	3.295	0.747	-0.107
528.9	3.792	3.076	0.865	-0.183
529.0	4.109	2.678	0.984	-0.287
529.1	4.237	2.155	1.096	-0.425
529.2	4.134	1.638	1.189	-0.599
529.3	3.856	1.243	1.250	-0.808
337π/2	3.654	1.088	1.264	-0.943
529.4	3.505	1.010	1.262	-1.043
529.5	3.162	0.915	1.216	-1.286
529.6	2.863	0.915	1.108	-1.516
529.7	2.618	0.971	0.947	-1.711
529.8	2.421	1.058	0.749	-1.855
529.9	2.265	1.160	0.537	-1.945
530.0	2.142	1.269	0.328	-1.984
530.1	2.044	1.380	0.134	-1.981
530.2	1.968	1.491	-0.037	-1.948
530.3	1.907	1.601	-0.184	-1.896
530.4	1.861	1.710	-0.308	-1.832
530.5	1.827	1.819	-0.412	-1.762
530.6	1.803	1.928	-0.498	-1.691
530.7	1.790	2.038	-0.571	-1.620
530.8	1.786	2.149	-0.632	-1.551
530.9	1.793	2.264	-0.684	-1.485
169π	1.797	2.298	-0.698	-1.466
531.0	1.811	2.382	-0.730	-1.422
531.1	1.842	2.504	-0.771	-1.361
531.2	1.890	2.631	-0.809	-1.303
531.3	1.958	2.764	-0.844	-1.246
531.4	2.053	2.900	-0.878	-1.189
531.5	2.181	3.039	-0.911	-1.133
531.6	2.354	3.171	-0.943	-1.074
531.7	2.582	3.286	-0.975	-1.013
531.8	2.877	3.356	-1.007	-0.948
531.9	3.237	3.343	-1.037	-0.876
532.0	3.635	3.193	-1.065	-0.798
532.1	3.996	2.867	-1.087	-0.710
532.2	4.211	2.385	-1.101	-0.613
532.3	4.201	1.850	-1.102	-0.507
532.4	3.983	1.397	-1.086	-0.394
339π/2	3.653	1.098	-1.049	-0.279
532.5	3.652	1.098	-1.049	-0.279
532.6	3.301	0.952	-0.988	-0.169
532.7	2.983	0.918	-0.906	-0.074
532.8	2.716	0.955	-0.808	0.001
532.9	2.500	1.031	-0.702	0.052

k_0h	ADMITTANCE		RAD.EFF.LENGTH	
533.0	2.329	1.128	-0.597	0.079
533.1	2.193	1.234	-0.501	0.087
533.2	2.086	1.344	-0.416	0.082
533.3	2.001	1.455	-0.343	0.069
533.4	1.934	1.566	-0.282	0.053
533.5	1.882	1.675	-0.230	0.037
533.6	1.843	1.784	-0.184	0.023
533.7	1.816	1.893	-0.143	0.012
533.8	1.798	2.002	-0.105	0.005
533.9	1.790	2.113	-0.067	0.001
534.0	1.792	2.226	-0.029	-0.001
170π	1.801	2.308	-0.000	-0.000
534.1	1.806	2.342	0.012	0.001
534.2	1.832	2.462	0.056	0.003
534.3	1.872	2.587	0.105	0.007
534.4	1.932	2.718	0.159	0.011
534.5	2.015	2.853	0.220	0.013
534.6	2.128	2.991	0.289	0.013
534.7	2.281	3.126	0.366	0.008
534.8	2.485	3.250	0.452	-0.005
534.9	2.752	3.341	0.548	-0.028
535.0	3.086	3.368	0.653	-0.065
535.1	3.473	3.280	0.766	-0.121
535.2	3.860	3.028	0.884	-0.201
535.3	4.147	2.603	1.002	-0.311
535.4	4.232	2.075	1.111	-0.456
535.5	4.091	1.577	1.199	-0.638
535.6	3.795	1.213	1.250	-0.852
341π/2	3.651	1.109	1.257	-0.949
535.7	3.445	1.008	1.251	-1.089
535.8	3.110	0.933	1.192	-1.330
535.9	2.822	0.945	1.073	-1.553
536.0	2.586	1.007	0.903	-1.738
536.1	2.398	1.097	0.702	-1.870
536.2	2.249	1.200	0.490	-1.948
536.3	2.130	1.309	0.284	-1.978
536.4	2.037	1.420	0.096	-1.968
536.5	1.963	1.530	-0.069	-1.931
536.6	1.906	1.640	-0.210	-1.876
536.7	1.862	1.749	-0.329	-1.811
536.8	1.829	1.858	-0.428	-1.741
536.9	1.808	1.967	-0.511	-1.670
537.0	1.796	2.077	-0.581	-1.600
537.1	1.794	2.189	-0.640	-1.533
537.2	1.803	2.303	-0.690	-1.468
171π	1.805	2.318	-0.696	-1.460
537.3	1.823	2.422	-0.735	-1.406
537.4	1.858	2.545	-0.775	-1.346
537.5	1.909	2.673	-0.812	-1.289
537.6	1.982	2.906	-0.847	-1.232
537.7	2.082	2.943	-0.881	-1.176
537.8	2.218	3.080	-0.914	-1.120
537.9	2.399	3.210	-0.946	-1.062

OMEGA = 20.0

k_0h	ADMITTANCE		RAD.EFF.LENGTH	
538.0	2.639	3.317	-0.979	-1.000
538.1	2.945	3.374	-1.010	-0.934
538.2	3.313	3.339	-1.040	-0.862
538.3	3.709	3.159	-1.067	-0.782
538.4	4.048	2.802	-1.089	-0.693
538.5	4.224	2.304	-1.101	-0.594
538.6	4.172	1.779	-1.100	-0.486
538.7	3.928	1.354	-1.080	-0.372
$343\pi/2$	3.650	1.120	-1.047	-0.277
538.8	3.591	1.086	-1.039	-0.258
538.9	3.245	0.963	-0.974	-0.150
539.0	2.937	0.944	-0.888	-0.058
539.1	2.680	0.989	-0.788	0.012
539.2	2.474	1.069	-0.682	0.058
539.3	2.309	1.167	-0.579	0.082
539.4	2.179	1.274	-0.484	0.087
539.5	2.076	1.384	-0.402	0.080
539.6	1.995	1.495	-0.332	0.066
539.7	1.931	1.605	-0.272	0.050
539.8	1.882	1.714	-0.221	0.035
539.9	1.845	1.823	-0.177	0.021
540.0	1.819	1.931	-0.137	0.011
540.1	1.803	2.041	-0.099	0.004
540.2	1.797	2.152	-0.061	0.000
540.3	1.802	2.265	-0.022	-0.001
172π	1.809	2.328	-0.000	-0.000
540.4	1.817	2.382	0.019	0.001
540.5	1.846	2.503	0.064	0.004
540.6	1.890	2.629	0.114	0.008
540.7	1.953	2.760	0.169	0.011
540.8	2.041	2.895	0.231	0.013
540.9	2.161	3.032	0.301	0.012
541.0	2.323	3.166	0.380	0.005
541.1	2.537	3.284	0.467	-0.009
541.2	2.815	3.366	0.565	-0.034
541.3	3.160	3.374	0.671	-0.075
541.4	3.550	3.259	0.785	-0.135
541.5	3.923	2.975	0.904	-0.221
541.6	4.176	2.526	1.020	-0.337
541.7	4.219	1.998	1.125	-0.489
541.8	4.045	1.522	1.207	-0.677
541.9	3.735	1.188	1.249	-0.896
$345\pi/2$	3.649	1.131	1.251	-0.954
542.0	3.385	1.011	1.238	-1.135
542.1	3.060	0.954	1.166	-1.373
542.2	2.782	0.976	1.036	-1.589
542.3	2.556	1.044	0.859	-1.762
542.4	2.376	1.136	0.655	-1.883
542.5	2.233	1.240	0.444	-1.950
542.6	2.120	1.349	0.243	-1.970
542.7	2.030	1.460	0.060	-1.954
542.8	1.959	1.570	-0.100	-1.913
542.9	1.905	1.679	-0.235	-1.856

k_0h	ADMITTANCE		RAD.EFF.LENGTH	
543.0	1.863	1.788	-0.349	-1.790
543.1	1.832	1.896	-0.444	-1.720
543.2	1.812	2.005	-0.523	-1.650
543.3	1.802	2.116	-0.590	-1.581
543.4	1.802	2.228	-0.647	-1.515
173π	1.813	2.338	-0.694	-1.454
543.5	1.813	2.343	-0.696	-1.451
543.6	1.836	2.462	-0.740	-1.390
543.7	1.874	2.586	-0.779	-1.331
543.8	1.929	2.714	-0.815	-1.275
543.9	2.006	2.848	-0.850	-1.219
544.0	2.112	2.985	-0.884	-1.164
544.1	2.256	3.121	-0.917	-1.107
544.2	2.447	3.247	-0.949	-1.049
544.3	2.698	3.346	-0.982	-0.987
544.4	3.015	3.389	-1.014	-0.920
544.5	3.390	3.330	-1.044	-0.847
544.6	3.779	3.119	-1.070	-0.766
544.7	4.094	2.733	-1.091	-0.675
544.8	4.228	2.224	-1.101	-0.574
544.9	4.137	1.713	-1.097	-0.465
545.0	3.871	1.317	-1.074	-0.351
$347\pi/2$	3.647	1.142	-1.046	-0.275
545.1	3.530	1.079	-1.028	-0.237
545.2	3.190	0.977	-0.959	-0.132
545.3	2.893	0.972	-0.870	-0.043
545.4	2.646	1.023	-0.768	0.023
545.5	2.449	1.107	-0.662	0.064
545.6	2.291	1.207	-0.560	0.084
545.7	2.167	1.314	-0.468	0.086
545.8	2.068	1.424	-0.388	0.078
545.9	1.990	1.534	-0.320	0.063
546.0	1.929	1.644	-0.263	0.047
546.1	1.882	1.753	-0.213	0.032
546.2	1.847	1.862	-0.170	0.019
546.3	1.823	1.970	-0.130	0.009
546.4	1.809	2.080	-0.092	0.003
546.5	1.805	2.191	-0.054	-0.000
546.6	1.811	2.305	-0.015	-0.000
174π	1.817	2.348	-0.000	-0.000
546.7	1.829	2.422	0.027	0.001
546.8	1.861	2.544	0.072	0.005
546.9	1.908	2.670	0.123	0.008
547.0	1.976	2.802	0.179	0.011
547.1	2.069	2.937	0.243	0.013
547.2	2.196	3.074	0.314	0.011
547.3	2.366	3.205	0.394	0.003
547.4	2.591	3.317	0.483	-0.013
547.5	2.881	3.387	0.582	-0.041
547.6	3.234	3.375	0.690	-0.085
547.7	3.624	3.232	0.805	-0.150
547.8	3.981	2.917	0.923	-0.241
547.9	4.198	2.448	1.038	-0.364

k_0h	ADMITTANCE		RAD.EFF.LENGTH	
548.0	4.199	1.924	1.139	-0.523
548.1	3.994	1.472	1.213	-0.717
$349\pi/2$	3.646	1.153	1.245	-0.959
548.2	3.674	1.170	1.245	-0.941
548.3	3.328	1.017	1.222	-1.180
548.4	3.011	0.977	1.138	-1.414
548.5	2.745	1.008	0.997	-1.622
548.6	2.528	1.081	0.814	-1.784
548.7	2.355	1.175	0.608	-1.893
548.8	2.218	1.280	0.399	-1.949
548.9	2.109	1.389	0.202	-1.961
549.0	2.024	1.499	0.025	-1.939
549.1	1.956	1.609	-0.129	-1.894
549.2	1.904	1.718	-0.259	-1.835
549.3	1.864	1.827	-0.368	-1.769
549.4	1.836	1.935	-0.459	-1.699
549.5	1.817	2.044	-0.535	-1.630
549.6	1.809	2.155	-0.599	-1.562
549.7	1.811	2.267	-0.654	-1.497
175π	1.820	2.358	-0.692	-1.447
549.8	1.824	2.383	-0.702	-1.434
549.9	1.850	2.503	-0.744	-1.374
550.0	1.890	2.627	-0.783	-1.317
550.1	1.949	2.756	-0.819	-1.261
550.2	2.031	2.890	-0.853	-1.206
550.3	2.144	3.027	-0.887	-1.151
550.4	2.295	3.161	-0.920	-1.095
550.5	2.496	3.283	-0.952	-1.036
550.6	2.758	3.373	-0.985	-0.974
550.7	3.086	3.399	-1.017	-0.907
550.8	3.466	3.316	-1.047	-0.832
550.9	3.846	3.073	-1.073	-0.749
551.0	4.132	2.661	-1.092	-0.657
551.1	4.225	2.145	-1.100	-0.555
551.2	4.097	1.653	-1.093	-0.444
551.3	3.813	1.286	-1.066	-0.329
$351\pi/2$	3.645	1.163	-1.044	-0.272
551.4	3.470	1.076	-1.016	-0.216
551.5	3.138	0.995	-0.943	-0.113
551.6	2.851	1.001	-0.851	-0.029
551.7	2.615	1.059	-0.748	0.032
551.8	2.425	1.146	-0.642	0.069
551.9	2.274	1.247	-0.542	0.085
552.0	2.155	1.354	-0.453	0.085
552.1	2.060	1.464	-0.375	0.075
552.2	1.986	1.574	-0.309	0.060
552.3	1.927	1.683	-0.254	0.044
552.4	1.883	1.792	-0.206	0.029
552.5	1.850	1.900	-0.163	0.017
552.6	1.827	2.009	-0.124	0.008
552.7	1.815	2.119	-0.086	0.002
552.8	1.813	2.230	-0.048	-0.000
552.9	1.821	2.345	-0.008	-0.000
176π	1.824	2.368	-0.000	-0.000

k_0h	ADMITTANCE		RAD.EFF.LENGTH	
553.0	1.842	2.462	0.034	0.002
553.1	1.876	2.585	0.080	0.005
553.2	1.927	2.712	0.132	0.009
553.3	1.999	2.844	0.189	0.012
553.4	2.098	2.979	0.254	0.012
553.5	2.232	3.115	0.327	0.009
553.6	2.411	3.243	0.408	-0.000
553.7	2.647	3.349	0.499	-0.019
553.8	2.948	3.406	0.600	-0.049
553.9	3.309	3.373	0.709	-0.097
554.0	3.697	3.199	0.824	-0.166
554.1	4.033	2.855	0.942	-0.263
554.2	4.212	2.370	1.054	-0.393
554.3	4.172	1.854	1.150	-0.558
554.4	3.941	1.429	1.217	-0.758
$353\pi/2$	3.644	1.174	1.239	-0.965
554.5	3.613	1.156	1.239	-0.986
554.6	3.272	1.027	1.203	-1.225
554.7	2.965	1.002	1.108	-1.454
554.8	2.709	1.042	0.957	-1.652
554.9	2.501	1.119	0.769	-1.803
555.0	2.335	1.215	0.562	-1.900
555.1	2.204	1.320	0.355	-1.947
555.2	2.100	1.429	0.163	-1.951
555.3	2.018	1.539	-0.009	-1.923
555.4	1.953	1.648	-0.156	-1.875
555.5	1.903	1.757	-0.281	-1.815
555.6	1.865	1.866	-0.386	-1.748
555.7	1.839	1.974	-0.473	-1.679
555.8	1.823	2.083	-0.546	-1.610
555.9	1.816	2.194	-0.608	-1.544
556.0	1.820	2.307	-0.661	-1.479
177π	1.828	2.378	-0.690	-1.441
556.1	1.836	2.423	-0.707	-1.418
556.2	1.864	2.543	-0.749	-1.359
556.3	1.908	2.668	-0.786	-1.303
556.4	1.970	2.798	-0.822	-1.247
556.5	2.058	2.932	-0.856	-1.193
556.6	2.176	3.068	-0.890	-1.138
556.7	2.336	3.200	-0.923	-1.082
556.8	2.547	3.317	-0.956	-1.024
556.9	2.820	3.397	-0.989	-0.961
557.0	3.158	3.406	-1.020	-0.893
557.1	3.541	3.296	-1.050	-0.817
557.2	3.908	3.022	-1.075	-0.733
557.3	4.163	2.587	-1.093	-0.639
557.4	4.214	2.070	-1.099	-0.535
557.5	4.052	1.597	-1.089	-0.423
557.6	3.754	1.261	-1.058	-0.306
$355\pi/2$	3.642	1.185	-1.043	-0.270
557.7	3.411	1.077	-1.004	-0.196
557.8	3.088	1.014	-0.926	-0.096
557.9	2.811	1.031	-0.832	-0.016

k_0h	ADMITTANCE		RAD.EFF.LENGTH	
558.0	2.584	1.095	-0.728	0.041
558.1	2.402	1.184	-0.623	0.073
558.2	2.258	1.287	-0.525	0.086
558.3	2.143	1.394	-0.438	0.083
558.4	2.053	1.504	-0.363	0.072
558.5	1.981	1.614	-0.299	0.057
558.6	1.926	1.722	-0.245	0.041
558.7	1.883	1.831	-0.198	0.027
558.8	1.852	1.939	-0.156	0.015
558.9	1.832	2.048	-0.117	0.007
559.0	1.821	2.158	-0.080	0.002
559.1	1.821	2.270	-0.041	-0.000
178π	1.832	2.388	-0.000	-0.000
559.2	1.832	2.384	-0.001	-0.000
559.3	1.854	2.503	0.042	0.002
559.4	1.892	2.626	0.089	0.006
559.5	1.946	2.753	0.141	0.009
559.6	2.023	2.886	0.200	0.011
559.7	2.128	3.021	0.266	0.011
559.8	2.269	3.155	0.340	0.007
559.9	2.458	3.280	0.423	-0.004
560.0	2.705	3.378	0.515	-0.024
560.1	3.016	3.421	0.617	-0.058
560.2	3.384	3.365	0.727	-0.109
560.3	3.766	3.161	0.843	-0.183
560.4	4.078	2.788	0.960	-0.286
560.5	4.219	2.292	1.070	-0.422
560.6	4.139	1.788	1.161	-0.594
560.7	3.886	1.392	1.219	-0.800
$357\pi/2$	3.641	1.196	1.232	-0.970
560.8	3.553	1.148	1.230	-1.030
560.9	3.218	1.040	1.182	-1.268
561.0	2.921	1.028	1.075	-1.492
561.1	2.675	1.076	0.916	-1.681
561.2	2.476	1.157	0.723	-1.820
561.3	2.317	1.254	0.516	-1.906
561.4	2.191	1.360	0.313	-1.942
561.5	2.091	1.469	0.125	-1.939
561.6	2.012	1.579	-0.041	-1.907
561.7	1.951	1.688	-0.183	-1.856
561.8	1.903	1.796	-0.302	-1.794
561.9	1.867	1.904	-0.403	-1.727
562.0	1.843	2.013	-0.486	-1.659
562.1	1.828	2.122	-0.556	-1.591
562.2	1.824	2.233	-0.616	-1.526
562.3	1.830	2.346	-0.667	-1.462
179π	1.836	2.399	-0.688	-1.435
562.4	1.847	2.463	-0.712	-1.402
562.5	1.878	2.584	-0.753	-1.344
562.6	1.925	2.710	-0.790	-1.289
562.7	1.992	2.840	-0.825	-1.234
562.8	2.085	2.974	-0.859	-1.180
562.9	2.211	3.109	-0.893	-1.126

k_0h	ADMITTANCE		RAD.EFF.LENGTH	
563.0	2.379	3.239	-0.926	-1.070
563.1	2.600	3.350	-0.959	-1.011
563.2	2.884	3.419	-0.992	-0.948
563.3	3.231	3.409	-1.024	-0.879
563.4	3.614	3.271	-1.053	-0.802
563.5	3.966	2.967	-1.078	-0.716
563.6	4.186	2.511	-1.094	-0.620
563.7	4.196	1.997	-1.098	-0.514
563.8	4.004	1.548	-1.084	-0.401
563.9	3.694	1.241	-1.049	-0.286
$359\pi/2$	3.640	1.207	-1.041	-0.268
564.0	3.354	1.082	-0.990	-0.176
564.1	3.040	1.036	-0.909	-0.079
564.2	2.773	1.063	-0.812	-0.003
564.3	2.555	1.132	-0.708	0.048
564.4	2.381	1.223	-0.604	0.077
564.5	2.243	1.327	-0.508	0.086
564.6	2.133	1.434	-0.423	0.082
564.7	2.046	1.544	-0.350	0.070
564.8	1.978	1.653	-0.289	0.054
564.9	1.924	1.762	-0.236	0.039
565.0	1.884	1.870	-0.191	0.025
565.1	1.855	1.978	-0.149	0.014
565.2	1.837	2.087	-0.111	0.006
565.3	1.828	2.197	-0.073	0.001
565.4	1.830	2.309	-0.035	-0.000
180π	1.840	2.409	-0.000	-0.000
565.5	1.842	2.424	0.006	0.000
565.6	1.868	2.543	0.049	0.003
565.7	1.908	2.667	0.097	0.006
565.8	1.966	2.795	0.151	0.009
565.9	2.048	2.928	0.210	0.011
566.0	2.159	3.063	0.278	0.011
566.1	2.308	3.195	0.353	0.005
566.2	2.506	3.316	0.438	-0.007
566.3	2.764	3.405	0.532	-0.030
566.4	3.086	3.432	0.635	-0.067
566.5	3.458	3.352	0.746	-0.122
566.6	3.832	3.118	0.863	-0.201
566.7	4.117	2.719	0.978	-0.310
566.8	4.217	2.215	1.085	-0.453
566.9	4.102	1.728	1.170	-0.632
567.0	3.829	1.360	1.219	-0.843
$361\pi/2$	3.639	1.218	1.226	-0.975
567.1	3.494	1.144	1.219	-1.075
567.2	3.166	1.056	1.159	-1.311
567.3	2.879	1.057	1.041	-1.528
567.4	2.642	1.111	0.874	-1.707
567.5	2.451	1.195	0.677	-1.834
567.6	2.299	1.294	0.471	-1.909
567.7	2.179	1.400	0.271	-1.936
567.8	2.083	1.509	0.089	-1.926
567.9	2.007	1.618	-0.071	-1.890

k_0h	ADMITTANCE		RAD.EFF.LENGTH	
568.0	1.948	1.727	-0.208	-1.837
568.1	1.903	1.835	-0.323	-1.774
568.2	1.869	1.943	-0.419	-1.707
568.3	1.847	2.052	-0.499	-1.639
568.4	1.834	2.161	-0.566	-1.572
568.5	1.831	2.272	-0.624	-1.508
568.6	1.840	2.386	-0.673	-1.446
181π	1.844	2.419	-0.686	-1.429
568.7	1.860	2.503	-0.717	-1.387
568.8	1.893	2.625	-0.757	-1.330
568.9	1.944	2.751	-0.794	-1.275
569.0	2.015	2.882	-0.829	-1.221
569.1	2.113	3.016	-0.862	-1.168
569.2	2.246	3.150	-0.896	-1.114
569.3	2.423	3.277	-0.929	-1.058
569.4	2.655	3.381	-0.963	-0.998
569.5	2.950	3.438	-0.996	-0.935
569.6	3.304	3.407	-1.027	-0.864
569.7	3.685	3.240	-1.056	-0.786
569.8	4.018	2.907	-1.080	-0.699
569.9	4.201	2.435	-1.094	-0.601
570.0	4.171	1.928	-1.096	-0.494
570.1	3.953	1.504	-1.078	-0.380
$363\pi/2$	3.638	1.229	-1.040	-0.266
570.2	3.634	1.227	-1.039	-0.265
570.3	3.298	1.091	-0.976	-0.157
570.4	2.993	1.060	-0.892	-0.064
570.5	2.737	1.095	-0.793	0.008
570.6	2.528	1.169	-0.688	0.055
570.7	2.361	1.263	-0.585	0.079
570.8	2.228	1.366	-0.491	0.086
570.9	2.123	1.474	-0.409	0.080
571.0	2.040	1.583	-0.339	0.067
571.1	1.974	1.692	-0.279	0.051
571.2	1.924	1.801	-0.228	0.036
571.3	1.885	1.909	-0.183	0.023
571.4	1.858	2.017	-0.143	0.012
571.5	1.842	2.126	-0.105	0.005
571.6	1.835	2.236	-0.067	0.001
571.7	1.839	2.348	-0.028	-0.000
182π	1.848	2.429	-0.000	-0.000
571.8	1.854	2.464	0.013	0.000
571.9	1.882	2.584	0.057	0.003
572.0	1.925	2.708	0.106	0.006
572.1	1.987	2.837	0.160	0.009
572.2	2.074	2.970	0.221	0.011
572.3	2.191	3.104	0.290	0.009
572.4	2.349	3.235	0.367	0.003
572.5	2.557	3.350	0.453	-0.011
572.6	2.825	3.429	0.549	-0.037
572.7	3.156	3.439	0.653	-0.076
572.8	3.532	3.334	0.765	-0.136
572.9	3.894	3.070	0.882	-0.220

k_0h	ADMITTANCE		RAD.EFF.LENGTH	
573.0	4.149	2.647	0.996	-0.335
573.1	4.208	2.141	1.099	-0.485
573.2	4.060	1.673	1.177	-0.670
573.3	3.772	1.334	1.217	-0.886
$365\pi/2$	3.636	1.240	1.219	-0.980
573.4	3.436	1.144	1.205	-1.120
573.5	3.116	1.075	1.133	-1.353
573.6	2.839	1.086	1.004	-1.562
573.7	2.612	1.147	0.831	-1.730
573.8	2.429	1.233	0.631	-1.846
573.9	2.283	1.334	0.426	-1.910
574.0	2.167	1.440	0.231	-1.929
574.1	2.075	1.549	0.054	-1.913
574.2	2.003	1.658	-0.100	-1.872
574.3	1.946	1.766	-0.232	-1.817
574.4	1.903	1.874	-0.342	-1.754
574.5	1.872	1.982	-0.434	-1.687
574.6	1.851	2.090	-0.511	-1.620
574.7	1.840	2.200	-0.576	-1.554
574.8	1.840	2.311	-0.631	-1.490
574.9	1.850	2.426	-0.679	-1.430
183π	1.852	2.439	-0.684	-1.423
575.0	1.872	2.544	-0.722	-1.371
575.1	1.909	2.666	-0.761	-1.315
575.2	1.963	2.793	-0.797	-1.261
575.3	2.039	2.924	-0.832	-1.208
575.4	2.143	3.058	-0.866	-1.155
575.5	2.283	3.191	-0.899	-1.101
575.6	2.469	3.314	-0.933	-1.045
575.7	2.711	3.410	-0.966	-0.986
575.8	3.017	3.453	-0.999	-0.921
575.9	3.378	3.400	-1.031	-0.850
576.0	3.754	3.204	-1.059	-0.770
576.1	4.064	2.843	-1.082	-0.681
576.2	4.209	2.359	-1.095	-0.581
576.3	4.141	1.863	-1.093	-0.473
576.4	3.900	1.466	-1.072	-0.358
$367\pi/2$	3.635	1.251	-1.038	-0.264
576.5	3.575	1.217	-1.028	-0.244
576.6	3.245	1.103	-0.961	-0.138
576.7	2.949	1.086	-0.874	-0.048
576.8	2.702	1.129	-0.773	0.018
576.9	2.502	1.206	-0.666	0.061
577.0	2.342	1.302	-0.567	0.082
577.1	2.215	1.406	-0.475	0.085
577.2	2.114	1.514	-0.395	0.077
577.3	2.034	1.623	-0.327	0.064
577.4	1.972	1.732	-0.269	0.048
577.5	1.923	1.840	-0.220	0.033
577.6	1.887	1.948	-0.176	0.020
577.7	1.862	2.056	-0.136	0.011
577.8	1.847	2.165	-0.098	0.004
577.9	1.842	2.275	-0.061	0.000

OMEGA = 20.0

k_0h	ADMITTANCE		RAD.EFF.LENGTH	
578.0	1.848	2.388	-0.022	-0.000
184π	1.856	2.449	-0.000	-0.000
578.1	1.865	2.504	0.020	0.001
578.2	1.896	2.625	0.065	0.004
578.3	1.942	2.750	0.115	0.007
578.4	2.009	2.879	0.170	0.010
578.5	2.101	3.012	0.232	0.011
578.6	2.225	3.146	0.302	0.008
578.7	2.391	3.273	0.381	0.000
578.8	2.609	3.383	0.468	-0.016
578.9	2.888	3.451	0.566	-0.044
579.0	3.228	3.443	0.671	-0.087
579.1	3.604	3.310	0.784	-0.151
579.2	3.951	3.017	0.900	-0.241
579.3	4.173	2.574	1.013	-0.362
579.4	4.193	2.069	1.111	-0.518
579.5	4.014	1.623	1.183	-0.710
579.6	3.713	1.313	1.213	-0.929
369π/2	3.634	1.262	1.213	-0.985
579.7	3.379	1.148	1.189	-1.164
579.8	3.067	1.096	1.106	-1.392
579.9	2.801	1.117	0.967	-1.594
580.0	2.582	1.183	0.787	-1.751
580.1	2.407	1.272	0.586	-1.855
580.2	2.267	1.373	0.383	-1.909
580.3	2.156	1.480	0.192	-1.920
580.4	2.068	1.589	0.020	-1.898
580.5	1.999	1.697	-0.128	-1.855
580.6	1.945	1.805	-0.254	-1.798
580.7	1.904	1.913	-0.360	-1.734
580.8	1.875	2.021	-0.448	-1.667
580.9	1.856	2.129	-0.522	-1.601
581.0	1.847	2.239	-0.584	-1.536
581.1	1.848	2.351	-0.638	-1.473
185π	1.859	2.460	-0.682	-1.417
581.2	1.860	2.466	-0.685	-1.414
581.3	1.885	2.584	-0.727	-1.356
581.4	1.925	2.707	-0.765	-1.301
581.5	1.983	2.835	-0.801	-1.248
581.6	2.064	2.966	-0.835	-1.195
581.7	2.173	3.100	-0.869	-1.143
581.8	2.321	3.231	-0.902	-1.089
581.9	2.516	3.349	-0.936	-1.033
582.0	2.770	3.437	-0.970	-0.973
582.1	3.085	3.465	-1.003	-0.908
582.2	3.451	3.388	-1.034	-0.835
582.3	3.819	3.163	-1.062	-0.754
582.4	4.103	2.776	-1.084	-0.663
582.5	4.209	2.285	-1.094	-0.562
582.6	4.106	1.803	-1.090	-0.451
582.7	3.845	1.434	-1.065	-0.336
371π/2	3.633	1.273	-1.036	-0.262
582.8	3.517	1.212	-1.017	-0.223
582.9	3.193	1.118	-0.946	-0.120

k_0h	ADMITTANCE		RAD.EFF.LENGTH	
583.0	2.907	1.113	-0.855	-0.034
583.1	2.670	1.163	-0.753	0.028
583.2	2.478	1.244	-0.649	0.066
583.3	2.324	1.342	-0.549	0.083
583.4	2.202	1.446	-0.460	0.084
583.5	2.106	1.554	-0.382	0.075
583.6	2.029	1.663	-0.316	0.061
583.7	1.969	1.771	-0.260	0.045
583.8	1.923	1.879	-0.212	0.031
583.9	1.889	1.986	-0.169	0.019
584.0	1.866	2.095	-0.130	0.009
584.1	1.853	2.204	-0.092	0.003
584.2	1.850	2.314	-0.054	0.000
584.3	1.858	2.428	-0.015	-0.000
186π	1.863	2.470	-0.000	-0.000
584.4	1.877	2.545	0.027	0.001
584.5	1.911	2.666	0.073	0.004
584.6	1.961	2.791	0.123	0.007
584.7	2.031	2.921	0.180	0.010
584.8	2.129	3.054	0.243	0.010
584.9	2.260	3.186	0.315	0.007
585.0	2.434	3.311	0.395	-0.003
585.1	2.663	3.414	0.484	-0.021
585.2	2.952	3.470	0.583	-0.051
585.3	3.300	3.442	0.690	-0.098
585.4	3.674	3.281	0.803	-0.167
585.5	4.003	2.959	0.919	-0.262
585.6	4.190	2.500	1.029	-0.390
585.7	4.170	2.001	1.123	-0.553
585.8	3.965	1.579	1.187	-0.750
373π/2	3.632	1.284	1.206	-0.989
585.9	3.655	1.298	1.207	-0.973
586.0	3.324	1.156	1.171	-1.207
586.1	3.021	1.119	1.075	-1.431
586.2	2.764	1.150	0.927	-1.623
586.3	2.555	1.220	0.743	-1.769
586.4	2.387	1.311	0.541	-1.862
586.5	2.253	1.413	0.340	-1.906
586.6	2.146	1.520	0.154	-1.909
586.7	2.062	1.628	-0.012	-1.883
586.8	1.995	1.737	-0.155	-1.836
586.9	1.944	1.845	-0.276	-1.778
587.0	1.905	1.952	-0.377	-1.714
587.1	1.878	2.060	-0.462	-1.648
587.2	1.860	2.168	-0.533	-1.582
587.3	1.853	2.278	-0.593	-1.518
587.4	1.857	2.391	-0.645	-1.457
187π	1.867	2.480	-0.681	-1.411
587.5	1.871	2.506	-0.690	-1.398
587.6	1.899	2.625	-0.731	-1.342
587.7	1.942	2.748	-0.769	-1.288
587.8	2.003	2.876	-0.804	-1.235
587.9	2.089	3.008	-0.838	-1.183

k_0h	ADMITTANCE		RAD.EFF.LENGTH	
588.0	2.205	3.141	-0.872	-1.131
588.1	2.361	3.270	-0.906	-1.077
588.2	2.566	3.384	-0.940	-1.021
588.3	2.830	3.462	-0.973	-0.960
588.4	3.155	3.473	-1.007	-0.894
588.5	3.523	3.372	-1.038	-0.820
588.6	3.880	3.117	-1.065	-0.738
588.7	4.135	2.707	-1.085	-0.645
588.8	4.202	2.212	-1.094	-0.542
588.9	4.066	1.748	-1.086	-0.430
589.0	3.789	1.407	-1.057	-0.315
375π/2	3.630	1.295	-1.035	-0.259
589.1	3.460	1.211	-1.005	-0.203
589.2	3.143	1.136	-0.929	-0.102
589.3	2.867	1.142	-0.836	-0.021
589.4	2.639	1.199	-0.733	0.037
589.5	2.454	1.283	-0.629	0.071
589.6	2.307	1.381	-0.532	0.084
589.7	2.190	1.486	-0.445	0.083
589.8	2.098	1.594	-0.370	0.073
589.9	2.024	1.702	-0.306	0.058
590.0	1.967	1.810	-0.251	0.043
590.1	1.923	1.918	-0.204	0.028
590.2	1.891	2.025	-0.162	0.017
590.3	1.870	2.133	-0.123	0.008
590.4	1.859	2.243	-0.086	0.003
590.5	1.858	2.354	-0.048	-0.000
590.6	1.868	2.468	-0.008	-0.000
188π	1.871	2.490	-0.000	-0.000
590.7	1.890	2.585	0.035	0.002
590.8	1.926	2.707	0.081	0.004
590.9	1.980	2.833	0.133	0.007
591.0	2.055	2.963	0.190	0.009
591.1	2.158	3.095	0.255	0.009
591.2	2.296	3.227	0.328	0.005
591.3	2.479	3.348	0.409	-0.006
591.4	2.718	3.443	0.500	-0.026
591.5	3.018	3.486	0.600	-0.060
591.6	3.372	3.436	0.708	-0.110
591.7	3.741	3.248	0.822	-0.183
591.8	4.049	2.898	0.937	-0.285
591.9	4.199	2.426	1.044	-0.419
592.0	4.142	1.937	1.133	-0.588
592.1	3.913	1.541	1.189	-0.791
377π/2	3.629	1.306	1.200	-0.994
592.2	3.597	1.287	1.198	-1.017
592.3	3.271	1.167	1.150	-1.249
592.4	2.977	1.144	1.043	-1.467
592.5	2.730	1.183	0.887	-1.650
592.6	2.528	1.257	0.698	-1.785
592.7	2.367	1.350	0.497	-1.867
592.8	2.239	1.453	0.299	-1.902
592.9	2.137	1.560	0.117	-1.898

k_0h	ADMITTANCE		RAD.EFF.LENGTH	
593.0	2.056	1.668	-0.043	-1.867
593.1	1.992	1.776	-0.180	-1.818
593.2	1.943	1.884	-0.296	-1.758
593.3	1.907	1.991	-0.393	-1.694
593.4	1.881	2.099	-0.475	-1.628
593.5	1.866	2.207	-0.543	-1.563
593.6	1.860	2.318	-0.601	-1.500
593.7	1.866	2.430	-0.651	-1.440
189π	1.875	2.500	-0.679	-1.405
593.8	1.883	2.546	-0.695	-1.383
593.9	1.913	2.666	-0.735	-1.327
594.0	1.959	2.790	-0.772	-1.274
594.1	2.025	2.918	-0.807	-1.222
594.2	2.116	3.050	-0.842	-1.171
594.3	2.238	3.182	-0.875	-1.119
594.4	2.402	3.308	-0.909	-1.065
594.5	2.617	3.416	-0.943	-1.008
594.6	2.891	3.484	-0.977	-0.947
594.7	3.225	3.477	-1.010	-0.880
594.8	3.594	3.350	-1.041	-0.805
594.9	3.937	3.066	-1.068	-0.721
595.0	4.160	2.637	-1.086	-0.626
595.1	4.189	2.142	-1.092	-0.521
595.2	4.022	1.699	-1.081	-0.409
595.3	3.732	1.386	-1.048	-0.293
379π/2	3.628	1.317	-1.033	-0.257
595.4	3.404	1.214	-0.991	-0.183
595.5	3.094	1.156	-0.912	-0.085
595.6	2.828	1.173	-0.817	-0.008
595.7	2.609	1.235	-0.713	0.045
595.8	2.432	1.321	-0.610	0.074
595.9	2.291	1.421	-0.515	0.084
596.0	2.179	1.526	-0.430	0.081
596.1	2.090	1.634	-0.357	0.070
596.2	2.020	1.742	-0.295	0.055
596.3	1.965	1.850	-0.243	0.040
596.4	1.924	1.957	-0.197	0.026
596.5	1.894	2.064	-0.156	0.015
596.6	1.874	2.172	-0.117	0.007
596.7	1.865	2.282	-0.079	0.002
596.8	1.866	2.393	-0.041	-0.000
190π	1.878	2.511	-0.000	-0.000
596.9	1.878	2.508	-0.001	-0.000
597.0	1.903	2.626	0.042	0.002
597.1	1.942	2.748	0.089	0.005
597.2	1.999	2.874	0.142	0.008
597.3	2.079	3.005	0.201	0.009
597.4	2.188	3.137	0.267	0.008
597.5	2.334	3.266	0.341	0.003
597.6	2.526	3.383	0.424	-0.010
597.7	2.775	3.471	0.516	-0.032
597.8	3.085	3.498	0.617	-0.068
597.9	3.443	3.426	0.727	-0.123

k_0h	ADMITTANCE		RAD.EFF.LENGTH	
598.0	3.806	3.209	0.841	-0.201
598.1	4.088	2.834	0.954	-0.308
598.2	4.201	2.354	1.058	-0.449
598.3	4.109	1.878	1.141	-0.625
598.4	3.860	1.508	1.189	-0.832
381π/2	3.627	1.328	1.193	-0.999
598.5	3.540	1.281	1.187	-1.060
598.6	3.219	1.181	1.127	-1.291
598.7	2.934	1.171	1.010	-1.502
598.8	2.697	1.217	0.846	-1.675
598.9	2.504	1.295	0.654	-1.798
599.0	2.349	1.390	0.453	-1.870
599.1	2.226	1.493	0.259	-1.896
599.2	2.128	1.600	0.082	-1.886
599.3	2.050	1.708	-0.072	-1.850
599.4	1.990	1.815	-0.205	-1.799
599.5	1.943	1.923	-0.316	-1.739
599.6	1.908	2.030	-0.409	-1.674
599.7	1.884	2.138	-0.487	-1.609
599.8	1.871	2.247	-0.552	-1.545
599.9	1.868	2.357	-0.608	-1.483
600.0	1.875	2.470	-0.657	-1.424
191π	1.882	2.521	-0.677	-1.399
600.1	1.894	2.586	-0.700	-1.368
600.2	1.928	2.707	-0.739	-1.313
600.3	1.977	2.831	-0.776	-1.261
600.4	2.047	2.960	-0.811	-1.209
600.5	2.143	3.092	-0.845	-1.158
600.6	2.273	3.223	-0.879	-1.107
600.7	2.445	3.346	-0.913	-1.053
600.8	2.670	3.448	-0.947	-0.996
600.9	2.954	3.504	-0.981	-0.934
601.0	3.295	3.477	-1.014	-0.866
601.1	3.663	3.323	-1.045	-0.790
601.2	3.988	3.011	-1.070	-0.704
601.3	4.178	2.565	-1.087	-0.607
601.4	4.169	2.075	-1.091	-0.500
601.5	3.975	1.654	-1.076	-0.387
601.6	3.675	1.369	-1.039	-0.272
383π/2	3.626	1.339	-1.031	-0.255
601.7	3.349	1.221	-0.978	-0.163
601.8	3.048	1.178	-0.895	-0.069
601.9	2.791	1.204	-0.797	0.003
602.0	2.581	1.271	-0.694	0.052
602.1	2.412	1.360	-0.592	0.077
602.2	2.276	1.461	-0.498	0.084
602.3	2.169	1.566	-0.416	0.079
602.4	2.083	1.674	-0.345	0.067
602.5	2.016	1.781	-0.286	0.052
602.6	1.964	1.889	-0.234	0.037
602.7	1.925	1.996	-0.189	0.024
602.8	1.896	2.103	-0.149	0.013
602.9	1.879	2.212	-0.111	0.006

k_0h	ADMITTANCE		RAD.EFF.LENGTH	
603.0	1.871	2.321	-0.073	0.001
603.1	1.874	2.433	-0.035	-0.000
192π	1.886	2.531	-0.000	-0.000
603.2	1.889	2.548	0.006	0.000
603.3	1.916	2.666	0.050	0.002
603.4	1.958	2.789	0.098	0.005
603.5	2.020	2.916	0.151	0.008
603.6	2.104	3.046	0.211	0.009
603.7	2.219	3.178	0.278	0.007
603.8	2.373	3.305	0.354	0.000
603.9	2.575	3.418	0.439	-0.014
604.0	2.834	3.495	0.532	-0.039
604.1	3.153	3.507	0.635	-0.078
604.2	3.514	3.410	0.745	-0.137
604.3	3.866	3.165	0.859	-0.220
604.4	4.121	2.767	0.971	-0.333
604.5	4.196	2.282	1.072	-0.480
604.6	4.071	1.823	1.148	-0.663
604.7	3.805	1.480	1.187	-0.874
385π/2	3.625	1.350	1.186	-1.003
604.8	3.483	1.279	1.173	-1.103
604.9	3.169	1.198	1.102	-1.331
605.0	2.894	1.199	0.974	-1.535
605.1	2.665	1.251	0.804	-1.697
605.2	2.480	1.333	0.610	-1.809
605.3	2.332	1.429	0.410	-1.870
605.4	2.214	1.533	0.220	-1.888
605.5	2.120	1.640	0.049	-1.872
605.6	2.045	1.747	-0.101	-1.833
605.7	1.987	1.855	-0.228	-1.780
605.8	1.943	1.962	-0.334	-1.719
605.9	1.910	2.069	-0.424	-1.655
606.0	1.888	2.177	-0.498	-1.591
606.1	1.877	2.286	-0.562	-1.527
606.2	1.875	2.396	-0.616	-1.467
606.3	1.885	2.510	-0.663	-1.408
193π	1.890	2.541	-0.675	-1.393
606.4	1.907	2.627	-0.705	-1.353
606.5	1.943	2.748	-0.744	-1.299
606.6	1.996	2.873	-0.780	-1.248
606.7	2.070	3.002	-0.814	-1.197
606.8	2.172	3.133	-0.848	-1.146
606.9	2.309	3.263	-0.882	-1.095
607.0	2.490	3.383	-0.916	-1.041
607.1	2.724	3.477	-0.951	-0.984
607.2	3.019	3.520	-0.985	-0.921
607.3	3.366	3.472	-1.018	-0.852
607.4	3.729	3.291	-1.048	-0.774
607.5	4.034	2.953	-1.073	-0.686
607.6	4.189	2.493	-1.087	-0.588
607.7	4.143	2.012	-1.088	-0.480
607.8	3.925	1.615	-1.070	-0.365
387π/2	3.624	1.361	-1.029	-0.253
607.9	3.618	1.357	-1.028	-0.251

k_0h	ADMITTANCE		RAD.EFF.LENGTH	
608.0	3.296	1.231	-0.963	-0.144
608.1	3.004	1.203	-0.877	-0.054
608.2	2.756	1.237	-0.778	0.014
608.3	2.554	1.308	-0.674	0.058
608.4	2.392	1.399	-0.573	0.079
608.5	2.262	1.500	-0.482	0.084
608.6	2.159	1.606	-0.402	0.077
608.7	2.077	1.713	-0.334	0.064
608.8	2.013	1.821	-0.276	0.049
608.9	1.963	1.928	-0.226	0.035
609.0	1.926	2.035	-0.182	0.022
609.1	1.900	2.142	-0.142	0.012
609.2	1.884	2.251	-0.104	0.005
609.3	1.878	2.361	-0.067	0.001
609.4	1.883	2.473	-0.028	-0.000
194π	1.893	2.552	-0.000	-0.000
609.5	1.900	2.588	0.013	0.000
609.6	1.930	2.707	0.057	0.003
609.7	1.975	2.830	0.106	0.005
609.8	2.041	2.958	0.161	0.008
609.9	2.131	3.088	0.222	0.008
610.0	2.252	3.219	0.291	0.006
610.1	2.413	3.344	0.368	-0.002
610.2	2.625	3.450	0.454	-0.018
610.3	2.895	3.518	0.549	-0.046
610.4	3.222	3.512	0.653	-0.088
610.5	3.584	3.390	0.764	-0.151
610.6	3.922	3.116	0.877	-0.240
610.7	4.148	2.699	0.987	-0.359
610.8	4.185	2.214	1.084	-0.513
610.9	4.030	1.774	1.154	-0.701
611.0	3.750	1.458	1.182	-0.917
389π/2	3.623	1.372	1.179	-1.007
611.1	3.428	1.281	1.157	-1.146
611.2	3.121	1.217	1.074	-1.369
611.3	2.855	1.229	0.937	-1.565
611.4	2.635	1.287	0.761	-1.717
611.5	2.458	1.371	0.565	-1.818
611.6	2.315	1.469	0.368	-1.869
611.7	2.202	1.573	0.182	-1.879
611.8	2.112	1.679	0.016	-1.858
611.9	2.041	1.787	-0.128	-1.816
612.0	1.985	1.894	-0.250	-1.761
612.1	1.943	2.001	-0.352	-1.700
612.2	1.912	2.108	-0.438	-1.636
612.3	1.893	2.216	-0.509	-1.572
612.4	1.883	2.325	-0.570	-1.510
612.5	1.883	2.436	-0.623	-1.450
612.6	1.895	2.550	-0.669	-1.393
195π	1.897	2.562	-0.673	-1.387
612.7	1.919	2.667	-0.710	-1.338
612.8	1.958	2.789	-0.748	-1.286
612.9	2.015	2.914	-0.783	-1.235

k_0h	ADMITTANCE		RAD.EFF.LENGTH	
613.0	2.094	3.044	-0.818	-1.185
613.1	2.202	3.174	-0.852	-1.134
613.2	2.346	3.302	-0.886	-1.083
613.3	2.536	3.418	-0.920	-1.029
613.4	2.781	3.504	-0.955	-0.971
613.5	3.085	3.533	-0.989	-0.908
613.6	3.436	3.463	-1.022	-0.837
613.7	3.793	3.254	-1.051	-0.758
613.8	4.074	2.891	-1.075	-0.668
613.9	4.193	2.422	-1.088	-0.568
614.0	4.112	1.953	-1.085	-0.458
614.1	3.874	1.582	-1.063	-0.343
391π/2	3.621	1.383	-1.027	-0.251
614.2	3.561	1.350	-1.017	-0.230
614.3	3.245	1.244	-0.948	-0.126
614.4	2.961	1.229	-0.859	-0.039
614.5	2.723	1.270	-0.758	0.024
614.6	2.529	1.345	-0.654	0.063
614.7	2.373	1.438	-0.556	0.081
614.8	2.249	1.540	-0.467	0.083
614.9	2.150	1.646	-0.389	0.075
615.0	2.071	1.753	-0.323	0.062
615.1	2.010	1.860	-0.267	0.046
615.2	1.962	1.967	-0.218	0.032
615.3	1.927	2.074	-0.175	0.020
615.4	1.903	2.182	-0.136	0.010
615.5	1.889	2.290	-0.098	0.004
615.6	1.885	2.400	-0.060	0.001
615.7	1.893	2.513	-0.021	-0.000
196π	1.901	2.572	-0.000	-0.000
615.8	1.911	2.628	0.020	0.001
615.9	1.944	2.748	0.065	0.003
616.0	1.993	2.872	0.115	0.006
616.1	2.063	2.999	0.171	0.008
616.2	2.158	3.130	0.233	0.008
616.3	2.286	3.259	0.303	0.004
616.4	2.456	3.381	0.381	-0.005
616.5	2.677	3.482	0.469	-0.023
616.6	2.957	3.537	0.566	-0.053
616.7	3.291	3.513	0.671	-0.099
616.8	3.652	3.365	0.782	-0.167
616.9	3.974	3.064	0.895	-0.261
617.0	4.167	2.629	1.003	-0.386
617.1	4.167	2.148	1.095	-0.547
617.2	3.985	1.729	1.157	-0.740
617.3	3.694	1.441	1.176	-0.959
393π/2	3.620	1.394	1.172	-1.012
617.4	3.373	1.287	1.139	-1.188
617.5	3.075	1.238	1.045	-1.407
617.6	2.818	1.260	0.899	-1.594
617.7	2.607	1.323	0.718	-1.735
617.8	2.436	1.410	0.522	-1.824
617.9	2.300	1.508	0.327	-1.866

OMEGA = 20.0

k_0h	ADMITTANCE		RAD.EFF.LENGTH	
618.0	2.191	1.613	0.146	-1.869
618.1	2.105	1.719	-0.015	-1.843
618.2	2.037	1.826	-0.153	-1.798
618.3	1.984	1.933	-0.270	-1.742
618.4	1.944	2.040	-0.369	-1.681
618.5	1.915	2.147	-0.451	-1.617
618.6	1.897	2.255	-0.520	-1.554
618.7	1.889	2.364	-0.579	-1.493
618.8	1.892	2.476	-0.629	-1.434
197π	1.905	2.583	-0.671	-1.381
618.9	1.906	2.590	-0.674	-1.378
619.0	1.933	2.708	-0.714	-1.324
619.1	1.975	2.830	-0.751	-1.272
619.2	2.035	2.956	-0.787	-1.222
619.3	2.119	3.085	-0.821	-1.172
619.4	2.233	3.216	-0.855	-1.122
619.5	2.384	3.341	-0.889	-1.071
619.6	2.583	3.452	-0.924	-1.017
619.7	2.838	3.529	-0.958	-0.959
619.8	3.151	3.542	-0.993	-0.894
619.9	3.506	3.449	-1.026	-0.823
620.0	3.852	3.212	-1.054	-0.741
620.1	4.108	2.827	-1.076	-0.650
620.2	4.190	2.353	-1.087	-0.548
620.3	4.076	1.898	-1.082	-0.437
620.4	3.821	1.554	-1.055	-0.322
395π/2	3.619	1.405	-1.026	-0.248
620.5	3.505	1.347	-1.005	-0.209
620.6	3.195	1.260	-0.932	-0.108
620.7	2.921	1.256	-0.840	-0.026
620.8	2.691	1.305	-0.738	0.033
620.9	2.505	1.383	-0.635	0.068
621.0	2.356	1.478	-0.538	0.082
621.1	2.236	1.580	-0.451	0.082
621.2	2.141	1.686	-0.376	0.073
621.3	2.066	1.793	-0.312	0.059
621.4	2.007	1.900	-0.258	0.044
621.5	1.962	2.006	-0.210	0.030
621.6	1.929	2.113	-0.168	0.018
621.7	1.907	2.221	-0.129	0.009
621.8	1.895	2.329	-0.092	0.003
621.9	1.893	2.440	-0.054	0.000
622.0	1.902	2.552	-0.014	-0.000
198π	1.908	2.593	-0.000	-0.000
622.1	1.924	2.669	0.028	0.001
622.2	1.959	2.789	0.073	0.003
622.3	2.012	2.913	0.124	0.006
622.4	2.085	3.041	0.181	0.008
622.5	2.186	3.171	0.244	0.007
622.6	2.321	3.300	0.316	0.002
622.7	2.499	3.418	0.395	-0.009
622.8	2.730	3.511	0.485	-0.029
622.9	3.020	3.554	0.583	-0.061

k_0h	ADMITTANCE		RAD.EFF.LENGTH	
623.0	3.360	3.509	0.689	-0.111
623.1	3.717	3.335	0.800	-0.183
623.2	4.020	3.007	0.913	-0.283
623.3	4.179	2.560	1.018	-0.415
623.4	4.143	2.086	1.104	-0.582
623.5	3.937	1.690	1.159	-0.780
397π/2	3.618	1.416	1.165	-1.016
623.6	3.638	1.428	1.167	-1.002
623.7	3.321	1.296	1.119	-1.229
623.8	3.030	1.262	1.013	-1.442
623.9	2.783	1.292	0.860	-1.620
624.0	2.580	1.360	0.675	-1.750
624.1	2.416	1.448	0.479	-1.829
624.2	2.285	1.548	0.287	-1.862
624.3	2.181	1.653	0.111	-1.858
624.4	2.098	1.759	-0.045	-1.828
624.5	2.033	1.866	-0.178	-1.781
624.6	1.983	1.973	-0.290	-1.723
624.7	1.945	2.079	-0.384	-1.662
624.8	1.918	2.186	-0.463	-1.598
624.9	1.902	2.294	-0.530	-1.536
625.0	1.896	2.404	-0.586	-1.476
625.1	1.901	2.516	-0.636	-1.418
199π	1.912	2.604	-0.669	-1.376
625.2	1.917	2.630	-0.679	-1.363
625.3	1.946	2.749	-0.719	-1.310
625.4	1.992	2.871	-0.755	-1.259
625.5	2.056	2.998	-0.790	-1.210
625.6	2.145	3.127	-0.824	-1.160
625.7	2.265	3.256	-0.859	-1.111
625.8	2.424	3.380	-0.893	-1.059
625.9	2.633	3.485	-0.928	-1.005
626.0	2.898	3.552	-0.962	-0.946
626.1	3.219	3.548	-0.997	-0.881
626.2	3.574	3.431	-1.029	-0.808
626.3	3.908	3.166	-1.057	-0.725
626.4	4.135	2.760	-1.078	-0.631
626.5	4.180	2.285	-1.086	-0.527
626.6	4.036	1.849	-1.077	-0.415
626.7	3.766	1.531	-1.047	-0.300
399π/2	3.617	1.428	-1.024	-0.246
626.8	3.451	1.348	-0.992	-0.189
626.9	3.147	1.278	-0.915	-0.091
627.0	2.882	1.285	-0.821	-0.013
627.1	2.661	1.340	-0.718	0.041
627.2	2.482	1.421	-0.616	0.071
627.3	2.339	1.517	-0.521	0.083
627.4	2.224	1.620	-0.437	0.081
627.5	2.133	1.726	-0.364	0.070
627.6	2.061	1.832	-0.302	0.056
627.7	2.005	1.939	-0.249	0.041
627.8	1.962	2.046	-0.203	0.027
627.9	1.931	2.152	-0.161	0.016

k_0h	ADMITTANCE		RAD.EFF.LENGTH	
628.0	1.911	2.260	-0.123	0.008
628.1	1.901	2.369	-0.085	0.003
628.2	1.901	2.479	-0.047	0.000
628.3	1.912	2.593	-0.008	-0.000
200π	1.916	2.614	-0.000	-0.000
628.4	1.936	2.709	0.035	0.001
628.5	1.975	2.830	0.082	0.004
628.6	2.031	2.955	0.133	0.006
628.7	2.109	3.083	0.191	0.007
628.8	2.216	3.213	0.256	0.006
628.9	2.358	3.339	0.328	0.000
629.0	2.545	3.453	0.410	-0.012
629.1	2.786	3.538	0.500	-0.034
629.2	3.084	3.567	0.600	-0.070
629.3	3.429	3.501	0.707	-0.124
629.4	3.780	3.300	0.819	-0.201
629.5	4.060	2.948	0.930	-0.306
629.6	4.184	2.490	1.032	-0.445
629.7	4.114	2.027	1.113	-0.617
629.8	3.887	1.656	1.158	-0.821
401π/2	3.616	1.439	1.158	-1.020
629.9	3.582	1.420	1.155	-1.044
630.0	3.270	1.308	1.096	-1.269
630.1	2.988	1.287	0.980	-1.475
630.2	2.749	1.325	0.820	-1.643
630.3	2.554	1.397	0.632	-1.762
630.4	2.397	1.487	0.436	-1.831
630.5	2.272	1.588	0.248	-1.856
630.6	2.172	1.693	0.077	-1.846
630.7	2.092	1.799	-0.073	-1.812
630.8	2.030	1.906	-0.201	-1.762
630.9	1.982	2.012	-0.309	-1.704
631.0	1.946	2.119	-0.399	-1.643
631.1	1.921	2.226	-0.475	-1.580
631.2	1.907	2.334	-0.539	-1.519
631.3	1.903	2.443	-0.594	-1.459
631.4	1.910	2.555	-0.642	-1.403
201π	1.919	2.624	-0.668	-1.370
631.5	1.928	2.671	-0.684	-1.348
631.6	1.960	2.790	-0.723	-1.297
631.7	2.009	2.913	-0.759	-1.246
631.8	2.078	3.039	-0.794	-1.197
631.9	2.172	3.169	-0.828	-1.149
632.0	2.298	3.297	-0.862	-1.099
632.1	2.466	3.417	-0.897	-1.047
632.2	2.684	3.516	-0.931	-0.993
632.3	2.958	3.572	-0.967	-0.933
632.4	3.286	3.549	-1.001	-0.867
632.5	3.641	3.407	-1.033	-0.792
632.6	3.959	3.116	-1.060	-0.708
632.7	4.155	2.693	-1.079	-0.612
632.8	4.164	2.221	-1.085	-0.507
632.9	3.993	1.804	-1.072	-0.394

k_0h	ADMITTANCE		RAD.EFF.LENGTH	
633.0	3.712	1.512	-1.037	-0.279
403π/2	3.615	1.450	-1.022	-0.244
633.1	3.397	1.353	-0.978	-0.170
633.2	3.101	1.299	-0.898	-0.075
633.3	2.844	1.316	-0.802	-0.001
633.4	2.632	1.376	-0.699	0.048
633.5	2.461	1.460	-0.598	0.075
633.6	2.323	1.557	-0.505	0.083
633.7	2.213	1.660	-0.423	0.079
633.8	2.126	1.766	-0.352	0.068
633.9	2.057	1.872	-0.292	0.053
634.0	2.003	1.979	-0.241	0.038
634.1	1.963	2.085	-0.196	0.025
634.2	1.933	2.192	-0.155	0.015
634.3	1.915	2.299	-0.116	0.007
634.4	1.907	2.408	-0.079	0.002
634.5	1.909	2.519	-0.041	-0.000
202π	1.923	2.635	-0.000	-0.000
634.6	1.923	2.633	-0.001	-0.000
634.7	1.949	2.750	0.042	0.002
634.8	1.991	2.871	0.090	0.004
634.9	2.051	2.996	0.142	0.006
635.0	2.134	3.125	0.201	0.007
635.1	2.246	3.254	0.267	0.005
635.2	2.396	3.378	0.341	-0.002
635.3	2.592	3.487	0.424	-0.016
635.4	2.842	3.564	0.516	-0.041
635.5	3.150	3.577	0.617	-0.079
635.6	3.497	3.489	0.725	-0.137
635.7	3.839	3.260	0.837	-0.219
635.8	4.094	2.886	0.946	-0.331
635.9	4.183	2.423	1.045	-0.475
636.0	4.080	1.973	1.119	-0.654
636.1	3.835	1.627	1.156	-0.862
405π/2	3.614	1.461	1.151	-1.024
636.2	3.527	1.416	1.142	-1.086
636.3	3.220	1.323	1.071	-1.308
636.4	2.947	1.314	0.945	-1.507
636.5	2.717	1.359	0.779	-1.665
636.6	2.530	1.434	0.589	-1.773
636.7	2.379	1.527	0.395	-1.832
636.8	2.259	1.628	0.210	-1.848
636.9	2.163	1.732	0.044	-1.833
637.0	2.087	1.839	-0.101	-1.795
637.1	2.027	1.945	-0.224	-1.744
637.2	1.981	2.051	-0.327	-1.686
637.3	1.948	2.158	-0.414	-1.624
637.4	1.925	2.265	-0.487	-1.562
637.5	1.912	2.373	-0.548	-1.501
637.6	1.910	2.483	-0.601	-1.443
637.7	1.919	2.596	-0.647	-1.387
203π	1.927	2.645	-0.666	-1.364
637.8	1.940	2.711	-0.689	-1.334
637.9	1.975	2.831	-0.727	-1.283

k_0h	ADMITTANCE		RAD.EFF.LENGTH	
638.0	2.027	2.954	-0.763	-1.234
638.1	2.100	3.081	-0.797	-1.185
638.2	2.200	3.210	-0.831	-1.137
638.3	2.333	3.337	-0.866	-1.087
638.4	2.509	3.454	-0.900	-1.036
638.5	2.736	3.546	-0.935	-0.981
638.6	3.020	3.589	-0.971	-0.920
638.7	3.354	3.547	-1.005	-0.853
638.8	3.705	3.379	-1.037	-0.777
638.9	4.006	3.062	-1.063	-0.690
639.0	4.169	2.626	-1.080	-0.593
639.1	4.142	2.159	-1.083	-0.486
639.2	3.947	1.765	-1.067	-0.372
639.3	3.657	1.499	-1.027	-0.258
407π/2	3.613	1.472	-1.020	-0.242
639.4	3.345	1.361	-0.964	-0.150
639.5	3.056	1.322	-0.880	-0.059
639.6	2.809	1.347	-0.782	0.009
639.7	2.605	1.412	-0.679	0.054
639.8	2.440	1.498	-0.580	0.077
639.9	2.308	1.596	-0.489	0.083
640.0	2.203	1.700	-0.409	0.077
640.1	2.119	1.805	-0.341	0.065
640.2	2.053	1.912	-0.282	0.050
640.3	2.002	2.018	-0.232	0.036
640.4	1.963	2.124	-0.188	0.023
640.5	1.936	2.231	-0.148	0.013
640.6	1.920	2.339	-0.110	0.006
640.7	1.913	2.448	-0.073	0.002
640.8	1.918	2.559	-0.034	-0.000
204π	1.930	2.656	-0.000	-0.000
640.9	1.933	2.673	0.006	0.000
641.0	1.963	2.791	0.050	0.002
641.1	2.007	2.913	0.098	0.004
641.2	2.071	3.038	0.152	0.006
641.3	2.159	3.166	0.212	0.006
641.4	2.278	3.294	0.279	0.003
641.5	2.435	3.416	0.355	-0.005
641.6	2.640	3.520	0.439	-0.021
641.7	2.901	3.587	0.532	-0.048
641.8	3.216	3.584	0.634	-0.090
641.9	3.565	3.472	0.743	-0.152
642.0	3.894	3.216	0.854	-0.239
642.1	4.122	2.822	0.962	-0.356
642.2	4.175	2.357	1.056	-0.507
642.3	4.043	1.924	1.124	-0.692
642.4	3.782	1.604	1.151	-0.903
409π/2	3.612	1.484	1.144	-1.028
642.5	3.473	1.416	1.126	-1.128
642.6	3.172	1.340	1.044	-1.346
642.7	2.908	1.343	0.909	-1.536
642.8	2.687	1.393	0.737	-1.683
642.9	2.507	1.472	0.546	-1.781

k_0h	ADMITTANCE		RAD.EFF.LENGTH	
643.0	2.362	1.566	0.354	-1.830
643.1	2.247	1.668	0.174	-1.840
643.2	2.155	1.772	0.013	-1.819
643.3	2.082	1.878	-0.127	-1.779
643.4	2.025	1.985	-0.245	-1.726
643.5	1.981	2.091	-0.344	-1.667
643.6	1.949	2.197	-0.427	-1.605
643.7	1.929	2.304	-0.497	-1.544
643.8	1.918	2.413	-0.557	-1.485
643.9	1.918	2.523	-0.608	-1.427
644.0	1.929	2.636	-0.653	-1.373
205π	1.934	2.666	-0.664	-1.358
644.1	1.952	2.752	-0.693	-1.320
644.2	1.990	2.872	-0.731	-1.270
644.3	2.046	2.996	-0.766	-1.221
644.4	2.124	3.123	-0.801	-1.173
644.5	2.229	3.251	-0.835	-1.125
644.6	2.369	3.376	-0.869	-1.076
644.7	2.554	3.489	-0.904	-1.024
644.8	2.791	3.573	-0.939	-0.968
644.9	3.084	3.603	-0.975	-0.907
645.0	3.422	3.540	-1.009	-0.838
645.1	3.767	3.346	-1.040	-0.761
645.2	4.046	3.005	-1.065	-0.673
645.3	4.175	2.558	-1.080	-0.574
645.4	4.115	2.102	-1.080	-0.465
645.5	3.899	1.731	-1.060	-0.350
411π/2	3.611	1.495	-1.018	-0.240
645.6	3.602	1.490	-1.016	-0.237
645.7	3.294	1.372	-0.949	-0.132
645.8	3.013	1.346	-0.862	-0.045
645.9	2.775	1.380	-0.763	0.019
646.0	2.579	1.449	-0.660	0.060
646.1	2.421	1.537	-0.562	0.079
646.2	2.294	1.636	-0.473	0.082
646.3	2.193	1.740	-0.396	0.075
646.4	2.113	1.845	-0.330	0.062
646.5	2.050	1.951	-0.273	0.048
646.6	2.001	2.057	-0.224	0.033
646.7	1.964	2.164	-0.181	0.021
646.8	1.939	2.270	-0.142	0.012
646.9	1.924	2.378	-0.104	0.005
647.0	1.920	2.487	-0.066	0.001
647.1	1.926	2.599	-0.028	-0.000
206π	1.937	2.677	-0.000	-0.000
647.2	1.945	2.714	0.013	0.000
647.3	1.977	2.832	0.058	0.002
647.4	2.025	2.954	0.107	0.004
647.5	2.093	3.080	0.161	0.006
647.6	2.186	3.208	0.223	0.006
647.7	2.311	3.335	0.291	0.002
647.8	2.476	3.454	0.368	-0.008
647.9	2.690	3.552	0.454	-0.025

k_0h	ADMITTANCE		RAD.EFF.LENGTH	
648.0	2.960	3.607	0.549	-0.055
648.1	3.282	3.586	0.652	-0.100
648.2	3.630	3.450	0.761	-0.167
648.3	3.945	3.168	0.872	-0.259
648.4	4.143	2.757	0.977	-0.383
648.5	4.161	2.293	1.067	-0.540
648.6	4.001	1.880	1.127	-0.730
648.7	3.729	1.585	1.145	-0.945
413π/2	3.610	1.506	1.137	-1.032
648.8	3.420	1.420	1.108	-1.168
648.9	3.126	1.360	1.015	-1.381
649.0	2.870	1.372	0.872	-1.563
649.1	2.658	1.429	0.695	-1.700
649.2	2.485	1.510	0.504	-1.787
649.3	2.346	1.606	0.314	-1.827
649.4	2.235	1.707	0.139	-1.830
649.5	2.147	1.812	-0.017	-1.805
649.6	2.077	1.918	-0.151	-1.762
649.7	2.023	2.024	-0.265	-1.708
649.8	1.981	2.130	-0.360	-1.648
649.9	1.952	2.237	-0.440	-1.587
650.0	1.933	2.344	-0.507	-1.527
650.1	1.924	2.452	-0.565	-1.468
650.2	1.926	2.563	-0.614	-1.412
207π	1.941	2.687	-0.662	-1.353
650.3	1.939	2.676	-0.658	-1.358
650.4	1.965	2.793	-0.698	-1.307
650.5	2.006	2.913	-0.735	-1.257
650.6	2.066	3.037	-0.770	-1.209
650.7	2.148	3.164	-0.804	-1.161
650.8	2.259	3.292	-0.839	-1.114
650.9	2.407	3.415	-0.873	-1.064
651.0	2.600	3.523	-0.908	-1.012
651.1	2.846	3.599	-0.944	-0.956
651.2	3.148	3.613	-0.979	-0.894
651.3	3.489	3.529	-1.013	-0.824
651.4	3.826	3.308	-1.043	-0.745
651.5	4.081	2.945	-1.067	-0.654
651.6	4.176	2.492	-1.080	-0.554
651.7	4.084	2.048	-1.077	-0.443
651.8	3.849	1.701	-1.053	-0.329
415π/2	3.609	1.517	-1.016	-0.237
651.9	3.548	1.485	-1.005	-0.216
652.0	3.245	1.386	-0.933	-0.114
652.1	2.972	1.373	-0.843	-0.031
652.2	2.743	1.413	-0.743	0.029
652.3	2.555	1.486	-0.641	0.065
652.4	2.403	1.576	-0.544	0.080
652.5	2.281	1.676	-0.458	0.081
652.6	2.184	1.780	-0.383	0.073
652.7	2.107	1.885	-0.319	0.059
652.8	2.047	1.991	-0.264	0.045
652.9	2.000	2.097	-0.217	0.031

k_0h	ADMITTANCE		RAD.EFF.LENGTH	
653.0	1.966	2.203	-0.174	0.019
653.1	1.943	2.310	-0.135	0.010
653.2	1.930	2.417	-0.098	0.004
653.3	1.927	2.527	-0.060	0.001
653.4	1.936	2.639	-0.021	-0.000
208π	1.945	2.698	-0.000	-0.000
653.5	1.956	2.754	0.021	0.001
653.6	1.991	2.873	0.066	0.002
653.7	2.043	2.995	0.116	0.005
653.8	2.115	3.121	0.171	0.006
653.9	2.213	3.249	0.234	0.005
654.0	2.345	3.375	0.303	0.000
654.1	2.518	3.490	0.382	-0.011
654.2	2.742	3.581	0.469	-0.031
654.3	3.021	3.624	0.565	-0.063
654.4	3.349	3.585	0.669	-0.112
654.5	3.694	3.423	0.779	-0.183
654.6	3.991	3.116	0.889	-0.281
654.7	4.158	2.692	0.991	-0.411
654.8	4.141	2.233	1.076	-0.574
654.9	3.957	1.840	1.129	-0.769
655.0	3.675	1.570	1.136	-0.986
417π/2	3.608	1.529	1.130	-1.035
655.1	3.368	1.427	1.088	-1.208
655.2	3.082	1.382	0.984	-1.415
655.3	2.834	1.403	0.834	-1.588
655.4	2.630	1.465	0.653	-1.714
655.5	2.464	1.549	0.462	-1.791
655.6	2.331	1.645	0.276	-1.823
655.7	2.225	1.747	0.105	-1.819
655.8	2.140	1.852	-0.046	-1.790
655.9	2.073	1.958	-0.175	-1.744
656.0	2.021	2.064	-0.284	-1.689
656.1	1.982	2.170	-0.376	-1.630
656.2	1.954	2.276	-0.452	-1.569
656.3	1.937	2.383	-0.517	-1.509
656.4	1.930	2.492	-0.573	-1.452
656.5	1.934	2.603	-0.621	-1.396
209π	1.948	2.708	-0.661	-1.347
656.6	1.950	2.716	-0.663	-1.344
656.7	1.979	2.833	-0.702	-1.293
656.8	2.023	2.954	-0.739	-1.244
656.9	2.086	3.079	-0.774	-1.197
657.0	2.173	3.206	-0.808	-1.150
657.1	2.290	3.333	-0.842	-1.102
657.2	2.446	3.453	-0.877	-1.053
657.3	2.648	3.556	-0.912	-1.000
657.4	2.904	3.622	-0.948	-0.944
657.5	3.213	3.621	-0.983	-0.880
657.6	3.555	3.513	-1.017	-0.809
657.7	3.881	3.266	-1.047	-0.728
657.8	4.109	2.884	-1.069	-0.636
657.9	4.169	2.428	-1.079	-0.533

OMEGA = 20.0

k_0h	ADMITTANCE		RAD.EFF.LENGTH	
658.0	4.048	1.999	-1.073	-0.422
658.1	3.798	1.677	-1.044	-0.307
419π/2	3.607	1.540	-1.014	-0.235
658.2	3.495	1.484	-0.992	-0.196
658.3	3.197	1.403	-0.917	-0.097
658.4	2.933	1.400	-0.824	-0.018
658.5	2.712	1.448	-0.723	0.037
658.6	2.531	1.524	-0.622	0.069
658.7	2.385	1.616	-0.528	0.081
658.8	2.269	1.716	-0.443	0.080
658.9	2.176	1.820	-0.371	0.070
659.0	2.102	1.925	-0.308	0.057
659.1	2.044	2.031	-0.255	0.042
659.2	2.000	2.136	-0.209	0.029
659.3	1.968	2.242	-0.167	0.018
659.4	1.946	2.349	-0.129	0.009
659.5	1.935	2.457	-0.091	0.004
659.6	1.935	2.567	-0.053	0.001
659.7	1.945	2.679	-0.014	-0.000
210π	1.952	2.719	-0.000	-0.000
659.8	1.969	2.795	0.028	0.001
659.9	2.006	2.914	0.074	0.003
660.0	2.061	3.037	0.125	0.005
660.1	2.138	3.163	0.181	0.006
660.2	2.242	3.290	0.245	0.004
660.3	2.381	3.414	0.316	-0.002
660.4	2.562	3.525	0.396	-0.014
660.5	2.795	3.609	0.485	-0.036
660.6	3.083	3.639	0.582	-0.072
660.7	3.415	3.579	0.687	-0.124
660.8	3.755	3.392	0.796	-0.200
660.9	4.032	3.062	0.905	-0.304
661.0	4.166	2.626	1.005	-0.440
661.1	4.116	2.176	1.084	-0.609
661.2	3.910	1.805	1.128	-0.809
421π/2	3.606	1.551	1.123	-1.039
661.3	3.622	1.561	1.124	-1.027
661.4	3.318	1.437	1.065	-1.247
661.5	3.039	1.406	0.951	-1.448
661.6	2.800	1.435	0.794	-1.611
661.7	2.604	1.501	0.612	-1.726
661.8	2.445	1.588	0.421	-1.793
661.9	2.317	1.685	0.238	-1.817
662.0	2.215	1.787	0.072	-1.807
662.1	2.133	1.892	-0.074	-1.774
662.2	2.069	1.998	-0.198	-1.727
662.3	2.020	2.103	-0.303	-1.671
662.4	1.983	2.209	-0.390	-1.612
662.5	1.957	2.315	-0.464	-1.551
662.6	1.942	2.423	-0.526	-1.492
662.7	1.937	2.532	-0.580	-1.436
662.8	1.943	2.643	-0.627	-1.381
211π	1.955	2.729	-0.659	-1.342
662.9	1.961	2.757	-0.668	-1.329

k_0h	ADMITTANCE		RAD.EFF.LENGTH	
663.0	1.992	2.874	-0.707	-1.280
663.1	2.040	2.996	-0.743	-1.232
663.2	2.107	3.121	-0.777	-1.185
663.3	2.199	3.247	-0.811	-1.138
663.4	2.323	3.373	-0.846	-1.091
663.5	2.486	3.491	-0.881	-1.041
663.6	2.697	3.588	-0.916	-0.989
663.7	2.962	3.643	-0.952	-0.931
663.8	3.278	3.624	-0.987	-0.867
663.9	3.620	3.493	-1.021	-0.794
664.0	3.932	3.220	-1.050	-0.711
664.1	4.132	2.821	-1.070	-0.617
664.2	4.157	2.365	-1.078	-0.513
664.3	4.008	1.955	-1.068	-0.400
664.4	3.745	1.657	-1.035	-0.285
423π/2	3.605	1.563	-1.012	-0.233
664.5	3.442	1.487	-0.979	-0.176
664.6	3.151	1.422	-0.900	-0.081
664.7	2.896	1.430	-0.805	-0.006
664.8	2.683	1.483	-0.704	0.044
664.9	2.509	1.562	-0.604	0.072
665.0	2.369	1.655	-0.511	0.082
665.1	2.257	1.756	-0.429	0.079
665.2	2.168	1.860	-0.359	0.068
665.3	2.097	1.965	-0.298	0.054
665.4	2.042	2.070	-0.247	0.040
665.5	2.000	2.176	-0.202	0.027
665.6	1.970	2.282	-0.161	0.016
665.7	1.950	2.389	-0.122	0.008
665.8	1.941	2.497	-0.085	0.003
665.9	1.943	2.607	-0.047	0.000
666.0	1.955	2.720	-0.007	-0.000
212π	1.959	2.740	-0.000	-0.000
666.1	1.981	2.836	0.035	0.001
666.2	2.022	2.955	0.082	0.003
666.3	2.081	3.079	0.134	0.005
666.4	2.162	3.205	0.191	0.005
666.5	2.272	3.331	0.256	0.003
666.6	2.418	3.453	0.329	-0.004
666.7	2.608	3.560	0.410	-0.018
666.8	2.850	3.635	0.500	-0.043
666.9	3.146	3.650	0.599	-0.081
667.0	3.481	3.569	0.705	-0.138
667.1	3.813	3.356	0.814	-0.218
667.2	4.067	3.004	0.921	-0.328
667.3	4.168	2.562	1.017	-0.470
667.4	4.086	2.123	1.090	-0.645
667.5	3.862	1.775	1.125	-0.849
425π/2	3.604	1.574	1.115	-1.042
667.6	3.568	1.555	1.111	-1.068
667.7	3.269	1.451	1.041	-1.285
667.8	2.998	1.432	0.918	-1.478
667.9	2.768	1.469	0.755	-1.631

k_0h	ADMITTANCE		RAD.EFF.LENGTH	
668.0	2.579	1.538	0.570	-1.736
668.1	2.426	1.627	0.381	-1.793
668.2	2.303	1.725	0.202	-1.809
668.3	2.205	1.827	0.040	-1.794
668.4	2.127	1.932	-0.100	-1.758
668.5	2.066	2.037	-0.219	-1.709
668.6	2.019	2.143	-0.320	-1.653
668.7	1.984	2.249	-0.404	-1.593
668.8	1.960	2.355	-0.475	-1.534
668.9	1.947	2.462	-0.535	-1.476
669.0	1.944	2.571	-0.587	-1.420
669.1	1.952	2.683	-0.632	-1.367
213π	1.963	2.750	-0.657	-1.336
669.2	1.972	2.798	-0.673	-1.316
669.3	2.007	2.916	-0.711	-1.267
669.4	2.058	3.037	-0.746	-1.220
669.5	2.129	3.162	-0.781	-1.173
669.6	2.227	3.289	-0.815	-1.127
669.7	2.357	3.413	-0.850	-1.079
669.8	2.528	3.527	-0.885	-1.030
669.9	2.748	3.617	-0.920	-0.977
670.0	3.022	3.660	-0.956	-0.918
670.1	3.344	3.623	-0.992	-0.853
670.2	3.683	3.468	-1.025	-0.779
670.3	3.978	3.171	-1.053	-0.694
670.4	4.148	2.757	-1.072	-0.598
670.5	4.139	2.306	-1.077	-0.492
670.6	3.966	1.915	-1.063	-0.379
670.7	3.693	1.642	-1.026	-0.264
427π/2	3.603	1.585	-1.010	-0.231
670.8	3.391	1.494	-0.965	-0.157
670.9	3.106	1.443	-0.882	-0.065
671.0	2.860	1.460	-0.786	0.005
671.1	2.655	1.518	-0.684	0.051
671.2	2.488	1.600	-0.586	0.075
671.3	2.353	1.695	-0.495	0.082
671.4	2.246	1.796	-0.415	0.077
671.5	2.160	1.900	-0.347	0.065
671.6	2.093	2.005	-0.289	0.051
671.7	2.040	2.110	-0.238	0.037
671.8	2.000	2.215	-0.194	0.024
671.9	1.972	2.321	-0.154	0.014
672.0	1.954	2.428	-0.116	0.007
672.1	1.947	2.536	-0.079	0.002
672.2	1.951	2.647	-0.040	0.000
214π	1.966	2.761	-0.000	-0.000
672.3	1.966	2.760	-0.000	-0.000
672.4	1.994	2.877	0.043	0.001
672.5	2.038	2.997	0.090	0.003
672.6	2.101	3.120	0.143	0.005
672.7	2.187	3.246	0.202	0.005
672.8	2.303	3.372	0.268	0.001
672.9	2.456	3.491	0.342	-0.007

k_0h	ADMITTANCE		RAD.EFF.LENGTH	
673.0	2.655	3.593	0.424	-0.023
673.1	2.906	3.658	0.516	-0.049
673.2	3.210	3.658	0.616	-0.091
673.3	3.546	3.555	0.722	-0.152
673.4	3.868	3.317	0.831	-0.237
673.5	4.097	2.945	0.937	-0.353
673.6	4.164	2.499	1.029	-0.501
673.7	4.052	2.074	1.095	-0.682
673.8	3.812	1.751	1.121	-0.889
429π/2	3.602	1.597	1.108	-1.046
673.9	3.516	1.553	1.095	-1.108
674.0	3.221	1.467	1.014	-1.321
674.1	2.958	1.459	0.882	-1.506
674.2	2.737	1.502	0.714	-1.649
674.3	2.555	1.576	0.528	-1.744
674.4	2.408	1.666	0.342	-1.792
674.5	2.290	1.765	0.166	-1.801
674.6	2.196	1.867	0.010	-1.781
674.7	2.122	1.972	-0.125	-1.742
674.8	2.063	2.077	-0.240	-1.691
674.9	2.018	2.182	-0.336	-1.635
675.0	1.985	2.288	-0.417	-1.576
675.1	1.964	2.394	-0.486	-1.517
675.2	1.952	2.502	-0.544	-1.460
675.3	1.951	2.611	-0.594	-1.405
675.4	1.962	2.723	-0.638	-1.352
215π	1.970	2.772	-0.655	-1.331
675.5	1.984	2.838	-0.678	-1.302
675.6	2.022	2.957	-0.715	-1.254
675.7	2.076	3.079	-0.750	-1.207
675.8	2.152	3.204	-0.784	-1.162
675.9	2.255	3.330	-0.819	-1.115
676.0	2.392	3.452	-0.853	-1.068
676.1	2.571	3.562	-0.889	-1.018
676.2	2.800	3.645	-0.925	-0.965
676.3	3.083	3.675	-0.961	-0.905
676.4	3.409	3.619	-0.996	-0.839
676.5	3.743	3.438	-1.028	-0.763
676.6	4.019	3.118	-1.055	-0.676
676.7	4.157	2.694	-1.072	-0.579
676.8	4.116	2.250	-1.074	-0.471
676.9	3.921	1.880	-1.056	-0.357
677.0	3.640	1.632	-1.015	-0.243
431π/2	3.601	1.608	-1.008	-0.229
677.1	3.341	1.503	-0.950	-0.138
677.2	3.064	1.467	-0.864	-0.050
677.3	2.825	1.492	-0.767	0.015
677.4	2.628	1.555	-0.665	0.056
677.5	2.468	1.639	-0.568	0.077
677.6	2.339	1.734	-0.479	0.081
677.7	2.236	1.836	-0.402	0.075
677.8	2.153	1.940	-0.336	0.063
677.9	2.089	2.044	-0.279	0.049

k_0h	ADMITTANCE		RAD.EFF.LENGTH	
678.0	2.038	2.150	-0.230	0.035
678.1	2.001	2.255	-0.187	0.023
678.2	1.975	2.361	-0.147	0.013
678.3	1.959	2.468	-0.110	0.006
678.4	1.954	2.576	-0.072	0.002
678.5	1.959	2.687	-0.034	0.000
216π	1.973	2.782	-0.000	-0.000
678.6	1.977	2.801	0.007	0.000
678.7	2.008	2.918	0.051	0.001
678.8	2.055	3.038	0.099	0.003
678.9	2.122	3.162	0.152	0.004
679.0	2.213	3.288	0.212	0.004
679.1	2.335	3.412	0.280	-0.000
679.2	2.496	3.528	0.355	-0.010
679.3	2.703	3.624	0.439	-0.027
679.4	2.964	3.679	0.532	-0.057
679.5	3.274	3.662	0.633	-0.101
679.6	3.610	3.536	0.740	-0.167
679.7	3.918	3.273	0.848	-0.257
679.8	4.120	2.884	0.951	-0.379
679.9	4.153	2.438	1.039	-0.533
680.0	4.015	2.030	1.097	-0.719
680.1	3.761	1.730	1.114	-0.929
433π/2	3.600	1.620	1.101	-1.049
680.2	3.464	1.555	1.077	-1.148
680.3	3.175	1.485	0.986	-1.355
680.4	2.921	1.488	0.846	-1.532
680.5	2.707	1.537	0.674	-1.665
680.6	2.533	1.614	0.487	-1.749
680.7	2.392	1.705	0.303	-1.789
680.8	2.278	1.805	0.132	-1.791
680.9	2.188	1.907	-0.019	-1.767
681.0	2.117	2.012	-0.149	-1.726
681.1	2.061	2.117	-0.260	-1.674
681.2	2.018	2.222	-0.352	-1.617
681.3	1.987	2.328	-0.430	-1.558
681.4	1.967	2.434	-0.496	-1.500
681.5	1.958	2.542	-0.552	-1.443
681.6	1.959	2.651	-0.600	-1.389
681.7	1.972	2.764	-0.643	-1.338
217π	1.977	2.793	-0.654	-1.325
681.8	1.997	2.879	-0.682	-1.289
681.9	2.037	2.998	-0.719	-1.241
682.0	2.095	3.121	-0.754	-1.195
682.1	2.176	3.246	-0.788	-1.150
682.2	2.284	3.371	-0.822	-1.104
682.3	2.428	3.491	-0.857	-1.057
682.4	2.616	3.597	-0.893	-1.007
682.5	2.854	3.671	-0.929	-0.953
682.6	3.144	3.687	-0.965	-0.892
682.7	3.474	3.610	-1.000	-0.824
682.8	3.800	3.405	-1.032	-0.747
682.9	4.054	3.063	-1.058	-0.658

k_0h	ADMITTANCE		RAD.EFF.LENGTH	
683.0	4.160	2.631	-1.072	-0.559
683.1	4.088	2.198	-1.071	-0.449
683.2	3.874	1.850	-1.049	-0.335
435π/2	3.599	1.631	-1.006	-0.226
683.3	3.588	1.625	-1.004	-0.222
683.4	3.292	1.516	-0.934	-0.120
683.5	3.023	1.492	-0.846	-0.036
683.6	2.793	1.524	-0.747	0.024
683.7	2.603	1.591	-0.646	0.061
683.8	2.449	1.678	-0.550	0.078
683.9	2.325	1.774	-0.464	0.080
684.0	2.226	1.876	-0.389	0.073
684.1	2.147	1.980	-0.325	0.060
684.2	2.085	2.084	-0.270	0.046
684.3	2.037	2.189	-0.223	0.032
684.4	2.002	2.295	-0.180	0.021
684.5	1.977	2.401	-0.141	0.012
684.6	1.964	2.508	-0.103	0.005
684.7	1.961	2.616	-0.066	0.001
684.8	1.968	2.727	-0.027	-0.000
218π	1.980	2.804	-0.000	-0.000
684.9	1.988	2.841	0.014	0.000
685.0	2.022	2.959	0.058	0.002
685.1	2.072	3.080	0.107	0.003
685.2	2.143	3.204	0.162	0.004
685.3	2.240	3.329	0.223	0.003
685.4	2.368	3.452	0.292	-0.002
685.5	2.537	3.565	0.368	-0.013
685.6	2.753	3.654	0.454	-0.033
685.7	3.022	3.697	0.548	-0.064
685.8	3.338	3.663	0.650	-0.113
685.9	3.672	3.513	0.757	-0.182
686.0	3.964	3.226	0.865	-0.279
686.1	4.137	2.823	0.965	-0.406
686.2	4.137	2.379	1.048	-0.566
686.3	3.974	1.990	1.098	-0.757
686.4	3.710	1.715	1.105	-0.969
437π/2	3.598	1.643	1.093	-1.052
686.5	3.413	1.561	1.057	-1.187
686.6	3.131	1.505	0.955	-1.388
686.7	2.885	1.518	0.809	-1.556
686.8	2.679	1.573	0.633	-1.679
686.9	2.511	1.652	0.447	-1.753
687.0	2.376	1.745	0.266	-1.784
687.1	2.267	1.845	0.099	-1.780
687.2	2.180	1.948	-0.047	-1.753
687.3	2.112	2.052	-0.172	-1.709
687.4	2.059	2.157	-0.278	-1.656
687.5	2.018	2.262	-0.367	-1.599
687.6	1.989	2.367	-0.442	-1.540
687.7	1.971	2.474	-0.505	-1.483
687.8	1.964	2.582	-0.559	-1.428
687.9	1.967	2.692	-0.606	-1.375

k_0h	ADMITTANCE		RAD.EFF.LENGTH	
219π	1.984	2.814	-0.652	-1.320
688.0	1.982	2.804	-0.648	-1.324
688.1	2.010	2.920	-0.687	-1.276
688.2	2.053	3.040	-0.723	-1.229
688.3	2.115	3.162	-0.758	-1.184
688.4	2.200	3.287	-0.792	-1.139
688.5	2.315	3.411	-0.826	-1.093
688.6	2.466	3.529	-0.861	-1.046
688.7	2.662	3.630	-0.897	-0.995
688.8	2.909	3.695	-0.933	-0.940
688.9	3.207	3.696	-0.969	-0.879
689.0	3.538	3.597	-1.004	-0.810
689.1	3.854	3.367	-1.036	-0.730
689.2	4.084	3.006	-1.060	-0.640
689.3	4.157	2.569	-1.072	-0.538
689.4	4.056	2.149	-1.068	-0.428
689.5	3.826	1.824	-1.041	-0.313
439π/2	3.597	1.654	-1.003	-0.224
689.6	3.536	1.623	-0.991	-0.202
689.7	3.245	1.531	-0.918	-0.103
689.8	2.983	1.518	-0.827	-0.023
689.9	2.761	1.558	-0.728	0.033
690.0	2.579	1.629	-0.627	0.066
690.1	2.431	1.717	-0.534	0.079
690.2	2.312	1.814	-0.450	0.079
690.3	2.217	1.916	-0.377	0.071
690.4	2.141	2.020	-0.315	0.057
690.5	2.082	2.124	-0.261	0.043
690.6	2.036	2.229	-0.215	0.030
690.7	2.003	2.334	-0.173	0.019
690.8	1.981	2.440	-0.134	0.010
690.9	1.969	2.547	-0.097	0.004
691.0	1.968	2.656	-0.060	0.001
691.1	1.978	2.768	-0.021	-0.000
220π	1.987	2.825	-0.000	-0.000
691.2	2.000	2.882	0.021	0.000
691.3	2.037	3.000	0.066	0.002
691.4	2.091	3.121	0.116	0.003
691.5	2.166	3.245	0.172	0.004
691.6	2.268	3.370	0.234	0.002
691.7	2.403	3.491	0.304	-0.004
691.8	2.579	3.600	0.382	-0.017
691.9	2.805	3.682	0.469	-0.038
692.0	3.082	3.713	0.564	-0.073
692.1	3.402	3.659	0.667	-0.125
692.2	3.731	3.485	0.774	-0.199
692.3	4.005	3.175	0.881	-0.301
692.4	4.147	2.761	0.978	-0.434
692.5	4.116	2.324	1.055	-0.600
692.6	3.931	1.955	1.098	-0.796
692.7	3.658	1.703	1.094	-1.009
441π/2	3.596	1.666	1.086	-1.056
692.8	3.363	1.569	1.035	-1.224
692.9	3.088	1.528	0.924	-1.420

k_0h	ADMITTANCE		RAD.EFF.LENGTH	
693.0	2.850	1.549	0.770	-1.578
693.1	2.652	1.609	0.592	-1.690
693.2	2.491	1.690	0.407	-1.755
693.3	2.361	1.784	0.229	-1.778
693.4	2.256	1.885	0.068	-1.769
693.5	2.173	1.988	-0.074	-1.737
693.6	2.108	2.092	-0.194	-1.692
693.7	2.057	2.197	-0.296	-1.638
693.8	2.019	2.301	-0.381	-1.581
693.9	1.992	2.407	-0.453	-1.523
694.0	1.976	2.514	-0.514	-1.467
694.1	1.970	2.622	-0.566	-1.412
694.2	1.976	2.732	-0.612	-1.360
221π	1.991	2.836	-0.650	-1.314
694.3	1.993	2.845	-0.653	-1.310
694.4	2.023	2.961	-0.691	-1.263
694.5	2.070	3.081	-0.727	-1.217
694.6	2.136	3.204	-0.761	-1.172
694.7	2.226	3.329	-0.795	-1.127
694.8	2.346	3.452	-0.830	-1.082
694.9	2.505	3.566	-0.865	-1.035
695.0	2.709	3.661	-0.901	-0.984
695.1	2.965	3.716	-0.938	-0.928
695.2	3.270	3.701	-0.974	-0.866
695.3	3.600	3.580	-1.009	-0.795
695.4	3.905	3.325	-1.039	-0.714
695.5	4.108	2.947	-1.061	-0.621
695.6	4.148	2.510	-1.071	-0.518
695.7	4.021	2.105	-1.063	-0.406
695.8	3.776	1.804	-1.033	-0.292
443π/2	3.595	1.677	-1.001	-0.222
695.9	3.485	1.624	-0.978	-0.182
696.0	3.199	1.548	-0.901	-0.087
696.1	2.945	1.547	-0.808	-0.011
696.2	2.731	1.592	-0.708	0.040
696.3	2.556	1.666	-0.609	0.069
696.4	2.414	1.756	-0.517	0.080
696.5	2.300	1.854	-0.435	0.078
696.6	2.208	1.956	-0.365	0.068
696.7	2.136	2.060	-0.305	0.055
696.8	2.079	2.164	-0.253	0.041
696.9	2.036	2.269	-0.207	0.028
697.0	2.005	2.374	-0.166	0.017
697.1	1.984	2.480	-0.128	0.009
697.2	1.974	2.587	-0.091	0.004
697.3	1.975	2.697	-0.053	0.001
697.4	1.987	2.808	-0.014	-0.000
222π	1.994	2.847	-0.000	-0.000
697.5	2.012	2.923	0.028	0.001
697.6	2.052	3.041	0.074	0.002
697.7	2.110	3.163	0.125	0.003
697.8	2.189	3.287	0.182	0.003
697.9	2.296	3.411	0.245	0.001

OMEGA = 20.0

k_0h	ADMITTANCE		RAD.EFF.LENGTH	
698.0	2.438	3.530	0.316	-0.006
698.1	2.623	3.634	0.396	-0.020
698.2	2.857	3.708	0.484	-0.044
698.3	3.143	3.725	0.581	-0.082
698.4	3.466	3.652	0.684	-0.138
698.5	3.788	3.453	0.791	-0.217
698.6	4.041	3.122	0.896	-0.324
698.7	4.152	2.700	0.990	-0.464
698.8	4.090	2.272	1.061	-0.635
698.9	3.885	1.925	1.095	-0.835
$445\pi/2$	3.594	1.689	1.078	-1.059
699.0	3.607	1.696	1.080	-1.049
699.1	3.315	1.581	1.011	-1.261
699.2	3.047	1.552	0.891	-1.449
699.3	2.817	1.581	0.732	-1.598
699.4	2.626	1.645	0.552	-1.700
699.5	2.472	1.729	0.368	-1.755
699.6	2.347	1.824	0.194	-1.771
699.7	2.247	1.925	0.037	-1.757
699.8	2.167	2.028	-0.099	-1.722
699.9	2.104	2.132	-0.215	-1.675
700.0	2.056	2.236	-0.313	-1.620
700.1	2.019	2.341	-0.395	-1.564
700.2	1.995	2.447	-0.464	-1.506
700.3	1.981	2.553	-0.523	-1.451
700.4	1.977	2.662	-0.573	-1.397
700.5	1.984	2.772	-0.618	-1.346
223π	1.998	2.857	-0.648	-1.309
700.6	2.004	2.886	-0.658	-1.297
700.7	2.037	3.003	-0.695	-1.250
700.8	2.087	3.123	-0.731	-1.205
700.9	2.157	3.246	-0.765	-1.161
701.0	2.252	3.370	-0.799	-1.116
701.1	2.379	3.491	-0.834	-1.071
701.2	2.545	3.603	-0.869	-1.023
701.3	2.758	3.691	-0.905	-0.972
701.4	3.023	3.735	-0.942	-0.916
701.5	3.333	3.702	-0.978	-0.852
701.6	3.661	3.558	-1.013	-0.780
701.7	3.950	3.280	-1.042	-0.697
701.8	4.126	2.888	-1.063	-0.602
701.9	4.134	2.452	-1.070	-0.497
702.0	3.982	2.065	-1.058	-0.385
702.1	3.726	1.787	-1.023	-0.271
$447\pi/2$	3.593	1.700	-0.999	-0.220
702.2	3.434	1.629	-0.964	-0.163
702.3	3.155	1.568	-0.884	-0.071
702.4	2.909	1.576	-0.789	-0.000
702.5	2.703	1.627	-0.689	0.047
702.6	2.534	1.704	-0.591	0.072
702.7	2.398	1.795	-0.501	0.080
702.8	2.288	1.894	-0.422	0.076
702.9	2.200	1.996	-0.353	0.066

k_0h	ADMITTANCE		RAD.EFF.LENGTH	
703.0	2.131	2.100	-0.295	0.052
703.1	2.077	2.204	-0.244	0.038
703.2	2.036	2.309	-0.200	0.026
703.3	2.007	2.414	-0.160	0.016
703.4	1.988	2.520	-0.122	0.008
703.5	1.980	2.627	-0.084	0.003
703.6	1.983	2.737	-0.047	0.001
703.7	1.998	2.849	-0.007	-0.000
224π	2.001	2.868	-0.000	-0.000
703.8	2.025	2.964	0.036	0.001
703.9	2.068	3.083	0.082	0.002
704.0	2.129	3.205	0.134	0.003
704.1	2.213	3.329	0.192	0.003
704.2	2.327	3.451	0.256	-0.001
704.3	2.476	3.568	0.329	-0.009
704.4	2.669	3.667	0.410	-0.025
704.5	2.912	3.732	0.499	-0.051
704.6	3.204	3.734	0.597	-0.092
704.7	3.529	3.640	0.701	-0.152
704.8	3.841	3.418	0.808	-0.236
704.9	4.071	3.067	0.911	-0.349
705.0	4.151	2.640	1.001	-0.494
705.1	4.060	2.224	1.065	-0.671
705.2	3.838	1.899	1.090	-0.874
$449\pi/2$	3.592	1.712	1.071	-1.061
705.3	3.555	1.693	1.065	-1.088
705.4	3.268	1.595	0.985	-1.295
705.5	3.007	1.579	0.856	-1.476
705.6	2.785	1.614	0.692	-1.615
705.7	2.602	1.682	0.511	-1.707
705.8	2.453	1.768	0.330	-1.753
705.9	2.333	1.864	0.160	-1.762
706.0	2.237	1.965	0.008	-1.744
706.1	2.161	2.068	-0.124	-1.706
706.2	2.101	2.172	-0.235	-1.657
706.3	2.055	2.276	-0.329	-1.603
706.4	2.021	2.381	-0.408	-1.546
706.5	1.998	2.487	-0.474	-1.490
706.6	1.986	2.593	-0.531	-1.435
706.7	1.984	2.702	-0.580	-1.382
706.8	1.994	2.813	-0.623	-1.332
225π	2.005	2.879	-0.647	-1.304
706.9	2.015	2.927	-0.663	-1.284
707.0	2.052	3.044	-0.699	-1.238
707.1	2.105	3.164	-0.734	-1.193
707.2	2.179	3.287	-0.769	-1.149
707.3	2.280	3.411	-0.803	-1.105
707.4	2.413	3.531	-0.838	-1.060
707.5	2.587	3.638	-0.873	-1.012
707.6	2.809	3.719	-0.910	-0.960
707.7	3.082	3.751	-0.947	-0.903
707.8	3.396	3.700	-0.983	-0.838
707.9	3.720	3.532	-1.017	-0.764

k_0h	ADMITTANCE		RAD.EFF.LENGTH	
708.0	3.992	3.232	-1.045	-0.679
708.1	4.138	2.828	-1.064	-0.583
708.2	4.115	2.398	-1.068	-0.476
708.3	3.940	2.030	-1.052	-0.363
708.4	3.675	1.775	-1.013	-0.249
$451\pi/2$	3.591	1.724	-0.997	-0.218
708.5	3.385	1.637	-0.950	-0.144
708.6	3.112	1.590	-0.866	-0.056
708.7	2.874	1.607	-0.770	0.010
708.8	2.676	1.663	-0.670	0.053
708.9	2.514	1.743	-0.573	0.075
709.0	2.382	1.835	-0.486	0.080
709.1	2.277	1.934	-0.408	0.075
709.2	2.193	2.036	-0.342	0.063
709.3	2.127	2.140	-0.285	0.050
709.4	2.075	2.244	-0.236	0.036
709.5	2.036	2.349	-0.193	0.024
709.6	2.009	2.454	-0.153	0.014
709.7	1.993	2.560	-0.115	0.007
709.8	1.987	2.668	-0.078	0.003
709.9	1.991	2.777	-0.040	0.000
226π	2.008	2.890	-0.000	-0.000
710.0	2.008	2.890	0.000	0.000
710.1	2.038	3.006	0.043	0.001
710.2	2.084	3.125	0.091	0.002
710.3	2.150	3.246	0.143	0.003
710.4	2.239	3.370	0.202	0.002
710.5	2.358	3.492	0.268	-0.002
710.6	2.514	3.605	0.342	-0.012
710.7	2.715	3.699	0.424	-0.029
710.8	2.967	3.754	0.515	-0.058
710.9	3.266	3.740	0.614	-0.102
711.0	3.591	3.624	0.718	-0.166
711.1	3.891	3.378	0.824	-0.255
711.2	4.096	3.011	0.925	-0.374
711.3	4.144	2.581	1.011	-0.526
711.4	4.026	2.181	1.068	-0.708
711.5	3.790	1.877	1.083	-0.913
$453\pi/2$	3.590	1.735	1.063	-1.064
711.6	3.505	1.693	1.047	-1.127
711.7	3.222	1.612	0.957	-1.329
711.8	2.969	1.606	0.821	-1.501
711.9	2.755	1.648	0.653	-1.630
712.0	2.579	1.720	0.472	-1.712
712.1	2.436	1.807	0.292	-1.750
712.2	2.321	1.904	0.127	-1.753
712.3	2.228	2.005	-0.020	-1.730
712.4	2.155	2.108	-0.147	-1.690
712.5	2.098	2.212	-0.254	-1.640
712.6	2.054	2.316	-0.344	-1.585
712.7	2.022	2.421	-0.420	-1.529
712.8	2.001	2.527	-0.484	-1.473
712.9	1.991	2.634	-0.539	-1.419

k_0h	ADMITTANCE		RAD.EFF.LENGTH	
713.0	1.991	2.742	-0.586	-1.367
713.1	2.003	2.854	-0.629	-1.318
227π	2.012	2.901	-0.645	-1.298
713.2	2.028	2.968	-0.667	-1.271
713.3	2.067	3.085	-0.703	-1.226
713.4	2.124	3.206	-0.738	-1.182
713.5	2.202	3.329	-0.772	-1.138
713.6	2.308	3.452	-0.807	-1.094
713.7	2.449	3.569	-0.842	-1.049
713.8	2.631	3.673	-0.877	-1.001
713.9	2.861	3.746	-0.914	-0.949
714.0	3.141	3.764	-0.951	-0.890
714.1	3.459	3.694	-0.987	-0.824
714.2	3.776	3.502	-1.020	-0.748
714.3	4.028	3.181	-1.047	-0.661
714.4	4.144	2.769	-1.064	-0.563
714.5	4.091	2.347	-1.065	-0.455
714.6	3.896	2.000	-1.045	-0.341
714.7	3.625	1.767	-1.002	-0.229
$455\pi/2$	3.589	1.747	-0.995	-0.215
714.8	3.337	1.647	-0.934	-0.126
714.9	3.071	1.614	-0.848	-0.042
715.0	2.841	1.639	-0.751	0.020
715.1	2.650	1.700	-0.651	0.058
715.2	2.494	1.782	-0.556	0.076
715.3	2.368	1.875	-0.470	0.079
715.4	2.267	1.974	-0.396	0.073
715.5	2.186	2.077	-0.331	0.061
715.6	2.123	2.180	-0.276	0.047
715.7	2.074	2.284	-0.228	0.034
715.8	2.037	2.388	-0.186	0.022
715.9	2.012	2.494	-0.147	0.013
716.0	1.997	2.600	-0.109	0.006
716.1	1.993	2.708	-0.072	0.002
716.2	2.000	2.818	-0.034	0.000
228π	2.015	2.912	-0.000	-0.000
716.3	2.019	2.931	0.007	0.000
716.4	2.052	3.047	0.051	0.001
716.5	2.101	3.166	0.099	0.002
716.6	2.171	3.288	0.153	0.003
716.7	2.265	3.411	0.213	0.001
716.8	2.390	3.531	0.280	-0.004
716.9	2.554	3.642	0.355	-0.015
717.0	2.764	3.729	0.439	-0.035
717.1	3.024	3.773	0.531	-0.066
717.2	3.328	3.743	0.630	-0.113
717.3	3.651	3.604	0.735	-0.182
717.4	3.937	3.335	0.840	-0.276
717.5	4.115	2.953	0.939	-0.401
717.6	4.131	2.525	1.019	-0.558
717.7	3.988	2.141	1.068	-0.745
717.8	3.741	1.861	1.074	-0.952
$457\pi/2$	3.588	1.759	1.056	-1.067
717.9	3.455	1.697	1.027	-1.164

k_0h	ADMITTANCE		RAD.EFF.LENGTH	
718.0	3.178	1.632	0.928	-1.361
718.1	2.933	1.635	0.784	-1.524
718.2	2.727	1.683	0.613	-1.643
718.3	2.557	1.758	0.432	-1.715
718.4	2.419	1.847	0.256	-1.746
718.5	2.309	1.944	0.095	-1.743
718.6	2.220	2.045	-0.047	-1.716
718.7	2.150	2.148	-0.169	-1.674
718.8	2.095	2.252	-0.272	-1.623
718.9	2.054	2.356	-0.359	-1.568
719.0	2.024	2.461	-0.432	-1.512
719.1	2.005	2.567	-0.493	-1.457
719.2	1.997	2.674	-0.546	-1.404
719.3	1.999	2.783	-0.592	-1.353
719.4	2.013	2.894	-0.634	-1.305
229π	2.019	2.922	-0.643	-1.293
719.5	2.040	3.009	-0.672	-1.258
719.6	2.082	3.127	-0.707	-1.214
719.7	2.143	3.248	-0.742	-1.170
719.8	2.226	3.370	-0.776	-1.127
719.9	2.338	3.492	-0.810	-1.083
720.0	2.485	3.607	-0.846	-1.038
720.1	2.675	3.706	-0.882	-0.990
720.2	2.914	3.770	-0.919	-0.937
720.3	3.201	3.774	-0.956	-0.877
720.4	3.520	3.683	-0.992	-0.810
720.5	3.829	3.469	-1.024	-0.732
720.6	4.059	3.128	-1.050	-0.643
720.7	4.144	2.710	-1.064	-0.543
720.8	4.062	2.300	-1.062	-0.434
720.9	3.850	1.973	-1.038	-0.320
$459\pi/2$	3.587	1.770	-0.992	-0.213
721.0	3.574	1.763	-0.990	-0.208
721.1	3.290	1.661	-0.919	-0.109
721.2	3.031	1.639	-0.830	-0.029
721.3	2.809	1.672	-0.732	0.028
721.4	2.625	1.736	-0.633	0.063
721.5	2.475	1.820	-0.539	0.078
721.6	2.354	1.915	-0.456	0.078
721.7	2.257	2.015	-0.383	0.071
721.8	2.180	2.117	-0.321	0.058
721.9	2.119	2.220	-0.267	0.044
722.0	2.072	2.324	-0.221	0.031
722.1	2.038	2.428	-0.179	0.020
722.2	2.015	2.534	-0.140	0.011
722.3	2.002	2.640	-0.103	0.005
722.4	2.000	2.748	-0.065	0.002
722.5	2.009	2.859	-0.027	0.000
230π	2.022	2.933	-0.000	-0.000
722.6	2.031	2.972	0.014	0.000
722.7	2.066	3.088	0.059	0.001
722.8	2.119	3.208	0.108	0.002
722.9	2.193	3.330	0.162	0.002

k_0h	ADMITTANCE		RAD.EFF.LENGTH	
723.0	2.292	3.452	0.223	0.000
723.1	2.424	3.571	0.292	-0.006
723.2	2.595	3.677	0.368	-0.019
723.3	2.813	3.758	0.453	-0.040
723.4	3.081	3.789	0.547	-0.074
723.5	3.390	3.741	0.647	-0.125
723.6	3.708	3.580	0.752	-0.198
723.7	3.978	3.289	0.856	-0.298
723.8	4.128	2.895	0.951	-0.428
723.9	4.113	2.472	1.026	-0.591
724.0	3.948	2.106	1.067	-0.782
724.1	3.692	1.848	1.063	-0.991
$461\pi/2$	3.586	1.782	1.048	-1.070
724.2	3.406	1.704	1.006	-1.200
724.3	3.135	1.653	0.897	-1.391
724.4	2.898	1.666	0.747	-1.545
724.5	2.699	1.719	0.574	-1.654
724.6	2.536	1.796	0.394	-1.717
724.7	2.404	1.887	0.221	-1.740
724.8	2.298	1.984	0.064	-1.731
724.9	2.213	2.086	-0.073	-1.701
725.0	2.146	2.188	-0.190	-1.657
725.1	2.093	2.292	-0.289	-1.606
725.2	2.054	2.396	-0.373	-1.551
725.3	2.026	2.501	-0.443	-1.496
725.4	2.009	2.607	-0.502	-1.441
725.5	2.003	2.714	-0.553	-1.389
725.6	2.007	2.823	-0.598	-1.339
231π	2.025	2.944	-0.642	-1.288
725.7	2.024	2.935	-0.639	-1.291
725.8	2.053	3.050	-0.676	-1.246
725.9	2.099	3.169	-0.711	-1.202
726.0	2.163	3.290	-0.746	-1.159
726.1	2.251	3.412	-0.780	-1.116
726.2	2.369	3.532	-0.814	-1.073
726.3	2.523	3.645	-0.850	-1.027
726.4	2.721	3.738	-0.886	-0.978
726.5	2.969	3.792	-0.923	-0.924
726.6	3.262	3.780	-0.960	-0.864
726.7	3.581	3.669	-0.996	-0.795
726.8	3.878	3.431	-1.028	-0.716
726.9	4.084	3.074	-1.052	-0.625
727.0	4.138	2.653	-1.063	-0.523
727.1	4.030	2.256	-1.057	-0.412
727.2	3.803	1.952	-1.029	-0.298
$463\pi/2$	3.585	1.794	-0.990	-0.211
727.3	3.524	1.763	-0.977	-0.189
727.4	3.245	1.677	-0.902	-0.092
727.5	2.993	1.667	-0.811	-0.016
727.6	2.779	1.705	-0.712	0.036
727.7	2.602	1.774	-0.614	0.066
727.8	2.458	1.860	-0.523	0.078
727.9	2.342	1.955	-0.442	0.077

k_0h	ADMITTANCE		RAD.EFF.LENGTH	
728.0	2.248	2.055	-0.371	0.068
728.1	2.174	2.157	-0.311	0.056
728.2	2.116	2.260	-0.259	0.042
728.3	2.072	2.364	-0.213	0.029
728.4	2.039	2.469	-0.172	0.019
728.5	2.018	2.574	-0.134	0.010
728.6	2.007	2.680	-0.097	0.005
728.7	2.007	2.789	-0.059	0.001
728.8	2.019	2.899	-0.020	0.000
232π	2.029	2.955	-0.000	-0.000
728.9	2.043	3.013	0.021	0.000
729.0	2.081	3.130	0.067	0.001
729.1	2.138	3.250	0.116	0.002
729.2	2.216	3.372	0.172	0.002
729.3	2.320	3.493	0.234	-0.001
729.4	2.459	3.609	0.304	-0.008
729.5	2.638	3.712	0.382	-0.022
729.6	2.864	3.784	0.468	-0.046
729.7	3.140	3.803	0.563	-0.083
729.8	3.451	3.736	0.664	-0.138
729.9	3.764	3.552	0.769	-0.215
730.0	4.015	3.240	0.871	-0.321
730.1	4.135	2.837	0.963	-0.457
730.2	4.091	2.422	1.032	-0.625
730.3	3.906	2.075	1.065	-0.820
730.4	3.642	1.839	1.050	-1.030
$465\pi/2$	3.584	1.806	1.040	-1.072
730.5	3.359	1.714	0.982	-1.236
730.6	3.094	1.676	0.864	-1.419
730.7	2.865	1.697	0.709	-1.564
730.8	2.673	1.755	0.534	-1.663
730.9	2.516	1.835	0.356	-1.717
731.0	2.389	1.926	0.187	-1.733
731.1	2.287	2.025	0.035	-1.719
731.2	2.206	2.126	-0.098	-1.686
731.3	2.141	2.229	-0.211	-1.641
731.4	2.091	2.332	-0.306	-1.589
731.5	2.054	2.436	-0.386	-1.534
731.6	2.028	2.541	-0.453	-1.479
731.7	2.013	2.647	-0.510	-1.426
731.8	2.005	2.754	-0.560	-1.374
731.9	2.016	2.864	-0.604	-1.325
233π	2.032	2.966	-0.640	-1.282
732.0	2.034	2.976	-0.643	-1.278
732.1	2.067	3.092	-0.680	-1.234
732.2	2.116	3.210	-0.715	-1.190
732.3	2.184	3.331	-0.749	-1.148
732.4	2.277	3.453	-0.783	-1.105
732.5	2.401	3.572	-0.818	-1.062
732.6	2.562	3.681	-0.854	-1.016
732.7	2.769	3.768	-0.890	-0.967
732.8	3.024	3.812	-0.928	-0.912
732.9	3.323	3.784	-0.965	-0.851

k_0h	ADMITTANCE		RAD.EFF.LENGTH	
733.0	3.640	3.650	-1.000	-0.780
733.1	3.924	3.390	-1.031	-0.699
733.2	4.104	3.018	-1.053	-0.606
733.3	4.127	2.598	-1.062	-0.502
733.4	3.994	2.217	-1.053	-0.390
733.5	3.756	1.934	-1.020	-0.277
$467\pi/2$	3.584	1.817	-0.988	-0.209
733.6	3.475	1.766	-0.963	-0.169
733.7	3.201	1.696	-0.885	-0.077
733.8	2.957	1.695	-0.792	-0.005
733.9	2.750	1.740	-0.693	0.043
734.0	2.580	1.812	-0.596	0.070
734.1	2.441	1.899	-0.507	0.079
734.2	2.329	1.995	-0.428	0.076
734.3	2.240	2.095	-0.360	0.066
734.4	2.169	2.197	-0.301	0.053
734.5	2.113	2.301	-0.250	0.040
734.6	2.071	2.404	-0.206	0.027
734.7	2.041	2.509	-0.165	0.017
734.8	2.021	2.614	-0.127	0.009
734.9	2.013	2.721	-0.090	0.004
735.0	2.015	2.829	-0.053	0.001
735.1	2.028	2.940	-0.013	0.000
234π	2.036	2.977	-0.000	-0.000
735.2	2.055	3.054	0.029	0.000
735.3	2.097	3.171	0.075	0.001
735.4	2.157	3.291	0.125	0.002
735.5	2.239	3.413	0.182	0.001
735.6	2.350	3.534	0.245	-0.002
735.7	2.495	3.648	0.316	-0.011
735.8	2.682	3.745	0.395	-0.027
735.9	2.917	3.809	0.483	-0.053
736.0	3.199	3.814	0.579	-0.093
736.1	3.512	3.727	0.680	-0.151
736.2	3.816	3.520	0.785	-0.234
736.3	4.046	3.189	0.885	-0.345
736.4	4.136	2.780	0.973	-0.487
736.5	4.064	2.375	1.036	-0.660
736.6	3.862	2.048	1.060	-0.858
$469\pi/2$	3.583	1.829	1.033	-1.075
736.7	3.592	1.835	1.035	-1.067
736.8	3.312	1.727	0.957	-1.269
736.9	3.055	1.701	0.831	-1.445
737.0	2.833	1.729	0.671	-1.581
737.1	2.648	1.792	0.495	-1.670
737.2	2.497	1.873	0.319	-1.715
737.3	2.375	1.966	0.154	-1.724
737.4	2.277	2.065	0.006	-1.707
737.5	2.199	2.166	-0.122	-1.671
737.6	2.138	2.269	-0.230	-1.624
737.7	2.090	2.372	-0.321	-1.572
737.8	2.055	2.476	-0.398	-1.517
737.9	2.031	2.581	-0.463	-1.463

OMEGA = 20.0

k_0h	ADMITTANCE		RAD.EFF.LENGTH	
738.0	2.018	2.687	-0.519	-1.410
738.1	2.016	2.795	-0.567	-1.360
738.2	2.025	2.905	-0.609	-1.311
235π	2.039	2.988	-0.638	-1.277
738.3	2.046	3.018	-0.648	-1.266
738.4	2.081	3.133	-0.684	-1.221
738.5	2.133	3.252	-0.719	-1.179
738.6	2.206	3.373	-0.753	-1.137
738.7	2.304	3.494	-0.787	-1.094
738.8	2.434	3.612	-0.822	-1.051
738.9	2.603	3.717	-0.858	-1.005
739.0	2.818	3.797	-0.895	-0.955
739.1	3.081	3.829	-0.932	-0.900
739.2	3.384	3.783	-0.969	-0.837
739.3	3.697	3.628	-1.004	-0.765
739.4	3.965	3.346	-1.034	-0.681
739.5	4.118	2.962	-1.054	-0.587
739.6	4.111	2.546	-1.060	-0.481
739.7	3.956	2.181	-1.047	-0.369
739.8	3.707	1.921	-1.010	-0.256
471π/2	3.582	1.841	-0.985	-0.207
739.9	3.427	1.773	-0.949	-0.150
740.0	3.158	1.716	-0.868	-0.061
740.1	2.922	1.725	-0.773	0.006
740.2	2.722	1.775	-0.674	0.049
740.3	2.558	1.850	-0.579	0.072
740.4	2.425	1.939	-0.491	0.079
740.5	2.318	2.035	-0.414	0.074
740.6	2.232	2.136	-0.348	0.064
740.7	2.164	2.238	-0.291	0.051
740.8	2.111	2.341	-0.242	0.037
740.9	2.071	2.445	-0.199	0.025
741.0	2.043	2.549	-0.159	0.015
741.1	2.025	2.654	-0.121	0.008
741.2	2.019	2.761	-0.084	0.003
741.3	2.023	2.870	-0.046	0.001
741.4	2.039	2.981	-0.007	-0.000
236π	2.042	2.999	-0.000	-0.000
741.5	2.068	3.096	0.036	0.000
741.6	2.113	3.213	0.083	0.001
741.7	2.177	3.333	0.134	0.002
741.8	2.264	3.455	0.192	0.000
741.9	2.380	3.574	0.257	-0.004
742.0	2.532	3.685	0.329	-0.014
742.1	2.727	3.777	0.409	-0.031
742.2	2.970	3.831	0.498	-0.060
742.3	3.258	3.821	0.595	-0.103
742.4	3.572	3.714	0.697	-0.166
742.5	3.865	3.484	0.801	-0.253
742.6	4.072	3.137	0.899	-0.369
742.7	4.132	2.725	0.982	-0.518
742.8	4.033	2.332	1.038	-0.696
742.9	3.816	2.026	1.053	-0.896
473π/2	3.581	1.853	1.025	-1.077

k_0h	ADMITTANCE		RAD.EFF.LENGTH	
743.0	3.543	1.834	1.017	-1.105
743.1	3.267	1.743	0.929	-1.302
743.2	3.017	1.728	0.796	-1.470
743.3	2.802	1.763	0.633	-1.595
743.4	2.624	1.829	0.457	-1.675
743.5	2.479	1.913	0.283	-1.712
743.6	2.362	2.006	0.122	-1.715
743.7	2.268	2.105	-0.021	-1.693
743.8	2.193	2.207	-0.144	-1.655
743.9	2.134	2.309	-0.249	-1.607
744.0	2.089	2.413	-0.336	-1.555
744.1	2.056	2.517	-0.410	-1.501
744.2	2.034	2.622	-0.473	-1.447
744.3	2.023	2.728	-0.526	-1.395
744.4	2.023	2.836	-0.573	-1.345
744.5	2.034	2.946	-0.615	-1.298
237π	2.046	3.010	-0.637	-1.272
744.6	2.057	3.059	-0.653	-1.253
744.7	2.096	3.175	-0.688	-1.210
744.8	2.151	3.294	-0.723	-1.167
744.9	2.228	3.415	-0.757	-1.126
745.0	2.332	3.535	-0.791	-1.084
745.1	2.468	3.650	-0.826	-1.040
745.2	2.645	3.751	-0.862	-0.994
745.3	2.868	3.823	-0.899	-0.944
745.4	3.138	3.843	-0.937	-0.887
745.5	3.444	3.779	-0.974	-0.823
745.6	3.752	3.601	-1.008	-0.749
745.7	4.002	3.299	-1.037	-0.664
745.8	4.126	2.906	-1.055	-0.567
745.9	4.090	2.497	-1.058	-0.460
746.0	3.915	2.150	-1.040	-0.347
746.1	3.659	1.912	-0.999	-0.235
475π/2	3.580	1.865	-0.983	-0.205
746.2	3.380	1.782	-0.934	-0.132
746.3	3.117	1.739	-0.850	-0.047
746.4	2.888	1.756	-0.754	0.015
746.5	2.696	1.811	-0.656	0.055
746.6	2.538	1.888	-0.562	0.074
746.7	2.410	1.979	-0.476	0.078
746.8	2.307	2.076	-0.402	0.072
746.9	2.225	2.176	-0.337	0.061
747.0	2.160	2.278	-0.282	0.048
747.1	2.109	2.381	-0.234	0.035
747.2	2.071	2.485	-0.192	0.023
747.3	2.045	2.589	-0.152	0.014
747.4	2.030	2.695	-0.115	0.007
747.5	2.025	2.802	-0.078	0.003
747.6	2.031	2.911	-0.040	0.001
238π	2.049	3.021	-0.000	-0.000
747.7	2.049	3.023	0.000	0.000
747.8	2.081	3.137	0.044	0.000
747.9	2.130	3.255	0.091	0.001

k_0h	ADMITTANCE		RAD.EFF.LENGTH	
748.0	2.197	3.375	0.144	0.001
748.1	2.289	3.496	0.202	-0.000
748.2	2.411	3.614	0.268	-0.006
748.3	2.571	3.722	0.342	-0.017
748.4	2.774	3.807	0.424	-0.036
748.5	3.025	3.851	0.514	-0.067
748.6	3.318	3.825	0.611	-0.114
748.7	3.630	3.697	0.713	-0.181
748.8	3.911	3.445	0.816	-0.273
748.9	4.093	3.083	0.912	-0.395
749.0	4.123	2.671	0.990	-0.549
749.1	3.999	2.293	1.039	-0.732
749.2	3.769	2.008	1.044	-0.934
477π/2	3.579	1.877	1.017	-1.079
749.3	3.495	1.836	0.998	-1.141
749.4	3.223	1.761	0.901	-1.333
749.5	2.980	1.756	0.761	-1.492
749.6	2.773	1.797	0.594	-1.608
749.7	2.602	1.866	0.419	-1.678
749.8	2.462	1.952	0.248	-1.708
749.9	2.350	2.047	0.091	-1.705
750.0	2.259	2.146	-0.047	-1.680
750.1	2.188	2.247	-0.166	-1.639
750.2	2.131	2.350	-0.266	-1.591
750.3	2.089	2.453	-0.350	-1.538
750.4	2.058	2.557	-0.421	-1.484
750.5	2.038	2.662	-0.482	-1.431
750.6	2.029	2.768	-0.534	-1.380
750.7	2.030	2.876	-0.579	-1.331
750.8	2.044	2.987	-0.620	-1.285
239π	2.053	3.033	-0.635	-1.267
750.9	2.070	3.100	-0.657	-1.241
751.0	2.111	3.217	-0.692	-1.198
751.1	2.170	3.336	-0.727	-1.156
751.2	2.252	3.456	-0.761	-1.115
751.3	2.361	3.576	-0.795	-1.073
751.4	2.504	3.688	-0.830	-1.030
751.5	2.688	3.785	-0.867	-0.983
751.6	2.919	3.848	-0.904	-0.932
751.7	3.196	3.854	-0.942	-0.874
751.8	3.504	3.771	-0.979	-0.809
751.9	3.804	3.571	-1.012	-0.733
752.0	4.033	3.251	-1.039	-0.646
752.1	4.129	2.851	-1.055	-0.547
752.2	4.065	2.450	-1.055	-0.439
752.3	3.872	2.124	-1.033	-0.325
752.4	3.610	1.907	-0.987	-0.215
479π/2	3.578	1.889	-0.981	-0.202
752.5	3.334	1.794	-0.918	-0.115
752.6	3.078	1.764	-0.832	-0.034
752.7	2.856	1.788	-0.735	0.024
752.8	2.671	1.847	-0.637	0.059
752.9	2.519	1.927	-0.545	0.076

k_0h	ADMITTANCE		RAD.EFF.LENGTH	
753.0	2.396	2.019	-0.462	0.078
753.1	2.297	2.116	-0.389	0.071
753.2	2.218	2.217	-0.327	0.059
753.3	2.156	.319	-0.273	0.046
753.4	2.108	2.422	-0.226	0.033
753.5	2.072	2.525	-0.185	0.022
753.6	2.048	2.629	-0.146	0.013
753.7	2.034	2.735	-0.108	0.006
753.8	2.031	2.842	-0.071	0.002
753.9	2.040	2.952	-0.033	0.000
240π	2.056	3.044	-0.000	-0.000
754.0	2.061	3.064	0.007	0.000
754.1	2.095	3.179	0.051	0.001
754.2	2.147	3.297	0.099	0.001
754.3	2.219	3.417	0.153	0.001
754.4	2.316	3.537	0.213	-0.002
754.5	2.444	3.653	0.280	-0.008
754.6	2.611	3.757	0.355	-0.020
754.7	2.822	3.836	0.438	-0.042
754.8	3.080	3.869	0.529	-0.075
754.9	3.378	3.826	0.627	-0.125
755.0	3.686	3.676	0.730	-0.197
755.1	3.952	3.403	0.831	-0.295
755.2	4.108	3.029	0.924	-0.422
755.3	4.108	2.620	0.997	-0.582
755.4	3.963	2.257	1.037	-0.768
755.5	3.722	1.995	1.033	-0.972
481π/2	3.577	1.901	1.010	-1.081
755.6	3.447	1.842	0.977	-1.176
755.7	3.181	1.781	0.870	-1.362
755.8	2.945	1.785	0.725	-1.512
755.9	2.745	1.832	0.556	-1.618
756.0	2.580	1.904	0.381	-1.679
756.1	2.446	1.992	0.213	-1.702
756.2	2.338	2.087	0.061	-1.694
756.3	2.251	2.186	-0.072	-1.665
756.4	2.183	2.288	-0.187	-1.623
756.5	2.129	2.390	-0.283	-1.574
756.6	2.088	2.493	-0.364	-1.521
756.7	2.059	2.597	-0.432	-1.468
756.8	2.042	2.702	-0.490	-1.416
756.9	2.034	2.809	-0.541	-1.366
757.0	2.038	2.917	-0.585	-1.318
757.1	2.054	3.028	-0.625	-1.272
241π	2.059	3.055	-0.634	-1.262
757.2	2.082	3.142	-0.661	-1.228
757.3	2.127	3.258	-0.696	-1.186
757.4	2.190	3.378	-0.730	-1.145
757.5	2.276	3.498	-0.764	-1.104
757.6	2.391	3.616	-0.799	-1.063
757.7	2.541	3.726	-0.834	-1.019
757.8	2.733	3.817	-0.871	-0.972
757.9	2.972	3.871	-0.909	-0.920

k_0h	ADMITTANCE		RAD.EFF.LENGTH	
758.0	3.255	3.863	-0.946	-0.861
758.1	3.563	3.760	-0.983	-0.794
758.2	3.853	3.538	-1.016	-0.717
758.3	4.060	3.200	-1.042	-0.627
758.4	4.126	2.797	-1.055	-0.527
758.5	4.036	2.408	-1.051	-0.417
758.6	3.828	2.101	-1.025	-0.304
483π/2	3.576	1.913	-0.978	-0.200
758.7	3.562	1.905	-0.975	-0.195
758.8	3.289	1.809	-0.902	-0.098
758.9	3.040	1.790	-0.813	-0.021
759.0	2.825	1.821	-0.716	0.032
759.1	2.647	1.884	-0.619	0.063
759.2	2.501	1.966	-0.528	0.077
759.3	2.383	2.059	-0.447	0.077
759.4	2.288	2.157	-0.377	0.068
759.5	2.212	2.257	-0.317	0.056
759.6	2.152	2.359	-0.265	0.043
759.7	2.107	2.462	-0.219	0.031
759.8	2.073	2.566	-0.178	0.020
759.9	2.051	2.670	-0.139	0.012
760.0	2.039	2.776	-0.102	0.006
760.1	2.038	2.883	-0.065	0.002
760.2	2.049	2.993	-0.026	0.000
242π	2.063	3.066	-0.000	0.000
760.3	2.072	3.105	0.015	0.000
760.4	2.110	3.221	0.059	0.001
760.5	2.165	3.339	0.108	0.001
760.6	2.241	3.459	0.163	0.000
760.7	2.343	3.578	0.223	-0.003
760.8	2.478	3.692	0.292	-0.010
760.9	2.652	3.792	0.368	-0.024
761.0	2.871	3.863	0.452	-0.048
761.1	3.137	3.884	0.545	-0.084
761.2	3.437	3.823	0.644	-0.138
761.3	3.741	3.652	0.746	-0.214
761.4	3.989	3.359	0.846	-0.317
761.5	4.117	2.975	0.935	-0.450
761.6	4.089	2.571	1.002	-0.615
761.7	3.923	2.226	1.034	-0.805
761.8	3.674	1.985	1.020	-1.009
485π/2	3.576	1.925	1.002	-1.083
761.9	3.400	1.851	0.954	-1.210
762.0	3.140	1.803	0.839	-1.389
762.1	2.911	1.816	0.688	-1.530
762.2	2.718	1.868	0.518	-1.627
762.3	2.560	1.943	0.344	-1.679
762.4	2.431	2.032	0.180	-1.695
762.5	2.327	2.127	0.032	-1.683
762.6	2.244	2.227	-0.097	-1.651
762.7	2.178	2.328	-0.206	-1.607
762.8	2.127	2.431	-0.299	-1.557
762.9	2.088	2.534	-0.377	-1.505

k_0h	ADMITTANCE		RAD.EFF.LENGTH	
763.0	2.062	2.638	-0.443	-1.452
763.1	2.046	2.743	-0.499	-1.401
763.2	2.040	2.849	-0.547	-1.351
763.3	2.046	2.958	-0.590	-1.304
243π	2.066	3.077	-0.632	-1.256
763.4	2.064	3.069	-0.629	-1.259
763.5	2.096	3.183	-0.666	-1.217
763.6	2.143	3.300	-0.700	-1.175
763.7	2.210	3.419	-0.734	-1.134
763.8	2.301	3.539	-0.768	-1.094
763.9	2.422	3.656	-0.803	-1.052
764.0	2.579	3.763	-0.839	-1.008
764.1	2.779	3.847	-0.876	-0.961
764.2	3.025	3.892	-0.913	-0.908
764.3	3.313	3.868	-0.951	-0.848
764.4	3.620	3.744	-0.987	-0.779
764.5	3.898	3.501	-1.019	-0.700
764.6	4.081	3.148	-1.043	-0.609
764.7	4.118	2.744	-1.054	-0.506
764.8	4.004	2.369	-1.046	-0.396
764.9	3.782	2.083	-1.016	-0.283
487π/2	3.575	1.937	-0.976	-0.198
765.0	3.514	1.907	-0.962	-0.175
765.1	3.245	1.827	-0.886	-0.082
765.2	3.003	1.817	-0.795	-0.010
765.3	2.796	1.855	-0.697	0.039
765.4	2.624	1.922	-0.601	0.067
765.5	2.484	2.006	-0.512	0.077
765.6	2.370	2.099	-0.434	0.075
765.7	2.279	2.197	-0.365	0.066
765.8	2.206	2.298	-0.307	0.054
765.9	2.149	2.400	-0.256	0.041
766.0	2.106	2.503	-0.211	0.029
766.1	2.074	2.606	-0.171	0.018
766.2	2.054	2.710	-0.133	0.010
766.3	2.044	2.816	-0.096	0.005
766.4	2.046	2.924	-0.059	0.002
766.5	2.059	3.034	-0.020	0.000
244π	2.069	3.088	-0.000	-0.000
766.6	2.084	3.147	0.022	0.000
766.7	2.125	3.262	0.067	0.001
766.8	2.184	3.381	0.117	0.001
766.9	2.264	3.500	0.172	-0.000
767.0	2.372	3.618	0.234	-0.004
767.1	2.513	3.730	0.304	-0.013
767.2	2.694	3.825	0.381	-0.028
767.3	2.922	3.888	0.467	-0.054
767.4	3.194	3.896	0.560	-0.093
767.5	3.496	3.816	0.660	-0.151
767.6	3.792	3.623	0.762	-0.232
767.7	4.021	3.312	0.860	-0.340
767.8	4.121	2.921	0.945	-0.479
767.9	4.066	2.526	1.006	-0.649

k_0h	ADMITTANCE		RAD.EFF.LENGTH	
768.0	3.882	2.200	1.029	-0.842
768.1	3.627	1.979	1.005	-1.046
489π/2	3.574	1.949	0.994	-1.085
768.2	3.355	1.862	0.929	-1.243
768.3	3.100	1.827	0.806	-1.414
768.4	2.879	1.848	0.651	-1.546
768.5	2.693	1.904	0.480	-1.633
768.6	2.540	1.982	0.309	-1.678
768.7	2.417	2.072	0.148	-1.687
768.8	2.317	2.168	0.005	-1.670
768.9	2.237	2.268	-0.120	-1.636
769.0	2.174	2.369	-0.225	-1.591
769.1	2.125	2.471	-0.314	-1.541
769.2	2.089	2.574	-0.389	-1.489
769.3	2.064	2.678	-0.452	-1.437
769.4	2.050	2.784	-0.507	-1.386
769.5	2.047	2.890	-0.554	-1.337
769.6	2.055	2.999	-0.596	-1.291
245π	2.073	3.100	-0.630	-1.251
769.7	2.075	3.111	-0.634	-1.247
769.8	2.110	3.225	-0.670	-1.205
769.9	2.160	3.342	-0.704	-1.164
770.0	2.232	3.461	-0.738	-1.124
770.1	2.327	3.580	-0.772	-1.083
770.2	2.454	3.695	-0.807	-1.042
770.3	2.618	3.798	-0.843	-0.997
770.4	2.826	3.877	-0.880	-0.949
770.5	3.080	3.910	-0.918	-0.896
770.6	3.372	3.869	-0.956	-0.835
770.7	3.676	3.725	-0.992	-0.764
770.8	3.939	3.461	-1.023	-0.683
770.9	4.097	3.096	-1.045	-0.590
771.0	4.105	2.694	-1.053	-0.485
771.1	3.969	2.334	-1.041	-0.374
771.2	3.736	2.069	-1.006	-0.262
491π/2	3.573	1.961	-0.973	-0.196
771.3	3.467	1.912	-0.948	-0.156
771.4	3.203	1.846	-0.868	-0.067
771.5	2.968	1.847	-0.776	0.001
771.6	2.768	1.890	-0.678	0.046
771.7	2.602	1.960	-0.584	0.070
771.8	2.467	2.045	-0.497	0.077
771.9	2.358	2.139	-0.420	0.074
772.0	2.270	2.238	-0.354	0.064
772.1	2.201	2.338	-0.297	0.051
772.2	2.147	2.440	-0.248	0.038
772.3	2.105	2.543	-0.204	0.027
772.4	2.076	2.647	-0.164	0.017
772.5	2.058	2.751	-0.127	0.009
772.6	2.050	2.857	-0.090	0.004
772.7	2.053	2.965	-0.052	0.001
772.8	2.068	3.075	-0.013	0.000
246π	2.076	3.111	-0.000	0.000
772.9	2.097	3.188	0.029	0.000

k_0h	ADMITTANCE		RAD.EFF.LENGTH	
773.0	2.141	3.304	0.075	0.001
773.1	2.203	3.423	0.126	0.001
773.2	2.288	3.542	0.182	-0.001
773.3	2.401	3.659	0.245	-0.006
773.4	2.549	3.768	0.316	-0.016
773.5	2.738	3.858	0.395	-0.033
773.6	2.973	3.912	0.482	-0.061
773.7	3.251	3.905	0.576	-0.103
773.8	3.554	3.806	0.676	-0.165
773.9	3.840	3.591	0.777	-0.251
774.0	4.048	3.263	0.873	-0.364
774.1	4.120	2.868	0.954	-0.509
774.2	4.039	2.484	1.008	-0.683
774.3	3.839	2.177	1.023	-0.879
493π/2	3.572	1.973	0.986	-1.087
774.4	3.579	1.977	0.988	-1.082
774.5	3.310	1.877	0.902	-1.274
774.6	3.062	1.853	0.773	-1.438
774.7	2.848	1.881	0.614	-1.560
774.8	2.669	1.941	0.442	-1.638
774.9	2.522	2.021	0.273	-1.674
775.0	2.403	2.112	0.117	-1.678
775.1	2.307	2.208	-0.022	-1.657
775.2	2.230	2.308	-0.141	-1.621
775.3	2.170	2.410	-0.243	-1.575
775.4	2.124	2.512	-0.329	-1.525
775.5	2.090	2.615	-0.401	-1.473
775.6	2.067	2.719	-0.462	-1.421
775.7	2.055	2.824	-0.514	-1.371
775.8	2.054	2.931	-0.560	-1.324
775.9	2.064	3.040	-0.601	-1.278
247π	2.079	3.122	-0.629	-1.246
776.0	2.087	3.152	-0.638	-1.235
776.1	2.124	3.267	-0.674	-1.193
776.2	2.178	3.384	-0.708	-1.153
776.3	2.254	3.503	-0.742	-1.113
776.4	2.355	3.621	-0.776	-1.073
776.5	2.487	3.734	-0.811	-1.031
776.6	2.659	3.833	-0.847	-0.987
776.7	2.874	3.904	-0.885	-0.938
776.8	3.135	3.925	-0.923	-0.884
776.9	3.431	3.867	-0.961	-0.821
777.0	3.729	3.702	-0.996	-0.749
777.1	3.976	3.418	-1.026	-0.665
777.2	4.108	3.043	-1.046	-0.570
777.3	4.088	2.646	-1.050	-0.464
777.4	3.931	2.303	-1.035	-0.352
777.5	3.690	2.059	-0.996	-0.241
495π/2	3.571	1.985	-0.971	-0.194
777.6	3.420	1.920	-0.933	-0.138
777.7	3.162	1.868	-0.851	-0.053
777.8	2.934	1.877	-0.757	0.011
777.9	2.741	1.926	-0.660	0.051

OMEGA = 20.0

k_0h	ADMITTANCE		RAD.EFF.LENGTH	
778.0	2.581	1.998	-0.567	0.072
778.1	2.452	2.085	-0.482	0.077
778.2	2.347	2.180	-0.407	0.072
778.3	2.263	2.278	-0.343	0.062
778.4	2.196	2.379	-0.288	0.049
778.5	2.144	2.481	-0.240	0.036
778.6	2.105	2.584	-0.197	0.025
778.7	2.078	2.687	-0.158	0.015
778.8	2.062	2.792	-0.120	0.008
778.9	2.056	2.898	-0.083	0.004
779.0	2.061	3.006	-0.046	0.001
779.1	2.079	3.117	-0.006	0.000
248π	2.083	3.134	-0.000	0.000
779.2	2.110	3.230	0.036	0.000
779.3	2.157	3.346	0.083	0.000
779.4	2.223	3.465	0.135	0.000
779.5	2.313	3.583	0.192	-0.002
779.6	2.432	3.699	0.257	-0.008
779.7	2.587	3.804	0.328	-0.019
779.8	2.784	3.888	0.408	-0.038
779.9	3.026	3.933	0.497	-0.068
780.0	3.309	3.910	0.592	-0.114
780.1	3.611	3.792	0.692	-0.180
780.2	3.885	3.556	0.792	-0.270
780.3	4.070	3.213	0.886	-0.390
780.4	4.113	2.817	0.962	-0.540
780.5	4.008	2.445	1.009	-0.718
780.6	3.795	2.158	1.014	-0.916
497π/2	3.570	1.998	0.978	-1.089
780.7	3.532	1.978	0.969	-1.117
780.8	3.267	1.893	0.874	-1.304
780.9	3.025	1.880	0.738	-1.459
781.0	2.818	1.914	0.576	-1.572
781.1	2.646	1.978	0.405	-1.641
781.2	2.505	2.060	0.239	-1.670
781.3	2.390	2.152	0.087	-1.668
781.4	2.298	2.249	-0.047	-1.644
781.5	2.224	2.349	-0.162	-1.606
781.6	2.167	2.450	-0.260	-1.559
781.7	2.123	2.553	-0.342	-1.508
781.8	2.091	2.656	-0.412	-1.457
781.9	2.070	2.760	-0.470	-1.406
782.0	2.060	2.865	-0.521	-1.357
782.1	2.061	2.972	-0.566	-1.310
782.2	2.073	3.082	-0.606	-1.265
249π	2.086	3.145	-0.627	-1.241
782.3	2.099	3.194	-0.643	-1.223
782.4	2.139	3.309	-0.678	-1.182
782.5	2.197	3.426	-0.712	-1.142
782.6	2.276	3.545	-0.745	-1.103
782.7	2.383	3.662	-0.780	-1.062
782.8	2.522	3.772	-0.815	-1.021
782.9	2.701	3.867	-0.852	-0.976

k_0h	ADMITTANCE		RAD.EFF.LENGTH	
783.0	2.924	3.929	-0.889	-0.927
783.1	3.191	3.938	-0.928	-0.871
783.2	3.488	3.862	-0.965	-0.807
783.3	3.780	3.675	-1.000	-0.733
783.4	4.008	3.373	-1.028	-0.648
783.5	4.113	2.991	-1.046	-0.550
783.6	4.066	2.602	-1.048	-0.443
783.7	3.891	2.276	-1.028	-0.331
783.8	3.643	2.052	-0.984	-0.220
499π/2	3.569	2.010	-0.968	-0.192
783.9	3.375	1.931	-0.918	-0.121
784.0	3.122	1.891	-0.833	-0.039
784.1	2.901	1.908	-0.738	0.020
784.2	2.715	1.962	-0.641	0.056
784.3	2.562	2.037	-0.550	0.073
784.4	2.437	2.125	-0.467	0.076
784.5	2.336	2.220	-0.395	0.070
784.6	2.256	2.319	-0.333	0.059
784.7	2.192	2.420	-0.279	0.047
784.8	2.142	2.522	-0.232	0.034
784.9	2.106	2.624	-0.190	0.023
785.0	2.080	2.728	-0.151	0.014
785.1	2.065	2.833	-0.114	0.007
785.2	2.062	2.939	-0.077	0.003
785.3	2.070	3.047	-0.039	0.001
250π	2.089	3.156	-0.000	0.000
785.4	2.090	3.158	0.001	-0.000
785.5	2.124	3.272	0.044	0.000
785.6	2.174	3.388	0.091	0.000
785.7	2.244	3.506	0.144	-0.000
785.8	2.339	3.624	0.202	-0.003
785.9	2.464	3.738	0.268	-0.010
786.0	2.626	3.840	0.341	-0.022
786.1	2.830	3.918	0.422	-0.043
786.2	3.079	3.951	0.512	-0.076
786.3	3.367	3.913	0.608	-0.125
786.4	3.665	3.774	0.708	-0.195
786.5	3.926	3.518	0.807	-0.291
786.6	4.087	3.163	0.897	-0.416
786.7	4.102	2.768	0.968	-0.572
786.8	3.974	2.410	1.007	-0.754
786.9	3.750	2.144	1.003	-0.952
501π/2	3.569	2.022	0.971	-1.091
787.0	3.485	1.983	0.948	-1.151
787.1	3.224	1.912	0.845	-1.332
787.2	2.990	1.909	0.703	-1.478
787.3	2.790	1.949	0.539	-1.582
787.4	2.624	2.016	0.369	-1.642
787.5	2.488	2.100	0.206	-1.664
787.6	2.378	2.193	0.058	-1.657
787.7	2.289	2.290	-0.072	-1.630
787.8	2.219	2.390	-0.183	-1.590
787.9	2.164	2.491	-0.276	-1.543

k_0h	ADMITTANCE		RAD.EFF.LENGTH	
788.0	2.122	2.593	-0.355	-1.492
788.1	2.092	2.696	-0.422	-1.441
788.2	2.074	2.801	-0.479	-1.391
788.3	2.065	2.906	-0.528	-1.343
788.4	2.068	3.014	-0.572	-1.297
788.5	2.083	3.123	-0.611	-1.253
251π	2.093	3.168	-0.625	-1.236
788.6	2.111	3.236	-0.647	-1.211
788.7	2.154	3.351	-0.682	-1.171
788.8	2.216	3.468	-0.715	-1.131
788.9	2.300	3.586	-0.749	-1.092
789.0	2.412	3.702	-0.784	-1.052
789.1	2.558	3.810	-0.819	-1.010
789.2	2.744	3.899	-0.856	-0.965
789.3	2.975	3.953	-0.894	-0.915
789.4	3.248	3.947	-0.933	-0.858
789.5	3.545	3.853	-0.970	-0.793
789.6	3.828	3.645	-1.004	-0.717
789.7	4.036	3.327	-1.031	-0.629
789.8	4.113	2.940	-1.046	-0.530
789.9	4.040	2.560	-1.044	-0.422
790.0	3.849	2.253	-1.020	-0.309
790.1	3.596	2.050	-0.972	-0.201
503π/2	3.568	2.034	-0.966	-0.190
790.2	3.331	1.945	-0.902	-0.104
790.3	3.084	1.916	-0.815	-0.027
790.4	2.870	1.941	-0.719	0.028
790.5	2.691	1.998	-0.623	0.060
790.6	2.543	2.076	-0.533	0.075
790.7	2.423	2.166	-0.453	0.076
790.8	2.326	2.261	-0.383	0.068
790.9	2.249	2.360	-0.322	0.057
791.0	2.188	2.461	-0.270	0.044
791.1	2.141	2.563	-0.224	0.032
791.2	2.106	2.665	-0.183	0.021
791.3	2.083	2.769	-0.145	0.013
791.4	2.070	2.874	-0.108	0.007
791.5	2.069	2.980	-0.071	0.003
791.6	2.079	3.089	-0.033	0.001
252π	2.096	3.179	-0.000	0.000
791.7	2.101	3.200	0.008	-0.000
791.8	2.138	3.314	0.052	0.000
791.9	2.191	3.430	0.100	0.000
792.0	2.266	3.548	0.153	-0.001
792.1	2.366	3.665	0.213	-0.004
792.2	2.497	3.777	0.280	-0.012
792.3	2.666	3.875	0.354	-0.026
792.4	2.878	3.945	0.436	-0.049
792.5	3.134	3.967	0.527	-0.085
792.6	3.424	3.912	0.623	-0.138
792.7	3.718	3.753	0.723	-0.212
792.8	3.963	3.478	0.821	-0.313
792.9	4.098	3.112	0.908	-0.443

k_0h	ADMITTANCE		RAD.EFF.LENGTH	
793.0	4.086	2.721	0.973	-0.604
793.1	3.938	2.379	1.004	-0.790
793.2	3.704	2.133	0.990	-0.989
505π/2	3.567	2.047	0.963	-1.092
793.3	3.440	1.990	0.925	-1.184
793.4	3.184	1.933	0.814	-1.358
793.5	2.956	1.939	0.667	-1.496
793.6	2.763	1.984	0.502	-1.590
793.7	2.603	2.055	0.334	-1.642
793.8	2.472	2.140	0.174	-1.658
793.9	2.366	2.233	0.030	-1.646
794.0	2.281	2.331	-0.095	-1.616
794.1	2.214	2.431	-0.202	-1.574
794.2	2.162	2.532	-0.292	-1.527
794.3	2.122	2.634	-0.368	-1.476
794.4	2.094	2.737	-0.432	-1.425
794.5	2.077	2.842	-0.487	-1.376
794.6	2.071	2.947	-0.535	-1.329
794.7	2.076	3.055	-0.577	-1.283
794.8	2.093	3.165	-0.615	-1.240
253π	2.099	3.191	-0.624	-1.231
794.9	2.124	3.278	-0.651	-1.199
795.0	2.170	3.393	-0.686	-1.160
795.1	2.236	3.510	-0.719	-1.121
795.2	2.325	3.628	-0.753	-1.082
795.3	2.442	3.742	-0.788	-1.042
795.4	2.595	3.847	-0.824	-1.000
795.5	2.788	3.930	-0.861	-0.954
795.6	3.026	3.974	-0.899	-0.903
795.7	3.304	3.954	-0.937	-0.845
795.8	3.601	3.840	-0.975	-0.778
795.9	3.873	3.612	-1.008	-0.700
796.0	4.059	3.279	-1.033	-0.611
796.1	4.108	2.890	-1.045	-0.510
796.2	4.011	2.522	-1.040	-0.400
796.3	3.806	2.234	-1.011	-0.288
507π/2	3.566	2.059	-0.963	-0.188
796.4	3.549	2.050	-0.959	-0.181
796.5	3.287	1.961	-0.885	-0.088
796.6	3.047	1.943	-0.796	-0.015
796.7	2.840	1.974	-0.700	0.035
796.8	2.667	2.036	-0.606	0.064
796.9	2.525	2.116	-0.518	0.075
797.0	2.410	2.206	-0.439	0.075
797.1	2.317	2.302	-0.371	0.066
797.2	2.243	2.401	-0.312	0.055
797.3	2.184	2.502	-0.262	0.042
797.4	2.140	2.604	-0.217	0.030
797.5	2.107	2.706	-0.176	0.020
797.6	2.086	2.810	-0.138	0.012
797.7	2.075	2.915	-0.102	0.006
797.8	2.076	3.021	-0.065	0.002
797.9	2.088	3.130	-0.026	0.000
254π	2.102	3.202	-0.000	0.000

k_0h	ADMITTANCE		RAD.EFF.LENGTH	
798.0	2.113	3.242	0.015	-0.000
798.1	2.152	3.356	0.059	0.000
798.2	2.210	3.473	0.108	-0.000
798.3	2.288	3.590	0.163	-0.002
798.4	2.393	3.706	0.223	-0.006
798.5	2.531	3.815	0.291	-0.014
798.6	2.707	3.909	0.367	-0.030
798.7	2.926	3.971	0.451	-0.055
798.8	3.189	3.980	0.542	-0.094
798.9	3.481	3.908	0.639	-0.151
799.0	3.769	3.728	0.738	-0.230
799.1	3.996	3.435	0.834	-0.336
799.2	4.105	3.062	0.918	-0.472
799.3	4.066	2.677	0.977	-0.637
799.4	3.899	2.353	0.999	-0.825
799.5	3.658	2.126	0.975	-1.024
509π/2	3.565	2.071	0.955	-1.094
799.6	3.395	2.001	0.901	-1.216
799.7	3.144	1.956	0.782	-1.383
799.8	2.923	1.970	0.631	-1.511
799.9	2.737	2.020	0.465	-1.596
800.0	2.583	2.093	0.299	-1.640
800.1	2.457	2.180	0.143	-1.650
800.2	2.356	2.274	0.004	-1.634
800.3	2.274	2.372	-0.117	-1.602
800.4	2.209	2.472	-0.220	-1.559
800.5	2.159	2.573	-0.307	-1.510
800.6	2.122	2.675	-0.380	-1.460
800.7	2.096	2.778	-0.442	-1.410
800.8	2.081	2.883	-0.495	-1.362
800.9	2.077	2.988	-0.541	-1.315
801.0	2.085	3.096	-0.582	-1.271
255π	2.106	3.214	-0.622	-1.226
801.1	2.104	3.207	-0.620	-1.228
801.2	2.137	3.320	-0.655	-1.188
801.3	2.187	3.435	-0.689	-1.149
801.4	2.257	3.552	-0.723	-1.110
801.5	2.350	3.669	-0.757	-1.071
801.6	2.473	3.782	-0.792	-1.032
801.7	2.633	3.883	-0.828	-0.989
801.8	2.834	3.960	-0.865	-0.943
801.9	3.079	3.993	-0.904	-0.891
802.0	3.361	3.958	-0.942	-0.832
802.1	3.655	3.824	-0.979	-0.763
802.2	3.914	3.576	-1.011	-0.684
802.3	4.076	3.230	-1.034	-0.592
802.4	4.098	2.842	-1.044	-0.489
802.5	3.979	2.487	-1.035	-0.379
802.6	3.762	2.219	-1.002	-0.267
511π/2	3.564	2.084	-0.961	-0.185
802.7	3.504	2.054	-0.946	-0.162
802.8	3.246	1.979	-0.868	-0.073
802.9	3.012	1.972	-0.778	-0.004

k_0h	ADMITTANCE		RAD.EFF.LENGTH	
803.0	2.812	2.009	-0.682	0.042
803.1	2.645	2.074	-0.588	0.067
803.2	2.508	2.155	-0.502	0.076
803.3	2.397	2.247	-0.426	0.073
803.4	2.308	2.343	-0.360	0.064
803.5	2.237	2.442	-0.303	0.052
803.6	2.181	2.543	-0.253	0.040
803.7	2.139	2.645	-0.210	0.028
803.8	2.109	2.747	-0.170	0.018
803.9	2.089	2.851	-0.132	0.011
804.0	2.081	2.956	-0.095	0.005
804.1	2.083	3.063	-0.058	0.002
804.2	2.096	3.172	-0.019	0.000
256π	2.109	3.225	-0.000	-0.000
804.3	2.125	3.284	0.022	-0.000
804.4	2.168	3.398	0.067	-0.000
804.5	2.229	3.515	0.117	-0.000
804.6	2.312	3.632	0.172	-0.002
804.7	2.422	3.747	0.234	-0.007
804.8	2.566	3.853	0.303	-0.017
804.9	2.749	3.941	0.380	-0.034
805.0	2.976	3.995	0.465	-0.062
805.1	3.244	3.991	0.557	-0.104
805.2	3.537	3.900	0.655	-0.164
805.3	3.816	3.700	0.753	-0.248
805.4	4.024	3.390	0.847	-0.359
805.5	4.106	3.012	0.926	-0.501
805.6	4.042	2.636	0.979	-0.670
805.7	3.859	2.330	0.993	-0.861
805.8	3.612	2.123	0.959	-1.059
513π/2	3.564	2.096	0.947	-1.095
805.9	3.351	2.014	0.875	-1.246
806.0	3.106	1.981	0.750	-1.405
806.1	2.892	2.002	0.595	-1.525
806.2	2.712	2.057	0.429	-1.601
806.3	2.564	2.132	0.265	-1.637
806.4	2.443	2.220	0.113	-1.641
806.5	2.345	2.315	-0.022	-1.622
806.6	2.267	2.413	-0.139	-1.587
806.7	2.205	2.513	-0.237	-1.543
806.8	2.158	2.614	-0.321	-1.495
806.9	2.123	2.716	-0.391	-1.445
807.0	2.099	2.819	-0.451	-1.395
807.1	2.086	2.924	-0.502	-1.347
807.2	2.084	3.030	-0.547	-1.302
807.3	2.093	3.138	-0.587	-1.258
257π	2.112	3.237	-0.621	-1.221
807.4	2.115	3.249	-0.624	-1.216
807.5	2.151	3.362	-0.659	-1.177
807.6	2.205	3.477	-0.693	-1.138
807.7	2.278	3.594	-0.727	-1.100
807.8	2.377	3.710	-0.761	-1.061
807.9	2.506	3.821	-0.796	-1.021

k_0h	ADMITTANCE		RAD.EFF.LENGTH	
808.0	2.672	3.918	-0.832	-0.979
808.1	2.881	3.988	-0.870	-0.932
808.2	3.133	4.010	-0.909	-0.879
808.3	3.418	3.958	-0.947	-0.818
808.4	3.707	3.804	-0.983	-0.748
808.5	3.951	3.537	-1.014	-0.666
808.6	4.089	3.181	-1.036	-0.573
808.7	4.083	2.797	-1.042	-0.468
808.8	3.944	2.457	-1.029	-0.357
808.9	3.718	2.208	-0.992	-0.246
515π/2	3.563	2.108	-0.958	-0.183
809.0	3.458	2.061	-0.931	-0.144
809.1	3.205	2.000	-0.851	-0.058
809.2	2.978	2.001	-0.759	0.006
809.3	2.785	2.044	-0.663	0.048
809.4	2.624	2.112	-0.571	0.069
809.5	2.492	2.195	-0.487	0.076
809.6	2.386	2.287	-0.413	0.072
809.7	2.300	2.384	-0.349	0.062
809.8	2.232	2.483	-0.294	0.050
809.9	2.179	2.584	-0.245	0.037
810.0	2.139	2.686	-0.202	0.026
810.1	2.110	2.788	-0.163	0.017
810.2	2.093	2.892	-0.126	0.010
810.3	2.086	2.997	-0.089	0.005
810.4	2.091	3.104	-0.052	0.002
810.5	2.108	3.214	-0.013	0.000
258π	2.115	3.248	-0.000	0.000
810.6	2.138	3.326	0.029	-0.000
810.7	2.184	3.441	0.075	-0.000
810.8	2.248	3.557	0.126	-0.001
810.9	2.336	3.673	0.182	-0.003
811.0	2.452	3.787	0.245	-0.009
811.1	2.602	3.890	0.315	-0.020
811.2	2.793	3.973	0.394	-0.039
811.3	3.027	4.017	0.480	-0.069
811.4	3.300	3.999	0.573	-0.114
811.5	3.592	3.889	0.670	-0.179
811.6	3.860	3.668	0.768	-0.267
811.7	4.047	3.344	0.859	-0.384
811.8	4.102	2.963	0.933	-0.531
811.9	4.014	2.599	0.979	-0.704
812.0	3.817	2.311	0.984	-0.897
517π/2	3.562	2.121	0.939	-1.096
812.1	3.566	2.123	0.940	-1.093
812.2	3.308	2.029	0.848	-1.275
812.3	3.069	2.008	0.716	-1.426
812.4	2.862	2.035	0.559	-1.536
812.5	2.689	2.094	0.393	-1.603
812.6	2.546	2.172	0.232	-1.633
812.7	2.430	2.261	0.084	-1.631
812.8	2.336	2.356	-0.047	-1.609
812.9	2.261	2.454	-0.159	-1.572

k_0h	ADMITTANCE		RAD.EFF.LENGTH	
813.0	2.202	2.554	-0.254	-1.527
813.1	2.156	2.655	-0.334	-1.479
813.2	2.123	2.757	-0.402	-1.429
813.3	2.102	2.860	-0.460	-1.380
813.4	2.091	2.965	-0.509	-1.333
813.5	2.091	3.071	-0.553	-1.288
813.6	2.102	3.180	-0.592	-1.246
259π	2.119	3.260	-0.619	-1.216
813.7	2.127	3.291	-0.629	-1.205
813.8	2.166	3.404	-0.663	-1.166
813.9	2.223	3.520	-0.697	-1.127
814.0	2.300	3.636	-0.731	-1.090
814.1	2.404	3.751	-0.765	-1.051
814.2	2.540	3.859	-0.800	-1.011
814.3	2.713	3.952	-0.837	-0.968
814.4	2.929	4.014	-0.875	-0.920
814.5	3.187	4.024	-0.914	-0.867
814.6	3.474	3.955	-0.952	-0.805
814.7	3.757	3.781	-0.987	-0.732
814.8	3.984	3.496	-1.017	-0.649
814.9	4.096	3.132	-1.036	-0.553
815.0	4.065	2.753	-1.039	-0.447
815.1	3.907	2.430	-1.022	-0.336
815.2	3.673	2.201	-0.981	-0.226
519π/2	3.561	2.133	-0.955	-0.181
815.3	3.414	2.071	-0.916	-0.127
815.4	3.165	2.023	-0.834	-0.045
815.5	2.945	2.032	-0.740	0.015
815.6	2.759	2.080	-0.645	0.053
815.7	2.604	2.151	-0.555	0.071
815.8	2.477	2.235	-0.472	0.075
815.9	2.375	2.328	-0.400	0.070
816.0	2.292	2.425	-0.338	0.060
816.1	2.227	2.524	-0.284	0.048
816.2	2.176	2.625	-0.237	0.035
816.3	2.139	2.727	-0.195	0.024
816.4	2.112	2.829	-0.157	0.015
816.5	2.097	2.933	-0.120	0.009
816.6	2.092	3.039	-0.083	0.004
816.7	2.099	3.146	-0.045	0.001
816.8	2.118	3.256	-0.006	0.000
260π	2.122	3.271	-0.000	0.000
816.9	2.151	3.368	0.037	-0.000
817.0	2.200	3.483	0.083	-0.000
817.1	2.269	3.599	0.135	-0.001
817.2	2.361	3.715	0.192	-0.005
817.3	2.483	3.826	0.256	-0.011
817.4	2.640	3.926	0.328	-0.024
817.5	2.838	4.002	0.407	-0.045
817.6	3.079	4.037	0.494	-0.077
817.7	3.356	4.003	0.588	-0.125
817.8	3.645	3.874	0.685	-0.194
817.9	3.901	3.634	0.782	-0.287

OMEGA = 20.0

k_0h	ADMITTANCE		RAD.EFF.LENGTH	
818.0	4.066	3.297	0.870	-0.410
818.1	4.094	2.917	0.940	-0.561
818.2	3.983	2.565	0.978	-0.739
818.3	3.775	2.295	0.973	-0.932
521π/2	3.560	2.146	0.931	-1.098
818.4	3.521	2.127	0.920	-1.126
818.5	3.266	2.047	0.819	-1.302
818.6	3.034	2.036	0.682	-1.445
818.7	2.834	2.069	0.523	-1.546
818.8	2.666	2.132	0.358	-1.605
818.9	2.529	2.212	0.200	-1.627
819.0	2.417	2.301	0.056	-1.621
819.1	2.327	2.397	-0.070	-1.595
819.2	2.255	2.495	-0.178	-1.557
819.3	2.198	2.595	-0.270	-1.511
819.4	2.156	2.696	-0.347	-1.463
819.5	2.125	2.799	-0.412	-1.414
819.6	2.105	2.902	-0.468	-1.366
819.7	2.096	3.006	-0.516	-1.320
819.8	2.098	3.113	-0.559	-1.275
819.9	2.112	3.222	-0.597	-1.233
261π	2.125	3.283	-0.617	-1.211
820.0	2.139	3.333	-0.633	-1.193
820.1	2.181	3.446	-0.667	-1.155
820.2	2.241	3.562	-0.701	-1.117
820.3	2.323	3.678	-0.735	-1.079
820.4	2.432	3.792	-0.769	-1.041
820.5	2.574	3.897	-0.805	-1.001
820.6	2.755	3.984	-0.842	-0.957
820.7	2.978	4.038	-0.880	-0.909
820.8	3.241	4.035	-0.919	-0.854
820.9	3.529	3.948	-0.957	-0.790
821.0	3.804	3.755	-0.991	-0.716
821.1	4.012	3.454	-1.020	-0.631
821.2	4.098	3.084	-1.036	-0.533
821.3	4.042	2.713	-1.036	-0.426
821.4	3.868	2.407	-1.014	-0.315
821.5	3.628	2.197	-0.969	-0.206
523π/2	3.560	2.158	-0.953	-0.179
821.6	3.370	2.084	-0.901	-0.110
821.7	3.127	2.047	-0.816	-0.032
821.8	2.914	2.064	-0.722	0.024
821.9	2.734	2.116	-0.627	0.057
822.0	2.585	2.190	-0.538	0.073
822.1	2.463	2.276	-0.458	0.075
822.2	2.364	2.369	-0.388	0.068
822.3	2.285	2.466	-0.328	0.058
822.4	2.223	2.566	-0.276	0.045
822.5	2.175	2.666	-0.230	0.033
822.6	2.139	2.768	-0.189	0.023
822.7	2.115	2.871	-0.150	0.014
822.8	2.101	2.975	-0.113	0.008
822.9	2.099	3.080	-0.077	0.003
823.0	2.108	3.188	-0.039	0.001
262π	2.128	3.295	-0.000	0.000
823.1	2.129	3.298	0.001	-0.000
823.2	2.165	3.410	0.044	-0.000
823.3	2.217	3.525	0.091	-0.001
823.4	2.290	3.641	0.144	-0.002
823.5	2.387	3.756	0.202	-0.006
823.6	2.515	3.865	0.268	-0.014
823.7	2.679	3.961	0.340	-0.027
823.8	2.884	4.031	0.421	-0.050
823.9	3.131	4.054	0.509	-0.086
824.0	3.411	4.004	0.603	-0.137
824.1	3.697	3.856	0.701	-0.210
824.2	3.938	3.597	0.796	-0.309
824.3	4.079	3.250	0.881	-0.436
824.4	4.081	2.872	0.944	-0.593
824.5	3.950	2.534	0.974	-0.774
824.6	3.731	2.284	0.961	-0.967
525π/2	3.559	2.171	0.923	-1.099
824.7	3.476	2.133	0.898	-1.158
824.8	3.226	2.068	0.790	-1.327
824.9	3.000	2.065	0.647	-1.461
825.0	2.806	2.104	0.487	-1.553
825.1	2.645	2.170	0.323	-1.604
825.2	2.512	2.252	0.168	-1.620
825.3	2.405	2.342	0.029	-1.610
825.4	2.318	2.438	-0.093	-1.581
825.5	2.250	2.537	-0.197	-1.542
825.6	2.196	2.637	-0.285	-1.496
825.7	2.155	2.738	-0.359	-1.447
825.8	2.126	2.840	-0.422	-1.399
825.9	2.108	2.943	-0.476	-1.351
826.0	2.101	3.048	-0.523	-1.306
826.1	2.106	3.155	-0.564	-1.263
826.2	2.122	3.264	-0.602	-1.221
263π	2.132	3.307	-0.616	-1.206
826.3	2.152	3.375	-0.637	-1.182
826.4	2.197	3.489	-0.671	-1.144
826.5	2.261	3.604	-0.705	-1.107
826.6	2.347	3.720	-0.738	-1.069
826.7	2.462	3.832	-0.773	-1.031
826.8	2.610	3.934	-0.809	-0.991
826.9	2.798	4.016	-0.846	-0.947
827.0	3.028	4.060	-0.885	-0.897
827.1	3.296	4.044	-0.924	-0.841
827.2	3.583	3.939	-0.961	-0.776
827.3	3.848	3.725	-0.995	-0.700
827.4	4.036	3.410	-1.022	-0.612
827.5	4.096	3.037	-1.036	-0.513
827.6	4.016	2.676	-1.032	-0.405
827.7	3.828	2.388	-1.006	-0.293
527π/2	3.558	2.184	-0.950	-0.177
827.8	3.583	2.197	-0.956	-0.187
827.9	3.328	2.099	-0.884	-0.094
828.0	3.091	2.073	-0.797	-0.020
828.1	2.884	2.097	-0.703	0.031
828.2	2.710	2.153	-0.610	0.061
828.3	2.566	2.229	-0.522	0.073
828.4	2.449	2.316	-0.445	0.074
828.5	2.355	2.410	-0.377	0.066
828.6	2.279	2.508	-0.318	0.055
828.7	2.219	2.607	-0.267	0.043
828.8	2.173	2.708	-0.222	0.031
828.9	2.140	2.809	-0.182	0.021
829.0	2.117	2.912	-0.144	0.013
829.1	2.106	3.016	-0.107	0.007
829.2	2.106	3.122	-0.070	0.003
829.3	2.117	3.230	-0.032	0.001
264π	2.135	3.318	-0.000	0.000
829.4	2.141	3.340	0.008	-0.000
829.5	2.179	3.453	0.052	-0.000
829.6	2.235	3.568	0.100	-0.001
829.7	2.312	3.683	0.153	-0.003
829.8	2.414	3.797	0.213	-0.007
829.9	2.548	3.904	0.279	-0.016
830.0	2.719	3.995	0.353	-0.032
830.1	2.931	4.057	0.435	-0.057
830.2	3.184	4.068	0.524	-0.095
830.3	3.466	4.002	0.618	-0.150
830.4	3.746	3.835	0.715	-0.227
830.5	3.971	3.558	0.809	-0.331
830.6	4.087	3.203	0.890	-0.464
830.7	4.063	2.830	0.947	-0.625
830.8	3.914	2.507	0.969	-0.808
830.9	3.687	2.276	0.946	-1.002
529π/2	3.557	2.196	0.915	-1.100
831.0	3.432	2.142	0.874	-1.189
831.1	3.186	2.090	0.759	-1.351
831.2	2.967	2.095	0.612	-1.476
831.3	2.780	2.140	0.451	-1.559
831.4	2.624	2.209	0.290	-1.603
831.5	2.497	2.292	0.138	-1.613
831.6	2.394	2.383	0.003	-1.598
831.7	2.311	2.479	-0.115	-1.567
831.8	2.245	2.578	-0.215	-1.526
831.9	2.193	2.678	-0.299	-1.480
832.0	2.155	2.779	-0.371	-1.432
832.1	2.128	2.881	-0.431	-1.384
832.2	2.112	2.985	-0.483	-1.337
832.3	2.107	3.090	-0.529	-1.293
832.4	2.114	3.196	-0.569	-1.250
832.5	2.132	3.306	-0.606	-1.210
265π	2.138	3.330	-0.614	-1.201
832.6	2.165	3.417	-0.641	-1.171
832.7	2.213	3.531	-0.675	-1.133
832.8	2.281	3.646	-0.708	-1.096
832.9	2.372	3.761	-0.742	-1.059
833.0	2.492	3.872	-0.777	-1.021
833.1	2.647	3.970	-0.813	-0.980
833.2	2.842	4.046	-0.851	-0.936
833.3	3.079	4.080	-0.890	-0.886
833.4	3.351	4.049	-0.928	-0.828
833.5	3.635	3.925	-0.966	-0.761
833.6	3.889	3.693	-0.999	-0.683
833.7	4.055	3.365	-1.024	-0.594
833.8	4.089	2.991	-1.035	-0.492
833.9	3.987	2.642	-1.027	-0.383
834.0	3.786	2.372	-0.997	-0.273
531π/2	3.556	2.209	-0.947	-0.175
834.1	3.538	2.200	-0.943	-0.168
834.2	3.286	2.116	-0.868	-0.078
834.3	3.055	2.100	-0.779	-0.009
834.4	2.855	2.131	-0.685	0.038
834.5	2.687	2.191	-0.592	0.064
834.6	2.549	2.269	-0.507	0.074
834.7	2.436	2.357	-0.431	0.073
834.8	2.345	2.451	-0.365	0.064
834.9	2.273	2.549	-0.308	0.053
835.0	2.216	2.649	-0.259	0.041
835.1	2.172	2.749	-0.215	0.029
835.2	2.141	2.851	-0.175	0.020
835.3	2.120	2.954	-0.137	0.012
835.4	2.111	3.058	-0.101	0.006
835.5	2.113	3.164	-0.064	0.003
835.6	2.126	3.272	-0.026	0.001
266π	2.141	3.342	-0.000	0.000
835.7	2.153	3.382	0.015	-0.000
835.8	2.194	3.495	0.060	-0.001
835.9	2.254	3.610	0.108	-0.001
836.0	2.335	3.725	0.163	-0.004
836.1	2.443	3.838	0.223	-0.009
836.2	2.582	3.942	0.291	-0.019
836.3	2.760	4.028	0.366	-0.036
836.4	2.979	4.082	0.449	-0.063
836.5	3.238	4.080	0.539	-0.104
836.6	3.521	3.997	0.634	-0.163
836.7	3.793	3.810	0.730	-0.245
836.8	4.000	3.518	0.821	-0.354
836.9	4.091	3.156	0.898	-0.492
837.0	4.042	2.790	0.949	-0.657
837.1	3.876	2.484	0.963	-0.843
837.2	3.643	2.272	0.930	-1.036
533π/2	3.555	2.222	0.907	-1.101
837.3	3.389	2.155	0.849	-1.218
837.4	3.148	2.114	0.727	-1.373
837.5	2.935	2.127	0.577	-1.489
837.6	2.755	2.176	0.416	-1.564
837.7	2.605	2.248	0.257	-1.599
837.8	2.482	2.332	0.109	-1.604
837.9	2.383	2.424	-0.022	-1.586
838.0	2.303	2.521	-0.135	-1.553
838.1	2.240	2.619	-0.232	-1.511
838.2	2.191	2.719	-0.313	-1.465
838.3	2.155	2.821	-0.382	-1.417
838.4	2.130	2.923	-0.440	-1.369
838.5	2.116	3.026	-0.491	-1.323
838.6	2.114	3.131	-0.535	-1.280
838.7	2.122	3.238	-0.574	-1.238
267π	2.145	3.354	-0.613	-1.196
838.8	2.143	3.348	-0.611	-1.198
838.9	2.178	3.460	-0.645	-1.160
839.0	2.230	3.574	-0.679	-1.123
839.1	2.302	3.689	-0.712	-1.086
839.2	2.398	3.803	-0.746	-1.049
839.3	2.524	3.911	-0.781	-1.011
839.4	2.685	4.006	-0.818	-0.970
839.5	2.887	4.075	-0.856	-0.925
839.6	3.130	4.098	-0.895	-0.874
839.7	3.405	4.051	-0.933	-0.815
839.8	3.686	3.909	-0.970	-0.746
839.9	3.926	3.658	-1.002	-0.666
840.0	4.069	3.319	-1.025	-0.574
840.1	4.077	2.948	-1.033	-0.472
840.2	3.955	2.612	-1.022	-0.362
840.3	3.743	2.360	-0.987	-0.252
535π/2	3.555	2.234	-0.945	-0.173
840.4	3.494	2.206	-0.929	-0.150
840.5	3.246	2.136	-0.851	-0.064
840.6	3.021	2.129	-0.761	0.001
840.7	2.827	2.166	-0.667	0.044
840.8	2.666	2.229	-0.575	0.067
840.9	2.532	2.309	-0.492	0.074
841.0	2.424	2.398	-0.418	0.071
841.1	2.337	2.493	-0.354	0.062
841.2	2.267	2.591	-0.299	0.051
841.3	2.213	2.690	-0.251	0.039
841.4	2.171	2.791	-0.208	0.027
841.5	2.142	2.892	-0.168	0.018
841.6	2.124	2.995	-0.131	0.011
841.7	2.116	3.099	-0.095	0.006
841.8	2.120	3.206	-0.058	0.002
841.9	2.136	3.314	-0.019	0.000
268π	2.148	3.366	-0.000	0.000
842.0	2.165	3.425	0.022	-0.000
842.1	2.210	3.538	0.067	-0.001
842.2	2.273	3.652	0.117	-0.002
842.3	2.359	3.767	0.172	-0.005
842.4	2.472	3.878	0.234	-0.011
842.5	2.618	3.979	0.303	-0.022
842.6	2.802	4.060	0.379	-0.041
842.7	3.028	4.105	0.463	-0.070
842.8	3.292	4.090	0.553	-0.114
842.9	3.574	3.989	0.649	-0.177
843.0	3.836	3.782	0.744	-0.264
843.1	4.024	3.475	0.833	-0.378
843.2	4.089	3.110	0.905	-0.521
843.3	4.018	2.754	0.949	-0.690
843.4	3.837	2.465	0.954	-0.878
843.5	3.598	2.271	0.912	-1.069
537π/2	3.554	2.247	0.899	-1.101
843.6	3.347	2.169	0.822	-1.246
843.7	3.112	2.139	0.694	-1.393
843.8	2.905	2.160	0.542	-1.500
843.9	2.731	2.213	0.381	-1.566
844.0	2.587	2.287	0.225	-1.595
844.1	2.469	2.373	0.081	-1.595
844.2	2.373	2.466	-0.046	-1.573
844.3	2.297	2.562	-0.155	-1.539
844.4	2.236	2.661	-0.248	-1.496
844.5	2.190	2.761	-0.326	-1.449
844.6	2.155	2.862	-0.392	-1.402
844.7	2.133	2.964	-0.449	-1.355
844.8	2.121	3.068	-0.498	-1.310
844.9	2.120	3.173	-0.540	-1.267
845.0	2.131	3.281	-0.579	-1.226
269π	2.151	3.377	-0.611	-1.191
845.1	2.155	3.390	-0.615	-1.187
845.2	2.193	3.502	-0.649	-1.149
845.3	2.248	3.616	-0.683	-1.112
845.4	2.324	3.731	-0.716	-1.076
845.5	2.425	3.844	-0.750	-1.039
845.6	2.557	3.949	-0.786	-1.001
845.7	2.725	4.040	-0.822	-0.960
845.8	2.934	4.101	-0.860	-0.914
845.9	3.182	4.114	-0.900	-0.862
846.0	3.459	4.051	-0.938	-0.801
846.1	3.735	3.889	-0.975	-0.731
846.2	3.959	3.621	-1.005	-0.649
846.3	4.078	3.273	-1.026	-0.555
846.4	4.061	2.906	-1.031	-0.451
846.5	3.920	2.585	-1.015	-0.341
846.6	3.700	2.352	-0.976	-0.232
539π/2	3.553	2.260	-0.942	-0.171
846.7	3.450	2.215	-0.914	-0.132
846.8	3.207	2.158	-0.834	-0.050
846.9	2.988	2.160	-0.742	0.011
847.0	2.801	2.201	-0.649	0.049
847.1	2.645	2.268	-0.559	0.069
847.2	2.517	2.349	-0.477	0.074
847.3	2.413	2.439	-0.406	0.070
847.4	2.329	2.534	-0.344	0.060
847.5	2.262	2.632	-0.290	0.048
847.6	2.210	2.732	-0.243	0.037
847.7	2.171	2.832	-0.201	0.026
847.8	2.144	2.934	-0.162	0.017
847.9	2.127	3.037	-0.125	0.010
848.0	2.122	3.141	-0.088	0.005
848.1	2.128	3.248	-0.051	0.002
848.2	2.146	3.356	-0.012	0.000
270π	2.154	3.389	-0.000	0.000
848.3	2.178	3.467	0.030	-0.000
848.4	2.226	3.580	0.075	-0.001
848.5	2.293	3.695	0.126	-0.002
848.6	2.383	3.808	0.182	-0.006
848.7	2.502	3.918	0.245	-0.013
848.8	2.654	4.015	0.315	-0.025
848.9	2.846	4.091	0.392	-0.046
849.0	3.078	4.125	0.477	-0.078
849.1	3.346	4.096	0.568	-0.125
849.2	3.626	3.977	0.663	-0.192
849.3	3.877	3.752	0.758	-0.284
849.4	4.044	3.432	0.843	-0.403
849.5	4.084	3.066	0.911	-0.551
849.6	3.990	2.720	0.948	-0.724
849.7	3.797	2.450	0.944	-0.912
541π/2	3.552	2.273	0.891	-1.102
849.8	3.554	2.274	0.892	-1.101
849.9	3.306	2.186	0.795	-1.272
850.0	3.076	2.166	0.661	-1.411
850.1	2.876	2.194	0.507	-1.509
850.2	2.708	2.251	0.347	-1.567
850.3	2.569	2.327	0.193	-1.590
850.4	2.455	2.414	0.054	-1.585
850.5	2.364	2.507	-0.069	-1.560
850.6	2.290	2.604	-0.174	-1.524
850.7	2.233	2.703	-0.263	-1.481
850.8	2.188	2.803	-0.339	-1.434
850.9	2.156	2.904	-0.402	-1.387
851.0	2.136	3.006	-0.457	-1.341
851.1	2.126	3.110	-0.504	-1.296
851.2	2.127	3.215	-0.546	-1.254
851.3	2.140	3.323	-0.584	-1.214
271π	2.157	3.401	-0.610	-1.186
851.4	2.166	3.433	-0.619	-1.175
851.5	2.207	3.545	-0.653	-1.138
851.6	2.266	3.659	-0.687	-1.102
851.7	2.346	3.773	-0.720	-1.066
851.8	2.452	3.884	-0.754	-1.030
851.9	2.590	3.988	-0.790	-0.991
852.0	2.765	4.073	-0.827	-0.949
852.1	2.981	4.127	-0.865	-0.902
852.2	3.235	4.126	-0.905	-0.849
852.3	3.513	4.047	-0.943	-0.787
852.4	3.781	3.866	-0.979	-0.715
852.5	3.988	3.582	-1.008	-0.631
852.6	4.083	3.228	-1.026	-0.535
852.7	4.042	2.868	-1.028	-0.430
852.8	3.884	2.563	-1.008	-0.319
852.9	3.657	2.348	-0.965	-0.212
543π/2	3.552	2.286	-0.939	-0.169
853.0	3.408	2.226	-0.899	-0.115
853.1	3.169	2.181	-0.816	-0.037
853.2	2.957	2.191	-0.724	0.019
853.3	2.776	2.237	-0.631	0.054
853.4	2.625	2.307	-0.543	0.070
853.5	2.502	2.390	-0.463	0.074
853.6	2.402	2.481	-0.394	0.068
853.7	2.321	2.576	-0.333	0.058
853.8	2.257	2.674	-0.281	0.046
853.9	2.208	2.773	-0.235	0.035
854.0	2.171	2.874	-0.194	0.024
854.1	2.146	2.976	-0.155	0.015
854.2	2.132	3.079	-0.119	0.009
854.3	2.128	3.183	-0.082	0.004
854.4	2.136	3.290	-0.045	0.002
854.5	2.157	3.399	-0.005	0.000
272π	2.161	3.413	-0.000	0.000
854.6	2.192	3.510	0.037	-0.001
854.7	2.243	3.623	0.083	-0.001
854.8	2.314	3.737	0.135	-0.003
854.9	2.409	3.850	0.192	-0.007
855.0	2.533	3.957	0.256	-0.015
855.1	2.692	4.051	0.327	-0.029
855.2	2.890	4.119	0.405	-0.052
855.3	3.129	4.144	0.491	-0.086
855.4	3.399	4.099	0.583	-0.137
855.5	3.676	3.962	0.678	-0.208
855.6	3.914	3.719	0.771	-0.304
855.7	4.059	3.388	0.853	-0.429
855.8	4.073	3.023	0.915	-0.581
855.9	3.959	2.691	0.945	-0.757
856.0	3.755	2.438	0.931	-0.946
545π/2	3.551	2.299	0.883	-1.103
856.1	3.511	2.279	0.870	-1.131
856.2	3.266	2.205	0.766	-1.296
856.3	3.042	2.195	0.628	-1.427
856.4	2.848	2.228	0.472	-1.517
856.5	2.686	2.289	0.314	-1.567
856.6	2.552	2.367	0.163	-1.583
856.7	2.443	2.455	0.027	-1.574
856.8	2.355	2.549	-0.091	-1.547
856.9	2.285	2.646	-0.192	-1.509
857.0	2.229	2.744	-0.278	-1.465
857.1	2.187	2.844	-0.351	-1.419
857.2	2.158	2.946	-0.412	-1.372
857.3	2.139	3.048	-0.465	-1.327
857.4	2.131	3.152	-0.511	-1.283
857.5	2.135	3.257	-0.551	-1.242
857.6	2.150	3.365	-0.589	-1.202
273π	2.164	3.425	-0.608	-1.181
857.7	2.179	3.475	-0.624	-1.164
857.8	2.223	3.588	-0.657	-1.128
857.9	2.285	3.701	-0.690	-1.092

OMEGA = 20.0

k_0h	ADMITTANCE		RAD.EFF.LENGTH	
858.0	2.370	3.815	-0.724	-1.056
858.1	2.481	3.925	-0.758	-1.020
858.2	2.625	4.025	-0.794	-0.981
858.3	2.807	4.105	-0.832	-0.939
858.4	3.029	4.150	-0.870	-0.891
858.5	3.288	4.136	-0.910	-0.837
858.6	3.565	4.039	-0.948	-0.773
858.7	3.824	3.840	-0.983	-0.699
858.8	4.013	3.541	-1.010	-0.613
858.9	4.083	3.184	-1.026	-0.515
859.0	4.018	2.832	-1.024	-0.409
859.1	3.846	2.543	-1.000	-0.298
859.2	3.613	2.347	-0.952	-0.192
547π/2	3.550	2.312	-0.936	-0.167
859.3	3.366	2.241	-0.883	-0.099
859.4	3.132	2.206	-0.798	-0.025
859.5	2.926	2.224	-0.706	0.027
859.6	2.752	2.274	-0.613	0.058
859.7	2.607	2.346	-0.527	0.072
859.8	2.488	2.430	-0.450	0.073
859.9	2.392	2.522	-0.382	0.066
860.0	2.314	2.618	-0.323	0.056
860.1	2.253	2.716	-0.272	0.044
860.2	2.206	2.815	-0.228	0.033
860.3	2.171	2.916	-0.187	0.022
860.4	2.148	3.018	-0.149	0.014
860.5	2.136	3.121	-0.113	0.008
860.6	2.135	3.225	-0.076	0.004
860.7	2.145	3.332	-0.038	0.001
274π	2.167	3.437	-0.000	0.000
860.8	2.168	3.441	0.001	-0.000
860.9	2.206	3.553	0.044	-0.001
861.0	2.260	3.666	0.092	-0.002
861.1	2.335	3.779	0.144	-0.004
861.2	2.435	3.891	0.202	-0.009
861.3	2.565	3.996	0.267	-0.018
861.4	2.731	4.086	0.339	-0.033
861.5	2.936	4.147	0.419	-0.058
861.6	3.180	4.160	0.506	-0.095
861.7	3.453	4.100	0.598	-0.149
861.8	3.724	3.943	0.692	-0.225
861.9	3.947	3.683	0.783	-0.326
862.0	4.069	3.344	0.862	-0.455
862.1	4.059	2.983	0.918	-0.612
862.2	3.926	2.664	0.940	-0.791
862.3	3.713	2.429	0.917	-0.979
549π/2	3.549	2.325	0.875	-1.103
862.4	3.468	2.288	0.847	-1.161
862.5	3.227	2.227	0.736	-1.319
862.6	3.009	2.225	0.594	-1.441
862.7	2.822	2.264	0.438	-1.522
862.8	2.665	2.328	0.281	-1.565
862.9	2.536	2.407	0.134	-1.576

k_0h	ADMITTANCE		RAD.EFF.LENGTH	
863.0	2.431	2.496	0.002	-1.563
863.1	2.347	2.590	-0.112	-1.533
863.2	2.279	2.687	-0.210	-1.494
863.3	2.227	2.786	-0.292	-1.450
863.4	2.187	2.886	-0.362	-1.404
863.5	2.159	2.987	-0.421	-1.358
863.6	2.143	3.090	-0.472	-1.313
863.7	2.137	3.194	-0.517	-1.270
863.8	2.142	3.300	-0.557	-1.229
863.9	2.160	3.408	-0.593	-1.191
275π	2.170	3.449	-0.606	-1.176
864.0	2.191	3.518	-0.628	-1.153
864.1	2.239	3.630	-0.661	-1.117
864.2	2.305	3.744	-0.694	-1.082
864.3	2.394	3.857	-0.728	-1.047
864.4	2.511	3.965	-0.763	-1.010
864.5	2.661	4.061	-0.799	-0.971
864.6	2.850	4.136	-0.836	-0.928
864.7	3.078	4.171	-0.875	-0.880
864.8	3.341	4.144	-0.915	-0.824
864.9	3.616	4.029	-0.952	-0.759
865.0	3.865	3.811	-0.986	-0.683
865.1	4.033	3.500	-1.012	-0.594
865.2	4.078	3.141	-1.025	-0.495
865.3	3.992	2.799	-1.019	-0.387
865.4	3.807	2.528	-0.991	-0.278
551π/2	3.548	2.338	-0.934	-0.165
865.5	3.570	2.349	-0.939	-0.174
865.6	3.325	2.257	-0.867	-0.084
865.7	3.097	2.233	-0.780	-0.014
865.8	2.897	2.257	-0.687	0.034
865.9	2.728	2.312	-0.596	0.061
866.0	2.589	2.386	-0.512	0.072
866.1	2.474	2.471	-0.436	0.072
866.2	2.382	2.564	-0.370	0.065
866.3	2.308	2.660	-0.314	0.054
866.4	2.249	2.758	-0.264	0.042
866.5	2.205	2.857	-0.220	0.031
866.6	2.172	2.958	-0.180	0.021
866.7	2.151	3.060	-0.143	0.013
866.8	2.141	3.163	-0.106	0.007
866.9	2.142	3.268	-0.070	0.003
867.0	2.154	3.375	-0.032	0.001
276π	2.173	3.462	-0.000	0.000
867.1	2.180	3.484	0.008	-0.000
867.2	2.220	3.595	0.052	-0.001
867.3	2.278	3.708	0.100	-0.002
867.4	2.357	3.822	0.153	-0.005
867.5	2.462	3.932	0.212	-0.010
867.6	2.598	4.034	0.278	-0.020
867.7	2.770	4.119	0.352	-0.037
867.8	2.982	4.172	0.432	-0.064
867.9	3.232	4.173	0.520	-0.104

k_0h	ADMITTANCE		RAD.EFF.LENGTH	
868.0	3.505	4.097	0.613	-0.162
868.1	3.770	3.922	0.706	-0.242
868.2	3.976	3.646	0.795	-0.348
868.3	4.075	3.301	0.870	-0.483
868.4	4.041	2.945	0.920	-0.644
868.5	3.891	2.641	0.933	-0.825
868.6	3.670	2.425	0.901	-1.012
553π/2	3.548	2.351	0.867	-1.104
868.7	3.426	2.299	0.823	-1.189
868.8	3.189	2.250	0.705	-1.340
868.9	2.977	2.256	0.560	-1.453
869.0	2.796	2.300	0.403	-1.526
869.1	2.645	2.367	0.249	-1.562
869.2	2.521	2.448	0.105	-1.568
869.3	2.420	2.538	-0.022	-1.551
869.4	2.339	2.632	-0.132	-1.520
869.5	2.274	2.729	-0.226	-1.480
869.6	2.224	2.828	-0.306	-1.435
869.7	2.187	2.928	-0.373	-1.389
869.8	2.161	3.029	-0.430	-1.344
869.9	2.146	3.132	-0.479	-1.300
870.0	2.143	3.236	-0.523	-1.257
870.1	2.151	3.342	-0.562	-1.217
870.2	2.171	3.450	-0.598	-1.179
277π	2.177	3.474	-0.605	-1.171
870.3	2.205	3.561	-0.632	-1.143
870.4	2.255	3.673	-0.665	-1.107
870.5	2.325	3.786	-0.698	-1.072
870.6	2.419	3.898	-0.732	-1.037
870.7	2.542	4.004	-0.767	-1.000
870.8	2.698	4.097	-0.803	-0.961
870.9	2.893	4.165	-0.841	-0.917
871.0	3.128	4.190	-0.880	-0.868
871.1	3.394	4.148	-0.920	-0.811
871.2	3.666	4.015	-0.957	-0.744
871.3	3.902	3.780	-0.990	-0.666
871.4	4.049	3.458	-1.014	-0.576
871.5	4.069	3.100	-1.023	-0.474
871.6	3.963	2.769	-1.014	-0.366
871.7	3.766	2.516	-0.981	-0.257
555π/2	3.547	2.364	-0.931	-0.163
871.8	3.527	2.354	-0.925	-0.155
871.9	3.285	2.276	-0.850	-0.069
872.0	3.063	2.262	-0.762	-0.004
872.1	2.869	2.292	-0.669	0.040
872.2	2.706	2.350	-0.579	0.064
872.3	2.572	2.426	-0.497	0.073
872.4	2.462	2.513	-0.423	0.071
872.5	2.373	2.605	-0.359	0.063
872.6	2.302	2.701	-0.304	0.052
872.7	2.246	2.800	-0.256	0.040
872.8	2.204	2.899	-0.213	0.029
872.9	2.173	3.000	-0.174	0.019

k_0h	ADMITTANCE		RAD.EFF.LENGTH	
873.0	2.154	3.102	-0.136	0.012
873.1	2.146	3.205	-0.100	0.007
873.2	2.149	3.310	-0.063	0.003
873.3	2.164	3.417	-0.025	0.001
278π	2.180	3.486	-0.000	0.000
873.4	2.192	3.527	0.016	-0.000
873.5	2.235	3.638	0.060	-0.001
873.6	2.297	3.751	0.108	-0.003
873.7	2.381	3.864	0.162	-0.006
873.8	2.491	3.972	0.223	-0.012
873.9	2.633	4.072	0.290	-0.023
874.0	2.811	4.151	0.364	-0.042
874.1	3.030	4.196	0.446	-0.071
874.2	3.284	4.184	0.534	-0.114
874.3	3.557	4.091	0.627	-0.176
874.4	3.813	3.898	0.720	-0.261
874.5	4.001	3.608	0.806	-0.372
874.6	4.076	3.258	0.877	-0.511
874.7	4.019	2.910	0.920	-0.676
874.8	3.854	2.622	0.924	-0.858
874.9	3.628	2.423	0.884	-1.044
557π/2	3.546	2.377	0.859	-1.104
875.0	3.384	2.313	0.797	-1.216
875.1	3.152	2.275	0.673	-1.359
875.2	2.947	2.289	0.526	-1.464
875.3	2.772	2.337	0.370	-1.529
875.4	2.626	2.406	0.218	-1.558
875.5	2.507	2.489	0.078	-1.558
875.6	2.410	2.579	-0.045	-1.539
875.7	2.332	2.674	-0.152	-1.505
875.8	2.270	2.771	-0.242	-1.465
875.9	2.222	2.870	-0.318	-1.420
876.0	2.187	2.970	-0.383	-1.375
876.1	2.163	3.072	-0.438	-1.330
876.2	2.151	3.174	-0.486	-1.286
876.3	2.149	3.279	-0.528	-1.245
876.4	2.159	3.385	-0.566	-1.206
279π	2.183	3.498	-0.603	-1.167
876.5	2.182	3.493	-0.602	-1.168
876.6	2.219	3.604	-0.636	-1.132
876.7	2.272	3.716	-0.669	-1.097
876.8	2.347	3.829	-0.702	-1.062
876.9	2.445	3.939	-0.736	-1.027
877.0	2.573	4.043	-0.771	-0.990
877.1	2.736	4.132	-0.808	-0.950
877.2	2.938	4.193	-0.846	-0.906
877.3	3.179	4.207	-0.885	-0.856
877.4	3.446	4.149	-0.924	-0.797
877.5	3.713	3.998	-0.961	-0.729
877.6	3.935	3.746	-0.993	-0.649
877.7	4.060	3.416	-1.015	-0.556
877.8	4.056	3.060	-1.021	-0.454
877.9	3.931	2.743	-1.008	-0.345

k_0h	ADMITTANCE		RAD.EFF.LENGTH	
878.0	3.725	2.507	-0.971	-0.237
559π/2	3.545	2.390	-0.928	-0.161
878.1	3.485	2.362	-0.911	-0.138
878.2	3.246	2.297	-0.833	-0.055
878.3	3.030	2.291	-0.744	0.006
878.4	2.842	2.327	-0.651	0.046
878.5	2.685	2.389	-0.563	0.066
878.6	2.556	2.467	-0.482	0.073
878.7	2.450	2.554	-0.411	0.069
878.8	2.365	2.647	-0.349	0.061
878.9	2.296	2.743	-0.295	0.049
879.0	2.243	2.842	-0.248	0.038
879.1	2.203	2.941	-0.206	0.027
879.2	2.175	3.042	-0.167	0.018
879.3	2.158	3.144	-0.130	0.011
879.4	2.151	3.247	-0.094	0.006
879.5	2.156	3.353	-0.057	0.003
879.6	2.174	3.460	-0.019	0.001
280π	2.186	3.510	-0.000	0.000
879.7	2.205	3.570	0.023	-0.001
879.8	2.251	3.681	0.068	-0.002
879.9	2.317	3.794	0.117	-0.003
880.0	2.405	3.906	0.172	-0.007
880.1	2.520	4.013	0.233	-0.014
880.2	2.668	4.108	0.301	-0.027
880.3	2.854	4.182	0.377	-0.047
880.4	3.078	4.218	0.460	-0.079
880.5	3.336	4.192	0.549	-0.125
880.6	3.607	4.082	0.641	-0.191
880.7	3.853	3.871	0.733	-0.280
880.8	4.022	3.568	0.817	-0.396
880.9	4.072	3.216	0.882	-0.540
881.0	3.994	2.878	0.918	-0.708
881.1	3.816	2.606	0.914	-0.891
881.2	3.585	2.425	0.864	-1.075
561π/2	3.545	2.403	0.851	-1.104
881.3	3.343	2.329	0.770	-1.241
881.4	3.117	2.301	0.641	-1.376
881.5	2.918	2.322	0.492	-1.473
881.6	2.749	2.374	0.337	-1.530
881.7	2.608	2.446	0.188	-1.553
881.8	2.493	2.530	0.052	-1.549
881.9	2.400	2.621	-0.068	-1.526
882.0	2.325	2.716	-0.170	-1.491
882.1	2.266	2.814	-0.257	-1.450
882.2	2.221	2.912	-0.330	-1.405
882.3	2.188	3.013	-0.393	-1.360
882.4	2.166	3.114	-0.446	-1.316
882.5	2.155	3.217	-0.493	-1.273
882.6	2.156	3.321	-0.534	-1.233
882.7	2.168	3.428	-0.571	-1.194
281π	2.189	3.523	-0.602	-1.162
882.8	2.193	3.536	-0.606	-1.157
882.9	2.233	3.647	-0.639	-1.122

k_0h	ADMITTANCE		RAD.EFF.LENGTH	
883.0	2.290	3.759	-0.672	-1.087
883.1	2.369	3.871	-0.706	-1.053
883.2	2.472	3.980	-0.740	-1.017
883.3	2.606	4.082	-0.775	-0.980
883.4	2.776	4.166	-0.812	-0.940
883.5	2.984	4.219	-0.851	-0.895
883.6	3.229	4.221	-0.890	-0.844
883.7	3.498	4.148	-0.929	-0.784
883.8	3.759	3.979	-0.966	-0.713
883.9	3.965	3.711	-0.996	-0.631
884.0	4.066	3.373	-1.015	-0.537
884.1	4.039	3.023	-1.018	-0.433
884.2	3.897	2.721	-1.001	-0.324
884.3	3.683	2.502	-0.960	-0.217
563π/2	3.544	2.416	-0.925	-0.159
884.4	3.443	2.373	-0.896	-0.121
884.5	3.209	2.320	-0.815	-0.042
884.6	2.998	2.322	-0.725	0.015
884.7	2.817	2.363	-0.634	0.050
884.8	2.665	2.428	-0.547	0.068
884.9	2.540	2.507	-0.468	0.072
885.0	2.439	2.596	-0.399	0.068
885.1	2.357	2.689	-0.338	0.059
885.2	2.291	2.786	-0.286	0.047
885.3	2.241	2.884	-0.240	0.036
885.4	2.203	2.983	-0.199	0.025
885.5	2.176	3.084	-0.161	0.017
885.6	2.161	3.186	-0.124	0.010
885.7	2.157	3.290	-0.088	0.005
885.8	2.164	3.395	-0.051	0.002
885.9	2.184	3.503	-0.012	0.000
282π	2.192	3.535	-0.000	0.000
886.0	2.218	3.613	0.030	-0.001
886.1	2.268	3.724	0.075	-0.002
886.2	2.337	3.837	0.126	-0.004
886.3	2.429	3.947	0.182	-0.008
886.4	2.550	4.052	0.244	-0.016
886.5	2.704	4.144	0.313	-0.030
886.6	2.897	4.212	0.390	-0.053
886.7	3.127	4.237	0.474	-0.087
886.8	3.388	4.197	0.563	-0.136
886.9	3.656	4.070	0.656	-0.206
887.0	3.890	3.841	0.746	-0.300
887.1	4.038	3.528	0.826	-0.421
887.2	4.064	3.176	0.886	-0.569
887.3	3.966	2.849	0.915	-0.741
887.4	3.777	2.594	0.902	-0.924
565π/2	3.543	2.430	0.843	-1.104
887.5	3.543	2.429	0.843	-1.104
887.6	3.304	2.347	0.742	-1.265
887.7	3.083	2.329	0.609	-1.392
887.8	2.890	2.356	0.458	-1.480
887.9	2.726	2.412	0.304	-1.530

k_0h	ADMITTANCE		RAD.EFF.LENGTH	
888.0	2.591	2.486	0.158	-1.546
888.1	2.481	2.572	0.026	-1.538
888.2	2.391	2.663	-0.089	-1.513
888.3	2.319	2.758	-0.188	-1.477
888.4	2.263	2.856	-0.271	-1.435
888.5	2.220	2.955	-0.342	-1.391
888.6	2.189	3.055	-0.402	-1.346
888.7	2.169	3.156	-0.454	-1.302
888.8	2.160	3.259	-0.499	-1.260
888.9	2.163	3.364	-0.539	-1.221
889.0	2.177	3.470	-0.576	-1.183
283π	2.196	3.547	-0.600	-1.157
889.1	2.205	3.579	-0.610	-1.146
889.2	2.248	3.690	-0.643	-1.111
889.3	2.309	3.802	-0.676	-1.077
889.4	2.391	3.913	-0.710	-1.043
889.5	2.500	4.021	-0.744	-1.008
889.6	2.640	4.119	-0.780	-0.971
889.7	2.816	4.198	-0.817	-0.930
889.8	3.031	4.243	-0.856	-0.884
889.9	3.280	4.232	-0.895	-0.831
890.0	3.549	4.143	-0.934	-0.770
890.1	3.801	3.956	-0.970	-0.697
890.2	3.990	3.674	-0.998	-0.613
890.3	4.068	3.332	-1.015	-0.517
890.4	4.019	2.989	-1.015	-0.412
890.5	3.862	2.701	-0.993	-0.303
890.6	3.642	2.500	-0.948	-0.198
567π/2	3.542	2.443	-0.922	-0.157
890.7	3.402	2.386	-0.881	-0.105
890.8	3.172	2.344	-0.798	-0.030
890.9	2.967	2.355	-0.707	0.023
891.0	2.792	2.400	-0.616	0.055
891.1	2.646	2.467	-0.531	0.069
891.2	2.526	2.548	-0.454	0.072
891.3	2.428	2.638	-0.387	0.066
891.4	2.349	2.731	-0.328	0.056
891.5	2.287	2.828	-0.277	0.045
891.6	2.238	2.926	-0.233	0.034
891.7	2.203	3.026	-0.192	0.024
891.8	2.179	3.127	-0.154	0.016
891.9	2.165	3.229	-0.118	0.009
892.0	2.163	3.333	-0.082	0.005
892.1	2.173	3.438	-0.044	0.002
892.2	2.195	3.546	-0.005	0.000
284π	2.199	3.559	-0.000	0.000
892.3	2.231	3.656	0.037	-0.001
892.4	2.285	3.767	0.084	-0.002
892.5	2.358	3.879	0.135	-0.005
892.6	2.455	3.989	0.192	-0.010
892.7	2.581	4.091	0.255	-0.019
892.8	2.742	4.179	0.325	-0.034
892.9	2.941	4.240	0.403	-0.059

k_0h	ADMITTANCE		RAD.EFF.LENGTH	
893.0	3.177	4.254	0.488	-0.095
893.1	3.440	4.200	0.578	-0.148
893.2	3.703	4.054	0.669	-0.222
893.3	3.923	3.810	0.758	-0.321
893.4	4.050	3.487	0.835	-0.447
893.5	4.053	3.138	0.889	-0.600
893.6	3.936	2.823	0.910	-0.773
893.7	3.737	2.585	0.888	-0.956
569π/2	3.542	2.456	0.835	-1.104
893.8	3.501	2.437	0.821	-1.133
893.9	3.265	2.368	0.713	-1.287
894.0	3.050	2.359	0.576	-1.406
894.1	2.863	2.391	0.424	-1.486
894.2	2.705	2.451	0.272	-1.528
894.3	2.575	2.527	0.130	-1.539
894.4	2.468	2.613	0.002	-1.527
894.5	2.382	2.705	-0.109	-1.500
894.6	2.313	2.801	-0.204	-1.463
894.7	2.260	2.898	-0.285	-1.420
894.8	2.219	2.997	-0.353	-1.376
894.9	2.190	3.097	-0.411	-1.332
895.0	2.172	3.199	-0.461	-1.289
895.1	2.166	3.302	-0.505	-1.248
895.2	2.170	3.407	-0.544	-1.209
895.3	2.187	3.513	-0.580	-1.171
285π	2.202	3.572	-0.599	-1.152
895.4	2.218	3.622	-0.614	-1.136
895.5	2.264	3.733	-0.647	-1.101
895.6	2.328	3.845	-0.680	-1.067
895.7	2.415	3.955	-0.713	-1.033
895.8	2.529	4.061	-0.748	-0.998
895.9	2.675	4.156	-0.784	-0.961
896.0	2.857	4.229	-0.822	-0.919
896.1	3.078	4.265	-0.861	-0.873
896.2	3.332	4.241	-0.900	-0.819
896.3	3.598	4.135	-0.939	-0.755
896.4	3.841	3.931	-0.973	-0.681
896.5	4.011	3.636	-1.000	-0.595
896.6	4.066	3.292	-1.014	-0.497
896.7	3.995	2.957	-1.010	-0.391
896.8	3.825	2.686	-0.984	-0.282
896.9	3.600	2.501	-0.935	-0.179
571π/2	3.541	2.470	-0.919	-0.155
897.0	3.362	2.402	-0.865	-0.089
897.1	3.137	2.371	-0.780	-0.019
897.2	2.938	2.388	-0.689	0.030
897.3	2.769	2.437	-0.599	0.058
897.4	2.628	2.507	-0.516	0.070
897.5	2.512	2.590	-0.441	0.071
897.6	2.418	2.680	-0.375	0.065
897.7	2.342	2.774	-0.319	0.054
897.8	2.283	2.870	-0.269	0.043
897.9	2.237	2.969	-0.225	0.032

OMEGA = 20.0

k_0h	ADMITTANCE		RAD.EFF.LENGTH	
898.0	2.203	3.068	-0.185	0.022
898.1	2.181	3.169	-0.148	0.014
898.2	2.170	3.271	-0.112	0.008
898.3	2.170	3.375	-0.075	0.004
898.4	2.182	3.481	-0.038	0.002
286π	2.205	3.584	-0.000	0.000
898.5	2.206	3.589	0.002	-0.000
898.6	2.246	3.699	0.045	-0.001
898.7	2.302	3.810	0.092	-0.003
898.8	2.380	3.922	0.144	-0.006
898.9	2.482	4.030	0.202	-0.011
899.0	2.614	4.130	0.266	-0.022
899.1	2.781	4.213	0.338	-0.038
899.2	2.986	4.266	0.416	-0.065
899.3	3.227	4.269	0.502	-0.104
899.4	3.491	4.199	0.592	-0.161
899.5	3.748	4.036	0.683	-0.239
899.6	3.953	3.776	0.769	-0.343
899.7	4.058	3.447	0.842	-0.473
899.8	4.037	3.102	0.890	-0.630
899.9	3.903	2.801	0.903	-0.806
900.0	3.696	2.580	0.873	-0.988
573π/2	3.540	2.483	0.827	-1.104
900.1	3.460	2.448	0.797	-1.160
900.2	3.228	2.390	0.683	-1.307
900.3	3.018	2.390	0.543	-1.418
900.4	2.837	2.427	0.391	-1.489
900.5	2.685	2.490	0.241	-1.525
900.6	2.559	2.568	0.102	-1.531
900.7	2.457	2.655	-0.022	-1.516
900.8	2.374	2.747	-0.129	-1.486
900.9	2.308	2.843	-0.220	-1.448
901.0	2.257	2.941	-0.298	-1.406
901.1	2.218	3.040	-0.364	-1.362
901.2	2.192	3.140	-0.420	-1.318
901.3	2.176	3.241	-0.468	-1.276
901.4	2.172	3.345	-0.511	-1.235
901.5	2.178	3.450	-0.549	-1.197
901.6	2.198	3.557	-0.584	-1.160
287π	2.208	3.597	-0.597	-1.147
901.7	2.231	3.666	-0.618	-1.125
901.8	2.280	3.776	-0.651	-1.091
901.9	2.348	3.888	-0.684	-1.058
902.0	2.440	3.997	-0.717	-1.024
902.1	2.559	4.101	-0.752	-0.989
902.2	2.711	4.192	-0.789	-0.951
902.3	2.900	4.259	-0.827	-0.909
902.4	3.126	4.285	-0.866	-0.861
902.5	3.383	4.248	-0.905	-0.806
902.6	3.646	4.125	-0.943	-0.741
902.7	3.878	3.903	-0.977	-0.664
902.8	4.028	3.598	-1.002	-0.576
902.9	4.059	3.253	-1.013	-0.477

k_0h	ADMITTANCE		RAD.EFF.LENGTH	
903.0	3.969	2.929	-1.005	-0.370
903.1	3.786	2.674	-0.975	-0.262
575π/2	3.539	2.496	-0.916	-0.153
903.2	3.558	2.506	-0.921	-0.161
903.3	3.322	2.420	-0.848	-0.075
903.4	3.103	2.398	-0.762	-0.008
903.5	2.910	2.422	-0.671	0.036
903.6	2.746	2.475	-0.583	0.061
903.7	2.611	2.548	-0.501	0.071
903.8	2.499	2.631	-0.428	0.070
903.9	2.409	2.722	-0.364	0.063
904.0	2.336	2.816	-0.309	0.052
904.1	2.279	2.913	-0.261	0.041
904.2	2.235	3.011	-0.218	0.030
904.3	2.204	3.111	-0.179	0.021
904.4	2.184	3.212	-0.142	0.013
904.5	2.175	3.314	-0.105	0.008
904.6	2.177	3.418	-0.069	0.004
904.7	2.191	3.524	-0.031	0.001
288π	2.211	3.609	-0.000	0.000
904.8	2.218	3.632	0.009	-0.000
904.9	2.261	3.742	0.052	-0.002
905.0	2.321	3.854	0.100	-0.003
905.1	2.402	3.964	0.153	-0.007
905.2	2.509	4.071	0.212	-0.013
905.3	2.647	4.168	0.277	-0.025
905.4	2.820	4.246	0.350	-0.043
905.5	3.032	4.291	0.430	-0.072
905.6	3.277	4.282	0.516	-0.114
905.7	3.541	4.196	0.606	-0.175
905.8	3.790	4.015	0.696	-0.257
905.9	3.979	3.741	0.780	-0.365
906.0	4.061	3.407	0.848	-0.501
906.1	4.018	3.068	0.890	-0.661
906.2	3.869	2.782	0.895	-0.838
906.3	3.655	2.578	0.856	-1.019
577π/2	3.539	2.510	0.819	-1.104
906.4	3.419	2.461	0.772	-1.186
906.5	3.191	2.415	0.652	-1.325
906.6	2.987	2.422	0.510	-1.428
906.7	2.812	2.464	0.359	-1.492
906.8	2.666	2.530	0.211	-1.521
906.9	2.545	2.609	0.076	-1.522
907.0	2.446	2.697	-0.044	-1.504
907.1	2.367	2.790	-0.148	-1.473
907.2	2.303	2.886	-0.236	-1.434
907.3	2.255	2.983	-0.310	-1.391
907.4	2.218	3.082	-0.374	-1.348
907.5	2.194	3.183	-0.428	-1.304
907.6	2.180	3.284	-0.475	-1.263
907.7	2.178	3.387	-0.516	-1.223
907.8	2.187	3.493	-0.554	-1.185
907.9	2.208	3.600	-0.589	-1.149
289π	2.214	3.622	-0.595	-1.142

k_0h	ADMITTANCE		RAD.EFF.LENGTH	
908.0	2.244	3.709	-0.622	-1.115
908.1	2.297	3.820	-0.655	-1.081
908.2	2.369	3.930	-0.688	-1.048
908.3	2.465	4.039	-0.721	-1.014
908.4	2.590	4.141	-0.756	-0.979
908.5	2.748	4.228	-0.793	-0.941
908.6	2.943	4.288	-0.831	-0.898
908.7	3.175	4.303	-0.871	-0.849
908.8	3.433	4.251	-0.910	-0.793
908.9	3.693	4.111	-0.948	-0.726
909.0	3.911	3.874	-0.980	-0.647
909.1	4.041	3.559	-1.003	-0.557
909.2	4.049	3.216	-1.011	-0.456
909.3	3.940	2.903	-0.999	-0.349
909.4	3.747	2.665	-0.965	-0.242
579π/2	3.538	2.523	-0.913	-0.151
909.5	3.517	2.513	-0.907	-0.143
909.6	3.284	2.440	-0.831	-0.061
909.7	3.070	2.428	-0.744	0.001
909.8	2.883	2.457	-0.654	0.042
909.9	2.725	2.514	-0.566	0.064
910.0	2.594	2.588	-0.486	0.071
910.1	2.487	2.673	-0.415	0.069
910.2	2.400	2.764	-0.354	0.061
910.3	2.330	2.859	-0.300	0.050
910.4	2.276	2.956	-0.253	0.039
910.5	2.235	3.054	-0.211	0.028
910.6	2.205	3.154	-0.172	0.019
910.7	2.187	3.255	-0.135	0.012
910.8	2.180	3.357	-0.099	0.007
910.9	2.184	3.461	-0.063	0.003
911.0	2.201	3.568	-0.025	0.001
290π	2.218	3.634	-0.000	0.000
911.1	2.231	3.676	0.016	-0.001
911.2	2.276	3.786	0.060	-0.002
911.3	2.340	3.897	0.108	-0.004
911.4	2.425	4.006	0.162	-0.008
911.5	2.538	4.111	0.222	-0.015
911.6	2.682	4.205	0.289	-0.028
911.7	2.861	4.278	0.362	-0.048
911.8	3.078	4.314	0.443	-0.079
911.9	3.327	4.292	0.530	-0.125
912.0	3.589	4.190	0.620	-0.189
912.1	3.830	3.992	0.709	-0.276
912.2	4.000	3.705	0.790	-0.389
912.3	4.060	3.368	0.854	-0.529
912.4	3.996	3.037	0.889	-0.692
912.5	3.833	2.766	0.885	-0.870
912.6	3.614	2.579	0.837	-1.048
581π/2	3.537	2.537	0.812	-1.104
912.7	3.379	2.476	0.746	-1.210
912.8	3.156	2.441	0.621	-1.342
912.9	2.958	2.455	0.477	-1.436

k_0h	ADMITTANCE		RAD.EFF.LENGTH	
913.0	2.789	2.502	0.327	-1.493
913.1	2.647	2.570	0.182	-1.516
913.2	2.531	2.651	0.050	-1.513
913.3	2.436	2.739	-0.066	-1.492
913.4	2.360	2.832	-0.166	-1.459
913.5	2.299	2.928	-0.250	-1.419
913.6	2.253	3.026	-0.322	-1.377
913.7	2.219	3.125	-0.383	-1.334
913.8	2.196	3.225	-0.435	-1.291
913.9	2.185	3.327	-0.481	-1.250
914.0	2.184	3.431	-0.521	-1.211
914.1	2.195	3.536	-0.558	-1.174
291π	2.221	3.647	-0.594	-1.138
914.2	2.220	3.643	-0.593	-1.139
914.3	2.258	3.752	-0.626	-1.105
914.4	2.314	3.863	-0.659	-1.072
914.5	2.390	3.973	-0.692	-1.039
914.6	2.491	4.080	-0.725	-1.005
914.7	2.621	4.179	-0.761	-0.969
914.8	2.786	4.262	-0.798	-0.931
914.9	2.987	4.315	-0.836	-0.887
915.0	3.224	4.319	-0.876	-0.837
915.1	3.483	4.252	-0.915	-0.779
915.2	3.737	4.094	-0.952	-0.710
915.3	3.941	3.842	-0.983	-0.630
915.4	4.049	3.520	-1.004	-0.538
915.5	4.035	3.181	-1.008	-0.436
915.6	3.908	2.881	-0.993	-0.328
915.7	3.708	2.659	-0.954	-0.222
583π/2	3.536	2.550	-0.910	-0.149
915.8	3.476	2.523	-0.893	-0.126
915.9	3.247	2.462	-0.814	-0.048
916.0	3.038	2.458	-0.726	0.010
916.1	2.857	2.493	-0.636	0.047
916.2	2.704	2.553	-0.550	0.066
916.3	2.578	2.629	-0.472	0.071
916.4	2.475	2.715	-0.403	0.068
916.5	2.392	2.807	-0.343	0.059
916.6	2.325	2.901	-0.291	0.048
916.7	2.273	2.998	-0.245	0.037
916.8	2.234	3.097	-0.204	0.027
916.9	2.207	3.196	-0.166	0.018
917.0	2.191	3.298	-0.129	0.011
917.1	2.186	3.400	-0.093	0.006
917.2	2.192	3.505	-0.056	0.003
917.3	2.211	3.611	-0.018	0.001
292π	2.224	3.660	-0.000	0.000
917.4	2.243	3.719	0.023	-0.001
917.5	2.292	3.829	0.068	-0.002
917.6	2.359	3.940	0.117	-0.005
917.7	2.450	4.048	0.172	-0.009
917.8	2.567	4.151	0.232	-0.018
917.9	2.717	4.242	0.300	-0.031

k_0h	ADMITTANCE		RAD.EFF.LENGTH	
918.0	2.903	4.308	0.375	-0.054
918.1	3.126	4.334	0.456	-0.087
918.2	3.377	4.299	0.543	-0.136
918.3	3.637	4.180	0.633	-0.204
918.4	3.866	3.966	0.721	-0.295
918.5	4.018	3.668	0.799	-0.413
918.6	4.054	3.330	0.857	-0.557
918.7	3.971	3.009	0.885	-0.724
918.8	3.796	2.754	0.873	-0.902
918.9	3.573	2.583	0.817	-1.077
585π/2	3.536	2.564	0.804	-1.103
919.0	3.340	2.494	0.719	-1.233
919.1	3.122	2.468	0.590	-1.357
919.2	2.930	2.489	0.444	-1.443
919.3	2.766	2.540	0.295	-1.492
919.4	2.630	2.610	0.154	-1.510
919.5	2.518	2.692	0.025	-1.503
919.6	2.426	2.781	-0.087	-1.479
919.7	2.353	2.875	-0.183	-1.445
919.8	2.295	2.971	-0.264	-1.405
919.9	2.251	3.069	-0.333	-1.362
920.0	2.219	3.168	-0.392	-1.320
920.1	2.199	3.268	-0.443	-1.278
920.2	2.189	3.370	-0.487	-1.238
920.3	2.191	3.474	-0.527	-1.199
920.4	2.205	3.579	-0.563	-1.163
293π	2.227	3.672	-0.592	-1.133
920.5	2.231	3.687	-0.597	-1.128
920.6	2.273	3.796	-0.630	-1.095
920.7	2.332	3.906	-0.662	-1.062
920.8	2.413	4.016	-0.695	-1.029
920.9	2.518	4.121	-0.730	-0.995
921.0	2.654	4.218	-0.765	-0.960
921.1	2.825	4.295	-0.803	-0.920
921.2	3.033	4.340	-0.841	-0.876
921.3	3.273	4.332	-0.881	-0.825
921.4	3.533	4.250	-0.920	-0.765
921.5	3.779	4.075	-0.956	-0.695
921.6	3.967	3.809	-0.986	-0.612
921.7	4.053	3.482	-1.004	-0.518
921.8	4.017	3.148	-1.005	-0.415
921.9	3.875	2.862	-0.985	-0.307
922.0	3.667	2.657	-0.942	-0.203
587π/2	3.535	2.578	-0.907	-0.148
922.1	3.436	2.536	-0.878	-0.110
922.2	3.210	2.486	-0.797	-0.035
922.3	3.007	2.490	-0.708	0.018
922.4	2.832	2.530	-0.619	0.051
922.5	2.685	2.593	-0.535	0.067
922.6	2.563	2.671	-0.459	0.071
922.7	2.464	2.757	-0.392	0.066
922.8	2.384	2.849	-0.333	0.057
922.9	2.320	2.944	-0.282	0.046

k_0h	ADMITTANCE		RAD.EFF.LENGTH	
923.0	2.271	3.041	-0.237	0.035
923.1	2.234	3.140	-0.197	0.025
923.2	2.209	3.239	-0.159	0.017
923.3	2.195	3.341	-0.123	0.010
923.4	2.192	3.443	-0.087	0.006
923.5	2.200	3.548	-0.050	0.003
923.6	2.222	3.655	-0.011	0.000
294π	2.230	3.685	-0.000	0.000
923.7	2.257	3.763	0.030	-0.001
923.8	2.309	3.873	0.076	-0.003
923.9	2.380	3.983	0.126	-0.006
924.0	2.475	4.090	0.181	-0.011
924.1	2.598	4.191	0.243	-0.020
924.2	2.753	4.277	0.312	-0.035
924.3	2.946	4.337	0.388	-0.059
924.4	3.173	4.353	0.470	-0.095
924.5	3.427	4.304	0.557	-0.147
924.6	3.683	4.168	0.647	-0.219
924.7	3.900	3.938	0.732	-0.315
924.8	4.031	3.631	0.807	-0.438
924.9	4.045	3.294	0.860	-0.587
925.0	3.943	2.984	0.880	-0.755
925.1	3.757	2.745	0.860	-0.933
589π/2	3.534	2.591	0.796	-1.103
925.2	3.532	2.590	0.795	-1.105
925.3	3.302	2.514	0.690	-1.254
925.4	3.089	2.497	0.558	-1.370
925.5	2.903	2.524	0.412	-1.449
925.6	2.744	2.578	0.264	-1.491
925.7	2.613	2.651	0.126	-1.502
925.8	2.505	2.734	0.002	-1.492
925.9	2.417	2.824	-0.107	-1.466
926.0	2.347	2.918	-0.199	-1.431
926.1	2.292	3.014	-0.278	-1.391
926.2	2.250	3.112	-0.344	-1.348
926.3	2.220	3.211	-0.401	-1.306
926.4	2.202	3.311	-0.450	-1.265
926.5	2.194	3.413	-0.493	-1.225
926.6	2.198	3.517	-0.532	-1.188
926.7	2.214	3.623	-0.567	-1.152
295π	2.233	3.698	-0.591	-1.128
926.8	2.243	3.730	-0.601	-1.118
926.9	2.288	3.840	-0.634	-1.085
927.0	2.351	3.949	-0.666	-1.052
927.1	2.436	4.058	-0.699	-1.020
927.2	2.547	4.162	-0.734	-0.986
927.3	2.688	4.255	-0.770	-0.950
927.4	2.865	4.327	-0.807	-0.910
927.5	3.078	4.363	-0.846	-0.865
927.6	3.323	4.343	-0.886	-0.813
927.7	3.581	4.245	-0.925	-0.751
927.8	3.818	4.053	-0.960	-0.679
927.9	3.989	3.775	-0.988	-0.594

k_0h	ADMITTANCE		RAD.EFF.LENGTH	
928.0	4.053	3.444	-1.003	-0.498
928.1	3.996	3.118	-1.001	-0.394
928.2	3.840	2.847	-0.977	-0.287
928.3	3.627	2.658	-0.930	-0.184
591π/2	3.534	2.605	-0.904	-0.146
928.4	3.396	2.551	-0.862	-0.095
928.5	3.175	2.512	-0.780	-0.024
928.6	2.978	2.523	-0.691	0.026
928.7	2.808	2.567	-0.602	0.055
928.8	2.666	2.633	-0.520	0.068
928.9	2.549	2.712	-0.445	0.070
929.0	2.454	2.800	-0.380	0.064
929.1	2.377	2.892	-0.323	0.055
929.2	2.316	2.987	-0.274	0.044
929.3	2.269	3.084	-0.230	0.033
929.4	2.234	3.183	-0.190	0.024
929.5	2.211	3.283	-0.153	0.016
929.6	2.199	3.384	-0.117	0.010
929.7	2.198	3.487	-0.081	0.005
929.8	2.209	3.591	-0.044	0.002
929.9	2.233	3.698	-0.005	0.000
296π	2.236	3.710	-0.000	0.000
930.0	2.271	3.807	0.037	-0.001
930.1	2.326	3.916	0.084	-0.003
930.2	2.401	4.026	0.134	-0.007
930.3	2.501	4.132	0.191	-0.012
930.4	2.629	4.230	0.254	-0.023
930.5	2.791	4.312	0.324	-0.040
930.6	2.989	4.364	0.400	-0.066
930.7	3.222	4.369	0.483	-0.104
930.8	3.477	4.305	0.571	-0.160
930.9	3.727	4.153	0.660	-0.236
931.0	3.930	3.908	0.743	-0.337
931.1	4.040	3.594	0.814	-0.464
931.2	4.032	3.260	0.861	-0.616
931.3	3.913	2.963	0.874	-0.787
931.4	3.719	2.739	0.845	-0.964
593π/2	3.533	2.619	0.788	-1.102
931.5	3.492	2.600	0.772	-1.131
931.6	3.265	2.536	0.661	-1.274
931.7	3.057	2.528	0.526	-1.382
931.8	2.877	2.560	0.379	-1.452
931.9	2.724	2.618	0.234	-1.488
932.0	2.597	2.692	0.099	-1.495
932.1	2.493	2.776	-0.021	-1.481
932.2	2.409	2.867	-0.126	-1.453
932.3	2.342	2.961	-0.215	-1.417
932.4	2.289	3.057	-0.290	-1.376
932.5	2.249	3.155	-0.355	-1.334
932.6	2.222	3.254	-0.409	-1.292
932.7	2.205	3.355	-0.457	-1.252
932.8	2.200	3.457	-0.499	-1.213
932.9	2.206	3.561	-0.537	-1.176

k_0h	ADMITTANCE		RAD.EFF.LENGTH	
933.0	2.224	3.666	-0.572	-1.141
297π	2.239	3.723	-0.589	-1.123
933.1	2.256	3.774	-0.605	-1.108
933.2	2.304	3.883	-0.637	-1.075
933.3	2.371	3.993	-0.670	-1.043
933.4	2.460	4.100	-0.703	-1.010
933.5	2.576	4.202	-0.738	-0.977
933.6	2.723	4.292	-0.774	-0.940
933.7	2.906	4.358	-0.812	-0.900
933.8	3.125	4.384	-0.851	-0.854
933.9	3.372	4.351	-0.891	-0.800
934.0	3.628	4.237	-0.930	-0.737
934.1	3.855	4.029	-0.964	-0.662
934.2	4.007	3.739	-0.990	-0.576
934.3	4.048	3.408	-1.002	-0.478
934.4	3.972	3.090	-0.996	-0.373
934.5	3.804	2.835	-0.968	-0.266
934.6	3.587	2.662	-0.917	-0.166
595π/2	3.532	2.633	-0.901	-0.144
934.7	3.358	2.569	-0.846	-0.080
934.8	3.141	2.540	-0.762	-0.013
934.9	2.949	2.557	-0.673	0.032
935.0	2.786	2.605	-0.586	0.058
935.1	2.649	2.674	-0.505	0.069
935.2	2.536	2.754	-0.432	0.069
935.3	2.444	2.842	-0.369	0.063
935.4	2.370	2.935	-0.314	0.053
935.5	2.312	3.030	-0.266	0.042
935.6	2.267	3.127	-0.223	0.031
935.7	2.234	3.226	-0.184	0.022
935.8	2.213	3.326	-0.147	0.015
935.9	2.204	3.427	-0.111	0.009
936.0	2.205	3.530	-0.075	0.005
936.1	2.218	3.635	-0.037	0.002
298π	2.242	3.736	-0.000	0.000
936.2	2.244	3.742	0.002	-0.000
936.3	2.285	3.850	0.045	-0.002
936.4	2.344	3.960	0.092	-0.004
936.5	2.423	4.068	0.143	-0.008
936.6	2.528	4.173	0.201	-0.014
936.7	2.662	4.268	0.265	-0.026
936.8	2.829	4.345	0.336	-0.044
936.9	3.033	4.390	0.413	-0.072
937.0	3.270	4.383	0.497	-0.114
937.1	3.525	4.304	0.585	-0.173
937.2	3.768	4.135	0.672	-0.253
937.3	3.956	3.877	0.754	-0.359
937.4	4.045	3.557	0.820	-0.490
937.5	4.015	3.228	0.861	-0.646
937.6	3.881	2.944	0.866	-0.818
937.7	3.679	2.737	0.828	-0.993
597π/2	3.531	2.647	0.780	-1.102
937.8	3.452	2.613	0.747	-1.156
937.9	3.229	2.559	0.632	-1.291

OMEGA = 20.0

k_0h	ADMITTANCE		RAD.EFF.LENGTH	
938.0	3.027	2.559	0.494	-1.392
938.1	2.852	2.596	0.348	-1.455
938.2	2.704	2.657	0.205	-1.484
938.3	2.582	2.734	0.074	-1.486
938.4	2.482	2.819	-0.043	-1.469
938.5	2.401	2.910	-0.144	-1.440
938.6	2.337	3.004	-0.230	-1.403
938.7	2.286	3.100	-0.303	-1.362
938.8	2.249	3.198	-0.364	-1.320
938.9	2.224	3.297	-0.417	-1.279
939.0	2.209	3.398	-0.463	-1.239
939.1	2.206	3.500	-0.504	-1.201
939.2	2.214	3.604	-0.541	-1.165
939.3	2.235	3.710	-0.576	-1.131
299π	2.246	3.749	-0.588	-1.118
939.4	2.269	3.818	-0.609	-1.098
939.5	2.321	3.927	-0.641	-1.065
939.6	2.391	4.036	-0.674	-1.034
939.7	2.485	4.142	-0.707	-1.001
939.8	2.606	4.242	-0.742	-0.967
939.9	2.759	4.327	-0.779	-0.930
940.0	2.948	4.387	-0.817	-0.889
940.1	3.172	4.404	-0.857	-0.842
940.2	3.421	4.357	-0.896	-0.787
940.3	3.673	4.226	-0.934	-0.722
940.4	3.888	4.003	-0.967	-0.646
940.5	4.021	3.704	-0.991	-0.557
940.6	4.040	3.373	-1.000	-0.458
940.7	3.946	3.066	-0.990	-0.352
940.8	3.767	2.826	-0.958	-0.246
599π/2	3.531	2.661	-0.898	-0.142
940.9	3.547	2.669	-0.903	-0.148
941.0	3.320	2.588	-0.829	-0.066
941.1	3.108	2.568	-0.744	-0.003
941.2	2.922	2.592	-0.655	0.038
941.3	2.764	2.644	-0.569	0.061
941.4	2.632	2.715	-0.490	0.070
941.5	2.523	2.796	-0.420	0.068
941.6	2.435	2.885	-0.358	0.061
941.7	2.364	2.978	-0.305	0.051
941.8	2.308	3.073	-0.258	0.040
941.9	2.266	3.171	-0.216	0.030
942.0	2.235	3.269	-0.177	0.021
942.1	2.216	3.369	-0.140	0.014
942.2	2.209	3.471	-0.105	0.008
942.3	2.212	3.574	-0.068	0.004
942.4	2.227	3.679	-0.031	0.001
300π	2.249	3.762	-0.000	0.000
942.5	2.256	3.786	0.009	-0.000
942.6	2.300	3.894	0.052	-0.002
942.7	2.362	4.003	0.100	-0.005
942.8	2.446	4.111	0.152	-0.009
942.9	2.555	4.214	0.211	-0.016

k_0h	ADMITTANCE		RAD.EFF.LENGTH	
943.0	2.695	4.306	0.276	-0.029
943.1	2.869	4.377	0.348	-0.049
943.2	3.079	4.414	0.426	-0.080
943.3	3.319	4.395	0.510	-0.124
943.4	3.573	4.301	0.598	-0.187
943.5	3.807	4.115	0.684	-0.271
943.6	3.978	3.844	0.763	-0.381
943.7	4.046	3.521	0.825	-0.518
943.8	3.996	3.199	0.859	-0.676
943.9	3.847	2.929	0.856	-0.849
944.0	3.640	2.738	0.810	-1.022
601π/2	3.530	2.675	0.772	-1.101
944.1	3.413	2.627	0.722	-1.179
944.2	3.194	2.585	0.602	-1.308
944.3	2.997	2.592	0.462	-1.400
944.4	2.828	2.634	0.317	-1.455
944.5	2.685	2.698	0.177	-1.479
944.6	2.568	2.775	0.049	-1.477
944.7	2.471	2.861	-0.064	-1.457
944.8	2.394	2.953	-0.161	-1.426
944.9	2.332	3.047	-0.244	-1.389
945.0	2.284	3.144	-0.314	-1.348
945.1	2.249	3.242	-0.374	-1.307
945.2	2.226	3.341	-0.425	-1.266
945.3	2.213	3.442	-0.470	-1.227
945.4	2.212	3.544	-0.510	-1.189
945.5	2.222	3.648	-0.546	-1.154
945.6	2.246	3.754	-0.580	-1.120
301π	2.252	3.775	-0.586	-1.114
945.7	2.283	3.862	-0.612	-1.088
945.8	2.338	3.971	-0.645	-1.056
945.9	2.412	4.079	-0.678	-1.024
946.0	2.510	4.184	-0.711	-0.992
946.1	2.637	4.281	-0.746	-0.958
946.2	2.796	4.362	-0.783	-0.920
946.3	2.991	4.415	-0.822	-0.879
946.4	3.219	4.421	-0.862	-0.830
946.5	3.470	4.360	-0.901	-0.774
946.6	3.716	4.213	-0.938	-0.707
946.7	3.918	3.975	-0.970	-0.628
946.8	4.031	3.668	-0.992	-0.538
946.9	4.028	3.340	-0.998	-0.438
947.0	3.917	3.045	-0.984	-0.331
947.1	3.729	2.820	-0.947	-0.227
603π/2	3.529	2.688	-0.895	-0.140
947.2	3.507	2.678	-0.889	-0.131
947.3	3.283	2.610	-0.813	-0.053
947.4	3.077	2.599	-0.727	0.006
947.5	2.896	2.628	-0.638	0.043
947.6	2.743	2.583	-0.554	0.063
947.7	2.616	2.756	-0.476	0.070
947.8	2.511	2.839	-0.408	0.067
947.9	2.426	2.928	-0.348	0.059

k_0h	ADMITTANCE		RAD.EFF.LENGTH	
948.0	2.358	3.021	-0.296	0.049
948.1	2.305	3.117	-0.250	0.038
948.2	2.265	3.214	-0.209	0.028
948.3	2.237	3.313	-0.170	0.020
948.4	2.220	3.413	-0.134	0.013
948.5	2.214	3.514	-0.098	0.007
948.6	2.219	3.618	-0.062	0.004
948.7	2.237	3.723	-0.024	0.001
302π	2.255	3.788	-0.000	0.000
948.8	2.269	3.830	0.016	-0.001
948.9	2.316	3.938	0.060	-0.003
949.0	2.382	4.047	0.108	-0.005
949.1	2.470	4.153	0.162	-0.010
949.2	2.584	4.254	0.221	-0.019
949.3	2.729	4.343	0.287	-0.033
949.4	2.909	4.408	0.360	-0.054
949.5	3.124	4.435	0.439	-0.087
949.6	3.367	4.404	0.524	-0.135
949.7	3.619	4.294	0.611	-0.201
949.8	3.843	4.093	0.696	-0.290
949.9	3.997	3.811	0.772	-0.405
950.0	4.042	3.486	0.829	-0.545
950.1	3.973	3.172	0.856	-0.707
950.2	3.812	2.917	0.844	-0.880
950.3	3.600	2.741	0.790	-1.050
605π/2	3.529	2.703	0.764	-1.100
950.4	3.375	2.645	0.696	-1.201
950.5	3.160	2.612	0.571	-1.322
950.6	2.969	2.626	0.430	-1.406
950.7	2.805	2.672	0.286	-1.455
950.8	2.667	2.738	0.149	-1.473
950.9	2.554	2.817	0.025	-1.467
951.0	2.461	2.904	-0.084	-1.445
951.1	2.387	2.996	-0.178	-1.413
951.2	2.328	3.091	-0.258	-1.375
951.3	2.283	3.187	-0.325	-1.334
951.4	2.250	3.285	-0.383	-1.293
951.5	2.228	3.384	-0.432	-1.253
951.6	2.218	3.485	-0.476	-1.215
951.7	2.219	3.588	-0.515	-1.178
951.8	2.231	3.692	-0.550	-1.143
303π	2.258	3.801	-0.585	-1.109
951.9	2.257	3.798	-0.584	-1.110
952.0	2.298	3.906	-0.616	-1.078
952.1	2.355	4.014	-0.649	-1.046
952.2	2.434	4.122	-0.681	-1.015
952.3	2.537	4.226	-0.715	-0.983
952.4	2.669	4.320	-0.751	-0.948
952.5	2.834	4.396	-0.788	-0.910
952.6	3.034	4.441	-0.827	-0.868
952.7	3.267	4.436	-0.867	-0.818
952.8	3.518	4.360	-0.906	-0.760
952.9	3.757	4.197	-0.943	-0.692

k_0h	ADMITTANCE		RAD.EFF.LENGTH	
953.0	3.945	3.945	-0.973	-0.611
953.1	4.037	3.633	-0.992	-0.519
953.2	4.013	3.309	-0.995	-0.417
953.3	3.886	3.026	-0.977	-0.311
953.4	3.691	2.818	-0.936	-0.208
607π/2	3.528	2.717	-0.892	-0.138
953.5	3.468	2.690	-0.874	-0.115
953.6	3.247	2.633	-0.796	-0.040
953.7	3.046	2.630	-0.709	0.014
953.8	2.871	2.664	-0.621	0.048
953.9	2.723	2.723	-0.538	0.065
954.0	2.600	2.797	-0.463	0.069
954.1	2.500	2.881	-0.396	0.066
954.2	2.418	2.971	-0.338	0.057
954.3	2.353	3.065	-0.287	0.047
954.4	2.302	3.160	-0.242	0.036
954.5	2.264	3.258	-0.202	0.027
954.6	2.238	3.356	-0.164	0.018
954.7	2.223	3.457	-0.128	0.012
954.8	2.220	3.558	-0.092	0.007
954.9	2.227	3.662	-0.056	0.003
955.0	2.248	3.767	-0.018	0.001
304π	2.261	3.814	-0.000	0.000
955.1	2.282	3.874	0.023	-0.001
955.2	2.332	3.982	0.068	-0.003
955.3	2.402	4.090	0.117	-0.006
955.4	2.494	4.196	0.171	-0.012
955.5	2.614	4.294	0.231	-0.021
955.6	2.765	4.379	0.298	-0.036
955.7	2.951	4.438	0.372	-0.060
955.8	3.170	4.455	0.452	-0.096
955.9	3.416	4.411	0.537	-0.146
956.0	3.663	4.285	0.624	-0.217
956.1	3.877	4.068	0.707	-0.310
956.2	4.011	3.777	0.780	-0.429
956.3	4.035	3.452	0.831	-0.573
956.4	3.948	3.148	0.851	-0.737
956.5	3.776	2.908	0.831	-0.910
956.6	3.561	2.748	0.769	-1.076
609π/2	3.527	2.731	0.756	-1.100
956.7	3.337	2.664	0.668	-1.221
956.8	3.127	2.641	0.540	-1.335
956.9	2.942	2.661	0.399	-1.412
957.0	2.783	2.711	0.257	-1.453
957.1	2.650	2.779	0.122	-1.466
957.2	2.541	2.860	0.002	-1.457
957.3	2.452	2.947	-0.104	-1.433
957.4	2.381	3.039	-0.194	-1.399
957.5	2.324	3.134	-0.270	-1.361
957.6	2.281	3.231	-0.336	-1.320
957.7	2.250	3.329	-0.391	-1.280
957.8	2.231	3.428	-0.439	-1.240
957.9	2.223	3.529	-0.482	-1.203

k_0h	ADMITTANCE		RAD.EFF.LENGTH	
958.0	2.226	3.632	-0.520	-1.167
958.1	2.241	3.736	-0.555	-1.132
305π	2.264	3.827	-0.583	-1.104
958.2	2.269	3.842	-0.588	-1.100
958.3	2.312	3.950	-0.620	-1.068
958.4	2.374	4.058	-0.652	-1.037
958.5	2.456	4.165	-0.685	-1.006
958.6	2.564	4.267	-0.719	-0.973
958.7	2.702	4.358	-0.755	-0.939
958.8	2.873	4.429	-0.793	-0.900
958.9	3.079	4.465	-0.832	-0.857
959.0	3.315	4.448	-0.872	-0.806
959.1	3.564	4.357	-0.911	-0.746
959.2	3.796	4.178	-0.947	-0.676
959.3	3.967	3.914	-0.975	-0.593
959.4	4.038	3.598	-0.992	-0.499
959.5	3.995	3.280	-0.991	-0.396
959.6	3.853	3.011	-0.969	-0.290
959.7	3.652	2.818	-0.924	-0.189
611π/2	3.527	2.745	-0.889	-0.136
959.8	3.429	2.705	-0.858	-0.100
959.9	3.212	2.659	-0.778	-0.029
960.0	3.016	2.663	-0.691	0.021
960.1	2.847	2.702	-0.605	0.052
960.2	2.704	2.763	-0.523	0.066
960.3	2.586	2.839	-0.449	0.069
960.4	2.489	2.924	-0.385	0.064
960.5	2.411	3.015	-0.328	0.055
960.6	2.348	3.108	-0.279	0.045
960.7	2.300	3.204	-0.235	0.035
960.8	2.264	3.301	-0.195	0.025
960.9	2.240	3.400	-0.158	0.017
961.0	2.227	3.500	-0.122	0.011
961.1	2.226	3.602	-0.086	0.006
961.2	2.236	3.706	-0.050	0.003
961.3	2.258	3.811	-0.011	0.001
306π	2.267	3.840	-0.000	0.000
961.4	2.296	3.918	0.030	-0.001
961.5	2.349	4.026	0.076	-0.004
961.6	2.423	4.134	0.125	-0.007
961.7	2.519	4.238	0.181	-0.013
961.8	2.644	4.334	0.242	-0.024
961.9	2.801	4.414	0.310	-0.041
962.0	2.993	4.466	0.385	-0.066
962.1	3.217	4.473	0.465	-0.104
962.2	3.463	4.415	0.550	-0.158
962.3	3.706	4.273	0.636	-0.233
962.4	3.907	4.042	0.718	-0.331
962.5	4.022	3.743	0.786	-0.454
962.6	4.025	3.420	0.832	-0.602
962.7	3.920	3.127	0.845	-0.768
962.8	3.739	2.902	0.817	-0.939
613π/2	3.526	2.759	0.748	-1.099
962.9	3.522	2.757	0.747	-1.101

k_0h	ADMITTANCE		RAD.EFF.LENGTH	
963.0	3.301	2.685	0.640	-1.240
963.1	3.096	2.671	0.510	-1.346
963.2	2.915	2.697	0.368	-1.415
963.3	2.762	2.750	0.228	-1.451
963.4	2.634	2.821	0.097	-1.458
963.5	2.529	2.902	-0.020	-1.446
963.6	2.443	2.991	-0.122	-1.420
963.7	2.375	3.083	-0.209	-1.386
963.8	2.321	3.178	-0.283	-1.347
963.9	2.280	3.274	-0.346	-1.307
964.0	2.252	3.373	-0.399	-1.267
964.1	2.234	3.472	-0.446	-1.228
964.2	2.228	3.573	-0.487	-1.191
964.3	2.233	3.676	-0.524	-1.155
964.4	2.250	3.780	-0.559	-1.122
307π	2.270	3.854	-0.582	-1.099
964.5	2.281	3.887	-0.592	-1.090
964.6	2.328	3.994	-0.624	-1.058
964.7	2.393	4.102	-0.656	-1.028
964.8	2.480	4.208	-0.689	-0.997
964.9	2.592	4.308	-0.724	-0.964
965.0	2.735	4.395	-0.760	-0.929
965.1	2.912	4.460	-0.798	-0.890
965.2	3.124	4.488	-0.837	-0.846
965.3	3.363	4.458	-0.877	-0.794
965.4	3.610	4.352	-0.915	-0.732
965.5	3.832	4.157	-0.950	-0.660
965.6	3.986	3.883	-0.977	-0.575
965.7	4.036	3.564	-0.991	-0.479
965.8	3.973	3.254	-0.986	-0.376
965.9	3.819	2.999	-0.960	-0.270
966.0	3.613	2.822	-0.911	-0.171
615π/2	3.525	2.773	-0.886	-0.135
966.1	3.391	2.722	-0.843	-0.085
966.2	3.179	2.686	-0.761	-0.018
966.3	2.988	2.697	-0.674	0.028
966.4	2.824	2.740	-0.588	0.055
966.5	2.686	2.804	-0.508	0.067
966.6	2.572	2.882	-0.436	0.068
966.7	2.479	2.967	-0.373	0.063
966.8	2.404	3.058	-0.319	0.054
966.9	2.344	3.152	-0.270	0.043
967.0	2.298	3.248	-0.227	0.033
967.1	2.265	3.345	-0.188	0.024
967.2	2.243	3.444	-0.151	0.016
967.3	2.232	3.544	-0.116	0.010
967.4	2.232	3.646	-0.080	0.006
967.5	2.245	3.750	-0.043	0.002
967.6	2.270	3.856	-0.004	0.000
308π	2.273	3.867	-0.000	0.000
967.7	2.310	3.962	0.038	-0.002
967.8	2.367	4.070	0.083	-0.004
967.9	2.444	4.177	0.134	-0.008

k_0h	ADMITTANCE		RAD.EFF.LENGTH	
968.0	2.546	4.279	0.190	-0.015
968.1	2.676	4.373	0.253	-0.027
968.2	2.838	4.448	0.321	-0.045
968.3	3.036	4.493	0.397	-0.073
968.4	3.264	4.489	0.478	-0.114
968.5	3.510	4.416	0.564	-0.171
968.6	3.747	4.258	0.649	-0.249
968.7	3.933	4.014	0.728	-0.352
968.8	4.028	3.709	0.792	-0.480
968.9	4.011	3.390	0.832	-0.631
969.0	3.890	3.109	0.837	-0.798
969.1	3.702	2.899	0.800	-0.968
617π/2	3.524	2.788	0.740	-1.097
969.2	3.483	2.769	0.723	-1.125
969.3	3.265	2.709	0.612	-1.257
969.4	3.065	2.702	0.479	-1.355
969.5	2.890	2.733	0.337	-1.417
969.6	2.742	2.790	0.199	-1.447
969.7	2.619	2.863	0.072	-1.450
969.8	2.517	2.945	-0.042	-1.435
969.9	2.435	3.034	-0.140	-1.407
970.0	2.369	3.127	-0.224	-1.372
970.1	2.318	3.222	-0.295	-1.333
970.2	2.280	3.318	-0.355	-1.293
970.3	2.253	3.416	-0.407	-1.254
970.4	2.238	3.516	-0.452	-1.216
970.5	2.233	3.617	-0.492	-1.179
970.6	2.241	3.720	-0.529	-1.144
970.7	2.261	3.825	-0.563	-1.111
309π	2.276	3.880	-0.580	-1.095
970.8	2.294	3.931	-0.595	-1.080
970.9	2.344	4.038	-0.628	-1.049
971.0	2.413	4.146	-0.660	-1.018
971.1	2.504	4.250	-0.693	-0.987
971.2	2.622	4.348	-0.728	-0.955
971.3	2.770	4.432	-0.764	-0.919
971.4	2.953	4.490	-0.802	-0.880
971.5	3.169	4.508	-0.842	-0.834
971.6	3.410	4.466	-0.882	-0.781
971.7	3.654	4.344	-0.920	-0.718
971.8	3.865	4.135	-0.954	-0.643
971.9	4.001	3.851	-0.978	-0.556
972.0	4.030	3.532	-0.989	-0.459
972.1	3.949	3.231	-0.981	-0.355
972.2	3.784	2.990	-0.950	-0.250
972.3	3.574	2.828	-0.898	-0.153
619π/2	3.524	2.802	-0.883	-0.133
972.4	3.354	2.741	-0.827	-0.071
972.5	3.146	2.714	-0.744	-0.008
972.6	2.961	2.731	-0.657	0.034
972.7	2.802	2.779	-0.572	0.058
972.8	2.669	2.845	-0.494	0.068
972.9	2.559	2.924	-0.424	0.067

k_0h	ADMITTANCE		RAD.EFF.LENGTH	
973.0	2.469	3.011	-0.363	0.061
973.1	2.397	3.102	-0.309	0.052
973.2	2.340	3.196	-0.262	0.041
973.3	2.297	3.292	-0.220	0.031
973.4	2.265	3.389	-0.182	0.022
973.5	2.245	3.488	-0.145	0.015
973.6	2.237	3.589	-0.110	0.009
973.7	2.239	3.691	-0.074	0.005
973.8	2.254	3.794	-0.037	0.002
310π	2.279	3.893	-0.000	0.000
973.9	2.282	3.900	0.003	-0.000
974.0	2.324	4.007	0.045	-0.002
974.1	2.385	4.114	0.092	-0.005
974.2	2.466	4.220	0.143	-0.010
974.3	2.573	4.321	0.200	-0.017
974.4	2.708	4.411	0.263	-0.030
974.5	2.877	4.481	0.333	-0.050
974.6	3.079	4.518	0.410	-0.080
974.7	3.311	4.502	0.491	-0.123
974.8	3.556	4.415	0.576	-0.185
974.9	3.785	4.242	0.660	-0.267
975.0	3.956	3.985	0.737	-0.374
975.1	4.031	3.676	0.797	-0.506
975.2	3.993	3.363	0.830	-0.660
975.3	3.859	3.095	0.827	-0.828
975.4	3.664	2.900	0.783	-0.995
621π/2	3.523	2.816	0.733	-1.096
975.5	3.445	2.783	0.698	-1.148
975.6	3.230	2.734	0.583	-1.273
975.7	3.035	2.735	0.448	-1.363
975.8	2.866	2.771	0.307	-1.418
975.9	2.723	2.830	0.172	-1.442
976.0	2.604	2.905	0.048	-1.441
976.1	2.506	2.988	-0.062	-1.423
976.2	2.427	3.078	-0.157	-1.394
976.3	2.364	3.170	-0.237	-1.358
976.4	2.316	3.266	-0.306	-1.320
976.5	2.279	3.362	-0.364	-1.280
976.6	2.255	3.461	-0.415	-1.241
976.7	2.242	3.560	-0.459	-1.204
976.8	2.239	3.662	-0.498	-1.168
976.9	2.249	3.765	-0.533	-1.134
977.0	2.271	3.869	-0.567	-1.101
311π	2.282	3.907	-0.578	-1.090
977.1	2.308	3.976	-0.599	-1.070
977.2	2.361	4.083	-0.631	-1.040
977.3	2.433	4.189	-0.664	-1.009
977.4	2.529	4.292	-0.697	-0.978
977.5	2.652	4.388	-0.732	-0.945
977.6	2.806	4.467	-0.769	-0.910
977.7	2.994	4.519	-0.807	-0.869
977.8	3.215	4.527	-0.847	-0.823
977.9	3.457	4.471	-0.887	-0.768

k_0h	ADMITTANCE		RAD.EFF.LENGTH	
978.0	3.696	4.334	-0.924	-0.703
978.1	3.895	4.110	-0.957	-0.626
978.2	4.012	3.818	-0.979	-0.538
978.3	4.021	3.501	-0.987	-0.439
978.4	3.923	3.211	-0.975	-0.335
978.5	3.748	2.985	-0.940	-0.231
623π/2	3.522	2.831	-0.880	-0.131
978.6	3.536	2.837	-0.884	-0.136
978.7	3.318	2.762	-0.810	-0.058
978.8	3.114	2.744	-0.726	0.001
978.9	2.934	2.767	-0.640	0.039
979.0	2.781	2.818	-0.556	0.060
979.1	2.652	2.887	-0.480	0.068
979.2	2.546	2.967	-0.412	0.066
979.3	2.460	3.054	-0.352	0.059
979.4	2.391	3.146	-0.300	0.050
979.5	2.337	3.240	-0.254	0.039
979.6	2.295	3.336	-0.213	0.029
979.7	2.266	3.433	-0.175	0.021
979.8	2.248	3.532	-0.139	0.014
979.9	2.242	3.633	-0.104	0.008
980.0	2.246	3.735	-0.068	0.005
980.1	2.263	3.839	-0.030	0.002
312π	2.285	3.920	-0.000	0.000
980.2	2.294	3.945	0.009	-0.000
980.3	2.340	4.051	0.052	-0.003
980.4	2.404	4.158	0.100	-0.006
980.5	2.489	4.263	0.152	-0.011
980.6	2.601	4.362	0.210	-0.020
980.7	2.742	4.449	0.274	-0.033
980.8	2.916	4.513	0.345	-0.055
980.9	3.123	4.541	0.422	-0.087
981.0	3.358	4.514	0.504	-0.134
981.1	3.601	4.411	0.589	-0.199
981.2	3.821	4.223	0.672	-0.285
981.3	3.976	3.955	0.745	-0.397
981.4	4.030	3.643	0.800	-0.533
981.5	3.973	3.337	0.827	-0.689
981.6	3.826	3.083	0.816	-0.857
981.7	3.626	2.903	0.764	-1.022
625π/2	3.522	2.845	0.725	-1.095
981.8	3.407	2.800	0.673	-1.169
981.9	3.197	2.761	0.553	-1.287
982.0	3.007	2.768	0.417	-1.370
982.1	2.843	2.809	0.278	-1.418
982.2	2.705	2.871	0.145	-1.436
982.3	2.590	2.947	0.024	-1.432
982.4	2.496	3.032	-0.082	-1.411
982.5	2.420	3.122	-0.173	-1.381
982.6	2.360	3.215	-0.251	-1.345
982.7	2.314	3.310	-0.317	-1.306
982.8	2.280	3.407	-0.373	-1.267
982.9	2.257	3.505	-0.422	-1.229

k_0h	ADMITTANCE		RAD.EFF.LENGTH	
983.0	2.246	3.605	-0.464	-1.192
983.1	2.246	3.706	-0.503	-1.156
983.2	2.258	3.809	-0.538	-1.123
983.3	2.282	3.914	-0.571	-1.091
313π	2.289	3.934	-0.577	-1.085
983.4	2.322	4.020	-0.603	-1.060
983.5	2.378	4.127	-0.635	-1.030
983.6	2.454	4.233	-0.668	-1.000
983.7	2.555	4.334	-0.701	-0.969
983.8	2.683	4.427	-0.737	-0.936
983.9	2.843	4.502	-0.774	-0.900
984.0	3.037	4.546	-0.812	-0.859
984.1	3.261	4.543	-0.852	-0.811
984.2	3.503	4.474	-0.892	-0.754
984.3	3.737	4.321	-0.929	-0.688
984.4	3.922	4.084	-0.959	-0.609
984.5	4.020	3.786	-0.979	-0.519
984.6	4.008	3.472	-0.984	-0.419
984.7	3.894	3.193	-0.968	-0.314
984.8	3.712	2.982	-0.929	-0.212
627π/2	3.521	2.860	-0.876	-0.129
984.9	3.498	2.849	-0.869	-0.120
985.0	3.282	2.785	-0.793	-0.045
985.1	3.083	2.775	-0.709	0.009
985.2	2.909	2.804	-0.623	0.044
985.3	2.761	2.858	-0.541	0.062
985.4	2.637	2.929	-0.466	0.068
985.5	2.535	3.010	-0.400	0.065
985.6	2.452	3.098	-0.342	0.058
985.7	2.386	3.190	-0.292	0.048
985.8	2.334	3.284	-0.247	0.038
985.9	2.295	3.380	-0.206	0.028
986.0	2.268	3.478	-0.169	0.020
986.1	2.252	3.577	-0.133	0.013
986.2	2.247	3.677	-0.098	0.008
986.3	2.254	3.780	-0.061	0.004
986.4	2.273	3.884	-0.024	0.001
314π	2.292	3.947	-0.000	0.000
986.5	2.307	3.989	0.016	-0.001
986.6	2.356	4.096	0.060	-0.003
986.7	2.423	4.202	0.108	-0.007
986.8	2.513	4.306	0.161	-0.012
986.9	2.629	4.402	0.220	-0.022
987.0	2.776	4.485	0.285	-0.037
987.1	2.956	4.544	0.357	-0.061
987.2	3.168	4.562	0.435	-0.095
987.3	3.405	4.522	0.517	-0.145
987.4	3.645	4.405	0.601	-0.214
987.5	3.854	4.201	0.682	-0.305
987.6	3.991	3.925	0.752	-0.420
987.7	4.025	3.612	0.802	-0.560
987.8	3.950	3.315	0.822	-0.719
987.9	3.792	3.074	0.803	-0.886

k_0h	ADMITTANCE		RAD.EFF.LENGTH	
988.0	3.587	2.910	0.743	-1.047
629π/2	3.520	2.874	0.717	-1.094
988.1	3.370	2.819	0.647	-1.189
988.2	3.164	2.789	0.523	-1.299
988.3	2.979	2.803	0.387	-1.374
988.4	2.821	2.848	0.249	-1.416
988.5	2.687	2.913	0.119	-1.429
988.6	2.577	2.990	0.002	-1.422
988.7	2.486	3.075	-0.101	-1.399
988.8	2.414	3.166	-0.188	-1.368
988.9	2.356	3.259	-0.263	-1.331
989.0	2.312	3.354	-0.327	-1.293
989.1	2.280	3.451	-0.381	-1.254
989.2	2.260	3.549	-0.429	-1.216
989.3	2.251	3.649	-0.470	-1.180
989.4	2.253	3.751	-0.508	-1.145
989.5	2.267	3.854	-0.542	-1.113
315π	2.295	3.961	-0.575	-1.081
989.6	2.294	3.959	-0.575	-1.081
989.7	2.336	4.065	-0.607	-1.051
989.8	2.396	4.171	-0.639	-1.021
989.9	2.476	4.276	-0.671	-0.991
990.0	2.581	4.376	-0.705	-0.960
990.1	2.715	4.465	-0.741	-0.927
990.2	2.880	4.535	-0.778	-0.890
990.3	3.080	4.572	-0.817	-0.848
990.4	3.307	4.558	-0.857	-0.799
990.5	3.549	4.474	-0.896	-0.741
990.6	3.775	4.306	-0.933	-0.672
990.7	3.945	4.057	-0.962	-0.591
990.8	4.023	3.754	-0.979	-0.499
990.9	3.992	3.446	-0.980	-0.398
991.0	3.864	3.179	-0.960	-0.294
991.1	3.675	2.983	-0.917	-0.193
631π/2	3.520	2.889	-0.873	-0.128
991.2	3.460	2.863	-0.854	-0.105
991.3	3.248	2.810	-0.777	-0.034
991.4	3.054	2.808	-0.692	0.017
991.5	2.885	2.841	-0.606	0.048
991.6	2.741	2.899	-0.526	0.064
991.7	2.622	2.971	-0.453	0.068
991.8	2.524	3.054	-0.389	0.064
991.9	2.444	3.142	-0.332	0.056
992.0	2.380	3.234	-0.283	0.046
992.1	2.331	3.328	-0.239	0.036
992.2	2.294	3.425	-0.200	0.026
992.3	2.269	3.522	-0.162	0.018
992.4	2.256	3.621	-0.127	0.012
992.5	2.253	3.722	-0.091	0.007
992.6	2.262	3.825	-0.055	0.004
992.7	2.284	3.929	-0.017	0.001
316π	2.298	3.974	-0.000	0.000
992.8	2.320	4.034	0.023	-0.001
992.9	2.372	4.140	0.068	-0.004

k_0h	ADMITTANCE		RAD.EFF.LENGTH	
993.0	2.444	4.246	0.116	-0.008
993.1	2.538	4.348	0.170	-0.014
993.2	2.659	4.442	0.230	-0.025
993.3	2.811	4.521	0.296	-0.041
993.4	2.996	4.573	0.369	-0.067
993.5	3.213	4.582	0.447	-0.104
993.6	3.451	4.528	0.530	-0.157
993.7	3.687	4.396	0.614	-0.229
993.8	3.884	4.179	0.692	-0.325
993.9	4.003	3.894	0.759	-0.445
994.0	4.016	3.583	0.803	-0.588
994.1	3.925	3.295	0.816	-0.748
994.2	3.757	3.069	0.789	-0.914
994.3	3.549	2.919	0.722	-1.072
633π/2	3.519	2.903	0.709	-1.092
994.4	3.334	2.840	0.619	-1.207
994.5	3.132	2.819	0.493	-1.310
994.6	2.953	2.839	0.357	-1.378
994.7	2.799	2.888	0.221	-1.413
994.8	2.671	2.955	0.094	-1.422
994.9	2.564	3.033	-0.020	-1.411
995.0	2.477	3.119	-0.119	-1.387
995.1	2.407	3.210	-0.203	-1.354
995.2	2.352	3.303	-0.275	-1.318
995.3	2.311	3.398	-0.337	-1.279
995.4	2.281	3.495	-0.389	-1.241
995.5	2.263	3.594	-0.435	-1.204
995.6	2.256	3.694	-0.476	-1.168
995.7	2.260	3.796	-0.512	-1.134
995.8	2.276	3.899	-0.546	-1.102
317π	2.301	3.988	-0.574	-1.076
995.9	2.306	4.004	-0.579	-1.071
996.0	2.351	4.110	-0.611	-1.042
996.1	2.415	4.216	-0.642	-1.012
996.2	2.499	4.319	-0.675	-0.982
996.3	2.609	4.417	-0.709	-0.951
996.4	2.748	4.503	-0.745	-0.917
996.5	2.919	4.567	-0.783	-0.880
996.6	3.123	4.595	-0.822	-0.837
996.7	3.353	4.570	-0.862	-0.787
996.8	3.593	4.471	-0.901	-0.727
996.9	3.810	4.288	-0.936	-0.656
997.0	3.965	4.028	-0.964	-0.573
997.1	4.023	3.723	-0.978	-0.480
997.2	3.973	3.421	-0.976	-0.378
997.3	3.832	3.167	-0.951	-0.274
997.4	3.637	2.986	-0.905	-0.175
635π/2	3.518	2.918	-0.870	-0.126
997.5	3.423	2.880	-0.839	-0.090
997.6	3.214	2.837	-0.760	-0.023
997.7	3.025	2.842	-0.674	0.024
997.8	2.861	2.880	-0.590	0.052
997.9	2.723	2.940	-0.511	0.065

k_0h	ADMITTANCE		RAD.EFF.LENGTH	
[illegible]	[illegible]	[illegible]	[illegible]	[illegible]
998.1	2.513	3.097	-0.378	0.063
998.2	2.437	3.186	-0.323	0.054
998.3	2.376	3.278	-0.275	0.044
998.4	2.329	3.373	-0.232	0.034
998.5	2.294	3.469	-0.193	0.025
998.6	2.272	3.567	-0.156	0.017
998.7	2.260	3.666	-0.121	0.011
998.8	2.259	3.767	-0.085	0.007
998.9	2.271	3.869	-0.049	0.003
999.0	2.295	3.974	-0.011	0.001
318π	2.304	4.001	-0.000	0.000
999.1	2.334	4.079	0.031	-0.002
999.2	2.389	4.185	0.075	-0.005
999.3	2.465	4.290	0.125	-0.009
999.4	2.563	4.390	0.180	-0.016
999.5	2.690	4.482	0.240	-0.028
999.6	2.847	4.556	0.308	-0.046
999.7	3.038	4.601	0.381	-0.073
999.8	3.258	4.599	0.460	-0.113
999.9	3.496	4.532	0.543	-0.169
1000.0	3.726	4.385	0.625	-0.246
1000.1	3.911	4.154	0.702	-0.345
1000.2	4.011	3.863	0.764	-0.469
1000.3	4.004	3.555	0.803	-0.616
1000.4	3.898	3.278	0.808	-0.777
1000.5	3.721	3.066	0.773	-0.942
637π/2	3.518	2.933	0.701	-1.091
1000.6	3.512	2.930	0.699	-1.095
1000.7	3.299	2.863	0.592	-1.223
1000.8	3.102	2.850	0.464	-1.319
1000.9	2.928	2.876	0.327	-1.380
1001.0	2.779	2.928	0.194	-1.410
1001.1	2.655	2.997	0.070	-1.414
1001.2	2.552	3.077	-0.040	-1.400
1001.3	2.459	3.163	-0.136	-1.374
1001.4	2.402	3.254	-0.217	-1.341
1001.5	2.349	3.348	-0.287	-1.304
1001.6	2.310	3.443	-0.346	-1.266
1001.7	2.282	3.540	-0.397	-1.228
1001.8	2.266	3.639	-0.441	-1.192
1001.9	2.261	3.739	-0.481	-1.157
1002.0	2.267	3.841	-0.517	-1.124
1002.1	2.286	3.944	-0.550	-1.092
319π	2.307	4.015	-0.572	-1.071
1002.2	2.319	4.049	-0.582	-1.062
1002.3	2.367	4.154	-0.614	-1.032
1002.4	2.434	4.260	-0.646	-1.003
1002.5	2.523	4.362	-0.679	-0.973
1002.6	2.637	4.458	-0.714	-0.942
1002.7	2.781	4.540	-0.750	-0.908
1002.8	2.958	4.598	-0.788	-0.870
1002.9	3.167	4.617	-0.827	-0.826

k_0h	ADMITTANCE		RAD.EFF.LENGTH	
1003.0	[illegible]	[illegible]	-0.867	-0.774
1003.1	3.636	4.466	[illegible]	[illegible]
1003.2	3.843	4.269	-0.940	-0.640
1003.3	3.981	4.000	-0.965	-0.555
1003.4	4.019	3.693	-0.977	-0.460
1003.5	3.951	3.399	-0.970	-0.357
1003.6	3.799	3.159	-0.942	-0.254
1003.7	3.600	2.992	-0.892	-0.158
639π/2	3.517	2.948	-0.867	-0.124
1003.8	3.386	2.898	-0.823	-0.076
1003.9	3.182	2.865	-0.742	-0.013
1004.0	2.998	2.876	-0.657	0.030
1004.1	2.839	2.919	-0.574	0.055
1004.2	2.705	2.981	-0.497	0.066
1004.3	2.594	3.057	-0.428	0.067
1004.4	2.503	3.141	-0.367	0.061
1004.5	2.430	3.230	-0.314	0.052
1004.6	2.372	3.323	-0.267	0.042
1004.7	2.327	3.418	-0.225	0.032
1004.8	2.295	3.514	-0.186	0.024
1004.9	2.274	3.612	-0.150	0.016
1005.0	2.264	3.711	-0.115	0.010
1005.1	2.265	3.812	-0.079	0.006
1005.2	2.280	3.915	-0.043	0.003
320π	2.310	4.029	-0.000	0.000
1005.3	2.306	4.019	-0.004	0.000
1005.4	2.348	4.124	0.038	-0.002
1005.5	2.407	4.230	0.083	-0.005
1005.6	2.485	4.333	0.134	-0.010
1005.7	2.590	4.432	0.189	-0.018
1005.8	2.721	4.521	0.251	-0.031
1005.9	2.884	4.590	0.319	-0.051
1006.0	3.080	4.627	0.393	-0.080
1006.1	3.304	4.614	0.473	-0.123
1006.2	3.541	4.533	0.555	-0.182
1006.3	3.764	4.371	0.636	-0.263
1006.4	3.934	4.129	0.711	-0.367
1006.5	4.015	3.833	0.769	-0.495
1006.6	3.989	3.529	0.801	-0.644
1006.7	3.869	3.264	0.798	-0.806
1006.8	3.685	3.067	0.756	-0.968
641π/2	3.516	2.963	0.694	-1.089
1006.9	3.475	2.944	0.675	-1.117
1007.0	3.265	2.888	0.564	-1.238
1007.1	3.072	2.882	0.434	-1.327
1007.2	2.903	2.913	0.298	-1.381
1007.3	2.760	2.968	0.167	-1.405
1007.4	2.640	3.039	0.046	-1.405
1007.5	2.541	3.120	-0.060	-1.389
1007.6	2.461	3.208	-0.152	-1.362
1007.7	2.396	3.299	-0.231	-1.328
1007.8	2.347	3.392	-0.298	-1.291
1007.9	2.309	3.488	-0.355	-1.253

TABLE 2.10

FAR FIELD OF LONG DIPOLE ANTENNAS IN AIR

a/λ = 0.007022

Θ	$k_0h = 2\pi$			$k_0h = 2\pi + \pi/2$		
		$E(\Theta)/V$			$E(\Theta)/V$	
	RE	IM	ABSVAL	RE	IM	ABSVAL
0	0.0000	0.0000	0.0000	0.0000	0.0000	0.0000
2	-0.0023	0.0014	0.0027	0.0129	0.0260	0.0290
4	-0.0050	0.0029	0.0058	0.0272	0.0586	0.0646
6	-0.0078	0.0046	0.0091	0.0423	0.0948	0.1038
8	-0.0105	0.0064	0.0123	0.0581	0.1338	0.1459
10	-0.0128	0.0084	0.0153	0.0747	0.1753	0.1905
12	-0.0145	0.0106	0.0180	0.0922	0.2188	0.2374
14	-0.0155	0.0131	0.0204	0.1106	0.2639	0.2862
16	-0.0156	0.0159	0.0223	0.1301	0.3103	0.3365
18	-0.0145	0.0191	0.0240	0.1505	0.3573	0.3877
20	-0.0120	0.0227	0.0257	0.1718	0.4043	0.4393
22	-0.0080	0.0268	0.0280	0.1937	0.4503	0.4902
24	-0.0022	0.0314	0.0315	0.2162	0.4942	0.5394
26	0.0054	0.0366	0.0370	0.2386	0.5349	0.5857
28	0.0151	0.0423	0.0449	0.2606	0.5709	0.6276
30	0.0268	0.0485	0.0555	0.2816	0.6006	0.6634
32	0.0407	0.0553	0.0687	0.3010	0.6224	0.6913
34	0.0567	0.0626	0.0844	0.3178	0.6345	0.7097
36	0.0746	0.0702	0.1024	0.3315	0.6354	0.7167
38	0.0942	0.0782	0.1224	0.3412	0.6237	0.7109
40	0.1154	0.0863	0.1441	0.3462	0.5982	0.6911
42	0.1375	0.0944	0.1668	0.3458	0.5583	0.6567
44	0.1603	0.1024	0.1902	0.3396	0.5039	0.6076
46	0.1830	0.1099	0.2134	0.3275	0.4355	0.5449
48	0.2049	0.1167	0.2358	0.3094	0.3548	0.4707
50	0.2255	0.1226	0.2567	0.2857	0.2638	0.3889
52	0.2439	0.1274	0.2751	0.2574	0.1658	0.3062
54	0.2593	0.1308	0.2904	0.2253	0.0647	0.2344
56	0.2710	0.1325	0.3017	0.1911	-0.0349	0.1942
58	0.2784	0.1325	0.3083	0.1563	-0.1280	0.2020
60	0.2809	0.1305	0.3098	0.1228	-0.2096	0.2429
62	0.2783	0.1266	0.3057	0.0926	-0.2747	0.2899
64	0.2703	0.1206	0.2960	0.0674	-0.3192	0.3262
66	0.2570	0.1128	0.2807	0.0486	-0.3398	0.3432
68	0.2388	0.1033	0.2602	0.0376	-0.3346	0.3367
70	0.2163	0.0924	0.2352	0.0347	-0.3033	0.3053
72	0.1903	0.0804	0.2066	0.0401	-0.2476	0.2509
74	0.1619	0.0677	0.1755	0.0530	-0.1708	0.1788
76	0.1323	0.0549	0.1433	0.0722	-0.0779	0.1062
78	0.1029	0.0423	0.1112	0.0958	0.0247	0.0989
80	0.0749	0.0307	0.0810	0.1216	0.1296	0.1777
82	0.0499	0.0203	0.0538	0.1472	0.2291	0.2724
84	0.0289	0.0117	0.0312	0.1702	0.3158	0.3588
86	0.0131	0.0053	0.0142	0.1884	0.3830	0.4268
88	0.0033	0.0013	0.0036	0.2000	0.4256	0.4702
90	0.0000	-0.0000	0.0000	0.2040	0.4401	0.4851

TABLE 2.10

FAR FIELD OF LONG DIPOLE ANTENNAS IN AIR

$a/\lambda = 0.007022$

	$k_0h = 3\pi$			$k_0h = 3\pi + \pi/2$		
Θ	E(Θ)/V			E(Θ)/V		
	RE	IM	ABSVAL	RE	IM	ABSVAL
0	0.0000	0.0000	0.0000	0.0000	0.0000	0.0000
2	-0.0035	0.0023	0.0042	0.0161	0.0353	0.0388
4	-0.0076	0.0051	0.0092	0.0340	0.0794	0.0864
6	-0.0118	0.0080	0.0142	0.0529	0.1284	0.1389
8	-0.0154	0.0113	0.0191	0.0731	0.1813	0.1954
10	-0.0181	0.0150	0.0235	0.0946	0.2373	0.2554
12	-0.0195	0.0191	0.0273	0.1176	0.2958	0.3183
14	-0.0191	0.0239	0.0306	0.1423	0.3559	0.3833
16	-0.0165	0.0293	0.0337	0.1687	0.4166	0.4494
18	-0.0113	0.0356	0.0373	0.1965	0.4764	0.5153
20	-0.0030	0.0426	0.0427	0.2255	0.5336	0.5793
22	0.0086	0.0506	0.0513	0.2551	0.5859	0.6390
24	0.0238	0.0594	0.0639	0.2844	0.6308	0.6919
26	0.0427	0.0690	0.0811	0.3124	0.6653	0.7350
28	0.0653	0.0793	0.1027	0.3379	0.6864	0.7651
30	0.0913	0.0900	0.1282	0.3594	0.6911	0.7789
32	0.1203	0.1010	0.1571	0.3754	0.6764	0.7736
34	0.1515	0.1118	0.1883	0.3845	0.6404	0.7470
36	0.1839	0.1220	0.2206	0.3852	0.5818	0.6978
38	0.2161	0.1311	0.2527	0.3767	0.5008	0.6267
40	0.2465	0.1385	0.2828	0.3585	0.3994	0.5367
42	0.2726	0.1438	0.3091	0.3308	0.2813	0.4342
44	0.2955	0.1463	0.3298	0.2940	0.1525	0.3317
46	0.3105	0.1458	0.3430	0.2517	0.0207	0.2526
48	0.3171	0.1417	0.3473	0.2049	-0.1050	0.2302
50	0.3142	0.1341	0.3416	0.1574	-0.2144	0.2660
52	0.3013	0.1229	0.3254	0.1131	-0.2978	0.3185
54	0.2786	0.1087	0.2990	0.0756	-0.3467	0.3548
56	0.2470	0.0920	0.2636	0.0483	-0.3551	0.3584
58	0.2085	0.0738	0.2211	0.0336	-0.3208	0.3225
60	0.1655	0.0551	0.1744	0.0326	-0.2459	0.2480
62	0.1213	0.0373	0.1269	0.0450	-0.1373	0.1445
64	0.0794	0.0218	0.0823	0.0685	-0.0065	0.0688
66	0.0434	0.0096	0.0445	0.0996	0.1316	0.1650
68	0.0167	0.0020	0.0168	0.1336	0.2601	0.2924
70	0.0018	-0.0003	0.0018	0.1654	0.3627	0.3986
72	0.0001	0.0029	0.0029	0.1902	0.4253	0.4659
74	0.0120	0.0115	0.0166	0.2040	0.4389	0.4840
76	0.0362	0.0248	0.0439	0.2050	0.4007	0.4501
78	0.0702	0.0415	0.0816	0.1930	0.3154	0.3698
80	0.1103	0.0602	0.1257	0.1703	0.1943	0.2584
82	0.1522	0.0790	0.1715	0.1408	0.0545	0.1510
84	0.1910	0.0960	0.2138	0.1097	-0.0840	0.1381
86	0.2225	0.1096	0.2480	0.0825	-0.2008	0.2171
88	0.2430	0.1183	0.2703	0.0640	-0.2787	0.2860
90	0.2501	0.1214	0.2779	0.0574	-0.3060	0.3114

TABLE 2.10

FAR FIELD OF LONG DIPOLE ANTENNAS IN AIR

$a/\lambda = 0.007022$

Θ	$k_0h = 4\pi$			$k_0h = 4\pi + \pi/2$		
	$E(\Theta)/V$			$E(\Theta)/V$		
	RE	IM	ABSVAL	RE	IM	ABSVAL
0	0.0000	0.0000	0.0000	0.0000	0.0000	0.0000
2	-0.0047	0.0034	0.0058	0.0191	0.0444	0.0484
4	-0.0102	0.0073	0.0126	0.0404	0.0998	0.1077
6	-0.0155	0.0117	0.0195	0.0631	0.1614	0.1733
8	-0.0198	0.0167	0.0259	0.0876	0.2278	0.2441
10	-0.0223	0.0223	0.0315	0.1142	0.2979	0.3190
12	-0.0223	0.0287	0.0364	0.1432	0.3706	0.3973
14	-0.0192	0.0362	0.0410	0.1746	0.4442	0.4773
16	-0.0120	0.0448	0.0464	0.2084	0.5166	0.5571
18	-0.0003	0.0546	0.0546	0.2440	0.5851	0.6339
20	0.0166	0.0656	0.0676	0.2805	0.6460	0.7043
22	0.0390	0.0777	0.0869	0.3167	0.6952	0.7639
24	0.0670	0.0907	0.1127	0.3507	0.7278	0.8079
26	0.1002	0.1042	0.1446	0.3805	0.7392	0.8314
28	0.1378	0.1178	0.1813	0.4036	0.7246	0.8294
30	0.1785	0.1308	0.2213	0.4175	0.6806	0.7984
32	0.2204	0.1423	0.2623	0.4200	0.6051	0.7366
34	0.2609	0.1515	0.3017	0.4092	0.4989	0.6452
36	0.2971	0.1574	0.3362	0.3844	0.3657	0.5306
38	0.3260	0.1590	0.3627	0.3461	0.2130	0.4064
40	0.3443	0.1558	0.3779	0.2961	0.0521	0.3007
42	0.3495	0.1473	0.3793	0.2383	-0.1027	0.2595
44	0.3398	0.1335	0.3651	0.1776	-0.2354	0.2949
46	0.3146	0.1150	0.3350	0.1203	-0.3299	0.3511
48	0.2752	0.0929	0.2905	0.0726	-0.3728	0.3798
50	0.2246	0.0690	0.2349	0.0403	-0.3562	0.3584
52	0.1674	0.0453	0.1734	0.0273	-0.2796	0.2810
54	0.1097	0.0245	0.1124	0.0346	-0.1521	0.1560
56	0.0585	0.0087	0.0591	0.0602	0.0084	0.0608
58	0.0201	0.0001	0.0201	0.0989	0.1765	0.2023
60	0.0000	-0.0000	0.0000	0.1428	0.3235	0.3536
62	0.0012	0.0086	0.0087	0.1830	0.4220	0.4600
64	0.0236	0.0249	0.0343	0.2111	0.4521	0.4989
66	0.0638	0.0469	0.0791	0.2212	0.4060	0.4624
68	0.1153	0.0712	0.1355	0.2113	0.2912	0.3598
70	0.1693	0.0943	0.1938	0.1842	0.1297	0.2253
72	0.2162	0.1124	0.2437	0.1463	-0.0448	0.1530
74	0.2474	0.1224	0.2760	0.1072	-0.1947	0.2223
76	0.2567	0.1224	0.2844	0.0763	-0.2862	0.2962
78	0.2421	0.1122	0.2669	0.0612	-0.2976	0.3038
80	0.2059	0.0934	0.2261	0.0652	-0.2255	0.2348
82	0.1547	0.0691	0.1695	0.0863	-0.0862	0.1220
84	0.0983	0.0434	0.1075	0.1181	0.0879	0.1473
86	0.0476	0.0208	0.0520	0.1512	0.2551	0.2966
88	0.0125	0.0055	0.0137	0.1759	0.3751	0.4143
90	0.0000	-0.0000	0.0000	0.1850	0.4187	0.4577

TABLE 2.10

FAR FIELD OF LONG DIPOLE ANTENNAS IN AIR

$a/\lambda = 0.007022$

Θ	$k_0h = 5\pi$			$k_0h = 5\pi + \pi/2$		
	E(Θ)/V			E(Θ)/V		
	RE	IM	ABSVAL	RE	IM	ABSVAL
0	0.0000	0.0000	0.0000	0.0000	0.0000	0.0000
2	-0.0059	0.0045	0.0074	0.0220	0.0534	0.0578
4	-0.0128	0.C098	0.0161	0.0465	0.1200	0.1287
6	-0.0191	0.0157	0.0247	0.0730	0.1940	0.2073
8	-0.0236	0.0224	0.0326	0.1019	0.2736	0.2920
10	-0.0253	0.0302	0.0394	0.1339	0.3574	0.3816
12	-0.0231	0.0392	0.0455	0.1691	0.4433	0.4745
14	-0.0157	0.0497	0.0522	0.2077	0.5287	0.5680
16	-0.0022	0.0618	0.0618	0.2492	0.6098	0.6588
18	0.0183	0.0754	0.0776	0.2925	0.6818	0.7418
20	0.0463	0.0904	0.1015	0.3358	0.7386	0.8113
22	0.0819	0.1063	0.1342	0.3766	0.7736	0.8604
24	0.1244	0.1226	0.1746	0.4117	0.7798	0.8818
26	0.1723	0.1383	0.2209	0.4377	0.7508	0.8691
28	0.2231	0.1523	0.2701	0.4506	0.6822	0.8176
30	0.2733	0.1631	0.3183	0.4474	0.5727	0.7268
32	0.3186	0.1695	0.3608	0.4258	0.4258	0.6022
34	0.3541	0.1700	0.3927	0.3855	0.2505	0.4597
36	0.3749	0.1636	0.4090	0.3284	0.0620	0.3342
38	0.3770	0.1500	0.4057	0.2594	-0.1192	0.2854
40	0.3580	0.1295	0.3807	0.1857	-0.2694	0.3272
42	0.3179	0.1035	0.3343	0.1166	-0.3656	0.3837
44	0.2600	0.0745	0.2705	0.0618	-0.3896	0.3946
46	0.1907	0.0457	0.1961	0.0296	-0.3336	0.3350
48	0.1193	0.0210	0.1211	0.0248	-0.2039	0.2054
50	0.0565	0.0042	0.0567	0.0472	-0.0223	0.0522
52	0.0129	-0.0019	0.0131	0.0910	0.1756	0.1978
54	-0.0034	0.0042	0.0054	0.1452	0.3467	0.3759
56	0.0109	0.0217	0.0243	0.1958	0.4500	0.4908
58	0.0532	0.0478	0.0715	0.2297	0.4577	0.5121
60	0.1144	0.0774	0.1381	0.2377	0.3645	0.4351
62	0.1806	0.1044	0.2086	0.2177	0.1922	0.2904
64	0.2360	0.1226	0.2659	0.1761	-0.0138	0.1767
66	0.2661	0.1276	0.2952	0.1258	-0.1951	0.2321
68	0.2626	0.1178	0.2878	0.0824	-0.2973	0.3085
70	0.2254	0.0949	0.2446	0.0594	-0.2877	0.2937
72	0.1635	0.0642	0.1757	0.0632	-0.1673	0.1788
74	0.0933	0.0332	0.0990	0.0906	0.0267	0.0945
76	0.0339	0.0097	0.0353	0.1303	0.2311	0.2653
78	0.0022	-0.0001	0.0022	0.1664	0.3773	0.4123
80	0.0074	0.0067	0.0099	0.1848	0.4145	0.4538
82	0.0479	0.0282	0.0556	0.1787	0.3288	0.3742
84	0.1117	0.0587	0.1262	0.1511	0.1493	0.2125
86	0.1796	0.0898	0.2009	0.1139	-0.0611	0.1293
88	0.2310	0.1129	0.2571	0.0827	-0.2277	0.2422
90	0.2501	0.1214	0.2779	0.0706	-0.2908	0.2992

TABLE 2.10

FAR FIELD OF LONG DIPOLE ANTENNAS IN AIR

$a/\lambda = 0.007022$

Θ	$k_0h = 6\pi$			$k_0h = 6\pi + \pi/2$		
	E(Θ)/V			E(Θ)/V		
	RE	IM	ABSVAL	RE	IM	ABSVAL
0	0.0000	0.0000	0.0000	0.0000	0.0000	0.0000
2	-0.0072	0.0056	0.0091	0.0248	0.0623	0.0670
4	-0.0153	0.0123	0.0196	0.0525	0.1399	0.1494
6	-0.0224	0.0198	0.0299	0.0827	0.2262	0.2408
8	-0.0269	0.0285	0.0392	0.1163	0.3188	0.3394
10	-0.0271	0.0387	0.0472	0.1538	0.4157	0.4433
12	-0.0216	0.0505	0.0550	0.1957	0.5139	0.5499
14	-0.0087	0.0643	0.0649	0.2418	0.6089	0.6552
16	0.0129	0.0800	0.0811	0.2911	0.6950	0.7535
18	0.0440	0.0974	0.1069	0.3415	0.7646	0.8374
20	0.0850	0.1160	0.1438	0.3900	0.8087	0.8979
22	0.1350	0.1348	0.1908	0.4323	0.8180	0.9252
24	0.1918	0.1525	0.2451	0.4637	0.7838	0.9107
26	0.2518	0.1674	0.3024	0.4789	0.7000	0.8481
28	0.3098	0.1776	0.3571	0.4734	0.5657	0.7376
30	0.3592	0.1810	0.4023	0.4441	0.3868	0.5890
32	0.3931	0.1761	0.4307	0.3912	0.1783	0.4299
34	0.4048	0.1619	0.4359	0.3183	-0.0362	0.3203
36	0.3897	0.1387	0.4137	0.2335	-0.2262	0.3251
38	0.3469	0.1083	0.3634	0.1484	-0.3591	0.3885
40	0.2799	0.0740	0.2895	0.0768	-0.4069	0.4141
42	0.1975	0.0408	0.2017	0.0314	-0.3551	0.3565
44	0.1134	0.0141	0.1143	0.0208	-0.2094	0.2104
46	0.0433	-0.0010	0.0433	0.0461	0.0008	0.0461
48	0.0020	-0.0008	0.0022	0.0995	0.2251	0.2461
50	-0.0007	0.0150	0.0150	0.1650	0.4030	0.4355
52	0.0363	0.0435	0.0567	0.2223	0.4810	0.5299
54	0.1035	0.0780	0.1296	0.2531	0.4310	0.4998
56	0.1821	0.1095	0.2125	0.2469	0.2641	0.3615
58	0.2484	0.1291	0.2800	0.2063	0.0326	0.2088
60	0.2809	0.1305	0.3098	0.1461	-0.1838	0.2348
62	0.2676	0.1124	0.2902	0.0890	-0.3049	0.3176
64	0.2112	0.0796	0.2257	0.0566	-0.2821	0.2877
66	0.1298	0.0420	0.1364	0.0602	-0.1212	0.1353
68	0.0515	0.0118	0.0528	0.0956	0.1142	0.1489
70	0.0045	-0.0005	0.0046	0.1446	0.3257	0.3564
72	0.0067	0.0096	0.0117	0.1833	0.4208	0.4589
74	0.0571	0.0388	0.0690	0.1930	0.3550	0.4041
76	0.1359	0.0764	0.1559	0.1702	0.1560	0.2309
78	0.2109	0.1084	0.2371	0.1276	-0.0856	0.1536
80	0.2510	0.1223	0.2792	0.0881	-0.2566	0.2713
82	0.2388	0.1124	0.2639	0.0723	-0.2747	0.2841
84	0.1791	0.0824	0.1971	0.0875	-0.1303	0.1569
86	0.0968	0.0439	0.1063	0.1240	0.1071	0.1638
88	0.0271	0.0122	0.0297	0.1602	0.3214	0.3591
90	0.0000	-0.0000	0.0000	0.1751	0.4069	0.4430

TABLE 2.10

FAR FIELD OF LONG DIPOLE ANTENNAS IN AIR

$a/\lambda = 0.007022$

θ	$k_0h = 7\pi$			$k_0h = 7\pi + \pi/2$		
	E(Θ)/V			E(Θ)/V		
	RE	IM	ABSVAL	RE	IM	ABSVAL
0	0.0000	0.0000	0.0000	0.0000	0.0000	0.0000
2	-0.0084	0.0068	0.0108	0.0275	0.0710	0.0762
4	-0.0178	0.0149	0.0232	0.0583	0.1596	0.1699
6	-0.0255	0.0241	0.0351	0.0923	0.2580	0.2741
8	-0.0295	0.0349	0.0457	0.1306	0.3635	0.3862
10	-0.0277	0.0476	0.0551	0.1741	0.4730	0.5040
12	-0.0179	0.0625	0.0650	0.2230	0.5821	0.6234
14	0.0018	0.0797	0.0797	0.2768	0.6845	0.7384
16	0.0330	0.0991	0.1044	0.3336	0.7714	0.8405
18	0.0764	0.1200	0.1423	0.3902	0.8321	0.9190
20	0.1315	0.1414	0.1930	0.4416	0.8544	0.9617
22	0.1957	0.1615	0.2537	0.4816	0.8266	0.9566
24	0.2645	0.1780	0.3188	0.5034	0.7400	0.8950
26	0.3310	0.1885	0.3809	0.5005	0.5923	0.7754
28	0.3865	0.1903	0.4308	0.4687	0.3908	0.6102
30	0.4213	0.1813	0.4587	0.4076	0.1550	0.4360
32	0.4271	0.1609	0.4564	0.3222	-0.0833	0.3328
34	0.3985	0.1300	0.4192	0.2238	-0.2831	0.3609
36	0.3362	0.0920	0.3486	0.1286	-0.4026	0.4226
38	0.2481	0.0528	0.2536	0.0551	-0.4100	0.4137
40	0.1497	0.0195	0.1510	0.0192	-0.2964	0.2970
42	0.0619	-0.0006	0.0619	0.0287	-0.0849	0.0896
44	0.0060	-0.0022	0.0064	0.079[illegible]	0.1684	0.1862
46	-0.0027	0.0159	0.0162	0.1537	0.3861	0.4156
48	0.0388	0.0495	0.0629	0.2253	0.4931	0.5421
50	0.1180	0.0891	0.1478	0.2672	0.4456	0.5195
52	0.2080	0.1221	0.2411	0.2628	0.2549	0.3661
54	0.2758	0.1368	0.3078	0.2138	-0.0078	0.2139
56	0.2940	0.1269	0.3202	0.1411	-0.2339	0.2731
58	0.2529	0.0948	0.2701	0.0769	-0.3217	0.3307
60	0.1671	0.0519	0.1750	0.0496	-0.2266	0.2320
62	0.0716	0.0148	0.0731	0.0698	0.0096	0.0704
64	0.0080	-0.0009	0.0081	0.1234	0.2707	0.2975
66	0.0056	0.0120	0.0132	0.1789	0.4203	0.4568
68	0.0654	0.0481	0.0812	0.2044	0.3750	0.4271
70	0.1581	0.0910	0.1824	0.1860	0.1563	0.2429
72	0.2361	0.1202	0.2649	0.1367	-0.1139	0.1779
74	0.2580	0.1208	0.2849	0.0880	-0.2787	0.2923
76	0.2111	0.0920	0.2303	0.0703	-0.2384	0.2486
78	0.1196	0.0479	0.1288	0.0923	-0.0156	0.0936
80	0.0332	0.0109	0.0350	0.1365	0.2537	0.2881
82	-0.0001	0.0005	0.0005	0.1705	0.4010	0.4357
84	0.0383	0.0222	0.0442	0.1704	0.3321	0.3732
86	0.1265	0.0646	0.1421	0.1374	0.0899	0.1642
88	0.2139	0.1049	0.2383	0.0966	-0.1706	0.1960
90	0.2501	0.1214	0.2779	0.0785	-0.2813	0.2920

TABLE 2.10

FAR FIELD OF LONG DIPOLE ANTENNAS IN AIR

$a/\lambda = 0.007022$

Θ	$k_0h = 8\pi$			$k_0h = 8\pi + \pi/2$		
	E(Θ)/V			E(Θ)/V		
	RE	IM	ABSVAL	RE	IM	ABSVAL
0	0.0000	0.0000	0.0000	0.0000	0.0000	0.0000
2	-0.0096	0.0080	0.0125	0.0302	0.0798	0.0853
4	-0.0202	0.0176	0.0267	0.0640	0.1792	0.1903
6	-0.0284	0.0286	0.0403	0.1019	0.2896	0.3070
8	-0.0316	0.0416	0.0523	0.1451	0.4076	0.4327
10	-0.0271	0.0570	0.0631	0.1947	0.5291	0.5638
12	-0.0121	0.0750	0.0760	0.2509	0.6479	0.6948
14	0.0157	0.0957	0.0970	0.3125	0.7549	0.8170
16	0.0579	0.1186	0.1320	0.3765	0.8381	0.9188
18	0.1147	0.1425	0.1829	0.4377	0.8829	0.9855
20	0.1840	0.1654	0.2474	0.4891	0.8742	1.0017
22	0.2608	0.1848	0.3196	0.5221	0.7992	0.9546
24	0.3370	0.1972	0.3905	0.5280	0.6520	0.8390
26	0.4018	0.1993	0.4485	0.5003	0.4386	0.6653
28	0.4428	0.1886	0.4813	0.4369	0.1806	0.4727
30	0.4487	0.1641	0.4778	0.3430	-0.0835	0.3531
32	0.4129	0.1275	0.4321	0.2323	-0.3026	0.3815
34	0.3369	0.0837	0.3471	0.1255	-0.4240	0.4422
36	0.2326	0.0407	0.2362	0.0464	-0.4098	0.4124
38	0.1224	0.0082	0.1227	0.0150	-0.2548	0.2553
40	0.0343	-0.0050	0.0347	0.0389	0.0011	0.0390
42	-0.0057	0.0056	0.0080	0.1084	0.2764	0.2969
44	0.0159	0.0379	0.0411	0.1961	0.4697	0.5090
46	0.0921	0.0818	0.1232	0.2655	0.4993	0.5654
48	0.1939	0.1215	0.2288	0.2856	0.3443	0.4473
50	0.2787	0.1410	0.3123	0.2467	0.0671	0.2557
52	0.3073	0.1308	0.3340	0.1670	-0.2033	0.2631
54	0.2632	0.0937	0.2794	0.0865	-0.3285	0.3397
56	0.1644	0.0451	0.1704	0.0466	-0.2368	0.2414
58	0.0583	0.0075	0.0588	0.0664	0.0258	0.0712
60	0.0000	-0.0000	0.0000	0.1296	0.3082	0.3343
62	0.0214	0.0270	0.0345	0.1935	0.4362	0.4772
64	0.1103	0.0744	0.1331	0.2154	0.3240	0.3891
66	0.2134	0.1155	0.2426	0.1820	0.0407	0.1865
68	0.2660	0.1260	0.2943	0.1188	-0.2235	0.2531
70	0.2331	0.0984	0.2531	0.0723	-0.2827	0.2918
72	0.1343	0.0490	0.1430	0.0748	-0.0915	0.1182
74	0.0342	0.0083	0.0352	0.1198	0.2118	0.2434
76	0.0002	0.0025	0.0025	0.1678	0.3995	0.4333
78	0.0559	0.0357	0.0663	0.1786	0.3258	0.3716
80	0.1617	0.0860	0.1832	0.1451	0.0460	0.1522
82	0.2418	0.1193	0.2696	0.0984	-0.2204	0.2414
84	0.2374	0.1126	0.2628	0.0797	-0.2604	0.2723
86	0.1513	0.0701	0.1668	0.1043	-0.0409	0.1120
88	0.0464	0.0212	0.0510	0.1476	0.2608	0.2997
90	0.0000	-0.0000	0.0000	0.1686	0.3990	0.4332

TABLE 2.10

FAR FIELD OF LONG DIPOLE ANTENNAS IN AIR

$a/\lambda = 0.007022$

Θ	$k_0h = 9\pi$			$k_0h = 9\pi + \pi/2$		
	$E(\Theta)/V$			$E(\Theta)/V$		
	RE	IM	ABSVAL	RE	IM	ABSVAL
0	0.0000	0.0000	0.0000	0.0000	0.0000	0.0000
2	-0.0108	0.0092	0.0142	0.0327	0.0884	0.0943
4	-0.0225	0.0203	0.0303	0.0697	0.1987	0.2105
6	-0.0311	0.0332	0.0455	0.1114	0.3210	0.3398
8	-0.0331	0.0485	0.0587	0.1598	0.4513	0.4787
10	-0.0253	0.0667	0.0713	0.2158	0.5840	0.6227
12	-0.0041	0.0880	0.0881	0.2795	0.7109	0.7639
14	0.0329	0.1122	0.1169	0.3488	0.8196	0.8908
16	0.0873	0.1383	0.1635	0.4191	0.8942	0.9875
18	0.1580	0.1643	0.2280	0.4832	0.9161	1.0357
20	0.2407	0.1873	0.3050	0.5311	0.8677	1.0173
22	0.3270	0.2034	0.3851	0.5519	0.7375	0.9211
24	0.4042	0.2084	0.4547	0.5358	0.5263	0.7511
26	0.4571	0.1988	0.4985	0.4779	0.2541	0.5412
28	0.4711	0.1729	0.5018	0.3813	-0.0376	0.3832
30	0.4364	0.1326	0.4561	0.2604	-0.2884	0.3885
32	0.3535	0.0838	0.3633	0.1394	-0.4329	0.4548
34	0.2369	0.0367	0.2397	0.0483	-0.4221	0.4249
36	0.1149	0.0034	0.1149	0.0129	-0.2484	0.2487
38	0.0232	-0.0057	0.0239	0.0431	0.0376	0.0572
40	-0.0069	0.0137	0.0153	0.1255	0.3313	0.3543
42	0.0365	0.0557	0.0666	0.2235	0.5082	0.5552
44	0.1362	0.1046	0.1717	0.2906	0.4803	0.5613
46	0.2475	0.1393	0.2840	0.2922	0.2507	0.3850
48	0.3148	0.1427	0.3456	0.2263	-0.0684	0.2364
50	0.2999	0.1109	0.3198	0.1287	-0.3014	0.3278
52	0.2064	0.0582	0.2145	0.0555	-0.3063	0.3113
54	0.0839	0.0120	0.0847	0.0491	-0.0730	0.0880
56	0.0040	-0.0014	0.0042	0.1089	0.2498	0.2725
58	0.0166	0.0269	0.0316	0.1882	0.4386	0.4773
60	0.1134	0.0796	0.1385	0.2269	0.3519	0.4187
62	0.2270	0.1223	0.2579	0.1963	0.0486	0.2023
64	0.2746	0.1251	0.3017	0.1231	-0.2373	0.2674
66	0.2181	0.0845	0.2340	0.0683	-0.2733	0.2817
68	0.0985	0.0290	0.1027	0.0754	-0.0252	0.0795
70	0.0082	-0.0003	0.0082	0.1321	0.2962	0.3244
72	0.0196	0.0193	0.0275	0.1821	0.4068	0.4457
74	0.1230	0.0727	0.1429	0.1782	0.2034	0.2704
76	0.2305	0.1172	0.2585	0.1275	-0.1293	0.1816
78	0.2484	0.1157	0.2740	0.0827	-0.2793	0.2913
80	0.1599	0.0688	0.1740	0.0879	-0.1019	0.1345
82	0.0425	0.0157	0.0453	0.1333	0.2327	0.2682
84	0.0012	0.0015	0.0020	0.1673	0.3974	0.4312
86	0.0732	0.0386	0.0827	0.1531	0.2287	0.2752
88	0.1926	0.0949	0.2147	0.1081	-0.1068	0.1519
90	0.2501	0.1214	0.2779	0.0839	-0.2745	0.2871

TABLE 2.10

FAR FIELD OF LONG DIPOLE ANTENNAS IN AIR

$a/\lambda = 0.007022$

	$k_0h = 10\pi$			$k_0h = 10\pi + \pi/2$		
Θ	$E(\Theta)/V$			$E(\Theta)/V$		
	RE	IM	ABSVAL	RE	IM	ABSVAL
0	0.0000	0.0000	0.0000	0.0000	0.0000	0.0000
2	-0.0120	0.0105	0.0159	0.0352	0.0970	0.1032
4	-0.0248	0.0231	0.0339	0.0752	0.2180	0.2306
6	-0.0335	0.0379	0.0506	0.1210	0.3521	0.3724
8	-0.0340	0.0556	0.0652	0.1746	0.4944	0.5243
10	-0.0222	0.0767	0.0799	0.2374	0.6377	0.6805
12	0.0060	0.1013	0.1015	0.3087	0.7710	0.8305
14	0.0533	0.1289	0.1395	0.3855	0.8783	0.9591
16	0.1207	0.1577	0.1986	0.4611	0.9390	1.0461
18	0.2053	0.1849	0.2763	0.5256	0.9310	1.0691
20	0.2997	0.2061	0.3637	0.5660	0.8354	1.0091
22	0.3907	0.2164	0.4466	0.5694	0.6449	0.8603
24	0.4609	0.2110	0.5069	0.5261	0.3721	0.6444
26	0.4914	0.1870	0.5258	0.4352	0.0568	0.4389
28	0.4677	0.1454	0.4897	0.3086	-0.2354	0.3881
30	0.3864	0.0926	0.3974	0.1724	-0.4256	0.4592
32	0.2619	0.0401	0.2649	0.0623	-0.4473	0.4516
34	0.1265	0.0028	0.1265	0.0120	-0.2803	0.2806
36	0.0239	-0.0064	0.0247	0.0382	0.0236	0.0449
38	-0.0071	0.0173	0.0187	0.1276	0.3424	0.3654
40	0.0473	0.0657	0.0809	0.2363	0.5261	0.5767
42	0.1626	0.1180	0.2009	0.3070	0.4721	0.5631
44	0.2797	0.1483	0.3165	0.2985	0.1975	0.3579
46	0.3314	0.1387	0.3593	0.2141	-0.1443	0.2582
48	0.2824	0.0919	0.2970	0.1052	-0.3384	0.3544
50	0.1575	0.0332	0.1610	0.0422	-0.2479	0.2514
52	0.0349	-0.0012	0.0349	0.0653	0.0708	0.0963
54	-0.0018	0.0118	0.0120	0.1509	0.3808	0.4096
56	0.0739	0.0642	0.0979	0.2263	0.4326	0.4882
58	0.2034	0.1177	0.2350	0.2273	0.1754	0.2871
60	0.2809	0.1305	0.3098	0.1550	-0.1731	0.2324
62	0.2384	0.0906	0.2550	0.0771	-0.2976	0.3074
64	0.1094	0.0296	0.1133	0.0648	-0.0768	0.1005
66	0.0075	-0.0010	0.0076	0.1234	0.2785	0.3046
68	0.0262	0.0264	0.0372	0.1866	0.4118	0.4521
70	0.1470	0.0871	0.1709	0.1869	0.1815	0.2605
72	0.2508	0.1243	0.2799	0.1276	-0.1724	0.2145
74	0.2319	0.1015	0.2531	0.0780	-0.2672	0.2784
76	0.1078	0.0400	0.1149	0.0922	0.0040	0.0923
78	0.0065	0.0006	0.0065	0.1468	0.3379	0.3685
80	0.0346	0.0231	0.0416	0.1722	0.3525	0.3923
82	0.1615	0.0847	0.1824	0.1386	0.0288	0.1416
84	0.2492	0.1215	0.2772	0.0921	-0.2555	0.2716
86	0.2009	0.0946	0.2221	0.0927	-0.1646	0.1889
88	0.0695	0.0322	0.0766	0.1370	0.1942	0.2377
90	0.0000	-0.0000	0.0000	0.1639	0.3931	0.4259

TABLE 2.10

FAR FIELD OF LONG DIPOLE ANTENNAS IN AIR

$a/\lambda = 0.007022$

Θ	$k_0h = 11\pi$ E(Θ)/V RE	IM	ABSVAL	$k_0h = 11\pi + \pi/2$ E(Θ)/V RE	IM	ABSVAL
0	0.0000	0.0000	0.0000	0.0000	0.0000	0.0000
2	-0.0132	0.0117	0.0177	0.0377	0.1056	0.1121
4	-0.0270	0.0260	0.0374	0.0808	0.2372	0.2506
6	-0.0357	0.0428	0.0557	0.1306	0.3831	0.4047
8	-0.0343	0.0630	0.0717	0.1897	0.5370	0.5696
10	-0.0178	0.0870	0.0889	0.2593	0.6901	0.7372
12	0.0182	0.1150	0.1164	0.3384	0.8280	0.8945
14	0.0767	0.1457	0.1647	0.4222	0.9304	1.0217
16	0.1576	0.1766	0.2367	0.5019	0.9721	1.0940
18	0.2555	0.2036	0.3268	0.5640	0.9274	1.0855
20	0.3588	0.2211	0.4214	0.5929	0.7785	0.9786
22	0.4488	0.2230	0.5011	0.5738	0.5261	0.7785
24	0.5032	0.2046	0.5432	0.4994	0.2005	0.5381
26	0.5012	0.1652	0.5277	0.3759	-0.1339	0.3990
28	0.4328	0.1101	0.4465	0.2273	-0.3876	0.4493
30	0.3074	0.0515	0.3117	0.0931	-0.4726	0.4817
32	0.1584	0.0066	0.1586	0.0167	-0.3441	0.3445
34	0.0366	-0.0081	0.0375	0.0257	-0.0398	0.0473
36	-0.0086	0.0149	0.0172	0.1137	0.3111	0.3312
38	0.0444	0.0670	0.0804	0.2339	0.5311	0.5804
40	0.1693	0.1236	0.2096	0.3172	0.4883	0.5823
42	0.2954	0.1534	0.3328	0.3105	0.1940	0.3661
44	0.3420	0.1365	0.3683	0.2150	0.1706	0.2750
46	0.2725	0.0803	0.2841	0.0961	-0.3511	0.3640
48	0.1285	0.0196	0.1299	0.0374	-0.2045	0.2079
50	0.0123	-0.0031	0.0126	0.0802	0.1612	0.1800
52	0.0135	0.0301	0.0330	0.1817	0.4393	0.4754
54	0.1308	0.0936	0.1609	0.2467	0.3759	0.4496
56	0.2605	0.1341	0.2930	0.2140	0.0213	0.2150
58	0.2807	0.1144	0.3031	0.1180	-0.2805	0.3043
60	0.1679	0.0502	0.1752	0.0573	-0.2173	0.2247
62	0.0305	0.0016	0.0306	0.0910	0.1488	0.1744
64	0.0083	0.0165	0.0185	0.1728	0.4145	0.4490
66	0.1240	0.0802	0.1477	0.2039	0.2721	0.3400
68	0.2491	0.1255	0.2790	0.1499	-0.1155	0.1892
70	0.2400	0.1021	0.2608	0.0817	-0.2798	0.2915
72	0.1047	0.0348	0.1104	0.0837	-0.0142	0.0849
74	0.0021	-0.0004	0.0021	0.1452	0.3457	0.3750
76	0.0553	0.0374	0.0667	0.1780	0.3324	0.3771
78	0.1980	0.1038	0.2236	0.1393	-0.0393	0.1447
80	0.2506	0.1186	0.2772	0.0883	-0.2727	0.2866
82	0.1452	0.0632	0.1583	0.0977	-0.0503	0.1098
84	0.0160	0.0054	0.0169	0.1479	0.3239	0.3560
86	0.0297	0.0167	0.0341	0.1606	0.3320	0.3688
88	0.1681	0.0832	0.1876	0.1180	-0.0380	0.1239
90	0.2501	0.1214	0.2779	0.0880	-0.2693	0.2833

TABLE 2.10

FAR FIELD OF LONG DIPOLE ANTENNAS IN AIR

$a/\lambda = 0.007022$

θ	$k_0h = 12\pi$ E(θ)/V RE	IM	ABSVAL	$k_0h = 12\pi + \pi/2$ E(θ)/V RE	IM	ABSVAL
0	0.0000	0.0000	0.0000	0.0000	0.0000	0.0000
2	-0.0145	0.0130	0.0194	0.0402	0.1141	0.1210
4	-0.0291	0.0289	0.0410	0.0863	0.2564	0.2705
6	-0.0376	0.0477	0.0608	0.1403	0.4138	0.4369
8	-0.0339	0.0705	0.0782	0.2050	0.5791	0.6144
10	-0.0123	0.0976	0.0984	0.2817	0.7411	0.7928
12	0.0324	0.1289	0.1329	0.3685	0.8816	0.9556
14	0.1029	0.1624	0.1923	0.4589	0.9756	1.0781
16	0.1975	0.1947	0.2773	0.5409	0.9929	1.1307
18	0.3075	0.2201	0.3781	0.5978	0.9055	1.0851
20	0.4158	0.2317	0.4760	0.6107	0.6991	0.9283
22	0.4982	0.2228	0.5457	0.5647	0.3874	0.6848
24	0.5278	0.1899	0.5610	0.4572	0.0237	0.4578
26	0.4854	0.1359	0.5041	0.3052	-0.2998	0.4278
28	0.3710	0.0718	0.3779	0.1473	-0.4749	0.4972
30	0.2131	0.0167	0.2137	0.0352	-0.4225	0.4240
32	0.0662	-0.0090	0.0668	0.0118	-0.1475	0.1480
34	-0.0074	0.0073	0.0104	0.0854	0.2317	0.2469
36	0.0288	0.0598	0.0663	0.2141	0.5151	0.5579
38	0.1560	0.1219	0.1980	0.3190	0.5258	0.6150
40	0.2964	0.1569	0.3353	0.3288	0.2404	0.4073
42	0.3530	0.1395	0.3796	0.2331	-0.1532	0.2789
44	0.2779	0.0779	0.2887	0.1008	-0.3579	0.3718
46	0.1207	0.0143	0.1215	0.0350	-0.1980	0.2010
48	0.0045	-0.0019	0.0049	0.0861	0.1968	0.2148
50	0.0285	0.0437	0.0522	0.1992	0.4618	0.5029
52	0.1699	0.1116	0.2032	0.2584	0.3287	0.4181
54	0.2891	0.1375	0.3201	0.2024	-0.0742	0.2155
56	0.2599	0.0936	0.2762	0.0941	-0.3109	0.3248
58	0.1084	0.0227	0.1108	0.0549	-0.1070	0.1203
60	0.0000	-0.0000	0.0000	0.1230	0.3000	0.3242
62	0.0586	0.0512	0.0778	0.2047	0.4059	0.4546
64	0.2123	0.1173	0.2425	0.1932	0.0694	0.2053
66	0.2676	0.1183	0.2926	0.1087	-0.2680	0.2892
68	0.1506	0.0513	0.1591	0.0691	-0.1449	0.1605
70	0.0127	0.0002	0.0127	0.1229	0.2690	0.2958
72	0.0383	0.0312	0.0494	0.1822	0.3793	0.4208
74	0.1908	0.1033	0.2170	0.1577	0.0212	0.1591
76	0.2540	0.1184	0.2803	0.0932	-0.2715	0.2871
78	0.1358	0.0547	0.1464	0.0908	-0.0527	0.1050
80	0.0065	0.0009	0.0067	0.1466	0.3408	0.3710
82	0.0558	0.0331	0.0649	0.1634	0.2929	0.3354
84	0.2094	0.1053	0.2344	0.1160	-0.1218	0.1682
86	0.2362	0.1128	0.2618	0.0892	-0.2431	0.2589
88	0.0952	0.0445	0.1051	0.1278	0.1239	0.1780
90	0.0000	-0.0000	0.0000	0.1603	0.3885	0.4203

TABLE 2.10

FAR FIELD OF LONG DIPOLE ANTENNAS IN AIR

$a/\lambda = 0.007022$

θ	$k_0h = 13\pi$ E(Θ)/V RE	IM	ABSVAL	$k_0h = 13\pi + \pi/2$ E(Θ)/V RE	IM	ABSVAL
0	0.0000	0.0000	0.0000	0.0000	0.0000	0.0000
2	-0.0156	0.0143	0.0212	0.0426	0.1226	0.1298
4	-0.0312	0.0318	0.0446	0.0917	0.2754	0.2903
6	-0.0393	0.0528	0.0658	0.1500	0.4444	0.4690
8	-0.0330	0.0782	0.0849	0.2206	0.6207	0.6587
10	-0.0055	0.1084	0.1086	0.3046	0.7906	0.8472
12	0.0486	0.1429	0.1509	0.3990	0.9317	1.0135
14	0.1317	0.1789	0.2222	0.4951	1.0136	1.1281
16	0.2398	0.2117	0.3199	0.5777	1.0014	1.1561
18	0.3599	0.2340	0.4293	0.6260	0.8660	1.0685
20	0.4689	0.2376	0.5256	0.6187	0.6001	0.8619
22	0.5362	0.2160	0.5781	0.5425	0.2358	0.5915
24	0.5330	0.1681	0.5588	0.4024	-0.1456	0.4279
26	0.4456	0.1019	0.4571	0.2296	-0.4248	0.4829
28	0.2906	0.0360	0.2928	0.0785	-0.4865	0.4928
30	0.1197	-0.0054	0.1198	0.0079	-0.2851	0.2852
32	0.0058	-0.0030	0.0065	0.0485	0.0981	0.1095
34	0.0068	0.0444	0.0449	0.1751	0.4583	0.4906
36	0.1232	0.1121	0.1666	0.3062	0.5662	0.6437
38	0.2801	0.1580	0.3216	0.3485	0.3309	0.4806
40	0.3630	0.1480	0.3920	0.2662	-0.0885	0.2805
42	0.3000	0.0850	0.3118	0.1212	-0.3569	0.3769
44	0.1338	0.0156	0.1347	0.0246	-0.2314	0.2339
46	0.0049	-0.0021	0.0054	0.0801	0.1814	0.1983
48	0.0326	0.0490	0.0588	0.2035	0.4698	0.5120
50	0.1880	0.1202	0.2231	0.2680	0.3178	0.4157
52	0.3028	0.1373	0.3325	0.2005	-0.1136	0.2305
54	0.2429	0.0789	0.2554	0.0828	-0.3138	0.3246
56	0.0730	0.0089	0.0735	0.0579	-0.0277	0.0641
58	-0.0016	0.0097	0.0099	0.1478	0.3756	0.4037
60	0.1128	0.0809	0.1388	0.2211	0.3446	0.4094
62	0.2612	0.1305	0.2920	0.1743	-0.0831	0.1931
64	0.2352	0.0908	0.2521	0.0814	-0.2841	0.2955
66	0.0683	0.0147	0.0699	0.0810	0.0476	0.0939
68	0.0031	0.0095	0.0100	0.1622	0.3978	0.4296
70	0.1359	0.0829	0.1592	0.1871	0.2054	0.2779
72	0.2594	0.1248	0.2878	0.1190	-0.2180	0.2484
74	0.1762	0.0704	0.1897	0.0785	-0.1659	0.1836
76	0.0185	0.0033	0.0188	0.1291	0.2740	0.3029
78	0.0430	0.0295	0.0522	0.1713	0.3407	0.3813
80	0.2076	0.1068	0.2335	0.1296	-0.0844	0.1547
82	0.2340	0.1092	0.2583	0.0876	-0.2469	0.2619
84	0.0765	0.0322	0.0829	0.1224	0.1449	0.1896
86	0.0043	0.0031	0.0053	0.1602	0.3825	0.4147
88	0.1415	0.0704	0.1581	0.1264	0.0333	0.1307
90	0.2501	0.1214	0.2779	0.0913	-0.2651	0.2804

TABLE 2.10

FAR FIELD OF LONG DIPOLE ANTENNAS IN AIR

$a/\lambda = 0.007022$

	$k_0h = 14\pi$			$k_0h = 14\pi + \pi/2$		
Θ	E(Θ)/V			E(Θ)/V		
	RE	IM	ABSVAL	RE	IM	ABSVAL
0	0.0000	0.0000	0.0000	0.0000	0.0000	0.0000
2	-0.0168	0.0156	0.0229	0.0450	0.1311	0.1386
4	-0.0333	0.0348	0.0481	0.0972	0.2944	0.3100
6	-0.0408	0.0579	0.0708	0.1598	0.4747	0.5009
8	-0.0315	0.0860	0.0916	0.2364	0.6617	0.7027
10	0.0024	0.1194	0.1194	0.3278	0.8385	0.9003
12	0.0667	0.1569	0.1705	0.4297	0.9780	1.0682
14	0.1628	0.1950	0.2540	0.5307	1.0441	1.1712
16	0.2839	0.2272	0.3636	0.6117	0.9973	1.1700
18	0.4116	0.2448	0.4789	0.6481	0.8097	1.0371
20	0.5159	0.2385	0.5684	0.6168	0.4849	0.7846
22	0.5609	0.2028	0.5965	0.5081	0.0790	0.5142
24	0.5182	0.1407	0.5370	0.3384	-0.2956	0.4493
26	0.3857	0.0670	0.3915	0.1559	-0.4972	0.5211
28	0.2021	0.0076	0.2023	0.0293	-0.4218	0.4228
30	0.0436	-0.0109	0.0449	0.0156	-0.0848	0.0862
32	-0.0094	0.0233	0.0251	0.1189	0.3382	0.3585
34	0.0753	0.0928	0.1195	0.2706	0.5777	0.6379
36	0.2417	0.1534	0.2863	0.3593	0.4485	0.5746
38	0.3639	0.1594	0.3973	0.3109	0.0311	0.3124
40	0.3347	0.1014	0.3498	0.1610	-0.3294	0.3666
42	0.1689	0.0238	0.1706	0.0415	-0.2943	0.2972
44	0.0144	-0.0044	0.0151	0.0630	0.1140	0.1302
46	0.0238	0.0451	0.0510	0.1930	0.4640	0.5025
48	0.1853	0.1216	0.2216	0.2767	0.3487	0.4451
50	0.3104	0.1390	0.3401	0.2118	-0.1039	0.2359
52	0.2411	0.0737	0.2521	0.0824	-0.3177	0.3282
54	0.0593	0.0036	0.0594	0.0580	0.0007	0.0580
56	0.0042	0.0189	0.0194	0.1612	0.4053	0.4362
58	0.1500	0.0992	0.1798	0.2302	0.2951	0.3743
60	0.2809	0.1305	0.3098	0.1603	-0.1665	0.2311
62	0.1946	0.0650	0.2052	0.0678	-0.2448	0.2540
64	0.0224	0.0002	0.0224	0.0998	0.1917	0.2161
66	0.0396	0.0371	0.0543	0.1882	0.4004	0.4424
68	0.2147	0.1161	0.2441	0.1736	0.0132	0.1741
70	0.2460	0.1057	0.2678	0.0896	-0.2766	0.2908
72	0.0764	0.0223	0.0795	0.0922	0.0634	0.1119
74	0.0064	0.0094	0.0113	0.1644	0.3957	0.4285
76	0.1610	0.0895	0.1842	0.1599	0.0847	0.1810
78	0.2525	0.1187	0.2790	0.0947	-0.2671	0.2834
80	0.1080	0.0433	0.1164	0.1028	0.0321	0.1077
82	0.0005	0.0014	0.0015	0.1580	0.3863	0.4174
84	0.1348	0.0707	0.1522	0.1403	0.0832	0.1631
86	0.2505	0.1212	0.2783	0.0931	-0.2635	0.2794
88	0.1224	0.0576	0.1352	0.1201	0.0524	0.1311
90	0.0000	-0.0000	0.0000	0.1574	0.3847	0.4157

TABLE 2.10

FAR FIELD OF LONG DIPOLE ANTENNAS IN AIR

$a/\lambda = 0.007022$

θ	$k_0h = 15\pi$			$k_0h = 15\pi + \pi/2$		
	E(Θ)/V			E(Θ)/V		
	RE	IM	ABSVAL	RE	IM	ABSVAL
0	0.0000	0.0000	0.0000	0.0000	0.0000	0.0000
2	-0.0180	0.0169	0.0247	0.0473	0.1395	0.1473
4	-0.0352	0.0378	0.0517	0.1026	0.3133	0.3297
6	-0.0420	0.0632	0.0759	0.1697	0.5049	0.5326
8	-0.0293	0.0940	0.0985	0.2524	0.7022	0.7462
10	0.0115	0.1306	0.1311	0.3513	0.8847	0.9519
12	0.0866	0.1710	0.1916	0.4605	1.0204	1.1195
14	0.1960	0.2105	0.2876	0.5654	1.0668	1.2074
16	0.3291	0.2411	0.4080	0.6426	0.9809	1.1727
18	0.4614	0.2523	0.5258	0.6636	0.7380	0.9925
20	0.5554	0.2344	0.6028	0.6047	0.3577	0.7026
22	0.5710	0.1841	0.6000	0.4630	-0.0751	0.4690
24	0.4846	0.1099	0.4969	0.2696	-0.4157	0.4954
26	0.3116	0.0348	0.3135	0.0911	-0.5104	0.5184
28	0.1175	-0.0094	0.1178	0.0058	-0.2899	0.2900
30	-0.0021	0.0014	0.0025	0.0568	0.1433	0.1542
32	0.0241	0.0638	0.0682	0.2079	0.5198	0.5598
34	0.1790	0.1383	0.2262	0.3463	0.5576	0.6564
36	0.3431	0.1687	0.3823	0.3559	0.2043	0.4104
38	0.3715	0.1255	0.3921	0.2219	-0.2458	0.3311
40	0.2262	0.0415	0.2300	0.0657	-0.3598	0.3658
42	0.0403	-0.0058	0.0407	0.0411	-0.0063	0.0415
44	0.0066	0.0328	0.0335	0.1651	0.4285	0.4592
46	0.1612	0.1156	0.1984	0.2801	0.4119	0.4981
48	0.3121	0.1438	0.3436	0.2362	-0.0437	0.2402
50	0.2572	0.0786	0.2689	0.0939	-0.3260	0.3392
52	0.0641	0.0032	0.0642	0.0527	-0.0248	0.0583
54	0.0052	0.0219	0.0225	0.1617	0.4088	0.4396
56	0.1655	0.1074	0.1973	0.2379	0.2851	0.3713
58	0.2887	0.1283	0.3159	0.1569	-0.1974	0.2522
60	0.1683	0.0492	0.1754	0.0620	-0.2114	0.2203
62	0.0046	-0.0014	0.0048	0.1141	0.2695	0.2926
64	0.0811	0.0622	0.1022	0.2027	0.3597	0.4129
66	0.2555	0.1277	0.2856	0.1579	-0.1141	0.1948
68	0.2000	0.0756	0.2138	0.0749	-0.2302	0.2421
70	0.0180	0.0011	0.0181	0.1155	0.2425	0.2686
72	0.0618	0.0449	0.0764	0.1820	0.3382	0.3841
74	0.2409	0.1216	0.2699	0.1339	-0.1429	0.1958
76	0.1900	0.0794	0.2060	0.0828	-0.1819	0.1998
78	0.0126	0.0020	0.0128	0.1351	0.3039	0.3326
80	0.0773	0.0461	0.0900	0.1629	0.2545	0.3021
82	0.2454	0.1207	0.2735	0.1064	-0.2209	0.2452
84	0.1574	0.0708	0.1726	0.1018	-0.0644	0.1204
86	0.0018	0.0004	0.0019	0.1533	0.3722	0.4025
88	0.1141	0.0572	0.1276	0.1334	0.1042	0.1693
90	0.2501	0.1214	0.2779	0.0939	-0.2617	0.2780

TABLE 2.10

FAR FIELD OF LONG DIPOLE ANTENNAS IN AIR

$a/\lambda = 0.007022$

Θ	$k_0h = 16\pi$			$k_0h = 16\pi + \pi/2$		
	E(Θ)/V			E(Θ)/V		
	RE	IM	ABSVAL	RE	IM	ABSVAL
0	0.0000	0.0000	0.0000	0.0000	0.0000	0.0000
2	-0.0192	0.0182	0.0265	0.0496	0.1479	0.1560
4	-0.0371	0.0409	0.0553	0.1081	0.3321	0.3493
6	-0.0430	0.0685	0.0809	0.1797	0.5349	0.5643
8	-0.0265	0.1022	0.1055	0.2687	0.7420	0.7892
10	0.0218	0.1418	0.1435	0.3752	0.9292	1.0021
12	0.1082	0.1849	0.2142	0.4913	1.0587	1.1671
14	0.2310	0.2253	0.3226	0.5989	1.0817	1.2364
16	0.3748	0.2531	0.4523	0.6699	0.9525	1.1644
18	0.5080	0.2563	0.5690	0.6720	0.6527	0.9368
20	0.5858	0.2255	0.6277	0.5830	0.2229	0.6242
22	0.5660	0.1608	0.5884	0.4094	-0.2186	0.4641
24	0.4346	0.0781	0.4416	0.2005	-0.4975	0.5363
26	0.2308	0.0086	0.2309	0.0411	-0.4635	0.4653
28	0.0478	-0.0126	0.0495	0.0109	-0.1089	0.1094
30	-0.0092	0.0293	0.0307	0.1247	0.3598	0.3808
32	0.0989	0.1094	0.1475	0.2964	0.6033	0.6722
34	0.2887	0.1680	0.3340	0.3820	0.4049	0.5566
36	0.3921	0.1525	0.4207	0.2977	-0.0806	0.3084
38	0.2997	0.0709	0.3080	0.1186	-0.3826	0.4005
40	0.0932	-0.0003	0.0932	0.0278	-0.1663	0.1686
42	-0.0066	0.0148	0.0162	0.1195	0.3286	0.3496
44	0.1161	0.0999	0.1532	0.2681	0.4802	0.5500
46	0.3005	0.1487	0.3352	0.2694	0.0725	0.2790
48	0.2881	0.0932	0.3028	0.1220	-0.3202	0.3427
50	0.0885	0.0076	0.0888	0.0450	-0.1006	0.1102
52	-0.0002	0.0173	0.0174	0.1478	0.3849	0.4123
54	0.1593	0.1071	0.1920	0.2451	0.3191	0.4024
56	0.2943	0.1294	0.3215	0.1665	-0.1880	0.2511
58	0.1633	0.0440	0.1691	0.0599	-0.2104	0.2188
60	0.0000	-0.0000	0.0000	0.1187	0.2946	0.3176
62	0.1074	0.0773	0.1323	0.2113	0.3322	0.3937
64	0.2714	0.1286	0.3004	0.1488	-0.1754	0.2300
66	0.1597	0.0525	0.1681	0.0694	-0.1681	0.1818
68	-0.0004	0.0002	0.0004	0.1354	0.3362	0.3624
70	0.1249	0.0783	0.1474	0.1867	0.2281	0.2948
72	0.2612	0.1219	0.2883	0.1109	-0.2489	0.2726
74	0.1069	0.0368	0.1131	0.0890	-0.0046	0.0892
76	0.0055	0.0078	0.0095	0.1608	0.3899	0.4218
78	0.1841	0.0983	0.2087	0.1463	0.0032	0.1464
80	0.2294	0.1048	0.2522	0.0894	-0.2390	0.2552
82	0.0353	0.0122	0.0373	0.1302	0.2471	0.2793
84	0.0563	0.0320	0.0647	0.1552	0.2747	0.3155
86	0.2410	0.1182	0.2685	0.1025	-0.2232	0.2456
88	0.1496	0.0708	0.1655	0.1138	-0.0171	0.1150
90	0.0000	-0.0000	0.0000	0.1550	0.3815	0.4118

TABLE 2.10

FAR FIELD OF LONG DIPOLE ANTENNAS IN AIR

$a/\lambda = 0.007022$

	$k_0h = 17\pi$			$k_0h = 17\pi + \pi/2$		
Θ	E(Θ)/V			E(Θ)/V		
	RE	IM	ABSVAL	RE	IM	ABSVAL
0	0.0000	0.0000	0.0000	0.0000	0.0000	0.0000
2	-0.0204	0.0196	0.0283	0.0520	0.1562	0.1647
4	-0.0390	0.0440	0.0588	0.1135	0.3509	0.3688
6	-0.0437	0.0739	0.0859	0.1897	0.5647	0.5957
8	-0.0231	0.1104	0.1128	0.2853	0.7812	0.8317
10	0.0332	0.1532	0.1567	0.3993	0.9720	1.0508
12	0.1315	0.1987	0.2382	0.5219	1.0927	1.2110
14	0.2674	0.2392	0.3588	0.6309	1.0887	1.2583
16	0.4203	0.2630	0.4958	0.6931	0.9124	1.1458
18	0.5504	0.2568	0.6073	0.6732	0.5556	0.8729
20	0.6061	0.2122	0.6421	0.5522	0.0852	0.5587
22	0.5460	0.1343	0.5623	0.3498	-0.3444	0.4909
24	0.3718	0.0477	0.3748	0.1359	-0.5354	0.5524
26	0.1512	-0.0087	0.1514	0.0105	-0.3616	0.3618
28	0.0025	-0.0016	0.0030	0.0438	0.0971	0.1066
30	0.0234	0.0678	0.0717	0.2078	0.5273	0.5668
32	0.1984	0.1496	0.2484	0.3657	0.5711	0.6782
34	0.3736	0.1732	0.4118	0.3682	0.1634	0.4028
36	0.3714	0.1106	0.3875	0.2043	-0.3074	0.3691
38	0.1790	0.0192	0.1801	0.0437	-0.3226	0.3255
40	0.0042	-0.0022	0.0048	0.0659	0.1501	0.1639
42	0.0571	0.0729	0.0925	0.2294	0.5050	0.5546
44	0.2633	0.1476	0.3019	[illegible]	0.3389	0.3829
46	0.3230	0.1159	0.3432	0.1708	-0.2644	0.3148
48	0.1368	0.0201	0.1383	0.0445	-0.2107	0.2154
50	-0.0049	0.0072	0.0087	0.1184	0.3145	0.3360
52	0.1310	0.0979	0.1635	0.2466	0.3853	0.4574
54	0.2976	0.1347	0.3267	0.1899	-0.1339	0.2324
56	0.1807	0.0488	0.1872	0.0625	-0.2436	0.2515
58	0.0009	-0.0005	0.0010	0.1114	0.2751	0.2968
60	0.1124	0.0818	0.1390	0.2171	0.3396	0.4031
62	0.2779	0.1282	0.3060	0.1503	-0.1892	0.2416
64	0.1397	0.0407	0.1454	0.0667	-0.1374	0.1528
66	-0.0010	0.0057	0.0058	0.1470	0.3725	0.4005
68	0.1680	0.0990	0.1950	0.1878	0.1450	0.2373
70	0.2512	0.1091	0.2739	0.0967	-0.2725	0.2891
72	0.0507	0.0120	0.0521	0.1008	0.1385	0.1713
74	0.0459	0.0347	0.0575	0.1745	0.3505	0.3916
76	0.2439	0.1219	0.2727	0.1260	-0.1709	0.2124
78	0.1517	0.0619	0.1639	0.0905	-0.0870	0.1256
80	0.0001	0.0015	0.0015	0.1535	0.3805	0.4103
82	0.1710	0.0896	0.1931	0.1373	0.0046	0.1373
84	0.2250	0.1054	0.2485	0.0937	-0.2181	0.2374
86	0.0227	0.0090	0.0244	0.1420	0.3041	0.3356
88	0.0873	0.0441	0.0978	0.1391	0.1716	0.2209
90	0.2501	0.1214	0.2779	0.0961	-0.2587	0.2760

TABLE 2.10

FAR FIELD OF LONG DIPOLE ANTENNAS IN AIR

$a/\lambda = 0.007022$

Θ	$k_0h = 18\pi$			$k_0h = 18\pi + \pi/2$		
	E(Θ)/V			E(Θ)/V		
	RE	IM	ABSVAL	RE	IM	ABSVAL
0	0.0000	0.0000	0.0000	0.0000	0.0000	0.0000
2	-0.0215	0.0210	0.0300	0.0542	0.1646	0.1733
4	-0.0407	0.0472	0.0623	0.1190	0.3696	0.3883
6	-0.0442	0.0794	0.0909	0.1999	0.5943	0.6270
8	-0.0191	0.1188	0.1203	0.3020	0.8198	0.8737
10	0.0458	0.1645	0.1708	0.4237	1.0128	1.0978
12	0.1563	0.2122	0.2635	0.5523	1.1223	1.2509
14	0.3050	0.2522	0.3958	0.6612	1.0876	1.2728
16	0.4649	0.2707	0.5380	0.7120	0.8612	1.1174
18	0.5876	0.2537	0.6400	0.6671	0.4492	0.8042
20	0.6154	0.1948	0.6455	0.5133	-0.0506	0.5158
22	0.5121	0.1060	0.5230	0.2870	-0.4462	0.5306
24	0.3008	0.0210	0.3015	0.0802	-0.5271	0.5331
26	0.0809	-0.0155	0.0823	0.0023	-0.2153	0.2153
28	-0.0123	0.0222	0.0254	0.1004	0.3007	0.3170
30	0.0902	0.1099	0.1422	0.2924	0.6174	0.6831
32	0.3003	0.1751	0.3476	0.4009	0.4313	0.5888
34	0.4099	0.1524	0.4374	0.3089	-0.0989	0.3244
36	0.2883	0.0577	0.2940	0.1076	-0.3977	0.4120
38	0.0609	-0.0071	0.0613	0.0280	-0.0928	0.0969
40	0.0053	0.0370	0.0374	0.1603	0.4260	0.4552
42	0.1922	0.1323	0.2333	0.3046	0.4161	0.5157
44	0.3413	0.1407	0.3692	0.2368	-0.1204	0.2657
46	0.2097	0.0451	0.2145	0.0667	-0.3136	0.3206
48	0.0074	-0.0029	0.0079	0.0781	0.1741	0.1908
50	0.0828	0.0779	0.1137	0.2317	0.4501	0.5062
52	0.2888	0.1405	0.3212	0.2245	-0.0207	0.2255
54	0.2189	0.0642	0.2282	0.0762	-0.2913	0.3011
56	0.0097	-0.0026	0.0100	0.0924	0.2048	0.2246
58	0.0956	0.0759	0.1221	0.2181	0.3793	0.4376
60	0.2809	0.1305	0.3098	0.1639	-0.1619	0.2304
62	0.1426	0.0394	0.1479	0.0644	-0.1524	0.1655
64	-0.0001	0.0085	0.0085	0.1486	0.3785	0.4066
66	0.1875	0.1083	0.2165	0.1917	0.1155	0.2238
68	0.2393	0.0981	0.2586	0.0903	-0.2704	0.2851
70	0.0241	0.0024	0.0242	0.1093	0.2162	0.2422
72	0.0889	0.0594	0.1069	0.1810	0.2850	0.3376
74	0.2590	0.1223	0.2864	0.1112	-0.2446	0.2686
76	0.0829	0.0279	0.0875	0.0995	0.0717	0.1226
78	0.0317	0.0235	0.0394	0.1655	0.3536	0.3904
80	0.2400	0.1196	0.2681	0.1175	-0.1849	0.2191
82	0.1352	0.0579	0.1470	0.1023	-0.0218	0.1046
84	0.0067	0.0053	0.0085	0.1559	0.3746	0.4058
86	0.2097	0.1044	0.2343	0.1150	-0.1308	0.1742
88	0.1758	0.0835	0.1946	0.1088	-0.0818	0.1361
90	0.0000	-0.0000	0.0000	0.1529	0.3787	0.4085

TABLE 2.10

FAR FIELD OF LONG DIPOLE ANTENNAS IN AIR

$a/\lambda = 0.007022$

Θ	$k_0h = 19\pi$ E(Θ)/V RE	IM	ABSVAL	$k_0h = 19\pi + \pi/2$ E(Θ)/V RE	IM	ABSVAL
0	0.0000	0.0000	0.0000	0.0000	0.0000	0.0000
2	-0.0227	0.0223	0.0318	0.0565	0.1729	0.1819
4	-0.0424	0.0504	0.0659	0.1244	0.3883	0.4077
6	-0.0444	0.0850	0.0959	0.2102	0.6237	0.6582
8	-0.0145	0.1272	0.1281	0.3191	0.8577	0.9151
10	0.0594	0.1760	0.1857	0.4483	1.0516	1.1432
12	0.1824	0.2254	0.2899	0.5824	1.1475	1.2868
14	0.3435	0.2640	0.4332	0.6896	1.0785	1.2801
16	0.5080	0.2761	0.5782	0.7262	0.7998	1.0803
18	0.6186	0.2470	0.6661	0.6538	0.3358	0.7350
20	0.6136	0.1741	0.6378	0.4676	-0.1798	0.5010
22	0.4661	0.0774	0.4725	0.2241	-0.5190	0.5654
24	0.2267	0.0002	0.2267	0.0372	-0.4734	0.4749
26	0.0268	-0.0109	0.0289	0.0172	-0.0392	0.0428
28	0.0052	0.0556	0.0558	0.1733	0.4749	0.5055
30	0.1795	0.1483	0.2328	0.3643	0.6150	0.7148
32	0.3820	0.1802	0.4224	0.3945	0.2155	0.4496
34	0.3874	0.1114	0.4031	0.2204	-0.3084	0.3791
36	0.1717	0.0126	0.1722	0.0402	-0.3215	0.3240
38	-0.0042	0.0035	0.0055	0.0777	0.2079	0.2219
40	0.0958	0.0969	0.1363	0.2634	0.5203	0.5832
42	0.3157	0.1560	0.3521	0.3002	0.1186	0.3228
44	0.2931	0.0842	0.3050	0.1259	-0.3366	0.3594
46	0.0567	-0.0028	0.0568	0.0439	-0.0357	0.0566
48	0.0273	0.0472	0.0545	0.1895	0.4544	0.4923
50	0.2523	0.1392	0.2882	0.2591	0.1556	0.3022
52	0.2699	0.0899	0.2845	0.1105	-0.3090	0.3282
54	0.0377	-0.0015	0.0377	0.0667	0.0725	0.0985
56	0.0599	0.0595	0.0844	0.2069	0.4257	0.4733
58	0.2760	0.1343	0.3069	0.1897	-0.0827	0.2070
60	0.1687	0.0484	0.1755	0.0653	-0.2071	0.2171
62	-0.0022	0.0059	0.0063	0.1387	0.3577	0.3836
64	0.1853	0.1091	0.2150	0.2004	0.1436	0.2465
66	0.2395	0.0954	0.2578	0.0910	-0.2730	0.2877
68	0.0162	-0.0000	0.0162	0.1104	0.2365	0.2610
70	0.1139	0.0736	0.1356	0.1858	0.2494	0.3110
72	0.2563	0.1157	0.2812	0.1036	-0.2640	0.2835
74	0.0440	0.0107	0.0453	0.1075	0.1710	0.2020
76	0.0778	0.0502	0.0926	0.1703	0.2777	0.3258
78	0.2547	0.1210	0.2820	0.1039	-0.2539	0.2743
80	0.0592	0.0209	0.0628	0.1158	0.1634	0.2002
82	0.0644	0.0380	0.0748	0.1569	0.2663	0.3091
84	0.2513	0.1215	0.2791	0.1000	-0.2542	0.2732
86	0.0630	0.0273	0.0687	0.1289	0.1918	0.2311
88	0.0622	0.0318	0.0699	0.1433	0.2328	0.2733
90	0.2501	0.1214	0.2779	0.0980	-0.2561	0.2743

TABLE 2.10

FAR FIELD OF LONG DIPOLE ANTENNAS IN AIR

$a/\lambda = 0.007022$

	$k_0h = 20\pi$			$k_0h = 20\pi + \pi/2$		
Θ	$E(\Theta)/V$			$E(\Theta)/V$		
	RE	IM	ABSVAL	RE	IM	ABSVAL
0	0.0000	0.0000	0.0000	0.0000	0.0000	0.0000
2	-0.0238	0.0237	0.0336	0.0588	0.1812	0.1905
4	-0.0441	0.0536	0.0694	0.1299	0.4069	0.4271
6	-0.0444	0.0906	0.1009	0.2206	0.6530	0.6892
8	-0.0094	0.1358	0.1361	0.3363	0.8948	0.9560
10	0.0741	0.1874	0.2015	0.4731	1.0884	1.1868
12	0.2099	0.2382	0.3175	0.6120	1.1680	1.3186
14	0.3824	0.2747	0.4708	0.7157	1.0616	1.2803
16	0.5489	0.2791	0.6158	0.7357	0.7291	1.0358
18	0.6429	0.2371	0.6852	0.6335	0.2180	0.6700
20	0.6005	0.1509	0.6192	0.4165	-0.2980	0.5121
22	0.4102	0.0500	0.4133	0.1641	-0.5593	0.5828
24	0.1550	-0.0133	0.1555	0.0097	-0.3786	0.3787
26	-0.0055	0.0046	0.0072	0.0536	0.1488	0.1582
28	0.0528	0.0939	0.1077	0.2532	0.5969	0.6484
30	0.2757	0.1761	0.3271	0.4114	0.5212	0.6640
32	0.4252	0.1636	0.4556	0.3482	-0.0277	0.3493
34	0.3122	0.0620	0.3183	0.1269	-0.4070	0.4264
36	0.0622	-0.0087	0.0629	0.0248	-0.1063	0.1091
38	0.0114	0.0464	0.0478	0.1718	0.4512	0.4829
40	0.2292	0.1463	0.2719	0.3233	0.3859	0.5034
42	0.3514	0.1292	0.3744	0.2192	-0.2037	0.2993
44	0.1539	0.0195	0.1551	0.0457	-0.2542	0.2582
46	-0.0056	0.0133	0.0144	0.1208	0.3315	0.3528
48	0.1771	0.1207	0.2143	0.2705	0.3557	0.4469
50	0.3119	0.1213	0.3347	0.1707	-0.2337	0.2894
52	0.0986	0.0115	0.0993	0.0495	-0.1106	0.1211
54	0.0174	0.0338	0.0380	0.1741	0.4265	0.4607
56	0.2496	0.1335	0.2831	0.2220	0.0623	0.2306
58	0.2153	0.0685	0.2259	0.0782	-0.2730	0.2840
60	0.0000	-0.0000	0.0000	0.1156	0.2906	0.3127
62	0.1607	0.1014	0.1900	0.2112	0.2243	0.3081
64	0.2539	0.1021	0.2737	0.1012	-0.2785	0.2963
66	0.0210	0.0002	0.0210	0.1021	0.2053	0.2293
68	0.1160	0.0766	0.1390	0.1914	0.2614	0.3240
70	0.2556	0.1123	0.2792	0.1033	-0.2670	0.2862
72	0.0291	0.0045	0.0295	0.1097	0.2080	0.2351
74	0.1094	0.0679	0.1288	0.1737	0.2252	0.2844
76	0.2462	0.1118	0.2704	0.0965	-0.2605	0.2778
78	0.0206	0.0044	0.0210	0.1265	0.2700	0.2982
80	0.1284	0.0716	0.1470	0.1532	0.1312	0.2017
82	0.2285	0.1057	0.2518	0.0943	-0.2275	0.2462
84	0.0066	0.0016	0.0067	0.1437	0.3437	0.3725
86	0.1624	0.0824	0.1821	0.1281	-0.0041	0.1281
88	0.1994	0.0952	0.2210	0.1053	-0.1387	0.1742
90	0.0000	-0.0000	0.0000	0.1512	0.3763	0.4056

e. MEASURED ADMITTANCES OF MONOPOLES; COMPARISON OF THEORY WITH EXPERIMENT

Table 2.11 lists measured values of the admittance of electrically thick monopoles when driven from a coaxial line that penetrates a large metal ground plane. The monopole with radius a is the extension above the ground plane of the inner conductor of the coaxial line; the inner radius of the outer conductor of the coaxial line is b. Measured admittances are given for monopoles with open tubular ends, flat metallically closed ends, and hemispherically capped ends for b/a quite near one. In all cases the length h is measured vertically from the ground plane to the end of the circular tube. The hemispherical cap is mounted on the tube so that with it the axial length is $h' = h + a$.

Measured values of the admittances of electrically thin monopoles with hemispherical caps are given in Table 2.12 for a wide range of lengths and ratios of b/a. For purposes of comparison theoretical values of the admittance of open-ended tubular monopoles are given in Table 2.13 for the same values of a/λ and b/a as some of those included in Table 2.12.

The measurement of the admittances listed in Tables 2.11 and 2.12 was carried out with great care.[14,15] Nevertheless, difficulties were encountered in obtaining highly accurate values, especially for the thicker antennas, owing primarily to the problem of establishing a reference point at the end of the transmission line when the monopole is replaced by a short-circuiting plug. A completely satisfactory short circuit was not obtained in every case, with consequent small shifts in the entire susceptance curve. The experimental errors introduced in this way are small but comparable in magnitude with the capacitive end correction or difference between the true apparent admittance terminating the line and that calculated under the assumption that only the TEM mode exists on the line even at its end. It follows that they are sufficiently accurate for most applications but not for determining or studying the capacitive end effect.

A comparison of theoretically and experimentally determined admittances based on Tables 2.11 and 2.12 is given in Figs. 2.5a,b,c, which show circular graphs of the admittance calculated with the TEM assumption and corresponding measured values. The agreement is satisfactory for many purposes but not within the accuracy required for determining the capacitive end correction. A clear indication of this fact is given in Fig. 2.6, in which the graphs of the conductance and susceptance of full-wave dipole antennas are shown as a function of the thickness ratio a/λ. The theoretical curve for the conductance (which is independent of capacitive end effects) is in very good agreement with measured points. On the other hand, the measured susceptances depart as much from the true theoretical curve as does the approximate theoretical value obtained with the TEM mode assumption. The difference $B - B_{\text{TEM}} = -\omega C_T$ is also shown in Fig. 2.6.

A comparison of theoretical and experimental susceptances for an electrically quite thin antenna is shown in Fig. 2.7. The theoretical curves (broken lines) were computed using the TEM approximation. However, when a/λ is small and b/a not too large, the error involved is quite small. The known correction when $(b/a - 1) \ll 1$ is indicated in the figure. Although applicable only for $b/a = 1.189$, the general order of magnitude of the correction should not differ greatly for $b/a = 3$ or 5.3. It should account for the small difference between theoretical and experimental curves for the susceptance with $b/a = 3.0$.

REFERENCES FOR SECTION 2

1. See, for example, King, R. W. P., H. R. Mimno, and A. H. Wing, Jr., *Transmission Lines, Antennas, and Wave Guides*, 2nd ed., Dover Publications, New York (1965), p. 40.
2. See, for example, King, R. W. P., *Transmission-Line Theory*, 2nd ed., Dover Publications, New York (1965), Ch. II.
3. Wu, T. T., and R. W. P. King, "Driving-point and Input Admittance of Linear Antennas," *J. Appl. Phys.*, **30**, 74 (1959).
4. Duncan, R. H., "Theory of the Infinite Cylindrical Antenna Including the Feed-point Singularity in Antenna Current," *J. Res. NBS*, **66D**, 181 (1962).
5. King, R. W. P., and C. W. Harrison, Jr., "Determination of Admittance and Effective Length of Cylindrical Antennas," *Radio Science*, **1** (New Series), 835 (1966).
6. King, R. W. P., *Transmission-Line Theory*, 2nd ed., Dover Publications, New York (1965), Ch. V.
7. King, R. W. P., R. B. Mack, and S. S. Sandler, *Arrays of Cylindrical Dipoles*, Cambridge University Press, Cambridge (1968), Ch. 7.
8. King, R. W. P., *Theory of Linear Antennas*, Harvard University Press, Cambridge, Mass. (1956), p. 214 and Fig. 34.8a.
9. Chang, D. C., and T. T. Wu, "A Note on the Theory of End-Corrections for Thick Monopoles." *Radio Science*, **3** (New Series), 639 (1968).
10. King, R. W. P., and T. T. Wu, "On the Admittance of a Monopole Driven from a Coaxial Line," *Trans. IEEE*, **AP-17**, 814 (November 1969).
11. See, for example, King, R. W. P., *Theory of Linear Antennas*, Harvard University Press, Cambridge, Mass. (1956), Ch. II.
12. Wu, T. T., "Theory of the Dipole Antenna and the Two-Wire Transmission Line," *J. Math. Phys.*, **2**, 550 (1961).
13. Collin, R. E., and F. J. Zucker, Eds., *Antenna Theory*, McGraw-Hill Book Co., Inc., New York (1969), Part 1, Ch. 8.

14. Hartig, E. O., "Circular Apertures and Their Effects on Half-Dipole Impedances," Doctoral Dissertation, Harvard University, Cambridge, Mass. (June 1950).

15. Holly, S., "Experimental Study of Electrically Thick Monopole Antennas," Doctoral Dissertation, Harvard University, Cambridge, Mass. (June 1969).

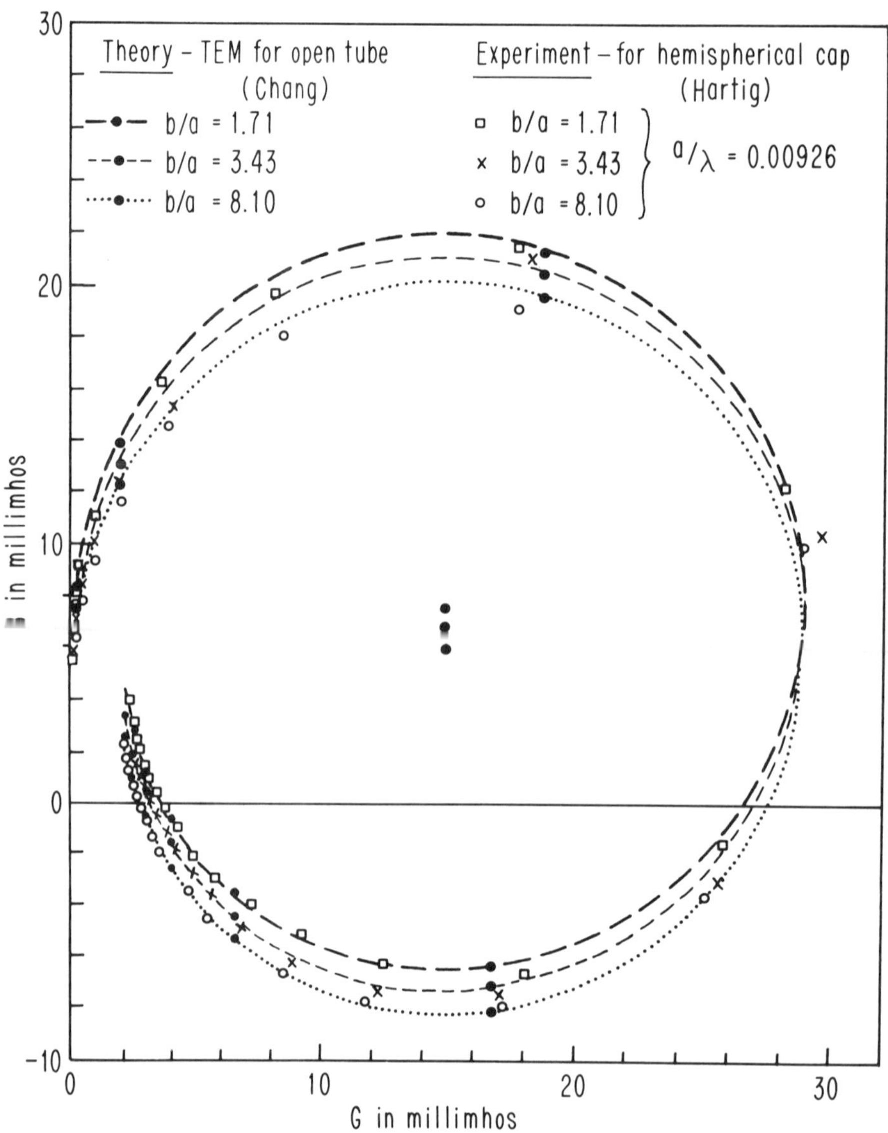

Fig. 2.5a. Measured and theoretical circular graphs of the admittance of a monopole with $a/\lambda = 0.00926$.

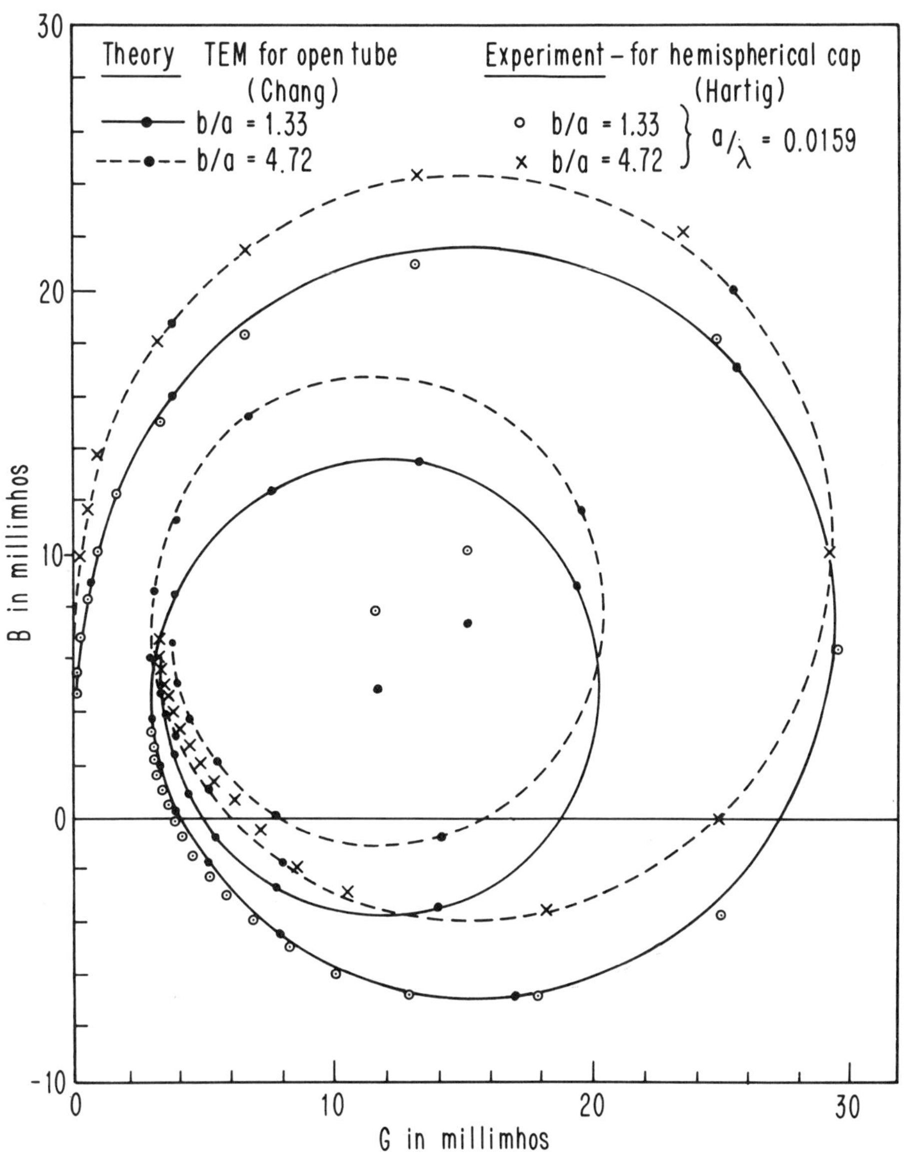

Fig. 2.5b. Measured and theoretical circular graphs of the admittance of a monopole with $a/\lambda = 0.0159$.

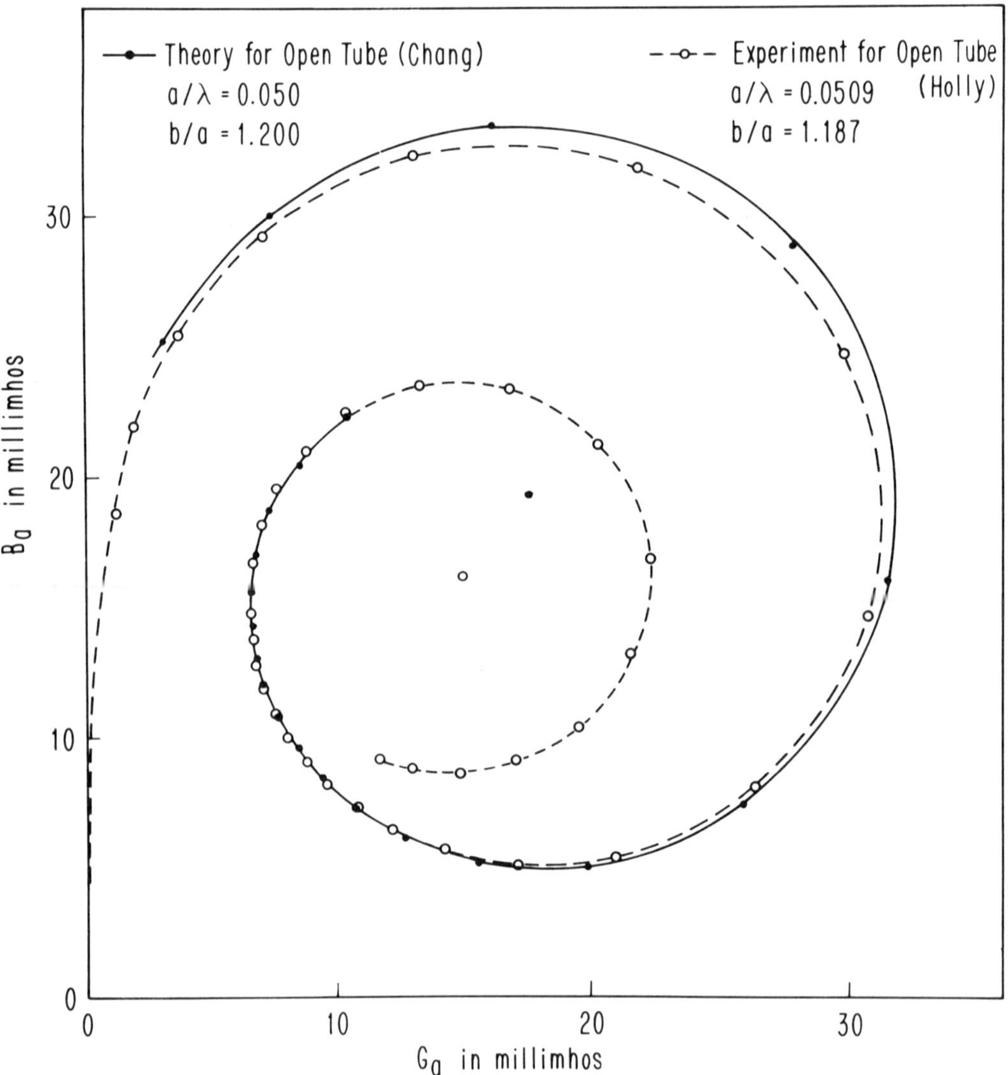

Fig. 2.5c. Measured and theoretical circular graphs of the admittance of a monopole with $a/\lambda = 0.05$.

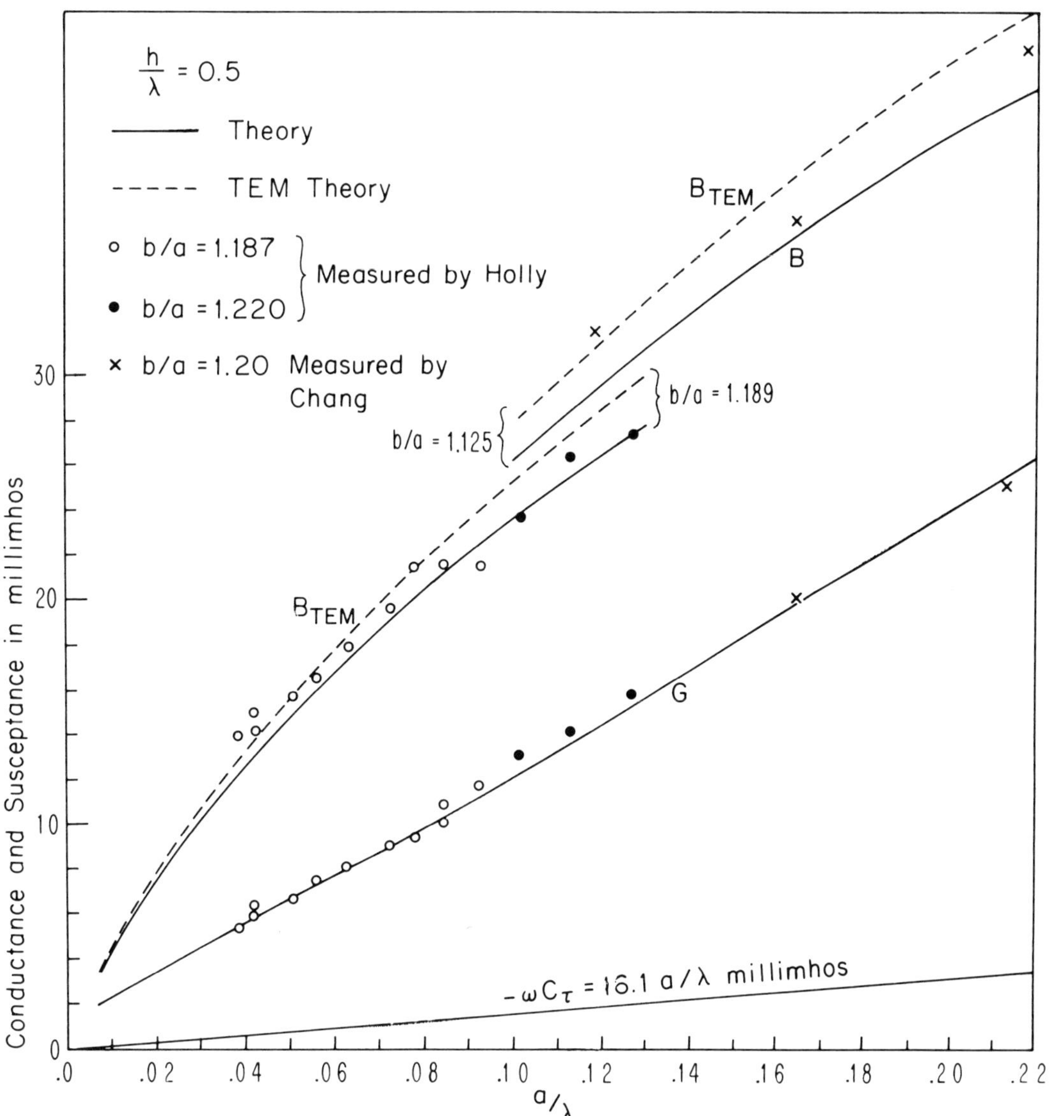

Fig. 2.6. Measured and theoretical conductance and susceptance of a monopole with $h/\lambda = 0.5$.

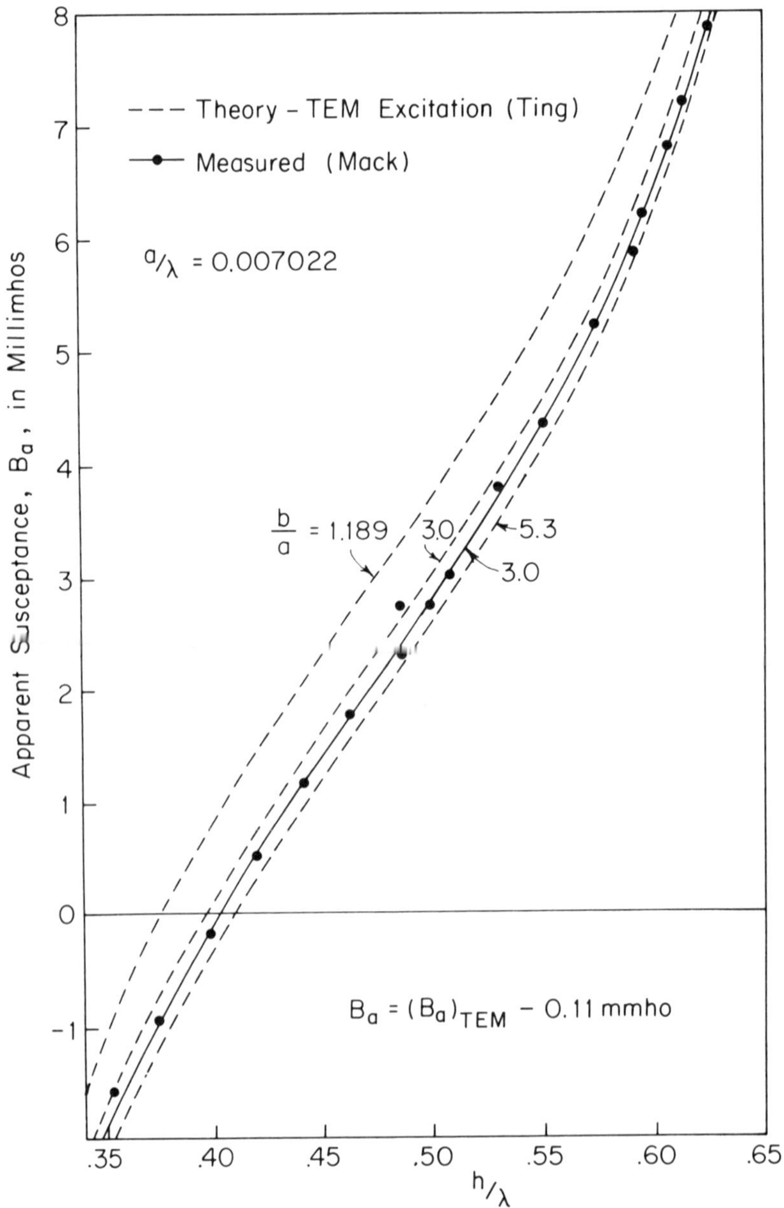

Fig. 2.7. Measured and theoretical susceptance of a monopole with $a/\lambda = 0.007022$; the formula for B_a is valid only when $(b/a - 1) \ll 1$.

TABLE 2.11

MEASURED ADMITTANCE OF TUBULAR, FLAT-TOPPED, AND HEMISPHERICALLY CAPPED MONOPOLES DRIVEN FROM COAXIAL LINE*

Y = G + jB (millimhos)

	$\frac{a}{\lambda} = 0.0423$; $\frac{b}{a} = 1.187$; $Z_c = 10.27$ ohms			$\frac{a}{\lambda} = 0.0509$; $\frac{b}{a} = 1.187$; $Z_c = 10.274$ ohms		
h/λ	Open	Flat	Capped	Open	Flat	Capped
.10	2.7 + j22.6	2.8 + j23.2	2.2 + j20.4	3.7 + j25.5	4.1 + j26.0	2.3 + j22.4
.12	5.0 + j26.2	5.6 + j26.9	3.5 + j23.7	7.2 + j29.3	7.6 + j29.7	4.3 + j26.1
.14	10.3 + j29.9	10.9 + j30.3	6.5 + j27.4	13.1 + j32.3	14.1 + j32.5	7.9 + j29.9
.16	18.9 + j30.7	19.8 + j30.6	12.4 + j30.5	21.9 + j31.8	23.1 + j31.1	14.5 + j32.5
.18	28.2 + j24.6	28.9 + j23.3	21.5 + j29.7	29.8 + j24.6	30.2 + j23.8	23.9 + j30.6
.20	30.1 + j13.9	29.8 + j12.8	29.6 + j21.8	30.8 + j14.6	30.5 + j13.9	30.0 + j22.5
.22	25.2 + j6.5	24.8 + j6.0	29.3 + j11.6	26.4 + j8.1	26.0 + j7.7	29.8 + j13.6
.24	20.2 + j3.7	19.6 + j3.7	23.8 + j5.6	21.0 + j5.4	20.7 + j5.4	25.5 + j8.0
.26	16.1 + j3.2	15.7 + j3.4	19.0 + j3.8	17.1 + j5.1	16.8 + j5.1	20.5 + j5.7
.28	13.2 + j3.8	12.9 + j3.9	15.3 + j3.7	14.3 + j5.7	14.0 + j5.8	16.8 + j5.3
.30	11.2 + j4.6	10.8 + j4.8	12.6 + j4.2	12.2 + j6.5	12.1 + j6.6	14.0 + j5.8
.32	9.6 + j5.6	9.5 + j5.7	10.8 + j5.0	10.9 + j7.4	10.7 + j7.5	12.2 + j6.6
.34	8.6 + j6.6	8.3 + j6.8	9.4 + j6.0	9.6 + j8.2	9.7 + j8.4	10.7 + j7.5
.36	7.7 + j7.7	7.6 + j7.7	8.4 + j6.9	8.8 + j9.1	8.7 + j9.2	9.7 + j8.5
.38	7.1 + j8.6	7.0 + j8.7	7.8 + j7.8	8.1 + j10.0	8.1 + j10.0	8.8 + j9.3
.40	6.7 + j9.6	6.6 + j9.6	7.2 + j8.8	7.6 + j11.0	7.6 + j11.0	8.1 + j10.2
.42	6.3 + j10.5	6.3 + j10.5	6.7 + j9.7	7.2 + j11.9	7.2 + j12.0	7.7 + j11.1
.44	6.1 + j11.3	6.0 + j11.5	6.5 + j10.7	6.9 + j12.8	6.9 + j12.8	7.3 + j12.0
.46	5.9 + j12.3	5.7 + j12.4	6.2 + j11.6	6.8 + j13.8	6.8 + j13.9	7.0 + j12.9
.48	5.7 + j13.3	5.7 + j13.4	6.1 + j12.5	6.7 + j14.8	6.7 + j14.9	6.8 + j13.9
.50	6.0 + j14.2	5.8 + j14.3	5.9 + j13.3	6.7 + j15.8	6.7 + j16.0	6.7 + j14.9
.52	6.0 + j15.3	5.9 + j15.4	6.0 + j14.5	6.8 + j16.9	6.9 + j17.1	6.9 + j15.9
.54	6.2 + j16.5	6.3 + j16.7	6.1 + j15.6	7.2 + j18.2	7.3 + j18.3	7.0 + j17.1
.56	6.7 + j17.9	6.8 + j18.0	6.4 + j16.8	7.7 + j19.6	7.9 + j19.9	7.4 + j18.3
.58	7.5 + j19.3	7.8 + j19.5	7.1 + j18.2	8.9 + j21.0	9.1 + j21.2	8.1 + j19.7
.60	9.2 + j20.7	9.3 + j21.8	8.0 + j19.6	10.5 + j22.5	10.9 + j22.7	9.2 + j21.2
.62	11.0 + j22.0	11.7 + j22.1	9.8 + j21.1	13.4 + j23.5	13.8 + j23.5	11.2 + j22.5
.64	14.8 + j22.4	15.4 + j22.3	12.4 + j22.2	16.9 + j23.3	17.4 + j23.2	14.1 + j23.4
.66	19.0 + j20.8	19.2 + j20.5	16.1 + j22.2	20.4 + j21.2	20.9 + j20.9	17.6 + j22.9
.68	21.6 + j17.0	21.7 + j16.4	20.0 + j20.0	22.6 + j17.2	22.4 + j16.8	20.9 + j20.4
.70	21.6 + j12.2	21.2 + j11.8	21.7 + j15.8	21.6 + j13.2	21.4 + j12.8	22.1 + j16.5
.72	19.2 + j8.7	18.7 + j8.5	20.9 + j11.5	19.5 + j10.4	19.1 + j10.3	21.3 + j12.6
.74	16.4 + j7.1	16.1 + j7.0	18.3 + j8.6	17.1 + j9.1	16.6 + j9.0	18.9 + j10.1

*Measured by S. Holly.

MEASURED ADMITTANCE OF TUBULAR, FLAT-TOPPED, AND HEMISPHERICALLY CAPPED MONOPOLES DRIVEN FROM COAXIAL LINE*

Y = G + jB (millimhos)

	$\frac{a}{\lambda}$ = 0.0635; $\frac{b}{a}$ = 1.187; Z_c = 10.27 ohms			$\frac{a}{\lambda}$ = 0.0847; $\frac{b}{a}$ = 1.187; Z_c = 10.27 ohms		
h/λ	Open	Flat	Capped	Open	Flat	Capped
.10	5.8 + j29.4	6.0 + j30.2	3.9 + j24.7	9.5 + j35.2	11.0 + j35.7	4.6 + j28.1
.12	10.2 + j33.0	11.1 + j33.6	5.3 + j28.7	15.9 + j38.1	18.3 + j37.9	8.0 + j32.6
.14	17.4 + j34.8	19.1 + j34.6	9.7 + j32.3	24.4 + j37.4	26.4 + j36.1	13.1 + j35.8
.16	26.3 + j31.6	27.3 + j30.5	16.5 + j34.1	31.8 + j31.9	32.6 + j29.3	20.8 + j36.7
.18	31.2 + j23.3	31.3 + j21.6	25.0 + j31.5	34.0 + j23.5	33.6 + j21.0	28.2 + j33.2
.20	30.2 + j14.8	29.3 + j13.7	30.2 + j23.8	32.2 + j16.1	30.9 + j14.9	32.5 + j25.9
.22	26.8 + j9.7	24.6 + j9.3	29.8 + j15.5	28.3 + j12.1	26.8 + j11.6	32.1 + j18.7
.24	21.6 + j7.7	20.6 + j7.5	25.9 + j10.2	24.2 + j10.2	23.4 + j10.3	29.0 + j13.8
.26	17.9 + j7.3	17.4 + j7.4	21.4 + j8.1	21.0 + j9.8	20.0 + j10.3	27.0 + j11.4
.28	15.8 + j7.7	15.1 + j7.7	18.1 + j7.5	18.4 + j10.3	17.8 + j10.6	21.9 + j10.6
.30	13.6 + j8.4	12.9 + j8.6	15.3 + j7.8	16.4 + j10.9	16.0 + j11.3	19.2 + j10.6
.32	12.0 + j9.3	11.4 + j9.5	13.5 + j8.5	14.8 + j11.6	14.5 + j12.1	17.1 + j11.1
.34	10.8 + j10.3	10.3 + j10.4	12.2 + j9.4	13.4 + j12.5	13.2 + j13.1	15.5 + j11.9
.36	9.9 + j11.2	9.7 + j11.3	10.8 + j10.3	12.3 + j13.4	12.5 + j14.2	14.1 + j12.8
.38	9.4 + j12.2	8.7 + j12.3	9.9 + j11.2	11.7 + j14.4	11.9 + j15.1	13.0 + j13.7
.40	9.1 + j12.8	8.4 + j13.2	9.3 + j12.1	11.0 + j15.3	11.3 + j16.0	12.3 + j14.6
.42	8.6 + j13.5	8.0 + j14.2	8.5 + j13.2	10.4 + j16.4	10.7 + j17.1	11.6 + j15.5
.44	8.3 + j14.5	7.7 + j15.1	8.1 + j14.1	10.0 + j17.7	10.5 + j18.3	11.2 + j16.6
.46	8.0 + j15.7	7.9 + j16.2	7.9 + j15.0	9.8 + j19.0	10.4 + j19.4	11.0 + j17.5
.48	8.1 + j16.7	8.1 + j17.4	7.7 + j16.1	9.8 + j20.3	10.6 + j20.6	10.6 + j18.8
.50	8.1 + j17.9	8.5 + j18.3	7.8 + j17.1	10.2 + j21.6	10.7 + j21.9	10.4 + j19.8
.52	8.1 + j19.1	8.4 + j19.1	7.9 + j18.5	10.8 + j23.0	11.2 + j23.2	10.7 + j21.1
.54	8.7 + j20.5	8.9 + j20.7	8.6 + j19.6	11.7 + j24.4	12.4 + j24.6	11.1 + j22.4
.56	9.5 + j21.8	9.7 + j22.1	9.3 + j20.3	13.0 + j25.8	13.7 + j25.8	12.1 + j23.6
.58	11.0 + j23.4	11.2 + j23.6	9.5 + j21.5	15.0 + j26.8	15.8 + j26.8	12.9 + j24.9
.60	13.0 + j24.5	13.5 + j24.6	10.5 + j22.9	17.6 + j27.4	18.6 + j27.1	14.8 + j26.2
.62	15.9 + j24.9	16.5 + j24.8	12.8 + j24.2	20.6 + j26.8	21.6 + j26.1	16.9 + j26.6
.64	19.0 + j24.0	19.8 + j23.5	15.6 + j24.6	23.6 + j24.9	24.1 + j24.0	19.9 + j26.4
.66	21.9 + j21.0	22.5 + j20.3	18.9 + j23.6	25.0 + j21.8	25.3 + j20.6	22.8 + j24.9
.68	23.1 + j17.2	23.0 + j16.5	21.5 + j20.9	25.5 + j18.4	24.9 + j17.2	24.5 + j22.3
.70	22.0 + j13.2	21.1 + j13.2	22.4 + j17.3	23.9 + j15.6	23.4 + j14.9	25.0 + j19.1
.72	19.7 + j11.5	19.0 + j11.3	21.4 + j13.9	21.8 + j13.9	21.4 + j13.5	23.9 + j16.3
.74	17.3 + j10.6	17.1 + j10.5	19.3 + j11.7	19.9 + j13.0	19.4 + j12.9	22.4 + j14.5

*Measured by S. Holly.

MEASURED ADMITTANCE OF TUBULAR, FLAT-TOPPED, AND HEMISPHERICALLY CAPPED MONOPOLES DRIVEN FROM COAXIAL LINE*

$Y = G + jB$ (millimhos)

	$\frac{a}{\lambda} = 0.1129$; $\frac{b}{a} = 1.220$; $Z_c = 11.84$ ohms			$\frac{a}{\lambda} = 0.1270$; $\frac{b}{a} = 1.220$; $Z_c = 11.84$ ohms		
h/λ	Open	Flat	Capped	Open	Flat	Capped
.10	14.7 + j40.5	18.2 + j40.8	6.6 + j29.5	16.9 + j42.2	17.6 + j40.8	6.2 + j28.2
.12	22.1 + j42.3	25.9 + j41.1	10.8 + j34.5	24.9 + j42.4	27.2 + j39.8	11.0 + j35.1
.14	30.3 + j40.4	33.3 + j37.4	16.6 + j37.8	32.3 + j38.8	32.9 + j34.9	13.1 + j40.9
.16	36.8 + j34.6	37.0 + j30.8	21.2 + j43.3	37.1 + j32.3	35.8 + j28.3	22.9 + j40.1
.18	38.1 + j27.0	36.5 + j24.3	30.5 + j36.5	37.5 + j25.0	34.8 + j22.6	26.9 + j37.7
.20	36.5 + j20.4	34.0 + j19.0	34.3 + j32.9	35.1 + j19.9	32.3 + j18.0	30.8 + j32.0
.22	32.9 + j16.7	30.3 + j15.7	35.1 + j24.4	31.8 + j16.9	29.4 + j15.7	33.6 + j25.4
.24	27.9 + j14.5	26.4 + j14.3	33.1 + j19.6	28.2 + j15.2	27.1 + j15.2	31.7 + j19.8
.26	24.6 + j13.8	23.6 + j13.9	30.1 + j16.8	25.4 + j14.6	24.6 + j15.1	29.7 + j17.8
.28	22.2 + j13.9	20.9 + j14.1	27.0 + j15.1	22.8 + j14.6	23.0 + j15.4	26.5 + j15.9
.30	20.1 + j14.4	19.2 + j14.9	24.7 + j14.4	21.2 + j15.0	20.8 + j15.9	24.2 + j15.4
.32	18.3 + j15.4	18.6 + j15.6	22.3 + j14.4	19.5 + j15.9	19.4 + j16.8	22.4 + j15.5
.34	17.1 + j16.3	16.8 + j16.7	20.7 + j14.9	18.4 + j16.9	17.7 + j18.0	21.5 + j16.8
.36	16.1 + j17.3	15.7 + j17.7	19.0 + j15.6	17.1 + j18.0	17.0 + j18.9	20.3 + j17.5
.38	15.1 + j18.4	15.0 + j18.9	17.9 + j16.4	16.7 + j18.7	16.5 + j19.6	19.4 + j18.0
.40	14.8 + j19.5	14.5 + j20.1	16.9 + j17.3	16.1 + j20.1	15.9 + j20.9	18.5 + j19.3
.42	14.0 + j20.8	14.4 + j21.4	16.1 + j18.5	15.7 + j21.5	15.8 + j22.5	17.7 + j20.0
.44	13.5 + j22.1	14.2 + j22.7	15.8 + j19.6	15.6 + j22.8	15.9 + j23.8	16.8 + j21.1
.46	13.4 + j23.4	14.0 + j24.2	15.5 + j20.8	15.7 + j24.2	15.9 + j25.3	16.5 + j22.3
.48	13.7 + j24.9	14.3 + j25.4	15.0 + j22.1	15.7 + j25.8	16.1 + j26.7	16.4 + j23.4
.50	14.2 + j26.4	15.2 + j27.0	15.1 + j23.4	15.8 + j27.3	16.8 + j28.1	16.5 + j24.4
.52	14.8 + j27.9	16.2 + j28.8	15.1 + j24.9	16.5 + j28.7	17.5 + j29.3	16.5 + j25.8
.54	16.1 + j29.4	17.3 + j30.2	15.7 + j26.5	17.8 + j30.2	18.6 + j30.5	16.9 + j27.2
.56	17.9 + j30.8	19.5 + j31.3	16.6 + j28.0	19.6 + j31.3	20.7 + j31.3	17.0 + j28.4
.58	20.2 + j31.8	22.0 + j32.0	17.9 + j29.4	21.5 + j31.9	23.0 + j31.6	19.3 + j29.3
.60	22.7 + j32.1	24.7 + j32.0	19.7 + j30.2	24.1 + j31.7	25.4 + j30.8	20.7 + j30.1
.62	25.8 + j31.6	27.3 + j30.7	22.2 + j30.7	26.5 + j30.6	27.5 + j29.2	22.8 + j30.3
.64	27.9 + j29.4	28.5 + j28.6	24.4 + j30.3	28.4 + j28.1	28.6 + j27.0	24.7 + j29.6
.66	29.6 + j26.8	29.1 + j25.7	26.9 + j29.2	28.9 + j26.0	28.6 + j24.7	26.3 + j28.3
.68	28.7 + j23.7	28.5 + j22.8	28.3 + j27.2	28.8 + j23.5	28.5 + j22.9	27.5 + j26.4
.70	27.9 + j21.2	26.6 + j20.8	29.0 + j24.8	27.7 + j21.7	26.9 + j21.1	27.9 + j24.2
.72	25.7 + j19.6	25.0 + j19.6	28.8 + j22.5	26.1 + j20.5	25.4 + j20.1	27.2 + j22.4
.74	24.0 + j18.6	23.8 + j18.7	27.3 + j20.6	24.4 + j19.7	24.3 + j20.1	25.9 + j21.1

*Measured by S. Holly.

TABLE 2.12

MEASURED ADMITTANCE OF HEMISPHERICALLY CAPPED MONOPOLE*

Y = G + jB (millimhos); h' = h + a

a/λ = 0.002980

k_0h'	h'/λ	b/a = 2.21	b/a = 5.32	b/a = 7.09	b/a = 10.64	b/a = 25.11
0.105	.017	0.00 + j1.24	0.00 + j0.91	0.00 + j0.88	0.00 + j0.76	0.00 + j1.14
0.209	.033	0.00 + j1.97	0.00 + j1.64	0.00 + j1.57	0.00 + j1.32	0.00 + j1.45
0.314	.050	0.00 + j2.63	0.00 + j2.32	0.00 + j2.22	0.00 + j1.99	0.00 + j2.00
0.419	.067	0.00 + j3.29	0.00 + j2.94	0.00 + j2.87	0.00 + j2.65	0.00 + j2.61
0.524	.083	0.05 + j4.02	0.04 + j3.68	0.04 + j3.53	0.04 + j3.36	0.04 + j3.34
0.628	.100	0.10 + j4.72	0.10 + j4.38	0.10 + j4.27	0.10 + j4.20	0.10 + j4.05
0.732	.117	0.19 + j5.30	0.17 + j5.28	0.20 + j5.26	0.22 + j5.15	0.24 + j4.98
0.838	.133	0.41 + j6.82	0.36 + j6.45	0.36 + j6.32	0.40 + j6.16	0.42 + j5.99
0.942	.150	0.83 + j8.39	0.76 + j8.00	0.76 + j7.97	0.82 + j7.74	0.84 + j7.67
1.047	.167	1.69 + j10.41	1.57 + j9.98	1.53 + j9.94	1.59 + j9.77	1.66 + j9.70
1.152	.183	3.57 + j13.47	3.44 + j12.96	3.30 + j12.78	3.46 + j12.71	3.62 + j12.64
1.256	.200	8.50 + j17.34	8.13 + j16.66	8.05 + j16.45	8.13 + j16.32	8.68 + j16.34
1.361	.217	20.86 + j16.87	19.59 + j17.01	20.32 + j17.29	20.32 + j17.01	20.99 + j15.84
1.413	.225	27.47 + j9.33	27.03 + j9.81	27.73 + j10.16	27.64 + j9.04	--- ---
1.466	.233	27.55 - j0.18	27.75 - j0.24	27.97 - j0.46	27.65 - j0.55	26.88 - j1.65
1.518	.242	22.32 - j6.46	23.24 - j6.51	23.15 - j6.91	23.12 - j6.77	--- ---
1.571	.250	17.90 - j8.49	18.62 - j9.22	18.89 - j9.52	17.55 - j8.99	17.14 - j9.66
1.623	.258	13.23 - j8.76	--- ---	--- ---	--- ---	--- ---
1.675	.267	10.20 - j8.56	10.31 - j8.88	10.43 - j9.00	10.30 - j9.14	9.68 - j9.23
1.780	.283	6.82 - j6.78	6.78 - j7.00	6.80 - j7.15	6.73 - j7.24	6.32 - j7.46
1.885	.300	4.92 - j5.37	5.00 - j5.52	5.03 - j5.74	4.95 - j5.88	4.68 - j6.02
1.989	.317	3.69 - j3.93	3.87 - j4.22	3.96 - j4.57	3.86 - j4.69	3.74 - j4.45
2.094	.333	3.02 - j2.98	3.23 - j3.31	3.28 - j3.56	3.12 - j3.69	3.02 - j3.78
2.199	.350	2.65 - j2.27	2.74 - j2.60	2.80 - j2.81	2.66 - j2.96	2.62 - j3.07
2.303	.367	2.25 - j1.57	2.39 - j1.89	2.40 - j2.19	2.34 - j2.30	2.26 - j2.36
2.408	.383	2.08 - j1.03	2.14 - j1.37	2.16 - j1.57	2.17 - j1.69	2.00 - j1.81
2.513	.400	1.86 - j0.55	1.92 - j0.85	1.96 - j1.05	1.92 - j1.16	1.81 - j1.31
2.618	.417	1.75 - j0.13	1.79 - j0.37	1.82 - j0.61	1.87 - j0.77	1.69 - j0.86
2.670	.425	1.72 + j0.11	1.78 - j0.17	1.76 - j0.41	--- ---	--- ---
2.722	.433	1.64 + j0.29	1.72 + j0.06	1.70 - j0.15	1.70 - j0.26	1.54 - j0.45
2.774	.441	--- ---	1.62 + j0.26	1.66 + j0.05	1.64 - j0.10	--- ---
2.827	.450	1.56 + j0.71	1.54 + j0.44	1.60 + j0.25	1.60 + j0.13	1.45 - j0.10
2.880	.458	--- ---	--- ---	--- ---	1.56 + j0.33	1.44 + j0.10
2.932	.467	1.52 + j1.10	1.50 + j0.84	1.52 + j0.64	1.52 + j0.49	1.39 + j0.31
3.036	.483	1.42 + j1.56	1.49 + j1.25	1.46 + j1.04	1.44 + j0.89	1.36 + j0.70
3.141	.500	1.41 + j1.89	1.43 + j1.60	1.43 + j1.39	1.38 + j1.24	1.30 + j1.05
3.246	.517	--- ---	1.42 + j2.03	1.39 + j1.79	1.37 + j1.65	1.26 + j1.47
3.351	.533	--- ---	1.39 + j2.42	--- ---	1.35 + j2.07	1.24 + j1.88
3.456	.550	--- ---	--- ---	--- ---	--- ---	1.25 + j2.31

*From data of E. O. Hartig.

MEASURED ADMITTANCE OF HEMISPHERICALLY CAPPED MONOPOLE*

$Y = G + jB$ (millimhos); $h' = h + a$

$a/\lambda = 0.003970$

k_0h'	h'/λ	b/a = 1.67	b/a = 4.00	b/a = 5.33	b/a = 8.00	b/a = 18.88
0.105	.017	0.00 + j1.44	0.00 + j1.02	0.00 + j0.99	0.00 + j0.91	0.00 + j1.04
0.209	.033	0.00 + j2.26	0.00 + j1.91	0.00 + j1.75	0.00 + j1.58	0.00 + j1.64
0.314	.050	0.00 + j3.00	0.00 + j2.66	0.00 + j2.51	0.00 + j2.34	0.00 + j2.25
0.419	.067	0.03 + j3.71	0.02 + j3.36	0.02 + j3.32	0.02 + j3.23	0.03 + j2.92
0.524	.083	0.10 + j4.51	0.04 + j4.15	0.10 + j3.99	0.04 + j3.86	0.10 + j3.68
0.628	.100	0.13 + j5.41	0.10 + j5.02	0.13 + j4.85	0.11 + j4.71	0.16 + j4.52
0.732	.117	0.25 + j6.40	0.22 + j6.02	0.27 + j5.88	0.22 + j5.64	0.29 + j5.51
0.838	.133	0.46 + j7.66	0.46 + j7.31	0.48 + j7.19	0.48 + j6.93	0.54 + j6.79
0.942	.150	0.98 + j9.37	0.94 + j8.98	0.97 + j8.87	0.96 + j8.60	1.03 + j8.43
1.047	.167	1.99 + j11.59	1.94 + j11.21	1.97 + j11.10	2.20 + j10.71	2.07 + j10.67
1.152	.183	4.28 + j14.67	4.11 + j14.30	4.25 + j14.17	4.22 + j13.87	4.98 + j14.79
1.256	.200	10.15 + j18.13	9.99 + j17.96	10.19 + j17.76	9.84 + j17.50	10.36 + j17.24
1.361	.217	22.73 + j15.54	22.87 + j15.89	22.80 + j16.05	22.87 + j15.90	23.33 + j14.79
1.413	.225	27.40 + j8.11	28.02 + j7.67	28.44 + j7.59	28.07 + j7.12	--- ---
1.466	.233	26.99 − j0.35	26.97 − j0.74	27.20 − j0.93	27.25 − j1.56	26.30 − j2.49
1.518	.242	22.02 − j6.06	22.55 − j6.95	22.96 − j7.01	22.59 − j7.32	--- ---
1.571	.250	17.23 − j7.78	17.44 − j9.23	17.65 − j8.95	17.35 − j9.49	16.52 − j9.48
1.623	.258	10.93 − j7.85	10.63 − j8.30	10.76 − j8.58	10.67 − j8.95	10.12 − j8.84
1.675	.267	7.19 − j6.16	7.08 − j6.71	6.83 − j7.02	7.19 − j7.22	6.93 − j7.43
1.780	.283	5.32 − j4.81	5.35 − j5.39	5.43 − j5.57	5.31 − j5.75	5.08 − j5.93
1.885	.300	4.23 − j3.60	4.28 − j4.17	4.30 − j4.40	4.23 − j4.56	4.05 − j4.78
1.989	.317	3.47 − j2.64	3.52 − j3.23	3.52 − j3.43	3.45 − j3.62	3.32 − j3.80
2.094	.333	3.02 − j1.88	3.00 − j2.43	3.00 − j2.62	2.97 − j2.83	2.85 − j3.00
2.199	.350	2.69 − j1.25	2.63 − j1.76	2.64 − j1.99	2.64 − j2.16	2.54 − j2.36
2.303	.367	2.36 − j0.65	2.36 − j1.20	2.35 − j1.41	2.37 − j1.54	2.25 − j1.77
2.408	.383	2.16 − j0.13	2.12 − j0.68	2.16 − j0.88	2.06 − j1.02	2.04 − j1.27
2.513	.400	2.10 + j0.11	2.03 − j0.44	2.10 − j0.62	--- ---	--- ---
2.618	.417	2.00 + j0.31	1.98 − j0.24	2.00 − j0.38	1.98 − j0.60	1.86 − j0.79
2.670	.425	1.92 + j0.55	1.92 + j0.00	1.94 − j0.15	1.88 − j0.16	--- ---
2.722	.433	1.86 + j0.73	1.86 + j0.21	1.86 + j0.10	--- ---	1.76 − j0.37
2.774	.441	--- ---	--- ---	--- ---	1.80 + j0.10	1.70 − j0.16
2.827	.450	1.76 + j1.18	1.76 + j0.61	1.78 + j0.49	1.76 + j0.28	1.67 + j0.05
2.880	.458	--- ---	--- ---	--- ---	1.72 + j0.49	--- ---
2.932	.467	1.69 + j1.56	1.69 + j1.01	1.68 + j0.89	1.68 + j0.73	1.59 + j0.46
3.036	.483	1.65 + j1.99	1.63 + j1.44	1.63 + j1.30	1.62 + j1.11	1.51 + j0.84
3.141	.500	1.58 + j2.39	1.58 + j1.85	1.59 + j1.70	1.55 + j1.54	1.47 + j1.25
3.246	.517	--- ---	1.50 + j2.30	1.54 + j2.13	1.53 + j1.94	1.43 + j1.67
3.351	.533	--- ---	--- ---	--- ---	1.52 + j2.37	1.42 + j2.13
3.456	.550	--- ---	--- ---	--- ---	--- ---	1.44 + j2.60

*From data of E. O. Hartig.

MEASURED ADMITTANCE OF HEMISPHERICALLY CAPPED MONOPOLE*

Y = G + jB (millimhos); h' = h + a

a/λ = 0.009260

k_0h'	h'/λ	b/a = 1.71	b/a = 2.28	b/a = 3.43	b/a = 8.10
0.105	.017	0.00 + j2.31	0.00 + j2.07	0.00 + j1.75	0.00 + j1.57
0.209	.033	0.01 + j3.53	0.01 + j3.25	0.01 + j2.84	0.00 + j2.41
0.314	.050	0.02 + j4.54	0.03 + j4.21	0.03 + j3.82	0.00 + j3.29
0.419	.067	0.10 + j5.58	0.10 + j5.21	0.06 + j4.80	0.05 + j4.21
0.524	.083	0.16 + j6.65	0.11 + j6.29	0.12 + j5.83	0.15 + j5.20
0.628	.100	0.28 + j7.91	0.23 + j7.46	0.27 + j7.06	0.30 + j6.33
0.732	.117	0.54 + j9.25	0.51 + j8.87	0.53 + j8.46	0.58 + j7.71
0.838	.133	1.03 + j11.01	0.98 + j10.60	1.01 + j10.07	1.06 + j9.38
0.942	.150	1.99 + j13.31	1.96 + j12.87	1.99 + j12.35	2.01 + j11.65
1.047	.167	3.78 + j16.21	3.99 + j15.79	4.06 + j15.26	3.95 + j14.56
1.152	.183	8.27 + j19.60	8.48 + j19.20	8.38 + j18.48	8.51 + j17.92
1.256	.200	17.81 + j21.36	17.37 + j20.21	18.23 + j20.92	17.92 + j18.90
1.361	.217	28.17 + j12.16	28.24 + j10.93	29.62 + j10.30	28.21 + j9.88
1.466	.233	25.82 − j1.77	25.45 − j1.91	25.62 − j3.10	25.23 − j3.71
1.571	.250	18.10 − j6.71	17.61 − j6.82	17.20 − j7.41	17.25 − j7.86
1.675	.267	12.54 − j6.26	12.27 − j6.68	12.30 − j7.36	11.79 − j7.97
1.780	.283	9.01 − j5.08	8.89 − j5.48	8.90 − j6.24	8.61 − j6.73
1.885	.300	7.12 − j4.02	6.94 − j4.24	6.84 − j4.93	6.74 − j5.47
1.989	.317	5.75 − j2.95	5.64 − j3.19	5.62 − j3.67	5.52 − j4.60
2.094	.333	4.87 − j2.02	4.77 − j2.24	4.83 − j2.71	4.82 − j3.53
2.199	.350	4.29 − j1.00	4.21 − j1.39	4.22 − j1.85	4.07 − j2.75
2.303	.367	3.78 − j0.25	3.77 − j0.72	3.83 − j1.17	3.56 − j2.01
2.408	.383	3.37 + j0.36	3.39 − j0.05	3.39 − j0.53	3.21 − j1.42
2.513	.400	3.17 + j0.96	3.08 + j0.52	3.14 + j0.06	3.04 − j0.83
2.618	.417	2.94 + j1.44	2.90 + j1.04	2.92 + j0.56	2.79 − j0.35
2.722	.433	2.73 + j2.03	2.77 + j1.56	2.75 + j1.08	2.61 + j0.20
2.827	.450	2.56 + j2.52	2.61 + j2.07	2.60 + j1.56	2.49 + j0.67
2.932	.467	2.50 + j2.99	2.46 + j2.56	2.49 + j2.09	2.39 + j1.18
3.036	.483	2.39 + j3.41	2.41 + j3.05	2.40 + j2.56	2.35 + j1.65
3.141	.500	2.33 + j3.97	2.31 + j3.59	2.36 + j3.03	2.23 + j2.15

*From data of E. O. Hartig.

MEASURED ADMITTANCE OF HEMISPHERICALLY CAPPED MONOPOLE*

$Y = G + jB$ (millimhos); $h' = h + a$

$a/\lambda = 0.015900$

k_0h'	h'/λ	b/a = 1.33	b/a = 2.00	b/a = 4.72
0.105	.017	0.00 + j4.04	0.00 + j2.71	0.00 + j2.13
0.209	.033	0.00 + j5.81	0.00 + j4.30	0.01 + j3.24
0.314	.050	0.10 + j7.16	0.06 + j5.60	0.10 + j4.35
0.419	.067	0.12 + j8.48	0.13 + j6.90	0.15 + j5.50
0.524	.083	0.23 + j9.89	0.25 + j8.26	0.28 + j6.82
0.628	.100	0.49 + j11.68	0.48 + j9.81	0.50 + j8.34
0.732	.117	0.97 + j13.74	0.95 + j11.62	0.98 + j10.06
0.838	.133	1.77 + j15.35	1.92 + j13.83	1.77 + j12.25
0.942	.150	3.35 + j18.11	3.40 + j16.55	3.40 + j15.02
1.047	.167	6.73 + j21.41	6.83 + j19.74	6.68 + j18.24
1.152	.183	13.38 + j24.22	13.12 + j22.53	13.25 + j20.95
1.256	.200	23.52 + j22.08	22.63 + j20.45	24.86 + j18.02
1.361	.217	29.20 + j9.99	29.49 + j9.90	29.58 + j6.29
1.466	.233	25.23 - j0.03	25.60 - j0.24	24.98 - j3.74
1.571	.250	18.33 - j3.48	18.21 - j4.15	17.86 - j6.74
1.675	.267	13.66 - j3.38	13.84 - j4.29	13.04 - j6.78
1.780	.283	10.63 - j2.74	10.37 - j3.63	10.20 - j5.96
1.885	.300	8.63 - j1.83	8.62 - j2.81	8.28 - j4.88
1.989	.317	7.25 - j0.45	7.17 - j1.99	6.91 - j3.87
2.094	.333	6.30 + j0.69	6.14 - j1.16	5.99 - j2.96
2.199	.350	5.48 + j1.45	5.39 - j0.31	5.24 - j2.22
2.303	.367	4.95 + j2.12	4.87 + j0.44	4.59 - j1.42
2.408	.383	4.51 + j2.81	4.46 + j1.09	4.22 - j0.75
2.513	.400	4.17 + j3.40	4.11 + j1.74	3.93 - j0.10
2.618	.417	3.86 + j3.99	3.76 + j2.33	3.67 + j0.48
2.722	.433	3.63 + j4.59	3.66 + j2.88	3.45 + j1.05
2.827	.450	3.51 + j5.10	3.45 + j3.42	3.29 + j1.63
2.932	.467	3.40 + j5.65	3.33 + j4.02	3.17 + j2.20
3.036	.483	3.26 + j6.26	3.23 + j4.61	3.08 + j2.63
3.141	.500	3.23 + j6.81	3.17 + j5.30	3.01 + j3.33

*From data of E. O. Hartig.

TABLE 2.13

ADMITTANCE OF TUBULAR MONOPOLES*

Y = G + jB (millimhos)

$a/\lambda = 0.00926$			B			
$k_0a = 0.0582$			b/a (Z_c in ohms)			
h/λ	k_0h	G	1.71 (32.2)	2.28 (49.4)	3.43 (73.9)	8.1 (125.4)
.10	0.628	0.28	8.30	7.91	7.47	6.63
.15	0.942	2.04	13.87	13.48	13.04	12.17
.20	1.257	18.88	21.15	20.77	20.31	19.32
.25	1.571	16.84	-6.41	-6.80	-7.25	-8.22
.30	1.885	6.67	-3.60	-3.99	-4.44	-5.36
.35	2.199	4.06	-0.78	-1.16	-1.61	-2.52
.40	2.513	3.02	1.15	0.77	0.32	-0.59
.45	2.827	2.51	2.70	2.32	1.87	0.96
.50	3.142	2.26	4.18	3.80	3.35	2.44
.55	3.456	2.25	5.86	5.48	5.03	4.12
.60	3.770	2.73	8.16	7.79	7.34	6.43
.65	4.084	5.17	11.89	11.51	11.06	10.15
.70	4.398	16.65	12.32	11.95	11.50	10.57
.75	4.712	13.84	-2.98	-3.36	-3.80	-4.73
.80	5.026	6.63	-2.06	-2.43	-2.88	-3.79
.85	5.341	4.34	-0.03	-0.34	-0.78	-1.70
.90	5.655	3.41	1.62	1.26	0.81	-0.97
.95	5.969	2.95	2.97	2.60	2.16	1.25
1.00	6.283	2.74	4.28	3.92	3.47	2.56

*Formulation of D. C. Chang; programmed by A. O. Aronson.

ADMITTANCE OF TUBULAR MONOPOLES*

Y = G + jB (millimhos)

a/λ = 0.0159			B		
k_0a = 0.0998			b/a (Z_c in ohms)		
h/λ	k_0h	G	1.33 (17.1)	2.0 (41.5)	4.72 (93)
.10	0.628	0.54	11.81	10.47	8.92
.15	0.942	3.78	18.78	17.43	15.96
.20	1.257	25.58	19.98	18.59	17.03
.25	1.571	17.06	-3.92	-5.13	-6.82
.30	1.885	7.97	-1.62	-2.83	-4.48
.35	2.199	5.15	1.09	-0.11	-1.75
.40	2.513	3.94	3.13	1.93	0.30
.45	2.827	3.33	4.86	3.66	2.03
.50	3.142	3.05	6.57	5.38	3.74
.55	3.456	3.11	8.57	7.38	5.75
.60	3.770	3.94	11.33	10.14	8.51
.65	4.084	7.79	15.30	14.12	12.49
.70	4.398	19.56	11.60	10.43	8.78
.75	4.712	14.09	-0.70	-1.87	-3.52
.80	5.026	7.80	0.10	-1.07	-2.70
.85	5.341	5.45	2.05	0.89	-0.74
.90	5.655	4.41	3.70	2.54	0.92
.95	5.969	3.88	5.16	4.02	2.39
1.00	6.283	3.67	6.66	5.52	3.89

*Formulation of D. C. Chang; programmed by A. O. Aronson.

3. Imperfectly Conducting Dipoles

For many purposes the conductivity σ of copper is sufficiently high so that no significant differences exist between the distributions of current and the admittances of copper antennas and the physically unavailable but mathematically convenient perfectly conducting antennas. The discussion and the tables in Section 2 assume $\sigma = \infty$ or $z^i = r^i + jx^i = 0$, where z^i is the internal impedance per unit length. In this section the properties of cylindrical antennas with nonzero values of z^i are considered. The basis of the results presented is given in three papers.[1,2,3] The impedance per unit length of the cylindrical antenna is conveniently expressed in terms of the dimensionless real parameter

$$\Phi_i = 2\lambda r^i/\zeta_0 \tag{3.1}$$

where r^i is the resistance per unit length, λ is the wavelength in air, and $\zeta_0 \doteq 120\pi$ ohms. Typical graphs of the distributions of current along cylindrical antennas that are a half wavelength and a full wavelength long when perfectly conducting are shown in Figs. 3.1 and 3.2, with Φ_i as the parameter. Typical graphs of the admittance of resistive dipoles are shown in Fig. 3.3, and graphs of the impedance in Fig. 3.4. The efficiency is shown in Fig. 3.5. Tables 3.1 and 3.2 give numerical values of the admittance and the impedance, respectively, of resistive dipoles for a wide range of values of h/λ and Φ_i.

REFERENCES FOR SECTION 3

1. King, R. W. P., and T. T. Wu, "The Imperfectly Conducting Cylindrical Transmitting Antenna," *Trans. IEEE*, **AP-14**, 524 (September 1966).
2. King, R. W. P., C. W. Harrison, Jr., and E. A. Aronson, "The Imperfectly Conducting Cylindrical Transmitting Antenna: Numerical Results," *Trans. IEEE*, **AP-14**, 535 (September 1966).
3. Shen, L. C., "An Experimental Study of Imperfectly Conducting Dipoles," *Trans. IEEE*, **AP-15**, 782 (November 1967).

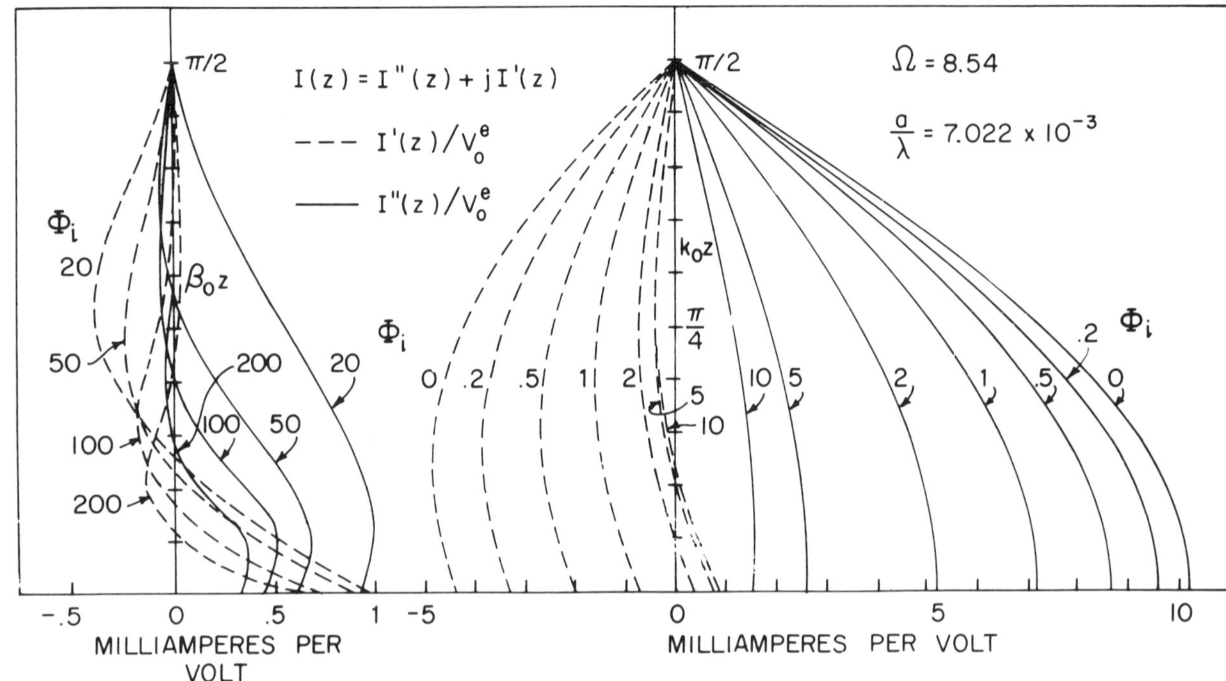

Fig. 3.1. Distribution of current along imperfectly conducting half-wave dipoles; the parameter is $\Phi_i = 2\lambda r^i/\zeta_0$.

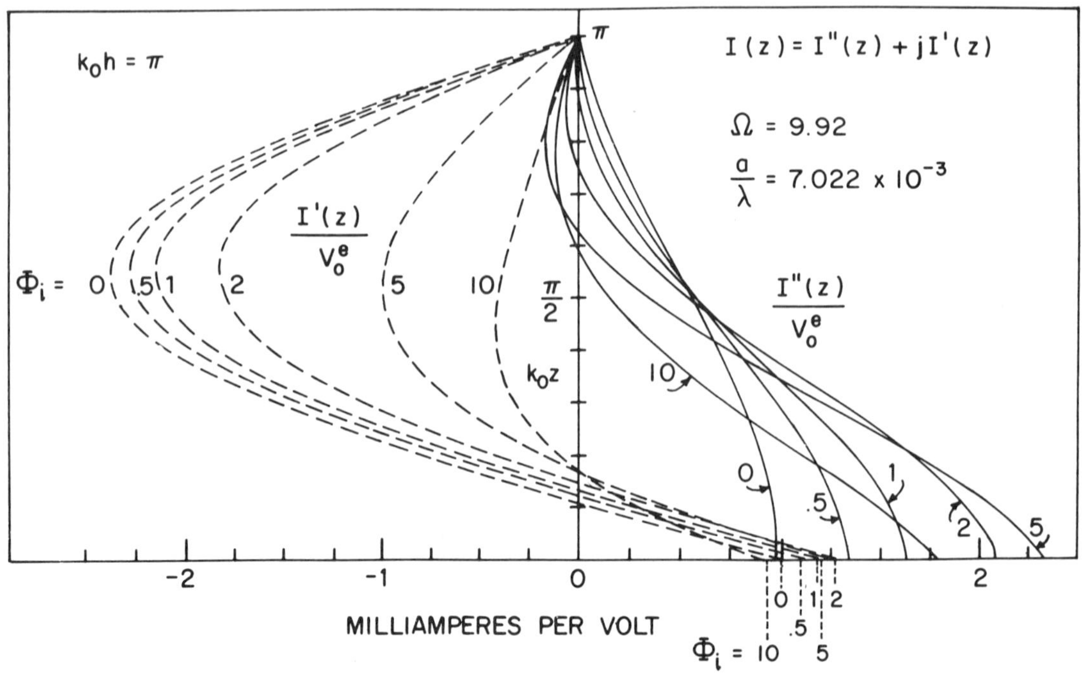

Fig. 3.2. Distribution of current along imperfectly conducting full-wave dipoles; the parameter is $\Phi_i = 2\lambda r^i/\zeta_0$.

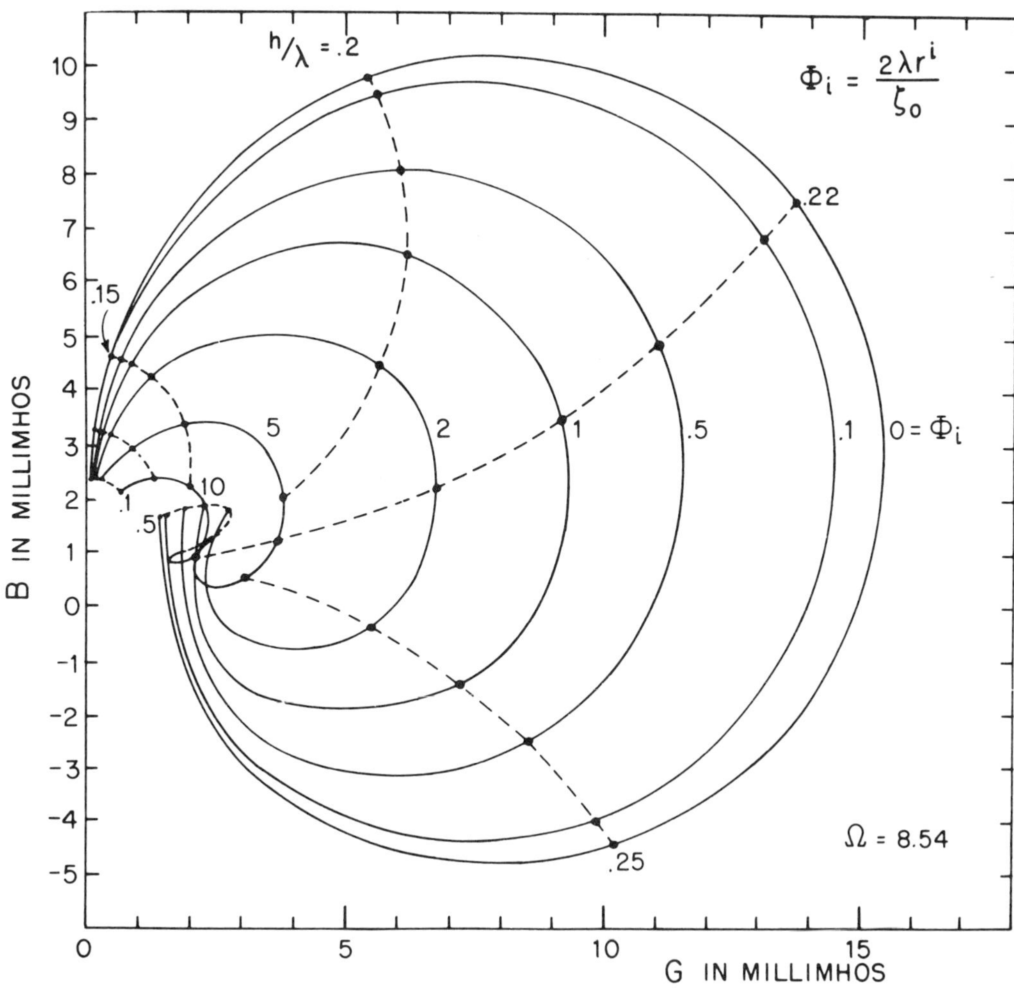

Fig. 3.3. Admittance $Y = G + jB$ of an imperfectly conducting dipole.

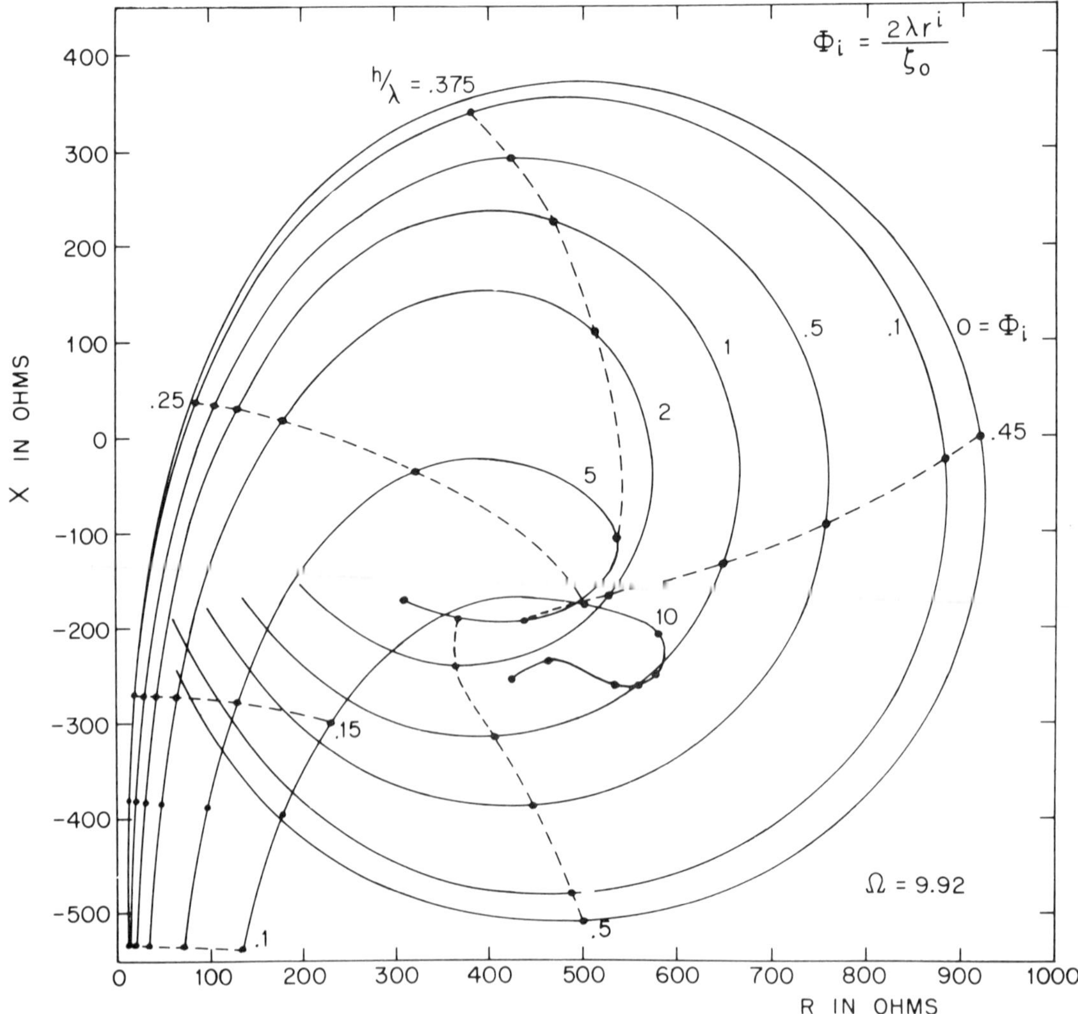

Fig. 3.4. Impedance $Z = R + jX$ of an imperfectly conducting dipole.

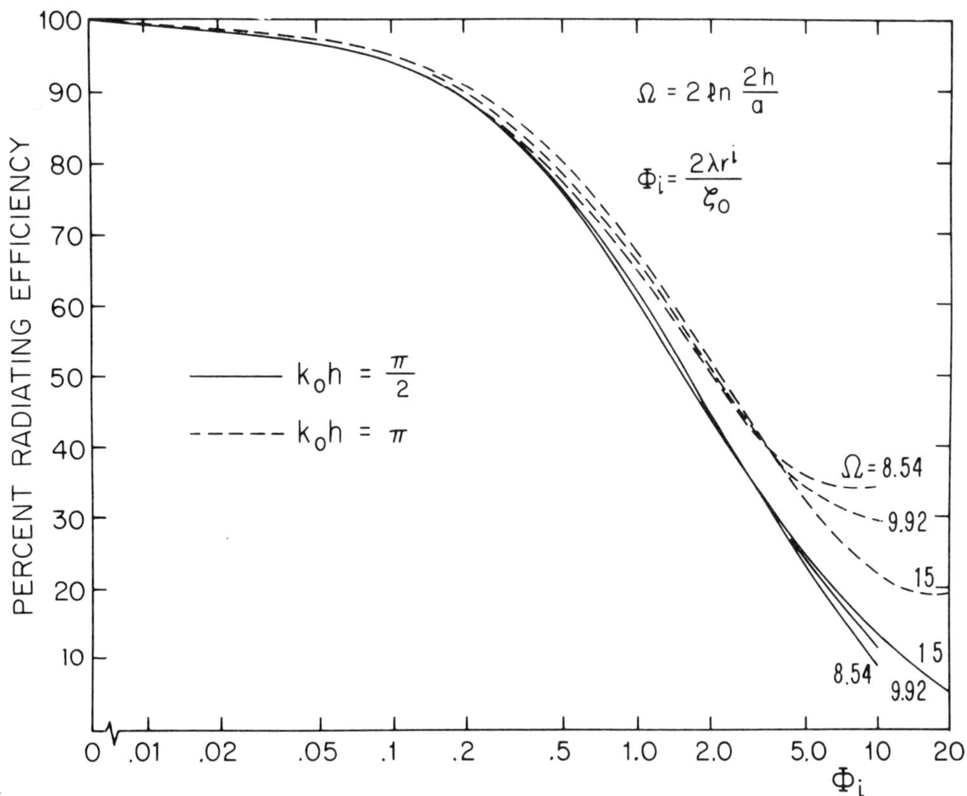

Fig. 3.5. Radiating efficiency of imperfectly conducting half-wave and full-wave dipoles as a function of $\Phi_i = 2\lambda r^i/\zeta_0$.

Table 3.1

Admittance of Resistive Antenna, Y = G + jB (millimhos) Ω = 8.54

$\frac{2\lambda r^i}{\zeta_o}$	$\frac{h}{\lambda}$ = .125	.250	.375
0	0.14+j3.29	10.21−j4.45	1.93−j1.31
0.01	0.15+j3.29	10.13−j4.45	1.93−j1.31
0.02	0.15+j3.29	10.09−j4.39	1.94−j1.30
0.05	0.15+j3.29	9.99−j4.24	1.95−j1.27
0.1	0.16+j3.29	9.81−j4.00	1.98−j1.24
0.2	0.18+j3.29	9.47−j3.56	2.02−j1.16
0.5	0.23+j3.28	8.52−j2.53	2.13−j0.94
1.0	0.32+j3.26	7.23−j1.44	2.24−j0.60
2	0.49+j3.21	5.46−j0.37	2.31−j0.06
5	0.93+j2.96	3.06+j0.52	2.08+j0.69
10	1.33+j2.38	1.74+j0.81	1.65+j0.87
20	1.37+j1.48	0.98+j0.97	1.18+j0.75

Admittance of Resistive Antenna, Y = G + jB (millimhos) Ω = 9.92

$\frac{2\lambda r^i}{\zeta_o}$	$\frac{h}{\lambda}$ = .125	.250	.375	.500	.625
0	0.09+j2.61	10.01−j4.86	1.41−j1.34	0.97+j1.00	1.23+j4.78
0.01	0.09+j2.61	10.03−j4.76	1.41−j1.33	0.98+j1.00	1.25+j4.77
0.02	0.09+j2.61	9.99−j4.70	1.42−j1.33	0.99+j1.00	1.28+j4.76
0.05	0.10+j2.61	9.89−j4.55	1.43−j1.31	1.01+j1.01	1.35+j4.74
0.1	0.10+j2.61	9.73−j4.30	1.45−j1.29	1.04+j1.02	1.47+j4.71
0.2	0.11+j2.61	9.40−j3.86	1.49−j1.24	1.10+j1.05	1.70+j4.62
0.5	0.15+j2.60	8.49−j2.80	1.60−j1.09	1.28+j1.11	2.29+j4.28
1.0	0.20+j2.59	7.22−j1.69	1.73−j0.84	1.54+j1.19	2.89+j3.59
2	0.31+j2.57	5.47−j0.58	1.86−j0.40	1.91+j1.26	3.14+j2.47
5	0.61+j2.43	3.09+j0.33	1.79+j0.36	2.14+j1.11	[illegible].45+j1.36
10	0.95+j2.09	1.77+j0.62	1.47+j0.69	1.71+j0.87	1.74+j1.04
20	1.12+j1.42	0.98+j0.75	1.08+j0.66		

Admittance of Resistive Antenna, Y = G + jB (millimhos) Ω = 15

$\frac{2\lambda r^i}{\zeta_o}$	$\frac{h}{\lambda}$ = .125	.250	.375	.500	.625
0	0.03+j1.46	10.09−j5.29	0.58−j1.08	0.36+j0.27	0.42+j2.13
0.01	0.03+j1.46	10.00−j5.25	0.58−j1.08	0.37+j0.27	0.43+j2.12
0.02	0.03+j1.46	9.97−j5.20	0.59−j1.08	0.37+j0.27	0.43+j2.12
0.05	0.03+j1.46	9.87−j5.04	0.59−j1.07	0.37+j0.28	0.45+j2.12
0.1	0.03+j1.46	9.71−j4.79	0.60−j1.06	0.38+j0.28	0.48+j2.12
0.2	0.04+j1.46	9.40−j4.33	0.63−j1.05	0.40+j0.29	0.54+j2.11
0.5	0.05+j1.46	8.51−j3.25	0.69−j1.00	0.46+j0.31	0.72+j2.06
1.0	0.06+j1.46	7.26−j2.09	0.79−j0.92	0.55+j0.34	0.98+j1.96
2	0.10+j1.45	5.51−j0.94	0.93−j0.74	0.72+j0.40	1.33+j1.67
5	0.20+j1.43	3.13+j0.01	1.11−j0.25	1.02+j0.53	1.55+j0.99
10	0.35+j1.35	1.80+j0.30	1.04+j0.20	1.12+j0.57	1.28+j0.66
20	0.55+j1.13	0.99+j0.40	0.82+j0.42		

TABLE 3.2

Impedance of Resistive Antenna, Z = R + jX (ohms) Ω = 8.54

$\frac{2\lambda r^i}{\zeta_0}$	$\frac{h}{\lambda}$ = .125	.250	.375
0	13.3 - j303.0	82.3 + j35.8	354.4 + j241.3
0.01	13.5 - j303.0	82.8 + j36.3	355.3 + j239.9
0.02	13.7 - j303.0	83.3 + j36.3	356.3 + j238.5
0.05	14.2 - j303.0	84.8 + j36.0	359.1 + j234.4
0.1	15.0 - j303.0	87.4 + j35.6	363.7 + j227.6
0.2	16.7 - j303.0	92.5 + j34.8	372.2 + j213.9
0.5	21.7 - j303.1	107.8 + j32.0	393.3 + j173.6
1.0	30.0 - j303.4	133.1 + j26.5	416.0 + j111.1
2.0	46.8 - j304.1	182.5 + j12.3	432.5 + j10.9
5.0	96.7 - j307.9	317.7 - j53.9	433.3 - j143.5
10	179.1 - j319.9	472.6 - j221.2	472.8 - j249.8
20	337.4 - j364.7	516.0 - j508.9	606.9 - j384.9

Impedance of Resistive Antenna, Z = R + jX (ohms) Ω = 9.92

$\frac{2\lambda r^i}{\zeta_0}$	$\frac{h}{\lambda}$ = .125	.250	.375	.500	.625
0	13.3 - j383.0	80.8 + j39.2	372.8 + j354.3	502.0 - j512.9	50.4 - j196.4
0.01	13.5 - j383.0	81.4 + j38.6	374.1 + j353.0	500.7 - j509.5	51.5 - j196.2
0.02	13.7 - j383.0	81.9 + j38.6	375.3 + j351.8	499.4 - j506.3	52.5 - j195.9
0.05	14.2 - j383.0	83.4 + j38.4	379.1 + j348.1	495.6 - j496.7	55.5 - j195.0
0.1	15.0 - j383.0	86.0 + j38.0	385.2 + j341.9	489.3 - j481.5	60.5 - j193.5
0.2	16.7 - j383.1	91.1 + j37.4	397.0 + j329.3	477.3 - j453.8	70.2 - j190.5
0.5	21.7 - j383.2	106.2 + j35.1	428.3 + j290.8	445.5 - j387.7	97.2 - j181.8
1.0	30.1 - j383.4	131.4 + j30.7	468.7 + j226.8	406.2 - j315.3	136.1 - j169.2
2.0	46.9 - j383.9	180.8 + j19.2	514.5 + j110.1	365.1 - j240.9	196.8 - j154.9
5.0	97.1 - j387.0	320.2 - j34.6	537.7 - j109.8	368.2 - j191.7	311.3 - j173.0
10	180.1 - j396.6	504.1 - j176.6	557.8 - j261.0	464.1 - j235.3	424.1 - j253.4
20	341.8 - j432.5	643.5 - j491.2	671.7 - j412.5		

Impedance of Resistive Antenna, Z = R + jX (ohms) Ω = 15

$\frac{2\lambda r^i}{\zeta_0}$	$\frac{h}{\lambda}$ = .125	.250	.375	.500	.625
0	13.4 - j685.1	77.8 + j40.8	386.8 + j718.6	1758.9 - j1323.3	89.4 - j452.8
0.01	13.5 - j685.1	78.4 + j41.2	388.4 + j717.8	1753.2 - j1314.7	90.7 - j452.4
0.02	13.7 - j685.1	78.9 + j41.1	390.1 + j717.0	1747.5 - j1306.3	92.0 - j452.0
0.05	14.2 - j685.1	80.4 + j41.0	395.0 + j714.7	1730.6 - j1281.6	95.9 - j450.9
0.1	15.1 - j685.1	82.8 + j40.8	403.2 + j710.7	1702.9 - j1242.3	102.2 - j448.9
0.2	16.8 - j685.1	87.8 + j40.4	419.3 + j702.6	1649.5 - j1170.4	114.7 - j444.9
0.5	21.8 - j685.2	102.6 + j39.2	466.1 + j676.7	1505.5 - j996.9	150.5 - j432.0
1.0	30.3 - j685.3	127.3 + j36.6	538.2 + j628.9	1314.3 - j803.0	204.4 - j409.4
2.0	47.2 - j685.6	176.2 + j30.1	659.5 + j521.6	1062.6 - j594.4	292.6 - j365.9
5.0	97.9 - j687.3	319.7 - j1.4	860.0 + j192.9	769.5 - j399.2	459.0 - j293.2
10	182.1 - j692.9	540.1 - j89.2	929.2 - j176.0	707.5 - j361.4	616.9 - j316.9
20	349.0 - j713.6	871.7 - j353.2	969.7 - j491.6		

4. The Circular Loop Antenna

The circular loop antenna consists of a ring of radius b made of highly conducting metal with circular cross section of radius $a \ll b$. It is driven at $\Phi = 0$ by a delta-function generator with voltage V, as shown in Fig. 4.1. It is assumed that the loop is electrically thin so that $|ka| \ll 1$, where $k = \beta - j\alpha$, as defined in (1.6), is the complex wave number characteristic of the isotropic, homogeneous medium in which the loop is immersed. The normalized current in the loop as given by King *et al.*[1] based on the earlier work of Wu,[2] Storer,[3] and Hallen[4] is

$$\frac{I(\Phi)}{V\Delta} = \frac{-j(1 - j\alpha/\beta)}{\pi\zeta_0}\left[\frac{1}{a_0} + 2\sum_{1}^{\infty}\frac{\cos n\Phi}{a_n}\right] \qquad (4.1)$$

where $\zeta_0 \doteq 120\pi$ ohms and the normalizing factor Δ is defined in (1.15). Formulas for the coefficients $(1/a_i)$ of the Fourier series are to be found in the literature,[5] as are graphical representations.[6] The normalized admittance Y/Δ is obtained from (4.1) by setting $\Phi = 0$.

Although the loop is assumed to be driven by a delta-function generator with its infinite capacitance, the associated infinite charging currents at $\Phi = 0$ are highly localized when $|ka|$ is sufficiently small, so that a very large number of terms in the Fourier series (4.1) are required before the presence of this current can be observed. Since relatively few terms are required to describe the rest of the current correctly,[5] (4.1) can be used effectively to determine $I(\Phi)/\Delta$ and Y/Δ for loops driven from two-wire or shielded-pair lines and half loops driven from coaxial lines over ground planes if corrections for the junction region are made in a manner closely paralleling that described for dipoles and monopoles.

Distributions of normalized current along circular loops in air over a range of sizes are shown in Fig. 4.2 in real and imaginary parts, magnitudes, and phases. Similar graphs for loops in dissipative media are shown in Fig. 4.3 for two sizes. Typical graphs of the normalized admittance are shown in Fig. 4.4. Complete numerical values of the admittance for six values of $\Omega = 2\ln(2\pi b/a)$ in the range $\beta b \leq 2.5$ with α/β extending from 0 (air) to 1 are given in Tables 4.1 to 4.6.

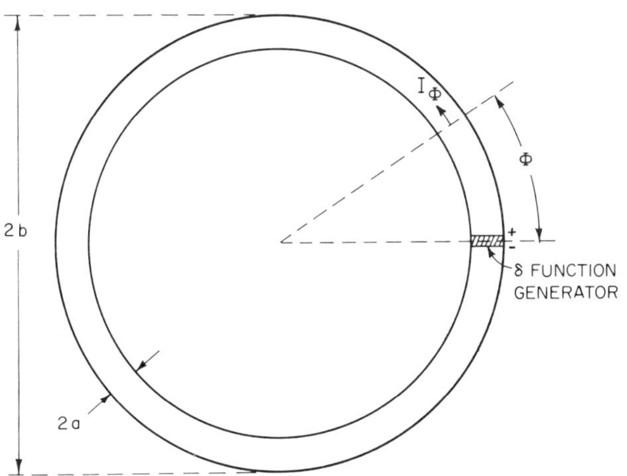

Fig. 4.1. Circular loop antenna.

REFERENCES FOR SECTION 4

1. King, R. W. P., C. W. Harrison, Jr., and D. G. Tingley, "The Current in Bare Circular Loop Antennas in a Dissipative Medium," *Trans. IEEE*, **AP-13**, 529 (1965).
2. Wu, T. T., "Theory of Thin Circular Loop Antennas," *J. Math. Phys.*, **3**, 1301 (1962).
3. Storer, J. E., "Impedance of Thin-Wire Loop Antennas," *Trans. AIEE (Communication and Electronics)*, **75**, 607 (1956).
4. Hallen, E., "Theoretical Investigations into Transmitting and Receiving Qualities of Antennae," *Nova Acta Regiae Soc. Sci. Upsaliensis*, **4**, 1 (1938).
5. King, R. W. P., C. W. Harrison, Jr., and D. G. Tingley, "The Admittance of Bare Circular Loop Antennas in a Dissipative Medium," *Trans. IEEE*, **AP-12**, 434 (1964).
6. Harrison, C. W., and R. W. P. King, "Folded Dipoles and Loops," *Trans. IEEE*, **AP-9**, 171 (1961).

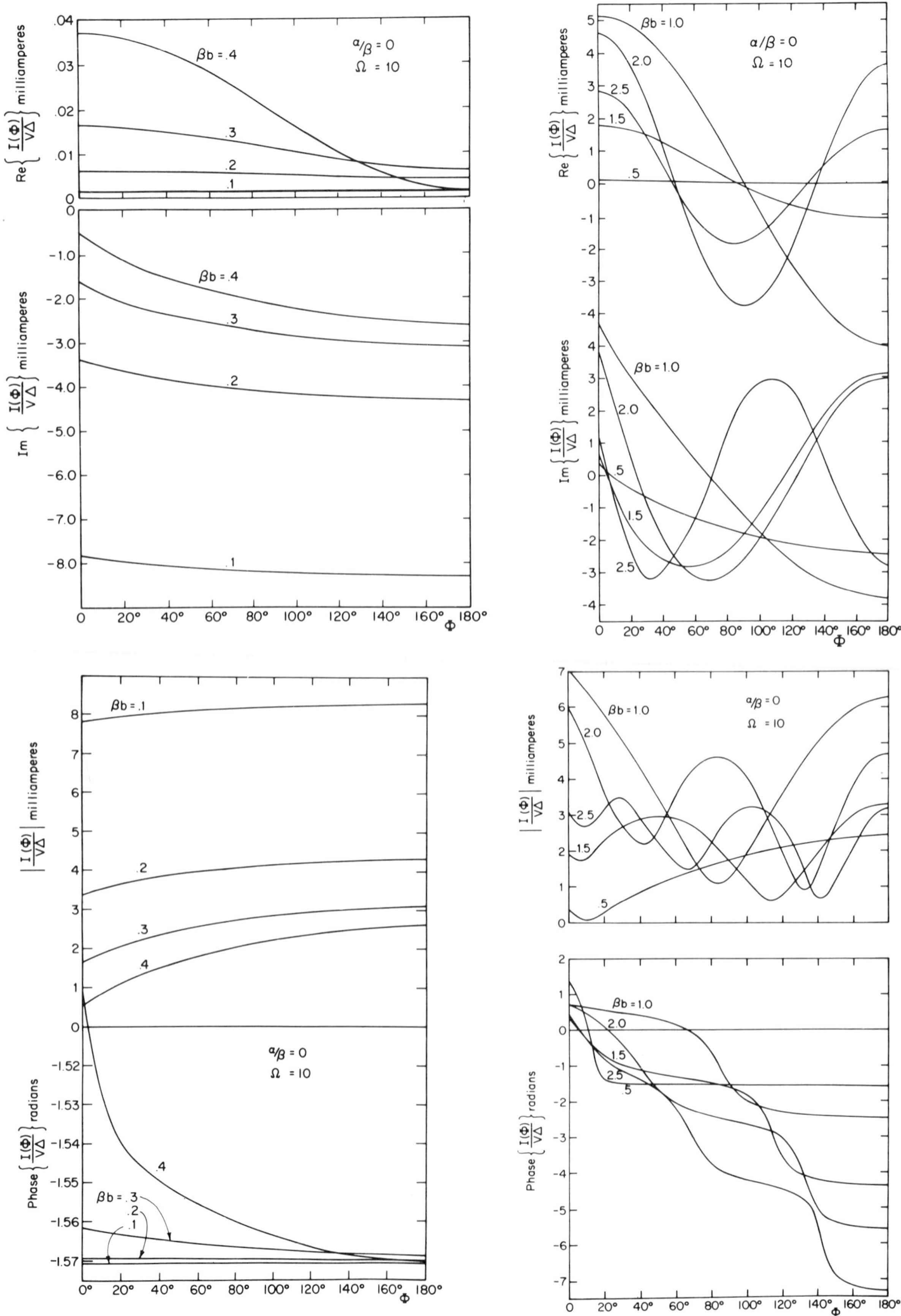

Fig. 4.2. Distribution of current around circular loops in air.

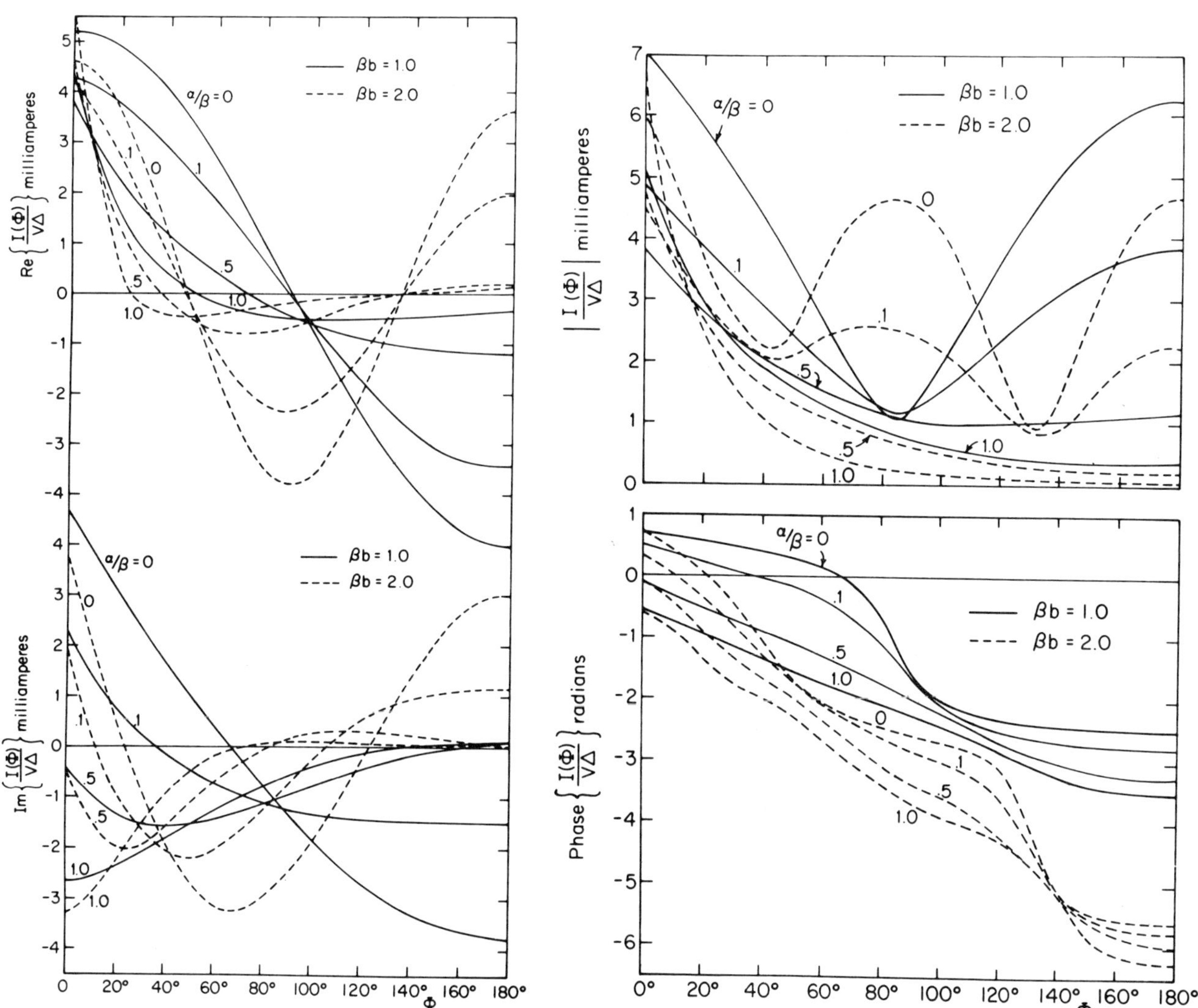

Fig. 4.3. Distribution of current along circular loops in dissipative media.

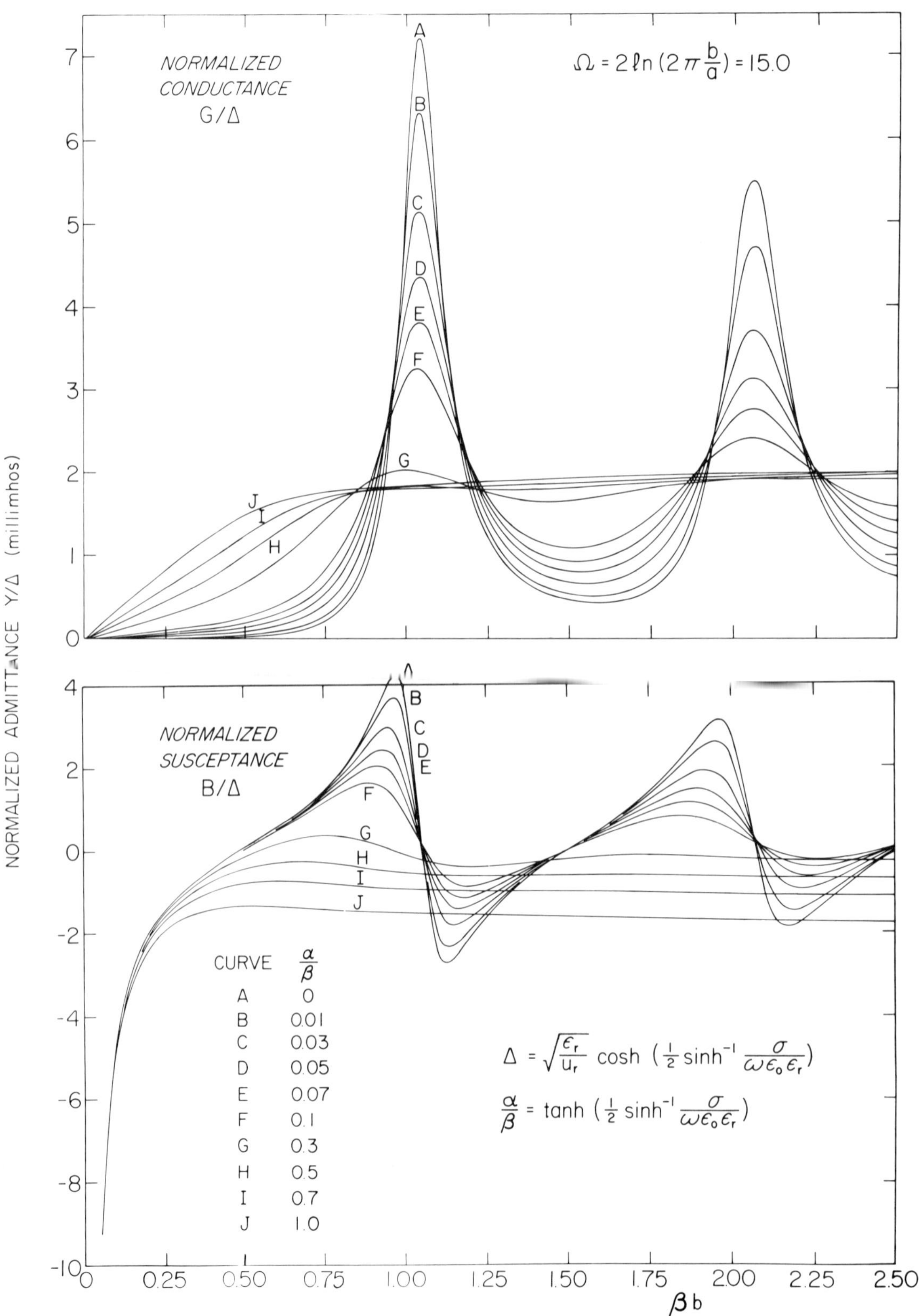

Fig. 4.4. Normalized admittance of circular loops in dissipative media.

TABLE 4.1

Normalized Admittance Y/Δ of Loop Antennas in Dissipative Media

$\Omega = 10$

βb	$\frac{\alpha}{\beta} = 0.00$		$\frac{\alpha}{\beta} = 0.01$		$\frac{\alpha}{\beta} = 0.03$		$\frac{\alpha}{\beta} = 0.05$		$\frac{\alpha}{\beta} = 0.07$	
	Y/Δ		Y/Δ		Y/Δ		Y/Δ		Y/Δ	
0.05	0.0003	-16.1742	0.0038	-16.1742	0.0108	-16.1744	0.0178	-16.1747	0.0248	-16.1751
0.10	0.0014	-7.8236	0.0084	-7.8237	0.0225	-7.8241	0.0366	-7.8247	0.0507	-7.8257
0.15	0.0032	-4.9201	0.0139	-4.9203	0.0353	-4.9209	0.0567	-4.9220	0.0781	-4.9236
0.20	0.0061	-3.3750	0.0206	-3.3752	0.0496	-3.3763	0.0787	-3.3779	0.1076	-3.3802
0.25	0.0102	-2.3691	0.0288	-2.3696	0.0660	-2.3711	0.1031	-2.3735	0.1401	-2.3767
0.30	0.0162	-1.6283	0.0392	-1.6290	0.0851	-1.6313	0.1309	-1.6346	0.1766	-1.6391
0.35	0.0248	-1.0333	0.0526	-1.0345	0.1081	-1.0378	0.1634	-1.0426	0.2186	-1.0487
0.40	0.0370	-0.5233	0.0702	-0.5250	0.1364	-0.5300	0.2024	-0.5367	0.2680	-0.5452
0.45	0.0546	-0.0624	0.0939	-0.0651	0.1723	-0.0723	0.2502	-0.0819	0.3276	-0.0938
0.50	0.0801	0.3730	0.1266	0.3688	0.2191	0.3580	0.3107	0.3442	0.4013	0.3274
0.55	0.1174	0.8006	0.1724	0.7939	0.2814	0.7776	0.3889	0.7574	0.4947	0.7335
0.60	0.1728	1.2351	0.2380	1.2247	0.3666	1.1997	0.4926	1.1698	0.6158	1.1350
0.65	0.2561	1.6908	0.3337	1.6740	0.4857	1.6354	0.6333	1.5903	0.7761	1.5393
0.70	0.3837	2.1816	0.4765	2.1545	0.6563	2.0937	0.8282	2.0249	0.9921	1.9491
0.75	0.5830	2.7214	0.6938	2.6768	0.9051	2.5796	1.1027	2.4733	1.2868	2.3599
0.80	0.9009	3.3195	1.0313	3.2450	1.2738	3.0877	1.4929	2.9229	1.6899	2.7536
0.85	1.4161	3.9671	1.5620	3.8414	1.8222	3.5873	2.0441	3.3345	2.2330	3.0870
0.90	2.2513	4.5965	2.3902	4.3886	2.6194	3.9908	2.7957	3.6205	2.9313	3.2790
0.95	3.5409	4.9903	3.6073	4.6755	3.6909	4.1131	3.7295	3.6295	3.7408	3.2122
1.00	5.2227	4.6875	5.1043	4.3157	4.8814	3.6941	4.6828	3.1964	4.5101	2.7892
1.05	6.6359	3.2896	6.3011	3.0461	5.7564	2.6416	5.3366	2.3176	5.0076	2.0502
1.10	6.8730	1.3221	6.5054	1.3362	5.9058	1.3233	5.4445	1.2783	5.0849	1.2165
1.15	6.0398	-0.1400	5.8108	0.0327	5.4080	0.2831	5.0746	0.4422	4.8014	0.5392
1.20	4.9127	-0.7848	4.8271	-0.5874	4.6528	-0.2698	4.4874	-0.0365	4.3390	0.1313
1.25	3.9355	-0.8825	3.9399	-0.7158	3.9256	-0.4304	3.8932	-0.2040	3.8535	-0.0287
1.30	3.1945	-0.7081	3.2478	-0.5803	3.3294	-0.3531	3.3857	-0.1640	3.4253	-0.0112
1.35	2.6537	-0.4132	2.7340	-0.3188	2.8724	-0.1475	2.9860	-0.0016	3.0804	0.1186
1.40	2.2627	-0.0671	2.3595	0.0008	2.5343	0.1243	2.6865	0.2295	2.8192	0.3155
1.45	1.9826	0.3014	2.0916	0.3479	2.2929	0.4306	2.4731	0.4985	2.6341	0.5512
1.50	1.7878	0.6819	1.9076	0.7098	2.1311	0.7558	2.3338	0.7885	2.5167	0.8082
1.55	1.6637	1.0722	1.7942	1.0822	2.0383	1.0923	2.2603	1.0898	2.4608	1.0758
1.60	1.6039	1.4734	1.7457	1.4644	2.0098	1.4359	2.2484	1.3956	2.4626	1.3455
1.65	1.6093	1.8871	1.7628	1.8558	2.0457	1.7829	2.2976	1.6998	2.5206	1.6099
1.70	1.6877	2.3137	1.8524	2.2545	2.1504	2.1276	2.4098	1.9945	2.6346	1.8597
1.75	1.8540	2.7491	2.0269	2.6535	2.3315	2.4594	2.5880	2.2679	2.8042	2.0835
1.80	2.1306	3.1797	2.3038	3.0368	2.5975	2.7605	2.8340	2.5028	3.0263	2.2664
1.85	2.5444	3.5744	2.7011	3.3725	2.9531	3.0020	3.1440	2.6759	3.2927	2.3907
1.90	3.1165	3.8733	3.2271	3.6068	3.3902	3.1435	3.5030	2.7596	3.5864	2.4389
1.95	3.8359	3.9815	3.8587	3.6637	3.8765	3.1400	3.8793	2.7291	3.8802	2.3989
2.00	4.6171	3.7906	4.5146	3.4690	4.3470	2.9598	4.2248	2.5745	4.1397	2.2712
2.05	5.2801	3.2530	5.0531	3.0048	4.7145	2.6125	4.4854	2.3133	4.3309	2.0738
2.10	5.6233	2.4710	5.3316	2.3600	4.9046	2.1627	4.6209	1.9919	4.4316	1.8404
2.15	5.5686	1.6718	5.2987	1.7030	4.8948	1.7088	4.6219	1.6713	4.4387	1.6107
2.20	5.2098	1.0560	5.0200	1.1830	4.7214	1.3374	4.5110	1.4048	4.3674	1.4188
2.25	4.7102	0.6964	4.6146	0.8631	4.4510	1.0927	4.3285	1.2227	4.2438	1.2853
2.30	4.1983	0.5656	4.1841	0.7327	4.1477	0.9786	4.1155	1.1316	4.0952	1.2167
2.35	3.7412	0.6027	3.7886	0.7500	3.8563	0.9755	3.9044	1.1231	3.9448	1.2091
2.40	3.3624	0.7520	3.4543	0.8714	3.6021	1.0578	3.7163	1.1816	3.8092	1.2532
2.45	3.0648	0.9739	3.1888	1.0632	3.3970	1.2017	3.5633	1.2908	3.6990	1.3374
2.50	2.8436	1.2429	2.9916	1.3016	3.2453	1.3880	3.4515	1.4360	3.6204	1.4506

βb	$\frac{\alpha}{\beta} = 0.10$		$\frac{\alpha}{\beta} = 0.30$		$\frac{\alpha}{\beta} = 0.50$		$\frac{\alpha}{\beta} = 0.70$		$\frac{\alpha}{\beta} = 1.00$	
	Y/Δ		Y/Δ		Y/Δ		Y/Δ		Y/Δ	
0.05	0.0353	-16.1760	0.1052	-16.1903	0.1750	-16.2186	0.2446	-16.2609	0.3485	-16.3505
0.10	0.0718	-7.8276	0.2123	-7.8572	0.3520	-7.9152	0.4905	-8.0016	0.6954	-8.1837
0.15	0.1102	-4.9267	0.3228	-4.9733	0.5328	-5.0639	0.7392	-5.1977	1.0399	-5.4773
0.20	0.1511	-3.3848	0.4382	-3.4511	0.7192	-3.5784	0.9916	-3.7645	1.3802	-4.1482
0.25	0.1956	-2.3831	0.5604	-2.4731	0.9127	-2.6427	1.2483	-2.8872	1.7135	-3.3821
0.30	0.2450	-1.6478	0.6914	-1.7670	1.1149	-1.9863	1.5087	-2.2964	2.0360	-2.9087
0.35	0.3009	-1.0606	0.8334	-1.2165	1.3268	-1.4950	1.7717	-1.8784	2.3432	-2.6117
0.40	0.3657	-0.5613	0.9892	-0.7645	1.5488	-1.1134	2.0348	-1.5774	2.6307	-2.4314
0.45	0.4424	-0.1159	1.1614	-0.3806	1.7806	-0.8128	2.2946	-1.3632	2.8944	-2.3322
0.50	0.5353	0.2969	1.3530	-0.0485	2.0207	-0.5776	2.5466	-1.2173	3.1316	-2.2904
0.55	0.6501	0.6908	1.5664	0.2395	2.2659	-0.3994	2.7858	-1.1265	3.3411	-2.2880
0.60	0.7951	1.0747	1.8033	0.4853	2.5114	-0.2731	3.0076	-1.0804	3.5237	-2.3117
0.65	0.9811	1.4529	2.0634	0.6864	2.7509	-0.1947	3.2078	-1.0692	3.6816	-2.3511
0.70	1.2227	1.8246	2.3432	0.8372	2.9772	-0.1596	3.3840	-1.0835	3.8180	-2.3985
0.75	1.5384	2.1802	2.6350	0.9311	3.1830	-0.1616	3.5356	-1.1147	3.9368	-2.4487
0.80	1.9479	2.4969	2.9256	0.9628	3.3624	-0.1923	3.6639	-1.1552	4.0416	-2.4985
0.85	2.4647	2.7317	3.1976	0.9319	3.5120	-0.2425	3.7715	-1.1986	4.1357	-2.5462
0.90	3.0788	2.8191	3.4323	0.8458	3.6316	-0.3021	3.8622	-1.2408	4.2219	-2.5909
0.95	3.7301	2.6862	3.6142	0.7204	3.7235	-0.3628	3.9396	-1.2788	4.3023	-2.6327
1.00	4.2963	2.2997	3.7357	0.5770	3.7927	-0.4177	4.0076	-1.3114	4.3784	-2.6718
1.05	4.6365	1.7229	3.7992	0.4377	3.8447	-0.4627	4.0692	-1.3384	4.4513	-2.7087
1.10	4.6827	1.1086	3.8155	0.3199	3.8853	-0.4961	4.1269	-1.3602	4.5218	-2.7437
1.15	4.4843	0.6080	3.7994	0.2339	3.9197	-0.5179	4.1823	-1.3779	4.5902	-2.7774
1.20	4.1542	0.2923	3.7664	0.1828	3.9519	-0.5295	4.2367	-1.3923	4.6568	-2.8099
1.25	3.7946	0.1555	3.7293	0.1644	3.9846	-0.5328	4.2907	-1.4044	4.7218	-2.8417
1.30	3.4663	0.1568	3.6975	0.1731	4.0195	-0.5302	4.3446	-1.4152	4.7852	-2.8728
1.35	3.1955	0.2528	3.6769	0.2024	4.0575	-0.5238	4.3983	-1.4251	4.8472	-2.9034
1.40	2.9886	0.4090	3.6706	0.2454	4.0987	-0.5153	4.4519	-1.4348	4.9077	-2.9336
1.45	2.8440	0.6017	3.6795	0.2962	4.1428	-0.5064	4.5051	-1.4446	4.9669	-2.9633
1.50	2.7573	0.8144	3.7030	0.3496	4.1893	-0.4981	4.5578	-1.4546	5.0246	-2.9927
1.55	2.7244	1.0356	3.7396	0.4014	4.2373	-0.4911	4.6098	-1.4650	5.0811	-3.0217
1.60	2.7418	1.2561	3.7869	0.4484	4.2862	-0.4858	4.6609	-1.4757	5.1363	-3.0503
1.65	2.8070	1.4673	3.8422	0.4885	4.3352	-0.4823	4.7111	-1.4866	5.1903	-3.0785
1.70	2.9171	1.6602	3.9028	0.5203	4.3837	-0.4806	4.7602	-1.4979	5.2431	-3.1064
1.75	3.0684	1.8250	3.9657	0.5434	4.4315	-0.4804	4.8084	-1.5092	5.2949	-3.1339
1.80	3.2541	1.9509	4.0284	0.5582	4.4780	-0.4815	4.8555	-1.5207	5.3455	-3.1610
1.85	3.4638	2.0275	4.0887	0.5656	4.5232	-0.4835	4.9017	-1.5322	5.3952	-3.1878
1.90	3.6820	2.0471	4.1451	0.5671	4.5671	-0.4861	4.9469	-1.5438	5.4438	-3.2142
1.95	3.8893	2.0078	4.1964	0.5645	4.6096	-0.4891	4.9913	-1.5552	5.4916	-3.2403
2.00	4.0653	1.9156	4.2423	0.5594	4.6509	-0.4922	5.0348	-1.5667	5.5384	-3.2661
2.05	4.1927	1.7855	4.2829	0.5537	4.6910	-0.4953	5.0775	-1.5780	5.5844	-3.2915
2.10	4.2625	1.6384	4.3189	0.5486	4.7302	-0.4983	5.1194	-1.5893	5.6295	-3.3166
2.15	4.2753	1.4968	4.3510	0.5451	4.7685	-0.5012	5.1606	-1.6005	5.6739	-3.3414
2.20	4.2408	1.3793	4.3804	0.5440	4.8062	-0.5039	5.2012	-1.6116	5.7174	-3.3659
2.25	4.1736	1.2975	4.4080	0.5453	4.8433	-0.5065	5.2411	-1.6227	5.7602	-3.3900
2.30	4.0896	1.2560	4.4349	0.5491	4.8799	-0.5089	5.2804	-1.6337	5.8023	-3.4139
2.35	4.0029	1.2535	4.4617	0.5551	4.9160	-0.5114	5.3190	-1.6447	5.8437	-3.4375
2.40	3.9247	1.2851	4.4891	0.5628	4.9517	-0.5138	5.3571	-1.6556	5.8844	-3.4607
2.45	3.8627	1.3440	4.5175	0.5717	4.9870	-0.5163	5.3945	-1.6665	5.9245	-3.4837
2.50	3.8218	1.4225	4.5471	0.5811	5.0219	-0.5188	5.4315	-1.6773	5.9639	-3.5064

TABLE 4.2

Normalized Admittance Y/Δ of Loop Antennas in Dissipative Media

$\Omega = 11$

βb	$\frac{\alpha}{\beta} = 0.00$		$\frac{\alpha}{\beta} = 0.01$		$\frac{\alpha}{\beta} = 0.03$		$\frac{\alpha}{\beta} = 0.05$		$\frac{\alpha}{\beta} = 0.07$	
	Y/Δ		Y/Δ		Y/Δ		Y/Δ		Y/Δ	
0.05	0.0002	-14.0300	0.0031	-14.0300	0.0089	-14.0302	0.0146	-14.0304	0.0204	-14.0308
0.10	0.0010	-6.7984	0.0068	-6.7985	0.0184	-6.7988	0.0300	-6.7993	0.0416	-6.8001
0.15	0.0024	-4.2891	0.0112	-4.2892	0.0288	-4.2897	0.0465	-4.2906	0.0641	-4.2919
0.20	0.0046	-2.9573	0.0165	-2.9576	0.0405	-2.9584	0.0644	-2.9597	0.0883	-2.9616
0.25	0.0077	-2.0931	0.0230	-2.0935	0.0537	-2.0947	0.0843	-2.0966	0.1149	-2.0992
0.30	0.0122	-1.4585	0.0312	-1.4590	0.0692	-1.4609	0.1071	-1.4636	0.1449	-1.4672
0.35	0.0187	-0.9503	0.0417	-0.9511	0.0877	-0.9538	0.1336	-0.9576	0.1793	-0.9626
0.40	0.0279	-0.5154	0.0555	-0.5168	0.1106	-0.5207	0.1654	-0.5261	0.2200	-0.5330
0.45	0.0413	-0.1228	0.0741	-0.1249	0.1395	-0.1306	0.2045	-0.1384	0.2691	-0.1481
0.50	0.0606	0.2483	0.0996	0.2450	0.1771	0.2365	0.2540	0.2253	0.3300	0.2116
0.55	0.0890	0.6135	0.1354	0.6084	0.2274	0.5955	0.3181	0.5792	0.4075	0.5595
0.60	0.1314	0.9867	0.1869	0.9785	0.2963	0.9587	0.4036	0.9344	0.5086	0.9057
0.65	0.1958	1.3813	0.2625	1.3682	0.3934	1.3372	0.5206	1.3002	0.6437	1.2577
0.70	0.2957	1.8124	0.3767	1.7908	0.5341	1.7413	0.6847	1.6840	0.8283	1.6198
0.75	0.4550	2.2968	0.5541	2.2605	0.7434	2.1796	0.9205	2.0890	1.0853	1.9908
0.80	0.7169	2.8520	0.8379	2.7892	1.0628	2.6537	1.2655	2.5085	1.4469	2.3570
0.85	1.1622	3.4865	1.3052	3.3748	1.5590	3.1444	1.7735	2.9108	1.9532	2.6796
0.90	1.9374	4.1612	2.0856	3.9612	2.3258	3.5737	2.5042	3.2098	2.6350	2.8737
0.95	3.2622	4.6662	3.3433	4.3304	3.4364	3.7337	3.4678	3.2271	3.4625	2.7970
1.00	5.1932	4.3834	5.0403	3.9540	4.7489	3.2605	4.4884	2.7285	4.2623	2.3094
1.05	6.8232	2.5879	6.3748	2.3528	5.6666	1.9755	5.1373	1.6845	4.7312	1.4514
1.10	6.7120	0.1681	6.2925	0.2928	5.6112	0.4414	5.0905	0.5083	4.6867	0.5287
1.15	5.3779	-1.2319	5.1883	-0.9740	4.8322	-0.5849	4.5203	-0.3207	4.2556	-0.1428
1.20	4.0411	-1.5838	4.0110	-1.3585	3.9190	-0.9827	3.8074	-0.6941	3.6936	-0.4765
1.25	3.0663	-1.4420	3.1097	-1.2781	3.1599	-0.9866	3.1756	-0.7445	3.1707	-0.5487
1.30	2.4031	-1.1269	2.4775	-1.0124	2.5967	-0.8009	2.6839	-0.6166	2.7473	-0.4611
1.35	1.9519	-0.7638	2.0401	-0.6842	2.1941	-0.5346	2.3218	-0.4011	2.4275	-0.2863
1.40	1.6408	-0.3945	1.7368	-0.3397	1.9114	-0.2367	2.0642	-0.1450	2.1975	-0.0667
1.45	1.4256	-0.0298	1.5278	0.0063	1.7179	0.0725	1.8891	0.1291	2.0422	0.1749
1.50	1.2804	0.3306	1.3894	0.3513	1.5941	0.3860	1.7808	0.4111	1.9497	0.4266
1.55	1.1914	0.6913	1.3085	0.6976	1.5289	0.7021	1.7302	0.6964	1.9125	0.6814
1.60	1.1531	1.0581	1.2801	1.0490	1.5178	1.0211	1.7333	0.9823	1.9270	0.9348
1.65	1.1669	1.4370	1.3055	1.4093	1.5618	1.3432	1.7904	1.2661	1.9924	1.1818
1.70	1.2412	1.8330	1.3927	1.7811	1.6671	1.6663	1.9053	1.5426	2.1102	1.4155
1.75	1.3928	2.2484	1.5567	2.1629	1.8446	1.9837	2.0841	1.8024	2.2824	1.6256
1.80	1.6496	2.6778	1.8213	2.5447	2.1085	2.2804	2.3332	2.0290	2.5094	1.7968
1.85	2.0527	3.0985	2.2174	2.8991	2.4729	2.5267	2.6546	2.1965	2.7859	1.9087
1.90	2.6507	3.4504	2.7748	3.1669	2.9409	2.6739	3.0377	2.2698	3.0959	1.9384
1.95	3.4702	3.6083	3.4925	3.2432	3.4849	2.6559	3.4490	2.2115	3.4087	1.8672
2.00	4.4330	3.3816	4.2812	2.9952	4.0239	2.4143	3.8269	2.0010	3.6804	1.6919
2.05	5.2599	2.6373	4.9228	2.3608	4.4283	1.9503	4.0950	1.6577	3.8656	1.4352
2.10	5.5806	1.5398	5.1709	1.4821	4.5834	1.3631	4.1968	1.2493	3.9357	1.1432
2.15	5.2983	0.5189	4.9680	0.6572	4.4695	0.8083	4.1275	0.8645	3.8917	0.8703
2.20	4.6658	-0.1268	4.4801	0.1015	4.1672	0.4039	3.9317	0.5719	3.7609	0.6582
2.25	3.9731	-0.3895	3.9146	-0.1562	3.7873	0.1840	3.6726	0.3983	3.5819	0.5261
2.30	3.3679	-0.3895	3.3961	-0.1889	3.4122	0.1236	3.4036	0.3368	3.3899	0.4735
2.35	2.8872	-0.2379	2.9690	-0.0792	3.0846	0.1794	3.1588	0.3649	3.2101	0.4884
2.40	2.5221	-0.0053	2.6370	0.1138	2.8196	0.3125	2.9548	0.4581	3.0575	0.5555
[illegible]	[illegible]	[illegible]	[illegible]	[illegible]	2.6184	0.4946	2.7980	0.5957	2.9397	0.6599
2.50	2.0636	0.5696	2.2157	[illegible]	[illegible]	[illegible]	[illegible]	0.7619	2.8597	0.7889

βb	$\frac{\alpha}{\beta} = 0.10$		$\frac{\alpha}{\beta} = 0.30$		$\frac{\alpha}{\beta} = 0.50$		$\frac{\alpha}{\beta} = 0.70$		$\frac{\alpha}{\beta} = 1.00$	
	Y/Δ		Y/Δ		Y/Δ		Y/Δ		Y/Δ	
0.05	0.0290	-14.0315	0.0865	-14.0432	0.1438	-14.0665	0.2010	-14.1012	0.2864	-14.1749
0.10	0.0590	-6.8017	0.1745	-6.8259	0.2894	-6.8737	0.4032	-6.9448	0.5717	-7.0946
0.15	0.0904	-4.2945	0.2654	-4.3327	0.4383	-4.4074	0.6080	-4.5177	0.8550	-4.7481
0.20	0.1241	-2.9653	0.3607	-3.0200	0.5921	-3.1250	0.8161	-3.2788	1.1349	-3.5958
0.25	0.1607	-2.1045	0.4617	-2.1788	0.7522	-2.3193	1.0282	-2.5220	1.4088	-2.9319
0.30	0.2014	-1.4744	0.5705	-1.5730	0.9201	-1.7555	1.2438	-2.0136	1.6736	-2.5221
0.35	0.2476	-0.9724	0.6890	-1.1020	1.0967	-1.3348	1.4619	-1.6552	1.9250	-2.2659
0.40	0.3013	-0.5462	0.8196	-0.7159	1.2826	-1.0092	1.6806	-1.3988	2.1590	-2.1114
0.45	0.3650	-0.1662	0.9649	-0.3885	1.4776	-0.7539	1.8965	-1.2182	2.3717	-2.0278
0.50	0.4425	0.1864	1.1277	-0.1055	1.6805	-0.5559	2.1057	-1.0976	2.5606	-1.9943
0.55	0.5389	0.5241	1.3106	0.1397	1.8884	-0.4081	2.3035	-1.0259	2.7246	-1.9950
0.60	0.6615	0.8552	1.5155	0.3484	2.0972	-0.3065	2.4854	-0.9938	2.8642	-2.0178
0.65	0.8204	1.1846	1.7427	0.5178	2.3008	-0.2482	2.6475	-0.9925	2.9817	-2.0536
0.70	1.0301	1.5127	1.9896	0.6419	2.4922	-0.2293	2.7874	-1.0137	3.0802	-2.0955
0.75	1.3096	1.8326	2.2492	0.7133	2.6643	-0.2441	2.9045	-1.0491	3.1631	-2.1388
0.80	1.6821	2.1242	2.5088	0.7257	2.8112	-0.2850	3.0001	-1.0917	3.2341	-2.1810
0.85	2.1681	2.3451	2.7508	0.6777	2.9297	-0.3428	3.0769	-1.1357	3.2962	-2.2205
0.90	2.7666	2.4230	2.9555	0.5761	3.0194	-0.4080	3.1386	-1.1772	3.3518	-2.2569
0.95	3.4193	2.2667	3.1066	0.4372	3.0832	-0.4720	3.1888	-1.2138	3.4030	-2.2903
1.00	3.9829	1.8251	3.1964	0.2838	3.1260	-0.5287	3.2313	-1.2446	3.4511	-2.3212
1.05	4.2803	1.1739	3.2285	0.1389	3.1537	-0.5744	3.2688	-1.2696	3.4969	-2.3499
1.10	4.2368	0.5116	3.2155	0.0202	3.1722	-0.6077	3.3037	-1.2896	3.5410	-2.3771
1.15	3.9389	0.0172	3.1740	-0.0631	3.1863	-0.6293	3.3374	-1.3055	3.5837	-2.4031
1.20	3.5375	-0.2519	3.1201	-0.1096	3.1999	-0.6408	3.3709	-1.3184	3.6252	-2.4283
1.25	3.1447	-0.3308	3.0667	-0.1233	3.2154	-0.6446	3.4046	-1.3294	3.6655	-2.4528
1.30	2.8124	-0.2801	3.0223	-0.1110	3.2342	-0.6430	3.4388	-1.3392	3.7048	-2.4768
1.35	2.5540	-0.1504	2.9919	-0.0801	3.2567	-0.6383	3.4733	-1.3486	3.7429	-2.5005
1.40	2.3661	0.0241	2.9775	-0.0376	3.2830	-0.6321	3.5079	-1.3579	3.7800	-2.5239
1.45	2.2412	0.2224	2.9792	0.0106	3.3125	-0.6261	3.5424	-1.3674	3.8161	-2.5470
1.50	2.1716	0.4317	2.9959	0.0595	3.3445	-0.6211	3.5766	-1.3773	3.8511	-2.5698
1.55	2.1519	0.6436	3.0255	0.1053	3.3780	-0.6178	3.6103	-1.3876	3.8851	-2.5923
1.60	2.1787	0.8512	3.0654	0.1450	3.4124	-0.6164	3.6433	-1.3983	3.9181	-2.6145
1.65	2.2498	1.0481	3.1128	0.1767	3.4469	-0.6170	3.6755	-1.4093	3.9502	-2.6365
1.70	2.3637	1.2265	3.1647	0.1994	3.4809	-0.6194	3.7069	-1.4205	3.9814	-2.6582
1.75	2.5180	1.3770	3.2182	0.2130	3.5139	-0.6233	3.7375	-1.4319	4.0118	-2.6797
1.80	2.7078	1.4881	3.2707	0.2179	3.5458	-0.6283	3.7673	-1.4434	4.0414	-2.7008
1.85	2.9231	1.5474	3.3199	0.2155	3.5764	-0.6342	3.7963	-1.4549	4.0702	-2.7217
1.90	3.1478	1.5448	3.3644	0.2075	3.6056	-0.6405	3.8245	-1.4663	4.0982	-2.7424
1.95	3.3593	1.4761	3.4031	0.1957	3.6336	-0.6470	3.8520	-1.4778	4.1256	-2.7627
2.00	3.5325	1.3480	3.4359	0.1822	3.6604	-0.6535	3.8789	-1.4891	4.1522	-2.7829
2.05	3.6463	1.1791	3.4631	0.1687	3.6863	-0.6598	3.9052	-1.5004	4.1782	-2.8028
2.10	3.6902	0.9967	3.4856	0.1567	3.7115	-0.6660	3.9310	-1.5116	4.2036	-2.8225
2.15	3.6681	0.8288	3.5044	0.1470	3.7360	-0.6718	3.9562	-1.5227	4.2284	-2.8419
2.20	3.5949	0.6970	3.5207	0.1403	3.7599	-0.6775	3.9808	-1.5337	4.2526	-2.8612
2.25	3.4912	0.6124	3.5357	0.1367	3.7835	-0.6829	4.0051	-1.5448	4.2762	-2.8802
2.30	3.3766	0.5765	3.5504	0.1360	3.8068	-0.6882	4.0288	-1.5557	4.2993	-2.8990
2.35	3.2669	0.5843	3.5658	0.1377	3.8298	-0.6935	4.0521	-1.5666	4.3219	-2.9176
2.40	3.1729	0.6277	3.5822	0.1411	3.8526	-0.6987	4.0750	-1.5775	4.3439	-2.9360
2.45	3.1013	0.6975	3.6001	0.1457	3.8751	-0.7039	4.0975	-1.5884	4.3655	-2.9542
2.50	3.0555	0.7850	3.6196	0.1508	3.8973	-0.7092	4.1195	-1.5992	4.3866	-2.9722

TABLE 4.3

Normalized Admittance Y/Δ of Loop Antennas in Dissipative Media
$\Omega = 12$

βb	$\frac{\alpha}{\beta} = 0.00$		$\frac{\alpha}{\beta} = 0.01$		$\frac{\alpha}{\beta} = 0.03$		$\frac{\alpha}{\beta} = 0.05$		$\frac{\alpha}{\beta} = 0.07$	
	Y/Δ		Y/Δ		Y/Δ		Y/Δ		Y/Δ	
0.05	0.0002	-12.3838	0.0026	-12.3839	0.0075	-12.3840	0.0124	-12.3842	0.0173	-12.3845
0.10	0.0008	-6.0079	0.0057	-6.0079	0.0156	-6.0082	0.0254	-6.0086	0.0353	-6.0093
0.15	0.0019	-3.7985	0.0094	-3.7986	0.0244	-3.7991	0.0393	-3.7998	0.0543	-3.8009
0.20	0.0036	-2.6282	0.0137	-2.6284	0.0341	-2.6291	0.0545	-2.6302	0.0748	-2.6317
0.25	0.0060	-1.8703	0.0191	-1.8706	0.0452	-1.8716	0.0713	-1.8732	0.0973	-1.8754
0.30	0.0095	-1.3150	0.0258	-1.3154	0.0581	-1.3169	0.0905	-1.3192	0.1227	-1.3222
0.35	0.0146	-0.8711	0.0343	-0.8717	0.0736	-0.8739	0.1128	-0.8771	0.1519	-0.8813
0.40	0.0218	-0.4917	0.0455	-0.4928	0.0926	-0.4960	0.1396	-0.5005	0.1864	-0.5063
0.45	0.0323	-0.1495	0.0605	-0.1511	0.1167	-0.1558	0.1726	-0.1622	0.2281	-0.1704
0.50	0.0474	0.1742	0.0811	0.1716	0.1479	0.1646	0.2142	0.1554	0.2799	0.1438
0.55	0.0698	0.4934	0.1100	0.4893	0.1896	0.4788	0.2683	0.4652	0.3459	0.4486
0.60	0.1033	0.8207	0.1516	0.8142	0.2470	0.7980	0.3407	0.7777	0.4324	0.7534
0.65	0.1544	1.1692	0.2131	1.1587	0.3283	1.1332	0.4403	1.1020	0.5489	1.0657
0.70	0.2347	1.5538	0.3068	1.5363	0.4470	1.4951	0.5814	1.4463	0.7096	1.3909
0.75	0.3646	1.9931	0.4544	1.9631	0.6262	1.8945	0.7871	1.8161	0.9369	1.7297
0.80	0.5833	2.5099	0.6961	2.4564	0.9060	2.3385	1.0952	2.2094	1.2640	2.0727
0.85	0.9693	3.1257	1.1088	3.0265	1.3561	2.8175	1.5639	2.6016	1.7363	2.3854
0.90	1.6821	3.8294	1.8378	3.6393	2.0877	3.2657	2.2692	2.9111	2.3977	2.5825
0.95	3.0140	4.4372	3.1124	4.0862	3.2198	3.4637	3.2489	2.9404	3.2328	2.5022
1.00	5.1747	4.1923	4.9896	3.7064	4.6349	2.9477	4.3203	2.3893	4.0503	1.9646
1.05	6.9797	1.9691	6.4182	1.7662	5.5604	1.4470	4.9408	1.2056	4.4768	1.0151
1.10	6.3872	-0.7830	5.9627	-0.5402	5.2650	-0.2246	4.7283	-0.0478	4.3118	0.0497
1.15	4.6500	-1.9270	4.5259	-1.6197	4.2584	-1.1427	3.9992	-0.8072	3.7666	-0.5726
1.20	3.2716	-1.9702	3.2919	-1.7477	3.2807	-1.3632	3.2274	-1.0562	3.1547	-0.8170
1.25	2.3889	-1.6551	2.4583	-1.5089	2.5554	-1.2396	2.6099	-1.0065	2.6355	-0.8115
1.30	1.8314	-1.2655	1.9157	-1.1693	2.0552	-0.9858	2.1616	-0.8194	2.2416	-0.6740
1.35	1.4688	-0.8811	1.5575	-0.8168	1.7149	-0.6922	1.8473	-0.5770	1.9579	-0.4744
1.40	1.2260	-0.5180	1.3170	-0.4750	1.4842	-0.3920	1.6321	-0.3156	1.7618	-0.2482
1.45	1.0616	-0.1741	1.1555	-0.1464	1.3314	-0.0948	1.4909	-0.0495	1.6344	-0.0119
1.50	0.9530	0.1578	1.0514	0.1730	1.2373	0.1983	1.4079	0.2164	1.5630	0.2270
1.55	0.8886	0.4858	0.9936	0.4894	1.1921	0.4896	1.3745	0.4812	1.5403	0.4650
1.60	0.8642	0.8179	0.9780	0.8090	1.1922	0.7821	1.3873	0.7450	1.5630	0.6997
1.65	0.8814	1.1624	1.0067	1.1381	1.2395	1.0783	1.4476	1.0071	1.6315	0.9283
1.70	0.9487	1.5270	1.0880	1.4818	1.3411	1.3786	1.5605	1.2645	1.7485	1.1455
1.75	1.0832	1.9184	1.2381	1.8427	1.5097	1.6793	1.7343	1.5096	1.9179	1.3419
1.80	1.3152	2.3383	1.4838	2.2166	1.7638	1.9683	1.9788	1.7271	2.1428	1.5025
1.85	1.6944	2.7739	1.8656	2.5823	2.1251	2.2172	2.3013	1.8899	2.4208	1.6049
1.90	2.2933	3.1723	2.4329	2.8810	2.6079	2.3718	2.6959	1.9569	2.7374	1.6218
1.95	3.1836	3.3899	3.2132	2.9853	3.1923	2.3471	3.1289	1.8794	3.0591	1.5283
2.00	4.3153	3.1453	4.1192	2.6969	3.7835	2.0555	3.5260	1.6252	3.3335	1.3182
2.05	5.2921	2.1830	4.8422	1.8938	4.2051	1.4892	3.7880	1.2175	3.5046	1.0191
2.10	5.5121	0.7726	5.0070	0.8014	4.3002	0.7930	3.8449	0.7482	3.5399	0.6908
2.15	4.9312	-0.3661	4.5938	-0.1131	4.0706	0.1875	3.7043	0.3340	3.4480	0.3999
2.20	4.0586	-0.9108	3.9217	-0.6096	3.6575	-0.1953	3.4383	0.0500	3.2701	0.1915
2.25	3.2688	-1.0093	3.2661	-0.7494	3.2090	-0.3536	3.1297	-0.0900	3.0553	0.0786
2.30	2.6622	-0.8742	2.7322	-0.6745	2.8088	-0.3473	2.8363	-0.1105	2.8422	0.0514
2.35	2.2212	-0.6356	2.3283	-0.4892	2.4846	-0.2378	2.5863	-0.0458	2.6541	0.0908
2.40	1.9063	-0.3555	2.0330	-0.2514	2.2373	-0.0685	2.3889	0.0746	2.5021	0.1774
2.45	1.6846	-0.0599	1.8233	0.0107	2.0584	0.1342	2.2442	0.2295	2.3899	0.2952
2.50	1.5338	0.2419	1.6816	0.2847	1.9391	0.3555	2.1489	0.4047	2.3176	0.4320

βb	$\frac{\alpha}{\beta} = 0.10$		$\frac{\alpha}{\beta} = 0.30$		$\frac{\alpha}{\beta} = 0.50$		$\frac{\alpha}{\beta} = 0.70$		$\frac{\alpha}{\beta} = 1.00$	
	Y/Δ		Y/Δ		Y/Δ		Y/Δ		Y/Δ	
0.05	0.0246	-12.3851	0.0734	-12.3951	0.1221	-12.4148	0.1707	-12.4443	0.2433	-12.5069
0.10	0.0500	-6.0107	0.1482	-6.0312	0.2458	-6.0718	0.3426	-6.1323	0.4857	-6.2596
0.15	0.0767	-3.8031	0.2256	-3.8356	0.3725	-3.8991	0.5168	-3.9930	0.7265	-4.1891
0.20	0.1053	-2.6349	0.3066	-2.6814	0.5035	-2.7709	0.6940	-2.9021	0.9644	-3.1724
0.25	0.1364	-1.8799	0.3929	-1.9431	0.6403	-2.0632	0.8749	-2.2365	1.1973	-2.5867
0.30	0.1710	-1.3283	0.4860	-1.4125	0.7840	-1.5689	1.0591	-1.7901	1.4220	-2.2256
0.35	0.2103	-0.8895	0.5877	-1.0005	0.9356	-1.2007	1.2458	-1.4763	1.6350	-2.0003
0.40	0.2561	-0.5175	0.7003	-0.6632	1.0958	-0.9165	1.4331	-1.2528	1.8323	-1.8652
0.45	0.3105	-0.1858	0.8262	-0.3774	1.2644	-0.6944	1.6182	-1.0966	2.0105	-1.7931
0.50	0.3770	0.1224	0.9680	-0.1304	1.4405	-0.5232	1.7973	-0.9940	2.1670	-1.7652
0.55	0.4599	0.4184	1.1283	0.0835	1.6215	-0.3969	1.9661	-0.9351	2.3008	-1.7675
0.60	0.5660	0.7100	1.3092	0.2652	1.8035	-0.3123	2.1203	-0.9117	2.4126	-1.7890
0.65	0.7047	1.0025	1.5114	0.4119	1.9810	-0.2670	2.2562	-0.9161	2.5043	-1.8214
0.70	0.8897	1.2971	1.7328	0.5174	2.1471	-0.2578	2.3715	-0.9402	2.5789	-1.8585
0.75	1.1403	1.5888	1.9670	0.5739	2.2951	-0.2796	2.4657	-0.9767	2.6398	-1.8962
0.80	1.4814	1.8602	2.2021	0.5746	2.4191	-0.3251	2.5400	-1.0187	2.6903	-1.9322
0.85	1.9387	2.0703	2.4203	0.5172	2.5159	-0.3855	2.5972	-1.0612	2.7332	-1.9653
0.90	2.5192	2.1430	2.6016	0.4083	2.5856	-0.4514	2.6408	-1.1004	2.7709	-1.9953
0.95	3.1671	1.9724	2.7293	0.2644	2.6309	-0.5149	2.6744	-1.1343	2.8051	-2.0225
1.00	3.7210	1.4915	2.7960	0.1090	2.6570	-0.5700	2.7015	-1.1622	2.8369	-2.0471
1.05	3.9720	0.7912	2.8065	-0.0341	2.6699	-0.6135	2.7248	-1.1845	2.8671	-2.0699
1.10	3.8479	0.1131	2.7747	-0.1480	2.6752	-0.6445	2.7463	-1.2019	2.8961	-2.0913
1.15	3.4768	-0.3481	2.7184	-0.2245	2.6776	-0.6640	2.7673	-1.2155	2.9242	-2.1117
1.20	3.0383	-0.5602	2.6537	-0.2641	2.6806	-0.6740	2.7887	-1.2264	2.9514	-2.1313
1.25	2.6421	-0.5859	2.5928	-0.2721	2.6864	-0.6768	2.8107	-1.2356	2.9779	-2.1505
1.30	2.3258	-0.4980	2.5435	-0.2559	2.6960	-0.6749	2.8334	-1.2439	3.0035	-2.1693
1.35	2.0906	-0.3481	2.5095	-0.2232	2.7098	-0.6703	2.8567	-1.2519	3.0283	-2.1878
1.40	1.9261	-0.1669	2.4921	-0.1809	2.7275	-0.6649	2.8803	-1.2600	3.0523	-2.2060
1.45	1.8213	0.0286	2.4909	-0.1347	2.7484	-0.6599	2.9039	-1.2684	3.0755	-2.2241
1.50	1.7673	0.2287	2.5043	-0.0892	2.7717	-0.6563	2.9273	-1.2772	3.0979	-2.2419
1.55	1.7583	0.4277	2.5300	-0.0479	2.7966	-0.6545	2.9503	-1.2865	3.1195	-2.2595
1.60	1.7913	0.6207	2.5653	-0.0133	2.8222	-0.6548	2.9728	-1.2961	3.1403	-2.2769
1.65	1.8649	0.8028	2.6074	0.0129	2.8477	-0.6570	2.9946	-1.3061	3.1604	-2.2941
1.70	1.9786	0.9673	2.6532	0.0299	2.8728	-0.6610	3.0157	-1.3162	3.1798	-2.3110
1.75	2.1314	1.1049	2.6999	0.0376	2.8968	-0.6665	3.0361	-1.3265	3.1986	-2.3277
1.80	2.3197	1.2040	2.7450	0.0369	2.9197	-0.6729	3.0559	-1.3368	3.2167	-2.3442
1.85	2.5346	1.2506	2.7862	0.0291	2.9413	-0.6800	3.0749	-1.3472	3.2343	-2.3604
1.90	2.7594	1.2325	2.8223	0.0160	2.9616	-0.6875	3.0934	-1.3574	3.2513	-2.3765
1.95	2.9691	1.1438	2.8522	-0.0002	2.9808	-0.6950	3.1113	-1.3676	3.2677	-2.3923
2.00	3.1348	0.9918	2.8760	-0.0175	2.9990	-0.7024	3.1287	-1.3778	3.2836	-2.4079
2.05	3.2323	0.7986	2.8943	-0.0343	3.0163	-0.7095	3.1457	-1.3878	3.2990	-2.4234
2.10	3.2512	0.5972	2.9079	-0.0490	3.0330	-0.7164	3.1622	-1.3977	3.3139	-2.4386
2.15	3.1994	0.4199	2.9181	-0.0608	3.0492	-0.7229	3.1783	-1.4076	3.3284	-2.4537
2.20	3.0976	0.2890	2.9263	-0.0692	3.0651	-0.7291	3.1940	-1.4174	3.3424	-2.4686
2.25	2.9705	0.2132	2.9336	-0.0743	3.0807	-0.7352	3.2094	-1.4272	3.3560	-2.4833
2.30	2.8399	0.1900	2.9410	-0.0765	3.0961	-0.7410	3.2244	-1.4369	3.3692	-2.4979
2.35	2.7212	0.2109	2.9495	-0.0762	3.1113	-0.7468	3.2390	-1.4465	3.3820	-2.5123
2.40	2.6238	0.2655	2.9594	-0.0742	3.1264	-0.7525	3.2534	-1.4561	3.3944	-2.5265
2.45	2.5523	0.3436	2.9711	-0.0712	3.1413	-0.7583	3.2674	-1.4657	3.4064	-2.5406
2.50	2.5088	0.4362	2.9845	-0.0679	3.1561	-0.7641	3.2811	-1.4753	3.4180	-2.5545

TABLE 4.4

Normalized Admittance Y/Δ of Loop Antennas in Dissipative Media

$\Omega = 15$

βb	$\frac{\alpha}{\beta} = 0.00$		$\frac{\alpha}{\beta} = 0.01$		$\frac{\alpha}{\beta} = 0.03$		$\frac{\alpha}{\beta} = 0.05$		$\frac{\alpha}{\beta} = 0.07$	
	Y/Δ		Y/Δ		Y/Δ		Y/Δ		Y/Δ	
0.05	0.0001	-9.1554	0.0018	-9.1554	0.0052	-9.1555	0.0086	-9.1556	0.0119	-9.1558
0.10	0.0004	-4.4503	0.0038	-4.4503	0.0107	-4.4505	0.0175	-4.4508	0.0243	-4.4512
0.15	0.0010	-2.8236	0.0062	-2.8236	0.0166	-2.8239	0.0270	-2.8244	0.0374	-2.8251
0.20	0.0019	-1.9645	0.0090	-1.9646	0.0232	-1.9651	0.0373	-1.9658	0.0514	-1.9669
0.25	0.0033	-1.4101	0.0124	-1.4103	0.0306	-1.4109	0.0487	-1.4120	0.0669	-1.4135
0.30	0.0052	-1.0053	0.0165	-1.0056	0.0391	-1.0065	0.0617	-1.0080	0.0842	-1.0101
0.35	0.0080	-0.6827	0.0218	-0.6831	0.0493	-0.6845	0.0768	-0.6866	0.1042	-0.6894
0.40	0.0120	-0.4077	0.0286	-0.4083	0.0618	-0.4103	0.0949	-0.4132	0.1278	-0.4171
0.45	0.0178	-0.1597	0.0377	-0.1606	0.0775	-0.1635	0.1170	-0.1677	0.1563	-0.1732
0.50	0.0262	0.0751	0.0501	0.0736	0.0977	0.0693	0.1450	0.0633	0.1918	0.0556
0.55	0.0386	0.3076	0.0674	0.3052	0.1247	0.2987	0.1814	0.2899	0.2373	0.2787
0.60	0.0574	0.5477	0.0925	0.5439	0.1619	0.5339	0.2302	0.5206	0.2972	0.5041
0.65	0.0865	0.8065	0.1298	0.8003	0.2150	0.7843	0.2981	0.7637	0.3789	0.7388
0.70	0.1330	1.0977	0.1874	1.0872	0.2937	1.0609	0.3960	1.0280	0.4938	0.9891
0.75	0.2105	1.4408	0.2807	1.4221	0.4158	1.3769	0.5429	1.3221	0.6614	1.2594
0.80	0.3471	1.8645	0.4403	1.8295	0.6149	1.7474	0.7729	1.6523	0.9137	1.5476
0.85	0.6069	2.4125	0.7334	2.3416	0.9588	2.1830	1.1478	2.0103	1.3024	1.8316
0.90	1.1506	3.1369	1.3176	2.9793	1.5834	2.6539	1.7708	2.3338	1.8960	2.0330
0.95	2.4073	3.9786	2.5577	3.6081	2.7152	2.9469	2.7497	2.4001	2.7177	1.9565
1.00	5.1475	3.9093	4.8714	3.2605	4.3462	2.3426	3.9015	1.7417	3.5390	1.3264
1.05	7.1521	0.2902	6.3132	0.3242	5.1358	0.3176	4.3563	0.2758	3.8075	0.2255
1.10	4.8807	-2.5090	4.6364	-2.0294	4.1444	-1.3637	3.7130	-0.9550	3.3577	-0.6988
1.15	2.8139	-2.5757	2.8608	-2.2810	2.8541	-1.7777	2.7712	-1.3890	2.6579	-1.0977
1.20	1.7662	-2.0714	1.8597	-1.9144	1.9844	-1.6150	2.0448	-1.3498	2.0621	-1.1262
1.25	1.2225	-1.5815	1.3131	-1.4928	1.4598	-1.3141	1.5655	-1.1434	1.6381	-0.9888
1.30	0.9134	-1.1718	0.9954	-1.1182	1.1394	-1.0073	1.2576	-0.8973	1.3524	-0.7937
1.35	0.7239	-0.8280	0.7994	-0.7940	0.9380	-0.7234	1.0592	-0.6526	1.1638	-0.5850
1.40	0.6016	-0.5295	0.6738	-0.5077	0.8094	-0.4632	0.9325	-0.4194	1.0427	-0.3783
1.45	0.5214	-0.2599	0.5930	-0.2466	0.7293	-0.2210	0.8555	-0.1977	0.9707	-0.1777
1.50	0.4702	-0.0066	0.5439	-0.0001	0.6848	0.0099	0.8163	0.0156	0.9372	0.0169
1.55	0.4422	0.2409	0.5203	0.2409	0.6699	0.2358	0.8092	0.2241	0.9371	0.2067
1.60	0.4354	0.4918	0.5209	0.4846	0.6835	0.4623	0.8332	0.4310	0.9690	0.3927
1.65	0.4520	0.7555	0.5481	0.7390	0.7283	0.6947	0.8909	0.6387	1.0348	0.5747
1.70	0.4988	1.0428	0.6095	1.0125	0.8123	0.9369	0.9889	0.8473	1.1394	0.7503
1.75	0.5901	1.3663	0.7201	1.3139	0.9497	1.1905	1.1384	1.0532	1.2898	0.9126
1.80	0.7540	1.7408	0.9083	1.6506	1.1639	1.4509	1.3551	1.2448	1.4937	1.0479
1.85	1.0470	2.1790	1.2257	2.0204	1.4899	1.6972	1.6560	1.3957	1.7552	1.1318
1.90	1.5822	2.6677	1.7624	2.3840	1.9692	1.8714	2.0488	1.4550	2.0643	1.1287
1.95	2.5642	3.0700	2.6377	2.5848	2.6124	1.8484	2.5029	1.3478	2.3833	1.0005
2.00	4.1394	2.8425	3.8244	2.2215	3.2960	1.4524	2.9132	1.0125	2.6400	0.7342
2.05	5.4573	1.1474	4.6704	0.9285	3.6910	0.6556	3.1192	0.4895	2.7553	0.3745
2.10	4.9449	-0.9671	4.3613	-0.5862	3.5472	-0.2063	3.0348	-0.0492	2.6983	0.0167
2.15	3.5420	-1.8088	3.3844	-1.3482	3.0341	-0.7565	2.7364	-0.4326	2.5100	-0.2522
2.20	2.4554	-1.7836	2.5010	-1.4552	2.4674	-0.9489	2.3691	-0.6141	2.2650	-0.3999
2.25	1.7788	-1.4964	1.8880	-1.2847	2.0019	-0.9170	2.0317	-0.6384	2.0233	-0.4406
2.30	1.3627	-1.1686	1.4859	-1.0324	1.6593	-0.7778	1.7602	-0.5666	1.8160	-0.4052
2.35	1.0987	-0.8571	1.2218	-0.7680	1.4176	-0.5948	1.5564	-0.4432	1.6527	-0.3222
2.40	0.9260	-0.5702	1.0466	-0.5119	1.2515	-0.3969	1.4112	-0.2944	1.5330	-0.2116
[illegible]	[illegible]	[illegible]	[illegible]	[illegible]	1.1417	-0.1956	1.3145	-0.1340	1.4528	-0.0861
2.50	0.7387	-0.0499	0.8592	-0.0298	[illegible]	[illegible]	[illegible]	[illegible]	1.4081	0.0461

βb	$\frac{\alpha}{\beta} = 0.10$		$\frac{\alpha}{\beta} = 0.30$		$\frac{\alpha}{\beta} = 0.50$		$\frac{\alpha}{\beta} = 0.70$		$\frac{\alpha}{\beta} = 1.00$	
	Y/Δ		Y/Δ		Y/Δ		Y/Δ		Y/Δ	
0.05	0.0170	-9.1563	0.0508	-9.1632	0.0845	-9.1768	0.1182	-9.1972	0.1685	-9.2406
0.10	0.0346	-4.4522	0.1026	-4.4664	0.1702	-4.4945	0.2373	-4.5363	0.3364	-4.6246
0.15	0.0529	-2.8266	0.1562	-2.8491	0.2581	-2.8932	0.3582	-2.9584	0.5033	-3.0947
0.20	0.0726	-1.9690	0.2126	-2.0012	0.3494	-2.0635	0.4815	-2.1549	0.6684	-2.3434
0.25	0.0940	-1.4165	0.2728	-1.4604	0.4449	-1.5443	0.6077	-1.6657	0.8298	-1.9108
0.30	0.1179	-1.0142	0.3380	-1.0728	0.5458	-1.1828	0.7367	-1.3386	0.9854	-1.6446
0.35	0.1451	-0.6950	0.4098	-0.7726	0.6529	-0.9143	0.8678	-1.1096	1.1322	-1.4793
0.40	0.1769	-0.4248	0.4898	-0.5273	0.7669	-0.7078	0.9997	-0.9477	1.2671	-1.3812
0.45	0.2148	-0.1837	0.5801	-0.3194	0.8876	-0.5475	1.1302	-0.8363	1.3871	-1.3300
0.50	0.2612	0.0409	0.6829	-0.1398	1.0146	-0.4252	1.2562	-0.7654	1.4903	-1.3118
0.55	0.3195	0.2579	0.8006	0.0158	1.1460	-0.3370	1.3742	-0.7277	1.5756	-1.3157
0.60	0.3949	0.4740	0.9352	0.1478	1.2785	-0.2811	1.4804	-0.7171	1.6438	-1.3333
0.65	0.4949	0.6941	1.0880	0.2532	1.4076	-0.2560	1.5718	-0.7276	1.6965	-1.3579
0.70	0.6314	0.9212	1.2580	0.3263	1.5273	-0.2596	1.6464	-0.7527	1.7362	-1.3849
0.75	0.8221	1.1537	1.4403	0.3592	1.6316	-0.2879	1.7038	-0.7862	1.7659	-1.4114
0.80	1.0933	1.3800	1.6242	0.3444	1.7154	-0.3347	1.7453	-0.8227	1.7882	-1.4355
0.85	1.4784	1.5656	1.7931	0.2788	1.7758	-0.3920	1.7734	-0.8579	1.8056	-1.4568
0.90	2.0007	1.6315	1.9274	0.1682	1.8133	-0.4517	1.7913	-0.8891	1.8197	-1.4753
0.95	2.6147	1.4445	2.0108	0.0294	1.8308	-0.5068	1.8022	-0.9149	1.8319	-1.4913
1.00	3.1199	0.9046	2.0375	-0.1139	1.8335	-0.5526	1.8090	-0.9351	1.8431	-1.5053
1.05	3.2410	0.1508	2.0148	-0.2388	1.8268	-0.5869	1.8139	-0.9503	1.8536	-1.5178
1.10	2.9489	-0.4763	1.9589	-0.3309	1.8159	-0.6097	1.8184	-0.9613	1.8637	-1.5294
1.15	2.4815	-0.7973	1.8878	-0.3859	1.8046	-0.6224	1.8235	-0.9694	1.8736	-1.5402
1.20	2.0429	-0.8667	1.8163	-0.4074	1.7956	-0.6273	1.8296	-0.9755	1.8832	-1.5507
1.25	1.7024	-0.7948	1.7540	-0.4024	1.7904	-0.6267	1.8368	-0.9805	1.8925	-1.5608
1.30	1.4588	-0.6578	1.7059	-0.3792	1.7894	-0.6228	1.8450	-0.9850	1.9015	-1.5708
1.35	1.2923	-0.4951	1.6737	-0.3447	1.7926	-0.6175	1.8540	-0.9895	1.9100	-1.5806
1.40	1.1848	-0.3251	1.6572	-0.3050	1.7995	-0.6123	1.8634	-0.9942	1.9182	-1.5904
1.45	1.1229	-0.1554	1.6550	-0.2645	1.8092	-0.6080	1.8730	-0.9994	1.9259	-1.6000
1.50	1.0983	0.0107	1.6651	-0.2268	1.8210	-0.6054	1.8826	-1.0051	1.9332	-1.6095
1.55	1.1065	0.1716	1.6852	-0.1944	1.8340	-0.6047	1.8919	-1.0112	1.9400	-1.6188
1.60	1.1458	0.3258	1.7129	-0.1692	1.8473	-0.6059	1.9009	-1.0176	1.9465	-1.6280
1.65	1.2166	0.4707	1.7453	-0.1522	1.8605	-0.6089	1.9094	-1.0242	1.9525	-1.6371
1.70	1.3203	0.6016	1.7800	-0.1439	1.8731	-0.6134	1.9175	-1.0310	1.9582	-1.6460
1.75	1.4586	0.7104	1.8143	-0.1439	1.8847	-0.6190	1.9251	-1.0379	1.9636	-1.6547
1.80	1.6306	0.7851	1.8459	-0.1513	1.8952	-0.6254	1.9322	-1.0447	1.9686	-1.6634
1.85	1.8298	0.8097	1.8731	-0.1645	1.9046	-0.6321	1.9389	-1.0515	1.9733	-1.6718
1.90	2.0394	0.7681	1.8948	-0.1817	1.9130	-0.6389	1.9453	-1.0582	1.9778	-1.6802
1.95	2.2303	0.6512	1.9104	-0.2007	1.9204	-0.6456	1.9514	-1.0649	1.9819	-1.6883
2.00	2.3663	0.4680	1.9203	-0.2197	1.9272	-0.6520	1.9572	-1.0714	1.9858	-1.6964
2.05	2.4187	0.2503	1.9253	-0.2371	1.9333	-0.6580	1.9627	-1.0778	1.9894	-1.7043
2.10	2.3825	0.0418	1.9265	-0.2516	1.9391	-0.6637	1.9680	-1.0842	1.9928	-1.7122
2.15	2.2778	-0.1205	1.9252	-0.2627	1.9447	-0.6690	1.9731	-1.0905	1.9960	-1.7199
2.20	2.1369	-0.2198	1.9228	-0.2703	1.9501	-0.6740	1.9780	-1.0967	1.9989	-1.7274
2.25	1.9897	-0.2578	1.9204	-0.2746	1.9555	-0.6787	1.9827	-1.1028	2.0016	-1.7349
2.30	1.8558	-0.2461	1.9189	-0.2762	1.9609	-0.6833	1.9873	-1.1089	2.0041	-1.7423
2.35	1.7454	-0.1988	1.9189	-0.2756	1.9663	-0.6879	1.9917	-1.1150	2.0064	-1.7496
2.40	1.6623	-0.1280	1.9209	-0.2738	1.9716	-0.6924	1.9959	-1.1210	2.0086	-1.7567
2.45	1.6070	-0.0433	1.9247	-0.2714	1.9770	-0.6969	2.0000	-1.1270	2.0105	-1.7638
2.50	1.5782	0.0482	1.9304	-0.2691	1.9823	-0.7015	2.0039	-1.1330	2.0123	-1.7708

TABLE 4.5

Normalized Admittance Y/Δ of Loop Antennas in Dissipative Media
$\Omega = 17$

βb	$\frac{\alpha}{\beta} = 0.00$		$\frac{\alpha}{\beta} = 0.01$		$\frac{\alpha}{\beta} = 0.03$		$\frac{\alpha}{\beta} = 0.05$		$\frac{\alpha}{\beta} = 0.07$	
	Y/Δ		Y/Δ		Y/Δ		Y/Δ		Y/Δ	
0.05	0.0001	-7.5205	0.0014	-7.5205	0.0041	-7.5206	0.0068	-7.5207	0.0095	-7.5209
0.10	0.0003	-3.6585	0.0030	-3.6586	0.0085	-3.6587	0.0139	-3.6589	0.0194	-3.6593
0.15	0.0007	-2.3246	0.0048	-2.3247	0.0131	-2.3249	0.0214	-2.3253	0.0297	-2.3258
0.20	0.0013	-1.6211	0.0070	-1.6211	0.0183	-1.6215	0.0296	-1.6221	0.0409	-1.6229
0.25	0.0022	-1.1677	0.0095	-1.1678	0.0241	-1.1683	0.0386	-1.1692	0.0531	-1.1703
0.30	0.0035	-0.8372	0.0126	-0.8373	0.0307	-0.8381	0.0488	-0.8392	0.0668	-0.8408
0.35	0.0054	-0.5741	0.0165	-0.5744	0.0386	-0.5754	0.0606	-0.5770	0.0826	-0.5792
0.40	0.0081	-0.3500	0.0215	-0.3504	0.0482	-0.3519	0.0748	-0.3541	0.1013	-0.3572
0.45	0.0120	-0.1479	0.0281	-0.1486	0.0602	-0.1507	0.0921	-0.1539	0.1238	-0.1582
0.50	0.0177	0.0435	0.0371	0.0424	0.0756	0.0393	0.1139	0.0347	0.1518	0.0287
0.55	0.0262	0.2333	0.0496	0.2317	0.0962	0.2269	0.1422	0.2201	0.1877	0.2115
0.60	0.0391	0.4302	0.0677	0.4275	0.1245	0.4201	0.1804	0.4099	0.2352	0.3971
0.65	0.0591	0.6436	0.0947	0.6392	0.1649	0.6273	0.2336	0.6115	0.3003	0.5919
0.70	0.0913	0.8859	0.1366	0.8784	0.2253	0.8588	0.3108	0.8332	0.3928	0.8023
0.75	0.1459	1.1755	0.2053	1.1621	0.3200	1.1277	0.4284	1.0845	0.5296	1.0338
0.80	0.2444	1.5418	0.3254	1.5158	0.4781	1.4519	0.6167	1.3748	0.7405	1.2878
0.85	0.4389	2.0347	0.5542	1.9795	0.7611	1.8503	0.9351	1.7039	1.0771	1.5490
0.90	0.8742	2.7367	1.0412	2.6034	1.3082	2.3164	1.4950	2.0251	1.6171	1.7480
0.95	2.0206	3.6945	2.2073	3.3276	2.4020	2.6627	2.4446	2.1152	2.4077	1.6792
1.00	5.1373	3.8032	4.7883	3.0269	4.1369	2.0120	3.6118	1.4047	3.2029	1.0128
1.05	6.9020	-0.9520	5.9742	-0.6197	4.7075	-0.2997	3.8984	-0.1729	3.3449	-0.1219
1.10	3.6382	-2.9701	3.5808	-2.4677	3.3293	-1.7242	3.0334	-1.2433	2.7606	-0.9319
1.15	1.8759	-2.4507	1.9839	-2.2219	2.0919	-1.7993	2.1019	-1.4479	2.0591	-1.1711
1.20	1.1348	-1.8366	1.2374	-1.7272	1.3914	-1.5038	1.4866	-1.2916	1.5375	-1.1033
1.25	0.7756	-1.3653	0.8601	-1.3063	1.0036	-1.1803	1.1141	-1.0522	1.1957	-0.9305
1.30	0.5773	-1.0030	0.6487	-0.9682	0.7775	-0.8925	0.8868	-0.8132	0.9773	-0.7352
1.35	0.4575	-0.7108	0.5211	-0.6892	0.6396	-0.6421	0.7455	-0.5927	0.8386	-0.5436
1.40	0.3810	-0.4625	0.4406	-0.4488	0.5539	-0.4200	0.6582	-0.3905	0.7528	-0.3620
1.45	0.3313	-0.2407	0.3899	-0.2326	0.5025	-0.2169	0.6077	-0.2025	0.7047	-0.1900
1.50	0.3002	-0.0334	0.3603	-0.0298	0.4761	-0.0250	0.5850	-0.0236	0.6860	-0.0255
1.55	0.2840	0.1691	0.3479	0.1684	0.4709	0.1623	0.5863	0.1507	0.6929	0.1342
1.60	0.2817	0.3752	0.3521	0.3695	0.4866	0.3509	0.6114	0.3240	0.7250	0.2907
1.65	0.2953	0.5939	0.3753	0.5814	0.5263	0.5460	0.6632	0.4994	0.7848	0.4449
1.70	0.3301	0.8357	0.4239	0.8130	0.5972	0.7528	0.7486	0.6782	0.8776	0.5953
1.75	0.3976	1.1146	0.5111	1.0749	0.7129	0.9750	0.8790	0.8586	1.0112	0.7365
1.80	0.5216	1.4501	0.6624	1.3791	0.8971	1.2119	1.0715	1.0317	1.1952	0.8562
1.85	0.7542	1.8676	0.9297	1.7349	1.1889	1.4489	1.3475	1.1732	1.4365	0.9302
1.90	1.2167	2.3858	1.4178	2.1237	1.6440	1.6321	1.7225	1.2303	1.7285	0.9202
1.95	2.1930	2.9156	2.3105	2.3957	2.2976	1.6184	2.1705	1.1143	2.0322	0.7817
2.00	4.0749	2.7502	3.6688	1.9994	3.0131	1.1685	2.5679	0.7461	2.2652	0.4992
2.05	5.5169	0.3773	4.4957	0.3382	3.3397	0.2537	2.7168	0.1851	2.3382	0.1304
2.10	4.1584	-1.8458	3.7161	-1.2369	3.0115	-0.6120	2.5427	-0.3395	2.2308	-0.2100
2.15	2.5313	-2.1189	2.5417	-1.6614	2.3870	-1.0227	2.1843	-0.6509	2.0087	-0.4353
2.20	1.6193	-1.7831	1.7409	-1.5201	1.8305	-1.0720	1.8125	-0.7496	1.7580	-0.5324
2.25	1.1321	-1.3982	1.2637	-1.2461	1.4308	-0.9550	1.5070	-0.7141	1.5328	-0.5325
2.30	0.8541	-1.0616	0.9759	-0.9692	1.1608	-0.7809	1.2798	-0.6108	1.3525	-0.4726
2.35	0.6847	-0.7746	0.7960	-0.7164	0.9810	-0.5939	1.1189	-0.4780	1.2181	-0.3796
2.40	0.5766	-0.5244	0.6806	-0.4873	0.8628	-0.4091	1.0096	-0.3342	1.1243	-0.2701
2.45	0.5066	-0.2986	0.6070	-0.2760	0.7881	-0.2299	0.9405	-0.1876	1.0648	-0.1532
2.50	0.4627	-0.0871	0.5630	-0.0756	0.7464	-0.0556	0.9040	-0.0414	1.0351	-0.0341

βb	$\frac{\alpha}{\beta} = 0.10$		$\frac{\alpha}{\beta} = 0.30$		$\frac{\alpha}{\beta} = 0.50$		$\frac{\alpha}{\beta} = 0.70$		$\frac{\alpha}{\beta} = 1.00$	
	Y/Δ		Y/Δ		Y/Δ		Y/Δ		Y/Δ	
0.05	0.0136	-7.5212	0.0405	-7.5267	0.0675	-7.5376	0.0943	-7.5539	0.1345	-7.5885
0.10	0.0275	-3.6600	0.0819	-3.6714	0.1359	-3.6938	0.1894	-3.7272	0.2686	-3.7977
0.15	0.0422	-2.3270	0.1247	-2.3449	0.2061	-2.3801	0.2860	-2.4322	0.4019	-2.5412
0.20	0.0578	-1.6246	0.1697	-1.6502	0.2791	-1.7001	0.3847	-1.7733	0.5338	-1.9242
0.25	0.0749	-1.1727	0.2179	-1.2077	0.3557	-1.2750	0.4859	-1.3724	0.6629	-1.5691
0.30	0.0939	-0.8441	0.2703	-0.8909	0.4368	-0.9792	0.5894	-1.1046	0.7871	-1.3507
0.35	0.1155	-0.5836	0.3281	-0.6458	0.5232	-0.7599	0.6948	-0.9175	0.9041	-1.2153
0.40	0.1408	-0.3632	0.3927	-0.4455	0.6153	-0.5915	0.8010	-0.7858	1.0111	-1.1355
0.45	0.1709	-0.1665	0.4660	-0.2758	0.7133	-0.4611	0.9062	-0.6957	1.1057	-1.0943
0.50	0.2080	0.0171	0.5498	-0.1291	0.8167	-0.3622	1.0076	-0.6393	1.1860	-1.0803
0.55	0.2547	0.1950	0.6464	-0.0019	0.9240	-0.2916	1.1022	-0.6106	1.2513	-1.0843
0.60	0.3153	0.3732	0.7577	0.1060	1.0325	-0.2481	1.1867	-0.6045	1.3021	-1.0993
0.65	0.3964	0.5562	0.8850	0.1918	1.1380	-0.2308	1.2585	-0.6159	1.3399	-1.1197
0.70	0.5082	0.7474	1.0278	0.2501	1.2354	-0.2380	1.3157	-0.6392	1.3670	-1.1415
0.75	0.6670	0.9467	1.1819	0.2736	1.3192	-0.2661	1.3582	-0.6692	1.3859	-1.1624
0.80	0.8980	1.1458	1.3378	0.2548	1.3848	-0.3098	1.3871	-0.7009	1.3990	-1.1810
0.85	1.2365	1.3150	1.4801	0.1906	1.4297	-0.3618	1.4047	-0.7309	1.4082	-1.1969
0.90	1.7134	1.3777	1.5900	0.0862	1.4545	-0.4148	1.4140	-0.7569	1.4152	-1.2103
0.95	2.2912	1.1906	1.6524	-0.0421	1.4622	-0.4627	1.4178	-0.7778	1.4209	-1.2216
1.00	2.7504	0.6387	1.6628	-0.1712	1.4576	-0.5015	1.4186	-0.7936	1.4260	-1.2312
1.05	2.7882	-0.1033	1.6296	-0.2800	1.4458	-0.5296	1.4183	-0.8050	1.4308	-1.2396
1.10	2.4265	-0.6519	1.5692	-0.3564	1.4312	-0.5473	1.4181	-0.8129	1.4356	-1.2472
1.15	1.9561	-0.8737	1.4987	-0.3982	1.4172	-0.5562	1.4187	-0.8183	1.4403	-1.2543
1.20	1.5609	-0.8741	1.4311	-0.4103	1.4059	-0.5584	1.4206	-0.8220	1.4449	-1.2612
1.25	1.2759	-0.7706	1.3741	-0.4000	1.3984	-0.5562	1.4236	-0.8250	1.4494	-1.2678
1.30	1.0821	-0.6282	1.3313	-0.3751	1.3951	-0.5515	1.4276	-0.8277	1.4537	-1.2743
1.35	0.9550	-0.4755	1.3033	-0.3419	1.3956	-0.5459	1.4323	-0.8305	1.4577	-1.2808
1.40	0.8763	-0.3237	1.2894	-0.3054	1.3994	-0.5406	1.4376	-0.8336	1.4615	-1.2872
1.45	0.8341	-0.1764	1.2881	-0.2693	1.4057	-0.5365	1.4430	-0.8372	1.4650	-1.2935
1.50	0.8213	-0.0345	1.2973	-0.2366	1.4137	-0.5340	1.4484	-0.8412	1.4682	-1.2997
1.55	0.8347	0.1019	1.3150	-0.2092	1.4227	-0.5334	1.4537	-0.8455	1.4712	-1.3059
1.60	0.8733	0.2323	1.3388	-0.1886	1.4319	-0.5345	1.4586	-0.8502	1.4738	-1.3119
1.65	0.9381	0.3550	1.3664	-0.1756	1.4409	-0.5372	1.4632	-0.8551	1.4762	-1.3179
1.70	1.0315	0.4662	1.3954	-0.1705	1.4493	-0.5411	1.4675	-0.8601	1.4784	-1.3237
1.75	1.1559	0.5586	1.4236	-0.1728	1.4567	-0.5460	1.4714	-0.8651	1.4803	-1.3294
1.80	1.3119	0.6208	1.4489	-0.1815	1.4632	-0.5515	1.4749	-0.8701	1.4820	-1.3350
1.85	1.4943	0.6363	1.4699	-0.1951	1.4687	-0.5572	1.4782	-0.8751	1.4836	-1.3405
1.90	1.6870	0.5876	1.4855	-0.2119	1.4733	-0.5628	1.4812	-0.8800	1.4849	-1.3459
1.95	1.8595	0.4646	1.4956	-0.2299	1.4772	-0.5683	1.4839	-0.8848	1.4861	-1.3512
2.00	1.9729	0.2780	1.5005	-0.2474	1.4805	-0.5735	1.4865	-0.8894	1.4871	-1.3564
2.05	1.9987	0.0641	1.5011	-0.2629	1.4834	-0.5783	1.4889	-0.8940	1.4879	-1.3615
2.10	1.9374	-0.1293	1.4985	-0.2754	1.4860	-0.5827	1.4912	-0.8986	1.4886	-1.3665
2.15	1.8164	-0.2673	1.4941	-0.2847	1.4885	-0.5868	1.4934	-0.9030	1.4891	-1.3715
2.20	1.6716	-0.3397	1.4890	-0.2906	1.4910	-0.5906	1.4954	-0.9074	1.4895	-1.3763
2.25	1.5312	-0.3549	1.4843	-0.2935	1.4934	-0.5942	1.4973	-0.9118	1.4897	-1.3811
2.30	1.4106	-0.3277	1.4807	-0.2940	1.4959	-0.5977	1.4991	-0.9161	1.4898	-1.3858
2.35	1.3159	-0.2728	1.4788	-0.2928	1.4985	-0.6011	1.5009	-0.9204	1.4898	-1.3904
2.40	1.2478	-0.2012	1.4788	-0.2906	1.5011	-0.6045	1.5025	-0.9247	1.4897	-1.3950
2.45	1.2051	-0.1207	1.4806	-0.2881	1.5037	-0.6080	1.5040	-0.9289	1.4895	-1.3994
2.50	1.1857	-0.0369	1.4841	-0.2858	1.5064	-0.6115	1.5054	-0.9331	1.4891	-1.4039

TABLE 4.6

Normalized Admittance Y/Δ of Loop Antennas in Dissipative Media
Ω = 20

βb	$\frac{\alpha}{\beta}$ = 0.00		$\frac{\alpha}{\beta}$ = 0.01		$\frac{\alpha}{\beta}$ = 0.03		$\frac{\alpha}{\beta}$ = 0.05		$\frac{\alpha}{\beta}$ = 0.07	
	Y/Δ		Y/Δ		Y/Δ		Y/Δ		Y/Δ	
0.05	0.0001	-6.3809	0.0012	-6.3810	0.0034	-6.3810	0.0057	-6.3811	0.0079	-6.3812
0.10	0.0002	-3.1057	0.0025	-3.1058	0.0070	-3.1059	0.0116	-3.1061	0.0161	-3.1064
0.15	0.0005	-1.9752	0.0040	-1.9752	0.0109	-1.9754	0.0178	-1.9757	0.0247	-1.9762
0.20	0.0009	-1.3794	0.0057	-1.3794	0.0151	-1.3797	0.0245	-1.3802	0.0339	-1.3809
0.25	0.0016	-0.9958	0.0077	-0.9959	0.0198	-0.9963	0.0319	-0.9970	0.0441	-0.9979
0.30	0.0026	-0.7164	0.0101	-0.7165	0.0252	-0.7171	0.0403	-0.7180	0.0554	-0.7193
0.35	0.0039	-0.4942	0.0132	-0.4944	0.0316	-0.4952	0.0501	-0.4965	0.0685	-0.4983
0.40	0.0059	-0.3050	0.0170	-0.3053	0.0394	-0.3065	0.0616	-0.3083	0.0838	-0.3108
0.45	0.0087	-0.1344	0.0222	-0.1349	0.0490	-0.1366	0.0758	-0.1392	0.1024	-0.1427
0.50	0.0128	0.0272	0.0291	0.0264	0.0614	0.0239	0.0936	0.0202	0.1255	0.0153
0.55	0.0190	0.1877	0.0387	0.1865	0.0779	0.1828	0.1168	0.1774	0.1552	0.1703
0.60	0.0283	0.3546	0.0525	0.3526	0.1005	0.3468	0.1479	0.3386	0.1944	0.3282
0.65	0.0429	0.5362	0.0731	0.5329	0.1329	0.5236	0.1914	0.5109	0.2484	0.4949
0.70	0.0666	0.7437	0.1054	0.7381	0.1814	0.7226	0.2550	0.7019	0.3256	0.6764
0.75	0.1070	0.9941	0.1585	0.9838	0.2581	0.9566	0.3526	0.9212	0.4409	0.8789
0.80	0.1812	1.3157	0.2527	1.2956	0.3879	1.2440	0.5113	1.1796	0.6216	1.1055
0.85	0.3315	1.7603	0.4367	1.7162	0.6265	1.6085	0.7868	1.4822	0.9175	1.3460
0.90	0.6846	2.4269	0.8473	2.3137	1.1092	2.0598	1.2922	1.7947	1.4103	1.5394
0.95	1.7135	3.4487	1.9280	3.0949	2.1525	2.4405	2.2026	1.9014	2.1640	1.4783
1.00	5.1313	3.7410	4.7106	2.8443	3.9455	1.7582	3.3597	1.1603	2.9227	0.7969
1.05	6.3542	-1.9606	5.4858	-1.3368	4.2674	-0.7195	3.4877	-0.4540	2.9593	-0.3265
1.10	2.6887	-2.9849	2.7570	-2.5378	2.6900	-1.8264	2.5089	-1.3393	2.3089	-1.0138
1.15	1.3056	-2.2037	1.4322	-2.0315	1.5893	-1.6909	1.6491	-1.3889	1.6490	-1.1405
1.20	0.7783	-1.5977	0.8759	-1.5196	1.0309	-1.3508	1.1363	-1.1811	1.2011	-1.0242
1.25	0.5302	-1.1739	0.6054	-1.1327	0.7367	-1.0402	0.8419	-0.9415	0.9227	-0.8442
1.30	0.3949	-0.8597	0.4565	-0.8356	0.5696	-0.7810	0.6677	-0.7213	0.7506	-0.6605
1.35	0.3135	-0.6105	0.3676	-0.5956	0.4695	-0.5622	0.5619	-0.5257	0.6441	-0.4884
1.40	0.2618	-0.4004	0.3122	-0.3912	0.4086	-0.3711	0.4981	-0.3500	0.5801	-0.3291
1.45	0.2285	-0.2136	0.2778	-0.2083	0.3730	-0.1979	0.4627	-0.1883	0.5460	-0.1802
1.50	0.2078	-0.0391	0.2584	-0.0370	0.3563	-0.0347	0.4489	-0.0353	0.5351	-0.0386
1.55	0.1975	0.1315	0.2514	0.1305	0.3556	0.1246	0.4539	0.1136	0.5449	0.0984
1.60	0.1970	0.3059	0.2566	0.3012	0.3712	0.2855	0.4779	0.2622	0.5754	0.2331
1.65	0.2079	0.4921	0.2763	0.4824	0.4060	0.4533	0.5242	0.4136	0.6293	0.3664
1.70	0.2345	0.7002	0.3158	0.6826	0.4666	0.6333	0.5991	0.5697	0.7119	0.4977
1.75	0.2859	0.9443	0.3862	0.9131	0.5655	0.8302	0.7137	0.7299	0.8311	0.6225
1.80	0.3820	1.2453	0.5103	1.1983	0.7255	1.0461	0.8853	0.8871	0.9972	0.7298
1.85	0.5685	1.6361	0.7369	1.5247	0.9868	1.2714	1.1376	1.0198	1.2189	0.7965
1.90	0.9635	2.1591	1.1755	1.9215	1.4131	1.4580	1.4911	1.0752	1.4921	0.7832
1.95	1.8979	2.7845	2.0572	2.2496	2.0605	1.4539	1.9246	0.9568	1.7776	0.6421
2.00	4.0386	2.7033	3.5455	1.8339	2.7850	0.9691	2.3015	0.5732	1.9871	0.3566
2.05	5.4367	-0.3557	4.2701	-0.1374	3.0304	-0.0134	2.3988	0.0037	2.0269	-0.0050
2.10	3.3406	-2.2772	3.0998	-1.5752	2.5661	-0.8202	2.1653	-0.4822	1.8894	-0.3170
2.15	1.8152	-2.0877	1.9261	-1.6954	1.9127	-1.0959	1.7683	-0.7240	1.6572	-0.5011
2.20	1.1145	-1.6268	1.2608	-1.4267	1.4092	-1.0537	1.4391	-0.7647	1.4179	-0.5612
2.25	0.7683	-1.2369	0.8977	-1.1272	1.0770	-0.8998	1.1726	-0.6975	1.2160	-0.5378
2.30	0.5773	-0.9277	0.6881	-0.8629	0.8640	-0.7211	0.9841	-0.5843	1.0616	-0.4678
2.35	0.4628	-0.6754	0.5603	-0.6353	0.7270	-0.5454	0.8555	-0.4551	0.9506	-0.3749
2.40	0.3906	-0.4603	0.4799	-0.4352	0.6394	-0.3792	0.7710	-0.3226	0.8757	-0.2722
2.45	0.3442	-0.2685	0.4296	-0.2534	0.5858	-0.2217	0.7196	-0.1915	0.8302	-0.1661
2.50	0.3156	-0.0895	0.4006	-0.0524	[illegible]	[illegible]	[illegible]	[illegible]	[illegible]	[illegible]

βb	$\frac{\alpha}{\beta}$ = 0.10		$\frac{\alpha}{\beta}$ = 0.30		$\frac{\alpha}{\beta}$ = 0.50		$\frac{\alpha}{\beta}$ = 0.70		$\frac{\alpha}{\beta}$ = 1.00	
	Y/Δ		Y/Δ		Y/Δ		Y/Δ		Y/Δ	
0.05	0.0113	-6.3815	0.0338	-6.3861	0.0562	-6.3952	0.0786	-6.4087	0.1120	-6.4375
0.10	0.0229	-3.1070	0.0682	-3.1164	0.1132	-3.1351	0.1578	-3.1629	0.2237	-3.2217
0.15	0.0351	-1.9772	0.1038	-1.9920	0.1717	-2.0214	0.2383	-2.0648	0.3349	-2.1557
0.20	0.0480	-1.3823	0.1414	-1.4036	0.2326	-1.4452	0.3207	-1.5062	0.4448	-1.6322
0.25	0.0622	-0.9999	0.1816	-1.0290	0.2966	-1.0852	0.4051	-1.1666	0.5524	-1.3310
0.30	0.0780	-0.7220	0.2254	-0.7611	0.3645	-0.8349	0.4917	-0.9399	0.6560	-1.1458
0.35	0.0959	-0.5020	0.2738	-0.5538	0.4369	-0.6495	0.5800	-0.7817	0.7533	-1.0313
0.40	0.1169	-0.3157	0.3281	-0.3845	0.5143	-0.5072	0.6690	-0.6706	0.8421	-0.9639
0.45	0.1420	-0.1495	0.3897	-0.2410	0.5968	-0.3972	0.7571	-0.5950	0.9202	-0.9296
0.50	0.1728	0.0058	0.4606	-0.1169	0.6842	-0.3141	0.8421	-0.5481	0.9859	-0.9182
0.55	0.2117	0.1567	0.5425	-0.0093	0.7750	-0.2552	0.9211	-0.5250	1.0386	-0.9221
0.60	0.2624	0.3084	0.6375	0.0820	0.8669	-0.2197	0.9914	-0.5214	1.0788	-0.9351
0.65	0.3305	0.4651	0.7467	0.1545	0.9563	-0.2069	1.0504	-0.5326	1.1079	-0.9523
0.70	0.4251	0.6303	0.8698	0.2031	1.0384	-0.2155	1.0967	-0.5540	1.1279	-0.9706
0.75	0.5609	0.8048	1.0034	0.2209	1.1084	-0.2423	1.1300	-0.5807	1.1409	-0.9877
0.80	0.7619	0.9825	1.1388	0.2009	1.1621	-0.2824	1.1516	-0.6086	1.1492	-1.0027
0.85	1.0632	1.1375	1.2616	0.1399	1.1974	-0.3293	1.1635	-0.6345	1.1544	-1.0153
0.90	1.5000	1.1974	1.3544	0.0430	1.2149	-0.3765	1.1683	-0.6566	1.1578	-1.0256
0.95	2.0413	1.0150	1.4031	-0.0744	1.2174	-0.4185	1.1686	-0.6739	1.1604	-1.0340
1.00	2.4573	0.4673	1.4043	-0.1905	1.2096	-0.4518	1.1667	-0.6868	1.1626	-1.0410
1.05	2.4343	-0.2422	1.3669	-0.2858	1.1958	-0.4752	1.1641	-0.6956	1.1648	-1.0470
1.10	2.0421	-0.7142	1.3069	-0.3501	1.1803	-0.4893	1.1619	-0.7015	1.1670	-1.0523
1.15	1.5945	-0.8641	1.2401	-0.3827	1.1658	-0.4956	1.1606	-0.7052	1.1693	-1.0573
1.20	1.2458	-0.8260	1.1779	-0.3889	1.1542	-0.4963	1.1605	-0.7076	1.1716	-1.0620
1.25	1.0062	-0.7116	1.1267	-0.3761	1.1463	-0.4932	1.1616	-0.7093	1.1738	-1.0667
1.30	0.8488	-0.5744	1.0889	-0.3513	1.1422	-0.4882	1.1636	-0.7109	1.1759	-1.0712
1.35	0.7483	-0.4351	1.0648	-0.3201	1.1416	-0.4826	1.1663	-0.7127	1.1779	-1.0758
1.40	0.6881	-0.3003	1.0532	-0.2868	1.1439	-0.4775	1.1695	-0.7148	1.1797	-1.0803
1.45	0.6577	-0.1716	1.0526	-0.2546	1.1485	-0.4736	1.1729	-0.7174	1.1813	-1.0847
1.50	0.6514	-0.0487	1.0613	-0.2258	1.1545	-0.4712	1.1762	-0.7203	1.1827	-1.0891
1.55	0.6666	0.0690	1.0771	-0.2022	1.1613	-0.4706	1.1795	-0.7236	1.1839	-1.0935
1.60	0.7030	0.1815	1.0981	-0.1848	1.1683	-0.4715	1.1825	-0.7272	1.1849	-1.0977
1.65	0.7619	0.2876	1.1222	-0.1744	1.1749	-0.4738	1.1852	-0.7309	1.1857	-1.1019
1.70	0.8458	0.3841	1.1472	-0.1710	1.1810	-0.4772	1.1876	-0.7347	1.1863	-1.1060
1.75	0.9578	0.4645	1.1711	-0.1742	1.1863	-0.4814	1.1897	-0.7386	1.1868	-1.1100
1.80	1.0993	0.5178	1.1921	-0.1831	1.1906	-0.4861	1.1916	-0.7424	1.1871	-1.1139
1.85	1.2660	0.5280	1.2090	-0.1963	1.1941	-0.4909	1.1932	-0.7462	1.1873	-1.1177
1.90	1.4425	0.4769	1.2210	-0.2120	1.1969	-0.4956	1.1946	-0.7499	1.1873	-1.1215
1.95	1.5982	0.3543	1.2279	-0.2284	1.1990	-0.5002	1.1959	-0.7535	1.1873	-1.1252
2.00	1.6933	0.1719	1.2301	-0.2441	1.2006	-0.5044	1.1970	-0.7571	1.1871	-1.1288
2.05	1.7011	-0.0310	1.2286	-0.2577	1.2020	-0.5083	1.1981	-0.7605	1.1868	-1.1324
2.10	1.6264	-0.2059	1.2244	-0.2686	1.2031	-0.5118	1.1990	-0.7639	1.1864	-1.1358
2.15	1.5014	-0.3217	1.2188	-0.2762	1.2042	-0.5150	1.1998	-0.7672	1.1859	-1.1392
2.20	1.3621	-0.3741	1.2128	-0.2808	1.2053	-0.5180	1.2006	-0.7705	1.1853	-1.1426
2.25	1.2334	-0.3749	1.2074	-0.2827	1.2064	-0.5208	1.2013	-0.7738	1.1846	-1.1459
2.30	1.1271	-0.3404	1.2033	-0.2825	1.2076	-0.5235	1.2019	-0.7770	1.1838	-1.1491
2.35	1.0463	-0.2841	1.2007	-0.2809	1.2089	-0.5261	1.2025	-0.7802	1.1829	-1.1523
2.40	0.9902	-0.2157	1.1999	-0.2784	1.2102	-0.5287	1.2029	-0.7834	1.1820	-1.1554
2.45	0.9565	-0.1415	1.2008	-0.2758	1.2116	-0.5314	1.2033	-0.7865	1.1809	-1.1585
2.50	0.9433	-0.0659	1.2033	-0.2736	1.2129	-0.5341	1.2037	-0.7897	1.1798	-1.1615

5. Broadside and Endfire Arrays

An array of $N = 7$ identical dipole antennas, each of half-length h and radius a and separated from its adjacent neighbors by the distance b is shown in Fig. 5.1. Each element is driven by a generator at its center that maintains either specified voltages V_i or specified currents $I_i(0)$ at the N driving points in the plane $z = 0$. A useful two-term expression of the current in each element is[1–4]

$$I_i(z) = jA_i \sin k_0(h - |z|) + B_i(\cos k_0 z - \cos k_0 h) \qquad i = 1, 2, \ldots, N \tag{5.1}$$

where

$$A_i = c_1 V_i, \qquad c_1 = j2\pi/\zeta_0 \Psi_{dR} \cos k_0 h \tag{5.2}$$

and the complex coefficients B_i are related to the A_i through the matrix equation

$$[\Phi_u]\{B\} = [\Phi_v]\{jA\} \tag{5.3}$$

Note that $\zeta_0 \doteq 120\pi$ ohms. The B and jA matrices are columnar; the Φ matrices are $N \times N$, with the elements given by

$$\begin{aligned}\Phi_{kiu} &= \Psi_{kidu} \cos k_0 h - \Psi_{kiu}(h)\\ \Phi_{kiv} &= \Psi_{kiv}(h) - (1 - \delta_{ik})\Psi_{kidv} \cos k_0 h\\ &\quad - j\delta_{ik}\Psi_{kidI} \cos k_0 h\end{aligned} \tag{5.4}$$

where $\delta_{ik} = 0$, $i \neq k$; $\delta_{ik} = 1$, $i = k$. The parameters Ψ_{dR}, Ψ_{dI}, Ψ_{du}, Ψ_{dv}, $\Psi_u(h)$, and $\Psi_v(h)$ are defined on page 141 of Reference 1. Extensive tables of Ψ_{dR}, Φ_{kiu}, and Φ_{kiv} are given in Appendix III of Reference 1.

The driving-point admittances $Y_i = I_i(0)/V_i$ and impedances $Z_i = 1/Y_i$ may be obtained from the following matrix equations:

$$\{I(0)\} = c_1(1 - \cos k_0 h)[\Phi_w][\Phi_u]^{-1}\{V\} \tag{5.5}$$

$$\{V\} = \frac{[\Phi_w]^{-1}[\Phi_u]\{I(0)\}}{c_1(1 - \cos k_0 h)} \tag{5.6}$$

Formula (5.5) is used to determine the Y_i when the driving voltages V_i, $i = 1, 2, \ldots, N$, are specified; formula (5.6) is used to determine the Z_i when the driving-point currents are specified.

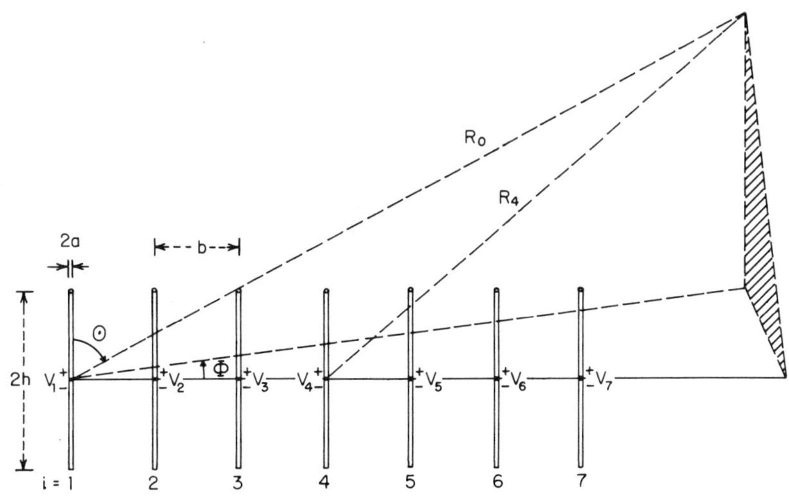

Fig. 5.1. Curtain array: seven identical elements.

The distribution of current (5.1) has been shown to be a very good approximation for elements up to $h = 5\lambda/8$, except that very near the driving point at $z = 0$ the trigonometric functions cannot accurately approximate the sharp rise in current that obtains there. Account is easily taken of this relatively minor correction by adding an empirically determined lumped susceptance B_T to the self-admittance of each element.[5] When $a/\lambda = 0.007022, B_T = 0.72$ millimhos. This correction may be added to B_i in $Y_i = G_i + jB_i$ when the Y_i are determined from (5.5). When the Z_i are obtained from (5.6), it is necessary to obtain $Y_i = Z_i^{-1} = G_i + jB_i$ and to add B_T to B_i. The resulting corrected admittances are then inverted to obtain the corrected impedances.

The corrected driving-point admittances and impedances of broadside arrays of $N = 2$ to 50 elements are given in Table 5.1 successively for $h/\lambda = 0.25$, $b/\lambda = 0.25$; $h/\lambda = 0.25$, $b/\lambda = 0.5$; $h/\lambda = 0.5$, $b/\lambda = 0.25$; and $h/\lambda = 0.5$, $b/\lambda = 0.5$, with $a/\lambda = 0.007022$ when the driving-point currents are specified and when the driving voltages are assigned. For the bidirectional broadside array all V_i or $I_i(0)$ are equal and in phase. The corrected driving-point admittances and impedances of unilateral endfire arrays of $N = 2$ to 50 elements are given in Table 5.2 for $h/\lambda = 0.25$, $b/\lambda = 0.25$ and $h/\lambda = 0.5$, $b/\lambda = 0.25$. In this case all V_i or $I_i(0)$ are equal in magnitude but with progressive phase shifts of one quarter period from element to element.

The self-admittances Y_{ii} and mutual admittances Y_{ik}, $k \neq i$, are the coefficients in the relations

$$I_i(0) = \sum_{i=1}^{N} \sum_{k=1}^{N} V_i Y_{ik} \tag{5.8}$$

They are given for arrays of $N = 2$ to 50 elements in Table 5.3 successively for $h/\lambda = 0.25$, $b/\lambda = 0.25$; $h/\lambda = 0.25$, $b/\lambda = 0.5$; $h/\lambda = 0.5$, $b/\lambda = 0.25$; and $h/\lambda = 0.5$, $b/\lambda = 0.5$. Note that the self- and mutual admittances of a given array are independent of the amplitudes and phases of the driving-point voltages and currents.

Similarly, the self-impedances Z_{ii} and mutual impedances Z_{ik}, $k \neq i$, are the coefficients in the relations

$$V_i = \sum_{i=1}^{N} \sum_{k=1}^{N} I_i(0) Z_{ik} \tag{5.9}$$

They are given in Table 5.4 for the same values of h/λ and b/λ as the admittances in Table 5.3. A knowledge of the self- and mutual admittances is convenient to determine the driving-point admittances for arrays with arbitrarily specified driving-point voltages. Likewise, a knowledge of self- and mutual impedances is useful in determining driving-point impedances for arrays with arbitrarily specified driving-point currents.

The radiation field of a curtain array of N parallel, identical, nonstaggered, equally-spaced elements such as that shown in Fig. 5.1 is given by

$$E_\theta(\theta, \Phi) = \frac{j\zeta_0 k_0 \sin\theta}{2\pi} \sum_{i=1}^{N} \int_{-h}^{h} I_i(z') \frac{e^{-jk_0(R_i - z'\cos\theta)}}{R_i}\, dz' \tag{5.10}$$

where

$$R_i = R_0 + \frac{k_0 b}{2}(N - 2i + 1)\cos\Phi\sin\theta \tag{5.11}$$

is the distance from the center of element i to the observation point and R_0 is the distance to this point from the center of element 1. For use in (5.10) the current $I_i(z)$ as given in (5.1) is conveniently rewritten in the equivalent form

$$I_i(z) = -\frac{j2\pi V_i}{\zeta_0 \Psi_{dR}} \times \left[\frac{\sin k_0(h - |z|) + \sin k_0|z| - \sin k_0 h}{1 - \cos k_0 h}\right] + I_i(0)\left[\frac{\cos k_0 z - \cos k_0 h}{1 - \cos k_0 h}\right] \tag{5.12}$$

The integration yields

$$E_\theta(\theta, \Phi) = -\frac{V_1}{\Psi_{dR}} \frac{e^{-jk_0 R_0}}{R_0} f(\theta, \Phi) \tag{5.13}$$

where

$$f(\theta, \Phi) = \sum_{i=1}^{N} \left\{-\frac{V_i}{V_1}\left[\frac{F(\theta, k_0 h) + H(\theta, k_0 h)}{1 - \cos k_0 h}\right] - \frac{j\zeta_0 \Psi_{dR}}{2\pi} \frac{I_i(0)}{V_1}\left[\frac{G(\theta, k_0 h)}{1 - \cos k_0 h}\right]\right\} \times \exp[-j(k_0 b/2)(N - 2i + 1)\cos\Phi\sin\theta] \tag{5.14}$$

The functions $F(\theta, k_0 h)$, $G(\theta, k_0 h)$, and $H(\theta, k_0 h)$ are defined on page 60 of Reference 1. In the equatorial plane $\theta = \pi/2$ and

$$f(\pi/2, \Phi) = \sum_{i=1}^{N} \left\{ \frac{V_i}{V_1}\left[-2 + \frac{k_0 h \sin k_0 h}{1 - \cos k_0 h} \right] - \frac{j\zeta_0 \Psi_{dR}}{2\pi} \times \frac{I_i(0)}{V_1}\left[\frac{\sin k_0 h - k_0 h \cos k_0 h}{1 - \cos k_0 h} \right] \right\} \times \exp[-j(k_0 b/2)(N - 2i + 1)\cos\Phi] \tag{5.15}$$

The far-field function $f(\pi/2, \Phi)/N$ is given in Table 5.5 for the several broadside arrays, and in Table 5.6 for the endfire arrays for which driving-point admittances and impedances are given in Tables 5.1 and 5.2. Note that $f(\theta, \Phi)$ is the field pattern referred to the driving voltage V_1. It differs from the field pattern referred to $I_1(0) = V_1 Y_1$ by the complex factor Y_1, the admittance of element 1. On the left of each page are the data for radiation patterns with the normalized driving voltages V_i/V_1 specified (denoted by VBR for broadside, VEN for endfire excitation), and on the right the data for radiation patterns with the normalized driving-point currents $I_i(0)/V_1$ specified (denoted by IBR for broadside, IEN for endfire excitation).

Typical far-field patterns are shown in Fig. 5.2 for broadside arrays, and in Figs. 5.3 and 5.4 for unilateral endfire arrays of 20 elements. The quantity represented is $f(\pi/2, \Phi)/N$ as a function of Φ. Graphs are shown for specified driving voltages V_i/V_1 and specified driving-point currents $I_i(0)/V_1$ with $h/\lambda = 0.25$ and 0.5 and with the distance between elements $b/\lambda = 0.25$. Note that conventional patterns obtained with the assumption that the distributions of current along all elements are identical correspond closely to the graphs for $h/\lambda = 0.25$ when the driving-point currents are specified, but with sharp nulls instead of deep minima. According to conventional theory the horizontal pattern of an array is independent of the length of the identical elements and the same when currents or voltages are specified. Hence, it predicts identical patterns for the four cases in Fig. 5.2 and for the four cases in Figs. 5.3 and 5.4.

REFERENCES FOR SECTION 5

1. King, R. W. P., R. B. Mack, and S. S. Sandler, *Arrays of Cylindrical Dipoles*, Cambridge University Press, Cambridge (1968), Ch. 5.
2. King, R., "Linear Arrays: Currents, Impedances and Fields, I," *Trans. IEEE*, **AP-7**, S440 (1959).
3. King, R. W. P., and S. S. Sandler, "Linear Arrays: Currents, Impedances and Fields, II," *Electromagnetic Theory and Antennas*, E. C. Jordan, ed., Macmillan Co., New York (1963), p. 1307.
4. King, R. W. P., and S. S. Sandler, "Theory of Broadside Arrays," and "Theory of Endfire Arrays," *Trans. IEEE*, **AP-12**, 269, 276 (1964).
5. King, R. W. P., R. B. Mack, and S. S. Sandler, *Op. cit.*, pp. 64, 88.

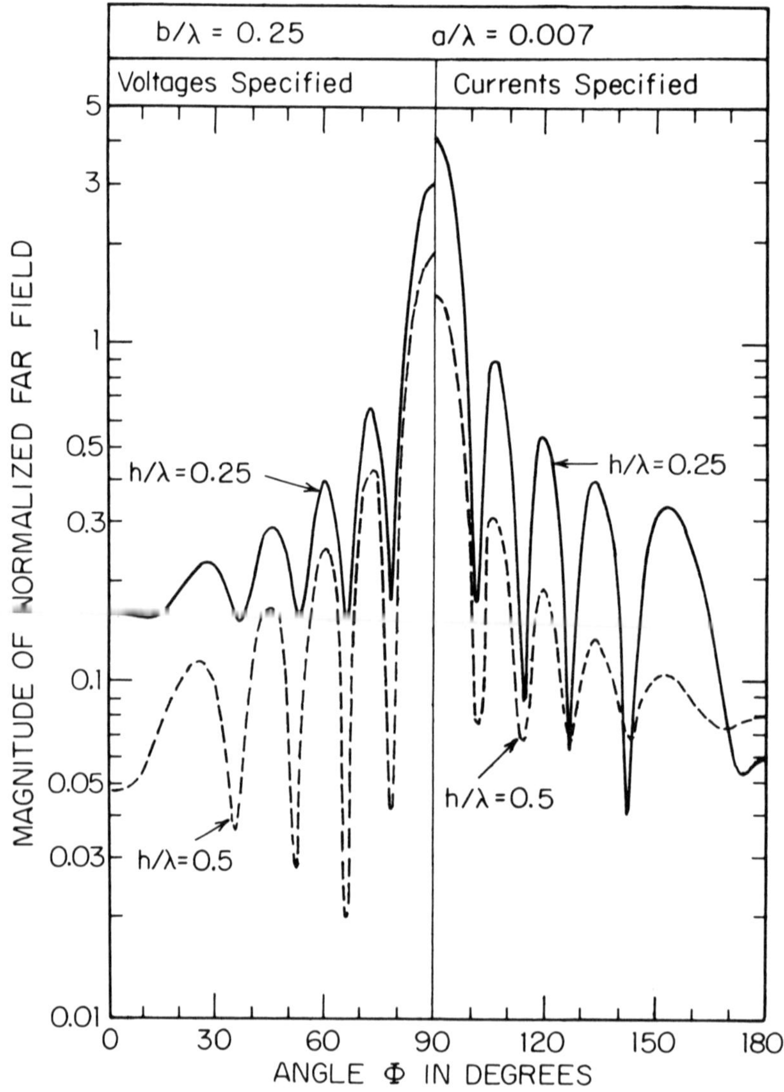

Fig. 5.2. Horizontal field patterns of 20-element broadside arrays.

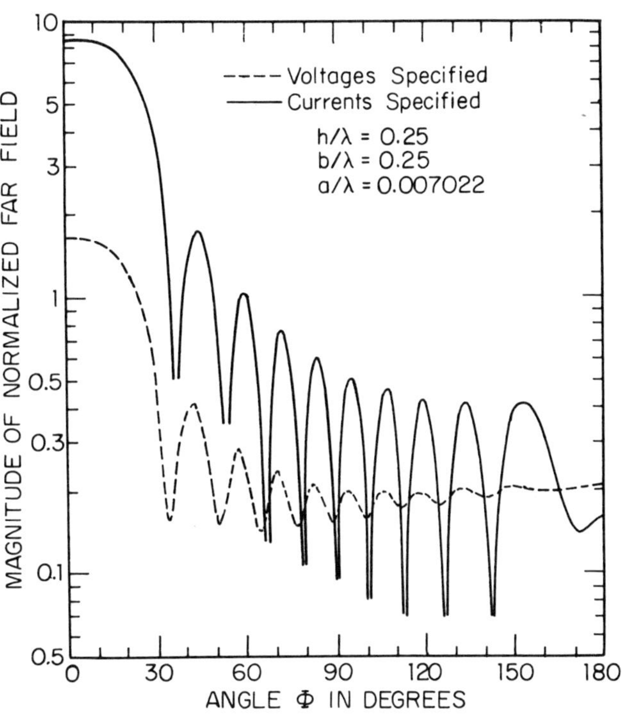

Fig. 5.3. Horizontal field patterns of 20-element unilateral endfire arrays; $h/\lambda = 0.25$.

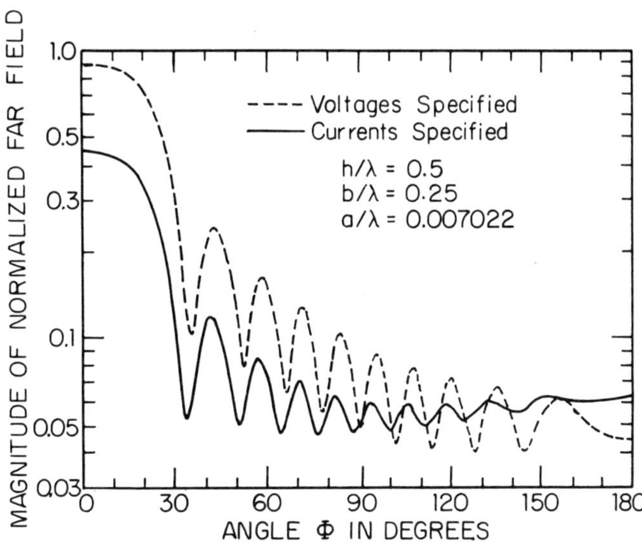

Fig. 5.4. Horizontal field patterns of 20-element unilateral endfire arrays; $h/\lambda = 0.5$.

TABLE 5.1

DRIVING-POINT ADMITTANCES AND IMPEDANCES OF BROADSIDE ARRAYS

BR

DRIVING POINT ADMITTANCES AND IMPEDANCES
FOR SPECIFIED BASE VOLTAGES

H/LMDA=0.2500 B/LMDA= 0.250 OMEGA= 8.53

2 ELEMENT ARRAY

ELEMENT	VO		YO		ZO	
	RE	IM	RE	IM	RE	IM
1	1.00	0.0	8.365	0.990	117.89	-13.95
2	1.00	0.0	8.365	0.990	117.89	-13.95

3 ELEMENT ARRAY

ELEMENT	VO		YO		ZO	
	RE	IM	RE	IM	RE	IM
1	1.00	0.0	8.852	0.458	112.67	-5.83
2	1.00	0.0	6.875	6.490	76.91	-72.61
3	1.00	0.0	8.852	0.458	112.67	-5.83

4 ELEMENT ARRAY

ELEMENT	VO		YO		ZO	
	RE	IM	RE	IM	RE	IM
1	1.00	0.0	8.664	0.243	115.33	-3.23
2	1.00	0.0	7.512	5.900	82.33	-64.66
3	1.00	0.0	7.512	5.900	82.33	-64.66
4	1.00	0.0	8.664	0.243	115.33	-3.23

5 ELEMENT ARRAY

ELEMENT	VO		YO		ZO	
	RE	IM	RE	IM	RE	IM
1	1.00	0.0	8.534	0.353	116.98	-4.84
2	1.00	0.0	7.300	5.583	86.43	-66.11
3	1.00	0.0	8.140	5.342	85.87	-56.36
4	1.00	0.0	7.300	5.583	86.43	-66.11
5	1.00	0.0	8.534	0.353	116.98	-4.84

6 ELEMENT ARRAY

ELEMENT	VO		YO		ZO	
	RE	IM	RE	IM	RE	IM
1	1.00	0.0	8.610	0.443	115.84	-5.97
2	1.00	0.0	7.095	5.713	85.51	-68.85
3	1.00	0.0	7.940	5.024	89.94	-56.91
4	1.00	0.0	7.940	5.024	89.94	-56.91
5	1.00	0.0	7.095	5.713	85.51	-68.85
6	1.00	0.0	8.610	0.443	115.84	-5.97

7 ELEMENT ARRAY

ELEMENT	VO		YO		ZO	
	RE	IM	RE	IM	RE	IM
1	1.00	0.0	8.677	0.385	115.02	-5.10
2	1.00	0.0	7.189	5.858	83.60	-68.12
3	1.00	0.0	7.737	5.147	89.60	-59.61
4	1.00	0.0	7.733	4.702	94.41	-57.40
5	1.00	0.0	7.737	5.147	89.60	-59.61
6	1.00	0.0	7.189	5.858	83.60	-68.12
7	1.00	0.0	8.677	0.385	115.02	-5.10

8 ELEMENT ARRAY

ELEMENT	VO		YO		ZO	
	RE	IM	RE	IM	RE	IM
1	1.00	0.0	8.631	0.333	115.69	-4.47
2	1.00	0.0	7.298	5.786	84.14	-66.71
3	1.00	0.0	7.825	5.289	87.72	-59.30
4	1.00	0.0	7.531	4.829	94.09	-60.34
5	1.00	0.0	7.531	4.829	94.09	-60.34
6	1.00	0.0	7.825	5.289	87.72	-59.30
7	1.00	0.0	7.298	5.786	84.14	-66.71
8	1.00	0.0	8.631	0.333	115.69	-4.47

DRIVING POINT ADMITTANCES AND IMPEDANCES
FOR SPECIFIED BASE CURRENTS

H/LMDA=0.2500 B/LMDA= 0.250 OMEGA= 8.53

2 ELEMENT ARRAY

ELEMENT	IZ(0)		YO		ZO	
	RE	IM	RE	IM	RE	IM
1	1.00	0.0	8.365	0.990	119.42	-3.85
2	1.00	0.0	8.365	0.990	119.42	-3.85

3 ELEMENT ARRAY

ELEMENT	IZ(0)		YO		ZO	
	RE	IM	RE	IM	RE	IM
1	1.00	0.0	9.254	3.794	97.32	-32.33
2	1.00	0.0	6.052	2.265	155.13	-39.60
3	1.00	0.0	9.254	3.794	97.32	-32.33

4 ELEMENT ARRAY

ELEMENT	IZ(0)		YO		ZO	
	RE	IM	RE	IM	RE	IM
1	1.00	0.0	12.853	3.735	73.75	-17.30
2	1.00	0.0	5.887	3.710	135.02	-68.58
3	1.00	0.0	5.887	3.710	135.02	-68.58
4	1.00	0.0	12.853	3.735	73.75	-17.30

5 ELEMENT ARRAY

ELEMENT	IZ(0)		YO		ZO	
	RE	IM	RE	IM	RE	IM
1	[illegible]	[illegible]	[illegible]	0.359	85.86	2.66
2	1.00	0.0	7.254	4.242	111.56	[illegible]
3	1.00	0.0	5.030	5.091	113.26	-98.43
4	1.00	0.0	7.254	4.242	111.56	-54.16
5	1.00	0.0	11.636	0.359	85.86	2.66

6 ELEMENT ARRAY

ELEMENT	IZ(0)		YO		ZO	
	RE	IM	RE	IM	RE	IM
1	1.00	0.0	9.689	1.431	102.66	-7.53
2	1.00	0.0	7.525	2.866	122.89	-35.05
3	1.00	0.0	6.006	6.261	89.94	-82.98
4	1.00	0.0	6.006	6.261	89.94	-82.98
5	1.00	0.0	7.525	2.866	122.89	-35.05
6	1.00	0.0	9.689	1.431	102.66	-7.53

7 ELEMENT ARRAY

ELEMENT	IZ(0)		YO		ZO	
	RE	IM	RE	IM	RE	IM
1	1.00	0.0	10.058	3.083	94.22	-22.14
2	1.00	0.0	6.534	2.788	139.11	-44.03
3	1.00	0.0	7.062	5.157	101.53	-63.78
4	1.00	0.0	7.305	8.160	67.20	-68.43
5	1.00	0.0	7.062	5.157	101.53	-63.78
6	1.00	0.0	6.534	2.788	139.11	-44.03
7	1.00	0.0	10.058	3.083	94.22	-22.14

8 ELEMENT ARRAY

ELEMENT	IZ(0)		YO		ZO	
	RE	IM	RE	IM	RE	IM
1	1.00	0.0	11.953	2.921	80.92	-14.90
2	1.00	0.0	6.343	3.540	131.65	-58.52
3	1.00	0.0	6.117	4.499	118.31	-73.10
4	1.00	0.0	9.165	6.488	78.16	-49.18
5	1.00	0.0	9.165	6.488	78.16	-49.18
6	1.00	0.0	6.117	4.499	118.31	-73.10
7	1.00	0.0	6.343	3.540	131.65	-58.52
8	1.00	0.0	11.953	2.921	80.92	-14.90

BR

DRIVING POINT ADMITTANCES AND IMPEDANCES
FOR SPECIFIED BASE VOLTAGES

H/LMDA=0.2500 B/LMDA= 0.250 OMEGA= 8.53

9 ELEMENT ARRAY

ELEMENT	VO RE	VO IM	YO RE	YO IM	ZO RE	ZO IM
1	1.00	0.0	8.589	0.371	116.21	-5.02
2	1.00	0.0	7.240	5.700	85.27	-67.13
3	1.00	0.0	7.932	5.223	87.94	-57.91
4	1.00	0.0	7.620	4.972	92.05	-60.06
5	1.00	0.0	7.329	4.954	93.66	-63.30
6	1.00	0.0	7.620	4.972	92.05	-60.06
7	1.00	0.0	7.932	5.223	87.94	-57.91
8	1.00	0.0	7.240	5.700	85.27	-67.13
9	1.00	0.0	8.589	0.371	116.21	-5.02

10 ELEMENT ARRAY

ELEMENT	VO RE	VO IM	YO RE	YO IM	ZO RE	ZO IM
1	1.00	0.0	8.621	0.406	115.74	-5.45
2	1.00	0.0	7.170	5.747	84.91	-68.07
3	1.00	0.0	7.879	5.138	89.05	-58.07
4	1.00	0.0	7.728	4.904	92.25	-58.54
5	1.00	0.0	7.417	5.097	91.58	-62.93
6	1.00	0.0	7.417	5.097	91.58	-62.93
7	1.00	0.0	7.728	4.904	92.25	-58.54
8	1.00	0.0	7.879	5.138	89.05	-58.07
9	1.00	0.0	7.170	5.747	84.91	-68.07
10	1.00	0.0	8.621	0.406	115.74	-5.45

12 ELEMENT ARRAY

ELEMENT	VO RE	VO IM	YO RE	YO IM	ZO RE	ZO IM
1	1.00	0.0	8.626	0.353	115.73	-4.73
2	1.00	0.0	7.261	5.771	84.40	-67.09
3	1.00	0.0	7.847	5.240	88.14	-58.86
4	1.00	0.0	7.604	4.864	93.33	-59.70
5	1.00	0.0	7.471	4.944	93.08	-61.60
6	1.00	0.0	7.614	5.172	89.87	-61.05
7	1.00	0.0	7.614	5.172	89.87	-61.05
8	1.00	0.0	7.471	4.944	93.08	-61.60
9	1.00	0.0	7.604	4.864	93.33	-59.70
10	1.00	0.0	7.847	5.240	88.14	-58.86
11	1.00	0.0	7.261	5.771	84.40	-67.09
12	1.00	0.0	8.626	0.353	115.73	-4.73

14 ELEMENT ARRAY

ELEMENT	VO RE	VO IM	YO RE	YO IM	ZO RE	ZO IM
1	1.00	0.0	8.623	0.394	115.73	-5.29
2	1.00	0.0	7.191	5.755	84.77	-67.84
3	1.00	0.0	7.868	5.165	88.82	-58.31
4	1.00	0.0	7.692	4.889	92.60	-58.86
5	1.00	0.0	7.439	5.047	92.05	-62.45
6	1.00	0.0	7.490	5.132	90.86	-62.26
7	1.00	0.0	7.668	5.019	91.30	-59.76
8	1.00	0.0	7.668	5.019	91.30	-59.76
9	1.00	0.0	7.490	5.132	90.86	-62.26
10	1.00	0.0	7.439	5.047	92.05	-62.45
11	1.00	0.0	7.692	4.889	92.60	-58.86
12	1.00	0.0	7.868	5.165	88.82	-58.31
13	1.00	0.0	7.191	5.755	84.77	-67.84
14	1.00	0.0	8.623	0.394	115.73	-5.29

16 ELEMENT ARRAY

ELEMENT	VO RE	VO IM	YO RE	YO IM	ZO RE	ZO IM
1	1.00	0.0	8.625	0.361	115.74	-4.84
2	1.00	0.0	7.247	5.767	84.49	-67.24
3	1.00	0.0	7.853	5.224	88.28	-58.73
4	1.00	0.0	7.625	4.871	93.14	-59.51
5	1.00	0.0	7.460	4.971	92.83	-61.85
6	1.00	0.0	7.578	5.157	90.19	-61.38
7	1.00	0.0	7.636	5.122	90.32	-60.59
8	1.00	0.0	7.544	4.979	92.34	-60.94
9	1.00	0.0	7.544	4.979	92.34	-60.94
10	1.00	0.0	7.636	5.122	90.32	-60.59
11	1.00	0.0	7.578	5.157	90.19	-61.38
12	1.00	0.0	7.460	4.971	92.83	-61.85
13	1.00	0.0	7.625	4.871	93.14	-59.51
14	1.00	0.0	7.853	5.224	88.28	-58.73
15	1.00	0.0	7.247	5.767	84.49	-67.24
16	1.00	0.0	8.625	0.361	115.74	-4.84

DRIVING POINT ADMITTANCES AND IMPEDANCES
FOR SPECIFIED BASE CURRENTS

H/LMDA=0.2500 B/LMDA= 0.250 OMEGA= 8.53

9 ELEMENT ARRAY

ELEMENT	IZ(0) RE	IZ(0) IM	YO RE	YO IM	ZO RE	ZO IM
1	1.00	0.0	11.410	1.070	87.56	-2.68
2	1.00	0.0	7.083	3.817	118.52	-51.82
3	1.00	0.0	5.540	5.147	110.16	-88.04
4	1.00	0.0	7.645	5.424	94.88	-58.38
5	1.00	0.0	10.060	4.030	89.69	-29.51
6	1.00	0.0	7.645	5.424	94.88	-58.38
7	1.00	0.0	5.540	5.147	110.16	-88.04
8	1.00	0.0	7.083	3.817	118.52	-51.82
9	1.00	0.0	11.410	1.070	87.56	-2.68

10 ELEMENT ARRAY

ELEMENT	IZ(0) RE	IZ(0) IM	YO RE	YO IM	ZO RE	ZO IM
1	1.00	0.0	10.059	1.620	98.63	-8.82
2	1.00	0.0	7.273	3.058	124.62	-40.06
3	1.00	0.0	6.089	5.793	96.94	-80.76
4	1.00	0.0	6.754	6.385	86.91	-72.91
5	1.00	0.0	8.279	3.766	106.39	-39.14
6	1.00	0.0	8.279	3.766	106.39	-39.14
7	1.00	0.0	6.754	6.385	86.91	-72.91
8	1.00	0.0	6.089	5.793	96.94	-80.76
9	1.00	0.0	7.273	3.058	124.62	-40.06
10	1.00	0.0	10.059	1.620	98.63	-8.82

12 ELEMENT ARRAY

ELEMENT	IZ(0) RE	IZ(0) IM	YO RE	YO IM	ZO RE	ZO IM
1	1.00	0.0	11.643	2.664	83.56	-13.95
2	1.00	0.0	6.493	3.480	130.45	-55.45
3	1.00	0.0	6.137	4.721	114.34	-74.55
4	1.00	0.0	8.562	6.526	80.00	-54.26
5	1.00	0.0	9.021	5.676	85.15	-46.78
6	1.00	0.0	6.689	4.380	115.05	-62.94
7	1.00	0.0	6.689	4.380	115.05	-62.94
8	1.00	0.0	9.021	5.676	85.15	-46.78
9	1.00	0.0	8.562	6.526	80.00	-54.26
10	1.00	0.0	6.137	4.721	114.34	-74.55
11	1.00	0.0	6.493	3.480	130.45	-55.45
12	1.00	0.0	11.643	2.664	83.56	-13.95

14 ELEMENT ARRAY

ELEMENT	IZ(0) RE	IZ(0) IM	YO RE	YO IM	ZO RE	ZO IM
1	1.00	0.0	10.237	1.721	96.76	-9.46
2	1.00	0.0	7.165	3.128	125.40	-42.15
3	1.00	0.0	6.113	5.624	99.52	-79.85
4	1.00	0.0	7.013	6.426	85.80	-69.81
5	1.00	0.0	8.441	4.057	102.45	-40.50
6	1.00	0.0	7.924	3.960	108.12	-44.22
7	1.00	0.0	6.801	5.838	93.87	-70.64
8	1.00	0.0	6.801	5.838	93.87	-70.64
9	1.00	0.0	7.924	3.960	108.12	-44.22
10	1.00	0.0	8.441	4.057	102.45	-40.50
11	1.00	0.0	7.013	6.426	85.80	-69.81
12	1.00	0.0	6.113	5.624	99.52	-79.85
13	1.00	0.0	7.165	3.128	125.40	-42.15
14	1.00	0.0	10.237	1.721	96.76	-9.46

16 ELEMENT ARRAY

ELEMENT	IZ(0) RE	IZ(0) IM	YO RE	YO IM	ZO RE	ZO IM
1	1.00	0.0	11.483	2.536	84.96	-13.43
2	1.00	0.0	6.569	3.448	129.84	-53.92
3	1.00	0.0	6.142	4.827	112.50	-75.24
4	1.00	0.0	8.324	6.526	80.82	-56.37
5	1.00	0.0	8.956	5.397	87.73	-45.81
6	1.00	0.0	6.881	4.335	113.88	-59.83
7	1.00	0.0	6.738	4.624	111.11	-64.38
8	1.00	0.0	8.481	5.778	86.98	-51.88
9	1.00	0.0	8.481	5.778	86.98	-51.88
10	1.00	0.0	6.738	4.624	111.11	-64.38
11	1.00	0.0	6.881	4.335	113.88	-59.83
12	1.00	0.0	8.956	5.397	87.73	-45.81
13	1.00	0.0	8.324	6.526	80.82	-56.37
14	1.00	0.0	6.142	4.827	112.50	-75.24
15	1.00	0.0	6.569	3.448	129.84	-53.92
16	1.00	0.0	11.483	2.536	84.96	-13.43

BR

DRIVING POINT ADMITTANCES AND IMPEDANCES FOR SPECIFIED BASE VOLTAGES

H/LMDA=0.2500 B/LMDA= 0.250 OMEGA= 8.53

18 ELEMENT ARRAY

ELEMENT	VO RE	VO IM	YO RE	YO IM	ZO RE	ZO IM
1	1.00	0.0	8.624	0.388	115.72	-5.21
2	1.00	0.0	7.201	5.758	84.71	-67.73
3	1.00	0.0	7.864	5.176	88.72	-58.39
4	1.00	0.0	7.678	4.885	92.72	-58.98
5	1.00	0.0	7.445	5.030	92.22	-62.31
6	1.00	0.0	7.511	5.139	90.68	-62.05
7	1.00	0.0	7.657	5.046	91.05	-60.01
8	1.00	0.0	7.632	5.004	91.63	-60.09
9	1.00	0.0	7.512	5.082	91.33	-61.78
10	1.00	0.0	7.512	5.082	91.33	-61.78
11	1.00	0.0	7.632	5.004	91.63	-60.09
12	1.00	0.0	7.657	5.046	91.05	-60.01
13	1.00	0.0	7.511	5.139	90.68	-62.05
14	1.00	0.0	7.445	5.030	92.22	-62.31
15	1.00	0.0	7.678	4.885	92.72	-58.98
16	1.00	0.0	7.864	5.176	88.72	-58.39
17	1.00	0.0	7.201	5.758	84.71	-67.73
18	1.00	0.0	8.624	0.388	115.72	-5.21

20 ELEMENT ARRAY

ELEMENT	VO RE	VO IM	YO RE	YO IM	ZO RE	ZO IM
1	1.00	0.0	8.625	0.365	115.74	-4.90
2	1.00	0.0	7.240	5.765	84.53	-67.31
3	1.00	0.0	7.855	5.216	88.35	-58.67
4	1.00	0.0	7.634	4.874	93.05	-59.42
5	1.00	0.0	7.457	4.982	92.72	-61.95
6	1.00	0.0	7.564	5.153	90.30	-61.51
7	1.00	0.0	7.642	5.106	90.47	-60.45
8	1.00	0.0	7.565	4.987	92.15	-60.75
9	1.00	0.0	7.533	5.006	92.09	-61.19
10	1.00	0.0	7.600	5.108	90.64	-60.92
11	1.00	0.0	7.600	5.108	90.64	-60.92
12	1.00	0.0	7.533	5.006	92.09	-61.19
13	1.00	0.0	7.565	4.987	92.15	-60.75
14	1.00	0.0	7.642	5.106	90.47	-60.45
15	1.00	0.0	7.564	5.153	90.30	-61.51
16	1.00	0.0	7.457	4.982	92.72	-61.95
17	1.00	0.0	7.634	4.874	93.05	-59.42
18	1.00	0.0	7.855	5.216	88.35	-58.67
19	1.00	0.0	7.240	5.765	84.53	-67.31
20	1.00	0.0	8.625	0.365	115.74	-4.90

22 ELEMENT ARRAY

ELEMENT	VO RE	VO IM	YO RE	YO IM	ZO RE	ZO IM
1	1.00	0.0	8.624	0.385	115.72	-5.17
2	1.00	0.0	7.206	5.759	84.68	-67.67
3	1.00	0.0	7.863	5.182	88.67	-58.44
4	1.00	0.0	7.671	4.882	92.77	-59.05
5	1.00	0.0	7.448	5.022	92.30	-62.24
6	1.00	0.0	7.520	5.142	90.61	-61.96
7	1.00	0.0	7.653	5.058	90.95	-60.10
8	1.00	0.0	7.618	5.000	91.75	-60.21
9	1.00	0.0	7.518	5.065	91.49	-61.64
10	1.00	0.0	7.533	5.090	91.14	-61.58
11	1.00	0.0	7.621	5.031	91.38	-60.33
12	1.00	0.0	7.621	5.031	91.38	-60.33
13	1.00	0.0	7.533	5.090	91.14	-61.58
14	1.00	0.0	7.518	5.065	91.49	-61.64
15	1.00	0.0	7.618	5.000	91.75	-60.21
16	1.00	0.0	7.653	5.058	90.95	-60.10
17	1.00	0.0	7.520	5.142	90.61	-61.96
18	1.00	0.0	7.448	5.022	92.30	-62.24
19	1.00	0.0	7.671	4.882	92.77	-59.05
20	1.00	0.0	7.863	5.182	88.67	-58.44
21	1.00	0.0	7.206	5.759	84.68	-67.67
22	1.00	0.0	8.624	0.385	115.72	-5.17

24 ELEMENT ARRAY

ELEMENT	VO RE	VO IM	YO RE	YO IM	ZO RE	ZO IM
1	1.00	0.0	8.624	0.368	115.74	-4.93
2	1.00	0.0	7.235	5.764	84.55	-67.36
3	1.00	0.0	7.856	5.211	88.39	-58.63
4	1.00	0.0	7.639	4.876	93.01	-59.37
5	1.00	0.0	7.455	4.988	92.66	-62.00
6	1.00	0.0	7.557	5.151	90.35	-61.58
7	1.00	0.0	7.644	5.098	90.55	-60.39
8	1.00	0.0	7.574	4.990	92.07	-60.66
9	1.00	0.0	7.529	5.017	91.98	-61.29
10	1.00	0.0	7.586	5.103	90.75	-61.05
11	1.00	0.0	7.606	5.091	90.80	-60.78
12	1.00	0.0	7.554	5.014	91.90	-60.99
13	1.00	0.0	7.554	5.014	91.90	-60.99
14	1.00	0.0	7.606	5.091	90.80	-60.78
15	1.00	0.0	7.586	5.103	90.75	-61.05
16	1.00	0.0	7.529	5.017	91.98	-61.29
17	1.00	0.0	7.574	4.990	92.07	-60.66
18	1.00	0.0	7.644	5.098	90.55	-60.39
19	1.00	0.0	7.557	5.151	90.35	-61.58
20	1.00	0.0	7.455	4.988	92.66	-62.00
21	1.00	0.0	7.639	4.876	93.01	-59.37
22	1.00	0.0	7.856	5.211	88.39	-58.63
23	1.00	0.0	7.235	5.764	84.55	-67.36
24	1.00	0.0	8.624	0.368	115.74	-4.93

DRIVING POINT ADMITTANCES AND IMPEDANCES FOR SPECIFIED BASE CURRENTS

H/LMDA=0.2500 B/LMDA= 0.250 OMEGA= 8.53

18 ELEMENT ARRAY

ELEMENT	IZ(0) RE	IZ(0) IM	YO RE	YO IM	ZO RE	ZO IM
1	1.00	0.0	10.343	1.784	95.67	-9.84
2	1.00	0.0	7.104	3.165	125.85	-43.32
3	1.00	0.0	6.124	5.536	100.90	-79.35
4	1.00	0.0	7.147	6.445	85.22	-68.27
5	1.00	0.0	8.513	4.203	100.63	-41.17
6	1.00	0.0	7.775	4.028	108.90	-46.34
7	1.00	0.0	6.810	5.644	96.43	-69.72
8	1.00	0.0	7.047	5.850	92.75	-67.53
9	1.00	0.0	8.055	4.245	104.19	-45.59
10	1.00	0.0	8.055	4.245	104.19	-45.59
11	1.00	0.0	7.047	5.850	92.75	-67.53
12	1.00	0.0	6.810	5.644	96.43	-69.72
13	1.00	0.0	7.775	4.028	108.90	-46.34
14	1.00	0.0	8.513	4.203	100.63	-41.17
15	1.00	0.0	7.147	6.445	85.22	-68.27
16	1.00	0.0	6.124	5.536	100.90	-79.35
17	1.00	0.0	7.104	3.165	125.85	-43.32
18	1.00	0.0	10.343	1.784	95.67	-9.84

20 ELEMENT ARRAY

ELEMENT	IZ(0) RE	IZ(0) IM	YO RE	YO IM	ZO RE	ZO IM
1	1.00	0.0	11.384	2.458	85.84	-13.11
2	1.00	0.0	6.616	3.428	129.46	-53.00
3	1.00	0.0	6.143	4.891	111.43	-75.66
4	1.00	0.0	8.194	6.522	81.29	-57.55
5	1.00	0.0	8.920	5.254	89.09	-45.29
6	1.00	0.0	6.980	4.311	113.28	-58.28
7	1.00	0.0	6.755	4.742	109.29	-65.08
8	1.00	0.0	8.264	5.804	87.79	-54.00
9	1.00	0.0	8.440	5.519	89.54	-50.91
10	1.00	0.0	6.940	4.587	109.95	-61.26
11	1.00	0.0	6.940	4.587	109.95	-61.26
12	1.00	0.0	8.440	5.519	89.54	-50.91
13	1.00	0.0	8.264	5.804	87.79	-54.00
14	1.00	0.0	6.755	4.742	109.29	-65.08
15	1.00	0.0	6.980	4.311	113.28	-58.28
16	1.00	0.0	8.920	5.254	89.09	-45.29
17	1.00	0.0	8.194	6.522	81.29	-57.55
18	1.00	0.0	6.143	4.891	111.43	-75.66
19	1.00	0.0	6.616	3.428	129.46	-53.00
20	1.00	0.0	11.384	2.458	85.84	-13.11

[illegible] ELEMENT ARRAY

ELEMENT	IZ(0) RE	IZ(0) IM	YO RE	YO IM	ZO RE	ZO IM
1	1.00	0.0	10.415	1.828	94.94	-10.10
2	1.00	0.0	7.065	3.188	126.15	-44.07
3	1.00	0.0	6.130	5.480	101.77	-79.03
4	1.00	0.0	7.232	6.457	84.87	-67.33
5	1.00	0.0	8.553	4.291	99.56	-41.57
6	1.00	0.0	7.692	4.063	109.35	-47.52
7	1.00	0.0	6.813	5.542	97.80	-69.21
8	1.00	0.0	7.174	5.855	92.17	-65.97
9	1.00	0.0	8.111	4.385	102.38	-46.26
10	1.00	0.0	7.895	4.309	104.98	-47.72
11	1.00	0.0	7.050	5.646	95.31	-66.60
12	1.00	0.0	7.050	5.646	95.31	-66.60
13	1.00	0.0	7.895	4.309	104.98	-47.72
14	1.00	0.0	8.111	4.385	102.38	-46.26
15	1.00	0.0	7.174	5.855	92.17	-65.97
16	1.00	0.0	6.813	5.542	97.80	-69.21
17	1.00	0.0	7.692	4.063	109.35	-47.52
18	1.00	0.0	8.553	4.291	99.56	-41.57
19	1.00	0.0	7.232	6.457	84.87	-67.33
20	1.00	0.0	6.130	5.480	101.77	-79.03
21	1.00	0.0	7.065	3.188	126.15	-44.07
22	1.00	0.0	10.415	1.828	94.94	-10.10

24 ELEMENT ARRAY

ELEMENT	IZ(0) RE	IZ(0) IM	YO RE	YO IM	ZO RE	ZO IM
1	1.00	0.0	11.316	2.406	86.45	-12.88
2	1.00	0.0	6.648	3.414	129.20	-52.37
3	1.00	0.0	6.143	4.933	110.71	-75.93
4	1.00	0.0	8.112	6.518	81.60	-58.32
5	1.00	0.0	8.895	5.165	89.95	-44.95
6	1.00	0.0	7.041	4.296	112.91	-57.34
7	1.00	0.0	6.763	4.813	108.23	-65.49
8	1.00	0.0	8.145	5.814	88.25	-55.19
9	1.00	0.0	8.415	5.385	90.90	-50.39
10	1.00	0.0	7.045	4.566	109.35	-59.71
11	1.00	0.0	6.962	4.709	108.14	-61.95
12	1.00	0.0	8.231	5.553	90.35	-53.04
13	1.00	0.0	8.231	5.553	90.35	-53.04
14	1.00	0.0	6.962	4.709	108.14	-61.95
15	1.00	0.0	7.045	4.566	109.35	-59.71
16	1.00	0.0	8.415	5.385	90.90	-50.39
17	1.00	0.0	8.145	5.814	88.25	-55.19
18	1.00	0.0	6.763	4.813	108.23	-65.49
19	1.00	0.0	7.041	4.296	112.91	-57.34
20	1.00	0.0	8.895	5.165	89.95	-44.95
21	1.00	0.0	8.112	6.518	81.60	-58.32
22	1.00	0.0	6.143	4.933	110.71	-75.93
23	1.00	0.0	6.648	3.414	129.20	-52.37
24	1.00	0.0	11.316	2.406	86.45	-12.88

BR

DRIVING POINT ADMITTANCES AND IMPEDANCES
FOR SPECIFIED BASE VOLTAGES

H/LMDA=0.2500 B/LMDA= 0.250 OMEGA= 8.53

26 ELEMENT ARRAY

ELEMENT	VO RE	VO IM	YO RE	YO IM	ZO RE	ZO IM
1	1.00	0.0	8.624	0.383	115.72	-5.14
2	1.00	0.0	7.210	5.760	84.66	-67.64
3	1.00	0.0	7.862	5.185	88.64	-58.46
4	1.00	0.0	7.667	4.881	92.81	-59.08
5	1.00	0.0	7.449	5.018	92.34	-62.21
6	1.00	0.0	7.525	5.144	90.57	-61.91
7	1.00	0.0	7.651	5.064	90.89	-60.15
8	1.00	0.0	7.611	4.998	91.80	-60.28
9	1.00	0.0	7.520	5.057	91.57	-61.58
10	1.00	0.0	7.542	5.093	91.07	-61.49
11	1.00	0.0	7.617	5.043	91.28	-60.43
12	1.00	0.0	7.607	5.027	91.50	-60.46
13	1.00	0.0	7.539	5.073	91.30	-61.44
14	1.00	0.0	7.539	5.073	91.30	-61.44
15	1.00	0.0	7.607	5.027	91.50	-60.46
16	1.00	0.0	7.617	5.043	91.28	-60.43
17	1.00	0.0	7.542	5.093	91.07	-61.49
18	1.00	0.0	7.520	5.057	91.57	-61.58
19	1.00	0.0	7.611	4.998	91.80	-60.28
20	1.00	0.0	7.651	5.064	90.89	-60.15
21	1.00	0.0	7.525	5.144	90.57	-61.91
22	1.00	0.0	7.449	5.018	92.34	-62.21
23	1.00	0.0	7.667	4.881	92.81	-59.08
24	1.00	0.0	7.862	5.185	88.64	-58.46
25	1.00	0.0	7.210	5.760	84.66	-67.64
26	1.00	0.0	8.624	0.383	115.72	-5.14

28 ELEMENT ARRAY

ELEMENT	VO RE	VO IM	YO RE	YO IM	ZO RE	ZO IM
1	1.00	0.0	8.624	0.369	115.74	-4.95
2	1.00	0.0	7.233	5.764	84.56	-67.39
3	1.00	0.0	7.857	5.208	88.42	-58.61
4	1.00	0.0	7.642	4.877	92.99	-59.34
5	1.00	0.0	7.454	4.992	92.62	-62.02
6	1.00	0.0	7.553	5.149	90.38	-61.62
7	1.00	0.0	7.645	5.093	90.59	-60.35
8	1.00	0.0	7.579	4.991	92.03	-60.60
9	1.00	0.0	7.528	5.023	91.92	-61.33
10	1.00	0.0	7.579	5.101	90.81	-61.11
11	1.00	0.0	7.608	5.083	90.88	-60.71
12	1.00	0.0	7.563	5.017	91.82	-60.90
13	1.00	0.0	7.550	5.025	91.79	-61.09
14	1.00	0.0	7.592	5.086	90.91	-60.91
15	1.00	0.0	7.592	5.086	90.91	-60.91
16	1.00	0.0	7.550	5.025	91.79	-61.09
17	1.00	0.0	7.563	5.017	91.82	-60.90
18	1.00	0.0	7.608	5.083	90.88	-60.71
19	1.00	0.0	7.579	5.101	90.81	-61.11
20	1.00	0.0	7.528	5.023	91.92	-61.33
21	1.00	0.0	7.579	4.991	92.03	-60.60
22	1.00	0.0	7.645	5.093	90.59	-60.35
23	1.00	0.0	7.553	5.149	90.38	-61.62
24	1.00	0.0	7.454	4.992	92.62	-62.02
25	1.00	0.0	7.642	4.877	92.99	-59.34
26	1.00	0.0	7.857	5.208	88.42	-58.61
27	1.00	0.0	7.233	5.764	84.56	-67.39
28	1.00	0.0	8.624	0.369	115.74	-4.95

30 ELEMENT ARRAY

ELEMENT	VO RE	VO IM	YO RE	YO IM	ZO RE	ZO IM
1	1.00	0.0	8.624	0.382	115.72	-5.12
2	1.00	0.0	7.212	5.760	84.65	-67.61
3	1.00	0.0	7.861	5.188	88.62	-58.48
4	1.00	0.0	7.665	4.881	92.83	-59.11
5	1.00	0.0	7.450	5.015	92.37	-62.18
6	1.00	0.0	7.528	5.145	90.54	-61.88
7	1.00	0.0	7.651	5.067	90.85	-60.18
8	1.00	0.0	7.607	4.997	91.83	-60.32
9	1.00	0.0	7.522	5.053	91.61	-61.54
10	1.00	0.0	7.547	5.094	91.03	-61.44
11	1.00	0.0	7.616	5.049	91.22	-60.47
12	1.00	0.0	7.601	5.025	91.55	-60.53
13	1.00	0.0	7.541	5.065	91.38	-61.38
14	1.00	0.0	7.548	5.076	91.23	-61.35
15	1.00	0.0	7.604	5.038	91.39	-60.55
16	1.00	0.0	7.604	5.038	91.39	-60.55
17	1.00	0.0	7.548	5.076	91.23	-61.35
18	1.00	0.0	7.541	5.065	91.38	-61.38
19	1.00	0.0	7.601	5.025	91.55	-60.53
20	1.00	0.0	7.616	5.049	91.22	-60.47
21	1.00	0.0	7.547	5.094	91.03	-61.44
22	1.00	0.0	7.522	5.053	91.61	-61.54
23	1.00	0.0	7.607	4.997	91.83	-60.32
24	1.00	0.0	7.651	5.067	90.85	-60.18
25	1.00	0.0	7.528	5.145	90.54	-61.88
26	1.00	0.0	7.450	5.015	92.37	-62.18
27	1.00	0.0	7.665	4.881	92.83	-59.11
28	1.00	0.0	7.861	5.188	88.62	-58.48
29	1.00	0.0	7.212	5.760	84.65	-67.61
30	1.00	0.0	8.624	0.382	115.72	-5.12

DRIVING POINT ADMITTANCES AND IMPEDANCES
FOR SPECIFIED BASE CURRENTS

H/LMDA=0.2500 B/LMDA= 0.250 OMEGA= 8.53

26 ELEMENT ARRAY

ELEMENT	IZ(O) RE	IZ(O) IM	YO RE	YO IM	ZO RE	ZO IM
1	1.00	0.0	10.467	1.860	94.42	-10.29
2	1.00	0.0	7.037	3.204	126.36	-44.61
3	1.00	0.0	6.134	5.442	102.37	-78.80
4	1.00	0.0	7.289	6.465	84.62	-66.69
5	1.00	0.0	8.580	4.351	98.85	-41.83
6	1.00	0.0	7.639	4.084	109.64	-48.29
7	1.00	0.0	6.814	5.478	98.65	-68.89
8	1.00	0.0	7.253	5.857	91.81	-65.03
9	1.00	0.0	8.143	4.470	101.32	-46.66
10	1.00	0.0	7.805	4.341	105.43	-48.91
11	1.00	0.0	7.050	5.539	96.67	-66.09
12	1.00	0.0	7.174	5.646	94.73	-65.04
13	1.00	0.0	7.945	4.447	103.16	-48.39
14	1.00	0.0	7.945	4.447	103.16	-48.39
15	1.00	0.0	7.174	5.646	94.73	-65.04
16	1.00	0.0	7.050	5.539	96.67	-66.09
17	1.00	0.0	7.805	4.341	105.43	-48.91
18	1.00	0.0	8.143	4.470	101.32	-46.66
19	1.00	0.0	7.253	5.857	91.81	-65.03
20	1.00	0.0	6.814	5.478	98.65	-68.89
21	1.00	0.0	7.639	4.084	109.64	-48.29
22	1.00	0.0	8.580	4.351	98.85	-41.83
23	1.00	0.0	7.289	6.465	84.62	-66.69
24	1.00	0.0	6.134	5.442	102.37	-78.80
25	1.00	0.0	7.037	3.204	126.36	-44.61
26	1.00	0.0	10.467	1.860	94.42	-10.29

28 ELEMENT ARRAY

ELEMENT	IZ(O) RE	IZ(O) IM	YO RE	YO IM	ZO RE	ZO IM
1	1.00	0.0	11.266	2.368	86.91	-12.71
2	1.00	0.0	6.671	3.404	129.02	-51.91
3	1.00	0.0	6.143	4.964	110.20	-76.13
4	1.00	0.0	8.054	6.515	81.81	-58.86
5	1.00	0.0	8.878	5.104	90.55	-44.72
6	1.00	0.0	7.082	4.285	112.66	-56.70
7	1.00	0.0	6.768	4.860	107.52	-65.77
8	1.00	0.0	8.070	5.820	88.56	-55.96
9	1.00	0.0	8.399	5.302	91.76	-50.05
10	1.00	0.0	7.109	4.553	108.98	-58.76
11	1.00	0.0	6.973	4.782	107.07	-62.37
12	1.00	0.0	8.117	5.568	90.81	-54.24
13	1.00	0.0	8.211	5.422	91.71	-52.52
14	1.00	0.0	7.069	4.690	107.54	-60.40
15	1.00	0.0	7.069	4.690	107.54	-60.40
16	1.00	0.0	8.211	5.422	91.71	-52.52
17	1.00	0.0	8.117	5.568	90.81	-54.24
18	1.00	0.0	6.973	4.782	107.07	-62.37
19	1.00	0.0	7.109	4.553	108.98	-58.76
20	1.00	0.0	8.399	5.302	91.76	-50.05
21	1.00	0.0	8.070	5.820	88.56	-55.96
22	1.00	0.0	6.768	4.860	107.52	-65.77
23	1.00	0.0	7.082	4.285	112.66	-56.70
24	1.00	0.0	8.878	5.104	90.55	-44.72
25	1.00	0.0	8.054	6.515	81.81	-58.86
26	1.00	0.0	6.143	4.964	110.20	-76.13
27	1.00	0.0	6.671	3.404	129.02	-51.91
28	1.00	0.0	11.266	2.368	86.91	-12.71

30 ELEMENT ARRAY

ELEMENT	IZ(O) RE	IZ(O) IM	YO RE	YO IM	ZO RE	ZO IM
1	1.00	0.0	10.506	1.885	94.02	-10.43
2	1.00	0.0	7.016	3.216	126.52	-45.00
3	1.00	0.0	6.137	5.414	102.81	-78.64
4	1.00	0.0	7.332	6.471	84.44	-66.23
5	1.00	0.0	8.598	4.395	98.34	-42.03
6	1.00	0.0	7.601	4.099	109.85	-48.83
7	1.00	0.0	6.814	5.434	99.25	-68.66
8	1.00	0.0	7.307	5.859	91.57	-64.39
9	1.00	0.0	8.163	4.528	100.61	-46.93
10	1.00	0.0	7.748	4.361	105.72	-49.68
11	1.00	0.0	7.049	5.473	97.53	-65.76
12	1.00	0.0	7.251	5.645	94.37	-64.10
13	1.00	0.0	7.973	4.530	102.10	-48.79
14	1.00	0.0	7.853	4.478	103.61	-49.58
15	1.00	0.0	7.172	5.537	96.09	-64.53
16	1.00	0.0	7.172	5.537	96.09	-64.53
17	1.00	0.0	7.853	4.478	103.61	-49.58
18	1.00	0.0	7.973	4.530	102.10	-48.79
19	1.00	0.0	7.251	5.645	94.37	-64.10
20	1.00	0.0	7.049	5.473	97.53	-65.76
21	1.00	0.0	7.748	4.361	105.72	-49.68
22	1.00	0.0	8.163	4.528	100.61	-46.93
23	1.00	0.0	7.307	5.859	91.57	-64.39
24	1.00	0.0	6.814	5.434	99.25	-68.66
25	1.00	0.0	7.601	4.099	109.85	-48.83
26	1.00	0.0	8.598	4.395	98.34	-42.03
27	1.00	0.0	7.332	6.471	84.44	-66.23
28	1.00	0.0	6.137	5.414	102.81	-78.64
29	1.00	0.0	7.016	3.216	126.52	-45.00
30	1.00	0.0	10.506	1.885	94.02	-10.43

BR

DRIVING POINT ADMITTANCES AND IMPEDANCES
FOR SPECIFIED BASE VOLTAGES

H/LMDA=0.2500 B/LMDA= 0.250 OMEGA= 8.53

35 ELEMENT ARRAY

ELEMENT	VO RE	VO IM	YO RE	YO IM	ZO RE	ZO IM
1	1.00	0.0	8.629	0.376	115.67	-5.03
2	1.00	0.0	7.221	5.770	84.52	-67.54
3	1.00	0.0	7.851	5.196	88.57	-58.62
4	1.00	0.0	7.656	4.871	92.99	-59.16
5	1.00	0.0	7.460	5.005	92.43	-62.02
6	1.00	0.0	7.539	5.156	90.37	-61.81
7	1.00	0.0	7.639	5.078	90.79	-60.35
8	1.00	0.0	7.596	4.984	92.03	-60.39
9	1.00	0.0	7.535	5.041	91.68	-61.34
10	1.00	0.0	7.560	5.108	90.81	-61.36
11	1.00	0.0	7.601	5.062	91.14	-60.70
12	1.00	0.0	7.586	5.009	91.80	-60.62
13	1.00	0.0	7.558	5.049	91.48	-61.11
14	1.00	0.0	7.565	5.095	90.94	-61.24
15	1.00	0.0	7.584	5.057	91.28	-60.86
16	1.00	0.0	7.583	5.016	91.73	-60.68
17	1.00	0.0	7.572	5.053	91.37	-60.98
18	1.00	0.0	7.567	5.091	90.97	-61.21
19	1.00	0.0	7.572	5.053	91.37	-60.98
20	1.00	0.0	7.583	5.016	91.73	-60.68
21	1.00	0.0	7.584	5.057	91.28	-60.86
22	1.00	0.0	7.565	5.095	90.94	-61.24
23	1.00	0.0	7.558	5.049	91.48	-61.11
24	1.00	0.0	7.586	5.009	91.80	-60.62
25	1.00	0.0	7.601	5.062	91.14	-60.70
26	1.00	0.0	7.560	5.108	90.81	-61.36
27	1.00	0.0	7.535	5.041	91.68	-61.34
28	1.00	0.0	7.596	4.984	92.03	-60.39
29	1.00	0.0	7.639	5.078	90.79	-60.35
30	1.00	0.0	7.539	5.156	90.37	-61.81
31	1.00	0.0	7.460	5.005	92.43	-62.02
32	1.00	0.0	7.656	4.871	92.99	-59.16
33	1.00	0.0	7.851	5.196	88.57	-58.62
34	1.00	0.0	7.221	5.770	84.52	-67.54
35	1.00	0.0	8.629	0.376	115.67	-5.03

40 ELEMENT ARRAY

ELEMENT	VO RE	VO IM	YO RE	YO IM	ZO RE	ZO IM
1	1.00	0.0	8.624	0.372	115.74	-4.99
[illegible]	1.00	0.0	7.228	5.763	84.58	-67.43
3	1.00	0.0	[illegible]	[illegible]	[illegible]	[illegible]
4	1.00	0.0	7.647	4.878	92.95	-59.29
5	1.00	0.0	7.453	4.997	92.56	-62.06
6	1.00	0.0	7.548	5.148	90.42	-61.68
7	1.00	0.0	7.647	5.087	90.65	-60.31
8	1.00	0.0	7.586	4.993	91.98	-60.54
9	1.00	0.0	7.526	5.031	91.84	-61.39
10	1.00	0.0	7.571	5.099	90.87	-61.20
11	1.00	0.0	7.611	5.074	90.97	-60.64
12	1.00	0.0	7.574	5.019	91.74	-60.80
13	1.00	0.0	7.547	5.037	91.67	-61.18
14	1.00	0.0	7.579	5.082	91.02	-61.04
15	1.00	0.0	7.597	5.071	91.06	-60.78
16	1.00	0.0	7.568	5.030	91.65	-60.91
17	1.00	0.0	7.557	5.037	91.62	-61.07
18	1.00	0.0	7.584	5.075	91.07	-60.95
19	1.00	0.0	7.589	5.072	91.09	-60.87
20	1.00	0.0	7.563	5.035	91.61	-60.99
21	1.00	0.0	7.563	5.035	91.61	-60.99
22	1.00	0.0	7.589	5.072	91.09	-60.87
23	1.00	0.0	7.584	5.075	91.07	-60.95
24	1.00	0.0	7.557	5.037	91.62	-61.07
25	1.00	0.0	7.568	5.030	91.65	-60.91
26	1.00	0.0	7.597	5.071	91.06	-60.78
27	1.00	0.0	7.579	5.082	91.02	-61.04
28	1.00	0.0	7.547	5.037	91.67	-61.18
29	1.00	0.0	7.574	5.019	91.74	-60.80
30	1.00	0.0	7.611	5.074	90.97	-60.64
31	1.00	0.0	7.571	5.099	90.87	-61.20
32	1.00	0.0	7.526	5.031	91.84	-61.39
33	1.00	0.0	7.586	4.993	91.98	-60.54
34	1.00	0.0	7.647	5.087	90.65	-60.31
35	1.00	0.0	7.548	5.148	90.42	-61.68
36	1.00	0.0	7.453	4.997	92.56	-62.06
37	1.00	0.0	7.647	4.878	92.95	-59.29
38	1.00	0.0	7.858	5.204	88.46	-58.58
39	1.00	0.0	7.228	5.763	84.58	-67.43
40	1.00	0.0	8.624	0.372	115.74	-4.99

DRIVING POINT ADMITTANCES AND IMPEDANCES
FOR SPECIFIED BASE CURRENTS

H/LMDA=0.2500 B/LMDA= 0.250 OMEGA= 8.53

35 ELEMENT ARRAY

ELEMENT	IZ(O) RE	IZ(O) IM	YO RE	YO IM	ZO RE	ZO IM
1	1.00	0.0	10.654	2.420	91.53	-14.61
2	1.00	0.0	6.765	3.169	130.69	-47.30
3	1.00	0.0	6.341	5.190	105.35	-74.27
4	1.00	0.0	7.644	6.809	80.04	-63.75
5	1.00	0.0	8.446	4.835	95.68	-46.62
6	1.00	0.0	7.265	3.986	114.50	-51.48
7	1.00	0.0	7.044	5.125	102.06	-63.82
8	1.00	0.0	7.671	6.170	86.64	-61.55
9	1.00	0.0	7.976	4.974	97.61	-52.06
10	1.00	0.0	7.351	4.214	110.97	-52.74
11	1.00	0.0	7.308	5.093	100.76	-60.30
12	1.00	0.0	7.671	5.973	88.75	-60.78
13	1.00	0.0	7.759	5.021	98.59	-54.65
14	1.00	0.0	7.380	4.302	109.67	-53.22
15	1.00	0.0	7.471	5.070	99.96	-58.20
16	1.00	0.0	7.669	5.903	89.51	-60.49
17	1.00	0.0	7.609	5.048	99.29	-56.49
18	1.00	0.0	7.388	4.326	109.31	-53.36
19	1.00	0.0	7.609	5.048	99.29	-56.49
20	1.00	0.0	7.669	5.903	89.51	-60.49
21	1.00	0.0	7.471	5.070	99.96	-58.20
22	1.00	0.0	7.380	4.302	109.67	-53.22
23	1.00	0.0	7.759	5.021	98.59	-54.65
24	1.00	0.0	7.671	5.973	88.75	-60.78
25	1.00	0.0	7.308	5.093	100.76	-60.30
26	1.00	0.0	7.351	4.214	110.97	-52.74
27	1.00	0.0	7.976	4.974	97.61	-52.06
28	1.00	0.0	7.671	6.170	86.64	-61.55
29	1.00	0.0	7.044	5.125	102.06	-63.82
30	1.00	0.0	7.265	3.986	114.50	-51.48
31	1.00	0.0	8.446	4.835	95.68	-46.62
32	1.00	0.0	7.644	6.809	80.04	-63.75
33	1.00	0.0	6.341	5.190	105.35	-74.27
34	1.00	0.0	6.765	3.169	130.69	-47.30
35	1.00	0.0	10.654	2.420	91.53	-14.61

40 ELEMENT ARRAY

ELEMENT	IZ(O) RE	IZ(O) IM	YO RE	YO IM	ZO RE	ZO IM
1	1.00	0.0	11.172	2.298	87.76	-12.40
2	1.00	0.0	6.715	3.384	128.67	-51.06
3	1.00	0.0	6.142	5.020	109.26	-76.49
4	[illegible]	[illegible]	[illegible]	6.508	82.20	-59.82
5	1.00	0.0	8.847	4.999	91.60	[illegible]
6	1.00	0.0	7.155	4.265	112.22	-55.60
7	1.00	0.0	6.776	4.941	106.33	-66.24
8	1.00	0.0	7.948	5.827	89.06	-57.22
9	1.00	0.0	8.371	5.171	93.13	-49.52
10	1.00	0.0	7.210	4.531	108.40	-57.30
11	1.00	0.0	6.988	4.893	105.48	-62.99
12	1.00	0.0	7.955	5.585	91.49	-55.96
13	1.00	0.0	8.180	5.246	93.59	-51.78
14	1.00	0.0	7.214	4.663	106.73	-58.34
15	1.00	0.0	7.095	4.851	105.27	-61.28
16	1.00	0.0	7.977	5.456	92.69	-55.03
17	1.00	0.0	8.078	5.306	93.62	-53.15
18	1.00	0.0	7.195	4.745	105.86	-59.22
19	1.00	0.0	7.157	4.804	105.40	-60.14
20	1.00	0.0	8.016	5.371	93.33	-54.15
21	1.00	0.0	8.016	5.371	93.33	-54.15
22	1.00	0.0	7.157	4.804	105.40	-60.14
23	1.00	0.0	7.195	4.745	105.86	-59.22
24	1.00	0.0	8.078	5.306	93.62	-53.15
25	1.00	0.0	7.977	5.456	92.69	-55.03
26	1.00	0.0	7.095	4.851	105.27	-61.28
27	1.00	0.0	7.214	4.663	106.73	-58.34
28	1.00	0.0	8.180	5.246	93.59	-51.78
29	1.00	0.0	7.955	5.585	91.49	-55.96
30	1.00	0.0	6.988	4.893	105.48	-62.99
31	1.00	0.0	7.210	4.531	108.40	-57.30
32	1.00	0.0	8.371	5.171	93.13	-49.52
33	1.00	0.0	7.948	5.827	89.06	-57.22
34	1.00	0.0	6.776	4.941	106.33	-66.24
35	1.00	0.0	7.155	4.265	112.22	-55.60
36	1.00	0.0	8.847	4.999	91.60	-44.31
37	1.00	0.0	7.953	6.508	82.20	-59.82
38	1.00	0.0	6.142	5.020	109.26	-76.49
39	1.00	0.0	6.715	3.384	128.67	-51.06
40	1.00	0.0	11.172	2.298	87.76	-12.40

BR

DRIVING POINT ADMITTANCES AND IMPEDANCES
FOR SPECIFIED BASE VOLTAGES

H/LMDA=0.2500 B/LMDA= 0.250 OMEGA= 8.53

45 ELEMENT ARRAY

ELEMENT	VO		YO		ZO	
	RE	IM	RE	IM	RE	IM
1	1.00	0.0	8.621	0.376	115.78	-5.05
2	1.00	0.0	7.222	5.756	84.67	-67.48
3	1.00	0.0	7.865	5.198	88.49	-58.49
4	1.00	0.0	7.653	4.885	92.84	-59.25
5	1.00	0.0	7.446	5.003	92.53	-62.17
6	1.00	0.0	7.541	5.141	90.53	-61.72
7	1.00	0.0	7.654	5.080	90.69	-60.19
8	1.00	0.0	7.593	5.001	91.86	-60.50
9	1.00	0.0	7.518	5.038	91.80	-61.52
10	1.00	0.0	7.563	5.090	91.00	-61.24
11	1.00	0.0	7.620	5.066	91.01	-60.51
12	1.00	0.0	7.582	5.029	91.60	-60.75
13	1.00	0.0	7.537	5.045	91.62	-61.33
14	1.00	0.0	7.570	5.072	91.17	-61.09
15	1.00	0.0	7.607	5.061	91.12	-60.62
16	1.00	0.0	7.578	5.041	91.48	-60.85
17	1.00	0.0	7.545	5.048	91.55	-61.25
18	1.00	0.0	7.572	5.063	91.26	-61.02
19	1.00	0.0	7.602	5.060	91.16	-60.67
20	1.00	0.0	7.576	5.049	91.40	-60.91
21	1.00	0.0	7.548	5.049	91.53	-61.22
22	1.00	0.0	7.574	5.056	91.33	-60.96
23	1.00	0.0	7.601	5.059	91.17	-60.68
24	1.00	0.0	7.574	5.056	91.33	-60.96
25	1.00	0.0	7.548	5.049	91.53	-61.22
26	1.00	0.0	7.576	5.049	91.40	-60.91
27	1.00	0.0	7.602	5.060	91.16	-60.67
28	1.00	0.0	7.572	5.063	91.26	-61.02
29	1.00	0.0	7.545	5.048	91.55	-61.25
30	1.00	0.0	7.578	5.041	91.48	-60.85
31	1.00	0.0	7.607	5.061	91.12	-60.62
32	1.00	0.0	7.570	5.072	91.17	-61.09
33	1.00	0.0	7.537	5.045	91.62	-61.33
34	1.00	0.0	7.582	5.029	91.60	-60.75
35	1.00	0.0	7.620	5.066	91.01	-60.51
36	1.00	0.0	7.563	5.090	91.00	-61.24
37	1.00	0.0	7.518	5.038	91.80	-61.52
38	1.00	0.0	7.593	5.001	91.86	-60.50
39	1.00	0.0	7.654	5.080	90.69	-60.19
40	1.00	0.0	7.541	5.141	90.53	-61.72
41	1.00	0.0	7.446	5.003	92.53	-62.17
42	1.00	0.0	7.653	4.885	92.84	-59.25
43	1.00	0.0	7.865	5.198	88.49	-58.49
44	1.00	0.0	7.222	5.756	84.67	-67.48
45	1.00	0.0	8.621	0.376	115.78	-5.05

50 ELEMENT ARRAY

ELEMENT	VO		YO		ZO	
	RE	IM	RE	IM	RE	IM
1	1.00	0.0	8.624	0.379	115.73	-5.08
2	1.00	0.0	7.217	5.761	84.63	-67.56
3	1.00	0.0	7.860	5.193	88.57	-58.52
4	1.00	0.0	7.659	4.880	92.87	-59.17
5	1.00	0.0	7.451	5.009	92.44	-62.14
6	1.00	0.0	7.535	5.146	90.50	-61.81
7	1.00	0.0	7.649	5.074	90.79	-60.23
8	1.00	0.0	7.600	4.995	91.89	-60.39
9	1.00	0.0	7.523	5.044	91.69	-61.48
10	1.00	0.0	7.556	5.096	90.96	-61.35
11	1.00	0.0	7.613	5.058	91.12	-60.55
12	1.00	0.0	7.590	5.022	91.64	-60.63
13	1.00	0.0	7.544	5.053	91.50	-61.29
14	1.00	0.0	7.561	5.080	91.13	-61.22
15	1.00	0.0	7.600	5.053	91.25	-60.67
16	1.00	0.0	7.587	5.033	91.52	-60.72
17	1.00	0.0	7.553	5.057	91.41	-61.20
18	1.00	0.0	7.563	5.071	91.21	-61.16
19	1.00	0.0	7.593	5.050	91.31	-60.72
20	1.00	0.0	7.587	5.040	91.45	-60.75
21	1.00	0.0	7.558	5.060	91.36	-61.16
22	1.00	0.0	7.562	5.066	91.27	-61.14
23	1.00	0.0	7.590	5.047	91.36	-60.75
24	1.00	0.0	7.588	5.044	91.41	-60.76
25	1.00	0.0	7.561	5.063	91.31	-61.14
26	1.00	0.0	7.561	5.063	91.31	-61.14
27	1.00	0.0	7.588	5.044	91.41	-60.76
28	1.00	0.0	7.590	5.047	91.36	-60.75
29	1.00	0.0	7.562	5.066	91.27	-61.14
30	1.00	0.0	7.558	5.060	91.36	-61.16
31	1.00	0.0	7.587	5.040	91.45	-60.75
32	1.00	0.0	7.593	5.050	91.31	-60.72
33	1.00	0.0	7.563	5.071	91.21	-61.16
34	1.00	0.0	7.553	5.057	91.41	-61.20
35	1.00	0.0	7.587	5.033	91.52	-60.72
36	1.00	0.0	7.600	5.053	91.25	-60.67
37	1.00	0.0	7.561	5.080	91.13	-61.22
38	1.00	0.0	7.544	5.053	91.50	-61.29
39	1.00	0.0	7.590	5.022	91.64	-60.63
40	1.00	0.0	7.613	5.058	91.12	-60.55
41	1.00	0.0	7.556	5.096	90.96	-61.35
42	1.00	0.0	7.523	5.044	91.69	-61.48
43	1.00	0.0	7.600	4.995	91.89	-60.39
44	1.00	0.0	7.649	5.074	90.79	-60.23
45	1.00	0.0	7.535	5.146	90.50	-61.81
46	1.00	0.0	7.451	5.009	92.44	-62.14
47	1.00	0.0	7.659	4.880	92.87	-59.17
48	1.00	0.0	7.860	5.193	88.57	-58.52
49	1.00	0.0	7.217	5.761	84.63	-67.56
50	1.00	0.0	8.624	0.379	115.73	-5.08

DRIVING POINT ADMITTANCES AND IMPEDANCES
FOR SPECIFIED BASE CURRENTS

H/LMDA=0.2500 B/LMDA= 0.250 OMEGA= 8.53

45 ELEMENT ARRAY

ELEMENT	IZ(0)		YO		ZO	
	RE	IM	RE	IM	RE	IM
1	1.00	0.0	11.029	1.820	89.78	-8.95
2	1.00	0.0	6.916	3.438	125.25	-49.21
3	1.00	0.0	5.988	5.190	107.24	-80.05
4	1.00	0.0	7.677	6.250	85.76	-61.78
5	1.00	0.0	8.985	4.616	93.68	-40.62
6	1.00	0.0	7.411	4.377	108.51	-53.55
7	1.00	0.0	6.609	5.166	104.17	-70.07
8	1.00	0.0	7.641	5.602	92.93	-59.38
9	1.00	0.0	8.539	4.796	95.38	-45.52
10	1.00	0.0	7.497	4.674	104.35	-55.04
11	1.00	0.0	6.809	5.155	103.12	-67.17
12	1.00	0.0	7.617	5.362	95.73	-58.34
13	1.00	0.0	8.371	4.850	96.08	-47.40
14	1.00	0.0	7.534	4.832	102.27	-55.82
15	1.00	0.0	6.900	5.149	102.64	-65.89
16	1.00	0.0	7.600	5.223	97.39	-57.71
17	1.00	0.0	8.293	4.873	96.41	-48.28
18	1.00	0.0	7.555	4.941	100.87	-56.36
19	1.00	0.0	6.942	5.146	102.41	-65.30
20	1.00	0.0	7.585	5.121	98.63	-57.23
21	1.00	0.0	8.260	4.882	96.55	-48.65
22	1.00	0.0	7.571	5.032	99.73	-56.80
23	1.00	0.0	6.955	5.145	102.35	-65.12
24	1.00	0.0	7.571	5.032	99.73	-56.80
25	1.00	0.0	8.260	4.882	96.55	-48.65
26	1.00	0.0	7.585	5.121	98.63	-57.23
27	1.00	0.0	6.942	5.146	102.41	-65.30
28	1.00	0.0	7.555	4.941	100.87	-56.36
29	1.00	0.0	8.293	4.873	96.41	-48.28
30	1.00	0.0	7.600	5.223	97.39	-57.71
31	1.00	0.0	6.900	5.149	102.64	-65.89
32	1.00	0.0	7.534	4.832	102.27	-55.82
33	1.00	0.0	8.371	4.850	96.08	-47.40
34	1.00	0.0	7.617	5.362	95.73	-58.34
35	1.00	0.0	6.809	5.155	103.12	-67.17
36	1.00	0.0	7.497	4.674	104.35	-55.04
37	1.00	0.0	8.539	4.796	95.38	-45.52
38	1.00	0.0	7.641	5.602	92.93	-59.38
39	1.00	0.0	6.609	5.166	104.17	-70.07
40	1.00	0.0	7.411	4.377	108.51	-53.55
41	1.00	0.0	8.985	4.616	93.68	-40.62
42	1.00	0.0	7.677	6.250	85.76	-61.78
43	1.00	0.0	5.988	5.190	107.24	-80.05
44	1.00	0.0	6.916	3.438	125.25	-49.21
45	1.00	0.0	11.029	1.820	89.78	-8.95

50 ELEMENT ARRAY

ELEMENT	IZ(0)		YO		ZO	
	RE	IM	RE	IM	RE	IM
1	1.00	0.0	10.617	1.956	92.93	-10.82
2	1.00	0.0	6.960	3.247	126.95	-46.09
3	1.00	0.0	6.143	5.339	103.99	-78.19
4	1.00	0.0	7.445	6.485	83.97	-65.02
5	1.00	0.0	8.645	4.509	97.03	-42.52
6	1.00	0.0	7.508	4.133	110.38	-50.18
7	1.00	0.0	6.814	5.328	100.70	-68.11
8	1.00	0.0	7.439	5.861	90.98	-62.87
9	1.00	0.0	8.209	4.664	98.97	-47.56
10	1.00	0.0	7.620	4.402	106.39	-51.41
11	1.00	0.0	7.044	5.331	99.39	-65.05
12	1.00	0.0	7.417	5.642	93.61	-62.12
13	1.00	0.0	8.027	4.703	99.97	-49.61
14	1.00	0.0	7.678	4.531	104.50	-51.87
15	1.00	0.0	7.164	5.341	98.58	-63.58
16	1.00	0.0	7.392	5.529	95.05	-61.84
17	1.00	0.0	7.923	4.712	100.66	-50.71
18	1.00	0.0	7.720	4.607	103.34	-52.02
19	1.00	0.0	7.240	5.357	97.95	-62.74
20	1.00	0.0	7.365	5.460	96.02	-61.79
21	1.00	0.0	7.854	4.704	101.26	-51.37
22	1.00	0.0	7.760	4.655	102.51	-51.98
23	1.00	0.0	7.292	5.380	97.36	-62.22
24	1.00	0.0	7.332	5.413	96.75	-61.92
25	1.00	0.0	7.803	4.686	101.85	-51.76
26	1.00	0.0	7.803	4.686	101.85	-51.76
27	1.00	0.0	7.332	5.413	96.75	-61.92
28	1.00	0.0	7.292	5.380	97.36	-62.22
29	1.00	0.0	7.760	4.655	102.51	-51.98
30	1.00	0.0	7.854	4.704	101.26	-51.37
31	1.00	0.0	7.365	5.460	96.02	-61.79
32	1.00	0.0	7.240	5.357	97.95	-62.74
33	1.00	0.0	7.720	4.607	103.34	-52.02
34	1.00	0.0	7.923	4.712	100.66	-50.71
35	1.00	0.0	7.392	5.529	95.05	-61.84
36	1.00	0.0	7.164	5.341	98.58	-63.58
37	1.00	0.0	7.678	4.531	104.50	-51.87
38	1.00	0.0	8.027	4.703	99.97	-49.61
39	1.00	0.0	7.417	5.642	93.61	-62.12
40	1.00	0.0	7.044	5.331	99.39	-65.05
41	1.00	0.0	7.620	4.402	106.39	-51.41
42	1.00	0.0	8.209	4.664	98.97	-47.56
43	1.00	0.0	7.439	5.861	90.98	-62.87
44	1.00	0.0	6.814	5.328	100.70	-68.11
45	1.00	0.0	7.508	4.133	110.38	-50.18
46	1.00	0.0	8.645	4.509	97.03	-42.52
47	1.00	0.0	7.445	6.485	83.97	-65.02
48	1.00	0.0	6.143	5.339	103.99	-78.19
49	1.00	0.0	6.960	3.247	126.95	-46.09
50	1.00	0.0	10.617	1.956	92.93	-10.82

8R

DRIVING POINT ADMITTANCES AND IMPEDANCES FOR SPECIFIED BASE VOLTAGES

H/LMDA=0.2500 B/LMDA= 0.500 OMEGA= 8.53

2 ELEMENT ARRAY

ELEMENT	VO		YO		ZO	
	RE	IM	RE	IM	RE	IM
1	1.00	0.0	15.494	-0.719	64.40	2.99
2	1.00	0.0	15.494	-0.719	64.40	2.99

3 ELEMENT ARRAY

ELEMENT	VO		YO		ZO	
	RE	IM	RE	IM	RE	IM
1	1.00	0.0	13.879	-0.878	71.76	4.54
2	1.00	0.0	19.920	1.386	49.96	-3.48
3	1.00	0.0	13.879	-0.878	71.76	4.54

4 ELEMENT ARRAY

ELEMENT	VO		YO		ZO	
	RE	IM	RE	IM	RE	IM
1	1.00	0.0	14.798	-0.795	67.38	3.62
2	1.00	0.0	18.446	1.400	53.90	-4.09
3	1.00	0.0	18.446	1.400	53.90	-4.09
4	1.00	0.0	14.798	-0.795	67.38	3.62

5 ELEMENT ARRAY

ELEMENT	VO		YO		ZO	
	RE	IM	RE	IM	RE	IM
1	1.00	0.0	14.206	-0.840	70.15	4.15
2	1.00	0.0	19.178	1.339	51.89	-3.62
3	1.00	0.0	17.085	1.443	58.12	-4.91
4	1.00	0.0	19.178	1.339	51.89	-3.62
5	1.00	0.0	14.206	-0.840	70.15	4.15

6 ELEMENT ARRAY

ELEMENT	VO		YO		ZO	
	RE	IM	RE	IM	RE	IM
1	1.00	0.0	14.659	-0.803	68.02	3.72
2	1.00	0.0	18.683	1.389	53.23	-3.96
3	1.00	0.0	17.798	1.376	55.85	-4.32
4	1.00	0.0	17.798	1.376	55.85	-4.32
5	1.00	0.0	18.683	1.389	53.23	-3.96
6	1.00	0.0	14.659	-0.803	68.02	3.72

7 ELEMENT ARRAY

ELEMENT	VO		YO		ZO	
	RE	IM	RE	IM	RE	IM
1	1.00	0.0	14.309	-0.829	69.65	4.03
2	1.00	0.0	19.037	1.346	52.27	-3.69
3	1.00	0.0	17.346	1.435	57.26	-4.74
4	1.00	0.0	18.478	1.303	53.85	-3.80
5	1.00	0.0	17.346	1.435	57.26	-4.74
6	1.00	0.0	19.037	1.346	52.27	-3.69
7	1.00	0.0	14.309	-0.829	69.65	4.03

8 ELEMENT ARRAY

ELEMENT	VO		YO		ZO	
	RE	IM	RE	IM	RE	IM
1	1.00	0.0	14.602	-0.806	68.28	3.77
2	1.00	0.0	18.755	1.382	53.03	-3.91
3	1.00	0.0	17.681	1.386	56.21	-4.41
4	1.00	0.0	18.030	1.363	55.15	-4.17
5	1.00	0.0	18.030	1.363	55.15	-4.17
6	1.00	0.0	17.681	1.386	56.21	-4.41
7	1.00	0.0	18.755	1.382	53.03	-3.91
8	1.00	0.0	14.602	-0.806	68.28	3.77

DRIVING POINT ADMITTANCES AND IMPEDANCES FOR SPECIFIED BASE CURRENTS

H/LMDA=0.2500 B/LMDA= 0.500 OMEGA= 8.53

2 ELEMENT ARRAY

ELEMENT	IZ(0)		YO		ZO	
	RE	IM	RE	IM	RE	IM
1	1.00	0.0	15.494	-0.719	63.99	5.94
2	1.00	0.0	15.494	-0.719	63.99	5.94

3 ELEMENT ARRAY

ELEMENT	IZ(0)		YO		ZO	
	RE	IM	RE	IM	RE	IM
1	1.00	0.0	12.215	-3.528	73.03	25.40
2	1.00	0.0	17.151	10.535	43.92	-25.14
3	1.00	0.0	12.215	-3.528	73.03	25.40

4 ELEMENT ARRAY

ELEMENT	IZ(0)		YO		ZO	
	RE	IM	RE	IM	RE	IM
1	1.00	0.0	14.419	-1.715	67.43	11.39
2	1.00	0.0	18.597	2.664	53.19	-5.56
3	1.00	0.0	18.597	2.664	53.19	-5.56
4	1.00	0.0	14.419	-1.715	67.43	11.39

5 ELEMENT ARRAY

ELEMENT	IZ(0)		YO		ZO	
	RE	IM	RE	IM	RE	IM
1	1.00	0.0	12.697	-3.285	71.63	22.59
2	1.00	0.0	17.978	8.373	47.09	-20.04
3	1.00	0.0	15.111	-2.753	62.86	14.45
4	1.00	0.0	17.978	8.373	47.09	-20.04
5	1.00	0.0	12.697	-3.285	71.63	22.59

6 ELEMENT ARRAY

ELEMENT	IZ(0)		YO		ZO	
	RE	IM	RE	IM	RE	IM
1	1.00	0.0	14.088	-2.037	68.36	13.38
2	1.00	0.0	18.875	3.900	51.52	-8.68
3	1.00	0.0	17.636	0.746	56.70	-0.08
4	1.00	0.0	17.636	0.746	56.70	-0.08
5	1.00	0.0	18.875	3.900	51.52	-8.68
6	1.00	0.0	14.088	-2.037	68.36	13.38

7 ELEMENT ARRAY

ELEMENT	IZ(0)		YO		ZO	
	RE	IM	RE	IM	RE	IM
1	1.00	0.0	12.898	-3.150	71.12	21.34
2	1.00	0.0	18.232	7.641	47.94	-18.20
3	1.00	0.0	15.747	-2.240	61.34	11.53
4	1.00	0.0	18.268	6.107	50.36	-14.85
5	1.00	0.0	15.747	-2.240	61.34	11.53
6	1.00	0.0	18.232	7.641	47.94	-18.20
7	1.00	0.0	12.898	-3.150	71.12	21.34

8 ELEMENT ARRAY

ELEMENT	IZ(0)		YO		ZO	
	RE	IM	RE	IM	RE	IM
1	1.00	0.0	13.926	-2.198	68.79	14.41
2	1.00	0.0	18.909	4.471	50.88	-10.09
3	1.00	0.0	17.316	0.137	57.68	1.94
4	1.00	0.0	18.115	1.779	55.01	-3.22
5	1.00	0.0	18.115	1.779	55.01	-3.22
6	1.00	0.0	17.316	0.137	57.68	1.94
7	1.00	0.0	18.909	4.471	50.88	-10.09
8	1.00	0.0	13.926	-2.198	68.79	14.41

8R

DRIVING POINT ADMITTANCES AND IMPEDANCES FOR SPECIFIED BASE VOLTAGES

H/LMDA=0.2500 B/LMDA= 0.500 OMEGA= 8.53

DRIVING POINT ADMITTANCES AND IMPEDANCES FOR SPECIFIED BASE CURRENTS

H/LMDA=0.2500 B/LMDA= 0.500 OMEGA= 8.53

9 ELEMENT ARRAY (specified base voltages)

ELEMENT	VO RE	VO IM	YO RE	YO IM	ZO RE	ZO IM
1	1.00	0.0	14.358	-0.824	69.42	3.98
2	1.0C	0.0	18.981	1.350	52.42	-3.73
3	1.0C	0.0	17.425	1.428	57.01	-4.67
4	1.00	0.0	18.349	1.312	54.22	-3.88
5	1.00	0.0	17.598	1.425	56.45	-4.57
6	1.00	0.0	18.349	1.312	54.22	-3.88
7	1.00	0.0	17.425	1.428	57.01	-4.67
8	1.00	0.0	18.981	1.350	52.42	-3.73
9	1.00	0.0	14.358	-0.824	69.42	3.98

9 ELEMENT ARRAY (specified base currents)

ELEMENT	IZ(0) RE	IZ(0) IM	YO RE	YO IM	ZO RE	ZO IM
1	1.00	0.0	13.008	-3.067	70.87	20.63
2	1.0C	0.0	18.356	7.273	48.32	-17.25
3	1.00	0.0	16.002	-1.972	60.77	10.22
4	1.00	0.0	18.341	5.358	51.25	-12.96
5	1.00	0.0	16.392	-1.629	59.78	8.57
6	1.00	0.0	18.341	5.358	51.25	-12.96
7	1.00	0.0	16.002	-1.972	60.77	10.22
8	1.00	0.0	18.356	7.273	48.32	-17.25
9	1.00	0.0	13.008	-3.067	70.87	20.63

10 ELEMENT ARRAY (specified base voltages)

ELEMENT	VO RE	VO IM	YO RE	YO IM	ZO RE	ZO IM
1	1.00	0.0	14.571	-C.807	68.42	3.79
2	1.00	0.0	18.788	1.378	52.94	-3.88
3	1.00	0.0	17.636	1.392	56.35	-4.45
4	1.00	0.0	18.099	1.355	54.94	-4.11
5	1.00	0.0	17.914	1.374	55.49	-4.26
6	1.00	0.0	17.914	1.374	55.49	-4.26
7	1.0C	0.0	18.099	1.355	54.94	-4.11
8	1.00	0.0	17.636	1.392	56.35	-4.45
9	1.C0	0.0	18.788	1.378	52.94	-3.88
10	1.00	0.0	14.571	-0.807	68.42	3.79

1C ELEMENT ARRAY (specified base currents)

ELEMENT	IZ(0) RE	IZ(0) IM	YO RE	YO IM	ZO RE	ZO IM
1	1.00	0.0	13.829	-2.294	69.03	15.04
2	1.00	0.0	18.903	4.797	50.55	-10.90
3	1.00	0.0	17.155	-0.164	58.14	3.00
4	1.00	0.0	18.260	2.279	54.37	-4.64
5	1.00	0.0	17.849	1.098	56.00	-1.19
6	1.C0	0.0	17.849	1.098	56.00	-1.19
7	1.00	0.0	18.260	2.279	54.37	-4.64
8	1.00	0.0	17.155	-0.164	58.14	3.00
9	1.00	0.0	18.903	4.797	50.55	-10.90
10	1.00	0.0	13.829	-2.294	69.03	15.04

12 ELEMENT ARRAY (specified base voltages)

ELEMENT	VO RE	VO IM	YO RE	YO IM	ZO RE	ZO IM
1	1.C0	0.0	14.553	-C.8C8	68.51	3.81
2	1.00	0.0	18.807	1.375	52.89	-3.87
3	1.CC	0.0	17.612	1.396	56.42	-4.47
4	1.00	0.0	18.130	1.351	54.85	-4.09
5	1.00	0.0	17.870	1.380	55.63	-4.30
6	1.00	0.0	17.983	1.366	55.29	-4.20
7	1.00	0.0	17.983	1.366	55.29	-4.20
8	1.C0	0.0	17.870	1.380	55.63	-4.30
9	1.00	0.0	18.130	1.351	54.85	-4.09
10	1.00	0.0	17.612	1.396	56.42	-4.47
11	1.00	0.0	18.807	1.375	52.89	-3.87
12	1.00	0.0	14.553	-C.8C8	68.51	3.81

12 ELEMENT ARRAY (specified base currents)

ELEMENT	IZ(0) RE	IZ(0) IM	YO RE	YO IM	ZO RE	ZO IM
1	1.00	0.0	13.765	-2.358	69.19	15.47
2	1.0C	0.0	18.889	5.008	50.35	-11.43
3	1.00	0.0	17.056	-0.345	58.40	3.65
4	1.C0	0.0	18.323	2.571	54.02	-5.46
5	1.00	0.0	17.711	0.760	56.46	-0.13
6	1.00	0.0	18.026	1.571	55.35	-2.61
7	1.00	0.0	18.026	1.571	55.35	-2.61
8	1.00	0.0	17.711	C.760	56.46	-0.13
9	1.C0	0.0	18.323	2.571	54.02	-5.46
10	1.00	0.0	17.056	-C.345	58.40	3.65
11	1.00	0.0	18.889	5.008	50.35	-11.43
12	1.0C	0.0	13.765	-2.358	69.19	15.47

14 ELEMENT ARRAY (specified base voltages)

ELEMENT	VO RE	VO IM	YO RE	YO IM	ZO RE	ZO IM
1	1.0C	0.0	14.540	-0.809	68.56	3.81
2	1.00	0.0	18.819	1.373	52.86	-3.86
3	1.0C	0.0	17.599	1.398	56.47	-4.49
4	1.C0	0.0	18.147	1.348	54.80	-4.07
5	1.00	0.0	17.848	1.384	55.69	-4.32
6	1.00	0.0	18.013	1.361	55.20	-4.17
7	1.00	0.0	17.938	1.372	55.42	-4.24
8	1.00	0.0	17.938	1.372	55.42	-4.24
9	1.C0	0.0	18.013	1.361	55.20	-4.17
10	1.00	0.0	17.848	1.384	55.69	-4.32
11	1.00	0.0	18.147	1.348	54.80	-4.07
12	1.0C	0.0	17.599	1.398	56.47	-4.49
13	1.0C	0.0	18.819	1.373	52.86	-3.86
14	1.00	0.0	14.540	-0.809	68.56	3.81

14 ELEMENT ARRAY (specified base currents)

ELEMENT	IZ(0) RE	IZ(0) IM	YO RE	YO IM	ZO RE	ZO IM
1	1.00	0.0	13.719	-2.405	69.30	15.78
2	1.00	0.0	18.875	5.156	50.21	-11.80
3	1.00	0.0	16.990	-0.466	58.57	4.09
4	1.C0	0.0	18.355	2.764	53.81	-5.99
5	1.00	0.0	17.626	0.556	56.73	0.53
6	1.00	0.0	18.109	1.850	55.01	-3.43
7	1.C0	0.0	17.903	1.218	55.81	-1.55
8	1.00	0.0	17.903	1.218	55.81	-1.55
9	1.00	0.0	18.109	1.850	55.01	-3.43
10	1.00	0.0	17.626	C.556	56.73	0.53
11	1.00	0.0	18.355	2.764	53.81	-5.99
12	1.0C	0.0	16.990	-C.466	58.57	4.09
13	1.00	0.0	18.875	5.156	50.21	-11.80
14	1.00	0.0	13.719	-2.405	69.30	15.78

16 ELEMENT ARRAY (specified base voltages)

ELEMENT	VO RE	VO IM	YO RE	YO IM	ZO RE	ZO IM
1	1.C0	0.0	14.531	-C.810	68.61	3.82
2	1.0C	0.0	18.827	1.372	52.83	-3.85
3	1.0C	0.0	17.589	1.400	56.49	-4.50
4	1.00	0.0	18.158	1.346	54.77	-4.06
5	1.00	0.0	17.834	1.386	55.74	-4.33
6	1.00	0.0	18.030	1.358	55.15	-4.15
7	1.00	0.0	17.916	1.376	55.49	-4.26
8	1.00	0.0	17.969	1.367	55.33	-4.21
9	1.00	0.0	17.969	1.367	55.33	-4.21
10	1.00	0.0	17.916	1.376	55.49	-4.26
11	1.00	0.0	18.030	1.358	55.15	-4.15
12	1.00	0.0	17.834	1.386	55.74	-4.33
13	1.0C	0.0	18.158	1.346	54.77	-4.06
14	1.00	0.0	17.589	1.400	56.49	-4.50
15	1.0C	0.0	18.827	1.372	52.83	-3.85
16	1.00	0.0	14.531	-0.810	68.61	3.82

16 ELEMENT ARRAY (specified base currents)

ELEMENT	IZ(0) RE	IZ(0) IM	YO RE	YO IM	ZO RE	ZO IM
1	1.00	0.0	13.684	-2.440	69.38	16.02
2	1.00	0.0	18.863	5.265	50.11	-12.07
3	1.00	0.0	16.942	-C.553	58.70	4.41
4	1.00	0.0	18.375	2.901	53.67	-6.37
5	1.00	0.0	17.568	0.419	56.90	0.97
6	1.00	0.0	18.155	2.035	54.79	-3.97
7	1.00	0.0	17.827	1.005	56.08	-0.90
8	1.00	0.0	17.995	1.490	55.47	-2.37
9	1.00	0.0	17.995	1.490	55.47	-2.37
10	1.00	0.0	17.827	1.005	56.08	-0.90
11	1.00	0.0	18.155	2.035	54.79	-3.97
12	1.00	0.0	17.568	0.419	56.90	0.97
13	1.0C	0.0	18.375	2.901	53.67	-6.37
14	1.00	0.0	16.942	-C.553	58.70	4.41
15	1.00	0.0	18.863	5.265	50.11	-12.07
16	1.00	0.0	13.684	-2.440	69.38	16.02

BR

DRIVING POINT ADMITTANCES AND IMPEDANCES
FOR SPECIFIED BASE VOLTAGES

H/LMDA=C.2500 B/LMDA= 0.500 CMEGA= 8.53

1E ELEMENT ARRAY

ELEMENT	VO		YO		ZO	
	RE	IM	RE	IM	RE	IM
1	1.00	0.0	14.524	-0.810	68.64	3.83
2	1.00	0.0	18.833	1.371	52.82	-3.84
3	1.CC	0.0	17.583	1.401	56.52	-4.50
4	1.00	0.0	18.165	1.344	54.75	-4.05
5	1.C0	0.0	17.826	1.388	55.76	-4.34
6	1.00	0.0	18.041	1.356	55.12	-4.14
7	1.00	0.0	17.903	1.378	55.53	-4.27
8	1.00	0.0	17.986	1.364	55.28	-4.19
9	1.00	0.0	17.947	1.371	55.40	-4.23
10	1.00	0.0	17.947	1.371	55.40	-4.23
11	1.00	0.0	17.986	1.364	55.28	-4.19
12	1.00	0.0	17.903	1.378	55.53	-4.27
13	1.00	0.0	18.041	1.356	55.12	-4.14
14	1.0C	0.0	17.826	1.388	55.76	-4.34
15	1.00	0.0	18.165	1.344	54.75	-4.05
16	1.00	0.0	17.583	1.401	56.52	-4.50
17	1.00	0.0	18.833	1.371	52.82	-3.84
18	1.00	0.0	14.524	-0.810	68.64	3.83

20 ELEMENT ARRAY

ELEMENT	VO		YO		ZO	
	RE	IM	RE	IM	RE	IM
1	1.00	0.0	14.519	-0.810	68.66	3.83
2	1.0C	0.0	18.838	1.370	52.80	-3.84
3	1.0C	0.0	17.578	1.402	56.53	-4.51
4	1.00	0.0	18.171	1.343	54.73	-4.05
5	1.00	0.0	17.819	1.389	55.78	-4.35
6	1.00	0.0	18.048	1.355	55.10	-4.14
7	1.00	0.0	17.894	1.380	55.55	-4.28
8	1.C0	0.0	17.997	1.362	55.25	-4.18
9	1.00	0.0	17.933	1.373	55.44	-4.25
10	1.00	0.0	17.964	1.368	55.35	-4.21
11	1.0C	0.0	17.964	1.368	55.35	-4.21
12	1.00	0.0	17.933	1.373	55.44	-4.25
13	1.0C	0.0	17.997	1.362	55.25	-4.18
14	1.00	0.0	17.894	1.380	55.55	-4.28
15	1.00	0.0	18.048	1.355	55.10	-4.14
16	1.00	0.0	17.819	1.389	55.78	-4.35
17	1.00	0.0	18.171	1.343	54.73	-4.05
18	1.00	0.0	17.578	1.402	56.53	-4.51
19	1.00	0.0	18.838	1.370	52.80	-3.84
20	1.00	0.0	14.519	-0.810	68.66	3.83

22 ELEMENT ARRAY

ELEMENT	VO		YO		ZO	
	RE	IM	RE	IM	RE	IM
1	1.00	0.0	14.515	-C.810	68.68	3.83
2	1.00	0.0	18.842	1.370	52.79	-3.84
3	1.0C	0.0	17.574	1.403	56.54	-4.51
4	1.0C	0.0	18.175	1.342	54.72	-4.04
5	1.C0	0.0	17.815	1.390	55.79	-4.35
6	1.00	0.0	18.053	1.353	55.08	-4.13
7	1.C0	0.0	17.888	1.381	55.57	-4.29
8	1.00	0.0	18.004	1.361	55.23	-4.17
9	1.C0	0.0	17.925	1.375	55.46	-4.26
10	1.00	0.0	17.974	1.366	55.32	-4.20
11	1.00	0.0	17.950	1.370	55.39	-4.23
12	1.00	0.0	17.950	1.370	55.39	-4.23
13	1.00	0.0	17.974	1.366	55.32	-4.20
14	1.00	0.0	17.925	1.375	55.46	-4.26
15	1.00	0.0	18.004	1.361	55.23	-4.17
16	1.00	0.0	17.888	1.381	55.57	-4.29
17	1.00	0.0	18.053	1.353	55.08	-4.13
18	1.00	0.0	17.815	1.390	55.79	-4.35
19	1.00	0.0	18.175	1.342	54.72	-4.04
20	1.C0	0.0	17.574	1.403	56.54	-4.51
21	1.C0	0.0	18.842	1.370	52.79	-3.84
22	1.00	0.0	14.515	-C.810	68.68	3.83

24 ELEMENT ARRAY

ELEMENT	VO		YO		ZO	
	RE	IM	RE	IM	RE	IM
1	1.C0	0.0	14.511	-C.811	68.70	3.84
2	1.CC	0.0	18.845	1.369	52.79	-3.83
3	1.CC	0.0	17.571	1.403	56.55	-4.52
4	1.00	0.0	18.178	1.341	54.71	-4.04
5	1.00	0.0	17.811	1.391	55.80	-4.36
6	1.C0	0.0	18.057	1.352	55.07	-4.12
7	1.00	0.0	17.884	1.382	55.58	-4.30
8	1.0C	0.0	18.009	1.359	55.21	-4.17
9	1.00	0.0	17.919	1.377	55.48	-4.26
10	1.00	0.0	17.981	1.364	55.30	-4.20
11	1.00	0.0	17.942	1.372	55.41	-4.24
12	1.0C	0.0	17.961	1.368	55.36	-4.22
13	1.0C	0.0	17.961	1.368	55.36	-4.22
14	1.00	0.0	17.942	1.372	55.41	-4.24
15	1.00	0.0	17.981	1.364	55.30	-4.20
16	1.00	0.0	17.919	1.377	55.48	-4.26
17	1.00	0.0	18.009	1.359	55.21	-4.17
18	1.00	0.0	17.884	1.382	55.58	-4.30
19	1.0C	0.0	18.057	1.352	55.07	-4.12
20	1.00	0.0	17.811	1.391	55.80	-4.36
21	1.00	0.0	18.178	1.341	54.71	-4.04
22	1.CC	0.0	17.571	1.403	56.55	-4.52
23	1.CC	0.0	18.845	1.369	52.79	-3.83
24	1.C0	0.0	14.5!1	-0.811	68.70	3.84

DRIVING POINT ADMITTANCES AND IMPEDANCES
FOR SPECIFIED BASE CURRENTS

H/LMDA=0.250C B/LMDA= C.5CC UMEGA= 8.53

18 ELEMENT ARRAY

ELEMENT	IZ(0)		YO		ZO	
	RE	IM	RE	IM	RE	IM
1	1.C0	0.0	13.657	-2.467	69.44	16.21
2	1.00	0.0	18.852	5.35C	50.03	-12.29
3	1.00	0.0	16.905	-0.618	58.79	4.65
4	1.00	0.0	18.387	3.003	53.56	-6.65
5	1.CC	0.0	17.526	C.321	57.03	1.30
6	1.0C	0.0	18.184	2.167	54.65	-4.35
7	1.CC	0.0	17.774	C.862	56.26	-0.45
8	1.00	0.0	18.048	1.671	55.25	-2.91
9	1.00	0.0	17.924	1.272	55.74	-1.72
10	1.00	0.0	17.924	1.272	55.74	-1.72
11	1.00	0.0	18.048	1.671	55.25	-2.91
12	1.00	0.0	17.774	0.862	56.26	-0.45
13	1.0C	0.0	18.184	2.167	54.65	-4.35
14	1.00	0.0	17.526	0.321	57.03	1.30
15	1.00	0.0	18.387	3.003	53.56	-6.65
16	1.00	0.0	16.905	-0.618	58.79	4.65
17	1.00	0.0	18.852	5.350	50.03	-12.29
18	1.00	0.0	13.657	-2.467	69.44	16.21

2C ELEMENT ARRAY

ELEMENT	IZ(0)		YO		ZO	
	RE	IM	RE	IM	RE	IM
1	1.00	0.0	13.634	-2.489	69.49	16.36
2	1.00	0.0	18.842	5.418	49.97	-12.46
3	1.00	0.0	16.876	-0.669	58.86	4.85
4	1.C0	0.0	18.395	3.082	53.48	-6.87
5	1.00	0.0	17.493	0.246	57.12	1.55
6	1.00	0.0	18.205	2.266	54.54	-4.63
7	1.00	0.0	17.735	0.758	56.38	-0.12
8	1.C0	0.0	18.083	1.8C0	55.10	-3.29
9	1.00	0.0	17.875	1.126	55.91	-1.27
10	1.00	0.0	17.981	1.450	55.52	-2.25
11	1.00	0.0	17.981	1.450	55.52	-2.25
12	1.00	0.0	17.875	1.126	55.91	-1.27
13	1.0C	0.0	18.083	1.800	55.10	-3.29
14	1.00	0.0	17.735	0.758	56.38	-0.12
15	1.0C	0.0	18.205	2.266	54.54	-4.63
16	1.00	0.0	17.493	0.246	57.12	1.55
17	1.00	0.0	18.395	3.082	53.48	-6.87
18	1.00	0.0	16.876	-0.669	58.86	4.85
19	1.00	0.0	18.842	5.418	49.97	-12.46
20	1.00	0.0	13.634	-2.489	69.49	16.36

22 ELEMENT ARRAY

ELEMENT	IZ(0)		YO		ZO	
	RE	IM	RE	IM	RE	IM
1	1.00	0.0	13.616	-2.507	69.54	16.48
2	1.0C	0.0	18.834	5.473	49.92	-12.60
3	1.CC	0.0	16.852	-C.711	58.91	5.00
4	1.00	0.0	18.401	3.146	53.41	-7.04
5	1.00	0.0	17.468	0.188	57.20	1.74
6	1.00	0.0	18.219	2.343	54.46	-4.85
7	1.C0	0.0	17.706	0.680	56.48	0.13
8	1.00	0.0	18.107	1.897	55.00	-3.58
9	1.00	0.0	17.839	1.020	56.04	-0.94
10	1.00	0.0	18.019	1.578	55.37	-2.64
11	1.00	0.0	17.935	1.302	55.70	-1.81
12	1.00	0.0	17.935	1.302	55.70	-1.81
13	1.0C	0.0	18.019	1.578	55.37	-2.64
14	1.00	0.0	17.839	1.020	56.04	-0.94
15	1.00	0.0	18.107	1.897	55.00	-3.58
16	1.00	0.0	17.706	C.680	56.48	0.13
17	1.00	0.0	18.219	2.343	54.46	-4.85
18	1.00	0.0	17.468	0.188	57.20	1.74
19	1.CC	0.0	18.401	3.146	53.41	-7.04
20	1.00	0.0	16.852	-0.711	58.91	5.00
21	1.00	0.0	18.834	5.473	49.92	-12.60
22	1.C0	0.0	13.616	-2.507	69.54	16.48

24 ELEMENT ARRAY

ELEMENT	IZ(0)		YO		ZO	
	RE	IM	RE	IM	RE	IM
1	1.00	0.0	13.601	-2.522	69.57	16.59
2	1.00	0.0	18.826	5.519	49.88	-12.71
3	1.00	0.0	16.833	-C.745	58.96	5.13
4	1.00	0.0	18.406	3.198	53.36	-7.18
5	1.C0	0.0	17.447	0.141	57.25	1.90
6	1.00	0.0	18.23C	2.405	54.39	-5.03
7	1.00	0.0	17.682	0.619	56.55	0.32
8	1.C0	0.0	18.124	1.973	54.91	-3.80
9	1.00	0.0	17.811	C.940	56.14	-0.69
10	1.00	0.0	18.045	1.673	55.26	-2.92
11	1.00	0.0	17.90C	1.194	55.83	-1.48
12	1.00	0.0	17.974	1.428	55.55	-2.19
13	1.0C	0.0	17.974	1.428	55.55	-2.19
14	1.00	0.0	17.900	1.194	55.83	-1.48
15	1.00	0.0	18.045	1.673	55.26	-2.92
16	1.00	0.0	17.811	C.940	56.14	-0.69
17	1.00	0.0	18.124	1.973	54.91	-3.80
18	1.00	0.0	17.682	0.619	56.55	0.32
19	1.00	0.0	18.230	2.405	54.39	-5.03
20	1.C0	0.0	17.447	0.141	57.25	1.90
21	1.C0	0.0	18.406	3.198	53.36	-7.18
22	1.CC	0.0	16.833	-C.745	58.96	5.13
23	1.00	0.0	18.826	5.519	49.88	-12.71
24	1.00	0.0	13.601	-?.522	69.57	16.59

BR

DRIVING POINT ADMITTANCES AND IMPEDANCES
FOR SPECIFIED BASE VOLTAGES

H/LMDA=0.2500 B/LMDA= 0.500 OMEGA= 8.53

26 ELEMENT ARRAY

ELEMENT	V0 RE	V0 IM	Y0 RE	Y0 IM	Z0 RE	Z0 IM
1	1.00	0.0	14.509	-0.811	68.71	3.84
2	1.00	0.0	18.847	1.369	52.78	-3.83
3	1.00	0.0	17.569	1.404	56.56	-4.52
4	1.00	0.0	18.181	1.341	54.71	-4.03
5	1.00	0.0	17.809	1.392	55.81	-4.36
6	1.00	0.0	18.060	1.352	55.06	-4.12
7	1.00	0.0	17.880	1.383	55.59	-4.30
8	1.00	0.0	18.013	1.358	55.20	-4.16
9	1.00	0.0	17.914	1.378	55.49	-4.27
10	1.00	0.0	17.986	1.363	55.28	-4.19
11	1.00	0.0	17.936	1.374	55.43	-4.24
12	1.00	0.0	17.968	1.367	55.33	-4.21
13	1.00	0.0	17.952	1.370	55.38	-4.23
14	1.00	0.0	17.952	1.370	55.38	-4.23
15	1.00	0.0	17.968	1.367	55.33	-4.21
16	1.00	0.0	17.936	1.374	55.43	-4.24
17	1.00	0.0	17.986	1.363	55.28	-4.19
18	1.00	0.0	17.914	1.378	55.49	-4.27
19	1.00	0.0	18.013	1.358	55.20	-4.16
20	1.00	0.0	17.880	1.383	55.59	-4.30
21	1.00	0.0	18.060	1.352	55.06	-4.12
22	1.00	0.0	17.809	1.392	55.81	-4.36
23	1.00	0.0	18.181	1.341	54.71	-4.03
24	1.00	0.0	17.569	1.404	56.56	-4.52
25	1.00	0.0	18.847	1.369	52.78	-3.83
26	1.00	0.0	14.509	-0.811	68.71	3.84

28 ELEMENT ARRAY

ELEMENT	V0 RE	V0 IM	Y0 RE	Y0 IM	Z0 RE	Z0 IM
1	1.00	0.0	14.506	-0.811	68.72	3.84
2	1.00	0.0	18.849	1.368	52.78	-3.83
3	1.00	0.0	17.567	1.404	56.56	-4.52
4	1.00	0.0	18.183	1.340	54.70	-4.03
5	1.00	0.0	17.806	1.392	55.82	-4.36
6	1.00	0.0	18.063	1.351	55.05	-4.12
7	1.00	0.0	17.878	1.384	55.60	-4.30
8	1.00	0.0	18.016	1.358	55.19	-4.16
9	1.00	0.0	17.911	1.378	55.50	-4.27
10	1.00	0.0	17.990	1.362	55.27	-4.18
11	1.00	0.0	17.931	1.375	55.44	-4.25
12	1.00	0.0	17.973	1.366	55.32	-4.20
13	1.00	0.0	17.946	1.371	55.40	-4.23
14	1.00	0.0	17.959	1.369	55.36	-4.22
15	1.00	0.0	17.959	1.369	55.36	-4.22
16	1.00	0.0	17.946	1.371	55.40	-4.23
17	1.00	0.0	17.973	1.366	55.32	-4.20
18	1.00	0.0	17.931	1.375	55.44	-4.25
19	1.00	0.0	17.990	1.362	55.27	-4.18
20	1.00	0.0	17.911	1.378	55.50	-4.27
21	1.00	0.0	18.016	1.358	55.19	-4.16
22	1.00	0.0	17.878	1.384	55.60	-4.30
23	1.00	0.0	18.063	1.351	55.05	-4.12
24	1.00	0.0	17.806	1.392	55.82	-4.36
25	1.00	0.0	18.183	1.340	54.70	-4.03
26	1.00	0.0	17.567	1.404	56.56	-4.52
27	1.00	0.0	18.849	1.368	52.78	-3.83
28	1.00	0.0	14.506	-0.811	68.72	3.84

30 ELEMENT ARRAY

ELEMENT	V0 RE	V0 IM	Y0 RE	Y0 IM	Z0 RE	Z0 IM
1	1.00	0.0	14.504	-0.811	68.73	3.84
2	1.00	0.0	18.851	1.368	52.77	-3.83
3	1.00	0.0	17.565	1.405	56.57	-4.52
4	1.00	0.0	18.184	1.340	54.70	-4.03
5	1.00	0.0	17.805	1.393	55.82	-4.37
6	1.00	0.0	18.065	1.351	55.05	-4.12
7	1.00	0.0	17.875	1.384	55.61	-4.31
8	1.00	0.0	18.018	1.357	55.19	-4.16
9	1.00	0.0	17.909	1.379	55.51	-4.27
10	1.00	0.0	17.993	1.361	55.26	-4.18
11	1.00	0.0	17.928	1.375	55.45	-4.25
12	1.00	0.0	17.977	1.365	55.31	-4.20
13	1.00	0.0	17.942	1.372	55.41	-4.24
14	1.00	0.0	17.965	1.367	55.34	-4.21
15	1.00	0.0	17.954	1.370	55.38	-4.23
16	1.00	0.0	17.954	1.370	55.38	-4.23
17	1.00	0.0	17.965	1.367	55.34	-4.21
18	1.00	0.0	17.942	1.372	55.41	-4.24
19	1.00	0.0	17.977	1.365	55.31	-4.20
20	1.00	0.0	17.928	1.375	55.45	-4.25
21	1.00	0.0	17.993	1.361	55.26	-4.18
22	1.00	0.0	17.909	1.379	55.51	-4.27
23	1.00	0.0	18.018	1.357	55.19	-4.16
24	1.00	0.0	17.875	1.384	55.61	-4.31
25	1.00	0.0	18.065	1.351	55.05	-4.12
26	1.00	0.0	17.805	1.393	55.82	-4.37
27	1.00	0.0	18.184	1.340	54.70	-4.03
28	1.00	0.0	17.565	1.405	56.57	-4.52
29	1.00	0.0	18.851	1.368	52.77	-3.83
30	1.00	0.0	14.504	-0.811	68.73	3.84

DRIVING POINT ADMITTANCES AND IMPEDANCES
FOR SPECIFIED BASE CURRENTS

H/LMDA=0.2500 B/LMDA= 0.500 OMEGA= 8.53

26 ELEMENT ARRAY

ELEMENT	IZ(0) RE	IZ(0) IM	Y0 RE	Y0 IM	Z0 RE	Z0 IM
1	1.00	0.0	13.588	-2.535	69.60	16.67
2	1.00	0.0	18.820	5.558	49.84	-12.81
3	1.00	0.0	16.817	-0.773	59.00	5.24
4	1.00	0.0	18.409	3.242	53.32	-7.30
5	1.00	0.0	17.429	0.102	57.30	2.03
6	1.00	0.0	18.239	2.456	54.34	-5.17
7	1.00	0.0	17.663	0.569	56.61	0.48
8	1.00	0.0	18.138	2.034	54.85	-3.97
9	1.00	0.0	17.789	0.877	56.21	-0.49
10	1.00	0.0	18.064	1.748	55.18	-3.14
11	1.00	0.0	17.874	1.113	55.92	-1.23
12	1.00	0.0	18.001	1.522	55.44	-2.47
13	1.00	0.0	17.941	1.319	55.68	-1.86
14	1.00	0.0	17.941	1.319	55.68	-1.86
15	1.00	0.0	18.001	1.522	55.44	-2.47
16	1.00	0.0	17.874	1.113	55.92	-1.23
17	1.00	0.0	18.064	1.748	55.18	-3.14
18	1.00	0.0	17.789	0.877	56.21	-0.49
19	1.00	0.0	18.138	2.034	54.85	-3.97
20	1.00	0.0	17.663	0.569	56.61	0.48
21	1.00	0.0	18.239	2.456	54.34	-5.17
22	1.00	0.0	17.429	0.102	57.30	2.03
23	1.00	0.0	18.409	3.242	53.32	-7.30
24	1.00	0.0	16.817	-0.773	59.00	5.24
25	1.00	0.0	18.820	5.558	49.84	-12.81
26	1.00	0.0	13.588	-2.535	69.60	16.67

28 ELEMENT ARRAY

ELEMENT	IZ(0) RE	IZ(0) IM	Y0 RE	Y0 IM	Z0 RE	Z0 IM
1	1.00	0.0	13.577	-2.547	69.62	16.75
2	1.00	0.0	18.814	5.591	49.81	-12.90
3	1.00	0.0	16.803	-0.798	59.03	5.33
4	1.00	0.0	18.412	3.279	53.28	-7.41
5	1.00	0.0	17.415	0.070	57.34	2.14
6	1.00	0.0	18.246	2.499	54.29	-5.29
7	1.00	0.0	17.647	0.528	56.66	0.62
8	1.00	0.0	18.148	2.084	54.79	-4.12
9	1.00	0.0	17.771	0.826	56.27	-0.33
10	1.00	0.0	18.079	1.809	55.11	-3.32
11	1.00	0.0	17.853	1.049	56.00	-1.03
12	1.00	0.0	18.022	1.597	55.36	-2.69
13	1.00	0.0	17.915	1.237	55.77	-1.61
14	1.00	0.0	17.969	1.413	55.57	-2.14
15	1.00	0.0	17.969	1.413	55.57	-2.14
16	1.00	0.0	17.915	1.237	55.77	-1.61
17	1.00	0.0	18.022	1.597	55.36	-2.69
18	1.00	0.0	17.853	1.049	56.00	-1.03
19	1.00	0.0	18.079	1.809	55.11	-3.32
20	1.00	0.0	17.771	0.826	56.27	-0.33
21	1.00	0.0	18.148	2.084	54.79	-4.12
22	1.00	0.0	17.647	0.528	56.66	0.62
23	1.00	0.0	18.246	2.499	54.29	-5.29
24	1.00	0.0	17.415	0.070	57.34	2.14
25	1.00	0.0	18.412	3.279	53.28	-7.41
26	1.00	0.0	16.803	-0.798	59.03	5.33
27	1.00	0.0	18.814	5.591	49.81	-12.90
28	1.00	0.0	13.577	-2.547	69.62	16.75

30 ELEMENT ARRAY

ELEMENT	IZ(0) RE	IZ(0) IM	Y0 RE	Y0 IM	Z0 RE	Z0 IM
1	1.00	0.0	13.567	-2.556	69.65	16.82
2	1.00	0.0	18.809	5.620	49.79	-12.97
3	1.00	0.0	16.791	-0.819	59.06	5.41
4	1.00	0.0	18.414	3.311	53.25	-7.49
5	1.00	0.0	17.402	0.042	57.38	2.24
6	1.00	0.0	18.251	2.535	54.25	-5.40
7	1.00	0.0	17.633	0.494	56.70	0.73
8	1.00	0.0	18.157	2.126	54.75	-4.24
9	1.00	0.0	17.756	0.784	56.32	-0.20
10	1.00	0.0	18.091	1.858	55.06	-3.46
11	1.00	0.0	17.835	0.997	56.06	-0.87
12	1.00	0.0	18.038	1.657	55.29	-2.87
13	1.00	0.0	17.895	1.172	55.85	-1.41
14	1.00	0.0	17.991	1.487	55.48	-2.37
15	1.00	0.0	17.945	1.331	55.66	-1.89
16	1.00	0.0	17.945	1.331	55.66	-1.89
17	1.00	0.0	17.991	1.487	55.48	-2.37
18	1.00	0.0	17.895	1.172	55.85	-1.41
19	1.00	0.0	18.038	1.657	55.29	-2.87
20	1.00	0.0	17.835	0.997	56.06	-0.87
21	1.00	0.0	18.091	1.858	55.06	-3.46
22	1.00	0.0	17.756	0.784	56.32	-0.20
23	1.00	0.0	18.157	2.126	54.75	-4.24
24	1.00	0.0	17.633	0.494	56.70	0.73
25	1.00	0.0	18.251	2.535	54.25	-5.40
26	1.00	0.0	17.402	0.042	57.38	2.24
27	1.00	0.0	18.414	3.311	53.25	-7.49
28	1.00	0.0	16.791	-0.819	59.06	5.41
29	1.00	0.0	18.809	5.620	49.79	-12.97
30	1.00	0.0	13.567	-2.556	69.65	16.82

BR

DRIVING POINT ADMITTANCES AND IMPEDANCES
FOR SPECIFIED BASE VOLTAGES

H/LMDA=0.2500 B/LMDA= 0.500 OMEGA= 8.53

35 ELEMENT ARRAY

ELEMENT	VO		YO		ZO	
	RE	IM	RE	IM	RE	IM
1	1.00	0.0	14.458	-0.814	68.95	3.88
2	1.00	0.0	18.886	1.361	52.67	-3.80
3	1.00	0.0	17.532	1.413	56.67	-4.57
4	1.00	0.0	18.217	1.331	54.60	-3.99
5	1.00	0.0	17.772	1.402	55.92	-4.41
6	1.00	0.0	18.098	1.341	54.95	-4.07
7	1.00	0.0	17.841	1.394	55.71	-4.35
8	1.00	0.0	18.053	1.347	55.08	-4.11
9	1.00	0.0	17.872	1.390	55.62	-4.33
10	1.00	0.0	18.032	1.350	55.15	-4.13
11	1.00	0.0	17.888	1.387	55.57	-4.31
12	1.00	0.0	18.020	1.352	55.18	-4.14
13	1.00	0.0	17.897	1.386	55.54	-4.30
14	1.00	0.0	18.013	1.353	55.20	-4.15
15	1.00	0.0	17.902	1.385	55.53	-4.30
16	1.00	0.0	18.010	1.354	55.21	-4.15
17	1.00	0.0	17.904	1.384	55.52	-4.29
18	1.00	0.0	18.008	1.354	55.22	-4.15
19	1.00	0.0	17.904	1.384	55.52	-4.29
20	1.00	0.0	18.010	1.354	55.21	-4.15
21	1.00	0.0	17.902	1.385	55.53	-4.30
22	1.00	0.0	18.013	1.353	55.20	-4.15
23	1.00	0.0	17.897	1.386	55.54	-4.30
24	1.00	0.0	18.020	1.352	55.18	-4.14
25	1.00	0.0	17.888	1.387	55.57	-4.31
26	1.00	0.0	18.032	1.350	55.15	-4.13
27	1.00	0.0	17.872	1.390	55.62	-4.33
28	1.00	0.0	18.053	1.347	55.08	-4.11
29	1.00	0.0	17.841	1.394	55.71	-4.35
30	1.00	0.0	18.098	1.341	54.95	-4.07
31	1.00	0.0	17.772	1.402	55.92	-4.41
32	1.00	0.0	18.217	1.331	54.60	-3.99
33	1.00	0.0	17.532	1.413	56.67	-4.57
34	1.00	0.0	18.886	1.361	52.67	-3.80
35	1.00	0.0	14.458	-0.814	68.95	3.88

40 ELEMENT ARRAY

ELEMENT	VO		YO		ZO	
	RE	IM	RE	IM	RE	IM
1	1.00	0.0	14.498	-0.811	68.76	3.85
2	1.00	0.0	18.856	1.367	52.76	-3.82
3	1.00	0.0	17.560	1.406	56.58	-4.53
4	1.00	0.0	18.190	1.339	54.68	-4.02
5	1.00	0.0	17.799	1.394	55.84	-4.37
6	1.00	0.0	18.071	1.349	55.03	-4.11
7	1.00	0.0	17.869	1.38[illegible]	[illegible]	[illegible]
8	1.00	0.0	18.025	1.355	55.17	-4.15
9	1.00	0.0	17.901	1.381	55.53	-4.28
10	1.00	0.0	18.001	1.359	55.24	-4.17
11	1.00	0.0	17.919	1.378	55.48	-4.27
12	1.00	0.0	17.987	1.362	55.28	-4.19
13	1.00	0.0	17.931	1.375	55.44	-4.25
14	1.00	0.0	17.977	1.364	55.31	-4.20
15	1.00	0.0	17.939	1.373	55.42	-4.24
16	1.00	0.0	17.969	1.366	55.33	-4.21
17	1.00	0.0	17.946	1.372	55.40	-4.23
18	1.00	0.0	17.963	1.368	55.35	-4.21
19	1.00	0.0	17.952	1.370	55.38	-4.23
20	1.00	0.0	17.958	1.369	55.37	-4.22
21	1.00	0.0	17.958	1.369	55.37	-4.22
22	1.00	0.0	17.952	1.370	55.38	-4.23
23	1.00	0.0	17.963	1.368	55.35	-4.21
24	1.00	0.0	17.946	1.372	55.40	-4.23
25	1.00	0.0	17.969	1.366	55.33	-4.21
26	1.00	0.0	17.939	1.373	55.42	-4.24
27	1.00	0.0	17.977	1.364	55.31	-4.20
28	1.00	0.0	17.931	1.375	55.44	-4.25
29	1.00	0.0	17.987	1.362	55.28	-4.19
30	1.00	0.0	17.919	1.378	55.48	-4.27
31	1.00	0.0	18.001	1.359	55.24	-4.17
32	1.00	0.0	17.901	1.381	55.53	-4.28
33	1.00	0.0	18.025	1.355	55.17	-4.15
34	1.00	0.0	17.869	1.386	55.63	-4.31
35	1.00	0.0	18.071	1.349	55.03	-4.11
36	1.00	0.0	17.799	1.394	55.84	-4.37
37	1.00	0.0	18.190	1.339	54.68	-4.02
38	1.00	0.0	17.560	1.406	56.58	-4.53
39	1.00	0.0	18.856	1.367	52.76	-3.82
40	1.00	0.0	14.498	-0.811	68.76	3.85

DRIVING POINT ADMITTANCES AND IMPEDANCES
FOR SPECIFIED BASE CURRENTS

H/LMDA=0.2500 B/LMDA= 0.500 OMEGA= 8.53

35 ELEMENT ARRAY

ELEMENT	IZ(0)		YO		ZO	
	RE	IM	RE	IM	RE	IM
1	1.00	0.0	13.291	-2.818	70.26	18.70
2	1.00	0.0	18.637	6.401	49.09	-14.97
3	1.00	0.0	16.462	-1.345	59.80	7.50
4	1.00	0.0	18.434	4.120	52.46	-9.68
5	1.00	0.0	17.078	-0.604	58.21	4.51
6	1.00	0.0	18.344	3.387	53.39	-7.76
7	1.00	0.0	17.305	-0.239	57.61	3.19
8	1.00	0.0	18.295	3.034	53.80	-6.80
9	1.00	0.0	17.417	-0.031	57.31	2.47
10	1.00	0.0	18.266	2.833	54.02	-6.25
11	1.00	0.0	17.479	0.097	57.14	2.04
12	1.00	0.0	18.248	2.711	54.16	-5.91
13	1.00	0.0	17.516	0.176	57.03	1.77
14	1.00	0.0	18.236	2.637	54.24	-5.70
15	1.00	0.0	17.537	0.224	56.98	1.61
16	1.00	0.0	18.230	2.597	54.28	-5.59
17	1.00	0.0	17.547	0.245	56.95	1.54
18	1.00	0.0	18.228	2.585	54.29	-5.55
19	1.00	0.0	17.547	0.245	56.95	1.54
20	1.00	0.0	18.230	2.597	54.28	-5.59
21	1.00	0.0	17.537	0.224	56.98	1.61
22	1.00	0.0	18.236	2.637	54.24	-5.70
23	1.00	0.0	17.516	0.176	57.03	1.77
24	1.00	0.0	18.248	2.711	54.16	-5.91
25	1.00	0.0	17.479	0.097	57.14	2.04
26	1.00	0.0	18.266	2.833	54.02	-6.25
27	1.00	0.0	17.417	-0.031	57.31	2.47
28	1.00	0.0	18.295	3.034	53.80	-6.80
29	1.00	0.0	17.305	-0.239	57.61	3.19
30	1.00	0.0	18.344	3.387	53.39	-7.76
31	1.00	0.0	17.078	-0.604	58.21	4.51
32	1.00	0.0	18.434	4.120	52.46	-9.68
33	1.00	0.0	16.462	-1.345	59.80	7.50
34	1.00	0.0	18.637	6.401	49.09	-14.97
35	1.00	0.0	13.291	-2.818	70.26	18.70

40 ELEMENT ARRAY

ELEMENT	IZ(0)		YO		ZO	
	RE	IM	RE	IM	RE	IM
1	1.00	0.0	13.532	-2.591	69.72	17.06
2	1.00	0.0	18.790	5.723	49.70	-13.23
3	1.00	0.0	16.748	-0.891	59.16	5.69
4	1.00	0.0	18.421	3.421	53.14	-7.79
5	1.00	0.0	17.359	-0.052	57.49	2.56
6	1.00	0.0	18.269	2.657	54.13	-5.74
7	1.00	0.0	17.587	0.381	56.84	1.10
8	1.00	0.0	18.184	2.2[illegible]4	[illegible]	[illegible]
9	1.00	0.0	17.706	0.649	56.48	0.23
10	1.00	0.0	18.126	2.016	54.89	-3.92
11	1.00	0.0	17.780	0.838	56.24	-0.37
12	1.00	0.0	18.083	1.839	55.09	-3.41
13	1.00	0.0	17.832	0.982	56.07	-0.83
14	1.00	0.0	18.048	1.702	55.24	-3.01
15	1.00	0.0	17.873	1.101	55.93	-1.19
16	1.00	0.0	18.018	1.588	55.37	-2.67
17	1.00	0.0	17.906	1.205	55.81	-1.51
18	1.00	0.0	17.990	1.487	55.49	-2.36
19	1.00	0.0	17.936	1.300	55.70	-1.80
20	1.00	0.0	17.963	1.392	55.59	-2.08
21	1.00	0.0	17.963	1.392	55.59	-2.08
22	1.00	0.0	17.936	1.300	55.70	-1.80
23	1.00	0.0	17.990	1.487	55.49	-2.36
24	1.00	0.0	17.906	1.205	55.81	-1.51
25	1.00	0.0	18.018	1.588	55.37	-2.67
26	1.00	0.0	17.873	1.101	55.93	-1.19
27	1.00	0.0	18.048	1.702	55.24	-3.01
28	1.00	0.0	17.832	0.982	56.07	-0.83
29	1.00	0.0	18.083	1.839	55.09	-3.41
30	1.00	0.0	17.780	0.838	56.24	-0.37
31	1.00	0.0	18.126	2.016	54.89	-3.92
32	1.00	0.0	17.706	0.649	56.48	0.23
33	1.00	0.0	18.184	2.264	54.60	-4.64
34	1.00	0.0	17.587	0.381	56.84	1.10
35	1.00	0.0	18.269	2.657	54.13	-5.74
36	1.00	0.0	17.359	-0.052	57.49	2.56
37	1.00	0.0	18.421	3.421	53.14	-7.79
38	1.00	0.0	16.748	-0.891	59.16	5.69
39	1.00	0.0	18.790	5.723	49.70	-13.23
40	1.00	0.0	13.532	-2.591	69.72	17.06

8R

DRIVING POINT ADMITTANCES AND IMPEDANCES
FOR SPECIFIED BASE VOLTAGES

H/LMDA=0.2500 B/LMDA= 0.500 OMEGA= 8.53

45 ELEMENT ARRAY

ELEMENT	VO		YO		ZO	
	RE	IM	RE	IM	RE	IM
1	1.00	0.0	14.464	-0.813	68.92	3.87
2	1.00	0.0	18.882	1.362	52.69	-3.80
3	1.00	0.0	17.536	1.412	56.66	-4.56
4	1.00	0.0	18.213	1.332	54.62	-3.99
5	1.00	0.0	17.776	1.401	55.91	-4.41
6	1.00	0.0	18.093	1.342	54.97	-4.08
7	1.00	0.0	17.846	1.393	55.70	-4.35
8	1.00	0.0	18.048	1.348	55.10	-4.12
9	1.00	0.0	17.877	1.389	55.60	-4.32
10	1.00	0.0	18.026	1.352	55.17	-4.14
11	1.00	0.0	17.894	1.386	55.55	-4.30
12	1.00	0.0	18.013	1.354	55.20	-4.15
13	1.00	0.0	17.904	1.384	55.52	-4.29
14	1.00	0.0	18.005	1.355	55.23	-4.16
15	1.00	0.0	17.910	1.383	55.50	-4.28
16	1.00	0.0	18.000	1.356	55.24	-4.16
17	1.00	0.0	17.915	1.382	55.49	-4.28
18	1.00	0.0	17.996	1.357	55.25	-4.17
19	1.00	0.0	17.917	1.381	55.48	-4.28
20	1.00	0.0	17.994	1.357	55.26	-4.17
21	1.00	0.0	17.919	1.381	55.48	-4.28
22	1.00	0.0	17.993	1.358	55.26	-4.17
23	1.00	0.0	17.919	1.381	55.48	-4.28
24	1.00	0.0	17.993	1.358	55.26	-4.17
25	1.00	0.0	17.919	1.381	55.48	-4.28
26	1.00	0.0	17.994	1.357	55.26	-4.17
27	1.00	0.0	17.917	1.381	55.48	-4.28
28	1.00	0.0	17.996	1.357	55.25	-4.17
29	1.00	0.0	17.915	1.382	55.49	-4.28
30	1.00	0.0	18.000	1.356	55.24	-4.16
31	1.00	0.0	17.910	1.383	55.50	-4.28
32	1.00	0.0	18.005	1.355	55.23	-4.16
33	1.00	0.0	17.904	1.384	55.52	-4.29
34	1.00	0.0	18.013	1.354	55.20	-4.15
35	1.00	0.0	17.894	1.386	55.55	-4.30
36	1.00	0.0	18.026	1.352	55.17	-4.14
37	1.00	0.0	17.877	1.389	55.60	-4.32
38	1.00	0.0	18.048	1.348	55.10	-4.12
39	1.00	0.0	17.846	1.393	55.70	-4.35
40	1.00	0.0	18.093	1.342	54.97	-4.08
41	1.00	0.0	17.776	1.401	55.91	-4.41
42	1.00	0.0	18.213	1.332	54.62	-3.99
43	1.00	0.0	17.536	1.412	56.66	-4.56
44	1.00	0.0	18.882	1.362	52.69	-3.80
45	1.00	0.0	14.464	-0.813	68.92	3.87

50 ELEMENT ARRAY

ELEMENT	VO		YO		ZO	
	RE	IM	RE	IM	RE	IM
1	1.00	0.0	14.494	-0.812	68.78	3.85
2	1.00	0.0	18.859	1.366	52.75	-3.82
3	1.00	0.0	17.557	1.407	56.59	-4.53
4	1.00	0.0	18.193	1.338	54.67	-4.02
5	1.00	0.0	17.796	1.395	55.85	-4.38
6	1.00	0.0	18.074	1.348	55.02	-4.10
7	1.00	0.0	17.866	1.387	55.64	-4.32
8	1.00	0.0	18.028	1.354	55.16	-4.14
9	1.00	0.0	17.897	1.382	55.54	-4.29
10	1.00	0.0	18.005	1.358	55.23	-4.17
11	1.00	0.0	17.915	1.379	55.49	-4.27
12	1.00	0.0	17.991	1.361	55.27	-4.18
13	1.00	0.0	17.926	1.377	55.46	-4.26
14	1.00	0.0	17.982	1.363	55.29	-4.19
15	1.00	0.0	17.934	1.375	55.43	-4.25
16	1.00	0.0	17.975	1.364	55.31	-4.20
17	1.00	0.0	17.940	1.374	55.42	-4.24
18	1.00	0.0	17.970	1.366	55.33	-4.20
19	1.00	0.0	17.945	1.372	55.40	-4.24
20	1.00	0.0	17.966	1.367	55.34	-4.21
21	1.00	0.0	17.949	1.371	55.39	-4.23
22	1.00	0.0	17.962	1.368	55.35	-4.21
23	1.00	0.0	17.952	1.370	55.38	-4.23
24	1.00	0.0	17.959	1.369	55.36	-4.22
25	1.00	0.0	17.955	1.370	55.37	-4.22
26	1.00	0.0	17.955	1.370	55.37	-4.22
27	1.00	0.0	17.959	1.369	55.36	-4.22
28	1.00	0.0	17.952	1.370	55.38	-4.23
29	1.00	0.0	17.962	1.368	55.35	-4.21
30	1.00	0.0	17.949	1.371	55.39	-4.23
31	1.00	0.0	17.966	1.367	55.34	-4.21
32	1.00	0.0	17.945	1.372	55.40	-4.24
33	1.00	0.0	17.970	1.366	55.33	-4.20
34	1.00	0.0	17.940	1.374	55.42	-4.24
35	1.00	0.0	17.975	1.364	55.31	-4.20
36	1.00	0.0	17.934	1.375	55.43	-4.25
37	1.00	0.0	17.982	1.363	55.29	-4.19
38	1.00	0.0	17.926	1.377	55.46	-4.26
39	1.00	0.0	17.991	1.361	55.27	-4.18
40	1.00	0.0	17.915	1.379	55.49	-4.27
41	1.00	0.0	18.005	1.358	55.23	-4.17
42	1.00	0.0	17.897	1.382	55.54	-4.29
43	1.00	0.0	18.028	1.354	55.16	-4.14
44	1.00	0.0	17.866	1.387	55.64	-4.32
45	1.00	0.0	18.074	1.348	55.02	-4.10
46	1.00	0.0	17.796	1.395	55.85	-4.38
47	1.00	0.0	18.193	1.338	54.67	-4.02
48	1.00	0.0	17.557	1.407	56.59	-4.53
49	1.00	0.0	18.859	1.366	52.75	-3.82
50	1.00	0.0	14.494	-0.812	68.78	3.85

DRIVING POINT ADMITTANCES AND IMPEDANCES
FOR SPECIFIED BASE CURRENTS

H/LMDA=0.2500 B/LMDA= 0.500 OMEGA= 8.53

45 ELEMENT ARRAY

ELEMENT	IZ(0)		YO		ZO	
	RE	IM	RE	IM	RE	IM
1	1.00	0.0	13.313	-2.796	70.22	18.54
2	1.00	0.0	18.657	6.335	49.15	-14.79
3	1.00	0.0	16.490	-1.300	59.75	7.32
4	1.00	0.0	18.438	4.047	52.53	-9.48
5	1.00	0.0	17.107	-0.545	58.14	4.30
6	1.00	0.0	18.341	3.306	53.46	-7.54
7	1.00	0.0	17.336	-0.169	57.53	2.95
8	1.00	0.0	18.289	2.943	53.88	-6.55
9	1.00	0.0	17.451	0.050	57.22	2.20
10	1.00	0.0	18.256	2.729	54.12	-5.96
11	1.00	0.0	17.518	0.191	57.03	1.72
12	1.00	0.0	18.234	2.592	54.27	-5.57
13	1.00	0.0	17.560	0.286	56.91	1.41
14	1.00	0.0	18.219	2.500	54.37	-5.31
15	1.00	0.0	17.588	0.351	56.83	1.19
16	1.00	0.0	18.209	2.437	54.44	-5.13
17	1.00	0.0	17.607	0.396	56.78	1.04
18	1.00	0.0	18.201	2.395	54.48	-5.01
19	1.00	0.0	17.619	0.426	56.74	0.95
20	1.00	0.0	18.197	2.369	54.51	-4.94
21	1.00	0.0	17.626	0.442	56.72	0.89
22	1.00	0.0	18.195	2.356	54.52	-4.90
23	1.00	0.0	17.628	0.448	56.71	0.88
24	1.00	0.0	18.195	2.356	54.52	-4.90
25	1.00	0.0	17.626	0.442	56.72	0.89
26	1.00	0.0	18.197	2.369	54.51	-4.94
27	1.00	0.0	17.619	0.426	56.74	0.95
28	1.00	0.0	18.201	2.395	54.48	-5.01
29	1.00	0.0	17.607	0.396	56.78	1.04
30	1.00	0.0	18.209	2.437	54.44	-5.13
31	1.00	0.0	17.588	0.351	56.83	1.19
32	1.00	0.0	18.219	2.500	54.37	-5.31
33	1.00	0.0	17.560	0.286	56.91	1.41
34	1.00	0.0	18.234	2.592	54.27	-5.57
35	1.00	0.0	17.518	0.191	57.03	1.72
36	1.00	0.0	18.256	2.729	54.12	-5.96
37	1.00	0.0	17.451	0.050	57.22	2.20
38	1.00	0.0	18.289	2.943	53.88	-6.55
39	1.00	0.0	17.336	-0.169	57.53	2.95
40	1.00	0.0	18.341	3.306	53.46	-7.54
41	1.00	0.0	17.107	-0.545	58.14	4.30
42	1.00	0.0	18.438	4.047	52.53	-9.48
43	1.00	0.0	16.490	-1.300	59.75	7.32
44	1.00	0.0	18.657	6.335	49.15	-14.79
45	1.00	0.0	13.313	-2.796	70.22	18.54

50 ELEMENT ARRAY

ELEMENT	IZ(0)		YO		ZO	
	RE	IM	RE	IM	RE	IM
1	1.00	0.0	13.510	-2.612	69.77	17.21
2	1.00	0.0	18.779	5.785	49.64	-13.39
3	1.00	0.0	16.722	-0.935	59.22	5.86
4	1.00	0.0	18.425	3.487	53.08	-7.97
5	1.00	0.0	17.333	-0.107	57.56	2.75
6	1.00	0.0	18.279	2.728	54.05	-5.94
7	1.00	0.0	17.561	0.317	56.92	1.31
8	1.00	0.0	18.198	2.341	54.52	-4.86
9	1.00	0.0	17.678	0.577	56.56	0.46
10	1.00	0.0	18.144	2.101	54.80	-4.17
11	1.00	0.0	17.750	0.755	56.34	-0.11
12	1.00	0.0	18.105	1.934	54.99	-3.69
13	1.00	0.0	17.800	0.889	56.17	-0.53
14	1.00	0.0	18.074	1.807	55.13	-3.32
15	1.00	0.0	17.837	0.994	56.05	-0.86
16	1.00	0.0	18.049	1.706	55.24	-3.02
17	1.00	0.0	17.867	1.082	55.95	-1.13
18	1.00	0.0	18.026	1.621	55.34	-2.77
19	1.00	0.0	17.892	1.159	55.86	-1.37
20	1.00	0.0	18.006	1.547	55.42	-2.54
21	1.00	0.0	17.913	1.228	55.78	-1.58
22	1.00	0.0	17.988	1.479	55.49	-2.34
23	1.00	0.0	17.933	1.292	55.71	-1.78
24	1.00	0.0	17.970	1.416	55.57	-2.15
25	1.00	0.0	17.952	1.354	55.64	-1.96
26	1.00	0.0	17.952	1.354	55.64	-1.96
27	1.00	0.0	17.970	1.416	55.57	-2.15
28	1.00	0.0	17.933	1.292	55.71	-1.78
29	1.00	0.0	17.988	1.479	55.49	-2.34
30	1.00	0.0	17.913	1.228	55.78	-1.58
31	1.00	0.0	18.006	1.547	55.42	-2.54
32	1.00	0.0	17.892	1.159	55.86	-1.37
33	1.00	0.0	18.026	1.621	55.34	-2.77
34	1.00	0.0	17.867	1.082	55.95	-1.13
35	1.00	0.0	18.049	1.706	55.24	-3.02
36	1.00	0.0	17.837	0.994	56.05	-0.86
37	1.00	0.0	18.074	1.807	55.13	-3.32
38	1.00	0.0	17.800	0.889	56.17	-0.53
39	1.00	0.0	18.105	1.934	54.99	-3.69
40	1.00	0.0	17.750	0.755	56.34	-0.11
41	1.00	0.0	18.144	2.101	54.80	-4.17
42	1.00	0.0	17.678	0.577	56.56	0.46
43	1.00	0.0	18.198	2.341	54.52	-4.86
44	1.00	0.0	17.561	0.317	56.92	1.31
45	1.00	0.0	18.279	2.728	54.05	-5.94
46	1.00	0.0	17.333	-0.107	57.56	2.75
47	1.00	0.0	18.425	3.487	53.08	-7.97
48	1.00	0.0	16.722	-0.935	59.22	5.86
49	1.00	0.0	18.779	5.785	49.64	-13.39
50	1.00	0.0	13.510	-2.612	69.77	17.21

BR

DRIVING POINT ADMITTANCES AND IMPEDANCES
FOR SPECIFIED BASE VOLTAGES

H/LMDA=0.5000 B/LMDA= 0.250 OMEGA= 9.92

2 ELEMENT ARRAY

ELEMENT	VO		YO		ZO	
	RE	IM	RE	IM	RE	IM
1	1.00	0.0	1.636	1.959	251.07	-300.76
2	1.00	0.0	1.636	1.959	251.07	-300.76

3 ELEMENT ARRAY

ELEMENT	VO		YO		ZO	
	RE	IM	RE	IM	RE	IM
1	1.00	0.0	1.842	1.700	293.12	-270.66
2	1.00	0.0	2.399	2.279	219.10	-208.11
3	1.00	0.0	1.842	1.700	293.12	-270.66

4 ELEMENT ARRAY

ELEMENT	VO		YO		ZO	
	RE	IM	RE	IM	RE	IM
1	1.00	0.0	1.708	1.572	316.97	-291.80
2	1.00	0.0	2.620	1.962	244.53	-183.14
3	1.00	0.0	2.620	1.962	244.53	-183.14
4	1.00	0.0	1.708	1.572	316.97	-291.80

5 ELEMENT ARRAY

ELEMENT	VO		YO		ZO	
	RE	IM	RE	IM	RE	IM
1	1.00	0.0	1.625	1.649	303.25	-307.71
2	1.00	0.0	2.465	1.821	262.43	-193.93
3	1.00	0.0	2.819	1.664	263.05	-155.32
4	1.00	0.0	2.465	1.821	262.43	-193.93
5	1.00	0.0	1.625	1.649	303.25	-307.71

6 ELEMENT ARRAY

ELEMENT	VO		YO		ZO	
	RE	IM	RE	IM	RE	IM
1	1.00	0.0	1.674	1.708	292.68	-298.54
2	1.00	0.0	2.367	1.912	255.71	-206.51
3	1.00	0.0	2.675	1.528	281.87	-161.04
4	1.00	0.0	2.675	1.528	281.87	-161.04
5	1.00	0.0	2.367	1.912	255.71	-206.51
6	1.00	0.0	1.674	1.708	292.68	-298.54

7 ELEMENT ARRAY

ELEMENT	VO		YO		ZO	
	RE	IM	RE	IM	RE	IM
1	1.00	0.0	1.720	1.671	299.17	-290.63
2	1.00	0.0	2.430	1.983	247.04	-201.63
3	1.00	0.0	2.580	1.615	278.50	-174.30
4	1.00	0.0	2.527	1.385	304.38	-166.81
5	1.00	0.0	2.580	1.615	278.50	-174.30
6	1.00	0.0	2.430	1.983	247.04	-201.63
7	1.00	0.0	1.720	1.671	299.17	-290.63

8 ELEMENT ARRAY

ELEMENT	VO		YO		ZO	
	RE	IM	RE	IM	RE	IM
1	1.00	0.0	1.689	1.636	305.54	-295.85
2	1.00	0.0	2.483	1.935	250.62	-195.30
3	1.00	0.0	2.640	1.681	269.49	-171.59
4	1.00	0.0	2.430	1.475	300.75	-182.54
5	1.00	0.0	2.430	1.475	300.75	-182.54
6	1.00	0.0	2.640	1.681	269.48	-171.59
7	1.00	0.0	2.483	1.935	250.62	-195.30
8	1.00	0.0	1.689	1.636	305.54	-295.85

DRIVING POINT ADMITTANCES AND IMPEDANCES
FOR SPECIFIED BASE CURRENTS

H/LMDA=0.5000 B/LMDA= 0.250 OMEGA= 9.92

2 ELEMENT ARRAY

ELEMENT	IZ(0)		YO		ZO	
	RE	IM	RE	IM	RE	IM
1	1.00	0.0	1.636	1.959	388.37	-294.28
2	1.00	0.0	1.636	1.959	388.37	-294.28

3 ELEMENT ARRAY

ELEMENT	IZ(0)		YO		ZO	
	RE	IM	RE	IM	RE	IM
1	1.00	0.0	1.442	1.849	429.88	-336.48
2	1.00	0.0	3.358	1.037	295.19	-27.89
3	1.00	0.0	1.442	1.849	429.88	-336.48

4 ELEMENT ARRAY

ELEMENT	IZ(0)		YO		ZO	
	RE	IM	RE	IM	RE	IM
1	1.00	0.0	1.429	1.893	418.22	-343.28
2	1.00	0.0	2.787	1.329	342.44	-74.82
3	1.00	0.0	2.787	1.329	342.44	-74.82
4	1.00	0.0	1.429	1.893	418.22	-343.28

5 ELEMENT ARRAY

ELEMENT	IZ(0)		YO		ZO	
	RE	IM	RE	IM	RE	IM
1	1.00	0.0	1.450	1.907	413.00	-337.93
2	1.00	0.0	2.841	1.466	329.27	-86.50
3	1.00	0.0	2.347	1.439	389.57	-119.26
4	1.00	0.0	2.841	1.466	329.27	-86.50
5	1.00	0.0	1.450	1.907	413.00	-337.93

6 ELEMENT ARRAY

ELEMENT	IZ(0)		YO		ZO	
	RE	IM	RE	IM	RE	IM
1	1.00	0.0	1.461	1.890	417.04	-334.08
2	1.00	0.0	2.939	1.452	320.36	-79.81
3	1.00	0.0	2.365	1.542	377.33	-131.12
4	1.00	0.0	2.365	1.542	377.33	-131.12
5	1.00	0.0	2.939	1.452	320.36	-79.81
6	1.00	0.0	1.461	1.890	417.04	-334.08

7 ELEMENT ARRAY

ELEMENT	IZ(0)		YO		ZO	
	RE	IM	RE	IM	RE	IM
1	1.00	0.0	1.448	1.883	419.88	-337.27
2	1.00	0.0	2.924	1.378	325.57	-73.25
3	1.00	0.0	2.434	1.546	368.40	-125.04
4	1.00	0.0	2.376	1.552	364.76	-143.00
5	1.00	0.0	2.434	1.546	368.40	-125.04
6	1.00	0.0	2.924	1.378	325.57	-73.25
7	1.00	0.0	1.448	1.883	419.88	-337.27

8 ELEMENT ARRAY

ELEMENT	IZ(0)		YO		ZO	
	RE	IM	RE	IM	RE	IM
1	1.00	0.0	1.442	1.893	417.33	-339.47
2	1.00	0.0	2.868	1.391	330.51	-77.35
3	1.00	0.0	2.435	1.494	373.06	-118.53
4	1.00	0.0	2.448	1.661	355.90	-136.76
5	1.00	0.0	2.448	1.661	355.90	-136.76
6	1.00	0.0	2.435	1.494	373.06	-118.53
7	1.00	0.0	2.868	1.391	330.51	-77.35
8	1.00	0.0	1.442	1.893	417.33	-339.47

BR

DRIVING POINT ADMITTANCES AND IMPEDANCES
FOR SPECIFIED BASE VOLTAGES

H/LMDA=0.5000 B/LMDA= 0.250 OMEGA= 9.92

9 ELEMENT ARRAY

ELEMENT	VO RE	VO IM	YO RE	YO IM	ZO RE	ZO IM
1	1.00	0.0	1.661	1.661	301.05	-300.99
2	1.00	0.0	2.444	1.894	255.67	-198.11
3	1.00	0.0	2.689	1.636	271.39	-165.14
4	1.00	0.0	2.491	1.543	290.17	-179.68
5	1.00	0.0	2.336	1.564	295.61	-197.85
6	1.00	0.0	2.491	1.543	290.17	-179.68
7	1.00	0.0	2.689	1.636	271.39	-165.14
8	1.00	0.0	2.444	1.894	255.67	-198.11
9	1.00	0.0	1.661	1.661	301.05	-300.99

10 ELEMENT ARRAY

ELEMENT	VO RE	VO IM	YO RE	YO IM	ZO RE	ZO IM
1	1.00	0.0	1.683	1.684	296.86	-297.16
2	1.00	0.0	2.410	1.926	253.24	-202.30
3	1.00	0.0	2.653	1.598	276.55	-166.56
4	1.00	0.0	2.542	1.497	292.09	-172.02
5	1.00	0.0	2.396	1.630	285.27	-194.11
6	1.00	0.0	2.396	1.630	285.27	-194.11
7	1.00	0.0	2.542	1.497	292.09	-172.02
8	1.00	0.0	2.653	1.598	276.55	-166.56
9	1.00	0.0	2.410	1.926	253.24	-202.30
10	1.00	0.0	1.683	1.684	296.86	-297.16

12 ELEMENT ARRAY

ELEMENT	VO RE	VO IM	YO RE	YO IM	ZO RE	ZO IM
1	1.00	0.0	1.686	1.648	303.30	-296.46
2	1.00	0.0	2.462	1.930	251.61	-197.19
3	1.00	0.0	2.648	1.654	271.66	-169.67
4	1.00	0.0	2.473	1.488	296.92	-178.65
5	1.00	0.0	2.410	1.546	293.96	-188.61
6	1.00	0.0	2.507	1.653	278.03	-183.28
7	1.00	0.0	2.507	1.653	278.03	-183.28
8	1.00	0.0	2.410	1.546	293.96	-188.61
9	1.00	0.0	2.473	1.488	296.92	-178.65
10	1.00	0.0	2.648	1.654	271.66	-169.67
11	1.00	0.0	2.462	1.930	251.61	-197.19
12	1.00	0.0	1.686	1.648	303.30	-296.46

14 ELEMENT ARRAY

ELEMENT	VO RE	VO IM	YO RE	YO IM	ZO RE	ZO IM
1	1.00	0.0	1.684	1.677	298.22	-296.90
2	1.00	0.0	2.422	1.928	252.75	-201.15
3	1.00	0.0	2.650	1.612	275.39	-167.53
4	1.00	0.0	2.522	1.492	293.70	-173.71
5	1.00	0.0	2.404	1.603	287.94	-191.97
6	1.00	0.0	2.438	1.644	281.95	-190.08
7	1.00	0.0	2.521	1.568	286.02	-177.92
8	1.00	0.0	2.521	1.568	286.02	-177.92
9	1.00	0.0	2.438	1.644	281.95	-190.08
10	1.00	0.0	2.404	1.603	287.94	-191.97
11	1.00	0.0	2.522	1.492	293.70	-173.71
12	1.00	0.0	2.650	1.612	275.39	-167.53
13	1.00	0.0	2.422	1.928	252.75	-201.15
14	1.00	0.0	1.684	1.677	298.22	-296.90

16 ELEMENT ARRAY

ELEMENT	VO RE	VO IM	YO RE	YO IM	ZO RE	ZO IM
1	1.00	0.0	1.685	1.653	302.36	-296.62
2	1.00	0.0	2.454	1.929	251.89	-197.94
3	1.00	0.0	2.649	1.645	272.42	-169.15
4	1.00	0.0	2.484	1.490	296.06	-177.56
5	1.00	0.0	2.407	1.561	292.51	-189.67
6	1.00	0.0	2.488	1.648	279.39	-185.07
7	1.00	0.0	2.515	1.625	280.49	-181.23
8	1.00	0.0	2.452	1.559	290.39	-184.64
9	1.00	0.0	2.452	1.559	290.39	-184.64
10	1.00	0.0	2.515	1.625	280.49	-181.23
11	1.00	0.0	2.488	1.648	279.39	-185.07
12	1.00	0.0	2.407	1.561	292.51	-189.67
13	1.00	0.0	2.484	1.490	296.06	-177.56
14	1.00	0.0	2.649	1.645	272.42	-169.15
15	1.00	0.0	2.454	1.929	251.89	-197.94
16	1.00	0.0	1.685	1.653	302.36	-296.62

DRIVING POINT ADMITTANCES AND IMPEDANCES
FOR SPECIFIED BASE CURRENTS

H/LMDA=0.5000 B/LMDA= 0.250 OMEGA= 9.92

9 ELEMENT ARRAY

ELEMENT	IZ(O) RE	IZ(O) IM	YO RE	YO IM	ZO RE	ZO IM
1	1.00	0.0	1.450	1.898	415.53	-337.39
2	1.00	0.0	2.878	1.435	327.23	-81.29
3	1.00	0.0	2.395	1.494	378.00	-122.15
4	1.00	0.0	2.453	1.606	360.63	-130.26
5	1.00	0.0	2.524	1.670	347.08	-130.65
6	1.00	0.0	2.453	1.606	360.63	-130.26
7	1.00	0.0	2.395	1.494	378.00	-122.15
8	1.00	0.0	2.878	1.435	327.23	-81.29
9	1.00	0.0	1.450	1.898	415.53	-337.39

10 ELEMENT ARRAY

ELEMENT	IZ(O) RE	IZ(O) IM	YO RE	YO IM	ZO RE	ZO IM
1	1.00	0.0	1.454	1.891	417.28	-335.88
2	1.00	0.0	2.915	1.427	323.96	-78.58
3	1.00	0.0	2.395	1.525	375.11	-126.11
4	1.00	0.0	2.411	1.604	365.60	-133.96
5	1.00	0.0	2.528	1.612	351.75	-124.13
6	1.00	0.0	2.528	1.612	351.75	-124.13
7	1.00	0.0	2.411	1.604	365.60	-133.96
8	1.00	0.0	2.395	1.525	375.11	-126.11
9	1.00	0.0	2.915	1.427	323.96	-78.58
10	1.00	0.0	1.454	1.891	417.28	-335.88

12 ELEMENT ARRAY

ELEMENT	IZ(O) RE	IZ(O) IM	YO RE	YO IM	ZO RE	ZO IM
1	1.00	0.0	1.445	1.893	417.26	-338.49
2	1.00	0.0	2.881	1.402	328.62	-77.81
3	1.00	0.0	2.421	1.503	373.86	-120.96
4	1.00	0.0	2.437	1.639	359.34	-135.46
5	1.00	0.0	2.482	1.645	353.75	-131.75
6	1.00	0.0	2.487	1.555	361.42	-121.30
7	1.00	0.0	2.487	1.555	361.42	-121.30
8	1.00	0.0	2.482	1.645	353.75	-131.75
9	1.00	0.0	2.437	1.639	359.34	-135.46
10	1.00	0.0	2.421	1.503	373.86	-120.96
11	1.00	0.0	2.881	1.402	328.62	-77.81
12	1.00	0.0	1.445	1.893	417.26	-338.49

14 ELEMENT ARRAY

ELEMENT	IZ(O) RE	IZ(O) IM	YO RE	YO IM	ZO RE	ZO IM
1	1.00	0.0	1.452	1.891	417.32	-336.48
2	1.00	0.0	2.907	1.421	325.09	-78.33
3	1.00	0.0	2.402	1.520	374.71	-124.75
4	1.00	0.0	2.418	1.614	363.78	-134.49
5	1.00	0.0	2.513	1.622	352.54	-126.58
6	1.00	0.0	2.515	1.589	355.19	-122.81
7	1.00	0.0	2.444	1.587	363.42	-128.93
8	1.00	0.0	2.444	1.587	363.42	-128.93
9	1.00	0.0	2.515	1.589	355.19	-122.81
10	1.00	0.0	2.513	1.622	352.54	-126.58
11	1.00	0.0	2.418	1.614	363.78	-134.49
12	1.00	0.0	2.402	1.520	374.71	-124.75
13	1.00	0.0	2.907	1.421	325.09	-78.33
14	1.00	0.0	1.452	1.891	417.32	-336.48

16 ELEMENT ARRAY

ELEMENT	IZ(O) RE	IZ(O) IM	YO RE	YO IM	ZO RE	ZO IM
1	1.00	0.0	1.447	1.892	417.24	-338.09
2	1.00	0.0	2.887	1.406	327.90	-77.96
3	1.00	0.0	2.417	1.507	374.09	-121.81
4	1.00	0.0	2.432	1.632	360.42	-135.17
5	1.00	0.0	2.491	1.639	353.37	-130.38
6	1.00	0.0	2.494	1.565	359.62	-121.83
7	1.00	0.0	2.472	1.565	362.21	-123.74
8	1.00	0.0	2.470	1.622	357.19	-130.44
9	1.00	0.0	2.470	1.622	357.19	-130.44
10	1.00	0.0	2.472	1.565	362.21	-123.74
11	1.00	0.0	2.494	1.565	359.62	-121.83
12	1.00	0.0	2.491	1.639	353.37	-130.38
13	1.00	0.0	2.432	1.632	360.42	-135.17
14	1.00	0.0	2.417	1.507	374.09	-121.81
15	1.00	0.0	2.887	1.406	327.90	-77.96
16	1.00	0.0	1.447	1.892	417.24	-338.09

BR

DRIVING POINT ADMITTANCES AND IMPEDANCES
FOR SPECIFIED BASE VOLTAGES

H/LMDA=0.5000 B/LMDA= 0.250 OMEGA= 9.92

18 ELEMENT ARRAY

ELEMENT	VO RE	VO IM	YO RE	YO IM	ZO RE	ZO IM
1	1.00	0.0	1.685	1.673	298.89	-296.81
2	1.00	0.0	2.428	1.928	252.56	-200.61
3	1.00	0.0	2.650	1.618	274.86	-167.88
4	1.00	0.0	2.515	1.491	294.23	-174.42
5	1.00	0.0	2.406	1.594	288.87	-191.40
6	1.00	0.0	2.450	1.646	281.24	-188.96
7	1.00	0.0	2.518	1.583	284.70	-178.95
8	1.00	0.0	2.501	1.563	287.51	-179.66
9	1.00	0.0	2.447	1.616	284.54	-187.98
10	1.00	0.0	2.447	1.616	284.54	-187.98
11	1.00	0.0	2.501	1.563	287.51	-179.66
12	1.00	0.0	2.518	1.583	284.70	-178.95
13	1.00	0.0	2.450	1.646	281.24	-188.96
14	1.00	0.0	2.406	1.594	288.87	-191.40
15	1.00	0.0	2.515	1.491	294.23	-174.42
16	1.00	0.0	2.650	1.618	274.86	-167.88
17	1.00	0.0	2.428	1.928	252.56	-200.61
18	1.00	0.0	1.685	1.673	298.89	-296.81

20 ELEMENT ARRAY

ELEMENT	VO RE	VO IM	YO RE	YO IM	ZO RE	ZO IM
1	1.00	0.0	1.685	1.656	301.86	-296.68
2	1.00	0.0	2.450	1.928	252.02	-198.34
3	1.00	0.0	2.650	1.641	272.80	-168.92
4	1.00	0.0	2.489	1.490	295.71	-177.04
5	1.00	0.0	2.406	1.567	291.86	-190.05
6	1.00	0.0	2.480	1.647	279.81	-185.80
7	1.00	0.0	2.517	1.616	281.35	-180.68
8	1.00	0.0	2.464	1.561	289.61	-183.52
9	1.00	0.0	2.449	1.574	288.98	-185.68
10	1.00	0.0	2.496	1.620	281.88	-183.00
11	1.00	0.0	2.496	1.620	281.88	-183.00
12	1.00	0.0	2.449	1.574	288.98	-185.68
13	1.00	0.0	2.464	1.561	289.61	-183.52
14	1.00	0.0	2.517	1.616	281.35	-180.68
15	1.00	0.0	2.480	1.647	279.81	-185.80
16	1.00	0.0	2.406	1.567	291.86	-190.05
17	1.00	0.0	2.489	1.490	295.71	-177.04
18	1.00	0.0	2.650	1.641	272.80	-168.92
19	1.00	0.0	2.450	1.928	252.02	-198.34
20	1.00	0.0	1.685	1.656	301.86	-296.68

22 ELEMENT ARRAY

ELEMENT	VO RE	VO IM	YO RE	YO IM	ZO RE	ZO IM
1	1.00	0.0	1.685	1.671	299.27	-296.77
2	1.00	0.0	2.431	1.929	252.47	-200.30
3	1.00	0.0	2.650	1.622	274.58	-168.05
4	1.00	0.0	2.511	1.491	294.48	-174.80
5	1.00	0.0	2.406	1.590	289.33	-191.15
6	1.00	0.0	2.455	1.647	280.95	-188.43
7	1.00	0.0	2.517	1.589	284.11	-179.32
8	1.00	0.0	2.494	1.562	287.99	-180.37
9	1.00	0.0	2.448	1.607	285.44	-187.42
10	1.00	0.0	2.458	1.618	283.81	-186.87
11	1.00	0.0	2.498	1.578	286.17	-180.70
12	1.00	0.0	2.498	1.578	286.17	-180.70
13	1.00	0.0	2.458	1.618	283.81	-186.87
14	1.00	0.0	2.448	1.607	285.44	-187.42
15	1.00	0.0	2.494	1.562	287.99	-180.37
16	1.00	0.0	2.517	1.589	284.11	-179.32
17	1.00	0.0	2.455	1.647	280.95	-188.43
18	1.00	0.0	2.406	1.590	289.33	-191.15
19	1.00	0.0	2.511	1.491	294.48	-174.80
20	1.00	0.0	2.650	1.622	274.58	-168.05
21	1.00	0.0	2.431	1.929	252.47	-200.30
22	1.00	0.0	1.685	1.671	299.27	-296.77

24 ELEMENT ARRAY

ELEMENT	VO RE	VO IM	YO RE	YO IM	ZO RE	ZO IM
1	1.00	0.0	1.685	1.658	301.55	-296.71
2	1.00	0.0	2.448	1.928	252.09	-198.58
3	1.00	0.0	2.650	1.638	273.03	-168.80
4	1.00	0.0	2.492	1.491	295.52	-176.74
5	1.00	0.0	2.406	1.570	291.51	-190.23
6	1.00	0.0	2.476	1.647	280.01	-186.19
7	1.00	0.0	2.517	1.612	281.78	-180.43
8	1.00	0.0	2.469	1.562	289.29	-183.00
9	1.00	0.0	2.449	1.580	288.35	-186.06
10	1.00	0.0	2.488	1.619	282.32	-183.72
11	1.00	0.0	2.497	1.611	282.76	-182.44
12	1.00	0.0	2.461	1.576	288.22	-184.57
13	1.00	0.0	2.461	1.576	288.22	-184.57
14	1.00	0.0	2.497	1.611	282.76	-182.44
15	1.00	0.0	2.488	1.619	282.32	-183.72
16	1.00	0.0	2.449	1.580	288.35	-186.06
17	1.00	0.0	2.469	1.562	289.29	-183.00
18	1.00	0.0	2.517	1.612	281.78	-180.43
19	1.00	0.0	2.476	1.647	280.01	-186.19
20	1.00	0.0	2.406	1.570	291.51	-190.23
21	1.00	0.0	2.492	1.491	295.52	-176.74
22	1.00	0.0	2.650	1.638	273.03	-168.80
23	1.00	0.0	2.448	1.928	252.09	-198.58
24	1.00	0.0	1.685	1.658	301.55	-296.71

DRIVING POINT ADMITTANCES AND IMPEDANCES
FOR SPECIFIED BASE CURRENTS

H/LMDA=0.5000 B/LMDA= 0.250 OMEGA= 9.92

18 ELEMENT ARRAY

ELEMENT	IZ(O) RE	IZ(O) IM	YO RE	YO IM	ZO RE	ZO IM
1	1.00	0.0	1.451	1.891	417.32	-336.76
2	1.00	0.0	2.904	1.418	325.59	-78.23
3	1.00	0.0	2.405	1.518	374.56	-124.18
4	1.00	0.0	2.421	1.618	363.09	-134.67
5	1.00	0.0	2.508	1.626	352.77	-127.43
6	1.00	0.0	2.510	1.583	356.26	-122.51
7	1.00	0.0	2.452	1.582	363.03	-127.56
8	1.00	0.0	2.451	1.598	361.61	-129.47
9	1.00	0.0	2.500	1.600	355.98	-125.25
10	1.00	0.0	2.500	1.600	355.98	-125.25
11	1.00	0.0	2.451	1.598	361.61	-129.47
12	1.00	0.0	2.452	1.582	363.03	-127.56
13	1.00	0.0	2.510	1.583	356.26	-122.51
14	1.00	0.0	2.508	1.626	352.77	-127.43
15	1.00	0.0	2.421	1.618	363.09	-134.67
16	1.00	0.0	2.405	1.518	374.56	-124.18
17	1.00	0.0	2.904	1.418	325.59	-78.23
18	1.00	0.0	1.451	1.891	417.32	-336.76

20 ELEMENT ARRAY

ELEMENT	IZ(O) RE	IZ(O) IM	YO RE	YO IM	ZO RE	ZO IM
1	1.00	0.0	1.447	1.892	417.24	-337.89
2	1.00	0.0	2.889	1.408	327.53	-78.03
3	1.00	0.0	2.415	1.509	374.19	-122.21
4	1.00	0.0	2.431	1.630	360.90	-135.06
5	1.00	0.0	2.494	1.637	353.22	-129.82
6	1.00	0.0	2.497	1.569	358.93	-122.01
7	1.00	0.0	2.468	1.568	362.44	-124.59
8	1.00	0.0	2.466	1.616	358.26	-130.15
9	1.00	0.0	2.478	1.617	356.80	-129.07
10	1.00	0.0	2.480	1.575	360.41	-124.27
11	1.00	0.0	2.480	1.575	360.41	-124.27
12	1.00	0.0	2.478	1.617	356.80	-129.07
13	1.00	0.0	2.466	1.616	358.26	-130.15
14	1.00	0.0	2.468	1.568	362.44	-124.59
15	1.00	0.0	2.497	1.569	358.93	-122.01
16	1.00	0.0	2.494	1.637	353.22	-129.82
17	1.00	0.0	2.431	1.630	360.90	-135.06
18	1.00	0.0	2.415	1.509	374.19	-122.21
19	1.00	0.0	2.889	1.408	327.53	-78.03
20	1.00	0.0	1.447	1.892	417.24	-337.89

22 ELEMENT ARRAY

ELEMENT	IZ(O) RE	IZ(O) IM	YO RE	YO IM	ZO RE	ZO IM
1	1.00	0.0	1.451	1.891	417.32	-336.92
2	1.00	0.0	2.902	1.416	325.87	-78.19
3	1.00	0.0	2.407	1.516	374.49	-123.89
4	1.00	0.0	2.422	1.620	362.74	-134.75
5	1.00	0.0	2.505	1.628	352.87	-127.83
6	1.00	0.0	2.508	1.580	356.74	-122.39
7	1.00	0.0	2.455	1.579	362.88	-126.99
8	1.00	0.0	2.454	1.602	360.92	-129.65
9	1.00	0.0	2.495	1.603	356.21	-126.11
10	1.00	0.0	2.495	1.593	357.05	-124.96
11	1.00	0.0	2.459	1.592	361.22	-128.09
12	1.00	0.0	2.459	1.592	361.22	-128.09
13	1.00	0.0	2.495	1.593	357.05	-124.96
14	1.00	0.0	2.495	1.603	356.21	-126.11
15	1.00	0.0	2.454	1.602	360.92	-129.65
16	1.00	0.0	2.455	1.579	362.88	-126.99
17	1.00	0.0	2.508	1.580	356.74	-122.39
18	1.00	0.0	2.505	1.628	352.87	-127.83
19	1.00	0.0	2.422	1.620	362.74	-134.75
20	1.00	0.0	2.407	1.516	374.49	-123.89
21	1.00	0.0	2.902	1.416	325.87	-78.19
22	1.00	0.0	1.451	1.891	417.32	-336.92

24 ELEMENT ARRAY

ELEMENT	IZ(O) RE	IZ(O) IM	YO RE	YO IM	ZO RE	ZO IM
1	1.00	0.0	1.448	1.892	417.25	-337.77
2	1.00	0.0	2.891	1.409	327.32	-78.06
3	1.00	0.0	2.414	1.510	374.24	-122.43
4	1.00	0.0	2.429	1.628	361.16	-135.00
5	1.00	0.0	2.496	1.635	353.15	-129.52
6	1.00	0.0	2.499	1.571	358.58	-122.09
7	1.00	0.0	2.465	1.570	362.53	-125.00
8	1.00	0.0	2.464	1.613	358.73	-130.03
9	1.00	0.0	2.482	1.614	356.66	-128.50
10	1.00	0.0	2.483	1.579	359.71	-124.45
11	1.00	0.0	2.475	1.579	360.63	-125.13
12	1.00	0.0	2.474	1.610	357.87	-128.78
13	1.00	0.0	2.474	1.610	357.87	-128.78
14	1.00	0.0	2.475	1.579	360.63	-125.13
15	1.00	0.0	2.483	1.579	359.71	-124.45
16	1.00	0.0	2.482	1.614	356.66	-128.50
17	1.00	0.0	2.464	1.613	358.73	-130.03
18	1.00	0.0	2.465	1.570	362.53	-125.00
19	1.00	0.0	2.499	1.571	358.58	-122.09
20	1.00	0.0	2.496	1.635	353.15	-129.52
21	1.00	0.0	2.429	1.628	361.16	-135.00
22	1.00	0.0	2.414	1.510	374.24	-122.43
23	1.00	0.0	2.891	1.409	327.32	-78.06
24	1.00	0.0	1.448	1.892	417.25	-337.77

BR

DRIVING POINT ADMITTANCES AND IMPEDANCES
FOR SPECIFIED BASE VOLTAGES

H/LMDA=0.5000 B/LMDA= 0.250 OMEGA= 9.92

26 ELEMENT ARRAY

ELEMENT	VO		YO		ZO	
	RE	IM	RE	IM	RE	IM
1	1.00	0.0	1.685	1.669	299.52	-296.75
2	1.00	0.0	2.433	1.929	252.41	-200.11
3	1.00	0.0	2.649	1.624	274.40	-168.15
4	1.00	0.0	2.509	1.490	294.62	-175.03
5	1.00	0.0	2.406	1.587	289.61	-191.02
6	1.00	0.0	2.458	1.647	280.81	-188.14
7	1.00	0.0	2.517	1.592	283.78	-179.51
8	1.00	0.0	2.490	1.562	288.21	-180.76
9	1.00	0.0	2.448	1.603	285.90	-187.17
10	1.00	0.0	2.463	1.619	283.51	-186.34
11	1.00	0.0	2.498	1.584	285.57	-181.08
12	1.00	0.0	2.491	1.577	286.63	-181.42
13	1.00	0.0	2.459	1.609	284.70	-186.31
14	1.00	0.0	2.459	1.609	284.70	-186.31
15	1.00	0.0	2.491	1.577	286.63	-181.42
16	1.00	0.0	2.498	1.584	285.57	-181.08
17	1.00	0.0	2.463	1.619	283.51	-186.34
18	1.00	0.0	2.448	1.603	285.90	-187.17
19	1.00	0.0	2.490	1.562	288.21	-180.76
20	1.00	0.0	2.517	1.592	283.78	-179.51
21	1.00	0.0	2.458	1.647	280.81	-188.14
22	1.00	0.0	2.406	1.587	289.61	-191.02
23	1.00	0.0	2.509	1.490	294.62	-175.03
24	1.00	0.0	2.649	1.624	274.40	-168.15
25	1.00	0.0	2.433	1.929	252.41	-200.11
26	1.00	0.0	1.685	1.669	299.52	-296.75

28 ELEMENT ARRAY

ELEMENT	VO		YO		ZO	
	RE	IM	RE	IM	RE	IM
1	1.00	0.0	1.685	1.659	301.35	-296.73
2	1.00	0.0	2.446	1.928	252.14	-198.73
3	1.00	0.0	2.650	1.637	273.17	-168.72
4	1.00	0.0	2.494	1.491	295.41	-176.55
5	1.00	0.0	2.406	1.572	291.29	-190.34
6	1.00	0.0	2.474	1.646	280.12	-186.42
7	1.00	0.0	2.517	1.609	282.03	-180.29
8	1.00	0.0	2.472	1.562	289.13	-182.70
9	1.00	0.0	2.448	1.583	288.01	-186.24
10	1.00	0.0	2.484	1.619	282.53	-184.11
11	1.00	0.0	2.497	1.607	283.20	-182.19
12	1.00	0.0	2.466	1.576	287.91	-184.04
13	1.00	0.0	2.460	1.582	287.59	-184.95
14	1.00	0.0	2.490	1.610	283.20	-183.16
15	1.00	0.0	2.490	1.610	283.20	-183.16
16	1.00	0.0	2.460	1.582	287.59	-184.95
17	1.00	0.0	2.466	1.576	287.91	-184.04
18	1.00	0.0	2.497	1.607	283.20	-182.19
19	1.00	0.0	2.484	1.619	282.53	-184.11
20	1.00	0.0	2.448	1.583	288.01	-186.24
21	1.00	0.0	2.472	1.562	289.13	-182.70
22	1.00	0.0	2.517	1.609	282.03	-180.29
23	1.00	0.0	2.474	1.646	280.12	-186.42
24	1.00	0.0	2.406	1.572	291.29	-190.34
25	1.00	0.0	2.494	1.491	295.41	-176.55
26	1.00	0.0	2.650	1.637	273.17	-168.72
27	1.00	0.0	2.446	1.928	252.14	-198.73
28	1.00	0.0	1.685	1.659	301.35	-296.73

30 ELEMENT ARRAY

ELEMENT	VO		YO		ZO	
	RE	IM	RE	IM	RE	IM
1	1.00	0.0	1.685	1.668	299.69	-296.74
2	1.00	0.0	2.434	1.929	252.38	-199.98
3	1.00	0.0	2.649	1.625	274.28	-168.22
4	1.00	0.0	2.507	1.490	294.71	-175.18
5	1.00	0.0	2.406	1.585	289.78	-190.93
6	1.00	0.0	2.460	1.647	280.72	-187.95
7	1.00	0.0	2.517	1.594	283.58	-179.62
8	1.00	0.0	2.488	1.562	288.34	-180.99
9	1.00	0.0	2.449	1.600	286.16	-187.03
10	1.00	0.0	2.466	1.619	283.36	-186.04
11	1.00	0.0	2.497	1.587	285.24	-181.26
12	1.00	0.0	2.487	1.576	286.86	-181.80
13	1.00	0.0	2.460	1.605	285.15	-186.06
14	1.00	0.0	2.464	1.610	284.41	-185.78
15	1.00	0.0	2.490	1.583	286.03	-181.80
16	1.00	0.0	2.490	1.583	286.03	-181.80
17	1.00	0.0	2.464	1.610	284.41	-185.78
18	1.00	0.0	2.460	1.605	285.15	-186.06
19	1.00	0.0	2.487	1.576	286.86	-181.80
20	1.00	0.0	2.497	1.587	285.24	-181.26
21	1.00	0.0	2.466	1.619	283.36	-186.04
22	1.00	0.0	2.449	1.600	286.16	-187.03
23	1.00	0.0	2.488	1.562	288.34	-180.99
24	1.00	0.0	2.517	1.594	283.58	-179.62
25	1.00	0.0	2.460	1.647	280.72	-187.95
26	1.00	0.0	2.406	1.585	289.78	-190.93
27	1.00	0.0	2.507	1.490	294.71	-175.18
28	1.00	0.0	2.649	1.625	274.28	-168.22
29	1.00	0.0	2.434	1.929	252.38	-199.98
30	1.00	0.0	1.685	1.668	299.69	-296.74

DRIVING POINT ADMITTANCES AND IMPEDANCES
FOR SPECIFIED BASE CURRENTS

H/LMDA=0.5000 B/LMDA= 0.250 OMEGA= 9.92

26 ELEMENT ARRAY

ELEMENT	IZ(O)		YO		ZO	
	RE	IM	RE	IM	RE	IM
1	1.00	0.0	1.450	1.891	417.32	-337.02
2	1.00	0.0	2.900	1.415	326.04	-78.16
3	1.00	0.0	2.408	1.515	374.45	-123.71
4	1.00	0.0	2.423	1.621	362.54	-134.80
5	1.00	0.0	2.504	1.629	352.92	-128.06
6	1.00	0.0	2.507	1.579	357.00	-122.34
7	1.00	0.0	2.457	1.578	362.81	-126.69
8	1.00	0.0	2.456	1.603	360.57	-129.73
9	1.00	0.0	2.492	1.605	356.31	-126.51
10	1.00	0.0	2.493	1.590	357.53	-124.84
11	1.00	0.0	2.462	1.590	361.08	-127.52
12	1.00	0.0	2.462	1.596	360.53	-128.27
13	1.00	0.0	2.490	1.597	357.28	-125.81
14	1.00	0.0	2.490	1.597	357.28	-125.81
15	1.00	0.0	2.462	1.596	360.53	-128.27
16	1.00	0.0	2.462	1.590	361.08	-127.52
17	1.00	0.0	2.493	1.590	357.53	-124.84
18	1.00	0.0	2.492	1.605	356.31	-126.51
19	1.00	0.0	2.456	1.603	360.57	-129.73
20	1.00	0.0	2.457	1.578	362.81	-126.69
21	1.00	0.0	2.507	1.579	357.00	-122.34
22	1.00	0.0	2.504	1.629	352.92	-128.06
23	1.00	0.0	2.423	1.621	362.54	-134.80
24	1.00	0.0	2.408	1.515	374.45	-123.71
25	1.00	0.0	2.900	1.415	326.04	-78.16
26	1.00	0.0	1.450	1.891	417.32	-337.02

28 ELEMENT ARRAY

ELEMENT	IZ(O)		YO		ZO	
	RE	IM	RE	IM	RE	IM
1	1.00	0.0	1.448	1.892	417.25	-337.69
2	1.00	0.0	2.892	1.410	327.18	-78.08
3	1.00	0.0	2.413	1.510	374.27	-122.58
4	1.00	0.0	2.429	1.627	361.32	-134.97
5	1.00	0.0	2.497	1.635	353.12	-129.34
6	1.00	0.0	2.500	1.572	358.38	-122.13
7	1.00	0.0	2.464	1.571	362.58	-125.23
8	1.00	0.0	2.463	1.612	359.00	-129.98
9	1.00	0.0	2.483	1.613	356.59	-128.20
10	1.00	0.0	2.484	1.581	359.37	-124.53
11	1.00	0.0	2.473	1.580	360.73	-125.53
12	1.00	0.0	2.472	1.608	358.35	-128.66
13	1.00	0.0	2.477	1.608	357.73	-128.21
14	1.00	0.0	2.478	1.583	359.94	-125.31
15	1.00	0.0	2.478	1.583	359.94	-125.31
16	1.00	0.0	2.477	1.608	357.73	-128.21
17	1.00	0.0	2.472	1.608	358.35	-128.66
18	1.00	0.0	2.473	1.580	360.73	-125.53
19	1.00	0.0	2.484	1.581	359.37	-124.53
20	1.00	0.0	2.483	1.613	356.59	-128.20
21	1.00	0.0	2.463	1.612	359.00	-129.98
22	1.00	0.0	2.464	1.571	362.58	-125.23
23	1.00	0.0	2.500	1.572	358.38	-122.13
24	1.00	0.0	2.497	1.635	353.12	-129.34
25	1.00	0.0	2.429	1.627	361.32	-134.97
26	1.00	0.0	2.413	1.510	374.27	-122.58
27	1.00	0.0	2.892	1.410	327.18	-78.08
28	1.00	0.0	1.448	1.892	417.25	-337.69

30 ELEMENT ARRAY

ELEMENT	IZ(O)		YO		ZO	
	RE	IM	RE	IM	RE	IM
1	1.00	0.0	1.450	1.891	417.32	-337.08
2	1.00	0.0	2.900	1.415	326.15	-78.15
3	1.00	0.0	2.408	1.515	374.42	-123.59
4	1.00	0.0	2.424	1.622	362.41	-134.82
5	1.00	0.0	2.503	1.629	352.95	-128.20
6	1.00	0.0	2.506	1.578	357.17	-122.30
7	1.00	0.0	2.458	1.577	362.77	-126.51
8	1.00	0.0	2.456	1.605	360.37	-129.77
9	1.00	0.0	2.491	1.606	356.36	-126.74
10	1.00	0.0	2.492	1.589	357.79	-124.78
11	1.00	0.0	2.464	1.588	361.01	-127.22
12	1.00	0.0	2.464	1.598	360.18	-128.35
13	1.00	0.0	2.488	1.599	357.38	-126.22
14	1.00	0.0	2.488	1.594	357.76	-125.69
15	1.00	0.0	2.465	1.594	360.38	-127.70
16	1.00	0.0	2.465	1.594	360.38	-127.70
17	1.00	0.0	2.488	1.594	357.76	-125.69
18	1.00	0.0	2.488	1.599	357.38	-126.22
19	1.00	0.0	2.464	1.598	360.18	-128.35
20	1.00	0.0	2.464	1.588	361.01	-127.22
21	1.00	0.0	2.492	1.589	357.79	-124.78
22	1.00	0.0	2.491	1.606	356.36	-126.74
23	1.00	0.0	2.456	1.605	360.37	-129.77
24	1.00	0.0	2.458	1.577	362.77	-126.51
25	1.00	0.0	2.506	1.578	357.17	-122.30
26	1.00	0.0	2.503	1.629	352.95	-128.20
27	1.00	0.0	2.424	1.622	362.41	-134.82
28	1.00	0.0	2.408	1.515	374.42	-123.59
29	1.00	0.0	2.900	1.415	326.15	-78.15
30	1.00	0.0	1.450	1.891	417.32	-337.08

BR

DRIVING POINT ADMITTANCES AND IMPEDANCES
FOR SPECIFIED BASE VOLTAGES

H/LMDA=0.5000 B/LMDA= 0.250 OMEGA= 9.92

35 ELEMENT ARRAY

ELEMENT	VO RE	VO IM	YO RE	YO IM	ZO RE	ZO IM
1	1.00	0.0	1.688	1.664	300.48	-296.09
2	1.00	0.0	2.440	1.933	251.78	-199.47
3	1.00	0.0	2.645	1.631	273.94	-168.88
4	1.00	0.0	2.501	1.486	295.56	-175.57
5	1.00	0.0	2.411	1.579	290.29	-190.09
6	1.00	0.0	2.467	1.652	279.87	-187.43
7	1.00	0.0	2.511	1.601	283.09	-180.51
8	1.00	0.0	2.480	1.556	289.33	-181.52
9	1.00	0.0	2.454	1.592	286.76	-186.01
10	1.00	0.0	2.475	1.625	282.30	-185.41
11	1.00	0.0	2.491	1.596	284.59	-182.40
12	1.00	0.0	2.477	1.569	288.10	-182.50
13	1.00	0.0	2.467	1.594	285.94	-184.75
14	1.00	0.0	2.476	1.618	283.01	-184.93
15	1.00	0.0	2.481	1.595	285.14	-183.31
16	1.00	0.0	2.476	1.573	287.69	-182.77
17	1.00	0.0	2.475	1.595	285.51	-183.99
18	1.00	0.0	2.476	1.616	283.20	-184.82
19	1.00	0.0	2.475	1.595	285.51	-183.99
20	1.00	0.0	2.476	1.573	287.69	-182.77
21	1.00	0.0	2.481	1.595	285.14	-183.31
22	1.00	0.0	2.476	1.618	283.01	-184.93
23	1.00	0.0	2.467	1.594	285.94	-184.75
24	1.00	0.0	2.477	1.569	288.10	-182.50
25	1.00	0.0	2.491	1.596	284.59	-182.40
26	1.00	0.0	2.475	1.625	282.30	-185.41
27	1.00	0.0	2.454	1.592	286.76	-186.01
28	1.00	0.0	2.480	1.556	289.33	-181.52
29	1.00	0.0	2.511	1.601	283.09	-180.51
30	1.00	0.0	2.467	1.652	279.87	-187.43
31	1.00	0.0	2.411	1.579	290.29	-190.09
32	1.00	0.0	2.501	1.486	295.56	-175.57
33	1.00	0.0	2.645	1.631	273.94	-168.88
34	1.00	0.0	2.440	1.933	251.78	-199.47
35	1.00	0.0	1.688	1.664	300.48	-296.09

DRIVING POINT ADMITTANCES AND IMPEDANCES
FOR SPECIFIED BASE CURRENTS

H/LMDA=0.5000 B/LMDA= 0.250 OMEGA= 9.92

35 ELEMENT ARRAY

ELEMENT	IZ(O) RE	IZ(O) IM	YO RE	YO IM	ZO RE	ZO IM
1	1.00	0.0	1.449	1.891	417.51	-337.40
2	1.00	0.0	2.898	1.410	326.62	-77.73
3	1.00	0.0	2.412	1.514	373.97	-123.16
4	1.00	0.0	2.424	1.626	361.94	-135.29
5	1.00	0.0	2.498	1.630	353.45	-128.69
6	1.00	0.0	2.505	1.573	357.68	-121.78
7	1.00	0.0	2.463	1.576	362.23	-125.97
8	1.00	0.0	2.457	1.610	359.79	-130.35
9	1.00	0.0	2.485	1.607	356.97	-127.35
10	1.00	0.0	2.491	1.583	358.45	-124.13
11	1.00	0.0	2.471	1.588	360.31	-126.52
12	1.00	0.0	2.464	1.605	359.43	-129.10
13	1.00	0.0	2.480	1.600	358.17	-127.03
14	1.00	0.0	2.487	1.586	358.65	-124.84
15	1.00	0.0	2.474	1.592	359.46	-126.73
16	1.00	0.0	2.467	1.604	359.32	-128.71
17	1.00	0.0	2.477	1.596	358.85	-126.87
18	1.00	0.0	2.486	1.586	358.70	-125.01
19	1.00	0.0	2.477	1.596	358.85	-126.87
20	1.00	0.0	2.467	1.604	359.32	-128.71
21	1.00	0.0	2.474	1.592	359.46	-126.73
22	1.00	0.0	2.487	1.586	358.65	-124.84
23	1.00	0.0	2.480	1.600	358.17	-127.03
24	1.00	0.0	2.464	1.605	359.43	-129.10
25	1.00	0.0	2.471	1.588	360.31	-126.52
26	1.00	0.0	2.491	1.583	358.45	-124.13
27	1.00	0.0	2.485	1.607	356.97	-127.35
28	1.00	0.0	2.457	1.610	359.79	-130.35
29	1.00	0.0	2.463	1.576	362.23	-125.97
30	1.00	0.0	2.505	1.573	357.68	-121.78
31	1.00	0.0	2.498	1.630	353.45	-128.69
32	1.00	0.0	2.424	1.626	361.94	-135.29
33	1.00	0.0	2.412	1.514	373.97	-123.16
34	1.00	0.0	2.898	1.410	326.62	-77.73
35	1.00	0.0	1.449	1.891	417.51	-337.40

40 ELEMENT ARRAY

ELEMENT	VO RE	VO IM	YO RE	YO IM	ZO RE	ZO IM
1	1.00	0.0	1.685	1.661	301.01	-296.74
2	1.00	0.0	2.444	1.928	252.20	-198.98
3	1.00	[illegible]	2.650	1.634	273.40	-168.60
4	1.00	0.0	2.497	[illegible]	[illegible]	-176.26
5	1.00	0.0	2.406	1.575	290.96	-190.49
6	1.00	0.0	2.471	1.646	280.28	-186.76
7	1.00	0.0	2.517	1.606	282.39	-180.12
8	1.00	0.0	2.476	1.562	288.92	-182.29
9	1.00	0.0	2.448	1.588	287.55	-186.47
10	1.00	0.0	2.480	1.619	282.77	-184.60
11	1.00	0.0	2.498	1.601	283.73	-181.92
12	1.00	0.0	2.472	1.577	287.58	-183.43
13	1.00	0.0	2.459	1.589	286.90	-185.31
14	1.00	0.0	2.482	1.610	283.60	-183.93
15	1.00	0.0	2.490	1.602	284.07	-182.70
16	1.00	0.0	2.470	1.583	287.03	-183.93
17	1.00	0.0	2.465	1.588	286.74	-184.71
18	1.00	0.0	2.484	1.605	283.97	-183.53
19	1.00	0.0	2.486	1.603	284.11	-183.16
20	1.00	0.0	2.468	1.586	286.79	-184.31
21	1.00	0.0	2.468	1.586	286.79	-184.31
22	1.00	0.0	2.486	1.603	284.11	-183.16
23	1.00	0.0	2.484	1.605	283.97	-183.53
24	1.00	0.0	2.465	1.588	286.74	-184.71
25	1.00	0.0	2.470	1.583	287.03	-183.93
26	1.00	0.0	2.490	1.602	284.07	-182.70
27	1.00	0.0	2.482	1.610	283.60	-183.93
28	1.00	0.0	2.459	1.589	286.90	-185.31
29	1.00	0.0	2.472	1.577	287.58	-183.43
30	1.00	0.0	2.498	1.601	283.73	-181.92
31	1.00	0.0	2.480	1.619	282.77	-184.60
32	1.00	0.0	2.448	1.588	287.55	-186.47
33	1.00	0.0	2.476	1.562	288.92	-182.29
34	1.00	0.0	2.517	1.606	282.39	-180.12
35	1.00	0.0	2.471	1.646	280.28	-186.76
36	1.00	0.0	2.406	1.575	290.96	-190.49
37	1.00	0.0	2.497	1.491	295.24	-176.26
38	1.00	0.0	2.650	1.634	273.40	-168.60
39	1.00	0.0	2.444	1.928	252.20	-198.98
40	1.00	0.0	1.685	1.661	301.01	-296.74

40 ELEMENT ARRAY

ELEMENT	IZ(O) RE	IZ(O) IM	YO RE	YO IM	ZO RE	ZO IM
1	1.00	0.0	1.449	1.892	417.26	-337.56
2	1.00	0.0	2.893	1.411	326.96	-78.10
3	1.00	0.0	2.412	1.511	374.31	-122.80
4	1.00	0.0	2.428	1.626	361.56	-134.92
5	1.00	0.0	2.498	1.633	353.07	-129.08
6	1.00	0.0	[illegible]	1.573	358.09	-122.18
7	1.00	0.0	2.462	1.572	[illegible]	-125.54
8	1.00	0.0	2.461	1.610	359.34	-129.91
9	1.00	0.0	2.485	1.611	356.51	-127.82
10	1.00	0.0	2.486	1.583	358.94	-124.62
11	1.00	0.0	2.470	1.583	360.83	-126.00
12	1.00	0.0	2.470	1.605	358.88	-128.55
13	1.00	0.0	2.481	1.605	357.60	-127.61
14	1.00	0.0	2.481	1.586	359.26	-125.46
15	1.00	0.0	2.474	1.586	360.12	-126.08
16	1.00	0.0	2.473	1.603	358.63	-128.00
17	1.00	0.0	2.478	1.603	358.10	-127.61
18	1.00	0.0	2.478	1.587	359.49	-125.83
19	1.00	0.0	2.476	1.587	359.74	-126.02
20	1.00	0.0	2.476	1.602	358.40	-127.74
21	1.00	0.0	2.476	1.602	358.40	-127.74
22	1.00	0.0	2.476	1.587	359.74	-126.02
23	1.00	0.0	2.478	1.587	359.49	-125.83
24	1.00	0.0	2.478	1.603	358.10	-127.61
25	1.00	0.0	2.473	1.603	358.63	-128.00
26	1.00	0.0	2.474	1.586	360.12	-126.08
27	1.00	0.0	2.481	1.586	359.26	-125.46
28	1.00	0.0	2.481	1.605	357.60	-127.61
29	1.00	0.0	2.470	1.605	358.88	-128.55
30	1.00	0.0	2.470	1.583	360.83	-126.00
31	1.00	0.0	2.486	1.583	358.94	-124.62
32	1.00	0.0	2.465	1.611	356.51	-127.82
33	1.00	0.0	2.461	1.610	359.34	-129.91
34	1.00	0.0	2.462	1.572	362.64	-125.54
35	1.00	0.0	2.501	1.573	358.09	-122.18
36	1.00	0.0	2.498	1.633	353.07	-129.08
37	1.00	0.0	2.428	1.626	361.56	-134.92
38	1.00	0.0	2.412	1.511	374.31	-122.80
39	1.00	0.0	2.893	1.411	326.96	-78.10
40	1.00	0.0	1.449	1.892	417.26	-337.56

BR

DRIVING POINT ADMITTANCES AND IMPEDANCES
FOR SPECIFIED BASE VOLTAGES

H/LMDA=0.5000 B/LMDA= 0.250 OMEGA= 9.92

45 ELEMENT ARRAY

ELEMENT	V0 RE	V0 IM	Y0 RE	Y0 IM	Z0 RE	Z0 IM
1	1.00	0.0	1.682	1.664	300.46	-297.19
2	1.00	0.0	2.440	1.925	252.61	-199.33
3	1.00	0.0	2.653	1.630	273.63	-168.15
4	1.00	0.0	2.501	1.494	294.68	-176.00
5	1.00	0.0	2.403	1.579	290.63	-191.05
6	1.00	0.0	2.466	1.643	280.83	-187.09
7	1.00	0.0	2.521	1.601	282.70	-179.55
8	1.00	0.0	2.481	1.566	288.29	-181.96
9	1.00	0.0	2.444	1.593	287.19	-187.10
10	1.00	0.0	2.474	1.615	283.42	-184.97
11	1.00	0.0	2.502	1.596	284.11	-181.24
12	1.00	0.0	2.477	1.581	286.86	-183.03
13	1.00	0.0	2.455	1.595	286.47	-186.06
14	1.00	0.0	2.476	1.605	284.38	-184.38
15	1.00	0.0	2.495	1.595	284.53	-181.88
16	1.00	0.0	2.477	1.588	286.16	-183.44
17	1.00	0.0	2.459	1.595	286.21	-185.63
18	1.00	0.0	2.476	1.600	284.93	-184.08
19	1.00	0.0	2.492	1.595	284.69	-182.16
20	1.00	0.0	2.477	1.592	285.70	-183.68
21	1.00	0.0	2.461	1.595	286.11	-185.46
22	1.00	0.0	2.476	1.596	285.33	-183.87
23	1.00	0.0	2.491	1.595	284.73	-182.23
24	1.00	0.0	2.476	1.596	285.33	-183.87
25	1.00	0.0	2.461	1.595	286.11	-185.46
26	1.00	0.0	2.477	1.592	285.70	-183.68
27	1.00	0.0	2.492	1.595	284.69	-182.16
28	1.00	0.0	2.476	1.600	284.93	-184.08
29	1.00	0.0	2.459	1.595	286.21	-185.63
30	1.00	0.0	2.477	1.588	286.16	-183.44
31	1.00	0.0	2.495	1.595	284.53	-181.88
32	1.00	0.0	2.476	1.605	284.38	-184.38
33	1.00	0.0	2.455	1.595	286.47	-186.06
34	1.00	0.0	2.477	1.581	286.86	-183.03
35	1.00	0.0	2.502	1.596	284.11	-181.24
36	1.00	0.0	2.474	1.615	283.42	-184.97
37	1.00	0.0	2.444	1.593	287.19	-187.10
38	1.00	0.0	2.481	1.566	288.29	-181.96
39	1.00	0.0	2.521	1.601	282.70	-179.55
40	1.00	0.0	2.466	1.643	280.83	-187.09
41	1.00	0.0	2.403	1.579	290.63	-191.05
42	1.00	0.0	2.501	1.494	294.68	-176.00
43	1.00	0.0	2.653	1.630	273.63	-168.15
44	1.00	0.0	2.440	1.925	252.61	-199.33
45	1.00	0.0	1.682	1.664	300.46	-297.19

50 ELEMENT ARRAY

ELEMENT	V0 RE	V0 IM	Y0 RE	Y0 IM	Z0 RE	Z0 IM
1	1.00	0.0	1.685	1.666	300.08	-296.72
2	1.00	0.0	2.437	1.929	252.31	-199.68
3	1.00	0.0	2.649	1.628	274.01	-168.35
4	1.00	0.0	2.504	1.490	294.90	-175.51
5	1.00	0.0	2.406	1.582	290.16	-190.77
6	1.00	0.0	2.463	1.647	280.55	-187.56
7	1.00	0.0	2.517	1.598	283.18	-179.81
8	1.00	0.0	2.484	1.562	288.57	-181.44
9	1.00	0.0	2.449	1.596	286.66	-186.80
10	1.00	0.0	2.471	1.619	283.11	-185.51
11	1.00	0.0	2.497	1.593	284.67	-181.55
12	1.00	0.0	2.481	1.576	287.18	-182.44
13	1.00	0.0	2.460	1.598	285.86	-185.71
14	1.00	0.0	2.472	1.610	284.02	-185.01
15	1.00	0.0	2.490	1.591	285.18	-182.24
16	1.00	0.0	2.481	1.582	286.54	-182.75
17	1.00	0.0	2.465	1.599	285.49	-185.22
18	1.00	0.0	2.472	1.606	284.50	-184.83
19	1.00	0.0	2.486	1.590	285.47	-182.59
20	1.00	0.0	2.481	1.586	286.17	-182.86
21	1.00	0.0	2.468	1.600	285.25	-184.96
22	1.00	0.0	2.471	1.603	284.81	-184.78
23	1.00	0.0	2.484	1.589	285.69	-182.78
24	1.00	0.0	2.482	1.588	285.91	-182.86
25	1.00	0.0	2.470	1.602	285.04	-184.83
26	1.00	0.0	2.470	1.602	285.04	-184.83
27	1.00	0.0	2.482	1.588	285.91	-182.86
28	1.00	0.0	2.484	1.589	285.69	-182.78
29	1.00	0.0	2.471	1.603	284.81	-184.78
30	1.00	0.0	2.468	1.600	285.25	-184.96
31	1.00	0.0	2.481	1.586	286.17	-182.86
32	1.00	0.0	2.486	1.590	285.47	-182.59
33	1.00	0.0	2.472	1.606	284.50	-184.83
34	1.00	0.0	2.465	1.599	285.49	-185.22
35	1.00	0.0	2.481	1.582	286.54	-182.75
36	1.00	0.0	2.490	1.591	285.18	-182.24
37	1.00	0.0	2.472	1.610	284.02	-185.01
38	1.00	0.0	2.460	1.598	285.86	-185.71
39	1.00	0.0	2.481	1.576	287.18	-182.44
40	1.00	0.0	2.497	1.593	284.67	-181.55
41	1.00	0.0	2.471	1.619	283.11	-185.51
42	1.00	0.0	2.449	1.596	286.66	-186.80
43	1.00	0.0	2.484	1.562	288.57	-181.44
44	1.00	0.0	2.517	1.598	283.18	-179.81
45	1.00	0.0	2.463	1.647	280.55	-187.56
46	1.00	0.0	2.406	1.582	290.16	-190.77
47	1.00	0.0	2.504	1.490	294.90	-175.51
48	1.00	0.0	2.649	1.628	274.01	-168.35
49	1.00	0.0	2.437	1.929	252.31	-199.68
50	1.00	0.0	1.685	1.666	300.08	-296.72

DRIVING POINT ADMITTANCES AND IMPEDANCES
FOR SPECIFIED BASE CURRENTS

H/LMDA=0.5000 B/LMDA= 0.250 OMEGA= 9.92

45 ELEMENT ARRAY

ELEMENT	IZ(0) RE	IZ(0) IM	Y0 RE	Y0 IM	Z0 RE	Z0 IM
1	1.00	0.0	1.449	1.892	417.12	-337.35
2	1.00	0.0	2.895	1.415	326.65	-78.39
3	1.00	0.0	2.409	1.512	374.62	-123.08
4	1.00	0.0	2.428	1.623	361.87	-134.61
5	1.00	0.0	2.502	1.633	352.75	-128.77
6	1.00	0.0	2.502	1.577	357.77	-122.52
7	1.00	0.0	2.459	1.573	362.99	-125.88
8	1.00	0.0	2.461	1.606	359.69	-129.55
9	1.00	0.0	2.489	1.611	356.14	-127.46
10	1.00	0.0	2.487	1.587	358.56	-125.01
11	1.00	0.0	2.466	1.583	361.24	-126.40
12	1.00	0.0	2.469	1.601	359.30	-128.12
13	1.00	0.0	2.485	1.605	357.15	-127.17
14	1.00	0.0	2.481	1.591	358.80	-125.92
15	1.00	0.0	2.469	1.587	360.61	-126.57
16	1.00	0.0	2.473	1.598	359.14	-127.48
17	1.00	0.0	2.483	1.602	357.55	-127.07
18	1.00	0.0	2.479	1.593	358.91	-126.41
19	1.00	0.0	2.470	1.588	360.35	-126.63
20	1.00	0.0	2.475	1.596	359.05	-127.08
21	1.00	0.0	2.483	1.602	357.70	-127.03
22	1.00	0.0	2.477	1.595	358.99	-126.76
23	1.00	0.0	2.470	1.588	360.28	-126.65
24	1.00	0.0	2.477	1.595	358.99	-126.76
25	1.00	0.0	2.483	1.602	357.70	-127.03
26	1.00	0.0	2.475	1.596	359.05	-127.08
27	1.00	0.0	2.470	1.588	360.35	-126.63
28	1.00	0.0	2.479	1.593	358.91	-126.41
29	1.00	0.0	2.483	1.602	357.55	-127.07
30	1.00	0.0	2.473	1.598	359.14	-127.48
31	1.00	0.0	2.469	1.587	360.61	-126.57
32	1.00	0.0	2.481	1.591	358.80	-125.92
33	1.00	0.0	2.485	1.605	357.15	-127.17
34	1.00	0.0	2.469	1.601	359.30	-128.12
35	1.00	0.0	2.466	1.583	361.24	-126.40
36	1.00	0.0	2.487	1.587	358.56	-125.01
37	1.00	0.0	2.489	1.611	356.14	-127.46
38	1.00	0.0	2.461	1.606	359.69	-129.55
39	1.00	0.0	2.459	1.573	362.99	-125.88
40	1.00	0.0	2.502	1.577	357.77	-122.52
41	1.00	0.0	2.502	1.633	352.75	-128.77
42	1.00	0.0	2.428	1.623	361.87	-134.61
43	1.00	0.0	2.409	1.512	374.62	-123.08
44	1.00	0.0	2.895	1.415	326.65	-78.39
45	1.00	0.0	1.449	1.892	417.12	-337.35

50 ELEMENT ARRAY

ELEMENT	IZ(0) RE	IZ(0) IM	Y0 RE	Y0 IM	Z0 RE	Z0 IM
1	1.00	0.0	1.450	1.891	417.31	-337.23
2	1.00	0.0	2.898	1.414	326.41	-78.12
3	1.00	0.0	2.410	1.514	374.38	-123.34
4	1.00	0.0	2.425	1.623	362.13	-134.87
5	1.00	0.0	2.501	1.631	353.00	-128.50
6	1.00	0.0	2.504	1.576	357.49	-122.25
7	1.00	0.0	2.459	1.575	362.71	-126.16
8	1.00	0.0	2.458	1.607	359.99	-129.84
9	1.00	0.0	2.489	1.608	356.44	-127.15
10	1.00	0.0	2.490	1.587	358.24	-124.70
11	1.00	0.0	2.467	1.586	360.91	-126.73
12	1.00	0.0	2.466	1.601	359.64	-128.46
13	1.00	0.0	2.485	1.601	357.50	-126.81
14	1.00	0.0	2.485	1.591	358.42	-125.56
15	1.00	0.0	2.469	1.590	360.23	-126.96
16	1.00	0.0	2.469	1.598	359.55	-127.88
17	1.00	0.0	2.483	1.598	357.97	-126.64
18	1.00	0.0	2.483	1.593	358.46	-125.97
19	1.00	0.0	2.470	1.592	359.89	-127.10
20	1.00	0.0	2.470	1.596	359.55	-127.57
21	1.00	0.0	2.482	1.597	358.22	-126.51
22	1.00	0.0	2.482	1.594	358.44	-126.22
23	1.00	0.0	2.471	1.594	359.71	-127.23
24	1.00	0.0	2.471	1.595	359.60	-127.37
25	1.00	0.0	2.482	1.595	358.36	-126.38
26	1.00	0.0	2.482	1.595	358.36	-126.38
27	1.00	0.0	2.471	1.595	359.60	-127.37
28	1.00	0.0	2.471	1.594	359.71	-127.23
29	1.00	0.0	2.482	1.594	358.44	-126.22
30	1.00	0.0	2.482	1.597	358.22	-126.51
31	1.00	0.0	2.470	1.596	359.55	-127.57
32	1.00	0.0	2.470	1.592	359.89	-127.10
33	1.00	0.0	2.483	1.593	358.46	-125.97
34	1.00	0.0	2.483	1.598	357.97	-126.64
35	1.00	0.0	2.469	1.598	359.55	-127.88
36	1.00	0.0	2.469	1.590	360.23	-126.96
37	1.00	0.0	2.485	1.591	358.42	-125.56
38	1.00	0.0	2.485	1.601	357.50	-126.81
39	1.00	0.0	2.466	1.601	359.64	-128.46
40	1.00	0.0	2.467	1.586	360.91	-126.73
41	1.00	0.0	2.490	1.587	358.24	-124.70
42	1.00	0.0	2.489	1.608	356.44	-127.15
43	1.00	0.0	2.458	1.607	359.99	-129.84
44	1.00	0.0	2.459	1.575	362.71	-126.16
45	1.00	0.0	2.504	1.576	357.49	-122.25
46	1.00	0.0	2.501	1.631	353.00	-128.50
47	1.00	0.0	2.425	1.623	362.13	-134.87
48	1.00	0.0	2.410	1.514	374.38	-123.34
49	1.00	0.0	2.898	1.414	326.41	-78.12
50	1.00	0.0	1.450	1.891	417.31	-337.23

BR

DRIVING POINT ADMITTANCES AND IMPEDANCES
FOR SPECIFIED BASE VOLTAGES

H/LMDA=0.5000 B/LMDA= 0.500 OMEGA= 9.92

2 ELEMENT ARRAY

ELEMENT	V0 RE	V0 IM	Y0 RE	Y0 IM	Z0 RE	Z0 IM
1	1.00	0.0	1.135	1.179	423.84	-440.27
2	1.00	0.0	1.135	1.179	423.84	-440.27

3 ELEMENT ARRAY

ELEMENT	V0 RE	V0 IM	Y0 RE	Y0 IM	Z0 RE	Z0 IM
1	1.00	0.0	1.039	1.412	338.01	-459.51
2	1.00	0.0	1.226	0.725	604.30	-357.33
3	1.00	0.0	1.039	1.412	338.01	-459.51

4 ELEMENT ARRAY

ELEMENT	V0 RE	V0 IM	Y0 RE	Y0 IM	Z0 RE	Z0 IM
1	1.00	0.0	1.121	1.266	391.95	-442.77
2	1.00	0.0	1.131	0.945	520.68	-434.89
3	1.00	0.0	1.131	0.945	520.68	-434.89
4	1.00	0.0	1.121	1.266	391.95	-442.77

5 ELEMENT ARRAY

ELEMENT	V0 RE	V0 IM	Y0 RE	Y0 IM	Z0 RE	Z0 IM
1	1.00	0.0	1.061	1.362	355.85	-456.97
2	1.00	0.0	1.202	0.820	567.85	-387.21
3	1.00	0.0	1.048	1.151	432.61	-475.11
4	1.00	0.0	1.202	0.820	567.85	-387.21
5	1.00	0.0	1.061	1.362	355.85	-456.97

6 ELEMENT ARRAY

ELEMENT	V0 RE	V0 IM	Y0 RE	Y0 IM	Z0 RE	Z0 IM
1	1.00	0.0	1.110	1.290	383.37	-445.48
2	1.00	0.0	1.145	0.905	537.42	-424.72
3	1.00	0.0	1.117	1.028	484.72	-445.95
4	1.00	0.0	1.117	1.028	484.72	-445.95
5	1.00	0.0	1.145	0.905	537.42	-424.72
6	1.00	0.0	1.110	1.290	383.37	-445.48

7 ELEMENT ARRAY

ELEMENT	V0 RE	V0 IM	Y0 RE	Y0 IM	Z0 RE	Z0 IM
1	1.00	0.0	1.071	1.346	361.99	-454.82
2	1.00	0.0	1.188	0.843	559.60	-397.11
3	1.00	0.0	1.066	1.108	450.87	-468.86
4	1.00	0.0	1.182	0.909	531.49	-408.75
5	1.00	0.0	1.066	1.108	450.87	-468.86
6	1.00	0.0	1.188	0.843	559.60	-397.11
7	1.00	0.0	1.071	1.346	361.99	-454.82

8 ELEMENT ARRAY

ELEMENT	V0 RE	V0 IM	Y0 RE	Y0 IM	Z0 RE	Z0 IM
1	1.00	0.0	1.104	1.299	379.86	-446.91
2	1.00	0.0	1.153	0.892	542.41	-419.87
3	1.00	0.0	1.107	1.049	476.02	-451.00
4	1.00	0.0	1.131	0.989	501.13	-438.09
5	1.00	0.0	1.131	0.989	501.13	-438.09
6	1.00	0.0	1.107	1.049	476.02	-451.00
7	1.00	0.0	1.153	0.892	542.41	-419.87
8	1.00	0.0	1.104	1.299	379.86	-446.91

DRIVING POINT ADMITTANCES AND IMPEDANCES
FOR SPECIFIED BASE CURRENTS

H/LMDA=0.5000 B/LMDA= 0.500 OMEGA= 9.92

2 ELEMENT ARRAY

ELEMENT	IZ(0) RE	IZ(0) IM	Y0 RE	Y0 IM	Z0 RE	Z0 IM
1	1.00	0.0	1.135	1.179	757.37	-306.21
2	1.00	0.0	1.135	1.179	757.37	-306.21

3 ELEMENT ARRAY

ELEMENT	IZ(0) RE	IZ(0) IM	Y0 RE	Y0 IM	Z0 RE	Z0 IM
1	1.00	0.0	1.214	1.328	658.44	-329.59
2	1.00	0.0	1.004	0.917	959.02	-187.97
3	1.00	0.0	1.214	1.328	658.44	-329.59

4 ELEMENT ARRAY

ELEMENT	IZ(0) RE	IZ(0) IM	Y0 RE	Y0 IM	Z0 RE	Z0 IM
1	1.00	0.0	1.168	1.241	713.86	-318.51
2	1.00	0.0	1.094	0.974	866.84	-201.45
3	1.00	0.0	1.094	0.974	866.84	-201.45
4	1.00	0.0	1.168	1.241	713.86	-318.51

5 ELEMENT ARRAY

ELEMENT	IZ(0) RE	IZ(0) IM	Y0 RE	Y0 IM	Z0 RE	Z0 IM
1	1.00	0.0	1.198	1.292	679.57	-324.60
2	1.00	0.0	1.047	0.950	910.95	-199.84
3	1.00	0.0	1.190	1.041	783.33	-211.49
4	1.00	0.0	1.047	0.950	910.95	-199.84
5	1.00	0.0	1.198	1.292	679.57	-324.60

6 ELEMENT ARRAY

ELEMENT	IZ(0) RE	IZ(0) IM	Y0 RE	Y0 IM	Z0 RE	Z0 IM
1	1.00	0.0	1.176	1.253	705.55	-319.77
2	1.00	0.0	1.078	0.965	882.02	-200.24
3	1.00	0.0	1.137	1.010	825.94	-210.58
4	1.00	0.0	1.137	1.010	825.94	-210.58
5	1.00	0.0	1.078	0.965	882.02	-200.24
6	1.00	0.0	1.176	1.253	705.55	-319.77

7 ELEMENT ARRAY

ELEMENT	IZ(0) RE	IZ(0) IM	Y0 RE	Y0 IM	Z0 RE	Z0 IM
1	1.00	0.0	1.193	1.283	685.71	-323.32
2	1.00	0.0	1.056	0.955	902.33	-200.44
3	1.00	0.0	1.169	1.027	800.16	-209.87
4	1.00	0.0	1.090	0.985	866.20	-210.34
5	1.00	0.0	1.169	1.027	800.16	-209.87
6	1.00	0.0	1.056	0.955	902.33	-200.44
7	1.00	0.0	1.193	1.283	685.71	-323.32

8 ELEMENT ARRAY

ELEMENT	IZ(0) RE	IZ(0) IM	Y0 RE	Y0 IM	Z0 RE	Z0 IM
1	1.00	0.0	1.179	1.258	702.30	-320.24
2	1.00	0.0	1.074	0.963	886.20	-200.23
3	1.00	0.0	1.145	1.015	818.98	-210.70
4	1.00	0.0	1.120	0.999	840.77	-209.51
5	1.00	0.0	1.120	0.999	840.77	-209.51
6	1.00	0.0	1.145	1.015	818.98	-210.70
7	1.00	0.0	1.074	0.963	886.20	-200.23
8	1.00	0.0	1.179	1.258	702.30	-320.24

BR

DRIVING POINT ADMITTANCES AND IMPEDANCES
FOR SPECIFIED BASE VOLTAGES

H/LMDA=0.5000 B/LMDA= 0.500 OMEGA= 9.92

9 ELEMENT ARRAY

ELEMENT	VO		YO		ZO	
	RE	IM	RE	IM	RE	IM
1	1.00	0.0	1.077	1.338	364.90	-453.59
2	1.00	0.0	1.182	0.853	556.46	-401.48
3	1.00	0.0	1.074	1.095	456.71	-465.30
4	1.00	0.0	1.170	0.932	523.10	-416.45
5	1.00	0.0	1.082	1.067	468.72	-462.01
6	1.00	0.0	1.170	0.932	523.10	-416.45
7	1.00	0.0	1.074	1.095	456.71	-465.30
8	1.00	0.0	1.182	0.853	556.46	-401.48
9	1.00	0.0	1.077	1.338	364.90	-453.59

10 ELEMENT ARRAY

ELEMENT	VO		YO		ZO	
	RE	IM	RE	IM	RE	IM
1	1.00	0.0	1.101	1.304	378.00	-447.73
2	1.00	0.0	1.157	0.887	544.61	-417.38
3	1.00	0.0	1.102	1.057	472.66	-453.38
4	1.00	0.0	1.138	0.976	506.09	-434.17
5	1.00	0.0	1.121	1.010	492.56	-443.65
6	1.00	0.0	1.121	1.010	492.56	-443.65
7	1.00	0.0	1.138	0.976	506.09	-434.17
8	1.00	0.0	1.102	1.057	472.66	-453.38
9	1.00	0.0	1.157	0.887	544.61	-417.38
10	1.00	0.0	1.101	1.304	378.00	-447.73

12 ELEMENT ARRAY

ELEMENT	VO		YO		ZO	
	RE	IM	RE	IM	RE	IM
1	1.00	0.0	1.099	1.307	376.85	-448.24
2	1.00	0.0	1.159	0.883	545.84	-415.90
3	1.00	0.0	1.099	1.061	470.93	-454.66
4	1.00	0.0	1.142	0.971	508.28	-432.17
5	1.00	0.0	1.116	1.018	489.30	-446.19
6	1.00	0.0	1.128	0.997	497.58	-439.93
7	1.00	0.0	1.128	0.997	497.58	-439.93
8	1.00	0.0	1.116	1.018	489.30	-446.19
9	1.00	0.0	1.142	0.971	508.28	-432.17
10	1.00	0.0	1.099	1.061	470.93	-454.66
11	1.00	0.0	1.159	0.883	545.84	-415.90
12	1.00	0.0	1.099	1.307	376.85	-448.24

14 ELEMENT ARRAY

ELEMENT	VO		YO		ZO	
	RE	IM	RE	IM	RE	IM
1	1.00	0.0	1.098	1.309	376.07	-448.58
2	1.00	0.0	1.161	0.881	546.62	-414.94
3	1.00	0.0	1.097	1.064	469.87	-455.46
4	1.00	0.0	1.144	0.968	509.49	-430.99
5	1.00	0.0	1.113	1.022	487.64	-447.54
6	1.00	0.0	1.132	0.992	499.79	-438.03
7	1.00	0.0	1.123	1.005	494.36	-442.53
8	1.00	0.0	1.123	1.005	494.36	-442.53
9	1.00	0.0	1.132	0.992	499.79	-438.03
10	1.00	0.0	1.113	1.022	487.64	-447.54
11	1.00	0.0	1.144	0.968	509.49	-430.99
12	1.00	0.0	1.097	1.064	469.87	-455.46
13	1.00	0.0	1.161	0.881	546.62	-414.94
14	1.00	0.0	1.098	1.309	376.07	-448.58

16 ELEMENT ARRAY

ELEMENT	VO		YO		ZO	
	RE	IM	RE	IM	RE	IM
1	1.00	0.0	1.097	1.311	375.51	-448.83
2	1.00	0.0	1.162	0.880	547.16	-414.27
3	1.00	0.0	1.096	1.065	469.17	-455.99
4	1.00	0.0	1.145	0.966	510.26	-430.23
5	1.00	0.0	1.111	1.024	486.63	-448.37
6	1.00	0.0	1.134	0.989	501.02	-436.92
7	1.00	0.0	1.120	1.009	492.72	-443.91
8	1.00	0.0	1.127	1.000	496.59	-440.66
9	1.00	0.0	1.127	1.000	496.59	-440.66
10	1.00	0.0	1.120	1.009	492.72	-443.91
11	1.00	0.0	1.134	0.989	501.02	-436.92
12	1.00	0.0	1.111	1.024	486.63	-448.37
13	1.00	0.0	1.145	0.966	510.26	-430.23
14	1.00	0.0	1.096	1.065	469.17	-455.99
15	1.00	0.0	1.162	0.880	547.16	-414.27
16	1.00	0.0	1.097	1.311	375.51	-448.83

DRIVING POINT ADMITTANCES AND IMPEDANCES
FOR SPECIFIED BASE CURRENTS

H/LMDA=0.5000 B/LMDA= 0.500 OMEGA= 9.92

9 ELEMENT ARRAY

ELEMENT	IZ(O)		YO		ZO	
	RE	IM	RE	IM	RE	IM
1	1.00	0.0	1.191	1.278	688.57	-322.73
2	1.00	0.0	1.060	0.956	899.05	-200.55
3	1.00	0.0	1.164	1.023	804.77	-209.80
4	1.00	0.0	1.099	0.990	858.45	-210.63
5	1.00	0.0	1.150	1.014	816.31	-208.46
6	1.00	0.0	1.099	0.990	858.45	-210.63
7	1.00	0.0	1.164	1.023	804.77	-209.80
8	1.00	0.0	1.060	0.956	899.05	-200.55
9	1.00	0.0	1.191	1.278	688.57	-322.73

10 ELEMENT ARRAY

ELEMENT	IZ(O)		YO		ZO	
	RE	IM	RE	IM	RE	IM
1	1.00	0.0	1.180	1.260	700.58	-320.49
2	1.00	0.0	1.072	0.962	888.10	-200.27
3	1.00	0.0	1.148	1.016	816.44	-210.60
4	1.00	0.0	1.115	0.997	844.67	-209.59
5	1.00	0.0	1.128	1.004	833.88	-209.61
6	1.00	0.0	1.128	1.004	833.88	-209.61
7	1.00	0.0	1.115	0.997	844.67	-209.59
8	1.00	0.0	1.148	1.016	816.44	-210.60
9	1.00	0.0	1.072	0.962	888.10	-200.27
10	1.00	0.0	1.180	1.260	700.58	-320.49

12 ELEMENT ARRAY

ELEMENT	IZ(O)		YO		ZO	
	RE	IM	RE	IM	RE	IM
1	1.00	0.0	1.181	1.261	699.53	-320.64
2	1.00	0.0	1.070	0.961	889.18	-200.31
3	1.00	0.0	1.150	1.017	815.14	-210.52
4	1.00	0.0	1.113	0.996	846.40	-209.70
5	1.00	0.0	1.131	1.005	831.39	-209.50
6	1.00	0.0	1.123	1.001	837.76	-209.70
7	1.00	0.0	1.123	1.001	837.76	-209.70
8	1.00	0.0	1.131	1.005	831.39	-209.50
9	1.00	0.0	1.113	0.996	846.40	-209.70
10	1.00	0.0	1.150	1.017	815.14	-210.52
11	1.00	0.0	1.070	0.961	889.18	-200.31
12	1.00	0.0	1.181	1.261	699.53	-320.64

14 ELEMENT ARRAY

ELEMENT	IZ(O)		YO		ZO	
	RE	IM	RE	IM	RE	IM
1	1.00	0.0	1.182	1.263	698.82	-320.75
2	1.00	0.0	1.070	0.961	889.86	-200.34
3	1.00	0.0	1.151	1.017	814.37	-210.46
4	1.00	0.0	1.112	0.995	847.35	-209.78
5	1.00	0.0	1.133	1.006	830.15	-209.40
6	1.00	0.0	1.121	1.000	839.46	-209.81
7	1.00	0.0	1.126	1.003	835.29	-209.58
8	1.00	0.0	1.126	1.003	835.29	-209.58
9	1.00	0.0	1.121	1.000	839.46	-209.81
10	1.00	0.0	1.133	1.006	830.15	-209.40
11	1.00	0.0	1.112	0.995	847.35	-209.78
12	1.00	0.0	1.151	1.017	814.37	-210.46
13	1.00	0.0	1.070	0.961	889.86	-200.34
14	1.00	0.0	1.182	1.263	698.82	-320.75

16 ELEMENT ARRAY

ELEMENT	IZ(O)		YO		ZO	
	RE	IM	RE	IM	RE	IM
1	1.00	0.0	1.182	1.263	698.32	-320.83
2	1.00	0.0	1.069	0.961	890.33	-200.36
3	1.00	0.0	1.152	1.018	813.86	-210.42
4	1.00	0.0	1.111	0.995	847.95	-209.84
5	1.00	0.0	1.133	1.006	829.42	-209.33
6	1.00	0.0	1.120	1.000	840.40	-209.89
7	1.00	0.0	1.128	1.003	834.06	-209.48
8	1.00	0.0	1.124	1.002	836.99	-209.69
9	1.00	0.0	1.124	1.002	836.99	-209.69
10	1.00	0.0	1.128	1.003	834.06	-209.48
11	1.00	0.0	1.120	1.000	840.40	-209.89
12	1.00	0.0	1.133	1.006	829.42	-209.33
13	1.00	0.0	1.111	0.995	847.95	-209.84
14	1.00	0.0	1.152	1.018	813.86	-210.42
15	1.00	0.0	1.069	0.961	890.33	-200.36
16	1.00	0.0	1.182	1.263	698.32	-320.83

BR

DRIVING POINT ADMITTANCES AND IMPEDANCES
FOR SPECIFIED BASE VOLTAGES

H/LMDA=0.5000 B/LMDA= 0.500 OMEGA= 9.92

18 ELEMENT ARRAY

ELEMENT	VO		YO		ZO	
	RE	IM	RE	IM	RE	IM
1	1.00	0.0	1.096	1.312	375.10	-449.02
2	1.00	0.0	1.162	0.878	547.55	-413.78
3	1.00	0.0	1.095	1.066	468.67	-456.37
4	1.00	0.0	1.146	0.964	510.79	-429.70
5	1.00	0.0	1.110	1.026	485.97	-448.92
6	1.00	0.0	1.135	0.987	501.79	-436.21
7	1.00	0.0	1.119	1.012	491.73	-444.74
8	1.00	0.0	1.129	0.997	497.82	-439.58
9	1.00	0.0	1.124	1.004	494.96	-442.05
10	1.00	0.0	1.124	1.004	494.96	-442.05
11	1.00	0.0	1.129	0.997	497.82	-439.58
12	1.00	0.0	1.119	1.012	491.73	-444.74
13	1.00	0.0	1.135	0.987	501.79	-436.21
14	1.00	0.0	1.110	1.026	485.97	-448.92
15	1.00	0.0	1.146	0.964	510.79	-429.70
16	1.00	0.0	1.095	1.066	468.67	-456.37
17	1.00	0.0	1.162	0.878	547.55	-413.78
18	1.00	0.0	1.096	1.312	375.10	-449.02

20 ELEMENT ARRAY

ELEMENT	VO		YO		ZO	
	RE	IM	RE	IM	RE	IM
1	1.00	0.0	1.095	1.313	374.77	-449.16
2	1.00	0.0	1.163	0.878	547.85	-413.40
3	1.00	0.0	1.095	1.067	468.30	-456.65
4	1.00	0.0	1.147	0.963	511.17	-429.32
5	1.00	0.0	1.109	1.027	485.49	-449.31
6	1.00	0.0	1.136	0.985	502.32	-435.72
7	1.00	0.0	1.117	1.013	491.08	-445.30
8	1.00	0.0	1.130	0.995	498.60	-438.88
9	1.00	0.0	1.122	1.006	493.98	-442.90
10	1.00	0.0	1.126	1.001	496.20	-440.98
11	1.00	0.0	1.126	1.001	496.20	-440.98
12	1.00	0.0	1.122	1.006	493.98	-442.90
13	1.00	0.0	1.130	0.995	498.60	-438.88
14	1.00	0.0	1.117	1.013	491.08	-445.30
15	1.00	0.0	1.136	0.985	502.32	-435.72
16	1.00	0.0	1.109	1.027	485.49	-449.31
17	1.00	0.0	1.147	0.963	511.17	-429.32
18	1.00	0.0	1.095	1.067	468.30	-456.65
19	1.00	0.0	1.163	0.878	547.85	-413.40
20	1.00	0.0	1.095	1.313	374.77	-449.16

[illegible] ELEMENT ARRAY

ELEMENT	VO		YO		ZO	
	RE	IM	RE	IM	RE	IM
1	1.00	0.0	1.095	1.313	374.51	-449.28
2	1.00	0.0	1.164	0.877	548.08	-413.10
3	1.00	0.0	1.094	1.068	468.01	-456.87
4	1.00	0.0	1.148	0.963	511.46	-429.02
5	1.00	0.0	1.109	1.028	485.14	-449.60
6	1.00	0.0	1.137	0.984	502.71	-435.36
7	1.00	0.0	1.117	1.014	490.62	-445.69
8	1.00	0.0	1.131	0.993	499.13	-438.40
9	1.00	0.0	1.121	1.008	493.34	-443.46
10	1.00	0.0	1.127	0.999	496.98	-440.30
11	1.00	0.0	1.124	1.003	495.23	-441.83
12	1.00	0.0	1.124	1.003	495.23	-441.83
13	1.00	0.0	1.127	0.999	496.98	-440.30
14	1.00	0.0	1.121	1.008	493.34	-443.46
15	1.00	0.0	1.131	0.993	499.13	-438.40
16	1.00	0.0	1.117	1.014	490.62	-445.69
17	1.00	0.0	1.137	0.984	502.71	-435.36
18	1.00	0.0	1.109	1.028	485.14	-449.60
19	1.00	0.0	1.148	0.963	511.46	-429.02
20	1.00	0.0	1.094	1.068	468.01	-456.87
21	1.00	0.0	1.164	0.877	548.08	-413.10
22	1.00	0.0	1.095	1.313	374.51	-449.28

24 ELEMENT ARRAY

ELEMENT	VO		YO		ZO	
	RE	IM	RE	IM	RE	IM
1	1.00	0.0	1.094	1.314	374.30	-449.37
2	1.00	0.0	1.164	0.876	548.27	-412.86
3	1.00	0.0	1.094	1.069	467.78	-457.04
4	1.00	0.0	1.148	0.962	511.69	-428.79
5	1.00	0.0	1.108	1.028	484.87	-449.82
6	1.00	0.0	1.137	0.984	503.00	-435.09
7	1.00	0.0	1.116	1.015	490.27	-445.98
8	1.00	0.0	1.132	0.992	499.52	-438.06
9	1.00	0.0	1.120	1.009	492.88	-443.85
10	1.00	0.0	1.128	0.997	497.51	-439.82
11	1.00	0.0	1.123	1.005	494.58	-442.39
12	1.00	0.0	1.126	1.001	496.01	-441.15
13	1.00	0.0	1.126	1.001	496.01	-441.15
14	1.00	0.0	1.123	1.005	494.58	-442.39
15	1.00	0.0	1.128	0.997	497.51	-439.82
16	1.00	0.0	1.120	1.009	492.88	-443.85
17	1.00	0.0	1.132	0.992	499.52	-438.06
18	1.00	0.0	1.116	1.015	490.27	-445.98
19	1.00	0.0	1.137	0.984	503.00	-435.09
20	1.00	0.0	1.108	1.028	484.87	-449.82
21	1.00	0.0	1.148	0.962	511.69	-428.79
22	1.00	0.0	1.094	1.069	467.78	-457.04
23	1.00	0.0	1.164	0.876	548.27	-412.86
24	1.00	0.0	1.094	1.314	374.30	-449.37

DRIVING POINT ADMITTANCES AND IMPEDANCES
FOR SPECIFIED BASE CURRENTS

H/LMDA=0.5000 B/LMDA= 0.500 OMEGA= 9.92

18 ELEMENT ARRAY

ELEMENT	IZ(0)		YO		ZO	
	RE	IM	RE	IM	RE	IM
1	1.00	0.0	1.183	1.264	697.94	-320.89
2	1.00	0.0	1.069	0.960	890.67	-200.38
3	1.00	0.0	1.152	1.018	813.51	-210.38
4	1.00	0.0	1.111	0.995	848.35	-209.88
5	1.00	0.0	1.134	1.006	828.95	-209.28
6	1.00	0.0	1.119	0.999	840.97	-209.95
7	1.00	0.0	1.129	1.004	833.34	-209.41
8	1.00	0.0	1.123	1.001	837.91	-209.78
9	1.00	0.0	1.126	1.002	835.76	-209.59
10	1.00	0.0	1.126	1.002	835.76	-209.59
11	1.00	0.0	1.123	1.001	837.91	-209.78
12	1.00	0.0	1.129	1.004	833.34	-209.41
13	1.00	0.0	1.119	0.999	840.97	-209.95
14	1.00	0.0	1.134	1.006	828.95	-209.28
15	1.00	0.0	1.111	0.995	848.35	-209.88
16	1.00	0.0	1.152	1.018	813.51	-210.38
17	1.00	0.0	1.069	0.960	890.67	-200.38
18	1.00	0.0	1.183	1.264	697.94	-320.89

20 ELEMENT ARRAY

ELEMENT	IZ(0)		YO		ZO	
	RE	IM	RE	IM	RE	IM
1	1.00	0.0	1.183	1.264	697.65	-320.93
2	1.00	0.0	1.068	0.960	890.93	-200.39
3	1.00	0.0	1.153	1.018	813.25	-210.35
4	1.00	0.0	1.110	0.995	848.64	-209.91
5	1.00	0.0	1.134	1.006	828.61	-209.24
6	1.00	0.0	1.119	0.999	841.36	-210.00
7	1.00	0.0	1.129	1.004	832.87	-209.36
8	1.00	0.0	1.122	1.001	838.49	-209.84
9	1.00	0.0	1.127	1.003	835.04	-209.52
10	1.00	0.0	1.125	1.002	836.68	-209.68
11	1.00	0.0	1.125	1.002	836.68	-209.68
12	1.00	0.0	1.127	1.003	835.04	-209.52
13	1.00	0.0	1.122	1.001	838.49	-209.84
14	1.00	0.0	1.129	1.004	832.87	-209.36
15	1.00	0.0	1.119	0.999	841.36	-210.00
16	1.00	0.0	1.134	1.006	828.61	-209.24
17	1.00	0.0	1.110	0.995	848.64	-209.91
18	1.00	0.0	1.153	1.018	813.25	-210.35
19	1.00	0.0	1.068	0.960	890.93	-200.39
20	1.00	0.0	1.183	1.264	697.65	-320.93

22 ELEMENT ARRAY

ELEMENT	IZ(0)		YO		ZO	
	RE	IM	RE	IM	RE	IM
1	1.00	0.0	1.183	1.265	697.42	-320.97
2	1.00	0.0	1.068	0.960	891.13	-200.41
3	1.00	0.0	1.153	1.018	813.04	-210.33
4	1.00	0.0	1.110	0.995	848.86	-209.94
5	1.00	0.0	1.135	1.007	828.37	-209.21
6	1.00	0.0	1.118	0.999	841.64	-210.04
7	1.00	0.0	1.130	1.004	832.55	-209.31
8	1.00	0.0	1.122	1.001	838.87	-209.89
9	1.00	0.0	1.127	1.003	834.58	-209.47
10	1.00	0.0	1.124	1.002	837.25	-209.74
11	1.00	0.0	1.125	1.002	835.97	-209.61
12	1.00	0.0	1.125	1.002	835.97	-209.61
13	1.00	0.0	1.124	1.002	837.25	-209.74
14	1.00	0.0	1.127	1.003	834.58	-209.47
15	1.00	0.0	1.122	1.001	838.87	-209.89
16	1.00	0.0	1.130	1.004	832.55	-209.31
17	1.00	0.0	1.118	0.999	841.64	-210.04
18	1.00	0.0	1.135	1.007	828.37	-209.21
19	1.00	0.0	1.110	0.995	848.86	-209.94
20	1.00	0.0	1.153	1.018	813.04	-210.33
21	1.00	0.0	1.068	0.960	891.13	-200.41
22	1.00	0.0	1.183	1.265	697.42	-320.97

24 ELEMENT ARRAY

ELEMENT	IZ(0)		YO		ZO	
	RE	IM	RE	IM	RE	IM
1	1.00	0.0	1.183	1.265	697.23	-321.00
2	1.00	0.0	1.068	0.960	891.29	-200.41
3	1.00	0.0	1.153	1.018	812.88	-210.31
4	1.00	0.0	1.110	0.994	849.03	-209.96
5	1.00	0.0	1.135	1.007	828.19	-209.18
6	1.00	0.0	1.118	0.999	841.85	-210.07
7	1.00	0.0	1.130	1.004	832.31	-209.28
8	1.00	0.0	1.122	1.001	839.14	-209.92
9	1.00	0.0	1.128	1.003	834.26	-209.42
10	1.00	0.0	1.123	1.001	837.64	-209.79
11	1.00	0.0	1.126	1.002	835.50	-209.55
12	1.00	0.0	1.125	1.002	836.54	-209.67
13	1.00	0.0	1.125	1.002	836.54	-209.67
14	1.00	0.0	1.126	1.002	835.50	-209.55
15	1.00	0.0	1.123	1.001	837.64	-209.79
16	1.00	0.0	1.128	1.003	834.26	-209.42
17	1.00	0.0	1.122	1.001	839.14	-209.92
18	1.00	0.0	1.130	1.004	832.31	-209.28
19	1.00	0.0	1.118	0.999	841.85	-210.07
20	1.00	0.0	1.135	1.007	828.19	-209.18
21	1.00	0.0	1.110	0.994	849.03	-209.96
22	1.00	0.0	1.153	1.018	812.88	-210.31
23	1.00	0.0	1.068	0.960	891.29	-200.41
24	1.00	0.0	1.183	1.265	697.23	-321.00

BR

DRIVING POINT ADMITTANCES AND IMPEDANCES
FOR SPECIFIED BASE VOLTAGES

H/LMDA=0.5000 B/LMDA= 0.500 OMEGA= 9.92

26 ELEMENT ARRAY

ELEMENT	VO RE	VO IM	YO RE	YO IM	ZO RE	ZO IM
1	1.00	0.0	1.094	1.314	374.12	-449.45
2	1.00	0.0	1.164	0.876	548.42	-412.66
3	1.00	0.0	1.093	1.069	467.60	-457.18
4	1.00	0.0	1.148	0.962	511.87	-428.61
5	1.00	0.0	1.108	1.029	484.66	-450.00
6	1.00	0.0	1.138	0.983	503.23	-434.87
7	1.00	0.0	1.116	1.016	490.01	-446.20
8	1.00	0.0	1.132	0.992	499.81	-437.79
9	1.00	0.0	1.120	1.010	492.54	-444.14
10	1.00	0.0	1.129	0.996	497.90	-439.48
11	1.00	0.0	1.122	1.006	494.13	-442.78
12	1.00	0.0	1.127	1.000	496.54	-440.68
13	1.00	0.0	1.125	1.003	495.37	-441.71
14	1.00	0.0	1.125	1.003	495.37	-441.71
15	1.00	0.0	1.127	1.000	496.54	-440.68
16	1.00	0.0	1.122	1.006	494.13	-442.78
17	1.00	0.0	1.129	0.996	497.90	-439.48
18	1.00	0.0	1.120	1.010	492.54	-444.14
19	1.00	0.0	1.132	0.992	499.81	-437.79
20	1.00	0.0	1.116	1.016	490.01	-446.20
21	1.00	0.0	1.138	0.983	503.23	-434.87
22	1.00	0.0	1.108	1.029	484.66	-450.00
23	1.00	0.0	1.148	0.962	511.87	-428.61
24	1.00	0.0	1.093	1.069	467.60	-457.18
25	1.00	0.0	1.164	0.876	548.42	-412.66
26	1.00	0.0	1.094	1.314	374.12	-449.45

28 ELEMENT ARRAY

ELEMENT	VO RE	VO IM	YO RE	YO IM	ZO RE	ZO IM
1	1.00	0.0	1.094	1.315	373.97	-449.51
2	1.00	0.0	1.164	0.876	548.55	-412.50
3	1.00	0.0	1.093	1.069	467.44	-457.30
4	1.00	0.0	1.149	0.961	512.02	-428.46
5	1.00	0.0	1.108	1.029	484.48	-450.14
6	1.00	0.0	1.138	0.983	503.41	-434.70
7	1.00	0.0	1.115	1.016	489.80	-446.38
8	1.00	0.0	1.133	0.991	500.04	-437.59
9	1.00	0.0	1.119	1.010	492.28	-444.36
10	1.00	0.0	1.129	0.996	498.19	-439.22
11	1.00	0.0	1.122	1.007	493.79	-443.07
12	1.00	0.0	1.127	0.999	496.93	-440.33
13	1.00	0.0	1.124	1.004	494.91	-442.10
14	1.00	0.0	1.125	1.001	495.90	-441.24
15	1.00	0.0	1.125	1.001	495.90	-441.24
16	1.00	0.0	1.124	1.004	494.91	-442.10
17	1.00	0.0	1.127	0.999	496.93	-440.33
18	1.00	0.0	1.122	1.007	493.79	-443.07
19	1.00	0.0	1.129	0.996	498.19	-439.22
20	1.00	0.0	1.119	1.010	492.28	-444.36
21	1.00	0.0	1.133	0.991	500.04	-437.59
22	1.00	0.0	1.115	1.016	489.80	-446.38
23	1.00	0.0	1.138	0.983	503.41	-434.70
24	1.00	0.0	1.108	1.029	484.48	-450.14
25	1.00	0.0	1.149	0.961	512.02	-428.46
26	1.00	0.0	1.093	1.069	467.44	-457.30
27	1.00	0.0	1.164	0.876	548.55	-412.50
28	1.00	0.0	1.094	1.315	373.97	-449.51

30 ELEMENT ARRAY

ELEMENT	VO RE	VO IM	YO RE	YO IM	ZO RE	ZO IM
1	1.00	0.0	1.094	1.315	373.85	-449.57
2	1.00	0.0	1.165	0.875	548.66	-412.36
3	1.00	0.0	1.093	1.070	467.31	-457.40
4	1.00	0.0	1.149	0.961	512.15	-428.33
5	1.00	0.0	1.108	1.030	484.34	-450.26
6	1.00	0.0	1.138	0.982	503.56	-434.57
7	1.00	0.0	1.115	1.017	489.63	-446.52
8	1.00	0.0	1.133	0.991	500.22	-437.42
9	1.00	0.0	1.119	1.011	492.07	-444.54
10	1.00	0.0	1.130	0.995	498.42	-439.02
11	1.00	0.0	1.121	1.007	493.53	-443.30
12	1.00	0.0	1.128	0.998	497.22	-440.08
13	1.00	0.0	1.123	1.005	494.58	-442.39
14	1.00	0.0	1.126	1.000	496.29	-440.90
15	1.00	0.0	1.125	1.003	495.45	-441.64
16	1.00	0.0	1.125	1.003	495.45	-441.64
17	1.00	0.0	1.126	1.000	496.29	-440.90
18	1.00	0.0	1.123	1.005	494.58	-442.39
19	1.00	0.0	1.128	0.998	497.22	-440.08
20	1.00	0.0	1.121	1.007	493.53	-443.30
21	1.00	0.0	1.130	0.995	498.42	-439.02
22	1.00	0.0	1.119	1.011	492.07	-444.54
23	1.00	0.0	1.133	0.991	500.22	-437.42
24	1.00	0.0	1.115	1.017	489.63	-446.52
25	1.00	0.0	1.138	0.982	503.56	-434.57
26	1.00	0.0	1.108	1.030	484.34	-450.26
27	1.00	0.0	1.149	0.961	512.15	-428.33
28	1.00	0.0	1.093	1.070	467.31	-457.40
29	1.00	0.0	1.165	0.875	548.66	-412.36
30	1.00	0.0	1.094	1.315	373.85	-449.57

DRIVING POINT ADMITTANCES AND IMPEDANCES
FOR SPECIFIED BASE CURRENTS

H/LMDA=0.5000 B/LMDA= 0.500 OMEGA= 9.92

26 ELEMENT ARRAY

ELEMENT	IZ(0) RE	IZ(0) IM	YO RE	YO IM	ZO RE	ZO IM
1	1.00	0.0	1.184	1.265	697.07	-321.03
2	1.00	0.0	1.068	0.960	891.42	-200.42
3	1.00	0.0	1.153	1.018	812.76	-210.30
4	1.00	0.0	1.110	0.994	849.16	-209.98
5	1.00	0.0	1.135	1.007	828.04	-209.16
6	1.00	0.0	1.118	0.999	842.01	-210.09
7	1.00	0.0	1.130	1.004	832.13	-209.25
8	1.00	0.0	1.121	1.000	839.35	-209.95
9	1.00	0.0	1.128	1.003	834.02	-209.39
10	1.00	0.0	1.123	1.001	837.91	-209.83
11	1.00	0.0	1.126	1.003	835.18	-209.51
12	1.00	0.0	1.124	1.002	836.92	-209.72
13	1.00	0.0	1.125	1.002	836.07	-209.61
14	1.00	0.0	1.125	1.002	836.07	-209.61
15	1.00	0.0	1.124	1.002	836.92	-209.72
16	1.00	0.0	1.126	1.003	835.18	-209.51
17	1.00	0.0	1.123	1.001	837.91	-209.83
18	1.00	0.0	1.128	1.003	834.02	-209.39
19	1.00	0.0	1.121	1.000	839.35	-209.95
20	1.00	0.0	1.130	1.004	832.13	-209.25
21	1.00	0.0	1.118	0.999	842.01	-210.09
22	1.00	0.0	1.135	1.007	828.04	-209.16
23	1.00	0.0	1.110	0.994	849.16	-209.98
24	1.00	0.0	1.153	1.018	812.76	-210.30
25	1.00	0.0	1.068	0.960	891.42	-200.42
26	1.00	0.0	1.184	1.265	697.07	-321.03

28 ELEMENT ARRAY

ELEMENT	IZ(0) RE	IZ(0) IM	YO RE	YO IM	ZO RE	ZO IM
1	1.00	0.0	1.184	1.265	696.94	-321.05
2	1.00	0.0	1.068	0.960	891.53	-200.43
3	1.00	0.0	1.153	1.018	812.65	-210.29
4	1.00	0.0	1.110	0.994	849.27	-210.00
5	1.00	0.0	1.135	1.007	827.92	-209.14
6	1.00	0.0	1.118	0.999	842.14	-210.11
7	1.00	0.0	1.130	1.004	831.99	-209.23
8	1.00	0.0	1.121	1.000	839.51	-209.98
9	1.00	0.0	1.128	1.003	833.84	-209.36
10	1.00	0.0	1.123	1.001	838.11	-209.86
11	1.00	0.0	1.127	1.003	834.95	-209.47
12	1.00	0.0	1.124	1.002	837.19	-209.76
13	1.00	0.0	1.126	1.002	835.75	-209.57
14	1.00	0.0	1.125	1.002	836.46	-209.66
15	1.00	0.0	1.125	1.002	836.46	-209.66
16	1.00	0.0	1.126	1.002	835.75	-209.57
17	1.00	0.0	1.124	1.002	837.19	-209.76
18	1.00	0.0	1.127	1.003	834.95	-209.47
19	1.00	0.0	1.123	1.001	838.11	-209.86
20	1.00	0.0	1.128	1.003	833.84	-209.36
21	1.00	0.0	1.121	1.000	839.51	-209.98
22	1.00	0.0	1.130	1.004	831.99	-209.23
23	1.00	0.0	1.118	0.999	842.14	-210.11
24	1.00	0.0	1.135	1.007	827.92	-209.14
25	1.00	0.0	1.110	0.994	849.27	-210.00
26	1.00	0.0	1.153	1.018	812.65	-210.29
27	1.00	0.0	1.068	0.960	891.53	-200.43
28	1.00	0.0	1.184	1.265	696.94	-321.05

30 ELEMENT ARRAY

ELEMENT	IZ(0) RE	IZ(0) IM	YO RE	YO IM	ZO RE	ZO IM
1	1.00	0.0	1.184	1.265	696.83	-321.06
2	1.00	0.0	1.068	0.960	891.62	-200.44
3	1.00	0.0	1.153	1.018	812.56	-210.28
4	1.00	0.0	1.110	0.994	849.36	-210.01
5	1.00	0.0	1.136	1.007	827.83	-209.13
6	1.00	0.0	1.118	0.999	842.24	-210.13
7	1.00	0.0	1.131	1.004	831.88	-209.21
8	1.00	0.0	1.121	1.000	839.63	-210.00
9	1.00	0.0	1.128	1.003	833.70	-209.34
10	1.00	0.0	1.123	1.001	838.27	-209.88
11	1.00	0.0	1.127	1.003	834.77	-209.45
12	1.00	0.0	1.124	1.002	837.39	-209.79
13	1.00	0.0	1.126	1.002	835.52	-209.54
14	1.00	0.0	1.125	1.002	836.73	-209.70
15	1.00	0.0	1.125	1.002	836.14	-209.62
16	1.00	0.0	1.125	1.002	836.14	-209.62
17	1.00	0.0	1.125	1.002	836.73	-209.70
18	1.00	0.0	1.126	1.002	835.52	-209.54
19	1.00	0.0	1.124	1.002	837.39	-209.79
20	1.00	0.0	1.127	1.003	834.77	-209.45
21	1.00	0.0	1.123	1.001	838.27	-209.88
22	1.00	0.0	1.128	1.003	833.70	-209.34
23	1.00	0.0	1.121	1.000	839.63	-210.00
24	1.00	0.0	1.131	1.004	831.88	-209.21
25	1.00	0.0	1.118	0.999	842.24	-210.13
26	1.00	0.0	1.136	1.007	827.83	-209.13
27	1.00	0.0	1.110	0.994	849.36	-210.01
28	1.00	0.0	1.153	1.018	812.56	-210.28
29	1.00	0.0	1.068	0.960	891.62	-200.44
30	1.00	0.0	1.184	1.265	696.83	-321.06

BR

DRIVING POINT ADMITTANCES AND IMPEDANCES
FOR SPECIFIED BASE VOLTAGES

H/LMDA=0.5000 B/LMDA= 0.500 OMEGA= 9.92

35 ELEMENT ARRAY

ELEMENT	VO		YO		ZO	
	RE	IM	RE	IM	RE	IM
1	1.00	0.0	1.088	1.323	370.91	-450.83
2	1.00	0.0	1.170	0.869	551.06	-409.27
3	1.00	0.0	1.088	1.076	464.59	-459.41
4	1.00	0.0	1.153	0.954	514.61	-425.82
5	1.00	0.0	1.103	1.036	481.61	-452.42
6	1.00	0.0	1.143	0.976	506.19	-432.11
7	1.00	0.0	1.110	1.024	486.77	-448.83
8	1.00	0.0	1.138	0.983	503.08	-434.86
9	1.00	0.0	1.114	1.018	488.99	-447.05
10	1.00	0.0	1.135	0.987	501.56	-436.26
11	1.00	0.0	1.116	1.016	490.16	-446.07
12	1.00	0.0	1.134	0.989	500.71	-437.05
13	1.00	0.0	1.117	1.014	490.82	-445.50
14	1.00	0.0	1.133	0.991	500.23	-437.50
15	1.00	0.0	1.118	1.013	491.18	-445.18
16	1.00	0.0	1.132	0.991	499.98	-437.73
17	1.00	0.0	1.118	1.013	491.35	-445.04
18	1.00	0.0	1.132	0.991	499.90	-437.81
19	1.00	0.0	1.118	1.013	491.35	-445.04
20	1.00	0.0	1.132	0.991	499.98	-437.73
21	1.00	0.0	1.118	1.013	491.18	-445.18
22	1.00	0.0	1.133	0.991	500.23	-437.50
23	1.00	0.0	1.117	1.014	490.82	-445.50
24	1.00	0.0	1.134	0.989	500.71	-437.05
25	1.00	0.0	1.116	1.016	490.16	-446.07
26	1.00	0.0	1.135	0.987	501.56	-436.26
27	1.00	0.0	1.114	1.018	488.99	-447.05
28	1.00	0.0	1.138	0.983	503.08	-434.86
29	1.00	0.0	1.110	1.024	486.77	-448.83
30	1.00	0.0	1.143	0.976	506.19	-432.11
31	1.00	0.0	1.103	1.036	481.61	-452.42
32	1.00	0.0	1.153	0.954	514.61	-425.82
33	1.00	0.0	1.088	1.076	464.59	-459.41
34	1.00	0.0	1.170	0.869	551.06	-409.27
35	1.00	0.0	1.088	1.323	370.91	-450.83

40 ELEMENT ARRAY

ELEMENT	VO		YO		ZO	
	RE	IM	RE	IM	RE	IM
1	1.00	0.0	1.093	1.316	373.42	-449.75
2	1.00	0.0	1.165	0.874	549.03	-411.89
3	1.00	0.0	1.092	1.071	466.89	-457.72
4	1.00	0.0	[illegible]	[illegible]	[illegible]	-427.92
5	1.00	0.0	1.107	1.031	483.88	-450.63
6	1.00	0.0	1.139	0.981	504.02	-434.13
7	1.00	0.0	1.114	1.018	489.11	-446.95
8	1.00	0.0	1.134	0.989	500.76	-436.93
9	1.00	0.0	1.118	1.012	491.47	-445.04
10	1.00	0.0	1.131	0.994	499.06	-438.44
11	1.00	0.0	1.120	1.009	492.81	-443.91
12	1.00	0.0	1.129	0.996	498.00	-439.39
13	1.00	0.0	1.122	1.007	493.70	-443.14
14	1.00	0.0	1.128	0.998	497.26	-440.05
15	1.00	0.0	1.123	1.005	494.35	-442.58
16	1.00	0.0	1.127	0.999	496.69	-440.56
17	1.00	0.0	1.124	1.004	494.88	-442.13
18	1.00	0.0	1.126	1.001	496.20	-440.98
19	1.00	0.0	1.125	1.003	495.34	-441.73
20	1.00	0.0	1.125	1.002	495.77	-441.36
21	1.00	0.0	1.125	1.002	495.77	-441.36
22	1.00	0.0	1.125	1.003	495.34	-441.73
23	1.00	0.0	1.126	1.001	496.20	-440.98
24	1.00	0.0	1.124	1.004	494.88	-442.13
25	1.00	0.0	1.127	0.999	496.69	-440.56
26	1.00	0.0	1.123	1.005	494.35	-442.58
27	1.00	0.0	1.128	0.998	497.26	-440.05
28	1.00	0.0	1.122	1.007	493.70	-443.14
29	1.00	0.0	1.129	0.996	498.00	-439.39
30	1.00	0.0	1.120	1.009	492.81	-443.91
31	1.00	0.0	1.131	0.994	499.06	-438.44
32	1.00	0.0	1.118	1.012	491.47	-445.04
33	1.00	0.0	1.134	0.989	500.76	-436.93
34	1.00	0.0	1.114	1.018	489.11	-446.95
35	1.00	0.0	1.139	0.981	504.02	-434.13
36	1.00	0.0	1.107	1.031	483.88	-450.63
37	1.00	0.0	1.150	0.960	512.55	-427.92
38	1.00	0.0	1.092	1.071	466.89	-457.72
39	1.00	0.0	1.165	0.874	549.03	-411.89
40	1.00	0.0	1.093	1.316	373.42	-449.75

DRIVING POINT ADMITTANCES AND IMPEDANCES
FOR SPECIFIED BASE CURRENTS

H/LMDA=0.5000 B/LMDA= 0.500 OMEGA= 9.92

35 ELEMENT ARRAY

ELEMENT	IZ(O)		YO		ZO	
	RE	IM	RE	IM	RE	IM
1	1.00	0.0	1.186	1.269	694.30	-321.52
2	1.00	0.0	1.065	0.959	893.63	-200.56
3	1.00	0.0	1.156	1.019	810.77	-210.04
4	1.00	0.0	1.107	0.994	851.09	-210.29
5	1.00	0.0	1.138	1.008	826.10	-208.82
6	1.00	0.0	1.115	0.998	843.99	-210.46
7	1.00	0.0	1.133	1.005	830.09	-208.86
8	1.00	0.0	1.118	1.000	841.48	-210.37
9	1.00	0.0	1.131	1.004	831.79	-208.95
10	1.00	0.0	1.120	1.000	840.27	-210.29
11	1.00	0.0	1.130	1.004	832.67	-209.02
12	1.00	0.0	1.121	1.001	839.62	-210.24
13	1.00	0.0	1.129	1.003	833.16	-209.06
14	1.00	0.0	1.121	1.001	839.25	-210.20
15	1.00	0.0	1.129	1.003	833.43	-209.09
16	1.00	0.0	1.121	1.001	839.05	-210.19
17	1.00	0.0	1.129	1.003	833.55	-209.10
18	1.00	0.0	1.122	1.001	839.00	-210.18
19	1.00	0.0	1.129	1.003	833.55	-209.10
20	1.00	0.0	1.121	1.001	839.05	-210.19
21	1.00	0.0	1.129	1.003	833.43	-209.09
22	1.00	0.0	1.121	1.001	839.25	-210.20
23	1.00	0.0	1.129	1.003	833.16	-209.06
24	1.00	0.0	1.121	1.001	839.62	-210.24
25	1.00	0.0	1.130	1.004	832.67	-209.02
26	1.00	0.0	1.120	1.000	840.27	-210.29
27	1.00	0.0	1.131	1.004	831.79	-208.95
28	1.00	0.0	1.118	1.000	841.48	-210.37
29	1.00	0.0	1.133	1.005	830.09	-208.86
30	1.00	0.0	1.115	0.998	843.99	-210.46
31	1.00	0.0	1.138	1.008	826.10	-208.82
32	1.00	0.0	1.107	0.994	851.09	-210.29
33	1.00	0.0	1.156	1.019	810.77	-210.04
34	1.00	0.0	1.065	0.959	893.63	-200.56
35	1.00	0.0	1.186	1.269	694.30	-321.52

40 ELEMENT ARRAY

ELEMENT	IZ(O)		YO		ZO	
	RE	IM	RE	IM	RE	IM
1	1.00	0.0	1.184	1.266	696.46	-321.13
2	1.00	0.0	1.067	0.960	891.93	-200.46
3	1.00	C.0	1.154	1.019	812.28	-210.24
4	1.00	0.0	1.109	0.994	849.65	-210.05
5	[illegible]	[illegible]	1.136	1.007	827.53	-209.08
6	1.00	0.0	1.117	0.999	[illegible]	[illegible]
7	1.00	0.0	1.131	1.004	831.54	-209.16
8	1.00	0.0	1.120	1.000	839.99	-210.06
9	1.00	0.0	1.129	1.003	833.31	-209.27
10	1.00	0.0	1.122	1.001	838.70	-209.96
11	1.00	0.0	1.128	1.003	834.30	-209.36
12	1.00	0.0	1.123	1.001	837.92	-209.88
13	1.00	0.0	1.127	1.003	834.94	-209.44
14	1.00	0.0	1.124	1.002	837.38	-209.81
15	1.00	C.0	1.126	1.002	835.39	-209.50
16	1.00	0.0	1.124	1.002	836.98	-209.75
17	1.00	0.0	1.126	1.002	835.75	-209.55
18	1.00	0.0	1.125	1.002	836.65	-209.70
19	1.00	0.0	1.125	1.002	836.06	-209.60
20	1.00	0.0	1.125	1.002	836.35	-209.65
21	1.00	0.0	1.125	1.002	836.35	-209.65
22	1.00	0.0	1.125	1.002	836.06	-209.60
23	1.00	0.0	1.125	1.002	836.65	-209.70
24	1.00	0.0	1.126	1.002	835.75	-209.55
25	1.00	0.0	1.124	1.002	836.98	-209.75
26	1.00	0.0	1.126	1.002	835.39	-209.50
27	1.00	0.0	1.124	1.002	837.38	-209.81
28	1.00	0.0	1.127	1.003	834.94	-209.44
29	1.00	0.0	1.123	1.001	837.92	-209.88
30	1.00	0.0	1.128	1.003	834.30	-209.36
31	1.00	0.0	1.122	1.001	838.70	-209.96
32	1.00	0.0	1.129	1.003	833.31	-209.27
33	1.00	0.0	1.120	1.000	839.99	-210.06
34	1.00	0.0	1.131	1.004	831.54	-209.16
35	1.00	0.0	1.117	0.999	842.56	-210.18
36	1.00	0.0	1.136	1.007	827.53	-209.08
37	1.00	0.0	1.109	0.994	849.65	-210.05
38	1.00	0.0	1.154	1.019	812.28	-210.24
39	1.00	0.0	1.067	0.960	891.93	-200.46
40	1.00	0.0	1.184	1.266	696.46	-321.13

BR

DRIVING POINT ADMITTANCES AND IMPEDANCES
FOR SPECIFIED BASE VOLTAGES

H/LMDA=0.5000 B/LMDA= 0.500 OMEGA= 9.92

45 ELEMENT ARRAY

ELEMENT	VO		YO		ZO	
	RE	IM	RE	IM	RE	IM
1	1.00	0.0	1.089	1.322	371.27	-450.66
2	1.00	0.0	1.169	0.869	550.77	-409.67
3	1.00	0.0	1.089	1.075	464.93	-459.14
4	1.00	0.0	1.153	0.955	514.29	-426.17
5	1.00	0.0	1.104	1.035	481.97	-452.12
6	1.00	0.0	1.142	0.976	505.84	-432.46
7	1.00	0.0	1.111	1.023	487.17	-448.50
8	1.00	0.0	1.137	0.984	502.68	-435.24
9	1.00	0.0	1.115	1.017	489.44	-446.67
10	1.00	0.0	1.134	0.988	501.10	-436.69
11	1.00	0.0	1.117	1.014	490.67	-445.63
12	1.00	0.0	1.133	0.991	500.18	-437.54
13	1.00	0.0	1.118	1.012	491.41	-444.99
14	1.00	0.0	1.132	0.992	499.60	-438.08
15	1.00	0.0	1.119	1.011	491.88	-444.58
16	1.00	0.0	1.131	0.993	499.23	-438.42
17	1.00	0.0	1.119	1.011	492.19	-444.31
18	1.00	0.0	1.130	0.994	498.98	-438.65
19	1.00	0.0	1.120	1.010	492.39	-444.14
20	1.00	0.0	1.130	0.994	498.84	-438.78
21	1.00	0.0	1.120	1.010	492.50	-444.04
22	1.00	0.0	1.130	0.994	498.77	-438.85
23	1.00	0.0	1.120	1.010	492.53	-444.01
24	1.00	0.0	1.130	0.994	498.77	-438.85
25	1.00	0.0	1.120	1.010	492.50	-444.04
26	1.00	0.0	1.130	0.994	498.84	-438.78
27	1.00	0.0	1.120	1.010	492.39	-444.14
28	1.00	0.0	1.130	0.994	498.98	-438.65
29	1.00	0.0	1.119	1.011	492.19	-444.31
30	1.00	0.0	1.131	0.993	499.23	-438.42
31	1.00	0.0	1.119	1.011	491.88	-444.58
32	1.00	0.0	1.132	0.992	499.60	-438.08
33	1.00	0.0	1.118	1.012	491.41	-444.99
34	1.00	0.0	1.133	0.991	500.18	-437.54
35	1.00	0.0	1.117	1.014	490.67	-445.63
36	1.00	0.0	1.134	0.988	501.10	-436.69
37	1.00	0.0	1.115	1.017	489.44	-446.67
38	1.00	0.0	1.137	0.984	502.68	-435.24
39	1.00	0.0	1.111	1.023	487.17	-448.50
40	1.00	0.0	1.142	0.976	505.84	-432.46
41	1.00	0.0	1.104	1.035	481.97	-452.12
42	1.00	0.0	1.153	0.955	514.29	-426.17
43	1.00	0.0	1.089	1.075	464.93	-459.14
44	1.00	0.0	1.169	0.869	550.77	-409.67
45	1.00	0.0	1.089	1.322	371.27	-450.66

50 ELEMENT ARRAY

ELEMENT	VO		YO		ZO	
	RE	IM	RE	IM	RE	IM
1	1.00	0.0	1.092	1.317	373.17	-449.86
2	1.00	0.0	1.166	0.874	549.23	-411.63
3	1.00	0.0	1.092	1.071	466.65	-457.89
4	1.00	0.0	1.150	0.959	512.77	-427.70
5	1.00	0.0	1.106	1.031	483.64	-450.82
6	1.00	0.0	1.139	0.980	504.26	-433.91
7	1.00	0.0	1.114	1.019	488.85	-447.16
8	1.00	0.0	1.134	0.989	501.03	-436.70
9	1.00	0.0	1.118	1.013	491.18	-445.28
10	1.00	0.0	1.131	0.993	499.36	-438.18
11	1.00	0.0	1.120	1.010	492.49	-444.18
12	1.00	0.0	1.130	0.995	498.35	-439.09
13	1.00	0.0	1.121	1.008	493.33	-443.45
14	1.00	0.0	1.128	0.997	497.65	-439.71
15	1.00	0.0	1.122	1.006	493.93	-442.94
16	1.00	0.0	1.128	0.998	497.14	-440.17
17	1.00	0.0	1.123	1.005	494.38	-442.55
18	1.00	0.0	1.127	0.999	496.74	-440.52
19	1.00	0.0	1.124	1.004	494.74	-442.24
20	1.00	0.0	1.126	1.000	496.41	-440.81
21	1.00	0.0	1.124	1.004	495.06	-441.97
22	1.00	0.0	1.126	1.001	496.12	-441.06
23	1.00	0.0	1.125	1.003	495.33	-441.73
24	1.00	0.0	1.125	1.002	495.85	-441.29
25	1.00	0.0	1.125	1.002	495.59	-441.51
26	1.00	0.0	1.125	1.002	495.59	-441.51
27	1.00	0.0	1.125	1.002	495.85	-441.29
28	1.00	0.0	1.125	1.003	495.33	-441.73
29	1.00	0.0	1.126	1.001	496.12	-441.06
30	1.00	0.0	1.124	1.004	495.06	-441.97
31	1.00	0.0	1.126	1.000	496.41	-440.81
32	1.00	0.0	1.124	1.004	494.74	-442.24
33	1.00	0.0	1.127	0.999	496.74	-440.52
34	1.00	0.0	1.123	1.005	494.38	-442.55
35	1.00	0.0	1.128	0.998	497.14	-440.17
36	1.00	0.0	1.122	1.006	493.93	-442.94
37	1.00	0.0	1.128	0.997	497.65	-439.71
38	1.00	0.0	1.121	1.008	493.33	-443.45
39	1.00	0.0	1.130	0.995	498.35	-439.09
40	1.00	0.0	1.120	1.010	492.49	-444.18
41	1.00	0.0	1.131	0.993	499.36	-438.18
42	1.00	0.0	1.118	1.013	491.18	-445.28
43	1.00	0.0	1.134	0.989	501.03	-436.70
44	1.00	0.0	1.114	1.019	488.85	-447.16
45	1.00	0.0	1.139	0.980	504.26	-433.91
46	1.00	0.0	1.106	1.031	483.64	-450.82
47	1.00	0.0	1.150	0.959	512.77	-427.70
48	1.00	0.0	1.092	1.071	466.65	-457.89
49	1.00	0.0	1.166	0.874	549.23	-411.63
50	1.00	0.0	1.092	1.317	373.17	-449.86

DRIVING POINT ADMITTANCES AND IMPEDANCES
FOR SPECIFIED BASE CURRENTS

H/LMDA=0.5000 B/LMDA= 0.500 OMEGA= 9.92

45 ELEMENT ARRAY

ELEMENT	IZ(O)		YO		ZO	
	RE	IM	RE	IM	RE	IM
1	1.00	0.0	1.186	1.269	694.62	-321.45
2	1.00	0.0	1.066	0.959	893.36	-200.55
3	1.00	0.0	1.156	1.019	811.02	-210.06
4	1.00	0.0	1.108	0.994	850.85	-210.26
5	1.00	0.0	1.137	1.007	826.35	-208.85
6	1.00	0.0	1.116	0.998	843.73	-210.42
7	1.00	0.0	1.133	1.005	830.35	-208.90
8	1.00	0.0	1.119	1.000	841.20	-210.32
9	1.00	0.0	1.130	1.004	832.08	-209.00
10	1.00	0.0	1.120	1.000	839.95	-210.24
11	1.00	0.0	1.129	1.003	833.01	-209.07
12	1.00	0.0	1.121	1.001	839.25	-210.18
13	1.00	0.0	1.129	1.003	833.56	-209.12
14	1.00	0.0	1.122	1.001	838.81	-210.14
15	1.00	0.0	1.128	1.003	833.91	-209.16
16	1.00	0.0	1.122	1.001	838.53	-210.10
17	1.00	0.0	1.128	1.003	834.13	-209.19
18	1.00	0.0	1.122	1.001	838.36	-210.08
19	1.00	0.0	1.128	1.003	834.27	-209.20
20	1.00	0.0	1.122	1.001	838.25	-210.07
21	1.00	0.0	1.128	1.003	834.35	-209.21
22	1.00	0.0	1.123	1.001	838.20	-210.06
23	1.00	0.0	1.128	1.003	834.37	-209.22
24	1.00	0.0	1.123	1.001	838.20	-210.06
25	1.00	0.0	1.128	1.003	834.35	-209.21
26	1.00	0.0	1.122	1.001	838.25	-210.07
27	1.00	0.0	1.128	1.003	834.27	-209.20
28	1.00	0.0	1.122	1.001	838.36	-210.08
29	1.00	0.0	1.128	1.003	834.13	-209.19
30	1.00	0.0	1.122	1.001	838.53	-210.10
31	1.00	0.0	1.128	1.003	833.91	-209.16
32	1.00	0.0	1.122	1.001	838.81	-210.14
33	1.00	0.0	1.129	1.003	833.56	-209.12
34	1.00	0.0	1.121	1.001	839.25	-210.18
35	1.00	0.0	1.129	1.003	833.01	-209.07
36	1.00	0.0	1.120	1.000	839.95	-210.24
37	1.00	0.0	1.130	1.004	832.08	-209.00
38	1.00	0.0	1.119	1.000	841.20	-210.32
39	1.00	0.0	1.133	1.005	830.35	-208.90
40	1.00	0.0	1.116	0.998	843.73	-210.42
41	1.00	0.0	1.137	1.007	826.35	-208.85
42	1.00	0.0	1.108	0.994	850.85	-210.26
43	1.00	0.0	1.156	1.019	811.02	-210.06
44	1.00	0.0	1.066	0.959	893.36	-200.55
45	1.00	0.0	1.186	1.269	694.62	-321.45

50 ELEMENT ARRAY

ELEMENT	IZ(O)		YO		ZO	
	RE	IM	RE	IM	RE	IM
1	1.00	0.0	1.184	1.266	696.25	-321.16
2	1.00	0.0	1.067	0.960	892.10	-200.47
3	1.00	0.0	1.154	1.019	812.12	-210.22
4	1.00	0.0	1.109	0.994	849.80	-210.08
5	1.00	0.0	1.136	1.007	827.37	-209.05
6	1.00	0.0	1.117	0.999	842.72	-210.21
7	1.00	0.0	1.131	1.005	831.37	-209.12
8	1.00	0.0	1.120	1.000	840.17	-210.10
9	1.00	0.0	1.129	1.004	833.13	-209.23
10	1.00	0.0	1.122	1.001	838.89	-210.00
11	1.00	0.0	1.128	1.003	834.09	-209.32
12	1.00	0.0	1.123	1.001	838.14	-209.92
13	1.00	0.0	1.127	1.003	834.70	-209.39
14	1.00	0.0	1.123	1.001	837.64	-209.86
15	1.00	0.0	1.127	1.003	835.12	-209.45
16	1.00	0.0	1.124	1.002	837.28	-209.81
17	1.00	0.0	1.126	1.002	835.43	-209.49
18	1.00	0.0	1.124	1.002	837.00	-209.77
19	1.00	0.0	1.126	1.002	835.68	-209.53
20	1.00	0.0	1.124	1.002	836.78	-209.73
21	1.00	0.0	1.126	1.002	835.89	-209.57
22	1.00	0.0	1.125	1.002	836.58	-209.69
23	1.00	0.0	1.125	1.002	836.07	-209.60
24	1.00	0.0	1.125	1.002	836.41	-209.66
25	1.00	0.0	1.125	1.002	836.24	-209.63
26	1.00	0.0	1.125	1.002	836.24	-209.63
27	1.00	0.0	1.125	1.002	836.41	-209.66
28	1.00	0.0	1.125	1.002	836.07	-209.60
29	1.00	0.0	1.125	1.002	836.58	-209.69
30	1.00	0.0	1.126	1.002	835.89	-209.57
31	1.00	0.0	1.124	1.002	836.78	-209.73
32	1.00	0.0	1.126	1.002	835.68	-209.53
33	1.00	0.0	1.124	1.002	837.00	-209.77
34	1.00	0.0	1.126	1.002	835.43	-209.49
35	1.00	0.0	1.124	1.002	837.28	-209.81
36	1.00	0.0	1.127	1.003	835.12	-209.45
37	1.00	0.0	1.123	1.001	837.64	-209.86
38	1.00	0.0	1.127	1.003	834.70	-209.39
39	1.00	0.0	1.123	1.001	838.14	-209.92
40	1.00	0.0	1.128	1.003	834.09	-209.32
41	1.00	0.0	1.122	1.001	838.89	-210.00
42	1.00	0.0	1.129	1.004	833.13	-209.23
43	1.00	0.0	1.120	1.000	840.17	-210.10
44	1.00	0.0	1.131	1.005	831.37	-209.12
45	1.00	0.0	1.117	0.999	842.72	-210.21
46	1.00	0.0	1.136	1.007	827.37	-209.05
47	1.00	0.0	1.109	0.994	849.80	-210.08
48	1.00	0.0	1.154	1.019	812.12	-210.22
49	1.00	0.0	1.067	0.960	892.10	-200.47
50	1.00	0.0	1.184	1.266	696.25	-321.16

TABLE 5.2

DRIVING-POINT ADMITTANCES AND IMPEDANCES OF UNILATERAL ENDFIRE ARRAYS

EN

DRIVING POINT ADMITTANCES AND IMPEDANCES
FOR SPECIFIED BASE VOLTAGES

H/LMDA=0.2500 B/LMDA= 0.250 OMEGA= 8.53

2 ELEMENT ARRAY

ELEMENT	VO		YO		ZO	
	RE	IM	RE	IM	RE	IM
1	1.00	0.0	12.277	-5.809	66.55	31.49
2	0.0	-1.00	1.566	-2.922	142.48	265.89

3 ELEMENT ARRAY

ELEMENT	VO		YO		ZO	
	RE	IM	RE	IM	RE	IM
1	1.00	0.0	12.391	-5.508	67.39	29.96
2	0.0	-1.00	3.186	-4.749	97.41	145.21
3	-1.00	0.0	1.146	-1.839	244.16	391.72

4 ELEMENT ARRAY

ELEMENT	VO		YO		ZO	
	RE	IM	RE	IM	RE	IM
1	1.00	0.0	12.363	-5.618	67.05	30.47
2	0.0	-1.00	3.283	-4.414	108.47	145.87
3	-1.00	0.0	2.538	-3.479	136.87	187.63
4	0.0	1.00	0.923	-1.426	319.71	494.27

5 ELEMENT ARRAY

ELEMENT	VO		YO		ZO	
	RE	IM	RE	IM	RE	IM
1	1.00	0.0	12.379	-5.557	67.23	30.18
2	0.0	-1.00	3.260	-4.554	103.94	145.18
3	-1.00	0.0	[illegible]	[illegible]	[illegible]	186.86
4	0.0	1.00	2.168	-2.960	161.03	219.85
5	1.00	0.0	0.785	-1.179	391.30	587.93

6 ELEMENT ARRAY

ELEMENT	VO		YO		ZO	
	RE	IM	RE	IM	RE	IM
1	1.00	0.0	12.368	-5.596	67.12	30.37
2	0.0	-1.00	3.273	-4.472	106.59	145.61
3	-1.00	0.0	2.609	-3.304	147.19	186.44
4	0.0	1.00	2.260	-2.675	184.27	218.14
5	1.00	0.0	1.932	-2.636	180.84	246.77
6	0.0	-1.00	0.691	-1.006	463.72	675.15

7 ELEMENT ARRAY

ELEMENT	VO		YO		ZO	
	RE	IM	RE	IM	RE	IM
1	1.00	0.0	12.376	-5.569	67.20	30.23
2	0.0	-1.00	3.264	-4.527	104.80	145.35
3	-1.00	0.0	2.620	-3.228	151.60	186.77
4	0.0	1.00	2.238	-2.798	174.34	217.93
5	1.00	0.0	2.022	-2.366	208.74	244.23
6	0.0	-1.00	1.765	-2.407	198.10	270.14
7	-1.00	0.0	0.623	-0.877	538.09	757.89

8 ELEMENT ARRAY

ELEMENT	VO		YO		ZO	
	RE	IM	RE	IM	RE	IM
1	1.00	0.0	12.369	-5.589	67.14	30.34
2	0.0	-1.00	3.271	-4.487	106.10	145.52
3	-1.00	0.0	2.612	-3.279	148.60	186.59
4	0.0	1.00	2.251	-2.724	180.26	218.18
5	1.00	0.0	2.001	-2.484	196.71	244.18
6	0.0	-1.00	1.855	-2.148	230.34	266.66
7	-1.00	0.0	1.640	-2.232	213.73	290.91
8	0.0	1.00	0.570	-0.775	615.74	836.95

DRIVING POINT ADMITTANCES AND IMPEDANCES
FOR SPECIFIED BASE CURRENTS

H/LMDA=0.2500 B/LMDA= 0.250 OMEGA= 8.53

2 ELEMENT ARRAY

ELEMENT	IZ(0)		YO		ZO	
	RE	IM	RE	IM	RE	IM
1	1.00	0.0	24.745	3.437	39.93	-4.39
2	0.0	-1.00	5.988	-3.089	118.89	75.63

3 ELEMENT ARRAY

ELEMENT	IZ(0)		YO		ZO	
	RE	IM	RE	IM	RE	IM
1	1.00	0.0	12.749	-4.641	66.65	28.03
2	0.0	-1.00	10.810	-4.251	76.36	35.12
3	-1.00	0.0	4.501	-2.687	141.24	106.91

4 ELEMENT ARRAY

ELEMENT	IZ(0)		YO		ZO	
	RE	IM	RE	IM	RE	IM
1	1.00	0.0	21.246	0.040	47.02	1.50
2	0.0	-1.00	6.807	-3.772	102.34	67.54
3	-1.00	0.0	7.125	-4.128	95.93	65.28
4	0.0	1.00	3.750	-2.443	155.80	131.43

5 ELEMENT ARRAY

ELEMENT	IZ(0)		YO		ZO	
	RE	IM	RE	IM	RE	IM
1	1.00	0.0	13.887	-4.595	62.81	24.04
2	0.0	-1.00	9.636	-4.017	83.58	41.09
3	-1.00	0.0	4.941	-3.219	123.74	98.64
4	0.0	1.00	5.462	-3.732	110.01	89.66
5	1.00	0.0	3.271	-2.249	[illegible]	[illegible]

6 ELEMENT ARRAY

ELEMENT	IZ(0)		YO		ZO	
	RE	IM	RE	IM	RE	IM
1	1.00	0.0	20.081	-1.041	49.42	4.33
2	0.0	-1.00	7.172	-3.904	98.50	63.50
3	-1.00	0.0	6.543	-3.791	103.59	71.42
4	0.0	1.00	4.036	-2.883	137.87	123.08
5	1.00	0.0	4.542	-3.420	120.24	109.61
6	0.0	-1.00	2.944	-2.105	176.84	169.72

7 ELEMENT ARRAY

ELEMENT	IZ(0)		YO		ZO	
	RE	IM	RE	IM	RE	IM
1	1.00	0.0	14.488	-4.475	61.16	21.93
2	0.0	-1.00	9.223	-3.996	85.95	43.95
3	-1.00	0.0	5.148	-3.345	119.64	94.47
4	0.0	1.00	5.106	-3.434	117.85	95.86
5	1.00	0.0	3.478	-2.629	149.17	143.68
6	0.0	-1.00	3.937	-3.157	128.94	126.98
7	-1.00	0.0	2.697	-1.982	185.04	185.38

8 ELEMENT ARRAY

ELEMENT	IZ(0)		YO		ZO	
	RE	IM	RE	IM	RE	IM
1	1.00	0.0	19.481	-1.576	50.63	5.97
2	0.0	-1.00	7.365	-3.946	96.89	61.38
3	-1.00	0.0	6.319	-3.708	106.13	74.37
4	0.0	1.00	4.177	-2.994	133.69	118.87
5	1.00	0.0	4.291	-3.156	128.34	115.92
6	0.0	-1.00	3.102	-2.443	158.06	161.16
7	-1.00	0.0	3.517	-2.952	136.03	142.05
8	0.0	1.00	2.508	-1.883	191.94	199.24

EN

DRIVING POINT ADMITTANCES AND IMPEDANCES
FOR SPECIFIED BASE VOLTAGES

H/LMDA=0.2500 B/LMDA= 0.250 OMEGA= 8.53

9 ELEMENT ARRAY

ELEMENT	VO RE	VO IM	YO RE	YO IM	ZO RE	ZO IM
1	1.00	0.0	12.375	-5.573	67.18	30.25
2	0.0	-1.00	3.265	-4.517	105.10	145.40
3	-1.00	0.0	2.618	-3.242	150.79	186.70
4	0.0	1.00	2.242	-2.774	176.20	218.06
5	1.00	0.0	2.013	-2.412	203.92	244.38
6	0.0	-1.00	1.833	-2.261	216.32	266.86
7	-1.00	0.0	1.729	-1.982	249.93	286.49
8	0.0	1.00	1.542	-2.093	228.12	309.74
9	1.00	0.0	0.528	-0.692	697.13	912.99

10 ELEMENT ARRAY

ELEMENT	VO RE	VO IM	YO RE	YO IM	ZO RE	ZO IM
1	1.00	0.0	12.370	-5.586	67.15	30.32
2	0.0	-1.00	3.270	-4.493	105.90	145.48
3	-1.00	0.0	2.613	-3.271	149.10	186.63
4	0.0	1.00	2.249	-2.737	179.18	218.12
5	1.00	0.0	2.004	-2.461	198.96	244.30
6	0.0	-1.00	1.846	-2.192	224.76	266.96
7	-1.00	0.0	1.707	-2.092	234.11	286.95
8	0.0	1.00	1.630	-1.850	268.09	304.29
9	1.00	0.0	1.462	-1.978	241.61	327.01
10	0.0	-1.00	0.494	-0.622	783.05	986.13

12 ELEMENT ARRAY

ELEMENT	VO RE	VO IM	YO RE	YO IM	ZO RE	ZO IM
1	1.00	0.0	12.371	-5.584	67.15	30.31
2	0.0	-1.00	3.270	-4.496	105.79	145.47
3	-1.00	0.0	2.613	-3.266	149.34	186.65
4	0.0	1.00	2.248	-2.743	178.72	218.10
5	1.00	0.0	2.005	-2.453	199.78	244.34
6	0.0	-1.00	1.844	-2.205	223.23	266.93
7	-1.00	0.0	1.711	-2.071	237.05	287.01
8	0.0	1.00	1.620	-1.892	261.18	304.91
9	1.00	0.0	1.527	-1.847	265.96	321.64
10	0.0	-1.00	1.482	-1.650	301.28	335.45
11	-1.00	0.0	1.338	-1.797	266.49	358.03
12	0.0	1.00	0.440	-0.510	970.08	1123.97

14 ELEMENT ARRAY

ELEMENT	VO RE	VO IM	YO RE	YO IM	ZO RE	ZO IM
1	1.00	0.0	12.371	-5.583	67.15	30.31
2	0.0	-1.00	3.270	-4.498	105.73	145.46
3	-1.00	0.0	2.614	-3.264	149.48	186.66
4	0.0	1.00	2.247	-2.746	178.48	218.09
5	1.00	0.0	2.006	-2.449	200.17	244.36
6	0.0	-1.00	1.843	-2.210	222.57	266.92
7	-1.00	0.0	1.712	-2.064	238.13	287.04
8	0.0	1.00	1.618	-1.903	259.26	304.94
9	1.00	0.0	1.531	-1.827	269.51	321.62
10	0.0	-1.00	1.472	-1.690	293.16	336.43
11	-1.00	0.0	1.402	-1.672	294.46	351.19
12	0.0	1.00	1.375	-1.502	331.49	362.22
13	1.00	0.0	1.246	-1.659	289.34	385.47
14	0.0	-1.00	0.400	-0.423	1180.29	1248.97

16 ELEMENT ARRAY

ELEMENT	VO RE	VO IM	YO RE	YO IM	ZO RE	ZO IM
1	1.00	0.0	12.371	-5.583	67.16	30.31
2	0.0	-1.00	3.269	-4.499	105.69	145.46
3	-1.00	0.0	2.614	-3.263	149.56	186.67
4	0.0	1.00	2.247	-2.748	178.33	218.09
5	1.00	0.0	2.007	-2.447	200.40	244.36
6	0.0	-1.00	1.842	-2.213	222.23	266.92
7	-1.00	0.0	1.713	-2.060	238.66	287.04
8	0.0	1.00	1.617	-1.909	258.43	304.95
9	1.00	0.0	1.532	-1.819	270.82	321.62
10	0.0	-1.00	1.470	-1.701	290.90	336.53
11	-1.00	0.0	1.406	-1.653	298.56	351.06
12	0.0	1.00	1.365	-1.540	322.27	363.61
13	1.00	0.0	1.309	-1.539	320.68	377.10
14	0.0	-1.00	1.293	-1.387	359.59	385.78
15	-1.00	0.0	1.173	-1.549	310.72	410.23
16	0.0	1.00	0.368	-0.353	1416.43	1357.66

DRIVING POINT ADMITTANCES AND IMPEDANCES
FOR SPECIFIED BASE CURRENTS

H/LMDA=0.2500 B/LMDA= 0.250 OMEGA= 8.53

9 ELEMENT ARRAY

ELEMENT	IZ(0) RE	IZ(0) IM	YO RE	YO IM	ZO RE	ZO IM
1	1.00	0.0	14.861	-4.369	60.23	20.62
2	0.0	-1.00	9.011	-3.995	87.12	45.59
3	-1.00	0.0	5.259	-3.396	117.91	92.28
4	0.0	1.00	4.962	-3.352	120.43	98.84
5	1.00	0.0	3.584	-2.729	144.85	139.39
6	0.0	-1.00	3.748	-2.923	137.19	133.35
7	-1.00	0.0	2.824	-2.288	165.91	176.73
8	0.0	1.00	3.198	-2.776	142.44	155.71
9	1.00	0.0	2.354	-1.796	198.33	211.96

10 ELEMENT ARRAY

ELEMENT	IZ(0) RE	IZ(0) IM	YO RE	YO IM	ZO RE	ZO IM
1	1.00	0.0	19.109	-1.899	51.37	7.04
2	0.0	-1.00	7.486	-3.964	95.99	60.07
3	-1.00	0.0	6.201	-3.672	107.39	76.06
4	0.0	1.00	4.254	-3.041	131.95	116.66
5	1.00	0.0	4.184	-3.079	131.01	118.96
6	0.0	-1.00	3.188	-2.533	153.66	156.83
7	-1.00	0.0	3.366	-2.740	144.45	148.50
8	0.0	1.00	2.612	-2.164	172.54	190.52
9	1.00	0.0	2.953	-2.632	147.96	167.96
10	0.0	-1.00	2.228	-1.722	203.92	223.51

12 ELEMENT ARRAY

ELEMENT	IZ(0) RE	IZ(0) IM	YO RE	YO IM	ZO RE	ZO IM
1	1.00	0.0	18.852	-2.117	51.87	7.81
2	0.0	-1.00	7.569	-3.974	95.42	59.17
3	-1.00	0.0	6.127	-3.652	108.15	77.17
4	0.0	1.00	4.301	-3.066	130.99	115.31
5	1.00	0.0	4.125	-3.043	132.32	120.69
6	0.0	-1.00	3.234	-2.573	151.83	154.56
7	-1.00	0.0	3.297	-2.675	147.23	151.61
8	0.0	1.00	2.673	-2.241	167.97	186.09
9	1.00	0.0	2.848	-2.453	156.65	174.53
10	0.0	-1.00	2.302	-1.965	184.02	214.66
11	-1.00	0.0	2.591	-2.397	157.69	189.72
12	0.0	1.00	2.029	-1.596	213.98	244.26

14 ELEMENT ARRAY

ELEMENT	IZ(0) RE	IZ(0) IM	YO RE	YO IM	ZO RE	ZO IM
1	1.00	0.0	18.663	-2.275	52.24	8.38
2	0.0	-1.00	7.630	-3.979	95.02	58.51
3	-1.00	0.0	6.076	-3.639	108.66	77.95
4	0.0	1.00	4.334	-3.082	130.38	114.38
5	1.00	0.0	4.088	-3.022	133.11	121.83
6	0.0	-1.00	3.264	-2.595	150.82	153.17
7	-1.00	0.0	3.258	-2.643	148.59	153.39
8	0.0	1.00	2.707	-2.276	166.06	183.77
9	1.00	0.0	2.798	-2.396	159.52	177.69
10	0.0	-1.00	2.350	-2.034	179.31	210.15
11	-1.00	0.0	2.512	-2.241	166.60	196.39
12	0.0	1.00	2.085	-1.814	193.65	235.30
13	1.00	0.0	2.336	-2.215	165.98	208.59
14	0.0	-1.00	1.879	-1.495	222.70	262.49

16 ELEMENT ARRAY

ELEMENT	IZ(0) RE	IZ(0) IM	YO RE	YO IM	ZO RE	ZO IM
1	1.00	0.0	18.517	-2.395	52.52	8.83
2	0.0	-1.00	7.678	-3.982	94.72	58.01
3	-1.00	0.0	6.038	-3.630	109.03	78.54
4	0.0	1.00	4.358	-3.094	129.95	113.70
5	1.00	0.0	4.062	-3.008	133.64	122.63
6	0.0	-1.00	3.284	-2.609	150.18	152.22
7	-1.00	0.0	3.234	-2.625	149.41	154.54
8	0.0	1.00	2.728	-2.295	165.02	182.35
9	1.00	0.0	2.769	-2.369	160.92	179.51
10	0.0	-1.00	2.376	-2.064	177.35	207.79
11	-1.00	0.0	2.472	-2.190	169.55	199.60
12	0.0	1.00	2.124	-1.876	188.82	230.73
13	1.00	0.0	2.273	-2.076	175.09	215.34
14	0.0	-1.00	1.923	-1.693	202.01	253.43
15	-1.00	0.0	2.145	-2.069	173.27	225.32
16	0.0	1.00	1.761	-1.411	230.47	278.79

EN

DRIVING POINT ADMITTANCES AND IMPEDANCES
FOR SPECIFIED BASE VOLTAGES

H/LMDA=0.2500 B/LMDA= 0.250 OMEGA= 8.53

18 ELEMENT ARRAY

ELEMENT	VO		YO		ZO	
	RE	IM	RE	IM	RE	IM
1	1.00	0.0	12.372	-5.582	67.16	30.30
2	0.0	-1.00	3.269	-4.500	105.66	145.45
3	-1.00	0.0	2.614	-3.262	149.62	186.67
4	0.0	1.00	2.247	-2.749	178.24	218.09
5	1.00	0.0	2.007	-2.445	200.54	244.37
6	0.0	-1.00	1.842	-2.214	222.02	266.92
7	-1.00	0.0	1.713	-2.058	238.96	287.04
8	0.0	1.00	1.617	-1.911	257.99	304.95
9	1.00	0.0	1.533	-1.816	271.46	321.61
10	0.0	-1.00	1.469	-1.706	289.91	336.56
11	-1.00	0.0	1.407	-1.646	300.09	351.02
12	0.0	1.00	1.363	-1.551	319.68	363.78
13	1.00	0.0	1.313	-1.521	325.31	376.86
14	0.0	-1.00	1.283	-1.424	349.33	387.60
15	-1.00	0.0	1.236	-1.433	345.23	400.29
16	0.0	1.00	1.227	-1.293	386.08	406.85
17	1.00	0.0	1.115	-1.458	330.96	432.88
18	0.0	-1.00	0.342	-0.294	1680.36	1444.75

20 ELEMENT ARRAY

ELEMENT	VO		YO		ZO	
	RE	IM	RE	IM	RE	IM
1	1.00	0.0	12.372	-5.582	67.16	30.30
2	0.0	-1.00	3.269	-4.501	105.64	145.45
3	-1.00	0.0	2.614	-3.261	149.66	186.67
4	0.0	1.00	2.247	-2.750	178.17	218.09
5	1.00	0.0	2.007	-2.444	200.64	244.37
6	0.0	-1.00	1.842	-2.216	221.88	266.93
7	-1.00	0.0	1.713	-2.056	239.15	287.04
8	0.0	1.00	1.617	-1.913	257.72	304.96
9	1.00	0.0	1.533	-1.814	271.83	321.60
10	0.0	-1.00	1.469	-1.708	289.39	336.59
11	-1.00	0.0	1.408	-1.642	300.84	350.99
12	0.0	1.00	1.362	-1.556	318.56	363.85
13	1.00	0.0	1.314	-1.514	327.02	376.77
14	0.0	-1.00	1.281	-1.434	346.44	387.86
15	-1.00	0.0	1.239	-1.415	350.35	399.91
16	0.0	1.00	1.217	-1.329	374.85	409.13
17	1.00	0.0	1.176	-1.345	368.49	421.33
18	0.0	-1.00	1.173	-1.215	411.30	425.92
19	-1.00	0.0	1.066	-1.381	350.30	453.83
20	0.0	1.00	0.321	-0.244	1972.61	1503.32

22 ELEMENT ARRAY

ELEMENT	VO		YO		ZO	
	RE	IM	RE	IM	RE	IM
1	1.00	0.0	12.372	-5.582	67.16	30.30
2	0.0	-1.00	3.269	-4.501	105.62	145.45
3	-1.00	0.0	2.615	-3.261	149.69	186.67
4	0.0	1.00	2.246	-2.750	178.12	218.09
5	1.00	0.0	2.007	-2.444	200.70	244.37
6	0.0	-1.00	1.841	-2.216	221.78	266.93
7	-1.00	0.0	1.713	-2.055	239.28	287.04
8	0.0	1.00	1.616	-1.914	257.55	304.97
9	1.00	0.0	1.533	-1.812	272.06	321.59
10	0.0	-1.00	1.468	-1.710	289.07	336.60
11	-1.00	0.0	1.408	-1.640	301.27	350.97
12	0.0	1.00	1.362	-1.558	317.95	363.89
13	1.00	0.0	1.315	-1.510	327.87	376.72
14	0.0	-1.00	1.280	-1.439	345.18	387.97
15	-1.00	0.0	1.241	-1.408	352.25	399.77
16	0.0	1.00	1.215	-1.339	371.67	409.49
17	1.00	0.0	1.180	-1.327	374.08	420.79
18	0.0	-1.00	1.163	-1.249	399.14	428.68
19	-1.00	0.0	1.127	-1.270	390.73	440.62
20	0.0	1.00	1.128	-1.148	435.50	443.33
21	1.00	0.0	1.024	-1.314	368.92	473.37
22	0.0	-1.00	0.302	-0.201	2291.68	1524.96

24 ELEMENT ARRAY

ELEMENT	VO		YO		ZO	
	RE	IM	RE	IM	RE	IM
1	1.00	0.0	12.372	-5.581	67.16	30.30
2	0.0	-1.00	3.269	-4.502	105.61	145.45
3	-1.00	0.0	2.615	-3.260	149.71	186.67
4	0.0	1.00	2.246	-2.751	178.09	218.09
5	1.00	0.0	2.007	-2.443	200.75	244.36
6	0.0	-1.00	1.841	-2.217	221.71	266.93
7	-1.00	0.0	1.714	-2.055	239.37	287.03
8	0.0	1.00	1.616	-1.915	257.43	304.98
9	1.00	0.0	1.533	-1.812	272.22	321.58
10	0.0	-1.00	1.468	-1.711	288.86	336.61
11	-1.00	0.0	1.408	-1.639	301.54	350.96
12	0.0	1.00	1.361	-1.560	317.58	363.92
13	1.00	0.0	1.315	-1.508	328.36	376.69
14	0.0	-1.00	1.279	-1.441	344.50	388.03
15	-1.00	0.0	1.241	-1.405	353.20	399.70
16	0.0	1.00	1.214	-1.343	370.28	409.64
17	1.00	0.0	1.181	-1.321	376.16	420.60
18	0.0	-1.00	1.161	-1.259	395.69	429.15
19	-1.00	0.0	1.131	-1.254	396.77	439.92
20	0.0	1.00	1.118	-1.182	422.45	446.60
21	1.00	0.0	1.084	-1.206	412.12	458.44
22	0.0	-1.00	1.089	-1.090	458.83	459.34
23	-1.00	0.0	0.988	-1.256	386.93	491.70
24	0.0	1.00	0.287	-0.163	2633.29	1500.35

DRIVING POINT ADMITTANCES AND IMPEDANCES
FOR SPECIFIED BASE CURRENTS

H/LMDA=0.2500 B/LMDA= 0.250 OMEGA= 8.53

18 ELEMENT ARRAY

ELEMENT	IZ(O)		YO		ZO	
	RE	IM	RE	IM	RE	IM
1	1.00	0.0	18.401	-2.489	52.74	9.20
2	0.0	-1.00	7.715	-3.984	94.49	57.61
3	-1.00	0.0	6.009	-3.623	109.31	79.00
4	0.0	1.00	4.377	-3.102	129.63	113.19
5	1.00	0.0	4.043	-2.998	134.02	123.23
6	0.0	-1.00	3.300	-2.619	149.73	151.52
7	-1.00	0.0	3.216	-2.612	149.95	155.36
8	0.0	1.00	2.743	-2.308	164.35	181.38
9	1.00	0.0	2.750	-2.352	161.76	180.68
10	0.0	-1.00	2.393	-2.082	176.27	206.34
11	-1.00	0.0	2.449	-2.165	170.99	201.44
12	0.0	1.00	2.146	-1.904	186.80	228.33
13	1.00	0.0	2.240	-2.029	178.10	218.60
14	0.0	-1.00	1.956	-1.750	197.08	248.80
15	-1.00	0.0	2.093	-1.942	182.54	232.14
16	0.0	1.00	1.797	-1.593	209.45	269.65
17	1.00	0.0	1.995	-1.948	179.79	240.39
18	0.0	-1.00	1.666	-1.339	237.49	293.58

20 ELEMENT ARRAY

ELEMENT	IZ(O)		YO		ZO	
	RE	IM	RE	IM	RE	IM
1	1.00	0.0	18.305	-2.566	52.92	9.50
2	0.0	-1.00	7.746	-3.985	94.30	57.28
3	-1.00	0.0	5.986	-3.618	109.54	79.38
4	0.0	1.00	4.392	-3.108	129.39	112.77
5	1.00	0.0	4.028	-2.990	134.31	123.70
6	0.0	-1.00	3.311	-2.627	149.40	150.99
7	-1.00	0.0	3.204	-2.603	150.35	155.97
8	0.0	1.00	2.754	-2.316	163.89	180.67
9	1.00	0.0	2.737	-2.341	162.32	181.52
10	0.0	-1.00	2.405	-2.093	175.58	205.35
11	-1.00	0.0	2.434	-2.150	171.85	202.64
12	0.0	1.00	2.161	-1.919	185.70	226.86
13	1.00	0.0	2.221	-2.007	179.58	220.46
14	0.0	-1.00	1.975	-1.775	195.02	246.37
15	-1.00	0.0	2.065	-1.899	185.62	235.44
16	0.0	1.00	1.825	-1.646	204.42	264.97
17	1.00	0.0	1.952	-1.830	189.23	247.28
18	0.0	-1.00	1.694	-1.509	216.17	284.37
19	-1.00	0.0	1.874	-1.845	185.74	254.15
20	0.0	1.00	1.586	-1.277	243.92	307.15

22 ELEMENT ARRAY

ELEMENT	IZ(O)		YO		ZO	
	RE	IM	RE	IM	RE	IM
1	1.00	0.0	[illegible]	-2.630	53.08	9.76
2	0.0	-1.00	7.772	[illegible]	[illegible]	57.01
3	-1.00	0.0	5.967	-3.613	109.72	[illegible]
4	0.0	1.00	4.404	-3.113	129.19	112.44
5	1.00	0.0	4.016	-2.984	134.54	124.08
6	0.0	-1.00	3.320	-2.632	149.14	150.57
7	-1.00	0.0	3.194	-2.597	150.65	156.45
8	0.0	1.00	2.763	-2.323	163.55	180.13
9	1.00	0.0	2.728	-2.333	162.73	182.14
10	0.0	-1.00	2.414	-2.101	175.11	204.63
11	-1.00	0.0	2.424	-2.140	172.43	203.48
12	0.0	1.00	2.170	-1.930	185.00	225.86
13	1.00	0.0	2.209	-1.993	180.46	221.68
14	0.0	-1.00	1.987	-1.790	193.89	244.87
15	-1.00	0.0	2.049	-1.878	187.13	237.33
16	0.0	1.00	1.842	-1.670	202.32	262.50
17	1.00	0.0	1.927	-1.791	192.36	250.62
18	0.0	-1.00	1.720	-1.558	211.06	279.64
19	-1.00	0.0	1.837	-1.736	195.31	261.09
20	0.0	1.00	1.609	-1.436	222.34	297.87
21	1.00	0.0	1.774	-1.756	191.21	266.82
22	0.0	-1.00	1.518	-1.222	249.88	319.70

24 ELEMENT ARRAY

ELEMENT	IZ(O)		YO		ZO	
	RE	IM	RE	IM	RE	IM
1	1.00	0.0	18.156	-2.684	53.21	9.98
2	0.0	-1.00	7.794	-3.986	94.02	56.77
3	-1.00	0.0	5.951	-3.610	109.87	79.94
4	0.0	1.00	4.415	-3.117	129.03	112.16
5	1.00	0.0	4.007	-2.979	134.73	124.39
6	0.0	-1.00	3.328	-2.637	148.94	150.23
7	-1.00	0.0	3.186	-2.591	150.88	156.84
8	0.0	1.00	2.770	-2.328	163.28	179.70
9	1.00	0.0	2.720	-2.327	163.04	182.62
10	0.0	-1.00	2.421	-2.107	174.75	204.09
11	-1.00	0.0	2.416	-2.133	172.84	204.12
12	0.0	1.00	2.178	-1.937	184.51	225.13
13	1.00	0.0	2.200	-1.984	181.05	222.53
14	0.0	-1.00	1.996	-1.800	193.17	243.86
15	-1.00	0.0	2.038	-1.866	188.03	238.56
16	0.0	1.00	1.853	-1.684	201.16	260.99
17	1.00	0.0	1.913	-1.771	193.90	252.53
18	0.0	-1.00	1.734	-1.581	208.92	277.15
19	-1.00	0.0	1.815	-1.699	198.49	264.46
20	0.0	1.00	1.632	-1.483	217.15	293.09
21	1.00	0.0	1.742	-1.654	200.91	273.82
22	0.0	-1.00	1.537	-1.372	228.04	310.36
23	-1.00	0.0	1.690	-1.679	196.29	278.59
24	0.0	1.00	1.459	-1.173	255.44	331.40

EN

DRIVING POINT ADMITTANCES AND IMPEDANCES
FOR SPECIFIED BASE VOLTAGES

H/LMDA=0.2500 B/LMDA= 0.250 OMEGA= 8.53

26 ELEMENT ARRAY

ELEMENT	VO RE	VO IM	YO RE	YO IM	ZO RE	ZO IM
1	1.00	0.0	12.372	-5.581	67.16	30.30
2	0.0	-1.00	3.269	-4.502	105.60	145.45
3	-1.00	0.0	2.615	-3.260	149.73	186.67
4	0.0	1.00	2.246	-2.751	178.06	218.09
5	1.00	0.0	2.007	-2.443	200.79	244.36
6	0.0	-1.00	1.841	-2.217	221.66	266.93
7	-1.00	0.0	1.714	-2.054	239.44	287.03
8	0.0	1.00	1.616	-1.915	257.34	304.98
9	1.00	0.0	1.534	-1.811	272.34	321.58
10	0.0	-1.00	1.468	-1.712	288.71	336.63
11	-1.00	0.0	1.409	-1.638	301.73	350.94
12	0.0	1.00	1.361	-1.561	317.34	363.94
13	1.00	0.0	1.315	-1.507	328.68	376.66
14	0.0	-1.00	1.279	-1.443	344.09	388.07
15	-1.00	0.0	1.242	-1.403	353.74	399.65
16	0.0	1.00	1.214	-1.346	369.53	409.72
17	1.00	0.0	1.182	-1.318	377.20	420.50
18	0.0	-1.00	1.160	-1.264	394.18	429.35
19	-1.00	0.0	1.132	-1.247	399.02	439.66
20	0.0	1.00	1.116	-1.191	418.74	447.18
21	1.00	0.0	1.088	-1.190	418.60	457.57
22	0.0	-1.00	1.079	-1.123	444.94	463.13
23	-1.00	0.0	1.048	-1.150	432.81	475.02
24	0.0	1.00	1.054	-1.038	481.44	474.13
25	1.00	0.0	0.957	-1.204	404.42	508.99
26	0.0	-1.00	0.273	-0.130	2989.62	1420.18

28 ELEMENT ARRAY

ELEMENT	VO RE	VO IM	YO RE	YO IM	ZO RE	ZO IM
1	1.00	0.0	12.372	-5.581	67.16	30.30
2	0.0	-1.00	3.269	-4.502	105.60	145.45
3	-1.00	0.0	2.615	-3.260	149.74	186.67
4	0.0	1.00	2.246	-2.752	178.04	218.09
5	1.00	0.0	2.007	-2.443	200.82	244.36
6	0.0	-1.00	1.841	-2.218	221.62	266.94
7	-1.00	0.0	1.714	-2.054	239.49	287.03
8	0.0	1.00	1.616	-1.916	257.27	304.98
9	1.00	0.0	1.534	-1.810	272.42	321.57
10	0.0	-1.00	1.468	-1.712	288.60	336.63
11	-1.00	0.0	1.409	-1.638	301.87	350.93
12	0.0	1.00	1.361	-1.562	317.17	363.95
13	1.00	0.0	1.315	-1.506	328.89	376.64
14	0.0	-1.00	1.279	-1.444	343.82	388.10
15	-1.00	0.0	1.242	-1.402	354.09	399.61
16	0.0	1.00	1.214	-1.347	369.08	409.78
17	1.00	0.0	1.182	-1.316	377.79	420.43
18	0.0	-1.00	1.160	-1.266	393.37	429.45
19	-1.00	0.0	1.133	-1.244	400.14	439.53
20	0.0	1.00	1.115	-1.196	417.12	447.42
21	1.00	0.0	1.090	-1.184	421.01	457.25
22	0.0	-1.00	1.077	-1.132	440.98	463.83
23	-1.00	0.0	1.052	-1.134	439.71	473.96
24	0.0	1.00	1.045	-1.071	466.73	478.47
25	1.00	0.0	1.016	-1.100	452.91	490.51
26	0.0	-1.00	1.024	-0.993	503.42	487.85
27	-1.00	0.0	0.929	-1.158	421.48	525.37
28	0.0	1.00	0.261	-0.099	3348.90	1276.60

30 ELEMENT ARRAY

ELEMENT	VO RE	VO IM	YO RE	YO IM	ZO RE	ZO IM
1	1.00	0.0	12.372	-5.581	67.16	30.30
2	0.0	-1.00	3.269	-4.502	105.59	145.45
3	-1.00	0.0	2.615	-3.259	149.75	186.67
4	0.0	1.00	2.246	-2.752	178.02	218.09
5	1.00	0.0	2.007	-2.442	200.85	244.36
6	0.0	-1.00	1.841	-2.218	221.59	266.94
7	-1.00	0.0	1.714	-2.054	239.53	287.03
8	0.0	1.00	1.616	-1.916	257.22	304.99
9	1.00	0.0	1.534	-1.810	272.49	321.57
10	0.0	-1.00	1.468	-1.713	288.52	336.64
11	-1.00	0.0	1.409	-1.637	301.97	350.93
12	0.0	1.00	1.361	-1.562	317.05	363.97
13	1.00	0.0	1.316	-1.506	329.05	376.63
14	0.0	-1.00	1.279	-1.444	343.63	388.12
15	-1.00	0.0	1.242	-1.401	354.33	399.59
16	0.0	1.00	1.213	-1.348	368.78	409.81
17	1.00	0.0	1.183	-1.315	378.18	420.39
18	0.0	-1.00	1.159	-1.268	392.87	429.52
19	-1.00	0.0	1.133	-1.242	400.79	439.45
20	0.0	1.00	1.114	-1.198	416.24	447.56
21	1.00	0.0	1.090	-1.181	422.22	457.08
22	0.0	-1.00	1.076	-1.137	439.24	464.13
23	-1.00	0.0	1.053	-1.128	442.28	473.57
24	0.0	1.00	1.043	-1.080	462.52	479.30
25	1.00	0.0	1.020	-1.084	460.20	489.26
26	0.0	-1.00	1.015	-1.025	487.92	492.76
27	-1.00	0.0	0.988	-1.056	472.48	505.05
28	0.0	1.00	0.998	-0.952	524.84	500.63
29	1.00	0.0	0.904	-1.116	438.15	540.95
30	0.0	-1.00	0.250	-0.072	3695.59	1065.04

DRIVING POINT ADMITTANCES AND IMPEDANCES
FOR SPECIFIED BASE CURRENTS

H/LMDA=0.2500 B/LMDA= 0.250 OMEGA= 8.53

26 ELEMENT ARRAY

ELEMENT	IZ(O) RE	IZ(O) IM	YO RE	YO IM	ZO RE	ZO IM
1	1.00	0.0	18.097	-2.730	53.32	10.16
2	0.0	-1.00	7.813	-3.987	93.91	56.57
3	-1.00	0.0	5.937	-3.607	110.00	80.17
4	0.0	1.00	4.423	-3.121	128.89	111.92
5	1.00	0.0	3.999	-2.975	134.88	124.66
6	0.0	-1.00	3.334	-2.641	148.77	149.95
7	-1.00	0.0	3.179	-2.587	151.07	157.15
8	0.0	1.00	2.775	-2.332	163.07	179.35
9	1.00	0.0	2.714	-2.322	163.28	183.01
10	0.0	-1.00	2.426	-2.112	174.48	203.65
11	-1.00	0.0	2.410	-2.128	173.16	204.61
12	0.0	1.00	2.183	-1.943	184.15	224.57
13	1.00	0.0	2.193	-1.977	181.48	223.17
14	0.0	-1.00	2.002	-1.806	192.67	243.12
15	-1.00	0.0	2.030	-1.857	188.63	239.43
16	0.0	1.00	1.860	-1.693	200.43	259.96
17	1.00	0.0	1.903	-1.759	194.81	253.77
18	0.0	-1.00	1.744	-1.594	207.75	275.62
19	-1.00	0.0	1.802	-1.680	200.06	266.40
20	0.0	1.00	1.645	-1.504	214.97	290.58
21	1.00	0.0	1.722	-1.619	204.14	277.22
22	0.0	-1.00	1.558	-1.417	222.78	305.55
23	-1.00	0.0	1.661	-1.582	206.11	285.64
24	0.0	1.00	1.476	-1.316	233.37	322.01
25	1.00	0.0	1.618	-1.610	201.05	289.60
26	0.0	-1.00	1.408	-1.129	260.66	342.38

28 ELEMENT ARRAY

ELEMENT	IZ(O) RE	IZ(O) IM	YO RE	YO IM	ZO RE	ZO IM
1	1.00	0.0	18.045	-2.770	53.42	10.33
2	0.0	-1.00	7.830	-3.987	93.81	56.40
3	-1.00	0.0	5.926	-3.604	110.12	80.36
4	0.0	1.00	4.431	-3.124	128.77	111.72
5	1.00	0.0	3.992	-2.972	135.02	124.88
6	0.0	-1.00	3.340	-2.644	148.63	149.70
7	-1.00	0.0	3.174	-2.584	151.23	157.42
8	0.0	1.00	2.780	-2.336	162.90	179.06
9	1.00	0.0	2.709	-2.319	163.47	183.34
10	0.0	-1.00	2.431	-2.115	174.27	203.30
11	-1.00	0.0	2.405	-2.124	173.40	205.00
12	0.0	1.00	2.188	-1.947	183.87	224.13
13	1.00	0.0	2.188	-1.972	181.80	223.67
14	0.0	-1.00	2.007	-1.812	192.30	242.56
15	-1.00	0.0	2.025	-1.851	189.06	240.07
16	0.0	1.00	1.866	-1.699	199.92	259.22
17	1.00	0.0	1.897	-1.751	195.43	254.65
18	0.0	-1.00	1.751	-1.602	207.00	274.58
19	-1.00	0.0	1.794	-1.669	200.99	267.65
20	0.0	1.00	1.654	-1.516	213.78	289.03
21	1.00	0.0	1.711	-1.601	205.73	279.17
22	0.0	-1.00	1.570	-1.437	220.58	303.01
23	-1.00	0.0	1.643	-1.549	209.39	289.07
24	0.0	1.00	1.494	-1.358	228.04	317.15
25	1.00	0.0	1.592	-1.518	210.99	296.69
26	0.0	-1.00	1.422	-1.266	238.37	332.93
27	-1.00	0.0	1.555	-1.549	205.53	299.94
28	0.0	1.00	1.362	-1.089	265.59	352.71

30 ELEMENT ARRAY

ELEMENT	IZ(O) RE	IZ(O) IM	YO RE	YO IM	ZO RE	ZO IM
1	1.00	0.0	18.000	-2.805	53.50	10.48
2	0.0	-1.00	7.844	-3.987	93.73	56.25
3	-1.00	0.0	5.915	-3.602	110.21	80.53
4	0.0	1.00	4.437	-3.127	128.67	111.54
5	1.00	0.0	3.986	-2.969	135.13	125.07
6	0.0	-1.00	3.345	-2.647	148.51	149.50
7	-1.00	0.0	3.169	-2.580	151.37	157.65
8	0.0	1.00	2.784	-2.339	162.75	178.82
9	1.00	0.0	2.705	-2.315	163.64	183.60
10	0.0	-1.00	2.434	-2.119	174.09	203.00
11	-1.00	0.0	2.401	-2.120	173.60	205.33
12	0.0	1.00	2.192	-1.950	183.65	223.77
13	1.00	0.0	2.184	-1.968	182.05	224.07
14	0.0	-1.00	2.011	-1.816	192.02	242.11
15	-1.00	0.0	2.020	-1.846	189.39	240.58
16	0.0	1.00	1.870	-1.704	199.55	258.65
17	1.00	0.0	1.892	-1.746	195.86	255.30
18	0.0	-1.00	1.756	-1.608	206.48	273.82
19	-1.00	0.0	1.788	-1.662	201.61	268.54
20	0.0	1.00	1.660	-1.524	213.02	287.98
21	1.00	0.0	1.703	-1.591	206.67	280.44
22	0.0	-1.00	1.578	-1.449	219.37	301.45
23	-1.00	0.0	1.633	-1.532	211.00	291.04
24	0.0	1.00	1.506	-1.378	225.81	314.60
25	1.00	0.0	1.576	-1.487	214.30	300.15
26	0.0	-1.00	1.439	-1.306	232.98	328.04
27	-1.00	0.0	1.531	-1.461	215.57	307.08
28	0.0	1.00	1.374	-1.220	243.10	343.22
29	1.00	0.0	1.499	-1.493	209.77	309.71
30	0.0	-1.00	1.322	-1.053	270.27	362.50

EN

DRIVING POINT ADMITTANCES AND IMPEDANCES
FOR SPECIFIED BASE VOLTAGES

H/LMDA=0.2500 B/LMDA= 0.250 OMEGA= 8.53

35 ELEMENT ARRAY

ELEMENT	VO		YO		ZO	
	RE	IM	RE	IM	RE	IM
1	1.00	0.0	12.373	-5.580	67.16	30.29
2	0.0	-1.00	3.268	-4.505	105.50	145.45
3	-1.00	0.0	2.616	-3.257	149.91	186.65
4	0.0	1.00	2.245	-2.755	177.78	218.12
5	1.00	0.0	2.008	-2.439	201.16	244.32
6	0.0	-1.00	1.840	-2.221	221.19	266.99
7	-1.00	0.0	1.715	-2.050	240.02	286.96
8	0.0	1.00	1.615	-1.920	256.63	305.07
9	1.00	0.0	1.535	-1.806	273.18	321.47
10	0.0	-1.00	1.466	-1.717	287.71	336.77
11	-1.00	0.0	1.410	-1.633	302.91	350.78
12	0.0	1.00	1.359	-1.567	315.96	364.15
13	1.00	0.0	1.317	-1.501	330.30	376.42
14	0.0	-1.00	1.277	-1.450	342.17	388.37
15	-1.00	0.0	1.244	-1.395	356.00	399.29
16	0.0	1.00	1.211	-1.355	366.85	410.17
17	1.00	0.0	1.185	-1.308	380.41	419.97
18	0.0	-1.00	1.157	-1.275	390.28	430.02
19	-1.00	0.0	1.135	-1.234	403.81	438.86
20	0.0	1.00	1.112	-1.207	412.68	448.27
21	1.00	0.0	1.093	-1.170	426.44	456.24
22	0.0	-1.00	1.072	-1.149	434.16	465.18
23	-1.00	0.0	1.057	-1.113	448.47	472.31
24	0.0	1.00	1.038	-1.098	454.77	480.96
25	1.00	0.0	1.026	-1.063	470.14	487.21
26	0.0	-1.00	1.008	-1.053	474.45	495.78
27	-1.00	0.0	0.998	-1.017	491.75	501.01
28	0.0	1.00	0.980	-1.014	492.93	509.86
29	1.00	0.0	0.973	-0.973	513.96	513.66
30	0.0	-1.00	0.955	-0.981	509.27	523.53
31	-1.00	0.0	0.953	-0.928	538.73	524.63
32	0.0	1.00	0.929	-0.962	519.53	537.89
33	1.00	0.0	0.942	-0.864	576.37	529.05
34	0.0	-1.00	0.852	-1.027	478.46	576.88
35	-1.00	0.0	0.227	-0.013	4389.03	258.69

40 ELEMENT ARRAY

ELEMENT	VO		YO		ZO	
	RE	IM	RE	IM	RE	IM
1	1.00	0.0	12.372	-5.581	67.16	30.29
2	0.0	-1.00	3.268	-4.503	105.57	145.45
3	-1.00	0.0	2.615	-3.259	149.79	186.67
4	[illegible]	1.00	2.246	-2.752	177.97	218.10
5	1.00	0.0	2.008	-2.442	[illegible]	[illegible]
6	0.0	-1.00	1.841	-2.219	221.50	266.95
7	-1.00	0.0	1.714	-2.053	239.65	287.02
8	0.0	1.00	1.616	-1.917	257.07	305.00
9	1.00	0.0	1.534	-1.809	272.67	321.55
10	0.0	-1.00	1.467	-1.714	288.31	336.66
11	-1.00	0.0	1.409	-1.636	302.23	350.90
12	0.0	1.00	1.360	-1.563	316.73	364.00
13	1.00	0.0	1.316	-1.504	329.42	376.58
14	0.0	-1.00	1.278	-1.446	343.18	388.18
15	-1.00	0.0	1.243	-1.399	354.87	399.51
16	0.0	1.00	1.213	-1.350	368.13	409.91
17	1.00	0.0	1.183	-1.312	378.96	420.27
18	0.0	-1.00	1.159	-1.270	391.91	429.67
19	-1.00	0.0	1.134	-1.239	401.96	439.27
20	0.0	1.00	1.113	-1.202	414.77	447.80
21	1.00	0.0	1.092	-1.176	424.05	456.79
22	0.0	-1.00	1.074	-1.142	436.88	464.55
23	-1.00	0.0	1.055	-1.121	445.35	473.05
24	0.0	1.00	1.040	-1.090	458.38	480.09
25	1.00	0.0	1.023	-1.072	465.94	488.24
26	0.0	-1.00	1.010	-1.042	479.39	494.57
27	-1.00	0.0	0.994	-1.029	485.86	502.50
28	0.0	1.00	0.984	-1.000	500.07	508.09
29	1.00	0.0	0.969	-0.990	505.08	515.95
30	0.0	-1.00	0.960	-0.960	520.58	520.71
31	-1.00	0.0	0.946	-0.955	523.51	528.74
32	0.0	1.00	0.939	-0.924	541.22	532.41
33	1.00	0.0	0.924	-0.925	540.80	541.03
34	0.0	-1.00	0.920	-0.888	562.68	543.08
35	-1.00	0.0	0.904	-0.899	555.89	553.20
36	0.0	1.00	0.904	-0.850	587.08	552.00
37	1.00	0.0	0.882	-0.886	564.45	566.59
38	0.0	-1.00	0.897	-0.793	625.54	553.26
39	-1.00	0.0	0.810	-0.954	517.21	609.26
40	0.0	1.00	0.209	0.035	4658.96	-774.73

DRIVING POINT ADMITTANCES AND IMPEDANCES
FOR SPECIFIED BASE CURRENTS

H/LMDA=0.2500 B/LMDA= 0.250 OMEGA= 8.53

35 ELEMENT ARRAY

ELEMENT	IZ(0)		YO		ZO	
	RE	IM	RE	IM	RE	IM
1	1.00	0.0	16.065	-3.853	57.58	16.39
2	0.0	-1.00	8.468	-4.017	89.95	50.31
3	-1.00	0.0	5.552	-3.491	114.35	86.73
4	0.0	1.00	4.683	-3.235	124.63	105.28
5	1.00	0.0	3.793	-2.860	139.44	131.60
6	0.0	-1.00	3.500	-2.747	144.22	142.87
7	-1.00	0.0	3.034	-2.483	155.90	164.54
8	0.0	1.00	2.901	-2.430	158.19	171.78
9	1.00	0.0	2.598	-2.225	168.43	190.93
10	0.0	-1.00	2.531	-2.204	169.22	195.49
11	-1.00	0.0	2.310	-2.035	178.72	213.16
12	0.0	1.00	2.277	-2.033	178.42	215.71
13	1.00	0.0	2.102	-1.886	187.54	232.49
14	0.0	-1.00	2.089	-1.896	186.36	233.39
15	-1.00	0.0	1.943	-1.764	195.35	249.71
16	0.0	1.00	1.944	-1.785	193.37	249.14
17	1.00	0.0	1.818	-1.663	202.40	265.31
18	0.0	-1.00	1.828	-1.691	199.65	263.34
19	-1.00	0.0	1.715	-1.575	208.89	279.63
20	0.0	1.00	1.733	-1.612	205.32	276.26
21	1.00	0.0	1.628	-1.499	214.96	292.96
22	0.0	-1.00	1.654	-1.543	210.47	288.07
23	-1.00	0.0	1.554	-1.431	220.72	305.51
24	0.0	1.00	1.586	-1.484	215.13	298.88
25	1.00	0.0	1.489	-1.369	226.30	317.47
26	0.0	-1.00	1.529	-1.433	219.28	308.73
27	-1.00	0.0	1.431	-1.311	231.84	329.07
28	0.0	1.00	1.481	-1.390	222.82	317.57
29	1.00	0.0	1.378	-1.255	237.61	340.62
30	0.0	-1.00	1.440	-1.357	225.43	325.13
31	-1.00	0.0	1.327	-1.196	244.23	352.68
32	0.0	1.00	1.409	-1.341	226.06	330.68
33	1.00	0.0	1.276	-1.123	253.95	366.68
34	0.0	-1.00	1.385	-1.376	219.54	332.05
35	-1.00	0.0	1.237	-0.974	281.07	384.95

40 ELEMENT ARRAY

ELEMENT	IZ(0)		YO		ZO	
	RE	IM	RE	IM	RE	IM
1	1.00	0.0	17.832	-2.932	53.82	11.02
2	0.0	-1.00	7.898	-3.987	93.42	55.68
3	-1.00	0.0	5.879	-3.594	110.56	81.13
4	0.0	1.00	4.461	-3.136	128.32	110.90
5	1.00	0.0	3.965	-2.959	135.52	125.76
6	0.0	-1.00	[illegible]	[illegible]	148.10	148.77
7	-1.00	0.0	3.153	-2.570	151.82	158.43
8	0.0	1.00	2.797	-2.348	162.27	177.99
9	1.00	0.0	2.692	-2.305	164.16	184.50
10	0.0	-1.00	2.446	-2.128	173.52	202.04
11	-1.00	0.0	2.389	-2.109	174.23	206.37
12	0.0	1.00	2.203	-1.961	182.97	222.65
13	1.00	0.0	2.172	-1.957	182.79	225.30
14	0.0	-1.00	2.023	-1.827	191.20	240.78
15	-1.00	0.0	2.007	-1.833	190.29	242.04
16	0.0	1.00	1.882	-1.717	198.55	257.04
17	1.00	0.0	1.878	-1.731	196.99	257.08
18	0.0	-1.00	1.769	-1.623	205.22	271.85
19	-1.00	0.0	1.773	-1.644	203.05	270.75
20	0.0	1.00	1.675	-1.542	211.38	285.49
21	1.00	0.0	1.685	-1.569	208.59	283.28
22	0.0	-1.00	1.596	-1.472	217.12	298.19
23	-1.00	0.0	1.612	-1.504	213.68	294.84
24	0.0	1.00	1.527	-1.408	222.55	310.11
25	1.00	0.0	1.548	-1.446	218.37	305.54
26	0.0	-1.00	1.468	-1.351	227.73	321.42
27	-1.00	0.0	1.493	-1.396	222.69	315.46
28	0.0	1.00	1.414	-1.299	232.75	332.23
29	1.00	0.0	1.446	-1.351	226.64	324.64
30	0.0	-1.00	1.367	-1.250	237.69	342.69
31	-1.00	0.0	1.404	-1.312	230.17	333.07
32	0.0	1.00	1.323	-1.204	242.68	352.98
33	1.00	0.0	1.368	-1.279	233.15	340.64
34	0.0	-1.00	1.281	-1.158	247.98	363.37
35	-1.00	0.0	1.338	-1.254	235.26	347.07
36	0.0	1.00	1.241	-1.108	254.21	374.42
37	1.00	0.0	1.315	-1.244	235.39	351.58
38	0.0	-1.00	1.200	-1.044	263.64	387.54
39	-1.00	0.0	1.297	-1.280	228.30	351.99
40	0.0	1.00	1.170	-0.909	290.79	405.03

EN

DRIVING POINT ADMITTANCES AND IMPEDANCES
FOR SPECIFIED BASE VOLTAGES

H/LMDA=0.2500 B/LMDA= 0.250 OMEGA= 8.53

45 ELEMENT ARRAY

ELEMENT	VO RE	VO IM	YO RE	YO IM	ZO RE	ZO IM
1	1.00	0.0	12.373	-5.580	67.16	30.29
2	0.0	-1.00	3.268	-4.505	105.51	145.45
3	-1.00	0.0	2.616	-3.257	149.89	186.65
4	0.0	1.00	2.245	-2.754	177.82	218.12
5	1.00	0.0	2.008	-2.440	201.11	244.33
6	0.0	-1.00	1.840	-2.221	221.25	266.99
7	-1.00	0.0	1.715	-2.051	239.94	286.97
8	0.0	1.00	1.615	-1.919	256.73	305.06
9	1.00	0.0	1.535	-1.807	273.06	321.48
10	0.0	-1.00	1.467	-1.716	287.85	336.75
11	-1.00	0.0	1.410	-1.634	302.75	350.80
12	0.0	1.00	1.360	-1.566	316.14	364.12
13	1.00	0.0	1.317	-1.502	330.08	376.45
14	0.0	-1.00	1.277	-1.449	342.43	388.34
15	-1.00	0.0	1.244	-1.396	355.70	399.33
16	0.0	1.00	1.212	-1.353	367.20	410.12
17	1.00	0.0	1.184	-1.309	380.00	420.04
18	0.0	-1.00	1.158	-1.274	390.76	429.94
19	-1.00	0.0	1.135	-1.235	403.25	438.96
20	0.0	1.00	1.112	-1.206	413.35	448.15
21	1.00	0.0	1.093	-1.172	425.64	456.39
22	0.0	-1.00	1.073	-1.147	435.11	465.00
23	-1.00	0.0	1.057	-1.116	447.32	472.54
24	0.0	1.00	1.039	-1.095	456.16	480.68
25	1.00	0.0	1.025	-1.067	468.42	487.57
26	0.0	-1.00	1.009	-1.048	476.59	495.34
27	-1.00	0.0	0.996	-1.022	489.03	501.62
28	0.0	1.00	0.982	-1.007	496.45	509.11
29	1.00	0.0	0.971	-0.982	509.23	514.78
30	0.0	-1.00	0.958	-0.969	515.78	522.09
31	-1.00	0.0	0.949	-0.945	529.12	527.12
32	0.0	1.00	0.936	-0.935	534.56	534.38
33	1.00	0.0	0.928	-0.911	548.81	538.70
34	0.0	-1.00	0.916	-0.904	552.76	546.07
35	-1.00	0.0	0.909	-0.879	568.47	549.52
36	0.0	1.00	0.897	-0.877	570.23	557.28
37	1.00	0.0	0.892	-0.849	588.40	559.56
38	0.0	-1.00	0.880	-0.852	586.56	568.18
39	-1.00	0.0	0.877	-0.819	609.31	568.58
40	0.0	1.00	0.863	-0.832	600.62	579.19
41	1.00	0.0	0.865	-0.786	633.49	575.73
42	0.0	-1.00	0.844	-0.822	607.70	591.93
43	-1.00	0.0	0.860	-0.734	672.79	574.01
44	0.0	1.00	0.775	-0.892	554.76	638.75
45	1.00	0.0	0.194	0.075	4481.70	-1739.73

DRIVING POINT ADMITTANCES AND IMPEDANCES
FOR SPECIFIED BASE CURRENTS

H/LMDA=0.2500 B/LMDA= 0.250 OMEGA= 8.53

45 ELEMENT ARRAY

ELEMENT	IZ(0) RE	IZ(0) IM	YO RE	YO IM	ZO RE	ZO IM
1	1.00	0.0	16.188	-3.787	57.33	15.96
2	0.0	-1.00	8.421	-4.019	90.19	50.76
3	-1.00	0.0	5.577	-3.497	114.08	86.25
4	0.0	1.00	4.663	-3.228	124.90	105.77
5	1.00	0.0	3.808	-2.867	139.14	131.07
6	0.0	-1.00	3.486	-2.739	144.53	143.42
7	-1.00	0.0	3.045	-2.490	155.56	163.95
8	0.0	1.00	2.890	-2.423	158.54	172.40
9	1.00	0.0	2.608	-2.233	168.05	190.27
10	0.0	-1.00	2.522	-2.197	169.63	196.19
11	-1.00	0.0	2.318	-2.042	178.28	212.41
12	0.0	1.00	2.268	-2.025	178.89	216.50
13	1.00	0.0	2.110	-1.893	187.03	231.64
14	0.0	-1.00	2.081	-1.888	186.91	234.31
15	-1.00	0.0	1.952	-1.772	194.75	248.72
16	0.0	1.00	1.935	-1.776	194.02	250.20
17	1.00	0.0	1.826	-1.671	201.69	264.15
18	0.0	-1.00	1.819	-1.682	200.42	264.59
19	-1.00	0.0	1.723	-1.585	208.04	278.27
20	0.0	1.00	1.723	-1.601	206.26	277.75
21	1.00	0.0	1.638	-1.510	213.92	291.32
22	0.0	-1.00	1.643	-1.530	211.64	289.88
23	-1.00	0.0	1.564	-1.444	219.41	303.49
24	0.0	1.00	1.574	-1.468	216.62	301.14
25	1.00	0.0	1.501	-1.385	224.59	314.92
26	0.0	-1.00	1.515	-1.413	221.26	311.63
27	-1.00	0.0	1.445	-1.331	229.52	325.74
28	0.0	1.00	1.463	-1.364	225.59	321.44
29	1.00	0.0	1.396	-1.283	234.25	336.05
30	0.0	-1.00	1.417	-1.320	229.62	330.62
31	-1.00	0.0	1.351	-1.238	238.83	345.94
32	0.0	1.00	1.377	-1.281	233.36	339.21
33	1.00	0.0	1.311	-1.196	243.32	355.51
34	0.0	-1.00	1.341	-1.246	236.80	347.20
35	-1.00	0.0	1.274	-1.156	247.80	364.88
36	0.0	1.00	1.309	-1.215	239.87	354.57
37	1.00	0.0	1.239	-1.117	252.39	374.19
38	0.0	-1.00	1.281	-1.189	242.44	361.18
39	-1.00	0.0	1.206	-1.078	257.34	383.71
40	0.0	1.00	1.258	-1.169	244.16	366.73
41	1.00	0.0	1.173	-1.035	263.28	393.98
42	0.0	-1.00	1.240	-1.163	243.90	370.44
43	-1.00	0.0	1.138	-0.977	272.51	406.40
44	0.0	1.00	1.226	-1.199	236.33	370.07
45	1.00	0.0	1.114	-0.853	299.73	423.30

50 ELEMENT ARRAY

ELEMENT	VO RE	VO IM	YO RE	YO IM	ZO RE	ZO IM
1	1.00	0.0	12.372	-5.581	67.16	30.29
2	0.0	-1.00	3.268	-4.503	105.56	145.45
3	-1.00	0.0	2.615	-3.259	149.80	186.67
4	0.0	1.00	2.246	-2.753	177.94	218.10
5	1.00	0.0	2.008	-2.441	200.95	244.35
6	0.0	-1.00	1.841	-2.219	221.45	266.95
7	-1.00	0.0	1.714	-2.053	239.70	287.01
8	0.0	1.00	1.616	-1.917	257.01	305.01
9	1.00	0.0	1.534	-1.809	272.74	321.54
10	0.0	-1.00	1.467	-1.714	288.22	336.68
11	-1.00	0.0	1.409	-1.636	302.33	350.88
12	0.0	1.00	1.360	-1.564	316.61	364.02
13	1.00	0.0	1.316	-1.504	329.56	376.56
14	0.0	-1.00	1.278	-1.447	343.01	388.20
15	-1.00	0.0	1.243	-1.399	355.05	399.48
16	0.0	1.00	1.213	-1.351	367.91	409.94
17	1.00	0.0	1.184	-1.312	379.21	420.23
18	0.0	-1.00	1.159	-1.271	391.63	429.72
19	-1.00	0.0	1.134	-1.238	402.29	439.21
20	0.0	1.00	1.113	-1.203	414.40	447.87
21	1.00	0.0	1.092	-1.175	424.48	456.70
22	0.0	-1.00	1.074	-1.144	436.38	464.65
23	-1.00	0.0	1.055	-1.119	445.93	472.93
24	0.0	1.00	1.040	-1.091	457.70	480.24
25	1.00	0.0	1.023	-1.070	466.73	488.07
26	0.0	-1.00	1.010	-1.044	478.46	494.79
27	-1.00	0.0	0.995	-1.026	486.96	502.24
28	0.0	1.00	0.983	-1.002	498.74	508.42
29	1.00	0.0	0.970	-0.987	506.68	515.57
30	0.0	-1.00	0.959	-0.964	518.62	521.21
31	-1.00	0.0	0.947	-0.951	525.93	528.14
32	0.0	1.00	0.938	-0.929	538.15	533.24
33	1.00	0.0	0.926	-0.918	544.75	540.02
34	0.0	-1.00	0.918	-0.897	557.40	544.56
35	-1.00	0.0	0.907	-0.888	563.12	551.30
36	0.0	1.00	0.900	-0.867	576.45	555.22
37	1.00	0.0	0.889	-0.860	581.04	562.04
38	0.0	-1.00	0.883	-0.839	595.40	565.24
39	-1.00	0.0	0.873	-0.835	598.42	572.32
40	0.0	1.00	0.868	-0.812	614.40	574.60
41	1.00	0.0	0.857	-0.812	615.11	582.26
42	0.0	-1.00	0.854	-0.786	633.77	583.23
43	-1.00	0.0	0.843	-0.791	630.67	592.03
44	0.0	1.00	0.842	-0.760	654.27	590.85
45	1.00	0.0	0.829	-0.775	643.84	602.12
46	0.0	-1.00	0.831	-0.731	678.32	596.41
47	-1.00	0.0	0.812	-0.769	649.60	614.46
48	0.0	1.00	0.829	-0.683	718.45	591.82
49	1.00	0.0	0.746	-0.840	591.36	665.81
50	0.0	-1.00	0.181	0.110	4028.79	-2446.82

50 ELEMENT ARRAY

ELEMENT	IZ(0) RE	IZ(0) IM	YO RE	YO IM	ZO RE	ZO IM
1	1.00	0.0	17.724	-3.012	54.03	11.38
2	0.0	-1.00	7.933	-3.987	93.23	55.32
3	-1.00	0.0	5.856	-3.589	110.78	81.52
4	0.0	1.00	4.476	-3.141	128.10	110.51
5	1.00	0.0	3.952	-2.953	135.76	126.18
6	0.0	-1.00	3.371	-2.662	147.85	148.34
7	-1.00	0.0	3.144	-2.564	152.08	158.89
8	0.0	1.00	2.805	-2.354	162.00	177.51
9	1.00	0.0	2.684	-2.299	164.46	185.01
10	0.0	-1.00	2.453	-2.134	173.21	201.51
11	-1.00	0.0	2.382	-2.103	174.56	206.94
12	0.0	1.00	2.209	-1.966	182.63	222.05
13	1.00	0.0	2.165	-1.951	183.17	225.93
14	0.0	-1.00	2.029	-1.833	190.81	240.10
15	-1.00	0.0	2.001	-1.827	190.72	242.76
16	0.0	1.00	1.888	-1.723	198.09	256.28
17	1.00	0.0	1.872	-1.724	197.48	257.90
18	0.0	-1.00	1.775	-1.630	204.69	270.98
19	-1.00	0.0	1.766	-1.637	203.62	271.69
20	0.0	1.00	1.681	-1.549	210.76	284.48
21	1.00	0.0	1.679	-1.561	209.26	284.37
22	0.0	-1.00	1.602	-1.479	216.39	297.01
23	-1.00	0.0	1.604	-1.495	214.48	296.12
24	0.0	1.00	1.535	-1.417	221.67	308.72
25	1.00	0.0	1.540	-1.436	219.34	307.06
26	0.0	-1.00	1.476	-1.362	226.66	319.74
27	-1.00	0.0	1.485	-1.384	223.89	317.31
28	0.0	1.00	1.423	-1.311	231.40	330.17
29	1.00	0.0	1.435	-1.337	228.17	326.94
30	0.0	-1.00	1.377	-1.265	235.93	340.10
31	-1.00	0.0	1.392	-1.294	232.20	336.02
32	0.0	1.00	1.335	-1.223	240.30	349.59
33	1.00	0.0	1.353	-1.256	235.99	344.57
34	0.0	-1.00	1.297	-1.183	244.53	358.73
35	-1.00	0.0	1.318	-1.220	239.55	352.64
36	0.0	1.00	1.263	-1.146	248.68	367.58
37	1.00	0.0	1.287	-1.188	242.87	360.23
38	0.0	-1.00	1.230	-1.111	252.78	376.22
39	-1.00	0.0	1.259	-1.160	245.92	367.33
40	0.0	1.00	1.200	-1.078	256.92	384.74
41	1.00	0.0	1.233	-1.134	248.65	373.87
42	0.0	-1.00	1.172	-1.044	261.20	393.29
43	-1.00	0.0	1.211	-1.113	250.89	379.73
44	0.0	1.00	1.144	-1.010	265.88	402.12
45	1.00	0.0	1.193	-1.098	252.29	384.59
46	0.0	-1.00	1.116	-0.972	271.59	411.77
47	-1.00	0.0	1.178	-1.095	251.71	387.63
48	0.0	1.00	1.087	-0.920	280.67	423.65
49	1.00	0.0	1.167	-1.131	243.74	386.63
50	0.0	-1.00	1.067	-0.805	307.98	440.06

EN

DRIVING POINT ADMITTANCES AND IMPEDANCES
FOR SPECIFIED BASE VOLTAGES

H/LMDA=0.5000 B/LMDA= 0.250 OMEGA= 9.92

2 ELEMENT ARRAY

ELEMENT	VO		YO		ZO	
	RE	IM	RE	IM	RE	IM
1	1.00	0.0	1.110	1.267	391.27	-446.47
2	0.0	-1.00	0.943	2.485	133.50	-351.75

3 ELEMENT ARRAY

ELEMENT	VO		YO		ZO	
	RE	IM	RE	IM	RE	IM
1	1.00	0.0	1.018	1.410	336.59	-466.32
2	0.0	-1.00	1.034	2.071	193.01	-386.56
3	-1.00	0.0	0.827	2.777	98.48	-330.72

4 ELEMENT ARRAY

ELEMENT	VO		YO		ZO	
	RE	IM	RE	IM	RE	IM
1	1.00	0.0	1.068	1.347	361.36	-455.70
2	0.0	-1.00	0.941	2.217	162.23	-382.27
3	-1.00	0.0	0.894	2.418	134.56	-363.82
4	0.0	1.00	0.735	2.934	80.38	-320.75

5 ELEMENT ARRAY

ELEMENT	VO		YO		ZO	
	RE	IM	RE	IM	RE	IM
1	1.00	0.0	1.039	1.385	346.51	-461.95
2	0.0	-1.00	0.992	2.147	177.34	-383.77
3	-1.00	0.0	0.813	2.547	113.76	-356.25
4	0.0	1.00	0.792	2.601	107.15	-351.88
5	1.00	0.0	0.671	3.028	69.73	-314.85

6 ELEMENT ARRAY

ELEMENT	VO		YO		ZO	
	RE	IM	RE	IM	RE	IM
1	1.00	0.0	1.059	1.358	357.17	-458.04
2	0.0	-1.00	0.962	2.192	167.83	-382.57
3	-1.00	0.0	0.858	2.485	124.13	-359.54
4	0.0	1.00	0.715	2.724	90.17	-343.43
5	1.00	0.0	0.716	2.716	90.80	-344.29
6	0.0	-1.00	0.622	3.092	62.54	-310.81

7 ELEMENT ARRAY

ELEMENT	VO		YO		ZO	
	RE	IM	RE	IM	RE	IM
1	1.00	0.0	1.044	1.379	348.91	-460.97
2	0.0	-1.00	0.983	2.159	174.71	-383.73
3	-1.00	0.0	0.832	2.526	117.61	-357.20
4	0.0	1.00	0.758	2.663	98.85	-347.33
5	1.00	0.0	0.645	2.832	76.44	-335.66
6	0.0	-1.00	0.661	2.795	80.14	-338.87
7	-1.00	0.0	0.586	3.140	57.45	-307.77

8 ELEMENT ARRAY

ELEMENT	VO		YO		ZO	
	RE	IM	RE	IM	RE	IM
1	1.00	0.0	1.056	1.362	355.55	-458.61
2	0.0	-1.00	0.966	2.185	169.30	-382.79
3	-1.00	0.0	0.850	2.495	122.33	-359.11
4	0.0	1.00	0.732	2.704	93.31	-344.57
5	1.00	0.0	0.684	2.775	83.82	-339.75
6	0.0	-1.00	0.593	2.907	67.34	-330.25
7	-1.00	0.0	0.619	2.854	72.53	-334.65
8	0.0	1.00	0.558	3.177	53.58	-305.33

DRIVING POINT ADMITTANCES AND IMPEDANCES
FOR SPECIFIED BASE CURRENTS

H/LMDA=0.5000 B/LMDA= 0.250 OMEGA= 9.92

2 ELEMENT ARRAY

ELEMENT	IZ(O)		YO		ZO	
	RE	IM	RE	IM	RE	IM
1	1.00	0.0	0.849	1.540	609.56	-588.51
2	0.0	-1.00	0.343	2.597	94.13	-515.47

3 ELEMENT ARRAY

ELEMENT	IZ(O)		YO		ZO	
	RE	IM	RE	IM	RE	IM
1	1.00	0.0	0.868	1.529	616.84	-574.32
2	0.0	-1.00	0.422	2.303	157.27	-589.82
3	-1.00	0.0	0.304	3.010	56.97	-429.19

4 ELEMENT ARRAY

ELEMENT	IZ(O)		YO		ZO	
	RE	IM	RE	IM	RE	IM
1	1.00	0.0	0.863	1.532	614.87	-578.28
2	0.0	-1.00	0.456	2.329	162.99	-575.32
3	-1.00	0.0	0.438	2.654	111.46	-491.72
4	0.0	1.00	0.308	3.131	52.07	-408.10

5 ELEMENT ARRAY

ELEMENT	IZ(O)		YO		ZO	
	RE	IM	RE	IM	RE	IM
1	1.00	0.0	0.867	1.530	615.51	-575.31
2	0.0	-1.00	0.445	2.318	161.64	-580.94
3	-1.00	0.0	0.478	2.689	116.35	-479.52
4	0.0	1.00	0.438	2.773	99.53	-465.93
5	1.00	0.0	0.287	3.202	46.00	-397.50

6 ELEMENT ARRAY

ELEMENT	IZ(O)		YO		ZO	
	RE	IM	RE	IM	RE	IM
1	1.00	0.0	0.864	1.531	615.23	-577.48
2	0.0	-1.00	0.451	2.327	161.80	-576.71
3	-1.00	0.0	0.465	2.673	115.34	-484.56
4	0.0	1.00	0.478	2.808	104.17	-455.09
5	1.00	0.0	0.418	2.852	88.53	-451.77
6	0.0	-1.00	0.272	3.257	41.75	-389.74

7 ELEMENT ARRAY

ELEMENT	IZ(O)		YO		ZO	
	RE	IM	RE	IM	RE	IM
1	1.00	0.0	0.866	1.531	615.32	-575.89
2	0.0	-1.00	0.447	2.320	161.95	-579.80
3	-1.00	0.0	0.472	2.687	115.33	-480.67
4	0.0	1.00	0.464	2.790	103.17	-459.92
5	1.00	0.0	0.455	2.887	92.90	-442.02
6	0.0	-1.00	0.401	2.913	80.64	-441.30
7	-1.00	0.0	0.260	3.299	38.73	-383.88

8 ELEMENT ARRAY

ELEMENT	IZ(O)		YO		ZO	
	RE	IM	RE	IM	RE	IM
1	1.00	0.0	0.865	1.531	615.32	-577.09
2	0.0	-1.00	0.450	2.326	161.67	-577.52
3	-1.00	0.0	0.468	2.676	115.60	-483.52
4	0.0	1.00	0.472	2.806	103.20	-456.12
5	1.00	0.0	0.442	2.868	91.95	-446.63
6	0.0	-1.00	0.437	2.947	84.80	-432.36
7	-1.00	0.0	0.387	2.961	74.82	-433.25
8	0.0	1.00	0.251	3.333	36.44	-379.24

EN

DRIVING POINT ADMITTANCES AND IMPEDANCES
FOR SPECIFIED BASE VOLTAGES

H/LMDA=0.5000 B/LMDA= 0.250 OMEGA= 9.92

9 ELEMENT ARRAY

ELEMENT	VO		YO		ZO	
	RE	IM	RE	IM	RE	IM
1	1.00	0.0	1.046	1.376	350.09	-460.62
2	0.0	-1.00	0.980	2.164	173.70	-383.53
3	-1.00	0.0	0.836	2.519	118.60	-357.59
4	0.0	1.00	0.751	2.673	97.37	-346.78
5	1.00	0.0	0.661	2.814	79.11	-336.84
6	0.0	-1.00	0.631	2.851	73.98	-334.41
7	-1.00	0.0	0.554	2.963	60.95	-326.15
8	0.0	1.00	0.586	2.901	66.86	-331.20
9	1.00	0.0	0.535	3.208	50.55	-303.29

10 ELEMENT ARRAY

ELEMENT	VO		YO		ZO	
	RE	IM	RE	IM	RE	IM
1	1.00	0.0	1.054	1.364	354.66	-458.89
2	0.0	-1.00	0.968	2.181	170.03	-382.96
3	-1.00	0.0	0.848	2.500	121.65	-358.80
4	0.0	1.00	0.736	2.698	94.13	-345.02
5	1.00	0.0	0.678	2.784	82.57	-339.15
6	0.0	-1.00	0.608	2.889	69.72	-331.44
7	-1.00	0.0	0.590	2.908	66.97	-330.30
8	0.0	1.00	0.523	3.006	56.15	-322.85
9	1.00	0.0	0.559	2.940	62.39	-328.28
10	0.0	-1.00	0.516	3.234	48.07	-301.53

12 ELEMENT ARRAY

ELEMENT	VO		YO		ZO	
	RE	IM	RE	IM	RE	IM
1	1.00	0.0	1.053	1.366	354.13	-459.08
2	0.0	-1.00	0.970	2.179	170.47	-383.05
3	-1.00	0.0	0.846	2.502	121.26	-358.62
4	0.0	1.00	0.738	2.694	94.55	-345.29
5	1.00	0.0	0.676	2.788	82.10	-338.78
6	0.0	-1.00	0.611	2.883	70.32	-331.91
7	-1.00	0.0	0.584	2.916	65.98	-329.69
8	0.0	1.00	0.537	2.990	58.14	-323.99
9	1.00	0.0	0.531	2.991	57.60	-324.15
10	0.0	-1.00	0.478	3.073	49.37	-317.72
11	-1.00	0.0	0.517	3.002	55.77	-323.53
12	0.0	1.00	0.485	3.276	44.25	-298.65

14 ELEMENT ARRAY

ELEMENT	VO		YO		ZO	
	RE	IM	RE	IM	RE	IM
1	1.00	0.0	1.053	1.367	353.79	-459.21
2	0.0	-1.00	0.971	2.178	170.75	-383.10
3	-1.00	0.0	0.845	2.504	121.02	-358.52
4	0.0	1.00	0.739	2.692	94.81	-345.44
5	1.00	0.0	0.674	2.791	81.81	-338.57
6	0.0	-1.00	0.612	2.880	70.64	-332.19
7	-1.00	0.0	0.582	2.921	65.60	-329.32
8	0.0	1.00	0.539	2.985	58.62	-324.45
9	1.00	0.0	0.526	2.998	56.78	-323.55
10	0.0	-1.00	0.490	3.058	51.12	-318.80
11	-1.00	0.0	0.491	3.050	51.50	-319.61
12	0.0	1.00	0.445	3.123	44.73	-313.84
13	1.0C	0.0	0.487	3.049	51.03	-319.81
14	0.0	-1.00	0.462	3.310	41.39	-296.35

16 ELEMENT ARRAY

ELEMENT	VO		YO		ZO	
	RE	IM	RE	IM	RE	IM
1	1.00	0.0	1.052	1.367	353.56	-459.30
2	0.0	-1.00	0.971	2.177	170.94	-383.12
3	-1.00	0.0	0.845	2.505	120.86	-358.46
4	0.0	1.00	0.740	2.691	94.98	-345.53
5	1.00	0.0	0.673	2.792	81.63	-338.46
6	0.0	-1.00	0.614	2.878	70.85	-332.35
7	-1.00	0.0	0.581	2.923	65.37	-329.10
8	0.0	1.00	0.541	2.981	58.89	-324.74
9	1.00	0.0	0.525	3.003	56.46	-323.19
10	0.0	-1.00	0.493	3.053	51.53	-319.26
11	-1.00	0.0	0.487	3.057	50.78	-319.04
12	0.0	1.00	0.457	3.109	46.31	-314.87
13	1.00	0.0	0.461	3.095	47.11	-316.06
14	0.0	-1.00	0.420	3.162	41.28	-310.76
15	-1.00	0.0	0.462	3.088	47.41	-316.78
16	0.0	1.00	0.444	3.337	39.15	-294.46

DRIVING POINT ADMITTANCES AND IMPEDANCES
FOR SPECIFIED BASE CURRENTS

H/LMDA=0.5000 B/LMDA= 0.250 OMEGA= 9.92

9 ELEMENT ARRAY

ELEMENT	IZ(0)		YO		ZO	
	RE	IM	RE	IM	RE	IM
1	1.00	0.0	0.866	1.531	615.28	-576.15
2	0.0	-1.00	0.448	2.321	161.98	-579.25
3	-1.00	0.0	0.470	2.685	115.23	-481.42
4	0.0	1.00	0.467	2.794	103.45	-458.93
5	1.00	0.0	0.449	2.884	91.98	-442.95
6	0.0	-1.00	0.424	2.928	83.88	-436.82
7	-1.00	0.0	0.422	2.995	78.79	-424.99
8	0.0	1.00	0.375	3.001	70.26	-426.82
9	1.00	0.0	0.243	3.361	34.61	-375.45

10 ELEMENT ARRAY

ELEMENT	IZ(0)		YO		ZO	
	RE	IM	RE	IM	RE	IM
1	1.00	0.0	0.865	1.531	615.33	-576.90
2	0.0	-1.00	0.449	2.325	161.69	-577.89
3	-1.00	0.0	0.469	2.678	115.61	-483.02
4	0.0	1.00	0.470	2.803	103.09	-456.85
5	1.00	0.0	0.445	2.872	92.23	-445.68
6	0.0	-1.00	0.431	2.945	83.92	-433.21
7	-1.00	0.0	0.409	2.975	77.91	-429.31
8	0.0	1.00	0.409	3.034	74.07	-419.11
9	1.00	0.0	0.365	3.035	66.54	-421.50
10	0.0	-1.00	0.237	3.385	33.10	-372.26

12 ELEMENT ARRAY

ELEMENT	IZ(0)		YO		ZO	
	RE	IM	RE	IM	RE	IM
1	1.00	0.0	0.865	1.531	615.32	-576.81
2	0.0	-1.00	0.449	2.324	161.72	-578.08
3	-1.00	0.0	0.469	2.679	115.57	-482.78
4	0.0	1.00	0.469	2.801	103.12	-457.19
5	1.00	0.0	0.446	2.874	92.24	-445.20
6	0.0	-1.00	0.429	2.942	83.81	-433.89
7	-1.00	0.0	0.412	2.979	78.19	-428.42
8	0.0	1.00	0.403	3.031	73.26	-419.85
9	1.00	0.0	0.386	3.046	69.38	-418.38
10	0.0	-1.00	0.388	3.095	66.99	-410.19
11	-1.00	0.0	0.349	3.089	60.79	-413.14
12	0.0	1.00	0.227	3.425	30.75	-367.17

14 ELEMENT ARRAY

ELEMENT	IZ(0)		YO		ZO	
	RE	IM	RE	IM	RE	IM
1	1.00	0.0	0.865	1.531	615.32	-576.75
2	0.0	-1.00	0.449	2.324	161.74	-578.18
3	-1.00	0.0	0.469	2.680	115.54	-482.65
4	0.0	1.00	0.469	2.801	103.16	-457.36
5	1.00	0.0	0.446	2.875	92.20	-444.97
6	0.0	-1.00	0.429	2.940	83.83	-434.22
7	-1.00	0.0	0.413	2.981	78.21	-427.96
8	0.0	1.00	0.401	3.028	73.14	-420.50
9	1.00	0.0	0.389	3.050	69.67	-417.54
10	0.0	-1.00	0.382	3.092	66.23	-410.85
11	-1.00	0.0	0.367	3.099	63.44	-410.63
12	0.0	1.00	0.371	3.141	61.85	-403.63
13	1.00	0.0	0.335	3.132	56.50	-406.78
14	0.0	-1.00	0.218	3.456	28.99	-363.23

16 ELEMENT ARRAY

ELEMENT	IZ(0)		YO		ZO	
	RE	IM	RE	IM	RE	IM
1	1.00	0.0	0.865	1.531	615.32	-576.71
2	0.0	-1.00	0.449	2.324	161.76	-578.25
3	-1.00	0.0	0.469	2.680	115.51	-482.57
4	0.0	1.00	0.469	2.800	103.18	-457.45
5	1.00	0.0	0.447	2.875	92.17	-444.85
6	0.0	-1.00	0.429	2.939	83.87	-434.38
7	-1.00	0.0	0.413	2.982	78.17	-427.74
8	0.0	1.00	0.401	3.027	73.17	-420.82
9	1.00	0.0	0.390	3.052	69.69	-417.09
10	0.0	-1.00	0.381	3.089	66.11	-411.48
11	-1.00	0.0	0.371	3.102	63.73	-409.82
12	0.0	1.00	0.366	3.139	61.13	-404.22
13	1.00	0.0	0.353	3.139	58.99	-404.75
14	0.0	-1.00	0.357	3.177	57.90	-398.53
15	-1.00	0.0	0.324	3.166	53.14	-401.72
16	0.0	1.00	0.212	3.481	27.60	-360.06

EN

DRIVING POINT ADMITTANCES AND IMPEDANCES
FOR SPECIFIED BASE VOLTAGES

H/LMDA=0.5000 B/LMDA= 0.250 OMEGA= 9.92

18 ELEMENT ARRAY

ELEMENT	VO RE	VO IM	YO RE	YO IM	ZO RE	ZO IM
1	1.00	0.0	1.052	1.368	353.39	-459.37
2	0.0	-1.00	0.972	2.176	171.07	-383.14
3	-1.00	0.0	0.844	2.506	120.75	-358.43
4	0.0	1.00	0.740	2.690	95.10	-345.58
5	1.00	0.0	0.673	2.793	81.51	-338.38
6	0.0	-1.00	0.614	2.877	70.98	-332.45
7	-1.00	0.0	0.580	2.925	65.21	-328.98
8	0.0	1.00	0.542	2.979	59.07	-324.90
9	1.00	0.0	0.523	3.005	56.25	-322.98
10	0.0	-1.00	0.494	3.050	51.76	-319.53
11	-1.00	0.0	0.485	3.061	50.50	-318.68
12	0.0	1.00	0.459	3.104	46.66	-315.31
13	1.00	0.0	0.457	3.102	46.47	-315.51
14	0.0	-1.00	0.432	3.149	42.74	-311.74
15	-1.00	0.0	0.438	3.132	43.77	-313.18
16	0.0	1.00	0.400	3.194	38.61	-308.24
17	1.00	0.0	0.442	3.120	44.55	-314.25
18	0.0	-1.00	0.428	3.360	37.33	-292.86

20 ELEMENT ARRAY

ELEMENT	VO RE	VO IM	YO RE	YO IM	ZO RE	ZO IM
1	1.00	0.0	1.052	1.368	353.27	-459.42
2	0.0	-1.00	0.972	2.176	171.17	-383.15
3	-1.00	0.0	0.844	2.506	120.67	-358.40
4	0.0	1.00	0.741	2.689	95.18	-345.62
5	1.00	0.0	0.672	2.794	81.42	-338.34
6	0.0	-1.00	0.615	2.876	71.08	-332.51
7	-1.00	0.0	0.579	2.926	65.11	-328.91
8	0.0	1.00	0.542	2.978	59.19	-324.99
9	1.00	0.0	0.523	3.007	56.12	-322.86
10	0.0	-1.00	0.495	3.048	51.92	-319.69
11	-1.00	0.0	0.484	3.064	50.32	-318.47
12	0.0	1.00	0.460	3.100	46.87	-315.58
13	1.00	0.0	0.456	3.106	46.23	-315.16
14	0.0	-1.00	0.434	3.144	43.05	-312.16
15	-1.00	0.0	0.434	3.139	43.19	-312.65
16	0.0	1.00	0.411	3.181	39.96	-309.17
17	1.00	0.0	0.418	3.162	41.12	-310.78
18	0.0	-1.00	0.384	3.221	36.45	-306.11
19	-1.00	0.0	0.426	3.147	42.20	-312.09
20	0.0	1.00	0.415	3.380	35.82	-291.49

22 ELEMENT ARRAY

ELEMENT	VO RE	VO IM	YO RE	YO IM	ZO RE	ZO IM
1	1.00	0.0	1.052	1.368	353.17	-459.46
2	0.0	-1.00	0.972	2.175	171.24	-383.16
3	-1.00	0.0	0.844	2.506	120.61	-358.38
4	0.0	1.00	0.741	2.689	95.24	-345.65
5	1.00	0.0	0.672	2.794	81.36	-338.30
6	0.0	-1.00	0.615	2.875	71.15	-332.55
7	-1.00	0.0	0.579	2.926	65.03	-328.85
8	0.0	1.00	0.543	2.977	59.27	-325.06
9	1.00	0.0	0.522	3.007	56.02	-322.78
10	0.0	-1.00	0.496	3.046	52.03	-319.79
11	-1.00	0.0	0.483	3.065	50.19	-318.35
12	0.0	1.00	0.461	3.098	47.01	-315.74
13	1.00	0.0	0.455	3.109	46.06	-314.95
14	0.0	-1.00	0.435	3.141	43.24	-312.43
15	-1.00	0.0	0.432	3.142	42.97	-312.30
16	0.0	1.00	0.413	3.176	40.24	-309.58
17	1.00	0.0	0.415	3.169	40.59	-310.27
18	0.0	-1.00	0.394	3.209	37.72	-307.00
19	-1.00	0.0	0.402	3.188	38.94	-308.73
20	0.0	1.00	0.370	3.244	34.66	-304.29
21	1.00	0.0	0.411	3.170	40.24	-310.22
22	0.0	-1.00	0.404	3.397	34.53	-290.30

24 ELEMENT ARRAY

ELEMENT	VO RE	VO IM	YO RE	YO IM	ZO RE	ZO IM
1	1.00	0.0	1.051	1.368	353.10	-459.49
2	0.0	-1.00	0.972	2.175	171.30	-383.17
3	-1.00	0.0	0.843	2.507	120.56	-358.36
4	0.0	1.00	0.741	2.689	95.29	-345.67
5	1.00	0.0	0.672	2.795	81.31	-338.28
6	0.0	-1.00	0.615	2.875	71.20	-332.58
7	-1.00	0.0	0.578	2.927	64.98	-328.82
8	0.0	1.00	0.543	2.977	59.33	-325.10
9	1.00	0.0	0.522	3.008	55.96	-322.73
10	0.0	-1.00	0.496	3.046	52.11	-319.85
11	-1.00	0.0	0.483	3.066	50.11	-318.27
12	0.0	1.00	0.462	3.097	47.11	-315.83
13	1.00	0.0	0.454	3.110	45.94	-314.83
14	0.0	-1.00	0.436	3.139	43.37	-312.59
15	-1.00	0.0	0.431	3.145	42.82	-312.10
16	0.0	1.00	0.414	3.173	40.41	-309.85
17	1.00	0.0	0.413	3.173	40.39	-309.93
18	0.0	-1.00	0.396	3.204	37.98	-307.41
19	-1.00	0.0	0.399	3.195	38.46	-308.23
20	0.0	1.00	0.380	3.232	35.87	-305.15
21	1.00	0.0	0.388	3.211	37.12	-306.96
22	0.0	-1.00	0.357	3.264	33.15	-302.70
23	-1.00	0.0	0.399	3.191	38.57	-308.57
24	0.0	1.00	0.394	3.412	33.42	-289.24

DRIVING POINT ADMITTANCES AND IMPEDANCES
FOR SPECIFIED BASE CURRENTS

H/LMDA=0.5000 B/LMDA= 0.250 OMEGA= 9.92

18 ELEMENT ARRAY

ELEMENT	IZ(O) RE	IZ(O) IM	YO RE	YO IM	ZO RE	ZO IM
1	1.00	0.0	0.865	1.531	615.32	-576.69
2	0.0	-1.00	0.449	2.324	161.77	-578.30
3	-1.00	0.0	0.469	2.680	115.50	-482.52
4	0.0	1.00	0.469	2.800	103.20	-457.52
5	1.00	0.0	0.447	2.876	92.15	-444.77
6	0.0	-1.00	0.428	2.939	83.90	-434.48
7	-1.00	0.0	0.414	2.983	78.14	-427.62
8	0.0	1.00	0.401	3.026	73.20	-420.98
9	1.00	0.0	0.390	3.054	69.65	-416.87
10	0.0	-1.00	0.380	3.087	66.13	-411.78
11	-1.00	0.0	0.371	3.105	63.75	-409.38
12	0.0	1.00	0.364	3.135	61.00	-404.84
13	1.00	0.0	0.356	3.143	59.29	-403.96
14	0.0	-1.00	0.352	3.175	57.21	-399.07
15	-1.00	0.0	0.340	3.172	55.50	-400.07
16	0.0	1.00	0.345	3.208	54.74	-394.40
17	1.00	0.0	0.314	3.196	50.41	-397.56
18	0.0	-1.00	0.206	3.502	26.48	-357.44

20 ELEMENT ARRAY

ELEMENT	IZ(O) RE	IZ(O) IM	YO RE	YO IM	ZO RE	ZO IM
1	1.00	0.0	0.865	1.531	615.32	-576.67
2	0.0	-1.00	0.449	2.324	161.78	-578.33
3	-1.00	0.0	0.469	2.680	115.49	-482.49
4	0.0	1.00	0.469	2.800	103.21	-457.56
5	1.00	0.0	0.447	2.876	92.13	-444.72
6	0.0	-1.00	0.428	2.939	83.92	-434.54
7	-1.00	0.0	0.414	2.983	78.12	-427.54
8	0.0	1.00	0.401	3.025	73.23	-421.07
9	1.00	0.0	0.390	3.054	69.62	-416.75
10	0.0	-1.00	0.380	3.086	66.17	-411.94
11	-1.00	0.0	0.372	3.106	63.71	-409.17
12	0.0	1.00	0.364	3.134	61.03	-405.14
13	1.00	0.0	0.356	3.146	59.30	-403.54
14	0.0	-1.00	0.350	3.172	57.08	-399.67
15	-1.00	0.0	0.343	3.176	55.80	-399.31
16	0.0	1.00	0.340	3.206	54.07	-394.90
17	1.00	0.0	0.330	3.200	52.66	-396.22
18	0.0	-1.00	0.335	3.233	52.14	-390.96
19	-1.00	0.0	0.306	3.220	48.15	-394.04
20	0.0	1.00	0.201	3.521	25.55	-355.21

22 ELEMENT ARRAY

ELEMENT	IZ(O) RE	IZ(O) IM	YO RE	YO IM	ZO RE	ZO IM
1	1.00	0.0	0.865	1.531	615.31	-576.65
2	0.0	-1.00	0.449	2.324	161.78	-578.36
3	-1.00	0.0	0.469	2.680	115.48	-482.46
4	0.0	1.00	0.469	2.800	103.22	-457.59
5	1.00	0.0	0.447	2.876	92.12	-444.69
6	0.0	-1.00	0.428	2.938	83.93	-434.58
7	-1.00	0.0	0.414	2.984	78.10	-427.49
8	0.0	1.00	0.401	3.025	73.25	-421.13
9	1.00	0.0	0.390	3.055	69.59	-416.68
10	0.0	-1.00	0.380	3.086	66.20	-412.03
11	-1.00	0.0	0.372	3.107	63.68	-409.06
12	0.0	1.00	0.364	3.133	61.07	-405.29
13	1.00	0.0	0.357	3.147	59.27	-403.33
14	0.0	-1.00	0.350	3.170	57.11	-399.97
15	-1.00	0.0	0.344	3.179	55.82	-398.89
16	0.0	1.00	0.339	3.202	53.95	-395.49
17	1.00	0.0	0.333	3.204	52.97	-395.48
18	0.0	-1.00	0.330	3.231	51.49	-391.43
19	-1.00	0.0	0.320	3.224	50.30	-392.98
20	0.0	1.00	0.326	3.255	49.94	-388.03
21	1.00	0.0	0.298	3.242	46.22	-391.02
22	0.0	-1.00	0.197	3.537	24.76	-353.28

24 ELEMENT ARRAY

ELEMENT	IZ(O) RE	IZ(O) IM	YO RE	YO IM	ZO RE	ZO IM
1	1.00	0.0	0.865	1.531	615.31	-576.64
2	0.0	-1.00	0.449	2.324	161.79	-578.37
3	-1.00	0.0	0.469	2.680	115.47	-482.44
4	0.0	1.00	0.469	2.799	103.23	-457.61
5	1.00	0.0	0.447	2.876	92.11	-444.66
6	0.0	-1.00	0.428	2.938	83.94	-434.61
7	-1.00	0.0	0.414	2.984	78.09	-427.46
8	0.0	1.00	0.401	3.025	73.26	-421.17
9	1.00	0.0	0.390	3.055	69.58	-416.63
10	0.0	-1.00	0.380	3.086	66.22	-412.09
11	-1.00	0.0	0.372	3.107	63.66	-408.99
12	0.0	1.00	0.364	3.132	61.10	-405.38
13	1.00	0.0	0.357	3.148	59.23	-403.21
14	0.0	-1.00	0.350	3.169	57.15	-400.12
15	-1.00	0.0	0.344	3.180	55.78	-398.68
16	0.0	1.00	0.338	3.201	53.98	-395.78
17	1.00	0.0	0.333	3.207	52.98	-395.07
18	0.0	-1.00	0.329	3.228	51.37	-392.01
19	-1.00	0.0	0.324	3.228	50.61	-392.25
20	0.0	1.00	0.322	3.253	49.32	-388.47
21	1.00	0.0	0.312	3.244	48.30	-390.19
22	0.0	-1.00	0.318	3.274	48.06	-385.50
23	-1.00	0.0	0.292	3.261	44.56	-388.38
24	0.0	1.00	0.194	3.551	24.08	-351.60

EN

DRIVING POINT ADMITTANCES AND IMPEDANCES
FOR SPECIFIED BASE VOLTAGES

H/LMDA=0.5000 B/LMDA= 0.250 OMEGA= 9.92

26 ELEMENT ARRAY

ELEMENT	VO		YO		ZO	
	RE	IM	RE	IM	RE	IM
1	1.00	0.0	1.051	1.368	353.04	-459.52
2	0.0	-1.00	0.973	2.175	171.35	-383.17
3	-1.00	0.0	0.843	2.507	120.52	-358.35
4	0.0	1.00	0.741	2.688	95.32	-345.69
5	1.00	0.0	0.672	2.795	81.28	-338.26
6	0.0	-1.00	0.616	2.875	71.23	-332.60
7	-1.00	0.0	0.578	2.927	64.94	-328.79
8	0.0	1.00	0.544	2.976	59.37	-325.13
9	1.00	0.0	0.521	3.009	55.91	-322.69
10	0.0	-1.00	0.497	3.045	52.16	-319.89
11	-1.00	0.0	0.482	3.067	50.04	-318.22
12	0.0	1.00	0.463	3.097	47.18	-315.89
13	1.00	0.0	0.453	3.111	45.86	-314.76
14	0.0	-1.00	0.436	3.137	43.47	-312.68
15	-1.00	0.0	0.431	3.146	42.71	-311.98
16	0.0	1.00	0.415	3.172	40.54	-310.00
17	1.00	0.0	0.413	3.175	40.24	-309.72
18	0.0	-1.00	0.397	3.201	38.14	-307.67
19	-1.00	0.0	0.398	3.198	38.27	-307.90
20	0.0	1.00	0.381	3.228	36.10	-305.54
21	1.00	0.0	0.385	3.217	36.67	-306.48
22	0.0	-1.00	0.368	3.253	34.30	-303.53
23	-1.00	0.0	0.376	3.231	35.57	-305.40
24	0.0	1.00	0.347	3.282	31.85	-301.30
25	1.00	0.0	0.388	3.209	37.13	-307.10
26	0.0	-1.00	0.386	3.425	32.45	-288.29

28 ELEMENT ARRAY

ELEMENT	VO		YO		ZO	
	RE	IM	RE	IM	RE	IM
1	1.00	0.0	1.051	1.369	352.99	-459.54
2	0.0	-1.00	0.973	2.175	171.39	-383.18
3	-1.00	0.0	0.843	2.507	120.49	-358.34
4	0.0	1.00	0.741	2.688	95.35	-345.70
5	1.00	0.0	0.671	2.795	81.25	-338.24
6	0.0	-1.00	0.616	2.874	71.26	-332.62
7	-1.00	0.0	0.578	2.928	64.91	-328.77
8	0.0	1.00	0.544	2.976	59.41	-325.15
9	1.00	0.0	0.521	3.009	55.87	-322.66
10	0.0	-1.00	0.497	3.045	52.20	-319.93
11	-1.00	0.0	0.482	3.067	50.00	-318.19
12	0.0	1.00	0.463	3.096	47.23	-315.94
13	1.00	0.0	0.453	3.112	45.80	-314.71
14	0.0	-1.00	0.437	3.137	43.53	-312.74
15	-1.00	0.0	0.430	3.147	42.63	-311.91
16	0.0	1.00	0.415	3.170	40.63	-310.09
17	1.00	0.0	0.412	3.177	40.14	-309.61
18	0.0	-1.00	0.398	3.199	38.26	-307.82
19	-1.00	0.0	0.397	3.201	38.14	-307.70
20	0.0	1.00	0.382	3.225	36.25	-305.80
21	1.00	0.0	0.384	3.221	36.50	-306.15
22	0.0	-1.00	0.369	3.248	34.52	-303.92
23	-1.00	0.0	0.373	3.236	35.15	-304.93
24	0.0	1.00	0.357	3.271	32.96	-302.10
25	1.00	0.0	0.366	3.248	34.23	-304.01
26	0.0	-1.00	0.338	3.298	30.72	-300.05
27	-1.00	0.0	0.378	3.226	35.86	-305.79
28	0.0	1.00	0.378	3.438	31.60	-287.43

30 ELEMENT ARRAY

ELEMENT	VO		YO		ZO	
	RE	IM	RE	IM	RE	IM
1	1.00	0.0	1.051	1.369	352.95	-459.55
2	0.0	-1.00	0.973	2.175	171.42	-383.18
3	-1.00	0.0	0.843	2.507	120.47	-358.33
4	0.0	1.00	0.742	2.688	95.38	-345.71
5	1.00	0.0	0.671	2.795	81.22	-338.23
6	0.0	-1.00	0.616	2.874	71.29	-332.64
7	-1.00	0.0	0.578	2.928	64.88	-328.76
8	0.0	1.00	0.544	2.976	59.43	-325.17
9	1.00	0.0	0.521	3.009	55.84	-322.64
10	0.0	-1.00	0.497	3.044	52.23	-319.95
11	-1.00	0.0	0.482	3.067	49.96	-318.16
12	0.0	1.00	0.463	3.096	47.27	-315.97
13	1.00	0.0	0.453	3.112	45.76	-314.67
14	0.0	-1.00	0.437	3.136	43.58	-312.79
15	-1.00	0.0	0.430	3.148	42.57	-311.86
16	0.0	1.00	0.416	3.170	40.69	-310.16
17	1.00	0.0	0.411	3.177	40.06	-309.53
18	0.0	-1.00	0.398	3.198	38.34	-307.91
19	-1.00	0.0	0.396	3.202	38.04	-307.59
20	0.0	1.00	0.383	3.223	36.37	-305.95
21	1.00	0.0	0.383	3.223	36.37	-305.95
22	0.0	-1.00	0.370	3.245	34.66	-304.17
23	-1.00	0.0	0.372	3.240	34.99	-304.61
24	0.0	1.00	0.358	3.267	33.16	-302.48
25	1.00	0.0	0.363	3.254	33.84	-303.56
26	0.0	-1.00	0.347	3.288	31.79	-300.82
27	-1.00	0.0	0.356	3.264	33.05	-302.77
28	0.0	1.00	0.329	3.313	29.73	-298.93
29	1.00	0.0	0.370	3.241	34.74	-304.60
30	0.0	-1.00	0.371	3.449	30.83	-286.66

DRIVING POINT ADMITTANCES AND IMPEDANCES
FOR SPECIFIED BASE CURRENTS

H/LMDA=0.5000 B/LMDA= 0.250 OMEGA= 9.92

26 ELEMENT ARRAY

ELEMENT	IZ(O)		YO		ZO	
	RE	IM	RE	IM	RE	IM
1	1.00	0.0	0.865	1.531	615.31	-576.63
2	0.0	-1.00	0.449	2.323	161.79	-578.39
3	-1.00	0.0	0.469	2.681	115.47	-482.42
4	0.0	1.00	0.469	2.799	103.24	-457.63
5	1.00	0.0	0.447	2.876	92.11	-444.64
6	0.0	-1.00	0.428	2.938	83.95	-434.63
7	-1.00	0.0	0.414	2.984	78.08	-427.43
8	0.0	1.00	0.401	3.024	73.27	-421.20
9	1.00	0.0	0.390	3.055	69.57	-416.60
10	0.0	-1.00	0.380	3.085	66.23	-412.13
11	-1.00	0.0	0.372	3.108	63.64	-408.94
12	0.0	1.00	0.364	3.132	61.11	-405.43
13	1.00	0.0	0.357	3.148	59.21	-403.15
14	0.0	-1.00	0.350	3.169	57.18	-400.20
15	-1.00	0.0	0.344	3.181	55.74	-398.57
16	0.0	1.00	0.338	3.200	54.02	-395.93
17	1.00	0.0	0.334	3.208	52.94	-394.86
18	0.0	-1.00	0.328	3.226	51.40	-392.29
19	-1.00	0.0	0.324	3.230	50.62	-391.84
20	0.0	1.00	0.320	3.250	49.20	-389.04
21	1.00	0.0	0.315	3.248	48.60	-389.48
22	0.0	-1.00	0.314	3.273	47.46	-385.91
23	-1.00	0.0	0.305	3.262	46.56	-387.76
24	0.0	1.00	0.311	3.291	46.42	-383.27
25	1.00	0.0	0.286	3.278	43.11	-386.05
26	0.0	-1.00	0.191	3.564	23.48	-350.10

28 ELEMENT ARRAY

ELEMENT	IZ(O)		YO		ZO	
	RE	IM	RE	IM	RE	IM
1	1.00	0.0	0.865	1.531	615.31	-576.63
2	0.0	-1.00	0.449	2.323	161.80	-578.40
3	-1.00	0.0	0.469	2.681	115.46	-482.41
4	0.0	1.00	0.469	2.799	103.24	-457.64
5	1.00	0.0	0.447	2.877	92.10	-444.63
6	0.0	-1.00	0.428	2.938	83.95	-434.65
7	-1.00	0.0	0.414	2.984	78.07	-427.41
8	0.0	1.00	0.401	3.024	73.28	-421.22
9	1.00	0.0	0.390	3.055	69.56	-416.58
10	0.0	-1.00	0.380	3.085	66.24	-412.15
11	-1.00	0.0	0.372	3.108	63.63	-408.91
12	0.0	1.00	0.364	3.131	61.13	-405.47
13	1.00	0.0	0.357	3.148	59.20	-403.10
14	0.0	-1.00	0.350	3.168	57.19	-400.26
15	-1.00	0.0	0.344	3.181	55.72	-398.50
16	0.0	1.00	0.338	3.199	54.04	-396.01
17	1.00	0.0	0.334	3.209	52.91	-394.75
18	0.0	-1.00	0.328	3.225	51.44	-392.44
19	-1.00	0.0	0.324	3.231	50.58	-391.64
20	0.0	1.00	0.320	3.248	49.23	-389.33
21	1.00	0.0	0.316	3.251	48.62	-389.08
22	0.0	-1.00	0.312	3.269	47.34	-386.47
23	-1.00	0.0	0.308	3.266	46.87	-387.06
24	0.0	1.00	0.307	3.290	45.84	-383.66
25	1.00	0.0	0.299	3.278	45.04	-385.61
26	0.0	-1.00	0.305	3.307	44.98	-381.29
27	-1.00	0.0	0.280	3.294	41.83	-383.96
28	0.0	1.00	0.188	3.575	22.96	-348.76

30 ELEMENT ARRAY

ELEMENT	IZ(O)		YO		ZO	
	RE	IM	RE	IM	RE	IM
1	1.00	0.0	0.865	1.531	615.31	-576.62
2	0.0	-1.00	0.449	2.323	161.80	-578.41
3	-1.00	0.0	0.469	2.681	115.46	-482.40
4	0.0	1.00	0.469	2.799	103.25	-457.66
5	1.00	0.0	0.447	2.877	92.10	-444.61
6	0.0	-1.00	0.428	2.938	83.96	-434.66
7	-1.00	0.0	0.414	2.984	78.07	-427.40
8	0.0	1.00	0.401	3.024	73.29	-421.23
9	1.00	0.0	0.390	3.056	69.55	-416.56
10	0.0	-1.00	0.380	3.085	66.25	-412.18
11	-1.00	0.0	0.372	3.108	63.62	-408.88
12	0.0	1.00	0.364	3.131	61.14	-405.50
13	1.00	0.0	0.357	3.149	59.18	-403.07
14	0.0	-1.00	0.350	3.168	57.21	-400.30
15	-1.00	0.0	0.344	3.182	55.71	-398.46
16	0.0	1.00	0.338	3.199	54.06	-396.07
17	1.00	0.0	0.334	3.209	52.89	-394.69
18	0.0	-1.00	0.328	3.225	51.46	-392.52
19	-1.00	0.0	0.324	3.232	50.55	-391.54
20	0.0	1.00	0.320	3.247	49.26	-389.47
21	1.00	0.0	0.316	3.252	48.58	-388.88
22	0.0	-1.00	0.312	3.267	47.36	-386.75
23	-1.00	0.0	0.309	3.269	46.88	-386.66
24	0.0	1.00	0.305	3.286	45.71	-384.22
25	1.00	0.0	0.302	3.282	45.35	-384.91
26	0.0	-1.00	0.301	3.305	44.41	-381.66
27	-1.00	0.0	0.293	3.293	43.70	-383.68
28	0.0	1.00	0.299	3.320	43.70	-379.51
29	1.00	0.0	0.276	3.308	40.68	-382.08
30	0.0	-1.00	0.185	3.585	22.49	-347.55

EN

DRIVING POINT ADMITTANCES AND IMPEDANCES
FOR SPECIFIED BASE VOLTAGES

H/LMDA=0.5000 B/LMDA= 0.250 OMEGA= 9.92

35 ELEMENT ARRAY

ELEMENT	VO		YO		ZO	
	RE	IM	RE	IM	RE	IM
1	1.00	0.0	1.050	1.370	352.26	-459.82
2	0.0	-1.00	0.974	2.172	171.92	-383.25
3	-1.00	0.0	0.841	2.510	120.10	-358.21
4	0.0	1.00	0.743	2.686	95.73	-345.87
5	1.00	0.0	0.670	2.798	80.90	-338.05
6	0.0	-1.00	0.618	2.872	71.62	-332.83
7	-1.00	0.0	0.576	2.931	64.56	-328.54
8	0.0	1.00	0.546	2.973	59.76	-325.40
9	1.00	0.0	0.519	3.012	55.51	-322.40
10	0.0	-1.00	0.499	3.041	52.58	-320.21
11	-1.00	0.0	0.479	3.071	49.61	-317.88
12	0.0	1.00	0.466	3.092	47.64	-316.26
13	1.00	0.0	0.450	3.116	45.37	-314.35
14	0.0	-1.00	0.440	3.132	43.99	-313.13
15	-1.00	0.0	0.427	3.153	42.14	-311.49
16	0.0	1.00	0.419	3.164	41.16	-310.56
17	1.00	0.0	0.407	3.183	39.57	-309.09
18	0.0	-1.00	0.402	3.192	38.88	-308.40
19	-1.00	0.0	0.391	3.209	37.45	-307.05
20	0.0	1.00	0.388	3.215	37.00	-306.56
21	1.00	0.0	0.378	3.232	35.67	-305.26
22	0.0	-1.00	0.376	3.235	35.43	-304.97
23	-1.00	0.0	0.366	3.252	34.13	-303.67
24	0.0	1.00	0.365	3.253	34.11	-303.60
25	1.00	0.0	0.355	3.270	32.78	-302.22
26	0.0	-1.00	0.356	3.268	32.98	-302.42
27	-1.00	0.0	0.345	3.288	31.56	-300.85
28	0.0	1.00	0.348	3.280	32.02	-301.44
29	1.00	0.0	0.336	3.305	30.45	-299.50
30	0.0	-1.00	0.341	3.290	31.19	-300.69
31	-1.00	0.0	0.328	3.322	29.41	-298.14
32	0.0	1.00	0.337	3.297	30.65	-300.14
33	1.00	0.0	0.312	3.343	27.68	-296.54
34	0.0	-1.00	0.351	3.273	32.42	-302.06
35	-1.00	0.0	0.356	3.472	29.22	-284.98

DRIVING POINT ADMITTANCES AND IMPEDANCES
FOR SPECIFIED BASE CURRENTS

H/LMDA=0.5000 B/LMDA= 0.250 OMEGA= 9.92

35 ELEMENT ARRAY

ELEMENT	IZ(0)		YO		ZO	
	RE	IM	RE	IM	RE	IM
1	1.00	0.0	0.865	1.531	615.30	-576.53
2	0.0	-1.00	0.448	2.323	161.85	-578.56
3	-1.00	0.0	0.469	2.681	115.40	-482.26
4	0.0	1.00	0.469	2.798	103.31	-457.81
5	1.00	0.0	0.447	2.877	92.03	-444.45
6	0.0	-1.00	0.428	2.937	84.03	-434.83
7	-1.00	0.0	0.414	2.985	77.99	-427.22
8	0.0	1.00	0.401	3.023	73.36	-421.43
9	1.00	0.0	0.390	3.057	69.47	-416.35
10	0.0	-1.00	0.380	3.084	66.34	-412.39
11	-1.00	0.0	0.371	3.109	63.53	-408.65
12	0.0	1.00	0.364	3.130	61.24	-405.75
13	1.00	0.0	0.356	3.150	59.07	-402.80
14	0.0	-1.00	0.350	3.166	57.33	-400.58
15	-1.00	0.0	0.344	3.184	55.58	-398.15
16	0.0	1.00	0.339	3.196	54.20	-396.41
17	1.00	0.0	0.333	3.211	52.74	-394.31
18	0.0	-1.00	0.329	3.222	51.63	-392.94
19	-1.00	0.0	0.324	3.235	50.37	-391.07
20	0.0	1.00	0.320	3.244	49.47	-389.99
21	1.00	0.0	0.316	3.256	48.34	-388.27
22	0.0	-1.00	0.313	3.263	47.62	-387.46
23	-1.00	0.0	0.309	3.275	46.59	-385.81
24	0.0	1.00	0.306	3.279	46.03	-385.26
25	1.00	0.0	0.302	3.291	45.04	-383.60
26	0.0	-1.00	0.300	3.294	44.63	-383.35
27	-1.00	0.0	0.296	3.307	43.66	-381.56
28	0.0	1.00	0.294	3.306	43.40	-381.73
29	1.00	0.0	0.291	3.322	42.43	-379.57
30	0.0	-1.00	0.288	3.316	42.25	-380.48
31	-1.00	0.0	0.288	3.338	41.47	-377.48
32	0.0	1.00	0.281	3.324	40.91	-379.64
33	1.00	0.0	0.287	3.350	41.02	-375.75
34	0.0	-1.00	0.265	3.338	38.28	-378.08
35	-1.00	0.0	0.180	3.608	21.52	-344.97

40 ELEMENT ARRAY

ELEMENT	VO		YO		ZO	
	RE	IM	RE	IM	RE	IM
1	1.00	0.0	1.051	1.369	352.82	-459.61
2	0.0	-1.00	0.973	2.174	171.52	-383.19
3	-1.00	0.0	0.843	2.508	120.39	-358.31
4	0.0	1.00	0.742	2.687	95.45	-345.74
5	1.00	0.0	0.671	2.796	81.15	-338.19
[illegible]	[illegible]	1.00	[illegible]	2.874	71.36	-332.68
7	-1.00	0.0	0.577	2.928	[illegible]	[illegible]
8	0.0	1.00	0.544	2.975	59.51	-325.22
9	1.00	0.0	0.520	3.010	55.76	-322.59
10	0.0	-1.00	0.498	3.044	52.32	-320.01
11	-1.00	0.0	0.481	3.068	49.87	-318.09
12	0.0	1.00	0.464	3.095	47.37	-316.05
13	1.00	0.0	0.452	3.113	45.65	-314.58
14	0.0	-1.00	0.438	3.135	43.70	-312.89
15	-1.00	0.0	0.429	3.149	42.44	-311.74
16	0.0	1.00	0.417	3.168	40.84	-310.28
17	1.00	0.0	0.410	3.179	39.90	-309.38
18	0.0	-1.00	0.400	3.196	38.52	-308.09
19	-1.00	0.0	0.394	3.205	37.83	-307.38
20	0.0	1.00	0.385	3.220	36.60	-306.20
21	1.00	0.0	0.381	3.227	36.10	-305.65
22	0.0	-1.00	0.372	3.241	34.97	-304.54
23	-1.00	0.0	0.370	3.246	34.63	-304.14
24	0.0	1.00	0.361	3.260	33.56	-303.07
25	1.00	0.0	0.360	3.263	33.37	-302.81
26	0.0	-1.00	0.351	3.277	32.32	-301.74
27	-1.00	0.0	0.351	3.278	32.28	-301.64
28	0.0	1.00	0.342	3.292	31.22	-300.52
29	1.00	0.0	0.343	3.291	31.33	-300.60
30	0.0	-1.00	0.334	3.307	30.21	-299.38
31	-1.00	0.0	0.336	3.302	30.51	-299.70
32	0.0	1.00	0.326	3.321	29.27	-298.28
33	1.00	0.0	0.330	3.312	29.80	-298.96
34	0.0	-1.00	0.319	3.335	28.41	-297.15
35	-1.00	0.0	0.324	3.319	29.17	-298.41
36	0.0	1.00	0.312	3.349	27.59	-295.99
37	1.00	0.0	0.321	3.325	28.78	-298.01
38	0.0	-1.00	0.298	3.368	26.08	-294.59
39	-1.00	0.0	0.336	3.299	30.59	-299.97
40	0.0	1.00	0.344	3.492	27.94	-283.60

40 ELEMENT ARRAY

ELEMENT	IZ(0)		YO		ZO	
	RE	IM	RE	IM	RE	IM
1	1.00	0.0	0.865	1.531	615.31	-576.60
2	0.0	-1.00	0.448	2.323	161.81	-578.44
3	-1.00	0.0	0.469	2.681	115.45	-482.37
4	0.0	1.00	0.469	2.799	103.26	-457.69
5	1.00	0.0	0.447	2.877	92.08	-444.58
6	0.0	-1.00	0.428	2.938	83.97	-434.70
7	-1.00	0.0	0.414	2.984	78.05	-427.36
8	0.0	1.00	[illegible]	[illegible]	73.30	-421.28
9	1.00	0.0	0.390	3.056	69.53	[illegible]
10	0.0	-1.00	0.380	3.085	66.27	-412.23
11	-1.00	0.0	0.372	3.108	63.60	-408.82
12	0.0	1.00	0.364	3.131	61.17	-405.57
13	1.00	0.0	0.357	3.149	59.15	-402.99
14	0.0	-1.00	0.350	3.168	57.24	-400.38
15	-1.00	0.0	0.344	3.182	55.67	-398.36
16	0.0	1.00	0.338	3.198	54.10	-396.18
17	1.00	0.0	0.333	3.210	52.84	-394.55
18	0.0	-1.00	0.328	3.224	51.52	-392.68
19	-1.00	0.0	0.324	3.233	50.48	-391.35
20	0.0	1.00	0.320	3.246	49.34	-389.69
21	1.00	0.0	0.316	3.254	48.48	-388.61
22	0.0	-1.00	0.312	3.265	47.47	-387.09
23	-1.00	0.0	0.309	3.272	46.76	-386.22
24	0.0	1.00	0.305	3.282	45.84	-384.80
25	1.00	0.0	0.303	3.288	45.25	-384.11
26	0.0	-1.00	0.299	3.298	44.40	-382.76
27	-1.00	0.0	0.297	3.302	43.92	-382.25
28	0.0	1.00	0.293	3.312	43.11	-380.90
29	1.00	0.0	0.291	3.315	42.75	-380.60
30	0.0	-1.00	0.288	3.325	41.94	-379.19
31	-1.00	0.0	0.287	3.326	41.70	-379.14
32	0.0	1.00	0.283	3.338	40.87	-377.56
33	1.00	0.0	0.282	3.336	40.75	-377.90
34	0.0	-1.00	0.279	3.350	39.91	-375.93
35	-1.00	0.0	0.277	3.343	39.84	-376.96
36	0.0	1.00	0.277	3.364	39.17	-374.15
37	1.00	0.0	0.270	3.349	38.71	-376.39
38	0.0	-1.00	0.277	3.374	38.88	-372.71
39	-1.00	0.0	0.256	3.363	36.37	-374.81
40	0.0	1.00	0.176	3.626	20.75	-342.86

EN

DRIVING POINT ADMITTANCES AND IMPEDANCES
FOR SPECIFIED BASE VOLTAGES

H/LMDA=0.5000 B/LMDA= 0.250 OMEGA= 9.92

45 ELEMENT ARRAY

ELEMENT	V0		Y0		Z0	
	RE	IM	RE	IM	RE	IM
1	1.00	0.0	1.050	1.370	352.36	-459.79
2	0.0	-1.00	0.974	2.172	171.85	-383.24
3	-1.00	0.0	0.842	2.509	120.15	-358.22
4	0.0	1.00	0.743	2.686	95.68	-345.85
5	1.00	0.0	0.670	2.798	80.94	-338.07
6	0.0	-1.00	0.618	2.872	71.57	-332.81
7	-1.00	0.0	0.576	2.930	64.61	-328.57
8	0.0	1.00	0.546	2.973	59.71	-325.37
9	1.00	0.0	0.519	3.012	55.56	-322.44
10	0.0	-1.00	0.499	3.042	52.52	-320.17
11	-1.00	0.0	0.480	3.070	49.67	-317.93
12	0.0	1.00	0.465	3.092	47.58	-316.22
13	1.00	0.0	0.450	3.116	45.44	-314.41
14	0.0	-1.00	0.439	3.133	43.92	-313.07
15	-1.00	0.0	0.427	3.152	42.22	-311.55
16	0.0	1.00	0.419	3.165	41.07	-310.49
17	1.00	0.0	0.408	3.182	39.66	-309.17
18	0.0	-1.00	0.402	3.193	38.78	-308.31
19	-1.00	0.0	0.392	3.208	37.56	-307.14
20	0.0	1.00	0.387	3.217	36.88	-306.45
21	1.00	0.0	0.379	3.230	35.80	-305.38
22	0.0	-1.00	0.375	3.237	35.28	-304.83
23	-1.00	0.0	0.367	3.250	34.30	-303.84
24	0.0	1.00	0.364	3.255	33.91	-303.40
25	1.00	0.0	0.356	3.267	33.00	-302.46
26	0.0	-1.00	0.354	3.271	32.72	-302.13
27	-1.00	0.0	0.347	3.283	31.85	-301.21
28	0.0	1.00	0.346	3.286	31.68	-300.99
29	1.00	0.0	0.339	3.298	30.82	-300.08
30	0.0	-1.00	0.338	3.299	30.77	-299.97
31	-1.00	0.0	0.331	3.311	29.89	-299.03
32	0.0	1.00	0.332	3.311	29.96	-299.05
33	1.00	0.0	0.324	3.324	29.04	-298.06
34	0.0	-1.00	0.326	3.321	29.24	-298.23
35	-1.00	0.0	0.317	3.335	28.25	-297.12
36	0.0	1.00	0.320	3.330	28.62	-297.52
37	1.00	0.0	0.311	3.347	27.50	-296.19
38	0.0	-1.00	0.316	3.338	28.07	-296.94
39	-1.00	0.0	0.305	3.360	26.80	-295.23
40	0.0	1.00	0.311	3.344	27.57	-296.52
41	1.00	0.0	0.300	3.372	26.14	-294.21
42	0.0	-1.00	0.308	3.347	27.28	-296.23
43	-1.00	0.0	0.287	3.389	24.78	-292.96
44	0.0	1.00	0.324	3.322	29.10	-298.21
45	1.00	0.0	0.334	3.509	26.88	-282.43

DRIVING POINT ADMITTANCES AND IMPEDANCES
FOR SPECIFIED BASE CURRENTS

H/LMDA=0.5000 B/LMDA= 0.250 OMEGA= 9.92

45 ELEMENT ARRAY

ELEMENT	IZ(0)		Y0		Z0	
	RE	IM	RE	IM	RE	IM
1	1.00	0.0	0.865	1.531	615.30	-576.54
2	0.0	-1.00	0.448	2.323	161.84	-578.54
3	-1.00	0.0	0.469	2.681	115.40	-482.28
4	0.0	1.00	0.469	2.799	103.30	-457.79
5	1.00	0.0	0.447	2.877	92.04	-444.48
6	0.0	-1.00	0.428	2.937	84.02	-434.81
7	-1.00	0.0	0.414	2.985	78.00	-427.25
8	0.0	1.00	0.401	3.023	73.35	-421.40
9	1.00	0.0	0.390	3.057	69.48	-416.39
10	0.0	-1.00	0.380	3.084	66.32	-412.36
11	-1.00	0.0	0.371	3.109	63.54	-408.69
12	0.0	1.00	0.364	3.130	61.22	-405.71
13	1.00	0.0	0.356	3.150	59.09	-402.85
14	0.0	-1.00	0.350	3.167	57.31	-400.53
15	-1.00	0.0	0.344	3.183	55.60	-398.20
16	0.0	1.00	0.339	3.197	54.18	-396.35
17	1.00	0.0	0.333	3.211	52.76	-394.38
18	0.0	-1.00	0.329	3.222	51.60	-392.86
19	-1.00	0.0	0.324	3.235	50.40	-391.16
20	0.0	1.00	0.320	3.244	49.43	-389.90
21	1.00	0.0	0.316	3.255	48.39	-388.39
22	0.0	-1.00	0.312	3.263	47.57	-387.33
23	-1.00	0.0	0.309	3.274	46.65	-385.96
24	0.0	1.00	0.306	3.280	45.96	-385.08
25	1.00	0.0	0.302	3.290	45.12	-383.82
26	0.0	-1.00	0.299	3.296	44.54	-383.08
27	-1.00	0.0	0.296	3.305	43.77	-381.89
28	0.0	1.00	0.294	3.309	43.28	-381.30
29	1.00	0.0	0.291	3.318	42.56	-380.15
30	0.0	-1.00	0.289	3.322	42.15	-379.69
31	-1.00	0.0	0.286	3.330	41.46	-378.56
32	0.0	1.00	0.284	3.333	41.14	-378.24
33	1.00	0.0	0.281	3.342	40.46	-377.08
34	0.0	-1.00	0.280	3.343	40.22	-376.94
35	-1.00	0.0	0.277	3.353	39.53	-375.69
36	0.0	1.00	0.276	3.352	39.40	-375.78
37	1.00	0.0	0.273	3.363	38.67	-374.35
38	0.0	-1.00	0.272	3.360	38.65	-374.79
39	-1.00	0.0	0.270	3.374	37.88	-372.96
40	0.0	1.00	0.268	3.366	37.90	-374.08
41	1.00	0.0	0.268	3.386	37.31	-371.40
42	0.0	-1.00	0.262	3.370	36.92	-373.69
43	-1.00	0.0	0.268	3.395	37.13	-370.17
44	0.0	1.00	0.249	3.384	34.79	-372.08
45	1.00	0.0	0.172	3.642	20.11	-341.09

50 ELEMENT ARRAY

ELEMENT	V0		Y0		Z0	
	RE	IM	RE	IM	RE	IM
1	1.00	0.0	1.051	1.369	352.75	-459.64
2	0.0	-1.00	0.973	2.174	171.57	-383.20
3	-1.00	0.0	0.842	2.508	120.36	-358.30
4	0.0	1.00	0.742	2.687	95.49	-345.76
5	1.00	0.0	0.671	2.796	81.12	-338.17
6	0.0	-1.00	0.617	2.873	71.40	-332.70
7	-1.00	0.0	0.577	2.929	64.77	-328.69
8	0.0	1.00	0.545	2.975	59.55	-325.25
9	1.00	0.0	0.520	3.010	55.72	-322.56
10	0.0	-1.00	0.498	3.043	52.36	-320.04
11	-1.00	0.0	0.481	3.069	49.83	-318.06
12	0.0	1.00	0.464	3.094	47.41	-316.08
13	1.00	0.0	0.451	3.114	45.61	-314.55
14	0.0	-1.00	0.438	3.134	43.75	-312.92
15	-1.00	0.0	0.428	3.150	42.40	-311.70
16	0.0	1.00	0.417	3.167	40.89	-310.33
17	1.00	0.0	0.410	3.180	39.85	-309.34
18	0.0	-1.00	0.400	3.195	38.58	-308.14
19	-1.00	0.0	0.394	3.205	37.76	-307.33
20	0.0	1.00	0.385	3.219	36.67	-306.26
21	1.00	0.0	0.380	3.228	36.02	-305.58
22	0.0	-1.00	0.373	3.240	35.06	-304.62
23	-1.00	0.0	0.369	3.247	34.54	-304.06
24	0.0	1.00	0.362	3.258	33.66	-303.16
25	1.00	0.0	0.359	3.264	33.26	-302.71
26	0.0	-1.00	0.352	3.275	32.45	-301.86
27	-1.00	0.0	0.350	3.280	32.14	-301.49
28	0.0	1.00	0.343	3.290	31.37	-300.69
29	1.00	0.0	0.342	3.293	31.15	-300.40
30	0.0	-1.00	0.335	3.304	30.41	-299.61
31	-1.00	0.0	0.334	3.306	30.27	-299.41
32	0.0	1.00	0.328	3.316	29.55	-298.63
33	1.00	0.0	0.328	3.318	29.49	-298.52
34	0.0	-1.00	0.321	3.328	28.75	-297.72
35	-1.00	0.0	0.322	3.328	28.79	-297.70
36	0.0	1.00	0.315	3.339	28.02	-296.86
37	1.00	0.0	0.316	3.338	28.16	-296.95
38	0.0	-1.00	0.309	3.349	27.34	-296.05
39	-1.00	0.0	0.312	3.346	27.60	-296.29
40	0.0	1.00	0.304	3.359	26.70	-295.26
41	1.00	0.0	0.307	3.354	27.10	-295.71
42	0.0	-1.00	0.298	3.370	26.08	-294.46
43	-1.00	0.0	0.303	3.360	26.66	-295.25
44	0.0	1.00	0.294	3.380	25.49	-293.61
45	1.00	0.0	0.300	3.364	26.26	-294.93
46	0.0	-1.00	0.289	3.392	24.94	-292.70
47	-1.00	0.0	0.297	3.367	26.04	-294.72
48	0.0	1.00	0.277	3.407	23.70	-291.57
49	1.00	0.0	0.314	3.341	27.86	-296.70
50	0.0	-1.00	0.325	3.523	25.99	-281.42

50 ELEMENT ARRAY

ELEMENT	IZ(0)		Y0		Z0	
	RE	IM	RE	IM	RE	IM
1	1.00	0.0	0.865	1.531	615.31	-576.59
2	0.0	-1.00	0.448	2.323	161.81	-578.46
3	-1.00	0.0	0.469	2.681	115.44	-482.36
4	0.0	1.00	0.469	2.799	103.27	-457.71
5	1.00	0.0	0.447	2.877	92.07	-444.56
6	0.0	-1.00	0.428	2.938	83.98	-434.72
7	-1.00	0.0	0.414	2.985	78.04	-427.34
8	0.0	1.00	0.401	3.024	73.31	-421.30
9	1.00	0.0	0.390	3.056	69.52	-416.48
10	0.0	-1.00	0.380	3.085	66.28	-412.26
11	-1.00	0.0	0.372	3.108	63.59	-408.79
12	0.0	1.00	0.364	3.131	61.18	-405.60
13	1.00	0.0	0.357	3.149	59.14	-402.96
14	0.0	-1.00	0.350	3.167	57.26	-400.42
15	-1.00	0.0	0.344	3.182	55.65	-398.32
16	0.0	1.00	0.338	3.198	54.12	-396.22
17	1.00	0.0	0.333	3.210	52.82	-394.51
18	0.0	-1.00	0.329	3.223	51.54	-392.72
19	-1.00	0.0	0.324	3.234	50.46	-391.30
20	0.0	1.00	0.320	3.245	49.36	-389.74
21	1.00	0.0	0.316	3.254	48.46	-388.55
22	0.0	-1.00	0.312	3.265	47.50	-387.16
23	-1.00	0.0	0.309	3.272	46.73	-386.14
24	0.0	1.00	0.305	3.282	45.87	-384.89
25	1.00	0.0	0.302	3.288	45.21	-384.02
26	0.0	-1.00	0.299	3.297	44.44	-382.86
27	-1.00	0.0	0.297	3.303	43.87	-382.13
28	0.0	1.00	0.294	3.311	43.17	-381.05
29	1.00	0.0	0.291	3.316	42.68	-380.42
30	0.0	-1.00	0.288	3.324	42.02	-379.40
31	-1.00	0.0	0.286	3.328	41.60	-378.88
32	0.0	1.00	0.284	3.336	40.98	-377.89
33	1.00	0.0	0.282	3.339	40.63	-377.47
34	0.0	-1.00	0.279	3.346	40.03	-376.50
35	-1.00	0.0	0.278	3.349	39.75	-376.19
36	0.0	1.00	0.275	3.356	39.16	-375.20
37	1.00	0.0	0.274	3.358	38.94	-375.02
38	0.0	-1.00	0.271	3.366	38.35	-373.99
39	-1.00	0.0	0.270	3.366	38.20	-373.95
40	0.0	1.00	0.268	3.375	37.59	-372.83
41	1.00	0.0	0.267	3.374	37.53	-373.00
42	0.0	-1.00	0.264	3.384	36.88	-371.68
43	-1.00	0.0	0.264	3.381	36.92	-372.21
44	0.0	1.00	0.261	3.394	36.22	-370.48
45	1.00	0.0	0.260	3.385	36.29	-371.65
46	0.0	-1.00	0.260	3.404	35.75	-369.08
47	-1.00	0.0	0.254	3.388	35.41	-371.40
48	0.0	1.00	0.261	3.412	35.66	-368.00
49	1.00	0.0	0.243	3.403	33.46	-369.74
50	0.0	-1.00	0.169	3.655	19.58	-339.58

TABLE 5.3

VEN

SELF AND MUTUAL ADMITTANCES

H/LMDA=0.2500 B/LMDA= 0.250 OMEGA= 8.53

2 ELEMENT ARRAY

I, K	YIK IN MMHO		I, K	YIK IN MMHO		I, K	YIK IN MMHO		I, K	YIK IN MMHO	
	RE	IM		RE	IM		RE	IM		RE	IM
1, 1	6.9214	-4.3657	1, 2	1.4437	5.3557						

VEN

SELF AND MUTUAL ADMITTANCES

H/LMDA=0.2500 B/LMDA= 0.250 OMEGA= 8.53

3 ELEMENT ARRAY

I, K	YIK IN MMHO		I, K	YIK IN MMHO		I, K	YIK IN MMHO		I, K	YIK IN MMHO	
	RE	IM		RE	IM		RE	IM		RE	IM
1, 1	6.8929	-4.4191	1, 2	1.8349	5.6226	1, 3	0.1242	-0.7456			
2, 2	3.1860	-4.7492									

VEN

SELF AND MUTUAL ADMITTANCES

H/LMDA=0.2500 B/LMDA= 0.250 OMEGA= 8.53

4 ELEMENT ARRAY

I, K	YIK IN MMHO		I, K	YIK IN MMHO		I, K	YIK IN MMHO		I, K	YIK IN MMHO	
	RE	IM		RE	IM		RE	IM		RE	IM
1, 1	6.8950	-4.4095	1, 2	1.8060	5.6300	1, 3	0.2524	-0.8875	1, 4	-0.2896	-0.0900
2, 2	3.1618	-4.8365	2, 3	2.2834	5.9996						

VEN

SELF AND MUTUAL ADMITTANCES

H/LMDA=0.2500 B/LMDA= 0.250 OMEGA= [illegible]

5 ELEMENT ARRAY

I, K	YIK IN MMHO		I, K	YIK IN MMHO		I, K	YIK IN MMHO		I, K	YIK IN MMHO	
	RE	IM		RE	IM		RE	IM		RE	IM
1, 1	6.8943	-4.4130	1, 2	1.8140	5.6292	1, 3	0.2565	-0.8702	1, 4	-0.3752	-0.1681
1, 5	-0.0560	0.1749									
2, 2	3.1616	-4.8185	2, 3	2.2439	6.0044	2, 4	0.4472	-1.0613			
3, 3	3.1401	-4.9213									

VEN

SELF AND MUTUAL ADMITTANCES

H/LMDA=0.2500 B/LMDA= 0.250 OMEGA= 8.53

6 ELEMENT ARRAY

I, K	YIK IN MMHO		I, K	YIK IN MMHO		I, K	YIK IN MMHO		I, K	YIK IN MMHO	
	RE	IM		RE	IM		RE	IM		RE	IM
1, 1	6.8947	-4.4114	1, 2	1.8105	5.6295	1, 3	0.2560	-0.8756	1, 4	-0.3632	-0.1711
1, 5	-0.1093	0.2345	1, 6	0.1211	0.0375						
2, 2	3.1618	-4.8256	2, 3	2.2550	6.0049	2, 4	0.4500	-1.0362	2, 5	-0.4827	-0.2924
3, 3	3.1392	-4.9041	3, 4	2.2051	6.0083						

VEN

SELF AND MUTUAL ADMITTANCES

H/LMDA=0.2500 B/LMDA= 0.250 OMEGA= 8.53

7 ELEMENT ARRAY

I, K	YIK IN MMHO		I, K	YIK IN MMHO		I, K	YIK IN MMHO		I, K	YIK IN MMHO	
	RE	IM		RE	IM		RE	IM		RE	IM
1, 1	6.8944	-4.4123	1, 2	1.8124	5.6293	1, 3	0.2562	-0.8731	1, 4	-0.3673	-0.1707
1, 5	-0.1116	0.2256	1, 6	0.1661	0.0768	1, 7	0.0270	-0.0907			
2, 2	3.1617	-4.8220	2, 3	2.2500	6.0046	2, 4	0.4502	-1.0442	2, 5	-0.4650	-0.2944
2, 6	-0.1969	0.3108									
3, 3	3.1398	-4.9109	3, 4	2.2160	6.0090	3, 5	0.4519	-1.0120			
4, 4	3.1385	-4.8866									

VEN

SELF AND MUTUAL ADMITTANCES

H/LMDA=0.2500 B/LMDA= 0.250 OMEGA= 8.53

8 ELEMENT ARRAY

I, K	YIK IN MMHO RE	IM	I, K	YIK IN MMHO RE	IM	I, K	YIK IN MMHO RE	IM	I, K	YIK IN MMHO RE	IM
1, 1	6.8946	-4.4117	1, 2	1.8113	5.6295	1, 3	0.2561	-0.8745	1, 4	-0.3653	-0.1709
1, 5	-0.1113	0.2288	1, 6	0.1591	0.0787	1, 7	0.0574	-0.1262	1, 8	-0.0714	-0.0205
2, 2	3.1618	-4.8241	2, 3	2.2527	6.0048	2, 4	0.4501	-1.0404	2, 5	-0.4710	-0.2946
2, 6	-0.1985	0.2974	2, 7	0.2243	0.1428						
3, 3	3.1395	-4.9075	3, 4	2.2112	6.0086	3, 5	0.4524	-1.0196	3, 6	-0.4478	-0.2957
4, 4	3.1390	-4.8935	4, 5	2.2269	6.0096						

VEN

SELF AND MUTUAL ADMITTANCES

H/LMDA=0.2500 B/LMDA= 0.250 OMEGA= 8.53

9 ELEMENT ARRAY

I, K	YIK IN MMHO RE	IM	I, K	YIK IN MMHO RE	IM	I, K	YIK IN MMHO RE	IM	I, K	YIK IN MMHO RE	IM
1, 1	6.8945	-4.4121	1, 2	1.8120	5.6294	1, 3	0.2562	-0.8736	1, 4	-0.3664	-0.1708
1, 5	-0.1114	0.2272	1, 6	0.1617	0.0783	1, 7	0.0590	-0.1206	1, 8	-0.1005	-0.0450
1, 9	-0.0161	0.0582									
2, 2	3.1618	-4.8228	2, 3	2.2511	6.0047	2, 4	0.4501	-1.0425	2, 5	-0.4680	-0.2944
2, 6	-0.1986	0.3022	2, 7	0.2137	0.1442	2, 8	0.1096	-0.1728			
3, 3	3.1396	-4.9094	3, 4	2.2137	6.0088	3, 5	0.4521	-1.0160	3, 6	-0.4536	-0.2961
3, 7	-0.1996	0.2844									
4, 4	3.1387	-4.8901	4, 5	2.2221	6.0093	4, 6	0.4528	-1.0273			
5, 5	3.1395	-4.9003									

VEN

SELF AND MUTUAL ADMITTANCES

H/LMDA=0.2500 B/LMDA= 0.250 OMEGA= 8.53

10 ELEMENT ARRAY

I, K	YIK IN MMHO RE	IM	I, K	YIK IN MMHO RE	IM	I, K	YIK IN MMHO RE	IM	I, K	YIK IN MMHO RE	IM
1, 1	6.8946	-4.4118	1, 2	1.8115	5.6294	1, 3	0.2561	-0.8742	1, 4	-0.3657	-0.1709
1, 5	-0.1113	0.2281	1, 6	0.1603	0.0785	1, 7	0.0588	-0.1227	1, 8	-0.0958	-0.0463
1, 9	-0.0364	0.0828	1,10	0.0488	0.0131						
2, 2	3.1618	-4.8237	2, 3	2.2521	6.0047	2, 4	0.4501	-1.0412	2, 5	-0.4697	-0.2945
2, 6	-0.1985	0.2997	2, 7	0.2176	0.1443	2, 8	0.1108	-0.1641	2, 9	-0.1391	-0.0875
3, 3	3.1396	-4.9082	3, 4	2.2122	6.0087	3, 5	0.4522	-1.0180	3, 6	-0.4508	-0.2959
3, 7	-0.1998	0.2890	3, 8	0.2034	0.1451						
4, 4	3.1389	-4.8920	4, 5	2.2247	6.0094	4, 6	0.4526	-1.0237	4, 7	-0.4595	-0.2964
5, 5	3.1393	-4.8969	5, 6	2.2172	6.0089						

VEN

SELF AND MUTUAL ADMITTANCES

H/LMDA=0.2500 B/LMDA= 0.250 OMEGA= 8.53

12 ELEMENT ARRAY

I, K	YIK IN MMHO RE	IM	I, K	YIK IN MMHO RE	IM	I, K	YIK IN MMHO RE	IM	I, K	YIK IN MMHO RE	IM
1, 1	6.8946	-4.4119	1, 2	1.8116	5.6294	1, 3	0.2561	-0.8741	1, 4	-0.3658	-0.1708
1, 5	-0.1113	0.2279	1, 6	0.1606	0.0785	1, 7	0.0588	-0.1223	1, 8	-0.0967	-0.0462
1, 9	-0.0373	0.0804	1,10	0.0664	0.0313	1,11	0.0256	-0.0601	1,12	-0.0362	-0.0091
2, 2	3.1618	-4.8235	2, 3	2.2519	6.0047	2, 4	0.4501	-1.0415	2, 5	-0.4694	-0.2945
2, 6	-0.1986	0.3002	2, 7	0.2168	0.1442	2, 8	0.1108	-0.1656	2, 9	-0.1345	-0.0886
2,10	-0.0729	0.1091	2,11	0.0980	0.0606						
3, 3	3.1396	-4.9085	3, 4	2.2125	6.0087	3, 5	0.4522	-1.0176	3, 6	-0.4514	-0.2959
3, 7	-0.1998	0.2881	3, 8	0.2052	0.1452	3, 9	0.1117	-0.1588	3,10	-0.1247	-0.0893
4, 4	3.1388	-4.8916	4, 5	2.2241	6.0094	4, 6	0.4526	-1.0244	4, 7	-0.4583	-0.2963
4, 8	-0.1999	0.2913	4, 9	0.2110	0.1454						
5, 5	3.1393	-4.8977	5, 6	2.2183	6.0090	5, 7	0.4524	-1.0221	5, 8	-0.4537	-0.2960
6, 6	3.1391	-4.8955	6, 7	2.2224	6.0093						

VEN

SELF AND MUTUAL ADMITTANCES

H/LMDA=0.2500 B/LMDA= 0.250 OMEGA= 8.53

14 ELEMENT ARRAY

I, K	YIK IN MMHO		I, K	YIK IN MMHO		I, K	YIK IN MMHO		I, K	YIK IN MMHO	
	RE	IM		RE	IM		RE	IM		RE	IM
1, 1	6.8945	-4.4119	1, 2	1.8116	5.6294	1, 3	0.2561	-0.8740	1, 4	-0.3659	-0.1708
1, 5	-0.1113	0.2279	1, 6	0.1607	0.0785	1, 7	0.0588	-0.1222	1, 8	-0.0969	-0.0462
1, 9	-0.0374	0.0800	1,10	0.0670	0.0311	1,11	0.0263	-0.0583	1,12	-0.0498	-0.0229
1,13	-0.0192	0.0464	1,14	0.0284	0.0068						
2, 2	3.1618	-4.8234	2, 3	2.2518	6.0047	2, 4	0.4501	-1.0416	2, 5	-0.4692	-0.2945
2, 6	-0.1986	0.3004	2, 7	0.2165	0.1442	2, 8	0.1108	-0.1660	2, 9	-0.1340	-0.0886
2,10	-0.0729	0.1102	2,11	0.0947	0.0614	2,12	0.0527	-0.0799	2,13	-0.0744	-0.0451
3, 3	3.1396	-4.9086	3, 4	2.2126	6.0087	3, 5	0.4522	-1.0174	3, 6	-0.4516	-0.2959
3, 7	-0.1998	0.2878	3, 8	0.2056	0.1452	3, 9	0.1117	-0.1582	3,10	-0.1259	-0.0893
3,11	-0.0736	0.1054	3,12	0.0873	0.0620						
4, 4	3.1388	-4.8915	4, 5	2.2239	6.0094	4, 6	0.4526	-1.0247	4, 7	-0.4579	-0.2963
4, 8	-0.2000	0.2918	4, 9	0.2102	0.1454	4,10	0.1117	-0.1603	4,11	-0.1301	-0.0894
5, 5	3.1394	-4.8979	5, 6	2.2186	6.0090	5, 7	0.4524	-1.0216	5, 8	-0.4543	-0.2961
5, 9	-0.1999	0.2903	5,10	0.2070	0.1452						
6, 6	3.1391	-4.8951	6, 7	2.2219	6.0092	6, 8	0.4525	-1.0228	6, 9	-0.4570	-0.2962
7, 7	3.1392	-4.8962	7, 8	2.2193	6.0091						

VEN

SELF AND MUTUAL ADMITTANCES

H/LMDA=0.2500 B/LMDA= 0.250 OMEGA= 8.53

16 ELEMENT ARRAY

I, K	YIK IN MMHO		I, K	YIK IN MMHO		I, K	YIK IN MMHO		I, K	YIK IN MMHO	
	RE	IM		RE	IM		RE	IM		RE	IM
1, 1	6.8945	-4.4119	1, 2	1.8117	5.6294	1, 3	0.2561	-0.8740	1, 4	-0.3659	-0.1708
1, 5	-0.1113	0.2278	1, 6	0.1608	0.0785	1, 7	0.0588	-0.1221	1, 8	-0.0970	-0.0462
1, 9	-0.0374	0.0799	1,10	0.0671	0.0311	1,11	0.0263	-0.0580	1,12	-0.0502	-0.0228
1,13	-0.0198	0.0450	1,14	0.0393	0.0176	1,15	0.0151	-0.0374	1,16	-0.0231	-0.0053
2, 2	3.1618	-4.8234	2, 3	2.2518	6.0047	2, 4	0.4501	-1.0416	2, 5	-0.4692	-0.2945
2, 6	-0.1986	0.3005	2, 7	0.2164	0.1442	2, 8	0.1108	-0.1661	2, 9	-0.1338	-0.0885
2,10	-0.0729	0.1105	2,11	0.0943	0.0614	2,12	0.0527	-0.0808	2,13	-0.0718	-0.0458
2,14	-0.0404	0.0622	2,15	0.0592	0.0353						
3, 3	3.1396	-4.9086	3, 4	2.2127	6.0087	3, 5	0.4522	-1.0174	3, 6	-0.4517	-0.2960
3, 7	-0.1997	0.2877	3, 8	0.2057	0.1452	3, 9	0.1116	-0.1580	3,10	-0.1262	-0.0893
3,11	-0.0736	0.1049	3,12	0.0882	0.0620	3,13	0.0533	-0.0771	3,14	-0.0659	-0.0463
4, 4	3.1388	-4.8914	4, 5	[illegible]	[illegible]	4, 6	0.4527	-1.0248	4, 7	-0.4578	-0.2963
4, 8	-0.2000	0.2920	4, 9	0.2099	0.1454	4,10	0.1118	[illegible]	4,11	-0.1296	-0.0894
4,12	-0.0736	0.1065	4,13	0.0915	0.0621						
5, 5	3.1394	-4.8980	5, 6	2.2187	6.0090	5, 7	0.4524	-1.0215	5, 8	-0.4545	-0.2961
5, 9	-0.1998	0.2900	5,10	0.2074	0.1453	5,11	0.1117	-0.1597	5,12	-0.1271	-0.0893
6, 6	3.1391	-4.8949	6, 7	2.2217	6.0092	6, 8	0.4525	-1.0231	6, 9	-0.4567	-0.2962
6,10	-0.1999	0.2908	6,11	0.2094	0.1453						
7, 7	3.1392	-4.8964	7, 8	2.2196	6.0091	7, 9	0.4525	-1.0224	7,10	-0.4549	-0.2961
8, 8	3.1392	-4.8958	8, 9	2.2213	6.0092						

VEN

SELF AND MUTUAL ADMITTANCES

H/LMDA=0.2500 B/LMDA= 0.250 OMEGA= 8.53

18 ELEMENT ARRAY

I, K	YIK IN MMHO		I, K	YIK IN MMHO		I, K	YIK IN MMHO		I, K	YIK IN MMHO	
	RE	IM		RE	IM		RE	IM		RE	IM
1, 1	6.8945	-4.4119	1, 2	1.8117	5.6294	1, 3	0.2561	-0.8740	1, 4	-0.3659	-0.1708
1, 5	-0.1114	0.2278	1, 6	0.1608	0.0785	1, 7	0.0588	-0.1221	1, 8	-0.0970	-0.0462
1, 9	-0.0374	0.0799	1,10	0.0672	0.0311	1,11	0.0264	-0.0579	1,12	-0.0504	-0.0228
1,13	-0.0198	0.0448	1,14	0.0397	0.0176	1,15	0.0156	-0.0362	1,16	-0.0321	-0.0141
1,17	-0.0122	0.0310	1,18	0.0193	0.0043						
2, 2	3.1618	-4.8234	2, 3	2.2518	6.0047	2, 4	0.4501	-1.0417	2, 5	-0.4691	-0.2945
2, 6	-0.1986	0.3005	2, 7	0.2164	0.1442	2, 8	0.1108	-0.1662	2, 9	-0.1337	-0.0885
2,10	-0.0729	0.1106	2,11	0.0941	0.0614	2,12	0.0527	-0.0810	2,13	-0.0715	-0.0458
2,14	-0.0404	0.0628	2,15	0.0571	0.0359	2,16	0.0323	-0.0503	2,17	-0.0488	-0.0287
3, 3	3.1396	-4.9087	3, 4	2.2127	6.0087	3, 5	0.4522	-1.0173	3, 6	-0.4517	-0.2960
3, 7	-0.1997	0.2876	3, 8	0.2058	0.1452	3, 9	0.1116	-0.1579	3,10	-0.1263	-0.0893
3,11	-0.0736	0.1048	3,12	0.0884	0.0620	3,13	0.0532	-0.0768	3,14	-0.0666	-0.0463
3,15	-0.0409	0.0599	3,16	0.0523	0.0364						
4, 4	3.1388	-4.8914	4, 5	2.2238	6.0094	4, 6	0.4527	-1.0249	4, 7	-0.4577	-0.2963
4, 8	-0.2000	0.2920	4, 9	0.2098	0.1454	4,10	0.1118	-0.1608	4,11	-0.1294	-0.0894
4,12	-0.0737	0.1068	4,13	0.0910	0.0621	4,14	0.0533	-0.0780	4,15	-0.0693	-0.0464
5, 5	3.1394	-4.8981	5, 6	2.2188	6.0090	5, 7	0.4524	-1.0214	5, 8	-0.4546	-0.2961
5, 9	-0.1998	0.2899	5,10	0.2075	0.1453	5,11	0.1117	-0.1595	5,12	-0.1274	-0.0893
5,13	-0.0736	0.1060	5,14	0.0891	0.0620						
6, 6	3.1391	-4.8948	6, 7	2.2216	6.0092	6, 8	0.4525	-1.0232	6, 9	-0.4566	-0.2962
6,10	-0.1999	0.2910	6,11	0.2091	0.1453	6,12	0.1117	-0.1600	6,13	-0.1290	-0.0894
7, 7	3.1392	-4.8965	7, 8	2.2197	6.0091	7, 9	0.4525	-1.0223	7,10	-0.4551	-0.2961
7,11	-0.1999	0.2905	7,12	0.2078	0.1453						
8, 8	3.1392	-4.8956	8, 9	2.2211	6.0092	8,10	0.4525	-1.0227	8,11	-0.4564	-0.2962
9, 9	3.1392	-4.8960	9,10	2.2199	6.0091						

VEN

SELF AND MUTUAL ADMITTANCES

H/LMDA=0.2500 B/LMDA= 0.250 OMEGA= 8.53

20 ELEMENT ARRAY

I, K	YIK IN MMHO RE	IM	I, K	YIK IN MMHO RE	IM	I, K	YIK IN MMHO RE	IM	I, K	YIK IN MMHO RE	IM
1, 1	6.8945	-4.4119	1, 2	1.8117	5.6294	1, 3	0.2561	-0.8740	1, 4	-0.3659	-0.1708
1, 5	-0.1114	0.2278	1, 6	0.1608	0.0785	1, 7	0.0588	-0.1221	1, 8	-0.0970	-0.0462
1, 9	-0.0374	0.0798	1,10	0.0672	0.0311	1,11	0.0264	-0.0579	1,12	-0.0504	-0.0227
1,13	-0.0198	0.0447	1,14	0.0398	0.0175	1,15	0.0156	-0.0361	1,16	-0.0324	-0.0141
1,17	-0.0127	0.0301	1,18	0.0270	0.0117	1,19	0.0102	-0.0264	1,20	-0.0165	-0.0035
2, 2	3.1618	-4.8233	2, 3	2.2518	6.0047	2, 4	0.4501	-1.0417	2, 5	-0.4691	-0.2945
2, 6	-0.1986	0.3005	2, 7	0.2163	0.1442	2, 8	0.1108	-0.1662	2, 9	-0.1337	-0.0885
2,10	-0.0729	0.1106	2,11	0.0941	0.0614	2,12	0.0527	-0.0811	2,13	-0.0713	-0.0458
2,14	-0.0404	0.0630	2,15	0.0568	0.0359	2,16	0.0322	-0.0509	2,17	-0.0470	-0.0291
2,18	-0.0266	0.0420	2,19	0.0412	0.0239						
3, 3	3.1396	-4.9087	3, 4	2.2127	6.0087	3, 5	0.4522	-1.0173	3, 6	-0.4517	-0.2960
3, 7	-0.1997	0.2876	3, 8	0.2058	0.1452	3, 9	0.1116	-0.1579	3,10	-0.1264	-0.0893
3,11	-0.0736	0.1047	3,12	0.0885	0.0620	3,13	0.0532	-0.0766	3,14	-0.0668	-0.0463
3,15	-0.0409	0.0596	3,16	0.0529	0.0364	3,17	0.0327	-0.0485	3,18	-0.0430	-0.0296
4, 4	3.1388	-4.8913	4, 5	2.2238	6.0094	4, 6	0.4527	-1.0249	4, 7	-0.4577	-0.2963
4, 8	-0.2000	0.2921	4, 9	0.2098	0.1454	4,10	0.1118	-0.1609	4,11	-0.1293	-0.0894
4,12	-0.0737	0.1069	4,13	0.0909	0.0621	4,14	0.0533	-0.0782	4,15	-0.0690	-0.0464
4,16	-0.0409	0.0606	4,17	0.0551	0.0364						
5, 5	3.1394	-4.8981	5, 6	2.2188	6.0090	5, 7	0.4524	-1.0214	5, 8	-0.4546	-0.2961
5, 9	-0.1998	0.2899	5,10	0.2076	0.1453	5,11	0.1117	-0.1594	5,12	-0.1275	-0.0893
5,13	-0.0736	0.1058	5,14	0.0893	0.0620	5,15	0.0532	-0.0776	5,16	-0.0674	-0.0463
6, 6	3.1391	-4.8948	6, 7	2.2215	6.0092	6, 8	0.4525	-1.0232	6, 9	-0.4565	-0.2962
6,10	-0.1999	0.2911	6,11	0.2090	0.1453	6,12	0.1117	-0.1602	6,13	-0.1288	-0.0894
6,14	-0.0736	0.1063	6,15	0.0906	0.0621						
7, 7	3.1392	-4.8966	7, 8	2.2198	6.0091	7, 9	0.4525	-1.0222	7,10	-0.4552	-0.2961
7,11	-0.1999	0.2904	7,12	0.2079	0.1453	7,13	0.1117	-0.1598	7,14	-0.1277	-0.0893
8, 8	3.1392	-4.8956	8, 9	2.2210	6.0092	8,10	0.4525	-1.0228	8,11	-0.4562	-0.2962
8,12	-0.1999	0.2907	8,13	0.2089	0.1453						
9, 9	3.1392	-4.8961	9,10	2.2201	6.0091	9,11	0.4525	-1.0225	9,12	-0.4553	-0.2961
10,10	3.1392	-4.8959	10,11	2.2209	6.0091						

VEN

SELF AND MUTUAL ADMITTANCES

H/LMDA=0.2500 B/LMDA= 0.250 OMEGA= 8.53

22 ELEMENT ARRAY

I, K	YIK IN MMHO RE	IM	I, K	YIK IN MMHO RE	IM	I, K	YIK IN MMHO RE	IM	I, K	YIK IN MMHO RE	IM
1, 1	6.8945	-4.4119	1, 2	1.8117	5.6294	1, 3	0.2561	-0.8740	1, 4	-0.3659	-0.1708
1, 5	-0.1114	0.2278	1, 6	0.1608	0.0785	1, 7	0.0588	-0.1220	1, 8	-0.0970	-0.0462
1, 9	-0.0374	0.0798	1,10	0.0673	0.0311	1,11	0.0264	-0.0579	1,12	-0.0504	-0.0227
1,13	-0.0198	0.0447	1,14	0.0398	0.0175	1,15	0.0156	-0.0360	1,16	-0.0325	-0.0141
1,17	-0.0127	0.0299	1,18	0.0273	0.0116	1,19	0.0106	-0.0255	1,20	-0.0231	-0.0098
1,21	-0.0086	0.0228	1,22	0.0143	0.0029						
2, 2	3.1618	-4.8233	2, 3	2.2517	6.0047	2, 4	0.4501	-1.0417	2, 5	-0.4691	-0.2945
2, 6	-0.1986	0.3006	2, 7	0.2163	0.1442	2, 8	0.1108	-0.1662	2, 9	-0.1336	-0.0885
2,10	-0.0729	0.1107	2,11	0.0940	0.0614	2,12	0.0527	-0.0811	2,13	-0.0713	-0.0458
2,14	-0.0404	0.0631	2,15	0.0567	0.0359	2,16	0.0322	-0.0510	2,17	-0.0468	-0.0291
2,18	-0.0265	0.0424	2,19	0.0397	0.0243	2,20	0.0224	-0.0358	2,21	-0.0355	-0.0203
3, 3	3.1396	-4.9087	3, 4	2.2127	6.0087	3, 5	0.4522	-1.0173	3, 6	-0.4517	-0.2960
3, 7	-0.1997	0.2876	3, 8	0.2058	0.1452	3, 9	0.1116	-0.1578	3,10	-0.1264	-0.0893
3,11	-0.0736	0.1047	3,12	0.0886	0.0620	3,13	0.0532	-0.0766	3,14	-0.0669	-0.0463
3,15	-0.0408	0.0595	3,16	0.0531	0.0364	3,17	0.0327	-0.0483	3,18	-0.0435	-0.0295
3,19	-0.0269	0.0405	3,20	0.0363	0.0247						
4, 4	3.1388	-4.8913	4, 5	2.2237	6.0094	4, 6	0.4527	-1.0249	4, 7	-0.4577	-0.2963
4, 8	-0.2000	0.2921	4, 9	0.2098	0.1454	4,10	0.1118	-0.1609	4,11	-0.1293	-0.0894
4,12	-0.0737	0.1069	4,13	0.0908	0.0621	4,14	0.0533	-0.0783	4,15	-0.0688	-0.0464
4,16	-0.0409	0.0608	4,17	0.0548	0.0364	4,18	0.0327	-0.0491	4,19	-0.0454	-0.0296
5, 5	3.1394	-4.8981	5, 6	2.2188	6.0090	5, 7	0.4524	-1.0213	5, 8	-0.4547	-0.2961
5, 9	-0.1998	0.2898	5,10	0.2076	0.1453	5,11	0.1117	-0.1593	5,12	-0.1276	-0.0893
5,13	-0.0736	0.1058	5,14	0.0894	0.0620	5,15	0.0532	-0.0775	5,16	-0.0676	-0.0463
5,17	-0.0408	0.0603	5,18	0.0535	0.0364						
6, 6	3.1391	-4.8948	6, 7	2.2215	6.0092	6, 8	0.4525	-1.0233	6, 9	-0.4565	-0.2962
6,10	-0.1999	0.2911	6,11	0.2090	0.1453	6,12	0.1117	-0.1602	6,13	-0.1287	-0.0894
6,14	-0.0736	0.1064	6,15	0.0905	0.0621	6,16	0.0532	-0.0778	6,17	-0.0686	-0.0464
7, 7	3.1392	-4.8966	7, 8	2.2198	6.0091	7, 9	0.4525	-1.0221	7,10	-0.4552	-0.2961
7,11	-0.1999	0.2904	7,12	0.2080	0.1453	7,13	0.1117	-0.1597	7,14	-0.1278	-0.0894
7,15	-0.0736	0.1061	7,16	0.0896	0.0620						
8, 8	3.1392	-4.8955	8, 9	2.2210	6.0092	8,10	0.4525	-1.0228	8,11	-0.4562	-0.2962
8,12	-0.1999	0.2908	8,13	0.2088	0.1453	8,14	0.1117	-0.1600	8,15	-0.1286	-0.0894
9, 9	3.1392	-4.8962	9,10	2.2201	6.0091	9,11	0.4525	-1.0224	9,12	-0.4554	-0.2961
9,13	-0.1999	0.2906	9,14	0.2081	0.1453						
10,10	3.1392	-4.8958	10,11	2.2208	6.0091	10,12	0.4525	-1.0226	10,13	-0.4561	-0.2962
11,11	3.1392	-4.8960	11,12	2.2202	6.0091						

VEN

SELF AND MUTUAL ADMITTANCES

H/LMDA=0.2500 B/LMDA= 0.250 OMEGA= 8.53

24 ELEMENT ARRAY

I, K	YIK IN MMHO RE	IM	I, K	YIK IN MMHO RE	IM	I, K	YIK IN MMHO RE	IM	I, K	YIK IN MMHO RE	IM
1, 1	6.8945	-4.4119	1, 2	1.8117	5.6294	1, 3	0.2561	-0.8740	1, 4	-0.3660	-0.1708
1, 5	-0.1114	0.2278	1, 6	0.1608	0.0785	1, 7	0.0588	-0.1220	1, 8	-0.0970	-0.0462
1, 9	-0.0374	0.0798	1,10	0.0673	0.0311	1,11	0.0264	-0.0579	1,12	-0.0505	-0.0227
1,13	-0.0198	0.0447	1,14	0.0398	0.0175	1,15	0.0156	-0.0360	1,16	-0.0326	-0.0141
1,17	-0.0127	0.0299	1,18	0.0273	0.0116	1,19	0.0106	-0.0254	1,20	-0.0234	-0.0098
1,21	-0.0090	0.0221	1,22	0.0202	0.0084	1,23	0.0074	-0.0200	1,24	-0.0126	-0.0025
2, 2	3.1618	-4.8233	2, 3	2.2517	6.0047	2, 4	0.4501	-1.0417	2, 5	-0.4691	-0.2945
2, 6	-0.1986	0.3006	2, 7	0.2163	0.1442	2, 8	0.1108	-0.1662	2, 9	-0.1336	-0.0885
2,10	-0.0729	0.1107	2,11	0.0940	0.0614	2,12	0.0527	-0.0811	2,13	-0.0713	-0.0458
2,14	-0.0404	0.0631	2,15	0.0567	0.0359	2,16	0.0322	-0.0511	2,17	-0.0467	-0.0291
2,18	-0.0265	0.0426	2,19	0.0395	0.0243	2,20	0.0224	-0.0362	2,21	-0.0342	-0.0207
2,22	-0.0192	0.0310	2,23	0.0310	0.0175						
3, 3	3.1396	-4.9087	3, 4	2.2128	6.0087	3, 5	0.4522	-1.0173	3, 6	-0.4518	-0.2960
3, 7	-0.1997	0.2876	3, 8	0.2059	0.1452	3, 9	0.1116	-0.1578	3,10	-0.1264	-0.0893
3,11	-0.0736	0.1047	3,12	0.0886	0.0620	3,13	0.0532	-0.0766	3,14	-0.0669	-0.0463
3,15	-0.0408	0.0595	3,16	0.0531	0.0364	3,17	0.0327	-0.0482	3,18	-0.0436	-0.0296
3,19	-0.0269	0.0402	3,20	0.0367	0.0246	3,21	0.0227	-0.0345	3,22	-0.0312	-0.0210
4, 4	3.1388	-4.8913	4, 5	2.2237	6.0094	4, 6	0.4527	-1.0249	4, 7	-0.4577	-0.2963
4, 8	-0.2000	0.2921	4, 9	0.2097	0.1454	4,10	0.1118	-0.1609	4,11	-0.1292	-0.0894
4,12	-0.0737	0.1069	4,13	0.0908	0.0621	4,14	0.0533	-0.0783	4,15	-0.0688	-0.0464
4,16	-0.0409	0.0609	4,17	0.0547	0.0364	4,18	0.0327	-0.0492	4,19	-0.0451	-0.0296
4,20	-0.0269	0.0409	4,21	0.0383	0.0247						
5, 5	3.1394	-4.8981	5, 6	2.2188	6.0090	5, 7	0.4524	-1.0213	5, 8	-0.4547	-0.2961
5, 9	-0.1998	0.2898	5,10	0.2076	0.1453	5,11	0.1117	-0.1593	5,12	-0.1276	-0.0893
5,13	-0.0736	0.1057	5,14	0.0895	0.0620	5,15	0.0532	-0.0774	5,16	-0.0676	-0.0463
5,17	-0.0408	0.0602	5,18	0.0537	0.0364	5,19	0.0327	-0.0488	5,20	-0.0440	-0.0295
6, 6	3.1391	-4.8948	6, 7	2.2215	6.0092	6, 8	0.4525	-1.0233	6, 9	-0.4564	-0.2962
6,10	-0.1999	0.2911	6,11	0.2090	0.1453	6,12	0.1117	-0.1603	6,13	-0.1287	-0.0894
6,14	-0.0736	0.1064	6,15	0.0904	0.0621	6,16	0.0532	-0.0779	6,17	-0.0685	-0.0464
6,18	-0.0409	0.0605	6,19	0.0546	0.0364						
7, 7	3.1392	-4.8966	7, 8	2.2199	6.0091	7, 9	0.4525	-1.0221	7,10	-0.4553	-0.2961
7,11	-0.1999	0.2903	7,12	0.2080	0.1453	7,13	0.1117	-0.1597	7,14	-0.1279	-0.0894
7,15	-0.0736	0.1060	7,16	0.0897	0.0620	7,17	0.0532	-0.0777	7,18	-0.0677	-0.0464
8, 8	3.1392	-4.8955	8, 9	2.2210	6.0092	8,10	0.4525	-1.0229	8,11	-0.4561	-0.2962
8,12	-0.1999	0.2908	8,13	0.2088	0.1453	8,14	0.1117	-0.1600	8,15	-0.1286	-0.0894
8,16	-0.0736	0.1062	8,17	0.0903	0.0620						
9, 9	3.1392	-4.8962	9,10	2.2201	6.0091	9,11	0.4525	-1.0224	9,12	-0.4555	-0.2961
9,13	-0.1999	0.2905	9,14	0.2082	0.1453	9,15	0.1117	-0.1599	9,16	-0.1280	-0.0894
10,10	3.1392	-4.8958	10,11	2.2208	6.0091	10,12	0.4525	-1.0227	10,13	-0.4560	-0.2962
10,14	[illegible]	[illegible]	10,15	0.2087	0.1453						
11,11	3.1392	-4.8960	11,12	2.2202	6.0091	11,13	0.4525	-1.0225	11,14	-0.4555	-0.2962
12,12	3.1392	-4.8959	12,13	2.2208	6.0091						

VEN

SELF AND MUTUAL ADMITTANCES

H/LMDA=0.2500 B/LMDA= 0.250 OMEGA= 8.53

26 ELEMENT ARRAY

I, K	YIK IN MMHO RE	IM	I, K	YIK IN MMHO RE	IM	I, K	YIK IN MMHO RE	IM	I, K	YIK IN MMHO RE	IM
1, 1	6.8945	-4.4119	1, 2	1.8117	5.6294	1, 3	0.2561	-0.8740	1, 4	-0.3660	-0.1708
1, 5	-0.1114	0.2278	1, 6	0.1608	0.0784	1, 7	0.0588	-0.1220	1, 8	-0.0970	-0.0462
1, 9	-0.0374	0.0798	1,10	0.0673	0.0311	1,11	0.0264	-0.0579	1,12	-0.0505	-0.0227
1,13	-0.0198	0.0447	1,14	0.0398	0.0175	1,15	0.0156	-0.0360	1,16	-0.0326	-0.0141
1,17	-0.0127	0.0299	1,18	0.0274	0.0116	1,19	0.0106	-0.0254	1,20	-0.0234	-0.0098
1,21	-0.0090	0.0220	1,22	0.0204	0.0084	1,23	0.0078	-0.0194	1,24	-0.0178	-0.0073
1,25	-0.0065	0.0178	1,26	0.0112	0.0022						
2, 2	3.1618	-4.8233	2, 3	2.2517	6.0047	2, 4	0.4501	-1.0417	2, 5	-0.4691	-0.2945
2, 6	-0.1986	0.3006	2, 7	0.2163	0.1442	2, 8	0.1108	-0.1663	2, 9	-0.1336	-0.0885
2,10	-0.0729	0.1107	2,11	0.0940	0.0614	2,12	0.0527	-0.0812	2,13	-0.0712	-0.0458
2,14	-0.0404	0.0631	2,15	0.0567	0.0359	2,16	0.0322	-0.0511	2,17	-0.0467	-0.0291
2,18	-0.0265	0.0426	2,19	0.0394	0.0243	2,20	0.0224	-0.0363	2,21	-0.0340	-0.0207
2,22	-0.0192	0.0314	2,23	0.0299	0.0179	2,24	0.0167	-0.0273	2,25	-0.0274	-0.0153
3, 3	3.1396	-4.9087	3, 4	2.2128	6.0087	3, 5	0.4522	-1.0173	3, 6	-0.4518	-0.2960
3, 7	-0.1997	0.2876	3, 8	0.2059	0.1452	3, 9	0.1116	-0.1578	3,10	-0.1264	-0.0893
3,11	-0.0736	0.1046	3,12	0.0886	0.0620	3,13	0.0532	-0.0765	3,14	-0.0670	-0.0463
3,15	-0.0408	0.0595	3,16	0.0532	0.0364	3,17	0.0327	-0.0481	3,18	-0.0437	-0.0296
3,19	-0.0269	0.0402	3,20	0.0368	0.0247	3,21	0.0227	-0.0343	3,22	-0.0315	-0.0210
3,23	-0.0195	0.0299	3,24	0.0272	0.0182						
4, 4	3.1388	-4.8913	4, 5	2.2237	6.0094	4, 6	0.4527	-1.0249	4, 7	-0.4576	-0.2963
4, 8	-0.2000	0.2921	4, 9	0.2097	0.1454	4,10	0.1118	-0.1609	4,11	-0.1292	-0.0894
4,12	-0.0737	0.1070	4,13	0.0908	0.0621	4,14	0.0533	-0.0783	4,15	-0.0688	-0.0464
4,16	-0.0409	0.0609	4,17	0.0547	0.0364	4,18	0.0327	-0.0493	4,19	-0.0450	-0.0296
4,20	-0.0269	0.0410	4,21	0.0381	0.0246	4,22	0.0227	-0.0349	4,23	-0.0330	-0.0210
5, 5	3.1394	-4.8981	5, 6	2.2188	6.0090	5, 7	0.4524	-1.0213	5, 8	-0.4547	-0.2961
5, 9	-0.1998	0.2898	5,10	0.2077	0.1453	5,11	0.1117	-0.1593	5,12	-0.1276	-0.0893
5,13	-0.0736	0.1057	5,14	0.0895	0.0620	5,15	0.0532	-0.0774	5,16	-0.0677	-0.0463
5,17	-0.0408	0.0601	5,18	0.0537	0.0364	5,19	0.0327	-0.0487	5,20	-0.0441	-0.0295
5,21	-0.0269	0.0407	5,22	0.0371	0.0246						
6, 6	3.1391	-4.8948	6, 7	2.2215	6.0092	6, 8	0.4525	-1.0233	6, 9	-0.4564	-0.2962
6,10	-0.1999	0.2912	6,11	0.2089	0.1453	6,12	0.1117	-0.1603	6,13	-0.1287	-0.0894
6,14	-0.0736	0.1064	6,15	0.0904	0.0621	6,16	0.0532	-0.0779	6,17	-0.0684	-0.0464
6,18	-0.0409	0.0605	6,19	0.0545	0.0364	6,20	0.0327	-0.0490	6,21	-0.0449	-0.0296

7, 7	3.1392	-4.8967	7, 8	2.2199	6.0091	7, 9	0.4525	-1.0221	7,10	-0.4553	-0.2961
7,11	-0.1999	0.2903	7,12	0.2081	0.1453	7,13	0.1117	-0.1597	7,14	-0.1279	-0.0894
7,15	-0.0736	0.1060	7,16	0.0897	0.0620	7,17	0.0532	-0.0776	7,18	-0.0678	-0.0464
7,19	-0.0409	0.0604	7,20	0.0538	0.0364						
8, 8	3.1392	-4.8955	8, 9	2.2210	6.0092	8,10	0.4525	-1.0229	8,11	-0.4561	-0.2962
8,12	-0.1999	0.2909	8,13	0.2087	0.1453	8,14	0.1117	-0.1600	8,15	-0.1285	-0.0894
8,16	-0.0736	0.1063	8,17	0.0903	0.0620	8,18	0.0532	-0.0778	8,19	-0.0684	-0.0464
9, 9	3.1392	-4.8962	9,10	2.2202	6.0091	9,11	0.4525	-1.0224	9,12	-0.4555	-0.2961
9,13	-0.1999	0.2905	9,14	0.2082	0.1453	9,15	0.1117	-0.1598	9,16	-0.1280	-0.0894
9,17	-0.0736	0.1061	9,18	0.0898	0.0620						
10,10	3.1392	-4.8958	10,11	2.2208	6.0091	10,12	0.4525	-1.0227	10,13	-0.4560	-0.2962
10,14	-0.1999	0.2907	10,15	0.2087	0.1453	10,16	0.1117	-0.1599	10,17	-0.1285	-0.0894
11,11	3.1392	-4.8961	11,12	2.2203	6.0091	11,13	0.4525	-1.0225	11,14	-0.4555	-0.2962
11,15	-0.1999	0.2906	11,16	0.2082	0.1453						
12,12	3.1392	-4.8959	12,13	2.2207	6.0091	12,14	0.4525	-1.0226	12,15	-0.4560	-0.2962
13,13	3.1392	-4.8960	13,14	2.2203	6.0091						

VEN

SELF AND MUTUAL ADMITTANCES

H/LMDA=0.2500 B/LMDA= 0.250 OMEGA= 8.53

28 ELEMENT ARRAY

I, K	YIK IN MMHO RE	IM	I, K	YIK IN MMHO RE	IM	I, K	YIK IN MMHO RE	IM	I, K	YIK IN MMHO RE	IM
1, 1	6.8945	-4.4119	1, 2	1.8117	5.6294	1, 3	0.2561	-0.8740	1, 4	-0.3660	-0.1708
1, 5	-0.1114	0.2278	1, 6	0.1608	0.0784	1, 7	0.0588	-0.1220	1, 8	-0.0970	-0.0462
1, 9	-0.0374	0.0798	1,10	0.0673	0.0311	1,11	0.0264	-0.0578	1,12	-0.0505	-0.0227
1,13	-0.0198	0.0446	1,14	0.0398	0.0175	1,15	0.0156	-0.0359	1,16	-0.0326	-0.0141
1,17	-0.0127	0.0298	1,18	0.0274	0.0116	1,19	0.0106	-0.0254	1,20	-0.0235	-0.0098
1,21	-0.0090	0.0220	1,22	0.0204	0.0084	1,23	0.0078	-0.0193	1,24	-0.0180	-0.0073
1,25	-0.0068	0.0172	1,26	0.0159	0.0065	1,27	0.0058	-0.0159	1,28	-0.0101	-0.0019
2, 2	3.1618	-4.8233	2, 3	2.2517	6.0047	2, 4	0.4501	-1.0417	2, 5	-0.4691	-0.2945
2, 6	-0.1986	0.3006	2, 7	0.2163	0.1442	2, 8	0.1108	-0.1663	2, 9	-0.1336	-0.0885
2,10	-0.0729	0.1107	2,11	0.0940	0.0614	2,12	0.0527	-0.0812	2,13	-0.0712	-0.0458
2,14	-0.0404	0.0631	2,15	0.0567	0.0359	2,16	0.0322	-0.0511	2,17	-0.0467	-0.0291
2,18	-0.0265	0.0426	2,19	0.0394	0.0243	2,20	0.0224	-0.0363	2,21	-0.0339	-0.0207
2,22	-0.0192	0.0315	2,23	0.0297	0.0179	2,24	0.0167	-0.0276	2,25	-0.0265	-0.0156
2,26	-0.0147	0.0243	2,27	0.0246	0.0136						
3, 3	3.1396	-4.9087	3, 4	2.2128	6.0087	3, 5	0.4522	-1.0173	3, 6	-0.4518	-0.2960
3, 7	-0.1997	0.2875	3, 8	0.2059	0.1452	3, 9	0.1116	-0.1578	3,10	-0.1264	-0.0893
3,11	-0.0736	0.1046	3,12	0.0886	0.0620	3,13	0.0532	-0.0765	3,14	-0.0670	-0.0463
3,15	-0.0408	0.0595	3,16	0.0532	0.0364	3,17	0.0327	-0.0481	3,18	-0.0437	-0.0296
3,19	-0.0269	0.0401	3,20	0.0368	0.0247	3,21	0.0227	-0.0342	3,22	-0.0316	-0.0210
3,23	-0.0195	0.0298	3,24	0.0276	0.0182	3,25	0.0170	-0.0263	3,26	-0.0241	-0.0159
4, 4	3.1388	-4.8913	4, 5	2.2237	6.0094	4, 6	0.4527	-1.0249	4, 7	-0.4576	-0.2963
4, 8	-0.2000	0.2921	4, 9	0.2097	0.1454	4,10	0.1118	-0.1610	4,11	-0.1292	-0.0894
4,12	-0.0737	0.1070	4,13	0.0908	0.0621	4,14	0.0533	-0.0783	4,15	-0.0687	-0.0464
4,16	-0.0409	0.0609	4,17	0.0547	0.0364	4,18	0.0327	-0.0493	4,19	-0.0450	-0.0296
4,20	-0.0269	0.0411	4,21	0.0380	0.0246	4,22	0.0227	-0.0350	4,23	-0.0328	-0.0210
4,24	-0.0195	0.0303	4,25	0.0288	0.0181						
5, 5	3.1394	-4.8981	5, 6	2.2189	6.0090	5, 7	0.4524	-1.0213	5, 8	-0.4547	-0.2961
5, 9	-0.1998	0.2898	5,10	0.2077	0.1453	5,11	0.1117	-0.1593	5,12	-0.1277	-0.0893
5,13	-0.0736	0.1057	5,14	0.0895	0.0620	5,15	0.0532	-0.0774	5,16	-0.0677	-0.0463
5,17	-0.0408	0.0601	5,18	0.0538	0.0364	5,19	0.0327	-0.0487	5,20	-0.0442	-0.0295
5,21	-0.0269	0.0406	5,22	0.0372	0.0246	5,23	0.0227	-0.0347	5,24	-0.0319	-0.0210
6, 6	3.1391	-4.8948	6, 7	2.2215	6.0092	6, 8	0.4525	-1.0233	6, 9	-0.4564	-0.2962
6,10	-0.1999	0.2912	6,11	0.2089	0.1453	6,12	0.1117	-0.1603	6,13	-0.1287	-0.0894
6,14	-0.0736	0.1065	6,15	0.0903	0.0621	6,16	0.0532	-0.0779	6,17	-0.0684	-0.0464
6,18	-0.0409	0.0606	6,19	0.0544	0.0364	6,20	0.0327	-0.0490	6,21	-0.0448	-0.0295
6,22	-0.0269	0.0408	6,23	0.0379	0.0246						
7, 7	3.1392	-4.8967	7, 8	2.2199	6.0091	7, 9	0.4525	-1.0221	7,10	-0.4553	-0.2961
7,11	-0.1999	0.2903	7,12	0.2081	0.1453	7,13	0.1117	-0.1596	7,14	-0.1279	-0.0894
7,15	-0.0736	0.1060	7,16	0.0897	0.0620	7,17	0.0532	-0.0776	7,18	-0.0679	-0.0464
7,19	-0.0409	0.0603	7,20	0.0539	0.0364	7,21	0.0327	-0.0489	7,22	-0.0442	-0.0295
8, 8	3.1392	-4.8955	8, 9	2.2209	6.0092	8,10	0.4525	-1.0229	8,11	-0.4561	-0.2962
8,12	-0.1999	0.2909	8,13	0.2087	0.1453	8,14	0.1117	-0.1601	8,15	-0.1285	-0.0894
8,16	-0.0736	0.1063	8,17	0.0902	0.0620	8,18	0.0532	-0.0778	8,19	-0.0683	-0.0464
8,20	-0.0409	0.0604	8,21	0.0544	0.0364						
9, 9	3.1392	-4.8963	9,10	2.2202	6.0091	9,11	0.4525	-1.0224	9,12	-0.4555	-0.2961
9,13	-0.1999	0.2905	9,14	0.2082	0.1453	9,15	0.1117	-0.1598	9,16	-0.1280	-0.0894
9,17	-0.0736	0.1061	9,18	0.0898	0.0620	9,19	0.0532	-0.0777	9,20	-0.0679	-0.0464
10,10	3.1392	-4.8957	10,11	2.2208	6.0091	10,12	0.4525	-1.0227	10,13	-0.4560	-0.2962
10,14	-0.1999	0.2907	10,15	0.2086	0.1453	10,16	0.1117	-0.1600	10,17	-0.1284	-0.0894
10,18	-0.0736	0.1062	10,19	0.0902	0.0620						
11,11	3.1392	-4.8961	11,12	2.2203	6.0091	11,13	0.4525	-1.0225	11,14	-0.4556	-0.2962
11,15	-0.1999	0.2906	11,16	0.2083	0.1453	11,17	0.1117	-0.1599	11,18	-0.1281	-0.0894
12,12	3.1392	-4.8959	12,13	2.2207	6.0091	12,14	0.4525	-1.0226	12,15	-0.4559	-0.2961
12,16	-0.1999	0.2907	12,17	0.2086	0.1453						
13,13	3.1392	-4.8960	13,14	2.2203	6.0091	13,15	0.4525	-1.0226	13,16	-0.4556	-0.2962
14,14	3.1392	-4.8959	14,15	2.2207	6.0091						

VEN

SELF AND MUTUAL ADMITTANCES

H/LMDA=0.2500 B/LMDA= 0.250 OMEGA= 8.53

30 ELEMENT ARRAY

I, K	YIK IN MMHO RE	IM	I, K	YIK IN MMHO RE	IM	I, K	YIK IN MMHO RE	IM	I, K	YIK IN MMHO RE	IM
1, 1	6.8945	-4.4119	1, 2	1.8117	5.6294	1, 3	0.2561	-0.8740	1, 4	-0.3660	-0.1708
1, 5	-0.1114	0.2278	1, 6	0.1608	0.0784	1, 7	0.0588	-0.1220	1, 8	-0.0970	-0.0462
1, 9	-0.0374	0.0798	1,10	0.0673	0.0311	1,11	0.0264	-0.0578	1,12	-0.0505	-0.0227
1,13	-0.0198	0.0446	1,14	0.0398	0.0175	1,15	0.0156	-0.0359	1,16	-0.0326	-0.0140
1,17	-0.0127	0.0298	1,18	0.0274	0.0116	1,19	0.0106	-0.0254	1,20	-0.0235	-0.0098
1,21	-0.0090	0.0219	1,22	0.0204	0.0084	1,23	0.0078	-0.0193	1,24	-0.0180	-0.0073
1,25	-0.0068	0.0171	1,26	0.0160	0.0064	1,27	0.0060	-0.0154	1,28	-0.0143	-0.0057
1,29	-0.0051	0.0144	1,30	0.0091	0.0017						
2, 2	3.1618	-4.8233	2, 3	2.2517	6.0047	2, 4	0.4501	-1.0417	2, 5	-0.4691	-0.2945
2, 6	-0.1986	0.3006	2, 7	0.2163	0.1442	2, 8	0.1108	-0.1663	2, 9	-0.1336	-0.0885
2,10	-0.0729	0.1107	2,11	0.0940	0.0614	2,12	0.0527	-0.0812	2,13	-0.0712	-0.0458
2,14	-0.0404	0.0632	2,15	0.0567	0.0359	2,16	0.0322	-0.0512	2,17	-0.0466	-0.0291
2,18	-0.0265	0.0427	2,19	0.0394	0.0243	2,20	0.0224	-0.0364	2,21	-0.0339	-0.0207
2,22	-0.0192	0.0315	2,23	0.0297	0.0179	2,24	0.0167	-0.0277	2,25	-0.0263	-0.0157
2,26	-0.0147	0.0246	2,27	0.0237	0.0139	2,28	0.0131	-0.0219	2,29	-0.0222	-0.0121
3, 3	3.1396	-4.9087	3, 4	2.2128	6.0087	3, 5	0.4522	-1.0173	3, 6	-0.4518	-0.2960
3, 7	-0.1997	0.2875	3, 8	0.2059	0.1452	3, 9	0.1116	-0.1578	3,10	-0.1264	-0.0893
3,11	-0.0736	0.1046	3,12	0.0886	0.0620	3,13	0.0532	-0.0765	3,14	-0.0670	-0.0463
3,15	-0.0408	0.0594	3,16	0.0532	0.0364	3,17	0.0327	-0.0481	3,18	-0.0437	-0.0296
3,19	-0.0269	0.0401	3,20	0.0369	0.0247	3,21	0.0227	-0.0342	3,22	-0.0317	-0.0210
3,23	-0.0195	0.0297	3,24	0.0276	0.0182	3,25	0.0170	-0.0262	3,26	-0.0244	-0.0159
3,27	-0.0150	0.0234	3,28	0.0215	0.0141						
4, 4	3.1388	-4.8913	4, 5	2.2237	6.0094	4, 6	0.4527	-1.0249	4, 7	-0.4576	-0.2963
4, 8	-0.2000	0.2921	4, 9	0.2097	0.1454	4,10	0.1118	-0.1610	4,11	-0.1292	-0.0894
4,12	-0.0737	0.1070	4,13	0.0908	0.0621	4,14	0.0533	-0.0783	4,15	-0.0687	-0.0464
4,16	-0.0409	0.0609	4,17	0.0547	0.0364	4,18	0.0327	-0.0493	4,19	-0.0450	-0.0296
4,20	-0.0269	0.0411	4,21	0.0380	0.0246	4,22	0.0227	-0.0350	4,23	-0.0327	-0.0210
4,24	-0.0195	0.0304	4,25	0.0287	0.0181	4,26	0.0170	-0.0266	4,27	-0.0255	-0.0159
5, 5	3.1394	-4.8981	5, 6	2.2189	6.0090	5, 7	0.4524	-1.0213	5, 8	-0.4547	-0.2961
5, 9	-0.1998	0.2898	5,10	0.2077	0.1453	5,11	0.1117	-0.1593	5,12	-0.1277	-0.0893
5,13	-0.0736	0.1057	5,14	0.0895	0.0620	5,15	0.0532	-0.0773	5,16	-0.0677	-0.0463
5,17	-0.0408	0.0601	5,18	0.0538	0.0364	5,19	0.0327	-0.0487	5,20	-0.0442	-0.0295
5,21	-0.0269	0.0406	5,22	0.0373	0.0246	5,23	0.0227	-0.0346	5,24	-0.0320	-0.0210
5,25	-0.0195	0.0301	5,26	0.0279	0.0182						
6, 6	3.1391	-4.8947	6, 7	2.2215	6.0092	6, 8	0.4525	-1.0233	6, 9	-0.4564	-0.2962
6,10	-0.1999	0.2912	6,11	0.2089	0.1453	6,12	0.1117	-0.1603	6,13	-0.1286	-0.0894
6,14	-0.0736	0.1065	6,15	0.0903	0.0621	6,16	0.0532	-0.0780	6,17	-0.0684	-0.0464
6,18	-0.0409	0.0606	6,19	0.0544	0.0364	6,20	0.0327	-0.0490	6,21	-0.0448	-0.0295
6,22	-0.0269	0.0409	6,23	0.0378	0.0246	6,24	0.0227	-0.0348	6,25	-0.0326	-0.0210
7, 7	3.1392	-4.8967	7, 8	2.2199	6.0091	7, 9	0.4525	-1.0221	7,10	-0.4553	-0.2961
7,11	-0.1999	0.2903	7,12	0.2081	0.1453	7,13	0.1117	-0.1596	7,14	-0.1280	-0.0894
7,15	-0.0736	0.1060	7,16	0.0898	0.0620	7,17	0.0532	-0.0776	7,18	-0.0679	-0.0464
7,19	-0.0409	0.0603	7,20	0.0539	0.0364	7,21	0.0327	-0.0488	7,22	-0.0443	-0.0296
7,23	-0.0269	0.0401	7,24	0.0373	0.0246						
8, 8	3.1392	-4.8955	8, 9	2.2209	6.0092	8,10	0.4525	-1.0229	8,11	-0.4561	-0.2962
8,12	-0.1999	0.2909	8,13	0.2087	0.1453	8,14	0.1117	-0.1601	8,15	-0.1285	-0.0894
8,16	-0.0736	0.1063	8,17	0.0902	0.0620	8,18	0.0532	-0.0778	8,19	-0.0683	-0.0464
8,20	-0.0409	0.0605	8,21	0.0543	0.0364	8,22	0.0327	-0.0489	8,23	-0.0447	-0.0295
9, 9	3.1392	-4.8963	9,10	2.2202	6.0091	9,11	0.4525	-1.0223	9,12	-0.4555	-0.2961
9,13	-0.1999	0.2905	9,14	0.2082	0.1453	9,15	0.1117	-0.1598	9,16	-0.1281	-0.0894
9,17	-0.0736	0.1061	9,18	0.0898	0.0620	9,19	0.0532	-0.0777	9,20	-0.0679	-0.0464
9,21	-0.0409	0.0604	9,22	0.0539	0.0364						
10,10	3.1392	-4.8957	10,11	2.2208	6.0091	10,12	0.4525	-1.0227	10,13	-0.4560	-0.2962
10,14	-0.1999	0.2908	10,15	0.2086	0.1453	10,16	0.1117	-0.1600	10,17	-0.1284	-0.0894
10,18	-0.0736	0.1062	10,19	0.0902	0.0620	10,20	0.0532	-0.0777	10,21	-0.0683	-0.0464
11,11	3.1392	-4.8961	11,12	2.2203	6.0091	11,13	0.4525	-1.0225	11,14	-0.4556	-0.2962
11,15	-0.1999	0.2906	11,16	0.2083	0.1453	11,17	0.1117	-0.1599	11,18	-0.1281	-0.0894
11,19	-0.0736	0.1062	11,20	0.0899	0.0620						
12,12	3.1392	-4.8958	12,13	2.2207	6.0091	12,14	0.4525	-1.0226	12,15	-0.4559	-0.2961
12,16	-0.1999	0.2907	12,17	0.2086	0.1453	12,18	0.1117	-0.1599	12,19	-0.1284	-0.0894
13,13	3.1392	-4.8960	13,14	2.2204	6.0091	13,15	0.4525	-1.0225	13,16	-0.4556	-0.2962
13,17	-0.1999	0.2906	13,18	0.2083	0.1453						
14,14	3.1392	-4.8959	14,15	2.2206	6.0091	14,16	0.4525	-1.0226	14,17	-0.4559	-0.2961
15,15	3.1392	-4.8960	15,16	2.2204	6.0091						

VEN

SELF AND MUTUAL ADMITTANCES

H/LMDA=0.2500 B/LMDA= 0.250 OMEGA= 8.53

35 ELEMENT ARRAY

I, K	YIK IN MMHO RE	IM	I, K	YIK IN MMHO RE	IM	I, K	YIK IN MMHO RE	IM	I, K	YIK IN MMHO RE	IM
1, 1	6.8945	-4.4119	1, 2	1.8117	5.6294	1, 3	0.2561	-0.8740	1, 4	-0.3660	-0.1708
1, 5	-0.1114	0.2278	1, 6	0.1608	0.0784	1, 7	0.0588	-0.1220	1, 8	-0.0971	-0.0461
1, 9	-0.0374	0.0798	1,10	0.0673	0.0311	1,11	0.0264	-0.0578	1,12	-0.0505	-0.0227
1,13	-0.0199	0.0446	1,14	0.0399	0.0175	1,15	0.0156	-0.0359	1,16	-0.0326	-0.0140
1,17	-0.0127	0.0298	1,18	0.0274	0.0116	1,19	0.0106	-0.0253	1,20	-0.0235	-0.0098
1,21	-0.0090	0.0219	1,22	0.0205	0.0084	1,23	0.0078	-0.0192	1,24	-0.0181	-0.0073
1,25	-0.0068	0.0170	1,26	0.0162	0.0064	1,27	0.0060	-0.0153	1,28	-0.0146	-0.0057
1,29	-0.0054	0.0138	1,30	0.0133	0.0051	1,31	0.0049	-0.0125	1,32	-0.0122	-0.0046
1,33	-0.0044	0.0114	1,34	0.0115	0.0040	1,35	0.0013	-0.0074			
2, 2	3.1618	-4.8233	2, 3	2.2517	6.0047	2, 4	0.4501	-1.0417	2, 5	-0.4690	-0.2945
2, 6	-0.1986	0.3006	2, 7	0.2163	0.1442	2, 8	0.1108	-0.1663	2, 9	-0.1336	-0.0885
2,10	-0.0729	0.1107	2,11	0.0939	0.0614	2,12	0.0527	-0.0812	2,13	-0.0712	-0.0458
2,14	-0.0404	0.0632	2,15	0.0566	0.0359	2,16	0.0322	-0.0512	2,17	-0.0466	-0.0292
2,18	-0.0265	0.0427	2,19	0.0393	0.0243	2,20	0.0224	-0.0364	2,21	-0.0338	-0.0207
2,22	-0.0192	0.0316	2,23	0.0296	0.0179	2,24	0.0167	-0.0278	2,25	-0.0262	-0.0157
2,26	-0.0147	0.0248	2,27	0.0234	0.0139	2,28	0.0131	-0.0223	2,29	-0.0211	-0.0124
2,30	-0.0118	0.0203	2,31	0.0191	0.0112	2,32	0.0106	-0.0186	2,33	-0.0173	-0.0102
2,34	-0.0095	0.0177									
3, 3	3.1396	-4.9087	3, 4	2.2128	6.0087	3, 5	0.4522	-1.0172	3, 6	-0.4518	-0.2960
3, 7	-0.1997	0.2875	3, 8	0.2059	0.1452	3, 9	0.1116	-0.1578	3,10	-0.1265	-0.0893
3,11	-0.0736	0.1046	3,12	0.0886	0.0620	3,13	0.0532	-0.0765	3,14	-0.0670	-0.0463
3,15	-0.0408	0.0594	3,16	0.0532	0.0364	3,17	0.0327	-0.0481	3,18	-0.0438	-0.0296
3,19	-0.0269	0.0400	3,20	0.0369	0.0247	3,21	0.0227	-0.0341	3,22	-0.0318	-0.0210
3,23	-0.0195	0.0296	3,24	0.0278	0.0182	3,25	0.0170	-0.0260	3,26	-0.0246	-0.0159
3,27	-0.0150	0.0231	3,28	0.0220	0.0141	3,29	0.0133	-0.0208	3,30	-0.0199	-0.0126
3,31	-0.0120	0.0187	3,32	0.0182	0.0114	3,33	0.0109	-0.0169			
4, 4	3.1388	-4.8913	4, 5	2.2237	6.0094	4, 6	0.4527	-1.0250	4, 7	-0.4576	-0.2963
4, 8	-0.2000	0.2922	4, 9	0.2097	0.1454	4,10	0.1118	-0.1610	4,11	-0.1292	-0.0894
4,12	-0.0737	0.1070	4,13	0.0907	0.0621	4,14	0.0533	-0.0784	4,15	-0.0687	-0.0464
4,16	-0.0409	0.0610	4,17	0.0546	0.0364	4,18	0.0327	-0.0494	4,19	-0.0449	-0.0296
4,20	-0.0269	0.0412	4,21	0.0379	0.0247	4,22	0.0227	-0.0351	4,23	-0.0326	-0.0210
4,24	-0.0195	0.0305	4,25	0.0285	0.0182	4,26	0.0170	-0.0269	4,27	-0.0252	-0.0159
4,28	-0.0150	0.0239	4,29	0.0225	0.0141	4,30	0.0133	-0.0215	4,31	-0.0203	-0.0126
4,32	-0.0120	0.0196									
5, 5	3.1394	-4.8982	5, 6	2.2189	6.0090	5, 7	0.4524	-1.0213	5, 8	-0.4547	-0.2961
5, 9	-0.1998	0.2898	5,10	0.2077	0.1453	5,11	0.1117	-0.1592	5,12	-0.1277	-0.0893
5,13	-0.0736	0.1057	5,14	0.0896	0.0620	5,15	0.0532	-0.0773	5,16	-0.0678	-0.0463
5,17	-0.0408	0.0601	5,18	0.0538	0.0364	5,19	0.0327	-0.0486	5,20	-0.0443	-0.0295
5,21	-0.0269	0.0405	5,22	0.0374	0.0246	5,23	0.0227	-0.0345	5,24	-0.0321	-0.0210
5,25	-0.0195	0.0299	5,26	0.0281	0.0181	5,27	0.0170	-0.0263	5,28	-0.0249	-0.0159
5,29	-0.0150	0.0234	5,30	0.0223	0.0141	5,31	0.0133	-0.0209			
6, 6	3.1391	-4.8947	6, 7	2.2214	6.0092	6, 8	0.4525	-1.0234	6, 9	-0.4564	-0.2962
6,10	-0.1999	0.2912	6,11	0.2089	0.1453	6,12	0.1117	-0.1603	6,13	-0.1286	-0.0894
6,14	-0.0736	0.1065	6,15	0.0903	0.0621	6,16	0.0532	-0.0780	6,17	-0.0683	-0.0464
6,18	-0.0409	0.0607	6,19	0.0543	0.0364	6,20	0.0327	-0.0491	6,21	-0.0447	-0.0296
6,22	-0.0269	0.0410	6,23	0.0377	0.0246	6,24	0.0227	-0.0350	6,25	-0.0324	-0.0210
6,26	-0.0195	0.0303	6,27	0.0283	0.0182	6,28	0.0170	-0.0267	6,29	-0.0250	-0.0159
6,30	-0.0150	0.0238									
7, 7	3.1392	-4.8967	7, 8	2.2199	6.0091	7, 9	0.4525	-1.0221	7,10	-0.4553	-0.2961
7,11	-0.1999	0.2903	7,12	0.2081	0.1453	7,13	0.1117	-0.1596	7,14	-0.1280	-0.0894
7,15	-0.0736	0.1059	7,16	0.0898	0.0620	7,17	0.0532	-0.0775	7,18	-0.0679	-0.0464
7,19	-0.0409	0.0602	7,20	0.0540	0.0364	7,21	0.0327	-0.0487	7,22	-0.0444	-0.0295
7,23	-0.0269	0.0406	7,24	0.0375	0.0246	7,25	0.0227	-0.0346	7,26	-0.0323	-0.0210
7,27	-0.0195	0.0300	7,28	0.0282	0.0181	7,29	0.0170	-0.0264			
8, 8	3.1392	-4.8954	8, 9	2.2209	6.0092	8,10	0.4525	-1.0229	8,11	-0.4560	-0.2962
8,12	-0.1999	0.2909	8,13	0.2087	0.1453	8,14	0.1117	-0.1601	8,15	-0.1284	-0.0894
8,16	-0.0736	0.1064	8,17	0.0901	0.0621	8,18	0.0532	-0.0779	8,19	-0.0682	-0.0464
8,20	-0.0409	0.0606	8,21	0.0542	0.0364	8,22	0.0327	-0.0490	8,23	-0.0446	-0.0296
8,24	-0.0269	0.0409	8,25	0.0376	0.0246	8,26	0.0227	-0.0349	8,27	-0.0323	-0.0210
8,28	-0.0195	0.0303									
9, 9	3.1392	-4.8963	9,10	2.2202	6.0091	9,11	0.4525	-1.0223	9,12	-0.4556	-0.2961
9,13	-0.1999	0.2904	9,14	0.2083	0.1453	9,15	0.1117	-0.1597	9,16	-0.1281	-0.0894
9,17	-0.0736	0.1060	9,18	0.0899	0.0620	9,19	0.0532	-0.0776	9,20	-0.0680	-0.0464
9,21	-0.0409	0.0603	9,22	0.0541	0.0364	9,23	0.0327	-0.0488	9,24	-0.0445	-0.0295
9,25	-0.0269	0.0406	9,26	0.0375	0.0246	9,27	0.0227	-0.0346			
10,10	3.1392	-4.8957	10,11	2.2207	6.0091	10,12	0.4525	-1.0228	10,13	-0.4559	-0.2962
10,14	-0.1999	0.2908	10,15	0.2085	0.1453	10,16	0.1117	-0.1600	10,17	-0.1283	-0.0894
10,18	-0.0736	0.1063	10,19	0.0901	0.0620	10,20	0.0532	-0.0778	10,21	-0.0682	-0.0464
10,22	-0.0409	0.0605	10,23	0.0542	0.0364	10,24	0.0327	-0.0490	10,25	-0.0445	-0.0295
10,26	-0.0269	0.0409									
11,11	3.1392	-4.8961	11,12	2.2204	6.0091	11,13	0.4525	-1.0224	11,14	-0.4556	-0.2961
11,15	-0.1999	0.2905	11,16	0.2083	0.1453	11,17	0.1117	-0.1598	11,18	-0.1282	-0.0894
11,19	-0.0736	0.1061	11,20	0.0900	0.0620	11,21	0.0532	-0.0776	11,22	-0.0681	-0.0464
11,23	-0.0409	0.0603	11,24	0.0541	0.0364	11,25	0.0327	-0.0488			
12,12	3.1392	-4.8958	12,13	2.2206	6.0091	12,14	0.4525	-1.0227	12,15	-0.4558	-0.2962
12,16	-0.1999	0.2908	12,17	0.2085	0.1453	12,18	0.1117	-0.1600	12,19	-0.1283	-0.0894
12,20	-0.0736	0.1063	12,21	0.0900	0.0620	12,22	0.0532	-0.0778	12,23	-0.0681	-0.0464
12,24	-0.0408	0.0605									
13,13	3.1392	-4.8961	13,14	2.2204	6.0091	13,15	0.4525	-1.0225	13,16	-0.4557	-0.2962
13,17	-0.1999	0.2905	13,18	0.2084	0.1453	13,19	0.1117	-0.1598	13,20	-0.1282	-0.0894
13,21	-0.0736	0.1061	13,22	0.0900	0.0620	13,23	0.0532	-0.0776			
14,14	3.1392	-4.8958	14,15	2.2206	6.0091	14,16	0.4525	-1.0227	14,17	-0.4558	-0.2962
14,18	-0.1999	0.2907	14,19	0.2085	0.1453	14,20	0.1117	-0.1600	14,21	-0.1283	-0.0894
14,22	-0.0736	0.1063									
15,15	3.1392	-4.8960	15,16	2.2205	6.0091	15,17	0.4525	-1.0225	15,18	-0.4557	-0.2962
15,19	-0.1999	0.2906	15,20	0.2084	0.1453	15,21	0.1117	-0.1598			
16,16	3.1392	-4.8959	16,17	2.2205	6.0091	16,18	0.4525	-1.0227	16,19	-0.4558	-0.2962
16,20	-0.1999	0.2907									
17,17	3.1392	-4.8960	17,18	2.2205	6.0091	17,19	0.4525	-1.0225			
18,18	3.1392	-4.8959									

VEN

SELF AND MUTUAL ADMITTANCES

H/LMDA=0.2500 B/LMDA= 0.250 OMEGA= 8.53

40 ELEMENT ARRAY

I, K	YIK IN MMHO RE	IM	I, K	YIK IN MMHO RE	IM	I, K	YIK IN MMHO RE	IM	I, K	YIK IN MMHO RE	IM
1, 1	6.8945	-4.4119	1, 2	1.8117	5.6294	1, 3	0.2561	-0.8740	1, 4	-0.3660	-0.1708
1, 5	-0.1114	0.2278	1, 6	0.1608	0.0784	1, 7	0.0588	-0.1220	1, 8	-0.0970	-0.0462
1, 9	-0.0374	0.0798	1,10	0.0673	0.0311	1,11	0.0264	-0.0578	1,12	-0.0505	-0.0227
1,13	-0.0199	0.0446	1,14	0.0399	0.0175	1,15	0.0156	-0.0359	1,16	-0.0326	-0.0140
1,17	-0.0127	0.0298	1,18	0.0274	0.0116	1,19	0.0106	-0.0253	1,20	-0.0235	-0.0098
1,21	-0.0090	0.0219	1,22	0.0205	0.0084	1,23	0.0078	-0.0192	1,24	-0.0181	-0.0073
1,25	-0.0068	0.0171	1,26	0.0161	0.0064	1,27	0.0060	-0.0153	1,28	-0.0145	-0.0057
1,29	-0.0054	0.0139	1,30	0.0132	0.0051	1,31	0.0048	-0.0126	1,32	-0.0120	-0.0046
1,33	-0.0044	0.0116	1,34	0.0110	0.0042	1,35	0.0040	-0.0107	1,36	-0.0102	-0.0039
1,37	-0.0037	0.0100	1,38	0.0093	0.0036	1,39	0.0032	-0.0096	1,40	-0.0061	-0.0010
2, 2	3.1618	-4.8233	2, 3	2.2517	6.0047	2, 4	0.4501	-1.0417	2, 5	-0.4691	-0.2945
2, 6	-0.1986	0.3006	2, 7	0.2163	0.1442	2, 8	0.1108	-0.1663	2, 9	-0.1336	-0.0885
2,10	-0.0729	0.1107	2,11	0.0940	0.0614	2,12	0.0527	-0.0812	2,13	-0.0712	-0.0458
2,14	-0.0404	0.0632	2,15	0.0566	0.0359	2,16	0.0322	-0.0512	2,17	-0.0466	-0.0292
2,18	-0.0265	0.0427	2,19	0.0394	0.0243	2,20	0.0224	-0.0364	2,21	-0.0339	-0.0207
2,22	-0.0192	0.0316	2,23	0.0296	0.0179	2,24	0.0167	-0.0278	2,25	-0.0262	-0.0157
2,26	-0.0147	0.0247	2,27	0.0235	0.0139	2,28	0.0131	-0.0222	2,29	-0.0212	-0.0124
2,30	-0.0118	0.0201	2,31	0.0193	0.0112	2,32	0.0107	-0.0184	2,33	-0.0177	-0.0102
2,34	-0.0097	0.0168	2,35	0.0163	0.0093	2,36	0.0089	-0.0155	2,37	-0.0152	-0.0085
2,38	-0.0082	0.0142	2,39	0.0146	0.0077						
3, 3	3.1396	-4.9087	3, 4	2.2128	6.0087	3, 5	0.4522	-1.0172	3, 6	-0.4518	-0.2960
3, 7	-0.1997	0.2875	3, 8	0.2059	0.1452	3, 9	0.1116	-0.1578	3,10	-0.1264	-0.0893
3,11	-0.0736	0.1046	3,12	0.0886	0.0620	3,13	0.0532	-0.0765	3,14	-0.0670	-0.0463
3,15	-0.0408	0.0594	3,16	0.0532	0.0364	3,17	0.0327	-0.0481	3,18	-0.0437	-0.0296
3,19	-0.0269	0.0401	3,20	0.0369	0.0247	3,21	0.0227	-0.0342	3,22	-0.0317	-0.0210
3,23	-0.0195	0.0296	3,24	0.0277	0.0182	3,25	0.0170	-0.0261	3,26	-0.0245	-0.0159
3,27	-0.0150	0.0232	3,28	0.0219	0.0141	3,29	0.0133	-0.0209	3,30	-0.0198	-0.0126
3,31	-0.0120	0.0189	3,32	0.0180	0.0114	3,33	0.0109	-0.0173	3,34	-0.0164	-0.0104
3,35	-0.0099	0.0159	3,36	0.0151	0.0095	3,37	0.0091	-0.0147	3,38	-0.0138	-0.0087
4, 4	3.1388	-4.8913	4, 5	2.2237	6.0094	4, 6	0.4527	-1.0249	4, 7	-0.4576	-0.2963
4, 8	-0.2000	0.2922	4, 9	0.2097	0.1454	4,10	0.1118	-0.1610	4,11	-0.1292	-0.0894
4,12	-0.0737	0.1070	4,13	0.0907	0.0621	4,14	0.0533	-0.0784	4,15	-0.0687	-0.0464
4,16	-0.0409	0.0609	4,17	0.0546	0.0364	4,18	0.0327	-0.0493	4,19	-0.0449	-0.0296
4,20	-0.0269	0.0412	4,21	0.0379	0.0246	4,22	0.0227	-0.0351	4,23	-0.0327	-0.0210
4,24	-0.0195	0.0304	4,25	0.0285	0.0181	4,26	0.0170	-0.0268	4,27	-0.0253	-0.0159
4,28	-0.0150	0.0238	4,29	0.0226	0.0141	4,30	0.0133	-0.0214	4,31	-0.0204	-0.0126
4,32	-0.0120	0.0194	4,33	0.0186	0.0114	4,34	0.0108	-0.0177	4,35	-0.0171	-0.0103
4,36	-0.0099	0.0162	4,37	0.0158	0.0095						
5, 5	3.1394	-4.8981	5, 6	2.2189	6.0090	5, 7	0.4524	-1.0213	5, 8	-0.4547	-0.2961
5, 9	-0.1998	0.2898	5,10	0.2077	0.1453	5,11	0.1117	-0.1593	5,12	-0.1277	-0.0893
5,13	-0.0736	0.1057	5,14	0.0895	0.0620	5,15	0.0532	-0.0773	5,16	-0.0677	-0.0463
5,17	-0.0408	0.0601	5,18	0.0538	0.0364	5,19	0.0327	-0.0486	5,20	-0.0442	-0.0295
5,21	-0.0269	0.0405	5,22	0.0373	0.0246	5,23	0.0227	-0.0346	5,24	-0.0321	-0.0210
5,25	-0.0195	0.0300	5,26	0.0280	0.0182	5,27	0.0170	-0.0264	5,28	-0.0248	-0.0159
5,29	-0.0150	[illegible]	[illegible]	[illegible]	[illegible]	5,31	0.0133	-0.0211	5,32	-0.0200	-0.0126
5,33	-0.0120	0.0191	5,34	0.0182	0.0114	[illegible]	[illegible]	[illegible]	[illegible]	-0.0166	-0.0104
6, 6	3.1391	-4.8947	6, 7	2.2215	6.0092	6, 8	0.4525	-1.0233	6, 9	-0.4564	-0.2962
6,10	-0.1999	0.2912	6,11	0.2089	0.1453	6,12	0.1117	-0.1603	6,13	-0.1286	-0.0894
6,14	-0.0736	0.1065	6,15	0.0903	0.0621	6,16	0.0532	-0.0780	6,17	-0.0684	-0.0464
6,18	-0.0409	0.0606	6,19	0.0543	0.0364	6,20	0.0327	-0.0491	6,21	-0.0447	-0.0295
6,22	-0.0269	0.0409	6,23	0.0377	0.0246	6,24	0.0227	-0.0349	6,25	-0.0325	-0.0210
6,26	-0.0195	0.0303	6,27	0.0284	0.0181	6,28	0.0170	-0.0266	6,29	-0.0251	-0.0159
6,30	-0.0150	0.0237	6,31	0.0225	0.0141	6,32	0.0133	-0.0213	6,33	-0.0203	-0.0126
6,34	-0.0120	0.0193	6,35	0.0185	0.0114						
7, 7	3.1392	-4.8967	7, 8	2.2199	6.0091	7, 9	0.4525	-1.0221	7,10	-0.4553	-0.2961
7,11	-0.1999	0.2903	7,12	0.2081	0.1453	7,13	0.1117	-0.1596	7,14	-0.1280	-0.0894
7,15	-0.0736	0.1059	7,16	0.0898	0.0620	7,17	0.0532	-0.0775	7,18	-0.0679	-0.0464
7,19	-0.0409	0.0603	7,20	0.0540	0.0364	7,21	0.0327	-0.0488	7,22	-0.0444	-0.0296
7,23	-0.0269	0.0407	7,24	0.0374	0.0246	7,25	0.0227	-0.0347	7,26	-0.0322	-0.0210
7,27	-0.0195	0.0301	7,28	0.0281	0.0182	7,29	0.0170	-0.0265	7,30	-0.0249	-0.0159
7,31	-0.0150	0.0236	7,32	0.0222	0.0141	7,33	0.0133	-0.0212	7,34	-0.0200	-0.0126
8, 8	3.1392	-4.8955	8, 9	2.2209	6.0092	8,10	0.4525	-1.0229	8,11	-0.4561	-0.2962
8,12	-0.1999	0.2909	8,13	0.2087	0.1453	8,14	0.1117	-0.1601	8,15	-0.1285	-0.0894
8,16	-0.0736	0.1063	8,17	0.0902	0.0620	8,18	0.0532	-0.0779	8,19	-0.0683	-0.0464
8,20	-0.0409	0.0605	8,21	0.0543	0.0364	8,22	0.0327	-0.0490	8,23	-0.0446	-0.0295
8,24	-0.0269	0.0409	8,25	0.0377	0.0246	8,26	0.0227	-0.0348	8,27	-0.0324	-0.0210
8,28	-0.0195	0.0302	8,29	0.0284	0.0181	8,30	0.0170	-0.0266	8,31	-0.0251	-0.0159
8,32	-0.0150	0.0236	8,33	0.0225	0.0141						
9, 9	3.1392	-4.8963	9,10	2.2202	6.0091	9,11	0.4525	-1.0223	9,12	-0.4555	-0.2961
9,13	-0.1999	0.2905	9,14	0.2082	0.1453	9,15	0.1117	-0.1598	9,16	-0.1281	-0.0894
9,17	-0.0736	0.1061	9,18	0.0899	0.0620	9,19	0.0532	-0.0776	9,20	-0.0680	-0.0464
9,21	-0.0409	0.0603	9,22	0.0540	0.0364	9,23	0.0327	-0.0488	9,24	-0.0444	-0.0296
9,25	-0.0269	0.0407	9,26	0.0375	0.0246	9,27	0.0227	-0.0347	9,28	-0.0322	-0.0210
9,29	-0.0195	0.0301	9,30	0.0281	0.0182	9,31	0.0170	-0.0265	9,32	-0.0249	-0.0159
10,10	3.1392	-4.8957	10,11	2.2207	6.0091	10,12	0.4525	-1.0227	10,13	-0.4559	-0.2962
10,14	-0.1999	0.2908	10,15	0.2086	0.1453	10,16	0.1117	-0.1600	10,17	-0.1284	-0.0894
10,18	-0.0736	0.1063	10,19	0.0901	0.0620	10,20	0.0532	-0.0778	10,21	-0.0682	-0.0464
10,22	-0.0409	0.0605	10,23	0.0542	0.0364	10,24	0.0327	-0.0490	10,25	-0.0446	-0.0295
10,26	-0.0269	0.0408	10,27	0.0376	0.0246	10,28	0.0227	-0.0348	10,29	-0.0324	-0.0210
10,30	-0.0195	0.0302	10,31	0.0283	0.0181						
11,11	3.1392	-4.8961	11,12	2.2203	6.0091	11,13	0.4525	-1.0224	11,14	-0.4556	-0.2962
11,15	-0.1999	0.2905	11,16	0.2083	0.1453	11,17	0.1117	-0.1598	11,18	-0.1281	-0.0894
11,19	-0.0736	0.1061	11,20	0.0899	0.0620	11,21	0.0532	-0.0777	11,22	-0.0680	-0.0464
11,23	-0.0409	0.0604	11,24	0.0540	0.0364	11,25	0.0327	-0.0489	11,26	-0.0444	-0.0296
11,27	-0.0269	0.0408	11,28	0.0375	0.0246	11,29	0.0227	-0.0348	11,30	-0.0322	-0.0210
12,12	3.1392	-4.8958	12,13	2.2206	6.0091	12,14	0.4525	-1.0227	12,15	-0.4559	-0.2961
12,16	-0.1999	0.2907	12,17	0.2085	0.1453	12,18	0.1117	-0.1600	12,19	-0.1283	-0.0894
12,20	-0.0736	0.1062	12,21	0.0901	0.0620	12,22	0.0532	-0.0778	12,23	-0.0682	-0.0464
12,24	-0.0409	0.0604	12,25	0.0542	0.0364	12,26	0.0327	-0.0489	12,27	-0.0446	-0.0295
12,28	-0.0269	0.0408	12,29	0.0376	0.0246						
13,13	3.1392	-4.8960	13,14	2.2204	6.0091	13,15	0.4525	-1.0225	13,16	-0.4557	-0.2962
13,17	-0.1999	0.2906	13,18	0.2083	0.1453	13,19	0.1117	-0.1599	13,20	-0.1282	-0.0894
13,21	-0.0736	0.1061	13,22	0.0899	0.0621	13,23	0.0532	-0.0777	13,24	-0.0680	-0.0464
13,25	-0.0409	0.0604	13,26	0.0541	0.0364	13,27	0.0327	-0.0489	13,28	-0.0444	-0.0296

14,14	3.1392	-4.8959	14,15	2.2206	6.0091	14,16	0.4525	-1.0226	14,17	-0.4558	-0.2961
14,18	-0.1999	0.2907	14,19	0.2085	0.1453	14,20	0.1117	-0.1599	14,21	-0.1283	-0.0894
14,22	-0.0736	0.1062	14,23	0.0901	0.0620	14,24	0.0532	-0.0777	14,25	-0.0682	-0.0464
14,26	-0.0409	0.0604	14,27	0.0542	0.0364						
15,15	3.1392	-4.8960	15,16	2.2204	6.0091	15,17	0.4525	-1.0225	15,18	-0.4557	-0.2962
15,19	-0.1999	0.2906	15,20	0.2084	0.1453	15,21	0.1117	-0.1599	15,22	-0.1282	-0.0894
15,23	-0.0736	0.1062	15,24	0.0899	0.0621	15,25	0.0532	-0.0777	15,26	-0.0680	-0.0464
16,16	3.1392	-4.8959	16,17	2.2206	6.0091	16,18	0.4525	-1.0226	16,19	-0.4558	-0.2961
16,2C	-0.1999	0.2907	16,21	0.2085	0.1453	16,22	0.1117	-0.1599	16,23	-0.1283	-0.0894
16,24	-0.0736	0.1062	16,25	0.0901	0.0620						
17,17	3.1392	-4.8960	17,18	2.2204	6.0091	17,19	0.4525	-1.0225	17,20	-0.4557	-0.2962
17,21	-0.1999	0.2906	17,22	0.2084	0.1453	17,23	0.1117	-0.1599	17,24	-0.1282	-0.0894
18,18	3.1392	-4.8959	18,19	2.2206	6.0091	18,20	0.4525	-1.0226	18,21	-0.4558	-0.2961
18,22	-0.1999	0.2907	18,23	0.2085	0.1453						
19,19	3.1392	-4.8960	19,20	2.2204	6.0091	19,21	0.4525	-1.0226	19,22	-0.4557	-0.2962
2C,2C	3.1392	-4.8959	20,21	2.2206	6.0091						

VEN

SELF AND MUTUAL ADMITTANCES

H/LMDA=0.2500 B/LMDA= 0.250 OMEGA= 8.53

45 ELEMENT ARRAY

I, K	YIK IN MMHO		I, K	YIK IN MMHO		I, K	YIK IN MMHO		I, K	YIK IN MMHO	
	RE	IM		RE	IM		RE	IM		RE	IM
1, 1	6.8945	-4.4119	1, 2	1.8117	5.6294	1, 3	0.2561	-0.8740	1, 4	-0.3660	-0.1708
1, 5	-0.1114	0.2278	1, 6	0.1608	0.0784	1, 7	0.0588	-0.1220	1, 8	-0.0970	-0.0461
1, 9	-0.0374	0.0798	1,10	0.0673	0.0311	1,11	0.0264	-0.0578	1,12	-0.0505	-0.0227
1,13	-0.0199	0.0446	1,14	0.0399	0.0175	1,15	0.0156	-0.0359	1,16	-0.0326	-0.0140
1,17	-0.0127	0.0298	1,18	0.0274	0.0116	1,19	0.0106	-0.0253	1,20	-0.0235	-0.0098
1,21	-0.0090	0.0219	1,22	0.0205	0.0084	1,23	0.0078	-0.0192	1,24	-0.0181	-0.0073
1,25	-0.0068	0.0171	1,26	0.0161	0.0064	1,27	0.0060	-0.0153	1,28	-0.0145	-0.0057
1,29	-0.0054	0.0138	1,30	0.0132	0.0051	1,31	0.0049	-0.0126	1,32	-0.0121	-0.0046
1,33	-0.0044	0.0115	1,34	0.0111	0.0042	1,35	0.0040	-0.0106	1,36	-0.0103	-0.0038
1,37	-0.0037	0.0098	1,38	0.0095	0.0035	1,39	0.0034	-0.0091	1,40	-0.0089	-0.0032
1,41	-0.0031	0.0085	1,42	0.0084	0.0030	1,43	0.0029	-0.0079	1,44	-0.0081	-0.0027
1,45	-0.0008	0.0052									
2, 2	3.1618	-4.8233	2, 3	2.2517	6.0047	2, 4	0.4501	-1.0417	2, 5	-0.4691	-0.2945
2, 6	-0.1986	0.3006	2, 7	0.2163	0.1442	2, 8	0.1108	-0.1663	2, 9	-0.1336	-0.0885
2,10	-0.0729	0.1107	2,11	0.0940	0.0614	2,12	0.0527	-0.0812	2,13	-0.0712	-0.0458
2,14	-0.0404	0.0632	2,15	0.0566	0.0359	2,16	0.0322	-0.0512	2,17	-0.0466	-0.0292
2,18	-0.0265	0.0427	2,19	0.0393	0.0243	2,20	0.0224	-0.0364	2,21	-0.0339	-0.0207
2,22	-0.0192	0.0316	2,23	0.0296	0.0179	2,24	0.0167	-0.0278	2,25	-0.0262	-0.0157
2,26	-0.0147	0.0248	2,27	0.0234	0.0139	2,28	0.0131	-0.0223	2,29	-0.0211	-0.0124
2,30	-0.0118	0.0202	2,31	0.0192	0.0112	2,32	0.0107	-0.0184	2,33	-0.0176	-0.0102
2,34	-0.0097	0.0169	2,35	0.0162	0.0093	2,36	0.0089	-0.0156	2,37	-0.0150	-0.0085
2,38	-0.0082	0.0145	2,39	0.0139	0.0079	2,40	0.0076	-0.0135	2,41	-0.0129	-0.0073
2,42	-0.0070	0.0127	2,43	0.0120	0.0068	2,44	0.0064	-0.0124			
3, 3	3.1396	-4.9087	3, 4	2.2128	6.0087	3, 5	0.4522	-1.0172	3, 6	-0.4518	-0.2960
3, 7	-0.1997	0.2875	3, 8	0.2059	0.1452	3, 9	0.1116	-0.1578	3,10	-0.1264	-0.0893
3,11	-0.0736	0.1046	3,12	0.0886	0.0620	3,13	0.0532	-0.0765	3,14	-0.0670	-0.0463
3,15	-0.0408	0.0594	3,16	0.0532	0.0364	3,17	0.0327	-0.0481	3,18	-0.0438	-0.0296
3,19	-0.0269	0.0401	3,20	0.0369	0.0247	3,21	0.0227	-0.0341	3,22	-0.0317	-0.0210
3,23	-0.0195	0.0296	3,24	0.0277	0.0182	3,25	0.0170	-0.0260	3,26	-0.0246	-0.0159
3,27	-0.0150	0.0232	3,28	0.0220	0.0141	3,29	0.0133	-0.0208	3,30	-0.0198	-0.0126
3,31	-0.0120	0.0189	3,32	0.0180	0.0114	3,33	0.0109	-0.0172	3,34	-0.0165	-0.0104
3,35	-0.0099	0.0158	3,36	0.0152	0.0095	3,37	0.0091	-0.0145	3,38	-0.0141	-0.0087
3,39	-0.0084	0.0135	3,4C	0.0131	0.0080	3,41	0.0077	-0.0125	3,42	-0.0123	-0.0075
3,43	-0.0072	0.0115									
4, 4	3.1388	-4.8913	4, 5	2.2237	6.0094	4, 6	0.4527	-1.0250	4, 7	-0.4576	-0.2963
4, 8	-0.2000	0.2922	4, 9	0.2097	0.1454	4,10	0.1118	-0.1610	4,11	-0.1292	-0.0894
4,12	-0.0737	0.1070	4,13	0.0907	0.0621	4,14	0.0533	-0.0784	4,15	-0.0687	-0.0464
4,16	-0.0409	0.0610	4,17	0.0546	0.0364	4,18	0.0327	-0.0494	4,19	-0.0449	-0.0296
4,2C	-0.0269	0.0412	4,21	0.0379	0.0247	4,22	0.0227	-0.0351	4,23	-0.0326	-0.0210
4,24	-0.0195	0.0305	4,25	0.0285	0.0182	4,26	0.0170	-0.0268	4,27	-0.0252	-0.0159
4,28	-0.0150	0.0239	4,29	0.0226	0.0141	4,30	0.0133	-0.0215	4,31	-0.0204	-0.0126
4,32	-0.0120	0.0194	4,33	0.0185	0.0114	4,34	0.0108	-0.0178	4,35	-0.0169	-0.0103
4,36	-0.0099	0.0163	4,37	0.0156	0.0095	4,38	0.0091	-0.0151	4,39	-0.0144	-0.0087
4,40	-0.0083	0.0140	4,41	0.0133	0.0080	4,42	0.0077	-0.0131			

5, 5	3.1394	-4.8982	5, 6	2.2189	6.0090	5, 7	0.4524	-1.0213	5, 8	-0.4547	-0.2961
5, 9	-0.1998	0.2898	5,10	0.2077	0.1453	5,11	0.1117	-0.1593	5,12	-0.1277	-0.0893
5,13	-0.0736	0.1057	5,14	0.0896	0.0620	5,15	0.0532	-0.0773	5,16	-0.0677	-0.0463
5,17	-0.0408	0.0601	5,18	0.0538	0.0364	5,19	0.0327	-0.0486	5,20	-0.0443	-0.0295
5,21	-0.0269	0.0405	5,22	0.0373	0.0246	5,23	0.0227	-0.0345	5,24	-0.0321	-0.0210
5,25	-0.0195	0.0300	5,26	0.0281	0.0181	5,27	0.0170	-0.0263	5,28	-0.0248	-0.0159
5,29	-0.0150	0.0234	5,30	0.0222	0.0141	5,31	0.0133	-0.0211	5,32	-0.0201	-0.0126
5,33	-0.0120	0.0191	5,34	0.0182	0.0114	5,35	0.0109	-0.0174	5,36	-0.0167	-0.0103
5,37	-0.0099	0.0160	5,38	0.0154	0.0095	5,39	0.0091	-0.0147	5,40	-0.0143	-0.0087
5,41	-0.0084	0.0136									
6, 6	3.1391	-4.8947	6, 7	2.2215	6.0092	6, 8	0.4525	-1.0233	6, 9	-0.4564	-0.2962
6,10	-0.1999	0.2912	6,11	0.2089	0.1453	6,12	0.1117	-0.1603	6,13	-0.1286	-0.0894
6,14	-0.0736	0.1065	6,15	0.0903	0.0621	6,16	0.0532	-0.0780	6,17	-0.0684	-0.0464
6,18	-0.0409	0.0606	6,19	0.0543	0.0364	6,20	0.0327	-0.0491	6,21	-0.0447	-0.0296
6,22	-0.0269	0.0410	6,23	0.0377	0.0246	6,24	0.0227	-0.0349	6,25	-0.0324	-0.0210
6,26	-0.0195	0.0303	6,27	0.0284	0.0182	6,28	0.0170	-0.0267	6,29	-0.0251	-0.0159
6,30	-0.0150	0.0237	6,31	0.0225	0.0141	6,32	0.0133	-0.0213	6,33	-0.0203	-0.0126
6,34	-0.0120	0.0193	6,35	0.0184	0.0114	6,36	0.0108	-0.0177	6,37	-0.0168	-0.0103
6,38	-0.0099	0.0162	6,39	0.0155	0.0095	6,40	0.0091	-0.0150			
7, 7	3.1392	-4.8967	7, 8	2.2199	6.0091	7, 9	0.4525	-1.0221	7,10	-0.4553	-0.2961
7,11	-0.1999	0.2903	7,12	0.2081	0.1453	7,13	0.1117	-0.1596	7,14	-0.1280	-0.0894
7,15	-0.0736	0.1059	7,16	0.0898	0.0620	7,17	0.0532	-0.0775	7,18	-0.0679	-0.0464
7,19	-0.0409	0.0602	7,20	0.0540	0.0364	7,21	0.0327	-0.0488	7,22	-0.0444	-0.0295
7,23	-0.0269	0.0406	7,24	0.0374	0.0246	7,25	0.0227	-0.0346	7,26	-0.0322	-0.0210
7,27	-0.0195	0.0300	7,28	0.0282	0.0182	7,29	0.0170	-0.0264	7,30	-0.0249	-0.0159
7,31	-0.0150	0.0235	7,32	0.0223	0.0141	7,33	0.0133	-0.0211	7,34	-0.0201	-0.0126
7,35	-0.0120	0.0191	7,36	0.0183	0.0114	7,37	0.0109	-0.0174	7,38	-0.0168	-0.0103
7,39	-0.0099	0.0160									
8, 8	3.1392	-4.8955	8, 9	2.2209	6.0092	8,10	0.4525	-1.0229	8,11	-0.4561	-0.2962
8,12	-0.1999	0.2909	8,13	0.2087	0.1453	8,14	0.1117	-0.1601	8,15	-0.1284	-0.0894
8,16	-0.0736	0.1064	8,17	0.0902	0.0620	8,18	0.0532	-0.0779	8,19	-0.0682	-0.0464
8,20	-0.0409	0.0605	8,21	0.0542	0.0364	8,22	0.0327	-0.0490	8,23	-0.0446	-0.0295
8,24	-0.0269	0.0409	8,25	0.0376	0.0246	8,26	0.0227	-0.0349	8,27	-0.0324	-0.0210
8,28	-0.0195	0.0303	8,29	0.0283	0.0181	8,30	0.0170	-0.0266	8,31	-0.0250	-0.0159
8,32	-0.0150	0.0237	8,33	0.0224	0.0141	8,34	0.0133	-0.0213	8,35	-0.0202	-0.0126
8,36	-0.0120	0.0193	8,37	0.0184	0.0114	8,38	0.0108	-0.0176			
9, 9	3.1392	-4.8963	9,10	2.2202	6.0091	9,11	0.4525	-1.0223	9,12	-0.4555	-0.2961
9,13	-0.1999	0.2904	9,14	0.2083	0.1453	9,15	0.1117	-0.1597	9,16	-0.1281	-0.0894
9,17	-0.0736	0.1060	9,18	0.0899	0.0620	9,19	0.0532	-0.0776	9,20	-0.0680	-0.0464
9,21	-0.0409	0.0603	9,22	0.0540	0.0364	9,23	0.0327	-0.0488	9,24	-0.0444	-0.0295
9,25	-0.0269	0.0407	9,26	0.0375	0.0246	9,27	0.0227	-0.0347	9,28	-0.0323	-0.0210
9,29	-0.0195	0.0301	9,30	0.0282	0.0182	9,31	0.0170	-0.0265	9,32	-0.0250	-0.0159
9,33	-0.0150	0.0235	9,34	0.0223	0.0141	9,35	0.0133	-0.0211	9,36	-0.0202	-0.0126
9,37	-0.0120	0.0191									
10,10	3.1392	-4.8957	10,11	2.2207	6.0091	10,12	0.4525	-1.0228	10,13	-0.4559	-0.2962
10,14	-0.1999	0.2908	10,15	0.2086	0.1453	10,16	0.1117	-0.1600	10,17	-0.1284	-0.0894
10,18	-0.0736	0.1063	10,19	0.0901	0.0620	10,20	0.0532	-0.0778	10,21	-0.0682	-0.0464
10,22	-0.0409	0.0605	10,23	0.0542	0.0364	10,24	0.0327	-0.0490	10,25	-0.0446	-0.0295
10,26	-0.0269	0.0408	10,27	0.0376	0.0246	10,28	0.0227	-0.0348	10,29	-0.0324	-0.0210
10,30	-0.0195	0.0302	10,31	0.0283	0.0181	10,32	0.0170	-0.0266	10,33	-0.0250	-0.0159
10,34	-0.0150	0.0237	10,35	0.0224	0.0141	10,36	0.0133	-0.0213			
11,11	3.1392	-4.8961	11,12	2.2204	6.0091	11,13	0.4525	-1.0224	11,14	-0.4556	-0.2962
[illegible]	[illegible]	0.2905	11,16	0.2083	0.1453	11,17	0.1117	-0.1598	11,18	-0.1282	-0.0894
11,19	-0.0736	[illegible]	[illegible]	0.0899	0.0620	11,21	0.0532	-0.0776	11,22	-0.0681	-0.0464
11,23	-0.0409	0.0603	11,24	0.0541	[illegible]	11,25	0.0327	-0.0488	11,26	-0.0445	-0.0296
11,27	-0.0269	0.0407	11,28	0.0375	0.0246	11,29	[illegible]	[illegible]	11,30	-0.0323	-0.0210
11,31	-0.0195	0.0301	11,32	0.0282	0.0182	11,33	0.0170	-0.0265	11,34	[illegible]	-0.0159
11,35	-0.0150	0.0235									
12,12	3.1392	-4.8958	12,13	2.2206	6.0091	12,14	0.4525	-1.0227	12,15	-0.4558	-0.2962
12,16	-0.1999	0.2907	12,17	0.2085	0.1453	12,18	0.1117	-0.1600	12,19	-0.1283	-0.0894
12,20	-0.0736	0.1063	12,21	0.0901	0.0620	12,22	0.0532	-0.0778	12,23	-0.0681	-0.0464
12,24	-0.0409	0.0605	12,25	0.0542	0.0364	12,26	0.0327	-0.0490	12,27	-0.0445	-0.0295
12,28	-0.0269	0.0408	12,29	0.0376	0.0246	12,30	0.0227	-0.0348	12,31	-0.0323	-0.0210
12,32	-0.0195	0.0302	12,33	0.0282	0.0181	12,34	0.0170	-0.0266			
13,13	3.1392	-4.8961	13,14	2.2204	6.0091	13,15	0.4525	-1.0225	13,16	-0.4557	-0.2962
13,17	-0.1999	0.2906	13,18	0.2084	0.1453	13,19	0.1117	-0.1598	13,20	-0.1282	-0.0894
13,21	-0.0736	0.1061	13,22	0.0900	0.0620	13,23	0.0532	-0.0777	13,24	-0.0681	-0.0464
13,25	-0.0409	0.0603	13,26	0.0541	0.0364	13,27	0.0327	-0.0488	13,28	-0.0445	-0.0295
13,29	-0.0269	0.0407	13,30	0.0375	0.0246	13,31	0.0227	-0.0347	13,32	-0.0323	-0.0210
13,33	-0.0195	0.0301									
14,14	3.1392	-4.8958	14,15	2.2206	6.0091	14,16	0.4525	-1.0227	14,17	-0.4558	-0.2962
14,18	-0.1999	0.2907	14,19	0.2085	0.1453	14,20	0.1117	-0.1600	14,21	-0.1283	-0.0894
14,22	-0.0736	0.1062	14,23	0.0900	0.0620	14,24	0.0532	-0.0778	14,25	-0.0681	-0.0464
14,26	-0.0409	0.0605	14,27	0.0541	0.0364	14,28	0.0327	-0.0490	14,29	-0.0445	-0.0295
14,30	-0.0269	0.0408	14,31	0.0376	0.0246	14,32	0.0227	-0.0348			
15,15	3.1392	-4.8960	15,16	2.2204	6.0091	15,17	0.4525	-1.0225	15,18	-0.4557	-0.2962
15,19	-0.1999	0.2906	15,20	0.2084	0.1453	15,21	0.1117	-0.1598	15,22	-0.1282	-0.0894
15,23	-0.0736	0.1061	15,24	0.0900	0.0620	15,25	0.0532	-0.0777	15,26	-0.0681	-0.0464
15,27	-0.0409	0.0604	15,28	0.0541	0.0364	15,29	0.0327	-0.0489	15,30	-0.0445	-0.0295
15,31	-0.0269	0.0407									
16,16	3.1392	-4.8959	16,17	2.2206	6.0091	16,18	0.4525	-1.0226	16,19	-0.4558	-0.2962
16,20	-0.1999	0.2907	16,21	0.2085	0.1453	16,22	0.1117	-0.1600	16,23	-0.1283	-0.0894
16,24	-0.0736	0.1062	16,25	0.0900	0.0620	16,26	0.0532	-0.0778	16,27	-0.0681	-0.0464
16,28	-0.0409	0.0605	16,29	0.0541	0.0364	16,30	0.0327	-0.0490			
17,17	3.1392	-4.8960	17,18	2.2205	6.0091	17,19	0.4525	-1.0225	17,20	-0.4557	-0.2962
17,21	-0.1999	0.2906	17,22	0.2084	0.1453	17,23	0.1117	-0.1599	17,24	-0.1282	-0.0894
17,25	-0.0736	0.1061	17,26	0.0900	0.0620	17,27	0.0532	-0.0777	17,28	-0.0681	-0.0464
17,29	-0.0409	0.0604									
18,18	3.1392	-4.8959	18,19	2.2205	6.0091	18,20	0.4525	-1.0226	18,21	-0.4558	-0.2962
18,22	-0.1999	0.2907	18,23	0.2084	0.1453	18,24	0.1117	-0.1599	18,25	-0.1283	-0.0894
18,26	-0.0736	0.1062	18,27	0.0900	0.0620	18,28	0.0532	-0.0778			
19,19	3.1392	-4.8960	19,20	2.2205	6.0091	19,21	0.4525	-1.0225	19,22	-0.4557	-0.2962
19,23	-0.1999	0.2906	19,24	0.2084	0.1453	19,25	0.1117	-0.1599	19,26	-0.1282	-0.0894
19,27	-0.0736	0.1061									
20,20	3.1392	-4.8959	20,21	2.2205	6.0091	20,22	0.4525	-1.0226	20,23	-0.4558	-0.2962
20,24	-0.1999	0.2907	20,25	0.2084	0.1453	20,26	0.1117	-0.1599			
21,21	3.1392	-4.8960	21,22	2.2205	6.0091	21,23	0.4525	-1.0225	21,24	-0.4558	-0.2962
21,25	-0.1999	0.2906									
22,22	3.1392	-4.8959	22,23	2.2205	6.0091	22,24	0.4525	-1.0226			
23,23	3.1392	-4.8960									

VEN

SELF AND MUTUAL ADMITTANCES

H/LMDA=0.2500 B/LMDA= 0.250 OMEGA= 8.53

50 ELEMENT ARRAY

I, K	YIK IN MMHO RE	IM	I, K	YIK IN MMHO RE	IM	I, K	YIK IN MMHO RE	IM	I, K	YIK IN MMHO RE	IM
1, 1	6.8945	-4.4119	1, 2	1.8117	5.6294	1, 3	0.2561	-0.8740	1, 4	-0.3660	-0.1708
1, 5	-0.1114	0.2278	1, 6	0.1608	0.0784	1, 7	0.0588	-0.1220	1, 8	-0.0970	-0.0462
1, 9	-0.0374	0.0798	1,10	0.0673	0.0311	1,11	0.0264	-0.0578	1,12	-0.0505	-0.0227
1,13	-0.0199	0.0446	1,14	0.0399	0.0175	1,15	0.0156	-0.0359	1,16	-0.0326	-0.0140
1,17	-0.0127	0.0298	1,18	0.0274	0.0116	1,19	0.0106	-0.0253	1,20	-0.0235	-0.0098
1,21	-0.0090	0.0219	1,22	0.0205	0.0084	1,23	0.0078	-0.0192	1,24	-0.0181	-0.0073
1,25	-0.0068	0.0171	1,26	0.0161	0.0064	1,27	0.0060	-0.0153	1,28	-0.0145	-0.0057
1,29	-0.0054	0.0138	1,30	0.0132	0.0051	1,31	0.0048	-0.0126	1,32	-0.0121	-0.0046
1,33	-0.0044	0.0116	1,34	0.0111	0.0042	1,35	0.0040	-0.0107	1,36	-0.0102	-0.0038
1,37	-0.0037	0.0099	1,38	0.0095	0.0035	1,39	0.0034	-0.0092	1,40	-0.0088	-0.0033
1,41	-0.0031	0.0086	1,42	0.0083	0.0030	1,43	0.0029	-0.0080	1,44	-0.0077	-0.0028
1,45	-0.0027	0.0076	1,46	0.0073	0.0026	1,47	0.0025	-0.0072	1,48	-0.0068	-0.0025
1,49	-0.0022	0.0070	1,50	0.0045	0.0007						
2, 2	3.1618	-4.8233	2, 3	2.2517	6.0047	2, 4	0.4501	-1.0417	2, 5	-0.4691	-0.2945
2, 6	-0.1986	0.3006	2, 7	0.2163	0.1442	2, 8	0.1108	-0.1663	2, 9	-0.1336	-0.0885
2,10	-0.0729	0.1107	2,11	0.0940	0.0614	2,12	0.0527	-0.0812	2,13	-0.0712	-0.0458
2,14	-0.0404	0.0632	2,15	0.0566	0.0359	2,16	0.0322	-0.0512	2,17	-0.0466	-0.0292
2,18	-0.0265	0.0427	2,19	0.0393	0.0243	2,20	0.0224	-0.0364	2,21	-0.0339	-0.0207
2,22	-0.0192	0.0316	2,23	0.0296	0.0179	2,24	0.0167	-0.0278	2,25	-0.0262	-0.0157
2,26	-0.0147	0.0248	2,27	0.0235	0.0139	2,28	0.0131	-0.0222	2,29	-0.0212	-0.0124
2,30	-0.0118	0.0202	2,31	0.0193	0.0112	2,32	0.0107	-0.0184	2,33	-0.0176	-0.0102
2,34	-0.0097	0.0169	2,35	0.0162	0.0093	2,36	0.0089	-0.0156	2,37	-0.0150	-0.0085
2,38	-0.0082	0.0144	2,39	0.0140	0.0079	2,40	0.0076	-0.0134	2,41	-0.0130	-0.0073
2,42	-0.0070	0.0126	2,43	0.0122	0.0068	2,44	0.0066	-0.0118	2,45	-0.0115	-0.0063
2,46	-0.0061	0.0110	2,47	0.0109	0.0059	2,48	0.0058	-0.0103	2,49	-0.0107	-0.0054
3, 3	3.1396	-4.9087	3, 4	2.2128	6.0087	3, 5	0.4522	-1.0172	3, 6	-0.4518	-0.2960
3, 7	-0.1997	0.2875	3, 8	0.2059	0.1452	3, 9	0.1116	-0.1578	3,10	-0.1264	-0.0893
3,11	-0.0736	0.1046	3,12	0.0886	0.0620	3,13	0.0532	-0.0765	3,14	-0.0670	-0.0463
3,15	-0.0408	0.0594	3,16	0.0532	0.0364	3,17	0.0327	-0.0481	3,18	-0.0437	-0.0296
3,19	-0.0269	0.0401	3,20	0.0369	0.0247	3,21	0.0227	-0.0342	3,22	-0.0317	-0.0210
3,23	-0.0195	0.0296	3,24	0.0277	0.0182	3,25	0.0170	-0.0261	3,26	-0.0245	-0.0159
3,27	-0.0150	0.0232	3,28	0.0219	0.0141	3,29	0.0133	-0.0208	3,30	-0.0198	-0.0126
3,31	-0.0120	0.0189	3,32	0.0180	0.0114	3,33	0.0109	-0.0172	3,34	-0.0165	-0.0104
3,35	-0.0099	0.0158	3,36	0.0152	0.0095	3,37	0.0091	-0.0146	3,38	-0.0140	-0.0087
3,39	-0.0084	0.0135	3,40	0.0130	0.0080	3,41	0.0077	-0.0126	3,42	-0.0121	-0.0075
3,43	-0.0072	0.0118	3,44	0.0113	0.0069	3,45	0.0067	-0.0111	3,46	-0.0106	-0.0065
3,47	-0.0063	0.0105	3,48	0.0099	0.0061						
4, 4	3.1388	-4.8913	4, 5	2.2237	6.0094	4, 6	0.4527	-1.0249	4, 7	-0.4576	-0.2963
4, 8	-0.2000	0.2922	4, 9	0.2097	0.1454	4,10	0.1118	-0.1610	4,11	-0.1292	-0.0894
4,12	-0.0737	0.1070	4,13	0.0907	0.0621	4,14	0.0533	-0.0784	4,15	-0.0687	-0.0464
4,16	-0.0409	0.0609	4,17	0.0546	0.0364	4,18	0.0327	-0.0494	4,19	-0.0449	-0.0296
4,20	-0.0269	0.0412	4,21	0.0379	0.0246	4,22	0.0227	-0.0351	4,23	-0.0326	-0.0210
4,24	-0.0195	0.0305	4,25	0.0285	0.0181	4,26	0.0170	-0.0268	4,27	-0.0253	-0.0159
4,28	-0.0150	0.0239	4,29	0.0226	0.0141	4,30	0.0133	-0.0214	4,31	-0.0204	-0.0126
4,32	-0.0120	0.0194	4,33	0.0186	0.0114	4,34	0.0108	-0.0177	4,35	-0.0170	-0.0103
4,36	-0.0099	0.0163	4,37	0.0156	0.0095	4,38	0.0091	-0.0150	4,39	-0.0145	-0.0087
4,40	-0.0083	0.0139	4,41	0.0135	0.0080	4,42	0.0077	-0.0129	4,43	-0.0126	-0.0074
4,44	-0.0072	0.0121	4,45	0.0118	0.0069	4,46	0.0067	-0.0113	4,47	-0.0111	-0.0065
5, 5	3.1394	-4.8981	5, 6	2.2189	6.0090	5, 7	0.4524	-1.0213	5, 8	-0.4547	-0.2961
5, 9	-0.1998	0.2898	5,10	0.2077	0.1453	5,11	0.1117	-0.1593	5,12	-0.1277	-0.0893
5,13	-0.0736	0.1057	5,14	0.0896	0.0620	5,15	0.0532	-0.0773	5,16	-0.0677	-0.0463
5,17	-0.0408	0.0601	5,18	0.0538	0.0364	5,19	0.0327	-0.0486	5,20	-0.0442	-0.0295
5,21	-0.0269	0.0405	5,22	0.0373	0.0246	5,23	0.0227	-0.0345	5,24	-0.0321	-0.0210
5,25	-0.0195	0.0300	5,26	0.0281	0.0182	5,27	0.0170	-0.0264	5,28	-0.0248	-0.0159
5,29	-0.0150	0.0235	5,30	0.0222	0.0141	5,31	0.0133	-0.0211	5,32	-0.0200	-0.0126
5,33	-0.0120	0.0191	5,34	0.0182	0.0114	5,35	0.0108	-0.0174	5,36	-0.0167	-0.0104
5,37	-0.0099	0.0160	5,38	0.0153	0.0095	5,39	0.0091	-0.0148	5,40	-0.0142	-0.0087
5,41	-0.0083	0.0137	5,42	0.0132	0.0080	5,43	0.0077	-0.0128	5,44	-0.0123	-0.0074
5,45	-0.0072	0.0120	5,46	0.0114	0.0069						
6, 6	3.1391	-4.8947	6, 7	2.2215	6.0092	6, 8	0.4525	-1.0233	6, 9	-0.4564	-0.2962
6,10	-0.1999	0.2912	6,11	0.2089	0.1453	6,12	0.1117	-0.1603	6,13	-0.1286	-0.0894
6,14	-0.0736	0.1065	6,15	0.0903	0.0621	6,16	0.0532	-0.0780	6,17	-0.0684	-0.0464
6,18	-0.0409	0.0606	6,19	0.0543	0.0364	6,20	0.0327	-0.0491	6,21	-0.0447	-0.0296
6,22	-0.0269	0.0409	6,23	0.0377	0.0246	6,24	0.0227	-0.0349	6,25	-0.0325	-0.0210
6,26	-0.0195	0.0303	6,27	0.0284	0.0181	6,28	0.0170	-0.0266	6,29	-0.0251	-0.0159
6,30	-0.0150	0.0237	6,31	0.0225	0.0141	6,32	0.0133	-0.0213	6,33	-0.0203	-0.0126
6,34	-0.0120	0.0193	6,35	0.0185	0.0114	6,36	0.0108	-0.0176	6,37	-0.0169	-0.0103
6,38	-0.0099	0.0162	6,39	0.0156	0.0095	6,40	0.0091	-0.0149	6,41	-0.0144	-0.0087
6,42	-0.0083	0.0138	6,43	0.0134	0.0080	6,44	0.0077	-0.0128	6,45	-0.0125	-0.0074
7, 7	3.1392	-4.8967	7, 8	2.2199	6.0091	7, 9	0.4525	-1.0221	7,10	-0.4553	-0.2961
7,11	-0.1999	0.2903	7,12	0.2081	0.1453	7,13	0.1117	-0.1596	7,14	-0.1280	-0.0894
7,15	-0.0736	0.1059	7,16	0.0898	0.0620	7,17	0.0532	-0.0775	7,18	-0.0679	-0.0464
7,19	-0.0409	0.0602	7,20	0.0540	0.0364	7,21	0.0327	-0.0488	7,22	-0.0444	-0.0296
7,23	-0.0269	0.0407	7,24	0.0374	0.0246	7,25	0.0227	-0.0347	7,26	-0.0322	-0.0210
7,27	-0.0195	0.0301	7,28	0.0281	0.0182	7,29	0.0170	-0.0265	7,30	-0.0249	-0.0159
7,31	-0.0150	0.0235	7,32	0.0223	0.0141	7,33	0.0133	-0.0212	7,34	-0.0201	-0.0126
7,35	-0.0120	0.0192	7,36	0.0183	0.0114	7,37	0.0108	-0.0175	7,38	-0.0167	-0.0104
7,39	-0.0099	0.0161	7,40	0.0154	0.0095	7,41	0.0091	-0.0148	7,42	-0.0142	-0.0087
7,43	-0.0083	0.0138	7,44	0.0132	0.0080						
8, 8	3.1392	-4.8955	8, 9	2.2209	6.0092	8,10	0.4525	-1.0229	8,11	-0.4561	-0.2962
8,12	-0.1999	0.2909	8,13	0.2087	0.1453	8,14	0.1117	-0.1601	8,15	-0.1284	-0.0894
8,16	-0.0736	0.1063	8,17	0.0902	0.0620	8,18	0.0532	-0.0779	8,19	-0.0682	-0.0464
8,20	-0.0409	0.0605	8,21	0.0542	0.0364	8,22	0.0327	-0.0490	8,23	-0.0446	-0.0295
8,24	-0.0269	0.0409	8,25	0.0377	0.0246	8,26	0.0227	-0.0348	8,27	-0.0324	-0.0210
8,28	-0.0195	0.0302	8,29	0.0283	0.0181	8,30	0.0170	-0.0266	8,31	-0.0251	-0.0159
8,32	-0.0150	0.0237	8,33	0.0224	0.0141	8,34	0.0133	-0.0213	8,35	-0.0203	-0.0126
8,36	-0.0120	0.0193	8,37	0.0184	0.0114	8,38	0.0108	-0.0176	8,39	-0.0169	-0.0103
8,40	-0.0099	0.0161	8,41	0.0155	0.0095	8,42	0.0091	-0.0149	8,43	-0.0144	-0.0087
9, 9	3.1392	-4.8963	9,10	2.2202	6.0091	9,11	0.4525	-1.0223	9,12	-0.4555	-0.2961
9,13	-0.1999	0.2905	9,14	0.2083	0.1453	9,15	0.1117	-0.1597	9,16	-0.1281	-0.0894
9,17	-0.0736	0.1060	9,18	0.0899	0.0620	9,19	0.0532	-0.0776	9,20	-0.0680	-0.0464
9,21	-0.0409	0.0603	9,22	0.0540	0.0364	9,23	0.0327	-0.0488	9,24	-0.0444	-0.0296
9,25	-0.0269	0.0407	9,26	0.0375	0.0246	9,27	0.0227	-0.0347	9,28	-0.0322	-0.0210
9,29	-0.0195	0.0301	9,30	0.0282	0.0182	9,31	0.0170	-0.0265	9,32	-0.0249	-0.0159
9,33	-0.0150	0.0236	9,34	0.0223	0.0141	9,35	0.0133	-0.0212	9,36	-0.0201	-0.0126
9,37	-0.0120	0.0192	9,38	0.0183	0.0114	9,39	0.0108	-0.0175	9,40	-0.0167	-0.0104
9,41	-0.0099	0.0161	9,42	0.0154	0.0095						

10,10	3.1392	-4.8957	10,11	2.2207	6.0091	10,12	0.4525	-1.0228	10,13	-0.4559	-0.2962
10,14	-0.1999	0.2908	10,15	0.2086	0.1453	10,16	0.1117	-0.1600	10,17	-0.1284	-0.0894
10,18	-0.0736	0.1063	10,19	0.0901	0.0620	10,20	0.0532	-0.0778	10,21	-0.0682	-0.0464
10,22	-0.0409	0.0605	10,23	0.0542	0.0364	10,24	0.0327	-0.0490	10,25	-0.0446	-0.0295
10,26	-0.0269	0.0408	10,27	0.0376	0.0246	10,28	0.0227	-0.0348	10,29	-0.0324	-0.0210
10,30	-0.0195	0.0302	10,31	0.0283	0.0181	10,32	0.0170	-0.0266	10,33	-0.0251	-0.0159
10,34	-0.0150	0.0236	10,35	0.0224	0.0141	10,36	0.0133	-0.0212	10,37	-0.0202	-0.0126
10,38	-0.0120	0.0192	10,39	0.0184	0.0114	10,40	0.0108	-0.0175	10,41	-0.0169	-0.0103
11,11	3.1392	-4.8961	11,12	2.2203	6.0091	11,13	0.4525	-1.0224	11,14	-0.4556	-0.2962
11,15	-0.1999	0.2905	11,16	0.2083	0.1453	11,17	0.1117	-0.1598	11,18	-0.1282	-0.0894
11,19	-0.0736	0.1061	11,20	0.0899	0.0620	11,21	0.0532	-0.0777	11,22	-0.0680	-0.0464
11,23	-0.0409	0.0604	11,24	0.0541	0.0364	11,25	0.0327	-0.0489	11,26	-0.0445	-0.0296
11,27	-0.0269	0.0407	11,28	0.0375	0.0246	11,29	0.0227	-0.0347	11,30	-0.0323	-0.0210
11,31	-0.0195	0.0301	11,32	0.0282	0.0182	11,33	0.0170	-0.0265	11,34	-0.0249	-0.0159
11,35	-0.0150	0.0236	11,36	0.0223	0.0141	11,37	0.0133	-0.0212	11,38	-0.0201	-0.0126
11,39	-0.0120	0.0192	11,40	0.0183	0.0114						
12,12	3.1392	-4.8958	12,13	2.2206	6.0091	12,14	0.4525	-1.0227	12,15	-0.4559	-0.2962
12,16	-0.1999	0.2907	12,17	0.2085	0.1453	12,18	0.1117	-0.1600	12,19	-0.1283	-0.0894
12,20	-0.0736	0.1062	12,21	0.0901	0.0620	12,22	0.0532	-0.0778	12,23	-0.0682	-0.0464
12,24	-0.0409	0.0605	12,25	0.0542	0.0364	12,26	0.0327	-0.0489	12,27	-0.0446	-0.0295
12,28	-0.0269	0.0408	12,29	0.0376	0.0246	12,30	0.0227	-0.0348	12,31	-0.0324	-0.0210
12,32	-0.0195	0.0302	12,33	0.0283	0.0181	12,34	0.0170	-0.0265	12,35	-0.0250	-0.0159
12,36	-0.0150	0.0236	12,37	0.0224	0.0141	12,38	0.0133	-0.0212	12,39	-0.0202	-0.0126
13,13	3.1392	-4.8961	13,14	2.2204	6.0091	13,15	0.4525	-1.0225	13,16	-0.4557	-0.2962
13,17	-0.1999	0.2906	13,18	0.2084	0.1453	13,19	0.1117	-0.1598	13,20	-0.1282	-0.0894
13,21	-0.0736	0.1061	13,22	0.0899	0.0620	13,23	0.0532	-0.0777	13,24	-0.0681	-0.0464
13,25	-0.0409	0.0604	13,26	0.0541	0.0364	13,27	0.0327	-0.0489	13,28	-0.0445	-0.0296
13,29	-0.0269	0.0408	13,30	0.0375	0.0246	13,31	0.0227	-0.0347	13,32	-0.0323	-0.0210
13,33	-0.0195	0.0301	13,34	0.0282	0.0182	13,35	0.0170	-0.0265	13,36	-0.0250	-0.0159
13,37	-0.0150	0.0236	13,38	0.0223	0.0141						
14,14	3.1392	-4.8959	14,15	2.2206	6.0091	14,16	0.4525	-1.0226	14,17	-0.4558	-0.2961
14,18	-0.1999	0.2907	14,19	0.2085	0.1453	14,20	0.1117	-0.1599	14,21	-0.1283	-0.0894
14,22	-0.0736	0.1062	14,23	0.0901	0.0620	14,24	0.0532	-0.0778	14,25	-0.0682	-0.0464
14,26	-0.0409	0.0604	14,27	0.0542	0.0364	14,28	0.0327	-0.0489	14,29	-0.0446	-0.0295
14,30	-0.0269	0.0408	14,31	0.0376	0.0246	14,32	0.0227	-0.0348	14,33	-0.0324	-0.0210
14,34	-0.0195	0.0302	14,35	0.0283	0.0181	14,36	0.0170	-0.0265	14,37	-0.0250	-0.0159
15,15	3.1392	-4.8960	15,16	2.2204	6.0091	15,17	0.4525	-1.0225	15,18	-0.4557	-0.2962
15,19	-0.1999	0.2906	15,20	0.2084	0.1453	15,21	0.1117	-0.1599	15,22	-0.1282	-0.0894
15,23	-0.0736	0.1061	15,24	0.0900	0.0620	15,25	0.0532	-0.0777	15,26	-0.0681	-0.0464
15,27	-0.0409	0.0604	15,28	0.0541	0.0364	15,29	0.0327	-0.0489	15,30	-0.0445	-0.0296
15,31	-0.0269	0.0408	15,32	0.0375	0.0246	15,33	0.0227	-0.0348	15,34	-0.0323	-0.0210
15,35	-0.0195	0.0302	15,36	0.0282	0.0182						
16,16	3.1392	-4.8959	16,17	2.2206	6.0091	16,18	0.4525	-1.0226	16,19	-0.4558	-0.2961
16,20	-0.1999	0.2907	16,21	0.2085	0.1453	16,22	0.1117	-0.1599	16,23	-0.1283	-0.0894
16,24	-0.0736	0.1062	16,25	0.0900	0.0620	16,26	0.0532	-0.0777	16,27	-0.0681	-0.0464
16,28	-0.0409	0.0604	16,29	0.0542	0.0364	16,30	0.0327	-0.0489	16,31	-0.0446	-0.0295
16,32	-0.0269	0.0408	16,33	0.0376	0.0246	16,34	0.0227	-0.0348	16,35	-0.0324	-0.0210
17,17	3.1392	-4.8960	17,18	2.2205	6.0091	17,19	0.4525	-1.0225	17,20	-0.4557	-0.2962
17,21	-0.1999	0.2906	17,22	0.2084	0.1453	17,23	0.1117	-0.1599	17,24	-0.1282	-0.0894
17,25	-0.0736	0.1062	17,26	0.0900	0.0620	17,27	0.0532	-0.0777	17,28	-0.0681	-0.0464
17,29	-0.0409	0.0604	17,30	0.0541	0.0364	17,31	0.0327	-0.0489	17,32	-0.0445	-0.0296
17,33	-0.0269	0.0408	17,34	0.0375	0.0246						
18,18	3.1392	-4.8959	18,19	2.2206	6.0091	[illegible]	[illegible]	-1.0226	18,21	-0.4558	-0.2961
18,22	-0.1999	0.2907	18,23	0.2085	0.1453	18,24	0.1117	-0.1599	18,25	[illegible]	[illegible]
18,26	-0.0736	0.1062	18,27	0.0900	0.0620	18,28	0.0532	-0.0777	18,29	-0.0681	-0.0464
18,30	-0.0409	0.0604	18,31	0.0542	0.0364	18,32	0.0327	-0.0489	18,33	-0.0446	-0.0295
19,19	3.1392	-4.8960	19,20	2.2205	6.0091	19,21	0.4525	-1.0225	19,22	-0.4557	-0.2962
19,23	-0.1999	0.2906	19,24	0.2084	0.1453	19,25	0.1117	-0.1599	19,26	-0.1282	-0.0894
19,27	-0.0736	0.1062	19,28	0.0900	0.0620	19,29	0.0532	-0.0777	19,30	-0.0681	-0.0464
19,31	-0.0409	0.0604	19,32	0.0541	0.0364						
20,20	3.1392	-4.8959	20,21	2.2205	6.0091	20,22	0.4525	-1.0226	20,23	-0.4558	-0.2961
20,24	-0.1999	0.2907	20,25	0.2085	0.1453	20,26	0.1117	-0.1599	20,27	-0.1283	-0.0894
20,28	-0.0736	0.1062	20,29	0.0900	0.0620	20,30	0.0532	-0.0777	20,31	-0.0681	-0.0464
21,21	3.1392	-4.8960	21,22	2.2205	6.0091	21,23	0.4525	-1.0226	21,24	-0.4557	-0.2962
21,25	-0.1999	0.2906	21,26	0.2084	0.1453	21,27	0.1117	-0.1599	21,28	-0.1282	-0.0894
21,29	-0.0736	0.1062	21,30	0.0900	0.0620						
22,22	3.1392	-4.8959	22,23	2.2205	6.0091	22,24	0.4525	-1.0226	22,25	-0.4558	-0.2961
22,26	-0.1999	0.2907	22,27	0.2085	0.1453	22,28	0.1117	-0.1599	22,29	-0.1283	-0.0894
23,23	3.1392	-4.8960	23,24	2.2205	6.0091	23,25	0.4525	-1.0226	23,26	-0.4557	-0.2962
23,27	-0.1999	0.2906	23,28	0.2084	0.1453						
24,24	3.1392	-4.8959	24,25	2.2205	6.0091	24,26	0.4525	-1.0226	24,27	-0.4558	-0.2961
25,25	3.1392	-4.8959	25,26	2.2205	6.0091						

VBR

SELF AND MUTUAL ADMITTANCES

H/LMDA=0.2500 B/LMDA= 0.500 OMEGA= 8.53

2 ELEMENT ARRAY

I, K	YIK IN MMHO RE	IM	I, K	YIK IN MMHO RE	IM	I, K	YIK IN MMHO RE	IM	I, K	YIK IN MMHO RE	IM
1, 1	11.1314	-2.1963	1, 2	4.3628	1.4772						

VBR

SELF AND MUTUAL ADMITTANCES

H/LMDA=0.2500 B/LMDA= 0.500 OMEGA= 8.53

3 ELEMENT ARRAY

I, K	YIK IN MMHO RE	IM	I, K	YIK IN MMHO RE	IM	I, K	YIK IN MMHO RE	IM	I, K	YIK IN MMHO RE	IM
1, 1	11.2842	-2.1582	1, 2	3.9677	1.2124	1, 3	-1.3727	0.0673			
2, 2	11.9878	-1.0387									

VBR

SELF AND MUTUAL ADMITTANCES

H/LMDA=0.2500 B/LMDA= 0.500 OMEGA= 8.53

4 ELEMENT ARRAY

I, K	YIK IN MMHO RE	IM	I, K	YIK IN MMHO RE	IM	I, K	YIK IN MMHO RE	IM	I, K	YIK IN MMHO RE	IM
1, 1	11.3286	-2.1493	1, 2	3.8957	1.2055	1, 3	-1.1621	0.2021	1, 4	0.7359	-0.0530
2, 2	12.1034	-1.0391	2, 3	3.6094	1.0320						

VBR

SELF AND MUTUAL ADMITTANCES

H/LMDA=0.2500 B/LMDA= 0.500 OMEGA= 8.53

5 ELEMENT ARRAY

I, K	YIK IN MMHO RE	IM	I, K	YIK IN MMHO RE	IM	I, K	YIK IN MMHO RE	IM	I, K	YIK IN MMHO RE	IM
1, 1	11.3486	-2.1455	1, 2	3.8705	1.2047	1, 3	-1.1157	0.2059	1, 4	0.5955	-0.1423
1, 5	-0.4934	0.0372									
2, 2	12.1342	-1.0432	2, 3	3.5520	1.0364	2, 4	-0.9722	0.2835			
3, 3	12.2101	-1.0418									

VBR

SELF AND MUTUAL ADMITTANCES

H/LMDA=0.2500 B/LMDA= 0.500 OMEGA= 8.53

6 ELEMENT ARRAY

I, K	YIK IN MMHO RE	IM	I, K	YIK IN MMHO RE	IM	I, K	YIK IN MMHO RE	IM	I, K	YIK IN MMHO RE	IM
1, 1	11.3596	-2.1434	1, 2	3.8583	1.2047	1, 3	-1.0980	0.2061	1, 4	0.5615	-0.1449
1, 5	-0.3893	0.1032	1, 6	0.3666	-0.0282						
2, 2	12.1474	-1.0455	2, 3	3.5330	1.C397	2, 4	-0.9351	0.2796	2, 5	0.4697	-0.1926
3, 3	12.2376	-1.0464	3, 4	3.4986	1.0417						

VBR

SELF AND MUTUAL ADMITTANCES

H/LMDA=0.2500 B/LMDA= 0.500 OMEGA= 8.53

7 ELEMENT ARRAY

I, K	YIK IN MMHO RE	IM	I, K	YIK IN MMHO RE	IM	I, K	YIK IN MMHO RE	IM	I, K	YIK IN MMHO RE	IM
1, 1	11.3664	-2.1421	1, 2	3.8512	1.2047	1, 3	-1.0889	0.2059	1, 4	0.5479	-0.1449
1, 5	-0.3627	C.1052	1, 6	0.2847	-0.0801	1, 7	-0.2894	0.0226			
2, 2	12.1545	-1.0471	2, 3	3.5239	1.0417	2, 4	-0.9215	0.2769	2, 5	0.4428	-0.1893
2, 6	-0.2967	0.1387									
3, 3	12.2491	-1.0491	3, 4	3.4812	1.0453	3, 5	-0.9007	0.2751			
4, 4	12.2638	-1.0511									

VBR

SELF AND MUTUAL ADMITTANCES

H/LMDA=0.2500 B/LMDA= 0.500 OMEGA= 8.53

8 ELEMENT ARRAY

I, K	YIK IN MMHO RE	IM	I, K	YIK IN MMHO RE	IM	I, K	YIK IN MMHO RE	IM	I, K	YIK IN MMHO RE	IM
1, 1	11.3711	-2.1413	1, 2	3.8466	1.2049	1, 3	-1.0835	0.2057	1, 4	0.5406	-0.1447
1, 5	-0.3517	0.1052	1, 6	0.2629	-0.0817	1, 7	-0.2221	0.0651	1, 8	0.2377	-0.0187
2, 2	12.1588	-1.0481	2, 3	3.5188	1.0430	2, 4	-0.9147	0.2753	2, 5	0.4323	-0.1871
2, 6	-0.2757	0.1359	2, 7	0.2119	-0.1071						
3, 3	12.2552	-1.0507	3, 4	3.4731	1.0474	3, 5	-0.8883	0.2722	3, 6	0.4179	-0.1856
4, 4	12.2746	-1.0538	4, 5	3.4646	1.0489						

VBR

SELF AND MUTUAL ADMITTANCES

H/LMDA=0.2500 B/LMDA= 0.500 OMEGA= 8.53

9 ELEMENT ARRAY

I, K	YIK IN MMHO RE	IM	I, K	YIK IN MMHO RE	IM	I, K	YIK IN MMHO RE	IM	I, K	YIK IN MMHO RE	IM
1, 1	11.3744	-2.1407	1, 2	3.8434	1.2050	1, 3	-1.0799	0.2055	1, 4	0.5362	-0.1445
1, 5	-0.3457	0.1050	1, 6	0.2537	-0.0817	1, 7	-0.2038	0.0664	1, 8	0.1809	-0.0546
1, 9	-0.2009	0.0160									
2, 2	12.1617	-1.0488	2, 3	3.5156	1.0439	2, 4	-0.9107	0.2742	2, 5	0.4268	-0.1857
2, 6	-0.2673	0.1340	2, 7	0.1949	-0.1047	2, 8	-0.1626	0.0866			
3, 3	12.2588	-1.0518	3, 4	3.4686	1.0487	3, 5	-0.8821	0.2705	3, 6	0.4084	-0.1832
3, 7	-0.2565	0.1327									
4, 4	12.2802	-1.0555	4, 5	3.4569	1.0510	4, 6	-0.8764	0.2694			
5, 5	12.2850	-1.0565									

VBR

SELF AND MUTUAL ADMITTANCES

H/LMDA=0.2500 B/LMDA= 0.500 OMEGA= 8.53

10 ELEMENT ARRAY

I, K	YIK IN MMHO RE	IM	I, K	YIK IN MMHO RE	IM	I, K	YIK IN MMHO RE	IM	I, K	YIK IN MMHO RE	IM
1, 1	11.3768	-2.1402	1, 2	3.8412	1.2051	1, 3	-1.0774	0.2053	1, 4	0.5332	-0.1443
1, 5	-0.3419	0.1047	1, 6	0.2485	-0.0815	1, 7	-0.1959	0.0663	1, 8	0.1651	-0.0557
1, 9	-0.1518	0.0469	1,10	0.1734	-0.0139						
2, 2	12.1637	-1.0493	2, 3	3.5134	1.0446	2, 4	-0.9081	0.2734	2, 5	0.4235	-0.1848
2, 6	-0.2627	0.1328	2, 7	0.1879	-0.1030	2, 8	-0.1484	0.0845	2, 9	0.1308	-0.0724
3, 3	12.2612	-1.0526	3, 4	3.4658	1.0496	3, 5	-0.8785	0.2694	3, 6	0.4034	-0.1818
3, 7	-0.2488	0.1307	3, 8	0.1794	-0.1019						
4, 4	12.2836	-1.0566	4, 5	3.4527	1.0523	4, 6	-0.8705	0.2677	4, 7	0.3992	-0.1809
5, 5	12.2904	-1.0581	5, 6	3.4495	1.0531						

VBR

SELF AND MUTUAL ADMITTANCES

H/LMDA=0.2500 B/LMDA= 0.500 OMEGA= 8.53

12 ELEMENT ARRAY

I, K	YIK IN MMHO RE	IM	I, K	YIK IN MMHO RE	IM	I, K	YIK IN MMHO RE	IM	I, K	YIK IN MMHO RE	IM
1, 1	11.3802	-2.1396	1, 2	3.8381	1.2052	1, 3	-1.0742	0.2050	1, 4	0.5295	-0.1440
1, 5	-0.3375	0.1044	1, 6	0.2430	-0.0811	1, 7	-0.1885	0.0660	1, 8	0.1542	-0.0555
1, 9	-0.1319	0.0478	1,10	0.1181	-0.0418	1,11	-0.1139	0.0364	1,12	0.1353	-0.0111
2, 2	12.1663	-1.0501	2, 3	3.5106	1.0454	2, 4	-0.9049	0.2724	2, 5	0.4198	-0.1836
2, 6	-0.2580	0.1314	2, 7	0.1816	-0.1012	2, 8	-0.1390	0.0821	2, 9	0.1133	-0.0692
2,10	-0.0980	0.0602	2,11	0.0926	-0.0540						
3, 3	12.2641	-1.0536	3, 4	3.4625	1.0508	3, 5	-0.8746	0.2680	3, 6	0.3985	-0.1801
3, 7	-0.2422	0.1286	3, 8	0.1694	-0.0991	3, 9	-0.1299	0.0806	3,10	0.1074	-0.0683
4, 4	12.2873	-1.0579	4, 5	3.4482	1.0539	4, 6	-0.8650	0.2657	4, 7	0.3917	-0.1784
4, 8	-0.2375	0.1275	4, 9	0.1668	-0.0985						
5, 5	12.2957	-1.0600	5, 6	3.4428	1.0553	5, 7	-0.8616	0.2648	5, 8	0.3900	-0.1780
6, 6	12.2988	-1.0609	6, 7	3.4414	1.0558						

VBR

SELF AND MUTUAL ADMITTANCES

H/LMDA=0.2500 B/LMDA= 0.500 OMEGA= 8.53

14 ELEMENT ARRAY

I, K	YIK IN MMHO RE	IM	I, K	YIK IN MMHO RE	IM	I, K	YIK IN MMHO RE	IM	I, K	YIK IN MMHO RE	IM
1, 1	11.3825	-2.1393	1, 2	3.8361	1.2053	1, 3	-1.0722	0.2048	1, 4	0.5274	-0.1437
1, 5	-0.3351	0.1041	1, 6	0.2402	-0.0808	1, 7	-0.1851	0.0657	1, 8	0.1499	-0.0552
1, 9	-0.1260	0.0475	1,10	0.1094	-0.0416	1,11	-0.0979	0.0371	1,12	0.0909	-0.0333
1,13	-0.0904	0.0296	1,14	0.1103	-0.0092						
2, 2	12.1679	-1.0506	2, 3	3.5090	1.0460	2, 4	-0.9032	0.2718	2, 5	0.4178	-0.1829
2, 6	-0.2557	0.1306	2, 7	0.1787	-0.1003	2, 8	-0.1354	0.0810	2, 9	0.1084	-0.0678
2,10	-0.0907	0.0583	2,11	0.0790	-0.0514	2,12	-0.0719	0.0464	2,13	0.0708	-0.0428
3, 3	12.2657	-1.0542	3, 4	3.4607	1.0515	3, 5	-0.8726	0.2672	3, 6	0.3961	-0.1792
3, 7	-0.2393	0.1276	3, 8	0.1658	-0.0978	3, 9	-0.1250	0.0789	3,10	0.1000	-0.0661
3,11	-0.0841	0.0571	3,12	0.0746	-0.0507						
4, 4	12.2892	-1.0587	4, 5	3.4460	1.0548	4, 6	-0.8624	0.2647	4, 7	0.3886	-0.1773
4, 8	-0.2335	0.1260	4, 9	0.1614	-0.0966	4,10	-0.1218	0.0781	4,11	0.0981	-0.0656
5, 5	12.2982	-1.0610	5, 6	3.4399	1.0565	5, 7	-0.8580	0.2635	5, 8	0.3855	-0.1763
5, 9	-0.2314	0.1254	5,10	0.1603	-0.0963						
6, 6	12.3022	-1.0622	6, 7	3.4372	1.0573	6, 8	-0.8563	0.2629	6, 9	0.3846	-0.1760
7, 7	12.3038	-1.0627	7, 8	3.4364	1.0576						

VBR

SELF AND MUTUAL ADMITTANCES

H/LMDA=0.2500 B/LMDA= 0.500 OMEGA= 8.53

16 ELEMENT ARRAY

I, K	YIK IN MMHO RE	IM	I, K	YIK IN MMHO RE	IM	I, K	YIK IN MMHO RE	IM	I, K	YIK IN MMHO RE	IM
1, 1	11.3840	-2.1390	1, 2	3.8348	1.2054	1, 3	-1.0709	0.2047	1, 4	0.5260	-0.1436
1, 5	-0.3336	0.1039	1, 6	0.2385	-0.0806	1, 7	-0.1831	0.0654	1, 8	0.1475	-0.0549
1, 9	-0.1232	0.0472	1,10	0.1058	-0.0414	1,11	-0.0930	0.0368	1,12	0.0837	-0.0331
1,13	-0.0771	0.0301	1,14	0.0734	-0.0276	1,15	-0.0745	0.0248	1,16	0.0927	-0.0079
2, 2	12.1690	-1.0509	2, 3	3.5080	1.0464	2, 4	-0.9020	0.2713	2, 5	0.4166	-0.1824
2, 6	-0.2543	0.1301	2, 7	0.1772	-0.0997	2, 8	-0.1335	0.0803	2, 9	0.1062	-0.0670
2,10	-0.0878	0.0574	2,11	0.0750	-0.0502	2,12	-0.0660	0.0448	2,13	0.0597	-0.0406
2,14	-0.0561	0.0375	2,15	0.0568	-0.0352						
3, 3	12.2668	-1.0546	3, 4	3.4596	1.0519	3, 5	-0.8714	0.2667	3, 6	0.3948	-0.1787
3, 7	-0.2377	0.1269	3, 8	0.1640	-0.0971	3, 9	-0.1227	0.0780	3,10	0.0971	-0.0651
3,11	-0.0802	0.0558	3,12	0.0687	-0.0489	3,13	-0.0608	0.0438	3,14	0.0562	-0.0400
4, 4	12.2904	-1.0592	4, 5	3.4447	1.0553	4, 6	-0.8609	0.2641	4, 7	0.3869	-0.1766
4, 8	-0.2316	0.1252	4, 9	0.1591	-0.0957	4,10	-0.1188	0.0769	4,11	0.0940	-0.0642
4,12	-0.0779	0.0551	4,13	0.0672	-0.0485						
5, 5	12.2996	-1.0616	5, 6	3.4383	1.0572	5, 7	-0.8562	0.2627	5, 8	0.3833	-0.1754
5, 9	-0.2288	0.1243	5,10	0.1570	-0.0950	5,11	-0.1173	0.0764	5,12	0.0932	-0.0639
6, 6	12.3040	-1.0630	6, 7	3.4352	1.0582	6, 8	-0.8539	0.2619	6, 9	0.3817	-0.1749
6,10	-0.2277	0.1239	6,11	0.1563	-0.0948						
7, 7	12.3062	-1.0637	7, 8	3.4337	1.0587	7, 9	-0.8529	0.2616	7,10	0.3812	-0.1747
8, 8	12.3071	-1.0640	8, 9	3.4332	1.0589						

VBR

SELF AND MUTUAL ADMITTANCES

H/LMDA=0.2500 B/LMDA= 0.500 OMEGA= 8.53

18 ELEMENT ARRAY

I, K	YIK IN MMHO RE	IM	I, K	YIK IN MMHO RE	IM	I, K	YIK IN MMHO RE	IM	I, K	YIK IN MMHO RE	IM
1, 1	11.3851	-2.1388	1, 2	3.8339	1.2055	1, 3	-1.0699	0.2046	1, 4	0.5250	-0.1435
1, 5	-0.3325	0.1038	1, 6	0.2373	-0.0804	1, 7	-0.1818	0.0653	1, 8	0.1461	-0.0547
1, 9	-0.1215	0.0470	1,10	0.1038	-0.0411	1,11	-0.0906	0.0366	1,12	0.0806	-0.0329
1,13	-0.0730	0.0299	1,14	0.0672	-0.0274	1,15	-0.0632	0.0253	1,16	0.0612	-0.0234
1,17	-0.0631	0.0214	1,18	0.0797	-0.0069						
2, 2	12.1697	-1.0511	2, 3	3.5072	1.0466	2, 4	-0.9013	0.2711	2, 5	0.4158	-0.1821
2, 6	-0.2534	0.1297	2, 7	0.1762	-0.0993	2, 8	-0.1324	0.0799	2, 9	0.1049	-0.0665
2,10	-0.0863	0.0568	2,11	0.0731	-0.0496	2,12	-0.0635	0.0440	2,13	0.0564	-0.0396
2,14	-0.0512	0.0361	2,15	0.0476	-0.0334	2,16	-0.0457	0.0313	2,17	0.0472	-0.0299
3, 3	12.2675	-1.0549	3, 4	3.4588	1.0523	3, 5	-0.8706	0.2664	3, 6	0.3939	-0.1783
3, 7	-0.2368	0.1265	3, 8	0.1629	-0.0966	3, 9	-0.1214	0.0775	3,10	0.0956	-0.0644
3,11	-0.0784	0.0550	3,12	0.0663	-0.0480	3,13	-0.0576	0.0426	3,14	0.0514	-0.0385
3,15	-0.0470	0.0352	3,16	0.0447	-0.0328						
4, 4	12.2912	-1.0595	4, 5	3.4439	1.0557	4, 6	-0.8600	0.2637	4, 7	0.3859	-0.1761
4, 8	-0.2305	0.1247	4, 9	0.1577	-0.0951	4,10	-0.1172	0.0762	4,11	0.0921	-0.0634
4,12	-0.0754	0.0541	4,13	0.0639	-0.0473	4,14	-0.0558	0.0421	4,15	0.0502	-0.0381
5, 5	12.3005	-1.0620	5, 6	3.4374	1.0576	5, 7	-0.8552	0.2622	5, 8	0.3821	-0.1749
5, 9	-0.2274	0.1237	5,10	0.1553	-0.0943	5,11	-0.1152	0.0756	5,12	0.0905	-0.0628
5,13	-0.0743	0.0537	5,14	0.0633	-0.0471						
6, 6	12.3050	-1.0634	6, 7	3.4340	1.0587	6, 8	-0.8526	0.2613	6, 9	0.3801	-0.1742
6,10	-0.2259	0.1232	6,11	0.1541	-0.0938	6,12	-0.1144	0.0753	6,13	0.0901	-0.0627
7, 7	12.3075	-1.0643	7, 8	3.4322	1.0594	7, 9	-0.8512	0.2608	7,10	0.3791	-0.1738
7,11	-0.2252	0.1229	7,12	0.1538	-0.0937						
8, 8	12.3088	-1.0648	8, 9	3.4313	1.0597	8,10	-0.8506	0.2606	8,11	0.3788	-0.1737
9, 9	12.3093	-1.0650	9,10	3.4310	1.0598						

VBR

SELF AND MUTUAL ADMITTANCES

H/LMDA=0.2500 B/LMDA= 0.500 OMEGA= 8.53

20 ELEMENT ARRAY

I, K	YIK IN MMHO		I, K	YIK IN MMHO		I, K	YIK IN MMHO		I, K	YIK IN MMHO	
	RE	IM		RE	IM		RE	IM		RE	IM
1, 1	11.3860	-2.1387	1, 2	3.8332	1.2055	1, 3	-1.0693	0.2045	1, 4	0.5243	-0.1434
1, 5	-0.3318	0.1037	1, 6	0.2365	-C.0803	1, 7	-0.1810	0.0651	1, 8	0.1451	-0.0546
1, 9	-0.1204	C.C469	1,10	0.1026	-0.0410	1,11	-0.0892	0.0364	1,12	0.0789	-0.0327
1,13	-0.07C9	0.0297	1,14	0.C646	-0.0272	1,15	-0.0596	0.0251	1,16	0.0558	-0.0233
1,17	-0.0533	0.0217	1,18	0.0523	-0.02C3	1,19	-0.0546	0.0187	1,20	0.0697	-0.0061
2, 2	12.1703	-1.0513	2, 3	3.5C67	1.0468	2, 4	-0.9007	0.2708	2, 5	0.4152	-0.1819
2, 6	-0.2528	0.1295	2, 7	0.1755	-0.0991	2, 8	-0.1316	0.0796	2, 9	0.1040	-0.0662
2,1C	-0.C853	0.0564	2,11	0.072C	-0.0491	2,12	-0.0622	0.0435	2,13	0.0548	-0.0390
2,14	-0.0491	0.0354	2,15	0.0447	-0.0325	2,16	-0.0415	0.0301	2,17	0.0393	-0.0282
2,18	-0.0384	0.0268	2,19	0.0402	-0.0259						
3, 3	12.268C	-1.0551	3, 4	3.4583	1.0525	3, 5	-0.8700	0.2661	3, 6	0.3933	-0.1780
3, 7	-0.2361	0.1262	3, 8	0.1621	-0.0963	3, 9	-0.1206	0.0771	3,10	0.0947	-0.0640
3,11	-0.C773	0.0546	3,12	0.065C	-0.0475	3,13	-0.0561	0.0420	3,14	0.0494	-0.0377
3,15	-0.0443	0.0343	3,16	0.0406	-0.0315	3,17	-0.0380	0.0294	3,18	0.0368	-0.0278
4, 4	12.2917	-1.0598	4, 5	3.4434	1.0560	4, 6	-0.8594	0.2634	4, 7	0.3853	-0.1758
4, 8	-0.2297	0.1244	4, 9	0.1569	-0.0947	4,10	-0.1163	0.0758	4,11	0.0910	-0.0629
4,12	-0.0741	0.0536	4,13	0.0623	-0.0466	4,14	-0.0537	0.0413	4,15	0.0474	-0.0371
4,16	-0.0428	0.0338	4,17	C.C396	-0.0312						
5, 5	12.3011	-1.0623	5, 6	3.4367	1.0579	5, 7	-0.8545	0.2619	5, 8	0.3814	-0.1745
5, 9	-0.2266	0.1233	5,1C	0.1543	-0.0938	5,11	-0.1141	0.0750	5,12	0.0892	-0.0622
5,13	-0.0726	0.0530	5,14	0.0611	-0.0462	5,15	-0.0529	0.0409	5,16	0.0469	-0.0369
6, 6	12.3057	-1.0638	6, 7	3.4333	1.C591	6, 8	-0.8518	0.2609	6, 9	0.3792	-0.1737
6,1C	-0.2248	0.1227	6,11	0.1529	-0.0933	6,12	-0.1129	0.0746	6,13	0.0883	-0.0619
6,14	-0.072C	0.0528	6,15	C.06C8	-0.0460						
7, 7	12.3083	-1.0647	7, 8	3.4313	1.0598	7, 9	-0.8502	0.2603	7,10	0.378C	-0.1733
7,11	-0.2239	0.1223	7,12	0.1522	-0.0930	7,13	-0.1124	0.0744	7,14	0.0880	-0.0618
8, 8	12.3C98	-1.0652	8, 9	3.4302	1.0602	8,10	-0.8494	0.260C	8,11	0.3774	-0.1730
8,12	-0.2234	0.1221	8,13	0.1519	-0.0929						
9, 9	12.3106	-1.0656	9,10	3.4296	1.0605	9,11	-0.8490	0.2599	9,12	0.3772	-0.1729
1C,1C	12.3110	-1.0657	10,11	3.4294	1.0605						

VBR

SELF AND MUTUAL ADMITTANCES

H/LMDA=0.2900 [illegible] OMEGA= 8.53

22 ELEMENT ARRAY

I, K	YIK IN MMHO		I, K	YIK IN MMHO		I, K	YIK IN MMHO		I, K	YIK IN MMHO	
	RE	IM		RE	IM		RE	IM		RE	IM
1, 1	11.3866	-2.1385	1, 2	3.8326	1.2056	1, 3	-1.0687	0.2044	1, 4	0.5238	-0.1433
1, 5	-0.3312	0.1036	1, 6	0.2359	-C.0802	1, 7	-0.1803	0.0650	1, 8	0.1444	-0.0545
1, 9	-0.1197	0.0467	1,1C	0.1017	-0.0409	1,11	-0.0883	0.0362	1,12	0.0778	-0.0325
1,13	-0.C696	0.0295	1,14	0.0631	-0.0270	1,15	-0.0578	0.0249	1,16	0.0535	-0.0231
1,17	-0.0501	0.0215	1,18	0.0476	-0.0202	1,19	-0.0459	0.0190	1,20	0.0455	-0.0180
1,21	-0.048C	0.0166	1,22	0.C618	-0.0055						
2, 2	12.17C7	-1.0515	2, 3	3.5063	1.0470	2, 4	-0.9003	0.2707	2, 5	0.4148	-0.1817
2, 6	-0.2524	0.1293	2, 7	0.175C	-0.0989	2, 8	-0.1311	0.0793	2, 9	0.1034	-0.0659
2,1C	-0.0847	0.0562	2,11	0.0713	-0.0488	2,12	-0.0614	0.0431	2,13	0.0538	-0.0386
2,14	-0.0479	0.0349	2,15	0.0433	-0.0320	2,16	-0.0396	0.0295	2,17	0.0368	-0.0274
2,18	-0.0346	0.0257	2,19	0.0333	-0.C244	2,20	-0.0329	0.0234	2,21	0.0350	-0.0227
3, 3	12.2684	-1.0553	3, 4	3.4579	1.0527	3, 5	-0.8696	0.2660	3, 6	0.3929	-0.1778
3, 7	-0.2357	0.1260	3, 8	0.1616	-0.0960	3, 9	-0.1201	0.0769	3,10	0.0941	-0.0637
3,11	-0.0766	0.0542	3,12	0.0643	-0.0471	3,13	-0.0552	0.0416	3,14	0.0483	-0.0372
3,15	-0.C43C	0.0337	3,16	0.C388	-C.C308	3,17	-0.0357	0.0285	3,18	0.0333	-0.C266
3,19	-0.0317	0.0251	3,20	0.0312	-0.C240						
4, 4	12.2921	-1.0600	4, 5	3.4429	1.0562	4, 6	-0.8590	0.2632	4, 7	0.3848	-0.1756
4, 8	-0.2292	0.1241	4, 9	0.1564	-0.0944	4,10	-0.1156	0.0755	4,11	0.0903	-0.0625
4,12	-0.C733	0.0532	4,13	0.0614	-0.0462	4,14	-0.0526	0.0408	4,15	0.0461	-0.0365
4,16	-0.0411	0.0331	4,17	0.C372	-C.C303	4,18	-0.0344	0.0281	4,19	0.0325	-0.0264
5, 5	12.3015	-1.0625	5, 6	3.4363	1.0581	5, 7	-0.8540	0.2616	5, 8	0.3808	-0.1742
5, 9	-0.2260	0.1230	5,1C	0.1537	-0.0935	5,11	-0.1134	0.0747	5,12	0.0884	-0.0618
5,13	-0.C717	0.0526	5,14	0.0600	-0.0457	5,15	-0.0514	0.0403	5,16	0.0451	-0.0361
5,17	-0.04C3	0.0328	5,18	0.0368	-C.0302						
6, 6	12.3062	-1.0640	6, 7	3.4328	1.C593	6, 8	-0.8512	0.2606	6, 9	0.3786	-0.1734
6,1C	-0.2241	0.1223	6,11	0.1521	-0.0929	6,12	-0.1121	0.0742	6,13	0.0873	-0.0614
6,14	-0.C708	0.0522	6,15	0.0593	-0.0454	6,16	-0.0509	0.0401	6,17	0.0448	-0.0360
7, 7	12.3088	-1.0650	7, 8	3.43C7	1.0601	7, 9	-0.8496	0.2600	7,10	0.3773	-0.1729
7,11	-0.2230	0.1219	7,12	0.1512	-0.0926	7,13	-0.1114	0.0739	7,14	0.0867	-0.0612
7,15	-0.C7C4	0.0521	7,16	0.C591	-0.0453						
8, 8	12.3104	-1.0656	8, 9	3.4295	1.0606	8,10	-0.8486	0.2596	8,11	0.3765	-0.1726
8,12	-0.2224	0.1216	8,13	0.1508	-0.0924	8,14	-0.1110	0.0738	8,15	0.0866	-0.0611
9, 9	12.3114	-1.0660	9,10	3.4287	1.0609	9,11	-0.848C	0.2594	9,12	0.3761	-0.1724
9,13	-0.2221	0.1215	9,14	0.1506	-0.0923						
1C,10	12.3119	-1.0662	10,11	3.4283	1.0611	10,12	-C.8477	0.2593	10,13	0.3759	-0.1724
11,11	12.3122	-1.0663	11,12	3.4282	1.0611						

VBR

SELF AND MUTUAL ADMITTANCES

H/LMDA=0.2500 B/LMDA= 0.500 OMEGA= 8.53

24 ELEMENT ARRAY

I, K	YIK IN MMHO RE	IM	I, K	YIK IN MMHO RE	IM	I, K	YIK IN MMHO RE	IM	I, K	YIK IN MMHO RE	IM
1, 1	11.3872	-2.1385	1, 2	3.8322	1.2056	1, 3	-1.0683	0.2044	1, 4	0.5233	-0.1432
1, 5	-0.3308	0.1035	1, 6	0.2355	-0.0802	1, 7	-0.1798	0.0650	1, 8	0.1439	-0.0544
1, 9	-0.1191	0.0466	1,10	0.1011	-0.0408	1,11	-0.0876	0.0361	1,12	0.0771	-0.0324
1,13	-0.0688	0.0294	1,14	0.0621	-0.0269	1,15	-0.0567	0.0247	1,16	0.0522	-0.0229
1,17	-0.0485	0.0214	1,18	0.0455	-0.0200	1,19	-0.0431	0.0188	1,20	0.0413	-0.0178
1,21	-0.0402	0.0169	1,22	0.0402	-0.0160	1,23	-0.0427	0.0149	1,24	0.0555	-0.0050
2, 2	12.1711	-1.0516	2, 3	3.5060	1.0471	2, 4	-0.9000	0.2705	2, 5	0.4144	-0.1816
2, 6	-0.2520	0.1291	2, 7	0.1746	-0.0987	2, 8	-0.1307	0.0792	2, 9	0.1030	-0.0657
2,10	-0.0842	0.0560	2,11	0.0708	-0.0486	2,12	-0.0608	0.0429	2,13	0.0532	-0.0383
2,14	-0.0472	0.0347	2,15	0.0425	-0.0316	2,16	-0.0386	0.0291	2,17	0.0355	-0.0270
2,18	-0.0330	0.0252	2,19	0.0311	-0.0237	2,20	-0.0296	0.0224	2,21	0.0288	-0.0214
2,22	-0.0288	0.0207	2,23	0.0308	-0.0203						
3, 3	12.2687	-1.0554	3, 4	3.4576	1.0528	3, 5	-0.8693	0.2658	3, 6	0.3926	-0.1776
3, 7	-0.2353	0.1258	3, 8	0.1613	-0.0959	3, 9	-0.1197	0.0767	3,10	0.0937	-0.0635
3,11	-0.0761	0.0540	3,12	0.0637	-0.0469	3,13	-0.0546	0.0413	3,14	0.0476	-0.0369
3,15	-0.0422	0.0333	3,16	0.0379	-0.0304	3,17	-0.0345	0.0280	3,18	0.0318	-0.0260
3,19	-0.0296	0.0243	3,20	0.0280	-0.0229	3,21	-0.0271	0.0218	3,22	0.0269	-0.0211
4, 4	12.2924	-1.0601	4, 5	3.4426	1.0563	4, 6	-0.8587	0.2630	4, 7	0.3845	-0.1754
4, 8	-0.2289	0.1239	4, 9	0.1560	-0.0942	4,10	-0.1152	0.0753	4,11	0.0898	-0.0623
4,12	-0.0728	0.0529	4,13	0.0608	-0.0459	4,14	-0.0520	0.0404	4,15	0.0453	-0.0361
4,16	-0.0401	0.0326	4,17	0.0361	-0.0298	4,18	-0.0329	0.0275	4,19	0.0304	-0.0255
4,20	-0.0286	0.0240	4,21	0.0274	-0.0227						
5, 5	12.3018	-1.0627	5, 6	3.4360	1.0583	5, 7	-0.8536	0.2614	5, 8	0.3805	-0.1740
5, 9	-0.2256	0.1228	5,10	0.1532	-0.0933	5,11	-0.1129	0.0744	5,12	0.0878	-0.0616
5,13	-0.0711	0.0523	5,14	0.0593	-0.0453	5,15	-0.0506	0.0399	5,16	0.0441	-0.0357
5,17	-0.0391	0.0323	5,18	0.0352	-0.0295	5,19	-0.0323	0.0272	5,20	0.0300	-0.0254
6, 6	12.3066	-1.0642	6, 7	3.4324	1.0595	6, 8	-0.8508	0.2604	6, 9	0.3782	-0.1732
6,10	-0.2237	0.1221	6,11	0.1516	-0.0926	6,12	-0.1115	0.0739	6,13	0.0866	-0.0611
6,14	-0.0701	0.0519	6,15	0.0584	-0.0450	6,16	-0.0499	0.0396	6,17	0.0435	-0.0355
6,18	-0.0387	0.0321	6,19	0.0350	-0.0294						
7, 7	12.3092	-1.0652	7, 8	3.4303	1.0603	7, 9	-0.8491	0.2598	7,10	0.3768	-0.1726
7,11	-0.2225	0.1216	7,12	0.1506	-0.0922	7,13	-0.1107	0.0736	7,14	0.0859	-0.0608
7,15	-0.0695	0.0516	7,16	0.0580	-0.0448	7,17	-0.0496	0.0395	7,18	0.0434	-0.0354
8, 8	12.3109	-1.0658	8, 9	3.4290	1.0608	8,10	-0.8481	0.2593	8,11	0.3759	-0.1723
8,12	-0.2218	0.1213	8,13	0.1500	-0.0920	8,14	-0.1102	0.0734	8,15	0.0856	-0.0607
8,16	-0.0692	0.0515	8,17	0.0578	-0.0447						
9, 9	12.3119	-1.0662	9,10	3.4282	1.0612	9,11	-0.8474	0.2591	9,12	0.3754	-0.1721
9,13	-0.2214	0.1211	9,14	0.1497	-0.0918	9,15	-0.1100	0.0733	9,16	0.0855	-0.0606
10,10	12.3125	-1.0665	10,11	3.4277	1.0614	10,12	-0.8470	0.2589	10,13	0.3751	-0.1719
10,14	-0.2212	0.1210	10,15	0.1496	-0.0918						
11,11	12.3129	-1.0667	11,12	3.4274	1.0615	11,13	-0.8468	0.2588	11,14	0.3750	-0.1719
12,12	12.3131	-1.0668	12,13	3.4273	1.0616						

VBR

SELF AND MUTUAL ADMITTANCES

H/LMDA=0.2500 B/LMDA= 0.500 OMEGA= 8.53

26 ELEMENT ARRAY

I, K	YIK IN MMHO RE	IM	I, K	YIK IN MMHO RE	IM	I, K	YIK IN MMHO RE	IM	I, K	YIK IN MMHO RE	IM
1, 1	11.3876	-2.1384	1, 2	3.8318	1.2057	1, 3	-1.0680	0.2043	1, 4	0.5230	-0.1432
1, 5	-0.3304	0.1035	1, 6	0.2351	-0.0801	1, 7	-0.1795	0.0649	1, 8	0.1435	-0.0543
1, 9	-0.1187	0.0466	1,10	0.1007	-0.0407	1,11	-0.0871	0.0360	1,12	0.0766	-0.0323
1,13	-0.0682	0.0293	1,14	0.0614	-0.0268	1,15	-0.0559	0.0246	1,16	0.0513	-0.0228
1,17	-0.0475	0.0212	1,18	0.0443	-0.0199	1,19	-0.0416	0.0187	1,20	0.0394	-0.0176
1,21	-0.0377	0.0167	1,22	0.0364	-0.0159	1,23	-0.0357	0.0152	1,24	0.0360	-0.0145
1,25	-0.0385	0.0136	1,26	0.0502	-0.0046						
2, 2	12.1713	-1.0517	2, 3	3.5057	1.0472	2, 4	-0.8997	0.2704	2, 5	0.4142	-0.1814
2, 6	-0.2517	0.1290	2, 7	0.1744	-0.0986	2, 8	-0.1304	0.0790	2, 9	0.1027	-0.0656
2,10	-0.0839	0.0558	2,11	0.0704	-0.0484	2,12	-0.0604	0.0427	2,13	0.0527	-0.0381
2,14	-0.0467	0.0344	2,15	0.0419	-0.0314	2,16	-0.0380	0.0288	2,17	0.0348	-0.0267
2,18	-0.0321	0.0248	2,19	0.0300	-0.0233	2,20	-0.0282	0.0219	2,21	0.0268	-0.0208
2,22	-0.0258	0.0198	2,23	0.0253	-0.0191	2,24	-0.0255	0.0185	2,25	0.0275	-0.0183
3, 3	12.2689	-1.0555	3, 4	3.4574	1.0529	3, 5	-0.8691	0.2657	3, 6	0.3923	-0.1775
3, 7	-0.2350	0.1257	3, 8	0.1610	-0.0957	3, 9	-0.1194	0.0766	3,10	0.0933	-0.0634
3,11	-0.0758	0.0538	3,12	0.0633	-0.0467	3,13	-0.0541	0.0411	3,14	0.0471	-0.0367
3,15	-0.0416	0.0331	3,16	0.0373	-0.0301	3,17	-0.0338	0.0277	3,18	0.0309	-0.0256
3,19	-0.0286	0.0239	3,20	0.0267	-0.0224	3,21	-0.0252	0.0211	3,22	0.0241	-0.0201
3,23	-0.0235	0.0193	3,24	0.0236	-0.0187						
4, 4	12.2927	-1.0602	4, 5	3.4424	1.0564	4, 6	-0.8584	0.2629	4, 7	0.3842	-0.1752
4, 8	-0.2286	0.1238	4, 9	0.1557	-0.0941	4,10	-0.1149	0.0751	4,11	0.0895	-0.0621
4,12	-0.0724	0.0527	4,13	0.0604	-0.0457	4,14	-0.0515	0.0402	4,15	0.0448	-0.0359
4,16	-0.0395	0.0324	4,17	0.0354	-0.0295	4,18	-0.0320	0.0271	4,19	0.0294	-0.0251
4,20	-0.0272	0.0234	4,21	0.0255	-0.0220	4,22	-0.0243	0.0208	4,23	0.0235	-0.0199
5, 5	12.3021	-1.0628	5, 6	3.4357	1.0584	5, 7	-0.8534	0.2613	5, 8	0.3802	-0.1739
5, 9	-0.2253	0.1226	5,10	0.1529	-0.0931	5,11	-0.1125	0.0743	5,12	0.0874	-0.0614
5,13	-0.0706	0.0520	5,14	0.0588	-0.0451	5,15	-0.0501	0.0397	5,16	0.0435	-0.0354
5,17	-0.0384	0.0319	5,18	0.0344	-0.0291	5,19	-0.0312	0.0268	5,20	0.0287	-0.0248
5,21	-0.0267	0.0232	5,22	0.0252	-0.0219						
6, 6	12.3068	-1.0643	6, 7	3.4321	1.0597	6, 8	-0.8505	0.2603	6, 9	0.3779	-0.1730
6,10	-0.2233	0.1219	6,11	0.1512	-0.0924	6,12	-0.1111	0.0737	6,13	0.0862	-0.0609
6,14	-0.0696	0.0516	6,15	0.0578	-0.0447	6,16	-0.0493	0.0393	6,17	0.0428	-0.0351
6,18	-0.0378	0.0317	6,19	0.0339	-0.0289	6,20	-0.0308	0.0266	6,21	0.0285	-0.0247

I, K	RE	IM	I, K	RE	IM	I, K	RE	IM	I, K	RE	IM
7, 7	12.3095	-1.0653	7, 8	3.4300	1.0605	7, 9	-0.8488	0.2596	7,10	0.3764	-0.1725
7,11	-0.2221	0.1214	7,12	0.1502	-0.0920	7,13	-0.1102	0.0733	7,14	0.0854	-0.0605
7,15	-0.0689	0.0513	7,16	0.0573	-0.0444	7,17	-0.0488	0.0391	7,18	0.0424	-0.0349
7,19	-0.0375	0.0316	7,20	0.0338	-0.0288						
8, 8	12.3112	-1.0660	8, 9	3.4287	1.0610	8,10	-0.8477	0.2591	8,11	0.3755	-0.1721
8,12	-0.2213	0.1211	8,13	0.1496	-0.0917	8,14	-0.1097	0.0731	8,15	0.0850	-0.0603
8,16	-0.0685	0.0512	8,17	0.057C	-0.0443	8,18	-0.0486	0.0390	8,19	0.0423	-0.0349
9, 9	12.3123	-1.0664	9,10	3.4278	1.0614	9,11	-0.8470	0.2588	9,12	0.3749	-0.1718
9,13	-0.2208	C.1208	9,14	0.1491	-0.0915	9,15	-0.1093	0.0729	9,16	0.0847	-0.0602
9,17	-0.0683	0.0511	9,18	0.0569	-0.0443						
10,10	12.3130	-1.0668	10,11	3.4272	1.0617	10,12	-0.8465	0.2586	10,13	0.3745	-0.1716
10,14	-0.2205	C.1207	10,15	0.1489	-0.0914	10,16	-0.1092	0.0729	10,17	0.0846	-0.0602
11,11	12.3134	-1.0670	11,12	3.4268	1.0618	11,13	-0.8462	0.2585	11,14	0.3743	-0.1715
11,15	-0.2204	0.1206	11,16	0.1488	-0.0914						
12,12	12.3137	-1.0671	12,13	3.4266	1.0619	12,14	-0.8461	0.2584	12,15	0.3742	-0.1715
13,13	12.3139	-1.0672	13,14	3.4265	1.0620						

VBR

SELF AND MUTUAL ADMITTANCES

H/LMDA=C.2500 B/LMDA= 0.500 OMEGA= 8.53

28 ELEMENT ARRAY

I, K	YIK IN MMHO		I, K	YIK IN MMHO		I, K	YIK IN MMHO		I, K	YIK IN MMHO	
	RE	IM		RE	IM		RE	IM		RE	IM
1, 1	11.3880	-2.1383	1, 2	3.8315	1.2057	1, 3	-1.0677	0.2043	1, 4	0.5227	-0.1431
1, 5	-0.3301	0.1034	1, 6	0.2348	-0.0800	1, 7	-0.1792	0.0648	1, 8	0.1432	-0.0543
1, 9	-0.1184	0.C465	1,1C	0.1003	-0.0406	1,11	-0.0867	0.0360	1,12	0.0761	-0.0323
1,13	-0.0677	0.0292	1,14	0.061C	-0.0267	1,15	-0.0554	0.0245	1,16	0.0507	-0.0227
1,17	-0.0468	C.0211	1,18	0.0435	-0.0198	1,19	-0.0407	0.0186	1,20	0.0383	-0.0175
1,21	-0.0364	0.0166	1,22	0.C347	-0.0158	1,23	-0.0334	0.0150	1,24	0.0325	-0.0144
1,25	-0.0321	0.0138	1,26	0.0325	-0.0132	1,27	-0.0349	0.0124	1,28	0.0458	-0.0042
2, 2	12.1716	-1.0518	2, 3	3.5055	1.0473	2, 4	-0.8995	0.2703	2, 5	0.4140	-0.1813
2, 6	-0.2515	0.1289	2, 7	0.1741	-0.0985	2, 8	-0.1302	0.0789	2, 9	0.1024	-0.0655
2,1C	-0.0836	0.0557	2,11	0.0701	-0.0483	2,12	-0.06C1	0.0426	2,13	0.0524	-0.0380
2,14	-0.0463	0.0343	2,15	C.C415	-0.0312	2,16	-0.0375	0.0286	2,17	0.0343	-0.0264
2,18	-0.0315	0.0246	2,19	0.0293	-0.0230	2,20	-0.0274	0.0216	2,21	0.0258	-0.0204
2,22	-0.0245	0.C194	2,23	0.0235	-0.0185	2,24	-0.0228	0.0177	2,25	0.0225	-0.0172
2,26	-0.0228	0.0168	2,27	0.0248	-0.C166						
3, 3	12.2691	-1.0556	3, 4	3.4572	1.0530	3, 5	-0.8689	0.2656	3, 6	0.3921	-0.1774
3, 7	-0.2348	0.1256	3, 8	0.1608	-0.0956	3, 9	-0.1192	0.0764	3,10	0.0931	-0.0633
3,11	-0.0755	0.0537	3,12	0.0631	-0.0465	3,13	-0.0538	0.0410	3,14	0.0468	-0.0365
3,15	-0.0413	0.0329	3,16	0.C365	-0.C299	3,17	-0.0333	0.0275	3,18	0.0304	-0.0254
3,19	-0.0279	0.0236	3,20	0.0259	-0.0221	3,21	-0.0243	0.0207	3,22	0.0229	-0.0196
[illegible]	[illegible]	0.0187	3,24	0.0211	-0.0179	3,25	-0.0208	0.0173	3,26	0.0210	-0.0169
4, 4	12.2928	-1.0603	4, 5	3.4422	1.0565	4, 6	-0.8582	[illegible]	[illegible]	[illegible]	-0.1751
4, 8	-0.2284	0.1237	4, 9	0.1554	-0.0940	4,10	-0.1147	0.0750	4,11	0.0892	[illegible]
4,12	-0.0722	0.0526	4,13	0.0601	-0.0455	4,14	-0.0512	0.0400	4,15	0.0444	-0.0357
4,16	-0.0391	0.C321	4,17	0.0349	-0.C292	4,18	-0.0315	0.0268	4,19	0.0287	-0.0248
4,2C	-0.0265	0.0231	4,21	0.0246	-0.0216	4,22	-0.0231	0.0203	4,23	0.0219	-0.C193
4,24	-0.0211	0.0184	4,25	0.0206	-0.0177						
5, 5	12.3023	-1.0629	5, 6	3.4355	1.0586	5, 7	-0.8531	0.2612	5, 8	0.3800	-0.1738
5, 9	-0.2250	0.1225	5,10	0.1526	-0.0930	5,11	-0.1123	0.0741	5,12	0.0871	-0.0612
5,13	-0.07C3	0.0519	5,14	0.0584	-0.0449	5,15	-0.0497	0.0395	5,16	0.0431	-0.0352
5,17	-0.0379	0.0317	5,18	0.0338	-0.0288	5,19	-0.0306	0.0264	5,20	0.0279	-0.0245
5,21	-0.0257	0.0228	5,22	0.024C	-0.0214	5,23	-0.0226	0.0202	5,24	0.0216	-0.0192
6, 6	12.3070	-1.0645	6, 7	3.4319	1.C598	6, 8	-C.8503	0.2601	6, 9	0.3776	-0.1729
6,1C	-0.2231	0.1217	6,11	0.1510	-0.0923	6,12	-0.1108	0.0735	6,13	0.0859	-0.0607
6,14	-0.0692	0.0514	6,15	0.C575	-0.0445	6,16	-0.0488	0.0391	6,17	0.0423	-0.0348
6,18	-0.0372	0.0314	6,19	0.0332	-0.0286	6,20	-0.0301	0.0262	6,21	0.0275	-0.0243
6,22	-0.0254	0.C226	6,23	0.0238	-0.0213						
7, 7	12.3097	-1.0655	7, 8	3.4298	1.0606	7, 9	-0.8485	0.2595	7,10	0.3762	-0.1723
7,11	-0.2218	0.1212	7,12	0.1499	-0.0918	7,13	-0.1099	0.0731	7,14	0.0851	-0.0603
7,15	-0.0685	0.0511	7,16	0.0568	-0.0442	7,17	-0.0483	0.0389	7,18	0.0418	-0.0346
7,19	-0.0369	0.0312	7,20	0.0329	-0.0284	7,21	-0.0298	0.0261	7,22	0.0274	-0.0242
8, 8	12.3114	-1.0661	8, 9	3.4284	1.0612	8,10	-0.8474	0.259C	8,11	0.3752	-0.1719
8,12	-0.2210	0.1209	8,13	0.1492	-0.0915	8,14	-0.1093	0.0729	8,15	0.0845	-0.0601
8,16	-0.0681	0.0509	8,17	0.0565	-C.0440	8,18	-0.0480	0.0387	8,19	0.0416	-0.0345
8,2C	-0.0367	0.0312	8,21	0.0328	-0.0284						
9, 9	12.3125	-1.0666	9,10	3.4275	1.0616	9,11	-0.8466	0.2586	9,12	0.3745	-0.1716
9,13	-0.2205	0.1206	9,14	0.1487	-0.0913	9,15	-0.1089	0.0727	9,16	0.0842	-0.0600
9,17	-0.0678	0.0508	9,18	0.0563	-0.0439	9,19	-0.0478	0.0387	9,20	0.0415	-0.0345
10,10	12.3133	-1.0669	10,11	3.4269	1.0619	10,12	-0.8461	0.2584	10,13	0.3741	-0.1714
10,14	-0.2201	0.1205	10,15	0.1484	-0.0912	10,16	-0.1086	0.0726	10,17	0.0840	-0.0599
1C,18	-0.C677	C.0507	10,19	0.0562	-0.0439						
11,11	12.3138	-1.0672	11,12	3.4264	1.0621	11,13	-0.8458	0.2582	11,14	0.3738	-0.1713
11,15	-0.2199	0.1204	11,16	0.1483	-0.0911	11,17	-0.1085	0.0725	11,18	0.0840	-0.0598
12,12	12.3141	-1.0673	12,13	3.4262	1.0622	12,14	-0.8456	0.2581	12,15	0.3737	-0.1712
12,16	-0.2198	0.1203	12,17	0.1482	-0.0911						
13,13	12.3143	-1.0674	13,14	3.4260	1.0623	13,15	-0.8455	0.2581	13,16	0.3736	-0.1712
14,14	12.3144	-1.0675	14,15	3.426C	1.0623						

VBR

SELF AND MUTUAL ADMITTANCES

H/LMDA=0.2500 B/LMDA= 0.500 OMEGA= 8.53

30 ELEMENT ARRAY

I, K	YIK IN MMHO		I, K	YIK IN MMHO		I, K	YIK IN MMHO		I, K	YIK IN MMHO	
	RE	IM		RE	IM		RE	IM		RE	IM
1, 1	11.3883	-2.1383	1, 2	3.8313	1.2057	1, 3	-1.0675	0.2043	1, 4	0.5225	-0.1431
1, 5	-0.3299	0.1034	1, 6	0.2346	-0.0800	1, 7	-0.1789	0.0648	1, 8	0.1430	-0.0542
1, 9	-0.1181	0.0465	1,10	0.1000	-0.0406	1,11	-0.0864	0.0359	1,12	0.0758	-0.0322
1,13	-0.0674	0.0292	1,14	0.0606	-0.0266	1,15	-0.0550	0.0245	1,16	0.0503	-0.0226
1,17	-0.0463	0.0211	1,18	0.0430	-0.0197	1,19	-0.0401	0.0185	1,20	0.0376	-0.0174
1,21	-0.0355	0.0165	1,22	0.0337	-0.0156	1,23	-0.0322	0.0149	1,24	0.0309	-0.0142
1,25	-0.0300	0.0136	1,26	0.0293	-0.0131	1,27	-0.0291	0.0126	1,28	0.0296	-0.0121
1,29	-0.0319	0.0114	1,30	0.0421	-0.0039						
2, 2	12.1717	-1.0518	2, 3	3.5053	1.0474	2, 4	-0.8994	0.2703	2, 5	0.4138	-0.1813
2, 6	-0.2514	0.1288	2, 7	0.1740	-0.0984	2, 8	-0.1300	0.0788	2, 9	0.1022	-0.0654
2,10	-0.0834	0.0556	2,11	0.0699	-0.0482	2,12	-0.0599	0.0425	2,13	0.0521	-0.0379
2,14	-0.0461	0.0342	2,15	0.0412	-0.0311	2,16	-0.0372	0.0285	2,17	0.0339	-0.0263
2,18	-0.0311	0.0244	2,19	0.0288	-0.0228	2,20	-0.0268	0.0214	2,21	0.0252	-0.0201
2,22	-0.0238	0.0191	2,23	0.0226	-0.0181	2,24	-0.0216	0.0173	2,25	0.0209	-0.0166
2,26	-0.0204	0.0160	2,27	0.0202	-0.0156	2,28	-0.0206	0.0153	2,29	0.0225	-0.0152
3, 3	12.2693	-1.0557	3, 4	3.4570	1.0531	3, 5	-0.8687	0.2655	3, 6	0.3920	-0.1773
3, 7	-0.2347	0.1255	3, 8	0.1606	-0.0955	3, 9	-0.1190	0.0763	3,10	0.0929	-0.0632
3,11	-0.0753	0.0536	3,12	0.0628	-0.0464	3,13	-0.0536	0.0408	3,14	0.0465	-0.0364
3,15	-0.0410	0.0328	3,16	0.0366	-0.0298	3,17	-0.0330	0.0273	3,18	0.0300	-0.0252
3,19	-0.0275	0.0234	3,20	0.0254	-0.0218	3,21	-0.0237	0.0205	3,22	0.0222	-0.0193
3,23	-0.0210	0.0183	3,24	0.0200	-0.0174	3,25	-0.0193	0.0167	3,26	0.0187	-0.0161
3,27	-0.0185	0.0156	3,28	0.0189	-0.0153						
4, 4	12.2930	-1.0604	4, 5	3.4420	1.0566	4, 6	-0.8581	0.2627	4, 7	0.3838	-0.1750
4, 8	-0.2282	0.1236	4, 9	0.1553	-0.0939	4,10	-0.1145	0.0749	4,11	0.0890	-0.0619
4,12	-0.0719	0.0525	4,13	0.0598	-0.0454	4,14	-0.0509	0.0399	4,15	0.0441	-0.0355
4,16	-0.0388	0.0320	4,17	0.0346	-0.0291	4,18	-0.0311	0.0266	4,19	0.0283	-0.0246
4,20	-0.0260	0.0228	4,21	0.0240	-0.0213	4,22	-0.0224	0.0200	4,23	0.0211	-0.0189
4,24	-0.0200	0.0179	4,25	0.0191	-0.0171	4,26	-0.0185	0.0164	4,27	0.0183	-0.0159
5, 5	12.3024	-1.0630	5, 6	3.4353	1.0586	5, 7	-0.8530	0.2611	5, 8	0.3798	-0.1737
5, 9	-0.2249	0.1224	5,10	0.1525	-0.0928	5,11	-0.1121	0.0740	5,12	0.0869	-0.0611
5,13	-0.0701	0.0517	5,14	0.0582	-0.0447	5,15	-0.0494	0.0393	5,16	0.0428	-0.0350
5,17	-0.0376	0.0315	5,18	0.0335	-0.0286	5,19	-0.0301	0.0262	5,20	0.0274	-0.0242
5,21	-0.0252	0.0225	5,22	0.0233	-0.0210	5,23	-0.0218	0.0198	5,24	0.0205	-0.0187
5,25	-0.0196	0.0178	5,26	0.0189	-0.0170						
6, 6	12.3072	-1.0646	6, 7	3.4317	1.0599	6, 8	-0.8501	0.2600	6, 9	0.3774	-0.1728
6,10	-0.2229	0.1216	6,11	0.1508	-0.0922	6,12	-0.1106	0.0734	6,13	0.0856	-0.0606
6,14	-0.0690	0.0513	6,15	0.0572	-0.0443	6,16	-0.0485	0.0389	6,17	0.0420	-0.0346
6,18	-0.0369	0.0312	6,19	0.0328	-0.0283	6,20	-0.0296	0.0260	6,21	0.0269	-0.0240
6,22	-0.0247	0.0223	6,23	0.0229	-0.0209	6,24	-0.0215	0.0197	6,25	0.0204	-0.0186
7, 7	12.3099	-1.0656	7, 8	3.4296	1.0607	7, 9	-0.8484	0.2593	7,10	0.3760	-0.1722
7,11	-0.2216	0.1211	7,12	0.1497	-0.0917	7,13	-0.1096	0.0730	7,14	0.0848	-0.0602
7,15	-0.0682	0.0510	7,16	0.0565	-0.0440	7,17	-0.0479	0.0387	7,18	0.0415	-0.0344
7,19	-0.0364	0.0310	7,20	0.0324	-0.0282	7,21	-0.0292	0.0258	7,22	0.0266	-0.0239
7,23	-0.0245	0.0222	7,24	0.0228	-0.0208						
8, 8	12.3116	-1.0662	8, 9	3.4282	1.0613	8,10	-0.8472	0.2589	8,11	0.3750	-0.1718
8,12	-0.2208	0.1207	8,13	0.1489	-0.0914	8,14	-0.1090	0.0727	8,15	0.0842	-0.0599
8,16	-0.0677	0.0507	8,17	0.0561	-0.0438	8,18	-0.0476	0.0385	8,19	0.0411	-0.0343
8,20	-0.0361	0.0309	8,21	0.0322	-0.0281	8,22	-0.0291	0.0258	8,23	0.0265	-0.0238
9, 9	12.3127	-1.0667	9,10	3.4273	1.0617	9,11	-0.8464	0.2585	9,12	0.3743	-0.1715
9,13	-0.2202	0.1205	9,14	0.1484	-0.0912	9,15	-0.1086	0.0725	9,16	0.0839	-0.0598
9,17	-0.0674	0.0506	9,18	0.0558	-0.0437	9,19	-0.0473	0.0384	9,20	0.0410	-0.0342
9,21	-0.0360	0.0308	9,22	0.0321	-0.0280						
10,10	12.3135	-1.0671	10,11	3.4266	1.0620	10,12	-0.8459	0.2582	10,13	0.3738	-0.1712
10,14	-0.2198	0.1203	10,15	0.1481	-0.0910	10,16	-0.1083	0.0724	10,17	0.0836	-0.0596
10,18	-0.0672	0.0505	10,19	0.0557	-0.0436	10,20	-0.0472	0.0383	10,21	0.0409	-0.0342
11,11	12.3141	-1.0673	11,12	3.4262	1.0622	11,13	-0.8455	0.2581	11,14	0.3735	-0.1711
11,15	-0.2195	0.1202	11,16	0.1479	-0.0909	11,17	-0.1081	0.0723	11,18	0.0835	-0.0596
11,19	-0.0671	0.0504	11,20	0.0556	-0.0436						
12,12	12.3144	-1.0675	12,13	3.4258	1.0624	12,14	-0.8452	0.2579	12,15	0.3733	-0.1710
12,16	-0.2194	0.1201	12,17	0.1477	-0.0908	12,18	-0.1080	0.0722	12,19	0.0834	-0.0596
13,13	12.3147	-1.0676	13,14	3.4256	1.0625	13,15	-0.8450	0.2579	13,16	0.3732	-0.1709
13,17	-0.2193	0.1200	13,18	0.1477	-0.0908						
14,14	12.3149	-1.0677	14,15	3.4255	1.0625	14,16	-0.8450	0.2578	14,17	0.3731	-0.1709
15,15	12.3149	-1.0677	15,16	3.4255	1.0625						

VBR

SELF AND MUTUAL ADMITTANCES

H/LMDA=0.2500 B/LMDA= 0.500 OMEGA= 8.53

35 ELEMENT ARRAY

I, K	YIK IN MMHO		I, K	YIK IN MMHO		I, K	YIK IN MMHO		I, K	YIK IN MMHO	
	RE	IM		RE	IM		RE	IM		RE	IM
1, 1	11.3889	-2.1382	1, 2	3.8308	1.2058	1, 3	-1.0670	0.2042	1, 4	0.5221	-0.1430
1, 5	-0.3295	0.1033	1, 6	0.2341	-0.0799	1, 7	-0.1785	0.0647	1, 8	0.1425	-0.0541
1, 9	-0.1176	0.0464	1,10	0.0995	-0.0405	1,11	-0.0859	0.0358	1,12	0.0753	-0.0321
1,13	-0.0668	0.0290	1,14	0.0599	-0.0265	1,15	-0.0543	0.0243	1,16	0.0495	-0.0225
1,17	-0.0455	0.0209	1,18	0.0421	-0.0195	1,19	-0.0392	0.0183	1,20	0.0366	-0.0172
1,21	-0.0344	0.0163	1,22	0.0324	-0.0154	1,23	-0.0307	0.0147	1,24	0.0292	-0.0140
1,25	-0.0278	0.0133	1,26	0.0267	-0.0128	1,27	-0.0257	0.0123	1,28	0.0249	-0.0118
1,29	-0.0242	0.0114	1,30	0.0237	-0.0110	1,31	-0.0234	0.0106	1,32	0.0234	-0.0103
1,33	-0.0240	0.0100	1,34	0.0262	-0.0095	1,35	-0.0349	0.0033			
2, 2	12.1721	-1.0520	2, 3	3.5050	1.0475	2, 4	-0.8990	0.2701	2, 5	0.4135	-0.1811
2, 6	-0.2510	0.1287	2, 7	0.1736	-0.0982	2, 8	-0.1297	0.0787	2, 9	0.1019	-0.0652
2,10	-0.0830	0.0554	2,11	0.0695	-0.0480	2,12	-0.0594	0.0423	2,13	0.0517	-0.0377
2,14	-0.0456	0.0339	2,15	0.0407	-0.0308	2,16	-0.0367	0.0282	2,17	0.0333	-0.0260
2,18	-0.0305	0.0241	2,19	0.0281	-0.0225	2,20	-0.0261	0.0210	2,21	0.0243	-0.0198
2,22	-0.0228	0.0186	2,23	0.0215	-0.0177	2,24	-0.0203	0.0168	2,25	0.0193	-0.0160
2,26	-0.0184	0.0153	2,27	0.0177	-0.0147	2,28	-0.0171	0.0141	2,29	0.0166	-0.0136
2,30	-0.0162	0.0132	2,31	0.0160	-0.0129	2,32	-0.0161	0.0126	2,33	0.0166	-0.0125
2,34	-0.0183	0.0125									
3, 3	12.2696	-1.0558	3, 4	3.4567	1.0532	3, 5	-0.8684	0.2654	3, 6	0.3917	-0.1772
3, 7	-0.2344	0.1253	3, 8	0.1603	-0.0953	3, 9	-0.1186	0.0762	3,10	0.0926	-0.0630
3,11	-0.0750	0.0534	3,12	0.0625	-0.0462	3,13	-0.0532	0.0406	3,14	0.0461	-0.0362
3,15	-0.0405	0.0325	3,16	0.0361	-0.0295	3,17	-0.0324	0.0270	3,18	0.0294	-0.0249
3,19	-0.0269	0.0231	3,20	0.0247	-0.0215	3,21	-0.0229	0.0201	3,22	0.0214	-0.0189
3,23	-0.0200	0.0178	3,24	0.0188	-0.0169	3,25	-0.0178	0.0160	3,26	0.0170	-0.0153
3,27	-0.0162	0.0146	3,28	0.0156	-0.0141	3,29	-0.0151	0.0136	3,30	0.0147	-0.0131
3,31	-0.0145	0.0128	3,32	0.0145	-0.0125	3,33	-0.0150	0.0124			
4, 4	12.2933	-1.0606	4, 5	3.4418	1.0568	4, 6	-0.8578	0.2626	4, 7	0.3835	-0.1749
4, 8	-0.2279	0.1234	4, 9	0.1549	-0.0937	4,10	-0.1141	0.0747	4,11	0.0887	-0.0617
4,12	-0.0716	0.0523	4,13	0.0595	-0.0452	4,14	-0.0505	0.0397	4,15	0.0437	-0.0353
4,16	-0.0383	0.0317	4,17	0.0340	-0.0288	4,18	-0.0306	0.0263	4,19	0.0277	-0.0242
4,20	-0.0253	0.0224	4,21	0.0233	-0.0209	4,22	-0.0215	0.0196	4,23	0.0201	-0.0184
4,24	-0.0188	0.0174	4,25	0.0177	-0.0164	4,26	-0.0168	0.0156	4,27	0.0160	-0.0149
4,28	-0.0153	0.0143	4,29	0.0148	-0.0138	4,30	-0.0144	0.0133	4,31	0.0141	-0.0129
4,32	-0.0141	0.0127									
5, 5	12.3027	-1.0632	5, 6	3.4351	1.0588	5, 7	-0.8527	0.2609	5, 8	0.3795	-0.1735
5, 9	-0.2245	0.1222	5,10	0.1521	-0.0926	5,11	-0.1117	0.0738	5,12	0.0866	-0.0609
5,13	-0.0697	0.0515	5,14	0.0578	-0.0445	5,15	-0.0490	0.0391	5,16	0.0423	-0.0347
5,17	-0.0371	0.0312	5,18	0.0329	-0.0283	5,19	-0.0295	0.0259	5,20	0.0267	-0.0238
5,21	-0.0244	0.0221	5,22	0.0224	-0.0206	5,23	-0.0208	0.0192	5,24	0.0194	-0.0181
5,25	-0.0181	0.0171	5,26	0.0171	-0.0162	5,27	-0.0162	0.0154	5,28	0.0155	-0.0147
5,29	-0.0149	0.0141	5,30	0.0145	-0.0136	5,31	-0.0142	0.0132			
6, 6	12.3075	-1.0647	6, 7	3.4314	1.0601	6, 8	-0.8498	0.2599	6, 9	0.3771	-0.1726
6,10	-0.2226	0.1214	6,11	0.1504	-0.0920	6,12	-0.1102	0.0732	6,13	0.0852	-0.0603
6,14	-0.0685	0.0510	6,15	0.0567	-0.0441	6,16	-0.0480	0.0386	6,17	0.0414	-0.0343
6,18	-0.0363	0.0309	6,19	0.0322	-0.0280	6,20	-0.0289	0.0256	6,21	0.0261	-0.0236
6,22	-0.0230	[illegible]	[illegible]	0.0219	-0.0203	6,24	-0.0203	0.0190	6,25	0.0189	-0.0179
6,26	-0.0178	0.0169	6,27	0.0168	-0.0161	[illegible]	[illegible]	[illegible]	[illegible]	0.0153	-0.0147
6,30	-0.0148	0.0141									
7, 7	12.3102	-1.0657	7, 8	3.4293	1.0609	7, 9	-0.8480	0.2591	7,10	0.3756	-0.1720
7,11	-0.2213	0.1209	7,12	0.1493	-0.0915	7,13	-0.1092	0.0728	7,14	0.0844	-0.0599
7,15	-0.0678	0.0507	7,16	0.0560	-0.0437	7,17	-0.0474	0.0384	7,18	0.0409	-0.0341
7,19	-0.0358	0.0306	7,20	0.0317	-0.0278	7,21	-0.0284	0.0254	7,22	0.0257	-0.0234
7,23	-0.0235	0.0217	7,24	0.0216	-0.0202	7,25	-0.0200	0.0189	7,26	0.0187	-0.0178
7,27	-0.0176	0.0168	7,28	0.0166	-0.0160	7,29	-0.0159	0.0153			
8, 8	12.3119	-1.0664	8, 9	3.4279	1.0615	8,10	-0.8468	0.2586	8,11	0.3746	-0.1715
8,12	-0.2204	0.1205	8,13	0.1485	-0.0911	8,14	-0.1086	0.0724	8,15	0.0838	-0.0596
8,16	-0.0672	0.0504	8,17	0.0555	-0.0435	8,18	-0.0470	0.0381	8,19	0.0405	-0.0339
8,20	-0.0354	0.0304	8,21	0.0314	-0.0276	8,22	-0.0281	0.0253	8,23	0.0255	-0.0233
8,24	-0.0233	0.0216	8,25	0.0214	-0.0201	8,26	-0.0199	0.0188	8,27	0.0186	-0.0177
8,28	-0.0175	0.0168									
9, 9	12.3131	-1.0669	9,10	3.4269	1.0619	9,11	-0.8460	0.2583	9,12	0.3739	-0.1712
9,13	-0.2198	0.1202	9,14	0.1480	-0.0909	9,15	-0.1081	0.0722	9,16	0.0833	-0.0594
9,17	-0.0668	0.0502	9,18	0.0552	-0.0433	9,19	-0.0467	0.0380	9,20	0.0402	-0.0338
9,21	-0.0352	0.0303	9,22	0.0312	-0.0275	9,23	-0.0279	0.0252	9,24	0.0253	-0.0232
9,25	-0.0231	0.0215	9,26	0.0213	-0.0201	9,27	-0.0198	0.0188			
10,10	12.3139	-1.0673	10,11	3.4262	1.0623	10,12	-0.8454	0.2580	10,13	0.3734	-0.1710
10,14	-0.2193	0.1200	10,15	0.1476	-0.0907	10,16	-0.1077	0.0720	10,17	0.0830	-0.0593
10,18	-0.0665	0.0501	10,19	0.0549	-0.0432	10,20	-0.0464	0.0379	10,21	0.0400	-0.0337
10,22	-0.0350	0.0302	10,23	0.0310	-0.0274	10,24	-0.0278	0.0251	10,25	0.0252	-0.0231
10,26	-0.0231	0.0215									
11,11	12.3145	-1.0676	11,12	3.4257	1.0625	11,13	-0.8450	0.2578	11,14	0.3730	-0.1708
11,15	-0.2190	0.1198	11,16	0.1473	-0.0905	11,17	-0.1075	0.0719	11,18	0.0828	-0.0592
11,19	-0.0663	0.0500	11,20	0.0548	-0.0431	11,21	-0.0463	0.0378	11,22	0.0399	-0.0336
11,23	-0.0349	0.0302	11,24	0.0310	-0.0274	11,25	-0.0278	0.0251			
12,12	12.3149	-1.0678	12,13	3.4253	1.0627	12,14	-0.8447	0.2576	12,15	0.3727	-0.1706
12,16	-0.2187	0.1197	12,17	0.1471	-0.0904	12,18	-0.1073	0.0718	12,19	0.0826	-0.0591
12,20	-0.0662	0.0499	12,21	0.0547	-0.0431	12,22	-0.0462	0.0378	12,23	0.0398	-0.0336
12,24	-0.0349	0.0302									
13,13	12.3152	-1.0679	13,14	3.4251	1.0628	13,15	-0.8444	0.2575	13,16	0.3725	-0.1705
13,17	-0.2186	0.1196	13,18	0.1469	-0.0904	13,19	-0.1072	0.0717	13,20	0.0825	-0.0590
13,21	-0.0661	0.0499	13,22	0.0546	-0.0430	13,23	-0.0462	0.0378			
14,14	12.3155	-1.0681	14,15	3.4249	1.0629	14,16	-0.8443	0.2574	14,17	0.3724	-0.1705
14,18	-0.2185	0.1196	14,19	0.1468	-0.0903	14,20	-0.1071	0.0717	14,21	0.0825	-0.0590
14,22	-0.0661	0.0499									
15,15	12.3156	-1.0681	15,16	3.4247	1.0630	15,17	-0.8442	0.2574	15,18	0.3723	-0.1704
15,19	-0.2184	0.1195	15,20	0.1468	-0.0903	15,21	-0.1071	0.0717			
16,16	12.3157	-1.0682	16,17	3.4246	1.0630	16,18	-0.8441	0.2573	16,19	0.3722	-0.1704
16,20	-0.2184	0.1195									
17,17	12.3158	-1.0682	17,18	3.4246	1.0631	17,19	-0.8441	0.2573			
18,18	12.3158	-1.0682									

VBR

SELF AND MUTUAL ADMITTANCES

H/LMDA=0.2500 B/LMDA= 0.500 OMEGA= 8.53

40 ELEMENT ARRAY

I, K	YIK IN MMHO		I, K	YIK IN MMHO		I, K	YIK IN MMHO		I, K	YIK IN MMHO	
	RE	IM		RE	IM		RE	IM		RE	IM
1, 1	11.3893	-2.1381	1, 2	3.8305	1.2058	1, 3	-1.0667	0.2042	1, 4	0.5218	-0.1430
1, 5	-0.3292	0.1033	1, 6	0.2338	-0.0799	1, 7	-0.1782	0.0647	1, 8	0.1422	-0.0541
1, 9	-0.1173	0.0463	1,10	0.0992	-0.0404	1,11	-0.0855	0.0357	1,12	0.0749	-0.0320
1,13	-0.0664	0.0290	1,14	0.0596	-0.0264	1,15	-0.0539	0.0242	1,16	0.0491	-0.0224
1,17	-0.0451	0.0208	1,18	0.0416	-0.0194	1,19	-0.0386	0.0182	1,20	0.0360	-0.0171
1,21	-0.0338	0.0162	1,22	0.0317	-0.0153	1,23	-0.0300	0.0145	1,24	0.0284	-0.0138
1,25	-0.0270	0.0132	1,26	0.0257	-0.0126	1,27	-0.0246	0.0121	1,28	0.0236	-0.0116
1,29	-0.0227	0.0112	1,30	0.0219	-0.0108	1,31	-0.0212	0.0104	1,32	0.0206	-0.0101
1,33	-0.0201	0.0097	1,34	0.0197	-0.0095	1,35	-0.0194	0.0092	1,36	0.0194	-0.0090
1,37	-0.0195	0.0087	1,38	0.0202	-0.0085	1,39	-0.0221	0.0081	1,40	0.0297	-0.0029
2, 2	12.1723	-1.0521	2, 3	3.5048	1.0476	2, 4	-0.8988	0.2700	2, 5	0.4133	-0.1810
2, 6	-0.2508	0.1286	2, 7	0.1734	-0.0981	2, 8	-0.1294	0.0786	2, 9	0.1017	-0.0651
2,10	-0.0828	0.0553	2,11	0.0693	-0.0479	2,12	-0.0592	0.0421	2,13	0.0514	-0.0375
2,14	-0.0453	0.0338	2,15	0.0404	-0.0307	2,16	-0.0363	0.0281	2,17	0.0330	-0.0258
2,18	-0.0302	0.0239	2,19	0.0277	-0.0223	2,20	-0.0257	0.0208	2,21	0.0239	-0.0196
2,22	-0.0223	0.0184	2,23	0.0209	-0.0174	2,24	-0.0197	0.0165	2,25	0.0187	-0.0157
2,26	-0.0177	0.0150	2,27	0.0169	-0.0143	2,28	-0.0161	0.0137	2,29	0.0155	-0.0132
2,30	-0.0149	0.0127	2,31	0.0144	-0.0123	2,32	-0.0140	0.0119	2,33	0.0136	-0.0115
2,34	-0.0133	0.0112	2,35	0.0131	-0.0109	2,36	-0.0131	0.0107	2,37	0.0132	-0.0106
2,38	-0.0138	0.0105	2,39	0.0153	-0.0106						
3, 3	12.2698	-1.0559	3, 4	3.4565	1.0533	3, 5	-0.8682	0.2653	3, 6	0.3915	-0.1771
3, 7	-0.2342	0.1252	3, 8	0.1601	-0.0952	3, 9	-0.1184	0.0760	3,10	0.0923	-0.0628
3,11	-0.0748	0.0533	3,12	0.0622	-0.0461	3,13	-0.0529	0.0405	3,14	0.0458	-0.0360
3,15	-0.0403	0.0324	3,16	0.0358	-0.0294	3,17	-0.0321	0.0268	3,18	0.0291	-0.0247
3,19	-0.0265	0.0229	3,20	0.0244	-0.0213	3,21	-0.0225	0.0199	3,22	0.0209	-0.0187
3,23	-0.0195	0.0176	3,24	0.0183	-0.0166	3,25	-0.0173	0.0157	3,26	0.0163	-0.0150
3,27	-0.0155	0.0143	3,28	0.0147	-0.0137	3,29	-0.0141	0.0131	3,30	0.0135	-0.0126
3,31	-0.0130	0.0121	3,32	0.0126	-0.0117	3,33	-0.0123	0.0114	3,34	0.0120	-0.0110
3,35	-0.0118	0.0108	3,36	0.0118	-0.0106	3,37	-0.0119	0.0104	3,38	0.0124	-0.0104
4, 4	12.2935	-1.0607	4, 5	3.4416	1.0569	4, 6	-0.8576	0.2625	4, 7	0.3833	-0.1748
4, 8	-0.2277	0.1233	4, 9	0.1547	-0.0936	4,10	-0.1139	0.0746	4,11	0.0885	-0.0615
4,12	-0.0713	0.0521	4,13	0.0592	-0.0450	4,14	-0.0502	0.0395	4,15	0.0434	-0.0351
4,16	-0.0380	0.0316	4,17	0.0337	-0.0286	4,18	-0.0302	0.0261	4,19	0.0274	-0.0240
4,20	-0.0249	0.0222	4,21	0.0229	-0.0207	4,22	-0.0211	0.0193	4,23	0.0196	-0.0181
4,24	-0.0183	0.0171	4,25	0.0172	-0.0161	4,26	-0.0161	0.0153	4,27	0.0153	-0.0146
4,28	-0.0145	0.0139	4,29	0.0138	-0.0133	4,30	-0.0132	0.0127	4,31	0.0127	-0.0123
4,32	-0.0123	0.0118	4,33	0.0119	-0.0114	4,34	-0.0116	0.0111	4,35	0.0114	-0.0108
4,36	-0.0113	0.0106	4,37	0.0114	-0.0105						
5, 5	12.3029	-1.0633	5, 6	3.4349	1.0589	5, 7	-0.8525	0.2608	5, 8	0.3793	-0.1734
5, 9	-0.2243	0.1221	5,10	0.1519	-0.0925	5,11	-0.1115	0.0737	5,12	0.0863	-0.0607
5,13	-0.0695	0.0514	5,14	0.0575	-0.0443	5,15	-0.0487	0.0389	5,16	0.0420	-0.0346
5,17	-0.0368	0.0310	5,18	0.0326	-0.0281	5,19	-0.0292	0.0257	5,20	0.0264	-0.0236
5,21	-0.0240	0.0219	5,22	0.0220	-0.0203	5,23	-0.0203	0.0190	5,24	0.0189	-0.0178
5,25	-0.0176	0.0168	5,26	0.0165	-0.0159	5,27	-0.0155	0.0150	5,28	0.0147	-0.0143
5,29	-0.0139	0.0137	5,30	0.0133	-0.0131	5,31	-0.0127	0.0125	5,32	0.0123	-0.0121
5,33	-0.0119	0.0117	5,34	0.0116	-0.0113	5,35	-0.0113	0.0110	5,36	0.0113	-0.0108
6, 6	12.3077	-1.0648	6, 7	3.4313	1.0602	6, 8	-0.8496	0.2597	6, 9	0.3769	-0.1725
6,10	-0.2223	0.1213	6,11	0.1502	-0.0918	6,12	-0.1100	0.0730	6,13	0.0850	-0.0602
6,14	-0.0683	0.0509	6,15	0.0565	-0.0439	6,16	-0.0478	0.0385	6,17	0.0411	-0.0342
6,18	-0.0360	0.0307	6,19	0.0318	-0.0278	6,20	-0.0285	0.0254	6,21	0.0257	-0.0233
6,22	-0.0234	0.0216	6,23	0.0215	-0.0201	6,24	-0.0198	0.0187	6,25	0.0184	-0.0176
6,26	-0.0171	0.0166	6,27	0.0161	-0.0157	6,28	-0.0151	0.0149	6,29	0.0143	-0.0141
6,30	-0.0136	0.0135	6,31	0.0130	-0.0129	6,32	-0.0125	0.0124	6,33	0.0121	-0.0120
6,34	-0.0117	0.0116	6,35	0.0115	-0.0113						
7, 7	12.3104	-1.0659	7, 8	3.4291	1.0610	7, 9	-0.8478	0.2590	7,10	0.3754	-0.1718
7,11	-0.2210	0.1208	7,12	0.1491	-0.0913	7,13	-0.1090	0.0726	7,14	0.0841	-0.0598
7,15	-0.0675	0.0505	7,16	0.0557	-0.0436	7,17	-0.0471	0.0382	7,18	0.0405	-0.0339
7,19	-0.0354	0.0304	7,20	0.0313	-0.0276	7,21	-0.0280	0.0252	7,22	0.0253	-0.0231
7,23	-0.0230	0.0214	7,24	0.0211	-0.0199	7,25	-0.0195	0.0186	7,26	0.0181	-0.0174
7,27	-0.0169	0.0164	7,28	0.0158	-0.0155	7,29	-0.0149	0.0147	7,30	0.0141	-0.0140
7,31	-0.0134	0.0134	7,32	0.0129	-0.0129	7,33	-0.0124	0.0124	7,34	0.0120	-0.0120
8, 8	12.3121	-1.0666	8, 9	3.4276	1.0616	8,10	-0.8466	0.2585	8,11	0.3744	-0.1714
8,12	-0.2202	0.1204	8,13	0.1483	-0.0910	8,14	-0.1083	0.0723	8,15	0.0835	-0.0595
8,16	-0.0669	0.0502	8,17	0.0552	-0.0433	8,18	-0.0466	0.0379	8,19	0.0401	-0.0337
8,20	-0.0350	0.0302	8,21	0.0310	-0.0274	8,22	-0.0277	0.0250	8,23	0.0250	-0.0230
8,24	-0.0227	0.0212	8,25	0.0208	-0.0198	8,26	-0.0192	0.0185	8,27	0.0178	-0.0173
8,28	-0.0167	0.0163	8,29	0.0156	-0.0155	8,30	-0.0148	0.0147	8,31	0.0140	-0.0140
8,32	-0.0134	0.0134	8,33	0.0128	-0.0128						
9, 9	12.3133	-1.0671	9,10	3.4267	1.0621	9,11	-0.8458	0.2581	9,12	0.3736	-0.1711
9,13	-0.2195	0.1200	9,14	0.1477	-0.0907	9,15	-0.1078	0.0720	9,16	0.0830	-0.0593
9,17	-0.0665	0.0500	9,18	0.0549	-0.0431	9,19	-0.0463	0.0378	9,20	0.0398	-0.0335
9,21	-0.0347	0.0301	9,22	0.0307	-0.0272	9,23	-0.0275	0.0249	9,24	0.0248	-0.0229
9,25	-0.0225	0.0211	9,26	0.0207	-0.0197	9,27	-0.0191	0.0184	9,28	0.0177	-0.0173
9,29	-0.0165	0.0163	9,30	0.0155	-0.0154	9,31	-0.0147	0.0146	9,32	0.0140	-0.0140
10,10	12.3141	-1.0674	10,11	3.4260	1.0624	10,12	-0.8452	0.2578	10,13	0.3731	-0.1708
10,14	-0.2190	0.1198	10,15	0.1473	-0.0905	10,16	-0.1074	0.0718	10,17	0.0827	-0.0591
10,18	-0.0662	0.0499	10,19	0.0546	-0.0430	10,20	-0.0460	0.0376	10,21	0.0396	-0.0334
10,22	-0.0345	0.0300	10,23	0.0305	-0.0271	10,24	-0.0273	0.0248	10,25	0.0246	-0.0228
10,26	-0.0224	0.0211	10,27	0.0205	-0.0196	10,28	-0.0190	0.0183	10,29	0.0176	-0.0172
10,30	-0.0165	0.0162	10,31	0.0155	-0.0154						
11,11	12.3147	-1.0677	11,12	3.4254	1.0627	11,13	-0.8447	0.2576	11,14	0.3727	-0.1706
11,15	-0.2187	0.1196	11,16	0.1470	-0.0903	11,17	-0.1071	0.0717	11,18	0.0824	-0.0589
11,19	-0.0660	0.0498	11,20	0.0544	-0.0429	11,21	-0.0459	0.0375	11,22	0.0394	-0.0333
11,23	-0.0344	0.0299	11,24	0.0304	-0.0271	11,25	-0.0272	0.0247	11,26	0.0245	-0.0227
11,27	-0.0223	0.0210	11,28	0.0205	-0.0196	11,29	-0.0189	0.0183	11,30	0.0176	-0.0172
12,12	12.3152	-1.0680	12,13	3.4251	1.0629	12,14	-0.8444	0.2574	12,15	0.3724	-0.1704
12,16	-0.2184	0.1195	12,17	0.1467	-0.0902	12,18	-0.1069	0.0716	12,19	0.0822	-0.0588
12,20	-0.0658	0.0497	12,21	0.0542	-0.0428	12,22	-0.0457	0.0375	12,23	0.0393	-0.0332
12,24	-0.0343	0.0298	12,25	0.0303	-0.0270	12,26	-0.0271	0.0247	12,27	0.0245	-0.0227
12,28	-0.0223	0.0210	12,29	0.0205	-0.0196						
13,13	12.3155	-1.0681	13,14	3.4248	1.0630	13,15	-0.8441	0.2573	13,16	0.3722	-0.1703
13,17	-0.2182	0.1194	13,18	0.1465	-0.0901	13,19	-0.1067	0.0715	13,20	0.0821	-0.0588
13,21	-0.0657	0.0496	13,22	0.0541	-0.0427	13,23	-0.0456	0.0374	13,24	0.0392	-0.0332
13,25	-0.0342	0.0298	13,26	0.0303	-0.0270	13,27	-0.0271	0.0247	13,28	0.0244	-0.0227

I, K	RE	IM	I, K	RE	IM	I, K	RE	IM	I, K	RE	IM
14,14	12.3158	-1.0683	14,15	3.4245	1.0631	14,16	-0.8439	0.2572	14,17	0.3720	-0.1702
14,18	-0.2180	0.1193	14,19	0.1464	-0.0900	14,20	-0.1066	0.0714	14,21	0.0820	-0.0587
14,22	-0.0656	0.0495	14,23	0.0540	-0.0427	14,24	-0.0456	0.0374	14,25	0.0392	-0.0332
14,26	-0.0342	0.0298	14,27	0.0302	-0.0270						
15,15	12.3160	-1.0684	15,16	3.4244	1.0632	15,17	-0.8438	0.2571	15,18	0.3719	-0.1701
15,19	-0.2179	0.1192	15,20	0.1463	-0.0900	15,21	-0.1065	0.0714	15,22	0.0819	-0.0587
15,23	-0.0655	0.0495	15,24	0.0540	-0.0427	15,25	-0.0455	0.0374	15,26	0.0391	-0.0332
16,16	12.3161	-1.0685	16,17	3.4242	1.0633	16,18	-0.8436	0.2570	16,19	0.3718	-0.1701
16,20	-0.2178	0.1192	16,21	0.1462	-0.0899	16,22	-0.1065	0.0713	16,23	0.0819	-0.0586
16,24	-0.0655	0.0495	16,25	0.0540	-0.0427						
17,17	12.3162	-1.0685	17,18	3.4241	1.0633	17,19	-0.8436	0.2570	17,20	0.3717	-0.1701
17,21	-0.2178	0.1192	17,22	0.1462	-0.0899	17,23	-0.1065	0.0713	17,24	0.0819	-0.0586
18,18	12.3163	-1.0686	18,19	3.4241	1.0634	18,20	-0.8435	0.2570	18,21	0.3717	-0.1700
18,22	-0.2178	0.1191	18,23	0.1462	-0.0899						
19,19	12.3164	-1.0686	19,20	3.4240	1.0634	19,21	-0.8435	0.2569	19,22	0.3716	-0.1700
20,20	12.3164	-1.0686	20,21	3.4240	1.0634						

VBR

SELF AND MUTUAL ADMITTANCES

H/LMDA=0.2500 B/LMDA= 0.500 OMEGA= 8.53

45 ELEMENT ARRAY

I, K	YIK IN MMHO RE	IM	I, K	YIK IN MMHO RE	IM	I, K	YIK IN MMHO RE	IM	I, K	YIK IN MMHO RE	IM
1, 1	11.3896	-2.1381	1, 2	3.8302	1.2058	1, 3	-1.0665	0.2041	1, 4	0.5216	-0.1429
1, 5	[illegible]	[illegible]	1, 6	0.2336	-0.0798	1, 7	-0.1779	0.0646	1, 8	0.1420	-0.0540
1, 9	-0.1171	0.0462	1,10	0.0990	-0.0403	1,11	[illegible]	[illegible]	1,12	0.0747	-0.0320
1,13	-0.0662	0.0289	1,14	0.0593	-0.0263	1,15	-0.0536	0.0242	1,16	0.0488	-0.0223
1,17	-0.0448	0.0207	1,18	0.0413	-0.0194	1,19	-0.0383	0.0181	1,20	0.0357	-0.0170
1,21	-0.0334	0.0161	1,22	0.0314	-0.0152	1,23	-0.0296	0.0144	1,24	0.0279	-0.0137
1,25	-0.0265	0.0131	1,26	0.0252	-0.0125	1,27	-0.0240	0.0120	1,28	0.0230	-0.0115
1,29	-0.0220	0.0111	1,30	0.0212	-0.0106	1,31	-0.0204	0.0103	1,32	0.0197	-0.0099
1,33	-0.0190	0.0096	1,34	0.0184	-0.0093	1,35	-0.0179	0.0090	1,36	0.0175	-0.0087
1,37	-0.0171	0.0085	1,38	0.0168	-0.0083	1,39	-0.0166	0.0081	1,40	0.0164	-0.0079
1,41	-0.0165	0.0077	1,42	0.0167	-0.0075	1,43	-0.0173	0.0074	1,44	0.0191	-0.0071
1,45	-0.0258	0.0026									
2, 2	12.1725	-1.0521	2, 3	3.5046	1.0477	2, 4	-0.8987	0.2699	2, 5	0.4131	-0.1809
2, 6	-0.2507	0.1285	2, 7	0.1732	-0.0980	2, 8	-0.1293	0.0785	2, 9	0.1015	-0.0650
2,10	-0.0826	0.0552	2,11	0.0691	-0.0478	2,12	-0.0590	0.0420	2,13	0.0513	-0.0374
2,14	-0.0451	0.0337	2,15	0.0402	-0.0306	2,16	-0.0361	0.0280	2,17	0.0328	-0.0257
2,18	-0.0299	0.0238	2,19	0.0275	-0.0222	2,20	-0.0254	0.0207	2,21	0.0236	-0.0194
2,22	-0.0220	0.0183	2,23	0.0206	-0.0173	2,24	-0.0194	0.0164	2,25	0.0183	-0.0155
2,26	-0.0173	0.0148	2,27	0.0165	-0.0141	2,28	-0.0157	0.0135	2,29	0.0150	-0.0130
2,30	-0.0144	0.0124	2,31	0.0138	-0.0120	2,32	-0.0133	0.0116	2,33	0.0128	-0.0112
2,34	-0.0124	0.0108	2,35	0.0120	-0.0105	2,36	-0.0117	0.0102	2,37	0.0114	-0.0099
2,38	-0.0112	0.0097	2,39	0.0111	-0.0095	2,40	-0.0110	0.0093	2,41	0.0110	-0.0092
2,42	-0.0112	0.0091	2,43	0.0117	-0.0091	2,44	-0.0131	0.0092			
3, 3	12.2699	-1.0560	3, 4	3.4564	1.0534	3, 5	-0.8681	0.2652	3, 6	0.3913	-0.1770
3, 7	-0.2340	0.1251	3, 8	0.1599	-0.0951	3, 9	-0.1183	0.0760	3,10	0.0922	-0.0628
3,11	-0.0746	0.0532	3,12	0.0621	-0.0460	3,13	-0.0528	0.0404	3,14	0.0457	-0.0359
3,15	-0.0401	0.0323	3,16	0.0356	-0.0293	3,17	-0.0319	0.0267	3,18	0.0289	-0.0246
3,19	-0.0263	0.0227	3,20	0.0241	-0.0211	3,21	-0.0223	0.0197	3,22	0.0207	-0.0185
3,23	-0.0193	0.0174	3,24	0.0180	-0.0164	3,25	-0.0169	0.0156	3,26	0.0160	-0.0148
3,27	-0.0151	0.0141	3,28	0.0143	-0.0134	3,29	-0.0137	0.0129	3,30	0.0130	-0.0123
3,31	-0.0125	0.0118	3,32	0.0120	-0.0114	3,33	-0.0115	0.0110	3,34	0.0112	-0.0106
3,35	-0.0108	0.0103	3,36	0.0105	-0.0100	3,37	-0.0102	0.0097	3,38	0.0100	-0.0095
3,39	-0.0099	0.0093	3,40	0.0098	-0.0091	3,41	-0.0098	0.0090	3,42	0.0100	-0.0089
3,43	-0.0105	0.0089									
4, 4	12.2936	-1.0607	4, 5	3.4414	1.0570	4, 6	-0.8575	0.2624	4, 7	0.3832	-0.1747
4, 8	-0.2276	0.1232	4, 9	0.1546	-0.0935	4,10	-0.1138	0.0745	4,11	0.0883	-0.0615
4,12	-0.0712	0.0520	4,13	0.0591	-0.0449	4,14	-0.0501	0.0394	4,15	0.0432	-0.0350
4,16	-0.0379	0.0314	4,17	0.0336	-0.0285	4,18	-0.0301	0.0260	4,19	0.0272	-0.0239
4,20	-0.0247	0.0221	4,21	0.0226	-0.0205	4,22	-0.0209	0.0192	4,23	0.0193	-0.0180
4,24	-0.0180	0.0169	4,25	0.0169	-0.0160	4,26	-0.0158	0.0151	4,27	0.0149	-0.0144
4,28	-0.0141	0.0137	4,29	0.0134	-0.0131	4,30	-0.0127	0.0125	4,31	0.0122	-0.0120
4,32	-0.0117	0.0115	4,33	0.0112	-0.0111	4,34	-0.0108	0.0107	4,35	0.0104	-0.0104
4,36	-0.0101	0.0100	4,37	0.0099	-0.0097	4,38	-0.0097	0.0095	4,39	0.0095	-0.0093
4,40	-0.0094	0.0091	4,41	0.0094	-0.0090	4,42	-0.0096	0.0089			

5, 5	12.3030	-1.0633	5, 6	3.4347	1.0590	5, 7	-0.8524	0.2607	5, 8	0.3791	-0.1733
5, 9	-0.2242	0.1220	5,10	0.1518	-0.0924	5,11	-0.1113	0.0736	5,12	0.0862	-0.0606
5,13	-0.0693	0.0513	5,14	0.0574	-0.0442	5,15	-0.0486	0.0388	5,16	0.0418	-0.0344
5,17	-0.0366	0.0309	5,18	0.0324	-0.0280	5,19	-0.0290	0.0256	5,20	0.0262	-0.0235
5,21	-0.0238	0.0217	5,22	0.0218	-0.0202	5,23	-0.0201	0.0188	5,24	0.0186	-0.0176
5,25	-0.0173	0.0166	5,26	0.0162	-0.0157	5,27	-0.0152	0.0148	5,28	0.0143	-0.0141
5,29	-0.0135	0.0134	5,30	0.0128	-0.0128	5,31	-0.0122	0.0123	5,32	0.0117	-0.0118
5,33	-0.0112	0.0113	5,34	0.0108	-0.0109	5,35	-0.0104	0.0105	5,36	0.0101	-0.0102
5,37	-0.0098	0.0099	5,38	0.0095	-0.0096	5,39	-0.0094	0.0094	5,40	0.0093	-0.0092
5,41	-0.0093	0.0090									
6, 6	12.3078	-1.0649	6, 7	3.4311	1.0603	6, 8	-0.8495	0.2597	6, 9	0.3768	-0.1724
6,10	-0.2222	0.1212	6,11	0.1501	-0.0917	6,12	-0.1098	0.0729	6,13	0.0848	-0.0601
6,14	-0.0681	0.0508	6,15	0.0563	-0.0438	6,16	-0.0476	0.0384	6,17	0.0410	-0.0341
6,18	-0.0358	0.0306	6,19	0.0316	-0.0277	6,20	-0.0283	0.0252	6,21	0.0255	-0.0232
6,22	-0.0232	0.0214	6,23	0.0212	-0.0199	6,24	-0.0195	0.0186	6,25	0.0181	-0.0174
6,26	-0.0168	0.0164	6,27	0.0157	-0.0155	6,28	-0.0148	0.0146	6,29	0.0139	-0.0139
6,30	-0.0132	0.0132	6,31	0.0125	-0.0126	6,32	-0.0119	0.0121	6,33	0.0114	-0.0116
6,34	-0.0109	0.0112	6,35	0.0105	-0.0108	6,36	-0.0101	0.0104	6,37	0.0098	-0.0101
6,38	-0.0096	0.0098	6,39	0.0094	-0.0096	6,40	-0.0093	0.0093			
7, 7	12.3106	-1.0659	7, 8	3.4289	1.0611	7, 9	-0.8477	0.2589	7,10	0.3753	-0.1718
7,11	-0.2209	0.1207	7,12	0.1489	-0.0912	7,13	-0.1088	0.0725	7,14	0.0840	-0.0597
7,15	-0.0673	0.0504	7,16	0.0556	-0.0434	7,17	-0.0469	0.0380	7,18	0.0404	-0.0338
7,19	-0.0352	0.0303	7,20	0.0311	-0.0274	7,21	-0.0278	0.0250	7,22	0.0251	-0.0230
7,23	-0.0228	0.0212	7,24	0.0208	-0.0197	7,25	-0.0192	0.0184	7,26	0.0177	-0.0172
7,27	-0.0165	0.0162	7,28	0.0154	-0.0153	7,29	-0.0145	0.0145	7,30	0.0136	-0.0138
7,31	-0.0129	0.0131	7,32	0.0123	-0.0125	7,33	-0.0117	0.0120	7,34	0.0112	-0.0115
7,35	-0.0107	0.0111	7,36	0.0104	-0.0107	7,37	-0.0100	0.0104	7,38	0.0098	-0.0100
7,39	-0.0095	0.0098									
8, 8	12.3123	-1.0667	8, 9	3.4275	1.0617	8,10	-0.8465	0.2584	8,11	0.3742	-0.1713
8,12	-0.2200	0.1202	8,13	0.1481	-0.0909	8,14	-0.1081	0.0722	8,15	0.0833	-0.0594
8,16	-0.0667	0.0501	8,17	0.0550	-0.0432	8,18	-0.0464	0.0378	8,19	0.0399	-0.0335
8,20	-0.0348	0.0301	8,21	0.0307	-0.0272	8,22	-0.0274	0.0248	8,23	0.0247	-0.0228
8,24	-0.0225	0.0211	8,25	0.0205	-0.0196	8,26	-0.0189	0.0183	8,27	0.0175	-0.0171
8,28	-0.0163	0.0161	8,29	0.0152	-0.0152	8,30	-0.0143	0.0144	8,31	0.0135	-0.0137
8,32	-0.0127	0.0130	8,33	0.0121	-0.0125	8,34	-0.0116	0.0119	8,35	0.0111	-0.0115
8,36	-0.0106	0.0110	8,37	0.0103	-0.0107	8,38	-0.0100	0.0103			
9, 9	12.3135	-1.0672	9,10	3.4265	1.0622	9,11	-0.8456	0.2580	9,12	0.3735	-0.1710
9,13	-0.2193	0.1199	9,14	0.1475	-0.0906	9,15	-0.1076	0.0719	9,16	0.0828	-0.0591
9,17	-0.0663	0.0499	9,18	0.0547	-0.0430	9,19	-0.0461	0.0376	9,20	0.0396	-0.0334
9,21	-0.0345	0.0299	9,22	0.0305	-0.0271	9,23	-0.0272	0.0247	9,24	0.0245	-0.0227
9,25	-0.0222	0.0210	9,26	0.0203	-0.0195	9,27	-0.0187	0.0182	9,28	0.0173	-0.0170
9,29	-0.0161	0.0160	9,30	0.0151	-0.0151	9,31	-0.0141	0.0143	9,32	0.0133	-0.0136
9,33	-0.0126	0.0130	9,34	0.0120	-0.0124	9,35	-0.0115	0.0119	9,36	0.0110	-0.0114
9,37	-0.0106	0.0110									
10,10	12.3143	-1.0675	10,11	3.4258	1.0625	10,12	-0.8450	0.2577	10,13	0.3729	-0.1707
10,14	-0.2189	0.1197	10,15	0.1471	-0.0904	10,16	-0.1072	0.0717	10,17	0.0825	-0.0589
10,18	-0.0660	0.0497	10,19	0.0544	-0.0428	10,20	-0.0458	0.0375	10,21	0.0393	-0.0332
10,22	-0.0343	0.0298	10,23	0.0303	-0.0270	10,24	-0.0270	0.0246	10,25	0.0243	-0.0226
10,26	-0.0221	0.0209	10,27	0.0202	-0.0194	10,28	-0.0186	0.0181	10,29	0.0172	-0.0169
10,30	-0.0160	0.0160	10,31	0.0150	-0.0151	10,32	-0.0141	0.0143	10,33	0.0133	-0.0136
10,34	-0.0126	0.0129	10,35	0.0120	-0.0124	10,36	-0.0115	0.0119			
11,11	12.3149	-1.0678	11,12	3.4253	1.0628	11,13	-0.8445	0.2575	11,14	0.3725	-0.1705
11,15	-0.2185	0.1195	11,16	0.1468	-0.0902	11,17	-0.1069	0.0716	11,18	0.0822	-0.0588
11,19	-0.0657	0.0496	11,20	0.0541	-0.0427	11,21	-0.0456	0.0374	11,22	0.0391	-0.0331
11,23	-0.0341	0.0297	11,24	0.0301	-0.0269	11,25	-0.0268	0.0245	11,26	0.0242	-0.0225
11,27	-0.0219	0.0208	11,28	0.0201	-0.0193	11,29	-0.0185	0.0180	11,30	0.0171	-0.0169
11,31	-0.0159	0.0159	11,32	0.0149	-0.0150	11,33	-0.0140	0.0143	11,34	0.0132	-0.0136
11,35	-0.0126	0.0129									
12,12	12.3154	-1.0681	12,13	3.4249	1.0630	12,14	-0.8442	0.2573	12,15	0.3722	-0.1703
12,16	-0.2182	0.1194	12,17	0.1465	-0.0901	12,18	-0.1067	0.0714	12,19	0.0820	-0.0587
12,20	-0.0655	0.0495	12,21	0.0540	-0.0426	12,22	-0.0454	0.0373	12,23	0.0390	-0.0331
12,24	-0.0340	0.0296	12,25	0.0300	-0.0268	12,26	-0.0267	0.0244	12,27	0.0241	-0.0225
12,28	-0.0219	0.0207	12,29	0.0200	-0.0193	12,30	-0.0184	0.0180	12,31	0.0170	-0.0169
12,32	-0.0159	0.0159	12,33	0.0149	-0.0150	12,34	-0.0140	0.0142			
13,13	12.3157	-1.0683	13,14	3.4246	1.0631	13,15	-0.8439	0.2572	13,16	0.3720	-0.1702
13,17	-0.2180	0.1192	13,18	0.1463	-0.0900	13,19	-0.1065	0.0713	13,20	0.0818	-0.0586
13,21	-0.0654	0.0494	13,22	0.0538	-0.0425	13,23	-0.0453	0.0372	13,24	0.0389	-0.0330
13,25	-0.0339	0.0296	13,26	0.0299	-0.0268	13,27	-0.0266	0.0244	13,28	0.0240	-0.0224
13,29	-0.0218	0.0207	13,30	0.0199	-0.0192	13,31	-0.0184	0.0180	13,32	0.0170	-0.0169
13,33	-0.0159	0.0159									
14,14	12.3160	-1.0684	14,15	3.4243	1.0633	14,16	-0.8437	0.2570	14,17	0.3718	-0.1701
14,18	-0.2178	0.1191	14,19	0.1462	-0.0899	14,20	-0.1064	0.0713	14,21	0.0817	-0.0585
14,22	-0.0653	0.0494	14,23	0.0537	-0.0425	14,24	-0.0452	0.0372	14,25	0.0388	-0.0329
14,26	-0.0338	0.0295	14,27	0.0298	-0.0267	14,28	-0.0266	0.0244	14,29	0.0240	-0.0224
14,30	-0.0218	0.0207	14,31	0.0199	-0.0192	14,32	-0.0183	0.0180			
15,15	12.3162	-1.0685	15,16	3.4241	1.0634	15,17	-0.8435	0.2569	15,18	0.3716	-0.1700
15,19	-0.2177	0.1191	15,20	0.1460	-0.0898	15,21	-0.1062	0.0712	15,22	0.0816	-0.0585
15,23	-0.0652	0.0493	15,24	0.0536	-0.0424	15,25	-0.0452	0.0371	15,26	0.0387	-0.0329
15,27	-0.0337	0.0295	15,28	0.0298	-0.0267	15,29	-0.0266	0.0244	15,30	0.0239	-0.0224
15,31	-0.0217	0.0207									
16,16	12.3164	-1.0686	16,17	3.4240	1.0635	16,18	-0.8434	0.2569	16,19	0.3715	-0.1699
16,20	-0.2176	0.1190	16,21	0.1459	-0.0897	16,22	-0.1062	0.0711	16,23	0.0815	-0.0584
16,24	-0.0651	0.0493	16,25	0.0536	-0.0424	16,26	-0.0451	0.0371	16,27	0.0387	-0.0329
16,28	-0.0337	0.0295	16,29	0.0298	-0.0267	16,30	-0.0266	0.0243			
17,17	12.3165	-1.0687	17,18	3.4239	1.0635	17,19	-0.8433	0.2568	17,20	0.3714	-0.1699
17,21	-0.2175	0.1190	17,22	0.1459	-0.0897	17,23	-0.1061	0.0711	17,24	0.0815	-0.0584
17,25	-0.0651	0.0492	17,26	0.0535	-0.0424	17,27	-0.0451	0.0371	17,28	0.0387	-0.0329
17,29	-0.0337	0.0295									
18,18	12.3166	-1.0687	18,19	3.4238	1.0636	18,20	-0.8432	0.2568	18,21	0.3713	-0.1698
18,22	-0.2174	0.1189	18,23	0.1458	-0.0897	18,24	-0.1061	0.0711	18,25	0.0814	-0.0584
18,26	-0.0651	0.0492	18,27	0.0535	-0.0424	18,28	-0.0451	0.0371			
19,19	12.3167	-1.0688	19,20	3.4237	1.0636	19,21	-0.8431	0.2567	19,22	0.3713	-0.1698
19,23	-0.2174	0.1189	19,24	0.1458	-0.0897	19,25	-0.1060	0.0711	19,26	0.0814	-0.0584
19,27	-0.0650	0.0492									
20,20	12.3167	-1.0688	20,21	3.4236	1.0636	20,22	-0.8431	0.2567	20,23	0.3712	-0.1698
20,24	-0.2174	0.1189	20,25	0.1458	-0.0896	20,26	-0.1060	0.0711			
21,21	12.3168	-1.0689	21,22	3.4236	1.0637	21,23	-0.8431	0.2567	21,24	0.3712	-0.1698
21,25	-0.2173	0.1189									
22,22	12.3168	-1.0689	22,23	3.4236	1.0637	22,24	-0.8431	0.2567			
23,23	12.3168	-1.0689									

VBR

SELF AND MUTUAL ADMITTANCES

H/LMDA=0.2500 B/LMDA= 0.500 OMEGA= 8.53

50 ELEMENT ARRAY

I, K	YIK IN MMHO RE	IM	I, K	YIK IN MMHO RE	IM	I, K	YIK IN MMHO RE	IM	I, K	YIK IN MMHO RE	IM
1, 1	11.3898	-2.1381	1, 2	3.8300	1.2058	1, 3	-1.0663	0.2041	1, 4	0.5214	-0.1429
1, 5	-0.3288	0.1032	1, 6	0.2335	-0.0798	1, 7	-0.1778	0.0646	1, 8	0.1418	-0.0540
1, 9	-0.1169	0.0462	1,10	0.0588	-0.0403	1,11	-0.0851	0.0356	1,12	0.0745	-0.0319
1,13	-0.0660	0.0288	1,14	0.0591	-0.0263	1,15	-0.0534	0.0241	1,16	0.0486	-0.0223
1,17	-0.0446	0.0207	1,18	0.0411	-0.0193	1,19	-0.0381	0.0181	1,20	0.0354	-0.0170
1,21	-0.0331	0.0160	1,22	0.0311	-0.0152	1,23	-0.0293	0.0144	1,24	0.0277	-0.0137
1,25	-0.0262	0.0130	1,26	0.0249	-0.0124	1,27	-0.0237	0.0119	1,28	0.0226	-0.0114
1,29	-0.0216	0.0110	1,30	0.0207	-0.0106	1,31	-0.0199	0.0102	1,32	0.0192	-0.0098
1,33	-0.0185	0.0095	1,34	0.0179	-0.0092	1,35	-0.0173	0.0089	1,36	0.0168	-0.0086
1,37	-0.0163	0.0084	1,38	0.0159	-0.0081	1,39	-0.0155	0.0079	1,40	0.0151	-0.0077
1,41	-0.0148	0.0075	1,42	0.0146	-0.0073	1,43	-0.0144	0.0072	1,44	0.0142	-0.0070
1,45	-0.0142	0.0069	1,46	0.0143	-0.0068	1,47	-0.0145	0.0066	1,48	0.0152	-0.0065
1,49	-0.0168	0.0063	1,50	0.0227	-0.0023						
2, 2	12.1727	-1.0522	2, 3	3.5045	1.0478	2, 4	-0.8986	0.2699	2, 5	0.4130	-0.1809
2, 6	-0.2505	0.1284	2, 7	0.1731	-0.0980	2, 8	-0.1292	0.0784	2, 9	0.1014	-0.0649
2,10	-0.0825	0.0551	2,11	0.0690	-0.0477	2,12	-0.0589	0.0420	2,13	0.0511	-0.0374
2,14	-0.0450	0.0336	2,15	0.0401	-0.0305	2,16	-0.0360	0.0279	2,17	0.0326	-0.0257
2,18	-0.0298	0.0237	2,19	0.0273	-0.0221	2,20	-0.0253	0.0206	2,21	0.0234	-0.0193
2,22	-0.0218	0.0182	2,23	0.0204	-0.0172	2,24	-0.0192	0.0163	2,25	0.0181	-0.0154
2,26	-0.0171	0.0147	2,27	0.0162	-0.0140	2,28	-0.0154	0.0134	2,29	0.0147	-0.0128
2,30	-0.0141	0.0123	2,31	0.0135	-0.0118	2,32	-0.0129	0.0114	2,33	0.0124	-0.0110
2,34	-0.0120	0.0106	2,35	0.0116	-0.0103	2,36	-0.0112	0.0099	2,37	0.0109	-0.0096
2,38	-0.0105	0.0094	2,39	0.0103	-0.0091	2,40	-0.0100	0.0089	2,41	0.0098	-0.0087
2,42	-0.0097	0.0085	2,43	0.0095	-0.0083	2,44	-0.0095	0.0082	2,45	0.0094	-0.0081
2,46	-0.0095	0.0080	2,47	0.0097	-0.0079	2,48	-0.0102	0.0080	2,49	0.0115	-0.0081
3, 3	12.2700	-1.0561	3, 4	3.4563	1.0535	3, 5	-0.8680	0.2651	3, 6	0.3912	-0.1769
3, 7	-0.2339	0.1251	3, 8	0.1598	-0.0951	3, 9	-0.1182	0.0759	3,10	0.0921	-0.0627
3,11	-0.0745	0.0531	3,12	0.0620	-0.0459	3,13	-0.0527	0.0403	3,14	0.0455	-0.0358
3,15	-0.0400	0.0322	3,16	0.0355	-0.0292	3,17	-0.0318	0.0267	3,18	0.0287	-0.0245
3,19	-0.0262	0.0227	3,20	0.0240	-0.0210	3,21	-0.0221	0.0196	3,22	0.0205	-0.0184
3,23	-0.0191	0.0173	3,24	0.0178	-0.0163	3,25	-0.0167	0.0155	3,26	0.0158	-0.0147
3,27	-0.0149	0.0140	3,28	0.0141	-0.0133	3,29	-0.0134	0.0127	3,30	0.0128	-0.0122
3,31	-0.0122	0.0117	3,32	0.0117	-0.0112	3,33	-0.0112	0.0108	3,34	0.0108	-0.0104
3,35	-0.0104	0.0101	3,36	0.0100	-0.0098	3,37	-0.0097	0.0095	3,38	0.0094	-0.0092
3,39	-0.0092	0.0089	3,40	0.0089	-0.0087	3,41	-0.0087	0.0085	3,42	0.0086	-0.0083
3,43	-0.0085	0.0081	3,44	0.0084	-0.0080	3,45	-0.0084	0.0079	3,46	0.0084	-0.0078
3,47	-0.0086	0.0077	3,48	0.0090	-0.0078						
4, 4	12.2937	-1.0608	4, 5	3.4413	1.0570	4, 6	-0.8574	0.2623	4, 7	0.3831	-0.1746
4, 8	-0.2274	0.1231	4, 9	0.1545	-0.0934	4,10	-0.1137	0.0744	4,11	0.0882	-0.0614
4,12	-0.0711	0.0520	4,13	0.0589	-0.0449	4,14	-0.0500	0.0393	4,15	0.0431	-0.0349
4,16	-0.0377	0.0314	4,17	0.0334	-0.0284	4,18	-0.0299	0.0259	4,19	0.0270	-0.0238
4,20	-0.0246	0.0220	4,21	0.0225	-0.0205	4,22	-0.0207	0.0191	4,23	0.0192	-0.0179
4,24	-0.0178	0.0168	4,25	0.0167	-0.0159	4,26	-0.0156	0.0150	4,27	0.0147	-0.0142
4,28	-0.0139	0.0135	4,29	0.0132	-0.0129	4,30	-0.0125	0.0123	4,31	0.0119	-0.0118
4,32	-0.0114	0.0113	4,33	0.0109	-0.0109	4,34	-0.0104	0.0105	4,35	0.0100	-0.0101
4,36	-0.0097	0.0098	4,37	0.0093	-0.0095	4,38	-0.0091	0.0092	4,39	0.0088	-0.0089
4,40	-0.0086	0.0087	4,41	0.0084	-0.0085	4,42	-0.0082	0.0083	4,43	0.0081	-0.0081
4,44	-0.0080	0.0079	4,45	0.0080	-0.0078	4,46	-0.0080	0.0077	4,47	0.0082	-0.0077
5, 5	12.3031	-1.0634	5, 6	3.4346	1.0591	5, 7	-0.8523	0.2607	5, 8	0.3790	-0.1732
5, 9	-0.2241	0.1219	5,10	0.1517	-0.0924	5,11	-0.1112	0.0735	5,12	0.0861	-0.0606
5,13	-0.0692	0.0512	5,14	0.0573	-0.0442	5,15	-0.0484	0.0387	5,16	0.0417	-0.0344
5,17	-0.0365	0.0308	5,18	0.0323	-0.0279	5,19	-0.0288	0.0255	5,20	0.0260	-0.0234
5,21	-0.0236	0.0216	5,22	0.0216	-0.0201	5,23	-0.0199	0.0187	5,24	0.0184	-0.0175
5,25	-0.0171	0.0165	5,26	0.0160	-0.0155	5,27	-0.0150	0.0147	5,28	0.0141	-0.0140
5,29	-0.0133	0.0133	5,30	0.0126	-0.0127	5,31	-0.0120	0.0121	5,32	0.0114	-0.0116
5,33	-0.0109	0.0111	5,34	0.0104	-0.0107	5,35	-0.0100	0.0103	5,36	0.0096	-0.0099
5,37	-0.0093	0.0096	5,38	0.0090	-0.0093	5,39	-0.0087	0.0090	5,40	0.0085	-0.0088
5,41	-0.0082	0.0085	5,42	0.0081	-0.0083	5,43	-0.0079	0.0081	5,44	0.0078	-0.0080
5,45	-0.0078	0.0079	5,46	0.0078	-0.0078						
6, 6	12.3079	-1.0650	6, 7	3.4310	1.0603	6, 8	-0.8494	0.2596	6, 9	0.3767	-0.1723
6,10	-0.2221	0.1211	6,11	0.1499	-0.0917	6,12	-0.1097	0.0729	6,13	0.0847	-0.0600
6,14	-0.0680	0.0507	6,15	0.0562	-0.0437	6,16	-0.0475	0.0383	6,17	0.0408	-0.0340
6,18	-0.0357	0.0305	6,19	0.0315	-0.0276	6,20	-0.0281	0.0251	6,21	0.0254	-0.0231
6,22	-0.0230	0.0213	6,23	0.0211	-0.0198	6,24	-0.0194	0.0185	6,25	0.0179	-0.0173
6,26	-0.0166	0.0163	6,27	0.0155	-0.0153	6,28	-0.0145	0.0145	6,29	0.0137	-0.0138
6,30	-0.0129	0.0131	6,31	0.0122	-0.0125	6,32	-0.0116	0.0119	6,33	0.0111	-0.0114
6,34	-0.0106	0.0110	6,35	0.0101	-0.0105	6,36	-0.0097	0.0102	6,37	0.0093	-0.0098
6,38	-0.0090	0.0095	6,39	0.0087	-0.0092	6,40	-0.0085	0.0089	6,41	0.0083	-0.0087
6,42	-0.0081	0.0085	6,43	0.0079	-0.0083	6,44	-0.0078	0.0081	6,45	0.0078	-0.0080
7, 7	12.3106	-1.0660	7, 8	3.4288	1.0612	7, 9	-0.8476	0.2589	7,10	0.3751	-0.1717
7,11	-0.2208	0.1206	7,12	0.1488	-0.0912	7,13	-0.1087	0.0724	7,14	0.0838	-0.0596
7,15	-0.0672	0.0503	7,16	0.0554	-0.0434	7,17	-0.0468	0.0380	7,18	0.0402	-0.0337
7,19	-0.0351	0.0302	7,20	0.0310	-0.0273	7,21	-0.0277	0.0249	7,22	0.0249	-0.0229
7,23	-0.0226	0.0211	7,24	0.0206	-0.0196	7,25	-0.0190	0.0183	7,26	0.0175	-0.0171
7,27	-0.0163	0.0161	7,28	0.0152	-0.0152	7,29	-0.0142	0.0143	7,30	0.0134	-0.0136
7,31	-0.0126	0.0129	7,32	0.0120	-0.0123	7,33	-0.0114	0.0118	7,34	0.0108	-0.0113
7,35	-0.0103	0.0109	7,36	0.0099	-0.0104	7,37	-0.0095	0.0101	7,38	0.0092	-0.0097
7,39	-0.0089	0.0094	7,40	0.0086	-0.0091	7,41	-0.0083	0.0089	7,42	0.0081	-0.0086
7,43	-0.0080	0.0084	7,44	0.0079	-0.0082						
8, 8	12.3124	-1.0667	8, 9	3.4274	1.0618	8,10	-0.8464	0.2583	8,11	0.3741	-0.1712
8,12	-0.2199	0.1202	8,13	0.1480	-0.0908	8,14	-0.1080	0.0721	8,15	0.0832	-0.0593
8,16	-0.0666	0.0500	8,17	0.0549	-0.0431	8,18	-0.0463	0.0377	8,19	0.0398	-0.0334
8,20	-0.0347	0.0300	8,21	0.0306	-0.0271	8,22	-0.0273	0.0247	8,23	0.0246	-0.0227
8,24	-0.0223	0.0210	8,25	0.0204	-0.0194	8,26	-0.0187	0.0181	8,27	0.0173	-0.0170
8,28	-0.0161	0.0160	8,29	0.0150	-0.0150	8,30	-0.0140	0.0142	8,31	0.0132	-0.0135
8,32	-0.0124	0.0128	8,33	0.0118	-0.0123	8,34	-0.0112	0.0117	8,35	0.0107	-0.0112
8,36	-0.0102	0.0108	8,37	0.0098	-0.0104	8,38	-0.0094	0.0100	8,39	0.0091	-0.0097
8,40	-0.0088	0.0094	8,41	0.0085	-0.0091	8,42	-0.0083	0.0088	8,43	0.0081	-0.0086
9, 9	12.3136	-1.0672	9,10	3.4264	1.0622	9,11	-0.8455	0.2579	9,12	0.3734	-0.1709
9,13	-0.2192	0.1199	9,14	0.1474	-0.0905	9,15	-0.1075	0.0718	9,16	0.0827	-0.0590
9,17	-0.0662	0.0498	9,18	0.0545	-0.0429	9,19	-0.0459	0.0375	9,20	0.0394	-0.0333
9,21	-0.0344	0.0298	9,22	0.0303	-0.0270	9,23	-0.0270	0.0246	9,24	0.0243	-0.0226
9,25	-0.0220	0.0208	9,26	0.0201	-0.0193	9,27	-0.0185	0.0180	9,28	0.0171	-0.0169
9,29	-0.0159	0.0159	9,30	0.0148	-0.0150	9,31	-0.0139	0.0141	9,32	0.0130	-0.0134
9,33	-0.0123	0.0128	9,34	0.0117	-0.0122	9,35	-0.0111	0.0116	9,36	0.0106	-0.0112
9,37	-0.0101	0.0107	9,38	0.0097	-0.0103	9,39	-0.0093	0.0100	9,40	0.0090	-0.0096
9,41	-0.0087	0.0093	9,42	0.0085	-0.0091						

10,10	12.3144	-1.0676	10,11	3.4257	1.0626	10,12	-0.8449	0.2576	10,13	0.3728	-0.1706
10,14	-0.2187	0.1196	10,15	0.1470	-0.0903	10,16	-0.1071	0.0716	10,17	0.0824	-0.0589
10,18	-0.0659	0.0496	10,19	0.0542	-0.0427	10,20	-0.0457	0.0374	10,21	0.0392	-0.0331
10,22	-0.0341	0.0297	10,23	0.0301	-0.0268	10,24	-0.0268	0.0245	10,25	0.0241	-0.0225
10,26	-0.0219	0.0207	10,27	0.0200	-0.0192	10,28	-0.0183	0.0179	10,29	0.0169	-0.0168
10,30	-0.0157	0.0158	10,31	0.0147	-0.0149	10,32	-0.0137	0.0141	10,33	0.0129	-0.0134
10,34	-0.0122	0.0127	10,35	0.0116	-0.0121	10,36	-0.0110	0.0116	10,37	0.0105	-0.0111
10,38	-0.0100	0.0107	10,39	0.0096	-0.0103	10,40	-0.0093	0.0099	10,41	0.0090	-0.0096
11,11	12.3150	-1.0679	11,12	3.4252	1.0628	11,13	-0.8444	0.2574	11,14	0.3724	-0.1704
11,15	-0.2184	0.1194	11,16	0.1466	-0.0901	11,17	-0.1068	0.0715	11,18	0.0821	-0.0587
11,19	-0.0656	0.0495	11,20	0.0540	-0.0426	11,21	-0.0454	0.0373	11,22	0.0390	-0.0330
11,23	-0.0339	0.0296	11,24	0.0299	-0.0268	11,25	-0.0266	0.0244	11,26	0.0240	-0.0224
11,27	-0.0217	0.0206	11,28	0.0198	-0.0192	11,29	-0.0182	0.0179	11,30	0.0168	-0.0167
11,31	-0.0156	0.0157	11,32	0.0146	-0.0148	11,33	-0.0137	0.0140	11,34	0.0129	-0.0133
11,35	-0.0121	0.0127	11,36	0.0115	-0.0121	11,37	-0.0109	0.0116	11,38	0.0104	-0.0111
11,39	-0.0100	0.0107	11,40	0.0096	-0.0103						
12,12	12.3155	-1.0682	12,13	3.4248	1.0631	12,14	-0.8441	0.2572	12,15	0.3721	-0.1702
12,16	-0.2181	0.1193	12,17	0.1464	-0.0900	12,18	-0.1065	0.0713	12,19	0.0819	-0.0586
12,20	-0.0654	0.0494	12,21	0.0538	-0.0425	12,22	-0.0453	0.0372	12,23	0.0388	-0.0329
12,24	-0.0338	0.0295	12,25	0.0298	-0.0267	12,26	-0.0265	0.0243	12,27	0.0238	-0.0223
12,28	-0.0216	0.0206	12,29	0.0197	-0.0191	12,30	0.0181	0.0178	12,31	0.0167	-0.0167
12,32	-0.0155	0.0157	12,33	0.0145	-0.0148	12,34	-0.0136	0.0140	12,35	0.0128	-0.0133
12,36	-0.0121	0.0127	12,37	0.0115	-0.0121	12,38	-0.0109	0.0116	12,39	0.0104	-0.0111
13,13	12.3158	-1.0683	13,14	3.4244	1.0632	13,15	-0.8438	0.2571	13,16	0.3718	-0.1701
13,17	-0.2178	0.1191	13,18	0.1462	-0.0899	13,19	-0.1063	0.0712	13,20	0.0817	-0.0585
13,21	-0.0652	0.0493	13,22	0.0536	-0.0424	13,23	-0.0451	0.0371	13,24	0.0387	-0.0329
13,25	-0.0337	0.0294	13,26	0.0297	-0.0266	13,27	-0.0264	0.0242	13,28	0.0238	-0.0222
13,29	-0.0215	0.0205	13,30	0.0197	-0.0191	13,31	-0.0181	0.0178	13,32	0.0167	-0.0166
13,33	-0.0155	0.0156	13,34	0.0145	-0.0148	13,35	-0.0136	0.0140	13,36	0.0128	-0.0133
13,37	-0.0121	0.0126	13,38	0.0115	-0.0121						
14,14	12.3161	-1.0685	14,15	3.4242	1.0634	14,16	-0.8436	0.2569	14,17	0.3716	-0.1700
14,18	-0.2177	0.1190	14,19	0.1460	-0.0898	14,20	-0.1062	0.0711	14,21	0.0815	-0.0584
14,22	-0.0651	0.0492	14,23	0.0535	-0.0424	14,24	-0.0450	0.0370	14,25	0.0386	-0.0328
14,26	-0.0336	0.0294	14,27	0.0296	-0.0266	14,28	-0.0263	0.0242	14,29	0.0237	-0.0222
14,30	-0.0215	0.0205	14,31	0.0196	-0.0190	14,32	-0.0180	0.0177	14,33	0.0166	-0.0166
14,34	-0.0155	0.0156	14,35	0.0144	-0.0147	14,36	-0.0135	0.0140	14,37	0.0128	-0.0133
15,15	12.3163	-1.0686	15,16	3.4240	1.0635	15,17	-0.8434	0.2568	15,18	0.3715	-0.1699
15,19	-0.2175	0.1190	15,20	0.1459	-0.0897	15,21	-0.1061	0.0711	15,22	0.0814	-0.0583
15,23	-0.0650	0.0492	15,24	0.0534	-0.0423	15,25	-0.0449	0.0370	15,26	0.0385	-0.0328
15,27	-0.0335	0.0293	15,28	0.0295	-0.0265	15,29	-0.0263	0.0242	15,30	0.0236	-0.0222
15,31	-0.0214	0.0205	15,32	0.0196	-0.0190	15,33	-0.0180	0.0177	15,34	0.0166	-0.0166
15,35	-0.0154	0.0156	15,36	0.0144	-0.0147						
16,16	12.3165	-1.0687	16,17	3.4238	1.0636	16,18	-0.8432	0.2568	16,19	0.3713	-0.1698
16,20	-0.2174	0.1189	16,21	0.1458	-0.0896	16,22	-0.1060	0.0710	16,23	0.0813	-0.0583
16,24	-0.0649	0.0491	16,25	0.0534	-0.0423	16,26	-0.0449	0.0369	16,27	0.0384	-0.0327
16,28	-0.0334	0.0293	16,29	0.0295	-0.0265	16,30	-0.0262	0.0241	16,31	0.0236	-0.0222
16,32	-0.0214	0.0205	16,33	0.0195	-0.0190	16,34	-0.0180	0.0177	16,35	0.0166	-0.0166
17,17	12.3167	-1.0688	17,18	3.4237	1.0636	17,19	-0.8431	0.2567	17,20	0.3712	-0.1697
17,21	-0.2173	0.1188	17,22	0.1457	-0.0896	17,23	-0.1059	0.0710	17,24	0.0813	-0.0582
17,25	-0.0648	0.0491	17,26	0.0533	-0.0422	17,27	-0.0448	0.0369	17,28	0.0384	-0.0327
17.29	-0.0334	0.0293	17,30	0.0294	-0.0265	17,31	-0.0262	0.0241	17,32	0.0236	-0.0221
17.33	-0.0214	0.0204	17,34	0.0195	-0.0190						
18,18	12.3168	-1.0689	18,19	3.4236	1.0637	18,20	-0.8430	0.2566	18,21	0.3711	-0.1697
18,22	-0.2172	0.1188	18,23	0.1456	-0.0895	18,24	-0.1058	0.0709	18,25	0.0812	-0.0582
18,26	-0.0648	0.0491	18,27	0.0532	-0.0422	18,28	-0.0448	0.0369	18,29	0.0384	-0.0327
18,30	-0.0334	0.0293	18,31	0.0294	-0.0265	18,32	-0.0262	0.0241	18,33	0.0236	-0.0221
19,19	12.3169	-1.0689	19,20	3.4235	1.0638	19,21	-0.8429	0.2566	19,22	0.3711	-0.1696
19,23	-0.2171	0.1187	19,24	0.1455	-0.0895	19,25	-0.1058	0.0709	19,26	0.0812	-0.0582
19,27	-0.0648	0.0490	19,28	0.0532	-0.0422	19,29	-0.0448	0.0369	19,30	0.0383	-0.0327
19,31	-0.0334	0.0293	19,32	0.0294	-0.0265						
20,20	12.3170	-1.0690	20,21	3.4234	1.0638	20,22	-0.8429	0.2565	20,23	0.3710	-0.1696
20,24	-0.2171	0.1187	20,25	0.1455	-0.0895	20,26	-0.1057	0.0709	20,27	0.0811	-0.0582
20,28	-0.0647	0.0490	20,29	0.0532	-0.0422	20,30	-0.0447	0.0369	20,31	0.0383	-0.0327
21,21	12.3170	-1.0690	21,22	3.4234	1.0638	21,23	-0.8428	0.2565	21,24	0.3710	-0.1696
21,25	-0.2171	0.1187	21,26	0.1455	-0.0894	21,27	-0.1057	0.0709	21,28	0.0811	-0.0582
21,29	-0.0647	0.0490	21,30	0.0532	-0.0422						
22,22	12.3171	-1.0690	22,23	3.4233	1.0639	22,24	-0.8428	0.2565	22,25	0.3709	-0.1696
22,26	-0.2170	0.1187	22,27	0.1455	-0.0894	22,28	-0.1057	0.0709	22,29	0.0811	-0.0582
23,23	12.3171	-1.0691	23,24	3.4233	1.0639	23,25	-0.8428	0.2565	23,26	0.3709	-0.1696
23,27	-0.2170	0.1187	23,28	0.1454	-0.0894						
24,24	12.3171	-1.0691	24,25	3.4233	1.0639	24,26	-0.8428	0.2565	24,27	0.3709	-0.1696
25,25	12.3171	-1.0691	25,26	3.4233	1.0639						

VBR

SELF AND MUTUAL ADMITTANCES

H/LMDA=0.5000 B/LMDA= 0.250 OMEGA= 9.92

2 ELEMENT ARRAY

I, K	YIK IN MMHO		I, K	YIK IN MMHO		I, K	YIK IN MMHO		I, K	YIK IN MMHO	
	RE	IM		RE	IM		RE	IM		RE	IM
1, 1	1.0267	1.8759	1, 2	0.6091	0.0835						

VBR

SELF AND MUTUAL ADMITTANCES

H/LMDA=0.5000 B/LMDA= 0.250 OMEGA= 9.92

3 ELEMENT ARRAY

I, K	YIK IN MMHO		I, K	YIK IN MMHO		I, K	YIK IN MMHO		I, K	YIK IN MMHO	
	RE	IM		RE	IM		RE	IM		RE	IM
1, 1	1.0400	1.8494	1, 2	0.6838	0.0953	1, 3	0.1177	-0.2443			
2, 2	1.0339	2.0707									

VBR

SELF AND MUTUAL ADMITTANCES

H/LMDA=0.5000 B/LMDA= 0.250 OMEGA= 9.92

4 ELEMENT ARRAY

I, K	YIK IN MMHO		I, K	YIK IN MMHO		I, K	YIK IN MMHO		I, K	YIK IN MMHO	
	RE	IM		RE	IM		RE	IM		RE	IM
1, 1	1.0333	1.8560	1, 2	0.6680	0.0835	1, 3	0.1315	-0.2844	1, 4	-0.1252	-0.0831
2, 2	1.0514	2.0348	2, 3	0.7676	0.1165						

VBR

SELF AND MUTUAL ADMITTANCES

H/LMDA=0.5000 B/LMDA= 0.250 OMEGA= 9.92

5 ELEMENT ARRAY

I, K	YIK IN MMHO		I, K	YIK IN MMHO		I, K	YIK IN MMHO		I, K	YIK IN MMHO	
	RE	IM		RE	IM		RE	IM		RE	IM
1, 1	1.0365	1.8539	1, 2	0.6722	0.0890	1, 3	0.1233	-0.2758	1, 4	-0.1489	-0.0953
1, 5	-0.0584	0.0767									
2, 2	1.0423	2.0442	2, 3	0.7495	0.1035	2, 4	0.1502	-0.3298			
3, 3	1.0667	2.0011									

VBR

SELF AND MUTUAL ADMITTANCES

H/LMDA=0.5000 B/LMDA= 0.250 OMEGA= 9.92

6 ELEMENT ARRAY

I, K	YIK IN MMHO		I, K	YIK IN MMHO		I, K	YIK IN MMHO		I, K	YIK IN MMHO	
	RE	IM		RE	IM		RE	IM		RE	IM
1, 1	1.0349	1.8548	1, 2	0.6707	0.0864	1, 3	0.1270	-0.2783	1, 4	-0.1436	-0.0893
1, 5	-0.0675	0.0919	1, 6	0.0529	0.0425						
2, 2	1.0466	2.0411	2, 3	0.7545	0.1096	2, 4	0.1404	-0.3194	2, 5	-0.1774	-0.1091
3, 3	1.0586	2.0098	3, 4	0.7318	0.0922						

VBR

SELF AND MUTUAL ADMITTANCES

H/LMDA=0.5000 B/LMDA= 0.250 OMEGA= 9.92

7 ELEMENT ARRAY

I, K	YIK IN MMHO		I, K	YIK IN MMHO		I, K	YIK IN MMHO		I, K	YIK IN MMHO	
	RE	IM		RE	IM		RE	IM		RE	IM
1, 1	1.0357	1.8543	1, 2	0.6714	0.0877	1, 3	0.1252	-0.2773	1, 4	-0.1453	-0.0920
1, 5	-0.0632	0.0882	1, 6	0.0635	0.0492	1, 7	0.0323	-0.0396			
2, 2	1.0444	2.0424	2, 3	0.7527	0.1068	2, 4	0.1448	-0.3225	2, 5	-0.1708	-0.1022
2, 6	-0.0778	0.1117									
3, 3	1.0625	2.0070	3, 4	0.7366	0.0981	3, 5	0.1325	-0.3086			
4, 4	1.0500	2.0188									

VBR

SELF AND MUTUAL ADMITTANCES

H/LMDA=0.5000 B/LMDA= 0.250 OMEGA= 9.92

8 ELEMENT ARRAY

I, K	YIK IN MMHO		I, K	YIK IN MMHO		I, K	YIK IN MMHO		I, K	YIK IN MMHO	
	RE	IM		RE	IM		RE	IM		RE	IM
1, 1	1.0352	1.8546	1, 2	0.6710	0.0870	1, 3	0.1261	-0.2778	1, 4	-0.1445	-0.0907
1, 5	-0.0652	0.0895	1, 6	0.0607	0.0460	1, 7	0.0372	-0.0477	1, 8	-0.0314	-0.0255
2, 2	1.0456	2.0417	2, 3	0.7535	0.1082	2, 4	0.1427	-0.3212	2, 5	-0.1730	-0.1053
2, 6	-0.0728	0.1071	2, 7	0.0784	0.0570						
3, 3	1.0605	2.0082	3, 4	0.7349	0.0953	3, 5	0.1366	-0.3115	3, 6	-0.1634	-0.0964
4, 4	1.0541	2.0159	4, 5	0.7420	0.1039						

VBR

SELF AND MUTUAL ADMITTANCES

H/LMDA=0.5000 B/LMDA= 0.250 OMEGA= 9.92

9 ELEMENT ARRAY

I, K	YIK IN MMHO		I, K	YIK IN MMHO		I, K	YIK IN MMHO		I, K	YIK IN MMHO	
	RE	IM		RE	IM		RE	IM		RE	IM
1, 1	1.0356	1.8544	1, 2	0.6713	0.0875	1, 3	0.1256	-0.2775	1, 4	-0.1450	-0.0914
1, 5	-0.0642	0.0889	1, 6	0.0618	0.0476	1, 7	0.0347	-0.0454	1, 8	-0.0379	-0.0292
1, 9	-0.0207	0.0259									
2, 2	1.0449	2.0421	2, 3	0.7530	0.1074	2, 4	0.1438	-0.3219	2, 5	-0.1720	-0.1037
2, 6	-0.0751	0.1088	2, 7	0.0748	0.0533	2, 8	0.0432	-0.0595			
3, 3	1.0616	2.0076	3, 4	0.7357	0.0967	3, 5	0.1346	-0.3103	3, 6	-0.1654	-0.0993
3, 7	-0.0684	0.1015									
4, 4	1.0521	2.0171	4, 5	0.7403	0.1011	4, 6	0.1408	-0.3152			
5, 5	1.0581	2.0130									

VBR

SELF AND MUTUAL ADMITTANCES

H/LMDA=0.5000 B/LMDA= 0.250 OMEGA= 9.92

10 ELEMENT ARRAY

I, K	YIK IN MMHO		I, K	YIK IN MMHO		I, K	YIK IN MMHO		I, K	YIK IN MMHO	
	RE	IM		RE	IM		RE	IM		RE	IM
1, 1	1.0353	1.8545	1, 2	0.6711	0.0872	1, 3	0.1260	-0.2777	1, 4	-0.1447	-0.0909
1, 5	-0.0648	0.0893	1, 6	0.0613	0.0467	1, 7	0.0360	-0.0463	1, 8	-0.0360	-0.0272
1, 9	-0.0236	0.0314	1,10	0.0220	0.0174						
2, 2	1.0454	2.0418	2, 3	0.7534	0.1079	2, 4	0.1432	-0.3215	2, 5	-0.1726	-0.1046
2, 6	-0.0739	0.1080	2, 7	0.0763	0.0551	2, 8	0.0404	-0.0565	2, 9	-0.0477	-0.0340
3, 3	1.0610	2.0079	3, 4	0.7352	0.0959	3, 5	0.1357	-0.3110	3, 6	-0.1645	-0.0978
3, 7	-0.0706	0.1031	3, 8	0.0704	0.0498						
4, 4	1.0532	2.0165	4, 5	0.7411	0.1025	4, 6	0.1388	-0.3140	4, 7	-0.1682	-0.1025
5, 5	1.0561	2.0142	5, 6	0.7383	0.0982						

VBR

SELF AND MUTUAL ADMITTANCES

H/LMDA=0.5000 B/LMDA= 0.250 OMEGA= 9.92

12 ELEMENT ARRAY

I, K	YIK IN MMHO		I, K	YIK IN MMHO		I, K	YIK IN MMHO		I, K	YIK IN MMHO	
	RE	IM		RE	IM		RE	IM		RE	IM
1, 1	1.0354	1.8545	1, 2	0.6711	0.0872	1, 3	0.1259	-0.2777	1, 4	-0.1447	-0.0910
1, 5	-0.0647	0.0892	1, 6	0.0614	0.0469	1, 7	0.0357	-0.0461	1, 8	-0.0363	-0.0276
1, 9	-0.0229	0.0304	1,10	0.0252	0.0183	1,11	0.0168	-0.0231	1,12	-0.0167	-0.0129
2, 2	1.0453	2.0419	2, 3	0.7533	0.1078	2, 4	0.1433	-0.3216	2, 5	-0.1725	-0.1044
2, 6	-0.0741	0.1081	2, 7	0.0760	0.0547	2, 8	0.0410	-0.0571	2, 9	-0.0462	-0.0329
2,10	-0.0257	0.0374	2,11	0.0338	0.0231						
3, 3	1.0611	2.0079	3, 4	0.7353	0.0961	3, 5	0.1355	-0.3108	3, 6	-0.1647	-0.0982
3, 7	-0.0701	0.1028	3, 8	0.0711	0.0506	3, 9	0.0390	-0.0540	3,10	-0.0421	-0.0294
4, 4	1.0529	2.0166	4, 5	0.7409	0.1022	4, 6	0.1392	-0.3142	4, 7	-0.1678	-0.1018
4, 8	-0.0719	0.1045	4, 9	0.0736	0.0536						
5, 5	1.0566	2.0140	5, 6	0.7386	0.0988	5, 7	0.1377	-0.3132	5, 8	-0.1661	-0.0993
6, 6	1.0552	2.0148	6, 7	0.7400	0.1011						

VBR

SELF AND MUTUAL ADMITTANCES

H/LMDA=0.5000 B/LMDA= 0.250 OMEGA= 9.92

14 ELEMENT ARRAY

I, K	YIK IN MMHO RE	IM	I, K	YIK IN MMHO RE	IM	I, K	YIK IN MMHO RE	IM	I, K	YIK IN MMHO RE	IM
1, 1	1.0354	1.8545	1, 2	0.6712	0.0872	1, 3	0.1259	-0.2776	1, 4	-0.1448	-0.0911
1, 5	-0.0646	0.0892	1, 6	0.0614	0.0470	1, 7	0.0356	-0.0461	1, 8	-0.0364	-0.0278
1, 9	-0.0227	0.0303	1,10	0.0255	0.0186	1,11	0.0163	-0.0224	1,12	-0.0191	-0.0135
1,13	-0.0128	0.0181	1,14	0.0133	0.0101						
2, 2	1.0453	2.0419	2, 3	0.7533	0.1078	2, 4	0.1434	-0.3216	2, 5	-0.1724	-0.1044
2, 6	-0.0742	0.1082	2, 7	0.0759	0.0546	2, 8	0.0412	-0.0573	2, 9	-0.0460	-0.0327
2,10	-0.0262	0.0379	2,11	0.0328	0.0224	2,12	0.0183	-0.0277	2,13	-0.0260	-0.0171
3, 3	1.0611	2.0078	3, 4	0.7354	0.0961	3, 5	0.1354	-0.3108	3, 6	-0.1647	-0.0983
3, 7	-0.0700	0.1027	3, 8	0.0712	0.0508	3, 9	0.0387	-0.0538	3,10	-0.0425	-0.0299
3,11	-0.0248	0.0357	3,12	0.0295	0.0198						
4, 4	1.0529	2.0166	4, 5	0.7409	0.1021	4, 6	0.1393	-0.3143	4, 7	-0.1676	-0.1016
4, 8	-0.0721	0.1047	4, 9	0.0733	0.0532	4,10	0.0398	-0.0550	4,11	-0.0444	-0.0320
5, 5	1.0567	2.0139	5, 6	0.7387	0.0990	5, 7	0.1374	-0.3131	5, 8	-0.1662	-0.0996
5, 9	-0.0712	0.1040	5,10	0.0721	0.0514						
6, 6	1.0550	2.0149	6, 7	0.7398	0.1008	6, 8	0.1382	-0.3135	6, 9	-0.1672	-0.1011
7, 7	1.0556	2.0145	7, 8	0.7390	0.0995						

VBR

SELF AND MUTUAL ADMITTANCES

H/LMDA=0.5000 B/LMDA= 0.250 OMEGA= 9.92

16 ELEMENT ARRAY

I, K	YIK IN MMHO RE	IM	I, K	YIK IN MMHO RE	IM	I, K	YIK IN MMHO RE	IM	I, K	YIK IN MMHO RE	IM
1, 1	1.0354	1.8545	1, 2	0.6712	0.0872	1, 3	0.1258	-0.2776	1, 4	-0.1448	-0.0911
1, 5	-0.0646	0.0892	1, 6	0.0614	0.0470	1, 7	0.0356	-0.0461	1, 8	-0.0365	-0.0278
1, 9	-0.0226	0.0302	1,10	0.0255	0.0187	1,11	0.0161	-0.0223	1,12	-0.0194	-0.0137
1,13	-0.0124	0.0175	1,14	0.0153	0.0105	1,15	0.0102	-0.0148	1,16	-0.0110	-0.0082
2, 2	1.0452	2.0419	2, 3	0.7532	0.1078	2, 4	0.1434	-0.3216	2, 5	-0.1724	-0.1043
2, 6	-0.0743	0.1082	2, 7	0.0759	0.0545	2, 8	0.0412	-0.0573	2, 9	-0.0459	-0.0326
2,10	-0.0263	0.0380	2,11	0.0326	0.0222	2,12	0.0186	-0.0280	2,13	-0.0251	-0.0166
2,14	-0.0140	0.0217	2,15	0.0208	0.0134						
3, 3	1.0611	2.0078	3, 4	0.7354	0.0962	3, 5	0.1354	-0.3108	3, 6	-0.1648	-0.0983
3, 7	-0.0699	0.1027	3, 8	0.0713	0.0509	3, 9	0.0386	-0.0537	3,10	-0.0426	-0.0300
3,11	-0.0245	0.0355	3,12	0.0299	0.0201	3,13	0.0176	-0.0263	3,14	-0.0225	-0.0145
4, 4	1.0528	2.0166	4, 5	0.7408	0.1020	4, 6	[illegible]	[illegible]	4, 7	-0.1676	-0.1016
4, 8	-0.0722	0.1047	4, 9	0.0732	0.0531	4,10	0.0400	-0.0551	4,11	-0.0442	[illegible]
4,12	-0.0254	0.0363	4,13	0.0314	0.0217						
5, 5	1.0567	2.0139	5, 6	0.7388	0.0990	5, 7	0.1374	-0.3130	5, 8	-0.1663	-0.0997
5, 9	-0.0710	0.1039	5,10	0.0722	0.0516	5,11	0.0393	-0.0546	5,12	-0.0432	-0.0304
6, 6	1.0549	2.0149	6, 7	0.7398	0.1007	6, 8	0.1384	-0.3135	6, 9	-0.1671	-0.1009
6,10	-0.0716	0.1042	6,11	0.0729	0.0527						
7, 7	1.0558	2.0145	7, 8	0.7391	0.0996	7, 9	0.1379	-0.3133	7,10	-0.1665	-0.1000
8, 8	1.0554	2.0146	8, 9	0.7397	0.1005						

VBR

SELF AND MUTUAL ADMITTANCES

H/LMDA=0.5000 B/LMDA= 0.250 OMEGA= 9.92

18 ELEMENT ARRAY

I, K	YIK IN MMHO RE	IM	I, K	YIK IN MMHO RE	IM	I, K	YIK IN MMHO RE	IM	I, K	YIK IN MMHO RE	IM
1, 1	1.0354	1.8545	1, 2	0.6712	0.0873	1, 3	0.1258	-0.2776	1, 4	-0.1448	-0.0911
1, 5	-0.0646	0.0892	1, 6	0.0614	0.0470	1, 7	0.0356	-0.0460	1, 8	-0.0365	-0.0278
1, 9	-0.0226	0.0302	1,10	0.0256	0.0187	1,11	0.0161	-0.0222	1,12	-0.0194	-0.0138
1,13	-0.0123	0.0174	1,14	0.0155	0.0107	1,15	0.0099	-0.0142	1,16	-0.0126	-0.0085
1,17	-0.0084	0.0124	1,18	0.0093	0.0069						
2, 2	1.0452	2.0419	2, 3	0.7532	0.1077	2, 4	0.1434	-0.3216	2, 5	-0.1724	-0.1043
2, 6	-0.0743	0.1082	2, 7	0.0758	0.0545	2, 8	0.0413	-0.0573	2, 9	-0.0459	-0.0326
2,10	-0.0263	0.0380	2,11	0.0325	0.0221	2,12	0.0187	-0.0281	2,13	-0.0249	-0.0165
2,14	-0.0142	0.0220	2,15	0.0201	0.0130	2,16	0.0111	-0.0177	2,17	-0.0173	-0.0109
3, 3	1.0612	2.0078	3, 4	0.7354	0.0962	3, 5	0.1353	-0.3108	3, 6	-0.1648	-0.0983
3, 7	-0.0699	0.1027	3, 8	0.0713	0.0509	3, 9	0.0385	-0.0537	3,10	-0.0427	-0.0301
3,11	-0.0245	0.0354	3,12	0.0300	0.0202	3,13	0.0174	-0.0262	3,14	-0.0228	-0.0148
3,15	-0.0134	0.0206	3,16	0.0180	0.0113						
4, 4	1.0528	2.0167	4, 5	0.7408	0.1020	4, 6	0.1394	-0.3144	4, 7	-0.1676	-0.1015
4, 8	-0.0723	0.1047	4, 9	0.0732	0.0530	4,10	0.0401	-0.0551	4,11	-0.0441	-0.0317
4,12	-0.0255	0.0364	4,13	0.0312	0.0215	4,14	0.0181	-0.0268	4,15	-0.0240	-0.0161
5, 5	1.0568	2.0138	5, 6	0.7388	0.0991	5, 7	0.1373	-0.3130	5, 8	-0.1663	-0.0997
5, 9	-0.0710	0.1039	5,10	0.0722	0.0517	5,11	0.0392	-0.0546	5,12	-0.0433	-0.0306
5,13	-0.0250	0.0361	5,14	0.0304	0.0205						
6, 6	1.0548	2.0149	6, 7	0.7398	0.1007	6, 8	0.1384	-0.3136	6, 9	-0.1670	-0.1008
6,10	-0.0716	0.1043	6,11	0.0728	0.0526	6,12	0.0396	-0.0548	6,13	-0.0439	-0.0314
7, 7	1.0558	2.0144	7, 8	0.7392	0.0997	7, 9	0.1379	-0.3133	7,10	-0.1665	-0.1001
7,11	-0.0714	0.1041	7,12	0.0724	0.0519						
8, 8	1.0553	2.0147	8, 9	0.7396	0.1004	8,10	0.1381	-0.3134	8,11	-0.1670	-0.1007
9, 9	1.0555	2.0146	9,10	0.7392	0.0998						

VBR

SELF AND MUTUAL ADMITTANCES

H/LMDA=0.5000 B/LMDA= 0.250 OMEGA= 9.92

20 ELEMENT ARRAY

I, K	YIK IN MMHO		I, K	YIK IN MMHO		I, K	YIK IN MMHO		I, K	YIK IN MMHO	
	RE	IM		RE	IM		RE	IM		RE	IM
1, 1	1.0354	1.8545	1, 2	0.6712	0.0873	1, 3	0.1258	-0.2776	1, 4	-0.1448	-0.0911
1, 5	-0.0646	0.0892	1, 6	0.0615	0.0470	1, 7	0.0356	-0.0460	1, 8	-0.0365	-0.0278
1, 9	-0.0226	0.0302	1,10	0.0256	0.0188	1,11	0.0160	-0.0222	1,12	-0.0195	-0.0138
1,13	-0.0122	0.0174	1,14	0.0155	0.0108	1,15	0.0098	-0.0142	1,16	-0.0128	-0.0087
1,17	-0.0082	0.0119	1,18	0.0106	0.0071	1,19	0.0071	-0.0106	1,20	-0.0080	-0.0059
2, 2	1.0452	2.0420	2, 3	0.7532	0.1077	2, 4	0.1434	-0.3216	2, 5	-0.1724	-0.1043
2, 6	-0.0743	0.1082	2, 7	0.0758	0.0545	2, 8	0.0413	-0.0573	2, 9	-0.0459	-0.0325
2,10	-0.0264	0.0380	2,11	0.0325	0.0221	2,12	0.0188	-0.0281	2,13	-0.0249	-0.0164
2,14	-0.0143	0.0221	2,15	0.0200	0.0129	2,16	0.0113	-0.0179	2,17	-0.0167	-0.0106
2,18	-0.0092	0.0148	2,19	0.0147	0.0091						
3, 3	1.0612	2.0078	3, 4	0.7354	0.0962	3, 5	0.1353	-0.3108	3, 6	-0.1648	-0.0983
3, 7	-0.0699	0.1026	3, 8	0.0713	0.0509	3, 9	0.0385	-0.0537	3,10	-0.0427	-0.0301
3,11	-0.0244	0.0354	3,12	0.0300	0.0203	3,13	0.0173	-0.0261	3,14	-0.0229	-0.0149
3,15	-0.0132	0.0205	3,16	0.0182	0.0115	3,17	0.0107	-0.0168	3,18	-0.0148	-0.0092
4, 4	1.0528	2.0167	4, 5	0.7408	0.1020	4, 6	0.1394	-0.3144	4, 7	-0.1676	-0.1015
4, 8	-0.0723	0.1048	4, 9	0.0732	0.0530	4,10	0.0401	-0.0552	4,11	-0.0441	-0.0316
4,12	-0.0256	0.0365	4,13	0.0312	0.0215	4,14	0.0182	-0.0269	4,15	-0.0239	-0.0160
4,16	-0.0138	0.0210	4,17	0.0192	0.0126						
5, 5	1.0568	2.0138	5, 6	0.7388	0.0991	5, 7	0.1373	-0.3130	5, 8	-0.1664	-0.0998
5, 9	-0.0710	0.1039	5,10	0.0723	0.0517	5,11	0.0392	-0.0545	5,12	-0.0434	-0.0306
5,13	-0.0249	0.0360	5,14	0.0305	0.0206	5,15	0.0178	-0.0266	5,16	-0.0232	-0.0151
6, 6	1.0548	2.0150	6, 7	0.7397	0.1007	6, 8	0.1384	-0.3136	6, 9	-0.1670	-0.1008
6,10	-0.0717	0.1043	6,11	0.0728	0.0525	6,12	0.0397	-0.0548	6,13	-0.0438	-0.0313
6,14	-0.0252	0.0362	6,15	0.0310	0.0213						
7, 7	1.0558	2.0144	7, 8	0.7392	0.0997	7, 9	0.1378	-0.3133	7,10	-0.1666	-0.1001
7,11	-0.0713	0.1041	7,12	0.0724	0.0520	7,13	0.0395	-0.0547	7,14	-0.0435	-0.0308
8, 8	1.0553	2.0147	8, 9	0.7396	0.1003	8,10	0.1382	-0.3134	8,11	-0.1669	-0.1006
8,12	-0.0715	0.1042	8,13	0.0727	0.0524						
9, 9	1.0556	2.0145	9,10	0.7393	0.0999	9,11	0.1380	-0.3134	9,12	-0.1666	-0.1002
10,10	1.0555	2.0146	10,11	0.7396	0.1003						

VBR

SELF AND MUTUAL ADMITTANCES

H/LMDA=0.5000 B/LMDA= 0.250 OMEGA= 9.92

22 ELEMENT ARRAY

I, K	YIK IN MMHO		I, K	YIK IN MMHO		I, K	YIK IN MMHO		I, K	YIK IN MMHO	
	RE	IM		RE	IM		RE	IM		RE	IM
1, 1	1.0354	1.8545	1, 2	0.6712	0.0873	1, 3	0.1258	-0.2776	1, 4	-0.1448	-0.0911
1, 5	-0.0646	0.0891	1, 6	0.0615	0.0470	1, 7	0.0356	-0.0460	1, 8	-0.0365	-0.0278
1, 9	-0.0226	0.0302	1,10	0.0256	0.0188	1,11	0.0160	-0.0222	1,12	-0.0195	-0.0138
1,13	-0.0122	0.0174	1,14	0.0155	0.0108	1,15	0.0098	-0.0141	1,16	-0.0128	-0.0087
1,17	-0.0081	0.0118	1,18	0.0108	0.0072	1,19	0.0069	-0.0102	1,20	-0.0092	-0.0060
1,21	-0.0061	0.0092	1,22	0.0070	0.0051						
2, 2	1.0452	2.0420	2, 3	0.7532	0.1077	2, 4	0.1434	-0.3217	2, 5	-0.1724	-0.1043
2, 6	-0.0743	0.1082	2, 7	0.0758	0.0545	2, 8	0.0413	-0.0573	2, 9	-0.0459	-0.0325
2,10	-0.0264	0.0381	2,11	0.0325	0.0221	2,12	0.0188	-0.0282	2,13	-0.0249	-0.0164
2,14	-0.0143	0.0221	2,15	0.0200	0.0128	2,16	0.0114	-0.0180	2,17	-0.0166	-0.0105
2,18	-0.0093	0.0150	2,19	0.0141	0.0088	2,20	0.0077	-0.0127	2,21	-0.0127	-0.0078
3, 3	1.0612	2.0078	3, 4	0.7354	0.0962	3, 5	0.1353	-0.3107	3, 6	-0.1648	-0.0983
3, 7	-0.0699	0.1026	3, 8	0.0713	0.0509	3, 9	0.0385	-0.0537	3,10	-0.0427	-0.0301
3,11	-0.0244	0.0354	3,12	0.0300	0.0203	3,13	0.0173	-0.0261	3,14	-0.0229	-0.0149
3,15	-0.0132	0.0204	3,16	0.0183	0.0116	3,17	0.0106	-0.0167	3,18	-0.0150	-0.0093
3,19	-0.0088	0.0140	3,20	0.0125	0.0076						
4, 4	1.0528	2.0167	4, 5	0.7408	0.1020	4, 6	0.1394	-0.3144	4, 7	-0.1676	-0.1015
4, 8	-0.0723	0.1048	4, 9	0.0731	0.0530	4,10	0.0401	-0.0552	4,11	-0.0441	-0.0316
4,12	-0.0256	0.0365	4,13	0.0311	0.0214	4,14	0.0182	-0.0270	4,15	-0.0238	-0.0159
4,16	-0.0139	0.0211	4,17	0.0191	0.0125	4,18	0.0110	-0.0171	4,19	-0.0159	-0.0103
5, 5	1.0568	2.0138	5, 6	0.7388	0.0991	5, 7	0.1373	-0.3130	5, 8	-0.1664	-0.0998
5, 9	-0.0709	0.1039	5,10	0.0723	0.0517	5,11	0.0392	-0.0545	5,12	-0.0434	-0.0307
5,13	-0.0249	0.0360	5,14	0.0306	0.0207	5,15	0.0177	-0.0266	5,16	-0.0233	-0.0152
5,17	-0.0135	0.0209	5,18	0.0186	0.0118						
6, 6	1.0548	2.0150	6, 7	0.7397	0.1006	6, 8	0.1385	-0.3136	6, 9	-0.1670	-0.1008
6,10	-0.0717	0.1043	6,11	0.0728	0.0525	6,12	0.0397	-0.0548	6,13	-0.0438	-0.0313
6,14	-0.0253	0.0362	6,15	0.0309	0.0212	6,16	0.0179	-0.0267	6,17	-0.0237	-0.0158
7, 7	1.0559	2.0144	7, 8	0.7392	0.0997	7, 9	0.1378	-0.3133	7,10	-0.1666	-0.1001
7,11	-0.0713	0.1041	7,12	0.0724	0.0520	7,13	0.0394	-0.0547	7,14	-0.0435	-0.0308
7,15	-0.0251	0.0361	7,16	0.0306	0.0208						
8, 8	1.0553	2.0147	8, 9	0.7396	0.1003	8,10	0.1382	-0.3135	8,11	-0.1669	-0.1006
8,12	-0.0715	0.1042	8,13	0.0727	0.0524	8,14	0.0396	-0.0547	8,15	-0.0438	-0.0312
9, 9	1.0556	2.0145	9,10	0.7393	0.0999	9,11	0.1380	-0.3133	9,12	-0.1666	-0.1002
9,13	-0.0714	0.1041	9,14	0.0724	0.0520						
10,10	1.0554	2.0146	10,11	0.7395	0.1002	10,12	0.1381	-0.3134	10,13	-0.1669	-0.1006
11,11	1.0555	2.0146	11,12	0.7393	0.0999						

VBR

SELF AND MUTUAL ADMITTANCES

H/LMDA=0.5000   B/LMDA= 0.250   OMEGA= 9.92

24 ELEMENT ARRAY

I, K	YIK IN MMHO RE	IM	I, K	YIK IN MMHO RE	IM	I, K	YIK IN MMHO RE	IM	I, K	YIK IN MMHO RE	IM
1, 1	1.0354	1.8545	1, 2	0.6712	0.0873	1, 3	0.1258	-0.2776	1, 4	-0.1448	-0.0911
1, 5	-0.0646	0.0891	1, 6	0.0615	0.0470	1, 7	0.0356	-0.0460	1, 8	-0.0365	-0.0278
1, 9	-0.0226	0.0302	1,10	0.0256	0.0188	1,11	0.0160	-0.0222	1,12	-0.0195	-0.0139
1,13	-0.0122	0.0174	1,14	0.0155	0.0108	1,15	0.0097	-0.0141	1,16	-0.0128	-0.0088
1,17	-0.0080	0.0118	1,18	0.0108	0.0073	1,19	0.0068	-0.0101	1,20	-0.0093	-0.0062
1,21	-0.0059	0.0089	1,22	0.0080	0.0052	1,23	0.0053	-0.0081	1,24	-0.0062	-0.0045
2, 2	1.0452	2.0420	2, 3	0.7532	0.1077	2, 4	0.1434	-0.3217	2, 5	-0.1724	-0.1043
2, 6	-0.0743	0.1082	2, 7	0.0758	0.0545	2, 8	0.0413	-0.0574	2, 9	-0.0459	-0.0325
2,10	-0.0264	0.0381	2,11	0.0325	0.0221	2,12	0.0188	-0.0282	2,13	-0.0249	-0.0164
2,14	-0.0143	0.0221	2,15	0.0199	0.0128	2,16	0.0114	-0.0180	2,17	-0.0165	-0.0104
2,18	-0.0094	0.0151	2,19	0.0140	0.0087	2,20	0.0079	-0.0129	2,21	-0.0122	-0.0075
2,22	-0.0066	0.0110	2,23	0.0111	0.0067						
3, 3	1.0612	2.0078	3, 4	0.7354	0.0962	3, 5	0.1353	-0.3107	3, 6	-0.1648	-0.0983
3, 7	-0.0699	0.1026	3, 8	0.0713	0.0509	3, 9	0.0385	-0.0537	3,10	-0.0427	-0.0301
3,11	-0.0244	0.0354	3,12	0.0300	0.0203	3,13	0.0173	-0.0261	3,14	-0.0229	-0.0149
3,15	-0.0132	0.0204	3,16	0.0183	0.0116	3,17	0.0105	-0.0166	3,18	-0.0151	-0.0094
3,19	-0.0087	0.0140	3,20	0.0127	0.0078	3,21	0.0074	-0.0120	3,22	-0.0108	-0.0065
4, 4	1.0528	2.0167	4, 5	0.7408	0.1020	4, 6	0.1394	-0.3144	4, 7	-0.1676	-0.1015
4, 8	-0.0723	0.1048	4, 9	0.0731	0.0530	4,10	0.0401	-0.0552	4,11	-0.0441	-0.0316
4,12	-0.0256	0.0365	4,13	0.0311	0.0214	4,14	0.0182	-0.0270	4,15	-0.0238	-0.0159
4,16	-0.0139	0.0211	4,17	0.0191	0.0125	4,18	0.0111	-0.0172	4,19	-0.0158	-0.0102
4,20	-0.0091	0.0144	4,21	0.0135	0.0086						
5, 5	1.0568	2.0138	5, 6	0.7388	0.0991	5, 7	0.1373	-0.3130	5, 8	-0.1664	-0.0998
5, 9	-0.0709	0.1038	5,10	0.0723	0.0518	5,11	0.0392	-0.0545	5,12	-0.0434	-0.0307
5,13	-0.0249	0.0360	5,14	0.0306	0.0207	5,15	0.0177	-0.0266	5,16	-0.0233	-0.0152
5,17	-0.0135	0.0208	5,18	0.0186	0.0119	5,19	0.0108	-0.0170	5,20	-0.0153	-0.0096
6, 6	1.0548	2.0150	6, 7	0.7397	0.1006	6, 8	0.1385	-0.3136	6, 9	-0.1670	-0.1008
6,10	-0.0717	0.1043	6,11	0.0728	0.0525	6,12	0.0397	-0.0548	6,13	-0.0438	-0.0313
6,14	-0.0253	0.0362	6,15	0.0309	0.0212	6,16	0.0180	-0.0268	6,17	-0.0236	-0.0157
6,18	-0.0137	0.0209	6,19	0.0190	0.0123						
7, 7	1.0559	2.0144	7, 8	0.7392	0.0997	7, 9	0.1378	-0.3133	7,10	-0.1666	-0.1002
7,11	-0.0713	0.1040	7,12	0.0724	0.0520	7,13	0.0394	-0.0547	7,14	-0.0435	-0.0309
7,15	-0.0251	0.0361	7,16	0.0307	0.0208	7,17	0.0178	-0.0267	7,18	-0.0234	-0.0153
8, 8	1.0553	2.0147	8, 9	0.7396	0.1003	8,10	0.1382	-0.3135	8,11	-0.1669	-0.1006
8,12	-0.0715	0.1042	8,13	0.0727	0.0524	8,14	0.0396	-0.0548	8,15	-0.0437	-0.0312
8,16	-0.0252	0.0362	8,17	0.0309	0.0211						
9, 9	1.0556	2.0145	9,10	0.7393	0.0999	9,11	0.1380	-0.3133	9,12	-0.1666	-0.1003
9,13	-0.0714	0.1041	9,14	0.0725	0.0521	9,15	0.0395	-0.0547	9,16	-0.0435	-0.0309
10,10	1.0554	2.0147	10,11	0.7395	0.1002	10,12	0.1381	-0.3134	10,13	-0.1669	-0.1005
10,14	-0.0714	[illegible]	10,15	0.0727	0.0523						
11,11	1.0555	2.0146	11,12	0.7393	0.1000	11,13	0.1380	-0.3134	11,14	-0.1667	-0.1003
12,12	1.0555	2.0146	12,13	0.7395	0.1002						

VBR

SELF AND MUTUAL ADMITTANCES

H/LMDA=0.5000   B/LMDA= 0.250   OMEGA= 9.92

26 ELEMENT ARRAY

I, K	YIK IN MMHO RE	IM	I, K	YIK IN MMHO RE	IM	I, K	YIK IN MMHO RE	IM	I, K	YIK IN MMHO RE	IM
1, 1	1.0354	1.8545	1, 2	0.6712	0.0873	1, 3	0.1258	-0.2776	1, 4	-0.1448	-0.0911
1, 5	-0.0646	0.0891	1, 6	0.0615	0.0470	1, 7	0.0356	-0.0460	1, 8	-0.0365	-0.0278
1, 9	-0.0226	0.0302	1,10	0.0256	0.0188	1,11	0.0160	-0.0222	1,12	-0.0195	-0.0139
1,13	-0.0122	0.0174	1,14	0.0156	0.0108	1,15	0.0097	-0.0141	1,16	-0.0128	-0.0088
1,17	-0.0080	0.0116	1,18	0.0109	0.0073	1,19	0.0068	-0.0101	1,20	-0.0094	-0.0062
1,21	-0.0058	0.0088	1,22	0.0082	0.0053	1,23	0.0051	-0.0078	1,24	-0.0071	-0.0046
1,25	-0.0047	0.0072	1,26	0.0056	0.0040						
2, 2	1.0452	2.0420	2, 3	0.7532	0.1077	2, 4	0.1434	-0.3217	2, 5	-0.1724	-0.1043
2, 6	-0.0743	0.1082	2, 7	0.0758	0.0545	2, 8	0.0413	-0.0574	2, 9	-0.0459	-0.0325
2,10	-0.0264	0.0381	2,11	0.0325	0.0221	2,12	0.0188	-0.0282	2,13	-0.0249	-0.0164
2,14	-0.0144	0.0221	2,15	0.0199	0.0128	2,16	0.0115	-0.0180	2,17	-0.0165	-0.0104
2,18	-0.0094	0.0151	2,19	0.0140	0.0087	2,20	0.0079	-0.0129	2,21	-0.0121	-0.0074
2,22	-0.0068	0.0112	2,23	0.0107	0.0065	2,24	0.0058	-0.0097	2,25	-0.0099	-0.0059
3, 3	1.0612	2.0078	3, 4	0.7354	0.0962	3, 5	0.1353	-0.3107	3, 6	-0.1648	-0.0984
3, 7	-0.0699	0.1026	3, 8	0.0713	0.0509	3, 9	0.0385	-0.0537	3,10	-0.0427	-0.0301
3,11	-0.0244	0.0354	3,12	0.0300	0.0203	3,13	0.0173	-0.0261	3,14	-0.0229	-0.0149
3,15	-0.0132	0.0204	3,16	0.0183	0.0117	3,17	0.0105	-0.0166	3,18	-0.0151	-0.0094
3,19	-0.0086	0.0139	3,20	0.0128	0.0078	3,21	0.0073	-0.0119	3,22	-0.0110	-0.0066
3,23	-0.0063	0.0105	3,24	0.0095	0.0056						
4, 4	1.0528	2.0167	4, 5	0.7408	0.1020	4, 6	0.1395	-0.3144	4, 7	-0.1676	-0.1015
4, 8	-0.0723	0.1048	4, 9	0.0731	0.0530	4,10	0.0401	-0.0552	4,11	-0.0441	-0.0316
4,12	-0.0256	0.0365	4,13	0.0311	0.0214	4,14	0.0182	-0.0270	4,15	-0.0238	-0.0159
4,16	-0.0139	0.0211	4,17	0.0191	0.0124	4,18	0.0111	-0.0172	4,19	-0.0158	-0.0101
4,20	-0.0091	0.0144	4,21	0.0134	0.0085	4,22	0.0076	-0.0123	4,23	-0.0117	-0.0073
5, 5	1.0568	2.0138	5, 6	0.7388	0.0991	5, 7	0.1373	-0.3129	5, 8	-0.1664	-0.0998
5, 9	-0.0709	0.1038	5,10	0.0723	0.0518	5,11	0.0392	-0.0545	5,12	-0.0434	-0.0307
5,13	-0.0249	0.0360	5,14	0.0306	0.0207	5,15	0.0177	-0.0266	5,16	-0.0233	-0.0153
5,17	-0.0135	0.0208	5,18	0.0186	0.0119	5,19	0.0108	-0.0170	5,20	-0.0154	-0.0096
5,21	-0.0089	0.0142	5,22	0.0130	0.0080						
6, 6	1.0548	2.0150	6, 7	0.7397	0.1006	6, 8	0.1385	-0.3136	6, 9	-0.1670	-0.1008
6,10	-0.0717	0.1043	6,11	0.0728	0.0525	6,12	0.0397	-0.0549	6,13	-0.0438	-0.0313
6,14	-0.0253	0.0363	6,15	0.0309	0.0212	6,16	0.0180	-0.0268	6,17	-0.0236	-0.0157
6,18	-0.0137	0.0210	6,19	0.0189	0.0123	6,20	0.0109	-0.0171	6,21	-0.0157	-0.0100

I, K	RE	IM	I, K	RE	IM	I, K	RE	IM	I, K	RE	IM
7, 7	1.0559	2.0144	7, 8	0.7392	0.0997	7, 9	0.1378	-0.3132	7,10	-0.1666	-0.1002
7,11	-0.0713	0.1040	7,12	0.0724	0.0520	7,13	0.0394	-0.0547	7,14	-0.0435	-0.0309
7,15	-0.0251	0.0361	7,16	0.0307	0.0208	7,17	0.0178	-0.0267	7,18	-0.0234	-0.0154
7,19	-0.0136	0.0209	7,20	0.0187	0.0120						
8, 8	1.0553	2.0147	8, 9	0.7396	0.1003	8,10	0.1382	-0.3135	8,11	-0.1669	-0.1006
8,12	-0.0715	0.1042	8,13	0.0727	0.0524	8,14	0.0396	-0.0548	8,15	-0.0437	-0.0312
8,16	-0.0252	0.0362	8,17	0.0309	0.0211	8,18	0.0179	-0.0267	8,19	-0.0236	-0.0156
9, 9	1.0556	2.0145	9,10	0.7393	0.0999	9,11	0.1379	-0.3133	9,12	-0.1667	-0.1003
9,13	-0.0714	0.1041	9,14	0.0725	0.0521	9,15	0.0395	-0.0547	9,16	-0.0436	-0.0309
9,17	-0.0251	0.0362	9,18	0.0307	0.0209						
10,10	1.0554	2.0147	10,11	0.7395	0.1002	10,12	0.1381	-0.3134	10,13	-0.1668	-0.1005
10,14	-0.0715	0.1042	10,15	0.0727	0.0523	10,16	0.0395	-0.0547	10,17	-0.0437	-0.0311
11,11	1.0555	2.0146	11,12	0.7393	0.1000	11,13	0.1380	-0.3134	11,14	-0.1667	-0.1003
11,15	-0.0714	0.1041	11,16	0.0725	0.0521						
12,12	1.0555	2.0146	12,13	0.7395	0.1002	12,14	0.1381	-0.3134	12,15	-0.1668	-0.1005
13,13	1.0555	2.0146	13,14	0.7393	0.1000						

VBR

SELF AND MUTUAL ADMITTANCES

H/LMDA=0.5000 B/LMDA= 0.250 OMEGA= 9.92

28 ELEMENT ARRAY

I, K	YIK IN MMHO		I, K	YIK IN MMHO		I, K	YIK IN MMHO		I, K	YIK IN MMHO	
	RE	IM		RE	IM		RE	IM		RE	IM
1, 1	1.0354	1.8545	1, 2	0.6712	0.0873	1, 3	0.1258	-0.2776	1, 4	-0.1448	-0.0911
1, 5	-0.0646	0.0891	1, 6	0.0615	0.0470	1, 7	0.0356	-0.0460	1, 8	-0.0365	-0.0278
1, 9	-0.0226	0.0302	1,10	0.0256	0.0188	1,11	0.0160	-0.0222	1,12	-0.0195	-0.0139
1,13	-0.0122	0.0173	1,14	0.0156	0.0108	1,15	0.0097	-0.0141	1,16	-0.0128	-0.0088
1,17	-0.0080	0.0118	1,18	0.0109	0.0073	1,19	0.0068	-0.0101	1,20	-0.0094	-0.0062
1,21	-0.0058	0.0088	1,22	0.0082	0.0054	1,23	0.0051	-0.0078	1,24	-0.0072	-0.0047
1,25	-0.0045	0.0070	1,26	0.0064	0.0041	1,27	0.0042	-0.0065	1,28	-0.0050	-0.0035
2, 2	1.0452	2.0420	2, 3	0.7532	0.1077	2, 4	0.1434	-0.3217	2, 5	-0.1724	-0.1043
2, 6	-0.0743	0.1083	2, 7	0.0758	0.0545	2, 8	0.0413	-0.0574	2, 9	-0.0459	-0.0325
2,10	-0.0264	0.0381	2,11	0.0325	0.0221	2,12	0.0188	-0.0282	2,13	-0.0248	-0.0163
2,14	-0.0144	0.0221	2,15	0.0199	0.0128	2,16	0.0115	-0.0180	2,17	-0.0165	-0.0104
2,18	-0.0094	0.0151	2,19	0.0140	0.0087	2,20	0.0079	-0.0129	2,21	-0.0121	-0.0074
2,22	-0.0068	0.0112	2,23	0.0106	0.0064	2,24	0.0059	-0.0099	2,25	-0.0095	-0.0057
2,26	-0.0051	0.0087	2,27	0.0089	0.0052						
3, 3	1.0612	2.0078	3, 4	0.7354	0.0962	3, 5	0.1353	-0.3107	3, 6	-0.1648	-0.0984
3, 7	-0.0699	0.1026	3, 8	0.0713	0.0509	3, 9	0.0385	-0.0537	3,10	-0.0427	-0.0301
3,11	-0.0244	0.0354	3,12	0.0301	0.0203	3,13	0.0173	-0.0261	3,14	-0.0229	-0.0150
3,15	-0.0132	0.0204	3,16	0.0183	0.0117	3,17	0.0105	-0.0166	3,18	-0.0151	-0.0094
3,19	-0.0086	0.0139	3,20	0.0128	0.0078	3,21	0.0073	-0.0119	3,22	-0.0110	-0.0067
3,23	-0.0063	0.0104	3,24	0.0096	0.0057	3,25	0.0055	-0.0092	3,26	-0.0084	-0.0049
4, 4	1.0528	2.0167	4, 5	0.7408	0.1020	4, 6	0.1395	-0.3144	4, 7	-0.1676	-0.1015
4, 8	-0.0723	0.1048	4, 9	0.0731	0.0530	4,10	0.0401	-0.0552	4,11	-0.0441	-0.0316
4,12	-0.0256	0.0365	4,13	0.0311	0.0214	4,14	0.0182	-0.0270	4,15	-0.0238	-0.0159
4,16	-0.0139	0.0212	4,17	0.0191	0.0124	4,18	0.0111	-0.0172	4,19	-0.0158	-0.0101
4,20	-0.0091	0.0144	4,21	0.0134	0.0084	4,22	0.0077	-0.0123	4,23	-0.0116	-0.0072
4,24	-0.0066	0.0107	4,25	0.0102	0.0063						
5, 5	1.0568	2.0136	5, 6	0.7388	0.0991	5, 7	0.1373	-0.3129	5, 8	-0.1664	-0.0998
5, 9	-0.0709	0.1038	5,10	0.0723	0.0518	5,11	0.0392	-0.0545	5,12	-0.0434	-0.0307
5,13	-0.0249	0.0360	5,14	0.0306	0.0207	5,15	0.0177	-0.0266	5,16	-0.0233	-0.0153
5,17	-0.0135	0.0208	5,18	0.0187	0.0119	5,19	0.0107	-0.0170	5,20	-0.0154	-0.0096
5,21	-0.0088	0.0142	5,22	0.0130	0.0080	5,23	0.0075	-0.0122	5,24	-0.0112	-0.0068
6, 6	1.0548	2.0150	6, 7	0.7397	0.1006	6, 8	0.1385	-0.3136	6, 9	-0.1670	-0.1008
6,10	-0.0717	0.1043	6,11	0.0727	0.0525	6,12	0.0397	-0.0549	6,13	-0.0438	-0.0312
6,14	-0.0253	0.0363	6,15	0.0309	0.0212	6,16	0.0180	-0.0268	6,17	-0.0236	-0.0157
6,18	-0.0137	0.0210	6,19	0.0189	0.0123	6,20	0.0109	-0.0171	6,21	-0.0157	-0.0100
6,22	-0.0090	0.0143	6,23	0.0133	0.0084						
7, 7	1.0559	2.0144	7, 8	0.7392	0.0998	7, 9	0.1378	-0.3132	7,10	-0.1666	-0.1002
7,11	-0.0712	0.1040	7,12	0.0725	0.0520	7,13	0.0394	-0.0546	7,14	-0.0435	-0.0309
7,15	-0.0251	0.0361	7,16	0.0307	0.0208	7,17	0.0178	-0.0267	7,18	-0.0234	-0.0154
7,19	-0.0136	0.0209	7,20	0.0187	0.0120	7,21	0.0108	-0.0170	7,22	-0.0154	-0.0097
8, 8	1.0553	2.0147	8, 9	0.7396	0.1003	8,10	0.1382	-0.3135	8,11	-0.1669	-0.1006
8,12	-0.0715	0.1042	8,13	0.0727	0.0523	8,14	0.0396	-0.0548	8,15	-0.0437	-0.0311
8,16	-0.0252	0.0362	8,17	0.0309	0.0211	8,18	0.0179	-0.0267	8,19	-0.0236	-0.0156
8,20	-0.0137	0.0209	8,21	0.0189	0.0122						
9, 9	1.0556	2.0145	9,10	0.7393	0.0999	9,11	0.1379	-0.3133	9,12	-0.1667	-0.1003
9,13	-0.0714	0.1041	9,14	0.0725	0.0521	9,15	0.0395	-0.0547	9,16	-0.0436	-0.0309
9,17	-0.0251	0.0361	9,18	0.0307	0.0209	9,19	0.0179	-0.0267	9,20	-0.0234	-0.0154
10,10	1.0554	2.0147	10,11	0.7395	0.1002	10,12	0.1381	-0.3134	10,13	-0.1668	-0.1005
10,14	-0.0715	0.1042	10,15	0.0726	0.0523	10,16	0.0396	-0.0547	10,17	-0.0437	-0.0311
10,18	-0.0252	0.0362	10,19	0.0308	0.0211						
11,11	1.0556	2.0146	11,12	0.7393	0.1000	11,13	0.1380	-0.3134	11,14	-0.1667	-0.1003
11,15	-0.0714	0.1041	11,16	0.0725	0.0521	11,17	0.0395	-0.0547	11,18	-0.0436	-0.0309
12,12	1.0554	2.0146	12,13	0.7395	0.1002	12,14	0.1381	-0.3134	12,15	-0.1668	-0.1005
12,16	-0.0714	0.1041	12,17	0.0726	0.0523						
13,13	1.0555	2.0146	13,14	0.7393	0.1000	13,15	0.1380	-0.3134	13,16	-0.1667	-0.1003
14,14	1.0555	2.0146	14,15	0.7395	0.1002						

VBR

SELF AND MUTUAL ADMITTANCES

H/LMDA=0.5000 B/LMDA= 0.250 OMEGA= 9.92

30 ELEMENT ARRAY

I, K	YIK IN MMHO RE	IM	I, K	YIK IN MMHO RE	IM	I, K	YIK IN MMHO RE	IM	I, K	YIK IN MMHO RE	IM
1, 1	1.0354	1.8545	1, 2	0.6712	0.0873	1, 3	0.1258	-0.2776	1, 4	-0.1448	-0.0911
1, 5	-0.0646	0.0891	1, 6	0.0615	0.0470	1, 7	0.0355	-0.0460	1, 8	-0.0365	-0.0278
1, 9	-0.0226	0.0302	1,10	0.0256	0.0188	1,11	0.0160	-0.0222	1,12	-0.0195	-0.0139
1,13	-0.0122	0.0173	1,14	0.0156	0.0108	1,15	0.0097	-0.0141	1,16	-0.0128	-0.0088
1,17	-0.0080	0.0118	1,18	0.0109	0.0073	1,19	0.0068	-0.0101	1,20	-0.0094	-0.0062
1,21	-0.0058	0.0088	1,22	0.0082	0.0054	1,23	0.0051	-0.0078	1,24	-0.0073	-0.0047
1,25	-0.0045	0.0069	1,26	0.0065	0.0041	1,27	0.0041	-0.0063	1,28	-0.0058	-0.0036
1,29	-0.0038	0.0059	1,30	0.0046	0.0032						
2, 2	1.0452	2.0420	2, 3	0.7532	0.1077	2, 4	0.1434	-0.3217	2, 5	-0.1724	-0.1043
2, 6	-0.0743	0.1083	2, 7	0.0758	0.0545	2, 8	0.0413	-0.0574	2, 9	-0.0459	-0.0325
2,10	-0.0264	0.0381	2,11	0.0325	0.0221	2,12	0.0188	-0.0282	2,13	-0.0248	-0.0163
2,14	-0.0144	0.0221	2,15	0.0199	0.0128	2,16	0.0115	-0.0180	2,17	-0.0165	-0.0104
2,18	-0.0094	0.0151	2,19	0.0140	0.0087	2,20	0.0080	-0.0129	2,21	-0.0121	-0.0074
2,22	-0.0068	0.0113	2,23	0.0106	0.0064	2,24	0.0059	-0.0099	2,25	-0.0094	-0.0057
2,26	-0.0052	0.0088	2,27	0.0085	0.0051	2,28	0.0045	-0.0078	2,29	-0.0080	-0.0047
3, 3	1.0612	2.0078	3, 4	0.7354	0.0962	3, 5	0.1353	-0.3107	3, 6	-0.1648	-0.0984
3, 7	-0.0699	0.1026	3, 8	0.0713	0.0509	3, 9	0.0385	-0.0537	3,10	-0.0427	-0.0301
3,11	-0.0244	0.0354	3,12	0.0301	0.0203	3,13	0.0173	-0.0261	3,14	-0.0229	-0.0150
3,15	-0.0132	0.0204	3,16	0.0183	0.0117	3,17	0.0105	-0.0166	3,18	-0.0151	-0.0094
3,19	-0.0086	0.0139	3,20	0.0128	0.0079	3,21	0.0073	-0.0119	3,22	-0.0110	-0.0067
3,23	-0.0062	0.0104	3,24	0.0097	0.0057	3,25	0.0055	-0.0092	3,26	-0.0085	-0.0050
3,27	-0.0049	0.0082	3,28	0.0075	0.0044						
4, 4	1.0528	2.0167	4, 5	0.7408	0.1020	4, 6	0.1395	-0.3144	4, 7	-0.1676	-0.1015
4, 6	-0.0723	0.1048	4, 9	0.0731	0.0530	4,10	0.0401	-0.0552	4,11	-0.0441	-0.0316
4,12	-0.0256	0.0365	4,13	0.0311	0.0214	4,14	0.0182	-0.0270	4,15	-0.0238	-0.0159
4,16	-0.0139	0.0212	4,17	0.0191	0.0124	4,18	0.0111	-0.0172	4,19	-0.0158	-0.0101
4,20	-0.0092	0.0144	4,21	0.0134	0.0084	4,22	0.0077	-0.0124	4,23	-0.0116	-0.0072
4,24	-0.0066	0.0107	4,25	0.0102	0.0063	4,26	0.0057	-0.0094	4,27	-0.0091	-0.0056
5, 5	1.0568	2.0138	5, 6	0.7388	0.0991	5, 7	0.1373	-0.3129	5, 8	-0.1664	-0.0998
5, 9	-0.0709	0.1038	5,10	0.0723	0.0518	5,11	0.0392	-0.0545	5,12	-0.0434	-0.0307
5,13	-0.0249	0.0360	5,14	0.0306	0.0207	5,15	0.0177	-0.0266	5,16	-0.0233	-0.0153
5,17	-0.0134	0.0208	5,18	0.0187	0.0119	5,19	0.0107	-0.0169	5,20	-0.0154	-0.0097
5,21	-0.0088	0.0142	5,22	0.0130	0.0080	5,23	0.0075	-0.0122	5,24	-0.0112	-0.0068
5,25	-0.0064	0.0106	5,26	0.0098	0.0058						
6, 6	1.0548	2.0150	6, 7	0.7397	0.1006	6, 8	0.1385	-0.3136	6, 9	-0.1670	-0.1008
6,10	-0.0717	0.1043	6,11	0.0727	0.0525	6,12	0.0398	-0.0549	6,13	-0.0438	-0.0312
6,14	-0.0253	0.0363	6,15	0.0309	0.0212	6,16	0.0180	-0.0268	6,17	-0.0236	-0.0157
6,18	-0.0137	0.0210	6,19	0.0189	0.0123	6,20	0.0110	-0.0171	6,21	-0.0156	-0.0100
6,22	-0.0090	0.0143	6,23	0.0133	0.0083	6,24	0.0076	-0.0122	6,25	-0.0115	-0.0071
7, 7	1.0559	2.0144	7, 8	0.7392	0.0998	7, 9	0.1378	-0.3132	7,10	-0.1666	-0.1002
7,11	-0.0712	0.1040	7,12	0.0725	0.0520	7,13	0.0394	-0.0546	7,14	-0.0435	-0.0309
7,15	-0.0251	0.0361	7,16	0.0307	0.0209	7,17	0.0178	-0.0266	7,18	-0.0234	-0.0154
7,19	-0.0136	[illegible]	[illegible]	0.0187	0.0120	7,21	0.0108	-0.0170	7,22	-0.0155	-0.0097
7,23	-0.0089	0.0143	7,24	0.0131	0.0081						
8, 8	1.0553	2.0147	8, 9	0.7396	0.1003	8,10	0.1382	-0.3135	8,11	-0.1669	-0.1006
8,12	-0.0715	0.1042	8,13	0.0727	0.0523	8,14	0.0396	-0.0548	8,15	-0.0437	-0.0311
8,16	-0.0252	0.0362	8,17	0.0308	0.0211	8,18	0.0179	-0.0267	8,19	-0.0236	-0.0156
8,20	-0.0137	0.0209	8,21	0.0189	0.0122	8,22	0.0109	-0.0170	8,23	-0.0156	-0.0099
9, 9	1.0556	2.0145	9,10	0.7393	0.0999	9,11	0.1379	-0.3133	9,12	-0.1667	-0.1003
9,13	-0.0713	0.1041	9,14	0.0725	0.0521	9,15	0.0395	-0.0547	9,16	-0.0436	-0.0309
9,17	-0.0251	0.0361	9,18	0.0307	0.0209	9,19	0.0179	-0.0267	9,20	-0.0234	-0.0154
9,21	-0.0136	0.0209	9,22	0.0187	0.0120						
10,10	1.0554	2.0147	10,11	0.7395	0.1002	10,12	0.1381	-0.3134	10,13	-0.1668	-0.1005
10,14	-0.0715	0.1042	10,15	0.0726	0.0523	10,16	0.0396	-0.0548	10,17	-0.0437	-0.0311
10,18	-0.0252	0.0362	10,19	0.0308	0.0211	10,20	0.0179	-0.0267	10,21	-0.0236	-0.0156
11,11	1.0556	2.0145	11,12	0.7393	0.1000	11,13	0.1380	-0.3134	11,14	-0.1667	-0.1003
11,15	-0.0714	0.1041	11,16	0.0725	0.0521	11,17	0.0395	-0.0547	11,18	-0.0436	-0.0310
11,19	-0.0252	0.0362	11,20	0.0307	0.0209						
12,12	1.0554	2.0146	12,13	0.7395	0.1002	12,14	0.1381	-0.3134	12,15	-0.1668	-0.1005
12,16	-0.0714	0.1042	12,17	0.0726	0.0523	12,18	0.0395	-0.0547	12,19	-0.0437	-0.0311
13,13	1.0555	2.0146	13,14	0.7394	0.1000	13,15	0.1380	-0.3134	13,16	-0.1667	-0.1003
13,17	-0.0714	0.1041	13,18	0.0725	0.0521						
14,14	1.0555	2.0146	14,15	0.7395	0.1001	14,16	0.1381	-0.3134	14,17	-0.1668	-0.1005
15,15	1.0555	2.0146	15,16	0.7394	0.1000						

VBR

SELF AND MUTUAL ADMITTANCES

H/LMDA=0.5000 B/LMDA= 0.250 OMEGA= 9.92

35 ELEMENT ARRAY

I, K	YIK IN MMHO		I, K	YIK IN MMHO		I, K	YIK IN MMHO		I, K	YIK IN MMHO	
	RE	IM		RE	IM		RE	IM		RE	IM
1, 1	1.0354	1.8545	1, 2	0.6712	0.0873	1, 3	0.1258	-0.2776	1, 4	-0.1448	-0.0911
1, 5	-0.0646	0.0891	1, 6	0.0615	0.0471	1, 7	0.0355	-0.0460	1, 8	-0.0365	-0.0279
1, 9	-0.0226	0.0301	1,10	0.0256	0.0188	1,11	0.0160	-0.0222	1,12	-0.0195	-0.0139
1,13	-0.0122	0.0173	1,14	0.0156	0.0108	1,15	0.0097	-0.0141	1,16	-0.0129	-0.0088
1,17	-0.0080	0.0118	1,18	0.0109	0.0073	1,19	0.0067	-0.0101	1,20	-0.0094	-0.0062
1,21	-0.0058	0.0088	1,22	0.0082	0.0054	1,23	0.0050	-0.0077	1,24	-0.0073	-0.0047
1,25	-0.0044	0.0069	1,26	0.0066	0.0042	1,27	0.0040	-0.0062	1,28	-0.0059	-0.0038
1,29	-0.0035	0.0056	1,30	0.0054	0.0034	1,31	0.0032	-0.0051	1,32	-0.0050	-0.0032
1,33	-0.0029	0.0046	1,34	0.0048	0.0030	1,35	0.0026	-0.0037			
2, 2	1.0452	2.0420	2, 3	0.7532	0.1077	2, 4	0.1434	-0.3217	2, 5	-0.1723	-0.1043
2, 6	-0.0743	0.1083	2, 7	0.0758	0.0544	2, 8	0.0413	-0.0574	2, 9	-0.0459	-0.0325
2,10	-0.0264	0.0381	2,11	0.0324	0.0220	2,12	0.0188	-0.0282	2,13	-0.0248	-0.0163
2,14	-0.0144	0.0221	2,15	0.0199	0.0128	2,16	0.0115	-0.0181	2,17	-0.0165	-0.0104
2,18	-0.0095	0.0151	2,19	0.0140	0.0087	2,20	0.0080	-0.0130	2,21	-0.0121	-0.0074
2,22	-0.0069	0.0113	2,23	0.0106	0.0064	2,24	0.0060	-0.0100	2,25	-0.0094	-0.0056
2,26	-0.0053	0.0089	2,27	0.0084	0.0049	2,28	0.0047	-0.0080	2,29	-0.0076	-0.0044
2,30	-0.0043	0.0073	2,31	0.0069	0.0040	2,32	0.0039	-0.0067	2,33	-0.0062	-0.0035
2,34	-0.0037	0.0064									
3, 3	1.0612	2.0078	3, 4	0.7354	0.0962	3, 5	0.1353	-0.3107	3, 6	-0.1648	-0.0984
3, 7	-0.0698	0.1026	3, 8	0.0713	0.0509	3, 9	0.0385	-0.0537	3,10	-0.0427	-0.0301
3,11	-0.0244	0.0354	3,12	0.0301	0.0203	3,13	0.0173	-0.0261	3,14	-0.0229	-0.0150
3,15	-0.0131	0.0204	3,16	0.0183	0.0117	3,17	0.0105	-0.0166	3,18	-0.0151	-0.0095
3,19	-0.0086	0.0139	3,20	0.0128	0.0079	3,21	0.0072	-0.0119	3,22	-0.0111	-0.0067
3,23	-0.0062	0.0103	3,24	0.0097	0.0058	3,25	0.0054	-0.0091	3,26	-0.0086	-0.0051
3,27	-0.0047	0.0081	3,28	0.0077	0.0045	3,29	0.0042	-0.0073	3,30	-0.0070	-0.0041
3,31	-0.0038	0.0066	3,32	0.0064	0.0037	3,33	0.0033	-0.0059			
4, 4	1.0528	2.0167	4, 5	0.7408	0.1020	4, 6	0.1395	-0.3144	4, 7	-0.1675	-0.1015
4, 8	-0.0723	0.1048	4, 9	0.0731	0.0529	4,10	0.0402	-0.0552	4,11	-0.0440	-0.0316
4,12	-0.0256	0.0365	4,13	0.0311	0.0214	4,14	0.0183	-0.0270	4,15	-0.0238	-0.0158
4,16	-0.0140	0.0212	4,17	0.0190	0.0124	4,18	0.0111	-0.0173	4,19	-0.0157	-0.0101
4,20	-0.0092	0.0145	4,21	0.0133	0.0084	4,22	0.0078	-0.0124	4,23	-0.0115	-0.0071
4,24	-0.0067	0.0108	4,25	0.0101	0.0062	4,26	0.0058	-0.0095	4,27	-0.0090	-0.0054
4,28	-0.0051	0.0085	4,29	0.0080	0.0048	4,30	0.0046	-0.0077	4,31	-0.0072	-0.0043
4,32	-0.0042	0.0070									
5, 5	1.0568	2.0138	5, 6	0.7388	0.0991	5, 7	0.1373	-0.3129	5, 8	-0.1664	-0.0998
5, 9	-0.0709	0.1038	5,10	0.0723	0.0518	5,11	0.0391	-0.0545	5,12	-0.0434	-0.0307
5,13	-0.0249	0.0360	5,14	0.0306	0.0207	5,15	0.0176	-0.0265	5,16	-0.0234	-0.0153
5,17	-0.0134	0.0208	5,18	0.0187	0.0119	5,19	0.0107	-0.0169	5,20	-0.0155	-0.0097
5,21	-0.0088	0.0142	5,22	0.0131	0.0081	5,23	0.0074	-0.0121	5,24	-0.0113	-0.0069
5,25	-0.0063	0.0105	5,26	0.0099	0.0059	5,27	0.0055	-0.0093	5,28	-0.0088	-0.0052
5,29	-0.0048	0.0083	5,30	0.0079	0.0046	5,31	0.0043	-0.0074			
6, 6	1.0548	2.0150	6, 7	0.7397	0.1006	6, 8	0.1385	-0.3136	6, 9	-0.1670	-0.1008
6,10	-0.0717	0.1043	6,11	0.0727	0.0525	6,12	0.0398	-0.0549	6,13	-0.0438	-0.0312
6,14	-0.0254	0.0363	6,15	0.0309	0.0211	6,16	0.0180	-0.0268	6,17	-0.0236	-0.0156
6,18	-0.0138	0.0210	6,19	0.0189	0.0122	6,20	0.0110	-0.0171	6,21	-0.0156	-0.0099
6,22	-0.0091	0.0144	6,23	0.0132	0.0083	6,24	0.0076	-0.0123	6,25	-0.0114	-0.0070
6,26	-0.0066	0.0107	6,27	0.0100	0.0061	6,28	0.0057	-0.0095	6,29	-0.0089	-0.0053
6,30	-0.0051	0.0085									
7, 7	1.0559	2.0144	7, 8	0.7392	0.0998	7, 9	0.1378	-0.3132	7,10	-0.1666	-0.1002
7,11	-0.0712	0.1040	7,12	0.0725	0.0520	7,13	0.0394	-0.0546	7,14	-0.0436	-0.0309
7,15	-0.0250	0.0361	7,16	0.0307	0.0209	7,17	0.0178	-0.0266	7,18	-0.0234	-0.0154
7,19	-0.0135	0.0209	7,20	0.0188	0.0120	7,21	0.0108	-0.0170	7,22	-0.0155	-0.0098
7,23	-0.0089	0.0142	7,24	0.0131	0.0081	7,25	0.0075	-0.0122	7,26	-0.0114	-0.0069
7,27	-0.0064	0.0106	7,28	0.0100	0.0060	7,29	0.0056	-0.0093			
8, 8	1.0552	2.0148	8, 9	0.7395	0.1003	8,10	0.1382	-0.3135	8,11	-0.1668	-0.1005
8,12	-0.0716	0.1042	8,13	0.0726	0.0523	8,14	0.0396	-0.0548	8,15	-0.0437	-0.0311
8,16	-0.0253	0.0362	8,17	0.0308	0.0210	8,18	0.0180	-0.0268	8,19	-0.0235	-0.0156
8,20	-0.0137	0.0210	8,21	0.0188	0.0122	8,22	0.0109	-0.0171	8,23	-0.0156	-0.0099
8,24	-0.0090	0.0143	8,25	0.0132	0.0082	8,26	0.0076	-0.0123	8,27	-0.0114	-0.0070
8,28	-0.0065	0.0107									
9, 9	1.0557	2.0145	9,10	0.7393	0.0999	9,11	0.1379	-0.3133	9,12	-0.1667	-0.1003
9,13	-0.0713	0.1041	9,14	0.0725	0.0521	9,15	0.0394	-0.0547	9,16	-0.0436	-0.0310
9,17	-0.0251	0.0361	9,18	0.0307	0.0209	9,19	0.0178	-0.0266	9,20	-0.0235	-0.0155
9,21	-0.0136	0.0209	9,22	0.0188	0.0121	9,23	0.0108	-0.0170	9,24	-0.0155	-0.0098
9,25	-0.0089	0.0142	9,26	0.0132	0.0082	9,27	0.0075	-0.0122			
10,10	1.0554	2.0147	10,11	0.7395	0.1002	10,12	0.1381	-0.3135	10,13	-0.1668	-0.1005
10,14	-0.0715	0.1042	10,15	0.0726	0.0523	10,16	0.0396	-0.0548	10,17	-0.0437	-0.0311
10,18	-0.0252	0.0362	10,19	0.0308	0.0210	10,20	0.0179	-0.0267	10,21	-0.0235	-0.0155
10,22	-0.0137	0.0210	10,23	0.0188	0.0121	10,24	0.0109	-0.0171	10,25	-0.0156	-0.0098
10,26	-0.0090	0.0143									
11,11	1.0556	2.0145	11,12	0.7394	0.1000	11,13	0.1380	-0.3133	11,14	-0.1667	-0.1003
11,15	-0.0714	0.1041	11,16	0.0725	0.0522	11,17	0.0395	-0.0547	11,18	-0.0436	-0.0310
11,19	-0.0251	0.0361	11,20	0.0308	0.0210	11,21	0.0178	-0.0267	11,22	-0.0235	-0.0155
11,23	-0.0136	0.0209	11,24	0.0188	0.0121	11,25	0.0108	-0.0170			
12,12	1.0554	2.0147	12,13	0.7394	0.1001	12,14	0.1381	-0.3134	12,15	-0.1668	-0.1004
12,16	-0.0715	0.1042	12,17	0.0726	0.0522	12,18	0.0396	-0.0548	12,19	-0.0437	-0.0310
12,20	-0.0252	0.0362	12,21	0.0308	0.0210	12,22	0.0179	-0.0267	12,23	-0.0235	-0.0155
12,24	-0.0137	0.0210									
13,13	1.0555	2.0145	13,14	0.7394	0.1000	13,15	0.1380	-0.3133	13,16	-0.1667	-0.1004
13,17	-0.0714	0.1041	13,18	0.0726	0.0522	13,19	0.0395	-0.0547	13,20	-0.0436	-0.0310
13,21	-0.0251	0.0361	13,22	0.0308	0.0210	13,23	0.0178	-0.0267			
14,14	1.0554	2.0146	14,15	0.7394	0.1001	14,16	0.1381	-0.3134	14,17	-0.1668	-0.1004
14,18	-0.0715	0.1042	14,19	0.0726	0.0522	14,20	0.0396	-0.0548	14,21	-0.0436	-0.0310
14,22	-0.0252	0.0362									
15,15	1.0555	2.0146	15,16	0.7394	0.1001	15,17	0.1380	-0.3133	15,18	-0.1667	-0.1004
15,19	-0.0714	0.1041	15,20	0.0726	0.0522	15,21	0.0395	-0.0547			
16,16	1.0554	2.0146	16,17	0.7394	0.1001	16,18	0.1381	-0.3134	16,19	-0.1668	-0.1004
16,20	-0.0715	0.1042									
17,17	1.0555	2.0146	17,18	0.7394	0.1001	17,19	0.1380	-0.3134			
18,18	1.0554	2.0146									

VBR

SELF AND MUTUAL ADMITTANCES

H/LMDA=0.5000 B/LMDA= 0.250 OMEGA= 9.92

40 ELEMENT ARRAY

I, K	YIK IN MMHO RE	IM	I, K	YIK IN MMHO RE	IM	I, K	YIK IN MMHO RE	IM	I, K	YIK IN MMHO RE	IM
1, 1	1.0354	1.8545	1, 2	0.6712	0.0873	1, 3	0.1258	-0.2776	1, 4	-0.1448	-0.0911
1, 5	-0.0646	0.0891	1, 6	0.0615	0.0470	1, 7	0.0355	-0.0460	1, 8	-0.0365	-0.0279
1, 9	-0.0226	0.0302	1,10	0.0256	0.0188	1,11	0.0160	-0.0222	1,12	-0.0195	-0.0139
1,13	-0.0122	0.0173	1,14	0.0156	0.0108	1,15	0.0097	-0.0141	1,16	-0.0129	-0.0088
1,17	-0.0080	0.0118	1,18	0.0109	0.0073	1,19	0.0067	-0.0101	1,20	-0.0094	-0.0062
1,21	-0.0058	0.0088	1,22	0.0062	0.0054	1,23	0.0051	-0.0077	1,24	-0.0073	-0.0047
1,25	-0.0045	0.0069	1,26	0.0065	0.0042	1,27	0.0040	-0.0062	1,28	-0.0059	-0.0038
1,29	-0.0036	0.0056	1,30	0.0054	0.0034	1,31	0.0032	-0.0052	1,32	-0.0049	-0.0031
1,33	-0.0030	0.0048	1,34	0.0045	0.0028	1,35	0.0027	-0.0044	1,36	-0.0042	-0.0026
1,37	-0.0025	0.0041	1,38	0.0038	0.0023	1,39	0.0024	-0.0040	1,40	-0.0031	-0.0021
2, 2	1.0452	2.0420	2, 3	0.7532	0.1077	2, 4	0.1434	-0.3217	2, 5	-0.1724	-0.1043
2, 6	-0.0743	0.1083	2, 7	0.0758	0.0544	2, 8	0.0413	-0.0574	2, 9	-0.0459	-0.0325
2,10	-0.0264	0.0381	2,11	0.0325	0.0221	2,12	0.0188	-0.0282	2,13	-0.0248	-0.0163
2,14	-0.0144	0.0221	2,15	0.0199	0.0128	2,16	0.0115	-0.0180	2,17	-0.0165	-0.0104
2,18	-0.0095	0.0151	2,19	0.0140	0.0087	2,20	0.0080	-0.0130	2,21	-0.0121	-0.0074
2,22	-0.0068	0.0113	2,23	0.0106	0.0064	2,24	0.0060	-0.0100	2,25	-0.0094	-0.0056
2,26	-0.0053	0.0089	2,27	0.0084	0.0050	2,28	0.0047	-0.0080	2,29	-0.0076	-0.0045
2,30	-0.0042	0.0073	2,31	0.0070	0.0040	2,32	0.0038	-0.0066	2,33	-0.0064	-0.0037
2,34	-0.0035	0.0061	2,35	0.0059	0.0034	2,36	0.0032	-0.0056	2,37	-0.0055	-0.0031
2,38	-0.0029	0.0051	2,39	0.0053	0.0030						
3, 3	1.0612	2.0078	3, 4	0.7354	0.0962	3, 5	0.1353	-0.3107	3, 6	-0.1648	-0.0984
3, 7	-0.0699	0.1026	3, 8	0.0713	0.0509	3, 9	0.0385	-0.0537	3,10	-0.0427	-0.0301
3,11	-0.0244	0.0354	3,12	0.0301	0.0203	3,13	0.0173	-0.0261	3,14	-0.0229	-0.0150
3,15	-0.0132	0.0204	3,16	0.0183	0.0117	3,17	0.0105	-0.0166	3,18	-0.0151	-0.0095
3,19	-0.0086	0.0139	3,20	0.0128	0.0079	3,21	0.0073	-0.0119	3,22	-0.0111	-0.0067
3,23	-0.0062	0.0103	3,24	0.0097	0.0058	3,25	0.0054	-0.0091	3,26	-0.0086	-0.0051
3,27	-0.0048	0.0081	3,28	0.0077	0.0045	3,29	0.0043	-0.0073	3,30	-0.0070	-0.0040
3,31	-0.0038	0.0067	3,32	0.0063	0.0036	3,33	0.0035	-0.0061	3,34	-0.0058	-0.0033
3,35	-0.0032	0.0056	3,36	0.0053	0.0030	3,37	0.0030	-0.0052	3,38	-0.0048	-0.0027
4, 4	1.0528	2.0167	4, 5	0.7408	0.1020	4, 6	0.1395	-0.3144	4, 7	-0.1675	-0.1015
4, 8	-0.0723	0.1048	4, 9	0.0731	0.0529	4,10	0.0402	-0.0552	4,11	-0.0441	-0.0316
4,12	-0.0256	0.0365	4,13	0.0311	0.0214	4,14	0.0183	-0.0270	4,15	-0.0238	-0.0159
4,16	-0.0139	0.0212	4,17	0.0190	0.0124	4,18	0.0111	-0.0173	4,19	-0.0158	-0.0101
4,20	-0.0092	0.0145	4,21	0.0134	0.0084	4,22	0.0077	-0.0124	4,23	-0.0115	-0.0072
4,24	-0.0066	0.0108	4,25	0.0101	0.0062	4,26	0.0058	-0.0095	4,27	-0.0090	-0.0054
4,28	-0.0051	0.0085	4,29	0.0081	0.0048	4,30	0.0045	-0.0076	4,31	-0.0073	-0.0043
4,32	-0.0041	0.0069	4,33	0.0067	0.0039	4,34	0.0037	-0.0063	4,35	-0.0061	-0.0036
4,36	-0.0033	0.0058	4,37	0.0057	0.0033						
5, 5	1.0568	2.0138	5, 6	0.7388	0.0991	5, 7	0.1373	-0.3129	5, 8	-0.1664	-0.0998
5, 9	-0.0709	0.1038	5,10	0.0723	0.0518	5,11	0.0392	-0.0545	5,12	-0.0434	-0.0307
5,13	-0.0249	0.0360	5,14	0.0306	0.0207	5,15	0.0176	-0.0265	5,16	-0.0233	-0.0153
5,17	-0.0134	0.0208	5,18	0.0187	0.0119	5,19	0.0107	-0.0169	5,20	-0.0154	-0.0097
5,21	-0.0088	0.0142	5,22	0.0131	0.0081	5,23	0.0074	-0.0121	5,24	-0.0113	-0.0068
5,25	-0.0064	0.0106	5,26	0.0099	0.0059	5,27	0.0055	-0.0093	5,28	-0.0088	-0.0052
5,29	[illegible]	[illegible]	5,30	0.0079	0.0046	5,31	0.0044	-0.0075	5,32	-0.0071	-0.0041
5,33	-0.0039	0.0068	5,34	[illegible]	[illegible]	5,35	0.0036	-0.0062	5,36	-0.0059	-0.0033
6, 6	1.0548	2.0150	6, 7	0.7397	0.1006	6, 8	0.1385	-0.3136	6, 9	-0.1670	-0.1008
6,10	-0.0717	0.1043	6,11	0.0727	0.0525	6,12	0.0398	-0.0549	6,13	-0.0438	-0.0312
6,14	-0.0253	0.0363	6,15	0.0309	0.0212	6,16	0.0180	-0.0268	6,17	-0.0236	-0.0157
6,18	-0.0138	0.0210	6,19	0.0189	0.0122	6,20	0.0110	-0.0171	6,21	-0.0156	-0.0099
6,22	-0.0090	0.0143	6,23	0.0132	0.0083	6,24	0.0076	-0.0123	6,25	-0.0114	-0.0071
6,26	-0.0065	0.0107	6,27	0.0100	0.0061	6,28	0.0057	-0.0094	6,29	-0.0089	-0.0054
6,30	-0.0050	0.0084	6,31	0.0080	0.0048	6,32	0.0045	-0.0076	6,33	-0.0072	-0.0043
6,34	-0.0040	0.0068	6,35	0.0066	0.0039						
7, 7	1.0559	2.0144	7, 8	0.7392	0.0998	7, 9	0.1378	-0.3132	7,10	-0.1666	-0.1002
7,11	-0.0712	0.1040	7,12	0.0725	0.0520	7,13	0.0394	-0.0546	7,14	-0.0435	-0.0309
7,15	-0.0250	0.0361	7,16	0.0307	0.0209	7,17	0.0178	-0.0266	7,18	-0.0234	-0.0154
7,19	-0.0136	0.0209	7,20	0.0187	0.0120	7,21	0.0108	-0.0170	7,22	-0.0155	-0.0098
7,23	-0.0089	0.0142	7,24	0.0131	0.0081	7,25	0.0075	-0.0122	7,26	-0.0113	-0.0069
7,27	-0.0064	0.0106	7,28	0.0099	0.0060	7,29	0.0056	-0.0094	7,30	-0.0088	-0.0052
7,31	-0.0050	0.0084	7,32	0.0079	0.0046	7,33	0.0044	-0.0075	7,34	-0.0071	-0.0041
8, 8	1.0552	2.0148	8, 9	0.7395	0.1003	8,10	0.1382	-0.3135	8,11	-0.1669	-0.1006
8,12	-0.0716	0.1042	8,13	0.0727	0.0523	8,14	0.0396	-0.0548	8,15	-0.0437	-0.0311
8,16	-0.0252	0.0362	8,17	0.0308	0.0211	8,18	0.0179	-0.0267	8,19	-0.0236	-0.0156
8,20	-0.0137	0.0210	8,21	0.0189	0.0122	8,22	0.0109	-0.0171	8,23	-0.0156	-0.0099
8,24	-0.0090	0.0143	8,25	0.0132	0.0083	8,26	0.0076	-0.0122	8,27	-0.0114	-0.0070
8,28	-0.0065	0.0106	8,29	0.0100	0.0061	8,30	0.0057	-0.0094	8,31	-0.0089	-0.0053
8,32	-0.0050	0.0084	8,33	0.0080	0.0047						
9, 9	1.0556	2.0145	9,10	0.7393	0.0999	9,11	0.1379	-0.3133	9,12	-0.1667	-0.1003
9,13	-0.0713	0.1041	9,14	0.0725	0.0521	9,15	0.0395	-0.0547	9,16	-0.0436	-0.0310
9,17	-0.0251	0.0361	9,18	0.0307	0.0209	9,19	0.0178	-0.0267	9,20	-0.0235	-0.0154
9,21	-0.0136	0.0209	9,22	0.0188	0.0121	9,23	0.0108	-0.0170	9,24	-0.0155	-0.0098
9,25	-0.0089	0.0143	9,26	0.0131	0.0082	9,27	0.0075	-0.0122	9,28	-0.0113	-0.0069
9,29	-0.0065	0.0106	9,30	0.0099	0.0060	9,31	0.0056	-0.0094	9,32	-0.0088	-0.0052
10,10	1.0554	2.0147	10,11	0.7395	0.1002	10,12	0.1381	-0.3135	10,13	-0.1668	-0.1005
10,14	-0.0715	0.1042	10,15	0.0726	0.0523	10,16	0.0396	-0.0548	10,17	-0.0437	-0.0311
10,18	-0.0252	0.0362	10,19	0.0308	0.0210	10,20	0.0179	-0.0267	10,21	-0.0235	-0.0156
10,22	-0.0137	0.0209	10,23	0.0188	0.0122	10,24	0.0109	-0.0171	10,25	-0.0156	-0.0099
10,26	-0.0090	0.0143	10,27	0.0132	0.0082	10,28	0.0076	-0.0122	10,29	-0.0114	-0.0070
10,30	-0.0065	0.0106	10,31	0.0100	0.0061						
11,11	1.0556	2.0145	11,12	0.7393	0.1000	11,13	0.1380	-0.3133	11,14	-0.1667	-0.1003
11,15	-0.0714	0.1041	11,16	0.0725	0.0522	11,17	0.0395	-0.0547	11,18	-0.0436	-0.0310
11,19	-0.0251	0.0351	11,20	0.0307	0.0209	11,21	0.0179	-0.0267	11,22	-0.0235	-0.0155
11,23	-0.0136	0.0209	11,24	0.0188	0.0121	11,25	0.0109	-0.0170	11,26	-0.0155	-0.0098
11,27	-0.0089	0.0143	11,28	0.0131	0.0082	11,29	0.0075	-0.0122	11,30	-0.0113	-0.0069
12,12	1.0554	2.0146	12,13	0.7395	0.1001	12,14	0.1381	-0.3134	12,15	-0.1668	-0.1005
12,16	-0.0715	0.1042	12,17	0.0726	0.0523	12,18	0.0396	-0.0548	12,19	-0.0437	-0.0311
12,20	-0.0252	0.0362	12,21	0.0308	0.0210	12,22	0.0179	-0.0267	12,23	-0.0235	-0.0155
12,24	-0.0136	0.0209	12,25	0.0188	0.0122	12,26	0.0109	-0.0170	12,27	-0.0156	-0.0099
12,28	-0.0090	0.0143	12,29	0.0132	0.0082						
13,13	1.0555	2.0146	13,14	0.7394	0.1000	13,15	0.1380	-0.3134	13,16	-0.1667	-0.1004
13,17	-0.0714	0.1041	13,18	0.0725	0.0522	13,19	0.0395	-0.0547	13,20	-0.0436	-0.0310
13,21	-0.0251	0.0351	13,22	0.0307	0.0209	13,23	0.0179	-0.0267	13,24	-0.0235	-0.0155
13,25	-0.0136	0.0209	13,26	0.0188	0.0121	13,27	0.0109	-0.0170	13,28	-0.0155	-0.0098

I, K	RE	IM	I, K	RE	IM	I, K	RE	IM	I, K	RE	IM
14,14	1.0555	2.0146	14,15	0.7394	0.1001	14,16	0.1381	-0.3134	14,17	-0.1668	-0.1004
14,18	-0.0714	0.1042	14,19	0.0726	0.0522	14,20	0.0395	-0.0547	14,21	-0.0437	-0.0311
14,22	-0.0252	0.0362	14,23	0.0308	0.0210	14,24	0.0179	-0.0267	14,25	-0.0235	-0.0155
14,26	-0.0136	0.0209	14,27	0.0188	0.0121						
15,15	1.0555	2.0146	15,16	0.7394	0.1000	15,17	0.1380	-0.3134	15,18	-0.1667	-0.1004
15,19	-0.0714	0.1041	15,20	0.0725	0.0522	15,21	0.0395	-0.0547	15,22	-0.0436	-0.0310
15,23	-0.0252	0.0362	15,24	0.0307	0.0210	15,25	0.0179	-0.0267	15,26	-0.0235	-0.0155
16,16	1.0555	2.0146	16,17	0.7394	0.1001	16,18	0.1381	-0.3134	16,19	-0.1668	-0.1004
16,20	-0.0714	0.1042	16,21	0.0726	0.0522	16,22	0.0395	-0.0547	16,23	-0.0437	-0.0311
16,24	-0.0252	0.0362	16,25	0.0308	0.0210						
17,17	1.0555	2.0146	17,18	0.7394	0.1001	17,19	0.1380	-0.3134	17,20	-0.1667	-0.1004
17,21	-0.0714	0.1041	17,22	0.0725	0.0522	17,23	0.0395	-0.0547	17,24	-0.0436	-0.0310
18,18	1.0555	2.0146	18,19	0.7394	0.1001	18,20	0.1381	-0.3134	18,21	-0.1668	-0.1004
18,22	-0.0714	0.1041	18,23	0.0726	0.0522						
19,19	1.0555	2.0146	19,20	0.7394	0.1001	19,21	0.1380	-0.3134	19,22	-0.1667	-0.1004
20,20	1.0555	2.0146	20,21	0.7394	0.1001						

VBR

SELF AND MUTUAL ADMITTANCES

H/LMDA=0.5000 B/LMDA= 0.250 OMEGA= 9.92

45 ELEMENT ARRAY

I, K	YIK IN MMHO		I, K	YIK IN MMHO		I, K	YIK IN MMHO		I, K	YIK IN MMHO	
	RE	IM		RE	IM		RE	IM		RE	IM
1, 1	1.0354	1.8545	1, 2	0.6712	0.0973	1, 3	0.1258	-0.2776	1, 4	-0.1448	-0.0911
1, 5	-0.0646	0.0891	1, 6	0.0615	0.0471	1, 7	0.0355	-0.0460	1, 8	-0.0365	-0.0279
1, 9	-0.0226	0.0302	1,10	0.0256	0.0188	1,11	0.0160	-0.0222	1,12	-0.0195	-0.0139
1,13	-0.0122	0.0173	1,14	0.0156	0.0108	1,15	0.0097	-0.0141	1,16	-0.0129	-0.0088
1,17	-0.0080	0.0118	1,18	0.0109	0.0073	1,19	0.0067	-0.0101	1,20	-0.0094	-0.0062
1,21	-0.0058	0.0088	1,22	0.0082	0.0054	1,23	0.0050	-0.0077	1,24	-0.0073	-0.0047
1,25	-0.0044	0.0069	1,26	0.0065	0.0042	1,27	0.0040	-0.0062	1,28	-0.0059	-0.0038
1,29	-0.0036	0.0056	1,30	0.0054	0.0034	1,31	0.0032	-0.0051	1,32	-0.0049	-0.0031
1,33	-0.0029	0.0047	1,34	0.0046	0.0028	1,35	0.0027	-0.0044	1,36	-0.0042	-0.0026
1,37	-0.0025	0.0041	1,38	0.0039	0.0024	1,39	0.0023	-0.0038	1,40	-0.0037	-0.0022
1,41	-0.0021	0.0035	1,42	0.0035	0.0021	1,43	0.0019	-0.0032	1,44	-0.0034	-0.0020
1,45	-0.0018	0.0027									
2, 2	1.0452	2.0420	2, 3	0.7532	0.1077	2, 4	0.1434	-0.3217	2, 5	-0.1723	-0.1043
2, 6	-0.0743	0.1083	2, 7	0.0758	0.0544	2, 8	0.0413	-0.0574	2, 9	-0.0459	-0.0325
2,10	-0.0264	0.0381	2,11	0.0325	0.0220	2,12	0.0188	-0.0282	2,13	-0.0248	-0.0163
2,14	-0.0144	0.0221	2,15	0.0199	0.0128	2,16	0.0115	-0.0180	2,17	-0.0165	-0.0104
2,18	-0.0095	0.0151	2,19	0.0140	0.0087	2,20	0.0080	-0.0130	2,21	-0.0121	-0.0074
2,22	-0.0069	0.0113	2,23	0.0106	0.0064	2,24	0.0060	-0.0100	2,25	-0.0094	-0.0056
2,26	-0.0053	0.0089	2,27	0.0084	0.0050	2,28	0.0047	-0.0080	2,29	-0.0076	-0.0044
2,30	-0.0042	0.0073	2,31	0.0069	0.0040	2,32	0.0038	-0.0067	2,33	-0.0064	-0.0036
2,34	-0.0035	0.0061	2,35	0.0059	0.0033	2,36	0.0032	-0.0057	2,37	-0.0054	-0.0031
2,38	-0.0030	0.0053	2,39	0.0050	0.0028	2,40	0.0028	-0.0049	2,41	-0.0047	-0.0026
2,42	-0.0026	0.0046	2,43	0.0043	0.0024	2,44	0.0025	-0.0045			
3, 3	1.0612	2.0078	3, 4	0.7354	0.0962	3, 5	0.1353	-0.3107	3, 6	-0.1648	-0.0984
3, 7	-0.0698	0.1026	3, 8	0.0713	0.0509	3, 9	0.0385	-0.0537	3,10	-0.0427	-0.0301
3,11	-0.0244	0.0354	3,12	0.0301	0.0203	3,13	0.0173	-0.0261	3,14	-0.0229	-0.0150
3,15	-0.0131	0.0204	3,16	0.0183	0.0117	3,17	0.0105	-0.0166	3,18	-0.0151	-0.0095
3,19	-0.0086	0.0139	3,20	0.0128	0.0079	3,21	0.0072	-0.0119	3,22	-0.0111	-0.0067
3,23	-0.0062	0.0103	3,24	0.0097	0.0058	3,25	0.0054	-0.0091	3,26	-0.0086	-0.0051
3,27	-0.0048	0.0081	3,28	0.0077	0.0045	3,29	0.0042	-0.0073	3,30	-0.0070	-0.0040
3,31	-0.0038	0.0066	3,32	0.0064	0.0036	3,33	0.0034	-0.0061	3,34	-0.0058	-0.0033
3,35	-0.0031	0.0056	3,36	0.0054	0.0030	3,37	0.0029	-0.0051	3,38	-0.0050	-0.0028
3,39	-0.0026	0.0048	3,40	0.0046	0.0026	3,41	0.0024	-0.0044	3,42	-0.0044	-0.0024
3,43	-0.0022	0.0041									
4, 4	1.0528	2.0167	4, 5	0.7408	0.1020	4, 6	0.1395	-0.3144	4, 7	-0.1675	-0.1015
4, 8	-0.0723	0.1048	4, 9	0.0731	0.0529	4,10	0.0402	-0.0552	4,11	-0.0441	-0.0316
4,12	-0.0256	0.0365	4,13	0.0311	0.0214	4,14	0.0183	-0.0270	4,15	-0.0238	-0.0158
4,16	-0.0139	0.0212	4,17	0.0190	0.0124	4,18	0.0111	-0.0173	4,19	-0.0157	-0.0101
4,20	-0.0092	0.0145	4,21	0.0133	0.0084	4,22	0.0077	-0.0124	4,23	-0.0115	-0.0072
4,24	-0.0067	0.0108	4,25	0.0101	0.0062	4,26	0.0058	-0.0095	4,27	-0.0090	-0.0054
4,28	-0.0051	0.0085	4,29	0.0080	0.0048	4,30	0.0046	-0.0077	4,31	-0.0073	-0.0043
4,32	-0.0041	0.0070	4,33	0.0066	0.0039	4,34	0.0037	-0.0064	4,35	-0.0061	-0.0035
4,36	-0.0034	0.0059	4,37	0.0056	0.0032	4,38	0.0031	-0.0054	4,39	-0.0052	-0.0030
4,40	-0.0029	0.0050	4,41	0.0048	0.0027	4,42	0.0027	-0.0047			

5, 5	1.0568	2.0138	5, 6	0.7388	0.0991	5, 7	0.1373	-0.3129	5, 8	-0.1664	-0.0998
5, 9	-0.0709	0.1038	5,10	0.0723	0.0518	5,11	0.0392	-0.0545	5,12	-0.0434	-0.0307
5,13	-0.0249	0.0360	5,14	0.0306	0.0207	5,15	0.0176	-0.0265	5,16	-0.0234	-0.0153
5,17	-0.0134	0.0208	5,18	0.0187	0.0119	5,19	0.0107	-0.0169	5,20	-0.0154	-0.0097
5,21	-0.0088	0.0142	5,22	0.0131	0.0081	5,23	0.0074	-0.0121	5,24	-0.0113	-0.0069
5,25	-0.0064	0.0105	5,26	0.0099	0.0059	5,27	0.0055	-0.0093	5,28	-0.0088	-0.0052
5,29	-0.0049	0.0083	5,30	0.0079	0.0046	5,31	0.0043	-0.0075	5,32	-0.0071	-0.0041
5,33	-0.0039	0.0068	5,34	0.0065	0.0037	5,35	0.0035	-0.0062	5,36	-0.0060	-0.0034
5,37	-0.0032	0.0057	5,38	0.0055	0.0031	5,39	0.0029	-0.0052	5,40	-0.0051	-0.0029
5,41	-0.0027	0.0049									
6, 6	1.0548	2.0150	6, 7	0.7397	0.1006	6, 8	0.1385	-0.3136	6, 9	-0.1670	-0.1008
6,10	-0.0717	0.1043	6,11	0.0727	0.0525	6,12	0.0398	-0.0549	6,13	-0.0438	-0.0312
6,14	-0.0254	0.0363	6,15	0.0309	0.0211	6,16	0.0180	-0.0268	6,17	-0.0236	-0.0156
6,18	-0.0138	0.0210	6,19	0.0189	0.0122	6,20	0.0110	-0.0171	6,21	-0.0156	-0.0099
6,22	-0.0091	0.0143	6,23	0.0132	0.0083	6,24	0.0076	-0.0123	6,25	-0.0114	-0.0071
6,26	-0.0066	0.0107	6,27	0.0100	0.0061	6,28	0.0057	-0.0094	6,29	-0.0089	-0.0054
6,30	-0.0050	0.0084	6,31	0.0080	0.0047	6,32	0.0045	-0.0076	6,33	-0.0072	-0.0042
6,34	-0.0041	0.0069	6,35	0.0066	0.0038	6,36	0.0037	-0.0063	6,37	-0.0060	-0.0035
6,38	-0.0034	0.0058	6,39	0.0055	0.0032	6,40	0.0031	-0.0054			
7, 7	1.0559	2.0144	7, 8	0.7392	0.0998	7, 9	0.1378	-0.3132	7,10	-0.1666	-0.1002
7,11	-0.0712	0.1040	7,12	0.0725	0.0520	7,13	0.0394	-0.0546	7,14	-0.0436	-0.0309
7,15	-0.0250	0.0361	7,16	0.0307	C.0209	7,17	0.0178	-0.0266	7,18	-0.0234	-0.0154
7,19	-0.0135	0.0209	7,20	0.0188	0.0120	7,21	0.0108	-0.0170	7,22	-0.0155	-0.0098
7,23	-0.0089	0.0142	7,24	0.0131	0.0081	7,25	0.0075	-0.0122	7,26	-0.0113	-0.0069
7,27	-0.0064	0.0106	7,28	0.0099	0.0060	7,29	0.0056	-0.0093	7,30	-0.0088	-0.0053
7,31	-0.0049	0.0083	7,32	0.0079	0.0047	7,33	0.0044	-0.0075	7,34	-0.0072	-0.0042
7,35	-0.0039	0.0068	7,36	0.0065	0.0038	7,37	0.0036	-0.0062	7,38	-0.0060	-0.0034
7,39	-0.0032	0.0057									
8, 8	1.0552	2.0148	8, 9	0.7395	0.1003	8,10	0.1382	-0.3135	8,11	-0.1669	-0.1005
8,12	-0.0716	0.1042	8,13	0.0727	0.0523	8,14	0.0396	-0.0548	8,15	-0.0437	-0.0311
8,16	-0.0253	0.0362	8,17	0.0308	0.0211	8,18	0.0180	-0.0268	8,19	-0.0235	-0.0156
8,20	-0.0137	0.0210	8,21	C.0188	0.0122	8,22	0.0109	-0.0171	8,23	-0.0156	-0.0099
8,24	-0.0090	0.0143	8,25	0.0132	0.0082	8,26	0.0076	-0.0123	8,27	-0.0114	-0.0070
8,28	-0.0065	0.0107	8,29	0.0100	0.0061	8,30	0.0057	-0.0094	8,31	-0.0089	-0.0053
8,32	-0.0050	0.0084	8,33	0.0079	0.0047	8,34	0.0045	-0.0076	8,35	-0.0072	-0.0042
8,36	-0.0040	0.0069	8,37	0.0065	0.0038	8,38	0.0037	-0.0063			
9, 9	1.0556	2.0145	9,10	0.7393	0.0999	9,11	0.1379	-0.3133	9,12	-0.1667	-0.1003
9,13	-0.0713	0.1041	9,14	0.0725	0.0521	9,15	0.0395	-0.0547	9,16	-0.0436	-0.0310
9,17	-0.0251	0.0361	9,18	0.0307	0.0209	9,19	0.0178	-0.0267	9,20	-0.0235	-0.0155
9,21	-0.0136	0.0209	9,22	0.0188	0.0121	9,23	0.0108	-0.0170	9,24	-0.0155	-0.0098
9,25	-0.0089	0.0142	9,26	0.0132	0.0082	9,27	0.0075	-0.0122	9,28	-0.0114	-0.0070
9,29	-0.0064	0.0106	9,30	0.0100	0.0060	9,31	0.0056	-0.0093	9,32	-0.0088	-0.0053
9,33	-0.0049	0.0083	9,34	0.0079	0.0047	9,35	0.0044	-0.0075	9,36	-0.0072	-0.0042
9,37	-0.0039	0.0068									
10,10	1.0554	2.0147	10,11	0.7395	0.1002	10,12	0.1381	-0.3135	10,13	-0.1668	-0.1005
10,14	-0.0715	0.1042	10,15	0.0726	0.0523	10,16	0.0396	-0.0548	10,17	-0.0437	-0.0311
10,18	-0.0252	0.0362	10,19	0.0308	0.0210	10,20	0.0179	-0.0267	10,21	-0.0235	-0.0155
10,22	-0.0137	0.0210	10,23	0.0188	0.0121	10,24	0.0109	-0.0171	10,25	-0.0156	-0.0099
10,26	-0.0090	0.0143	10,27	0.0132	0.0082	10,28	0.0076	-0.0122	10,29	-0.0114	-0.0070
10,30	-0.0065	0.0107	10,31	0.0100	0.0060	10,32	0.0057	-0.0094	10,33	-0.0089	-0.0053
10,34	-0.0050	0.0084	10,35	0.0079	0.0047	10,36	0.0045	-0.0076			
11,11	1.0556	2.0145	11,12	0.7394	0.1000	11,13	0.1380	-0.3133	11,14	-0.1667	-0.1003
[illegible]	0.0716	0.1041	11,16	0.0725	0.0522	11,17	0.0395	-0.0547	11,18	-0.0436	-0.0310
11,19	-0.0251	0.0361	[illegible]	[illegible]	0.0209	11,21	0.0178	-0.0267	11,22	-0.0235	-0.0155
11,23	-0.0136	0.0209	11,24	0.0188	0.0121	[illegible]	[illegible]	-0.0170	11,26	-0.0155	-0.0098
11,27	-0.0089	0.0142	11,28	0.0132	0.0082	11,29	0.0075	-0.0122	11,30	[illegible]	0.0070
11,31	-0.0064	0.0106	11,32	0.0100	0.0060	11,33	0.0056	-0.0094	11,34	-0.0088	-0.0053
11,35	-0.0049	0.0083									
12,12	1.0554	2.0147	12,13	0.7395	0.1001	12,14	0.1381	-0.3134	12,15	-0.1668	-0.1004
12,16	-0.0715	0.1042	12,17	0.0726	0.0522	12,18	0.0396	-0.0548	12,19	-0.0437	-0.0311
12,20	-0.0252	0.0362	12,21	0.0308	0.0210	12,22	0.0179	-0.0267	12,23	-0.0235	-0.0155
12,24	-0.0137	0.0210	12,25	0.0188	0.0121	12,26	0.0109	-0.0171	12,27	-0.0156	-0.0098
12,28	-0.0090	0.0143	12,29	0.0132	0.0082	12,30	0.0076	-0.0122	12,31	-0.0114	-0.0070
12,32	-0.0065	0.0107	12,33	0.010C	0.0060	12,34	0.0057	-0.0094			
13,13	1.0555	2.0146	13,14	0.7394	0.1000	13,15	0.1380	-0.3133	13,16	-0.1667	-0.1004
13,17	-0.0714	0.1041	13,18	0.0726	0.0522	13,19	0.0395	-0.0547	13,20	-0.0436	-0.0310
13,21	-0.0251	0.0361	13,22	0.0308	0.0210	13,23	0.0179	-0.0267	13,24	-0.0235	-0.0155
13,25	-0.0136	0.0209	13,26	0.0188	0.0121	13,27	0.0108	-0.0170	13,28	-0.0155	-0.0098
13,29	-0.0089	0.0143	13,30	0.0132	0.0082	13,31	0.0075	-0.0122	13,32	-0.0114	-0.0070
13,33	-0.0064	0.0106									
14,14	1.0554	2.0146	14,15	0.7394	0.1001	14,16	0.1381	-0.3134	14,17	-0.1668	-0.1004
14,18	-0.0715	0.1042	14,19	0.0726	0.0522	14,20	0.0396	-0.0548	14,21	-0.0437	-0.0310
14,22	-0.0252	0.0362	14,23	0.0308	0.0210	14,24	0.0179	-0.0267	14,25	-0.0235	-0.0155
14,26	-0.0137	0.0209	14,27	0.0188	0.0121	14,28	0.0109	-0.0171	14,29	-0.0156	-0.0098
14,30	-0.0090	0.0143	14,31	0.0132	0.0082	14,32	0.0076	-0.0122			
15,15	1.0555	2.0146	15,16	0.7394	0.1001	15,17	0.1380	-0.3134	15,18	-0.1667	-0.1004
15,19	-0.0714	0.1041	15,20	0.0726	0.0522	15,21	0.0395	-0.0547	15,22	-0.0436	-0.0310
15,23	-0.0251	0.0361	15,24	0.0308	0.0210	15,25	0.0179	-0.0267	15,26	-0.0235	-0.0155
15,27	-0.0136	0.0209	15,28	0.0188	0.0121	15,29	0.0109	-0.0170	15,30	-0.0155	-0.0098
15,31	-0.0089	0.0143									
16,16	1.0555	2.0146	16,17	0.7394	0.1001	16,18	0.1381	-0.3134	16,19	-0.1668	-0.1004
16,20	-0.0715	0.1042	16,21	0.0726	0.0522	16,22	0.0396	-0.0548	16,23	-0.0436	-0.0310
16,24	-0.0252	0.0362	16,25	0.0308	0.0210	16,26	0.0179	-0.0267	16,27	-0.0235	-0.0155
16,28	-0.0137	0.0209	16,29	0.0188	0.0121	16,30	0.0109	-0.0171			
17,17	1.0555	2.0146	17,18	0.7394	0.1001	17,19	0.1380	-0.3134	17,20	-0.1667	-0.1004
17,21	-0.0714	0.1041	17,22	0.0726	0.0522	17,23	0.0395	-0.0547	17,24	-0.0436	-0.0310
17,25	-0.0251	0.0361	17,26	0.0308	0.0210	17,27	0.0179	-0.0267	17,28	-0.0235	-0.0155
17,29	-0.0136	0.0209									
18,18	1.0555	2.0146	18,19	0.7394	0.1001	18,20	0.1381	-0.3134	18,21	-0.1668	-0.1004
18,22	-0.0714	0.1042	18,23	0.0726	0.0522	18,24	0.0395	-0.0547	18,25	-0.0436	-0.0310
18,26	-0.0252	0.0362	18,27	0.0308	0.0210	18,28	0.0179	-0.0267			
19,19	1.0555	2.0146	19,20	0.7394	0.1001	19,21	0.1380	-0.3134	19,22	-0.1668	-0.1004
19,23	-0.0714	0.1041	19,24	0.0726	0.0522	19,25	0.0395	-0.0547	19,26	-0.0436	-0.0310
19,27	-0.0251	0.0361									
20,20	1.0555	2.0146	20,21	0.7394	0.1001	20,22	0.1381	-0.3134	20,23	-0.1668	-0.1004
20,24	-0.0714	0.1042	20,25	0.0726	0.0522	20,26	0.0395	-0.0547			
21,21	1.0555	2.0146	21,22	0.7394	0.1001	21,23	0.1380	-0.3134	21,24	-0.1668	-0.1004
21,25	-0.0714	0.1041									
22,22	1.0555	2.0146	22,23	0.7394	0.1001	22,24	0.1381	-0.3134			
23,23	1.0555	2.0146									

VBR

SELF AND MUTUAL ADMITTANCES

H/LMDA=0.5000 B/LMDA= 0.250 OMEGA= 9.92

50 ELEMENT ARRAY

I, K	YIK IN MMHO		I, K	YIK IN MMHO		I, K	YIK IN MMHO		I, K	YIK IN MMHO	
	RE	IM		RE	IM		RE	IM		RE	IM
1, 1	1.0354	1.8545	1, 2	0.6712	0.0873	1, 3	0.1258	-0.2776	1, 4	-0.1448	-0.0911
1, 5	-0.0646	0.0891	1, 6	0.0615	0.0470	1, 7	0.0355	-0.0460	1, 8	-0.0365	-0.0279
1, 9	-0.0226	0.0302	1,10	0.0256	0.0188	1,11	0.0160	-0.0222	1,12	-0.0195	-0.0139
1,13	-0.0122	0.0173	1,14	0.0156	0.0108	1,15	0.0097	-0.0141	1,16	-0.0129	-0.0088
1,17	-0.0080	0.0118	1,18	0.0109	0.0073	1,19	0.0067	-0.0101	1,20	-0.0094	-0.0062
1,21	-0.0058	0.0088	1,22	0.0082	0.0054	1,23	0.0050	-0.0077	1,24	-0.0073	-0.0047
1,25	-0.0045	0.0069	1,26	0.0065	0.0042	1,27	0.0040	-0.0062	1,28	-0.0059	-0.0038
1,29	-0.0036	0.0056	1,30	0.0054	0.0034	1,31	0.0032	-0.0052	1,32	-0.0049	-0.0031
1,33	-0.0030	0.0047	1,34	0.0045	0.0028	1,35	0.0027	-0.0044	1,36	-0.0042	-0.0026
1,37	-0.0025	0.0041	1,38	0.0039	0.0024	1,39	0.0023	-0.0038	1,40	-0.0037	-0.0022
1,41	-0.0022	0.0036	1,42	0.0034	0.0021	1,43	0.0020	-0.0033	1,44	-0.0032	-0.0019
1,45	-0.0019	0.0032	1,46	0.0030	0.0018	1,47	0.0018	-0.0030	1,48	-0.0028	-0.0017
1,49	-0.0017	0.0030	1,50	0.0023	0.0015						
2, 2	1.0452	2.0420	2, 3	0.7532	0.1077	2, 4	0.1434	-0.3217	2, 5	-0.1723	-0.1043
2, 6	-0.0743	0.1083	2, 7	0.0758	0.0544	2, 8	0.0413	-0.0574	2, 9	-0.0459	-0.0325
2,10	-0.0264	0.0381	2,11	0.0325	0.0221	2,12	0.0188	-0.0282	2,13	-0.0248	-0.0163
2,14	-0.0144	0.0221	2,15	0.0199	0.0128	2,16	0.0115	-0.0180	2,17	-0.0165	-0.0104
2,18	-0.0095	0.0151	2,19	0.0140	0.0087	2,20	0.0080	-0.0130	2,21	-0.0121	-0.0074
2,22	-0.0068	0.0113	2,23	0.0106	0.0064	2,24	0.0060	-0.0100	2,25	-0.0094	-0.0056
2,26	-0.0053	0.0089	2,27	0.0084	0.0050	2,28	0.0047	-0.0080	2,29	-0.0076	-0.0045
2,30	-0.0042	0.0073	2,31	0.0070	0.0040	2,32	0.0038	-0.0066	2,33	-0.0064	-0.0037
2,34	-0.0035	0.0061	2,35	0.0059	0.0033	2,36	0.0032	-0.0056	2,37	-0.0055	-0.0031
2,38	-0.0029	0.0052	2,39	0.0051	0.0028	2,40	0.0027	-0.0049	2,41	-0.0048	-0.0026
2,42	-0.0025	0.0046	2,43	0.0045	0.0025	2,44	0.0024	-0.0043	2,45	-0.0042	-0.0023
2,46	-0.0022	0.0040	2,47	0.0040	0.0022	2,48	0.0020	-0.0037	2,49	-0.0039	-0.0021
3, 3	1.0612	2.0078	3, 4	0.7354	0.0962	3, 5	0.1353	-0.3107	3, 6	-0.1648	-0.0984
3, 7	-0.0699	0.1026	3, 8	0.0713	0.0509	3, 9	0.0385	-0.0537	3,10	-0.0427	-0.0301
3,11	-0.0244	0.0354	3,12	0.0301	0.0203	3,13	0.0173	-0.0261	3,14	-0.0229	-0.0150
3,15	-0.0131	0.0204	3,16	0.0183	0.0117	3,17	0.0105	-0.0166	3,18	-0.0151	-0.0095
3,19	-0.0086	0.0139	3,20	0.0128	0.0079	3,21	0.0072	-0.0119	3,22	-0.0111	-0.0067
3,23	-0.0062	0.0103	3,24	0.0097	0.0058	3,25	0.0054	-0.0091	3,26	-0.0086	-0.0051
3,27	-0.0048	0.0081	3,28	0.0077	0.0045	3,29	0.0043	-0.0073	3,30	-0.0070	-0.0040
3,31	-0.0038	0.0066	3,32	0.0063	0.0036	3,33	0.0035	-0.0061	3,34	-0.0058	-0.0033
3,35	-0.0032	0.0056	3,36	0.0054	0.0030	3,37	0.0029	-0.0052	3,38	-0.0050	-0.0028
3,39	-0.0027	0.0048	3,40	0.0046	0.0025	3,41	0.0025	-0.0045	3,42	-0.0043	-0.0024
3,43	-0.0023	0.0042	3,44	0.0040	0.0022	3,45	0.0022	-0.0039	3,46	-0.0038	-0.0020
3,47	-0.0020	0.0037	3,48	0.0035	0.0019						
4, 4	1.0528	2.0167	4, 5	0.7408	0.1020	4, 6	0.1395	-0.3144	4, 7	-0.1675	-0.1015
4, 8	-0.0723	0.1046	4, 9	0.0731	0.0529	4,10	0.0402	-0.0552	4,11	-0.0441	-0.0316
4,12	-0.0256	0.0365	4,13	0.0311	0.0214	4,14	0.0183	-0.0270	4,15	-0.0238	-0.0158
4,16	-0.0139	0.0212	4,17	0.0190	0.0124	4,18	0.0111	-0.0173	4,19	-0.0158	-0.0101
4,20	-0.0092	0.0145	4,21	0.0134	0.0084	4,22	0.0077	-0.0124	4,23	-0.0115	-0.0072
4,24	-0.0066	0.0108	4,25	0.0101	0.0062	4,26	0.0058	-0.0095	4,27	-0.0090	-0.0054
4,28	-0.0051	0.0085	4,29	0.0081	0.0048	4,30	0.0046	-0.0076	4,31	-0.0073	-0.0043
4,32	-0.0041	0.0069	4,33	0.0066	0.0039	4,34	0.0037	-0.0063	4,35	-0.0061	-0.0036
4,36	-0.0034	0.0058	4,37	0.0056	0.0033	4,38	0.0031	-0.0054	4,39	-0.0052	-0.0030
4,40	-0.0029	0.0050	4,41	0.0049	0.0028	4,42	0.0026	-0.0047	4,43	-0.0045	-0.0026
4,44	-0.0024	0.0044	4,45	0.0043	0.0024	4,46	0.0023	-0.0041	4,47	-0.0040	-0.0023
5, 5	1.0568	2.0136	5, 6	0.7388	0.0991	5, 7	0.1373	-0.3129	5, 8	-0.1664	-0.0998
5, 9	-0.0709	0.1038	5,10	0.0723	0.0518	5,11	0.0392	-0.0545	5,12	-0.0434	-0.0307
5,13	-0.0249	0.0360	5,14	0.0306	0.0207	5,15	0.0176	-0.0265	5,16	-0.0234	-0.0153
5,17	-0.0134	0.0208	5,18	0.0187	0.0119	5,19	0.0107	-0.0169	5,20	-0.0154	-0.0097
5,21	-0.0088	0.0142	5,22	0.0131	0.0081	5,23	0.0074	-0.0121	5,24	-0.0113	-0.0069
5,25	-0.0064	0.0106	5,26	0.0099	0.0059	5,27	0.0055	-0.0093	5,28	-0.0088	-0.0052
5,29	-0.0049	0.0083	5,30	0.0079	0.0046	5,31	0.0044	-0.0075	5,32	-0.0071	-0.0041
5,33	-0.0039	0.0068	5,34	0.0065	0.0037	5,35	0.0035	-0.0062	5,36	-0.0059	-0.0034
5,37	-0.0032	0.0057	5,38	0.0055	0.0031	5,39	0.0030	-0.0053	5,40	-0.0051	-0.0028
5,41	-0.0027	0.0049	5,42	0.0047	0.0026	5,43	0.0025	-0.0046	5,44	-0.0044	-0.0024
5,45	-0.0024	0.0043	5,46	0.0041	0.0022						
6, 6	1.0548	2.0150	6, 7	0.7397	0.1006	6, 8	0.1385	-0.3136	6, 9	-0.1670	-0.1008
6,10	-0.0717	0.1043	6,11	0.0727	0.0525	6,12	0.0398	-0.0549	6,13	-0.0438	-0.0312
6,14	-0.0253	0.0363	6,15	0.0309	0.0211	6,16	0.0180	-0.0268	6,17	-0.0236	-0.0156
6,18	-0.0138	0.0210	6,19	0.0189	0.0122	6,20	0.0110	-0.0171	6,21	-0.0156	-0.0099
6,22	-0.0090	0.0143	6,23	0.0132	0.0083	6,24	0.0076	-0.0123	6,25	-0.0114	-0.0071
6,26	-0.0065	0.0107	6,27	0.0100	0.0061	6,28	0.0057	-0.0094	6,29	-0.0089	-0.0054
6,30	-0.0050	0.0084	6,31	0.0080	0.0048	6,32	0.0045	-0.0076	6,33	-0.0072	-0.0043
6,34	-0.0040	0.0069	6,35	0.0066	0.0038	6,36	0.0037	-0.0063	6,37	-0.0060	-0.0035
6,38	-0.0033	0.0058	6,39	0.0056	0.0032	6,40	0.0031	-0.0053	6,41	-0.0052	-0.0029
6,42	-0.0028	0.0050	6,43	0.0048	0.0027	6,44	0.0026	-0.0046	6,45	-0.0045	-0.0025
7, 7	1.0559	2.0144	7, 8	0.7392	0.0998	7, 9	0.1378	-0.3132	7,10	-0.1666	-0.1002
7,11	-0.0712	0.1040	7,12	0.0725	0.0520	7,13	0.0394	-0.0546	7,14	-0.0435	-0.0309
7,15	-0.0250	0.0361	7,16	0.0307	0.0209	7,17	0.0178	-0.0266	7,18	-0.0234	-0.0154
7,19	-0.0135	0.0209	7,20	0.0187	0.0120	7,21	0.0108	-0.0170	7,22	-0.0155	-0.0098
7,23	-0.0089	0.0142	7,24	0.0131	0.0081	7,25	0.0075	-0.0122	7,26	-0.0113	-0.0069
7,27	-0.0064	0.0106	7,28	0.0099	0.0060	7,29	0.0056	-0.0094	7,30	-0.0088	-0.0052
7,31	-0.0049	0.0083	7,32	0.0079	0.0046	7,33	0.0044	-0.0075	7,34	-0.0071	-0.0042
7,35	-0.0040	0.0068	7,36	0.0065	0.0037	7,37	0.0036	-0.0062	7,38	-0.0060	-0.0034
7,39	-0.0033	0.0057	7,40	0.0055	0.0031	7,41	0.0030	-0.0053	7,42	-0.0051	-0.0029
7,43	-0.0028	0.0049	7,44	0.0047	0.0026						
8, 8	1.0552	2.0148	8, 9	0.7395	0.1003	8,10	0.1382	-0.3135	8,11	-0.1669	-0.1005
8,12	-0.0716	0.1042	8,13	0.0727	0.0523	8,14	0.0396	-0.0548	8,15	-0.0437	-0.0311
8,16	-0.0252	0.0362	8,17	0.0308	0.0211	8,18	0.0180	-0.0267	8,19	-0.0236	-0.0156
8,20	-0.0137	0.0210	8,21	0.0189	0.0122	8,22	0.0109	-0.0171	8,23	-0.0156	-0.0099
8,24	-0.0090	0.0143	8,25	0.0132	0.0082	8,26	0.0076	-0.0122	8,27	-0.0114	-0.0070
8,28	-0.0065	0.0107	8,29	0.0100	0.0061	8,30	0.0057	-0.0094	8,31	-0.0089	-0.0053
8,32	-0.0050	0.0084	8,33	0.0080	0.0047	8,34	0.0045	-0.0076	8,35	-0.0072	-0.0042
8,36	-0.0040	0.0069	8,37	0.0066	0.0038	8,38	0.0036	-0.0063	8,39	-0.0060	-0.0035
8,40	-0.0033	0.0058	8,41	0.0056	0.0032	8,42	0.0030	-0.0053	8,43	-0.0052	-0.0029
9, 9	1.0556	2.0145	9,10	0.7393	0.0999	9,11	0.1379	-0.3133	9,12	-0.1667	-0.1003
9,13	-0.0713	0.1041	9,14	0.0725	0.0521	9,15	0.0395	-0.0547	9,16	-0.0436	-0.0310
9,17	-0.0251	0.0361	9,18	0.0307	0.0209	9,19	0.0178	-0.0267	9,20	-0.0235	-0.0155
9,21	-0.0136	0.0209	9,22	0.0188	0.0121	9,23	0.0108	-0.0170	9,24	-0.0155	-0.0098
9,25	-0.0089	0.0142	9,26	0.0131	0.0082	9,27	0.0075	-0.0122	9,28	-0.0113	-0.0069
9,29	-0.0064	0.0106	9,30	0.0099	0.0060	9,31	0.0056	-0.0094	9,32	-0.0088	-0.0053
9,33	-0.0050	0.0084	9,34	0.0079	0.0047	9,35	0.0044	-0.0075	9,36	-0.0071	-0.0042
9,37	-0.0040	0.0068	9,38	0.0065	0.0038	9,39	0.0036	-0.0062	9,40	-0.0060	-0.0034
9,41	-0.0033	0.0058	9,42	0.0055	0.0031						

10,10	1.0554	2.0147	10,11	0.7395	0.1002	10,12	0.1381	-0.3135	10,13	-0.1668	-0.1005
10,14	-0.0715	0.1042	10,15	0.0726	0.0523	10,16	0.0396	-0.0548	10,17	-0.0437	-0.0311
10,18	-0.0252	0.0362	10,19	0.0308	0.0210	10,20	0.0179	-0.0267	10,21	-0.0235	-0.0155
10,22	-0.0137	0.0209	10,23	0.0188	0.0122	10,24	0.0109	-0.0171	10,25	-0.0156	-0.0099
10,26	-0.0090	0.0143	10,27	0.0132	0.0082	10,28	0.0076	-0.0122	10,29	-0.0114	-0.0070
10,30	-0.0065	0.0106	10,31	0.0100	0.0061	10,32	0.0057	-0.0094	10,33	-0.0089	-0.0053
10,34	-0.0050	0.0084	10,35	0.0080	0.0047	10,36	0.0044	-0.0075	10,37	-0.0072	-0.0042
10,38	-0.0040	0.0068	10,39	0.0066	0.0038	10,40	0.0036	-0.0063	10,41	-0.0060	-0.0035
11,11	1.0556	2.0145	11,12	0.7393	0.1000	11,13	0.1380	-0.3133	11,14	-0.1667	-0.1003
11,15	-0.0714	0.1041	11,16	0.0725	0.0522	11,17	0.0395	-0.0547	11,18	-0.0436	-0.0310
11,19	-0.0251	0.0361	11,20	0.0307	0.0209	11,21	0.0179	-0.0267	11,22	-0.0235	-0.0155
11,23	-0.0136	0.0209	11,24	0.0188	0.0121	11,25	0.0109	-0.0170	11,26	-0.0155	-0.0098
11,27	-0.0089	0.0143	11,28	0.0131	0.0082	11,29	0.0075	-0.0122	11,30	-0.0114	-0.0070
11,31	-0.0065	0.0106	11,32	0.0099	0.0060	11,33	0.0056	-0.0094	11,34	-0.0088	-0.0053
11,35	-0.0050	0.0084	11,36	0.0079	0.0047	11,37	0.0044	-0.0075	11,38	-0.0072	-0.0042
11,39	-0.0040	0.0068	11,40	0.0065	0.0038						
12,12	1.0554	2.0147	12,13	0.7395	0.1001	12,14	0.1381	-0.3134	12,15	-0.1668	-0.1004
12,16	-0.0715	0.1042	12,17	0.0726	0.0523	12,18	0.0396	-0.0548	12,19	-0.0437	-0.0311
12,20	-0.0252	0.0362	12,21	0.0308	0.0210	12,22	0.0179	-0.0267	12,23	-0.0235	-0.0155
12,24	-0.0137	0.0209	12,25	0.0188	0.0121	12,26	0.0109	-0.0171	12,27	-0.0156	-0.0099
12,28	-0.0090	0.0143	12,29	0.0132	0.0082	12,30	0.0076	-0.0122	12,31	-0.0114	-0.0070
12,32	-0.0065	0.0106	12,33	0.0100	0.0061	12,34	0.0056	-0.0094	12,35	-0.0089	-0.0053
12,36	-0.0050	0.0084	12,37	0.0080	0.0047	12,38	0.0044	-0.0075	12,39	-0.0072	-0.0042
13,13	1.0555	2.0146	13,14	0.7394	0.1000	13,15	0.1380	-0.3134	13,16	-0.1667	-0.1004
13,17	-0.0714	0.1041	13,18	0.0725	0.0522	13,19	0.0395	-0.0547	13,20	-0.0436	-0.0310
13,21	-0.0251	0.0361	13,22	0.0308	0.0210	13,23	0.0179	-0.0267	13,24	-0.0235	-0.0155
13,25	-0.0136	0.0209	13,26	0.0188	0.0121	13,27	0.0109	-0.0170	13,28	-0.0155	-0.0098
13,29	-0.0089	0.0143	13,30	0.0132	0.0082	13,31	0.0075	-0.0122	13,32	-0.0114	-0.0070
13,33	-0.0065	0.0106	13,34	0.0100	0.0060	13,35	0.0056	-0.0094	13,36	-0.0088	-0.0053
13,37	-0.0050	0.0084	13,38	0.0079	0.0047						
14,14	1.0554	2.0146	14,15	0.7394	0.1001	14,16	0.1381	-0.3134	14,17	-0.1668	-0.1004
14,18	-0.0715	0.1042	14,19	0.0726	0.0522	14,20	0.0396	-0.0547	14,21	-0.0437	-0.0311
14,22	-0.0252	0.0362	14,23	0.0308	0.0210	14,24	0.0179	-0.0267	14,25	-0.0235	-0.0155
14,26	-0.0136	0.0209	14,27	0.0188	0.0121	14,28	0.0109	-0.0171	14,29	-0.0156	-0.0099
14,30	-0.0090	0.0143	14,31	0.0132	0.0082	14,32	0.0075	-0.0122	14,33	-0.0114	-0.0070
14,34	-0.0065	0.0106	14,35	0.0100	0.0061	14,36	0.0056	-0.0094	14,37	-0.0089	-0.0053
15,15	1.0555	2.0146	15,16	0.7394	0.1000	15,17	0.1380	-0.3134	15,18	-0.1667	-0.1004
15,19	-0.0714	0.1041	15,20	0.0726	0.0522	15,21	0.0395	-0.0547	15,22	-0.0436	-0.0310
15,23	-0.0252	0.0361	15,24	0.0308	0.0210	15,25	0.0179	-0.0267	15,26	-0.0235	-0.0155
15,27	-0.0136	0.0209	15,28	0.0188	0.0121	15,29	0.0109	-0.0170	15,30	-0.0155	-0.0098
15,31	-0.0089	0.0143	15,32	0.0132	0.0082	15,33	0.0075	-0.0122	15,34	-0.0114	-0.0070
15,35	-0.0065	0.0106	15,36	0.0100	0.0060						
16,16	1.0555	2.0146	16,17	0.7394	0.1001	16,18	0.1381	-0.3134	16,19	-0.1668	-0.1004
16,20	-0.0714	0.1042	16,21	0.0726	0.0522	16,22	0.0395	-0.0547	16,23	-0.0437	-0.0310
16,24	-0.0252	0.0362	16,25	0.0308	0.0210	16,26	0.0179	-0.0267	16,27	-0.0235	-0.0155
16,28	-0.0136	0.0209	16,29	0.0188	0.0121	16,30	0.0109	-0.0170	16,31	-0.0156	-0.0098
16,32	-0.0090	0.0143	16,33	0.0132	0.0082	16,34	0.0075	-0.0122	16,35	-0.0114	-0.0070
17,17	1.0555	2.0146	17,18	0.7394	0.1001	17,19	0.1380	-0.3134	17,20	-0.1667	-0.1004
17,21	-0.0714	0.1041	17,22	0.0726	0.0522	17,23	0.0395	-0.0547	17,24	-0.0436	-0.0310
17,25	-0.0252	0.0362	17,26	0.0308	0.0210	17,27	0.0179	-0.0267	17,28	-0.0235	-0.0155
17,29	-0.0136	0.0209	17,30	0.0188	0.0121	17,31	0.0109	-0.0170	17,32	-0.0155	-0.0098
17,33	-0.0089	0.0143	17,34	0.0132	0.0082						
18,18	1.0555	2.0146	18,19	0.7394	0.1001	18,20	0.1381	-0.3134	18,21	-0.1668	-0.1004
18,22	-0.0714	0.1042	18,23	0.0726	0.0522	18,24	0.0395	-0.0547	18,25	-0.0437	-0.0310
18,26	-0.0252	0.0362	18,27	0.0308	0.0210	18,28	0.0179	-0.0267	18,29	-0.0235	-0.0155
18,30	-0.0136	0.0209	18,31	0.0188	0.0121	18,32	0.0109	-0.0170	18,33	-0.0156	-0.0098
19,19	1.0555	2.0146	19,20	0.7394	0.1001	19,21	0.1380	-0.3134	19,22	-0.1667	-0.1004
19,23	-0.0714	0.1041	19,24	0.0726	0.0522	19,25	0.0395	-0.0547	19,26	-0.0436	-0.0310
19,27	-0.0252	0.0362	19,28	0.0308	0.0210	19,29	0.0179	-0.0267	19,30	-0.0235	-0.0155
19,31	-0.0136	0.0209	19,32	0.0188	0.0121						
20,20	1.0555	2.0146	20,21	0.7394	0.1001	20,22	0.1381	-0.3134	20,23	-0.1668	-0.1004
20,24	-0.0714	0.1041	20,25	0.0726	0.0522	20,26	0.0395	-0.0547	20,27	-0.0437	-0.0310
20,28	-0.0252	0.0362	20,29	0.0308	0.0210	20,30	0.0179	-0.0267	20,31	-0.0235	-0.0155
21,21	1.0555	2.0146	21,22	0.7394	0.1001	21,23	0.1380	-0.3134	21,24	-0.1667	-0.1004
21,25	-0.0714	0.1041	21,26	0.0726	0.0522	21,27	0.0395	-0.0547	21,28	-0.0436	-0.0310
21,29	-0.0252	0.0362	21,30	0.0308	0.0210						
22,22	1.0555	2.0146	22,23	0.7394	0.1001	22,24	0.1381	-0.3134	22,25	-0.1668	-0.1004
22,26	-0.0714	0.1041	22,27	0.0726	0.0522	22,28	0.0395	-0.0547	22,29	-0.0437	-0.0310
23,23	1.0555	2.0146	23,24	0.7394	0.1001	23,25	0.1380	-0.3134	23,26	-0.1667	-0.1004
23,27	-0.0714	0.1041	23,28	0.0726	0.0522						
24,24	1.0555	2.0146	24,25	0.7394	0.1001	24,26	0.1380	-0.3134	24,27	-0.1668	-0.1004
25,25	1.0555	2.0146	25,26	0.7394	0.1001						

VBR

SELF AND MUTUAL ADMITTANCES

H/LMDA=0.5000 B/LMDA= 0.500 OMEGA= 9.92

2 ELEMENT ARRAY

I, K	YIK IN MMHO RE	IM	I, K	YIK IN MMHO RE	IM	I, K	YIK IN MMHO RE	IM	I, K	YIK IN MMHO RE	IM
1, 1	1.0286	1.6363	1, 2	0.1062	-0.4574						

VBR

SELF AND MUTUAL ADMITTANCES

H/LMDA=0.5000 B/LMDA= 0.500 OMEGA= 9.92

3 ELEMENT ARRAY

I, K	YIK IN MMHO RE	IM	I, K	YIK IN MMHO RE	IM	I, K	YIK IN MMHO RE	IM	I, K	YIK IN MMHO RE	IM
1, 1	1.0390	1.6183	1, 2	0.0939	-0.4214	1, 3	-0.0941	0.2153			
2, 2	1.0342	1.5662									

VBR

SELF AND MUTUAL ADMITTANCES

H/LMDA=0.5000 B/LMDA= 0.500 OMEGA= 9.92

4 ELEMENT ARRAY

I, K	YIK IN MMHO RE	IM	I, K	YIK IN MMHO RE	IM	I, K	YIK IN MMHO RE	IM	I, K	YIK IN MMHO RE	IM
1, 1	1.0444	1.6126	1, 2	0.0871	-0.4125	1, 3	-0.0844	0.1952	1, 4	0.0739	-0.1291
2, 2	1.0426	1.5515	2, 3	0.0854	-0.3887						

VBR

SELF AND MUTUAL ADMITTANCES

H/LMDA=0.5000 B/LMDA= 0.500 OMEGA= 9.92

5 ELEMENT ARRAY

I, K	YIK IN MMHO RE	IM	I, K	YIK IN MMHO RE	IM	I, K	YIK IN MMHO RE	IM	I, K	YIK IN MMHO RE	IM
1, 1	1.0472	1.6102	1, 2	0.0836	-0.4092	1, 3	-0.0798	0.1896	1, 4	0.0668	-0.1160
1, 5	-0.0571	0.0877									
2, 2	1.0471	1.5470	2, 3	0.0795	-0.3810	2, 4	-0.0770	0.1755			
3, 3	1.0502	1.5376									

VBR

SELF AND MUTUAL ADMITTANCES

H/LMDA=0.5000 B/LMDA= 0.500 OMEGA= 9.92

6 ELEMENT ARRAY

I, K	YIK IN MMHO RE	IM	I, K	YIK IN MMHO RE	IM	I, K	YIK IN MMHO RE	IM	I, K	YIK IN MMHO RE	IM
1, 1	1.0489	1.6090	1, 2	0.0817	-0.4077	1, 3	-0.0773	0.1873	1, 4	0.0634	-0.1120
1, 5	-0.0518	0.0781	1, 6	0.0450	-0.0651						
2, 2	1.0494	1.5452	2, 3	0.0765	-0.3783	2, 4	-0.0728	0.1707	2, 5	0.0608	-0.1026
3, 3	1.0542	1.5334	3, 4	0.0739	-0.3730						

VBR

SELF AND MUTUAL ADMITTANCES

H/LMDA=0.5000 B/LMDA= 0.500 OMEGA= 9.92

7 ELEMENT ARRAY

I, K	YIK IN MMHO RE	IM	I, K	YIK IN MMHO RE	IM	I, K	YIK IN MMHO RE	IM	I, K	YIK IN MMHO RE	IM
1, 1	1.0499	1.6082	1, 2	0.0806	-0.4068	1, 3	-0.0758	0.1861	1, 4	0.0615	-0.1102
1, 5	-0.0491	0.0749	1, 6	0.0409	-0.0576	1, 7	-0.0365	0.0513			
2, 2	1.0507	1.5442	2, 3	0.0749	-0.3770	2, 4	-0.0707	0.1687	2, 5	0.0578	-0.0991
2, 6	-0.0470	0.0682									
3, 3	1.0563	1.5317	3, 4	0.0712	-0.3705	3, 5	-0.0687	0.1654			
4, 4	1.0580	1.5293									

VBR

SELF AND MUTUAL ADMITTANCES

H/LMDA=0.5000 B/LMDA= 0.500 OMEGA= 9.92

8 ELEMENT ARRAY

I, K	YIK IN MMHO RE	IM	I, K	YIK IN MMHO RE	IM	I, K	YIK IN MMHO RE	IM	I, K	YIK IN MMHO RE	IM
1, 1	1.0506	1.6076	1, 2	0.0798	-0.4062	1, 3	-0.0749	0.1854	1, 4	0.0603	-0.1092
1, 5	-0.0476	0.0735	1, 6	0.0387	-0.0550	1, 7	-0.0331	0.0452	1, 8	0.0304	-0.0422
2, 2	1.0515	1.5436	2, 3	0.0739	-0.3763	2, 4	-0.0694	0.1677	2, 5	0.0561	-0.0976
2, 6	-0.0446	0.0655	2, 7	0.0369	-0.0498						
3, 3	1.0575	1.5309	3, 4	0.0697	-0.3693	3, 5	-0.0666	0.1635	3, 6	0.0545	-0.0953
4, 4	1.0600	1.5277	4, 5	0.0684	-0.3678						

VBR

SELF AND MUTUAL ADMITTANCES

H/LMDA=0.5000 B/LMDA= 0.500 OMEGA= 9.92

9 ELEMENT ARRAY

I, K	YIK IN MMHO RE	IM	I, K	YIK IN MMHO RE	IM	I, K	YIK IN MMHO RE	IM	I, K	YIK IN MMHO RE	IM
1, 1	1.0511	1.6073	1, 2	0.0793	-0.4058	1, 3	-0.0743	0.1850	1, 4	0.0596	-0.1087
1, 5	-0.0467	0.0727	1, 6	0.0374	-0.0537	1, 7	-0.0313	0.0430	1, 8	0.0276	-0.0370
1, 9	-0.0259	0.0358									
2, 2	1.0521	1.5432	2, 3	0.0733	-0.3759	2, 4	-0.0686	0.1672	2, 5	0.0551	-0.0968
2, 6	-0.0433	0.0642	2, 7	0.0350	-0.0475	2, 8	-0.0299	0.0388			
3, 3	1.0582	1.5303	3, 4	0.0688	-0.3686	3, 5	-0.0655	0.1626	3, 6	0.0530	-0.0939
3, 7	-0.0420	0.0625									
4, 4	1.0611	1.5269	4, 5	0.0670	-0.3667	4, 6	-0.0645	0.1617			
5, 5	1.0619	1.5261									

VBR

SELF AND MUTUAL ADMITTANCES

H/LMDA=0.5000 B/LMDA= 0.500 OMEGA= 9.92

10 ELEMENT ARRAY

I, K	YIK IN MMHO RE	IM	I, K	YIK IN MMHO RE	IM	I, K	YIK IN MMHO RE	IM	I, K	YIK IN MMHO RE	IM
1, 1	1.0515	1.6070	1, 2	0.0789	-0.4055	1, 3	-0.0739	0.1846	1, 4	0.0591	-0.1083
1, 5	-0.0460	0.0722	1, 6	0.0366	-0.0530	1, 7	-0.0303	0.0419	1, 8	0.0261	-0.0351
1, 9	-0.0236	0.0313	1,10	0.0225	-0.0310						
2, 2	1.0525	1.5430	2, 3	0.0728	-0.3756	2, 4	-0.0681	0.1668	2, 5	0.0544	-0.0964
2, 6	-0.0425	0.0636	2, 7	0.0339	-0.0465	2, 8	-0.0283	0.0369	2, 9	0.0248	-0.0316
3, 3	1.0587	1.5300	3, 4	0.0682	-0.3682	3, 5	-0.0648	0.1621	3, 6	0.0520	-0.0931
3, 7	-0.0408	0.0613	3, 8	0.0329	-0.0451						
4, 4	1.0618	1.5264	4, 5	0.0661	-0.3660	4, 6	-0.0634	0.1608	4, 7	0.0513	-0.0924
5, 5	1.0630	1.5253	5, 6	0.0655	-0.3655						

VBR

SELF AND MUTUAL ADMITTANCES

H/LMDA=0.5000 B/LMDA= 0.500 OMEGA= 9.92

12 ELEMENT ARRAY

I, K	YIK IN MMHO RE	IM	I, K	YIK IN MMHO RE	IM	I, K	YIK IN MMHO RE	IM	I, K	YIK IN MMHO RE	IM
1, 1	1.0520	1.6066	1, 2	0.0783	-0.4052	1, 3	-0.0733	0.1842	1, 4	0.0584	-0.1078
1, 5	-0.0453	0.0716	1, 6	0.0357	-0.0523	1, 7	-0.0291	0.0409	1, 8	0.0245	-0.0336
1, 9	-0.0214	0.0287	1,10	0.0193	-0.0255	1,11	-0.0180	0.0237	1,12	0.0177	-0.0243
2, 2	1.0530	1.5426	2, 3	0.0722	-0.3752	2, 4	-0.0675	0.1663	2, 5	0.0537	-0.0958
2, 6	-0.0415	0.0629	2, 7	0.0327	-0.0455	2, 8	-0.0267	0.0355	2, 9	0.0226	-0.0292
2,10	-0.0199	0.0252	2,11	0.0183	-0.0229						
3, 3	1.0593	1.5296	3, 4	0.0675	-0.3677	3, 5	-0.0640	0.1615	3, 6	0.0510	-0.0924
3, 7	-0.0395	0.0603	3, 8	0.0312	-0.0436	3, 9	-0.0255	0.0340	3,10	0.0219	-0.0282
4, 4	1.0626	1.5258	4, 5	0.0652	-0.3654	4, 6	-0.0623	0.1599	4, 7	0.0498	-0.0913
4, 8	-0.0387	0.0595	4, 9	0.0306	-0.0431						
5, 5	1.0641	1.5245	5, 6	0.0642	-0.3645	5, 7	-0.0616	0.1594	5, 8	0.0494	-0.0910
6, 6	1.0646	1.5240	6, 7	0.0639	-0.3643						

VBR

SELF AND MUTUAL ADMITTANCES

H/LMDA=0.5000 B/LMDA= 0.500 OMEGA= 9.92

14 ELEMENT ARRAY

I, K	YIK IN MMHO RE	IM	I, K	YIK IN MMHO RE	IM	I, K	YIK IN MMHO RE	IM	I, K	YIK IN MMHO RE	IM
1, 1	1.0523	1.6063	1, 2	0.0780	-0.4049	1, 3	-0.0730	0.1840	1, 4	0.0580	-0.1075
1, 5	-0.0449	0.0713	1, 6	0.0352	-0.0519	1, 7	-0.0285	0.0404	1, 8	0.0238	-0.0330
1, 9	-0.0204	0.0279	1,10	0.0180	-0.0243	1,11	-0.0163	0.0217	1,12	0.0151	-0.0199
1,13	-0.0145	0.0190	1,14	0.0145	-0.0200						
2, 2	1.0533	1.5424	2, 3	0.0719	-0.3749	2, 4	-0.0671	0.1661	2, 5	0.0533	-0.0955
2, 6	-0.0410	0.0625	2, 7	0.0321	-0.0451	2, 8	-0.0260	0.0349	2, 9	0.0217	-0.0284
2,10	-0.0187	0.0240	2,11	0.0166	-0.0210	2,12	-0.0152	0.0189	2,13	0.0144	-0.0178
3, 3	1.0597	1.5293	3, 4	0.0671	-0.3674	3, 5	-0.0635	0.1612	3, 6	0.0505	-0.0920
3, 7	-0.0389	0.0599	3, 8	0.0304	-0.0430	3, 9	-0.0246	0.0332	3,10	0.0206	-0.0270
3,11	-0.0179	0.0229	3,12	0.0160	-0.0202						
4, 4	1.0630	1.5255	4, 5	0.0647	-0.3650	4, 6	-0.0617	0.1595	4, 7	0.0492	-0.0908
4, 8	-0.0378	0.0589	4, 9	0.0296	-0.0423	4,10	-0.0240	0.0327	4,11	0.0202	-0.0267
5, 5	1.0646	1.5241	5, 6	0.0636	-0.3640	5, 7	-0.0609	0.1588	5, 8	0.0485	-0.0903
5, 9	-0.0374	0.0586	5,10	0.0293	-0.0421						
6, 6	1.0654	1.5235	6, 7	0.0630	-0.3636	6, 8	-0.0605	0.1586	6, 9	0.0483	-0.0901
7, 7	1.0657	1.5232	7, 8	0.0629	-0.3635						

VBR

SELF AND MUTUAL ADMITTANCES

H/LMDA=0.5000 B/LMDA= 0.500 OMEGA= 9.92

16 ELEMENT ARRAY

I, K	YIK IN MMHO RE	IM	I, K	YIK IN MMHO RE	IM	I, K	YIK IN MMHO RE	IM	I, K	YIK IN MMHO RE	IM
1, 1	1.0526	1.6061	1, 2	0.0778	-0.4047	1, 3	-0.0727	0.1838	1, 4	0.0578	-0.1073
1, 5	-0.0446	0.0711	1, 6	0.0349	-0.0517	1, 7	-0.0281	0.0401	1, 8	0.0234	-0.0327
1, 9	-0.0199	0.0275	1,10	0.0174	-0.0238	1,11	-0.0155	0.0210	1,12	0.0141	-0.0189
1,13	-0.0130	0.0173	1,14	0.0124	-0.0162	1,15	-0.0120	0.0158	1,16	0.0122	-0.0169
2, 2	1.0536	1.5422	2, 3	0.0717	-0.3748	2, 4	-0.0669	0.1659	2, 5	0.0530	-0.0953
2, 6	-0.0407	0.0623	2, 7	0.0318	-0.0449	2, 8	-0.0256	0.0346	2, 9	0.0212	-0.0281
2,10	-0.0181	0.0236	2,11	0.0159	-0.0204	2,12	-0.0142	0.0180	2,13	0.0130	-0.0163
2,14	-0.0121	0.0151	2,15	0.0117	-0.0145						
3, 3	1.0599	1.5292	3, 4	0.0669	-0.3673	3, 5	-0.0633	0.1610	3, 6	0.0502	-0.0918
3, 7	-0.0386	0.0596	3, 8	0.0300	-0.0427	3, 9	-0.0241	0.0329	3,10	0.0200	-0.0266
3,11	-0.0171	0.0223	3,12	0.0150	-0.0193	3,13	-0.0135	0.0171	3,14	0.0125	-0.0157
4, 4	1.0633	1.5253	4, 5	0.0644	-0.3648	4, 6	-0.0614	0.1593	4, 7	0.0488	-0.0905
4, 8	-0.0374	0.0586	4, 9	0.0291	-0.0419	4,10	-0.0234	0.0322	4,11	0.0194	-0.0260
4,12	-0.0166	0.0219	4,13	0.0147	-0.0191						
5, 5	1.0649	1.5238	5, 6	0.0632	-0.3638	5, 7	-0.0605	0.1585	5, 8	0.0481	-0.0899
5, 9	-0.0368	0.0581	5,10	0.0287	-0.0415	5,11	-0.0230	0.0319	5,12	0.0192	-0.0259
6, 6	1.0658	1.5232	6, 7	0.0626	-0.3633	6, 8	-0.0600	0.1582	6, 9	0.0477	-0.0897
6,10	-0.0366	0.0580	6,11	0.0285	-0.0415						
7, 7	1.0662	1.5228	7, 8	0.0623	-0.3631	7, 9	-0.0598	0.1580	7,10	0.0476	-0.0896
8, 8	1.0664	1.5227	8, 9	0.0622	-0.3630						

VBR

SELF AND MUTUAL ADMITTANCES

H/LMDA=0.5000 B/LMDA= 0.500 OMEGA= 9.92

18 ELEMENT ARRAY

I, K	YIK IN MMHO RE	IM	I, K	YIK IN MMHO RE	IM	I, K	YIK IN MMHO RE	IM	I, K	YIK IN MMHO RE	IM
1, 1	1.0528	1.6060	1, 2	0.0776	-0.4046	1, 3	-0.0726	0.1836	1, 4	0.0576	-0.1072
1, 5	-0.0444	0.0709	1, 6	0.0347	-0.0515	1, 7	-0.0279	0.0400	1, 8	0.0231	-0.0325
1, 9	-0.0197	0.0273	1,10	0.0171	-0.0235	1,11	-0.0151	0.0207	1,12	0.0136	-0.0185
1,13	-0.0124	0.0167	1,14	0.0115	-0.0154	1,15	-0.0108	0.0143	1,16	0.0104	-0.0137
1,17	-0.0103	0.0135	1,18	0.0105	-0.0146						
2, 2	1.0537	1.5421	2, 3	0.0715	-0.3746	2, 4	-0.0667	0.1658	2, 5	0.0528	-0.0952
2, 6	-0.0405	0.0622	2, 7	0.0316	-0.0447	2, 8	-0.0253	0.0345	2, 9	0.0210	-0.0279
2,10	-0.0178	0.0233	2,11	0.0155	-0.0201	2,12	-0.0137	0.0176	2,13	0.0123	-0.0157
2,14	-0.0113	0.0143	2,15	0.0106	-0.0132	2,16	-0.0101	0.0125	2,17	0.0099	-0.0122
3, 3	1.0601	1.5290	3, 4	0.0667	-0.3671	3, 5	-0.0631	0.1609	3, 6	0.0500	-0.0917
3, 7	-0.0383	0.0595	3, 8	0.0298	-0.0425	3, 9	-0.0238	0.0326	3,10	0.0197	-0.0263
3,11	-0.0167	0.0220	3,12	0.0145	-0.0189	3,13	-0.0129	0.0166	3,14	0.0117	-0.0149
3,15	-0.0108	0.0136	3,16	0.0102	-0.0127						
4, 4	1.0635	1.5252	4, 5	0.0642	-0.3647	4, 6	-0.0612	0.1591	4, 7	0.0486	-0.0903
4, 8	-0.0372	0.0584	4, 9	0.0288	-0.0417	4,10	-0.0230	0.0319	4,11	0.0190	-0.0257
4,12	-0.0161	0.0215	4,13	0.0140	-0.0185	4,14	-0.0125	0.0163	4,15	0.0114	-0.0147
5, 5	1.0651	1.5237	5, 6	0.0630	-0.3636	5, 7	-0.0602	0.1583	5, 8	0.0478	-0.0897
5, 9	-0.0365	0.0579	5,10	0.0283	-0.0413	5,11	-0.0226	0.0316	5,12	0.0187	-0.0255
5,13	-0.0159	0.0213	5,14	0.0139	-0.0184						
6, 6	1.0660	1.5230	6, 7	0.0623	-0.3631	6, 8	-0.0597	0.1579	6, 9	0.0474	-0.0894
6,10	-0.0362	0.0577	6,11	0.0280	-0.0411	6,12	-0.0224	0.0315	6,13	0.0186	-0.0254
7, 7	1.0665	1.5226	7, 8	0.0619	-0.3628	7, 9	-0.0594	0.1577	7,10	0.0472	-0.0892
7,11	-0.0361	0.0576	7,12	0.0280	-0.0410						
8, 8	1.0668	1.5224	8, 9	0.0617	-0.3627	8,10	-0.0593	0.1576	8,11	0.0471	-0.0892
9, 9	1.0669	1.5223	9,10	0.0617	-0.3626						

VBR

SELF AND MUTUAL ADMITTANCES

H/LMDA=0.5000 B/LMDA= 0.500 OMEGA= 9.92

20 ELEMENT ARRAY

I, K	YIK IN MMHO RE	IM	I, K	YIK IN MMHO RE	IM	I, K	YIK IN MMHO RE	IM	I, K	YIK IN MMHO RE	IM
1, 1	1.0529	1.6059	1, 2	0.0775	-0.4045	1, 3	-0.0724	0.1835	1, 4	0.0575	-0.1071
1, 5	-0.0442	0.0708	1, 6	0.0345	-0.0514	1, 7	-0.0277	0.0398	1, 8	0.0229	-0.0323
1, 9	-0.0195	0.0271	1,10	0.0168	-0.0233	1,11	-0.0148	0.0205	1,12	0.0133	-0.0182
1,13	-0.0120	0.0164	1,14	0.0110	-0.0150	1,15	-0.0102	0.0138	1,16	0.0096	-0.0129
1,17	-0.0092	0.0122	1,18	0.0089	-0.0118	1,19	-0.0089	0.0118	1,20	0.0092	-0.0128
2, 2	1.0539	1.5420	2, 3	0.0714	-0.3746	2, 4	-0.0666	0.1657	2, 5	0.0527	-0.0951
2, 6	-0.0404	0.0621	2, 7	0.0314	-0.0446	2, 8	-0.0252	0.0343	2, 9	0.0208	-0.0277
2,10	-0.0176	0.0232	2,11	0.0152	-0.0199	2,12	-0.0134	0.0174	2,13	0.0120	-0.0155
2,14	-0.0109	0.0140	2,15	0.0100	-0.0128	2,16	-0.0094	0.0118	2,17	0.0089	-0.0111
2,18	-0.0086	0.0106	2,19	0.0085	-0.0105						
3, 3	1.0602	1.5290	3, 4	0.0666	-0.3670	3, 5	-0.0629	0.1608	3, 6	0.0499	-0.0916
3, 7	-0.0382	0.0593	3, 8	0.0296	-0.0424	3, 9	-0.0237	0.0325	3,10	0.0195	-0.0262
3,11	-0.0165	0.0218	3,12	0.0142	-0.0187	3,13	-0.0126	0.0164	3,14	0.0112	-0.0146
3,15	-0.0102	0.0131	3,16	0.0095	-0.0120	3,17	-0.0089	0.0112	3,18	0.0085	-0.0107
4, 4	1.0636	1.5251	4, 5	0.0641	-0.3646	4, 6	-0.0611	0.1590	4, 7	0.0484	-0.0902
4, 8	-0.0370	0.0583	4, 9	0.0286	-0.0415	4,10	-0.0228	0.0318	4,11	0.0188	-0.0255
4,12	-0.0159	0.0213	4,13	0.0137	-0.0182	4,14	-0.0121	0.0160	4,15	0.0109	-0.0142
4,16	-0.0099	0.0129	4,17	0.0092	-0.0119						
5, 5	1.0653	1.5236	5, 6	0.0628	-0.3635	5, 7	-0.0601	0.1582	5, 8	0.0476	-0.0896
5, 9	-0.0363	0.0578	5,10	0.0281	-0.0411	5,11	-0.0224	0.0314	5,12	0.0184	-0.0252
5,13	-0.0155	0.0210	5,14	0.0134	-0.0180	5,15	-0.0119	0.0158	5,16	0.0107	-0.0141
6, 6	1.0662	1.5229	6, 7	0.0621	-0.3630	6, 8	-0.0595	0.1578	6, 9	0.0472	-0.0892
6,10	-0.0360	0.0575	6,11	0.0278	-0.0409	6,12	-0.0221	0.0312	6,13	0.0182	-0.0251
6,14	-0.0154	0.0209	6,15	0.0133	-0.0180						
7, 7	1.0667	1.5225	7, 8	0.0617	-0.3627	7, 9	-0.0592	0.1575	7,10	0.0469	-0.0890
7,11	-0.0358	0.0573	7,12	0.0276	-0.0408	7,13	-0.0220	0.0311	7,14	0.0181	-0.0251
8, 8	1.0670	1.5222	8, 9	0.0615	-0.3625	8,10	-0.0590	0.1574	8,11	0.0468	-0.0889
8,12	-0.0357	0.0573	8,13	0.0275	-0.0407						
9, 9	1.0672	1.5221	9,10	0.0614	-0.3624	9,11	-0.0589	0.1573	9,12	0.0467	-0.0889
10,10	1.0672	1.5220	10,11	0.0613	-0.3623						

VBR

SELF AND MUTUAL ADMITTANCES

H/LMDA=0.5000 B/LMDA= 0.500 OMEGA= 9.92

22 ELEMENT ARRAY

I, K	YIK IN MMHO RE	IM	I, K	YIK IN MMHO RE	IM	I, K	YIK IN MMHO RE	IM	I, K	YIK IN MMHO RE	IM
1, 1	1.0530	1.6058	1, 2	0.0774	-0.4044	1, 3	-0.0723	0.1835	1, 4	0.0574	-0.1070
1, 5	-0.0441	0.0707	1, 6	0.0344	-0.0513	1, 7	-0.0276	0.0397	1, 8	0.0228	-0.0322
1, 9	-0.0193	0.0270	1,10	0.0167	-0.0232	1,11	-0.0147	0.0203	1,12	0.0131	-0.0181
1,13	-0.0118	0.0162	1,14	0.0108	-0.0148	1,15	-0.0099	0.0136	1,16	0.0092	-0.0126
1,17	-0.0087	0.0118	1,18	0.0083	-0.0111	1,19	-0.0080	0.0106	1,20	0.0078	-0.0103
1,21	-0.0079	0.0104	1,22	0.0082	-0.0114						
2, 2	1.0539	1.5419	2, 3	0.0713	-0.3745	2, 4	-0.0665	0.1656	2, 5	0.0526	-0.0950
2, 6	-0.0403	0.0620	2, 7	0.0313	-0.0445	2, 8	-0.0251	0.0342	2, 9	0.0206	-0.0276
2,10	-0.0175	0.0231	2,11	0.0151	-0.0198	2,12	-0.0132	0.0173	2,13	0.0118	-0.0153
2,14	-0.0107	0.0138	2,15	0.0097	-0.0125	2,16	-0.0090	0.0115	2,17	0.0084	-0.0107
2,18	-0.0079	0.0100	2,19	0.0076	-0.0095	2,20	-0.0074	0.0092	2,21	0.0074	-0.0092
3, 3	1.0603	1.5289	3, 4	0.0665	-0.3670	3, 5	-0.0628	0.1607	3, 6	0.0498	-0.0915
3, 7	-0.0381	0.0593	3, 8	0.0295	-0.0423	3, 9	-0.0235	0.0324	3,10	0.0193	-0.0261
3,11	-0.0163	0.0217	3,12	0.0141	-0.0186	3,13	-0.0123	0.0162	3,14	0.0110	-0.0144
3,15	-0.0099	0.0129	3,16	0.0091	-0.0118	3,17	-0.0084	0.0108	3,18	0.0079	-0.0101
3,19	-0.0075	0.0095	3,20	0.0073	-0.0092						
4, 4	1.0637	1.5250	4, 5	0.0640	-0.3645	4, 6	-0.0610	0.1589	4, 7	0.0483	-0.0901
4, 8	-0.0369	0.0582	4, 9	0.0285	-0.0414	4,10	-0.0227	0.0316	4,11	0.0186	-0.0254
4,12	-0.0157	0.0212	4,13	0.0135	-0.0181	4,14	-0.0119	0.0158	4,15	0.0106	-0.0140
4,16	-0.0096	0.0126	4,17	0.0088	-0.0115	4,18	-0.0082	0.0106	4,19	0.0077	-0.0099
5, 5	1.0654	1.5235	5, 6	0.0627	-0.3634	5, 7	-0.0600	0.1581	5, 8	0.0475	-0.0895
5, 9	-0.0362	0.0577	5,10	0.0279	-0.0410	5,11	-0.0222	0.0313	5,12	0.0182	-0.0251
5,13	-0.0153	0.0209	5,14	0.0132	-0.0178	5,15	-0.0116	0.0156	5,16	0.0104	-0.0138
5,17	-0.0094	0.0125	5,18	0.0087	-0.0114						
6, 6	1.0663	1.5228	6, 7	0.0620	-0.3629	6, 8	-0.0594	0.1577	6, 9	0.0470	-0.0891
6,10	-0.0358	0.0574	6,11	0.0276	-0.0407	6,12	-0.0219	0.0311	6,13	0.0179	-0.0249
6,14	-0.0151	0.0207	6,15	0.0130	-0.0177	6,16	-0.0115	0.0155	6,17	0.0103	-0.0138
7, 7	1.0668	1.5224	7, 8	0.0616	-0.3625	7, 9	-0.0590	0.1574	7,10	0.0467	-0.0889
7,11	-0.0356	0.0572	7,12	0.0274	-0.0406	7,13	-0.0217	0.0309	7,14	0.0178	-0.0248
7,15	-0.0150	0.0207	7,16	0.0130	-0.0177						
8, 8	1.0671	1.5221	8, 9	0.0613	-0.3623	8,10	-0.0588	0.1573	8,11	0.0466	-0.0888
8,12	-0.0354	0.0571	8,13	0.0273	-0.0405	8,14	-0.0217	0.0309	8,15	0.0178	-0.0248
9, 9	1.0673	1.5219	9,10	0.0612	-0.3622	9,11	-0.0587	0.1572	9,12	0.0465	-0.0887
9,13	-0.0354	0.0570	9,14	0.0272	-0.0405						
10,10	1.0675	1.5219	10,11	0.0611	-0.3622	10,12	-0.0586	0.1571	10,13	0.0464	-0.0887
11,11	1.0675	1.5218	11,12	0.0610	-0.3621						

VBR

SELF AND MUTUAL ADMITTANCES

H/LMDA=0.5000 B/LMDA= 0.500 OMEGA= 9.92

24 ELEMENT ARRAY

I, K	YIK IN MMHO		I, K	YIK IN MMHO		I, K	YIK IN MMHO		I, K	YIK IN MMHO	
	RE	IM		RE	IM		RE	IM		RE	IM
1, 1	1.0531	1.6057	1, 2	0.0773	-0.4044	1, 3	-0.0723	0.1834	1, 4	0.0573	-0.1069
1, 5	-0.0441	0.0707	1, 6	0.0343	-0.0512	1, 7	-0.0275	0.0397	1, 8	0.0227	-0.0322
1, 9	-0.0192	0.0269	1,10	0.0166	-0.0231	1,11	-0.0145	0.0202	1,12	0.0129	-0.0179
1,13	-0.0117	0.0161	1,14	0.0106	-0.0146	1,15	-0.0097	0.0134	1,16	0.0090	-0.0124
1,17	-0.0084	0.0115	1,18	0.0079	-0.0108	1,19	-0.0075	0.0102	1,20	0.0072	-0.0097
1,21	-0.0070	0.0094	1,22	0.0069	-0.0092	1,23	-0.0070	0.0093	1,24	0.0073	-0.0103
2, 2	1.0540	1.5419	2, 3	0.0712	-0.3744	2, 4	-0.0664	0.1656	2, 5	0.0525	-0.0950
2, 6	-0.0402	0.0619	2, 7	0.0312	-0.0445	2, 8	-0.0250	0.0342	2, 9	0.0206	-0.0276
2,10	-0.0173	0.0230	2,11	0.0150	-0.0197	2,12	-0.0131	0.0172	2,13	0.0117	-0.0152
2,14	-0.0105	0.0137	2,15	0.0095	-0.0124	2,16	-0.0088	0.0113	2,17	0.0081	-0.0105
2,18	-0.0076	0.0098	2,19	0.0072	-0.0092	2,20	-0.0069	0.0087	2,21	0.0066	-0.0083
2,22	-0.0065	0.0081	2,23	0.0066	-0.0082						
3, 3	1.0604	1.5288	3, 4	0.0664	-0.3669	3, 5	-0.0628	0.1606	3, 6	0.0497	-0.0914
3, 7	-0.0380	0.0592	3, 8	0.0294	-0.0422	3, 9	-0.0234	0.0323	3,10	0.0192	-0.0260
3,11	-0.0162	0.0216	3,12	0.0139	-0.0185	3,13	-0.0122	0.0161	3,14	0.0109	-0.0143
3,15	-0.0098	0.0128	3,16	0.0089	-0.0116	3,17	-0.0082	0.0106	3,18	0.0076	-0.0098
3,19	-0.0071	0.0092	3,20	0.0068	-0.0086	3,21	-0.0065	0.0082	3,22	0.0064	-0.0080
4, 4	1.0637	1.5250	4, 5	0.0639	-0.3644	4, 6	-0.0609	0.1589	4, 7	0.0482	-0.0901
4, 8	-0.0368	0.0581	4, 9	0.0284	-0.0413	4,10	-0.0226	0.0316	4,11	0.0185	-0.0253
4,12	-0.0155	0.0211	4,13	0.0134	-0.0180	4,14	-0.0117	0.0156	4,15	0.0104	-0.0138
4,16	-0.0094	0.0124	4,17	0.0085	-0.0113	4,18	-0.0079	0.0103	4,19	0.0073	-0.0096
4,20	-0.0069	0.0090	4,21	0.0066	-0.0085						
5, 5	1.0654	1.5235	5, 6	0.0626	-0.3634	5, 7	-0.0599	0.1581	5, 8	0.0474	-0.0894
5, 9	-0.0361	0.0576	5,10	0.0278	-0.0409	5,11	-0.0221	0.0312	5,12	0.0181	-0.0250
5,13	-0.0152	0.0208	5,14	0.0130	-0.0177	5,15	-0.0114	0.0154	5,16	0.0101	-0.0136
5,17	-0.0091	0.0122	5,18	0.0083	-0.0111	5,19	-0.0077	0.0102	5,20	0.0072	-0.0095
6, 6	1.0664	1.5227	6, 7	0.0619	-0.3628	6, 8	-0.0593	0.1576	6, 9	0.0469	-0.0890
6,10	-0.0357	0.0573	6,11	0.0275	-0.0406	6,12	-0.0218	0.0310	6,13	0.0178	-0.0248
6,14	-0.0149	0.0206	6,15	0.0128	-0.0176	6,16	-0.0112	0.0153	6,17	0.0100	-0.0136
6,18	-0.0090	0.0122	6,19	0.0083	-0.0111						
7, 7	1.0669	1.5223	7, 8	0.0615	-0.3625	7, 9	-0.0589	0.1573	7,10	0.0466	-0.0888
7,11	-0.0354	0.0571	7,12	0.0272	-0.0405	7,13	-0.0216	0.0308	7,14	0.0176	-0.0247
7,15	-0.0148	0.0205	7,16	0.0127	-0.0175	7,17	-0.0112	0.0152	7,18	0.0100	-0.0135
8, 8	1.0673	1.5220	8, 9	0.0612	-0.3623	8,10	-0.0587	0.1572	8,11	0.0464	-0.0887
8,12	-0.0353	0.0570	8,13	0.0271	-0.0404	8,14	-0.0215	0.0307	8,15	0.0176	-0.0246
8,16	-0.0148	0.0205	8,17	0.0127	-0.0175						
9, 9	1.0675	1.5218	9,10	0.0610	-0.3621	9,11	-0.0586	0.1571	9,12	0.0463	-0.0886
9,13	-0.0352	0.0569	9,14	0.0270	-0.0403	9,15	-0.0214	0.0307	9,16	0.0175	-0.0246
10,10	1.0676	1.5217	10,11	0.0609	-0.3620	10,12	-0.0585	0.1570	10,13	0.0462	-0.0885
10,14	-0.0351	0.0569	10,15	0.0270	-0.0403						
11,11	1.0677	1.5217	11,12	0.0608	-0.3620	11,13	-0.0584	0.1570	11,14	0.0462	-0.0885
12,12	1.0677	1.5216	12,13	0.0608	-0.3620						

VBR

SELF AND MUTUAL ADMITTANCES

H/LMDA=0.5000 B/LMDA= 0.500 OMEGA= 9.92

26 ELEMENT ARRAY

I, K	YIK IN MMHO		I, K	YIK IN MMHO		I, K	YIK IN MMHO		I, K	YIK IN MMHO	
	RE	IM		RE	IM		RE	IM		RE	IM
1, 1	1.0531	1.6057	1, 2	0.0772	-0.4043	1, 3	-0.0722	0.1834	1, 4	0.0572	-0.1069
1, 5	-0.0440	0.0706	1, 6	0.0343	-0.0512	1, 7	-0.0275	0.0396	1, 8	0.0226	-0.0321
1, 9	-0.0191	0.0269	1,10	0.0165	-0.0231	1,11	-0.0145	0.0201	1,12	0.0128	-0.0179
1,13	-0.0115	0.0160	1,14	0.0105	-0.0145	1,15	-0.0096	0.0133	1,16	0.0089	-0.0123
1,17	-0.0082	0.0114	1,18	0.0077	-0.0106	1,19	-0.0073	0.0100	1,20	0.0069	-0.0094
1,21	-0.0066	0.0090	1,22	0.0064	-0.0086	1,23	-0.0063	0.0084	1,24	0.0062	-0.0082
1,25	-0.0063	0.0084	1,26	0.0066	-0.0093						
2, 2	1.0541	1.5419	2, 3	0.0712	-0.3744	2, 4	-0.0663	0.1655	2, 5	0.0524	-0.0949
2, 6	-0.0402	0.0619	2, 7	0.0312	-0.0444	2, 8	-0.0249	0.0341	2, 9	0.0205	-0.0275
2,10	-0.0173	0.0229	2,11	0.0149	-0.0196	2,12	-0.0130	0.0171	2,13	0.0116	-0.0151
2,14	-0.0104	0.0136	2,15	0.0094	-0.0123	2,16	-0.0086	0.0112	2,17	0.0080	-0.0104
2,18	-0.0074	0.0096	2,19	0.0070	-0.0090	2,20	-0.0066	0.0085	2,21	0.0063	-0.0080
2,22	-0.0060	0.0077	2,23	0.0059	-0.0074	2,24	-0.0058	0.0073	2,25	0.0059	-0.0074
3, 3	1.0604	1.5288	3, 4	0.0663	-0.3669	3, 5	-0.0627	0.1606	3, 6	0.0497	-0.0914
3, 7	-0.0379	0.0592	3, 8	0.0293	-0.0422	3, 9	-0.0234	0.0323	3,10	0.0192	-0.0259
3,11	-0.0161	0.0216	3,12	0.0139	-0.0184	3,13	-0.0121	0.0160	3,14	0.0107	-0.0142
3,15	-0.0096	0.0127	3,16	0.0088	-0.0115	3,17	-0.0080	0.0105	3,18	0.0074	-0.0097
3,19	-0.0069	0.0090	3,20	0.0065	-0.0084	3,21	-0.0062	0.0079	3,22	0.0059	-0.0075
3,23	-0.0057	0.0073	3,24	0.0057	-0.0071						
4, 4	1.0638	1.5249	4, 5	0.0639	-0.3644	4, 6	-0.0608	0.1588	4, 7	0.0482	-0.0900
4, 8	-0.0367	0.0581	4, 9	0.0283	-0.0413	4,10	-0.0225	0.0315	4,11	0.0184	-0.0253
4,12	-0.0155	0.0210	4,13	0.0133	-0.0179	4,14	-0.0116	0.0156	4,15	0.0103	-0.0137
4,16	-0.0092	0.0123	4,17	0.0084	-0.0111	4,18	-0.0077	0.0102	4,19	0.0071	-0.0094
4,20	-0.0066	0.0087	4,21	0.0063	-0.0082	4,22	-0.0060	0.0078	4,23	0.0058	-0.0074
5, 5	1.0655	1.5234	5, 6	0.0626	-0.3633	5, 7	-0.0598	0.1580	5, 8	0.0473	-0.0894
5, 9	-0.0360	0.0575	5,10	0.0277	-0.0408	5,11	-0.0220	0.0311	5,12	0.0180	-0.0249
5,13	-0.0151	0.0207	5,14	0.0129	-0.0176	5,15	-0.0113	0.0153	5,16	0.0100	-0.0135
5,17	-0.0090	0.0121	5,18	0.0082	-0.0110	5,19	-0.0075	0.0100	5,20	0.0069	-0.0093
5,21	-0.0065	0.0086	5,22	0.0062	-0.0081						
6, 6	1.0664	1.5227	6, 7	0.0619	-0.3627	6, 8	-0.0592	0.1576	6, 9	0.0468	-0.0890
6,10	-0.0356	0.0572	6,11	0.0274	-0.0406	6,12	-0.0217	0.0309	6,13	0.0177	-0.0247
6,14	-0.0148	0.0205	6,15	0.0127	-0.0175	6,16	-0.0111	0.0152	6,17	0.0098	-0.0134
6,18	-0.0088	0.0120	6,19	0.0080	-0.0109	6,20	-0.0074	0.0100	6,21	0.0069	-0.0092

I, K	RE	IM	I, K	RE	IM	I, K	RE	IM	I, K	RE	IM
7, 7	1.0670	1.5222	7, 8	0.0614	-0.3624	7, 9	-0.0589	0.1573	7,10	0.0465	-0.0888
7,11	-0.0353	0.0570	7,12	0.0272	-0.0404	7,13	-0.0215	0.0307	7,14	0.0175	-0.0246
7,15	-0.0147	0.0204	7,16	0.0126	-0.0174	7,17	-0.0110	0.0151	7,18	0.0097	-0.0134
7,19	-0.0088	0.0120	7,20	0.0080	-0.0109						
8, 8	1.0673	1.5220	8, 9	0.0611	-0.3622	8,10	-0.0586	0.1571	8,11	0.0463	-0.0886
8,12	-0.0352	0.0569	8,13	0.0270	-0.0403	8,14	-0.0213	0.0306	8,15	0.0174	-0.0245
8,16	-0.0146	0.0203	8,17	0.0125	-0.0173	8,18	-0.0109	0.0151	8,19	0.0097	-0.0133
9, 9	1.0676	1.5218	9,10	0.0609	-0.3620	9,11	-0.0585	0.1570	9,12	0.0462	-0.0885
9,13	-0.0351	0.0568	9,14	0.0269	-0.0402	9,15	-0.0213	0.0306	9,16	0.0174	-0.0245
9,17	-0.0145	0.0203	9,18	0.0125	-0.0173						
10,10	1.0677	1.5217	10,11	0.0608	-0.3619	10,12	-0.0583	0.1569	10,13	0.0461	-0.0884
10,14	-0.0350	0.0567	10,15	0.0269	-0.0402	10,16	-0.0212	0.0306	10,17	0.0173	-0.0245
11,11	1.0678	1.5216	11,12	0.0607	-0.3619	11,13	-0.0583	0.1568	11,14	0.0461	-0.0884
11,15	-0.0350	0.0567	11,16	0.0268	-0.0402						
12,12	1.0679	1.5215	12,13	0.0607	-0.3618	12,14	-0.0583	0.1568	12,15	0.0460	-0.0884
13,13	1.0679	1.5215	13,14	0.0607	-0.3618						

VBR

SELF AND MUTUAL ADMITTANCES

H/LMDA=0.5000 B/LMDA= 0.500 OMEGA= 9.92

28 ELEMENT ARRAY

I, K	YIK IN MMHO		I, K	YIK IN MMHO		I, K	YIK IN MMHO		I, K	YIK IN MMHO	
	RE	IM		RE	IM		RE	IM		RE	IM
1, 1	1.0532	1.6056	1, 2	0.0772	-0.4043	1, 3	-0.0721	0.1833	1, 4	0.0572	-0.1068
1, 5	-0.0439	0.0706	1, 6	0.0342	-0.0511	1, 7	-0.0274	0.0396	1, 8	0.0226	-0.0321
1, 9	-0.0191	0.0268	1,10	0.0164	-0.0230	1,11	-0.0144	0.0201	1,12	0.0128	-0.0178
1,13	-0.0115	0.0160	1,14	0.0104	-0.0145	1,15	-0.0095	0.0132	1,16	0.0087	-0.0122
1,17	-0.0081	0.0113	1,18	0.0076	-0.0105	1,19	-0.0071	0.0099	1,20	0.0067	-0.0093
1,21	-0.0064	0.0088	1,22	0.0061	-0.0084	1,23	-0.0059	0.0080	1,24	0.0057	-0.0077
1,25	-0.0056	0.0075	1,26	0.0056	-0.0075	1,27	-0.0057	0.0077	1,28	0.0061	-0.0085
2, 2	1.0541	1.5418	2, 3	0.0711	-0.3744	2, 4	-0.0663	0.1655	2, 5	0.0524	-0.0949
2, 6	-0.0401	0.0619	2, 7	0.0311	-0.0444	2, 8	-0.0248	0.0341	2, 9	0.0204	-0.0275
2,10	-0.0172	0.0229	2,11	0.0148	-0.0196	2,12	-0.0129	0.0170	2,13	0.0115	-0.0151
2,14	-0.0103	0.0135	2,15	0.0093	-0.0122	2,16	-0.0085	0.0112	2,17	0.0079	-0.0103
2,18	-0.0073	0.0095	2,19	0.0068	-0.0089	2,20	-0.0064	0.0083	2,21	0.0061	-0.0078
2,22	-0.0058	0.0074	2,23	0.0056	-0.0071	2,24	-0.0054	0.0068	2,25	0.0053	-0.0066
2,26	-0.0053	0.0066	2,27	0.0053	-0.0067						
3, 3	1.0605	1.5288	3, 4	0.0663	-0.3668	3, 5	-0.0627	0.1606	3, 6	0.0496	-0.0913
3, 7	-0.0379	0.0591	3, 8	0.0293	-0.0422	3, 9	-0.0233	0.0322	3,10	0.0191	-0.0259
3,11	-0.0161	0.0215	3,12	0.0138	-0.0184	3,13	-0.0120	0.0160	3,14	0.0107	-0.0141
3,15	-0.0096	0.0126	3,16	0.0087	-0.0114	3,17	-0.0079	0.0104	3,18	0.0073	-0.0096
3,19	-0.0068	0.0089	3,20	0.0063	-0.0083	3,21	-0.0060	0.0078	3,22	0.0057	-0.0073
3,23	[illegible]	0.0070	3,24	0.0052	-0.0067	3,25	-0.0051	0.0065	3,26	0.0051	-0.0064
4, 4	1.0638	1.5249	4, 5	0.0638	-0.3644	4, 6	[illegible]	0.1588	4, 7	0.0481	-0.0900
4, 8	-0.0367	0.0580	4, 9	0.0283	-0.0412	4,10	-0.0224	0.0315	4,11	[illegible]	[illegible]
4,12	-0.0154	0.0209	4,13	0.0132	-0.0178	4,14	-0.0115	0.0155	4,15	0.0102	-0.0137
4,16	-0.0091	0.0122	4,17	0.0083	-0.0111	4,18	-0.0076	0.0101	4,19	0.0070	-0.0093
4,20	-0.0065	0.0086	4,21	0.0061	-0.0080	4,22	-0.0057	0.0075	4,23	0.0054	-0.0071
4,24	-0.0052	0.0068	4,25	0.0051	-0.0066						
5, 5	1.0655	1.5234	5, 6	0.0625	-0.3633	5, 7	-0.0598	0.1580	5, 8	0.0473	-0.0893
5, 9	-0.0360	0.0575	5,10	0.0277	-0.0408	5,11	-0.0219	0.0311	5,12	0.0179	-0.0249
5,13	-0.0150	0.0206	5,14	0.0128	-0.0176	5,15	-0.0112	0.0152	5,16	0.0099	-0.0135
5,17	-0.0089	0.0120	5,18	0.0080	-0.0109	5,19	-0.0073	0.0099	5,20	0.0068	-0.0091
5,21	-0.0063	0.0085	5,22	0.0059	-0.0079	5,23	-0.0056	0.0075	5,24	0.0054	-0.0071
6, 6	1.0665	1.5226	6, 7	0.0618	-0.3627	6, 8	-0.0592	0.1575	6, 9	0.0468	-0.0889
6,10	-0.0356	0.0572	6,11	0.0273	-0.0405	6,12	-0.0216	0.0308	6,13	0.0176	-0.0247
6,14	-0.0147	0.0204	6,15	0.0126	-0.0174	6,16	-0.0110	0.0151	6,17	0.0097	-0.0133
6,18	-0.0087	0.0119	6,19	0.0079	-0.0108	6,20	-0.0072	0.0098	6,21	0.0067	-0.0091
6,22	-0.0062	0.0084	6,23	0.0059	-0.0079						
7, 7	1.0670	1.5222	7, 8	0.0614	-0.3624	7, 9	-0.0588	0.1572	7,10	0.0465	-0.0887
7,11	-0.0353	0.0570	7,12	0.0271	-0.0403	7,13	-0.0214	0.0307	7,14	0.0174	-0.0245
7,15	-0.0146	0.0203	7,16	0.0125	-0.0173	7,17	-0.0109	0.0150	7,18	0.0096	-0.0132
7,19	-0.0086	0.0118	7,20	0.0078	-0.0107	7,21	-0.0072	0.0098	7,22	0.0066	-0.0091
8, 8	1.0674	1.5219	8, 9	0.0611	-0.3621	8,10	-0.0585	0.1570	8,11	0.0463	-0.0885
8,12	-0.0351	0.0568	8,13	0.0269	-0.0402	8,14	-0.0213	0.0306	8,15	0.0173	-0.0244
8,16	-0.0145	0.0202	8,17	0.0124	-0.0172	8,18	-0.0108	0.0150	8,19	0.0096	-0.0132
8,20	-0.0086	0.0118	8,21	0.0078	-0.0107						
9, 9	1.0676	1.5217	9,10	0.0609	-0.3620	9,11	-0.0584	0.1569	9,12	0.0461	-0.0884
9,13	-0.0350	0.0567	9,14	0.0268	-0.0401	9,15	-0.0212	0.0305	9,16	0.0172	-0.0244
9,17	-0.0144	0.0202	9,18	0.0123	-0.0172	9,19	-0.0108	0.0149	9,20	0.0095	-0.0132
10,10	1.0678	1.5216	10,11	0.0607	-0.3619	10,12	-0.0583	0.1568	10,13	0.0460	-0.0884
10,14	-0.0349	0.0567	10,15	0.0267	-0.0401	10,16	-0.0211	0.0305	10,17	0.0172	-0.0243
10,18	-0.0144	0.0202	10,19	0.0123	-0.0172						
11,11	1.0679	1.5215	11,12	0.0606	-0.3618	11,13	-0.0582	0.1568	11,14	0.0459	-0.0883
11,15	-0.0348	0.0566	11,16	0.0267	-0.0400	11,17	-0.0211	0.0304	11,18	0.0172	-0.0243
12,12	1.0680	1.5214	12,13	0.0606	-0.3618	12,14	-0.0581	0.1567	12,15	0.0459	-0.0883
12,16	-0.0348	0.0566	12,17	0.0267	-0.0400						
13,13	1.0680	1.5214	13,14	0.0605	-0.3617	13,15	-0.0581	0.1567	13,16	0.0459	-0.0883
14,14	1.0681	1.5214	14,15	0.0605	-0.3617						

VBR

SELF AND MUTUAL ADMITTANCES

H/LMDA=0.5000 B/LMDA= 0.500 OMEGA= 9.92

30 ELEMENT ARRAY

I, K	YIK IN MMHO		I, K	YIK IN MMHO		I, K	YIK IN MMHO		I, K	YIK IN MMHO	
	RE	IM		RE	IM		RE	IM		RE	IM
1, 1	1.0533	1.6056	1, 2	0.0771	-0.4042	1, 3	-0.0721	0.1833	1, 4	0.0571	-0.1068
1, 5	-0.0439	0.0705	1, 6	0.0342	-0.0511	1, 7	-0.0274	0.0395	1, 8	0.0225	-0.0320
1, 9	-0.0190	0.0268	1,10	0.0164	-0.0230	1,11	-0.0143	0.0200	1,12	0.0127	-0.0177
1,13	-0.0114	0.0159	1,14	0.0103	-0.0144	1,15	-0.0094	0.0131	1,16	0.0087	-0.0121
1,17	-0.0080	0.0112	1,18	0.0075	-0.0104	1,19	-0.0070	0.0098	1,20	0.0066	-0.0092
1,21	-0.0062	0.0087	1,22	0.0059	-0.0082	1,23	-0.0057	0.0078	1,24	0.0055	-0.0075
1,25	-0.0053	0.0072	1,26	0.0052	-0.0070	1,27	-0.0051	0.0069	1,28	0.0051	-0.0068
1,29	-0.0053	0.0070	1,30	0.0056	-0.0079						
2, 2	1.0542	1.5418	2, 3	0.0711	-0.3743	2, 4	-0.0663	0.1655	2, 5	0.0524	-0.0949
2, 6	-0.0401	0.0618	2, 7	0.0311	-0.0443	2, 8	-0.0248	0.0340	2, 9	0.0204	-0.0274
2,10	-0.0172	0.0229	2,11	0.0147	-0.0195	2,12	-0.0129	0.0170	2,13	0.0114	-0.0150
2,14	-0.0102	0.0135	2,15	0.0093	-0.0122	2,16	-0.0085	0.0111	2,17	0.0078	-0.0102
2,18	-0.0072	0.0094	2,19	0.0067	-0.0088	2,20	-0.0063	0.0082	2,21	0.0059	-0.0077
2,22	-0.0056	0.0073	2,23	0.0054	-0.0069	2,24	-0.0051	0.0066	2,25	0.0050	-0.0064
2,26	-0.0048	0.0062	2,27	0.0048	-0.0060	2,28	-0.0048	0.0060	2,29	0.0049	-0.0061
3, 3	1.0605	1.5287	3, 4	0.0663	-0.3668	3, 5	-0.0626	0.1605	3, 6	0.0496	-0.0913
3, 7	-0.0378	0.0591	3, 8	0.0292	-0.0421	3, 9	-0.0233	0.0322	3,10	0.0191	-0.0259
3,11	-0.0160	0.0215	3,12	0.0137	-0.0183	3,13	-0.0120	0.0159	3,14	0.0106	-0.0141
3,15	-0.0095	0.0126	3,16	0.0086	-0.0114	3,17	-0.0078	0.0103	3,18	0.0072	-0.0095
3,19	-0.0067	0.0088	3,20	0.0062	-0.0082	3,21	-0.0058	0.0077	3,22	0.0055	-0.0072
3,23	-0.0052	0.0068	3,24	0.0050	-0.0065	3,25	-0.0048	0.0062	3,26	0.0047	-0.0060
3,27	-0.0046	0.0058	3,28	0.0046	-0.0058						
4, 4	1.0639	1.5249	4, 5	0.0638	-0.3643	4, 6	-0.0607	0.1588	4, 7	0.0481	-0.0899
4, 8	-0.0366	0.0580	4, 9	0.0282	-0.0412	4,10	-0.0224	0.0314	4,11	0.0183	-0.0252
4,12	-0.0153	0.0209	4,13	0.0131	-0.0178	4,14	-0.0114	0.0154	4,15	0.0101	-0.0136
4,16	-0.0091	0.0122	4,17	0.0082	-0.0110	4,18	-0.0075	0.0100	4,19	0.0069	-0.0092
4,20	-0.0064	0.0085	4,21	0.0059	-0.0079	4,22	-0.0056	0.0074	4,23	0.0053	-0.0070
4,24	-0.0050	0.0066	4,25	0.0048	-0.0063	4,26	-0.0047	0.0061	4,27	0.0046	-0.0059
5, 5	1.0656	1.5233	5, 6	0.0625	-0.3632	5, 7	-0.0597	0.1579	5, 8	0.0472	-0.0893
5, 9	-0.0359	0.0574	5,10	0.0276	-0.0407	5,11	-0.0219	0.0310	5,12	0.0178	-0.0248
5,13	-0.0149	0.0206	5,14	0.0128	-0.0175	5,15	-0.0111	0.0152	5,16	0.0098	-0.0134
5,17	-0.0088	0.0120	5,18	0.0079	-0.0108	5,19	-0.0072	0.0098	5,20	0.0067	-0.0090
5,21	-0.0062	0.0084	5,22	0.0058	-0.0078	5,23	-0.0054	0.0073	5,24	0.0051	-0.0069
5,25	-0.0049	0.0066	5,26	0.0047	-0.0063						
6, 6	1.0665	1.5226	6, 7	0.0618	-0.3627	6, 8	-0.0591	0.1575	6, 9	0.0467	-0.0889
6,10	-0.0355	0.0571	6,11	0.0273	-0.0405	6,12	-0.0215	0.0308	6,13	0.0176	-0.0246
6,14	-0.0147	0.0204	6,15	0.0125	-0.0173	6,16	-0.0109	0.0150	6,17	0.0096	-0.0133
6,18	-0.0086	0.0118	6,19	0.0078	-0.0107	6,20	-0.0071	0.0097	6,21	0.0065	-0.0090
6,22	-0.0061	0.0083	6,23	0.0057	-0.0077	6,24	-0.0054	0.0073	6,25	0.0051	-0.0069
7, 7	1.0671	1.5221	7, 8	0.0613	-0.3623	7, 9	-0.0588	0.1572	7,10	0.0464	-0.0887
7,11	-0.0352	0.0569	7,12	0.0270	-0.0403	7,13	-0.0213	0.0306	7,14	0.0174	-0.0245
7,15	-0.0145	0.0203	7,16	0.0124	-0.0172	7,17	-0.0108	0.0149	7,18	0.0095	-0.0132
7,19	-0.0085	0.0118	7,20	0.0077	-0.0106	7,21	-0.0070	0.0097	7,22	0.0065	-0.0089
7,23	-0.0060	0.0083	7,24	0.0056	-0.0077						
8, 8	1.0674	1.5219	8, 9	0.0610	-0.3621	8,10	-0.0585	0.1570	8,11	0.0462	-0.0885
8,12	-0.0350	0.0568	8,13	0.0269	-0.0402	8,14	-0.0212	0.0305	8,15	0.0172	-0.0244
8,16	-0.0144	0.0202	8,17	0.0123	-0.0171	8,18	-0.0107	0.0149	8,19	0.0094	-0.0131
8,20	-0.0084	0.0117	8,21	0.0076	-0.0106	8,22	-0.0070	0.0097	8,23	0.0065	-0.0089
9, 9	1.0677	1.5217	9,10	0.0608	-0.3619	9,11	-0.0583	0.1569	9,12	0.0460	-0.0884
9,13	-0.0349	0.0567	9,14	0.0267	-0.0401	9,15	-0.0211	0.0304	9,16	0.0172	-0.0243
9,17	-0.0143	0.0201	9,18	0.0122	-0.0171	9,19	-0.0106	0.0148	9,20	0.0094	-0.0131
9,21	-0.0084	0.0117	9,22	0.0076	-0.0106						
10,10	1.0679	1.5215	10,11	0.0607	-0.3618	10,12	-0.0582	0.1568	10,13	0.0459	-0.0883
10,14	-0.0348	0.0566	10,15	0.0267	-0.0400	10,16	-0.0210	0.0304	10,17	0.0171	-0.0243
10,18	-0.0143	0.0201	10,19	0.0122	-0.0171	10,20	-0.0106	0.0148	10,21	0.0094	-0.0131
11,11	1.0680	1.5215	11,12	0.0606	-0.3617	11,13	-0.0581	0.1567	11,14	0.0459	-0.0882
11,15	-0.0348	0.0565	11,16	0.0266	-0.0400	11,17	-0.0210	0.0304	11,18	0.0171	-0.0242
11,19	-0.0143	0.0201	11,20	0.0122	-0.0171						
12,12	1.0681	1.5214	12,13	0.0605	-0.3617	12,14	-0.0581	0.1567	12,15	0.0458	-0.0882
12,16	-0.0347	0.0565	12,17	0.0266	-0.0399	12,18	-0.0210	0.0303	12,19	0.0171	-0.0242
13,13	1.0681	1.5213	13,14	0.0604	-0.3616	13,15	-0.0580	0.1566	13,16	0.0458	-0.0882
13,17	-0.0347	0.0565	13,18	0.0266	-0.0399						
14,14	1.0682	1.5213	14,15	0.0604	-0.3616	14,16	-0.0580	0.1566	14,17	0.0458	-0.0882
15,15	1.0682	1.5213	15,16	0.0604	-0.3616						

VBR

SELF AND MUTUAL ADMITTANCES

H/LMDA=0.5000 B/LMDA= 0.500 OMEGA= 9.92

35 ELEMENT ARRAY

I, K	YIK IN MMHO RE	IM	I, K	YIK IN MMHO RE	IM	I, K	YIK IN MMHO RE	IM	I, K	YIK IN MMHO RE	IM
1, 1	1.0533	1.6055	1, 2	0.0771	-0.4042	1, 3	-0.0720	0.1832	1, 4	0.0571	-0.1067
1, 5	-0.0438	0.0705	1, 6	0.0341	-0.0510	1, 7	-0.0273	0.0395	1, 8	0.0224	-0.0319
1, 9	-0.0189	0.0267	1,10	0.0163	-0.0229	1,11	-0.0142	0.0200	1,12	0.0126	-0.0177
1,13	-0.0113	0.0158	1,14	0.0102	-0.0143	1,15	-0.0093	0.0130	1,16	0.0085	-0.0120
1,17	-0.0079	0.0111	1,18	0.0073	-0.0103	1,19	-0.0068	0.0096	1,20	0.0064	-0.0090
1,21	-0.0060	0.0085	1,22	0.0057	-0.0080	1,23	-0.0054	0.0076	1,24	0.0052	-0.0073
1,25	-0.0049	0.0069	1,26	0.0047	-0.0066	1,27	-0.0046	0.0064	1,28	0.0044	-0.0062
1,29	-0.0043	0.0060	1,30	0.0042	-0.0058	1,31	-0.0042	0.0057	1,32	0.0041	-0.0056
1,33	-0.0042	0.0056	1,34	0.0043	-0.0058	1,35	-0.0046	0.0065			
2, 2	1.0542	1.5417	2, 3	0.0710	-0.3743	2, 4	-0.0662	0.1654	2, 5	0.0523	-0.0948
2, 6	-0.0400	0.0618	2, 7	0.0310	-0.0443	2, 8	-0.0247	0.0340	2, 9	0.0203	-0.0274
2,10	-0.0171	0.0228	2,11	0.0147	-0.0195	2,12	-0.0128	0.0169	2,13	0.0113	-0.0150
2,14	-0.0101	0.0134	2,15	0.0091	-0.0121	2,16	-0.0083	0.0110	2,17	0.0076	-0.0101
2,18	-0.0071	0.0093	2,19	0.0066	-0.0087	2,20	-0.0061	0.0081	2,21	0.0057	-0.0076
2,22	-0.0054	0.0071	2,23	0.0051	-0.0067	2,24	-0.0049	0.0064	2,25	0.0046	-0.0061
2,26	-0.0044	0.0058	2,27	0.0043	-0.0056	2,28	-0.0041	0.0054	2,29	0.0040	-0.0052
2,30	-0.0039	0.0050	2,31	0.0039	-0.0049	2,32	-0.0038	0.0049	2,33	0.0039	-0.0049
2,34	-0.0040	0.0050									
3, 3	1.0606	1.5287	3, 4	0.0662	-0.3668	3, 5	-0.0626	0.1605	3, 6	0.0495	-0.0913
3, 7	-0.0378	0.0590	3, 8	0.0292	-0.0421	3, 9	-0.0232	0.0321	3,10	0.0190	-0.0258
3,11	-0.0159	0.0214	3,12	0.0136	-0.0182	3,13	-0.0119	0.0158	3,14	0.0105	-0.0140
3,15	-0.0094	0.0125	3,16	0.0085	-0.0113	3,17	-0.0077	0.0102	3,18	0.0071	-0.0094
3,19	-0.0065	0.0087	3,20	0.0061	-0.0081	3,21	-0.0057	0.0075	3,22	0.0053	-0.0070
3,23	-0.0050	0.0066	3,24	0.0047	-0.0063	3,25	-0.0045	0.0060	3,26	0.0043	-0.0057
3,27	-0.0041	0.0054	3,28	0.0040	-0.0052	3,29	-0.0038	0.0050	3,30	0.0037	-0.0049
3,31	-0.0037	0.0047	3,32	0.0037	-0.0047	3,33	-0.0037	0.0047			
4, 4	1.0640	1.5248	4, 5	0.0637	-0.3643	4, 6	-0.0607	0.1587	4, 7	0.0480	-0.0899
4, 8	-0.0366	0.0579	4, 9	0.0281	-0.0411	4,10	-0.0223	0.0314	4,11	0.0182	-0.0251
4,12	-0.0152	0.0208	4,13	0.0130	-0.0177	4,14	-0.0113	0.0154	4,15	0.0100	-0.0135
4,16	-0.0089	0.0121	4,17	0.0081	-0.0109	4,18	-0.0073	0.0099	4,19	0.0067	-0.0091
4,20	-0.0062	0.0084	4,21	0.0058	-0.0078	4,22	-0.0054	0.0073	4,23	0.0050	-0.0068
4,24	-0.0047	0.0064	4,25	0.0045	-0.0061	4,26	-0.0043	0.0058	4,27	0.0041	-0.0055
4,28	-0.0039	0.0053	4,29	0.0038	-0.0051	4,30	-0.0037	0.0049	4,31	0.0036	-0.0047
4,32	-0.0036	0.0047									
5, 5	1.0657	1.5233	5, 6	0.0624	-0.3632	5, 7	-0.0597	0.1579	5, 8	0.0472	-0.0892
5, 9	-0.0359	0.0574	5,10	0.0275	-0.0407	5,11	-0.0218	0.0310	5,12	0.0178	-0.0247
5,13	-0.0148	0.0205	5,14	0.0127	-0.0174	5,15	-0.0110	0.0151	5,16	0.0097	-0.0133
5,17	-0.0087	0.0119	5,18	0.0078	-0.0107	5,19	-0.0071	0.0097	5,20	0.0065	-0.0089
5,21	-0.0060	0.0082	5,22	0.0056	-0.0076	5,23	-0.0052	0.0071	5,24	0.0049	-0.0067
5,25	-0.0046	0.0063	5,26	0.0044	-0.0060	5,27	-0.0042	0.0057	5,28	0.0040	-0.0054
5,29	-0.0038	0.0052	5,30	0.0037	-0.0050	5,31	-0.0036	0.0048			
6, 6	1.0666	1.5225	6, 7	0.0617	-0.3626	6, 8	-0.0591	0.1574	6, 9	0.0467	-0.0888
6,10	-0.0354	0.0570	6,11	0.0272	-0.0404	6,12	-0.0215	0.0307	6,13	0.0175	-0.0245
6,14	-0.0146	0.0203	6,15	0.0124	-0.0172	6,16	-0.0108	0.0149	6,17	0.0095	-0.0131
6,18	-0.0085	[illegible]	6,19	0.0076	-0.0106	6,20	-0.0069	0.0096	6,21	0.0064	-0.0088
6,22	-0.0059	0.0081	6,23	0.0054	[illegible]	[illegible]	[illegible]	0.0070	6,25	0.0048	-0.0066
6,26	-0.0045	0.0062	6,27	0.0043	-0.0059	6,28	-0.0041	0.0056	[illegible]	[illegible]	-0.0054
6,30	-0.0038	0.0052									
7, 7	1.0672	1.5221	7, 8	0.0612	-0.3623	7, 9	-0.0587	0.1571	7,10	0.0463	-0.0886
7,11	-0.0351	0.0568	7,12	0.0269	-0.0402	7,13	-0.0212	0.0305	7,14	0.0173	-0.0244
7,15	-0.0144	0.0202	7,16	0.0123	-0.0171	7,17	-0.0107	0.0148	7,18	0.0094	-0.0130
7,19	-0.0084	0.0116	7,20	0.0075	-0.0105	7,21	-0.0068	0.0095	7,22	0.0063	-0.0087
7,23	-0.0058	0.0081	7,24	0.0054	-0.0075	7,25	-0.0050	0.0070	7,26	0.0047	-0.0066
7,27	-0.0045	0.0062	7,28	0.0042	-0.0059	7,29	-0.0041	0.0056			
8, 8	1.0675	1.5218	8, 9	0.0609	-0.3620	8,10	-0.0584	0.1569	8,11	0.0461	-0.0884
8,12	-0.0349	0.0567	8,13	0.0268	-0.0401	8,14	-0.0211	0.0304	8,15	0.0171	-0.0243
8,16	-0.0143	0.0201	8,17	0.0122	-0.0170	8,18	-0.0105	0.0148	8,19	0.0093	-0.0130
8,20	-0.0083	0.0116	8,21	0.0074	-0.0104	8,22	-0.0068	0.0095	8,23	0.0062	-0.0087
8,24	-0.0057	0.0080	8,25	0.0053	-0.0075	8,26	-0.0050	0.0070	8,27	0.0047	-0.0065
8,28	-0.0044	0.0062									
9, 9	1.0678	1.5216	9,10	0.0607	-0.3619	9,11	-0.0582	0.1568	9,12	0.0459	-0.0883
9,13	-0.0348	0.0566	9,14	0.0266	-0.0400	9,15	-0.0210	0.0303	9,16	0.0170	-0.0242
9,17	-0.0142	0.0200	9,18	0.0121	-0.0170	9,19	-0.0105	0.0147	9,20	0.0092	-0.0129
9,21	-0.0082	0.0115	9,22	0.0074	-0.0104	9,23	-0.0067	0.0094	9,24	0.0062	-0.0087
9,25	-0.0057	0.0080	9,26	0.0053	-0.0074	9,27	-0.0050	0.0070			
10,10	1.0679	1.5215	10,11	0.0606	-0.3617	10,12	-0.0581	0.1567	10,13	0.0458	-0.0882
10,14	-0.0347	0.0565	10,15	0.0265	-0.0399	10,16	-0.0209	0.0303	10,17	0.0169	-0.0241
10,18	-0.0141	0.0199	10,19	0.0120	-0.0169	10,20	-0.0104	0.0147	10,21	0.0092	-0.0129
10,22	-0.0082	0.0115	10,23	0.0074	-0.0104	10,24	-0.0067	0.0094	10,25	0.0061	-0.0087
10,26	-0.0057	0.0080									
11,11	1.0681	1.5214	11,12	0.0604	-0.3616	11,13	-0.0580	0.1566	11,14	0.0457	-0.0881
11,15	-0.0346	0.0564	11,16	0.0265	-0.0399	11,17	-0.0208	0.0302	11,18	0.0169	-0.0241
11,19	-0.0141	0.0199	11,20	0.0120	-0.0169	11,21	-0.0104	0.0146	11,22	0.0091	-0.0129
11,23	-0.0082	0.0115	11,24	0.0073	-0.0104	11,25	-0.0067	0.0094			
12,12	1.0682	1.5213	12,13	0.0604	-0.3616	12,14	-0.0579	0.1565	12,15	0.0457	-0.0881
12,16	-0.0346	0.0564	12,17	0.0264	-0.0398	12,18	-0.0208	0.0302	12,19	0.0169	-0.0241
12,20	-0.0140	0.0199	12,21	0.0120	-0.0169	12,22	-0.0104	0.0146	12,23	0.0091	-0.0129
12,24	-0.0081	0.0115									
13,13	1.0682	1.5212	13,14	0.0603	-0.3615	13,15	-0.0579	0.1565	13,16	0.0456	-0.0880
13,17	-0.0345	0.0564	13,18	0.0264	-0.0398	13,19	-0.0208	0.0302	13,20	0.0168	-0.0241
13,21	-0.0140	0.0199	13,22	0.0119	-0.0169	13,23	-0.0104	0.0146			
14,14	1.0683	1.5212	14,15	0.0603	-0.3615	14,16	-0.0578	0.1565	14,17	0.0456	-0.0880
14,18	-0.0345	0.0563	14,19	0.0264	-0.0398	14,20	-0.0207	0.0302	14,21	0.0168	-0.0240
14,22	-0.0140	0.0199									
15,15	1.0683	1.5212	15,16	0.0602	-0.3615	15,17	-0.0578	0.1564	15,18	0.0456	-0.0880
15,19	-0.0345	0.0563	15,20	0.0264	-0.0398	15,21	-0.0207	0.0301			
16,16	1.0684	1.5211	16,17	0.0602	-0.3615	16,18	-0.0578	0.1564	16,19	0.0456	-0.0880
16,20	-0.0345	0.0563									
17,17	1.0684	1.5211	17,18	0.0602	-0.3614	17,19	-0.0578	0.1564			
18,18	1.0684	1.5211									

VBR

SELF AND MUTUAL ADMITTANCES

H/LMDA=0.5000 B/LMDA= 0.500 OMEGA= 9.92

40 ELEMENT ARRAY

I, K	YIK IN MMHO RE	IM	I, K	YIK IN MMHO RE	IM	I, K	YIK IN MMHO RE	IM	I, K	YIK IN MMHO RE	IM
1, 1	1.0534	1.6054	1, 2	0.0770	-0.4041	1, 3	-0.0720	0.1832	1, 4	0.0570	-0.1067
1, 5	-0.0438	0.0704	1, 6	0.0340	-0.0510	1, 7	-0.0272	0.0394	1, 8	0.0224	-0.0319
1, 9	-0.0189	0.0267	1,10	0.0162	-0.0228	1,11	-0.0142	0.0199	1,12	0.0125	-0.0176
1,13	-0.0112	0.0158	1,14	0.0101	-0.0142	1,15	-0.0092	0.0130	1,16	0.0084	-0.0119
1,17	-0.0078	0.0110	1,18	0.0072	-0.0102	1,19	-0.0067	0.0095	1,20	0.0063	-0.0089
1,21	-0.0059	0.0084	1,22	0.0056	-0.0079	1,23	-0.0053	0.0075	1,24	0.0050	-0.0071
1,25	-0.0048	0.0068	1,26	0.0046	-0.0065	1,27	-0.0044	0.0062	1,28	0.0042	-0.0060
1,29	-0.0040	0.0057	1,30	0.0039	-0.0055	1,31	-0.0038	0.0053	1,32	0.0037	-0.0052
1,33	-0.0036	0.0050	1,34	0.0035	-0.0049	1,35	-0.0035	0.0048	1,36	0.0035	-0.0047
1,37	-0.0035	0.0047	1,38	0.0035	-0.0048	1,39	-0.0037	0.0050	1,40	0.0039	-0.0056
2, 2	1.0543	1.5417	2, 3	0.0709	-0.3742	2, 4	-0.0661	0.1654	2, 5	0.0522	-0.0948
2, 6	-0.0399	0.0617	2, 7	0.0309	-0.0442	2, 8	-0.0247	0.0339	2, 9	0.0202	-0.0273
2,10	-0.0170	0.0227	2,11	0.0146	-0.0194	2,12	-0.0127	0.0169	2,13	0.0113	-0.0149
2,14	-0.0101	0.0133	2,15	0.0091	-0.0120	2,16	-0.0083	0.0109	2,17	0.0076	-0.0100
2,18	-0.0070	0.0093	2,19	0.0065	-0.0086	2,20	-0.0060	0.0080	2,21	0.0056	-0.0075
2,22	-0.0053	0.0071	2,23	0.0050	-0.0067	2,24	-0.0047	0.0063	2,25	0.0045	-0.0060
2,26	-0.0043	0.0057	2,27	0.0041	-0.0054	2,28	-0.0039	0.0052	2,29	0.0038	-0.0050
2,30	-0.0036	0.0048	2,31	0.0035	-0.0046	2,32	-0.0034	0.0045	2,33	0.0033	-0.0044
2,34	-0.0033	0.0042	2,35	0.0032	-0.0042	2,36	-0.0032	0.0041	2,37	0.0032	-0.0041
2,38	-0.0032	0.0041	2,39	0.0034	-0.0043						
3, 3	1.0606	1.5286	3, 4	0.0661	-0.3667	3, 5	-0.0625	0.1604	3, 6	0.0494	-0.0912
3, 7	-0.0377	0.0590	3, 8	0.0291	-0.0420	3, 9	-0.0231	0.0321	3,10	0.0189	-0.0257
3,11	-0.0159	0.0214	3,12	0.0136	-0.0182	3,13	-0.0118	0.0158	3,14	0.0104	-0.0139
3,15	-0.0093	0.0124	3,16	0.0084	-0.0112	3,17	-0.0076	0.0102	3,18	0.0070	-0.0093
3,19	-0.0064	0.0086	3,20	0.0060	-0.0080	3,21	-0.0056	0.0074	3,22	0.0052	-0.0070
3,23	-0.0049	0.0065	3,24	0.0046	-0.0062	3,25	-0.0044	0.0058	3,26	0.0041	-0.0055
3,27	-0.0039	0.0053	3,28	0.0038	-0.0050	3,29	-0.0036	0.0048	3,30	0.0035	-0.0046
3,31	-0.0034	0.0045	3,32	0.0033	-0.0043	3,33	-0.0032	0.0042	3,34	0.0031	-0.0041
3,35	-0.0030	0.0040	3,36	0.0030	-0.0039	3,37	-0.0030	0.0039	3,38	0.0031	-0.0039
4, 4	1.0640	1.5248	4, 5	0.0637	-0.3642	4, 6	-0.0606	0.1587	4, 7	0.0479	-0.0898
4, 8	-0.0365	0.0579	4, 9	0.0281	-0.0411	4,10	-0.0223	0.0313	4,11	0.0182	-0.0251
4,12	-0.0152	0.0208	4,13	0.0130	-0.0177	4,14	-0.0113	0.0153	4,15	0.0099	-0.0135
4,16	-0.0089	0.0120	4,17	0.0080	-0.0108	4,18	-0.0073	0.0098	4,19	0.0066	-0.0090
4,20	-0.0061	0.0083	4,21	0.0057	-0.0077	4,22	-0.0053	0.0072	4,23	0.0049	-0.0067
4,24	-0.0046	0.0063	4,25	0.0044	-0.0060	4,26	-0.0041	0.0056	4,27	0.0039	-0.0054
4,28	-0.0037	0.0051	4,29	0.0036	-0.0049	4,30	-0.0034	0.0047	4,31	0.0033	-0.0045
4,32	-0.0032	0.0043	4,33	0.0031	-0.0042	4,34	-0.0030	0.0041	4,35	0.0030	-0.0040
4,36	-0.0029	0.0039	4,37	0.0029	-0.0038						
5, 5	1.0657	1.5232	5, 6	0.0624	-0.3631	5, 7	-0.0596	0.1578	5, 8	0.0471	-0.0892
5, 9	-0.0358	0.0573	5,10	0.0275	-0.0406	5,11	-0.0217	0.0309	5,12	0.0177	-0.0247
5,13	-0.0148	0.0204	5,14	0.0126	-0.0174	5,15	-0.0109	0.0150	5,16	0.0096	-0.0132
5,17	-0.0086	0.0118	5,18	0.0077	-0.0106	5,19	-0.0070	0.0097	5,20	0.0064	-0.0088
5,21	-0.0059	0.0081	5,22	0.0055	-0.0075	5,23	-0.0051	0.0070	5,24	0.0048	-0.0066
5,25	-0.0045	0.0062	5,26	0.0042	-0.0058	5,27	-0.0040	0.0055	5,28	0.0038	-0.0052
5,29	-0.0036	0.0050	5,30	0.0035	-0.0048	5,31	-0.0033	0.0046	5,32	0.0032	-0.0044
5,33	-0.0031	0.0043	5,34	0.0030	-0.0041	5,35	-0.0030	0.0040	5,36	0.0029	-0.0039
6, 6	1.0666	1.5225	6, 7	0.0616	-0.3626	6, 8	-0.0590	0.1574	6, 9	0.0466	-0.0888
6,10	-0.0354	0.0570	6,11	0.0271	-0.0403	6,12	-0.0214	0.0307	6,13	0.0174	-0.0245
6,14	-0.0145	0.0202	6,15	0.0124	-0.0172	6,16	-0.0107	0.0149	6,17	0.0094	-0.0131
6,18	-0.0084	0.0117	6,19	0.0076	-0.0105	6,20	-0.0069	0.0095	6,21	0.0063	-0.0087
6,22	-0.0058	0.0080	6,23	0.0053	-0.0074	6,24	-0.0050	0.0069	6,25	0.0046	-0.0065
6,26	-0.0044	0.0061	6,27	0.0041	-0.0058	6,28	-0.0039	0.0055	6,29	0.0037	-0.0052
6,30	-0.0035	0.0049	6,31	0.0034	-0.0047	6,32	-0.0033	0.0045	6,33	0.0032	-0.0044
6,34	-0.0031	0.0042	6,35	0.0030	-0.0041						
7, 7	1.0672	1.5220	7, 8	0.0612	-0.3622	7, 9	-0.0586	0.1571	7,10	0.0463	-0.0885
7,11	-0.0351	0.0568	7,12	0.0269	-0.0402	7,13	-0.0212	0.0305	7,14	0.0172	-0.0243
7,15	-0.0143	0.0201	7,16	0.0122	-0.0171	7,17	-0.0106	0.0148	7,18	0.0093	-0.0130
7,19	-0.0083	0.0116	7,20	0.0074	-0.0104	7,21	-0.0067	0.0094	7,22	0.0062	-0.0086
7,23	-0.0057	0.0080	7,24	0.0052	-0.0074	7,25	-0.0049	0.0069	7,26	0.0046	-0.0064
7,27	-0.0043	0.0061	7,28	0.0040	-0.0057	7,29	-0.0038	0.0054	7,30	0.0037	-0.0051
7,31	-0.0035	0.0049	7,32	0.0034	-0.0047	7,33	-0.0032	0.0045	7,34	0.0031	-0.0044
8, 8	1.0676	1.5218	8, 9	0.0609	-0.3620	8,10	-0.0584	0.1569	8,11	0.0460	-0.0884
8,12	-0.0349	0.0566	8,13	0.0267	-0.0400	8,14	-0.0210	0.0304	8,15	0.0171	-0.0242
8,16	-0.0142	0.0200	8,17	0.0121	-0.0170	8,18	-0.0105	0.0147	8,19	0.0092	-0.0129
8,20	-0.0082	0.0115	8,21	0.0073	-0.0103	8,22	-0.0067	0.0094	8,23	0.0061	-0.0086
8,24	-0.0056	0.0079	8,25	0.0052	-0.0073	8,26	-0.0048	0.0068	8,27	0.0045	-0.0064
8,28	-0.0042	0.0060	8,29	0.0040	-0.0057	8,30	-0.0038	0.0054	8,31	0.0036	-0.0051
8,32	-0.0035	0.0049	8,33	0.0033	-0.0047						
9, 9	1.0678	1.5216	9,10	0.0607	-0.3618	9,11	-0.0582	0.1567	9,12	0.0459	-0.0882
9,13	-0.0347	0.0565	9,14	0.0266	-0.0399	9,15	-0.0209	0.0303	9,16	0.0169	-0.0241
9,17	-0.0141	0.0199	9,18	0.0120	-0.0169	9,19	-0.0104	0.0146	9,20	0.0091	-0.0128
9,21	-0.0081	0.0114	9,22	0.0073	-0.0103	9,23	-0.0066	0.0093	9,24	0.0060	-0.0086
9,25	-0.0056	0.0079	9,26	0.0051	-0.0073	9,27	-0.0048	0.0068	9,28	0.0045	-0.0064
9,29	-0.0042	0.0060	9,30	0.0040	-0.0057	9,31	-0.0038	0.0054	9,32	0.0036	-0.0051
10,10	1.0680	1.5214	10,11	0.0605	-0.3617	10,12	-0.0580	0.1566	10,13	0.0458	-0.0881
10,14	-0.0346	0.0564	10,15	0.0265	-0.0398	10,16	-0.0208	0.0302	10,17	0.0169	-0.0241
10,18	-0.0140	0.0199	10,19	0.0119	-0.0168	10,20	-0.0103	0.0146	10,21	0.0091	-0.0128
10,22	-0.0081	0.0114	10,23	0.0072	-0.0103	10,24	-0.0066	0.0093	10,25	0.0060	-0.0085
10,26	-0.0055	0.0079	10,27	0.0051	-0.0073	10,28	-0.0048	0.0068	10,29	0.0045	-0.0064
10,30	-0.0042	0.0060	10,31	0.0040	-0.0057						
11,11	1.0681	1.5213	11,12	0.0604	-0.3616	11,13	-0.0579	0.1565	11,14	0.0457	-0.0881
11,15	-0.0345	0.0564	11,16	0.0264	-0.0398	11,17	-0.0207	0.0301	11,18	0.0168	-0.0240
11,19	-0.0140	0.0198	11,20	0.0119	-0.0168	11,21	-0.0103	0.0145	11,22	0.0090	-0.0128
11,23	-0.0080	0.0114	11,24	0.0072	-0.0102	11,25	-0.0065	0.0093	11,26	0.0060	-0.0085
11,27	-0.0055	0.0078	11,28	0.0051	-0.0073	11,29	-0.0048	0.0068	11,30	0.0045	-0.0064
12,12	1.0682	1.5212	12,13	0.0603	-0.3615	12,14	-0.0578	0.1565	12,15	0.0456	-0.0880
12,16	-0.0345	0.0563	12,17	0.0263	-0.0397	12,18	-0.0207	0.0301	12,19	0.0168	-0.0240
12,20	-0.0139	0.0198	12,21	0.0118	-0.0168	12,22	-0.0102	0.0145	12,23	0.0090	-0.0128
12,24	-0.0080	0.0114	12,25	0.0072	-0.0102	12,26	-0.0065	0.0093	12,27	0.0060	-0.0085
12,28	-0.0055	0.0078	12,29	0.0051	-0.0073						
13,13	1.0683	1.5212	13,14	0.0602	-0.3615	13,15	-0.0578	0.1564	13,16	0.0455	-0.0880
13,17	-0.0344	0.0563	13,18	0.0263	-0.0397	13,19	-0.0207	0.0301	13,20	0.0167	-0.0240
13,21	-0.0139	0.0198	13,22	0.0118	-0.0168	13,23	-0.0102	0.0145	13,24	0.0090	-0.0127
13,25	-0.0080	0.0113	13,26	0.0072	-0.0102	13,27	-0.0065	0.0093	13,28	0.0060	-0.0085

14,14	1.0684	1.5211	14,15	0.0602	-0.3614	14,16	-0.0577	0.1564	14,17	0.0455	-0.0879
14,18	-0.0344	0.0562	14,19	0.0263	-0.0397	14,20	-0.0206	0.0301	14,21	0.0167	-0.0239
14,22	-0.0139	0.0198	14,23	0.0118	-0.0167	14,24	-0.0102	0.0145	14,25	0.0090	-0.0127
14,26	-0.0080	0.0113	14,27	0.0072	-0.0102						
15,15	1.0684	1.5211	15,16	0.0601	-0.3614	15,17	-0.0577	0.1564	15,18	0.0455	-0.0879
15,19	-0.0344	0.0562	15,20	0.0262	-0.0397	15,21	-0.0206	0.0300	15,22	0.0167	-0.0239
15,23	-0.0139	0.0197	15,24	0.0118	-0.0167	15,25	-0.0102	0.0145	15,26	0.0090	-0.0127
16,16	1.0685	1.5210	16,17	0.0601	-0.3614	16,18	-0.0577	0.1563	16,19	0.0454	-0.0879
16,20	-0.0343	0.0562	16,21	0.0262	-0.0396	16,22	-0.0206	0.0300	16,23	0.0167	-0.0239
16,24	-0.0139	0.0197	16,25	0.0118	-0.0167						
17,17	1.0685	1.5210	17,18	0.0601	-0.3613	17,19	-0.0577	0.1563	17,20	0.0454	-0.0879
17,21	-0.0343	0.0562	17,22	0.0262	-0.0396	17,23	-0.0206	0.0300	17,24	0.0167	-0.0239
18,18	1.0685	1.5210	18,19	0.0601	-0.3613	18,20	-0.0576	0.1563	18,21	0.0454	-0.0879
18,22	-0.0343	0.0562	18,23	0.0262	-0.0396						
19,19	1.0685	1.5210	19,20	0.0600	-0.3613	19,21	-0.0576	0.1563	19,22	0.0454	-0.0879
20,20	1.0685	1.5210	20,21	0.0600	-0.3613						

VBR

SELF AND MUTUAL ADMITTANCES

H/LMDA=0.5000   B/LMDA= 0.500   OMEGA= 9.92

45 ELEMENT ARRAY

I, K	YIK IN MMHO		I, K	YIK IN MMHO		I, K	YIK IN MMHO		I, K	YIK IN MMHO	
	RE	IM		RE	IM		RE	IM		RE	IM
1, 1	1.0535	1.6054	1, 2	0.0710	[illegible]	1, 3	-0.0719	0.1831	1, 4	0.0570	-0.1066
1, 5	-0.0437	0.0704	1, 6	0.0340	-0.0509	1, 7	-0.0272	0.0394	1, 8	[illegible]	0.0310
1, 9	-0.0188	0.0266	1,10	0.0162	-0.0228	1,11	-0.0141	0.0199	1,12	0.0125	-0.0176
1,13	-0.0112	0.0157	1,14	0.0101	-0.0142	1,15	-0.0092	0.0129	1,16	0.0084	-0.0119
1,17	-0.0077	0.0109	1,18	0.0072	-0.0102	1,19	-0.0067	0.0095	1,20	0.0062	-0.0089
1,21	-0.0059	0.0083	1,22	0.0055	-0.0079	1,23	-0.0052	0.0074	1,24	0.0049	-0.0071
1,25	-0.0047	0.0067	1,26	0.0045	-0.0064	1,27	-0.0043	0.0061	1,28	0.0041	-0.0059
1,29	-0.0039	0.0056	1,30	0.0038	-0.0054	1,31	-0.0036	0.0052	1,32	0.0035	-0.0050
1,33	-0.0034	0.0049	1,34	0.0033	-0.0047	1,35	-0.0032	0.0046	1,36	0.0031	-0.0045
1,37	-0.0031	0.0043	1,38	0.0030	-0.0042	1,39	-0.0030	0.0042	1,40	0.0030	-0.0041
1,41	-0.0029	0.0041	1,42	0.0030	-0.0041	1,43	-0.0030	0.0041	1,44	0.0032	-0.0043
1,45	-0.0034	0.0049									
2, 2	1.0543	1.5417	2, 3	0.0709	-0.3742	2, 4	-0.0661	0.1653	2, 5	0.0522	-0.0947
2, 6	-0.0399	0.0617	2, 7	0.0309	-0.0442	2, 8	-0.0246	0.0339	2, 9	0.0202	-0.0273
2,10	-0.0170	0.0227	2,11	0.0146	-0.0194	2,12	-0.0127	0.0168	2,13	0.0112	-0.0149
2,14	-0.0100	0.0133	2,15	0.0090	-0.0120	2,16	-0.0082	0.0109	2,17	0.0075	-0.0100
2,18	-0.0069	0.0092	2,19	0.0064	-0.0085	2,20	-0.0060	0.0080	2,21	0.0056	-0.0074
2,22	-0.0052	0.0070	2,23	0.0049	-0.0066	2,24	-0.0047	0.0062	2,25	0.0044	-0.0059
2,26	-0.0042	0.0056	2,27	0.0040	-0.0054	2,28	-0.0038	0.0051	2,29	0.0037	-0.0049
2,30	-0.0035	0.0047	2,31	0.0034	-0.0045	2,32	-0.0033	0.0044	2,33	0.0032	-0.0042
2,34	-0.0031	0.0041	2,35	0.0030	-0.0040	2,36	-0.0029	0.0038	2,37	0.0028	-0.0037
2,38	-0.0028	0.0037	2,39	0.0027	-0.0036	2,40	-0.0027	0.0035	2,41	0.0027	-0.0035
2,42	-0.0027	0.0035	2,43	0.0028	-0.0035	2,44	-0.0029	0.0037			
3, 3	1.0607	1.5286	3, 4	0.0661	-0.3667	3, 5	-0.0625	0.1604	3, 6	0.0494	-0.0912
3, 7	-0.0377	0.0590	3, 8	0.0291	-0.0420	3, 9	-0.0231	0.0321	3,10	0.0189	-0.0257
3,11	-0.0158	0.0213	3,12	0.0135	-0.0182	3,13	-0.0118	0.0158	3,14	0.0104	-0.0139
3,15	-0.0093	0.0124	3,16	0.0084	-0.0112	3,17	-0.0076	0.0101	3,18	0.0069	-0.0093
3,19	-0.0064	0.0086	3,20	0.0059	-0.0079	3,21	-0.0055	0.0074	3,22	0.0051	-0.0069
3,23	-0.0048	0.0065	3,24	0.0045	-0.0061	3,25	-0.0043	0.0058	3,26	0.0041	-0.0055
3,27	-0.0039	0.0052	3,28	0.0037	-0.0050	3,29	-0.0035	0.0048	3,30	0.0034	-0.0046
3,31	-0.0032	0.0044	3,32	0.0031	-0.0042	3,33	-0.0030	0.0041	3,34	0.0029	-0.0039
3,35	-0.0028	0.0038	3,36	0.0027	-0.0037	3,37	-0.0027	0.0036	3,38	0.0026	-0.0035
3,39	-0.0026	0.0034	3,40	0.0026	-0.0033	3,41	-0.0025	0.0033	3,42	0.0026	-0.0033
3,43	-0.0026	0.0033									
4, 4	1.0640	1.5247	4, 5	0.0636	-0.3642	4, 6	-0.0606	0.1586	4, 7	0.0479	-0.0898
4, 8	-0.0365	0.0579	4, 9	0.0281	-0.0411	4,10	-0.0222	0.0313	4,11	0.0181	-0.0250
4,12	-0.0152	0.0207	4,13	0.0129	-0.0176	4,14	-0.0112	0.0153	4,15	0.0099	-0.0134
4,16	-0.0088	0.0120	4,17	0.0079	-0.0108	4,18	-0.0072	0.0098	4,19	0.0066	-0.0090
4,20	-0.0061	0.0083	4,21	0.0056	-0.0077	4,22	-0.0052	0.0071	4,23	0.0049	-0.0067
4,24	-0.0046	0.0063	4,25	0.0043	-0.0059	4,26	-0.0041	0.0056	4,27	0.0038	-0.0053
4,28	-0.0036	0.0050	4,29	0.0035	-0.0048	4,30	-0.0033	0.0046	4,31	0.0032	-0.0044
4,32	-0.0031	0.0042	4,33	0.0029	-0.0041	4,34	-0.0028	0.0039	4,35	0.0028	-0.0038
4,36	-0.0027	0.0037	4,37	0.0026	-0.0036	4,38	-0.0026	0.0035	4,39	0.0025	-0.0034
4,40	-0.0025	0.0033	4,41	0.0025	-0.0033	4,42	-0.0025	0.0033			

5, 5	1.0657	1.5232	5, 6	0.0623	-0.3631	5, 7	-0.0596	0.1578	5, 8	0.0471	-0.0892
5, 9	-0.0358	0.0573	5,10	0.0275	-0.0406	5,11	-0.0217	0.0309	5,12	0.0177	-0.0247
5,13	-0.0147	0.0204	5,14	0.0126	-0.0173	5,15	-0.0109	0.0150	5,16	0.0096	-0.0132
5,17	-0.0085	0.0118	5,18	0.0077	-0.0106	5,19	-0.0070	0.0096	5,20	0.0064	-0.0088
5,21	-0.0058	0.0081	5,22	0.0054	-0.0075	5,23	-0.0050	0.0070	5,24	0.0047	-0.0065
5,25	-0.0044	0.0061	5,26	0.0041	-0.0058	5,27	-0.0039	0.0055	5,28	0.0037	-0.0052
5,29	-0.0035	0.0049	5,30	0.0034	-0.0047	5,31	-0.0032	0.0045	5,32	0.0031	-0.0043
5,33	-0.0030	0.0041	5,34	0.0028	-0.0040	5,35	-0.0028	0.0038	5,36	0.0027	-0.0037
5,37	-0.0026	0.0036	5,38	0.0025	-0.0035	5,39	-0.0025	0.0034	5,40	0.0025	-0.0033
5,41	-0.0024	0.0033									
6, 6	1.0667	1.5225	6, 7	0.0616	-0.3625	6, 8	-0.0590	0.1573	6, 9	0.0466	-0.0888
6,10	-0.0353	0.0570	6,11	0.0271	-0.0403	6,12	-0.0214	0.0306	6,13	0.0174	-0.0244
6,14	-0.0145	0.0202	6,15	0.0123	-0.0171	6,16	-0.0107	0.0148	6,17	0.0094	-0.0130
6,18	-0.0084	0.0116	6,19	0.0075	-0.0105	6,20	-0.0068	0.0095	6,21	0.0062	-0.0087
6,22	-0.0057	0.0080	6,23	0.0053	-0.0074	6,24	-0.0049	0.0069	6,25	0.0046	-0.0064
6,26	-0.0043	0.0060	6,27	0.0040	-0.0057	6,28	-0.0038	0.0054	6,29	0.0036	-0.0051
6,30	-0.0034	0.0048	6,31	0.0033	-0.0046	6,32	-0.0031	0.0044	6,33	0.0030	-0.0042
6,34	-0.0029	0.0041	6,35	0.0028	-0.0039	6,36	-0.0027	0.0038	6,37	0.0026	-0.0037
6,38	-0.0026	0.0036	6,39	0.0025	-0.0035	6,40	-0.0025	0.0034			
7, 7	1.0672	1.5220	7, 8	0.0611	-0.3622	7, 9	-0.0586	0.1570	7,10	0.0462	-0.0885
7,11	-0.0350	0.0568	7,12	0.0268	-0.0401	7,13	-0.0211	0.0304	7,14	0.0172	-0.0243
7,15	-0.0143	0.0201	7,16	0.0122	-0.0170	7,17	-0.0105	0.0147	7,18	0.0092	-0.0129
7,19	-0.0082	0.0115	7,20	0.0074	-0.0104	7,21	-0.0067	0.0094	7,22	0.0061	-0.0086
7,23	-0.0056	0.0079	7,24	0.0052	-0.0073	7,25	-0.0048	0.0068	7,26	0.0045	-0.0064
7,27	-0.0042	0.0060	7,28	0.0040	-0.0056	7,29	-0.0037	0.0053	7,30	0.0035	-0.0050
7,31	-0.0034	0.0048	7,32	0.0032	-0.0046	7,33	-0.0031	0.0044	7,34	0.0030	-0.0042
7,35	-0.0028	0.0040	7,36	0.0027	-0.0039	7,37	-0.0027	0.0038	7,38	0.0026	-0.0037
7,39	-0.0025	0.0036									
8, 8	1.0676	1.5217	8, 9	0.0608	-0.3619	8,10	-0.0583	0.1568	8,11	0.0460	-0.0883
8,12	-0.0348	0.0566	8,13	0.0266	-0.0400	8,14	-0.0210	0.0303	8,15	0.0170	-0.0242
8,16	-0.0142	0.0200	8,17	0.0120	-0.0169	8,18	-0.0104	0.0146	8,19	0.0091	-0.0129
8,20	-0.0081	0.0114	8,21	0.0073	-0.0103	8,22	-0.0066	0.0093	8,23	0.0060	-0.0085
8,24	-0.0055	0.0079	8,25	0.0051	-0.0073	8,26	-0.0047	0.0068	8,27	0.0044	-0.0063
8,28	-0.0041	0.0059	8,29	0.0039	-0.0056	8,30	-0.0037	0.0053	8,31	0.0035	-0.0050
8,32	-0.0033	0.0048	8,33	0.0032	-0.0046	8,34	-0.0030	0.0044	8,35	0.0029	-0.0042
8,36	-0.0028	0.0040	8,37	0.0027	-0.0039	8,38	-0.0027	0.0038			
9, 9	1.0679	1.5215	9,10	0.0606	-0.3618	9,11	-0.0581	0.1567	9,12	0.0458	-0.0882
9,13	-0.0347	0.0565	9,14	0.0265	-0.0399	9,15	-0.0209	0.0302	9,16	0.0169	-0.0241
9,17	-0.0141	0.0199	9,18	0.0119	-0.0169	9,19	-0.0103	0.0146	9,20	0.0091	-0.0128
9,21	-0.0081	0.0114	9,22	0.0072	-0.0102	9,23	-0.0065	0.0093	9,24	0.0060	-0.0085
9,25	-0.0055	0.0078	9,26	0.0051	-0.0072	9,27	-0.0047	0.0067	9,28	0.0044	-0.0063
9,29	-0.0041	0.0059	9,30	0.0039	-0.0056	9,31	-0.0037	0.0053	9,32	0.0035	-0.0050
9,33	-0.0033	0.0048	9,34	0.0032	-0.0045	9,35	-0.0030	0.0043	9,36	0.0029	-0.0042
9,37	-0.0028	0.0040									
10,10	1.0680	1.5214	10,11	0.0605	-0.3616	10,12	-0.0580	0.1566	10,13	0.0457	-0.0881
10,14	-0.0346	0.0564	10,15	0.0264	-0.0398	10,16	-0.0208	0.0302	10,17	0.0168	-0.0240
10,18	-0.0140	0.0198	10,19	0.0119	-0.0168	10,20	-0.0103	0.0145	10,21	0.0090	-0.0128
10,22	-0.0080	0.0113	10,23	0.0072	-0.0102	10,24	-0.0065	0.0092	10,25	0.0059	-0.0085
10,26	-0.0054	0.0078	10,27	0.0050	-0.0072	10,28	-0.0047	0.0067	10,29	0.0044	-0.0063
10,30	-0.0041	0.0059	10,31	0.0038	-0.0055	10,32	-0.0036	0.0052	10,33	0.0034	-0.0050
10,34	-0.0033	0.0047	10,35	0.0031	-0.0045	10,36	-0.0030	0.0043			
11,11	1.0682	1.5213	11,12	0.0603	-0.3616	11,13	-0.0579	0.1565	11,14	0.0456	-0.0880
11,15	-0.0345	0.0563	11,16	0.0263	-0.0397	11,17	-0.0207	0.0301	11,18	0.0168	-0.0240
11,19	-0.0139	0.0198	11,20	0.0118	-0.0168	11,21	-0.0102	0.0145	11,22	0.0090	-0.0127
11,23	-0.0080	0.0113	11,24	0.0071	-0.0102	11,25	-0.0065	0.0092	11,26	0.0059	-0.0084
11,27	-0.0054	0.0078	11,28	0.0050	-0.0072	11,29	-0.0046	0.0067	11,30	0.0043	-0.0062
11,31	-0.0041	0.0059	11,32	0.0038	-0.0055	11,33	-0.0036	0.0052	11,34	0.0034	-0.0050
11,35	-0.0033	0.0047									
12,12	1.0683	1.5212	12,13	0.0602	-0.3615	12,14	-0.0578	0.1564	12,15	0.0456	-0.0880
12,16	-0.0344	0.0563	12,17	0.0263	-0.0397	12,18	-0.0206	0.0301	12,19	0.0167	-0.0239
12,20	-0.0139	0.0197	12,21	0.0118	-0.0167	12,22	-0.0102	0.0144	12,23	0.0089	-0.0127
12,24	-0.0079	0.0113	12,25	0.0071	-0.0101	12,26	-0.0064	0.0092	12,27	0.0059	-0.0084
12,28	-0.0054	0.0077	12,29	0.0050	-0.0072	12,30	-0.0046	0.0067	12,31	0.0043	-0.0062
12,32	-0.0041	0.0059	12,33	0.0038	-0.0055	12,34	-0.0036	0.0052			
13,13	1.0684	1.5211	13,14	0.0602	-0.3614	13,15	-0.0577	0.1564	13,16	0.0455	-0.0879
13,17	-0.0344	0.0562	13,18	0.0262	-0.0396	13,19	-0.0206	0.0300	13,20	0.0167	-0.0239
13,21	-0.0138	0.0197	13,22	0.0117	-0.0167	13,23	-0.0101	0.0144	13,24	0.0089	-0.0127
13,25	-0.0079	0.0113	13,26	0.0071	-0.0101	13,27	-0.0064	0.0092	13,28	0.0058	-0.0084
13,29	-0.0054	0.0077	13,30	0.0050	-0.0072	13,31	-0.0046	0.0067	13,32	0.0043	-0.0062
13,33	-0.0040	0.0059									
14,14	1.0684	1.5211	14,15	0.0601	-0.3614	14,16	-0.0577	0.1563	14,17	0.0454	-0.0879
14,18	-0.0343	0.0562	14,19	0.0262	-0.0396	14,20	-0.0206	0.0300	14,21	0.0166	-0.0239
14,22	-0.0138	0.0197	14,23	0.0117	-0.0167	14,24	-0.0101	0.0144	14,25	0.0089	-0.0126
14,26	-0.0079	0.0112	14,27	0.0071	-0.0101	14,28	-0.0064	0.0092	14,29	0.0058	-0.0084
14,30	-0.0054	0.0077	14,31	0.0050	-0.0072	14,32	-0.0046	0.0067			
15,15	1.0685	1.5210	15,16	0.0601	-0.3613	15,17	-0.0576	0.1563	15,18	0.0454	-0.0879
15,19	-0.0343	0.0562	15,20	0.0262	-0.0396	15,21	-0.0205	0.0300	15,22	0.0166	-0.0239
15,23	-0.0138	0.0197	15,24	0.0117	-0.0167	15,25	-0.0101	0.0144	15,26	0.0089	-0.0126
15,27	-0.0079	0.0112	15,28	0.0071	-0.0101	15,29	-0.0064	0.0092	15,30	0.0058	-0.0084
15,31	-0.0054	0.0077									
16,16	1.0685	1.5210	16,17	0.0600	-0.3613	16,18	-0.0576	0.1563	16,19	0.0454	-0.0878
16,20	-0.0343	0.0561	16,21	0.0261	-0.0396	16,22	-0.0205	0.0300	16,23	0.0166	-0.0238
16,24	-0.0138	0.0197	16,25	0.0117	-0.0166	16,26	-0.0101	0.0144	16,27	0.0089	-0.0126
16,28	-0.0079	0.0112	16,29	0.0070	-0.0101	16,30	-0.0064	0.0092			
17,17	1.0686	1.5210	17,18	0.0600	-0.3613	17,19	-0.0576	0.1563	17,20	0.0454	-0.0878
17,21	-0.0343	0.0561	17,22	0.0261	-0.0396	17,23	-0.0205	0.0299	17,24	0.0166	-0.0238
17,25	-0.0138	0.0197	17,26	0.0117	-0.0166	17,27	-0.0101	0.0144	17,28	0.0088	-0.0126
17,29	-0.0079	0.0112									
18,18	1.0686	1.5209	18,19	0.0600	-0.3613	18,20	-0.0576	0.1562	18,21	0.0453	-0.0878
18,22	-0.0342	0.0561	18,23	0.0261	-0.0395	18,24	-0.0205	0.0299	18,25	0.0166	-0.0238
18,26	-0.0138	0.0196	18,27	0.0117	-0.0166	18,28	-0.0101	0.0144			
19,19	1.0686	1.5209	19,20	0.0600	-0.3612	19,21	-0.0575	0.1562	19,22	0.0453	-0.0878
19,23	-0.0342	0.0561	19,24	0.0261	-0.0395	19,25	-0.0205	0.0299	19,26	0.0166	-0.0238
19,27	-0.0138	0.0196									
20,20	1.0686	1.5209	20,21	0.0600	-0.3612	20,22	-0.0575	0.1562	20,23	0.0453	-0.0878
20,24	-0.0342	0.0561	20,25	0.0261	-0.0395	20,26	-0.0205	0.0299			
21,21	1.0686	1.5209	21,22	0.0599	-0.3612	21,23	-0.0575	0.1562	21,24	0.0453	-0.0878
21,25	-0.0342	0.0561									
22,22	1.0686	1.5209	22,23	0.0599	-0.3612	22,24	-0.0575	0.1562			
23,23	1.0686	1.5209									

VBR

SELF AND MUTUAL ADMITTANCES

H/LMDA=0.5000 B/LMDA= 0.500 OMEGA= 9.92

50 ELEMENT ARRAY

I, K	YIK IN MMHO		I, K	YIK IN MMHO		I, K	YIK IN MMHO		I, K	YIK IN MMHO	
	RE	IM		RE	IM		RE	IM		RE	IM
1, 1	1.0535	1.6054	1, 2	0.0769	-0.4041	1, 3	-0.0719	0.1831	1, 4	0.0569	-0.1066
1, 5	-0.0437	0.0703	1, 6	0.0340	-0.0509	1, 7	-0.0271	0.0393	1, 8	0.0223	-0.0318
1, 9	-0.0188	0.0266	1,10	0.0161	-0.0228	1,11	-0.0141	0.0198	1,12	0.0124	-0.0175
1,13	-0.0111	0.0157	1,14	0.0100	-0.0142	1,15	-0.0091	0.0129	1,16	0.0084	-0.0118
1,17	-0.0077	0.0109	1,18	0.0071	-0.0101	1,19	-0.0066	0.0094	1,20	0.0062	-0.0088
1,21	-0.0058	0.0083	1,22	0.0055	-0.0078	1,23	-0.0052	0.0074	1,24	0.0049	-0.0070
1,25	-0.0046	0.0067	1,26	0.0044	-0.0063	1,27	-0.0042	0.0061	1,28	0.0040	-0.0058
1,29	-0.0039	0.0056	1,30	0.0037	-0.0053	1,31	-0.0036	0.0051	1,32	0.0034	-0.0050
1,33	-0.0033	0.0048	1,34	0.0032	-0.0046	1,35	-0.0031	0.0045	1,36	0.0030	-0.0043
1,37	-0.0029	0.0042	1,38	0.0029	-0.0041	1,39	-0.0028	0.0040	1,40	0.0027	-0.0039
1,41	-0.0027	0.0038	1,42	0.0026	-0.0037	1,43	-0.0026	0.0037	1,44	0.0026	-0.0036
1,45	-0.0026	0.0036	1,46	0.0026	-0.0035	1,47	-0.0026	0.0036	1,48	0.0027	-0.0036
1,49	-0.0028	0.0038	1,50	0.0030	-0.0043						
2, 2	1.0544	1.5416	2, 3	0.0709	-0.3742	2, 4	-0.0661	0.1653	2, 5	0.0522	-0.0947
2, 6	-0.0399	0.0617	2, 7	0.0309	-0.0442	2, 8	-0.0246	0.0339	2, 9	0.0202	-0.0273
2,10	-0.0169	0.0227	2,11	0.0145	-0.0194	2,12	-0.0127	0.0168	2,13	0.0112	-0.0148
2,14	-0.0100	0.0133	2,15	0.0090	-0.0120	2,16	-0.0082	0.0109	2,17	0.0075	-0.0100
2,18	-0.0069	0.0092	2,19	0.0064	-0.0085	2,20	-0.0059	0.0079	2,21	0.0055	-0.0074
2,22	-0.0052	0.0070	2,23	0.0049	-0.0066	2,24	-0.0046	0.0062	2,25	0.0044	-0.0059
2,26	-0.0041	0.0056	2,27	0.0039	-0.0053	2,28	-0.0038	0.0051	2,29	0.0036	-0.0049
2,30	-0.0034	0.0047	2,31	0.0033	-0.0045	2,32	-0.0032	0.0043	2,33	0.0031	-0.0041
2,34	-0.0030	0.0040	2,35	0.0029	-0.0039	2,36	-0.0028	0.0038	2,37	0.0027	-0.0036
2,38	-0.0026	0.0035	2,39	0.0026	-0.0034	2,40	-0.0025	0.0034	2,41	0.0025	-0.0033
2,42	-0.0024	0.0032	2,43	0.0024	-0.0031	2,44	-0.0023	0.0031	2,45	0.0023	-0.0031
2,46	-0.0023	0.0030	2,47	0.0024	-0.0030	2,48	-0.0024	0.0031	2,49	0.0025	-0.0032
3, 3	1.0607	1.5286	3, 4	0.0661	-0.3667	3, 5	-0.0625	0.1604	3, 6	0.0494	-0.0912
3, 7	-0.0377	0.0589	3, 8	0.0291	-0.0420	3, 9	-0.0231	0.0320	3,10	0.0189	-0.0257
3,11	-0.0158	0.0213	3,12	0.0135	-0.0181	3,13	-0.0118	0.0157	3,14	0.0104	-0.0139
3,15	-0.0092	0.0124	3,16	0.0083	-0.0111	3,17	-0.0076	0.0101	3,18	0.0069	-0.0093
3,19	-0.0064	0.0085	3,20	0.0059	-0.0079	3,21	-0.0055	0.0074	3,22	0.0051	-0.0069
3,23	-0.0048	0.0065	3,24	0.0045	-0.0061	3,25	-0.0042	0.0057	3,26	0.0040	-0.0054
3,27	-0.0038	0.0052	3,28	0.0036	-0.0049	3,29	-0.0035	0.0047	3,30	0.0033	-0.0045
3,31	-0.0032	0.0043	3,32	0.0030	-0.0041	3,33	-0.0029	0.0040	3,34	0.0028	-0.0038
3,35	-0.0027	0.0037	3,36	0.0026	-0.0036	3,37	-0.0026	0.0035	3,38	0.0025	-0.0034
3,39	-0.0024	0.0033	3,40	0.0024	-0.0032	3,41	-0.0023	0.0031	3,42	0.0023	-0.0030
3,43	-0.0022	0.0030	3,44	0.0022	-0.0029	3,45	-0.0022	0.0029	3,46	0.0022	-0.0029
3,47	-0.0022	0.0029	3,48	0.0023	-0.0029						
4, 4	1.0641	1.5247	4, 5	0.0636	-0.3642	4, 6	-0.0606	0.1586	4, 7	0.0479	-0.0898
4, 8	-0.0364	0.0578	4, 9	0.0280	-0.0411	4,10	-0.0222	0.0313	4,11	0.0181	-0.0250
4,12	-0.0151	0.0207	4,13	0.0129	-0.0176	4,14	-0.0112	0.0153	4,15	0.0099	-0.0134
4,16	-0.0088	0.0120	4,17	0.0079	-0.0108	4,18	-0.0072	0.0098	4,19	0.0065	-0.0089
4,20	-0.0060	0.0082	4,21	0.0056	-0.0076	4,22	-0.0052	0.0071	4,23	0.0048	-0.0066
4,24	-0.0045	0.0062	4,25	0.0042	-0.0059	4,26	-0.0040	0.0055	4,27	0.0038	-0.0052
4,28	-0.0036	0.0050	4,29	0.0034	-0.0047	4,30	-0.0033	0.0045	4,31	0.0031	-0.0043
4,32	-0.0030	0.0041	4,33	0.0029	-0.0040	4,34	-0.0028	0.0038	4,35	0.0027	-0.0037
4,36	[illegible]	[illegible]	4,37	[illegible]	-0.0035	4,38	-0.0024	0.0033	4,39	0.0024	-0.0032
4,40	-0.0023	0.0032	4,41	0.0022	[illegible]	[illegible]	[illegible]	[illegible]	4,43	0.0022	-0.0029
4,44	-0.0021	0.0029	4,45	0.0021	-0.0028	4,46	-0.0021	0.0028	[illegible]	[illegible]	[illegible]
5, 5	1.0658	1.5232	5, 6	0.0623	-0.3631	5, 7	-0.0595	0.1578	5, 8	0.0471	-0.0891
5, 9	-0.0357	0.0573	5,10	0.0274	-0.0406	5,11	-0.0217	0.0308	5,12	0.0176	-0.0246
5,13	-0.0147	0.0204	5,14	0.0125	-0.0173	5,15	-0.0109	0.0150	5,16	0.0096	-0.0132
5,17	-0.0085	0.0117	5,18	0.0076	-0.0106	5,19	-0.0069	0.0096	5,20	0.0063	-0.0088
5,21	-0.0058	0.0081	5,22	0.0054	-0.0075	5,23	-0.0050	0.0069	5,24	0.0046	-0.0065
5,25	-0.0043	0.0061	5,26	0.0041	-0.0057	5,27	-0.0039	0.0054	5,28	0.0036	-0.0051
5,29	-0.0035	0.0049	5,30	0.0033	-0.0046	5,31	-0.0031	0.0044	5,32	0.0030	-0.0042
5,33	-0.0029	0.0041	5,34	0.0028	-0.0039	5,35	-0.0027	0.0037	5,36	0.0026	-0.0036
5,37	-0.0025	0.0035	5,38	0.0024	-0.0034	5,39	-0.0023	0.0033	5,40	0.0023	-0.0032
5,41	-0.0022	0.0031	5,42	0.0022	-0.0030	5,43	-0.0021	0.0030	5,44	0.0021	-0.0029
5,45	-0.0021	0.0029	5,46	0.0021	-0.0028						
6, 6	1.0667	1.5224	6, 7	0.0616	-0.3625	6, 8	-0.0589	0.1573	6, 9	0.0465	-0.0887
6,10	-0.0353	0.0569	6,11	0.0271	-0.0403	6,12	-0.0213	0.0306	6,13	0.0173	-0.0244
6,14	-0.0144	0.0202	6,15	0.0123	-0.0171	6,16	-0.0107	0.0148	6,17	0.0094	-0.0130
6,18	-0.0083	0.0116	6,19	0.0075	-0.0104	6,20	-0.0068	0.0095	6,21	0.0062	-0.0086
6,22	-0.0057	0.0080	6,23	0.0052	-0.0074	6,24	-0.0049	0.0068	6,25	0.0045	-0.0064
6,26	-0.0042	0.0060	6,27	0.0040	-0.0056	6,28	-0.0037	0.0053	6,29	0.0035	-0.0050
6,30	-0.0034	0.0048	6,31	0.0032	-0.0046	6,32	-0.0031	0.0044	6,33	0.0029	-0.0042
6,34	-0.0028	0.0040	6,35	0.0027	-0.0038	6,36	-0.0026	0.0037	6,37	0.0025	-0.0036
6,38	-0.0024	0.0034	6,39	0.0023	-0.0033	6,40	-0.0023	0.0032	6,41	0.0022	-0.0031
6,42	-0.0022	0.0031	6,43	0.0021	-0.0030	6,44	-0.0021	0.0029	6,45	0.0021	-0.0029
7, 7	1.0673	1.5220	7, 8	0.0611	-0.3622	7, 9	-0.0586	0.1570	7,10	0.0462	-0.0885
7,11	-0.0350	0.0567	7,12	0.0268	-0.0401	7,13	-0.0211	0.0304	7,14	0.0171	-0.0243
7,15	-0.0143	0.0200	7,16	0.0121	-0.0170	7,17	-0.0105	0.0147	7,18	0.0092	-0.0129
7,19	-0.0082	0.0115	7,20	0.0073	-0.0103	7,21	-0.0066	0.0094	7,22	0.0061	-0.0086
7,23	-0.0056	0.0079	7,24	0.0051	-0.0073	7,25	-0.0048	0.0068	7,26	0.0044	-0.0063
7,27	-0.0042	0.0059	7,28	0.0039	-0.0056	7,29	-0.0037	0.0053	7,30	0.0035	-0.0050
7,31	-0.0033	0.0047	7,32	0.0031	-0.0045	7,33	-0.0030	0.0043	7,34	0.0029	-0.0041
7,35	-0.0027	0.0040	7,36	0.0026	-0.0038	7,37	-0.0025	0.0037	7,38	0.0025	-0.0035
7,39	-0.0024	0.0034	7,40	0.0023	-0.0033	7,41	-0.0022	0.0032	7,42	0.0022	-0.0031
7,43	-0.0022	0.0031	7,44	0.0021	-0.0030						
8, 8	1.0676	1.5217	8, 9	0.0608	-0.3619	8,10	-0.0583	0.1568	8,11	0.0460	-0.0883
8,12	-0.0348	0.0566	8,13	0.0266	-0.0400	8,14	-0.0209	0.0303	8,15	0.0170	-0.0241
8,16	-0.0141	0.0199	8,17	0.0120	-0.0169	8,18	-0.0104	0.0146	8,19	0.0091	-0.0128
8,20	-0.0081	0.0114	8,21	0.0073	-0.0103	8,22	-0.0066	0.0093	8,23	0.0060	-0.0085
8,24	-0.0055	0.0078	8,25	0.0051	-0.0072	8,26	-0.0047	0.0067	8,27	0.0044	-0.0063
8,28	-0.0041	0.0059	8,29	0.0038	-0.0055	8,30	-0.0036	0.0052	8,31	0.0034	-0.0050
8,32	-0.0033	0.0047	8,33	0.0031	-0.0045	8,34	-0.0030	0.0043	8,35	0.0028	-0.0041
8,36	-0.0027	0.0039	8,37	0.0026	-0.0038	8,38	-0.0025	0.0036	8,39	0.0024	-0.0035
8,40	-0.0024	0.0034	8,41	0.0023	-0.0033	8,42	-0.0022	0.0032	8,43	0.0022	-0.0031
9, 9	1.0679	1.5215	9,10	0.0606	-0.3617	9,11	-0.0581	0.1567	9,12	0.0458	-0.0882
9,13	-0.0347	0.0565	9,14	0.0265	-0.0399	9,15	-0.0208	0.0302	9,16	0.0169	-0.0241
9,17	-0.0140	0.0199	9,18	0.0119	-0.0168	9,19	-0.0103	0.0145	9,20	0.0090	-0.0128
9,21	-0.0080	0.0113	9,22	0.0072	-0.0102	9,23	-0.0065	0.0092	9,24	0.0059	-0.0084
9,25	-0.0054	0.0078	9,26	0.0050	-0.0072	9,27	-0.0046	0.0067	9,28	0.0043	-0.0062
9,29	-0.0040	0.0059	9,30	0.0038	-0.0055	9,31	-0.0036	0.0052	9,32	0.0034	-0.0049
9,33	-0.0032	0.0047	9,34	0.0031	-0.0045	9,35	-0.0029	0.0043	9,36	0.0028	-0.0041
9,37	-0.0027	0.0039	9,38	0.0026	-0.0038	9,39	-0.0025	0.0036	9,40	0.0024	-0.0035
9,41	-0.0023	0.0034	9,42	0.0023	-0.0033						

10,10	1.0681	1.5214	10,11	0.0604	-0.3616	10,12	-0.0580	0.1566	10,13	0.0457	-0.0881
10,14	-0.0345	0.0564	10,15	0.0264	-0.0398	10,16	-0.0207	0.0301	10,17	0.0168	-0.0240
10,18	-0.0139	0.0198	10,19	0.0118	-0.0168	10,20	-0.0102	0.0145	10,21	0.0090	-0.0127
10,22	-0.0080	0.0113	10,23	0.0071	-0.0102	10,24	-0.0064	0.0092	10,25	0.0059	-0.0084
10,26	-0.0054	0.0077	10,27	0.0050	-0.0072	10,28	-0.0046	0.0067	10,29	0.0043	-0.0062
10,30	-0.0040	0.0058	10,31	0.0038	-0.0055	10,32	-0.0036	0.0052	10,33	0.0034	-0.0049
10,34	-0.0032	0.0047	10,35	0.0030	-0.0044	10,36	-0.0029	0.0042	10,37	0.0028	-0.0041
10,38	-0.0027	0.0039	10,39	0.0026	-0.0038	10,40	-0.0025	0.0036	10,41	0.0024	-0.0035
11,11	1.0682	1.5212	11,12	0.0603	-0.3615	11,13	-0.0578	0.1565	11,14	0.0456	-0.0880
11,15	-0.0345	0.0563	11,16	0.0263	-0.0397	11,17	-0.0207	0.0301	11,18	0.0167	-0.0239
11,19	-0.0139	0.0197	11,20	0.0118	-0.0167	11,21	-0.0102	0.0144	11,22	0.0089	-0.0127
11,23	-0.0079	0.0113	11,24	0.0071	-0.0101	11,25	-0.0064	0.0092	11,26	0.0058	-0.0084
11,27	-0.0053	0.0077	11,28	0.0049	-0.0071	11,29	-0.0046	0.0066	11,30	0.0043	-0.0062
11,31	-0.0040	0.0058	11,32	0.0037	-0.0055	11,33	-0.0035	0.0052	11,34	0.0033	-0.0049
11,35	-0.0032	0.0046	11,36	0.0030	-0.0044	11,37	-0.0029	0.0042	11,38	0.0028	-0.0041
11,39	-0.0027	0.0039	11,40	0.0026	-0.0038						
12,12	1.0683	1.5212	12,13	0.0602	-0.3614	12,14	-0.0578	0.1564	12,15	0.0455	-0.0879
12,16	-0.0344	0.0562	12,17	0.0262	-0.0397	12,18	-0.0206	0.0300	12,19	0.0167	-0.0239
12,20	-0.0138	0.0197	12,21	0.0117	-0.0167	12,22	-0.0101	0.0144	12,23	0.0089	-0.0126
12,24	-0.0079	0.0112	12,25	0.0071	-0.0101	12,26	-0.0064	0.0092	12,27	0.0058	-0.0084
12,28	-0.0053	0.0077	12,29	0.0049	-0.0071	12,30	-0.0046	0.0066	12,31	0.0042	-0.0062
12,32	-0.0040	0.0058	12,33	0.0037	-0.0055	12,34	-0.0035	0.0051	12,35	0.0033	-0.0049
12,36	-0.0032	0.0046	12,37	0.0030	-0.0044	12,38	-0.0029	0.0042	12,39	0.0028	-0.0041
13,13	1.0684	1.5211	13,14	0.0601	-0.3614	13,15	-0.0577	0.1563	13,16	0.0455	-0.0879
13,17	-0.0343	0.0562	13,18	0.0262	-0.0396	13,19	-0.0206	0.0300	13,20	0.0166	-0.0239
13,21	-0.0138	0.0197	13,22	0.0117	-0.0167	13,23	-0.0101	0.0144	13,24	0.0088	-0.0126
13,25	-0.0078	0.0112	13,26	0.0070	-0.0101	13,27	-0.0063	0.0091	13,28	0.0058	-0.0083
13,29	-0.0053	0.0077	13,30	0.0049	-0.0071	13,31	-0.0045	0.0066	13,32	0.0042	-0.0062
13,33	-0.0040	0.0058	13,34	0.0037	-0.0054	13,35	-0.0035	0.0051	13,36	0.0033	-0.0049
13,37	-0.0032	0.0046	13,38	0.0030	-0.0044						
14,14	1.0685	1.5210	14,15	0.0601	-0.3613	14,16	-0.0576	0.1563	14,17	0.0454	-0.0878
14,18	-0.0343	0.0562	14,19	0.0261	-0.0396	14,20	-0.0205	0.0300	14,21	0.0166	-0.0238
14,22	-0.0138	0.0196	14,23	0.0117	-0.0166	14,24	-0.0101	0.0144	14,25	0.0088	-0.0126
14,26	-0.0078	0.0112	14,27	0.0070	-0.0101	14,28	-0.0063	0.0091	14,29	0.0058	-0.0083
14,30	-0.0053	0.0077	14,31	0.0049	-0.0071	14,32	-0.0045	0.0066	14,33	0.0042	-0.0062
14,34	-0.0039	0.0058	14,35	0.0037	-0.0054	14,36	-0.0035	0.0051	14,37	0.0033	-0.0049
15,15	1.0685	1.5210	15,16	0.0600	-0.3613	15,17	-0.0576	0.1563	15,18	0.0454	-0.0878
15,19	-0.0343	0.0561	15,20	0.0261	-0.0396	15,21	-0.0205	0.0299	15,22	0.0166	-0.0238
15,23	-0.0137	0.0196	15,24	0.0116	-0.0166	15,25	-0.0101	0.0143	15,26	0.0088	-0.0126
15,27	-0.0078	0.0112	15,28	0.0070	-0.0100	15,29	-0.0063	0.0091	15,30	0.0058	-0.0083
15,31	-0.0053	0.0076	15,32	0.0049	-0.0071	15,33	-0.0045	0.0066	15,34	0.0042	-0.0061
15,35	-0.0039	0.0058	15,36	0.0037	-0.0054						
16,16	1.0686	1.5210	16,17	0.0600	-0.3613	16,18	-0.0576	0.1562	16,19	0.0453	-0.0878
16,20	-0.0342	0.0561	16,21	0.0261	-0.0395	16,22	-0.0205	0.0299	16,23	0.0165	-0.0238
16,24	-0.0137	0.0196	16,25	0.0116	-0.0166	16,26	-0.0100	0.0143	16,27	0.0088	-0.0126
16,28	-0.0078	0.0112	16,29	0.0070	-0.0100	16,30	-0.0063	0.0091	16,31	0.0057	-0.0083
16,32	-0.0053	0.0076	16,33	0.0049	-0.0071	16,34	-0.0045	0.0066	16,35	0.0042	-0.0061
17,17	1.0686	1.5209	17,18	0.0600	-0.3612	17,19	-0.0575	0.1562	17,20	0.0453	-0.0878
17,21	-0.0342	0.0561	17,22	0.0261	-0.0395	17,23	-0.0204	0.0299	17,24	0.0165	-0.0238
17,25	-0.0137	0.0196	17,26	0.0116	-0.0166	17,27	-0.0100	0.0143	17,28	0.0088	-0.0126
17,29	-0.0078	0.0112	17,30	0.0070	-0.0100	17,31	-0.0063	0.0091	17,32	0.0057	-0.0083
17,33	-0.0053	0.0076	17,34	0.0049	-0.0071						
18,18	1.0686	1.5209	18,19	0.0599	-0.3612	18,20	-0.0575	0.1562	18,21	0.0453	-0.0878
18,22	-0.0342	0.0561	18,23	0.0261	-0.0395	18,24	-0.0204	0.0299	18,25	0.0165	-0.0238
18,26	-0.0137	0.0196	18,27	0.0116	-0.0166	18,28	-0.0100	0.0143	18,29	0.0088	-0.0126
18,30	-0.0078	0.0112	18,31	0.0070	-0.0100	18,32	-0.0063	0.0091	18,33	0.0057	-0.0083
19,19	1.0687	1.5209	19,20	0.0599	-0.3612	19,21	-0.0575	0.1562	19,22	0.0453	-0.0877
19,23	-0.0342	0.0561	19,24	0.0260	-0.0395	19,25	-0.0204	0.0299	19,26	0.0165	-0.0238
19,27	-0.0137	0.0196	19,28	0.0116	-0.0166	19,29	-0.0100	0.0143	19,30	0.0088	-0.0125
19,31	-0.0078	0.0112	19,32	0.0070	-0.0100						
20,20	1.0687	1.5209	20,21	0.0599	-0.3612	20,22	-0.0575	0.1562	20,23	0.0453	-0.0877
20,24	-0.0342	0.0560	20,25	0.0260	-0.0395	20,26	-0.0204	0.0299	20,27	0.0165	-0.0237
20,28	-0.0137	0.0196	20,29	0.0116	-0.0166	20,30	-0.0100	0.0143	20,31	0.0088	-0.0125
21,21	1.0687	1.5209	21,22	0.0599	-0.3612	21,23	-0.0575	0.1562	21,24	0.0452	-0.0877
21,25	-0.0341	0.0560	21,26	0.0260	-0.0395	21,27	-0.0204	0.0299	21,28	0.0165	-0.0237
21,29	-0.0137	0.0196	21,30	0.0116	-0.0166						
22,22	1.0687	1.5208	22,23	0.0599	-0.3612	22,24	-0.0575	0.1561	22,25	0.0452	-0.0877
22,26	-0.0341	0.0560	22,27	0.0260	-0.0395	22,28	-0.0204	0.0299	22,29	0.0165	-0.0237
23,23	1.0687	1.5208	23,24	0.0599	-0.3612	23,25	-0.0575	0.1561	23,26	0.0452	-0.0877
23,27	-0.0341	0.0560	23,28	0.0260	-0.0395						
24,24	1.0687	1.5208	24,25	0.0599	-0.3612	24,26	-0.0575	0.1561	24,27	0.0452	-0.0877
25,25	1.0687	1.5208	25,26	0.0599	-0.3612						

TABLE 5.4

IEN

SELF AND MUTUAL IMPEDANCES

H/LMDA=0.2500 B/LMDA= 0.250 OMEGA= 8.53

2 ELEMENT ARRAY

I, K	ZIK IN OHMS RE	IM	I, K	ZIK IN OHMS RE	IM	I, K	ZIK IN OHMS RE	IM	I, K	ZIK IN OHMS RE	IM
1, 1	79.409	35.625	1, 2	40.010	-39.479						

IEN

SELF AND MUTUAL IMPEDANCES

H/LMDA=0.2500 B/LMDA= 0.250 OMEGA= 8.53

3 ELEMENT ARRAY

I, K	ZIK IN OHMS RE	IM	I, K	ZIK IN OHMS RE	IM	I, K	ZIK IN OHMS RE	IM	I, K	ZIK IN OHMS RE	IM
1, 1	80.912	36.217	1, 2	39.443	-37.292	1, 3	-23.033	-31.254			
2, 2	76.363	35.116									

IEN

SELF AND MUTUAL IMPEDANCES

H/LMDA=0.2500 B/LMDA= 0.250 OMEGA= 8.53

4 ELEMENT ARRAY

I, K	ZIK IN OHMS RE	IM	I, K	ZIK IN OHMS RE	IM	I, K	ZIK IN OHMS RE	IM	I, K	ZIK IN OHMS RE	IM
1, 1	80.076	35.756	1, 2	39.985	-38.367	1, 3	-21.335	-30.711	1, 4	-24.978	16.025
2, 2	77.726	35.739	2, 3	38.788	-35.220						

IEN

SELF AND MUTUAL IMPEDANCES

H/LMDA=0.2500 B/LMDA= 0.250 [illegible]

5 ELEMENT ARRAY

I, K	ZIK IN OHMS RE	IM	I, K	ZIK IN OHMS RE	IM	I, K	ZIK IN OHMS RE	IM	I, K	ZIK IN OHMS RE	IM
1, 1	80.612	36.103	1, 2	39.578	-37.734	1, 3	-22.225	-31.188	1, 4	-24.485	14.670
1, 5	12.376	20.810									
2, 2	76.979	35.268	2, 3	39.339	-36.269	2, 4	-19.814	-30.107			
3, 3	79.170	36.359									

IEN

SELF AND MUTUAL IMPEDANCES

H/LMDA=0.2500 B/LMDA= 0.250 OMEGA= 8.53

6 ELEMENT ARRAY

I, K	ZIK IN OHMS RE	IM	I, K	ZIK IN OHMS RE	IM	I, K	ZIK IN OHMS RE	IM	I, K	ZIK IN OHMS RE	IM
1, 1	80.235	35.836	1, 2	39.886	-38.156	1, 3	-21.674	-30.828	1, 4	-24.906	15.404
1, 5	11.220	20.362	1, 6	17.901	-10.149						
2, 2	77.453	35.618	2, 3	38.928	-35.651	2, 4	-20.639	-30.588	2, 5	-23.946	13.436
3, 3	78.370	35.886	3, 4	39.881	-37.320						

IEN

SELF AND MUTUAL IMPEDANCES

H/LMDA=0.2500 B/LMDA= 0.250 OMEGA= 8.53

7 ELEMENT ARRAY

I, K	ZIK IN OHMS RE	IM	I, K	ZIK IN OHMS RE	IM	I, K	ZIK IN OHMS RE	IM	I, K	ZIK IN OHMS RE	IM
1, 1	80.517	36.048	1, 2	39.646	-37.851	1, 3	-22.054	-31.104	1, 4	-24.586	14.938
1, 5	11.866	20.742	1, 6	17.495	-9.151	1, 7	-8.661	-15.763			
2, 2	77.122	35.350	2, 3	39.238	-36.063	2, 4	-20.134	-30.228	2, 5	-24.374	14.138
2, 6	10.191	19.884									
3, 3	78.882	36.240	3, 4	39.472	-36.693	3, 5	-21.493	-31.062			

IEN

SELF AND MUTUAL IMPEDANCES

H/LMDA=0.2500 B/LMDA= 0.250 OMEGA= 8.53

8 ELEMENT ARRAY

I, K	ZIK IN OHMS RE	IM	I, K	ZIK IN OHMS RE	IM	I, K	ZIK IN OHMS RE	IM	I, K	ZIK IN OHMS RE	IM
1, 1	80.296	35.875	1, 2	39.839	-38.084	1, 3	-21.773	-30.885	1, 4	-24.834	15.266
1, 5	11.446	20.450	1, 6	17.839	-9.718	1, 7	-7.769	-15.391	1, 8	-14.123	7.587
2, 2	77.368	35.562	2, 3	38.996	-35.765	2, 4	-20.480	-30.503	2, 5	-24.051	13.694
2, 6	10.792	20.266	2, 7	17.063	-8.253						
3, 3	78.523	35.967	3, 4	39.781	-37.111	3, 5	-20.958	-30.699	3, 6	-24.791	14.848
4, 4	78.091	35.766	4, 5	39.065	-36.074						

IEN

SELF AND MUTUAL IMPEDANCES

H/LMDA=0.2500 B/LMDA= 0.250 OMEGA= 8.53

9 ELEMENT ARRAY

I, K	ZIK IN OHMS RE	IM	I, K	ZIK IN OHMS RE	IM	I, K	ZIK IN OHMS RE	IM	I, K	ZIK IN OHMS RE	IM
1, 1	80.475	36.019	1, 2	39.680	-37.899	1, 3	-21.992	-31.063	1, 4	-24.636	15.020
1, 5	11.747	20.679	1, 6	17.573	-9.345	1, 7	-8.284	-15.707	1, 8	-13.781	6.786
1, 9	6.780	12.825									
2, 2	77.177	35.389	2, 3	39.190	-35.992	2, 4	-20.225	-30.286	2, 5	-24.301	14.006
2, 6	10.404	19.975	2, 7	17.411	-8.790	2, 8	-6.977	-14.999			
3, 3	78.791	36.183	3, 4	39.540	-36.809	3, 5	-21.326	-30.977	3, 6	-24.467	14.390
3, 7	11.413	20.640									
4, 4	77.752	35.498	4, 5	39.374	-36.488	4, 6	-20.446	-30.344			
5, 5	78.596	36.120									

IEN

SELF AND MUTUAL IMPEDANCES

H/LMDA=0.2500 B/LMDA= 0.250 OMEGA= 8.53

10 ELEMENT ARRAY

I, K	ZIK IN OHMS RE	IM	I, K	ZIK IN OHMS RE	IM	I, K	ZIK IN OHMS RE	IM	I, K	ZIK IN OHMS RE	IM
1, 1	80.326	35.897	1, 2	39.813	-38.050	1, 3	-21.815	-30.915	1, 4	-24.798	15.214
1, 5	11.518	20.495	1, 6	17.782	-9.615	1, 7	-7.940	-15.460	1, 8	-14.072	7.253
1, 9	6.046	12.506	1,10	11.768	-6.147						
2, 2	77.331	35.533	2, 3	39.031	-35.812	2, 4	-20.423	-30.462	2, 5	-24.102	13.771
2, 6	10.681	20.202	2, 7	17.144	-8.437	2, 8	-7.457	-15.317	2, 9	-13.421	6.069
3, 3	78.581	36.007	3, 4	39.733	-37.040	3, 5	-21.054	-30.757	3, 6	-24.718	14.712
3, 7	11.003	20.346	3, 8	17.749	-9.336						
4, 4	78.005	35.710	4, 5	39.133	-36.189	4, 6	-20.799	-30.619	4, 7	-24.150	13.944
5, 5	78.242	35.848	5, 6	39.682	-36.904						

IEN

SELF AND MUTUAL IMPEDANCES

H/LMDA=0.2500 B/LMDA= 0.250 OMEGA= 8.53

12 ELEMENT ARRAY

I, K	ZIK IN OHMS RE	IM	I, K	ZIK IN OHMS RE	IM	I, K	ZIK IN OHMS RE	IM	I, K	ZIK IN OHMS RE	IM
1, 1	80.344	35.910	1, 2	39.798	-38.032	1, 3	-21.838	-30.933	1, 4	-24.778	15.187
1, 5	11.550	20.519	1, 6	17.754	-9.574	1, 7	-7.996	-15.497	1, 8	-14.026	7.170
1, 9	6.185	12.564	1,10	11.725	-5.872	1,11	-5.010	-10.612	1,12	-10.149	5.219
2, 2	77.312	35.516	2, 3	39.050	-35.836	2, 4	-20.395	-30.439	2, 5	-24.128	13.806
2, 6	10.637	20.170	2, 7	17.184	-8.496	2, 8	-7.370	-15.266	2, 9	-13.486	6.214
2,10	5.797	12.447	2,11	11.161	-4.869						
3, 3	78.610	36.030	3, 4	39.708	-37.006	3, 5	-21.095	-30.788	3, 6	-24.681	14.660
3, 7	11.073	20.391	3, 8	17.692	-9.235	3, 9	-7.616	-15.378	3,10	-14.000	6.959
4, 4	77.967	35.681	4, 5	39.168	-36.236	4, 6	-20.740	-30.578	4, 7	-24.201	14.022
4, 8	10.892	20.288	4, 9	17.219	-8.616						
5, 5	78.300	35.888	5, 6	39.635	-36.833	5, 7	-20.893	-30.675	5, 8	-24.647	14.578
6, 6	78.155	35.792	6, 7	39.201	-36.304						

IEN

SELF AND MUTUAL IMPEDANCES

H/LMDA=0.2500 B/LMDA= 0.250 OMEGA= 8.53

14 ELEMENT ARRAY

I, K	ZIK IN OHMS RE	IM	I, K	ZIK IN OHMS RE	IM	I, K	ZIK IN OHMS RE	IM	I, K	ZIK IN OHMS RE	IM
1, 1	80.355	35.919	1, 2	39.788	-38.020	1, 3	-21.851	-30.944	1, 4	-24.765	15.172
1, 5	11.568	20.533	1, 6	17.737	-9.552	1, 7	-8.023	-15.516	1, 8	-14.002	7.136
1, 9	6.232	12.594	1,10	11.686	-5.802	1,11	-5.127	-10.661	1,12	-10.112	4.983
1,13	4.314	9.266	1,14	8.962	-4.568						
2, 2	77.300	35.505	2, 3	39.062	-35.850	2, 4	-20.379	-30.426	2, 5	-24.144	13.825
2, 6	10.615	20.152	2, 7	17.206	-8.524	2, 8	-7.335	-15.239	2, 9	-13.519	6.262
2,10	5.727	12.404	2,11	11.216	-4.989	2,12	-4.802	-10.563	2,13	-9.614	4.109
3, 3	78.626	36.043	3, 4	39.693	-36.988	3, 5	-21.117	-30.805	3, 6	-24.661	14.635
3, 7	11.105	20.415	3, 8	17.663	-9.195	3, 9	-7.671	-15.415	3,10	-13.953	6.878
3,11	5.927	12.497	3,12	11.665	-5.630						
4, 4	77.947	35.664	4, 5	39.187	-36.260	4, 6	-20.712	-30.556	4, 7	-24.227	14.057
4, 8	10.848	20.255	4, 9	17.259	-8.675	4,10	-7.530	-15.332	4,11	-13.546	6.354
5, 5	78.328	35.910	5, 6	39.609	-36.800	5, 7	-20.934	-30.705	5, 8	-24.611	14.527
5, 9	10.961	20.330	5,10	17.637	-9.136						
6, 6	78.116	35.763	6, 7	39.235	-36.351	6, 8	-20.835	-30.635	6, 9	-24.250	14.100
7, 7	78.212	35.832	7, 8	39.588	-36.762						

IEN

SELF AND MUTUAL IMPEDANCES

H/LMDA=0.2500 B/LMDA= 0.250 OMEGA= 8.53

16 ELEMENT ARRAY

I, K	ZIK IN OHMS RE	IM	I, K	ZIK IN OHMS RE	IM	I, K	ZIK IN OHMS RE	IM	I, K	ZIK IN OHMS RE	IM
1, 1	80.362	35.925	1, 2	39.781	-38.013	1, 3	-21.860	-30.952	1, 4	-24.757	15.163
1, 5	11.579	20.543	1, 6	17.726	-9.539	1, 7	-8.038	-15.529	1, 8	-13.987	7.118
1, 9	6.254	12.611	1,10	11.665	-5.773	1,11	-5.167	-10.687	1,12	-10.078	4.923
1,13	4.417	9.309	1,14	8.930	-4.361	1,15	-3.814	-8.256	1,16	-8.052	4.084
2, 2	77.292	35.498	2, 3	39.071	-35.859	2, 4	-20.370	-30.417	2, 5	-24.154	13.836
2, 6	10.603	20.140	2, 7	17.219	-8.539	2, 8	-7.317	-15.224	2, 9	-13.537	6.285
2,10	5.697	12.382	2,11	11.244	-5.030	2,12	-4.741	-10.526	2,13	-9.662	4.212
2,14	4.134	9.224	2,15	8.483	-3.581						
3, 3	78.637	36.052	3, 4	39.683	-36.977	3, 5	-21.130	-30.816	3, 6	-24.649	14.620
3, 7	11.123	20.430	3, 8	17.646	-9.174	3, 9	-7.697	-15.435	3,10	-13.929	6.844
3,11	[illegible]	12.527	3,12	11.625	-5.562	3,13	-4.911	-10.604	3,14	-10.061	4.777
4, 4	77.934	35.653	4, 5	39.199	-36.274	4, 6	-20.690	[illegible]	4, 7	[illegible]	14.076
4, 8	10.826	20.237	4, 9	17.281	-8.703	4,10	-7.495	-15.306	4,11	-13.579	6.402
4,12	5.856	12.459	4,13	11.267	-5.105						
5, 5	78.344	35.924	5, 6	39.594	-36.782	5, 7	-20.956	-30.723	5, 8	-24.590	14.502
5, 9	10.993	20.354	5,10	17.608	-9.096	5,11	-7.585	-15.366	5,12	-13.909	6.798
6, 6	78.096	35.746	6, 7	39.255	-36.375	6, 8	-20.806	-30.613	6, 9	-24.277	14.135
6,10	10.917	20.299	6,11	17.298	-8.735						
7, 7	78.240	35.854	7, 8	39.562	-36.729	7, 9	-20.875	-30.665	7,10	-24.575	14.476
8, 8	78.174	35.803	8, 9	39.269	-36.398						

IEN

SELF AND MUTUAL IMPEDANCES

H/LMDA=0.2500 B/LMDA= 0.250 OMEGA= 8.53

18 ELEMENT ARRAY

I, K	ZIK IN OHMS RE	IM	I, K	ZIK IN OHMS RE	IM	I, K	ZIK IN OHMS RE	IM	I, K	ZIK IN OHMS RE	IM
1, 1	80.367	35.930	1, 2	39.776	-38.007	1, 3	-21.867	-30.957	1, 4	-24.751	15.156
1, 5	11.587	20.549	1, 6	17.719	-9.531	1, 7	-8.048	-15.537	1, 8	-13.978	7.107
1, 9	6.267	12.622	1,10	11.652	-5.757	1,11	-5.187	-10.702	1,12	-10.060	4.898
1,13	4.452	9.332	1,14	8.900	-4.308	1,15	-3.905	-8.294	1,16	-8.024	3.899
1,17	3.435	7.467	1,18	7.329	-3.709						
2, 2	77.286	35.493	2, 3	39.076	-35.865	2, 4	-20.363	-30.410	2, 5	-24.161	13.843
2, 6	10.594	20.133	2, 7	17.228	-8.549	2, 8	-7.306	-15.214	2, 9	-13.549	6.298
2,10	5.682	12.368	2,11	11.260	-5.049	2,12	-4.716	-10.506	2,13	-9.686	4.247
2,14	4.082	9.191	2,15	8.525	-3.672	2,16	-3.654	-8.219	2,17	-7.617	3.193
3, 3	78.644	36.059	3, 4	39.676	-36.969	3, 5	-21.138	-30.824	3, 6	-24.640	14.610
3, 7	11.133	20.439	3, 8	17.635	-9.161	3, 9	-7.712	-15.447	3,10	-13.914	6.827
3,11	5.995	12.544	3,12	11.604	-5.534	3,13	-4.950	-10.631	3,14	-10.027	4.719
3,15	4.230	9.260	3,16	8.886	-4.181						
4, 4	77.926	35.646	4, 5	39.208	-36.283	4, 6	-20.686	-30.533	4, 7	-24.253	14.087
4, 8	10.813	20.226	4, 9	17.294	-8.719	4,10	-7.476	-15.291	4,11	-13.598	6.425
4,12	5.826	12.437	4,13	11.295	-5.146	4,14	-4.851	-10.572	4,15	-9.705	4.311
5, 5	78.355	35.933	5, 6	39.584	-36.771	5, 7	-20.969	-30.734	5, 8	-24.578	14.487
5, 9	11.010	20.368	5,10	17.591	-9.075	5,11	-7.610	-15.386	5,12	-13.885	6.765
5,13	5.902	12.487	5,14	11.588	-5.496						
6, 6	78.084	35.735	6, 7	39.267	-36.389	6, 8	-20.790	-30.599	6, 9	-24.292	14.154
6,10	10.894	20.281	6,11	17.320	-8.763	6,12	-7.549	-15.341	6,13	-13.611	6.451
7, 7	78.257	35.868	7, 8	39.547	-36.711	7, 9	-20.897	-30.682	7,10	-24.555	14.451
7,11	10.948	20.322	7,12	17.580	-9.056						
8, 8	78.153	35.786	8, 9	39.289	-36.422	8,10	-20.847	-30.643	8,11	-24.303	14.170
9, 9	78.202	35.825	9,10	39.537	-36.696						

IEN

SELF AND MUTUAL IMPEDANCES

H/LMDA=0.2500 B/LMDA= 0.250 OMEGA= 8.53

20 ELEMENT ARRAY

I, K	ZIK IN OHMS RE	IM	I, K	ZIK IN OHMS RE	IM	I, K	ZIK IN OHMS RE	IM	I, K	ZIK IN OHMS RE	IM
1, 1	80.371	35.933	1, 2	39.773	-38.003	1, 3	-21.871	-30.961	1, 4	-24.747	15.151
1, 5	11.592	20.554	1, 6	17.714	-9.525	1, 7	-8.054	-15.543	1, 8	-13.972	7.100
1, 9	6.276	12.629	1,10	11.644	-5.748	1,11	-5.198	-10.712	1,12	-10.049	4.884
1,13	4.470	9.345	1,14	8.884	-4.285	1,15	-3.937	-8.314	1,16	-7.997	3.851
1,17	3.517	7.501	1,18	7.304	-3.541	1,19	-3.137	-6.832	1,20	-6.740	3.409
2, 2	77.282	35.489	2, 3	39.081	-35.870	2, 4	-20.358	-30.406	2, 5	-24.166	13.849
2, 6	10.589	20.127	2, 7	17.234	-8.555	2, 8	-7.299	-15.207	2, 9	-13.556	6.306
2,10	5.672	12.360	2,11	11.270	-5.061	2,12	-4.702	-10.495	2,13	-9.701	4.264
2,14	4.059	9.174	2,15	8.547	-3.703	2,16	-3.607	-8.190	2,17	-7.655	3.275
2,18	3.291	7.434	2,19	6.931	-2.894						
3, 3	78.649	36.063	3, 4	39.671	-36.964	3, 5	-21.144	-30.829	3, 6	-24.634	14.604
3, 7	11.141	20.446	3, 8	17.628	-9.153	3, 9	-7.721	-15.455	3,10	-13.905	6.816
3,11	6.007	12.555	3,12	11.592	-5.519	3,13	-4.969	-10.646	3,14	-10.008	4.694
3,15	4.264	9.283	3,16	8.855	-4.129	3,17	-3.739	-8.250	3,18	-7.984	3.738
4, 4	77.921	35.641	4, 5	39.213	-36.289	4, 6	-20.680	-30.526	4, 7	-24.260	14.095
4, 8	10.804	20.218	4, 9	17.303	-8.728	4,10	-7.465	-15.281	4,11	-13.609	6.439
4,12	5.810	12.423	4,13	11.311	-5.166	4,14	-4.825	-10.553	4,15	-9.730	4.347
4,16	4.177	9.232	4,17	8.563	-3.760						
5, 5	78.362	35.939	5, 6	39.577	-36.763	5, 7	-20.977	-30.741	5, 8	-24.570	14.477
5, 9	11.021	20.378	5,10	17.581	-9.063	5,11	-7.625	-15.398	5,12	-13.870	6.748
5,13	5.923	12.504	5,14	11.567	-5.468	5,15	-4.890	-10.596	5,16	-9.994	4.661
6, 6	78.075	35.727	6, 7	39.275	-36.398	6, 8	-20.780	-30.590	6, 9	-24.303	14.166
6,10	10.881	20.269	6,11	17.333	-8.778	6,12	-7.530	-15.326	6,13	-13.630	6.474
6,14	5.871	12.466	6,15	11.322	-5.187						
7, 7	78.267	35.877	7, 8	39.537	-36.700	7, 9	-20.910	-30.693	7,10	-24.542	14.436
7,11	10.966	20.337	7,12	17.563	-9.035	7,13	-7.575	-15.360	7,14	-13.861	6.732
8, 8	78.141	35.775	8, 9	39.301	-36.436	8,10	-20.831	-30.629	8,11	-24.319	14.189
8,12	10.926	20.305	8,13	17.341	-8.791						
9, 9	78.218	35.838	9,10	39.521	-36.678	9,11	-20.868	-30.660	9,12	-24.535	14.425
10,10	78.181	35.808	10,11	39.308	-36.446						

IEN

SELF AND MUTUAL IMPEDANCES

H/LMDA=0.2500 B/LMDA= 0.250 OMEGA= 8.53

22 ELEMENT ARRAY

I, K	ZIK IN OHMS RE	IM	I, K	ZIK IN OHMS RE	IM	I, K	ZIK IN OHMS RE	IM	I, K	ZIK IN OHMS RE	IM
1, 1	80.375	35.936	1, 2	39.770	-38.000	1, 3	-21.874	-30.964	1, 4	-24.743	15.148
1, 5	11.596	20.557	1, 6	17.710	-9.521	1, 7	-8.059	-15.547	1, 8	-13.967	7.095
1, 9	6.281	12.634	1,10	11.639	-5.741	1,11	-5.206	-10.718	1,12	-10.042	4.875
1,13	4.480	9.353	1,14	8.874	-4.273	1,15	-3.952	-8.326	1,16	-7.982	3.831
1,17	3.546	7.520	1,18	7.279	-3.498	1,19	-3.212	-6.863	1,20	-6.717	3.255
1,21	2.896	6.310	1,22	6.250	-3.163						
2, 2	77.279	35.486	2, 3	39.084	-35.873	2, 4	-20.355	-30.402	2, 5	-24.170	13.853
2, 6	10.584	20.123	2, 7	17.238	-8.560	2, 8	-7.294	-15.202	2, 9	-13.562	6.312
2,10	5.666	12.354	2,11	11.277	-5.068	2,12	-4.694	-10.487	2,13	-9.709	4.274
2,14	4.047	9.164	2,15	8.560	-3.719	2,16	-3.587	-8.175	2,17	-7.675	3.303
2,18	3.249	7.408	2,19	6.965	-2.968	2,20	-3.006	-6.803	2,21	-6.372	2.656
3, 3	78.653	36.067	3, 4	39.667	-36.960	3, 5	-21.149	-30.834	3, 6	-24.630	14.599
3, 7	11.146	20.450	3, 8	17.623	-9.148	3, 9	-7.728	-15.461	3,10	-13.899	6.809
3,11	6.015	12.562	3,12	11.584	-5.509	3,13	-4.981	-10.655	3,14	-9.997	4.680
3,15	4.281	9.296	3,16	8.839	-4.106	3,17	-3.770	-8.271	3,18	-7.957	3.691
3,19	3.368	7.463	3,20	7.268	-3.395						
4, 4	77.917	35.637	4, 5	39.218	-36.294	4, 6	-20.675	-30.522	4, 7	-24.265	14.100
4, 8	10.799	20.213	4, 9	17.309	-8.735	4,10	-7.458	-15.274	4,11	-13.617	6.447
4,12	5.801	12.415	4,13	11.321	-5.178	4,14	-4.811	-10.541	4,15	-9.744	4.364
4,16	4.154	9.215	4,17	8.585	-3.791	4,18	-3.692	-8.226	4,19	-7.689	3.353
5, 5	78.367	35.944	5, 6	39.572	-36.758	5, 7	-20.983	-30.747	5, 8	-24.564	14.471
5, 9	11.028	20.384	5,10	17.573	-9.055	5,11	-7.634	-15.407	5,12	-13.861	6.737
5,13	5.936	12.515	5,14	11.554	-5.452	5,15	-4.909	-10.611	5,16	-9.976	4.636
5,17	4.211	9.253	5,18	8.827	-4.078						
6, 6	78.070	35.722	6, 7	39.281	-36.404	6, 8	-20.773	-30.584	6, 9	-24.310	14.173
6,10	10.873	20.262	6,11	17.341	-8.788	6,12	-7.519	-15.316	6,13	-13.641	6.487
6,14	5.855	12.453	6,15	11.338	-5.207	6,16	-4.864	-10.578	6,17	-9.753	4.383
7, 7	78.274	35.883	7, 8	39.530	-36.692	7, 9	-20.918	-30.701	7,10	-24.534	14.426
7,11	10.977	20.346	7,12	17.552	-9.023	7,13	-7.589	-15.373	7,14	-13.846	6.715
7,15	5.893	12.482	7,16	11.547	-5.439						
8, 8	78.132	35.767	8, 9	39.309	-36.445	8,10	-20.820	-30.620	8,11	-24.329	14.201
8,12	10.913	20.293	8,13	17.354	-8.806	8,14	-7.556	-15.346	8,15	-13.648	6.497
9, 9	78.228	35.848	9,10	39.511	-36.667	9,11	-20.881	-30.671	9,12	-24.522	14.410
9,13	10.943	20.319	9,14	17.546	-9.014						
10,10	78.169	35.797	10,11	39.320	-36.460	10,12	-20.852	-30.646	10,13	-24.334	14.208
11,11	78.197	35.821	11,12	39.506	-36.660						

IEN

SELF AND MUTUAL IMPEDANCES

H/LMDA=0.2500 B/LMDA= 0.250 OMEGA= 8.53

24 ELEMENT ARRAY

I, K	ZIK IN OHMS RE	IM	I, K	ZIK IN OHMS RE	IM	I, K	ZIK IN OHMS RE	IM	I, K	ZIK IN OHMS RE	IM
1, 1	80.377	35.938	1, 2	39.767	-37.998	1, 3	-21.877	-30.967	1, 4	-24.741	15.145
1, 5	11.599	20.560	1, 6	17.707	-9.518	1, 7	-8.063	-15.550	1, 8	-13.964	7.091
1, 9	6.286	12.638	1,10	11.634	-5.737	1,11	-5.211	-10.723	1,12	-10.036	4.870
1,13	4.487	9.359	1,14	8.867	-4.265	1,15	-3.962	-8.334	1,16	-7.973	3.819
1,17	3.560	7.531	1,18	7.266	-3.479	1,19	-3.238	-6.881	1,20	-6.695	3.215
1,21	2.965	6.338	1,22	6.229	-3.020	1,23	-2.697	-5.871	1,24	-5.835	2.957
2, 2	77.277	35.483	2, 3	39.087	-35.876	2, 4	-20.352	-30.399	2, 5	-24.173	13.856
2, 6	10.581	20.120	2, 7	17.241	-8.563	2, 8	-7.290	-15.199	2, 9	-13.566	6.316
2,10	5.661	12.350	2,11	11.282	-5.073	2,12	-4.688	-10.482	2,13	-9.715	4.281
2,14	4.039	9.157	2,15	8.568	-3.728	2,16	-3.576	-8.165	2,17	-7.686	3.317
2,18	3.230	7.394	2,19	6.983	-2.994	2,20	-2.967	-6.779	2,21	-6.403	2.725
2,22	2.775	6.283	2,23	5.907	-2.462						
3, 3	78.656	36.070	3, 4	39.664	-36.957	3, 5	-21.152	-30.837	3, 6	-24.627	14.596
3, 7	11.150	20.454	3, 8	17.619	-9.143	3, 9	-7.732	-15.465	3,10	-13.894	6.804
3,11	6.021	12.567	3,12	11.578	-5.503	3,13	-4.988	-10.662	3,14	-9.990	4.672
3,15	4.291	9.305	3,16	8.829	-4.094	3,17	-3.786	-8.283	3,18	-7.942	3.671
3,19	3.396	7.482	3,20	7.243	-3.352	3,21	-3.076	-6.829	3,22	-6.685	3.121
4, 4	77.914	35.634	4, 5	39.221	-36.297	4, 6	-20.671	-30.518	4, 7	-24.269	14.104
4, 8	10.794	20.208	4, 9	17.313	-8.739	4,10	-7.453	-15.269	4,11	-13.622	6.453
4,12	5.794	12.409	4,13	11.327	-5.185	4,14	-4.802	-10.534	4,15	-9.753	4.375
4,16	4.142	9.204	4,17	8.598	-3.807	4,18	-3.672	-8.210	4,19	-7.708	3.382
4,20	3.325	7.440	4,21	6.995	-3.040						
5, 5	78.370	35.947	5, 6	39.568	-36.754	5, 7	-20.987	-30.751	5, 8	-24.559	14.466
5, 9	11.033	20.389	5,10	17.568	-9.049	5,11	-7.641	-15.412	5,12	-13.855	6.730
5,13	5.944	12.522	5,14	11.546	-5.443	5,15	-4.920	-10.621	5,16	-9.965	4.623
5,17	4.228	9.266	5,18	8.810	-4.056	5,19	-3.723	-8.244	5,20	-7.931	3.645
6, 6	78.066	35.718	6, 7	39.285	-36.409	6, 8	-20.768	-30.579	6, 9	-24.315	14.179
6,10	10.867	20.256	6,11	17.348	-8.795	6,12	-7.512	-15.309	6,13	-13.649	6.496
6,14	5.846	12.444	6,15	11.348	-5.219	6,16	-4.850	-10.567	6,17	-9.767	4.400
6,18	4.188	9.237	6,19	8.606	-3.823						
7, 7	78.279	35.888	7, 8	39.525	-36.687	7, 9	-20.924	-30.706	7,10	-24.528	14.420
7,11	10.984	20.353	7,12	17.545	-9.015	7,13	-7.599	-15.381	7,14	-13.837	6.704
7,15	5.906	12.493	7,16	11.534	-5.424	7,17	-4.883	-10.592	7,18	-9.958	4.612
8, 8	78.127	35.762	8, 9	39.315	-36.451	8,10	-20.814	-30.613	8,11	-24.336	14.209
8,12	10.904	20.285	8,13	17.363	-8.816	8,14	-7.545	-15.336	8,15	-13.659	6.510
8,16	5.877	12.470	8,17	11.353	-5.227						
9, 9	78.235	35.854	9,10	39.505	-36.659	9,11	-20.890	-30.679	9,12	-24.514	14.401
9,13	10.954	20.328	9,14	17.536	-9.002	9,15	-7.571	-15.357	9,16	-13.832	6.697
10,10	78.160	35.789	10,11	39.328	-36.469	10,12	-20.842	-30.637	10,13	-24.345	14.220
10,14	10.930	20.307	10,15	17.367	-8.822						
11,11	78.207	35.830	11,12	39.496	-36.648	11,13	-20.865	-30.657	11,14	-24.510	14.396
12,12	78.185	35.810	12,13	39.332	-36.474						

IEN

SELF AND MUTUAL IMPEDANCES

H/LMDA=0.2500 B/LMDA= 0.250 OMEGA= 8.53

26 ELEMENT ARRAY

I, K	ZIK IN OHMS RE	IM	I, K	ZIK IN OHMS RE	IM	I, K	ZIK IN OHMS RE	IM	I, K	ZIK IN OHMS RE	IM
1, 1	80.379	35.940	1, 2	39.766	-37.996	1, 3	-21.879	-30.969	1, 4	-24.739	15.143
1, 5	11.602	20.562	1, 6	17.705	-9.516	1, 7	-8.065	-15.552	1, 8	-13.961	7.088
1, 9	6.289	12.641	1,10	11.631	-5.733	1,11	-5.215	-10.726	1,12	-10.033	4.865
1,13	4.491	9.363	1,14	8.863	-4.260	1,15	-3.968	-8.339	1,16	-7.967	3.812
1,17	3.569	7.538	1,18	7.258	-3.469	1,19	-3.252	-6.891	1,20	-6.682	3.198
1,21	2.990	6.354	1,22	6.208	-2.983	1,23	-2.761	-5.898	1,24	-5.815	2.824
1,25	2.529	5.497	1,26	5.478	-2.782						
2, 2	77.275	35.481	2, 3	39.089	-35.878	2, 4	-20.350	-30.397	2, 5	-24.175	13.858
2, 6	10.579	20.118	2, 7	17.244	-8.566	2, 8	-7.288	-15.196	2, 9	-13.569	6.319
2,10	5.658	12.346	2,11	11.285	-5.077	2,12	-4.684	-10.478	2,13	-9.720	4.285
2,14	4.034	9.152	2,15	8.573	-3.734	2,16	-3.569	-8.159	2,17	-7.693	3.325
2,18	3.220	7.386	2,19	6.993	-3.007	2,20	-2.950	-6.766	2,21	-6.420	2.748
2,22	2.739	6.261	2,23	5.936	-2.526	2,24	-2.585	-5.847	2,25	-5.514	2.301
3, 3	78.658	36.072	3, 4	39.662	-36.954	3, 5	-21.154	-30.839	3, 6	-24.624	14.593
3, 7	11.153	20.457	3, 8	17.616	-9.140	3, 9	-7.736	-15.468	3,10	-13.891	6.800
3,11	6.025	12.571	3,12	11.574	-5.498	3,13	-4.993	-10.666	3,14	-9.985	4.666
3,15	4.298	9.311	3,16	8.822	-4.087	3,17	-3.795	-8.291	3,18	-7.933	3.660
3,19	3.410	7.492	3,20	7.230	-3.334	3,21	-3.102	-6.846	3,22	-6.662	3.082
3,23	2.840	6.307	3,24	6.199	-2.896						
4, 4	77.911	35.631	4, 5	39.223	-36.300	4, 6	-20.668	-30.516	4, 7	-24.272	14.107
4, 8	10.791	20.205	4, 9	17.316	-8.743	4,10	-7.449	-15.266	4,11	-13.626	6.457
4,12	5.790	12.404	4,13	11.332	-5.190	4,14	-4.797	-10.528	4,15	-9.759	4.381
4,16	4.134	9.197	4,17	8.606	-3.816	4,18	-3.661	-8.201	4,19	-7.720	3.396
4,20	3.307	7.426	4,21	7.013	-3.066	4,22	-3.037	-6.809	4,23	-6.431	2.790
5, 5	78.373	35.950	5, 6	39.565	-36.751	5, 7	-20.991	-30.754	5, 8	-24.556	14.463
5, 9	11.037	20.393	5,10	17.564	-9.045	5,11	-7.645	-15.417	5,12	-13.850	6.725
5,13	5.950	12.527	5,14	11.541	-5.437	5,15	-4.927	-10.627	5,16	-9.957	4.615
5,17	4.238	9.275	5,18	8.800	-4.044	5,19	-3.738	-8.256	5,20	-7.916	3.625
5,21	3.353	7.457	5,22	7.220	-3.311						
6, 6	78.063	35.715	6, 7	39.288	-36.412	6, 8	-20.765	-30.576	6, 9	-24.318	14.183
6,10	10.863	20.252	6,11	17.352	-8.799	6,12	-7.507	-15.304	6,13	-13.654	6.501
6,14	5.839	12.438	6,15	11.355	-5.226	6,16	-4.841	-10.559	6,17	-9.776	4.410
6,18	4.176	9.227	6,19	8.619	-3.839	6,20	-3.702	-8.230	6,21	-7.727	3.410

I, K	RE	IM	I, K	RE	IM	I, K	RE	IM	I, K	RE	IM
7, 7	78.282	35.891	7, 8	39.521	-36.683	7, 9	-20.928	-30.710	7,10	-24.524	14.415
7,11	10.989	20.357	7,12	17.540	-9.009	7,13	-7.605	-15.386	7,14	-13.831	6.697
7,15	5.914	12.500	7,16	11.526	-5.415	7,17	-4.894	-10.602	7,18	-9.947	4.598
7,19	4.205	9.249	7,20	8.794	-4.034						
8, 8	78.122	35.758	8, 9	39.319	-36.456	8,10	-20.809	-30.609	8,11	-24.341	14.214
8,12	10.898	20.280	8,13	17.369	-8.823	8,14	-7.537	-15.329	8,15	-13.667	6.519
8,16	5.867	12.461	8,17	11.363	-5.239	8,18	-4.869	-10.581	8,19	-9.781	4.418
9, 9	78.240	35.859	9,10	39.500	-36.654	9,11	-20.895	-30.684	9,12	-24.508	14.395
9,13	10.961	20.335	9,14	17.528	-8.994	9,15	-7.580	-15.366	9,16	-13.823	6.687
9,17	5.890	12.480	9,18	11.522	-5.409						
10,10	78.155	35.784	10,11	39.334	-36.476	10,12	-20.835	-30.631	10,13	-24.352	14.227
10,14	10.921	20.300	10,15	17.376	-8.832	10,16	-7.559	-15.348	10,17	-13.670	6.524
11,11	78.214	35.837	11,12	39.489	-36.641	11,13	-20.873	-30.665	11,14	-24.501	14.386
11,15	10.941	20.317	11,16	17.525	-8.990						
12,12	78.176	35.803	12,13	39.341	-36.484	12,14	-20.855	-30.648	12,15	-24.355	14.231
13,13	78.195	35.819	13,14	39.486	-36.637						

IEN

SELF AND MUTUAL IMPEDANCES

H/LMDA=0.2500 B/LMDA= 0.250 OMEGA= 8.53

28 ELEMENT ARRAY

I, K	ZIK IN OHMS		I, K	ZIK IN OHMS		I, K	ZIK IN OHMS		I, K	ZIK IN OHMS	
	RE	IM		RE	IM		RE	IM		RE	IM
1, 1	80.381	35.941	1, 2	39.764	-37.994	1, 3	-21.881	-30.970	1, 4	-24.737	15.141
1, 5	11.604	20.564	1, 6	17.703	-9.514	1, 7	-8.067	-15.554	1, 8	-13.959	7.086
1, 9	6.291	12.643	1,10	11.629	-5.731	1,11	-5.218	-10.729	1,12	-10.030	4.862
1,13	4.495	9.367	1,14	8.859	-4.256	1,15	-3.972	-8.343	1,16	-7.962	3.807
1,17	3.574	7.543	1,18	7.252	-3.462	1,19	-3.260	-6.897	1,20	-6.675	3.188
1,21	3.002	6.363	1,22	6.197	-2.967	1,23	-2.784	-5.912	1,24	-5.796	2.789
1,25	2.589	5.522	1,26	5.460	-2.657	1,27	-2.385	-5.174	1,28	-5.168	2.631
2, 2	77.274	35.480	2, 3	39.090	-35.879	2, 4	-20.348	-30.396	2, 5	-24.177	13.860
2, 6	10.577	20.116	2, 7	17.246	-8.568	2, 8	-7.285	-15.194	2, 9	-13.571	6.321
2,10	5.655	12.344	2,11	11.288	-5.080	2,12	-4.681	-10.475	2,13	-9.723	4.289
2,14	4.030	9.148	2,15	8.577	-3.738	2,16	-3.565	-8.154	2,17	-7.698	3.330
2,18	3.214	7.380	2,19	7.000	-3.014	2,20	-2.941	-6.759	2,21	-6.429	2.760
2,22	2.724	6.249	2,23	5.951	-2.547	2,24	-2.551	-5.827	2,25	-5.541	2.360
2,26	2.424	5.475	2,27	5.177	-2.164						
3, 3	78.660	36.074	3, 4	39.660	-36.953	3, 5	-21.157	-30.841	3, 6	-24.622	14.591
3, 7	11.155	20.459	3, 8	17.613	-9.138	3, 9	-7.738	-15.471	3,10	-13.888	6.797
3,11	6.028	12.574	3,12	11.571	-5.495	3,13	-4.997	-10.670	3,14	-9.981	4.662
3,15	4.303	9.315	3,16	8.818	-4.081	3,17	-3.801	-8.296	3,18	-7.927	3.653
3,19	3.418	7.500	3,20	7.222	-3.324	3,21	-3.115	-6.856	3,22	-6.650	3.065
3,23	2.864	6.323	3,24	6.178	-2.860	3,25	-2.645	-5.868	3,26	-5.788	2.708
4, 4	77.909	35.629	4, 5	39.226	-36.302	4, 6	-20.666	-30.514	4, 7	-24.274	14.109
4, 8	10.789	20.203	4, 9	17.319	-8.746	4,10	-7.446	-15.263	4,11	-13.629	6.460
4,12	5.786	12.401	4,13	11.336	-5.194	4,14	-4.793	-10.524	4,15	-9.763	4.386
4,16	4.129	9.192	4,17	8.611	-3.822	4,18	-3.654	-8.195	4,19	-7.727	3.404
4,20	3.296	7.418	4,21	7.024	-3.079	4,22	-3.020	-6.796	4,23	-6.448	2.814
4,24	2.804	6.288	4,25	5.962	-2.586						
5, 5	78.375	35.952	5, 6	39.563	-36.748	5, 7	-20.993	-30.757	5, 8	-24.553	14.460
5, 9	11.040	20.395	5,10	17.561	-9.042	5,11	-7.649	-15.420	5,12	-13.846	6.722
5,13	5.954	12.531	5,14	11.536	-5.432	5,15	-4.932	-10.632	5,16	-9.952	4.609
5,17	4.245	9.280	5,18	8.794	-4.036	5,19	-3.747	-8.264	5,20	-7.907	3.614
5,21	3.367	7.468	5,22	7.207	-3.292	5,23	-3.063	-6.824	5,24	-6.641	3.043
6, 6	78.060	35.713	6, 7	39.291	-36.415	6, 8	-20.762	-30.573	6, 9	-24.321	14.186
6,10	10.859	20.249	6,11	17.355	-8.803	6,12	-7.503	-15.301	6,13	-13.658	6.506
6,14	5.835	12.434	6,15	11.359	-5.231	6,16	-4.835	-10.554	6,17	-9.782	4.417
6,18	4.168	9.220	6,19	8.627	-3.848	6,20	-3.691	-8.221	6,21	-7.739	3.424
6,22	3.334	7.444	6,23	7.031	-3.092						
7, 7	78.285	35.894	7, 8	39.518	-36.680	7, 9	-20.932	-30.714	7,10	-24.520	14.412
7,11	10.993	20.361	7,12	17.536	-9.005	7,13	-7.609	-15.391	7,14	-13.826	6.692
7,15	5.920	12.505	7,16	11.520	-5.409	7,17	-4.901	-10.608	7,18	-9.940	4.590
7,19	4.215	9.258	7,20	8.784	-4.022	7,21	-3.718	-8.241	7,22	-7.902	3.605
8, 8	78.119	35.755	8, 9	39.322	-36.459	8,10	-20.805	-30.605	8,11	-24.345	14.218
8,12	10.894	20.276	8,13	17.373	-8.827	8,14	-7.532	-15.324	8,15	-13.672	6.525
8,16	5.861	12.455	8,17	11.370	-5.246	8,18	-4.860	-10.574	8,19	-9.790	4.428
8,20	4.193	9.240	8,21	8.631	-3.854						
9, 9	78.244	35.862	9,10	39.496	-36.650	9,11	-20.900	-30.688	9,12	-24.504	14.390
9,13	10.966	20.339	9,14	17.523	-8.989	9,15	-7.586	-15.371	9,16	-13.817	6.680
9,17	5.898	12.487	9,18	11.514	-5.400	9,19	-4.880	-10.590	9,20	-9.936	4.585
10,10	78.150	35.780	10,11	39.339	-36.480	10,12	-20.830	-30.626	10,13	-24.357	14.233
10,14	10.915	20.294	10,15	17.382	-8.838	10,16	-7.552	-15.341	10,17	-13.678	6.532
10,18	5.880	12.471	10,19	11.373	-5.250						
11,11	78.219	35.841	11,12	39.484	-36.635	11,13	-20.879	-30.670	11,14	-24.495	14.380
11,15	10.948	20.323	11,16	17.518	-8.982	11,17	-7.569	-15.356	11,18	-13.814	6.676
12,12	78.171	35.798	12,13	39.347	-36.490	12,14	-20.848	-30.642	12,15	-24.362	14.239
12,16	10.932	20.309	12,17	17.385	-8.842						
13,13	78.202	35.826	13,14	39.479	-36.629	13,15	-20.863	-30.656	13,16	-24.493	14.377
14,14	78.187	35.812	14,15	39.349	-36.493						

IEN

SELF AND MUTUAL IMPEDANCES

H/LMDA=0.2500   B/LMDA= 0.250   OMEGA= 8.53

30 ELEMENT ARRAY

I, K	ZIK IN OHMS RE	IM	I, K	ZIK IN OHMS RE	IM	I, K	ZIK IN OHMS RE	IM	I, K	ZIK IN OHMS RE	IM
1, 1	80.382	35.943	1, 2	39.763	-37.993	1, 3	-21.882	-30.972	1, 4	-24.736	15.140
1, 5	11.605	20.566	1, 6	17.701	-9.512	1, 7	-8.069	-15.556	1, 8	-13.957	7.084
1, 9	6.293	12.645	1,10	11.627	-5.729	1,11	-5.220	-10.731	1,12	-10.028	4.860
1,13	4.498	9.369	1,14	8.857	-4.253	1,15	-3.975	-8.346	1,16	-7.959	3.804
1,17	3.578	7.546	1,18	7.248	-3.457	1,19	-3.265	-6.902	1,20	-6.669	3.182
1,21	3.009	6.370	1,22	6.189	-2.958	1,23	-2.795	-5.921	1,24	-5.785	2.774
1,25	2.611	5.536	1,26	5.442	-2.624	1,27	-2.442	-5.197	1,28	-5.152	2.513
1,29	2.261	4.892	1,30	4.896	-2.499						
2, 2	77.273	35.479	2, 3	39.092	-35.881	2, 4	-20.347	-30.394	2, 5	-24.179	13.861
2, 6	10.575	20.114	2, 7	17.248	-8.570	2, 8	-7.284	-15.192	2, 9	-13.573	6.323
2,10	5.653	12.342	2,11	11.290	-5.082	2,12	-4.678	-10.473	2,13	-9.726	4.291
2,14	4.028	9.146	2,15	8.580	-3.741	2,16	-3.561	-8.151	2,17	-7.702	3.334
2,18	3.209	7.376	2,19	7.005	-3.019	2,20	-2.935	-6.753	2,21	-6.435	2.767
2,22	2.715	6.242	2,23	5.960	-2.558	2,24	-2.537	-5.816	2,25	-5.555	2.380
2,26	2.393	5.456	2,27	5.202	-2.219	2,28	-2.287	-5.153	2,29	-4.884	2.046
3, 3	78.661	36.075	3, 4	39.659	-36.951	3, 5	-21.158	-30.843	3, 6	-24.620	14.589
3, 7	11.157	20.461	3, 8	17.612	-9.136	3, 9	-7.740	-15.473	3,10	-13.886	6.795
3,11	6.031	12.576	3,12	11.568	-5.492	3,13	-4.999	-10.673	3,14	-9.978	4.659
3,15	4.306	9.318	3,16	8.814	-4.078	3,17	-3.805	-8.300	3,18	-7.922	3.648
3,19	3.424	7.504	3,20	7.216	-3.317	3,21	-3.123	-6.863	3,22	-6.642	3.055
3,23	2.876	6.332	3,24	6.167	-2.844	3,25	-2.667	-5.883	3,26	-5.769	2.674
3,27	2.481	5.495	3,28	5.435	-2.548						
4, 4	77.908	35.628	4, 5	39.227	-36.304	4, 6	-20.665	-30.512	4, 7	-24.276	14.111
4, 8	10.787	20.201	4, 9	17.321	-8.748	4,10	-7.444	-15.261	4,11	-13.631	6.462
4,12	5.784	12.399	4,13	11.338	-5.197	4,14	-4.789	-10.521	4,15	-9.766	4.389
4,16	4.125	9.189	4,17	8.615	-3.826	4,18	-3.649	-8.190	4,19	-7.732	3.410
4,20	3.290	7.412	4,21	7.031	-3.086	4,22	-3.011	-6.788	4,23	-6.457	2.826
4,24	2.788	6.276	4,25	5.977	-2.608	4,26	-2.611	-5.851	4,27	-5.565	2.416
5, 5	78.377	35.954	5, 6	39.561	-36.746	5, 7	-20.995	-30.758	5, 8	-24.551	14.458
5, 9	11.042	20.398	5,10	17.559	-9.039	5,11	-7.651	-15.422	5,12	-13.844	6.719
5,13	5.957	12.534	5,14	11.533	-5.429	5,15	-4.936	-10.635	5,16	-9.949	4.605
5,17	4.249	9.285	5,18	8.789	-4.031	5,19	-3.753	-8.269	5,20	-7.901	3.607
5,21	3.376	7.475	5,22	7.198	-3.282	5,23	-3.076	-6.834	5,24	-6.628	3.026
5,25	2.828	6.302	5,26	6.159	-2.824						
6, 6	78.058	35.711	6, 7	39.293	-36.417	6, 8	-20.760	-30.571	6, 9	-24.324	14.188
6,10	10.857	20.247	6,11	17.358	-8.806	6,12	-7.500	-15.298	6,13	-13.661	6.509
6,14	5.831	12.431	6,15	11.363	-5.235	6,16	-4.831	-10.550	6,17	-9.787	4.422
6,18	4.163	9.215	6,19	8.632	-3.854	6,20	-3.684	-8.215	6,21	-7.746	3.433
6,22	3.324	7.436	6,23	7.041	-3.105	6,24	-3.045	-6.812	6,25	-6.463	2.838
7, 7	78.288	35.896	7, 8	39.516	-36.677	7, 9	-20.934	-30.716	7,10	-24.518	14.409
7,11	10.995	20.364	7,12	17.533	-9.002	7,13	-7.613	-15.394	7,14	-13.823	6.689
7,15	5.924	12.509	7,16	11.516	-5.404	7,17	-4.906	-10.613	7,18	-9.934	4.584
7,19	[illegible]	[illegible]	7,20	8.778	-4.014	7,21	-3.727	-8.249	7,22	-7.893	3.594
7,23	3.348	7.455	7,24	7.194	-3.214						
8, 8	78.117	35.752	8, 9	39.325	-36.462	8,10	-20.802	-30.603	8,11	-24.348	14.221
8,12	10.891	20.273	8,13	17.377	-8.831	8,14	-7.528	-15.320	8,15	-13.676	6.529
8,16	5.856	12.451	8,17	11.375	-5.251	8,18	-4.854	-10.568	8,19	-9.796	4.435
8,20	4.185	9.233	8,21	8.639	-3.864	8,22	-3.706	-8.233	8,23	-7.750	3.438
9, 9	78.247	35.865	9,10	39.493	-36.647	9,11	-20.903	-30.691	9,12	-24.500	14.387
9,13	10.970	20.343	9,14	17.519	-8.984	9,15	-7.591	-15.375	9,16	-13.812	6.675
9,17	5.903	12.492	9,18	11.508	-5.394	9,19	-4.887	-10.597	9,20	-9.929	4.577
9,21	4.203	9.248	9,22	8.775	-4.010						
10,10	78.147	35.777	10,11	39.342	-36.484	10,12	-20.826	-30.623	10,13	-24.360	14.237
10,14	10.911	20.290	10,15	17.386	-8.843	10,16	-7.547	-15.336	10,17	-13.683	6.538
10,18	5.873	12.465	10,19	11.380	-5.258	10,20	-4.871	-10.583	10,21	-9.799	4.438
11,11	78.223	35.845	11,12	39.481	-36.631	11,13	-20.883	-30.674	11,14	-24.491	14.375
11,15	10.953	20.328	11,16	17.513	-8.976	11,17	-7.575	-15.361	11,18	-13.808	6.669
11,19	5.888	12.478	11,20	11.506	-5.391						
12,12	78.167	35.794	12,13	39.351	-36.494	12,14	-20.843	-30.637	12,15	-24.367	14.244
12,16	10.926	20.303	12,17	17.391	-8.848	12,18	-7.561	-15.349	12,19	-13.685	6.541
13,13	78.207	35.830	13,14	39.474	-36.624	13,15	-20.869	-30.661	13,16	-24.487	14.371
13,17	10.939	20.315	13,18	17.511	-8.974						
14,14	78.181	35.807	14,15	39.355	-36.499	14,16	-20.856	-30.649	14,17	-24.369	14.247
15,15	78.194	35.818	15,16	39.472	-36.622						

IEN

SELF AND MUTUAL IMPEDANCES

H/LMDA=0.2500 B/LMDA= 0.250 OMEGA= 8.53

35 ELEMENT ARRAY

I, K	ZIK IN OHMS		I, K	ZIK IN OHMS		I, K	ZIK IN OHMS		I, K	ZIK IN OHMS	
	RE	IM		RE	IM		RE	IM		RE	IM
1, 1	80.404	35.962	1, 2	39.744	-37.972	1, 3	-21.905	-30.992	1, 4	-24.716	15.118
1, 5	11.628	20.587	1, 6	17.680	-9.489	1, 7	-8.094	-15.578	1, 8	-13.935	7.059
1, 9	6.319	12.669	1,10	11.603	-5.702	1,11	-5.248	-10.756	1,12	-10.001	4.831
1,13	4.527	9.396	1,14	8.828	-4.222	1,15	-4.008	-8.376	1,16	-7.928	3.770
1,17	3.614	7.579	1,18	7.214	-3.420	1,19	-3.305	-6.938	1,20	-6.631	3.140
1,21	3.055	6.410	1,22	6.146	-2.909	1,23	-2.848	-5.968	1,24	-5.734	2.716
1,25	2.676	5.592	1,26	5.380	-2.550	1,27	-2.529	-5.268	1,28	-5.071	2.406
1,29	2.405	4.987	1,30	4.797	-2.275	1,31	-2.302	-4.742	1,32	-4.549	2.152
1,33	2.226	4.531	1,34	4.320	-2.011	1,35	-2.232	-4.341			
2, 2	77.253	35.459	2, 3	39.112	-35.902	2, 4	-20.326	-30.374	2, 5	-24.200	13.883
2, 6	10.553	20.092	2, 7	17.271	-8.593	2, 8	-7.260	-15.169	2, 9	-13.597	6.348
2,10	5.628	12.317	2,11	11.316	-5.108	2,12	-4.651	-10.446	2,13	-9.753	4.320
2,14	3.998	9.117	2,15	8.610	-3.772	2,16	-3.529	-8.120	2,17	-7.735	3.368
2,18	3.174	7.341	2,19	7.041	-3.057	2,20	-2.895	-6.714	2,21	-6.477	2.810
2,22	2.669	6.198	2,23	6.008	-2.609	2,24	-2.482	-5.764	2,25	-5.612	2.442
2,26	2.322	5.392	2,27	5.274	-2.302	2,28	-2.184	-5.070	2,29	-4.982	2.184
2,30	2.061	4.786	2,31	4.729	-2.087	2,32	-1.945	-4.531	2,33	-4.512	2.016
2,34	1.812	4.295									
3, 3	78.684	36.096	3, 4	39.637	-36.928	3, 5	-21.182	-30.865	3, 6	-24.598	14.565
3, 7	11.182	20.484	3, 8	17.588	-9.111	3, 9	-7.767	-15.498	3,10	-13.860	6.768
3,11	6.059	12.603	3,12	11.540	-5.463	3,13	-5.030	-10.701	3,14	-9.948	4.627
3,15	4.339	9.349	3,16	8.781	-4.043	3,17	-3.842	-8.335	3,18	-7.886	3.610
3,19	3.465	7.543	3,20	7.176	-3.274	3,21	-3.169	-6.906	3,22	-6.596	3.006
3,23	2.930	6.382	3,24	6.113	-2.784	3,25	-2.734	-5.943	3,26	-5.703	2.598
3,27	2.570	5.569	3,28	5.349	-2.438	3,29	-2.433	-5.249	3,30	-5.039	2.296
3,31	2.319	4.973	3,32	4.762	-2.163	3,33	-2.236	-4.736			
4, 4	77.885	35.606	4, 5	39.250	-36.327	4, 6	-20.641	-30.489	4, 7	-24.300	14.136
4, 8	10.762	20.177	4, 9	17.347	-8.774	4,10	-7.417	-15.235	4,11	-13.658	6.490
4,12	5.755	12.370	4,13	11.368	-5.227	4,14	-4.758	-10.491	4,15	-9.798	4.422
4,16	4.091	9.156	4,17	8.650	-3.862	4,18	-3.611	-8.154	4,19	-7.771	3.450
4,20	3.248	7.371	4,21	7.074	-3.132	4,22	-2.962	-6.741	4,23	-6.508	2.879
4,24	2.730	6.222	4,25	6.037	-2.674	4,26	-2.536	-5.784	4,27	-5.641	2.505
4,28	2.372	5.410	4,29	5.302	-2.363	4,30	-2.227	-5.084	4,31	-5.012	2.248
4,32	2.092	4.794									
5, 5	78.402	35.977	5, 6	39.537	-36.721	5, 7	-21.022	-30.783	5, 8	-24.526	14.432
5, 9	11.070	20.424	5,10	17.532	-9.011	5,11	-7.681	-15.450	5,12	-13.815	6.688
5,13	5.989	12.564	5,14	11.502	-5.396	5,15	-4.971	-10.668	5,16	-9.914	4.569
5,17	4.288	9.321	5,18	8.751	-3.991	5,19	-3.796	-8.310	5,20	-7.859	3.562
5,21	3.424	7.521	5,22	7.150	-3.230	5,23	-3.132	-6.886	5,24	-6.572	2.964
5,25	2.898	6.365	5,26	6.089	-2.745	5,27	-2.705	-5.928	5,28	-5.679	2.559
5,29	2.546	5.558	5,30	5.324	-2.397	5,31	-2.416	-5.242			
6, 6	78.033	35.686	6, 7	39.318	-36.443	6, 8	-20.733	-30.545	6, 9	-24.351	14.216
6,10	10.829	20.219	6,11	17.387	-8.835	6,12	-7.469	-15.268	6,13	-13.692	6.541
6,14	5.798	12.398	6,15	11.397	-5.270	6,16	-4.795	-10.515	6,17	-9.824	4.460
6,18	4.123	9.176	6,19	8.673	-3.897	6,20	-3.639	-8.171	6,21	-7.792	3.481
6,22	3.272	7.386	6,23	7.094	-3.161	6,24	-2.983	-6.754	6,25	-6.527	2.908
6,26	2.748	6.233	6,27	6.057	-2.702	6,28	-2.551	-5.793	6,29	-5.661	2.534
6,30	2.382	5.415									
7, 7	78.315	35.922	7, 8	39.489	-36.649	7, 9	-20.963	-30.744	7,10	-24.489	14.379
7,11	11.027	20.394	7,12	17.502	-8.970	7,13	-7.646	-15.426	7,14	-13.789	6.654
7,15	5.960	12.544	7,16	11.480	-5.366	7,17	-4.946	-10.651	7,18	-9.894	4.542
7,19	4.267	9.306	7,20	8.733	-3.967	7,21	-3.778	-8.297	7,22	-7.842	3.540
7,23	3.408	7.510	7,24	7.133	-3.209	7,25	-3.119	-6.878	7,26	-6.555	2.943
7,27	2.887	6.358	7,28	6.072	-2.723	7,29	-2.698	-5.924			
8, 8	78.089	35.725	8, 9	39.354	-36.492	8,10	-20.772	-30.573	8,11	-24.378	14.253
8,12	10.858	20.241	8,13	17.410	-8.865	8,14	-7.493	-15.286	8,15	-13.712	6.566
8,16	5.818	12.413	8,17	11.414	-5.292	8,18	-4.812	-10.527	8,19	-9.840	4.480
8,20	4.137	9.187	8,21	8.688	-3.915	8,22	-3.651	-8.180	8,23	-7.807	3.499
8,24	3.282	7.394	8,25	7.109	-3.179	8,26	-2.991	-6.760	8,27	-6.542	2.926
8,28	2.753	6.236									
9, 9	78.278	35.894	9,10	39.462	-36.615	9,11	-20.936	-30.723	9,12	-24.468	14.352
9,13	11.006	20.377	9,14	17.484	-8.948	9,15	-7.629	-15.413	9,16	-13.773	6.635
9,17	5.946	12.533	9,18	11.465	-5.349	9,19	-4.935	-10.642	9,20	-9.881	4.526
9,21	4.257	9.299	9,22	8.720	-3.951	9,23	-3.770	-8.291	9,24	-7.829	3.525
9,25	3.402	7.506	9,26	7.121	-3.193	9,27	-3.116	-6.875			
10,10	78.115	35.746	10,11	39.375	-36.517	10,12	-20.792	-30.589	10,13	-24.396	14.273
10,14	10.874	20.254	10,15	17.425	-8.883	10,16	-7.506	-15.296	10,17	-13.725	6.581
10,18	5.828	12.422	10,19	11.426	-5.306	10,20	-4.820	-10.534	10,21	-9.851	4.493
10,22	4.143	9.192	10,23	8.700	-3.928	10,24	-3.656	-8.184	10,25	-7.818	3.512
10,26	3.285	7.396									
11,11	78.258	35.879	11,12	39.445	-36.595	11,13	-20.922	-30.711	11,14	-24.453	14.336
11,15	10.994	20.368	11,16	17.471	-8.933	11,17	-7.620	-15.405	11,18	-13.762	6.622
11,19	5.939	12.527	11,20	11.454	-5.337	11,21	-4.929	-10.637	11,22	-9.870	4.515
11,23	4.253	9.295	11,24	8.709	-3.940	11,25	-3.767	-8.289			
12,12	78.129	35.757	12,13	39.389	-36.534	12,14	-20.803	-30.598	12,15	-24.408	14.287
12,16	10.882	20.261	12,17	17.436	-8.895	12,18	-7.512	-15.302	12,19	-13.735	6.593
12,20	5.833	12.426	12,21	11.436	-5.317	12,22	-4.823	-10.537	12,23	-9.861	4.504
12,24	4.145	9.193									
13,13	78.248	35.870	13,14	39.433	-36.582	13,15	-20.914	-30.704	13,16	-24.443	14.324
13,17	10.988	20.362	13,18	17.461	-8.922	13,19	-7.616	-15.401	13,20	-13.753	6.611
13,21	5.936	12.524	13,22	11.445	-5.327	13,23	-4.927	-10.636			
14,14	78.137	35.764	14,15	39.399	-36.545	14,16	-20.808	-30.603	14,17	-24.417	14.297
14,18	10.886	20.264	14,19	17.445	-8.905	14,20	-7.515	-15.304	14,21	-13.744	6.602
14,22	5.834	12.427									
15,15	78.242	35.865	15,16	39.424	-36.571	15,17	-20.909	-30.700	15,18	-24.434	14.315
15,19	10.985	20.360	15,20	17.453	-8.913	15,21	-7.614	-15.400			
16,16	78.141	35.768	16,17	39.408	-36.554	16,18	-20.811	-30.606	16,19	-24.426	14.306
16,20	10.888	20.266									
17,17	78.239	35.862	17,18	39.416	-36.563	17,19	-20.908	-30.699			
18,18	78.142	35.769									

IEN

SELF AND MUTUAL IMPEDANCES

H/LMDA=0.2500 B/LMDA= 0.250 OMEGA= 8.53

40 ELEMENT ARRAY

I, K	ZIK IN OHMS		I, K	ZIK IN OHMS		I, K	ZIK IN OHMS		I, K	ZIK IN OHMS	
	RE	IM		RE	IM		RE	IM		RE	IM
1, 1	80.386	35.947	1, 2	39.759	-37.989	1, 3	-21.887	-30.976	1, 4	-24.732	15.135
1, 5	11.610	20.570	1, 6	17.697	-9.507	1, 7	-8.074	-15.561	1, 8	-13.952	7.079
1, 9	6.299	12.650	1,10	11.621	-5.723	1,11	-5.226	-10.737	1,12	-10.021	4.853
1,13	4.505	9.376	1,14	8.849	-4.245	1,15	-3.984	-8.354	1,16	-7.951	3.795
1,17	3.588	7.555	1,18	7.238	-3.447	1,19	-3.277	-6.912	1,20	-6.658	3.169
1,21	3.024	6.383	1,22	6.175	-2.941	1,23	-2.815	-5.937	1,24	-5.767	2.752
1,25	2.638	5.557	1,26	5.416	-2.591	1,27	-2.486	-5.229	1,28	-5.113	2.453
1,29	2.353	4.941	1,30	4.847	-2.334	1,31	-2.236	-4.686	1,32	-4.612	2.230
1,33	2.130	4.459	1,34	4.404	-2.140	1,35	-2.032	-4.252	1,36	-4.221	2.065
1,37	1.937	4.063	1,38	4.060	-2.011	1,39	-1.822	-3.883	1,40	-3.913	2.028
2, 2	77.269	35.475	2, 3	39.096	-35.885	2, 4	-20.343	-30.390	2, 5	-24.183	13.866
2, 6	10.571	20.109	2, 7	17.253	-8.575	2, 8	-7.278	-15.187	2, 9	-13.579	6.329
2,10	5.648	12.336	2,11	11.297	-5.088	2,12	-4.672	-10.466	2,13	-9.733	4.298
2,14	4.020	9.138	2,15	8.588	-3.749	2,16	-3.553	-8.142	2,17	-7.711	3.344
2,18	3.199	7.366	2,19	7.016	-3.031	2,20	-2.922	-6.741	2,21	-6.449	2.781
2,22	2.699	6.227	2,23	5.977	-2.577	2,24	-2.515	-5.796	2,25	-5.578	2.406
2,26	2.361	5.429	2,27	5.234	-2.260	2,28	-2.229	-5.113	2,29	-4.935	2.134
2,30	2.116	4.837	2,31	4.672	-2.024	2,32	-2.019	-4.595	2,33	-4.438	1.925
2,34	1.935	4.382	2,35	4.226	-1.833	2,36	-1.865	-4.193	2,37	-4.032	1.744
2,38	1.815	4.030	2,39	3.848	-1.637						
3, 3	78.666	36.080	3, 4	39.654	-36.946	3, 5	-21.163	-30.848	3, 6	-24.615	14.584
3, 7	11.162	20.466	3, 8	17.606	-9.130	3, 9	-7.746	-15.479	3,10	-13.880	6.789
3,11	6.037	12.583	3,12	11.561	-5.485	3,13	-5.007	-10.680	3,14	-9.970	4.651
3,15	4.315	9.326	3,16	8.805	-4.068	3,17	-3.815	-8.310	3,18	-7.912	3.637
3,19	3.436	7.516	3,20	7.203	-3.304	3,21	-3.138	-6.877	3,22	-6.626	3.038
3,23	2.896	6.350	3,24	6.147	-2.821	3,25	-2.695	-5.907	3,26	-5.741	2.640
3,27	2.526	5.529	3,28	5.393	-2.486	3,29	-2.379	-5.201	3,30	-5.092	2.355
3,31	2.252	4.914	3,32	4.828	-2.242	3,33	-2.138	-4.660	3,34	-4.596	2.145
3,35	2.034	4.431	3,36	4.393	-2.065	3,37	-1.932	-4.222	3,38	-4.216	2.006
4, 4	77.903	35.623	4, 5	39.232	-36.309	4, 6	-20.659	-30.507	4, 7	-24.281	14.117
4, 8	10.781	20.195	4, 9	17.327	-8.754	4,10	-7.438	-15.254	4,11	-13.638	6.469
4,12	5.777	12.391	4,13	11.346	-5.204	4,14	-4.781	-10.513	4,15	-9.775	4.398
4,16	4.116	9.180	4,17	8.625	-3.837	4,18	-3.638	-8.179	4,19	-7.744	3.422
4,20	3.276	7.399	4,21	7.045	-3.101	4,22	-2.993	-6.772	4,23	-6.475	2.846
4,24	2.765	6.256	4,25	6.001	-2.636	4,26	-2.577	-5.823	4,27	-5.599	2.461
4,28	2.419	5.455	4,29	5.254	-2.311	4,30	-2.285	-5.137	4,31	-4.953	2.181
4,32	2.170	4.861	4,33	4.687	-2.066	4,34	-2.073	-4.620	4,35	-4.450	1.962
4,36	1.991	4.408	4,37	4.233	-1.861						
5, 5	78.383	35.959	5, 6	39.556	-36.741	5, 7	-21.001	-30.764	5, 8	-24.545	14.452
5, 9	11.049	20.404	5,10	17.552	-9.033	5,11	-7.658	-15.429	5,12	-13.836	6.711
5,13	5.965	12.542	5,14	11.525	-5.420	5,15	-4.945	-10.644	5,16	-9.939	4.595
5,17	4.260	9.295	5,18	8.778	-4.019	5,19	-3.766	-8.282	5,20	-7.888	3.593
5,21	3.392	7.490	5,22	7.182	-3.264	5,23	-3.097	-6.853	5,24	-6.607	3.002
5,25	2.858	6.327	5,26	6.129	-2.787	5,27	-2.659	-5.886	5,28	-5.725	2.609
5,29	2.491	5.508	5,30	5.379	-2.459	5,31	-2.346	-5.181	5,32	-5.080	2.330
5,33	2.210	[illegible]	[illegible]	4.819	-2.221	5,35	-2.102	-4.638	5,36	-4.591	2.131
6, 6	78.053	35.705	6, 7	39.299	-36.423	6, 8	-20.754	-30.565	6, 9	-24.330	14.195
6,10	10.850	20.240	6,11	17.365	-8.813	6,12	-7.492	-15.290	6,13	-13.669	6.517
6,14	5.822	12.422	6,15	11.372	-5.245	6,16	-4.821	-10.540	6,17	-9.797	4.433
6,18	4.151	9.203	6,19	8.645	-3.867	6,20	-3.669	-8.201	6,21	-7.761	3.449
6,22	3.306	7.419	6,23	7.060	-3.126	6,24	-3.020	-6.790	6,25	-6.489	2.868
6,26	2.791	6.273	6,27	6.013	-2.656	6,28	-2.601	-5.840	6,29	-5.610	2.479
6,30	2.443	5.471	6,31	5.263	-2.327	6,32	-2.309	-5.154	6,33	-4.960	2.194
6,34	2.196	4.879	6,35	4.692	-2.075						
7, 7	78.294	35.902	7, 8	39.509	-36.671	7, 9	-20.941	-30.723	7,10	-24.511	14.402
7,11	11.003	20.371	7,12	17.525	-8.994	7,13	-7.622	-15.402	7,14	-13.814	6.679
7,15	5.934	12.519	7,16	11.506	-5.393	7,17	-4.918	-10.624	7,18	-9.922	4.572
7,19	4.236	9.277	7,20	8.763	-3.999	7,21	-3.744	-8.265	7,22	-7.875	3.575
7,23	3.371	7.475	7,24	7.170	-3.248	7,25	-3.077	-6.838	7,26	-6.597	2.987
7,27	2.839	6.313	7,28	6.120	-2.775	7,29	-2.641	-5.872	7,30	-5.718	2.598
7,31	2.473	5.494	7,32	5.374	-2.450	7,33	-2.326	-5.167	7,34	-5.076	2.325
8, 8	78.110	35.746	8, 9	39.332	-36.469	8,10	-20.795	-30.595	8,11	-24.356	14.229
8,12	10.882	20.264	8,13	17.386	-8.840	8,14	-7.519	-15.311	8,15	-13.686	6.539
8,16	5.845	12.440	8,17	11.387	-5.263	8,18	-4.841	-10.556	8,19	-9.810	4.449
8,20	4.169	9.218	8,21	8.656	-3.882	8,22	-3.686	-8.214	8,23	-7.771	3.462
8,24	3.321	7.431	8,25	7.069	-3.137	8,26	-3.035	-6.802	8,27	-6.496	2.878
8,28	2.805	6.285	8,29	6.019	-2.664	8,30	-2.616	-5.851	8,31	-5.615	2.485
8,32	2.458	5.483	8,33	5.266	-2.331						
9, 9	78.254	35.872	9,10	39.485	-36.639	9,11	-20.912	-30.700	9,12	-24.491	14.378
9,13	10.980	20.352	9,14	17.509	-8.974	9,15	-7.602	-15.386	9,16	-13.801	6.663
9,17	5.916	12.504	9,18	11.495	-5.380	9,19	-4.902	-10.611	9,20	-9.913	4.560
9,21	4.222	9.265	9,22	8.755	-3.988	9,23	-3.731	-8.254	9,24	-7.868	3.566
9,25	3.359	7.464	9,26	7.164	-3.240	9,27	-3.065	-6.828	9,28	-6.592	2.981
9,29	2.827	6.303	9,30	6.117	-2.770	9,31	-2.628	-5.861	9,32	-5.716	2.595
10,10	78.139	35.769	10,11	39.351	-36.492	10,12	-20.817	-30.613	10,13	-24.370	14.247
10,14	10.900	20.280	10,15	17.398	-8.855	10,16	-7.534	-15.324	10,17	-13.696	6.551
10,18	5.859	12.452	10,19	11.395	-5.274	10,20	-4.854	-10.567	10,21	-9.817	4.458
10,22	4.181	9.228	10,23	8.662	-3.889	10,24	-3.697	-8.224	10,25	-7.776	3.468
10,26	3.332	7.440	10,27	7.073	-3.142	10,28	-3.046	-6.811	10,29	-6.499	2.881
10,30	2.816	6.294	10,31	6.021	-2.667						
11,11	78.232	35.854	11,12	39.471	-36.621	11,13	-20.894	-30.685	11,14	-24.480	14.364
11,15	10.965	20.340	11,16	17.500	-8.963	11,17	-7.589	-15.375	11,18	-13.793	6.654
11,19	5.905	12.494	11,20	11.488	-5.372	11,21	-4.892	-10.602	11,22	-9.907	4.553
11,23	4.212	9.256	11,24	8.751	-3.983	11,25	-3.721	-8.246	11,26	-7.864	3.561
11,27	3.349	7.456	11,28	7.162	-3.237	11,29	-3.055	-6.819	11,30	-6.591	2.979
12,12	78.156	35.783	12,13	39.362	-36.506	12,14	-20.831	-30.626	12,15	-24.379	14.257
12,16	10.912	20.290	12,17	17.405	-8.863	12,18	-7.545	-15.334	12,19	-13.702	6.558
12,20	5.869	12.460	12,21	11.400	-5.279	12,22	-4.863	-10.575	12,23	-9.821	4.462
12,24	4.189	9.235	12,25	8.665	-3.893	12,26	-3.706	-8.231	12,27	-7.778	3.471
12,28	3.341	7.448	12,29	7.074	-3.144						
13,13	78.219	35.842	13,14	39.462	-36.611	13,15	-20.883	-30.674	13,16	-24.473	14.356
13,17	10.955	20.330	13,18	17.494	-8.957	13,19	-7.580	-15.367	13,20	-13.788	6.649
13,21	5.896	12.487	13,22	11.484	-5.368	13,23	-4.884	-10.595	13,24	-9.905	4.550
13,25	4.204	9.249	13,26	8.749	-3.980	13,27	-3.713	-8.238	13,28	-7.863	3.560

I, K	RE	IM	I, K	RE	IM	I, K	RE	IM	I, K	RE	IM
14,14	78.168	35.794	14,15	39.369	-36.513	14,16	-20.841	-30.635	14,17	-24.385	14.263
14,18	10.921	20.298	14,19	17.409	-8.868	14,20	-7.553	-15.341	14,21	-13.705	6.562
14,22	5.876	12.467	14,23	11.403	-5.282	14,24	-4.870	-10.582	14,25	-9.823	4.465
14,26	4.197	9.242	14,27	8.666	-3.894						
15,15	78.209	35.833	15,16	39.457	-36.605	15,17	-20.874	-30.666	15,18	-24.469	14.351
15,19	10.947	20.323	15,20	17.491	-8.953	15,21	-7.573	-15.360	15,22	-13.786	6.646
15,23	5.889	12.480	15,24	11.483	-5.366	15,25	-4.877	-10.588	15,26	-9.904	4.549
16,16	78.176	35.802	16,17	39.373	-36.518	16,18	-20.849	-30.642	16,19	-24.388	14.267
16,20	10.928	20.305	16,21	17.412	-8.870	16,22	-7.560	-15.348	16,23	-13.707	6.564
16,24	5.883	12.474	16,25	11.403	-5.283						
17,17	78.201	35.826	17,18	39.454	-36.602	17,19	-20.867	-30.660	17,20	-24.467	14.349
17,21	10.941	20.317	17,22	17.490	-8.951	17,23	-7.566	-15.354	17,24	-13.785	6.645
18,18	78.183	35.808	18,19	39.375	-36.520	18,20	-20.855	-30.648	18,21	-24.389	14.268
18,22	10.935	20.311	18,23	17.412	-8.871						
19,19	78.195	35.820	19,20	39.452	-36.600	19,21	-20.861	-30.654	19,22	-24.466	14.348
20,20	78.189	35.814	20,21	39.376	-36.521						

IEN

SELF AND MUTUAL IMPEDANCES

H/LMDA=0.2500 B/LMDA= 0.250 OMEGA= 8.53

45 ELEMENT ARRAY

I, K	ZIK IN OHMS		I, K	ZIK IN OHMS		I, K	ZIK IN OHMS		I, K	ZIK IN OHMS	
	RE	IM		RE	IM		RE	IM		RE	IM
1, 1	80.401	35.959	1, 2	39.746	-37.975	1, 3	-21.902	-30.989	1, 4	-24.718	15.121
1, 5	11.625	20.584	1, 6	17.683	-9.492	1, 7	-8.090	-15.575	1, 8	-13.938	7.063
1, 9	6.315	12.665	1,10	11.606	-5.706	1,11	-5.244	-10.752	1,12	-10.006	4.836
1,13	4.523	9.392	1,14	8.833	-4.227	1,15	-4.003	-8.371	1,16	-7.934	3.776
1,17	3.608	7.573	1,18	7.220	-3.426	1,19	-3.298	-6.932	1,20	-6.638	3.147
1,21	3.047	6.403	1,22	6.154	-2.918	1,23	-2.839	-5.959	1,24	-5.744	2.726
1,25	2.664	5.581	1,26	5.391	-2.563	1,27	-2.515	-5.255	1,28	-5.085	2.422
1,29	2.385	4.970	1,30	4.816	-2.299	1,31	-2.272	-4.719	1,32	-4.577	2.190
1,33	2.173	4.497	1,34	4.363	-2.093	1,35	-2.085	-4.298	1,36	-4.170	2.005
1,37	2.008	4.119	1,38	3.995	-1.923	1,39	-1.940	-3.959	1,40	-3.833	1.847
1,41	1.882	3.816	1,42	3.682	-1.770	1,43	-1.843	-3.690	1,44	-3.537	1.674
1,45	1.867	3.572									
2, 2	77.256	35.462	2, 3	39.109	-35.899	2, 4	-20.329	-30.377	2, 5	-24.197	13.880
2, 6	10.556	20.095	2, 7	17.267	-8.589	2, 8	-7.264	-15.172	2, 9	-13.593	6.344
2,10	5.632	12.321	2,11	11.312	-5.104	2,12	-4.656	-10.450	2,13	-9.749	4.315
2,14	4.003	9.122	2,15	8.605	-3.767	2,16	-3.534	-8.125	2,17	-7.729	3.362
2,18	3.180	7.347	2,19	7.035	-3.051	2,20	-2.902	-6.721	2,21	-6.470	2.802
2,22	2.677	6.206	2,23	6.000	-2.600	2,24	-2.491	-5.773	2,25	-5.602	2.431
2,26	2.335	5.404	2,27	5.261	-2.288	2,28	-2.200	-5.085	2,29	-4.965	2.165
2,30	2.084	4.806	2,31	4.706	-2.058	2,32	-1.981	-4.559	2,33	-4.476	1.965
2,34	1.890	4.340	2,35	4.272	-1.883	2,36	-1.808	-4.142	2,37	-4.090	1.811
2,38	1.732	3.963	2,39	3.927	-1.748	2,40	-1.661	-3.799	2,41	-3.781	1.696
2,42	1.590	3.645	2,43	3.653	-1.660	2,44	-1.500	-3.497			
3, 3	78.681	36.093	3, 4	39.640	-36.932	3, 5	-21.179	-30.862	3, 6	-24.601	14.569
3, 7	11.178	20.481	3, 8	17.591	-9.114	3, 9	-7.763	-15.494	3,10	-13.864	6.772
3,11	6.055	12.599	3,12	11.545	-5.468	3,13	-5.025	-10.697	3,14	-9.953	4.632
3,15	4.334	9.344	3,16	8.787	-4.049	3,17	-3.835	-8.329	3,18	-7.893	3.616
3,19	3.458	7.536	3,20	7.183	-3.282	3,21	-3.161	-6.898	3,22	-6.604	3.014
3,23	2.921	6.373	3,24	6.123	-2.795	3,25	-2.722	-5.932	3,26	-5.715	2.612
3,27	2.555	5.556	3,28	5.364	-2.455	3,29	-2.412	-5.232	3,30	-5.059	2.320
3,31	2.289	4.949	3,32	4.791	-2.202	3,33	-2.182	-4.700	3,34	-4.553	2.097
3,35	2.087	4.479	3,36	4.340	-2.003	3,37	-2.005	-4.282	3,38	-4.147	1.918
3,39	1.932	4.106	3,40	3.971	-1.837	3,41	-1.872	-3.949	3,42	-3.807	1.757
3,43	1.830	3.812									
4, 4	77.889	35.609	4, 5	39.246	-36.324	4, 6	-20.644	-30.492	4, 7	-24.296	14.132
4, 8	10.765	20.180	4, 9	17.343	-8.770	4,10	-7.421	-15.239	4,11	-13.654	6.486
4,12	5.760	12.375	4,13	11.363	-5.222	4,14	-4.763	-10.496	4,15	-9.793	4.417
4,16	4.097	9.161	4,17	8.644	-3.857	4,18	-3.617	-8.160	4,19	-7.764	3.443
4,20	3.255	7.378	4,21	7.066	-3.124	4,22	-2.970	-6.749	4,23	-6.498	2.870
4,24	2.740	6.231	4,25	6.026	-2.663	4,26	-2.549	-5.797	4,27	-5.627	2.490
4,28	2.389	5.425	4,29	5.285	-2.343	4,30	-2.251	-5.105	4,31	-4.988	2.218
4,32	2.131	4.824	4,33	4.727	-2.109	4,34	-2.026	-4.576	4,35	-4.497	2.014
4,36	1.931	4.355	4,37	4.293	-1.931	4,38	-1.846	-4.155	4,39	-4.112	1.859
4,40	1.766	3.974	4,41	3.950	-1.799	4,42	-1.686	-3.805			

5, 5	78.399	35.974	5, 6	39.541	-36.725	5, 7	-21.018	-30.779	5, 8	-24.530	14.436
5, 9	11.066	20.420	5,10	17.536	-9.015	5,11	-7.676	-15.446	5,12	-13.819	6.693
5,13	5.984	12.559	5,14	11.507	-5.401	5,15	-4.965	-10.663	5,16	-9.920	4.575
5,17	4.281	9.315	5,18	8.758	-3.998	5,19	-3.789	-8.303	5,20	-7.866	3.570
5,21	3.415	7.512	5,22	7.159	-3.239	5,23	-3.122	-6.877	5,24	-6.582	2.975
5,25	2.885	6.353	5,26	6.102	-2.759	5,27	-2.689	-5.914	5,28	-5.695	2.577
5,29	2.525	5.540	5,30	5.346	-2.423	5,31	-2.384	-5.217	5,32	-5.041	2.289
5,33	2.263	4.935	5,34	4.774	-2.172	5,35	-2.157	-4.688	5,36	-4.536	2.068
5,37	2.066	4.468	5,38	4.322	-1.973	5,39	-1.986	-4.274	5,40	-4.128	1.886
5,41	1.919	4.101									
6, 6	78.037	35.690	6, 7	39.314	-36.439	6, 8	-20.737	-30.549	6, 9	-24.346	14.212
6,10	10.833	20.223	6,11	17.382	-8.831	6,12	-7.474	-15.273	6,13	-13.687	6.536
6,14	5.804	12.404	6,15	11.391	-5.264	6,16	-4.801	-10.521	6,17	-9.817	4.453
6,18	4.130	9.183	6,19	8.666	-3.889	6,20	-3.647	-8.179	6,21	-7.784	3.473
6,22	3.281	7.395	6,23	7.085	-3.151	6,24	-2.994	-6.765	6,25	-6.515	2.895
6,26	2.762	6.245	6,27	6.042	-2.687	6,28	-2.570	-5.809	6,29	-5.642	2.513
6,30	2.407	5.437	6,31	5.299	-2.365	6,32	-2.268	-5.115	6,33	-5.002	2.239
6,34	2.147	4.833	6,35	4.741	-2.130	6,36	-2.039	-4.584	6,37	-4.512	2.035
6,38	1.943	4.361	6,39	4.309	-1.954	6,40	-1.853	-4.159			
7, 7	78.311	35.918	7, 8	39.493	-36.654	7, 9	-20.959	-30.739	7,10	-24.494	14.384
7,11	11.022	20.389	7,12	17.507	-8.975	7,13	-7.641	-15.421	7,14	-13.795	6.660
7,15	5.954	12.538	7,16	11.486	-5.373	7,17	-4.939	-10.645	7,18	-9.901	4.549
7,19	4.259	9.299	7,20	8.741	-3.975	7,21	-3.769	-8.288	7,22	-7.851	3.549
7,23	3.398	7.500	7,24	7.144	-3.220	7,25	-3.106	-6.865	7,26	-6.569	2.957
7,27	2.871	6.343	7,28	6.089	-2.742	7,29	-2.676	-5.905	7,30	-5.683	2.561
7,31	2.512	5.532	7,32	5.334	-2.407	7,33	-2.373	-5.210	7,34	-5.030	2.273
7,35	2.253	4.929	7,36	4.762	-2.156	7,37	-2.150	-4.683	7,38	-4.523	2.051
7,39	2.060	4.465									
8, 8	78.093	35.729	8, 9	39.349	-36.487	8,10	-20.777	-30.578	8,11	-24.373	14.247
8,12	10.863	20.246	8,13	17.404	-8.859	8,14	-7.499	-15.292	8,15	-13.706	6.560
8,16	5.824	12.420	8,17	11.407	-5.285	8,18	-4.819	-10.534	8,19	-9.832	4.472
8,20	4.145	9.195	8,21	8.679	-3.906	8,22	-3.661	-8.190	8,23	-7.796	3.488
8,24	3.294	7.405	8,25	7.096	-3.166	8,26	-3.005	-6.773	8,27	-6.526	2.909
8,28	2.772	6.253	8,29	6.053	-2.700	8,30	-2.579	-5.816	8,31	-5.652	2.525
8,32	2.415	5.443	8,33	5.309	-2.378	8,34	-2.275	-5.120	8,35	-5.012	2.252
8,36	2.152	4.837	8,37	4.752	-2.143	8,38	-2.043	-4.586			
9, 9	78.273	35.890	9,10	39.467	-36.620	9,11	-20.931	-30.718	9,12	-24.473	14.358
9,13	11.000	20.371	9,14	17.490	-8.954	9,15	-7.623	-15.406	9,16	-13.780	6.641
9,17	5.939	12.526	9,18	11.473	-5.357	9,19	-4.926	-10.634	9,20	-9.890	4.535
9,21	4.247	9.289	9,22	8.730	-3.962	9,23	-3.758	-8.280	9,24	-7.841	3.537
9,25	3.388	7.493	9,26	7.135	-3.209	9,27	-3.098	-6.859	9,28	-6.560	2.946
9,29	2.863	6.338	9,30	6.080	-2.731	9,31	-2.670	-5.900	9,32	-5.674	2.551
9,33	2.507	5.528	9,34	5.325	-2.396	9,35	-2.369	-5.207	9,36	-5.020	2.262
9,37	2.251	4.927									
10,10	78.120	35.751	10,11	39.369	-36.512	10,12	-20.797	-30.595	10,13	-24.390	14.267
10,14	10.880	20.260	10,15	17.418	-8.876	10,16	-7.513	-15.303	10,17	-13.718	6.574
10,18	5.836	12.429	10,19	11.418	-5.298	10,20	-4.829	-10.543	10,21	-9.842	4.483
10,22	4.154	9.202	10,23	8.688	-3.917	10,24	-3.668	-8.196	10,25	-7.805	3.498
10,26	3.301	7.410	10,27	7.104	-3.175	10,28	-3.011	-6.778	10,29	-6.534	2.918
10,30	2.777	6.257	10,31	6.060	-2.709	10,32	-2.583	-5.819	10,33	-5.660	2.534
10,34	2.419	5.445	10,35	5.317	-2.387	10,36	-2.277	-5.121			
11,11	78.253	35.873	11,12	39.451	-36.601	11,13	-20.915	-30.705	11,14	-24.460	14.343
11,15	10.987	20.361	11,16	17.479	-8.941	11,17	-7.612	-15.397	11,18	-13.770	6.630
11,19	[illegible]	[illegible]	11,20	11.464	-5.346	11,21	-4.919	-10.627	11,22	-9.881	4.526
11,23	4.241	9.284	11,24	[illegible]	[illegible]	11,25	-3.753	-8.275	11,26	-7.833	3.529
11,27	3.383	7.488	11,28	7.128	-3.201	11,29	[illegible]	[illegible]	11,30	-6.553	2.938
11,31	2.860	6.335	11,32	6.073	-2.723	11,33	-2.667	-5.898	11,34	[illegible]	[illegible]
11,35	2.506	5.527									
12,12	78.135	35.764	12,13	39.382	-36.527	12,14	-20.809	-30.605	12,15	-24.401	14.280
12,16	10.890	20.268	12,17	17.428	-8.887	12,18	-7.521	-15.311	12,19	-13.726	6.583
12,20	5.843	12.435	12,21	11.426	-5.306	12,22	-4.835	-10.548	12,23	-9.849	4.491
12,24	4.159	9.207	12,25	8.695	-3.924	12,26	-3.673	-8.200	12,27	-7.811	3.505
12,28	3.304	7.414	12,29	7.111	-3.182	12,30	-3.014	-6.781	12,31	-6.541	2.925
12,32	2.779	6.259	12,33	6.067	-2.716	12,34	-2.584	-5.820			
13,13	78.241	35.863	13,14	39.441	-36.589	13,15	-20.906	-30.696	13,16	-24.450	14.332
13,17	10.980	20.354	13,18	17.470	-8.932	13,19	-7.606	-15.392	13,20	-13.763	6.622
13,21	5.924	12.513	13,22	11.457	-5.339	13,23	-4.914	-10.623	13,24	-9.875	4.519
13,25	4.237	9.280	13,26	8.716	-3.946	13,27	-3.749	-8.273	13,28	-7.827	3.522
13,29	3.381	7.486	13,30	7.122	-3.194	13,31	-3.092	-6.854	13,32	-6.546	2.932
13,33	2.859	6.334									
14,14	78.145	35.772	14,15	39.391	-36.537	14,16	-20.817	-30.612	14,17	-24.408	14.288
14,18	10.896	20.274	14,19	17.435	-8.894	14,20	-7.526	-15.315	14,21	-13.733	6.590
14,22	5.847	12.439	14,23	11.432	-5.313	14,24	-4.838	-10.551	14,25	-9.855	4.498
14,26	4.162	9.209	14,27	8.701	-3.930	14,28	-3.675	-8.202	14,29	-7.817	3.511
14,30	3.306	7.415	14,31	7.116	-3.188	14,32	-3.015	-6.781			
15,15	78.233	35.856	15,16	39.433	-36.580	15,17	-20.900	-30.691	15,18	-24.444	14.325
15,19	10.975	20.349	15,20	17.464	-8.925	15,21	-7.602	-15.388	15,22	-13.757	6.616
15,23	5.921	12.510	15,24	11.451	-5.333	15,25	-4.911	-10.621	15,26	-9.869	4.513
15,27	4.235	9.278	15,28	8.711	-3.941	15,29	-3.748	-8.271	15,30	-7.822	3.517
15,31	3.380	7.485									
16,16	78.151	35.777	16,17	39.398	-36.544	16,18	-20.822	-30.616	16,19	-24.415	14.295
16,20	10.900	20.277	16,21	17.440	-8.900	16,22	-7.529	-15.318	16,23	-13.738	6.596
16,24	5.850	12.442	16,25	11.437	-5.318	16,26	-4.840	-10.553	16,27	-9.860	4.503
16,28	4.164	9.210	16,29	8.706	-3.936	16,30	-3.676	-8.202			
17,17	78.229	35.852	17,18	39.427	-36.574	17,19	-20.896	-30.688	17,20	-24.438	14.319
17,21	10.972	20.347	17,22	17.459	-8.919	17,23	-7.600	-15.386	17,24	-13.752	6.610
17,25	5.919	12.509	17,26	11.446	-5.328	17,27	-4.910	-10.620	17,28	-9.865	4.508
17,29	4.234	9.278									
18,18	78.154	35.781	18,19	39.404	-36.550	18,20	-20.824	-30.619	18,21	-24.420	14.300
18,22	10.902	20.279	18,23	17.445	-8.905	18,24	-7.531	-15.319	18,25	-13.743	6.601
18,26	5.851	12.443	18,27	11.442	-5.323	18,28	-4.841	-10.553			
19,19	78.226	35.849	19,20	39.422	-36.569	19,21	-20.894	-30.686	19,22	-24.433	14.314
19,23	10.970	20.345	19,24	17.454	-8.915	19,25	-7.598	-15.385	19,26	-13.747	6.606
19,27	5.919	12.508									
20,20	78.156	35.783	20,21	39.408	-36.555	20,22	-20.826	-30.620	20,23	-24.424	14.305
20,24	10.903	20.280	20,25	17.450	-8.910	20,26	-7.531	-15.320			
21,21	78.224	35.848	21,22	39.417	-36.564	21,23	-20.893	-30.685	21,24	-24.429	14.309
21,25	10.970	20.345									
22,22	78.157	35.784	22,23	39.413	-36.560	22,24	-20.827	-30.621			
23,23	78.224	35.847									

IEN

SELF AND MUTUAL IMPEDANCES

H/LMDA=0.2500 B/LMDA= 0.250 OMEGA= 8.53

50 ELEMENT ARRAY

I, K	ZIK IN OHMS		I, K	ZIK IN OHMS		I, K	ZIK IN OHMS		I, K	ZIK IN OHMS	
	RE	IM		RE	IM		RE	IM		RE	IM
1, 1	80.388	35.948	1, 2	39.757	-37.987	1, 3	-21.889	-30.978	1, 4	-24.729	15.133
1, 5	11.612	20.572	1, 6	17.694	-9.505	1, 7	-8.077	-15.563	1, 8	-13.950	7.076
1, 9	6.302	12.653	1,10	11.619	-5.720	1,11	-5.229	-10.740	1,12	-10.018	4.850
1,13	4.508	9.379	1,14	8.846	-4.242	1,15	-3.987	-8.357	1,16	-7.948	3.791
1,17	3.592	7.559	1,18	7.234	-3.443	1,19	-3.281	-6.916	1,20	-6.653	3.164
1,21	3.029	6.387	1,22	6.170	-2.936	1,23	-2.820	-5.942	1,24	-5.761	2.746
1,25	2.644	5.563	1,26	5.410	-2.584	1,27	-2.493	-5.236	1,28	-5.105	2.445
1,29	2.362	4.949	1,30	4.838	-2.324	1,31	-2.247	-4.696	1,32	-4.601	2.217
1,33	2.145	4.471	1,34	4.390	-2.123	1,35	-2.053	-4.269	1,36	-4.200	2.039
1,37	1.971	4.087	1,38	4.030	-1.963	1,39	-1.896	-3.921	1,40	-3.875	1.896
1,41	1.827	3.770	1,42	3.734	-1.835	1,43	-1.762	-3.630	1,44	-3.607	1.782
1,45	1.700	3.499	1,46	3.491	-1.737	1,47	-1.636	-3.376	1,48	-3.390	1.707
1,49	1.554	3.256	1,50	3.294	-1.735						
2, 2	77.267	35.472	2, 3	39.098	-35.887	2, 4	-20.341	-30.388	2, 5	-24.186	13.868
2, 6	10.568	20.107	2, 7	17.255	-8.577	2, 8	-7.276	-15.184	2, 9	-13.581	6.331
2,10	5.645	12.333	2,11	11.299	-5.091	2,12	-4.669	-10.463	2,13	-9.736	4.301
2,14	4.017	9.135	2,15	8.592	-3.752	2,16	-3.549	-8.139	2,17	-7.715	3.347
2,18	3.195	7.362	2,19	7.020	-3.035	2,20	-2.918	-6.737	2,21	-6.453	2.786
2,22	2.694	6.222	2,23	5.982	-2.582	2,24	-2.510	-5.791	2,25	-5.584	2.412
2,26	2.354	5.423	2,27	5.241	-2.267	2,28	-2.221	-5.105	2,29	-4.944	2.143
2,30	2.107	4.828	2,31	4.682	-2.034	2,32	-2.006	-4.584	2,33	-4.451	1.939
2,34	1.918	4.367	2,35	4.244	-1.853	2,36	-1.840	-4.173	2,37	-4.057	1.777
2,38	1.770	3.998	2,39	3.888	-1.707	2,40	-1.707	-3.841	2,41	-3.734	1.643
2,42	1.652	3.698	2,43	3.592	-1.584	2,44	-1.603	-3.569	2,45	-3.460	1.526
2,46	1.562	3.452	2,47	3.335	-1.467	2,48	-1.535	-3.350	2,49	-3.213	1.390
3, 3	78.668	36.082	3, 4	39.652	-36.944	3, 5	-21.166	-30.850	3, 6	-24.613	14.582
3, 7	11.165	20.469	3, 8	17.603	-9.128	3, 9	-7.749	-15.482	3,10	-13.877	6.786
3,11	6.040	12.586	3,12	11.558	-5.482	3,13	-5.010	-10.683	3,14	-9.967	4.647
3,15	4.318	9.330	3,16	8.801	-4.065	3,17	-3.819	-8.314	3,18	-7.908	3.633
3,19	3.441	7.520	3,20	7.199	-3.299	3,21	-3.143	-6.881	3,22	-6.621	3.033
3,23	2.902	6.355	3,24	6.141	-2.815	3,25	-2.702	-5.913	3,26	-5.734	2.633
3,27	2.533	5.536	3,28	5.385	-2.478	3,29	-2.389	-5.210	3,30	-5.082	2.345
3,31	2.263	4.925	3,32	4.816	-2.229	3,33	-2.153	-4.673	3,34	-4.581	2.128
3,35	2.055	4.449	3,36	4.371	-2.038	3,37	-1.968	-4.248	3,38	-4.183	1.958
3,39	1.888	4.067	3,40	4.014	-1.886	3,41	-1.816	-3.901	3,42	-3.861	1.823
3,43	1.748	3.749	3,44	3.722	-1.767	3,45	-1.684	-3.608	3,46	-3.598	1.720
3,47	1.618	3.475	3,48	3.488	-1.689						
4, 4	77.901	35.621	4, 5	39.235	-36.311	4, 6	-20.657	-30.504	4, 7	-24.284	14.119
4, 8	10.778	20.193	4, 9	17.330	-8.757	4,10	-7.435	-15.252	4,11	-13.641	6.472
4,12	5.774	12.388	4,13	11.349	-5.208	4,14	-4.778	-10.510	4,15	-9.778	4.402
4,16	4.112	9.176	4,17	8.629	-3.841	4,18	-3.634	-8.175	4,19	-7.748	3.426
4,20	3.272	7.394	4,21	7.050	-3.106	4,22	-2.988	-6.767	4,23	-6.481	2.851
4,24	2.759	6.250	4,25	6.007	-2.643	4,26	-2.570	-5.816	4,27	-5.607	2.468
4,28	2.411	5.447	4,29	5.263	-2.320	4,30	-2.275	-5.128	4,31	-4.964	2.193
4,32	2.157	4.849	4,33	4.701	-2.081	4,34	-2.055	-4.604	4,35	-4.468	1.983
4,36	1.965	4.386	4,37	4.260	-1.895	4,38	-1.885	-4.192	4,39	-4.072	1.816
4,40	1.814	4.017	4,41	3.902	-1.744	4,42	-1.751	-3.859	4,43	-3.746	1.677
4,44	1.695	3.717	4,45	3.602	-1.614	4,46	-1.649	-3.590	4,47	-3.466	1.549
5, 5	78.385	35.962	5, 6	39.553	-36.738	5, 7	-21.004	-30.767	5, 8	-24.542	14.449
5, 9	11.052	20.407	5,10	17.549	-9.030	5,11	-7.662	-15.432	5,12	-13.833	6.708
5,13	5.968	12.545	5,14	11.521	-5.417	5,15	-4.949	-10.648	5,16	-9.935	4.591
5,17	4.264	9.299	5,18	8.773	-4.015	5,19	-3.771	-8.286	5,20	-7.883	3.588
5,21	3.397	7.495	5,22	7.176	-3.258	5,23	-3.103	-6.859	5,24	-6.601	2.995
5,25	2.864	6.334	5,26	6.122	-2.780	5,27	-2.667	-5.893	5,28	-5.717	2.600
5,29	2.500	5.517	5,30	5.369	-2.448	5,31	-2.358	-5.192	5,32	-5.067	2.317
5,33	2.234	4.908	5,34	4.802	-2.203	5,35	-2.125	-4.657	5,36	-4.568	2.103
5,37	2.028	4.434	5,38	4.360	-2.014	5,39	-1.941	-4.233	5,40	-4.173	1.936
5,41	1.862	4.051	5,42	4.005	-1.867	5,43	-1.789	-3.885	5,44	-3.853	1.806
5,45	1.719	3.732	5,46	3.718	-1.755						
6, 6	78.050	35.702	6, 7	39.302	-36.426	6, 8	-20.751	-30.562	6, 9	-24.333	14.198
6,10	10.847	20.237	6,11	17.368	-8.816	6,12	-7.489	-15.287	6,13	-13.673	6.520
6,14	5.819	12.418	6,15	11.376	-5.248	6,16	-4.817	-10.536	6,17	-9.802	4.437
6,18	4.146	9.199	6,19	8.650	-3.872	6,20	-3.664	-8.196	6,21	-7.767	3.454
6,22	3.300	7.413	6,23	7.066	-3.132	6,24	-3.014	-6.784	6,25	-6.496	2.875
6,26	2.783	6.266	6,27	6.021	-2.664	6,28	-2.593	-5.831	6,29	-5.619	2.489
6,30	2.432	5.461	6,31	5.274	-2.339	6,32	-2.295	-5.141	6,33	-4.974	2.210
6,34	2.177	4.862	6,35	4.711	-2.097	6,36	-2.074	-4.617	6,37	-4.477	1.998
6,38	1.983	4.399	6,39	4.268	-1.909	6,40	-1.903	-4.204	6,41	-4.079	1.828
6,42	1.833	4.030	6,43	3.907	-1.754	6,44	-1.771	-3.873	6,45	-3.749	1.684
7, 7	78.297	35.905	7, 8	39.506	-36.668	7, 9	-20.944	-30.726	7,10	-24.507	14.399
7,11	11.007	20.375	7,12	17.521	-8.990	7,13	-7.625	-15.406	7,14	-13.810	6.676
7,15	5.938	12.523	7,16	11.501	-5.389	7,17	-4.922	-10.628	7,18	-9.918	4.567
7,19	4.241	9.282	7,20	8.758	-3.993	7,21	-3.750	-8.271	7,22	-7.869	3.569
7,23	3.377	7.481	7,24	7.164	-3.241	7,25	-3.084	-6.845	7,26	-6.589	2.979
7,27	2.847	6.321	7,28	6.112	-2.765	7,29	-2.651	-5.881	7,30	-5.707	2.587
7,31	2.485	5.506	7,32	5.361	-2.435	7,33	-2.343	-5.181	7,34	-5.059	2.305
7,35	2.219	4.898	7,36	4.795	-2.192	7,37	-2.111	-4.647	7,38	-4.562	2.093
7,39	2.014	4.423	7,40	4.354	-2.006	7,41	-1.926	-4.222	7,42	-4.169	1.929
7,43	1.846	4.040	7,44	4.002	-1.862						
8, 8	78.107	35.743	8, 9	39.335	-36.472	8,10	-20.792	-30.592	8,11	-24.359	14.232
8,12	10.879	20.261	8,13	17.389	-8.844	8,14	-7.515	-15.307	8,15	-13.690	6.543
8,16	5.841	12.436	8,17	11.391	-5.268	8,18	-4.836	-10.551	8,19	-9.815	4.454
8,20	4.164	9.212	8,21	8.661	-3.887	8,22	-3.680	-8.208	8,23	-7.777	3.468
8,24	3.315	7.425	8,25	7.076	-3.144	8,26	-3.028	-6.795	8,27	-6.504	2.886
8,28	2.796	6.276	8,29	6.029	-2.675	8,30	-2.605	-5.841	8,31	-5.626	2.498
8,32	2.444	5.470	8,33	5.281	-2.348	8,34	-2.307	-5.150	8,35	-4.980	2.218
8,36	2.188	4.871	8,37	4.716	-2.105	8,38	-2.085	-4.625	8,39	-4.481	2.004
8,40	1.995	4.408	8,41	4.271	-1.914	8,42	-1.915	-4.214	8,43	-4.081	1.831
9, 9	78.258	35.876	9,10	39.482	-36.635	9,11	-20.915	-30.703	9,12	-24.488	14.374
9,13	10.983	20.356	9,14	17.505	-8.970	9,15	-7.606	-15.390	9,16	-13.796	6.659
9,17	5.921	12.509	9,18	11.490	-5.375	9,19	-4.907	-10.616	9,20	-9.908	4.554
9,21	4.227	9.271	9,22	8.749	-3.982	9,23	-3.737	-8.261	9,24	-7.861	3.559
9,25	3.366	7.472	9,26	7.157	-3.232	9,27	-3.074	-6.836	9,28	-6.583	2.971
9,29	2.837	6.313	9,30	6.106	-2.758	9,31	-2.641	-5.874	9,32	-5.702	2.580
9,33	2.476	5.498	9,34	5.356	-2.429	9,35	-2.334	-5.174	9,36	-5.055	2.300
9,37	2.210	4.890	9,38	4.792	-2.188	9,39	-2.101	-4.639	9,40	-4.559	2.090
9,41	2.004	4.415	9,42	4.353	-2.004						

10,10	78.135	35.765	10,11	39.354	-36.496	10,12	-20.813	-30.610	10,13	-24.374	14.251
10,14	10.896	20.276	10,15	17.402	-8.859	10,16	-7.530	-15.320	10,17	-13.701	6.556
10,18	5.854	12.447	10,19	11.400	-5.279	10,20	-4.848	-10.561	10,21	-9.823	4.464
10,22	4.174	9.221	10,23	8.668	-3.896	10,24	-3.690	-8.217	10,25	-7.783	3.476
10,26	3.324	7.432	10,27	7.081	-3.151	10,28	-3.036	-6.802	10,29	-6.509	2.892
10,30	2.804	6.283	10,31	6.033	-2.680	10,32	-2.613	-5.847	10,33	-5.630	2.503
10,34	2.452	5.477	10,35	5.284	-2.352	10,36	-2.314	-5.157	10,37	-4.983	2.222
10,38	2.196	4.878	10,39	4.718	-2.107	10,40	-2.093	-4.633	10,41	-4.482	2.006
11,11	78.236	35.858	11,12	39.467	-36.617	11,13	-20.898	-30.689	11,14	-24.476	14.360
11,15	10.970	20.344	11,16	17.495	-8.959	11,17	-7.594	-15.380	11,18	-13.788	6.649
11,19	5.910	12.500	11,20	11.482	-5.366	11,21	-4.898	-10.608	11,22	-9.901	4.547
11,23	4.219	9.263	11,24	8.743	-3.975	11,25	-3.729	-8.253	11,26	-7.856	3.553
11,27	3.359	7.465	11,28	7.152	-3.227	11,29	-3.067	-6.830	11,30	-6.579	2.966
11,31	2.830	6.307	11,32	6.102	-2.754	11,33	-2.635	-5.868	11,34	-5.699	2.576
11,35	2.469	5.492	11,36	5.353	-2.426	11,37	-2.327	-5.168	11,38	-5.054	2.298
11,39	2.203	4.884	11,40	4.791	-2.186						
12,12	78.152	35.779	12,13	39.366	-36.510	12,14	-20.827	-30.621	12,15	-24.384	14.262
12,16	10.908	20.285	12,17	17.410	-8.868	12,18	-7.540	-15.329	12,19	-13.708	6.564
12,20	5.863	12.454	12,21	11.406	-5.286	12,22	-4.856	-10.568	12,23	-9.828	4.470
12,24	4.182	9.228	12,25	8.673	-3.901	12,26	-3.697	-8.223	12,27	-7.787	3.480
12,28	3.330	7.438	12,29	7.085	-3.155	12,30	-3.042	-6.808	12,31	-6.512	2.896
12,32	2.810	6.288	12,33	6.036	-2.683	12,34	-2.619	-5.853	12,35	-5.632	2.505
12,36	2.458	5.482	12,37	5.286	-2.354	12,38	-2.321	-5.162	12,39	-4.984	2.223
13,13	78.223	35.846	13,14	39.457	-36.607	13,15	-20.887	-30.679	13,16	-24.468	14.351
13,17	10.960	20.336	13,18	17.489	-8.951	13,19	-7.586	-15.372	13,20	-13.782	6.642
13,21	5.903	12.493	13,22	11.478	-5.361	13,23	-4.891	-10.602	13,24	-9.897	4.542
13,25	4.213	9.257	13,26	8.740	-3.971	13,27	-3.723	-8.248	13,28	-7.853	3.549
13,29	3.353	7.460	13,30	7.149	-3.223	13,31	-3.061	-6.825	13,32	-6.577	2.964
13,33	2.825	6.302	13,34	6.100	-2.752	13,35	-2.629	-5.862	13,36	-5.698	2.575
13,37	2.463	5.487	13,38	5.353	-2.425						
14,14	78.163	35.789	14,15	39.374	-36.518	14,16	-20.836	-30.630	14,17	-24.390	14.269
14,18	10.915	20.293	14,19	17.415	-8.874	14,20	-7.547	-15.335	14,21	-13.712	6.569
14,22	5.869	12.460	14,23	11.410	-5.290	14,24	-4.862	-10.574	14,25	-9.832	4.473
14,26	4.187	9.233	14,27	8.676	-3.904	14,28	-3.702	-8.228	14,29	-7.790	3.483
14,30	3.335	7.443	14,31	7.087	-3.158	14,32	-3.048	-6.812	14,33	-6.514	2.898
14,34	2.815	6.293	14,35	6.037	-2.685	14,36	-2.624	-5.858	14,37	-5.633	2.506
15,15	78.214	35.838	15,16	39.451	-36.600	15,17	-20.880	-30.672	15,18	-24.463	14.345
15,19	10.954	20.329	15,20	17.484	-8.946	15,21	-7.580	-15.367	15,22	-13.778	6.638
15,23	5.898	12.488	15,24	11.474	-5.357	15,25	-4.886	-10.597	15,26	-9.894	4.539
15,27	4.208	9.253	15,28	8.737	-3.969	15,29	-3.719	-8.243	15,30	-7.851	3.547
15,31	3.348	7.455	15,32	7.148	-3.222	15,33	-3.056	-6.821	15,34	-6.576	2.963
15,35	2.820	6.297	15,36	6.100	-2.751						
16,16	78.170	35.796	16,17	39.379	-36.524	16,18	-20.842	-30.636	16,19	-24.394	14.274
16,20	10.921	20.298	16,21	17.419	-8.878	16,22	-7.552	-15.340	16,23	-13.715	6.572
16,24	5.874	12.465	16,25	11.413	-5.293	16,26	-4.867	-10.578	16,27	-9.834	4.476
16,28	4.192	9.237	16,29	8.678	-3.906	16,30	-3.707	-8.232	16,31	-7.791	3.485
16,32	3.340	7.447	16,33	7.088	-3.159	16,34	-3.052	-6.817	16,35	-6.515	2.898
17,17	78.208	35.832	17,18	39.447	-36.595	17,19	-20.874	-30.667	17,20	-24.459	14.341
17,21	10.948	20.325	17,22	17.482	-8.943	17,23	-7.575	-15.362	17,24	-13.776	6.636
17,25	5.893	12.483	17,26	11.472	-5.355	17,27	-4.882	-10.593	17,28	-9.892	4.537
17,29	4.20[illegible]	[illegible]	[illegible]	[illegible]	-3.967	17,31	-3.715	-8.239	17,32	-7.850	3.546
17,33	3.344	7.451	17,34	7.147	-3.221						
18,18	78.176	35.801	18,19	39.382	-36.528	18,20	-20.847	-30.641	18,21	-24.397	14.277
18,22	10.926	20.303	18,23	17.421	-8.880	18,24	-7.556	-15.344	18,25	-13.717	6.574
18,26	5.878	12.469	18,27	11.414	-5.294	18,28	-4.871	-10.582	18,29	-9.835	4.477
18,30	4.196	9.241	18,31	8.678	-3.907	18,32	-3.711	-8.236	18,33	-7.792	3.485
19,19	78.203	35.827	19,20	39.444	-36.592	19,21	-20.870	-30.662	19,22	-24.457	14.339
19,23	10.944	20.320	19,24	17.480	-8.941	19,25	-7.571	-15.358	19,26	-13.775	6.634
19,27	5.889	12.480	19,28	11.471	-5.354	19,29	-4.878	-10.589	19,30	-9.892	4.536
19,31	4.200	9.245	19,32	8.736	-3.967						
20,20	78.181	35.806	20,21	39.384	-36.530	20,22	-20.852	-30.645	20,23	-24.399	14.279
20,24	10.930	20.306	20,25	17.423	-8.882	20,26	-7.560	-15.348	20,27	-13.718	6.575
20,28	5.882	12.473	20,29	11.415	-5.295	20,30	-4.874	-10.586	20,31	-9.835	4.477
21,21	78.198	35.823	21,22	39.442	-36.590	21,23	-20.866	-30.658	21,24	-24.456	14.338
21,25	10.940	20.317	21,26	17.479	-8.940	21,27	-7.567	-15.355	21,28	-13.774	6.633
21,29	5.886	12.476	21,30	11.471	-5.353						
22,22	78.184	35.810	22,23	39.386	-36.532	22,24	-20.855	-30.648	22,25	-24.400	14.280
22,26	10.934	20.310	22,27	17.423	-8.883	22,28	-7.564	-15.351	22,29	-13.719	6.576
23,23	78.195	35.820	23,24	39.441	-36.589	23,25	-20.862	-30.655	23,26	-24.455	14.337
23,27	10.937	20.313	23,28	17.478	-8.940						
24,24	78.188	35.813	24,25	39.387	-36.532	24,26	-20.859	-30.652	24,27	-24.400	14.280
25,25	78.191	35.816	25,26	39.441	-36.589						

IBR

SELF AND MUTUAL IMPEDANCES

H/LMDA=C.25C0   B/LMDA= 0.500   OMEGA= 8.53

2 ELEMENT ARRAY

I, K	ZIK IN OHMS RE	IM	I, K	ZIK IN OHMS RE	IM	I, K	ZIK IN OHMS RE	IM	I, K	ZIK IN OHMS RE	IM
1, 1	83.967	36.707	1, 2	-19.978	-30.764						

IBR

SELF AND MUTUAL IMPEDANCES

H/LMDA=0.2500   B/LMDA= 0.500   OMEGA= 8.53

3 ELEMENT ARRAY

I, K	ZIK IN OHMS RE	IM	I, K	ZIK IN OHMS RE	IM	I, K	ZIK IN OHMS RE	IM	I, K	ZIK IN OHMS RE	IM
1, 1	84.342	37.060	1, 2	-20.703	-31.276	1, 3	9.394	19.618			
2, 2	85.303	37.397									

IBR

SELF AND MUTUAL IMPEDANCES

H/LMDA=0.2500   B/LMDA= 0.500   OMEGA= 8.53

4 ELEMENT ARRAY

I, K	ZIK IN OHMS RE	IM	I, K	ZIK IN OHMS RE	IM	I, K	ZIK IN OHMS RE	IM	I, K	ZIK IN OHMS RE	IM
1, 1	84.515	37.259	1, 2	-20.969	-31.547	1, 3	9.899	20.016	1, 4	-6.016	-14.342
2, 2	85.7C8	37.758	2, 3	-21.449	-31.787						

IBR

SELF AND MUTUAL IMPEDANCES

H/LMDA=0.2500   B/LMDA= 0.500   OMEGA= 8.53

5 ELEMENT ARRAY

I, K	ZIK IN OHMS RE	IM	I, K	ZIK IN OHMS RE	IM	I, K	ZIK IN OHMS RE	IM	I, K	ZIK IN OHMS RE	IM
1, 1	84.615	37.388	1, 2	-21.107	-31.710	1, 3	10.105	20.236	1, 4	-6.406	-14.667
1, 5	4.427	11.347									
2, 2	85.897	37.964	2, 3	-21.731	-32.064	2, 4	10.419	20.409			
3, 3	86.126	38.126									

IBR

SELF AND MUTUAL IMPEDANCES

H/LMDA=0.2500   B/LMDA= 0.500   OMEGA= 8.53

6 ELEMENT ARRAY

I, K	ZIK IN OHMS RE	IM	I, K	ZIK IN OHMS RE	IM	I, K	ZIK IN OHMS RE	IM	I, K	ZIK IN OHMS RE	IM
1, 1	84.680	37.477	1, 2	-21.191	-31.820	1, 3	10.219	20.375	1, 4	-6.575	-14.852
1, 5	4.745	11.622	1, 6	-3.513	-9.421						
2, 2	86.0C7	38.097	2, 3	-21.879	-32.232	2, 4	10.637	20.634	2, 5	-6.808	-14.986
3, 3	86.323	38.336	3, 4	-22.021	-32.343						

IBR

SELF AND MUTUAL IMPEDANCES

H/LMDA=C.25CC   B/LMDA= 0.500   CMEGA= 8.53

7 ELEMENT ARRAY

I, K	ZIK IN OHMS RE	IM	I, K	ZIK IN OHMS RE	IM	I, K	ZIK IN OHMS RE	IM	I, K	ZIK IN OHMS RE	IM
1, 1	84.726	37.543	1, 2	-21.249	-31.898	1, 3	10.292	20.470	1, 4	-6.672	-14.972
1, 5	4.889	11.783	1, 6	-3.784	-9.659	1, 7	2.923	8.076			
2, 2	86.079	38.190	2, 3	-21.970	-32.344	2, 4	10.758	20.775	2, 5	-6.986	-15.175
2, 6	5.075	11.891									
3, 3	86.438	38.472	3, 4	-22.174	-32.514	3, 5	10.861	20.859			
4, 4	86.524	38.549									

IBR

SELF AND MUTUAL IMPEDANCES

H/LMDA=0.2500 B/LMDA= 0.500 UMEGA= 8.53

8 ELEMENT ARRAY

I, K	ZIK IN OHMS		I, K	ZIK IN OHMS		I, K	ZIK IN OHMS		I, K	ZIK IN OHMS	
	RE	IM		RE	IM		RE	IM		RE	IM
1, 1	84.761	37.595	1, 2	-21.291	-31.958	1, 3	10.343	20.540	1, 4	-6.736	-15.057
1, 5	4.974	11.889	1, 6	-3.909	-9.801	1, 7	3.159	8.287	1, 8	-2.511	-7.082
2, 2	86.131	38.259	2, 3	-22.032	-32.425	2, 4	10.836	20.873	2, 5	-7.089	-15.298
2, 6	5.226	12.055	2, 7	-4.064	-9.892						
3, 3	86.514	38.567	3, 4	-22.268	-32.629	3, 5	10.985	21.003	3, 6	-7.170	-15.365
4, 4	86.642	38.687	4, 5	-22.329	-32.686						

IBR

SELF AND MUTUAL IMPEDANCES

H/LMDA=C.25CO B/LMDA= 0.500 UMEGA= 8.53

9 ELEMENT ARRAY

I, K	ZIK IN OHMS		I, K	ZIK IN OHMS		I, K	ZIK IN OHMS		I, K	ZIK IN OHMS	
	RE	IM		RE	IM		RE	IM		RE	IM
1, 1	84.788	37.635	1, 2	-21.324	-32.004	1, 3	10.382	20.594	1, 4	-6.782	-15.120
1, 5	5.031	11.965	1, 6	-3.984	-9.897	1, 7	3.270	8.414	1, 8	-2.720	-7.271
1, 9	2.2C7	6.317									
2, 2	86.170	38.312	2, 3	-22.078	-32.487	2, 4	10.891	20.945	2, 5	-7.158	-15.384
2, 6	5.316	12.163	2, 7	-4.196	-10.037	2, 8	3.403	8.492			
3, 3	86.568	38.637	3, 4	-22.334	-32.711	3, 5	11.066	21.103	3, 6	-7.276	-15.489
3, 7	5.382	12.219									
4, 4	86.721	38.784	4, 5	-22.427	-32.802	4, 6	11.113	21.148			
5, 5	86.763	38.826									

IBR

SELF AND MUTUAL IMPEDANCES

H/LMCA=0.2500 B/LMDA= 0.500 UMEGA= 8.53

[illegible] ELEMENT ARRAY

I, K	ZIK IN OHMS		I, K	ZIK IN OHMS		I, K	ZIK IN OHMS		I, K	ZIK IN OHMS	
	RE	IM		RE	IM		RE	IM		RE	IM
1, 1	84.810	37.669	1, 2	-21.350	-32.042	1, 3	10.411	20.637	1, 4	-6.817	-15.169
1, 5	5.073	12.023	1, 6	-4.036	-9.966	1, 7	3.338	8.501	1, 8	-2.820	-7.387
1, 9	2.395	6.488	1,10	-1.973	-5.709						
2, 2	86.200	38.355	2, 3	-22.113	-32.535	2, 4	10.932	21.000	2, 5	-7.207	-15.449
2, 6	5.377	12.241	2, 7	-4.276	-10.134	2, 8	3.520	8.622	2, 9	-2.937	-7.456
3, 3	86.609	38.692	3, 4	-22.382	-32.774	3, 5	11.123	21.176	3, 6	-7.347	-15.577
3, 7	5.474	12.329	3, 8	-4.332	-10.182						
4, 4	86.777	38.855	4, 5	-22.493	-32.886	4, 6	11.196	21.249	4, 7	-7.385	-15.614
5, 5	86.843	38.924	5, 6	-22.525	-32.919						

IBR

SELF AND MUTUAL IMPEDANCES

H/LMDA=0.2500 B/LMDA= 0.500 UMEGA= 8.53

12 ELEMENT ARRAY

I, K	ZIK IN OHMS		I, K	ZIK IN OHMS		I, K	ZIK IN OHMS		I, K	ZIK IN OHMS	
	RE	IM		RE	IM		RE	IM		RE	IM
1, 1	84.843	37.721	1, 2	-21.388	-32.100	1, 3	10.455	20.701	1, 4	-6.868	-15.241
1, 5	5.131	12.105	1, 6	-4.1C4	-10.061	1, 7	3.421	8.613	1, 8	-2.926	-7.525
1, 9	2.543	6.668	1,10	-2.228	-5.964	1,11	1.945	5.358	1,12	-1.636	-4.801
2, 2	86.245	38.419	2, 3	-22.165	-32.607	2, 4	1C.990	21.081	2, 5	-7.274	-15.540
2, 6	5.456	12.347	2, 7	-4.371	-10.259	2, 8	3.642	8.775	2, 9	-3.108	-7.655
2,1C	2.687	6.764	2,11	-2.322	-6.C19						
3, 3	86.667	38.772	3, 4	-22.448	-32.864	3, 5	11.200	21.278	3, 6	-7.437	-15.695
3, 7	5.584	12.468	3, 8	-4.471	-10.352	3, 9	3.715	8.842	3,10	-3.152	-7.691
4, 4	86.852	38.956	4, 5	-22.581	-33.000	4, 6	11.299	21.380	4, 7	-7.509	-15.770
4, 8	5.634	12.520	4, 9	-4.499	-1C.380						
5, 5	86.943	39.053	5, 6	-22.644	-33.069	5, 7	11.340	21.425	5, 8	-7.531	-15.793
6, 6	86.982	39.096	6, 7	-22.663	-33.090						

IBR

SELF AND MUTUAL IMPEDANCES

H/LMDA=0.2500 B/LMDA= 0.500 OMEGA= 8.53

14 ELEMENT ARRAY

I, K	ZIK IN OHMS		I, K	ZIK IN OHMS		I, K	ZIK IN OHMS		I, K	ZIK IN OHMS	
	RE	IM		RE	IM		RE	IM		RE	IM
1, 1	84.867	37.759	1, 2	-21.416	-32.142	1, 3	10.487	20.747	1, 4	-6.902	-15.292
1, 5	5.169	12.161	1, 6	-4.148	-10.123	1, 7	3.471	8.684	1, 8	-2.985	-7.607
1, 9	2.615	6.765	1,1C	-2.319	-6.083	1,11	2.072	5.513	1,12	-1.855	-5.021
1,13	1.647	4.579	1,14	-1.405	-4.155						
2, 2	86.277	38.465	2, 3	-22.200	-32.657	2, 4	11.030	21.137	2, 5	-7.318	-15.602
2, 6	5.506	12.416	2, 7	-4.429	-10.337	2, 8	3.710	8.865	2, 9	-3.190	-7.761
2,10	2.79C	6.894	2,11	-2.467	-6.188	2,12	2.191	5.592	2,13	-1.934	-5.C68
3, 3	86.7C7	38.827	3, 4	-22.492	-32.925	3, 5	11.250	21.345	3, 6	-7.493	-15.770
3, 7	5.648	12.554	3, 8	-4.546	-10.451	3, 9	3.806	8.958	3,10	-3.267	-7.834
3,11	2.849	6.947	3,12	-2.503	-6.218						
4, 4	86.902	39.023	4, 5	-22.636	-33.075	4, 6	11.361	21.464	4, 7	-7.581	-15.865
4, 8	5.718	12.628	4, 9	-4.600	-10.508	4,10	3.845	8.998	4,11	-3.289	-7.856
5, 5	87.005	39.136	5, 6	-22.714	-33.161	5, 7	11.420	21.529	5, 8	-7.624	-15.913
5, 9	5.748	12.661	5,10	-4.616	-10.525						
6, 6	87.061	39.199	6, 7	-22.753	-33.206	6, 8	11.446	21.559	6, 9	-7.638	-15.928
7, 7	87.C86	39.228	7, 8	-22.766	-33.221						

IBR

SELF AND MUTUAL IMPEDANCES

H/LMDA=0.2500 B/LMDA= 0.500 OMEGA= 8.53

16 ELEMENT ARRAY

I, K	ZIK IN OHMS		I, K	ZIK IN OHMS		I, K	ZIK IN OHMS		I, K	ZIK IN OHMS	
	RE	IM		RE	IM		RE	IM		RE	IM
1, 1	84.886	37.789	1, 2	-21.437	-32.174	1, 3	10.510	20.782	1, 4	-6.928	-15.330
1, 5	5.198	12.202	1, 6	-4.179	-10.169	1, 7	3.506	8.734	1, 8	-3.024	-7.663
1, 9	2.659	6.828	1,10	-2.371	-6.155	1,11	2.135	5.599	1,12	-1.935	-5.126
1,13	1.759	4.716	1,14	-1.597	-4.349	1,15	1.435	4.008	1,16	-1.236	-3.671
2, 2	86.301	38.500	2, 3	-22.226	-32.695	2, 4	11.059	21.178	2, 5	-7.350	-15.647
2, 6	5.541	12.465	2, 7	-4.468	-10.392	2, 8	3.754	8.926	2, 9	-3.240	-7.829
2,10	2.849	6.973	2,11	-2.538	-6.281	2,12	2.281	5.706	2,13	-2.060	-5.215
2,14	1.861	4.784	2,15	-1.666	-4.389						
3, 3	86.736	38.869	3, 4	-22.524	-32.969	3, 5	11.285	21.394	3, 6	-7.531	-15.824
3, 7	5.691	12.613	3, 8	-4.595	-10.517	3, 9	3.862	9.033	3,10	-3.332	-7.920
3,11	2.927	7.048	3,12	-2.602	-6.342	3,13	2.330	5.751	3,14	-2.090	-5.241
4, 4	86.937	39.072	4, 5	-22.675	-33.128	4, 6	11.404	21.521	4, 7	-7.628	-15.929
4, 8	5.771	12.700	4, 9	-4.661	-10.588	4,10	3.916	9.091	4,11	-3.375	-7.965
4,12	2.958	7.081	4,13	-2.620	-6.359						
5, 5	87.048	39.194	5, 6	-22.761	-33.224	5, 7	11.472	21.599	5, 8	-7.683	-15.991
5, 9	5.815	12.749	5,10	-4.694	-10.626	5,11	3.939	9.117	5,12	-3.388	-7.979
6, 6	87.113	39.269	6, 7	-22.811	-33.283	6, 8	11.511	21.644	6, 9	-7.712	-16.025
6,10	5.834	12.772	6,11	-4.705	-10.638						
7, 7	87.150	39.313	7, 8	-22.838	-33.314	7, 9	11.528	21.665	7,10	-7.721	-16.035
8, 8	87.167	39.333	8, 9	-22.846	-33.325						

IBR

SELF AND MUTUAL IMPEDANCES

H/LMDA=0.2500 B/LMDA= 0.500 OMEGA= 8.53

18 ELEMENT ARRAY

I, K	ZIK IN OHMS		I, K	ZIK IN OHMS		I, K	ZIK IN OHMS		I, K	ZIK IN OHMS	
	RE	IM		RE	IM		RE	IM		RE	IM
1, 1	84.901	37.812	1, 2	-21.454	-32.200	1, 3	10.528	20.810	1, 4	-6.948	-15.360
1, 5	5.219	12.234	1, 6	-4.2C3	-10.203	1, 7	3.531	8.772	1, 8	-3.052	-7.704
1, 9	2.690	6.873	1,10	-2.406	-6.206	1,11	2.175	5.655	1,12	-1.981	-5.191
1,13	1.815	4.792	1,14	-1.668	-4.443	1,15	1.535	4.130	1,16	-1.408	-3.844
1,17	1.276	3.571	1,18	-1.108	-3.293						
2, 2	86.319	38.528	2, 3	-22.247	-32.725	2, 4	11.081	21.210	2, 5	-7.374	-15.681
2, 6	5.567	12.502	2, 7	-4.497	-10.432	2, 8	3.785	8.970	2, 9	-3.275	-7.878
2,1C	2.888	7.027	2,11	-2.583	-6.342	2,12	2.333	5.777	2,13	-2.123	-5.298
2,14	1.941	4.885	2,15	-1.777	-4.520	2,16	1.624	4.190	2,17	-1.469	-3.879
3, 3	86.758	38.901	3, 4	-22.549	-33.004	3, 5	11.311	21.431	3, 6	-7.560	-15.864
3, 7	5.722	12.656	3, 8	-4.629	-10.564	3, 9	3.900	9.085	3,10	-3.375	-7.978
3,11	2.976	7.114	3,12	-2.659	-6.417	3,13	2.399	5.840	3,14	-2.178	-5.349
3,15	1.983	4.923	3,16	-1.803	-4.542						
4, 4	86.963	39.109	4, 5	-22.703	-33.167	4, 6	11.435	21.564	4, 7	-7.662	-15.975
4, 8	5.809	12.750	4, 9	-4.703	-10.645	4,10	3.963	9.153	4,11	-3.428	-8.036
4,12	3.020	7.162	4,13	-2.695	-6.455	4,14	2.425	5.867	4,15	-2.193	-5.365
5, 5	87.079	39.236	5, 6	-22.795	-33.270	5, 7	11.509	21.649	5, 8	-7.724	-16.C46
5, 9	5.860	12.809	5,10	-4.745	-10.693	5,11	3.997	9.193	5,12	-3.455	-8.066
5,13	3.04C	7.183	5,14	-2.706	-6.467						
6, 6	87.150	39.319	6, 7	-22.851	-33.337	6, 8	11.555	21.704	6, 9	-7.761	-16.090
6,1C	5.889	12.844	6,11	-4.768	-10.720	6,12	4.013	9.211	6,13	-3.464	-8.076
7, 7	87.194	39.371	7, 8	-22.886	-33.379	7, 9	11.582	21.736	7,10	-7.781	-16.114
7,11	5.903	12.861	7,12	-4.775	-10.728						
8, 8	87.220	39.403	8, 9	-22.905	-33.402	8,10	11.595	21.752	8,11	-7.788	-16.122
9, 9	87.233	39.418	9,10	-22.911	-33.410						

IER

SELF AND MUTUAL IMPEDANCES

H/LMDA=C.25C0 B/LMDA= 0.500 UMEGA= 8.53

20 ELEMENT ARRAY

I, K	ZIK IN CHMS		I, K	ZIK IN OHMS		I, K	ZIK IN OHMS		I, K	ZIK IN OHMS	
	RE	IM		RE	IM		RE	IM		RE	IM
1, 1	84.913	37.832	1, 2	-21.467	-32.221	1, 3	10.543	20.832	1, 4	-6.964	-15.384
1, 5	5.236	12.260	1, 6	-4.221	-10.230	1, 7	3.551	8.801	1, 8	-3.073	-7.735
1, 9	2.714	6.907	1,10	-2.432	-6.243	1,11	2.203	5.696	1,12	-2.013	-5.237
1,13	1.851	4.844	1,14	-1.711	-4.502	1,15	1.586	4.200	1,16	-1.472	-3.929
1,17	1.366	3.682	1,18	-1.262	-3.450	1,19	1.152	3.225	1,20	-1.006	-2.990
2, 2	86.335	38.550	2, 3	-22.263	-32.749	2, 4	11.099	21.235	2, 5	-7.394	-15.709
2, 6	5.588	12.531	2, 7	-4.519	-10.463	2, 8	3.809	9.004	2, 9	-3.301	-7.915
2,10	2.917	7.067	2,11	-2.615	-6.386	2,12	2.369	5.826	2,13	-2.164	-5.353
2,14	1.988	4.948	2,15	-1.834	-4.595	2,16	1.696	4.281	2,17	-1.569	-3.997
2,18	1.446	3.734	2,19	-1.317	-3.481						
3, 3	86.776	38.926	3, 4	-22.568	-33.031	3, 5	11.332	21.460	3, 6	-7.583	-15.895
3, 7	5.747	12.690	3, 8	-4.656	-10.600	3, 9	3.929	9.124	3,10	-3.406	-8.020
3,11	3.011	7.161	3,12	-2.698	-6.469	3,13	2.443	5.899	3,14	-2.229	-5.417
3,15	2.045	5.003	3,16	-1.882	-4.640	3,17	1.733	4.315	3,18	-1.592	-4.017
4, 4	86.984	39.138	4, 5	-22.726	-33.158	4, 6	11.459	21.598	4, 7	-7.688	-16.011
4, 8	5.837	12.789	4, 9	-4.733	-10.686	4,10	3.996	9.199	4,11	-3.466	-8.086
4,12	3.062	7.218	4,13	-2.742	-6.518	4,14	2.481	5.939	4,15	-2.260	-5.450
4,16	2.068	5.026	4,17	-1.895	-4.653						
5, 5	87.103	39.269	5, 6	-22.820	-33.306	5, 7	11.537	21.687	5, 8	-7.754	-16.087
5, 9	5.893	12.854	5,10	-4.781	-10.742	5,11	4.037	9.246	5,12	-3.500	-8.126
5,13	3.091	7.250	5,14	-2.765	-6.544	5,15	2.497	5.958	5,16	-2.269	-5.460
6, 6	87.178	39.356	6, 7	-22.881	-33.378	6, 8	11.588	21.748	6, 9	-7.796	-16.138
6,1C	5.928	12.896	6,11	-4.811	-10.777	6,12	4.061	9.275	6,13	-3.519	-8.148
6,14	3.104	7.266	6,15	-2.772	-6.552						
7, 7	87.227	39.415	7, 8	-22.921	-33.426	7, 9	11.620	21.787	7,10	-7.823	-16.170
7,11	5.949	12.922	7,12	-4.827	-10.797	7,13	4.072	9.288	7,14	-3.525	-8.155
8, 8	87.258	39.454	8, 9	-22.947	-33.457	8,10	11.640	21.812	8,11	-7.838	-16.188
8,12	5.959	12.934	8,13	-4.832	-10.803						
9, 9	87.278	39.478	9,10	-22.961	-33.475	9,11	11.650	21.823	9,12	-7.843	-16.194
1C,1C	87.287	39.490	1C,11	-22.966	-33.480						

SELF AND MUTUAL IMPEDANCES

H/LMDA=0.2500 B/LMDA= 0.500 UMEGA= [illegible]

22 ELEMENT ARRAY

I, K	ZIK IN OHMS		I, K	ZIK IN OHMS		I, K	ZIK IN OHMS		I, K	ZIK IN OHMS	
	RE	IM		RE	IM		RE	IM		RE	IM
1, 1	84.923	37.848	1, 2	-21.479	-32.238	1, 3	10.555	20.850	1, 4	-6.977	-15.403
1, 5	5.250	12.280	1, 6	-4.236	-10.253	1, 7	3.567	8.824	1, 8	-3.090	-7.760
1, 9	2.732	6.934	1,10	-2.452	-6.272	1,11	2.225	5.728	1,12	-2.037	-5.271
1,13	1.878	4.882	1,14	-1.740	-4.544	1,15	1.619	4.247	1,16	-1.511	-3.984
1,17	1.413	3.746	1,18	-1.322	-3.528	1,19	1.235	3.326	1,20	-1.147	-3.134
1,21	1.052	2.944	1,22	-0.923	-2.741						
2, 2	86.347	38.569	2, 3	-22.277	-32.768	2, 4	11.113	21.256	2, 5	-7.409	-15.731
2, 6	5.604	12.555	2, 7	-4.537	-10.488	2, 8	3.828	9.031	2, 9	-3.322	-7.943
2,10	2.939	7.098	2,11	-2.639	-6.420	2,12	2.395	5.862	2,13	-2.193	-5.393
2,14	2.020	4.993	2,15	-1.871	-4.645	2,16	1.739	4.339	2,17	-1.621	-4.066
2,18	1.512	3.818	2,19	-1.409	-3.589	2,20	1.307	3.374	2,21	-1.197	-3.162
3, 3	86.791	38.947	3, 4	-22.584	-33.C53	3, 5	11.349	21.484	3, 6	-7.600	-15.920
3, 7	5.766	12.716	3, 8	-4.676	-10.629	3, 9	3.951	9.154	3,10	-3.430	-8.053
3,11	3.036	7.197	3,12	-2.726	-6.508	3,13	2.474	5.941	3,14	-2.264	-5.465
3,15	2.065	5.056	3,16	-1.928	-4.701	3,17	1.790	4.387	3,18	-1.663	-4.105
3,19	1.545	3.848	3,20	-1.430	-3.607						
4, 4	87.001	39.161	4, 5	-22.744	-33.223	4, 6	11.478	21.624	4, 7	-7.709	-16.C39
4, 8	5.859	12.819	4, 9	-4.757	-10.719	4,10	4.022	9.234	4,11	-3.493	-8.124
4,12	3.093	7.259	4,13	-2.776	-6.563	4,14	2.518	5.99C	4,15	-2.302	-5.506
4,16	2.118	5.092	4,17	-1.955	-4.730	4,18	1.810	4.408	4,19	-1.675	-4.117
5, 5	87.122	39.296	5, 6	-22.841	-33.334	5, 7	11.559	21.717	5, 8	-7.778	-16.119
5, 9	5.918	12.888	5,10	-4.809	-10.779	5,11	4.067	9.286	5,12	-3.533	-8.170
5,13	3.127	7.299	5,14	-2.8C6	-6.597	5,15	2.543	6.018	5,16	-2.322	-5.529
5,17	2.132	5.1C8	5,18	-1.964	-4.738						
6, 6	87.200	39.386	6, 7	-22.905	-33.410	6, 8	11.613	21.782	6, 9	-7.823	-16.174
6,1C	5.957	12.936	6,11	-4.842	-10.820	6,12	4.096	9.321	6,13	-3.557	-8.199
6,14	3.147	7.322	6,15	-2.821	-6.616	6,16	2.554	6.031	6,17	-2.328	-5.536
7, 7	87.252	39.449	7, 8	-22.948	-33.462	7, 9	11.649	21.826	7,10	-7.854	-16.212
7,11	5.983	12.968	7,12	-4.864	-10.846	7,13	4.114	9.343	7,14	-3.571	-8.215
7,15	3.156	7.334	7,16	-2.826	-6.622						
8, 8	87.287	39.493	8, 9	-22.978	-33.499	8,10	11.674	21.857	8,11	-7.874	-16.237
8,12	5.999	12.987	8,13	-4.876	-10.861	8,14	4.122	9.353	8,15	-3.575	-8.220
9, 9	87.311	39.523	9,1C	-22.997	-33.523	9,11	11.689	21.875	9,12	-7.886	-16.251
9,13	6.007	12.997	9,14	-4.880	-10.866						
10,10	87.326	39.541	1C,11	-23.008	-33.536	10,12	11.696	21.884	10,13	-7.889	-16.255
11,11	87.333	39.550	11,12	-23.C11	-33.541						

IBR

SELF AND MUTUAL IMPEDANCES

H/LMDA=0.2500 B/LMDA= 0.500 OMEGA= 8.53

24 ELEMENT ARRAY

I, K	ZIK IN OHMS RE	IM	I, K	ZIK IN OHMS RE	IM	I, K	ZIK IN OHMS RE	IM	I, K	ZIK IN OHMS RE	IM
1, 1	84.931	37.862	1, 2	-21.488	-32.253	1, 3	10.565	20.866	1, 4	-6.988	-15.420
1, 5	5.262	12.298	1, 6	-4.249	-10.271	1, 7	3.580	8.844	1, 8	-3.105	-7.781
1, 9	2.747	6.956	1,10	-2.468	-6.295	1,11	2.242	5.753	1,12	-2.055	-5.298
1,13	1.898	4.911	1,14	-1.762	-4.576	1,15	1.644	4.282	1,16	-1.539	-4.022
1,17	1.444	3.789	1,18	-1.358	-3.578	1,19	1.278	3.385	1,20	-1.202	-3.206
1,21	1.128	3.037	1,22	-1.053	-2.874	1,23	0.970	2.710	1,24	-0.854	-2.533
2, 2	86.358	38.584	2, 3	-22.288	-32.785	2, 4	11.126	21.274	2, 5	-7.422	-15.749
2, 6	5.618	12.574	2, 7	-4.551	-10.509	2, 8	3.844	9.053	2, 9	-3.338	-7.967
2,10	2.957	7.123	2,11	-2.658	-6.447	2,12	2.416	5.891	2,13	-2.215	-5.424
2,14	2.045	5.027	2,15	-1.898	-4.682	2,16	1.770	4.380	2,17	-1.655	-4.112
2,18	1.552	3.871	2,19	-1.456	-3.652	2,20	1.367	3.450	2,21	-1.281	-3.261
2,22	1.194	3.080	2,23	-1.099	-2.900						
3, 3	86.803	38.965	3, 4	-22.597	-33.072	3, 5	11.363	21.503	3, 6	-7.615	-15.941
3, 7	5.781	12.738	3, 8	-4.693	-10.652	3, 9	3.969	9.179	3,10	-3.449	-8.080
3,11	3.057	7.225	3,12	-2.748	-6.539	3,13	2.498	5.974	3,14	-2.290	-5.500
3,15	2.114	5.096	3,16	-1.961	-4.745	3,17	1.826	4.436	3,18	-1.706	-4.162
3,19	1.597	3.914	3,20	-1.495	-3.688	3,21	1.397	3.477	3,22	-1.300	-3.277
4, 4	87.015	39.181	4, 5	-22.759	-33.244	4, 6	11.494	21.646	4, 7	-7.726	-16.062
4, 8	5.876	12.844	4, 9	-4.776	-10.745	4,10	4.042	9.261	4,11	-3.515	-8.153
4,12	3.116	7.291	4,13	-2.802	-6.598	4,14	2.546	6.027	4,15	-2.333	-5.548
4,16	2.152	5.138	4,17	-1.995	-4.781	4,18	1.856	4.467	4,19	-1.730	-4.187
4,20	1.615	3.933	4,21	-1.506	-3.698						
5, 5	87.138	39.318	5, 6	-22.858	-33.357	5, 7	11.577	21.742	5, 8	-7.797	-16.145
5, 9	5.938	12.916	5,10	-4.830	-10.808	5,11	4.090	9.318	5,12	-3.558	-8.203
5,13	3.154	7.335	5,14	-2.835	-6.637	5,15	2.576	6.062	5,16	-2.359	-5.577
5,17	2.174	5.162	5,18	-2.012	-4.801	5,19	1.868	4.482	5,20	-1.738	-4.195
6, 6	87.217	39.411	6, 7	-22.924	-33.435	6, 8	11.633	21.809	6, 9	-7.845	-16.203
6,10	5.980	12.967	6,11	-4.867	-10.853	6,12	4.123	9.357	6,13	-3.586	-8.237
6,14	3.178	7.364	6,15	-2.856	-6.662	6,16	2.593	6.082	6,17	-2.373	-5.594
6,18	2.184	5.174	6,19	-2.018	-4.807						
7, 7	87.272	39.476	7, 8	-22.970	-33.491	7, 9	11.672	21.857	7,10	-7.879	-16.245
7,11	6.010	13.003	7,12	-4.892	-10.884	7,13	4.144	9.383	7,14	-3.604	-8.260
7,15	3.193	7.383	7,16	-2.868	-6.676	7,17	2.601	6.092	7,18	-2.377	-5.599
8, 8	87.310	39.524	8, 9	-23.002	-33.531	8,10	11.700	21.891	8,11	-7.902	-16.274
8,12	6.029	13.027	8,13	-4.909	-10.904	8,14	4.157	9.400	8,15	-3.614	-8.272
8,16	3.200	7.391	8,17	-2.872	-6.681						
9, 9	87.337	39.557	9,10	-23.025	-33.560	9,11	11.718	21.915	9,12	-7.918	-16.293
9,13	6.042	13.043	9,14	-4.918	-10.916	9,15	4.164	9.408	9,16	-3.618	-8.276
10,10	87.356	39.581	10,11	-23.040	-33.578	10,12	11.730	21.930	10,13	-7.926	-16.304
10,14	6.048	13.050	10,15	-4.921	-10.920						
11,11	87.367	39.595	11,12	-23.048	-33.589	11,13	11.736	21.937	11,14	-7.929	-16.308
12,12	87.373	39.602	12,13	-23.051	-33.593						

IBR

SELF AND MUTUAL IMPEDANCES

H/LMDA=0.2500 B/LMDA= 0.500 OMEGA= 8.53

26 ELEMENT ARRAY

I, K	ZIK IN OHMS RE	IM	I, K	ZIK IN OHMS RE	IM	I, K	ZIK IN OHMS RE	IM	I, K	ZIK IN OHMS RE	IM
1, 1	84.939	37.874	1, 2	-21.496	-32.265	1, 3	10.574	20.879	1, 4	-6.997	-15.434
1, 5	5.272	12.313	1, 6	-4.259	-10.286	1, 7	3.591	8.860	1, 8	-3.116	-7.798
1, 9	2.760	6.974	1,10	-2.481	-6.315	1,11	2.256	5.773	1,12	-2.071	-5.320
1,13	1.914	4.934	1,14	-1.780	-4.601	1,15	1.663	4.309	1,16	-1.559	-4.052
1,17	1.467	3.822	1,18	-1.383	-3.614	1,19	1.307	3.426	1,20	-1.236	-3.253
1,21	1.169	3.092	1,22	-1.105	-2.941	1,23	1.041	2.797	1,24	-0.976	-2.656
1,25	0.901	2.513	1,26	-0.796	-2.357						
2, 2	86.367	38.598	2, 3	-22.298	-32.799	2, 4	11.136	21.288	2, 5	-7.433	-15.765
2, 6	5.630	12.591	2, 7	-4.564	-10.526	2, 8	3.857	9.071	2, 9	-3.352	-7.986
2,10	2.971	7.143	2,11	-2.673	-6.468	2,12	2.432	5.914	2,13	-2.233	-5.449
2,14	2.064	5.053	2,15	-1.919	-4.711	2,16	1.792	4.411	2,17	-1.680	-4.146
2,18	1.580	3.909	2,19	-1.488	-3.695	2,20	1.404	3.500	2,21	-1.326	-3.319
2,22	1.251	3.151	2,23	-1.177	-2.992	2,24	1.102	2.837	2,25	-1.018	-2.680
3, 3	86.814	38.979	3, 4	-22.608	-33.087	3, 5	11.374	21.520	3, 6	-7.628	-15.958
3, 7	5.795	12.756	3, 8	-4.707	-10.671	3, 9	3.983	9.199	3,10	-3.465	-8.101
3,11	3.074	7.247	3,12	-2.766	-6.563	3,13	2.517	6.000	3,14	-2.311	-5.528
3,15	2.136	5.126	3,16	-1.985	-4.778	3,17	1.853	4.472	3,18	-1.736	-4.202
3,19	1.631	3.959	3,20	-1.534	-3.740	3,21	1.445	3.538	3,22	-1.360	-3.352
3,23	1.278	3.176	3,24	-1.195	-3.006						
4, 4	87.027	39.197	4, 5	-22.771	-33.261	4, 6	11.507	21.664	4, 7	-7.740	-16.081
4, 8	5.891	12.864	4, 9	-4.792	-10.766	4,10	4.059	9.284	4,11	-3.533	-8.177
4,12	3.135	7.316	4,13	-2.822	-6.625	4,14	2.568	6.057	4,15	-2.357	-5.580
4,16	2.178	5.172	4,17	-2.023	-4.820	4,18	1.887	4.510	4,19	-1.766	-4.235
4,20	1.657	3.987	4,21	-1.556	-3.763	4,22	1.462	3.555	4,23	-1.370	-3.361
5, 5	87.151	39.336	5, 6	-22.872	-33.376	5, 7	11.592	21.762	5, 8	-7.812	-16.166
5, 9	5.955	12.938	5,10	-4.848	-10.832	5,11	4.109	9.343	5,12	-3.578	-8.230
5,13	3.175	7.364	5,14	-2.858	-6.668	5,15	2.601	6.055	5,16	-2.386	-5.614
5,17	2.204	5.202	5,18	-2.046	-4.846	5,19	1.907	4.532	5,20	-1.782	-4.252
5,21	1.668	4.000	5,22	-1.563	-3.770						
6, 6	87.232	39.431	6, 7	-22.939	-33.456	6, 8	11.649	21.831	6, 9	-7.862	-16.227
6,10	5.999	12.992	6,11	-4.887	-10.879	6,12	4.144	9.385	6,13	-3.609	-8.268
6,14	3.203	7.397	6,15	-2.883	-6.697	6,16	2.622	6.121	6,17	-2.405	-5.636
6,18	2.219	5.221	6,19	-2.058	-4.860	6,20	1.916	4.542	6,21	-1.787	-4.258

I, K	RE	IM	I, K	RE	IM	I, K	RE	IM	I, K	RE	IM
7, 7	87.288	39.499	7, 8	-22.987	-33.515	7, 9	11.691	21.882	7,10	-7.898	-16.271
7,11	6.030	13.031	7,12	-4.915	-10.914	7,13	4.168	9.415	7,14	-3.630	-8.294
7,15	3.221	7.420	7,16	-2.898	-6.716	7,17	2.635	6.136	7,18	-2.415	-5.648
7,19	2.226	5.229	7,20	-2.062	-4.864						
8, 8	87.328	39.548	8, 9	-23.022	-33.557	8,10	11.720	21.919	8,11	-7.924	-16.303
8,12	6.053	13.059	8,13	-4.934	-10.938	8,14	4.185	9.436	8,15	-3.644	-8.311
8,16	3.233	7.434	8,17	-2.907	-6.727	8,18	2.641	6.144	8,19	-2.418	-5.652
9, 9	87.358	39.585	9,10	-23.047	-33.589	9,11	11.742	21.946	9,12	-7.942	-16.326
9,13	6.068	13.078	9,14	-4.947	-10.954	9,15	4.195	9.449	9,16	-3.652	-8.321
9,17	3.238	7.441	9,18	-2.910	-6.731						
10,10	87.379	39.612	10,11	-23.064	-33.611	10,12	11.757	21.965	10,13	-7.955	-16.342
10,14	6.078	13.090	10,15	-4.954	-10.963	10,16	4.200	9.455	10,17	-3.654	-8.324
11,11	87.393	39.630	11,12	-23.076	-33.627	11,13	11.766	21.977	11,14	-7.962	-16.351
11,15	6.083	13.096	11,16	-4.957	-10.966						
12,12	87.403	39.642	12,13	-23.083	-33.635	12,14	11.771	21.982	12,15	-7.964	-16.354
13,13	87.407	39.648	13,14	-23.085	-33.638						

IBR

SELF AND MUTUAL IMPEDANCES

H/LMDA=0.2500 B/LMDA= 0.500 OMEGA= 8.53

28 ELEMENT ARRAY

I, K	ZIK IN OHMS RE	IM	I, K	ZIK IN OHMS RE	IM	I, K	ZIK IN OHMS RE	IM	I, K	ZIK IN OHMS RE	IM
1, 1	84.945	37.884	1, 2	-21.504	-32.276	1, 3	10.582	20.891	1, 4	-7.005	-15.446
1, 5	5.281	12.325	1, 6	-4.268	-10.300	1, 7	3.601	8.874	1, 8	-3.126	-7.813
1, 9	2.770	6.990	1,10	-2.492	-6.331	1,11	2.268	5.790	1,12	-2.083	-5.338
1,13	1.927	4.953	1,14	-1.794	-4.621	1,15	1.678	4.331	1,16	-1.576	-4.075
1,17	1.485	3.847	1,18	-1.403	-3.642	1,19	1.328	3.456	1,20	-1.260	-3.286
1,21	1.156	3.130	1,22	-1.136	-2.985	1,23	1.079	2.848	1,24	-1.023	-2.719
1,25	0.968	2.595	1,26	-0.910	-2.472	1,27	0.843	2.345	1,28	-0.747	-2.205
2, 2	86.375	38.609	2, 3	-22.307	-32.811	2, 4	11.145	21.301	2, 5	-7.442	-15.778
2, 6	5.640	12.605	2, 7	-4.574	-10.541	2, 8	3.868	9.086	2, 9	-3.364	-8.002
2,10	2.984	7.160	2,11	-2.686	-6.486	2,12	2.446	5.933	2,13	-2.247	-5.469
2,14	2.080	5.075	2,15	-1.936	-4.734	2,16	1.810	4.436	2,17	-1.700	-4.173
2,18	1.601	3.938	2,19	-1.512	-3.727	2,20	1.430	3.535	2,21	-1.355	-3.360
2,22	1.285	3.197	2,23	-1.219	-3.046	2,24	1.154	2.903	2,25	-1.091	-2.766
2,26	1.024	2.631	2,27	-0.949	-2.494						
3, 3	86.823	38.992	3, 4	-22.617	-33.101	3, 5	11.385	21.534	3, 6	-7.638	-15.973
3, 7	5.806	12.772	3, 8	-4.718	-10.687	3, 9	3.996	9.216	3,10	-3.478	-8.119
3,11	3.087	7.266	3,12	-2.781	-6.583	3,13	2.533	6.022	3,14	-2.327	-5.551
3,15	2.154	5.150	3,16	-2.004	-4.804	3,17	1.874	4.501	3,18	-1.759	-4.233
3,19	1.656	3.993	3,20	-1.562	-3.777	3,21	1.477	3.581	3,22	-1.397	-3.400
[illegible]	[illegible]	[illegible]	3,24	-1.251	-3.076	3,25	1.180	2.926	3,26	-1.107	-2.779
4, 4	87.037	39.211	4, 5	-22.782	-33.276	4, 6	11.518	21.680	[illegible]	[illegible]	[illegible]
4, 8	5.904	12.881	4, 9	-4.805	-10.784	4,10	4.073	9.303	4,11	-3.547	-8.197
4,12	3.150	7.337	4,13	-2.838	-6.648	4,14	2.586	6.081	4,15	-2.376	-5.605
4,16	2.198	5.200	4,17	-2.045	-4.849	4,18	1.911	4.542	4,19	-1.793	-4.270
4,20	1.686	4.027	4,21	-1.590	-3.807	4,22	1.501	3.606	4,23	-1.417	-3.421
4,24	1.338	3.248	4,25	-1.260	-3.084						
5, 5	87.162	39.351	5, 6	-22.884	-33.392	5, 7	11.604	21.779	5, 8	-7.825	-16.184
5, 9	5.969	12.957	5,10	-4.862	-10.852	5,11	4.124	9.364	5,12	-3.594	-8.252
5,13	3.193	7.387	5,14	-2.877	-6.693	5,15	2.621	6.122	5,16	-2.408	-5.643
5,17	2.227	5.233	5,18	-2.071	-4.879	5,19	1.935	4.569	5,20	-1.813	-4.293
5,21	1.704	4.046	5,22	-1.604	-3.823	5,23	1.511	3.618	5,24	-1.423	-3.427
6, 6	87.245	39.448	6, 7	-22.953	-33.474	6, 8	11.663	21.850	6, 9	-7.877	-16.246
6,10	6.014	13.012	6,11	-4.903	-10.901	6,12	4.161	9.408	6,13	-3.627	-8.292
6,14	3.222	7.423	6,15	-2.903	-6.725	6,16	2.644	6.151	6,17	-2.429	-5.668
6,18	2.246	5.256	6,19	-2.087	-4.899	6,20	1.948	4.585	6,21	-1.824	-4.306
6,22	1.712	4.056	6,23	-1.608	-3.828						
7, 7	87.302	39.517	7, 8	-23.002	-33.534	7, 9	11.706	21.902	7,10	-7.914	-16.293
7,11	6.047	13.053	7,12	-4.933	-10.938	7,13	4.187	9.441	7,14	-3.650	-8.321
7,15	3.243	7.449	7,16	-2.922	-6.748	7,17	2.661	6.170	7,18	-2.443	-5.685
7,19	2.257	5.270	7,20	-2.097	-4.910	7,21	1.955	4.593	7,22	-1.828	-4.310
8, 8	87.344	39.569	8, 9	-23.038	-33.579	8,10	11.737	21.942	8,11	-7.942	-16.327
8,12	6.072	13.084	8,13	-4.954	-10.965	8,14	4.206	9.465	8,15	-3.667	-8.342
8,16	3.258	7.467	8,17	-2.934	-6.763	8,18	2.671	6.183	8,19	-2.451	-5.695
8,20	2.263	5.277	8,21	-2.100	-4.913						
9, 9	87.375	39.608	9,10	-23.064	-33.613	9,11	11.761	21.971	9,12	-7.963	-16.353
9,13	6.090	13.106	9,14	-4.970	-10.984	9,15	4.219	9.481	9,16	-3.678	-8.356
9,17	3.267	7.478	9,18	-2.941	-6.772	9,19	2.675	6.189	9,20	-2.453	-5.698
10,10	87.398	39.637	10,11	-23.084	-33.638	10,12	11.778	21.993	10,13	-7.977	-16.372
10,14	6.102	13.122	10,15	-4.980	-10.997	10,16	4.228	9.492	10,17	-3.684	-8.364
10,18	3.271	7.484	10,19	-2.943	-6.774						
11,11	87.415	39.658	11,12	-23.099	-33.656	11,13	11.790	22.008	11,14	-7.987	-16.384
11,15	6.110	13.132	11,16	-4.986	-11.005	11,17	4.232	9.497	11,18	-3.687	-8.366
12,12	87.426	39.673	12,13	-23.108	-33.669	12,14	11.798	22.018	12,15	-7.993	-16.391
12,16	6.114	13.137	12,17	-4.988	-11.007						
13,13	87.434	39.683	13,14	-23.114	-33.676	13,15	11.801	22.023	13,16	-7.995	-16.394
14,14	87.438	39.688	14,15	-23.116	-33.678						

IBR

SELF AND MUTUAL IMPEDANCES

H/LMDA=0.2500 B/LMDA= 0.500 OMEGA= 8.53

30 ELEMENT ARRAY

I, K	ZIK IN OHMS		I, K	ZIK IN OHMS		I, K	ZIK IN OHMS		I, K	ZIK IN OHMS	
	RE	IM		RE	IM		RE	IM		RE	IM
1, 1	84.951	37.893	1, 2	-21.510	-32.286	1, 3	10.589	20.901	1, 4	-7.013	-15.456
1, 5	5.288	12.336	1, 6	-4.276	-10.311	1, 7	3.609	8.886	1, 8	-3.135	-7.825
1, 9	2.779	7.003	1,10	-2.502	-6.345	1,11	2.278	5.805	1,12	-2.094	-5.353
1,13	1.938	4.970	1,14	-1.806	-4.638	1,15	1.690	4.349	1,16	-1.589	-4.095
1,17	1.499	3.868	1,18	-1.418	-3.664	1,19	1.345	3.480	1,20	-1.278	-3.312
1,21	1.216	3.159	1,22	-1.159	-3.016	1,23	1.104	2.884	1,24	-1.053	-2.760
1,25	1.003	2.643	1,26	-0.955	-2.530	1,27	0.905	2.421	1,28	-0.853	-2.312
1,29	0.793	2.199	1,30	-0.704	-2.072						
2, 2	86.382	38.619	2, 3	-22.314	-32.822	2, 4	11.153	21.312	2, 5	-7.451	-15.790
2, 6	5.648	12.617	2, 7	-4.583	-10.554	2, 8	3.877	9.100	2, 9	-3.373	-8.016
2,10	2.994	7.175	2,11	-2.697	-6.502	2,12	2.458	5.949	2,13	-2.259	-5.486
2,14	2.092	5.093	2,15	-1.949	-4.753	2,16	1.825	4.457	2,17	-1.715	-4.195
2,18	1.618	3.962	2,19	-1.530	-3.752	2,20	1.450	3.563	2,21	-1.377	-3.390
2,22	1.310	3.231	2,23	-1.247	-3.083	2,24	1.187	2.946	2,25	-1.130	-2.817
2,26	1.074	2.693	2,27	-1.017	-2.574	2,28	0.958	2.455	2,29	-0.890	-2.333
3, 3	86.831	39.004	3, 4	-22.626	-33.113	3, 5	11.393	21.546	3, 6	-7.648	-15.985
3, 7	5.816	12.785	3, 8	-4.729	-10.701	3, 9	4.006	9.231	3,10	-3.489	-8.134
3,11	3.099	7.283	3,12	-2.793	-6.600	3,13	2.546	6.039	3,14	-2.341	-5.570
3,15	2.168	5.170	3,16	-2.020	-4.825	3,17	1.891	4.523	3,18	-1.777	-4.257
3,19	1.675	4.020	3,20	-1.584	-3.806	3,21	1.500	3.612	3,22	-1.424	-3.435
3,23	1.352	3.272	3,24	-1.285	-3.121	3,25	1.222	2.979	3,26	-1.159	-2.844
3,27	1.097	2.714	3,28	-1.033	-2.586						
4, 4	87.046	39.224	4, 5	-22.791	-33.289	4, 6	11.528	21.693	4, 7	-7.762	-16.112
4, 8	5.914	12.896	4, 9	-4.816	-10.799	4,10	4.084	9.319	4,11	-3.560	-8.214
4,12	3.163	7.355	4,13	-2.852	-6.666	4,14	2.600	6.100	4,15	-2.391	-5.626
4,16	2.215	5.222	4,17	-2.063	-4.873	4,18	1.930	4.568	4,19	-1.813	-4.298
4,20	1.709	4.057	4,21	-1.615	-3.840	4,22	1.528	3.643	4,23	-1.449	-3.462
4,24	1.374	3.296	4,25	-1.304	-3.140	4,26	1.236	2.993	4,27	-1.168	-2.852
5, 5	87.172	39.365	5, 6	-22.894	-33.406	5, 7	11.615	21.794	5, 8	-7.837	-16.199
5, 9	5.980	12.973	5,10	-4.875	-10.869	5,11	4.137	9.382	5,12	-3.608	-8.271
5,13	3.207	7.407	5,14	-2.892	-6.714	5,15	2.637	6.144	5,16	-2.425	-5.666
5,17	2.246	5.258	5,18	-2.091	-4.906	5,19	1.956	4.598	5,20	-1.837	-4.325
5,21	1.730	4.081	5,22	-1.633	-3.861	5,23	1.544	3.661	5,24	-1.462	-3.477
5,25	1.384	3.306	5,26	-1.309	-3.146						
6, 6	87.255	39.463	6, 7	-22.964	-33.490	6, 8	11.675	21.866	6, 9	-7.889	-16.263
6,10	6.027	13.030	6,11	-4.917	-10.920	6,12	4.175	9.428	6,13	-3.642	-8.313
6,14	3.238	7.445	6,15	-2.921	-6.748	6,16	2.663	6.175	6,17	-2.448	-5.694
6,18	2.267	5.284	6,19	-2.110	-4.929	6,20	1.973	4.618	6,21	-1.852	-4.342
6,22	1.742	4.096	6,23	-1.643	-3.873	6,24	1.552	3.669	6,25	-1.466	-3.481
7, 7	87.314	39.533	7, 8	-23.014	-33.551	7, 9	11.719	21.920	7,10	-7.928	-16.311
7,11	6.062	13.073	7,12	-4.948	-10.958	7,13	4.203	9.462	7,14	-3.667	-8.344
7,15	3.261	7.473	7,16	-2.941	-6.773	7,17	2.681	6.198	7,18	-2.465	-5.714
7,19	2.281	5.301	7,20	-2.123	-4.944	7,21	1.983	4.630	7,22	-1.860	-4.352
7,23	1.748	4.103	7,24	-1.646	-3.876						
8, 8	87.357	39.586	8, 9	-23.051	-33.597	8,10	11.752	21.961	8,11	-7.957	-16.347
8,12	6.088	13.105	8,13	-4.971	-10.987	8,14	4.224	9.488	8,15	-3.686	-8.367
8,16	3.278	7.493	8,17	-2.956	-6.791	8,18	2.694	6.213	8,19	-2.476	-5.728
8,20	2.290	5.312	8,21	-2.130	-4.952	8,22	1.988	4.636	8,23	-1.863	-4.355
9, 9	87.389	39.627	9,10	-23.080	-33.633	9,11	11.777	21.992	9,12	-7.979	-16.375
9,13	6.107	13.130	9,14	-4.988	-11.009	9,15	4.239	9.507	9,16	-3.699	-8.384
9,17	3.289	7.508	9,18	-2.965	-6.804	9,19	2.702	6.223	9,20	-2.482	-5.735
9,21	2.294	5.318	9,22	-2.132	-4.955						
10,10	87.413	39.658	10,11	-23.101	-33.660	10,12	11.795	22.016	10,13	-7.996	-16.396
10,14	6.122	13.148	10,15	-5.001	-11.025	10,16	4.250	9.521	10,17	-3.708	-8.395
10,18	3.296	7.517	10,19	-2.971	-6.811	10,20	2.705	6.228	10,21	-2.484	-5.738
11,11	87.432	39.681	11,12	-23.117	-33.680	11,13	11.809	22.034	11,14	-8.008	-16.411
11,15	6.132	13.161	11,16	-5.009	-11.035	11,17	4.257	9.529	11,18	-3.713	-8.401
11,19	3.300	7.521	11,20	-2.973	-6.813						
12,12	87.446	39.699	12,13	-23.129	-33.695	12,14	11.819	22.046	12,15	-8.016	-16.422
12,16	6.138	13.169	12,17	-5.014	-11.042	12,18	4.260	9.534	12,19	-3.715	-8.404
13,13	87.456	39.711	13,14	-23.137	-33.706	13,15	11.825	22.055	13,16	-8.020	-16.428
13,17	6.141	13.173	13,18	-5.016	-11.044						
14,14	87.462	39.720	14,15	-23.141	-33.712	14,16	11.829	22.059	14,17	-8.022	-16.430
15,15	87.465	39.723	15,16	-23.143	-33.714						

IBR

SELF AND MUTUAL IMPEDANCES

H/LMDA=0.2500 B/LMDA= 0.500 OMEGA= 8.53

35 ELEMENT ARRAY

I, K	ZIK IN OHMS		I, K	ZIK IN OHMS		I, K	ZIK IN OHMS		I, K	ZIK IN OHMS	
	RE	IM		RE	IM		RE	IM		RE	IM
1, 1	84.963	37.912	1, 2	-21.523	-32.306	1, 3	10.603	20.922	1, 4	-7.027	-15.478
1, 5	5.303	12.359	1, 6	-4.292	-10.335	1, 7	3.626	8.911	1, 8	-3.152	-7.851
1, 9	2.797	7.029	1,10	-2.521	-6.372	1,11	2.298	5.834	1,12	-2.114	-5.383
1,13	1.960	5.001	1,14	-1.828	-4.671	1,15	1.714	4.384	1,16	-1.614	-4.131
1,17	1.525	3.906	1,18	-1.446	-3.705	1,19	1.375	3.523	1,20	-1.310	-3.358
1,21	1.250	3.208	1,22	-1.196	-3.069	1,23	1.145	2.942	1,24	-1.098	-2.823
1,25	1.053	2.713	1,26	-1.011	-2.609	1,27	0.971	2.512	1,28	-0.933	-2.419
1,29	0.895	2.330	1,30	-0.858	-2.245	1,31	0.822	2.162	1,32	-0.783	-2.080
1,33	0.742	1.996	1,34	-0.692	-1.908	1,35	0.617	1.805			
2, 2	86.396	38.640	2, 3	-22.329	-32.844	2, 4	11.169	21.335	2, 5	-7.467	-15.814
2, 6	5.666	12.641	2, 7	-4.601	-10.579	2, 8	3.896	9.126	2, 9	-3.393	-8.044
2,10	3.015	7.204	2,11	-2.719	-6.532	2,12	2.480	5.980	2,13	-2.283	-5.519
2,14	2.117	5.127	2,15	-1.975	-4.789	2,16	1.852	4.495	2,17	-1.744	-4.235
2,18	1.648	4.004	2,19	-1.563	-3.797	2,20	1.485	3.611	2,21	-1.415	-3.441
2,22	1.350	3.286	2,23	-1.291	-3.144	2,24	1.236	3.012	2,25	-1.184	-2.890
2,26	1.135	2.776	2,27	-1.089	-2.669	2,28	1.045	2.567	2,29	-1.002	-2.471
2,30	0.960	2.378	2,31	-0.918	-2.288	2,32	0.875	2.199	2,33	-0.829	-2.109
2,34	0.774	2.014									
3, 3	86.847	39.026	3, 4	-22.643	-33.136	3, 5	11.411	21.571	3, 6	-7.666	-16.011
3, 7	5.835	12.812	3, 8	-4.749	-10.729	3, 9	4.027	9.260	3,10	-3.511	-8.165
3,11	3.122	7.314	3,12	-2.817	-6.633	3,13	2.571	6.074	3,14	-2.368	-5.606
3,15	2.196	5.208	3,16	-2.049	-4.865	3,17	1.922	4.565	3,18	-1.809	-4.301
3,19	1.710	4.067	3,20	-1.621	-3.856	3,21	1.540	3.666	3,22	-1.467	-3.493
3,23	1.399	3.336	3,24	-1.337	-3.190	3,25	1.279	3.056	3,26	-1.225	-2.931
3,27	1.174	2.813	3,28	-1.125	-2.703	3,29	1.077	2.598	3,30	-1.031	-2.498
3,31	0.986	2.401	3,32	-0.939	-2.306	3,33	0.888	2.210			
4, 4	87.064	39.248	4, 5	-22.810	-33.315	4, 6	11.548	21.720	4, 7	-7.782	-16.139
4, 8	5.936	12.924	4, 9	-4.838	-10.829	4,10	4.108	9.350	4,11	-3.584	-8.247
4,12	3.189	7.389	4,13	-2.879	-6.702	4,14	2.628	6.138	4,15	-2.420	-5.665
4,16	2.245	5.263	4,17	-2.095	-4.917	4,18	1.965	4.614	4,19	-1.850	-4.347
4,20	1.748	4.109	4,21	-1.657	-3.896	4,22	1.574	3.703	4,23	-1.498	-3.528
4,24	1.429	3.368	4,25	-1.365	-3.220	4,26	1.305	3.083	4,27	-1.248	-2.956
4,28	1.194	2.836	4,29	-1.143	-2.722	4,30	1.093	2.614	4,31	-1.043	-2.510
4,32	0.993	2.408									
5, 5	87.192	39.392	5, 6	-22.914	-33.434	5, 7	11.636	21.822	5, 8	-7.859	-16.229
5, 9	6.004	13.004	5,10	-4.899	-10.901	5,11	4.163	9.416	5,12	-3.634	-8.306
5,13	3.235	7.444	5,14	-2.921	-6.752	5,15	2.667	6.185	5,16	-2.457	-5.709
5,17	2.280	5.304	5,18	-2.127	-4.954	5,19	1.995	4.649	5,20	-1.878	-4.379
5,21	1.774	4.139	5,22	-1.681	-3.924	5,23	1.596	3.729	5,24	-1.519	-3.552
5,25	1.448	3.390	5,26	-1.382	-3.240	5,27	1.320	3.100	5,28	-1.261	-2.970
5,29	1.205	2.848	5,30	-1.151	-2.731	5,31	1.098	2.619			
6, 6	87.277	39.491	6, 7	-22.986	-33.520	6, 8	11.698	21.897	6, 9	-7.914	-16.296
6,10	6.053	13.064	6,11	-4.943	-10.955	6,12	4.203	9.465	6,13	-3.671	-8.351
6,14	3.269	7.485	6,15	-2.952	-6.791	6,16	2.696	6.220	6,17	-2.484	-5.741
6,18	2.304	[illegible]	6,19	-2.150	-4.982	6,20	2.016	4.675	6,21	-1.898	-4.403
6,22	1.792	4.161	6,23	-1.691	[illegible]	[illegible]	1.611	3.747	6,25	-1.532	-3.568
6,26	1.460	3.404	6,27	-1.392	-3.251	6,28	1.328	3.110	[illegible]	[illegible]	-2.977
6,30	1.209	2.851									
7, 7	87.337	39.565	7, 8	-23.038	-33.583	7, 9	11.744	21.954	7,10	-7.955	-16.346
7,11	6.089	13.109	7,12	-4.977	-10.996	7,13	4.233	9.502	7,14	-3.699	-8.386
7,15	3.294	7.517	7,16	-2.976	-6.820	7,17	2.718	6.247	7,18	-2.504	-5.766
7,19	2.323	5.357	7,20	-2.167	-5.003	7,21	2.031	4.694	7,22	-1.912	-4.420
7,23	1.805	4.177	7,24	-1.709	-3.958	7,25	1.621	3.759	7,26	-1.541	-3.573
7,27	1.466	3.411	7,28	-1.397	-3.257	7,29	1.331	3.113			
8, 8	87.382	39.620	8, 9	-23.078	-33.633	8,10	11.779	21.997	8,11	-7.986	-16.385
8,12	6.118	13.145	8,13	-5.002	-11.029	8,14	4.257	9.532	8,15	-3.721	-8.413
8,16	3.314	7.542	8,17	-2.994	-6.842	8,18	2.735	6.267	8,19	-2.519	-5.785
8,20	2.337	5.374	8,21	-2.180	-5.018	8,22	2.043	4.707	8,23	-1.921	-4.432
8,24	1.813	4.187	8,25	-1.716	-3.966	8,26	1.627	3.766	8,27	-1.545	-3.583
8,28	1.468	3.414									
9, 9	87.417	39.663	9,10	-23.108	-33.671	9,11	11.807	22.032	9,12	-8.011	-16.417
9,13	6.140	13.173	9,14	-5.023	-11.054	9,15	4.275	9.555	9,16	-3.737	-8.434
9,17	3.329	7.561	9,18	-3.008	-6.860	9,19	2.747	6.283	9,20	-2.530	-5.799
9,21	2.347	5.386	9,22	-2.188	-5.029	9,23	2.050	4.717	9,24	-1.928	-4.440
9,25	1.818	4.193	9,26	-1.719	-3.970	9,27	1.628	3.768			
10,10	87.444	39.697	10,11	-23.132	-33.701	10,12	11.828	22.059	10,13	-8.030	-16.441
10,14	6.158	13.195	10,15	-5.038	-11.075	10,16	4.290	9.573	10,17	-3.750	-8.450
10,18	3.341	7.576	10,19	-3.018	-6.873	10,20	2.756	6.295	10,21	-2.538	-5.809
10,22	2.353	5.395	10,23	-2.194	-5.036	10,24	2.054	4.722	10,25	-1.931	-4.444
10,26	1.820	4.195									
11,11	87.465	39.725	11,12	-23.151	-33.726	11,13	11.845	22.081	11,14	-8.045	-16.461
11,15	6.171	13.213	11,16	-5.051	-11.090	11,17	4.301	9.587	11,18	-3.760	-8.463
11,19	3.349	7.586	11,20	-3.026	-6.882	11,21	2.762	6.302	11,22	-2.543	-5.815
11,23	2.357	5.400	11,24	-2.197	-5.039	11,25	2.056	4.724			
12,12	87.482	39.746	12,13	-23.166	-33.745	12,14	11.858	22.098	12,15	-8.057	-16.476
12,16	6.182	13.226	12,17	-5.060	-11.102	12,18	4.309	9.598	12,19	-3.767	-8.471
12,20	3.355	7.594	12,21	-3.030	-6.888	12,22	2.766	6.307	12,23	-2.546	-5.819
12,24	2.358	5.401									
13,13	87.495	39.763	13,14	-23.178	-33.760	13,15	11.869	22.111	13,16	-8.066	-16.487
13,17	6.190	13.236	13,18	-5.067	-11.111	13,19	4.314	9.605	13,20	-3.771	-8.477
13,21	3.359	7.598	13,22	-3.033	-6.891	13,23	2.767	6.309			
14,14	87.505	39.776	14,15	-23.187	-33.771	14,16	11.876	22.121	14,17	-8.073	-16.496
14,18	6.195	13.243	14,19	-5.071	-11.116	14,20	4.317	9.609	14,21	-3.773	-8.480
14,22	3.360	7.600									
15,15	87.513	39.786	15,16	-23.193	-33.779	15,17	11.882	22.128	15,18	-8.077	-16.501
15,19	6.198	13.247	15,20	-5.073	-11.119	15,21	4.318	9.610			
16,16	87.518	39.793	16,17	-23.197	-33.785	16,18	11.885	22.132	16,19	-8.079	-16.504
16,20	6.199	13.249									
17,17	87.521	39.797	17,18	-23.199	-33.787	17,19	11.886	22.133			
18,18	87.522	39.798									

IBR

SELF AND MUTUAL IMPEDANCES

H/LMDA=0.2500 B/LMDA= 0.500 OMEGA= 8.53

40 ELEMENT ARRAY

I, K	ZIK IN OHMS		I, K	ZIK IN OHMS		I, K	ZIK IN OHMS		I, K	ZIK IN OHMS	
	RE	IM		RE	IM		RE	IM		RE	IM
1, 1	84.972	37.926	1, 2	-21.533	-32.321	1, 3	10.613	20.937	1, 4	-7.038	-15.494
1, 5	5.315	12.376	1, 6	-4.304	-10.352	1, 7	3.639	8.929	1, 8	-3.166	-7.570
1, 9	2.811	7.049	1,10	-2.535	-6.392	1,11	2.313	5.855	1,12	-2.130	-5.405
1,13	1.976	5.023	1,14	-1.845	-4.695	1,15	1.731	4.408	1,16	-1.632	-4.156
1,17	1.544	3.933	1,18	-1.466	-3.732	1,19	1.395	3.552	1,20	-1.331	-3.389
1,21	1.273	3.240	1,22	-1.219	-3.103	1,23	1.170	2.978	1,24	-1.124	-2.861
1,25	1.082	2.753	1,26	-1.042	-2.653	1,27	1.004	2.558	1,28	-0.969	-2.470
1,29	0.935	2.386	1,30	-0.903	-2.307	1,31	0.872	2.232	1,32	-0.842	-2.160
1,33	0.812	2.090	1,34	-0.783	-2.023	1,35	0.754	1.958	1,36	-0.725	-1.893
1,37	0.694	1.828	1,38	-0.659	-1.761	1,39	0.617	1.689	1,40	-0.552	-1.603
2, 2	86.407	38.656	2, 3	-22.341	-32.860	2, 4	11.181	21.352	2, 5	-7.480	-15.831
2, 6	5.679	12.660	2, 7	-4.615	-10.598	2, 8	3.911	9.146	2, 9	-3.408	-8.064
2,10	3.030	7.225	2,11	-2.735	-6.554	2,12	2.497	6.003	2,13	-2.301	-5.543
2,14	2.135	5.152	2,15	-1.994	-4.815	2,16	1.872	4.521	2,17	-1.765	-4.263
2,18	1.670	4.033	2,19	-1.585	-3.828	2,20	1.509	3.643	2,21	-1.439	-3.475
2,22	1.376	3.322	2,23	-1.318	-3.181	2,24	1.265	3.052	2,25	-1.215	-2.933
2,26	1.169	2.821	2,27	-1.125	-2.718	2,28	1.084	2.620	2,29	-1.045	-2.529
2,30	1.008	2.443	2,31	-0.973	-2.361	2,32	0.939	2.282	2,33	-0.905	-2.207
2,34	0.872	2.135	2,35	-0.840	-2.064	2,36	0.807	1.995	2,37	-0.772	-1.925
2,38	0.734	1.854	2,39	-0.687	-1.777						
3, 3	86.859	39.044	3, 4	-22.656	-33.154	3, 5	11.425	21.589	3, 6	-7.680	-16.030
3, 7	5.850	12.832	3, 8	-4.764	-10.750	3, 9	4.043	9.281	3,10	-3.528	-8.187
3,11	3.139	7.337	3,12	-2.835	-6.656	3,13	2.589	6.098	3,14	-2.387	-5.631
3,15	2.216	5.234	3,16	-2.070	-4.892	3,17	1.943	4.594	3,18	-1.832	-4.332
3,19	1.734	4.098	3,20	-1.645	-3.889	3,21	1.566	3.701	3,22	-1.494	-3.530
3,23	1.428	3.374	3,24	-1.368	-3.232	3,25	1.312	3.100	3,26	-1.260	-2.978
3,27	1.212	2.864	3,28	-1.166	-2.758	3,29	1.123	2.659	3,30	-1.082	-2.565
3,31	1.043	2.477	3,32	-1.006	-2.392	3,33	0.969	2.312	3,34	-0.933	-2.234
3,35	0.898	2.159	3,36	-0.862	-2.084	3,37	0.824	2.010	3,38	-0.783	-1.934
4, 4	87.077	39.267	4, 5	-22.824	-33.334	4, 6	11.563	21.740	4, 7	-7.797	-16.160
4, 8	5.952	12.946	4, 9	-4.855	-10.852	4,10	4.125	9.373	4,11	-3.602	-8.270
4,12	3.207	7.414	4,13	-2.898	-6.728	4,14	2.648	6.164	4,15	-2.441	-5.693
4,16	2.267	5.292	4,17	-2.118	-4.947	4,18	1.988	4.645	4,19	-1.875	-4.380
4,20	1.774	4.144	4,21	-1.684	-3.932	4,22	1.602	3.742	4,23	-1.529	-3.569
4,24	1.461	3.411	4,25	-1.399	-3.266	4,26	1.341	3.132	4,27	-1.288	-3.008
4,28	1.238	2.893	4,29	-1.191	-2.785	4,30	1.146	2.684	4,31	-1.104	-2.589
4,32	1.063	2.498	4,33	-1.024	-2.411	4,34	0.985	2.328	4,35	-0.947	-2.248
4,36	0.908	2.169	4,37	-0.868	-2.090						
5, 5	87.207	39.412	5, 6	-22.930	-33.455	5, 7	11.652	21.844	5, 8	-7.876	-16.251
5, 9	6.021	13.027	5,10	-4.917	-10.925	5,11	4.181	9.440	5,12	-3.654	-8.332
5,13	3.255	7.470	5,14	-2.942	-6.780	5,15	2.689	6.213	5,16	-2.480	-5.739
5,17	2.303	5.335	5,18	-2.152	-4.987	5,19	2.020	4.683	5,20	-1.905	-4.415
5,21	1.802	4.177	5,22	-1.711	-3.964	5,23	1.628	3.771	5,24	-1.552	-3.597
5,25	1.484	3.437	5,26	-1.420	-3.290	5,27	1.361	3.155	5,28	-1.307	-3.030
5,29	1.255	2.913	5,30	-1.207	-2.804	5,31	1.161	2.701	5,32	-1.117	-2.603
5,33	1.074	2.510	5,34	-1.033	-2.422	5,35	0.992	2.336	5,36	-0.951	-2.252
6, 6	87.293	39.513	6, 7	-23.003	-33.542	6, 8	11.716	21.920	6, 9	-7.932	-16.319
6,10	6.071	13.088	6,11	-4.963	-10.981	6,12	4.223	9.491	6,13	-3.692	-8.379
6,14	3.290	7.514	6,15	-2.975	-6.820	6,16	2.720	6.251	6,17	-2.508	-5.774
6,18	2.330	5.368	6,19	-2.177	-5.017	6,20	2.044	4.711	6,21	-1.927	-4.442
6,22	1.823	4.202	6,23	-1.730	-3.987	6,24	1.646	3.793	6,25	-1.570	-3.617
6,26	1.500	3.456	6,27	-1.435	-3.308	6,28	1.375	3.172	6,29	-1.319	-3.045
6,30	1.266	2.926	6,31	-1.217	-2.815	6,32	1.169	2.711	6,33	-1.124	-2.611
6,34	1.079	2.516	6,35	-1.036	-2.425						
7, 7	87.355	39.588	7, 8	-23.056	-33.607	7, 9	11.763	21.978	7,10	-7.974	-16.372
7,11	6.109	13.136	7,12	-4.997	-11.024	7,13	4.255	9.531	7,14	-3.722	-8.415
7,15	3.318	7.548	7,16	-3.000	-6.852	7,17	2.743	6.280	7,18	-2.531	-5.801
7,19	2.351	5.393	7,20	-2.196	-5.041	7,21	2.062	4.734	7,22	-1.944	-4.463
7,23	1.839	4.222	7,24	-1.745	-4.005	7,25	1.660	3.810	7,26	-1.582	-3.633
7,27	1.511	3.470	7,28	-1.445	-3.321	7,29	1.384	3.183	7,30	-1.327	-3.055
7,31	1.273	2.935	7,32	-1.222	-2.822	7,33	1.173	2.715	7,34	-1.126	-2.614
8, 8	87.401	39.645	8, 9	-23.097	-33.658	8,10	11.799	22.024	8,11	-8.007	-16.413
8,12	6.139	13.173	8,13	-5.025	-11.058	8,14	4.280	9.562	8,15	-3.745	-8.444
8,16	3.339	7.575	8,17	-3.020	-6.877	8,18	2.762	6.303	8,19	-2.548	-5.823
8,20	2.367	5.413	8,21	-2.211	-5.060	8,22	2.076	4.751	8,23	-1.957	-4.479
8,24	1.851	4.236	8,25	-1.756	-4.019	8,26	1.670	3.822	8,27	-1.591	-3.644
8,28	1.519	3.480	8,29	-1.452	-3.329	8,30	1.390	3.190	8,31	-1.332	-3.060
8,32	1.277	2.939	8,33	-1.224	-2.824						
9, 9	87.437	39.690	9,10	-23.129	-33.698	9,11	11.828	22.060	9,12	-8.033	-16.446
9,13	6.163	13.204	9,14	-5.047	-11.086	9,15	4.300	9.588	9,16	-3.763	-8.468
9,17	3.357	7.597	9,18	-3.036	-6.897	9,19	2.777	6.322	9,20	-2.562	-5.840
9,21	2.379	5.429	9,22	-2.223	-5.075	9,23	2.087	4.765	9,24	-1.967	-4.491
9,25	1.860	4.247	9,26	-1.764	-4.029	9,27	1.677	3.831	9,28	-1.597	-3.651
9,29	1.524	3.486	9,30	-1.456	-3.334	9,31	1.393	3.193	9,32	-1.333	-3.062
10,10	87.465	39.726	10,11	-23.155	-33.731	10,12	11.851	22.090	10,13	-8.054	-16.473
10,14	6.183	13.228	10,15	-5.064	-11.108	10,16	4.317	9.609	10,17	-3.778	-8.487
10,18	3.370	7.614	10,19	-3.049	-6.913	10,20	2.789	6.337	10,21	-2.572	-5.854
10,22	2.389	5.442	10,23	-2.232	-5.086	10,24	2.095	4.775	10,25	-1.974	-4.500
10,26	1.866	4.255	10,27	-1.769	-4.036	10,28	1.681	3.837	10,29	-1.601	-3.655
10,30	1.527	3.489	10,31	-1.458	-3.336						
11,11	87.488	39.755	11,12	-23.175	-33.757	11,13	11.870	22.114	11,14	-8.071	-16.495
11,15	6.198	13.246	11,16	-5.079	-11.127	11,17	4.330	9.625	11,18	-3.791	-8.503
11,19	3.382	7.628	11,20	-3.059	-6.926	11,21	2.798	6.349	11,22	-2.581	-5.864
11,23	2.397	5.451	11,24	-2.239	-5.094	11,25	2.101	4.782	11,26	-1.979	-4.506
11,27	1.870	4.260	11,28	-1.772	-4.040	11,29	1.683	3.839	11,30	-1.602	-3.657
12,12	87.507	39.779	12,13	-23.192	-33.779	12,14	11.885	22.133	12,15	-8.085	-16.512
12,16	6.211	13.264	12,17	-5.090	-11.142	12,18	4.340	9.639	12,19	-3.800	-8.515
12,20	3.390	7.639	12,21	-3.067	-6.936	12,22	2.805	6.357	12,23	-2.587	-5.872
12,24	2.402	5.458	12,25	-2.243	-5.100	12,26	2.104	4.787	12,27	-1.982	-4.510
12,28	1.872	4.263	12,29	-1.773	-4.041						
13,13	87.522	39.798	13,14	-23.206	-33.796	13,15	11.898	22.149	13,16	-8.096	-16.527
13,17	6.221	13.278	13,18	-5.100	-11.154	13,19	4.349	9.650	13,20	-3.808	-8.524
13,21	3.397	7.648	13,22	-3.073	-6.944	13,23	2.810	6.364	13,24	-2.591	-5.877
13,25	2.406	5.462	13,26	-2.246	-5.103	13,27	2.106	4.789	13,28	-1.983	-4.511

14,14	87.534	39.814	14,15	-23.217	-33.811	14,16	11.908	22.162	14,17	-8.105	-16.539
14,18	6.229	13.288	14,19	-5.107	-11.163	14,20	4.355	9.658	14,21	-3.813	-8.532
14,22	3.402	7.654	14,23	-3.077	-6.949	14,24	2.813	6.368	14,25	-2.594	-5.881
14,26	2.407	5.465	14,27	-2.247	-5.104						
15,15	87.544	39.827	15,16	-23.226	-33.822	15,17	11.916	22.172	15,18	-8.112	-16.548
15,19	6.236	13.296	15,20	-5.112	-11.170	15,21	4.360	9.664	15,22	-3.817	-8.537
15,23	3.405	7.658	15,24	-3.079	-6.952	15,25	2.815	6.370	15,26	-2.594	-5.882
16,16	87.552	39.837	16,17	-23.233	-33.831	16,18	11.922	22.180	16,19	-8.118	-16.554
16,20	6.240	13.302	16,21	-5.116	-11.175	16,22	4.363	9.668	16,23	-3.819	-8.540
16,24	3.406	7.660	16,25	-3.080	-6.953						
17,17	87.556	39.845	17,18	-23.238	-33.838	17,19	11.926	22.186	17,20	-8.121	-16.559
17,21	6.243	13.306	17,22	-5.118	-11.178	17,23	4.364	9.670	17,24	-3.820	-8.541
18,18	87.563	39.851	18,19	-23.242	-33.843	18,20	11.929	22.190	18,21	-8.124	-16.562
18,22	6.244	13.307	18,23	-5.119	-11.178						
19,19	87.565	39.854	19,20	-23.244	-33.845	19,21	11.931	22.192	19,22	-8.124	-16.563
20,20	87.567	39.856	20,21	-23.245	-33.846						

IBR

SELF AND MUTUAL IMPEDANCES

H/LMDA=0.2500 B/LMDA= 0.500 OMEGA= 8.53

45 ELEMENT ARRAY

I, K	ZIK IN OHMS		I, K	ZIK IN OHMS		I, K	ZIK IN OHMS		I, K	ZIK IN OHMS	
	RE	IM		RE	IM		RE	IM		RE	IM
1, 1	84.980	[illegible]	[illegible]	[illegible]	-32.333	1, 3	10.622	20.950	1, 4	-7.048	-15.508
1, 5	5.325	12.389	1, 6	-4.314	-10.380	[illegible]	[illegible]	[illegible]	1, 8	-3.176	-7.885
1, 9	2.822	7.064	1,10	-2.546	-6.408	1,11	2.324	5.871	[illegible]	[illegible]	[illegible]
1,13	1.988	5.041	1,14	-1.858	-4.712	1,15	1.744	4.427	1,16	-1.646	-4.175
1,17	1.558	3.952	1,18	-1.480	-3.753	1,19	1.410	3.573	1,20	-1.347	-3.411
1,21	1.289	3.263	1,22	-1.237	-3.127	1,23	1.188	3.003	1,24	-1.143	-2.887
1,25	1.101	2.781	1,26	-1.062	-2.682	1,27	1.026	2.589	1,28	-0.992	-2.502
1,29	0.959	2.420	1,30	-0.928	-2.343	1,31	0.899	2.271	1,32	-0.871	-2.202
1,33	0.845	2.136	1,34	-0.819	-2.073	1,35	0.794	2.013	1,36	-0.770	-1.955
1,37	0.746	1.899	1,38	-0.722	-1.845	1,39	0.699	1.792	1,40	-0.675	-1.740
1,41	0.651	1.687	1,42	-0.625	-1.634	1,43	0.595	1.578	1,44	-0.559	-1.518
1,45	0.502	1.445									
2, 2	86.416	38.669	2, 3	-22.351	-32.874	2, 4	11.191	21.366	2, 5	-7.491	-15.846
2, 6	5.690	12.675	2, 7	-4.626	-10.613	2, 8	3.922	9.162	2, 9	-3.420	-8.080
2,10	3.042	7.242	2,11	-2.747	-6.571	2,12	2.510	6.021	2,13	-2.314	-5.561
2,14	2.149	5.170	2,15	-2.008	-4.834	2,16	1.887	4.541	2,17	-1.780	-4.283
2,18	1.686	4.055	2,19	-1.601	-3.850	2,20	1.526	3.666	2,21	-1.457	-3.499
2,22	1.395	3.347	2,23	-1.338	-3.208	2,24	1.285	3.079	2,25	-1.236	-2.961
2,26	1.191	2.852	2,27	-1.149	-2.749	2,28	1.109	2.654	2,29	-1.072	-2.565
2,30	1.036	2.481	2,31	-1.003	-2.401	2,32	0.971	2.326	2,33	-0.941	-2.255
2,34	0.911	2.187	2,35	-0.883	-2.122	2,36	0.856	2.060	2,37	-0.829	-2.000
2,38	0.802	1.941	2,39	-0.776	-1.884	2,40	0.749	1.828	2,41	-0.722	-1.772
2,42	0.693	1.716	2,43	-0.661	-1.657	2,44	0.621	1.592			
3, 3	86.869	39.057	3, 4	-22.666	-33.168	3, 5	11.436	21.604	3, 6	-7.692	-16.046
3, 7	5.861	12.847	3, 8	-4.776	-10.766	3, 9	4.056	9.298	3,10	-3.540	-8.204
3,11	3.152	7.354	3,12	-2.849	-6.675	3,13	2.604	6.117	3,14	-2.401	-5.651
3,15	2.231	5.254	3,16	-2.085	-4.913	3,17	1.959	4.616	3,18	-1.849	-4.354
3,19	1.751	4.121	3,20	-1.663	-3.913	3,21	1.585	3.726	3,22	-1.514	-3.556
3,23	1.449	3.402	3,24	-1.389	-3.260	3,25	1.334	3.130	3,26	-1.284	-3.009
3,27	1.236	2.897	3,28	-1.192	-2.793	3,29	1.151	2.696	3,30	-1.112	-2.605
3,31	1.075	2.519	3,32	-1.040	-2.438	3,33	1.006	2.361	3,34	-0.974	-2.288
3,35	0.943	2.218	3,36	-0.914	-2.152	3,37	0.884	2.087	3,38	-0.856	-2.025
3,39	0.827	1.965	3,40	-0.798	-1.905	3,41	0.769	1.846	3,42	-0.738	-1.786
3,43	0.703	1.724									
4, 4	87.068	39.282	4, 5	-22.836	-33.349	4, 6	11.574	21.756	4, 7	-7.810	-16.176
4, 8	5.964	12.962	4, 9	-4.868	-10.869	4,10	4.138	9.391	4,11	-3.616	-8.289
4,12	3.222	7.433	4,13	-2.913	-6.747	4,14	2.663	6.184	4,15	-2.457	-5.714
4,16	2.283	5.314	4,17	-2.135	-4.969	4,18	2.006	4.668	4,19	-1.893	-4.403
4,20	1.793	4.168	4,21	-1.703	-3.958	4,22	1.623	3.769	4,23	-1.550	-3.597
4,24	1.483	3.440	4,25	-1.422	-3.297	4,26	1.366	3.165	4,27	-1.314	-3.043
4,28	1.265	2.929	4,29	-1.220	-2.824	4,30	1.177	2.725	4,31	-1.137	-2.632
4,32	1.099	2.545	4,33	-1.062	-2.463	4,34	1.028	2.384	4,35	-0.994	-2.310
4,36	0.962	2.239	4,37	-0.931	-2.170	4,38	0.900	2.104	4,39	-0.870	-2.040
4,40	0.839	1.977	4,41	-0.808	-1.914	4,42	0.775	1.851			

5, 5	87.219	39.427	5, 6	-22.942	-33.471	5, 7	11.665	21.861	5, 8	-7.889	-16.269
5, 9	6.034	13.045	5,10	-4.931	-10.943	5,11	4.196	9.459	5,12	-3.668	-8.351
5,13	3.270	7.490	5,14	-2.958	-6.801	5,15	2.705	6.235	5,16	-2.497	-5.761
5,17	2.321	5.358	5,18	-2.170	-5.010	5,19	2.039	4.707	5,20	-1.924	-4.441
5,21	1.823	4.204	5,22	-1.732	-3.991	5,23	1.650	3.800	5,24	-1.576	-3.627
5,25	1.508	3.469	5,26	-1.446	-3.324	5,27	1.388	3.190	5,28	-1.335	-3.067
5,29	1.285	2.952	5,30	-1.239	-2.846	5,31	1.195	2.746	5,32	-1.154	-2.652
5,33	1.115	2.563	5,34	-1.077	-2.479	5,35	1.042	2.400	5,36	-1.007	-2.324
5,37	0.973	2.251	5,38	-0.941	-2.181	5,39	0.908	2.113	5,40	-0.876	-2.046
5,41	0.843	1.980									
6, 6	87.306	39.530	6, 7	-23.016	-33.559	6, 8	11.729	21.936	6, 9	-7.946	-16.338
6,10	6.086	13.107	6,11	-4.977	-11.000	6,12	4.238	9.511	6,13	-3.708	-8.399
6,14	3.307	7.535	6,15	-2.992	-6.842	6,16	2.737	6.273	6,17	-2.526	-5.797
6,18	2.349	5.392	6,19	-2.196	-5.043	6,20	2.064	4.738	6,21	-1.948	-4.469
6,22	1.845	4.231	6,23	-1.753	-4.017	6,24	1.670	3.825	6,25	-1.595	-3.650
6,26	1.526	3.491	6,27	-1.463	-3.345	6,28	1.404	3.210	6,29	-1.350	-3.085
6,30	1.300	2.970	6,31	-1.252	-2.862	6,32	1.208	2.761	6,33	-1.166	-2.666
6,34	1.125	2.576	6,35	-1.087	-2.491	6,36	1.050	2.410	6,37	-1.014	-2.333
6,38	0.980	2.258	6,39	-0.945	-2.186	6,40	0.911	2.116			
7, 7	87.368	39.605	7, 8	-23.070	-33.626	7, 9	11.777	21.997	7,10	-7.989	-16.391
7,11	6.125	13.156	7,12	-5.013	-11.044	7,13	4.271	9.552	7,14	-3.738	-8.437
7,15	3.335	7.570	7,16	-3.019	-6.875	7,17	2.762	6.304	7,18	-2.550	-5.826
7,19	2.371	5.419	7,20	-2.217	-5.068	7,21	2.084	4.762	7,22	-1.967	-4.492
7,23	1.863	4.252	7,24	-1.770	-4.038	7,25	1.686	3.844	7,26	-1.610	-3.669
7,27	1.540	3.508	7,28	-1.476	-3.360	7,29	1.417	3.225	7,30	-1.362	-3.099
7,31	1.310	2.983	7,32	-1.262	-2.874	7,33	1.217	2.771	7,34	-1.174	-2.675
7,35	1.133	2.584	7,36	-1.093	-2.498	7,37	1.055	2.416	7,38	-1.018	-2.337
7,39	0.982	2.260									
8, 8	87.415	39.664	8, 9	-23.112	-33.677	8,10	11.815	22.044	8,11	-8.023	-16.433
8,12	6.156	13.195	8,13	-5.042	-11.080	8,14	4.298	9.585	8,15	-3.763	-8.468
8,16	3.358	7.599	8,17	-3.040	-6.902	8,18	2.782	6.329	8,19	-2.569	-5.850
8,20	2.388	5.441	8,21	-2.234	-5.089	8,22	2.099	4.782	8,23	-1.981	-4.511
8,24	1.877	4.270	8,25	-1.783	-4.054	8,26	1.698	3.859	8,27	-1.621	-3.682
8,28	1.551	3.521	8,29	-1.486	-3.373	8,30	1.426	3.236	8,31	-1.370	-3.110
8,32	1.318	2.992	8,33	-1.269	-2.882	8,34	1.223	2.779	8,35	-1.179	-2.681
8,36	1.137	2.589	8,37	-1.096	-2.501	8,38	1.057	2.417			
9, 9	87.452	39.710	9,10	-23.145	-33.719	9,11	11.845	22.081	9,12	-8.050	-16.468
9,13	6.181	13.226	9,14	-5.065	-11.109	9,15	4.319	9.612	9,16	-3.783	-8.493
9,17	3.377	7.622	9,18	-3.057	-6.924	9,19	2.798	6.350	9,20	-2.584	-5.869
9,21	2.403	5.459	9,22	-2.247	-5.106	9,23	2.112	4.797	9,24	-1.993	-4.525
9,25	1.888	4.283	9,26	-1.793	-4.066	9,27	1.707	3.871	9,28	-1.630	-3.693
9,29	1.559	3.531	9,30	-1.493	-3.382	9,31	1.433	3.244	9,32	-1.376	-3.117
9,33	1.323	2.998	9,34	-1.273	-2.887	9,35	1.226	2.783	9,36	-1.181	-2.684
9,37	1.138	2.591									
10,10	87.481	39.747	10,11	-23.172	-33.753	10,12	11.869	22.112	10,13	-8.072	-16.496
10,14	6.201	13.252	10,15	-5.084	-11.133	10,16	4.337	9.634	10,17	-3.799	-8.514
10,18	3.392	7.642	10,19	-3.071	-6.942	10,20	2.812	6.367	10,21	-2.596	-5.884
10,22	2.414	5.474	10,23	-2.258	-5.119	10,24	2.122	4.810	10,25	-2.002	-4.537
10,26	1.896	4.294	10,27	-1.801	-4.077	10,28	1.715	3.880	10,29	-1.636	-3.701
10,30	1.565	3.538	10,31	-1.498	-3.388	10,32	1.437	3.250	10,33	-1.380	-3.122
10,34	1.326	3.002	10,35	-1.275	-2.890	10,36	1.227	2.784			
11,11	87.505	39.777	11,12	-23.193	-33.780	11,13	11.889	22.138	11,14	-8.090	-16.519
11,15	6.218	13.274	11,16	-5.099	-11.153	11,17	4.351	9.653	11,18	-3.813	-8.531
11,19	3.404	7.658	11,20	-3.083	-6.957	11,21	2.823	6.380	11,22	-2.606	-5.897
11,23	2.424	5.486	11,24	-2.267	-5.130	11,25	2.130	4.820	11,26	-2.010	-4.546
11,27	1.903	4.303	11,28	-1.807	-4.084	11,29	1.720	3.887	11,30	-1.641	-3.707
11,31	1.569	3.543	11,32	-1.502	-3.393	11,33	1.440	3.253	11,34	-1.382	-3.124
11,35	1.327	3.003									
12,12	87.525	39.803	12,13	-23.211	-33.803	12,14	11.905	22.159	12,15	-8.106	-16.539
12,16	6.232	13.292	12,17	-5.112	-11.170	12,18	4.363	9.668	12,19	-3.824	-8.545
12,20	3.415	7.671	12,21	-3.093	-6.969	12,22	2.831	6.392	12,23	-2.615	-5.908
12,24	2.431	5.495	12,25	-2.274	-5.139	12,26	2.136	4.828	12,27	-2.015	-4.553
12,28	1.908	4.309	12,29	-1.811	-4.090	12,30	1.724	3.891	12,31	-1.644	-3.711
12,32	1.571	3.546	12,33	-1.504	-3.395	12,34	1.441	3.254			
13,13	87.542	39.824	13,14	-23.226	-33.823	13,15	11.919	22.176	13,16	-8.118	-16.555
13,17	6.244	13.307	13,18	-5.123	-11.184	13,19	4.373	9.681	13,20	-3.833	-8.557
13,21	3.423	7.682	13,22	-3.100	-6.979	13,23	2.839	6.401	13,24	-2.621	-5.916
13,25	2.437	5.503	13,26	-2.279	-5.146	13,27	2.141	4.834	13,28	-2.020	-4.559
13,29	1.911	4.313	13,30	-1.814	-4.093	13,31	1.726	3.894	13,32	-1.646	-3.713
13,33	1.572	3.547									
14,14	87.555	39.841	14,15	-23.239	-33.839	14,16	11.931	22.191	14,17	-8.129	-16.569
14,18	6.254	13.319	14,19	-5.132	-11.195	14,20	4.381	9.692	14,21	-3.840	-8.567
14,22	3.430	7.691	14,23	-3.107	-6.987	14,24	2.844	6.408	14,25	-2.626	-5.922
14,26	2.441	5.509	14,27	-2.283	-5.151	14,28	2.144	4.838	14,29	-2.022	-4.562
14,30	1.913	4.316	14,31	-1.816	-4.095	14,32	1.727	3.895			
15,15	87.567	39.856	15,16	-23.249	-33.852	15,17	11.940	22.204	15,18	-8.138	-16.580
15,19	6.262	13.329	15,20	-5.139	-11.205	15,21	4.388	9.700	15,22	-3.846	-8.574
15,23	3.435	7.697	15,24	-3.111	-6.993	15,25	2.848	6.413	15,26	-2.630	-5.927
15,27	2.445	5.512	15,28	-2.285	-5.154	15,29	2.146	4.841	15,30	-2.023	-4.564
15,31	1.914	4.317									
16,16	87.576	39.868	16,17	-23.258	-33.863	16,18	11.948	22.214	16,19	-8.145	-16.589
16,20	6.268	13.338	16,21	-5.145	-11.212	16,22	4.393	9.707	16,23	-3.851	-8.580
16,24	3.440	7.703	16,25	-3.115	-6.998	16,26	2.851	6.417	16,27	-2.632	-5.930
16,28	2.446	5.515	16,29	-2.286	-5.156	16,30	2.147	4.841			
17,17	87.584	39.878	17,18	-23.265	-33.872	17,19	11.954	22.222	17,20	-8.151	-16.597
17,21	6.273	13.344	17,22	-5.150	-11.218	17,23	4.397	9.712	17,24	-3.855	-8.585
17,25	3.442	7.706	17,26	-3.117	-7.000	17,27	2.853	6.419	17,28	-2.633	-5.932
17,29	2.447	5.516									
18,18	87.590	39.887	18,19	-23.271	-33.880	18,20	11.959	22.228	18,21	-8.155	-16.602
18,22	6.277	13.349	18,23	-5.153	-11.222	18,24	4.400	9.715	18,25	-3.857	-8.588
18,26	3.444	7.708	18,27	-3.118	-7.002	18,28	2.854	6.420			
19,19	87.595	39.893	19,20	-23.275	-33.885	19,21	11.963	22.233	19,22	-8.158	-16.606
19,23	6.280	13.353	19,24	-5.155	-11.225	19,25	4.402	9.718	19,26	-3.858	-8.589
19,27	3.445	7.709									
20,20	87.599	39.898	20,21	-23.278	-33.890	20,22	11.966	22.237	20,23	-8.160	-16.609
20,24	6.281	13.355	20,25	-5.156	-11.226	20,26	4.402	9.718			
21,21	87.602	39.901	21,22	-23.281	-33.892	21,23	11.968	22.239	21,24	-8.161	-16.611
21,25	6.282	13.356									
22,22	87.604	39.903	22,23	-23.282	-33.894	22,24	11.968	22.240			
23,23	87.604	39.904									

IBR

SELF AND MUTUAL IMPEDANCES

H/LMDA=0.2500 B/LMDA= 0.500 OMEGA= 8.53

50 ELEMENT ARRAY

I, K	ZIK IN OHMS RE	IM	I, K	ZIK IN OHMS RE	IM	I, K	ZIK IN OHMS RE	IM	I, K	ZIK IN OHMS RE	IM
1, 1	84.986	37.948	1, 2	-21.548	-32.343	1, 3	10.629	20.961	1, 4	-7.055	-15.518
1, 5	5.332	12.400	1, 6	-4.322	-10.378	1, 7	3.657	8.955	1, 8	-3.185	-7.897
1, 9	2.831	7.076	1,10	-2.555	-6.421	1,11	2.333	5.884	1,12	-2.151	-5.435
1,13	1.998	5.054	1,14	-1.868	-4.727	1,15	1.755	4.441	1,16	-1.656	-4.190
1,17	1.569	3.968	1,18	-1.492	-3.769	1,19	1.422	3.590	1,20	-1.359	-3.428
1,21	1.302	3.280	1,22	-1.249	-3.145	1,23	1.201	3.021	1,24	-1.157	-2.907
1,25	1.116	2.801	1,26	-1.077	-2.703	1,27	1.042	2.611	1,28	-1.008	-2.525
1,29	0.976	2.444	1,30	-0.946	-2.368	1,31	0.918	2.297	1,32	-0.891	-2.229
1,33	0.866	2.165	1,34	-0.841	-2.104	1,35	0.818	2.046	1,36	-0.795	-1.991
1,37	0.773	1.938	1,38	-0.752	-1.887	1,39	0.732	1.837	1,40	-0.712	-1.790
1,41	0.692	1.744	1,42	-0.672	-1.699	1,43	0.653	1.655	1,44	-0.633	-1.611
1,45	0.613	1.568	1,46	-0.592	-1.524	1,47	0.570	1.480	1,48	-0.544	-1.433
1,49	0.512	1.380	1,50	-0.461	-1.317						
2, 2	86.423	38.679	2, 3	-22.358	-32.884	2, 4	11.199	21.377	2, 5	-7.499	-15.857
2, 6	5.699	12.686	2, 7	-4.635	-10.626	2, 8	3.931	9.174	2, 9	-3.430	-8.093
2,10	3.052	7.255	2,11	-2.757	-6.584	2,12	2.520	6.035	2,13	-2.325	-5.575
2,14	2.160	5.185	2,15	-2.020	-4.849	2,16	1.898	4.557	2,17	-1.792	-4.299
2,18	1.698	4.071	2,19	-1.614	-3.867	2,20	1.539	3.683	2,21	-1.471	-3.517
2,22	1.409	3.366	2,23	-1.352	-3.227	2,24	1.300	3.100	2,25	-1.252	-2.982
2,26	1.207	2.873	2,27	-1.165	-2.772	2,28	1.127	2.678	2,29	-1.090	-2.589
2,30	1.056	2.507	2,31	-1.023	-2.429	2,32	0.992	2.355	2,33	-0.963	-2.285
2,34	0.935	2.219	2,35	-0.909	-2.157	2,36	0.883	2.097	2,37	-0.858	-2.039
2,38	0.834	1.984	2,39	-0.811	-1.932	2,40	0.789	1.880	2,41	-0.767	-1.831
2,42	0.745	1.783	2,43	-0.723	-1.736	2,44	0.701	1.690	2,45	-0.679	-1.644
2,46	0.655	1.597	2,47	-0.631	-1.550	2,48	0.602	1.500	2,49	-0.567	-1.445
3, 3	86.878	39.068	3, 4	-22.675	-33.180	3, 5	11.445	21.616	3, 6	-7.701	-16.058
3, 7	5.871	12.860	3, 8	-4.786	-10.779	3, 9	4.066	9.311	3,10	-3.551	-8.217
3,11	3.163	7.368	3,12	-2.859	-6.689	3,13	2.615	6.132	3,14	-2.413	-5.666
3,15	2.243	5.270	3,16	-2.098	-4.929	3,17	1.972	4.632	3,18	-1.862	-4.371
3,19	1.764	4.139	3,20	-1.677	-3.932	3,21	1.599	3.745	3,22	-1.528	-3.576
3,23	1.464	3.422	3,24	-1.405	-3.281	3,25	1.351	3.151	3,26	-1.301	-3.032
3,27	1.254	2.921	3,28	-1.211	-2.818	3,29	1.170	2.722	3,30	-1.132	-2.632
3,31	1.096	2.547	3,32	-1.062	-2.468	3,33	1.030	2.393	3,34	-1.000	-2.322
3,35	0.970	2.254	3,36	-0.942	-2.190	3,37	0.916	2.129	3,38	-0.890	-2.070
3,39	0.864	2.014	3,40	-0.840	-1.959	3,41	0.816	1.907	3,42	-0.792	-1.856
3,43	0.769	1.806	3,44	-0.745	-1.757	3,45	0.721	1.708	3,46	-0.696	-1.660
3,47	0.670	1.610	3,48	-0.640	-1.557						
4, 4	87.097	39.294	4, 5	-22.845	-33.361	4, 6	11.584	21.768	4, 7	-7.819	-16.189
4, 8	5.974	12.976	4, 9	-4.878	-10.882	4,10	4.149	9.405	4,11	-3.627	-8.303
4,12	3.233	7.447	4,13	-2.924	-6.762	4,14	2.675	6.200	4,15	-2.469	-5.730
4,16	2.296	5.330	4,17	-2.148	-4.986	4,18	2.019	4.686	4,19	-1.907	-4.422
4,20	1.807	4.187	4,21	-1.718	-3.977	4,22	1.638	3.789	4,23	-1.566	-3.618
4,24	1.500	3.462	4,25	-1.439	-3.319	4,26	1.384	3.188	4,27	-1.332	-3.067
4,28	1.285	2.955	4,29	-1.240	-2.850	4,30	1.198	2.753	4,31	-1.159	-2.662
4,32	1.122	2.576	4,33	-1.087	-2.495	4,34	1.054	2.419	4,35	-1.023	-2.347
[illegible]	0.992	2.278	4,37	-0.963	-2.213	4,38	0.935	2.150	4,39	-0.908	-2.090
4,40	0.882	[illegible]	[illegible]	[illegible]	-1.977	4,42	0.831	1.923	4,43	-0.806	-1.870
4,44	0.781	1.819	4,45	-0.756	[illegible]	[illegible]	0.730	1.716	4,47	-0.702	-1.664
5, 5	87.228	39.440	5, 6	-22.952	-33.484	5, 7	11.675	21.874	5, 8	-7.899	-16.282
5, 9	6.045	13.059	5,10	-4.942	-10.958	5,11	4.207	9.474	5,12	-3.680	-8.366
5,13	3.282	7.506	5,14	-2.970	-6.817	5,15	2.718	6.251	5,16	-2.510	-5.778
5,17	2.334	5.375	5,18	-2.184	-5.028	5,19	2.054	4.726	5,20	-1.939	-4.460
5,21	1.838	4.224	5,22	-1.748	-4.012	5,23	1.666	3.822	5,24	-1.593	-3.649
5,25	1.526	3.492	5,26	-1.464	-3.348	5,27	1.407	3.215	5,28	-1.355	-3.093
5,29	1.306	2.980	5,30	-1.261	-2.874	5,31	1.218	2.776	5,32	-1.178	-2.683
5,33	1.140	2.597	5,34	-1.105	-2.515	5,35	1.071	2.438	5,36	-1.038	-2.365
5,37	1.007	2.295	5,38	-0.977	-2.228	5,39	0.949	2.165	5,40	-0.921	-2.104
5,41	0.893	2.045	5,42	-0.867	-1.988	5,43	0.840	1.933	5,44	-0.814	-1.878
5,45	0.787	1.825	5,46	-0.759	-1.771						
6, 6	87.316	39.543	6, 7	-23.026	-33.573	6, 8	11.740	21.952	6, 9	-7.957	-16.352
6,10	6.097	13.122	6,11	-4.989	-11.015	6,12	4.250	9.527	6,13	-3.720	-8.415
6,14	3.320	7.552	6,15	-3.005	-6.859	6,16	2.751	6.291	6,17	-2.541	-5.815
6,18	2.363	5.411	6,19	-2.211	-5.062	6,20	2.080	4.758	6,21	-1.964	-4.490
6,22	1.862	4.252	6,23	-1.770	-4.039	6,24	1.688	3.848	6,25	-1.613	-3.674
6,26	1.545	3.515	6,27	-1.483	-3.370	6,28	1.425	3.237	6,29	-1.372	-3.114
6,30	1.322	2.999	6,31	-1.276	-2.893	6,32	1.233	2.793	6,33	-1.192	-2.700
6,34	1.154	2.612	6,35	-1.117	-2.530	6,36	1.082	2.452	6,37	-1.049	-2.378
6,38	1.017	2.307	6,39	-0.987	-2.240	6,40	0.957	2.175	6,41	-0.928	-2.113
6,42	0.900	2.053	6,43	-0.872	-1.994	6,44	0.844	1.937	6,45	-0.816	-1.881
7, 7	87.379	39.619	7, 8	-23.081	-33.640	7, 9	11.789	22.012	7,10	-8.001	-16.406
7,11	6.137	13.171	7,12	-5.026	-11.061	7,13	4.284	9.569	7,14	-3.752	-8.454
7,15	3.349	7.588	7,16	-3.033	-6.893	7,17	2.777	6.323	7,18	-2.565	-5.845
7,19	2.386	5.439	7,20	-2.233	-5.089	7,21	2.100	4.783	7,22	-1.984	-4.514
7,23	1.880	4.275	7,24	-1.788	-4.061	7,25	1.705	3.868	7,26	-1.629	-3.693
7,27	1.560	3.534	7,28	-1.497	-3.388	7,29	1.439	3.254	7,30	-1.385	-3.130
7,31	1.335	3.014	7,32	-1.288	-2.907	7,33	1.244	2.807	7,34	-1.203	-2.713
7,35	1.164	2.624	7,36	-1.126	-2.541	7,37	1.091	2.462	7,38	-1.057	-2.387
7,39	1.025	2.315	7,40	-0.993	-2.247	7,41	0.962	2.181	7,42	-0.933	-2.118
7,43	0.903	2.056	7,44	-0.874	-1.996						
8, 8	87.427	39.678	8, 9	-23.124	-33.693	8,10	11.827	22.060	8,11	-8.035	-16.450
8,12	6.169	13.211	8,13	-5.055	-11.097	8,14	4.311	9.603	8,15	-3.777	-8.486
8,16	3.373	7.618	8,17	-3.055	-6.921	8,18	2.798	6.349	8,19	-2.585	-5.870
8,20	2.405	5.462	8,21	-2.251	-5.111	8,22	2.117	4.804	8,23	-2.000	-4.534
8,24	1.895	4.294	8,25	-1.802	-4.079	8,26	1.718	3.885	8,27	-1.642	-3.709
8,28	1.573	3.549	8,29	-1.509	-3.402	8,30	1.450	3.267	8,31	-1.395	-3.142
8,32	1.345	3.026	8,33	-1.297	-2.918	8,34	1.253	2.817	8,35	-1.211	-2.722
8,36	1.171	2.633	8,37	-1.133	-2.549	8,38	1.097	2.469	8,39	-1.062	-2.393
8,40	1.029	2.321	8,41	-0.997	-2.251	8,42	0.965	2.184	8,43	-0.934	-2.119
9, 9	87.464	39.725	9,10	-23.158	-33.735	9,11	11.857	22.098	9,12	-8.063	-16.485
9,13	6.194	13.244	9,14	-5.079	-11.127	9,15	4.334	9.631	9,16	-3.798	-8.512
9,17	3.392	7.642	9,18	-3.073	-6.944	9,19	2.815	6.371	9,20	-2.601	-5.890
9,21	2.420	5.482	9,22	-2.265	-5.129	9,23	2.131	4.821	9,24	-2.013	-4.550
9,25	1.908	4.309	9,26	-1.814	-4.093	9,27	1.729	3.899	9,28	-1.653	-3.722
9,29	1.582	3.561	9,30	-1.518	-3.414	9,31	1.459	3.278	9,32	-1.404	-3.152
9,33	1.352	3.036	9,34	-1.304	-2.927	9,35	1.259	2.825	9,36	-1.216	-2.729
9,37	1.176	2.639	9,38	-1.137	-2.554	9,39	1.101	2.474	9,40	-1.065	-2.397
9,41	1.031	2.323	9,42	-0.998	-2.252						

10,10	87.494	39.764	10,11	-23.185	-33.770	10,12	11.882	22.130	10,13	-8.086	-16.514
10,14	6.216	13.271	10,15	-5.099	-11.152	10,16	4.352	9.654	10,17	-3.815	-8.534
10,18	3.408	7.663	10,19	-3.088	-6.963	10,20	2.829	6.389	10,21	-2.614	-5.907
10,22	2.433	5.498	10,23	-2.277	-5.144	10,24	2.142	4.835	10,25	-2.023	-4.564
10,26	1.918	4.322	10,27	-1.823	-4.105	10,28	1.738	3.910	10,29	-1.661	-3.733
10,30	1.590	3.571	10,31	-1.525	-3.423	10,32	1.465	3.286	10,33	-1.410	-3.160
10,34	1.358	3.043	10,35	-1.309	-2.933	10,36	1.264	2.831	10,37	-1.220	-2.734
10,38	1.179	2.643	10,39	-1.140	-2.558	10,40	1.103	2.476	10,41	-1.066	-2.398
11,11	87.519	39.795	11,12	-23.207	-33.798	11,13	11.903	22.156	11,14	-8.105	-16.538
11,15	6.234	13.293	11,16	-5.115	-11.173	11,17	4.367	9.674	11,18	-3.829	-8.552
11,19	3.422	7.680	11,20	-3.101	-6.980	11,21	2.841	6.404	11,22	-2.626	-5.922
11,23	2.443	5.511	11,24	-2.287	-5.157	11,25	2.151	4.847	11,26	-2.032	-4.575
11,27	1.926	4.332	11,28	-1.831	-4.115	11,29	1.745	3.919	11,30	-1.668	-3.741
11,31	1.596	3.579	11,32	-1.531	-3.430	11,33	1.471	3.293	11,34	-1.414	-3.166
11,35	1.362	3.048	11,36	-1.313	-2.937	11,37	1.266	2.834	11,38	-1.223	-2.737
11,39	1.181	2.645	11,40	-1.141	-2.559						
12,12	87.540	39.821	12,13	-23.226	-33.822	12,14	11.921	22.178	12,15	-8.121	-16.559
12,16	6.248	13.312	12,17	-5.129	-11.191	12,18	4.381	9.690	12,19	-3.842	-8.568
12,20	3.433	7.694	12,21	-3.112	-6.993	12,22	2.851	6.417	12,23	-2.635	-5.934
12,24	2.452	5.522	12,25	-2.296	-5.167	12,26	2.159	4.857	12,27	-2.039	-4.584
12,28	1.933	4.341	12,29	-1.837	-4.123	12,30	1.751	3.926	12,31	-1.673	-3.748
12,32	1.601	3.585	12,33	-1.535	-3.435	12,34	1.474	3.297	12,35	-1.418	-3.170
12,36	1.365	3.051	12,37	-1.315	-2.940	12,38	1.268	2.836	12,39	-1.223	-2.738
13,13	87.557	39.843	13,14	-23.242	-33.843	13,15	11.935	22.197	13,16	-8.135	-16.576
13,17	6.261	13.329	13,18	-5.141	-11.206	13,19	4.392	9.704	13,20	-3.852	-8.581
13,21	3.443	7.707	13,22	-3.121	-7.005	13,23	2.860	6.428	13,24	-2.643	-5.944
13,25	2.460	5.532	13,26	-2.302	-5.176	13,27	2.165	4.865	13,28	-2.045	-4.591
13,29	1.938	4.347	13,30	-1.842	-4.129	13,31	1.755	3.932	13,32	-1.677	-3.753
13,33	1.605	3.589	13,34	-1.538	-3.439	13,35	1.477	3.300	13,36	-1.419	-3.172
13,37	1.366	3.053	13,38	-1.316	-2.941						
14,14	87.572	39.862	14,15	-23.256	-33.860	14,16	11.948	22.213	14,17	-8.147	-16.591
14,18	6.272	13.342	14,19	-5.151	-11.219	14,20	4.401	9.716	14,21	-3.861	-8.592
14,22	3.451	7.717	14,23	-3.128	-7.014	14,24	2.867	6.436	14,25	-2.649	-5.952
14,26	2.466	5.539	14,27	-2.308	-5.183	14,28	2.170	4.872	14,29	-2.050	-4.597
14,30	1.942	4.353	14,31	-1.846	-4.134	14,32	1.759	3.936	14,33	-1.679	-3.756
14,34	1.607	3.592	14,35	-1.540	-3.441	14,36	1.478	3.302	14,37	-1.420	-3.173
15,15	87.584	39.878	15,16	-23.267	-33.875	15,17	11.958	22.227	15,18	-8.156	-16.604
15,19	6.281	13.354	15,20	-5.160	-11.230	15,21	4.409	9.726	15,22	-3.868	-8.602
15,23	3.458	7.726	15,24	-3.135	-7.022	15,25	2.872	6.444	15,26	-2.655	-5.959
15,27	2.470	5.545	15,28	-2.312	-5.188	15,29	2.174	4.877	15,30	-2.053	-4.601
15,31	1.945	4.356	15,32	-1.848	-4.137	15,33	1.761	3.938	15,34	-1.681	-3.758
15,35	1.608	3.593	15,36	-1.540	-3.442						
16,16	87.595	39.891	16,17	-23.277	-33.887	16,18	11.967	22.238	16,19	-8.165	-16.615
16,20	6.289	13.364	16,21	-5.167	-11.239	16,22	4.415	9.735	16,23	-3.874	-8.609
16,24	3.463	7.733	16,25	-3.140	-7.029	16,26	2.877	6.450	16,27	-2.659	-5.964
16,28	2.474	5.550	16,29	-2.316	-5.193	16,30	2.177	4.880	16,31	-2.055	-4.604
16,32	1.947	4.359	16,33	-1.850	-4.139	16,34	1.762	3.940	16,35	-1.682	-3.759
17,17	87.604	39.903	17,18	-23.285	-33.898	17,19	11.975	22.248	17,20	-8.172	-16.624
17,21	6.295	13.372	17,22	-5.173	-11.247	17,23	4.421	9.742	17,24	-3.879	-8.616
17,25	3.468	7.739	17,26	-3.144	-7.034	17,27	2.881	6.454	17,28	-2.662	-5.968
17,29	2.477	5.554	17,30	-2.318	-5.196	17,31	2.179	4.883	17,32	-2.057	-4.606
17,33	1.948	4.360	17,34	-1.850	-4.139						
18,18	87.611	39.913	18,19	-23.292	-33.907	18,20	11.981	22.256	18,21	-8.178	-16.631
18,22	6.301	13.379	18,23	-5.177	-11.253	18,24	4.425	9.747	18,25	-3.883	-8.621
18,26	3.471	7.743	18,27	-3.147	-7.038	18,28	2.883	6.458	18,29	-2.664	-5.971
18,30	2.479	5.556	18,31	-2.319	-5.197	18,32	2.180	4.884	18,33	-2.057	-4.607
19,19	87.617	39.921	19,20	-23.298	-33.914	19,21	11.987	22.263	19,22	-8.182	-16.637
19,23	6.305	13.385	19,24	-5.181	-11.258	19,25	4.429	9.752	19,26	-3.886	-8.625
19,27	3.474	7.746	19,28	-3.149	-7.041	19,29	2.885	6.460	19,30	-2.666	-5.973
19,31	2.480	5.557	19,32	-2.320	-5.198						
20,20	87.623	39.928	20,21	-23.303	-33.920	20,22	11.991	22.269	20,23	-8.186	-16.642
20,24	6.308	13.389	20,25	-5.184	-11.262	20,26	4.431	9.755	20,27	-3.888	-8.627
20,28	3.476	7.748	20,29	-3.150	-7.042	20,30	2.886	6.461	20,31	-2.666	-5.973
21,21	87.627	39.933	21,22	-23.306	-33.925	21,23	11.994	22.273	21,24	-8.189	-16.646
21,25	6.311	13.392	21,26	-5.186	-11.264	21,27	4.433	9.757	21,28	-3.889	-8.629
21,29	3.476	7.750	21,30	-3.151	-7.043						
22,22	87.630	39.937	22,23	-23.309	-33.929	22,24	11.997	22.276	22,25	-8.191	-16.648
22,26	6.312	13.394	22,27	-5.188	-11.266	22,28	4.434	9.758	22,29	-3.890	-8.629
23,23	87.633	39.941	23,24	-23.311	-33.932	23,25	11.998	22.278	23,26	-8.192	-16.650
23,27	6.313	13.395	23,28	-5.188	-11.267						
24,24	87.634	39.943	24,25	-23.313	-33.933	24,26	11.999	22.279	24,27	-8.193	-16.651
25,25	87.635	39.944	25,26	-23.313	-33.934						

IBR

SELF AND MUTUAL IMPEDANCES

H/LMDA=0.5000 B/LMDA= 0.250 OMEGA= 9.92

2 ELEMENT ARRAY

I, K	ZIK IN OHMS		I, K	ZIK IN OHMS		I, K	ZIK IN OHMS		I, K	ZIK IN OHMS	
	RE	IM		RE	IM		RE	IM		RE	IM
1, 1	351.847	-551.993	1, 2	36.520	257.716						

IBR

SELF AND MUTUAL IMPEDANCES

H/LMDA=0.5000 B/LMDA= 0.250 OMEGA= 9.92

3 ELEMENT ARRAY

I, K	ZIK IN OHMS		I, K	ZIK IN OHMS		I, K	ZIK IN OHMS		I, K	ZIK IN OHMS	
	RE	IM		RE	IM		RE	IM		RE	IM
1, 1	347.113	-559.084	1, 2	72.564	279.936	1, 3	10.204	-57.327			
2, 2	157.270	-589.815									

IBR

SELF AND MUTUAL IMPEDANCES

H/LMDA=0.5000 B/LMDA= 0.250 OMEGA= 9.92

4 ELEMENT ARRAY

I, K	ZIK IN OHMS		I, K	ZIK IN OHMS		I, K	ZIK IN OHMS		I, K	ZIK IN OHMS	
	RE	IM		RE	IM		RE	IM		RE	IM
1, 1	347.458	-558.803	1, 2	70.932	281.269	1, 3	13.991	-65.614	1, 4	-14.156	-0.133
2, 2	151.754	-599.569	2, 3	108.975	308.280						

IBR

SELF AND MUTUAL IMPEDANCES

H/LMDA=0.5000 B/LMDA= 0.250 OMEGA= 9.92

5 ELEMENT ARRAY

I, K	ZIK IN OHMS		I, K	ZIK IN OHMS		I, K	ZIK IN OHMS		I, K	ZIK IN OHMS	
	RE	IM		RE	IM		RE	IM		RE	IM
1, 1	347.420	-558.939	1, 2	71.221	281.164	1, 3	14.354	-64.548	1, 4	-17.685	-3.590
1, 5	-2.310	7.986									
2, 2	152.095	-598.923	2, 3	106.586	309.485	2, 4	21.509	-75.486			
3, 3	146.566	-609.136									

IBR

SELF AND MUTUAL IMPEDANCES

H/LMDA=0.5000 B/LMDA= 0.250 OMEGA= 9.92

6 ELEMENT ARRAY

I, K	ZIK IN OHMS		I, K	ZIK IN OHMS		I, K	ZIK IN OHMS		I, K	ZIK IN OHMS	
	RE	IM		RE	IM		RE	IM		RE	IM
1, 1	347.441	-558.870	1, 2	71.085	281.186	1, 3	14.283	-64.743	1, 4	-16.943	-3.866
1, 5	-4.671	10.521	1, 6	5.842	1.689						
2, 2	152.098	-599.208	2, 3	107.009	309.405	2, 4	21.824	-73.876	2, 5	-22.797	-9.633
3, 3	146.849	-608.523	3, 4	104.060	310.842						

IBR

SELF AND MUTUAL IMPEDANCES

H/LMDA=0.5000 B/LMDA= 0.250 OMEGA= 9.92

7 ELEMENT ARRAY

I, K	ZIK IN OHMS		I, K	ZIK IN OHMS		I, K	ZIK IN OHMS		I, K	ZIK IN OHMS	
	RE	IM		RE	IM		RE	IM		RE	IM
1, 1	347.428	-558.907	1, 2	71.157	281.170	1, 3	14.299	-64.648	1, 4	-17.083	-3.806
1, 5	-4.897	9.992	1, 6	7.767	3.317	1, 7	1.208	-4.383			
2, 2	152.103	-599.060	2, 3	106.813	309.402	2, 4	21.757	-74.176	2, 5	-21.660	-9.890
2, 6	-9.050	14.335									
3, 3	146.874	-608.788	3, 4	104.474	310.782	3, 5	22.298	-72.104			

IBR

SELF AND MUTUAL IMPEDANCES

H/LMDA=0.5000 B/LMDA= 0.250 OMEGA= 9.92

8 ELEMENT ARRAY

I, K	ZIK IN OHMS		I, K	ZIK IN OHMS		I, K	ZIK IN OHMS		I, K	ZIK IN OHMS	
	RE	IM		RE	IM		RE	IM		RE	IM
1, 1	347.437	-558.885	1, 2	71.115	281.181	1, 3	14.288	-64.702	1, 4	-17.009	-3.822
1, 5	-4.847	10.099	1, 6	7.361	3.506	1, 7	2.425	-5.920	1, 8	-3.439	-0.930
2, 2	152.096	-599.143	2, 3	106.920	309.399	2, 4	21.761	-74.030	2, 5	-21.881	-9.832
2, 6	-9.272	13.497	2, 7	10.705	6.591						
3, 3	146.866	-608.650	3, 4	104.285	310.771	3, 5	22.249	-72.396	3, 6	-20.365	-10.252
4, 4	147.163	-608.174	4, 5	104.901	310.690						

IBR

SELF AND MUTUAL IMPEDANCES

H/LMDA=0.5000 B/LMDA= 0.250 OMEGA= 9.92

9 ELEMENT ARRAY

I, K	ZIK IN OHMS		I, K	ZIK IN OHMS		I, K	ZIK IN OHMS		I, K	ZIK IN OHMS	
	RE	IM		RE	IM		RE	IM		RE	IM
1, 1	347.430	-558.899	1, 2	71.142	281.173	1, 3	14.296	-64.669	1, 4	-17.053	-3.811
1, 5	-4.861	10.039	1, 6	7.447	3.463	1, 7	2.586	-5.594	1, 8	-4.712	-1.891
1, 9	-0.743	2.797									
2, 2	152.102	-599.091	2, 3	106.856	309.403	2, 4	21.755	-74.114	2, 5	-21.766	-9.838
2, 6	-9.222	13.670	2, 7	10.048	6.782	2, 8	5.007	-8.270			
3, 3	146.867	-608.728	3, 4	104.388	310.773	3, 5	22.244	-72.256	3, 6	-20.580	-10.208
3, 7	-9.556	12.522									
4, 4	147.158	-608.033	4, 5	104.711	310.680	4, 6	22.177	-72.708			
5, 5	147.186	-608.440									

IBR

SELF AND MUTUAL IMPEDANCES

H/LMDA=0.5000 B/LMDA= 0.250 OMEGA= 9.92

10 ELEMENT ARRAY

I, K	ZIK IN OHMS		I, K	ZIK IN OHMS		I, K	ZIK IN OHMS		I, K	ZIK IN OHMS	
	RE	IM		RE	IM		RE	IM		RE	IM
1, 1	347.435	-558.889	1, 2	71.123	281.179	1, 3	14.290	-64.691	1, 4	-17.025	-3.819
1, 5	-4.851	10.075	1, 6	7.397	3.476	1, 7	2.550	-5.665	1, 8	-4.441	-2.031
1, 9	-1.530	3.875	1,10	2.335	0.608						
2, 2	152.097	-599.126	2, 3	106.898	309.400	2, 4	21.761	-74.062	2, 5	-21.835	-9.832
2, 6	-9.228	13.576	2, 7	10.190	6.739	2, 8	5.174	-7.734	2, 9	-6.652	-3.999
3, 3	146.868	-608.679	3, 4	104.327	310.774	3, 5	22.244	-72.336	3, 6	-20.470	-10.207
3, 7	-9.519	12.690	3, 8	9.276	7.011						
4, 4	147.157	-608.112	4, 5	104.814	310.681	4, 6	22.174	-72.567	4, 7	-20.816	-10.151
5, 5	147.179	-608.301	5, 6	104.515	310.666						

IBR

SELF AND MUTUAL IMPEDANCES

H/LMDA=0.5000 B/LMDA= 0.250 OMEGA= 9.92

12 ELEMENT ARRAY

I, K	ZIK IN OHMS		I, K	ZIK IN OHMS		I, K	ZIK IN OHMS		I, K	ZIK IN OHMS	
	RE	IM		RE	IM		RE	IM		RE	IM
1, 1	347.434	-558.891	1, 2	71.127	281.178	1, 3	14.291	-64.686	1, 4	-17.031	-3.817
1, 5	-4.853	10.068	1, 6	7.408	3.473	1, 7	2.553	-5.650	1, 8	-4.465	-2.009
1, 9	-1.623	3.698	1,10	3.065	1.377	1,11	1.074	-2.801	1,12	-1.724	-0.431
2, 2	152.098	-599.119	2, 3	106.890	309.400	2, 4	21.760	-74.072	2, 5	-21.821	-9.833
2, 6	-9.227	13.596	2, 7	10.160	6.740	2, 8	5.142	-7.784	2, 9	-6.304	-4.111
2,10	-3.422	5.130	2,11	4.676	2.773						
3, 3	146.868	-608.689	3, 4	104.339	310.774	3, 5	22.244	-72.320	3, 6	-20.494	-10.208
3, 7	-9.518	12.655	3, 8	9.338	6.979	3, 9	5.334	-7.217	3,10	-5.672	-4.306
4, 4	147.157	-608.096	4, 5	104.793	310.680	4, 6	22.175	-72.598	4, 7	-20.771	-10.149
4, 8	-9.474	12.785	4, 9	9.566	6.939						
5, 5	147.180	-608.330	5, 6	104.557	310.669	5, 7	22.165	-72.500	5, 8	-20.587	-10.145
6, 6	147.173	-608.240	6, 7	104.724	310.670						

IBR

SELF AND MUTUAL IMPEDANCES

H/LMDA=0.5000 B/LMDA= 0.250 OMEGA= 9.92

14 ELEMENT ARRAY

I, K	ZIK IN OHMS		I, K	ZIK IN OHMS		I, K	ZIK IN OHMS		I, K	ZIK IN OHMS	
	RE	IM		RE	IM		RE	IM		RE	IM
1, 1	347.434	-558.892	1, 2	71.128	281.177	1, 3	14.292	-64.684	1, 4	-17.033	-3.816
1, 5	-4.854	10.065	1, 6	7.412	3.472	1, 7	2.555	-5.644	1, 8	-4.472	-2.007
1, 9	-1.626	3.686	1,10	3.083	1.360	1,11	1.147	-2.668	1,12	-2.286	-1.010
1,13	-0.804	2.153	1,14	1.344	0.324						
2, 2	152.098	-599.116	2, 3	106.886	309.400	2, 4	21.759	-74.077	2, 5	-21.815	-9.834
2, 6	-9.227	13.604	2, 7	10.150	6.741	2, 8	5.141	-7.799	2, 9	-6.282	-4.113
2,10	-3.398	5.166	2,11	4.419	2.861	2,12	2.480	-3.740	2,13	-3.531	-2.066
3, 3	146.868	-608.693	3, 4	104.344	310.773	3, 5	22.245	-72.313	3, 6	-20.502	-10.208
3, 7	-9.518	12.644	3, 8	9.354	6.979	3, 9	5.334	-7.193	3,10	-5.714	-4.282
3,11	-3.539	4.762	3,12	3.947	3.010						
4, 4	147.158	-608.090	4, 5	104.785	310.680	4, 6	22.175	-72.608	4, 7	-20.758	-10.150
4, 8	-9.474	12.805	4, 9	9.538	6.939	4,10	5.304	-7.279	4,11	-5.880	-4.252
5, 5	147.180	-608.340	5, 6	104.569	310.668	5, 7	22.165	-72.483	5, 8	-20.611	-10.145
5, 9	-9.467	12.745	5,10	9.409	6.938						
6, 6	147.173	-608.224	6, 7	104.702	310.670	6, 8	22.169	-72.533	6, 9	-20.722	-10.145
7, 7	147.174	-608.269	7, 8	104.600	310.672						

IBR

SELF AND MUTUAL IMPEDANCES

H/LMDA=0.5000 B/LMDA= 0.250 OMEGA= 9.92

16 ELEMENT ARRAY

I, K	ZIK IN OHMS		I, K	ZIK IN OHMS		I, K	ZIK IN OHMS		I, K	ZIK IN OHMS	
	RE	IM		RE	IM		RE	IM		RE	IM
1, 1	347.434	-558.892	1, 2	71.129	281.177	1, 3	14.292	-64.683	1, 4	-17.034	-3.816
1, 5	-4.855	10.064	1, 6	7.414	3.472	1, 7	2.555	-5.642	1, 8	-4.475	-2.006
1, 9	-1.627	3.682	1,10	3.069	1.358	1,11	1.150	-2.659	1,12	-2.300	-0.996
1,13	-0.864	2.048	1,14	1.795	0.779	1,15	0.629	-1.726	1,16	-1.088	-0.253
2, 2	152.098	-599.115	2, 3	106.884	309.401	2, 4	21.759	-74.079	2, 5	-21.813	-9.834
2, 6	-9.226	13.607	2, 7	10.146	6.742	2, 8	5.140	-7.804	2, 9	-6.275	-4.113
2,10	-3.397	5.177	2,11	4.403	2.362	2,12	2.460	-3.767	2,13	-3.331	-2.137
2,14	-1.902	2.893	2,15	2.798	1.615						
3, 3	146.867	-608.695	3, 4	104.346	310.773	3, 5	22.245	-72.310	3, 6	-20.506	-10.207
[illegible]	[illegible]	[illegible]	[illegible]	9.360	6.979	3, 9	5.334	-7.185	3,10	-5.726	-4.282
3,11	-3.539	4.745	3,12	[illegible]	[illegible]	[illegible]	[illegible]	-3.461	3,14	-2.962	-2.257
4, 4	147.158	-608.087	4, 5	104.782	310.681	4, 6	22.175	-72.612	4, 7	-20.752	-10.150
4, 8	-9.474	12.812	4, 9	9.528	6.940	4,10	5.304	-7.293	4,11	-5.860	-4.252
4,12	-3.516	4.807	4,13	4.107	2.967						
5, 5	147.180	-608.344	5, 6	104.574	310.668	5, 7	22.165	-72.477	5, 8	-20.620	-10.145
5, 9	-9.468	12.734	5,10	9.425	6.938	5,11	5.299	-7.252	5,12	-5.763	-4.252
6, 6	147.173	-608.218	6, 7	104.694	310.670	6, 8	22.169	-72.543	6, 9	-20.709	-10.145
6,10	-9.471	12.766	6,11	9.506	6.937						
7, 7	147.174	-608.279	7, 8	104.613	310.672	7, 9	22.168	-72.516	7,10	-20.637	-10.148
8, 8	147.174	-608.253	8, 9	104.680	310.669						

IBR

SELF AND MUTUAL IMPEDANCES

H/LMDA=0.5000 B/LMDA= 0.250 OMEGA= 9.92

18 ELEMENT ARRAY

I, K	ZIK IN OHMS		I, K	ZIK IN OHMS		I, K	ZIK IN OHMS		I, K	ZIK IN OHMS	
	RE	IM		RE	IM		RE	IM		RE	IM
1, 1	347.434	-558.893	1, 2	71.130	281.177	1, 3	14.292	-64.683	1, 4	-17.035	-3.816
1, 5	-4.855	10.063	1, 6	7.415	3.471	1, 7	2.556	-5.641	1, 8	-4.477	-2.005
1, 9	-1.628	3.680	1,10	3.091	1.358	1,11	1.151	-2.656	1,12	-2.304	-0.995
1,13	-0.866	2.041	1,14	1.805	0.768	1,15	0.679	-1.640	1,16	-1.461	-0.624
1,17	-0.509	1.427	1,18	0.906	0.204						
2, 2	152.099	-599.114	2, 3	106.883	309.401	2, 4	21.759	-74.080	2, 5	-21.811	-9.834
2, 6	-9.226	13.608	2, 7	10.144	6.742	2, 8	5.140	-7.807	2, 9	-6.272	-4.114
2,10	-3.396	5.181	2,11	4.397	2.863	2,12	2.460	-3.776	2,13	-3.319	-2.138
2,14	-1.886	2.914	2,15	2.637	1.674	2,16	1.519	-2.331	2,17	-2.294	-1.306
3, 3	146.867	-608.696	3, 4	104.348	310.773	3, 5	22.245	-72.309	3, 6	-20.508	-10.207
3, 7	-9.518	12.637	3, 8	9.363	6.979	3, 9	5.335	-7.182	3,10	-5.730	-4.282
3,11	-3.539	4.739	3,12	3.988	2.991	3,13	2.570	-3.447	3,14	-2.986	-2.242
3,15	-1.975	2.671	3,16	2.336	1.773						
4, 4	147.158	-608.085	4, 5	104.780	310.681	4, 6	22.174	-72.614	4, 7	-20.750	-10.150
4, 8	-9.474	12.815	4, 9	9.524	6.940	4,10	5.303	-7.299	4,11	-5.853	-4.252
4,12	-3.516	4.818	4,13	4.091	2.968	4,14	2.552	-3.495	4,15	-3.089	-2.222
5, 5	147.180	-608.346	5, 6	104.577	310.668	5, 7	22.165	-72.474	5, 8	-20.623	-10.145
5, 9	-9.468	12.729	5,10	9.431	6.938	5,11	5.299	-7.244	5,12	-5.775	-4.252
5,13	-3.512	4.787	5,14	4.015	2.968						
6, 6	147.173	-608.215	6, 7	104.691	310.670	6, 8	22.168	-72.547	6, 9	-20.703	-10.145
6,10	-9.470	12.773	6,11	9.497	6.937	6,12	5.301	-7.267	6,13	-5.837	-4.250
7, 7	147.174	-608.283	7, 8	104.618	310.672	7, 9	22.168	-72.510	7,10	-20.646	-10.147
7,11	-9.469	12.755	7,12	9.443	6.939						
8, 8	147.174	-608.247	8, 9	104.672	310.669	8,10	22.168	-72.526	8,11	-20.695	-10.145
9, 9	147.174	-608.263	9,10	104.625	310.672						

IBR

SELF AND MUTUAL IMPEDANCES

H/LMDA=0.5000 B/LMDA= 0.250 OMEGA= 9.92

20 ELEMENT ARRAY

I, K	ZIK IN OHMS RE	IM	I, K	ZIK IN OHMS RE	IM	I, K	ZIK IN OHMS RE	IM	I, K	ZIK IN OHMS RE	IM
1, 1	347.433	-558.893	1, 2	71.130	281.177	1, 3	14.292	-64.683	1, 4	-17.035	-3.816
1, 5	-4.855	10.062	1, 6	7.415	3.471	1, 7	2.556	-5.640	1, 8	-4.478	-2.005
1, 9	-1.628	3.680	1,10	3.092	1.357	1,11	1.151	-2.654	1,12	-2.306	-0.994
1,13	-0.866	2.038	1,14	1.809	0.767	1,15	0.680	-1.635	1,16	-1.470	-0.614
1,17	-0.551	1.356	1,18	1.223	0.514	1,19	0.421	-1.207	1,20	-0.771	-0.168
2, 2	152.099	-599.113	2, 3	106.882	309.401	2, 4	21.759	-74.081	2, 5	-21.811	-9.834
2, 6	-9.226	13.609	2, 7	10.143	6.742	2, 8	5.139	-7.808	2, 9	-6.270	-4.114
2,10	-3.396	5.183	2,11	4.394	2.863	2,12	2.459	-3.779	2,13	-3.314	-2.139
2,14	-1.886	2.921	2,15	2.626	1.675	2,16	1.505	-2.349	2,17	-2.160	-1.356
2,18	-1.248	1.936	2,19	1.930	1.084						
3, 3	146.867	-608.697	3, 4	104.348	310.773	3, 5	22.245	-72.308	3, 6	-20.509	-10.207
3, 7	-9.518	12.636	3, 8	9.364	6.979	3, 9	5.335	-7.180	3,10	-5.732	-4.282
3,11	-3.539	4.736	3,12	3.991	2.991	3,13	2.571	-3.442	3,14	-2.993	-2.242
3,15	-1.976	2.660	3,16	2.356	1.761	3,17	1.580	-2.150	3,18	-1.910	-1.441
4, 4	147.158	-608.084	4, 5	104.779	310.681	4, 6	22.174	-72.615	4, 7	-20.749	-10.150
4, 8	-9.473	12.817	4, 9	9.522	6.940	4,10	5.303	-7.301	4,11	-5.849	-4.253
4,12	-3.515	4.822	4,13	4.086	2.968	4,14	2.552	-3.503	4,15	-3.077	-2.222
4,16	-1.961	2.698	4,17	2.441	1.743						
5, 5	147.180	-608.347	5, 6	104.578	310.668	5, 7	22.165	-72.472	5, 8	-20.625	-10.145
5, 9	-9.468	12.727	5,10	9.434	6.937	5,11	5.299	-7.240	5,12	-5.779	-4.251
5,13	-3.512	4.781	5,14	4.024	2.968	5,15	2.549	-3.479	5,16	-3.014	-2.223
6, 6	147.173	-608.213	6, 7	104.689	310.671	6, 8	22.168	-72.549	6, 9	-20.701	-10.146
6,10	-9.470	12.776	6,11	9.493	6.937	6,12	5.301	-7.272	6,13	-5.830	-4.251
6,14	-3.514	4.798	6,15	4.074	2.966						
7, 7	147.174	-608.285	7, 8	104.620	310.672	7, 9	22.168	-72.507	7,10	-20.649	-10.147
7,11	-9.470	12.750	7,12	9.449	6.939	7,13	5.300	-7.259	7,14	-5.787	-4.253
8, 8	147.175	-608.244	8, 9	104.669	310.670	8,10	22.168	-72.530	8,11	-20.689	-10.145
8,12	-9.470	12.762	8,13	9.487	6.937						
9, 9	147.174	-608.267	9,10	104.630	310.672	9,11	22.168	-72.520	9,12	-20.654	-10.147
10,10	147.174	-608.257	10,11	104.665	310.670						

IBR

SELF AND MUTUAL IMPEDANCES

H/LMDA=0.5000 B/LMDA= 0.250 OMEGA= 9.92

22 ELEMENT ARRAY

I, K	ZIK IN OHMS RE	IM	I, K	ZIK IN OHMS RE	IM	I, K	ZIK IN OHMS RE	IM	I, K	ZIK IN OHMS RE	IM
1, 1	347.433	-558.893	1, 2	71.130	281.177	1, 3	14.293	-64.682	1, 4	-17.036	-3.816
1, 5	-4.855	10.062	1, 6	7.416	3.471	1, 7	2.556	-5.640	1, 8	-4.478	-2.005
1, 9	-1.628	3.679	1,10	3.093	1.357	1,11	1.151	-2.654	1,12	-2.307	-0.994
1,13	-0.867	2.037	1,14	1.811	0.766	1,15	0.681	-1.633	1,16	-1.473	-0.613
1,17	-0.552	1.351	1,18	1.230	0.505	1,19	0.458	-1.147	1,20	-1.044	-0.432
1,21	-0.356	1.041	1,22	0.668	0.141						
2, 2	152.099	-599.113	2, 3	106.882	309.401	2, 4	21.759	-74.081	2, 5	-21.810	-9.835
2, 6	-9.226	13.610	2, 7	10.142	6.742	2, 8	5.139	-7.809	2, 9	-6.269	-4.114
2,10	-3.396	5.184	2,11	4.393	2.864	2,12	2.459	-3.781	2,13	-3.312	-2.139
2,14	-1.885	2.924	2,15	2.622	1.675	2,16	1.504	-2.354	2,17	-2.152	-1.357
2,18	-1.237	1.950	2,19	1.816	1.128	2,20	1.050	-1.644	2,21	-1.656	-0.919
3, 3	146.867	-608.697	3, 4	104.349	310.773	3, 5	22.245	-72.308	3, 6	-20.509	-10.207
3, 7	-9.519	12.635	3, 8	9.365	6.979	3, 9	5.335	-7.179	3,10	-5.733	-4.282
3,11	-3.539	4.735	3,12	3.993	2.991	3,13	2.571	-3.440	3,14	-2.996	-2.242
3,15	-1.976	2.656	3,16	2.362	1.760	3,17	1.580	-2.141	3,18	-1.926	-1.430
3,19	-1.301	1.783	3,20	1.603	1.202						
4, 4	147.158	-608.084	4, 5	104.778	310.681	4, 6	22.174	-72.616	4, 7	-20.748	-10.150
4, 8	-9.473	12.818	4, 9	9.521	6.940	4,10	5.303	-7.302	4,11	-5.848	-4.253
4,12	-3.515	4.824	4,13	4.083	2.968	4,14	2.552	-3.507	4,15	-3.072	-2.223
4,16	-1.960	2.705	4,17	2.431	1.744	4,18	1.567	-2.172	4,19	-1.998	-1.415
5, 5	147.180	-608.348	5, 6	104.579	310.668	5, 7	22.165	-72.471	5, 8	-20.626	-10.145
5, 9	-9.468	12.726	5,10	9.435	6.937	5,11	5.300	-7.238	5,12	-5.781	-4.251
5,13	-3.512	4.778	5,14	4.027	2.968	5,15	2.549	-3.474	5,16	-3.021	-2.223
5,17	-1.958	2.686	5,18	2.378	1.745						
6, 6	147.174	-608.212	6, 7	104.688	310.671	6, 8	22.168	-72.551	6, 9	-20.699	-10.146
6,10	-9.470	12.778	6,11	9.491	6.937	6,12	5.301	-7.274	6,13	-5.827	-4.251
6,14	-3.514	4.802	6,15	4.068	2.967	6,16	2.551	-3.488	6,17	-3.063	-2.221
7, 7	147.174	-608.286	7, 8	104.621	310.672	7, 9	22.168	-72.505	7,10	-20.651	-10.147
7,11	-9.470	12.748	7,12	9.451	6.939	7,13	5.301	-7.255	7,14	-5.792	-4.252
7,15	-3.513	4.792	7,16	4.034	2.969						
8, 8	147.175	-608.243	8, 9	104.667	310.670	8,10	22.168	-72.532	8,11	-20.687	-10.145
8,12	-9.470	12.765	8,13	9.483	6.937	8,14	5.301	-7.264	8,15	-5.823	-4.250
9, 9	147.174	-608.269	9,10	104.633	310.672	9,11	22.168	-72.517	9,12	-20.657	-10.147
9,13	-9.470	12.757	9,14	9.455	6.939						
10,10	147.174	-608.254	10,11	104.661	310.670	10,12	22.168	-72.524	10,13	-20.684	-10.145
11,11	147.174	-608.261	11,12	104.635	310.672						

IBR

SELF AND MUTUAL IMPEDANCES

H/LMDA=0.5000 B/LMDA= 0.250 OMEGA= 9.92

24 ELEMENT ARRAY

I, K	ZIK IN OHMS RE	IM	I, K	ZIK IN OHMS RE	IM	I, K	ZIK IN OHMS RE	IM	I, K	ZIK IN OHMS RE	IM
1, 1	347.433	-558.893	1, 2	71.130	281.177	1, 3	14.293	-64.682	1, 4	-17.036	-3.816
1, 5	-4.855	10.062	1, 6	7.416	3.471	1, 7	2.556	-5.640	1, 8	-4.478	-2.005
1, 9	-1.628	3.679	1,10	3.093	1.357	1,11	1.152	-2.653	1,12	-2.308	-0.994
1,13	-0.867	2.036	1,14	1.812	0.766	1,15	0.681	-1.632	1,16	-1.475	-0.613
1,17	-0.553	1.349	1,18	1.233	0.505	1,19	0.459	-1.143	1,20	-1.050	-0.425
1,21	-0.388	0.988	1,22	0.907	0.370	1,23	0.306	-0.910	1,24	-0.586	-0.121
2, 2	152.099	-599.112	2, 3	106.882	309.401	2, 4	21.759	-74.082	2, 5	-21.810	-9.835
2, 6	-9.226	13.610	2, 7	10.142	6.742	2, 8	5.139	-7.809	2, 9	-6.269	-4.114
2,10	-3.396	5.185	2,11	4.392	2.864	2,12	2.459	-3.782	2,13	-3.311	-2.139
2,14	-1.885	2.925	2,15	2.620	1.675	2,16	1.504	-2.357	2,17	-2.148	-1.358
2,18	-1.236	1.955	2,19	1.809	1.129	2,20	1.040	-1.657	2,21	-1.558	-0.957
2,22	-0.900	1.422	2,23	1.444	0.792						
3, 3	146.867	-608.698	3, 4	104.349	310.773	3, 5	22.245	-72.307	3, 6	-20.510	-10.207
3, 7	-9.519	12.635	3, 8	9.365	6.979	3, 9	5.335	-7.178	3,10	-5.734	-4.281
3,11	-3.539	4.734	3,12	3.994	2.991	3,13	2.571	-3.439	3,14	-2.998	-2.241
3,15	-1.976	2.654	3,16	2.364	1.760	3,17	1.580	-2.137	3,18	-1.931	-1.430
3,19	-1.301	1.776	3,20	1.616	1.192	3,21	1.096	-1.514	3,22	-1.373	-1.022
4, 4	147.158	-608.083	4, 5	104.778	310.681	4, 6	22.174	-72.617	4, 7	-20.747	-10.150
4, 8	-9.473	12.818	4, 9	9.521	6.940	4,10	5.303	-7.303	4,11	-5.847	-4.253
4,12	-3.515	4.825	4,13	4.082	2.968	4,14	2.552	-3.508	4,15	-3.070	-2.223
4,16	-1.960	2.708	4,17	2.427	1.744	4,18	1.567	-2.177	4,19	-1.989	-1.416
4,20	-1.290	1.802	4,21	1.678	1.179						
5, 5	147.180	-608.348	5, 6	104.579	310.668	5, 7	22.165	-72.471	5, 8	-20.627	-10.145
5, 9	-9.468	12.725	5,10	9.436	6.937	5,11	5.300	-7.238	5,12	-5.783	-4.251
5,13	-3.513	4.776	5,14	4.029	2.968	5,15	2.550	-3.472	5,16	-3.024	-2.223
5,17	-1.959	2.682	5,18	2.384	1.745	5,19	1.565	-2.161	5,20	-1.945	-1.417
6, 6	147.174	-608.212	6, 7	104.687	310.671	6, 8	22.168	-72.551	6, 9	-20.699	-10.146
6,10	-9.470	12.779	6,11	9.490	6.937	6,12	5.301	-7.276	6,13	-5.825	-4.251
6,14	-3.513	4.804	6,15	4.066	2.967	6,16	2.550	-3.491	6,17	-3.058	-2.221
6,18	-1.960	2.693	6,19	2.420	1.743						
7, 7	147.174	-608.287	7, 8	104.622	310.672	7, 9	22.168	-72.504	7,10	-20.652	-10.147
7,11	-9.470	12.747	7,12	9.453	6.939	7,13	5.301	-7.253	7,14	-5.794	-4.252
7,15	-3.513	4.789	7,16	4.037	2.969	7,17	2.550	-3.483	7,18	-3.029	-2.223
8, 8	147.175	-608.242	8, 9	104.666	310.670	8,10	22.167	-72.533	8,11	-20.686	-10.145
8,12	-9.469	12.766	8,13	9.481	6.937	8,14	5.300	-7.266	8,15	-5.820	-4.251
8,16	-3.513	4.796	8,17	4.063	2.967						
9, 9	147.174	-608.270	9,10	104.634	310.672	9,11	22.168	-72.515	9,12	-20.659	-10.147
9,13	-9.470	12.755	9,14	9.457	6.939	9,15	5.301	-7.260	9,16	-5.797	-4.252
10,10	147.175	-608.252	10,11	104.659	310.670	10,12	22.168	-72.526	10,13	-20.682	-10.145
10,14	-9.470	[illegible]	[illegible]	[illegible]	[illegible]						
11,11	147.174	-608.263	11,12	104.638	310.672	11,13	22.168	-72.521	11,14	-20.661	-10.147
12,12	147.174	-608.258	12,13	104.658	310.670						

IBR

SELF AND MUTUAL IMPEDANCES

H/LMDA=0.5000 B/LMDA= 0.250 OMEGA= 9.92

26 ELEMENT ARRAY

I, K	ZIK IN OHMS RE	IM	I, K	ZIK IN OHMS RE	IM	I, K	ZIK IN OHMS RE	IM	I, K	ZIK IN OHMS RE	IM
1, 1	347.433	-558.893	1, 2	71.131	281.176	1, 3	14.293	-64.682	1, 4	-17.036	-3.816
1, 5	-4.855	10.062	1, 6	7.416	3.471	1, 7	2.556	-5.640	1, 8	-4.478	-2.005
1, 9	-1.628	3.679	1,10	3.093	1.357	1,11	1.152	-2.653	1,12	-2.308	-0.994
1,13	-0.867	2.036	1,14	1.812	0.766	1,15	0.682	-1.631	1,16	-1.475	-0.613
1,17	-0.553	1.348	1,18	1.234	0.504	1,19	0.460	-1.141	1,20	-1.053	-0.424
1,21	-0.389	0.985	1,22	0.912	0.364	1,23	0.335	-0.865	1,24	-0.798	-0.321
1,25	-0.266	0.806	1,26	0.521	0.104						
2, 2	152.099	-599.112	2, 3	106.882	309.401	2, 4	21.758	-74.082	2, 5	-21.809	-9.835
2, 6	-9.226	13.611	2, 7	10.141	6.742	2, 8	5.139	-7.810	2, 9	-6.268	-4.114
2,10	-3.396	5.185	2,11	4.392	2.864	2,12	2.459	-3.782	2,13	-3.310	-2.140
2,14	-1.885	2.926	2,15	2.620	1.676	2,16	1.504	-2.358	2,17	-2.147	-1.358
2,18	-1.236	1.957	2,19	1.806	1.129	2,20	1.039	-1.661	2,21	-1.552	-0.958
2,22	-0.890	1.432	2,23	1.358	0.826	2,24	0.782	-1.247	2,25	-1.275	-0.691
3, 3	146.867	-608.698	3, 4	104.350	310.773	3, 5	22.245	-72.307	3, 6	-20.510	-10.207
3, 7	-9.519	12.635	3, 8	9.366	6.979	3, 9	5.335	-7.178	3,10	-5.735	-4.281
3,11	-3.539	4.733	3,12	3.994	2.991	3,13	2.571	-3.438	3,14	-2.998	-2.241
3,15	-1.976	2.653	3,16	2.365	1.760	3,17	1.580	-2.136	3,18	-1.933	-1.430
3,19	-1.301	1.773	3,20	1.621	1.192	3,21	1.096	-1.507	3,22	-1.385	-1.014
3,23	-0.940	1.308	3,24	1.195	0.884						
4, 4	147.158	-608.083	4, 5	104.777	310.681	4, 6	22.174	-72.617	4, 7	-20.747	-10.150
4, 8	-9.473	12.819	4, 9	9.520	6.940	4,10	5.303	-7.304	4,11	-5.846	-4.253
4,12	-3.515	4.825	4,13	4.081	2.969	4,14	2.551	-3.509	4,15	-3.069	-2.223
4,16	-1.960	2.709	4,17	2.425	1.744	4,18	1.567	-2.180	4,19	-1.986	-1.416
4,20	-1.290	1.807	4,21	1.671	1.179	4,22	1.086	-1.529	4,23	-1.439	-1.002
5, 5	147.180	-608.349	5, 6	104.580	310.668	5, 7	22.166	-72.470	5, 8	-20.627	-10.145
5, 9	-9.468	12.725	5,10	9.437	6.937	5,11	5.300	-7.237	5,12	-5.783	-4.251
5,13	-3.513	4.776	5,14	4.030	2.968	5,15	2.550	-3.471	5,16	-3.026	-2.222
5,17	-1.959	2.680	5,18	2.387	1.744	5,19	1.565	-2.158	5,20	-1.950	-1.417
5,21	-1.288	1.793	5,22	1.632	1.181						
6, 6	147.174	-608.211	6, 7	104.687	310.671	6, 8	22.168	-72.552	6, 9	-20.698	-10.146
6,10	-9.470	12.779	6,11	9.489	6.938	6,12	5.301	-7.277	6,13	-5.824	-4.251
6,14	-3.513	4.805	6,15	4.065	2.967	6,16	2.550	-3.493	6,17	-3.056	-2.222
6,18	-1.959	2.696	6,19	2.416	1.743	6,20	1.566	-2.167	6,21	-1.980	-1.415

7, 7	147.174	-608.287	7, 8	104.623	310.672	7, 9	22.168	-72.504	7,10	-20.653	-10.147
7,11	-9.470	12.746	7,12	9.454	6.939	7,13	5.301	-7.252	7,14	-5.795	-4.252
7,15	-3.513	4.788	7,16	4.039	2.968	7,17	2.550	-3.481	7,18	-3.032	-2.223
7,19	-1.959	2.689	7,20	2.391	1.745						
8, 8	147.175	-608.241	8, 9	104.665	310.670	8,10	22.167	-72.534	8,11	-20.685	-10.145
8,12	-9.469	12.767	8,13	9.480	6.937	8,14	5.300	-7.267	8,15	-5.818	-4.251
8,16	-3.513	4.798	8,17	4.060	2.967	8,18	2.550	-3.486	8,19	-3.054	-2.221
9, 9	147.174	-608.271	9,10	104.635	310.672	9,11	22.168	-72.514	9,12	-20.660	-10.147
9,13	-9.470	12.754	9,14	9.459	6.939	9,15	5.301	-7.259	9,16	-5.799	-4.252
9,17	-3.513	4.793	9,18	4.041	2.969						
10,10	147.175	-608.251	10,11	104.658	310.670	10,12	22.168	-72.527	10,13	-20.680	-10.145
10,14	-9.470	12.762	10,15	9.477	6.937	10,16	5.301	-7.263	10,17	-5.817	-4.251
11,11	147.174	-608.264	11,12	104.639	310.672	11,13	22.168	-72.519	11,14	-20.663	-10.147
11,15	-9.470	12.758	11,16	9.460	6.939						
12,12	147.174	-608.256	12,13	104.656	310.670	12,14	22.168	-72.523	12,15	-20.679	-10.145
13,13	147.174	-608.260	13,14	104.640	310.672						

IBR

SELF AND MUTUAL IMPEDANCES

H/LMDA=0.5000 B/LMDA= 0.250 OMEGA= 9.92

28 ELEMENT ARRAY

I, K	ZIK IN OHMS		I, K	ZIK IN OHMS		I, K	ZIK IN OHMS		I, K	ZIK IN OHMS	
	RE	IM		RE	IM		RE	IM		RE	IM
1, 1	347.433	-558.893	1, 2	71.131	281.176	1, 3	14.293	-64.682	1, 4	-17.036	-3.815
1, 5	-4.855	10.062	1, 6	7.416	3.471	1, 7	2.556	-5.639	1, 8	-4.479	-2.005
1, 9	-1.628	3.678	1,10	3.093	1.357	1,11	1.152	-2.653	1,12	-2.308	-0.994
1,13	-0.867	2.036	1,14	1.812	0.766	1,15	0.682	-1.631	1,16	-1.476	-0.613
1,17	-0.553	1.348	1,18	1.234	0.504	1,19	0.460	-1.140	1,20	-1.054	-0.424
1,21	-0.390	0.983	1,22	0.914	0.363	1,23	0.336	-0.861	1,24	-0.803	-0.315
1,25	-0.292	0.765	1,26	0.710	0.282	1,27	0.234	-0.721	1,28	-0.467	-0.091
2, 2	152.099	-599.112	2, 3	106.881	309.401	2, 4	21.758	-74.082	2, 5	-21.809	-9.835
2, 6	-9.226	13.611	2, 7	10.141	6.742	2, 8	5.139	-7.810	2, 9	-6.268	-4.114
2,10	-3.395	5.186	2,11	4.391	2.864	2,12	2.459	-3.783	2,13	-3.310	-2.140
2,14	-1.885	2.927	2,15	2.619	1.676	2,16	1.504	-2.359	2,17	-2.146	-1.358
2,18	-1.236	1.958	2,19	1.805	1.129	2,20	1.039	-1.663	2,21	-1.549	-0.958
2,22	-0.890	1.436	2,23	1.352	0.827	2,24	0.774	-1.257	2,25	-1.199	-0.722
2,26	-0.688	1.107	2,27	1.138	0.611						
3, 3	146.867	-608.698	3, 4	104.350	310.773	3, 5	22.245	-72.307	3, 6	-20.510	-10.207
3, 7	-9.519	12.634	3, 8	9.366	6.979	3, 9	5.335	-7.178	3,10	-5.735	-4.281
3,11	-3.539	4.733	3,12	3.995	2.991	3,13	2.571	-3.438	3,14	-2.999	-2.241
3,15	-1.976	2.653	3,16	2.366	1.760	3,17	1.580	-2.135	3,18	-1.934	-1.430
3,19	-1.301	1.771	3,20	1.623	1.192	3,21	1.096	-1.504	3,22	-1.389	-1.014
3,23	-0.940	1.302	3,24	1.206	0.876	3,25	0.818	-1.147	3,26	-1.054	-0.774
4, 4	147.158	-608.083	4, 5	104.777	310.681	4, 6	22.174	-72.617	4, 7	-20.747	-10.150
4, 8	-9.473	12.819	4, 9	9.520	6.940	4,10	5.303	-7.304	4,11	-5.846	-4.253
4,12	-3.515	4.826	4,13	4.081	2.969	4,14	2.551	-3.510	4,15	-3.068	-2.223
4,16	-1.960	2.710	4,17	2.424	1.744	4,18	1.567	-2.181	4,19	-1.985	-1.416
4,20	-1.290	1.809	4,21	1.669	1.180	4,22	1.086	-1.534	4,23	-1.433	-1.003
4,24	-0.931	1.322	4,25	1.253	0.866						
5, 5	147.180	-608.349	5, 6	104.580	310.668	5, 7	22.166	-72.470	5, 8	-20.627	-10.145
5, 9	-9.468	12.724	5,10	9.437	6.937	5,11	5.300	-7.237	5,12	-5.784	-4.251
5,13	-3.513	4.775	5,14	4.031	2.968	5,15	2.550	-3.470	5,16	-3.027	-2.222
5,17	-1.959	2.679	5,18	2.388	1.744	5,19	1.566	-2.156	5,20	-1.952	-1.417
5,21	-1.289	1.790	5,22	1.637	1.180	5,23	1.085	-1.522	5,24	-1.398	-1.004
6, 6	147.174	-608.211	6, 7	104.687	310.671	6, 8	22.168	-72.552	6, 9	-20.698	-10.146
6,10	-9.470	12.780	6,11	9.488	6.938	6,12	5.301	-7.277	6,13	-5.824	-4.251
6,14	-3.513	4.806	6,15	4.064	2.967	6,16	2.550	-3.494	6,17	-3.055	-2.222
6,18	-1.959	2.697	6,19	2.414	1.744	6,20	1.566	-2.170	6,21	-1.977	-1.415
6,22	-1.289	1.798	6,23	1.663	1.179						
7, 7	147.174	-608.288	7, 8	104.623	310.671	7, 9	22.168	-72.503	7,10	-20.653	-10.147
7,11	-9.470	12.746	7,12	9.454	6.939	7,13	5.301	-7.252	7,14	-5.796	-4.252
7,15	-3.514	4.787	7,16	4.040	2.968	7,17	2.550	-3.480	7,18	-3.034	-2.223
7,19	-1.959	2.687	7,20	2.394	1.745	7,21	1.566	-2.164	7,22	-1.955	-1.417
8, 8	147.175	-608.241	8, 9	104.665	310.670	8,10	22.167	-72.534	8,11	-20.684	-10.145
8,12	-9.469	12.768	8,13	9.479	6.937	8,14	5.300	-7.268	8,15	-5.817	-4.251
8,16	-3.513	4.799	8,17	4.059	2.967	8,18	2.550	-3.488	8,19	-3.052	-2.222
8,20	-1.959	2.691	8,21	2.412	1.743						
9, 9	147.174	-608.271	9,10	104.635	310.672	9,11	22.168	-72.514	9,12	-20.661	-10.147
9,13	-9.470	12.753	9,14	9.460	6.939	9,15	5.301	-7.258	9,16	-5.800	-4.252
9,17	-3.513	4.792	9,18	4.043	2.968	9,19	2.550	-3.484	9,20	-3.035	-2.223
10,10	147.175	-608.251	10,11	104.658	310.670	10,12	22.167	-72.528	10,13	-20.679	-10.145
10,14	-9.469	12.763	10,15	9.476	6.937	10,16	5.300	-7.264	10,17	-5.815	-4.251
10,18	-3.513	4.795	10,19	4.058	2.967						
11,11	147.174	-608.265	11,12	104.640	310.672	11,13	22.168	-72.518	11,14	-20.664	-10.147
11,15	-9.470	12.757	11,16	9.462	6.939	11,17	5.301	-7.261	11,18	-5.801	-4.252
12,12	147.174	-608.255	12,13	104.655	310.670	12,14	22.168	-72.524	12,15	-20.678	-10.145
12,16	-9.470	12.760	12,17	9.475	6.937						
13,13	147.174	-608.261	13,14	104.641	310.671	13,15	22.168	-72.521	13,16	-20.665	-10.147
14,14	147.174	-608.258	14,15	104.654	310.670						

IBR

SELF AND MUTUAL IMPEDANCES

H/LMDA=0.5000 B/LMDA= 0.250 OMEGA= 9.92

30 ELEMENT ARRAY

I, K	ZIK IN OHMS RE	IM	I, K	ZIK IN OHMS RE	IM	I, K	ZIK IN OHMS RE	IM	I, K	ZIK IN OHMS RE	IM
1, 1	347.433	-558.893	1, 2	71.131	281.176	1, 3	14.293	-64.682	1, 4	-17.036	-3.815
1, 5	-4.855	10.062	1, 6	7.416	3.471	1, 7	2.556	-5.639	1, 8	-4.479	-2.005
1, 9	-1.629	3.678	1,10	3.094	1.357	1,11	1.152	-2.653	1,12	-2.308	-0.993
1,13	-0.867	2.036	1,14	1.813	0.766	1,15	0.682	-1.630	1,16	-1.476	-0.613
1,17	-0.553	1.347	1,18	1.235	0.504	1,19	0.460	-1.140	1,20	-1.054	-0.424
1,21	-0.390	0.983	1,22	0.915	0.363	1,23	0.336	-0.860	1,24	-0.805	-0.315
1,25	-0.293	0.763	1,26	0.714	0.277	1,27	0.258	-0.685	1,28	-0.638	-0.250
1,29	-0.208	0.650	1,30	0.422	0.081						
2, 2	152.099	-599.112	2, 3	106.881	309.401	2, 4	21.758	-74.082	2, 5	-21.809	-9.835
2, 6	-9.226	13.611	2, 7	10.141	6.742	2, 8	5.139	-7.810	2, 9	-6.268	-4.114
2,10	-3.395	5.186	2,11	4.391	2.864	2,12	2.459	-3.783	2,13	-3.309	-2.140
2,14	-1.885	2.927	2,15	2.619	1.676	2,16	1.504	-2.359	2,17	-2.145	-1.358
2,18	-1.236	1.959	2,19	1.804	1.130	2,20	1.039	-1.664	2,21	-1.548	-0.959
2,22	-0.890	1.438	2,23	1.350	0.827	2,24	0.773	-1.260	2,25	-1.194	-0.723
2,26	-0.681	1.115	2,27	1.069	0.639	2,28	0.612	-0.992	2,29	-1.024	-0.544
3, 3	146.867	-608.698	3, 4	104.350	310.773	3, 5	22.245	-72.307	3, 6	-20.510	-10.207
3, 7	-9.519	12.634	3, 8	9.366	6.979	3, 9	5.335	-7.178	3,10	-5.735	-4.281
3,11	-3.539	4.733	3,12	3.995	2.991	3,13	2.571	-3.438	3,14	-2.999	-2.241
3,15	-1.976	2.653	3,16	2.367	1.760	3,17	1.580	-2.134	3,18	-1.935	-1.430
3,19	-1.301	1.770	3,20	1.624	1.192	3,21	1.096	-1.503	3,22	-1.390	-1.014
3,23	-0.940	1.300	3,24	1.209	0.876	3,25	0.818	-1.142	3,26	-1.063	-0.768
3,27	-0.720	1.018	3,28	0.940	0.686						
4, 4	147.158	-608.083	4, 5	104.777	310.681	4, 6	22.174	-72.617	4, 7	-20.747	-10.150
4, 8	-9.473	12.819	4, 9	9.520	6.940	4,10	5.303	-7.304	4,11	-5.846	-4.253
4,12	-3.515	4.826	4,13	4.080	2.969	4,14	2.551	-3.510	4,15	-3.068	-2.223
4,16	-1.960	2.710	4,17	2.424	1.744	4,18	1.567	-2.182	4,19	-1.984	-1.416
4,20	-1.290	1.810	4,21	1.667	1.180	4,22	1.086	-1.535	4,23	-1.430	-1.003
4,24	-0.931	1.326	4,25	1.248	0.866	4,26	0.810	-1.159	4,27	-1.106	-0.758
5, 5	147.180	-608.349	5, 6	104.580	310.668	5, 7	22.166	-72.470	5, 8	-20.628	-10.145
5, 9	-9.468	12.724	5,10	9.437	6.937	5,11	5.300	-7.236	5,12	-5.784	-4.251
5,13	-3.513	4.775	5,14	4.031	2.967	5,15	2.550	-3.470	5,16	-3.027	-2.222
5,17	-1.959	2.678	5,18	2.389	1.744	5,19	1.566	-2.155	5,20	-1.953	-1.416
5,21	-1.289	1.789	5,22	1.639	1.180	5,23	1.085	-1.519	5,24	-1.402	-1.004
5,25	-0.930	1.315	5,26	1.218	0.867						
6, 6	147.174	-608.211	6, 7	104.687	310.671	6, 8	22.168	-72.552	6, 9	-20.698	-10.146
6,10	-9.470	12.780	6,11	9.488	6.938	6,12	5.300	-7.277	6,13	-5.823	-4.251
6,14	-3.513	4.806	6,15	4.063	2.967	6,16	2.550	-3.494	6,17	-3.055	-2.222
6,18	-1.959	2.698	6,19	2.413	1.744	6,20	1.566	-2.171	6,21	-1.975	-1.416
6,22	-1.289	1.800	6,23	1.661	1.179	6,24	1.086	-1.526	6,25	-1.426	-1.002
7, 7	147.174	-608.288	7, 8	104.623	310.671	7, 9	22.168	-72.503	7,10	-20.653	-10.147
7,11	-9.470	12.745	7,12	9.455	6.939	7,13	5.301	-7.251	7,14	-5.796	-4.252
7,15	-3.514	4.786	7,16	4.041	2.968	7,17	2.550	-3.479	7,18	-3.035	-2.223
7,19	-1.959	2.686	7,20	2.395	1.745	7,21	1.566	-2.162	7,22	-1.957	-1.417
7,23	-1.289	[illegible]	[illegible]	[illegible]	1.181						
8, 8	147.175	-608.241	8, 9	104.665	310.670	8,10	22.167	-72.535	8,11	-20.684	-10.145
8,12	-9.469	12.768	8,13	9.479	6.937	8,14	5.300	-7.269	8,15	-5.817	-4.251
8,16	-3.513	4.799	8,17	4.058	2.967	8,18	2.550	-3.489	8,19	-3.051	-2.222
8,20	-1.959	2.693	8,21	2.410	1.743	8,22	1.566	-2.166	8,23	-1.974	-1.415
9, 9	147.174	-608.272	9,10	104.635	310.672	9,11	22.168	-72.513	9,12	-20.661	-10.147
9,13	-9.470	12.753	9,14	9.460	6.939	9,15	5.301	-7.257	9,16	-5.801	-4.252
9,17	-3.514	4.791	9,18	4.044	2.968	9,19	2.550	-3.483	9,20	-3.037	-2.223
9,21	-1.959	2.690	9,22	2.396	1.745						
10,10	147.175	-608.250	10,11	104.657	310.670	10,12	22.167	-72.528	10,13	-20.679	-10.145
10,14	-9.469	12.763	10,15	9.475	6.937	10,16	5.300	-7.265	10,17	-5.814	-4.251
10,18	-3.513	4.796	10,19	4.056	2.967	10,20	2.550	-3.486	10,21	-3.050	-2.222
11,11	147.174	-608.265	11,12	104.640	310.671	11,13	22.168	-72.518	11,14	-20.665	-10.147
11,15	-9.470	12.756	11,16	9.462	6.939	11,17	5.301	-7.260	11,18	-5.802	-4.252
11,19	-3.513	4.794	11,20	4.044	2.968						
12,12	147.175	-608.255	12,13	104.654	310.670	12,14	22.168	-72.525	12,15	-20.677	-10.145
12,16	-9.470	12.761	12,17	9.474	6.937	12,18	5.301	-7.262	12,19	-5.813	-4.251
13,13	147.174	-608.262	13,14	104.642	310.671	13,15	22.168	-72.520	13,16	-20.666	-10.147
13,17	-9.470	12.758	13,18	9.463	6.939						
14,14	147.174	-608.258	14,15	104.653	310.670	14,16	22.168	-72.523	14,17	-20.676	-10.145
15,15	147.174	-608.260	15,16	104.643	310.671						

IBR

SELF AND MUTUAL IMPEDANCES

H/LMDA=0.5000 B/LMDA= 0.250 OMEGA= 9.92

35 ELEMENT ARRAY

I, K	ZIK IN OHMS		I, K	ZIK IN OHMS		I, K	ZIK IN OHMS		I, K	ZIK IN OHMS	
	RE	IM		RE	IM		RE	IM		RE	IM
1, 1	347.433	-558.893	1, 2	71.131	281.176	1, 3	14.293	-64.682	1, 4	-17.036	-3.815
1, 5	-4.855	10.061	1, 6	7.417	3.471	1, 7	2.556	-5.639	1, 8	-4.479	-2.005
1, 9	-1.629	3.678	1,10	3.094	1.356	1,11	1.152	-2.652	1,12	-2.309	-0.993
1,13	-0.868	2.035	1,14	1.813	0.765	1,15	0.682	-1.630	1,16	-1.477	-0.612
1,17	-0.554	1.346	1,18	1.236	0.503	1,19	0.461	-1.139	1,20	-1.056	-0.423
1,21	-0.391	0.981	1,22	0.917	0.362	1,23	0.337	-0.858	1,24	-0.807	-0.314
1,25	-0.294	0.759	1,26	0.719	0.275	1,27	0.260	-0.679	1,28	-0.646	-0.244
1,29	-0.232	0.611	1,30	0.586	0.218	1,31	0.209	-0.554	1,32	-0.536	-0.196
1,33	-0.192	0.504	1,34	0.518	0.160	1,35	0.061	-0.337			
2, 2	152.099	-599.111	2, 3	106.881	309.401	2, 4	21.758	-74.083	2, 5	-21.808	-9.835
2, 6	-9.225	13.612	2, 7	10.140	6.743	2, 8	5.139	-7.811	2, 9	-6.267	-4.115
2,10	-3.395	5.187	2,11	4.390	2.864	2,12	2.458	-3.784	2,13	-3.308	-2.140
2,14	-1.884	2.928	2,15	2.617	1.676	2,16	1.503	-2.361	2,17	-2.144	-1.359
2,18	-1.235	1.961	2,19	1.802	1.130	2,20	1.038	-1.666	2,21	-1.545	-0.959
2,22	-0.889	1.441	2,23	1.346	0.828	2,24	0.772	-1.265	2,25	-1.187	-0.724
2,26	-0.679	1.123	2,27	1.059	0.640	2,28	0.604	-1.008	2,29	-0.952	-0.572
2,30	-0.541	0.913	2,31	0.862	0.515	2,32	0.488	-0.835	2,33	-0.781	-0.472
2,34	-0.423	0.813									
3, 3	146.867	-608.699	3, 4	104.351	310.773	3, 5	22.245	-72.306	3, 6	-20.511	-10.207
3, 7	-9.519	12.633	3, 8	9.367	6.979	3, 9	5.335	-7.177	3,10	-5.736	-4.281
3,11	-3.540	4.732	3,12	3.996	2.991	3,13	2.571	-3.436	3,14	-3.001	-2.241
3,15	-1.976	2.651	3,16	2.368	1.760	3,17	1.581	-2.132	3,18	-1.937	-1.430
3,19	-1.302	1.768	3,20	1.626	1.191	3,21	1.097	-1.500	3,22	-1.394	-1.013
3,23	-0.941	1.296	3,24	1.214	0.876	3,25	0.819	-1.136	3,26	-1.071	-0.767
3,27	-0.721	1.007	3,28	0.956	0.679	3,29	0.642	-0.901	3,30	-0.863	-0.607
3,31	-0.577	0.812	3,32	0.786	0.547	3,33	0.526	-0.733			
4, 4	147.158	-608.082	4, 5	104.776	310.681	4, 6	22.174	-72.618	4, 7	-20.746	-10.151
4, 8	-9.473	12.820	4, 9	9.519	6.941	4,10	5.303	-7.305	4,11	-5.845	-4.253
4,12	-3.515	4.827	4,13	4.079	2.969	4,14	2.551	-3.511	4,15	-3.066	-2.223
4,16	-1.960	2.712	4,17	2.422	1.745	4,18	1.566	-2.184	4,19	-1.981	-1.417
4,20	-1.289	1.812	4,21	1.664	1.180	4,22	1.085	-1.539	4,23	-1.426	-1.003
4,24	-0.931	1.331	4,25	1.242	0.867	4,26	0.810	-1.168	4,27	-1.095	-0.759
4,28	-0.713	1.037	4,29	0.975	0.672	4,30	0.634	-0.931	4,31	-0.876	-0.601
4,32	-0.569	0.846									
5, 5	147.180	-608.350	5, 6	104.581	310.668	5, 7	22.166	-72.469	5, 8	-20.629	-10.144
5, 9	-9.469	12.723	5,10	9.438	6.937	5,11	5.300	-7.235	5,12	-5.785	-4.251
5,13	-3.513	4.773	5,14	4.033	2.967	5,15	2.550	-3.468	5,16	-3.029	-2.222
5,17	-1.959	2.676	5,18	2.391	1.744	5,19	1.566	-2.153	5,20	-1.956	-1.416
5,21	-1.289	1.785	5,22	1.642	1.180	5,23	1.086	-1.515	5,24	-1.408	-1.003
5,25	-0.931	1.308	5,26	1.226	0.867	5,27	0.810	-1.146	5,28	-1.083	-0.759
5,29	-0.714	1.016	5,30	0.968	0.671	5,31	0.635	-0.908			
6, 6	147.174	-608.210	6, 7	104.685	310.671	6, 8	22.168	-72.553	6, 9	-20.696	-10.146
6,10	-9.469	12.781	6,11	9.487	6.938	6,12	5.300	-7.279	6,13	-5.822	-4.251
6,14	-3.513	4.808	6,15	4.062	2.967	6,16	2.550	-3.496	6,17	-3.052	-2.222
6,18	-1.959	2.700	6,19	2.411	1.744	6,20	1.565	-2.174	6,21	-1.972	-1.416
6,22	-1.289	1.804	6,23	1.656	1.180	6,24	1.085	-1.532	6,25	-1.419	-1.003
6,26	-0.930	1.325	6,27	1.235	0.867	6,28	0.809	-1.163	6,29	-1.088	-0.759
6,30	-0.712	1.034									
7, 7	147.174	-608.289	7, 8	104.624	310.671	7, 9	22.168	-72.502	7,10	-20.655	-10.146
7,11	-9.470	12.744	7,12	9.456	6.938	7,13	5.301	-7.250	7,14	-5.798	-4.252
7,15	-3.514	4.784	7,16	4.043	2.968	7,17	2.551	-3.477	7,18	-3.037	-2.223
7,19	-1.960	2.683	7,20	2.398	1.744	7,21	1.566	-2.159	7,22	-1.962	-1.416
7,23	-1.290	1.790	7,24	1.648	1.180	7,25	1.086	-1.519	7,26	-1.413	-1.003
7,27	-0.931	1.312	7,28	1.231	0.867	7,29	0.810	-1.149			
8, 8	147.175	-608.239	8, 9	104.663	310.670	8,10	22.167	-72.536	8,11	-20.682	-10.145
8,12	-9.469	12.770	8,13	9.477	6.937	8,14	5.300	-7.271	8,15	-5.814	-4.251
8,16	-3.513	4.802	8,17	4.056	2.967	8,18	2.550	-3.491	8,19	-3.048	-2.222
8,20	-1.959	2.696	8,21	2.406	1.744	8,22	1.565	-2.171	8,23	-1.968	-1.416
8,24	-1.288	1.802	8,25	1.652	1.180	8,26	1.085	-1.530	8,27	-1.415	-1.003
8,28	-0.930	1.324									
9, 9	147.173	-608.273	9,10	104.637	310.671	9,11	22.169	-72.512	9,12	-20.663	-10.146
9,13	-9.470	12.751	9,14	9.462	6.938	9,15	5.301	-7.255	9,16	-5.803	-4.252
9,17	-3.514	4.788	9,18	4.047	2.968	9,19	2.551	-3.480	9,20	-3.040	-2.223
9,21	-1.960	2.685	9,22	2.401	1.744	9,23	1.566	-2.160	9,24	-1.964	-1.416
9,25	-1.290	1.791	9,26	1.650	1.180	9,27	1.086	-1.520			
10,10	147.175	-608.249	10,11	104.656	310.670	10,12	22.167	-72.530	10,13	-20.677	-10.146
10,14	-9.469	12.766	10,15	9.473	6.938	10,16	5.300	-7.267	10,17	-5.811	-4.251
10,18	-3.513	4.799	10,19	4.053	2.967	10,20	2.550	-3.490	10,21	-3.045	-2.222
10,22	-1.959	2.695	10,23	2.404	1.744	10,24	1.565	-2.170	10,25	-1.966	-1.416
10,26	-1.288	1.801									
11,11	147.174	-608.267	11,12	104.642	310.671	11,13	22.168	-72.516	11,14	-20.667	-10.146
11,15	-9.470	12.754	11,16	9.465	6.938	11,17	5.301	-7.257	11,18	-5.805	-4.252
11,19	-3.514	4.790	11,20	4.049	2.968	11,21	2.551	-3.481	11,22	-3.042	-2.222
11,23	-1.960	2.686	11,24	2.403	1.744	11,25	1.566	-2.161			
12,12	147.175	-608.253	12,13	104.652	310.670	12,14	22.167	-72.527	12,15	-20.674	-10.146
12,16	-9.469	12.764	12,17	9.471	6.938	12,18	5.300	-7.266	12,19	-5.809	-4.251
12,20	-3.513	4.798	12,21	4.051	2.968	12,22	2.550	-3.489	12,23	-3.044	-2.222
12,24	-1.959	2.694									
13,13	147.174	-608.264	13,14	104.645	310.671	13,15	22.168	-72.518	13,16	-20.669	-10.146
13,17	-9.470	12.755	13,18	9.467	6.938	13,19	5.301	-7.258	13,20	-5.807	-4.252
13,21	-3.514	4.791	13,22	4.050	2.968	13,23	2.551	-3.481			
14,14	147.175	-608.255	14,15	104.650	310.671	14,16	22.167	-72.526	14,17	-20.673	-10.146
14,18	-9.469	12.763	14,19	9.469	6.938	14,20	5.300	-7.265	14,21	-5.808	-4.251
14,22	-3.513	4.798									
15,15	147.174	-608.263	15,16	104.646	310.671	15,17	22.168	-72.518	15,18	-20.670	-10.146
15,19	-9.470	12.756	15,20	9.468	6.938	15,21	5.301	-7.258			
16,16	147.175	-608.255	16,17	104.649	310.671	16,18	22.167	-72.525	16,19	-20.671	-10.146
16,20	-9.469	12.762									
17,17	147.174	-608.262	17,18	104.648	310.671	17,19	22.168	-72.519			
18,18	147.175	-608.256									

IBR

SELF AND MUTUAL IMPEDANCES

H/LMDA=0.5000 B/LMDA= 0.250 OMEGA= 9.92

40 ELEMENT ARRAY

I, K	ZIK IN OHMS		I, K	ZIK IN OHMS		I, K	ZIK IN OHMS		I, K	ZIK IN OHMS	
	RE	IM		RE	IM		RE	IM		RE	IM
1, 1	347.433	-558.893	1, 2	71.131	281.176	1, 3	14.293	-64.682	1, 4	-17.036	-3.815
1, 5	-4.855	10.061	1, 6	7.416	3.471	1, 7	2.556	-5.639	1, 8	-4.479	-2.005
1, 9	-1.629	3.678	1,10	3.094	1.357	1,11	1.152	-2.652	1,12	-2.309	-0.993
1,13	-0.867	2.035	1,14	1.813	0.766	1,15	0.682	-1.630	1,16	-1.476	-0.612
1,17	-0.553	1.347	1,18	1.235	0.504	1,19	0.460	-1.139	1,20	-1.055	-0.423
1,21	-0.391	0.982	1,22	0.916	0.362	1,23	0.336	-0.859	1,24	-0.806	-0.314
1,25	-0.294	0.760	1,26	0.717	0.276	1,27	0.259	-0.680	1,28	-0.644	-0.245
1,29	-0.231	0.614	1,30	0.583	0.219	1,31	0.207	-0.558	1,32	-0.532	-0.198
1,33	-0.187	0.511	1,34	0.487	0.179	1,35	0.170	-0.470	1,36	-0.448	-0.164
1,37	-0.155	0.436	1,38	0.413	0.154	1,39	0.128	-0.427	1,40	-0.278	-0.048
2, 2	152.099	-599.112	2, 3	106.881	309.401	2, 4	21.758	-74.082	2, 5	-21.809	-9.835
2, 6	-9.226	13.611	2, 7	10.141	6.742	2, 8	5.139	-7.810	2, 9	-6.268	-4.115
2,10	-3.395	5.186	2,11	4.391	2.864	2,12	2.458	-3.783	2,13	-3.309	-2.140
2,14	-1.885	2.928	2,15	2.618	1.676	2,16	1.503	-2.360	2,17	-2.145	-1.358
2,18	-1.235	1.960	2,19	1.803	1.130	2,20	1.039	-1.665	2,21	-1.546	-0.959
2,22	-0.889	1.440	2,23	1.347	0.827	2,24	0.773	-1.263	2,25	-1.190	-0.723
2,26	-0.680	1.121	2,27	1.062	0.640	2,28	0.604	-1.005	2,29	-0.956	-0.571
2,30	-0.542	0.908	2,31	0.868	0.514	2,32	0.489	-0.826	2,33	-0.793	-0.466
2,34	-0.445	0.756	2,35	0.730	0.424	2,36	0.408	-0.694	2,37	-0.677	-0.389
2,38	-0.379	0.638	2,39	0.668	0.341						
3, 3	146.867	-608.698	3, 4	104.350	310.773	3, 5	22.245	-72.306	3, 6	-20.511	-10.207
3, 7	-9.519	12.634	3, 8	9.367	6.979	3, 9	5.335	-7.177	3,10	-5.736	-4.281
3,11	-3.539	4.732	3,12	3.996	2.991	3,13	2.571	-3.437	3,14	-3.000	-2.241
3,15	-1.976	2.652	3,16	2.367	1.760	3,17	1.580	-2.133	3,18	-1.936	-1.430
3,19	-1.301	1.769	3,20	1.625	1.192	3,21	1.096	-1.502	3,22	-1.392	-1.013
3,23	-0.940	1.298	3,24	1.212	0.876	3,25	0.818	-1.138	3,26	-1.069	-0.767
3,27	-0.721	1.010	3,28	0.953	0.680	3,29	0.642	-0.905	3,30	-0.858	-0.608
3,31	-0.576	0.819	3,32	0.777	0.548	3,33	0.521	-0.746	3,34	-0.709	-0.497
3,35	-0.474	0.684	3,36	0.649	0.454	3,37	0.434	-0.633	3,38	-0.595	-0.421
4, 4	147.158	-608.083	4, 5	104.777	310.681	4, 6	22.174	-72.618	4, 7	-20.746	-10.150
4, 8	-9.473	12.820	4, 9	9.519	6.941	4,10	5.303	-7.305	4,11	-5.845	-4.253
4,12	-3.515	4.827	4,13	4.080	2.969	4,14	2.551	-3.511	4,15	-3.067	-2.223
4,16	-1.960	2.711	4,17	2.423	1.745	4,18	1.566	-2.183	4,19	-1.983	-1.416
4,20	-1.289	1.811	4,21	1.666	1.180	4,22	1.086	-1.538	4,23	-1.428	-1.003
4,24	-0.931	1.329	4,25	1.244	0.867	4,26	0.810	-1.165	4,27	-1.098	-0.759
4,28	-0.713	1.034	4,29	0.980	0.672	4,30	0.635	-0.926	4,31	-0.882	-0.600
4,32	-0.570	0.837	4,33	0.801	0.541	4,34	0.515	-0.761	4,35	-0.733	-0.490
4,36	-0.469	0.695	4,37	0.677	0.447						
5, 5	147.180	-608.349	5, 6	104.580	310.668	5, 7	22.166	-72.470	5, 8	-20.628	-10.144
5, 9	-9.468	12.724	5,10	9.438	6.937	5,11	5.300	-7.236	5,12	-5.785	-4.251
5,13	-3.513	4.774	5,14	4.032	2.967	5,15	2.550	-3.469	5,16	-3.028	-2.222
5,17	-1.959	2.677	5,18	2.390	1.744	5,19	1.566	-2.154	5,20	-1.954	-1.416
5,21	-1.289	1.787	5,22	1.641	1.180	5,23	1.085	-1.517	5,24	-1.406	-1.003
5,25	-0.931	1.311	5,26	1.224	0.867	5,27	0.810	-1.149	5,28	-1.079	-0.759
5,29	-0.713	1.020	5,30	0.962	0.672	5,31	0.634	-0.915	5,32	-0.866	-0.601
5,33	[illegible]	[illegible]	[illegible]	0.784	0.542	5,35	0.515	-0.755	5,36	-0.714	-0.492
6, 6	147.174	-608.211	6, 7	104.686	310.671	6, 8	22.168	-72.553	6, 9	-20.697	-10.146
6,10	-9.470	12.781	6,11	9.488	6.938	6,12	5.300	-7.278	6,13	-5.823	-4.251
6,14	-3.513	4.807	6,15	4.063	2.967	6,16	2.550	-3.495	6,17	-3.054	-2.222
6,18	-1.959	2.699	6,19	2.412	1.744	6,20	1.566	-2.172	6,21	-1.973	-1.416
6,22	-1.289	1.802	6,23	1.658	1.179	6,24	1.085	-1.530	6,25	-1.421	-1.003
6,26	-0.931	1.322	6,27	1.238	0.866	6,28	0.810	-1.159	6,29	-1.093	-0.758
6,30	-0.713	1.028	6,31	0.975	0.671	6,32	0.634	-0.921	6,33	-0.879	-0.600
6,34	-0.570	0.831	6,35	0.799	0.540						
7, 7	147.174	-608.288	7, 8	104.624	310.671	7, 9	22.168	-72.502	7,10	-20.654	-10.147
7,11	-9.470	12.745	7,12	9.455	6.938	7,13	5.301	-7.251	7,14	-5.797	-4.252
7,15	-3.514	4.785	7,16	4.042	2.968	7,17	2.551	-3.478	7,18	-3.036	-2.223
7,19	-1.959	2.684	7,20	2.397	1.745	7,21	1.566	-2.160	7,22	-1.960	-1.417
7,23	-1.289	1.792	7,24	1.645	1.180	7,25	1.086	-1.521	7,26	-1.410	-1.004
7,27	-0.931	1.315	7,28	1.227	0.867	7,29	0.810	-1.153	7,30	-1.082	-0.759
7,31	-0.713	1.024	7,32	0.965	0.672	7,33	0.634	-0.918	7,34	-0.867	-0.601
8, 8	147.175	-608.240	8, 9	104.664	310.670	8,10	22.167	-72.535	8,11	-20.683	-10.145
8,12	-9.469	12.769	8,13	9.478	6.937	8,14	5.300	-7.270	8,15	-5.816	-4.251
8,16	-3.513	4.800	8,17	4.057	2.967	8,18	2.550	-3.490	8,19	-3.049	-2.222
8,20	-1.959	2.695	8,21	2.408	1.744	8,22	1.566	-2.169	8,23	-1.970	-1.416
8,24	-1.289	1.799	8,25	1.655	1.179	8,26	1.085	-1.527	8,27	-1.419	-1.003
8,28	-0.931	1.320	8,29	1.236	0.866	8,30	0.810	-1.157	8,31	-1.091	-0.758
8,32	-0.713	1.026	8,33	0.974	0.671						
9, 9	147.174	-608.272	9,10	104.636	310.672	9,11	22.168	-72.513	9,12	-20.662	-10.147
9,13	-9.470	12.752	9,14	9.461	6.939	9,15	5.301	-7.256	9,16	-5.802	-4.252
9,17	-3.514	4.790	9,18	4.045	2.968	9,19	2.551	-3.481	9,20	-3.039	-2.223
9,21	-1.959	2.687	9,22	2.399	1.745	9,23	1.566	-2.163	9,24	-1.962	-1.417
9,25	-1.289	1.794	9,26	1.647	1.180	9,27	1.086	-1.523	9,28	-1.411	-1.004
9,29	-0.931	1.317	9,30	1.228	0.867	9,31	0.810	-1.155	9,32	-1.083	-0.759
10,10	147.175	-608.250	10,11	104.657	310.670	10,12	22.167	-72.529	10,13	-20.678	-10.145
10,14	-9.469	12.764	10,15	9.474	6.937	10,16	5.300	-7.266	10,17	-5.813	-4.251
10,18	-3.513	4.798	10,19	4.055	2.967	10,20	2.550	-3.488	10,21	-3.047	-2.222
10,22	-1.959	2.693	10,23	2.407	1.744	10,24	1.566	-2.167	10,25	-1.969	-1.416
10,26	-1.289	1.798	10,27	1.654	1.179	10,28	1.085	-1.526	10,29	-1.418	-1.003
10,30	-0.931	1.318	10,31	1.236	0.866						
11,11	147.174	-608.266	11,12	104.641	310.671	11,13	22.168	-72.517	11,14	-20.666	-10.147
11,15	-9.470	12.755	11,16	9.464	6.938	11,17	5.301	-7.258	11,18	-5.804	-4.252
11,19	-3.514	4.792	11,20	4.047	2.968	11,21	2.550	-3.483	11,22	-3.040	-2.223
11,23	-1.959	2.689	11,24	2.400	1.745	11,25	1.566	-2.164	11,26	-1.963	-1.417
11,27	-1.289	1.795	11,28	1.648	1.180	11,29	1.085	-1.524	11,30	-1.411	-1.004
12,12	147.175	-608.254	12,13	104.653	310.670	12,14	22.167	-72.526	12,15	-20.676	-10.146
12,16	-9.469	12.762	12,17	9.472	6.937	12,18	5.300	-7.264	12,19	-5.811	-4.251
12,20	-3.513	4.796	12,21	4.053	2.967	12,22	2.550	-3.486	12,23	-3.046	-2.222
12,24	-1.959	2.691	12,25	2.406	1.744	12,26	1.566	-2.166	12,27	-1.969	-1.416
12,28	-1.289	1.796	12,29	1.654	1.179						
13,13	147.174	-608.263	13,14	104.643	310.671	13,15	22.168	-72.519	13,16	-20.667	-10.146
13,17	-9.470	12.757	13,18	9.465	6.938	13,19	5.301	-7.260	13,20	-5.805	-4.252
13,21	-3.513	4.793	13,22	4.048	2.968	13,23	2.550	-3.484	13,24	-3.041	-2.223
13,25	-1.959	2.690	13,26	2.400	1.745	13,27	1.566	-2.165	13,28	-1.963	-1.417

14,14	147.175	-608.256	14,15	104.652	310.670	14,16	22.168	-72.524	14,17	-20.674	-10.146
14,18	-9.469	12.761	14,19	9.471	6.938	14,20	5.300	-7.263	14,21	-5.810	-4.251
14,22	-3.513	4.795	14,23	4.053	2.967	14,24	2.550	-3.486	14,25	-3.046	-2.222
14,26	-1.959	2.691	14,27	2.406	1.744						
15,15	147.174	-608.261	15,16	104.645	310.671	15,17	22.168	-72.520	15,18	-20.668	-10.146
15,19	-9.470	12.758	15,20	9.466	6.938	15,21	5.301	-7.261	15,22	-5.805	-4.252
15,23	-3.513	4.794	15,24	4.048	2.968	15,25	2.550	-3.485	15,26	-3.041	-2.223
16,16	147.174	-608.257	16,17	104.651	310.670	16,18	22.168	-72.523	16,19	-20.674	-10.146
16,20	-9.470	12.760	16,21	9.471	6.938	16,22	5.301	-7.262	16,23	-5.810	-4.251
16,24	-3.513	4.794	16,25	4.053	2.967						
17,17	147.174	-608.260	17,18	104.645	310.671	17,19	22.168	-72.521	17,20	-20.669	-10.146
17,21	-9.470	12.758	17,22	9.466	6.938	17,23	5.301	-7.262	17,24	-5.805	-4.252
18,18	147.174	-608.258	18,19	104.650	310.670	18,20	22.168	-72.523	18,21	-20.673	-10.146
18,22	-9.470	12.759	18,23	9.471	6.938						
19,19	147.174	-608.260	19,20	104.646	310.671	19,21	22.168	-72.522	19,22	-20.669	-10.146
20,20	147.174	-608.259	20,21	104.650	310.670						

IBR

SELF AND MUTUAL IMPEDANCES

H/LMDA=0.5000 B/LMDA= 0.250 OMEGA= 9.92

45 ELEMENT ARRAY

I, K	ZIK IN OHMS		I, K	ZIK IN OHMS		I, K	ZIK IN OHMS		I, K	ZIK IN OHMS	
	RE	IM		RE	IM		RE	IM		RE	IM
1, 1	347.433	-558.893	1, 2	71.131	281.176	1, 3	14.293	-64.682	1, 4	-17.036	-3.815
1, 5	-4.855	10.061	1, 6	7.416	3.471	1, 7	2.556	-5.639	1, 8	-4.479	-2.005
1, 9	-1.629	3.678	1,10	3.094	1.357	1,11	1.152	-2.652	1,12	-2.309	-0.993
1,13	-0.867	2.035	1,14	1.813	0.766	1,15	0.682	-1.630	1,16	-1.477	-0.612
1,17	-0.554	1.347	1,18	1.236	0.503	1,19	0.461	-1.139	1,20	-1.055	-0.423
1,21	-0.391	0.981	1,22	0.917	0.362	1,23	0.337	-0.858	1,24	-0.807	-0.314
1,25	-0.294	0.760	1,26	0.718	0.276	1,27	0.260	-0.679	1,28	-0.645	-0.244
1,29	-0.231	0.613	1,30	0.584	0.219	1,31	0.208	-0.557	1,32	-0.533	-0.197
1,33	-0.188	0.509	1,34	0.489	0.179	1,35	0.171	-0.468	1,36	-0.451	-0.163
1,37	-0.157	0.432	1,38	0.418	0.149	1,39	0.144	-0.401	1,40	-0.390	-0.137
1,41	-0.133	0.373	1,42	0.365	0.127	1,43	0.126	-0.347	1,44	-0.360	-0.105
1,45	-0.039	0.236									
2, 2	152.099	-599.111	2, 3	106.881	309.401	2, 4	21.758	-74.083	2, 5	-21.809	-9.835
2, 6	-9.225	13.612	2, 7	10.140	6.743	2, 8	5.139	-7.811	2, 9	-6.267	-4.115
2,10	-3.395	5.187	2,11	4.390	2.864	2,12	2.458	-3.784	2,13	-3.308	-2.140
2,14	-1.884	2.928	2,15	2.617	1.676	2,16	1.503	-2.361	2,17	-2.144	-1.358
2,18	-1.235	1.961	2,19	1.802	1.130	2,20	1.039	-1.666	2,21	-1.545	-0.959
2,22	-0.889	1.441	2,23	1.347	0.828	2,24	0.773	-1.264	2,25	-1.189	-0.724
2,26	-0.680	1.122	2,27	1.060	0.640	2,28	0.604	-1.006	2,29	-0.955	-0.571
2,30	-0.541	0.909	2,31	0.866	0.514	2,32	0.489	-0.828	2,33	-0.791	-0.466
2,34	-0.445	0.759	2,35	0.726	0.425	2,36	0.407	-0.700	2,37	-0.670	-0.390
2,38	-0.374	0.648	2,39	0.621	0.360	2,40	0.345	-0.604	2,41	-0.577	-0.333
2,42	-0.319	0.566	2,43	0.535	0.313	2,44	0.282	-0.564			
3, 3	146.867	-608.699	3, 4	104.350	310.773	3, 5	22.245	-72.306	3, 6	-20.511	-10.207
3, 7	-9.519	12.634	3, 8	9.367	6.979	3, 9	5.335	-7.177	3,10	-5.736	-4.281
3,11	-3.540	4.732	3,12	3.996	2.991	3,13	2.571	-3.437	3,14	-3.000	-2.241
3,15	-1.976	2.651	3,16	2.368	1.760	3,17	1.580	-2.133	3,18	-1.936	-1.430
3,19	-1.302	1.769	3,20	1.626	1.192	3,21	1.096	-1.501	3,22	-1.393	-1.013
3,23	-0.940	1.297	3,24	1.213	0.876	3,25	0.819	-1.137	3,26	-1.070	-0.767
3,27	-0.721	1.009	3,28	0.955	0.679	3,29	0.642	-0.904	3,30	-0.859	-0.607
3,31	-0.576	0.817	3,32	0.780	0.547	3,33	0.521	-0.743	3,34	-0.712	-0.497
3,35	-0.474	0.680	3,36	0.655	0.453	3,37	0.435	-0.626	3,38	-0.605	-0.416
3,39	-0.400	0.579	3,40	0.562	0.384	3,41	0.370	-0.537	3,42	-0.526	-0.355
3,43	-0.347	0.497									
4, 4	147.158	-608.082	4, 5	104.776	310.681	4, 6	22.174	-72.618	4, 7	-20.746	-10.151
4, 8	-9.473	12.820	4, 9	9.519	6.941	4,10	5.303	-7.305	4,11	-5.845	-4.253
4,12	-3.515	4.827	4,13	4.079	2.969	4,14	2.551	-3.511	4,15	-3.066	-2.223
4,16	-1.960	2.712	4,17	2.422	1.745	4,18	1.566	-2.183	4,19	-1.982	-1.417
4,20	-1.289	1.812	4,21	1.665	1.180	4,22	1.086	-1.538	4,23	-1.427	-1.003
4,24	-0.931	1.330	4,25	1.243	0.867	4,26	0.810	-1.166	4,27	-1.097	-0.759
4,28	-0.713	1.035	4,29	0.978	0.672	4,30	0.634	-0.928	4,31	-0.880	-0.601
4,32	-0.569	0.839	4,33	0.798	0.541	4,34	0.515	-0.764	4,35	-0.729	-0.491
4,36	-0.468	0.700	4,37	0.669	0.448	4,38	0.429	-0.646	4,39	-0.617	-0.412
4,40	-0.394	0.599	4,41	0.571	0.380	4,42	0.364	-0.559			

5, 5	147.180	-608.350	5, 6	104.581	310.668	5, 7	22.166	-72.469	5, 8	-20.628	-10.144
5, 9	-9.469	12.723	5,10	9.438	6.937	5,11	5.300	-7.235	5,12	-5.785	-4.251
5,13	-3.513	4.774	5,14	4.032	2.967	5,15	2.550	-3.468	5,16	-3.029	-2.222
5,17	-1.959	2.676	5,18	2.391	1.744	5,19	1.566	-2.153	5,20	-1.955	-1.416
5,21	-1.289	1.786	5,22	1.642	1.180	5,23	1.086	-1.516	5,24	-1.407	-1.003
5,25	-0.931	1.310	5,26	1.225	0.867	5,27	0.810	-1.148	5,28	-1.081	-0.759
5,29	-0.713	1.019	5,30	0.964	0.672	5,31	0.635	-0.913	5,32	-0.868	-0.600
5,33	-0.570	0.824	5,34	0.788	0.541	5,35	0.515	-0.750	5,36	-0.720	-0.491
5,37	-0.469	0.687	5,38	0.662	0.448	5,39	0.429	-0.632	5,40	-0.612	-0.411
5,41	-0.395	0.583									
6, 6	147.174	-608.210	6, 7	104.686	310.671	6, 8	22.168	-72.553	6, 9	-20.697	-10.146
6,10	-9.469	12.781	6,11	9.487	6.938	6,12	5.300	-7.278	6,13	-5.822	-4.251
6,14	-3.513	4.808	6,15	4.062	2.967	6,16	2.550	-3.496	6,17	-3.053	-2.222
6,18	-1.959	2.700	6,19	2.411	1.744	6,20	1.566	-2.173	6,21	-1.973	-1.416
6,22	-1.289	1.803	6,23	1.657	1.180	6,24	1.085	-1.531	6,25	-1.420	-1.003
6,26	-0.930	1.324	6,27	1.237	0.866	6,28	0.809	-1.161	6,29	-1.091	-0.759
6,30	-0.713	1.030	6,31	0.973	0.672	6,32	0.634	-0.924	6,33	-0.876	-0.600
6,34	-0.569	0.835	6,35	0.794	0.541	6,36	0.514	-0.761	6,37	-0.725	-0.491
6,38	-0.468	0.698	6,39	0.665	0.448	6,40	0.428	-0.644			
7, 7	147.174	-608.289	7, 8	104.624	310.671	7, 9	22.168	-72.502	7,10	-20.654	-10.146
7,11	-9.470	12.744	7,12	9.456	6.938	7,13	5.301	-7.250	7,14	-5.798	-4.252
7,15	-3.514	4.785	7,16	4.042	2.968	7,17	2.551	-3.477	7,18	-3.037	-2.223
7,19	-1.960	2.684	7,20	2.397	1.744	7,21	1.566	-2.159	7,22	-1.961	-1.416
7,23	-1.289	1.791	7,24	1.647	1.180	7,25	1.086	-1.520	7,26	-1.411	-1.003
7,27	-0.931	1.313	7,28	1.229	0.867	7,29	0.810	-1.151	7,30	-1.084	-0.759
7,31	-0.714	1.021	7,32	0.967	0.672	7,33	0.635	-0.915	7,34	-0.871	-0.601
7,35	-0.570	0.827	7,36	0.791	0.541	7,37	0.515	-0.752	7,38	-0.723	-0.491
7,39	-0.469	0.688									
8, 8	147.175	-608.240	8, 9	104.664	310.670	8,10	22.167	-72.536	8,11	-20.683	-10.145
8,12	-9.469	12.770	8,13	9.477	6.937	8,14	5.300	-7.270	8,15	-5.815	-4.251
8,16	-3.513	4.801	8,17	4.056	2.967	8,18	2.550	-3.491	8,19	-3.048	-2.222
8,20	-1.959	2.696	8,21	2.407	1.744	8,22	1.565	-2.170	8,23	-1.969	-1.416
8,24	-1.289	1.800	8,25	1.654	1.180	8,26	1.085	-1.528	8,27	-1.417	-1.003
8,28	-0.930	1.321	8,29	1.234	0.866	8,30	0.809	-1.159	8,31	-1.089	-0.759
8,32	-0.713	1.029	8,33	0.971	0.672	8,34	0.634	-0.922	8,35	-0.874	-0.600
8,36	-0.569	0.834	8,37	0.792	0.541	8,38	0.514	-0.760			
9, 9	147.173	-608.273	9,10	104.637	310.671	9,11	22.169	-72.512	9,12	-20.663	-10.147
9,13	-9.470	12.751	9,14	9.462	6.938	9,15	5.301	-7.255	9,16	-5.802	-4.252
9,17	-3.514	4.789	9,18	4.046	2.968	9,19	2.551	-3.480	9,20	-3.040	-2.223
9,21	-1.960	2.686	9,22	2.400	1.744	9,23	1.566	-2.162	9,24	-1.963	-1.416
9,25	-1.289	1.793	9,26	1.649	1.180	9,27	1.086	-1.522	9,28	-1.413	-1.003
9,29	-0.931	1.315	9,30	1.230	0.867	9,31	0.810	-1.153	9,32	-1.086	-0.759
9,33	-0.714	1.022	9,34	0.969	0.672	9,35	0.635	-0.916	9,36	-0.872	-0.600
9,37	-0.570	0.827									
10,10	147.175	-608.249	10,11	104.656	310.670	10,12	22.167	-72.530	10,13	-20.677	-10.146
10,14	-9.469	12.765	10,15	9.473	6.938	10,16	5.300	-7.267	10,17	-5.812	-4.251
10,18	-3.513	4.798	10,19	4.054	2.967	10,20	2.550	-3.489	10,21	-3.046	-2.222
10,22	-1.959	2.694	10,23	2.405	1.744	10,24	1.566	-2.168	10,25	-1.968	-1.416
10,26	-1.289	1.799	10,27	1.652	1.180	10,28	1.085	-1.527	10,29	-1.416	-1.003
10,30	-0.930	1.320	10,31	1.233	0.867	10,32	0.809	-1.158	10,33	-1.088	-0.759
10,34	-0.713	1.028	10,35	0.970	0.672	10,36	0.634	-0.922			
11,11	147.174	-608.267	11,12	104.642	310.671	11,13	22.168	-72.516	11,14	-20.666	-10.146
[illegible]	[illegible]	12.756	11,16	9.465	6.938	11,17	5.301	-7.258	11,18	-5.805	-4.252
11,19	-3.514	4.791	11,20	4.048	[illegible]	[illegible]	[illegible]	[illegible]	11,22	-3.041	-2.223
11,23	-1.960	2.688	11,24	2.401	1.744	11,25	1.566	-2.163	11,26	-1.964	-1.416
11,27	-1.289	1.794	11,28	1.650	1.180	11,29	1.086	-1.522	11,30	-1.414	-1.003
11,31	-0.931	1.315	11,32	1.231	0.867	11,33	0.810	-1.153	11,34	-1.087	-0.759
11,35	-0.714	1.023									
12,12	147.175	-608.253	12,13	104.653	310.670	12,14	22.167	-72.527	12,15	-20.675	-10.146
12,16	-9.469	12.763	12,17	9.471	6.938	12,18	5.300	-7.265	12,19	-5.810	-4.251
12,20	-3.513	4.797	12,21	4.052	2.967	12,22	2.550	-3.488	12,23	-3.045	-2.222
12,24	-1.959	2.693	12,25	2.404	1.744	12,26	1.566	-2.168	12,27	-1.967	-1.416
12,28	-1.289	1.798	12,29	1.652	1.180	12,30	1.085	-1.527	12,31	-1.415	-1.003
12,32	-0.930	1.320	12,33	1.232	0.867	12,34	0.809	-1.158			
13,13	147.174	-608.264	13,14	104.644	310.671	13,15	22.168	-72.518	13,16	-20.668	-10.146
13,17	-9.470	12.756	13,18	9.466	6.938	13,19	5.301	-7.259	13,20	-5.806	-4.252
13,21	-3.514	4.792	13,22	4.049	2.968	13,23	2.551	-3.483	13,24	-3.042	-2.222
13,25	-1.959	2.688	13,26	2.402	1.744	13,27	1.566	-2.163	13,28	-1.965	-1.416
13,29	-1.289	1.794	13,30	1.650	1.180	13,31	1.086	-1.523	13,32	-1.414	-1.003
13,33	-0.931	1.316									
14,14	147.175	-608.255	14,15	104.651	310.671	14,16	22.167	-72.525	14,17	-20.673	-10.146
14,18	-9.469	12.762	14,19	9.470	6.938	14,20	5.300	-7.264	14,21	-5.809	-4.251
14,22	-3.513	4.796	14,23	4.052	2.968	14,24	2.550	-3.487	14,25	-3.044	-2.222
14,26	-1.959	2.692	14,27	2.404	1.744	14,28	1.566	-2.167	14,29	-1.966	-1.416
14,30	-1.289	1.798	14,31	1.651	1.180	14,32	1.085	-1.527			
15,15	147.174	-608.262	15,16	104.646	310.671	15,17	22.168	-72.519	15,18	-20.669	-10.146
15,19	-9.470	12.757	15,20	9.467	6.938	15,21	5.301	-7.260	15,22	-5.807	-4.252
15,23	-3.514	4.792	15,24	4.050	2.968	15,25	2.551	-3.483	15,26	-3.043	-2.222
15,27	-1.959	2.689	15,28	2.403	1.744	15,29	1.566	-2.163	15,30	-1.966	-1.416
15,31	-1.289	1.794									
16,16	147.175	-608.256	16,17	104.650	310.671	16,18	22.168	-72.524	16,19	-20.672	-10.146
16,20	-9.469	12.761	16,21	9.469	6.938	16,22	5.300	-7.264	16,23	-5.809	-4.251
16,24	-3.513	4.796	16,25	4.051	2.968	16,26	2.550	-3.487	16,27	-3.044	-2.222
16,28	-1.959	2.692	16,29	2.403	1.744	16,30	1.566	-2.167			
17,17	147.174	-608.261	17,18	104.646	310.671	17,19	22.168	-72.520	17,20	-20.670	-10.146
17,21	-9.470	12.757	17,22	9.468	6.938	17,23	5.301	-7.260	17,24	-5.807	-4.252
17,25	-3.514	4.792	17,26	4.050	2.968	17,27	2.551	-3.483	17,28	-3.043	-2.222
17,29	-1.959	2.689									
18,18	147.175	-608.257	18,19	104.649	310.671	18,20	22.168	-72.524	18,21	-20.672	-10.146
18,22	-9.469	12.761	18,23	9.469	6.938	18,24	5.300	-7.263	18,25	-5.808	-4.251
18,26	-3.513	4.796	18,27	4.050	2.968	18,28	2.550	-3.487			
19,19	147.174	-608.261	19,20	104.647	310.671	19,21	22.168	-72.520	19,22	-20.671	-10.146
19,23	-9.470	12.757	19,24	9.468	6.938	19,25	5.301	-7.260	19,26	-5.808	-4.251
19,27	-3.514	4.792									
20,20	147.174	-608.257	20,21	104.648	310.671	20,22	22.168	-72.524	20,23	-20.671	-10.146
20,24	-9.469	12.761	20,25	9.468	6.938	20,26	5.300	-7.263			
21,21	147.174	-608.261	21,22	104.648	310.671	21,23	22.168	-72.520	21,24	-20.671	-10.146
21,25	-9.470	12.757									
22,22	147.174	-608.257	22,23	104.648	310.671	22,24	22.168	-72.524			
23,23	147.174	-608.261									

IBR

SELF AND MUTUAL IMPEDANCES

H/LMDA=0.5000 B/LMDA= 0.250 OMEGA= 9.92

50 ELEMENT ARRAY

I, K	ZIK IN OHMS RE	IM	I, K	ZIK IN OHMS RE	IM	I, K	ZIK IN OHMS RE	IM	I, K	ZIK IN OHMS RE	IM
1, 1	347.433	-558.893	1, 2	71.131	281.176	1, 3	14.293	-64.682	1, 4	-17.036	-3.815
1, 5	-4.855	10.061	1, 6	7.416	3.471	1, 7	2.556	-5.639	1, 8	-4.479	-2.005
1, 9	-1.629	3.678	1,10	3.094	1.357	1,11	1.152	-2.652	1,12	-2.309	-0.993
1,13	-0.867	2.035	1,14	1.813	0.766	1,15	0.682	-1.630	1,16	-1.476	-0.612
1,17	-0.554	1.347	1,18	1.235	0.504	1,19	0.460	-1.139	1,20	-1.055	-0.423
1,21	-0.391	0.982	1,22	0.916	0.362	1,23	0.337	-0.859	1,24	-0.806	-0.314
1,25	-0.294	0.760	1,26	0.718	0.276	1,27	0.259	-0.680	1,28	-0.645	-0.245
1,29	-0.231	0.613	1,30	0.584	0.219	1,31	0.207	-0.557	1,32	-0.532	-0.197
1,33	-0.188	0.510	1,34	0.488	0.179	1,35	0.171	-0.469	1,36	-0.450	-0.163
1,37	-0.156	0.434	1,38	0.417	0.150	1,39	0.144	-0.403	1,40	-0.388	-0.138
1,41	-0.132	0.375	1,42	0.362	0.128	1,43	0.123	-0.351	1,44	-0.339	-0.119
1,45	-0.114	0.330	1,46	0.317	0.111	1,47	0.106	-0.312	1,48	-0.298	-0.106
1,49	-0.088	0.310	1,50	0.203	0.032						
2, 2	152.099	-599.112	2, 3	106.881	309.401	2, 4	21.758	-74.082	2, 5	-21.809	-9.835
2, 6	-9.225	13.611	2, 7	10.141	6.742	2, 8	5.139	-7.810	2, 9	-6.267	-4.115
2,10	-3.395	5.186	2,11	4.391	2.864	2,12	2.458	-3.783	2,13	-3.309	-2.140
2,14	-1.885	2.928	2,15	2.618	1.676	2,16	1.503	-2.360	2,17	-2.144	-1.358
2,18	-1.235	1.960	2,19	1.803	1.130	2,20	1.039	-1.665	2,21	-1.546	-0.959
2,22	-0.889	1.440	2,23	1.347	0.827	2,24	0.773	-1.263	2,25	-1.189	-0.724
2,26	-0.680	1.121	2,27	1.061	0.640	2,28	0.604	-1.005	2,29	-0.955	-0.571
2,30	-0.542	0.909	2,31	0.867	0.514	2,32	0.489	-0.827	2,33	-0.792	-0.466
2,34	-0.445	0.758	2,35	0.728	0.425	2,36	0.407	-0.698	2,37	-0.672	-0.390
2,38	-0.374	0.646	2,39	0.624	0.359	2,40	0.346	-0.600	2,41	-0.581	-0.332
2,42	-0.321	0.560	2,43	0.544	0.309	2,44	0.299	-0.524	2,45	-0.511	-0.288
2,46	-0.279	0.490	2,47	0.483	0.269	2,48	0.265	-0.459	2,49	-0.485	-0.239
3, 3	146.867	-608.698	3, 4	104.350	310.773	3, 5	22.245	-72.306	3, 6	-20.511	-10.207
3, 7	-9.519	12.634	3, 8	9.367	6.979	3, 9	5.335	-7.177	3,10	-5.736	-4.281
3,11	-3.539	4.732	3,12	3.996	2.991	3,13	2.571	-3.437	3,14	-3.000	-2.241
3,15	-1.976	2.652	3,16	2.368	1.760	3,17	1.580	-2.133	3,18	-1.936	-1.430
3,19	-1.301	1.769	3,20	1.625	1.192	3,21	1.096	-1.501	3,22	-1.392	-1.013
3,23	-0.940	1.297	3,24	1.212	0.876	3,25	0.818	-1.138	3,26	-1.070	-0.767
3,27	-0.721	1.009	3,28	0.954	0.679	3,29	0.642	-0.905	3,30	-0.858	-0.607
3,31	-0.576	0.818	3,32	0.779	0.547	3,33	0.521	-0.744	3,34	-0.711	-0.497
3,35	-0.474	0.682	3,36	0.653	0.454	3,37	0.434	-0.628	3,38	-0.603	-0.417
3,39	-0.400	0.582	3,40	0.559	0.384	3,41	0.369	-0.541	3,42	-0.520	-0.356
3,43	-0.343	0.505	3,44	0.486	0.331	3,45	0.319	-0.474	3,46	-0.454	-0.309
3,47	-0.298	0.448	3,48	0.424	0.292						
4, 4	147.158	-608.082	4, 5	104.777	310.681	4, 6	22.174	-72.618	4, 7	-20.746	-10.150
4, 8	-9.473	12.820	4, 9	9.519	6.941	4,10	5.303	-7.305	4,11	-5.845	-4.253
4,12	-3.515	4.827	4,13	4.079	2.969	4,14	2.551	-3.511	4,15	-3.067	-2.223
4,16	-1.960	2.711	4,17	2.423	1.745	4,18	1.566	-2.183	4,19	-1.982	-1.416
4,20	-1.289	1.811	4,21	1.665	1.180	4,22	1.086	-1.538	4,23	-1.427	-1.003
4,24	-0.931	1.329	4,25	1.243	0.867	4,26	0.810	-1.166	4,27	-1.097	-0.759
4,28	-0.713	1.034	4,29	0.979	0.672	4,30	0.634	-0.927	4,31	-0.881	-0.600
4,32	-0.569	0.838	4,33	0.800	0.541	4,34	0.515	-0.763	4,35	-0.730	-0.491
4,36	-0.469	0.699	4,37	0.671	0.448	4,38	0.429	-0.643	4,39	-0.620	-0.411
4,40	-0.395	0.595	4,41	0.576	0.379	4,42	0.365	-0.553	4,43	-0.537	-0.351
4,44	-0.339	0.516	4,45	0.503	0.326	4,46	0.316	-0.481	4,47	-0.474	-0.304
5, 5	147.180	-608.350	5, 6	104.580	310.668	5, 7	22.166	-72.469	5, 8	-20.628	-10.144
5, 9	-9.468	12.723	5,10	9.438	6.937	5,11	5.300	-7.236	5,12	-5.785	-4.251
5,13	-3.513	4.774	5,14	4.032	2.967	5,15	2.550	-3.469	5,16	-3.028	-2.222
5,17	-1.959	2.677	5,18	2.390	1.744	5,19	1.566	-2.154	5,20	-1.955	-1.416
5,21	-1.289	1.786	5,22	1.641	1.180	5,23	1.085	-1.516	5,24	-1.406	-1.003
5,25	-0.931	1.310	5,26	1.224	0.867	5,27	0.810	-1.149	5,28	-1.080	-0.759
5,29	-0.713	1.019	5,30	0.963	0.672	5,31	0.635	-0.914	5,32	-0.867	-0.601
5,33	-0.569	0.826	5,34	0.786	0.541	5,35	0.515	-0.752	5,36	-0.718	-0.491
5,37	-0.469	0.689	5,38	0.659	0.448	5,39	0.429	-0.635	5,40	-0.609	-0.412
5,41	-0.395	0.588	5,42	0.564	0.380	5,43	0.365	-0.547	5,44	-0.525	-0.352
5,45	-0.338	0.511	5,46	0.489	0.327						
6, 6	147.174	-608.210	6, 7	104.686	310.671	6, 8	22.168	-72.553	6, 9	-20.697	-10.146
6,10	-9.470	12.781	6,11	9.487	6.938	6,12	5.300	-7.278	6,13	-5.823	-4.251
6,14	-3.513	4.807	6,15	4.062	2.967	6,16	2.550	-3.496	6,17	-3.053	-2.222
6,18	-1.959	2.699	6,19	2.412	1.744	6,20	1.566	-2.173	6,21	-1.973	-1.416
6,22	-1.289	1.803	6,23	1.657	1.180	6,24	1.085	-1.530	6,25	-1.421	-1.003
6,26	-0.930	1.323	6,27	1.237	0.866	6,28	0.810	-1.160	6,29	-1.092	-0.759
6,30	-0.713	1.029	6,31	0.974	0.672	6,32	0.634	-0.922	6,33	-0.877	-0.600
6,34	-0.569	0.834	6,35	0.796	0.541	6,36	0.515	-0.759	6,37	-0.727	-0.491
6,38	-0.469	0.695	6,39	0.668	0.448	6,40	0.429	-0.640	6,41	-0.617	-0.411
6,42	-0.395	0.592	6,43	0.573	0.379	6,44	0.365	-0.550	6,45	-0.535	-0.351
7, 7	147.174	-608.288	7, 8	104.624	310.671	7, 9	22.168	-72.502	7,10	-20.654	-10.147
7,11	-9.470	12.745	7,12	9.455	6.938	7,13	5.301	-7.251	7,14	-5.798	-4.252
7,15	-3.514	4.785	7,16	4.042	2.968	7,17	2.551	-3.478	7,18	-3.036	-2.223
7,19	-1.959	2.684	7,20	2.397	1.745	7,21	1.566	-2.160	7,22	-1.960	-1.417
7,23	-1.289	1.792	7,24	1.646	1.180	7,25	1.086	-1.521	7,26	-1.410	-1.003
7,27	-0.931	1.314	7,28	1.228	0.867	7,29	0.810	-1.152	7,30	-1.083	-0.759
7,31	-0.713	1.023	7,32	0.966	0.672	7,33	0.635	-0.916	7,34	-0.869	-0.601
7,35	-0.570	0.828	7,36	0.789	0.541	7,37	0.515	-0.754	7,38	-0.720	-0.491
7,39	-0.469	0.691	7,40	0.661	0.448	7,41	0.429	-0.637	7,42	-0.610	-0.412
7,43	-0.395	0.590	7,44	0.565	0.380						
8, 8	147.175	-608.240	8, 9	104.664	310.670	8,10	22.167	-72.536	8,11	-20.683	-10.145
8,12	-9.469	12.769	8,13	9.478	6.937	8,14	5.300	-7.270	8,15	-5.815	-4.251
8,16	-3.513	4.801	8,17	4.057	2.967	8,18	2.550	-3.490	8,19	-3.049	-2.222
8,20	-1.959	2.695	8,21	2.408	1.744	8,22	1.566	-2.169	8,23	-1.970	-1.416
8,24	-1.289	1.800	8,25	1.655	1.179	8,26	1.085	-1.528	8,27	-1.418	-1.003
8,28	-0.930	1.320	8,29	1.235	0.866	8,30	0.810	-1.158	8,31	-1.090	-0.758
8,32	-0.713	1.027	8,33	0.972	0.672	8,34	0.634	-0.921	8,35	-0.875	-0.600
8,36	-0.569	0.832	8,37	0.794	0.541	8,38	0.515	-0.757	8,39	-0.726	-0.491
8,40	-0.469	0.693	8,41	0.667	0.448	8,42	0.429	-0.638	8,43	-0.617	-0.411
9, 9	147.173	-608.272	9,10	104.636	310.672	9,11	22.168	-72.512	9,12	-20.662	-10.147
9,13	-9.470	12.752	9,14	9.461	6.938	9,15	5.301	-7.256	9,16	-5.802	-4.252
9,17	-3.514	4.789	9,18	4.045	2.968	9,19	2.551	-3.481	9,20	-3.039	-2.223
9,21	-1.959	2.687	9,22	2.399	1.745	9,23	1.566	-2.162	9,24	-1.962	-1.417
9,25	-1.289	1.794	9,26	1.648	1.180	9,27	1.086	-1.522	9,28	-1.412	-1.003
9,29	-0.931	1.316	9,30	1.229	0.867	9,31	0.810	-1.154	9,32	-1.085	-0.759
9,33	-0.713	1.024	9,34	0.967	0.672	9,35	0.635	-0.918	9,36	-0.870	-0.601
9,37	-0.570	0.830	9,38	0.789	0.541	9,39	0.515	-0.755	9,40	-0.721	-0.491
9,41	-0.469	0.692	9,42	0.662	0.449						

10,10	147.175	-608.249	10,11	104.656	310.670	10,12	22.167	-72.529	10,13	-20.678	-10.145
10,14	-9.469	12.765	10,15	9.474	6.937	10,16	5.300	-7.266	10,17	-5.812	-4.251
10,18	-3.513	4.798	10,19	4.054	2.967	10,20	2.550	-3.488	10,21	-3.047	-2.222
10,22	-1.959	2.693	10,23	2.406	1.744	10,24	1.566	-2.168	10,25	-1.968	-1.416
10,26	-1.289	1.798	10,27	1.653	1.180	10,28	1.085	-1.526	10,29	-1.417	-1.003
10,30	-0.931	1.319	10,31	1.234	0.866	10,32	0.810	-1.157	10,33	-1.089	-0.759
10,34	-0.713	1.026	10,35	0.972	0.672	10,36	0.634	-0.920	10,37	-0.875	-0.600
10,38	-0.569	0.831	10,39	0.794	0.541	10,40	0.515	-0.756	10,41	-0.726	-0.490
11,11	147.174	-608.266	11,12	104.641	310.671	11,13	22.168	-72.517	11,14	-20.666	-10.147
11,15	-9.470	12.755	11,16	9.464	6.938	11,17	5.301	-7.258	11,18	-5.804	-4.252
11,19	-3.514	4.791	11,20	4.047	2.968	11,21	2.551	-3.482	11,22	-3.040	-2.223
11,23	-1.959	2.688	11,24	2.400	1.744	11,25	1.566	-2.163	11,26	-1.963	-1.416
11,27	-1.289	1.795	11,28	1.649	1.180	11,29	1.086	-1.523	11,30	-1.413	-1.003
11,31	-0.931	1.317	11,32	1.230	0.867	11,33	0.810	-1.155	11,34	-1.085	-0.759
11,35	-0.713	1.025	11,36	0.968	0.672	11,37	0.635	-0.918	11,38	-0.871	-0.601
11,39	-0.569	0.830	11,40	0.790	0.541						
12,12	147.175	-608.254	12,13	104.653	310.670	12,14	22.167	-72.526	12,15	-20.675	-10.146
12,16	-9.469	12.762	12,17	9.472	6.938	12,18	5.300	-7.264	12,19	-5.811	-4.251
12,20	-3.513	4.797	12,21	4.053	2.967	12,22	2.550	-3.487	12,23	-3.046	-2.222
12,24	-1.959	2.692	12,25	2.405	1.744	12,26	1.566	-2.167	12,27	-1.968	-1.416
12,28	-1.289	1.797	12,29	1.653	1.180	12,30	1.085	-1.526	12,31	-1.416	-1.003
12,32	-0.931	1.319	12,33	1.234	0.866	12,34	0.810	-1.156	12,35	-1.089	-0.759
12,36	-0.713	1.026	12,37	0.971	0.672	12,38	0.634	-0.919	12,39	-0.875	-0.600
13,13	147.174	-608.263	13,14	104.644	310.671	13,15	22.168	-72.519	13,16	-20.668	-10.146
13,17	-9.470	12.756	13,18	9.465	6.938	13,19	5.301	-7.259	13,20	-5.805	-4.252
13,21	-3.514	4.792	13,22	4.048	2.968	13,23	2.550	-3.483	13,24	-3.041	-2.223
13,25	-1.959	2.689	13,26	2.401	1.744	13,27	1.566	-2.164	13,28	-1.964	-1.416
13,29	-1.289	1.795	13,30	1.649	1.180	13,31	1.086	-1.524	13,32	-1.413	-1.003
13,33	-0.931	1.317	13,34	1.230	0.867	13,35	0.810	-1.155	13,36	-1.085	-0.759
13,37	-0.713	1.025	13,38	0.968	0.672						
14,14	147.175	-608.256	14,15	104.651	310.670	14,16	22.168	-72.525	14,17	-20.674	-10.146
14,18	-9.469	12.761	14,19	9.471	6.938	14,20	5.300	-7.263	14,21	-5.810	-4.251
14,22	-3.513	4.796	14,23	4.052	2.967	14,24	2.550	-3.486	14,25	-3.045	-2.222
14,26	-1.959	2.691	14,27	2.405	1.744	14,28	1.566	-2.166	14,29	-1.967	-1.416
14,30	-1.289	1.797	14,31	1.652	1.180	14,32	1.085	-1.525	14,33	-1.416	-1.003
14,34	-0.931	1.318	14,35	1.234	0.866	14,36	0.810	-1.156	14,37	-1.089	-0.759
15,15	147.174	-608.262	15,16	104.645	310.671	15,17	22.168	-72.520	15,18	-20.669	-10.146
15,19	-9.470	12.757	15,20	9.466	6.938	15,21	5.301	-7.260	15,22	-5.806	-4.252
15,23	-3.513	4.793	15,24	4.049	2.968	15,25	2.550	-3.484	15,26	-3.042	-2.223
15,27	-1.959	2.690	15,28	2.401	1.744	15,29	1.566	-2.165	15,30	-1.964	-1.416
15,31	-1.289	1.796	15,32	1.649	1.180	15,33	1.085	-1.524	15,34	-1.413	-1.003
15,35	-0.931	1.318	15,36	1.230	0.867						
16,16	147.174	-608.257	16,17	104.650	310.670	16,18	22.168	-72.524	16,19	-20.673	-10.146
16,20	-9.470	12.760	16,21	9.470	6.938	16,22	5.300	-7.263	16,23	-5.809	-4.251
16,24	-3.513	4.795	16,25	4.052	2.967	16,26	2.550	-3.486	16,27	-3.045	-2.222
16,28	-1.959	2.691	16,29	2.404	1.744	16,30	1.566	-2.166	16,31	-1.967	-1.416
16,32	-1.289	1.796	16,33	1.652	1.180	16,34	1.085	-1.525	16,35	-1.416	-1.003
17,17	147.174	-608.261	17,18	104.646	310.671	17,19	22.168	-72.521	17,20	-20.669	-10.146
17,21	-9.470	12.758	17,22	9.467	6.938	17,23	5.301	-7.261	17,24	-5.806	-4.252
17,25	-3.513	4.793	17,26	4.049	2.968	17,27	2.550	-3.484	17,28	-3.042	-2.223
17,29	-1.959	2.690	17,30	2.402	1.744	17,31	1.566	-2.165	17,32	-1.964	-1.416
17,33	[illegible]	[illegible]	17,34	1.649	1.180						
18,18	147.174	-608.258	18,19	104.650	310.671	18,20	22.168	-72.523	18,21	-20.673	[illegible]
18,22	-9.470	12.760	18,23	9.470	6.938	18,24	5.300	-7.262	18,25	-5.809	-4.251
18,26	-3.513	4.795	18,27	4.052	2.967	18,28	2.550	-3.485	18,29	-3.045	-2.222
18,30	-1.959	2.691	18,31	2.404	1.744	18,32	1.566	-2.165	18,33	-1.967	-1.416
19,19	147.174	-608.260	19,20	104.646	310.671	19,21	22.168	-72.521	19,22	-20.670	-10.146
19,23	-9.470	12.758	19,24	9.467	6.938	19,25	5.301	-7.261	19,26	-5.806	-4.252
19,27	-3.513	4.794	19,28	4.049	2.968	19,29	2.550	-3.485	19,30	-3.042	-2.223
19,31	-1.959	2.690	19,32	2.402	1.744						
20,20	147.174	-608.258	20,21	104.649	310.671	20,22	22.168	-72.523	20,23	-20.672	-10.146
20,24	-9.470	12.759	20,25	9.470	6.938	20,26	5.301	-7.262	20,27	-5.809	-4.251
20,28	-3.513	4.794	20,29	4.052	2.967	20,30	2.550	-3.485	20,31	-3.045	-2.222
21,21	147.174	-608.260	21,22	104.646	310.671	21,23	22.168	-72.522	21,24	-20.670	-10.146
21,25	-9.470	12.759	21,26	9.467	6.938	21,27	5.301	-7.261	21,28	-5.807	-4.252
21,29	-3.513	4.794	21,30	4.049	2.968						
22,22	147.174	-608.259	22,23	104.649	310.671	22,24	22.168	-72.522	22,25	-20.672	-10.146
22,26	-9.470	12.759	22,27	9.470	6.938	22,28	5.301	-7.262	22,29	-5.809	-4.251
23,23	147.174	-608.259	23,24	104.647	310.671	23,25	22.168	-72.522	23,26	-20.670	-10.146
23,27	-9.470	12.759	23,28	9.467	6.938						
24,24	147.174	-608.259	24,25	104.649	310.671	24,26	22.168	-72.522	24,27	-20.672	-10.146
25,25	147.174	-608.259	25,26	104.647	310.671						

IBR

SELF AND MUTUAL IMPEDANCES

H/LMDA=0.5000 B/LMDA= 0.500 OMEGA= 9.92

2 ELEMENT ARRAY

I, K	ZIK IN OHMS RE	IM	I, K	ZIK IN OHMS RE	IM	I, K	ZIK IN OHMS RE	IM	I, K	ZIK IN OHMS RE	IM
1, 1	547.138	-403.985	1, 2	210.229	97.772						

IBR

SELF AND MUTUAL IMPEDANCES

H/LMDA=0.5000 B/LMDA= 0.500 OMEGA= 9.92

3 ELEMENT ARRAY

I, K	ZIK IN OHMS RE	IM	I, K	ZIK IN OHMS RE	IM	I, K	ZIK IN OHMS RE	IM	I, K	ZIK IN OHMS RE	IM
1, 1	559.862	-398.523	1, 2	186.039	77.212	1, 3	-87.459	-8.283			
2, 2	588.032	-344.127									

IBR

SELF AND MUTUAL IMPEDANCES

H/LMDA=0.5000 B/LMDA= 0.500 OMEGA= 9.92

4 ELEMENT ARRAY

I, K	ZIK IN OHMS RE	IM	I, K	ZIK IN OHMS RE	IM	I, K	ZIK IN OHMS RE	IM	I, K	ZIK IN OHMS RE	IM
1, 1	563.300	-397.415	1, 2	180.268	75.719	1, 3	-75.314	1.303	1, 4	45.604	1.881
2, 2	597.950	-342.275	2, 3	163.875	63.297						

IBR

SELF AND MUTUAL IMPEDANCES

H/LMDA=0.5000 B/LMDA= 0.500 OMEGA= 9.92

5 ELEMENT ARRAY

I, K	ZIK IN OHMS RE	IM	I, K	ZIK IN OHMS RE	IM	I, K	ZIK IN OHMS RE	IM	I, K	ZIK IN OHMS RE	IM
1, 1	564.729	-396.974	1, 2	178.368	75.422	1, 3	-71.790	2.142	1, 4	37.910	-4.133
1, 5	-29.649	-1.061									
2, 2	600.432	-342.235	2, 3	159.221	62.855	2, 4	-63.636	7.233			
3, 3	606.988	-340.831									

IBR

SELF AND MUTUAL IMPEDANCES

H/LMDA=0.5000 B/LMDA= 0.500 OMEGA= 9.92

6 ELEMENT ARRAY

I, K	ZIK IN OHMS RE	IM	I, K	ZIK IN OHMS RE	IM	I, K	ZIK IN OHMS RE	IM	I, K	ZIK IN OHMS RE	IM
1, 1	565.492	-396.741	1, 2	177.485	75.311	1, 3	-70.492	2.308	1, 4	35.391	-4.716
1, 5	-24.072	3.278	1, 6	21.750	0.786						
2, 2	601.414	-342.276	2, 3	157.778	62.911	2, 4	-60.788	7.406	2, 5	30.414	-7.693
3, 3	609.154	-340.901	3, 4	154.761	62.623						

IBR

SELF AND MUTUAL IMPEDANCES

H/LMDA=0.5000 B/LMDA= 0.500 OMEGA= 9.92

7 ELEMENT ARRAY

I, K	ZIK IN OHMS RE	IM	I, K	ZIK IN OHMS RE	IM	I, K	ZIK IN OHMS RE	IM	I, K	ZIK IN OHMS RE	IM
1, 1	565.960	-396.599	1, 2	176.984	75.256	1, 3	-69.853	2.367	1, 4	34.404	-4.833
1, 5	-22.121	3.724	1, 6	17.402	-2.592	1, 7	-17.066	-0.639			
2, 2	601.922	-342.313	2, 3	157.132	62.971	2, 4	-59.785	7.340	2, 5	28.392	-7.776
2, 6	-18.628	5.782									
3, 3	609.981	-340.990	3, 4	153.476	62.725	3, 5	-58.008	7.450			
4, 4	611.201	-340.998									

IBR

SELF AND MUTUAL IMPEDANCES

H/LMDA=0.5000 B/LMDA= 0.500 OMEGA= 9.92

8 ELEMENT ARRAY

I, K	ZIK IN OHMS		I, K	ZIK IN OHMS		I, K	ZIK IN OHMS		I, K	ZIK IN OHMS	
	RE	IM		RE	IM		RE	IM		RE	IM
1, 1	566.272	-396.503	1, 2	176.666	75.223	1, 3	-69.478	2.394	1, 4	33.899	-4.874
1, 5	-21.326	3.816	1, 6	15.818	-2.952	1, 7	-13.523	2.115	1, 8	13.969	0.544
2, 2	602.227	-342.342	2, 3	156.774	63.015	2, 4	-59.303	7.285	2, 5	27.627	-7.714
2, 6	-17.072	5.826	2, 7	13.162	-4.506						
3, 3	610.399	-341.053	3, 4	152.913	62.805	3, 5	-57.114	7.354	3, 6	26.408	-7.771
4, 4	611.970	-341.096	4, 5	152.221	62.835						

IBR

SELF AND MUTUAL IMPEDANCES

H/LMDA=0.5000 B/LMDA= 0.500 OMEGA= 9.92

9 ELEMENT ARRAY

I, K	ZIK IN OHMS		I, K	ZIK IN OHMS		I, K	ZIK IN OHMS		I, K	ZIK IN OHMS	
	RE	IM		RE	IM		RE	IM		RE	IM
1, 1	566.494	-396.435	1, 2	176.448	75.201	1, 3	-69.234	2.410	1, 4	33.596	-4.893
1, 5	-20.909	3.848	1, 6	15.154	-3.028	1, 7	-12.195	2.416	1, 8	10.993	-1.772
1, 9	-11.775	-0.474									
2, 2	602.428	-342.363	2, 3	156.551	63.046	2, 4	-59.025	7.245	2, 5	27.245	-7.666
2, 6	-16.458	5.771	2, 7	11.905	-4.531	2, 8	-10.071	3.653			
3, 3	610.646	-341.097	3, 4	152.606	62.861	3, 5	-56.691	7.284	3, 6	25.729	-7.685
3, 7	-15.545	5.804									
4, 4	612.353	-341.165	4, 5	151.693	62.921	4, 6	-56.228	7.256			
5, 5	612.709	-341.199									

IBR

SELF AND MUTUAL IMPEDANCES

H/LMDA=0.5000 B/LMDA= 0.500 OMEGA= 9.92

10 ELEMENT ARRAY

I, K	ZIK IN OHMS		I, K	ZIK IN OHMS		I, K	ZIK IN OHMS		I, K	ZIK IN OHMS	
	RE	IM		RE	IM		RE	IM		RE	IM
1, 1	566.658	-396.384	1, 2	176.291	75.186	1, 3	-69.064	2.419	1, 4	33.395	-4.903
1, 5	-20.654	3.862	1, 6	14.800	-3.054	1, 7	-11.627	2.481	1, 8	9.853	-2.031
1, 9	-9.218	1.518	1,10	10.144	0.419						
2, 2	602.569	-342.379	2, 3	156.400	63.070	2, 4	-58.847	7.216	2, 5	27.018	-7.630
2, 6	-16.142	5.728	2, 7	11.394	-4.482	2, 8	-9.021	3.669	2, 9	8.095	-3.052
3, 3	610.806	-341.129	3, 4	152.417	62.900	3, 5	-56.450	7.236	3, 6	25.393	-7.624
3, 7	-15.002	5.729	3, 8	10.672	-4.504						
4, 4	612.577	-341.212	4, 5	151.408	62.979	4, 6	-55.831	7.183	4, 7	25.053	-7.600
5, 5	613.074	-341.269	5, 6	151.175	63.006						

IBR

SELF AND MUTUAL IMPEDANCES

H/LMDA=0.5000 B/LMDA= 0.500 OMEGA= 9.92

12 ELEMENT ARRAY

I, K	ZIK IN OHMS		I, K	ZIK IN OHMS		I, K	ZIK IN OHMS		I, K	ZIK IN OHMS	
	RE	IM		RE	IM		RE	IM		RE	IM
1, 1	566.882	-396.314	1, 2	176.081	75.167	1, 3	-68.845	2.429	1, 4	33.149	-4.912
1, 5	-20.360	3.872	1, 6	14.430	-3.071	1, 7	-11.126	2.512	1, 8	9.087	-2.106
1, 9	-7.785	1.793	1,10	7.026	-1.524	1,11	-6.905	1.171	1,12	7.892	0.338
2, 2	602.751	-342.401	2, 3	156.211	63.101	2, 4	-58.635	7.178	2, 5	26.765	-7.585
2, 6	-15.824	5.673	2, 7	10.961	-4.416	2, 8	-8.354	3.591	2, 9	6.820	-3.022
2,10	-5.949	2.615	2,11	5.734	-2.272						
3, 3	610.998	-341.171	3, 4	152.201	62.949	3, 5	-56.192	7.177	3, 6	25.068	-7.553
3, 7	-14.559	5.639	3, 8	9.988	-4.391	3, 9	-7.612	3.580	3,10	6.320	-3.033
4, 4	612.820	-341.271	4, 5	151.117	63.049	4, 6	-55.464	7.098	4, 7	24.553	-7.494
4, 8	-14.200	5.600	4, 9	9.772	-4.373						
5, 5	613.422	-341.351	5, 6	150.736	63.106	5, 7	-55.222	7.061	5, 8	24.424	-7.477
6, 6	613.634	-341.388	6, 7	150.634	63.123						

IBR

SELF AND MUTUAL IMPEDANCES

H/LMDA=0.5000 B/LMDA= 0.500 OMEGA= 9.92

14 ELEMENT ARRAY

I, K	ZIK IN OHMS RE	IM	I, K	ZIK IN OHMS RE	IM	I, K	ZIK IN OHMS RE	IM	I, K	ZIK IN OHMS RE	IM
1, 1	567.028	-396.269	1, 2	175.949	75.155	1, 3	-68.713	2.435	1, 4	33.005	-4.915
1, 5	-20.198	3.876	1, 6	14.241	-3.075	1, 7	-10.897	2.519	1, 8	8.796	-2.118
1, 9	-7.387	1.817	1,10	6.416	-1.583	1,11	-5.761	1.390	1,12	5.399	-1.210
1,13	-5.476	0.948	1,14	6.418	0.281						
2, 2	602.862	-342.416	2, 3	156.101	63.122	2, 4	-58.516	7.155	2, 5	26.631	-7.558
2, 6	-15.667	5.642	2, 7	10.770	-4.379	2, 8	-8.111	3.546	2, 9	6.486	-2.968
2,10	-5.433	2.550	2,11	4.746	-2.241	2,12	-4.362	2.008	2,13	4.388	-1.794
3, 3	611.107	-341.197	3, 4	152.084	62.979	3, 5	-56.060	7.143	3, 6	24.914	-7.513
3, 7	-14.371	5.592	3, 8	9.749	-4.334	3, 9	-7.283	3.509	3,10	5.811	-2.941
3,11	-4.894	2.538	3,12	4.370	-2.247						
4, 4	612.946	-341.305	4, 5	150.976	63.088	4, 6	-55.298	7.052	4, 7	24.352	-7.440
4, 8	-13.942	5.535	4, 9	9.419	-4.291	4,10	-7.039	3.480	4,11	5.657	-2.928
5, 5	613.582	-341.396	5, 6	150.549	63.158	5, 7	-54.995	7.000	5, 8	24.133	-7.402
5, 9	-13.791	5.510	5,10	9.333	-4.279						
6, 6	613.853	-341.448	6, 7	150.367	63.194	6, 8	-54.878	6.977	6, 9	24.072	-7.391
7, 7	613.960	-341.471	7, 8	150.315	63.205						

IBR

SELF AND MUTUAL IMPEDANCES

H/LMDA=0.5000 B/LMDA= 0.500 OMEGA= 9.92

16 ELEMENT ARRAY

I, K	ZIK IN OHMS RE	IM	I, K	ZIK IN OHMS RE	IM	I, K	ZIK IN OHMS RE	IM	I, K	ZIK IN OHMS RE	IM
1, 1	567.128	-396.238	1, 2	175.860	75.148	1, 3	-68.625	2.438	1, 4	32.913	-4.917
1, 5	-20.097	3.877	1, 6	14.128	-3.076	1, 7	-10.767	2.520	1, 8	8.642	-2.120
1, 9	-7.200	1.822	1,10	6.177	-1.592	1,11	-5.433	1.409	1,12	4.895	-1.259
1,13	-4.529	1.128	1,14	4.352	-1.000	1,15	-4.512	0.793	1,16	5.384	0.238
2, 2	602.936	-342.426	2, 3	156.029	63.136	2, 4	-58.440	7.139	2, 5	26.549	-7.540
2, 6	-15.575	5.622	2, 7	10.665	-4.356	2, 8	-7.986	3.520	2, 9	6.334	-2.937
2,10	-5.237	2.513	2,11	4.477	-2.196	2,12	-3.944	1.953	2,13	3.587	-1.765
2,14	-3.407	1.619	2,15	3.527	-1.475						
3, 3	611.176	-341.214	3, 4	152.012	62.999	3, 5	-55.982	7.121	3, 6	24.826	-7.488
3, 7	-14.271	5.564	3, 8	9.629	-4.301	3, 9	-7.137	3.471	3,10	5.623	-2.895
3,11	-4.635	2.479	3,12	3.968	-2.172	3,13	-3.526	1.941	3,14	3.288	-1.769
4, 4	613.021	-341.327	4, 5	150.894	63.113	4, 6	-55.207	7.024	4, 7	24.246	-7.408
4, 8	-13.818	5.498	4, 9	9.266	-4.247	4,10	-6.843	3.427	4,11	5.387	-2.861
4,12	-4.454	2.456	4,13	3.850	-2.161						
5, 5	613.671	-341.424	5, 6	150.449	63.189	5, 7	-54.880	6.965	5, 8	23.997	-7.361
5, 9	-13.625	5.461	5,10	9.119	-4.220	5,11	-6.737	3.408	5,12	5.324	-2.852
6, 6	613.965	-341.483	6, 7	150.238	63.234	6, 8	-54.725	6.931	6, 9	23.885	-7.337
6,10	-13.548	5.445	6,11	9.076	-4.212						
7, 7	614.109	-341.516	7, 8	150.139	63.257	7, 9	-54.661	6.916	7,10	23.851	-7.329
8, 8	614.170	-341.531	8, 9	150.109	63.264						

IBR

SELF AND MUTUAL IMPEDANCES

H/LMDA=0.5000 B/LMDA= 0.500 OMEGA= 9.92

18 ELEMENT ARRAY

I, K	ZIK IN OHMS RE	IM	I, K	ZIK IN OHMS RE	IM	I, K	ZIK IN OHMS RE	IM	I, K	ZIK IN OHMS RE	IM
1, 1	567.201	-396.215	1, 2	175.796	75.142	1, 3	-68.563	2.440	1, 4	32.848	-4.917
1, 5	-20.029	3.877	1, 6	14.054	-3.076	1, 7	-10.684	2.520	1, 8	8.548	-2.120
1, 9	-7.091	1.822	1,10	6.047	-1.593	1,11	-5.275	1.413	1,12	4.692	-1.266
1,13	-4.250	1.144	1,14	3.924	-1.040	1,15	-3.706	0.946	1,16	3.627	-0.849
1,17	-3.820	0.681	1,18	4.621	0.205						
2, 2	602.988	-342.434	2, 3	155.979	63.146	2, 4	-58.389	7.127	2, 5	26.495	-7.527
2, 6	-15.515	5.608	2, 7	10.599	-4.341	2, 8	-7.911	3.503	2, 9	6.247	-2.917
2,10	-5.134	2.491	2,11	4.350	-2.170	2,12	-3.781	1.922	2,13	3.362	-1.727
2,14	-3.058	1.572	2,15	2.856	-1.448	2,16	-2.776	1.350	2,17	2.933	-1.247
3, 3	611.222	-341.226	3, 4	151.964	63.013	3, 5	-55.931	7.106	3, 6	24.770	-7.472
3, 7	-14.209	5.546	3, 8	9.559	-4.281	3, 9	-7.056	3.447	3,10	5.527	-2.867
3,11	-4.516	2.447	3,12	3.815	-2.132	3,13	-3.314	1.891	3,14	2.958	-1.705
3,15	-2.720	1.560	3,16	2.610	-1.450						
4, 4	613.070	-341.342	4, 5	150.842	63.130	4, 6	-55.150	7.006	4, 7	24.183	-7.388
4, 8	-13.746	5.475	4, 9	9.183	-4.221	4,10	-6.744	3.397	4,11	5.265	-2.825
4,12	-4.297	2.412	4,13	3.633	-2.105	4,14	-3.172	1.872	4,15	2.863	-1.695
5, 5	613.727	-341.442	5, 6	150.388	63.209	5, 7	-54.813	6.942	5, 8	23.921	-7.336
5, 9	-13.537	5.433	5,10	9.014	-4.186	5,11	-6.607	3.369	5,12	5.157	-2.804
5,13	-4.216	2.397	5,14	3.584	-2.098						
6, 6	614.031	-341.505	6, 7	150.165	63.258	6, 8	-54.642	6.903	6, 9	23.788	-7.305
6,10	-13.433	5.408	6,11	8.935	-4.168	6,12	-6.551	3.357	6,13	5.125	-2.798
7, 7	614.191	-341.543	7, 8	150.046	63.288	7, 9	-54.553	6.881	7,10	23.723	-7.289
7,11	-13.389	5.397	7,12	8.911	-4.163						
8, 8	614.275	-341.566	8, 9	149.987	63.304	8,10	-54.514	6.870	8,11	23.703	-7.284
9, 9	614.312	-341.576	9,10	149.968	63.309						

IBR

SELF AND MUTUAL IMPEDANCES

H/LMDA=0.5000 B/LMDA= 0.500 OMEGA= 9.92

20 ELEMENT ARRAY

I, K	ZIK IN OHMS		I, K	ZIK IN OHMS		I, K	ZIK IN OHMS		I, K	ZIK IN OHMS	
	RE	IM		RE	IM		RE	IM		RE	IM
1, 1	567.256	-396.198	1, 2	175.748	75.138	1, 3	-68.517	2.441	1, 4	32.802	-4.918
1, 5	-19.980	3.877	1, 6	14.001	-3.075	1, 7	-10.627	2.519	1, 8	8.485	-2.119
1, 9	-7.021	1.822	1,10	5.967	-1.593	1,11	-5.182	1.413	1,12	4.581	-1.267
1,13	-4.114	1.147	1,14	3.748	-1.047	1,15	-3.465	0.960	1,16	3.255	-0.884
1,17	-3.122	0.813	1,18	3.096	-0.736	1,19	-3.302	0.595	1,20	4.038	0.179
2, 2	603.027	-342.440	2, 3	155.943	63.153	2, 4	-58.353	7.119	2, 5	26.456	-7.518
2, 6	-15.474	5.598	2, 7	10.554	-4.330	2, 8	-7.862	3.490	2, 9	6.192	-2.904
2,10	-5.070	2.476	2,11	4.276	-2.153	2,12	-3.693	1.902	2,13	3.254	-1.704
2,14	-2.918	1.544	2,15	2.664	-1.414	2,16	-2.477	1.308	2,17	2.358	-1.223
2,18	-2.330	1.154	2,19	2.500	-1.078						
3, 3	611.256	-341.235	3, 4	151.930	63.023	3, 5	-55.896	7.095	3, 6	24.732	-7.460
3, 7	-14.167	5.533	3, 8	9.513	-4.266	3, 9	-7.005	3.431	3,10	5.468	-2.849
3,11	-4.448	2.427	3,12	3.733	-2.109	3,13	-3.214	1.863	3,14	2.829	-1.670
3,15	-2.541	1.517	3,16	2.331	-1.394	3,17	-2.194	1.297	3,18	2.149	-1.224
4, 4	613.105	-341.353	4, 5	150.806	63.142	4, 6	-55.112	6.993	4, 7	24.141	-7.374
4, 8	-13.700	5.460	4, 9	9.131	-4.203	4,10	-6.685	3.377	4,11	5.197	-2.803
4,12	-4.215	2.386	4,13	3.532	-2.074	4,14	-3.042	1.834	4,15	2.683	-1.648
4,16	-2.424	1.501	4,17	2.252	-1.386						
5, 5	613.764	-341.455	5, 6	150.348	63.223	5, 7	-54.769	6.927	5, 8	23.873	-7.319
5, 9	-13.483	5.414	5,10	8.952	-4.165	5,11	-6.536	3.344	5,12	5.071	-2.775
5,13	-4.111	2.364	5,14	3.448	-2.057	5,15	-2.978	1.822	5,16	2.643	-1.642
6, 6	614.074	-341.520	6, 7	150.118	63.275	6, 8	-54.591	6.885	6, 9	23.730	-7.285
6,10	-13.367	5.385	6,11	8.858	-4.142	6,12	-6.460	3.326	6,13	5.012	-2.761
6,14	-4.068	2.354	6,15	3.423	-2.052						
7, 7	614.242	-341.562	7, 8	149.990	63.308	7, 9	-54.490	6.858	7,10	23.651	-7.263
7,11	-13.305	5.369	7,12	8.811	-4.129	7,13	-6.427	3.318	7,14	4.993	-2.757
8, 8	614.337	-341.588	8, 9	149.917	63.328	8,10	-54.435	6.843	8,11	23.610	-7.252
8,12	-13.277	5.361	8,13	8.796	-4.126						
9, 9	614.390	-341.603	9,10	149.879	63.339	9,11	-54.410	6.836	9,12	23.597	-7.248
10,10	614.414	-341.610	10,11	149.867	63.343						

IBR

SELF AND MUTUAL IMPEDANCES

H/LMDA=[illegible].5000 B/LMDA= 0.500 OMEGA= 9.92

22 ELEMENT ARRAY

I, K	ZIK IN OHMS		I, K	ZIK IN OHMS		I, K	ZIK IN OHMS		I, K	ZIK IN OHMS	
	RE	IM		RE	IM		RE	IM		RE	IM
1, 1	567.299	-396.184	1, 2	175.711	75.136	1, 3	-68.482	2.442	1, 4	32.767	-4.918
1, 5	-19.944	3.876	1, 6	13.963	-3.075	1, 7	-10.586	2.519	1, 8	8.441	-2.119
1, 9	-6.972	1.821	1,10	5.913	-1.592	1,11	-5.120	1.412	1,12	4.511	-1.267
1,13	-4.032	1.147	1,14	3.651	-1.047	1,15	-3.345	0.962	1,16	3.101	-0.889
1,17	-2.909	0.825	1,18	2.769	-0.767	1,19	-2.687	0.711	1,20	2.694	-0.649
1,21	-2.901	0.528	1,22	3.577	0.158						
2, 2	603.056	-342.444	2, 3	155.915	63.159	2, 4	-58.325	7.112	2, 5	26.428	-7.511
2, 6	-15.444	5.590	2, 7	10.522	-4.321	2, 8	-7.827	3.481	2, 9	6.153	-2.894
2,10	-5.028	2.465	2,11	4.229	-2.141	2,12	-3.638	1.889	2,13	3.190	-1.689
2,14	-2.842	1.527	2,15	2.570	-1.394	2,16	-2.355	1.284	2,17	2.190	-1.193
2,18	-2.069	1.117	2,19	1.998	-1.055	2,20	-2.001	1.005	2,21	2.172	-0.947
3, 3	611.281	-341.243	3, 4	151.905	63.030	3, 5	-55.870	7.086	3, 6	24.705	-7.451
3, 7	-14.138	5.523	3, 8	9.482	-4.256	3, 9	-6.970	3.420	3,10	5.429	-2.837
3,11	-4.404	2.413	3,12	3.684	-2.093	3,13	-3.156	1.846	3,14	2.759	-1.650
3,15	-2.455	1.492	3,16	2.220	-1.364	3,17	-2.040	1.259	3,18	1.909	-1.174
3,19	-1.828	1.106	3,20	1.818	-1.055						
4, 4	613.130	-341.362	4, 5	150.780	63.151	4, 6	-55.084	6.983	4, 7	24.112	-7.363
4, 8	-13.668	5.448	4, 9	9.097	-4.191	4,10	-6.647	3.363	4,11	5.153	-2.787
4,12	-4.165	2.369	4,13	3.474	-2.055	4,14	-2.972	1.812	4,15	2.597	-1.621
4,16	-2.313	1.468	4,17	2.099	-1.344	4,18	-1.942	1.245	4,19	1.842	-1.168
5, 5	613.791	-341.465	5, 6	150.320	63.233	5, 7	-54.739	6.916	5, 8	23.840	-7.307
5, 9	-13.447	5.401	5,10	8.912	-4.150	5,11	-6.491	3.328	5,12	5.020	-2.757
5,13	-4.051	2.343	5,14	3.376	-2.032	5,15	-2.889	1.794	5,16	2.529	-1.606
5,17	-2.261	1.458	5,18	2.066	-1.340						
6, 6	614.104	-341.531	6, 7	150.087	63.287	6, 8	-54.556	6.872	6, 9	23.693	-7.270
6,10	-13.325	5.370	6,11	8.811	-4.124	6,12	-6.406	3.306	6,13	4.949	-2.739
6,14	-3.992	2.328	6,15	3.329	-2.021	6,16	-2.854	1.786	6,17	2.508	-1.602
7, 7	614.275	-341.574	7, 8	149.954	63.322	7, 9	-54.450	6.843	7,10	23.606	-7.247
7,11	-13.255	5.350	7,12	8.753	-4.108	7,13	-6.359	3.293	7,14	4.913	-2.729
7,15	-3.966	2.321	7,16	3.315	-2.018						
8, 8	614.377	-341.603	8, 9	149.874	63.345	8,10	-54.386	6.824	8,11	23.556	-7.232
8,12	-13.215	5.338	8,13	8.723	-4.099	8,14	-6.338	3.288	8,15	4.901	-2.726
9, 9	614.438	-341.621	9,10	149.826	63.359	9,11	-54.350	6.813	9,12	23.528	-7.224
9,13	-13.196	5.333	9,14	8.713	-4.097						
10,10	614.474	-341.632	10,11	149.800	63.367	10,12	-54.333	6.808	10,13	23.520	-7.221
11,11	614.490	-341.637	11,12	149.792	63.370						

IBR

SELF AND MUTUAL IMPEDANCES

H/LMDA=0.5000 B/LMDA= 0.500 OMEGA= 9.92

24 ELEMENT ARRAY

I, K	ZIK IN OHMS RE	IM	I, K	ZIK IN OHMS RE	IM	I, K	ZIK IN OHMS RE	IM	I, K	ZIK IN OHMS RE	IM
1, 1	567.334	-396.174	1, 2	175.681	75.133	1, 3	-68.454	2.442	1, 4	32.739	-4.917
1, 5	-19.915	3.876	1, 6	13.933	-3.074	1, 7	-10.555	2.518	1, 8	8.407	-2.118
1, 9	-6.936	1.820	1,10	5.873	-1.591	1,11	-5.077	1.411	1,12	4.463	-1.266
1,13	-3.978	1.146	1,14	3.588	-1.047	1,15	-3.272	0.962	1,16	3.014	-0.890
1,17	-2.803	0.827	1,18	2.632	-0.771	1,19	-2.498	0.722	1,20	2.402	-0.676
1,21	-2.352	0.631	1,22	2.378	-0.579	1,23	-2.582	0.474	1,24	3.206	0.140
2, 2	603.079	-342.448	2, 3	155.894	63.164	2, 4	-58.304	7.107	2, 5	26.406	-7.505
2, 6	-15.422	5.584	2, 7	10.498	-4.315	2, 8	-7.801	3.475	2, 9	6.126	-2.887
2,10	-4.998	2.457	2,11	4.196	-2.132	2,12	-3.601	1.879	2,13	3.148	-1.678
2,14	-2.794	1.515	2,15	2.513	-1.380	2,16	-2.288	1.268	2,17	2.107	-1.174
2,18	-1.962	1.095	2,19	1.850	-1.028	2,20	-1.770	0.972	2,21	1.728	-0.926
2,22	-1.747	0.888	2,23	1.915	-0.843						
3, 3	611.301	-341.248	3, 4	151.885	63.036	3, 5	-55.850	7.080	3, 6	24.684	-7.444
3, 7	-14.116	5.516	3, 8	9.458	-4.248	3, 9	-6.945	3.411	3,10	5.402	-2.828
3,11	-4.374	2.403	3,12	3.650	-2.082	3,13	-3.118	1.833	3,14	2.716	-1.636
3,15	-2.404	1.476	3,16	2.159	-1.346	3,17	-1.965	1.238	3,18	1.811	-1.148
3,19	-1.693	1.073	3,20	1.607	-1.011	3,21	-1.559	0.962	3,22	1.569	-0.926
4, 4	613.149	-341.368	4, 5	150.761	63.158	4, 6	-55.064	6.976	4, 7	24.091	-7.355
4, 8	-13.646	5.440	4, 9	9.072	-4.182	4,10	-6.620	3.354	4,11	5.124	-2.777
4,12	-4.132	2.357	4,13	3.437	-2.041	4,14	-2.929	1.797	4,15	2.547	-1.604
4,16	-2.253	1.448	4,17	2.025	-1.321	4,18	-1.846	1.216	4,19	1.708	-1.130
4,20	-1.608	1.060	4,21	1.549	-1.005						
5, 5	613.811	-341.472	5, 6	150.300	63.241	5, 7	-54.717	6.907	5, 8	23.817	-7.298
5, 9	-13.422	5.391	5,10	8.885	-4.140	5,11	-6.461	3.317	5,12	4.986	-2.744
5,13	-4.013	2.329	5,14	3.333	-2.017	5,15	-2.838	1.775	5,16	2.467	-1.585
5,17	-2.185	1.432	5,18	1.967	-1.308	5,19	-1.801	1.208	5,20	1.680	-1.126
6, 6	614.125	-341.540	6, 7	150.064	63.296	6, 8	-54.532	6.862	6, 9	23.667	-7.260
6,10	-13.297	5.359	6,11	8.780	-4.112	6,12	-6.371	3.293	6,13	4.909	-2.724
6,14	-3.946	2.311	6,15	3.276	-2.001	6,16	-2.790	1.763	6,17	2.429	-1.575
6,18	-2.156	1.425	6,19	1.950	-1.305						
7, 7	614.299	-341.584	7, 8	149.928	63.332	7, 9	-54.422	6.832	7,10	23.576	-7.235
7,11	-13.221	5.337	7,12	8.716	-4.094	7,13	-6.317	3.277	7,14	4.864	-2.711
7,15	-3.910	2.301	7,16	3.247	-1.993	7,17	-2.769	1.757	7,18	2.417	-1.573
8, 8	614.404	-341.614	8, 9	149.844	63.357	8,10	-54.354	6.812	8,11	23.520	-7.218
8,12	-13.175	5.323	8,13	8.678	-4.082	8,14	-6.287	3.268	8,15	4.841	-2.704
8,16	-3.893	2.296	8,17	3.237	-1.991						
9, 9	614.470	-341.634	9,10	149.791	63.373	9,11	-54.311	6.798	9,12	23.486	-7.207
9,13	-13.148	5.314	9,14	8.657	-4.076	9,15	-6.272	3.264	9,16	4.833	-2.702
10,10	614.512	-341.647	10,11	149.758	63.383	10,12	-54.286	6.790	10,13	23.467	-7.201
10,14	-13.135	5.310	10,15	8.651	-4.074						
11,11	614.536	-341.655	11,12	149.741	63.389	11,13	-54.274	6.786	11,14	23.461	-7.199
12,12	614.548	-341.659	12,13	149.735	63.391						

IBR

SELF AND MUTUAL IMPEDANCES

H/LMDA=0.5000 B/LMDA= 0.500 OMEGA= 9.92

26 ELEMENT ARRAY

I, K	ZIK IN OHMS RE	IM	I, K	ZIK IN OHMS RE	IM	I, K	ZIK IN OHMS RE	IM	I, K	ZIK IN OHMS RE	IM
1, 1	567.362	-396.165	1, 2	175.657	75.132	1, 3	-68.432	2.442	1, 4	32.717	-4.917
1, 5	-19.893	3.875	1, 6	13.910	-3.073	1, 7	-10.531	2.517	1, 8	8.382	-2.117
1, 9	-6.908	1.819	1,10	5.844	-1.590	1,11	-5.045	1.410	1,12	4.428	-1.265
1,13	-3.939	1.145	1,14	3.545	-1.046	1,15	-3.224	0.961	1,16	2.958	-0.889
1,17	-2.737	0.826	1,18	2.554	-0.772	1,19	-2.402	0.723	1,20	2.278	-0.680
1,21	-2.182	0.640	1,22	2.115	-0.604	1,23	-2.087	0.567	1,24	2.125	-0.523
1,25	-2.322	0.430	1,26	2.900	0.126						
2, 2	603.098	-342.451	2, 3	155.877	63.167	2, 4	-58.287	7.103	2, 5	26.389	-7.501
2, 6	-15.404	5.579	2, 7	10.479	-4.310	2, 8	-7.782	3.469	2, 9	6.105	-2.881
2,10	-4.975	2.451	2,11	4.171	-2.125	2,12	-3.575	1.872	2,13	3.119	-1.670
2,14	-2.761	1.506	2,15	2.476	-1.371	2,16	-2.245	1.258	2,17	2.057	-1.162
2,18	-1.902	1.081	2,19	1.776	-1.011	2,20	-1.674	0.952	2,21	1.595	-0.901
2,22	-1.541	0.858	2,23	1.518	-0.823	2,24	-1.548	0.795	2,25	1.710	-0.759
3, 3	611.316	-341.253	3, 4	151.870	63.041	3, 5	-55.835	7.075	3, 6	24.669	-7.439
3, 7	-14.100	5.510	3, 8	9.441	-4.242	3, 9	-6.926	3.405	3,10	5.382	-2.821
3,11	-4.353	2.395	3,12	3.626	-2.074	3,13	-3.092	1.824	3,14	2.686	-1.626
3,15	-2.370	1.465	3,16	2.120	-1.333	3,17	-1.920	1.223	3,18	1.757	-1.131
3,19	-1.626	1.053	3,20	1.521	-0.987	3,21	-1.439	0.932	3,22	1.382	-0.886
3,23	-1.355	0.849	3,24	1.377	-0.823						
4, 4	613.163	-341.373	4, 5	150.746	63.163	4, 6	-55.049	6.970	4, 7	24.075	-7.349
4, 8	-13.628	5.433	4, 9	9.054	-4.175	4,10	-6.600	3.346	4,11	5.102	-2.769
4,12	-4.109	2.348	4,13	3.411	-2.032	4,14	-2.900	1.786	4,15	2.515	-1.592
4,16	-2.216	1.434	4,17	1.981	-1.305	4,18	-1.793	1.199	4,19	1.643	-1.109
4,20	-1.524	1.034	4,21	1.432	-0.972	4,22	-1.366	0.921	4,23	1.331	-0.880
5, 5	613.826	-341.478	5, 6	150.284	63.247	5, 7	-54.701	6.901	5, 8	23.800	-7.291
5, 9	-13.403	5.383	5,10	8.865	-4.132	5,11	-6.439	3.308	5,12	4.963	-2.735
5,13	-3.987	2.319	5,14	3.303	-2.005	5,15	-2.805	1.763	5,16	2.429	-1.571
5,17	-2.140	1.416	5,18	1.914	-1.289	5,19	-1.735	1.185	5,20	1.594	-1.098
5,21	-1.485	1.027	5,22	1.407	-0.968						
6, 6	614.141	-341.546	6, 7	150.047	63.303	6, 8	-54.514	6.855	6, 9	23.648	-7.252
6,10	-13.276	5.350	6,11	8.757	-4.103	6,12	-6.346	3.283	6,13	4.882	-2.713
6,14	-3.916	2.299	6,15	3.242	-1.988	6,16	-2.751	1.748	6,17	2.383	-1.558
6,18	-2.101	1.405	6,19	1.881	-1.281	6,20	-1.711	1.179	6,21	1.579	-1.096

I, K	RE	IM	I, K	RE	IM	I, K	RE	IM	I, K	RE	IM
7, 7	614.317	-341.591	7, 8	149.909	63.340	7, 9	-54.402	6.824	7,10	23.555	-7.226
7,11	-13.198	5.327	7,12	8.691	-4.083	7,13	-6.289	3.266	7,14	4.833	-2.698
7,15	-3.874	2.286	7,16	3.206	-1.978	7,17	-2.721	1.739	7,18	2.359	-1.551
7,19	-2.083	1.401	7,20	1.871	-1.279						
8, 8	614.424	-341.622	8, 9	149.823	63.365	8,10	-54.331	6.802	8,11	23.495	-7.207
8,12	-13.148	5.312	8,13	8.648	-4.070	8,14	-6.253	3.255	8,15	4.803	-2.689
8,16	-3.849	2.279	8,17	3.187	-1.972	8,18	-2.707	1.735	8,19	2.351	-1.549
9, 9	614.493	-341.643	9,10	149.767	63.383	9,11	-54.285	6.787	9,12	23.457	-7.195
9,13	-13.116	5.301	9,14	8.622	-4.061	9,15	-6.232	3.248	9,16	4.787	-2.684
9,17	-3.838	2.275	9,18	3.180	-1.970						
10,10	614.538	-341.658	10,11	149.730	63.395	10,12	-54.255	6.777	10,13	23.433	-7.187
10,14	-13.097	5.295	10,15	8.607	-4.057	10,16	-6.222	3.245	10,17	4.781	-2.682
11,11	614.567	-341.668	11,12	149.707	63.403	11,13	-54.237	6.771	11,14	23.419	-7.182
11,15	-13.088	5.292	11,16	8.603	-4.055						
12,12	614.585	-341.674	12,13	149.694	63.407	12,14	-54.228	6.768	12,15	23.415	-7.181
13,13	614.593	-341.677	13,14	149.690	63.409						

IBR

SELF AND MUTUAL IMPEDANCES

H/LMDA=0.5000 B/LMDA= 0.500 OMEGA= 9.92

28 ELEMENT ARRAY

I, K	ZIK IN OHMS		I, K	ZIK IN OHMS		I, K	ZIK IN OHMS		I, K	ZIK IN OHMS	
	RE	IM		RE	IM		RE	IM		RE	IM
1, 1	567.385	-396.158	1, 2	175.638	75.130	1, 3	-68.414	2.443	1, 4	32.699	-4.917
1, 5	-19.875	3.875	1, 6	13.891	-3.073	1, 7	-10.511	2.516	1, 8	8.361	-2.116
1, 9	-6.887	1.818	1,10	5.821	-1.589	1,11	-5.021	1.409	1,12	4.401	-1.264
1,13	-3.910	1.144	1,14	3.514	-1.045	1,15	-3.189	0.960	1,16	2.919	-0.888
1,17	-2.693	0.826	1,18	2.503	-0.771	1,19	-2.343	0.723	1,20	2.207	-0.680
1,21	-2.094	0.642	1,22	2.003	-0.607	1,23	-1.932	0.575	1,24	1.886	-0.545
1,25	-1.872	0.514	1,26	1.917	-0.476	1,27	-2.107	0.393	1,28	2.645	0.114
2, 2	603.113	-342.454	2, 3	155.863	63.171	2, 4	-58.273	7.100	2, 5	26.375	-7.497
2, 6	-15.390	5.575	2, 7	10.465	-4.306	2, 8	-7.766	3.465	2, 9	6.089	-2.876
2,10	-4.958	2.446	2,11	4.153	-2.120	2,12	-3.555	1.866	2,13	3.097	-1.664
2,14	-2.738	1.499	2,15	2.450	-1.363	2,16	-2.216	1.250	2,17	2.023	-1.153
2,18	-1.864	1.071	2,19	1.730	-1.000	2,20	-1.620	0.939	2,21	1.528	-0.886
2,22	-1.454	0.840	2,23	1.398	-0.801	2,24	-1.361	0.767	2,25	1.350	-0.740
2,26	-1.386	0.718	2,27	1.542	-0.689						
3, 3	611.328	-341.256	3, 4	151.858	63.045	3, 5	-55.822	7.071	3, 6	24.656	-7.434
3, 7	-14.087	5.505	3, 8	9.427	-4.237	3, 9	-6.912	3.400	3,10	5.367	-2.815
3,11	-4.336	2.389	3,12	3.608	-2.067	3,13	-3.072	1.817	3,14	2.665	-1.618
3,15	-2.347	1.457	3,16	2.094	-1.324	3,17	-1.890	1.213	3,18	1.723	-1.120
3,19	[illegible]	1.810	3,20	1.472	-0.972	3,21	-1.379	0.914	3,22	1.304	-0.864
3,23	-1.247	0.822	3,24	1.209	[illegible]	[illegible]	-1.106	0.759	3,26	1.224	-0.740
4, 4	613.175	-341.378	4, 5	150.734	63.168	4, 6	-55.036	6.965	4, 7	24.062	-7.344
4, 8	-13.615	5.428	4, 9	9.040	-4.170	4,10	-6.585	3.340	4,11	5.086	-2.762
4,12	-4.092	2.342	4,13	3.392	-2.024	4,14	-2.880	1.778	4,15	2.492	-1.583
4,16	-2.190	1.425	4,17	1.952	-1.295	4,18	-1.760	1.186	4,19	1.604	-1.095
4,20	-1.477	1.018	4,21	1.373	-0.953	4,22	-1.290	0.897	4,23	1.226	-0.850
4,24	-1.182	0.812	4,25	1.163	-0.782						
5, 5	613.838	-341.482	5, 6	150.272	63.252	5, 7	-54.688	6.896	5, 8	23.786	-7.286
5, 9	-13.389	5.378	5,10	8.850	-4.126	5,11	-6.423	3.302	5,12	4.945	-2.728
5,13	-3.968	2.311	5,14	3.283	-1.997	5,15	-2.782	1.754	5,16	2.404	-1.561
5,17	-2.111	1.405	5,18	1.880	-1.277	5,19	-1.696	1.171	5,20	1.547	-1.081
5,21	-1.427	1.006	5,22	1.331	-0.943	5,23	-1.256	0.890	5,24	1.204	-0.847
6, 6	614.154	-341.551	6, 7	150.034	63.308	6, 8	-54.501	6.849	6, 9	23.633	-7.246
6,10	-13.261	5.344	6,11	8.741	-4.096	6,12	-6.328	3.275	6,13	4.863	-2.705
6,14	-3.895	2.290	6,15	3.218	-1.979	6,16	-2.725	1.737	6,17	2.353	-1.546
6,18	-2.066	1.392	6,19	1.841	-1.266	6,20	-1.662	1.161	6,21	1.519	-1.074
6,22	-1.405	1.001	6,23	1.318	-0.941						
7, 7	614.331	-341.597	7, 8	149.895	63.346	7, 9	-54.387	6.817	7,10	23.539	-7.219
7,11	-13.181	5.320	7,12	8.672	-4.075	7,13	-6.269	3.257	7,14	4.811	-2.689
7,15	-3.850	2.276	7,16	3.178	-1.966	7,17	-2.690	1.727	7,18	2.323	-1.537
7,19	-2.041	1.385	7,20	1.821	-1.260	7,21	-1.647	1.157	7,22	1.510	-1.072
8, 8	614.439	-341.628	8, 9	149.807	63.372	8,10	-54.315	6.795	8,11	23.477	-7.200
8,12	-13.129	5.303	8,13	8.627	-4.061	8,14	-6.230	3.245	8,15	4.777	-2.678
8,16	-3.821	2.267	8,17	3.154	-1.959	8,18	-2.670	1.720	8,19	2.307	-1.532
8,20	-2.029	1.381	8,21	1.814	-1.258						
9, 9	614.509	-341.650	9,10	149.749	63.391	9,11	-54.266	6.779	9,12	23.436	-7.186
9,13	-13.094	5.292	9,14	8.597	-4.051	9,15	-6.205	3.236	9,16	4.756	-2.671
9,17	-3.804	2.261	9,18	3.141	-1.954	9,19	-2.660	1.717	9,20	2.302	-1.531
10,10	614.557	-341.666	10,11	149.710	63.404	10,12	-54.233	6.768	10,13	23.409	-7.177
10,14	-13.071	5.284	10,15	8.579	-4.045	10,16	-6.190	3.231	10,17	4.745	-2.667
10,18	-3.795	2.259	10,19	3.136	-1.953						
11,11	614.589	-341.677	11,12	149.684	63.413	11,13	-54.211	6.760	11,14	23.391	-7.171
11,15	-13.057	5.279	11,16	8.568	-4.041	11,17	-6.182	3.229	11,18	4.741	-2.666
12,12	614.610	-341.684	12,13	149.667	63.419	12,14	-54.198	6.756	12,15	23.381	-7.167
12,16	-13.050	5.277	12,17	8.564	-4.040						
13,13	614.623	-341.689	13,14	149.657	63.422	13,15	-54.192	6.754	13,16	23.378	-7.166
14,14	614.629	-341.691	14,15	149.654	63.424						

IBR

SELF AND MUTUAL IMPEDANCES

H/LMDA=0.5000 B/LMDA= 0.500 OMEGA= 9.92

30 ELEMENT ARRAY

I, K	ZIK IN OHMS		I, K	ZIK IN OHMS		I, K	ZIK IN OHMS		I, K	ZIK IN OHMS	
	RE	IM		RE	IM		RE	IM		RE	IM
1, 1	567.404	-396.152	1, 2	175.621	75.129	1, 3	-68.399	2.443	1, 4	32.685	-4.917
1, 5	-19.860	3.874	1, 6	13.876	-3.072	1, 7	-10.495	2.516	1, 8	8.345	-2.115
1, 9	-6.870	1.817	1,10	5.803	-1.588	1,11	-5.002	1.408	1,12	4.381	-1.263
1,13	-3.888	1.143	1,14	3.490	-1.044	1,15	-3.163	0.959	1,16	2.891	-0.887
1,17	-2.662	0.825	1,18	2.468	-0.770	1,19	-2.303	0.722	1,20	2.161	-0.679
1,21	-2.041	0.641	1,22	1.938	-0.607	1,23	-1.852	0.576	1,24	1.783	-0.548
1,25	-1.730	0.521	1,26	1.698	-0.496	1,27	-1.695	0.469	1,28	1.744	-0.436
1,29	-1.926	0.362	1,30	2.428	0.103						
2, 2	603.126	-342.456	2, 3	155.851	63.173	2, 4	-58.262	7.097	2, 5	26.364	-7.494
2, 6	-15.378	5.572	2, 7	10.453	-4.302	2, 8	-7.754	3.461	2, 9	6.076	-2.872
2,10	-4.945	2.442	2,11	4.138	-2.116	2,12	-3.539	1.862	2,13	3.080	-1.659
2,14	-2.720	1.494	2,15	2.430	-1.358	2,16	-2.194	1.244	2,17	1.999	-1.147
2,18	-1.837	1.064	2,19	1.700	-0.992	2,20	-1.585	0.930	2,21	1.487	-0.876
2,22	-1.405	0.828	2,23	1.337	-0.787	2,24	-1.282	0.751	2,25	1.241	-0.720
2,26	-1.216	0.693	2,27	1.214	-0.671	2,28	-1.253	0.655	2,29	1.402	-0.631
3, 3	611.339	-341.260	3, 4	151.848	63.048	3, 5	-55.812	7.067	3, 6	24.646	-7.431
3, 7	-14.076	5.502	3, 8	9.416	-4.233	3, 9	-6.900	3.395	3,10	5.355	-2.811
3,11	-4.323	2.384	3,12	3.595	-2.062	3,13	-3.058	1.812	3,14	2.649	-1.613
3,15	-2.330	1.451	3,16	2.075	-1.317	3,17	-1.869	1.206	3,18	1.699	-1.112
3,19	-1.559	1.031	3,20	1.441	-0.962	3,21	-1.343	0.902	3,22	1.261	-0.850
3,23	-1.193	0.805	3,24	1.138	-0.767	3,25	-1.097	0.734	3,26	1.071	-0.706
3,27	-1.066	0.685	3,28	1.099	-0.671						
4, 4	613.185	-341.381	4, 5	150.724	63.171	4, 6	-55.027	6.962	4, 7	24.052	-7.340
4, 8	-13.605	5.424	4, 9	9.029	-4.165	4,10	-6.574	3.335	4,11	5.074	-2.757
4,12	-4.078	2.336	4,13	3.378	-2.019	4,14	-2.864	1.772	4,15	2.475	-1.576
4,16	-2.172	1.418	4,17	1.931	-1.287	4,18	-1.737	1.178	4,19	1.578	-1.086
4,20	-1.447	1.007	4,21	1.338	-0.940	4,22	-1.248	0.882	4,23	1.174	-0.833
4,24	-1.114	0.790	4,25	1.069	-0.754	4,26	-1.039	0.724	4,27	1.030	-0.702
5, 5	613.848	-341.486	5, 6	150.262	63.256	5, 7	-54.678	6.891	5, 8	23.776	-7.281
5, 9	-13.378	5.373	5,10	8.839	-4.121	5,11	-6.411	3.296	5,12	4.932	-2.722
5,13	-3.954	2.305	5,14	3.267	-1.991	5,15	-2.765	1.747	5,16	2.385	-1.553
5,17	-2.091	1.396	5,18	1.857	-1.268	5,19	-1.670	1.160	5,20	1.517	-1.070
5,21	-1.391	0.993	5,22	1.288	-0.928	5,23	-1.204	0.872	5,24	1.136	-0.824
5,25	-1.084	0.784	5,26	1.049	-0.751						
6, 6	614.164	-341.555	6, 7	150.024	63.313	6, 8	-54.490	6.845	6, 9	23.622	-7.241
6,10	-13.249	5.338	6,11	8.728	-4.091	6,12	-6.315	3.270	6,13	4.848	-2.699
6,14	-3.879	2.284	6,15	3.201	-1.971	6,16	-2.706	1.729	6,17	2.332	-1.538
6,18	-2.043	1.383	6,19	1.814	-1.255	6,20	-1.631	1.149	6,21	1.483	-1.060
6,22	-1.362	0.985	6,23	1.264	-0.921	6,24	-1.185	0.867	6,25	1.125	-0.822
7, 7	614.341	-341.602	7, 8	149.884	63.351	7, 9	-54.376	6.812	7,10	23.527	-7.214
7,11	-13.168	5.314	7,12	8.658	-4.069	7,13	-6.254	3.251	7,14	4.795	-2.682
7,15	-3.832	2.269	7,16	3.159	-1.958	7,17	-2.668	1.718	7,18	2.299	-1.527
7,19	-2.014	1.373	7,20	1.789	-1.247	7,21	-1.610	1.143	7,22	1.465	-1.055
7,23	-1.349	0.982	7,24	1.256	-0.920						
8, 8	614.451	-341.633	8, 9	149.795	63.377	8,10	-54.302	6.789	8,11	23.464	-7.194
8,12	-13.114	5.297	8,13	8.612	-4.054	8,14	-6.213	3.238	8,15	4.759	-2.670
8,16	-3.801	2.258	8,17	3.132	-1.949	8,18	-2.645	1.710	8,19	2.278	-1.520
8,20	-1.997	1.368	8,21	1.775	-1.243	8,22	-1.600	1.140	8,23	1.460	-1.054
9, 9	614.522	-341.656	9,10	149.736	63.397	9,11	-54.252	6.773	9,12	23.421	-7.179
9,13	-13.077	5.285	9,14	8.580	-4.043	9,15	-6.186	3.228	9,16	4.735	-2.662
9,17	-3.780	2.251	9,18	3.114	-1.943	9,19	-2.630	1.705	9,20	2.267	-1.517
9,21	-1.988	1.365	9,22	1.771	-1.242						
10,10	614.571	-341.672	10,11	149.695	63.411	10,12	-54.217	6.761	10,13	23.392	-7.169
10,14	-13.052	5.276	10,15	8.558	-4.036	10,16	-6.167	3.222	10,17	4.720	-2.657
10,18	-3.768	2.247	10,19	3.105	-1.940	10,20	-2.623	1.703	10,21	2.263	-1.516
11,11	614.605	-341.684	11,12	149.667	63.421	11,13	-54.193	6.752	11,14	23.371	-7.162
11,15	-13.035	5.270	11,16	8.544	-4.031	11,17	-6.156	3.218	11,18	4.711	-2.654
11,19	-3.762	2.245	11,20	3.101	-1.939						
12,12	614.628	-341.692	12,13	149.647	63.428	12,14	-54.177	6.747	12,15	23.358	-7.157
12,16	-13.025	5.266	12,17	8.536	-4.028	12,18	-6.151	3.216	12,19	4.708	-2.653
13,13	614.644	-341.698	13,14	149.634	63.432	13,15	-54.167	6.743	13,16	23.351	-7.154
13,17	-13.020	5.264	13,18	8.534	-4.027						
14,14	614.654	-341.702	14,15	149.627	63.435	14,16	-54.162	6.741	14,17	23.348	-7.154
15,15	614.659	-341.704	15,16	149.625	63.436						

IBR

SELF AND MUTUAL IMPEDANCES

H/LMDA=0.5000 B/LMDA= 0.500 OMEGA= 9.92

35 ELEMENT ARRAY

I, K	ZIK IN OHMS RE	IM	I, K	ZIK IN OHMS RE	IM	I, K	ZIK IN OHMS RE	IM	I, K	ZIK IN OHMS RE	IM
1, 1	567.442	-396.140	1, 2	175.590	75.127	1, 3	-68.370	2.443	1, 4	32.657	-4.916
1, 5	-19.832	3.873	1, 6	13.848	-3.071	1, 7	-10.467	2.514	1, 8	8.315	-2.114
1, 9	-6.839	1.815	1,10	5.771	-1.586	1,11	-4.968	1.406	1,12	4.345	-1.261
1,13	-3.850	1.141	1,14	3.450	-1.041	1,15	-3.120	0.957	1,16	2.844	-0.885
1,17	-2.612	0.822	1,18	2.413	-0.768	1,19	-2.242	0.720	1,20	2.095	-0.677
1,21	-1.966	0.639	1,22	1.854	-0.605	1,23	-1.755	0.574	1,24	1.669	-0.547
1,25	-1.594	0.521	1,26	1.529	-0.498	1,27	-1.473	0.477	1,28	1.426	-0.457
1,29	-1.390	0.439	1,30	1.364	-0.421	1,31	-1.353	0.404	1,32	1.364	-0.385
1,33	-1.417	0.361	1,34	1.580	-0.301	1,35	-2.009	-0.083			
2, 2	603.150	-342.460	2, 3	155.830	63.178	2, 4	-58.241	7.091	2, 5	26.343	-7.488
2, 6	-15.357	5.566	2, 7	10.431	-4.296	2, 8	-7.732	3.455	2, 9	6.053	-2.865
2,10	-4.920	2.434	2,11	4.113	-2.108	2,12	-3.513	1.854	2,13	3.052	-1.650
2,14	-2.690	1.485	2,15	2.398	-1.348	2,16	-2.160	1.233	2,17	1.962	-1.136
2,18	-1.796	1.052	2,19	1.655	-0.979	2,20	-1.535	0.916	2,21	1.431	-0.860
2,22	-1.341	0.811	2,23	1.264	-0.768	2,24	-1.196	0.729	2,25	1.138	-0.694
2,26	-1.087	0.664	2,27	1.044	-0.636	2,28	-1.009	0.612	2,29	0.981	-0.590
2,30	-0.962	0.571	2,31	0.955	-0.556	2,32	-0.964	0.543	2,33	1.006	-0.534
2,34	-1.137	0.519									
3, 3	611.358	-341.266	3, 4	151.830	63.055	3, 5	-55.794	7.061	3, 6	24.627	-7.424
3, 7	-14.057	5.494	3, 8	9.397	-4.225	3, 9	-6.880	3.388	3,10	5.333	-2.803
3,11	-4.301	2.376	3,12	3.571	-2.053	3,13	-3.033	1.802	3,14	2.622	-1.603
3,15	-2.301	1.440	3,16	2.045	-1.306	3,17	-1.836	1.194	3,18	1.663	-1.098
3,19	-1.519	1.017	3,20	1.397	-0.946	3,21	-1.294	0.885	3,22	1.205	-0.831
3,23	-1.128	0.783	3,24	1.062	-0.742	3,25	-1.006	0.705	3,26	0.957	-0.672
3,27	-0.916	0.643	3,28	0.882	-0.617	3,29	-0.855	0.595	3,30	0.836	-0.576
3,31	-0.828	0.560	3,32	0.835	-0.549	3,33	-0.871	0.542			
4, 4	613.202	-341.388	4, 5	150.707	63.178	4, 6	-55.009	6.954	4, 7	24.034	-7.333
4, 8	-13.586	5.416	4, 9	9.009	-4.157	4,10	-6.553	3.327	4,11	5.053	-2.748
4,12	-4.056	2.327	4,13	3.354	-2.009	4,14	-2.839	1.762	4,15	2.448	-1.565
4,16	-2.143	1.406	4,17	1.899	-1.274	4,18	-1.702	1.164	4,19	1.540	-1.071
4,20	-1.405	0.991	4,21	1.291	-0.922	4,22	-1.194	0.862	4,23	1.112	-0.810
4,24	-1.041	0.764	4,25	0.981	-0.723	4,26	-0.929	0.688	4,27	0.886	-0.656
4,28	-0.850	0.629	4,29	0.821	-0.605	4,30	-0.801	0.585	4,31	0.790	-0.568
4,32	-0.795	0.556									
5, 5	613.865	-341.493	5, 6	150.244	63.264	5, 7	-54.660	6.884	5, 8	23.757	-7.273
5, 9	-13.359	5.365	5,10	8.818	-4.112	5,11	-6.389	3.287	5,12	4.910	-2.713
5,13	-3.930	2.295	5,14	3.242	-1.980	5,15	-2.738	1.735	5,16	2.356	-1.541
5,17	-2.059	1.383	5,18	1.823	-1.253	5,19	-1.632	1.145	5,20	1.475	-1.053
5,21	-1.345	0.974	5,22	1.235	-0.907	5,23	-1.143	0.848	5,24	1.064	-0.797
5,25	-0.997	0.752	5,26	0.940	-0.713	5,27	-0.892	0.678	5,28	0.852	-0.648
5,29	-0.821	0.622	5,30	0.798	-0.600	5,31	-0.785	0.582			
6, 6	614.182	-341.563	6, 7	150.006	63.321	6, 8	-54.471	6.836	6, 9	23.602	-7.233
6,10	-13.229	5.329	6,11	8.707	-4.081	6,12	-6.293	3.260	6,13	4.825	-2.688
[illegible]	[illegible]	2.272	6,15	3.174	-1.959	6,16	-2.677	1.716	6,17	2.300	-1.524
6,18	-2.008	1.361	[illegible]	[illegible]	-1.239	6,20	-1.589	1.131	6,21	1.436	-1.041
6,22	-1.308	0.963	6,23	1.202	-0.896	6,24	[illegible]	[illegible]	6,25	1.036	-0.788
6,26	-0.972	0.745	6,27	0.918	-0.706	6,28	-0.874	0.673	6,29	0.838	[illegible]
6,30	-0.812	0.620									
7, 7	614.360	-341.610	7, 8	149.865	63.359	7, 9	-54.356	6.803	7,10	23.506	-7.204
7,11	-13.146	5.304	7,12	8.635	-4.059	7,13	-6.230	3.240	7,14	4.769	-2.670
7,15	-3.804	2.256	7,16	3.129	-1.945	7,17	-2.636	1.703	7,18	2.264	-1.511
7,19	-1.975	1.356	7,20	1.746	-1.229	7,21	-1.561	1.122	7,22	1.410	-1.032
7,23	-1.285	0.956	7,24	1.181	-0.890	7,25	-1.094	0.833	7,26	1.021	-0.784
7,27	-0.959	0.741	7,28	0.909	-0.704	7,29	-0.868	0.672			
8, 8	614.470	-341.642	8, 9	149.775	63.387	8,10	-54.280	6.779	8,11	23.441	-7.183
8,12	-13.091	5.286	8,13	8.587	-4.043	8,14	-6.186	3.225	8,15	4.730	-2.657
8,16	-3.770	2.244	8,17	3.098	-1.934	8,18	-2.608	1.693	8,19	2.238	-1.503
8,20	-1.952	1.348	8,21	1.725	-1.222	8,22	-1.543	1.116	8,23	1.394	-1.027
8,24	-1.271	0.951	8,25	1.169	-0.886	8,26	-1.084	0.830	8,27	1.013	-0.781
8,28	-0.955	0.740									
9, 9	614.543	-341.666	9,10	149.714	63.407	9,11	-54.228	6.762	9,12	23.397	-7.168
9,13	-13.051	5.273	9,14	8.552	-4.030	9,15	-6.156	3.214	9,16	4.703	-2.647
9,17	-3.745	2.236	9,18	3.076	-1.926	9,19	-2.589	1.686	9,20	2.221	-1.496
9,21	-1.936	1.343	9,22	1.712	-1.217	9,23	-1.531	1.112	9,24	1.384	-1.023
9,25	-1.263	0.948	9,26	1.163	-0.884	9,27	-1.081	0.829			
10,10	614.594	-341.683	10,11	149.671	63.422	10,12	-54.192	6.749	10,13	23.364	-7.157
10,14	-13.023	5.262	10,15	8.527	-4.021	10,16	-6.134	3.206	10,17	4.684	-2.640
10,18	-3.728	2.229	10,19	3.061	-1.920	10,20	-2.575	1.681	10,21	2.209	-1.492
10,22	-1.926	1.339	10,23	1.703	-1.214	10,24	-1.524	1.109	10,25	1.379	-1.022
10,26	-1.261	0.947									
11,11	614.630	-341.696	11,12	149.640	63.433	11,13	-54.165	6.739	11,14	23.341	-7.148
11,15	-13.003	5.255	11,16	8.509	-4.015	11,17	-6.118	3.200	11,18	4.670	-2.635
11,19	-3.716	2.225	11,20	3.050	-1.916	11,21	-2.566	1.678	11,22	2.202	-1.489
11,23	-1.921	1.337	11,24	1.699	-1.212	11,25	-1.522	1.109			
12,12	614.657	-341.706	12,13	149.617	63.442	12,14	-54.145	6.732	12,15	23.324	-7.141
12,16	-12.988	5.249	12,17	8.496	-4.010	12,18	-6.107	3.196	12,19	4.660	-2.631
12,20	-3.708	2.222	12,21	3.044	-1.914	12,22	-2.561	1.676	12,23	2.198	-1.488
12,24	-1.919	1.336									
13,13	614.676	-341.713	13,14	149.600	63.448	13,15	-54.131	6.726	13,16	23.311	-7.136
13,17	-12.977	5.245	13,18	8.487	-4.006	13,19	-6.099	3.193	13,20	4.654	-2.629
13,21	-3.703	2.220	13,22	3.041	-1.913	13,23	-2.559	1.676			
14,14	614.690	-341.719	14,15	149.588	63.453	14,16	-54.120	6.722	14,17	23.303	-7.133
14,18	-12.970	5.242	14,19	8.481	-4.004	14,20	-6.095	3.191	14,21	4.651	-2.628
14,22	-3.702	2.219									
15,15	614.700	-341.723	15,16	149.580	63.456	15,17	-54.113	6.719	15,18	23.297	-7.131
15,19	-12.966	5.240	15,20	8.479	-4.003	15,21	-6.094	3.191			
16,16	614.707	-341.725	16,17	149.574	63.458	16,18	-54.109	6.718	16,19	23.295	-7.130
16,20	-12.965	5.240									
17,17	614.711	-341.727	17,18	149.572	63.459	17,19	-54.108	6.717			
18,18	614.712	-341.728									

IBR

SELF AND MUTUAL IMPEDANCES

H/LMDA=0.5000 B/LMDA= 0.500 OMEGA= 9.92

40 ELEMENT ARRAY

I, K	ZIK IN OHMS		I, K	ZIK IN OHMS		I, K	ZIK IN OHMS		I, K	ZIK IN OHMS	
	RE	IM		RE	IM		RE	IM		RE	IM
1, 1	567.468	-396.133	1, 2	175.568	75.126	1, 3	-68.351	2.443	1, 4	32.638	-4.916
1, 5	-19.813	3.873	1, 6	13.829	-3.070	1, 7	-10.447	2.513	1, 8	8.295	-2.112
1, 9	-6.818	1.814	1,10	5.750	-1.585	1,11	-4.946	1.405	1,12	4.322	-1.259
1,13	-3.827	1.139	1,14	3.425	-1.040	1,15	-3.094	0.955	1,16	2.817	-0.883
1,17	-2.582	0.820	1,18	2.382	-0.766	1,19	-2.209	0.718	1,20	2.059	-0.675
1,21	-1.928	0.637	1,22	1.812	-0.603	1,23	-1.710	0.572	1,24	1.619	-0.544
1,25	-1.539	0.519	1,26	1.467	-0.496	1,27	-1.403	0.475	1,28	1.345	-0.456
1,29	-1.295	0.438	1,30	1.250	-0.421	1,31	-1.211	0.406	1,32	1.179	-0.391
1,33	-1.152	0.378	1,34	1.132	-0.365	1,35	-1.120	0.352	1,36	1.119	-0.340
1,37	-1.136	0.326	1,38	1.189	-0.307	1,39	-1.334	0.258	1,40	1.707	0.069
2, 2	603.167	-342.463	2, 3	155.815	63.182	2, 4	-58.227	7.087	2, 5	26.329	-7.484
2, 6	-15.343	5.562	2, 7	10.417	-4.292	2, 8	-7.717	3.450	2, 9	6.038	-2.861
2,10	-4.905	2.429	2,11	4.097	-2.103	2,12	-3.496	1.848	2,13	3.035	-1.645
2,14	-2.671	1.479	2,15	2.379	-1.342	2,16	-2.139	1.227	2,17	1.940	-1.129
2,18	-1.773	1.045	2,19	1.631	-0.972	2,20	-1.508	0.908	2,21	1.403	-0.852
2,22	-1.311	0.802	2,23	1.230	-0.758	2,24	-1.159	0.718	2,25	1.096	-0.683
2,26	-1.041	0.651	2,27	0.992	-0.622	2,28	-0.948	0.595	2,29	0.910	-0.572
2,30	-0.877	0.550	2,31	0.848	-0.531	2,32	-0.823	0.513	2,33	0.803	-0.498
2,34	-0.789	0.484	2,35	0.780	-0.472	2,36	-0.781	0.462	2,37	0.795	-0.454
2,38	-0.836	0.449	2,39	0.953	-0.439						
3, 3	611.371	-341.270	3, 4	151.817	63.059	3, 5	-55.782	7.056	3, 6	24.615	-7.419
3, 7	-14.045	5.490	3, 8	9.384	-4.220	3, 9	-6.867	3.382	3,10	5.320	-2.797
3,11	-4.287	2.370	3,12	3.557	-2.047	3,13	-3.018	1.796	3,14	2.606	-1.596
3,15	-2.284	1.433	3,16	2.027	-1.299	3,17	-1.817	1.186	3,18	1.643	-1.090
3,19	-1.498	1.008	3,20	1.374	-0.937	3,21	-1.269	0.875	3,22	1.178	-0.821
3,23	-1.099	0.772	3,24	1.030	-0.730	3,25	-0.970	0.691	3,26	0.917	-0.657
3,27	-0.870	0.626	3,28	0.829	-0.599	3,29	-0.793	0.574	3,30	0.761	-0.551
3,31	-0.734	0.531	3,32	0.711	-0.513	3,33	-0.692	0.497	3,34	0.679	-0.483
3,35	-0.670	0.471	3,36	0.670	-0.462	3,37	-0.681	0.456	3,38	0.717	-0.453
4, 4	613.214	-341.392	4, 5	150.695	63.183	4, 6	-54.997	6.950	4, 7	24.022	-7.328
4, 8	-13.574	5.411	4, 9	8.997	-4.152	4,10	-6.541	3.321	4,11	5.039	-2.743
4,12	-4.042	2.321	4,13	3.340	-2.002	4,14	-2.824	1.755	4,15	2.432	-1.558
4,16	-2.126	1.398	4,17	1.882	-1.266	4,18	-1.683	1.156	4,19	1.520	-1.062
4,20	-1.383	0.981	4,21	1.267	-0.912	4,22	-1.169	0.851	4,23	1.084	-0.798
4,24	-1.011	0.751	4,25	0.947	-0.710	4,26	-0.891	0.672	4,27	0.842	-0.639
4,28	-0.799	0.610	4,29	0.762	-0.583	4,30	-0.729	0.559	4,31	0.701	-0.538
4,32	-0.677	0.518	4,33	0.658	-0.502	4,34	-0.643	0.487	4,35	0.634	-0.475
4,36	-0.633	0.465	4,37	0.643	-0.459						
5, 5	613.876	-341.498	5, 6	150.233	63.269	5, 7	-54.648	6.879	5, 8	23.745	-7.268
5, 9	-13.347	5.359	5,10	8.806	-4.106	5,11	-6.376	3.281	5,12	4.896	-2.706
5,13	-3.916	2.288	5,14	3.227	-1.973	5,15	-2.722	1.728	5,16	2.340	-1.533
5,17	-2.042	1.375	5,18	1.804	-1.245	5,19	-1.612	1.136	5,20	1.454	-1.043
5,21	-1.321	0.964	5,22	1.210	-0.895	5,23	-1.115	0.836	5,24	1.034	-0.784
5,25	-0.963	0.738	5,26	0.902	-0.697	5,27	-0.849	0.661	5,28	0.803	-0.628
5,29	-0.763	0.599	5,30	0.727	-0.573	5,31	-0.697	0.550	5,32	0.672	-0.530
5,33	-0.651	0.512	5,34	0.635	-0.496	5,35	-0.625	0.483	5,36	0.622	-0.473
6, 6	614.193	-341.568	6, 7	149.994	63.326	6, 8	-54.459	6.831	6, 9	23.590	-7.227
6,10	-13.216	5.323	6,11	8.694	-4.075	6,12	-6.279	3.253	6,13	4.810	-2.681
6,14	-3.839	2.265	6,15	3.158	-1.952	6,16	-2.660	1.709	6,17	2.283	-1.515
6,18	-1.990	1.359	6,19	1.756	-1.230	6,20	-1.568	1.121	6,21	1.412	-1.030
6,22	-1.283	0.952	6,23	1.174	-0.884	6,24	-1.082	0.825	6,25	1.003	-0.774
6,26	-0.934	0.728	6,27	0.875	-0.688	6,28	-0.824	0.653	6,29	0.780	-0.621
6,30	-0.741	0.593	6,31	0.708	-0.568	6,32	-0.680	0.545	6,33	0.657	-0.526
6,34	-0.639	0.509	6,35	0.628	-0.495						
7, 7	614.372	-341.615	7, 8	149.853	63.365	7, 9	-54.343	6.797	7,10	23.493	-7.198
7,11	-13.133	5.298	7,12	8.622	-4.052	7,13	-6.215	3.233	7,14	4.754	-2.663
7,15	-3.788	2.249	7,16	3.112	-1.937	7,17	-2.618	1.695	7,18	2.245	-1.503
7,19	-1.955	1.347	7,20	1.724	-1.218	7,21	-1.538	1.111	7,22	1.385	-1.020
7,23	-1.258	0.943	7,24	1.151	-0.876	7,25	-1.060	0.818	7,26	0.982	-0.767
7,27	-0.916	0.722	7,28	0.858	-0.682	7,29	-0.809	0.647	7,30	0.766	-0.616
7,31	-0.729	0.589	7,32	0.698	-0.564	7,33	-0.673	0.543	7,34	0.653	-0.525
8, 8	614.483	-341.648	8, 9	149.762	63.393	8,10	-54.267	6.773	8,11	23.428	-7.177
8,12	-13.076	5.279	8,13	8.572	-4.035	8,14	-6.171	3.218	8,15	4.714	-2.649
8,16	-3.753	2.236	8,17	3.080	-1.925	8,18	-2.589	1.684	8,19	2.218	-1.493
8,20	-1.930	1.338	8,21	1.701	-1.210	8,22	-1.517	1.104	8,23	1.366	-1.013
8,24	-1.240	0.936	8,25	1.134	-0.870	8,26	-1.045	0.812	8,27	0.969	-0.762
8,28	-0.904	0.718	8,29	0.848	-0.679	8,30	-0.799	0.644	8,31	0.758	-0.614
8,32	-0.723	0.587	8,33	0.695	-0.564						
9, 9	614.556	-341.672	9,10	149.700	63.414	9,11	-54.215	6.755	9,12	23.382	-7.161
9,13	-13.036	5.265	9,14	8.536	-4.023	9,15	-6.139	3.206	9,16	4.686	-2.639
9,17	-3.727	2.227	9,18	3.057	-1.916	9,19	-2.567	1.676	9,20	2.198	-1.486
9,21	-1.912	1.331	9,22	1.685	-1.204	9,23	-1.502	1.098	9,24	1.352	-1.008
9,25	-1.228	0.932	9,26	1.123	-0.866	9,27	-1.035	0.809	9,28	0.960	-0.759
9,29	-0.896	0.715	9,30	0.842	-0.677	9,31	-0.795	0.643	9,32	0.756	-0.613
10,10	614.608	-341.690	10,11	149.656	63.429	10,12	-54.177	6.742	10,13	23.349	-7.149
10,14	-13.007	5.254	10,15	8.510	-4.013	10,16	-6.116	3.197	10,17	4.664	-2.631
10,18	-3.708	2.219	10,19	3.039	-1.910	10,20	-2.552	1.670	10,21	2.184	-1.480
10,22	-1.899	1.326	10,23	1.673	-1.200	10,24	-1.491	1.094	10,25	1.342	-1.005
10,26	-1.219	0.929	10,27	1.116	-0.863	10,28	-1.029	0.806	10,29	0.955	-0.757
10,30	-0.893	0.714	10,31	0.839	-0.676						
11,11	614.645	-341.703	11,12	149.624	63.441	11,13	-54.148	6.731	11,14	23.324	-7.139
11,15	-12.985	5.246	11,16	8.490	-4.005	11,17	-6.098	3.190	11,18	4.648	-2.624
11,19	-3.693	2.214	11,20	3.026	-1.905	11,21	-2.540	1.665	11,22	2.173	-1.476
11,23	-1.889	1.322	11,24	1.665	-1.196	11,25	-1.484	1.091	11,26	1.336	-1.002
11,27	-1.214	0.927	11,28	1.112	-0.862	11,29	-1.026	0.805	11,30	0.953	-0.756

12,12	614.673	-341.714	12,13	149.600	63.450	12,14	-54.127	6.723	12,15	23.305	-7.132
12,16	-12.968	5.239	12,17	8.475	-3.999	12,18	-6.085	3.185	12,19	4.636	-2.620
12,20	-3.682	2.209	12,21	3.017	-1.901	12,22	-2.531	1.662	12,23	2.166	-1.473
12,24	-1.883	1.320	12,25	1.659	-1.194	12,26	-1.479	1.089	12,27	1.332	-1.001
12,28	-1.211	0.926	12,29	1.110	-0.861						
13,13	614.694	-341.722	13,14	149.582	63.457	13,15	-54.111	6.716	13,16	23.291	-7.126
13,17	-12.955	5.234	13,18	8.464	-3.995	13,19	-6.075	3.181	13,20	4.627	-2.616
13,21	-3.675	2.206	13,22	3.010	-1.898	13,23	-2.525	1.660	13,24	2.161	-1.471
13,25	-1.879	1.318	13,26	1.656	-1.193	13,27	-1.477	1.088	13,28	1.331	-1.000
14,14	614.710	-341.728	14,15	149.568	63.463	14,16	-54.099	6.711	14,17	23.280	-7.122
14,18	-12.946	5.230	14,19	8.456	-3.991	14,20	-6.067	3.178	14,21	4.621	-2.613
14,22	-3.669	2.204	14,23	3.005	-1.896	14,24	-2.522	1.658	14,25	2.158	-1.470
14,26	-1.877	1.317	14,27	1.655	-1.192						
15,15	614.722	-341.733	15,16	149.557	63.467	15,17	-54.090	6.707	15,18	23.272	-7.118
15,19	-12.939	5.227	15,20	8.449	-3.989	15,21	-6.062	3.176	15,22	4.617	-2.611
15,23	-3.665	2.202	15,24	3.002	-1.895	15,25	-2.520	1.657	15,26	2.157	-1.469
16,16	614.731	-341.737	16,17	149.549	63.471	16,18	-54.083	6.705	16,19	23.266	-7.116
16,20	-12.934	5.225	16,21	8.445	-3.987	16,22	-6.059	3.174	16,23	4.614	-2.610
16,24	-3.664	2.202	16,25	3.001	-1.895						
17,17	614.738	-341.740	17,18	149.543	63.473	17,19	-54.078	6.702	17,20	23.262	-7.114
17,21	-12.930	5.224	17,22	8.443	-3.986	17,23	-6.057	3.174	17,24	4.613	-2.610
18,18	614.743	-341.742	18,19	149.539	63.475	18,20	-54.075	6.701	18,21	23.260	-7.113
18,22	-12.929	5.223	18,23	8.442	-3.986						
19,19	614.746	-341.744	19,20	149.537	63.476	19,21	-54.073	6.700	19,22	23.259	-7.113
20,20	614.747	-341.744	20,21	149.536	63.476						

IBR

SELF AND MUTUAL IMPEDANCES

H/LMDA=0.5000 B/LMDA= 0.500 OMEGA= 9.92

45 ELEMENT ARRAY

I, K	ZIK IN OHMS		I, K	ZIK IN OHMS		I, K	ZIK IN OHMS		I, K	ZIK IN OHMS	
	RE	IM		RE	IM		RE	IM		RE	IM
1, 1	567.487	-396.127	1, 2	175.552	75.125	1, 3	-68.336	2.443	1, 4	32.624	-4.915
1, 5	-19.800	3.872	1, 6	13.815	-3.069	1, 7	-10.433	2.512	1, 8	8.281	-2.111
1, 9	-6.804	1.813	1,10	5.735	-1.584	1,11	-4.931	1.403	1,12	4.307	-1.258
1,13	-3.811	1.138	1,14	3.408	-1.038	1,15	-3.076	0.954	1,16	2.799	-0.881
1,17	-2.564	0.819	1,18	2.362	-0.764	1,19	-2.188	0.716	1,20	2.037	-0.673
1,21	-1.904	0.635	1,22	1.787	-0.601	1,23	-1.684	0.570	1,24	1.591	-0.543
1,25	-1.508	0.517	1,26	1.434	-0.494	1,27	-1.367	0.473	1,28	1.307	-0.454
1,29	-1.252	0.436	1,30	1.203	-0.419	1,31	-1.158	0.404	1,32	1.118	-0.389
1,33	-1.082	0.376	1,34	1.050	-0.364	1,35	-1.022	0.352	1,36	0.997	-0.341
1,37	-0.977	0.331	1,38	0.961	-0.321	1,39	-0.950	0.311	1,40	0.945	-0.302
1,41	-0.950	0.293	1,42	0.970	-0.282	1,43	-1.020	0.267	1,44	1.151	-0.225
1,45	-1.480	-0.058									
2, 2	603.179	-342.465	2, 3	155.804	63.185	2, 4	-58.217	7.084	2, 5	26.319	-7.481
2, 6	-15.333	5.559	2, 7	10.407	-4.288	2, 8	-7.707	3.447	2, 9	6.027	-2.857
2,10	-4.894	2.426	2,11	4.086	-2.099	2,12	-3.485	1.844	2,13	3.023	-1.641
2,14	-2.659	1.475	2,15	2.366	-1.338	2,16	-2.126	1.222	2,17	1.926	-1.124
2,18	-1.758	1.040	2,19	1.615	-0.967	2,20	-1.492	0.903	2,21	1.386	-0.846
2,22	-1.292	0.796	2,23	1.211	-0.752	2,24	-1.138	0.712	2,25	1.074	-0.676
2,26	-1.017	0.643	2,27	0.966	-0.614	2,28	-0.920	0.587	2,29	0.879	-0.562
2,30	-0.842	0.540	2,31	0.808	-0.520	2,32	-0.778	0.501	2,33	0.752	-0.484
2,34	-0.728	0.468	2,35	0.707	-0.453	2,36	-0.690	0.440	2,37	0.675	-0.428
2,38	-0.663	0.418	2,39	0.656	-0.408	2,40	-0.653	0.400	2,41	0.658	-0.394
2,42	-0.673	0.389	2,43	0.713	-0.387	2,44	-0.817	0.380			
3, 3	611.380	-341.273	3, 4	151.808	63.062	3, 5	-55.773	7.053	3, 6	24.606	-7.416
3, 7	-14.036	5.486	3, 8	9.375	-4.217	3, 9	-6.858	3.379	3,10	5.311	-2.793
3,11	-4.277	2.366	3,12	3.547	-2.043	3,13	-3.007	1.792	3,14	2.596	-1.592
3,15	-2.273	1.429	3,16	2.015	-1.294	3,17	-1.805	1.181	3,18	1.631	-1.085
3,19	-1.484	1.003	3,20	1.360	-0.931	3,21	-1.254	0.869	3,22	1.162	-0.814
3,23	-1.082	0.766	3,24	1.012	-0.722	3,25	-0.951	0.684	3,26	0.896	-0.649
3,27	-0.847	0.618	3,28	0.804	-0.589	3,29	-0.765	0.563	3,30	0.731	-0.540
3,31	-0.700	0.519	3,32	0.672	-0.499	3,33	-0.648	0.481	3,34	0.626	-0.465
3,35	-0.607	0.450	3,36	0.590	-0.437	3,37	-0.577	0.425	3,38	0.566	-0.414
3,39	-0.559	0.405	3,40	0.556	-0.397	3,41	-0.560	0.392	3,42	0.573	-0.388
3,43	-0.607	0.388									
4, 4	613.223	-341.396	4, 5	150.687	63.186	4, 6	-54.989	6.946	4, 7	24.014	-7.324
4, 8	-13.565	5.407	4, 9	8.988	-4.148	4,10	-6.532	3.317	4,11	5.030	-2.738
4,12	-4.033	2.317	4,13	3.330	-1.998	4,14	-2.814	1.751	4,15	2.422	-1.553
4,16	-2.115	1.393	4,17	1.870	-1.261	4,18	-1.671	1.150	4,19	1.507	-1.056
4,20	-1.370	0.976	4,21	1.253	-0.906	4,22	-1.154	0.845	4,23	1.068	-0.791
4,24	-0.994	0.744	4,25	0.928	-0.702	4,26	-0.871	0.664	4,27	0.821	-0.630
4,28	-0.776	0.600	4,29	0.736	-0.572	4,30	-0.701	0.547	4,31	0.669	-0.525
4,32	-0.641	0.504	4,33	0.616	-0.485	4,34	-0.594	0.468	4,35	0.574	-0.453
4,36	-0.558	0.439	4,37	0.544	-0.426	4,38	-0.533	0.415	4,39	0.525	-0.406
4,40	-0.522	0.398	4,41	0.524	-0.392	4,42	-0.537	0.389			

5, 5	613.885	-341.502	5, 6	150.225	63.272	5, 7	-54.640	6.875	5, 8	23.737	-7.264
5, 9	-13.338	5.355	5,10	8.797	-4.102	5,11	-6.368	3.277	5,12	4.887	-2.702
5,13	-3.906	2.284	5,14	3.217	-1.968	5,15	-2.712	1.723	5,16	2.329	-1.528
5,17	-2.031	1.370	5,18	1.793	-1.239	5,19	-1.600	1.130	5,20	1.441	-1.037
5,21	-1.308	0.958	5,22	1.195	-0.889	5,23	-1.100	0.829	5,24	1.017	-0.776
5,25	-0.946	0.730	5,26	0.883	-0.688	5,27	-0.828	0.651	5,28	0.780	-0.618
5,29	-0.737	0.588	5,30	0.699	-0.561	5,31	-0.666	0.537	5,32	0.636	-0.515
5,33	-0.609	0.495	5,34	0.586	-0.477	5,35	-0.566	0.460	5,36	0.548	-0.445
5,37	-0.533	0.432	5,38	0.522	-0.421	5,39	-0.514	0.411	5,40	0.509	-0.402
5,41	-0.511	0.396									
6, 6	614.201	-341.572	6, 7	149.986	63.330	6, 8	-54.451	6.827	6, 9	23.582	-7.223
6,10	-13.208	5.319	6,11	8.685	-4.070	6,12	-6.270	3.249	6,13	4.801	-2.677
6,14	-3.829	2.260	6,15	3.148	-1.947	6,16	-2.650	1.704	6,17	2.272	-1.510
6,18	-1.978	1.353	6,19	1.744	-1.224	6,20	-1.555	1.115	6,21	1.399	-1.023
6,22	-1.269	0.945	6,23	1.159	-0.877	6,24	-1.065	0.817	6,25	0.985	-0.765
6,26	-0.915	0.719	6,27	0.855	-0.679	6,28	-0.801	0.642	6,29	0.755	-0.610
6,30	-0.713	0.580	6,31	0.677	-0.554	6,32	-0.645	0.530	6,33	0.616	-0.508
6,34	-0.591	0.489	6,35	0.569	-0.471	6,36	-0.550	0.455	6,37	0.534	-0.441
6,38	-0.522	0.429	6,39	0.512	-0.418	6,40	-0.507	0.409			
7, 7	614.380	-341.619	7, 8	149.844	63.369	7, 9	-54.335	6.793	7,10	23.484	-7.194
7,11	-13.124	5.294	7,12	8.612	-4.048	7,13	-6.206	3.228	7,14	4.744	-2.658
7,15	-3.778	2.243	7,16	3.102	-1.931	7,17	-2.607	1.689	7,18	2.233	-1.497
7,19	-1.942	1.341	7,20	1.711	-1.212	7,21	-1.524	1.104	7,22	1.370	-1.013
7,23	-1.242	0.935	7,24	1.134	-0.868	7,25	-1.042	0.809	7,26	0.963	-0.758
7,27	-0.895	0.712	7,28	0.835	-0.672	7,29	-0.783	0.636	7,30	0.738	-0.604
7,31	-0.698	0.575	7,32	0.662	-0.549	7,33	-0.631	0.525	7,34	0.604	-0.504
7,35	-0.580	0.485	7,36	0.559	-0.468	7,37	-0.542	0.453	7,38	0.528	-0.440
7,39	-0.518	0.428									
8, 8	614.491	-341.652	8, 9	149.753	63.397	8,10	-54.259	6.769	8,11	23.419	-7.172
8,12	-13.067	5.275	8,13	8.562	-4.031	8,14	-6.161	3.213	8,15	4.704	-2.644
8,16	-3.742	2.231	8,17	3.069	-1.919	8,18	-2.577	1.678	8,19	2.205	-1.487
8,20	-1.917	1.331	8,21	1.688	-1.203	8,22	-1.502	1.096	8,23	1.350	-1.006
8,24	-1.223	0.928	8,25	1.116	-0.861	8,26	-1.026	0.803	8,27	0.948	-0.752
8,28	-0.880	0.707	8,29	0.822	-0.667	8,30	-0.771	0.631	8,31	0.726	-0.599
8,32	-0.687	0.571	8,33	0.653	-0.545	8,34	-0.623	0.522	8,35	0.596	-0.502
8,36	-0.574	0.483	8,37	0.555	-0.467	8,38	-0.539	0.452			
9, 9	614.565	-341.676	9,10	149.691	63.418	9,11	-54.205	6.751	9,12	23.373	-7.156
9,13	-13.027	5.260	9,14	8.526	-4.018	9,15	-6.129	3.201	9,16	4.675	-2.633
9,17	-3.715	2.221	9,18	3.045	-1.910	9,19	-2.555	1.670	9,20	2.185	-1.479
9,21	-1.898	1.324	9,22	1.670	-1.197	9,23	-1.486	1.090	9,24	1.335	-1.000
9,25	-1.209	0.923	9,26	1.103	-0.856	9,27	-1.014	0.798	9,28	0.937	-0.747
9,29	-0.870	0.703	9,30	0.813	-0.663	9,31	-0.762	0.628	9,32	0.719	-0.596
9,33	-0.680	0.568	9,34	0.647	-0.543	9,35	-0.618	0.521	9,36	0.593	-0.500
9,37	-0.572	0.483									
10,10	614.617	-341.694	10,11	149.647	63.434	10,12	-54.167	6.737	10,13	23.339	-7.144
10,14	-12.997	5.249	10,15	8.499	-4.007	10,16	-6.105	3.192	10,17	4.653	-2.625
10,18	-3.695	2.213	10,19	3.026	-1.903	10,20	-2.538	1.663	10,21	2.169	-1.473
10,22	-1.884	1.318	10,23	1.657	-1.191	10,24	-1.474	1.085	10,25	1.324	-0.995
10,26	-1.199	0.919	10,27	1.094	-0.852	10,28	-1.005	0.795	10,29	0.929	-0.744
10,30	-0.863	0.700	10,31	0.806	-0.661	10,32	-0.757	0.626	10,33	0.714	-0.595
10,34	-0.676	0.567	10,35	0.644	-0.542	10,36	-0.616	0.520			
11,11	614.655	-341.708	11,12	149.614	63.446	11,13	-54.138	6.726	11,14	23.314	-7.134
11,15	-12.974	5.240	11,16	8.479	-3.999	11,17	-6.086	3.184	11,18	4.636	-2.618
11,19	-3.680	2.207	11,20	3.012	-1.897	11,21	-2.525	1.658	11,22	2.158	-1.468
11,23	-1.873	1.314	11,24	1.647	-1.187	11,25	-1.465	1.081	11,26	1.315	-0.992
11,27	-1.191	0.915	11,28	1.087	-0.849	11,29	-0.999	0.792	11,30	0.923	-0.742
11,31	-0.858	0.698	11,32	0.802	-0.659	11,33	-0.753	0.625	11,34	0.711	-0.594
11,35	-0.675	0.567									
12,12	614.683	-341.719	12,13	149.590	63.456	12,14	-54.116	6.717	12,15	23.294	-7.126
12,16	-12.956	5.233	12,17	8.463	-3.993	12,18	-6.072	3.178	12,19	4.623	-2.613
12,20	-3.668	2.202	12,21	3.001	-1.893	12,22	-2.515	1.654	12,23	2.148	-1.464
12,24	-1.864	1.310	12,25	1.639	-1.184	12,26	-1.458	1.079	12,27	1.309	-0.989
12,28	-1.186	0.913	12,29	1.082	-0.848	12,30	-0.994	0.791	12,31	0.919	-0.741
12,32	-0.855	0.697	12,33	0.800	-0.658	12,34	-0.752	0.624			
13,13	614.704	-341.728	13,14	149.570	63.463	13,15	-54.099	6.710	13,16	23.279	-7.120
13,17	-12.943	5.227	13,18	8.451	-3.988	13,19	-6.061	3.174	13,20	4.613	-2.608
13,21	-3.659	2.198	13,22	2.993	-1.889	13,23	-2.508	1.651	13,24	2.142	-1.461
13,25	-1.858	1.308	13,26	1.634	-1.182	13,27	-1.453	1.076	13,28	1.305	-0.988
13,29	-1.182	0.912	13,30	1.079	-0.846	13,31	-0.992	0.790	13,32	0.918	-0.740
13,33	-0.854	0.697									
14,14	614.721	-341.734	14,15	149.556	63.469	14,16	-54.086	6.705	14,17	23.267	-7.115
14,18	-12.932	5.223	14,19	8.441	-3.984	14,20	-6.052	3.170	14,21	4.605	-2.605
14,22	-3.652	2.195	14,23	2.986	-1.887	14,24	-2.502	1.648	14,25	2.136	-1.459
14,26	-1.854	1.306	14,27	1.630	-1.180	14,28	-1.449	1.075	14,29	1.302	-0.986
14,30	-1.180	0.911	14,31	1.077	-0.846	14,32	-0.991	0.789			
15,15	614.734	-341.740	15,16	149.544	63.474	15,17	-54.076	6.700	15,18	23.258	-7.111
15,19	-12.924	5.219	15,20	8.433	-3.980	15,21	-6.045	3.167	15,22	4.599	-2.602
15,23	-3.646	2.192	15,24	2.982	-1.884	15,25	-2.497	1.646	15,26	2.133	-1.457
15,27	-1.850	1.305	15,28	1.627	-1.179	15,29	-1.447	1.074	15,30	1.300	-0.986
15,31	-1.179	0.910									
16,16	614.745	-341.744	16,17	149.535	63.478	16,18	-54.068	6.697	16,19	23.250	-7.108
16,20	-12.917	5.216	16,21	8.428	-3.978	16,22	-6.040	3.165	16,23	4.594	-2.600
16,24	-3.642	2.191	16,25	2.978	-1.883	16,26	-2.495	1.645	16,27	2.130	-1.456
16,28	-1.849	1.304	16,29	1.626	-1.178	16,30	-1.447	1.074			
17,17	614.753	-341.748	17,18	149.527	63.481	17,19	-54.061	6.694	17,20	23.244	-7.105
17,21	-12.912	5.214	17,22	8.423	-3.976	17,23	-6.036	3.163	17,24	4.591	-2.599
17,25	-3.639	2.190	17,26	2.976	-1.882	17,27	-2.493	1.644	17,28	2.129	-1.456
17,29	-1.848	1.304									
18,18	614.759	-341.751	18,19	149.522	63.484	18,20	-54.056	6.692	18,21	23.240	-7.103
18,22	-12.908	5.212	18,23	8.420	-3.974	18,24	-6.033	3.162	18,25	4.589	-2.598
18,26	-3.638	2.189	18,27	2.975	-1.882	18,28	-2.492	1.644			
19,19	614.764	-341.753	19,20	149.518	63.486	19,21	-54.052	6.690	19,22	23.237	-7.102
19,23	-12.906	5.211	19,24	8.418	-3.974	19,25	-6.032	3.161	19,26	4.588	-2.597
19,27	-3.637	2.189									
20,20	614.768	-341.755	20,21	149.515	63.487	20,22	-54.050	6.689	20,23	23.235	-7.101
20,24	-12.904	5.211	20,25	8.417	-3.973	20,26	-6.031	3.161			
21,21	614.770	-341.756	21,22	149.513	63.488	21,23	-54.049	6.688	21,24	23.234	-7.101
21,25	-12.904	5.210									
22,22	614.772	-341.756	22,23	149.512	63.489	22,24	-54.048	6.688			
23,23	614.772	-341.757									

IBR

SELF AND MUTUAL IMPEDANCES

H/LMDA=0.5000 B/LMDA= 0.500 OMEGA= 9.92

50 ELEMENT ARRAY

I, K	ZIK IN OHMS RE	IM	I, K	ZIK IN OHMS RE	IM	I, K	ZIK IN OHMS RE	IM	I, K	ZIK IN OHMS RE	IM
1, 1	567.502	-396.122	1, 2	175.540	75.124	1, 3	-68.325	2.443	1, 4	32.613	-4.915
1, 5	-19.789	3.872	1, 6	13.805	-3.069	1, 7	-10.423	2.512	1, 8	8.271	-2.111
1, 9	-6.794	1.812	1,10	5.724	-1.583	1,11	-4.920	1.402	1,12	4.295	-1.257
1,13	-3.799	1.137	1,14	3.396	-1.037	1,15	-3.064	0.953	1,16	2.786	-0.880
1,17	-2.550	0.818	1,18	2.349	-0.763	1,19	-2.174	0.715	1,20	2.022	-0.672
1,21	-1.889	0.634	1,22	1.771	-0.600	1,23	-1.666	0.569	1,24	1.573	-0.541
1,25	-1.489	0.516	1,26	1.414	-0.493	1,27	-1.346	0.471	1,28	1.284	-0.452
1,29	-1.228	0.434	1,30	1.177	-0.417	1,31	-1.130	0.402	1,32	1.087	-0.388
1,33	-1.049	0.374	1,34	1.013	-0.362	1,35	-0.981	0.350	1,36	0.951	-0.339
1,37	-0.924	0.329	1,38	0.900	-0.319	1,39	-0.879	0.310	1,40	0.860	-0.302
1,41	-0.844	0.294	1,42	0.831	-0.286	1,43	-0.821	0.278	1,44	0.816	-0.271
1,45	-0.815	0.264	1,46	0.823	-0.257	1,47	-0.844	0.248	1,48	0.891	-0.235
1,49	-1.009	0.199	1,50	1.304	0.049						
2, 2	603.188	-342.467	2, 3	155.796	63.187	2, 4	-58.209	7.082	2, 5	26.311	-7.479
2, 6	-15.326	5.556	2, 7	10.399	-4.286	2, 8	-7.699	3.444	2, 9	6.019	-2.855
2,10	-4.886	2.423	2,11	4.078	-2.096	2,12	-3.476	1.842	2,13	3.014	-1.638
2,14	-2.650	1.472	2,15	2.357	-1.335	2,16	-2.117	1.219	2,17	1.917	-1.121
2,18	-1.748	1.036	2,19	1.605	-0.963	2,20	-1.481	0.899	2,21	1.374	-0.842
2,22	-1.281	0.792	2,23	1.198	-0.747	2,24	-1.125	0.707	2,25	1.060	-0.671
2,26	-1.002	0.638	2,27	0.950	-0.608	2,28	-0.903	0.581	2,29	0.861	-0.557
2,30	-0.822	0.534	2,31	0.788	-0.513	2,32	-0.756	0.494	2,33	0.727	-0.476
2,34	-0.701	0.460	2,35	0.677	-0.445	2,36	-0.656	0.431	2,37	0.636	-0.418
2,38	-0.619	0.406	2,39	0.603	-0.395	2,40	-0.590	0.384	2,41	0.578	-0.375
2,42	-0.569	0.367	2,43	0.563	-0.359	2,44	-0.559	0.353	2,45	0.560	-0.347
2,46	-0.566	0.343	2,47	0.582	-0.340	2,48	-0.619	0.339	2,49	0.713	-0.334
3, 3	611.387	-341.275	3, 4	151.802	63.065	3, 5	-55.766	7.050	3, 6	24.600	-7.413
3, 7	-14.029	5.483	3, 8	9.368	-4.214	3, 9	-6.851	3.376	3,10	5.304	-2.790
3,11	-4.271	2.363	3,12	3.540	-2.040	3,13	-3.000	1.789	3,14	2.588	-1.588
3,15	-2.266	1.426	3,16	2.007	-1.291	3,17	-1.797	1.178	3,18	1.622	-1.081
3,19	-1.475	0.999	3,20	1.351	-0.927	3,21	-1.244	0.865	3,22	1.152	-0.810
3,23	-1.072	0.761	3,24	1.001	-0.718	3,25	-0.939	0.679	3,26	0.883	-0.644
3,27	-0.834	0.612	3,28	0.790	-0.583	3,29	-0.750	0.557	3,30	0.714	-0.533
3,31	-0.682	0.511	3,32	0.653	-0.491	3,33	-0.626	0.473	3,34	0.602	-0.456
3,35	-0.581	0.440	3,36	0.561	-0.426	3,37	-0.543	0.413	3,38	0.527	-0.401
3,39	-0.513	0.389	3,40	0.501	-0.379	3,41	-0.491	0.370	3,42	0.483	-0.362
3,43	-0.476	0.354	3,44	0.473	-0.348	3,45	-0.473	0.343	3,46	0.479	-0.339
3,47	-0.493	0.338	3,48	0.525	-0.339						
4, 4	613.229	-341.398	4, 5	150.680	63.189	4, 6	-54.983	6.943	4, 7	24.007	-7.322
4, 8	-13.559	5.404	4, 9	8.982	-4.145	4,10	-6.525	3.315	4,11	5.024	-2.735
4,12	-4.026	2.313	4,13	3.324	-1.995	4,14	-2.807	1.747	4,15	2.414	-1.550
4,16	-2.108	1.390	4,17	1.863	-1.257	4,18	-1.663	1.146	4,19	1.499	-1.052
4,20	-1.361	0.971	4,21	1.244	-0.901	4,22	-1.144	0.840	4,23	1.058	-0.787
4,24	-0.983	0.739	4,25	0.917	-0.697	4,26	-0.859	0.659	4,27	0.808	-0.625
4,28	-0.762	0.594	4,29	0.722	-0.566	4,30	-0.685	0.540	4,31	0.652	-0.517
4,32	-0.622	0.496	4,33	0.596	-0.477	4,34	-0.571	0.459	4,35	0.550	-0.443
4,36	-0.530	0.428	4,37	0.512	-0.414	4,38	-0.496	0.401	4,39	0.482	-0.389
4,40	[illegible]	0.379	4,41	0.460	-0.369	4,42	-0.451	0.361	4,43	0.445	-0.353
4,44	-0.442	0.347	4,45	[illegible]	0.362	4,46	-0.446	0.338	4,47	0.459	-0.337
5, 5	613.891	-341.504	5, 6	150.219	63.275	5, 7	-54.634	6.872	5, 8	23.731	-7.261
5, 9	-13.332	5.352	5,10	8.791	-4.099	5,11	-6.361	3.274	5,12	4.881	-2.699
5,13	-3.900	2.280	5,14	3.211	-1.965	5,15	-2.705	1.720	5,16	2.322	-1.524
5,17	-2.023	1.366	5,18	1.785	-1.235	5,19	-1.592	1.126	5,20	1.432	-1.033
5,21	-1.299	0.953	5,22	1.186	-0.884	5,23	-1.090	0.824	5,24	1.007	-0.771
5,25	-0.935	0.724	5,26	0.872	-0.683	5,27	-0.816	0.645	5,28	0.767	-0.612
5,29	-0.723	0.582	5,30	0.684	-0.554	5,31	-0.649	0.529	5,32	0.618	-0.507
5,33	-0.590	0.486	5,34	0.565	-0.467	5,35	-0.542	0.450	5,36	0.521	-0.434
5,37	-0.503	0.419	5,38	0.486	-0.406	5,39	-0.472	0.394	5,40	0.459	-0.383
5,41	-0.448	0.373	5,42	0.439	-0.364	5,43	-0.433	0.356	5,44	0.429	-0.349
5,45	-0.428	0.344	5,46	0.432	-0.340						
6, 6	614.207	-341.575	6, 7	149.980	63.333	6, 8	-54.445	6.824	6, 9	23.576	-7.220
6,10	-13.202	5.316	6,11	8.679	-4.067	6,12	-6.264	3.245	6,13	4.794	-2.673
6,14	-3.823	2.257	6,15	3.141	-1.943	6,16	-2.642	1.700	6,17	2.264	-1.506
6,18	-1.970	1.349	6,19	1.736	-1.220	6,20	-1.546	1.111	6,21	1.390	-1.019
6,22	-1.260	0.940	6,23	1.149	-0.872	6,24	-1.055	0.812	6,25	0.974	-0.760
6,26	-0.904	0.714	6,27	0.843	-0.673	6,28	-0.789	0.636	6,29	0.741	-0.603
6,30	-0.698	0.573	6,31	0.661	-0.546	6,32	-0.627	0.522	6,33	0.597	-0.499
6,34	-0.570	0.479	6,35	0.545	-0.461	6,36	-0.523	0.444	6,37	0.504	-0.428
6,38	-0.486	0.414	6,39	0.471	-0.401	6,40	-0.458	0.389	6,41	0.446	-0.379
6,42	-0.437	0.369	6,43	0.429	-0.361	6,44	-0.425	0.354	6,45	0.424	-0.348
7, 7	614.386	-341.622	7, 8	149.839	63.372	7, 9	-54.329	6.790	7,10	23.478	-7.191
7,11	-13.118	5.290	7,12	8.606	-4.044	7,13	-6.199	3.225	7,14	4.737	-2.655
7,15	-3.771	2.240	7,16	3.095	-1.928	7,17	-2.600	1.685	7,18	2.225	-1.493
7,19	-1.934	1.337	7,20	1.703	-1.208	7,21	-1.515	1.100	7,22	1.361	-1.009
7,23	-1.233	0.930	7,24	1.124	-0.863	7,25	-1.031	0.804	7,26	0.952	-0.752
7,27	-0.883	0.706	7,28	0.822	-0.665	7,29	-0.770	0.629	7,30	0.723	-0.596
7,31	-0.681	0.567	7,32	0.645	-0.540	7,33	-0.612	0.516	7,34	0.582	-0.494
7,35	-0.556	0.474	7,36	0.533	-0.456	7,37	-0.512	0.439	7,38	0.493	-0.424
7,39	-0.476	0.410	7,40	0.462	-0.398	7,41	-0.449	0.387	7,42	0.439	-0.376
7,43	-0.431	0.368	7,44	0.426	-0.360						
8, 8	614.497	-341.655	8, 9	149.747	63.400	8,10	-54.253	6.766	8,11	23.413	-7.169
8,12	-13.061	5.271	8,13	8.556	-4.027	8,14	-6.154	3.209	8,15	4.697	-2.640
8,16	-3.735	2.227	8,17	3.061	-1.915	8,18	-2.569	1.674	8,19	2.197	-1.482
8,20	-1.908	1.327	8,21	1.679	-1.199	8,22	-1.493	1.091	8,23	1.340	-1.001
8,24	-1.213	0.923	8,25	1.106	-0.856	8,26	-1.014	0.797	8,27	0.935	-0.746
8,28	-0.867	0.700	8,29	0.808	-0.660	8,30	-0.756	0.624	8,31	0.710	-0.591
8,32	-0.669	0.562	8,33	0.633	-0.536	8,34	-0.601	0.512	8,35	0.572	-0.490
8,36	-0.547	0.471	8,37	0.524	-0.453	8,38	-0.504	0.436	8,39	0.486	-0.422
8,40	-0.470	0.408	8,41	0.457	-0.396	8,42	-0.446	0.385	8,43	0.437	-0.376
9, 9	614.571	-341.680	9,10	149.685	63.421	9,11	-54.199	6.747	9,12	23.366	-7.153
9,13	-13.020	5.257	9,14	8.519	-4.014	9,15	-6.122	3.197	9,16	4.667	-2.629
9,17	-3.708	2.217	9,18	3.037	-1.906	9,19	-2.547	1.665	9,20	2.176	-1.474
9,21	-1.889	1.319	9,22	1.661	-1.192	9,23	-1.476	1.085	9,24	1.325	-0.994
9,25	-1.198	0.917	9,26	1.092	-0.850	9,27	-1.001	0.792	9,28	0.923	-0.741
9,29	-0.856	0.696	9,30	0.797	-0.655	9,31	-0.746	0.620	9,32	0.701	-0.588
9,33	-0.661	0.559	9,34	0.625	-0.533	9,35	-0.594	0.509	9,36	0.565	-0.488
9,37	-0.541	0.468	9,38	0.518	-0.451	9,39	-0.499	0.435	9,40	0.482	-0.420
9,41	-0.467	0.407	9,42	0.455	-0.396						

10,10	614.623	-341.698	10,11	149.641	63.437	10,12	-54.160	6.733	10,13	23.332	-7.140
10,14	-12.990	5.245	10,15	8.492	-4.003	10,16	-6.097	3.188	10,17	4.645	-2.621
10,18	-3.687	2.209	10,19	3.018	-1.899	10,20	-2.529	1.658	10,21	2.160	-1.468
10,22	-1.874	1.313	10,23	1.647	-1.186	10,24	-1.463	1.080	10,25	1.313	-0.989
10,26	-1.187	0.912	10,27	1.081	-0.846	10,28	-0.991	0.788	10,29	0.914	-0.737
10,30	-0.847	0.692	10,31	0.789	-0.652	10,32	-0.738	0.617	10,33	0.694	-0.585
10,34	-0.654	0.556	10,35	0.619	-0.530	10,36	-0.588	0.507	10,37	0.561	-0.486
10,38	-0.537	0.467	10,39	0.515	-0.449	10,40	-0.497	0.434	10,41	0.480	-0.420
11,11	614.661	-341.712	11,12	149.608	63.450	11,13	-54.131	6.722	11,14	23.306	-7.130
11,15	-12.967	5.236	11,16	8.471	-3.995	11,17	-6.078	3.180	11,18	4.628	-2.614
11,19	-3.671	2.202	11,20	3.003	-1.893	11,21	-2.516	1.653	11,22	2.148	-1.463
11,23	-1.862	1.308	11,24	1.636	-1.182	11,25	-1.453	1.075	11,26	1.303	-0.986
11,27	-1.178	0.909	11,28	1.073	-0.842	11,29	-0.984	0.785	11,30	0.907	-0.734
11,31	-0.841	0.689	11,32	0.783	-0.650	11,33	-0.733	0.614	11,34	0.689	-0.583
11,35	-0.650	0.554	11,36	0.615	-0.529	11,37	-0.585	0.506	11,38	0.558	-0.485
11,39	-0.535	0.466	11,40	0.514	-0.449						
12,12	614.690	-341.723	12,13	149.582	63.459	12,14	-54.109	6.713	12,15	23.286	-7.122
12,16	-12.949	5.229	12,17	8.455	-3.989	12,18	-6.063	3.174	12,19	4.614	-2.608
12,20	-3.659	2.197	12,21	2.992	-1.888	12,22	-2.505	1.648	12,23	2.138	-1.458
12,24	-1.853	1.305	12,25	1.628	-1.178	12,26	-1.445	1.072	12,27	1.296	-0.982
12,28	-1.172	0.906	12,29	1.067	-0.840	12,30	-0.978	0.782	12,31	0.902	-0.732
12,32	-0.836	0.687	12,33	0.779	-0.648	12,34	-0.729	0.613	12,35	0.685	-0.581
12,36	-0.647	0.553	12,37	0.613	-0.528	12,38	-0.583	0.505	12,39	0.557	-0.485
13,13	614.712	-341.731	13,14	149.563	63.467	13,15	-54.092	6.706	13,16	23.271	-7.116
13,17	-12.934	5.223	13,18	8.442	-3.983	13,19	-6.052	3.169	13,20	4.603	-2.603
13,21	-3.649	2.193	13,22	2.983	-1.884	13,23	-2.497	1.645	13,24	2.130	-1.455
13,25	-1.846	1.301	13,26	1.621	-1.175	13,27	-1.439	1.069	13,28	1.290	-0.980
13,29	-1.166	0.904	13,30	1.062	-0.838	13,31	-0.974	0.780	13,32	0.898	-0.730
13,33	-0.833	0.686	13,34	0.776	-0.647	13,35	-0.727	0.612	13,36	0.684	-0.581
13,37	-0.646	0.553	13,38	0.612	-0.528						
14,14	614.729	-341.739	14,15	149.548	63.474	14,16	-54.078	6.700	14,17	23.258	-7.110
14,18	-12.923	5.218	14,19	8.432	-3.979	14,20	-6.042	3.165	14,21	4.595	-2.599
14,22	-3.641	2.189	14,23	2.975	-1.881	14,24	-2.490	1.642	14,25	2.124	-1.452
14,26	-1.840	1.299	14,27	1.616	-1.173	14,28	-1.435	1.067	14,29	1.286	-0.978
14,30	-1.163	0.902	14,31	1.059	-0.836	14,32	-0.971	0.779	14,33	0.895	-0.729
14,34	-0.831	0.685	14,35	0.774	-0.646	14,36	-0.726	0.611	14,37	0.683	-0.580
15,15	614.743	-341.744	15,16	149.535	63.479	15,17	-54.067	6.696	15,18	23.248	-7.106
15,19	-12.914	5.214	15,20	8.423	-3.975	15,21	-6.035	3.161	15,22	4.588	-2.596
15,23	-3.635	2.186	15,24	2.970	-1.878	15,25	-2.485	1.639	15,26	2.119	-1.450
15,27	-1.836	1.297	15,28	1.612	-1.171	15,29	-1.431	1.066	15,30	1.283	-0.977
15,31	-1.160	0.901	15,32	1.056	-0.835	15,33	-0.969	0.778	15,34	0.894	-0.728
15,35	-0.830	0.685	15,36	0.774	-0.646						
16,16	614.754	-341.749	16,17	149.526	63.483	16,18	-54.058	6.692	16,19	23.240	-7.102
16,20	-12.907	5.211	16,21	8.417	-3.972	16,22	-6.029	3.159	16,23	4.582	-2.594
16,24	-3.630	2.184	16,25	2.965	-1.876	16,26	-2.481	1.638	16,27	2.115	-1.448
16,28	-1.833	1.295	16,29	1.609	-1.170	16,30	-1.428	1.065	16,31	1.281	-0.976
16,32	-1.158	0.900	16,33	1.055	-0.835	16,34	-0.968	0.778	16,35	0.893	-0.728
17,17	614.762	-341.753	17,18	149.518	63.487	17,19	-54.051	6.688	17,20	23.234	-7.099
17,21	-12.901	5.208	17,22	8.411	-3.970	17,23	-6.024	3.156	17,24	4.578	-2.592
17,25	-3.626	2.182	17,26	2.961	-1.874	17,27	-2.477	1.636	17,28	2.113	-1.447
17,29	-1.830	1.294	17,30	1.607	-1.169	17,31	-1.427	1.064	17,32	1.279	-0.975
17,33	-1.157	0.899	17,34	1.055	-0.834						
18,18	614.769	-341.756	18,19	149.511	63.490	18,20	-54.045	6.686	18,21	23.229	-7.097
18,22	-12.896	5.206	18,23	8.407	-3.968	18,24	-6.020	3.155	18,25	4.574	-2.590
18,26	-3.623	2.181	18,27	2.959	-1.873	18,28	-2.475	1.635	18,29	2.111	-1.446
18,30	-1.829	1.294	18,31	1.606	-1.168	18,32	-1.426	1.063	18,33	1.279	-0.975
19,19	614.775	-341.759	19,20	149.506	63.492	19,21	-54.040	6.684	19,22	23.224	-7.095
19,23	-12.892	5.204	19,24	8.404	-3.966	19,25	-6.017	3.153	19,26	4.572	-2.589
19,27	-3.621	2.180	19,28	2.957	-1.872	19,29	-2.474	1.634	19,30	2.110	-1.446
19,31	-1.828	1.293	19,32	1.605	-1.168						
20,20	614.780	-341.761	20,21	149.502	63.494	20,22	-54.037	6.682	20,23	23.221	-7.094
20,24	-12.890	5.203	20,25	8.401	-3.965	20,26	-6.015	3.152	20,27	4.570	-2.588
20,28	-3.619	2.179	20,29	2.956	-1.872	20,30	-2.473	1.634	20,31	2.109	-1.446
21,21	614.783	-341.763	21,22	145.499	63.495	21,23	-54.034	6.681	21,24	23.219	-7.092
21,25	-12.887	5.202	21,26	8.400	-3.964	21,27	-6.013	3.152	21,28	4.569	-2.588
21,29	-3.618	2.179	21,30	2.956	-1.872						
22,22	614.786	-341.764	22,23	149.496	63.497	22,24	-54.032	6.680	22,25	23.217	-7.092
22,26	-12.886	5.201	22,27	8.399	-3.964	22,28	-6.013	3.151	22,29	4.569	-2.588
23,23	614.788	-341.765	23,24	149.495	63.497	23,25	-54.031	6.679	23,26	23.216	-7.091
23,27	-12.885	5.201	23,28	8.398	-3.964						
24,24	614.790	-341.766	24,25	149.494	63.498	24,26	-54.030	6.679	24,27	23.216	-7.091
25,25	614.790	-341.766	25,26	149.493	63.498						

TABLE 5.5

RADIATION PATTERNS OF BROADSIDE ARRAYS

VBR

RADIATION PATTERN FOR THETA=PI/2

H/LMDA=0.2500 B/LMDA= 0.250 OMEGA= 8.53

2 ELEMENT ARRAY

PHI DEG	ER(THETA,PHI) RE	IM	ABSVAL	PHI DEG	ER(THETA,PHI) RE	IM	ABSVAL
0	-0.23228	-2.20669	2.21888	90	-0.32849	-3.12073	3.13797
5	-0.23297	-2.21328	2.22550	95	-0.32772	-3.11343	3.13062
10	-0.23503	-2.23287	2.24520	100	-0.32544	-3.09176	3.10884
15	-0.23841	-2.26495	2.27746	105	-0.32173	-3.05648	3.07336
20	-0.24301	-2.30870	2.32145	110	-0.31671	-3.00882	3.02544
25	-0.24873	-2.36296	2.37601	115	-0.31056	-2.95040	2.96670
30	-0.25539	-2.42626	2.43966	120	-0.30348	-2.88318	2.89911
35	-0.26282	-2.49685	2.51064	125	-0.29572	-2.80940	2.82492
40	-0.27081	-2.57274	2.58695	130	-0.28751	-2.73142	2.74651
45	-0.27912	-2.65172	2.66637	135	-0.27912	-2.65172	2.66637
50	-0.28751	-2.73142	2.74651	140	-0.27081	-2.57274	2.58695
55	-0.29572	-2.80940	2.82491	145	-0.26282	-2.49685	2.51064
60	-0.30348	-2.88318	2.89911	150	-0.25539	-2.42626	2.43966
65	-0.31056	-2.95040	2.96669	155	-0.24873	-2.36296	2.37601
70	-0.31671	-3.00882	3.02544	160	-0.24301	-2.30870	2.32145
75	-0.32173	-3.05648	3.07336	165	-0.23841	-2.26495	2.27746
80	-0.32544	-3.09175	3.10884	170	-0.23503	-2.23287	2.24520
85	-0.32772	-3.11343	3.13062	175	-0.23297	-2.21328	2.22550
				180	-0.23228	-2.20669	2.21888

IBR

RADIATION PATTERN FOR THETA=PI/2

H/LMDA=0.2500 B/LMDA= 0.250 OMEGA= 8.53

2 ELEMENT ARRAY

PHI DEG	ER(THETA,PHI) RE	IM	ABSVAL	PHI DEG	ER(THETA,PHI) RE	IM	ABSVAL
0	-0.23228	-2.20669	2.21888	90	-0.32849	-3.12073	3.13797
5	-0.23297	-2.21328	2.22550	95	-0.32772	-3.11343	3.13062
10	-0.23503	-2.23287	2.24520	100	-0.32544	-3.09176	3.10884
15	-0.23841	-2.26495	2.27746	105	-0.32173	-3.05648	3.07336
20	-0.24301	-2.30870	2.32145	110	-0.31671	-3.00882	3.02544
25	-0.24873	-2.36296	2.37601	115	-0.31056	-2.95040	2.96670
30	-0.25539	-2.42626	2.43966	120	-0.30348	-2.88318	2.89911
35	-0.26282	-2.49685	2.51064	125	-0.29572	-2.80940	2.82492
40	-0.27081	-2.57274	2.58695	130	-0.28751	-2.73142	2.74651
45	-0.27912	-2.65172	2.66637	135	-0.27912	-2.65172	2.66637
50	-0.28751	-2.73142	2.74651	140	-0.27081	-2.57274	2.58695
55	-0.29572	-2.80940	2.82491	145	-0.26282	-2.49685	2.51064
60	-0.30348	-2.88318	2.89911	150	-0.25539	-2.42626	2.43966
65	-0.31056	-2.95040	2.96669	155	-0.24873	-2.36296	2.37601
70	-0.31671	-3.00882	3.02544	160	-0.24301	-2.30870	2.32145
75	-0.32173	-3.05648	3.07336	165	-0.23841	-2.26495	2.27746
80	-0.32544	-3.09175	3.10884	170	-0.23503	-2.23287	2.24520
85	-0.32772	-3.11343	3.13062	175	-0.23297	-2.21328	2.22550
				180	-0.23228	-2.20669	2.21888

VBR

RADIATION PATTERN FOR THETA=PI/2

H/LMDA=0.2500 B/LMDA= 0.250 OMEGA= 8.53

3 ELEMENT ARRAY

PHI DEG	ER(THETA,PHI) RE	IM	ABSVAL	PHI DEG	ER(THETA,PHI) RE	IM	ABSVAL
0	0.57451	-0.85491	1.03002	90	0.22321	-3.05648	3.06462
5	0.57241	-0.86807	1.03981	95	0.22650	-3.03588	3.04432
10	0.56613	-0.90745	1.06956	100	0.23620	-2.97509	2.98445
15	0.55572	-0.97269	1.12024	105	0.25184	-2.87703	2.88803
20	0.54128	-1.06316	1.19302	110	0.27270	-2.74633	2.75984
25	0.52300	-1.17775	1.28865	115	0.29782	-2.58893	2.60600
30	0.50112	-1.31481	1.40707	120	0.32610	-2.41166	2.43361
35	0.47605	-1.47195	1.54701	125	0.35640	-2.22176	2.25016
40	0.44829	-1.64589	1.70585	130	0.38757	-2.02642	2.06315
45	0.41853	[illegible]	[illegible]	135	0.41853	-1.83245	1.87963
50	0.38757	-2.02642	2.06315	140	0.44829	[illegible]	[illegible]
55	0.35641	-2.22175	2.25016	145	0.47605	-1.47195	1.54701
60	0.32610	-2.41166	2.43360	150	0.50112	-1.31481	1.40708
65	0.29782	-2.58893	2.60600	155	0.52299	-1.17775	1.28865
70	0.27270	-2.74633	2.75984	160	0.54128	-1.06316	1.19302
75	0.25184	-2.87703	2.88803	165	0.55572	-0.97269	1.12025
80	0.23620	-2.97509	2.98445	170	0.56613	-0.90745	1.06956
85	0.22650	-3.03588	3.04432	175	0.57241	-0.86807	1.03981
				180	0.57451	-0.85491	1.03002

IBR

RADIATION PATTERN FOR THETA=PI/2

H/LMDA=0.2500 B/LMDA= 0.250 OMEGA= 8.53

3 ELEMENT ARRAY

PHI DEG	ER(THETA,PHI) RE	IM	ABSVAL	PHI DEG	ER(THETA,PHI) RE	IM	ABSVAL
0	0.15947	-1.16657	1.17742	90	0.63788	-3.46813	3.52630
5	0.16233	-1.18033	1.19144	95	0.63340	-3.44659	3.50431
10	0.17089	-1.22149	1.23339	100	0.62019	-3.38304	3.43941
15	0.18507	-1.28970	1.30291	105	0.59888	-3.28053	3.33474
20	0.20472	-1.38427	1.39933	110	0.57048	-3.14389	3.19523
25	0.22963	-1.50407	1.52150	115	0.53628	-2.97934	3.02722
30	0.25941	-1.64736	1.66766	120	0.49776	-2.79402	2.83801
35	0.29356	-1.81163	1.83526	125	0.45649	-2.59549	2.63533
40	0.33135	-1.99347	2.02082	130	0.41404	-2.39129	2.42687
45	0.37189	-2.18850	2.21987	135	0.37189	-2.18850	2.21987
50	0.41404	-2.39129	2.42686	140	0.33135	-1.99347	2.02082
55	0.45649	[illegible]	[illegible]	145	0.29356	-1.81163	1.83526
60	0.49776	-2.79402	2.83801	150	0.25941	[illegible]	[illegible]
65	0.53628	-2.97934	3.02722	155	0.22963	-1.50408	1.52150
70	0.57048	-3.14389	3.19523	160	0.20472	-1.38428	1.39933
75	0.59888	-3.28052	3.33474	165	0.18507	-1.28970	1.30291
80	0.62019	-3.38304	3.43941	170	0.17089	-1.22149	1.23339
85	0.63340	-3.44659	3.50431	175	0.16233	-1.18033	1.19144
				180	0.15947	-1.16657	1.17742

VBR

RADIATION PATTERN FOR THETA=PI/2

H/LMDA=0.2500 B/LMDA= 0.250 OMEGA= 8.53

4 ELEMENT ARRAY

PHI DEG	ER(THETA,PHI) RE	IM	ABSVAL	PHI DEG	ER(THETA,PHI) RE	IM	ABSVAL
0	0.74618	0.15186	0.76147	90	0.44808	-3.01738	3.05047
5	0.74583	0.13861	0.75860	95	0.45269	-2.98015	3.01433
10	0.74466	0.09848	0.75115	100	0.46615	-2.87098	2.90858
15	0.74230	0.03037	0.74292	105	0.48732	-2.69723	2.74090
20	0.73818	-0.06730	0.74124	110	0.51448	-2.47017	2.52317
25	0.73160	-0.19626	0.75747	115	0.54553	-2.20374	2.27026
30	0.72181	-0.35796	0.80570	120	0.57827	-1.91309	1.99858
35	0.70814	-0.55295	0.89845	125	0.61062	-1.61315	1.72485
40	0.69011	-0.78028	1.04168	130	0.64082	-1.31738	1.46497
45	0.66758	-1.03695	1.23326	135	0.66758	-1.03696	1.23326
50	0.64082	-1.31738	1.46497	140	0.69011	-0.78029	1.04168
55	0.61062	-1.61314	1.72484	145	0.70814	-0.55295	0.89845
60	0.57827	-1.91309	1.99857	150	0.72181	-0.35796	0.80570
65	0.54553	-2.20373	2.27025	155	0.73160	-0.19626	0.75747
70	0.51448	-2.47017	2.52317	160	0.73818	-0.06730	0.74124
75	0.48732	-2.69723	2.74090	165	0.74230	0.03037	0.74292
80	0.46615	-2.87098	2.90858	170	0.74466	0.09848	0.75115
85	0.45269	-2.98014	3.01433	175	0.74583	0.13861	0.75860
				180	0.74618	0.15186	0.76147

IBR

RADIATION PATTERN FOR THETA=PI/2

H/LMDA=0.2500 B/LMDA= 0.250 OMEGA= 8.53

4 ELEMENT ARRAY

PHI DEG	ER(THETA,PHI) RE	IM	ABSVAL	PHI DEG	ER(THETA,PHI) RE	IM	ABSVAL
0	-0.14296	0.07198	0.16006	90	0.49323	-4.69309	4.71894
5	-0.14044	0.05187	0.14972	95	0.48559	-4.63734	4.66270
10	-0.13279	-0.00903	0.13309	100	0.46318	-4.47389	4.49780
15	-0.11974	-0.11229	0.16415	105	0.42756	-4.21368	4.23532
20	-0.10091	-0.26022	0.27910	110	0.38109	-3.87352	3.89222
25	-0.07587	-0.45532	0.46159	115	0.32669	-3.47422	3.48954
30	-0.04425	-0.69959	0.70099	120	0.26751	-3.03837	3.05012
35	-0.00582	-0.99376	0.99378	125	0.20666	-2.58828	2.59651
40	0.03930	-1.33629	1.33686	130	0.14693	-2.14409	2.14911
45	0.09057	-1.72254	1.72492	135	0.09057	-1.72255	1.72493
50	0.14692	-2.14408	2.14911	140	0.03930	-1.33629	1.33686
55	0.20666	-2.58827	2.59651	145	-0.00582	-0.99376	0.99378
60	0.26751	-3.03836	3.05012	150	-0.04425	-0.69960	0.70099
65	0.32669	-3.47421	3.48953	155	-0.07587	-0.45532	0.46160
70	0.38109	-3.87352	3.89222	160	-0.10091	-0.26022	0.27910
75	0.42756	-4.21368	4.23531	165	-0.11974	-0.11229	0.16416
80	0.46318	-4.47389	4.49780	170	-0.13279	-0.00903	0.13309
85	0.48559	-4.63734	4.66269	175	-0.14044	0.05187	0.14972
				180	-0.14296	0.07198	0.16006

VBR

RADIATION PATTERN FOR THETA=PI/2

H/LMDA=0.2500 B/LMDA= 0.250 OMEGA= 8.53

5 ELEMENT ARRAY

PHI	ER(THETA,PHI)			PHI	ER(THETA,PHI)		
DEG	RE	IM	ABSVAL	DEG	RE	IM	ABSVAL
0	0.48548	0.66613	0.82427	90	0.58651	-2.97004	3.02740
5	0.48878	0.65952	0.82090	95	0.58977	-2.91241	2.97153
10	0.49844	0.63868	0.81016	100	0.59891	-2.74494	2.80951
15	0.51382	0.60056	0.79037	105	0.61219	-2.48299	2.55734
20	0.53383	0.54030	0.75954	110	0.62710	-2.14954	2.23915
25	0.55698	0.45162	0.71707	115	0.64083	-1.77181	1.88414
30	0.58144	0.32744	0.66730	120	0.65074	-1.37756	1.52353
35	0.60517	0.16077	0.62616	125	0.65484	-0.99187	1.18853
40	0.62605	-0.05399	0.62838	130	0.65202	-0.63468	0.90992
45	0.64217	-0.31966	0.71733	135	0.64217	-0.31966	0.71733
50	0.65202	-0.63468	0.90992	140	0.62605	-0.05399	0.62838
55	0.65484	-0.99186	1.18853	145	0.60517	0.16077	0.62616
60	0.65074	-1.37756	1.52352	150	0.58144	0.32743	0.66730
65	0.64083	-1.77180	1.88413	155	0.55698	0.45162	0.71707
70	0.62710	-2.14954	2.23914	160	0.53383	0.54030	0.75954
75	0.61219	-2.48298	2.55734	165	0.51382	0.60056	0.79037
80	0.59891	-2.74493	2.80951	170	0.49844	0.63868	0.81016
85	0.58977	-2.91241	2.97153	175	0.48878	0.65952	0.82090
				180	0.48548	0.66613	0.82427

IBR

RADIATION PATTERN FOR THETA=PI/2

H/LMDA=0.2500 B/LMDA= 0.250 OMEGA= 8.53

5 ELEMENT ARRAY

PHI	ER(THETA,PHI)			PHI	ER(THETA,PHI)		
DEG	RE	IM	ABSVAL	DEG	RE	IM	ABSVAL
0	0.08851	0.97004	0.97407	90	-0.63587	-4.12412	4.17286
5	0.08686	0.96022	0.96414	95	-0.62491	-4.04427	4.09227
10	0.08173	0.92937	0.93296	100	-0.59303	-3.81215	3.85800
15	0.07260	0.87336	0.87637	105	-0.54306	-3.44890	3.49140
20	0.05862	0.78561	0.78779	110	-0.47923	-2.98615	3.02436
25	0.03873	0.65761	0.65875	115	-0.40659	-2.46137	2.49472
30	0.01173	0.47980	0.47995	120	-0.33029	-1.91283	1.94114
35	-0.02353	0.24284	0.24397	125	-0.25505	-1.37518	1.39863
40	-0.06792	-0.06075	0.09112	130	-0.18466	-0.87607	0.89532
45	-0.12178	-0.43453	0.45127	135	-0.12178	-0.43453	0.45127
50	-0.18466	-0.87607	0.89531	140	-0.06792	-0.06075	0.09112
55	-0.25505	-1.37517	1.39862	145	-0.02353	0.24283	0.24397
60	-0.33029	-1.91283	1.94113	150	0.01173	0.47980	0.47994
65	-0.40659	-2.46136	2.49471	155	0.03873	0.65761	0.65875
70	-0.47923	-2.98615	3.02436	160	0.05862	0.78561	0.78779
75	-0.54306	-3.44890	3.49139	165	0.07260	0.87336	0.87637
80	-0.59303	-3.81214	3.85799	170	0.08173	0.92937	0.93296
85	-0.62491	-4.04427	4.09226	175	0.08686	0.96022	0.96414
				180	0.08851	0.97004	0.97407

VBR

RADIATION PATTERN FOR THETA=PI/2

H/LMDA=0.2500 B/LMDA= 0.250 OMEGA= 8.53

6 ELEMENT ARRAY

PHI	ER(THETA,PHI)			PHI	ER(THETA,PHI)		
DEG	RE	IM	ABSVAL	DEG	RE	IM	ABSVAL
0	0.06489	0.68279	0.68587	90	0.69244	-2.94032	3.02076
2	0.06581	0.68337	0.68653	92	0.69234	-2.92693	3.00770
4	0.06856	0.68507	0.68849	94	0.69203	-2.88703	2.96881
6	0.07314	0.68780	0.69168	96	0.69144	-2.82143	2.90492
8	0.07954	0.69138	0.69594	98	0.69049	-2.73145	2.81738
10	0.08775	0.69558	0.70109	100	0.68905	-2.61889	2.70802
12	0.09774	0.70008	0.70687	102	0.68696	-2.48595	2.57912
14	0.10949	0.70450	0.71296	104	0.68408	-2.33516	2.43329
16	0.12297	0.70840	0.71899	106	0.68021	-2.16933	2.27348
18	0.13812	0.71122	0.72451	108	0.67519	-1.99145	2.10280
20	0.15490	0.71237	0.72901	110	0.66885	-1.80460	1.92457
22	0.17323	0.71116	0.73195	112	0.66104	-1.61188	1.74216
24	0.19302	0.70683	0.73271	114	0.65163	-1.41629	1.55901
26	0.21419	0.69856	0.73066	116	0.64053	-1.22073	1.37857
28	0.23659	0.68547	0.72515	118	0.62766	-1.02786	1.20435
30	0.26011	0.66662	0.71556	120	0.61301	-0.84011	1.03999
32	0.28457	0.64104	0.70136	122	0.59659	-0.65960	0.88938
34	0.30982	0.60774	0.68216	124	0.57845	-0.48813	0.75688
36	0.33566	0.56575	0.65783	126	0.55867	-0.32714	0.64741
38	0.36189	0.51410	0.62870	128	0.53739	-0.17776	0.56603
40	0.38828	0.45190	0.59580	130	0.51476	-0.04077	0.51637
42	0.41462	0.37833	0.56128	132	0.49096	0.08337	0.49798
44	0.44066	0.29270	0.52901	134	0.46618	0.19449	0.50513
46	0.46618	0.19449	0.50513	136	0.44066	0.29270	0.52901
48	0.49096	0.08337	0.49798	138	0.41462	0.37832	0.56128
50	0.51476	-0.04077	0.51637	140	0.38828	0.45190	0.59579
52	0.53739	-0.17776	0.56603	142	0.36189	0.51410	0.62870
54	0.55867	-0.32714	0.64741	144	0.33566	0.56575	0.65783
56	0.57845	-0.48812	0.75688	146	0.30982	0.60774	0.68216
58	0.59659	-0.65960	0.88938	148	0.28457	0.64104	0.70136
60	0.61301	-0.84011	1.03999	150	0.26011	0.66662	0.71556
62	0.62766	-1.02786	1.20435	152	0.23659	0.68547	0.72515
64	0.64053	-1.22072	1.37856	154	0.21419	0.69856	0.73066
66	0.65163	-1.41628	1.55900	156	0.19302	0.70683	0.73271
68	0.66104	-1.61187	1.74216	158	0.17323	0.71116	0.73195
70	0.66885	-1.80460	1.92456	160	0.15490	0.71237	0.72901
72	0.67519	-1.99145	2.10279	162	0.13812	0.71122	0.72451
74	0.68021	-2.16933	2.27347	164	0.12297	0.70840	0.71899
76	0.68408	-2.33515	2.43329	166	0.10949	0.70450	0.71296
78	0.68696	-2.48594	2.57911	168	0.09774	0.70008	0.70687
80	0.68905	-2.61889	2.70802	170	0.08775	0.69558	0.70109
82	0.69049	-2.73145	2.81738	172	0.07954	0.69138	0.69594
84	0.69144	-2.82143	2.90492	174	0.07314	0.68780	0.69168
86	0.69203	-2.88703	2.96881	176	0.06856	0.68507	0.68849
88	0.69234	-2.92693	3.00770	178	0.06581	0.68337	0.68653
				180	0.06489	0.68279	0.68587

IBR

RADIATION PATTERN FOR THETA=PI/2

H/LMDA=0.2500 B/LMDA= 0.250 OMEGA= 8.53

6 ELEMENT ARRAY

PHI	ER(THETA,PHI)			PHI	ER(THETA,PHI)		
DEG	RE	IM	ABSVAL	DEG	RE	IM	ABSVAL
0	0.06752	0.90129	0.90382	90	-0.18496	-3.47253	3.47745
2	0.06751	0.90177	0.90429	92	-0.18414	-3.45687	3.46177
4	0.06748	0.90316	0.90568	94	-0.18170	-3.41021	3.41505
6	0.06742	0.90535	0.90786	96	-0.17768	-3.33348	3.33821
8	0.06733	0.90814	0.91063	98	-0.17217	-3.22819	3.23278
10	0.06718	0.91125	0.91372	100	-0.16525	-3.09642	3.10083
12	0.06696	0.91433	0.91677	102	-0.15706	-2.94070	2.94489
14	0.06665	0.91692	0.91934	104	-0.14775	-2.76395	2.76789
16	0.06622	0.91851	0.92089	106	-0.13747	-2.56941	2.57308
18	0.06564	0.91847	0.92081	108	-0.12640	-2.36052	2.36390
20	0.06488	0.91611	0.91840	110	-0.11471	-2.14085	2.14392
22	0.06389	0.91063	0.91287	112	-0.10258	-1.91397	1.91671
24	0.06263	0.90117	0.90335	114	-0.09020	-1.68337	1.68579
26	0.06107	0.88679	0.88889	116	-0.07772	-1.45240	1.45448
28	0.05914	0.86649	0.86850	118	-0.06532	-1.22417	1.22591
30	0.05680	0.83919	0.84111	120	-0.05313	-1.00150	1.00291
32	0.05400	0.80380	0.80561	122	-0.04129	-0.78687	0.78795
34	0.05068	0.75920	0.76089	124	-0.02991	-0.58239	0.58316
36	0.04681	0.70429	0.70585	126	-0.01908	-0.38978	0.39025
38	0.04232	0.63800	0.63940	128	-0.00888	-0.21039	0.21058
40	0.03717	0.55932	0.56056	130	0.00063	-0.04517	0.04517
42	0.03133	0.46737	0.46842	132	0.00942	0.10532	0.10574
44	0.02477	0.36139	0.36223	134	0.01747	0.24081	0.24144
46	0.01747	0.24081	0.24145	136	0.02477	0.36139	0.36223
48	0.00942	0.10532	0.10574	138	0.03133	0.46737	0.46842
50	0.00063	-0.04516	0.04517	140	0.03717	0.55932	0.56056
52	-0.00888	-0.21039	0.21058	142	0.04232	0.63800	0.63940
54	-0.01908	-0.38978	0.39025	144	0.04681	0.70429	0.70584
56	-0.02991	-0.58238	0.58315	146	0.05068	0.75920	0.76089
58	-0.04129	-0.78686	0.78795	148	0.05400	0.80380	0.80561
60	-0.05313	-1.00149	1.00290	150	0.05680	0.83919	0.84111
62	-0.06532	-1.22416	1.22591	152	0.05914	0.86648	0.86850
64	-0.07772	-1.45239	1.45447	154	0.06107	0.88679	0.88889
66	-0.09019	-1.68336	1.68577	156	0.06263	0.90117	0.90335
68	-0.10258	-1.91396	1.91671	158	0.06389	0.91063	0.91287
70	-0.11470	-2.14085	2.14392	160	0.06488	0.91611	0.91840
72	-0.12639	-2.36052	2.36390	162	0.06564	0.91847	0.92081
74	-0.13747	-2.56940	2.57307	164	0.06622	0.91851	0.92089
76	-0.14775	-2.76394	2.76788	166	0.06665	0.91692	0.91934
78	-0.15706	-2.94069	2.94488	168	0.06696	0.91433	0.91677
80	-0.16525	-3.09642	3.10082	170	0.06718	0.91125	0.91372
82	-0.17217	-3.22819	3.23278	172	0.06733	0.90814	0.91063
84	-0.17768	-3.33347	3.33821	174	0.06742	0.90535	0.90786
86	-0.18170	-3.41021	3.41504	176	0.06748	0.90316	0.90568
88	-0.18414	-3.45687	3.46177	178	0.06751	0.90177	0.90429
				180	0.06752	0.90129	0.90382

VBR

RADIATION PATTERN FOR THETA=PI/2

H/LMDA=0.2500 B/LMDA= 0.250 OMEGA= 8.53

7 ELEMENT ARRAY

PHI DEG	ER(THETA,PHI) RE	IM	ABSVAL	PHI DEG	ER(THETA,PHI) RE	IM	ABSVAL
0	-0.27411	0.35409	0.44779	90	0.76686	-2.92794	3.02670
2	-0.27332	0.35596	0.44879	92	0.76592	-2.90962	3.00875
4	-0.27094	0.36153	0.45179	94	0.76308	-2.85515	2.95536
6	-0.26695	0.37076	0.45686	96	0.75828	-2.76595	2.86801
8	-0.26131	0.38353	0.46409	98	0.75142	-2.64435	2.74904
10	-0.25397	0.39970	0.47356	100	0.74238	-2.49348	2.60164
12	-0.24485	0.41906	0.48535	102	0.73102	-2.31711	2.42969
14	-0.23388	0.44134	0.49948	104	0.71721	-2.11958	2.23763
16	-0.22098	0.46618	0.51590	106	0.70081	-1.90556	2.03035
18	-0.20607	0.49315	0.53447	108	0.68171	-1.67995	1.81300
20	-0.18907	0.52170	0.55491	110	0.65984	-1.44765	1.59094
22	-0.16990	0.55119	0.57678	112	0.63517	-1.21346	1.36965
24	-0.14851	0.58083	0.59951	114	0.60773	-0.98187	1.15473
26	-0.12483	0.60971	0.62236	116	0.57758	-0.75699	0.95218
28	-0.09886	0.63678	0.64441	118	0.54489	-0.54245	0.76887
30	-0.07060	0.66085	0.66461	120	0.50984	-0.34127	0.61352
32	-0.04009	0.68058	0.68175	122	0.47270	-0.15591	0.49775
34	-0.00739	0.69450	0.69454	124	0.43376	0.01185	0.43392
36	0.02737	0.70105	0.70158	126	0.39338	0.16080	0.42497
38	0.06403	0.69856	0.70148	128	0.35192	0.29033	0.45622
40	0.10239	0.68531	0.69291	130	0.30977	0.40038	0.50623
42	0.14221	0.65958	0.67474	132	0.26735	0.49137	0.55939
44	0.18319	0.61970	0.64621	134	0.22503	0.56409	0.60732
46	0.22503	0.56409	0.60732	136	0.18319	0.61970	0.64621
48	0.26735	0.49137	0.55939	138	0.14221	0.65958	0.67474
50	0.30977	0.40038	0.50623	140	0.10239	0.68531	0.69292
52	0.35192	0.29033	0.45622	142	0.06403	0.69856	0.70148
54	0.39338	0.16080	0.42497	144	0.02737	0.70105	0.70158
56	0.43376	0.01185	0.43392	146	-0.00739	0.69450	0.69454
58	0.47270	-0.15590	0.49774	148	-0.04009	0.68058	0.68176
60	0.50984	-0.34127	0.61352	150	-0.07060	0.66085	0.66461
62	0.54489	-0.54244	0.76886	152	-0.09886	0.63678	0.64441
64	0.57758	-0.75699	0.95217	154	-0.12483	0.60971	0.62236
66	0.60773	-0.98186	1.15472	156	-0.14851	0.58083	0.59951
68	0.63517	-1.21346	1.36964	158	-0.16990	0.55119	0.57678
70	0.65984	-1.44765	1.59094	160	-0.18907	0.52170	0.55491
72	0.68171	-1.67994	1.81299	162	-0.20607	0.49315	0.53447
74	0.70081	-1.90555	2.03034	164	-0.22098	0.46618	0.51590
76	0.71721	-2.11957	2.23762	166	-0.23388	0.44134	0.49948
78	0.73102	-2.31711	2.42969	168	-0.24485	0.41906	0.48535
80	0.74238	-2.49347	2.60164	170	-0.25397	0.39970	0.47356
82	0.75142	-2.64435	2.74904	172	-0.26131	0.38353	0.46409
84	0.75828	-2.76595	2.86801	174	-0.26695	0.37076	0.45686
86	0.76308	-2.85515	2.95536	176	-0.27094	0.36153	0.45179
88	0.76592	-2.90962	3.00874	178	-0.27332	0.35596	0.44879
				180	-0.27411	0.35409	0.44779

IBR

RADIATION PATTERN FOR THETA=PI/2

H/LMDA=0.2500 B/LMDA= 0.250 OMEGA= 8.53

7 ELEMENT ARRAY

PHI DEG	ER(THETA,PHI) RE	IM	ABSVAL	PHI DEG	ER(THETA,PHI) RE	IM	ABSVAL
0	0.00703	0.55450	0.55455	90	0.37965	-3.65648	3.67614
2	0.00676	0.55660	0.55664	92	0.37733	-3.63413	3.65367
4	0.00596	0.56286	0.56290	94	0.37046	-3.56764	3.58682
6	0.00463	0.57322	0.57323	96	0.35921	-3.45874	3.47734
8	0.00278	0.58752	0.58753	98	0.34390	-3.31020	3.32801
10	0.00044	0.60558	0.60558	100	0.32492	-3.12575	3.14259
12	-0.00239	0.62711	0.62711	102	0.30278	-2.90995	2.92566
14	-0.00567	0.65177	0.65179	104	0.27804	-2.66797	2.68242
16	-0.00936	0.67909	0.67916	106	0.25131	-2.40544	2.41853
18	-0.01342	0.70854	0.70866	108	0.22323	-2.12822	2.13989
20	-0.01778	0.73941	0.73962	110	0.19444	-1.84224	1.85247
22	-0.02236	0.77089	0.77121	112	0.16555	-1.55327	1.56206
24	-0.02707	0.80202	0.80248	114	0.13714	-1.26674	1.27415
26	-0.03180	0.83168	0.83228	116	0.10974	-0.98765	0.99373
28	-0.03642	0.85857	0.85934	118	0.08379	-0.72041	0.72527
30	-0.04078	0.88125	0.88219	120	0.05969	-0.46874	0.47253
32	-0.04470	0.89811	0.89922	122	0.03773	-0.23566	0.23866
34	-0.04800	0.90739	0.90866	124	0.01813	-0.02343	0.02962
36	-0.05046	0.90720	0.90860	126	0.00102	0.16643	0.16643
38	-0.05188	0.89558	0.89708	128	-0.01355	0.33307	0.33335
40	-0.05201	0.87050	0.87205	130	-0.02559	0.47631	0.47700
42	-0.05063	0.82993	0.83148	132	-0.03517	0.59655	0.59759
44	-0.04751	0.77193	0.77339	134	-0.04242	0.69466	0.69596
46	-0.04242	0.69466	0.69596	136	-0.04751	0.77193	0.77339
48	-0.03517	0.59655	0.59759	138	-0.05063	0.82993	0.83148
50	-0.02559	0.47632	0.47700	140	-0.05201	0.87050	0.87205
52	-0.01355	0.33307	0.33335	142	-0.05188	0.89558	0.89708
54	0.00102	0.16643	0.16643	144	-0.05046	0.90720	0.90861
56	0.01813	-0.02342	0.02962	146	-0.04800	0.90739	0.90866
58	0.03773	-0.23566	0.23866	148	-0.04470	0.89811	0.89922
60	0.05969	-0.46874	0.47253	150	-0.04078	0.88125	0.88219
62	0.08379	-0.72041	0.72527	152	-0.03642	0.85857	0.85934
64	0.10974	-0.98765	0.99373	154	-0.03180	0.83168	0.83229
66	0.13714	-1.26673	1.27413	156	-0.02707	0.80202	0.80248
68	0.16555	-1.55326	1.56206	158	-0.02236	0.77089	0.77122
70	0.19444	-1.84224	1.85247	160	-0.01778	0.73941	0.73962
72	0.22323	-2.12821	2.13989	162	-0.01342	0.70854	0.70866
74	0.25131	-2.40543	2.41852	164	-0.00936	0.67909	0.67916
76	0.27804	-2.66796	2.68241	166	-0.00567	0.65177	0.65179
78	0.30278	-2.90994	2.92565	168	-0.00239	0.62711	0.62711
80	0.32492	-3.12575	3.14259	170	0.00044	0.60558	0.60558
82	0.34390	-3.31020	3.32801	172	0.00278	0.58752	0.58753
84	0.35921	-3.45874	3.47734	174	0.00463	0.57322	0.57323
86	0.37046	-3.56764	3.58682	176	0.00596	0.56286	0.56290
88	0.37733	-3.63413	3.65367	178	0.00676	0.55660	0.55664
				180	0.00703	0.55450	0.55455

VBR

RADIATION PATTERN FOR THETA=PI/2

H/LMDA=0.2500 B/LMDA= 0.250 OMEGA= 8.53

8 ELEMENT ARRAY

PHI DEG	ER(THETA,PHI) RE	IM	ABSVAL	PHI DEG	ER(THETA,PHI) RE	IM	ABSVAL
0	-0.38993	-0.06853	0.39591	90	0.81660	-2.91767	3.02979
2	-0.38979	-0.06645	0.39541	92	0.81462	-2.89379	3.00627
4	-0.38935	-0.06019	0.39398	94	0.80869	-2.82294	2.93649
6	-0.38856	-0.04973	0.39173	96	0.79874	-2.70745	2.82281
8	-0.38733	-0.03499	0.38890	98	0.78473	-2.55110	2.66907
10	-0.38553	-0.01591	0.38586	100	0.76658	-2.35892	2.48035
12	-0.38301	0.00757	0.38308	102	0.74423	-2.13694	2.26283
14	-0.37958	0.03552	0.38124	104	0.71765	-1.89193	2.02346
16	-0.37502	0.06795	0.38113	106	0.68685	-1.63107	1.76979
18	-0.36910	0.10481	0.38370	108	0.65191	-1.36169	1.50970
20	-0.36156	0.14599	0.38992	110	0.61297	-1.09094	1.25135
22	-0.35211	0.19125	0.40070	112	0.57028	-0.82551	1.00334
24	-0.34048	0.24021	0.41669	114	0.52416	-0.57143	0.77541
26	-0.32639	0.29234	0.43817	116	0.47501	-0.33385	0.58060
28	-0.30956	0.34687	0.46492	118	0.42334	-0.11701	0.43921
30	-0.28974	0.40284	0.49621	120	0.36970	0.07596	0.37742
32	-0.26670	0.45900	0.53086	122	0.31471	0.24296	0.39758
34	-0.24026	0.51385	0.56725	124	0.25901	0.38289	0.46227
36	-0.21029	0.56561	0.60344	126	0.20326	0.49563	0.53570
38	-0.17672	0.61222	0.63721	128	0.14812	0.58187	0.60042
40	-0.13956	0.65137	0.66615	130	0.09420	0.64298	0.64985
42	-0.09888	0.68053	0.68768	132	0.04206	0.68093	0.68223
44	-0.05486	0.69702	0.69918	134	-0.00777	0.69807	0.69812
46	-0.00777	0.69807	0.69812	136	-0.05486	0.69702	0.69918
48	0.04206	0.68093	0.68223	138	-0.09888	0.68053	0.68768
50	0.09419	0.64298	0.64985	140	-0.13956	0.65137	0.66615
52	0.14812	0.58187	0.60042	142	-0.17672	0.61222	0.63722
54	0.20326	0.49563	0.53570	144	-0.21029	0.56561	0.60344
56	0.25901	0.38290	0.46227	146	-0.24026	0.51385	0.56725
58	0.31471	0.24296	0.39758	148	-0.26670	0.45900	0.53086
60	0.36970	0.07597	0.37742	150	-0.28974	0.40284	0.49621
62	0.42334	-0.11701	0.43921	152	-0.30956	0.34687	0.46492
64	0.47501	-0.33385	0.58059	154	-0.32639	0.29234	0.43817
66	0.52415	-0.57141	0.77540	156	-0.34048	0.24021	0.41669
68	0.57028	-0.82551	1.00333	158	-0.35211	0.19125	0.40070
70	0.61297	-1.09094	1.25135	160	-0.36156	0.14599	0.38992
72	0.65191	-1.36169	1.50969	162	-0.36910	0.10481	0.38370
74	0.68685	-1.63106	1.76978	164	-0.37502	0.06795	0.38113
76	0.71765	-1.89191	2.02345	166	-0.37958	0.03552	0.38124
78	0.74423	-2.13694	2.26282	168	-0.38301	0.00758	0.38308
80	0.76658	-2.35892	2.48035	170	-0.38553	-0.01591	0.38586
82	0.78473	-2.55110	2.66907	172	-0.38733	-0.03499	0.38890
84	0.79874	-2.70745	2.82281	174	-0.38856	-0.04973	0.39173
86	0.80869	-2.82293	2.93648	176	-0.38935	-0.06019	0.39398
88	0.81462	-2.89379	3.00627	178	-0.38979	-0.06645	0.39541
				180	-0.38993	-0.06853	0.39591

IBR

RADIATION PATTERN FOR THETA=PI/2

H/LMDA=0.2500 B/LMDA= 0.250 OMEGA= 8.53

8 ELEMENT ARRAY

PHI DEG	ER(THETA,PHI) RE	IM	ABSVAL	PHI DEG	ER(THETA,PHI) RE	IM	ABSVAL
0	0.09369	-0.04607	0.10441	90	0.25024	-4.30468	4.31194
2	0.09349	-0.04304	0.10292	92	0.24807	-4.27023	4.27743
4	0.09288	-0.03395	0.09889	94	0.24165	-4.16803	4.17503
6	0.09185	-0.01875	0.09375	96	0.23122	-4.00141	4.00808
8	0.09039	0.00263	0.09043	98	0.21715	-3.77571	3.78195
10	0.08847	0.03024	0.09349	100	0.19997	-3.49812	3.50383
12	0.08606	0.06416	0.10735	102	0.18028	-3.17722	3.18233
14	0.08314	0.10442	0.13348	104	0.15876	-2.82266	2.82712
16	0.07967	0.15099	0.17072	106	0.13614	-2.44472	2.44851
18	0.07564	0.20375	0.21734	108	0.11313	-2.05384	2.05696
20	0.07101	0.26247	0.27190	110	0.09043	-1.66027	1.66273
22	0.06578	0.32675	0.33330	112	0.06869	-1.27361	1.27546
24	0.05995	0.39598	0.40049	114	0.04846	-0.90252	0.90382
26	0.05354	0.46931	0.47235	116	0.03021	-0.55446	0.55529
28	0.04658	0.54561	0.54759	118	0.01432	-0.23556	0.23600
30	0.03914	0.62340	0.62463	120	0.00102	0.04958	0.04959
32	0.03131	0.70087	0.70157	122	-0.00954	0.29783	0.29798
34	0.02323	0.77581	0.77615	124	-0.01734	0.50751	0.50781
36	0.01503	0.84563	0.84576	126	-0.02243	0.67829	0.67866
38	0.00693	0.90737	0.90740	128	-0.02497	0.81103	0.81141
40	-0.00085	0.95771	0.95771	130	-0.02516	0.90757	0.90792
42	-0.00805	0.99306	0.99309	132	-0.02326	0.97061	0.97089
44	-0.01439	1.00959	1.00969	134	-0.01957	1.00339	1.00358
46	-0.01957	1.00339	1.00358	136	-0.01439	1.00959	1.00969
48	-0.02326	0.97061	0.97089	138	-0.00805	0.99306	0.99309
50	-0.02516	0.90758	0.90793	140	-0.00085	0.95771	0.95772
52	-0.02497	0.81103	0.81141	142	0.00693	0.90737	0.90740
54	-0.02243	0.67829	0.67866	144	0.01503	0.84563	0.84577
56	-0.01734	0.50751	0.50781	146	0.02323	0.77581	0.77615
58	-0.00954	0.29783	0.29798	148	0.03131	0.70087	0.70157
60	0.00102	0.04958	0.04959	150	0.03914	0.62340	0.62463
62	0.01432	-0.23556	0.23599	152	0.04658	0.54561	0.54759
64	0.03021	-0.55446	0.55528	154	0.05354	0.46931	0.47236
66	0.04846	-0.90250	0.90380	156	0.05995	0.39598	0.40049
68	0.06869	-1.27360	1.27545	158	0.06578	0.32675	0.33330
70	0.09043	-1.66027	1.66273	160	0.07101	0.26247	0.27191
72	0.11313	-2.05384	2.05695	162	0.07564	0.20375	0.21734
74	0.13614	-2.44470	2.44849	164	0.07967	0.15099	0.17072
76	0.15876	-2.82265	2.82711	166	0.08314	0.10442	0.13348
78	0.18028	-3.17721	3.18232	168	0.08606	0.06417	0.10735
80	0.19997	-3.49811	3.50382	170	0.08847	0.03025	0.09349
82	0.21715	-3.77571	3.78195	172	0.09039	0.00263	0.09043
84	0.23122	-4.00140	4.00807	174	0.09185	-0.01875	0.09375
86	0.24165	-4.16802	4.17502	176	0.09288	-0.03395	0.09889
88	0.24807	-4.27023	4.27743	178	0.09349	-0.04304	0.10292
				180	0.09369	-0.04607	0.10441

VBR

RADIATION PATTERN FOR THETA=PI/2

H/LMDA=0.2500 B/LMDA= 0.250 OMEGA= 8.53

9 ELEMENT ARRAY

PHI DEG	ER(THETA,PHI) RE	IM	ABSVAL	PHI DEG	ER(THETA,PHI) RE	IM	ABSVAL
0	-0.27444	-0.35831	0.45134	90	0.85604	-2.90534	3.02882
2	-0.27511	-0.35719	0.45085	92	0.85268	-2.87529	2.99905
4	-0.27708	-0.35377	0.44936	94	0.84260	-2.78638	2.91099
6	-0.28031	-0.34793	0.44680	96	0.82585	-2.64222	2.76827
8	-0.28471	-0.33942	0.44302	98	0.80248	-2.44859	2.57674
10	-0.29014	-0.32795	0.43787	100	0.77263	-2.21314	2.34413
12	-0.29643	-0.31313	0.43119	102	0.73648	-1.94490	2.07967
14	-0.30338	-0.29452	0.42283	104	0.69430	-1.65382	1.79365
16	-0.31072	-0.27163	0.41271	106	0.64648	-1.35026	1.49704
18	-0.31816	-0.24396	0.40093	108	0.59351	-1.04442	1.20128
20	-0.32535	-0.21102	0.38779	110	0.53600	-0.74594	0.91854
22	-0.33189	-0.17235	0.37397	112	0.47468	-0.46340	0.66338
24	-0.33735	-0.12761	0.36067	114	0.41041	-0.20407	0.45835
26	-0.34125	-0.07655	0.34973	116	0.34411	0.02635	0.34512
28	-0.34310	-0.01916	0.34363	118	0.27678	0.22382	0.35595
30	-0.34236	0.04435	0.34522	120	0.20943	0.38598	0.43914
32	-0.33850	0.11345	0.35701	122	0.14309	0.51206	0.53168
34	-0.33100	0.18723	0.38028	124	0.07875	0.60273	0.60785
36	-0.31934	0.26436	0.41457	126	0.01729	0.65989	0.66012
38	-0.30308	0.34304	0.45775	128	-0.04046	0.68640	0.68759
40	-0.28181	0.42097	0.50659	130	-0.09384	0.68585	0.69224
42	-0.25524	0.49535	0.55724	132	-0.14231	0.66226	0.67738
44	-0.22315	0.56288	0.60550	134	-0.18548	0.61986	0.64702
46	-0.18548	0.61986	0.64702	136	-0.22315	0.56288	0.60550
48	-0.14231	0.66226	0.67738	138	-0.25524	0.49535	0.55724
50	-0.09384	0.68585	0.69224	140	-0.28181	0.42098	0.50660
52	-0.04046	0.68640	0.68760	142	-0.30308	0.34305	0.45775
54	0.01729	0.65989	0.66012	144	-0.31934	0.26436	0.41457
56	0.07874	0.60273	0.60786	146	-0.33100	0.18723	0.38028
58	0.14309	0.51206	0.53168	148	-0.33850	0.11345	0.35701
60	0.20943	0.38598	0.43914	150	-0.34236	0.04435	0.34522
62	0.27678	0.22383	0.35595	152	-0.34310	-0.01916	0.34363
64	0.34411	0.02636	0.34512	154	-0.34125	-0.07655	0.34973
66	0.41041	-0.20406	0.45834	156	-0.33735	-0.12761	0.36067
68	0.47468	-0.46340	0.66337	158	-0.33189	-0.17235	0.37397
70	0.53599	-0.74593	0.91854	160	-0.32535	-0.21102	0.38779
72	0.59351	-1.04442	1.20127	162	-0.31816	-0.24396	0.40093
74	0.64648	-1.35024	1.49703	164	-0.31072	-0.27163	0.41271
76	0.69430	-1.65381	1.79364	166	-0.30338	-0.29452	0.42283
78	0.73648	-1.94489	2.07966	168	-0.29643	-0.31313	0.43119
80	0.77263	-2.21313	2.34413	170	-0.29014	-0.32795	0.43787
82	0.80248	-2.44859	2.57673	172	-0.28471	-0.33942	0.44302
84	0.82585	-2.64221	2.76827	174	-0.28031	-0.34793	0.44680
86	0.84260	-2.78637	2.91099	176	-0.27708	-0.35377	0.44936
88	0.85268	-2.87529	2.99905	178	-0.27511	-0.35719	0.45085
				180	-0.27444	-0.35831	0.45134

IBR

RADIATION PATTERN FOR THETA=PI/2

H/LMDA=0.2500 B/LMDA= 0.250 OMEGA= 8.53

9 ELEMENT ARRAY

PHI DEG	ER(THETA,PHI) RE	IM	ABSVAL	PHI DEG	ER(THETA,PHI) RE	IM	ABSVAL
0	-0.00736	-0.55052	0.55057	90	-0.37291	-4.03704	4.05423
2	-0.00714	-0.54880	0.54884	92	-0.36927	-3.99588	4.01290
4	-0.00648	-0.54357	0.54360	94	-0.35849	-3.87407	3.89062
6	-0.00536	-0.53464	0.53467	96	-0.34097	-3.67652	3.69230
8	-0.00378	-0.52173	0.52174	98	-0.31737	-3.41111	3.42584
10	-0.00170	-0.50441	0.50441	100	-0.28854	-3.08825	3.10170
12	0.00089	-0.48218	0.48218	102	-0.25551	-2.72023	2.73221
14	0.00402	-0.45445	0.45447	104	-0.21941	-2.32065	2.33100
16	0.00774	-0.42061	0.42068	106	-0.18143	-1.90362	1.91225
18	0.01205	-0.37998	0.38017	108	-0.14275	-1.48309	1.48994
20	0.01698	-0.33195	0.33239	110	-0.10450	-1.07223	1.07731
22	0.02255	-0.27595	0.27686	112	-0.06772	-0.68282	0.68617
24	0.02874	-0.21153	0.21348	114	-0.03328	-0.32484	0.32654
26	0.03554	-0.13847	0.14296	116	-0.00191	-0.00615	0.00644
28	0.04290	-0.05679	0.07117	118	0.02584	0.26761	0.26885
30	0.05073	0.03313	0.06059	120	0.04962	0.49311	0.49560
32	0.05893	0.13049	0.14318	122	0.06924	0.66919	0.67277
34	0.06733	0.23397	0.24346	124	0.08468	0.79666	0.80115
36	0.07573	0.34166	0.34995	126	0.09608	0.87800	0.88324
38	0.08389	0.45104	0.45877	128	0.10368	0.91703	0.92287
40	0.09150	0.55888	0.56632	130	0.10782	0.91855	0.92486
42	0.09821	0.66129	0.66855	132	0.10889	0.88802	0.89467
44	0.10364	0.75376	0.76085	134	0.10735	0.83117	0.83807
46	0.10735	0.83117	0.83807	136	0.10364	0.75376	0.76085
48	0.10889	0.88802	0.89467	138	0.09821	0.66130	0.66855
50	0.10782	0.91855	0.92486	140	0.09150	0.55888	0.56632
52	0.10368	0.91703	0.92287	142	0.08389	0.45104	0.45878
54	0.09608	0.87800	0.88324	144	0.07573	0.34167	0.34996
56	0.08468	0.79667	0.80115	146	0.06733	0.23397	0.24346
58	0.06924	0.66920	0.67277	148	0.05893	0.13049	0.14318
60	0.04962	0.49311	0.49560	150	0.05073	0.03313	0.06059
62	0.02584	0.26761	0.26886	152	0.04290	-0.05679	0.07117
64	-0.00191	-0.00614	0.00643	154	0.03554	-0.13847	0.14296
66	-0.03328	-0.32482	0.32652	156	0.02874	-0.21153	0.21348
68	-0.06772	-0.68281	0.68616	158	0.02255	-0.27594	0.27686
70	-0.10450	-1.07222	1.07730	160	0.01698	-0.33195	0.33238
72	-0.14275	-1.48308	1.48993	162	0.01205	-0.37998	0.38017
74	-0.18142	-1.90360	1.91223	164	0.00774	-0.42061	0.42068
76	-0.21940	-2.32063	2.33098	166	0.00402	-0.45445	0.45447
78	-0.25551	-2.72023	2.73220	168	0.00089	-0.48218	0.48218
80	-0.28854	-3.08824	3.10169	170	-0.00170	-0.50441	0.50441
82	-0.31737	-3.41111	3.42584	172	-0.00378	-0.52173	0.52174
84	-0.34097	-3.67651	3.69229	174	-0.00536	-0.53464	0.53467
86	-0.35849	-3.87406	3.89061	176	-0.00648	-0.54357	0.54360
88	-0.36927	-3.99588	4.01290	178	-0.00714	-0.54880	0.54884
				180	-0.00736	-0.55052	0.55057

VBR

RADIATION PATTERN FOR THETA=PI/2

H/LMDA=0.2500 B/LMDA= 0.250 OMEGA= 8.53

10 ELEMENT ARRAY

PHI DEG	ER(THETA,PHI) RE	IM	ABSVAL	PHI DEG	ER(THETA,PHI) RE	IM	ABSVAL
0	-0.03498	-0.40099	0.40251	90	0.89086	-2.89604	3.02996
2	-0.03610	-0.40146	0.40307	92	0.88572	-2.85908	2.99313
4	-0.03945	-0.40279	0.40472	94	0.87036	-2.75004	2.88448
6	-0.04502	-0.40483	0.40732	96	0.84500	-2.57428	2.70942
8	-0.05279	-0.40726	0.41067	98	0.80997	-2.34033	2.47653
10	-0.06271	-0.40969	0.41446	100	0.76580	-2.05934	2.19712
12	-0.07473	-0.41157	0.41830	102	0.71316	-1.74426	1.88442
14	-0.08874	-0.41225	0.42169	104	0.65290	-1.40910	1.55301
16	-0.10463	-0.41093	0.42404	106	0.58606	-1.06802	1.21825
18	-0.12220	-0.40672	0.42468	108	0.51384	-0.73450	0.89640
20	-0.14122	-0.39864	0.42291	110	0.43760	-0.42070	0.60703
22	-0.16139	-0.38559	0.41800	112	0.35882	-0.13678	0.38401
24	-0.18233	-0.36647	0.40932	114	0.27905	0.10942	0.29974
26	-0.20357	-0.34016	0.39642	116	0.19987	0.31262	0.37105
28	-0.22457	-0.30563	0.37927	118	0.12284	0.47009	0.48587
30	-0.24470	-0.26198	0.35849	120	0.04942	0.58152	0.58362
32	-0.26325	-0.20856	0.33585	122	-0.01906	0.64877	0.64905
34	-0.27942	-0.14505	0.31483	124	-0.08145	0.67542	0.68031
36	-0.29239	-0.07161	0.30103	126	-0.13686	0.66639	0.68030
38	-0.30128	0.01104	0.30148	128	-0.18462	0.62749	0.65408
40	-0.30521	0.10149	0.32165	130	-0.22432	0.56493	0.60783
42	-0.30335	0.19756	0.36201	132	-0.25583	0.48497	0.54831
44	-0.29492	0.29622	0.41800	134	-0.27925	0.39359	0.48259
46	-0.27925	0.39359	0.48259	136	-0.29492	0.29622	0.41800
48	-0.25583	0.48497	0.54831	138	-0.30335	0.19756	0.36201
50	-0.22432	0.56493	0.60783	140	-0.30521	0.10149	0.32165
52	-0.18462	0.62749	0.65408	142	-0.30128	0.01104	0.30148
54	-0.13686	0.66639	0.68030	144	-0.29239	-0.07161	0.30103
56	-0.08146	0.67542	0.68031	146	-0.27942	-0.14505	0.31483
58	-0.01906	0.64877	0.64905	148	-0.26325	-0.20855	0.33585
60	0.04942	0.58152	0.58362	150	-0.24470	-0.26198	0.35849
62	0.12284	0.47009	0.48587	152	-0.22457	-0.30563	0.37927
64	0.19987	0.31263	0.37106	154	-0.20357	-0.34016	0.39642
66	0.27905	0.10943	0.29974	156	-0.18233	-0.36647	0.40932
68	0.35882	-0.13677	0.38401	158	-0.16139	-0.38559	0.41800
70	0.43760	-0.42069	0.60703	160	-0.14122	-0.39864	0.42291
72	0.51384	-0.73450	0.89639	162	-0.12220	-0.40672	0.42469
74	0.58606	-1.06800	1.21823	164	-0.10463	-0.41093	0.42404
76	0.65290	-1.40909	1.55299	166	-0.08874	-0.41225	0.42169
78	0.71316	-1.74426	1.88441	168	-0.07473	-0.41157	0.41830
80	0.76580	-2.05933	2.19711	170	-0.06271	-0.40969	0.41446
82	0.80997	-2.34033	2.47653	172	-0.05279	-0.40726	0.41067
84	0.84500	-2.57427	2.70940	174	-0.04502	-0.40483	0.40732
86	0.87036	-2.75003	2.88447	176	-0.03945	-0.40279	0.40472
88	0.88572	-2.85908	2.99313	178	-0.03610	-0.40146	0.40307
				180	-0.03498	-0.40099	0.40251

IBR

RADIATION PATTERN FOR THETA=PI/2

H/LMDA=0.2500 B/LMDA= 0.250 OMEGA= 8.53

10 ELEMENT ARRAY

PHI DEG	ER(THETA,PHI) RE	IM	ABSVAL	PHI DEG	ER(THETA,PHI) RE	IM	ABSVAL
0	-0.03794	-0.57873	0.57997	90	-0.12631	-3.58349	3.58571
2	-0.03788	-0.57907	0.58031	92	-0.12468	-3.53838	3.54057
4	-0.03769	-0.58004	0.58126	94	-0.11985	-3.40528	3.40739
6	-0.03737	-0.58141	0.58261	96	-0.11205	-3.19067	3.19264
8	-0.03690	-0.58283	0.58399	98	-0.10163	-2.90489	2.90667
10	-0.03625	-0.58377	0.58490	100	-0.08905	-2.56141	2.56296
12	-0.03542	-0.58359	0.58466	102	-0.07487	-2.17597	2.17726
14	-0.03437	-0.58146	0.58248	104	-0.05968	-1.76555	1.76655
16	-0.03306	-0.57645	0.57739	106	-0.04408	-1.34736	1.34808
18	-0.03147	-0.56746	0.56833	108	-0.02867	-0.93782	0.93826
20	-0.02955	-0.55329	0.55408	110	-0.01399	-0.55178	0.55196
22	-0.02727	-0.53268	0.53338	112	-0.00050	-0.20171	0.20171
24	-0.02460	-0.50430	0.50490	114	0.01141	0.10274	0.10337
26	-0.02151	-0.46686	0.46736	116	0.02147	0.35496	0.35561
28	-0.01797	-0.41917	0.41955	118	0.02949	0.55145	0.55223
30	-0.01399	-0.36023	0.36050	120	0.03543	0.69164	0.69255
32	-0.00956	-0.28936	0.28952	122	0.03928	0.77759	0.77858
34	-0.00471	-0.20631	0.20636	124	0.04117	0.81348	0.81453
36	0.00050	-0.11139	0.11140	126	0.04126	0.80517	0.80622
38	0.00601	-0.00564	0.00824	128	0.03976	0.75957	0.76061
40	0.01170	0.10908	0.10971	130	0.03693	0.68417	0.68517
42	0.01743	0.22999	0.23064	132	0.03303	0.58658	0.58751
44	0.02304	0.35325	0.35400	134	0.02831	0.47406	0.47490
46	0.02831	0.47405	0.47490	136	0.02304	0.35325	0.35400
48	0.03303	0.58658	0.58751	138	0.01743	0.22999	0.23065
50	0.03693	0.68417	0.68517	140	0.01170	0.10909	0.10971
52	0.03976	0.75957	0.76061	142	0.00601	-0.00564	0.00824
54	0.04126	0.80517	0.80622	144	0.00051	-0.11139	0.11139
56	0.04117	0.81348	0.81453	146	-0.00471	-0.20631	0.20636
58	0.03928	0.77759	0.77858	148	-0.00956	-0.28936	0.28952
60	0.03543	0.69164	0.69255	150	-0.01398	-0.36023	0.36050
62	0.02949	0.55145	0.55224	152	-0.01797	-0.41917	0.41955
64	0.02147	0.35497	0.35561	154	-0.02151	-0.46686	0.46735
66	0.01141	0.10275	0.10338	156	-0.02460	-0.50430	0.50490
68	-0.00050	-0.20170	0.20170	158	-0.02727	-0.53268	0.53338
70	-0.01399	-0.55178	0.55195	160	-0.02955	-0.55329	0.55408
72	-0.02867	-0.93782	0.93825	162	-0.03147	-0.56746	0.56833
74	-0.04408	-1.34734	1.34806	164	-0.03306	-0.57645	0.57739
76	-0.05968	-1.76553	1.76654	166	-0.03437	-0.58146	0.58248
78	-0.07487	-2.17596	2.17725	168	-0.03542	-0.58359	0.58466
80	-0.08905	-2.56141	2.56295	170	-0.03625	-0.58377	0.58489
82	-0.10163	-2.90489	2.90666	172	-0.03690	-0.58283	0.58399
84	-0.11205	-3.19066	3.19263	174	-0.03737	-0.58141	0.58261
86	-0.11985	-3.40527	3.40738	176	-0.03769	-0.58004	0.58126
88	-0.12468	-3.53838	3.54056	178	-0.03788	-0.57907	0.58031
				180	-0.03794	-0.57873	0.57997

VBR

RADIATION PATTERN FOR THETA=PI/2

H/LMDA=0.2500 B/LMDA= 0.250 OMEGA= 8.53

12 ELEMENT ARRAY

PHI DEG	ER(THETA,PHI) RE	ER(THETA,PHI) IM	ABSVAL	PHI DEG	ER(THETA,PHI) RE	ER(THETA,PHI) IM	ABSVAL
0	0.26480	0.04308	0.26828	90	0.94022	-2.88636	3.03564
2	0.26463	0.04103	0.26780	92	0.93082	-2.83337	2.98235
4	0.26411	0.03487	0.26640	94	0.90291	-2.67807	2.82618
6	0.26313	0.02455	0.26427	96	0.85733	-2.43116	2.57789
8	0.26151	0.01001	0.26171	98	0.79548	-2.10937	2.25438
10	0.25904	-0.00881	0.25919	100	0.71928	-1.73396	1.87723
12	0.25542	-0.03193	0.25741	102	0.63114	-1.32883	1.47110
14	0.25031	-0.05931	0.25724	104	0.53386	-0.91848	1.06236
16	0.24331	-0.09079	0.25970	106	0.43058	-0.52605	0.67980
18	0.23403	-0.12606	0.26582	108	0.32464	-0.17163	0.36722
20	0.22202	-0.16457	0.27636	110	0.21946	0.12904	0.25458
22	0.20686	-0.20552	0.29160	112	0.11835	0.36539	0.38408
24	0.18819	-0.24776	0.31113	114	0.02441	0.53229	0.53285
26	0.16568	-0.28978	0.33381	116	-0.05966	0.62982	0.63264
28	0.13914	-0.32969	0.35785	118	-0.13165	0.66260	0.67555
30	0.10852	-0.36518	0.38097	120	-0.18997	0.63886	0.66651
32	0.07394	-0.39361	0.40050	122	-0.23368	0.56931	0.61540
34	0.03578	-0.41207	0.41362	124	-0.26249	0.46599	0.53483
36	-0.00535	-0.41751	0.41755	126	-0.27678	0.34117	0.43932
38	-0.04853	-0.40699	0.40987	128	-0.27748	0.20643	0.34585
40	-0.09257	-0.37789	0.38906	130	-0.26600	0.07196	0.27556
42	-0.13597	-0.32825	0.35530	132	-0.24408	-0.05392	0.24997
44	-0.17700	-0.25715	0.31218	134	-0.21373	-0.16499	0.27000
46	-0.21373	-0.16499	0.27000	136	-0.17700	-0.25715	0.31218
48	-0.24408	-0.05392	0.24997	138	-0.13597	-0.32825	0.35530
50	-0.26600	0.07196	0.27556	140	-0.09257	-0.37789	0.38906
52	-0.27748	0.20643	0.34585	142	-0.04853	-0.40699	0.40987
54	-0.27678	0.34117	0.43932	144	-0.00535	-0.41751	0.41755
56	-0.26249	0.46599	0.53483	146	0.03578	-0.41207	0.41362
58	-0.23368	0.56931	0.61540	148	0.07394	-0.39361	0.40050
60	-0.18997	0.63886	0.66650	150	0.10852	-0.36518	0.38097
62	-0.13165	0.66260	0.67555	152	0.13914	-0.32969	0.35785
64	-0.05966	0.62983	0.63264	154	0.16568	-0.28979	0.33381
66	0.02440	0.53230	0.53286	156	0.18819	-0.24776	0.31113
68	0.11835	0.36539	0.38408	158	0.20686	-0.20552	0.29160
70	0.21946	0.12904	0.25458	160	0.22202	-0.16457	0.27636
72	0.32464	-0.17162	0.36721	162	0.23403	-0.12606	0.26582
74	0.43058	-0.52603	0.67978	164	0.24331	-0.09079	0.25970
76	0.53386	-0.91846	1.06234	166	0.25031	-0.05931	0.25724
78	0.63114	-1.32882	1.47109	168	0.25542	-0.03193	0.25741
80	0.71928	-1.73395	1.87722	170	0.25904	-0.00881	0.25919
82	0.79548	-2.10936	2.25437	172	0.26151	0.01001	0.26171
84	0.85733	-2.43114	2.57788	174	0.26313	0.02455	0.26427
86	0.90291	-2.67806	2.82617	176	0.26411	0.03487	0.26640
88	0.93082	-2.83337	2.98235	178	0.26463	0.04103	0.26780
				180	0.26480	0.04308	0.26828

IBR

RADIATION PATTERN FOR THETA=PI/2

H/LMDA=0.2500 B/LMDA= 0.250 OMEGA= 8.53

12 ELEMENT ARRAY

PHI DEG	ER(THETA,PHI) RE	ER(THETA,PHI) IM	ABSVAL	PHI DEG	ER(THETA,PHI) RE	ER(THETA,PHI) IM	ABSVAL
0	-0.07311	0.03498	0.08105	90	0.17546	-4.17180	4.17549
2	-0.07298	0.03200	0.07969	92	0.17223	-4.09656	4.10018
4	-0.07258	0.02304	0.07615	94	0.16278	-3.87600	3.87942
6	-0.07189	0.00805	0.07234	96	0.14781	-3.52516	3.52825
8	-0.07088	-0.01301	0.07206	98	0.12843	-3.06752	3.07021
10	-0.06952	-0.04017	0.08029	100	0.10602	-2.53302	2.53523
12	-0.06776	-0.07341	0.09990	102	0.08211	-1.95527	1.95699
14	-0.06555	-0.11259	0.13028	104	0.05824	-1.36885	1.37009
16	-0.06286	-0.15740	0.16949	106	0.03583	-0.80649	0.80729
18	-0.05962	-0.20732	0.21572	108	0.01607	-0.29670	0.29713
20	-0.05583	-0.26146	0.26736	110	-0.00018	0.13804	0.13804
22	-0.05146	-0.31861	0.32274	112	-0.01241	0.48249	0.48265
24	-0.04653	-0.37704	0.37990	114	-0.02051	0.72899	0.72928
26	-0.04108	-0.43456	0.43650	116	-0.02467	0.87721	0.87756
28	-0.03520	-0.48843	0.48970	118	-0.02538	0.93323	0.93358
30	-0.02901	-0.53538	0.53617	120	-0.02333	0.90829	0.90859
32	-0.02268	-0.57168	0.57213	122	-0.01932	0.81718	0.81740
34	-0.01643	-0.59328	0.59350	124	-0.01418	0.67665	0.67680
36	-0.01051	-0.59599	0.59608	126	-0.00872	0.50390	0.50398
38	-0.00519	-0.57580	0.57583	128	-0.00364	0.31528	0.31530
40	-0.00076	-0.52925	0.52925	130	0.00053	0.12525	0.12525
42	0.00250	-0.45384	0.45384	132	0.00338	-0.05426	0.05436
44	0.00437	-0.34852	0.34854	134	0.00468	-0.21421	0.21426
46	0.00468	-0.21421	0.21427	136	0.00437	-0.34852	0.34854
48	0.00338	-0.05426	0.05437	138	0.00250	-0.45384	0.45384
50	0.00053	0.12525	0.12525	140	-0.00076	-0.52925	0.52925
52	-0.00364	0.31528	0.31530	142	-0.00519	-0.57580	0.57583
54	-0.00872	0.50390	0.50398	144	-0.01051	-0.59599	0.59608
56	-0.01418	0.67664	0.67679	146	-0.01643	-0.59328	0.59350
58	-0.01932	0.81717	0.81740	148	-0.02268	-0.57168	0.57213
60	-0.02333	0.90829	0.90859	150	-0.02901	-0.53538	0.53617
62	-0.02538	0.93323	0.93358	152	-0.03520	-0.48844	0.48970
64	-0.02467	0.87721	0.87756	154	-0.04108	-0.43457	0.43650
66	-0.02051	0.72900	0.72929	156	-0.04653	-0.37704	0.37990
68	-0.01241	0.48249	0.48265	158	-0.05146	-0.31861	0.32273
70	-0.00018	0.13805	0.13805	160	-0.05583	-0.26146	0.26736
72	0.01608	-0.29669	0.29713	162	-0.05962	-0.20732	0.21572
74	0.03584	-0.80647	0.80726	164	-0.06285	-0.15741	0.16949
76	0.05824	-1.36882	1.37006	166	-0.06555	-0.11259	0.13028
78	0.08211	-1.95526	1.95698	168	-0.06776	-0.07341	0.09990
80	0.10602	-2.53301	2.53522	170	-0.06952	-0.04017	0.08029
82	0.12843	-3.06751	3.07020	172	-0.07088	-0.01301	0.07206
84	0.14781	-3.52514	3.52823	174	-0.07189	0.00805	0.07234
86	0.16278	-3.87599	3.87940	176	-0.07258	0.02304	0.07615
88	0.17223	-4.09656	4.10018	178	-0.07298	0.03200	0.07969
				180	-0.07311	0.03498	0.08105

VBR

RADIATION PATTERN FOR THETA=PI/2

H/LMDA=0.2500 B/LMDA= 0.250 OMEGA= 8.53

14 ELEMENT ARRAY

PHI DEG	ER(THETA,PHI) RE	ER(THETA,PHI) IM	ABSVAL	PHI DEG	ER(THETA,PHI) RE	ER(THETA,PHI) IM	ABSVAL
0	0.02324	0.28322	0.28417	90	0.97570	-2.87636	3.03734
2	0.02445	0.28363	0.28468	92	0.96080	-2.80477	2.96477
4	0.02810	0.28478	0.28616	94	0.91685	-2.59663	2.75374
6	0.03414	0.28642	0.28845	96	0.84609	-2.27107	2.42355
8	0.04255	0.28814	0.29126	98	0.75210	-1.85738	2.00387
10	0.05325	0.28935	0.29421	100	0.63963	-1.39167	1.53162
12	0.06610	0.28932	0.29677	102	0.51434	-0.91266	1.04761
14	0.08092	0.28711	0.29829	104	0.38246	-0.45754	0.59634
16	0.09742	0.28166	0.29803	106	0.25044	-0.05824	0.25712
18	0.11523	0.27178	0.29520	108	0.12449	0.26143	0.28955
20	0.13383	0.25622	0.28906	110	0.01032	0.48736	0.48746
22	0.15255	0.23373	0.27911	112	-0.08735	0.61562	0.62178
24	0.17058	0.20319	0.26530	114	-0.16499	0.65158	0.67214
26	0.18696	0.16371	0.24851	116	-0.22043	0.60816	0.64688
28	0.20060	0.11485	0.23115	118	-0.25300	0.50350	0.56349
30	0.21029	0.05675	0.21781	120	-0.26341	0.35838	0.44477
32	0.21479	-0.00962	0.21501	122	-0.25363	0.19373	0.31916
34	0.21290	-0.08233	0.22827	124	-0.22665	0.02860	0.22845
36	0.20353	-0.15834	0.25787	126	-0.18615	-0.12138	0.22223
38	0.18585	-0.23344	0.29839	128	-0.13619	-0.24483	0.28016
40	0.15938	-0.30233	0.34177	130	-0.08089	-0.33491	0.34454
42	0.12417	-0.35882	0.37970	132	-0.02412	-0.38903	0.38978
44	0.08083	-0.39629	0.40445	134	0.03072	-0.40823	0.40938
46	0.03072	-0.40823	0.40938	136	0.08083	-0.39629	0.40445
48	-0.02412	-0.38903	0.38978	138	0.12416	-0.35882	0.37970
50	-0.08089	-0.33491	0.34454	140	0.15938	-0.30233	0.34177
52	-0.13619	-0.24483	0.28016	142	0.18585	-0.23344	0.29839
54	-0.18615	-0.12138	0.22223	144	0.20353	-0.15834	0.25787
56	-0.22665	0.02859	0.22845	146	0.21290	-0.08233	0.22827
58	-0.25363	0.19373	0.31915	148	0.21479	-0.00962	0.21501
60	-0.26341	0.35838	0.44477	150	0.21029	0.05675	0.21781
62	-0.25300	0.50350	0.56349	152	0.20060	0.11485	0.23115
64	-0.22044	0.60816	0.64687	154	0.18696	0.16371	0.24851
66	-0.16499	0.65158	0.67214	156	0.17058	0.20319	0.26530
68	-0.08736	0.61562	0.62179	158	0.15255	0.23373	0.27911
70	0.01032	0.48736	0.48747	160	0.13383	0.25622	0.28906
72	0.12449	0.26143	0.28956	162	0.11523	0.27178	0.29520
74	0.25043	-0.05822	0.25711	164	0.09743	0.28166	0.29803
76	0.38246	-0.45752	0.59632	166	0.08092	0.28711	0.29829
78	0.51434	-0.91265	1.04760	168	0.06610	0.28932	0.29677
80	0.63963	-1.39166	1.53162	170	0.05325	0.28935	0.29421
82	0.75210	-1.85737	2.00387	172	0.04255	0.28814	0.29126
84	0.84609	-2.27105	2.42353	174	0.03414	0.28642	0.28845
86	0.91685	-2.59662	2.75373	176	0.02810	0.28478	0.28616
88	0.96080	-2.80477	2.96477	178	0.02445	0.28363	0.28468
				180	0.02324	0.28322	0.28417

IBR

RADIATION PATTERN FOR THETA=PI/2

H/LMDA=0.2500 B/LMDA= 0.250 OMEGA= 8.53

14 ELEMENT ARRAY

PHI DEG	ER(THETA,PHI) RE	ER(THETA,PHI) IM	ABSVAL	PHI DEG	ER(THETA,PHI) RE	ER(THETA,PHI) IM	ABSVAL
0	0.02678	0.42971	0.43054	90	-0.09536	-3.63884	3.64009
2	0.02668	0.43000	0.43083	92	-0.09289	-3.54942	3.55064
4	0.02638	0.43077	0.43157	94	-0.08573	-3.28939	3.29050
6	0.02587	0.43169	0.43247	96	-0.07449	-2.88244	2.88340
8	0.02513	0.43225	0.43298	98	-0.06015	-2.36491	2.36567
10	0.02414	0.43168	0.43236	100	-0.04395	-1.78160	1.78214
12	0.02287	0.42905	0.42966	102	-0.02718	-1.18065	1.18096
14	0.02129	0.42318	0.42372	104	-0.01115	-0.60837	0.60847
16	0.01937	0.41275	0.41321	106	0.00299	-0.10465	0.10470
18	0.01708	0.39628	0.39665	108	0.01437	0.30059	0.30093
20	0.01441	0.37225	0.37253	110	0.02240	0.58944	0.58987
22	0.01133	0.33917	0.33936	112	0.02685	0.75658	0.75705
24	0.00787	0.29571	0.29581	114	0.02780	0.80824	0.80871
26	0.00403	0.24090	0.24093	116	0.02564	0.76010	0.76054
28	-0.00011	0.17433	0.17433	118	0.02094	0.63446	0.63481
30	-0.00449	0.09637	0.09647	120	0.01443	0.45699	0.45721
32	-0.00897	0.00841	0.01229	122	0.00688	0.25373	0.25382
34	-0.01340	-0.08694	0.08797	124	-0.00098	0.04851	0.04852
36	-0.01758	-0.18568	0.18651	126	-0.00847	-0.13895	0.13921
38	-0.02128	-0.28242	0.28322	128	-0.01507	-0.29415	0.29454
40	-0.02426	-0.37041	0.37120	130	-0.02039	-0.40818	0.40869
42	-0.02626	-0.44192	0.44270	132	-0.02419	-0.47742	0.47803
44	-0.02704	-0.48871	0.48945	134	-0.02640	-0.50279	0.50348
46	-0.02640	-0.50279	0.50348	136	-0.02704	-0.48871	0.48945
48	-0.02419	-0.47742	0.47804	138	-0.02626	-0.44192	0.44270
50	-0.02039	-0.40818	0.40869	140	-0.02426	-0.37041	0.37120
52	-0.01507	-0.29415	0.29454	142	-0.02128	-0.28242	0.28322
54	-0.00847	-0.13895	0.13921	144	-0.01758	-0.18569	0.18652
56	-0.00098	0.04851	0.04852	146	-0.01340	-0.08694	0.08796
58	0.00688	0.25372	0.25382	148	-0.00897	0.00841	0.01230
60	0.01443	0.45698	0.45721	150	-0.00449	0.09637	0.09647
62	0.02094	0.63446	0.63480	152	-0.00011	0.17433	0.17433
64	0.02564	0.76010	0.76053	154	0.00403	0.24089	0.24093
66	0.02780	0.80824	0.80871	156	0.00787	0.29571	0.29581
68	0.02685	0.75658	0.75706	158	0.01133	0.33917	0.33936
70	0.02240	0.58945	0.58987	160	0.01441	0.37225	0.37253
72	0.01437	0.30059	0.30094	162	0.01708	0.39628	0.39665
74	0.00300	-0.10463	0.10467	164	0.01937	0.41275	0.41320
76	-0.01115	-0.60834	0.60845	166	0.02129	0.42318	0.42371
78	-0.02718	-1.18064	1.18095	168	0.02287	0.42905	0.42966
80	-0.04394	-1.78159	1.78213	170	0.02414	0.43168	0.43236
82	-0.06015	-2.36490	2.36566	172	0.02513	0.43225	0.43298
84	-0.07449	-2.88242	2.88338	174	0.02587	0.43169	0.43247
86	-0.08572	-3.28937	3.29049	176	0.02638	0.43077	0.43157
88	-0.09289	-3.54942	3.55063	178	0.02668	0.43000	0.43083
				180	0.02678	0.42971	0.43054

VBR

RADIATION PATTERN FOR THETA=PI/2

H/LMDA=0.2500 B/LMDA= 0.250 OMEGA= 8.53

16 ELEMENT ARRAY

PHI	ER(THETA,PHI)			PHI	ER(THETA,PHI)		
DEG	RE	IM	ABSVAL	DEG	RE	IM	ABSVAL
0	-0.20075	-0.03103	0.20313	90	1.00209	-2.87101	3.04087
2	-0.20058	-0.02900	0.20266	92	0.98059	-2.77795	2.94594
4	-0.20001	-0.02291	0.20132	94	0.91768	-2.50989	2.67239
6	-0.19891	-0.01270	0.19931	96	0.81796	-2.09851	2.25229
8	-0.19701	0.00170	0.19701	98	0.68868	-1.59123	1.73387
10	-0.19398	0.02029	0.19504	100	0.53913	-1.04433	1.17528
12	-0.18940	0.04305	0.19423	102	0.37986	-0.51482	0.63979
14	-0.18278	0.06979	0.19565	104	0.22180	-0.05288	0.22802
16	-0.17358	0.10008	0.20037	106	0.07530	0.30417	0.31335
18	-0.16127	0.13323	0.20918	108	-0.05070	0.53555	0.53795
20	-0.14533	0.16813	0.22223	110	-0.14940	0.63791	0.65517
22	-0.12534	0.20321	0.23876	112	-0.21667	0.62361	0.66018
24	-0.10105	0.23642	0.25711	114	-0.25120	0.51702	0.57482
26	-0.07241	0.26518	0.27489	116	-0.25451	0.34958	0.43241
28	-0.03973	0.28646	0.28920	118	-0.23049	0.15469	0.27759
30	-0.00366	0.29696	0.29698	120	-0.18492	-0.03690	0.18857
32	0.03465	0.29336	0.29540	122	-0.12472	-0.20055	0.23617
34	0.07362	0.27266	0.28242	124	-0.05718	-0.31969	0.32476
36	0.11115	0.23270	0.25788	126	0.01074	-0.38632	0.38647
38	0.14480	0.17270	0.22537	128	0.07301	-0.40035	0.40695
40	0.17182	0.09380	0.19575	130	0.12493	-0.36793	0.38856
42	0.18944	-0.00045	0.18944	132	0.16337	-0.29939	0.34106
44	0.19510	-0.10376	0.22097	134	0.18678	-0.20716	0.27893
46	0.18678	-0.20715	0.27893	136	0.19510	-0.10376	0.22097
48	0.16337	-0.29939	0.34106	138	0.18944	-0.00045	0.18944
50	0.12493	-0.36792	0.38856	140	0.17182	0.09380	0.19575
52	0.07301	-0.40035	0.40695	142	0.14480	0.17270	0.22537
54	0.01074	-0.38632	0.38647	144	0.11115	0.23270	0.25788
56	-0.05718	-0.31969	0.32476	146	0.07362	0.27266	0.28242
58	-0.12472	-0.20056	0.23617	148	0.03465	0.29336	0.29540
60	-0.18492	-0.03691	0.18857	150	-0.00366	0.29696	0.29698
62	-0.23049	0.15469	0.27758	152	-0.03972	0.28646	0.28920
64	-0.25451	0.34957	0.43241	154	-0.07241	0.26518	0.27489
66	-0.25121	0.51701	0.57481	156	-0.10105	0.23642	0.25711
68	-0.21667	0.62361	0.66018	158	-0.12534	0.20321	0.23876
70	-0.14940	0.63791	0.65517	160	-0.14533	0.16813	0.22223
72	-0.05070	0.53556	0.53795	162	-0.16127	0.13323	0.20918
74	0.07530	0.30418	0.31336	164	-0.17358	0.10008	0.20037
76	0.22179	-0.05286	0.22801	166	-0.18277	0.06978	0.19564
78	0.37986	-0.51481	0.63978	168	-0.18940	0.04305	0.19423
80	0.53913	-1.04432	1.17527	170	-0.19398	0.02030	0.19504
82	0.68868	-1.59122	1.73386	172	-0.19701	0.00170	0.19701
84	0.81796	-2.09848	2.25226	174	-0.19891	-0.01270	0.19931
86	0.91768	-2.50987	2.67238	176	-0.20001	-0.02291	0.20132
88	0.98059	-2.77795	2.94594	178	-0.20058	-0.02900	0.20266
				180	-0.20075	-0.03103	0.20313

IBR

RADIATION PATTERN FOR THETA=PI/2

H/LMDA=0.2500 B/LMDA= 0.250 OMEGA= 8.53

16 ELEMENT ARRAY

PHI	ER(THETA,PHI)			PHI	ER(THETA,PHI)		
DEG	RE	IM	ABSVAL	DEG	RE	IM	ABSVAL
0	0.06118	-0.02867	0.06756	90	0.13868	-4.10330	4.10564
2	0.06108	-0.02570	0.06627	92	0.13428	-3.97222	3.97449
4	0.06079	-0.01677	0.06306	94	0.12161	-3.59451	3.59657
6	0.06028	-0.00185	0.06031	96	0.10225	-3.01435	3.01608
8	0.05950	0.01911	0.06249	98	0.07853	-2.29799	2.29933
10	0.05839	0.04607	0.07438	100	0.05316	-1.52411	1.52503
12	0.05690	0.07889	0.09727	102	0.02885	-0.77260	0.77313
14	0.05494	0.11719	0.12943	104	0.00790	-0.11405	0.11433
16	0.05245	0.16029	0.16866	106	-0.00809	0.39871	0.39879
18	0.04937	0.20707	0.21287	108	-0.01835	0.73581	0.73604
20	0.04566	0.25586	0.25990	110	-0.02297	0.89163	0.89192
22	0.04130	0.30436	0.30715	112	-0.02276	0.88256	0.88285
24	0.03634	0.34959	0.35147	114	-0.01901	0.74191	0.74216
26	0.03085	0.38789	0.38912	116	-0.01326	0.51318	0.51335
28	0.02498	0.41503	0.41578	118	-0.00699	0.24290	0.24300
30	0.01893	0.42646	0.42688	120	-0.00142	-0.02579	0.02582
32	0.01296	0.41766	0.41786	122	0.00261	-0.25792	0.25793
34	0.00737	0.38465	0.38472	124	0.00469	-0.42962	0.42965
36	0.00248	0.32473	0.32474	126	0.00488	-0.52892	0.52894
38	-0.00140	0.23714	0.23714	128	0.00352	-0.55482	0.55483
40	-0.00403	0.12386	0.12393	130	0.00120	-0.51517	0.51517
42	-0.00526	-0.00974	0.01107	132	-0.00141	-0.42387	0.42387
44	-0.00508	-0.15458	0.15466	134	-0.00368	-0.29788	0.29790
46	-0.00368	-0.29787	0.29790	136	-0.00508	-0.15458	0.15466
48	-0.00141	-0.42387	0.42387	138	-0.00526	-0.00975	0.01107
50	0.00120	-0.51517	0.51517	140	-0.00403	0.12386	0.12393
52	0.00352	-0.55481	0.55483	142	-0.00140	0.23714	0.23714
54	0.00488	-0.52892	0.52894	144	0.00248	0.32473	0.32474
56	0.00469	-0.42962	0.42965	146	0.00737	0.38465	0.38472
58	0.00261	-0.25792	0.25793	148	0.01296	0.41766	0.41786
60	-0.00142	-0.02579	0.02583	150	0.01893	0.42646	0.42688
62	-0.00699	0.24289	0.24299	152	0.02498	0.41503	0.41579
64	-0.01326	0.51317	0.51335	154	0.03085	0.38789	0.38912
66	-0.01901	0.74190	0.74215	156	0.03634	0.34959	0.35148
68	-0.02276	0.88256	0.88285	158	0.04130	0.30436	0.30715
70	-0.02297	0.89163	0.89193	160	0.04566	0.25586	0.25990
72	-0.01835	0.73582	0.73605	162	0.04937	0.20707	0.21287
74	-0.00809	0.39873	0.39881	164	0.05245	0.16029	0.16865
76	0.00790	-0.11403	0.11430	166	0.05494	0.11719	0.12943
78	0.02885	-0.77258	0.77312	168	0.05690	0.07889	0.09727
80	0.05317	-1.52409	1.52502	170	0.05839	0.04607	0.07438
82	0.07853	-2.29798	2.29932	172	0.05950	0.01911	0.06249
84	0.10226	-3.01432	3.01605	174	0.06028	-0.00185	0.06031
86	0.12161	-3.59448	3.59654	176	0.06079	-0.01677	0.06306
88	0.13428	-3.97221	3.97448	178	0.06108	-0.02570	0.06627
				180	0.06118	-0.02867	0.06756

VBR

RADIATION PATTERN FOR THETA=PI/2

H/LMDA=0.2500 B/LMDA= 0.250 OMEGA= 8.53

18 ELEMENT ARRAY

PHI	ER(THETA,PHI)			PHI	ER(THETA,PHI)		
DEG	RE	IM	ABSVAL	DEG	RE	IM	ABSVAL
0	-0.01712	-0.21869	0.21936	90	1.02281	-2.86528	3.04236
2	-0.01840	-0.21907	0.21984	92	0.99351	-2.74824	2.92231
4	-0.02222	-0.22009	0.22120	94	0.90854	-2.41463	2.57990
6	-0.02855	-0.22144	0.22327	96	0.77634	-1.91365	2.06513
8	-0.03732	-0.22260	0.22570	98	0.60985	-1.31709	1.45143
10	-0.04842	-0.22281	0.22801	100	0.42506	-0.70636	0.82439
12	-0.06164	-0.22111	0.22955	102	0.23912	-0.15831	0.28677
14	-0.07666	-0.21637	0.22955	104	0.06840	0.26700	0.27562
16	-0.09300	-0.20726	0.22717	106	-0.07332	0.53457	0.53958
18	-0.11001	-0.19241	0.22164	108	-0.17627	0.63747	0.66139
20	-0.12681	-0.17046	0.21245	110	-0.23561	0.59438	0.63938
22	-0.14231	-0.14027	0.19982	112	-0.25161	0.44310	0.50955
24	-0.15521	-0.10114	0.18526	114	-0.22919	0.23142	0.32570
26	-0.16406	-0.05303	0.17242	116	-0.17695	0.00778	0.17712
28	-0.16731	0.00313	0.16734	118	-0.10575	-0.18668	0.21455
30	-0.16348	0.06517	0.17599	120	-0.02715	-0.32350	0.32464
32	-0.15132	0.12951	0.19918	122	0.04803	-0.38900	0.39196
34	-0.13002	0.19113	0.23116	124	0.11094	-0.38357	0.39929
36	-0.09945	0.24371	0.26322	126	0.15543	-0.31871	0.35460
38	-0.06037	0.28015	0.28658	128	0.17841	-0.21295	0.27781
40	-0.01462	0.29332	0.29368	130	0.17969	-0.08742	0.19983
42	0.03485	0.27720	0.27938	132	0.16154	0.03785	0.16591
44	0.08400	0.22823	0.24320	134	0.12797	0.14669	0.19466
46	0.12797	0.14669	0.19467	136	0.08400	0.22823	0.24320
48	0.16153	0.03785	0.16591	138	0.03485	0.27720	0.27938
50	0.17969	-0.08742	0.19983	140	-0.01462	0.29332	0.29368
52	0.17841	-0.21295	0.27781	142	-0.06037	0.28015	0.28658
54	0.15543	-0.31871	0.35460	144	-0.09945	0.24371	0.26322
56	0.11094	-0.38357	0.39929	146	-0.13002	0.19113	0.23116
58	0.04803	-0.38901	0.39196	148	-0.15132	0.12951	0.19918
60	-0.02715	-0.32351	0.32464	150	-0.16348	0.06517	0.17599
62	-0.10575	-0.18669	0.21456	152	-0.16731	0.00313	0.16734
64	-0.17695	0.00778	0.17712	154	-0.16406	-0.05303	0.17242
66	-0.22919	0.23141	0.32569	156	-0.15522	-0.10114	0.18526
68	-0.25161	0.44310	0.50955	158	-0.14231	-0.14027	0.19982
70	-0.23561	0.59438	0.63938	160	-0.12681	-0.17046	0.21245
72	-0.17628	0.63747	0.66139	162	-0.11001	-0.19241	0.22164
74	-0.07332	0.53458	0.53959	164	-0.09300	-0.20726	0.22717
76	0.06839	0.26701	0.27563	166	-0.07666	-0.21637	0.22955
78	0.23912	-0.15830	0.28676	168	-0.06165	-0.22112	0.22955
80	0.42506	-0.70634	0.82438	170	-0.04843	-0.22280	0.22801
82	0.60985	-1.31708	1.45142	172	-0.03732	-0.22260	0.22570
84	0.77634	-1.91363	2.06511	174	-0.02855	-0.22144	0.22327
86	0.90854	-2.41461	2.57988	176	-0.02222	-0.22009	0.22120
88	0.99351	-2.74824	2.92230	178	-0.01840	-0.21907	0.21984
				180	-0.01712	-0.21869	0.21936

IBR

RADIATION PATTERN FOR THETA=PI/2

H/LMDA=0.2500 B/LMDA= 0.250 OMEGA= 8.53

18 ELEMENT ARRAY

PHI	ER(THETA,PHI)			PHI	ER(THETA,PHI)		
DEG	RE	IM	ABSVAL	DEG	RE	IM	ABSVAL
0	-0.02105	-0.34330	0.34395	90	-0.07603	-3.67258	3.67337
2	-0.02092	-0.34356	0.34420	92	-0.07276	-3.52426	3.52501
4	-0.02053	-0.34420	0.34481	94	-0.06344	-3.10130	3.10195
6	-0.01987	-0.34481	0.34538	96	-0.04942	-2.46562	2.46612
8	-0.01892	-0.34467	0.34519	98	-0.03269	-1.70762	1.70793
10	-0.01764	-0.34282	0.34328	100	-0.01556	-0.92995	0.93008
12	-0.01602	-0.33801	0.33839	102	-0.00021	-0.22978	0.22978
14	-0.01402	-0.32876	0.32906	104	0.01159	0.31664	0.31685
16	-0.01162	-0.31340	0.31361	106	0.01875	0.66443	0.66469
18	-0.00883	-0.29020	0.29033	108	0.02097	0.80392	0.80419
20	-0.00564	-0.25751	0.25757	110	0.01871	0.75779	0.75802
22	-0.00209	-0.21400	0.21401	112	0.01304	0.57297	0.57312
24	0.00173	-0.15892	0.15893	114	0.00537	0.30934	0.30938
26	0.00572	-0.09241	0.09259	116	-0.00275	0.02793	0.02807
28	0.00972	-0.01590	0.01864	118	-0.00995	-0.21903	0.21925
30	0.01352	0.06762	0.06896	120	-0.01516	-0.39499	0.39528
32	0.01687	0.15339	0.15432	122	-0.01775	-0.48191	0.48223
34	0.01950	0.23482	0.23563	124	-0.01758	-0.47945	0.47977
36	0.02113	0.30381	0.30454	126	-0.01487	-0.40146	0.40174
38	0.02151	0.35130	0.35196	128	-0.01017	-0.27082	0.27102
40	0.02044	0.36839	0.36896	130	-0.00422	-0.11403	0.11411
42	0.01784	0.34771	0.34817	132	0.00221	0.04360	0.04365
44	0.01375	0.28510	0.28543	134	0.00840	0.18135	0.18155
46	0.00840	0.18136	0.18155	136	0.01375	0.28510	0.28543
48	0.00221	0.04360	0.04366	138	0.01784	0.34771	0.34817
50	-0.00422	-0.11403	0.11411	140	0.02044	0.36840	0.36896
52	-0.01017	-0.27082	0.27101	142	0.02151	0.35130	0.35196
54	-0.01487	-0.40146	0.40174	144	0.02113	0.30381	0.30454
56	-0.01758	-0.47945	0.47977	146	0.01950	0.23483	0.23563
58	-0.01775	-0.48191	0.48223	148	0.01687	0.15339	0.15432
60	-0.01516	-0.39499	0.39528	150	0.01352	0.06763	0.06896
62	-0.00995	-0.21903	0.21926	152	0.00972	-0.01590	0.01864
64	-0.00275	0.02793	0.02806	154	0.00572	-0.09241	0.09259
66	0.00537	0.30932	0.30937	156	0.00173	-0.15891	0.15892
68	0.01304	0.57297	0.57312	158	-0.00209	-0.21400	0.21401
70	0.01871	0.75779	0.75802	160	-0.00564	-0.25751	0.25757
72	0.02097	0.80392	0.80419	162	-0.00883	-0.29020	0.29033
74	0.01875	0.66444	0.66470	164	-0.01162	-0.31340	0.31361
76	0.01160	0.31666	0.31687	166	-0.01402	-0.32876	0.32906
78	-0.00020	-0.22977	0.22977	168	-0.01602	-0.33802	0.33839
80	-0.01555	-0.92993	0.93006	170	-0.01764	-0.34282	0.34328
82	-0.03269	-1.70760	1.70792	172	-0.01892	-0.34467	0.34519
84	-0.04941	-2.46559	2.46609	174	-0.01987	-0.34481	0.34538
86	-0.06344	-3.10128	3.10192	176	-0.02053	-0.34420	0.34481
88	-0.07276	-3.52426	3.52501	178	-0.02092	-0.34356	0.34420
				180	-0.02105	-0.34330	0.34395

VBR

RADIATION PATTERN FOR THETA=PI/2

H/LMDA=0.2500   B/LMDA= 0.250   OMEGA= 8.53

20 ELEMENT ARRAY

PHI DEG	ER(THETA,PHI) RE	IM	ABSVAL	PHI DEG	ER(THETA,PHI) RE	IM	ABSVAL
0	0.16178	0.02408	0.16356	90	1.03921	-2.86188	3.04472
2	0.16160	0.02207	0.16310	92	1.00104	-2.71810	2.89657
4	0.16101	0.01603	0.16180	94	0.89144	-2.31304	2.47887
6	0.15980	0.00589	0.15991	96	0.72442	-1.71961	1.86597
8	0.15765	-0.00837	0.15787	98	0.52097	-1.04096	1.16404
10	0.15410	-0.02678	0.15641	100	0.30587	-0.38810	0.49414
12	0.14861	-0.04916	0.15653	102	0.10405	0.14312	0.17694
14	0.14057	-0.07514	0.15939	104	-0.06302	0.48980	0.49384
16	0.12932	-0.10399	0.16594	106	-0.18020	0.63108	0.65631
18	0.11427	-0.13449	0.17648	108	-0.24051	0.58702	0.63438
20	0.09493	-0.16489	0.19026	110	-0.24546	0.40888	0.47690
22	0.07106	-0.19278	0.20546	112	-0.20408	0.16405	0.26184
24	0.04278	-0.21515	0.21936	114	-0.13081	-0.08015	0.15341
26	0.01069	-0.22855	0.22880	116	-0.04271	-0.27004	0.27340
28	-0.02402	-0.22929	0.23055	118	0.04345	-0.37372	0.37624
30	-0.05951	-0.21403	0.22215	120	0.11371	-0.38331	0.39982
32	-0.09327	-0.18026	0.20296	122	0.15865	-0.31161	0.34967
34	-0.12224	-0.12722	0.17643	124	0.17415	-0.18530	0.25429
36	-0.14309	-0.05654	0.15386	126	0.16119	-0.03678	0.16533
38	-0.15254	0.02703	0.15492	128	0.12487	0.10334	0.16208
40	-0.14790	0.11537	0.18758	130	0.07292	0.21163	0.22384
42	-0.12763	0.19728	0.23496	132	0.01415	0.27460	0.27496
44	-0.09190	0.25954	0.27533	134	-0.04308	0.28888	0.29208
46	-0.04308	0.28888	0.29207	136	-0.09190	0.25954	0.27533
48	0.01415	0.27460	0.27496	138	-0.12762	0.19728	0.23496
50	0.07292	0.21163	0.22384	140	-0.14790	0.11537	0.18758
52	0.12487	0.10334	0.16208	142	-0.15254	0.02704	0.15492
54	0.16119	-0.03678	0.16533	144	-0.14309	-0.05654	0.15385
56	0.17415	-0.18530	0.25429	146	-0.12224	-0.12722	0.17643
58	0.15865	-0.31161	0.34967	148	-0.09327	-0.18026	0.20296
60	0.11371	-0.38331	0.39982	150	-0.05951	-0.21402	0.22214
62	0.04346	-0.37372	0.37624	152	-0.02402	-0.22930	0.23055
64	-0.04270	-0.27004	0.27340	154	0.01069	-0.22854	0.22879
66	-0.13081	-0.08016	0.15342	156	0.04278	-0.21515	0.21936
68	-0.20408	0.16405	0.26184	158	0.07106	-0.19278	0.20546
70	-0.24546	0.40887	0.47689	160	0.09493	-0.16489	0.19026
72	-0.24051	0.58701	0.63438	162	0.11427	-0.13449	0.17648
74	-0.18021	0.63108	0.65631	164	0.12932	-0.10399	0.16595
76	-0.06302	0.48982	0.49385	166	0.14057	-0.07514	0.15939
78	0.10405	0.14312	0.17695	168	0.14861	-0.04916	0.15653
80	0.30587	-0.38809	0.49413	170	0.15410	-0.02678	0.15641
82	0.52097	-1.04094	1.16403	172	0.15764	-0.00838	0.15787
84	0.72442	-1.71958	1.86594	174	0.15980	0.00589	0.15991
86	0.89144	-2.31301	2.47885	176	0.16101	0.01603	0.16180
88	1.00104	-2.71809	2.89657	178	0.16160	0.02207	0.16310
				180	0.16178	0.02408	0.16356

IBR

RADIATION PATTERN FOR THETA=PI/2

H/LMDA=0.2500   B/LMDA= 0.250   OMEGA= 8.53

20 ELEMENT ARRAY

PHI DEG	ER(THETA,PHI) RE	IM	ABSVAL	PHI DEG	ER(THETA,PHI) RE	IM	ABSVAL
0	-0.05320	0.02455	0.05859	90	0.11659	-4.06101	4.06268
2	-0.05313	0.02157	0.05734	92	0.11092	-3.85948	3.86107
4	-0.05291	0.01265	0.05440	94	0.09496	-3.29141	3.29278
6	-0.05250	-0.00228	0.05255	96	0.07164	-2.45816	2.45920
8	-0.05185	-0.02320	0.05680	98	0.04509	-1.50329	1.50396
10	-0.05087	-0.05003	0.07135	100	0.01964	-0.58164	0.58197
12	-0.04947	-0.08245	0.09615	102	-0.00104	0.17256	0.17256
14	-0.04755	-0.11980	0.12889	104	-0.01470	0.67049	0.67065
16	-0.04502	-0.16091	0.16709	106	-0.02074	0.88147	0.88172
18	-0.04180	-0.20393	0.20817	108	-0.02015	0.83179	0.83204
20	-0.03784	-0.24628	0.24917	110	-0.01501	0.59149	0.59168
22	-0.03314	-0.28448	0.28641	112	-0.00781	0.25366	0.25378
24	-0.02778	-0.31429	0.31551	114	-0.00086	-0.08793	0.08793
26	-0.02191	-0.33085	0.33158	116	0.00421	-0.35770	0.35772
28	-0.01577	-0.32918	0.32956	118	0.00668	-0.50987	0.50991
30	-0.00965	-0.30476	0.30491	120	0.00668	-0.53179	0.53183
32	-0.00394	-0.25443	0.25446	122	0.00497	-0.43971	0.43974
34	0.00097	-0.17749	0.17749	124	0.00259	-0.26957	0.26959
36	0.00474	-0.07665	0.07680	126	0.00054	-0.06580	0.06580
38	0.00709	0.04107	0.04167	128	-0.00048	0.12904	0.12904
40	0.00792	0.16404	0.16423	130	-0.00020	0.28189	0.28189
42	0.00733	0.27654	0.27664	132	0.00126	0.37316	0.37317
44	0.00565	0.36031	0.36035	134	0.00342	0.39725	0.39726
46	0.00342	0.39725	0.39726	136	0.00565	0.36031	0.36035
48	0.00126	0.37316	0.37317	138	0.00733	0.27654	0.27664
50	-0.00020	0.28189	0.28189	140	0.00792	0.16405	0.16424
52	-0.00048	0.12904	0.12904	142	0.00709	0.04107	0.04168
54	0.00054	-0.06580	0.06580	144	0.00474	-0.07665	0.07679
56	0.00259	-0.26957	0.26958	146	0.00097	-0.17748	0.17749
58	0.00497	-0.43971	0.43974	148	-0.00394	-0.25443	0.25446
60	0.00668	-0.53179	0.53183	150	-0.00965	-0.30475	0.30491
62	0.00668	-0.50987	0.50991	152	-0.01577	-0.32918	0.32956
64	0.00421	-0.35770	0.35772	154	-0.02191	-0.33085	0.33158
66	-0.00086	-0.08794	0.08794	156	-0.02778	-0.31428	0.31551
68	-0.00781	0.25366	0.25378	158	-0.03314	-0.28448	0.28641
70	-0.01501	0.59148	0.59167	160	-0.03784	-0.24628	0.24917
72	-0.02015	0.83179	0.83203	162	-0.04180	-0.20393	0.20817
74	-0.02074	0.88148	0.88172	164	-0.04502	-0.16091	0.16709
76	-0.01469	0.67051	0.67067	166	-0.04755	-0.11980	0.12890
78	-0.00104	0.17257	0.17257	168	-0.04947	-0.08245	0.09616
80	0.01965	-0.58163	0.58196	170	-0.05087	-0.05003	0.07135
82	0.04509	-1.50327	1.50395	172	-0.05185	-0.02320	0.05680
84	0.07165	-2.45811	2.45916	174	-0.05250	-0.00228	0.05255
86	0.09496	-3.29138	3.29275	176	-0.05291	0.01265	0.05440
88	0.11092	-3.85947	3.86106	178	-0.05313	0.02157	0.05734
				180	-0.05320	0.02455	0.05859

VBR

RADIATION PATTERN FOR THETA=PI/2

H/LMDA=0.2500   B/LMDA= 0.250   OMEGA= 8.53

22 ELEMENT ARRAY

PHI DEG	ER(THETA,PHI) RE	IM	ABSVAL	PHI DEG	ER(THETA,PHI) RE	IM	ABSVAL
0	0.01342	0.17800	0.17851	90	1.05278	-2.85818	3.04590
2	0.01473	0.17835	0.17896	92	1.00456	-2.68521	2.86697
4	0.01867	0.17928	0.18025	94	0.86770	-2.20425	2.36889
6	0.02519	0.18040	0.18215	96	0.66407	-1.51882	1.65765
8	0.03420	0.18106	0.18426	98	0.42544	-0.77047	0.88013
10	0.04553	0.18037	0.18603	100	0.18753	-0.10245	0.21369
12	0.05886	0.17717	0.18669	102	-0.01660	0.37504	0.37541
14	0.07372	0.17008	0.18537	104	-0.16231	0.60776	0.62906
16	0.08940	0.15764	0.18122	106	-0.23714	0.60371	0.64861
18	0.10492	0.13836	0.17364	108	-0.24211	0.42245	0.48691
20	0.11901	0.11101	0.16274	110	-0.19014	0.15283	0.24395
22	0.13012	0.07487	0.15012	112	-0.10238	-0.11362	0.15294
24	0.13652	0.03011	0.13980	114	-0.00322	-0.30537	0.30539
26	0.13641	-0.02186	0.13815	116	0.08471	-0.38458	0.39379
28	0.12817	-0.07804	0.15006	118	0.14459	-0.34958	0.37830
30	0.11064	-0.13370	0.17355	120	0.16777	-0.22788	0.28298
32	0.08347	-0.18243	0.20062	122	0.15412	-0.06328	0.16661
34	0.04745	-0.21669	0.22183	124	0.11069	0.09814	0.14794
36	0.00483	-0.22877	0.22882	126	0.04914	0.21955	0.22498
38	-0.04065	-0.21229	0.21615	128	-0.01724	0.27989	0.28042
40	-0.08384	-0.16401	0.18420	130	-0.07627	0.27533	0.28569
42	-0.11869	-0.08569	0.14639	132	-0.11886	0.21624	0.24675
44	-0.13911	0.01455	0.13987	134	-0.14007	0.12191	0.18569
46	-0.14007	0.12190	0.18569	136	-0.13911	0.01455	0.13987
48	-0.11887	0.21624	0.24675	138	-0.11869	-0.08569	0.14639
50	-0.07627	0.27532	0.28569	140	-0.08384	-0.16401	0.18419
52	-0.01725	0.27989	0.28042	142	-0.04065	-0.21229	0.21615
54	0.04914	0.21955	0.22498	144	0.00483	-0.22878	0.22883
56	0.11069	0.09815	0.14794	146	0.04745	-0.21669	0.22183
58	0.15412	-0.06328	0.16661	148	0.08347	-0.18243	0.20062
60	0.16777	-0.22788	0.28297	150	0.11064	-0.13370	0.17354
62	0.14459	-0.34958	0.37830	152	0.12817	-0.07805	0.15007
64	0.08471	-0.38458	0.39379	154	0.13641	-0.02186	0.13815
66	-0.00321	-0.30538	0.30539	156	0.13652	0.03011	0.13980
68	-0.10238	-0.11362	0.15294	158	0.13012	0.07487	0.15012
70	-0.19014	0.15282	0.24394	160	0.11901	0.11100	0.16274
72	-0.24211	0.42245	0.48691	162	0.10492	0.13836	0.17364
74	-0.23715	0.60370	0.64861	164	0.08940	0.15764	0.18122
76	-0.16231	0.60777	0.62907	166	0.07372	0.17008	0.18537
78	-0.01660	0.37505	0.37542	168	0.05886	0.17716	0.18669
80	0.18754	-0.10244	0.21369	170	0.04553	0.18037	0.18603
82	0.42545	-0.77046	0.88012	172	0.03420	0.18106	0.18426
84	0.66406	-1.51878	1.65761	174	0.02519	0.18040	0.18215
86	0.86770	-2.20422	2.36886	176	0.01867	0.17928	0.18025
88	1.00457	-2.68521	2.86697	178	0.01473	0.17835	0.17896
				180	0.01342	0.17800	0.17851

IBR

RADIATION PATTERN FOR THETA=PI/2

H/LMDA=0.2500   B/LMDA= 0.250   OMEGA= 8.53

22 ELEMENT ARRAY

PHI DEG	ER(THETA,PHI) RE	IM	ABSVAL	PHI DEG	ER(THETA,PHI) RE	IM	ABSVAL
0	0.01758	0.28670	0.28723	90	-0.06271	-3.69555	3.69608
2	0.01742	0.28693	0.28746	92	-0.05868	-3.47414	3.47464
4	0.01696	0.28748	0.28798	94	-0.04747	-2.85812	2.85851
6	0.01617	0.28782	0.28827	96	-0.03150	-1.97914	1.97939
8	0.01503	0.28707	0.28747	98	-0.01412	-1.01744	1.01754
10	0.01351	0.28403	0.28436	100	0.00125	-0.15582	0.15583
12	0.01159	0.27718	0.27742	102	0.01192	0.46441	0.46457
14	0.00925	0.26472	0.26488	104	0.01652	0.77277	0.77295
16	0.00648	0.24476	0.24484	106	0.01521	0.77776	0.77791
18	0.00331	0.21544	0.21546	108	0.00943	0.55383	0.55391
20	-0.00019	0.17528	0.17528	110	0.00148	0.21323	0.21323
22	-0.00393	0.12353	0.12359	112	-0.00620	-0.12746	0.12761
24	-0.00773	0.06062	0.06111	114	-0.01160	-0.37634	0.37652
26	-0.01136	-0.01137	0.01607	116	-0.01353	-0.48377	0.48396
28	-0.01457	-0.08830	0.08949	118	-0.01180	-0.44601	0.44617
30	-0.01702	-0.16381	0.16469	120	-0.00714	-0.29663	0.29672
32	-0.01842	-0.22949	0.23023	122	-0.00084	-0.09049	0.09049
34	-0.01847	-0.27557	0.27619	124	0.00559	0.11402	0.11416
36	-0.01699	-0.29223	0.29272	126	0.01075	0.26965	0.26986
38	-0.01394	-0.27152	0.27188	128	0.01370	0.34884	0.34911
40	-0.00947	-0.20967	0.20989	130	0.01400	0.34581	0.34609
42	-0.00398	-0.10941	0.10949	132	0.01174	0.27300	0.27326
44	0.00193	0.01839	0.01849	134	0.00746	0.15448	0.15466
46	0.00746	0.15448	0.15466	136	0.00193	0.01839	0.01849
48	0.01174	0.27300	0.27325	138	-0.00398	-0.10941	0.10949
50	0.01400	0.34581	0.34609	140	-0.00947	-0.20967	0.20988
52	0.01370	0.34884	0.34911	142	-0.01394	-0.27152	0.27187
54	0.01075	0.26965	0.26986	144	-0.01699	-0.29223	0.29273
56	0.00559	0.11403	0.11417	146	-0.01847	-0.27557	0.27619
58	-0.00084	-0.09049	0.09049	148	-0.01842	-0.22949	0.23022
60	-0.00714	-0.29663	0.29671	150	-0.01702	-0.16381	0.16469
62	-0.01180	-0.44601	0.44617	152	-0.01457	-0.08830	0.08949
64	-0.01353	-0.48377	0.48396	154	-0.01136	-0.01137	0.01607
66	-0.01160	-0.37635	0.37652	156	-0.00773	0.06062	0.06111
68	-0.00620	-0.12747	0.12762	158	-0.00393	0.12353	0.12359
70	0.00148	0.21322	0.21322	160	-0.00019	0.17528	0.17528
72	0.00943	0.55382	0.55390	162	0.00331	0.21544	0.21546
74	0.01521	0.77776	0.77790	164	0.00648	0.24475	0.24484
76	0.01652	0.77278	0.77295	166	0.00925	0.26472	0.26488
78	0.01192	0.46442	0.46458	168	0.01159	0.27718	0.27742
80	0.00126	-0.15581	0.15582	170	0.01351	0.28404	0.28436
82	-0.01411	-1.01743	1.01752	172	0.01503	0.28707	0.28746
84	-0.03149	-1.97909	1.97934	174	0.01617	0.28782	0.28827
86	-0.04746	-2.85808	2.85848	176	0.01696	0.28748	0.28798
88	-0.05868	-3.47413	3.47463	178	0.01742	0.28693	0.28746
				180	0.01758	0.28670	0.28723

VBR

RADIATION PATTERN FOR THETA=PI/2

H/LMDA=0.2500 B/LMDA= 0.250 OMEGA= 8.53

24 ELEMENT ARRAY

PHI	ER(THETA,PHI)			PHI	ER(THETA,PHI)		
DEG	RE	IM	ABSVAL	DEG	RE	IM	ABSVAL
0	-0.13554	-0.01958	0.13695	90	1.06396	-2.85583	3.04758
2	-0.13537	-0.01759	0.13650	92	1.00465	-2.65103	2.83501
4	-0.13475	-0.01159	0.13525	94	0.83841	-2.08965	2.25157
6	-0.13345	-0.00152	0.13346	96	0.59755	-1.31390	1.44340
8	-0.13105	0.01265	0.13166	98	0.32749	-0.51093	0.60688
10	-0.12701	0.03085	0.13070	100	0.07638	0.14327	0.16236
12	-0.12065	0.05282	0.13170	102	-0.11550	0.53267	0.54505
14	-0.11124	0.07793	0.13582	104	-0.22397	0.62703	0.66583
16	-0.09809	0.10505	0.14372	106	-0.24479	0.47882	0.53776
18	-0.08059	0.13241	0.15501	108	-0.19255	0.19550	0.27440
20	-0.05845	0.15752	0.16801	110	-0.09524	-0.10068	0.13859
22	-0.03179	0.17715	0.17998	112	0.01382	-0.31075	0.31105
24	-0.00139	0.18754	0.18755	114	0.10441	-0.38328	0.39725
26	0.03123	0.18477	0.18739	116	0.15588	-0.31974	0.35572
28	0.06368	0.16542	0.17725	118	0.16053	-0.16293	0.22872
30	0.09276	0.12743	0.15761	120	0.12326	0.02441	0.12566
32	0.11472	0.07118	0.13501	122	0.05837	0.18226	0.19138
34	0.12572	0.00037	0.12572	124	-0.01544	0.27008	0.27052
36	0.12249	-0.07731	0.14485	126	-0.08016	0.27417	0.28565
38	0.10317	-0.15044	0.18242	128	-0.12246	0.20570	0.23939
40	0.06816	-0.20498	0.21602	130	-0.13582	0.09242	0.16429
42	0.02076	-0.22690	0.22785	132	-0.12072	-0.03211	0.12492
44	-0.03268	-0.20562	0.20820	134	-0.08323	-0.13809	0.16123
46	-0.08323	-0.13809	0.16123	136	-0.03268	-0.20562	0.20820
48	-0.12072	-0.03211	0.12492	138	0.02076	-0.22690	0.22785
50	-0.13582	0.09242	0.16428	140	0.06816	-0.20498	0.21602
52	-0.12246	0.20570	0.23939	142	0.10317	-0.15043	0.18241
54	-0.08016	0.27417	0.28565	144	0.12249	-0.07732	0.14485
56	-0.01544	0.27008	0.27052	146	0.12572	0.00037	0.12572
58	0.05837	0.18226	0.19138	148	0.11472	0.07118	0.13501
60	0.12326	0.02441	0.12565	150	0.09276	0.12743	0.15761
62	0.16053	-0.16293	0.22872	152	0.06368	0.16541	0.17725
64	0.15588	-0.31974	0.35571	154	0.03123	0.18477	0.18739
66	0.10441	-0.38328	0.39725	156	-0.00139	0.18755	0.18755
68	0.01383	-0.31075	0.31106	158	-0.03179	0.17716	0.17999
70	-0.09524	-0.10068	0.13859	160	-0.05845	0.15752	0.16802
72	-0.19255	0.19549	0.27439	162	-0.08059	0.13242	0.15501
74	-0.24478	0.47881	0.53775	164	-0.09809	0.10505	0.14372
76	-0.22397	0.62703	0.66583	166	-0.11124	0.07793	0.13582
78	-0.11550	0.53268	0.54506	168	-0.12065	0.05282	0.13170
80	0.07638	0.14328	0.16237	170	-0.12701	0.03085	0.13070
82	0.32749	-0.51092	0.60686	172	-0.13105	0.01265	0.13166
84	0.59754	-1.31386	1.44336	174	-0.13345	-0.00152	0.13346
86	0.83840	-2.08961	2.25154	176	-0.13475	-0.01159	0.13525
88	1.00465	-2.65102	2.83500	178	-0.13537	-0.01759	0.13650
				180	-0.13554	-0.01958	0.13695

IBR

RADIATION PATTERN FOR THETA=PI/2

H/LMDA=0.2500 B/LMDA= 0.250 OMEGA= 8.53

24 ELEMENT ARRAY

PHI	ER(THETA,PHI)			PHI	ER(THETA,PHI)		
DEG	RE	IM	ABSVAL	DEG	RE	IM	ABSVAL
0	0.04740	-0.02161	0.05210	90	0.10177	-4.03208	4.03337
2	0.04735	-0.01863	0.05089	92	0.09474	-3.74599	3.74718
4	0.04719	-0.00969	0.04817	94	0.07550	-2.96116	2.96212
6	0.04685	0.00525	0.04714	96	0.04893	-1.87487	1.87551
8	0.04627	0.02617	0.05316	98	0.02141	-0.74710	0.74741
10	0.04535	0.05287	0.06965	100	-0.00116	0.17681	0.17681
12	0.04395	0.08485	0.09556	102	-0.01502	0.73383	0.73399
14	0.04195	0.12108	0.12814	104	-0.01933	0.87940	0.87961
16	0.03923	0.15982	0.16456	106	-0.01595	0.68381	0.68400
18	0.03571	0.19842	0.20161	108	-0.00836	0.29454	0.29466
20	0.03133	0.23327	0.23536	110	-0.00029	-0.11896	0.11896
22	0.02614	0.25977	0.26108	112	0.00548	-0.41803	0.41806
24	0.02027	0.27266	0.27341	114	0.00774	-0.52881	0.52886
26	0.01399	0.26657	0.26694	116	0.00684	-0.45041	0.45047
28	0.00766	0.23696	0.23709	118	0.00412	-0.23978	0.23982
30	0.00172	0.18132	0.18133	120	0.00120	0.01744	0.01748
32	-0.00334	0.10059	0.10064	122	-0.00073	0.23805	0.23805
34	-0.00707	0.00037	0.00708	124	-0.00120	0.36476	0.36476
36	-0.00915	-0.10827	0.10866	126	-0.00051	0.37683	0.37683
38	-0.00948	-0.20924	0.20945	128	0.00061	0.28795	0.28795
40	-0.00824	-0.28315	0.28327	130	0.00129	0.13522	0.13522
42	-0.00589	-0.31089	0.31095	132	0.00093	-0.03552	0.03553
44	-0.00310	-0.27855	0.27857	134	-0.00063	-0.18285	0.18285
46	-0.00063	-0.18285	0.18285	136	-0.00310	-0.27855	0.27857
48	0.00093	-0.03552	0.03553	138	-0.00589	-0.31089	0.31095
50	0.00129	0.13521	0.13522	140	-0.00824	-0.28315	0.28327
52	0.00061	0.28795	0.28795	142	-0.00948	-0.20924	0.20945
54	-0.00051	0.37683	0.37683	144	-0.00915	-0.10828	0.10866
56	-0.00120	0.36476	0.36476	146	-0.00707	0.00038	0.00708
58	-0.00073	0.23805	0.23805	148	-0.00334	0.10059	0.10065
60	0.00120	0.01744	0.01748	150	0.00171	0.18133	0.18133
62	0.00412	-0.23978	0.23982	152	0.00766	0.23696	0.23709
64	0.00684	-0.45041	0.45046	154	0.01399	0.26657	0.26694
66	0.00774	-0.52881	0.52887	156	0.02027	0.27266	0.27341
68	0.00548	-0.41803	0.41807	158	0.02614	0.25977	0.26109
70	-0.00028	-0.11896	0.11896	160	0.03133	0.23327	0.23536
72	-0.00836	0.29454	0.29465	162	0.03571	0.19843	0.20162
74	-0.01595	0.68380	0.68398	164	0.03923	0.15982	0.16456
76	-0.01933	0.87940	0.87961	166	0.04195	0.12108	0.12814
78	-0.01502	0.73384	0.73399	168	0.04395	0.08485	0.09556
80	-0.00115	0.17682	0.17683	170	0.04535	0.05287	0.06966
82	0.02142	-0.74708	0.74739	172	0.04627	0.02616	0.05316
84	0.04893	-1.87482	1.87545	174	0.04685	0.00525	0.04714
86	0.07550	-2.96111	2.96208	176	0.04719	-0.00969	0.04817
88	0.09474	-3.74597	3.74717	178	0.04735	-0.01863	0.05089
				180	0.04740	-0.02161	0.05210

VBR

RADIATION PATTERN FOR THETA=PI/2

H/LMDA=0.2500 B/LMDA= 0.250 OMEGA= 8.53

26 ELEMENT ARRAY

PHI	ER(THETA,PHI)			PHI	ER(THETA,PHI)		
DEG	RE	IM	ABSVAL	DEG	RE	IM	ABSVAL
0	-0.01096	-0.15002	0.15042	90	1.07354	-2.85324	3.04852
2	-0.01230	-0.15035	0.15085	92	1.00201	-2.61427	2.79972
4	-0.01633	-0.15120	0.15208	94	0.80431	-1.96937	2.12728
6	-0.02298	-0.15211	0.15384	96	0.52629	-1.10810	1.22673
8	-0.03214	-0.15233	0.15569	98	0.23004	-0.26938	0.35424
10	-0.04357	-0.15080	0.15696	100	-0.02298	0.34080	0.34157
12	-0.05685	-0.14617	0.15683	102	-0.18787	0.61360	0.64172
14	-0.07132	-0.13690	0.15436	104	-0.24645	0.56059	0.61237
16	-0.08598	-0.12140	0.14876	106	-0.20927	0.29179	0.35907
18	-0.09951	-0.09823	0.13983	108	-0.10929	-0.03772	0.11562
20	-0.11021	-0.06650	0.12872	110	0.00993	-0.28837	0.28854
22	-0.11615	-0.02626	0.11908	112	0.10775	-0.38054	0.39550
24	-0.11530	0.02101	0.11720	114	0.15715	-0.30940	0.34702
26	-0.10586	0.07196	0.12800	116	0.15012	-0.13050	0.19891
28	-0.08670	0.12123	0.14904	118	0.09664	0.07272	0.12094
30	-0.05779	0.16158	0.17161	120	0.01875	0.22407	0.22485
32	-0.02072	0.18475	0.18591	122	-0.05782	0.27942	0.28534
34	0.02106	0.18295	0.18415	124	-0.11155	0.23483	0.25998
36	0.06220	0.15098	0.16329	126	-0.13036	0.11939	0.17677
38	0.09603	0.08866	0.13070	128	-0.11322	-0.02106	0.11516
40	0.11553	0.00290	0.11557	130	-0.06845	-0.14171	0.15737
42	0.11498	-0.09161	0.14701	132	-0.00987	-0.21178	0.21201
44	0.09168	-0.17374	0.19644	134	0.04759	-0.22013	0.22521
46	0.04759	-0.22013	0.22522	136	0.09168	-0.17374	0.19644
48	-0.00987	-0.21178	0.21201	138	0.11498	-0.09161	0.14701
50	-0.06844	-0.14171	0.15738	140	0.11553	0.00290	0.11557
52	-0.11322	-0.02106	0.11516	142	0.09603	0.08867	0.13070
54	-0.13036	0.11939	0.17677	144	0.06220	0.15097	0.16329
56	-0.11155	0.23482	0.25997	146	0.02105	0.18295	0.18415
58	-0.05782	0.27942	0.28534	148	-0.02072	0.18476	0.18591
60	0.01875	0.22407	0.22485	150	-0.05779	0.16158	0.17161
62	0.09664	0.07272	0.12094	152	-0.08670	0.12123	0.14904
64	0.15012	-0.13049	0.19891	154	-0.10586	0.07197	0.12801
66	0.15715	-0.30939	0.34701	156	-0.11530	0.02101	0.11720
68	0.10775	-0.38054	0.39550	158	-0.11615	-0.02626	0.11908
70	0.00993	-0.28837	0.28854	160	-0.11021	-0.06650	0.12872
72	-0.10929	-0.03773	0.11562	162	-0.09951	-0.09823	0.13983
74	-0.20926	0.29178	0.35906	164	-0.08598	-0.12140	0.14876
76	-0.24646	0.56058	0.61237	166	-0.07132	-0.13690	0.15436
78	-0.18788	0.61360	0.64172	168	-0.05685	-0.14617	0.15684
80	-0.02298	0.34081	0.34158	170	-0.04357	-0.15080	0.15697
82	0.23004	-0.26936	0.35423	172	-0.03214	-0.15233	0.15569
84	0.52629	-1.10806	1.22669	174	-0.02298	-0.15211	0.15384
86	0.80430	-1.96933	2.12724	176	-0.01633	-0.15120	0.15208
88	1.00201	-2.61426	2.79971	178	-0.01230	-0.15035	0.15085
				180	-0.01096	-0.15002	0.15042

IBR

RADIATION PATTERN FOR THETA=PI/2

H/LMDA=0.2500 B/LMDA= 0.250 OMEGA= 8.53

26 ELEMENT ARRAY

PHI	ER(THETA,PHI)			PHI	ER(THETA,PHI)		
DEG	RE	IM	ABSVAL	DEG	RE	IM	ABSVAL
0	-0.01524	-0.24663	0.24710	90	-0.05293	-3.71232	3.71269
2	-0.01507	-0.24686	0.24732	92	-0.04818	-3.40415	3.40449
4	-0.01454	-0.24734	0.24777	94	-0.03539	-2.57189	2.57213
6	-0.01363	-0.24743	0.24781	96	-0.01839	-1.45856	1.45868
8	-0.01233	-0.24612	0.24642	98	-0.00205	-0.37097	0.37097
10	-0.01060	-0.24194	0.24217	100	0.00941	0.42537	0.42548
12	-0.00842	-0.23311	0.23327	102	0.01375	0.78865	0.78877
14	-0.00579	-0.21760	0.21768	104	0.01124	0.73110	0.73119
16	-0.00274	-0.19334	0.19336	106	0.00430	0.39171	0.39174
18	0.00068	-0.15855	0.15855	108	-0.00360	-0.03144	0.03164
20	0.00432	-0.11221	0.11229	110	-0.00923	-0.35847	0.35859
22	0.00798	-0.05462	0.05520	112	-0.01069	-0.48501	0.48513
24	0.01140	0.01199	0.01654	114	-0.00784	-0.40201	0.40209
26	0.01424	0.08297	0.08418	116	-0.00211	-0.17825	0.17826
28	0.01615	0.15103	0.15189	118	0.00425	0.08057	0.08068
30	0.01677	0.20657	0.20725	120	0.00901	0.27667	0.27682
32	0.01583	0.23879	0.23932	122	0.01066	0.35221	0.35237
34	0.01323	0.23770	0.23807	124	0.00886	0.30036	0.30049
36	0.00911	0.19688	0.19709	126	0.00430	0.15681	0.15687
38	0.00386	0.11655	0.11661	128	-0.00157	-0.02068	0.02074
40	-0.00181	0.00610	0.00636	130	-0.00711	-0.17477	0.17492
42	-0.00700	-0.11510	0.11532	132	-0.01092	-0.26552	0.26575
44	-0.01073	-0.21968	0.21995	134	-0.01219	-0.27777	0.27804
46	-0.01219	-0.27778	0.27804	136	-0.01073	-0.21968	0.21995
48	-0.01092	-0.26552	0.26575	138	-0.00700	-0.11511	0.11532
50	-0.00711	-0.17478	0.17492	140	-0.00181	0.00609	0.00636
52	-0.00157	-0.02068	0.02074	142	0.00386	0.11655	0.11661
54	0.00430	0.15681	0.15687	144	0.00911	0.19688	0.19709
56	0.00886	0.30036	0.30049	146	0.01323	0.23770	0.23807
58	0.01066	0.35221	0.35237	148	0.01583	0.23879	0.23932
60	0.00901	0.27667	0.27682	150	0.01677	0.20657	0.20725
62	0.00426	0.08057	0.08068	152	0.01615	0.15103	0.15189
64	-0.00211	-0.17825	0.17826	154	0.01424	0.08298	0.08419
66	-0.00784	-0.40200	0.40208	156	0.01140	0.01199	0.01655
68	-0.01069	-0.48501	0.48513	158	0.00798	-0.05462	0.05520
70	-0.00923	-0.35847	0.35859	160	0.00432	-0.11221	0.11229
72	-0.00360	-0.03144	0.03165	162	0.00068	-0.15854	0.15854
74	0.00430	0.39169	0.39172	164	-0.00274	-0.19334	0.19336
76	0.01124	0.73109	0.73118	166	-0.00579	-0.21760	0.21768
78	0.01375	0.78866	0.78878	168	-0.00842	-0.23312	0.23327
80	0.00942	0.42538	0.42549	170	-0.01060	-0.24194	0.24217
82	-0.00204	-0.37095	0.37095	172	-0.01233	-0.24612	0.24643
84	-0.01838	-1.45851	1.45862	174	-0.01363	-0.24743	0.24781
86	-0.03538	-2.57184	2.57208	176	-0.01454	-0.24734	0.24777
88	-0.04818	-3.40414	3.40448	178	-0.01507	-0.24686	0.24732
				180	-0.01524	-0.24663	0.24710

VBR

RADIATION PATTERN FOR THETA=PI/2

H/LMDA=0.2500 B/LMDA= 0.250 OMEGA= 8.53

28 ELEMENT ARRAY

PHI DEG	ER(THETA,PHI) RE	IM	ABSVAL	PHI DEG	ER(THETA,PHI) RE	IM	ABSVAL
0	0.11667	0.01645	0.11782	90	1.08164	-2.85152	3.04977
2	0.11649	0.01447	0.11739	92	0.99689	-2.57586	2.76203
4	0.11586	0.00850	0.11617	94	0.76616	-1.84443	1.99722
6	0.11447	-0.00151	0.11448	96	0.45219	-0.90377	1.01058
8	0.11183	-0.01559	0.11291	98	0.13648	-0.04995	0.14534
10	0.10730	-0.03359	0.11243	100	-0.10653	0.48719	0.49870
12	0.10009	-0.05510	0.11425	102	-0.23155	0.62300	0.66464
14	0.08939	-0.07920	0.11943	104	-0.23328	0.42931	0.48860
16	0.07448	-0.10433	0.12819	106	-0.14318	0.08112	0.16456
18	0.05488	-0.12810	0.13936	108	-0.01461	-0.23004	0.23050
20	0.03060	-0.14728	0.15043	110	0.09754	-0.37425	0.38675
22	0.00236	-0.15794	0.15796	112	0.15483	-0.32170	0.35702
24	-0.02819	-0.15590	0.15843	114	0.14511	-0.13260	0.19658
26	-0.05839	-0.13752	0.14941	116	0.08193	0.08734	0.11975
28	-0.08457	-0.10076	0.13155	118	-0.00464	0.24014	0.24018
30	-0.10248	-0.04648	0.11253	120	-0.08115	0.27378	0.28556
32	-0.10792	0.02037	0.10982	122	-0.12264	0.19330	0.22892
34	-0.09773	0.09025	0.13303	124	-0.11944	0.04685	0.12830
36	-0.07101	0.14966	0.16565	126	-0.07770	-0.10160	0.12790
38	-0.03007	0.18333	0.18578	128	-0.01461	-0.19904	0.19957
40	0.01895	0.17813	0.17913	130	0.04882	-0.21980	0.22515
42	0.06648	0.12802	0.14425	132	0.09465	-0.16822	0.19302
44	0.10116	0.03864	0.10829	134	0.11257	-0.07047	0.13281
46	0.11257	-0.07047	0.13281	136	0.10116	0.03864	0.10829
48	0.09465	-0.16822	0.19302	138	0.06648	0.12801	0.14425
50	0.04882	-0.21980	0.22516	140	0.01895	0.17813	0.17914
52	-0.01460	-0.19904	0.19957	142	-0.03007	0.18333	0.18578
54	-0.07770	-0.10160	0.12790	144	-0.07100	0.14967	0.16565
56	-0.11944	0.04684	0.12830	146	-0.09773	0.09025	0.13303
58	-0.12264	0.19330	0.22892	148	-0.10792	0.02037	0.10982
60	-0.08115	0.27378	0.28556	150	-0.10248	-0.04648	0.11253
62	-0.00464	0.24014	0.24019	152	-0.08457	-0.10075	0.13154
64	0.08193	0.08734	0.11975	154	-0.05839	-0.13752	0.14940
66	0.14511	-0.13259	0.19657	156	-0.02820	-0.15591	0.15843
68	0.15483	-0.32170	0.35702	158	0.00236	-0.15794	0.15795
70	0.09754	-0.37425	0.38676	160	0.03060	-0.14728	0.15042
72	-0.01460	-0.23004	0.23051	162	0.05488	-0.12810	0.13936
74	-0.14318	0.08110	0.16455	164	0.07448	-0.10433	0.12819
76	-0.23328	0.42930	0.48859	166	0.08939	-0.07920	0.11943
78	-0.23155	0.62300	0.66464	168	0.10009	-0.05510	0.11425
80	-0.10653	0.48720	0.49871	170	0.10730	-0.03359	0.11243
82	0.13648	-0.04994	0.14533	172	0.11183	-0.01559	0.11291
84	0.45218	-0.90373	1.01054	174	0.11447	-0.00151	0.11448
86	0.76615	-1.84438	1.99718	176	0.11586	0.00850	0.11617
88	0.99689	-2.57585	2.76202	178	0.11649	0.01447	0.11739
				180	0.11667	0.01645	0.11782

IBR

RADIATION PATTERN FOR THETA=PI/2

H/LMDA=0.2500 B/LMDA= 0.250 OMEGA= 8.53

28 ELEMENT ARRAY

PHI DEG	ER(THETA,PHI) RE	IM	ABSVAL	PHI DEG	ER(THETA,PHI) RE	IM	ABSVAL
0	-0.04297	0.01940	0.04715	90	0.09107	-4.01093	4.01196
2	-0.04293	0.01642	0.04597	92	0.08262	-3.62677	3.62771
4	-0.04281	0.00745	0.04345	94	0.06020	-2.60650	2.60720
6	-0.04252	-0.00751	0.04318	96	0.03130	-1.29158	1.29195
8	-0.04198	-0.02843	0.05070	98	0.00484	-0.09293	0.09306
10	-0.04106	-0.05500	0.06863	100	-0.01234	0.66866	0.66877
12	-0.03960	-0.08646	0.09510	102	-0.01780	0.87231	0.87249
14	-0.03745	-0.12138	0.12702	104	-0.01367	0.61376	0.61391
16	-0.03446	-0.15736	0.16109	106	-0.00490	0.13317	0.13326
18	-0.03053	-0.19090	0.19333	108	0.00329	-0.30383	0.30385
20	-0.02565	-0.21735	0.21885	110	0.00758	-0.51436	0.51442
22	-0.01993	-0.23116	0.23202	112	0.00736	-0.45273	0.45279
24	-0.01360	-0.22658	0.22699	114	0.00420	-0.19821	0.19826
26	-0.00707	-0.19872	0.19884	116	0.00048	0.10504	0.10504
28	-0.00085	-0.14508	0.14508	118	-0.00196	0.32110	0.32110
30	0.00445	-0.06733	0.06748	120	-0.00251	0.37559	0.37560
32	0.00826	0.02719	0.02842	122	-0.00172	0.27226	0.27227
34	0.01018	0.12487	0.12528	124	-0.00071	0.07503	0.07503
36	0.01007	0.20680	0.20704	126	-0.00039	-0.12895	0.12895
38	0.00816	0.25201	0.25214	128	-0.00102	-0.26605	0.26605
40	0.00507	0.24290	0.24295	130	-0.00218	-0.29931	0.29932
42	0.00164	0.17197	0.17198	132	-0.00304	-0.23284	0.23286
44	-0.00121	0.04802	0.04803	134	-0.00284	-0.10128	0.10132
46	-0.00284	-0.10128	0.10132	136	-0.00121	0.04802	0.04803
48	-0.00304	-0.23284	0.23286	138	0.00164	0.17197	0.17198
50	-0.00218	-0.29931	0.29932	140	0.00507	0.24290	0.24295
52	-0.00102	-0.26605	0.26605	142	0.00816	0.25202	0.25215
54	-0.00039	-0.12895	0.12895	144	0.01007	0.20680	0.20705
56	-0.00071	0.07502	0.07502	146	0.01018	0.12487	0.12528
58	-0.00172	0.27226	0.27227	148	0.00826	0.02719	0.02842
60	-0.00251	0.37559	0.37560	150	0.00445	-0.06733	0.06748
62	-0.00196	0.32110	0.32110	152	-0.00085	-0.14508	0.14508
64	0.00048	0.10505	0.10505	154	-0.00707	-0.19871	0.19884
66	0.00420	-0.19820	0.19824	156	-0.01360	-0.22658	0.22699
68	0.00736	-0.45272	0.45278	158	-0.01993	-0.23116	0.23201
70	0.00758	-0.51436	0.51442	160	-0.02565	-0.21734	0.21885
72	0.00329	-0.30384	0.30386	162	-0.03053	-0.19090	0.19332
74	-0.00491	0.13314	0.13323	164	-0.03446	-0.15736	0.16109
76	-0.01367	0.61374	0.61390	166	-0.03745	-0.12138	0.12702
78	-0.01780	0.87231	0.87249	168	-0.03960	-0.08646	0.09510
80	-0.01233	0.66867	0.66878	170	-0.04106	-0.05500	0.06863
82	0.00485	-0.09291	0.09304	172	-0.04198	-0.02843	0.05070
84	0.03131	-1.29151	1.29189	174	-0.04252	-0.00751	0.04318
86	0.06021	-2.60645	2.60714	176	-0.04281	0.00745	0.04345
88	0.08263	-3.62676	3.62770	178	-0.04293	0.01642	0.04597
				180	-0.04297	0.01940	0.04715

VBR

RADIATION PATTERN FOR THETA=PI/2

H/LMDA=0.2500 B/LMDA= 0.250 OMEGA= 8.53

30 ELEMENT ARRAY

PHI DEG	ER(THETA,PHI) RE	IM	ABSVAL	PHI DEG	ER(THETA,PHI) RE	IM	ABSVAL
0	0.00921	0.12960	0.12993	90	1.08876	-2.84961	3.05051
2	0.01058	0.12992	0.13035	92	0.98971	-2.53503	2.72138
4	0.01468	0.13070	0.13152	94	0.72448	-1.71547	1.86217
6	0.02143	0.13143	0.13317	96	0.37651	-0.70426	0.79859
8	0.03069	0.13123	0.13477	98	0.04923	0.14201	0.15031
10	0.04215	0.12889	0.13561	100	-0.17159	0.58008	0.60493
12	0.05525	0.12291	0.13476	102	-0.24645	0.56958	0.62061
14	0.06912	0.11159	0.13126	104	-0.19061	0.25857	0.32123
16	0.08248	0.09330	0.12453	106	-0.06110	-0.11504	0.13026
18	0.09362	0.06681	0.11501	108	0.07061	-0.34671	0.35383
20	0.10046	0.03178	0.10537	110	0.14766	-0.35005	0.37992
22	0.10077	-0.01062	0.10133	112	0.14677	-0.16979	0.22444
24	0.09250	-0.05723	0.10877	114	0.08103	0.06910	0.10649
26	0.07433	-0.10256	0.12666	116	-0.01209	0.23854	0.23884
28	0.04627	-0.13897	0.14647	118	-0.09004	0.26842	0.28312
30	0.01029	-0.15770	0.15804	120	-0.12308	0.16706	0.20750
32	-0.02935	-0.15076	0.15359	122	-0.10364	0.00088	0.10365
34	-0.06623	-0.11364	0.13153	124	-0.04542	-0.14731	0.15415
36	-0.09262	-0.04827	0.10444	126	0.02554	-0.21742	0.21891
38	-0.10117	0.03483	0.10700	128	0.08247	-0.19230	0.20924
40	-0.08716	0.11659	0.14557	130	0.10746	-0.09535	0.14366
42	-0.05087	0.17275	0.18008	132	0.09590	0.02750	0.09976
44	0.00096	0.18090	0.18090	134	0.05538	0.12976	0.14108
46	0.05538	0.12976	0.14108	136	0.00096	0.18090	0.18091
48	0.09590	0.02750	0.09976	138	-0.05087	0.17275	0.18008
50	0.10746	-0.09535	0.14366	140	-0.08716	0.11659	0.14557
52	0.08247	-0.19230	0.20924	142	-0.10117	0.03483	0.10700
54	0.02554	-0.21742	0.21891	144	-0.09262	-0.04827	0.10444
56	-0.04542	-0.14731	0.15415	146	-0.06623	-0.11364	0.13153
58	-0.10364	0.00087	0.10365	148	-0.02935	-0.15076	0.15359
60	-0.12308	0.16706	0.20750	150	0.01029	-0.15770	0.15804
62	-0.09004	0.26842	0.28312	152	0.04627	-0.13897	0.14647
64	-0.01209	0.23854	0.23884	154	0.07432	-0.10255	0.12666
66	0.08102	0.06911	0.10650	156	0.09250	-0.05723	0.10877
68	0.14677	-0.16979	0.22443	158	0.10077	-0.01061	0.10132
70	0.14766	-0.35005	0.37992	160	0.10046	0.03178	0.10537
72	0.07061	-0.34671	0.35383	162	0.09362	0.06681	0.11501
74	-0.06110	-0.11505	0.13027	164	0.08248	0.09330	0.12453
76	-0.19061	0.25855	0.32122	166	0.06912	0.11159	0.13126
78	-0.24645	0.56958	0.62061	168	0.05525	0.12291	0.13475
80	-0.17159	0.58009	0.60494	170	0.04215	0.12889	0.13561
82	0.04923	0.14203	0.15032	172	0.03069	0.13123	0.13477
84	0.37650	-0.70422	0.79855	174	0.02143	0.13143	0.13317
86	0.72447	-1.71542	1.86213	176	0.01468	0.13070	0.13152
88	0.98971	-2.53502	2.72137	178	0.01058	0.12992	0.13035
				180	0.00921	0.12960	0.12993

IBR

RADIATION PATTERN FOR THETA=PI/2

H/LMDA=0.2500 B/LMDA= 0.250 OMEGA= 8.53

30 ELEMENT ARRAY

PHI DEG	ER(THETA,PHI) RE	IM	ABSVAL	PHI DEG	ER(THETA,PHI) RE	IM	ABSVAL
0	0.01356	0.21675	0.21717	90	-0.04542	-3.72517	3.72545
2	0.01337	0.21697	0.21738	92	-0.04002	-3.31720	3.31744
4	0.01278	0.21737	0.21774	94	-0.02599	-2.25332	2.25347
6	0.01177	0.21725	0.21757	96	-0.00888	-0.93781	0.93786
8	0.01033	0.21538	0.21563	98	0.00500	0.16822	0.16829
10	0.00841	0.21010	0.21027	100	0.01137	0.74820	0.74829
12	0.00601	0.19936	0.19945	102	0.00965	0.74633	0.74639
14	0.00314	0.18092	0.18095	104	0.00275	0.35035	0.35036
16	-0.00014	0.15267	0.15267	106	-0.00470	-0.13392	0.13400
18	-0.00368	0.11311	0.11317	108	-0.00868	-0.44083	0.44092
20	-0.00728	0.06201	0.06244	110	-0.00754	-0.45500	0.45507
22	-0.01063	0.00119	0.01070	112	-0.00241	-0.22981	0.22982
24	-0.01337	-0.06485	0.06621	114	0.00377	0.07631	0.07641
26	-0.01507	-0.12855	0.12943	116	0.00787	0.29830	0.29840
28	-0.01535	-0.17966	0.18031	118	0.00805	0.34359	0.34368
30	-0.01395	-0.20658	0.20705	120	0.00443	0.21972	0.21977
32	-0.01080	-0.19895	0.19924	122	-0.00122	0.00969	0.00976
34	-0.00619	-0.15119	0.15132	124	-0.00640	-0.18081	0.18093
36	-0.00072	-0.06630	0.06631	126	-0.00904	-0.27375	0.27390
38	0.00467	0.04156	0.04182	128	-0.00819	-0.24541	0.24554
40	0.00886	0.14727	0.14754	130	-0.00428	-0.12418	0.12426
42	0.01086	0.21937	0.21964	132	0.00125	0.03182	0.03184
44	0.01006	0.22921	0.22943	134	0.00655	0.16291	0.16304
46	0.00655	0.16291	0.16304	136	0.01006	0.22921	0.22943
48	0.00125	0.03182	0.03185	138	0.01086	0.21937	0.21964
50	-0.00428	-0.12418	0.12426	140	0.00886	0.14728	0.14754
52	-0.00819	-0.24541	0.24554	142	0.00467	0.04156	0.04182
54	-0.00904	-0.27375	0.27390	144	-0.00072	-0.06630	0.06630
56	-0.00640	-0.18082	0.18093	146	-0.00619	-0.15119	0.15132
58	-0.00122	0.00968	0.00976	148	-0.01080	-0.19894	0.19924
60	0.00443	0.21972	0.21976	150	-0.01395	-0.20658	0.20705
62	0.00805	0.34359	0.34368	152	-0.01535	-0.17966	0.18031
64	0.00787	0.29830	0.29840	154	-0.01507	-0.12855	0.12943
66	0.00377	0.07633	0.07642	156	-0.01337	-0.06485	0.06621
68	-0.00241	-0.22981	0.22982	158	-0.01063	0.00119	0.01070
70	-0.00754	-0.45500	0.45506	160	-0.00728	0.06201	0.06244
72	-0.00868	-0.44084	0.44092	162	-0.00368	0.11311	0.11317
74	-0.00470	-0.13394	0.13402	164	-0.00014	0.15267	0.15267
76	0.00275	0.35033	0.35034	166	0.00314	0.18092	0.18095
78	0.00965	0.74632	0.74638	168	0.00601	0.19936	0.19945
80	0.01138	0.74821	0.74830	170	0.00841	0.21010	0.21027
82	0.00501	0.16823	0.16831	172	0.01033	0.21538	0.21563
84	-0.00887	-0.93776	0.93780	174	0.01177	0.21725	0.21757
86	-0.02598	-2.25326	2.25341	176	0.01278	0.21737	0.21774
88	-0.04001	-3.31718	3.31742	178	0.01337	0.21697	0.21738
				180	0.01356	0.21675	0.21717

VBR

RADIATION PATTERN FOR THETA=PI/2

H/LMDA=0.2500 B/LMDA= 0.250 OMEGA= 8.53

35 ELEMENT ARRAY

PHI	ER(THETA,PHI)			PHI	ER(THETA,PHI)		
DEG	RE	IM	ABSVAL	DEG	RE	IM	ABSVAL
0	0.06095	-0.06925	0.09225	90	1.10294	-2.84631	3.05253
1	0.06068	-0.06965	0.09237	91	1.06713	-2.73708	2.93775
2	0.05984	-0.07085	0.09274	92	0.96386	-2.42450	2.60906
3	0.05844	-0.07283	0.09337	93	0.80502	-1.95135	2.11088
4	0.05643	-0.07554	0.09429	94	0.60857	-1.38104	1.50918
5	0.05380	-0.07894	0.09553	95	0.39615	-0.78718	0.88124
6	0.05050	-0.08293	0.09710	96	0.19010	-0.24192	0.30767
7	0.04650	-0.08742	0.09902	97	0.01060	0.19507	0.19536
8	0.04175	-0.09228	0.10129	98	-0.12682	0.48501	0.50132
9	0.03622	-0.09735	0.10387	99	-0.21307	0.61436	0.65026
10	0.02988	-0.10241	0.10669	100	-0.24620	0.59477	0.64371
11	0.02273	-0.10726	0.10964	101	-0.23113	0.45882	0.51375
12	0.01476	-0.11159	0.11257	102	-0.17844	0.25256	0.30924
13	0.00602	-0.11512	0.11527	103	-0.10239	0.02652	0.10577
14	-0.00341	-0.11748	0.11753	104	-0.01862	-0.17319	0.17419
15	-0.01344	-0.11832	0.11908	105	0.05821	-0.31207	0.31746
16	-0.02389	-0.11724	0.11965	106	0.11638	-0.37163	0.38943
17	-0.03455	-0.11388	0.11901	107	0.14850	-0.35052	0.38068
18	-0.04515	-0.10788	0.11694	108	0.15213	-0.26250	0.30339
19	-0.05537	-0.09893	0.11337	109	0.12959	-0.13204	0.18501
20	-0.06481	-0.08682	0.10835	110	0.08711	0.01145	0.08786
21	-0.07307	-0.07147	0.10221	111	0.03343	0.13958	0.14352
22	-0.07969	-0.05294	0.09567	112	-0.02179	0.23000	0.23103
23	-0.08421	-0.03148	0.08991	113	-0.06955	0.26969	0.27852
24	-0.08619	-0.00761	0.08652	114	-0.10287	0.25622	0.27610
25	-0.08523	0.01795	0.08710	115	-0.11759	0.19691	0.22935
26	-0.08102	0.04418	0.09228	116	-0.11281	0.10646	0.15511
27	-0.07339	0.06983	0.10131	117	-0.09067	0.00349	0.09073
28	-0.06233	0.09346	0.11234	118	-0.05573	-0.09308	0.10849
29	-0.04803	0.11348	0.12322	119	-0.01410	-0.16732	0.16792
30	-0.03092	0.12824	0.13192	120	0.02763	-0.20862	0.21044
31	-0.01167	0.13621	0.13671	121	0.06339	-0.21284	0.22208
32	0.00880	0.13608	0.13636	122	0.08844	-0.18227	0.20259
33	0.02934	0.12698	0.13032	123	0.09993	-0.12449	0.15963
34	0.04864	0.10860	0.11899	124	0.09710	-0.05060	0.10949
35	0.06530	0.08138	0.10434	125	0.08119	0.02692	0.08554
36	0.07793	0.04659	0.09079	126	0.05509	0.09627	0.11092
37	0.08528	0.00638	0.08551	127	0.02277	0.14800	0.14974
38	0.08638	-0.03627	0.09369	128	-0.01133	0.17613	0.17650
39	0.08069	-0.07771	0.11203	129	-0.04288	0.17857	0.18364
40	0.06819	-0.11392	0.13277	130	-0.06824	0.15693	0.17113
41	0.04948	-0.14083	0.14927	131	-0.08480	0.11590	0.14361
42	0.02583	-0.15486	0.15700	132	-0.09118	0.06216	0.11035
43	-0.00088	-0.15337	0.15337	133	-0.08727	0.00334	0.08733
44	-0.02826	-0.13516	0.13808	134	-0.07410	-0.05312	0.09117
45	-0.05359	-0.10085	0.11421	135	-0.05359	-0.10085	0.11421
46	-0.07410	-0.05313	0.09118	136	-0.02826	-0.13516	0.13808
47	-0.08727	0.00333	0.08733	137	-0.00089	-0.15337	0.15337
48	-0.09118	0.06215	0.11035	138	0.02583	-0.15486	0.15700
49	-0.08480	0.11589	0.14360	139	0.04948	-0.14083	0.14927
50	-0.06824	0.15693	0.17113	140	0.06819	-0.11392	0.13277
51	-0.04288	0.17857	0.18364	141	0.08069	-0.07771	0.11203
52	-0.01133	0.17613	0.17650	142	0.08638	-0.03627	0.09369
53	0.02276	0.14800	0.14974	143	0.08528	0.00638	0.08551
54	0.05509	0.09627	0.11092	144	0.07793	0.04659	0.09079
55	0.08119	0.02693	0.08554	145	0.06530	0.08138	0.10434
56	0.09710	-0.05059	0.10949	146	0.04864	0.10859	0.11899
57	0.09993	-0.12448	0.15963	147	0.02934	0.12698	0.13032
58	0.08844	-0.18227	0.20259	148	0.00880	0.13608	0.13636
59	0.06339	-0.21284	0.22208	149	-0.01167	0.13621	0.13671
60	0.02763	-0.20862	0.21045	150	-0.03092	0.12824	0.13191
61	-0.01410	-0.16733	0.16792	151	-0.04803	0.11348	0.12323
62	-0.05572	-0.09309	0.10849	152	-0.06233	0.09347	0.11234
63	-0.09066	0.00348	0.09073	153	-0.07339	0.06984	0.10131
64	-0.11281	0.10645	0.15511	154	-0.08102	0.04418	0.09228
65	-0.11759	0.19691	0.22935	155	-0.08523	0.01795	0.08710
66	-0.10287	0.25622	0.27610	156	-0.08619	-0.00761	0.08652
67	-0.06956	0.26969	0.27852	157	-0.08421	-0.03148	0.08990
68	-0.02179	0.23000	0.23103	158	-0.07969	-0.05293	0.09567
69	0.03343	0.13958	0.14353	159	-0.07307	-0.07147	0.10221
70	0.08711	0.01145	0.08786	160	-0.06481	-0.08682	0.10835
71	0.12959	-0.13203	0.18500	161	-0.05537	-0.09893	0.11337
72	0.15213	-0.26248	0.30338	162	-0.04515	-0.10788	0.11694
73	0.14850	-0.35051	0.38067	163	-0.03455	-0.11388	0.11901
74	0.11638	-0.37164	0.38943	164	-0.02389	-0.11725	0.11965
75	0.05822	-0.31208	0.31747	165	-0.01344	-0.11832	0.11908
76	-0.01861	-0.17321	0.17420	166	-0.00342	-0.11748	0.11753
77	-0.10239	0.02650	0.10576	167	0.00602	-0.11512	0.11527
78	-0.17843	0.25254	0.30922	168	0.01477	-0.11160	0.11257
79	-0.23113	0.45881	0.51374	169	0.02273	-0.10726	0.10964
80	-0.24620	0.59477	0.64371	170	0.02989	-0.10242	0.10669
81	-0.21307	0.61436	0.65026	171	0.03622	-0.09735	0.10387
82	-0.12683	0.48503	0.50134	172	0.04175	-0.09229	0.10129
83	0.01059	0.19511	0.19540	173	0.04650	-0.08742	0.09902
84	0.19008	-0.24187	0.30763	174	0.05050	-0.08293	0.09710
85	0.39614	-0.78712	0.88118	175	0.05380	-0.07894	0.09553
86	0.60856	-1.38098	1.50912	176	0.05643	-0.07554	0.09429
87	0.80501	-1.95130	2.11083	177	0.05844	-0.07283	0.09337
88	0.96385	-2.42446	2.60902	178	0.05984	-0.07085	0.09274
89	1.06713	-2.73707	2.93774	179	0.06068	-0.06965	0.09237
				180	0.06095	-0.06925	0.09225

IBR

RADIATION PATTERN FOR THETA=PI/2

H/LMDA=0.2500 B/LMDA= 0.250 OMEGA= 8.53

35 ELEMENT ARRAY

PHI	ER(THETA,PHI)			PHI	ER(THETA,PHI)		
DEG	RE	IM	ABSVAL	DEG	RE	IM	ABSVAL
0	-0.02712	-0.12484	0.12775	90	0.13540	-3.79729	3.79970
1	-0.02707	-0.12538	0.12827	91	0.13040	-3.65267	3.65499
2	-0.02692	-0.12699	0.12981	92	0.11608	-3.23868	3.24076
3	-0.02666	-0.12964	0.13235	93	0.09436	-2.61165	2.61336
4	-0.02629	-0.13326	0.13583	94	0.06808	-1.85508	1.85633
5	-0.02581	-0.13776	0.14016	95	0.04053	-1.06603	1.06680
6	-0.02520	-0.14303	0.14524	96	0.01498	-0.33981	0.34014
7	-0.02446	-0.14891	0.15090	97	-0.00586	0.24452	0.24459
8	-0.02358	-0.15520	0.15698	98	-0.02021	0.63519	0.63551
9	-0.02255	-0.16165	0.16322	99	-0.02737	0.81360	0.81406
10	-0.02137	-0.16799	0.16934	100	-0.02776	0.79429	0.79478
11	-0.02003	-0.17387	0.17502	101	-0.02273	0.61951	0.61993
12	-0.01852	-0.17890	0.17986	102	-0.01424	0.34949	0.34978
13	-0.01686	-0.18265	0.18343	103	-0.00447	0.05070	0.05090
14	-0.01504	-0.18465	0.18526	104	0.00457	-0.21581	0.21585
15	-0.01308	-0.18439	0.18485	105	0.01132	-0.40394	0.40410
16	-0.01101	-0.18134	0.18168	106	0.01491	-0.48844	0.48866
17	-0.00885	-0.17503	0.17525	107	0.01516	-0.46652	0.46677
18	-0.00663	-0.16499	0.16512	108	0.01253	-0.35540	0.35562
19	-0.00439	-0.15084	0.15090	109	0.00793	-0.18657	0.18674
20	-0.00220	-0.13234	0.13236	110	0.00248	0.00162	0.00296
21	-0.00010	-0.10942	0.10942	111	-0.00269	0.17180	0.17182
22	0.00185	-0.08223	0.08225	112	-0.00669	0.29418	0.29426
23	0.00359	-0.05118	0.05130	113	-0.00897	0.35099	0.35110
24	0.00506	-0.01701	0.01774	114	-0.00937	0.33820	0.33833
25	0.00621	0.01922	0.02019	115	-0.00809	0.26467	0.26479
26	0.00700	0.05609	0.05653	116	-0.00559	0.14904	0.14915
27	0.00740	0.09189	0.09218	117	-0.00249	0.01538	0.01558
28	0.00742	0.12463	0.12485	118	0.00060	-0.11159	0.11159
29	0.00706	0.15218	0.15235	119	0.00315	-0.21081	0.21084
30	0.00635	0.17237	0.17248	120	0.00481	-0.26795	0.26799
31	0.00536	0.18316	0.18324	121	0.00547	-0.27705	0.27710
32	0.00417	0.18289	0.18294	122	0.00516	-0.24053	0.24059
33	0.00287	0.17048	0.17051	123	0.00412	-0.16783	0.16788
34	0.00157	0.14563	0.14564	124	0.00263	-0.07311	0.07316
35	0.00036	0.10904	0.10904	125	0.00103	0.02746	0.02748
36	-0.00067	0.06250	0.06251	126	-0.00040	0.11846	0.11846
37	-0.00143	0.00899	0.00910	127	-0.00146	0.18735	0.18736
38	-0.00189	-0.04750	0.04753	128	-0.00206	0.22602	0.22603
39	-0.00204	-0.10207	0.10209	129	-0.00219	0.23139	0.23140
40	-0.00191	-0.14940	0.14941	130	-0.00194	0.20524	0.20525
41	-0.00157	-0.18419	0.18420	131	-0.00146	0.15340	0.15341
42	-0.00111	-0.20181	0.20181	132	-0.00091	0.08443	0.08443
43	-0.00064	-0.19892	0.19893	133	-0.00043	0.00818	0.00819
44	-0.00027	-0.17414	0.17414	134	-0.00014	-0.06560	0.06560
45	-0.00008	-0.12847	0.12847	135	-0.00008	-0.12847	0.12847
46	-0.00014	-0.06560	0.06560	136	-0.00027	-0.17414	0.17414
47	-0.00043	0.00817	0.00818	137	-0.00064	-0.19893	0.19893
48	-0.00091	0.08442	0.08443	138	-0.00111	-0.20181	0.20182
49	-0.00146	0.15340	0.15341	139	-0.00157	-0.18419	0.18420
50	-0.00194	0.20525	0.20526	140	-0.00191	-0.14940	0.14941
51	-0.00219	0.23139	0.23140	141	-0.00204	-0.10207	0.10209
52	-0.00206	0.22602	0.22603	142	-0.00189	-0.04750	0.04753
53	-0.00146	0.18735	0.18736	143	-0.00143	0.00898	0.00909
54	-0.00040	0.11847	0.11847	144	-0.00067	0.06250	0.06250
55	0.00103	0.02747	0.02749	145	0.00036	0.10904	0.10904
56	0.00263	-0.07311	0.07316	146	0.00157	0.14563	0.14564
57	0.00412	-0.16783	0.16788	147	0.00287	0.17049	0.17051
58	0.00516	-0.24053	0.24058	148	0.00417	0.18289	0.18294
59	0.00547	-0.27705	0.27710	149	0.00536	0.18316	0.18324
60	0.00481	-0.26795	0.26800	150	0.00635	0.17236	0.17248
61	0.00315	-0.21082	0.21084	151	0.00706	0.15219	0.15235
62	0.00060	-0.11160	0.11161	152	0.00742	0.12463	0.12485
63	-0.00249	0.01537	0.01557	153	0.00740	0.09189	0.09219
64	-0.00559	0.14903	0.14914	154	0.00700	0.05609	0.05652
65	-0.00809	0.26466	0.26479	155	0.00621	0.01922	0.02020
66	-0.00937	0.33820	0.33833	156	0.00506	-0.01701	0.01775
67	-0.00897	0.35099	0.35110	157	0.00359	-0.05118	0.05130
68	-0.00669	0.29419	0.29427	158	0.00185	-0.08222	0.08224
69	-0.00269	0.17181	0.17183	159	-0.00010	-0.10942	0.10942
70	0.00248	0.00162	0.00297	160	-0.00220	-0.13234	0.13236
71	0.00793	-0.18656	0.18673	161	-0.00439	-0.15084	0.15090
72	0.01253	-0.35538	0.35561	162	-0.00663	-0.16499	0.16512
73	0.01516	-0.46652	0.46676	163	-0.00884	-0.17504	0.17526
74	0.01491	-0.48844	0.48867	164	-0.01101	-0.18135	0.18168
75	0.01132	-0.40395	0.40411	165	-0.01308	-0.18438	0.18484
76	0.00456	-0.21583	0.21587	166	-0.01504	-0.18465	0.18526
77	-0.00447	0.05068	0.05087	167	-0.01686	-0.18265	0.18343
78	-0.01424	0.34946	0.34975	168	-0.01852	-0.17891	0.17986
79	-0.02273	0.61950	0.61992	169	-0.02003	-0.17387	0.17502
80	-0.02776	0.79429	0.79478	170	-0.02137	-0.16799	0.16934
81	-0.02736	0.81361	0.81407	171	-0.02255	-0.16165	0.16322
82	-0.02021	0.63522	0.63554	172	-0.02358	-0.15520	0.15698
83	-0.00586	0.24456	0.24463	173	-0.02446	-0.14891	0.15090
84	0.01499	-0.33975	0.34008	174	-0.02520	-0.14303	0.14524
85	0.04054	-1.06596	1.06673	175	-0.02581	-0.13777	0.14016
86	0.06809	-1.85500	1.85625	176	-0.02629	-0.13326	0.13583
87	0.09437	-2.61159	2.61329	177	-0.02666	-0.12964	0.13235
88	0.11609	-3.23863	3.24071	178	-0.02692	-0.12699	0.12981
89	0.13040	-3.65266	3.65499	179	-0.02707	-0.12538	0.12827
				180	-0.02712	-0.12484	0.12775

VBR

RADIATION PATTERN FOR THETA=PI/2

H/LMDA=0.2500 B/LMDA= 0.250 OMEGA= 8.53

40 ELEMENT ARRAY

PHI DEG	ER(THETA,PHI) RE	IM	ABSVAL	PHI DEG	ER(THETA,PHI) RE	IM	ABSVAL
C	-0.08238	-0.01101	0.08311	90	1.11346	-2.84379	3.05400
1	-0.08233	-0.01052	0.08300	91	1.06539	-2.70193	2.90439
2	-0.08220	-0.00905	0.08270	92	0.92860	-2.30177	2.48203
3	-0.08194	-0.00659	0.08221	93	0.72395	-1.71414	1.86075
4	-0.08151	-0.00316	0.08157	94	0.48197	-1.04010	1.14634
5	-0.08085	0.00129	0.08086	95	0.23733	-0.38977	0.45634
6	-0.07987	0.00674	0.08015	96	0.02274	0.13997	0.14180
7	-0.07846	0.01316	0.07956	97	-0.13652	0.48383	0.50272
8	-0.07653	0.02054	0.07924	98	-0.22648	0.61857	0.65873
9	-0.07395	0.02881	0.07936	99	-0.24599	0.56372	0.61506
10	-0.07057	0.03786	0.08009	100	-0.20581	0.37296	0.42598
11	-0.06629	0.04755	0.08158	101	-0.12552	0.11894	0.17293
12	-0.06097	0.05766	0.08392	102	-0.02906	-0.12452	0.12787
13	-0.05451	0.06792	0.08709	103	0.06025	-0.29856	0.30458
14	-0.04684	0.07798	0.09096	104	0.12398	-0.37017	0.39038
15	-0.03791	0.08739	0.09526	105	0.15156	-0.33597	0.36857
16	-0.02777	0.09568	0.09963	106	0.14140	-0.21877	0.26049
17	-0.01648	0.10227	0.10359	107	0.10012	-0.05853	0.11597
18	-0.00424	0.10656	0.10665	108	0.04017	0.09945	0.10726
19	0.00868	0.10794	0.10829	109	-0.02341	0.21611	0.21737
20	0.02192	0.10580	0.10804	110	-0.07637	0.26707	0.27778
21	0.03499	0.09964	0.10561	111	-0.10822	0.24676	0.26945
22	0.04731	0.08911	0.10089	112	-0.11395	0.16733	0.20244
23	0.05825	0.07406	0.09422	113	-0.09451	0.05369	0.10869
24	0.06709	0.05462	0.08651	114	-0.05604	-0.06384	0.08495
25	0.07314	0.03128	0.07955	115	-0.00806	-0.15725	0.15746
26	0.07575	0.00493	0.07591	116	0.03877	-0.20699	0.21059
27	0.07438	-0.02313	0.07789	117	0.07497	-0.20548	0.21873
28	0.06869	-0.05121	0.08568	118	0.09396	-0.15742	0.18333
29	0.05862	-0.07725	0.09698	119	0.09313	-0.07737	0.12108
30	0.04442	-0.09906	0.10856	120	0.07394	0.01457	0.07536
31	0.02672	-0.11435	0.11744	121	0.04127	0.09774	0.10610
32	0.00655	-0.12112	0.12129	122	0.00216	0.15534	0.15536
33	-0.01470	-0.11780	0.11871	123	-0.03573	0.17749	0.18105
34	-0.03533	-0.10365	0.10950	124	-0.06555	0.16247	0.17520
35	-0.05346	-0.07893	0.09533	125	-0.08245	0.11612	0.14241
36	-0.06723	-0.04512	0.08096	126	-0.08424	0.04979	0.09785
37	-0.07495	-0.00497	0.07512	127	-0.07150	-0.02248	0.07495
38	-0.07540	0.03761	0.08426	128	-0.04721	-0.08691	0.09890
39	-0.06798	0.07784	0.10334	129	-0.01600	-0.13251	0.13348
40	-0.05295	0.11054	0.12257	130	0.01676	-0.15274	0.15365
41	-0.03149	0.13088	0.13462	131	0.04589	-0.14606	0.15310
42	-0.00573	0.13504	0.13516	132	0.06723	-0.11562	0.13375
43	0.02143	0.12101	0.12290	133	0.07809	-0.06814	0.10363
44	0.04662	0.08921	0.10066	134	0.07754	-0.01230	0.07851
45	0.06634	0.04282	0.07896	135	0.06634	0.04281	0.07896
46	0.07754	-0.01230	0.07851	136	0.04662	0.08921	0.10065
47	0.07809	-0.06813	0.10363	137	0.02144	0.12101	0.12289
48	0.06723	-0.11562	0.13375	138	-0.00573	0.13504	0.13516
49	0.04589	-0.14606	0.15310	139	[illegible]	[illegible]	[illegible]
50	0.01676	-0.15274	0.15366	140	-0.05295	0.11055	0.12257
51	-0.01600	-0.13251	0.13348	141	-0.06798	0.07783	0.10334
52	-0.04721	-0.08691	0.09890	142	-0.07540	0.03761	0.08426
53	-0.07150	-0.02249	0.07495	143	-0.07495	-0.00497	0.07512
54	-0.08424	0.04978	0.09785	144	-0.06723	-0.04512	0.08096
55	-0.08245	0.11612	0.14241	145	-0.05346	-0.07892	0.09533
56	-0.06555	0.16247	0.17520	146	-0.03533	-0.10365	0.10950
57	-0.03573	0.17749	0.18105	147	-0.01470	-0.11780	0.11871
58	0.00215	0.15535	0.15536	148	0.00655	-0.12112	0.12129
59	0.04127	0.09775	0.10610	149	0.02672	-0.11435	0.11744
60	0.07394	0.01457	0.07536	150	0.04442	-0.09906	0.10856
61	0.09313	-0.07737	0.12108	151	0.05862	-0.07726	0.09698
62	0.09396	-0.15742	0.18333	152	0.06869	-0.05121	0.08568
63	0.07497	-0.20548	0.21873	153	0.07438	-0.02313	0.07789
64	0.03878	-0.20699	0.21059	154	0.07575	0.00493	0.07591
65	-0.00806	-0.15726	0.15746	155	0.07314	0.03128	0.07955
66	-0.05603	-0.06385	0.08495	156	0.06709	0.05462	0.08651
67	-0.09450	0.05368	0.10868	157	0.05825	0.07405	0.09422
68	-0.11395	0.16731	0.20243	158	0.04731	0.08911	0.10090
69	-0.10822	0.24676	0.26945	159	0.03498	0.09965	0.10561
70	-0.07637	0.26707	0.27778	160	0.02192	0.10580	0.10804
71	-0.02341	0.21611	0.21737	161	0.00868	0.10794	0.10829
72	0.04016	0.09947	0.10727	162	-0.00424	0.10657	0.10665
73	0.10011	-0.05852	0.11596	163	-0.01648	0.10228	0.10360
74	0.14140	-0.21876	0.26048	164	-0.02776	0.09568	0.09963
75	0.15156	-0.33597	0.36857	165	-0.03791	0.08739	0.09526
76	0.12398	-0.37017	0.39039	166	-0.04684	0.07797	0.09096
77	0.06026	-0.29858	0.30460	167	-0.05451	0.06792	0.08709
78	-0.02905	-0.12454	0.12789	168	-0.06097	0.05766	0.08392
79	-0.12552	0.11893	0.17292	169	-0.06629	0.04755	0.08158
80	-0.20581	0.37296	0.42597	170	-0.07057	0.03786	0.08009
81	-0.24599	0.56372	0.61505	171	-0.07395	0.02881	0.07936
82	-0.22649	0.61858	0.65874	172	-0.07653	0.02054	0.07924
83	-0.13653	0.48385	0.50274	173	-0.07846	0.01316	0.07956
84	0.02273	0.14001	0.14185	174	-0.07987	0.00674	0.08015
85	0.23732	-0.38971	0.45628	175	-0.08085	0.00129	0.08086
86	0.48196	-1.04003	1.14628	176	-0.08151	-0.00316	0.08157
87	0.72394	-1.71408	1.86069	177	-0.08194	-0.00660	0.08221
88	0.92858	-2.30173	2.48198	178	-0.08220	-0.00905	0.08270
89	1.06539	-2.70192	2.90438	179	-0.08233	-0.01052	0.08300
				180	-0.08238	-0.01101	0.08311

IBR

RADIATION PATTERN FOR THETA=PI/2

H/LMDA=0.2500 B/LMDA= 0.250 OMEGA= 8.53

40 ELEMENT ARRAY

PHI DEG	ER(THETA,PHI) RE	IM	ABSVAL	PHI DEG	ER(THETA,PHI) RE	IM	ABSVAL
0	0.03415	-C.01511	0.03735	90	0.07145	-3.97142	3.97206
1	0.03415	-0.01436	0.03705	91	0.06804	-3.77467	3.77529
2	0.03415	-0.01211	0.03623	92	0.05842	-3.21950	3.22002
3	0.03413	-0.00834	0.03513	93	0.04424	-2.40357	2.40398
4	0.03409	-0.00308	0.03423	94	0.02790	-1.46645	1.46671
5	0.03401	0.00371	0.03421	95	0.01197	-0.56037	0.56050
6	0.03388	0.01199	0.03594	96	-0.00124	0.18034	0.18035
7	0.03367	0.02173	0.04007	97	-0.01019	0.66467	0.66475
8	0.C3335	0.03287	0.04683	98	-0.01427	0.85941	0.85953
9	0.03290	0.04530	0.05598	99	-0.01388	0.79034	0.79047
10	0.03228	0.05883	0.06710	100	-0.01019	0.53065	0.53075
11	0.03147	0.07324	0.07971	101	-0.00478	0.18016	0.18022
12	0.03042	0.08820	0.09330	102	0.00071	-0.15907	0.15907
13	0.02912	0.10328	0.10731	103	0.00496	-0.40496	0.40499
14	0.02753	0.11797	0.12114	104	0.00721	-0.51067	0.51072
15	0.02564	0.13162	0.13409	105	0.00730	-0.47004	0.47009
16	0.02344	0.14352	0.14542	106	0.00563	-0.31316	0.31321
17	0.02093	0.15285	0.15427	107	0.00292	-0.09420	0.09425
18	0.01813	0.15875	0.15978	108	0.00002	0.12468	0.12468
19	0.01508	0.16035	0.16106	109	-0.00236	0.28922	0.28923
20	0.01182	0.15683	0.15728	110	-0.00375	0.36492	0.36494
21	0.00843	0.14752	0.14777	111	-0.00403	0.34282	0.34284
22	0.00499	0.13193	0.13203	112	-0.00336	0.23839	0.23841
23	0.00162	0.10992	0.10994	113	-0.00211	0.08493	0.08496
24	-0.00158	0.08173	0.08175	114	-0.00070	-0.07633	0.07634
25	-0.00447	0.04812	0.04833	115	0.00052	-0.20680	0.20680
26	-0.00692	0.01037	0.01247	116	0.00132	-0.27904	0.27904
27	-0.00884	-0.02961	0.03090	117	0.00163	-0.28172	0.28172
28	-0.01012	-0.06942	0.07016	118	0.00154	-0.22029	0.22030
29	-0.01069	-0.10619	0.10673	119	0.00122	-0.11386	0.11386
30	-0.01054	-0.13683	0.13723	120	0.00086	0.01059	0.01062
31	-0.00969	-0.15819	0.15849	121	0.00059	0.12487	0.12487
32	-0.00822	-0.16752	0.16772	122	0.00047	0.20575	0.20575
33	-0.00624	-0.16277	0.16289	123	0.00047	0.23920	0.23920
34	-0.00394	-0.14296	0.14302	124	0.00049	0.22218	0.22218
35	-0.00150	-0.10860	0.10861	125	0.00043	0.16201	0.16201
36	0.00085	-0.06183	0.06184	126	0.00019	0.07368	0.07368
37	0.00290	-0.00656	0.00717	127	-0.00028	-0.02391	0.02392
38	0.00447	0.05177	0.05196	128	-0.00092	-0.11202	0.11203
39	0.00543	0.10655	0.10668	129	-0.00165	-0.17546	0.17547
40	0.00571	0.15070	0.15081	130	-0.00230	-0.20496	0.20497
41	0.00533	0.17768	0.17776	131	-0.00270	-0.19803	0.19804
42	0.00438	0.18240	0.18246	132	-0.00272	-0.15856	0.15859
43	0.00303	0.16232	0.16235	133	-0.00227	-0.09540	0.09543
44	0.00148	0.11822	0.11822	134	-0.00136	-0.02022	0.02027
45	-0.00005	0.05464	0.05464	135	-0.00005	0.05464	0.05464
46	-0.00136	-0.02022	0.02027	136	0.00148	0.11821	0.11822
47	-0.00227	-0.09540	0.09542	137	0.00303	0.16231	0.16234
48	-0.00272	-0.15856	0.15858	138	0.00438	0.18240	0.18246
49	-0.00270	-0.19802	0.19804	139	0.00533	0.17769	0.17777
50	-0.00230	-0.20496	[illegible]	140	[illegible]	[illegible]	[illegible]
51	-0.00165	-0.17546	0.17547	141	0.00543	0.10654	0.10668
52	-0.00092	-0.11202	0.11203	142	0.00447	0.05177	0.05197
53	-0.00028	-0.02392	0.02392	143	0.00290	-0.00656	0.00717
54	0.00019	0.07367	0.07367	144	0.00085	-0.06183	0.06184
55	0.00043	0.16201	0.16201	145	-0.00150	-0.10859	0.10860
56	0.00049	0.22218	0.22218	146	-0.00394	-0.14296	0.14302
57	0.00047	0.23920	0.23920	147	-0.00624	-0.16277	0.16289
58	0.00047	0.20576	0.20576	148	-0.00822	-0.16752	0.16772
59	0.00059	0.12488	0.12488	149	-0.00969	-0.15819	0.15849
60	0.00086	0.01059	0.01063	150	-0.01054	-0.13683	0.13723
61	0.00122	-0.11385	0.11386	151	-0.01069	-0.10620	0.10674
62	0.00154	-0.22028	0.22029	152	-0.01012	-0.06942	0.07016
63	0.00163	-0.28171	0.28172	153	-0.00884	-0.02961	0.03090
64	0.00132	-0.27904	0.27905	154	-0.00693	0.01037	0.01247
65	0.00052	-0.20681	0.20681	155	-0.00447	0.04812	0.04832
66	-0.00070	-0.07635	0.07635	156	-0.00158	0.08174	0.08175
67	-0.00211	0.08492	0.08494	157	0.00161	0.10992	0.10993
68	-0.00336	0.23837	0.23840	158	0.00499	0.13194	0.13203
69	-0.00403	0.34282	0.34284	159	0.00843	0.14753	0.14777
70	-0.00375	0.36492	0.36494	160	0.01182	0.15684	0.15728
71	-0.00236	0.28922	0.28923	161	0.01508	0.16035	0.16106
72	0.00002	0.12469	0.12469	162	0.01813	0.15875	0.15978
73	0.00292	-0.09418	0.09423	163	0.02093	0.15285	0.15428
74	0.00563	-0.31314	0.31319	164	0.02344	0.14352	0.14542
75	0.00731	-0.47003	0.47008	165	0.02564	0.13162	0.13410
76	0.00721	-0.51067	0.51072	166	0.02753	0.11796	0.12113
77	0.00496	-0.40497	0.40500	167	0.02912	0.10328	0.10731
78	0.00071	-0.15910	C.15910	168	0.03042	0.08819	0.09329
79	-0.00478	0.18014	0.18021	169	0.03147	0.07324	0.07971
80	-0.01019	0.53063	0.53073	170	0.03228	0.05883	0.06710
81	-0.01389	0.79034	0.79046	171	0.03290	0.04529	0.05598
82	-0.01427	0.85942	0.85954	172	0.03335	0.03287	0.04683
83	-0.01018	0.66470	0.66478	173	0.03367	0.02173	0.04007
84	-0.00123	0.18040	0.18040	174	0.03388	0.01199	0.03594
85	0.01198	-0.56029	0.56042	175	0.03401	0.00371	0.03421
86	0.02791	-1.46636	1.46663	176	0.03409	-0.00308	0.03423
87	0.04425	-2.40350	2.40390	177	0.03413	-0.00834	0.03513
88	0.05843	-3.21943	3.21996	178	0.03415	-0.01211	0.03623
89	0.06805	-3.77466	3.77527	179	0.03415	-0.01436	0.03705
				180	0.03415	-0.01511	0.03735

VBR

RADIATION PATTERN FOR THETA=PI/2

H/LMDA=0.2500 B/LMDA= 0.250 OMEGA= 8.53

45 ELEMENT ARRAY

PHI DEG	ER(THETA,PHI) RE	IM	ABSVAL	PHI DEG	ER(THETA,PHI) RE	IM	ABSVAL
0	0.05589	0.06742	0.08758	90	1.12169	-2.84174	3.05511
1	0.05612	0.06713	0.08750	91	1.05963	-2.66324	2.86630
2	0.05678	0.06624	0.08725	92	0.88571	-2.16792	2.34187
3	0.05785	0.06472	0.08680	93	0.63382	-1.46552	1.59671
4	0.05928	0.06250	0.08614	94	0.35155	-0.70603	0.78871
5	0.06101	0.05950	0.08522	95	0.08932	-0.04055	0.09809
6	0.06296	0.05563	0.08401	96	-0.11081	0.41625	0.43075
7	0.06501	0.05076	0.08248	97	-0.22372	0.61172	0.65135
8	0.06702	0.04479	0.08061	98	-0.24538	0.56225	0.61347
9	0.06883	0.03762	0.07844	99	-0.19167	0.34113	0.39129
10	0.07026	0.02917	0.07607	100	-0.09259	0.05221	0.10629
11	0.07109	0.01938	0.07369	101	0.01639	-0.20200	0.20266
12	0.07109	0.00829	0.07157	102	0.10344	-0.34800	0.36305
13	0.07001	-0.00402	0.07013	103	0.14766	-0.35742	0.38672
14	0.06762	-0.01737	0.06982	104	0.14275	-0.24809	0.28622
15	0.06369	-0.03145	0.07103	105	0.09662	-0.07173	0.12034
16	0.05803	-0.04582	0.07394	106	0.02747	0.10655	0.11003
17	0.05051	-0.05990	0.07835	107	-0.04248	0.22994	0.23383
18	0.04109	-0.07297	0.08374	108	-0.09337	0.26601	0.28193
19	0.02984	-0.08420	0.08933	109	-0.11284	0.21343	0.24142
20	0.01698	-0.09266	0.09421	110	-0.09849	0.09830	0.13915
21	0.00288	-0.09742	0.09746	111	-0.05741	-0.03714	0.06838
22	-0.01190	-0.09756	0.09828	112	-0.00331	-0.14928	0.14932
23	-0.02665	-0.09235	0.09612	113	0.04793	-0.20633	0.21183
24	-0.04048	-0.08134	0.09086	114	0.08264	-0.19636	0.21304
25	-0.05240	-0.06447	0.08308	115	0.09279	-0.12844	0.15845
26	-0.06140	-0.04219	0.07450	116	0.07750	-0.02764	0.08228
27	-0.06650	-0.01557	0.06830	117	0.04261	0.07412	0.08549
28	-0.06690	0.01369	0.06828	118	-0.00153	0.14782	0.14783
29	-0.06208	0.04329	0.07568	119	-0.04305	0.17519	0.18040
30	-0.05196	0.07044	0.08753	120	-0.07174	0.15247	0.16851
31	-0.03694	0.09214	0.09927	121	-0.08134	0.08984	0.12119
32	-0.01804	0.10545	0.10698	122	-0.07074	0.00704	0.07109
33	0.00318	0.10799	0.10803	123	-0.04370	-0.07283	0.08493
34	0.02465	0.09829	0.10133	124	-0.00751	-0.12965	0.12987
35	0.04398	0.07630	0.08807	125	0.02902	-0.15106	0.15382
36	0.05873	0.04361	0.07315	126	0.05770	-0.13452	0.14637
37	0.06673	0.00360	0.06683	127	0.07273	-0.08684	0.11328
38	0.06645	-0.03877	0.07693	128	0.07169	-0.02145	0.07483
39	0.05731	-0.07746	0.09636	129	0.05573	0.04552	0.07196
40	0.03996	-0.10617	0.11344	130	0.02896	0.09920	0.10334
41	0.01636	-0.11937	0.12049	131	-0.00272	0.12906	0.12909
42	-0.01035	-0.11347	0.11394	132	-0.03292	0.13069	0.13477
43	-0.03615	-0.08782	0.09497	133	-0.05608	0.10588	0.11982
44	-0.05678	-0.04530	0.07264	134	-0.06842	0.06155	0.09203
45	-0.06841	0.00759	0.06883	135	-0.06841	0.00759	0.06883
46	-0.06842	0.06155	0.09203	136	-0.05678	-0.04531	0.07264
47	-0.05608	0.10589	0.11982	137	-0.03615	-0.08782	0.09497
48	-0.03291	0.13068	0.13477	138	-0.01035	-0.11347	0.11394
49	-0.00272	0.12906	0.12909	139	0.01636	-0.11937	0.12049
50	0.02896	0.09920	0.10334	140	0.03996	-0.10617	0.11344
51	0.05573	0.04553	0.07196	141	0.05731	-0.07747	0.09636
52	0.07169	-0.02145	0.07483	142	0.06645	-0.03877	0.07693
53	0.07273	-0.08684	0.11327	143	0.06673	0.00360	0.06683
54	0.05770	-0.13451	0.14637	144	0.05873	0.04361	0.07315
55	0.02902	-0.15106	0.15382	145	0.04398	0.07630	0.08807
56	-0.00751	-0.12966	0.12987	146	0.02465	0.09829	0.10133
57	-0.04370	-0.07283	0.08493	147	0.00318	0.10799	0.10804
58	-0.07074	0.00703	0.07109	148	-0.01804	0.10545	0.10698
59	-0.08135	0.08983	0.12119	149	-0.03694	0.09214	0.09927
60	-0.07174	0.15247	0.16851	150	-0.05196	0.07044	0.08753
61	-0.04305	0.17518	0.18040	151	-0.06208	0.04329	0.07569
62	-0.00153	0.14782	0.14783	152	-0.06690	0.01369	0.06828
63	0.04260	0.07413	0.08550	153	-0.06650	-0.01557	0.06830
64	0.07750	-0.02763	0.08227	154	-0.06140	-0.04219	0.07450
65	0.09279	-0.12844	0.15845	155	-0.05241	-0.06447	0.08308
66	0.08264	-0.19635	0.21303	156	-0.04048	-0.08134	0.09086
67	0.04794	-0.20633	0.21183	157	-0.02665	-0.09236	0.09612
68	-0.00330	-0.14929	0.14933	158	-0.01191	-0.09755	0.09828
69	-0.05741	-0.03715	0.06838	159	0.00288	-0.09741	0.09745
70	-0.09848	0.09829	0.13914	160	0.01698	-0.09267	0.09421
71	-0.11284	0.21342	0.24142	161	0.02984	-0.08420	0.08933
72	-0.09338	0.26601	0.28193	162	0.04109	-0.07297	0.08375
73	-0.04249	0.22994	0.23384	163	0.05051	-0.05990	0.07836
74	0.02746	0.10657	0.11005	164	0.05803	-0.04582	0.07394
75	0.09662	-0.07172	0.12033	165	0.06369	-0.03145	0.07103
76	0.14274	-0.24807	0.28621	166	0.06762	-0.01737	0.06982
77	0.14766	-0.35741	0.38672	167	0.07001	-0.00403	0.07013
78	0.10345	-0.34801	0.36306	168	0.07109	0.00829	0.07157
79	0.01639	-0.20201	0.20267	169	0.07109	0.01938	0.07369
80	-0.09259	0.05220	0.10629	170	0.07026	0.02917	0.07607
81	-0.19167	0.34112	0.39128	171	0.06883	0.03762	0.07844
82	-0.24538	0.56224	0.61345	172	0.06702	0.04479	0.08061
83	-0.22373	0.61173	0.65136	173	0.06501	0.05076	0.08248
84	-0.11082	0.41629	0.43079	174	0.06296	0.05563	0.08401
85	0.08930	-0.04049	0.09805	175	0.06101	0.05950	0.08522
86	0.35154	-0.70596	0.78864	176	0.05928	0.06250	0.08614
87	0.63380	-1.46545	1.59664	177	0.05784	0.06472	0.08680
88	0.88570	-2.16786	2.34181	178	0.05678	0.06624	0.08725
89	1.05963	-2.66323	2.86629	179	0.05612	0.06713	0.08750
				180	0.05589	0.06742	0.08758

IBR

RADIATION PATTERN FOR THETA=PI/2

H/LMDA=0.2500 B/LMDA= 0.250 OMEGA= 8.53

45 ELEMENT ARRAY

PHI DEG	ER(THETA,PHI) RE	IM	ABSVAL	PHI DEG	ER(THETA,PHI) RE	IM	ABSVAL
0	-0.01270	0.12116	0.12182	90	-0.08905	-3.90045	3.90146
1	-0.01277	0.12068	0.12135	91	-0.08336	-3.65690	3.65784
2	-0.01297	0.11921	0.11991	92	-0.06757	-2.98079	2.98156
3	-0.01331	0.11670	0.11746	93	-0.04520	-2.02120	2.02170
4	-0.01378	0.11306	0.11390	94	-0.02107	-0.98201	0.98224
5	-0.01437	0.10817	0.10912	95	-0.00003	-0.06903	0.06903
6	-0.01507	0.10190	0.10300	96	0.01423	0.56101	0.56119
7	-0.01586	0.09407	0.09540	97	0.02001	0.83524	0.83548
8	-0.01673	0.08455	0.08619	98	0.01788	0.77451	0.77471
9	-0.01765	0.07318	0.07528	99	0.01025	0.47725	0.47736
10	-0.01859	0.05987	0.06269	100	0.00057	0.08409	0.08409
11	-0.01951	0.04456	0.04864	101	-0.00779	-0.26531	0.26543
12	-0.02037	0.02730	0.03406	102	-0.01236	-0.46991	0.47007
13	-0.02111	0.00823	0.02266	103	-0.01221	-0.48937	0.48952
14	-0.02168	-0.01236	0.02496	104	-0.00801	-0.34611	0.34620
15	-0.02202	-0.03399	0.04050	105	-0.00156	-0.10902	0.10903
16	-0.02206	-0.05604	0.06022	106	0.00482	0.13406	0.13414
17	-0.02174	-0.07764	0.08062	107	0.00911	0.30552	0.30566
18	-0.02099	-0.09775	0.09998	108	0.01012	0.36014	0.36029
19	-0.01977	-0.11517	0.11685	109	0.00780	0.29444	0.29454
20	-0.01804	-0.12857	0.12983	110	0.00313	0.14229	0.14233
21	-0.01580	-0.13661	0.13752	111	-0.00228	-0.04004	0.04011
22	-0.01306	-0.13803	0.13865	112	-0.00673	-0.19364	0.19376
23	-0.00988	-0.13179	0.13216	113	-0.00892	-0.27483	0.27497
24	-0.00635	-0.11727	0.11744	114	-0.00833	-0.26627	0.26640
25	-0.00261	-0.09441	0.09444	115	-0.00527	-0.17892	0.17899
26	0.00116	-0.06386	0.06387	116	-0.00076	-0.04549	0.04550
27	0.00476	-0.02713	0.02755	117	0.00386	0.09141	0.09149
28	0.00795	0.01339	0.01557	118	0.00729	0.19262	0.19275
29	0.01051	0.05447	0.05548	119	0.00862	0.23285	0.23301
30	0.01221	0.09228	0.09309	120	0.00758	0.20613	0.20627
31	0.01289	0.12264	0.12332	121	0.00452	0.12518	0.12526
32	0.01244	0.14155	0.14209	122	0.00032	0.01587	0.01587
33	0.01086	0.14572	0.14612	123	-0.00393	-0.09109	0.09117
34	0.00826	0.13319	0.13345	124	-0.00717	-0.16858	0.16873
35	0.00486	0.10392	0.10404	125	-0.00865	-0.19955	0.19974
36	0.00101	0.06012	0.06013	126	-0.00807	-0.18002	0.18020
37	-0.00288	0.00640	0.00702	127	-0.00561	-0.11852	0.11865
38	-0.00635	-0.05048	0.05088	128	-0.00188	-0.03258	0.03263
39	-0.00894	-0.10235	0.10274	129	0.00227	0.05641	0.05646
40	-0.01028	-0.14072	0.14110	130	0.00595	0.12851	0.12865
41	-0.01016	-0.15821	0.15854	131	0.00843	0.16943	0.16964
42	-0.00855	-0.15009	0.15033	132	0.00924	0.17295	0.17319
43	-0.00566	-0.11561	0.11574	133	0.00826	0.14115	0.14139
44	-0.00190	-0.05877	0.05880	134	0.00572	0.08300	0.08319
45	0.00212	0.01160	0.01180	135	0.00212	0.01160	0.01179
46	0.00572	0.08299	0.08319	136	-0.00190	-0.05878	0.05881
47	0.00826	0.14115	0.14139	137	-0.00566	-0.11560	0.11574
48	0.00924	0.17294	0.17319	138	-0.00855	-0.15009	0.15033
49	0.00843	0.16943	0.16964	139	-0.01016	-0.15821	0.15854
50	0.00595	0.12851	0.12865	140	-0.01028	-0.14072	0.14110
51	0.00227	0.05642	0.05646	141	-0.00894	-0.10236	0.10274
52	-0.00188	-0.03257	0.03263	142	-0.00635	-0.05048	0.05088
53	-0.00561	-0.11851	0.11864	143	-0.00289	0.00640	0.00702
54	-0.00807	-0.18001	0.18020	144	0.00101	0.06011	0.06012
55	-0.00865	-0.19955	0.19974	145	0.00486	0.10393	0.10404
56	-0.00717	-0.16858	0.16873	146	0.00826	0.13319	0.13345
57	-0.00393	-0.09109	0.09117	147	0.01086	0.14572	0.14612
58	0.00032	0.01586	0.01586	148	0.01244	0.14155	0.14210
59	0.00452	0.12517	0.12525	149	0.01289	0.12265	0.12332
60	0.00758	0.20612	0.20626	150	0.01221	0.09228	0.09309
61	0.00862	0.23284	0.23300	151	0.01051	0.05448	0.05548
62	0.00729	0.19262	0.19276	152	0.00795	0.01338	0.01557
63	0.00386	0.09143	0.09151	153	0.00476	-0.02713	0.02754
64	-0.00075	-0.04548	0.04549	154	0.00116	-0.06386	0.06387
65	-0.00527	-0.17890	0.17898	155	-0.00261	-0.09440	0.09444
66	-0.00833	-0.26626	0.26639	156	-0.00635	-0.11728	0.11745
67	-0.00892	-0.27483	0.27498	157	-0.00988	-0.13180	0.13217
68	-0.00673	-0.19365	0.19377	158	-0.01306	-0.13802	0.13864
69	-0.00228	-0.04005	0.04012	159	-0.01580	-0.13660	0.13751
70	0.00313	0.14228	0.14232	160	-0.01804	-0.12857	0.12983
71	0.00780	0.29443	0.29454	161	-0.01977	-0.11517	0.11685
72	0.01012	0.36014	0.36029	162	-0.02099	-0.09775	0.09998
73	0.00911	0.30553	0.30567	163	-0.02174	-0.07765	0.08063
74	0.00483	0.13408	0.13416	164	-0.02206	-0.05604	0.06023
75	-0.00156	-0.10900	0.10901	165	-0.02202	-0.03399	0.04050
76	-0.00801	-0.34609	0.34618	166	-0.02168	-0.01235	0.02496
77	-0.01221	-0.48936	0.48952	167	-0.02111	0.00823	0.02266
78	-0.01236	-0.46992	0.47008	168	-0.02037	0.02729	0.03406
79	-0.00779	-0.26532	0.26544	169	-0.01951	0.04456	0.04865
80	0.00056	0.08408	0.08408	170	-0.01859	0.05987	0.06269
81	0.01025	0.47724	0.47735	171	-0.01765	0.07318	0.07528
82	0.01788	0.77449	0.77470	172	-0.01673	0.08455	0.08618
83	0.02002	0.83525	0.83549	173	-0.01586	0.09407	0.09540
84	0.01424	0.56105	0.56123	174	-0.01507	0.10190	0.10300
85	-0.00002	-0.06895	0.06895	175	-0.01437	0.10817	0.10912
86	-0.02105	-0.98192	0.98215	176	-0.01378	0.11306	0.11389
87	-0.04519	-2.02111	2.02161	177	-0.01331	0.11670	0.11746
88	-0.06756	-2.98072	2.98148	178	-0.01297	0.11921	0.11991
89	-0.08336	-3.65688	3.65783	179	-0.01277	0.12068	0.12135
				180	-0.01270	0.12116	0.12182

VBR

RADIATION PATTERN FOR THETA=PI/2

H/LMDA=0.2500 B/LMDA= 0.250 OMEGA= 8.53

50 ELEMENT ARRAY

PHI DEG	ER(THETA,PHI) RE	IM	ABSVAL	PHI DEG	ER(THETA,PHI) RE	IM	ABSVAL
0	-0.00497	-0.07699	0.07715	90	1.12833	-2.84013	3.05606
1	-0.00533	-0.07707	0.07725	91	1.05059	-2.62106	2.82377
2	-0.00641	-0.07726	0.07753	92	0.83648	-2.02426	2.19028
3	-0.00820	-0.07753	0.07797	93	0.53769	-1.21111	1.32510
4	-0.01068	-0.07780	0.07853	94	0.22357	-0.39138	0.45074
5	-0.01386	-0.07792	0.07914	95	-0.03848	0.24308	0.24611
6	-0.01768	-0.07773	0.07971	96	-0.20113	0.57507	0.60923
7	-0.02209	-0.07706	0.08016	97	-0.24826	0.59036	0.64044
8	-0.02701	-0.07566	0.08033	98	-0.19615	0.36979	0.41860
9	-0.03233	-0.07329	0.08011	99	-0.08510	0.05079	0.09910
10	-0.03789	-0.06969	0.07933	100	0.03530	-0.22542	0.22817
11	-0.04350	-0.06458	0.07787	101	0.12205	-0.36139	0.38144
12	-0.04891	-0.05773	0.07566	102	0.15076	-0.32938	0.36225
13	-0.05381	-0.04893	0.07274	103	0.12044	-0.16942	0.20787
14	-0.05787	-0.03807	0.06927	104	0.05025	0.03637	0.06203
15	-0.06070	-0.02515	0.06570	105	-0.02970	0.20080	0.20298
16	-0.06190	-0.01033	0.06276	106	-0.08992	0.26503	0.27987
17	-0.06110	0.00604	0.06140	107	-0.11145	0.21664	0.24363
18	-0.05794	0.02338	0.06248	108	-0.09090	0.08796	0.12649
19	-0.05219	0.04087	0.06629	109	-0.03980	-0.06213	0.07378
20	-0.04372	0.05744	0.07219	110	0.02107	-0.17368	0.17496
21	-0.03264	0.07184	0.07891	111	0.06973	-0.20829	0.21965
22	-0.01924	0.08269	0.08490	112	0.09052	-0.16065	0.18440
23	-0.00414	0.08861	0.08871	113	0.07871	-0.05654	0.09691
24	0.01181	0.08836	0.08914	114	0.04088	0.06033	0.07288
25	0.02745	0.08108	0.08560	115	-0.00834	0.14623	0.14646
26	0.04146	0.06645	0.07832	116	-0.05190	0.17301	0.18063
27	0.05241	0.04492	0.06902	117	-0.07603	0.13631	0.15608
28	0.05895	0.01782	0.06158	118	-0.07435	0.05431	0.09207
29	0.05997	-0.01255	0.06127	119	-0.04912	-0.04109	0.06404
30	0.05482	-0.04306	0.06970	120	-0.00973	-0.11656	0.11697
31	0.04349	-0.06995	0.08237	121	0.03085	-0.14868	0.15184
32	0.02677	-0.08934	0.09326	122	0.06036	-0.13040	0.14369
33	0.00628	-0.09774	0.09794	123	0.07077	-0.07148	0.10059
34	-0.01559	-0.09277	0.09407	124	0.06023	0.00653	0.06058
35	-0.03592	-0.07381	0.08208	125	0.03297	0.07830	0.08496
36	-0.05162	-0.04244	0.06683	126	-0.00246	0.12283	0.12285
37	-0.05995	-0.00266	0.06001	127	-0.03596	0.12915	0.13406
38	-0.05902	0.03942	0.07098	128	-0.05879	0.09834	0.11457
39	-0.04831	0.07639	0.09038	129	-0.06565	0.04175	0.07780
40	-0.02897	0.10075	0.10484	130	-0.05573	-0.02337	0.06044
41	-0.00391	0.10659	0.10666	131	-0.03244	-0.07922	0.08560
42	0.02256	0.09104	0.09379	132	-0.00211	-0.11218	0.11220
43	0.04536	0.05553	0.07170	133	0.02782	-0.11579	0.11909
44	0.05962	0.00619	0.05994	134	0.05064	-0.09142	0.10451
45	0.06180	-0.04689	0.07757	135	0.06180	-0.04689	0.07757
46	0.05064	-0.09142	0.10451	136	0.05962	0.00618	0.05994
47	0.02782	-0.11579	0.11909	137	0.04536	0.05552	0.07170
48	[illegible]	-0.11218	0.11220	138	0.02256	0.09104	0.09379
49	-0.03244	-0.07921	0.08560	139	-0.00391	[illegible]	[illegible]
50	-0.05573	-0.02337	0.06044	140	-0.02897	0.10075	0.10484
51	-0.06565	0.04175	0.07780	141	-0.04831	0.07638	0.09038
52	-0.05879	0.09834	0.11457	142	-0.05903	0.03942	0.07098
53	-0.03597	0.12915	0.13406	143	-0.05995	-0.00266	0.06001
54	-0.00246	0.12283	0.12285	144	-0.05162	-0.04244	0.06683
55	0.03297	0.07830	0.08496	145	-0.03592	-0.07380	0.08208
56	0.06023	0.00654	0.06058	146	-0.01559	-0.09277	0.09407
57	0.07077	-0.07148	0.10058	147	0.00628	-0.09774	0.09794
58	0.06036	-0.13039	0.14369	148	0.02677	-0.08934	0.09327
59	0.03085	-0.14868	0.15185	149	0.04349	-0.06996	0.08237
60	-0.00973	-0.11657	0.11697	150	0.05481	-0.04306	0.06970
61	-0.04911	-0.04109	0.06404	151	0.05997	-0.01255	0.06127
62	-0.07435	0.05430	0.09206	152	0.05895	0.01782	0.06158
63	-0.07603	0.13631	0.15608	153	0.05241	0.04491	0.06902
64	-0.05190	0.17301	0.18063	154	0.04146	0.06645	0.07832
65	-0.00835	0.14623	0.14647	155	0.02745	0.08108	0.08560
66	0.04088	0.06034	0.07289	156	0.01181	0.08836	0.08915
67	0.07871	-0.05653	0.09690	157	-0.00413	0.08861	0.08870
68	0.09052	-0.16064	0.18439	158	-0.01924	0.08270	0.08490
69	0.06973	-0.20829	0.21965	159	-0.03264	0.07185	0.07892
70	0.02108	-0.17368	0.17496	160	-0.04372	0.05744	0.07219
71	-0.03979	-0.06213	0.07378	161	-0.05219	0.04087	0.06629
72	-0.09089	0.08795	0.12648	162	-0.05794	0.02338	0.06248
73	-0.11145	0.21663	0.24362	163	-0.06110	0.00604	0.06140
74	-0.08993	0.26503	0.27987	164	-0.06190	-0.01033	0.06276
75	-0.02970	0.20081	0.20299	165	-0.06070	-0.02515	0.06570
76	0.05025	0.03638	0.06203	166	-0.05787	-0.03807	0.06927
77	0.12044	-0.16941	0.20785	167	-0.05381	-0.04893	0.07273
78	0.15076	-0.32937	0.36224	168	-0.04891	-0.05773	0.07567
79	0.12205	-0.36140	0.38145	169	-0.04350	-0.06458	0.07787
80	0.03530	-0.22543	0.22818	170	-0.03789	-0.06969	0.07933
81	-0.08510	0.05078	0.09910	171	-0.03233	-0.07329	0.08011
82	-0.19615	0.36977	0.41857	172	-0.02701	-0.07566	0.08034
83	-0.24826	0.59035	0.64043	173	-0.02209	-0.07706	0.08016
84	-0.20114	0.57509	0.60925	174	-0.01768	-0.07773	0.07971
85	-0.03849	0.24313	0.24616	175	-0.01386	-0.07791	0.07914
86	0.22356	-0.39131	0.45066	176	-0.01069	-0.07780	0.07853
87	0.53767	-1.21103	1.32502	177	-0.00820	-0.07754	0.07797
88	0.83646	-2.02419	2.19021	178	-0.00641	-0.07726	0.07753
89	1.05059	-2.62105	2.82376	179	-0.00533	-0.07707	0.07725
				180	-0.00497	-0.07699	0.07715

IBR

RADIATION PATTERN FOR THETA=PI/2

H/LMDA=0.2500 B/LMDA= 0.250 OMEGA= 8.53

50 ELEMENT ARRAY

PHI DEG	ER(THETA,PHI) RE	IM	ABSVAL	PHI DEG	ER(THETA,PHI) RE	IM	ABSVAL
0	-0.00924	-0.13651	0.13682	90	-0.02413	-3.76162	3.76169
1	-0.00917	-0.13656	0.13687	91	-0.02207	-3.47300	3.47307
2	-0.00897	-0.13670	0.13699	92	-0.01648	-2.68640	2.68645
3	-0.00863	-0.13684	0.13711	93	-0.00899	-1.61355	1.61357
4	-0.00815	-0.13686	0.13711	94	-0.00165	-0.53000	0.53000
5	-0.00752	-0.13657	0.13677	95	0.00368	0.31165	0.31167
6	-0.00675	-0.13572	0.13589	96	0.00593	0.75618	0.75621
7	-0.00582	-0.13405	0.13418	97	0.00514	0.78320	0.78321
8	-0.00473	-0.13123	0.13131	98	0.00227	0.49726	0.49727
9	-0.00349	-0.12690	0.12694	99	-0.00115	0.07803	0.07804
10	-0.00210	-0.12068	0.12070	100	-0.00365	-0.28873	0.28875
11	-0.00059	-0.11220	0.11220	101	-0.00433	-0.47354	0.47356
12	0.00104	-0.10114	0.10114	102	-0.00310	-0.43778	0.43779
13	0.00274	-0.08723	0.08727	103	-0.00063	-0.23187	0.23187
14	0.00448	-0.07033	0.07047	104	0.00198	0.03774	0.03779
15	0.00619	-0.05048	0.05086	105	0.00365	0.25662	0.25665
16	0.00781	-0.02794	0.02901	106	0.00375	0.34632	0.34634
17	0.00925	-0.00324	0.00980	107	0.00233	0.28855	0.28856
18	0.01044	0.02277	0.02505	108	-0.00001	0.12368	0.12368
19	0.01129	0.04891	0.05020	109	-0.00232	-0.07235	0.07239
20	0.01171	0.07366	0.07459	110	-0.00369	-0.22105	0.22109
21	0.01164	0.09526	0.09597	111	-0.00363	-0.27107	0.27109
22	0.01102	0.11180	0.11234	112	-0.00219	-0.21375	0.21376
23	0.00983	0.12135	0.12175	113	0.00008	-0.08110	0.08110
24	0.00810	0.12226	0.12253	114	0.00234	0.07078	0.07082
25	0.00590	0.11335	0.11350	115	0.00376	0.18485	0.18489
26	0.00334	0.09422	0.09428	116	0.00386	0.22359	0.22362
27	0.00060	0.06553	0.06553	117	0.00259	0.17994	0.17996
28	-0.00212	0.02912	0.02920	118	0.00042	0.07627	0.07627
29	-0.00457	-0.01185	0.01270	119	-0.00194	-0.04673	0.04677
30	-0.00651	-0.05313	0.05353	120	-0.00370	-0.14590	0.14594
31	-0.00775	-0.08967	0.09000	121	-0.00429	-0.19025	0.19030
32	-0.00811	-0.11622	0.11650	122	-0.00352	-0.16973	0.16977
33	-0.00754	-0.12816	0.12838	123	-0.00164	-0.09605	0.09606
34	-0.00607	-0.12233	0.12248	124	0.00079	0.00357	0.00366
35	-0.00387	-0.09794	0.09801	125	0.00304	0.09655	0.09660
36	-0.00122	-0.05713	0.05714	126	0.00447	0.15549	0.15556
37	0.00152	-0.00523	0.00545	127	0.00467	0.16570	0.16576
38	0.00395	0.04969	0.04985	128	0.00359	0.12792	0.12797
39	0.00568	0.09791	0.09807	129	0.00151	0.05631	0.05633
40	0.00643	0.12963	0.12979	130	-0.00102	-0.02711	0.02713
41	0.00605	0.13717	0.13731	131	-0.00337	-0.09937	0.09942
42	0.00460	0.11694	0.11703	132	-0.00496	-0.14267	0.14276
43	0.00234	0.07092	0.07096	133	-0.00542	-0.14835	0.14845
44	-0.00030	0.00720	0.00721	134	-0.00464	-0.11784	0.11793
45	-0.00280	-0.06102	0.06108	135	-0.00280	-0.06102	0.06109
46	-0.00464	-0.11784	0.11793	136	-0.00030	0.00719	0.00720
47	-0.00542	-0.14835	0.14845	137	0.00234	0.07091	0.07095
48	-0.00496	-0.14267	0.14276	138	0.00460	0.11694	0.11703
49	-0.00337	-0.09937	0.09942	139	0.00605	0.13718	0.13731
50	-0.00102	-0.02712	0.02714	140	[illegible]	[illegible]	[illegible]
51	0.00151	0.05631	0.05633	141	0.00568	0.09790	0.09807
52	0.00359	0.12792	0.12797	142	0.00395	0.04970	0.04985
53	0.00467	0.16569	0.16576	143	0.00152	-0.00524	0.00545
54	0.00447	0.15549	0.15556	144	-0.00122	-0.05713	0.05714
55	0.00304	0.09655	0.09660	145	-0.00387	-0.09793	0.09800
56	0.00079	0.00358	0.00367	146	-0.00607	-0.12233	0.12248
57	-0.00164	-0.09604	0.09605	147	-0.00754	-0.12816	0.12838
58	-0.00352	-0.16973	0.16976	148	-0.00811	-0.11622	0.11651
59	-0.00429	-0.19025	0.19030	149	-0.00775	-0.08967	0.09000
60	-0.00370	-0.14590	0.14595	150	-0.00651	-0.05314	0.05353
61	-0.00194	-0.04674	0.04678	151	-0.00457	-0.01186	0.01271
62	0.00042	0.07626	0.07626	152	-0.00212	0.02912	0.02920
63	0.00259	0.17993	0.17995	153	0.00060	0.06552	0.06553
64	0.00386	0.22359	0.22362	154	0.00334	0.09422	0.09428
65	0.00376	0.18486	0.18490	155	0.00590	0.11335	0.11350
66	0.00234	0.07080	0.07083	156	0.00810	0.12226	0.12253
67	0.00009	-0.08109	0.08109	157	0.00983	0.12135	0.12175
68	-0.00219	-0.21374	0.21375	158	0.01102	0.11180	0.11234
69	-0.00363	-0.27107	0.27109	159	0.01164	0.09527	0.09598
70	-0.00369	-0.22106	0.22109	160	0.01171	0.07366	0.07459
71	-0.00232	-0.07236	0.07239	161	0.01129	0.04891	0.05020
72	-0.00001	0.12366	0.12366	162	0.01044	0.02278	0.02506
73	0.00233	0.28854	0.28855	163	0.00925	-0.00323	0.00980
74	0.00375	0.34632	0.34634	164	0.00781	-0.02794	0.02901
75	0.00365	0.25664	0.25666	165	0.00619	-0.05048	0.05086
76	0.00198	0.03776	0.03781	166	0.00448	-0.07033	0.07047
77	-0.00063	-0.23184	0.23184	167	0.00274	-0.08723	0.08727
78	-0.00310	-0.43777	0.43778	168	0.00104	-0.10114	0.10115
79	-0.00433	-0.47354	0.47356	169	-0.00059	-0.11220	0.11220
80	-0.00366	-0.28874	0.28876	170	-0.00210	-0.12068	0.12070
81	-0.00115	0.07801	0.07802	171	-0.00349	-0.12690	0.12695
82	0.00227	0.49723	0.49724	172	-0.00473	-0.13123	0.13132
83	0.00514	0.78318	0.78320	173	-0.00582	-0.13405	0.13418
84	0.00594	0.75620	0.75623	174	-0.00675	-0.13572	0.13589
85	0.00368	0.31171	0.31173	175	-0.00752	-0.13656	0.13677
86	-0.00164	-0.52991	0.52991	176	-0.00815	-0.13686	0.13710
87	-0.00897	-1.61345	1.61347	177	-0.00863	-0.13684	0.13711
88	-0.01647	-2.68631	2.68636	178	-0.00897	-0.13670	0.13699
89	-0.02206	-3.47298	3.47305	179	-0.00917	-0.13656	0.13687
				180	-0.00924	-0.13651	0.13682

VBR

RADIATION PATTERN FOR THETA=PI/2

H/LMDA=0.2500 B/LMDA= 0.500 OMEGA= 8.53

2 ELEMENT ARRAY

PHI DEG	ER(THETA,PHI) RE	IM	ABSVAL	PHI DEG	ER(THETA,PHI) RE	IM	ABSVAL
0	-0.00000	-0.00000	0.00000	90	-0.96610	-5.78030	5.86048
5	-0.00577	-0.03455	0.03503	95	-0.95706	-5.72622	5.80565
10	-0.02305	-0.13793	0.13985	100	-0.93038	-5.56660	5.64382
15	-0.05168	-0.30924	0.31353	105	-0.88736	-5.30915	5.38280
20	-0.09138	-0.54676	0.55434	110	-0.83000	-4.96599	5.03488
25	-0.14167	-0.84763	0.85939	115	-0.76093	-4.55273	4.61588
30	-0.20182	-1.20749	1.22424	120	-0.68314	-4.08730	4.14399
35	-0.27077	-1.62005	1.64252	125	-0.59980	-3.58870	3.63848
40	-0.34710	-2.07675	2.10556	130	-0.51409	-3.07585	3.11851
45	-0.42897	-2.56655	2.60215	135	-0.42897	-2.56656	2.60216
50	-0.51409	-3.07585	3.11851	140	-0.34710	-2.07675	2.10556
55	-0.59980	-3.58869	3.63847	145	-0.27077	-1.62005	1.64252
60	-0.68314	-4.08729	4.14399	150	-0.20182	-1.20749	1.22424
65	-0.76093	-4.55272	4.61588	155	-0.14167	-0.84764	0.85940
70	-0.83000	-4.96599	5.03487	160	-0.09138	-0.54676	0.55434
75	-0.88735	-5.30915	5.38279	165	-0.05168	-0.30924	0.31353
80	-0.93038	-5.56660	5.64382	170	-0.02305	-0.13793	0.13985
85	-0.95706	-5.72622	5.80565	175	-0.00577	-0.03455	0.03503
				180	-0.00000	-0.00000	0.00000

IBR

RADIATION PATTERN FOR THETA=PI/2

H/LMDA=0.2500 B/LMDA= 0.500 OMEGA= 8.53

2 ELEMENT ARRAY

PHI DEG	ER(THETA,PHI) RE	IM	ABSVAL	PHI DEG	ER(THETA,PHI) RE	IM	ABSVAL
0	-0.00000	-0.00000	0.00000	90	-0.96610	-5.78030	5.86048
5	-0.00577	-0.03455	0.03503	95	-0.95706	-5.72622	5.80565
10	-0.02305	-0.13793	0.13985	100	-0.93038	-5.56660	5.64382
15	-0.05168	-0.30924	0.31353	105	-0.88736	-5.30915	5.38280
20	-0.09138	-0.54676	0.55434	110	-0.83000	-4.96599	5.03488
25	-0.14167	-0.84763	0.85939	115	-0.76093	-4.55273	4.61588
30	-0.20182	-1.20749	1.22424	120	-0.68314	-4.08730	4.14399
35	-0.27077	-1.62005	1.64252	125	-0.59980	-3.58870	3.63848
40	-0.34710	-2.07675	2.10556	130	-0.51409	-3.07585	3.11851
45	-0.42897	-2.56655	2.60215	135	-0.42897	-2.56656	2.60216
50	-0.51409	-3.07585	3.11851	140	-0.34710	-2.07675	2.10556
55	-0.59980	-3.58869	3.63847	145	-0.27077	-1.62005	1.64252
60	-0.68314	-4.08729	4.14399	150	-0.20182	-1.20749	1.22424
65	-0.76093	-4.55272	4.61588	155	-0.14167	-0.84764	0.85940
70	-0.83000	-4.96599	5.03487	160	-0.09138	-0.54676	0.55434
75	-0.88735	-5.30915	5.38279	165	-0.05168	-0.30924	0.31353
80	-0.93038	-5.56660	5.64382	170	-0.02305	-0.13793	0.13985
85	-0.95706	-5.72622	5.80565	175	-0.00577	-0.03455	0.03503
				180	-0.00000	-0.00000	0.00000

VBR

RADIATION PATTERN FOR THETA=PI/2

H/LMDA=0.2500 B/LMDA= 0.500 OMEGA= 8.53

3 ELEMENT ARRAY

PHI DEG	ER(THETA,PHI) RE	IM	ABSVAL	PHI DEG	ER(THETA,PHI) RE	IM	ABSVAL
0	0.62344	0.97477	1.15709	90	-0.74389	-5.92897	5.97545
5	0.62339	0.97453	1.15686	95	-0.71842	-5.80038	5.84471
10	0.62266	0.97084	1.15336	100	-0.64466	-5.42794	5.46609
15	0.61953	0.95501	1.13836	105	-0.53007	-4.84939	4.87828
20	0.61121	0.91300	1.09870	110	-0.38578	-4.12083	4.13884
25	0.59404	0.82632	1.01768	115	-0.22479	-3.30802	3.31565
30	0.56378	0.67351	0.87832	120	-0.06023	-2.47711	2.47784
35	0.51604	0.43247	0.67329	125	0.09640	-1.68631	1.68906
40	0.44694	0.08362	0.45470	130	0.23627	-0.98008	1.00815
45	0.35387	-0.38630	0.52389	135	0.35387	-0.38631	0.52389
50	0.23627	-0.98007	1.00815	140	0.44694	0.08362	0.45470
55	0.09640	-1.68630	1.68905	145	0.51604	0.43247	0.67329
60	-0.06022	-2.47710	2.47783	150	0.56377	0.67350	0.87832
65	-0.22479	-3.30801	3.31564	155	0.59404	0.82632	1.01768
70	-0.38577	-4.12082	4.13884	160	0.61121	0.91300	1.09870
75	-0.53007	-4.84938	4.87827	165	0.61953	0.95501	1.13836
80	-0.64466	-5.42793	5.46608	170	0.62266	0.97084	1.15336
85	-0.71842	-5.80038	5.84470	175	0.62339	0.97453	1.15686
				180	0.62344	0.97477	1.15709

IBR

RADIATION PATTERN FOR THETA=PI/2

H/LMDA=0.2500 B/LMDA= 0.500 OMEGA= 8.53

3 ELEMENT ARRAY

PHI DEG	ER(THETA,PHI) RE	IM	ABSVAL	PHI DEG	ER(THETA,PHI) RE	IM	ABSVAL
0	0.75296	1.58960	1.75892	90	-1.93253	-4.48632	4.88485
5	0.75287	1.58939	1.75868	95	-1.88251	-4.37316	4.76113
10	0.75143	1.58614	1.75514	100	-1.73763	-4.04537	4.40277
15	0.74528	1.57222	1.73991	105	-1.51258	-3.53620	3.84612
20	0.72894	1.53524	1.69950	110	-1.22918	-2.89499	3.14513
25	0.69522	1.45895	1.61613	115	-0.91301	-2.17965	2.36315
30	0.63577	1.32446	1.46915	120	-0.58979	-1.44837	1.56384
35	0.54201	1.11233	1.23736	125	-0.28217	-0.75239	0.80356
40	0.40631	0.80531	0.90200	130	-0.00745	-0.13084	0.13105
45	0.22352	0.39173	0.45101	135	0.22352	0.39173	0.45101
50	-0.00745	-0.13084	0.13105	140	0.40631	0.80531	0.90200
55	-0.28217	-0.75238	0.80355	145	0.54201	1.11233	1.23736
60	-0.58978	-1.44836	1.56384	150	0.63577	1.32446	1.46915
65	-0.91300	-2.17964	2.36313	155	0.69522	1.45895	1.61612
70	-1.22918	-2.89499	3.14512	160	0.72894	1.53524	1.69950
75	-1.51258	-3.53619	3.84610	165	0.74528	1.57221	1.73991
80	-1.73763	-4.04537	4.40277	170	0.75143	1.58614	1.75514
85	-1.88251	-4.37315	4.76112	175	0.75287	1.58939	1.75868
				180	0.75296	1.58960	1.75892

VBR

RADIATION PATTERN FOR THETA=PI/2

H/LMDA=0.2500 B/LMDA= 0.500 OMEGA= 8.53

4 ELEMENT ARRAY

PHI DEG	ER(THETA,PHI) RE	IM	ABSVAL	PHI DEG	ER(THETA,PHI) RE	IM	ABSVAL
0	0.00000	0.00000	0.00000	90	-0.58484	-6.20101	6.22853
5	0.00839	0.02893	0.03012	95	-0.54268	-5.93927	5.96401
10	0.03347	0.11535	0.12010	100	-0.42423	-5.20007	5.21735
15	0.07480	0.25725	0.26791	105	-0.25155	-4.10972	4.11741
20	0.13110	0.44849	0.46726	110	-0.05499	-2.84303	2.84356
25	0.19958	0.67496	0.70385	115	0.13399	-1.58259	1.58825
30	0.27512	0.91046	0.95112	120	0.28954	-0.48112	0.56152
35	0.34966	1.11349	1.16710	125	0.39564	0.36278	0.53679
40	0.41212	1.22694	1.29431	130	0.44735	0.91197	1.01578
45	0.44922	1.18261	1.26506	135	0.44922	1.18261	1.26505
50	0.44735	0.91197	1.01578	140	0.41213	1.22694	1.29431
55	0.39564	0.36279	0.53679	145	0.34966	1.11349	1.16710
60	0.28954	-0.48111	0.56152	150	0.27512	0.91046	0.95112
65	0.13400	-1.58256	1.58823	155	0.19958	0.67496	0.70385
70	-0.05499	-2.84303	2.84356	160	0.13110	0.44849	0.46726
75	-0.25154	-4.10970	4.11739	165	0.07480	0.25726	0.26791
80	-0.42423	-5.20007	5.21734	170	0.03347	0.11535	0.12010
85	-0.54268	-5.93925	5.96399	175	0.00839	0.02893	0.03012
				180	0.00000	0.00000	0.00000

IBR

RADIATION PATTERN FOR THETA=PI/2

H/LMDA=0.2500 B/LMDA= 0.500 OMEGA= 8.53

4 ELEMENT ARRAY

PHI DEG	ER(THETA,PHI) RE	IM	ABSVAL	PHI DEG	ER(THETA,PHI) RE	IM	ABSVAL
0	0.00000	0.00000	0.00000	90	-1.28469	-5.33429	5.48681
5	0.00831	0.03242	0.03347	95	-1.22331	-5.08587	5.23092
10	0.03314	0.12929	0.13347	100	-1.05022	-4.38515	4.50916
15	0.07398	0.28854	0.29788	105	-0.79574	-3.35425	3.44735
20	0.12926	0.50398	0.52029	110	-0.50176	-2.16201	2.21947
25	0.19547	0.76150	0.78619	115	-0.21193	-0.98445	1.00701
30	0.26609	1.03504	1.06870	120	0.03741	0.03181	0.04911
35	0.33082	1.28341	1.32536	125	0.22310	0.79304	0.82382
40	0.37552	1.44989	1.49773	130	0.33684	1.26535	1.30941
45	0.38323	1.46669	1.51593	135	0.38323	1.46669	1.51592
50	0.33684	1.26535	1.30941	140	0.37552	1.44989	1.49773
55	0.22310	0.79305	0.82383	145	0.33082	1.28341	1.32537
60	0.03741	0.03182	0.04911	150	0.26609	1.03505	1.06870
65	-0.21193	-0.98443	1.00698	155	0.19547	0.76150	0.78619
70	-0.50176	-2.16201	2.21947	160	0.12926	0.50398	0.52029
75	-0.79573	-3.35423	3.44732	165	0.07398	0.28855	0.29788
80	-1.05022	-4.38515	4.50916	170	0.03314	0.12929	0.13347
85	-1.22331	-5.08587	5.23092	175	0.00831	0.03242	0.03347
				180	0.00000	0.00000	0.00000

VBR

RADIATION PATTERN FOR THETA=PI/2

H/LMDA=0.2500 B/LMDA= 0.500 OMEGA= 8.53

5 ELEMENT ARRAY

PHI	ER(THETA,PHI)			PHI	ER(THETA,PHI)		
DEG	RE	IM	ABSVAL	DEG	RE	IM	ABSVAL
0	-0.35709	-0.53276	0.64137	90	-0.51566	-6.25635	6.27757
5	-0.35698	-0.53236	0.64097	95	-0.45356	-5.83977	5.85735
10	-0.35534	-0.52637	0.63509	100	-0.28637	-4.69954	4.70825
15	-0.34831	-0.50075	0.60997	105	-0.06400	-3.12411	3.12477
20	-0.32982	-0.43360	0.54478	110	0.15138	-1.47900	1.48673
25	-0.29242	-0.29901	0.41823	115	0.30660	-0.08946	0.31938
30	-0.22897	-0.07478	0.24088	120	0.37256	0.84508	0.92356
35	-0.13534	0.24512	0.28000	125	0.34826	1.27819	1.32479
40	-0.01380	0.63491	0.63506	130	0.25479	1.28880	1.31374
45	0.12380	1.02307	1.03053	135	0.12380	1.02307	1.03054
50	0.25479	1.28880	1.31374	140	-0.01380	0.63491	0.63506
55	0.34826	1.27819	1.32479	145	-0.13534	0.24513	0.28001
60	0.37256	0.84509	0.92356	150	-0.22897	-0.07478	0.24087
65	0.30660	-0.08944	0.31938	155	-0.29242	-0.29901	0.41823
70	0.15138	-1.47899	1.48672	160	-0.32982	-0.43360	0.54478
75	-0.06400	-3.12408	3.12474	165	-0.34831	-0.50075	0.60997
80	-0.28636	-4.69953	4.70825	170	-0.35534	-0.52637	0.63509
85	-0.45356	-5.83975	5.85733	175	-0.35698	-0.53236	0.64097
				180	-0.35709	-0.53276	0.64137

IBR

RADIATION PATTERN FOR THETA=PI/2

H/LMDA=0.2500 B/LMDA= 0.500 OMEGA= 8.53

5 ELEMENT ARRAY

PHI	ER(THETA,PHI)			PHI	ER(THETA,PHI)		
DEG	RE	IM	ABSVAL	DEG	RE	IM	ABSVAL
0	-0.45509	-1.01755	1.11468	90	-1.82804	-4.65478	5.00087
5	-0.45492	-1.01714	1.11423	95	-1.68998	-4.30997	4.62946
10	-0.45237	-1.01099	1.10759	100	-1.31417	-3.37061	3.61774
15	-0.44146	-0.98470	1.07913	105	-0.80146	-2.08640	2.23503
20	-0.41280	-0.91569	1.00444	110	-0.27875	-0.77204	0.82082
25	-0.35512	-0.77683	0.85415	115	0.14261	0.29586	0.32844
30	-0.25816	-0.54369	0.60187	120	0.39701	0.95320	1.03257
35	-0.11748	-0.20615	0.23727	125	0.47353	1.17095	1.26308
40	0.05955	0.21695	0.22498	130	0.40535	1.02919	1.10614
45	0.24835	0.66452	0.70942	135	0.24835	0.66453	0.70942
50	0.40535	1.02919	1.10614	140	0.05956	0.21696	0.22498
55	0.47353	1.17095	1.26308	145	-0.11748	-0.20614	0.23727
60	0.39701	0.95320	1.03258	150	-0.25816	-0.54369	0.60187
65	0.14262	0.29587	0.32845	155	-0.35512	-0.77683	0.85415
70	-0.27875	-0.77203	0.82081	160	-0.41280	-0.91569	1.00444
75	-0.80145	-2.08638	2.23501	165	-0.44146	-0.98470	1.07913
80	-1.31417	-3.37060	3.61773	170	-0.45237	-1.01099	1.10759
85	-1.68997	-4.30996	4.62945	175	-0.45492	-1.01714	1.11423
				180	-0.45509	-1.01755	1.11468

VBR

RADIATION PATTERN FOR THETA=PI/2

H/LMDA=0.2500 B/LMDA= 0.500 OMEGA= 8.53

6 ELEMENT ARRAY

PHI	ER(THETA,PHI)			PHI	ER(THETA,PHI)		
DEG	RE	IM	ABSVAL	DEG	RE	IM	ABSVAL
0	-0.00000	-0.00000	0.00000	90	-0.45380	-6.35939	6.37556
2	-0.00148	-0.00417	0.00442	92	-0.44049	-6.25667	6.27215
4	-0.00591	-0.01668	0.01769	94	-0.40157	-5.95493	5.96845
6	-0.01328	-0.03748	0.03976	96	-0.33999	-5.47288	5.48343
8	-0.02358	-0.06652	0.07058	98	-0.26034	-4.83995	4.84695
10	-0.03676	-0.10362	0.10994	100	-0.16847	-4.09389	4.09735
12	-0.05273	-0.14846	0.15754	102	-0.07099	-3.27765	3.27842
14	-0.07137	-0.20050	0.21283	104	0.02535	-2.43609	2.43622
16	-0.09246	-0.25892	0.27493	106	0.11413	-1.61264	1.61667
18	-0.11570	-0.32247	0.34259	108	0.18983	-0.84629	0.86732
20	-0.14066	-0.38943	0.41405	110	0.24813	-0.16926	0.30036
22	-0.16679	-0.45753	0.48699	112	0.28621	0.39475	0.48759
24	-0.19337	-0.52389	0.55844	114	0.30285	0.83134	0.88478
26	-0.21952	-0.58498	0.62482	116	0.29836	1.13538	1.17392
28	-0.24421	-0.63667	0.68190	118	0.27443	1.31029	1.33872
30	-0.26623	-0.67429	0.72495	120	0.23388	1.36678	1.38665
[illegible]	[illegible]	[illegible]	[illegible]	[illegible]	[illegible]	[illegible]	[illegible]
34	-0.29695	-0.68718	0.74860	124	0.11780	1.19299	1.19879
36	-0.30286	-0.65247	0.71934	126	0.05051	1.00420	1.00547
38	-0.30071	-0.58455	0.65736	128	-0.01757	0.77638	0.77658
40	-0.28934	-0.48045	0.56085	130	-0.08287	0.52977	0.53621
42	-0.26795	-0.33903	0.43214	132	-0.14244	0.28213	0.31605
44	-0.23613	-0.16150	0.28608	134	-0.19403	0.04804	0.19989
46	-0.19403	0.04804	0.19989	136	-0.23613	-0.16150	0.28608
48	-0.14244	0.28213	0.31605	138	-0.26795	-0.33903	0.43214
50	-0.08287	0.52976	0.53620	140	-0.28934	-0.48045	0.56085
52	-0.01757	0.77638	0.77657	142	-0.30071	-0.58454	0.65736
54	0.05051	1.00420	1.00547	144	-0.30286	-0.65247	0.71934
56	0.11780	1.19299	1.19879	146	-0.29695	-0.68718	0.74860
58	0.18031	1.32106	1.33331	148	-0.28427	-0.69284	0.74889
60	0.23388	1.36678	1.38665	150	-0.26623	-0.67429	0.72495
62	0.27443	1.31029	1.33872	152	-0.24421	-0.63667	0.68190
64	0.29836	1.13538	1.17393	154	-0.21953	-0.58498	0.62482
66	0.30285	0.83135	0.88480	156	-0.19337	-0.52389	0.55844
68	0.28621	0.39476	0.48760	158	-0.16679	-0.45753	0.48698
70	0.24813	-0.16925	0.30036	160	-0.14066	-0.38943	0.41405
72	0.18983	-0.84627	0.86730	162	-0.11570	-0.32247	0.34259
74	0.11414	-1.61261	1.61664	164	-0.09246	-0.25892	0.27494
76	0.02535	-2.43606	2.43619	166	-0.07137	-0.20050	0.21283
78	-0.07099	-3.27764	3.27841	168	-0.05273	-0.14846	0.15755
80	-0.16847	-4.09387	4.09734	170	-0.03676	-0.10362	0.10995
82	-0.26033	-4.83994	4.84693	172	-0.02358	-0.06652	0.07057
84	-0.33999	-5.47285	5.48340	174	-0.01328	-0.03748	0.03976
86	-0.40157	-5.95491	5.96843	176	-0.00591	-0.01668	0.01769
88	-0.44049	-6.25667	6.27215	178	-0.00148	-0.00417	0.00442
				180	-0.00000	-0.00000	0.00000

IBR

RADIATION PATTERN FOR THETA=PI/2

H/LMDA=0.2500 B/LMDA= 0.500 OMEGA= 8.53

6 ELEMENT ARRAY

PHI	ER(THETA,PHI)			PHI	ER(THETA,PHI)		
DEG	RE	IM	ABSVAL	DEG	RE	IM	ABSVAL
0	-0.00000	-0.00000	0.00000	90	-1.38639	-5.19529	5.37709
2	-0.00149	-0.00512	0.00533	92	-1.36158	-5.10417	5.28265
4	-0.00596	-0.02046	0.02131	94	-1.28879	-4.83674	5.00550
6	-0.01339	-0.04599	0.04790	96	-1.17277	-4.41033	4.56359
8	-0.02377	-0.08163	0.08502	98	-1.02101	-3.85211	3.98512
10	-0.03705	-0.12720	0.13248	100	-0.84307	-3.19692	3.30622
12	-0.05312	-0.18235	0.18993	102	-0.64981	-2.48428	2.56786
14	-0.07182	-0.24652	0.25677	104	-0.45251	-1.75528	1.81267
16	-0.09291	-0.31882	0.33208	106	-0.26200	-1.04953	1.08173
18	-0.11601	-0.39795	0.41451	108	-0.08794	-0.40225	0.41176
20	-0.14061	-0.48209	0.50218	110	0.06184	0.15778	0.16947
22	-0.16604	-0.56886	0.59259	112	0.18175	0.60993	0.63644
24	-0.19143	-0.65519	0.68259	114	0.26861	0.94234	0.97987
26	-0.21573	-0.73738	0.76829	116	0.32165	1.15183	1.19590
28	-0.23770	-0.81101	0.84513	118	0.34231	1.24321	1.28947
30	-0.25591	-0.87114	0.90795	120	0.33384	1.22797	1.27254
32	-0.26885	-0.91234	0.95113	122	0.30087	1.12259	1.16221
34	-0.27492	-0.92906	0.96888	124	0.24891	[illegible]	[illegible]
36	-0.27258	-0.91588	0.95558	126	0.18383	0.72167	0.74472
38	-0.26045	-0.86801	0.90624	128	0.11141	0.46809	0.48116
40	-0.23744	-0.78175	0.81702	130	0.03699	0.20530	0.20860
42	-0.20296	-0.65511	0.68583	132	-0.03484	-0.05005	0.06098
44	-0.15701	-0.48835	0.51297	134	-0.10040	-0.28455	0.30174
46	-0.10040	-0.28455	0.30175	136	-0.15701	-0.48835	0.51297
48	-0.03484	-0.05005	0.06099	138	-0.20296	-0.65511	0.68583
50	0.03699	0.20529	0.20860	140	-0.23744	-0.78175	0.81702
52	0.11141	0.46808	0.48116	142	-0.26045	-0.86801	0.90624
54	0.18383	0.72167	0.74472	144	-0.27258	-0.91588	0.95558
56	0.24891	0.94679	0.97897	146	-0.27492	-0.92906	0.96888
58	0.30087	1.12259	1.16221	148	-0.26885	-0.91234	0.95113
60	0.33384	1.22796	1.27253	150	-0.25591	-0.87114	0.90795
62	0.34231	1.24321	1.28947	152	-0.23770	-0.81102	0.84513
64	0.32165	1.15183	1.19590	154	-0.21573	-0.73738	0.76829
66	0.26861	0.94235	0.97989	156	-0.19143	-0.65519	0.68259
68	0.18175	0.60994	0.63644	158	-0.16604	-0.56885	0.59259
70	0.06184	0.15779	0.16947	160	-0.14061	-0.48209	0.50218
72	-0.08794	-0.40224	0.41174	162	-0.11601	-0.39795	0.41451
74	-0.26200	-1.04950	1.08170	164	-0.09291	-0.31883	0.33209
76	-0.45250	-1.75525	1.81264	166	-0.07182	-0.24652	0.25677
78	-0.64981	-2.48426	2.56784	168	-0.05312	-0.18236	0.18993
80	-0.84307	-3.19691	3.30620	170	-0.03705	-0.12720	0.13249
82	-1.02101	-3.85210	3.98511	172	-0.02377	-0.08163	0.08502
84	-1.17277	-4.41030	4.56357	174	-0.01339	-0.04599	0.04790
86	-1.28878	-4.83673	5.00548	176	-0.00596	-0.02046	0.02131
88	-1.36158	-5.10417	5.28265	178	-0.00149	-0.00512	0.00533
				180	-0.00000	-0.00000	0.00000

VBR

RADIATION PATTERN FOR THETA=PI/2

H/LMDA=0.2500 B/LMDA= 0.500 OMEGA= 8.53

7 ELEMENT ARRAY

PHI	ER(THETA,PHI)			PHI	ER(THETA,PHI)		
DEG	RE	IM	ABSVAL	DEG	RE	IM	ABSVAL
0	0.24798	0.36022	0.43732	90	-0.42033	-6.38799	6.40180
2	0.24798	0.36021	0.43731	92	-0.40329	-6.24654	6.25955
4	0.24791	0.36000	0.43710	94	-0.35396	-5.83407	5.84480
6	0.24762	0.35911	0.43621	96	-0.27753	-5.18492	5.19234
8	0.24685	0.35674	0.43382	98	-0.18191	-4.35207	4.35587
10	0.24524	0.35174	0.42879	100	-0.07683	-3.40156	3.40242
12	0.24232	0.34273	0.41974	102	0.02731	-2.40541	2.40557
14	0.23755	0.32803	0.40501	104	0.12065	-1.43454	1.43960
16	0.23032	0.30582	0.38285	106	0.19487	-0.55209	0.58547
18	0.21999	0.27419	0.35153	108	0.24402	0.19189	0.31043
20	0.20588	0.23127	0.30963	110	0.26498	0.76381	0.80847
22	0.18739	0.17545	0.25670	112	0.25763	1.14808	1.17663
24	0.16399	0.10561	0.19505	114	0.22454	1.34636	1.36496
26	0.13533	0.02137	0.13701	116	0.17052	1.37530	1.38583
28	0.10131	-0.07663	0.12702	118	0.10189	1.26290	1.26701
30	0.06214	-0.18635	0.19644	120	0.02572	1.04436	1.04468
32	0.01846	-0.30418	0.30474	122	-0.05097	0.75775	0.75946
34	-0.02863	-0.42468	0.42565	124	-0.12184	0.44006	0.45662
36	-0.07754	-0.54057	0.54611	126	-0.18174	0.12415	0.22009
38	-0.12615	-0.64283	0.65509	128	-0.22702	-0.16348	0.27976
40	-0.17189	-0.72109	0.74130	130	-0.25565	-0.40382	0.47794
42	-0.21186	-0.76433	0.79315	132	-0.26712	-0.58549	0.64354
44	-0.24298	-0.76178	0.79959	134	-0.26227	-0.70424	0.75149
46	-0.26227	-0.70424	0.75149	136	-0.24298	-0.76178	0.79960
48	-0.26712	-0.58549	0.64355	138	-0.21186	-0.76433	0.79315
50	-0.25565	-0.40382	0.47794	140	-0.17189	-0.72110	0.74130
52	-0.22702	-0.16349	0.27976	142	-0.12615	-0.64284	0.65510
54	-0.18174	0.12415	0.22009	144	-0.07754	-0.54058	0.54611
56	-0.12184	0.44006	0.45661	146	-0.02863	-0.42469	0.42565
58	-0.05097	0.75774	0.75946	148	0.01846	-0.30418	0.30474
60	0.02572	1.04436	1.04468	150	0.06214	-0.18636	0.19644
62	0.10189	1.26290	1.26700	152	0.10130	-0.07663	0.12702
64	0.17052	1.37530	1.38583	154	0.13533	0.02136	0.13700
66	0.22454	1.34637	1.36497	156	0.16399	0.10561	0.19505
68	0.25763	1.14809	1.17664	158	0.18739	0.17545	0.25671
70	0.26498	0.76382	0.80848	160	0.20588	0.23126	0.30963
72	0.24402	0.19190	0.31044	162	0.21998	0.27418	0.35152
74	0.19487	-0.55205	0.58544	164	0.23032	0.30582	0.38285
76	0.12066	-1.43450	1.43956	166	0.23755	0.32803	0.40501
78	0.02731	-2.40540	2.40555	168	0.24232	0.34273	0.41974
80	-0.07683	-3.40154	3.40241	170	0.24524	0.35174	0.42879
82	-0.18191	-4.35205	4.35585	172	0.24685	0.35673	0.43382
84	-0.27753	-5.18488	5.19230	174	0.24762	0.35911	0.43621
86	-0.35396	-5.83405	5.84477	176	0.24791	0.36000	0.43710
88	-0.40329	-6.24653	6.25954	178	0.24798	0.36021	0.43731
				180	0.24798	0.36022	0.43732

IBR

RADIATION PATTERN FOR THETA=PI/2

H/LMDA=0.2500 B/LMDA= 0.500 OMEGA= 8.53

7 ELEMENT ARRAY

PHI	ER(THETA,PHI)			PHI	ER(THETA,PHI)		
DEG	RE	IM	ABSVAL	DEG	RE	IM	ABSVAL
0	0.32790	0.75178	0.82017	90	-1.77252	-4.72582	5.04730
2	0.32789	0.75176	0.82016	92	-1.72925	-4.61224	4.92576
4	0.32780	0.75153	0.81991	94	-1.60331	-4.28156	4.57190
6	0.32739	0.75053	0.81883	96	-1.40584	-3.76274	4.01679
8	0.32630	0.74783	0.81592	98	-1.15400	-3.10045	3.30824
10	0.32402	0.74218	0.80982	100	-0.86909	-2.35008	2.50563
12	0.31989	0.73196	0.79881	102	-0.57416	-1.57175	1.67333
14	0.31315	0.71530	0.78084	104	-0.29168	-0.82411	0.87421
16	0.30295	0.69007	0.75365	106	-0.04136	-0.15871	0.16401
18	0.28839	0.65407	0.71483	108	0.16164	0.38455	0.41714
20	0.26857	0.60508	0.66200	110	0.30773	0.78019	0.83869
22	0.24266	0.54109	0.59301	112	0.39332	1.01827	1.09159
24	0.21003	0.46056	0.50619	114	0.42050	1.10361	1.18101
26	0.17030	0.36264	0.40064	116	0.39621	1.05357	1.12561
28	0.12352	0.24750	0.27661	118	0.33091	0.89483	0.95406
30	0.07024	0.11664	0.13616	120	0.23716	0.65958	0.70092
32	0.01166	-0.02682	0.02924	122	0.12813	0.38172	0.40265
34	-0.05029	-0.17794	0.18491	124	0.01627	0.09347	0.09488
36	-0.11292	-0.32985	0.34864	126	-0.08771	-0.17733	0.19784
38	-0.17281	-0.47385	0.50437	128	-0.17551	-0.40899	0.44506
40	-0.22589	-0.59968	0.64082	130	-0.24162	-0.58684	0.63463
42	-0.26773	-0.69616	0.74587	132	-0.28324	-0.70325	0.75815
44	-0.29383	-0.75201	0.80738	134	-0.30007	-0.75698	0.81429
46	-0.30007	-0.75698	0.81429	136	-0.29383	-0.75201	0.80738
48	-0.28324	-0.70326	0.75815	138	-0.26773	-0.69616	0.74587
50	-0.24162	-0.58684	0.63464	140	-0.22589	-0.59968	0.64082
52	-0.17551	-0.40899	0.44506	142	-0.17281	-0.47385	0.50438
54	-0.08771	-0.17733	0.19784	144	-0.11292	-0.32986	0.34865
56	0.01627	0.09346	0.09487	146	-0.05029	-0.17794	0.18491
58	0.12813	0.38172	0.40265	148	0.01166	-0.02682	0.02924
60	0.23716	0.65958	0.70092	150	0.07024	0.11663	0.13615
62	0.33091	0.89483	0.95405	152	0.12351	0.24749	0.27660
64	0.39621	1.05357	1.12561	154	0.17030	0.36264	0.40063
66	0.42050	1.10361	1.18101	156	0.21003	0.46056	0.50619
68	0.39332	1.01828	1.09160	158	0.24266	0.54109	0.59301
70	0.30773	0.78020	0.83869	160	0.26857	0.60507	0.66200
72	0.16164	0.38456	0.41715	162	0.28839	0.65407	0.71482
74	-0.04135	-0.15868	0.16398	164	0.30295	0.69007	0.75365
76	-0.29167	-0.82408	0.87417	166	0.31315	0.71530	0.78084
78	-0.57415	-1.57173	1.67332	168	0.31989	0.73196	0.79881
80	-0.86908	-2.35006	2.50561	170	0.32402	0.74218	0.80983
82	-1.15400	-3.10043	3.30823	172	0.32630	0.74783	0.81592
84	-1.40583	-3.76271	4.01676	174	0.32739	0.75053	0.81883
86	-1.60330	-4.28154	4.57188	176	0.32780	0.75153	0.81991
88	-1.72925	-4.61224	4.92575	178	0.32789	0.75176	0.82016
				180	0.32790	0.75178	0.82017

VBR

RADIATION PATTERN FOR THETA=PI/2

H/LMDA=0.2500 B/LMDA= 0.500 OMEGA= 8.53

8 ELEMENT ARRAY

PHI	ER(THETA,PHI)			PHI	ER(THETA,PHI)		
DEG	RE	IM	ABSVAL	DEG	RE	IM	ABSVAL
0	0.00000	0.00001	0.00001	90	-0.38764	-6.44165	6.45330
2	0.00154	0.00389	0.00418	92	-0.36710	-6.25254	6.26331
4	0.00615	0.01551	0.01669	94	-0.30838	-5.70595	5.71427
6	0.01383	0.03485	0.03749	96	-0.21963	-4.86121	4.86617
8	0.02453	0.06178	0.06647	98	-0.11304	-3.80788	3.80956
10	0.03818	0.09605	0.10336	100	-0.00290	-2.65365	2.65365
12	0.05464	0.13715	0.14764	102	0.09655	-1.50985	1.51294
14	0.07364	0.18419	0.19837	104	0.17315	-0.47771	0.50813
16	0.09480	0.23578	0.25413	106	0.21849	0.36322	0.42387
18	0.11754	0.28989	0.31282	108	0.22886	0.96272	0.98954
20	0.14105	0.34375	0.37156	110	0.20540	1.30275	1.31884
22	0.16430	0.39372	0.42663	112	0.15361	1.39589	1.40432
24	0.18599	0.43544	0.47349	114	0.08219	1.27972	1.28236
26	0.20460	0.46376	0.50689	116	0.00160	1.00874	1.00874
28	0.21840	0.47318	0.52115	118	-0.07754	0.64529	0.64993
30	0.22561	0.45815	0.51069	120	-0.14578	0.25085	0.29013
32	0.22445	0.41381	0.47076	122	-0.19594	-0.12112	0.23036
34	0.21341	0.33666	0.39860	124	-0.22368	-0.42999	0.48469
36	0.19140	0.22555	0.29582	126	-0.22765	-0.65026	0.68896
38	0.15804	0.08257	0.17831	128	-0.20923	-0.77157	0.79943
40	0.11386	-0.08615	0.14278	130	-0.17194	-0.79674	0.81508
42	0.06048	-0.26986	0.27655	132	-0.12077	-0.73870	0.74851
44	0.00074	-0.45320	0.45320	134	-0.06134	-0.61680	0.61984
46	-0.06134	-0.61679	0.61984	136	0.00074	-0.45320	0.45320
48	-0.12077	-0.73870	0.74851	138	0.06047	-0.26987	0.27656
50	-0.17194	-0.79674	0.81508	140	0.11386	-0.08615	0.14278
52	-0.20923	-0.77157	0.79943	142	0.15804	0.08256	0.17831
54	-0.22765	-0.65026	0.68896	144	0.19140	0.22555	0.29582
56	-0.22368	-0.42999	0.48469	146	0.21341	0.33666	0.39860
58	-0.19594	-0.12113	0.23036	148	0.22445	0.41380	0.47076
60	-0.14578	0.25084	0.29013	150	0.22561	0.45815	0.51069
62	-0.07754	0.64528	0.64993	152	0.21840	0.47317	0.52115
64	0.00160	1.00873	1.00873	154	0.20460	0.46376	0.50689
66	0.08219	1.27971	1.28235	156	0.18599	0.43543	0.47349
68	0.15361	1.39589	1.40432	158	0.16430	0.39373	0.42664
70	0.20540	1.30275	1.31884	160	0.14105	0.34375	0.37156
72	0.22886	0.96272	0.98955	162	0.11754	0.28989	0.31282
74	0.21850	0.36325	0.42390	164	0.09480	0.23578	0.25412
76	0.17315	-0.47767	0.50809	166	0.07364	0.18419	0.19837
78	0.09655	-1.50983	1.51292	168	0.05464	0.13715	0.14763
80	-0.00289	-2.65363	2.65363	170	0.03818	0.09605	0.10336
82	-0.11304	-3.80787	3.80954	172	0.02453	0.06179	0.06648
84	-0.21963	-4.86116	4.86612	174	0.01383	0.03485	0.03749
86	-0.30837	-5.70592	5.71424	176	0.00615	0.01551	0.01669
88	-0.36710	-6.25253	6.26330	178	0.00154	0.00389	0.00418
				180	0.00000	0.00001	0.00001

IBR

RADIATION PATTERN FOR THETA=PI/2

H/LMDA=0.2500 B/LMDA= 0.500 OMEGA= 8.53

8 ELEMENT ARRAY

PHI	ER(THETA,PHI)			PHI	ER(THETA,PHI)		
DEG	RE	IM	ABSVAL	DEG	RE	IM	ABSVAL
0	0.00000	0.00001	0.00001	90	-1.43663	-5.12706	5.32454
2	0.00159	0.00511	0.00535	92	-1.39078	-4.96604	5.15712
4	0.00633	0.02039	0.02135	94	-1.25856	-4.50145	4.67408
6	0.01423	0.04581	0.04797	96	-1.05514	-3.78604	3.93032
8	0.02524	0.08125	0.08508	98	-0.80334	-2.89925	3.00849
10	0.03926	0.12640	0.13236	100	-0.53044	-1.93607	2.00742
12	0.05614	0.18070	0.18922	102	-0.26438	-0.99404	1.02860
14	0.07557	0.24316	0.25464	104	-0.03017	-0.16068	0.16349
16	0.09709	0.31224	0.32699	106	0.15308	0.49676	0.51981
18	0.12002	0.38568	0.40392	108	0.27408	0.93800	0.97722
20	0.14341	0.46037	0.48219	110	0.32997	1.15197	1.19830
22	0.16606	0.53230	0.55760	112	0.32578	1.15497	1.20003
24	0.18648	0.59650	0.62497	114	0.27279	0.98515	1.02222
26	0.20290	0.64719	0.67825	116	0.18632	0.69488	0.71942
28	0.21341	0.67800	0.71079	118	0.08323	0.34222	0.35220
30	0.21601	0.68243	0.71580	120	-0.02038	-0.01706	0.02658
32	0.20887	0.65448	0.68700	122	-0.11101	-0.33593	0.35380
34	0.19052	0.58941	0.61944	124	-0.17899	-0.58025	0.60723
36	0.16013	0.48474	0.51051	126	-0.21896	-0.73059	0.76270
38	0.11785	0.34121	0.36099	128	-0.22972	-0.78184	0.81489
40	0.06500	0.16366	0.17609	130	-0.21362	-0.74105	0.77123
42	0.00437	-0.03832	0.03857	132	-0.17556	-0.62427	0.64849
44	-0.05982	-0.25017	0.25722	134	-0.12199	-0.45289	0.46903
46	-0.12198	-0.45285	0.46903	136	-0.05982	-0.25017	0.25722
48	-0.17556	-0.62427	0.64848	138	0.00437	-0.03833	0.03857
50	-0.21362	-0.74105	0.77123	140	0.06500	0.16365	0.17609
52	-0.22972	-0.78184	0.81489	142	0.11784	0.34121	0.36098
54	-0.21896	-0.73059	0.76270	144	0.16013	0.48474	0.51050
56	-0.17900	-0.58025	0.60723	146	0.19052	0.58941	0.61944
58	-0.11101	-0.33594	0.35380	148	0.20887	0.65447	0.68700
60	-0.02038	-0.01707	0.02659	150	0.21601	0.68243	0.71580
62	0.08322	0.34222	0.35219	152	0.21341	0.67800	0.71079
64	0.18632	0.69487	0.71942	154	0.20290	0.64719	0.67825
66	0.27279	0.98514	1.02221	156	0.18648	0.59650	0.62497
68	0.32578	1.15497	1.20003	158	0.16607	0.53230	0.55761
70	0.32997	1.15197	1.19830	160	0.14341	0.46037	0.48219
72	0.27408	0.93800	0.97723	162	0.12002	0.38568	0.40392
74	0.15309	0.49678	0.51984	164	0.09709	0.31224	0.32698
76	-0.03016	-0.16065	0.16345	166	0.07557	0.24316	0.25464
78	-0.26437	-0.99402	1.02858	168	0.05614	0.18070	0.18922
80	-0.53043	-1.93606	2.00741	170	0.03926	0.12640	0.13236
82	-0.80333	-2.89923	3.00847	172	0.02524	0.08125	0.08508
84	-1.05513	-3.78600	3.93028	174	0.01423	0.04581	0.04797
86	-1.25855	-4.50142	4.67405	176	0.00633	0.02039	0.02135
88	-1.39078	-4.96603	5.15711	178	0.00159	0.00511	0.00535
				180	0.00000	0.00001	0.00001

VBR

RADIATION PATTERN FOR THETA=PI/2

H/LMDA=0.2500 B/LMDA= 0.500 OMEGA= 8.53

9 ELEMENT ARRAY

PHI DEG	ER(THETA,PHI) RE	IM	ABSVAL	PHI DEG	ER(THETA,PHI) RE	IM	ABSVAL
0	-0.18907	-0.26967	0.32935	90	-0.36792	-6.45913	6.46960
2	-0.18906	-0.26966	0.32933	92	-0.34332	-6.21840	6.22787
4	-0.18897	-0.26941	0.32907	94	-0.27390	-5.52944	5.53622
6	-0.18858	-0.26832	0.32796	96	-0.17200	-4.48605	4.48935
8	-0.18753	-0.26541	0.32498	98	-0.05540	-3.22643	3.22691
10	-0.18533	-0.25931	0.31873	100	0.05608	-1.90991	1.91073
12	-0.18137	-0.24832	0.30750	102	0.14439	-0.69101	0.70594
14	-0.17493	-0.23050	0.28936	104	0.19644	0.30370	0.36170
16	-0.16522	-0.20379	0.26235	106	0.20623	0.99233	1.01353
18	-0.15147	-0.16618	0.22486	108	0.17542	1.34528	1.35666
20	-0.13295	-0.11605	0.17647	110	0.11250	1.38329	1.38785
22	-0.10911	-0.05245	0.12106	112	0.03069	1.16754	1.16794
24	-0.07968	0.02446	0.08335	114	-0.05477	0.78435	0.78626
26	-0.04484	0.11298	0.12155	116	-0.12938	0.32810	0.35269
28	-0.00530	0.20940	0.20946	118	-0.18176	-0.11430	0.21471
30	0.03756	0.30782	0.31010	120	-0.20514	-0.47520	0.51759
32	0.08162	0.39995	0.40819	122	-0.19785	-0.71260	0.73956
34	0.12402	0.47549	0.49140	124	-0.16294	-0.81135	0.82755
36	0.16133	0.52277	0.54710	126	-0.10708	-0.77985	0.78717
38	0.18972	0.53011	0.56304	128	-0.03902	-0.64379	0.64497
40	0.20543	0.48746	0.52898	130	0.03193	-0.43860	0.43976
42	0.20523	0.38868	0.43953	132	0.09729	-0.20222	0.22441
44	0.18701	0.23376	0.29936	134	0.15042	0.03090	0.15356
46	0.15042	0.03091	0.15357	136	0.18701	0.23376	0.29936
48	0.09729	-0.20221	0.22440	138	0.20523	0.38868	0.43953
50	0.03193	-0.43860	0.43976	140	0.20543	0.48746	0.52898
52	-0.03902	-0.64379	0.64497	142	0.18972	0.53011	0.56303
54	-0.10708	-0.77985	0.78717	144	0.16133	0.52277	0.54710
56	-0.16294	-0.81135	0.82755	146	0.12402	0.47548	0.49139
58	-0.19785	-0.71260	0.73956	148	0.08162	0.39995	0.40820
60	-0.20514	-0.47521	0.51759	150	0.03756	0.30781	0.31009
62	-0.18176	-0.11431	0.21472	152	-0.00530	0.20940	0.20946
64	-0.12938	0.32809	0.35268	154	-0.04484	0.11297	0.12154
66	-0.05477	0.78433	0.78624	156	-0.07968	0.02446	0.08335
68	0.03068	1.16754	1.16794	158	-0.10911	-0.05245	0.12106
70	0.11250	1.38328	1.38785	160	-0.13295	-0.11606	0.17648
72	0.17542	1.34528	1.35667	162	-0.15147	-0.16618	0.22486
74	0.20623	0.99235	1.01356	164	-0.16522	-0.20378	0.26235
76	0.19645	0.30374	0.36173	166	-0.17493	-0.23050	0.28936
78	0.14439	-0.69099	0.70592	168	-0.18137	-0.24832	0.30750
80	0.05609	-1.90989	1.91071	170	-0.18533	-0.25931	0.31873
82	-0.05539	-3.22641	3.22688	172	-0.18753	-0.26541	0.32498
84	-0.17199	-4.48600	4.48930	174	-0.18858	-0.26832	0.32796
86	-0.27390	-5.52940	5.53617	176	-0.18897	-0.26941	0.32907
88	-0.34332	-6.21839	6.22786	178	-0.18906	-0.26966	0.32933
				180	-0.18907	-0.26967	0.32935

IBR

RADIATION PATTERN FOR THETA=PI/2

H/LMDA=0.2500 B/LMDA= 0.500 OMEGA= 8.53

9 ELEMENT ARRAY

PHI DEG	ER(THETA,PHI) RE	IM	ABSVAL	PHI DEG	ER(THETA,PHI) RE	IM	ABSVAL
0	-0.25748	-0.59791	0.65100	90	-1.73869	-4.76476	5.07208
2	-0.25747	-0.59789	0.65098	92	-1.66859	-4.57511	4.86988
4	-0.25735	-0.59759	0.65065	94	-1.46851	-4.03359	4.29259
6	-0.25682	-0.59627	0.64922	96	-1.16732	-3.21756	3.42277
8	-0.25541	-0.59272	0.64541	98	-0.80728	-2.24049	2.38149
10	-0.25244	-0.58528	0.63740	100	-0.43669	-1.23222	1.30731
12	-0.24709	-0.57187	0.62297	102	-0.10171	-0.31710	0.33301
14	-0.23840	-0.55010	0.59954	104	0.16089	0.40534	0.43610
16	-0.22533	-0.51737	0.56431	106	0.32879	0.87399	0.93379
18	-0.20685	-0.47110	0.51451	108	0.39622	1.07191	1.14280
20	-0.18204	-0.40905	0.44773	110	0.37295	1.02386	1.08967
22	-0.15026	-0.32965	0.36228	112	0.28078	0.78701	0.83559
24	-0.11128	-0.23244	0.25771	114	0.14838	0.43737	0.46185
26	-0.06552	-0.11861	0.13551	116	0.00576	0.05500	0.05530
28	-0.01422	0.00857	0.01660	118	-0.12073	-0.28918	0.31337
30	0.04046	0.14344	0.14904	120	-0.21191	-0.54300	0.58289
32	0.09529	0.27765	0.29355	122	-0.25756	-0.67781	0.72510
34	0.14611	0.40049	0.42631	124	-0.25638	-0.68861	0.73479
36	0.18802	0.49939	0.53361	126	-0.21453	-0.59032	0.62809
38	0.21581	0.56118	0.60124	128	-0.14329	-0.41171	0.43594
40	0.22464	0.57365	0.61606	130	-0.05639	-0.18837	0.19663
42	0.21083	0.52776	0.56832	132	0.03251	0.04392	0.05464
44	0.17276	0.41995	0.45409	134	0.11172	0.25426	0.27772
46	0.11172	0.25427	0.27773	136	0.17276	0.41995	0.45409
48	0.03251	0.04392	0.05465	138	0.21083	0.52776	0.56832
50	-0.05639	-0.18837	0.19663	140	0.22464	0.57365	0.61606
52	-0.14329	-0.41171	0.43593	142	0.21581	0.56117	0.60124
54	-0.21453	-0.59032	0.62809	144	0.18802	0.49940	0.53362
56	-0.25638	-0.68861	0.73478	146	0.14611	0.40049	0.42631
58	-0.25757	-0.67781	0.72510	148	0.09530	0.27766	0.29356
60	-0.21191	-0.54301	0.58289	150	0.04046	0.14343	0.14903
62	-0.12073	-0.28918	0.31337	152	-0.01422	0.00858	0.01660
64	0.00576	0.05499	0.05529	154	-0.06552	-0.11862	0.13551
66	0.14838	0.43735	0.46184	156	-0.11128	-0.23244	0.25771
68	0.28078	0.78700	0.83559	158	-0.15026	-0.32965	0.36228
70	0.37295	1.02386	1.08967	160	-0.18204	-0.40905	0.44773
72	0.39622	1.07191	1.14280	162	-0.20685	-0.47110	0.51451
74	0.32880	0.87400	0.93380	164	-0.22533	-0.51736	0.56430
76	0.16089	0.40537	0.43613	166	-0.23840	-0.55010	0.59954
78	-0.10171	-0.31708	0.33299	168	-0.24709	-0.57187	0.62297
80	-0.43668	-1.23220	1.30729	170	-0.25244	-0.58528	0.63740
82	-0.80726	-2.24047	2.38147	172	-0.25541	-0.59272	0.64541
84	-1.16730	-3.21752	3.42272	174	-0.25682	-0.59627	0.64922
86	-1.46850	-4.03355	4.29256	176	-0.25735	-0.59759	0.65065
88	-1.66859	-4.57510	4.86988	178	-0.25747	-0.59789	0.65098
				180	-0.25748	-0.59791	0.65100

[illegible]

[illegible]

H/LMDA=0.2500 B/LMDA= 0.500 OMEGA= 8.53

10 ELEMENT ARRAY

PHI DEG	ER(THETA,PHI) RE	IM	ABSVAL	PHI DEG	ER(THETA,PHI) RE	IM	ABSVAL
0	-0.00000	0.00000	0.00000	90	-0.34775	-6.49192	6.50122
2	-0.00157	-0.00367	0.00399	92	-0.31934	-6.19096	6.19919
4	-0.00628	-0.01467	0.01596	94	-0.24042	-5.33933	5.34474
6	-0.01411	-0.03297	0.03586	96	-0.12836	-4.07983	4.08185
8	-0.02501	-0.05839	0.06351	98	-0.00734	-2.61676	2.61677
10	-0.03885	-0.09056	0.09855	100	0.09743	-1.17422	1.17826
12	-0.05541	-0.12874	0.14016	102	0.16553	0.04764	0.17225
14	-0.07428	-0.17163	0.18702	104	0.18576	0.90800	0.92681
16	-0.09481	-0.21721	0.23700	106	0.15805	1.34546	1.35471
18	-0.11608	-0.26254	0.28706	108	0.09270	1.37979	1.38290
20	-0.13682	-0.30372	0.33312	110	0.00726	1.09715	1.09717
22	-0.15545	-0.33588	0.37011	112	-0.07792	0.62380	0.62865
24	-0.17007	-0.35333	0.39212	114	-0.14440	0.09587	0.17333
26	-0.17857	-0.35011	0.39302	116	-0.17928	-0.36803	0.40937
28	-0.17881	-0.32065	0.36714	118	-0.17732	-0.68571	0.70827
30	-0.16892	-0.26079	0.31072	120	-0.14106	-0.81983	0.83188
32	-0.14754	-0.16900	0.22434	122	-0.07938	-0.77567	0.77972
34	-0.11430	-0.04770	0.12385	124	-0.00496	-0.59118	0.59120
36	-0.07011	0.09564	0.11859	126	0.06862	-0.32287	0.33008
38	-0.01748	0.24786	0.24847	128	0.12948	-0.03137	0.13323
40	0.03936	0.39039	0.39237	130	0.16921	0.23050	0.28594
42	0.09466	0.50073	0.50960	132	0.18375	0.42538	0.46337
44	0.14164	0.55543	0.57320	134	0.17333	0.53425	0.56166
46	0.17333	0.53425	0.56167	136	0.14164	0.55543	0.57321
48	0.18375	0.42538	0.46338	138	0.09466	0.50073	0.50960
50	0.16921	0.23051	0.28595	140	0.03936	0.39038	0.39236
52	0.12948	-0.03137	0.13323	142	-0.01748	0.24786	0.24848
54	0.06862	-0.32287	0.33008	144	-0.07011	0.09565	0.11859
56	-0.00496	-0.59117	0.59119	146	-0.11430	-0.04770	0.12385
58	-0.07938	-0.77567	0.77972	148	-0.14754	-0.16900	0.22434
60	-0.14106	-0.81983	0.83188	150	-0.16892	-0.26079	0.31071
62	-0.17732	-0.68572	0.70827	152	-0.17881	-0.32065	0.36714
64	-0.17928	-0.36803	0.40938	154	-0.17857	-0.35011	0.39301
66	-0.14440	0.09585	0.17332	156	-0.17007	-0.35333	0.39213
68	-0.07793	0.62379	0.62864	158	-0.15545	-0.33587	0.37010
70	0.00726	1.09714	1.09716	160	-0.13682	-0.30372	0.33312
72	0.09269	1.37979	1.38290	162	-0.11608	-0.26254	0.28706
74	0.15805	1.34547	1.35472	164	-0.09481	-0.21721	0.23700
76	0.18576	0.90803	0.92684	166	-0.07428	-0.17163	0.18702
78	0.16554	0.04766	0.17226	168	-0.05541	-0.12874	0.14016
80	0.09744	-1.17420	1.17824	170	-0.03885	-0.09057	0.09855
82	-0.00733	-2.61673	2.61674	172	-0.02501	-0.05838	0.06351
84	-0.12835	-4.07976	4.08178	174	-0.01411	-0.03297	0.03586
86	-0.24042	-5.33928	5.34469	176	-0.00628	-0.01467	0.01596
88	-0.31934	-6.19095	6.19918	178	-0.00157	-0.00367	0.00399
				180	-0.00000	0.00000	0.00000

[illegible]

RADIATION PATTERN FOR THETA=PI/2

H/LMDA=0.2500 B/LMDA= 0.500 OMEGA= 8.53

10 ELEMENT ARRAY

PHI DEG	ER(THETA,PHI) RE	IM	ABSVAL	PHI DEG	ER(THETA,PHI) RE	IM	ABSVAL
0	-0.00000	0.00000	0.00000	90	-1.46673	-5.08627	5.29353
2	-0.00165	-0.00510	0.00536	92	-1.39383	-4.83679	5.03362
4	-0.00659	-0.02038	0.02142	94	-1.18821	-4.13263	4.30006
6	-0.01480	-0.04578	0.04812	96	-0.88619	-3.09700	3.22129
8	-0.02623	-0.08113	0.08526	98	-0.53940	-1.90521	1.98010
10	-0.04073	-0.12597	0.13239	100	-0.20386	-0.74790	0.77518
12	-0.05803	-0.17944	0.18859	102	0.07140	0.20751	0.21945
14	-0.07767	-0.24003	0.25228	104	0.25338	0.84729	0.88436
16	-0.09889	-0.30537	0.32098	106	0.32993	1.12810	1.17536
18	-0.12064	-0.37206	0.39113	108	0.30978	1.07718	1.12083
20	-0.14146	-0.43550	0.45790	110	0.21804	0.77773	0.80772
22	-0.15955	-0.48993	0.51525	112	0.08897	0.34465	0.35594
24	-0.17276	-0.52854	0.55606	114	-0.04219	-0.10290	0.11122
26	-0.17877	-0.54403	0.57265	116	-0.14634	-0.46546	0.48792
28	-0.17533	-0.52925	0.55754	118	-0.20516	-0.67920	0.70951
30	-0.16052	-0.47836	0.50457	120	-0.21280	-0.72195	0.75266
32	-0.13325	-0.38807	0.41031	122	-0.17473	-0.60959	0.63414
34	-0.09365	-0.25920	0.27559	124	-0.10449	-0.38564	0.39954
36	-0.04349	-0.09787	0.10710	126	-0.01960	-0.10751	0.10928
38	0.01350	0.08352	0.08460	128	0.06257	0.16704	0.17837
40	0.07167	0.26639	0.27587	130	0.12792	0.39061	0.41103
42	0.12379	0.42716	0.44474	132	0.16757	0.53254	0.55828
44	0.16184	0.53997	0.55370	134	0.17827	0.58078	0.60752
46	0.17827	0.58078	0.60752	136	0.16184	0.53997	0.56371
48	0.16757	0.53254	0.55829	138	0.12379	0.42716	0.44474
50	0.12792	0.39062	0.41103	140	0.07167	0.26639	0.27586
52	0.06257	0.16705	0.17838	142	0.01350	0.08352	0.08460
54	-0.01960	-0.10751	0.10928	144	-0.04349	-0.09787	0.10710
56	-0.10449	-0.38563	0.39954	146	-0.09365	-0.25920	0.27559
58	-0.17472	-0.60959	0.63413	148	-0.13325	-0.38807	0.41031
60	-0.21280	-0.72195	0.75266	150	-0.16052	-0.47835	0.50457
62	-0.20516	-0.67920	0.70951	152	-0.17533	-0.52925	0.55754
64	-0.14635	-0.46546	0.48793	154	-0.17877	-0.54403	0.57265
66	-0.04220	-0.10292	0.11124	156	-0.17276	-0.52855	0.55606
68	0.08897	0.34464	0.35594	158	-0.15955	-0.48992	0.51525
70	0.21804	0.77772	0.80771	160	-0.14146	-0.43550	0.45790
72	0.30978	1.07717	1.12083	162	-0.12064	-0.37206	0.39113
74	0.32994	1.12811	1.17536	164	-0.09890	-0.30537	0.32099
76	0.25338	0.84731	0.88439	166	-0.07767	-0.24003	0.25228
78	0.07141	0.20753	0.21947	168	-0.05803	-0.17944	0.18859
80	-0.20385	-0.74788	0.77516	170	-0.04073	-0.12597	0.13239
82	-0.53939	-1.90519	1.98007	172	-0.02622	-0.08113	0.08526
84	-0.89617	-3.09695	3.22124	174	-0.01480	-0.04578	0.04812
86	-1.18820	-4.13259	4.30002	176	-0.00659	-0.02038	0.02142
88	-1.39383	-4.83678	5.03360	178	-0.00165	-0.00510	0.00536
				180	-0.00000	0.00000	0.00000

VBR

RADIATION PATTERN FOR THETA=PI/2

H/LMDA=0.2500 B/LMDA= 0.500 OMEGA= 8.53

12 ELEMENT ARRAY

PHI DEG	ER(THETA,PHI) RE	IM	ABSVAL	PHI DEG	ER(THETA,PHI) RE	IM	ABSVAL
0	0.00000	0.00001	0.00001	90	-0.32108	-6.52578	6.53367
2	0.00159	0.00353	0.00387	92	-0.28422	-6.08834	6.09497
4	0.00635	0.01405	0.01542	94	-0.18544	-4.88256	4.88607
6	0.01426	0.03155	0.03463	96	-0.05597	-3.19607	3.19656
8	0.02526	0.05580	0.06125	98	0.06439	-1.41279	1.41426
10	0.03915	0.08628	0.09475	100	0.14082	0.09603	0.17045
12	0.05561	0.12199	0.13407	102	0.15474	1.06999	1.08113
14	0.07402	0.16115	0.17734	104	0.10889	1.41626	1.42044
16	0.09345	0.20105	0.22171	106	0.02482	1.21206	1.21231
18	0.11258	0.23783	0.26313	108	-0.06561	0.65806	0.66132
20	0.12964	0.26647	0.29633	110	-0.13149	0.00342	0.13153
22	0.14247	0.28102	0.31507	112	-0.15256	-0.53092	0.55241
24	0.14868	0.27521	0.31281	114	-0.12455	-0.80717	0.81672
26	0.14587	0.24332	0.28370	116	-0.05868	-0.79151	0.79368
28	0.13205	0.18161	0.22454	118	0.02341	-0.54131	0.54182
30	0.10609	0.08992	0.13907	120	0.09736	-0.16925	0.19526
32	0.06831	-0.02663	0.07332	122	0.14335	0.20013	0.24617
34	0.02095	-0.15653	0.15793	124	0.15105	0.46733	0.49114
36	-0.03157	-0.28167	0.28344	126	0.12106	0.57930	0.59181
38	-0.08278	-0.37894	0.38787	128	0.06321	0.53313	0.53686
40	-0.12480	-0.42368	0.44168	130	-0.00732	0.36559	0.36567
42	-0.14957	-0.39536	0.42270	132	-0.07440	0.13498	0.15413
44	-0.15058	-0.28423	0.32165	134	-0.12482	-0.09787	0.15861
46	-0.12482	-0.09787	0.15862	136	-0.15058	-0.28422	0.32165
48	-0.07440	0.13497	0.15412	138	-0.14957	-0.39536	0.42270
50	-0.00732	0.36559	0.36566	140	-0.12480	-0.42369	0.44169
52	0.06321	0.53313	0.53687	142	-0.08278	-0.37894	0.38788
54	0.12106	0.57930	0.59181	144	-0.03157	-0.28168	0.28345
56	0.15105	0.46734	0.49114	146	0.02095	-0.15653	0.15793
58	0.14335	0.20013	0.24617	148	0.06831	-0.02664	0.07332
60	0.09736	-0.16925	0.19525	150	0.10609	0.08992	0.13907
62	0.02341	-0.54131	0.54181	152	0.13205	0.18160	0.22454
64	-0.05868	-0.79151	0.79368	154	0.14587	0.24332	0.28370
66	-0.12454	-0.80717	0.81673	156	0.14868	0.27521	0.31280
68	-0.15256	-0.53093	0.55241	158	0.14247	0.28103	0.31508
70	-0.13149	0.00341	0.13153	160	0.12964	0.26647	0.29633
72	-0.06562	0.65805	0.66131	162	0.11258	0.23783	0.26313
74	0.02482	1.21204	1.21229	164	0.09346	0.20106	0.22172
76	0.10888	1.41626	1.42044	166	0.07402	0.16116	0.17735
78	0.15474	1.07001	1.08114	168	0.05561	0.12199	0.13407
80	0.14083	0.09606	0.17047	170	0.03915	0.08628	0.09474
82	0.06440	-1.41276	1.41423	172	0.02526	0.05580	0.06125
84	-0.05595	-3.19598	3.19647	174	0.01426	0.03155	0.03463
86	-0.18543	-4.88249	4.88601	176	0.00635	0.01405	0.01542
88	-0.28422	-6.08833	6.09496	178	0.00159	0.00353	0.00387
				180	0.00000	0.00001	0.00001

IBR

RADIATION PATTERN FOR THETA=PI/2

H/LMDA=0.2500 B/LMDA= 0.500 OMEGA= 8.53

12 ELEMENT ARRAY

PHI DEG	ER(THETA,PHI) RE	IM	ABSVAL	PHI DEG	ER(THETA,PHI) RE	IM	ABSVAL
0	0.00000	0.00001	0.00001	90	-1.48685	-5.05900	5.27297
2	0.00170	0.00512	0.00539	92	-1.38109	-4.70313	4.90172
4	0.00678	0.02041	0.02151	94	-1.09079	-3.72556	3.88196
6	0.01523	0.04583	0.04830	96	-0.68856	-2.36863	2.46668
8	0.02696	0.08112	0.08548	98	-0.27039	-0.95338	0.99099
10	0.04177	0.12565	0.13241	100	0.07259	0.21442	0.22637
12	0.05926	0.17819	0.18778	102	0.27917	0.92762	0.96872
14	0.07875	0.23659	0.24935	104	0.33178	1.12400	1.17195
16	0.09916	0.29756	0.31365	106	0.25584	0.88478	0.92103
18	0.11898	0.35639	0.37573	108	0.10628	0.39079	0.40498
20	0.13621	0.40688	0.42907	110	-0.05272	-0.14446	0.15378
22	0.14844	0.44161	0.46589	112	-0.16774	-0.54105	0.56646
24	0.15304	0.45249	0.47766	114	-0.20948	-0.69772	0.72849
26	0.14745	0.43174	0.45622	116	-0.17666	-0.60567	0.63090
28	0.12970	0.37349	0.39536	118	-0.09093	-0.33272	0.34492
30	0.09897	0.27554	0.29277	120	0.01400	0.01169	0.01824
32	0.05625	0.14139	0.15217	122	0.10444	0.31641	0.33320
34	0.00478	-0.01825	0.01886	124	0.15640	0.50044	0.52431
36	-0.04966	-0.18490	0.19146	126	0.16037	0.52949	0.55324
38	-0.09918	-0.33351	0.34795	128	0.12115	0.41598	0.43326
40	-0.13472	-0.43566	0.45602	130	0.05377	0.20646	0.21334
42	-0.14780	-0.46519	0.48811	132	-0.02232	-0.03709	0.04329
44	-0.13271	-0.40523	0.42640	134	-0.08884	-0.25543	0.27044
46	-0.08884	-0.25543	0.27044	136	-0.13271	-0.40522	0.42640
48	-0.02232	-0.03710	0.04330	138	-0.14780	-0.46519	0.48811
50	0.05377	0.20645	0.21334	140	-0.13473	-0.43567	0.45603
52	0.12115	0.41595	0.43327	142	-0.09918	-0.33351	0.34795
54	0.16037	0.52948	0.55324	144	-0.04966	-0.18491	0.19147
56	0.15640	0.50044	0.52431	146	0.00478	-0.01825	0.01886
58	0.10444	0.31641	0.33321	148	0.05624	0.14139	0.15216
60	0.01400	0.01165	0.01824	150	0.09897	0.27554	0.29277
62	-0.09092	-0.33272	0.34492	152	0.12970	0.37348	0.39536
64	-0.17666	-0.60566	0.63090	154	0.14745	0.43174	0.45623
66	-0.20948	-0.69772	0.72849	156	0.15303	0.45248	0.47766
68	-0.16774	-0.54106	0.56646	158	0.14844	0.44162	0.46590
70	-0.05273	-0.14447	0.15379	160	0.13621	0.40688	0.42907
72	0.10628	0.39078	0.40497	162	0.11898	0.35639	0.37573
74	0.25583	0.88476	0.92101	164	0.09917	0.29757	0.31366
76	0.33178	1.12400	1.17195	166	0.07875	0.23660	0.24937
78	0.27917	0.92763	0.96873	168	0.05926	0.17819	0.18778
80	0.07260	0.21444	0.22639	170	0.04177	0.12564	0.13241
82	-0.27038	-0.95336	0.99096	172	0.02696	0.08112	0.08549
84	-0.68853	-2.36857	2.46661	174	0.01523	0.04583	0.04830
86	-1.09077	-3.72550	3.88190	176	0.00678	0.02041	0.02151
88	-1.38108	-4.70312	4.90170	178	0.00170	0.00512	0.00539
				180	0.00000	0.00001	0.00001

VBR

RADIATION PATTERN FOR THETA=PI/2

H/LMDA=0.2500 B/LMDA= 0.500 OMEGA= 8.53

14 ELEMENT ARRAY

PHI DEG	ER(THETA,PHI) RE	IM	ABSVAL	PHI DEG	ER(THETA,PHI) RE	IM	ABSVAL
0	-0.00000	-0.00000	0.00000	90	-0.30200	-6.55014	6.55710
2	-0.00160	-0.00339	0.00375	92	-0.25617	-5.95257	5.95808
4	-0.00639	-0.01356	0.01500	94	-0.13851	-4.35634	4.35854
6	-0.01435	-0.03042	0.03363	96	0.00092	-2.27225	2.27225
8	-0.02538	-0.05370	0.05940	98	0.10528	-0.32507	0.34169
10	-0.03923	-0.08275	0.09158	100	0.13677	0.97965	0.98915
12	-0.05544	-0.11624	0.12879	102	0.09237	1.42316	1.42615
14	-0.07317	-0.15192	0.16862	104	0.00249	1.11187	1.11187
16	-0.09116	-0.18628	0.20739	106	-0.08548	0.38585	0.39520
18	-0.10764	-0.21453	0.24002	108	-0.12987	-0.34828	0.37170
20	-0.12032	-0.23074	0.26023	110	-0.11261	-0.78104	0.78912
22	-0.12660	-0.22832	0.26107	112	-0.04472	-0.79557	0.79683
24	-0.12381	-0.20116	0.23621	114	0.04158	-0.46911	0.47095
26	-0.10972	-0.14528	0.18206	116	0.10888	0.00081	0.10889
28	-0.08324	-0.06080	0.10308	118	0.13072	0.40025	0.42106
30	-0.04508	0.04596	0.06438	120	0.10097	0.58547	0.59411
32	0.00160	0.16127	0.16128	122	0.03379	0.52235	0.52344
34	0.05095	0.26420	0.26907	124	-0.04428	0.27515	0.27868
36	0.09485	0.32942	0.34280	126	-0.10558	-0.03858	0.11241
38	0.12411	0.33265	0.35505	128	-0.13115	-0.30129	0.32859
40	0.13055	0.25873	0.28980	130	-0.11564	-0.43569	0.45077
42	0.10961	0.10999	0.15528	132	-0.06688	-0.42087	0.42615
44	0.06276	-0.08779	0.10791	134	-0.00120	-0.28565	0.28566
46	-0.00120	-0.28565	0.28565	136	0.06276	-0.08779	0.10791
48	-0.06688	-0.42087	0.42615	138	0.10961	0.10999	0.15528
50	-0.11564	-0.43569	0.45078	140	0.13055	0.25873	0.28980
52	-0.13115	-0.30128	0.32859	142	0.12411	0.33265	0.35505
54	-0.10558	-0.03858	0.11241	144	0.09485	0.32941	0.34280
56	-0.04428	0.27514	0.27868	146	0.05095	0.26420	0.26907
58	0.03379	0.52235	0.52344	148	0.00160	0.16126	0.16127
60	0.10096	0.58547	0.59411	150	-0.04508	0.04596	0.06438
62	0.13072	0.40025	0.42106	152	-0.08324	-0.06080	0.10308
64	0.10889	0.00082	0.10889	154	-0.10972	-0.14527	0.18205
66	0.04158	-0.46909	0.47092	156	-0.12381	-0.20116	0.23621
68	-0.04472	-0.79556	0.79682	158	-0.12660	-0.22831	0.26107
70	-0.11261	-0.78105	0.78912	160	-0.12032	-0.23074	0.26023
72	-0.12987	-0.34829	0.37171	162	-0.10764	-0.21453	0.24002
74	-0.08548	0.38581	0.39517	164	-0.09116	-0.18627	0.20738
76	0.00249	1.11184	1.11184	166	-0.07317	-0.15190	0.16861
78	0.09237	1.42316	1.42615	168	-0.05544	-0.11624	0.12879
80	0.13677	0.97967	0.98917	170	-0.03923	-0.08276	0.09159
82	0.10529	-0.32503	0.34166	172	-0.02538	-0.05370	0.05940
84	0.00094	-2.27216	2.27216	174	-0.01435	-0.03042	0.03363
86	-0.13849	-4.35625	4.35845	176	-0.00639	-0.01356	0.01500
88	-0.25617	-5.95255	5.95806	178	-0.00160	-0.00339	0.00375
				180	-0.00000	-0.00000	0.00000

IBR

RADIATION PATTERN FOR THETA=PI/2

H/LMDA=0.2500 B/LMDA= 0.500 OMEGA= 8.53

14 ELEMENT ARRAY

PHI DEG	ER(THETA,PHI) RE	IM	ABSVAL	PHI DEG	ER(THETA,PHI) RE	IM	ABSVAL
0	-0.00000	-0.00000	0.00001	90	-1.50128	-5.03943	5.25830
2	-0.00174	-0.00512	0.00540	92	-1.35708	-4.56000	4.75765
4	-0.00694	-0.02046	0.02160	94	-0.97392	-3.28489	3.42623
6	-0.01557	-0.04591	0.04848	96	-0.47975	-1.63653	1.70540
8	-0.02753	-0.08114	0.08568	98	-0.02902	-0.12614	0.12943
10	-0.04254	-0.12531	0.13234	100	0.25704	0.84264	0.88097
12	-0.06004	-0.17677	0.18669	102	0.33191	1.11125	1.15976
14	-0.07911	-0.23268	0.24576	104	0.23105	0.79145	0.82449
16	-0.09829	-0.28862	0.30490	106	0.04544	0.18025	0.18589
18	-0.11558	-0.33850	0.35769	108	-0.12306	-0.38591	0.40505
20	-0.12843	-0.37462	0.39603	110	-0.20333	-0.66814	0.69839
22	-0.13393	-0.38825	0.41070	112	-0.17682	-0.59858	0.62415
24	-0.12921	-0.37081	0.39268	114	-0.07398	-0.27111	0.28102
26	-0.11205	-0.31577	0.33506	116	0.04827	0.13075	0.13937
28	-0.08159	-0.22097	0.23556	118	0.13518	0.42661	0.44752
30	-0.03923	-0.09119	0.09927	120	0.15587	0.51115	0.53439
32	0.01079	0.06007	0.06104	122	0.11100	0.38071	0.39657
34	0.06128	0.21025	0.21900	124	0.02645	0.11472	0.11773
36	0.10276	0.33009	0.34572	126	-0.06127	-0.17047	0.18115
38	0.12521	0.38907	0.40872	128	-0.12036	-0.37106	0.39009
40	0.12071	0.36379	0.38330	130	-0.13400	-0.42961	0.45002
42	0.08656	0.24757	0.26227	132	-0.10301	-0.34488	0.35994
44	0.02779	0.05812	0.06443	134	-0.04205	-0.16048	0.16590
46	-0.04205	-0.16048	0.16590	136	0.02779	0.05812	0.06442
48	-0.10301	-0.34488	0.35994	138	0.08656	0.24757	0.26227
50	-0.13400	-0.42961	0.45003	140	0.12071	0.36379	0.38329
52	-0.12036	-0.37105	0.39009	142	0.12521	0.38906	0.40871
54	-0.06127	-0.17047	0.18115	144	0.10276	0.33009	0.34572
56	0.02645	0.11472	0.11773	146	0.06128	0.21025	0.21900
58	0.11100	0.38071	0.39657	148	0.01079	0.06007	0.06103
60	0.15587	0.51115	0.53439	150	-0.03923	-0.09118	0.09926
62	0.13518	0.42661	0.44752	152	-0.08159	-0.22097	0.23555
64	0.04827	0.13075	0.13938	154	-0.11205	-0.31576	0.33505
66	-0.07397	-0.27109	0.28100	156	-0.12922	-0.37082	0.39268
68	-0.17681	-0.59857	0.62414	158	-0.13393	-0.38825	0.41070
70	-0.20333	-0.66814	0.69839	160	-0.12843	-0.37462	0.39603
72	-0.12306	-0.38592	0.40506	162	-0.11558	-0.33850	0.35769
74	0.04543	0.18022	0.18585	164	-0.09829	-0.28861	0.30489
76	0.23104	0.79143	0.82446	166	-0.07911	-0.23267	0.24575
78	0.33191	1.11125	1.15976	168	-0.06004	-0.17677	0.18669
80	0.25704	0.84265	0.88098	170	-0.04254	-0.12532	0.13234
82	-0.02900	-0.12611	0.12940	172	-0.02753	-0.08114	0.08568
84	-0.47972	-1.63645	1.70532	174	-0.01557	-0.04591	0.04848
86	-0.97390	-3.28482	3.42615	176	-0.00694	-0.02046	0.02160
88	-1.35707	-4.55998	4.75763	178	-0.00174	-0.00512	0.00540
				180	-0.00000	-0.00000	0.00001

VBR

RADIATION PATTERN FOR THETA=PI/2

H/LMDA=0.2500 B/LMDA= 0.500 OMEGA= 8.53

16 ELEMENT ARRAY

PHI DEG	ER(THETA,PHI) RE	IM	ABSVAL	PHI DEG	ER(THETA,PHI) RE	IM	ABSVAL
0	0.00000	0.00000	0.00000	90	-0.28767	-6.56849	6.57479
2	0.00161	0.00330	0.00367	92	-0.23242	-5.78830	5.79296
4	0.00642	0.01315	0.01463	94	-0.09749	-3.77948	3.78074
6	0.01440	0.02948	0.03280	96	0.04353	-1.36699	1.36769
8	0.02542	0.05195	0.05784	98	0.11862	0.54430	0.55707
10	0.03917	0.07973	0.08883	100	0.09896	1.39376	1.39727
12	0.05502	0.11118	0.12405	102	0.01225	1.17573	1.17580
14	0.07189	0.14343	0.16044	104	-0.07859	0.34845	0.35720
16	0.08816	0.17230	0.19355	106	-0.11559	-0.47147	0.48543
18	0.10158	0.19211	0.21732	108	-0.07912	-0.83794	0.84166
20	0.10943	0.19616	0.22462	110	0.00483	-0.65394	0.65395
22	0.10877	0.17780	0.20844	112	0.08449	-0.13674	0.16074
24	0.09705	0.13221	0.16400	114	0.11455	0.36795	0.38537
26	0.07282	0.05861	0.09348	116	0.08076	0.59399	0.59945
28	0.03675	-0.03708	0.05221	118	0.00437	0.47355	0.47357
30	-0.00768	-0.14077	0.14098	120	-0.07307	0.12765	0.14708
32	-0.05380	-0.23035	0.23655	122	-0.11318	-0.23525	0.26106
34	-0.09236	-0.27930	0.29418	124	-0.09911	-0.43921	0.45025
36	-0.11330	-0.26381	0.28711	126	-0.04061	-0.41623	0.41821
38	-0.10861	-0.17280	0.20410	128	0.03409	-0.21299	0.21571
40	-0.07575	-0.01753	0.07775	130	0.09325	0.05571	0.10862
42	-0.02035	0.16366	0.16492	132	0.11517	0.27123	0.29467
44	0.04326	0.31075	0.31375	134	0.09495	0.35993	0.37224
46	0.09495	0.35993	0.37224	136	0.04326	0.31075	0.31375
48	0.11517	0.27123	0.29467	138	-0.02035	0.16367	0.16493
50	0.09325	0.05571	0.10862	140	-0.07575	-0.01753	0.07775
52	0.03409	-0.21295	0.21570	142	-0.10861	-0.17280	0.20410
54	-0.04061	-0.41624	0.41821	144	-0.11330	-0.26381	0.28711
56	-0.09911	-0.43921	0.45025	146	-0.09236	-0.27930	0.29418
58	-0.11318	-0.23526	0.26107	148	-0.05380	-0.23036	0.23656
60	-0.07308	0.12763	0.14707	150	-0.00768	-0.14077	0.14098
62	0.00437	0.47354	0.47356	152	0.03674	-0.03709	0.05221
64	0.08076	0.59400	0.59946	154	0.07282	0.05861	0.09348
66	0.11455	0.36797	0.38539	156	0.09705	0.13220	0.16399
68	0.08449	-0.13673	0.16073	158	0.10877	0.17781	0.20844
70	0.00484	-0.65393	0.65395	160	0.10943	0.19616	0.22462
72	-0.07912	-0.83794	0.84167	162	0.10158	0.19211	0.21732
74	-0.11559	-0.47150	0.48546	164	0.08816	0.17231	0.19355
76	-0.07859	0.34841	0.35716	166	0.07189	0.14344	0.16045
78	0.01224	1.17572	1.17579	168	0.05502	0.11117	0.12404
80	0.09896	1.39376	1.39727	170	0.03917	0.07973	0.08884
82	0.11863	0.54432	0.55710	172	0.02542	0.05195	0.05784
84	0.04354	-1.36689	1.36758	174	0.01440	0.02948	0.03280
86	-0.09747	-3.77937	3.78062	176	0.00642	0.01315	0.01463
88	-0.23241	-5.78827	5.79294	178	0.00161	0.00330	0.00367
				180	0.00000	0.00000	0.00000

IBR

RADIATION PATTERN FOR THETA=PI/2

H/LMDA=0.2500 B/LMDA= 0.500 OMEGA= 8.53

16 ELEMENT ARRAY

PHI DEG	ER(THETA,PHI) RE	IM	ABSVAL	PHI DEG	ER(THETA,PHI) RE	IM	ABSVAL
0	0.00000	0.00000	0.00000	90	-1.51217	-5.02465	5.24726
2	0.00177	0.00514	0.00543	92	-1.32422	-4.40537	4.60009
4	0.00707	0.02050	0.02168	94	-0.84341	-2.81925	2.94270
6	0.01585	0.04598	0.04864	96	-0.27488	-0.93825	0.97769
8	0.02799	0.08115	0.08584	98	0.16031	0.51105	0.53560
10	0.04310	0.12492	0.13214	100	0.33220	1.09726	1.14645
12	0.06048	0.17516	0.18531	102	0.25169	0.84878	0.88531
14	0.07890	0.22818	0.24144	104	0.04190	0.16346	0.16874
16	0.09650	0.27847	0.29472	106	-0.14358	-0.45554	0.47763
18	0.11077	0.31851	0.33723	108	-0.20382	-0.67142	0.70167
20	0.11865	0.33923	0.35938	110	-0.13056	-0.44746	0.46612
22	0.11692	0.33118	0.35121	112	0.00841	0.00236	0.00873
24	0.10291	0.28656	0.30448	114	0.12234	0.38334	0.40239
26	0.07533	0.20201	0.21560	116	0.15135	0.49518	0.51780
28	0.03543	0.08177	0.08911	118	0.09232	0.31909	0.33218
30	-0.01228	-0.05996	0.06120	120	-0.01066	-0.00889	0.01388
32	-0.05978	-0.19849	0.20729	122	-0.09812	-0.29828	0.31400
34	-0.09650	-0.30181	0.31687	124	-0.12804	-0.40996	0.42949
36	-0.11170	-0.33789	0.35588	126	-0.09302	-0.31345	0.32696
38	-0.09798	-0.28565	0.30198	128	-0.01705	-0.07963	0.08143
40	-0.05516	-0.14659	0.15662	130	0.06144	0.17154	0.18221
42	0.00715	0.04759	0.04813	132	0.10908	0.33279	0.35021
44	0.07009	0.23740	0.24753	134	0.11038	0.35121	0.36815
46	0.11038	0.35121	0.36815	136	0.07009	0.23740	0.24753
48	0.10908	0.33279	0.35021	138	0.00716	0.04760	0.04814
50	0.06144	0.17154	0.18221	140	-0.05516	-0.14659	0.15663
52	-0.01705	-0.07962	0.08143	142	-0.09798	-0.28564	0.30198
54	-0.09302	-0.31346	0.32697	144	-0.11170	-0.33789	0.35588
56	-0.12805	-0.40996	0.42949	146	-0.09650	-0.30181	0.31687
58	-0.09812	-0.29828	0.31400	148	-0.05978	-0.19849	0.20730
60	-0.01066	-0.00890	0.01389	150	-0.01228	-0.05995	0.06120
62	0.09232	0.31908	0.33217	152	0.03543	0.08176	0.08911
64	0.15135	0.49519	0.51780	154	0.07533	0.20201	0.21560
66	0.12234	0.38336	0.40240	156	0.10290	0.28656	0.30448
68	0.00841	0.00237	0.00874	158	0.11693	0.33118	0.35122
70	-0.13056	-0.44746	0.46612	160	0.11865	0.33923	0.35938
72	-0.20382	-0.67142	0.70167	162	0.11077	0.31851	0.33723
74	-0.14358	-0.45556	0.47766	164	0.09651	0.27848	0.29472
76	0.04189	0.16342	0.16871	166	0.07890	0.22819	0.24145
78	0.25169	0.84877	0.88530	168	0.06048	0.17516	0.18531
80	0.33220	1.09727	1.14645	170	0.04310	0.12492	0.13215
82	0.16032	0.51107	0.53562	172	0.02799	0.08115	0.08584
84	-0.27485	-0.93817	0.97760	174	0.01585	0.04598	0.04864
86	-0.84338	-2.81916	2.94261	176	0.00707	0.02050	0.02168
88	-1.32421	-4.40535	4.60007	178	0.00177	0.00514	0.00543
				180	0.00000	0.00000	0.00000

VBR

RADIATION PATTERN FOR THETA=PI/2

H/LMDA=0.2500 B/LMDA= 0.500 OMEGA= 8.53

18 ELEMENT ARRAY

PHI DEG	ER(THETA,PHI) RE	IM	ABSVAL	PHI DEG	ER(THETA,PHI) RE	IM	ABSVAL
0	-0.00000	-0.00001	0.00001	90	-0.27652	-6.58282	6.58862
2	-0.00161	-0.00320	0.00358	92	-0.21146	-5.59881	5.60280
4	-0.00643	-0.01281	0.01433	94	-0.06142	-3.17018	3.17077
6	-0.01442	-0.02871	0.03213	96	0.07254	-0.53307	0.53798
8	-0.02541	-0.05046	0.05650	98	0.10914	1.12953	1.13479
10	-0.03901	-0.07708	0.08639	100	0.04398	1.35639	1.35710
12	-0.05442	-0.10654	0.11963	102	-0.05460	0.55323	0.55592
14	-0.07028	-0.13548	0.15263	104	-0.10323	-0.41575	0.42837
16	-0.08459	-0.15886	0.17998	106	-0.06666	-0.84336	0.84599
18	-0.09465	-0.17026	0.19480	108	0.02164	-0.57057	0.57098
20	-0.09737	-0.16263	0.18955	110	0.09175	0.05362	0.10627
22	-0.08976	-0.12994	0.15793	112	0.09249	0.52545	0.53353
24	-0.06980	-0.06969	0.09863	114	0.02649	0.55406	0.55469
26	-0.03753	0.01414	0.04011	116	-0.05711	0.20297	0.21085
28	0.00386	0.10879	0.10886	118	-0.10101	-0.23034	0.25152
30	0.04757	0.19246	0.19825	120	-0.07861	-0.45530	0.46204
32	0.08372	0.23792	0.25222	122	-0.00744	-0.36729	0.36736
34	0.10141	0.22064	0.24283	124	0.06703	-0.07010	0.09699
36	0.09239	0.13042	0.15983	126	0.10185	0.23105	0.25250
38	0.05519	-0.01748	0.05789	128	0.08057	0.36870	0.37740
40	-0.00175	-0.17817	0.17818	130	0.01815	0.29603	0.29659
42	-0.06032	-0.28659	0.29287	132	-0.05166	0.08343	0.09813
44	-0.09776	-0.28254	0.29897	134	-0.09597	-0.14467	0.17360
46	-0.09597	-0.14467	0.17361	136	-0.09776	-0.28254	0.29897
48	-0.05167	0.08343	0.09813	138	-0.06032	-0.28659	0.29287
50	0.01815	0.29603	0.29659	140	-0.00175	-0.17818	0.17819
52	0.08057	0.36871	0.37741	142	0.05519	-0.01749	0.05789
54	0.10185	0.23104	0.25250	144	0.09239	0.13041	0.15982
56	0.06703	-0.07010	0.09699	146	0.10141	0.22064	0.24283
58	-0.00744	-0.36728	0.36736	148	0.08372	0.23792	0.25222
60	-0.07861	-0.45530	0.46204	150	0.04758	0.19247	0.19826
62	-0.10101	-0.23035	0.25152	152	0.00386	0.10879	0.10885
64	-0.05711	0.20297	0.21085	154	-0.03753	0.01414	0.04011
66	0.02648	0.55405	0.55468	156	-0.06980	-0.06970	0.09864
68	0.09249	0.52545	0.53353	158	-0.08976	-0.12993	0.15792
70	0.09175	0.05362	0.10627	160	-0.09737	-0.16263	0.18955
72	0.02164	-0.57055	0.57096	162	-0.09465	-0.17026	0.19480
74	-0.06666	-0.84336	0.84599	164	-0.08459	-0.15886	0.17998
76	-0.10324	-0.41578	0.42841	166	-0.07028	-0.13547	0.15262
78	-0.05460	0.55322	0.55591	168	-0.05442	-0.10655	0.11965
80	0.04398	1.35638	1.35709	170	-0.03901	-0.07707	0.08638
82	0.10915	1.12955	1.13481	172	-0.02541	-0.05046	0.05650
84	0.07256	-0.53296	0.53788	174	-0.01442	-0.02871	0.03213
86	-0.06140	-3.17005	3.17065	176	-0.00643	-0.01281	0.01433
88	-0.21144	-5.59878	5.60277	178	-0.00161	-0.00320	0.00358
				180	-0.00000	-0.00001	0.00001

IBR

RADIATION PATTERN FOR THETA=PI/2

H/LMDA=0.2500 B/LMDA= 0.500 OMEGA= 8.53

18 ELEMENT ARRAY

PHI DEG	ER(THETA,PHI) RE	IM	ABSVAL	PHI DEG	ER(THETA,PHI) RE	IM	ABSVAL
0	-0.00000	-0.00001	0.00001	90	-1.52069	-5.01307	5.23864
2	-0.00180	-0.00514	0.00545	92	-1.28400	-4.23865	4.42886
4	-0.00718	-0.02055	0.02177	94	-0.70430	-2.33915	2.44288
6	-0.01610	-0.04608	0.04881	96	-0.08692	-0.30885	0.32085
8	-0.02836	-0.08116	0.08597	98	0.28273	0.91889	0.96140
10	-0.04352	-0.12446	0.13185	100	0.30524	1.01174	1.05678
12	-0.06065	-0.17328	0.18359	102	0.09572	0.33565	0.34903
14	-0.07823	-0.22312	0.23644	104	-0.12796	-0.40238	0.42224
16	-0.09395	-0.26717	0.28321	106	-0.20229	-0.66304	0.69322
18	-0.10479	-0.29658	0.31455	108	-0.10882	-0.37376	0.38928
20	-0.10730	-0.30130	0.31983	110	0.04917	0.13919	0.14762
22	-0.09824	-0.27194	0.28914	112	0.14526	0.46487	0.48703
24	-0.07557	-0.20293	0.21654	114	0.12158	0.40614	0.42395
26	-0.03977	-0.09617	0.10407	116	0.01404	0.06771	0.06915
28	0.00511	0.03561	0.03597	118	-0.09079	-0.27604	0.29058
30	0.05093	0.16767	0.17524	120	-0.12292	-0.39558	0.41424
32	0.08645	0.26636	0.28003	122	-0.07107	-0.24488	0.25499
34	0.10002	0.29745	0.31382	124	0.02148	0.04582	0.05060
36	0.08394	0.23945	0.25373	126	0.09361	0.28345	0.29851
38	0.03916	0.09751	0.10508	128	0.10573	0.33739	0.35357
40	-0.02191	-0.08853	0.09120	130	0.05797	0.20043	0.20864
42	-0.07705	-0.25001	0.26161	132	-0.01766	-0.03236	0.03687
44	-0.10161	-0.31285	0.32893	134	-0.08019	-0.23365	0.24702
46	-0.08019	-0.23365	0.24703	136	-0.10161	-0.31285	0.32894
48	-0.01766	-0.03237	0.03687	138	-0.07705	-0.25001	0.26161
50	0.05797	0.20043	0.20864	140	-0.02191	-0.08854	0.09121
52	0.10573	0.33739	0.35357	142	0.03916	0.09750	0.10507
54	0.09361	0.28345	0.29850	144	0.08394	0.23944	0.25373
56	0.02148	0.04582	0.05061	146	0.10002	0.29745	0.31382
58	-0.07107	-0.24488	0.25498	148	0.08645	0.26636	0.28003
60	-0.12292	-0.39558	0.41424	150	0.05094	0.16768	0.17525
62	-0.09079	-0.27604	0.29059	152	0.00511	0.03560	0.03597
64	0.01404	0.06770	0.06914	154	-0.03977	-0.09617	0.10407
66	0.12158	0.40613	0.42394	156	-0.07557	-0.20293	0.21655
68	0.14526	0.46487	0.48703	158	-0.09823	-0.27194	0.28914
70	0.04918	0.13919	0.14762	160	-0.10730	-0.30129	0.31983
72	-0.10882	-0.37375	0.38926	162	-0.10479	-0.29658	0.31455
74	-0.20229	-0.66304	0.69321	164	-0.09395	-0.26717	0.28321
76	-0.12797	-0.40241	0.42227	166	-0.07823	-0.22311	0.23643
78	0.09572	0.33563	0.34902	168	-0.06065	-0.17330	0.18361
80	0.30524	1.01173	1.05677	170	-0.04351	-0.12445	0.13184
82	0.28273	0.91890	0.96142	172	-0.02836	-0.08116	0.08597
84	-0.08689	-0.30877	0.32077	174	-0.01610	-0.04608	0.04881
86	-0.70426	-2.33906	2.44278	176	-0.00718	-0.02055	0.02177
88	-1.28399	-4.23862	4.42883	178	-0.00180	-0.00514	0.00545
				180	-0.00000	-0.00001	0.00001

VBR

RADIATION PATTERN FOR THETA=PI/2

H/LMDA=0.2500 B/LMDA= 0.500 OMEGA= 8.53

20 ELEMENT ARRAY

PHI DEG	ER(THETA,PHI) RE	IM	ABSVAL	PHI DEG	ER(THETA,PHI) RE	IM	ABSVAL
0	0.00000	-0.00001	0.00001	90	-0.26759	-6.59431	6.59973
2	0.00161	0.00315	0.00354	92	-0.19239	-5.38675	5.39018
4	0.00644	0.01252	0.01408	94	-0.02991	-2.54631	2.54649
6	0.01442	0.02800	0.03149	96	0.08878	0.18536	0.20552
8	0.02537	0.04912	0.05528	98	0.08307	1.40656	1.40901
10	0.03878	0.07468	0.08415	100	-0.01152	0.97224	0.97231
12	0.05367	0.10225	0.11548	102	-0.08787	-0.14970	0.17358
14	0.06840	0.12783	0.14498	104	-0.07181	-0.82128	0.82442
16	0.08054	0.14574	0.16651	106	0.01477	-0.60662	0.60680
18	0.08702	0.14893	0.17249	108	0.08533	0.09031	0.12425
20	0.08451	0.13034	0.15534	110	0.07480	0.56916	0.57405
22	0.07026	0.08535	0.11055	112	-0.00371	0.47265	0.47267
24	0.04332	0.01497	0.04584	114	-0.07792	-0.00607	0.07816
26	0.00586	-0.07108	0.07132	116	-0.08445	-0.40482	0.41354
28	-0.03591	-0.15290	0.15706	118	-0.02093	-0.41871	0.41923
30	-0.07201	-0.20332	0.21570	120	0.05848	-0.10243	0.11795
32	-0.09077	-0.19578	0.21580	122	0.09186	0.25000	0.26634
34	-0.08289	-0.11750	0.14380	124	0.05705	0.37468	0.37900
36	-0.04654	0.01749	0.04972	126	-0.01739	0.21997	0.22066
38	0.00884	0.16298	0.16322	128	-0.07882	-0.06513	0.10225
40	0.06298	0.25175	0.25951	130	-0.08882	-0.27539	0.28936
42	0.09139	0.22556	0.24338	132	-0.04509	-0.29301	0.29646
44	0.07724	0.07443	0.10727	134	0.02315	-0.13862	0.14054
46	0.02316	-0.13860	0.14053	136	0.07724	0.07443	0.10727
48	-0.04509	-0.29301	0.29646	138	0.09139	0.22557	0.24338
50	-0.08882	-0.27539	0.28936	140	0.06298	0.25174	0.25950
52	-0.07882	-0.06513	0.10225	142	0.00884	0.16298	0.16322
54	-0.01739	0.21996	0.22065	144	-0.04654	0.01750	0.04972
56	0.05704	0.37468	0.37899	146	-0.08289	-0.11750	0.14380
58	0.09186	0.25000	0.26635	148	-0.09077	-0.19578	0.21580
60	0.05848	-0.10242	0.11795	150	-0.07201	-0.20331	0.21568
62	-0.02093	-0.41870	0.41923	152	-0.03591	-0.15292	0.15708
64	-0.08445	-0.40482	0.41354	154	0.00586	-0.07108	0.07132
66	-0.07792	-0.00608	0.07816	156	0.04332	0.01496	0.04583
68	-0.00371	0.47265	0.47266	158	0.07026	0.08536	0.11055
70	0.07480	0.56916	0.57405	160	0.08451	0.13034	0.15534
72	0.08533	0.09033	0.12426	162	0.08702	0.14893	0.17249
74	0.01477	-0.60659	0.60677	164	0.08054	0.14574	0.16652
76	-0.07181	-0.82130	0.82443	166	0.06840	0.12784	0.14499
78	-0.08787	-0.14972	0.17360	168	0.05367	0.10223	0.11547
80	-0.01153	0.97222	0.97229	170	0.03878	0.07469	0.08416
82	0.08307	1.40656	1.40901	172	0.02537	0.04914	0.05530
84	0.08879	0.18545	0.20561	174	0.01442	0.02800	0.03149
86	-0.02989	-2.54617	2.54634	176	0.00644	0.01252	0.01408
88	-0.19238	-5.38670	5.39014	178	0.00161	0.00315	0.00354
				180	0.00000	-0.00001	0.00001

IBR

RADIATION PATTERN FOR THETA=PI/2

H/LMDA=0.2500 B/LMDA= 0.500 OMEGA= 8.53

20 ELEMENT ARRAY

PHI DEG	ER(THETA,PHI) RE	IM	ABSVAL	PHI DEG	ER(THETA,PHI) RE	IM	ABSVAL
0	-0.00000	-0.00000	0.00000	90	-1.52754	-5.00374	5.23171
2	0.00183	0.00517	0.00548	92	-1.23748	-4.05996	4.24437
4	0.00728	0.02061	0.02185	94	-0.56125	-1.85600	1.93900
6	0.01630	0.04614	0.04893	96	0.07357	0.22237	0.23422
8	0.02866	0.08112	0.08604	98	0.33367	1.08853	1.13853
10	0.04381	0.12390	0.13142	100	0.20329	0.67966	0.70941
12	0.06060	0.17118	0.18159	102	-0.06612	-0.19742	0.20819
14	0.07715	0.21745	0.23073	104	-0.20171	-0.65429	0.68468
16	0.09072	0.25475	0.27042	106	-0.12097	-0.40888	0.42640
18	0.09783	0.27301	0.29000	108	0.05373	0.15586	0.16487
20	0.09477	0.26156	0.27820	110	0.14799	0.47564	0.49813
22	0.07861	0.21216	0.22626	112	0.09407	0.31866	0.33226
24	0.04856	0.12304	0.13228	114	-0.03542	-0.09476	0.10117
26	0.00747	0.00322	0.00814	116	-0.11754	-0.37120	0.38936
28	-0.03724	-0.12484	0.13028	118	-0.09041	-0.30170	0.31496
30	-0.07416	-0.22729	0.23908	120	0.00861	0.00718	0.01121
32	-0.09059	-0.26707	0.28201	122	0.09141	0.27904	0.29363
34	-0.07747	-0.21888	0.23219	124	0.09637	0.31144	0.32601
36	-0.03505	-0.08597	0.09284	126	0.02912	0.11100	0.11476
38	0.02348	0.08979	0.09281	128	-0.05418	-0.15212	0.16148
40	0.07398	0.23485	0.24623	130	-0.09515	-0.29294	0.30801
42	0.09062	0.27291	0.28756	132	-0.07200	-0.23569	0.24645
44	0.06020	0.16839	0.17882	134	-0.00595	-0.03872	0.03918
46	-0.00594	-0.03871	0.03916	136	0.06020	0.16838	0.17882
48	-0.07200	-0.23569	0.24644	138	0.09062	0.27291	0.28756
50	-0.09515	-0.29294	0.30800	140	0.07398	0.23485	0.24622
52	-0.05418	-0.15212	0.16148	142	0.02348	0.08979	0.09281
54	0.02912	0.11099	0.11475	144	-0.03505	-0.08597	0.09284
56	0.09637	0.31143	0.32600	146	-0.07747	-0.21888	0.23219
58	0.09141	0.27904	0.29364	148	-0.09059	-0.26706	0.28201
60	0.00861	0.00719	0.01122	150	-0.07416	-0.22728	0.23907
62	-0.09040	-0.30170	0.31495	152	-0.03724	-0.12486	0.13030
64	-0.11754	-0.37119	0.38936	154	0.00747	0.00322	0.00814
66	-0.03543	-0.09478	0.10118	156	0.04856	0.12304	0.13227
68	0.09407	0.31865	0.33225	158	0.07861	0.21217	0.22626
70	0.14799	0.47564	0.49813	160	0.09477	0.26156	0.27820
72	0.05374	0.15588	0.16488	162	0.09783	0.27300	0.29000
74	-0.12096	-0.40885	0.42637	164	0.09072	0.25475	0.27042
76	-0.20171	-0.65430	0.68468	166	0.07715	0.21746	0.23074
78	-0.06612	-0.19743	0.20821	168	0.06059	0.17117	0.18158
80	0.20329	0.67965	0.70940	170	0.04381	0.12391	0.13143
82	0.33367	1.08854	1.13853	172	0.02867	0.08114	0.08605
84	0.07360	0.22244	0.23430	174	0.01630	0.04614	0.04893
86	-0.56120	-1.85589	1.93889	176	0.00728	0.02061	0.02185
88	-1.23747	-4.05993	4.24433	178	0.00183	0.00517	0.00548
				180	-0.00000	-0.00000	0.00000

VBR

RADIATION PATTERN FOR THETA=PI/2

H/LMDA=0.2500 B/LMDA= 0.500 OMEGA= 8.53

22 ELEMENT ARRAY

PHI DEG	ER(THETA,PHI) RE	IM	ABSVAL	PHI DEG	ER(THETA,PHI) RE	IM	ABSVAL
0	-0.00000	-0.00001	0.00001	90	-0.26027	-6.60373	6.60885
2	-0.00161	-0.00306	0.00346	92	-0.17468	-5.15446	5.15742
4	-0.00644	-0.01226	0.01385	94	-0.00283	-1.92517	1.92517
6	-0.01441	-0.02743	0.03098	96	0.09343	0.75517	0.76093
8	-0.02529	-0.04797	0.05423	98	0.04751	1.39272	1.39353
10	-0.03848	-0.07247	0.08205	100	-0.05391	0.40143	0.40503
12	-0.05280	-0.09815	0.11145	102	-0.08220	-0.66884	0.67387
14	-0.06627	-0.12043	0.13746	104	-0.01015	-0.74231	0.74238
16	-0.07610	-0.13288	0.15313	106	0.07197	-0.02471	0.07609
18	-0.07886	-0.12808	0.15041	108	0.06991	0.55958	0.56393
20	-0.07118	-0.09951	0.12235	110	-0.01033	0.44907	0.44919
22	-0.05093	-0.04473	0.06778	112	-0.07830	-0.09992	0.12695
24	-0.01873	0.03083	0.03608	114	-0.06262	-0.45653	0.46081
26	0.02060	0.11109	0.11298	116	0.01719	-0.30788	0.30836
28	0.05777	0.17027	0.17981	118	0.07904	0.11536	0.13984
30	0.08056	0.17971	0.19693	120	0.06436	0.37249	0.37801
32	0.07805	0.12128	0.14422	122	-0.00957	0.25802	0.25819
34	0.04646	0.00288	0.04655	124	-0.07404	-0.06834	0.10075
36	-0.00580	-0.13295	0.13307	126	-0.07477	-0.29780	0.30704
38	-0.05783	-0.21857	0.22609	128	-0.01503	-0.25964	0.26008
40	-0.08352	-0.19353	0.21078	130	0.05517	-0.02529	0.06069
42	-0.06556	-0.04898	0.08183	132	0.08383	0.20144	0.21818
44	-0.00926	0.14522	0.14551	134	0.05440	0.26296	0.26852
46	0.05440	0.26297	0.26853	136	-0.00926	0.14522	0.14551
48	0.08383	0.20144	0.21818	138	-0.06556	-0.04898	0.08183
50	0.05517	-0.02529	0.06069	140	-0.08352	-0.19353	0.21079
52	-0.01503	-0.25964	0.26007	142	-0.05783	-0.21857	0.22609
54	-0.07477	-0.29781	0.30705	144	-0.00580	-0.13296	0.13309
56	-0.07404	-0.06835	0.10076	146	0.04646	0.00288	0.04655
58	-0.00957	0.25801	0.25819	148	0.07805	0.12128	0.14422
60	0.06436	0.37249	0.37801	150	0.08056	0.17972	0.19694
62	0.07904	0.11537	0.13985	152	0.05777	0.17025	0.17979
64	0.01720	-0.30787	0.30835	154	0.02060	0.11108	0.11298
66	-0.06261	-0.45653	0.46081	156	-0.01873	0.03083	0.03607
68	-0.07830	-0.09993	0.12696	158	-0.05093	-0.04472	0.06778
70	-0.01033	0.44906	0.44918	160	-0.07118	-0.09951	0.12235
72	0.06991	0.55959	0.56394	162	-0.07886	-0.12808	0.15041
74	0.07197	-0.02467	0.07608	164	-0.07610	-0.13289	0.15314
76	-0.01015	-0.74229	0.74236	166	-0.06627	-0.12042	0.13745
78	-0.08221	-0.66885	0.67388	168	-0.05280	-0.09816	0.11146
80	-0.05392	0.40140	0.40501	170	-0.03848	-0.07246	0.08205
82	0.04751	1.39271	1.39352	172	-0.02529	-0.04795	0.05421
84	0.09344	0.75525	0.76101	174	-0.01441	-0.02743	0.03098
86	-0.00281	-1.92502	1.92502	176	-0.00644	-0.01226	0.01385
88	-0.17467	-5.15441	5.15737	178	-0.00161	-0.00306	0.00346
				180	-0.00000	-0.00001	0.00001

IBR

RADIATION PATTERN FOR THETA=PI/2

H/LMDA=0.2500 B/LMDA= 0.500 OMEGA= 8.53

22 ELEMENT ARRAY

PHI DEG	ER(THETA,PHI) RE	IM	ABSVAL	PHI DEG	ER(THETA,PHI) RE	IM	ABSVAL
0	-0.00000	-0.00001	0.00001	90	-1.53318	-4.99605	5.22601
2	-0.00184	-0.00517	0.00549	92	-1.18549	-3.86991	4.04741
4	-0.00736	-0.02065	0.02193	94	-0.41861	-1.38133	1.44337
6	-0.01648	-0.04624	0.04909	96	0.19892	0.63404	0.66451
8	-0.02892	-0.08110	0.08610	98	0.31843	1.04094	1.08856
10	-0.04399	-0.12325	0.13086	100	0.06524	0.22969	0.23878
12	-0.06034	-0.16878	0.17924	102	-0.17491	-0.55838	0.58513
14	-0.07572	-0.21122	0.22438	104	-0.16127	-0.53282	0.55669
16	-0.08692	-0.24133	0.25651	106	0.02392	0.05980	0.06441
18	-0.09007	-0.24805	0.26389	108	0.14524	0.46459	0.48676
20	-0.08143	-0.22078	0.23532	110	0.08807	0.29764	0.31040
22	-0.05876	-0.15349	0.16435	112	-0.05400	-0.15710	0.16612
24	-0.02308	-0.04976	0.05485	114	-0.11906	-0.38091	0.39908
26	0.01982	0.07284	0.07549	116	-0.05289	-0.18483	0.19225
28	0.05921	0.18254	0.19190	118	0.05913	0.17272	0.18256
30	0.08154	0.23998	0.25345	120	0.10147	0.32262	0.33820
32	0.07557	0.21354	0.22652	122	0.04301	0.15211	0.15807
34	0.03900	0.09893	0.10634	124	-0.05024	-0.14197	0.15060
36	-0.01643	-0.06641	0.06841	126	-0.09173	-0.28606	0.30041
38	-0.06601	-0.20774	0.21797	128	-0.05210	-0.17695	0.18446
40	-0.08210	-0.24418	0.25761	130	0.02748	0.06728	0.07267
42	-0.05086	-0.13894	0.14795	132	0.08116	0.24328	0.25646
44	0.01396	0.06070	0.06229	134	0.07240	0.23092	0.24200
46	0.07240	0.23092	0.24200	136	0.01396	0.06070	0.06229
48	0.08116	0.24328	0.25646	138	-0.05086	-0.13894	0.14795
50	0.02749	0.06728	0.07268	140	-0.08210	-0.24419	0.25762
52	-0.05210	-0.17694	0.18445	142	-0.06601	-0.20774	0.21797
54	-0.09173	-0.28606	0.30041	144	-0.01643	-0.06642	0.06842
56	-0.05025	-0.14198	0.15061	146	0.03900	0.09893	0.10634
58	0.04301	0.15211	0.15807	148	0.07557	0.21354	0.22652
60	0.10147	0.32262	0.33820	150	0.08155	0.23998	0.25346
62	0.05913	0.17273	0.18257	152	0.05921	0.18252	0.19188
64	-0.05289	-0.18482	0.19224	154	0.01982	0.07284	0.07549
66	-0.11906	-0.38091	0.39908	156	-0.02308	-0.04977	0.05486
68	-0.05400	-0.15711	0.16613	158	-0.05876	-0.15348	0.16435
70	0.08806	0.29764	0.31039	160	-0.08143	-0.22078	0.23532
72	0.14524	0.46459	0.48677	162	-0.09007	-0.24805	0.26389
74	0.02393	0.05983	0.06444	164	-0.08692	-0.24134	0.25651
76	-0.16127	-0.53280	0.55667	166	-0.07571	-0.21122	0.22438
78	-0.17491	-0.55839	0.58514	168	-0.06034	-0.16880	0.17926
80	0.06523	0.22968	0.23876	170	-0.04399	-0.12324	0.13086
82	0.31843	1.04093	1.08855	172	-0.02892	-0.08108	0.08608
84	0.19894	0.63409	0.66457	174	-0.01648	-0.04624	0.04909
86	-0.41857	-1.38121	1.44324	176	-0.00736	-0.02065	0.02193
88	-1.18547	-3.86987	4.04737	178	-0.00184	-0.00517	0.00549
				180	-0.00000	-0.00001	0.00001

VBR

RADIATION PATTERN FOR THETA=PI/2

H/LMDA=0.2500 B/LMDA= 0.500 OMEGA= 8.53

24 ELEMENT ARRAY

PHI DEG	ER(THETA,PHI) RE	IM	ABSVAL	PHI DEG	ER(THETA,PHI) RE	IM	ABSVAL
0	0.00000	-0.00000	0.00000	90	-0.25418	-6.61159	6.61648
2	0.00161	0.00303	0.00343	92	-0.15800	-4.90417	4.90671
4	0.00644	0.01204	0.01365	94	0.01987	-1.32315	1.32330
6	0.01439	0.02686	0.03048	96	0.08815	1.15607	1.15943
8	0.02520	0.04688	0.05322	98	0.00968	1.14106	1.14110
10	0.03814	0.07041	0.08008	100	-0.07489	-0.18107	0.19594
12	0.05182	0.09424	0.10755	102	-0.04645	-0.84912	0.85039
14	0.06394	0.11317	0.12998	104	0.04637	-0.30209	0.30563
16	0.07133	0.12023	0.13980	106	0.07319	0.47618	0.48177
18	0.07032	0.10783	0.12873	108	0.00124	0.50411	0.50411
20	0.05770	0.07047	0.09108	110	-0.07138	-0.08217	0.10884
22	0.03235	0.00868	0.03350	112	-0.05202	-0.46423	0.46714
24	-0.00304	-0.06690	0.06697	114	0.03179	-0.23684	0.23896
26	-0.04077	-0.13415	0.14021	116	0.07647	0.22865	0.24109
28	-0.06889	-0.16369	0.17760	118	0.03134	0.36924	0.37057
30	-0.07476	-0.13113	0.15094	120	-0.04875	0.08554	0.09845
32	-0.05141	-0.03424	0.06177	122	-0.07494	-0.25261	0.26349
34	-0.00408	0.09337	0.09345	124	-0.02247	-0.28911	0.28998
36	0.04758	0.18649	0.19246	126	0.05182	-0.03480	0.06242
38	0.07599	0.17906	0.19452	128	0.07551	0.22331	0.23573
40	0.06107	0.05326	0.08103	130	0.03019	0.24959	0.25141
42	0.00694	-0.12591	0.12610	132	-0.04028	0.05720	0.06996
44	-0.05361	-0.23120	0.23733	134	-0.07656	-0.16108	0.17835
46	-0.07656	-0.16108	0.17835	136	-0.05361	-0.23120	0.23733
48	-0.04028	0.05720	0.06996	138	0.00694	-0.12592	0.12611
50	0.03018	0.24959	0.25141	140	0.06107	0.05324	0.08102
52	0.07551	0.22331	0.23573	142	0.07599	0.17906	0.19452
54	0.05182	-0.03481	0.06243	144	0.04758	0.18648	0.19246
56	-0.02247	-0.28911	0.28998	146	-0.00408	0.09337	0.09346
58	-0.07494	-0.25261	0.26349	148	-0.05141	-0.03424	0.06177
60	-0.04875	0.08553	0.09845	150	-0.07476	-0.13112	0.15093
62	0.03134	0.36924	0.37057	152	-0.06889	-0.16371	0.17762
64	0.07647	0.22865	0.24110	154	-0.04078	-0.13416	0.14022
66	0.03180	-0.23681	0.23894	156	-0.00304	-0.06691	0.06698
68	-0.05202	-0.46423	0.46714	158	0.03235	0.00868	0.03349
70	-0.07138	-0.08218	0.10885	160	0.05770	0.07046	0.09107
72	0.00124	0.50411	0.50411	162	0.07032	0.10783	0.12873
74	0.07319	0.47620	0.48179	164	0.07133	0.12023	0.13980
76	0.04638	-0.30206	0.30560	166	0.06394	0.11318	0.12999
78	-0.04645	-0.84912	0.85039	168	0.05182	0.09423	0.10753
80	-0.07490	-0.18109	0.19597	170	0.03814	0.07042	0.08008
82	0.00967	1.14104	1.14108	172	0.02520	0.04690	0.05324
84	0.08816	1.15613	1.15948	174	0.01439	0.02686	0.03048
86	0.01989	-1.32299	1.32314	176	0.00644	0.01204	0.01365
88	-0.15798	-4.90411	4.90666	178	0.00161	0.00303	0.00343
				180	0.00000	-0.00000	0.00000

IBR

RADIATION PATTERN FOR THETA=PI/2

H/LMDA=0.2500 B/LMDA= 0.500 OMEGA= 8.53

24 ELEMENT ARRAY

PHI DEG	ER(THETA,PHI) RE	IM	ABSVAL	PHI DEG	ER(THETA,PHI) RE	IM	ABSVAL
0	0.00000	0.00000	0.00000	90	-1.53791	-4.98960	5.22123
2	0.00187	0.00519	0.00552	92	-1.12873	-3.66935	3.83903
4	0.00744	0.02070	0.02200	94	-0.28043	-0.92630	0.96782
6	0.01663	0.04628	0.04918	96	0.28462	0.91413	0.95741
8	0.02913	0.08100	0.08608	98	0.25053	0.82206	0.85939
10	0.04409	0.12250	0.13019	100	-0.06810	-0.20534	0.21633
12	0.05991	0.16615	0.17663	102	-0.20003	-0.64790	0.67807
14	0.07396	0.20444	0.21741	104	-0.04569	-0.16335	0.16962
16	0.08261	0.22701	0.24157	106	0.13008	0.41077	0.43088
18	0.08167	0.22206	0.23660	108	0.10576	0.35077	0.36636
20	0.06762	0.17975	0.19205	110	-0.04710	-0.13580	0.14374
22	0.03934	0.09746	0.10510	112	-0.11711	-0.37524	0.39309
24	0.00017	-0.01445	0.01445	114	-0.03353	-0.12266	0.12715
26	-0.04091	-0.12941	0.13573	116	0.07995	0.24340	0.25620
28	-0.07039	-0.20818	0.21976	118	0.08631	0.28101	0.29397
30	-0.07470	-0.21227	0.22503	120	-0.00722	-0.00601	0.00939
32	-0.04748	-0.12544	0.13412	122	-0.08415	-0.25808	0.27145
34	0.00334	0.02620	0.02641	124	-0.06380	-0.21074	0.22018
36	0.05434	0.17182	0.18021	126	0.02150	0.05045	0.05484
38	0.07576	0.22452	0.23696	128	0.07953	0.24163	0.25438
40	0.04996	0.13659	0.14544	130	0.05792	0.19001	0.19864
42	-0.01133	-0.05079	0.05204	132	-0.01523	-0.02989	0.03355
44	-0.06684	-0.21063	0.22098	134	-0.07147	-0.21064	0.22243
46	-0.07147	-0.21064	0.22243	136	-0.06684	-0.21063	0.22098
48	-0.01523	-0.02990	0.03355	138	-0.01133	-0.05079	0.05204
50	0.05792	0.19001	0.19864	140	0.04996	0.13658	0.14543
52	0.07953	0.24163	0.25438	142	0.07577	0.22453	0.23696
54	0.02150	0.05045	0.05484	144	0.05435	0.17182	0.18021
56	-0.06380	-0.21073	0.22018	146	0.00334	0.02620	0.02641
58	-0.08415	-0.25807	0.27145	148	-0.04748	-0.12544	0.13412
60	-0.00722	-0.00602	0.00940	150	-0.07470	-0.21226	0.22502
62	0.08631	0.28101	0.29397	152	-0.07040	-0.20819	0.21977
64	0.07995	0.24341	0.25620	154	-0.04091	-0.12943	0.13574
66	-0.03352	-0.12263	0.12713	156	0.00017	-0.01445	0.01446
68	-0.11711	-0.37524	0.39309	158	0.03934	0.09745	0.10509
70	-0.04710	-0.13581	0.14375	160	0.06762	0.17975	0.19205
72	0.10576	0.35076	0.36636	162	0.08167	0.22206	0.23660
74	0.13009	0.41078	0.43089	164	0.08261	0.22700	0.24157
76	-0.04568	-0.16332	0.16959	166	0.07396	0.20445	0.21742
78	-0.20003	-0.64790	0.67807	168	0.05991	0.16614	0.17661
80	-0.06810	-0.20536	0.21635	170	0.04409	0.12251	0.13020
82	0.25053	0.82204	0.85937	172	0.02913	0.08102	0.08609
84	0.28463	0.91417	0.95746	174	0.01663	0.04628	0.04918
86	-0.28038	-0.92618	0.96769	176	0.00744	0.02070	0.02200
88	-1.12871	-3.66930	3.83898	178	0.00187	0.00519	0.00552
				180	0.00000	0.00000	0.00000

VBR

RADIATION PATTERN FOR THETA=PI/2

H/LMDA=0.2500 B/LMDA= 0.500 OMEGA= 8.53

26 ELEMENT ARRAY

PHI DEG	ER(THETA,PHI) RE	IM	ABSVAL	PHI DEG	ER(THETA,PHI) RE	IM	ABSVAL
0	-0.00000	-0.00000	0.00000	90	-0.24902	-6.61826	6.62294
2	-0.00161	-0.00297	0.00338	92	-0.14213	-4.63803	4.64021
4	-0.00643	-0.01184	0.01347	94	0.03820	-0.75534	0.75630
6	-0.01437	-0.02642	0.03007	96	0.07498	1.38128	1.38331
8	-0.02509	-0.04591	0.05232	98	-0.02403	0.72998	0.73037
10	-0.03775	-0.06846	0.07818	100	-0.07258	-0.62544	0.62964
12	-0.05074	-0.09042	0.10369	102	0.00146	-0.67906	0.67906
14	-0.06143	-0.10604	0.12255	104	0.07065	0.22384	0.23473
16	-0.06629	-0.10779	0.12654	106	0.02665	0.59056	0.59116
18	-0.06152	-0.08825	0.10758	108	-0.05788	0.05207	0.07785
20	-0.04436	-0.04348	0.06212	110	-0.05401	-0.45224	0.45546
22	-0.01506	0.02224	0.02685	112	0.02941	-0.24451	0.24628
24	0.02126	0.09285	0.09525	114	0.06996	0.25995	0.26920
26	0.05406	0.14113	0.15113	116	0.01262	0.34180	0.34204
28	0.06578	0.13783	0.15449	118	-0.06136	-0.03221	0.06930
30	0.05778	0.06877	0.08982	120	-0.05449	-0.31515	0.31983
32	0.01823	-0.04653	0.04997	122	0.02108	-0.17466	0.17593
34	-0.03287	-0.15051	0.15405	124	0.07034	0.15639	0.17148
36	-0.06759	-0.17153	0.18437	126	0.03724	0.26949	0.27205
38	-0.06136	-0.07489	0.09681	128	-0.03682	0.07330	0.08203
40	-0.01295	0.09087	0.09178	130	-0.07087	-0.17831	0.19188
42	0.04664	0.20216	0.20747	132	-0.03329	-0.22394	0.22640
44	0.07076	0.14837	0.16438	134	0.03510	-0.05188	0.06263
46	0.03510	-0.05187	0.06263	136	0.07076	0.14837	0.16438
48	-0.03329	-0.22394	0.22640	138	0.04664	0.20216	0.20747
50	-0.07087	-0.17831	0.19188	140	-0.01295	0.09086	0.09178
52	-0.03682	0.07330	0.08202	142	-0.06136	-0.07488	0.09681
54	0.03724	0.26948	0.27204	144	-0.06759	-0.17154	0.18437
56	0.07034	0.15640	0.17149	146	-0.03287	-0.15051	0.15405
58	0.02108	-0.17465	0.17592	148	0.01823	-0.04653	0.04997
60	-0.05449	-0.31515	0.31983	150	0.05778	0.06878	0.08983
62	-0.06136	-0.03222	0.06931	152	0.06978	0.13782	0.15447
64	0.01262	0.34180	0.34204	154	0.05406	0.14112	0.15113
66	0.06996	0.25997	0.26922	156	0.02126	0.09284	0.09525
68	0.02941	-0.24450	0.24627	158	-0.01506	0.02224	0.02686
70	-0.05401	-0.45224	0.45546	160	-0.04436	-0.04349	0.06212
72	-0.05788	0.05206	0.07785	162	-0.06152	-0.08825	0.10757
74	0.02664	0.59055	0.59115	164	-0.06629	-0.10780	0.12655
76	0.07066	0.22388	0.23477	166	-0.06143	-0.10603	0.12254
78	0.00146	-0.67905	0.67905	168	-0.05074	-0.09044	0.10370
80	-0.07259	-0.62546	0.62966	170	-0.03775	-0.06846	0.07818
82	-0.02404	0.72995	0.73035	172	-0.02509	-0.04590	0.05231
84	0.07499	1.38130	1.38334	174	-0.01437	-0.02642	0.03007
86	0.03822	-0.75518	0.75614	176	-0.00643	-0.01184	0.01347
88	-0.14211	-4.63797	4.64015	178	-0.00161	-0.00297	0.00338
				180	-0.00000	-0.00000	0.00000

IBR

RADIATION PATTERN FOR THETA=PI/2

H/LMDA=0.2500 B/LMDA= 0.500 OMEGA= 8.53

26 ELEMENT ARRAY

PHI DEG	ER(THETA,PHI) RE	IM	ABSVAL	PHI DEG	ER(THETA,PHI) RE	IM	ABSVAL
0	-0.00000	-0.00000	0.00000	90	-1.54193	-4.98410	5.21716
2	-0.00188	-0.00520	0.00553	92	-1.06785	-3.45935	3.62041
4	-0.00751	-0.02075	0.02207	94	-0.15033	-0.50128	0.52333
6	-0.01678	-0.04637	0.04931	96	0.32944	1.06050	1.11049
8	-0.02930	-0.08091	0.08605	98	0.14908	0.49463	0.51661
10	-0.04411	-0.12165	0.12940	100	-0.16334	-0.51764	0.54280
12	-0.05932	-0.16325	0.17370	102	-0.14519	-0.47750	0.49908
14	-0.07192	-0.19716	0.20986	104	0.07601	0.23147	0.24363
16	-0.07786	-0.21190	0.22575	106	0.13569	0.44033	0.46077
18	-0.07280	-0.19535	0.20848	108	-0.01310	-0.02699	0.03000
20	-0.05367	-0.13923	0.14922	110	-0.11651	-0.37054	0.38842
22	-0.02095	-0.04552	0.05011	112	-0.03726	-0.13296	0.13808
24	0.01932	0.06768	0.07039	114	0.08344	0.25626	0.26950
26	0.05506	0.16523	0.17417	116	0.07352	0.24238	0.25329
28	0.07123	0.20425	0.21631	118	-0.03503	-0.09622	0.10240
30	0.05684	0.15477	0.16488	120	-0.08638	-0.27236	0.28573
32	0.01377	0.02523	0.02874	122	-0.02200	-0.08370	0.08655
34	-0.03851	-0.12506	0.13085	124	0.06461	0.19175	0.20235
36	-0.06906	-0.20552	0.21682	126	0.06725	0.21616	0.22638
38	-0.05413	-0.15029	0.15974	128	-0.00798	-0.00904	0.01206
40	0.00125	0.01924	0.01928	130	-0.06990	-0.20881	0.22021
42	0.05779	0.18210	0.19105	132	-0.05481	-0.17745	0.18572
44	0.06680	0.19579	0.20687	134	0.01434	0.02813	0.03157
46	0.01435	0.02814	0.03158	136	0.06680	0.19579	0.20687
48	-0.05480	-0.17744	0.18571	138	0.05779	0.18209	0.19104
50	-0.06990	-0.20881	0.22020	140	0.00125	0.01924	0.01928
52	-0.00798	-0.00904	0.01206	142	-0.05413	-0.15029	0.15974
54	0.06725	0.21616	0.22637	144	-0.06906	-0.20553	0.21682
56	0.06461	0.19176	0.20235	146	-0.03851	-0.12506	0.13085
58	-0.02199	-0.08370	0.08654	148	0.01377	0.02523	0.02874
60	-0.08639	-0.27236	0.28573	150	0.05684	0.15478	0.16488
62	-0.03504	-0.09623	0.10241	152	0.07123	0.20424	0.21630
64	0.07352	0.24238	0.25328	154	0.05506	0.16523	0.17416
66	0.08344	0.25627	0.26951	156	0.01932	0.06768	0.07038
68	-0.03725	-0.13295	0.13807	158	-0.02095	-0.04552	0.05011
70	-0.11651	-0.37054	0.38843	160	-0.05368	-0.13924	0.14922
72	-0.01311	-0.02700	0.03001	162	-0.07280	-0.19535	0.20848
74	0.13569	0.44032	0.46075	164	-0.07786	-0.21190	0.22575
76	0.07602	0.23150	0.24366	166	-0.07192	-0.19715	0.20986
78	-0.14519	-0.47749	0.49907	168	-0.05933	-0.16327	0.17371
80	-0.16335	-0.51765	0.54281	170	-0.04411	-0.12165	0.12940
82	0.14907	0.49461	0.51658	172	-0.02930	-0.08090	0.08604
84	0.32945	1.06052	1.11051	174	-0.01678	-0.04637	0.04931
86	-0.15028	-0.50116	0.52321	176	-0.00751	-0.02075	0.02207
88	-1.06783	-3.45930	3.62036	178	-0.00188	-0.00520	0.00553
				180	-0.00000	-0.00000	0.00000

VBR

RADIATION PATTERN FOR THETA=PI/2

H/LMDA=0.2500 B/LMDA= 0.500 OMEGA= 8.53

28 ELEMENT ARRAY

PHI DEG	ER(THETA,PHI) RE	IM	ABSVAL	PHI DEG	ER(THETA,PHI) RE	IM	ABSVAL
0	0.00000	0.00000	0.00000	90	-0.24459	-6.62398	6.62849
2	0.00161	0.00292	0.00333	92	-0.12694	-4.35821	4.36006
4	0.00642	0.01165	0.01331	94	0.05222	-0.23509	0.24082
6	0.01434	0.02595	0.02964	96	0.05625	1.43726	1.43836
8	0.02496	0.04499	0.05145	98	-0.04872	0.25012	0.25482
10	0.03733	0.06661	0.07636	100	-0.05103	-0.83777	0.83933
12	0.04958	0.08672	0.09989	102	0.04198	-0.27375	0.27695
14	0.05876	0.09898	0.11511	104	0.05536	0.56028	0.56301
16	0.06104	0.09557	0.11340	106	-0.03068	0.30253	0.30408
18	0.05261	0.06947	0.08714	108	-0.06176	-0.37750	0.38252
20	0.03143	0.01891	0.03668	110	0.01342	-0.32958	0.32986
22	-0.00051	-0.04768	0.04768	112	0.06557	0.22358	0.23299
24	-0.03544	-0.10871	0.11434	114	0.01027	0.34250	0.34266
26	-0.06041	-0.13392	0.14691	116	-0.06105	-0.07377	0.09576
28	-0.06185	-0.09843	0.11625	118	-0.03779	-0.32224	0.32445
30	-0.03383	-0.00415	0.03409	120	0.04179	-0.07340	0.08447
32	0.01401	0.10606	0.10698	122	0.06058	0.24691	0.25423
34	0.05594	0.16002	0.16951	124	-0.00488	0.19418	0.19424
36	0.06280	0.10296	0.12061	126	-0.06326	-0.10633	0.12373
38	0.02444	-0.04419	0.05050	128	-0.04044	-0.23976	0.24315
40	-0.03453	-0.17079	0.17424	130	0.03119	-0.07042	0.07702
42	-0.06585	-0.15166	0.16534	132	0.06585	0.16157	0.17448
44	-0.03759	0.02306	0.04410	134	0.02822	0.19336	0.19541
46	0.02822	0.19337	0.19541	136	-0.03759	0.02306	0.04410
48	0.06585	0.16158	0.17449	138	-0.06585	-0.15167	0.16535
50	0.03119	-0.07042	0.07702	140	-0.03453	-0.17080	0.17425
52	-0.04044	-0.23976	0.24315	142	0.02444	-0.04420	0.05051
54	-0.06326	-0.10634	0.12373	144	0.06280	0.10296	0.12060
56	-0.00488	0.19417	0.19423	146	0.05594	0.16002	0.16952
58	0.06058	0.24692	0.25424	148	0.01401	0.10606	0.10698
60	0.04179	-0.07340	0.08446	150	-0.03383	-0.00414	0.03408
62	-0.03779	-0.32223	0.32444	152	-0.06185	-0.09844	0.11626
64	-0.06105	-0.07377	0.09576	154	-0.06041	-0.13392	0.14691
66	0.01027	0.34249	0.34264	156	-0.03544	-0.10872	0.11435
68	0.06557	0.22359	0.23300	158	-0.00051	-0.04767	0.04767
70	0.01342	-0.32958	0.32985	160	0.03143	0.01890	0.03668
72	-0.06176	-0.37751	0.38252	162	0.05261	0.06947	0.08715
74	-0.03069	0.30249	0.30405	164	0.06104	0.09557	0.11340
76	0.05536	0.56029	0.56302	166	0.05876	0.09899	0.11512
78	0.04199	-0.27373	0.27693	168	0.04958	0.08670	0.09987
80	-0.05103	-0.83778	0.83933	170	0.03733	0.06660	0.07635
82	-0.04873	0.25008	0.25479	172	0.02496	0.04500	0.05146
84	0.05625	1.43726	1.43836	174	0.01434	0.02595	0.02964
86	0.05223	-0.23493	0.24066	176	0.00642	0.01165	0.01331
88	-0.12693	-4.35814	4.35998	178	0.00161	0.00292	0.00333
				180	0.00000	0.00000	0.00000

IBR

RADIATION PATTERN FOR THETA=PI/2

H/LMDA=0.2500 B/LMDA= 0.500 OMEGA= 8.53

28 ELEMENT ARRAY

PHI DEG	ER(THETA,PHI) RE	IM	ABSVAL	PHI DEG	ER(THETA,PHI) RE	IM	ABSVAL
0	0.00000	0.00000	0.00000	90	-1.54539	-4.97935	5.21365
2	0.00190	0.00521	0.00554	92	-1.00344	-3.24113	3.39290
4	0.00757	0.02079	0.02213	94	-0.03151	-0.11542	0.11965
6	0.01690	0.04639	0.04937	96	0.33530	1.08056	1.13139
8	0.02944	0.08076	0.08596	98	0.03554	0.12817	0.13301
10	0.04406	0.12071	0.12850	100	-0.20129	-0.64539	0.67605
12	0.05859	0.16012	0.17050	102	-0.04230	-0.14934	0.15521
14	0.06962	0.18938	0.20177	104	0.14195	0.45105	0.47286
16	0.07275	0.19612	0.20918	106	0.05159	0.17795	0.18527
18	0.06360	0.16830	0.17991	108	-0.10492	-0.32835	0.34470
20	0.03991	0.10001	0.10768	110	-0.06271	-0.21119	0.22030
22	0.00411	-0.00111	0.00425	112	0.07429	0.22633	0.23821
24	-0.03480	-0.10863	0.11407	114	0.07376	0.24224	0.25322
26	-0.06215	-0.18057	0.19097	116	-0.04255	-0.12172	0.12895
28	-0.06312	-0.17558	0.18658	118	-0.08062	-0.25748	0.26981
30	-0.03225	-0.08057	0.08679	120	0.00622	0.00519	0.00810
32	0.01812	0.06571	0.06816	122	0.07635	0.23536	0.24744
34	0.05871	0.17697	0.18646	124	0.03273	0.11447	0.11906
36	0.05930	0.16780	0.17797	126	-0.05261	-0.15240	0.16123
38	0.01398	0.02723	0.03060	128	-0.06324	-0.20074	0.21046
40	-0.04476	-0.14333	0.15015	130	0.00653	0.00551	0.00855
42	-0.06466	-0.19030	0.20098	132	0.06481	0.19242	0.20304
44	-0.02149	-0.05083	0.05519	134	0.04623	0.15022	0.15717
46	0.04624	0.15022	0.15718	136	-0.02149	-0.05083	0.05519
48	0.06481	0.19242	0.20304	138	-0.06466	-0.19030	0.20099
50	0.00653	0.00551	0.00855	140	-0.04477	-0.14334	0.15017
52	-0.06324	-0.20074	0.21046	142	0.01397	0.02722	0.03060
54	-0.05261	-0.15241	0.16123	144	0.05930	0.16779	0.17796
56	0.03273	0.11446	0.11905	146	0.05871	0.17698	0.18646
58	0.07635	0.23537	0.24744	148	0.01812	0.06571	0.06816
60	0.00622	0.00519	0.00810	150	-0.03225	-0.08057	0.08678
62	-0.08062	-0.25748	0.26981	152	-0.06312	-0.17559	0.18659
64	-0.04256	-0.12172	0.12895	154	-0.06215	-0.18058	0.19097
66	0.07376	0.24222	0.25320	156	-0.03480	-0.10864	0.11408
68	0.07429	0.22634	0.23822	158	0.00411	-0.00111	0.00425
70	-0.06271	-0.21118	0.22030	160	0.03991	0.10000	0.10767
72	-0.10492	-0.32835	0.34471	162	0.06360	0.16830	0.17991
74	0.05158	0.17792	0.18525	164	0.07275	0.19612	0.20918
76	0.14195	0.45105	0.47286	166	0.06963	0.18938	0.20178
78	-0.04229	-0.14932	0.15519	168	0.05859	0.16011	0.17049
80	-0.20129	-0.64539	0.67605	170	0.04405	0.12071	0.12850
82	0.03553	0.12815	0.13298	172	0.02944	0.08078	0.08597
84	0.33530	1.08055	1.13138	174	0.01690	0.04639	0.04937
86	-0.03147	-0.11531	0.11952	176	0.00757	0.02079	0.02213
88	-1.00342	-3.24108	3.39285	178	0.00190	0.00521	0.00554
				180	0.00000	0.00000	0.00000

VBR

RADIATION PATTERN FOR THETA=PI/2

H/LMDA=0.2500 B/LMDA= 0.500 OMEGA= 8.53

30 ELEMENT ARRAY

PHI DEG	ER(THETA,PHI) RE	IM	ABSVAL	PHI DEG	ER(THETA,PHI) RE	IM	ABSVAL
0	-0.00000	-0.00000	0.00000	90	-0.24076	-6.62894	6.63331
2	-0.00161	-0.00289	0.00331	92	-0.11235	-4.06686	4.06841
4	-0.00641	-0.01149	0.01316	94	0.06203	0.22630	0.23464
6	-0.01430	-0.02558	0.02931	96	0.03440	1.34244	1.34289
8	-0.02482	-0.04413	0.05064	98	-0.06158	-0.20968	0.21854
10	-0.03687	-0.06480	0.07456	100	-0.01852	-0.79484	0.79506
12	-0.04834	-0.08305	0.09610	102	0.06098	0.18270	0.19261
14	-0.05595	-0.09203	0.10770	104	0.01366	0.56411	0.56427
16	-0.05562	-0.08360	0.10042	106	-0.06021	-0.14590	0.15784
18	-0.04371	-0.05163	0.06765	108	-0.01493	-0.44305	0.44330
20	-0.01916	0.00306	0.01941	110	0.05876	0.09894	0.11507
22	0.01398	0.06735	0.06879	112	0.02149	0.37214	0.37276
24	0.04530	0.11496	0.12356	114	-0.05511	-0.04058	0.06843
26	0.06024	0.11503	0.12985	116	-0.03287	-0.32275	0.32442
28	0.04725	0.05197	0.07024	118	0.04643	-0.02976	0.05515
30	0.00755	-0.05255	0.05309	120	0.04724	0.27313	0.27719
32	-0.03904	-0.13548	0.14099	122	-0.02888	0.10819	0.11198
34	-0.06123	-0.12613	0.14021	124	-0.05929	-0.20250	0.21100
36	-0.03816	-0.00992	0.03943	126	-0.00013	-0.17955	0.17955
38	0.01731	0.12973	0.13088	128	0.05874	0.09546	0.11209
40	0.05904	0.15767	0.16836	130	0.03618	0.21151	0.21458
42	0.04460	0.02056	0.04911	132	-0.03343	0.04131	0.05314
44	-0.01673	-0.15594	0.15683	134	-0.06067	-0.16219	0.17316
46	-0.06067	-0.16218	0.17316	136	-0.01673	-0.15595	0.15684
48	-0.03343	0.04131	0.05314	138	0.04460	0.02055	0.04911
50	0.03618	0.21150	0.21458	140	0.05904	0.15766	0.16835
52	0.05874	0.09546	0.11209	142	0.01732	0.12974	0.13089
54	-0.00013	-0.17955	0.17955	144	-0.03816	-0.00992	0.03943
56	-0.05929	-0.20251	0.21101	146	-0.06123	-0.12613	0.14021
58	-0.02888	0.10818	0.11197	148	-0.03904	-0.13548	0.14099
60	0.04724	0.27313	0.27719	150	0.00755	-0.05255	0.05309
62	0.04644	-0.02975	0.05515	152	0.04725	0.05196	0.07023
64	-0.03287	-0.32274	0.32441	154	0.06024	0.11502	0.12984
66	-0.05511	-0.04061	0.06845	156	0.04530	0.11495	0.12355
68	0.02149	0.37214	0.37276	158	0.01398	0.06736	0.06879
70	0.05876	0.09895	0.11508	160	-0.01916	0.00307	0.01940
72	-0.01493	-0.44305	0.44330	162	-0.04371	-0.05163	0.06765
74	-0.06021	-0.14594	0.15787	164	-0.05562	-0.08361	0.10042
76	0.01366	0.56409	0.56426	166	-0.05595	-0.09202	0.10769
78	0.06098	0.18272	0.19262	168	-0.04834	-0.08307	0.09611
80	-0.01852	-0.79483	0.79505	170	-0.03687	-0.06481	0.07456
82	-0.06158	-0.20971	0.21857	172	-0.02482	-0.04413	0.05063
84	0.03439	1.34241	1.34285	174	-0.01430	-0.02558	0.02931
86	0.06204	0.22644	0.23479	176	-0.00641	-0.01149	0.01316
88	-0.11233	-4.06678	4.06833	178	-0.00161	-0.00289	0.00331
				180	-0.00000	-0.00000	0.00000

IBR

RADIATION PATTERN FOR THETA=PI/2

H/LMDA=0.2500 B/LMDA= 0.500 OMEGA= 8.53

30 ELEMENT ARRAY

PHI DEG	ER(THETA,PHI) RE	IM	ABSVAL	PHI DEG	ER(THETA,PHI) RE	IM	ABSVAL
0	-0.00000	-0.00001	0.00001	90	-1.54841	-4.97521	5.21059
2	-0.00192	-0.00523	0.00557	92	-0.93609	-3.01600	3.15793
4	-0.00763	-0.02084	0.02219	94	0.07336	0.22357	0.23530
6	-0.01702	-0.04646	0.04948	96	0.30689	0.99020	1.03667
8	-0.02955	-0.08060	0.08584	98	-0.06958	-0.21138	0.22254
10	-0.04393	-0.11966	0.12747	100	-0.17968	-0.58118	0.60832
12	-0.05772	-0.15674	0.16703	102	0.06358	0.19193	0.20219
14	-0.06709	-0.18118	0.19320	104	0.12630	0.40951	0.42854
16	-0.06733	-0.17982	0.19201	106	-0.05562	-0.16594	0.17501
18	-0.05422	-0.14123	0.15128	108	-0.09942	-0.32244	0.33742
20	-0.02662	-0.06275	0.06816	110	0.04558	0.13303	0.14062
22	0.01074	0.04138	0.04275	112	0.08578	0.27706	0.29003
24	0.04590	0.13667	0.14417	114	-0.03279	-0.09124	0.09695
26	0.06250	0.17699	0.18770	116	-0.07932	-0.25338	0.26551
28	0.04820	0.12895	0.13767	118	0.01606	0.03729	0.04060
30	0.00560	0.00328	0.00649	120	0.07521	0.23565	0.24736
32	-0.04209	-0.13071	0.13732	122	0.00575	0.03119	0.03171
34	-0.06085	-0.17546	0.18571	124	-0.06757	-0.20539	0.21622
36	-0.03145	-0.08053	0.08645	126	-0.03174	-0.10959	0.11410
38	0.02655	0.09008	0.09391	128	0.04899	0.14093	0.14920
40	0.06103	0.18170	0.19167	130	0.05555	0.17619	0.18474
42	0.03337	0.08802	0.09413	132	-0.01373	-0.02837	0.03152
44	-0.03258	-0.10909	0.11385	134	-0.06214	-0.18523	0.19537
46	-0.06214	-0.18522	0.19537	136	-0.03258	-0.10910	0.11386
48	-0.01373	-0.02838	0.03152	138	0.03336	0.08801	0.09412
50	0.05555	0.17619	0.18474	140	0.06102	0.18169	0.19167
52	0.04899	0.14093	0.14920	142	0.02655	0.09009	0.09392
54	-0.03174	-0.10960	0.11410	144	-0.03145	-0.08053	0.08646
56	-0.06757	-0.20540	0.21623	146	-0.06085	-0.17546	0.18571
58	0.00574	0.03119	0.03171	148	-0.04209	-0.13071	0.13732
60	0.07521	0.23565	0.24736	150	0.00560	0.00328	0.00649
62	0.01606	0.03730	0.04061	152	0.04819	0.12894	0.13766
64	-0.07932	-0.25338	0.26550	154	0.06250	0.17698	0.18770
66	-0.03280	-0.09126	0.09697	156	0.04589	0.13666	0.14416
68	0.08578	0.27705	0.29003	158	0.01074	0.04138	0.04275
70	0.04558	0.13304	0.14063	160	-0.02661	-0.06274	0.06815
72	-0.09941	-0.32243	0.33741	162	-0.05422	-0.14123	0.15128
74	-0.05563	-0.16596	0.17504	164	-0.06733	-0.17983	0.19202
76	0.12630	0.40949	0.42852	166	-0.06709	-0.18117	0.19319
78	0.06359	0.19195	0.20221	168	-0.05772	-0.15675	0.16704
80	-0.17968	-0.58117	0.60831	170	-0.04394	-0.11966	0.12747
82	-0.06959	-0.21140	0.22256	172	-0.02955	-0.08060	0.08584
84	0.30688	0.99017	1.03663	174	-0.01702	-0.04646	0.04948
86	0.07340	0.22368	0.23541	176	-0.00763	-0.02084	0.02219
88	-0.93606	-3.01594	3.15786	178	-0.00192	-0.00523	0.00557
				180	-0.00000	-0.00001	0.00001

VBR

RADIATION PATTERN FOR THETA=PI/2

H/LMDA=0.2500 B/LMDA= 0.500 OMEGA= 8.53

35 ELEMENT ARRAY

PHI DEG	ER(THETA,PHI) RE	IM	ABSVAL	PHI DEG	ER(THETA,PHI) RE	IM	ABSVAL
0	0.04419	0.05822	0.07309	90	-0.23320	-6.63834	6.64244
1	0.04419	0.05821	0.07309	91	-0.18714	-5.68241	5.68549
2	0.04417	0.05817	0.07304	92	-0.07825	-3.30245	3.30338
3	0.04406	0.05797	0.07281	93	0.02652	-0.66642	0.66695
4	0.04378	0.05743	0.07221	94	0.06974	1.06940	1.07167
5	0.04318	0.05631	0.07096	95	0.04154	1.39735	1.39796
6	0.04210	0.05429	0.06870	96	-0.01878	0.64713	0.64740
7	0.04034	0.05099	0.06502	97	-0.05643	-0.34887	0.35340
8	0.03769	0.04604	0.05950	98	-0.04285	-0.84366	0.84475
9	0.03393	0.03905	0.05173	99	0.00634	-0.61141	0.61144
10	0.02887	0.02970	0.04142	100	0.04827	0.02503	0.05437
11	0.02237	0.01780	0.02859	101	0.04904	0.52722	0.52949
12	0.01439	0.00339	0.01478	102	0.00976	0.55444	0.55453
13	0.00501	-0.01324	0.01415	103	-0.03664	0.16587	0.16987
14	-0.00549	-0.03136	0.03184	104	-0.05309	-0.29011	0.29493
15	-0.01663	-0.04983	0.05253	105	-0.02747	-0.47068	0.47148
16	-0.02768	-0.06702	0.07251	106	0.01926	-0.28198	0.28263
17	-0.03770	-0.08090	0.08925	107	0.05070	0.09105	0.10421
18	-0.04557	-0.08924	0.10021	108	0.04328	0.35612	0.35874
19	-0.05012	-0.08984	0.10288	109	0.00340	0.33563	0.33565
20	-0.05028	-0.08098	0.09531	110	-0.03870	0.07655	0.08578
21	-0.04528	-0.06184	0.07664	111	-0.05217	-0.21198	0.21830
22	-0.03492	-0.03308	0.04810	112	-0.02789	-0.32478	0.32597
23	-0.01975	0.00279	0.01995	113	0.01609	-0.20159	0.20223
24	-0.00123	0.04140	0.04142	114	0.04849	0.04976	0.06948
25	0.01830	0.07666	0.07882	115	0.04720	0.24669	0.25117
26	0.03582	0.10163	0.10776	116	0.01384	0.26269	0.26306
27	0.04806	0.10995	0.12000	117	-0.02855	0.10390	0.10775
28	0.05217	0.09735	0.11044	118	-0.05190	-0.11068	0.12225
29	0.04647	0.06344	0.07864	119	-0.04162	-0.23884	0.24244
30	0.03113	0.01297	0.03372	120	-0.00505	-0.20750	0.20756
31	0.00858	-0.04423	0.04506	121	0.03438	-0.04990	0.06060
32	-0.01663	-0.09451	0.09596	122	0.05260	0.12744	0.13787
33	-0.03855	-0.12355	0.12943	123	0.03933	0.21714	0.22067
34	-0.05124	-0.12049	0.13093	124	0.00324	0.17424	0.17427
35	-0.05050	-0.08232	0.09658	125	-0.03438	0.03442	0.04865
36	-0.03550	-0.01657	0.03918	126	-0.05252	-0.11592	0.12727
37	-0.00964	0.05876	0.05954	127	-0.04197	-0.19365	0.19814
38	0.01979	0.11941	0.12104	128	-0.00929	-0.16326	0.16352
39	0.04344	0.14249	0.14896	129	0.02784	-0.04965	0.05696
40	0.05289	0.11543	0.12697	130	0.05059	0.08232	0.09662
41	0.04392	0.04320	0.06160	131	0.04838	0.16588	0.17279
42	0.01870	-0.05047	0.05382	132	0.02328	0.16485	0.16649
43	-0.01407	-0.12943	0.13019	133	-0.01221	0.08696	0.08781
44	-0.04183	-0.15877	0.16419	134	-0.04179	-0.02578	0.04910
45	-0.05293	-0.12066	0.13176	135	-0.05293	-0.12065	0.13175
46	-0.04179	-0.02578	0.04910	136	-0.04183	-0.15876	0.16418
47	-0.01222	0.08695	0.08781	137	-0.01407	-0.12944	0.13020
48	0.02327	0.16484	0.16647	138	0.01870	-0.05046	0.05381
49	0.04838	0.18988	[illegible]	139	[illegible]	[illegible]	[illegible]
50	0.05059	0.08233	0.09663	140	0.05289	0.11543	0.12697
51	0.02784	-0.04965	0.05696	141	0.04344	0.14247	0.14894
52	-0.00929	-0.16327	0.16353	142	0.01979	0.11942	0.12105
53	-0.04197	-0.19364	0.19814	143	-0.00964	0.05877	0.05955
54	-0.05252	-0.11594	0.12728	144	-0.03550	-0.01655	0.03917
55	-0.03438	0.03441	0.04864	145	-0.05050	-0.08231	0.09657
56	0.00324	0.17424	0.17427	146	-0.05124	-0.12049	0.13093
57	0.03933	0.21715	0.22068	147	-0.03855	-0.12356	0.12943
58	0.05260	0.12744	0.13787	148	-0.01663	-0.09451	0.09597
59	0.03439	-0.04988	0.06059	149	0.00858	-0.04422	0.04504
60	-0.00505	-0.20750	0.20756	150	0.03113	0.01296	0.03372
61	-0.04161	-0.23885	0.24245	151	0.04647	0.06345	0.07865
62	-0.05191	-0.11070	0.12226	152	0.05217	0.09734	0.11043
63	-0.02855	0.10388	0.10774	153	0.04806	0.10996	0.12000
64	0.01384	0.26268	0.26304	154	0.03582	0.10163	0.10776
65	0.04720	0.24671	0.25118	155	0.01831	0.07667	0.07882
66	0.04849	0.04978	0.06950	156	-0.00123	0.04139	0.04141
67	0.01609	-0.20157	0.20221	157	-0.01975	0.00280	0.01995
68	-0.02788	-0.32478	0.32598	158	-0.03492	-0.03308	0.04810
69	-0.05217	-0.21199	0.21831	159	-0.04528	-0.06184	0.07664
70	-0.03870	0.07654	0.08577	160	-0.05028	-0.08097	0.09531
71	0.00340	0.33563	0.33565	161	-0.05012	-0.08984	0.10288
72	0.04328	0.35613	0.35875	162	-0.04557	-0.08923	0.10020
73	0.05070	0.09109	0.10424	163	-0.03770	-0.08090	0.08926
74	0.01927	-0.28194	0.28260	164	-0.02768	-0.06699	0.07249
75	-0.02747	-0.47068	0.47148	165	-0.01663	-0.04982	0.05253
76	-0.05309	-0.29014	0.29496	166	-0.00549	-0.03135	0.03183
77	-0.03665	0.16583	0.16983	167	0.00501	-0.01325	0.01416
78	0.00976	0.55442	0.55451	168	0.01439	0.00337	0.01478
79	0.04904	0.52723	0.52951	169	0.02237	0.01781	0.02859
80	0.04828	0.02505	0.05439	170	0.02887	0.02970	0.04142
81	0.00635	-0.61138	0.61142	171	0.03393	0.03905	0.05173
82	-0.04285	-0.84367	0.84476	172	0.03769	0.04604	0.05950
83	-0.05644	-0.34894	0.35348	173	0.04034	0.05099	0.06502
84	-0.01879	0.64703	0.64731	174	0.04210	0.05429	0.06870
85	0.04153	1.39732	1.39793	175	0.04318	0.05631	0.07096
86	0.06975	1.06950	1.07177	176	0.04378	0.05743	0.07222
87	0.02653	-0.66622	0.66675	177	0.04406	0.05797	0.07281
88	-0.07823	-3.30221	3.30313	178	0.04417	0.05817	0.07304
89	-0.18712	-5.68234	5.68542	179	0.04419	0.05821	0.07309
				180	0.04419	0.05822	0.07309

IBR

RADIATION PATTERN FOR THETA=PI/2

H/LMDA=0.2500 B/LMDA= 0.500 OMEGA= 8.53

35 ELEMENT ARRAY

PHI DEG	ER(THETA,PHI) RE	IM	ABSVAL	PHI DEG	ER(THETA,PHI) RE	IM	ABSVAL
0	0.07281	0.17029	0.18520	90	-1.63873	-4.86569	5.13424
1	0.07280	0.17029	0.18520	91	-1.39757	-4.15236	4.38124
2	0.07277	0.17021	0.18511	92	-0.79965	-2.38251	2.51312
3	0.07263	0.16985	0.18473	93	-0.14417	-0.43874	0.46182
4	0.07226	0.16891	0.18371	94	0.27658	0.81467	0.86034
5	0.07147	0.16692	0.18158	95	0.34096	1.01460	1.07035
6	0.07005	0.16333	0.17772	96	0.14222	0.43157	0.45440
7	0.06774	0.15749	0.17144	97	-0.10374	-0.29883	0.31632
8	0.06424	0.14868	0.16197	98	-0.21226	-0.62834	0.66323
9	0.05928	0.13618	0.14853	99	-0.13813	-0.41707	0.43935
10	0.05260	0.11936	0.13044	100	0.02578	0.06622	0.07106
11	0.04401	0.09774	0.10719	101	0.14119	0.41363	0.43706
12	0.03342	0.07118	0.07864	102	0.13065	0.39151	0.41274
13	0.02094	0.03994	0.04510	103	0.02222	0.07600	0.07918
14	0.00689	0.00492	0.00846	104	-0.08837	-0.25374	0.26869
15	-0.00814	-0.03231	0.03332	105	-0.11802	-0.34987	0.36924
16	-0.02324	-0.06940	0.07319	106	-0.05500	-0.17135	0.17996
17	-0.03723	-0.10327	0.10978	107	0.04253	0.11572	0.12329
18	-0.04872	-0.13034	0.13915	108	0.09833	0.28687	0.30325
19	-0.05622	-0.14684	0.15724	109	0.07576	0.22923	0.24143
20	-0.05839	-0.14938	0.16038	110	-0.00061	0.00874	0.00876
21	-0.05425	-0.13561	0.14606	111	-0.07011	-0.19901	0.21099
22	-0.04350	-0.10500	0.11366	112	-0.08335	-0.24643	0.26014
23	-0.02678	-0.05954	0.06528	113	-0.03592	-0.11457	0.12007
24	-0.00582	-0.00399	0.00705	114	0.03357	0.08829	0.09445
25	0.01657	0.05403	0.05652	115	0.07492	0.21552	0.22817
26	0.03678	0.10497	0.11123	116	0.06194	0.18626	0.19628
27	0.05097	0.13884	0.14790	117	0.00757	0.03268	0.03354
28	0.05581	0.14728	0.15750	118	-0.04875	-0.13342	0.14205
29	0.04945	0.12599	0.13534	119	-0.06983	-0.20227	0.21398
30	0.03228	0.07677	0.08328	120	-0.04445	-0.13657	0.14362
31	0.00732	0.00843	0.01117	121	0.00805	0.01241	0.01480
32	-0.01996	-0.06392	0.06696	122	0.05305	0.14613	0.15546
33	-0.04271	-0.12188	0.12915	123	0.06340	0.18391	0.19454
34	-0.05438	-0.14834	0.15799	124	0.03510	0.10962	0.11510
35	-0.05081	-0.13312	0.14249	125	-0.01278	-0.02577	0.02877
36	-0.03192	-0.07745	0.08377	126	-0.05141	-0.14052	0.14963
37	-0.00259	0.00420	0.00493	127	-0.05958	-0.17156	0.18161
38	0.02825	0.08678	0.09126	128	-0.03460	-0.10720	0.11265
39	0.05006	0.14181	0.15038	129	0.00814	0.01207	0.01456
40	0.05438	0.14738	0.15709	130	0.04522	0.12060	0.12880
41	0.03846	0.09787	0.10515	131	0.05806	0.16396	0.17394
42	0.00718	0.00870	0.01128	132	0.04169	0.12480	0.13158
43	-0.02777	-0.08668	0.09102	133	0.00569	0.02706	0.02765
44	-0.05194	-0.14855	0.15737	134	-0.03196	-0.08007	0.08622
45	-0.05411	-0.14762	0.15723	135	-0.05411	-0.14762	0.15722
46	-0.03196	-0.08007	0.08622	136	-0.05193	-0.14854	0.15736
47	0.00569	0.02705	0.02765	137	-0.02777	-0.08669	0.09103
48	0.04169	0.12478	0.13156	138	0.00718	0.00871	0.01129
49	0.05806	0.16396	0.17394	139	0.03846	0.09786	0.10515
50	[illegible]	[illegible]	[illegible]	140	0.05438	0.14738	0.15709
51	0.00814	0.01207	0.01456	141	0.05005	0.14179	0.15037
52	-0.03460	-0.10721	0.11265	142	0.02825	0.08678	0.09127
53	-0.05958	-0.17156	0.18161	143	-0.00258	0.00421	0.00494
54	-0.05142	-0.14053	0.14964	144	-0.03191	-0.07744	0.08376
55	-0.01278	-0.02578	0.02877	145	-0.05080	-0.13312	0.14248
56	0.03510	0.10961	0.11510	146	-0.05438	-0.14834	0.15799
57	0.06341	0.18392	0.19455	147	-0.04271	-0.12188	0.12915
58	0.05305	0.14613	0.15546	148	-0.01996	-0.06392	0.06697
59	0.00806	0.01243	0.01481	149	0.00733	0.00844	0.01118
60	-0.04445	-0.13656	0.14361	150	0.03228	0.07676	0.08327
61	-0.06983	-0.20227	0.21398	151	0.04945	0.12599	0.13535
62	-0.04876	-0.13343	0.14206	152	0.05581	0.14728	0.15750
63	0.00757	0.03267	0.03353	153	0.05097	0.13885	0.14791
64	0.06193	0.18624	0.19627	154	0.03678	0.10497	0.11123
65	0.07492	0.21552	0.22818	155	0.01658	0.05404	0.05652
66	0.03357	0.08830	0.09447	156	-0.00582	-0.00400	0.00706
67	-0.03592	-0.11455	0.12005	157	-0.02678	-0.05953	0.06527
68	-0.08335	-0.24642	0.26014	158	-0.04350	-0.10500	0.11365
69	-0.07011	-0.19901	0.21100	159	-0.05425	-0.13560	0.14605
70	-0.00062	0.00873	0.00875	160	-0.05839	-0.14938	0.16038
71	0.07576	0.22923	0.24142	161	-0.05622	-0.14684	0.15724
72	0.09833	0.28687	0.30326	162	-0.04871	-0.13033	0.13914
73	0.04254	0.11575	0.12332	163	-0.03723	-0.10327	0.10978
74	-0.05499	-0.17132	0.17993	164	-0.02323	-0.06938	0.07317
75	-0.11802	-0.34987	0.36924	165	-0.00814	-0.03231	0.03332
76	-0.08838	-0.25376	0.26871	166	0.00689	0.00493	0.00847
77	0.02221	0.07597	0.07915	167	0.02094	0.03994	0.04509
78	0.13064	0.39149	0.41271	168	0.03342	0.07117	0.07862
79	0.14119	0.41364	0.43707	169	0.04401	0.09775	0.10720
80	0.02579	0.06624	0.07108	170	0.05260	0.11936	0.13044
81	-0.13813	-0.41705	0.43933	171	0.05928	0.13618	0.14853
82	-0.21226	-0.62835	0.66323	172	0.06424	0.14868	0.16197
83	-0.10376	-0.29888	0.31638	173	0.06774	0.15749	0.17144
84	0.14219	0.43150	0.45432	174	0.07005	0.16333	0.17772
85	0.34095	1.01457	1.07032	175	0.07147	0.16692	0.18158
86	0.27661	0.81473	0.86041	176	0.07226	0.16891	0.18372
87	-0.14410	-0.43855	0.46166	177	0.07263	0.16985	0.18473
88	-0.79957	-2.38233	2.51293	178	0.07277	0.17021	0.18511
89	-1.39754	-4.15232	4.38119	179	0.07280	0.17029	0.18520
				180	0.07281	0.17029	0.18520

VBR

RADIATION PATTERN FOR THETA=PI/2

H/LMDA=0.2500 B/LMDA= 0.500 OMEGA= 8.53

40 ELEMENT ARRAY

PHI DEG	ER(THETA,PHI) RE	IM	ABSVAL	PHI DEG	ER(THETA,PHI) RE	IM	ABSVAL
0	0.00000	-0.00000	0.00000	90	-0.22733	-6.64633	6.65021
1	0.00040	0.00068	0.00079	91	-0.17110	-5.41067	5.41338
2	0.00160	0.00271	0.00315	92	-0.04761	-2.51314	2.51359
3	0.00359	0.00612	0.00709	93	0.04859	0.24584	0.25059
4	0.00636	0.01084	0.01257	94	0.05751	1.42848	1.42964
5	0.00989	0.01682	0.01951	95	0.00095	0.91917	0.91917
6	0.01409	0.02391	0.02775	96	-0.04810	-0.23554	0.24040
7	0.01885	0.03192	0.03707	97	-0.03921	-0.84637	0.84728
8	0.02399	0.04040	0.04699	98	0.01176	-0.52176	0.52189
9	0.02922	0.04884	0.05691	99	0.04745	0.21909	0.22417
10	0.03416	0.05646	0.06599	100	0.03121	0.60533	0.60613
11	0.03834	0.06232	0.07317	101	-0.01745	0.34787	0.34831
12	0.04119	0.06534	0.07724	102	-0.04681	-0.19718	0.20266
13	0.04209	0.06438	0.07692	103	-0.02735	-0.47111	0.47190
14	0.04045	0.05839	0.07103	104	0.01973	-0.25872	0.25947
15	0.03579	0.04668	0.05882	105	0.04630	0.17039	0.17656
16	0.02787	0.02912	0.04031	106	0.02635	0.38514	0.38604
17	0.01684	0.00648	0.01804	107	-0.01945	0.21307	0.21395
18	0.00334	-0.01941	0.01970	108	-0.04588	-0.13871	0.14610
19	-0.01142	-0.04557	0.04697	109	-0.02784	-0.32483	0.32602
20	-0.02566	-0.06799	0.07267	110	0.01661	-0.19358	0.19429
21	-0.03723	-0.08221	0.09025	111	0.04515	0.10151	0.11110
22	-0.04387	-0.08412	0.09487	112	0.03157	0.27794	0.27973
23	-0.04375	-0.07109	0.08347	113	-0.01084	0.19065	0.19096
24	-0.03593	-0.04311	0.05612	114	-0.04322	-0.05760	0.07201
25	-0.02095	-0.00377	0.02128	115	-0.03697	-0.23539	0.23827
26	-0.00105	0.03962	0.03963	116	0.00164	-0.19656	0.19656
27	0.01986	0.07699	0.07951	117	0.03860	0.00619	0.03910
28	0.03689	0.09759	0.10433	118	0.04264	0.18879	0.19355
29	0.04529	0.09319	0.10361	119	0.01115	0.20281	0.20312
30	0.04189	0.06159	0.07448	120	-0.02926	0.05152	0.05925
31	0.02653	0.00888	0.02798	121	-0.04593	-0.13052	0.13836
32	0.00277	-0.05044	0.05051	122	-0.02637	-0.19835	0.20010
33	-0.02255	-0.09665	0.09925	123	0.01337	-0.10956	0.11037
34	-0.04102	-0.11140	0.11871	124	0.04274	0.05600	0.07044
35	-0.04556	-0.08521	0.09662	125	0.04035	0.16990	0.17462
36	-0.03342	-0.02367	0.04095	126	0.00890	0.15425	0.15451
37	-0.00808	0.05163	0.05226	127	-0.02854	0.03081	0.04200
38	0.02107	0.10947	0.11148	128	-0.04605	-0.10646	0.11599
39	0.04192	0.12168	0.12870	129	-0.03258	-0.16405	0.16725
40	0.04472	0.07757	0.08954	130	0.00197	-0.11078	0.11079
41	0.02706	-0.00713	0.02798	131	0.03490	0.01006	0.03633
42	-0.00382	-0.09344	0.09352	132	0.04598	0.11764	0.12631
43	-0.03332	-0.13589	0.13992	133	0.02946	0.14809	0.15099
44	-0.04617	-0.10684	0.11639	134	-0.00391	0.09006	0.09015
45	-0.03466	-0.01576	0.03807	135	-0.03466	-0.01576	0.03807
46	-0.00392	0.09006	0.09015	136	-0.04617	-0.10684	0.11639
47	0.02946	0.14807	0.15098	137	-0.03332	-0.13589	0.13991
48	0.04598	0.11765	0.12632	138	-0.00382	-0.09344	0.09352
49	0.03491	0.01007	0.03633	139	0.02705	-0.00714	0.02798
50	0.00197	-0.11077	0.11078	140	0.04472	0.07756	0.08953
51	-0.03258	-0.16404	0.16724	141	0.04192	0.12167	0.12869
52	-0.04605	-0.10647	0.11600	142	0.02107	0.10947	0.11148
53	-0.02854	0.03080	0.04199	143	-0.00808	0.05163	0.05226
54	0.00889	0.15426	0.15451	144	-0.03342	-0.02368	0.04096
55	0.04034	0.16991	0.17463	145	-0.04556	-0.08521	0.09663
56	0.04274	0.05600	0.07044	146	-0.04102	-0.11141	0.11872
57	0.01338	-0.10955	0.11036	147	-0.02255	-0.09665	0.09925
58	-0.02637	-0.19836	0.20010	148	0.00277	-0.05042	0.05050
59	-0.04593	-0.13052	0.13836	149	0.02653	0.00887	0.02797
60	-0.02926	0.05151	0.05924	150	0.04189	0.06159	0.07448
61	0.01114	0.20280	0.20311	151	0.04529	0.09319	0.10361
62	0.04264	0.18880	0.19355	152	0.03689	0.09759	0.10433
63	0.03861	0.00620	0.03910	153	0.01986	0.07699	0.07951
64	0.00164	-0.19655	0.19656	154	-0.00105	0.03962	0.03963
65	-0.03696	-0.23540	0.23828	155	-0.02095	-0.00376	0.02128
66	-0.04322	-0.05762	0.07203	156	-0.03593	-0.04311	0.05612
67	-0.01085	0.19063	0.19094	157	-0.04375	-0.07108	0.08347
68	0.03157	0.27795	0.27974	158	-0.04387	-0.08412	0.09488
69	0.04515	0.10152	0.11110	159	-0.03723	-0.08220	0.09024
70	0.01661	-0.19358	0.19429	160	-0.02566	-0.06799	0.07267
71	-0.02784	-0.32483	0.32602	161	-0.01142	-0.04557	0.04698
72	-0.04588	-0.13874	0.14613	162	0.00333	-0.01941	0.01970
73	-0.01945	0.21303	0.21392	163	0.01684	0.00649	0.01804
74	0.02634	0.38514	0.38604	164	0.02787	0.02913	0.04032
75	0.04630	0.17042	0.17660	165	0.03579	0.04668	0.05882
76	0.01973	-0.25868	0.25943	166	0.04045	0.05840	0.07104
77	-0.02735	-0.47111	0.47191	167	0.04209	0.06438	0.07691
78	-0.04681	-0.19723	0.20271	168	0.04119	0.06533	0.07723
79	-0.01745	0.34785	0.34829	169	0.03834	0.06232	0.07317
80	0.03121	0.60533	0.60614	170	0.03416	0.05645	0.06598
81	0.04745	0.21912	0.22420	171	0.02922	0.04884	0.05691
82	0.01177	-0.52170	0.52183	172	0.02399	0.04041	0.04699
83	-0.03921	-0.84638	0.84729	173	0.01885	0.03192	0.03707
84	-0.04811	-0.23563	0.24049	174	0.01409	0.02391	0.02775
85	0.00093	0.91907	0.91907	175	0.00989	0.01683	0.01952
86	0.05751	1.42850	1.42966	176	0.00636	0.01085	0.01257
87	0.04860	0.24602	0.25078	177	0.00359	0.00613	0.00710
88	-0.04759	-2.51286	2.51331	178	0.00160	0.00271	0.00315
89	-0.17109	-5.41059	5.41329	179	0.00040	0.00068	0.00079
				180	0.00000	-0.00000	0.00000

IBR

RADIATION PATTERN FOR THETA=PI/2

H/LMDA=0.2500 B/LMDA= 0.500 OMEGA= 8.53

40 ELEMENT ARRAY

PHI DEG	ER(THETA,PHI) RE	IM	ABSVAL	PHI DEG	ER(THETA,PHI) RE	IM	ABSVAL
0	0.0	-0.00000	0.00000	90	-1.55913	-4.96045	5.19971
1	0.00049	0.00132	0.00141	91	-1.26377	-4.02397	4.21776
2	0.00197	0.00527	0.00563	92	-0.57457	-1.83680	1.92457
3	0.00444	0.01186	0.01266	93	0.07308	0.22350	0.23514
4	0.00787	0.02102	0.02245	94	0.33787	1.07342	1.12534
5	0.01223	0.03266	0.03487	95	0.20327	0.65403	0.68489
6	0.01743	0.04654	0.04970	96	-0.07057	-0.21559	0.22685
7	0.02335	0.06231	0.06654	97	-0.20152	-0.63955	0.67055
8	0.02973	0.07930	0.08469	98	-0.11000	-0.35736	0.37391
9	0.03627	0.09663	0.10321	99	0.06665	0.20316	0.21381
10	0.04250	0.11305	0.12078	100	0.14465	0.45888	0.48114
11	0.04786	0.12701	0.13572	101	0.06890	0.22685	0.23708
12	0.05167	0.13664	0.14608	102	-0.06156	-0.18696	0.19683
13	0.05319	0.13997	0.14973	103	-0.11305	-0.35847	0.37587
14	0.05171	0.13505	0.14461	104	-0.04783	-0.15995	0.16695
15	0.04666	0.12039	0.12911	105	0.05555	0.16767	0.17663
16	0.03773	0.09523	0.10243	106	0.09328	0.29543	0.30981
17	0.02508	0.06010	0.06512	107	0.03729	0.12641	0.13180
18	0.00944	0.01716	0.01958	108	-0.04867	-0.14544	0.15336
19	-0.00780	-0.02963	0.03064	109	-0.08028	-0.25346	0.26587
20	-0.02459	-0.07453	0.07848	110	-0.03345	-0.11386	0.11867
21	-0.03844	-0.11062	0.11711	111	0.04072	0.11962	0.12636
22	-0.04679	-0.13088	0.13899	112	0.07136	0.22389	0.23499
23	-0.04751	-0.12968	0.13811	113	0.03436	0.11595	0.12094
24	-0.03959	-0.10456	0.11181	114	-0.03117	-0.08869	0.09401
25	-0.02365	-0.05778	0.06244	115	-0.06445	-0.20011	0.21023
26	-0.00230	0.00288	0.00369	116	-0.03867	-0.12811	0.13382
27	0.02008	0.06478	0.06782	117	0.01930	0.05058	0.05413
28	0.03804	0.11256	0.11881	118	0.05742	0.17532	0.18448
29	0.04643	0.13205	0.13998	119	0.04490	0.14523	0.15201
30	0.04205	0.11506	0.12250	120	-0.00439	-0.00364	0.00570
31	0.02512	0.06329	0.06809	121	-0.04770	-0.14181	0.14962
32	-0.00006	-0.00992	0.00992	122	-0.05065	-0.15953	0.16737
33	-0.02573	-0.08183	0.08578	123	-0.01348	-0.05102	0.05277
34	-0.04292	-0.12694	0.13400	124	0.03258	0.09185	0.09746
35	-0.04464	-0.12630	0.13395	125	0.05210	0.15909	0.16741
36	-0.02911	-0.07652	0.08187	126	0.03220	0.10592	0.11070
37	-0.00143	0.00581	0.00598	127	-0.01036	-0.02165	0.02400
38	0.02744	0.08792	0.09210	128	-0.04411	-0.12896	0.13630
39	0.04473	0.13313	0.14044	129	-0.04622	-0.14336	0.15063
40	0.04165	0.11770	0.12485	130	-0.01707	-0.06092	0.06327
41	0.01839	0.04482	0.04845	131	0.02242	0.05849	0.06264
42	-0.01464	-0.05272	0.05472	132	0.04643	0.13659	0.14427
43	-0.04080	-0.12554	0.13200	133	0.04092	0.12762	0.13402
44	-0.04548	-0.13240	0.13999	134	0.01087	0.04226	0.04364
45	-0.02476	-0.06505	0.06960	135	-0.02476	-0.06505	0.06960
46	0.01087	0.04226	0.04363	136	-0.04548	-0.13240	0.14000
47	0.04092	0.12760	0.13400	137	-0.04080	-0.12554	0.13200
48	0.04643	0.13660	0.14428	138	-0.01464	-0.05272	0.05472
49	0.02243	0.05850	0.06265	139	0.01839	0.04482	0.04845
50	-0.01706	-0.06091	0.06326	140	0.04165	0.11769	0.12484
51	-0.04622	-0.14336	0.15062	141	0.04473	0.13312	0.14043
52	-0.04411	-0.12896	0.13630	142	0.02744	0.08793	0.09211
53	-0.01037	-0.02166	0.02401	143	-0.00143	0.00580	0.00598
54	0.03220	0.10592	0.11070	144	-0.02911	-0.07653	0.08188
55	0.05210	0.15910	0.16741	145	-0.04465	-0.12630	0.13396
56	0.03258	0.09185	0.09746	146	-0.04293	-0.12695	0.13401
57	-0.01348	-0.05101	0.05276	147	-0.02573	-0.08183	0.08578
58	-0.05065	-0.15953	0.16737	148	-0.00006	-0.00991	0.00991
59	-0.04770	-0.14181	0.14962	149	0.02511	0.06328	0.06808
60	-0.00439	-0.00366	0.00571	150	0.04205	0.11506	0.12250
61	0.04489	0.14522	0.15200	151	0.04643	0.13205	0.13998
62	0.05742	0.17532	0.18448	152	0.03804	0.11256	0.11881
63	0.01930	0.05059	0.05415	153	0.02008	0.06479	0.06782
64	-0.03867	-0.12810	0.13381	154	-0.00230	0.00288	0.00369
65	-0.06446	-0.20011	0.21024	155	-0.02365	-0.05778	0.06243
66	-0.03118	-0.08871	0.09403	156	-0.03959	-0.10456	0.11181
67	0.03435	0.11594	0.12092	157	-0.04751	-0.12968	0.13811
68	0.07136	0.22389	0.23499	158	-0.04679	-0.13088	0.13899
69	0.04072	0.11963	0.12637	159	-0.03844	-0.11062	0.11710
70	-0.03345	-0.11385	0.11866	160	-0.02459	-0.07453	0.07848
71	-0.08028	-0.25346	0.26587	161	-0.00780	-0.02964	0.03065
72	-0.04867	-0.14546	0.15338	162	0.00943	0.01715	0.01958
73	0.03728	0.12638	0.13177	163	0.02508	0.06011	0.06513
74	0.09328	0.29542	0.30980	164	0.03773	0.09523	0.10243
75	0.05556	0.16770	0.17666	165	0.04665	0.12039	0.12911
76	-0.04782	-0.15992	0.16692	166	0.05171	0.13506	0.14462
77	-0.11305	-0.35847	0.37587	167	0.05319	0.13997	0.14973
78	-0.06157	-0.18699	0.19687	168	0.05167	0.13663	0.14608
79	0.06890	0.22683	0.23707	169	0.04786	0.12700	0.13572
80	0.14465	0.45888	0.48114	170	0.04250	0.11305	0.12078
81	0.06666	0.20318	0.21383	171	0.03627	0.09663	0.10322
82	-0.10998	-0.35732	0.37386	172	0.02973	0.07930	0.08469
83	-0.20152	-0.63956	0.67056	173	0.02335	0.06231	0.06654
84	-0.07060	-0.21566	0.22692	174	0.01743	0.04654	0.04970
85	0.20324	0.65396	0.68481	175	0.01223	0.03267	0.03488
86	0.33788	1.07343	1.12535	176	0.00787	0.02103	0.02246
87	0.07313	0.22364	0.23529	177	0.00444	0.01187	0.01267
88	-0.57449	-1.83659	1.92434	178	0.00197	0.00527	0.00563
89	-1.26374	-4.02391	4.21769	179	0.00049	0.00132	0.00141
				180	0.0	-0.00000	0.00000

VBR

RADIATION PATTERN FOR THETA=PI/2

H/LMDA=0.2500 B/LMDA= 0.500 OMEGA= 8.53

45 ELEMENT ARRAY

PHI DEG	ER(THETA,PHI) RE	IM	ABSVAL	PHI DEG	ER(THETA,PHI) RE	IM	ABSVAL
0	-0.03383	-0.04404	0.05553	90	-0.22291	-6.65182	6.65556
1	-0.03383	-0.04404	0.05553	91	-0.15585	-5.10878	5.11115
2	-0.03380	-0.04398	0.05546	92	-0.02055	-1.73283	1.73295
3	-0.03366	-0.04374	0.05519	93	0.05870	0.91832	0.92020
4	-0.03329	-0.04308	0.05445	94	0.03310	1.33584	1.33625
5	-0.03253	-0.04172	0.05290	95	-0.03106	0.17948	0.18215
6	-0.03115	-0.03926	0.05011	96	-0.04378	-0.78821	0.78943
7	-0.02892	-0.03529	0.04563	97	0.00370	-0.59498	0.59499
8	-0.02560	-0.02940	0.03899	98	0.04297	0.22949	0.23348
9	-0.02097	-0.02125	0.02985	99	0.02342	0.60598	0.60643
10	-0.01488	-0.01065	0.01830	100	-0.02611	0.20160	0.20328
11	-0.00734	0.00229	0.00769	101	-0.04038	-0.36544	0.36767
12	0.00147	0.01707	0.01713	102	-0.00242	-0.40948	0.40949
13	0.01111	0.03266	0.03450	103	0.03797	0.03410	0.05104
14	0.02081	0.04747	0.05183	104	0.02992	0.37459	0.37578
15	0.02954	0.05939	0.06633	105	-0.01539	0.23136	0.23187
16	0.03603	0.06602	0.07521	106	-0.04124	-0.16394	0.16905
17	0.03895	0.06503	0.07580	107	-0.01644	-0.32258	0.32300
18	0.03715	0.05471	0.06613	108	0.02807	-0.09122	0.09544
19	0.03002	0.03474	0.04591	109	0.03871	0.22050	0.22387
20	0.01779	0.00676	0.01903	110	0.00353	0.25014	0.25016
21	0.00183	-0.02517	0.02524	111	-0.03558	-0.00533	0.03598
22	-0.01531	-0.05488	0.05697	112	-0.03346	-0.23160	0.23400
23	-0.03017	-0.07498	0.08082	113	0.00672	-0.18184	0.18197
24	-0.03902	-0.07868	0.08782	114	0.03916	0.06275	0.07396
25	-0.03891	-0.06219	0.07336	115	0.02813	0.21975	0.22154
26	-0.02874	-0.02698	0.03942	116	-0.01350	0.12988	0.13058
27	-0.01022	0.01910	0.02167	117	-0.04045	-0.09005	0.09872
28	0.01200	0.06261	0.06375	118	-0.02458	-0.20041	0.20191
29	0.03119	0.08806	0.09342	119	0.01670	-0.09797	0.09935
30	0.04051	0.08356	0.09286	120	0.04075	0.09602	0.10431
31	0.03564	0.04675	0.05879	121	0.02390	0.18193	0.18350
32	0.01716	-0.01152	0.02066	122	-0.01630	0.08532	0.08686
33	-0.00869	-0.06886	0.06941	123	-0.04058	-0.08653	0.09558
34	-0.03154	-0.09913	0.10403	124	-0.02649	-0.16657	0.16866
35	-0.04099	-0.08470	0.09410	125	0.01191	-0.08888	0.08967
36	-0.03162	-0.02747	0.04188	126	0.03931	0.06391	0.07503
37	-0.00664	0.04762	0.04808	127	0.03190	0.15141	0.15474
38	0.02223	0.10171	0.10411	128	-0.00285	0.10380	0.10384
39	0.03985	0.10186	0.10937	129	-0.03492	-0.02801	0.04476
40	0.03571	0.04222	0.05529	130	-0.03821	-0.12931	0.13484
41	0.01089	-0.04683	0.04808	131	-0.01114	-0.12188	0.12238
42	-0.02086	-0.11158	0.11352	132	0.02425	-0.02044	0.03172
43	-0.04012	-0.10686	0.11414	133	0.04109	0.09063	0.09951
44	-0.03379	-0.02907	0.04457	134	0.02787	0.12915	0.13212
45	-0.00481	0.07388	0.07404	135	-0.00481	0.07388	0.07403
46	0.02787	0.12914	0.13211	136	-0.03379	-0.02905	0.04456
47	0.04109	0.09064	0.09951	137	-0.04012	-0.10686	0.11414
48	[illegible]	[illegible]	0.03172	138	-0.02086	-0.11158	0.11351
49	-0.01114	-0.12187	0.12238	139	[illegible]	[illegible]	[illegible]
50	-0.03821	-0.12930	0.13483	140	0.03570	0.04221	0.05529
51	-0.03492	-0.02801	0.04477	141	0.03985	0.10184	0.10936
52	-0.00285	0.10379	0.10383	142	0.02223	0.10172	0.10412
53	0.03190	0.15142	0.15474	143	-0.00664	0.04763	0.04809
54	0.03931	0.06392	0.07504	144	-0.03162	-0.02744	0.04186
55	0.01192	-0.08887	0.08966	145	-0.04099	-0.08470	0.09410
56	-0.02649	-0.16656	0.16865	146	-0.03154	-0.09914	0.10404
57	-0.04058	-0.08653	0.09558	147	-0.00869	-0.06886	0.06941
58	-0.01630	0.08530	0.08684	148	0.01716	-0.01152	0.02067
59	0.02390	0.18193	0.18349	149	0.03564	0.04676	0.05880
60	0.04075	0.09602	0.10431	150	0.04051	0.08356	0.09286
61	0.01670	-0.09797	0.09938	151	0.03120	0.08806	0.09343
62	-0.02458	-0.20041	0.20191	152	0.01200	0.06260	0.06375
63	-0.04045	-0.09006	0.09873	153	-0.01022	0.01911	0.02168
64	-0.01351	0.12985	0.13055	154	-0.02874	-0.02698	0.03942
65	0.02813	0.21974	0.22154	155	-0.03891	-0.06219	0.07335
66	0.03917	0.06277	0.07398	156	-0.03902	-0.07869	0.08784
67	0.00673	-0.18183	0.18196	157	-0.03017	-0.07499	0.08083
68	-0.03345	-0.23161	0.23401	158	-0.01531	-0.05488	0.05697
69	-0.03558	-0.00534	0.03598	159	0.00183	-0.02517	0.02523
70	0.00352	0.25013	0.25015	160	0.01779	0.00677	0.01903
71	0.03871	0.22051	0.22388	161	0.03002	0.03474	0.04591
72	0.02807	-0.09118	0.09541	162	0.03715	0.05472	0.06614
73	-0.01643	-0.32257	0.32299	163	0.03895	0.06503	0.07580
74	-0.04124	-0.16397	0.16908	164	0.03603	0.06605	0.07524
75	-0.01540	0.23133	0.23184	165	0.02954	0.05940	0.06634
76	0.02991	0.37459	0.37579	166	0.02081	0.04747	0.05183
77	0.03797	0.03415	0.05107	167	0.01111	0.03265	0.03449
78	-0.00241	-0.40946	0.40947	168	0.00147	0.01705	0.01711
79	-0.04038	-0.36546	0.36768	169	-0.00734	0.00229	0.00769
80	-0.02612	0.20158	0.20326	170	-0.01488	-0.01065	0.01830
81	0.02342	0.60598	0.60643	171	-0.02097	-0.02125	0.02985
82	0.04298	0.22956	0.23355	172	-0.02560	-0.02940	0.03898
83	0.00371	-0.59491	0.59493	173	-0.02892	-0.03529	0.04563
84	-0.04378	-0.78825	0.78947	174	-0.03115	-0.03926	0.05011
85	-0.03107	0.17936	0.18203	175	-0.03253	-0.04171	0.05290
86	0.03309	1.33578	1.33619	176	-0.03329	-0.04308	0.05445
87	0.05871	0.91846	0.92034	177	-0.03366	-0.04374	0.05519
88	-0.02053	-1.73251	1.73264	178	-0.03380	-0.04398	0.05546
89	-0.15583	-5.10867	5.11105	179	-0.03383	-0.04404	0.05553
				180	-0.03383	-0.04404	0.05553

IBR

RADIATION PATTERN FOR THETA=PI/2

H/LMDA=0.2500 B/LMDA= 0.500 OMEGA= 8.53

45 ELEMENT ARRAY

PHI DEG	ER(THETA,PHI) RE	IM	ABSVAL	PHI DEG	ER(THETA,PHI) RE	IM	ABSVAL
0	-0.05803	-0.13484	0.14679	90	-1.63001	-4.87379	5.13914
1	-0.05802	-0.13483	0.14679	91	-1.24559	-3.72772	3.93031
2	-0.05798	-0.13473	0.14667	92	-0.40937	-1.23214	1.29836
3	-0.05780	-0.13427	0.14618	93	0.23605	0.69998	0.73871
4	-0.05731	-0.13303	0.14485	94	0.32208	0.96590	1.01818
5	-0.05628	-0.13045	0.14208	95	0.02888	0.09439	0.09870
6	-0.05443	-0.12581	0.13708	96	-0.19944	-0.59310	0.62574
7	-0.05143	-0.11829	0.12899	97	-0.13541	-0.41051	0.43227
8	-0.04696	-0.10708	0.11692	98	0.07027	0.20268	0.21452
9	-0.04071	-0.09143	0.10008	99	0.14847	0.44413	0.46829
10	-0.03247	-0.07085	0.07794	100	0.03529	0.11313	0.11851
11	-0.02221	-0.04528	0.05044	101	-0.10006	-0.29373	0.31030
12	-0.01014	-0.01532	0.01837	102	-0.09404	-0.28478	0.29990
13	0.00319	0.01759	0.01788	103	0.02371	0.06268	0.06702
14	0.01685	0.05102	0.05373	104	0.09718	0.28796	0.30392
15	0.02954	0.08158	0.08676	105	0.04539	0.14199	0.14906
16	0.03963	0.10513	0.11236	106	-0.05445	-0.15529	0.16456
17	0.04543	0.11738	0.12586	107	-0.07895	-0.23649	0.24932
18	0.04542	0.11457	0.12325	108	-0.00786	-0.03146	0.03242
19	0.03873	0.09471	0.10232	109	0.06574	0.19055	0.20157
20	0.02553	0.05855	0.06387	110	0.05634	0.17155	0.18056
21	0.00738	0.01047	0.01280	111	-0.01730	-0.04305	0.04640
22	-0.01273	-0.04143	0.04334	112	-0.06539	-0.19129	0.20216
23	-0.03065	-0.08625	0.09154	113	-0.03588	-0.11243	0.11802
24	-0.04192	-0.11248	0.12004	114	0.03178	0.08620	0.09187
25	-0.04296	-0.11123	0.11924	115	0.05980	0.17613	0.18600
26	-0.03244	-0.07978	0.08613	116	0.02065	0.06829	0.07135
27	-0.01227	-0.02415	0.02709	117	-0.03849	-0.10631	0.11306
28	0.01215	0.04064	0.04241	118	-0.05336	-0.15790	0.16667
29	0.03307	0.09379	0.09944	119	-0.01147	-0.04164	0.04320
30	0.04279	0.11529	0.12298	120	0.04017	0.11121	0.11824
31	0.03673	0.09414	0.10105	121	0.04855	0.14375	0.15173
32	0.01601	0.03505	0.03854	122	0.00809	0.03181	0.03282
33	-0.01181	-0.04033	0.04203	123	-0.03856	-0.10595	0.11275
34	-0.03498	-0.09989	0.10583	124	-0.04622	-0.13608	0.14372
35	-0.04241	-0.11461	0.12221	125	-0.01002	-0.03720	0.03853
36	-0.02936	-0.07379	0.07942	126	0.03409	0.09180	0.09793
37	-0.00107	0.00583	0.00593	127	0.04586	0.13314	0.14082
38	0.02836	0.08444	0.08908	128	0.01676	0.05590	0.05836
39	0.04260	0.11801	0.12547	129	-0.02585	-0.06648	0.07133
40	0.03239	0.08378	0.08982	130	-0.04542	-0.12888	0.13665
41	0.00237	-0.00217	0.00322	131	-0.02721	-0.08402	0.08832
42	-0.02979	-0.08914	0.09399	132	0.01218	0.02607	0.02878
43	-0.04322	-0.12026	0.12779	133	0.04106	0.11250	0.11976
44	-0.02771	-0.07074	0.07598	134	0.03828	0.11190	0.11827
45	0.00760	0.02992	0.03087	135	0.00760	0.02992	0.03088
46	0.03827	0.11185	0.11826	136	-0.02771	-0.07073	0.07596
47	0.04106	0.11251	0.11977	137	-0.04322	-0.12027	0.12780
48	0.01218	0.02606	0.02877	138	-0.02979	-0.08914	0.09398
49	-0.02721	-0.08402	0.08831	139	0.00237	-0.00218	0.00322
50	[illegible]	[illegible]	[illegible]	140	0.03239	0.08377	0.08981
51	-0.02586	-0.06645	[illegible]	141	[illegible]	[illegible]	[illegible]
52	0.01675	0.05589	0.05834	142	0.02836	0.08445	[illegible]
53	0.04586	0.13314	0.14082	143	-0.00107	0.00583	0.00593
54	0.03409	0.09180	0.09793	144	-0.02935	-0.07378	0.07940
55	-0.01002	-0.03720	0.03852	145	-0.04241	-0.11461	0.12220
56	-0.04622	-0.13607	0.14371	146	-0.03498	-0.09990	0.10585
57	-0.03856	-0.10595	0.11275	147	-0.01181	-0.04034	0.04203
58	0.00808	0.03179	0.03280	148	0.01601	0.03505	0.03853
59	0.04854	0.14374	0.15172	149	0.03673	0.09415	0.10106
60	0.04017	0.11121	0.11824	150	0.04279	0.11529	0.12298
61	-0.01147	-0.04164	0.04319	151	0.03307	0.09379	0.09945
62	-0.05336	-0.15789	0.16666	152	0.01215	0.04064	0.04241
63	-0.03849	-0.10631	0.11307	153	-0.01227	-0.02414	0.02708
64	0.02065	0.06827	0.07132	154	-0.03244	-0.07978	0.08612
65	0.05980	0.17612	0.18599	155	-0.04296	-0.11123	0.11923
66	0.03179	0.08622	0.09189	156	-0.04193	-0.11249	0.12005
67	-0.03588	-0.11242	0.11800	157	-0.03066	-0.08626	0.09155
68	-0.06540	-0.19130	0.20216	158	-0.01273	-0.04143	0.04334
69	-0.01731	-0.04306	0.04641	159	0.00738	0.01047	0.01281
70	0.05633	0.17154	0.18056	160	0.02553	0.05855	0.06388
71	0.06574	0.19055	0.20158	161	0.03873	0.09471	0.10232
72	-0.00785	-0.03143	0.03240	162	0.04542	0.11458	0.12325
73	-0.07895	-0.23648	0.24931	163	0.04543	0.11738	0.12586
74	-0.05446	-0.15531	0.16458	164	0.03964	0.10515	0.11238
75	0.04538	0.14196	0.14904	165	0.02954	0.08159	0.08677
76	0.09718	0.28796	0.30392	166	0.01685	0.05102	0.05373
77	0.02372	0.06272	0.06705	167	0.00319	0.01759	0.01787
78	-0.09403	-0.28476	0.29988	168	-0.01015	-0.01533	0.01839
79	-0.10006	-0.29374	0.31031	169	-0.02221	-0.04528	0.05043
80	0.03528	0.11312	0.11849	170	-0.03247	-0.07085	0.07793
81	0.14847	0.44413	0.46829	171	-0.04071	-0.09143	0.10008
82	0.07029	0.20273	0.21457	172	-0.04696	-0.10708	0.11692
83	-0.13539	-0.41046	0.43222	173	-0.05143	-0.11829	0.12899
84	-0.19946	-0.59313	0.62576	174	-0.05443	-0.12581	0.13708
85	0.02884	0.09430	0.09861	175	-0.05628	-0.13045	0.14208
86	0.32206	0.96585	1.01813	176	-0.05731	-0.13303	0.14485
87	0.23610	0.70008	0.73882	177	-0.05780	-0.13427	0.14618
88	-0.40928	-1.23192	1.29812	178	-0.05798	-0.13473	0.14667
89	-1.24555	-3.72764	3.93023	179	-0.05802	-0.13483	0.14679
				180	-0.05803	-0.13484	0.14679

VBR

RADIATION PATTERN FOR THETA=PI/2

H/LMDA=0.2500 B/LMDA= 0.500 OMEGA= 8.53

50 ELEMENT ARRAY

PHI DEG	ER(THETA,PHI) RE	IM	ABSVAL	PHI DEG	ER(THETA,PHI) RE	IM	ABSVAL
0	-0.00000	-0.00001	0.00001	90	-0.21926	-6.65677	6.66038
1	-0.00040	-0.00066	0.00077	91	-0.14097	-4.78120	4.78328
2	-0.00159	-0.00263	0.00307	92	0.00240	-0.99173	0.99174
3	-0.00356	-0.00588	0.00687	93	0.05789	1.31907	1.32034
4	-0.00631	-0.01039	0.01215	94	0.00486	0.90974	0.90975
5	-0.00977	-0.01606	0.01880	95	-0.04399	-0.48547	0.48746
6	-0.01384	-0.02268	0.02657	96	-0.01517	-0.77871	0.77886
7	-0.01833	-0.02990	0.03507	97	0.03517	0.06058	0.07005
8	-0.02297	-0.03718	0.04371	98	0.02609	0.60759	0.60815
9	-0.02733	-0.04373	0.05157	99	-0.02383	0.19074	0.19222
10	-0.03088	-0.04854	0.05753	100	-0.03461	-0.40979	0.41125
11	-0.03297	-0.05044	0.06026	101	0.00873	-0.32579	0.32591
12	-0.03292	-0.04822	0.05839	102	0.03772	0.20346	0.20693
13	-0.03010	-0.04086	0.05075	103	0.00860	0.36483	0.36493
14	-0.02410	-0.02788	0.03685	104	-0.03317	-0.01105	0.03496
15	-0.01491	-0.00968	0.01777	105	-0.02472	-0.32169	0.32264
16	-0.00311	0.01211	0.01250	106	0.02040	-0.14267	0.14412
17	0.01000	0.03449	0.03591	107	0.03516	0.21473	0.21759
18	0.02244	0.05324	0.05778	108	-0.00139	0.23455	0.23455
19	0.03173	0.06359	0.07107	109	-0.03573	-0.07143	0.07987
20	0.03544	0.06137	0.07087	110	-0.01897	-0.24922	0.24994
21	0.03181	0.04459	0.05478	111	0.02444	-0.07153	0.07559
22	0.02059	0.01493	0.02543	112	0.03366	0.18647	0.18948
23	0.00361	-0.02150	0.02180	113	-0.00340	0.17350	0.17353
24	-0.01514	-0.05465	0.05671	114	-0.03570	-0.06732	0.07620
25	-0.03020	-0.07302	0.07902	115	-0.02015	-0.20110	0.20211
26	-0.03623	-0.06775	0.07682	116	0.02189	-0.06546	0.06903
27	-0.03008	-0.03712	0.04778	117	0.03536	0.14333	0.14763
28	-0.01270	0.01043	0.01644	118	0.00361	0.15733	0.15737
29	0.01021	0.05711	0.05802	119	-0.03255	-0.02407	0.04048
30	0.02962	0.08186	0.08705	120	-0.02836	-0.16384	0.16628
31	0.03656	0.06993	0.07892	121	0.01020	-0.09827	0.09880
32	0.02668	0.02251	0.03491	122	0.03644	0.07869	0.08672
33	0.00343	-0.03985	0.04000	123	0.02002	0.15354	0.15484
34	-0.02218	-0.08444	0.08730	124	-0.01921	0.04970	0.05329
35	-0.03632	-0.08321	0.09080	125	-0.03671	-0.10309	0.10943
36	-0.03004	-0.03161	0.04361	126	-0.01389	-0.13454	0.13525
37	-0.00568	0.04321	0.04358	127	0.02365	-0.02081	0.03151
38	0.02271	0.09415	0.09685	128	0.03621	0.10799	0.11390
39	0.03682	0.08341	0.09117	129	0.01174	0.11954	0.12012
40	0.02616	0.01224	0.02888	130	-0.02420	0.01180	0.02692
41	-0.00316	-0.07230	0.07237	131	-0.03642	-0.10130	0.10764
42	-0.03061	-0.10597	0.11030	132	-0.01425	-0.11266	0.11356
43	-0.03516	-0.05667	0.06669	133	0.02074	-0.02016	0.02892
44	-0.01204	0.04256	0.04423	134	0.03696	0.08515	0.09282
45	0.02120	0.11109	0.11309	135	0.02120	0.11108	0.11308
46	0.03696	0.08516	0.09283	136	-0.01203	0.04256	0.04423
47	0.02074	-0.02016	0.02893	137	-0.03515	-0.05667	0.06669
48	-0.01425	-0.11266	0.11356	138	-0.03061	-0.10597	0.11030
49	-0.03642	-0.10129	0.10764	139	-0.00316	-0.07230	0.07237
50	-0.02420	0.01181	0.02693	140	0.02616	0.01223	0.02888
51	0.01174	0.11954	0.12012	141	0.03682	0.08339	0.09116
52	0.03621	0.10798	0.11389	142	0.02272	0.09415	0.09686
53	0.02365	-0.02081	0.03150	143	-0.00568	0.04321	0.04359
54	-0.01389	-0.13452	0.13523	144	-0.03004	-0.03163	0.04362
55	-0.03671	-0.10309	0.10943	145	-0.03632	-0.08322	0.09080
56	-0.01922	0.04968	0.05327	146	-0.02219	-0.08445	0.08732
57	0.02002	0.15355	0.15485	147	0.00342	-0.03985	0.04000
58	0.03644	0.07870	0.08673	148	0.02668	0.02251	0.03491
59	0.01020	-0.09824	0.09877	149	0.03656	0.06992	0.07890
60	-0.02836	-0.16385	0.16629	150	0.02962	0.08186	0.08705
61	-0.03255	-0.02408	0.04049	151	0.01021	0.05711	0.05802
62	0.00361	0.15732	0.15736	152	-0.01270	0.01042	0.01643
63	0.03536	0.14334	0.14764	153	-0.03008	-0.03711	0.04777
64	0.02190	-0.06545	0.06902	154	-0.03623	-0.06776	0.07683
65	-0.02015	-0.20110	0.20211	155	-0.03020	-0.07301	0.07901
66	-0.03570	-0.06735	0.07623	156	-0.01514	-0.05466	0.05671
67	-0.00340	0.17348	0.17352	157	0.00361	-0.02151	0.02181
68	0.03365	0.18648	0.18949	158	0.02059	0.01492	0.02543
69	0.02444	-0.07152	0.07558	159	0.03181	0.04460	0.05478
70	-0.01897	-0.24922	0.24994	160	0.03544	0.06137	0.07086
71	-0.03573	-0.07144	0.07988	161	0.03173	0.06358	0.07106
72	-0.00140	0.23453	0.23453	162	0.02244	0.05325	0.05778
73	0.03516	0.21475	0.21761	163	0.01000	0.03451	0.03593
74	0.02041	-0.14263	0.14409	164	-0.00311	0.01213	0.01252
75	-0.02472	-0.32170	0.32264	165	-0.01491	-0.00968	0.01777
76	-0.03318	-0.01109	0.03498	166	-0.02410	-0.02787	0.03685
77	0.00859	0.36481	0.36491	167	-0.03010	-0.04086	0.05075
78	0.03773	0.20351	0.20697	168	-0.03292	-0.04823	0.05839
79	0.00873	-0.32578	0.32589	169	-0.03297	-0.05045	0.06027
80	-0.03462	-0.40980	0.41126	170	-0.03088	-0.04855	0.05753
81	-0.02384	0.19071	0.19220	171	-0.02733	-0.04373	0.05157
82	0.02609	0.60759	0.60815	172	-0.02297	-0.03718	0.04370
83	0.03518	0.06066	0.07012	173	-0.01833	-0.02990	0.03507
84	-0.01516	-0.77867	0.77882	174	-0.01384	-0.02268	0.02657
85	-0.04399	-0.48556	0.48755	175	-0.00977	-0.01605	0.01878
86	0.00484	0.90961	0.90963	176	-0.00631	-0.01039	0.01215
87	0.05790	1.31914	1.32041	177	-0.00356	-0.00586	0.00686
88	0.00242	-0.99141	0.99142	178	-0.00159	-0.00263	0.00307
89	-0.14096	-4.78108	4.78315	179	-0.00040	-0.00066	0.00077
				180	-0.00000	-0.00001	0.00001

IBR

RADIATION PATTERN FOR THETA=PI/2

H/LMDA=0.2500 B/LMDA= 0.500 OMEGA= 8.53

50 ELEMENT ARRAY

PHI DEG	ER(THETA,PHI) RE	IM	ABSVAL	PHI DEG	ER(THETA,PHI) RE	IM	ABSVAL
0	-0.00000	-0.00001	0.00001	90	-1.56570	-4.95134	5.19299
1	-0.00051	-0.00134	0.00143	91	-1.11807	-3.53952	3.71191
2	-0.00203	-0.00534	0.00571	92	-0.21963	-0.70245	0.73598
3	-0.00455	-0.01197	0.01281	93	0.31554	0.99483	1.04367
4	-0.00805	-0.02118	0.02266	94	0.20389	0.65066	0.68186
5	-0.01247	-0.03281	0.03510	95	-0.12475	-0.38843	0.40797
6	-0.01768	-0.04649	0.04974	96	-0.17838	-0.56697	0.59437
7	-0.02345	-0.06163	0.06594	97	0.02715	0.07837	0.08293
8	-0.02945	-0.07729	0.08271	98	0.14432	0.45572	0.47803
9	-0.03516	-0.09210	0.09859	99	0.03260	0.11027	0.11499
10	-0.03993	-0.10431	0.11169	100	-0.10373	-0.32391	0.34012
11	-0.04297	-0.11180	0.11977	101	-0.06787	-0.21989	0.23013
12	-0.04343	-0.11230	0.12041	102	0.05957	0.18166	0.19118
13	-0.04055	-0.10377	0.11141	103	0.08286	0.26400	0.27670
14	-0.03378	-0.08486	0.09133	104	-0.01591	-0.04253	0.04541
15	-0.02309	-0.05553	0.06014	105	-0.07961	-0.24973	0.26212
16	-0.00911	-0.01770	0.01991	106	-0.02222	-0.07707	0.08021
17	0.00666	0.02447	0.02536	107	0.06065	0.18607	0.19570
18	0.02191	0.06457	0.06819	108	0.04935	0.15979	0.16724
19	0.03375	0.09471	0.10055	109	-0.03036	-0.08811	0.09319
20	0.03928	0.10714	0.11411	110	-0.06064	-0.19098	0.20038
21	0.03642	0.09646	0.10310	111	-0.00432	-0.02121	0.02164
22	0.02478	0.06216	0.06691	112	0.05359	0.16406	0.17260
23	0.00639	0.01035	0.01216	113	0.03429	0.11271	0.11781
24	-0.01425	-0.04611	0.04826	114	-0.02992	-0.08636	0.09140
25	-0.03102	-0.09013	0.09532	115	-0.05009	-0.15726	0.16504
26	-0.03794	-0.10558	0.11219	116	-0.00304	-0.01721	0.01748
27	-0.03158	-0.08387	0.08962	117	0.04525	0.13660	0.14390
28	-0.01312	-0.02957	0.03235	118	0.03284	0.10690	0.11183
29	0.01101	0.03824	0.03980	119	-0.02047	-0.05585	0.05948
30	0.03089	0.09137	0.09646	120	-0.04567	-0.14074	0.14796
31	0.03706	0.10401	0.11042	121	-0.01371	-0.04953	0.05139
32	0.02544	0.06649	0.07119	122	0.03337	0.09679	0.10239
33	0.00063	-0.00605	0.00609	123	0.03897	0.12226	0.12832
34	-0.02499	-0.07719	0.08114	124	-0.00095	0.00509	0.00517
35	-0.03695	-0.10633	0.11256	125	-0.03829	-0.11297	0.11929
36	-0.02721	-0.07287	0.07778	126	-0.03154	-0.10064	0.10547
37	-0.00037	0.00688	0.00689	127	0.00946	0.02078	0.02283
38	0.02721	0.08418	0.08847	128	0.03902	0.11559	0.12200
39	0.03686	0.10616	0.11237	129	0.02692	0.08680	0.09087
40	0.02062	0.05339	0.05723	130	-0.01226	-0.02911	0.03159
41	-0.01112	-0.04016	0.04167	131	-0.03816	-0.11239	0.11869
42	-0.03501	-0.10540	0.11106	132	-0.02644	-0.08470	0.08873
43	-0.03159	-0.08852	0.09399	133	0.00987	0.02155	0.02370
44	-0.00211	0.00182	0.00279	134	0.03625	0.10498	0.11107
45	0.02984	0.09317	0.09783	135	0.02983	0.09317	0.09783
46	0.03625	0.10499	0.11107	136	-0.00211	0.00183	0.00279
47	0.00987	0.02155	0.02370	137	-0.03159	-0.08852	0.09399
48	-0.02644	-0.08470	0.08873	138	-0.03501	-0.10540	0.11106
49	-0.03815	-0.11238	0.11868	139	-0.01113	-0.04016	0.04168
50	-0.01226	-0.02911	0.03158	140	0.02062	0.05338	0.05723
51	0.02691	0.08679	0.09087	141	0.03686	0.10614	0.11236
52	0.03902	0.11559	0.12200	142	0.02722	0.08418	0.08847
53	0.00946	0.02079	0.02284	143	-0.00037	0.00688	0.00689
54	-0.03153	-0.10063	0.10545	144	-0.02721	-0.07288	0.07780
55	-0.03829	-0.11297	0.11928	145	-0.03695	-0.10633	0.11257
56	-0.00095	0.00507	0.00516	146	-0.02499	-0.07720	0.08115
57	0.03897	0.12226	0.12832	147	0.00063	-0.00605	0.00609
58	0.03337	0.09680	0.10239	148	0.02544	0.06649	0.07119
59	-0.01370	-0.04950	0.05136	149	0.03705	0.10400	0.11041
60	-0.04567	-0.14074	0.14796	150	0.03089	0.09137	0.09645
61	-0.02047	-0.05586	0.05949	151	0.01101	0.03824	0.03980
62	0.03283	0.10688	0.11181	152	-0.01312	-0.02957	0.03235
63	0.04525	0.13660	0.14390	153	-0.03158	-0.08386	0.08961
64	-0.00303	-0.01720	0.01747	154	-0.03794	-0.10558	0.11220
65	-0.05009	-0.15726	0.16504	155	-0.03102	-0.09012	0.09531
66	-0.02992	-0.08638	0.09142	156	-0.01426	-0.04611	0.04827
67	0.03429	0.11265	0.11779	157	0.00639	0.01034	0.01215
68	0.05359	0.16407	0.17260	158	0.02478	0.06215	0.06691
69	-0.00432	-0.02120	0.02163	159	0.03642	0.09646	0.10311
70	-0.06064	-0.19098	0.20038	160	0.03928	0.10713	0.11411
71	-0.03036	-0.08812	0.09320	161	0.03375	0.09471	0.10055
72	0.04934	0.15977	0.16722	162	0.02191	0.06458	0.06820
73	0.06065	0.18608	0.19571	163	0.00667	0.02449	0.02538
74	-0.02221	-0.07704	0.08018	164	-0.00911	-0.01769	0.01990
75	-0.07962	-0.24974	0.26212	165	-0.02309	-0.05554	0.06015
76	-0.01592	-0.04257	0.04545	166	-0.03378	-0.08485	0.09133
77	0.08286	0.26399	0.27669	167	-0.04055	-0.10377	0.11141
78	0.05958	0.18169	0.19121	168	-0.04344	-0.11231	0.12041
79	-0.06787	-0.21988	0.23011	169	-0.04297	-0.11180	0.11977
80	-0.10373	-0.32392	0.34012	170	-0.03993	-0.10431	0.11169
81	0.03259	0.11025	0.11497	171	-0.03516	-0.09210	0.09858
82	0.14432	0.45572	0.47803	172	-0.02945	-0.07729	0.08271
83	0.02717	0.07842	0.08300	173	-0.02345	-0.06163	0.06594
84	-0.17836	-0.56693	0.59433	174	-0.01768	-0.04649	0.04974
85	-0.12478	-0.39850	0.40805	175	-0.01247	-0.03280	0.03509
86	0.20385	0.65057	0.68176	176	-0.00805	-0.02118	0.02266
87	0.31556	0.99488	1.04373	177	-0.00455	-0.01196	0.01280
88	-0.21954	-0.70221	0.73573	178	-0.00203	-0.00534	0.00571
89	-1.11803	-3.53942	3.71181	179	-0.00051	-0.00134	0.00143
				180	-0.00000	-0.00001	0.00001

VBR

RADIATION PATTERN FOR THETA=PI/2

H/LMDA=0.5000 B/LMDA= 0.250 OMEGA= 9.92

2 ELEMENT ARRAY

PHI DEG	ER(THETA,PHI) RE	IM	ABSVAL	PHI DEG	ER(THETA,PHI) RE	IM	ABSVAL
0	-0.94035	-0.62538	1.12931	90	-1.32985	-0.88442	1.59709
5	-0.94315	-0.62724	1.13268	95	-1.32673	-0.88235	1.59335
10	-0.95150	-0.63279	1.14271	100	-1.31750	-0.87620	1.58226
15	-0.96517	-0.64189	1.15913	105	-1.30247	-0.86621	1.56421
20	-0.98381	-0.65429	1.18152	110	-1.28216	-0.85270	1.53981
25	-1.00693	-0.66966	1.20928	115	-1.25726	-0.83614	1.50991
30	-1.03391	-0.68760	1.24168	120	-1.22862	-0.81709	1.47552
35	-1.06399	-0.70761	1.27781	125	-1.19718	-0.79618	1.43776
40	-1.09633	-0.72912	1.31664	130	-1.16395	-0.77409	1.39785
45	-1.12999	-0.75150	1.35706	135	-1.12999	-0.75150	1.35706
50	-1.16395	-0.77409	1.39785	140	-1.09633	-0.72912	1.31664
55	-1.19718	-0.79618	1.43775	145	-1.06399	-0.70761	1.27781
60	-1.22862	-0.81709	1.47552	150	-1.03391	-0.68760	1.24168
65	-1.25726	-0.83614	1.50992	155	-1.00693	-0.66966	1.20928
70	-1.28216	-0.85270	1.53981	160	-0.98381	-0.65429	1.18152
75	-1.30247	-0.86621	1.56420	165	-0.96517	-0.64189	1.15913
80	-1.31750	-0.87620	1.58226	170	-0.95150	-0.63279	1.14271
85	-1.32673	-0.88235	1.59335	175	-0.94315	-0.62724	1.13268
				180	-0.94035	-0.62538	1.12931

IBR

RADIATION PATTERN FOR THETA=PI/2

H/LMDA=0.5000 B/LMDA= 0.250 OMEGA= 9.92

2 ELEMENT ARRAY

PHI DEG	ER(THETA,PHI) RE	IM	ABSVAL	PHI DEG	ER(THETA,PHI) RE	IM	ABSVAL
0	-0.94035	-0.62538	1.12931	90	-1.32985	-0.88442	1.59709
5	-0.94315	-0.62724	1.13268	95	-1.32673	-0.88235	1.59335
10	-0.95150	-0.63279	1.14271	100	-1.31750	-0.87620	1.58226
15	-0.96517	-0.64189	1.15913	105	-1.30247	-0.86621	1.56421
20	-0.98381	-0.65429	1.18152	110	-1.28216	-0.85270	1.53981
25	-1.00693	-0.66966	1.20928	115	-1.25726	-0.83614	1.50991
30	-1.03391	-0.68760	1.24168	120	-1.22862	-0.81709	1.47552
35	-1.06399	-0.70761	1.27781	125	-1.19718	-0.79618	1.43776
40	-1.09633	-0.72912	1.31664	130	-1.16395	-0.77409	1.39785
45	-1.12999	-0.75150	1.35706	135	-1.12999	-0.75150	1.35706
50	-1.16395	-0.77409	1.39785	140	-1.09633	-0.72912	1.31664
55	-1.19718	-0.79618	1.43775	145	-1.06399	-0.70761	1.27781
60	-1.22862	-0.81709	1.47552	150	-1.03391	-0.68760	1.24168
65	-1.25726	-0.83614	1.50992	155	-1.00693	-0.66966	1.20928
70	-1.28216	-0.85270	1.53981	160	-0.98381	-0.65429	1.18152
75	-1.30247	-0.86621	1.56420	165	-0.96517	-0.64189	1.15913
80	-1.31750	-0.87620	1.58226	170	-0.95150	-0.63279	1.14271
85	-1.32673	-0.88235	1.59335	175	-0.94315	-0.62724	1.13268
				180	-0.94035	-0.62538	1.12931

VBR

RADIATION PATTERN FOR THETA=PI/2

H/LMDA=0.5000 B/LMDA= 0.250 OMEGA= 9.92

3 ELEMENT ARRAY

PHI DEG	ER(THETA,PHI) RE	IM	ABSVAL	PHI DEG	ER(THETA,PHI) RE	IM	ABSVAL
0	-0.38568	-0.43244	0.57944	90	-1.36561	-1.09623	1.75118
5	-0.39153	-0.43641	0.58631	95	-1.35645	-1.09002	1.74014
10	-0.40906	-0.44828	0.60687	100	-1.32939	-1.07169	1.70757
15	-0.43810	-0.46796	0.64103	105	-1.28574	-1.04213	1.65504
20	-0.47837	-0.49523	0.68854	110	-1.22756	-1.00272	1.58504
25	-0.52937	-0.52978	0.74894	115	-1.15750	-0.95526	1.50078
30	-0.59038	-0.57111	0.82141	120	-1.07860	-0.90181	1.40593
35	-0.66032	-0.61848	0.90474	125	-0.99407	-0.84456	1.30440
40	-0.73775	-0.67093	0.99721	130	-0.90713	-0.78566	1.20006
45	-0.82078	-0.72718	1.09657	135	-0.82079	-0.72718	1.09657
50	-0.90713	-0.78566	1.20006	140	-0.73775	-0.67093	0.99721
55	-0.99407	-0.84456	1.30439	145	-0.66032	-0.61848	0.90474
60	-1.07860	-0.90181	1.40593	150	-0.59038	-0.57111	0.82141
65	-1.15750	-0.95526	1.50078	155	-0.52938	-0.52978	0.74894
70	-1.22756	-1.00272	1.58504	160	-0.47837	-0.49523	0.68854
75	-1.28574	[illegible]	[illegible]	165	-0.43810	-0.46796	0.64103
80	-1.32939	-1.07169	1.70757	170	-0.40906	-0.44828	[illegible]
85	-1.35645	-1.09002	1.74014	175	-0.39153	-0.43641	0.58631
				180	-0.38568	-0.43244	0.57944

IBR

RADIATION PATTERN FOR THETA=PI/2

H/LMDA=0.5000 B/LMDA= 0.250 OMEGA= 9.92

3 ELEMENT ARRAY

PHI DEG	ER(THETA,PHI) RE	IM	ABSVAL	PHI DEG	ER(THETA,PHI) RE	IM	ABSVAL
0	-0.10138	-0.45535	0.46650	90	-1.02773	-0.97531	1.41685
5	-0.10692	-0.45846	0.47076	95	-1.01906	-0.97045	1.40721
10	-0.12348	-0.46776	0.48379	100	-0.99348	-0.95609	1.37881
15	-0.15094	-0.48317	0.50620	105	-0.95222	-0.93293	1.33307
20	-0.18900	-0.50454	0.53878	110	-0.89723	-0.90206	1.27229
25	-0.23722	-0.53160	0.58213	115	-0.83100	-0.86489	1.19941
30	-0.29489	-0.56397	0.63642	120	-0.75641	-0.82302	1.11782
35	-0.36101	-0.60108	0.70116	125	-0.67650	-0.77817	1.03112
40	-0.43420	-0.64216	0.77518	130	-0.59431	-0.73204	0.94291
45	-0.51269	-0.68622	0.85660	135	-0.51269	-0.68622	0.85660
50	-0.59431	-0.73204	0.94291	140	-0.43420	-0.64217	0.77518
55	-0.67650	-0.77817	1.03112	145	-0.36101	-0.60108	0.70116
60	-0.75641	-0.82302	1.11782	150	-0.29489	-0.56397	0.63642
65	-0.83100	-0.86489	1.19941	155	-0.23722	-0.53160	0.58213
70	-0.89723	-0.90206	1.27229	160	-0.18900	-0.50454	0.53878
75	-0.95222	-0.93293	1.33307	165	-0.15094	-0.48317	0.50620
80	-0.99348	-0.95609	1.37881	170	-0.12348	-0.46776	0.48379
85	-1.01906	-0.97045	[illegible]	175	-0.10692	-0.45846	0.47076
				180	-0.10138	-0.45535	0.46650

VBR

RADIATION PATTERN FOR THETA=PI/2

H/LMDA=0.5000 B/LMDA= 0.250 OMEGA= 9.92

4 ELEMENT ARRAY

PHI DEG	ER(THETA,PHI) RE	IM	ABSVAL	PHI DEG	ER(THETA,PHI) RE	IM	ABSVAL
0	0.07458	-0.17438	0.18966	90	-1.43384	-1.16991	1.85057
5	0.06827	-0.17882	0.19141	95	-1.41611	-1.15855	1.82965
10	0.04918	-0.19221	0.19841	100	-1.36415	-1.12523	1.76834
15	0.01678	-0.21484	0.21549	105	-1.28143	-1.07212	1.67078
20	-0.02969	-0.24705	0.24883	110	-1.17334	-1.00255	1.54332
25	-0.09106	-0.28925	0.30324	115	-1.04652	-0.92068	1.39386
30	-0.16800	-0.34169	0.38075	120	-0.90816	-0.83102	1.23100
35	-0.26079	-0.40435	0.48116	125	-0.76539	-0.73805	1.06327
40	-0.36898	-0.47678	0.60288	130	-0.62461	-0.64587	0.89849
45	-0.49114	-0.55790	0.74328	135	-0.49114	-0.55790	0.74328
50	-0.62461	-0.64587	0.89849	140	-0.36898	-0.47678	0.60289
55	-0.76539	-0.73805	1.06327	145	-0.26079	-0.40435	0.48116
60	-0.90816	-0.83102	1.23099	150	-0.16800	-0.34169	0.38075
65	-1.04651	-0.92067	1.39386	155	-0.09106	-0.28925	0.30324
70	-1.17334	-1.00255	1.54332	160	-0.02969	-0.24705	0.24883
75	-1.28143	-1.07211	1.67078	165	0.01678	-0.21484	0.21549
80	-1.36415	-1.12523	1.76834	170	0.04918	-0.19221	0.19841
85	-1.41611	-1.15855	1.82965	175	0.06827	-0.17882	0.19141
				180	0.07458	-0.17438	0.18966

IBR

RADIATION PATTERN FOR THETA=PI/2

H/LMDA=0.5000 B/LMDA= 0.250 OMEGA= 9.92

4 ELEMENT ARRAY

PHI DEG	ER(THETA,PHI) RE	IM	ABSVAL	PHI DEG	ER(THETA,PHI) RE	IM	ABSVAL
0	0.29915	-0.20835	0.36455	90	-0.94293	-1.06707	1.42400
5	0.29425	-0.21225	0.36281	95	-0.92798	-1.05736	1.40682
10	0.27937	-0.22401	0.35809	100	-0.88415	-1.02887	1.35657
15	0.25401	-0.24384	0.35211	105	-0.81447	-0.98344	1.27692
20	0.21739	-0.27203	0.34822	110	-0.72358	-0.92389	1.17352
25	0.16866	-0.30887	0.35192	115	-0.61720	-0.85374	1.05348
30	0.10706	-0.35453	0.37034	120	-0.50152	-0.77683	0.92466
35	0.03217	-0.40895	0.41022	125	-0.38262	-0.69697	0.79509
40	-0.05583	-0.47170	0.47499	130	-0.26592	-0.61764	0.67246
45	-0.15589	-0.54179	0.56377	135	-0.15589	-0.54179	0.56377
50	-0.26592	-0.61764	0.67245	140	-0.05583	-0.47170	0.47499
55	-0.38262	-0.69697	0.79509	145	0.03217	-0.40895	0.41022
60	-0.50152	-0.77683	0.92466	150	0.10706	-0.35453	0.37034
65	-0.61720	-0.85374	1.05347	155	0.16866	-0.30887	0.35192
70	-0.72358	-0.92389	1.17352	160	0.21739	-0.27203	0.34822
75	-0.81447	-0.98344	1.27691	165	0.25401	-0.24384	0.35210
80	-0.88415	-1.02887	1.35657	170	0.27937	-0.22401	0.35809
85	-0.92797	-1.05736	1.40682	175	0.29425	-0.21225	0.36281
				180	0.29915	-0.20835	0.36455

VBR

RADIATION PATTERN FOR THETA=PI/2

H/LMDA=0.5000 B/LMDA= 0.250 OMEGA= 9.92

5 ELEMENT ARRAY

PHI	ER(THETA,PHI)			PHI	ER(THETA,PHI)		
DEG	RE	IM	ABSVAL	DEG	RE	IM	ABSVAL
0	0.30128	0.04657	0.30486	90	-1.45886	-1.18925	1.88217
5	0.29788	0.04336	0.30102	95	-1.43128	-1.17118	1.84938
10	0.28720	0.03345	0.28914	100	-1.35112	-1.11854	1.75404
15	0.26780	0.01605	0.26828	105	-1.22567	-1.03591	1.60480
20	0.23742	-0.01013	0.23764	110	-1.06586	-0.93009	1.41461
25	0.19313	-0.04670	0.19870	115	-0.88461	-0.80924	1.19892
30	0.13163	-0.09545	0.16259	120	-0.69514	-0.68174	0.97365
35	0.04970	-0.15803	0.16566	125	-0.50942	-0.55526	0.75354
40	-0.05525	-0.23565	0.24204	130	-0.33699	-0.43607	0.55110
45	-0.18442	-0.32866	0.37687	135	-0.18442	-0.32866	0.37687
50	-0.33698	-0.43607	0.55110	140	-0.05525	-0.23566	0.24204
55	-0.50941	-0.55526	0.75353	145	0.04969	-0.15803	0.16566
60	-0.69514	-0.68174	0.97365	150	0.13163	-0.09545	0.16259
65	-0.88460	-0.80924	1.19891	155	0.19313	-0.04671	0.19870
70	-1.06586	-0.93009	1.41461	160	0.23742	-0.01014	0.23764
75	-1.22567	-1.03590	1.60479	165	0.26780	0.01604	0.26828
80	-1.35112	-1.11854	1.75404	170	0.28720	0.03345	0.28914
85	-1.43128	-1.17118	1.84938	175	0.29788	0.04336	0.30102
				180	0.30128	0.04657	0.30486

IBR

RADIATION PATTERN FOR THETA=PI/2

H/LMDA=0.5000 B/LMDA= 0.250 OMEGA= 9.92

5 ELEMENT ARRAY

PHI	ER(THETA,PHI)			PHI	ER(THETA,PHI)		
DEG	RE	IM	ABSVAL	DEG	RE	IM	ABSVAL
0	0.38906	0.04110	0.39123	90	-0.90513	-1.11214	1.43392
5	0.38778	0.03794	0.38964	95	-0.88295	-1.09553	1.40705
10	0.38349	0.02820	0.38453	100	-0.81860	-1.04716	1.32915
15	0.37485	0.01117	0.37502	105	-0.71829	-0.97117	1.20794
20	0.35971	-0.01425	0.36000	110	-0.59129	-0.87374	1.05501
25	0.33527	-0.04950	0.33890	115	-0.44851	-0.76229	0.88445
30	0.29830	-0.09613	0.31340	120	-0.30101	-0.64444	0.71127
35	0.24555	-0.15557	0.29068	125	-0.15864	-0.52721	0.55056
40	0.17424	-0.22882	0.28761	130	-0.02906	-0.41638	0.41739
45	0.08269	-0.31609	0.32672	135	0.08269	-0.31609	0.32672
50	-0.02906	-0.41638	0.41739	140	0.17424	-0.22882	0.28761
55	-0.15863	-0.52721	0.55056	145	0.24554	-0.15557	0.29068
60	-0.30101	-0.64443	0.71127	150	0.29830	-0.09613	0.31340
65	-0.44851	-0.76228	0.88444	155	0.33527	-0.04951	0.33890
70	-0.59129	-0.87374	1.05501	160	0.35971	-0.01425	0.36000
75	-0.71829	-0.97117	1.20793	165	0.37485	0.01117	0.37502
80	-0.81860	-1.04716	1.32915	170	0.38349	0.02820	0.38453
85	-0.88295	-1.09553	1.40705	175	0.38778	0.03794	0.38964
				180	0.38906	0.04110	0.39123

VBR

RADIATION PATTERN FOR THETA=PI/2

H/LMDA=0.5000 B/LMDA= 0.250 OMEGA= 9.92

6 ELEMENT ARRAY

PHI	ER(THETA,PHI)			PHI	ER(THETA,PHI)		
DEG	RE	IM	ABSVAL	DEG	RE	IM	ABSVAL
0	0.29662	0.17417	0.34398	90	-1.46153	-1.21043	1.89769
2	0.29681	0.17409	0.34410	92	-1.45523	-1.20598	1.88999
4	0.29737	0.17382	0.34444	94	-1.43644	-1.19271	1.86706
6	0.29824	0.17333	0.34495	96	-1.40555	-1.17086	1.82934
8	0.29935	0.17258	0.34554	98	-1.36317	-1.14086	1.77758
10	0.30060	0.17149	0.34607	100	-1.31013	-1.10325	1.71278
12	0.30182	0.16996	0.34638	102	-1.24745	-1.05872	1.63616
14	0.30286	0.16787	0.34627	104	-1.17631	-1.00806	1.54916
16	0.30349	0.16509	0.34549	106	-1.09802	-0.95214	1.45334
18	0.30348	0.16145	0.34375	108	-1.01396	-0.89190	1.35040
20	0.30253	0.15677	0.34074	110	-0.92557	-0.82830	1.24208
22	0.30034	0.15085	0.33609	112	-0.83429	-0.76232	1.13012
24	0.29656	0.14346	0.32943	114	-0.74153	-0.69492	1.01626
26	0.29080	0.13436	0.32034	116	-0.64863	-0.62702	0.90215
28	0.28267	0.12330	0.30840	118	-0.55686	-0.55949	0.78938
30	0.27174	0.11001	0.29317	120	-0.46733	-0.49312	0.67938
32	0.25757	0.09422	0.27426	122	-0.38104	-0.42861	0.57350
34	0.23970	0.07565	0.25136	124	-0.29886	-0.36657	0.47296
36	0.21770	0.05404	0.22431	126	-0.22147	-0.30751	0.37896
38	0.19113	0.02912	0.19333	128	-0.14940	-0.25183	0.29281
40	0.15958	0.00068	0.15958	130	-0.08304	-0.19983	0.21640
42	0.12270	-0.03150	0.12667	132	-0.02261	-0.15173	0.15340
44	0.08017	-0.06757	0.10485	134	0.03178	-0.10763	0.11222
46	0.03178	-0.10763	0.11222	136	0.08017	-0.06757	0.10485
48	-0.02261	-0.15173	0.15340	138	0.12270	-0.03150	0.12667
50	-0.08304	-0.19983	0.21639	140	0.15958	0.00068	0.15958
52	-0.14940	-0.25183	0.29281	142	0.19113	0.02912	0.19333
54	-0.22147	-0.30751	0.37896	144	0.21770	0.05403	0.22431
56	-0.29886	-0.36657	0.47296	146	0.23970	0.07565	0.25136
58	-0.38104	-0.42861	0.57350	148	0.25757	0.09422	0.27426
60	-0.46732	-0.49312	0.67938	150	0.27174	0.11001	0.29317
62	-0.55685	-0.55949	0.78938	152	0.28267	0.12330	0.30840
64	-0.64863	-0.62702	0.90215	154	0.29080	0.13436	0.32034
66	-0.74153	-0.69492	1.01625	156	0.29656	0.14346	0.32943
68	-0.83429	-0.76232	1.13012	158	0.30034	0.15085	0.33609
70	-0.92557	-0.82829	1.24207	160	0.30253	0.15677	0.34074
72	-1.01396	-0.89189	1.35040	162	0.30348	0.16145	0.34375
74	-1.09801	-0.95214	1.45334	164	0.30349	0.16509	0.34549
76	-1.17630	-1.00806	1.54915	166	0.30286	0.16787	0.34627
78	-1.24744	-1.05872	1.63616	168	0.30182	0.16996	0.34638
80	-1.31013	-1.10325	1.71277	170	0.30060	0.17149	0.34607
82	-1.36317	-1.14086	1.77758	172	0.29935	0.17258	0.34554
84	-1.40555	-1.17086	1.82934	174	0.29824	0.17333	0.34495
86	-1.43644	-1.19270	1.86706	176	0.29737	0.17382	0.34444
88	-1.45523	-1.20598	1.88999	178	0.29681	0.17409	0.34410
				180	0.29662	0.17417	0.34398

IBR

RADIATION PATTERN FOR THETA=PI/2

H/LMDA=0.5000 B/LMDA= 0.250 OMEGA= 9.92

6 ELEMENT ARRAY

PHI	ER(THETA,PHI)			PHI	ER(THETA,PHI)		
DEG	RE	IM	ABSVAL	DEG	RE	IM	ABSVAL
0	0.25478	0.19004	0.31785	90	-0.87793	-1.12854	1.42981
2	0.25529	0.18990	0.31817	92	-0.87300	-1.12444	1.42355
4	0.25683	0.18946	0.31915	94	-0.85834	-1.11221	1.40491
6	0.25935	0.18870	0.32073	96	-0.83426	-1.09208	1.37428
8	0.26278	0.18756	0.32285	98	-0.80128	-1.06443	1.33231
10	0.26704	0.18599	0.32543	100	-0.76011	-1.02974	1.27990
12	0.27200	0.18388	0.32833	102	-0.71161	-0.98864	1.21811
14	0.27752	0.18115	0.33141	104	-0.65677	-0.94184	1.14822
16	0.28340	0.17765	0.33448	106	-0.59670	-0.89013	1.07163
18	0.28945	0.17324	0.33733	108	-0.53254	-0.83436	0.98983
20	0.29542	0.16776	0.33973	110	-0.46551	-0.77541	0.90441
22	0.30102	0.16102	0.34138	112	-0.39680	-0.71416	0.81699
24	0.30596	0.15283	0.34201	114	-0.32755	-0.65149	0.72919
26	0.30990	0.14296	0.34129	116	-0.25888	-0.58823	0.64267
28	0.31246	0.13120	0.33889	118	-0.19179	-0.52518	0.55910
30	0.31327	0.11730	0.33451	120	-0.12719	-0.46306	0.48021
32	0.31190	0.10103	0.32786	122	-0.06585	-0.40252	0.40787
34	0.30794	0.08214	0.31871	124	-0.00843	-0.34413	0.34423
36	0.30096	0.06040	0.30697	126	0.04457	-0.28835	0.29178
38	0.29054	0.03559	0.29272	128	0.09277	-0.23558	0.25319
40	0.27628	0.00752	0.27638	130	0.13594	-0.18610	0.23047
42	0.25779	-0.02398	0.25890	132	0.17398	-0.14012	0.22339
44	0.23474	-0.05905	0.24205	134	0.20687	-0.09775	0.22880
46	0.20687	-0.09775	0.22880	136	0.23474	-0.05905	0.24205
48	0.17398	-0.14012	0.22339	138	0.25779	-0.02398	0.25890
50	0.13595	-0.18610	0.23047	140	0.27628	0.00752	0.27638
52	0.09277	-0.23558	0.25319	142	0.29054	0.03559	0.29272
54	0.04457	-0.28835	0.29178	144	0.30096	0.06040	0.30696
56	-0.00843	-0.34413	0.34423	146	0.30794	0.08214	0.31871
58	-0.06585	-0.40252	0.40787	148	0.31190	0.10103	0.32786
60	-0.12719	-0.46306	0.48021	150	0.31327	0.11730	0.33451
62	-0.19179	-0.52517	0.55910	152	0.31246	0.13120	0.33889
64	-0.25888	-0.58822	0.64267	154	0.30990	0.14296	0.34129
66	-0.32755	-0.65148	0.72919	156	0.30596	0.15283	0.34201
68	-0.39679	-0.71416	0.81699	158	0.30102	0.16102	0.34138
70	-0.46551	-0.77541	0.90441	160	0.29542	0.16776	0.33973
72	-0.53254	-0.83436	0.98983	162	0.28945	0.17324	0.33733
74	-0.59669	-0.89013	1.07162	164	0.28340	0.17765	0.33448
76	-0.65677	-0.94184	1.14822	166	0.27752	0.18115	0.33140
78	-0.71161	-0.98864	1.21811	168	0.27200	0.18388	0.32833
80	-0.76011	-1.02974	1.27989	170	0.26704	0.18599	0.32543
82	-0.80128	-1.06443	1.33231	172	0.26278	0.18756	0.32285
84	-0.83426	-1.09208	1.37428	174	0.25935	0.18870	0.32073
86	-0.85834	-1.11221	1,40491	176	0.25683	0.18946	0.31915
88	-0.87300	-1.12444	1.42355	178	0.25529	0.18990	0.31817
				180	0.25478	0.19004	0.31785

VBR

RADIATION PATTERN FOR THETA=PI/2

H/LMDA=0.5000 B/LMDA= 0.250 OMEGA= 9.92

7 ELEMENT ARRAY

PHI DEG	ER(THETA,PHI) RE	IM	ABSVAL	PHI DEG	ER(THETA,PHI) RE	IM	ABSVAL
0	0.14195	0.18016	0.22937	90	-1.46850	-1.23471	1.91859
2	0.14276	0.18054	0.23016	92	-1.45986	-1.22827	1.90784
4	0.14516	0.18168	0.23255	94	-1.43417	-1.20912	1.87585
6	0.14913	0.18354	0.23648	96	-1.39209	-1.17771	1.82343
8	0.15462	0.18608	0.24193	98	-1.33470	-1.13480	1.75191
10	0.16156	0.18923	0.24881	100	-1.26346	-1.08141	1.66306
12	0.16984	0.19291	0.25702	102	-1.18014	-1.01876	1.55904
14	0.17935	0.19701	0.26641	104	-1.08675	-0.94827	1.44230
16	0.18990	0.20138	0.27679	106	-0.98547	-0.87148	1.31553
18	0.20130	0.20585	0.28792	108	-0.87858	-0.78998	1.18152
20	0.21329	0.21023	0.29948	110	-0.76839	-0.70543	1.04310
22	0.22555	0.21428	0.31111	112	-0.65712	-0.61941	0.90303
24	0.23774	0.21771	0.32236	114	-0.54688	-0.53344	0.76396
26	0.24941	0.22021	0.33271	116	-0.43961	-0.44894	0.62833
28	0.26009	0.22142	0.34157	118	-0.33700	-0.36714	0.49836
30	0.26921	0.22095	0.34827	120	-0.24048	-0.28914	0.37608
32	0.27616	0.21835	0.35205	122	-0.15122	-0.21583	0.26354
34	0.28026	0.21318	0.35212	124	-0.07008	-0.14791	0.16367
36	0.28078	0.20492	0.34761	126	0.00237	-0.08586	0.08589
38	0.27697	0.19308	0.33762	128	0.06581	-0.03001	0.07233
40	0.26802	0.17713	0.32126	130	0.12020	0.01950	0.12177
42	0.25316	0.15657	0.29766	132	0.16570	0.06270	0.17717
44	0.23161	0.13092	0.26605	134	0.20267	0.09975	0.22589
46	0.20267	0.09975	0.22589	136	0.23161	0.13092	0.26605
48	0.16570	0.06270	0.17717	138	0.25316	0.15657	0.29766
50	0.12020	0.01950	0.12177	140	0.26802	0.17713	0.32126
52	0.06581	-0.03001	0.07233	142	0.27697	0.19308	0.33762
54	0.00237	-0.08586	0.08589	144	0.28078	0.20492	0.34761
56	-0.07008	-0.14790	0.16367	146	0.28026	0.21318	0.35212
58	-0.15122	-0.21583	0.26354	148	0.27616	0.21835	0.35205
60	-0.24048	-0.28914	0.37608	150	0.26921	0.22095	0.34827
62	-0.33699	-0.36714	0.49835	152	0.26009	0.22142	0.34157
64	-0.43960	-0.44893	0.62832	154	0.24941	0.22021	0.33271
66	-0.54688	-0.53344	0.76396	156	0.23774	0.21771	0.32236
68	-0.65711	-0.61941	0.90303	158	0.22555	0.21428	0.31111
70	-0.76839	-0.70543	1.04309	160	0.21329	0.21023	0.29948
72	-0.87858	-0.78998	1.18151	162	0.20130	0.20585	0.28792
74	-0.98547	-0.87147	1.31552	164	0.18990	0.20138	0.27679
76	-1.08674	-0.94827	1.44230	166	0.17935	0.19701	0.26641
78	-1.18014	-1.01876	1.55904	168	0.16984	0.19291	0.25702
80	-1.26346	-1.08141	1.66306	170	0.16156	0.18923	0.24881
82	-1.33470	-1.13480	1.75191	172	0.15462	0.18608	0.24193
84	-1.39209	-1.17771	1.82343	174	0.14913	0.18354	0.23648
86	-1.43417	-1.20912	1.87584	176	0.14516	0.18168	0.23255
88	-1.45986	-1.22827	1.90784	178	0.14276	0.18054	0.23016
				180	0.14195	0.18016	0.22937

IBR

RADIATION PATTERN FOR THETA=PI/2

H/LMDA=0.5000 B/LMDA= 0.250 OMEGA= 9.92

7 ELEMENT ARRAY

PHI DEG	ER(THETA,PHI) RE	IM	ABSVAL	PHI DEG	ER(THETA,PHI) RE	IM	ABSVAL
0	0.02980	0.20551	0.20766	90	-0.84691	-1.14183	1.42163
2	0.03072	0.20580	0.20808	92	-0.84048	-1.13599	1.41311
4	0.03349	0.20667	0.20937	94	-0.82138	-1.11860	1.38777
6	0.03810	0.20809	0.21155	96	-0.79014	-1.09007	1.34632
8	0.04451	0.21001	0.21467	98	-0.74766	-1.05106	1.28986
10	0.05270	0.21237	0.21881	100	-0.69512	-1.00249	1.21990
12	0.06260	0.21508	0.22400	102	-0.63395	-0.94543	1.13830
14	0.07415	0.21804	0.23030	104	-0.56579	-0.88115	1.04716
16	0.08724	0.22111	0.23770	106	-0.49241	-0.81101	0.94880
18	0.10174	0.22413	0.24614	108	-0.41563	-0.73645	0.84564
20	0.11748	0.22690	0.25551	110	-0.33728	-0.65891	0.74022
22	0.13425	0.22921	0.26563	112	-0.25913	-0.57984	0.63511
24	0.15178	0.23080	0.27623	114	-0.18282	-0.50059	0.53293
26	0.16976	0.23138	0.28697	116	-0.10983	-0.42244	0.43648
28	0.18780	0.23061	0.29741	118	-0.04145	-0.34651	0.34898
30	0.20548	0.22815	0.30704	120	0.02126	-0.27380	0.27463
32	0.22229	0.22360	0.31529	122	0.07751	-0.20514	0.21929
34	0.23767	0.21655	0.32153	124	0.12672	-0.14116	0.18969
36	0.25101	0.20657	0.32508	126	0.16857	-0.08235	0.18761
38	0.26164	0.19320	0.32524	128	0.20295	-0.02902	0.20502
40	0.26889	0.17600	0.32137	130	0.22997	0.01867	0.23073
42	0.27203	0.15453	0.31286	132	0.24991	0.06072	0.25718
44	0.27036	0.12839	0.29929	134	0.26319	0.09722	0.28057
46	0.26319	0.09722	0.28057	136	0.27036	0.12839	0.29929
48	0.24991	0.06072	0.25718	138	0.27203	0.15453	0.31286
50	0.22997	0.01867	0.23073	140	0.26889	0.17600	0.32137
52	0.20295	-0.02902	0.20502	142	0.26165	0.19320	0.32524
54	0.16857	-0.08235	0.18761	144	0.25101	0.20657	0.32508
56	0.12672	-0.14116	0.18969	146	0.23767	0.21656	0.32153
58	0.07751	-0.20514	0.21929	148	0.22229	0.22360	0.31529
60	0.02126	-0.27380	0.27463	150	0.20548	0.22815	0.30704
62	-0.04145	-0.34651	0.34898	152	0.18780	0.23061	0.29741
64	-0.10983	-0.42243	0.43648	154	0.16976	0.23138	0.28697
66	-0.18281	-0.50059	0.53293	156	0.15178	0.23080	0.27624
68	-0.25912	-0.57984	0.63510	158	0.13425	0.22921	0.26563
70	-0.33728	-0.65891	0.74022	160	0.11748	0.22690	0.25551
72	-0.41563	-0.73645	0.84564	162	0.10174	0.22413	0.24614
74	-0.49241	-0.81101	0.94879	164	0.08724	0.22111	0.23769
76	-0.56579	-0.88115	1.04716	166	0.07415	0.21804	0.23030
78	-0.63395	-0.94543	1.13830	168	0.06260	0.21508	0.22400
80	-0.69511	-1.00249	1.21990	170	0.05270	0.21237	0.21881
82	-0.74766	-1.05106	1.28985	172	0.04451	0.21001	0.21467
84	-0.79014	-1.09007	1.34631	174	0.03810	0.20809	0.21155
86	-0.82138	-1.11860	1.38777	176	0.03349	0.20667	0.20937
88	-0.84048	-1.13599	1.41311	178	0.03072	0.20580	0.20808
				180	0.02980	0.20551	0.20766

VBR

RADIATION PATTERN FOR THETA=PI/2

H/LMDA=0.5000 B/LMDA= 0.250 OMEGA= 9.92

8 ELEMENT ARRAY

PHI DEG	ER(THETA,PHI) RE	IM	ABSVAL	PHI DEG	ER(THETA,PHI) RE	IM	ABSVAL
0	-0.04831	0.09595	0.10743	90	-1.48010	-1.24924	1.93683
2	-0.04736	0.09658	0.10757	92	-1.46865	-1.24058	1.92249
4	-0.04451	0.09849	0.10808	94	-1.43467	-1.21484	1.87992
6	-0.03973	0.10166	0.10915	96	-1.37926	-1.17282	1.81049
8	-0.03301	0.10609	0.11111	98	-1.30423	-1.11575	1.71636
10	-0.02432	0.11176	0.11438	100	-1.21194	-1.04531	1.60046
12	-0.01363	0.11865	0.11943	102	-1.10527	-0.96352	1.46628
14	-0.00092	0.12673	0.12673	104	-0.98742	-0.87264	1.31776
16	0.01380	0.13593	0.13663	106	-0.86181	-0.77511	1.15910
18	0.03052	0.14617	0.14932	108	-0.73192	-0.67342	0.99459
20	0.04916	0.15733	0.16483	110	-0.60116	-0.57001	0.82843
22	0.06960	0.16924	0.18300	112	-0.47270	-0.46722	0.66464
24	0.09166	0.18171	0.20352	114	-0.34943	-0.36718	0.50688
26	0.11506	0.19446	0.22595	116	-0.23383	-0.27175	0.35850
28	0.13944	0.20716	0.24971	118	-0.12790	-0.18250	0.22286
30	0.16433	0.21940	0.27412	120	-0.03319	-0.10067	0.10600
32	0.18914	0.23070	0.29833	122	0.04930	-0.02715	0.05628
34	0.21315	0.24051	0.32137	124	0.11900	0.03749	0.12476
36	0.23551	0.24819	0.34215	126	0.17582	0.09301	0.19891
38	0.25525	0.25303	0.35941	128	0.22006	0.13944	0.26052
40	0.27127	0.25426	0.37180	130	0.25235	0.17706	0.30827
42	0.28239	0.25104	0.37784	132	0.27358	0.20634	0.34266
44	0.28734	0.24255	0.37603	134	0.28483	0.22792	0.36480
46	0.28483	0.22792	0.36480	136	0.28734	0.24255	0.37603
48	0.27358	0.20634	0.34266	138	0.28239	0.25104	0.37785
50	0.25235	0.17706	0.30827	140	0.27127	0.25426	0.37180
52	0.22006	0.13944	0.26052	142	0.25525	0.25303	0.35941
54	0.17582	0.09301	0.19891	144	0.23551	0.24819	0.34215
56	0.11900	0.03750	0.12477	146	0.21315	0.24051	0.32137
58	0.04930	-0.02715	0.05628	148	0.18914	0.23070	0.29833
60	-0.03318	-0.10067	0.10600	150	0.16433	0.21940	0.27412
62	-0.12790	-0.18250	0.22285	152	0.13944	0.20716	0.24971
64	-0.23382	-0.27174	0.35849	154	0.11506	0.19446	0.22595
66	-0.34943	-0.36717	0.50687	156	0.09166	0.18171	0.20352
68	-0.47270	-0.46722	0.66463	158	0.06960	0.16924	0.18300
70	-0.60115	-0.57001	0.82843	160	0.04916	0.15733	0.16483
72	-0.73192	-0.67342	0.99458	162	0.03052	0.14617	0.14932
74	-0.86181	-0.77511	1.15910	164	0.01380	0.13593	0.13663
76	-0.98741	-0.87264	1.31776	166	-0.00092	0.12673	0.12673
78	-1.10526	-0.96352	1.46628	168	-0.01363	0.11865	0.11943
80	-1.21194	-1.04531	1.60046	170	-0.02432	0.11176	0.11438
82	-1.30422	-1.11574	1.71636	172	-0.03301	0.10609	0.11111
84	-1.37926	-1.17281	1.81048	174	-0.03973	0.10166	0.10915
86	-1.43466	-1.21484	1.87992	176	-0.04451	0.09849	0.10808
88	-1.46865	-1.24058	1.92249	178	-0.04736	0.09658	0.10757
				180	-0.04831	0.09595	0.10743

IBR

RADIATION PATTERN FOR THETA=PI/2

H/LMDA=0.5000 B/LMDA= 0.250 OMEGA= 9.92

8 ELEMENT ARRAY

PHI DEG	ER(THETA,PHI) RE	IM	ABSVAL	PHI DEG	ER(THETA,PHI) RE	IM	ABSVAL
0	-0.15178	0.11405	0.18985	90	-0.82477	-1.16122	1.42431
2	-0.15100	0.11463	0.18958	92	-0.81665	-1.15330	1.41316
4	-0.14865	0.11640	0.18880	94	-0.79261	-1.12976	1.38007
6	-0.14471	0.11933	0.18756	96	-0.75350	-1.09132	1.32617
8	-0.13912	0.12341	0.18597	98	-0.70072	-1.03908	1.25327
10	-0.13181	0.12863	0.18417	100	-0.63615	-0.97454	1.16379
12	-0.12271	0.13495	0.18239	102	-0.56199	-0.89953	1.06066
14	-0.11173	0.14232	0.18094	104	-0.48074	-0.81609	0.94716
16	-0.09881	0.15067	0.18018	106	-0.39502	-0.72639	0.82685
18	-0.08387	0.15991	0.18057	108	-0.30749	-0.63269	0.70345
20	-0.06688	0.16991	0.18260	110	-0.22071	-0.53721	0.58078
22	-0.04782	0.18050	0.18673	112	-0.13704	-0.44206	0.46282
24	-0.02675	0.19148	0.19334	114	-0.05860	-0.34918	0.35407
26	-0.00377	0.20257	0.20261	116	0.01287	-0.26029	0.26061
28	0.02095	0.21346	0.21449	118	0.07599	-0.17682	0.19246
30	0.04712	0.22377	0.22868	120	0.12977	-0.09993	0.16379
32	0.07438	0.23303	0.24462	122	0.17366	-0.03047	0.17632
34	0.10224	0.24074	0.26155	124	0.20746	0.03101	0.20977
36	0.13008	0.24629	0.27853	126	0.23133	0.08426	0.24620
38	0.15720	0.24905	0.29451	128	0.24574	0.12924	0.27765
40	0.18274	0.24831	0.30831	130	0.25138	0.16618	0.30134
42	0.20577	0.24333	0.31867	132	0.24915	0.19547	0.31668
44	0.22524	0.23336	0.32433	134	0.24007	0.21765	0.32404
46	0.24007	0.21765	0.32404	136	0.22524	0.23336	0.32433
48	0.24915	0.19547	0.31668	138	0.20577	0.24333	0.31867
50	0.25138	0.16618	0.30134	140	0.18274	0.24831	0.30831
52	0.24574	0.12924	0.27765	142	0.15720	0.24905	0.29451
54	0.23133	0.08426	0.24620	144	0.13008	0.24629	0.27854
56	0.20746	0.03102	0.20977	146	0.10224	0.24074	0.26155
58	0.17366	-0.03047	0.17632	148	0.07438	0.23303	0.24462
60	0.12978	-0.09993	0.16379	150	0.04712	0.22377	0.22868
62	0.07599	-0.17682	0.19246	152	0.02095	0.21346	0.21449
64	0.01288	-0.26029	0.26061	154	-0.00377	0.20257	0.20261
66	-0.05859	-0.34918	0.35406	156	-0.02675	0.19148	0.19334
68	-0.13704	-0.44206	0.46281	158	-0.04782	0.18050	0.18673
70	-0.22070	-0.53721	0.58078	160	-0.06688	0.16991	0.18260
72	-0.30749	-0.63269	0.70345	162	-0.08387	0.15991	0.18057
74	-0.39502	-0.72639	0.82685	164	-0.09881	0.15067	0.18018
76	-0.48074	-0.81608	0.94715	166	-0.11173	0.14232	0.18094
78	-0.56199	-0.89953	1.06066	168	-0.12271	0.13495	0.18239
80	-0.63614	-0.97454	1.16379	170	-0.13181	0.12863	0.18417
82	-0.70072	-1.03907	1.25327	172	-0.13911	0.12341	0.18597
84	-0.75349	-1.09131	1.32617	174	-0.14471	0.11933	0.18756
86	-0.79260	-1.12976	1.38007	176	-0.14865	0.11640	0.18880
88	-0.81665	-1.15330	1.41316	178	-0.15100	0.11463	0.18958
				180	-0.15178	0.11405	0.18985

VBR

RADIATION PATTERN FOR THETA=PI/2

H/LMDA=0.5000 B/LMDA= 0.250 OMEGA= 9.92

9 ELEMENT ARRAY

PHI DEG	ER(THETA,PHI) RE	IM	ABSVAL	PHI DEG	ER(THETA,PHI) RE	IM	ABSVAL
0	-0.16859	-0.01686	0.16943	90	-1.48632	-1.25601	1.94594
2	-0.16804	-0.01630	0.16883	92	-1.47176	-1.24487	1.92764
4	-0.16640	-0.01462	0.16704	94	-1.42868	-1.21189	1.87345
6	-0.16358	-0.01178	0.16401	96	-1.35879	-1.15828	1.78548
8	-0.15950	-0.00775	0.15968	98	-1.26486	-1.08600	1.66711
10	-0.15400	-0.00246	0.15402	100	-1.15053	-0.99762	1.52281
12	-0.14693	0.00417	0.14699	102	-1.02011	-0.89624	1.35789
14	-0.13807	0.01221	0.13861	104	-0.87836	-0.78525	1.17819
16	-0.12723	0.02174	0.12907	106	-0.73023	-0.66824	0.98984
18	-0.11417	0.03283	0.11879	108	-0.58062	-0.54877	0.79892
20	-0.09868	0.04554	0.10868	110	-0.43415	-0.43026	0.61124
22	-0.08057	0.05988	0.10039	112	-0.29497	-0.31581	0.43213
24	-0.05971	0.07585	0.09653	114	-0.16657	-0.20810	0.26655
26	-0.03600	0.09337	0.10007	116	-0.05175	-0.10934	0.12096
28	-0.00946	0.11229	0.11269	118	0.04751	-0.02119	0.05203
30	0.01977	0.13237	0.13384	120	0.13002	0.05523	0.14126
32	0.05140	0.15327	0.16166	122	0.19533	0.11933	0.22890
34	0.08499	0.17453	0.19412	124	0.24371	0.17102	0.29773
36	0.11986	0.19554	0.22935	126	0.27598	0.21066	0.34719
38	0.15515	0.21557	0.26559	128	0.29345	0.23892	0.37842
40	0.18972	0.23372	0.30103	130	0.29779	0.25679	0.39322
42	0.22224	0.24899	0.33375	132	0.29086	0.26543	0.39377
44	0.25114	0.26021	0.36163	134	0.27465	0.26612	0.38243
46	0.27465	0.26612	0.38243	136	0.25114	0.26021	0.36163
48	0.29086	0.26543	0.39376	138	0.22224	0.24899	0.33375
50	0.29779	0.25679	0.39322	140	0.18972	0.23373	0.30103
52	0.29346	0.23892	0.37842	142	0.15515	0.21557	0.26559
54	0.27598	0.21066	0.34719	144	0.11986	0.19554	0.22935
56	0.24371	0.17103	0.29773	146	0.08499	0.17453	0.19412
58	0.19533	0.11933	0.22890	148	0.05140	0.15327	0.16166
60	0.13002	0.05523	0.14126	150	0.01977	0.13237	0.13384
62	0.04752	-0.02119	0.05203	152	-0.00946	0.11229	0.11269
64	-0.05174	-0.10933	0.12096	154	-0.03600	0.09337	0.10007
66	-0.16657	-0.20809	0.26655	156	-0.05971	0.07585	0.09653
68	-0.29496	-0.31580	0.43213	158	-0.08057	0.05988	0.10039
70	-0.43415	-0.43026	0.61124	160	-0.09868	0.04554	0.10868
72	-0.58062	-0.54877	0.79892	162	-0.11417	0.03283	0.11879
74	-0.73022	-0.66824	0.98983	164	-0.12723	0.02174	0.12907
76	-0.87835	-0.78524	1.17818	166	-0.13807	0.01221	0.13861
78	-1.02010	-0.89623	1.35788	168	-0.14693	0.00417	0.14699
80	-1.15053	-0.99762	1.52281	170	-0.15400	-0.00246	0.15402
82	-1.26486	-1.08600	1.66711	172	-0.15950	-0.00775	0.15968
84	-1.35879	-1.15828	1.78548	174	-0.16358	-0.01178	0.16401
86	-1.42868	-1.21189	1.87345	176	-0.16640	-0.01462	0.16704
88	-1.47176	-1.24487	1.92764	178	-0.16804	-0.01630	0.16883
				180	-0.16859	-0.01686	0.16943

IBR

RADIATIUN PATTERN FOR THETA=PI/2

H/LMDA=0.5000 B/LMDA= 0.250 OMEGA= 9.92

9 ELEMENT ARRAY

PHI DEG	ER(THETA,PHI) RE	IM	ABSVAL	PHI DEG	ER(THETA,PHI) RE	IM	ABSVAL
0	-0.21216	-0.01675	0.21282	90	-0.81549	-1.17567	1.43082
2	-0.21199	-0.01617	0.21261	92	-0.80549	-1.16536	1.41665
4	-0.21145	-0.01444	0.21195	94	-0.77594	-1.13482	1.37473
6	-0.21048	-0.01152	0.21080	96	-0.72814	-1.08514	1.30680
8	-0.20897	-0.00739	0.20910	98	-0.66422	-1.01813	1.21564
10	-0.20676	-0.00198	0.20677	100	-0.58693	-0.93615	1.10493
12	-0.20365	0.00476	0.20371	102	-0.49954	-0.84203	0.97906
14	-0.19942	0.01288	0.19984	104	-0.40563	-0.73888	0.84290
16	-0.19381	0.02246	0.19510	106	-0.30888	-0.63001	0.70165
18	-0.18653	0.03354	0.18952	108	-0.21288	-0.51868	0.56067
20	-0.17730	0.04616	0.18321	110	-0.12099	-0.40806	0.42562
22	-0.16582	0.06031	0.17644	112	-0.03611	-0.30100	0.30316
24	-0.15181	0.07595	0.16975	114	0.03934	-0.20002	0.20385
26	-0.13506	0.09300	0.16398	116	0.10356	-0.10716	0.14902
28	-0.11537	0.11128	0.16029	118	0.15540	-0.02401	0.15724
30	-0.09266	0.13055	0.16009	120	0.19429	0.04837	0.20022
32	-0.06694	0.15047	0.16469	122	0.22030	0.10939	0.24596
34	-0.03839	0.17057	0.17484	124	0.23397	0.15893	0.28284
36	-0.00731	0.19028	0.19042	126	0.23629	0.19725	0.30779
38	0.02578	0.20889	0.21047	128	0.22856	0.22494	0.32068
40	0.06017	0.22556	0.23345	130	0.21231	0.24288	0.32259
42	0.09494	0.23932	0.25747	132	0.18919	0.25211	0.31520
44	0.12895	0.24912	0.28051	134	0.16086	0.25378	0.30047
46	0.16086	0.25378	0.30047	136	0.12895	0.24912	0.28051
48	0.18919	0.25211	0.31520	138	0.09494	0.23932	0.25747
50	0.21231	0.24288	0.32259	140	0.06017	0.22556	0.23345
52	0.22856	0.22494	0.32068	142	0.02578	0.20889	0.21047
54	0.23629	0.19725	0.30779	144	-0.00731	0.19028	0.19042
56	0.23397	0.15893	0.28284	146	-0.03839	0.17057	0.17484
58	0.22030	0.10939	0.24597	148	-0.06694	0.15047	0.16469
60	0.19429	0.04837	0.20022	150	-0.09266	0.13055	0.16009
62	0.15540	-0.02401	0.15724	152	-0.11537	0.11128	0.16029
64	0.10356	-0.10716	0.14902	154	-0.13506	0.09300	0.16398
66	0.03934	-0.20001	0.20384	156	-0.15181	0.07595	0.16975
68	-0.03611	-0.30100	0.30316	158	-0.16581	0.06031	0.17644
70	-0.12099	-0.40806	0.42561	160	-0.17730	0.04616	0.18321
72	-0.21288	-0.51868	0.56067	162	-0.18653	0.03354	0.18952
74	-0.30887	-0.63000	0.70165	164	-0.19381	0.02246	0.19510
76	-0.40563	-0.73888	0.84290	166	-0.19942	0.01288	0.19984
78	-0.49954	-0.84203	0.97905	168	-0.20365	0.00476	0.20371
80	-0.58693	-0.93615	1.10493	170	-0.20676	-0.00198	0.20677
82	-0.66422	-1.01813	1.21564	172	-0.20897	-0.00739	0.20910
84	-0.72814	-1.08514	1.30680	174	-0.21048	-0.01152	0.21080
86	-0.77593	-1.13481	1.37473	176	-0.21145	-0.01444	0.21195
88	-0.80549	-1.16536	1.41665	178	-0.21199	-0.01617	0.21261
				180	-0.21216	-0.01675	0.21282

VBR

RADIATION PATTERN FOR THETA=PI/2

H/LMDA=0.5000 B/LMDA= 0.250 OMEGA= 9.92

10 ELEMENT ARRAY

PHI DEG	ER(THETA,PHI) RE	IM	ABSVAL	PHI DEG	ER(THETA,PHI) RE	IM	ABSVAL
0	-0.17387	-0.09892	0.20003	90	-1.48791	-1.26355	1.95203
2	-0.17403	-0.09873	0.20009	92	-1.46996	-1.24954	1.92928
4	-0.17451	-0.09817	0.20022	94	-1.41701	-1.20815	1.86213
6	-0.17520	-0.09716	0.20034	96	-1.33159	-1.14123	1.75372
8	-0.17600	-0.09563	0.20030	98	-1.21777	-1.05174	1.60907
10	-0.17669	-0.09346	0.19989	100	-1.08082	-0.94355	1.43474
12	-0.17705	-0.09050	0.19884	102	-0.92695	-0.82122	1.23840
14	-0.17677	-0.08655	0.19682	104	-0.76282	-0.68967	1.02837
16	-0.17550	-0.08140	0.19346	106	-0.59519	-0.55397	0.81310
18	-0.17284	-0.07483	0.18834	108	-0.43054	-0.41899	0.60077
20	-0.16834	-0.06658	0.18103	110	-0.27472	-0.28922	0.39890
22	-0.16152	-0.05640	0.17108	112	-0.13264	-0.16852	0.21446
24	-0.15190	-0.04404	0.15815	114	-0.00814	-0.05999	0.06054
26	-0.13899	-0.02928	0.14204	116	0.09614	0.03409	0.10201
28	-0.12235	-0.01195	0.12293	118	0.17879	0.11233	0.21115
30	-0.10161	0.00803	0.10193	120	0.23953	0.17415	0.29615
32	-0.07652	0.03067	0.08244	122	0.27910	0.21974	0.35522
34	-0.04699	0.05581	0.07296	124	0.29910	0.24992	0.38977
36	-0.01317	0.08316	0.08419	126	0.30174	0.26604	0.40228
38	0.02455	0.11219	0.11484	128	0.28970	0.26981	0.39588
40	0.06544	0.14217	0.15651	130	0.26585	0.26316	0.37407
42	0.10841	0.17212	0.20341	132	0.23311	0.24812	0.34045
44	0.15199	0.20079	0.25183	134	0.19431	0.22670	0.29857
46	0.19430	0.22669	0.29857	136	0.15199	0.20079	0.25183
48	0.23311	0.24812	0.34045	138	0.10841	0.17212	0.20342
50	0.26585	0.26316	0.37407	140	0.06544	0.14217	0.15651
52	0.28970	0.26981	0.39588	142	0.02455	0.11219	0.11485
54	0.30174	0.26604	0.40228	144	-0.01317	0.08316	0.08419
56	0.29910	0.24992	0.38977	146	-0.04699	0.05581	0.07296
58	0.27910	0.21974	0.35522	148	-0.07652	0.03067	0.08244
60	0.23953	0.17415	0.29615	150	-0.10161	0.00803	0.10193
62	0.17879	0.11233	0.21115	152	-0.12235	-0.01195	0.12293
64	0.09614	0.03410	0.10201	154	-0.13899	-0.02928	0.14204
66	-0.00814	-0.05999	0.06054	156	-0.15190	-0.04404	0.15815
68	-0.13264	-0.16852	0.21446	158	-0.16152	-0.05640	0.17108
70	-0.27471	-0.28922	0.39890	160	-0.16834	-0.06658	0.18103
72	-0.43054	-0.41899	0.60076	162	-0.17284	-0.07483	0.18834
74	-0.59518	-0.55397	0.81309	164	-0.17550	-0.08140	0.19346
76	-0.76281	-0.68967	1.02836	166	-0.17677	-0.08655	0.19682
78	-0.92695	-0.82122	1.23840	168	-0.17705	-0.09050	0.19884
80	-1.08082	-0.94355	1.43473	170	-0.17669	-0.09346	0.19989
82	-1.21776	-1.05173	1.60907	172	-0.17600	-0.09563	0.20030
84	-1.33159	-1.14122	1.75372	174	-0.17520	-0.09716	0.20034
86	-1.41701	-1.20814	1.86213	176	-0.17451	-0.09817	0.20022
88	-1.46996	-1.24954	1.92928	178	-0.17403	-0.09873	0.20009
				180	-0.17387	-0.09892	0.20003

IBR

RADIATION PATTERN FOR THETA=PI/2

H/LMDA=0.5000 B/LMDA= 0.250 OMEGA= 9.92

10 ELEMENT ARRAY

PHI DEG	ER(THETA,PHI) RE	IM	ABSVAL	PHI DEG	ER(THETA,PHI) RE	IM	ABSVAL
0	-0.14761	-0.11327	0.18607	90	-0.80771	-1.18029	1.43020
2	-0.14815	-0.11303	0.18634	92	-0.79565	-1.16732	1.41269
4	-0.14975	-0.11227	0.18716	94	-0.76015	-1.12901	1.36106
6	-0.15233	-0.11096	0.18846	96	-0.70310	-1.06704	1.27786
8	-0.15580	-0.10901	0.19014	98	-0.62755	-0.98412	1.16718
10	-0.15997	-0.10630	0.19207	100	-0.53745	-0.88382	1.03440
12	-0.16465	-0.10271	0.19406	102	-0.43736	-0.77028	0.88579
14	-0.16955	-0.09805	0.19586	104	-0.33220	-0.64806	0.72824
16	-0.17437	-0.09213	0.19721	106	-0.22685	-0.52181	0.56898
18	-0.17871	-0.08475	0.19779	108	-0.12591	-0.39602	0.41555
20	-0.18214	-0.07567	0.19723	110	-0.03344	-0.27485	0.27688
22	-0.18416	-0.06469	0.19519	112	0.04728	-0.16189	0.16866
24	-0.18427	-0.05157	0.19135	114	0.11385	-0.06005	0.12871
26	-0.18190	-0.03616	0.18546	116	0.16480	0.02852	0.16725
28	-0.17650	-0.01831	0.17744	118	0.19963	0.10247	0.22439
30	-0.16754	0.00202	0.16755	120	0.21868	0.16120	0.27167
32	-0.15455	0.02480	0.15653	122	0.22305	0.20481	0.30282
34	-0.13718	0.04984	0.14595	124	0.21440	0.23402	0.31738
36	-0.11521	0.07681	0.13847	126	0.19485	0.24999	0.31695
38	-0.08863	0.10519	0.13755	128	0.16672	0.25426	0.30405
40	-0.05767	0.13425	0.14611	130	0.13242	0.24860	0.28166
42	-0.02286	0.16304	0.16464	132	0.09430	0.23484	0.25306
44	0.01493	0.19038	0.19096	134	0.05451	0.21485	0.22165
46	0.05451	0.21485	0.22165	136	0.01493	0.19038	0.19096
48	0.09430	0.23484	0.25306	138	-0.02286	0.16304	0.16464
50	0.13242	0.24860	0.28166	140	-0.05767	0.13425	0.14611
52	0.16672	0.25427	0.30405	142	-0.08863	0.10519	0.13755
54	0.19485	0.24999	0.31695	144	-0.11521	0.07681	0.13847
56	0.21440	0.23402	0.31738	146	-0.13718	0.04984	0.14595
58	0.22305	0.20481	0.30282	148	-0.15455	0.02480	0.15653
60	0.21868	0.16120	0.27167	150	-0.16754	0.00202	0.16755
62	0.19963	0.10247	0.22439	152	-0.17650	-0.01831	0.17744
64	0.16480	0.02852	0.16725	154	-0.18190	-0.03616	0.18546
66	0.11385	-0.06005	0.12872	156	-0.18427	-0.05157	0.19135
68	0.04728	-0.16189	0.16866	158	-0.18416	-0.06468	0.19519
70	-0.03343	-0.27485	0.27688	160	-0.18214	-0.07567	0.19723
72	-0.12591	-0.39602	0.41555	162	-0.17871	-0.08475	0.19779
74	-0.22684	-0.52180	0.56898	164	-0.17437	-0.09213	0.19721
76	-0.33219	-0.64806	0.72824	166	-0.16955	-0.09805	0.19586
78	-0.43736	-0.77028	0.88579	168	-0.16465	-0.10271	0.19405
80	-0.53744	-0.88381	1.03439	170	-0.15997	-0.10630	0.19207
82	-0.62755	-0.98412	1.16718	172	-0.15580	-0.10901	0.19014
84	-0.70310	-1.06703	1.27785	174	-0.15233	-0.11096	0.18846
86	-0.76014	-1.12900	1.36105	176	-0.14975	-0.11227	0.18716
88	-0.79565	-1.16732	1.41269	178	-0.14815	-0.11303	0.18634
				180	-0.14761	-0.11327	0.18606

VBR

RADIATION PATTERN FOR THETA=PI/2

H/LMDA=0.5000 B/LMDA= 0.250 OMEGA= 9.92

12 ELEMENT ARRAY

PHI DEG	ER(THETA,PHI) RE	IM	ABSVAL	PHI DEG	ER(THETA,PHI) RE	IM	ABSVAL
0	0.03532	-0.06683	0.07559	90	-1.49550	-1.27837	1.96742
2	0.03439	-0.06743	0.07570	92	-1.46943	-1.25751	1.93405
4	0.03162	-0.06921	0.07609	94	-1.39298	-1.19627	1.83615
6	0.02697	-0.07216	0.07704	96	-1.27131	-1.09850	1.68015
8	0.02042	-0.07626	0.07894	98	-1.11245	-0.97028	1.47614
10	0.01196	-0.08144	0.08231	100	-0.92667	-0.81937	1.23697
12	0.00157	-0.08764	0.08765	102	-0.72548	-0.65457	0.97714
14	-0.01071	-0.09473	0.09533	104	-0.52076	-0.48501	0.71163
16	-0.02482	-0.10255	0.10551	106	-0.32372	-0.31947	0.45482
18	-0.04058	-0.11088	0.11807	108	-0.14419	-0.16577	0.21971
20	-0.05775	-0.11940	0.13263	110	0.01009	-0.03029	0.03193
22	-0.07594	-0.12772	0.14859	112	0.13382	0.08235	0.15713
24	-0.09462	-0.13536	0.16515	114	0.22431	0.16944	0.28111
26	-0.11308	-0.14174	0.18132	116	0.28135	0.23012	0.36347
28	-0.13044	-0.14616	0.19590	118	0.30691	0.26524	0.40564
30	-0.14565	-0.14786	0.20755	120	0.30473	0.27703	0.41183
32	-0.15750	-0.14600	0.21476	122	0.27976	0.26877	0.38795
34	-0.16466	-0.13976	0.21598	124	0.23764	0.24441	0.34089
36	-0.16577	-0.12832	0.20963	126	0.18416	0.20814	0.27791
38	-0.15951	-0.11099	0.19433	128	0.12481	0.16415	0.20621
40	-0.14476	-0.08730	0.16904	130	0.06449	0.11627	0.13296
42	-0.12069	-0.05706	0.13350	132	0.00721	0.06784	0.06822
44	-0.08698	-0.02052	0.08937	134	-0.04397	0.02157	0.04897
46	-0.04397	0.02157	0.04898	136	-0.08698	-0.02052	0.08937
48	0.00721	0.06784	0.06822	138	-0.12069	-0.05706	0.13350
50	0.06449	0.11627	0.13295	140	-0.14476	-0.08730	0.16904
52	0.12481	0.16415	0.20621	142	-0.15951	-0.11099	0.19433
54	0.18415	0.20814	0.27791	144	-0.16577	-0.12832	0.20963
56	0.23764	0.24441	0.34089	146	-0.16466	-0.13976	0.21598
58	0.27976	0.26877	0.38795	148	-0.15750	-0.14600	0.21476
60	0.30472	0.27703	0.41183	150	-0.14565	-0.14785	0.20755
62	0.30691	0.26524	0.40564	152	-0.13044	-0.14616	0.19590
64	0.28135	0.23012	0.36347	154	-0.11308	-0.14174	0.18132
66	0.22432	0.16944	0.28112	156	-0.09462	-0.13536	0.16515
68	0.13383	0.08235	0.15713	158	-0.07594	-0.12772	0.14859
70	0.01009	-0.03029	0.03193	160	-0.05775	-0.11940	0.13263
72	-0.14419	-0.16577	0.21970	162	-0.04058	-0.11088	0.11807
74	-0.32371	-0.31946	0.45481	164	-0.02482	-0.10255	0.10551
76	-0.52075	-0.48500	0.71162	166	-0.01071	-0.09473	0.09533
78	-0.72548	-0.65457	0.97713	168	0.00157	-0.08764	0.08765
80	-0.92666	-0.81937	1.23696	170	0.01196	-0.08144	0.08231
82	-1.11245	-0.97028	1.47614	172	0.02042	-0.07626	0.07894
84	-1.27130	-1.09849	1.68015	174	0.02697	-0.07216	0.07704
86	-1.39298	-1.19626	1.83614	176	0.03162	-0.06921	0.07609
88	-1.46943	-1.25751	1.93405	178	0.03439	-0.06743	0.07569
				180	0.03532	-0.06683	0.07559

IBR

RADIATION PATTERN FOR THETA=PI/2

H/LMDA=0.5000 B/LMDA= 0.250 OMEGA= 9.92

12 ELEMENT ARRAY

PHI DEG	ER(THETA,PHI) RE	IM	ABSVAL	PHI DEG	ER(THETA,PHI) RE	IM	ABSVAL
C	0.10158	-0.07895	0.12865	90	-0.78505	-1.19309	1.42821
2	0.10081	-0.07952	0.12840	92	-0.76846	-1.17382	1.40299
4	0.09849	-0.08124	0.12767	94	-0.71997	-1.11721	1.32911
6	0.09457	-0.08407	0.12654	96	-0.64324	-1.02681	1.21165
8	0.08898	-0.08799	0.12513	98	-0.54396	-0.90816	1.05861
10	0.08161	-0.09292	0.12367	100	-0.42934	-0.76838	0.88019
12	0.07235	-0.09878	0.12244	102	-0.30739	-0.61554	0.68802
14	0.06112	-0.10543	0.12187	104	-0.18624	-0.45802	0.49444
16	0.04784	-0.11269	0.12243	106	-0.07340	-0.30392	0.31266
18	0.03248	-0.12033	0.12464	108	0.02479	-0.16049	0.16239
20	0.01510	-0.12802	0.12891	110	0.10362	-0.03364	0.10894
22	-0.00416	-0.13538	0.13545	112	0.16023	0.07226	0.17577
24	-0.02500	-0.14193	0.14411	114	0.19369	0.15463	0.24784
26	-0.04699	-0.14709	0.15441	116	0.20486	0.21255	0.29521
28	-0.06952	-0.15022	0.16553	118	0.19609	0.24669	0.31513
30	-0.09178	-0.15060	0.17636	120	0.17088	0.25900	0.31029
32	-0.11278	-0.14746	0.18564	122	0.13342	0.25243	0.28552
34	-0.13136	-0.14004	0.19201	124	0.08818	0.23053	0.24681
36	-0.14622	-0.12763	0.19409	126	0.03952	0.19716	0.20108
38	-0.15603	-0.10964	0.19070	128	-0.00863	0.15616	0.15640
40	-0.15945	-0.08568	0.18102	130	-0.05300	0.11111	0.12310
42	-0.15532	-0.05567	0.16500	132	-0.09111	0.06514	0.11200
44	-0.14276	-0.01991	0.14414	134	-0.12131	0.02081	0.12308
46	-0.12131	0.02081	0.12308	136	-0.14276	-0.01991	0.14414
48	-0.09111	0.06514	0.11200	138	-0.15532	-0.05567	0.16500
50	-0.05300	0.11111	0.12310	140	-0.15945	-0.08568	0.18101
52	-0.00863	0.15616	0.15640	142	-0.15603	-0.10964	0.19070
54	0.03952	0.19716	0.20108	144	-0.14622	-0.12763	0.19409
56	0.08818	0.23053	0.24681	146	-0.13136	-0.14004	0.19201
58	0.13342	0.25243	0.28551	148	-0.11278	-0.14746	0.18564
60	0.17088	0.25900	0.31029	150	-0.09178	-0.15060	0.17636
62	0.19609	0.24669	0.31513	152	-0.06952	-0.15022	0.16553
64	0.20486	0.21255	0.29521	154	-0.04699	-0.14709	0.15441
66	0.19370	0.15463	0.24785	156	-0.02500	-0.14193	0.14411
68	0.16023	0.07226	0.17577	158	-0.00416	-0.13538	0.13545
70	0.10362	-0.03364	0.10894	160	0.01510	-0.12802	0.12891
72	0.02479	-0.16048	0.16239	162	0.03248	-0.12033	0.12464
74	-0.C7340	-0.30392	0.31265	164	0.04784	-0.11269	0.12243
76	-0.18624	-0.45801	0.49443	166	0.06112	-0.10543	0.12186
78	-0.30739	-0.61553	0.68802	168	0.07235	-0.09878	0.12244
80	-0.42933	-0.76838	0.88019	170	0.08161	-0.09292	0.12367
82	-0.54395	-0.90816	1.05860	172	0.08898	-0.08799	0.12513
84	-0.64323	-1.02680	1.21164	174	0.09457	-0.08407	0.12654
86	-0.71997	-1.11721	1.32910	176	0.09849	-0.08124	0.12767
88	-0.76846	-1.17382	1.40299	178	0.10081	-0.07952	0.12840
				180	0.10158	-0.07895	0.12865

VBR

[illegible] FOR THETA=PI/2

H/LMDA=0.5000 B/LMDA= 0.250 OMEGA= 9.92

14 ELEMENT ARRAY

PHI DEG	ER(THETA,PHI) RE	IM	ABSVAL	PHI DEG	ER(THETA,PHI) RE	IM	ABSVAL
0	0.12235	0.06856	0.14025	90	-1.49920	-1.28550	1.97487
2	0.12251	0.06833	0.14028	92	-1.46365	-1.25667	1.92912
4	0.12296	0.06762	0.14033	94	-1.36023	-1.17263	1.79591
6	0.12360	0.06634	0.14028	96	-1.19819	-1.04050	1.58692
8	0.12423	0.06440	0.13993	98	-0.99172	-0.87125	1.32007
10	0.12459	0.06162	0.13899	100	-0.75836	-0.67847	1.01756
12	0.12434	0.05781	0.13712	102	-0.51696	-0.47696	0.70337
14	0.12308	0.05274	0.13390	104	-0.28573	-0.28117	0.40087
16	0.12033	0.04615	0.12888	106	-0.08041	-0.10387	0.13136
18	0.11557	0.03779	0.12159	108	0.08708	0.04493	0.09798
20	0.10825	0.02741	0.11166	110	0.20947	0.15866	0.26277
22	0.09783	0.01482	0.09894	112	0.28436	0.23444	0.36854
24	0.08383	-0.00008	0.08383	114	0.31384	0.27287	0.41588
26	0.06592	-0.01729	0.06815	116	0.30371	0.27753	0.41142
28	0.04393	-0.03660	0.05718	118	0.26235	0.25422	0.36532
30	0.01800	-0.05758	0.06033	120	0.19951	0.21006	0.28971
32	-0.01135	-0.07955	0.08036	122	0.12516	0.15264	0.19739
34	-0.04318	-0.10149	0.11029	124	0.04847	0.08924	0.10156
36	-0.07602	-0.12208	0.14381	126	-0.02291	0.02625	0.03484
38	-0.10790	-0.13967	0.17649	128	-0.08329	-0.03124	0.08895
40	-0.13635	-0.15241	0.20450	130	-0.12908	-0.07962	0.15166
42	-0.15856	-0.15829	0.22404	132	-0.15873	-0.11676	0.19705
44	-0.17152	-0.15535	0.23142	134	-0.17238	-0.14189	0.22327
46	-0.17238	-0.14189	0.22327	136	-0.17152	-0.15535	0.23142
48	-0.15873	-0.11676	0.19705	138	-0.15856	-0.15829	0.22404
50	-0.12909	-0.07962	0.15166	140	-0.13635	-0.15241	0.20450
52	-0.08329	-0.03124	0.08895	142	-0.10789	-0.13967	0.17649
54	-0.02291	0.02625	0.03484	144	-0.07602	-0.12208	0.14381
56	0.04847	0.08924	0.10155	146	-0.04318	-0.10149	0.11029
58	0.12516	0.15264	0.19739	148	-0.01135	-0.07955	0.08036
60	0.19951	0.21006	0.28971	150	0.01800	-0.05758	0.06033
62	0.26235	0.25422	0.36532	152	0.04393	-0.03660	0.05717
64	0.30371	0.27753	0.41142	154	0.06592	-0.01729	0.06815
66	0.31384	0.27287	0.41588	156	0.08383	-0.00008	0.08383
68	0.28436	0.23444	0.36854	158	0.09783	0.01482	0.09894
70	0.20947	0.15866	0.26277	160	0.10825	0.02741	0.11166
72	0.08708	0.04493	0.09799	162	0.11557	0.03779	0.12159
74	-0.08041	-0.10386	0.13135	164	0.12033	0.04615	0.12888
76	-0.28572	-0.28116	0.40085	166	0.12308	0.05274	0.13390
78	-0.51695	-0.47695	0.70337	168	0.12434	0.05781	0.13712
80	-0.75835	-0.67847	1.01756	170	0.12459	0.06162	0.13899
82	-0.99172	-0.87124	1.32006	172	0.12423	0.06440	0.13993
84	-1.19818	-1.04050	1.58690	174	0.12360	0.06634	0.14028
86	-1.36023	-1.17263	1.79590	176	0.12296	0.06762	0.14033
88	-1.46365	-1.25667	1.92911	178	0.12251	0.06833	0.14028
				180	0.12235	0.06856	0.14025

IBR

RADIATION PATTERN FOR THETA=PI/2

H/LMDA=0.5000 B/LMDA= [illegible] OMEGA= [illegible]

14 ELEMENT ARRAY

PHI DEG	ER(THETA,PHI) RE	IM	ABSVAL	PHI DEG	ER(THETA,PHI) RE	IM	ABSVAL
0	0.10351	0.08071	0.13126	90	-0.77783	-1.20229	1.43197
2	0.10406	0.08041	0.13151	92	-0.75592	-1.17552	1.39759
4	0.10568	0.07950	0.13224	94	-0.69240	-1.09746	1.29762
6	0.10827	0.07790	0.13338	96	-0.59364	-0.97467	1.14122
8	0.11168	0.07551	0.13481	98	-0.46933	-0.81727	0.94245
10	0.11567	0.07216	0.13633	100	-0.33133	-0.63783	0.71875
12	0.11993	0.06769	0.13771	102	-0.19216	-0.45002	0.48933
14	0.12408	0.06186	0.13864	104	-0.06364	-0.26724	0.27471
16	0.12764	0.05445	0.13877	106	0.04443	-0.10138	0.11069
18	0.13007	0.04523	0.13771	108	0.12519	0.03822	0.13089
20	0.13074	0.03399	0.13509	110	0.17514	0.14536	0.22760
22	0.12901	0.02058	0.13064	112	0.19426	0.21722	0.29141
24	0.12420	0.00494	0.12430	114	0.18558	0.25424	0.31477
26	0.11569	-0.01288	0.11640	116	0.15447	0.25962	0.30210
28	0.10296	-0.03263	0.10801	118	0.10777	0.23864	0.26185
30	0.08568	-0.05385	0.10120	120	0.05287	0.19785	0.20479
32	0.06380	-0.07583	0.09910	122	-0.00319	0.14429	0.14432
34	0.03762	-0.09756	0.10456	124	-0.05435	0.08476	0.10069
36	0.00790	-0.11774	0.11800	126	-0.09602	0.02527	0.09929
38	-0.02411	-0.13479	0.13693	128	-0.12530	-0.02933	0.12869
40	-0.05666	-0.14693	0.15748	130	-0.14094	-0.07556	0.15992
42	-0.08749	-0.15230	0.17564	132	-0.14319	-0.11132	0.18138
44	-0.11402	-0.14909	0.18769	134	-0.13348	-0.13576	0.19039
46	-0.13348	-0.13576	0.19039	136	-0.11402	-0.14909	0.18769
48	-0.14319	-0.11132	0.18138	138	-0.08749	-0.15230	0.17564
50	-0.14094	-0.07556	0.15992	140	-0.05666	-0.14693	0.15748
52	-0.12530	-0.02933	0.12869	142	-0.02411	-0.13479	0.13693
54	-0.09602	0.02527	0.09929	144	0.00790	-0.11774	0.11800
56	-0.05435	0.08476	0.10069	146	0.03762	-0.09756	0.10456
58	-0.00319	0.14429	0.14432	148	0.06380	-0.07583	0.09910
60	0.05286	0.19785	0.20479	150	0.08568	-0.05385	0.10120
62	0.10777	0.23864	0.26184	152	0.10296	-0.03263	0.10801
64	0.15447	0.25962	0.30210	154	0.11569	-0.01288	0.11640
66	0.18558	0.25424	0.31477	156	0.12420	0.00494	0.12430
68	0.19426	0.21722	0.29141	158	0.12901	0.02058	0.13064
70	0.17514	0.14536	0.22760	160	0.13074	0.03399	0.13509
72	0.12519	0.03823	0.13089	162	0.13007	0.04523	0.13771
74	0.04444	-0.10137	0.11068	164	0.12764	0.05445	0.13877
76	-0.06364	-0.26723	0.27471	166	0.12408	0.06186	0.13864
78	-0.19216	-0.45001	0.48933	168	0.11993	0.06769	0.13771
80	-0.33133	-0.63783	0.71875	170	0.11567	0.07216	0.13633
82	-0.46933	-0.81727	0.94244	172	0.11168	0.07551	0.13481
84	-0.59364	-0.97466	1.14121	174	0.10827	0.07790	0.13338
86	-0.69240	-1.09745	1.29762	176	0.10568	0.07950	0.13224
88	-0.75592	-1.17552	1.39759	178	0.10406	0.08041	0.13151
				180	0.10351	0.08071	0.13126

VBR

RADIATION PATTERN FOR THETA=PI/2

H/LMDA=0.5000 B/LMDA= 0.250 OMEGA= 9.92

16 ELEMENT ARRAY

PHI	ER(THETA,PHI)			PHI	ER(THETA,PHI)		
DEG	RE	IM	ABSVAL	DEG	RE	IM	ABSVAL
0	-0.02789	0.05152	0.05859	90	-1.50326	-1.29326	1.98301
2	-0.02699	0.05209	0.05867	92	-1.45665	-1.25499	1.92271
4	-0.02426	0.05379	0.05901	94	-1.32224	-1.14443	1.74872
6	-0.01969	0.05660	0.05992	96	-1.11552	-0.97369	1.48069
8	-0.01325	0.06045	0.06189	98	-0.85969	-0.76106	1.14817
10	-0.00493	0.06526	0.06545	100	-0.58236	-0.52845	0.78639
12	0.00524	0.07089	0.07108	102	-0.31165	-0.29840	0.43147
14	0.01718	0.07713	0.07902	104	-0.07247	-0.09134	0.11660
16	0.03069	0.08369	0.08914	106	0.11637	0.07684	0.13945
18	0.04544	0.09017	0.10098	108	0.24401	0.19625	0.31314
20	0.06094	0.09608	0.11378	110	0.30805	0.26350	0.40538
22	0.07647	0.10078	0.12651	112	0.31374	0.28137	0.42143
24	0.09109	0.10353	0.13789	114	0.27229	0.25756	0.37481
26	0.10363	0.10349	0.14646	116	0.19851	0.20305	0.28397
28	0.11275	0.09978	0.15056	118	0.10840	0.13027	0.16947
30	0.11697	0.09154	0.14853	120	0.01687	0.05137	0.05407
32	0.11484	0.07804	0.13885	122	-0.06389	-0.02318	0.06796
34	0.10507	0.05883	0.12042	124	-0.12542	-0.08546	0.15177
36	0.08677	0.03387	0.09314	126	-0.16333	-0.13054	0.20909
38	0.05971	0.00367	0.05983	128	-0.17696	-0.15639	0.23616
40	0.02457	-0.03051	0.03918	130	-0.16868	-0.16351	0.23493
42	-0.01687	-0.06662	0.06872	132	-0.14297	-0.15430	0.21035
44	-0.06159	-0.10176	0.11895	134	-0.10537	-0.13235	0.16917
46	-0.10536	-0.13235	0.16917	136	-0.06159	-0.10176	0.11895
48	-0.14297	-0.15430	0.21035	138	-0.01687	-0.06662	0.06872
50	-0.16868	-0.16351	0.23493	140	0.02457	-0.03051	0.03918
52	-0.17696	-0.15639	0.23616	142	0.05971	0.00367	0.05983
54	-0.16333	-0.13054	0.20909	144	0.08677	0.03386	0.09314
56	-0.12543	-0.08547	0.15178	146	0.10507	0.05883	0.12042
58	-0.06389	-0.02318	0.06797	148	0.11484	0.07804	0.13885
60	0.01687	0.05137	0.05406	150	0.11697	0.09154	0.14853
62	0.10840	0.13027	0.16947	152	0.11275	0.09978	0.15056
64	0.19851	0.20305	0.28397	154	0.10363	0.10349	0.14646
66	0.27229	0.25756	0.37480	156	0.09109	0.10353	0.13790
68	0.31374	0.28137	0.42143	158	0.07647	0.10078	0.12651
70	0.30805	0.26350	0.40538	160	0.06094	0.09608	0.11378
72	0.24402	0.19625	0.31314	162	0.04544	0.09017	0.10098
74	0.11637	0.07685	0.13946	164	0.03069	0.08369	0.08914
76	-0.07246	-0.09133	0.11658	166	0.01718	0.07713	0.07902
78	-0.31164	-0.29840	0.43147	168	0.00524	0.07089	0.07108
80	-0.58236	-0.52844	0.78638	170	-0.00493	0.06526	0.06545
82	-0.85968	-0.76106	1.14816	172	-0.01325	0.06045	0.06189
84	-1.11551	-0.97369	1.48068	174	-0.01969	0.05660	0.05992
86	-1.32223	-1.14442	1.74871	176	-0.02426	0.05379	0.05901
88	-1.45664	-1.25499	1.92271	178	-0.02699	0.05209	0.05867
				180	-0.02789	0.05152	0.05859

IBR

RADIATION PATTERN FOR THETA=PI/2

H/LMDA=0.5000 B/LMDA= 0.250 OMEGA= 9.92

16 ELEMENT ARRAY

PHI	ER(THETA,PHI)			PHI	ER(THETA,PHI)		
DEG	RE	IM	ABSVAL	DEG	RE	IM	ABSVAL
0	-0.07631	0.06050	0.09739	90	-0.76506	-1.20910	1.43082
2	-0.07555	0.06107	0.09715	92	-0.73730	-1.17358	1.38597
4	-0.07326	0.06276	0.09646	94	-0.65762	-1.07093	1.25673
6	-0.06936	0.06552	0.09541	96	-0.53625	-0.91233	1.05826
8	-0.06375	0.06931	0.09417	98	-0.38841	-0.71463	0.81336
10	-0.05631	0.07400	0.09299	100	-0.23193	-0.49807	0.54942
12	-0.04691	0.07944	0.09225	102	-0.08453	-0.28353	0.29586
14	-0.03546	0.08539	0.09247	104	0.03864	-0.08994	0.09789
16	-0.02194	0.09156	0.09415	106	0.12702	0.06784	0.14400
18	-0.00640	0.09753	0.09774	108	0.17564	0.18050	0.25186
20	0.01093	0.10278	0.10336	110	0.18528	0.24472	0.30695
22	0.02963	0.10670	0.11074	112	0.16166	0.26285	0.30859
24	0.04906	0.10856	0.11913	114	0.11403	0.24188	0.26741
26	0.06831	0.10756	0.12742	116	0.05340	0.19187	0.19916
28	0.08622	0.10287	0.13422	118	-0.00926	0.12430	0.12465
30	0.10140	0.09372	0.13807	120	-0.06449	0.05049	0.08190
32	0.11229	0.07946	0.13756	122	-0.10540	-0.01972	0.10723
34	0.11733	0.05975	0.13167	124	-0.12816	-0.07883	0.15047
36	0.11510	0.03462	0.12019	126	-0.13198	-0.12204	0.17976
38	0.10451	0.00467	0.10462	128	-0.11862	-0.14728	0.18911
40	0.08514	-0.02880	0.08987	130	-0.09173	-0.15482	0.17995
42	0.05738	-0.06376	0.08577	132	-0.05598	-0.14677	0.15708
44	0.02269	-0.09743	0.10003	134	-0.01633	-0.12637	0.12742
46	-0.01633	-0.12637	0.12742	136	0.02269	-0.09743	0.10003
48	-0.05598	-0.14676	0.15708	138	0.05738	-0.06376	0.08577
50	-0.09172	-0.15482	0.17995	140	0.08514	-0.02880	0.08987
52	-0.11862	-0.14728	0.18911	142	0.10451	0.00467	0.10462
54	-0.13198	-0.12204	0.17976	144	0.11509	0.03462	0.12019
56	-0.12817	-0.07883	0.15047	146	0.11733	0.05975	0.13167
58	-0.10540	-0.01972	0.10723	148	0.11229	0.07946	0.13756
60	-0.06449	0.05049	0.08190	150	0.10139	0.09372	0.13807
62	-0.00926	0.12430	0.12465	152	0.08622	0.10287	0.13422
64	0.05340	0.19187	0.19916	154	0.06831	0.10756	0.12742
66	0.11403	0.24188	0.26741	156	0.04906	0.10856	0.11913
68	0.16166	0.26285	0.30859	158	0.02963	0.10670	0.11074
70	0.18528	0.24472	0.30695	160	0.01093	0.10278	0.10336
72	0.17565	0.18050	0.25186	162	-0.00640	0.09753	0.09774
74	0.12702	0.06785	0.14401	164	-0.02194	0.09156	0.09415
76	0.03865	-0.08994	0.09789	166	-0.03547	0.08539	0.09246
78	-0.08453	-0.28352	0.29586	168	-0.04691	0.07944	0.09225
80	-0.23192	-0.49806	0.54942	170	-0.05631	0.07400	0.09299
82	-0.38841	-0.71463	0.81336	172	-0.06375	0.06931	0.09417
84	-0.53625	-0.91232	1.05825	174	-0.06936	0.06552	0.09541
86	-0.65762	-1.07093	1.25672	176	-0.07326	0.06276	0.09646
88	-0.73730	-1.17358	1.38596	178	-0.07555	0.06107	0.09715
				180	-0.07631	0.06050	0.09739

VBR

RADIATION PATTERN FOR THETA=PI/2

H/LMDA=0.5000 B/LMDA= 0.250 OMEGA= 9.92

18 ELEMENT ARRAY

PHI	ER(THETA,PHI)			PHI	ER(THETA,PHI)		
DEG	RE	IM	ABSVAL	DEG	RE	IM	ABSVAL
0	-0.09415	-0.05227	0.10769	90	-1.50542	-1.29754	1.98744
2	-0.09431	-0.05201	0.10770	92	-1.44643	-1.24872	1.91088
4	-0.09476	-0.05120	0.10771	94	-1.27803	-1.10908	1.69217
6	-0.09536	-0.04975	0.10755	96	-1.02442	-0.89785	1.36219
8	-0.09585	-0.04752	0.10699	98	-0.72096	-0.64326	0.96622
10	-0.09593	-0.04431	0.10566	100	-0.40794	-0.37781	0.55602
12	-0.09514	-0.03989	0.10317	102	-0.12365	-0.13281	0.18146
14	-0.09299	-0.03400	0.09901	104	0.10158	0.06623	0.12127
16	-0.08891	-0.02638	0.09274	106	0.24951	0.20306	0.32169
18	-0.08227	-0.01679	0.08397	108	0.31551	0.27209	0.41663
20	-0.07251	-0.00507	0.07269	110	0.30757	0.27792	0.41453
22	-0.05915	0.00878	0.05980	112	0.24324	0.23318	0.33696
24	-0.04192	0.02460	0.04860	114	0.14532	0.15541	0.21277
26	-0.02084	0.04193	0.04683	116	0.03734	0.06355	0.07371
28	0.00358	0.06000	0.06011	118	-0.06031	-0.02511	0.06533
30	0.03032	0.07765	0.08336	120	-0.13302	-0.09727	0.16479
32	0.05770	0.09335	0.10974	122	-0.17316	-0.14478	0.22571
34	0.08343	0.10521	0.13427	124	-0.17982	-0.16478	0.24390
36	0.10464	0.11119	0.15268	126	-0.15755	-0.15902	0.22385
38	0.11820	0.10923	0.16094	128	-0.11446	-0.13254	0.17512
40	0.12099	0.09763	0.15547	130	-0.06015	-0.09222	0.11010
42	0.11052	0.07542	0.13381	132	-0.00397	-0.04538	0.04555
44	0.08548	0.04277	0.09558	134	0.04637	0.00135	0.04639
46	0.04637	0.00135	0.04639	136	0.08548	0.04277	0.09558
48	-0.00396	-0.04538	0.04555	138	0.11052	0.07542	0.13381
50	-0.06015	-0.09222	0.11010	140	0.12099	0.09763	0.15547
52	-0.11446	-0.13254	0.17512	142	0.11820	0.10923	0.16094
54	-0.15755	-0.15902	0.22385	144	0.10464	0.11119	0.15268
56	-0.17982	-0.16479	0.24390	146	0.08343	0.10521	0.13427
58	-0.17316	-0.14478	0.22571	148	0.05770	0.09335	0.10974
60	-0.13302	-0.09728	0.16480	150	0.03032	0.07765	0.08336
62	-0.06031	-0.02511	0.06533	152	0.00358	0.06000	0.06011
64	0.03733	0.06355	0.07371	154	-0.02084	0.04193	0.04683
66	0.14532	0.15540	0.21276	156	-0.04192	0.02460	0.04860
68	0.24324	0.23318	0.33696	158	-0.05915	0.00878	0.05980
70	0.30757	0.27792	0.41453	160	-0.07251	-0.00507	0.07269
72	0.31551	0.27209	0.41663	162	-0.08227	-0.01679	0.08397
74	0.24951	0.20306	0.32170	164	-0.08891	-0.02638	0.09274
76	0.10159	0.06623	0.12128	166	-0.09299	-0.03400	0.09902
78	-0.12364	-0.13281	0.18146	168	-0.09514	-0.03989	0.10317
80	-0.40793	-0.37780	0.55601	170	-0.09592	-0.04431	0.10566
82	-0.72095	-0.64326	0.96621	172	-0.09585	-0.04752	0.10699
84	-1.02440	-0.89784	1.36217	174	-0.09536	-0.04975	0.10755
86	-1.27802	-1.10908	1.69215	176	-0.09476	-0.05120	0.10771
88	-1.44642	-1.24872	1.91088	178	-0.09431	-0.05201	0.10770
				180	-0.09415	-0.05227	0.10769

IBR

RADIATION PATTERN FOR THETA=PI/2

H/LMDA=0.5000 B/LMDA= 0.250 OMEGA= 9.92

18 ELEMENT ARRAY

PHI	ER(THETA,PHI)			PHI	ER(THETA,PHI)		
DEG	RE	IM	ABSVAL	DEG	RE	IM	ABSVAL
0	-0.07955	-0.06270	0.10129	90	-0.76132	-1.21449	1.43339
2	-0.08011	-0.06237	0.10153	92	-0.72689	-1.16906	1.37661
4	-0.08173	-0.06136	0.10220	94	-0.62916	-1.03907	1.21470
6	-0.08431	-0.05957	0.10324	96	-0.48371	-0.84231	0.97132
8	-0.08764	-0.05687	0.10447	98	-0.31306	-0.60497	0.68117
10	-0.09140	-0.05308	0.10569	100	-0.14240	-0.35715	0.38449
12	-0.09520	-0.04796	0.10659	102	0.00513	-0.12801	0.12811
14	-0.09852	-0.04127	0.10681	104	0.11228	0.05868	0.12669
16	-0.10075	-0.03278	0.10595	106	0.17027	0.18764	0.25338
18	-0.10118	-0.02229	0.10361	108	0.17938	0.25347	0.31052
20	-0.09906	-0.00968	0.09953	110	0.14783	0.26020	0.29927
22	-0.09362	0.00501	0.09375	112	0.08933	0.21943	0.23692
24	-0.08418	0.02156	0.08690	114	0.01986	0.14734	0.14867
26	-0.07027	0.03948	0.08060	116	-0.04546	0.06158	0.07654
28	-0.05172	0.05795	0.07767	118	-0.09485	-0.02167	0.09729
30	-0.02883	0.07578	0.08108	120	-0.12134	-0.08984	0.15098
32	-0.00246	0.09147	0.09151	122	-0.12317	-0.13512	0.18283
34	0.02585	0.10319	0.10638	124	-0.10310	-0.15461	0.18583
36	0.05388	0.10897	0.12156	126	-0.06711	-0.14974	0.16409
38	0.07885	0.10689	0.13283	128	-0.02284	-0.12512	0.12718
40	0.09764	0.09540	0.13651	130	0.02204	-0.08712	0.08986
42	0.10716	0.07366	0.13004	132	0.06099	-0.04261	0.07440
44	0.10488	0.04199	0.11297	134	0.08940	0.00209	0.08942
46	0.08940	0.00210	0.08942	136	0.10488	0.04199	0.11297
48	0.06099	-0.04261	0.07440	138	0.10716	0.07366	0.13004
50	0.02204	-0.08712	0.08986	140	0.09764	0.09540	0.13651
52	-0.02283	-0.12512	0.12718	142	0.07886	0.10689	0.13283
54	-0.06711	-0.14974	0.16409	144	0.05388	0.10897	0.12156
56	-0.10310	-0.15461	0.18583	146	0.02585	0.10319	0.10638
58	-0.12317	-0.13512	0.18283	148	-0.00246	0.09147	0.09151
60	-0.12134	-0.08985	0.15098	150	-0.02883	0.07578	0.08108
62	-0.09485	-0.02167	0.09729	152	-0.05172	0.05795	0.07767
64	-0.04546	0.06158	0.07654	154	-0.07027	0.03948	0.08060
66	0.01986	0.14734	0.14867	156	-0.08418	0.02156	0.08690
68	0.08933	0.21943	0.23691	158	-0.09362	0.00501	0.09375
70	0.14783	0.26020	0.29927	160	-0.09906	-0.00968	0.09953
72	0.17938	0.25347	0.31052	162	-0.10118	-0.02229	0.10361
74	0.17027	0.18765	0.25339	164	-0.10075	-0.03278	0.10595
76	0.11229	0.05869	0.12670	166	-0.09852	-0.04127	0.10682
78	0.00513	-0.12801	0.12811	168	-0.09520	-0.04796	0.10659
80	-0.14240	-0.35715	0.38449	170	-0.09140	-0.05307	0.10569
82	-0.31306	-0.60496	0.68116	172	-0.08764	-0.05687	0.10447
84	-0.48370	-0.84230	0.97131	174	-0.08431	-0.05957	0.10324
86	-0.62915	-1.03906	1.21469	176	-0.08173	-0.06136	0.10220
88	-0.72689	-1.16906	1.37661	178	-0.08011	-0.06237	0.10153
				180	-0.07955	-0.06270	0.10129

VBR

RADIATION PATTERN FOR THETA=PI/2

H/LMDA=0.5000 B/LMDA= 0.250 OMEGA= 9.92

20 ELEMENT ARRAY

PHI DEG	ER(THETA,PHI) RE	IM	ABSVAL	PHI DEG	ER(THETA,PHI) RE	IM	ABSVAL
0	0.02310	-0.04202	0.04795	90	-1.50794	-1.30229	1.99245
2	0.02220	-0.04257	0.04801	92	-1.43502	-1.24152	1.89754
4	0.01951	-0.04422	0.04833	94	-1.22924	-1.06964	1.62947
6	0.01499	-0.04691	0.04925	96	-0.92672	-0.81568	1.23456
8	0.00863	-0.05057	0.05130	98	-0.57868	-0.52111	0.77873
10	0.00042	-0.05507	0.05507	100	-0.24055	-0.23126	0.33369
12	-0.00956	-0.06018	0.06094	102	0.03931	0.01352	0.04157
14	-0.02116	-0.06562	0.06894	104	0.22846	0.18510	0.29404
16	-0.03402	-0.07094	0.07868	106	0.31497	0.27145	0.41580
18	-0.04763	-0.07559	0.08935	108	0.30709	0.27695	0.41353
20	-0.06118	-0.07887	0.09981	110	0.22879	0.21921	0.31686
22	-0.07359	-0.07994	0.10866	112	0.11250	0.12371	0.16721
24	-0.08355	-0.07789	0.11422	114	-0.00869	0.01769	0.01971
26	-0.08950	-0.07179	0.11473	116	-0.10770	-0.07527	0.13140
28	-0.08985	-0.06086	0.10852	118	-0.16763	-0.13897	0.21774
30	-0.08314	-0.04461	0.09435	120	-0.18301	-0.16612	0.24716
32	-0.06836	-0.02303	0.07213	122	-0.15849	-0.15782	0.22367
34	-0.04526	0.00316	0.04537	124	-0.10574	-0.12154	0.16110
36	-0.01472	0.03235	0.03554	126	-0.03959	-0.06831	0.07895
38	0.02096	0.06194	0.06539	128	0.02554	-0.00994	0.02740
40	0.05803	0.08844	0.10578	130	0.07823	0.04325	0.08939
42	0.09136	0.10770	0.14123	132	0.11155	0.08393	0.13960
44	0.11503	0.11548	0.16299	134	0.12325	0.10820	0.16400
46	0.12325	0.10820	0.16400	136	0.11503	0.11548	0.16299
48	0.11155	0.08393	0.13960	138	0.09136	0.10770	0.14123
50	0.07823	0.04325	0.08939	140	0.05803	0.08844	0.10578
52	0.02554	-0.00994	0.02740	142	0.02097	0.06194	0.06539
54	-0.03959	-0.06831	0.07895	144	-0.01472	0.03235	0.03554
56	-0.10574	-0.12153	0.16109	146	-0.04526	0.00316	0.04537
58	-0.15849	-0.15782	0.22367	148	-0.06836	-0.02303	0.07213
60	-0.18301	-0.16612	0.24716	150	-0.08314	-0.04461	0.09435
62	-0.16763	-0.13897	0.21774	152	-0.08985	-0.06086	0.10852
64	-0.10770	-0.07527	0.13140	154	-0.08950	-0.07179	0.11473
66	-0.00870	0.01769	0.01971	156	-0.08355	-0.07788	0.11422
68	0.11250	0.12371	0.16721	158	-0.07359	-0.07994	0.10866
70	0.22879	0.21921	0.31685	160	-0.06118	-0.07887	0.09982
72	0.30709	0.27695	0.41353	162	-0.04763	-0.07559	0.08935
74	0.31497	0.27145	0.41580	164	-0.03403	-0.07094	0.07868
76	0.22847	0.18510	0.29404	166	-0.02116	-0.06562	0.06894
78	0.03931	0.01353	0.04158	168	-0.00956	-0.06018	0.06094
80	-0.24055	-0.23126	0.33368	170	0.00042	-0.05507	0.05507
82	-0.57867	-0.52110	0.77872	172	0.00863	-0.05057	0.05130
84	-0.92670	-0.81567	1.23454	174	0.01499	-0.04691	0.04925
86	-1.22923	-1.06963	1.62945	176	0.01951	-0.04422	0.04833
88	-1.43501	-1.24152	1.89753	178	0.02220	-0.04257	0.04801
				180	0.02309	-0.04202	0.04795

IBR

RADIATION PATTERN FOR THETA=PI/2

H/LMDA=0.5000 B/LMDA= 0.250 OMEGA= 9.92

20 ELEMENT ARRAY

PHI DEG	ER(THETA,PHI) RE	IM	ABSVAL	PHI DEG	ER(THETA,PHI) RE	IM	ABSVAL
0	0.06110	-0.04911	0.07839	90	-0.75301	-1.21872	1.43259
2	0.06035	-0.04967	0.07816	92	-0.71145	-1.16217	1.36265
4	0.05807	-0.05134	0.07751	94	-0.59494	-1.00219	1.16548
6	0.05418	-0.05405	0.07653	96	-0.42606	-0.76565	0.87621
8	0.04855	-0.05772	0.07542	98	-0.23640	-0.49095	0.54490
10	0.04103	-0.06218	0.07450	100	-0.05934	-0.22020	0.22805
12	0.03150	-0.06720	0.07422	102	0.07734	0.00908	0.07788
14	0.01988	-0.07245	0.07513	104	0.15689	0.17055	0.23174
16	0.00623	-0.07748	0.07773	106	0.17604	0.25276	0.30802
18	-0.00925	-0.08170	0.08222	108	0.14407	0.25946	0.29678
20	-0.02607	-0.08441	0.08834	110	0.07918	0.20676	0.22141
22	-0.04348	-0.08479	0.09529	112	0.00323	0.11821	0.11826
24	-0.06038	-0.08195	0.10179	114	-0.06354	0.01915	0.06636
26	-0.07535	-0.07503	0.10633	116	-0.10660	-0.06832	0.12661
28	-0.08671	-0.06332	0.10737	118	-0.11928	-0.12885	0.17559
30	-0.09268	-0.04642	0.10366	120	-0.10280	-0.15532	0.18626
32	-0.09158	-0.02443	0.09478	122	-0.06454	-0.14847	0.16189
34	-0.08216	0.00184	0.08218	124	-0.01531	-0.11503	0.11604
36	-0.06394	0.03076	0.07095	126	0.03349	-0.06522	0.07332
38	-0.03759	0.05974	0.07058	128	0.07239	-0.01010	0.07309
40	-0.00515	0.08538	0.08553	130	0.09527	0.04056	0.10354
42	0.02985	0.10372	0.10793	132	0.09996	0.07965	0.12781
44	0.06266	0.11079	0.12728	134	0.08778	0.10333	0.13558
46	0.08778	0.10333	0.13558	136	0.06266	0.11079	0.12728
48	0.09996	0.07965	0.12781	138	0.02985	0.10372	0.10793
50	0.09527	0.04055	0.10354	140	-0.00515	0.08538	0.08553
52	0.07239	-0.01010	0.07309	142	-0.03759	0.05974	0.07058
54	0.03349	-0.06522	0.07332	144	-0.06394	0.03076	0.07095
56	-0.01531	-0.11503	0.11604	146	-0.08216	0.00184	0.08218
58	-0.06454	-0.14847	0.16189	148	-0.09158	-0.02443	0.09478
60	-0.10280	-0.15532	0.18626	150	-0.09268	-0.04642	0.10365
62	-0.11929	-0.12885	0.17559	152	-0.08671	-0.06332	0.10737
64	-0.10660	-0.06832	0.12661	154	-0.07535	-0.07503	0.10633
66	-0.06354	0.01915	0.06636	156	-0.06038	-0.08195	0.10179
68	0.00323	0.11821	0.11826	158	-0.04348	-0.08479	0.09528
70	0.07918	0.20676	0.22140	160	-0.02607	-0.08441	0.08835
72	0.14407	0.25946	0.29678	162	-0.00925	-0.08170	0.08222
74	0.17604	0.25276	0.30802	164	0.00622	-0.07748	0.07773
76	0.15690	0.17056	0.23175	166	0.01988	-0.07245	0.07513
78	0.07735	0.00909	0.07788	168	0.03150	-0.06720	0.07422
80	-0.05934	-0.22019	0.22805	170	0.04103	-0.06218	0.07450
82	-0.23640	-0.49095	0.54490	172	0.04855	-0.05772	0.07542
84	-0.42605	-0.76564	0.87619	174	0.05418	-0.05405	0.07653
86	-0.59493	-1.00218	1.16547	176	0.05807	-0.05134	0.07751
88	-0.71145	-1.16217	1.36265	178	0.06035	-0.04967	0.07816
				180	0.06110	-0.04911	0.07839

VBR

RADIATION PATTERN FOR [illegible]

H/LMDA=0.5000 B/LMDA= 0.250 OMEGA= 9.92

22 ELEMENT ARRAY

PHI DEG	ER(THETA,PHI) RE	IM	ABSVAL	PHI DEG	ER(THETA,PHI) RE	IM	ABSVAL
0	0.07641	0.04214	0.08726	90	-1.50937	-1.30514	1.99539
2	0.07658	0.04186	0.08727	92	-1.42124	-1.23133	1.88045
4	0.07702	0.04098	0.08725	94	-1.17568	-1.02517	1.55987
6	0.07758	0.03940	0.08701	96	-0.82424	-0.72849	1.10003
8	0.07796	0.03695	0.08627	98	-0.43766	-0.39910	0.59231
10	0.07775	0.03341	0.08462	100	-0.08805	-0.09670	0.13078
12	0.07646	0.02851	0.08160	102	0.16826	0.13092	0.21319
14	0.07346	0.02198	0.07668	104	0.30226	0.25741	0.39701
16	0.06812	0.01358	0.06946	106	0.31563	0.28126	0.42277
18	0.05979	0.00316	0.05988	108	0.23602	0.22263	0.32445
20	0.04795	-0.00928	0.04884	110	0.10630	0.11539	0.15689
22	0.03232	-0.02347	0.03994	112	-0.02821	-0.00292	0.02836
24	0.01301	-0.03883	0.04095	114	-0.13101	-0.10034	0.16502
26	-0.00929	-0.05434	0.05513	116	-0.18143	-0.15667	0.23971
28	-0.03321	-0.06855	0.07617	118	-0.17627	-0.16568	0.24191
30	-0.05658	-0.07960	0.09766	120	-0.12680	-0.13354	0.18416
32	-0.07655	-0.08535	0.11465	122	-0.05290	-0.07468	0.09152
34	-0.08980	-0.08366	0.12273	124	0.02367	-0.00666	0.02459
36	-0.09301	-0.07280	0.11812	126	0.08483	0.05445	0.10080
38	-0.08356	-0.05195	0.09839	128	0.11954	0.09723	0.15409
40	-0.06030	-0.02177	0.06411	130	0.12472	0.11632	0.17055
42	-0.02443	0.01519	0.02877	132	0.10412	0.11206	0.15296
44	0.01998	0.05429	0.05785	134	0.06591	0.08905	0.11079
46	0.06591	0.08905	0.11079	136	0.01998	0.05429	0.05784
48	0.10412	0.11206	0.15296	138	-0.02443	0.01519	0.02877
50	0.12472	0.11632	0.17055	140	-0.06030	-0.02177	0.06411
52	0.11954	0.09723	0.15409	142	-0.08355	-0.05195	0.09839
54	0.08483	0.05445	0.10080	144	-0.09301	-0.07280	0.11812
56	0.02367	-0.00666	0.02459	146	-0.08980	-0.08366	0.12273
58	-0.05290	-0.07468	0.09152	148	-0.07655	-0.08535	0.11465
60	-0.12680	-0.13354	0.18416	150	-0.05658	-0.07960	0.09766
62	-0.17627	-0.16568	0.24191	152	-0.03321	-0.06855	0.07617
64	-0.18143	-0.15667	0.23972	154	-0.00929	-0.05434	0.05513
66	-0.13101	-0.10034	0.16503	156	0.01301	-0.03882	0.04095
68	-0.02821	-0.00293	0.02836	158	0.03232	-0.02347	0.03994
70	0.10629	0.11538	0.15688	160	0.04795	-0.00928	0.04884
72	0.23601	0.22262	0.32444	162	0.05979	0.00316	0.05988
74	0.31563	0.28126	0.42277	164	0.06812	0.01358	0.06946
76	0.30226	0.25741	0.39702	166	0.07346	0.02198	0.07668
78	0.16826	0.13092	0.21320	168	0.07645	0.02850	0.08160
80	-0.08805	-0.09669	0.13077	170	0.07775	0.03341	0.08463
82	-0.43765	-0.39909	0.59230	172	0.07796	0.03695	0.08627
84	-0.82422	-0.72848	1.10001	174	0.07758	0.03940	0.08701
86	-1.17566	-1.02516	1.55985	176	0.07702	0.04098	0.08725
88	-1.42123	-1.23133	1.88045	178	0.07658	0.04186	0.08727
				180	0.07641	0.04214	0.08726

IBR

RADIATION PATTERN FOR THETA=PI/2

H/LMDA=0.5000 B/LMDA= 0.250 OMEGA= [illegible]

22 ELEMENT ARRAY

PHI DEG	ER(THETA,PHI) RE	IM	ABSVAL	PHI DEG	ER(THETA,PHI) RE	IM	ABSVAL
0	0.06453	0.05127	0.08242	90	-0.75085	-1.22224	1.43445
2	0.06509	0.05092	0.08264	92	-0.70132	-1.15346	1.34993
4	0.06672	0.04983	0.08328	94	-0.56437	-0.96125	1.11468
6	0.06927	0.04791	0.08423	96	-0.37158	-0.68443	0.77880
8	0.07250	0.04497	0.08531	98	-0.16560	-0.37674	0.41153
10	0.07601	0.04082	0.08628	100	0.01144	-0.09373	0.09442
12	0.07932	0.03518	0.08677	102	0.12872	0.11999	0.17598
14	0.08177	0.02782	0.08638	104	0.17347	0.23961	0.29581
16	0.08263	0.01852	0.08468	106	0.15209	0.26339	0.30415
18	0.08105	0.00716	0.08136	108	0.08597	0.20979	0.22672
20	0.07618	-0.00620	0.07643	110	0.00372	0.11017	0.11023
22	0.06726	-0.02124	0.07054	112	-0.06745	-0.00050	0.06746
24	0.05380	-0.03731	0.06547	114	-0.10858	-0.09222	0.14246
26	0.03570	-0.05337	0.06421	116	-0.11243	-0.14587	0.18417
28	0.01351	-0.06792	0.06925	118	-0.08323	-0.15526	0.17617
30	-0.01146	-0.07913	0.07995	120	-0.03344	-0.12587	0.13023
32	-0.03707	-0.08490	0.09264	122	0.02112	-0.07099	0.07406
34	-0.06038	-0.08319	0.10279	124	0.06590	-0.00704	0.06628
36	-0.07794	-0.07240	0.10638	126	0.09107	0.05079	0.10427
38	-0.08630	-0.05188	0.10069	128	0.09300	0.09159	0.13053
40	-0.08268	-0.02236	0.08565	130	0.07396	0.11006	0.13260
42	-0.06584	0.01353	0.06721	132	0.04041	0.10630	0.11372
44	-0.03678	0.05123	0.06307	134	0.00077	0.08452	0.08452
46	0.00076	0.08451	0.08452	136	-0.03678	0.05123	0.06307
48	0.04041	0.10630	0.11372	138	-0.06584	0.01353	0.06721
50	0.07396	0.11006	0.13260	140	-0.08268	-0.02236	0.08565
52	0.09300	0.09159	0.13053	142	-0.08630	-0.05187	0.10069
54	0.09107	0.05079	0.10427	144	-0.07794	-0.07240	0.10638
56	0.06590	-0.00704	0.06628	146	-0.06038	-0.08319	0.10279
58	0.02113	-0.07099	0.07406	148	-0.03707	-0.08489	0.09264
60	-0.03343	-0.12587	0.13023	150	-0.01146	-0.07913	0.07995
62	-0.08323	-0.15526	0.17617	152	0.01351	-0.06792	0.06925
64	-0.11243	-0.14587	0.18417	154	0.03570	-0.05337	0.06421
66	-0.10858	-0.09223	0.14246	156	0.05380	-0.03731	0.06547
68	-0.06746	-0.00050	0.06746	158	0.06726	-0.02124	0.07054
70	0.00372	0.11016	0.11023	160	0.07618	-0.00620	0.07643
72	0.08597	0.20979	0.22672	162	0.08105	0.00716	0.08136
74	0.15208	0.26339	0.30415	164	0.08263	0.01852	0.08468
76	0.17347	0.23962	0.29582	166	0.08177	0.02782	0.08638
78	0.12873	0.11999	0.17598	168	0.07932	0.03518	0.08677
80	0.01145	-0.09372	0.09442	170	0.07601	0.04082	0.08628
82	-0.16560	-0.37674	0.41153	172	0.07249	0.04497	0.08531
84	-0.37157	-0.68442	0.77878	174	0.06927	0.04791	0.08423
86	-0.56436	-0.96124	1.11467	176	0.06672	0.04983	0.08328
88	-0.70132	-1.15345	1.34993	178	0.06509	0.05092	0.08264
				180	0.06453	0.05127	0.08242

VBR RADIATION PATTERN FOR THETA=PI/2

H/LMDA=0.5000 B/LMDA= 0.250 OMEGA= 9.92

24 ELEMENT ARRAY

PHI DEG	ER(THETA,PHI) RE	IM	ABSVAL	PHI DEG	ER(THETA,PHI) RE	IM	ABSVAL
0	-0.01973	0.03554	0.04065	90	-1.51108	-1.30835	1.99878
2	-0.01884	0.03608	0.04070	92	-1.40625	-1.22016	1.86181
4	-0.01618	0.03767	0.04100	94	-1.11826	-0.97719	1.48506
6	-0.01170	0.04028	0.04194	96	-0.71834	-0.63774	0.96059
8	-0.00540	0.04377	0.04410	98	-0.30062	-0.27947	0.41046
10	0.00271	0.04798	0.04805	100	0.04552	0.02279	0.05091
12	0.01251	0.05261	0.05408	102	0.25982	0.21686	0.33843
14	0.02374	0.05725	0.06198	104	0.32436	0.28457	0.43150
16	0.03588	0.06135	0.07107	106	0.26261	0.24202	0.35712
18	0.04816	0.06416	0.08022	108	0.12636	0.12993	0.18124
20	0.05948	0.06481	0.08797	110	-0.02377	-0.00157	0.02383
22	0.06842	0.06235	0.09257	112	-0.13722	-0.10836	0.17485
24	0.07331	0.05583	0.09215	114	-0.18578	-0.16318	0.24726
26	0.07246	0.04448	0.08502	116	-0.16698	-0.15957	0.23096
28	0.06440	0.02800	0.07022	118	-0.09912	-0.10916	0.14745
30	0.04830	0.00675	0.04876	120	-0.01133	-0.03443	0.03624
32	0.02443	-0.01797	0.03032	122	0.06735	0.03995	0.07830
34	-0.00543	-0.04371	0.04405	124	0.11622	0.09426	0.14964
36	-0.03776	-0.06697	0.07688	126	0.12693	0.11791	0.17325
38	-0.06741	-0.08346	0.10728	128	0.10297	0.11011	0.15076
40	-0.08819	-0.08880	0.12515	130	0.05607	0.07779	0.09589
42	-0.09407	-0.07952	0.12318	132	0.00126	0.03212	0.03214
44	-0.08077	-0.05431	0.09733	134	-0.04757	-0.01512	0.04991
46	-0.04757	-0.01512	0.04992	136	-0.08077	-0.05431	0.09733
48	0.00126	0.03212	0.03214	138	-0.09407	-0.07952	0.12318
50	0.05607	0.07779	0.09589	140	-0.08819	-0.08880	0.12515
52	0.10297	0.11011	0.15076	142	-0.06741	-0.08346	0.10728
54	0.12693	0.11791	0.17325	144	-0.03776	-0.06697	0.07688
56	0.11622	0.09426	0.14964	146	-0.00543	-0.04371	0.04404
58	0.06735	0.03995	0.07830	148	0.02443	-0.01796	0.03032
60	-0.01133	-0.03443	0.03624	150	0.04830	0.00675	0.04877
62	-0.09912	-0.10916	0.14745	152	0.06440	0.02800	0.07022
64	-0.16698	-0.15957	0.23096	154	0.07246	0.04448	0.08503
66	-0.18578	-0.16318	0.24727	156	0.07331	0.05583	0.09215
68	-0.13722	-0.10837	0.17485	158	0.06842	0.06235	0.09257
70	-0.02378	-0.00158	0.02383	160	0.05948	0.06481	0.08797
72	0.12636	0.12992	0.18124	162	0.04817	0.06416	0.08023
74	0.26260	0.24201	0.35712	164	0.03588	0.06134	0.07107
76	0.32436	0.28457	0.43150	166	0.02374	0.05725	0.06198
78	0.25983	0.21687	0.33844	168	0.01251	0.05261	0.05408
80	0.04553	0.02280	0.05091	170	0.00271	0.04798	0.04805
82	-0.30061	-0.27946	0.41045	172	-0.00540	0.04377	0.04410
84	-0.71831	-0.63773	0.96056	174	-0.01170	0.04028	0.04194
86	-1.11824	-0.97718	1.48504	176	-0.01618	0.03767	0.04100
88	-1.40624	-1.22015	1.86180	178	-0.01884	0.03607	0.04070
				180	-0.01973	0.03554	0.04065

IBR RADIATION PATTERN FOR THETA=PI/2

H/LMDA=0.5000 B/LMDA= 0.250 OMEGA= 9.92

24 ELEMENT ARRAY

PHI DEG	ER(THETA,PHI) RE	IM	ABSVAL	PHI DEG	ER(THETA,PHI) RE	IM	ABSVAL
0	-0.05094	0.04137	0.06562	90	-0.74496	-1.22513	1.43384
2	-0.05020	0.04192	0.06540	92	-0.68706	-1.14294	1.33356
4	-0.04793	0.04357	0.06477	94	-0.52938	-0.91642	1.05833
6	-0.04404	0.04624	0.06386	96	-0.31455	-0.59967	0.67716
8	-0.03838	0.04980	0.06287	98	-0.09786	-0.26485	0.28235
10	-0.03079	0.05403	0.06219	100	0.07023	0.01833	0.07258
12	-0.02113	0.05864	0.06233	102	0.15888	0.20107	0.25626
14	-0.00940	0.06315	0.06385	104	0.16389	0.26603	0.31247
16	0.00426	0.06700	0.06713	106	0.10549	0.22777	0.25101
18	0.01941	0.06941	0.07208	108	0.01849	0.12389	0.12526
20	0.03528	0.06953	0.07797	110	-0.06074	0.00100	0.06075
22	0.05070	0.06640	0.08354	112	-0.10582	-0.09958	0.14530
24	0.06410	0.05913	0.08720	114	-0.10675	-0.15202	0.18576
26	0.07359	0.04704	0.08734	116	-0.06993	-0.14984	0.16535
28	0.07722	0.02989	0.08281	118	-0.01299	-0.10348	0.10430
30	0.07323	0.00817	0.07369	120	0.04291	-0.03374	0.05459
32	0.06057	-0.01674	0.06284	122	0.08034	0.03628	0.08815
34	0.03931	-0.04238	0.05780	124	0.09020	0.08790	0.12595
36	0.01114	-0.06527	0.06622	126	0.07287	0.11088	0.13268
38	-0.02046	-0.08128	0.08381	128	0.03633	0.10411	0.11027
40	-0.05038	-0.08623	0.09987	130	-0.00756	0.07388	0.07427
42	-0.07262	-0.07695	0.10580	132	-0.04694	0.03060	0.05604
44	-0.08143	-0.05236	0.09681	134	-0.07298	-0.01456	0.07442
46	-0.07298	-0.01456	0.07442	136	-0.08143	-0.05236	0.09681
48	-0.04694	0.03060	0.05604	138	-0.07262	-0.07695	0.10580
50	-0.00757	0.07388	0.07427	140	-0.05038	-0.08623	0.09987
52	0.03633	0.10411	0.11027	142	-0.02046	-0.08127	0.08381
54	0.07287	0.11088	0.13268	144	0.01114	-0.06527	0.06622
56	0.09020	0.08790	0.12595	146	0.03931	-0.04238	0.05780
58	0.08034	0.03628	0.08815	148	0.06057	-0.01674	0.06284
60	0.04291	-0.03374	0.05459	150	0.07323	0.00817	0.07369
62	-0.01299	-0.10348	0.10429	152	0.07722	0.02989	0.08281
64	-0.06993	-0.14984	0.16535	154	0.07360	0.04704	0.08734
66	-0.10675	-0.15202	0.18576	156	0.06410	0.05913	0.08720
68	-0.10582	-0.09958	0.14530	158	0.05070	0.06640	0.08355
70	-0.06074	0.00100	0.06075	160	0.03528	0.06953	0.07797
72	0.01849	0.12389	0.12526	162	0.01941	0.06942	0.07208
74	0.10548	0.22777	0.25101	164	0.00426	0.06700	0.06713
76	0.16389	0.26604	0.31247	166	-0.00940	0.06315	0.06385
78	0.15888	0.20107	0.25627	168	-0.02113	0.05864	0.06233
80	0.07023	0.01834	0.07258	170	-0.03078	0.05403	0.06219
82	-0.09786	-0.26485	0.28235	172	-0.03838	0.04980	0.06287
84	-0.31454	-0.59966	0.67714	174	-0.04404	0.04624	0.06386
86	-0.52937	-0.91641	1.05832	176	-0.04793	0.04357	0.06477
88	-0.68706	-1.14294	1.33355	178	-0.05020	0.04192	0.06540
				180	-0.05094	0.04137	0.06562

VBR RADIATION PATTERN FOR THETA=PI/2

H/LMDA=0.5000 B/LMDA= 0.250 OMEGA= 9.92

26 ELEMENT ARRAY

PHI DEG	ER(THETA,PHI) RE	IM	ABSVAL	PHI DEG	ER(THETA,PHI) RE	IM	ABSVAL
0	-0.06423	-0.03526	0.07327	90	-1.51209	-1.31038	2.00088
2	-0.06440	-0.03496	0.07328	92	-1.38934	-1.20676	1.84026
4	-0.06485	-0.03403	0.07324	94	-1.05722	-0.92556	1.40512
6	-0.06537	-0.03234	0.07293	96	-0.61110	-0.54533	0.81905
8	-0.06563	-0.02971	0.07204	98	-0.17185	-0.16651	0.23929
10	-0.06516	-0.02588	0.07011	100	0.15502	0.12166	0.19706
12	-0.06336	-0.02058	0.06662	102	0.31171	0.26786	0.41099
14	-0.05955	-0.01354	0.06107	104	0.30039	0.26983	0.40378
16	-0.05304	-0.00458	0.05323	106	0.17332	0.16774	0.24119
18	-0.04320	0.00634	0.04366	108	0.00708	0.02428	0.02529
20	-0.02966	0.01896	0.03520	110	-0.12726	-0.09947	0.16152
22	-0.01247	0.03267	0.03496	112	-0.18668	-0.16333	0.24804
24	0.00768	0.04636	0.04699	114	-0.16458	-0.15657	0.22716
26	0.02927	0.05844	0.06536	116	-0.08486	-0.09490	0.12731
28	0.04990	0.06686	0.08343	118	0.01323	-0.00933	0.01619
30	0.06638	0.06936	0.09601	120	0.09210	0.06746	0.11416
32	0.07514	0.06383	0.09860	122	0.12824	0.11218	0.17038
34	0.07289	0.04896	0.08780	124	0.11700	0.11619	0.16489
36	0.05758	0.02483	0.06270	126	0.06988	0.08495	0.11000
38	0.02948	-0.00642	0.03017	128	0.00724	0.03316	0.03395
40	-0.00793	-0.04025	0.04102	130	-0.04996	-0.02168	0.05446
42	-0.04780	-0.07005	0.08480	132	-0.08656	-0.06483	0.10814
44	-0.08059	-0.08813	0.11942	134	-0.09611	-0.08760	0.13004
46	-0.09611	-0.08760	0.13004	136	-0.08059	-0.08813	0.11942
48	-0.08656	-0.06483	0.10814	138	-0.04780	-0.07005	0.08480
50	-0.04996	-0.02168	0.05446	140	-0.00793	-0.04025	0.04102
52	0.00724	0.03316	0.03395	142	0.02949	-0.00641	0.03018
54	0.06988	0.08495	0.11000	144	0.05757	0.02482	0.06270
56	0.11700	0.11619	0.16489	146	0.07289	0.04896	0.08780
58	0.12824	0.11218	0.17038	148	0.07515	0.06384	0.09860
60	0.09210	0.06747	0.11416	150	0.06638	0.06936	0.09601
62	0.01323	-0.00933	0.01619	152	0.04990	0.06686	0.08343
64	-0.08486	-0.09490	0.12730	154	0.02927	0.05844	0.06536
66	-0.16457	-0.15657	0.22715	156	0.00768	0.04636	0.04699
68	-0.18668	-0.16333	0.24804	158	-0.01246	0.03267	0.03497
70	-0.12726	-0.09947	0.16152	160	-0.02966	0.01896	0.03520
72	0.00707	0.02428	0.02529	162	-0.04320	0.00634	0.04366
74	0.17331	0.16773	0.24118	164	-0.05304	-0.00458	0.05323
76	0.30038	0.26983	0.40378	166	-0.05955	-0.01354	0.06107
78	0.31171	0.26786	0.41099	168	-0.06336	-0.02058	0.06662
80	0.15503	0.12167	0.19707	170	-0.06516	-0.02588	0.07011
82	-0.17185	-0.16650	0.23928	172	-0.06563	-0.02971	0.07204
84	-0.61108	-0.54532	0.81902	174	-0.06537	-0.03234	0.07293
86	-1.05720	-0.92555	1.40510	176	-0.06485	-0.03403	0.07324
88	-1.38934	-1.20676	1.84025	178	-0.06440	-0.03496	0.07328
				180	-0.06423	-0.03526	0.07327

IBR RADIATION PATTERN FOR THETA=PI/2

H/LMDA=0.5000 B/LMDA= 0.250 OMEGA= 9.92

26 ELEMENT ARRAY

PHI DEG	ER(THETA,PHI) RE	IM	ABSVAL	PHI DEG	ER(THETA,PHI) RE	IM	ABSVAL
0	-0.05424	-0.04337	0.06945	90	-0.74362	-1.22761	1.43527
2	-0.05481	-0.04300	0.06966	92	-0.67652	-1.13094	1.31784
4	-0.05643	-0.04186	0.07026	94	-0.49673	-0.86846	1.00048
6	-0.05896	-0.03982	0.07114	96	-0.26043	-0.51324	0.57554
8	-0.06207	-0.03669	0.07210	98	-0.03725	-0.15878	0.16309
10	-0.06532	-0.03224	0.07284	100	0.11503	0.11165	0.16031
12	-0.06811	-0.02617	0.07297	102	0.16936	0.24986	0.30185
14	-0.06968	-0.01827	0.07204	104	0.13498	0.25326	0.28699
16	-0.06915	-0.00836	0.06965	106	0.04887	0.15887	0.16622
18	-0.06558	0.00353	0.06567	108	-0.04199	0.02495	0.04885
20	-0.05810	0.01709	0.06056	110	-0.09927	-0.09138	0.13493
22	-0.04611	0.03163	0.05591	112	-0.10556	-0.15224	0.18525
24	-0.02946	0.04600	0.05463	114	-0.06672	-0.14706	0.16149
26	-0.00875	0.05856	0.05921	116	-0.00506	-0.09009	0.09023
28	0.01448	0.06725	0.06879	118	0.05220	-0.01010	0.05317
30	0.03771	0.06985	0.07938	120	0.08399	0.06226	0.10455
32	0.05750	0.06436	0.08630	122	0.08185	0.10487	0.13303
34	0.06999	0.04957	0.08576	124	0.05066	0.10927	0.12044
36	0.07166	0.02571	0.07614	126	0.00439	0.08027	0.08039
38	0.06035	-0.00498	0.06055	128	-0.04023	0.03150	0.05109
40	0.03630	-0.03801	0.05256	130	-0.06961	-0.02052	0.07258
42	0.00296	-0.06690	0.06696	132	-0.07691	-0.06173	0.09862
44	-0.03299	-0.08428	0.09050	134	-0.06263	-0.08365	0.10450
46	-0.06263	-0.08365	0.10450	136	-0.03299	-0.08428	0.09050
48	-0.07691	-0.06173	0.09862	138	0.00296	-0.06690	0.06696
50	-0.06962	-0.02052	0.07258	140	0.03630	-0.03801	0.05256
52	-0.04023	0.03150	0.05109	142	0.06035	-0.00498	0.06055
54	0.00439	0.08027	0.08039	144	0.07166	0.02571	0.07613
56	0.05066	0.10927	0.12044	146	0.06999	0.04957	0.08576
58	0.08185	0.10487	0.13303	148	0.05750	0.06436	0.08630
60	0.08399	0.06226	0.10455	150	0.03771	0.06985	0.07938
62	0.05220	-0.01010	0.05317	152	0.01448	0.06725	0.06879
64	-0.00506	-0.09009	0.09023	154	-0.00875	0.05856	0.05921
66	-0.06672	-0.14706	0.16149	156	-0.02946	0.04600	0.05463
68	-0.10556	-0.15224	0.18525	158	-0.04611	0.03163	0.05591
70	-0.09927	-0.09139	0.13493	160	-0.05810	0.01709	0.06056
72	-0.04200	0.02495	0.04885	162	-0.06558	0.00353	0.06567
74	0.04887	0.15886	0.16621	164	-0.06915	-0.00836	0.06965
76	0.13498	0.25326	0.28698	166	-0.06968	-0.01827	0.07204
78	0.16936	0.24986	0.30185	168	-0.06811	-0.02618	0.07297
80	0.11503	0.11166	0.16031	170	-0.06532	-0.03224	0.07284
82	-0.03725	-0.15877	0.16308	172	-0.06207	-0.03670	0.07210
84	-0.26042	-0.51323	0.57552	174	-0.05896	-0.03982	0.07114
86	-0.49672	-0.86845	1.00047	176	-0.05643	-0.04186	0.07026
88	-0.67652	-1.13093	1.31784	178	-0.05481	-0.04300	0.06966
				180	-0.05424	-0.04337	0.06945

VBR

RADIATION PATTERN FOR THETA=PI/2

H/LMDA=0.5000 B/LMDA= 0.250 OMEGA= 9.92

28 ELEMENT ARRAY

PHI DEG	ER(THETA,PHI) RE	IM	ABSVAL	PHI DEG	ER(THETA,PHI) RE	IM	ABSVAL
0	0.01724	-0.03082	0.03531	90	-1.51332	-1.31268	2.00332
2	0.01636	-0.03134	0.03536	92	-1.37124	-1.19235	1.81714
4	0.01372	-0.03291	0.03565	94	-0.99315	-0.87111	1.32106
6	0.00928	-0.03543	0.03663	96	-0.50370	-0.45225	0.67693
8	0.00302	-0.03878	0.03890	98	-0.05353	-0.06181	0.08177
10	-0.00501	-0.04271	0.04301	100	0.23855	0.19860	0.31040
12	-0.01462	-0.04687	0.04910	102	0.32568	0.28562	0.43318
14	-0.02543	-0.05073	0.05675	104	0.23987	0.22144	0.32646
16	-0.03676	-0.05359	0.06499	106	0.06673	0.07498	0.10037
18	-0.04757	-0.05457	0.07239	108	-0.09704	-0.07203	0.12085
20	-0.05642	-0.05267	0.07718	110	-0.18286	-0.15787	0.24158
22	-0.06161	-0.04690	0.07743	112	-0.17069	-0.15971	0.23376
24	-0.06131	-0.03647	0.07134	114	-0.08582	-0.09351	0.12693
26	-0.05396	-0.02110	0.05794	116	0.02183	0.00104	0.02185
28	-0.03872	-0.00129	0.03875	118	0.10347	0.08127	0.13157
30	-0.01607	0.02130	0.02668	120	0.13076	0.11862	0.17655
32	0.01173	0.04376	0.04530	122	0.10244	0.10632	0.14764
34	0.04040	0.06204	0.07403	124	0.03876	0.05701	0.06894
36	0.06398	0.07153	0.09597	126	-0.03093	-0.00614	0.03153
38	0.07588	0.06813	0.10198	128	-0.08097	-0.05987	0.10070
40	0.07066	0.04964	0.08635	130	-0.09759	-0.08863	0.13183
42	0.04622	0.01727	0.04934	132	-0.08084	-0.08779	0.11934
44	0.00578	-0.02334	0.02405	134	-0.04131	-0.06241	0.07484
46	-0.04131	-0.06241	0.07484	136	0.00578	-0.02334	0.02404
48	-0.08084	-0.08779	0.11934	138	0.04622	0.01727	0.04934
50	-0.09759	-0.08863	0.13183	140	0.07066	0.04964	0.08635
52	-0.08097	-0.05987	0.10070	142	0.07588	0.06813	0.10198
54	-0.03093	-0.00614	0.03153	144	0.06398	0.07153	0.09597
56	0.03876	0.05701	0.06894	146	0.04040	0.06204	0.07403
58	0.10244	0.10632	0.14764	148	0.01173	0.04376	0.04530
60	0.13076	0.11862	0.17655	150	-0.01607	0.02130	0.02668
62	0.10347	0.08127	0.13157	152	-0.03872	-0.00129	0.03875
64	0.02183	0.00104	0.02185	154	-0.05396	-0.02109	0.05794
66	-0.08582	-0.09351	0.12692	156	-0.06131	-0.03647	0.07134
68	-0.17069	-0.15971	0.23376	158	-0.06161	-0.04689	0.07742
70	-0.18286	-0.15787	0.24158	160	-0.05642	-0.05267	0.07718
72	-0.09704	-0.07203	0.12085	162	-0.04756	-0.05457	0.07239
74	0.06672	0.07497	0.10036	164	-0.03676	-0.05359	0.06499
76	0.23987	0.22143	0.32645	166	-0.02543	-0.05073	0.05675
78	0.32568	0.28562	0.43318	168	-0.01462	-0.04687	0.04910
80	0.23855	0.19861	0.31040	170	-0.00501	-0.04271	0.04301
82	-0.05352	-0.06181	0.08176	172	0.00302	-0.03878	0.03890
84	-0.50367	-0.45223	0.67690	174	0.00928	-0.03543	0.03662
86	-0.99313	-0.87110	1.32103	176	0.01372	-0.03291	0.03565
88	-1.37123	-1.19235	1.81713	178	0.01636	-0.03134	0.03536
				180	0.01724	-0.03082	0.03531

IBR

RADIATION PATTERN FOR THETA=PI/2

H/LMDA=0.5000 B/LMDA= 0.250 OMEGA= 9.92

28 ELEMENT ARRAY

PHI DEG	ER(THETA,PHI) RE	IM	ABSVAL	PHI DEG	ER(THETA,PHI) RE	IM	ABSVAL
0	0.04367	-0.03575	0.05644	90	-0.73919	-1.22971	1.43478
2	0.04294	-0.03630	0.05623	92	-0.66255	-1.11744	1.29909
4	0.04068	-0.03794	0.05562	94	-0.46080	-0.81758	0.93849
6	0.03679	-0.04056	0.05476	96	-0.20602	-0.42618	0.47336
8	0.03109	-0.04401	0.05389	98	0.01695	-0.06063	0.06296
10	0.02342	-0.04803	0.05343	100	0.14538	0.18417	0.23463
12	0.01366	-0.05220	0.05396	102	0.16111	0.26727	0.31207
14	0.00187	-0.05597	0.05600	104	0.09159	0.20876	0.22797
16	-0.01165	-0.05859	0.05974	106	-0.00860	0.07247	0.07298
18	-0.02622	-0.05918	0.06473	108	-0.08507	-0.06540	0.10730
20	-0.04071	-0.05673	0.06982	110	-0.10622	-0.14687	0.18125
22	-0.05352	-0.05027	0.07343	112	-0.07201	-0.14998	0.16637
24	-0.06264	-0.03910	0.07385	114	-0.00752	-0.08898	0.08930
26	-0.06594	-0.02301	0.06984	116	0.05307	-0.00066	0.05308
28	-0.06152	-0.00262	0.06157	118	0.08310	0.07501	0.11195
30	-0.04832	0.02033	0.05243	120	0.07350	0.11092	0.13306
32	-0.02676	0.04288	0.05055	122	0.03325	0.10021	0.10558
34	0.00081	0.06103	0.06104	124	-0.01758	0.05432	0.05710
36	0.02991	0.07031	0.07640	126	-0.05789	-0.00513	0.05812
38	0.05439	0.06682	0.08616	128	-0.07382	-0.05616	0.09276
40	0.06770	0.04863	0.08335	130	-0.06245	-0.08384	0.10454
42	0.06471	0.01713	0.06694	132	-0.03068	-0.08343	0.08889
44	0.04392	-0.02207	0.04915	134	0.00905	-0.05944	0.06013
46	0.00905	-0.05944	0.06013	136	0.04392	-0.02207	0.04915
48	-0.03067	-0.08343	0.08889	138	0.06471	0.01713	0.06694
50	-0.06245	-0.08384	0.10454	140	0.06770	0.04863	0.08335
52	-0.07382	-0.05616	0.09276	142	0.05439	0.06682	0.08616
54	-0.05789	-0.00513	0.05812	144	0.02991	0.07031	0.07640
56	-0.01758	0.05432	0.05710	146	0.00081	0.06103	0.06104
58	0.03325	0.10021	0.10558	148	-0.02676	0.04288	0.05055
60	0.07349	0.11092	0.13306	150	-0.04832	0.02033	0.05243
62	0.08310	0.07501	0.11195	152	-0.06152	-0.00262	0.06157
64	0.05307	-0.00066	0.05308	154	-0.06594	-0.02301	0.06984
66	-0.00751	-0.08898	0.08930	156	-0.06264	-0.03910	0.07385
68	-0.07200	-0.14998	0.16637	158	-0.05352	-0.05027	0.07343
70	-0.10621	-0.14687	0.18125	160	-0.04071	-0.05672	0.06982
72	-0.08507	-0.06540	0.10731	162	-0.02622	-0.05918	0.06473
74	-0.00860	0.07246	0.07297	164	-0.01165	-0.05859	0.05974
76	0.09159	0.20876	0.22797	166	0.00187	-0.05597	0.05600
78	0.16111	0.26727	0.31207	168	0.01366	-0.05220	0.05396
80	0.14538	0.18417	0.23463	170	0.02342	-0.04803	0.05343
82	0.01696	-0.06063	0.06295	172	0.03109	-0.04402	0.05389
84	-0.20601	-0.42616	0.47334	174	0.03679	-0.04056	0.05476
86	-0.46078	-0.81757	0.93847	176	0.04068	-0.03794	0.05562
88	-0.66254	-1.11744	1.29909	178	0.04294	-0.03630	0.05623
				180	0.04367	-0.03575	0.05644

VBR

RADIATION PATTERN FOR THETA=PI/2

H/LMDA=0.5000 B/LMDA= 0.250 OMEGA= 9.92

30 ELEMENT ARRAY

PHI DEG	ER(THETA,PHI) RE	IM	ABSVAL	PHI DEG	ER(THETA,PHI) RE	IM	ABSVAL
0	0.05537	0.03028	0.06311	90	-1.51408	-1.31421	2.00489
2	0.05554	0.02997	0.06311	92	-1.35149	-1.17617	1.79162
4	0.05599	0.02899	0.06305	94	-0.92652	-0.81413	1.23339
6	0.05647	0.02722	0.06269	96	-0.39820	-0.36050	0.53714
8	0.05662	0.02443	0.06166	98	0.05109	0.03121	0.05987
10	0.05587	0.02036	0.05947	100	0.29426	0.25117	0.38688
12	0.05358	0.01473	0.05557	102	0.30567	0.27268	0.40962
14	0.04899	0.00728	0.04953	104	0.15572	0.15019	0.21635
16	0.04139	-0.00208	0.04144	106	-0.03682	-0.01791	0.04095
18	0.03023	-0.01319	0.03298	108	-0.16614	-0.13950	0.21694
20	0.01540	-0.02551	0.02980	110	-0.18296	-0.16668	0.24750
22	-0.00259	-0.03800	0.03809	112	-0.10348	-0.10646	0.14846
24	-0.02231	-0.04902	0.05386	114	0.01370	-0.00474	0.01450
26	-0.04135	-0.05647	0.06999	116	0.10474	0.08329	0.13382
28	-0.05636	-0.05798	0.08085	118	0.13142	0.11996	0.17794
30	-0.06356	-0.05145	0.08177	120	0.09294	0.09790	0.13499
32	-0.05961	-0.03576	0.06951	122	0.01789	0.03645	0.04060
34	-0.04277	-0.01161	0.04432	124	-0.05488	-0.03243	0.06375
36	-0.01424	0.01785	0.02284	126	-0.09518	-0.08010	0.12440
38	0.02101	0.04686	0.05136	128	-0.09208	-0.09177	0.13000
40	0.05436	0.06778	0.08688	130	-0.05384	-0.06900	0.08752
42	0.07520	0.07293	0.10476	132	-0.00027	-0.02519	0.02519
44	0.07418	0.05741	0.09380	134	0.04735	0.02195	0.05220
46	0.04735	0.02195	0.05220	136	0.07418	0.05741	0.09380
48	-0.00027	-0.02519	0.02519	138	0.07520	0.07293	0.10476
50	-0.05384	-0.06900	0.08752	140	0.05436	0.06778	0.08688
52	-0.09208	-0.09177	0.13001	142	0.02101	0.04686	0.05136
54	-0.09518	-0.08010	0.12440	144	-0.01424	0.01785	0.02284
56	-0.05488	-0.03244	0.06375	146	-0.04277	-0.01161	0.04432
58	0.01789	0.03645	0.04060	148	-0.05961	-0.03576	0.06951
60	0.09294	0.09790	0.13499	150	-0.06356	-0.05145	0.08177
62	0.13142	0.11996	0.17794	152	-0.05636	-0.05798	0.08085
64	0.10474	0.08329	0.13382	154	-0.04135	-0.05647	0.06999
66	0.01371	-0.00473	0.01450	156	-0.02231	-0.04902	0.05386
68	-0.10347	-0.10646	0.14846	158	-0.00259	-0.03799	0.03808
70	-0.18296	-0.16668	0.24750	160	0.01540	-0.02551	0.02980
72	-0.16614	-0.13950	0.21694	162	0.03023	-0.01319	0.03299
74	-0.03683	-0.01792	0.04096	164	0.04139	-0.00208	0.04144
76	0.15571	0.15018	0.21634	166	0.04899	0.00728	0.04953
78	0.30567	0.27268	0.40962	168	0.05358	0.01472	0.05556
80	0.29427	0.25117	0.38688	170	0.05587	0.02036	0.05947
82	0.05110	0.03121	0.05988	172	0.05661	0.02443	0.06166
84	-0.39817	-0.36048	0.53711	174	0.05647	0.02722	0.06269
86	-0.92649	-0.81411	1.23335	176	0.05599	0.02899	0.06305
88	-1.35148	-1.17617	1.79161	178	0.05554	0.02997	0.06311
				180	0.05537	0.03028	0.06311

IBR

RADIATION PATTERN FOR THETA=PI/2

H/LMDA=0.5000 B/LMDA= 0.250 OMEGA= 9.92

30 ELEMENT ARRAY

PHI DEG	ER(THETA,PHI) RE	IM	ABSVAL	PHI DEG	ER(THETA,PHI) RE	IM	ABSVAL
0	0.04676	0.03758	0.05999	90	-0.73833	-1.23155	1.43591
2	0.04733	0.03720	0.06020	92	-0.65132	-1.10265	1.28064
4	0.04895	0.03601	0.06077	94	-0.42657	-0.76442	0.87539
6	0.05144	0.03388	0.06160	96	-0.15486	-0.34017	0.37376
8	0.05444	0.03059	0.06244	98	0.06288	0.02695	0.06841
10	0.05742	0.02587	0.06297	100	0.16200	0.23416	0.28474
12	0.05968	0.01944	0.06277	102	0.13870	0.25594	0.29110
14	0.06035	0.01109	0.06136	104	0.04239	0.14245	0.14863
16	0.05843	0.00076	0.05843	106	-0.05660	-0.01474	0.05849
18	0.05294	-0.01134	0.05414	108	-0.10409	-0.12943	0.16609
20	0.04308	-0.02460	0.04961	110	-0.08461	-0.15620	0.17764
22	0.02853	-0.03788	0.04742	112	-0.02073	-0.10089	0.10300
24	0.00974	-0.04951	0.05046	114	0.04622	-0.00596	0.04661
26	-0.01179	-0.05734	0.05854	116	0.08073	0.07695	0.11153
28	-0.03337	-0.05901	0.06780	118	0.06965	0.11220	0.13206
30	-0.05130	-0.05252	0.07341	120	0.02446	0.09230	0.09548
32	-0.06139	-0.03685	0.07160	122	-0.02885	0.03497	0.04534
34	-0.06002	-0.01282	0.06137	124	-0.06458	-0.02989	0.07116
36	-0.04542	0.01633	0.04827	126	-0.06884	-0.07517	0.10193
38	-0.01901	0.04487	0.04873	128	-0.04338	-0.08661	0.09686
40	0.01393	0.06532	0.06679	130	-0.00185	-0.06530	0.06532
42	0.04473	0.07034	0.08336	132	0.03797	-0.02373	0.04478
44	0.06336	0.05535	0.08413	134	0.06181	0.02130	0.06538
46	0.06181	0.02130	0.06538	136	0.06336	0.05536	0.08413
48	0.03797	-0.02373	0.04478	138	0.04473	0.07034	0.08336
50	-0.00185	-0.06530	0.06532	140	0.01393	0.06532	0.06679
52	-0.04338	-0.08661	0.09686	142	-0.01901	0.04487	0.04873
54	-0.06884	-0.07517	0.10193	144	-0.04542	0.01633	0.04827
56	-0.06458	-0.02989	0.07117	146	-0.06002	-0.01282	0.06137
58	-0.02885	0.03497	0.04534	148	-0.06139	-0.03685	0.07160
60	0.02446	0.09229	0.09548	150	-0.05130	-0.05252	0.07341
62	0.06965	0.11220	0.13206	152	-0.03337	-0.05901	0.06779
64	0.08073	0.07695	0.11153	154	-0.01179	-0.05734	0.05853
66	0.04623	-0.00596	0.04661	156	0.00974	-0.04951	0.05046
68	-0.02073	-0.10089	0.10300	158	0.02853	-0.03788	0.04742
70	-0.08460	-0.15620	0.17764	160	0.04308	-0.02460	0.04961
72	-0.10409	-0.12943	0.16609	162	0.05294	-0.01134	0.05414
74	-0.05661	-0.01475	0.05850	164	0.05843	0.00076	0.05843
76	0.04238	0.14245	0.14862	166	0.06035	0.01109	0.06136
78	0.13870	0.25594	0.29110	168	0.05968	0.01944	0.06277
80	0.16200	0.23416	0.28474	170	0.05742	0.02587	0.06297
82	0.06289	0.02695	0.06842	172	0.05444	0.03059	0.06244
84	-0.15485	-0.34016	0.37374	174	0.05144	0.03388	0.06160
86	-0.42655	-0.76441	0.87536	176	0.04895	0.03601	0.06077
88	-0.65131	-1.10265	1.28064	178	0.04733	0.03720	0.06020
				160	0.04676	0.03758	0.05999

VBR

RADIATION PATTERN FOR THETA=PI/2

H/LMDA=0.5000 B/LMDA= 0.250 OMEGA= 9.92

35 ELEMENT ARRAY

PHI DEG	ER(THETA,PHI) RE	IM	ABSVAL	PHI DEG	ER(THETA,PHI) RE	IM	ABSVAL
0	-0.02339	-0.03589	0.04284	90	-1.51591	-1.31789	2.00868
1	-0.02357	-0.03592	0.04297	91	-1.45916	-1.26947	1.93409
2	-0.02412	-0.03602	0.04336	92	-1.29659	-1.13066	1.72033
3	-0.02503	-0.03618	0.04400	93	-1.05003	-0.91970	1.39586
4	-0.02629	-0.03637	0.04488	94	-0.75186	-0.66376	1.00292
5	-0.02786	-0.03658	0.04598	95	-0.43979	-0.39463	0.59089
6	-0.02971	-0.03676	0.04726	96	-0.15104	-0.14386	0.20859
7	-0.03180	-0.03687	0.04869	97	0.08336	0.06189	0.10383
8	-0.03408	-0.03687	0.05021	98	0.24276	0.20451	0.31742
9	-0.03647	-0.03669	0.05173	99	0.31926	0.27642	0.42230
10	-0.03888	-0.03627	0.05317	100	0.31781	0.28088	0.42414
11	-0.04123	-0.03554	0.05443	101	0.25416	0.23040	0.34305
12	-0.04338	-0.03441	0.05536	102	0.15125	0.14389	0.20876
13	-0.04519	-0.03280	0.05584	103	0.03472	0.04297	0.05524
14	-0.04653	-0.03063	0.05571	104	-0.07157	-0.05175	0.08832
15	-0.04724	-0.02783	0.05483	105	-0.14924	-0.12385	0.19393
16	-0.04713	-0.02433	0.05304	106	-0.18773	-0.16323	0.24877
17	-0.04605	-0.02008	0.05024	107	-0.18506	-0.16696	0.24925
18	-0.04384	-0.01506	0.04636	108	-0.14700	-0.13882	0.20219
19	-0.04038	-0.00927	0.04143	109	-0.08499	-0.08773	0.12215
20	-0.03557	-0.00278	0.03567	110	-0.01340	-0.02558	0.02888
21	-0.02937	0.00432	0.02968	111	0.05346	0.03524	0.06403
22	-0.02181	0.01187	0.02483	112	0.10386	0.08404	0.13360
23	-0.01301	0.01964	0.02355	113	0.13040	0.11345	0.17285
24	-0.00318	0.02733	0.02752	114	0.13081	0.12030	0.17772
25	0.00734	0.03460	0.03537	115	0.10767	0.10563	0.15083
26	0.01812	0.04103	0.04485	116	0.06740	0.07399	0.10008
27	0.02862	0.04618	0.05433	117	0.01866	0.03229	0.03729
28	0.03821	0.04960	0.06261	118	-0.02940	-0.01166	0.03163
29	0.04621	0.05085	0.06871	119	-0.06872	-0.05052	0.08529
30	0.05193	0.04956	0.07178	120	-0.09360	-0.07852	0.12218
31	0.05471	0.04547	0.07114	121	-0.10131	-0.09224	0.13701
32	0.05403	0.03847	0.06632	122	-0.09216	-0.09082	0.12938
33	0.04955	0.02866	0.05724	123	-0.06911	-0.07578	0.10256
34	0.04118	0.01637	0.04431	124	-0.03697	-0.05052	0.06261
35	0.02918	0.00220	0.02926	125	-0.00142	-0.01958	0.01963
36	0.01415	-0.01298	0.01921	126	0.03196	0.01220	0.03421
37	-0.00290	-0.02809	0.02824	127	0.05852	0.04035	0.07108
38	-0.02066	-0.04186	0.04668	128	0.07508	0.06140	0.09699
39	-0.03750	-0.05293	0.06487	129	0.08020	0.07319	0.10858
40	-0.05168	-0.06003	0.07921	130	0.07420	0.07500	0.10550
41	-0.06148	-0.06205	0.08735	131	0.05883	0.06747	0.08951
42	-0.06543	-0.05828	0.08762	132	0.03687	0.05231	0.06399
43	-0.06252	-0.04850	0.07913	133	0.01164	0.03195	0.03401
44	-0.05242	-0.03314	0.06202	134	-0.01351	0.00917	0.01633
45	-0.03562	-0.01332	0.03803	135	-0.03562	-0.01332	0.03803
46	-0.01351	0.00917	0.01633	136	-0.05242	-0.03314	0.06202
47	0.01164	0.03195	0.03401	137	-0.06252	-0.04850	0.07913
48	0.03686	0.05231	0.06399	138	-0.06543	-0.05828	0.08762
49	0.05883	0.06747	0.08951	139	-0.06148	-0.06205	0.08735
50	0.07420	0.07500	0.10550	140	-0.05168	-0.06003	0.07921
51	0.08020	0.07319	0.10858	141	-0.03750	-0.05293	0.06487
52	0.07508	0.06140	0.09699	142	-0.02066	-0.04185	0.04668
53	0.05852	0.04035	0.07108	143	-0.00291	-0.02809	0.02824
54	0.03196	0.01220	0.03421	144	0.01415	-0.01298	0.01921
55	-0.00142	-0.01958	0.01963	145	0.02918	0.00220	0.02926
56	-0.03697	-0.05052	0.06260	146	0.04118	0.01637	0.04431
57	-0.06911	-0.07578	0.10256	147	0.04955	0.02866	0.05724
58	-0.09215	-0.09081	0.12938	148	0.05403	0.03847	0.06633
59	-0.10131	-0.09224	0.13701	149	0.05471	0.04547	0.07114
60	-0.09361	-0.07852	0.12218	150	0.05193	0.04956	0.07178
61	-0.06873	-0.05052	0.08530	151	0.04621	0.05085	0.06871
62	-0.02940	-0.01167	0.03163	152	0.03821	0.04960	0.06261
63	0.01865	0.03228	0.03728	153	0.02862	0.04618	0.05433
64	0.06739	0.07399	0.10008	154	0.01812	0.04103	0.04485
65	0.10767	0.10563	0.15083	155	0.00734	0.03460	0.03537
66	0.13081	0.12031	0.17772	156	-0.00318	0.02733	0.02752
67	0.13041	0.11345	0.17285	157	-0.01301	0.01964	0.02355
68	0.10386	0.08404	0.13361	158	-0.02181	0.01187	0.02483
69	0.05346	0.03524	0.06403	159	-0.02936	0.00433	0.02968
70	-0.01340	-0.02558	0.02888	160	-0.03557	-0.00278	0.03567
71	-0.08499	-0.08773	0.12215	161	-0.04038	-0.00928	0.04143
72	-0.14699	-0.13881	0.20218	162	-0.04384	-0.01506	0.04636
73	-0.18506	-0.16696	0.24924	163	-0.04605	-0.02008	0.05024
74	-0.18773	-0.16323	0.24877	164	-0.04713	-0.02433	0.05304
75	-0.14924	-0.12385	0.19394	165	-0.04723	-0.02783	0.05482
76	-0.07158	-0.05176	0.08833	166	-0.04653	-0.03063	0.05571
77	0.03471	0.04296	0.05523	167	-0.04519	-0.03280	0.05584
78	0.15124	0.14388	0.20875	168	-0.04338	-0.03441	0.05537
79	0.25416	0.23040	0.34305	169	-0.04123	-0.03554	0.05443
80	0.31781	0.28088	0.42414	170	-0.03888	-0.03627	0.05318
81	0.31927	0.27642	0.42231	171	-0.03647	-0.03669	0.05173
82	0.24277	0.20452	0.31743	172	-0.03408	-0.03687	0.05021
83	0.08338	0.06191	0.10385	173	-0.03180	-0.03687	0.04869
84	-0.15101	-0.14384	0.20855	174	-0.02971	-0.03676	0.04726
85	-0.43976	-0.39461	0.59085	175	-0.02786	-0.03658	0.04598
86	-0.75182	-0.66374	1.00289	176	-0.02629	-0.03637	0.04488
87	-1.05001	-0.91968	1.39582	177	-0.02503	-0.03618	0.04400
88	-1.29657	-1.13064	1.72031	178	-0.02412	-0.03602	0.04336
89	-1.45915	-1.26947	1.93408	179	-0.02357	-0.03592	0.04297
				180	-0.02339	-0.03589	0.04284

IBR

RADIATION PATTERN FOR THETA=PI/2

H/LMDA=0.5000 B/LMDA= 0.250 OMEGA= 9.92

35 ELEMENT ARRAY

PHI DEG	ER(THETA,PHI) RE	IM	ABSVAL	PHI DEG	ER(THETA,PHI) RE	IM	ABSVAL
0	-0.00347	-0.04322	0.04336	90	-0.73283	-1.23462	1.43573
1	-0.00370	-0.04325	0.04340	91	-0.70302	-1.18941	1.38164
2	-0.00439	-0.04333	0.04355	92	-0.61792	-1.05977	1.22676
3	-0.00553	-0.04345	0.04380	93	-0.48978	-0.86270	0.99204
4	-0.00712	-0.04360	0.04417	94	-0.33666	-0.62350	0.70858
5	-0.00915	-0.04373	0.04468	95	-0.17932	-0.37178	0.41277
6	-0.01161	-0.04382	0.04534	96	-0.03778	-0.13699	0.14211
7	-0.01447	-0.04383	0.04615	97	0.07186	0.05597	0.09109
8	-0.01770	-0.04368	0.04713	98	0.13980	0.19012	0.23599
9	-0.02125	-0.04333	0.04826	99	0.16361	0.25828	0.30574
10	-0.02507	-0.04269	0.04951	100	0.14803	0.26335	0.30210
11	-0.02909	-0.04171	0.05085	101	0.10348	0.21689	0.24031
12	-0.03319	-0.04028	0.05219	102	0.04371	0.13645	0.14328
13	-0.03726	-0.03833	0.05346	103	-0.01694	0.04218	0.04546
14	-0.04116	-0.03578	0.05454	104	-0.06610	-0.04664	0.08090
15	-0.04472	-0.03256	0.05532	105	-0.09533	-0.11462	0.14908
16	-0.04776	-0.02859	0.05566	106	-0.10116	-0.15219	0.18275
17	-0.05007	-0.02384	0.05546	107	-0.08514	-0.15647	0.17813
18	-0.05145	-0.01829	0.05460	108	-0.05287	-0.13084	0.14112
19	-0.05168	-0.01197	0.05304	109	-0.01251	-0.08352	0.08445
20	-0.05056	-0.00493	0.05080	110	0.02708	-0.02556	0.03724
21	-0.04793	0.00270	0.04801	111	0.05813	0.03146	0.06610
22	-0.04366	0.01076	0.04497	112	0.07530	0.07750	0.10806
23	-0.03770	0.01899	0.04222	113	0.07640	0.10559	0.13033
24	-0.03008	0.02710	0.04048	114	0.06253	0.11261	0.12881
25	-0.02092	0.03471	0.04053	115	0.03746	0.09944	0.10626
26	-0.01049	0.04141	0.04272	116	0.00673	0.07022	0.07054
27	0.00084	0.04677	0.04677	117	-0.02361	0.03137	0.03926
28	0.01255	0.05031	0.05186	118	-0.04807	-0.00982	0.04906
29	0.02404	0.05163	0.05696	119	-0.06270	-0.04644	0.07803
30	0.03459	0.05037	0.06110	120	-0.06558	-0.07303	0.09816
31	0.04344	0.04627	0.06347	121	-0.05698	-0.08630	0.10342
32	0.04984	0.03926	0.06345	122	-0.03907	-0.08535	0.09387
33	0.05313	0.02946	0.06075	123	-0.01539	-0.07154	0.07318
34	0.05277	0.01721	0.05551	124	0.00987	-0.04799	0.04899
35	0.04849	0.00315	0.04859	125	0.03263	-0.01894	0.03773
36	0.04028	-0.01188	0.04200	126	0.04953	0.01104	0.05074
37	0.02853	-0.02678	0.03913	127	0.05836	0.03772	0.06949
38	0.01398	-0.04031	0.04266	128	0.05832	0.05778	0.08210
39	-0.00226	-0.05116	0.05121	129	0.04992	0.06912	0.08527
40	-0.01877	-0.05809	0.06105	130	0.03484	0.07101	0.07910
41	-0.03392	-0.06006	0.06898	131	0.01551	0.06400	0.06585
42	-0.04606	-0.05640	0.07281	132	-0.00527	0.04966	0.04994
43	-0.05367	-0.04693	0.07129	133	-0.02476	0.03030	0.03913
44	-0.05559	-0.03210	0.06419	134	-0.04064	0.00853	0.04152
45	-0.05121	-0.01303	0.05284	135	-0.05121	-0.01303	0.05284
46	-0.04064	0.00853	0.04152	136	-0.05559	-0.03210	0.06419
47	-0.02477	0.03030	0.03913	137	-0.05367	-0.04693	0.07129
48	-0.00527	0.04966	0.04994	138	-0.04606	-0.05640	0.07282
49	0.01551	0.06400	0.06585	139	-0.03392	-0.06006	0.06898
50	0.03484	0.07101	0.07910	140	-0.01877	-0.05809	0.06105
51	0.04992	0.06912	0.08527	141	-0.00226	-0.05116	0.05121
52	0.05832	0.05778	0.08209	142	0.01398	-0.04031	0.04266
53	0.05836	0.03772	0.06950	143	0.02853	-0.02678	0.03913
54	0.04953	0.01104	0.05075	144	0.04028	-0.01188	0.04200
55	0.03263	-0.01894	0.03773	145	0.04849	0.00315	0.04859
56	0.00987	-0.04799	0.04899	146	0.05277	0.01721	0.05551
57	-0.01539	-0.07154	0.07317	147	0.05313	0.02946	0.06075
58	-0.03907	-0.08535	0.09387	148	0.04984	0.03926	0.06345
59	-0.05698	-0.08630	0.10342	149	0.04344	0.04627	0.06347
60	-0.06558	-0.07303	0.09816	150	0.03459	0.05037	0.06110
61	-0.06270	-0.04644	0.07803	151	0.02404	0.05163	0.05696
62	-0.04807	-0.00982	0.04907	152	0.01255	0.05031	0.05186
63	-0.02361	0.03137	0.03926	153	0.00084	0.04677	0.04677
64	0.00673	0.07022	0.07054	154	-0.01049	0.04141	0.04272
65	0.03746	0.09943	0.10626	155	-0.02092	0.03471	0.04053
66	0.06253	0.11261	0.12881	156	-0.03008	0.02709	0.04048
67	0.07640	0.10559	0.13033	157	-0.03770	0.01899	0.04221
68	0.07530	0.07751	0.10806	158	-0.04366	0.01076	0.04497
69	0.05813	0.03146	0.06610	159	-0.04793	0.00271	0.04801
70	0.02708	-0.02556	0.03724	160	-0.05056	-0.00493	0.05080
71	-0.01251	-0.08352	0.08445	161	-0.05168	-0.01197	0.05305
72	-0.05286	-0.13084	0.14111	162	-0.05145	-0.01829	0.05460
73	-0.08513	-0.15647	0.17813	163	-0.05007	-0.02384	0.05546
74	-0.10116	-0.15219	0.18275	164	-0.04776	-0.02859	0.05566
75	-0.09533	-0.11462	0.14909	165	-0.04472	-0.03255	0.05531
76	-0.06610	-0.04665	0.08091	166	-0.04116	-0.03578	0.05454
77	-0.01695	0.04217	0.04545	167	-0.03726	-0.03833	0.05346
78	0.04370	0.13644	0.14327	168	-0.03319	-0.04028	0.05219
79	0.10348	0.21689	0.24031	169	-0.02909	-0.04171	0.05085
80	0.14803	0.26335	0.30210	170	-0.02507	-0.04269	0.04951
81	0.16361	0.25828	0.30574	171	-0.02125	-0.04333	0.04826
82	0.13980	0.19013	0.23600	172	-0.01770	-0.04368	0.04713
83	0.07187	0.05599	0.09111	173	-0.01447	-0.04382	0.04615
84	-0.03777	-0.13697	0.14208	174	-0.01161	-0.04382	0.04534
85	-0.17930	-0.37176	0.41274	175	-0.00915	-0.04373	0.04468
86	-0.33664	-0.62348	0.70855	176	-0.00712	-0.04360	0.04417
87	-0.48976	-0.86269	0.99202	177	-0.00553	-0.04345	0.04380
88	-0.61791	-1.05976	1.22674	178	-0.00439	-0.04333	0.04355
89	-0.70302	-1.18941	1.38164	179	-0.00370	-0.04325	0.04340
				180	-0.00347	-0.04322	0.04336

VBR

RADIATION PATTERN FOR THETA=PI/2

H/LMDA=0.5000 B/LMDA= 0.250 OMEGA= 9.92

40 ELEMENT ARRAY

PHI DEG	ER(THETA,PHI) RE	IM	ABSVAL	PHI DEG	ER(THETA,PHI) RE	IM	ABSVAL
0	-0.01255	0.02211	0.02542	90	-1.51739	-1.32053	2.01153
1	-0.01233	0.02224	0.02543	91	-1.44333	-1.25713	1.91405
2	-0.01168	0.02262	0.02546	92	-1.23420	-1.07789	1.63863
3	-0.01060	0.02324	0.02555	93	-0.92634	-0.81343	1.23279
4	-0.00909	0.02410	0.02575	94	-0.57175	-0.50768	0.76461
5	-0.00712	0.02517	0.02616	95	-0.22734	-0.20897	0.30879
6	-0.00471	0.02643	0.02685	96	0.05640	0.03944	0.06882
7	-0.00185	0.02785	0.02791	97	0.24484	0.20732	0.32082
8	0.00144	0.02938	0.02942	98	0.32469	0.28224	0.43022
9	0.00513	0.03097	0.03139	99	0.30460	0.27036	0.40728
10	0.00920	0.03255	0.03382	100	0.21105	0.19317	0.28611
11	0.01357	0.03403	0.03663	101	0.08079	0.08149	0.11475
12	0.01816	0.03531	0.03970	102	-0.04807	-0.03207	0.05779
13	0.02284	0.03628	0.04287	103	-0.14438	-0.12004	0.18776
14	0.02748	0.03682	0.04594	104	-0.18967	-0.16524	0.25155
15	0.03187	0.03677	0.04866	105	-0.18037	-0.16303	0.24313
16	0.03582	0.03600	0.05079	106	-0.12647	-0.12047	0.17467
17	0.03907	0.03437	0.05204	107	-0.04721	-0.05312	0.07107
18	0.04135	0.03175	0.05213	108	0.03470	0.01976	0.03993
19	0.04241	0.02804	0.05084	109	0.09892	0.08007	0.12727
20	0.04197	0.02316	0.04794	110	0.13192	0.11493	0.17496
21	0.03982	0.01714	0.04335	111	0.12930	0.11892	0.17566
22	0.03580	0.01002	0.03718	112	0.09561	0.09440	0.13436
23	0.02985	0.00197	0.02992	113	0.04217	0.04998	0.06539
24	0.02203	-0.00673	0.02303	114	-0.01643	-0.00218	0.01657
25	0.01255	-0.01572	0.02011	115	-0.06606	-0.04956	0.08259
26	0.00178	-0.02451	0.02457	116	-0.09626	-0.08206	0.12648
27	-0.00969	-0.03252	0.03393	117	-0.10208	-0.09393	0.13872
28	-0.02113	-0.03913	0.04447	118	-0.08457	-0.08452	0.11956
29	-0.03168	-0.04367	0.05395	119	-0.04978	-0.05773	0.07623
30	-0.04038	-0.04553	0.06086	120	-0.00684	-0.02073	0.02183
31	-0.04628	-0.04422	0.06401	121	0.03438	0.01800	0.03881
32	-0.04852	-0.03943	0.06252	122	0.06542	0.05050	0.08265
33	-0.04649	-0.03115	0.05596	123	0.08084	0.07090	0.10753
34	-0.03992	-0.01968	0.04451	124	0.07893	0.07629	0.10977
35	-0.02899	-0.00575	0.02955	125	0.06159	0.06693	0.09096
36	-0.01441	0.00956	0.01729	126	0.03354	0.04581	0.05677
37	0.00257	0.02482	0.02495	127	0.00100	0.01772	0.01775
38	0.02019	0.03836	0.04335	128	-0.02963	-0.01179	0.03188
39	0.03637	0.04849	0.06061	129	-0.05299	-0.03753	0.06494
40	0.04891	0.05362	0.07258	130	-0.06561	-0.05554	0.08596
41	0.05579	0.05262	0.07669	131	-0.06626	-0.06354	0.09181
42	0.05550	0.04497	0.07143	132	-0.05587	-0.06113	0.08281
43	0.04739	0.03102	0.05664	133	-0.03708	-0.04957	0.06190
44	0.03188	0.01206	0.03408	134	-0.01358	-0.03136	0.03418
45	0.01062	-0.00972	0.01440	135	0.01062	-0.00972	0.01440
46	-0.01358	-0.03136	0.03418	136	0.03187	0.01206	0.03408
47	-0.03708	-0.04956	0.06190	137	0.04738	0.03102	0.05668
48	-0.05587	-0.06113	0.08281	138	0.05550	0.04497	0.07143
49	[illegible]	[illegible]	[illegible]	139	0.05579	0.05262	0.07669
50	-0.06561	-0.05554	0.08596	140	0.04891	0.05362	[illegible]
51	-0.05299	-0.03753	0.06493	141	0.03637	0.04849	0.06061
52	-0.02963	-0.01179	0.03188	142	0.02019	0.03837	0.04335
53	0.00100	0.01772	0.01775	143	0.00257	0.02482	0.02495
54	0.03354	0.04580	0.05677	144	-0.01441	0.00956	0.01729
55	0.06159	0.06693	0.09096	145	-0.02898	-0.00575	0.02955
56	0.07893	0.07629	0.10977	146	-0.03992	-0.01968	0.04451
57	0.08084	0.07090	0.10753	147	-0.04649	-0.03114	0.05596
58	0.06543	0.05050	0.08265	148	-0.04852	-0.03943	0.06252
59	0.03438	0.01800	0.03881	149	-0.04628	-0.04422	0.06401
60	-0.00684	-0.02073	0.02183	150	-0.04038	-0.04553	0.06086
61	-0.04978	-0.05773	0.07623	151	-0.03168	-0.04367	0.05395
62	-0.08457	-0.08451	0.11956	152	-0.02113	-0.03913	0.04447
63	-0.10208	-0.09393	0.13872	153	-0.00969	-0.03252	0.03394
64	-0.09626	-0.08206	0.12649	154	0.00178	-0.02451	0.02457
65	-0.06607	-0.04956	0.08259	155	0.01254	-0.01572	0.02011
66	-0.01644	-0.00218	0.01658	156	0.02203	-0.00673	0.02304
67	0.04216	0.04998	0.06539	157	0.02985	0.00197	0.02992
68	0.09560	0.09439	0.13435	158	0.03581	0.01002	0.03718
69	0.12930	0.11892	0.17566	159	0.03982	0.01714	0.04336
70	0.13192	0.11493	0.17496	160	0.04197	0.02317	0.04794
71	0.09892	0.08007	0.12727	161	0.04241	0.02804	0.05084
72	0.03471	0.01976	0.03994	162	0.04135	0.03175	0.05214
73	-0.04720	-0.05311	0.07106	163	0.03907	0.03437	0.05204
74	-0.12646	-0.12047	0.17466	164	0.03582	0.03600	0.05079
75	-0.18037	-0.16303	0.24312	165	0.03187	0.03677	0.04866
76	-0.18967	-0.16525	0.25156	166	0.02747	0.03681	0.04594
77	-0.14439	-0.12004	0.18777	167	0.02284	0.03628	0.04287
78	-0.04808	-0.03208	0.05780	168	0.01815	0.03531	0.03970
79	0.08079	0.08148	0.11474	169	0.01357	0.03402	0.03663
80	0.21104	0.19317	0.28610	170	0.00920	0.03255	0.03382
81	0.30460	0.27035	0.40727	171	0.00513	0.03097	0.03139
82	0.32469	0.28225	0.43022	172	0.00144	0.02938	0.02942
83	0.24485	0.20733	0.32084	173	-0.00185	0.02785	0.02791
84	0.05643	0.03946	0.06885	174	-0.00470	0.02643	0.02685
85	-0.22730	-0.20895	0.30875	175	-0.00712	0.02517	0.02616
86	-0.57171	-0.50765	0.76457	176	-0.00909	0.02410	0.02575
87	-0.92631	-0.81341	1.23275	177	-0.01060	0.02324	0.02555
88	-1.23418	-1.07787	1.63860	178	-0.01168	0.02262	0.02546
89	-1.44333	-1.25712	1.91404	179	-0.01233	0.02224	0.02543
				180	-0.01255	0.02211	0.02542

IBR

RADIATION PATTERN FOR THETA=PI/2

H/LMDA=0.5000 B/LMDA= 0.250 OMEGA= 9.92

40 ELEMENT ARRAY

PHI DEG	ER(THETA,PHI) RE	IM	ABSVAL	PHI DEG	ER(THETA,PHI) RE	IM	ABSVAL
0	-0.03058	0.02545	0.03979	90	-0.72877	-1.23796	1.43654
1	-0.03040	0.02559	0.03974	91	-0.69049	-1.17869	1.36605
2	-0.02985	0.02600	0.03959	92	-0.58281	-1.01112	1.16706
3	-0.02893	0.02668	0.03935	93	-0.42569	-0.76378	0.87440
4	-0.02761	0.02760	0.03904	94	-0.24738	-0.47765	0.53791
5	-0.02588	0.02875	0.03868	95	-0.07831	-0.19786	0.21279
6	-0.02369	0.03011	0.03831	96	0.05539	0.03516	0.06560
7	-0.02104	0.03162	0.03798	97	0.13695	0.19307	0.23671
8	-0.01788	0.03324	0.03775	98	0.16177	0.26414	0.30974
9	-0.01421	0.03492	0.03770	99	0.13718	0.25391	0.28860
10	-0.01000	0.03656	0.03791	100	0.07964	0.18233	0.19896
11	-0.00528	0.03809	0.03845	101	0.01002	0.07808	0.07872
12	-0.00008	0.03939	0.03939	102	-0.05155	-0.02834	0.05882
13	0.00554	0.04033	0.04071	103	-0.09017	-0.11122	0.14318
14	0.01149	0.04080	0.04239	104	-0.09895	-0.15432	0.18332
15	0.01763	0.04063	0.04429	105	-0.07954	-0.15308	0.17251
16	0.02378	0.03969	0.04627	106	-0.04055	-0.11392	0.12093
17	0.02971	0.03783	0.04810	107	0.00547	-0.05125	0.05154
18	0.03515	0.03492	0.04954	108	0.04567	0.01699	0.04873
19	0.03980	0.03087	0.05037	109	0.07013	0.07382	0.10182
20	0.04334	0.02562	0.05035	110	0.07404	0.10707	0.13017
21	0.04543	0.01919	0.04931	111	0.05826	0.11150	0.12581
22	0.04575	0.01166	0.04722	112	0.02854	0.08916	0.09362
23	0.04406	0.00321	0.04418	113	-0.00646	0.04796	0.04839
24	0.04018	-0.00589	0.04061	114	-0.03761	-0.00080	0.03762
25	0.03404	-0.01522	0.03729	115	-0.05755	-0.04538	0.07329
26	0.02574	-0.02431	0.03541	116	-0.06226	-0.07624	0.09843
27	0.01558	-0.03256	0.03610	117	-0.05171	-0.08789	0.10197
28	0.00404	-0.03934	0.03954	118	-0.02944	-0.07956	0.08483
29	-0.00819	-0.04398	0.04474	119	-0.00138	-0.05481	0.05483
30	-0.02026	-0.04590	0.05017	120	0.02571	-0.02027	0.03274
31	-0.03116	-0.04461	0.05441	121	0.04586	0.01611	0.04861
32	-0.03986	-0.03984	0.05635	122	0.05509	0.04683	0.07231
33	-0.04538	-0.03158	0.05529	123	0.05211	0.06631	0.08433
34	-0.04692	-0.02020	0.05108	124	0.03828	0.07171	0.08128
35	-0.04397	-0.00640	0.04443	125	0.01709	0.06318	0.06545
36	-0.03647	0.00872	0.03750	126	-0.00679	0.04347	0.04400
37	-0.02487	0.02374	0.03438	127	-0.02852	0.01706	0.03323
38	-0.01015	0.03703	0.03840	128	-0.04410	-0.01084	0.04541
39	0.00619	0.04694	0.04734	129	-0.05098	-0.03530	0.06201
40	0.02224	0.05194	0.05651	130	-0.04844	-0.05250	0.07144
41	0.03593	0.05097	0.06236	131	-0.03748	-0.06024	0.07095
42	0.04525	0.04356	0.06281	132	-0.02046	-0.05807	0.06157
43	0.04861	0.03008	0.05716	133	-0.00055	-0.04714	0.04714
44	0.04511	0.01182	0.04663	134	0.01887	-0.02980	0.03527
45	0.03481	-0.00909	0.03598	135	0.03481	-0.00909	0.03598
46	0.01887	-0.02980	0.03527	136	0.04511	0.01181	0.04663
47	-0.00055	-0.04714	0.04714	137	0.04861	0.03008	0.05716
48	-0.02045	-0.05807	0.06157	138	0.04525	0.04356	0.06281
49	-0.03748	-0.06024	0.07095	139	0.03593	0.05097	0.06236
[illegible]	[illegible]	[illegible]	0.07144	140	0.02224	0.05194	0.05651
51	-0.05098	-0.03530	0.06201	141	[illegible]	[illegible]	[illegible]
52	-0.04410	-0.01084	0.04541	142	-0.01014	0.03703	0.03840
53	-0.02853	0.01705	0.03323	143	-0.02487	0.02374	0.03438
54	-0.00679	0.04347	0.04399	144	-0.03647	0.00872	0.03750
55	0.01708	0.06318	0.06545	145	-0.04397	-0.00640	0.04443
56	0.03828	0.07171	0.08129	146	-0.04692	-0.02020	0.05108
57	0.05211	0.06631	0.08434	147	-0.04538	-0.03158	0.05529
58	0.05510	0.04684	0.07231	148	-0.03986	-0.03984	0.05635
59	0.04586	0.01611	0.04861	149	-0.03116	-0.04461	0.05441
60	0.02571	-0.02027	0.03274	150	-0.02026	-0.04590	0.05017
61	-0.00138	-0.05481	0.05483	151	-0.00819	-0.04398	0.04474
62	-0.02944	-0.07956	0.08483	152	0.00404	-0.03934	0.03954
63	-0.05171	-0.08789	0.10197	153	0.01558	-0.03256	0.03610
64	-0.06226	-0.07624	0.09843	154	0.02574	-0.02431	0.03541
65	-0.05755	-0.04538	0.07329	155	0.03404	-0.01522	0.03729
66	-0.03762	-0.00080	0.03763	156	0.04018	-0.00588	0.04061
67	-0.00647	0.04796	0.04839	157	0.04406	0.00321	0.04418
68	0.02854	0.08916	0.09361	158	0.04576	0.01166	0.04722
69	0.05826	0.11150	0.12581	159	0.04543	0.01920	0.04932
70	0.07404	0.10707	0.13017	160	0.04334	0.02563	0.05035
71	0.07014	0.07382	0.10182	161	0.03980	0.03087	0.05037
72	0.04567	0.01699	0.04873	162	0.03515	0.03492	0.04955
73	0.00548	-0.05124	0.05153	163	0.02971	0.03783	0.04810
74	-0.04054	-0.11392	0.12092	164	0.02378	0.03969	0.04627
75	-0.07954	-0.15308	0.17251	165	0.01763	0.04063	0.04429
76	-0.09895	-0.15432	0.18332	166	0.01149	0.04080	0.04238
77	-0.09017	-0.11122	0.14318	167	0.00554	0.04033	0.04071
78	-0.05155	-0.02835	0.05883	168	-0.00008	0.03939	0.03939
79	0.01002	0.07808	0.07872	169	-0.00528	0.03809	0.03845
80	0.07963	0.18233	0.19896	170	-0.01000	0.03656	0.03791
81	0.13718	0.25390	0.28859	171	-0.01421	0.03492	0.03770
82	0.16177	0.26414	0.30974	172	-0.01788	0.03324	0.03775
83	0.13696	0.19308	0.23672	173	-0.02104	0.03162	0.03798
84	0.05540	0.03518	0.06562	174	-0.02369	0.03011	0.03831
85	-0.07829	-0.19784	0.21277	175	-0.02588	0.02875	0.03868
86	-0.24736	-0.47763	0.53788	176	-0.02761	0.02760	0.03904
87	-0.42567	-0.76376	0.87437	177	-0.02893	0.02667	0.03935
88	-0.58280	-1.01110	1.16704	178	-0.02985	0.02600	0.03959
89	-0.69048	-1.17869	1.36604	179	-0.03040	0.02559	0.03973
				180	-0.03058	0.02545	0.03978

VBR

RADIATION PATTERN FOR THETA=PI/2

H/LMDA=0.5000 B/LMDA= 0.250 OMEGA= 9.92

45 ELEMENT ARRAY

PHI DEG	ER(THETA,PHI) RE	IM	ABSVAL	PHI DEG	ER(THETA,PHI) RE	IM	ABSVAL
0	0.03370	-0.00007	0.03370	90	-1.51845	-1.32249	2.01363
1	0.03358	-0.00022	0.03358	91	-1.42493	-1.24221	1.89037
2	0.03323	-0.00067	0.03323	92	-1.16506	-1.01885	1.54771
3	0.03261	-0.00141	0.03264	93	-0.79547	-0.70036	1.05985
4	0.03170	-0.00246	0.03180	94	-0.39380	-0.35269	0.52865
5	0.03047	-0.00382	0.03071	95	-0.03872	-0.04309	0.05793
6	0.02887	-0.00549	0.02939	96	0.20933	0.17612	0.27356
7	0.02684	-0.00747	0.02786	97	0.32145	0.27898	0.42562
8	0.02434	-0.00975	0.02622	98	0.30427	0.26962	0.40653
9	0.02131	-0.01232	0.02461	99	0.19415	0.17780	0.26327
10	0.01771	-0.01513	0.02329	100	0.04410	0.04822	0.06534
11	0.01352	-0.01814	0.02262	101	-0.09249	-0.07308	0.11787
12	0.00874	-0.02127	0.02300	102	-0.17611	-0.15091	0.23192
13	0.00340	-0.02442	0.02466	103	-0.18981	-0.16883	0.25402
14	-0.00242	-0.02746	0.02756	104	-0.14038	-0.13065	0.19178
15	-0.00858	-0.03022	0.03142	105	-0.05267	-0.05631	0.07710
16	-0.01492	-0.03252	0.03578	106	0.04057	0.02641	0.04841
17	-0.02117	-0.03415	0.04018	107	0.10952	0.09109	0.14245
18	-0.02703	-0.03487	0.04411	108	0.13588	0.12026	0.18145
19	-0.03213	-0.03445	0.04711	109	0.11671	0.10930	0.15990
20	-0.03608	-0.03270	0.04869	110	0.06327	0.06608	0.09148
21	-0.03846	-0.02943	0.04843	111	-0.00427	0.00707	0.00826
22	-0.03888	-0.02456	0.04599	112	-0.06395	-0.04862	0.08034
23	-0.03703	-0.01810	0.04121	113	-0.09876	-0.08501	0.13031
24	-0.03270	-0.01018	0.03425	114	-0.10099	-0.09352	0.13764
25	-0.02589	-0.00112	0.02591	115	-0.07329	-0.07443	0.10446
26	-0.01678	0.00862	0.01886	116	-0.02664	-0.03576	0.04459
27	-0.00585	0.01838	0.01929	117	0.02380	0.00994	0.02580
28	0.00617	0.02739	0.02808	118	0.06352	0.04951	0.08054
29	0.01827	0.03476	0.03927	119	0.08252	0.07287	0.11009
30	0.02926	0.03960	0.04924	120	0.07755	0.07526	0.10807
31	0.03786	0.04111	0.05589	121	0.05215	0.05791	0.07793
32	0.04284	0.03871	0.05774	122	0.01484	0.02695	0.03076
33	0.04323	0.03217	0.05388	123	-0.02367	-0.00873	0.02523
34	0.03845	0.02174	0.04417	124	-0.05352	-0.03999	0.06681
35	0.02859	0.00820	0.02974	125	-0.06808	-0.05973	0.09056
36	0.01441	-0.00714	0.01608	126	-0.06513	-0.06431	0.09153
37	-0.00253	-0.02246	0.02261	127	-0.04692	-0.05397	0.07152
38	-0.02005	-0.03569	0.04093	128	-0.01900	-0.03226	0.03744
39	-0.03550	-0.04470	0.05708	129	0.01146	-0.00485	0.01244
40	-0.04623	-0.04772	0.06644	130	0.03750	0.02199	0.04347
41	-0.04998	-0.04365	0.06636	131	0.05390	0.04279	0.06882
42	-0.04543	-0.03241	0.05581	132	0.05805	0.05391	0.07922
43	-0.03258	-0.01518	0.03595	133	0.05015	0.05402	0.07371
44	-0.01300	0.00566	0.01418	134	0.03283	0.04406	0.05495
45	0.01022	0.02671	0.02860	135	0.01022	0.02671	0.02859
46	0.03283	0.04406	0.05495	136	-0.01300	0.00566	0.01418
47	0.05015	0.05402	0.07371	137	-0.03258	-0.01518	0.03594
48	0.05804	0.05391	0.07922	138	-0.04543	-0.03241	0.05581
49	0.05390	0.04279	0.06882	139	-0.04998	-0.04365	0.06636
50	0.03750	0.02199	0.04347	140	-0.04623	-0.04772	0.06644
51	0.01146	-0.00485	0.01244	141	-0.03550	-0.04470	0.05708
52	-0.01900	-0.03226	0.03744	142	-0.02004	-0.03569	0.04093
53	-0.04692	-0.05397	0.07151	143	-0.00254	-0.02247	0.02261
54	-0.06513	-0.06431	0.09153	144	0.01441	-0.00714	0.01608
55	-0.06808	-0.05973	0.09057	145	0.02859	0.00820	0.02974
56	-0.05353	-0.03999	0.06682	146	0.03845	0.02174	0.04417
57	-0.02367	-0.00873	0.02523	147	0.04323	0.03217	0.05389
58	0.01483	0.02694	0.03076	148	0.04284	0.03871	0.05774
59	0.05214	0.05791	0.07792	149	0.03786	0.04111	0.05589
60	0.07755	0.07526	0.10807	150	0.02926	0.03960	0.04924
61	0.08252	0.07286	0.11009	151	0.01827	0.03477	0.03927
62	0.06352	0.04952	0.08054	152	0.00616	0.02739	0.02808
63	0.02381	0.00994	0.02580	153	-0.00585	0.01838	0.01929
64	-0.02664	-0.03575	0.04459	154	-0.01678	0.00862	0.01886
65	-0.07329	-0.07443	0.10446	155	-0.02588	-0.00112	0.02591
66	-0.10099	-0.09352	0.13764	156	-0.03270	-0.01019	0.03425
67	-0.09877	-0.08501	0.13032	157	-0.03703	-0.01810	0.04121
68	-0.06396	-0.04862	0.08034	158	-0.03888	-0.02456	0.04598
69	-0.00428	0.00706	0.00826	159	-0.03845	-0.02943	0.04842
70	0.06326	0.06608	0.09148	160	-0.03608	-0.03270	0.04869
71	0.11671	0.10929	0.15989	161	-0.03213	-0.03445	0.04711
72	0.13588	0.12026	0.18146	162	-0.02703	-0.03487	0.04412
73	0.10952	0.09110	0.14246	163	-0.02117	-0.03415	0.04018
74	0.04057	0.02642	0.04842	164	-0.01492	-0.03252	0.03578
75	-0.05266	-0.05630	0.07709	165	-0.00858	-0.03022	0.03142
76	-0.14038	-0.13065	0.19177	166	-0.00242	-0.02746	0.02756
77	-0.18980	-0.16882	0.25402	167	0.00340	-0.02442	0.02466
78	-0.17611	-0.15092	0.23193	168	0.00874	-0.02127	0.02300
79	-0.09249	-0.07308	0.11788	169	0.01352	-0.01814	0.02262
80	0.04410	0.04821	0.06534	170	0.01771	-0.01513	0.02329
81	0.19415	0.17780	0.26326	171	0.02131	-0.01232	0.02461
82	0.30426	0.26961	0.40653	172	0.02434	-0.00975	0.02622
83	0.32145	0.27898	0.42563	173	0.02684	-0.00747	0.02786
84	0.20935	0.17613	0.27358	174	0.02887	-0.00549	0.02939
85	-0.03868	-0.04307	0.05789	175	0.03047	-0.00382	0.03071
86	-0.39375	-0.35267	0.52860	176	0.03170	-0.00246	0.03180
87	-0.79543	-0.70034	1.05980	177	0.03261	-0.00141	0.03264
88	-1.16502	-1.01883	1.54767	178	0.03323	-0.00067	0.03323
89	-1.42492	-1.24220	1.89036	179	0.03358	-0.00022	0.03358
				180	0.03370	-0.00007	0.03370

IBR

RADIATION PATTERN FOR THETA=PI/2

H/LMDA=0.5000 B/LMDA= 0.250 OMEGA= 9.92

45 ELEMENT ARRAY

PHI DEG	ER(THETA,PHI) RE	IM	ABSVAL	PHI DEG	ER(THETA,PHI) RE	IM	ABSVAL
0	0.04097	0.00164	0.04100	90	-0.72648	-1.24050	1.43758
1	0.04094	0.00147	0.04097	91	-0.67869	-1.16538	1.34860
2	0.04086	0.00096	0.04087	92	-0.54651	-0.95636	1.10149
3	0.04069	0.00011	0.04069	93	-0.36046	-0.65818	0.75042
4	0.04041	-0.00108	0.04042	94	-0.16190	-0.33246	0.36979
5	0.03996	-0.00262	0.04005	95	0.00817	-0.04209	0.04287
6	0.03928	-0.00451	0.03954	96	0.11966	0.16395	0.20297
7	0.03829	-0.00675	0.03888	97	0.16032	0.26120	0.30648
8	0.03692	-0.00931	0.03808	98	0.13711	0.25332	0.28805
9	0.03507	-0.01218	0.03713	99	0.07218	0.16795	0.18280
10	0.03265	-0.01532	0.03607	100	-0.00511	0.04676	0.04704
11	0.02958	-0.01866	0.03498	101	-0.06739	-0.06714	0.09513
12	0.02579	-0.02212	0.03398	102	-0.09674	-0.14072	0.17077
13	0.02123	-0.02559	0.03325	103	-0.08856	-0.15834	0.18143
14	0.01588	-0.02893	0.03300	104	-0.05095	-0.12333	0.13344
15	0.00977	-0.03195	0.03342	105	-0.00033	-0.05411	0.05412
16	0.00301	-0.03447	0.03460	106	0.04484	0.02339	0.05057
17	-0.00426	-0.03626	0.03651	107	0.07020	0.08441	0.10979
18	-0.01180	-0.03707	0.03891	108	0.06951	0.11242	0.13217
19	-0.01930	-0.03668	0.04145	109	0.04565	0.10286	0.11254
20	-0.02636	-0.03488	0.04372	110	0.00863	0.06291	0.06350
21	-0.03254	-0.03149	0.04528	111	-0.02838	0.00782	0.02944
22	-0.03734	-0.02644	0.04575	112	-0.05355	-0.04452	0.06964
23	-0.04027	-0.01974	0.04485	113	-0.05988	-0.07908	0.09919
24	-0.04089	-0.01156	0.04250	114	-0.04689	-0.08764	0.09940
25	-0.03887	-0.00222	0.03894	115	-0.02007	-0.07029	0.07309
26	-0.03403	0.00779	0.03491	116	0.01141	-0.03436	0.03620
27	-0.02643	0.01781	0.03188	117	0.03776	0.00842	0.03869
28	-0.01641	0.02704	0.03163	118	0.05156	0.04573	0.06892
29	-0.00462	0.03459	0.03489	119	0.04965	0.06801	0.08420
30	0.00801	0.03956	0.04036	120	0.03358	0.07067	0.07824
31	0.02028	0.04117	0.04589	121	0.00876	0.05473	0.05542
32	0.03086	0.03886	0.04962	122	-0.01745	0.02584	0.03118
33	0.03840	0.03243	0.05026	123	-0.03791	-0.00767	0.03868
34	0.04174	0.02214	0.04725	124	-0.04759	-0.03721	0.06041
35	0.04012	0.00878	0.04107	125	-0.04463	-0.05601	0.07162
36	0.03332	-0.00632	0.03392	126	-0.03052	-0.06057	0.06782
37	0.02185	-0.02140	0.03058	127	-0.00938	-0.05100	0.05186
38	0.00698	-0.03439	0.03509	128	0.01332	-0.03061	0.03338
39	-0.00934	-0.04324	0.04424	129	0.03218	-0.00471	0.03253
40	-0.02471	-0.04623	0.05242	130	0.04316	0.02075	0.04789
41	-0.03655	-0.04233	0.05593	131	0.04428	0.04055	0.06005
42	-0.04264	-0.03151	0.05302	132	0.03589	0.05120	0.06253
43	-0.04152	-0.01491	0.04411	133	0.02028	0.05137	0.05523
44	-0.03288	0.00512	0.03328	134	0.00104	0.04190	0.04191
45	-0.01787	0.02531	0.03098	135	-0.01787	0.02531	0.03098
46	0.00104	0.04190	0.04191	136	-0.03288	0.00512	0.03328
47	0.02028	0.05137	0.05523	137	-0.04151	-0.01491	0.04411
48	0.03589	0.05120	0.06253	138	-0.04264	-0.03151	0.05302
49	0.04428	0.04055	0.06005	139	-0.03655	-0.04233	0.05593
50	0.04316	0.02075	0.04789	140	-0.02471	-0.04623	0.05242
51	0.03218	-0.00471	0.03253	141	-0.00935	-0.04324	0.04424
52	0.01332	-0.03061	0.03338	142	0.00698	-0.03439	0.03509
53	-0.00938	-0.05100	0.05186	143	0.02185	-0.02140	0.03058
54	-0.03052	-0.06057	0.06782	144	0.03332	-0.00633	0.03391
55	-0.04463	-0.05601	0.07162	145	0.04012	0.00878	0.04107
56	-0.04759	-0.03721	0.06041	146	0.04174	0.02214	0.04725
57	-0.03791	-0.00767	0.03868	147	0.03840	0.03244	0.05027
58	-0.01745	0.02584	0.03118	148	0.03086	0.03886	0.04962
59	0.00876	0.05472	0.05542	149	0.02028	0.04117	0.04589
60	0.03358	0.07067	0.07824	150	0.00801	0.03956	0.04036
61	0.04965	0.06801	0.08420	151	-0.00462	0.03459	0.03489
62	0.05156	0.04574	0.06893	152	-0.01641	0.02704	0.03163
63	0.03776	0.00843	0.03869	153	-0.02643	0.01781	0.03187
64	0.01141	-0.03436	0.03620	154	-0.03403	0.00779	0.03491
65	-0.02006	-0.07028	0.07309	155	-0.03887	-0.00222	0.03893
66	-0.04689	-0.08764	0.09940	156	-0.04090	-0.01156	0.04250
67	-0.05988	-0.07909	0.09920	157	-0.04027	-0.01975	0.04485
68	-0.05355	-0.04453	0.06964	158	-0.03734	-0.02644	0.04575
69	-0.02838	0.00781	0.02944	159	-0.03254	-0.03149	0.04528
70	0.00863	0.06291	0.06350	160	-0.02636	-0.03488	0.04372
71	0.04565	0.10286	0.11254	161	-0.01930	-0.03668	0.04145
72	0.06951	0.11242	0.13217	162	-0.01180	-0.03707	0.03891
73	0.07020	0.08441	0.10979	163	-0.00426	-0.03626	0.03651
74	0.04484	0.02340	0.05058	164	0.00301	-0.03447	0.03460
75	-0.00033	-0.05411	0.05411	165	0.00977	-0.03195	0.03341
76	-0.05095	-0.12333	0.13344	166	0.01588	-0.02893	0.03300
77	-0.08856	-0.15834	0.18143	167	0.02123	-0.02559	0.03325
78	-0.09674	-0.14073	0.17077	168	0.02579	-0.02212	0.03398
79	-0.06739	-0.06714	0.09513	169	0.02958	-0.01866	0.03498
80	-0.00512	0.04676	0.04704	170	0.03265	-0.01532	0.03607
81	0.07218	0.16795	0.18280	171	0.03507	-0.01218	0.03713
82	0.13710	0.25332	0.28804	172	0.03692	-0.00931	0.03808
83	0.16032	0.26121	0.30648	173	0.03829	-0.00675	0.03888
84	0.11967	0.16396	0.20299	174	0.03928	-0.00451	0.03954
85	0.00818	-0.04207	0.04285	175	0.03996	-0.00262	0.04004
86	-0.16188	-0.33244	0.36976	176	0.04041	-0.00108	0.04042
87	-0.36044	-0.65815	0.75039	177	0.04069	0.00011	0.04069
88	-0.54649	-0.95633	1.10146	178	0.04086	0.00096	0.04087
89	-0.67868	-1.16538	1.34860	179	0.04094	0.00147	0.04097
				180	0.04097	0.00164	0.04100

VBR

RADIATION PATTERN FOR THETA=PI/2

H/LMDA=0.5000 B/LMDA= 0.250 OMEGA= 9.92

50 ELEMENT ARRAY

PHI DEG	ER(THETA,PHI) RE	IM	ABSVAL	PHI DEG	ER(THETA,PHI) RE	IM	ABSVAL
0	-0.03263	-0.01764	0.03709	90	-1.51924	-1.32413	2.01530
1	-0.03268	-0.01756	0.03710	91	-1.40413	-1.22508	1.86344
2	-0.03281	-0.01729	0.03709	92	-1.09006	-0.95445	1.44887
3	-0.03301	-0.01684	0.03706	93	-0.66066	-0.58337	0.88136
4	-0.03324	-0.01618	0.03697	94	-0.22506	-0.20499	0.30442
5	-0.03344	-0.01528	0.03677	95	0.11616	0.09416	0.14954
6	-0.03354	-0.01410	0.03638	96	0.30033	0.25918	0.39671
7	-0.03345	-0.01260	0.03574	97	0.31772	0.27993	0.42345
8	-0.03307	-0.01074	0.03478	98	0.20818	0.18884	0.28107
9	-0.03230	-0.00849	0.03340	99	0.04195	0.04524	0.06170
10	-0.03101	-0.00582	0.03155	100	-0.10716	-0.08719	0.13815
11	-0.02908	-0.00270	0.02921	101	-0.18650	-0.16155	0.24674
12	-0.02640	0.00085	0.02642	102	-0.17862	-0.16043	0.24009
13	-0.02289	0.00481	0.02339	103	-0.10132	-0.09685	0.14016
14	-0.01847	0.00910	0.02059	104	0.00471	-0.00463	0.00660
15	-0.01315	0.01361	0.01892	105	0.09435	0.07707	0.12183
16	-0.00698	0.01815	0.01945	106	0.13543	0.11872	0.18010
17	-0.00011	0.02252	0.02252	107	0.11866	0.10981	0.16167
18	0.00720	0.02642	0.02738	108	0.05783	0.05997	0.08331
19	0.01462	0.02954	0.03296	109	-0.01856	-0.00730	0.01994
20	0.02169	0.03152	0.03826	110	-0.07981	-0.06505	0.10296
21	0.02788	0.03202	0.04245	111	-0.10470	-0.09306	0.14008
22	0.03258	0.03071	0.04477	112	-0.08800	-0.08426	0.12183
23	0.03521	0.02739	0.04461	113	-0.04027	-0.04563	0.06086
24	0.03525	0.02195	0.04152	114	0.01789	0.00610	0.01891
25	0.03233	0.01450	0.03543	115	0.06453	0.05144	0.08252
26	0.02632	0.00538	0.02687	116	0.08423	0.07524	0.11294
27	0.01744	-0.00480	0.01809	117	0.07271	0.07136	0.10188
28	0.00627	-0.01520	0.01645	118	0.03695	0.04363	0.05717
29	-0.00619	-0.02475	0.02551	119	-0.00864	0.00339	0.00929
30	-0.01859	-0.03226	0.03723	120	-0.04790	-0.03514	0.05940
31	-0.02934	-0.03658	0.04689	121	-0.06848	-0.05978	0.09090
32	-0.03679	-0.03675	0.05200	122	-0.06545	-0.06399	0.09153
33	-0.03953	-0.03225	0.05101	123	-0.04187	-0.04823	0.06386
34	-0.03662	-0.02313	0.04331	124	-0.00689	-0.01893	0.02015
35	-0.02790	-0.01018	0.02970	125	0.02782	0.01411	0.03119
36	-0.01420	0.00507	0.01508	126	0.05193	0.04106	0.06620
37	0.00265	0.02040	0.02057	127	0.05935	0.05484	0.08081
38	0.01996	0.03327	0.03880	128	0.04941	0.05279	0.07230
39	0.03454	0.04115	0.05373	129	0.02637	0.03677	0.04525
40	0.04326	0.04206	0.06034	130	-0.00238	0.01213	0.01236
41	0.04377	0.03512	0.05611	131	-0.02876	-0.01415	0.03205
42	0.03516	0.02088	0.04089	132	-0.04617	-0.03548	0.05823
43	0.01842	0.00156	0.01849	133	-0.05105	-0.04717	0.06950
44	-0.00344	-0.01921	0.01952	134	-0.04328	-0.04735	0.06415
45	-0.02580	-0.03697	0.04508	135	-0.02580	-0.03697	0.04509
46	-0.04328	-0.04735	0.06415	136	-0.00344	-0.01921	0.01952
47	-0.05105	-0.04717	0.06950	137	0.01842	0.00156	0.01849
48	-0.04617	-0.03548	0.05823	138	0.03516	0.02088	0.04089
49	[illegible]	[illegible]	[illegible]	139	[illegible]	[illegible]	[illegible]
50	-0.00239	0.01213	0.01236	140	0.04326	0.04207	0.06034
51	0.02637	0.03677	0.04525	141	0.03454	0.04115	0.05372
52	0.04941	0.05279	0.07230	142	0.01997	0.03327	0.03880
53	0.05935	0.05484	0.08081	143	0.00265	0.02040	0.02057
54	0.05194	0.04106	0.06620	144	-0.01420	0.00506	0.01508
55	0.02782	0.01411	0.03119	145	-0.02790	-0.01017	0.02970
56	-0.00689	-0.01893	0.02014	146	-0.03662	-0.02313	0.04331
57	-0.04186	-0.04822	0.06386	147	-0.03953	-0.03225	0.05101
58	-0.06545	-0.06399	0.09153	148	-0.03679	-0.03675	0.05200
59	-0.06848	-0.05978	0.09090	149	-0.02934	-0.03658	0.04689
60	-0.04790	-0.03514	0.05941	150	-0.01859	-0.03226	0.03723
61	-0.00865	0.00339	0.00929	151	-0.00619	-0.02475	0.02551
62	0.03694	0.04363	0.05717	152	0.00627	-0.01521	0.01645
63	0.07271	0.07136	0.10188	153	0.01744	-0.00480	0.01809
64	0.08423	0.07524	0.11294	154	0.02632	0.00538	0.02687
65	0.06453	0.05144	0.08253	155	0.03233	0.01450	0.03543
66	0.01790	0.00611	0.01891	156	0.03525	0.02195	0.04152
67	-0.04026	-0.04562	0.06085	157	0.03521	0.02739	0.04461
68	-0.08799	-0.08426	0.12183	158	0.03258	0.03071	0.04478
69	-0.10470	-0.09305	0.14008	159	0.02788	0.03202	0.04246
70	-0.07981	-0.06505	0.10296	160	0.02169	0.03152	0.03826
71	-0.01856	-0.00731	0.01994	161	0.01462	0.02954	0.03296
72	0.05782	0.05997	0.08330	162	0.00720	0.02642	0.02738
73	0.11865	0.10981	0.16167	163	-0.00011	0.02252	0.02252
74	0.13543	0.11872	0.18010	164	-0.00698	0.01815	0.01945
75	0.09436	0.07707	0.12184	165	-0.01315	0.01361	0.01892
76	0.00471	-0.00462	0.00660	166	-0.01847	0.00910	0.02059
77	-0.10131	-0.09684	0.14015	167	-0.02289	0.00481	0.02339
78	-0.17862	-0.16043	0.24009	168	-0.02641	0.00085	0.02642
79	-0.18650	-0.16155	0.24674	169	-0.02908	-0.00270	0.02921
80	-0.10716	-0.08720	0.13816	170	-0.03101	-0.00582	0.03155
81	0.04195	0.04523	0.06169	171	-0.03230	-0.00850	0.03340
82	0.20816	0.18884	0.28105	172	-0.03308	-0.01075	0.03478
83	0.31772	0.27993	0.42344	173	-0.03345	-0.01260	0.03574
84	0.30034	0.25919	0.39672	174	-0.03354	-0.01410	0.03638
85	0.11619	0.09418	0.14957	175	-0.03344	-0.01528	0.03676
86	-0.22502	-0.20497	0.30438	176	-0.03324	-0.01618	0.03697
87	-0.66062	-0.58334	0.88131	177	-0.03301	-0.01684	0.03706
88	-1.09002	-0.95443	1.44882	178	-0.03281	-0.01729	0.03709
89	-1.40412	-1.22507	1.86343	179	-0.03268	-0.01756	0.03710
				180	-0.03263	-0.01764	0.03709

IBR

RADIATION PATTERN FOR THETA=PI/2

H/LMDA=0.5000 B/LMDA= 0.250 OMEGA= 9.92

50 ELEMENT ARRAY

PHI DEG	ER(THETA,PHI) RE	IM	ABSVAL	PHI DEG	ER(THETA,PHI) RE	IM	ABSVAL
0	-0.02759	-0.02254	0.03563	90	-0.72462	-1.24179	1.43774
1	-0.02774	-0.02244	0.03568	91	-0.66634	-1.14910	1.32832
2	-0.02816	-0.02213	0.03581	92	-0.50818	-0.89582	1.02992
3	-0.02884	-0.02158	0.03603	93	-0.29455	-0.54836	0.62246
4	-0.02976	-0.02080	0.03630	94	-0.08258	-0.19378	0.21064
5	-0.03084	-0.01972	0.03661	95	0.07656	0.08697	0.11587
6	-0.03204	-0.01833	0.03691	96	0.15324	0.24239	0.28677
7	-0.03326	-0.01658	0.03716	97	0.14660	0.26275	0.30089
8	-0.03440	-0.01442	0.03730	98	0.08141	0.17812	0.19584
9	-0.03534	-0.01182	0.03727	99	-0.00363	0.04382	0.04397
10	-0.03594	-0.00876	0.03699	100	-0.07092	-0.08055	0.10732
11	-0.03603	-0.00520	0.03641	101	-0.09689	-0.15094	0.17936
12	-0.03546	-0.00118	0.03548	102	-0.07805	-0.15075	0.16975
13	-0.03406	0.00328	0.03421	103	-0.02903	-0.09184	0.09632
14	-0.03166	0.00809	0.03268	104	0.02585	-0.00564	0.02646
15	-0.02816	0.01310	0.03106	105	0.06338	0.07120	0.09532
16	-0.02347	0.01815	0.02967	106	0.07038	0.11089	0.13134
17	-0.01759	0.02299	0.02895	107	0.04751	0.10331	0.11372
18	-0.01061	0.02731	0.02930	108	0.00725	0.05718	0.05764
19	-0.00274	0.03078	0.03090	109	-0.03244	-0.00570	0.03293
20	0.00569	0.03303	0.03352	110	-0.05576	-0.06008	0.08197
21	0.01422	0.03370	0.03658	111	-0.05499	-0.08690	0.10284
22	0.02227	0.03247	0.03937	112	-0.03250	-0.07931	0.08571
23	0.02915	0.02912	0.04120	113	0.00119	-0.04357	0.04358
24	0.03417	0.02356	0.04151	114	0.03230	0.00478	0.03265
25	0.03665	0.01593	0.03997	115	0.04925	0.04748	0.06841
26	0.03606	0.00658	0.03666	116	0.04677	0.07024	0.08438
27	0.03209	-0.00386	0.03232	117	0.02714	0.06710	0.07238
28	0.02478	-0.01451	0.02872	118	-0.00138	0.04147	0.04150
29	0.01457	-0.02429	0.02833	119	-0.02805	0.00386	0.02831
30	0.00236	-0.03200	0.03209	120	-0.04367	-0.03240	0.05438
31	-0.01055	-0.03648	0.03798	121	-0.04358	-0.05582	0.07081
32	-0.02255	-0.03681	0.04316	122	-0.02878	-0.06012	0.06665
33	-0.03190	-0.03245	0.04550	123	-0.00506	-0.04558	0.04586
34	-0.03701	-0.02349	0.04383	124	0.01942	-0.01817	0.02659
35	-0.03675	-0.01073	0.03828	125	0.03686	0.01294	0.03907
36	-0.03071	0.00429	0.03101	126	0.04232	0.03846	0.05718
37	-0.01943	0.01940	0.02746	127	0.03483	0.05164	0.06229
38	-0.00446	0.03208	0.03239	128	0.01735	0.04986	0.05279
39	0.01174	0.03986	0.04156	129	-0.00455	0.03481	0.03511
40	0.02619	0.04085	0.04852	130	-0.02451	0.01151	0.02708
41	0.03590	0.03419	0.04957	131	-0.03723	-0.01344	0.03958
42	0.03854	0.02047	0.04364	132	-0.03980	-0.03375	0.05218
43	0.03305	0.00186	0.03310	133	-0.03215	-0.04492	0.05524
44	0.02010	-0.01812	0.02707	134	-0.01678	-0.04511	0.04813
45	0.00217	-0.03518	0.03524	135	0.00217	-0.03518	0.03524
46	-0.01678	-0.04511	0.04813	136	0.02010	-0.01813	0.02707
47	-0.03215	-0.04492	0.05524	137	0.03305	0.00186	0.03310
48	-0.03980	-0.03375	0.05218	138	0.03854	0.02047	0.04364
49	-0.03723	-0.01344	0.03958	139	0.03590	0.03419	0.04958
50	-0.02491	[illegible]	[illegible]	140	[illegible]	[illegible]	[illegible]
51	-0.00455	0.03481	0.03511	141	0.01174	0.03986	0.04155
52	0.01735	0.04986	0.05279	142	-0.00446	0.03208	0.03239
53	0.03483	0.05164	0.06229	143	-0.01943	0.01940	0.02745
54	0.04232	0.03846	0.05718	144	-0.03071	0.00429	0.03101
55	0.03687	0.01294	0.03907	145	-0.03675	-0.01072	0.03828
56	0.01942	-0.01817	0.02659	146	-0.03701	-0.02348	0.04383
57	-0.00505	-0.04558	0.04586	147	-0.03190	-0.03245	0.04550
58	-0.02878	-0.06011	0.06665	148	-0.02255	-0.03681	0.04316
59	-0.04358	-0.05582	0.07082	149	-0.01055	-0.03648	0.03798
60	-0.04367	-0.03240	0.05438	150	0.00236	-0.03200	0.03209
61	-0.02805	0.00386	0.02831	151	0.01457	-0.02429	0.02833
62	-0.00138	0.04147	0.04149	152	0.02478	-0.01452	0.02872
63	0.02713	0.06710	0.07238	153	0.03209	-0.00386	0.03232
64	0.04677	0.07024	0.08439	154	0.03606	0.00658	0.03666
65	0.04925	0.04748	0.06841	155	0.03665	0.01593	0.03997
66	0.03230	0.00478	0.03265	156	0.03417	0.02356	0.04151
67	0.00119	-0.04356	0.04358	157	0.02915	0.02911	0.04120
68	-0.03250	-0.07931	0.08571	158	0.02227	0.03247	0.03937
69	-0.05499	-0.08690	0.10284	159	0.01422	0.03370	0.03658
70	-0.05576	-0.06008	0.08197	160	0.00569	0.03303	0.03352
71	-0.03244	-0.00570	0.03293	161	-0.00274	0.03078	0.03090
72	0.00725	0.05718	0.05763	162	-0.01061	0.02731	0.02930
73	0.04751	0.10331	0.11371	163	-0.01759	0.02299	0.02895
74	0.07038	0.11090	0.13134	164	-0.02347	0.01815	0.02967
75	0.06338	0.07120	0.09533	165	-0.02816	0.01311	0.03106
76	0.02585	-0.00563	0.02646	166	-0.03166	0.00808	0.03268
77	-0.02902	-0.09183	0.09631	167	-0.03405	0.00328	0.03421
78	-0.07805	-0.15074	0.16975	168	-0.03546	-0.00118	0.03548
79	-0.09689	-0.15094	0.17936	169	-0.03603	-0.00520	0.03641
80	-0.07093	-0.08055	0.10733	170	-0.03594	-0.00876	0.03699
81	-0.00363	0.04382	0.04397	171	-0.03534	-0.01182	0.03727
82	0.08140	0.17811	0.19583	172	-0.03440	-0.01442	0.03730
83	0.14660	0.26275	0.30088	173	-0.03326	-0.01658	0.03716
84	0.15324	0.24240	0.28678	174	-0.03204	-0.01833	0.03691
85	0.07658	0.08699	0.11590	175	-0.03084	-0.01972	0.03661
86	-0.08255	-0.19375	0.21061	176	-0.02975	-0.02079	0.03630
87	-0.29453	-0.54833	0.62242	177	-0.02884	-0.02158	0.03603
88	-0.50816	-0.89579	1.02989	178	-0.02816	-0.02213	0.03581
89	-0.66633	-1.14910	1.32831	179	-0.02774	-0.02244	0.03568
				180	-0.02759	-0.02254	0.03563

VBR

RADIATION PATTERN FOR THETA=PI/2

H/LMDA=0.5000 B/LMDA= 0.500 OMEGA= 9.92

2 ELEMENT ARRAY

PHI DEG	ER(THETA,PHI) RE	IM	ABSVAL	PHI DEG	ER(THETA,PHI) RE	IM	ABSVAL
0	-0.00000	-0.00000	0.00000	90	-1.75191	-0.61360	1.85626
5	-0.01047	-0.00367	0.01110	95	-1.73552	-0.60786	1.83889
10	-0.04181	-0.01464	0.04430	100	-1.68714	-0.59091	1.78763
15	-0.09373	-0.03283	0.09931	105	-1.60911	-0.56358	1.70496
20	-0.16571	-0.05804	0.17558	110	-1.50511	-0.52716	1.59476
25	-0.25690	-0.08998	0.27221	115	-1.37986	-0.48329	1.46204
30	-0.36597	-0.12818	0.38777	120	-1.23879	-0.43388	1.31258
35	-0.49101	-0.17197	0.52026	125	-1.08767	-0.38095	1.15246
40	-0.62943	-0.22045	0.66692	130	-0.93224	-0.32651	0.98776
45	-0.77788	-0.27245	0.82421	135	-0.77788	-0.27245	0.82421
50	-0.93224	-0.32651	0.98776	140	-0.62943	-0.22045	0.66692
55	-1.08767	-0.38095	1.15246	145	-0.49101	-0.17197	0.52026
60	-1.23879	-0.43388	1.31257	150	-0.36597	-0.12818	0.38777
65	-1.37985	-0.48329	1.46204	155	-0.25690	-0.08998	0.27221
70	-1.50511	-0.52716	1.59475	160	-0.16571	-0.05804	0.17558
75	-1.60911	-0.56358	1.70495	165	-0.09373	-0.03283	0.09931
80	-1.68714	-0.59091	1.78763	170	-0.04181	-0.01464	0.04430
85	-1.73552	-0.60786	1.83889	175	-0.01047	-0.00367	0.01110
				180	-0.00000	-0.00000	0.00000

VBR

RADIATION PATTERN FOR THETA=PI/2

H/LMDA=0.5000 B/LMDA= 0.500 OMEGA= 9.92

3 ELEMENT ARRAY

PHI DEG	ER(THETA,PHI) RE	IM	ABSVAL	PHI DEG	ER(THETA,PHI) RE	IM	ABSVAL
0	0.41808	0.15345	0.44535	90	-1.74961	-0.59541	1.84814
5	0.41800	0.15342	0.44527	95	-1.70923	-0.58146	1.80543
10	0.41684	0.15303	0.44404	100	-1.59229	-0.54106	1.68170
15	0.41187	0.15131	0.43879	105	-1.41063	-0.47831	1.48952
20	0.39868	0.14675	0.42483	110	-1.18187	-0.39928	1.24749
25	0.37147	0.13735	0.39604	115	-0.92666	-0.31111	0.97749
30	0.32349	0.12077	0.34529	120	-0.66577	-0.22098	0.70148
35	0.24780	0.09463	0.26526	125	-0.41746	-0.13520	0.43881
40	0.13827	0.05679	0.14948	130	-0.19572	-0.05859	0.20430
45	-0.00928	0.00581	0.01095	135	-0.00928	0.00581	0.01095
50	-0.19572	-0.05859	0.20430	140	0.13827	0.05679	0.14947
55	-0.41746	-0.13520	0.43881	145	0.24780	0.09463	0.26525
60	-0.66576	-0.22098	0.70148	150	0.32348	0.12077	0.34529
65	-0.92666	-0.31111	0.97749	155	0.37147	0.13735	0.39604
70	-1.18187	-0.39928	1.24749	160	0.39868	0.14675	0.42483
75	-1.41063	-0.47830	1.48951	165	0.41187	0.15131	0.43879
80	-1.59229	-0.54106	1.68170	170	0.41684	0.15303	0.44404
85	-1.70923	-0.58146	1.80543	175	0.41800	0.15342	0.44527
				180	0.41808	0.15345	0.44535

VBR

RADIATION PATTERN FOR THETA=PI/2

H/LMDA=0.5000 B/LMDA= 0.500 OMEGA= 9.92

4 ELEMENT ARRAY

PHI DEG	ER(THETA,PHI) RE	IM	ABSVAL	PHI DEG	ER(THETA,PHI) RE	IM	ABSVAL
0	0.00000	0.0	0.00000	90	-1.79152	-0.60888	1.89216
5	0.00967	0.00361	0.01032	95	-1.71185	-0.58082	1.80770
10	0.03856	0.01438	0.04115	100	-1.48700	-0.50165	1.56934
15	0.08603	0.03209	0.09182	105	-1.15582	-0.38516	1.21830
20	0.15014	0.05603	0.16026	110	-0.77200	-0.25036	0.81158
25	0.22648	0.08464	0.24178	115	-0.39161	-0.11713	0.40876
30	0.30687	0.11497	0.32770	120	-0.06143	-0.00199	0.06146
35	0.37836	0.14239	0.40426	125	0.18852	0.08445	0.20657
40	0.42312	0.16052	0.45255	130	0.34717	0.13836	0.37372
45	0.41988	0.16174	0.44995	135	0.41988	0.16174	0.44995
50	0.34717	0.13836	0.37372	140	0.42313	0.16052	0.45255
55	0.18852	0.08445	0.20658	145	0.37836	0.14239	0.40426
60	-0.06143	-0.00199	0.06146	150	0.30687	0.11497	0.32770
65	-0.39161	-0.11713	0.40875	155	0.22648	0.08464	0.24178
70	-0.77200	-0.25036	0.81158	160	0.15014	0.05603	0.16026
75	-1.15581	-0.38515	1.21829	165	0.08603	0.03209	0.09182
80	-1.48700	-0.50165	1.56934	170	0.03856	0.01438	0.04115
85	-1.71185	-0.58082	1.80770	175	0.00967	0.00361	0.01032
				180	0.00000	0.0	0.00000

IBR

RADIATION PATTERN FOR THETA=PI/2

H/LMDA=0.5000 B/LMDA= 0.500 OMEGA= 9.92

2 ELEMENT ARRAY

PHI DEG	ER(THETA,PHI) RE	IM	ABSVAL	PHI DEG	ER(THETA,PHI) RE	IM	ABSVAL
0	-0.00000	-0.00000	0.00000	90	-1.75191	-0.61360	1.85626
5	-0.01047	-0.00367	0.01110	95	-1.73552	-0.60786	1.83889
10	-0.04181	-0.01464	0.04430	100	-1.68714	-0.59091	1.78763
15	-0.09373	-0.03283	0.09931	105	-1.60911	-0.56358	1.70496
20	-0.16571	-0.05804	0.17558	110	-1.50511	-0.52716	1.59476
25	-0.25690	-0.08998	0.27221	115	-1.37986	-0.48329	1.46204
30	-0.36597	-0.12818	0.38777	120	-1.23879	-0.43388	1.31258
35	-0.49101	-0.17197	0.52026	125	-1.08767	-0.38095	1.15246
40	-0.62943	-0.22045	0.66692	130	-0.93224	-0.32651	0.98776
45	-0.77788	-0.27245	0.82421	135	-0.77788	-0.27245	0.82421
50	-0.93224	-0.32651	0.98776	140	-0.62943	-0.22045	0.66692
55	-1.08767	-0.38095	1.15246	145	-0.49101	-0.17197	0.52026
60	-1.23879	-0.43388	1.31257	150	-0.36597	-0.12818	0.38777
65	-1.37985	-0.48329	1.46204	155	-0.25690	-0.08998	0.27221
70	-1.50511	-0.52716	1.59475	160	-0.16571	-0.05804	0.17558
75	-1.60911	-0.56358	1.70495	165	-0.09373	-0.03283	0.09931
80	-1.68714	-0.59091	1.78763	170	-0.04181	-0.01464	0.04430
85	-1.73552	-0.60786	1.83889	175	-0.01047	-0.00367	0.01110
				180	-0.00000	-0.00000	0.00000

IBR

RADIATION PATTERN FOR THETA=PI/2

H/LMDA=0.5000 B/LMDA= 0.500 OMEGA= 9.92

3 ELEMENT ARRAY

PHI DEG	ER(THETA,PHI) RE	IM	ABSVAL	PHI DEG	ER(THETA,PHI) RE	IM	ABSVAL
0	0.37114	-0.01760	0.37156	90	-1.85727	-0.89311	2.06085
5	0.37106	-0.01763	0.37148	95	-1.81577	-0.87680	2.01638
10	0.36987	-0.01810	0.37031	100	-1.69555	-0.82957	1.88761
15	0.36476	-0.02011	0.36532	105	-1.50880	-0.75620	1.68770
20	0.35120	-0.02543	0.35212	110	-1.27363	-0.66381	1.43624
25	0.32322	-0.03643	0.32527	115	-1.01128	-0.56073	1.15633
30	0.27390	-0.05581	0.27952	120	-0.74307	-0.45536	0.87149
35	0.19609	-0.08637	0.21427	125	-0.48781	-0.35507	0.60335
40	0.08349	-0.13061	0.15502	130	-0.25985	-0.26551	0.37151
45	-0.06819	-0.19021	0.20206	135	-0.06820	-0.19021	0.20206
50	-0.25985	-0.26551	0.37151	140	0.08349	-0.13061	0.15502
55	-0.48781	-0.35507	0.60335	145	0.19609	-0.08637	0.21427
60	-0.74307	-0.45535	0.87149	150	0.27390	-0.05581	0.27952
65	-1.01127	-0.56073	1.15632	155	0.32322	-0.03643	0.32527
70	-1.27363	-0.66381	1.43624	160	0.35120	-0.02543	0.35212
75	-1.50880	-0.75620	1.68770	165	0.36476	-0.02011	0.36531
80	-1.69555	-0.82957	1.88761	170	0.36987	-0.01810	0.37031
85	-1.81577	-0.87680	2.01638	175	0.37106	-0.01763	0.37148
				180	0.37114	-0.01760	0.37156

IBR

RADIATION PATTERN FOR THETA=PI/2

H/LMDA=0.5000 B/LMDA= 0.500 OMEGA= 9.92

4 ELEMENT ARRAY

PHI DEG	ER(THETA,PHI) RE	IM	ABSVAL	PHI DEG	ER(THETA,PHI) RE	IM	ABSVAL
0	0.00000	0.0	0.00000	90	-1.83587	-0.84816	2.02232
5	0.00957	0.00248	0.00988	95	-1.75529	-0.81692	1.93607
10	0.03814	0.00989	0.03940	100	-1.52783	-0.72851	1.69263
15	0.08510	0.02202	0.08790	105	-1.19267	-0.59758	1.33400
20	0.14848	0.03820	0.15331	110	-0.80403	-0.44441	0.91867
25	0.22386	0.05690	0.23097	115	-0.41847	-0.29028	0.50929
30	0.30300	0.07521	0.31220	120	-0.08323	-0.15309	0.17425
35	0.37291	0.08855	0.38327	125	0.17130	-0.04457	0.17700
40	0.41565	0.09057	0.42540	130	0.33387	0.03057	0.33527
45	0.40982	0.07375	0.41640	135	0.40982	0.07375	0.41640
50	0.33387	0.03057	0.33527	140	0.41565	0.09057	0.42540
55	0.17130	-0.04457	0.17700	145	0.37291	0.08855	0.38328
6C	-0.08323	-0.15309	0.17425	150	0.30301	0.07521	0.31220
65	-0.41846	-0.29028	0.50928	155	0.22386	0.05690	0.23097
70	-0.80402	-0.44441	0.91867	160	0.14848	0.03820	0.15331
75	-1.19266	-0.59758	1.33399	165	0.08510	0.02202	0.08790
80	-1.52783	-0.72851	1.69262	170	0.03814	0.00989	0.03940
85	-1.75528	-0.81692	1.93607	175	0.00957	0.00248	0.00988
				180	0.00000	0.0	0.00000

VBR

RADIATION PATTERN FOR THETA=PI/2

H/LMDA=0.5000 B/LMDA= 0.500 OMEGA= 9.92

5 ELEMENT ARRAY

PHI DEG	ER(THETA,PHI) RE	IM	ABSVAL	PHI DEG	ER(THETA,PHI) RE	IM	ABSVAL
0	-0.23608	-0.08275	0.25016	90	-1.79295	-0.60272	1.89155
5	-0.23595	-0.08270	0.25002	95	-1.66728	-0.55948	1.75865
10	-0.23396	-0.08200	0.24791	100	-1.32399	-0.44144	1.39565
15	-0.22544	-0.07900	0.23888	105	-0.85179	-0.27927	0.89641
20	-0.20311	-0.07113	0.21521	110	-0.36282	-0.11172	0.37963
25	-0.15828	-0.05531	0.16766	115	0.04367	0.02695	0.05132
30	-0.08330	-0.02884	0.08815	120	0.30767	0.11612	0.32885
35	0.02443	0.00926	0.02613	125	0.41661	0.15160	0.44334
40	0.15752	0.05647	0.16733	130	0.39659	0.14257	0.42143
45	0.29410	0.10525	0.31236	135	0.29410	0.10525	0.31237
50	0.39659	0.14257	0.42143	140	0.15752	0.05647	0.16734
55	0.41661	0.15160	0.44334	145	0.02443	0.00926	0.02613
60	0.30767	0.11612	0.32886	150	-0.08330	-0.02884	0.08815
65	0.04368	0.02695	0.05133	155	-0.15827	-0.05531	0.16766
70	-0.36281	-0.11172	0.37962	160	-0.20311	-0.07113	0.21521
75	-0.85178	-0.27927	0.89640	165	-0.22544	-0.07900	0.23888
80	-1.32399	-0.44144	1.39564	170	-0.23396	-0.08200	0.24791
85	-1.66728	-0.55948	1.75865	175	-0.23595	-0.08270	0.25002
				180	-0.23608	-0.08275	0.25016

IBR

RADIATION PATTERN FOR THETA=PI/2

H/LMDA=0.5000 B/LMDA= 0.500 OMEGA= 9.92

5 ELEMENT ARRAY

PHI DEG	ER(THETA,PHI) RE	IM	ABSVAL	PHI DEG	ER(THETA,PHI) RE	IM	ABSVAL
0	-0.19729	0.01800	0.19810	90	-1.87905	-0.95132	2.10614
5	-0.19715	0.01804	0.19798	95	-1.74885	-0.89537	1.96473
10	-0.19517	0.01862	0.19605	100	-1.39288	-0.74144	1.57793
15	-0.18666	0.02114	0.18785	105	-0.90240	-0.52626	1.04464
20	-0.16436	0.02770	0.16668	110	-0.39281	-0.29670	0.49227
25	-0.11962	0.04077	0.12638	115	0.03349	-0.09510	0.10082
30	-0.04490	0.06221	0.07673	120	0.31427	0.05160	0.31848
35	0.06217	0.09190	0.11096	125	0.43593	0.13543	0.45649
40	0.19380	0.12593	0.23111	130	0.42458	0.16431	0.45527
45	0.32742	0.15502	0.36227	135	0.32743	0.15502	0.36227
50	0.42458	0.16431	0.45527	140	0.19380	0.12593	0.23112
55	0.43593	0.13543	0.45649	145	0.06218	0.09190	0.11096
60	0.31428	0.05160	0.31848	150	-0.04490	0.06221	0.07673
65	0.03350	-0.09509	0.10082	155	-0.11962	0.04077	0.12638
70	-0.39281	-0.29670	0.49227	160	-0.16436	0.02770	0.16668
75	-0.90239	-0.52626	1.04463	165	-0.18666	0.02114	0.18785
80	-1.39288	-0.74144	1.57792	170	-0.19517	0.01862	0.19605
85	-1.74884	-0.89537	1.96472	175	-0.19715	0.01804	0.19798
				180	-0.19729	0.01800	0.19810

VBR

RADIATION PATTERN FOR THETA=PI/2

H/LMDA=0.5000 B/LMDA= 0.500 OMEGA= 9.92

6 ELEMENT ARRAY

PHI DEG	ER(THETA,PHI) RE	IM	ABSVAL	PHI DEG	ER(THETA,PHI) RE	IM	ABSVAL
0	-0.00000	-0.00000	0.00000	90	-1.80845	-0.60783	1.90787
2	-0.00147	-0.00056	0.00157	92	-1.77794	-0.59728	1.87558
4	-0.00586	-0.00223	0.00627	94	-1.68834	-0.56630	1.78078
6	-0.01317	-0.00501	0.01409	96	-1.54533	-0.51688	1.62948
8	-0.02338	-0.00889	0.02501	98	-1.35784	-0.45215	1.43114
10	-0.03642	-0.01385	0.03897	100	-1.13730	-0.37612	1.19788
12	-0.05220	-0.01985	0.05585	102	-0.89673	-0.29334	0.94349
14	-0.07054	-0.02683	0.07547	104	-0.64965	-0.20853	0.68230
16	-0.09116	-0.03470	0.09754	106	-0.40916	-0.12627	0.42820
18	-0.11367	-0.04329	0.12163	108	-0.18696	-0.05062	0.19369
20	-0.13751	-0.05241	0.14716	110	0.00737	0.01509	0.01679
22	-0.16194	-0.06179	0.17333	112	0.16683	0.06845	0.18033
24	-0.18602	-0.07109	0.19914	114	0.28731	0.10808	0.30696
26	-0.20861	-0.07989	0.22338	116	0.36749	0.13358	0.39102
28	-0.22835	-0.08770	0.24461	118	0.40870	0.14546	0.43382
30	-0.24373	-0.09396	0.26122	120	0.41445	0.14502	0.43909
32	-0.25310	-0.09806	0.27144	122	0.38990	0.13411	0.41232
34	-0.25478	-0.09939	0.27348	124	0.34131	0.11498	0.36015
36	-0.24712	-0.09734	0.26560	126	0.27541	0.09002	0.28975
38	-0.22872	-0.09138	0.24630	128	0.19892	0.06163	0.20825
40	-0.19853	-0.08110	0.21446	130	0.11809	0.03201	0.12235
42	-0.15604	-0.06631	0.16954	132	0.03832	0.00306	0.03844
44	-0.10147	-0.04704	0.11184	134	-0.03596	-0.02367	0.04305
46	-0.03596	-0.02367	0.04305	136	-0.10147	-0.04704	0.11184
48	0.03832	0.00306	0.03844	138	-0.15604	-0.06631	0.16954
50	0.11808	0.03201	0.12235	140	-0.19853	-0.08110	0.21446
52	0.19892	0.06163	0.20825	142	-0.22872	-0.09138	0.24630
54	0.27541	0.09002	0.28975	144	-0.24712	-0.09734	0.26560
56	0.34131	0.11498	0.36015	146	-0.25478	-0.09939	0.27348
58	0.38990	0.13411	0.41232	148	-0.25311	-0.09806	0.27144
60	0.41445	0.14502	0.43909	150	-0.24373	-0.09396	0.26122
62	0.40870	0.14546	0.43382	152	-0.22835	-0.08770	0.24461
64	0.36749	0.13358	0.39102	154	-0.20861	-0.07989	0.22338
66	0.28731	0.10808	0.30697	156	-0.18602	-0.07109	0.19914
68	0.16684	0.06845	0.18033	158	-0.16194	-0.06179	0.17333
70	0.00737	0.01509	0.01679	160	-0.13751	-0.05241	0.14716
72	-0.18695	-0.05062	0.19368	162	-0.11367	-0.04329	0.12163
74	-0.40915	-0.12627	0.42819	164	-0.09116	-0.03470	0.09754
76	-0.64964	-0.20853	0.68229	166	-0.07054	-0.02683	0.07547
78	-0.89672	-0.29334	0.94348	168	-0.05220	-0.01985	0.05585
80	-1.13730	-0.37612	1.19788	170	-0.03642	-0.01385	0.03897
82	-1.35783	-0.45215	1.43113	172	-0.02338	-0.00889	0.02501
84	-1.54532	-0.51688	1.62947	174	-0.01317	-0.00501	0.01409
86	-1.68833	-0.56629	1.78077	176	-0.00586	-0.00223	0.00627
88	-1.77794	-0.59728	1.87558	178	-0.00147	-0.00056	0.00157
				180	-0.00000	-0.00000	0.00000

IBR

RADIATION PATTERN FOR THETA=PI/2

H/LMDA=0.5000 B/LMDA= 0.500 OMEGA= 9.92

6 ELEMENT ARRAY

PHI DEG	ER(THETA,PHI) RE	IM	ABSVAL	PHI DEG	ER(THETA,PHI) RE	IM	ABSVAL
0	-0.00000	0.0	0.00000	90	-1.86331	-0.92049	2.07827
2	-0.00142	-0.00028	0.00144	92	-1.83207	-0.90710	2.04434
4	-0.00566	-0.00113	0.00577	94	-1.74035	-0.86772	1.94467
6	-0.01273	-0.00254	0.01298	96	-1.59392	-0.80464	1.78550
8	-0.02259	-0.00450	0.02303	98	-1.40186	-0.72150	1.57664
10	-0.03519	-0.00700	0.03588	100	-1.17583	-0.62297	1.33066
12	-0.05043	-0.01001	0.05142	102	-0.92905	-0.51436	1.06193
14	-0.06814	-0.01348	0.06946	104	-0.67533	-0.40128	0.78555
16	-0.08803	-0.01732	0.08972	106	-0.42801	-0.28917	0.51654
18	-0.10972	-0.02141	0.11179	108	-0.19904	-0.18301	0.27039
20	-0.13265	-0.02558	0.13509	110	0.00177	-0.08695	0.08697
22	-0.15607	-0.02962	0.15886	112	0.16724	-0.00415	0.16729
24	-0.17906	-0.03322	0.18212	114	0.29311	0.06333	0.29987
26	-0.20048	-0.03606	0.20369	116	0.37798	0.11457	0.39496
28	-0.21897	-0.03772	0.22220	118	0.42311	0.14968	0.44880
30	-0.23304	-0.03775	0.23607	120	0.43198	0.16966	0.46410
32	-0.24103	-0.03569	0.24366	122	0.40979	0.17624	0.44608
34	-0.24128	-0.03106	0.24327	124	0.36283	0.17160	0.40136
36	[illegible]	[illegible]	[illegible]	126	0.29790	0.15820	0.33730
38	-0.21232	-0.01249	0.21269	128	0.22182	0.13854	0.26153
40	-0.18070	0.00197	0.18071	130	0.14089	0.11503	0.18188
42	-0.13685	0.01991	0.13829	132	0.06063	0.08980	0.10834
44	-0.08105	0.04102	0.09084	134	-0.01447	0.06466	0.06626
46	-0.01448	0.06466	0.06626	136	-0.08105	0.04102	0.09084
48	0.06062	0.08979	0.10834	138	-0.13685	0.01991	0.13829
50	0.14089	0.11502	0.18188	140	-0.18070	0.00197	0.18071
52	0.22182	0.13854	0.26153	142	-0.21232	-0.01249	0.21269
54	0.29790	0.15820	0.33730	144	-0.23217	-0.02344	0.23335
56	0.36282	0.17160	0.40136	146	-0.24128	-0.03106	0.24327
58	0.40979	0.17624	0.44608	148	-0.24103	-0.03569	0.24366
60	0.43198	0.16966	0.46410	150	-0.23304	-0.03775	0.23607
62	0.42311	0.14968	0.44880	152	-0.21897	-0.03772	0.22220
64	0.37798	0.11457	0.39496	154	-0.20048	-0.03606	0.20369
66	0.29311	0.06333	0.29988	156	-0.17906	-0.03322	0.18212
68	0.16724	-0.00415	0.16729	158	-0.15607	-0.02962	0.15885
70	0.00177	-0.08695	0.08696	160	-0.13264	-0.02558	0.13509
72	-0.19904	-0.18301	0.27038	162	-0.10972	-0.02141	0.11179
74	-0.42800	-0.28917	0.51653	164	-0.08804	-0.01732	0.08972
76	-0.67532	-0.40127	0.78554	166	-0.06814	-0.01348	0.06946
78	-0.92904	-0.51436	1.06193	168	-0.05044	-0.01001	0.05142
80	-1.17582	-0.62296	1.33066	170	-0.03519	-0.00700	0.03588
82	-1.40186	-0.72150	1.57663	172	-0.02259	-0.00450	0.02303
84	-1.59391	-0.80464	1.78550	174	-0.01273	-0.00254	0.01298
86	-1.74034	-0.86771	1.94467	176	-0.00566	-0.00113	0.00577
88	-1.83207	-0.90710	2.04434	178	-0.00142	-0.00028	0.00144
				180	-0.00000	0.0	0.00000

VBR

RADIATION PATTERN FOR THETA=PI/2

H/LMDA=0.5000 B/LMDA= 0.500 OMEGA= 9.92

7 ELEMENT ARRAY

PHI	ER(THETA,PHI)			PHI	ER(THETA,PHI)		
DEG	RE	IM	ABSVAL	DEG	RE	IM	ABSVAL
0	0.16272	0.05520	0.17183	90	-1.80965	-0.60503	1.90811
2	0.16272	0.05520	0.17182	92	-1.76791	-0.59078	1.86400
4	0.16265	0.05517	0.17175	94	-1.64626	-0.54925	1.73547
6	0.16234	0.05506	0.17142	96	-1.45509	-0.48402	1.53348
8	0.16152	0.05477	0.17055	98	-1.21039	-0.40062	1.27497
10	0.15979	0.05415	0.16872	100	-0.93206	-0.30590	0.98097
12	0.15668	0.05304	0.16542	102	-0.64175	-0.20732	0.67440
14	0.15161	0.05123	0.16003	104	-0.36068	-0.11217	0.37772
16	0.14393	0.04848	0.15188	106	-0.10766	-0.02691	0.11097
18	0.13298	0.04457	0.14025	108	0.10257	0.04344	0.11139
20	0.11811	0.03926	0.12447	110	0.26034	0.09558	0.27733
22	0.09873	0.03232	0.10389	112	0.36145	0.12814	0.38349
24	0.07442	0.02362	0.07808	114	0.40697	0.14156	0.43088
26	0.04498	0.01308	0.04685	116	0.40245	0.13783	0.42540
28	0.01056	0.00075	0.01059	118	0.35687	0.12012	0.37654
30	-0.02825	-0.01318	0.03118	120	0.28127	0.09228	0.29602
32	-0.07035	-0.02831	0.07583	122	0.18745	0.05845	0.19635
34	-0.11403	-0.04405	0.12224	124	0.08676	0.02255	0.08965
36	-0.15697	-0.05958	0.16790	126	-0.01083	-0.01195	0.01613
38	-0.19626	-0.07387	0.20970	128	-0.09741	-0.04234	0.10621
40	-0.22854	-0.08574	0.24409	130	-0.16744	-0.06671	0.18024
42	-0.25017	-0.09388	0.26720	132	-0.21778	-0.08403	0.23343
44	-0.25755	-0.09702	0.27522	134	-0.24752	-0.09402	0.26477
46	-0.24752	-0.09402	0.26477	136	-0.25755	-0.09702	0.27522
48	-0.21778	-0.08403	0.23343	138	-0.25017	-0.09388	0.26720
50	-0.16744	-0.06671	0.18024	140	-0.22854	-0.08574	0.24409
52	-0.09741	-0.04234	0.10621	142	-0.19626	-0.07387	0.20971
54	-0.01083	-0.01195	0.01613	144	-0.15697	-0.05958	0.16790
56	0.08676	0.02255	0.08965	146	-0.11403	-0.04405	0.12225
58	0.18745	0.05845	0.19635	148	-0.07035	-0.02831	0.07584
60	0.28127	0.09228	0.29602	150	-0.02825	-0.01318	0.03118
62	0.35687	0.12012	0.37654	152	0.01056	0.00075	0.01059
64	0.40245	0.13783	0.42540	154	0.04498	0.01308	0.04685
66	0.40697	0.14156	0.43088	156	0.07442	0.02362	0.07808
68	0.36145	0.12814	0.38349	158	0.09873	0.03232	0.10389
70	0.26034	0.09558	0.27733	160	0.11811	0.03925	0.12446
72	0.10258	0.04344	0.11140	162	0.13298	0.04457	0.14025
74	-0.10765	-0.02690	0.11096	164	0.14393	0.04848	0.15188
76	-0.36067	-0.11216	0.37771	166	0.15161	0.05123	0.16003
78	-0.64174	-0.20731	0.67440	168	0.15668	0.05304	0.16542
80	-0.93205	-0.30590	0.98097	170	0.15980	0.05415	0.16872
82	-1.21039	-0.40062	1.27496	172	0.16152	0.05477	0.17055
84	-1.45508	-0.48402	1.53347	174	0.16234	0.05506	0.17142
86	-1.64625	-0.54925	1.73546	176	0.16265	0.05517	0.17175
88	-1.76790	-0.59078	1.86400	178	0.16272	0.05520	0.17182
				180	0.16272	0.05520	0.17183

IBR

RADIATION PATTERN FOR THETA=PI/2

H/LMDA=0.5000 B/LMDA= 0.500 OMEGA= 9.92

7 ELEMENT ARRAY

PHI	ER(THETA,PHI)			PHI	ER(THETA,PHI)		
DEG	RE	IM	ABSVAL	DEG	RE	IM	ABSVAL
0	0.13227	-0.01464	0.13308	90	-1.88895	-0.98015	2.12810
2	0.13227	-0.01464	0.13307	92	-1.84562	-0.96039	2.08054
4	0.13220	-0.01465	0.13301	94	-1.71935	-0.90267	1.94190
6	0.13190	-0.01473	0.13272	96	-1.52083	-0.81151	1.72379
8	0.13110	-0.01492	0.13194	98	-1.26655	-0.69389	1.44417
10	0.12941	-0.01533	0.13032	100	-0.97706	-0.55856	1.12545
12	0.12637	-0.01606	0.12739	102	-0.67471	-0.41511	0.79218
14	0.12141	-0.01725	0.12263	104	-0.38147	-0.27310	0.46915
16	0.11391	-0.01904	0.11549	106	-0.11680	-0.14116	0.18321
18	0.10323	-0.02158	0.10546	108	0.10396	-0.02632	0.10724
20	0.08872	-0.02500	0.09218	110	0.27067	0.06646	0.27871
22	0.06983	-0.02940	0.07577	112	0.37884	0.13452	0.40202
24	0.04616	-0.03484	0.05783	114	0.42938	0.17743	0.46459
26	0.01755	-0.04126	0.04484	116	0.42784	0.19675	0.47091
28	-0.01582	-0.04853	0.05105	118	0.38333	0.19557	0.43033
30	-0.05334	-0.05635	0.07759	120	0.30708	0.17804	0.35496
32	-0.09386	-0.06426	0.11374	122	0.21119	0.14881	0.25836
34	-0.13562	-0.07162	0.15337	124	0.10733	0.11256	0.15552
36	-0.17628	-0.07761	0.19261	126	0.00576	0.07354	0.07377
38	-0.21291	-0.08126	0.22789	128	-0.08529	0.03537	0.09234
40	-0.24210	-0.08148	0.25544	130	-0.16003	0.00077	0.16003
42	-0.26022	-0.07717	0.27142	132	-0.21509	-0.02841	0.21696
44	-0.26369	-0.06732	0.27215	134	-0.24938	-0.05119	0.25458
46	-0.24938	-0.05119	0.25458	136	-0.26369	-0.06732	0.27215
48	-0.21509	-0.02841	0.21696	138	-0.26022	-0.07717	0.27142
50	-0.16003	0.00077	0.16003	140	-0.24210	-0.08148	0.25544
52	-0.08529	0.03537	0.09234	142	-0.21291	-0.08126	0.22789
54	0.00576	0.07354	0.07377	144	-0.17629	-0.07761	0.19261
56	0.10732	0.11256	0.15552	146	-0.13562	-0.07162	0.15337
58	0.21119	0.14881	0.25835	148	-0.09386	-0.06426	0.11374
60	0.30708	0.17804	0.35496	150	-0.05334	-0.05635	0.07759
62	0.38332	0.19557	0.43033	152	-0.01583	-0.04853	0.05105
64	0.42784	0.19675	0.47091	154	0.01755	-0.04126	0.04484
66	0.42938	0.17744	0.46460	156	0.04616	-0.03484	0.05783
68	0.37884	0.13452	0.40202	158	0.06983	-0.02940	0.07577
70	0.27067	0.06646	0.27872	160	0.08872	-0.02500	0.09217
72	0.10396	-0.02632	0.10724	162	0.10323	-0.02158	0.10546
74	-0.11679	-0.14115	0.18320	164	0.11391	-0.01904	0.11549
76	-0.38145	-0.27309	0.46913	166	0.12141	-0.01725	0.12263
78	-0.67471	-0.41510	0.79218	168	0.12637	-0.01606	0.12739
80	-0.97705	-0.55855	1.12544	170	0.12941	-0.01533	0.13032
82	-1.26655	-0.69389	1.44417	172	0.13110	-0.01492	0.13194
84	-1.52082	-0.81151	1.72378	174	0.13190	-0.01473	0.13272
86	-1.71934	-0.90267	1.94189	176	0.13220	-0.01465	0.13301
88	-1.84562	-0.96039	2.08054	178	0.13227	-0.01464	0.13307
				180	0.13227	-0.01464	0.13308

VBR

RADIATION PATTERN FOR THETA=PI/2

H/LMDA=0.5000 B/LMDA= 0.500 OMEGA= 9.92

8 ELEMENT ARRAY

PHI	ER(THETA,PHI)			PHI	ER(THETA,PHI)		
DEG	RE	IM	ABSVAL	DEG	RE	IM	ABSVAL
0	0.00000	0.00000	0.00000	90	-1.81764	-0.60761	1.91651
2	0.00141	0.00054	0.00151	92	-1.76233	-0.58869	1.85805
4	0.00563	0.00215	0.00603	94	-1.60261	-0.53408	1.68926
6	0.01266	0.00483	0.01355	96	-1.35623	-0.44994	1.42891
8	0.02244	0.00857	0.02402	98	-1.04992	-0.34554	1.10532
10	0.03490	0.01332	0.03736	100	-0.71577	-0.23196	0.75242
12	0.04986	0.01904	0.05338	102	-0.38682	-0.12063	0.40519
14	0.06703	0.02561	0.07176	104	-0.09291	-0.02180	0.09544
16	0.08595	0.03285	0.09201	106	0.14273	0.05659	0.15354
18	0.10593	0.04053	0.11341	108	0.30584	0.10975	0.32494
20	0.12603	0.04829	0.13497	110	0.39187	0.13627	0.41488
22	0.14507	0.05569	0.15539	112	0.40542	0.13794	0.42825
24	0.16154	0.06218	0.17310	114	0.35856	0.11911	0.37782
26	0.17373	0.06713	0.18625	116	0.26823	0.08579	0.28162
28	0.17975	0.06983	0.19284	118	0.15356	0.04471	0.15994
30	0.17771	0.06958	0.19084	120	0.03313	0.00236	0.03322
32	0.16587	0.06572	0.17841	122	-0.07713	-0.03575	0.08501
34	0.14294	0.05775	0.15417	124	-0.16538	-0.06556	0.17790
36	0.10836	0.04543	0.11750	126	-0.22451	-0.08472	0.23996
38	0.06257	0.02888	0.06892	128	-0.25208	-0.09252	0.26853
40	0.00733	0.00870	0.01138	130	-0.24969	-0.08967	0.26530
42	-0.05412	-0.01398	0.05590	132	-0.22194	-0.07791	0.23522
44	-0.11704	-0.03749	0.12290	134	-0.17532	-0.05964	0.18519
46	-0.17532	-0.05964	0.18519	136	-0.11704	-0.03749	0.12290
48	-0.22194	-0.07791	0.23522	138	-0.05413	-0.01398	0.05590
50	-0.24969	-0.08967	0.26530	140	0.00733	0.00870	0.01138
52	-0.25208	-0.09252	0.26853	142	0.06257	0.02888	0.06892
54	-0.22451	-0.08472	0.23996	144	0.10836	0.04543	0.11750
56	-0.16538	-0.06556	0.17790	146	0.14294	0.05775	0.15417
58	-0.07713	-0.03575	0.08501	148	0.16587	0.06572	0.17841
60	0.03313	0.00236	0.03321	150	0.17771	0.06958	0.19084
62	0.15356	0.04470	0.15994	152	0.17975	0.06983	0.19284
64	0.26823	0.08579	0.28162	154	0.17373	0.06713	0.18625
66	0.35855	0.11911	0.37782	156	0.16154	0.06218	0.17310
68	0.40542	0.13794	0.42825	158	0.14507	0.05569	0.15539
70	0.39187	0.13627	0.41488	160	0.12603	0.04829	0.13497
72	0.30585	0.10975	0.32494	162	0.10593	0.04053	0.11341
74	0.14274	0.05659	0.15355	164	0.08595	0.03285	0.09201
76	-0.09290	-0.02180	0.09542	166	0.06703	0.02561	0.07176
78	-0.38681	-0.12063	0.40518	168	0.04986	0.01904	0.05338
80	-0.71576	-0.23196	0.75241	170	0.03490	0.01332	0.03736
82	-1.04992	-0.34553	1.10531	172	0.02244	0.00857	0.02402
84	-1.35621	-0.44994	1.42890	174	0.01266	0.00483	0.01355
86	-1.60260	-0.53408	1.68925	176	0.00563	0.00215	0.00603
88	-1.76233	-0.58869	1.85805	178	0.00141	0.00054	0.00151
				180	0.00000	0.00000	0.00000

IBR

RADIATION PATTERN FOR THETA=PI/2

H/LMDA=0.5000 B/LMDA= 0.500 OMEGA= 9.92

8 ELEMENT ARRAY

PHI	ER(THETA,PHI)			PHI	ER(THETA,PHI)		
DEG	RE	IM	ABSVAL	DEG	RE	IM	ABSVAL
0	0.00000	0.00000	0.00000	90	-1.87670	-0.95592	2.10613
2	0.00134	0.00021	0.00135	92	-1.81981	-0.93006	2.04370
4	0.00534	0.00085	0.00541	94	-1.65549	-0.85516	1.86331
6	0.01199	0.00191	0.01215	96	-1.40191	-0.73888	1.58470
8	0.02127	0.00338	0.02153	98	-1.08645	-0.59285	1.23767
10	0.03307	0.00523	0.03348	100	-0.74197	-0.43111	0.85813
12	0.04724	0.00744	0.04782	102	-0.40238	-0.26836	0.48366
14	0.06347	0.00991	0.06424	104	-0.09835	-0.11814	0.15371
16	0.08132	0.01252	0.08228	106	0.14622	0.00860	0.14647
18	0.10011	0.01510	0.10124	108	0.31650	0.10450	0.33331
20	0.11893	0.01740	0.12019	110	0.40761	0.16630	0.44023
22	0.13657	0.01910	0.13790	112	0.42405	0.19461	0.46657
24	0.15160	0.01983	0.15289	114	0.37794	0.19337	0.42454
26	0.16230	0.01915	0.16343	116	0.28650	0.16886	0.33256
28	0.16684	0.01661	0.16767	118	0.16918	0.12861	0.21252
30	0.16336	0.01179	0.16378	120	0.04499	0.08033	0.09207
32	0.15019	0.00437	0.15026	122	-0.06972	0.03100	0.07630
34	0.12611	-0.00583	0.12624	124	-0.16269	-0.01377	0.16327
36	0.09062	-0.01867	0.09252	126	-0.22647	-0.05010	0.23194
38	0.04426	-0.03371	0.05564	128	-0.25834	-0.07587	0.26925
40	-0.01113	-0.05004	0.05126	130	-0.25970	-0.09060	0.27505
42	-0.07221	-0.06631	0.09804	132	-0.23504	-0.09510	0.25355
44	-0.13416	-0.08070	0.15656	134	-0.19079	-0.09107	0.21141
46	-0.19078	-0.09107	0.21140	136	-0.13416	-0.08070	0.15656
48	-0.23504	-0.09510	0.25355	138	-0.07222	-0.06631	0.09804
50	-0.25970	-0.09060	0.27505	140	-0.01113	-0.05004	0.05126
52	-0.25834	-0.07587	0.26925	142	0.04426	-0.03371	0.05563
54	-0.22647	-0.05010	0.23194	144	0.09062	-0.01867	0.09252
56	-0.16269	-0.01377	0.16327	146	0.12611	-0.00583	0.12624
58	-0.06972	0.03100	0.07630	148	0.15019	0.00437	0.15025
60	0.04499	0.08033	0.09207	150	0.16336	0.01180	0.16378
62	0.16918	0.12861	0.21252	152	0.16684	0.01661	0.16767
64	0.28650	0.16886	0.33256	154	0.16230	0.01915	0.16343
66	0.37794	0.19337	0.42454	156	0.15160	0.01983	0.15289
68	0.42405	0.19461	0.46657	158	0.13658	0.01910	0.13791
70	0.40761	0.16630	0.44023	160	0.11893	0.01740	0.12019
72	0.31650	0.10451	0.33331	162	0.10011	0.01510	0.10124
74	0.14623	0.00860	0.14648	164	0.08132	0.01252	0.08228
76	-0.09833	-0.11813	0.15370	166	0.06347	0.00991	0.06424
78	-0.40237	-0.26835	0.48365	168	0.04724	0.00744	0.04782
80	-0.74197	-0.43111	0.85812	170	0.03307	0.00523	0.03348
82	-1.08644	-0.59285	1.23767	172	0.02127	0.00338	0.02153
84	-1.40189	-0.73888	1.58469	174	0.01199	0.00191	0.01215
86	-1.65548	-0.85515	1.86330	176	0.00534	0.00085	0.00541
88	-1.81981	-0.93006	2.04370	178	0.00134	0.00021	0.00135
				180	0.00000	0.00000	0.00000

VBR

RADIATION PATTERN FOR THETA=PI/2

H/LMDA=0.5000 B/LMDA= 0.500 OMEGA= 9.92

9 ELEMENT ARRAY

PHI DEG	ER(THETA,PHI) RE	IM	ABSVAL	PHI DEG	ER(THETA,PHI) RE	IM	ABSVAL
0	-0.12344	-0.04088	0.13004	90	-1.81850	-0.60606	1.91683
2	-0.12344	-0.04088	0.13003	92	-1.74842	-0.58227	1.84283
4	-0.12335	-0.04085	0.12993	94	-1.54808	-0.51430	1.63127
6	-0.12296	-0.04071	0.12952	96	-1.24540	-0.41173	1.31170
8	-0.12193	-0.04033	0.12843	98	-0.88144	-0.28864	0.92749
10	-0.11976	-0.03955	0.12612	100	-0.50335	-0.16118	0.52853
12	-0.11586	-0.03815	0.12198	102	-0.15662	-0.04489	0.16292
14	-0.10952	-0.03587	0.11525	104	0.12191	0.04773	0.13092
16	-0.10001	-0.03245	0.10515	106	0.30895	0.10884	0.32756
18	-0.08660	-0.02762	0.09090	108	0.39697	0.13607	0.41965
20	-0.06868	-0.02116	0.07186	110	0.39347	0.13222	0.41509
22	-0.04584	-0.01293	0.04763	112	0.31783	0.10412	0.33445
24	-0.01807	-0.00291	0.01831	114	0.19663	0.06106	0.20589
26	0.01415	0.00872	0.01662	116	0.05840	0.01292	0.05981
28	0.04967	0.02157	0.05415	118	-0.07105	-0.03142	0.07769
30	0.08659	0.03494	0.09337	120	-0.17210	-0.06529	0.18407
32	0.12220	0.04789	0.13124	122	-0.23319	-0.08485	0.24814
34	0.15307	0.05917	0.16411	124	-0.25101	-0.08913	0.26637
36	0.17528	0.06738	0.18778	126	-0.22939	-0.07964	0.24282
38	0.18478	0.07106	0.19798	128	-0.17722	-0.05962	0.18698
40	0.17800	0.06890	0.19087	130	-0.10614	-0.03325	0.11123
42	0.15249	0.06000	0.16387	132	-0.02823	-0.00481	0.02863
44	0.10773	0.04412	0.11641	134	0.04579	0.02193	0.05077
46	0.04579	0.02193	0.05077	136	0.10773	0.04412	0.11641
48	-0.02823	-0.00481	0.02863	138	0.15249	0.06000	0.16387
50	-0.10614	-0.03325	0.11123	140	0.17800	0.06890	0.19087
52	-0.17722	-0.05962	0.18698	142	0.18478	0.07106	0.19797
54	-0.22939	-0.07964	0.24282	144	0.17528	0.06738	0.18778
56	-0.25101	-0.08913	0.26637	146	0.15307	0.05917	0.16411
58	-0.23319	-0.08485	0.24814	148	0.12220	0.04789	0.13125
60	-0.17210	-0.06529	0.18407	150	0.08658	0.03494	0.09337
62	-0.07105	-0.03142	0.07769	152	0.04967	0.02157	0.05415
64	0.05839	0.01292	0.05980	154	0.01414	0.00872	0.01662
66	0.19662	0.06106	0.20588	156	-0.01807	-0.00291	0.01831
68	0.31783	0.10412	0.33445	158	-0.04584	-0.01293	0.04763
70	0.39347	0.13222	0.41509	160	-0.06868	-0.02116	0.07186
72	0.39697	0.13607	0.41965	162	-0.08660	-0.02762	0.09090
74	0.30896	0.10884	0.32757	164	-0.10001	-0.03245	0.10514
76	0.12192	0.04773	0.13093	166	-0.10952	-0.03587	0.11525
78	-0.15661	-0.04488	0.16292	168	-0.11586	-0.03815	0.12198
80	-0.50334	-0.16118	0.52852	170	-0.11976	-0.03955	0.12612
82	-0.88143	-0.28864	0.92749	172	-0.12193	-0.04033	0.12843
84	-1.24539	-0.41172	1.31168	174	-0.12296	-0.04071	0.12952
86	-1.54807	-0.51429	1.63126	176	-0.12335	-0.04085	0.12993
88	-1.74842	-0.58227	1.84282	178	-0.12344	-0.04088	0.13003
				180	-0.12344	-0.04088	0.13004

IBR

RADIATION PATTERN FOR THETA=PI/2

H/LMDA=0.5000 B/LMDA= 0.500 OMEGA= 9.92

9 ELEMENT ARRAY

PHI DEG	ER(THETA,PHI) RE	IM	ABSVAL	PHI DEG	ER(THETA,PHI) RE	IM	ABSVAL
0	-0.09863	0.01213	0.09937	90	-1.89455	-0.99679	2.14077
2	-0.09862	0.01213	0.09936	92	-1.82178	-0.96230	2.06031
4	-0.09853	0.01215	0.09928	94	-1.61370	-0.86333	1.83012
6	-0.09816	0.01222	0.09892	96	-1.29916	-0.71261	1.48177
8	-0.09717	0.01242	0.09796	98	-0.92063	-0.52898	1.06178
10	-0.09509	0.01284	0.09595	100	-0.52691	-0.33436	0.62404
12	-0.09134	0.01360	0.09235	102	-0.16516	-0.15032	0.22332
14	-0.08527	0.01482	0.08654	104	0.12631	0.00502	0.12641
16	-0.07615	0.01663	0.07794	106	0.32313	0.11925	0.34444
18	-0.06330	0.01916	0.06613	108	0.41718	0.18690	0.45713
20	-0.04614	0.02247	0.05132	110	0.41585	0.20922	0.46551
22	-0.02434	0.02657	0.03603	112	0.33889	0.19315	0.39007
24	0.00210	0.03134	0.03141	114	0.21359	0.14938	0.26065
26	0.03266	0.03653	0.04900	116	0.06939	0.09021	0.11381
28	0.06614	0.04169	0.07818	118	-0.06693	0.02747	0.07234
30	0.10062	0.04619	0.11072	120	-0.17484	-0.02903	0.17723
32	0.13339	0.04919	0.14217	122	-0.24204	-0.07249	0.25267
34	0.16102	0.04971	0.16852	124	-0.26471	-0.09947	0.28278
36	0.17964	0.04670	0.18561	126	-0.24634	-0.10957	0.26961
38	0.18526	0.03923	0.18937	128	-0.19575	-0.10480	0.22204
40	0.17444	0.02663	0.17647	130	-0.12467	-0.08872	0.15301
42	0.14491	0.00876	0.14518	132	-0.04537	-0.06554	0.07971
44	0.09637	-0.01376	0.09735	134	0.03114	-0.03938	0.05021
46	0.03114	-0.03938	0.05021	136	0.09637	-0.01376	0.09735
48	-0.04537	-0.06554	0.07970	138	0.14491	0.00876	0.14518
50	-0.12467	-0.08872	0.15301	140	0.17444	0.02663	0.17646
52	-0.19575	-0.10480	0.22204	142	0.18526	0.03923	0.18937
54	-0.24634	-0.10957	0.26961	144	0.17964	0.04670	0.18561
56	-0.26471	-0.09947	0.28278	146	0.16102	0.04971	0.16852
58	-0.24204	-0.07249	0.25267	148	0.13339	0.04919	0.14217
60	-0.17484	-0.02903	0.17723	150	0.10062	0.04619	0.11072
62	-0.06693	0.02747	0.07235	152	0.06614	0.04169	0.07818
64	0.06938	0.09021	0.11380	154	0.03266	0.03653	0.04899
66	0.21358	0.14938	0.26064	156	0.00210	0.03134	0.03141
68	0.33889	0.19315	0.39007	158	-0.02434	0.02657	0.03603
70	0.41584	0.20922	0.46551	160	-0.04614	0.02247	0.05132
72	0.41718	0.18690	0.45713	162	-0.06330	0.01916	0.06613
74	0.32314	0.11926	0.34445	164	-0.07614	0.01663	0.07794
76	0.12632	0.00502	0.12642	166	-0.08527	0.01482	0.08654
78	-0.16515	-0.15031	0.22331	168	-0.09134	0.01360	0.09235
80	-0.52690	-0.33435	0.62403	170	-0.09509	0.01284	0.09595
82	-0.92062	-0.52897	1.06177	172	-0.09717	0.01242	0.09796
84	-1.29915	-0.71260	1.48175	174	-0.09816	0.01222	0.09892
86	-1.61368	-0.86332	1.83011	176	-0.09853	0.01215	0.09928
88	-1.82178	-0.96230	2.06031	178	-0.09862	0.01213	0.09936
				180	-0.09863	0.01213	0.09937

VBR

RADIATION PATTERN FOR THETA=PI/2

H/LMDA=0.5000 B/LMDA= 0.500 OMEGA= 9.92

10 ELEMENT ARRAY

PHI DEG	ER(THETA,PHI) RE	IM	ABSVAL	PHI DEG	ER(THETA,PHI) RE	IM	ABSVAL
0	0.00000	0.00000	0.00000	90	-1.82335	-0.60759	1.92192
2	-0.00137	-0.00052	0.00146	92	-1.73625	-0.57800	1.82993
4	-0.00546	-0.00208	0.00584	94	-1.49010	-0.49446	1.57000
6	-0.01227	-0.00467	0.01313	96	-1.12709	-0.37147	1.18673
8	-0.02173	-0.00828	0.02325	98	-0.70742	-0.22972	0.74378
10	-0.03372	-0.01285	0.03609	100	-0.29683	-0.09172	0.31068
12	-0.04799	-0.01829	0.05135	102	0.04644	0.02268	0.05168
14	-0.06408	-0.02443	0.06858	104	0.28214	0.09990	0.29931
16	-0.08131	-0.03103	0.08703	106	0.39379	0.13463	0.41617
18	-0.09867	-0.03770	0.10563	108	0.38901	0.12986	0.41012
20	-0.11483	-0.04395	0.12295	110	0.29484	0.09521	0.30983
22	-0.12808	-0.04915	0.13719	112	0.14965	0.04411	0.15601
24	-0.13647	-0.05257	0.14625	114	-0.00596	-0.00937	0.01110
26	-0.13790	-0.05342	0.14789	116	-0.13735	-0.05341	0.14737
28	-0.13036	-0.05095	0.13997	118	-0.22138	-0.08026	0.23549
30	-0.11229	-0.04456	0.12081	120	-0.24868	-0.08705	0.26348
32	-0.08295	-0.03392	0.08962	122	-0.22271	-0.07536	0.23512
34	-0.04286	-0.01921	0.04697	124	-0.15653	-0.05005	0.16433
36	0.00577	-0.00116	0.00588	126	-0.06834	-0.01771	0.07059
38	0.05886	0.01875	0.06177	128	0.02296	0.01495	0.02739
40	0.11048	0.03841	0.11697	130	0.10135	0.04233	0.10983
42	0.15329	0.05514	0.16290	132	0.15597	0.06071	0.16737
44	0.17942	0.06605	0.19119	134	0.18184	0.06850	0.19432
46	0.18184	0.06850	0.19432	136	0.17942	0.06605	0.19119
48	0.15597	0.06071	0.16737	138	0.15329	0.05514	0.16290
50	0.10135	0.04233	0.10983	140	0.11048	0.03841	0.11697
52	0.02296	0.01495	0.02740	142	0.05886	0.01875	0.06178
54	-0.06834	-0.01771	0.07059	144	0.00577	-0.00116	0.00589
56	-0.15653	-0.05005	0.16433	146	-0.04286	-0.01921	0.04697
58	-0.22271	-0.07536	0.23512	148	-0.08295	-0.03392	0.08962
60	-0.24868	-0.08705	0.26348	150	-0.11229	-0.04456	0.12081
62	-0.22139	-0.08026	0.23549	152	-0.13036	-0.05095	0.13997
64	-0.13735	-0.05341	0.14737	154	-0.13790	-0.05342	0.14789
66	-0.00597	-0.00937	0.01111	156	-0.13647	-0.05257	0.14625
68	0.14965	0.04411	0.15601	158	-0.12808	-0.04915	0.13718
70	0.29483	0.09521	0.30982	160	-0.11483	-0.04395	0.12295
72	0.38901	0.12986	0.41011	162	-0.09867	-0.03770	0.10563
74	0.39379	0.13463	0.41617	164	-0.08131	-0.03103	0.08703
76	0.28215	0.09990	0.29931	166	-0.06408	-0.02443	0.06858
78	0.04645	0.02268	0.05169	168	-0.04799	-0.01829	0.05135
80	-0.29682	-0.09172	0.31067	170	-0.03372	-0.01285	0.03609
82	-0.70741	-0.22972	0.74378	172	-0.02173	-0.00828	0.02325
84	-1.12707	-0.37147	1.18671	174	-0.01227	-0.00467	0.01313
86	-1.49009	-0.49446	1.56998	176	-0.00546	-0.00208	0.00584
88	-1.73625	-0.57800	1.82993	178	-0.00137	-0.00052	0.00146
				180	0.00000	0.00000	0.00000

IBR

RADIATION PATTERN FOR THETA=PI/2

H/LMDA=0.5000 B/LMDA= 0.500 OMEGA= 9.92

10 ELEMENT ARRAY

PHI DEG	ER(THETA,PHI) RE	IM	ABSVAL	PHI DEG	ER(THETA,PHI) RE	IM	ABSVAL
0	0.00000	0.00000	0.00000	90	-1.88465	-0.97700	2.12284
2	-0.00127	-0.00016	0.00128	92	-1.79483	-0.93463	2.02360
4	-0.00509	-0.00066	0.00514	94	-1.54095	-0.81434	1.74289
6	-0.01144	-0.00148	0.01154	96	-1.16633	-0.63527	1.32812
8	-0.02027	-0.00262	0.02044	98	-0.73289	-0.42497	0.84719
10	-0.03145	-0.00404	0.03170	100	-0.30828	-0.21399	0.37526
12	-0.04472	-0.00568	0.04508	102	0.04746	-0.03015	0.05623
14	-0.05967	-0.00745	0.06013	104	0.29264	0.10615	0.31129
16	-0.07561	-0.00918	0.07617	106	0.40994	0.18475	0.44965
18	-0.09156	-0.01064	0.09218	108	0.40682	0.20629	0.45613
20	-0.10622	-0.01152	0.10685	110	0.31073	0.18050	0.35935
22	-0.11795	-0.01145	0.11851	112	0.16095	0.12307	0.20261
24	-0.12484	-0.01002	0.12524	114	-0.00080	0.05179	0.05180
26	-0.12487	-0.00680	0.12505	116	-0.13870	-0.01695	0.13973
28	-0.11612	-0.00147	0.11612	118	-0.22858	-0.07085	0.23931
30	-0.09712	0.00614	0.09731	120	-0.26030	-0.10300	0.27994
32	-0.06724	0.01590	0.06909	122	-0.23689	-0.11190	0.26199
34	-0.02714	0.02727	0.03847	124	-0.17130	-0.10054	0.19862
36	0.02090	0.03923	0.04445	126	-0.08189	-0.07475	0.11087
38	0.07268	0.05025	0.08836	128	0.01209	-0.04152	0.04325
40	0.12224	0.05834	0.13544	130	0.09417	-0.00748	0.09447
42	0.16221	0.06131	0.17341	132	0.15301	0.02217	0.15461
44	0.18482	0.05706	0.19343	134	0.18320	0.04412	0.18844
46	0.18320	0.04412	0.18844	136	0.18483	0.05706	0.19343
48	0.15301	0.02217	0.15461	138	0.16221	0.06130	0.17341
50	0.09417	-0.00748	0.09447	140	0.12223	0.05834	0.13544
52	0.01209	-0.04152	0.04325	142	0.07268	0.05025	0.08836
54	-0.08189	-0.07475	0.11087	144	0.02090	0.03923	0.04445
56	-0.17129	-0.10054	0.19862	146	-0.02714	0.02727	0.03847
58	-0.23689	-0.11190	0.26199	148	-0.06724	0.01590	0.06909
60	-0.26030	-0.10300	0.27994	150	-0.09712	0.00614	0.09731
62	-0.22858	-0.07085	0.23931	152	-0.11612	-0.00147	0.11612
64	-0.13870	-0.01695	0.13973	154	-0.12487	-0.00680	0.12505
66	-0.00080	0.05179	0.05180	156	-0.12484	-0.01002	0.12524
68	0.16095	0.12307	0.20261	158	-0.11795	-0.01145	0.11851
70	0.31073	0.18050	0.35935	160	-0.10622	-0.01152	0.10685
72	0.40682	0.20629	0.45613	162	-0.09156	-0.01064	0.09218
74	0.40994	0.18475	0.44965	164	-0.07561	-0.00918	0.07617
76	0.29264	0.10615	0.31130	166	-0.05967	-0.00745	0.06013
78	0.04747	-0.03014	0.05623	168	-0.04472	-0.00568	0.04508
80	-0.30827	-0.21398	0.37526	170	-0.03145	-0.00404	0.03170
82	-0.73288	-0.42497	0.84718	172	-0.02027	-0.00262	0.02044
84	-1.16631	-0.63527	1.32810	174	-0.01144	-0.00148	0.01154
86	-1.54093	-0.81433	1.74287	176	-0.00509	-0.00066	0.00514
88	-1.79483	-0.93463	2.02360	178	-0.00127	-0.00016	0.00128
				180	0.00000	0.00000	0.00000

VBR

RADIATION PATTERN FOR THETA=PI/2

H/LMDA=0.5000 B/LMDA= 0.500 OMEGA= 9.92

12 ELEMENT ARRAY

PHI DEG	ER(THETA,PHI) RE	IM	ABSVAL	PHI DEG	ER(THETA,PHI) RE	IM	ABSVAL
0	0.00000	0.00000	0.00000	90	-1.82724	-0.60762	1.92562
2	0.00134	0.00051	0.00143	92	-1.70158	-0.56515	1.79297
4	0.00533	0.00202	0.00570	94	-1.35580	-0.44840	1.42803
6	0.01196	0.00454	0.01280	96	-0.87407	-0.28615	0.91971
8	0.02117	0.00803	0.02264	98	-0.36826	-0.11655	0.38626
10	0.03276	0.01242	0.03503	100	0.05425	0.02395	0.05930
12	0.04638	0.01759	0.04961	102	0.31944	0.11049	0.33801
14	0.06142	0.02331	0.06570	104	0.40298	0.13535	0.42511
16	0.07693	0.02923	0.08230	106	0.33038	0.10784	0.34753
18	0.09156	0.03483	0.09796	108	0.16250	0.04923	0.16980
20	0.10353	0.03946	0.11079	110	-0.02699	-0.01513	0.03095
22	0.11074	0.04234	0.11856	112	-0.17443	-0.06368	0.18569
24	0.11094	0.04262	0.11885	114	-0.24197	-0.08397	0.25612
26	0.10202	0.03949	0.10939	116	-0.22312	-0.07449	0.23523
28	0.08247	0.03237	0.08859	118	-0.13811	-0.04288	0.14461
30	0.05197	0.02112	0.05609	120	-0.02256	-0.00188	0.02263
32	0.01193	0.00620	0.01345	122	0.08582	0.03534	0.09281
34	-0.03407	-0.01111	0.03583	124	0.15816	0.05893	0.16879
36	-0.08016	-0.02868	0.08514	126	0.18055	0.06445	0.19171
38	-0.11867	-0.04371	0.12646	128	0.15454	0.05290	0.16334
40	-0.14114	-0.05306	0.15079	130	0.09349	0.02940	0.09800
42	-0.14017	-0.05393	0.15019	132	0.01669	0.00099	0.01672
44	-0.11157	-0.04463	0.12017	134	-0.05659	-0.02541	0.06204
46	-0.05659	-0.02541	0.06204	136	-0.11157	-0.04463	0.12017
48	0.01669	0.00099	0.01672	138	-0.14017	-0.05393	0.15019
50	0.09349	0.02940	0.09800	140	-0.14114	-0.05306	0.15079
52	0.15454	0.05291	0.16334	142	-0.11867	-0.04371	0.12646
54	0.18055	0.06445	0.19171	144	-0.08016	-0.02868	0.08514
56	0.15817	0.05893	0.16879	146	-0.03407	-0.01111	0.03583
58	0.08582	0.03534	0.09281	148	0.01193	0.00620	0.01344
60	-0.02256	-0.00188	0.02263	150	0.05197	0.02112	0.05609
62	-0.13811	-0.04288	0.14461	152	0.08246	0.03237	0.08859
64	-0.22312	-0.07449	0.23522	154	0.10202	0.03949	0.10939
66	-0.24197	-0.08397	0.25612	156	0.11094	0.04262	0.11885
68	-0.17443	-0.06368	0.18569	158	0.11075	0.04234	0.11856
70	-0.02700	-0.01514	0.03095	160	0.10353	0.03946	0.11079
72	0.16250	0.04923	0.16979	162	0.09156	0.03483	0.09796
74	0.33037	0.10784	0.34753	164	0.07694	0.02923	0.08230
76	0.40298	0.13535	0.42511	166	0.06143	0.02331	0.06570
78	0.31944	0.11049	0.33801	168	0.04638	0.01759	0.04961
80	0.05426	0.02395	0.05931	170	0.03276	0.01242	0.03503
82	-0.36825	-0.11654	0.38625	172	0.02117	0.00803	0.02264
84	-0.87404	-0.28614	0.91969	174	0.01196	0.00454	0.01280
86	-1.35578	-0.44839	1.42800	176	0.00533	0.00202	0.00570
88	-1.70157	-0.56515	1.79297	178	0.00134	0.00051	0.00143
				180	0.00000	0.00000	0.00000

IBR

RADIATION PATTERN FOR THETA=PI/2

H/LMDA=0.5000 B/LMDA= 0.500 OMEGA= 9.92

12 ELEMENT ARRAY

PHI DEG	ER(THETA,PHI) RE	IM	ABSVAL	PHI DEG	ER(THETA,PHI) RE	IM	ABSVAL
0	0.00000	0.00000	0.00000	90	-1.88992	-0.99099	2.13397
2	0.00123	0.00013	0.00124	92	-1.76014	-0.92816	1.98987
4	0.00491	0.00052	0.00493	94	-1.40296	-0.75429	1.59288
6	0.01101	0.00117	0.01108	96	-0.90506	-0.50896	1.03835
8	0.01948	0.00207	0.01959	98	-0.38176	-0.24547	0.45387
10	0.03013	0.00316	0.03030	100	0.05608	-0.01633	0.05841
12	0.04263	0.00440	0.04286	102	0.33177	0.14016	0.36016
14	0.05638	0.00563	0.05666	104	0.41973	0.20798	0.46843
16	0.07046	0.00668	0.07078	106	0.34569	0.19445	0.39662
18	0.08357	0.00726	0.08388	108	0.17207	0.12471	0.21251
20	0.09400	0.00703	0.09427	110	-0.02521	0.03220	0.04089
22	0.09975	0.00559	0.09991	112	-0.18013	-0.05158	0.18737
24	0.09866	0.00258	0.09870	114	-0.25299	-0.10481	0.27384
26	0.08874	-0.00228	0.08877	116	-0.23626	-0.11894	0.26451
28	0.06864	-0.00906	0.06924	118	-0.15006	-0.09793	0.17919
30	0.03817	-0.01743	0.04196	120	-0.03065	-0.05425	0.06231
32	-0.00113	-0.02661	0.02663	122	0.08312	-0.00359	0.08319
34	-0.04559	-0.03524	0.05762	124	0.16114	0.04004	0.16604
36	-0.08931	-0.04147	0.09847	126	0.18837	0.06749	0.20009
38	-0.12466	-0.04319	0.13193	128	0.16559	0.07558	0.18202
40	-0.14336	-0.03844	0.14843	130	0.10581	0.06638	0.12491
42	-0.13828	-0.02600	0.14071	132	0.02834	0.04533	0.05346
44	-0.10566	-0.00608	0.10584	134	-0.04724	0.01914	0.05097
46	-0.04724	0.01914	0.05098	136	-0.10566	-0.00608	0.10583
48	0.02834	0.04533	0.05346	138	-0.13828	-0.02600	0.14071
50	0.10581	0.06637	0.12491	140	-0.14336	-0.03844	0.14843
52	0.16559	0.07558	0.18203	142	-0.12466	-0.04319	0.13193
54	0.18837	0.06749	0.20009	144	-0.08931	-0.04147	0.09847
56	0.16114	0.04004	0.16604	146	-0.04559	-0.03524	0.05762
58	0.08312	-0.00359	0.08320	148	-0.00113	-0.02661	0.02663
60	-0.03065	-0.05425	0.06231	150	0.03817	-0.01743	0.04196
62	-0.15006	-0.09793	0.17919	152	0.06864	-0.00906	0.06924
64	-0.23626	-0.11894	0.26451	154	0.08874	-0.00228	0.08877
66	-0.25299	-0.10481	0.27384	156	0.09866	0.00258	0.09869
68	-0.18013	-0.05158	0.18737	158	0.09976	0.00560	0.09991
70	-0.02522	0.03220	0.04090	160	0.09400	0.00703	0.09427
72	0.17206	0.12471	0.21251	162	0.08357	0.00726	0.08388
74	0.34568	0.19444	0.39662	164	0.07046	0.00668	0.07078
76	0.41973	0.20798	0.46843	166	0.05638	0.00563	0.05666
78	0.33178	0.14016	0.36017	168	0.04263	0.00440	0.04286
80	0.05609	-0.01633	0.05842	170	0.03013	0.00316	0.03030
82	-0.38175	-0.24547	0.45386	172	0.01948	0.00207	0.01959
84	-0.90503	-0.50894	1.03832	174	0.01101	0.00117	0.01108
86	-1.40294	-0.75428	1.59285	176	0.00491	0.00052	0.00493
88	-1.76014	-0.92816	1.98987	178	0.00123	0.00013	0.00124
				180	0.00000	0.00000	0.00000

VBR

RADIATION PATTERN FOR THETA=PI/2

H/LMDA=0.5000 B/LMDA= 0.500 OMEGA= 9.92

14 ELEMENT ARRAY

PHI DEG	ER(THETA,PHI) RE	IM	ABSVAL	PHI DEG	ER(THETA,PHI) RE	IM	ABSVAL
0	-0.00000	-0.00000	0.00000	90	-1.83005	-0.60767	1.92830
2	-0.00131	-0.00049	0.00140	92	-1.65934	-0.55017	1.74817
4	-0.00522	-0.00197	0.00558	94	-1.20434	-0.39715	1.26813
6	-0.01172	-0.00442	0.01252	96	-0.61331	-0.19902	0.64480
8	-0.02070	-0.00780	0.02212	98	-0.06663	-0.01692	0.06874
10	-0.03193	-0.01204	0.03412	100	0.29157	0.10064	0.30845
12	-0.04493	-0.01695	0.04803	102	0.40182	0.13430	0.42367
14	-0.05892	-0.02224	0.06298	104	0.29935	0.09692	0.31465
16	-0.07265	-0.02744	0.07766	106	0.08582	0.02370	0.08903
18	-0.08440	-0.03192	0.09023	108	-0.12051	-0.04506	0.12866
20	-0.09203	-0.03487	0.09842	110	-0.23268	-0.08039	0.24618
22	-0.09315	-0.03540	0.09965	112	-0.22182	-0.07352	0.23368
24	-0.08548	-0.03264	0.09150	114	-0.11577	-0.03503	0.12096
26	-0.06738	-0.02597	0.07221	116	0.02429	0.01343	0.02776
28	-0.03861	-0.01525	0.04151	118	0.13532	0.05021	0.14433
30	-0.00098	-0.00113	0.00150	120	0.17762	0.06219	0.18819
32	0.04111	0.01481	0.04370	122	0.14541	0.04824	0.15320
34	0.08071	0.03003	0.08612	124	0.06192	0.01724	0.06428
36	0.10912	0.04129	0.11667	126	-0.03518	-0.01724	0.03918
38	0.11766	0.04535	0.12610	128	-0.11030	-0.04274	0.11829
40	0.10028	0.03983	0.10790	130	-0.14174	-0.05199	0.15097
42	0.05643	0.02435	0.06146	132	-0.12588	-0.04422	0.13342
44	-0.00677	0.00130	0.00689	134	-0.07430	-0.02401	0.07808
46	-0.07430	-0.02401	0.07808	136	-0.00677	0.00130	0.00689
48	-0.12588	-0.04422	0.13342	138	0.05643	0.02435	0.06146
50	-0.14174	-0.05199	0.15097	140	0.10028	0.03983	0.10790
52	-0.11030	-0.04274	0.11829	142	0.11766	0.04535	0.12609
54	-0.03518	-0.01724	0.03918	144	0.10912	0.04129	0.11667
56	0.06192	0.01724	0.06427	146	0.08071	0.03003	0.08612
58	0.14541	0.04824	0.15320	148	0.04111	0.01481	0.04370
60	0.17762	0.06219	0.18819	150	-0.00098	-0.00113	0.00150
62	0.13532	0.05021	0.14433	152	-0.03860	-0.01525	0.04151
64	0.02430	0.01344	0.02776	154	-0.06738	-0.02597	0.07221
66	-0.11577	-0.03503	0.12095	156	-0.08548	-0.03264	0.09150
68	-0.22182	-0.07351	0.23368	158	-0.09315	-0.03540	0.09965
70	-0.23268	-0.08039	0.24618	160	-0.09203	-0.03487	0.09842
72	-0.12051	-0.04507	0.12866	162	-0.08440	-0.03192	0.09023
74	0.08581	0.02370	0.08902	164	-0.07264	-0.02744	0.07765
76	0.29934	0.09691	0.31464	166	-0.05892	-0.02224	0.06298
78	0.40182	0.13430	0.42367	168	-0.04493	-0.01695	0.04803
80	0.29158	0.10065	0.30846	170	-0.03193	-0.01204	0.03412
82	-0.06662	-0.01692	0.06873	172	-0.02070	-0.00780	0.02212
84	-0.61329	-0.19901	0.64477	174	-0.01172	-0.00442	0.01252
86	-1.20431	-0.39714	1.26810	176	-0.00522	-0.00197	0.00558
88	-1.65933	-0.55017	1.74816	178	-0.00131	-0.00049	0.00140
				180	-0.00000	-0.00000	0.00000

IBR

RADIATION PATTERN FOR THETA=PI/2

H/LMDA=0.5000 B/LMDA= 0.500 OMEGA= 9.92

14 ELEMENT ARRAY

PHI DEG	ER(THETA,PHI) RE	IM	ABSVAL	PHI DEG	ER(THETA,PHI) RE	IM	ABSVAL
0	-0.00000	0.00000	0.00000	90	-1.89365	-1.00093	2.14191
2	-0.00119	-0.00010	0.00119	92	-1.71719	-0.91388	1.94523
4	-0.00475	-0.00042	0.00477	94	-1.24674	-0.68019	1.42022
6	-0.01066	-0.00094	0.01070	96	-0.63527	-0.37162	0.73598
8	-0.01883	-0.00164	0.01890	98	-0.06904	-0.07705	0.10345
10	-0.02903	-0.00249	0.02914	100	0.30276	0.12955	0.32931
12	-0.04081	-0.00340	0.04095	102	0.41809	0.21284	0.46915
14	-0.05341	-0.00421	0.05357	104	0.31268	0.18283	0.36221
16	-0.06563	-0.00471	0.06580	106	0.09112	0.08392	0.12388
18	-0.07585	-0.00458	0.07599	108	-0.12426	-0.02700	0.12716
20	-0.08203	-0.00347	0.08211	110	-0.24290	-0.10317	0.26391
22	-0.08192	-0.00103	0.08192	112	-0.23390	-0.12238	0.26398
24	-0.07337	0.00302	0.07343	114	-0.12504	-0.08930	0.15365
26	-0.05494	0.00869	0.05563	116	0.02097	-0.02731	0.03443
28	-0.02654	0.01563	0.03080	118	0.13868	0.03460	0.14293
30	0.00987	0.02293	0.02497	120	0.18619	0.07386	0.20030
32	0.04986	0.02914	0.05775	122	0.15625	0.08092	0.17596
34	0.08648	0.03232	0.09233	124	0.07175	0.05957	0.09326
36	0.11124	0.03048	0.11534	126	-0.02903	0.02227	0.03659
38	0.11579	0.02209	0.11788	128	-0.10927	-0.01615	0.11045
40	0.09457	0.00692	0.09482	130	-0.14587	-0.04392	0.15234
42	0.04763	-0.01327	0.04945	132	-0.13406	-0.05528	0.14501
44	-0.01729	-0.03441	0.03851	134	-0.08469	-0.05056	0.09863
46	-0.08469	-0.05056	0.09863	136	-0.01729	-0.03441	0.03851
48	-0.13406	-0.05528	0.14502	138	0.04763	-0.01327	0.04945
50	-0.14588	-0.04392	0.15234	140	0.09457	0.00692	0.09482
52	-0.10927	-0.01615	0.11045	142	0.11579	0.02208	0.11788
54	-0.02903	0.02227	0.03659	144	0.11124	0.03048	0.11534
56	0.07175	0.05957	0.09325	146	0.08648	0.03232	0.09233
58	0.15625	0.08093	0.17596	148	0.04985	0.02914	0.05774
60	0.18619	0.07386	0.20030	150	0.00988	0.02293	0.02497
62	0.13868	0.03460	0.14294	152	-0.02654	0.01563	0.03080
64	0.02097	-0.02730	0.03443	154	-0.05494	0.00869	0.05562
66	-0.12503	-0.08930	0.15364	156	-0.07337	0.00302	0.07343
68	-0.23390	-0.12238	0.26398	158	-0.08191	-0.00103	0.08192
70	-0.24290	-0.10317	0.26391	160	-0.08203	-0.00347	0.08211
72	-0.12426	-0.02700	0.12716	162	-0.07585	-0.00458	0.07599
74	0.09111	0.08392	0.12387	164	-0.06563	-0.00471	0.06580
76	0.31267	0.18283	0.36220	166	-0.05340	-0.00421	0.05357
78	0.41809	0.21284	0.46915	168	-0.04081	-0.00340	0.04095
80	0.30276	0.12955	0.32931	170	-0.02903	-0.00249	0.02914
82	-0.06903	-0.07704	0.10344	172	-0.01883	-0.00164	0.01890
84	-0.63524	-0.37161	0.73595	174	-0.01066	-0.00094	0.01070
86	-1.24671	-0.68018	1.42019	176	-0.00475	-0.00042	0.00477
88	-1.71718	-0.91388	1.94522	178	-0.00119	-0.00010	0.00119
				180	-0.00000	0.00000	0.00000

VBR

RADIATION PATTERN FOR THETA=PI/2

H/LMDA=0.5000 B/LMDA= 0.500 OMEGA= 9.92

16 ELEMENT ARRAY

PHI	ER(THETA,PHI)			PHI	ER(THETA,PHI)		
DEG	RE	IM	ABSVAL	DEG	RE	IM	ABSVAL
0	0.00000	0.00000	0.00000	90	-1.83218	-0.60772	1.93034
2	0.00129	0.00048	0.00137	92	-1.61024	-0.53318	1.69622
4	0.00513	0.00192	0.00548	94	-1.04036	-0.34211	1.09517
6	0.01150	0.00431	0.01229	96	-0.36042	-0.11507	0.37834
8	0.02029	0.00760	0.02167	98	0.17058	0.06059	0.18102
10	0.03118	0.01169	0.03330	100	0.39555	0.13264	0.41720
12	0.04359	0.01634	0.04655	102	0.31895	0.10383	0.33542
14	0.05649	0.02119	0.06033	104	0.07753	0.02146	0.08045
16	0.06836	0.02566	0.07302	106	-0.15046	-0.05400	0.15986
18	0.07714	0.02898	0.08241	108	-0.24102	-0.08149	0.25442
20	0.08037	0.03023	0.08587	110	-0.17377	-0.05572	0.18248
22	0.07555	0.02847	0.08074	112	-0.01830	-0.00170	0.01838
24	0.06075	0.02297	0.06494	114	0.12289	0.04523	0.13095
26	0.03538	0.01347	0.03786	116	0.17555	0.06042	0.18565
28	0.00113	0.00059	0.00128	118	0.12638	0.04066	0.13276
30	-0.03741	-0.01403	0.03995	120	0.01708	0.00152	0.01715
32	-0.07278	-0.02762	0.07785	122	-0.08810	-0.03434	0.09456
34	-0.09568	-0.03672	0.10249	124	-0.13859	-0.04974	0.14725
36	-0.09731	-0.03800	0.10447	126	-0.11869	-0.04016	0.12530
38	-0.07272	-0.02948	0.07847	128	-0.04744	-0.01319	0.04924
40	-0.02417	-0.01180	0.02690	130	0.03727	0.01728	0.04108
42	0.03690	0.01104	0.03852	132	0.09893	0.03835	0.10610
44	0.09121	0.03203	0.09667	134	0.11699	0.04312	0.12469
46	0.11699	0.04312	0.12469	136	0.09121	0.03203	0.09667
48	0.09893	0.03835	0.10610	138	0.03691	0.01104	0.03852
50	0.03727	0.01728	0.04108	140	-0.02417	-0.01180	0.02690
52	-0.04744	-0.01319	0.04924	142	-0.07272	-0.02948	0.07846
54	-0.11869	-0.04016	0.12531	144	-0.09731	-0.03800	0.10447
56	-0.13860	-0.04974	0.14725	146	-0.09568	-0.03672	0.10249
58	-0.08810	-0.03434	0.09456	148	-0.07278	-0.02762	0.07785
60	0.01708	0.00152	0.01715	150	-0.03741	-0.01403	0.03995
62	0.12638	0.04066	0.13275	152	0.00113	0.00059	0.00127
64	0.17555	0.06042	0.18565	154	0.03538	0.01347	0.03785
66	0.12290	0.04523	0.13096	156	0.06074	0.02296	0.06494
68	-0.01829	-0.00170	0.01837	158	0.07556	0.02847	0.08074
70	-0.17377	-0.05572	0.18248	160	0.08037	0.03023	0.08587
72	-0.24102	-0.08149	0.25442	162	0.07714	0.02898	0.08241
74	-0.15047	-0.05400	0.15987	164	0.06836	0.02566	0.07302
76	0.07752	0.02146	0.08044	166	0.05649	0.02119	0.06034
78	0.31894	0.10383	0.33542	168	0.04359	0.01634	0.04655
80	0.39555	0.13264	0.41720	170	0.03118	0.01169	0.03330
82	0.17059	0.06059	0.18103	172	0.02029	0.00760	0.02167
84	-0.36039	-0.11506	0.37831	174	0.01150	0.00431	0.01229
86	-1.04033	-0.34209	1.09513	176	0.00513	0.00192	0.00548
88	-1.61024	-0.53318	1.69621	178	0.00129	0.00048	0.00137
				180	0.00000	0.00000	0.00000

IBR

RADIATION PATTERN FOR THETA=PI/2

H/LMDA=0.5000 B/LMDA= 0.500 OMEGA= 9.92

16 ELEMENT ARRAY

PHI	ER(THETA,PHI)			PHI	ER(THETA,PHI)		
DEG	RE	IM	ABSVAL	DEG	RE	IM	ABSVAL
0	0.00000	-0.00000	0.00000	90	-1.89644	-1.00837	2.14786
2	0.00116	0.00009	0.00116	92	-1.66689	-0.89349	1.89125
4	0.00462	0.00034	0.00464	94	-1.07728	-0.59593	1.23112
6	0.01037	0.00075	0.01039	96	-0.37333	-0.23352	0.44035
8	0.01828	0.00131	0.01832	98	0.17709	0.06235	0.18774
10	0.02806	0.00196	0.02813	100	0.41105	0.20636	0.45994
12	0.03917	0.00260	0.03926	102	0.33231	0.19079	0.38319
14	0.05062	0.00307	0.05071	104	0.08176	0.07580	0.11149
16	0.06098	0.00311	0.06106	106	-0.15613	-0.05098	0.16424
18	0.06830	0.00241	0.06834	108	-0.25207	-0.12056	0.27942
20	0.07027	0.00063	0.07027	110	-0.18371	-0.11039	0.21432
22	0.06457	-0.00252	0.06462	112	-0.02209	-0.04360	0.04888
24	0.04948	-0.00709	0.04998	114	0.12662	0.03359	0.13100
26	0.02459	-0.01278	0.02771	116	0.18447	0.08028	0.20118
28	-0.00825	-0.01873	0.02047	118	0.13605	0.07955	0.15760
30	-0.04443	-0.02352	0.05027	120	0.02322	0.04093	0.04705
32	-0.07658	-0.02526	0.08064	122	-0.08790	-0.01050	0.08852
34	-0.09569	-0.02206	0.09820	124	-0.14413	-0.04936	0.15234
36	-0.09345	-0.01276	0.09431	126	-0.12764	-0.06141	0.14164
38	-0.06555	0.00217	0.06559	128	-0.05651	-0.04688	0.07343
40	-0.01498	0.02000	0.02499	130	0.03107	-0.01668	0.03527
42	0.04625	0.03567	0.05841	132	0.09742	0.01485	0.09855
44	0.09857	0.04287	0.10749	134	0.12046	0.03637	0.12583
46	0.12046	0.03637	0.12583	136	0.09857	0.04287	0.10749
48	0.09743	0.01485	0.09855	138	0.04626	0.03567	0.05841
50	0.03107	-0.01668	0.03527	140	-0.01498	0.02000	0.02499
52	-0.05651	-0.04688	0.07343	142	-0.06555	0.00217	0.06559
54	-0.12764	-0.06141	0.14165	144	-0.09345	-0.01276	0.09431
56	-0.14413	-0.04936	0.15234	146	-0.09569	-0.02206	0.09820
58	-0.08790	-0.01050	0.08852	148	-0.07658	-0.02526	0.08064
60	0.02321	0.04093	0.04705	150	-0.04443	-0.02352	0.05027
62	0.13605	0.07955	0.15760	152	-0.00825	-0.01873	0.02047
64	0.18447	0.08028	0.20118	154	0.02459	-0.01278	0.02771
66	0.12663	0.03360	0.13101	156	0.04947	-0.00709	0.04998
68	-0.02209	-0.04360	0.04887	158	0.06457	-0.00252	0.06462
70	-0.18371	-0.11039	0.21432	160	0.07027	0.00063	0.07027
72	-0.25207	-0.12056	0.27942	162	0.06830	0.00241	0.06834
74	-0.15614	-0.05098	0.16425	164	0.06098	0.00311	0.06106
76	0.08175	0.07580	0.11148	166	0.05062	0.00307	0.05072
78	0.33231	0.19079	0.38318	168	0.03917	0.00260	0.03926
80	0.41105	0.20636	0.45994	170	0.02806	0.00196	0.02813
82	0.17710	0.06236	0.18775	172	0.01828	0.00131	0.01832
84	-0.37330	-0.23350	0.44032	174	0.01037	0.00075	0.01039
86	-1.07724	-0.59592	1.23108	176	0.00462	0.00034	0.00464
88	-1.66688	-0.89349	1.89124	178	0.00116	0.00009	0.00116
				180	0.00000	-0.00000	0.00000

VBR

RADIATION PATTERN FOR [illegible]

H/LMDA=0.5000 B/LMDA= 0.500 OMEGA= 9.92

18 ELEMENT ARRAY

PHI	ER(THETA,PHI)			PHI	ER(THETA,PHI)		
DEG	RE	IM	ABSVAL	DEG	RE	IM	ABSVAL
0	-0.00000	-0.00000	0.00000	90	-1.83385	-0.60776	1.93194
2	-0.00126	-0.00047	0.00135	92	-1.55489	-0.51428	1.63773
4	-0.00505	-0.00188	0.00539	94	-0.86861	-0.28476	0.91409
6	-0.01133	-0.00422	0.01209	96	-0.12948	-0.03885	0.13519
8	-0.01993	-0.00743	0.02127	98	0.32656	0.11076	0.34484
10	-0.03050	-0.01136	0.03255	100	0.37440	0.12332	0.39419
12	-0.04230	-0.01576	0.04514	102	0.13781	0.04216	0.14412
14	-0.05409	-0.02016	0.05773	104	-0.13291	-0.04772	0.14122
16	-0.06405	-0.02387	0.06835	106	-0.24011	-0.08067	0.25330
18	-0.06978	-0.02601	0.07447	108	-0.14828	-0.04680	0.15549
20	-0.06864	-0.02558	0.07325	110	0.03381	0.01532	0.03712
22	-C.05827	-0.02169	0.06218	112	0.16004	0.05602	0.16956
24	-0.03747	-0.01389	0.03996	114	0.15321	0.05052	0.16132
26	-0.00729	-0.00252	0.00771	116	0.04125	0.01028	0.04251
28	0.02815	0.01089	0.03019	118	-0.08366	-0.03208	0.08960
30	0.06142	0.02361	0.06580	120	-0.13814	-0.04837	0.14636
32	0.08286	0.03203	0.08883	122	-0.09846	-0.03184	0.10348
34	0.08325	0.03265	0.08943	124	-0.00272	0.00286	0.00395
36	0.05779	0.02355	0.06241	126	0.08472	0.03278	0.09084
38	0.01005	0.00587	0.01164	128	0.11572	0.04156	0.12296
40	-0.04614	-0.01544	0.04866	130	0.08154	0.02715	0.08594
42	-0.08923	-0.03248	0.09496	132	0.00890	-0.00008	0.00890
44	-0.09773	-0.03714	0.10455	134	-0.06150	-0.02525	0.06648
46	-0.06150	-0.02525	0.06648	136	-0.09773	-0.03714	0.10455
48	0.00889	-0.00008	0.00889	138	-0.08923	-0.03248	0.09496
50	0.08154	0.02715	0.08594	140	-0.04615	-0.01545	0.04866
52	0.11572	0.04156	0.12296	142	0.01005	0.00587	0.01164
54	0.08472	0.03278	0.09084	144	0.05779	0.02355	0.06241
56	-0.00271	0.00286	0.00394	146	0.08325	0.03265	0.08943
58	-0.09846	-0.03184	0.10348	148	0.08286	0.03203	0.08883
60	-0.13814	-0.04837	0.14636	150	0.06142	0.02361	0.06581
62	-0.08366	-0.03208	0.08960	152	0.02815	0.01089	0.03019
64	0.04125	0.01028	0.04251	154	-0.00729	-0.00252	0.00771
66	0.15321	0.05052	0.16132	156	-0.03747	-0.01389	0.03997
68	0.16004	0.05602	0.16956	158	-0.05827	-0.02169	0.06217
70	0.03381	0.01532	0.03712	160	-0.06864	-0.02558	0.07325
72	-0.14827	-0.04680	0.15548	162	-0.06978	-0.02601	0.07447
74	-0.24011	-0.08067	0.25330	164	-0.06405	-0.02387	0.06835
76	-0.13292	-0.04772	0.14123	166	-0.05409	-0.02016	0.05773
78	0.13781	0.04216	0.14411	168	-0.04230	-0.01576	0.04514
80	0.37440	0.12332	0.39419	170	-0.03050	-0.01136	0.03255
82	0.32657	0.11076	0.34484	172	-0.01993	-0.00743	0.02127
84	-0.12946	-0.03884	0.13516	174	-0.01133	-0.00422	0.01209
86	-0.86857	-0.28475	0.91405	176	-0.00505	-0.00188	0.00539
88	-1.55488	-0.51427	1.63772	178	-0.00126	-0.00047	0.00135
				180	-0.00000	-0.00000	0.00000

IBR

RADIATION PATTERN FOR THETA=PI/2

H/LMDA=0.5000 B/LMDA= 0.500 OMEGA= 9.92

18 ELEMENT ARRAY

PHI	ER(THETA,PHI)			PHI	ER(THETA,PHI)		
DEG	RE	IM	ABSVAL	DEG	RE	IM	ABSVAL
0	-0.00000	-0.00000	0.00000	90	-1.89860	-1.01413	2.15247
2	-0.00113	-0.00007	0.00113	92	-1.60994	-0.86802	1.82903
4	-0.00452	-0.00027	0.00452	94	-0.89958	-0.50488	1.03158
6	-0.01012	-0.00060	0.01014	96	-0.13400	-0.10367	0.16942
8	-0.01780	-0.00104	0.01783	98	0.33904	0.16058	0.37514
10	-0.02721	-0.00152	0.02725	100	0.38927	0.21296	0.44372
12	-0.03765	-0.00194	0.03770	102	0.14395	0.10375	0.17744
14	-0.04797	-0.00212	0.04802	104	-0.13796	-0.04473	0.14503
16	-0.05644	-0.00179	0.05647	106	-0.25086	-0.12421	0.27993
18	-0.06085	-0.00063	0.06085	108	-0.15648	-0.10014	0.18578
20	-0.05873	0.00165	0.05875	110	0.03342	-0.01316	0.03592
22	-0.04795	0.00519	0.04823	112	0.16697	0.06561	0.17940
24	-0.02755	0.00982	0.02925	114	0.16243	0.08691	0.18422
26	0.00128	0.01482	0.01487	116	0.04692	0.04884	0.06772
28	0.03435	0.01890	0.03921	118	-0.08468	-0.01336	0.08573
30	0.06436	0.02028	0.06748	120	-0.14491	-0.05755	0.15592
32	0.08203	0.01714	0.08380	122	-0.10703	-0.06137	0.12338
34	0.07872	0.00845	0.07918	124	-0.00860	-0.03027	0.03147
36	0.05041	-0.00508	0.05067	126	0.08428	0.01232	0.08517
38	0.00142	-0.02029	0.02034	128	0.12072	0.04219	0.12788
40	-0.05389	-0.03179	0.06257	130	0.08963	0.04712	0.10126
42	-0.09392	-0.03351	0.09972	132	0.01670	0.02976	0.03412
44	-0.09786	-0.02161	0.10022	134	-0.05690	0.00234	0.05695
46	-0.05690	0.00234	0.05695	136	-0.09786	-0.02161	0.10022
48	0.01670	0.02976	0.03412	138	-0.09392	-0.03351	0.09972
50	0.08963	0.04712	0.10126	140	-0.05389	-0.03179	0.06257
52	0.12072	0.04219	0.12788	142	0.00142	-0.02029	0.02034
54	0.08428	0.01232	0.08517	144	0.05041	-0.00508	0.05066
56	-0.00860	-0.03027	0.03147	146	0.07872	0.00845	0.07918
58	-0.10703	-0.06136	0.12338	148	0.08203	0.01714	0.08380
60	-0.14491	-0.05755	0.15592	150	0.06437	0.02028	0.06749
62	-0.08468	-0.01336	0.08573	152	0.03435	0.01890	0.03921
64	0.04691	0.04884	0.06772	154	0.00128	0.01482	0.01487
66	0.16243	0.08691	0.18422	156	-0.02755	0.00981	0.02925
68	0.16697	0.06561	0.17940	158	-0.04795	0.00519	0.04823
70	0.03342	-0.01316	0.03592	160	-0.05873	0.00165	0.05875
72	-0.15647	-0.10014	0.18577	162	-0.06085	-0.00063	0.06085
74	-0.25086	-0.12422	0.27993	164	-0.05644	-0.00179	0.05647
76	-0.13797	-0.04474	0.14504	166	-0.04797	-0.00212	0.04801
78	0.14395	0.10375	0.17744	168	-0.03765	-0.00195	0.03770
80	0.38927	0.21296	0.44372	170	-0.02720	-0.00152	0.02724
82	0.33904	0.16059	0.37515	172	-0.01780	-0.00104	0.01783
84	-0.13397	-0.10365	0.16939	174	-0.01012	-0.00060	0.01014
86	-0.89954	-0.50487	1.03153	176	-0.00452	-0.00027	0.00452
88	-1.60993	-0.86802	1.82902	178	-0.00113	-0.00007	0.00113
				180	-0.00000	-0.00000	0.00000

VBR

RADIATION PATTERN FOR THETA=PI/2

H/LMDA=0.5000 B/LMDA= 0.500 OMEGA= 9.92

20 ELEMENT ARRAY

PHI	ER(THETA,PHI)			PHI	ER(THETA,PHI)		
DEG	RE	IM	ABSVAL	DEG	RE	IM	ABSVAL
0	-0.00000	-0.00000	0.00000	90	-1.83519	-0.60780	1.93322
2	0.00125	0.00046	0.00134	92	-1.49381	-0.49360	1.57324
4	0.00499	0.00185	0.00532	94	-0.69382	-0.22663	0.72990
6	0.01116	0.00413	0.01190	96	0.06772	0.02587	0.07249
8	0.01960	0.00726	0.02090	98	0.39608	0.13215	0.41754
10	0.02986	0.01106	0.03184	100	0.25969	0.08367	0.27284
12	0.04105	0.01520	0.04377	102	-0.05814	-0.02279	0.06245
14	0.05169	0.01914	0.05512	104	-0.23546	-0.07945	0.24850
16	0.05967	0.02208	0.06362	106	-0.15971	-0.05091	0.16762
18	0.06234	0.02303	0.06646	108	0.04208	0.01761	0.04562
20	0.05698	0.02099	0.06072	110	0.16801	0.05772	0.17765
22	0.04165	0.01521	0.04434	112	0.12526	0.04010	0.13153
24	0.01635	0.00568	0.01731	114	-0.01861	-0.00989	0.02107
26	-0.01585	-0.00647	0.01712	116	-0.12614	-0.04477	0.13385
28	-0.04816	-0.01874	0.05168	118	-0.11556	-0.03809	0.12168
30	-0.07092	-0.02752	0.07607	120	-0.01374	-0.00126	0.01380
32	-0.07417	-0.02905	0.07965	122	0.08706	0.03281	0.09304
34	-0.05213	-0.02106	0.05623	124	0.11178	0.03885	0.11834
36	-0.00795	-0.00455	0.00916	126	0.05349	0.01610	0.05586
38	0.04395	0.01528	0.04653	128	-0.03600	-0.01597	0.03938
40	0.08091	0.03003	0.08631	130	-0.09341	-0.03498	0.09974
42	0.08150	0.03156	0.08739	132	-0.08725	-0.03061	0.09246
44	0.03880	0.01683	0.04229	134	-0.03009	-0.00844	0.03126
46	-0.03009	-0.00844	0.03125	136	0.03880	0.01683	0.04229
48	-0.08725	-0.03061	0.09246	138	0.08150	0.03156	0.08739
50	-0.09341	-0.03498	0.09974	140	0.08091	0.03003	0.08630
52	-0.03600	-0.01597	0.03938	142	0.04395	0.01528	0.04653
54	0.05349	0.01610	0.05586	144	-0.00795	-0.00455	0.00916
56	0.11178	0.03885	0.11834	146	-0.05213	-0.02106	0.05623
58	0.08706	0.03281	0.09304	148	-0.07417	-0.02905	0.07965
60	-0.01374	-0.00126	0.01380	150	-0.07091	-0.02752	0.07607
62	-0.11556	-0.03809	0.12168	152	-0.04817	-0.01874	0.05168
64	-0.12614	-0.04477	0.13385	154	-0.01585	-0.00647	0.01712
66	-0.01861	-0.00989	0.02108	156	0.01634	0.00568	0.01730
68	0.12526	0.04010	0.13152	158	0.04165	0.01521	0.04434
70	0.16801	0.05772	0.17765	160	0.05698	0.02099	0.06072
72	0.04209	0.01761	0.04562	162	0.06234	0.02303	0.06646
74	-0.15970	-0.05091	0.16762	164	0.05967	0.02208	0.06362
76	-0.23546	-0.07945	0.24850	166	0.05169	0.01914	0.05512
78	-0.05815	-0.02279	0.06246	168	0.04104	0.01520	0.04377
80	0.25969	0.08366	0.27283	170	0.02986	0.01106	0.03184
82	0.39608	0.13216	0.41755	172	0.01960	0.00726	0.02091
84	0.06774	0.02587	0.07252	174	0.01116	0.00413	0.01190
86	-0.69378	-0.22661	0.72985	176	0.00499	0.00185	0.00532
88	-1.49379	-0.49360	1.57323	178	0.00125	0.00046	0.00134
				180	-0.00000	-0.00000	0.00000

IBR

RADIATION PATTERN FOR THETA=PI/2

H/LMDA=0.5000 B/LMDA= 0.500 OMEGA= 9.92

20 ELEMENT ARRAY

PHI	ER(THETA,PHI)			PHI	ER(THETA,PHI)		
DEG	RE	IM	ABSVAL	DEG	RE	IM	ABSVAL
0	-0.00000	-0.00000	0.00000	90	-1.90032	-1.01873	2.15616
2	0.00111	0.00006	0.00111	92	-1.54695	-0.83821	1.75945
4	0.00442	0.00021	0.00443	94	-0.71863	-0.41017	0.82744
6	0.00989	0.00047	0.00991	96	0.07044	0.01037	0.07120
8	0.01736	0.00081	0.01738	98	0.41126	0.21210	0.46273
10	0.02642	0.00116	0.02644	100	0.27015	0.16259	0.31530
12	0.03622	0.00140	0.03625	102	-0.06015	-0.00446	0.06031
14	0.04539	0.00132	0.04541	104	-0.24550	-0.11916	0.27289
16	0.05197	0.00068	0.05197	106	-0.16772	-0.10416	0.19743
18	0.05350	-0.00084	0.05351	108	0.04265	-0.00527	0.04298
20	0.04751	-0.00344	0.04764	110	0.17571	0.07717	0.19191
22	0.03232	-0.00711	0.03309	112	0.13313	0.08034	0.15549
24	0.00814	-0.01135	0.01397	114	-0.01682	0.01834	0.02488
26	-0.02188	-0.01510	0.02659	116	-0.13138	-0.04738	0.13967
28	-0.05106	-0.01673	0.05373	118	-0.12356	-0.06607	0.14011
30	-0.07012	-0.01445	0.07159	120	-0.01868	-0.03285	0.03779
32	-0.06977	-0.00713	0.07013	122	0.08837	0.01874	0.09034
34	-0.04508	0.00464	0.04532	124	0.11823	0.04962	0.12822
36	-0.00003	0.01779	0.01779	126	0.06093	0.04320	0.07469
38	0.05047	0.02699	0.05723	128	-0.03190	0.01062	0.03362
40	0.08391	0.02641	0.08797	130	-0.09478	-0.02339	0.09762
42	0.07983	0.01312	0.08090	132	-0.09321	-0.03904	0.10106
44	0.03301	-0.00966	0.03440	134	-0.03767	-0.03159	0.04916
46	-0.03767	-0.03158	0.04916	136	0.03301	-0.00966	0.03440
48	-0.09321	-0.03904	0.10106	138	0.07983	0.01312	0.08090
50	-0.09478	-0.02339	0.09762	140	0.08391	0.02641	0.08796
52	-0.03190	0.01062	0.03362	142	0.05047	0.02699	0.05723
54	0.06093	0.04320	0.07469	144	-0.00003	0.01779	0.01779
56	0.11823	0.04962	0.12822	146	-0.04508	0.00464	0.04532
58	0.08838	0.01874	0.09034	148	-0.06977	-0.00713	0.07013
60	-0.01867	-0.03285	0.03778	150	-0.07012	-0.01445	0.07159
62	-0.12355	-0.06607	0.14011	152	-0.05107	-0.01673	0.05374
64	-0.13138	-0.04738	0.13967	154	-0.02188	-0.01510	0.02659
66	-0.01683	0.01834	0.02489	156	0.00813	-0.01136	0.01397
68	0.13313	0.08033	0.15549	158	0.03232	-0.00711	0.03310
70	0.17571	0.07717	0.19191	160	0.04751	-0.00344	0.04764
72	0.04266	-0.00526	0.04298	162	0.05350	-0.00084	0.05351
74	-0.16771	-0.10415	0.19742	164	0.05197	0.00068	0.05197
76	-0.24551	-0.11916	0.27290	166	0.04540	0.00133	0.04542
78	-0.06015	-0.00446	0.06032	168	0.03622	0.00139	0.03624
80	0.27014	0.16258	0.31530	170	0.02642	0.00116	0.02645
82	0.41126	0.21210	0.46273	172	0.01737	0.00081	0.01739
84	0.07047	0.01038	0.07123	174	0.00989	0.00047	0.00991
86	-0.71858	-0.41015	0.82740	176	0.00442	0.00021	0.00443
88	-1.54694	-0.83821	1.75943	178	0.00111	0.00006	0.00111
				180	-0.00000	-0.00000	0.00000

VBR

RADIATION PATTERN FOR THETA=PI/2

H/LMDA=0.5000 B/LMDA= 0.500 OMEGA= 9.92

22 ELEMENT ARRAY

PHI	ER(THETA,PHI)			PHI	ER(THETA,PHI)		
DEG	RE	IM	ABSVAL	DEG	RE	IM	ABSVAL
0	-0.00000	-0.00000	0.00000	90	-1.83629	-0.60783	1.93428
2	-0.00123	-0.00045	0.00131	92	-1.42753	-0.47130	1.50332
4	-0.00493	-0.00181	0.00525	94	-0.52065	-0.16921	0.54746
6	-0.01103	-0.00406	0.01175	96	0.22249	0.07630	0.23521
8	-0.01931	-0.00711	0.02058	98	0.38523	0.12707	0.40564
10	-0.02925	-0.01077	0.03117	100	0.09732	0.02915	0.10159
12	-0.03981	-0.01465	0.04242	102	-0.19672	-0.06741	0.20795
14	-0.04928	-0.01812	0.05250	104	-0.20159	-0.06567	0.21201
16	-0.05524	-0.02028	0.05885	106	0.00829	0.00601	0.01024
18	-0.05486	-0.02007	0.05841	108	0.16453	0.05638	0.17392
20	-0.04553	-0.01652	0.04843	110	0.11784	0.03746	0.12365
22	-0.02604	-0.00917	0.02761	112	-0.04349	-0.01783	0.04701
24	0.00205	0.00142	0.00249	114	-0.13514	-0.04653	0.14292
26	0.03332	0.01323	0.03585	116	-0.07792	-0.02410	0.08157
28	0.05865	0.02286	0.06294	118	0.04828	0.01959	0.05210
30	0.06738	0.02631	0.07233	120	0.11302	0.03957	0.11975
32	0.05192	0.02067	0.05588	122	0.06575	0.02048	0.06886
34	0.01328	0.00616	0.01464	124	-0.03529	-0.01536	0.03849
36	-0.03535	-0.01245	0.03748	126	-0.09577	-0.03478	0.10189
38	-0.07104	-0.02663	0.07586	128	-0.07144	-0.02370	0.07527
40	-0.07132	-0.02783	0.07656	130	0.00688	0.00533	0.00870
42	-0.02943	-0.01307	0.03221	132	0.07304	0.02822	0.07831
44	0.03513	0.01096	0.03680	134	0.08171	0.02946	0.08686
46	0.08171	0.02946	0.08686	136	0.03513	0.01096	0.03680
48	0.07304	0.02822	0.07831	138	-0.02943	-0.01307	0.03220
50	0.00688	0.00533	0.00870	140	-0.07132	-0.02783	0.07656
52	-0.07144	-0.02370	0.07526	142	-0.07103	-0.02663	0.07586
54	-0.09577	-0.03478	0.10189	144	-0.03536	-0.01245	0.03748
56	-0.03530	-0.01536	0.03850	146	0.01328	0.00616	0.01464
58	0.06575	0.02048	0.06886	148	0.05192	0.02067	0.05588
60	0.11302	0.03957	0.11975	150	0.06738	0.02631	0.07233
62	0.04828	0.01959	0.05210	152	0.05864	0.02286	0.06294
64	-0.07792	-0.02410	0.08156	154	0.03332	0.01323	0.03584
66	-0.13513	-0.04653	0.14292	156	0.00204	0.00142	0.00249
68	-0.04350	-0.01783	0.04701	158	-0.02604	-0.00917	0.02761
70	0.11783	0.03746	0.12365	160	-0.04553	-0.01652	0.04843
72	0.16453	0.05638	0.17392	162	-0.05486	-0.02007	0.05841
74	0.00830	0.00602	0.01025	164	-0.05524	-0.02028	0.05885
76	-0.20158	-0.06567	0.21201	166	-0.04928	-0.01812	0.05250
78	-0.19673	-0.06741	0.20795	168	-0.03981	-0.01465	0.04242
80	0.09731	0.02915	0.10158	170	-0.02925	-0.01077	0.03117
82	0.38523	0.12707	0.40564	172	-0.01930	-0.00711	0.02057
84	0.22251	0.07631	0.23524	174	-0.01103	-0.00406	0.01175
86	-0.52060	-0.16920	0.54741	176	-0.00493	-0.00181	0.00525
88	-1.42752	-0.47130	1.50330	178	-0.00123	-0.00045	0.00131
				180	-0.00000	-0.00000	0.00000

IBR

RADIATION PATTERN FOR THETA=PI/2

H/LMDA=0.5000 B/LMDA= 0.500 OMEGA= 9.92

22 ELEMENT ARRAY

PHI	ER(THETA,PHI)			PHI	ER(THETA,PHI)		
DEG	RE	IM	ABSVAL	DEG	RE	IM	ABSVAL
0	-0.00000	-0.00000	0.00000	90	-1.90172	-1.02249	2.15917
2	-0.00109	-0.00004	0.00109	92	-1.47850	-0.80463	1.68327
4	-0.00434	-0.00017	0.00434	94	-0.53927	-0.31473	0.62439
6	-0.00971	-0.00037	0.00972	96	0.23094	0.10278	0.25278
8	-0.01699	-0.00062	0.01700	98	0.40004	0.21801	0.45559
10	-0.02569	-0.00084	0.02570	100	0.10136	0.07839	0.12814
12	-0.03484	-0.00092	0.03485	102	-0.20469	-0.09144	0.22419
14	-0.04287	-0.00065	0.04288	104	-0.21070	-0.12050	0.24273
16	-0.04755	0.00025	0.04755	106	0.00768	-0.02103	0.02239
18	-0.04627	0.00203	0.04632	108	0.17189	0.07663	0.18820
20	-0.03672	0.00481	0.03704	110	0.12484	0.07766	0.14703
22	-0.01796	0.00834	0.01980	112	-0.04370	0.00237	0.04377
24	0.00831	0.01185	0.01448	114	-0.14180	-0.06157	0.15459
26	0.03672	0.01394	0.03928	116	-0.08443	-0.05663	0.10166
28	0.05849	0.01289	0.05989	118	0.04774	-0.00060	0.04775
30	0.06362	0.00734	0.06405	120	0.11861	0.04714	0.12764
32	0.04547	-0.00254	0.04554	122	0.07258	0.04710	0.08652
34	0.00594	-0.01418	0.01538	124	-0.03264	0.00815	0.03364
36	-0.04122	-0.02255	0.04698	126	-0.09912	-0.03146	0.10399
38	-0.07329	-0.02195	0.07651	128	-0.07822	-0.04144	0.08852
40	-0.06901	-0.00956	0.06966	130	0.00131	-0.02018	0.02022
42	-0.02347	0.01098	0.02591	132	0.07214	0.01157	0.07307
44	0.04197	0.02888	0.05095	134	0.08588	0.03105	0.09132
46	0.08589	0.03105	0.09133	136	0.04197	0.02888	0.05095
48	0.07214	0.01157	0.07307	138	-0.02347	0.01098	0.02591
50	0.00131	-0.02018	0.02022	140	-0.06901	-0.00956	0.06967
52	-0.07822	-0.04144	0.08851	142	-0.07329	-0.02195	0.07651
54	-0.09912	-0.03146	0.10399	144	-0.04122	-0.02255	0.04699
56	-0.03264	0.00815	0.03364	146	0.00594	-0.01418	0.01538
58	0.07258	0.04710	0.08652	148	0.04547	-0.00254	0.04554
60	0.11861	0.04714	0.12764	150	0.06363	0.00734	0.06405
62	0.04775	-0.00060	0.04775	152	0.05848	0.01288	0.05988
64	-0.08442	-0.05663	0.10166	154	0.03672	0.01394	0.03928
66	-0.14180	-0.06157	0.15459	156	0.00831	0.01185	0.01448
68	-0.04371	0.00237	0.04377	158	-0.01796	0.00835	0.01980
70	0.12484	0.07766	0.14703	160	-0.03672	0.00481	0.03704
72	0.17189	0.07664	0.18820	162	-0.04627	0.00204	0.04632
74	0.00769	-0.02102	0.02239	164	-0.04755	0.00025	0.04755
76	-0.21070	-0.12050	0.24272	166	-0.04287	-0.00065	0.04288
78	-0.20470	-0.09144	0.22419	168	-0.03485	-0.00092	0.03486
80	0.10135	0.07839	0.12813	170	-0.02568	-0.00084	0.02570
82	0.40004	0.21801	0.45558	172	-0.01698	-0.00061	0.01699
84	0.23096	0.10279	0.25280	174	-0.00971	-0.00037	0.00972
86	-0.53922	-0.31471	0.62434	176	-0.00434	-0.00017	0.00434
88	-1.47849	-0.80463	1.68326	178	-0.00109	-0.00004	0.00109
				180	-0.00000	-0.00000	0.00000

VBR

RADIATION PATTERN FOR THETA=PI/2

H/LMDA=0.5000 B/LMDA= 0.500 OMEGA= 9.92

24 ELEMENT ARRAY

PHI DEG	ER(THETA,PHI) RE	IM	ABSVAL	PHI DEG	ER(THETA,PHI) RE	IM	ABSVAL
0	-0.00000	-0.00000	0.00000	90	-1.83721	-0.60787	1.93516
2	0.00122	0.00045	0.00130	92	-1.35659	-0.44753	1.42850
4	0.00487	0.00179	0.00519	94	-0.35351	-0.11395	0.37142
6	0.01089	0.00399	0.01159	96	0.32968	0.11087	0.34782
8	0.01902	0.00697	0.02026	98	0.30976	0.10092	0.32579
10	0.02866	0.01049	0.03052	100	-0.06389	-0.02403	0.06826
12	0.03858	0.01411	0.04108	102	-0.23798	-0.07916	0.25080
14	0.04684	0.01710	0.04986	104	-0.07168	-0.02108	0.07471
16	0.05077	0.01848	0.05402	106	0.14294	0.04957	0.15129
18	0.04741	0.01714	0.05041	108	0.13527	0.04370	0.14216
20	0.03447	0.01223	0.03658	110	-0.03712	-0.01538	0.04018
22	0.01178	0.00368	0.01234	112	-0.13500	-0.04597	0.14262
24	-0.01723	-0.00723	0.01869	114	-0.05609	-0.01642	0.05844
26	-0.04480	-0.01763	0.04814	116	0.07761	0.02879	0.08277
28	-0.06015	-0.02346	0.06456	118	0.10480	0.03519	0.11055
30	-0.05360	-0.02106	0.05759	120	0.01149	0.00108	0.01154
32	-0.02271	-0.00942	0.02458	122	-0.08384	-0.03085	0.08934
34	0.02252	0.00786	0.02385	124	-0.08120	-0.02718	0.08563
36	0.06010	0.02261	0.06421	126	0.00320	0.00391	0.00505
38	0.06572	0.02562	0.07054	128	0.07651	0.02873	0.08173
40	0.02953	0.01279	0.03218	130	0.07237	0.02491	0.07654
42	-0.03061	-0.00974	0.03212	132	0.00489	-0.00068	0.00494
44	-0.07326	-0.02678	0.07801	134	-0.06088	-0.02394	0.06542
46	-0.06088	-0.02394	0.06542	136	-0.07326	-0.02678	0.07801
48	0.00489	-0.00068	0.00494	138	-0.03061	-0.00974	0.03212
50	0.07237	0.02492	0.07654	140	0.02952	0.01279	0.03218
52	0.07651	0.02873	0.08173	142	0.06572	0.02562	0.07054
54	0.00319	0.00391	0.00505	144	0.06010	0.02261	0.06421
56	-0.08120	-0.02718	0.08563	146	0.02252	0.00786	0.02386
58	-0.08384	-0.03085	0.08933	148	-0.02271	-0.00942	0.02458
60	0.01149	0.00108	0.01154	150	-0.05360	-0.02106	0.05759
62	0.10480	0.03519	0.11055	152	-0.06015	-0.02346	0.06456
64	0.07761	0.02879	0.08278	154	-0.04480	-0.01763	0.04814
66	-0.05608	-0.01642	0.05843	156	-0.01723	-0.00724	0.01869
68	-0.13500	-0.04597	0.14262	158	0.01178	0.00368	0.01234
70	-0.03712	-0.01538	0.04018	160	0.03447	0.01223	0.03657
72	0.13527	0.04370	0.14216	162	0.04741	0.01714	0.05041
74	0.14294	0.04957	0.15129	164	0.05076	0.01848	0.05402
76	-0.07167	-0.02107	0.07470	166	0.04684	0.01711	0.04987
78	-0.23798	-0.07916	0.25080	168	0.03858	0.01411	0.04108
80	-0.06390	-0.02404	0.06827	170	0.02866	0.01049	0.03052
82	0.30976	0.10092	0.32578	172	0.01903	0.00697	0.02026
84	0.32970	0.11087	0.34784	174	0.01089	0.00399	0.01159
86	-0.35347	-0.11394	0.37138	176	0.00487	0.00179	0.00519
88	-1.35657	-0.44753	1.42848	178	0.00122	0.00045	0.00130
				180	-0.00000	-0.00000	0.00000

IBR

RADIATION PATTERN FOR THETA=PI/2

H/LMDA=0.5C00 B/LMDA= 0.500 OMEGA= 9.92

24 ELEMENT ARRAY

PHI DEG	ER(THETA,PHI) RE	IM	ABSVAL	PHI DEG	ER(THETA,PHI) RE	IM	ABSVAL
0	-0.00000	-0.00000	0.00000	90	-1.90289	-1.02561	2.16168
2	0.00107	0.00003	0.00107	92	-1.40517	-0.76775	1.60123
4	0.00427	0.00013	0.00427	94	-0.36611	-0.22131	0.42780
6	0.00953	0.00027	0.00953	96	0.34211	0.16980	0.38193
8	0.01663	0.00045	0.01663	98	0.32170	0.18533	0.37126
10	0.02500	0.00057	0.02500	100	-0.06632	-0.01271	0.06752
12	0.03352	0.00052	0.03352	102	-0.24795	-0.12949	0.27972
14	0.04039	0.00006	0.04039	104	-0.07540	-0.06081	0.09687
16	0.04317	-0.00104	0.04318	106	0.14894	0.06190	0.16129
18	0.03920	-0.00300	0.03931	108	0.14243	0.08378	0.16524
20	0.02649	-0.00580	0.02712	110	-0.03744	0.00347	0.03760
22	0.00510	-0.00898	0.01033	112	-0.14175	-0.06535	0.15609
24	-0.02146	-0.01146	0.02433	114	-0.06112	-0.04794	0.07767
26	-0.04570	-0.01167	0.04716	116	0.07971	0.01952	0.08207
28	-0.05744	-0.00807	0.05801	118	0.11140	0.05502	0.12425
30	-0.04797	-0.00022	0.04797	120	0.01562	0.02743	0.03157
32	-0.01585	0.01010	0.01880	122	-0.08613	-0.02372	0.08934
34	0.02822	0.01848	0.03373	124	-0.08751	-0.04416	0.09803
36	0.06238	0.01927	0.06528	126	-0.00135	-0.02035	0.02040
38	0.06353	0.00911	0.06418	128	0.07753	0.01813	0.07962
40	0.02386	-0.00905	0.02552	130	0.07797	0.03556	0.08570
42	-0.03681	-0.02488	0.04442	132	0.01073	0.02184	0.02433
44	-0.07642	-0.02560	0.08060	134	-0.05895	-0.00631	0.05929
46	-0.05895	-0.00631	0.05929	136	-0.07642	-0.02560	0.08060
48	0.01072	0.02184	0.02433	138	-0.03681	-0.02488	0.04443
50	0.07797	0.03556	0.08570	140	0.02386	-0.00905	0.02552
52	0.07753	0.01813	0.07962	142	0.06353	0.00911	0.06418
54	-0.00136	-0.02035	0.02040	144	0.06238	0.01927	0.06528
56	-0.08751	-0.04416	0.09803	146	0.02822	0.01848	0.03373
58	-0.08613	-0.02372	0.08934	148	-0.01585	0.01010	0.01880
60	0.01562	0.02743	0.03157	150	-0.04797	-0.00022	0.04797
62	0.11140	0.05502	0.12425	152	-0.05745	-0.00807	0.05801
64	0.07971	0.01953	0.08207	154	-0.04570	-0.01167	0.04717
66	-0.06111	-0.04794	0.07767	156	-0.02146	-0.01146	0.02433
68	-0.14175	-0.06535	0.15609	158	0.00510	-0.00898	0.01033
70	-0.03745	0.00347	0.03761	160	0.02649	-0.00580	0.02712
72	0.14243	0.08378	0.16524	162	0.03920	-0.00300	0.03931
74	0.14895	0.06190	0.16130	164	0.04317	-0.00104	0.04318
76	-0.07539	-0.06081	0.09686	166	0.04039	0.00007	0.04039
78	-0.24795	-0.12949	0.27972	168	0.03351	0.00052	0.03352
80	-0.06633	-0.01271	0.06753	170	0.02500	0.00058	0.02501
82	0.32169	0.18533	0.37126	172	0.01663	0.00045	0.01664
84	0.34212	0.16981	0.38195	174	0.00953	0.00027	0.00953
86	-0.36606	-0.22129	0.42775	176	0.00427	0.00013	0.00427
88	-1.40515	-0.76774	1.60121	178	0.00107	0.00003	0.00107
				180	-0.00000	-0.00000	0.00000

VBR

RADIATION PATTERN FOR THETA=PI/2

H/LMDA=0.5000 B/LMDA= 0.500 OMEGA= 9.92

26 ELEMENT ARRAY

PHI DEG	ER(THETA,PHI) RE	IM	ABSVAL	PHI DEG	ER(THETA,PHI) RE	IM	ABSVAL
0	-0.00000	-0.00000	0.00000	90	-1.83799	-0.60789	1.93591
2	-0.00121	-0.00044	0.00129	92	-1.28153	-0.42246	1.34936
4	-0.00483	-0.00176	0.00514	94	-0.19648	-0.06216	0.20608
6	-0.01078	-0.00393	0.01147	96	0.38786	0.12918	0.40881
8	-0.01877	-0.00684	0.01997	98	0.19212	0.06128	0.20165
10	-0.02808	-0.01022	0.02989	100	-0.18318	-0.06260	0.19358
12	-0.03735	-0.01358	0.03974	102	-0.18247	-0.05903	0.19179
14	-0.04438	-0.01609	0.04720	104	0.07465	0.02736	0.07951
16	-0.04626	-0.01669	0.04917	106	0.16398	0.05422	0.17271
18	-0.04004	-0.01427	0.04250	108	0.00181	-0.00214	0.00280
20	-0.02395	-0.00818	0.02531	110	-0.13234	-0.04523	0.13986
22	0.00086	0.00114	0.00143	112	-0.05889	-0.01751	0.06143
24	0.02890	0.01166	0.03117	114	0.08443	0.03061	0.08980
26	0.05033	0.01970	0.05405	116	0.09352	0.03064	0.09841
28	0.05397	0.02105	0.05793	118	-0.02230	-0.01033	0.02458
30	0.03328	0.01321	0.03581	120	-0.09563	-0.03348	0.10132
32	-0.00671	-0.00205	0.00702	122	-0.04113	-0.01201	0.04285
34	-0.04696	-0.01767	0.05017	124	0.05756	0.02226	0.06171
36	-0.06163	-0.02390	0.06610	126	0.07943	0.02746	0.08405
38	-0.03527	-0.01464	0.03819	128	0.01028	0.00122	0.01035
40	0.01992	0.00604	0.02082	130	-0.06374	-0.02444	0.06826
42	0.06397	0.02349	0.06815	132	-0.06638	-0.02324	0.07034
44	0.05624	0.02214	0.06044	134	-0.00464	0.00043	0.00466
46	-0.00463	0.00043	0.00465	136	0.05624	0.02214	0.06044
48	-0.06638	-0.02324	0.07033	138	0.06397	0.02348	0.06815
50	-0.06374	-0.02444	0.06826	140	0.01992	0.00604	0.02082
52	0.01027	0.00122	0.01035	142	-0.03527	-0.01464	0.03819
54	0.07943	0.02746	0.08404	144	-0.06163	-0.02390	0.06610
56	0.05756	0.02226	0.06171	146	-0.04696	-0.01767	0.05017
58	-0.04113	-0.01201	0.04285	148	-0.00671	-0.00205	0.00702
60	-0.09563	-0.03348	0.10132	150	0.03328	0.01321	0.03581
62	-0.02230	-0.01033	0.02458	152	0.05397	0.02105	0.05793
64	0.09352	0.03064	0.09841	154	0.05033	0.01970	0.05405
66	0.08443	0.03061	0.08981	156	0.02890	0.01166	0.03116
68	-0.05888	-0.01751	0.06143	158	0.00086	0.00114	0.00143
70	-0.13234	-0.04523	0.13986	160	-0.02395	-0.00818	0.02531
72	0.00180	-0.00214	0.00280	162	-0.04004	-0.01427	0.04250
74	0.16398	0.05421	0.17271	164	-0.04626	-0.01669	0.04917
76	0.07466	0.02737	0.07952	166	-0.04437	-0.01609	0.04720
78	-0.18247	-0.05903	0.19178	168	-0.03736	-0.01358	0.03975
80	-0.18318	-0.06260	0.19358	170	-0.02808	-0.01022	0.02989
82	0.19211	0.06128	0.20165	172	-0.01876	-0.00683	0.01997
84	0.38787	0.12918	0.40882	174	-0.01078	-0.00393	0.01147
86	-0.19643	-0.06215	0.20603	176	-0.00483	-0.00176	0.00514
88	-1.28151	-0.42246	1.34934	178	-0.00121	-0.00044	0.00129
				180	-0.00000	-0.00000	0.00000

IBR

RADIATION PATTERN FOR THETA=PI/2

H/LMDA=0.[illegible] B/LMDA= [illegible] [illegible]

26 ELEMENT ARRAY

PHI DEG	ER(THETA,PHI) RE	IM	ABSVAL	PHI DEG	ER(THETA,PHI) RE	IM	ABSVAL
0	-0.00000	-0.00000	0.00000	90	-1.90387	-1.02825	2.16380
2	-0.00105	-0.00002	0.00105	92	-1.32753	-0.72800	1.51404
4	-0.00420	-0.00009	0.00420	94	-0.20339	-0.13242	0.24270
6	-0.00938	-0.00020	0.00938	96	0.40245	0.20980	0.45385
8	-0.01631	-0.00030	0.01631	98	0.19952	0.12547	0.23569
10	-0.02434	-0.00034	0.02434	100	-0.19046	-0.08645	0.20917
12	-0.03222	-0.00016	0.03222	102	-0.19034	-0.11249	0.22110
14	-0.03793	0.00044	0.03793	104	0.07742	0.02064	0.08012
16	-0.03884	0.00170	0.03888	106	0.17170	0.09189	0.19474
18	-0.03232	0.00376	0.03254	108	0.00302	0.02248	0.02268
20	-0.01693	0.00645	0.01811	110	-0.13859	-0.06272	0.15213
22	0.00603	0.00908	0.01089	112	-0.06350	-0.04835	0.07981
24	0.03112	0.01034	0.03280	114	0.08741	0.02656	0.09135
26	0.C4902	0.00865	0.04978	116	0.09980	0.05491	0.11390
28	0.04944	0.00299	0.04953	118	-0.02068	0.01026	0.02308
30	0.02695	-0.00586	0.02758	120	-0.10038	-0.03991	0.10803
32	-0.01251	-0.01442	0.01909	122	-0.04654	-0.03579	0.05871
34	-0.04979	-0.01730	0.05271	124	0.05747	0.00816	0.05804
36	-0.06013	-0.01025	0.06100	126	0.08458	0.03688	0.09227
38	-0.03017	0.00537	0.03064	128	0.01537	0.02321	0.02784
40	0.02574	0.02050	0.03291	130	-0.06351	-0.01137	0.06452
42	0.06685	0.02246	0.07053	132	-0.07110	-0.03023	0.07726
44	0.05415	0.00558	0.05444	134	-0.01024	-0.01961	0.02213
46	-0.01024	-0.01961	0.02212	136	0.05415	0.00558	0.05444
48	-0.C7110	-0.03023	0.07726	138	0.06685	0.02246	0.07052
50	-0.06351	-0.01137	0.06452	140	0.02574	0.02050	0.03290
52	0.01537	0.02321	0.02784	142	-0.03017	0.00537	0.03064
54	0.08458	0.03688	0.09227	144	-0.06013	-0.01025	0.06100
56	0.05747	0.00816	0.05804	146	-0.04979	-0.01730	0.05271
58	-0.04654	-0.03579	0.05871	148	-0.01251	-0.01442	0.01909
60	-0.10039	-0.03991	0.10803	150	0.02695	-0.00586	0.02758
62	-0.02068	0.01026	0.02309	152	0.04943	0.00299	0.04952
64	0.C9980	0.05491	0.11390	154	0.04902	0.00865	0.04978
66	0.08741	0.02656	0.09136	156	0.03112	0.01034	0.03280
68	-0.06349	-0.04835	0.07981	158	0.00603	0.00908	0.01090
70	-0.13860	-0.06272	0.15213	160	-0.01693	0.00645	0.01812
72	0.00302	0.02247	0.02268	162	-0.03232	0.00376	0.03254
74	0.17169	0.09189	0.19474	164	-0.03884	0.00170	0.03888
76	0.07743	0.02064	C.08013	166	-0.03792	0.00044	0.03793
78	-0.19034	-0.11249	0.22109	168	-0.03222	-0.00017	0.03222
80	-0.19047	-0.08646	0.20917	170	-0.02434	-0.00034	0.02435
82	0.19952	0.12546	0.23568	172	-0.01630	-0.00030	0.01630
84	0.40246	0.20981	0.45386	174	-0.00938	-0.00020	0.00938
86	-0.20334	-0.13240	0.24265	176	-0.00420	-0.00009	0.00420
88	-1.32751	-0.72799	1.51402	178	-0.00105	-0.00002	0.00105
				180	-0.00000	-0.00000	0.00000

VBR

RADIATION PATTERN FOR THETA=PI/2

H/LMDA=0.5000 B/LMDA= 0.500 OMEGA= 9.92

28 ELEMENT ARRAY

PHI DEG	ER(THETA,PHI) RE	IM	ABSVAL	PHI DEG	ER(THETA,PHI) RE	IM	ABSVAL
0	0.00000	0.00000	0.00000	90	-1.83866	-0.60792	1.93655
2	0.00120	0.00043	0.00127	92	-1.20290	-0.39626	1.26649
4	0.00478	0.00173	0.00509	94	-0.05315	-0.01501	0.05523
6	0.01066	0.00386	0.01134	96	0.39925	0.13206	0.42052
8	0.01851	0.00671	0.01969	98	0.05772	0.01659	0.06006
10	0.02752	0.00996	0.02927	100	-0.23606	-0.07879	0.24886
12	0.03612	0.01305	0.03841	102	-0.06515	-0.01922	0.06793
14	0.04188	0.01508	0.04451	104	0.16194	0.05491	0.17099
16	0.04173	0.01490	0.04431	106	0.07481	0.02276	0.07820
18	0.03283	0.01148	0.03478	108	-0.11405	-0.03970	0.12077
20	0.01413	0.00444	0.01481	110	-0.08524	-0.02682	0.08936
22	-0.01166	-0.00522	0.01278	112	0.07373	0.02692	0.07849
24	-0.03693	-0.01465	0.03973	114	0.09337	0.03051	0.09823
26	-0.05038	-0.01962	0.05407	116	-0.03308	-0.01372	0.03582
28	-0.04194	-0.01635	0.04502	118	-0.09380	-0.03201	0.09911
30	-0.01046	-0.00432	0.01132	120	-0.00987	-0.00094	0.00992
32	0.03052	0.01145	0.03260	122	0.07902	0.02853	0.08401
34	0.05560	0.02144	0.05959	124	0.04986	0.01567	0.05226
36	0.04292	0.01723	0.04625	126	-0.04288	-0.01723	0.04621
38	-0.00484	-0.00060	0.00488	128	-0.07236	-0.02543	0.07670
40	-0.05275	-0.01928	0.05616	130	-0.01072	-0.00169	0.01086
42	-0.05592	-0.02181	0.06002	132	0.05824	0.02244	0.06242
44	-0.00359	-0.00317	0.00479	134	0.05712	0.02006	0.06055
46	0.05713	0.02006	0.06055	136	-0.00359	-0.00317	0.00479
48	0.05824	0.02244	0.06242	138	-0.05592	-0.02181	0.06002
50	-0.01072	-0.00169	0.01085	140	-0.05275	-0.01928	0.05616
52	-0.07236	-0.02543	0.07670	142	-0.00485	-0.00060	0.00488
54	-0.04288	-0.01723	0.04621	144	0.04292	0.01723	0.04625
56	0.04985	0.01566	0.05226	146	0.05560	0.02144	0.05959
58	0.07902	0.02853	0.08401	148	0.03052	0.01145	0.03260
60	-0.00987	-0.00094	0.00991	150	-0.01046	-0.00432	0.01131
62	-0.09380	-0.03201	0.09911	152	-0.04195	-0.01635	0.04502
64	-0.03309	-0.01372	0.03582	154	-0.05038	-0.01963	0.05407
66	0.09337	0.03050	0.09823	156	-0.03693	-0.01465	0.03973
68	0.07374	0.02692	0.07850	158	-0.01166	-0.00522	0.01278
70	-0.08524	-0.02682	0.08936	160	0.01413	0.00444	0.01481
72	-0.11405	-0.03970	0.12077	162	0.03283	0.01148	0.03478
74	0.07480	0.02276	0.07819	164	0.04173	0.01490	0.04431
76	0.16194	0.05491	0.17099	166	0.04188	0.01508	0.04451
78	-0.06514	-0.01922	0.06792	168	0.03612	0.01305	0.03840
80	-0.23606	-0.07879	0.24886	170	0.02752	0.00996	0.02927
82	0.05771	0.01659	0.06005	172	0.01852	0.00671	0.01969
84	0.39925	0.13206	0.42052	174	0.01066	0.00386	0.01134
86	-0.05311	-0.01499	0.05519	176	0.00478	0.00173	0.00509
88	-1.20288	-0.39626	1.26647	178	0.00120	0.00043	0.00127
				180	0.00000	0.00000	0.00000

IBR

RADIATION PATTERN FOR THETA=PI/2

H/LMDA=0.5000 B/LMDA= 0.500 OMEGA= 9.92

28 ELEMENT ARRAY

PHI DEG	ER(THETA,PHI) RE	IM	ABSVAL	PHI DEG	ER(THETA,PHI) RE	IM	ABSVAL
0	0.00000	-0.00000	0.00000	90	-1.90471	-1.03050	2.16561
2	0.00104	0.00001	0.00104	92	-1.24616	-0.68577	1.42239
4	0.00414	0.00006	0.00414	94	-0.05485	-0.05029	0.07442
6	0.00922	0.00012	0.00922	96	0.41424	0.22329	0.47059
8	0.01600	0.00017	0.01600	98	0.05993	0.05219	0.07947
10	0.02371	0.00014	0.02372	100	-0.24556	-0.12640	0.27618
12	0.03095	-0.00014	0.03095	102	-0.06819	-0.05490	0.08755
14	0.03548	-0.00087	0.03549	104	0.16887	0.07995	0.18684
16	0.03456	-0.00225	0.03464	106	0.07894	0.05752	0.09768
18	0.02570	-0.00433	0.02606	108	-0.11894	-0.04828	0.12836
20	0.00816	-0.00679	0.01062	110	-0.09050	-0.05847	0.10774
22	-0.01528	-0.00872	0.01759	112	0.07625	0.02157	0.07924
24	-0.03727	-0.00867	0.03826	114	0.09930	0.05538	0.11370
26	-0.04729	-0.00525	0.04758	116	-0.03265	0.00289	0.03278
28	-0.03639	0.00176	0.03643	118	-0.09919	-0.04537	0.10908
30	-0.00455	0.01013	0.01110	120	-0.01342	-0.02354	0.02710
32	0.03422	0.01512	0.03741	122	0.08194	0.02657	0.08614
34	0.05523	0.01174	0.05646	124	0.05532	0.03509	0.06551
36	0.03866	-0.00082	0.03866	126	-0.04158	-0.00118	0.04160
38	-0.01044	-0.01567	0.01883	128	-0.07661	-0.03087	0.08259
40	-0.05591	-0.02047	0.05955	130	-0.01573	-0.02136	0.02652
42	-0.05430	-0.00751	0.05481	132	0.05768	0.00970	0.05849
44	0.00160	0.01538	0.01546	134	0.06146	0.02633	0.06686
46	0.06146	0.02633	0.06686	136	0.00160	0.01538	0.01546
48	0.05768	0.00970	0.05849	138	-0.05430	-0.00751	0.05482
50	-0.01573	-0.02136	0.02652	140	-0.05592	-0.02047	0.05955
52	-0.07661	-0.03087	0.08259	142	-0.01044	-0.01567	0.01883
54	-0.04158	-0.00118	0.04160	144	0.03865	-0.00082	0.03866
56	0.05532	0.03509	0.06551	146	0.05523	0.01174	0.05646
58	0.08194	0.02658	0.08615	148	0.03422	0.01512	0.03741
60	-0.01342	-0.02354	0.02710	150	-0.00454	0.01013	0.01110
62	-0.09919	-0.04537	0.10908	152	-0.03639	0.00176	0.03644
64	-0.03266	0.00289	0.03278	154	-0.04729	-0.00525	0.04758
66	0.09930	0.05538	0.11370	156	-0.03727	-0.00867	0.03827
68	0.07625	0.02157	0.07925	158	-0.01528	-0.00872	0.01759
70	-0.09049	-0.05847	0.10774	160	0.00816	-0.00679	0.01061
72	-0.11894	-0.04828	0.12837	162	0.02570	-0.00433	0.02606
74	0.07893	0.05752	0.09767	164	0.03456	-0.00225	0.03464
76	0.16888	0.07996	0.18685	166	0.03548	-0.00087	0.03549
78	-0.06819	-0.05490	0.08754	168	0.03094	-0.00015	0.03094
80	-0.24556	-0.12640	0.27618	170	0.02371	0.00013	0.02371
82	0.05992	0.05219	0.07946	172	0.01600	0.00018	0.01600
84	0.41424	0.22329	0.47059	174	0.00922	0.00012	0.00922
86	-0.05481	-0.05027	0.07437	176	0.00414	0.00006	0.00414
88	-1.24613	-0.68577	1.42237	178	0.00104	0.00001	0.00104
				180	0.00000	-0.00000	0.00000

VBR

RADIATION PATTERN FOR THETA=PI/2

H/LMDA=0.5000 B/LMDA= 0.500 OMEGA= 9.92

30 ELEMENT ARRAY

PHI DEG	ER(THETA,PHI) RE	IM	ABSVAL	PHI DEG	ER(THETA,PHI) RE	IM	ABSVAL
0	-0.00000	-0.00000	0.00000	90	-1.83924	-0.60794	1.93711
2	-0.00119	-0.00043	0.00127	92	-1.12128	-0.36911	1.18047
4	-0.00474	-0.00171	0.00504	94	0.07343	0.02653	0.07808
6	-0.01057	-0.00381	0.01123	96	0.36930	0.12139	0.38874
8	-0.01827	-0.00659	0.01942	98	-0.06889	-0.02505	0.07330
10	-0.02696	-0.00971	0.02865	100	-0.21793	-0.07147	0.22936
12	-0.03488	-0.01253	0.03706	102	0.06156	0.02267	0.06560
14	-0.03937	-0.01407	0.04180	104	0.15441	0.05058	0.16248
16	-0.03722	-0.01314	0.03947	106	-0.05163	-0.01946	0.05517
18	-0.02584	-0.00881	0.02730	108	-0.12179	-0.04000	0.12819
20	-0.00514	-0.00104	0.00524	110	0.03892	0.01535	0.04184
22	0.02044	0.00850	0.02214	112	0.10387	0.03447	0.10944
24	0.04134	0.01623	0.04441	114	-0.02281	-0.01007	0.02494
26	0.04568	0.01771	0.04899	116	-0.09294	-0.03148	0.09813
28	0.02628	0.01023	0.02820	118	0.00254	0.00322	0.00410
30	-0.01103	-0.00409	0.01177	120	0.08288	0.02902	0.08781
32	-0.04479	-0.01720	0.04798	122	0.02182	0.00542	0.02249
34	-0.04846	-0.01906	0.05207	124	-0.06708	-0.02469	0.07147
36	-0.01270	-0.00582	0.01397	126	-0.04706	-0.01505	0.04941
38	0.03767	0.01356	0.04004	128	0.03908	0.01580	0.04216
40	0.05576	0.02145	0.05974	130	0.06391	0.02254	0.06777
42	0.01674	0.00779	0.01847	132	0.00246	-0.00107	0.00268
44	-0.04448	-0.01543	0.04708	134	-0.05718	-0.02186	0.06122
46	-0.05718	-0.02186	0.06122	136	-0.04448	-0.01543	0.04708
48	0.00246	-0.00107	0.00268	138	0.01674	0.00779	0.01846
50	0.06391	0.02254	0.06777	140	0.05575	0.02145	0.05974
52	0.03908	0.01580	0.04216	142	0.03768	0.01356	0.04004
54	-0.04706	-0.01506	0.04941	144	-0.01270	-0.00582	0.01397
56	-0.06708	-0.02469	0.07148	146	-0.04846	-0.01906	0.05207
58	0.02182	0.00542	0.02248	148	-0.04479	-0.01720	0.04798
60	0.08288	0.02902	0.08781	150	-0.01103	-0.00409	0.01177
62	0.00254	0.00322	0.00411	152	0.02627	0.01022	0.02819
64	-0.09294	-0.03148	0.09813	154	0.04568	0.01771	0.04899
66	-0.02282	-0.01007	0.02495	156	0.04134	0.01623	0.04441
68	0.10387	0.03447	0.10944	158	0.02044	0.00850	0.02214
70	0.03892	0.01535	0.04184	160	-0.00514	-0.00104	0.00524
72	-0.12179	-0.04000	0.12819	162	-0.02584	-0.00881	0.02730
74	-0.05164	-0.01946	0.05518	164	-0.03722	-0.01314	0.03947
76	0.15441	0.05058	0.16248	166	-0.03936	-0.01407	0.04180
78	0.06157	0.02267	0.06561	168	-0.03488	-0.01253	0.03706
80	-0.21793	-0.07147	0.22935	170	-0.02696	-0.00971	0.02865
82	-0.06890	-0.02505	0.07331	172	-0.01827	-0.00659	0.01942
84	0.36929	0.12138	0.38873	174	-0.01057	-0.00381	0.01123
86	0.07348	0.02654	0.07812	176	-0.00474	-0.00171	0.00504
88	-1.12125	-0.36911	1.18044	178	-0.00119	-0.00043	0.00127
				180	-0.00000	-0.00000	0.00000

IBR

RADIATION PATTERN FOR THETA=PI/2

H/LMDA=0.5000 B/LMDA= 0.500 OMEGA= 9.92

30 ELEMENT ARRAY

PHI DEG	ER(THETA,PHI) RE	IM	ABSVAL	PHI DEG	ER(THETA,PHI) RE	IM	ABSVAL
0	-0.00000	-0.00000	0.00000	90	-1.90544	-1.03246	2.16718
2	-0.00103	-0.00001	0.00103	92	-1.16166	-0.64147	1.32700
4	-0.00409	-0.00003	0.00409	94	0.07635	0.02318	0.07979
6	-0.00910	-0.00006	0.00910	96	0.38315	0.21271	0.43824
8	-0.01571	-0.00006	0.01571	98	-0.07160	-0.02052	0.07448
10	-0.02310	0.00005	0.02310	100	-0.22679	-0.12679	0.25982
12	-0.02968	0.00042	0.02968	102	0.06389	0.01614	0.06590
14	-0.03305	0.00124	0.03308	104	0.16141	0.09032	0.18496
16	-0.03036	0.00270	0.03048	106	-0.05334	-0.01009	0.05428
18	-0.01938	0.00474	0.01995	108	-0.12800	-0.07040	0.14609
2C	-0.00027	0.00686	0.00687	110	0.03968	0.00239	0.03975
22	0.02254	0.00798	0.02391	112	0.10969	0.05740	0.12380
24	0.04003	0.00661	0.04057	114	-0.02221	0.00685	0.02324
26	0.04132	0.00179	0.04136	116	-0.09825	-0.04658	0.10873
28	0.02053	-0.00566	0.02129	118	0.00017	-0.01703	0.01703
30	-0.01562	-0.01217	0.01981	120	0.08702	0.03461	0.09365
32	-0.04588	-0.01256	0.04757	122	0.02601	0.02639	0.03706
34	-0.04540	-0.00383	0.04556	124	-0.06889	-0.01918	0.07151
36	-0.00742	0.01016	0.01258	126	-0.05219	-0.03138	0.06090
38	0.04146	0.01834	0.04534	128	0.03748	0.00034	0.03749
4C	0.05510	0.01057	0.05610	130	0.06778	0.02715	0.07301
42	0.01214	-0.00979	0.01560	132	0.00709	0.01704	0.01846
44	-0.04884	-0.02310	0.05402	134	-0.05697	-0.01120	0.05806
46	-0.05697	-0.01120	0.05806	136	-0.04884	-0.02310	0.05403
48	0.00709	0.01704	0.01846	138	0.01214	-0.00979	0.01560
50	0.06777	0.02715	0.07301	140	0.05509	0.01057	0.05610
52	0.03748	0.00034	0.03748	142	0.04147	0.01834	0.04534
54	-0.05220	-0.03139	0.06090	144	-0.00742	0.01016	0.01258
56	-0.06889	-0.01918	0.07151	146	-0.04540	-0.00383	0.04556
58	0.02601	0.02639	0.03705	148	-0.04588	-0.01256	0.04757
60	0.08702	0.03461	0.09365	150	-0.01563	-0.01217	0.01981
62	0.00017	-0.01702	0.01703	152	0.02052	-0.00566	0.02129
64	-0.09824	-0.04658	0.10873	154	0.04132	0.00179	0.04136
66	-0.02222	0.00684	0.02325	156	0.04002	0.00661	0.04057
68	0.10969	0.05740	0.12380	158	0.02255	0.00798	0.02392
70	0.03968	0.00239	0.03976	160	-0.00027	0.00686	0.00687
72	-0.12800	-0.07040	0.14608	162	-0.01938	0.00474	0.01995
74	-0.05335	-0.01010	0.05430	164	-0.03036	0.00270	0.03048
76	0.16140	0.09032	0.18495	166	-0.03305	0.00124	0.03308
78	0.06390	0.01614	0.06590	168	-0.02969	0.00041	0.02969
80	-0.22678	-0.12679	0.25982	170	-0.02310	0.00005	0.02310
82	-0.07161	-0.02052	0.07449	172	-0.01571	-0.00006	0.01571
84	0.38314	0.21271	0.43822	174	-0.00910	-0.00006	0.00910
86	0.07639	0.02320	0.07984	176	-0.00409	-0.00003	0.00409
88	-1.16163	-0.64146	1.32698	178	-0.00103	-0.00001	0.00103
				180	-0.00000	-0.00000	0.00000

VBR

RADIATION PATTERN FOR THETA=PI/2

H/LMDA=0.5000 B/LMDA= 0.500 OMEGA= 9.92

35 ELEMENT ARRAY

PHI DEG	ER(THETA,PHI) RE	IM	ABSVAL	PHI DEG	ER(THETA,PHI) RE	IM	ABSVAL
0	0.02822	0.00833	0.02942	90	-1.84029	-0.60791	1.93810
1	0.02822	0.00833	0.02942	91	-1.57278	-0.51901	1.65620
2	0.02820	0.00832	0.02940	92	-0.90795	-0.29836	0.95572
3	0.02812	0.00829	0.02932	93	-0.17482	-0.05569	0.18348
4	0.02790	0.00822	0.02909	94	0.30273	0.10130	0.31923
5	0.02744	0.00806	0.02860	95	0.38564	0.12703	0.40603
6	0.02662	0.00777	0.02773	96	0.17098	0.05473	0.17953
7	0.02528	0.00729	0.02631	97	-0.10542	-0.03663	0.11160
8	0.02327	0.00658	0.02418	98	-0.23617	-0.07849	0.24887
9	0.02041	0.00558	0.02116	99	-0.16358	-0.05281	0.17190
10	0.01659	0.00423	0.01712	100	0.01647	0.00743	0.01807
11	0.01171	0.00252	0.01198	101	0.15191	0.05138	0.16037
12	0.00576	0.00043	0.00578	102	0.15121	0.04943	0.15908
13	-0.00116	-0.00199	0.00230	103	0.03713	0.01042	0.03857
14	-0.00878	-0.00464	0.00993	104	-0.08885	-0.03112	0.09415
15	-0.01669	-0.00737	0.01824	105	-0.13205	-0.04396	0.13918
16	-0.02426	-0.00995	0.02622	106	-0.07166	-0.02231	0.07505
17	-0.03071	-0.01210	0.03301	107	0.03497	0.01360	0.03752
18	-0.03517	-0.01349	0.03767	108	0.10444	0.03572	0.11038
19	-0.03674	-0.01381	0.03925	109	0.09051	0.02938	0.09516
20	-0.03467	-0.01279	0.03696	110	0.01224	0.00215	0.01243
21	-0.02856	-0.01028	0.03035	111	-0.06774	-0.02429	0.07196
22	-0.01849	-0.00634	0.01954	112	-0.09261	-0.03116	0.09771
23	-0.00525	-0.00126	0.00540	113	-0.05021	-0.01540	0.05252
24	0.00966	0.00438	0.01061	114	0.02381	0.01001	0.02583
25	0.02403	0.00978	0.02594	115	0.07592	0.02672	0.08048
26	0.03527	0.01395	0.03792	116	0.07278	0.02407	0.07666
27	0.04085	0.01596	0.04386	117	0.02118	0.00538	0.02185
28	0.03890	0.01511	0.04174	118	-0.04094	-0.01576	0.04387
29	0.02885	0.01120	0.03095	119	-0.07254	-0.02540	0.07686
30	0.01194	0.00469	0.01283	120	-0.05593	-0.01820	0.05882
31	-0.00867	-0.00324	0.00926	121	-0.00527	0.00008	0.00527
32	-0.02828	-0.01081	0.03027	122	0.04570	0.01740	0.04890
33	-0.04161	-0.01604	0.04460	123	0.06624	0.02331	0.07022
34	-0.04429	-0.01726	0.04754	124	0.04651	0.01509	0.04889
35	-0.03443	-0.01370	0.03705	125	0.00091	-0.00153	0.00178
36	-0.01370	-0.00594	0.01493	126	-0.04283	-0.01656	0.04592
37	0.01238	0.00399	0.01300	127	-0.06066	-0.02175	0.06444
38	0.03565	0.01306	0.03796	128	-0.04485	-0.01493	0.04727
39	0.04777	0.01809	0.05108	129	-0.00620	-0.00051	0.00622
40	0.04331	0.01694	0.04651	130	0.03378	0.01361	0.03642
41	0.02244	0.00951	0.02438	131	0.05500	0.02037	0.05865
42	-0.00815	-0.00185	0.00836	132	0.04853	0.01698	0.05142
43	-0.03690	-0.01292	0.03910	133	0.01948	0.00571	0.02030
44	-0.05155	-0.01908	0.05497	134	-0.01715	-0.00775	0.01882
45	-0.04451	-0.01729	0.04775	135	-0.04451	-0.01729	0.04775
46	-0.01715	-0.00775	0.01882	136	-0.05155	-0.01908	0.05496
47	0.01948	0.00571	0.02030	137	-0.03690	-0.01292	0.03910
48	0.04853	0.01698	0.05141	138	-0.00815	-0.00185	0.00836
[illegible]	[illegible]	[illegible]	0.05865	139	0.02244	0.00951	0.02437
50	0.03378	0.01361	0.03642	140	0.04331	0.01694	0.04651
51	-0.00620	-0.00051	0.00622	141	0.04777	0.01809	0.05108
52	-0.04486	-0.01494	0.04728	142	0.03565	0.01306	0.03796
53	-0.06066	-0.02175	0.06444	143	0.01238	0.00399	0.01301
54	-0.04283	-0.01656	0.04592	144	-0.01369	-0.00594	0.01493
55	0.00091	-0.00153	0.00178	145	-0.03442	-0.01370	0.03705
56	0.04651	0.01509	0.04889	146	-0.04429	-0.01726	0.04754
57	0.06624	0.02331	0.07022	147	-0.04161	-0.01604	0.04460
58	0.04570	0.01740	0.04890	148	-0.02828	-0.01081	0.03027
59	-0.00527	0.00009	0.00527	149	-0.00867	-0.00324	0.00925
60	-0.05593	-0.01820	0.05882	150	0.01194	0.00469	0.01283
61	-0.07255	-0.02540	0.07686	151	0.02886	0.01120	0.03095
62	-0.04095	-0.01576	0.04388	152	0.03890	0.01511	0.04173
63	0.02117	0.00538	0.02185	153	0.04085	0.01596	0.04386
64	0.07277	0.02407	0.07665	154	0.03527	0.01395	0.03792
65	0.07592	0.02672	0.08048	155	0.02403	0.00978	0.02594
66	0.02382	0.01001	0.02584	156	0.00966	0.00438	0.01060
67	-0.05021	-0.01540	0.05251	157	-0.00524	-0.00126	0.00539
68	-0.09260	-0.03116	0.09771	158	-0.01849	-0.00634	0.01954
69	-0.06774	-0.02429	0.07196	159	-0.02856	-0.01028	0.03035
70	0.01224	0.00215	0.01243	160	-0.03467	-0.01279	0.03696
71	0.09051	0.02938	0.09516	161	-0.03674	-0.01381	0.03925
72	0.10444	0.03572	0.11038	162	-0.03516	-0.01349	0.03766
73	0.03498	0.01360	0.03753	163	-0.03071	-0.01210	0.03301
74	-0.07165	-0.02231	0.07504	164	-0.02425	-0.00995	0.02621
75	-0.13205	-0.04396	0.13918	165	-0.01668	-0.00737	0.01824
76	-0.08886	-0.03112	0.09415	166	-0.00878	-0.00464	0.00993
77	0.03712	0.01042	0.03856	167	-0.00116	-0.00199	0.00230
78	0.15120	0.04943	0.15908	168	0.00575	0.00043	0.00577
79	0.15192	0.05138	0.16037	169	0.01171	0.00252	0.01198
80	0.01648	0.00743	0.01807	170	0.01659	0.00423	0.01712
81	-0.16358	-0.05281	0.17189	171	0.02042	0.00558	0.02116
82	-0.23617	-0.07849	0.24887	172	0.02327	0.00658	0.02418
83	-0.10544	-0.03664	0.11162	173	0.02528	0.00729	0.02631
84	0.17096	0.05473	0.17950	174	0.02662	0.00777	0.02773
85	0.38564	0.12703	0.40602	175	0.02744	0.00806	0.02860
86	0.30275	0.10131	0.31925	176	0.02790	0.00822	0.02909
87	-0.17476	-0.05568	0.18342	177	0.02812	0.00829	0.02932
88	-0.90788	-0.29834	0.95565	178	0.02820	0.00832	0.02940
89	-1.57276	-0.51902	1.65619	179	0.02822	0.00833	0.02942
				180	0.02822	0.00833	0.02942

IBR

RADIATION PATTERN FOR THETA=PI/2

H/LMDA=0.5000 B/LMDA= 0.500 OMEGA= 9.92

35 ELEMENT ARRAY

PHI DEG	ER(THETA,PHI) RE	IM	ABSVAL	PHI DEG	ER(THETA,PHI) RE	IM	ABSVAL
C	0.02112	-0.00365	0.02143	90	-1.90970	-1.04220	2.17557
1	0.02111	-0.00365	0.02143	91	-1.63210	-0.89497	1.86138
2	0.02110	-0.00365	0.02141	92	-0.94218	-0.52708	1.07959
3	0.02103	-0.00366	0.02134	93	-0.18126	-0.11593	0.21516
4	0.02084	-0.00366	0.02116	94	0.31450	0.16077	0.35321
5	0.02044	-0.00368	0.02077	95	0.40061	0.22137	0.45770
6	0.01972	-0.00370	0.02007	96	0.17760	0.11120	0.20954
7	0.01856	-0.00374	0.01893	97	-0.10970	-0.04487	0.11853
8	0.01681	-0.00379	0.01723	98	-0.24578	-0.12988	0.27799
9	0.01433	-0.00386	0.01484	99	-0.17038	-0.10275	0.19897
10	0.01102	-0.00394	0.01170	100	0.01707	-0.00693	0.01842
11	0.00680	-0.00402	0.00790	101	0.15838	0.07625	0.17578
12	0.00168	-0.00409	0.00442	102	0.15797	0.08994	0.18178
13	-0.00424	-0.00411	0.00590	103	0.03913	0.03615	0.05327
14	-0.01071	-0.00404	0.01145	104	-0.09262	-0.03597	0.09936
15	-0.01733	-0.00383	0.01775	105	-0.13829	-0.07218	0.15599
16	-0.02354	-0.00340	0.02379	106	-0.07562	-0.05174	0.09163
17	-0.02863	-0.00270	0.02876	107	0.03599	0.00321	0.03613
18	-0.03181	-0.00168	0.03185	108	0.10938	0.04946	0.12005
19	-0.03229	-0.00030	0.03230	109	0.09564	0.05568	0.11067
20	-0.02946	0.00141	0.02949	110	0.01396	0.02230	0.02631
21	-0.02299	0.00334	0.02323	111	-0.07045	-0.02309	0.07414
22	-0.01308	0.00531	0.01411	112	-0.09770	-0.04817	0.10893
23	-0.00054	0.00703	0.00705	113	-0.05417	-0.03823	0.06630
24	0.01313	0.00818	0.01547	114	0.02357	-0.00365	0.02385
25	0.02584	0.00837	0.02716	115	0.07942	0.03022	0.08497
26	0.03516	0.00732	0.03591	116	0.07765	0.04145	0.08802
27	0.03883	0.00489	0.03914	117	0.02432	0.02514	0.03498
28	0.03526	0.00121	0.03528	118	-0.04142	-0.00564	0.04180
29	0.02416	-0.00325	0.02438	119	-0.07625	-0.03038	0.08208
30	0.00701	-0.00767	0.01039	120	-0.06059	-0.03459	0.06977
31	-0.01293	-0.01102	0.01699	121	-0.00816	-0.01777	0.01955
32	-0.03102	-0.01225	0.03335	122	0.04619	0.00812	0.04690
33	-0.04225	-0.01062	0.04357	123	0.06975	0.02731	0.07490
34	-0.04268	-0.00602	0.04310	124	0.05108	0.02963	0.05905
35	-0.03091	0.00078	0.03092	125	0.00410	0.01548	0.01601
36	-0.00914	0.00815	0.01224	126	-0.04263	-0.00588	0.04304
37	0.01680	0.01387	0.02179	127	-0.06351	-0.02250	0.06737
38	0.03874	0.01581	0.04184	128	-0.04929	-0.02632	0.05588
39	0.04862	0.01269	0.05025	129	-0.01010	-0.01674	0.01955
40	0.04165	0.00483	0.04193	130	0.03220	0.00019	0.03220
41	0.01876	-0.00557	0.01957	131	0.05642	0.01569	0.05856
42	-0.01267	-0.01490	0.01956	132	0.05228	0.02279	0.05703
43	-0.04073	-0.01929	0.04507	133	0.02396	0.01923	0.03072
44	-0.05332	-0.01624	0.05574	134	-0.01377	0.00771	0.01578
45	-0.04354	-0.00605	0.04396	135	-0.04354	-0.00605	0.04396
46	-0.01377	0.00771	0.01579	136	-0.05332	-0.01623	0.05573
47	0.02395	0.01923	0.03072	137	-0.04073	-0.01929	0.04507
48	0.05227	0.02279	0.05703	138	-0.01267	-0.01490	0.01956
49	0.05642	0.01569	0.05856	139	0.01876	-0.00557	0.01957
[illegible]	[illegible]	[illegible]	[illegible]	[illegible]	[illegible]	[illegible]	[illegible]
51	-0.01010	-0.01674	0.01955	141	0.04862	0.01268	0.05025
52	-0.04929	-0.02632	0.05588	142	0.03874	0.01581	0.04184
53	-0.06351	-0.02250	0.06737	143	0.01681	0.01387	0.02179
54	-0.04264	-0.00588	0.04304	144	-0.00913	0.00815	0.01224
55	0.00410	0.01548	0.01601	145	-0.03091	0.00078	0.03092
56	0.05108	0.02963	0.05905	146	-0.04268	-0.00602	0.04310
57	0.06975	0.02731	0.07491	147	-0.04225	-0.01062	0.04357
58	0.04619	0.00813	0.04690	148	-0.03102	-0.01225	0.03335
59	-0.00815	-0.01777	0.01955	149	-0.01293	-0.01102	0.01698
60	-0.06059	-0.03459	0.06977	150	0.00701	-0.00767	0.01039
61	-0.07625	-0.03038	0.08208	151	0.02416	-0.00325	0.02438
62	-0.04143	-0.00565	0.04181	152	0.03525	0.00121	0.03527
63	0.02432	0.02514	0.03498	153	0.03883	0.00489	0.03914
64	0.07765	0.04145	0.08801	154	0.03516	0.00732	0.03591
65	0.07942	0.03023	0.08498	155	0.02584	0.00837	0.02716
66	0.02358	-0.00364	0.02386	156	0.01313	0.00817	0.01546
67	-0.05416	-0.03823	0.06629	157	-0.00054	0.00703	0.00705
68	-0.09770	-0.04817	0.10893	158	-0.01308	0.00531	0.01411
69	-0.07045	-0.02309	0.07414	159	-0.02299	0.00334	0.02323
70	0.01396	0.02230	0.02631	160	-0.02946	0.00141	0.02949
71	0.09564	0.05568	0.11067	161	-0.03229	-0.00030	0.03229
72	0.10939	0.04947	0.12005	162	-0.03181	-0.00167	0.03185
73	0.03600	0.00321	0.03614	163	-0.02863	-0.00270	0.02876
74	-0.07561	-0.05174	0.09162	164	-0.02354	-0.00340	0.02378
75	-0.13829	-0.07218	0.15599	165	-0.01733	-0.00382	0.01775
76	-0.09263	-0.03598	0.09937	166	-0.01071	-0.00404	0.01144
77	0.03912	0.03614	0.05326	167	-0.00424	-0.00411	0.00591
78	0.15796	0.08994	0.18177	168	0.00167	-0.00409	0.00442
79	0.15838	0.07625	0.17578	169	0.00680	-0.00402	0.00790
80	0.01708	-0.00693	0.01843	170	0.01102	-0.00394	0.01170
81	-0.17038	-0.10275	0.19896	171	0.01433	-0.00386	0.01484
82	-0.24578	-0.12988	0.27799	172	0.01681	-0.00379	0.01723
83	-0.10973	-0.04488	0.11855	173	0.01856	-0.00374	0.01893
84	0.17758	0.11118	0.20951	174	0.01972	-0.00370	0.02007
85	0.40060	0.22137	0.45769	175	0.02044	-0.00368	0.02077
86	0.31453	0.16078	0.35324	176	0.02084	-0.00366	0.02116
87	-0.18120	-0.11590	0.21509	177	0.02103	-0.00366	0.02134
88	-0.94210	-0.52705	1.07951	178	0.02110	-0.00365	0.02141
89	-1.63208	-0.89497	1.86136	179	0.02111	-0.00365	0.02143
				180	0.02112	-0.00365	0.02143

VBR

RADIATION PATTERN FOR THETA=PI/2

H/LMDA=0.5000 B/LMDA= 0.500 OMEGA= 9.92

40 ELEMENT ARRAY

PHI DEG	ER(THETA,PHI) RE	IM	ABSVAL	PHI DEG	ER(THETA,PHI) RE	IM	ABSVAL
0	-0.00000	-0.00000	0.00000	90	-1.84129	-0.60802	1.93908
1	0.00029	0.00010	0.00031	91	-1.49617	-0.49346	1.57544
2	0.00115	0.00041	0.00122	92	-0.68858	-0.22576	0.72465
3	0.00258	0.00091	0.00274	93	0.07605	0.02679	0.08063
4	0.00458	0.00162	0.00486	94	0.39731	0.13153	0.41851
5	0.00711	0.00251	0.00754	95	0.24844	0.08072	0.26122
6	0.01012	0.00357	0.01073	96	-0.07317	-0.02584	0.07760
7	0.01352	0.00477	0.01434	97	-0.23625	-0.07839	0.24892
8	0.01715	0.00605	0.01819	98	-0.13840	-0.04439	0.14535
9	0.02080	0.00732	0.02206	99	0.06860	0.02433	0.07279
10	0.02417	0.00849	0.02562	100	0.16942	0.05631	0.17853
11	0.02688	0.00941	0.02848	101	0.09016	0.02845	0.09454
12	0.02849	0.00993	0.03017	102	-0.06258	-0.02236	0.06646
13	0.02854	0.00987	0.03020	103	-0.13229	-0.04406	0.13944
14	0.02660	0.00908	0.02811	104	-0.06548	-0.02030	0.06856
15	0.02236	0.00746	0.02357	105	0.05530	0.01999	0.05880
16	0.01570	0.00496	0.01646	106	0.10880	0.03637	0.11472
17	0.00687	0.00168	0.00707	107	0.05306	0.01624	0.05549
18	-0.00350	-0.00215	0.00411	108	-0.04677	-0.01724	0.04984
19	-0.01431	-0.00610	0.01556	109	-0.09278	-0.03124	0.09790
20	-0.02405	-0.00962	0.02590	110	-0.04821	-0.01475	0.05042
21	-0.03099	-0.01206	0.03325	111	0.03675	0.01401	0.03933
22	-0.03344	-0.01281	0.03581	112	0.08092	0.02757	0.08549
23	-0.03022	-0.01144	0.03231	113	0.04844	0.01504	0.05072
24	-0.02107	-0.00783	0.02248	114	-0.02475	-0.01010	0.02673
25	-0.00702	-0.00241	0.00743	115	-0.07067	-0.02454	0.07481
26	0.00944	0.00390	0.01021	116	-0.05181	-0.01654	0.05439
27	0.02470	0.00973	0.02654	117	0.01024	0.00526	0.01151
28	0.03466	0.01354	0.03721	118	0.05955	0.02129	0.06324
29	0.03593	0.01403	0.03857	119	0.05615	0.01856	0.05913
30	0.02706	0.01066	0.02908	120	0.00694	0.00068	0.00698
31	0.00956	0.00396	0.01035	121	-0.04495	-0.01683	0.04800
32	-0.01192	-0.00432	0.01268	122	-0.05833	-0.02007	0.06168
33	-0.03051	-0.01159	0.03264	123	-0.02570	-0.00750	0.02677
34	-0.03926	-0.01519	0.04210	124	0.02472	0.01023	0.02675
35	-0.03384	-0.01341	0.03640	125	0.05408	0.01951	0.05749
36	-0.01493	-0.00641	0.01625	126	0.04251	0.01409	0.04479
37	0.01100	0.00347	0.01154	127	0.00123	-0.00119	0.00171
38	0.03348	0.01230	0.03567	128	-0.03890	-0.01502	0.04170
39	0.04209	0.01609	0.04506	129	-0.05068	-0.01807	0.05381
40	0.03151	0.01265	0.03395	130	-0.02836	-0.00900	0.02976
41	0.00516	0.00303	0.00599	131	0.01119	0.00554	0.01249
42	-0.02501	-0.00846	0.02640	132	0.04192	0.01606	0.04489
43	-0.04351	-0.01602	0.04636	133	0.04572	0.01641	0.04858
44	-0.03928	-0.01528	0.04215	134	0.02233	0.00704	0.02341
45	-0.01281	-0.00603	0.01416	135	-0.01281	-0.00603	0.01416
46	0.02233	0.00704	0.02341	136	-0.03928	-0.01528	0.04215
47	0.04572	0.01641	0.04857	137	-0.04351	-0.01602	0.04636
48	0.04192	0.01606	0.04489	138	-0.02501	-0.00846	0.02640
49	0.01120	0.00554	0.01249	139	0.00516	0.00303	0.00599
50	-0.02836	-0.00900	0.02975	140	0.03150	0.01265	0.03395
51	-0.05068	-0.01807	0.05380	141	0.04208	0.01609	0.04506
52	-0.03890	-0.01502	0.04170	142	0.03348	0.01230	0.03567
53	0.00122	-0.00119	0.00171	143	0.01100	0.00347	0.01154
54	0.04251	0.01409	0.04479	144	-0.01494	-0.00641	0.01625
55	0.05408	0.01951	0.05749	145	-0.03384	-0.01341	0.03640
56	0.02472	0.01023	0.02675	146	-0.03926	-0.01519	0.04210
57	-0.02570	-0.00750	0.02677	147	-0.03051	-0.01159	0.03264
58	-0.05833	-0.02007	0.06168	148	-0.01192	-0.00432	0.01268
59	-0.04495	-0.01683	0.04800	149	0.00956	0.00396	0.01035
60	0.00694	0.00068	0.00697	150	0.02706	0.01066	0.02909
61	0.05614	0.01856	0.05913	151	0.03593	0.01403	0.03857
62	0.05955	0.02129	0.06324	152	0.03466	0.01354	0.03721
63	0.01024	0.00526	0.01152	153	0.02470	0.00973	0.02654
64	-0.05181	-0.01654	0.05438	154	0.00944	0.00390	0.01021
65	-0.07068	-0.02454	0.07482	155	-0.00702	-0.00241	0.00742
66	-0.02476	-0.01010	0.02674	156	-0.02107	-0.00783	0.02248
67	0.04843	0.01504	0.05071	157	-0.03022	-0.01143	0.03231
68	0.08092	0.02757	0.08549	158	-0.03344	-0.01282	0.03581
69	0.03675	0.01401	0.03933	159	-0.03099	-0.01206	0.03325
70	-0.04821	-0.01475	0.05042	160	-0.02405	-0.00962	0.02591
71	-0.09278	-0.03124	0.09790	161	-0.01431	-0.00610	0.01556
72	-0.04678	-0.01724	0.04985	162	-0.00350	-0.00215	0.00411
73	0.05305	0.01623	0.05548	163	0.00687	0.00168	0.00707
74	0.10880	0.03637	0.11472	164	0.01570	0.00496	0.01647
75	0.05531	0.02000	0.05881	165	0.02236	0.00746	0.02357
76	-0.06547	-0.02029	0.06855	166	0.02660	0.00909	0.02811
77	-0.13229	-0.04406	0.13944	167	0.02854	0.00987	0.03020
78	-0.06260	-0.02236	0.06647	168	0.02849	0.00993	0.03017
79	0.09016	0.02845	0.09454	169	0.02688	0.00941	0.02848
80	0.16942	0.05631	0.17853	170	0.02417	0.00849	0.02562
81	0.06861	0.02433	0.07280	171	0.02080	0.00732	0.02206
82	-0.13838	-0.04439	0.14533	172	0.01715	0.00605	0.01819
83	-0.23626	-0.07839	0.24892	173	0.01352	0.00477	0.01434
84	-0.07320	-0.02585	0.07762	174	0.01012	0.00357	0.01073
85	0.24841	0.08071	0.26119	175	0.00711	0.00251	0.00754
86	0.39731	0.13153	0.41852	176	0.00458	0.00162	0.00486
87	0.07610	0.02681	0.08068	177	0.00259	0.00091	0.00274
88	-0.68850	-0.22574	0.72457	178	0.00115	0.00041	0.00122
89	-1.49615	-0.49346	1.57542	179	0.00029	0.00010	0.00031
				180	-0.00000	-0.00000	0.00000

IBR

RADIATION PATTERN FOR THETA=PI/2

H/LMDA=0.5000 B/LMDA= 0.500 OMEGA= 9.92

40 ELEMENT ARRAY

PHI DEG	ER(THETA,PHI) RE	IM	ABSVAL	PHI DEG	ER(THETA,PHI) RE	IM	ABSVAL
0	-0.00000	-0.00000	0.00000	90	-1.90798	-1.03926	2.17266
1	0.00024	-0.00000	0.00024	91	-1.55035	-0.84922	1.76770
2	0.00097	-0.00002	0.00097	92	-0.71346	-0.40166	0.81875
3	0.00218	-0.00004	0.00218	93	0.07902	0.02943	0.08432
4	0.00387	-0.00007	0.00387	94	0.41202	0.22161	0.46784
5	0.00601	-0.00012	0.00601	95	0.25761	0.15079	0.29850
6	0.00854	-0.00018	0.00854	96	-0.07604	-0.02785	0.08098
7	0.01140	-0.00025	0.01141	97	-0.24536	-0.13036	0.27783
8	0.01444	-0.00036	0.01444	98	-0.14380	-0.08858	0.16889
9	0.01747	-0.00051	0.01748	99	0.07131	0.02530	0.07567
10	0.02022	-0.00071	0.02023	100	0.17631	0.09274	0.19921
11	0.02236	-0.00099	0.02238	101	0.09403	0.06147	0.11234
12	0.02350	-0.00136	0.02354	102	-0.06505	-0.02182	0.06861
13	0.02325	-0.00184	0.02332	103	-0.13803	-0.07170	0.15554
14	0.02122	-0.00241	0.02136	104	-0.06871	-0.04752	0.08354
15	0.01717	-0.00306	0.01744	105	0.05742	0.01747	0.06002
16	0.01103	-0.00373	0.01165	106	0.11385	0.05790	0.12773
17	0.00306	-0.00433	0.00530	107	0.05613	0.04014	0.06901
18	-0.00611	-0.00472	0.00772	108	-0.04839	-0.01224	0.04991
19	-0.01547	-0.00475	0.01618	109	-0.09734	-0.04770	0.10840
20	-0.02361	-0.00426	0.02399	110	-0.05143	-0.03650	0.06306
21	-0.02896	-0.00312	0.02913	111	0.03768	0.00608	0.03817
22	-0.03006	-0.00131	0.03009	112	0.08500	0.03910	0.09356
23	-0.02593	0.00107	0.02595	113	0.05200	0.03489	0.06262
24	-0.01648	0.00372	0.01690	114	-0.02475	0.00101	0.02477
25	-0.00288	0.00614	0.00678	115	-0.07411	-0.03069	0.08021
26	0.01242	0.00772	0.01462	116	-0.05572	-0.03387	0.06521
27	0.02594	0.00787	0.02710	117	0.00905	-0.00879	0.01262
28	0.03390	0.00618	0.03446	118	0.06197	0.02134	0.06554
29	0.03332	0.00267	0.03342	119	0.06017	0.03192	0.06811
30	0.02319	-0.00208	0.02329	120	0.00944	0.01652	0.01903
31	0.00539	-0.00688	0.00874	121	-0.04584	-0.01040	0.04701
32	-0.01532	-0.01022	0.01841	122	-0.06187	-0.02732	0.06763
33	-0.03218	-0.01067	0.03390	123	-0.02928	-0.02261	0.03699
34	-0.03869	-0.00753	0.03942	124	0.02364	-0.00183	0.02371
35	-0.03120	-0.00130	0.03123	125	0.05619	0.01865	0.05920
36	-0.01107	0.00615	0.01267	126	0.04634	0.02457	0.05245
37	0.01476	0.01200	0.01902	127	0.00423	0.01335	0.01400
38	0.03579	0.01348	0.03824	128	-0.03863	-0.00585	0.03907
39	0.04205	0.00926	0.04305	129	-0.05327	-0.01978	0.05682
40	0.02912	0.00039	0.02912	130	-0.03220	-0.02020	0.03801
41	0.00139	-0.00955	0.00966	131	0.00838	-0.00825	0.01176
42	-0.02854	-0.01578	0.03261	132	0.04171	0.00765	0.04241
43	-0.04521	-0.01454	0.04749	133	0.04821	0.01787	0.05141
44	-0.03834	-0.00546	0.03873	134	0.02614	0.01731	0.03135
45	-0.00965	0.00745	0.01220	135	-0.00965	0.00745	0.01219
46	0.02614	0.01731	0.03135	136	-0.03834	-0.00546	0.03873
47	0.04820	0.01787	0.05141	137	-0.04521	-0.01454	0.04749
48	0.04171	0.00765	0.04241	138	-0.02854	-0.01578	0.03261
49	0.00838	-0.00825	0.01176	139	0.00139	-0.00955	0.00966
50	-0.03220	-0.02020	0.03801	140	0.02911	0.00039	0.02912
51	-0.05327	-0.01978	0.05682	141	0.04204	0.00926	0.04305
52	-0.03863	-0.00585	0.03907	142	0.03579	0.01348	0.03824
53	0.00422	0.01335	0.01400	143	0.01476	0.01199	0.01902
54	0.04634	0.02457	0.05245	144	-0.01107	0.00615	0.01267
55	0.05619	0.01865	0.05921	145	-0.03120	-0.00130	0.03123
56	0.02364	-0.00183	0.02371	146	-0.03869	-0.00753	0.03942
57	-0.02928	-0.02261	0.03699	147	-0.03218	-0.01067	0.03390
58	-0.06187	-0.02732	0.06763	148	-0.01531	-0.01022	0.01841
59	-0.04584	-0.01040	0.04701	149	0.00538	-0.00688	0.00873
60	0.00943	0.01652	0.01903	150	0.02320	-0.00208	0.02329
61	0.06017	0.03191	0.06811	151	0.03332	0.00267	0.03342
62	0.06197	0.02134	0.06554	152	0.03390	0.00618	0.03446
63	0.00906	-0.00879	0.01262	153	0.02594	0.00787	0.02710
64	-0.05572	-0.03387	0.06521	154	0.01242	0.00772	0.01462
65	-0.07411	-0.03069	0.08021	155	-0.00288	0.00614	0.00678
66	-0.02476	0.00100	0.02478	156	-0.01648	0.00371	0.01690
67	0.05199	0.03489	0.06261	157	-0.02593	0.00107	0.02595
68	0.08500	0.03910	0.09356	158	-0.03006	-0.00131	0.03009
69	0.03768	0.00608	0.03817	159	-0.02896	-0.00312	0.02913
70	-0.05143	-0.03649	0.06306	160	-0.02361	-0.00426	0.02399
71	-0.09734	-0.04770	0.10840	161	-0.01547	-0.00475	0.01618
72	-0.04840	-0.01224	0.04992	162	-0.00611	-0.00472	0.00772
73	0.05612	0.04014	0.06899	163	0.00307	-0.00432	0.00530
74	0.11385	0.05790	0.12773	164	0.01104	-0.00373	0.01165
75	0.05743	0.01747	0.06003	165	0.01717	-0.00306	0.01744
76	-0.06870	-0.04751	0.08353	166	0.02122	-0.00241	0.02136
77	-0.13803	-0.07170	0.15554	167	0.02325	-0.00184	0.02332
78	-0.06506	-0.02183	0.06863	168	0.02350	-0.00136	0.02354
79	0.09403	0.06147	0.11234	169	0.02236	-0.00099	0.02238
80	0.17631	0.09274	0.19921	170	0.02022	-0.00071	0.02023
81	0.07132	0.02530	0.07568	171	0.01747	-0.00050	0.01748
82	-0.14378	-0.08857	0.16888	172	0.01444	-0.00036	0.01445
83	-0.24536	-0.13036	0.27784	173	0.01140	-0.00025	0.01141
84	-0.07607	-0.02787	0.08102	174	0.00854	-0.00018	0.00854
85	0.25758	0.15077	0.29847	175	0.00601	-0.00011	0.00601
86	0.41203	0.22161	0.46784	176	0.00387	-0.00007	0.00387
87	0.07907	0.02946	0.08438	177	0.00219	-0.00004	0.00219
88	-0.71337	-0.40162	0.81866	178	0.00097	-0.00002	0.00097
89	-1.55032	-0.84922	1.76767	179	0.00024	-0.00000	0.00024
				180	-0.00000	-0.00000	0.00000

VBR

RADIATION PATTERN FOR THETA=PI/2

H/LMDA=0.5000 B/LMDA= 0.500 OMEGA= 9.92

45 ELEMENT ARRAY

PHI DEG	ER(THETA,PHI) RE	IM	ABSVAL	PHI DEG	ER(THETA,PHI) RE	IM	ABSVAL
0	-0.02153	-0.00624	0.02241	90	-1.84191	-0.60800	1.93966
1	-0.02153	-0.00624	0.02241	91	-1.41156	-0.46531	1.48627
2	-0.02150	-0.00623	0.02239	92	-0.47235	-0.15436	0.49694
3	-0.02140	-0.00620	0.02228	93	0.25976	0.08689	0.27390
4	-0.02113	-0.00610	0.02199	94	0.36744	0.12078	0.38678
5	-0.02056	-0.00590	0.02139	95	0.04231	0.01242	0.04409
6	-0.01954	-0.00555	0.02031	96	-0.22153	-0.07381	0.23350
7	-0.01789	-0.00497	0.01857	97	-0.15984	-0.05174	0.16800
8	-0.01544	-0.00412	0.01598	98	0.07051	0.02473	0.07472
9	-0.01204	-0.00294	0.01239	99	0.16810	0.05555	0.17704
10	-0.00759	-0.00140	0.00772	100	0.04907	0.01478	0.05125
11	-0.00212	0.00049	0.00218	101	-0.10659	-0.03632	0.11261
12	0.00418	0.00265	0.00495	102	-0.11070	-0.03598	0.11640
13	0.01094	0.00495	0.01201	103	0.01692	0.00715	0.01837
14	0.01753	0.00716	0.01894	104	0.10680	0.03591	0.11268
15	0.02314	0.00898	0.02482	105	0.05895	0.01839	0.06175
16	0.02678	0.01006	0.02861	106	-0.05253	-0.01885	0.05581
17	0.02749	0.01006	0.02927	107	-0.08990	-0.02980	0.09471
18	0.02450	0.00871	0.02600	108	-0.01854	-0.00470	0.01913
19	0.01755	0.00593	0.01852	109	0.06716	0.02347	0.07114
20	0.00711	0.00192	0.00736	110	0.06760	0.02197	0.07108
21	-0.00542	-0.00279	0.00610	111	-0.00903	-0.00459	0.01012
22	-0.01777	-0.00735	0.01923	112	-0.06896	-0.02381	0.07295
23	-0.02708	-0.01071	0.02912	113	-0.04698	-0.01480	0.04926
24	-0.03061	-0.01184	0.03282	114	0.02523	0.01002	0.02715
25	-0.02655	-0.01010	0.02841	115	0.06457	0.02214	0.06826
26	-0.01494	-0.00554	0.01594	116	0.03148	0.00945	0.03287
27	0.00176	0.00092	0.00199	117	-0.03288	-0.01258	0.03520
28	0.01889	0.00748	0.02032	118	-0.05857	-0.02004	0.06190
29	0.03065	0.01198	0.03291	119	-0.02213	-0.00625	0.02299
30	0.03215	0.01255	0.03451	120	0.03463	0.01320	0.03706
31	0.02158	0.00851	0.02319	121	0.05345	0.01838	0.05652
32	0.00183	0.00093	0.00206	122	0.01870	0.00514	0.01939
33	-0.01970	-0.00740	0.02105	123	-0.03219	-0.01246	0.03452
34	-0.03357	-0.01290	0.03596	124	-0.04994	-0.01742	0.05289
35	-0.03247	-0.01274	0.03488	125	-0.02049	-0.00594	0.02133
36	-0.01542	-0.00644	0.01671	126	0.02607	0.01048	0.02810
37	0.01034	0.00337	0.01088	127	0.04722	0.01688	0.05015
38	0.03179	0.01181	0.03391	128	0.02641	0.00834	0.02770
39	0.03646	0.01408	0.03908	129	-0.01577	-0.00696	0.01724
40	0.02010	0.00835	0.02177	130	-0.04299	-0.01593	0.04584
41	-0.00919	-0.00259	0.00955	131	-0.03441	-0.01173	0.03636
42	-0.03403	-0.01233	0.03619	132	0.00070	0.00151	0.00167
43	-0.03775	-0.01446	0.04043	133	0.03371	0.01313	0.03618
44	-0.01607	-0.00698	0.01753	134	0.04039	0.01464	0.04296
45	0.01786	0.00561	0.01873	135	0.01786	0.00561	0.01873
46	0.04039	0.01464	0.04296	136	-0.01607	-0.00698	0.01752
47	0.03372	0.01313	0.03618	137	-0.03775	-0.01446	0.04043
48	0.00069	0.00151	0.00166	138	-0.03403	-0.01233	0.03619
[illegible]	[illegible]	[illegible]	[illegible]	[illegible]	[illegible]	[illegible]	[illegible]
50	-0.04299	-0.01592	0.04584	140	0.02010	0.00835	0.02177
51	-0.01577	-0.00696	0.01724	141	0.03645	0.01408	0.03908
52	0.02641	0.00834	0.02769	142	0.03179	0.01181	0.03391
53	0.04722	0.01689	0.05015	143	0.01035	0.00337	0.01088
54	0.02608	0.01048	0.02810	144	-0.01541	-0.00644	0.01671
55	-0.02049	-0.00594	0.02133	145	-0.03247	-0.01274	0.03488
56	-0.04994	-0.01742	0.05289	146	-0.03357	-0.01291	0.03597
57	-0.03219	-0.01246	0.03452	147	-0.01970	-0.00740	0.02105
58	0.01869	0.00514	0.01939	148	0.00183	0.00093	0.00206
59	0.05345	0.01838	0.05652	149	0.02158	0.00851	0.02320
60	0.03463	0.01320	0.03706	150	0.03215	0.01255	0.03451
61	-0.02213	-0.00625	0.02299	151	0.03066	0.01198	0.03291
62	-0.05857	-0.02004	0.06190	152	0.01889	0.00748	0.02032
63	-0.03288	-0.01258	0.03521	153	0.00177	0.00092	0.00199
64	0.03147	0.00944	0.03286	154	-0.01494	-0.00554	0.01594
65	0.06457	0.02214	0.06826	155	-0.02655	-0.01010	0.02840
66	0.02524	0.01002	0.02716	156	-0.03062	-0.01184	0.03283
67	-0.04698	-0.01480	0.04925	157	-0.02709	-0.01071	0.02913
68	-0.06896	-0.02381	0.07295	158	-0.01776	-0.00735	0.01923
69	-0.00903	-0.00459	0.01013	159	-0.00542	-0.00279	0.00610
70	0.06760	0.02197	0.07108	160	0.00711	0.00192	0.00737
71	0.06716	0.02347	0.07114	161	0.01755	0.00593	0.01852
72	-0.01853	-0.00470	0.01912	162	0.02450	0.00871	0.02600
73	-0.08989	-0.02980	0.09470	163	0.02749	0.01006	0.02927
74	-0.05254	-0.01885	0.05582	164	0.02679	0.01006	0.02861
75	0.05894	0.01838	0.06174	165	0.02314	0.00898	0.02482
76	0.10681	0.03591	0.11268	166	0.01753	0.00716	0.01894
77	0.01693	0.00716	0.01839	167	0.01094	0.00495	0.01201
78	-0.11069	-0.03598	0.11639	168	0.00418	0.00265	0.00495
79	-0.10659	-0.03632	0.11261	169	-0.00212	0.00049	0.00218
80	0.04906	0.01478	0.05124	170	-0.00759	-0.00140	0.00772
81	0.16810	0.05555	0.17704	171	-0.01204	-0.00294	0.01239
82	0.07053	0.02473	0.07474	172	-0.01544	-0.00412	0.01598
83	-0.15982	-0.05173	0.16799	173	-0.01789	-0.00497	0.01857
84	-0.22154	-0.07381	0.23351	174	-0.01954	-0.00555	0.02031
85	0.04227	0.01241	0.04406	175	-0.02056	-0.00590	0.02139
86	0.36742	0.12077	0.38676	176	-0.02113	-0.00610	0.02199
87	0.25980	0.08690	0.27395	177	-0.02140	-0.00620	0.02228
88	-0.47227	-0.15434	0.49685	178	-0.02150	-0.00623	0.02239
89	-1.41153	-0.46530	1.48624	179	-0.02153	-0.00624	0.02241
				180	-0.02153	-0.00624	0.02241

IBR

RADIATION PATTERN FOR THETA=PI/2

H/LMDA=0.5000 B/LMDA= 0.500 OMEGA= 9.92

45 ELEMENT ARRAY

PHI DEG	ER(THETA,PHI) RE	IM	ABSVAL	PHI DEG	ER(THETA,PHI) RE	IM	ABSVAL
0	-0.01597	0.00288	0.01622	90	-1.91092	-1.04588	2.17841
1	-0.01596	0.00288	0.01622	91	-1.46444	-0.80676	1.67195
2	-0.01594	0.00288	0.01620	92	-0.48995	-0.28095	0.56479
3	-0.01586	0.00288	0.01612	93	0.26973	0.13816	0.30306
4	-0.01563	0.00288	0.01589	94	0.38143	0.21284	0.43680
5	-0.01514	0.00289	0.01541	95	0.04381	0.03664	0.05711
6	-0.01427	0.00289	0.01456	96	-0.23024	-0.12023	0.25974
7	-0.01286	0.00290	0.01318	97	-0.16615	-0.09916	0.19349
8	-0.01076	0.00291	0.01115	98	0.07338	0.02817	0.07860
9	-0.00787	0.00290	0.00838	99	0.17503	0.09497	0.19913
10	-0.00409	0.00288	0.00500	100	0.05120	0.03940	0.06461
11	0.00052	0.00282	0.00286	101	-0.11108	-0.05134	0.12237
12	0.00581	0.00268	0.00639	102	-0.11565	-0.06748	0.13390
13	0.01140	0.00242	0.01165	103	0.01742	-0.00315	0.01771
14	0.01674	0.00200	0.01686	104	0.11163	0.05543	0.12463
15	0.02110	0.00135	0.02115	105	0.06205	0.04224	0.07506
16	0.02362	0.00046	0.02363	106	-0.05464	-0.01767	0.05742
17	0.02347	-0.00068	0.02348	107	-0.09440	-0.05013	0.10688
18	0.02002	-0.00202	0.02012	108	-0.02011	-0.02202	0.02983
19	0.01312	-0.00340	0.01356	109	0.07014	0.02776	0.07543
20	0.00332	-0.00460	0.00568	110	0.07153	0.04131	0.08260
21	-0.00801	-0.00533	0.00962	111	-0.00856	0.00779	0.01158
22	-0.01872	-0.00525	0.01944	112	-0.07233	-0.03098	0.07868
23	-0.02622	-0.00413	0.02654	113	-0.05037	-0.03253	0.05996
24	-0.02808	-0.00193	0.02815	114	0.02546	0.00089	0.02547
25	-0.02286	0.00111	0.02288	115	0.06805	0.03040	0.07453
26	-0.01092	0.00436	0.01176	116	0.03450	0.02564	0.04299
27	0.00515	0.00692	0.00862	117	-0.03339	-0.00509	0.03378
28	0.02075	0.00780	0.02217	118	-0.06200	-0.02811	0.06808
29	0.03046	0.00631	0.03111	119	-0.02502	-0.02124	0.03282
30	0.02995	0.00247	0.03005	120	0.03506	0.00591	0.03556
31	0.01806	-0.00279	0.01827	121	0.05673	0.02522	0.06208
32	-0.00183	-0.00766	0.00787	122	0.02171	0.01919	0.02898
33	-0.02220	-0.01001	0.02435	123	-0.03218	-0.00416	0.03245
34	-0.03395	-0.00833	0.03496	124	-0.05288	-0.02194	0.05726
35	-0.03058	-0.00267	0.03069	125	-0.02377	-0.01886	0.03035
36	-0.01203	0.00491	0.01299	126	0.02525	0.00037	0.02525
37	0.01375	0.01086	0.01753	127	0.04949	0.01783	0.05260
38	0.03363	0.01177	0.03563	128	0.02990	0.01918	0.03552
39	0.03584	0.00633	0.03640	129	-0.01383	0.00494	0.01468
40	0.01731	-0.00328	0.01762	130	-0.04406	-0.01210	0.04569
41	-0.01270	-0.01189	0.01740	131	-0.03765	-0.01856	0.04198
42	-0.03633	-0.01398	0.03893	132	-0.00235	-0.01072	0.01097
43	-0.03752	-0.00729	0.03822	133	0.03299	0.00424	0.03326
44	-0.01342	0.00487	0.01428	134	0.04245	0.01504	0.04504
45	0.02133	0.01479	0.02595	135	0.02133	0.01479	0.02595
46	0.04245	0.01504	0.04503	136	-0.01341	0.00488	0.01427
47	0.03299	0.00424	0.03326	137	-0.03752	-0.00729	0.03822
48	-0.00235	-0.01072	0.01098	138	-0.03633	-0.01398	0.03893
[illegible]	[illegible]	[illegible]	[illegible]	[illegible]	[illegible]	[illegible]	[illegible]
50	-0.04405	-0.01210	0.04568	140	0.01731	-0.00328	0.01762
51	-0.01383	0.00494	0.01469	141	0.03584	0.00633	0.03639
52	0.02990	0.01918	0.03552	142	0.03363	0.01177	0.03564
53	0.04949	0.01783	0.05261	143	0.01375	0.01087	0.01753
54	0.02525	0.00037	0.02525	144	-0.01202	0.00491	0.01299
55	-0.02377	-0.01886	0.03035	145	-0.03058	-0.00267	0.03069
56	-0.05288	-0.02194	0.05725	146	-0.03396	-0.00833	0.03496
57	-0.03218	-0.00416	0.03245	147	-0.02220	-0.01001	0.02435
58	0.02170	0.01919	0.02897	148	-0.00183	-0.00766	0.00787
59	0.05672	0.02522	0.06208	149	0.01806	-0.00279	0.01828
60	0.03506	0.00591	0.03556	150	0.02995	0.00247	0.03005
61	-0.02502	-0.02124	0.03282	151	0.03047	0.00631	0.03111
62	-0.06200	-0.02812	0.06808	152	0.02075	0.00780	0.02217
63	-0.03339	-0.00509	0.03378	153	0.00516	0.00692	0.00863
64	0.03449	0.02564	0.04298	154	-0.01092	0.00436	0.01175
65	0.06805	0.03040	0.07453	155	-0.02286	0.00111	0.02288
66	0.02546	0.00090	0.02548	156	-0.02808	-0.00193	0.02815
67	-0.05037	-0.03253	0.05996	157	-0.02622	-0.00414	0.02655
68	-0.07233	-0.03098	0.07869	158	-0.01872	-0.00525	0.01944
69	-0.00857	0.00779	0.01158	159	-0.00801	-0.00533	0.00961
70	0.07153	0.04130	0.08260	160	0.00332	-0.00460	0.00568
71	0.07014	0.02777	0.07543	161	0.01312	-0.00340	0.01356
72	-0.02010	-0.02202	0.02982	162	0.02002	-0.00201	0.02012
73	-0.09439	-0.05013	0.10688	163	0.02347	-0.00068	0.02348
74	-0.05465	-0.01767	0.05744	164	0.02363	0.00046	0.02363
75	0.06204	0.04223	0.07505	165	0.02111	0.00135	0.02115
76	0.11163	0.05544	0.12463	166	0.01674	0.00200	0.01686
77	0.01744	-0.00315	0.01772	167	0.01140	0.00242	0.01165
78	-0.11565	-0.06748	0.13389	168	0.00580	0.00267	0.00639
79	-0.11109	-0.05134	0.12238	169	0.00052	0.00282	0.00287
80	0.05120	0.03940	0.06460	170	-0.00409	0.00288	0.00500
81	0.17502	0.09497	0.19913	171	-0.00787	0.00290	0.00838
82	0.07340	0.02818	0.07862	172	-0.01076	0.00291	0.01115
83	-0.16613	-0.09915	0.19347	173	-0.01286	0.00290	0.01318
84	-0.23025	-0.12024	0.25976	174	-0.01427	0.00289	0.01456
85	0.04378	0.03662	0.05707	175	-0.01514	0.00289	0.01541
86	0.38141	0.21284	0.43678	176	-0.01563	0.00288	0.01589
87	0.26977	0.13818	0.30310	177	-0.01586	0.00288	0.01612
88	-0.48986	-0.28091	0.56469	178	-0.01594	0.00288	0.01620
89	-1.46441	-0.80674	1.67192	179	-0.01596	0.00288	0.01622
				180	-0.01597	0.00288	0.01622

VBR

RADIATION PATTERN FOR THETA=PI/2

H/LMDA=0.5000 B/LMDA= 0.500 OMEGA= 9.92

50 ELEMENT ARRAY

PHI	ER(THETA,PHI)			PHI	ER(THETA,PHI)		
DEG	RE	IM	ABSVAL	DEG	RE	IM	ABSVAL
0	-0.00000	-0.00000	0.00000	90	-1.84253	-0.60807	1.94027
1	-0.00028	-0.00010	0.00030	91	-1.32009	-0.43495	1.38990
2	-0.00113	-0.00039	0.00119	92	-0.26754	-0.08682	0.28127
3	-0.00252	-0.00088	0.00267	93	0.36788	0.12200	0.38758
4	-0.00446	-0.00155	0.00472	94	0.24673	0.08035	0.25948
5	-0.00690	-0.00240	0.00731	95	-0.13985	-0.04731	0.14764
6	-0.00976	-0.00339	0.01033	96	-0.21327	-0.06990	0.22443
7	-0.01290	-0.00447	0.01365	97	0.02337	0.00911	0.02508
8	-0.01610	-0.00557	0.01703	98	0.16913	0.05602	0.17817
9	-0.01903	-0.00656	0.02013	99	0.04665	0.01409	0.04873
10	-0.02130	-0.00731	0.02252	100	-0.11741	-0.03959	0.12391
11	-0.02242	-0.00764	0.02368	101	-0.08598	-0.02748	0.09026
12	-0.02188	-0.00736	0.02309	102	0.06243	0.02190	0.06616
13	-0.01925	-0.00634	0.02027	103	0.09988	0.03272	0.10510
14	-0.01429	-0.00447	0.01497	104	-0.00977	-0.00466	0.01083
15	-0.00708	-0.00182	0.00731	105	-0.09148	-0.03071	0.09650
16	0.00178	0.00142	0.00228	106	-0.03405	-0.01007	0.03551
17	0.01120	0.00480	0.01218	107	0.06489	0.02258	0.06871
18	0.01953	0.00774	0.02101	108	0.06252	0.02012	0.06568
19	0.02487	0.00954	0.02664	109	-0.02665	-0.01028	0.02856
20	0.02548	0.00955	0.02721	110	-0.07064	-0.02367	0.07450
21	0.02033	0.00743	0.02165	111	-0.01388	-0.00330	0.01426
22	0.00978	0.00331	0.01033	112	0.05697	0.01997	0.06037
23	-0.00413	-0.00200	0.00459	113	0.04548	0.01446	0.04772
24	-0.01776	-0.00714	0.01914	114	-0.02570	-0.01002	0.02758
25	-0.02667	-0.01044	0.02864	115	-0.05787	-0.01959	0.06110
26	-0.02711	-0.01050	0.02907	116	-0.01253	-0.00292	0.01287
27	-0.01777	-0.00683	0.01904	117	0.04587	0.01652	0.04876
28	-0.00102	-0.00035	0.00107	118	0.04246	0.01380	0.04465
29	0.01708	0.00664	0.01833	119	-0.01372	-0.00609	0.01501
30	0.02870	0.01117	0.03080	120	-0.04972	-0.01741	0.05268
31	0.02764	0.01085	0.02969	121	-0.02358	-0.00700	0.02459
32	0.01298	0.00529	0.01402	122	0.02914	0.01127	0.03124
33	-0.00911	-0.00322	0.00966	123	0.04504	0.01549	0.04762
34	-0.02735	-0.01040	0.02926	124	0.00855	0.00169	0.00872
35	-0.03085	-0.01205	0.03312	125	-0.03570	-0.01343	0.03814
36	-0.01605	-0.00667	0.01738	126	-0.03853	-0.01310	0.04070
37	0.00950	0.00305	0.00997	127	0.00022	0.00138	0.00139
38	0.03001	0.01119	0.03202	128	0.03692	0.01384	0.03943
39	0.03096	0.01210	0.03324	129	0.03402	0.01157	0.03593
40	0.00994	0.00455	0.01093	130	-0.00284	-0.00225	0.00362
41	-0.01937	-0.00661	0.02047	131	-0.03529	-0.01338	0.03774
42	-0.03511	-0.01315	0.03749	132	-0.03277	-0.01136	0.03469
43	-0.02357	-0.00956	0.02544	133	-0.00006	0.00111	0.00111
44	0.00792	0.00197	0.00817	134	0.03130	0.01216	0.03358
45	0.03405	0.01223	0.03618	135	0.03405	0.01223	0.03618
46	0.03130	0.01216	0.03358	136	0.00793	0.00197	0.00817
47	-0.00006	0.00111	0.00111	137	-0.02357	-0.00956	0.02543
48	-0.03277	-0.01136	0.03469	138	-0.03511	-0.01315	0.03749
49	-0.03529	-0.01338	0.03774	139	-0.01937	-0.00661	0.02047
50	-0.00284	-0.00225	0.00362	140	0.00994	0.00455	0.01093
51	0.03402	0.01157	0.03593	141	0.03096	0.01210	0.03324
52	0.03692	0.01384	0.03943	142	0.03001	0.01119	0.03203
53	0.00022	0.00138	0.00139	143	0.00950	0.00305	0.00998
54	-0.03853	-0.01310	0.04069	144	-0.01605	-0.00667	0.01738
55	-0.03570	-0.01343	0.03814	145	-0.03085	-0.01205	0.03312
56	0.00855	0.00169	0.00871	146	-0.02735	-0.01040	0.02926
57	0.04504	0.01549	0.04763	147	-0.00911	-0.00322	0.00966
58	0.02914	0.01127	0.03125	148	0.01298	0.00529	0.01402
59	-0.02357	-0.00700	0.02458	149	0.02763	0.01085	0.02969
60	-0.04972	-0.01741	0.05268	150	0.02870	0.01117	0.03080
61	-0.01372	-0.00609	0.01501	151	0.01709	0.00664	0.01833
62	0.04246	0.01379	0.04464	152	-0.00102	-0.00035	0.00108
63	0.04588	0.01652	0.04876	153	-0.01777	-0.00683	0.01904
64	-0.01253	-0.00292	0.01286	154	-0.02711	-0.01050	0.02908
65	-0.05787	-0.01959	0.06109	155	-0.02666	-0.01044	0.02863
66	-0.02571	-0.01003	0.02759	156	-0.01776	-0.00714	0.01914
67	0.04548	0.01446	0.04772	157	-0.00414	-0.00200	0.00460
68	0.05697	0.01997	0.06037	158	0.00978	0.00331	0.01033
69	-0.01387	-0.00330	0.01426	159	0.02034	0.00743	0.02165
70	-0.07064	-0.02367	0.07450	160	0.02548	0.00955	0.02721
71	-0.02665	-0.01028	0.02857	161	0.02487	0.00954	0.02664
72	0.06252	0.02011	0.06567	162	0.01953	0.00774	0.02101
73	0.06490	0.02259	0.06871	163	0.01121	0.00480	0.01219
74	-0.03404	-0.01007	0.03550	164	0.00179	0.00142	0.00228
75	-0.09148	-0.03071	0.09650	165	-0.00709	-0.00182	0.00731
76	-0.00979	-0.00466	0.01084	166	-0.01429	-0.00447	0.01497
77	0.09987	0.03272	0.10510	167	-0.01925	-0.00634	0.02027
78	0.06244	0.02190	0.06617	168	-0.02188	-0.00736	0.02309
79	-0.08597	-0.02748	0.09026	169	-0.02242	-0.00764	0.02368
80	-0.11741	-0.03959	0.12391	170	-0.02130	-0.00731	0.02252
81	0.04664	0.01409	0.04872	171	-0.01903	-0.00656	0.02013
82	0.16913	0.05602	0.17817	172	-0.01609	-0.00557	0.01703
83	0.02339	0.00912	0.02511	173	-0.01290	-0.00447	0.01365
84	-0.21325	-0.06989	0.22442	174	-0.00976	-0.00339	0.01033
85	-0.13988	-0.04732	0.14767	175	-0.00690	-0.00239	0.00730
86	0.24669	0.08034	0.25945	176	-0.00446	-0.00155	0.00472
87	0.36791	0.12200	0.38761	177	-0.00252	-0.00088	0.00267
88	-0.26745	-0.08680	0.28118	178	-0.00113	-0.00039	0.00119
89	-1.32005	-0.43495	1.38986	179	-0.00028	-0.00010	0.00030
				180	-0.00000	-0.00000	0.00000

IBR

RADIATION PATTERN FOR THETA=PI/2

H/LMDA=0.5000 B/LMDA= 0.500 OMEGA= 9.92

50 ELEMENT ARRAY

PHI	ER(THETA,PHI)			PHI	ER(THETA,PHI)		
DEG	RE	IM	ABSVAL	DEG	RE	IM	ABSVAL
0	-0.00000	-0.00000	0.00000	90	-1.90949	-1.04333	2.17593
1	-0.00024	0.00001	0.00024	91	-1.36803	-0.75311	1.56163
2	-0.00094	0.00003	0.00094	92	-0.27714	-0.16334	0.32169
3	-0.00210	0.00008	0.00210	93	0.38147	0.20359	0.43240
4	-0.00372	0.00014	0.00372	94	0.25577	0.14835	0.29568
5	-0.00575	0.00022	0.00575	95	-0.14519	-0.06987	0.16113
6	-0.00812	0.00032	0.00812	96	-0.22134	-0.12464	0.25403
7	-0.01071	0.00045	0.01072	97	0.02433	0.00200	0.02441
8	-0.01333	0.00062	0.01334	98	0.17578	0.09416	0.19941
9	-0.01569	0.00083	0.01571	99	0.04854	0.03688	0.06096
10	-0.01745	0.00110	0.01748	100	-0.12218	-0.05973	0.13599
11	-0.01818	0.00144	0.01824	101	-0.08966	-0.05585	0.10563
12	-0.01746	0.00183	0.01755	102	0.06495	0.02497	0.06958
13	-0.01491	0.00228	0.01508	103	0.10432	0.05847	0.11959
14	-0.01036	0.00272	0.01071	104	-0.00991	0.00590	0.01153
15	-0.00394	0.00309	0.00500	105	-0.09561	-0.04769	0.10684
16	0.00381	0.00328	0.00502	106	-0.03604	-0.02857	0.04599
17	0.01183	0.00314	0.01224	107	0.06767	0.02749	0.07304
18	0.01865	0.00256	0.01882	108	0.06592	0.03954	0.07687
19	0.02258	0.00143	0.02262	109	-0.02735	-0.00338	0.02755
20	0.02212	-0.00019	0.02212	110	-0.07427	-0.03725	0.08309
21	0.01651	-0.00212	0.01664	111	-0.01542	-0.01799	0.02369
22	0.00625	-0.00397	0.00740	112	0.05949	0.02319	0.06385
23	-0.00659	-0.00520	0.00840	113	0.04859	0.03017	0.05719
24	-0.01855	-0.00524	0.01928	114	-0.02606	-0.00251	0.02618
25	-0.02556	-0.00375	0.02584	115	-0.06113	-0.02910	0.06770
26	-0.02441	-0.00081	0.02443	116	-0.01449	-0.01669	0.02210
27	-0.01428	0.00287	0.01457	117	0.04761	0.01545	0.05005
28	0.00213	0.00606	0.00643	118	0.04568	0.02601	0.05257
29	0.01880	0.00731	0.02017	119	-0.01292	0.00431	0.01362
30	0.02834	0.00565	0.02889	120	-0.05223	-0.02079	0.05621
31	0.02533	0.00121	0.02536	121	-0.02642	-0.01966	0.03294
32	0.00969	-0.00438	0.01063	122	0.02919	0.00392	0.02945
33	-0.01193	-0.00846	0.01462	123	0.04787	0.02129	0.05240
34	-0.02837	-0.00854	0.02963	124	0.01104	0.01387	0.01773
35	-0.02956	-0.00385	0.02981	125	-0.03620	-0.00798	0.03707
36	-0.01311	0.00363	0.01360	126	-0.04146	-0.01975	0.04592
37	0.01249	0.00970	0.01582	127	-0.00213	-0.01024	0.01046
38	0.03134	0.01017	0.03295	128	0.03747	0.00888	0.03851
39	0.02983	0.00385	0.03008	129	0.03689	0.01782	0.04097
40	0.00703	-0.00581	0.00912	130	-0.00036	0.00900	0.00901
41	-0.02217	-0.01218	0.02530	131	-0.03547	-0.00754	0.03626
42	-0.03590	-0.00994	0.03725	132	-0.03541	-0.01597	0.03885
43	-0.02178	0.00034	0.02178	133	-0.00287	-0.00967	0.01009
44	0.01101	0.01128	0.01576	134	0.03067	0.00439	0.03098
45	0.03608	0.01370	0.03860	135	0.03608	0.01369	0.03859
46	0.03067	0.00439	0.03099	136	0.01101	0.01128	0.01576
47	-0.00287	-0.00967	0.01009	137	-0.02177	0.00034	0.02178
48	-0.03541	-0.01597	0.03885	138	-0.03590	-0.00993	0.03724
49	-0.03546	-0.00754	0.03626	139	-0.02217	-0.01218	0.02530
50	-0.00036	0.00900	0.00901	140	0.00703	-0.00581	0.00912
51	0.03689	0.01782	0.04097	141	0.02983	0.00385	0.03007
52	0.03747	0.00888	0.03851	142	0.03134	0.01017	0.03295
53	-0.00213	-0.01024	0.01046	143	0.01249	0.00970	0.01582
54	-0.04145	-0.01975	0.04592	144	-0.01311	0.00363	0.01361
55	-0.03620	-0.00798	0.03707	145	-0.02956	-0.00385	0.02981
56	0.01104	0.01386	0.01772	146	-0.02837	-0.00854	0.02963
57	0.04788	0.02129	0.05240	147	-0.01193	-0.00846	0.01462
58	0.02919	0.00392	0.02945	148	0.00969	-0.00438	0.01063
59	-0.02641	-0.01966	0.03293	149	0.02533	0.00121	0.02535
60	-0.05223	-0.02079	0.05622	150	0.02834	0.00565	0.02889
61	-0.01292	0.00431	0.01362	151	0.01880	0.00731	0.02017
62	0.04568	0.02601	0.05256	152	0.00213	0.00606	0.00642
63	0.04761	0.01545	0.05006	153	-0.01428	0.00287	0.01457
64	-0.01449	-0.01669	0.02210	154	-0.02441	-0.00082	0.02443
65	-0.06113	-0.02910	0.06770	155	-0.02556	-0.00375	0.02583
66	-0.02607	-0.00252	0.02619	156	-0.01855	-0.00524	0.01928
67	0.04858	0.03017	0.05719	157	-0.00660	-0.00520	0.00840
68	0.05949	0.02319	0.06385	158	0.00624	-0.00397	0.00740
69	-0.01541	-0.01798	0.02368	159	0.01651	-0.00212	0.01664
70	-0.07427	-0.03725	0.08309	160	0.02211	-0.00019	0.02212
71	-0.02735	-0.00339	0.02756	161	0.02257	0.00143	0.02262
72	0.06591	0.03954	0.07686	162	0.01865	0.00256	0.01882
73	0.06768	0.02750	0.07305	163	0.01183	0.00315	0.01225
74	-0.03603	-0.02856	0.04598	164	0.00381	0.00328	0.00503
75	-0.09561	-0.04769	0.10684	165	-0.00394	0.00309	0.00501
76	-0.00993	0.00589	0.01154	166	-0.01036	0.00272	0.01071
77	0.10431	0.05847	0.11958	167	-0.01491	0.00228	0.01508
78	0.06496	0.02497	0.06959	168	-0.01746	0.00183	0.01755
79	-0.08965	-0.05585	0.10562	169	-0.01818	0.00144	0.01824
80	-0.12218	-0.05973	0.13600	170	-0.01745	0.00110	0.01748
81	0.04853	0.03688	0.06095	171	-0.01569	0.00083	0.01571
82	0.17578	0.09416	0.19941	172	-0.01332	0.00062	0.01334
83	0.02435	0.00201	0.02443	173	-0.01071	0.00045	0.01072
84	-0.22133	-0.12464	0.25401	174	-0.00812	0.00032	0.00812
85	-0.14522	-0.06989	0.16116	175	-0.00574	0.00022	0.00575
86	0.25574	0.14834	0.29564	176	-0.00372	0.00014	0.00372
87	0.38149	0.20360	0.43242	177	-0.00210	0.00008	0.00210
88	-0.27704	-0.16330	0.32158	178	-0.00094	0.00003	0.00094
89	-1.36800	-0.75309	1.56159	179	-0.00024	0.00001	0.00024
				180	-0.00000	-0.00000	0.00000

TABLE 5.6

RADIATION PATTERNS OF UNILATERAL ENDFIRE ARRAYS

VEN

RADIATION PATTERN FOR THETA=PI/2

H/LMDA=0.2500 B/LMDA= 0.250 OMEGA= 8.53

2 ELEMENT ARRAY

PHI	ER(THETA,PHI)			PHI	ER(THETA,PHI)		
DEG	RE	IM	ABSVAL	DEG	RE	IM	ABSVAL
0	-3.47092	-0.18076	3.47563	90	-1.72461	-1.39612	2.21888
5	-3.46782	-0.18612	3.47281	95	-1.50279	-1.47086	2.10281
10	-3.45836	-0.20214	3.46426	100	-1.27570	-1.53822	1.99838
15	-3.44206	-0.22869	3.44965	105	-1.04632	-1.59762	1.90976
20	-3.41817	-0.26548	3.42846	110	-0.81772	-1.64875	1.84039
25	-3.38566	-0.31214	3.40002	115	-0.59293	-1.69156	1.79247
30	-3.34334	-0.36814	3.36355	120	-0.37487	-1.72630	1.76653
35	-3.28988	-0.43285	3.31824	125	-0.16623	-1.75341	1.76128
40	-3.22394	-0.50544	3.26332	130	0.03055	-1.77358	1.77384
45	-3.14419	-0.58497	3.19814	135	0.21334	-1.78763	1.80031
50	-3.04948	-0.67033	3.12229	140	0.38038	-1.79649	1.83632
55	-2.93889	-0.76026	3.03563	145	0.53021	-1.80119	1.87760
60	-2.81180	-0.85340	2.93846	150	0.66170	-1.80272	1.92033
65	-2.66803	-0.94827	2.83154	155	0.77398	-1.80210	1.96127
70	-2.50781	-1.04336	2.71619	160	0.86645	-1.80021	1.99787
75	-2.33189	-1.13713	2.59437	165	0.93870	-1.79786	2.02816
80	-2.14150	-1.22810	2.46865	170	0.99046	-1.79569	2.05074
85	-1.93836	-1.31484	2.34223	175	1.02157	-1.79419	2.06464
				180	1.03195	-1.79366	2.06933

IEN

RADIATION PATTERN FOR THETA=PI/2

H/LMDA=0.2500 B/LMDA= 0.250 OMEGA= 8.53

2 ELEMENT ARRAY

PHI	ER(THETA,PHI)			PHI	ER(THETA,PHI)		
DEG	RE	IM	ABSVAL	DEG	RE	IM	ABSVAL
0	-6.71073	-7.01047	9.70467	90	-4.79438	-4.53536	6.59966
5	-6.71091	-7.00865	9.70348	95	-4.46195	-4.15683	6.09822
10	-6.71108	-7.00285	9.69941	100	-4.11139	-3.76192	5.57275
15	-6.71019	-6.99200	9.69096	105	-3.74764	-3.35608	5.03072
20	-6.70650	-6.97436	9.67568	110	-3.37610	-2.94512	4.48016
25	-6.69768	-6.94764	9.65032	115	-3.00243	-2.53501	3.92949
30	-6.68092	-6.90904	9.61091	120	-2.63234	-2.13169	3.38723
35	-6.65300	-6.85543	9.55297	125	-2.27141	-1.74089	2.86181
40	-6.61047	-6.78346	9.47173	130	-1.92495	-1.36793	2.36150
45	-6.54981	-6.68978	9.36233	135	-1.59786	-1.01770	1.89443
50	-6.46761	-6.57121	9.22013	140	-1.29454	-0.69446	1.46905
55	-6.36073	-6.42490	9.04092	145	-1.01882	-0.40191	1.09523
60	-6.22652	-6.24855	8.82122	150	-0.77399	-0.14310	0.78710
65	-6.06295	-6.04060	8.55851	155	-0.56273	0.07948	0.56832
70	-5.86878	-5.80029	8.25142	160	-0.38721	0.26391	0.46859
75	-5.64370	-5.52787	7.89991	165	-0.24909	0.40870	0.47863
80	-5.38833	-5.22457	7.50535	170	-0.14961	0.51281	0.53419
85	-5.10435	-4.89265	7.07053	175	-0.08960	0.57555	0.58248
				180	-0.06955	0.59651	0.60055

VEN

RADIATION PATTERN FOR THETA=PI/2

H/LMDA=0.2500 B/LMDA= 0.250 OMEGA= 8.53

3 ELEMENT ARRAY

PHI	ER(THETA,PHI)			PHI	ER(THETA,PHI)		
DEG	RE	IM	ABSVAL	DEG	RE	IM	ABSVAL
0	-2.07961	2.20203	3.02881	90	-0.85255	-0.57521	1.02845
5	-2.08231	2.19364	3.02458	95	-0.61854	-0.75030	0.97239
10	-2.09002	2.16827	3.01157	100	-0.38218	-0.89497	0.97315
15	-2.10161	2.12524	2.98888	105	-0.14966	-1.00648	1.01755
20	-2.11523	2.06357	2.95508	110	0.07326	-1.08386	1.08633
25	-2.12833	1.98206	2.90832	115	0.28160	-1.12780	1.16243
30	-2.13780	1.87948	2.84651	120	0.47147	-1.14062	1.23421
35	-2.14004	1.75484	2.76753	125	0.64016	-1.12592	1.29518
40	-2.13117	1.60754	2.66947	130	0.78626	-1.08835	1.34265
45	-2.10719	1.43774	2.55095	135	0.90955	-1.03319	1.37650
50	[illegible]	[illegible]	[illegible]	140	1.01088	-0.96600	1.39821
55	-1.99920	1.03591	2.25165	145	1.09185	-0.89232	1.41010
60	-1.90924	0.80936	2.07371	150	1.15475	-0.81735	1.41475
65	-1.79287	0.57134	1.88171	155	1.20211	-0.74581	1.41467
70	-1.64978	0.32744	1.68196	160	1.23651	-0.68174	1.41199
75	-1.48105	0.08403	1.48343	165	1.26040	-0.62849	1.40841
80	-1.28919	-0.15206	1.29812	170	1.27586	-0.58863	1.40510
85	-1.07803	-0.37395	1.14105	175	1.28447	-0.56398	1.40283
				180	1.28723	-0.55565	1.40203

IEN

RADIATION PATTERN FOR THETA=PI/2

H/LMDA=0.2500 B/LMDA= 0.250 OMEGA= 8.53

3 ELEMENT ARRAY

PHI	ER(THETA,PHI)			PHI	ER(THETA,PHI)		
DEG	RE	IM	ABSVAL	DEG	RE	IM	ABSVAL
0	-4.84830	2.64883	5.52470	90	-1.39432	0.91954	1.67024
5	-4.84707	2.64932	5.52386	95	-0.95160	0.67090	1.16432
10	-4.84269	2.65038	5.52052	100	-0.52405	0.42712	0.67606
15	-4.83313	2.65087	5.51237	105	-0.12221	0.19436	0.22959
20	-4.81511	2.64889	5.49562	110	0.24505	-0.02203	0.24604
25	-4.78427	2.64191	5.46525	115	0.57101	-0.21778	0.61113
30	-4.73538	2.62687	5.41519	120	0.85147	-0.38993	0.93651
35	-4.66266	2.60034	5.33875	125	1.08477	-0.53692	1.21037
40	-4.56019	2.55872	5.22899	130	1.27169	-0.65853	1.43208
45	[illegible]	[illegible]	[illegible]	135	[illegible]	[illegible]	[illegible]
50	-4.24424	2.41650	4.88396	140	1.51969	-0.83071	1.73192
55	-4.02245	2.31027	4.63869	145	1.59111	-0.88603	1.82118
60	-3.75524	2.17824	4.34126	150	1.63577	-0.92494	1.87916
65	-3.44314	2.02005	3.99197	155	1.66019	-0.95077	1.91316
70	-3.08907	1.83669	3.59385	160	1.67056	-0.96677	1.93014
75	-2.69848	1.63059	3.15288	165	1.67243	-0.97587	1.93632
80	-2.27913	1.40555	2.67768	170	1.67034	-0.98052	1.93687
85	-1.84073	1.16657	2.17926	175	1.66770	-0.98256	1.9356[illegible]
				180	1.66658	-0.98311	1.93494

VEN

RADIATION PATTERN FOR THETA=PI/2

H/LMDA=0.2500 B/LMDA= 0.250 OMEGA= 8.53

4 ELEMENT ARRAY

PHI	ER(THETA,PHI)			PHI	ER(THETA,PHI)		
DEG	RE	IM	ABSVAL	DEG	RE	IM	ABSVAL
0	0.21853	2.73850	2.74721	90	-0.41960	-0.63766	0.76333
5	0.20896	2.73419	2.74216	95	-0.25868	-0.82392	0.86357
10	0.18039	2.72045	2.72642	100	-0.08199	-0.94627	0.94982
15	0.13328	2.69498	2.69827	105	0.10050	-1.00313	1.00815
20	0.06852	2.65411	2.65499	110	0.27902	-0.99762	1.03590
25	-0.01239	2.59311	2.59314	115	0.44485	-0.93699	1.03723
30	-0.10712	2.50661	2.50890	120	0.59115	-0.83150	1.02022
35	-0.21238	2.38915	2.39857	125	0.71337	-0.69319	0.99469
40	-0.32367	2.23591	2.25921	130	0.80940	-0.53451	0.96996
45	-0.43539	2.04347	2.08934	135	0.87945	-0.36721	0.95303
50	-0.54089	1.81064	1.88970	140	0.92560	-0.20147	0.94727
55	-0.63283	1.53913	1.66415	145	0.95138	-0.04542	0.95247
60	-0.70369	1.23410	1.42063	150	0.96115	0.09503	0.96584
65	-0.74647	0.90425	1.17255	155	0.95959	0.21603	0.98361
70	-0.75544	0.56153	0.94128	160	0.95126	0.31550	1.00221
75	-0.72689	0.22045	0.75959	165	0.94022	0.39258	1.01889
80	-0.65971	-0.10322	0.66773	170	0.92984	0.44725	1.03181
85	-0.55570	-0.39387	0.68113	175	0.92260	0.47982	1.03991
				180	0.92002	0.49064	1.04267

IEN

RADIATION PATTERN FOR THETA=PI/2

H/LMDA=0.2500 B/LMDA= 0.250 OMEGA= 8.53

4 ELEMENT ARRAY

PHI	ER(THETA,PHI)			PHI	ER(THETA,PHI)		
DEG	RE	IM	ABSVAL	DEG	RE	IM	ABSVAL
0	-5.17279	6.85201	8.58533	90	0.25791	0.01529	0.25836
5	-5.16926	6.85208	8.58325	95	0.65948	-0.60259	0.89332
10	-5.15730	6.85044	8.57475	100	0.96741	-1.10757	1.47058
15	-5.13296	6.84168	8.55312	105	1.17488	-1.48555	1.89399
20	-5.08986	6.81703	8.50756	110	1.28258	-1.73225	2.15539
25	-5.01961	6.76481	8.42372	115	1.29811	-1.85286	2.26234
30	-4.91249	6.67113	8.28471	120	1.23453	-1.86064	2.23295
35	-4.75820	6.52087	8.07231	125	1.10857	-1.77475	2.09253
40	-4.54702	6.29908	7.76877	130	0.93858	-1.61771	1.87027
45	-4.27103	5.99255	7.35883	135	0.74265	-1.41290	1.59619
50	-3.92551	5.59168	6.83202	140	0.53718	-1.18249	1.29878
55	-3.51013	5.09229	6.18485	145	0.33586	-0.94589	1.00375
60	-3.03003	4.49716	5.42269	150	0.14927	-0.71890	0.73424
65	-2.49628	3.81708	4.56087	155	-0.01514	-0.51351	0.51373
70	-1.92571	3.07094	3.62478	160	-0.15262	-0.33809	0.37094
75	-1.33999	2.28500	2.64892	165	-0.26058	-0.19805	0.32730
80	-0.76392	1.49092	1.67523	170	-0.33787	-0.09653	0.35139
85	-0.22325	0.72308	0.75676	175	-0.38422	-0.03516	0.38582
				180	-0.39965	-0.01464	0.39992

VEN

RADIATION PATTERN FOR THETA=PI/2

H/LMDA=0.2500 B/LMDA= 0.250 OMEGA= 8.53

5 ELEMENT ARRAY

PHI DEG	ER(THETA,PHI) RE	IM	ABSVAL	PHI DEG	ER(THETA,PHI) RE	IM	ABSVAL
0	1.99801	1.58337	2.54934	90	-0.48665	-0.66714	0.82578
5	1.98711	1.58791	2.54363	95	-0.33745	-0.77528	0.84554
10	1.95367	1.60056	2.52559	100	-0.14753	-0.80760	0.82097
15	1.89555	1.61843	2.49247	105	0.06082	-0.76714	0.76955
20	1.80963	1.63675	2.44002	110	0.26525	-0.66397	0.71500
25	1.69239	1.64896	2.36289	115	0.44643	-0.51329	0.68027
30	1.54078	1.64702	2.25537	120	0.59021	-0.33289	0.67762
35	1.35322	1.62193	2.11231	125	0.68880	-0.14057	0.70300
40	1.13073	1.56454	1.93038	130	0.74070	0.04805	0.74225
45	0.87793	1.46671	1.70938	135	0.74973	0.22096	0.78162
50	0.60368	1.32267	1.45392	140	0.72357	0.37039	0.81286
55	0.32126	1.13049	1.17525	145	0.67196	0.49273	0.83326
60	0.04762	0.89331	0.89458	150	0.60520	0.58784	0.84369
65	-0.19805	0.62001	0.65087	155	0.53295	0.65806	0.84680
70	-0.39666	0.32515	0.51289	160	0.46345	0.70722	0.84555
75	-0.53188	0.02783	0.53260	165	0.40326	0.73967	0.84245
80	-0.59282	-0.25035	0.64352	170	0.35713	0.75950	0.83928
85	-0.57608	-0.48805	0.75503	175	0.32827	0.77000	0.83705
				180	0.31845	0.77326	0.83627

IEN

RADIATION PATTERN FOR THETA=PI/2

H/LMDA=0.2500 B/LMDA= 0.250 OMEGA= 8.53

5 ELEMENT ARRAY

PHI DEG	ER(THETA,PHI) RE	IM	ABSVAL	PHI DEG	ER(THETA,PHI) RE	IM	ABSVAL
0	2.82077	5.41440	6.10512	90	-0.51327	-1.13362	1.24440
5	2.82219	5.41169	6.10337	95	-0.64717	-1.29446	1.44723
10	2.82525	5.40128	6.09556	100	-0.69465	-1.29692	1.47124
15	2.82631	5.37641	6.07402	105	-0.66261	-1.16128	1.33702
20	2.81946	5.32623	6.02645	110	-0.56507	-0.91944	1.07920
25	2.79683	5.23655	5.93665	115	-0.42067	-0.60956	0.74062
30	2.74910	5.09102	5.78585	120	-0.24961	-0.27038	0.36798
35	2.66626	4.87286	5.55462	125	-0.07098	0.06377	0.09542
40	2.53883	4.56717	5.22539	130	0.09948	0.36647	0.37973
45	2.35921	4.16365	4.78559	135	0.25052	0.62078	0.66942
50	2.12333	3.65947	4.23087	140	0.37575	0.81916	0.90123
55	1.83211	3.06176	3.56805	145	0.47315	0.96202	1.07207
60	1.49257	2.38906	2.81698	150	0.54412	1.05557	1.18756
65	1.11824	1.67141	2.01099	155	0.59233	1.10950	1.25771
70	0.72849	0.94827	1.19579	160	0.62260	1.13476	1.29434
75	0.34678	0.26468	0.43625	165	0.63994	1.14199	1.30907
80	-0.00197	-0.33430	0.33430	170	0.64880	1.14027	1.31193
85	-0.29468	-0.80955	0.86152	175	0.65267	1.13642	1.31051
				180	0.65372	1.13467	1.30951

VEN

RADIATION PATTERN FOR THETA=PI/2

H/LMDA=0.2500 B/LMDA= 0.250 OMEGA= 8.53

6 ELEMENT ARRAY

PHI DEG	ER(THETA,PHI) RE	IM	ABSVAL	PHI DEG	ER(THETA,PHI) RE	IM	ABSVAL
0	2.37616	-0.34205	2.40065	90	-0.52873	-0.43635	0.68553
2	2.37545	-0.34012	2.39967	92	-0.44747	-0.47798	0.65475
4	2.37327	-0.33433	2.39670	94	-0.35612	-0.50935	0.62149
6	2.36949	-0.32469	2.39163	96	-0.25722	-0.52974	0.58889
8	2.36389	-0.31121	2.38429	98	-0.15342	-0.53875	0.56017
10	2.35618	-0.29394	2.37444	100	-0.04740	-0.53626	0.53835
12	2.34597	-0.27291	2.36179	102	0.05824	-0.52244	0.52567
14	2.33281	-0.24820	2.34597	104	0.16100	-0.49773	0.52312
16	2.31618	-0.21989	2.32659	106	0.25858	-0.46284	0.53018
18	2.29551	-0.18812	2.30320	108	0.34895	-0.41870	0.54505
20	2.27016	-0.15304	2.27532	110	0.43037	-0.36642	0.56522
22	2.23949	-0.11489	2.24244	112	0.50142	-0.30725	0.58807
24	2.20282	-0.07392	2.20406	114	0.56106	-0.24257	0.61125
26	2.15948	-0.03046	2.15969	116	0.60858	-0.17377	0.63290
28	2.10879	0.01508	2.10884	118	0.64364	-0.10229	0.65171
30	2.05016	0.06225	2.05110	120	0.66622	-0.02951	0.66688
32	1.98303	0.11049	1.98610	122	0.67663	0.04323	0.67801
34	1.90693	0.15918	1.91356	124	0.67542	0.11473	0.68510
36	1.82154	0.20765	1.83334	126	0.66339	0.18389	0.68840
38	1.72665	0.25512	1.74540	128	0.64151	0.24975	0.68841
40	1.62224	0.30078	1.64989	130	0.61089	0.31153	0.68574
42	1.50847	0.34376	1.54714	132	0.57274	0.36861	0.68111
44	1.38573	0.38316	1.43773	134	0.52832	0.42053	0.67525
46	1.25463	0.41807	1.32245	136	0.47888	0.46699	0.66889
48	1.11604	0.44758	1.20245	138	0.42567	0.50786	0.66266
50	0.97108	0.47082	1.07919	140	0.36987	0.54312	0.65711
52	0.82108	0.48698	0.95463	142	0.31258	0.57291	0.65264
54	0.66767	0.49534	0.83135	144	0.25482	0.59744	0.64951
56	0.51264	0.49531	0.71283	146	0.19748	0.61703	0.64787
58	0.35799	0.48645	0.60398	148	0.14136	0.63208	0.64769
60	0.20588	0.46847	0.51172	150	0.08711	0.64300	0.64888
62	0.05854	0.44133	0.44519	152	0.03529	0.65029	0.65125
64	-0.08174	0.40516	0.41332	154	-0.01366	0.65443	0.65457
66	-0.21268	0.36035	0.41843	156	-0.05938	0.65592	0.65861
68	-0.33211	0.30752	0.45262	158	-0.10164	0.65526	0.66310
70	-0.43800	0.24753	0.50310	160	-0.14025	0.65292	0.66782
72	-0.52852	0.18146	0.55881	162	-0.17512	0.64936	0.67256
74	-0.60217	0.11059	0.61224	164	-0.20618	0.64500	0.67715
76	-0.65775	0.03639	0.65875	166	-0.23344	0.64022	0.68145
78	-0.69446	-0.03955	0.69558	168	-0.25692	0.63537	0.68535
80	-0.71193	-0.11553	0.72124	170	-0.27665	0.63075	0.68875
82	-0.71023	-0.18980	0.73516	172	-0.29270	0.62661	0.69160
84	-0.68989	-0.26064	0.73749	174	-0.30511	0.62318	0.69386
86	-0.65187	-0.32637	0.72901	176	-0.31393	0.62061	0.69549
88	-0.59756	-0.38541	0.71107	178	-0.31920	0.61902	0.69647
				180	-0.32096	0.61848	0.69680

IEN

RADIATION PATTERN FOR THETA=PI/2

H/LMDA=0.2500 B/LMDA= 0.250 OMEGA= 8.53

6 ELEMENT ARRAY

PHI DEG	ER(THETA,PHI) RE	IM	ABSVAL	PHI DEG	ER(THETA,PHI) RE	IM	ABSVAL
0	6.99617	4.57630	8.35996	90	-1.59919	-0.98051	1.87585
2	6.99622	4.57543	8.35953	92	-1.53673	-0.90883	1.78536
4	6.99627	4.57275	8.35810	94	-1.43730	-0.81708	1.65331
6	6.99597	4.56804	8.35527	96	-1.30533	-0.70862	1.48527
8	6.99478	4.56094	8.35040	98	-1.14581	-0.58709	1.28747
10	6.99192	4.55096	8.34256	100	-0.96419	-0.45628	1.06671
12	6.98641	4.53748	8.33059	102	-0.76612	-0.31999	0.83027
14	6.97707	4.51974	8.31310	104	-0.55733	-0.18194	0.58628
16	6.96253	4.49688	8.28847	106	-0.34345	-0.04567	0.34648
18	6.94121	4.46793	8.25487	108	-0.12984	0.08557	0.15551
20	6.91142	4.43184	8.21029	110	0.07850	0.20889	0.22316
22	6.87132	4.38748	8.15260	112	0.27713	0.32183	0.42470
24	6.81893	4.33369	8.07952	114	0.46218	0.42236	0.62610
26	6.75225	4.26928	7.98872	116	0.63048	0.50897	0.81028
28	6.66922	4.19309	7.87784	118	0.77951	0.58061	0.97198
30	6.56780	4.10397	7.74459	120	0.90752	0.63674	1.10861
32	6.44603	4.00089	7.58672	122	1.01342	0.67724	1.21888
34	6.30206	3.88290	7.40222	124	1.09680	0.70239	1.30243
36	6.13428	3.74926	7.18932	126	1.15784	0.71287	1.35970
38	5.94127	3.59939	6.94653	128	1.19725	0.70962	1.39175
40	5.72200	3.43299	6.67284	130	1.21622	0.69384	1.40021
42	5.47582	3.25004	6.36768	132	1.21628	0.66690	1.38711
44	5.20253	3.05084	6.03108	134	1.19925	0.63029	1.35479
46	4.90247	2.83605	5.66369	136	1.16715	0.58556	1.30580
48	4.57657	2.60672	5.26687	138	1.12212	0.53429	1.24283
50	4.22634	2.36430	4.84271	140	1.06634	0.47801	1.16858
52	3.85399	2.11062	4.39408	142	1.00197	0.41818	1.08573
54	3.46232	1.84794	3.92461	144	0.93111	0.35616	0.99690
56	3.05482	1.57886	3.43871	146	0.85572	0.29320	0.90456
58	2.63557	1.30634	2.94155	148	0.77764	0.23042	0.81106
60	2.20917	1.03362	2.43902	150	0.69852	0.16878	0.71862
62	1.78073	0.76416	1.93776	152	0.61983	0.10912	0.62936
64	1.35568	0.50153	1.44548	154	0.54287	0.05210	0.54536
66	0.93973	0.24941	0.97226	156	0.46873	-0.00170	0.46873
68	0.53864	0.01138	0.53876	158	0.39834	-0.05186	0.40170
70	0.15816	-0.20910	0.26217	160	0.33244	-0.09805	0.34660
72	-0.19620	-0.40882	0.45346	162	0.27166	-0.14004	0.30563
74	-0.51932	-0.58494	0.78220	164	0.21647	-0.17768	0.28005
76	-0.80660	-0.73506	1.09129	166	0.16722	-0.21088	0.26913
78	-1.05414	-0.85732	1.35875	168	0.12419	-0.23959	0.26987
80	-1.25884	-0.95044	1.57734	170	0.08757	-0.26382	0.27797
82	-1.41854	-1.01379	1.74357	172	0.05748	-0.28358	0.28935
84	-1.53205	-1.04744	1.85588	174	0.03401	-0.29890	0.30083
86	-1.59923	-1.05208	1.91427	176	0.01723	-0.30981	0.31029
88	-1.62098	-1.02911	1.92006	178	0.00714	-0.31635	0.31643
				180	0.00378	-0.31852	0.31855

VEN

RADIATION PATTERN FOR THETA=PI/2

H/LMDA=0.2500 B/LMDA= 0.250 OMEGA= 8.53

7 ELEMENT ARRAY

PHI DEG	ER(THETA,PHI) RE	IM	ABSVAL	PHI DEG	ER(THETA,PHI) RE	IM	ABSVAL
0	1.32752	-1.85813	2.28362	90	-0.35284	-0.27444	0.44701
2	1.32849	-1.85614	2.28258	92	-0.26042	-0.33982	0.42813
4	1.33137	-1.85015	2.27938	94	-0.16128	-0.39292	0.42473
6	1.33603	-1.84003	2.27391	96	-0.05867	-0.43198	0.43594
8	1.34225	-1.82559	2.26592	98	0.04414	-0.45576	0.45789
10	1.34974	-1.80657	2.25510	100	0.14397	-0.46365	0.48548
12	1.35810	-1.78266	2.24104	102	0.23786	-0.45561	0.51397
14	1.36686	-1.75348	2.22328	104	0.32317	-0.43222	0.53968
16	1.37546	-1.71863	2.20127	106	0.39766	-0.39456	0.56019
18	1.38326	-1.67770	2.17441	108	0.45957	-0.34417	0.57416
20	1.38954	-1.63025	2.14209	110	0.50761	-0.28298	0.58116
22	1.39351	-1.57591	2.10365	112	0.54103	-0.21316	0.58151
24	1.39430	-1.51432	2.05845	114	0.55959	-0.13709	0.57614
26	1.39101	-1.44520	2.00587	116	0.56352	-0.05719	0.56642
28	1.38271	-1.36839	1.94534	118	0.55349	0.02416	0.55402
30	1.36844	-1.28384	1.87640	120	0.53053	0.10467	0.54076
32	1.34727	-1.19168	1.79867	122	0.49599	0.18228	0.52843
34	1.31830	-1.09221	1.71196	124	0.45142	0.25519	0.51856
36	1.28072	-0.98594	1.61626	126	0.39854	0.32186	0.51228
38	1.23383	-0.87361	1.51180	128	0.33912	0.38113	0.51016
40	1.17707	-0.75622	1.39906	130	0.27495	0.43213	0.51219
42	1.11011	-0.63500	1.27889	132	0.20775	0.47435	0.51785
44	1.03281	-0.51141	1.15249	134	0.13912	0.50759	0.52631
46	0.94531	-0.38715	1.02152	136	0.07053	0.53191	0.53657
48	0.84808	-0.26415	0.88826	138	0.00326	0.54765	0.54766
50	0.74187	-0.14444	0.75580	140	-0.06161	0.55532	0.55873
52	0.62780	-0.03020	0.62853	142	-0.12321	0.55563	0.56912
54	0.50731	0.07635	0.51303	144	-0.18086	0.54938	0.57839
56	0.38218	0.17303	0.41953	146	-0.23409	0.53747	0.58624
58	0.25449	0.25775	0.36222	148	-0.28260	0.52084	0.59257
60	0.12658	0.32862	0.35216	150	-0.32625	0.50043	0.59739
62	0.00099	0.38405	0.38405	152	-0.36506	0.47717	0.60080
64	-0.11963	0.42278	0.43938	154	-0.39915	0.45195	0.60298
66	-0.23256	0.44400	0.50122	156	-0.42874	0.42559	0.60410
68	-0.33516	0.44740	0.55901	158	-0.45411	0.39886	0.60440
70	-0.42492	0.43318	0.60679	160	-0.47560	0.37244	0.60408
72	-0.49960	0.40212	0.64133	162	-0.49358	0.34694	0.60331
74	-0.55733	0.35556	0.66109	164	-0.50842	0.32289	0.60228
76	-0.59665	0.29537	0.66575	166	-0.52049	0.30073	0.60112
78	-0.61664	0.22387	0.65602	168	-0.53016	0.28084	0.59995
80	-0.61696	0.14381	0.63350	170	-0.53774	0.26354	0.59885
82	-0.59784	0.05824	0.60067	172	-0.54353	0.24906	0.59788
84	-0.56016	-0.02962	0.56095	174	-0.54777	0.23762	0.59709
86	-0.50536	-0.11647	0.51861	176	-0.55065	0.22934	0.59650
88	-0.43544	-0.19909	0.47879	178	-0.55232	0.22434	0.59615
				180	-0.55287	0.22266	0.59603

IEN

RADIATION PATTERN FOR THETA=PI/2

H/LMDA=0.2500 B/LMDA= 0.250 OMEGA= 8.53

7 ELEMENT ARRAY

PHI DEG	ER(THETA,PHI) RE	IM	ABSVAL	PHI DEG	ER(THETA,PHI) RE	IM	ABSVAL
0	5.76765	-2.91819	6.46387	90	-0.68770	0.45844	0.82650
2	5.76702	-2.91864	6.46350	92	-0.49436	0.36914	0.61697
4	5.76497	-2.91991	6.46226	94	-0.29312	0.27120	0.39933
6	5.76114	-2.92182	6.45971	96	-0.09130	0.16842	0.19158
8	5.75492	-2.92404	6.45516	98	0.10414	0.06457	0.12254
10	5.74542	-2.92613	6.44764	100	0.28687	-0.03674	0.28921
12	5.73156	-2.92752	6.43593	102	0.45135	-0.13223	0.47032
14	5.71202	-2.92752	6.41853	104	0.59305	-0.21901	0.63220
16	5.68528	-2.92531	6.39374	106	0.70853	-0.29469	0.76736
18	5.64965	-2.92000	6.35963	108	0.79548	-0.35742	0.87209
20	5.60326	-2.91057	6.31411	110	0.85277	-0.40595	0.94447
22	5.54416	-2.89593	6.25493	112	0.88037	-0.43958	0.98401
24	5.47030	-2.87492	6.17976	114	0.87926	-0.45819	0.99148
26	5.37963	-2.84636	6.08623	116	0.85130	-0.46217	0.96867
28	5.27011	-2.80904	5.97200	118	0.79912	-0.45234	0.91826
30	5.13983	-2.76177	5.83483	120	0.72586	-0.42993	0.84363
32	4.98707	-2.70341	5.67267	122	0.63508	-0.39642	0.74865
34	4.81035	-2.63293	5.48377	124	0.53054	-0.35352	0.63753
36	4.60856	-2.54943	5.26673	126	0.41603	-0.30305	0.51470
38	4.38104	-2.45222	5.02064	128	0.29525	-0.24686	0.38486
40	4.12761	-2.34082	4.74517	130	0.17169	-0.18677	0.25369
42	3.84877	-2.21507	4.44067	132	0.04847	-0.12451	0.13361
44	3.54562	-2.07511	4.10822	134	-0.07164	-0.06166	0.09452
46	3.22004	-1.92147	3.74976	136	-0.18635	0.00041	0.18635
48	2.87465	-1.75507	3.36806	138	-0.29380	0.06048	0.29996
50	2.51284	-1.57724	2.96682	140	-0.39262	0.11760	0.40985
52	2.13877	-1.38976	2.55064	142	-0.48187	0.17101	0.51131
54	1.75727	-1.19481	2.12499	144	-0.56103	0.22017	0.60269
56	1.37382	-0.99494	1.69625	146	-0.62996	0.26475	0.68333
58	0.99434	-0.79306	1.27187	148	-0.68880	0.30457	0.75313
60	0.62512	-0.59235	0.86120	150	-0.73798	0.33963	0.81238
62	0.27260	-0.39618	0.48090	152	-0.77813	0.37005	0.86164
64	-0.05686	-0.20798	0.21562	154	-0.81005	0.39606	0.90168
66	-0.35716	-0.03121	0.35852	156	-0.83461	0.41795	0.93341
68	-0.62272	0.13086	0.63632	158	-0.85277	0.43611	0.95781
70	-0.84867	0.27518	0.89217	160	-0.86550	0.45091	0.97592
72	-1.03110	0.39911	1.10565	162	-0.87376	0.46277	0.98874
74	-1.16723	0.50051	1.27002	164	-0.87847	0.47210	0.99729
76	-1.25552	0.57781	1.38210	166	-0.88048	0.47929	1.00248
78	-1.29581	0.63012	1.44089	168	-0.88058	0.48472	1.00518
80	-1.28930	0.65724	1.44715	170	-0.87947	0.48871	1.00613
82	-1.23859	0.65972	1.40333	172	-0.87773	0.49157	1.00601
84	-1.14757	0.63885	1.31341	174	-0.87587	0.49353	1.00534
86	-1.02129	0.59655	1.18276	176	-0.87426	0.49479	1.00456
88	-0.86575	0.53540	1.01793	178	-0.87318	0.49549	1.00397
				180	-0.87281	0.49572	1.00375

VEN

RADIATION PATTERN FOR THETA=PI/2

H/LMDA=0.2500 B/LMDA= 0.250 OMEGA= 8.53

8 ELEMENT ARRAY

PHI DEG	ER(THETA,PHI) RE	IM	ABSVAL	PHI DEG	ER(THETA,PHI) RE	IM	ABSVAL
0	-0.39940	-2.15163	2.18839	90	-0.22694	-0.32563	0.39691
2	-0.39716	-2.15092	2.18728	92	-0.15677	-0.40033	0.42993
4	-0.39045	-2.14871	2.18390	94	-0.07946	-0.45521	0.46209
6	-0.37926	-2.14479	2.17806	96	0.00174	-0.48831	0.48831
8	-0.36361	-2.13879	2.16948	98	0.08342	-0.49873	0.50566
10	-0.34353	-2.13021	2.15773	100	0.16219	-0.48665	0.51297
12	-0.31906	-2.11841	2.14231	102	0.23488	-0.45327	0.51052
14	-0.29027	-2.10266	2.12260	104	0.29864	-0.40070	0.49975
16	-0.25731	-2.08208	2.09792	106	0.35110	-0.33181	0.48308
18	-0.22036	-2.05575	2.06752	108	0.39044	-0.25004	0.46364
20	-0.17969	-2.02266	2.03062	110	0.41546	-0.15921	0.44492
22	-0.13565	-1.98178	1.98642	112	0.42562	-0.06329	0.43030
24	-0.08871	-1.93209	1.93412	114	0.42098	0.03378	0.42234
26	-0.03946	-1.87260	1.87302	116	0.40221	0.12832	0.42219
28	0.01138	-1.80245	1.80248	118	0.37047	0.21700	0.42934
30	0.06299	-1.72088	1.72203	120	0.32733	0.29699	0.44198
32	0.11436	-1.62737	1.63138	122	0.27468	0.36608	0.45767
34	0.16441	-1.52165	1.53050	124	0.21460	0.42263	0.47399
36	0.21191	-1.40375	1.41966	126	0.14926	0.46568	0.48902
38	0.25557	-1.27410	1.29948	128	0.08081	0.49485	0.50140
40	0.29406	-1.13353	1.17105	130	0.01131	0.51030	0.51043
42	0.32602	-0.98332	1.03595	132	-0.05738	0.51270	0.51590
44	0.35016	-0.82522	0.89644	134	-0.12361	0.50306	0.51802
46	0.36529	-0.66147	0.75564	136	-0.18604	0.48271	0.51732
48	0.37041	-0.49474	0.61804	138	-0.24356	0.45319	0.51449
50	0.36475	-0.32810	0.49061	140	-0.29542	0.41612	0.51032
52	0.34787	-0.16496	0.38500	142	-0.34112	0.37317	0.50558
54	0.31968	-0.00893	0.31981	144	-0.38041	0.32596	0.50097
56	0.28053	0.13627	0.31188	146	-0.41332	0.27604	0.49702
58	0.23121	0.26696	0.35317	148	-0.44003	0.22479	0.49412
60	0.17297	0.37970	0.41725	150	-0.46092	0.17346	0.49248
62	0.10755	0.47142	0.48354	152	-0.47648	0.12310	0.49212
64	0.03707	0.53961	0.54089	154	-0.48728	0.07458	0.49296
66	-0.03597	0.58248	0.58359	156	-0.49397	0.02861	0.49480
68	-0.10882	0.59905	0.60886	158	-0.49720	-0.01428	0.49740
70	-0.17856	0.58933	0.61579	160	-0.49764	-0.05370	0.50053
72	-0.24228	0.55430	0.60493	162	-0.49594	-0.08941	0.50393
74	-0.29716	0.49593	0.57815	164	-0.49269	-0.12126	0.50740
76	-0.34069	0.41718	0.53862	166	-0.48848	-0.14918	0.51075
78	-0.37074	0.32184	0.49095	168	-0.48379	-0.17318	0.51385
80	-0.38572	0.21440	0.44131	170	-0.47907	-0.19329	0.51659
82	-0.38471	0.09985	0.39746	172	-0.47469	-0.20960	0.51891
84	-0.36746	-0.01656	0.36784	174	-0.47097	-0.22218	0.52075
86	-0.33449	-0.12955	0.35870	176	-0.46815	-0.23110	0.52208
88	-0.28701	-0.23409	0.37037	178	-0.46638	-0.23642	0.52288
				180	-0.46578	-0.23820	0.52315

IEN

RADIATION PATTERN FOR THETA=PI/2

H/LMDA=0.2500 B/LMDA= 0.250 OMEGA= 8.53

8 ELEMENT ARRAY

PHI DEG	ER(THETA,PHI) RE	IM	ABSVAL	PHI DEG	ER(THETA,PHI) RE	IM	ABSVAL
0	4.27882	-7.12384	8.31008	90	0.16887	0.00537	0.16895
2	4.27759	-7.12393	8.30952	92	0.30673	-0.25384	0.39815
4	4.27380	-7.12397	8.30761	94	0.42304	-0.48781	0.64570
6	4.26708	-7.12337	8.30364	96	0.51367	-0.68739	0.85811
8	4.25683	-7.12111	8.29643	98	0.57592	-0.84547	1.02299
10	4.24222	-7.11578	8.28437	100	0.60856	-0.95729	1.13435
12	4.22219	-7.10561	8.26538	102	0.61181	-1.02046	1.18981
14	4.19551	-7.08841	8.23698	104	0.58725	-1.03495	1.18995
16	4.16074	-7.06172	8.19632	106	0.53763	-1.00296	1.13797
18	4.11631	-7.02273	8.14019	108	0.46665	-0.92859	1.03925
20	4.06053	-6.96839	8.06513	110	0.37874	-0.81754	0.90100
22	3.99163	-6.89546	7.96746	112	0.27872	-0.67665	0.73181
24	3.90786	-6.80058	7.84342	114	0.17158	-0.51354	0.54145
26	3.80749	-6.68036	7.68923	116	0.06218	-0.33607	0.34178
28	3.68894	-6.53152	7.50127	118	-0.04492	-0.15204	0.15854
30	3.55082	-6.35099	7.27622	120	-0.14572	0.03125	0.14903
32	3.39204	-6.13606	7.01122	122	-0.23683	0.20724	0.31470
34	3.21190	-5.88455	6.70405	124	-0.31564	0.37035	0.48661
36	3.01019	-5.59496	6.35333	126	-0.38029	0.51612	0.64109
38	2.78723	-5.26664	5.95870	128	-0.42969	0.64125	0.77190
40	2.54403	-4.89995	5.52101	130	-0.46346	0.74365	0.87625
42	2.28229	-4.49640	5.04247	132	-0.48189	0.82232	0.95312
44	2.00447	-4.05876	4.52674	134	-0.48579	0.87727	1.00280
46	1.71379	-3.59114	3.97911	136	-0.47641	0.90937	1.02661
48	1.41423	-3.09902	3.40646	138	-0.45529	0.92019	1.02667
50	1.11044	-2.58922	2.81730	140	-0.42423	0.91185	1.00570
52	0.80770	-2.06978	2.22179	142	-0.38507	0.88679	0.96679
54	0.51171	-1.54981	1.63210	144	-0.33970	0.84771	0.91324
56	0.22848	-1.03923	1.06405	146	-0.28993	0.79733	0.84841
58	-0.03594	-0.54848	0.54965	148	-0.23745	0.73838	0.77562
60	-0.27568	-0.08811	0.28942	150	-0.18379	0.67343	0.69806
62	-0.48528	0.33160	0.58775	152	-0.13027	0.60483	0.61870
64	-0.66001	0.70113	0.96291	154	-0.07802	0.53471	0.54037
66	-0.79607	1.01214	1.28769	156	-0.02796	0.46492	0.46576
68	-0.89084	1.25793	1.54142	158	0.01918	0.39702	0.39749
70	-0.94304	1.43377	1.71611	160	0.06286	0.33231	0.33821
72	-0.95287	1.53727	1.80864	162	0.10268	0.27184	0.29058
74	-0.92205	1.56854	1.81948	164	0.13840	0.21640	0.25687
76	-0.85380	1.53028	1.75235	166	0.16986	0.16662	0.23794
78	-0.75273	1.42773	1.61401	168	0.19702	0.12294	0.23223
80	-0.62465	1.26851	1.41396	170	0.21986	0.08565	0.23596
82	-0.47633	1.06226	1.16417	172	0.23843	0.05498	0.24469
84	-0.31518	0.82026	0.87873	174	0.25278	0.03104	0.25468
86	-0.14892	0.55487	0.57451	176	0.26298	0.01390	0.26335
88	0.01484	0.27897	0.27937	178	0.26908	0.00361	0.26910
				180	0.27111	0.00018	0.27111

VEN

RADIATION PATTERN FOR THETA=PI/2

H/LMDA=0.2500 B/LMDA= 0.250 OMEGA= 8.53

9 ELEMENT ARRAY

PHI DEG	ER(THETA,PHI) RE	IM	ABSVAL	PHI DEG	ER(THETA,PHI) RE	IM	ABSVAL
0	-1.75679	-1.16667	2.10889	90	-0.27519	-0.35888	0.45225
2	-1.75461	-1.16784	2.10773	92	-0.21408	-0.41015	0.46266
4	-1.74801	-1.17132	2.10417	94	-0.13949	-0.43826	0.45992
6	-1.73680	-1.17689	2.09799	96	-0.05600	-0.44222	0.44575
8	-1.72067	-1.18425	2.08882	98	0.03137	-0.42238	0.42354
10	-1.69919	-1.19294	2.07614	100	0.11749	-0.38040	0.39813
12	-1.67185	-1.20239	2.05932	102	0.19743	-0.31911	0.37525
14	-1.63806	-1.21188	2.03762	104	0.26678	-0.24225	0.36036
16	-1.59719	-1.22058	2.01018	106	0.32192	-0.15422	0.35695
18	-1.54858	-1.22752	1.97609	108	0.36015	-0.05981	0.36509
20	-1.49164	-1.23162	1.93439	110	0.37988	0.03612	0.38159
22	-1.42581	-1.23169	1.88415	112	0.38057	0.12894	0.40182
24	-1.35067	-1.22649	1.82444	114	0.36276	0.21446	0.42141
26	-1.26598	-1.21471	1.75448	116	0.32792	0.28918	0.43721
28	-1.17170	-1.19504	1.67362	118	0.27830	0.35034	0.44742
30	-1.06812	-1.16622	1.58144	120	0.21673	0.39609	0.45151
32	-0.95581	-1.12710	1.47781	122	0.14640	0.42546	0.44994
34	-0.83576	-1.07670	1.36300	124	0.07068	0.43829	0.44395
36	-0.70934	-1.01427	1.23771	126	-0.00715	0.43520	0.43526
38	-0.57835	-0.93942	1.10318	128	-0.08398	0.41744	0.42580
40	-0.44500	-0.85214	0.96134	130	-0.15710	0.38673	0.41742
42	-0.31189	-0.75290	0.81494	132	-0.22421	0.34515	0.41158
44	-0.18192	-0.64269	0.66795	134	-0.28355	0.29497	0.40916
46	-0.05824	-0.52310	0.52633	136	-0.33390	0.23850	0.41033
48	0.05588	-0.39626	0.40018	138	-0.37452	0.17802	0.41468
50	0.15723	-0.26491	0.30805	140	-0.40519	0.11563	0.42136
52	0.24275	-0.13228	0.27645	142	-0.42609	0.05319	0.42940
54	0.30980	-0.00203	0.30980	144	-0.43777	-0.00770	0.43783
56	0.35625	0.12189	0.37653	146	-0.44102	-0.06575	0.44590
58	0.38074	0.23543	0.44765	148	-0.43687	-0.11998	0.45305
60	0.38274	0.33458	0.50836	150	-0.42643	-0.16969	0.45896
62	0.36272	0.41566	0.55167	152	-0.41090	-0.21447	0.46351
64	0.32217	0.47551	0.57438	154	-0.39146	-0.25414	0.46672
66	0.26362	0.51173	0.57564	156	-0.36925	-0.28872	0.46873
68	0.19055	0.52284	0.55648	158	-0.34535	-0.31838	0.46972
70	0.10726	0.50846	0.51966	160	-0.32073	-0.34344	0.46991
72	0.01866	0.46944	0.46981	162	-0.29625	-0.36427	0.46953
74	-0.07000	0.40779	0.41376	164	-0.27265	-0.38131	0.46876
76	-0.15343	0.32675	0.36098	166	-0.25057	-0.39501	0.46778
78	-0.22661	0.23055	0.32327	168	-0.23052	-0.40583	0.46673
80	-0.28511	0.12429	0.31102	170	-0.21294	-0.41419	0.46572
82	-0.32538	0.01361	0.32566	172	-0.19814	-0.42047	0.46481
84	-0.34500	-0.09559	0.35799	174	-0.18640	-0.42499	0.46407
86	-0.34281	-0.19752	0.39564	176	-0.17788	-0.42803	0.46353
88	-0.31903	-0.28681	0.42900	178	-0.17273	-0.42978	0.46319
				180	-0.17100	-0.43035	0.46308

IEN

RADIATION PATTERN FOR THETA=PI/2

H/LMDA=0.2500 B/LMDA= 0.250 OMEGA= 8.53

9 ELEMENT ARRAY

PHI DEG	ER(THETA,PHI) RE	IM	ABSVAL	PHI DEG	ER(THETA,PHI) RE	IM	ABSVAL
0	-2.99144	-6.02062	6.72284	90	-0.27644	-0.70564	0.75785
2	-2.99212	-6.01973	6.72235	92	-0.35000	-0.79838	0.87173
4	-2.99405	-6.01686	6.72064	94	-0.39878	-0.83903	0.92898
6	-2.99690	-6.01136	6.71698	96	-0.42132	-0.82803	0.92906
8	-3.00013	-6.00213	6.71016	98	-0.41777	-0.76880	0.87497
10	-3.00298	-5.98769	6.69852	100	-0.38975	-0.66736	0.77284
12	-3.00447	-5.96614	6.67994	102	-0.34024	-0.53186	0.63138
14	-3.00345	-5.93520	6.65187	104	-0.27324	-0.37187	0.46146
16	-2.99853	-5.89229	6.61138	106	-0.19346	-0.19771	0.27662
18	-2.98819	-5.83453	6.55523	108	-0.10603	-0.01981	0.10786
20	-2.97072	-5.75882	6.47990	110	-0.01611	0.15198	0.15283
22	-2.94432	-5.66193	6.38173	112	0.07139	0.30892	0.31706
24	-2.90712	-5.54062	6.25698	114	0.15211	0.44379	0.46914
26	-2.85722	-5.39177	6.10204	116	0.22240	0.55119	0.59437
28	-2.79277	-5.21249	5.91351	118	0.27950	0.62763	0.68705
30	-2.71206	-5.00031	5.68844	120	0.32158	0.67157	0.74459
32	-2.61361	-4.75337	5.42453	122	0.34773	0.68327	0.76667
34	-2.49625	-4.47061	5.12031	124	0.35794	0.66455	0.75482
36	-2.35924	-4.15193	4.77540	126	0.35294	0.61852	0.71213
38	-2.20236	-3.79837	4.39067	128	0.33411	0.54922	0.64286
40	-2.02604	-3.41231	3.96847	130	0.30328	0.46128	0.55205
42	-1.83140	-2.99754	3.51273	132	0.26259	0.35963	0.44530
44	-1.62035	-2.55929	3.02910	134	0.21436	0.24913	0.32866
46	-1.39560	-2.10430	2.52502	136	0.16089	0.13439	0.20964
48	-1.16067	-1.64067	2.00972	138	0.10440	0.01954	0.10621
50	-0.91984	-1.17771	1.49436	140	0.04689	-0.09188	0.10315
52	-0.67805	-0.72564	0.99313	142	-0.00989	-0.19702	0.19726
54	-0.44075	-0.29526	0.53050	144	-0.06448	-0.29370	0.30070
56	-0.21370	0.10253	0.23702	146	-0.11575	-0.38046	0.39768
58	-0.00274	0.45723	0.45724	148	-0.16288	-0.45645	0.48465
60	0.18647	0.75926	0.78182	150	-0.20535	-0.52138	0.56037
62	0.34880	1.00056	1.05961	152	-0.24289	-0.57544	0.62460
64	0.47989	1.17507	1.26928	154	-0.27546	-0.61918	0.67769
66	0.57645	1.27923	1.40311	156	-0.30319	-0.65344	0.72035
68	0.63650	1.31227	1.45849	158	-0.32637	-0.67926	0.75360
70	0.65954	1.27643	1.43675	160	-0.34539	-0.69780	0.77860
72	0.64663	1.17698	1.34291	162	-0.36068	-0.71023	0.79657
74	0.60045	1.02206	1.18539	164	-0.37273	-0.71777	0.80878
76	0.52516	0.82229	0.97568	166	-0.38202	-0.72153	0.81642
78	0.42621	0.59031	0.72809	168	-0.38902	-0.72257	0.82064
80	0.31010	0.34006	0.46022	170	-0.39415	-0.72183	0.82243
82	0.18399	0.08599	0.20309	172	-0.39780	-0.72010	0.82267
84	0.05532	-0.15771	0.16713	174	-0.40030	-0.71805	0.82209
86	-0.06863	-0.37797	0.38415	176	-0.40189	-0.71622	0.82127
88	-0.18114	-0.56354	0.59194	178	-0.40277	-0.71497	0.82061
				180	-0.40306	-0.71453	0.82037

VEN

RADIATION PATTERN FOR THETA=PI/2

H/LMDA=0.2500 B/LMDA= 0.250 OMEGA= 8.53

10 ELEMENT ARRAY

PHI DEG	ER(THETA,PHI) RE	IM	ABSVAL	PHI DEG	ER(THETA,PHI) RE	IM	ABSVAL
0	-1.99514	0.43112	2.04118	90	-0.30830	-0.25839	0.40226
2	-1.99444	0.42862	2.03998	92	-0.22451	-0.29908	0.37397
4	-1.99223	0.42112	2.03625	94	-0.12563	-0.31988	0.34367
6	-1.98820	0.40862	2.02975	96	-0.01906	-0.31937	0.31993
8	-1.98179	0.39111	2.02001	98	0.08748	-0.29761	0.31020
10	-1.97228	0.36859	2.00642	100	0.18652	-0.25616	0.31687
12	-1.95873	0.34111	1.98821	102	0.27143	-0.19790	0.33591
14	-1.94005	0.30876	1.96447	104	0.33684	-0.12676	0.35991
16	-1.91501	0.27170	1.93419	106	0.37901	-0.04745	0.38197
18	-1.88225	0.23022	1.89627	108	0.39595	0.03498	0.39749
20	-1.84037	0.18470	1.84961	110	0.38749	0.11547	0.40433
22	-1.78796	0.13570	1.79310	112	0.35514	0.18930	0.40244
24	-1.72370	0.08396	1.72575	114	0.30187	0.25242	0.39350
26	-1.64640	0.03039	1.64668	116	0.23175	0.30166	0.38040
28	-1.55512	-0.02386	1.55530	118	0.14960	0.33486	0.36676
30	-1.44926	-0.07747	1.45133	120	0.06057	0.35094	0.35613
32	-1.32868	-0.12892	1.33491	122	-0.03022	0.34988	0.35118
34	-1.19376	-0.17653	1.20674	124	-0.11802	0.33255	0.35288
36	-1.04554	-0.21853	1.06814	126	-0.19872	0.30067	0.36040
38	-0.88577	-0.25312	0.92123	128	-0.26900	0.25649	0.37169
40	-0.71693	-0.27859	0.76915	130	-0.32648	0.20269	0.38428
42	-0.54225	-0.29337	0.61652	132	-0.36969	0.14210	0.39606
44	-0.36569	-0.29619	0.47059	134	-0.39805	0.07758	0.40554
46	-0.19181	-0.28620	0.34453	136	-0.41180	0.01178	0.41197
48	-0.02560	-0.26310	0.26434	138	-0.41182	-0.05291	0.41520
50	0.12768	-0.22720	0.26061	140	-0.39950	-0.11447	0.41558
52	0.26281	-0.17953	0.31828	142	-0.37660	-0.17132	0.41374
54	0.37494	-0.12189	0.39425	144	-0.34507	-0.22231	0.41048
56	0.45983	-0.05676	0.46332	146	-0.30693	-0.26673	0.40663
58	0.51430	0.01268	0.51446	148	-0.26414	-0.30423	0.40290
60	0.53643	0.08274	0.54278	150	-0.21856	-0.33482	0.39984
62	0.52587	0.14940	0.54668	152	-0.17183	-0.35879	0.39781
64	0.48395	0.20857	0.52699	154	-0.12536	-0.37663	0.39694
66	0.41380	0.25634	0.48676	156	-0.08031	-0.38899	0.39720
68	0.32018	0.28928	0.43151	158	-0.03760	-0.39664	0.39842
70	0.20935	0.30478	0.36975	160	0.00209	-0.40038	0.40038
72	0.08870	0.30122	0.31401	162	0.03827	-0.40101	0.40283
74	-0.03372	0.27824	0.28027	164	0.07065	-0.39931	0.40551
76	-0.14971	0.23678	0.28014	166	0.09906	-0.39603	0.40823
78	-0.25151	0.17913	0.30878	168	0.12347	-0.39182	0.41081
80	-0.33234	0.10881	0.34970	170	0.14389	-0.38727	0.41313
82	-0.38694	0.03032	0.38813	172	0.16041	-0.38287	0.41511
84	-0.41193	-0.05110	0.41509	174	0.17311	-0.37903	0.41669
86	-0.40612	-0.12990	0.42639	176	0.18210	-0.37606	0.41783
88	-0.37051	-0.20065	0.42136	178	0.18746	-0.37419	0.41852
				180	0.18924	-0.37356	0.41876

IEN

RADIATION PATTERN FOR THETA=PI/2

H/LMDA=0.2500 B/LMDA= 0.250 OMEGA= 8.53

10 ELEMENT ARRAY

PHI DEG	ER(THETA,PHI) RE	IM	ABSVAL	PHI DEG	ER(THETA,PHI) RE	IM	ABSVAL
0	-7.23732	-4.09980	8.31788	90	-0.97527	-0.52601	1.10808
2	-7.23744	-4.09820	8.31720	92	-0.91458	-0.45367	1.02092
4	-7.23747	-4.09322	8.31477	94	-0.79012	-0.35204	0.86500
6	-7.23645	-4.08431	8.30950	96	-0.61420	-0.23000	0.65585
8	-7.23276	-4.07056	8.29954	98	-0.40214	-0.09739	0.41376
10	-7.22417	-4.05074	8.28234	100	-0.17093	0.03572	0.17462
12	-7.20783	-4.02328	8.25467	102	0.06214	0.15990	0.17154
14	-7.18033	-3.98632	8.21266	104	0.28076	0.26695	0.38741
16	-7.13774	-3.93776	8.15189	106	0.47077	0.35049	0.58692
18	-7.07570	-3.87534	8.06745	108	0.62104	0.40625	0.74211
20	-6.98950	-3.79665	7.95410	110	0.72398	0.43222	0.84318
22	-6.87424	-3.69928	7.80639	112	0.77581	0.42862	0.88634
24	-6.72497	-3.58094	7.61894	114	0.77645	0.39766	0.87236
26	-6.53693	-3.43952	7.38659	116	0.72918	0.34316	0.80590
28	-6.30581	-3.27332	7.10477	118	0.64003	0.27017	0.69471
30	-6.02796	-3.08116	6.76977	120	0.51705	0.18440	0.54895
32	-5.70079	-2.86254	6.37912	122	0.36955	0.09182	0.38079
34	-5.32300	-2.61787	5.93191	124	0.20731	-0.00180	0.20732
36	-4.89496	-2.34854	5.42921	126	0.03988	-0.09127	0.09961
38	-4.41897	-2.05709	4.87431	128	-0.12400	-0.17223	0.21222
40	-3.89948	-1.74728	4.27304	130	-0.27686	-0.24129	0.36725
42	-3.34327	-1.42411	3.63394	132	-0.41278	-0.29617	0.50804
44	-2.75949	-1.09378	2.96836	134	-0.52748	-0.33560	0.62519
46	-2.15958	-0.76356	2.29059	136	-0.61837	-0.35931	0.71518
48	-1.55699	-0.44157	1.61839	138	-0.68441	-0.36785	0.77700
50	-0.96678	-0.13647	0.97636	140	-0.72591	-0.36244	0.81136
52	-0.40504	0.14293	0.42952	142	-0.74430	-0.34477	0.82027
54	0.11185	0.38814	0.40393	144	-0.74184	-0.31686	0.80667
56	0.56811	0.59149	0.82013	146	-0.72137	-0.28085	0.77411
58	0.94949	0.74668	1.20791	148	-0.68603	-0.23892	0.72645
60	1.24422	0.84920	1.50640	150	-0.63909	-0.19312	0.66763
62	1.44395	0.89685	1.69981	152	-0.58371	-0.14531	0.60153
64	1.54447	0.88996	1.78253	154	-0.52288	-0.09712	0.53183
66	1.54627	0.83160	1.75571	156	-0.45926	-0.04993	0.46197
68	1.45479	0.72750	1.62655	158	-0.39517	-0.00483	0.39520
70	1.28036	0.58590	1.40804	160	-0.33256	0.03736	0.33465
72	1.03773	0.41706	1.11840	162	-0.27299	0.07602	0.28338
74	0.74533	0.23269	0.78081	164	-0.21772	0.11077	0.24427
76	0.42412	0.04521	0.42652	166	-0.16765	0.14137	0.21930
78	0.09625	-0.13307	0.16423	168	-0.12347	0.16773	0.20827
80	-0.21633	-0.29078	0.36242	170	-0.08564	0.18984	0.20826
82	-0.49347	-0.41830	0.64690	172	-0.05444	0.20774	0.21476
84	-0.71813	-0.50838	0.87986	174	-0.03007	0.22154	0.22357
86	-0.87754	-0.55667	1.03921	176	-0.01262	0.23131	0.23165
88	-0.96399	-0.56193	1.11581	178	-0.00214	0.23714	0.23715
				180	0.00136	0.23907	0.23908

VEN

RADIATION PATTERN FOR THETA=PI/2

H/LMDA=0.2500 B/LMDA= 0.250 OMEGA= 8.53

12 ELEMENT ARRAY

PHI DEG	ER(THETA,PHI) RE	IM	ABSVAL	PHI DEG	ER(THETA,PHI) RE	IM	ABSVAL
0	0.45046	1.87798	1.93125	90	-0.15656	-0.21871	0.26897
2	0.44773	1.87729	1.92995	92	-0.08246	-0.28753	0.29912
4	0.43954	1.87510	1.92593	94	0.00046	-0.32518	0.32518
6	0.42586	1.87094	1.91880	96	0.08398	-0.32844	0.33901
8	0.40666	1.86409	1.90794	98	0.15990	-0.29777	0.33799
10	0.38193	1.85355	1.89249	100	0.22088	-0.23711	0.32405
12	0.35169	1.83804	1.87138	102	0.26119	-0.15330	0.30286
14	0.31605	1.81608	1.84338	104	0.27725	-0.05527	0.28270
16	0.27522	1.78601	1.80709	106	0.26790	0.04708	0.27201
18	0.22958	1.74606	1.76109	108	0.23444	0.14390	0.27508
20	0.17970	1.69441	1.70391	110	0.18034	0.22640	0.28945
22	0.12639	1.62932	1.63422	112	0.11078	0.28767	0.30826
24	0.07072	1.54924	1.55086	114	0.03205	0.32317	0.32476
26	0.01405	1.45295	1.45302	116	-0.04913	0.33100	0.33462
28	-0.04197	1.33970	1.34036	118	-0.12625	0.31180	0.33639
30	-0.09543	1.20942	1.21318	120	-0.19361	0.26846	0.33099
32	-0.14421	1.06282	1.07256	122	-0.24670	0.20560	0.32114
34	-0.18608	0.90159	0.92059	124	-0.28246	0.12900	0.31052
36	-0.21879	0.72843	0.76057	126	-0.29942	0.04489	0.30276
38	-0.24025	0.54715	0.59757	128	-0.29758	-0.04054	0.30033
40	-0.24872	0.36262	0.43972	130	-0.27827	-0.12173	0.30373
42	-0.24296	0.18066	0.30276	132	-0.24382	-0.19406	0.31162
44	-0.22248	0.00776	0.22261	134	-0.19727	-0.25407	0.32166
46	-0.18766	-0.14920	0.23974	136	-0.14203	-0.29958	0.33154
48	-0.13991	-0.28342	0.31607	138	-0.08160	-0.32961	0.33956
50	-0.08170	-0.38868	0.39717	140	-0.01926	-0.34427	0.34481
52	-0.01654	-0.45990	0.46020	142	0.04208	-0.34457	0.34713
54	0.05118	-0.49370	0.49635	144	0.10002	-0.33215	0.34688
56	0.11646	-0.48887	0.50255	146	0.15275	-0.30911	0.34479
58	0.17405	-0.44670	0.47941	148	0.19902	-0.27774	0.34169
60	0.21888	-0.37119	0.43092	150	0.23817	-0.24038	0.33839
62	0.24661	-0.26890	0.36486	152	0.27000	-0.19924	0.33555
64	0.25407	-0.14867	0.29437	154	0.29471	-0.15629	0.33358
66	0.23972	-0.02095	0.24063	156	0.31282	-0.11324	0.33268
68	0.20396	0.10299	0.22849	158	0.32507	-0.07146	0.33283
70	0.14925	0.21207	0.25933	160	0.33233	-0.03203	0.33387
72	0.08002	0.29649	0.30710	162	0.33553	0.00430	0.33556
74	0.00234	0.34867	0.34868	164	0.33561	0.03700	0.33764
76	-0.07657	0.36412	0.37209	166	0.33345	0.06578	0.33988
78	-0.14906	0.34196	0.37304	168	0.32989	0.09052	0.34209
80	-0.20786	0.28505	0.35279	170	0.32565	0.11120	0.34411
82	-0.24684	0.19979	0.31756	172	0.32133	0.12789	0.34585
84	-0.26176	0.09542	0.27861	174	0.31746	0.14070	0.34724
86	-0.25073	-0.01690	0.25130	176	0.31442	0.14975	0.34826
88	-0.21454	-0.12535	0.24848	178	0.31248	0.15513	0.34887
				180	0.31182	0.15691	0.34908

IEN

RADIATION PATTERN FOR THETA=PI/2

H/LMDA=0.2500 B/LMDA= 0.250 OMEGA= 8.53

12 ELEMENT ARRAY

PHI DEG	ER(THETA,PHI) RE	IM	ABSVAL	PHI DEG	ER(THETA,PHI) RE	IM	ABSVAL
0	-3.98014	7.33941	8.34916	90	0.13120	0.00357	0.13125
2	-3.97814	7.33957	8.34834	92	0.25501	-0.25618	0.36147
4	-3.97190	7.33957	8.34538	94	0.34559	-0.47560	0.58790
6	-3.96064	7.33799	8.33863	96	0.39540	-0.63424	0.74739
8	-3.94310	7.33249	8.32547	98	0.40157	-0.71910	0.82363
10	-3.91754	7.31979	8.30219	100	0.36594	-0.72557	0.81263
12	-3.88179	7.29574	8.26415	102	0.29456	-0.65739	0.72037
14	-3.83330	7.25541	8.20580	104	0.19674	-0.52561	0.56122
16	-3.76922	7.19310	8.12082	106	0.08382	-0.34694	0.35692
18	-3.68649	7.10260	8.00232	108	-0.03219	-0.14150	0.14511
20	-3.58201	6.97728	7.84304	110	-0.13988	0.06946	0.15617
22	-3.45279	6.81043	7.63569	112	-0.22959	0.26588	0.35129
24	-3.29615	6.59556	7.37333	114	-0.29423	0.43074	0.52164
26	-3.10993	6.32678	7.04981	116	-0.32967	0.55145	0.64248
28	-2.89281	5.99932	6.66034	118	-0.33486	0.62050	0.70509
30	-2.64450	5.60999	6.20205	120	-0.31159	0.63568	0.70794
32	-2.36608	5.15777	5.67459	122	-0.26392	0.59956	0.65508
34	-2.06016	4.64431	5.08074	124	-0.19758	0.51871	0.55507
36	-1.73112	4.07437	4.42688	126	-0.11922	0.40253	0.41981
38	-1.38513	3.45623	3.72345	128	-0.03565	0.26208	0.26449
40	-1.03018	2.80180	2.98519	130	0.04670	0.10886	0.11845
42	-0.67590	2.12661	2.23143	132	0.12233	-0.04617	0.13075
44	-0.33320	1.44940	1.48721	134	0.18693	-0.19342	0.26898
46	-0.01374	0.79148	0.79160	136	0.23756	-0.32518	0.40271
48	0.27064	0.17567	0.32266	138	0.27263	-0.43590	0.51413
50	0.50885	-0.37500	0.63211	140	0.29182	-0.52223	0.59823
52	0.69133	-0.83891	1.08707	142	0.29582	-0.58285	0.65362
54	0.81098	-1.19754	1.44630	144	0.28617	-0.61819	0.68121
56	0.86393	-1.43717	1.67685	146	0.26490	-0.63008	0.68350
58	0.85022	-1.55046	1.76828	148	0.23439	-0.62132	0.66406
60	0.77412	-1.53757	1.72145	150	0.19709	-0.59533	0.62711
62	0.64415	-1.40679	1.54725	152	0.15535	-0.55581	0.57711
64	0.47267	-1.17449	1.26604	154	0.11132	-0.50643	0.51852
66	0.27516	-0.86429	0.90703	156	0.06685	-0.45067	0.45560
68	0.06896	-0.50542	0.51010	158	0.02348	-0.39164	0.39234
70	-0.12806	-0.13055	0.18287	160	-0.01762	-0.33198	0.33245
72	-0.29913	0.22704	0.37553	162	-0.05557	-0.27392	0.27950
74	-0.43013	0.53637	0.68753	164	-0.08981	-0.21918	0.23686
76	-0.51092	0.77167	0.92548	166	-0.11998	-0.16907	0.20732
78	-0.53634	0.91476	1.06040	168	-0.14593	-0.12454	0.19185
80	-0.50670	0.95682	1.08270	170	-0.16763	-0.08624	0.18851
82	-0.42766	0.89911	0.99564	172	-0.18514	-0.05460	0.19303
84	-0.30962	0.75267	0.81386	174	-0.19859	-0.02984	0.20082
86	-0.16650	0.53696	0.56218	176	-0.20808	-0.01210	0.20843
88	-0.01417	0.27762	0.27798	178	-0.21372	-0.00146	0.21373
				180	-0.21560	0.00210	0.21561

VEN

RADIATION PATTERN FOR THETA=PI/2

H/LMDA=0.2500 B/LMDA= 0.250 OMEGA= 8.53

14 ELEMENT ARRAY

PHI DEG	ER(THETA,PHI) RE	IM	ABSVAL	PHI DEG	ER(THETA,PHI) RE	IM	ABSVAL
0	1.78595	-0.46298	1.84499	90	-0.21671	-0.18351	0.28397
2	1.78529	-0.46005	1.84361	92	-0.12992	-0.22146	0.25676
4	1.78310	-0.45122	1.83931	94	-0.02575	-0.22949	0.23093
6	1.77881	-0.43646	1.83158	96	0.08077	-0.20646	0.22170
8	1.77148	-0.41569	1.81960	98	0.17470	-0.15550	0.23388
10	1.75977	-0.38888	1.80223	100	0.24334	-0.08354	0.25728
12	1.74208	-0.35604	1.77809	102	0.27795	-0.00021	0.27795
14	1.71648	-0.31728	1.74555	104	0.27480	0.08357	0.28722
16	1.68086	-0.27288	1.70286	106	0.23539	0.15710	0.28300
18	1.63300	-0.22336	1.64821	108	0.16595	0.21138	0.26874
20	1.57071	-0.16949	1.57983	110	0.07629	0.24017	0.25200
22	1.49198	-0.11241	1.49621	112	-0.02180	0.24066	0.24165
24	1.39518	-0.05361	1.39621	114	-0.11623	0.21356	0.24314
26	1.27927	0.00502	1.27928	116	-0.19617	0.16279	0.25491
28	1.14409	0.06124	1.14572	118	-0.25327	0.09469	0.27040
30	0.99051	0.11252	0.99688	120	-0.28248	0.01710	0.28300
32	0.82070	0.15614	0.83542	122	-0.28221	-0.06176	0.28889
34	0.63827	0.18940	0.66578	124	-0.25418	-0.13418	0.28743
36	0.44828	0.20983	0.49496	126	-0.20279	-0.19382	0.28051
38	0.25723	0.21544	0.33553	128	-0.13427	-0.23613	0.27164
40	0.07277	0.20504	0.21757	130	-0.05580	-0.25868	0.26463
42	-0.09669	0.17850	0.20301	132	0.02538	-0.26102	0.26226
44	-0.24255	0.13698	0.27856	134	0.10273	-0.24450	0.26520
46	-0.35676	0.08303	0.36630	136	0.17087	-0.21181	0.27214
48	-0.43267	0.02063	0.43317	138	0.22596	-0.16654	0.28071
50	-0.46586	-0.04501	0.46802	140	0.26575	-0.11272	0.28867
52	-0.45487	-0.10781	0.46747	142	0.28946	-0.05437	0.29452
54	-0.40181	-0.16137	0.43301	144	0.29761	0.00482	0.29765
56	-0.31256	-0.19966	0.37088	146	0.29171	0.06177	0.29818
58	-0.19653	-0.21778	0.29334	148	0.27393	0.11403	0.29672
60	-0.06609	-0.21268	0.22271	150	0.24680	0.15997	0.29411
62	0.06460	-0.18379	0.19481	152	0.21295	0.19863	0.29121
64	0.18107	-0.13335	0.22487	154	0.17487	0.22970	0.28869
66	0.27015	-0.06642	0.27820	156	0.13482	0.25336	0.28700
68	0.32174	0.00950	0.32188	158	0.09469	0.27020	0.28631
70	0.33016	0.08531	0.34100	160	0.05598	0.28107	0.28659
72	0.29508	0.15136	0.33163	162	0.01982	0.28696	0.28765
74	0.22178	0.19879	0.29784	164	-0.01302	0.28894	0.28923
76	0.12061	0.22083	0.25162	166	-0.04206	0.28803	0.29109
78	0.00566	0.21386	0.21393	168	-0.06706	0.28522	0.29299
80	-0.10709	0.17812	0.20783	170	-0.08796	0.28136	0.29479
82	-0.20201	0.11785	0.23388	172	-0.10480	0.27720	0.29635
84	-0.26610	0.04085	0.26921	174	-0.11770	0.27335	0.29761
86	-0.29079	-0.04256	0.29389	176	-0.12679	0.27026	0.29853
88	-0.27325	-0.12097	0.29883	178	-0.13220	0.26828	0.29909
				180	-0.13399	0.26760	0.29927

IEN

RADIATION PATTERN FOR THETA=PI/2

H/LMDA=0.2500 B/LMDA= 0.250 OMEGA= 8.53

14 ELEMENT ARRAY

PHI DEG	ER(THETA,PHI) RE	IM	ABSVAL	PHI DEG	ER(THETA,PHI) RE	IM	ABSVAL
0	7.43231	3.89459	8.39089	90	-0.70688	-0.35622	0.79157
2	7.43251	3.89218	8.38995	92	-0.63868	-0.27772	0.69645
4	7.43245	3.88462	8.38639	94	-0.48290	-0.16388	0.50995
6	7.43018	3.87086	8.37802	96	-0.26617	-0.03265	0.26816
8	7.42246	3.84924	8.36119	98	-0.02214	0.09661	0.09912
10	7.40478	3.81745	8.33089	100	0.21369	0.20598	0.29681
12	7.37145	3.77262	8.28075	102	0.40927	0.28150	0.49673
14	7.31570	3.71141	8.20329	104	0.54019	0.31483	0.62524
16	7.22983	3.63016	8.09002	106	0.59256	0.30402	0.66600
18	7.10547	3.52502	7.93180	108	0.56404	0.25325	0.61829
20	6.93395	3.39221	7.71924	110	0.46317	0.17170	0.49397
22	6.70672	3.22830	7.44326	112	0.30718	0.07183	0.31546
24	6.41599	3.03053	7.09570	114	0.11884	-0.03261	0.12323
26	6.05534	2.79717	6.67018	116	-0.07706	-0.12855	0.14988
28	5.62056	2.52796	6.16290	118	-0.25707	-0.20518	0.32891
30	5.11051	2.22446	5.57364	120	-0.40192	-0.25507	0.47602
32	4.52785	1.89042	4.90664	122	-0.49826	-0.27468	0.56895
34	3.87985	1.53201	4.17136	124	-0.53951	-0.26428	0.60076
36	3.17883	1.15792	3.38316	126	-0.52568	-0.22739	0.57275
38	2.44239	0.77930	2.56370	128	-0.46245	-0.16990	0.49267
40	1.69315	0.40932	1.74193	130	-0.35966	-0.09910	0.37306
42	0.95797	0.06263	0.96001	132	-0.22962	-0.02266	0.23074
44	0.26660	-0.24566	0.36252	134	-0.08546	0.05226	0.10017
46	-0.35028	-0.50105	0.61135	136	0.06035	0.11961	0.13397
48	-0.86343	-0.69108	1.10594	138	0.19706	0.17491	0.26349
50	-1.24787	-0.80666	1.48589	140	0.31639	0.21533	0.38271
52	-1.48564	-0.84337	1.70833	142	0.41273	0.23970	0.47729
54	-1.56831	-0.80242	1.76167	144	0.48319	0.24826	0.54324
56	-1.49887	-0.69120	1.65056	146	0.52724	0.24236	0.58028
58	-1.29257	-0.52312	1.39441	148	0.54628	0.22414	0.59048
60	-0.97652	-0.31683	1.02663	150	0.54312	0.19615	0.57746
62	-0.58780	-0.09461	0.59537	152	0.52144	0.16111	0.54577
64	-0.17023	0.11971	0.20810	154	0.48534	0.12164	0.50035
66	0.23003	0.30334	0.38069	156	0.43894	0.08009	0.44619
68	0.56947	0.43715	0.71791	158	0.38609	0.03845	0.38800
70	0.81232	0.50800	0.95809	160	0.33024	-0.00169	0.33024
72	0.93497	0.51047	1.06525	162	0.27423	-0.03914	0.27701
74	0.92891	0.44762	1.03113	164	0.22039	-0.07310	0.23220
76	0.80189	0.33063	0.86737	166	0.17047	-0.10309	0.19921
78	0.57681	0.17736	0.60346	168	0.12574	-0.12884	0.18003
80	0.28861	0.00987	0.28878	170	0.08708	-0.15033	0.17373
82	-0.02069	-0.14862	0.15005	172	0.05503	-0.16761	0.17641
84	-0.30786	-0.27695	0.41410	174	0.02992	-0.18083	0.18329
86	-0.53453	-0.35897	0.64388	176	0.01192	-0.19012	0.19050
88	-0.67239	-0.38567	0.77515	178	0.00110	-0.19565	0.19565
				180	-0.00250	-0.19747	0.19749

VEN

RADIATION PATTERN FOR THETA=PI/2

H/LMDA=0.2500 B/LMDA= 0.250 OMEGA= 8.53

16 ELEMENT ARRAY

PHI	ER(THETA,PHI)			PHI	ER(THETA,PHI)		
DEG	RE	IM	ABSVAL	DEG	RE	IM	ABSVAL
0	-0.47142	-1.71113	1.77488	90	-0.11985	-0.16466	0.20366
2	-0.46829	-1.71049	1.77343	92	-0.04298	-0.22781	0.23183
4	-0.45887	-1.70831	1.76887	94	0.04226	-0.24772	0.25130
6	-0.44309	-1.70388	1.76054	96	0.12065	-0.22158	0.25230
8	-0.42084	-1.69599	1.74742	98	0.17828	-0.15512	0.23631
10	-0.39205	-1.68301	1.72807	100	0.20511	-0.06123	0.21405
12	-0.35670	-1.66290	1.70073	102	0.19669	0.04266	0.20127
14	-0.31494	-1.63331	1.66340	104	0.15482	0.13806	0.20744
16	-0.26712	-1.59164	1.61390	106	0.08695	0.20872	0.22611
18	-0.21391	-1.53524	1.55007	108	0.00464	0.24345	0.24350
20	-0.15638	-1.46156	1.46990	110	-0.07860	0.23768	0.25034
22	-0.09604	-1.36844	1.37181	112	-0.14972	0.19378	0.24488
24	-0.03490	-1.25438	1.25486	114	-0.19822	0.12010	0.23177
26	0.02457	-1.11883	1.11910	116	-0.21764	0.02908	0.21958
28	0.C7947	-0.96259	0.96586	118	-0.20622	-0.06511	0.21625
30	0.12663	-0.78803	0.79814	120	-0.16673	-0.14895	0.22358
32	0.16285	-0.59939	0.62111	122	-0.10563	-0.21154	0.23645
34	0.18518	-0.40279	0.44332	124	-0.03171	-0.24589	0.24792
36	0.19129	-0.20619	0.28126	126	0.04535	-0.24940	0.25349
38	0.17989	-0.01901	0.18089	128	0.11645	-0.22366	0.25216
40	0.15107	0.14848	0.21182	130	0.17412	-0.17366	0.24592
42	0.10663	0.28601	0.30524	132	0.21328	-0.10660	0.23843
44	0.05017	0.38444	0.38770	134	0.23143	-0.03065	0.23345
46	-0.01294	0.43692	0.43711	136	0.22858	0.04615	0.23320
48	-0.07596	0.44011	0.44661	138	0.20682	0.11683	0.23754
50	-0.13141	0.39509	0.41637	140	0.16969	0.17609	0.24454
52	-0.17198	0.30791	0.35268	142	0.12157	0.22052	0.25180
54	-0.19155	0.18945	0.26942	144	0.06705	0.24859	0.25748
56	-0.18632	0.05468	0.19417	146	0.01046	0.26045	0.26066
58	-0.15562	-0.07897	0.17451	148	-0.04453	0.25751	0.26133
60	-0.10250	-0.19372	0.21916	150	-0.09505	0.24207	0.26007
62	-0.03364	-0.27395	0.27600	152	-0.13917	0.21690	0.25771
64	0.04128	-0.30867	0.31142	154	-0.17579	0.18488	0.25512
66	0.11091	-0.29348	0.31374	156	-0.20461	0.14875	0.25297
68	0.16394	-0.23164	0.28378	158	-0.22590	0.11094	0.25167
70	0.19111	-0.13390	0.23335	160	-0.24036	0.07341	0.25132
72	0.18702	-0.01701	0.18779	162	-0.24900	0.03771	0.25184
74	0.15134	0.09891	0.18080	164	-0.25294	0.00492	0.25299
76	0.08928	0.19380	0.21337	166	-0.25336	-0.02428	0.25452
78	0.01095	0.25124	0.25148	168	-0.25136	-0.04949	0.25618
80	-0.07021	0.26152	0.27078	170	-0.24794	-0.07058	0.25779
82	-0.13981	0.22347	0.26360	172	-0.24398	-0.08756	0.25922
84	-0.18515	0.14479	0.23504	174	-0.24018	-0.10054	0.26037
86	-0.19766	0.04060	0.20178	176	-0.23708	-0.10967	0.26122
88	-0.17463	-0.06939	0.18791	178	-0.23506	-0.11509	0.26173
				180	-0.23437	-0.11688	0.26190

IEN

RADIATION PATTERN FOR THETA=PI/2

H/LMDA=0.2500 B/LMDA= 0.250 OMEGA= 8.53

16 ELEMENT ARRAY

PHI	ER(THETA,PHI)			PHI	ER(THETA,PHI)		
DEG	RE	IM	ABSVAL	DEG	RE	IM	ABSVAL
0	3.83052	-7.51768	8.43732	90	0.10939	0.00291	0.10943
2	3.82768	-7.51792	8.43625	92	0.22327	-0.25628	0.33989
4	3.81873	-7.51778	8.43206	94	0.28987	-0.45453	0.53909
6	3.80235	-7.51467	8.42188	96	0.29958	-0.55880	0.63404
8	3.77638	-7.50431	8.40093	98	0.25381	-0.55516	0.61042
10	3.73787	-7.48076	8.36262	100	0.16400	-0.45020	0.47915
12	3.68319	-7.43653	8.29867	102	0.04890	-0.26846	0.27287
14	3.60815	-7.36276	8.19933	104	-0.06951	-0.04655	0.08366
16	3.50820	-7.24946	8.05371	106	-0.17020	0.17430	0.24362
18	3.37873	-7.08594	7.85024	108	-0.23686	0.35611	0.42769
20	3.21538	-6.86139	7.57742	110	-0.26032	0.47044	0.53766
22	3.01452	-6.56561	7.22458	112	-0.23950	0.50250	0.55666
24	2.77373	-6.18998	6.78303	114	-0.18077	0.45224	0.48703
26	2.49235	-5.72851	6.24721	116	-0.09605	0.33281	0.34639
28	2.17205	-5.17907	5.61610	118	-0.00027	0.16695	0.16695
30	1.81730	-4.54460	4.89448	120	0.09144	-0.01775	0.09314
32	1.43578	-3.83417	4.09418	122	0.16611	-0.19374	0.25520
34	1.03851	-3.06376	3.23499	124	0.21468	-0.33762	0.40010
36	0.63965	-2.25650	2.34541	126	0.23275	-0.43300	0.49159
38	0.25595	-1.44228	1.46482	128	0.22056	-0.47174	0.52076
40	-0.09420	-0.65645	0.66318	130	0.18226	-0.45388	0.48911
42	-0.39231	0.06241	0.39724	132	0.12472	-0.38626	0.40589
44	-0.62148	0.67563	0.91799	134	0.05618	-0.28051	0.28608
46	-0.76849	1.14831	1.38174	136	-0.01503	-0.15081	0.15156
48	-0.82571	1.45355	1.67171	138	-0.08150	-0.01170	0.08234
50	-0.79281	1.57646	1.76458	140	-0.13748	0.12367	0.18492
52	-0.67768	1.51748	1.66192	142	-0.17915	0.24469	0.30327
54	-0.49658	1.29415	1.38615	144	-0.20471	0.34395	0.40026
56	-0.27301	0.94088	0.97969	146	-0.21411	0.41728	0.46900
58	-0.03550	0.50631	0.50756	148	-0.20869	0.46350	0.50831
60	0.18570	0.04840	0.19191	150	-0.19075	0.48385	0.52009
62	0.36247	-0.37258	0.51981	152	-0.16310	0.48130	0.50818
64	0.47282	-0.70199	0.84637	154	-0.12872	0.45989	0.47756
66	0.50431	-0.89864	1.03048	156	-0.09042	0.42413	0.43367
68	0.45616	-0.94126	1.04596	158	-0.05069	0.37854	0.38192
70	0.33963	-0.83197	0.89862	160	-0.01153	0.32725	0.32745
72	0.17647	-0.59632	0.62189	162	0.02548	0.27385	0.27503
74	-0.00458	-0.27927	0.27931	164	0.05930	0.22125	0.22906
76	-0.17258	0.06229	0.18348	166	0.08923	0.17173	0.19353
78	-0.29967	0.36921	0.47552	168	0.11495	0.12692	0.17124
80	-0.36592	0.59036	0.69456	170	0.13635	0.08796	0.16226
82	-0.36275	0.69149	0.78087	172	0.15351	0.05555	0.16325
84	-0.29426	0.66102	0.72355	174	0.16657	0.03012	0.16928
86	-0.17612	0.51141	0.54089	176	0.17573	0.01188	0.17614
88	-0.03237	0.27619	0.27808	178	0.18115	0.00092	0.18115
				180	0.18295	-0.00274	0.18297

VEN

RADIATION PATTERN FOR THETA=PI/2

H/LMDA=0.2500 B/LMDA= 0.250 OMEGA= 8.53

18 ELEMENT ARRAY

PHI	ER(THETA,PHI)			PHI	ER(THETA,PHI)		
DEG	RE	IM	ABSVAL	DEG	RE	IM	ABSVAL
0	-1.64870	0.47723	1.71638	90	-0.16675	-0.14227	0.21920
2	-1.64807	0.47392	1.71486	92	-0.07714	-0.17674	0.19284
4	-1.64590	0.46394	1.71004	94	0.02947	-0.17122	0.17374
6	-1.64130	0.44719	1.70113	96	0.12797	-0.12686	0.18019
8	-1.63281	0.42353	1.68684	98	0.19580	-0.05366	0.20302
10	-1.61844	0.39282	1.66543	100	0.21811	0.03196	0.22044
12	-1.59571	0.35504	1.63473	102	0.19095	0.11109	0.22091
14	-1.56178	0.31034	1.59232	104	0.12157	0.16663	0.20626
16	-1.51359	0.25921	1.53563	106	0.02628	0.18707	0.18891
18	-1.44806	0.20251	1.46215	108	-0.07378	0.16869	0.18412
20	-1.36237	0.14163	1.36971	110	-0.15750	0.11600	0.19561
22	-1.25435	0.07854	1.25681	112	-0.20840	0.04027	0.21225
24	-1.12286	0.01584	1.12297	114	-0.21759	-0.04324	0.22185
26	-0.96820	-0.04334	0.96917	116	-0.18502	-0.11861	0.21977
28	-0.79260	-0.09542	0.79833	118	-0.11859	-0.17243	0.20927
30	-0.60051	-0.13668	0.61586	120	-0.03184	-0.19611	0.19868
32	-0.39875	-0.16359	0.43100	122	0.05920	-0.18695	0.19610
34	-0.19646	-0.17331	0.26198	124	0.13922	-0.14797	0.20317
36	-0.00468	-0.16420	0.16426	126	0.19620	-0.08677	0.21453
38	0.16449	-0.13630	0.21363	128	0.22304	-0.01361	0.22345
40	0.29904	-0.09178	0.31281	130	0.21797	0.06053	0.22622
42	0.38857	-0.03504	0.39014	132	0.18408	0.12576	0.22294
44	0.42592	0.02739	0.42680	134	0.12802	0.17460	0.21650
46	0.40873	0.08744	0.41798	136	0.05845	0.20262	0.21088
48	0.34060	0.13642	0.36690	138	-0.01553	0.20854	0.20911
50	0.23148	0.16636	0.28506	140	-0.08568	0.19383	0.21192
52	0.09716	0.17144	0.19705	142	-0.14555	0.16198	0.21776
54	-0.04242	0.14932	0.15523	144	-0.19087	0.11768	0.22423
56	-0.16585	0.10214	0.19478	146	-0.21963	0.06603	0.22934
58	-0.25362	0.03668	0.25625	148	-0.23181	0.01185	0.23212
60	-0.29162	-0.03630	0.29387	150	-0.22892	-0.04076	0.23252
62	-0.27413	-0.10365	0.29307	152	-0.21346	-0.08868	0.23115
64	-0.20541	-0.15229	0.25570	154	-0.18849	-0.12988	0.22890
66	-0.09950	-0.17188	0.19860	156	-0.15711	-0.16331	0.22661
68	0.02204	-0.15730	0.15884	158	-0.12223	-0.18877	0.22489
70	0.13420	-0.11028	0.17370	160	-0.08633	-0.20674	0.22405
72	0.21367	-0.03957	0.21730	162	-0.05137	-0.21813	0.22409
74	0.24392	0.04047	0.24725	164	-0.01879	-0.22408	0.22487
76	0.21904	0.11267	0.24632	166	0.01046	-0.22587	0.22611
78	0.14537	0.16089	0.21684	168	0.03583	-0.22473	0.22756
80	0.04025	0.17377	0.17837	170	0.05708	-0.22179	0.22902
82	-0.07181	0.14765	0.16419	172	0.07418	-0.21806	0.23033
84	-0.16480	0.08773	0.18670	174	0.08724	-0.21432	0.23140
86	-0.21734	0.00716	0.21746	176	0.09641	-0.21122	0.23218
88	-0.21776	-0.07585	0.23059	178	0.10184	-0.20919	0.23266
				180	0.10364	-0.20848	0.23282

IEN

RADIATION PATTERN FOR THETA=PI/2

H/LMDA=0.2500 B/LMDA= 0.250 OMEGA= 8.53

18 ELEMENT ARRAY

PHI	ER(THETA,PHI)			PHI	ER(THETA,PHI)		
DEG	RE	IM	ABSVAL	DEG	RE	IM	ABSVAL
0	-7.59678	-3.78085	8.48562	90	-0.55685	-0.26842	0.61816
2	-7.59706	-3.77758	8.48442	92	-0.47807	-0.18298	0.51190
4	-7.59680	-3.76719	8.47956	94	-0.29160	-0.05941	0.29759
6	-7.59269	-3.74805	8.46739	96	-0.04640	0.07162	0.08534
8	-7.57926	-3.71747	8.44184	98	0.19782	0.17951	0.26713
10	-7.54893	-3.67180	8.39455	100	0.38521	0.24088	0.45432
12	-7.49218	-3.60654	8.31504	102	0.47640	0.24450	0.53547
14	-7.39777	-3.51658	8.19106	104	0.45637	0.19298	0.49549
16	-7.25321	-3.39651	8.00908	106	0.33590	0.10107	0.35078
18	-7.04534	-3.24097	7.75504	108	0.14703	-0.00870	0.14729
20	-6.76128	-3.04522	7.41541	110	-0.06591	-0.11169	0.12968
22	-6.38953	-2.80575	6.97841	112	-0.25705	-0.18676	0.31773
24	-5.92140	-2.52104	6.43573	114	-0.38885	-0.22043	0.44698
26	-5.35267	-2.19223	5.78420	116	-0.43902	-0.20882	0.48615
28	-4.68522	-1.82394	5.02773	118	-0.40326	-0.15743	0.43290
30	-3.92863	-1.42474	4.17899	120	-0.29401	-0.07883	0.30440
32	-3.10135	-1.00744	3.26087	122	-0.13597	0.01063	0.13638
34	-2.23123	-0.58891	2.30764	124	0.04017	0.09443	0.10262
36	-1.35503	-0.18935	1.36820	126	0.20402	0.15892	0.25862
38	-0.51662	0.16911	0.54359	128	0.33075	0.19533	0.38413
40	0.23621	0.46441	0.52103	130	0.40442	0.20050	0.45139
42	0.85628	0.67711	1.09165	132	0.41906	0.17650	0.45471
44	1.30263	0.79317	1.52511	134	0.37797	0.12944	0.39952
46	1.54658	0.80661	1.74429	136	0.29151	0.06781	0.29929
48	1.57724	0.72157	1.73446	138	0.17434	0.00078	0.17434
50	1.40550	0.55316	1.51044	140	0.04251	-0.06322	0.07618
52	1.06541	0.32682	1.11441	142	-0.08892	-0.11747	0.14733
54	0.61205	0.07587	0.61674	144	-0.20761	-0.15755	0.26062
56	0.11564	-0.16266	0.19957	146	-0.30488	-0.18132	0.35472
58	-0.34775	-0.35342	0.49582	148	-0.37585	-0.18873	0.42058
60	-0.70769	-0.46848	0.84871	150	-0.41916	-0.18133	0.45670
62	-0.91116	-0.49246	1.03572	152	-0.43624	-0.16171	0.46525
64	-0.93229	-0.42568	1.02488	154	-0.43049	-0.13298	0.45056
66	-0.77781	-0.28472	0.82828	156	-0.40643	-0.09835	0.41816
68	-0.48648	-0.09969	0.49659	158	-0.36900	-0.06076	0.37397
70	-0.12221	0.09134	0.15257	160	-0.32300	-0.02270	0.32380
72	0.23846	0.24988	0.34540	162	-0.27270	0.01388	0.27306
74	0.52195	0.34500	0.62567	164	-0.22171	0.04760	0.22676
76	0.67300	0.35986	0.76317	166	-0.17279	0.07758	0.18941
78	0.66641	0.29526	0.72889	168	-0.12804	0.10334	0.16454
80	0.51210	0.16932	0.53937	170	-0.08884	0.12474	0.15314
82	0.25226	0.01311	0.25260	172	-0.05613	0.14184	0.15254
84	-0.04897	-0.13666	0.14517	174	-0.03041	0.15480	0.15776
86	-0.32084	-0.24626	0.40445	176	-0.01194	0.16387	0.16430
88	-0.50251	-0.29263	0.58151	178	-0.00083	0.16922	0.16922
				180	0.00287	0.17098	0.17100

VEN

RADIATION PATTERN FOR THETA=PI/2

H/LMDA=0.2500 B/LMDA= 0.250 OMEGA= 8.53

20 ELEMENT ARRAY

PHI DEG	ER(THETA,PHI) RE	ER(THETA,PHI) IM	ABSVAL	PHI DEG	ER(THETA,PHI) RE	ER(THETA,PHI) IM	ABSVAL
0	0.48129	1.59554	1.66655	90	-0.09724	-0.13205	0.16399
2	0.47781	1.59494	1.66497	92	-0.01824	-0.18951	0.19039
4	0.46730	1.59277	1.65990	94	0.06658	-0.19152	0.20277
6	0.44962	1.58798	1.65040	96	0.13344	-0.13850	0.19232
8	0.42458	1.57883	1.63492	98	0.16372	-0.04656	0.17021
10	0.39200	1.56293	1.61134	100	0.14919	0.05762	0.15993
12	0.35183	1.53737	1.57712	102	0.09407	0.14482	0.17269
14	0.30427	1.49880	1.52937	104	0.01351	0.19159	0.19206
16	0.24992	1.44365	1.46513	106	-0.07098	0.18641	0.19946
18	0.18990	1.36848	1.38159	108	-0.13751	0.13215	0.19071
20	0.12598	1.27029	1.27652	110	-0.16963	0.04433	0.17533
22	0.06067	1.14708	1.14868	112	-0.16021	-0.05382	0.16901
24	-0.00278	0.99835	0.99835	114	-0.11259	-0.13791	0.17803
26	-0.06054	0.82567	0.82789	116	-0.03897	-0.18854	0.19253
28	-0.10837	0.63320	0.64241	118	0.04321	-0.19555	0.20026
30	-0.14213	0.42795	0.45093	120	0.11582	-0.15942	0.19705
32	-0.15825	0.21983	0.27087	122	0.16411	-0.09005	0.18719
34	-0.15447	0.02120	0.15592	124	0.17956	-0.00343	0.17959
36	-0.13041	-0.15405	0.20184	126	0.16091	0.08251	0.18084
38	-0.08820	-0.29201	0.30504	128	0.11349	0.15182	0.18955
40	-0.03266	-0.38063	0.38202	130	0.04728	0.19340	0.19909
42	0.02880	-0.41190	0.41291	132	-0.02562	0.20235	0.20396
44	0.08694	-0.38388	0.39360	134	-0.09345	0.17985	0.20268
46	0.13196	-0.30211	0.32967	136	-0.14680	0.13194	0.19738
48	0.15520	-0.17996	0.23764	138	-0.17966	0.06759	0.19195
50	0.15104	-0.03762	0.15566	140	-0.18978	-0.00323	0.18981
52	0.11855	0.10058	0.15547	142	-0.17827	-0.07134	0.19201
54	0.06240	0.21009	0.21916	144	-0.14873	-0.12948	0.19720
56	-0.00728	0.27091	0.27101	146	-0.10625	-0.17291	0.20294
58	-0.07644	0.27198	0.28252	148	-0.05639	-0.19949	0.20730
60	-0.12984	0.21430	0.25057	150	-0.00434	-0.20933	0.20938
62	-0.15457	0.11159	0.19064	152	0.04556	-0.20425	0.20927
64	-0.14354	-0.01201	0.14404	154	0.09016	-0.18709	0.20768
66	-0.09787	-0.12703	0.16036	156	0.12752	-0.16117	0.20552
68	-0.02747	-0.20562	0.20745	158	0.15683	-0.12980	0.20357
70	0.05074	-0.22864	0.23420	160	0.17816	-0.09593	0.20235
72	0.11679	-0.19099	0.22387	162	0.19225	-0.06199	0.20200
74	0.15287	-0.10349	0.18460	164	0.20023	-0.02979	0.20243
76	0.14840	0.00962	0.14871	166	0.20341	-0.00058	0.20341
78	0.10354	0.11700	0.15623	168	0.20316	0.02490	0.20468
80	0.02967	0.18884	0.19116	170	0.20074	0.04629	0.20600
82	-0.05324	0.20539	0.21218	172	0.19725	0.06350	0.20722
84	-0.12211	0.16269	0.20342	174	0.19360	0.07664	0.20822
86	-0.15726	0.07381	0.17372	176	0.19051	0.08584	0.20895
88	-0.14829	-0.03500	0.15236	178	0.18846	0.09128	0.20940
				180	0.18774	0.09308	0.20955

IEN

RADIATION PATTERN FOR THETA=PI/2

H/LMDA=0.2500 B/LMDA= 0.250 OMEGA= 8.53

20 ELEMENT ARRAY

PHI DEG	ER(THETA,PHI) RE	ER(THETA,PHI) IM	ABSVAL	PHI DEG	ER(THETA,PHI) RE	ER(THETA,PHI) IM	ABSVAL
0	-3.74133	7.67056	8.53434	90	0.09484	0.00255	0.09488
2	-3.73760	7.67088	8.53300	92	0.20011	-0.25475	0.32394
4	-3.72573	7.67047	8.52743	94	0.24256	-0.42562	0.48989
6	-3.70372	7.66520	8.51310	96	0.21359	-0.46611	0.51271
8	-3.66828	7.64827	8.48247	98	0.12538	-0.37180	0.39237
10	-3.61498	7.61025	8.42520	100	0.00612	-0.17666	0.17677
12	-3.53844	7.53932	8.32838	102	-0.10913	0.05929	0.12419
14	-3.43255	7.42165	8.17700	104	-0.18882	0.26872	0.32843
16	-3.29104	7.24204	7.95475	106	-0.21344	0.39621	0.45004
18	-3.10790	6.98481	7.64504	108	-0.17980	0.41244	0.44992
20	-2.87825	6.63518	7.23256	110	-0.10044	0.31973	0.33513
22	-2.59913	6.18090	6.70515	112	0.00124	0.14842	0.14843
24	-2.27055	5.61435	6.05610	114	0.09818	-0.05378	0.11195
26	-1.89638	4.93467	5.28651	116	0.16699	-0.23573	0.28889
28	-1.48525	4.15013	4.40789	118	0.19329	-0.35581	0.40492
30	-1.05095	3.27975	3.44401	120	0.17414	-0.39078	0.42782
32	-0.61246	2.35449	2.43285	122	0.11725	-0.33884	0.35855
34	-0.19304	1.41684	1.42993	124	0.03779	-0.21723	0.22049
36	0.18137	0.51874	0.54953	126	-0.04602	-0.05574	0.07228
38	0.48484	-0.28240	0.56109	128	-0.11726	0.11130	0.16167
40	0.69471	-0.92983	1.16069	130	-0.16359	0.25273	0.30105
42	0.79536	-1.37510	1.58855	132	-0.17895	0.34607	0.38959
44	0.78174	-1.58621	1.76838	134	-0.16364	0.38021	0.41393
46	0.66184	-1.55512	1.69009	136	-0.12316	0.35536	0.37609
48	0.45756	-1.30261	1.38063	138	-0.06622	0.28086	0.28856
50	0.20312	-0.87912	0.90228	140	-0.00271	0.17188	0.17190
52	-0.05906	-0.36027	0.36508	142	0.05815	0.04578	0.07401
54	-0.28480	0.16326	0.32828	144	0.10909	-0.08090	0.13581
56	-0.43595	0.60021	0.74182	146	0.14542	-0.19471	0.24302
58	-0.48797	0.87547	1.00227	148	0.16510	-0.28632	0.33051
60	-0.43559	0.94538	1.04091	150	0.16838	-0.35083	0.38915
62	-0.29483	0.80797	0.86009	152	0.15725	-0.38726	0.41797
64	-0.10040	0.50474	0.51463	154	0.13471	-0.39766	0.41985
66	0.10151	0.11265	0.15164	156	0.10418	-0.38607	0.39988
68	0.26373	-0.27284	0.37947	158	0.06904	-0.35760	0.36420
70	0.34934	-0.56019	0.66019	160	0.03224	-0.31753	0.31916
72	0.34105	-0.68428	0.76456	162	-0.00385	-0.27083	0.27086
74	0.24562	-0.62281	0.66949	164	-0.03749	-0.22173	0.22488
76	0.09213	-0.40184	0.41227	166	-0.06755	-0.17364	0.18631
78	-0.07592	-0.08846	0.11657	168	-0.09343	-0.12903	0.15931
80	-0.21250	0.22795	0.31164	170	-0.11488	-0.08970	0.14575
82	-0.28187	0.46067	0.54006	172	-0.13198	-0.05670	0.14364
84	-0.26823	0.54967	0.61163	174	-0.14490	-0.03070	0.14811
86	-0.17969	0.47769	0.51037	176	-0.15388	-0.01202	0.15435
88	-0.04513	0.27364	0.27734	178	-0.15916	-0.00080	0.15917
				180	-0.16091	0.00296	0.16094

VEN

RADIATION PATTERN FOR THETA=PI/2

H/LMDA=0.2500 B/LMDA= 0.250 OMEGA= 8.53

22 ELEMENT ARRAY

PHI DEG	ER(THETA,PHI) RE	ER(THETA,PHI) IM	ABSVAL	PHI DEG	ER(THETA,PHI) RE	ER(THETA,PHI) IM	ABSVAL
0	1.54954	-0.48413	1.62341	90	-0.13536	-0.11615	0.17837
2	1.54896	-0.48048	1.62177	92	-0.04325	-0.14674	0.15298
4	1.54678	-0.46947	1.61646	94	0.06282	-0.12759	0.14222
6	1.54178	-0.45089	1.60636	96	0.14601	-0.06504	0.15984
8	1.53192	-0.42451	1.58965	98	0.17823	0.01984	0.17933
10	1.51439	-0.39009	1.56382	100	0.14958	0.09872	0.17922
12	1.48579	-0.34757	1.52590	102	0.07100	0.14566	0.16204
14	1.44226	-0.29719	1.47256	104	-0.03039	0.14572	0.14885
16	1.37977	-0.23971	1.40044	106	-0.12109	0.09944	0.15668
18	1.29451	-0.17654	1.30650	108	-0.17252	0.02198	0.17391
20	1.18347	-0.10991	1.18856	110	-0.16987	-0.06241	0.18098
22	1.04500	-0.04292	1.04588	112	-0.11576	-0.12842	0.17289
24	0.87959	0.02046	0.87983	114	-0.02802	-0.15737	0.15985
26	0.69051	0.07567	0.69464	116	0.06687	-0.14222	0.15716
28	0.48432	0.11792	0.49847	118	0.14230	-0.08855	0.16760
30	0.27108	0.14284	0.30641	120	0.17898	-0.01191	0.17938
32	0.06405	0.14721	0.16053	122	0.16943	0.06738	0.18234
34	-0.12132	0.12984	0.17770	124	0.11849	0.12999	0.17589
36	-0.26903	0.09231	0.28443	126	0.04060	0.16225	0.16725
38	-0.36489	0.03941	0.36701	128	-0.04502	0.15874	0.16500
40	-0.39923	-0.02099	0.39978	130	-0.11958	0.12240	0.17111
42	-0.36950	-0.07870	0.37779	132	-0.16883	0.06262	0.18007
44	-0.28216	-0.12275	0.30770	134	-0.18539	-0.00773	0.18555
46	-0.15314	-0.14355	0.20990	136	-0.16909	-0.07546	0.18516
48	-0.00638	-0.13525	0.13540	138	-0.12565	-0.12968	0.18057
50	0.12978	-0.09775	0.16247	140	-0.06451	-0.16335	0.17562
52	0.22796	-0.03772	0.23106	142	0.00362	-0.17375	0.17379
54	0.26789	0.03204	0.26979	144	0.06883	-0.16206	0.17607
56	0.24173	0.09493	0.25971	146	0.12349	-0.13226	0.18095
58	0.15721	0.13454	0.20692	148	0.16281	-0.08995	0.18601
60	0.03680	0.13919	0.14397	150	0.18494	-0.04109	0.18945
62	-0.08711	0.10593	0.13714	152	0.19046	0.00885	0.19066
64	-0.18050	0.04243	0.18542	154	0.18165	0.05556	0.18995
66	-0.21736	-0.03414	0.22003	156	0.16178	0.09608	0.18816
68	-0.18780	-0.10147	0.21346	158	0.13443	0.12881	0.18618
70	-0.10175	-0.13865	0.17197	160	0.10298	0.15331	0.18468
72	0.01325	-0.13302	0.13368	162	0.07031	0.17003	0.18399
74	0.12025	-0.08504	0.14728	164	0.03866	0.18002	0.18412
76	0.18466	-0.00901	0.18488	166	0.00958	0.18462	0.18487
78	0.18578	0.07082	0.19882	168	-0.01596	0.18528	0.18597
80	0.12406	0.12796	0.17823	170	-0.03747	0.18339	0.18718
82	0.02127	0.14272	0.14430	172	-0.05479	0.18016	0.18831
84	-0.08682	0.10930	0.13958	174	-0.06797	0.17662	0.18925
86	-0.16282	0.03830	0.16727	176	-0.07721	0.17354	0.18995
88	-0.18081	-0.04648	0.18669	178	-0.08266	0.17148	0.19037
				180	-0.08446	0.17076	0.19051

IEN

RADIATION PATTERN FOR THETA=PI/2

H/LMDA=0.2500 B/LMDA= 0.250 OMEGA= 8.53

22 ELEMENT ARRAY

PHI DEG	ER(THETA,PHI) RE	ER(THETA,PHI) IM	ABSVAL	PHI DEG	ER(THETA,PHI) RE	ER(THETA,PHI) IM	ABSVAL
0	7.73978	3.70924	8.58270	90	-0.46074	-0.21502	0.50844
2	7.74014	3.70504	8.58121	92	-0.37013	-0.12282	0.38998
4	7.73955	3.69164	8.57489	94	-0.15632	0.00681	0.15647
6	7.73297	3.66663	8.55821	96	0.10075	0.12745	0.16246
8	7.71209	3.62609	8.52202	98	0.31099	0.19846	0.36892
10	7.66543	3.56474	8.45377	100	0.40567	0.19863	0.45168
12	7.57867	3.47622	8.33788	102	0.35938	0.13199	0.38285
14	7.43513	3.35345	8.15639	104	0.19539	0.02461	0.19693
16	7.21675	3.18927	7.89005	106	-0.02595	-0.08590	0.08973
18	6.90535	2.97721	7.51982	108	-0.23010	-0.16374	0.28241
20	6.48459	2.71253	7.02906	110	-0.35362	-0.18637	0.39973
22	5.94225	2.39335	6.40613	112	-0.36323	-0.15038	0.39313
24	5.27310	2.02195	5.64747	114	-0.26304	-0.07056	0.27234
26	4.48174	1.60591	4.76077	116	-0.08931	0.02657	0.09317
28	3.58537	1.15894	3.76803	118	0.10382	0.11199	0.15271
30	2.61551	0.70110	2.70785	120	0.26197	0.16290	0.30849
32	1.61837	0.25812	1.63882	122	0.34567	0.16827	0.38445
34	0.65285	-0.14048	0.66779	124	0.33889	0.13028	0.36307
36	-0.21406	-0.46444	0.51140	126	0.25001	0.06184	0.25754
38	-0.91476	-0.68695	1.14398	128	0.10602	-0.01824	0.10758
40	-1.39083	-0.78942	1.59924	130	-0.05676	-0.09083	0.10711
42	-1.60387	-0.76607	1.77743	132	-0.20261	-0.14097	0.24682
44	-1.54540	-0.62728	1.66786	134	-0.30413	-0.16052	0.34389
46	-1.24337	-0.40046	1.30626	136	-0.34651	-0.14872	0.37708
48	-0.76273	-0.12778	0.77336	138	-0.32811	-0.11091	0.34635
50	-0.19853	0.13984	0.24284	140	-0.25812	-0.05634	0.26420
52	0.33885	0.35165	0.48834	142	-0.15264	0.00444	0.15271
54	0.74427	0.46753	0.87893	144	-0.03037	0.06167	0.06874
56	0.94035	0.46743	1.05011	146	0.09113	0.10792	0.14125
58	0.89621	0.35729	0.96481	148	0.19787	0.13869	0.24164
60	0.63699	0.16896	0.65901	150	0.28071	0.15244	0.31943
62	0.23974	-0.04657	0.24422	152	0.33541	0.15006	0.36745
64	-0.18470	-0.23177	0.29637	154	0.36193	0.13414	0.38599
66	-0.52062	-0.33804	0.62074	156	0.36331	0.10815	0.37907
68	-0.67944	-0.33952	0.75955	158	0.34441	0.07577	0.35265
70	-0.62546	-0.24100	0.67028	160	0.31088	0.04043	0.31349
72	-0.38666	-0.07682	0.39421	162	0.26822	0.00494	0.26827
74	-0.04614	0.09957	0.10974	164	0.22135	-0.02860	0.22319
76	0.28416	0.23241	0.36710	166	0.17425	-0.05878	0.18390
78	0.49947	0.28149	0.57332	168	0.12995	-0.08480	0.15517
80	0.53608	0.23511	0.58537	170	0.09049	-0.10638	0.13966
82	0.39157	0.11343	0.40767	172	0.05726	-0.12350	0.13613
84	0.12396	-0.03905	0.12997	174	0.03103	-0.13639	0.13987
86	-0.16978	-0.16930	0.23977	176	0.01213	-0.14533	0.14583
88	-0.38827	-0.23421	0.45344	178	0.00078	-0.15057	0.15057
				180	-0.00300	-0.15229	0.15232

VEN

RADIATION PATTERN FOR THETA=PI/2

H/LMDA=0.2500 B/LMDA= 0.250 OMEGA= 8.53

24 ELEMENT ARRAY

PHI DEG	ER(THETA,PHI) RE	IM	ABSVAL	PHI DEG	ER(THETA,PHI) RE	IM	ABSVAL
0	-0.48610	-1.50920	1.58555	90	-0.08189	-0.11022	0.13731
2	-0.48229	-1.50864	1.58385	92	-0.00123	-0.16195	0.16196
4	-0.47079	-1.50646	1.57831	94	0.08063	-0.14635	0.16709
6	-0.45134	-1.50123	1.56761	96	0.13069	-0.07092	0.14869
8	-0.42364	-1.49059	1.54962	98	0.12893	0.03264	0.13299
10	-0.38741	-1.47131	1.52146	100	0.07617	0.12213	0.14393
12	-0.34256	-1.43949	1.47968	102	-0.00668	0.16230	0.16244
14	-0.28940	-1.39071	1.42050	104	-0.08744	0.13857	0.16385
16	-0.22887	-1.32051	1.34019	106	-0.13553	0.06168	0.14891
18	-0.16273	-1.22480	1.23556	108	-0.13357	-0.03790	0.13884
20	-0.09373	-1.10066	1.10464	110	-0.08317	-0.12270	0.14823
22	-0.02564	-0.94709	0.94744	112	-0.00325	-0.16266	0.16269
24	0.03683	-0.76592	0.76680	114	0.07788	-0.14536	0.16491
26	0.08842	-0.56253	0.56943	116	0.13303	-0.07890	0.15467
28	0.12389	-0.34637	0.36786	118	0.14517	0.01279	0.14574
30	0.13892	-0.13089	0.19087	120	0.11210	0.09943	0.14984
32	0.13111	0.06731	0.14738	122	0.04563	0.15479	0.16138
34	0.10094	0.23019	0.25135	124	-0.03355	0.16433	0.16772
36	0.05254	0.34096	0.34498	126	-0.10296	0.12801	0.16428
38	-0.00629	0.38731	0.38736	128	-0.14490	0.05839	0.15623
40	-0.06468	0.36480	0.37049	130	-0.15057	-0.02463	0.15257
42	-0.11051	0.27938	0.30044	132	-0.12104	-0.10017	0.15712
44	-0.13287	0.14826	0.19909	134	-0.06541	-0.15175	0.16525
46	-0.12500	-0.00166	0.12501	136	0.00267	-0.17044	0.17046
48	-0.08680	-0.13829	0.16327	138	0.06884	-0.15551	0.17006
50	-0.02605	-0.23105	0.23252	140	0.12125	-0.11300	0.16574
52	0.04246	-0.25863	0.26210	142	0.15244	-0.05310	0.16143
54	0.10008	-0.21550	0.23760	144	0.15987	0.01275	0.16038
56	0.12948	-0.11493	0.17313	146	0.14524	0.07427	0.16312
58	0.12028	0.01313	0.12099	148	0.11325	0.12384	0.16782
60	0.07341	0.12983	0.14915	150	0.07008	0.15718	0.17209
62	0.00226	0.19892	0.19893	152	0.02203	0.17310	0.17450
64	-0.07048	0.19881	0.21093	154	-0.02543	0.17292	0.17478
66	-0.11985	0.13070	0.17733	156	-0.06824	0.15955	0.17353
68	-0.12749	0.01935	0.12895	158	-0.10389	0.13666	0.17167
70	-0.08911	-0.09472	0.13005	160	-0.13132	0.10799	0.17002
72	-0.01736	-0.16954	0.17042	162	-0.15058	0.07685	0.16906
74	0.06173	-0.17749	0.18792	164	-0.16256	0.04589	0.16892
76	0.11799	-0.11643	0.16576	166	-0.16859	0.01704	0.16945
78	0.12889	-0.01110	0.12937	168	-0.17018	-0.00852	0.17039
80	0.08905	0.09612	0.13103	170	-0.16883	-0.03011	0.17149
82	0.01329	0.16213	0.16267	172	-0.16588	-0.04752	0.17255
84	-0.06858	0.16066	0.17468	174	-0.16245	-0.06078	0.17345
86	-0.12353	0.09319	0.15474	176	-0.15939	-0.07004	0.17410
88	-0.12888	-0.01167	0.12941	178	-0.15732	-0.07550	0.17450
				180	-0.15659	-0.07731	0.17464

IEN

RADIATION PATTERN FOR THETA=PI/2

H/LMDA=0.2500 B/LMDA= 0.250 OMEGA= 8.53

24 ELEMENT ARRAY

PHI DEG	ER(THETA,PHI) RE	IM	ABSVAL	PHI DEG	ER(THETA,PHI) RE	IM	ABSVAL
0	3.68273	-7.80506	8.63026	90	0.08432	0.00232	0.08436
2	3.67805	-7.80546	8.62863	92	0.18140	-0.25179	0.31033
4	3.66307	-7.80464	8.62152	94	0.19946	-0.38990	0.43796
6	3.63495	-7.79658	8.60229	96	0.13554	-0.36223	0.38675
8	3.58908	-7.77130	8.56006	98	0.01999	-0.18892	0.18997
10	3.51926	-7.71507	8.47983	100	-0.09793	0.05251	0.11112
12	3.41811	-7.61080	8.34313	102	-0.17126	0.26231	0.31327
14	3.27752	-7.43879	8.12882	104	-0.17383	0.35981	0.39960
16	3.08957	-7.17800	7.81467	106	-0.10891	0.31359	0.33196
18	2.84748	-6.80788	7.37939	108	-0.00538	0.14951	0.14960
20	2.54703	-6.31099	6.80558	110	0.09557	-0.06422	0.11514
22	2.18800	-5.67615	6.08326	112	0.15716	-0.24652	0.29235
24	1.77582	-4.90218	5.21391	114	0.15997	-0.33431	0.37061
26	1.32273	-4.00143	4.21438	116	0.10706	-0.30306	0.32141
28	0.84859	-3.00286	3.12046	118	0.02026	-0.17095	0.17214
30	0.38044	-1.95330	1.99000	120	-0.06952	0.01235	0.07061
32	-0.04913	-0.91637	0.91769	122	-0.13364	0.18580	0.22887
34	-0.40539	0.03224	0.40667	124	-0.15455	0.29782	0.33554
36	-0.65636	0.81317	1.04501	126	-0.12961	0.32055	0.34576
38	-0.77865	1.35519	1.56296	128	-0.06964	0.25455	0.26390
40	-0.76347	1.60910	1.78104	130	0.00628	0.12420	0.12435
42	-0.62115	1.56080	1.67986	132	0.07738	-0.03276	0.08403
44	-0.38261	1.24043	1.29810	134	0.12699	-0.17718	0.21799
46	-0.09668	0.72379	0.73022	136	0.14591	-0.27841	0.31432
48	0.17754	0.12354	0.21629	138	0.13322	-0.31971	0.34635
50	0.38233	-0.43026	0.57559	140	0.09483	-0.29925	0.31391
52	0.47469	-0.81738	0.94522	142	0.04082	-0.22731	0.23095
54	0.43810	-0.95662	1.05216	144	-0.01758	-0.12154	0.12281
56	0.28862	-0.82868	0.87750	146	-0.07040	-0.00194	0.07042
58	0.07250	-0.48451	0.48991	148	-0.11048	0.11333	0.15827
60	-0.14547	-0.03420	0.14944	150	-0.13404	0.21062	0.24965
62	-0.30042	0.38320	0.48693	152	-0.14045	0.28176	0.31483
64	-0.34756	0.64138	0.72950	154	-0.13149	0.32391	0.34958
66	-0.27715	0.66716	0.72244	156	-0.11042	0.33845	0.35601
68	-0.11854	0.46548	0.48033	158	-0.08115	0.32963	0.33947
70	0.06939	0.11836	0.13720	160	-0.04748	0.30311	0.30680
72	0.21877	-0.24366	0.32747	162	-0.01269	0.26491	0.26521
74	0.27742	-0.48882	0.56205	164	0.02071	0.22055	0.22152
76	0.22818	-0.53279	0.57959	166	0.05100	0.17465	0.18194
78	0.09512	-0.37000	0.38203	168	0.07722	0.13071	0.15181
80	-0.06627	-0.07438	0.09962	170	0.09892	0.09125	0.13458
82	-0.19149	0.23132	0.30030	172	0.11611	0.05782	0.12971
84	-0.23288	0.42555	0.48510	174	0.12901	0.03132	0.13275
86	-0.17825	0.43659	0.47158	176	0.13790	0.01225	0.13844
88	-0.05462	0.26981	0.27529	178	0.14310	0.00079	0.14310
				180	0.14481	-0.00306	0.14484

VEN

RADIATION PATTERN FOR THETA=PI/2

H/LMDA=0.2500 B/LMDA= 0.250 OMEGA= 8.53

26 ELEMENT ARRAY

PHI DEG	ER(THETA,PHI) RE	IM	ABSVAL	PHI DEG	ER(THETA,PHI) RE	IM	ABSVAL
0	-1.47343	0.48743	1.55196	90	-0.11383	-0.09814	0.15029
2	-1.47289	0.48347	1.55021	92	-0.01960	-0.12458	0.12611
4	-1.47070	0.47149	1.54443	94	0.08299	-0.09225	0.12409
6	-1.46522	0.45120	1.53311	96	0.14483	-0.01619	0.14574
8	-1.45375	0.42220	1.51381	98	0.13732	0.06803	0.15325
10	-1.43261	0.38417	1.48323	100	0.06530	0.12136	0.13781
12	-1.39737	0.33701	1.43743	102	-0.03647	0.11955	0.12499
14	-1.34308	0.28110	1.37218	104	-0.12078	0.06385	0.13662
16	-1.26483	0.21762	1.28342	106	-0.15012	-0.02042	0.15150
18	-1.15838	0.14870	1.16789	108	-0.11302	-0.09620	0.14841
20	-1.02106	0.07768	1.02401	110	-0.02744	-0.13136	0.13420
22	-0.85277	0.00907	0.85282	112	0.06902	-0.11217	0.13170
24	-0.65706	-0.05165	0.65909	114	0.13643	-0.04771	0.14453
26	-0.44189	-0.09858	0.45275	116	0.14908	0.03544	0.15323
28	-0.21999	-0.12623	0.25363	118	0.10434	0.10510	0.14809
30	-0.00834	-0.13070	0.13097	120	0.02141	0.13622	0.13789
32	0.17336	-0.11088	0.20579	122	-0.06818	0.11939	0.13749
34	0.30553	-0.06952	0.31334	124	-0.13336	0.06247	0.14727
36	0.37238	-0.01372	0.37263	126	-0.15414	-0.01415	0.15479
38	0.36610	0.04557	0.36893	128	-0.12676	-0.08576	0.15305
40	0.29036	0.09516	0.30556	130	-0.06256	-0.13170	0.14581
42	0.16200	0.12252	0.20311	132	0.01780	-0.14095	0.14207
44	0.00968	0.11918	0.11958	134	0.09179	-0.11370	0.14613
46	-0.13100	0.08396	0.15559	136	0.14156	-0.05947	0.15354
48	-0.22554	0.02467	0.22688	138	0.15774	0.00707	0.15790
50	-0.25007	-0.04265	0.25368	140	0.14008	0.07060	0.15687
52	-0.19939	-0.09759	0.22199	142	0.09566	0.11889	0.15260
54	-0.09051	-0.12162	0.15161	144	0.03570	0.14480	0.14914
56	0.04067	-0.10488	0.11249	146	-0.02776	0.14660	0.14920
58	0.14987	-0.05112	0.15835	148	-0.08444	0.12705	0.15255
60	0.19934	0.02188	0.20053	150	-0.12732	0.09177	0.15694
62	0.17213	0.08744	0.19307	152	-0.15305	0.04750	0.16025
64	0.07994	0.11966	0.14391	154	-0.16152	0.00068	0.16152
66	-0.03992	0.10423	0.11161	156	-0.15495	-0.04351	0.16094
68	-0.13834	0.04566	0.14568	158	-0.13689	-0.08155	0.15935
70	-0.17451	-0.03282	0.17757	160	-0.11132	-0.11162	0.15764
72	-0.13381	-0.09796	0.16583	162	-0.08195	-0.13330	0.15647
74	-0.03514	-0.12062	0.12563	164	-0.05185	-0.14722	0.15609
76	0.07651	-0.08942	0.11769	166	-0.02330	-0.15467	0.15642
78	0.15005	-0.01718	0.15103	168	0.00222	-0.15720	0.15722
80	0.15190	0.06372	0.16473	170	0.02389	-0.15641	0.15822
82	0.08209	0.11574	0.14190	172	0.04138	-0.15375	0.15922
84	-0.02544	0.11388	0.11668	174	0.05469	-0.15043	0.16007
86	-0.11890	0.05814	0.13236	176	0.06398	-0.14741	0.16070
88	-0.15367	-0.02572	0.15581	178	0.06946	-0.14533	0.16108
				180	0.07126	-0.14460	0.16121

IEN

RADIATION PATTERN FOR THETA=PI/2

H/LMDA=0.2500 B/LMDA= 0.250 OMEGA= 8.53

26 ELEMENT ARRAY

PHI DEG	ER(THETA,PHI) RE	IM	ABSVAL	PHI DEG	ER(THETA,PHI) RE	IM	ABSVAL
0	-7.86687	-3.66053	8.67682	90	-0.39378	-0.17920	0.43264
2	-7.86731	-3.65536	8.67504	92	-0.29083	-0.08074	0.30183
4	-7.86623	-3.63877	8.66708	94	-0.05463	0.05088	0.07465
6	-7.85652	-3.60743	8.64514	96	0.19584	0.15164	0.24768
8	-7.82638	-3.55599	8.59635	98	0.34174	0.17584	0.38432
10	-7.75963	-3.47729	8.50314	100	0.32046	0.11651	0.34099
12	-7.63618	-3.36286	8.34387	102	0.15107	0.00596	0.15119
14	-7.43313	-3.20361	8.09411	104	-0.08054	-0.10217	0.13010
16	-7.12640	-2.99089	7.72858	106	-0.26620	-0.15918	0.31017
18	-6.69328	-2.71786	7.22404	108	-0.32602	-0.14287	0.35595
20	-6.11588	-2.38127	6.56311	110	-0.24122	-0.06496	0.24982
22	-5.38530	-1.98326	5.73888	112	-0.05658	0.03730	0.06776
24	-4.50628	-1.53324	4.75998	114	0.14503	0.11966	0.18802
26	-3.50149	-1.04917	3.65530	116	0.28093	0.15001	0.31848
28	-2.41448	-0.55788	2.47809	118	0.30194	0.12005	0.32493
30	-1.30986	-0.09380	1.31321	120	0.20765	0.04558	0.21259
32	-0.26972	0.30424	0.40659	122	0.04043	-0.04258	0.05871
34	0.61505	0.59837	0.85810	124	-0.13571	-0.11203	0.17597
36	1.25810	0.75861	1.46912	126	-0.26070	-0.14056	0.29618
38	1.59445	0.77036	1.77080	128	-0.29818	-0.12228	0.32228
40	1.59769	0.64066	1.72135	130	-0.24357	-0.06691	0.25259
42	1.29298	0.40106	1.35375	132	-0.12035	0.00580	0.12049
44	0.76108	0.10513	0.76831	134	0.03181	0.07380	0.08037
46	0.12914	-0.18026	0.22175	136	0.17110	0.11954	0.20873
48	-0.45265	-0.38915	0.59693	138	0.26522	0.13388	0.29710
50	-0.84547	-0.47332	0.96894	140	0.29763	0.11680	0.31973
52	-0.95895	-0.41686	1.04564	142	0.26817	0.07552	0.27860
54	-0.77975	-0.24324	0.81681	144	0.18951	0.02122	0.19069
56	-0.38001	-0.01088	0.38017	146	0.08134	-0.03437	0.08830
58	0.10092	0.20345	0.22711	148	-0.03528	-0.08148	0.08879
60	0.49944	0.32930	0.59822	150	-0.14236	-0.11381	0.18226
62	0.68300	0.32739	0.75742	152	-0.22743	-0.12876	0.26135
64	0.59848	0.20493	0.63260	154	-0.28421	-0.12694	0.31127
66	0.29339	0.01411	0.29373	156	-0.31181	-0.11114	0.33102
68	-0.10116	-0.16736	0.19556	158	-0.31334	-0.08528	0.32474
70	-0.42211	-0.26708	0.49951	160	-0.29428	-0.05352	0.29910
72	-0.54146	-0.24805	0.59557	162	-0.26091	-0.01957	0.26165
74	-0.42011	-0.12433	0.43812	164	-0.21937	0.01362	0.21979
76	-0.12399	0.04518	0.13197	166	-0.17483	0.04403	0.18029
78	0.20652	0.18328	0.27612	168	-0.13141	0.07043	0.14910
80	0.42097	0.22952	0.47947	170	-0.09195	0.09231	0.13029
82	0.42783	0.16753	0.45946	172	-0.05832	0.10960	0.12415
84	0.23516	0.03154	0.23727	174	-0.03164	0.12250	0.12652
86	-0.05636	-0.11120	0.12467	176	-0.01239	0.13137	0.13195
88	-0.30432	-0.19330	0.36052	178	-0.00080	0.13654	0.13654
				180	0.00307	0.13824	0.13827

VEN

RADIATION PATTERN FOR THETA=PI/2

H/LMDA=0.2500 B/LMDA= 0.250 OMEGA= 8.53

28 ELEMENT ARRAY

PHI DEG	ER(THETA,PHI) RE	IM	ABSVAL	PHI DEG	ER(THETA,PHI) RE	IM	ABSVAL
0	0.48829	1.44142	1.52188	90	-0.07077	-0.09459	0.11813
2	0.48418	1.44089	1.52007	92	0.01113	-0.14053	0.14097
4	0.47174	1.43869	1.51405	94	0.08766	-0.10809	0.13916
6	0.45061	1.43294	1.50212	96	0.11711	-0.01635	0.11824
8	0.42034	1.42058	1.48147	98	0.08352	0.08351	0.11811
10	0.38052	1.39748	1.44836	100	0.00500	0.13726	0.13736
12	0.33106	1.35864	1.39839	102	-0.07669	0.11706	0.13995
14	0.27244	1.29858	1.32685	104	-0.11893	0.03519	0.12403
16	0.20609	1.21200	1.22939	106	-0.10048	-0.06438	0.11934
18	0.13459	1.09457	1.10282	108	-0.03146	-0.13047	0.13421
20	0.06194	0.94409	0.94612	110	0.05329	-0.13110	0.14152
22	-0.00659	0.76168	0.76171	112	0.11272	-0.06801	0.13165
24	-0.06473	0.55299	0.55677	114	0.11965	0.02675	0.12260
26	-0.10605	0.32895	0.34563	116	0.07238	0.10845	0.13039
28	-0.12505	0.10587	0.16385	118	-0.00659	0.14125	0.14141
30	-0.11865	-0.09576	0.15247	120	-0.08262	0.11322	0.14017
32	-0.08759	-0.25382	0.26851	122	-0.12480	0.03886	0.13071
34	-0.03737	-0.34860	0.35059	124	-0.11809	-0.04995	0.12822
36	0.02168	-0.36759	0.36822	126	-0.06741	-0.11870	0.13650
38	0.07582	-0.31020	0.31933	128	0.00678	-0.14376	0.14392
40	0.11075	-0.19074	0.22056	130	0.07792	-0.11942	0.14259
42	0.11559	-0.03794	0.12166	132	0.12339	-0.05708	0.13595
44	0.08688	0.10974	0.13996	134	0.13127	0.02125	0.13297
46	0.03119	0.21319	0.21545	136	0.10220	0.09175	0.13734
48	-0.03521	0.24397	0.24650	138	0.04699	0.13617	0.14405
50	-0.09039	0.19455	0.21452	140	-0.01837	0.14593	0.14708
52	-0.11398	0.08315	0.14108	142	-0.07796	0.12236	0.14509
54	-0.09533	-0.05003	0.10766	144	-0.11984	0.07423	0.14097
56	-0.03942	-0.15563	0.16054	146	-0.13799	0.01391	0.13869
58	0.03304	-0.19350	0.19630	148	-0.13218	-0.04626	0.14004
60	0.09238	-0.14971	0.17591	150	-0.10661	-0.09654	0.14382
62	0.11214	-0.04423	0.12055	152	-0.06786	-0.13109	0.14761
64	0.08173	0.07567	0.11138	154	-0.02305	-0.14797	0.14976
66	0.01320	0.15539	0.15594	156	0.02155	-0.14836	0.14992
68	-0.06232	0.15824	0.17007	158	0.06140	-0.13543	0.14870
70	-0.10822	0.08401	0.13700	160	0.09379	-0.11324	0.14704
72	-0.10064	-0.02950	0.10487	162	0.11775	-0.08586	0.14573
74	-0.04181	-0.12432	0.13117	164	0.13357	-0.05676	0.14513
76	0.03924	-0.15173	0.15672	166	0.14243	-0.02861	0.14528
78	0.10049	-0.09816	0.14048	168	0.14591	-0.00316	0.14594
80	0.10885	0.00679	0.10906	170	0.14568	0.01855	0.14686
82	0.05873	0.10580	0.12101	172	0.14331	0.03612	0.14779
84	-0.02375	0.14491	0.14685	174	0.14012	0.04949	0.14860
86	-0.09405	0.10336	0.13975	176	0.13713	0.05880	0.14921
88	-0.11347	0.00514	0.11359	178	0.13505	0.06429	0.14957
				180	0.13431	0.06609	0.14969

IEN

RADIATION PATTERN FOR THETA=PI/2

H/LMDA=0.2500 B/LMDA= 0.250 OMEGA= 8.53

28 ELEMENT ARRAY

PHI DEG	ER(THETA,PHI) RE	IM	ABSVAL	PHI DEG	ER(THETA,PHI) RE	IM	ABSVAL
0	-3.64173	7.92563	8.72226	90	0.07630	0.00214	0.07633
2	-3.63607	7.92610	8.72032	92	0.16526	-0.24752	0.29762
4	-3.61781	7.92473	8.71148	94	0.15914	-0.34852	0.38314
6	-3.58314	7.91319	8.68662	96	0.06648	-0.25377	0.26234
8	-3.52591	7.87773	8.63080	98	-0.05767	-0.02529	0.06297
10	-3.43793	7.79949	8.52358	100	-0.14466	0.20643	0.25207
12	-3.30962	7.65521	8.34002	102	-0.15022	0.31737	0.35112
14	-3.13088	7.41857	8.05218	104	-0.07618	0.25561	0.26672
16	-2.89250	7.06249	7.63187	106	0.03416	0.06300	0.07166
18	-2.58787	6.56241	7.05424	108	0.12210	-0.15438	0.19683
20	-2.21517	5.90079	6.30288	110	0.14483	-0.28586	0.32045
22	-1.77965	5.07245	5.37558	112	0.09536	-0.27187	0.28811
24	-1.29564	4.09026	4.29056	114	0.00221	-0.12790	0.12792
26	-0.78772	2.98980	3.09183	116	-0.08808	0.07085	0.11304
28	-0.29037	1.83191	1.85478	118	-0.13465	0.23088	0.26728
30	0.15475	0.70080	0.71769	120	-0.12017	0.28462	0.30895
32	0.50396	-0.30333	0.58821	122	-0.05532	0.21652	0.22348
34	0.71930	-1.07826	1.29616	124	0.02946	0.06274	0.06932
36	0.77756	-1.53898	1.72425	126	0.09925	-0.11067	0.14865
38	0.67869	-1.63922	1.77416	128	0.12902	-0.23779	0.27054
40	0.45091	-1.39009	1.46139	130	0.11153	-0.27734	0.29892
42	0.14960	-0.86873	0.88151	132	0.05697	-0.22322	0.23037
44	-0.15159	-0.21136	0.26010	134	-0.01356	-0.10061	0.10152
46	-0.37712	0.41226	0.55873	136	-0.07701	0.04792	0.09070
48	-0.47003	0.83990	0.96247	138	-0.11588	0.17859	0.21289
50	-0.41047	0.96345	1.04725	140	-0.12234	0.25939	0.28680
52	-0.22514	0.76595	0.79835	142	-0.09837	0.27627	0.29325
54	0.01812	0.33293	0.33342	144	-0.05306	0.23266	0.23863
56	0.23137	-0.17110	0.28777	146	0.00125	0.14464	0.14465
58	0.33785	-0.55846	0.65270	148	0.05269	0.03407	0.06275
60	0.30208	-0.68863	0.75198	150	0.09235	-0.07759	0.12061
62	0.14581	-0.52551	0.54536	152	0.11534	-0.17351	0.20835
64	-0.05932	-0.15481	0.16578	154	0.12070	-0.24345	0.27173
66	-0.22167	0.25075	0.33468	156	0.11043	-0.28370	0.30443
68	-0.27046	0.50757	0.57513	158	0.08829	-0.29576	0.30865
70	-0.18890	0.50562	0.53976	160	0.05867	-0.28447	0.29045
72	-0.02250	0.26070	0.26167	162	0.02571	-0.25628	0.25756
74	0.14312	-0.09395	0.17120	164	-0.00723	-0.21776	0.21788
76	0.22528	-0.37518	0.43762	166	-0.03773	-0.17480	0.17882
78	0.18643	-0.44386	0.48143	168	-0.06435	-0.13194	0.14680
80	0.05292	-0.27612	0.28115	170	-0.08640	-0.09259	0.12664
82	-0.10017	0.02689	0.10372	172	-0.10379	-0.05883	0.11930
84	-0.19036	0.29644	0.35230	174	-0.11673	-0.03192	0.12102
86	-0.17248	0.38932	0.42582	176	-0.12559	-0.01251	0.12621
88	-0.06190	0.26470	0.27184	178	-0.13073	-0.00081	0.13073
				180	-0.13241	0.00309	0.13245

VEN

RADIATION PATTERN FOR THETA=PI/2

H/LMDA=0.2500 B/LMDA= 0.250 OMEGA= 8.53

30 ELEMENT ARRAY

PHI DEG	ER(THETA,PHI) RE	IM	ABSVAL	PHI DEG	ER(THETA,PHI) RE	IM	ABSVAL
0	1.41253	-0.48878	1.49471	90	-0.09816	-0.08496	0.12982
2	1.41203	-0.48453	1.49285	92	-0.00222	-0.10709	0.10711
4	1.40980	-0.47164	1.48660	94	0.09409	-0.06249	0.11295
6	1.40376	-0.44969	1.47403	96	0.13030	0.02133	0.13203
8	1.39047	-0.41816	1.45198	98	0.08482	0.09266	0.12562
10	1.36529	-0.37656	1.41627	100	-0.01276	0.10786	0.10861
12	1.32267	-0.32481	1.36197	102	-0.10183	0.05800	0.11719
14	1.25661	-0.26351	1.28394	104	-0.12891	-0.02672	0.13165
16	1.16144	-0.19438	1.17760	106	-0.07941	-0.09627	0.12479
18	1.03289	-0.12053	1.03989	108	0.01595	-0.11085	0.11199
20	0.86942	-0.04663	0.87067	110	0.10159	-0.06316	0.11962
22	0.67368	0.02120	0.67402	112	0.13024	0.01935	0.13167
24	0.45383	0.07596	0.46014	114	0.08829	0.09170	0.12730
26	0.22415	0.11081	0.25004	116	0.00036	0.11658	0.11658
28	0.00467	0.12055	0.12064	118	-0.08701	0.08294	0.12020
30	-0.18092	0.10339	0.20838	120	-0.13092	0.00916	0.13124
32	-0.30908	0.06240	0.31531	122	-0.11266	-0.06868	0.13194
34	-0.36179	0.00627	0.36184	124	-0.04369	-0.11547	0.12346
36	-0.33235	-0.05155	0.33633	126	0.04326	-0.11264	0.12066
38	-0.23000	-0.09534	0.24898	128	0.11114	-0.06399	0.12824
40	-0.08115	-0.11138	0.13781	130	0.13453	0.00890	0.13482
42	0.07436	-0.09281	0.11893	132	0.10786	0.07768	0.13292
44	0.19260	-0.04342	0.19744	134	0.04428	0.11880	0.12678
46	0.23860	0.02160	0.23958	136	-0.03227	0.12099	0.12522
48	0.19929	0.07931	0.21449	138	-0.09712	0.08682	0.13027
50	0.09078	0.10690	0.14024	140	-0.13283	0.02923	0.13601
52	-0.04446	0.09136	0.10161	142	-0.13295	-0.03444	0.13734
54	-0.15203	0.03686	0.15644	144	-0.10141	-0.08818	0.13438
56	-0.18745	-0.03451	0.19060	146	-0.04905	-0.12127	0.13081
58	-0.13666	-0.09063	0.16398	148	0.01075	-0.12963	0.13007
60	-0.02481	-0.10386	0.10679	150	0.06591	-0.11508	0.13262
62	0.09316	-0.06568	0.11399	152	0.10796	-0.08337	0.13640
64	0.15822	0.00653	0.15835	154	0.13276	-0.04192	0.13922
66	0.13761	0.07597	0.15719	156	0.14010	0.00215	0.14012
68	0.04370	0.10486	0.11360	158	0.13251	0.04318	0.13936
70	-0.07080	0.07569	0.10364	160	0.11396	0.07755	0.13784
72	-0.14105	0.00318	0.14108	162	0.08877	0.10361	0.13644
74	-0.12710	-0.07186	0.14601	164	0.06082	0.12127	0.13567
76	-0.03828	-0.10518	0.11193	166	0.03315	0.13153	0.13564
78	0.07126	-0.07573	0.10399	168	0.00783	0.13595	0.13618
80	0.13463	0.00002	0.13463	170	-0.01392	0.13630	0.13701
82	0.11337	0.07648	0.13675	172	-0.03156	0.13423	0.13789
84	0.02172	0.10630	0.10850	174	-0.04497	0.13117	0.13866
86	-0.08202	0.07014	0.10792	176	-0.05432	0.12821	0.13924
88	-0.13235	-0.01026	0.13274	178	-0.05981	0.12613	0.13960
				180	-0.06162	0.12539	0.13971

IEN

RADIATION PATTERN FOR THETA=PI/2

H/LMDA=0.2500 B/LMDA= 0.250 OMEGA= 8.53

30 ELEMENT ARRAY

PHI DEG	ER(THETA,PHI) RE	IM	ABSVAL	PHI DEG	ER(THETA,PHI) RE	IM	ABSVAL
0	7.98166	3.62565	8.76653	90	-0.34438	-0.15353	0.37706
2	7.98216	3.61948	8.76444	92	-0.22900	-0.04943	0.23427
4	7.98045	3.59952	8.75467	94	0.02358	0.08000	0.08340
6	7.96692	3.56142	8.72671	96	0.24897	0.15305	0.29225
8	7.92568	3.49817	8.66335	98	0.30914	0.12773	0.33448
10	7.83499	3.40052	8.54111	100	0.17604	0.02513	0.17782
12	7.66820	3.25774	8.33152	102	-0.05870	-0.08791	0.10571
14	7.39547	3.05876	8.00307	104	-0.24815	-0.14281	0.28631
16	6.98677	2.79396	7.52471	106	-0.28234	-0.11047	0.30318
18	6.41610	2.45726	6.87055	108	-0.15007	-0.01506	0.15082
20	5.66728	2.04888	6.02627	110	0.06331	0.08483	0.10585
22	4.74050	1.57792	4.99622	112	0.23342	0.13289	0.26860
24	3.65906	1.06459	3.81078	114	0.26980	0.10619	0.28994
26	2.47420	0.54089	2.53263	116	0.16124	0.02400	0.16302
28	1.26636	0.04924	1.26732	118	-0.02704	-0.06724	0.07247
30	0.14019	-0.36216	0.38835	120	-0.19576	-0.12107	0.23017
32	-0.78813	-0.64702	1.01969	122	-0.26487	-0.11417	0.28843
34	-1.41133	-0.77128	1.60833	124	-0.20922	-0.05438	0.21617
36	-1.65665	-0.72389	1.80790	126	-0.06246	0.02729	0.06816
38	-1.51068	-0.52485	1.59926	128	0.10634	0.09401	0.14193
40	-1.03561	-0.22702	1.06020	130	0.22731	0.11973	0.25691
42	-0.36774	0.09159	0.37897	132	0.25806	0.09811	0.27608
44	0.30691	0.34514	0.46186	134	0.19516	0.04148	0.19952
46	0.79863	0.46413	0.92370	136	0.06835	-0.02708	0.07352
48	0.97112	0.41898	1.05765	138	-0.07610	-0.08396	0.11332
50	0.79034	0.23378	0.82419	140	-0.19371	-0.11300	0.22426
52	0.34392	-0.01788	0.34439	142	-0.25511	-0.10899	0.27742
54	-0.18385	-0.23724	0.30014	144	-0.25110	-0.07681	0.26258
56	-0.57923	-0.33837	0.67082	146	-0.19037	-0.02773	0.19238
58	-0.68625	-0.28499	0.74307	148	-0.09291	0.02508	0.09623
60	-0.47643	-0.10873	0.48868	150	0.01781	0.07030	0.07252
62	-0.06540	0.10290	0.12192	152	0.12102	0.10055	0.15734
64	0.33887	0.24675	0.41919	154	0.20226	0.11296	0.23166
66	0.53700	0.25494	0.59444	156	0.25447	0.10839	0.27659
68	0.43993	0.13027	0.45881	158	0.27699	0.09024	0.29131
70	0.11620	-0.05443	0.12831	160	0.27373	0.06300	0.28088
72	-0.24537	-0.19563	0.31381	162	0.25101	0.03118	0.25294
74	-0.44135	-0.21687	0.49175	164	0.21583	-0.00140	0.21584
76	-0.36950	-0.11225	0.38617	166	0.17456	-0.03197	0.17747
78	-0.08654	0.05096	0.10043	168	0.13244	-0.05879	0.14490
80	0.22859	0.17326	0.28683	170	0.09319	-0.08106	0.12351
82	0.38511	0.18349	0.42659	172	0.05931	-0.09857	0.11503
84	0.29662	0.08129	0.30756	174	0.03223	-0.11155	0.11611
86	0.03067	-0.06484	0.07172	176	0.01263	-0.12041	0.12107
88	-0.23886	-0.16242	0.28885	178	0.00085	-0.12552	0.12553
				180	-0.00310	-0.12720	0.12724

VEN

RADIATION PATTERN FOR THETA=PI/2

H/LMDA=0.2500 B/LMDA= 0.250 OMEGA= 8.53

35 ELEMENT ARRAY

PHI	ER(THETA,PHI)			PHI	ER(THETA,PHI)		
DEG	RE	IM	ABSVAL	DEG	RE	IM	ABSVAL
0	-0.60960	1.30108	1.43681	90	-0.06895	-0.06102	0.09208
1	-0.61034	1.30020	1.43632	91	-0.02123	-0.08755	0.09009
2	-0.61252	1.29751	1.43482	92	0.03093	-0.09435	0.09929
3	-0.61605	1.29286	1.43213	93	0.07601	-0.07986	0.11025
4	-0.62078	1.28600	1.42799	94	0.10414	-0.04733	0.11439
5	-0.62648	1.27659	1.42203	95	0.10928	-0.00407	0.10936
6	-0.63286	1.26422	1.41378	96	0.09048	0.04024	0.09903
7	-0.63957	1.24837	1.40267	97	0.05205	0.07574	0.09190
8	-0.64617	1.22849	1.38806	98	0.00249	0.09458	0.09461
9	-0.65215	1.20396	1.36924	99	-0.04739	0.09264	0.10405
10	-0.65693	1.17415	1.34543	100	-0.08686	0.07041	0.11181
11	-0.65985	1.13844	1.31584	101	-0.10762	0.03280	0.11250
12	-0.66022	1.09622	1.27968	102	-0.10547	-0.01201	0.10615
13	-0.65726	1.04696	1.23617	103	-0.08114	-0.05437	0.09767
14	-0.65019	0.99026	1.18463	104	-0.03999	-0.08530	0.09421
15	-0.63821	0.92585	1.12451	105	0.00926	-0.09837	0.09880
16	-0.62055	0.85370	1.05541	106	0.05638	-0.09097	0.10703
17	-0.59651	0.77402	0.97720	107	0.09185	-0.06480	0.11241
18	-0.56551	0.68731	0.89006	108	0.10871	-0.02536	0.11163
19	-0.52714	0.59446	0.79451	109	0.10390	0.01926	0.10567
20	-0.48120	0.49668	0.69155	110	0.07873	0.06012	0.09906
21	-0.42782	0.39561	0.58270	111	0.03844	0.08922	0.09715
22	-0.36743	0.29325	0.47011	112	-0.00899	0.10106	0.10146
23	-0.30088	0.19197	0.35691	113	-0.05448	0.09360	0.10830
24	-0.22944	0.09441	0.24811	114	-0.08960	0.06851	0.11279
25	-0.15482	0.00340	0.15486	115	-0.10813	0.03072	0.11241
26	-0.07915	-0.07816	0.11123	116	-0.10710	-0.01266	0.10784
27	-0.00493	-0.14747	0.14755	117	-0.08712	-0.05376	0.10237
28	0.06504	-0.20203	0.21224	118	-0.05213	-0.08536	0.10002
29	0.12782	-0.23981	0.27175	119	-0.00847	-0.10218	0.10253
30	0.18047	-0.25949	0.31607	120	0.03631	-0.10166	0.10795
31	0.22031	-0.26060	0.34125	121	0.07481	-0.08426	0.11268
32	0.24509	-0.24373	0.34565	122	0.10100	-0.05320	0.11415
33	0.25326	-0.21054	0.32935	123	0.11110	-0.01376	0.11195
34	0.24421	-0.16384	0.29408	124	0.10403	0.02770	0.10766
35	0.21840	-0.10745	0.24340	125	0.08140	0.06480	0.10404
36	0.17753	-0.04600	0.18339	126	0.04703	0.09213	0.10344
37	0.12451	0.01534	0.12545	127	0.00625	0.10602	0.10620
38	0.06342	0.07129	0.09542	128	-0.03501	0.10490	0.11059
39	-0.00077	0.11691	0.11692	129	-0.07110	0.08941	0.11423
40	-0.06257	0.14814	0.16081	130	-0.09744	0.06209	0.11554
41	-0.11635	0.16217	0.19959	131	-0.11101	0.02692	0.11423
42	-0.15703	0.15792	0.22270	132	-0.11063	-0.01137	0.11121
43	-0.18051	0.13617	0.22611	133	-0.09692	-0.04797	0.10814
44	-0.18430	0.09968	0.20953	134	-0.07208	-0.07861	0.10665
45	-0.16792	0.05294	0.17606	135	-0.03944	-0.09999	0.10749
46	-0.13307	0.00172	0.13308	136	-0.00294	-0.11015	0.11019
47	-0.08367	-0.04751	0.09622	137	0.03334	-0.10849	0.11350
48	-0.02552	-0.08846	0.09207	138	0.06570	-0.09574	0.11611
49	0.03428	-0.11576	0.12073	139	0.09115	-0.07370	0.11722
50	0.08816	-0.12583	0.15364	140	0.10764	-0.04497	0.11666
51	0.12907	-0.11741	0.17448	141	0.11417	-0.01254	0.11486
52	0.15145	-0.09189	0.17715	142	0.11076	0.02050	0.11264
53	0.15216	-0.05320	0.16119	143	0.09828	0.05133	0.11087
54	0.13101	-0.00735	0.13122	144	0.07831	0.07756	0.11022
55	0.09102	0.03843	0.09880	145	0.05287	0.09745	0.11087
56	0.03812	0.07679	0.08574	146	0.02420	0.10993	0.11257
57	-0.01962	0.10147	0.10335	147	-0.00546	0.11462	0.11475
58	-0.07316	0.10836	0.13075	148	-0.03409	0.11175	0.11683
59	-0.11391	0.09633	0.14918	149	-0.05994	0.10205	0.11835
60	-0.13516	0.06751	0.15108	150	-0.08167	0.08663	0.11906
61	-0.13331	0.02705	0.13602	151	-0.09841	0.06684	0.11896
62	-0.10863	-0.01772	0.11007	152	-0.10967	0.04411	0.11821
63	-0.06537	-0.05853	0.08775	153	-0.11537	0.01987	0.11707
64	-0.01116	-0.08772	0.08843	154	-0.11577	-0.00457	0.11586
65	0.04428	-0.09972	0.10911	155	-0.11136	-0.02808	0.11484
66	0.09080	-0.09216	0.12937	156	-0.10281	-0.04975	0.11421
67	0.11974	-0.06648	0.13696	157	-0.09091	-0.06888	0.11406
68	0.12564	-0.02773	0.12866	158	-0.07648	-0.08502	0.11436
69	0.10733	0.01633	0.10857	159	-0.06034	-0.09795	0.11504
70	0.06837	0.05674	0.08885	160	-0.04322	-0.10760	0.11596
71	0.01639	0.08515	0.08672	161	-0.02579	-0.11410	0.11698
72	-0.03827	0.09560	0.10297	162	-0.00862	-0.11766	0.11798
73	-0.08463	0.08580	0.12052	163	0.00786	-0.11861	0.11887
74	-0.11327	0.05775	0.12715	164	0.02330	-0.11730	0.11959
75	-0.11833	0.01737	0.11960	165	0.03746	-0.11413	0.12012
76	-0.09878	-0.02668	0.10232	166	0.05022	-0.10950	0.12046
77	-0.05872	-0.06484	0.08747	167	0.06150	-0.10377	0.12063
78	-0.00662	-0.08871	0.08896	168	0.07131	-0.09732	0.12065
79	0.04645	-0.09297	0.10393	169	0.07971	-0.09045	0.12056
80	0.08915	-0.07658	0.11752	170	0.08678	-0.08346	0.12040
81	0.11233	-0.04309	0.12031	171	0.09264	-0.07657	0.12018
82	0.11102	0.00007	0.11102	172	0.09742	-0.06998	0.11995
83	0.08558	0.04327	0.09590	173	0.10126	-0.06387	0.11972
84	0.04163	0.07677	0.08733	174	0.10428	-0.05837	0.11950
85	-0.01115	0.09297	0.09364	175	0.10661	-0.05357	0.11931
86	-0.06114	0.08814	0.10727	176	0.10835	-0.04955	0.11914
87	-0.09736	0.06331	0.11613	177	0.10961	-0.04637	0.11902
88	-0.11189	0.02405	0.11444	178	0.11045	-0.04408	0.11892
89	-0.10162	-0.02078	0.10372	179	0.11094	-0.04269	0.11887
				180	0.11110	-0.04223	0.11885

IEN

RADIATION PATTERN FOR THETA=PI/2

H/LMDA=0.2500 B/LMDA= 0.250 OMEGA= 8.53

35 ELEMENT ARRAY

PHI	ER(THETA,PHI)			PHI	ER(THETA,PHI)		
DEG	RE	IM	ABSVAL	DEG	RE	IM	ABSVAL
0	-7.31608	3.51118	8.11501	90	-0.15766	0.11979	0.19801
1	-7.31493	3.51241	8.11451	91	-0.03669	0.07566	0.08409
2	-7.31124	3.51600	8.11274	92	0.08927	0.01577	0.09065
3	-7.30425	3.52158	8.10886	93	0.19199	-0.04605	0.19744
4	-7.29270	3.52855	8.10149	94	0.24923	-0.09596	0.26706
5	-7.27486	3.53608	8.08872	95	0.24949	-0.12316	0.27824
6	-7.24850	3.54306	8.06809	96	0.19431	-0.12224	0.22956
7	-7.21097	3.54816	8.03664	97	0.09742	-0.09417	0.13550
8	-7.15915	3.54982	7.99091	98	-0.01864	-0.04593	0.04956
9	-7.08958	3.54621	7.92703	99	-0.12785	0.01133	0.12835
10	-6.99849	3.53532	7.84075	100	-0.20654	0.06478	0.21646
11	-6.88187	3.51491	7.72753	101	-0.23849	0.10291	0.25974
12	-6.73564	3.48263	7.58272	102	-0.21813	0.11787	0.24794
13	-6.55572	3.43596	7.40157	103	-0.15137	0.10710	0.18543
14	-6.33827	3.37242	7.17961	104	-0.05381	0.07364	0.09120
15	-6.07984	3.28954	6.91271	105	0.05303	0.02522	0.05872
16	-5.77768	3.18501	6.59741	106	0.14645	-0.02754	0.14902
17	-5.42990	3.05683	6.23121	107	0.20746	-0.07352	0.22011
18	-5.03586	2.90342	5.81289	108	0.22452	-0.10344	0.24720
19	-4.59636	2.72379	5.34280	109	0.19549	-0.11166	0.22513
20	-4.11393	2.51770	4.82319	110	0.12770	-0.09716	0.16046
21	-3.59314	2.28585	4.25861	111	0.03585	-0.06352	0.07294
22	-3.04066	2.03000	3.65602	112	-0.06121	-0.01793	0.06378
23	-2.46541	1.75311	3.02517	113	-0.14449	0.03037	0.14765
24	-1.87858	1.45943	2.37886	114	-0.19852	0.07204	0.21119
25	-1.29335	1.15448	1.73366	115	-0.21409	0.09939	0.23604
26	-0.72465	0.84508	1.11322	116	-0.18963	0.10780	0.21813
27	-0.18860	0.53910	0.57113	117	-0.13098	0.09632	0.16259
28	0.29821	0.24528	0.38613	118	-0.04992	0.06766	0.08408
29	0.71959	-0.02715	0.72011	119	0.03845	0.02745	0.04724
30	1.06077	-0.26894	1.09433	120	0.11851	-0.01694	0.11972
31	1.30953	-0.47141	1.39179	121	0.17695	-0.05779	0.18615
32	1.45732	-0.62699	1.58647	122	0.20483	-0.08840	0.22309
33	1.50029	-0.72998	1.66845	123	0.19876	-0.10412	0.22438
34	1.44018	-0.77707	1.63644	124	0.16105	-0.10296	0.19115
35	1.28478	-0.76789	1.49677	125	0.09887	-0.08571	0.13084
36	1.04810	-0.70541	1.26337	126	0.02271	-0.05558	0.06004
37	0.74987	-0.59603	0.95789	127	-0.05545	-0.01756	0.05816
38	0.41463	-0.44947	0.61151	128	-0.12415	0.02252	0.12618
39	0.07012	-0.27829	0.28699	129	-0.17403	0.05889	0.18372
40	-0.25480	-0.09710	0.27267	130	-0.19901	0.08668	0.21707
41	-0.53251	0.07849	0.53826	131	-0.19688	0.10251	0.22197
42	-0.73937	0.23326	0.77529	132	-0.16917	0.10482	0.19901
43	-0.85805	0.35368	0.92808	133	-0.12058	0.09389	0.15282
44	-0.87969	0.42945	0.97892	134	-0.05811	0.07162	0.09223
45	-0.80524	0.45459	0.92470	135	0.01009	0.04117	0.04239
46	-0.64581	0.42834	0.77495	136	0.07587	0.00640	0.07614
47	-0.42192	0.35542	0.55167	137	0.13201	-0.02863	0.13508
48	-0.16147	0.24576	0.29406	138	0.17299	-0.06019	0.18316
49	0.10318	0.11353	0.15341	139	0.19536	-0.08518	0.21312
50	0.33913	-0.02440	0.34001	140	0.19791	-0.10147	0.22241
51	0.51715	-0.15055	0.53862	141	0.18157	-0.10798	0.21125
52	0.61550	-0.24897	0.66395	142	0.14894	-0.10463	0.18202
53	0.62321	-0.30745	0.69492	143	0.10392	-0.09227	0.13897
54	0.54190	-0.31932	0.62898	144	0.05109	-0.07245	0.08865
55	0.38581	-0.28446	0.47934	145	-0.00473	-0.04721	0.04745
56	0.18009	-0.20947	0.27624	146	-0.05905	-0.01878	0.06196
57	-0.04280	-0.10687	0.11512	147	-0.10790	0.01056	0.10842
58	-0.24789	0.00676	0.24799	148	-0.14826	0.03874	0.15323
59	-0.40327	0.11328	0.41888	149	-0.17801	0.06400	0.18916
60	-0.48520	0.19586	0.52324	150	-0.19605	0.08497	0.21367
61	-0.48217	0.24172	0.53936	151	-0.20221	0.10076	0.22593
62	-0.39707	0.24436	0.46624	152	-0.19707	0.11089	0.22613
63	-0.24679	0.20469	0.32063	153	-0.18185	0.11530	0.21532
64	-0.05948	0.13092	0.14380	154	-0.15814	0.11423	0.19509
65	0.13045	0.03719	0.13565	155	-0.12786	0.10824	0.16753
66	0.28844	-0.05898	0.29441	156	-0.09292	0.09802	0.13506
67	0.38619	-0.13988	0.41074	157	-0.05520	0.08438	0.10083
68	0.40700	-0.19087	0.44953	158	-0.01649	0.06819	0.07016
69	0.34898	-0.20320	0.40383	159	0.02175	0.05029	0.05479
70	0.22544	-0.17565	0.28579	160	0.05827	0.03146	0.06622
71	0.06228	-0.11479	0.13060	161	0.09213	0.01238	0.09296
72	-0.10710	-0.03360	0.11225	162	0.12270	-0.00637	0.12287
73	-0.24859	0.05108	0.25378	163	0.14953	-0.02431	0.15149
74	-0.33422	0.12202	0.35580	164	0.17249	-0.04111	0.17732
75	-0.34791	0.16512	0.38511	165	0.19157	-0.05652	0.19973
76	-0.28873	0.17227	0.33622	166	0.20696	-0.07040	0.21860
77	-0.17096	0.14303	0.22290	167	0.21893	-0.08268	0.23402
78	-0.02099	0.08458	0.08714	168	0.22784	-0.09338	0.24623
79	0.12848	0.01017	0.12888	169	0.23411	-0.10256	0.25559
80	0.24542	-0.06383	0.25358	170	0.23815	-0.11030	0.26246
81	0.30548	-0.12147	0.32874	171	0.24039	-0.11674	0.26724
82	0.29714	-0.15073	0.33318	172	0.24124	-0.12202	0.27035
83	0.22401	-0.14601	0.26739	173	0.24108	-0.12627	0.27215
84	0.10384	-0.10931	0.15077	174	0.24024	-0.12963	0.27299
85	-0.03566	-0.04960	0.06109	175	0.23903	-0.13223	0.27317
86	-0.16315	0.01927	0.16429	176	0.23770	-0.13420	0.27296
87	-0.25069	0.08177	0.26368	177	0.23645	-0.13562	0.27258
88	-0.27987	0.12417	0.30618	178	0.23544	-0.13657	0.27218
89	-0.24576	0.13757	0.28164	179	0.23479	-0.13713	0.27190
				180	0.23457	-0.13730	0.27180

VEN

RADIATION PATTERN FOR THETA=PI/2

H/LMDA=0.2500 B/LMDA= 0.250 OMEGA= 8.53

40 ELEMENT ARRAY

PHI DEG	ER(THETA,PHI) RE	IM	ABSVAL	PHI DEG	ER(THETA,PHI) RE	IM	ABSVAL
0	-0.48817	-1.30102	1.38959	90	-0.05040	-0.06636	0.08333
1	-0.48695	-1.30093	1.38908	91	-0.01029	-0.09439	0.09495
2	-0.48326	-1.30060	1.38748	92	0.03296	-0.09476	0.10033
3	-0.47706	-1.29981	1.38459	93	0.06690	-0.06757	0.09508
4	-0.46825	-1.29820	1.38007	94	0.08174	-0.02089	0.08437
5	-0.45672	-1.29528	1.37344	95	0.07322	0.03161	0.07976
6	-0.44234	-1.29041	1.36412	96	0.04376	0.07475	0.08661
7	-0.42500	-1.28283	1.35140	97	0.00173	0.09621	0.09623
8	-0.40456	-1.27168	1.33448	98	-0.04093	0.09005	0.09892
9	-0.38094	-1.25595	1.31245	99	-0.07220	0.05827	0.09278
10	-0.35411	-1.23461	1.28439	100	-0.08333	0.01012	0.08394
11	-0.32409	-1.20657	1.24934	101	-0.07129	-0.04069	0.08208
12	-0.29103	-1.17073	1.20636	102	-0.03949	-0.07994	0.08917
13	-0.25517	-1.12605	1.15460	103	0.00323	-0.09688	0.09693
14	-0.21691	-1.07162	1.09335	104	0.04517	-0.08710	0.09812
15	-0.17681	-1.00673	1.02214	105	0.07502	-0.05361	0.09220
16	-0.13560	-0.93095	0.94078	106	0.08486	-0.00576	0.08506
17	-0.09418	-0.84422	0.84946	107	0.07225	0.04345	0.08431
18	-0.05360	-0.74696	0.74888	108	0.04069	0.08098	0.09063
19	-0.01506	-0.64012	0.64029	109	-0.00146	0.09718	0.09719
20	0.02014	-0.52529	0.52568	110	-0.04328	0.08819	0.09824
21	0.05071	-0.40473	0.40789	111	-0.07419	0.05674	0.09340
22	0.07537	-0.28133	0.29125	112	-0.08662	0.01112	0.08733
23	0.09300	-0.15864	0.18389	113	-0.07774	-0.03706	0.08612
24	0.10276	-0.04070	0.11052	114	-0.05001	-0.07596	0.09094
25	0.10414	0.06809	0.12442	115	-0.01035	-0.09637	0.09693
26	0.09713	0.16324	0.18996	116	0.03171	-0.09384	0.09905
27	0.08227	0.24046	0.25414	117	0.06641	-0.06945	0.09609
28	0.06066	0.29601	0.30216	118	0.08599	-0.02927	0.09083
29	0.03405	0.32709	0.32886	119	0.08637	0.01726	0.08808
30	0.00468	0.33223	0.33226	120	0.06787	0.05972	0.09040
31	-0.02480	0.31156	0.31255	121	0.03486	0.08900	0.09558
32	-0.05154	0.26710	0.27203	122	-0.00537	0.09924	0.09939
33	-0.07278	0.20279	0.21545	123	-0.04437	0.08885	0.09931
34	-0.08611	0.12438	0.15128	124	-0.07432	0.06051	0.09584
35	-0.08984	0.03912	0.09799	125	-0.08955	0.02041	0.09184
36	-0.08322	-0.04480	0.09451	126	-0.08754	-0.02333	0.09060
37	-0.06668	-0.11903	0.13644	127	-0.06919	-0.06231	0.09311
38	-0.04191	-0.17590	0.18082	128	-0.03833	-0.08950	0.09736
39	-0.01176	-0.20937	0.20970	129	-0.00085	-0.10044	0.10044
40	0.01999	-0.21590	0.21683	130	0.03652	-0.09381	0.10067
41	0.04909	-0.19506	0.20114	131	0.06752	-0.07139	0.09826
42	0.07136	-0.14985	0.16597	132	0.08730	-0.03751	0.09502
43	0.08333	-0.08652	0.12012	133	0.09316	0.00195	0.09318
44	0.08286	-0.01400	0.08403	134	0.08474	0.04065	0.09399
45	0.06961	0.05729	0.09015	135	0.06389	0.07283	0.09688
46	0.04525	0.11682	0.12528	136	0.03413	0.09411	0.10011
47	0.01341	0.15562	0.15620	137	-0.00001	0.10204	0.10204
48	-0.02082	0.16777	0.16906	138	-0.03376	[illegible]	[illegible]
49	-0.05159	0.15156	0.16010	139	-0.06278	0.07807	0.10018
50	-0.07335	0.11002	0.13223	140	-0.08373	0.05056	0.09781
51	-0.08188	0.05060	0.09625	141	-0.09452	0.01750	0.09613
52	-0.07519	-0.01587	0.07684	142	-0.09449	-0.01697	0.09600
53	-0.05413	-0.07711	0.09421	143	-0.08425	-0.04899	0.09745
54	-0.02240	-0.12159	0.12364	144	-0.06548	-0.07529	0.09978
55	0.01394	-0.14083	0.14152	145	-0.04058	-0.09360	0.10202
56	0.04760	-0.13120	0.13957	146	-0.01234	-0.10269	0.10342
57	0.07150	-0.09491	0.11883	147	0.01646	-0.10236	0.10367
58	0.08031	-0.03971	0.08959	148	0.04328	-0.09332	0.10287
59	0.07175	0.02250	0.07519	149	0.06608	-0.07699	0.10146
60	0.04729	0.07813	0.09133	150	0.08337	-0.05521	0.10000
61	0.01207	0.11489	0.11553	151	0.09433	-0.03006	0.09900
62	-0.02608	0.12459	0.12729	152	0.09870	-0.00359	0.09877
63	-0.05834	0.10517	0.12026	153	0.09679	0.02234	0.09933
64	-0.07695	0.06136	0.09842	154	0.08927	0.04615	0.10049
65	-0.07716	0.00378	0.07725	155	0.07710	0.06665	0.10191
66	-0.05855	-0.05354	0.07934	156	0.06142	0.08305	0.10329
67	-0.02537	-0.09647	0.09975	157	0.04338	0.09493	0.10437
68	0.01433	-0.11437	0.11526	158	0.02410	0.10222	0.10502
69	0.05060	-0.10283	0.11460	159	0.00458	0.10513	0.10523
70	0.07404	-0.06497	0.09850	160	-0.01433	0.10407	0.10505
71	0.07834	-0.01077	0.07908	161	-0.03197	0.09959	0.10460
72	0.06207	0.04541	0.07690	162	-0.04786	0.09232	0.10399
73	0.02923	0.08863	0.09332	163	-0.06171	0.08292	0.10336
74	-0.01162	0.10732	0.10794	164	-0.07337	0.07201	0.10280
75	-0.04952	0.09652	0.10848	165	-0.08283	0.06018	0.10238
76	-0.07409	0.05934	0.09493	166	-0.09018	0.04794	0.10213
77	-0.07840	0.00618	0.07865	167	-0.09559	0.03572	0.10205
78	-0.06101	-0.04809	0.07769	168	-0.09930	0.02387	0.10212
79	-0.02652	-0.08824	0.09213	169	-0.10153	0.01266	0.10231
80	0.01559	-0.10299	0.10416	170	-0.10257	0.00226	0.10259
81	0.05351	-0.08827	0.10322	171	-0.10266	-0.00718	0.10291
82	0.07643	-0.04845	0.09049	172	-0.10207	-0.01559	0.10325
83	0.07770	0.00495	0.07785	173	-0.10101	-0.02294	0.10358
84	0.05676	0.05647	0.08006	174	-0.09970	-0.02923	0.10389
85	0.01945	0.09122	0.09327	175	-0.09829	-0.03446	0.10416
86	-0.02363	0.09923	0.10201	176	-0.09695	-0.03869	0.10438
87	-0.06010	0.07832	0.09872	177	-0.09579	-0.04192	0.10456
88	-0.07940	0.03474	0.08666	178	-0.09489	-0.04421	0.10469
89	-0.07586	-0.01870	0.07813	179	-0.09433	-0.04557	0.10476
				180	-0.09414	-0.04602	0.10478

IEN

RADIATION PATTERN FOR THETA=PI/2

H/LMDA=0.2500 B/LMDA= 0.250 OMEGA= 8.53

40 ELEMENT ARRAY

PHI DEG	ER(THETA,PHI) RE	IM	ABSVAL	PHI DEG	ER(THETA,PHI) RE	IM	ABSVAL
0	3.57178	-8.22933	8.97104	90	0.06040	0.00178	0.06042
1	3.56961	-8.22960	8.97041	91	0.10894	-0.13423	0.17287
2	3.56296	-8.22998	8.96812	92	0.12459	-0.22764	0.25950
3	3.55133	-8.22939	8.96297	93	0.10375	-0.25254	0.27302
4	3.53393	-8.22600	8.95298	94	0.05344	-0.20361	0.21050
5	3.50967	-8.21722	8.93535	95	-0.01098	-0.09700	0.09762
6	3.47716	-8.19982	8.90661	96	-0.07057	0.03497	0.07876
7	3.43477	-8.16978	8.86244	97	-0.10838	0.15367	0.18804
8	3.38065	-8.12250	8.79795	98	-0.11421	0.22548	0.25276
9	3.31279	-8.05281	8.70760	99	-0.08734	0.23124	0.24719
10	3.22910	-7.95503	8.58543	100	-0.03636	0.17116	0.17498
11	3.12748	-7.82317	8.42515	101	0.02364	0.06407	0.06829
12	3.00593	-7.65106	8.22036	102	0.07557	-0.05865	0.09565
13	2.86269	-7.43264	7.96486	103	0.10519	-0.16233	0.19343
14	2.69639	-7.16225	7.65300	104	0.10501	-0.21883	0.24272
15	2.50622	-6.83500	7.28000	105	0.07601	-0.21402	0.22711
16	2.29207	-6.44715	6.84247	106	0.02699	-0.15099	0.15339
17	2.05477	-5.99656	6.33883	107	-0.02819	-0.04849	0.05609
18	1.79618	-5.48318	5.76988	108	-0.07458	0.06492	0.09888
19	1.51944	-4.90954	5.13929	109	-0.10020	0.15894	0.18788
20	1.22896	-4.28120	4.45410	110	-0.09903	0.20956	0.23178
21	0.93052	-3.60695	3.72504	111	-0.07228	0.20505	0.21742
22	0.63119	-2.89917	2.96708	112	-0.02766	0.14830	0.15086
23	0.33919	-2.17372	2.20002	113	0.02295	0.05524	0.05982
24	0.06358	-1.44971	1.45111	114	0.06675	-0.04984	0.08330
25	-0.18613	-0.74894	0.77173	115	0.09323	-0.14083	0.16889
26	-0.40054	-0.09496	0.41164	116	0.09662	-0.19625	0.21875
27	-0.57098	0.48810	0.75117	117	0.07696	-0.20415	0.21818
28	-0.69031	0.97722	1.19645	118	0.03973	-0.16431	0.16905
29	-0.75359	1.35211	1.54794	119	-0.00584	-0.08750	0.08769
30	-0.75884	1.59723	1.76833	120	-0.04912	0.00769	0.04972
31	-0.70760	1.70364	1.84475	121	-0.08059	0.09958	0.12811
32	-0.60528	1.67056	1.77683	122	-0.09388	0.16839	0.19279
33	-0.46117	1.50651	1.57551	123	-0.08688	0.20039	0.21842
34	-0.28815	1.22971	1.26302	124	-0.06190	0.19033	0.20015
35	-0.10187	0.86758	0.87354	125	-0.02482	0.14186	0.14401
36	0.08042	0.45517	0.46222	126	0.01643	0.06609	0.06810
37	0.24141	0.03255	0.24359	127	0.05359	-0.02122	0.05764
38	0.36547	-0.35860	0.51202	128	0.07978	-0.10312	0.13038
39	0.44052	-0.67930	0.80964	129	0.09060	-0.16476	0.18803
40	0.45959	-0.89758	1.00841	130	0.08483	-0.19601	0.21358
41	0.42198	-0.99252	1.07850	131	0.06427	-0.19271	0.20315
42	0.33381	-0.95745	1.01397	132	0.03316	-0.15700	0.16046
43	0.20756	-0.80145	0.82789	133	-0.00277	-0.09619	0.09623
44	0.06089	-0.54908	0.55245	134	-0.03748	-0.02112	0.04302
45	-0.08559	-0.23777	0.25271	135	-0.06557	0.05594	0.08619
46	-0.21107	0.08683	0.22823	136	-0.08313	0.12344	0.14882
47	-0.29762	0.37709	0.48039	137	-0.08813	0.17215	0.19340
48	[illegible]	[illegible]	[illegible]	138	-0.08037	0.19627	0.21217
49	-0.31313	0.69632	0.76348	139	-0.06219	0.19390	0.20363
50	-0.24246	0.68117	0.72304	140	-0.03605	0.16679	0.17064
51	-0.13421	0.55152	0.56761	141	-0.00596	0.11970	0.11985
52	-0.00803	0.33333	0.33343	142	0.02416	0.05936	0.06409
53	0.11322	0.06793	0.13203	143	0.05076	-0.00658	0.05119
54	0.20750	-0.19501	0.28475	144	0.07104	-0.07057	0.10013
55	0.25782	-0.40656	0.48142	145	0.08322	-0.12613	0.15111
56	0.25560	-0.52784	0.58647	146	0.08658	-0.16829	0.18926
57	0.20259	-0.53785	0.57474	147	0.08139	-0.19397	0.21036
58	0.11076	-0.43802	0.45181	148	0.06875	-0.20206	0.21343
59	0.00001	-0.25221	0.25221	149	0.05033	-0.19311	0.19956
60	-0.10581	-0.02207	0.10809	150	0.02814	-0.16916	0.17148
61	-0.18403	0.20162	0.27298	151	0.00428	-0.13320	0.13327
62	-0.21816	0.37002	0.42954	152	-0.01927	-0.08877	0.09084
63	-0.20163	0.44712	0.49048	153	-0.04084	-0.03958	0.05687
64	-0.13948	0.41803	0.44069	154	-0.05906	0.01080	0.06004
65	-0.04729	0.29260	0.29639	155	-0.07306	0.05929	0.09408
66	0.05244	0.10319	0.11575	156	-0.08233	0.10339	0.13217
67	0.13557	-0.10293	0.17022	157	-0.08678	0.14128	0.16580
68	0.18226	-0.27508	0.32998	158	-0.08659	0.17178	0.19237
69	0.18187	-0.37166	0.41377	159	-0.08222	0.19434	0.21102
70	0.13567	-0.37057	0.39462	160	-0.07429	0.20898	0.22179
71	0.05655	-0.27484	0.28060	161	-0.06349	0.21611	0.22525
72	-0.03430	-0.11178	0.11692	162	-0.05056	0.21650	0.22232
73	-0.11293	0.07428	0.13517	163	-0.03621	0.21106	0.21415
74	-0.15893	0.23365	0.28258	164	-0.02111	0.20086	0.20197
75	-0.16089	0.32463	0.36231	165	-0.00583	0.18701	0.18710
76	-0.11939	0.32462	0.34588	166	0.00918	0.17049	0.17074
77	-0.04679	0.23612	0.24071	167	0.02353	0.15229	0.15410
78	0.03623	0.08580	0.09314	168	0.03695	0.13324	0.13827
79	0.10653	-0.08316	0.13514	169	0.04923	0.11413	0.12429
80	0.14488	-0.22323	0.26612	170	0.06030	0.09545	0.11290
81	0.14138	-0.29595	0.32799	171	0.07008	0.07775	0.10467
82	0.09811	-0.28267	0.29921	172	0.07859	0.06135	0.09970
83	0.02834	-0.18947	0.19158	173	0.08584	0.04656	0.09765
84	-0.04739	-0.04506	0.06539	174	0.09193	0.03351	0.09785
85	-0.10729	0.10787	0.15214	175	0.09690	0.02239	0.09945
86	-0.13461	0.22518	0.26234	176	0.10084	0.01323	0.10170
87	-0.12236	0.27404	0.30012	177	0.10381	0.00609	0.10399
88	-0.07514	0.24221	0.25359	178	0.10590	0.00096	0.10590
89	-0.00751	0.14106	0.14126	179	0.10713	-0.00210	0.10715
				180	0.10753	-0.00312	0.10758

VEN

RADIATION PATTERN FOR THETA=PI/2

H/LMDA=0.2500 B/LMDA= 0.250 OMEGA= 8.53

45 ELEMENT ARRAY

PHI	ER(THETA,PHI)			PHI	ER(THETA,PHI)		
DEG	RE	IM	ABSVAL	DEG	RE	IM	ABSVAL
0	1.23467	0.54612	1.35006	90	-0.05607	-0.06754	0.08778
1	1.23370	0.54699	1.34952	91	-0.01799	-0.08585	0.08771
2	1.23071	0.54955	1.34783	92	0.02685	-0.07297	0.07776
3	1.22548	0.55365	1.34474	93	0.06194	-0.03376	0.07054
4	1.21759	0.55909	1.33982	94	0.07436	0.01747	0.07638
5	1.20655	0.56551	1.33250	95	0.05956	0.06217	0.08610
6	1.19167	0.57250	1.32206	96	0.02296	0.08435	0.08742
7	1.17219	0.57952	1.30762	97	-0.02212	0.07625	0.07940
8	1.14725	0.58592	1.28821	98	-0.05934	0.04101	0.07213
9	1.11596	0.59094	1.26277	99	-0.07530	-0.00864	0.07579
10	1.07741	0.59371	1.23016	100	-0.06435	-0.05504	0.08467
11	1.03074	0.59329	1.18929	101	-0.03046	-0.08195	0.08743
12	0.97523	0.58864	1.13911	102	0.01429	-0.08021	0.08147
13	0.91034	0.57872	1.07872	103	0.05416	-0.05074	0.07422
14	0.83582	0.56248	1.00747	104	0.07531	-0.00398	0.07542
15	0.75182	0.53897	0.92505	105	0.07056	0.04395	0.08313
16	0.65892	0.50738	0.83163	106	0.04171	0.07687	0.08746
17	0.55825	0.46715	0.72793	107	-0.00131	0.08399	0.08400
18	0.45157	0.41810	0.61540	108	-0.04399	0.06334	0.07711
19	0.34123	0.36046	0.49636	109	-0.07221	0.02207	0.07551
20	0.23022	0.29505	0.37424	110	-0.07689	-0.02614	0.08121
21	0.12206	0.22329	0.25447	111	-0.05678	-0.06579	0.08691
22	0.02066	0.14727	0.14871	112	-0.01858	-0.08456	0.08658
23	-0.06987	0.06974	0.09872	113	0.02550	-0.07698	0.08110
24	-0.14552	-0.00596	0.14564	114	0.06172	-0.04589	0.07691
25	-0.20270	-0.07608	0.21651	115	0.07917	-0.00112	0.07917
26	-0.23864	-0.13664	0.27499	116	0.07289	0.04374	0.08501
27	-0.25174	-0.18383	0.31172	117	0.04514	0.07559	0.08804
28	-0.24191	-0.21430	0.32318	118	0.00433	0.08559	0.08570
29	-0.21081	-0.22567	0.30882	119	-0.03767	0.07143	0.08075
30	-0.16193	-0.21685	0.27064	120	-0.06912	0.03762	0.07869
31	-0.10048	-0.18845	0.21356	121	-0.08162	-0.00615	0.08185
32	-0.03309	-0.14295	0.14673	122	-0.07223	-0.04800	0.08672
33	0.03282	-0.08474	0.09088	123	-0.04393	-0.07707	0.08871
34	0.08977	-0.01987	0.09194	124	-0.00446	-0.08635	0.08646
35	0.13113	0.04446	0.13846	125	0.03601	-0.07411	0.08240
36	0.15204	0.10065	0.18234	126	0.06751	-0.04402	0.08059
37	0.15022	0.14163	0.20646	127	0.08278	-0.00377	0.08286
38	0.12645	0.16194	0.20547	128	0.07874	0.03698	0.08699
39	0.08476	0.15867	0.17989	129	0.05690	0.06904	0.08947
40	0.03197	0.13216	0.13597	130	0.02266	0.08573	0.08867
41	-0.02317	0.08626	0.08932	131	-0.01619	0.08406	0.08561
42	-0.07134	0.02812	0.07668	132	-0.05142	0.06512	0.08297
43	-0.10423	-0.03283	0.10927	133	-0.07603	0.03339	0.08304
44	-0.11611	-0.08619	0.14460	134	-0.08562	-0.00440	0.08573
45	-0.10506	-0.12251	0.16138	135	-0.07898	-0.04090	0.08894
46	-0.07356	-0.13503	0.15377	136	-0.05802	-0.06953	0.09056
47	-0.02826	-0.12128	0.12453	137	-0.02707	-0.08565	0.08983
48	0.02122	-0.08384	0.08649	138	0.00813	-0.08714	0.08752
49	0.06418	-0.03023	0.07094	139	0.04158	-0.07451	0.08533
50	0.09114	0.02841	0.09546	140	0.06810	-0.05050	0.08478
51	0.09608	0.07945	0.12467	141	0.08401	-0.01935	0.08621
52	0.07801	0.11154	0.13611	142	0.08757	0.01403	0.08869
53	0.04145	0.11728	0.12439	143	0.07899	0.04488	0.09085
54	-0.00444	0.09526	0.09536	144	0.06012	0.06928	0.09173
55	-0.04796	0.05067	0.06977	145	0.03400	0.08456	0.09114
56	-0.07783	-0.00556	0.07802	146	0.00429	0.08947	0.08957
57	-0.08614	-0.05927	0.10456	147	-0.02535	0.08418	0.08792
58	-0.07070	-0.09659	0.11970	148	-0.05165	0.07001	0.08700
59	-0.03583	-0.10764	0.11345	149	-0.07211	0.04909	0.08723
60	0.00860	-0.08938	0.08979	150	-0.08514	0.02400	0.08846
61	0.04976	-0.04675	0.06828	151	-0.09011	-0.00258	0.09014
62	0.07556	0.00832	0.07602	152	-0.08725	-0.02821	0.09170
63	0.07829	0.06014	0.09872	153	-0.07747	-0.05086	0.09267
64	0.05708	0.09364	0.10967	154	-0.06216	-0.06904	0.09290
65	0.01843	0.09889	0.10059	155	-0.04297	-0.08187	0.09246
66	-0.02560	0.07427	0.07855	156	-0.02160	-0.08901	0.09159
67	-0.06101	0.02726	0.06682	157	0.00035	-0.09062	0.09062
68	-0.07637	-0.02761	0.08121	158	0.02150	-0.08720	0.08981
69	-0.06659	-0.07310	0.09888	159	0.04075	-0.07954	0.08937
70	-0.03480	-0.09472	0.10091	160	0.05733	-0.06855	0.08937
71	0.00846	-0.08551	0.08593	161	0.07077	-0.05519	0.08975
72	0.04863	-0.04846	0.06865	162	0.08089	-0.04038	0.09041
73	0.07196	0.00421	0.07208	163	0.08772	-0.02495	0.09120
74	0.07034	0.05495	0.08925	164	0.09148	-0.00958	0.09198
75	0.04420	0.08665	0.09728	165	0.09253	0.00515	0.09268
76	0.00259	0.08858	0.08862	166	0.09129	0.01885	0.09321
77	-0.03985	0.06008	0.07210	167	0.08820	0.03125	0.09357
78	-0.06803	0.01105	0.06892	168	0.08373	0.04221	0.09376
79	-0.07178	-0.04144	0.08288	169	0.07830	0.05167	0.09381
80	-0.04963	-0.07897	0.09327	170	0.07231	0.05966	0.09375
81	-0.00947	-0.08835	0.08886	171	0.06610	0.06628	0.09361
82	0.03416	-0.06632	0.07460	172	0.05995	0.07165	0.09342
83	0.06533	-0.02080	0.06857	173	0.05410	0.07592	0.09323
84	0.07258	0.03186	0.07926	174	0.04874	0.07924	0.09303
85	0.05312	0.07275	0.09008	175	0.04402	0.08176	0.09286
86	0.01404	0.08717	0.08829	176	0.04003	0.08362	0.09271
87	-0.03032	0.07002	0.07630	177	0.03686	0.08494	0.09259
88	-0.06356	0.02764	0.06931	178	0.03456	0.08581	0.09251
89	-0.07338	-0.02453	0.07737	179	0.03317	0.08630	0.09245
				180	0.03270	0.08646	0.09244

IEN

RADIATION PATTERN FOR THETA=PI/2

H/LMDA=0.2500 B/LMDA= 0.250 OMEGA= 8.53

45 ELEMENT ARRAY

PHI	ER(THETA,PHI)			PHI	ER(THETA,PHI)		
DEG	RE	IM	ABSVAL	DEG	RE	IM	ABSVAL
0	3.64117	7.55944	8.39067	90	-0.04849	-0.17832	0.18480
1	3.64290	7.55786	8.38999	91	-0.09478	-0.21436	0.23438
2	3.64787	7.55273	8.38753	92	-0.10519	-0.17126	0.20098
3	3.65547	7.54275	8.38185	93	-0.07692	-0.06686	0.10192
4	3.66466	7.52580	8.37062	94	-0.02132	0.05919	0.06292
5	3.67401	7.49889	8.35055	95	0.04062	0.16056	0.16562
6	3.68163	7.45829	8.31749	96	0.08627	0.20120	0.21891
7	3.68524	7.39949	8.26641	97	0.09955	0.16813	0.19539
8	3.68213	7.31732	8.19153	98	0.07658	0.07530	0.10740
9	3.66921	7.20610	8.08647	99	0.02660	-0.04234	0.05001
10	3.64306	7.05975	7.94430	100	-0.03175	-0.14213	0.14564
11	3.59999	6.87208	7.75792	101	-0.07751	-0.18918	0.20445
12	3.53617	6.63701	7.52027	102	-0.09485	-0.16842	0.19329
13	3.44774	6.34904	7.22477	103	-0.07848	-0.08908	0.11871
14	3.33105	6.00359	6.86579	104	-0.03505	0.01953	0.04012
15	3.18287	5.59756	6.43920	105	0.01964	0.11906	0.12066
16	3.00066	5.12978	5.94294	106	0.06650	0.17571	0.18787
17	2.78288	4.60166	5.37770	107	0.08983	0.17156	0.19366
18	2.52939	4.01773	4.74762	108	0.08244	0.10987	0.13736
19	2.24168	3.38599	4.06080	109	0.04769	0.01291	0.04941
20	1.92321	2.71826	3.32981	110	-0.00217	-0.08635	0.08638
21	1.57969	2.03035	2.57250	111	-0.05044	-0.15558	0.16355
22	1.21907	1.34166	1.81279	112	-0.08164	-0.17352	0.19176
23	0.85158	0.67479	1.08652	113	-0.08637	-0.13613	0.16122
24	0.48943	0.05443	0.49245	114	-0.06395	-0.05694	0.08563
25	0.14632	-0.49402	0.51523	115	-0.02214	0.03827	0.04421
26	-0.16330	-0.94624	0.96022	116	0.02577	0.12010	0.12284
27	-0.42528	-1.28116	1.34990	117	0.06521	0.16455	0.17699
28	-0.62689	-1.48322	1.61026	118	0.08486	0.15987	0.18100
29	-0.75821	-1.54459	1.72065	119	0.07967	0.10916	0.13514
30	-0.81339	-1.46696	1.67737	120	0.05191	0.02845	0.05920
31	-0.79179	-1.26274	1.49045	121	0.01008	-0.05871	0.05957
32	-0.69875	-0.95531	1.18358	122	-0.03393	-0.12830	0.13271
33	-0.54583	-0.57795	0.79496	123	-0.06826	-0.16239	0.17615
34	-0.35037	-0.17145	0.39007	124	-0.08425	-0.15344	0.17505
35	-0.13426	0.21955	0.25735	125	-0.07845	-0.10544	0.13142
36	0.07808	0.55132	0.55682	126	-0.05310	-0.03196	0.06198
37	0.26227	0.78636	0.82894	127	-0.01511	0.04801	0.05033
38	0.39699	0.89857	0.98236	128	0.02601	0.11515	0.11806
39	0.46692	0.87756	0.99404	129	0.06058	0.15442	0.16588
40	0.46521	0.73103	0.86650	130	0.08105	0.15816	0.17771
41	0.39496	0.48480	0.62532	131	0.08348	0.12714	0.15210
42	0.26931	0.17990	0.32386	132	0.06810	0.06958	0.09735
43	0.10992	-0.13319	0.17269	133	0.03881	-0.00135	0.03883
44	-0.05609	-0.40234	0.40623	134	0.00210	-0.07072	0.07075
45	-0.20046	-0.58264	0.61616	135	-0.03457	-0.12510	0.12979
46	-0.29868	-0.64479	0.71061	136	-0.06433	-0.15500	0.16782
47	-0.33458	-0.58108	0.67052	137	-0.08209	-0.15622	0.17647
48	-0.30367	-0.40781	0.50845	138	-0.08531	-0.13005	0.15554
49	-0.21439	-0.16277	0.26918	139	-0.07415	-0.08245	0.11089
50	-0.08683	0.10162	0.13367	140	-0.05115	-0.02234	0.05581
51	0.05106	0.32889	0.33283	141	-0.02049	0.04010	0.04503
52	0.16933	0.47076	0.50029	142	0.01283	0.09531	0.09617
53	0.24251	0.49807	0.55397	143	0.04390	0.13575	0.14267
54	0.25549	0.40796	0.48135	144	0.06859	0.15674	0.17109
55	0.20729	0.22521	0.30609	145	0.08404	0.15674	0.17785
56	0.11155	-0.00312	0.11159	146	0.08890	0.13715	0.16345
57	-0.00660	-0.21901	0.21911	147	0.08324	0.10163	0.13137
58	-0.11662	-0.36791	0.38595	148	0.06839	0.05538	0.08800
59	-0.19043	-0.41304	0.45482	149	0.04655	0.00422	0.04674
60	-0.20971	-0.34536	0.40405	150	0.02041	-0.04624	0.05054
61	-0.17105	-0.18647	0.25304	151	-0.00721	-0.09110	0.09138
62	-0.08694	0.01706	0.08860	152	-0.03375	-0.12670	0.13112
63	0.01759	0.20652	0.20727	153	-0.05701	-0.15069	0.16112
64	0.11200	0.32783	0.34643	154	-0.07546	-0.16213	0.17883
65	0.16914	0.34737	0.38636	155	-0.08812	-0.16124	0.18375
66	0.17324	0.26216	0.31423	156	-0.09464	-0.14922	0.17670
67	0.12472	0.10084	0.16038	157	-0.09519	-0.12794	0.15947
68	0.04002	-0.08495	0.09391	158	-0.09031	-0.09960	0.13445
69	-0.05345	-0.23665	0.24262	159	-0.08084	-0.06659	0.10474
70	-0.12600	-0.30723	0.33207	160	-0.06778	-0.03116	0.07460
71	-0.15510	-0.27630	0.31686	161	-0.05215	0.00468	0.05236
72	-0.13265	-0.15662	0.20525	162	-0.03498	0.03927	0.05259
73	-0.06752	0.01014	0.06828	163	-0.01713	0.07135	0.07338
74	0.01746	0.16740	0.16831	164	0.00060	0.10002	0.10002
75	0.09328	0.26267	0.27874	165	0.01762	0.12478	0.12601
76	0.13457	0.26536	0.29753	166	0.03348	0.14541	0.14922
77	0.12826	0.17701	0.21859	167	0.04787	0.16198	0.16890
78	0.07789	0.03050	0.08365	168	0.06065	0.17473	0.18496
79	0.00221	-0.12172	0.12174	169	0.07172	0.18404	0.19753
80	-0.07158	-0.22622	0.23728	170	0.08115	0.19041	0.20698
81	-0.11757	-0.24746	0.27397	171	0.08902	0.19432	0.21374
82	-0.12025	-0.18012	0.21657	172	0.09545	0.19629	0.21827
83	-0.07989	-0.05056	0.09454	173	0.10061	0.19681	0.22104
84	-0.01216	0.09318	0.09397	174	0.10467	0.19633	0.22249
85	0.05781	0.19902	0.20725	175	0.10779	0.19525	0.22302
86	0.10470	0.22978	0.25251	176	0.11012	0.19390	0.22299
87	0.11216	0.17632	0.20897	177	0.11179	0.19257	0.22266
88	0.07860	0.06030	0.09906	178	0.11290	0.19147	0.22227
89	0.01735	-0.07444	0.07643	179	0.11354	0.19074	0.22198
				180	0.11375	0.19049	0.22187

VEN

RADIATION PATTERN FOR THETA=PI/2

H/LMDA=0.2500 B/LMDA= 0.250 OMEGA= 8.53

50 ELEMENT ARRAY

PHI DEG	ER(THETA,PHI) RE	IM	ABSVAL	PHI DEG	ER(THETA,PHI) RE	IM	ABSVAL
0	-1.22355	0.48529	1.31628	90	-0.05796	-0.05082	0.07708
1	-1.22349	0.48392	1.31572	91	-0.01126	-0.06526	0.06622
2	-1.22321	0.47978	1.31394	92	0.04026	-0.05064	0.06469
3	-1.22240	0.47278	1.31064	93	0.07346	-0.01340	0.07467
4	-1.22053	0.46278	1.30532	94	0.07362	0.02995	0.07947
5	-1.21687	0.44961	1.29727	95	0.04090	0.06018	0.07276
6	-1.21051	0.43306	1.28564	96	-0.00989	0.06392	0.06468
7	-1.20035	0.41293	1.26939	97	-0.05608	0.03951	0.06860
8	-1.18512	0.38905	1.24735	98	-0.07733	-0.00232	0.07736
9	-1.16345	0.36131	1.21826	99	-0.06450	-0.04328	0.07767
10	-1.13388	0.32969	1.18084	100	-0.02353	-0.06558	0.06967
11	-1.09497	0.29432	1.13384	101	0.02753	-0.05966	0.06570
12	-1.04535	0.25551	1.07613	102	0.06652	-0.02814	0.07223
13	-0.98387	0.21377	1.00682	103	0.07689	0.01543	0.07842
14	-0.90970	0.16986	0.92542	104	0.05453	0.05261	0.07578
15	-0.82250	0.12482	0.83192	105	0.00927	0.06786	0.06849
16	-0.72258	0.07992	0.72699	106	-0.03972	0.05499	0.06783
17	-0.61100	0.03670	0.61210	107	-0.07212	0.01951	0.07471
18	-0.48976	-0.00315	0.48977	108	-0.07489	-0.02397	0.07864
19	-0.36184	-0.03786	0.36381	109	-0.04734	-0.05788	0.07478
20	-0.23120	-0.06570	0.24036	110	-0.00095	-0.06884	0.06885
21	-0.10274	-0.08516	0.13345	111	0.04565	-0.05279	0.06979
22	0.01795	-0.09511	0.09679	112	0.07428	-0.01627	0.07604
23	0.12490	-0.09501	0.15692	113	0.07423	0.02648	0.07881
24	0.21219	-0.08506	0.22860	114	0.04603	0.05928	0.07505
25	0.27457	-0.06632	0.28246	115	0.00076	0.07008	0.07009
26	0.30807	-0.04076	0.31076	116	-0.04459	0.05529	0.07103
27	0.31065	-0.01117	0.31086	117	-0.07362	0.02064	0.07646
28	0.28275	0.01900	0.28339	118	-0.07635	-0.02126	0.07925
29	0.22764	0.04601	0.23224	119	-0.05242	-0.05574	0.07652
30	0.15150	0.06626	0.16536	120	-0.01067	-0.07120	0.07199
31	0.06313	0.07680	0.09942	121	0.03443	-0.06287	0.07168
32	-0.02685	0.07586	0.08047	122	0.06794	-0.03400	0.07597
33	-0.10712	0.06324	0.12439	123	0.07936	0.00567	0.07957
34	-0.16716	0.04058	0.17201	124	0.06571	0.04343	0.07877
35	-0.19883	0.01130	0.19915	125	0.03193	0.06777	0.07492
36	-0.19786	-0.01981	0.19885	126	-0.01121	0.07177	0.07264
37	-0.16490	-0.04724	0.17154	127	-0.05073	0.05482	0.07469
38	-0.10587	-0.06578	0.12464	128	-0.07543	0.02238	0.07868
39	-0.03132	-0.07156	0.07811	129	-0.07892	-0.01607	0.08054
40	0.04505	-0.06299	0.07744	130	-0.06101	-0.04996	0.07886
41	0.10878	-0.04136	0.11637	131	-0.02723	-0.07057	0.07564
42	0.14748	-0.01085	0.14788	132	0.01317	-0.07311	0.07429
43	0.15354	0.02209	0.15512	133	0.04988	-0.05763	0.07622
44	0.12606	0.05000	0.13562	134	0.07421	-0.02853	0.07951
45	0.07152	0.06614	0.09742	135	0.08103	0.00690	0.08132
46	0.00274	0.06622	0.06628	136	0.06958	0.04044	0.08048
47	-0.06381	0.04970	0.08088	137	0.04322	0.06494	0.07801
48	-0.11181	0.02034	0.11364	138	0.00819	0.07569	0.07613
49	[illegible]	[illegible]	[illegible]	139	[illegible]	[illegible]	[illegible]
50	-0.11202	-0.04545	0.12089	140	-0.05819	0.05307	0.07875
51	-0.06497	-0.06380	0.09105	141	-0.07713	0.02540	0.08120
52	-0.00144	-0.06392	0.06394	142	-0.08203	-0.00638	0.08228
53	0.06033	-0.04528	0.07543	143	-0.07289	-0.03659	0.08156
54	0.10231	-0.01297	0.10312	144	-0.05207	-0.06041	0.07975
55	0.11208	0.02343	0.11450	145	-0.02356	-0.07452	0.07816
56	0.08691	0.05258	0.10158	146	0.00784	-0.07743	0.07782
57	0.03499	0.06495	0.07377	147	0.03747	-0.06947	0.07893
58	-0.02676	0.05607	0.06213	148	0.06144	-0.05247	0.08079
59	-0.07798	0.02841	0.08300	149	0.07710	-0.02925	0.08246
60	-0.10155	-0.00903	0.10196	150	0.08319	-0.00309	0.08325
61	-0.08960	-0.04347	0.09959	151	0.07978	0.02281	0.08298
62	-0.04652	-0.06269	0.07807	152	0.06806	0.04567	0.08196
63	0.01215	-0.05946	0.06069	153	0.04995	0.06344	0.08074
64	0.06510	-0.03450	0.07368	154	0.02777	0.07489	0.07987
65	0.09287	0.00335	0.09293	155	0.00390	0.07959	0.07968
66	0.08522	0.04010	0.09419	156	-0.01947	0.07781	0.08021
67	0.04528	0.06168	0.07652	157	-0.04056	0.07036	0.08121
68	-0.01142	0.05949	0.06057	158	-0.05809	0.05839	0.08237
69	-0.06262	0.03396	0.07123	159	-0.07128	0.04323	0.08336
70	-0.08809	-0.00516	0.08824	160	-0.07982	0.02619	0.08401
71	-0.07775	-0.04237	0.08855	161	-0.08381	0.00851	0.08425
72	-0.03597	-0.06255	0.07215	162	-0.08365	-0.00877	0.08411
73	0.01996	-0.05715	0.06053	163	-0.07994	-0.02486	0.08372
74	0.06681	-0.02805	0.07246	164	-0.07339	-0.03920	0.08320
75	0.08503	0.01290	0.08600	165	-0.06473	-0.05144	0.08268
76	0.06705	0.04860	0.08281	166	-0.05469	-0.06144	0.08225
77	0.02069	0.06380	0.06707	167	-0.04389	-0.06922	0.08197
78	-0.03402	0.05175	0.06194	168	-0.03289	-0.07494	0.08184
79	-0.07341	0.01735	0.07543	169	-0.02211	-0.07882	0.08187
80	-0.08041	-0.02474	0.08413	170	-0.01190	-0.08114	0.08201
81	-0.05210	-0.05623	0.07666	171	-0.00248	-0.08219	0.08223
82	-0.00111	-0.06322	0.06322	172	0.00598	-0.08227	0.08249
83	0.04993	-0.04241	0.06551	173	0.01342	-0.08167	0.08277
84	0.07836	-0.00276	0.07841	174	0.01978	-0.08064	0.08303
85	0.07163	0.03828	0.08122	175	0.02508	-0.07940	0.08327
86	0.03292	0.06249	0.07063	176	0.02934	-0.07814	0.08347
87	-0.02027	0.05898	0.06236	177	0.03260	-0.07701	0.08362
88	-0.06401	0.02914	0.07033	178	0.03490	-0.07611	0.08373
89	-0.07872	-0.01384	0.07993	179	0.03626	-0.07555	0.08380
				180	0.03671	-0.07536	0.08382

IEN

RADIATION PATTERN FOR THETA=PI/2

H/LMDA=0.2500 B/LMDA= 0.250 OMEGA= 8.53

50 ELEMENT ARRAY

PHI DEG	ER(THETA,PHI) RE	IM	ABSVAL	PHI DEG	ER(THETA,PHI) RE	IM	ABSVAL
0	-8.43765	-3.54251	9.15113	90	-0.21418	-0.08924	0.23203
1	-8.43796	-3.53965	9.15032	91	-0.16438	-0.03554	0.16818
2	-8.43835	-3.53085	9.14728	92	-0.04244	0.03256	0.05349
3	-8.43703	-3.51534	9.14008	93	0.09528	0.08432	0.12723
4	-8.43104	-3.49189	9.12556	94	0.18695	0.09715	0.21069
5	-8.41632	-3.45884	9.09934	95	0.19305	0.06644	0.20417
6	-8.38759	-3.41406	9.05580	96	0.11322	0.00706	0.11344
7	-8.33862	-3.35513	8.98830	97	-0.01485	-0.05394	0.05595
8	-8.26208	-3.27930	8.88909	98	-0.13332	-0.08963	0.16065
9	-8.14996	-3.18371	8.74973	99	-0.19042	-0.08509	0.20857
10	-7.99357	-3.06548	8.56121	100	-0.16290	-0.04348	0.16860
11	-7.78409	-2.92192	8.31442	101	-0.06510	0.01602	0.06704
12	-7.51286	-2.75078	8.00061	102	0.05871	0.06717	0.08921
13	-7.17205	-2.55048	7.61204	103	0.15462	0.08828	0.17805
14	-6.75525	-2.32044	7.14268	104	0.18252	0.07127	0.19594
15	-6.25836	-2.06137	6.58911	105	0.13257	0.02451	0.13481
16	-5.68033	-1.77560	5.95138	106	0.02799	-0.03146	0.04211
17	-5.02412	-1.46733	5.23401	107	-0.08614	-0.07308	0.11296
18	-4.29738	-1.14279	4.44673	108	-0.16253	-0.08367	0.18280
19	-3.51322	-0.81035	3.60547	109	-0.17117	-0.05993	0.18136
20	-2.69052	-0.48035	2.73306	110	-0.11063	-0.01255	0.11133
21	-1.85373	-0.16479	1.86104	111	-0.00713	0.03879	0.03944
22	-1.03235	0.12325	1.03968	112	0.09726	0.07374	0.12206
23	-0.25967	0.37047	0.45241	113	0.16197	0.07928	0.18033
24	0.42931	0.56441	0.70913	114	0.16344	0.05429	0.17222
25	1.00049	0.69476	1.21806	115	0.10316	0.00931	0.10358
26	1.42393	0.75466	1.61155	116	0.00556	-0.03818	0.03858
27	1.67733	0.74197	1.83411	117	-0.09240	-0.07067	0.11633
28	1.74952	0.66028	1.86997	118	-0.15543	-0.07699	0.17345
29	1.64325	0.51939	1.72338	119	-0.16240	-0.05586	0.17174
30	1.37699	0.33513	1.41719	120	-0.11283	-0.01572	0.11392
31	0.98509	0.12844	0.99342	121	-0.02561	0.02897	0.03866
32	0.51573	-0.07657	0.52138	122	0.06876	0.06301	0.09326
33	0.02680	-0.25522	0.25662	123	0.13908	0.07564	0.15831
34	-0.42038	-0.38539	0.57031	124	0.16371	0.06365	0.17565
35	-0.76880	-0.45087	0.89126	125	0.13665	0.03187	0.14031
36	-0.97408	-0.44416	1.07057	126	0.06817	-0.00908	0.06877
37	-1.01190	-0.36838	1.07687	127	-0.01959	-0.04645	0.05041
38	-0.88316	-0.23752	0.91454	128	-0.10026	-0.06940	0.12194
39	-0.61545	-0.07494	0.61999	129	-0.15119	-0.07202	0.16747
40	-0.25997	0.09012	0.27515	130	-0.15954	-0.05450	0.16859
41	0.11625	0.22752	0.25550	131	-0.12486	-0.02255	0.12688
42	0.44203	0.31188	0.54098	132	-0.05814	0.01468	0.05996
43	0.65568	0.32795	0.73312	133	0.02225	0.04736	0.05233
44	0.71790	0.27426	0.76851	134	0.09592	0.06758	0.11734
45	0.62100	0.16416	0.64233	135	0.14561	0.07109	0.16203
46	0.39194	0.02358	0.39265	136	0.16103	0.05792	0.17113
47	0.08743	-0.11441	0.14399	137	0.14046	0.03197	0.14405
48	[illegible]	[illegible]	[illegible]	138	[illegible]	[illegible]	[illegible]
49	-0.45072	-0.26023	0.52045	139	0.02207	-0.03198	0.03886
50	-0.55404	-0.23443	0.60159	140	-0.04922	-0.05638	0.07484
51	-0.50580	-0.14824	0.52708	141	-0.10993	-0.06924	0.12992
52	-0.32390	-0.02635	0.32497	142	-0.14948	-0.06883	0.16457
53	-0.06269	0.09665	0.11520	143	-0.16220	-0.05605	0.17162
54	0.20154	0.18585	0.27415	144	-0.14755	-0.03389	0.15139
55	0.39215	0.21615	0.44778	145	-0.10963	-0.00665	0.10983
56	0.45481	0.18002	0.48914	146	-0.05574	0.02100	0.05957
57	0.37449	0.09021	0.38521	147	0.00516	0.04485	0.04514
58	0.18073	-0.02381	0.18229	148	0.06415	0.06170	0.08900
59	-0.06116	-0.12496	0.13912	149	0.11361	0.06974	0.13331
60	-0.27085	-0.18050	0.32549	150	0.14810	0.06858	0.16320
61	-0.37952	-0.17308	0.41713	151	0.16471	0.05905	0.17498
62	-0.35335	-0.10681	0.36914	152	0.16307	0.04294	0.16863
63	-0.20545	-0.00617	0.20554	153	0.14486	0.02257	0.14661
64	0.00842	0.09243	0.09281	154	0.11328	0.00039	0.11329
65	0.20953	0.15363	0.25981	155	0.07245	-0.02123	0.07550
66	0.32490	0.15603	0.36042	156	0.02673	-0.04036	0.04841
67	0.31425	0.10023	0.32985	157	-0.01979	-0.05556	0.05898
68	0.18525	0.00874	0.18545	158	-0.06362	-0.06597	0.09165
69	-0.00940	-0.08257	0.08311	159	-0.10208	-0.07128	0.12450
70	-0.19238	-0.13832	0.23695	160	-0.13336	-0.07160	0.15137
71	-0.29221	-0.13754	0.32296	161	-0.15655	-0.06744	0.17046
72	-0.27157	-0.08198	0.28368	162	-0.17145	-0.05950	0.18148
73	-0.14217	0.00461	0.14224	163	-0.17850	-0.04865	0.18501
74	0.04031	0.08624	0.09519	164	-0.17854	-0.03576	0.18208
75	0.19937	0.12950	0.23774	165	-0.17264	-0.02164	0.17399
76	0.26966	0.11745	0.29413	166	-0.16209	-0.00706	0.16224
77	0.22419	0.05654	0.23121	167	-0.14811	0.00737	0.14830
78	0.08538	-0.02624	0.08932	168	-0.13189	0.02116	0.13357
79	-0.08509	-0.09511	0.12762	169	-0.11442	0.03399	0.11937
80	-0.21323	-0.12092	0.24513	170	-0.09668	0.04559	0.10689
81	-0.24486	-0.09367	0.26216	171	-0.07933	0.05588	0.09703
82	-0.16888	-0.02656	0.17096	172	-0.06295	0.06478	0.09033
83	-0.02142	0.05007	0.05446	173	-0.04794	0.07235	0.08679
84	0.13085	0.10238	0.16615	174	-0.03464	0.07862	0.08591
85	0.22070	0.10800	0.24570	175	-0.02320	0.08371	0.08686
86	0.21006	0.06568	0.22009	176	-0.01376	0.08769	0.08876
87	0.10640	-0.00454	0.10649	177	-0.00638	0.09068	0.09090
88	-0.04167	-0.07072	0.08208	178	-0.00111	0.09274	0.09274
89	-0.16696	-0.10352	0.19645	179	0.00206	0.09395	0.09397
				180	0.00312	0.09435	0.09440

VEN

RADIATION PATTERN FOR THETA=PI/2

H/LMDA=0.5000 B/LMDA= 0.250 OMEGA= 9.92

2 ELEMENT ARRAY

PHI	ER(THETA,PHI)			PHI	ER(THETA,PHI)		
DEG	RE	IM	ABSVAL	DEG	RE	IM	ABSVAL
0	-1.36480	0.57976	1.48284	90	-1.10713	0.22272	1.12931
5	-1.36540	0.57897	1.48307	95	-1.04825	0.18135	1.06382
10	-1.36710	0.57656	1.48371	100	-0.98496	0.13945	0.99478
15	-1.36969	0.57247	1.48451	105	-0.91819	0.09757	0.92336
20	-1.37279	0.56657	1.48511	110	-0.84901	0.05623	0.85087
25	-1.37588	0.55872	1.48500	115	-0.77852	0.01596	0.77869
30	-1.37836	0.54874	1.48357	120	-0.70791	-0.02277	0.70828
35	-1.37950	0.53644	1.48013	125	-0.63835	-0.05952	0.64112
40	-1.37854	0.52162	1.47392	130	-0.57096	-0.09391	0.57863
45	-1.37466	0.50411	1.46418	135	-0.50682	-0.12562	0.52216
50	-1.36707	0.48378	1.45015	140	-0.44691	-0.15440	0.47283
55	-1.35501	0.46051	1.43113	145	-0.39210	-0.18005	0.43146
60	-1.33780	0.43430	1.40653	150	-0.34315	-0.20244	0.39841
65	-1.31488	0.40516	1.37589	155	-0.30072	-0.22145	0.37346
70	-1.28585	0.37322	1.33892	160	-0.26532	-0.23704	0.35578
75	-1.25048	0.33869	1.29554	165	-0.23737	-0.24918	0.34415
80	-1.20875	0.30184	1.24586	170	-0.21719	-0.25785	0.33714
85	-1.16083	0.26304	1.19026	175	-0.20500	-0.26306	0.33350
				180	-0.20092	-0.26479	0.33239

IEN

RADIATION PATTERN FOR THETA=PI/2

H/LMDA=0.5000 B/LMDA= 0.250 OMEGA= 9.92

2 ELEMENT ARRAY

PHI	ER(THETA,PHI)			PHI	ER(THETA,PHI)		
DEG	RE	IM	ABSVAL	DEG	RE	IM	ABSVAL
0	-0.81871	0.67930	1.06383	90	-0.64741	0.05133	0.64945
5	-0.81899	0.67749	1.06289	95	-0.61099	-0.01099	0.61109
10	-0.81980	0.67202	1.06004	100	-0.57201	-0.07279	0.57662
15	-0.82100	0.66283	1.05517	105	-0.53104	-0.13331	0.54752
20	-0.82237	0.64981	1.04812	110	-0.48873	-0.19186	0.52504
25	-0.82361	0.63286	1.03868	115	-0.44575	-0.24780	0.51000
30	-0.82435	0.61182	1.02659	120	-0.40281	-0.30057	0.50259
35	-0.82418	0.58658	1.01160	125	-0.36059	-0.34973	0.50233
40	-0.82262	0.55701	0.99347	130	-0.31978	-0.39490	0.50814
45	-0.81923	0.52307	0.97198	135	-0.28100	-0.43584	0.51857
50	-0.81352	0.48475	0.94699	140	-0.24484	-0.47238	0.53206
55	-0.80506	0.44214	0.91848	145	-0.21180	-0.50444	0.54710
60	-0.79346	0.39541	0.88653	150	-0.18233	-0.53201	0.56239
65	-0.77840	0.34485	0.85137	155	-0.15681	-0.55513	0.57685
70	-0.75966	0.29083	0.81343	160	-0.13553	-0.57387	0.58966
75	-0.73713	0.23385	0.77333	165	-0.11875	-0.58832	0.60019
80	-0.71080	0.17449	0.73190	170	-0.10664	-0.59857	0.60799
85	-0.68081	0.11341	0.69019	175	-0.09932	-0.60468	0.61279
				180	-0.09688	-0.60672	0.61440

VEN

RADIATION PATTERN FOR THETA=PI/2

H/LMDA=0.5000 B/LMDA= 0.250 OMEGA= 9.92

3 ELEMENT ARRAY

PHI	ER(THETA,PHI)			PHI	ER(THETA,PHI)		
DEG	RE	IM	ABSVAL	DEG	RE	IM	ABSVAL
0	-0.51881	1.26142	1.36394	90	-0.43281	0.38888	0.58185
5	-0.52028	1.26120	1.36430	95	-0.38513	0.27481	0.47312
10	-0.52460	1.26036	1.36518	100	-0.33413	0.16435	0.37236
15	-0.53152	1.25838	1.36603	105	-0.28125	0.06022	0.28762
20	-0.54063	1.25441	1.36595	110	-0.22794	-0.03524	0.23065
25	-0.55136	1.24732	1.36374	115	-0.17561	-0.12027	0.21285
30	-0.56296	1.23575	1.35794	120	-0.12553	-0.19374	0.23085
35	-0.57456	1.21819	1.34689	125	-0.07873	-0.25517	0.26704
40	-0.58516	1.19311	1.32888	130	-0.03600	-0.30471	0.30683
45	-0.59368	1.15901	1.30221	135	0.00213	-0.34305	0.34305
50	-0.59899	1.11461	1.26536	140	0.03538	-0.37133	0.37301
55	-0.60000	1.05898	1.21714	145	0.06373	-0.39098	0.39614
60	-0.59571	0.99162	1.15680	150	0.08731	-0.40363	0.41297
65	-0.58532	0.91262	1.08419	155	0.10640	-0.41092	0.42448
70	-0.56824	0.82267	0.99984	160	0.12133	-0.41444	0.43183
75	-0.54420	0.72313	0.90502	165	0.13247	-0.41559	0.43619
80	-0.51327	0.61595	0.80177	170	0.14016	-0.41553	0.43853
85	-0.47588	0.50359	0.69287	175	0.14466	-0.41514	0.43962
				180	0.14614	-0.41495	0.43993

IEN

RADIATION PATTERN FOR THETA=PI/2

H/LMDA=0.5000 B/LMDA= 0.250 OMEGA= 9.92

3 ELEMENT ARRAY

PHI	ER(THETA,PHI)			PHI	ER(THETA,PHI)		
DEG	RE	IM	ABSVAL	DEG	RE	IM	ABSVAL
0	0.00492	0.90280	0.90281	90	-0.30217	0.04549	0.30557
5	0.00252	0.90148	0.90149	95	-0.28540	-0.03976	0.28816
10	-0.00465	0.89742	0.89743	100	-0.26163	-0.11876	0.28732
15	-0.01647	0.89023	0.89038	105	-0.23170	-0.18967	0.29944
20	-0.03271	0.87933	0.87994	110	-0.19674	-0.25114	0.31902
25	-0.05305	0.86395	0.86558	115	-0.15803	-0.30233	0.34114
30	-0.07702	0.84320	0.84671	120	-0.11697	-0.34297	0.36236
35	-0.10402	0.81613	0.82274	125	-0.07494	-0.37331	0.38076
40	-0.13327	0.78186	0.79314	130	-0.03327	-0.39412	0.39552
45	-0.16382	0.73961	0.75753	135	0.00688	-0.40650	0.40655
50	-0.19457	0.68886	0.71581	140	0.04452	-0.41184	0.41424
55	-0.22429	0.62941	0.66818	145	0.07885	-0.41171	0.41919
60	-0.25168	0.56150	0.61532	150	0.10928	-0.40769	0.42208
65	-0.27541	0.48579	0.55843	155	0.13540	-0.40133	0.42355
70	-0.29423	0.40347	0.49936	160	0.15695	-0.39404	0.42414
75	-0.30704	0.31614	0.44070	165	0.17377	-0.38704	0.42426
80	-0.31295	0.22583	0.38593	170	0.18580	-0.38132	0.42418
85	-0.31140	0.13481	0.33933	175	0.19303	-0.37760	0.42407
				180	0.19543	-0.37631	0.42403

VEN

RADIATION PATTERN FOR THETA=PI/2

H/LMDA=0.5000 B/LMDA= 0.250 OMEGA= 9.92

4 ELEMENT ARRAY

PHI	ER(THETA,PHI)			PHI	ER(THETA,PHI)		
DEG	RE	IM	ABSVAL	DEG	RE	IM	ABSVAL
0	0.48958	1.18513	1.28227	90	-0.17250	0.07336	0.18745
5	0.48785	1.18622	1.28262	95	-0.18979	-0.05250	0.19692
10	0.48255	1.18916	1.28333	100	-0.19423	-0.16082	0.25217
15	0.47330	1.19296	1.28342	105	-0.18651	-0.24803	0.31033
20	0.45957	1.19601	1.28127	110	-0.16812	-0.31227	0.35465
25	0.44067	1.19612	1.27471	115	-0.14119	-0.35344	0.38059
30	0.41588	1.19063	1.26117	120	-0.10819	-0.37299	0.38837
35	0.38458	1.17655	1.23781	125	-0.07170	-0.37370	0.38052
40	0.34637	1.15078	1.20178	130	-0.03412	-0.35917	0.36079
45	0.30120	1.11042	1.15055	135	0.00250	-0.33341	0.33342
50	0.24951	1.05305	1.08220	140	0.03653	-0.30046	0.30267
55	0.19233	0.97710	0.99585	145	0.06689	-0.26402	0.27236
60	0.13131	0.88219	0.89191	150	0.09293	-0.22728	0.24554
65	0.06868	0.76939	0.77245	155	0.11442	-0.19281	0.22421
70	0.00709	0.64128	0.64132	160	0.13142	-0.16257	0.20904
75	-0.05053	0.50196	0.50450	165	0.14416	-0.13792	0.19951
80	-0.10129	0.35675	0.37085	170	0.15296	-0.11978	0.19428
85	-0.14261	0.21176	0.25531	175	0.15811	-0.10871	0.19187
				180	0.15980	-0.10499	0.19120

IEN

RADIATION PATTERN FOR THETA=PI/2

H/LMDA=0.5000 B/LMDA= 0.250 OMEGA= 9.92

4 ELEMENT ARRAY

PHI	ER(THETA,PHI)			PHI	ER(THETA,PHI)		
DEG	RE	IM	ABSVAL	DEG	RE	IM	ABSVAL
0	0.59262	0.53972	0.80156	90	-0.22521	-0.05123	0.23096
5	0.58957	0.54085	0.80007	95	-0.22559	-0.12419	0.25752
10	0.58026	0.54404	0.79541	100	-0.20972	-0.18634	0.28055
15	0.56437	0.54867	0.78712	105	-0.17955	-0.23525	0.29594
20	0.54136	0.55376	0.77441	110	-0.13795	-0.26957	0.30282
25	0.51065	0.55793	0.75634	115	-0.08832	-0.28910	0.30229
30	0.47170	0.55951	0.73182	120	-0.03427	-0.29470	0.29668
35	0.42420	0.55660	0.69982	125	0.02080	-0.28808	0.28883
40	0.36821	0.54719	0.65954	130	0.07393	-0.27158	0.28146
45	0.30434	0.52934	0.61059	135	0.12285	-0.24785	0.27663
50	0.23390	0.50137	0.55324	140	0.16603	-0.21959	0.27529
55	0.15888	0.46212	0.48867	145	0.20266	-0.18935	0.27735
60	0.08204	0.41114	0.41925	150	0.23256	-0.15936	0.28192
65	0.00670	0.34887	0.34894	155	0.25605	-0.13144	0.28782
70	-0.06346	0.27675	0.28393	160	0.27375	-0.10702	0.29393
75	-0.12473	0.19719	0.23332	165	0.28643	-0.08714	0.29939
80	-0.17372	0.11342	0.20747	170	0.29484	-0.07250	0.30362
85	-0.20777	0.02927	0.20982	175	0.29960	-0.06355	0.30627
				180	0.30114	-0.06055	0.30717

VEN

RADIATION PATTERN FOR THETA=PI/2

H/LMDA=0.5000 B/LMDA= 0.250 OMEGA= 9.92

5 ELEMENT ARRAY

PHI DEG	ER(THETA,PHI) RE	IM	ABSVAL	PHI DEG	ER(THETA,PHI) RE	IM	ABSVAL
0	1.12321	0.46576	1.21595	90	-0.29777	-0.04787	0.30160
5	1.12256	0.46816	1.21627	95	-0.32933	-0.11579	0.34909
10	1.12012	0.47514	1.21673	100	-0.32598	-0.16697	0.36626
15	1.11442	0.48603	1.21579	105	-0.29229	-0.19935	0.35380
20	1.10311	0.49968	1.21100	110	-0.23536	-0.21284	0.31733
25	1.08311	0.51444	1.19907	115	-0.16367	-0.20916	0.26558
30	1.05087	0.52818	1.17614	120	-0.08580	-0.19138	0.20973
35	1.00279	0.53829	1.13813	125	-0.00934	-0.16335	0.16362
40	0.93563	0.54186	1.08121	130	0.05985	-0.12914	0.14233
45	0.84720	0.53589	1.00246	135	0.11804	-0.09247	0.14995
50	0.73696	0.51760	0.90056	140	0.16358	-0.05641	0.17303
55	0.60650	0.48487	0.77650	145	0.19656	-0.02322	0.19793
60	0.45997	0.43665	0.63422	150	0.21839	0.00571	0.21846
65	0.30399	0.37333	0.48144	155	0.23122	0.02971	0.23312
70	0.14723	0.29696	0.33146	160	0.23751	0.04865	0.24244
75	-0.00041	0.21125	0.21125	165	0.23963	0.06277	0.24771
80	-0.12905	0.12127	0.17709	170	0.23960	0.07244	0.25031
85	-0.23016	0.03291	0.23250	175	0.23895	0.07806	0.25138
				180	0.23863	0.07990	0.25166

IEN

RADIATION PATTERN FOR THETA=PI/2

H/LMDA=0.5000 B/LMDA= 0.250 OMEGA= 9.92

5 ELEMENT ARRAY

PHI DEG	ER(THETA,PHI) RE	IM	ABSVAL	PHI DEG	ER(THETA,PHI) RE	IM	ABSVAL
0	0.73239	-0.06115	0.73494	90	-0.24161	-0.04058	0.24499
5	0.73101	-0.05775	0.73329	95	-0.22954	-0.09532	0.24855
10	0.72650	-0.04762	0.72806	100	-0.19352	-0.14130	0.23961
15	0.71783	-0.03098	0.71850	105	-0.13904	-0.17460	0.22320
20	0.70335	-0.00828	0.70340	110	-0.07305	-0.19291	0.20628
25	0.68097	0.01971	0.68126	115	-0.00281	-0.19571	0.19573
30	0.64841	0.05176	0.65047	120	0.06500	-0.18414	0.19527
35	0.60351	0.08614	0.60962	125	0.12512	-0.16065	0.20362
40	0.54463	0.12050	0.55780	130	0.17410	-0.12849	0.21638
45	0.47114	0.15200	0.49505	135	0.21040	-0.09119	0.22931
50	0.38379	0.17745	0.42282	140	0.23417	-0.05208	0.23989
55	0.28506	0.19362	0.34460	145	0.24686	-0.01396	0.24726
60	0.17927	0.19767	0.26685	150	0.25075	0.02103	0.25163
65	0.07236	0.18762	0.20110	155	0.24844	0.05148	0.25372
70	-0.02860	0.16283	0.16532	160	0.24257	0.07659	0.25437
75	-0.11625	0.12424	0.17015	165	0.23546	0.09603	0.25429
80	-0.18389	0.07453	0.19842	170	0.22900	0.10977	0.25395
85	-0.22652	0.01787	0.22722	175	0.22458	0.11794	0.25366
				180	0.22301	0.12064	0.25355

VEN

RADIATION PATTERN FOR THETA=PI/2

H/LMDA=0.5000 B/LMDA= 0.250 OMEGA= 9.92

6 ELEMENT ARRAY

PHI DEG	ER(THETA,PHI) RE	IM	ABSVAL	PHI DEG	ER(THETA,PHI) RE	IM	ABSVAL
0	1.07555	-0.44703	1.16475	90	-0.33337	0.08488	0.34400
2	1.07577	-0.44661	1.16480	92	-0.32825	0.06269	0.33419
4	1.07643	-0.44537	1.16492	94	-0.31595	0.03975	0.31844
6	1.07746	-0.44326	1.16508	96	-0.29709	0.01661	0.29756
8	1.07878	-0.44026	1.16516	98	-0.27243	-0.00617	0.27250
10	1.08025	-0.43633	1.16504	100	-0.24283	-0.02808	0.24445
12	1.08171	-0.43139	1.16456	102	-0.20921	-0.04862	0.21478
14	1.08296	-0.42537	1.16351	104	-0.17255	-0.06738	0.18523
16	1.08378	-0.41821	1.16167	106	-0.13382	-0.08400	0.15800
18	1.08388	-0.40980	1.15877	108	-0.09401	-0.09819	0.13594
20	1.08299	-0.40007	1.15452	110	-0.05404	-0.10975	0.12234
22	1.08076	-0.38892	1.14861	112	-0.01478	-0.11855	0.11947
24	1.07687	-0.37628	1.14071	114	0.02300	-0.12453	0.12663
26	1.07093	-0.36205	1.13048	116	0.05861	-0.12772	0.14052
28	1.06258	-0.34619	1.11756	118	0.09150	-0.12820	0.15750
30	1.05144	-0.32864	1.10160	120	0.12121	-0.12612	0.17492
32	1.03712	-0.30938	1.08228	122	0.14741	-0.12169	0.19116
[illegible]	[illegible]	[illegible]	[illegible]	124	0.16988	-0.11513	0.20522
36	0.99749	-0.26574	1.03228	126	0.18854	-0.10671	0.21664
38	0.97154	-0.24146	1.00110	128	0.20337	-0.09672	0.22520
40	0.94113	-0.21567	0.96552	130	0.21447	-0.08545	0.23086
42	0.90605	-0.18851	0.92545	132	0.22201	-0.07320	0.23376
44	0.86617	-0.16017	0.88085	134	0.22622	-0.06024	0.23410
46	0.82144	-0.13090	0.83180	136	0.22738	-0.04686	0.23216
48	0.77191	-0.10095	0.77848	138	0.22583	-0.03329	0.22827
50	0.71772	-0.07067	0.72120	140	0.22189	-0.01976	0.22277
52	0.65915	-0.04039	0.66038	142	0.21594	-0.00647	0.21603
54	0.59655	-0.01052	0.59664	144	0.20831	0.00641	0.20841
56	0.53042	0.01853	0.53075	146	0.19937	0.01876	0.20025
58	0.46138	0.04634	0.46370	148	0.18944	0.03045	0.19187
60	0.39014	0.07246	0.39681	150	0.17884	0.04141	0.18357
62	0.31751	0.09648	0.33184	152	0.16785	0.05157	0.17560
64	0.24439	0.11797	0.27137	154	0.15674	0.06091	0.16816
66	0.17174	0.13656	0.21942	156	0.14572	0.06942	0.16141
68	0.10057	0.15192	0.18220	158	0.13502	0.07708	0.15547
70	0.03191	0.16377	0.16685	160	0.12479	0.08391	0.15038
72	-0.03324	0.17191	0.17509	162	0.11519	0.08994	0.14615
74	-0.09388	0.17620	0.19965	164	0.10634	0.09521	0.14273
76	-0.14912	0.17659	0.23113	166	0.09834	0.09973	0.14006
78	-0.19813	0.17313	0.26312	168	0.09127	0.10356	0.13804
80	-0.24023	0.16596	0.29198	170	0.08520	0.10673	0.13656
82	-0.27485	0.15529	0.31568	172	0.08017	0.10926	0.13552
84	-0.30160	0.14143	0.33312	174	0.07622	0.11121	0.13482
86	-0.32027	0.12477	0.34372	176	0.07339	0.11257	0.13438
88	-0.33081	0.10575	0.34730	178	0.07168	0.11339	0.13414
				180	0.07111	0.11365	0.13407

IEN

RADIATION PATTERN FOR THETA=PI/2

H/LMDA=0.5000 B/LMDA= 0.250 OMEGA= 9.92

6 ELEMENT ARRAY

PHI DEG	ER(THETA,PHI) RE	IM	ABSVAL	PHI DEG	ER(THETA,PHI) RE	IM	ABSVAL
0	0.42851	-0.53262	0.68360	90	-0.20111	0.01372	0.20157
2	0.42878	-0.53204	0.68332	92	-0.19177	-0.01221	0.19216
4	0.42958	-0.53030	0.68247	94	-0.17816	-0.03780	0.18213
6	0.43089	-0.52737	0.68102	96	-0.16067	-0.06240	0.17237
8	0.43264	-0.52322	0.67892	98	-0.13982	-0.08539	0.16383
10	0.43478	-0.51778	0.67611	100	-0.11615	-0.10620	0.15738
12	0.43719	-0.51100	0.67251	102	-0.09029	-0.12437	0.15368
14	0.43978	-0.50281	0.66800	104	-0.06286	-0.13949	0.15300
16	0.44241	-0.49312	0.66249	106	-0.03453	-0.15127	0.15517
18	0.44492	-0.48184	0.65584	108	-0.00593	-0.15953	0.15964
20	0.44714	-0.46890	0.64792	110	0.02233	-0.16416	0.16567
22	0.44888	-0.45420	0.63858	112	0.04968	-0.16517	0.17248
24	0.44994	-0.43769	0.62770	114	0.07561	-0.16266	0.17937
26	0.45009	-0.41929	0.61513	116	0.09968	-0.15681	0.18581
28	0.44911	-0.39897	0.60074	118	0.12152	-0.14788	0.19140
30	0.44677	-0.37672	0.58440	120	0.14084	-0.13618	0.19591
[illegible]	[illegible]	[illegible]	[illegible]	122	[illegible]	-0.12208	0.19923
34	0.43705	-0.32651	0.54555	124	0.17122	-0.10593	0.20134
36	0.42922	-0.29869	0.52292	126	0.18211	-0.08817	0.20233
38	0.41912	-0.26923	0.49814	128	0.19013	-0.06920	0.20233
40	0.40658	-0.23830	0.47127	130	0.19538	-0.04942	0.20153
42	0.39145	-0.20614	0.44241	132	0.19800	-0.02921	0.20014
44	0.37363	-0.17302	0.41174	134	0.19818	-0.00892	0.19838
46	0.35305	-0.13928	0.37953	136	0.19613	0.01114	0.19645
48	0.32970	-0.10529	0.34611	138	0.19212	0.03068	0.19455
50	0.30366	-0.07148	0.31196	140	0.18640	0.04948	0.19285
52	0.27503	-0.03829	0.27768	142	0.17924	0.06734	0.19148
54	0.24403	-0.00621	0.24410	144	0.17093	0.08411	0.19050
56	0.21091	0.02428	0.21230	146	0.16172	0.09968	0.18997
58	0.17602	0.05267	0.18373	148	0.15187	0.11400	0.18990
60	0.13977	0.07848	0.16029	150	0.14162	0.12702	0.19024
62	0.10263	0.10125	0.14417	152	0.13119	0.13875	0.19095
64	0.06514	0.12055	0.13703	154	0.12079	0.14920	0.19196
66	0.02787	0.13604	0.13887	156	0.11058	0.15843	0.19320
68	-0.00857	0.14741	0.14766	158	0.10074	0.16649	0.19460
70	-0.04357	0.15447	0.16050	160	0.09139	0.17347	0.19607
72	-0.07651	0.15709	0.17473	162	0.08265	0.17944	0.19756
74	-0.10678	0.15527	0.18844	164	0.07462	0.18448	0.19900
76	-0.13381	0.14910	0.20034	166	0.06738	0.18870	0.20036
78	-0.15712	0.13879	0.20964	168	0.06099	0.19215	0.20160
80	-0.17626	0.12466	0.21589	170	0.05551	0.19494	0.20269
82	-0.19089	0.10711	0.21889	172	0.05098	0.19712	0.20360
84	-0.20077	0.08663	0.21866	174	0.04743	0.19875	0.20433
86	-0.20576	0.06382	0.21543	176	0.04487	0.19988	0.20485
88	-0.20584	0.03929	0.20955	178	0.04334	0.20054	0.20517
				180	0.04282	0.20076	0.20528

VEN RADIATION PATTERN FOR THETA=PI/2

H/LMDA=0.5000 B/LMDA= 0.250 OMEGA= 9.92

7 ELEMENT ARRAY

PHI	ER(THETA,PHI)			PHI	ER(THETA,PHI)		
DEG	RE	IM	ABSVAL	DEG	RE	IM	ABSVAL
0	0.42547	-1.03807	1.12187	90	-0.18072	0.14350	0.23076
2	0.42597	-1.03790	1.12191	92	-0.16917	0.09955	0.19629
4	0.42748	-1.03738	1.12201	94	-0.15305	0.05442	0.16244
6	0.42996	-1.03644	1.12208	96	-0.13304	0.00964	0.13339
8	0.43336	-1.03494	1.12201	98	-0.10989	-0.03340	0.11486
10	0.43760	-1.03273	1.12162	100	-0.08444	-0.07340	0.11188
12	0.44259	-1.02960	1.12070	102	-0.05753	-0.10925	0.12348
14	0.44821	-1.02528	1.11897	104	-0.03003	-0.14005	0.14323
16	0.45432	-1.01949	1.11614	106	-0.00274	-0.16511	0.16513
18	0.46073	-1.01190	1.11185	108	0.02357	-0.18400	0.18550
20	0.46725	-1.00215	1.10572	110	0.04822	-0.19652	0.20235
22	0.47367	-0.98986	1.09736	112	0.07065	-0.20268	0.21464
24	0.47973	-0.97465	1.08632	114	0.09038	-0.20274	0.22198
26	0.48515	-0.95612	1.07217	116	0.10710	-0.19710	0.22432
28	0.48964	-0.93389	1.05447	118	0.12057	-0.18632	0.22193
30	0.49288	-0.90760	1.03280	120	0.13069	-0.17109	0.21530
32	0.49455	-0.87692	1.00677	122	0.13748	-0.15216	0.20507
34	0.49431	-0.84160	0.97603	124	0.14105	-0.13032	0.19204
36	0.49182	-0.80142	0.94030	126	0.14159	-0.10637	0.17709
38	0.48675	-0.75630	0.89940	128	0.13935	-0.08109	0.16122
40	0.47880	-0.70622	0.85323	130	0.13464	-0.05521	0.14552
42	0.46770	-0.65131	0.80184	132	0.12780	-0.02938	0.13113
44	0.45320	-0.59182	0.74542	134	0.11920	-0.00418	0.11927
46	0.43515	-0.52816	0.68433	136	0.10919	0.01990	0.11099
48	0.41343	-0.46086	0.61913	138	0.09814	0.04249	0.10694
50	0.38804	-0.39064	0.55061	140	0.08637	0.06329	0.10708
52	0.35904	-0.31835	0.47985	142	0.07420	0.08211	0.11066
54	0.32662	-0.24496	0.40827	144	0.06189	0.09883	0.11661
56	0.29106	-0.17159	0.33787	146	0.04969	0.11342	0.12383
58	0.25275	-0.09942	0.27160	148	0.03780	0.12592	0.13147
60	0.21221	-0.02972	0.21428	150	0.02639	0.13641	0.13894
62	0.17003	0.03625	0.17385	152	0.01559	0.14501	0.14585
64	0.12690	0.09721	0.15986	154	0.00549	0.15190	0.15200
66	0.08360	0.15198	0.17346	156	-0.00384	0.15724	0.15729
68	0.04095	0.19946	0.20362	158	-0.01236	0.16125	0.16172
70	-0.00022	0.23873	0.23873	160	-0.02004	0.16411	0.16533
72	-0.03905	0.26905	0.27187	162	-0.02689	0.16602	0.16819
74	-0.07473	0.28990	0.29938	164	-0.03292	0.16719	0.17040
76	-0.10651	0.30105	0.31934	166	-0.03814	0.16777	0.17205
78	-0.13372	0.30252	0.33076	168	-0.04259	0.16793	0.17325
80	-0.15581	0.29462	0.33328	170	-0.04629	0.16783	0.17409
82	-0.17236	0.27794	0.32705	172	-0.04926	0.16757	0.17466
84	-0.18314	0.25332	0.31259	174	-0.05155	0.16726	0.17502
86	-0.18804	0.22184	0.29081	176	-0.05316	0.16698	0.17524
88	-0.18715	0.18476	0.26299	178	-0.05413	0.16679	0.17536
				180	-0.05445	0.16673	0.17539

IEN RADIATION PATTERN FOR THETA=PI/2

H/LMDA=0.5000 B/LMDA= 0.250 OMEGA= 9.92

7 ELEMENT ARRAY

PHI	ER(THETA,PHI)			PHI	ER(THETA,PHI)		
DEG	RE	IM	ABSVAL	DEG	RE	IM	ABSVAL
0	-0.08427	-0.63980	0.64533	90	-0.13300	0.01147	0.13350
2	-0.08361	-0.63959	0.64503	92	-0.12623	-0.02143	0.12803
4	-0.08166	-0.63893	0.64413	94	-0.11543	-0.05293	0.12698
6	-0.07841	-0.63777	0.64257	96	-0.10097	-0.08200	0.13007
8	-0.07388	-0.63603	0.64031	98	-0.08333	-0.10771	0.13619
10	-0.06808	-0.63359	0.63724	100	-0.06312	-0.12931	0.14389
12	-0.06102	-0.63031	0.63326	102	-0.04098	-0.14620	0.15184
14	-0.05275	-0.62601	0.62823	104	-0.01764	-0.15798	0.15897
16	-0.04331	-0.62049	0.62200	106	0.00620	-0.16446	0.16457
18	-0.03275	-0.61354	0.61442	108	0.02980	-0.16561	0.16827
20	-0.02116	-0.60492	0.60529	110	0.05250	-0.16161	0.16992
22	-0.00864	-0.59438	0.59444	112	0.07367	-0.15280	0.16963
24	0.00469	-0.58168	0.58170	114	0.09279	-0.13965	0.16766
26	0.01868	-0.56657	0.56688	116	0.10941	-0.12274	0.16443
28	0.03314	-0.54884	0.54984	118	0.12320	-0.10274	0.16042
30	0.04787	-0.52828	0.53044	120	0.13393	-0.08036	0.15619
32	0.06262	-0.50472	0.50859	122	0.14149	-0.05632	0.15229
34	0.07712	-0.47806	0.48424	124	0.14586	-0.03134	0.14919
36	0.09108	-0.44825	0.45740	126	0.14712	-0.00609	0.14725
38	0.10417	-0.41529	0.42815	128	0.14544	0.01881	0.14665
40	0.11607	-0.37929	0.39665	130	0.14105	0.04281	0.14740
42	0.12642	-0.34044	0.36316	132	0.13422	0.06546	0.14934
44	0.13490	-0.29904	0.32806	134	0.12529	0.08640	0.15219
46	0.14118	-0.25549	0.29190	136	0.11460	0.10534	0.15566
48	0.14496	-0.21027	0.25539	138	0.10251	0.12210	0.15942
50	0.14598	-0.16399	0.21955	140	0.08935	0.13659	0.16322
52	0.14404	-0.11733	0.18578	142	0.07548	0.14878	0.16683
54	0.13900	-0.07107	0.15612	144	0.06120	0.15873	0.17012
56	0.13082	-0.02602	0.13338	146	0.04681	0.16653	0.17298
58	0.11953	0.01694	0.12073	148	0.03255	0.17234	0.17539
60	0.10527	0.05695	0.11969	150	0.01865	0.17634	0.17733
62	0.08828	0.09316	0.12835	152	0.00528	0.17875	0.17882
64	0.06892	0.12476	0.14253	154	-0.00741	0.17977	0.17993
66	0.04761	0.15104	0.15836	156	-0.01929	0.17966	0.18069
68	0.02491	0.17140	0.17320	158	-0.03029	0.17862	0.18117
70	0.00142	0.18541	0.18541	160	-0.04035	0.17689	0.18143
72	-0.02218	0.19278	0.19405	162	-0.04942	0.17467	0.18152
74	-0.04519	0.19343	0.19863	164	-0.05751	0.17215	0.18150
76	-0.06690	0.18748	0.19906	166	-0.06459	0.16952	0.18140
78	-0.08660	0.17526	0.19549	168	-0.07067	0.16692	0.18126
80	-0.10366	0.15730	0.18839	170	-0.07578	0.16450	0.18111
82	-0.11749	0.13432	0.17846	172	-0.07992	0.16236	0.18096
84	-0.12763	0.10720	0.16667	174	-0.08312	0.16060	0.18084
86	-0.13371	0.07694	0.15426	176	-0.08540	0.15929	0.18074
88	-0.13553	0.04464	0.14269	178	-0.08675	0.15849	0.18068
				180	-0.08721	0.15822	0.18066

VEN RADIATION PATTERN FOR THETA=PI/2

H/LMDA=0.5000 B/LMDA= 0.250 OMEGA= 9.92

8 ELEMENT ARRAY

PHI	ER(THETA,PHI)			PHI	ER(THETA,PHI)		
DEG	RE	IM	ABSVAL	DEG	RE	IM	ABSVAL
0	-0.42223	-1.00125	1.08664	90	-0.10018	0.03496	0.10611
2	-0.42170	-1.00151	1.08667	92	-0.10947	-0.01606	0.11064
4	-0.42008	-1.00224	1.08672	94	-0.11364	-0.06392	0.13038
6	-0.41733	-1.00337	1.08670	96	-0.11269	-0.10680	0.15525
8	-0.41340	-1.00472	1.08645	98	-0.10685	-0.14320	0.17867
10	-0.40820	-1.00609	1.08574	100	-0.09656	-0.17197	0.19723
12	-0.40162	-1.00719	1.08431	102	-0.08245	-0.19237	0.20929
14	-0.39355	-1.00768	1.08180	104	-0.06527	-0.20404	0.21423
16	-0.38385	-1.00716	1.07783	106	-0.04589	-0.20704	0.21207
18	-0.37239	-1.00519	1.07195	108	-0.02520	-0.20179	0.20335
20	-0.35903	-1.00127	1.06369	110	-0.00412	-0.18901	0.18905
22	-0.34366	-0.99486	1.05254	112	0.01650	-0.16970	0.17050
24	-0.32615	-0.98539	1.03796	114	0.03586	-0.14504	0.14940
26	-0.30643	-0.97228	1.01943	116	0.05329	-0.11631	0.12794
28	-0.28445	-0.95495	0.99642	118	0.06826	-0.08486	0.10891
30	-0.26023	-0.93284	0.96846	120	0.08037	-0.05200	0.09572
32	-0.23382	-0.90542	0.93512	122	0.08937	-0.01897	0.09136
34	-0.20536	-0.87224	0.89609	124	0.09516	0.01313	0.09607
36	-0.17505	-0.83294	0.85113	126	0.09778	0.04333	0.10695
38	-0.14318	-0.78728	0.80019	128	0.09736	0.07088	0.12043
40	-0.11012	-0.73518	0.74338	130	0.09416	0.09519	0.13389
42	-0.07633	-0.67673	0.68102	132	0.08849	0.11587	0.14580
44	-0.04236	-0.61221	0.61367	134	0.08072	0.13274	0.15535
46	-0.00882	-0.54215	0.54222	136	0.07124	0.14574	0.16222
48	0.02363	-0.46728	0.46787	138	0.06046	0.15497	0.16634
50	0.05426	-0.38858	0.39235	140	0.04876	0.16065	0.16789
52	0.08234	-0.30725	0.31809	142	0.03653	0.16309	0.16714
54	0.10717	-0.22470	0.24895	144	0.02409	0.16266	0.16444
56	0.12806	-0.14252	0.19160	146	0.01174	0.15977	0.16020
58	0.14442	-0.06240	0.15732	148	-0.00028	0.15483	0.15483
60	0.15574	0.01386	0.15636	150	-0.01177	0.14829	0.14875
62	0.16170	0.08449	0.18244	152	-0.02256	0.14054	0.14234
64	0.16209	0.14777	0.21934	154	-0.03257	0.13197	0.13593
66	0.15694	0.20213	0.25591	156	-0.04171	0.12293	0.12981
68	0.14644	0.24623	0.28648	158	-0.04994	0.11373	0.12421
70	0.13101	0.27901	0.30823	160	-0.05727	0.10464	0.11928
72	0.11126	0.29977	0.31975	162	-0.06371	0.09588	0.11512
74	0.08799	0.30823	0.32054	164	-0.06928	0.08766	0.11173
76	0.06214	0.30453	0.31081	166	-0.07403	0.08011	0.10908
78	0.03478	0.28926	0.29134	168	-0.07802	0.07338	0.10710
80	0.00703	0.26343	0.26352	170	-0.08128	0.06755	0.10568
82	-0.01998	0.22844	0.22931	172	-0.08387	0.06269	0.10471
84	-0.04514	0.18604	0.19143	174	-0.08584	0.05887	0.10408
86	-0.06744	0.13821	0.15379	176	-0.08721	0.05611	0.10370
88	-0.08601	0.08711	0.12242	178	-0.08803	0.05445	0.10351
				180	-0.08830	0.05389	0.10344

IEN RADIATION PATTERN FOR THETA=PI/2

H/LMDA=0.5000 B/LMDA= 0.250 OMEGA= 9.92

8 ELEMENT ARRAY

PHI	ER(THETA,PHI)			PHI	ER(THETA,PHI)		
DEG	RE	IM	ABSVAL	DEG	RE	IM	ABSVAL
0	-0.49292	-0.36485	0.61326	90	-0.11734	-0.02319	0.11961
2	-0.49228	-0.36519	0.61294	92	-0.11684	-0.05351	0.12851
4	-0.49034	-0.36620	0.61199	94	-0.11075	-0.08094	0.13718
6	-0.48706	-0.36782	0.61034	96	-0.09941	-0.10440	0.14416
8	-0.48237	-0.36998	0.60792	98	-0.08341	-0.12299	0.14860
10	-0.47619	-0.37257	0.60462	100	-0.06354	-0.13607	0.15017
12	-0.46839	-0.37543	0.60027	102	-0.04078	-0.14325	0.14894
14	-0.45884	-0.37838	0.59473	104	-0.01617	-0.14443	0.14534
16	-0.44742	-0.38121	0.58780	106	0.00916	-0.13976	0.14006
18	-0.43397	-0.38368	0.57926	108	0.03414	-0.12964	0.13406
20	-0.41837	-0.38551	0.56890	110	0.05772	-0.11466	0.12837
22	-0.40049	-0.38640	0.55650	112	0.07900	-0.09558	0.12401
24	-0.38024	-0.38603	0.54185	114	0.09721	-0.07331	0.12175
26	-0.35755	-0.38406	0.52473	116	0.11176	-0.04878	0.12194
28	-0.33243	-0.38014	0.50499	118	0.12226	-0.02297	0.12440
30	-0.30491	-0.37394	0.48249	120	0.12850	0.00318	0.12854
32	-0.27511	-0.36513	0.45717	122	0.13045	0.02879	0.13359
34	-0.24324	-0.35340	0.42902	124	0.12828	0.05308	0.13883
36	-0.20957	-0.33853	0.39815	126	0.12226	0.07541	0.14365
38	-0.17449	-0.32031	0.36475	128	0.11281	0.09525	0.14764
40	-0.13847	-0.29864	0.32917	130	0.10043	0.11222	0.15060
42	-0.10206	-0.27351	0.29193	132	0.08566	0.12611	0.15245
44	-0.06590	-0.24502	0.25373	134	0.06907	0.13681	0.15326
46	-0.03071	-0.21342	0.21561	136	0.05123	0.14434	0.15316
48	0.00276	-0.17905	0.17907	138	0.03268	0.14884	0.15238
50	0.03373	-0.14243	0.14637	140	0.01392	0.15051	0.15115
52	0.06142	-0.10420	0.12095	142	-0.00462	0.14963	0.14970
54	0.08511	-0.06511	0.10716	144	-0.02255	0.14652	0.14824
56	0.10414	-0.02606	0.10735	146	-0.03958	0.14152	0.14695
58	0.11795	0.01200	0.11856	148	-0.05549	0.13500	0.14596
60	0.12617	0.04807	0.13502	150	-0.07011	0.12730	0.14533
62	0.12858	0.08111	0.15202	152	-0.08334	0.11877	0.14509
64	0.12516	0.11014	0.16672	154	-0.09514	0.10972	0.14522
66	0.11614	0.13425	0.17751	156	-0.10553	0.10043	0.14568
68	0.10195	0.15266	0.18357	158	-0.11454	0.09117	0.14639
70	0.08326	0.16476	0.18460	160	-0.12225	0.08214	0.14728
72	0.06093	0.17016	0.18074	162	-0.12875	0.07354	0.14827
74	0.03598	0.16872	0.17251	164	-0.13416	0.06553	0.14931
76	0.00960	0.16055	0.16083	166	-0.13860	0.05822	0.15033
78	-0.01700	0.14604	0.14702	168	-0.14216	0.05173	0.15128
80	-0.04255	0.12584	0.13284	170	-0.14498	0.04613	0.15214
82	-0.06586	0.10084	0.12044	172	-0.14713	0.04148	0.15286
84	-0.08581	0.07214	0.11211	174	-0.14871	0.03782	0.15345
86	-0.10148	0.04100	0.10945	176	-0.14979	0.03519	0.15387
88	-0.11215	0.00876	0.11249	178	-0.15042	0.03360	0.15413
				180	-0.15062	0.03307	0.15421

VEN

RADIATION PATTERN FOR THETA=PI/2

H/LMDA=0.5000 B/LMDA= 0.250 OMEGA= 9.92

9 ELEMENT ARRAY

PHI	ER(THETA,PHI)			PHI	ER(THETA,PHI)		
DEG	RE	IM	ABSVAL	DEG	RE	IM	ABSVAL
0	-0.97941	-0.39514	1.05611	90	-0.16687	-0.01757	0.16779
2	-0.97918	-0.39575	1.05613	92	-0.18303	-0.04826	0.18928
4	-0.97846	-0.39756	1.05614	94	-0.18837	-0.07511	0.20279
6	-0.97712	-0.40052	1.05602	96	-0.18313	-0.09692	0.20720
8	-0.97498	-0.40455	1.05558	98	-0.16813	-0.11283	0.20248
10	-0.97178	-0.40953	1.05455	100	-0.14469	-0.12231	0.18945
12	-0.96716	-0.41532	1.05256	102	-0.11452	-0.12519	0.16967
14	-0.96074	-0.42170	1.04922	104	-0.07961	-0.12170	0.14542
16	-0.95205	-0.42846	1.04402	106	-0.04207	-0.11233	0.11994
18	-0.94059	-0.43531	1.03644	108	-0.00401	-0.09787	0.09795
20	-0.92582	-0.44193	1.02588	110	0.03259	-0.07932	0.08575
22	-0.90716	-0.44796	1.01173	112	0.06599	-0.05779	0.08771
24	-0.88407	-0.45299	0.99336	114	0.09474	-0.03445	0.10081
26	-0.85600	-0.45658	0.97015	116	0.11780	-0.01048	0.11826
28	-0.82248	-0.45826	0.94152	118	0.13448	0.01305	0.13511
30	-0.78309	-0.45754	0.90696	120	0.14450	0.03517	0.14872
32	-0.73756	-0.45394	0.86606	122	0.14793	0.05508	0.15785
34	-0.68576	-0.44698	0.81857	124	0.14514	0.07218	0.16210
36	-0.62771	-0.43620	0.76439	126	0.13679	0.08604	0.16160
38	-0.56370	-0.42124	0.70371	128	0.12368	0.09644	0.15684
40	-0.49423	-0.40179	0.63694	130	0.10676	0.10333	0.14857
42	-0.42006	-0.37766	0.56487	132	0.08700	0.10683	0.13777
44	-0.34223	-0.34882	0.48866	134	0.06539	0.10715	0.12553
46	-0.26203	-0.31537	0.41002	136	0.04286	0.10463	0.11307
48	-0.18102	-0.27762	0.33143	138	0.02022	0.09967	0.10170
50	-0.10096	-0.23607	0.25675	140	-0.00180	0.09270	0.09272
52	-0.02374	-0.19143	0.19290	142	-0.02263	0.08416	0.08715
54	0.04864	-0.14459	0.15255	144	-0.04184	0.07449	0.08544
56	0.11418	-0.09664	0.14959	146	-0.05913	0.06409	0.08720
58	0.17099	-0.04881	0.17782	148	-0.07432	0.05333	0.09147
60	0.21738	-0.00243	0.21740	150	-0.08734	0.04252	0.09714
62	0.25197	0.04112	0.25530	152	-0.09824	0.03193	0.10330
64	0.27377	0.08045	0.28534	154	-0.10710	0.02177	0.10929
66	0.28227	0.11429	0.30453	156	-0.11410	0.01221	0.11475
68	0.27753	0.14151	0.31152	158	-0.11943	0.00336	0.11947
70	0.26015	0.16121	0.30605	160	-0.12331	-0.00471	0.12340
72	0.23132	0.17277	0.28872	162	-0.12598	-0.01194	0.12654
74	0.19276	0.17594	0.26098	164	-0.12766	-0.01832	0.12897
76	0.14664	0.17081	0.22512	166	-0.12859	-0.02386	0.13078
78	0.09550	0.15786	0.18450	168	-0.12895	-0.02858	0.13208
80	0.04208	0.13794	0.14421	170	-0.12894	-0.03249	0.13297
82	-0.01079	0.11221	0.11273	172	-0.12872	-0.03563	0.13356
84	-0.06039	0.08212	0.10194	174	-0.12841	-0.03804	0.13392
86	-0.10425	0.04931	0.11533	176	-0.12811	-0.03974	0.13413
88	-0.14028	0.01550	0.14113	178	-0.12790	-0.04075	0.13424
				180	-0.12783	-0.04109	0.13427

IEN

RADIATION PATTERN FOR THETA=PI/2

H/LMDA=0.5000 B/LMDA= 0.250 OMEGA= 9.92

9 ELEMENT ARRAY

PHI	ER(THETA,PHI)			PHI	ER(THETA,PHI)		
DEG	RE	IM	ABSVAL	DEG	RE	IM	ABSVAL
0	-0.57973	0.09630	0.58767	90	-0.13259	-0.01944	0.13401
2	-0.57952	0.09556	0.58734	92	-0.12935	-0.04329	0.13641
4	-0.57886	0.09336	0.58634	94	-0.11854	-0.06478	0.13508
6	-0.57768	0.08968	0.58460	96	-0.10095	-0.08275	0.13054
8	-0.57585	0.08455	0.58203	98	-0.07779	-0.09625	0.12375
10	-0.57319	0.07797	0.57847	100	-0.05050	-0.10460	0.11615
12	-0.56947	0.06997	0.57375	102	-0.02074	-0.10743	0.10941
14	-0.56442	0.06060	0.56767	104	0.00978	-0.10467	0.10513
16	-0.55775	0.04991	0.55998	106	0.03937	-0.09658	0.10430
18	-0.54913	0.03799	0.55044	108	0.06649	-0.08368	0.10688
20	-0.53820	0.02496	0.53878	110	0.08983	-0.06673	0.11190
22	-0.52462	0.01099	0.52473	112	0.10834	-0.04666	0.11795
24	-0.50804	-0.00373	0.50805	114	0.12131	-0.02450	0.12376
26	-0.48816	-0.01895	0.48853	116	0.12836	-0.00135	0.12837
28	-0.46470	-0.03437	0.46597	118	0.12946	0.02173	0.13127
30	-0.43748	-0.04964	0.44029	120	0.12486	0.04376	0.13230
32	-0.40639	-0.06436	0.41146	122	0.11507	0.06388	0.13161
34	-0.37144	-0.07811	0.37956	124	0.10081	0.08139	0.12956
36	-0.33278	-0.09042	0.34484	126	0.08292	0.09576	0.12667
38	-0.29070	-0.10082	0.30769	128	0.06236	0.10666	0.12355
40	-0.24569	-0.10884	0.26872	130	0.04005	0.11393	0.12077
42	-0.19839	-0.11405	0.22883	132	0.01694	0.11760	0.11881
44	-0.14962	-0.11605	0.18935	134	-0.00615	0.11781	0.11797
46	-0.10036	-0.11455	0.15229	136	-0.02848	0.11484	0.11832
48	-0.05175	-0.10935	0.12098	138	-0.04943	0.10907	0.11975
50	-0.00501	-0.10040	0.10052	140	-0.06854	0.10092	0.12199
52	0.03856	-0.08780	0.09589	142	-0.08547	0.09085	0.12474
54	0.07766	-0.07183	0.10579	144	-0.10004	0.07933	0.12767
56	0.11107	-0.05295	0.12305	146	-0.11215	0.06680	0.13054
58	0.13769	-0.03180	0.14131	148	-0.12186	0.05370	0.13316
60	0.15661	-0.00918	0.15687	150	-0.12926	0.04040	0.13542
62	0.16721	0.01397	0.16779	152	-0.13454	0.02722	0.13727
64	0.16919	0.03661	0.17311	154	-0.13794	0.01444	0.13869
66	0.16264	0.05766	0.17256	156	-0.13970	0.00228	0.13972
68	0.14803	0.07607	0.16643	158	-0.14012	-0.00909	0.14042
70	0.12621	0.09084	0.15550	160	-0.13947	-0.01956	0.14083
72	0.09843	0.10114	0.14113	162	-0.13801	-0.02903	0.14103
74	0.06623	0.10633	0.12527	164	-0.13601	-0.03747	0.14108
76	0.03141	0.10604	0.11059	166	-0.13369	-0.04486	0.14102
78	-0.00409	0.10018	0.10026	168	-0.13127	-0.05119	0.14090
80	-0.03829	0.08897	0.09686	170	-0.12892	-0.05649	0.14075
82	-0.06931	0.07296	0.10063	172	-0.12679	-0.06077	0.14060
84	-0.09545	0.05296	0.10916	174	-0.12501	-0.06407	0.14047
86	-0.11531	0.03005	0.11916	176	-0.12368	-0.06640	0.14037
88	-0.12788	0.00546	0.12800	178	-0.12285	-0.06779	0.14031
				180	-0.12256	-0.06826	0.14029

VEN

RADIATION PATTERN FOR THETA=PI/2

H/LMDA=0.5000 B/LMDA= 0.250 OMEGA= 9.92

10 ELEMENT ARRAY

PHI	ER(THETA,PHI)			PHI	ER(THETA,PHI)		
DEG	RE	IM	ABSVAL	DEG	RE	IM	ABSVAL
0	-0.94679	0.40552	1.02998	90	-0.19313	0.05216	0.20005
2	-0.94707	0.40488	1.02999	92	-0.18889	0.02784	0.19093
4	-0.94787	0.40292	1.02995	94	-0.17225	0.00249	0.17227
6	-0.94904	0.39959	1.02973	96	-0.14493	-0.02213	0.14661
8	-0.95034	0.39479	1.02908	98	-0.10929	-0.04436	0.11795
10	-0.95146	0.38840	1.02768	100	-0.06813	-0.06277	0.09263
12	-0.95196	0.38026	1.02510	102	-0.02442	-0.07629	0.08010
14	-0.95134	0.37019	1.02083	104	0.01885	-0.08423	0.08631
16	-0.94901	0.35802	1.01430	106	0.05895	-0.08631	0.10452
18	-0.94430	0.34354	1.00485	108	0.09356	-0.08268	0.12486
20	-0.93647	0.32659	0.99178	110	0.12088	-0.07387	0.14167
22	-0.92474	0.30702	0.97438	112	0.13975	-0.06069	0.15236
24	-0.90833	0.28473	0.95191	114	0.14962	-0.04419	0.15601
26	-0.88643	0.25968	0.92368	116	0.15056	-0.02555	0.15272
28	-0.85830	0.23192	0.88908	118	0.14319	-0.00600	0.14331
30	-0.82326	0.20158	0.84758	120	0.12854	0.01332	0.12922
32	-0.78079	0.16892	0.79885	122	0.10797	0.03135	0.11243
34	-0.73053	0.13434	0.74278	124	0.08305	0.04723	0.09554
36	-0.67236	0.09836	0.67952	126	0.05537	0.06030	0.08187
38	-0.60645	0.06166	0.60958	128	0.02652	0.07012	0.07497
40	-0.53331	0.02501	0.53390	130	-0.00208	0.07649	0.07652
42	-0.45379	-0.01065	0.45392	132	-0.02921	0.07941	0.08461
44	-0.36915	-0.04433	0.37180	134	-0.05387	0.07904	0.09565
46	-0.28100	-0.07498	0.29083	136	-0.07534	0.07571	0.10681
48	-0.19135	-0.10155	0.21663	138	-0.09318	0.06981	0.11643
50	-0.10249	-0.12306	0.16015	140	-0.10716	0.06182	0.12371
52	-0.01694	-0.13865	0.13968	142	-0.11729	0.05224	0.12839
54	0.06266	-0.14764	0.16039	144	-0.12373	0.04156	0.13053
56	0.13368	-0.14962	0.20064	146	-0.12681	0.03024	0.13037
58	0.19367	-0.14447	0.24162	148	-0.12693	0.01870	0.12830
60	0.24049	-0.13243	0.27454	150	-0.12455	0.00728	0.12476
62	0.27248	-0.11411	0.29541	152	-0.12015	-0.00372	0.12021
64	0.28862	-0.09046	0.30246	154	-0.11422	-0.01408	0.11509
66	0.28861	-0.06277	0.29535	156	-0.10721	-0.02366	0.10979
68	0.27295	-0.03262	0.27489	158	-0.09953	-0.03234	0.10466
70	0.24297	-0.00173	0.24298	160	-0.09157	-0.04009	0.09996
72	0.20080	0.02807	0.20275	162	-0.08363	-0.04687	0.09587
74	0.14923	0.05501	0.15905	164	-0.07599	-0.05273	0.09249
76	0.09159	0.07746	0.11996	166	-0.06885	-0.05770	0.08983
78	0.03154	0.09411	0.09925	168	-0.06240	-0.06182	0.08784
80	-0.02716	0.10400	0.10749	170	-0.05677	-0.06518	0.08643
82	-0.08094	0.10665	0.13388	172	-0.05204	-0.06782	0.08549
84	-0.12660	0.10206	0.16261	174	-0.04831	-0.06981	0.08489
86	-0.16156	0.09075	0.18530	176	-0.04561	-0.07119	0.08455
88	-0.18403	0.07367	0.19823	178	-0.04398	-0.07201	0.08437
				180	-0.04343	-0.07228	0.08432

IEN

RADIATION PATTERN FOR THETA=PI/2

H/LMDA=0.5000 B/LMDA= 0.250 OMEGA= 9.92

10 ELEMENT ARRAY

PHI	ER(THETA,PHI)			PHI	ER(THETA,PHI)		
DEG	RE	IM	ABSVAL	DEG	RE	IM	ABSVAL
0	-0.32275	0.46407	0.56527	90	-0.11829	0.00707	0.11850
2	-0.32315	0.46338	0.56493	92	-0.10843	-0.01910	0.11010
4	-0.32433	0.46128	0.56389	94	-0.09128	-0.04385	0.10126
6	-0.32622	0.45771	0.56206	96	-0.06815	-0.06538	0.09444
8	-0.32871	0.45256	0.55934	98	-0.04075	-0.08216	0.09171
10	-0.33163	0.44569	0.55554	100	-0.01105	-0.09304	0.09369
12	-0.33478	0.43694	0.55045	102	0.01894	-0.09734	0.09917
14	-0.33789	0.42611	0.54382	104	0.04724	-0.09488	0.10599
16	-0.34068	0.41299	0.53537	106	0.07208	-0.08595	0.11218
18	-0.34278	0.39739	0.52480	108	0.09200	-0.07130	0.11640
20	-0.34381	0.37913	0.51180	110	0.10595	-0.05202	0.11804
22	-0.34335	0.35805	0.49607	112	0.11333	-0.02947	0.11710
24	-0.34095	0.33407	0.47734	114	0.11397	-0.00513	0.11408
26	-0.33616	0.30717	0.45536	116	0.10813	0.01947	0.10987
28	-0.32853	0.27742	0.42999	118	0.09647	0.04290	0.10558
30	-0.31764	0.24502	0.40116	120	0.07990	0.06390	0.10231
32	-0.30314	0.21027	0.36893	122	0.05955	0.08144	0.10089
34	-0.28476	0.17364	0.33353	124	0.03667	0.09479	0.10164
36	-0.26234	0.13574	0.29538	126	0.01250	0.10351	0.10427
38	-0.23587	0.09731	0.25516	128	-0.01176	0.10746	0.10811
40	-0.20552	0.05924	0.21389	130	-0.03504	0.10676	0.11237
42	-0.17166	0.02252	0.17313	132	-0.05646	0.10176	0.11637
44	-0.13486	-0.01179	0.13538	134	-0.07532	0.09297	0.11965
46	-0.09593	-0.04261	0.10497	136	-0.09113	0.08104	0.12195
48	-0.05589	-0.06887	0.08869	138	-0.10363	0.06668	0.12323
50	-0.01592	-0.08963	0.09103	140	-0.11272	0.05060	0.12356
52	0.02263	-0.10410	0.10653	142	-0.11850	0.03352	0.12315
54	0.05833	-0.11174	0.12605	144	-0.12116	0.01605	0.12222
56	0.08976	-0.11228	0.14375	146	-0.12103	-0.00122	0.12104
58	0.11555	-0.10584	0.15670	148	-0.11848	-0.01786	0.11982
60	0.13454	-0.09288	0.16349	150	-0.11394	-0.03349	0.11876
62	0.14579	-0.07427	0.16362	152	-0.10783	-0.04786	0.11798
64	0.14876	-0.05120	0.15733	154	-0.10058	-0.06081	0.11753
66	0.14331	-0.02521	0.14551	156	-0.09257	-0.07225	0.11743
68	0.12975	0.00199	0.12976	158	-0.08418	-0.08219	0.11765
70	0.10889	0.02854	0.11257	160	-0.07571	-0.09066	0.11811
72	0.08202	0.05259	0.09743	162	-0.06744	-0.09775	0.11876
74	0.05078	0.07248	0.08850	164	-0.05958	-0.10359	0.11950
76	0.01716	0.08677	0.08845	166	-0.05233	-0.10830	0.12029
78	-0.01669	0.09444	0.09591	168	-0.04583	-0.11204	0.12105
80	-0.04858	0.09495	0.10666	170	-0.04018	-0.11493	0.12175
82	-0.07644	0.08827	0.11677	172	-0.03547	-0.11711	0.12236
84	-0.09848	0.07492	0.12374	174	-0.03175	-0.11868	0.12285
86	-0.11329	0.05592	0.12634	176	-0.02908	-0.11973	0.12321
88	-0.12000	0.03272	0.12438	178	-0.02746	-0.12034	0.12343
				180	-0.02692	-0.12054	0.12350

VEN

RADIATION PATTERN FOR THETA=PI/2

H/LMDA=0.5000   B/LMDA= 0.250   OMEGA= 9.92

12 ELEMENT ARRAY

PHI	ER(THETA,PHI)			PHI	ER(THETA,PHI)		
DEG	RE	IM	ABSVAL	DEG	RE	IM	ABSVAL
0	0.39319	0.90460	0.98635	90	-0.07095	0.02307	0.07461
2	0.39244	0.90490	0.98633	92	-0.07955	-0.02715	0.08406
4	0.39017	0.90574	0.98620	94	-0.07986	-0.07227	0.10770
6	0.38629	0.90692	0.98576	96	-0.07214	-0.10821	0.13005
8	0.38068	0.90812	0.98468	98	-0.05750	-0.13198	0.14396
10	0.37314	0.90887	0.98249	100	-0.03765	-0.14198	0.14688
12	0.36348	0.90860	0.97861	102	-0.01478	-0.13801	0.13880
14	0.35146	0.90659	0.97233	104	0.00876	-0.12127	0.12159
16	0.33683	0.90203	0.96286	106	0.03068	-0.09407	0.09894
18	0.31935	0.89400	0.94932	108	0.04900	-0.05953	0.07711
20	0.29883	0.88151	0.93079	110	0.06220	-0.02124	0.06573
22	0.27513	0.86355	0.90632	112	0.06935	0.01720	0.07145
24	0.24819	0.83909	0.87503	114	0.07016	0.05247	0.08761
26	0.21809	0.80717	0.83612	116	0.06491	0.08184	0.10445
28	0.18503	0.76698	0.78898	118	0.05441	0.10335	0.11680
30	0.14940	0.71788	0.73326	120	0.03984	0.11592	0.12258
32	0.11179	0.65954	0.66895	122	0.02262	0.11930	0.12143
34	0.07297	0.59200	0.59648	124	0.00423	0.11401	0.11409
36	0.03393	0.51576	0.51688	126	-0.01391	0.10117	0.10212
38	-0.00416	0.43182	0.43184	128	-0.03054	0.08233	0.08782
40	-0.04000	0.34176	0.34409	130	-0.04468	0.05931	0.07425
42	-0.07220	0.24772	0.25802	132	-0.05562	0.03392	0.06515
44	-0.09936	0.15237	0.18191	134	-0.06300	0.00793	0.06350
46	-0.12020	0.05886	0.13384	136	-0.06671	-0.01715	0.06888
48	-0.13362	-0.02938	0.13681	138	-0.06690	-0.04009	0.07800
50	-0.13884	-0.10876	0.17636	140	-0.06393	-0.05999	0.08767
52	-0.13550	-0.17582	0.22198	142	-0.05827	-0.07629	0.09599
54	-0.12377	-0.22749	0.25898	144	-0.05049	-0.08873	0.10209
56	-0.10434	-0.26132	0.28138	146	-0.04116	-0.09734	0.10568
58	-0.07850	-0.27580	0.28676	148	-0.03086	-0.10233	0.10688
60	-0.04804	-0.27058	0.27481	150	-0.02011	-0.10409	0.10601
62	-0.01518	-0.24654	0.24700	152	-0.00934	-0.10309	0.10352
64	0.01757	-0.20590	0.20665	154	0.00108	-0.09987	0.09988
66	0.04766	-0.15212	0.15941	156	0.01089	-0.09495	0.09558
68	0.07268	-0.08967	0.11542	158	0.01989	-0.08886	0.09106
70	0.09058	-0.02374	0.09364	160	0.02796	-0.08205	0.08668
72	0.09993	0.04021	0.10772	162	0.03506	-0.07493	0.08273
74	0.10004	0.09688	0.13927	164	0.04117	-0.06786	0.07938
76	0.09107	0.14169	0.16843	166	0.04634	-0.06112	0.07670
78	0.07401	0.17114	0.18645	168	0.05060	-0.05493	0.07469
80	0.05065	0.18317	0.19005	170	0.05405	-0.04947	0.07327
82	0.02336	0.17741	0.17894	172	0.05674	-0.04486	0.07233
84	-0.00511	0.15514	0.15522	174	0.05875	-0.04119	0.07175
86	-0.03197	0.11917	0.12338	176	0.06014	-0.03854	0.07143
88	-0.05462	0.07353	0.09160	178	0.06095	-0.03693	0.07127
				180	0.06122	-0.03639	0.07122

IEN

RADIATION PATTERN FOR THETA=PI/2

H/LMDA=0.5000   B/LMDA= 0.250   OMEGA= 9.92

12 ELEMENT ARRAY

PHI	ER(THETA,PHI)			PHI	ER(THETA,PHI)		
DEG	RE	IM	ABSVAL	DEG	RE	IM	ABSVAL
0	0.44184	0.29241	0.52984	90	-0.07958	-0.01462	0.08092
2	0.44111	0.29286	0.52947	92	-0.07721	-0.04446	0.08910
4	0.43887	0.29420	0.52835	94	-0.06655	-0.06932	0.09609
6	0.43503	0.29632	0.52636	96	-0.04881	-0.08682	0.09960
8	0.42946	0.29907	0.52334	98	-0.02592	-0.09542	0.09888
10	0.42194	0.30226	0.51903	100	-0.00030	-0.09450	0.09450
12	0.41226	0.30559	0.51317	102	0.02540	-0.08440	0.08814
14	0.40013	0.30873	0.50539	104	0.04858	-0.06635	0.08224
16	0.38530	0.31127	0.49532	106	0.06702	-0.04230	0.07926
18	0.36750	0.31275	0.48257	108	0.07904	-0.01467	0.08039
20	0.34653	0.31267	0.46674	110	0.08366	0.01391	0.08481
22	0.32222	0.31047	0.44745	112	0.08065	0.04089	0.09043
24	0.29453	0.30558	0.42441	114	0.07052	0.06400	0.09523
26	0.26352	0.29746	0.39740	116	0.05437	0.08150	0.09797
28	0.22943	0.28560	0.36634	118	0.03377	0.09223	0.09822
30	0.19270	0.26956	0.33135	120	0.01057	0.09570	0.09628
32	0.15393	0.24905	0.29278	122	-0.01332	0.09203	0.09299
34	0.11396	0.22394	0.25127	124	-0.03609	0.08191	0.08951
36	0.07385	0.19435	0.20791	126	-0.05618	0.06644	0.08701
38	0.03479	0.16065	0.16437	128	-0.07237	0.04699	0.08629
40	-0.00185	0.12351	0.12352	130	-0.08386	0.02508	0.08753
42	-0.03467	0.08391	0.09079	132	-0.09027	0.00219	0.09030
44	-0.06228	0.04315	0.07577	134	-0.09159	-0.02030	0.09381
46	-0.08345	0.00278	0.08350	136	-0.08816	-0.04121	0.09731
48	-0.09718	-0.03543	0.10344	138	-0.08055	-0.05964	0.10023
50	-0.10282	-0.06964	0.12418	140	-0.06954	-0.07496	0.10225
52	-0.10019	-0.09801	0.14016	142	-0.05595	-0.08684	0.10330
54	-0.08966	-0.11891	0.14892	144	-0.04066	-0.09516	0.10348
56	-0.07215	-0.13101	0.14957	146	-0.02446	-0.10004	0.10299
58	-0.04914	-0.13352	0.14228	148	-0.00810	-0.10176	0.10209
60	-0.02260	-0.12623	0.12824	150	0.00783	-0.10073	0.10103
62	0.00517	-0.10966	0.10978	152	0.02286	-0.09739	0.10004
64	0.03169	-0.08506	0.09077	154	0.03664	-0.09225	0.09926
66	0.05451	-0.05437	0.07699	156	0.04897	-0.08579	0.09878
68	0.07150	-0.02011	0.07427	158	0.05973	-0.07848	0.09863
70	0.08101	0.01483	0.08235	160	0.06892	-0.07074	0.09876
72	0.08211	0.04740	0.09481	162	0.07660	-0.06291	0.09913
74	0.07469	0.07472	0.10565	164	0.08289	-0.05532	0.09965
76	0.05952	0.09434	0.11154	166	0.08791	-0.04819	0.10025
78	0.03817	0.10449	0.11124	168	0.09184	-0.04173	0.10088
80	0.01290	0.10430	0.10509	170	0.09484	-0.03607	0.10147
82	-0.01361	0.09391	0.09489	172	0.09706	-0.03134	0.10200
84	-0.03853	0.07447	0.08384	174	0.09864	-0.02759	0.10242
86	-0.05918	0.04801	0.07621	176	0.09968	-0.02488	0.10274
88	-0.07337	0.01727	0.07537	178	0.10028	-0.02324	0.10293
				180	0.10047	-0.02270	0.10300

VEN

RADIATION PATTERN FOR THETA=PI/2

H/LMDA=0.5000   B/LMDA= 0.250   OMEGA= 9.92

14 ELEMENT ARRAY

PHI	ER(THETA,PHI)			PHI	ER(THETA,PHI)		
DEG	RE	IM	ABSVAL	DEG	RE	IM	ABSVAL
0	0.87058	-0.38354	0.95132	90	-0.13521	0.03752	0.14032
2	0.87090	-0.38270	0.95127	92	-0.12982	0.01184	0.1303[illegible]
4	0.87177	-0.38013	0.95105	94	-0.10736	-0.01450	0.10834
6	0.87295	-0.37574	0.95037	96	-0.07168	-0.03788	0.0810[illegible]
8	0.87398	-0.36933	0.94882	98	-0.02818	-0.05526	0.06203
10	0.87428	-0.36069	0.94576	100	0.01700	-0.06452	0.06672
12	0.87307	-0.34954	0.94044	102	0.05787	-0.06480	0.0868[illegible]
14	0.86941	-0.33557	0.93192	104	0.08944	-0.05647	0.10577
16	0.86225	-0.31848	0.91919	106	0.10824	-0.04101	0.1157[illegible]
18	0.85041	-0.29801	0.90111	108	0.11270	-0.02078	0.1146[illegible]
20	0.83262	-0.27394	0.87653	110	0.10320	0.00143	0.10321
22	0.80764	-0.24619	0.84433	112	0.08180	0.02277	0.08491
24	0.77427	-0.21481	0.80351	114	0.05183	0.04070	0.0659[illegible]
26	0.73147	-0.18004	0.75330	116	0.01731	0.05333	0.05607
28	0.67849	-0.14237	0.69326	118	-0.01758	0.05953	0.06207
30	0.61496	-0.10254	0.62345	120	-0.04906	0.05902	0.07675
32	0.54104	-0.06158	0.54453	122	-0.07415	0.05228	0.09073
34	0.45753	-0.02077	0.45800	124	-0.09085	0.04046	0.09945
36	0.36595	0.01835	0.36641	126	-0.09831	0.02507	0.10145
38	0.26862	0.05407	0.27401	128	-0.09665	0.00783	0.09697
40	0.16862	0.08464	0.18867	130	-0.08690	-0.00954	0.08742
42	0.06972	0.10835	0.12885	132	-0.07067	-0.02556	0.07515
44	-0.02378	0.12375	0.12601	134	-0.04990	-0.03904	0.06336
46	-0.10735	0.12975	0.16841	136	-0.02666	-0.04918	0.05595
48	-0.17656	0.12587	0.21683	138	-0.00288	-0.05557	0.05565
50	-0.22747	0.11230	0.25368	140	0.01980	-0.05816	0.06144
52	-0.25710	0.09004	0.27242	142	0.04009	-0.05720	0.06985
54	-0.26383	0.06091	0.27077	144	0.05712	-0.05315	0.07803
56	-0.24770	0.02742	0.24922	146	0.07041	-0.04662	0.08444
58	-0.21068	-0.00735	0.21081	148	0.07983	-0.03826	0.08852
60	-0.15658	-0.04006	0.16162	150	0.08554	-0.02872	0.09023
62	-0.09090	-0.06744	0.11319	152	0.08789	-0.01861	0.08984
64	-0.02036	-0.08666	0.08902	154	0.08740	-0.00843	0.08781
66	0.04774	-0.09570	0.10695	156	0.08463	0.00140	0.08464
68	0.10632	-0.09365	0.14169	158	0.08016	0.01057	0.08085
70	0.14928	-0.08088	0.16978	160	0.07454	0.01890	0.07690
72	0.17233	-0.05903	0.18216	162	0.06827	0.02625	0.07314
74	0.17350	-0.03090	0.17623	164	0.06178	0.03260	0.06985
76	0.15348	-0.00008	0.15348	166	0.05542	0.03794	0.06716
78	0.11551	0.02949	0.11921	168	0.04948	0.04235	0.06513
80	0.06503	0.05404	0.08455	170	0.04418	0.04588	0.06369
82	0.00894	0.07050	0.07107	172	0.03967	0.04862	0.06275
84	-0.04532	0.07693	0.08929	174	0.03607	0.05065	0.06218
86	-0.09079	0.07276	0.11634	176	0.03345	0.05205	0.06187
88	-0.12190	0.05889	0.13538	178	0.03186	0.05286	0.06172
				180	0.03133	0.05313	0.06168

IEN

RADIATION PATTERN FOR THETA=PI/2

H/LMDA=0.5000   B/LMDA= 0.250   OMEGA= 9.92

14 ELEMENT ARRAY

PHI	ER(THETA,PHI)			PHI	ER(THETA,PHI)		
DEG	RE	IM	ABSVAL	DEG	RE	IM	ABSVAL
0	0.26928	-0.42401	0.50229	90	-0.08362	0.00453	0.08374
2	0.26978	-0.42323	0.50190	92	-0.07266	-0.02162	0.07580
4	0.27125	-0.42087	0.50071	94	-0.05188	-0.04464	0.06844
6	0.27358	-0.41679	0.49855	96	-0.02432	-0.06123	0.06589
8	0.27657	-0.41080	0.49522	98	0.00613	-0.06910	0.06937
10	0.27995	-0.40265	0.49041	100	0.03530	-0.06723	0.07594
12	0.28338	-0.39202	0.48372	102	0.05935	-0.05601	0.08161
14	0.28642	-0.37858	0.47472	104	0.07528	-0.03712	0.08393
16	0.28857	-0.36199	0.46294	106	0.08126	-0.01323	0.08233
18	0.28922	-0.34195	0.44786	108	0.07686	0.01243	0.07786
20	0.28776	-0.31820	0.42902	110	0.06300	0.03651	0.07282
22	0.28351	-0.29064	0.40601	112	0.04171	0.05606	0.06988
24	0.27579	-0.25928	0.37853	114	0.01579	0.06883	0.07062
26	0.26399	-0.22437	0.34645	116	-0.01159	0.07354	0.07444
28	0.24758	-0.18637	0.30989	118	-0.03733	0.06994	0.07927
30	0.22622	-0.14605	0.26927	120	-0.05875	0.05876	0.08309
32	0.19979	-0.10445	0.22544	122	-0.07390	0.04151	0.08476
34	0.16848	-0.06287	0.17983	124	-0.08161	0.02024	0.08408
36	0.13285	-0.02287	0.13481	126	-0.08163	-0.00279	0.08168
38	0.09388	0.01386	0.09490	128	-0.07444	-0.02535	0.07864
40	0.05295	0.04557	0.06986	130	-0.06119	-0.04549	0.07624
42	0.01182	0.07060	0.07158	132	-0.04338	-0.06169	0.07542
44	-0.02741	0.08755	0.09174	134	-0.02278	-0.07297	0.07644
46	-0.06248	0.09549	0.11411	136	-0.00110	-0.07888	0.07889
48	-0.09113	0.09403	0.13095	138	0.02008	-0.07946	0.08196
50	-0.11133	0.08355	0.13919	140	0.03945	-0.07518	0.08490
52	-0.12149	0.06519	0.13787	142	0.05605	-0.06677	0.08718
54	-0.12073	0.04086	0.12746	144	0.06928	-0.05517	0.08856
56	-0.10907	0.01313	0.10985	146	0.07886	-0.04135	0.08904
58	-0.08750	-0.01499	0.08878	148	0.08483	-0.02627	0.08880
60	-0.05808	-0.04035	0.07072	150	0.08742	-0.01080	0.08809
62	-0.02372	-0.06001	0.06453	152	0.08706	0.00434	0.08716
64	0.01197	-0.07164	0.07263	154	0.08424	0.01859	0.08627
66	0.04512	-0.07379	0.08649	156	0.07953	0.03157	0.08557
68	0.07199	-0.06621	0.09781	158	0.07347	0.04303	0.08514
70	0.08948	-0.04988	0.10244	160	0.06658	0.05288	0.08502
72	0.09552	-0.02698	0.09926	162	0.05930	0.06112	0.08516
74	0.08946	-0.00063	0.08946	164	0.05203	0.06784	0.08549
76	0.07213	0.02552	0.07651	166	0.04508	0.07318	0.08595
78	0.04585	0.04778	0.06622	168	0.03870	0.07733	0.08647
80	0.01414	0.06298	0.06455	170	0.03306	0.08045	0.08698
82	-0.01874	0.06894	0.07144	172	0.02832	0.08272	0.08744
84	-0.04837	0.06479	0.08086	174	0.02455	0.08432	0.08782
86	-0.07078	0.05115	0.08733	176	0.02182	0.08536	0.08810
88	-0.08304	0.03003	0.08830	178	0.02017	0.08594	0.08828
				180	0.01962	0.08613	0.08833

VEN

RADIATION PATTERN FOR THETA=PI/2

H/LMDA=0.5000 B/LMDA= 0.250 OMEGA= 9.92

16 ELEMENT ARRAY

PHI	ER(THETA,PHI)			PHI	ER(THETA,PHI)		
DEG	RE	IM	ABSVAL	DEG	RE	IM	ABSVAL
0	-0.37568	-0.84233	0.92231	90	-0.05516	0.01725	0.05779
2	-0.37475	-0.84267	0.92224	92	-0.06261	-0.03217	0.07039
4	-0.37191	-0.84357	0.92192	94	-0.05856	-0.07356	0.09402
6	-0.36702	-0.84472	0.92101	96	-0.04418	-0.10012	0.10943
8	-0.35985	-0.84555	0.91894	98	-0.02245	-0.10802	0.11032
10	-0.35012	-0.84530	0.91495	100	0.00248	-0.09691	0.09694
12	-0.33748	-0.84300	0.90805	102	0.02610	-0.06975	0.07448
14	-0.32156	-0.83748	0.89709	104	0.04439	-0.03206	0.05476
16	-0.30201	-0.82738	0.88078	106	0.05448	0.00927	0.05526
18	-0.27855	-0.81127	0.85775	108	0.05507	0.04723	0.07256
20	-0.25098	-0.78762	0.82664	110	0.04657	0.07595	0.08910
22	-0.21930	-0.75498	0.78618	112	0.03083	0.09152	0.09657
24	-0.18375	-0.71204	0.73537	114	0.01074	0.09243	0.09305
26	-0.14482	-0.65784	0.67359	116	-0.01039	0.07960	0.08027
28	-0.10338	-0.59186	0.60082	118	-0.02933	0.05589	0.06311
30	-0.06064	-0.51429	0.51785	120	-0.04351	0.02545	0.05040
32	-0.01817	-0.42611	0.42650	122	-0.05126	-0.00704	0.05174
34	0.02215	-0.32929	0.33003	124	-0.05199	-0.03719	0.06392
36	0.05824	-0.22680	0.23416	126	-0.04612	-0.06144	0.07682
38	0.08797	-0.12263	0.15092	128	-0.03486	-0.07740	0.08489
40	0.10932	-0.02161	0.11144	130	-0.01996	-0.08404	0.08638
42	0.12068	0.07084	0.13994	132	-0.00338	-0.08156	0.08163
44	0.12106	0.14918	0.19212	134	0.01299	-0.07115	0.07233
46	0.11032	0.20826	0.23568	136	0.02754	-0.05469	0.06123
48	0.08934	0.24397	0.25982	138	0.03911	-0.03435	0.05205
50	0.06010	0.25385	0.26087	140	0.04697	-0.01235	0.04857
52	0.02557	0.23761	0.23898	142	0.05090	0.00933	0.05175
54	-0.01049	0.19750	0.19778	144	0.05105	0.02913	0.05877
56	-0.04390	0.13834	0.14514	146	0.04787	0.04592	0.06633
58	-0.07062	0.06723	0.09751	148	0.04200	0.05906	0.07248
60	-0.08727	-0.00715	0.08756	150	0.03418	0.06835	0.07642
62	-0.09167	-0.07552	0.11877	152	0.02512	0.07390	0.07806
64	-0.08327	-0.12924	0.15375	154	0.01550	0.07610	0.07767
66	-0.06336	-0.16154	0.17352	156	0.00587	0.07549	0.07572
68	-0.03496	-0.16859	0.17218	158	-0.00335	0.07268	0.07276
70	-0.00244	-0.15023	0.15025	160	-0.01183	0.06828	0.06930
72	0.02910	-0.11009	0.11387	162	-0.01939	0.06287	0.06580
74	0.05469	-0.05512	0.07765	164	-0.02594	0.05697	0.06260
76	0.07029	0.00548	0.07050	166	-0.03147	0.05100	0.05993
78	0.07355	0.06174	0.09603	168	-0.03600	0.04531	0.05787
80	0.06422	0.10462	0.12276	170	-0.03961	0.04016	0.05641
82	0.04423	0.12751	0.13497	172	-0.04240	0.03574	0.05546
84	0.01736	0.12740	0.12858	174	-0.04446	0.03220	0.05489
86	-0.01148	0.10532	0.10595	176	-0.04586	0.02961	0.05459
88	-0.03713	0.06607	0.07579	178	-0.04668	0.02803	0.05445
				180	-0.04695	0.02751	0.05441

IEN

RADIATION PATTERN FOR THETA=PI/2

H/LMDA=0.5000 B/LMDA= 0.250 OMEGA= 9.92

16 ELEMENT ARRAY

PHI	ER(THETA,PHI)			PHI	ER(THETA,PHI)		
DEG	RE	IM	ABSVAL	DEG	RE	IM	ABSVAL
0	-0.40927	-0.25093	0.48007	90	-0.06028	-0.01057	0.06120
2	-0.40845	-0.25148	0.47966	92	-0.05623	-0.03975	0.06886
4	-0.40597	-0.25307	0.47839	94	-0.04153	-0.06128	0.07403
6	-0.40166	-0.25558	0.47608	96	-0.01902	-0.07147	0.07395
8	-0.39527	-0.25877	0.47244	98	0.00706	-0.06874	0.06910
10	-0.38648	-0.26231	0.46709	100	0.03188	-0.05389	0.06261
12	-0.37489	-0.26577	0.45954	102	0.05096	-0.02982	0.05905
14	-0.36011	-0.26862	0.44926	104	0.06100	-0.00093	0.06100
16	-0.34171	-0.27023	0.43565	106	0.06038	0.02775	0.06645
18	-0.31936	-0.26988	0.41812	108	0.04943	0.05145	0.07135
20	-0.29281	-0.26682	0.39614	110	0.03021	0.06650	0.07304
22	-0.26199	-0.26025	0.36929	112	0.00607	0.07082	0.07108
24	-0.22709	-0.24944	0.33733	114	-0.01901	0.06416	0.06692
26	-0.18856	-0.23372	0.30030	116	-0.04116	0.04799	0.06322
28	-0.14723	-0.21266	0.25865	118	-0.05718	0.02509	0.06244
30	-0.10428	-0.18609	0.21331	120	-0.06501	-0.00100	0.06502
32	-0.06124	-0.15422	0.16594	122	-0.06393	-0.02660	0.06924
34	-0.01997	-0.11775	0.11943	124	-0.05451	-0.04841	0.07291
36	0.01748	-0.07786	0.07980	126	-0.03837	-0.06398	0.07460
38	0.04901	-0.03631	0.06099	128	-0.01784	-0.07185	0.07403
40	0.07267	0.00469	0.07282	130	0.00446	-0.07169	0.07183
42	0.08693	0.04259	0.09680	132	0.02600	-0.06411	0.06918
44	0.09087	0.07465	0.11760	134	0.04462	-0.05045	0.06735
46	0.08444	0.09831	0.12959	136	0.05876	-0.03254	0.06717
48	0.06858	0.11143	0.13084	138	0.06753	-0.01234	0.06865
50	0.04528	0.11269	0.12145	140	0.07068	0.00825	0.07116
52	0.01744	0.10188	0.10336	142	0.06852	0.02759	0.07387
54	-0.01136	0.08005	0.08086	144	0.06178	0.04446	0.07611
56	-0.03728	0.04965	0.06209	146	0.05144	0.05803	0.07755
58	-0.05676	0.01430	0.05853	148	0.03862	0.06790	0.07811
60	-0.06698	-0.02155	0.07036	150	0.02438	0.07404	0.07795
62	-0.06645	-0.05315	0.08509	152	0.00969	0.07671	0.07732
64	-0.05526	-0.07612	0.09406	154	-0.00464	0.07634	0.07648
66	-0.03522	-0.08717	0.09401	156	-0.01800	0.07352	0.07569
68	-0.00963	-0.08467	0.08522	158	-0.03001	0.06884	0.07509
70	0.01717	-0.06906	0.07116	160	-0.04043	0.06290	0.07477
72	0.04052	-0.04285	0.05897	162	-0.04918	0.05625	0.07472
74	0.05630	-0.01033	0.05724	164	-0.05633	0.04937	0.07490
76	0.06164	0.02306	0.06581	166	-0.06200	0.04263	0.07524
78	0.05555	0.05169	0.07588	168	-0.06636	0.03635	0.07566
80	0.03914	0.07069	0.08080	170	-0.06961	0.03076	0.07610
82	0.01545	0.07686	0.07840	172	-0.07195	0.02601	0.07651
84	-0.01105	0.06926	0.07014	174	-0.07357	0.02224	0.07685
86	-0.03536	0.04942	0.06077	176	-0.07461	0.01949	0.07711
88	-0.05288	0.02107	0.05692	178	-0.07519	0.01783	0.07727
				180	-0.07537	0.01728	0.07733

[illegible]

[illegible]

H/LMDA=0.5000 B/LMDA= 0.250 OMEGA= 9.92

18 ELEMENT ARRAY

PHI	ER(THETA,PHI)			PHI	ER(THETA,PHI)		
DEG	RE	IM	ABSVAL	DEG	RE	IM	ABSVAL
0	-0.81835	0.36910	0.89774	90	-0.10374	0.02926	0.10779
2	-0.81870	0.36809	0.89764	92	-0.09657	0.00261	0.09661
4	-0.81963	0.36498	0.89722	94	-0.06831	-0.02362	0.07228
6	-0.82073	0.35961	0.89606	96	-0.02627	-0.04352	0.05083
8	-0.82132	0.35170	0.89345	98	0.01959	-0.05288	0.05639
10	-0.82045	0.34088	0.88845	100	0.05903	-0.05006	0.07740
12	-0.81692	0.32676	0.87985	102	0.08387	-0.03624	0.09137
14	-0.80930	0.30888	0.86625	104	0.08964	-0.01494	0.09088
16	-0.79597	0.28687	0.84608	106	0.07626	0.00887	0.07677
18	-0.77517	0.26041	0.81775	108	0.04767	0.03002	0.05633
20	-0.74518	0.22940	0.77969	110	0.01066	0.04424	0.04550
22	-0.70436	0.19397	0.73058	112	-0.02684	0.04899	0.05586
24	-0.65143	0.15459	0.66952	114	-0.05751	0.04387	0.07233
26	-0.58563	0.11212	0.59626	116	-0.07603	0.03044	0.08189
28	-0.50697	0.06790	0.51149	118	-0.07989	0.01172	0.08074
30	-0.41646	0.02371	0.41713	120	-0.06951	-0.00855	0.07003
32	-0.31630	-0.01828	0.31683	122	-0.04780	-0.02671	0.05476
34	-0.21000	-0.05560	0.21724	124	-0.01926	-0.03986	0.04427
36	-0.10231	-0.08574	0.13349	126	0.01101	-0.04622	0.04752
38	0.00094	-0.10634	0.10635	128	0.03830	-0.04533	0.05935
40	0.09325	-0.11559	0.14852	130	0.05896	-0.03790	0.07009
42	0.16809	-0.11253	0.20228	132	0.07081	-0.02551	0.07527
44	0.21963	-0.09733	0.24023	134	0.07325	-0.01024	0.07396
46	0.24365	-0.07155	0.25394	136	0.06697	0.00575	0.06722
48	0.23836	-0.03811	0.24138	138	0.05369	0.02051	0.05748
50	0.20502	-0.00112	0.20503	140	0.03567	0.03256	0.04830
52	0.14829	0.03455	0.15226	142	0.01530	0.04096	0.04373
54	0.07597	0.06388	0.09926	144	-0.00520	0.04534	0.04564
56	-0.00175	0.08255	0.08256	146	-0.02406	0.04581	0.05175
58	-0.07359	0.08764	0.11444	148	-0.04001	0.04285	0.05863
60	-0.12886	0.07834	0.15081	150	-0.05234	0.03715	0.06419
62	-0.15931	0.05624	0.16895	152	-0.06082	0.02951	0.06760
64	-0.16065	0.02523	0.16262	154	-0.06560	0.02073	0.06880
66	-0.13357	-0.00903	0.13387	156	-0.06713	0.01149	0.06811
68	-0.08384	-0.04018	0.09297	158	-0.06599	0.00239	0.06603
70	-0.02143	-0.06229	0.06588	160	-0.06283	-0.00617	0.06314
72	0.04133	-0.07119	0.08231	162	-0.05831	-0.01389	0.05995
74	0.09215	-0.06531	0.11295	164	-0.05302	-0.02061	0.05688
76	0.12130	-0.04619	0.12980	166	-0.04744	-0.02630	0.05424
78	0.12365	-0.01814	0.12498	168	-0.04200	-0.03095	0.05217
80	0.09980	0.01263	0.10059	170	-0.03700	-0.03465	0.05069
82	0.05588	0.03936	0.06835	172	-0.03267	-0.03749	0.04973
84	0.00220	0.05631	0.05635	174	-0.02917	-0.03957	0.04916
86	-0.04916	0.06002	0.07758	176	-0.02661	-0.04098	0.04886
88	-0.08705	0.05010	0.10043	178	-0.02505	-0.04180	0.04873
				180	-0.02452	-0.04206	0.04869

IEN

RADIATION PATTERN FOR THETA=PI/2

H/LMDA=0.5000 B/LMDA= 0.250 OMEGA= 9.92

18 ELEMENT ARRAY

PHI	ER(THETA,PHI)			PHI	ER(THETA,PHI)		
DEG	RE	IM	ABSVAL	DEG	RE	IM	ABSVAL
0	-0.23594	0.39680	0.46164	90	-0.06460	0.00324	0.06468
2	-0.23653	0.39595	0.46122	92	-0.05238	-0.02271	0.05709
4	-0.23824	0.39336	0.45988	94	-0.02838	-0.04325	0.05173
6	-0.24092	0.38882	0.45741	96	0.00171	-0.05357	0.05360
8	-0.24428	0.38203	0.45345	98	0.03098	-0.05133	0.05996
10	-0.24794	0.37259	0.44755	100	0.05290	-0.03718	0.06466
12	-0.25137	0.36004	0.43910	102	0.06280	-0.01450	0.06445
14	-0.25393	0.34387	0.42746	104	0.05885	0.01149	0.05996
16	-0.25488	0.32361	0.41193	106	0.04230	0.03498	0.05489
18	-0.25337	0.29890	0.39184	108	0.01702	0.05094	0.05371
20	-0.24852	0.26952	0.36661	110	-0.01146	0.05615	0.05731
22	-0.23944	0.23550	0.33585	112	-0.03730	0.04981	0.06223
24	-0.22534	0.19724	0.29947	114	-0.05553	0.03352	0.06486
26	-0.20560	0.15551	0.25779	116	-0.06300	0.01075	0.06392
28	-0.17996	0.11154	0.21172	118	-0.05882	-0.01397	0.06045
30	-0.14855	0.06705	0.16298	120	-0.04427	-0.03610	0.05713
32	-0.11209	0.02413	0.11466	122	-0.02240	-0.05189	0.05651
34	-0.07192	-0.01483	0.07343	124	0.00282	-0.05898	0.05905
36	-0.03004	-0.04736	0.05608	126	0.02722	-0.05671	0.06290
38	0.01096	-0.07117	0.07200	128	0.04720	-0.04594	0.06587
40	0.04807	-0.08448	0.09720	130	0.06024	-0.02874	0.06675
42	0.07815	-0.08637	0.11648	132	0.06508	-0.00791	0.06556
44	0.09831	-0.07700	0.12488	134	0.06177	0.01360	0.06325
46	0.10637	-0.05785	0.12109	136	0.05138	0.03313	0.06113
48	0.10130	-0.03168	0.10613	138	0.03572	0.04863	0.06034
50	0.08357	-0.00230	0.08360	140	0.01695	0.05881	0.06121
52	0.05538	0.02581	0.06110	142	-0.00280	0.06321	0.06328
54	0.02053	0.04822	0.05241	144	-0.02164	0.06204	0.06571
56	-0.01593	0.06124	0.06328	146	-0.03813	0.05604	0.06778
58	-0.04838	0.06264	0.07915	148	-0.05130	0.04629	0.06910
60	-0.07158	0.05222	0.08861	150	-0.06072	0.03399	0.06958
62	-0.08158	0.03196	0.08762	152	-0.06634	0.02032	0.06938
64	-0.07660	0.00584	0.07682	154	-0.06847	0.00632	0.06876
66	-0.05756	-0.02087	0.06122	156	-0.06762	-0.00716	0.06800
68	-0.02808	-0.04263	0.05104	158	-0.06441	-0.01953	0.06731
70	0.00606	-0.05478	0.05512	160	-0.05952	-0.03042	0.06684
72	0.03800	-0.05466	0.06657	162	-0.05357	-0.03965	0.06664
74	0.06117	-0.04225	0.07435	164	-0.04711	-0.04720	0.06669
76	0.07078	-0.02033	0.07364	166	-0.04062	-0.05318	0.06692
78	0.06487	0.00610	0.06516	168	-0.03447	-0.05775	0.06726
80	0.04487	0.03089	0.05447	170	-0.02893	-0.06114	0.06764
82	0.01533	0.04823	0.05061	172	-0.02420	-0.06356	0.06801
84	-0.01703	0.05400	0.05662	174	-0.02042	-0.06520	0.06832
86	-0.04489	0.04682	0.06486	176	-0.01767	-0.06625	0.06856
88	-0.06199	0.02843	0.06820	178	-0.01600	-0.06682	0.06871
				180	-0.01544	-0.06700	0.06876

VEN

RADIATION PATTERN FOR THETA=PI/2

H/LMDA=0.5000 B/LMDA= 0.250 OMEGA= 9.92

20 ELEMENT ARRAY

PHI DEG	ER(THETA,PHI) RE	IM	ABSVAL	PHI DEG	ER(THETA,PHI) RE	IM	ABSVAL
0	0.36347	0.79762	0.87654	90	-0.04522	0.01379	0.04728
2	0.36237	0.79799	0.87642	92	-0.05130	-0.03478	0.06198
4	0.35901	0.79894	0.87590	94	-0.04282	-0.07146	0.08331
6	0.35317	0.79998	0.87447	96	-0.02271	-0.08684	0.08977
8	0.34452	0.80029	0.87129	98	0.00296	-0.07787	0.07793
10	0.33263	0.79872	0.86522	100	0.02689	-0.04828	0.05527
12	0.31702	0.79384	0.85480	102	0.04264	-0.00713	0.04323
14	0.29718	0.78392	0.83836	104	0.04632	0.03405	0.05749
16	0.27268	0.76705	0.81408	106	0.03755	0.06453	0.07466
18	0.24327	0.74122	0.78012	108	0.01924	0.07715	0.07951
20	0.20890	0.70446	0.73478	110	-0.00343	0.06987	0.06996
22	0.16992	0.65508	0.67676	112	-0.02451	0.04579	0.05194
24	0.12712	0.59194	0.60544	114	-0.03895	0.01181	0.04070
26	0.08184	0.51469	0.52115	116	-0.04369	-0.02345	0.04959
28	0.03595	0.42413	0.42565	118	-0.03821	-0.05190	0.06445
30	-0.00814	0.32245	0.32255	120	-0.02440	-0.06784	0.07210
32	-0.04763	0.21342	0.21867	122	-0.00577	-0.06888	0.06912
34	-0.07957	0.10238	0.12967	124	0.01348	-0.05602	0.05762
36	-0.10123	-0.00398	0.10131	126	0.02948	-0.03295	0.04421
38	-0.11046	-0.09813	0.14775	128	0.03943	-0.00486	0.03973
40	-0.10619	-0.17247	0.20253	130	0.04202	0.02282	0.04781
42	-0.08877	-0.22037	0.23758	132	0.03742	0.04547	0.05889
44	-0.06029	-0.23736	0.24489	134	0.02701	0.06003	0.06583
46	-0.02449	-0.22213	0.22348	136	0.01291	0.06519	0.06646
48	0.01347	-0.17741	0.17792	138	-0.00248	0.06128	0.06133
50	0.04773	-0.11010	0.12000	140	-0.01698	0.04989	0.05270
52	0.07263	-0.03082	0.07890	142	-0.02887	0.03336	0.04411
54	0.08379	0.04750	0.09632	144	-0.03708	0.01423	0.03972
56	0.07912	0.11161	0.13681	146	-0.04120	-0.00513	0.04152
58	0.05947	0.15044	0.16177	148	-0.04136	-0.02280	0.04723
60	0.02874	0.15740	0.16000	150	-0.03809	-0.03751	0.05345
62	-0.00668	0.13206	0.13223	152	-0.03217	-0.04853	0.05823
64	-0.03917	0.08064	0.08965	154	-0.02449	-0.05572	0.06087
66	-0.06154	0.01498	0.06333	156	-0.01586	-0.05932	0.06140
68	-0.06878	-0.04987	0.08496	158	-0.00699	-0.05985	0.06025
70	-0.05941	-0.09907	0.11551	160	0.00154	-0.05795	0.05797
72	-0.03596	-0.12155	0.12676	162	0.00936	-0.05434	0.05514
74	-0.00456	-0.11285	0.11295	164	0.01624	-0.04966	0.05224
76	0.02666	-0.07633	0.08085	166	0.02206	-0.04448	0.04965
78	0.04961	-0.02234	0.05441	168	0.02684	-0.03929	0.04758
80	0.05845	0.03447	0.06786	170	0.03062	-0.03443	0.04608
82	0.05122	0.07905	0.09419	172	0.03351	-0.03019	0.04510
84	0.03035	0.10009	0.10459	174	0.03561	-0.02673	0.04453
86	0.00201	0.09298	0.09300	176	0.03703	-0.02420	0.04423
88	-0.02575	0.06091	0.06613	178	0.03785	-0.02265	0.04410
				180	0.03811	-0.02213	0.04407

IEN

RADIATION PATTERN FOR THETA=PI/2

H/LMDA=0.5000 B/LMDA= 0.250 OMEGA= 9.92

20 ELEMENT ARRAY

PHI DEG	ER(THETA,PHI) RE	IM	ABSVAL	PHI DEG	ER(THETA,PHI) RE	IM	ABSVAL
0	0.38606	0.22340	0.44604	90	-0.04855	-0.00823	0.04924
2	0.38518	0.22403	0.44560	92	-0.04292	-0.03663	0.05643
4	0.38248	0.22586	0.44419	94	-0.02471	-0.05426	0.05962
6	0.37772	0.22869	0.44156	96	0.00070	-0.05637	0.05638
8	0.37053	0.23221	0.43728	98	0.02594	-0.04272	0.04998
10	0.36044	0.23594	0.43079	100	0.04382	-0.01743	0.04716
12	0.34689	0.23928	0.42141	102	0.04940	0.01230	0.05091
14	0.32929	0.24148	0.40835	104	0.04130	0.03835	0.05635
16	0.30714	0.24167	0.39082	106	0.02194	0.05390	0.05819
18	0.28003	0.23887	0.36807	108	-0.00328	0.05523	0.05533
20	0.24782	0.23207	0.33951	110	-0.02769	0.04248	0.05070
22	0.21070	0.22031	0.30484	112	-0.04508	0.01930	0.04904
24	0.16932	0.20280	0.26420	114	-0.05136	-0.00827	0.05203
26	0.12488	0.17906	0.21830	116	-0.04537	-0.03360	0.05646
28	0.07910	0.14907	0.16876	118	-0.02892	-0.05099	0.05862
30	0.03428	0.11344	0.11851	120	-0.00609	-0.05701	0.05734
32	-0.00690	0.07353	0.07385	122	0.01794	-0.05096	0.05403
34	-0.04160	0.03147	0.05216	124	0.03817	-0.03473	0.05160
36	-0.06711	-0.00986	0.06783	126	0.05080	-0.01205	0.05221
38	-0.08131	-0.04705	0.09394	128	0.05388	0.01248	0.05531
40	-0.08307	-0.07655	0.11297	130	0.04745	0.03443	0.05863
42	-0.07264	-0.09510	0.11967	132	0.03322	0.05035	0.06032
44	-0.05185	-0.10038	0.11298	134	0.01401	0.05821	0.05988
46	-0.02410	-0.09154	0.09466	136	-0.00691	0.05755	0.05796
48	0.00595	-0.06966	0.06991	138	-0.02648	0.04921	0.05588
50	0.03306	-0.03790	0.05029	140	-0.04227	0.03502	0.05489
52	0.05227	-0.00132	0.05228	142	-0.05274	0.01727	0.05550
54	0.05993	0.03386	0.06883	144	-0.05729	-0.00167	0.05731
56	0.05455	0.06123	0.08201	146	-0.05613	-0.01976	0.05950
58	0.03735	0.07553	0.08426	148	-0.05008	-0.03541	0.06134
60	0.01219	0.07383	0.07483	150	-0.04034	-0.04764	0.06243
62	-0.01510	0.05646	0.05844	152	-0.02821	-0.05603	0.06273
64	-0.03798	0.02719	0.04671	154	-0.01495	-0.06061	0.06242
66	-0.05078	-0.00735	0.05131	156	-0.00163	-0.06176	0.06179
68	-0.05023	-0.03907	0.06364	158	0.01092	-0.06010	0.06109
70	-0.03642	-0.06032	0.07047	160	0.02217	-0.05632	0.06053
72	-0.01300	-0.06588	0.06715	162	0.03181	-0.05112	0.06021
74	0.01368	-0.05442	0.05611	164	0.03974	-0.04514	0.06014
76	0.03624	-0.02901	0.04642	166	0.04602	-0.03892	0.06027
78	0.04830	0.00350	0.04843	168	0.05081	-0.03291	0.06054
80	0.04641	0.03432	0.05772	170	0.05434	-0.02744	0.06087
82	0.03108	0.05504	0.06321	172	0.05683	-0.02273	0.06120
84	0.00675	0.06008	0.06046	174	0.05850	-0.01895	0.06149
86	-0.01948	0.04822	0.05200	176	0.05955	-0.01619	0.06171
88	-0.03990	0.02302	0.04607	178	0.06012	-0.01452	0.06185
				180	0.06031	-0.01395	0.06190

VEN

RADIATION PATTERN FOR THETA=PI/2

H/LMDA=0.5000 B/LMDA= 0.250 OMEGA= 9.92

22 ELEMENT ARRAY

PHI DEG	ER(THETA,PHI) RE	IM	ABSVAL	PHI DEG	ER(THETA,PHI) RE	IM	ABSVAL
0	0.77946	-0.35856	0.85798	90	-0.08402	0.02397	0.08737
2	0.77985	-0.35739	0.85784	92	-0.07483	-0.00338	0.07491
4	0.78081	-0.35378	0.85722	94	-0.04130	-0.02853	0.05020
6	0.78177	-0.34748	0.85552	96	0.00429	-0.04313	0.04335
8	0.78176	-0.33810	0.85174	98	0.04633	-0.04274	0.06304
10	0.77942	-0.32514	0.84452	100	0.07131	-0.02814	0.07666
12	0.77306	-0.30803	0.83217	102	0.07209	-0.00478	0.07225
14	0.76065	-0.28621	0.81272	104	0.04983	0.01934	0.05345
16	0.73999	-0.25924	0.78408	106	0.01287	0.03644	0.03865
18	0.70878	-0.22688	0.74421	108	-0.02643	0.04148	0.04918
20	0.66492	-0.18927	0.69133	110	-0.05601	0.03351	0.06527
22	0.60674	-0.14700	0.62429	112	-0.06772	0.01566	0.06950
24	0.53337	-0.10128	0.54290	114	-0.05936	-0.00627	0.05969
26	0.44514	-0.05397	0.44840	116	-0.03469	-0.02573	0.04319
28	0.34393	-0.00760	0.34401	118	-0.00163	-0.03744	0.03748
30	0.23344	0.03479	0.23602	120	0.03044	-0.03875	0.04927
32	0.11931	0.06983	0.13824	122	0.05343	-0.02998	0.06127
34	0.00886	0.09431	0.09472	124	0.06243	-0.01404	0.06399
36	-0.08939	0.10561	0.13836	126	0.05652	0.00475	0.05671
38	-0.16670	0.10234	0.19561	128	0.03833	0.02188	0.04413
40	-0.21542	0.08478	0.23150	130	0.01288	0.03373	0.03611
42	-0.23040	0.05530	0.23694	132	-0.01399	0.03831	0.04079
44	-0.21045	0.01829	0.21124	134	-0.03699	0.03537	0.05118
46	-0.15927	-0.02021	0.16054	136	-0.05241	0.02613	0.05856
48	-0.08569	-0.05340	0.10097	138	-0.05849	0.01281	0.05988
50	-0.00280	-0.07498	0.07503	140	-0.05538	-0.00202	0.05542
52	0.07401	-0.08056	0.10939	142	-0.04469	-0.01599	0.04747
54	0.12992	-0.06885	0.14704	144	-0.02887	-0.02726	0.03971
56	0.15393	-0.04236	0.15966	146	-0.01062	-0.03472	0.03630
58	0.14170	-0.00716	0.14188	148	0.00762	-0.03800	0.03875
60	0.09708	0.02830	0.10112	150	0.02391	-0.03734	0.04434
62	0.03176	0.05513	0.06362	152	0.03702	-0.03339	0.04985
64	-0.03727	0.06644	0.07618	154	0.04637	-0.02703	0.05367
66	-0.09193	0.05934	0.10941	156	0.05194	-0.01919	0.05537
68	-0.11797	0.03602	0.12334	158	0.05410	-0.01072	0.05515
70	-0.10914	0.00340	0.10919	160	0.05348	-0.00229	0.05353
72	-0.06924	-0.02880	0.07499	162	0.05078	0.00557	0.05109
74	-0.01116	-0.05091	0.05212	164	0.04673	0.01256	0.04839
76	0.04692	-0.05636	0.07333	166	0.04196	0.01852	0.04587
78	0.08714	-0.04386	0.09756	168	0.03702	0.02341	0.04380
80	0.09766	-0.01787	0.09928	170	0.03231	0.02727	0.04228
82	0.07637	0.01280	0.07744	172	0.02814	0.03021	0.04128
84	0.03151	0.03799	0.04935	174	0.02473	0.03233	0.04071
86	-0.02130	0.04952	0.05391	176	0.02222	0.03376	0.04041
88	-0.06439	0.04403	0.07800	178	0.02068	0.03458	0.04029
				180	0.02016	0.03484	0.04026

IEN

RADIATION PATTERN FOR THETA=PI/2

H/LMDA=0.5000 B/LMDA= 0.250 OMEGA= 9.92

22 ELEMENT ARRAY

PHI DEG	ER(THETA,PHI) RE	IM	ABSVAL	PHI DEG	ER(THETA,PHI) RE	IM	ABSVAL
0	0.21273	-0.37668	0.43260	90	-0.05261	0.00248	0.05266
2	0.21340	-0.37577	0.43214	92	-0.03909	-0.02312	0.04542
4	0.21533	-0.37297	0.43066	94	-0.01247	-0.04057	0.04245
6	0.21831	-0.36798	0.42787	96	0.01799	-0.04383	0.04738
8	0.22195	-0.36040	0.42326	98	0.04196	-0.03186	0.05269
10	0.22573	-0.34964	0.41618	100	0.05159	-0.00892	0.05235
12	0.22894	-0.33506	0.40581	102	0.04404	0.01709	0.04724
14	0.23071	-0.31601	0.39126	104	0.02222	0.03744	0.04354
16	0.23003	-0.29191	0.37165	106	-0.00647	0.04555	0.04601
18	0.22581	-0.26237	0.34616	108	-0.03281	0.03903	0.05098
20	0.21692	-0.22736	0.31424	110	-0.04874	0.02024	0.05278
22	0.20237	-0.18727	0.27573	112	-0.04985	-0.00476	0.05007
24	0.18142	-0.14311	0.23107	114	-0.03635	-0.02835	0.04610
26	0.15380	-0.09653	0.18158	116	-0.01268	-0.04374	0.04554
28	0.11987	-0.04987	0.12983	118	0.01421	-0.04690	0.04901
30	0.08085	-0.00600	0.08107	120	0.03701	-0.03746	0.05266
32	0.03887	0.03185	0.05025	122	0.05011	-0.01837	0.05338
34	-0.00305	0.06051	0.06059	124	0.05081	0.00515	0.05107
36	-0.04119	0.07738	0.08766	126	0.03966	0.02729	0.04814
38	-0.07155	0.08092	0.10802	128	0.01988	0.04311	0.04747
40	-0.09043	0.07120	0.11509	130	-0.00374	0.04958	0.04972
42	-0.09519	0.05017	0.10760	132	-0.02620	0.04600	0.05294
44	-0.08493	0.02170	0.08766	134	-0.04334	0.03377	0.05494
46	-0.06111	-0.00885	0.06175	136	-0.05255	0.01577	0.05487
48	-0.02763	-0.03548	0.04497	138	-0.05302	-0.00448	0.05321
50	0.00943	-0.05270	0.05354	140	-0.04552	-0.02360	0.05127
52	0.04283	-0.05681	0.07115	142	-0.03194	-0.03891	0.05034
54	0.06557	-0.04698	0.08066	144	-0.01478	-0.04876	0.05095
56	0.07257	-0.02568	0.07698	146	0.00345	-0.05255	0.05266
58	0.06212	0.00156	0.06214	148	0.02059	-0.05062	0.05465
60	0.03671	0.02745	0.04584	150	0.03504	-0.04393	0.05619
62	0.00277	0.04480	0.04489	152	0.04587	-0.03381	0.05698
64	-0.03075	0.04861	0.05752	154	0.05278	-0.02167	0.05705
66	-0.05465	0.03771	0.06640	156	0.05593	-0.00880	0.05662
68	-0.06219	0.01539	0.06406	158	0.05585	0.00375	0.05598
70	-0.05119	-0.01140	0.05244	160	0.05324	0.01525	0.05538
72	-0.02503	-0.03408	0.04229	162	0.04883	0.02523	0.05496
74	0.00816	-0.04522	0.04595	164	0.04336	0.03350	0.05480
76	0.03793	-0.04107	0.05591	166	0.03745	0.04007	0.05485
78	0.05477	-0.02298	0.05940	168	0.03160	0.04508	0.05505
80	0.05333	0.00288	0.05341	170	0.02619	0.04874	0.05534
82	0.03426	0.02757	0.04397	172	0.02151	0.05131	0.05564
84	0.00411	0.04247	0.04267	174	0.01774	0.05302	0.05590
86	-0.02684	0.04233	0.05012	176	0.01497	0.05408	0.05611
88	-0.04810	0.02722	0.05527	178	0.01330	0.05465	0.05624
				180	0.01273	0.05483	0.05629

VEN

RADIATION PATTERN FOR THETA=PI/2

H/LMDA=0.5000 B/LMDA= 0.250 OMEGA= 9.92

24 ELEMENT ARRAY

PHI	ER(THETA,PHI)			PHI	ER(THETA,PHI)		
DEG	RE	IM	ABSVAL	DEG	RE	IM	ABSVAL
0	-0.35423	-0.76336	0.84154	90	-0.03838	0.01150	0.04006
2	-0.35298	-0.76376	0.84138	92	-0.04298	-0.03612	0.05614
4	-0.34913	-0.76473	0.84066	94	-0.03022	-0.06723	0.07371
6	-0.34238	-0.76560	0.83867	96	-0.00589	-0.07047	0.07072
8	-0.33228	-0.76524	0.83426	98	0.01988	-0.04605	0.05016
10	-0.31824	-0.76206	0.82584	100	0.03688	-0.00490	0.03720
12	-0.29962	-0.75409	0.81143	102	0.03886	0.03626	0.05315
14	-0.27581	-0.73901	0.78880	104	0.02577	0.06183	0.06699
16	-0.24636	-0.71430	0.75559	106	0.00331	0.06318	0.06326
18	-0.21111	-0.67744	0.70958	108	-0.01966	0.04127	0.04571
20	-0.17040	-0.62622	0.64899	110	-0.03468	0.00546	0.03511
22	-0.12515	-0.55910	0.57293	112	-0.03679	-0.03073	0.04794
24	-0.07705	-0.47571	0.48191	114	-0.02596	-0.05483	0.06067
26	-0.02858	-0.37728	0.37837	116	-0.00663	-0.05961	0.05998
28	0.01705	-0.26710	0.26764	118	0.01431	-0.04491	0.04714
30	0.05616	-0.15065	0.16078	120	0.03005	-0.01680	0.03443
32	0.08501	-0.03563	0.09217	122	0.03610	0.01518	0.03916
34	0.10037	0.06866	0.12161	124	0.03132	0.04142	0.05193
36	0.10027	0.15231	0.18235	126	0.01786	0.05507	0.05789
38	0.08460	0.20632	0.22299	128	0.00000	0.05361	0.05361
40	0.05569	0.22439	0.23120	130	-0.01728	0.03880	0.04247
42	0.01827	0.20476	0.20557	132	-0.02980	0.01546	0.03357
44	-0.02086	0.15145	0.15288	134	-0.03510	-0.01027	0.03657
46	-0.05401	0.07460	0.09210	136	-0.03268	-0.03264	0.04619
48	-0.07412	-0.01059	0.07487	138	-0.02380	-0.04757	0.05319
50	-0.07656	-0.08644	0.11547	140	-0.01080	-0.05315	0.05423
52	-0.06058	-0.13655	0.14938	142	0.00355	-0.04958	0.04971
54	-0.03005	-0.14988	0.15286	144	0.01677	-0.03864	0.04212
56	0.00710	-0.12413	0.12433	146	0.02699	-0.02296	0.03544
58	0.04072	-0.06714	0.07852	148	0.03321	-0.00538	0.03364
60	0.06120	0.00457	0.06137	150	0.03520	0.01165	0.03708
62	0.06252	0.06999	0.09385	152	0.03338	0.02631	0.04250
64	0.04441	0.10979	0.11843	154	0.02855	0.03752	0.04715
66	0.01287	0.11246	0.11319	156	0.02167	0.04491	0.04987
68	-0.02154	0.07841	0.08132	158	0.01371	0.04866	0.05056
70	-0.04710	0.02024	0.05126	160	0.00549	0.04930	0.04960
72	-0.05505	-0.04133	0.06884	162	-0.00236	0.04753	0.04759
74	-0.04294	-0.08462	0.09489	164	-0.00944	0.04412	0.04511
76	-0.01564	-0.09484	0.09612	166	-0.01551	0.03976	0.04268
78	0.01629	-0.06956	0.07145	168	-0.02050	0.03507	0.04062
80	0.04066	-0.01967	0.04517	170	-0.02444	0.03051	0.03909
82	0.04833	0.03496	0.05965	172	-0.02743	0.02642	0.03809
84	0.03684	0.07323	0.08197	174	-0.02958	0.02305	0.03750
86	0.01133	0.08107	0.08186	176	-0.03101	0.02056	0.03721
88	-0.01762	0.05683	0.05950	178	-0.03183	0.01903	0.03709
				180	-0.03210	0.01852	0.03706

IEN

RADIATION PATTERN FOR THETA=PI/2

H/LMDA=0.5000 B/LMDA= 0.250 OMEGA= 9.92

24 ELEMENT ARRAY

PHI	ER(THETA,PHI)			PHI	ER(THETA,PHI)		
DEG	RE	IM	ABSVAL	DEG	RE	IM	ABSVAL
0	-0.36838	-0.20351	0.42086	90	-0.04066	-0.00671	0.04121
2	-0.36745	-0.20421	0.42038	92	-0.03352	-0.03424	0.04791
4	-0.36454	-0.20625	0.41884	94	-0.01241	-0.04750	0.04909
6	-0.35934	-0.20936	0.41588	96	0.01387	-0.04145	0.04371
8	-0.35135	-0.21312	0.41093	98	0.03448	-0.01889	0.03931
10	-0.33991	-0.21691	0.40322	100	0.04109	0.01087	0.04250
12	-0.32428	-0.21994	0.39183	102	0.03120	0.03600	0.04764
14	-0.30373	-0.22121	0.37575	104	0.00894	0.04690	0.04775
16	-0.27765	-0.21958	0.35398	106	-0.01684	0.03981	0.04323
18	-0.24568	-0.21382	0.32570	108	-0.03625	0.01792	0.04043
20	-0.20792	-0.20274	0.29041	110	-0.04219	-0.01031	0.04343
22	-0.16505	-0.18532	0.24817	112	-0.03281	-0.03455	0.04764
24	-0.11848	-0.16096	0.19986	114	-0.01184	-0.04653	0.04801
26	-0.07042	-0.12968	0.14757	116	0.01317	-0.04267	0.04466
28	-0.02384	-0.09237	0.09540	118	0.03379	-0.02492	0.04198
30	0.01778	-0.05097	0.05398	120	0.04356	0.00049	0.04357
32	0.05078	-0.00847	0.05148	122	0.03994	0.02545	0.04736
34	0.07196	0.03121	0.07843	124	0.02459	0.04271	0.04928
36	0.07915	0.06366	0.10157	126	0.00244	0.04789	0.04795
38	0.07193	0.08466	0.11109	128	-0.02017	0.04035	0.04512
40	0.05207	0.09102	0.10486	130	-0.03745	0.02284	0.04386
42	0.02362	0.08152	0.08487	132	-0.04555	0.00024	0.04555
44	-0.00751	0.05759	0.05808	134	-0.04329	-0.02198	0.04855
46	-0.03457	0.02360	0.04186	136	-0.03193	-0.03916	0.05053
48	-0.05138	-0.01360	0.05315	138	-0.01453	-0.04836	0.05050
50	-0.05395	-0.04585	0.07080	140	0.00515	-0.04866	0.04893
52	-0.04176	-0.06555	0.07772	142	0.02347	-0.04093	0.04718
54	-0.01829	-0.06769	0.07011	144	0.03766	-0.02726	0.04649
56	0.00961	-0.05160	0.05249	146	0.04613	-0.01034	0.04727
58	0.03347	-0.02167	0.03987	148	0.04844	0.00717	0.04897
60	0.04576	0.01350	0.04771	150	0.04515	0.02311	0.05072
62	0.04245	0.04326	0.06061	152	0.03744	0.03598	0.05193
64	0.02463	0.05826	0.06326	154	0.02678	0.04502	0.05239
66	-0.00149	0.05361	0.05363	156	0.01463	0.05012	0.05221
68	-0.02653	0.03091	0.04074	158	0.00227	0.05164	0.05169
70	-0.04122	-0.00192	0.04127	160	-0.00935	0.05022	0.05108
72	-0.03996	-0.03313	0.05191	162	-0.01961	0.04665	0.05061
74	-0.02318	-0.05134	0.05633	164	-0.02819	0.04173	0.05036
76	0.00258	-0.04985	0.04992	166	-0.03504	0.03614	0.05033
78	0.02705	-0.02937	0.03992	168	-0.04026	0.03046	0.05048
80	0.04031	0.00204	0.04037	170	-0.04406	0.02514	0.05072
82	0.03692	0.03196	0.04883	172	-0.04670	0.02049	0.05100
84	0.01824	0.04857	0.05189	174	-0.04844	0.01672	0.05125
86	-0.00797	0.04542	0.04611	176	-0.04951	0.01395	0.05144
88	-0.03078	0.02402	0.03904	178	-0.05008	0.01227	0.05156
				180	-0.05026	0.01171	0.05161

VEN

RADIATION PATTERN FOR THETA=PI/2

H/LMDA=0.5000 B/LMDA= 0.250 OMEGA= 9.92

26 ELEMENT ARRAY

PHI	ER(THETA,PHI)			PHI	ER(THETA,PHI)		
DEG	RE	IM	ABSVAL	DEG	RE	IM	ABSVAL
0	-0.74893	0.35036	0.82683	90	-0.07052	0.02029	0.07338
2	-0.74935	0.34903	0.82665	92	-0.05923	-0.00753	0.05971
4	-0.75033	0.34495	0.82582	94	-0.02123	-0.03076	0.03737
6	-0.75110	0.33776	0.82354	96	0.02471	-0.03876	0.04596
8	-0.75034	0.32693	0.81848	98	0.05722	-0.02848	0.06392
10	-0.74626	0.31181	0.80879	100	0.06230	-0.00545	0.06254
12	-0.73657	0.29168	0.79222	102	0.03919	0.01926	0.04366
14	-0.71865	0.26586	0.76625	104	-0.00028	0.03443	0.03443
16	-0.68967	0.23394	0.72827	106	-0.03788	0.03379	0.05076
18	-0.64691	0.19587	0.67591	108	-0.05758	0.01838	0.06045
20	-0.58812	0.15221	0.60750	110	-0.05218	-0.00441	0.05237
22	-0.51207	0.10432	0.52259	112	-0.02552	-0.02465	0.03548
24	-0.41907	0.05444	0.42259	114	0.01025	-0.03419	0.03569
26	-0.31153	0.00574	0.31158	116	0.04051	-0.02985	0.05032
28	-0.19439	-0.03790	0.19806	118	0.05417	-0.01414	0.05598
30	-0.07522	-0.07227	0.10431	120	0.04743	0.00637	0.04786
32	0.03622	-0.09344	0.10022	122	0.02425	0.02398	0.03410
34	0.12893	-0.09862	0.16232	124	-0.00617	0.03279	0.03337
36	0.19236	-0.08697	0.21111	126	-0.03331	0.03052	0.04517
38	0.21852	-0.06031	0.22669	128	-0.04899	0.01870	0.05244
40	0.20425	-0.02339	0.20558	130	-0.04962	0.00167	0.04965
42	0.15297	0.01653	0.15386	132	-0.03646	-0.01528	0.03953
44	0.07535	0.05088	0.09092	134	-0.01435	-0.02757	0.03108
46	-0.01183	0.07162	0.07259	136	0.01024	-0.03252	0.03409
48	-0.08868	0.07348	0.11517	138	0.03132	-0.02965	0.04313
50	-0.13684	0.05576	0.14776	140	0.04471	-0.02042	0.04915
52	-0.14457	0.02319	0.14642	142	0.04866	-0.00742	0.04922
54	-0.11082	-0.01479	0.11180	144	0.04367	0.00644	0.04414
56	-0.04658	-0.04658	0.06588	146	0.03184	0.01862	0.03689
58	0.02754	-0.06196	0.06781	148	0.01604	0.02739	0.03174
60	0.08736	-0.05584	0.10368	150	-0.00085	0.03196	0.03197
62	0.11318	-0.03047	0.11721	152	-0.01645	0.03234	0.03629
64	0.09705	0.00483	0.09717	154	-0.02917	0.02918	0.04126
66	0.04624	0.03675	0.05906	156	-0.03821	0.02342	0.04482
68	-0.01889	0.05301	0.05628	158	-0.04347	0.01610	0.04636
70	-0.07251	0.04742	0.08664	160	-0.04535	0.00817	0.04608
72	-0.09359	0.02268	0.09630	162	-0.04451	0.00038	0.04451
74	-0.07473	-0.01045	0.07545	164	-0.04175	-0.00675	0.04229
76	-0.02540	-0.03764	0.04541	166	-0.03782	-0.01292	0.03996
78	0.03203	-0.04720	0.05704	168	-0.03338	-0.01800	0.03793
80	0.07220	-0.03540	0.08042	170	-0.02897	-0.02202	0.03639
82	0.07809	-0.00818	0.07852	172	-0.02496	-0.02505	0.03536
84	0.04855	0.02160	0.05313	174	-0.02163	-0.02723	0.03477
86	-0.00148	0.04020	0.04022	176	-0.01916	-0.02867	0.03448
88	-0.04826	0.03940	0.06230	178	-0.01764	-0.02949	0.03436
				180	-0.01712	-0.02976	0.03433

IEN

RADIATION PATTERN FOR THETA=PI/2

H/LMDA=0.5000 B/LMDA= 0.250 OMEGA= 9.92

26 ELEMENT ARRAY

PHI	ER(THETA,PHI)			PHI	ER(THETA,PHI)		
DEG	RE	IM	ABSVAL	DEG	RE	IM	ABSVAL
0	-0.19544	0.36097	0.41048	90	-0.04435	0.00198	0.04440
2	-0.19618	0.36001	0.40999	92	-0.02956	-0.02313	0.03754
4	-0.19831	0.35700	0.40838	94	-0.00103	-0.03700	0.03702
6	-0.20154	0.35159	0.40526	96	0.02758	-0.03299	0.04300
8	-0.20540	0.34319	0.39996	98	0.04291	-0.01314	0.04488
10	-0.20918	0.33106	0.39161	100	0.03813	0.01299	0.04029
12	-0.21199	0.31435	0.37915	102	0.01592	0.03308	0.03671
14	-0.21270	0.29227	0.36147	104	-0.01326	0.03792	0.04017
16	-0.21063	0.26418	0.33750	106	-0.03620	0.02556	0.04432
18	-0.20265	0.22978	0.30637	108	-0.04308	0.00178	0.04311
20	-0.18929	0.18935	0.26774	110	-0.03143	-0.02281	0.03883
22	-0.16897	0.14391	0.22195	112	-0.00675	-0.03774	0.03834
24	-0.14130	0.09537	0.17047	114	0.02042	-0.03710	0.04235
26	-0.10669	0.04655	0.11640	116	0.03925	-0.02160	0.04481
28	-0.06668	0.00107	0.06669	118	0.04287	0.00225	0.04293
30	-0.02403	-0.03704	0.04416	120	0.03062	0.02521	0.03966
32	0.01734	-0.06389	0.06620	122	0.00763	0.03904	0.03978
34	0.05272	-0.07652	0.09292	124	-0.01769	0.03935	0.04315
36	0.07732	-0.07370	0.10682	126	-0.03692	0.02673	0.04558
38	0.08723	-0.05660	0.10398	128	-0.04446	0.00583	0.04484
40	0.08052	-0.02911	0.08562	130	-0.03883	-0.01659	0.04222
42	0.05822	0.00257	0.05827	132	-0.02253	-0.03409	0.04086
44	0.02470	0.03097	0.03961	134	-0.00066	-0.04235	0.04235
46	-0.01267	0.04902	0.05063	136	0.02092	-0.04000	0.04514
48	-0.04489	0.05204	0.06872	138	0.03724	-0.02848	0.04688
50	-0.06357	0.03937	0.07477	140	0.04531	-0.01111	0.04665
52	-0.06345	0.01502	0.06520	142	0.04435	0.00802	0.04507
54	-0.04443	-0.01311	0.04632	144	0.03555	0.02517	0.04356
56	-0.01227	-0.03551	0.03757	146	0.02132	0.03765	0.04326
58	0.02263	-0.04431	0.04975	148	0.00454	0.04407	0.04431
60	0.04843	-0.03628	0.06051	150	-0.01209	0.04435	0.04597
62	0.05596	-0.01446	0.05780	152	-0.02650	0.03935	0.04744
64	0.04244	0.01255	0.04425	154	-0.03742	0.03048	0.04826
66	0.01303	0.03375	0.03618	156	-0.04435	0.01932	0.04838
68	-0.02063	0.04023	0.04521	158	-0.04744	0.00734	0.04800
70	-0.04486	0.02915	0.05350	160	-0.04725	-0.00428	0.04744
72	-0.04960	0.00532	0.04989	162	-0.04455	-0.01474	0.04693
74	-0.03292	-0.02059	0.03883	164	-0.04020	-0.02360	0.04661
76	-0.00225	-0.03670	0.03677	166	-0.03495	-0.03071	0.04652
78	0.02869	-0.03550	0.04564	168	-0.02946	-0.03613	0.04662
80	0.04592	-0.01749	0.04914	170	-0.02423	-0.04007	0.04682
82	0.04169	0.00877	0.04260	172	-0.01961	-0.04279	0.04707
84	0.01819	0.03066	0.03565	174	-0.01585	-0.04457	0.04730
86	-0.01337	0.03757	0.03987	176	-0.01309	-0.04565	0.04749
88	-0.03806	0.02615	0.04618	178	-0.01140	-0.04622	0.04761
				180	-0.01084	-0.04640	0.04764

VEN

RADIATION PATTERN FOR THETA=PI/2

H/LMDA=0.5000 B/LMDA= 0.250 OMEGA= 9.92

28 ELEMENT ARRAY

PHI	ER(THETA,PHI)			PHI	ER(THETA,PHI)		
DEG	RE	IM	ABSVAL	DEG	RE	IM	ABSVAL
0	0.34687	0.73591	0.81356	90	-0.03337	0.00987	0.03480
2	0.34547	0.73634	0.81335	92	-0.03642	-0.03669	0.05170
4	0.34115	0.73732	0.81242	94	-0.01974	-0.06150	0.06459
6	0.33352	0.73798	0.80984	96	0.00693	-0.05254	0.05300
8	0.32198	0.73680	0.80408	98	0.02890	-0.01629	0.03317
10	0.30578	0.73175	0.79306	100	0.03471	0.02683	0.04387
12	0.28411	0.72023	0.77424	102	0.02204	0.05410	0.05841
14	0.25629	0.69929	0.74478	104	-0.00167	0.05245	0.05248
16	0.22190	0.66584	0.70184	106	-0.02374	0.02440	0.03404
18	0.18107	0.61696	0.64299	108	-0.03311	-0.01455	0.03616
20	0.13467	0.55051	0.56675	110	-0.02580	-0.04475	0.05165
22	0.08453	0.46566	0.47327	112	-0.00624	-0.05238	0.05275
24	0.03352	0.36364	0.36518	114	0.01568	-0.03539	0.03870
26	-0.01448	0.24833	0.24875	116	0.02974	-0.00313	0.02991
28	-0.05493	0.12659	0.13799	118	0.03016	0.02915	0.04194
30	-0.08324	0.00815	0.08364	120	0.01759	0.04776	0.05090
32	-0.09563	-0.09523	0.13496	122	-0.00190	0.04613	0.04617
34	-0.09018	-0.17144	0.19371	124	-0.02005	0.02649	0.03322
36	-0.06779	-0.21035	0.22100	126	-0.03001	-0.00226	0.03009
38	-0.03266	-0.20657	0.20914	128	-0.02874	-0.02899	0.04082
40	0.00784	-0.16183	0.16202	130	-0.01758	-0.04474	0.04807
42	0.04440	-0.08622	0.09698	132	-0.00102	-0.04543	0.04544
44	0.06781	0.00262	0.06786	134	0.01530	-0.03234	0.03577
46	0.07162	0.08277	0.10945	136	0.02661	-0.01067	0.02867
48	0.05451	0.13335	0.14406	138	0.03030	0.01279	0.03289
50	0.02144	0.14083	0.14245	140	0.02620	0.03196	0.04133
52	-0.01712	0.10410	0.10549	142	0.01613	0.04290	0.04584
54	-0.04816	0.03612	0.06020	144	0.00300	0.04434	0.04444
56	-0.06066	-0.03918	0.07222	146	-0.01016	0.03730	0.03866
58	-0.05000	-0.09476	0.10714	148	-0.02094	0.02430	0.03208
60	-0.02037	-0.11043	0.11229	150	-0.02785	0.00841	0.02909
62	0.01617	-0.08133	0.08293	152	-0.03044	-0.00752	0.03135
64	0.04428	-0.02079	0.04892	154	-0.02904	-0.02135	0.03605
66	0.05193	0.04460	0.06845	156	-0.02453	-0.03182	0.04018
68	0.03599	0.08611	0.09333	158	-0.01797	-0.03850	0.04248
70	0.00421	0.08594	0.08604	160	-0.01042	-0.04158	0.04287
72	-0.02812	0.04571	0.05367	162	-0.00275	-0.04168	0.04177
74	-0.04537	-0.01391	0.04746	164	0.00440	-0.03957	0.03981
76	-0.03942	-0.06323	0.07451	166	0.01065	-0.03607	0.03761
78	-0.01388	-0.07824	0.07946	168	0.01583	-0.03189	0.03560
80	0.01765	-0.05280	0.05568	170	0.01992	-0.02762	0.03406
82	0.03863	-0.00167	0.03866	172	0.02300	-0.02369	0.03302
84	0.03840	0.04758	0.06114	174	0.02520	-0.02040	0.03242
86	0.01779	0.06930	0.07155	176	0.02665	-0.01795	0.03213
88	-0.01146	0.05332	0.05454	178	0.02747	-0.01644	0.03201
				180	0.02774	-0.01593	0.03199

IEN

RADIATION PATTERN FOR THETA=PI/2

H/LMDA=0.5000 B/LMDA= 0.250 OMEGA= 9.92

28 ELEMENT ARRAY

PHI	ER(THETA,PHI)			PHI	ER(THETA,PHI)		
DEG	RE	IM	ABSVAL	DEG	RE	IM	ABSVAL
0	0.35429	0.18830	0.40122	90	-0.03498	-0.00565	0.03544
2	0.35331	0.18908	0.40072	92	-0.02641	-0.03220	0.04165
4	0.35021	0.19130	0.39905	94	-0.00305	-0.04079	0.04091
6	0.34458	0.19465	0.39575	96	0.02203	-0.02710	0.03493
8	0.33576	0.19859	0.39009	98	0.03497	0.00105	0.03498
10	0.32292	0.20234	0.38107	100	0.02878	0.02829	0.04036
12	0.30511	0.20488	0.36752	102	0.00702	0.04023	0.04083
14	0.28147	0.20497	0.34819	104	-0.01854	0.03100	0.03612
16	0.25134	0.20120	0.32195	106	-0.03445	0.00591	0.03495
18	0.21452	0.19211	0.28797	108	-0.03269	-0.02191	0.03936
20	0.17154	0.17640	0.24605	110	-0.01451	-0.03862	0.04125
22	0.12379	0.15319	0.19696	112	0.01082	-0.03650	0.03807
24	0.07376	0.12236	0.14288	114	0.03101	-0.01722	0.03547
26	0.02494	0.08485	0.08844	116	0.03685	0.00974	0.03812
28	-0.01843	0.04294	0.04673	118	0.02619	0.03213	0.04146
30	-0.05186	0.00029	0.05186	120	0.00424	0.04057	0.04079
32	-0.07149	-0.03830	0.08110	122	-0.01940	0.03224	0.03763
34	-0.07499	-0.06753	0.10092	124	-0.03522	0.01130	0.03699
36	-0.06255	-0.08264	0.10365	126	-0.03754	-0.01364	0.03994
38	-0.03738	-0.08070	0.08894	128	-0.02624	-0.03336	0.04245
40	-0.00561	-0.06182	0.06208	130	-0.00603	-0.04144	0.04188
42	0.02471	-0.02990	0.03879	132	0.01591	-0.03607	0.03943
44	0.04548	0.00756	0.04611	134	0.03268	-0.01990	0.03827
46	0.05089	0.04086	0.06526	136	0.03983	0.00157	0.03986
48	0.03952	0.06061	0.07235	138	0.03621	0.02215	0.04245
50	0.01534	0.06074	0.06264	140	0.02369	0.03680	0.04377
52	-0.01306	0.04105	0.04307	142	0.00605	0.04275	0.04318
54	-0.03516	0.00809	0.03608	144	-0.01238	0.03970	0.04158
56	-0.04242	-0.02641	0.04997	146	-0.02787	0.02930	0.04044
58	-0.03193	-0.04948	0.05889	148	-0.03801	0.01434	0.04062
60	-0.00812	-0.05202	0.05265	150	-0.04186	-0.00215	0.04191
62	0.01857	-0.03296	0.03783	152	-0.03976	-0.01754	0.04346
64	0.03602	-0.00040	0.03602	154	-0.03294	-0.03002	0.04456
66	0.03602	0.03127	0.04770	156	-0.02300	-0.03864	0.04497
68	0.01852	0.04761	0.05108	158	-0.01160	-0.04326	0.04479
70	-0.00788	0.04100	0.04175	160	-0.00010	-0.04430	0.04430
72	-0.02983	0.01473	0.03327	162	0.01049	-0.04250	0.04378
74	-0.03596	-0.01813	0.04027	164	0.01959	-0.03874	0.04341
76	-0.02297	-0.04098	0.04698	166	0.02693	-0.03386	0.04326
78	0.00223	-0.04218	0.04224	168	0.03255	-0.02857	0.04331
80	0.02601	-0.02129	0.03361	170	0.03663	-0.02343	0.04348
82	0.03529	0.01044	0.03680	172	0.03943	-0.01886	0.04370
84	0.02492	0.03592	0.04372	174	0.04125	-0.01510	0.04392
86	0.00067	0.04152	0.04152	176	0.04234	-0.01234	0.04410
88	-0.02389	0.02445	0.03419	178	0.04291	-0.01065	0.04421
				180	0.04308	-0.01009	0.04425

VEN

RADIATION PATTERN FOR THETA=PI/2

H/LMDA=0.5000 B/LMDA= 0.250 OMEGA= 9.92

30 ELEMENT ARRAY

PHI	ER(THETA,PHI)			PHI	ER(THETA,PHI)		
DEG	RE	IM	ABSVAL	DEG	RE	IM	ABSVAL
0	0.72406	-0.34369	0.80149	90	-0.06072	0.01758	0.06322
2	0.72451	-0.34223	0.80127	92	-0.04731	-0.01052	0.04846
4	0.72549	-0.33768	0.80023	94	-0.00584	-0.03106	0.03160
6	0.72601	-0.32962	0.79734	96	0.03721	-0.03176	0.04892
8	0.72439	-0.31736	0.79086	98	0.05584	-0.01303	0.05734
10	0.71830	-0.30006	0.77845	100	0.04024	0.01292	0.04226
12	0.70484	-0.27685	0.75727	102	0.00150	0.03020	0.03024
14	0.68074	-0.24701	0.72417	104	-0.03640	0.02895	0.04651
16	0.64262	-0.21019	0.67612	106	-0.05167	0.01079	0.05278
18	0.58747	-0.16669	0.61067	108	-0.03698	-0.01300	0.03920
20	0.51331	-0.11777	0.52665	110	-0.00236	-0.02870	0.02880
22	0.41992	-0.06579	0.42504	112	0.03209	-0.02806	0.04263
24	0.30967	-0.01432	0.31000	114	0.04808	-0.01232	0.04964
26	0.18816	0.03203	0.19086	116	0.03861	0.00941	0.03974
28	0.06438	0.06815	0.09375	118	0.01021	0.02572	0.02768
30	-0.04973	0.08930	0.10221	120	-0.02197	0.02892	0.03632
32	-0.14081	0.09222	0.16832	122	-0.04259	0.01832	0.04636
34	-0.19642	0.07641	0.21076	124	-0.04327	-0.00030	0.04327
36	-0.20824	0.04497	0.21304	126	-0.02536	-0.01823	0.03123
38	-0.17513	0.00479	0.17519	128	0.00219	-0.02803	0.02812
40	-0.10525	-0.03439	0.11072	130	0.02773	-0.02645	0.03832
42	-0.01597	-0.06213	0.06415	132	0.04198	-0.01500	0.04458
44	0.06917	-0.07033	0.09864	134	0.04105	0.00136	0.04107
46	0.12636	-0.05620	0.13829	136	0.02682	0.01671	0.03160
48	0.13898	-0.02411	0.14105	138	0.00511	0.02632	0.02682
50	0.10408	0.01500	0.10515	140	-0.01698	0.02798	0.03273
52	0.03506	0.04680	0.05848	142	-0.03353	0.02208	0.04015
54	-0.04172	0.05898	0.07225	144	-0.04121	0.01099	0.04265
56	-0.09617	0.04656	0.10685	146	-0.03950	-0.00211	0.03956
58	-0.10669	0.01484	0.10772	148	-0.03014	-0.01421	0.03332
60	-0.07013	-0.02204	0.07351	150	-0.01607	-0.02313	0.02816
62	-0.00438	-0.04707	0.04727	152	-0.00044	-0.02782	0.02783
64	0.05893	-0.04838	0.07625	154	0.01408	-0.02824	0.03156
66	0.08930	-0.02566	0.09291	156	0.02574	-0.02507	0.03594
68	0.07272	0.00915	0.07329	158	0.03372	-0.01937	0.03889
70	0.01947	0.03753	0.04228	160	0.03797	-0.01228	0.03990
72	-0.04084	0.04435	0.06029	162	0.03898	-0.00479	0.03928
74	-0.07520	0.02633	0.07967	164	0.03754	0.00234	0.03761
76	-0.06556	-0.00573	0.06581	166	0.03447	0.00866	0.03554
78	-0.01920	-0.03305	0.03823	168	0.03056	0.01392	0.03358
80	0.03565	-0.03982	0.05345	170	0.02644	0.01808	0.03203
82	0.06656	-0.02267	0.07031	172	0.02259	0.02121	0.03098
84	0.05632	0.00734	0.05680	174	0.01934	0.02343	0.03038
86	0.01290	0.03163	0.03416	176	0.01690	0.02489	0.03008
88	-0.03607	0.03560	0.05068	178	0.01540	0.02571	0.02997
				180	0.01489	0.02598	0.02994

IEN

RADIATION PATTERN FOR THETA=PI/2

H/LMDA=0.5000 B/LMDA= 0.250 OMEGA= 9.92

30 ELEMENT ARRAY

PHI	ER(THETA,PHI)			PHI	ER(THETA,PHI)		
DEG	RE	IM	ABSVAL	DEG	RE	IM	ABSVAL
0	0.18193	-0.34823	0.39289	90	-0.03833	0.00163	0.03836
2	0.18274	-0.34722	0.39237	92	-0.02231	-0.02288	0.03195
4	0.18505	-0.34403	0.39064	94	0.00736	-0.03279	0.03361
6	0.18851	-0.33818	0.38717	96	0.03206	-0.02189	0.03882
8	0.19252	-0.32895	0.38114	98	0.03667	0.00283	0.03678
10	0.19621	-0.31538	0.37143	100	0.01880	0.02583	0.03195
12	0.19846	-0.29645	0.35675	102	-0.01025	0.03296	0.03451
14	0.19788	-0.27121	0.33572	104	-0.03286	0.02006	0.03850
16	0.19293	-0.23903	0.30717	106	-0.03593	-0.00486	0.03625
18	0.18205	-0.19983	0.27033	108	-0.01822	-0.02700	0.03257
20	0.16396	-0.15442	0.22523	110	0.00963	-0.03369	0.03504
22	0.13791	-0.10467	0.17314	112	0.03191	-0.02154	0.03850
24	0.10416	-0.05368	0.11717	114	0.03681	0.00236	0.03688
26	0.06425	-0.00563	0.06449	116	0.02243	0.02503	0.03361
28	0.02131	0.03463	0.04066	118	-0.00318	0.03488	0.03502
30	-0.02007	0.06237	0.06552	120	-0.02699	0.02749	0.03852
32	-0.05431	0.07401	0.09180	122	-0.03786	0.00706	0.03851
34	-0.07583	0.06835	0.10209	124	-0.03153	-0.01650	0.03558
36	-0.08047	0.04740	0.09339	126	-0.01167	-0.03275	0.03477
38	-0.06699	0.01668	0.06903	128	0.01273	-0.03530	0.03753
40	-0.03819	-0.01554	0.04123	130	0.03183	-0.02393	0.03982
42	-0.00116	-0.04013	0.04015	132	0.03890	-0.00378	0.03908
44	0.03395	-0.04975	0.06023	134	0.03237	0.01743	0.03676
46	0.05661	-0.04153	0.07021	136	0.01545	0.03259	0.03607
48	0.05936	-0.01866	0.06223	138	-0.00576	0.03750	0.03794
50	0.04109	0.01006	0.04231	140	-0.02484	0.03164	0.04023
52	0.00835	0.03313	0.03416	142	-0.03697	0.01763	0.04096
54	-0.02622	0.04088	0.04856	144	-0.03997	-0.00025	0.03997
56	-0.04845	0.02995	0.05696	146	-0.03424	-0.01754	0.03847
58	-0.04872	0.00543	0.04902	148	-0.02206	-0.03075	0.03784
60	-0.02686	-0.02090	0.03403	150	-0.00655	-0.03797	0.03853
62	0.00690	-0.03587	0.03653	152	0.00923	-0.03887	0.03995
64	0.03611	-0.03173	0.04807	154	0.02290	-0.03429	0.04124
66	0.04602	-0.01064	0.04723	156	0.03303	-0.02581	0.04192
68	0.03152	0.01588	0.03529	158	0.03912	-0.01515	0.04195
70	0.00047	0.03290	0.03290	160	0.04138	-0.00390	0.04156
72	-0.02996	0.03055	0.04279	162	0.04049	0.00675	0.04105
74	-0.04261	0.01017	0.04381	164	0.03735	0.01604	0.04065
76	-0.03034	-0.01609	0.03434	166	0.03284	0.02361	0.04045
78	-0.00052	-0.03222	0.03223	168	0.02776	0.02942	0.04045
80	0.02892	-0.02823	0.04042	170	0.02273	0.03363	0.04059
82	0.04023	-0.00657	0.04077	172	0.01819	0.03651	0.04079
84	0.02672	0.01912	0.03286	174	0.01445	0.03836	0.04099
86	-0.00301	0.03254	0.03268	176	0.01169	0.03947	0.04116
88	-0.03034	0.02512	0.03939	178	0.01000	0.04004	0.04127
				180	0.00943	0.04021	0.04130

VEN

RADIATION PATTERN FOR THETA=PI/2

H/LMDA=0.5000 B/LMDA= 0.250 OMEGA= 9.92

35 ELEMENT ARRAY

PHI DEG	ER(THETA,PHI) RE	IM	ABSVAL	PHI DEG	ER(THETA,PHI) RE	IM	ABSVAL
0	-0.25577	0.73213	0.77552	90	-0.03588	0.02375	0.04303
1	-0.25614	0.73193	0.77546	91	-0.03125	0.00141	0.03129
2	-0.25725	0.73131	0.77524	92	-0.01995	-0.02070	0.02875
3	-0.25907	0.73018	0.77478	93	-0.00456	-0.03775	0.03803
4	-0.26154	0.72840	0.77393	94	0.01149	-0.04614	0.04755
5	-0.26459	0.72577	0.77249	95	0.02471	-0.04425	0.05069
6	-0.26811	0.72204	0.77021	96	0.03233	-0.03278	0.04604
7	-0.27197	0.71692	0.76677	97	0.03283	-0.01446	0.03588
8	-0.27601	0.71006	0.76182	98	0.02629	0.00656	0.02709
9	-0.28005	0.70107	0.75493	99	0.01426	0.02568	0.02938
10	-0.28387	0.68954	0.74568	100	-0.00055	0.03886	0.03887
11	-0.28721	0.67503	0.73359	101	-0.01495	0.04345	0.04595
12	-0.28978	0.65711	0.71817	102	-0.02593	0.03873	0.04661
13	-0.29129	0.63535	0.69894	103	-0.03129	0.02595	0.04065
14	-0.29139	0.60936	0.67545	104	-0.03007	0.00804	0.03113
15	-0.28973	0.57880	0.64727	105	-0.02270	-0.01115	0.02529
16	-0.28595	0.54343	0.61407	106	-0.01085	-0.02762	0.02967
17	-0.27970	0.50312	0.57564	107	0.00297	-0.03810	0.03821
18	-0.27064	0.45790	0.53191	108	0.01593	-0.04066	0.04367
19	-0.25850	0.40798	0.48298	109	0.02550	-0.03504	0.04334
20	-0.24305	0.35378	0.42922	110	0.02992	-0.02263	0.03751
21	-0.22416	0.29595	0.37126	111	0.02848	-0.00607	0.02912
22	-0.20183	0.23541	0.31009	112	0.02164	0.01132	0.02442
23	-0.17620	0.17332	0.24716	113	0.01086	0.02618	0.02834
24	-0.14758	0.11109	0.18472	114	-0.00173	0.03581	0.03585
25	-0.11644	0.05036	0.12687	115	-0.01377	0.03862	0.04101
26	-0.08349	-0.00709	0.08379	116	-0.02312	0.03435	0.04140
27	-0.04960	-0.05937	0.07737	117	-0.02820	0.02401	0.03704
28	-0.01584	-0.10462	0.10581	118	-0.02829	0.00966	0.02989
29	0.01659	-0.14106	0.14203	119	-0.02354	-0.00607	0.02431
30	0.04640	-0.16720	0.17352	120	-0.01493	-0.02043	0.02531
31	0.07225	-0.18191	0.19573	121	-0.00403	-0.03109	0.03135
32	0.09290	-0.18459	0.20665	122	0.00732	-0.03646	0.03718
33	0.10722	-0.17527	0.20547	123	0.01730	-0.03588	0.03983
34	0.11440	-0.15471	0.19241	124	0.02443	-0.02970	0.03846
35	0.11395	-0.12438	0.16869	125	0.02773	-0.01912	0.03368
36	0.10588	-0.08653	0.13674	126	0.02688	-0.00590	0.02752
37	0.09067	-0.04401	0.10079	127	0.02219	0.00790	0.02355
38	0.06938	-0.00017	0.06938	128	0.01447	0.02028	0.02491
39	0.04358	0.04144	0.06014	129	0.00492	0.02958	0.02998
40	0.01530	0.07731	0.07881	130	-0.00510	0.03470	0.03507
41	-0.01313	0.10430	0.10513	131	-0.01427	0.03516	0.03795
42	-0.03925	0.12001	0.12627	132	-0.02147	0.03116	0.03784
43	-0.06068	0.12303	0.13718	133	-0.02592	0.02340	0.03492
44	-0.07540	0.11317	0.13599	134	-0.02723	0.01304	0.03019
45	-0.08197	0.09161	0.12293	135	-0.02541	0.00141	0.02545
46	-0.07972	0.06077	0.10024	136	-0.02084	-0.01099	0.02315
47	-0.06892	0.02422	0.07305	137	-0.01414	-0.02023	0.02468
48	-0.05077	-0.01374	0.05260	138	-0.00614	[illegible]	[illegible]
49	-0.03730	[illegible]	[illegible]	139	0.00230	-0.03278	0.03286
50	-0.00154	-0.07596	0.07597	140	0.01034	-0.03428	0.03581
51	0.02360	-0.09251	0.09548	141	0.01727	-0.03257	0.03687
52	0.04487	-0.09616	0.10612	142	0.02254	-0.02802	0.03597
53	0.05954	-0.08655	0.10505	143	0.02582	-0.02122	0.03343
54	0.06569	-0.06515	0.09252	144	0.02698	-0.01289	0.02990
55	0.06255	-0.03513	0.07174	145	0.02607	-0.00381	0.02634
56	0.05062	-0.00096	0.05063	146	0.02330	0.00528	0.02389
57	0.03171	0.03220	0.04519	147	0.01902	0.01371	0.02344
58	0.00868	0.05931	0.05994	148	0.01361	0.02096	0.02499
59	-0.01494	0.07619	0.07764	149	0.00752	0.02665	0.02769
60	-0.03548	0.08027	0.08776	150	0.00117	0.03055	0.03058
61	-0.04971	0.07106	0.08672	151	-0.00506	0.03260	0.03299
62	-0.05540	0.05026	0.07480	152	-0.01085	0.03284	0.03459
63	-0.05171	0.02155	0.05602	153	-0.01594	0.03144	0.03525
64	-0.03938	-0.01002	0.04063	154	-0.02016	0.02863	0.03501
65	-0.02062	-0.03885	0.04398	155	-0.02340	0.02468	0.03401
66	0.00119	-0.05984	0.05985	156	-0.02564	0.01989	0.03245
67	0.02215	-0.06930	0.07275	157	-0.02689	0.01455	0.03058
68	0.03847	-0.06566	0.07610	158	-0.02722	0.00895	0.02866
69	0.04723	-0.04981	0.06864	159	-0.02674	0.00331	0.02694
70	0.04692	-0.02495	0.05314	160	-0.02555	-0.00215	0.02564
71	0.03772	0.00400	0.03793	161	-0.02380	-0.00729	0.02489
72	0.02155	0.03134	0.03803	162	-0.02161	-0.01199	0.02471
73	0.00163	0.05172	0.05174	163	-0.01912	-0.01616	0.02503
74	-0.01805	0.06120	0.06381	164	-0.01643	-0.01979	0.02572
75	-0.03362	0.05809	0.06711	165	-0.01365	-0.02285	0.02661
76	-0.04201	0.04323	0.06028	166	-0.01086	-0.02536	0.02759
77	-0.04166	0.01989	0.04616	167	-0.00815	-0.02736	0.02855
78	-0.03279	-0.00693	0.03351	168	-0.00555	-0.02890	0.02943
79	-0.01738	-0.03158	0.03605	169	-0.00312	-0.03003	0.03019
80	0.00125	-0.04893	0.04894	170	-0.00089	-0.03082	0.03083
81	0.01914	-0.05545	0.05866	171	0.00113	-0.03132	0.03134
82	0.03256	-0.04999	0.05966	172	0.00293	-0.03160	0.03173
83	0.03874	-0.03396	0.05152	173	0.00450	-0.03170	0.03202
84	0.03651	-0.01103	0.03815	174	0.00585	-0.03169	0.03223
85	0.02653	0.01369	0.02985	175	0.00697	-0.03161	0.03237
86	0.01108	0.03481	0.03653	176	0.00788	-0.03149	0.03246
87	-0.00640	0.04782	0.04825	177	0.00858	-0.03136	0.03251
88	-0.02209	0.05008	0.05474	178	0.00907	-0.03125	0.03254
89	-0.03264	0.04133	0.05267	179	0.00936	-0.03118	0.03256
				180	0.00946	-0.03116	0.03256

IEN

RADIATION PATTERN FOR THETA=PI/2

H/LMDA=0.5000 B/LMDA= 0.250 OMEGA= 9.92

35 ELEMENT ARRAY

PHI DEG	ER(THETA,PHI) RE	IM	ABSVAL	PHI DEG	ER(THETA,PHI) RE	IM	ABSVAL
0	0.11784	0.35628	0.37526	90	-0.02757	0.00078	0.02758
1	0.11749	0.35625	0.37513	91	-0.02275	-0.01463	0.02704
2	0.11646	0.35615	0.37471	92	-0.01288	-0.02663	0.02958
3	0.11471	0.35593	0.37396	93	-0.00016	-0.03258	0.03258
4	0.11226	0.35551	0.37281	94	0.01259	-0.03120	0.03365
5	0.10907	0.35478	0.37117	95	0.02257	-0.02287	0.03213
6	0.10512	0.35361	0.36890	96	0.02758	-0.00951	0.02918
7	0.10041	0.35183	0.36587	97	0.02656	0.00588	0.02720
8	0.09491	0.34924	0.36190	98	0.01974	0.01988	0.02801
9	0.08861	0.34562	0.35680	99	0.00864	0.02944	0.03068
10	0.08153	0.34076	0.35038	100	-0.00433	0.03251	0.03280
11	0.07369	0.33439	0.34241	101	-0.01636	0.02852	0.03288
12	0.06512	0.32626	0.33270	102	-0.02490	0.01840	0.03096
13	0.05591	0.31613	0.32104	103	-0.02817	0.00439	0.02851
14	0.04614	0.30378	0.30726	104	-0.02553	-0.01049	0.02760
15	0.03595	0.28899	0.29121	105	-0.01758	-0.02309	0.02902
16	0.02552	0.27161	0.27281	106	-0.00601	-0.03082	0.03140
17	0.01503	0.25158	0.25202	107	0.00678	-0.03215	0.03286
18	0.00475	0.22887	0.22892	108	0.01819	-0.02692	0.03249
19	-0.00506	0.20359	0.20365	109	0.02595	-0.01627	0.03063
20	-0.01410	0.17596	0.17652	110	0.02858	-0.00242	0.02869
21	-0.02207	0.14633	0.14798	111	0.02564	0.01184	0.02824
22	-0.02866	0.11519	0.11870	112	0.01777	0.02375	0.02966
23	-0.03357	0.08318	0.08970	113	0.00653	0.03105	0.03173
24	-0.03657	0.05109	0.06283	114	-0.00590	0.03246	0.03300
25	-0.03747	0.01980	0.04238	115	-0.01722	0.02783	0.03273
26	-0.03618	-0.00970	0.03746	116	-0.02540	0.01814	0.03121
27	-0.03271	-0.03636	0.04891	117	-0.02903	0.00522	0.02949
28	-0.02719	-0.05918	0.06513	118	-0.02757	-0.00856	0.02886
29	-0.01991	-0.07719	0.07972	119	-0.02137	-0.02078	0.02981
30	-0.01128	-0.08959	0.09030	120	-0.01158	-0.02941	0.03161
31	-0.00184	-0.09582	0.09584	121	0.00009	-0.03309	0.03309
32	0.00776	-0.09559	0.09590	122	0.01170	-0.03135	0.03346
33	0.01680	-0.08900	0.09057	123	0.02141	-0.02462	0.03262
34	0.02455	-0.07654	0.08038	124	0.02776	-0.01408	0.03113
35	0.03032	-0.05913	0.06645	125	0.02988	-0.00146	0.02991
36	0.03354	-0.03808	0.05075	126	0.02757	0.01128	0.02979
37	0.03383	-0.01504	0.03702	127	0.02129	0.02230	0.03083
38	0.03103	0.00810	0.03208	128	0.01205	0.03008	0.03241
39	0.02530	0.02940	0.03878	129	0.00122	0.03366	0.03369
40	0.01707	0.04696	0.04996	130	-0.00971	0.03271	0.03412
41	0.00708	0.05917	0.05959	131	-0.01930	0.02752	0.03361
42	-0.00367	0.06490	0.06500	132	-0.02641	0.01891	0.03248
43	-0.01405	0.06360	0.06514	133	-0.03025	0.00808	0.03131
44	-0.02289	0.05546	0.06000	134	-0.03050	-0.00359	0.03071
45	-0.02910	0.04141	0.05062	135	-0.02727	-0.01472	0.0309[illegible]
46	-0.03186	0.02309	0.03935	136	-0.02107	-0.02409	0.03201
47	-0.03073	0.00266	0.03084	137	[illegible]	[illegible]	[illegible]
48	[illegible]	-0.01138	0.03104	138	-0.00312	-0.03421	0.03435
49	-0.01735	-0.03452	0.03863	139	0.00667	-0.03422	0.03486
50	-0.00667	-0.04655	0.04702	140	0.01571	-0.03098	0.03473
51	0.00494	-0.05189	0.05213	141	0.02321	-0.02496	0.03409
52	0.01587	-0.04985	0.05232	142	0.02860	-0.01687	0.03320
53	0.02453	-0.04075	0.04756	143	0.03153	-0.00751	0.03241
54	0.02958	-0.02594	0.03934	144	0.03192	0.00229	0.03200
55	0.03018	-0.00764	0.03113	145	0.02988	0.01171	0.03210
56	0.02612	0.01134	0.02847	146	0.02573	0.02011	0.03265
57	0.01794	0.02804	0.03328	147	0.01989	0.02693	0.03348
58	0.00687	0.03980	0.04039	148	0.01287	0.03185	0.03435
59	-0.00533	0.04474	0.04505	149	0.00520	0.03469	0.03508
60	-0.01664	0.04205	0.04522	150	-0.00261	0.03543	0.03553
61	-0.02511	0.03223	0.04086	151	-0.01010	0.03421	0.03567
62	-0.02922	0.01702	0.03381	152	-0.01690	0.03125	0.03553
63	-0.02816	-0.00089	0.02817	153	-0.02271	0.02687	0.03518
64	-0.02204	-0.01826	0.02862	154	-0.02735	0.02141	0.03473
65	-0.01189	-0.03193	0.03407	155	-0.03072	0.01521	0.03428
66	0.00046	-0.03937	0.03937	156	-0.03280	0.00863	0.03392
67	0.01272	-0.03921	0.04122	157	-0.03364	0.00197	0.03369
68	0.02254	-0.03152	0.03875	158	-0.03334	-0.00450	0.03364
69	0.02800	-0.01782	0.03319	159	-0.03204	-0.01058	0.03374
70	0.02797	-0.00085	0.02799	160	-0.02991	-0.01611	0.03397
71	0.02239	0.01601	0.02753	161	-0.02711	-0.02098	0.03428
72	0.01232	0.02936	0.03184	162	-0.02382	-0.02513	0.03463
73	-0.00024	0.03649	0.03649	163	-0.02021	-0.02855	0.03498
74	-0.01275	0.03597	0.03816	164	-0.01641	-0.03126	0.03530
75	-0.02262	0.02795	0.03596	165	-0.01255	-0.03329	0.03558
76	-0.02776	0.01416	0.03116	166	-0.00875	-0.03471	0.03579
77	-0.02706	-0.00245	0.02717	167	-0.00507	-0.03560	0.03596
78	-0.02061	-0.01831	0.02757	168	-0.00161	-0.03603	0.03606
79	-0.00977	-0.03003	0.03158	169	0.00161	-0.03609	0.03613
80	0.00317	-0.03507	0.03522	170	0.00455	-0.03586	0.03615
81	0.01542	-0.03240	0.03588	171	0.00719	-0.03543	0.03615
82	0.02431	-0.02264	0.03322	172	0.00953	-0.03485	0.03613
83	0.02790	-0.00799	0.02902	173	0.01155	-0.03419	0.03609
84	0.02536	0.00828	0.02668	174	0.01328	-0.03352	0.03606
85	0.01725	0.02254	0.02838	175	0.01471	-0.03287	0.03602
86	0.00533	0.03163	0.03207	176	0.01587	-0.03229	0.03598
87	-0.00777	0.03356	0.03445	177	0.01675	-0.03181	0.03595
88	-0.01915	0.02795	0.03388	178	0.01738	-0.03144	0.03593
89	-0.02627	0.01613	0.03083	179	0.01775	-0.03122	0.03591
				180	0.01787	-0.03115	0.03591

VEN

RADIATION PATTERN FOR THETA=PI/2

H/LMDA=0.5000 B/LMDA= 0.250 OMEGA= 9.92

40 ELEMENT ARRAY

PHI DEG	ER(THETA,PHI) RE	IM	ABSVAL	PHI DEG	ER(THETA,PHI) RE	IM	ABSVAL
0	-0.33117	-0.67747	0.75408	90	-0.02406	0.00693	0.02504
1	-0.33072	-0.67761	0.75401	91	-0.02712	-0.01708	0.03205
2	-0.32937	-0.67798	0.75375	92	-0.02237	-0.03565	0.04209
3	-0.32708	-0.67847	0.75319	93	-0.01131	-0.04367	0.04511
4	-0.32375	-0.67892	0.75216	94	0.00276	-0.03912	0.03922
5	-0.31928	-0.67907	0.75039	95	0.01576	-0.02363	0.02841
6	-0.31355	-0.67862	0.74755	96	0.02404	-0.00185	0.02411
7	-0.30640	-0.67717	0.74326	97	0.02533	0.01994	0.03224
8	-0.29766	-0.67429	0.73706	98	0.01944	0.03563	0.04059
9	-0.28718	-0.66945	0.72845	99	0.00819	0.04102	0.04183
10	-0.27478	-0.66211	0.71686	100	-0.00515	0.03490	0.03528
11	-0.26032	-0.65166	0.70173	101	-0.01681	0.01928	0.02558
12	-0.24370	-0.63748	0.68247	102	-0.02361	-0.00127	0.02364
13	-0.22484	-0.61897	0.65854	103	-0.02378	-0.02101	0.03173
14	-0.20375	-0.59554	0.62943	104	-0.01746	-0.03462	0.03877
15	-0.18052	-0.56668	0.59474	105	-0.00651	-0.03863	0.03917
16	-0.15534	-0.53199	0.55421	106	0.00600	-0.03227	0.03283
17	-0.12849	-0.49124	0.50777	107	0.01670	-0.01755	0.02423
18	-0.10042	-0.44438	0.45559	108	0.02282	0.00143	0.02286
19	-0.07169	-0.39166	0.39816	109	0.02287	0.01962	0.03014
20	-0.04298	-0.33360	0.33636	110	0.01703	0.03240	0.03661
21	-0.01510	-0.27110	0.27152	111	0.00694	0.03672	0.03737
22	0.01104	-0.20540	0.20570	112	-0.00474	0.03179	0.03214
23	0.03449	-0.13813	0.14238	113	-0.01505	0.01915	0.02435
24	0.05430	-0.07127	0.08960	114	-0.02151	0.00213	0.02161
25	0.06957	-0.00708	0.06993	115	-0.02267	-0.01503	0.02720
26	0.07956	0.05200	0.09505	116	-0.01843	-0.02828	0.03375
27	0.08374	0.10346	0.13310	117	-0.00994	-0.03469	0.03609
28	0.08187	0.14489	0.16642	118	0.00069	-0.03307	0.03307
29	0.07409	0.17423	0.18932	119	0.01100	-0.02407	0.02647
30	0.06094	0.18989	0.19943	120	0.01871	-0.00998	0.02121
31	0.04340	0.19107	0.19594	121	0.02223	0.00596	0.02301
32	0.02286	0.17785	0.17931	122	0.02095	0.02030	0.02917
33	0.00104	0.15135	0.15135	123	0.01531	0.03013	0.03380
34	-0.02011	0.11379	0.11556	124	0.00665	0.03365	0.03430
35	-0.03861	0.06841	0.07855	125	-0.00321	0.03044	0.03060
36	-0.05260	0.01928	0.05602	126	-0.01227	0.02141	0.02468
37	-0.06060	-0.02899	0.06718	127	-0.01884	0.00854	0.02069
38	-0.06168	-0.07169	0.09457	128	-0.02181	-0.00562	0.02252
39	-0.05565	-0.10445	0.11835	129	-0.02078	-0.01845	0.02779
40	-0.04315	-0.12385	0.13115	130	-0.01611	-0.02780	0.03213
41	-0.02562	-0.12782	0.13036	131	-0.00877	-0.03222	0.03340
42	-0.00525	-0.11605	0.11617	132	-0.00010	-0.03125	0.03126
43	0.01532	-0.09015	0.09144	133	0.00842	-0.02533	0.02669
44	0.03330	-0.05359	0.06310	134	0.01547	-0.01563	0.02199
45	0.04612	-0.01138	0.04750	135	0.02005	-0.00383	0.02041
46	0.05188	0.03059	0.06023	136	0.02160	0.00825	0.02312
47	0.04966	0.06628	0.08282	137	0.02007	0.01888	0.02756
48	0.03974	0.09045	0.09880	138	0.01583	0.02672	0.03106
49	0.02367	0.09953	0.10230	139	0.00960	0.03090	0.03236
50	0.00404	0.09223	0.09232	140	0.00228	0.03114	0.03123
51	-0.01584	0.06996	0.07173	141	-0.00516	0.02767	0.02815
52	-0.03251	0.03664	0.04898	142	-0.01184	0.02113	0.02422
53	-0.04299	-0.00181	0.04303	143	-0.01706	0.01246	0.02112
54	-0.04534	-0.03848	0.05947	144	-0.02033	0.00274	0.02051
55	-0.03912	-0.06667	0.07730	145	-0.02146	-0.00697	0.02256
56	-0.02550	-0.08120	0.08511	146	-0.02047	-0.01573	0.02581
57	-0.00717	-0.07946	0.07978	147	-0.01761	-0.02279	0.02880
58	0.01220	-0.06202	0.06321	148	-0.01328	-0.02769	0.03070
59	0.02862	-0.03262	0.04340	149	-0.00795	-0.03018	0.03121
60	0.03867	0.00251	0.03875	150	-0.00213	-0.03028	0.03036
61	0.04020	0.03590	0.05390	151	0.00370	-0.02821	0.02845
62	0.03293	0.06040	0.06879	152	0.00910	-0.02430	0.02595
63	0.01850	0.07079	0.07316	153	0.01376	-0.01901	0.02347
64	0.00021	0.06498	0.06498	154	0.01744	-0.01280	0.02164
65	-0.01770	0.04455	0.04794	155	0.02001	-0.00614	0.02093
66	-0.03106	0.01446	0.03426	156	0.02144	0.00054	0.02145
67	-0.03674	-0.01807	0.04094	157	0.02177	0.00691	0.02283
68	-0.03342	-0.04524	0.05625	158	0.02109	0.01266	0.02460
69	-0.02200	-0.06058	0.06445	159	0.01956	0.01763	0.02633
70	-0.00539	-0.06053	0.06077	160	0.01734	0.02169	0.02777
71	0.01220	-0.04539	0.04700	161	0.01460	0.02480	0.02878
72	0.02628	-0.01925	0.03257	162	0.01152	0.02698	0.02934
73	0.03327	0.01102	0.03505	163	0.00826	0.02829	0.02947
74	0.03145	0.03754	0.04897	164	0.00494	0.02881	0.02923
75	0.02140	0.05349	0.05761	165	0.00168	0.02867	0.02872
76	0.00589	0.05486	0.05517	166	-0.00144	0.02798	0.02802
77	-0.01084	0.04159	0.04298	167	-0.00434	0.02687	0.02722
78	-0.02427	0.01754	0.02994	168	-0.00699	0.02544	0.02639
79	-0.03078	-0.01059	0.03255	169	-0.00936	0.02382	0.02560
80	-0.02870	-0.03504	0.04529	170	-0.01145	0.02209	0.02489
81	-0.01873	-0.04921	0.05265	171	-0.01326	0.02034	0.02428
82	-0.00378	-0.04940	0.04955	172	-0.01479	0.01863	0.02379
83	0.01191	-0.03588	0.03780	173	-0.01608	0.01703	0.02342
84	0.02390	-0.01271	0.02707	174	-0.01714	0.01556	0.02315
85	0.02890	0.01341	0.03186	175	-0.01798	0.01428	0.02296
86	0.02560	0.03508	0.04343	176	-0.01864	0.01320	0.02285
87	0.01510	0.04629	0.04869	177	-0.01913	0.01235	0.02277
88	0.00052	0.04414	0.04414	178	-0.01947	0.01173	0.02273
89	-0.01391	0.02954	0.03265	179	-0.01967	0.01136	0.02271
				180	-0.01974	0.01123	0.02271

IEN

RADIATION PATTERN FOR THETA=PI/2

H/LMDA=0.5000 B/LMDA= 0.250 OMEGA= 9.92

40 ELEMENT ARRAY

PHI DEG	ER(THETA,PHI) RE	IM	ABSVAL	PHI DEG	ER(THETA,PHI) RE	IM	ABSVAL
0	-0.32451	-0.15801	0.36093	90	-0.02468	-0.00381	0.02497
1	-0.32423	-0.15825	0.36079	91	-0.02163	-0.01801	0.02815
2	-0.32338	-0.15897	0.36034	92	-0.01225	-0.02692	0.02957
3	-0.32191	-0.16013	0.35954	93	0.00072	-0.02801	0.02802
4	-0.31974	-0.16167	0.35829	94	0.01347	-0.02105	0.02499
5	-0.31675	-0.16352	0.35647	95	0.02231	-0.00809	0.02373
6	-0.31280	-0.16558	0.35392	96	0.02467	0.00709	0.02567
7	-0.30773	-0.16771	0.35046	97	0.01990	0.02016	0.02832
8	-0.30136	-0.16976	0.34588	98	0.00940	0.02742	0.02898
9	-0.29349	-0.17155	0.33995	99	-0.00380	0.02689	0.02715
10	-0.28393	-0.17286	0.33241	100	-0.01592	0.01880	0.02463
11	-0.27250	-0.17347	0.32303	101	-0.02354	0.00551	0.02418
12	-0.25902	-0.17313	0.31155	102	-0.02456	-0.00924	0.02624
13	-0.24338	-0.17155	0.29777	103	-0.01873	-0.02136	0.02840
14	-0.22551	-0.16848	0.28150	104	-0.00771	-0.02756	0.02862
15	-0.20539	-0.16365	0.26262	105	0.00542	-0.02626	0.02682
16	-0.18312	-0.15682	0.24109	106	0.01709	-0.01790	0.02475
17	-0.15890	-0.14779	0.21700	107	0.02417	-0.00482	0.02465
18	-0.13304	-0.13642	0.19055	108	0.02483	0.00946	0.02657
19	-0.10598	-0.12267	0.16211	109	0.01895	0.02121	0.02844
20	-0.07830	-0.10659	0.13226	110	0.00815	0.02743	0.02861
21	-0.05070	-0.08839	0.10190	111	-0.00474	0.02664	0.02706
22	-0.02400	-0.06839	0.07248	112	-0.01643	0.01916	0.02524
23	0.00092	-0.04711	0.04712	113	-0.02402	0.00698	0.02501
24	0.02314	-0.02521	0.03422	114	-0.02569	-0.00685	0.02659
25	0.04177	-0.00351	0.04191	115	-0.02115	-0.01896	0.02840
26	0.05599	0.01708	0.05854	116	-0.01157	-0.02652	0.02893
27	0.06518	0.03555	0.07424	117	0.00071	-0.02786	0.02787
28	0.06893	0.05089	0.08568	118	0.01281	-0.02283	0.02618
29	0.06715	0.06217	0.09152	119	0.02199	-0.01269	0.02539
30	0.06014	0.06863	0.09125	120	0.02628	0.00018	0.02628
31	0.04855	0.06976	0.08500	121	0.02484	0.01294	0.02801
32	0.03345	0.06541	0.07347	122	0.01812	0.02289	0.02920
33	0.01621	0.05585	0.05816	123	0.00765	0.02803	0.02906
34	-0.00152	0.04182	0.04185	124	-0.00434	0.02746	0.02780
35	-0.01802	0.02449	0.03040	125	-0.01543	0.02146	0.02643
36	-0.03159	0.00543	0.03206	126	-0.02346	0.01136	0.02607
37	-0.04083	-0.01348	0.04300	127	-0.02702	-0.00081	0.02703
38	-0.04474	-0.03030	0.05403	128	-0.02556	-0.01275	0.02856
39	-0.04292	-0.04318	0.06088	129	-0.01952	-0.02231	0.02964
40	-0.03566	-0.05061	0.06191	130	-0.01010	-0.02791	0.02968
41	-0.02396	-0.05166	0.05694	131	0.00097	-0.02876	0.02877
42	-0.00942	-0.04614	0.04709	132	0.01180	-0.02488	0.02754
43	0.00589	-0.03471	0.03520	133	0.02066	-0.01710	0.02682
44	0.01974	-0.01885	0.02730	134	0.02625	-0.00676	0.02710
45	0.03006	-0.00075	0.03007	135	0.02785	0.00447	0.02821
46	0.03526	0.01700	0.03914	136	0.02541	0.01493	0.02947
47	0.03451	0.03171	0.04687	137	0.01945	0.02320	0.03027
48	0.02796	0.04110	0.04971	138	0.01093	0.02827	0.03031
49	0.01674	0.04362	0.04672	139	0.00110	0.02964	0.02966
50	0.00283	0.03881	0.03891	140	-0.00876	0.02734	0.02871
51	-0.01125	0.02744	0.02966	141	-0.01743	0.02183	0.02794
52	-0.02287	0.01148	0.02559	142	-0.02399	0.01393	0.02774
53	-0.02981	-0.00624	0.03046	143	-0.02782	0.00465	0.02821
54	-0.03072	-0.02243	0.03803	144	-0.02868	-0.00495	0.02910
55	-0.02539	-0.03402	0.04245	145	-0.02665	-0.01389	0.03005
56	-0.01491	-0.03873	0.04151	146	-0.02212	-0.02135	0.03075
57	-0.00148	-0.03560	0.03563	147	-0.01568	-0.02675	0.03101
58	0.01205	-0.02523	0.02796	148	-0.00801	-0.02977	0.03083
59	0.02271	-0.00977	0.02472	149	0.00015	-0.03034	0.03034
60	0.02812	0.00751	0.02910	150	0.00812	-0.02860	0.02973
61	0.02702	0.02287	0.03540	151	0.01532	-0.02487	0.02921
62	0.01966	0.03289	0.03831	152	0.02130	-0.01957	0.02893
63	0.00774	0.03530	0.03614	153	0.02578	-0.01320	0.02896
64	-0.00589	0.02954	0.03012	154	0.02860	-0.00624	0.02927
65	-0.01794	0.01695	0.02468	155	0.02975	0.00085	0.02976
66	-0.02541	0.00050	0.02541	156	0.02934	0.00766	0.03032
67	-0.02642	-0.01582	0.03079	157	0.02754	0.01387	0.03083
68	-0.02070	-0.02803	0.03484	158	0.02458	0.01925	0.03122
69	-0.00968	-0.03309	0.03447	159	0.02075	0.02366	0.03147
70	0.00377	-0.02973	0.02997	160	0.01629	0.02703	0.03155
71	0.01612	-0.01882	0.02478	161	0.01147	0.02935	0.03151
72	0.02408	-0.00318	0.02429	162	0.00651	0.03070	0.03138
73	0.02548	0.01309	0.02865	163	0.00160	0.03115	0.03119
74	0.01993	0.02571	0.03253	164	-0.00310	0.03083	0.03098
75	0.00893	0.03130	0.03255	165	-0.00747	0.02987	0.03079
76	-0.00449	0.02839	0.02874	166	-0.01145	0.02842	0.03064
77	-0.01659	0.01779	0.02433	167	-0.01499	0.02659	0.03053
78	-0.02396	0.00243	0.02408	168	-0.01808	0.02452	0.03047
79	-0.02450	-0.01344	0.02795	169	-0.02071	0.02232	0.03045
80	-0.01806	-0.02540	0.03117	170	-0.02291	0.02008	0.03047
81	-0.00645	-0.03013	0.03082	171	-0.02473	0.01789	0.03052
82	0.00698	-0.02633	0.02724	172	-0.02619	0.01580	0.03058
83	0.01835	-0.01512	0.02378	173	-0.02734	0.01387	0.03066
84	0.02436	0.00029	0.02436	174	-0.02824	0.01214	0.03074
85	0.02326	0.01548	0.02794	175	-0.02891	0.01064	0.03081
86	0.01537	0.02610	0.03029	176	-0.02941	0.00939	0.03087
87	0.00298	0.02914	0.02929	177	-0.02976	0.00841	0.03093
88	-0.01026	0.02378	0.02590	178	-0.02999	0.00770	0.03096
89	-0.02048	0.01161	0.02354	179	-0.03012	0.00727	0.03099
				180	-0.03016	0.00713	0.03100

VEN

RADIATION PATTERN FOR THETA=PI/2

H/LMDA=0.5000 B/LMDA= 0.250 OMEGA= 9.92

45 ELEMENT ARRAY

PHI DEG	ER(THETA,PHI) RE	IM	ABSVAL	PHI DEG	ER(THETA,PHI) RE	IM	ABSVAL
0	0.69719	0.23569	0.73595	90	-0.03345	-0.00016	0.03345
1	0.69695	0.23613	0.73586	91	-0.03729	-0.01595	0.04056
2	0.69619	0.23743	0.73556	92	-0.02765	-0.02571	0.03775
3	0.69477	0.23955	0.73491	93	-0.00832	-0.02609	0.02738
4	0.69248	0.24241	0.73368	94	0.01353	-0.01717	0.02186
5	0.68900	0.24588	0.73155	95	0.03002	-0.00233	0.03011
6	0.68394	0.24979	0.72813	96	0.03541	0.01302	0.03773
7	0.67685	0.25396	0.72292	97	0.02808	0.02346	0.03658
8	0.66719	0.25812	0.71538	98	0.01096	0.02541	0.02768
9	0.65439	0.26198	0.70488	99	-0.00964	0.01842	0.02079
10	0.63784	0.26520	0.69077	100	-0.02638	0.00513	0.02687
11	0.61692	0.26737	0.67237	101	-0.03352	-0.00968	0.03489
12	0.59107	0.26807	0.64902	102	-0.02885	-0.02086	0.03560
13	0.55978	0.26681	0.62011	103	-0.01433	-0.02467	0.02853
14	0.52264	0.26313	0.58514	104	0.00478	-0.02001	0.02058
15	0.47946	0.25653	0.54377	105	0.02185	-0.00869	0.02351
16	0.43025	0.24658	0.49590	106	0.03117	0.00533	0.03163
17	0.37532	0.23288	0.44170	107	0.02988	0.01732	0.03454
18	0.31533	0.21519	0.38176	108	0.01873	0.02340	0.02997
19	0.25132	0.19339	0.31711	109	0.00168	0.02176	0.02183
20	0.18472	0.16756	0.24940	110	-0.01555	0.01317	0.02037
21	0.11742	0.13806	0.18124	111	-0.02744	0.00055	0.02745
22	0.05160	0.10551	0.11745	112	-0.03041	-0.01201	0.03270
23	-0.01023	0.07083	0.07157	113	-0.02383	-0.02062	0.03151
24	-0.06543	0.03527	0.07433	114	-0.01005	-0.02279	0.02491
25	-0.11137	0.00033	0.11137	115	0.00649	-0.01806	0.01919
26	-0.14569	-0.03229	0.14923	116	0.02076	-0.00808	0.02227
27	-0.16650	-0.06076	0.17724	117	0.02863	0.00408	0.02892
28	-0.17265	-0.08331	0.19170	118	0.02806	0.01485	0.03175
29	-0.16396	-0.09838	0.19121	119	0.01954	0.02124	0.02886
30	-0.14138	-0.10481	0.17599	120	0.00575	0.02165	0.02240
31	-0.10711	-0.10201	0.14791	121	-0.00934	0.01618	0.01868
32	-0.06447	-0.09016	0.11084	122	-0.02162	0.00651	0.02258
33	-0.01780	-0.07027	0.07249	123	-0.02797	-0.00467	0.02836
34	0.02797	-0.04421	0.05232	124	-0.02700	-0.01446	0.03063
35	0.06777	-0.01465	0.06934	125	-0.01928	-0.02047	0.02812
36	0.09704	0.01519	0.09822	126	-0.00698	-0.02139	0.02250
37	0.11227	0.04188	0.11982	127	0.00674	-0.01721	0.01848
38	0.11165	0.06216	0.12779	128	0.01859	-0.00913	0.02071
39	0.09544	0.07339	0.12040	129	0.02597	0.00086	0.02598
40	0.06611	0.07407	0.09928	130	0.02744	0.01047	0.02937
41	0.02813	0.06408	0.06999	131	0.02298	0.01767	0.02899
42	-0.01252	0.04490	0.04661	132	0.01382	0.02107	0.02520
43	-0.04929	0.01946	0.05299	133	0.00206	0.02017	0.02028
44	-0.07605	-0.00816	0.07649	134	-0.00983	0.01537	0.01824
45	-0.08825	-0.03336	0.09435	135	-0.01958	0.00776	0.02106
46	-0.08385	-0.05178	0.09855	136	-0.02551	-0.00111	0.02553
47	-0.06380	-0.06016	0.08770	137	-0.02677	-0.00961	0.02844
48	-0.03214	-0.05698	0.06542	138	-0.02342	-0.01631	0.02856
49	0.00475	-0.04788	[illegible]	139	[illegible]	-0.02021	0.02596
50	0.03925	-0.02073	0.04438	140	-0.00677	-0.02086	0.02193
51	0.06411	0.00492	0.06430	141	0.00353	-0.01836	0.01870
52	0.07403	0.02865	0.07938	142	0.01304	-0.01327	0.01860
53	0.06696	0.04532	0.08086	143	0.02046	-0.00646	0.02146
54	0.04470	0.05127	0.06802	144	0.02494	0.00105	0.02497
55	0.01265	0.04523	0.04696	145	0.02611	0.00825	0.02738
56	-0.02137	0.02872	0.03579	146	0.02407	0.01430	0.02799
57	-0.04894	0.00580	0.04928	147	0.01931	0.01857	0.02679
58	-0.06318	-0.01780	0.06564	148	0.01258	0.02074	0.02425
59	-0.06059	-0.03614	0.07055	149	0.00476	0.02074	0.02128
60	-0.04206	-0.04454	0.06126	150	-0.00327	0.01875	0.01903
61	-0.01278	-0.04090	0.04285	151	-0.01071	0.01513	0.01854
62	0.01908	-0.02633	0.03252	152	-0.01695	0.01033	0.01985
63	0.04459	-0.00498	0.04486	153	-0.02158	0.00486	0.02212
64	0.05659	0.01711	0.05912	154	-0.02437	-0.00080	0.02438
65	0.05181	0.03365	0.06178	155	-0.02530	-0.00621	0.02605
66	0.03190	0.03992	0.05111	156	-0.02451	-0.01103	0.02688
67	0.00305	0.03423	0.03437	157	-0.02224	-0.01503	0.02685
68	-0.02587	0.01845	0.03178	158	-0.01880	-0.01806	0.02607
69	-0.04596	-0.00248	0.04602	159	-0.01453	-0.02007	0.02477
70	-0.05109	-0.02204	0.05564	160	-0.00976	-0.02108	0.02323
71	-0.03990	-0.03415	0.05252	161	-0.00480	-0.02118	0.02172
72	-0.01626	-0.03510	0.03868	162	0.00009	-0.02049	0.02049
73	0.01198	-0.02478	0.02752	163	0.00471	-0.01916	0.01973
74	0.03549	-0.00671	0.03612	164	0.00892	-0.01734	0.01950
75	0.04660	0.01307	0.04839	165	0.01261	-0.01517	0.01973
76	0.04181	0.02798	0.05031	166	0.01575	-0.01280	0.02030
77	0.02306	0.03315	0.04038	167	0.01833	-0.01034	0.02104
78	-0.00302	0.02701	0.02717	168	0.02037	-0.00788	0.02184
79	-0.02736	0.01184	0.02981	169	0.02191	-0.00552	0.02260
80	-0.04160	-0.00698	0.04218	170	0.02303	-0.00329	0.02326
81	-0.04100	-0.02293	0.04698	171	0.02378	-0.00125	0.02381
82	-0.02609	-0.03052	0.04015	172	0.02423	0.00058	0.02424
83	-0.00240	-0.02728	0.02738	173	0.02446	0.00219	0.02456
84	0.02152	-0.01455	0.02598	174	0.02453	0.00357	0.02479
85	0.03720	0.00299	0.03732	175	0.02449	0.00472	0.02494
86	0.03924	0.01906	0.04362	176	0.02439	0.00565	0.02504
87	0.02723	0.02800	0.03905	177	0.02427	0.00636	0.02509
88	0.00578	0.02678	0.02739	178	0.02417	0.00687	0.02512
89	-0.01721	0.01605	0.02353	179	0.02409	0.00716	0.02514
				180	0.02407	0.00726	0.02514

IEN

RADIATION PATTERN FOR THETA=PI/2

H/LMDA=0.5000 B/LMDA= 0.250 OMEGA= 9.92

45 ELEMENT ARRAY

PHI DEG	ER(THETA,PHI) RE	IM	ABSVAL	PHI DEG	ER(THETA,PHI) RE	IM	ABSVAL
0	0.32867	-0.11752	0.34905	90	-0.02582	-0.00289	0.02598
1	0.32865	-0.11713	0.34890	91	-0.02128	-0.01475	0.02589
2	0.32857	-0.11596	0.34843	92	-0.00897	-0.02124	0.02306
3	0.32835	-0.11398	0.34757	93	0.00654	-0.02002	0.02106
4	0.32787	-0.11118	0.34621	94	0.01955	-0.01154	0.02270
5	0.32697	-0.10753	0.34420	95	0.02535	0.00112	0.02537
6	0.32545	-0.10299	0.34136	96	0.02190	0.01337	0.02566
7	0.32305	-0.09752	0.33745	97	0.01054	0.02082	0.02333
8	0.31948	-0.09112	0.33222	98	-0.00456	0.02081	0.02131
9	0.31443	-0.08376	0.32540	99	-0.01794	0.01339	0.02239
10	0.30756	-0.07547	0.31669	100	-0.02485	0.00123	0.02488
11	0.29853	-0.06629	0.30580	101	-0.02290	-0.01136	0.02556
12	0.28700	-0.05632	0.29248	102	-0.01288	-0.01998	0.02377
13	0.27268	-0.04570	0.27648	103	0.00160	-0.02165	0.02171
14	0.25532	-0.03460	0.25766	104	0.01546	-0.01586	0.02215
15	0.23478	-0.02329	0.23594	105	0.02394	-0.00465	0.02439
16	0.21104	-0.01206	0.21139	106	0.02421	0.00813	0.02554
17	0.18422	-0.00128	0.18422	107	0.01630	0.01818	0.02441
18	0.15463	0.00865	0.15487	108	0.00296	0.02218	0.02238
19	0.12280	0.01732	0.12402	109	-0.01130	0.01890	0.02202
20	0.08948	0.02430	0.09272	110	-0.02182	0.00948	0.02379
21	0.05565	0.02921	0.06285	111	-0.02525	-0.00296	0.02542
22	0.02247	0.03174	0.03889	112	-0.02062	-0.01446	0.02518
23	-0.00872	0.03171	0.03289	113	-0.00952	-0.02142	0.02344
24	-0.03651	0.02908	0.04667	114	0.00448	-0.02178	0.02224
25	-0.05949	0.02399	0.06415	115	0.01706	-0.01554	0.02308
26	-0.07639	0.01681	0.07822	116	0.02447	-0.00467	0.02491
27	-0.08622	0.00812	0.08660	117	0.02462	0.00755	0.02575
28	-0.08837	-0.00131	0.08838	118	0.01763	0.01756	0.02488
29	-0.08279	-0.01058	0.08346	119	0.00564	0.02257	0.02326
30	-0.07004	-0.01870	0.07249	120	-0.00787	0.02129	0.02270
31	-0.05136	-0.02475	0.05701	121	-0.01916	0.01421	0.02385
32	-0.02860	-0.02793	0.03997	122	-0.02524	0.00335	0.02547
33	-0.00413	-0.02775	0.02805	123	-0.02466	-0.00834	0.02604
34	0.01936	-0.02408	0.03090	124	-0.01775	-0.01785	0.02518
35	0.03919	-0.01727	0.04282	125	-0.00642	-0.02287	0.02375
36	0.05298	-0.00810	0.05360	126	0.00641	-0.02227	0.02318
37	0.05903	0.00223	0.05907	127	0.01763	-0.01638	0.02406
38	0.05658	0.01223	0.05788	128	0.02465	-0.00671	0.02555
39	0.04604	0.02036	0.05034	129	0.02604	0.00442	0.02641
40	0.02902	0.02527	0.03848	130	0.02167	0.01452	0.02608
41	0.00820	0.02603	0.02729	131	0.01269	0.02144	0.02492
42	-0.01307	0.02235	0.02589	132	0.00116	0.02388	0.02391
43	-0.03122	0.01473	0.03452	133	-0.01050	0.02149	0.02392
44	-0.04313	0.00443	0.04336	134	-0.01999	0.01494	0.02495
45	-0.04668	-0.00672	0.04716	135	-0.02560	0.00562	0.02621
46	-0.04126	-0.01656	0.04446	136	-0.02648	-0.00463	0.02688
47	-0.02795	-0.02310	0.03626	137	-0.02268	-0.01300	[illegible]
48	[illegible]	[illegible]	0.02882	138	-0.01508	-0.02087	0.02575
49	0.01036	-0.02139	0.02377	139	-0.00512	-0.02428	0.02481
50	0.02733	-0.01325	0.03037	140	0.00550	-0.02385	0.02447
51	0.03772	-0.00213	0.03778	141	0.01514	-0.01983	0.02495
52	0.03924	0.00947	0.04036	142	0.02243	-0.01301	0.02593
53	0.03156	0.01884	0.03676	143	0.02649	-0.00449	0.02687
54	0.01656	0.02366	0.02889	144	0.02699	0.00450	0.02736
55	-0.00208	0.02265	0.02274	145	0.02407	0.01279	0.02726
56	-0.01971	0.01591	0.02533	146	0.01833	0.01941	0.02670
57	-0.03181	0.00507	0.03221	147	0.01061	0.02372	0.02598
58	-0.03526	-0.00739	0.03597	148	0.00191	0.02538	0.02546
59	-0.02917	-0.01735	0.03394	149	-0.00678	0.02443	0.02535
60	-0.01524	-0.02284	0.02746	150	-0.01460	0.02115	0.02570
61	0.00262	-0.02193	0.02209	151	-0.02090	0.01602	0.02633
62	0.01934	-0.01479	0.02435	152	-0.02523	0.00964	0.02701
63	0.03011	-0.00337	0.03029	153	-0.02742	0.00266	0.02755
64	0.03177	0.00904	0.03303	154	-0.02748	-0.00434	0.02782
65	0.02389	0.01872	0.03035	155	-0.02562	-0.01084	0.02782
66	0.00888	0.02271	0.02438	156	-0.02216	-0.01645	0.02759
67	-0.00858	0.01968	0.02147	157	-0.01748	-0.02090	0.02724
68	-0.02303	0.01050	0.02531	158	-0.01201	-0.02405	0.02688
69	-0.02990	-0.00200	0.02997	159	-0.00614	-0.02587	0.02659
70	-0.02702	-0.01384	0.03036	160	-0.00023	-0.02643	0.02643
71	-0.01538	-0.02118	0.02617	161	0.00543	-0.02586	0.02642
72	0.00115	-0.02154	0.02157	162	0.01061	-0.02433	0.02654
73	0.01704	-0.01475	0.02253	163	0.01517	-0.02205	0.02676
74	0.02694	-0.00301	0.02711	164	0.01901	-0.01921	0.02703
75	0.02752	0.00974	0.02920	165	0.02212	-0.01601	0.02730
76	0.01864	0.01915	0.02672	166	0.02449	-0.01261	0.02755
77	0.00342	0.02194	0.02221	167	0.02619	-0.00918	0.02776
78	-0.01280	0.01711	0.02137	168	0.02730	-0.00581	0.02791
79	-0.02431	0.00629	0.02511	169	0.02789	-0.00262	0.02802
80	-0.02706	-0.00674	0.02789	170	0.02808	0.00035	0.02808
81	-0.02013	-0.01737	0.02659	171	0.02794	0.00303	0.02810
82	-0.00607	-0.02181	0.02264	172	0.02757	0.00542	0.02810
83	0.01003	-0.01843	0.02099	173	0.02706	0.00750	0.02808
84	0.02233	-0.00844	0.02387	174	0.02647	0.00926	0.02805
85	0.02640	0.00460	0.02680	175	0.02588	0.01073	0.02802
86	0.02081	0.01597	0.02623	176	0.02533	0.01190	0.02798
87	0.00768	0.02154	0.02287	177	0.02485	0.01280	0.02795
88	-0.00815	0.01929	0.02094	178	0.02449	0.01343	0.02793
89	-0.02086	0.01002	0.02315	179	0.02427	0.01380	0.02792
				180	0.02419	0.01393	0.02792

VEN RADIATION PATTERN FOR THETA=PI/2

H/LMDA=0.5000 B/LMDA= 0.250 OMEGA= 9.92

50 ELEMENT ARRAY

PHI DEG	ER(THETA,PHI) RE	IM	ABSVAL	PHI DEG	ER(THETA,PHI) RE	IM	ABSVAL
0	-0.64427	0.32215	0.72032	90	-0.03567	0.01052	0.03719
1	-0.64443	0.32163	0.72023	91	-0.03085	-0.00411	0.03112
2	-0.64483	0.32006	0.71989	92	-0.01261	-0.01663	0.02087
3	-0.64532	0.31735	0.71914	93	0.01071	-0.02155	0.02407
4	-0.64565	0.31340	0.71770	94	0.02878	-0.01687	0.03336
5	-0.64546	0.30804	0.71519	95	0.03383	-0.00485	0.03418
6	-0.64426	0.30108	0.71114	96	0.02400	0.00907	0.02566
7	-0.64149	0.29230	0.70495	97	0.00397	0.01875	0.01916
8	-0.63651	0.28147	0.69597	98	-0.01732	0.02006	0.02650
9	-0.62856	0.26837	0.68345	99	-0.03064	0.01262	0.03314
10	-0.61686	0.25278	0.66665	100	-0.03050	-0.00015	0.03050
11	-0.60058	0.23457	0.64477	101	-0.01732	-0.01261	0.02143
12	-0.57893	0.21365	0.61709	102	0.00291	-0.01946	0.01968
13	-0.55114	0.19005	0.58299	103	0.02146	-0.01791	0.02795
14	-0.51661	0.16394	0.54200	104	0.03061	-0.00883	0.03185
15	-0.47494	0.13564	0.49393	105	0.02684	0.00380	0.02710
16	-0.42599	0.10566	0.43890	106	0.01209	0.01463	0.01898
17	-0.37004	0.07472	0.37750	107	-0.00726	0.01926	0.02058
18	-0.30778	0.04372	0.31087	108	-0.02326	0.01596	0.02821
19	-0.24045	0.01377	0.24084	109	-0.02961	0.00628	0.03027
20	-0.16983	-0.01392	0.17040	110	-0.02410	-0.00575	0.02477
21	-0.09827	-0.03804	0.10538	111	-0.00928	-0.01532	0.01791
22	-0.02860	-0.05734	0.06408	112	0.00880	-0.01877	0.02073
23	0.03598	-0.07068	0.07930	113	0.02315	-0.01495	0.02756
24	0.09210	-0.07721	0.12019	114	0.02848	-0.00553	0.02901
25	0.13652	-0.07653	0.15650	115	0.02315	0.00581	0.02386
26	0.16641	-0.06872	0.18004	116	0.00948	0.01484	0.01760
27	0.17975	-0.05453	0.18784	117	-0.00731	0.01835	0.01975
28	0.17570	-0.03535	0.17922	118	-0.02116	0.01528	0.02610
29	0.15482	-0.01317	0.15538	119	-0.02738	0.00692	0.02824
30	0.11935	0.00956	0.11974	120	-0.02413	-0.00369	0.02441
31	0.07312	0.03015	0.07909	121	-0.01288	-0.01289	0.01822
32	0.02137	0.04601	0.05073	122	0.00236	-0.01769	0.01785
33	-0.02973	0.05502	0.06253	123	0.01652	-0.01669	0.02348
34	-0.07378	0.05585	0.09253	124	0.02519	-0.01042	0.02726
35	-0.10500	0.04830	0.11558	125	0.02594	-0.00103	0.02596
36	-0.11914	0.03345	0.12375	126	0.01889	0.00851	0.02072
37	-0.11429	0.01360	0.11510	127	0.00645	0.01539	0.01669
38	-0.09140	-0.00798	0.09175	128	-0.00757	0.01772	0.01927
39	-0.05441	-0.02754	0.06098	129	-0.01918	0.01505	0.02438
40	-0.00978	-0.04149	0.04262	130	-0.02535	0.00831	0.02668
41	0.03441	-0.04713	0.05836	131	-0.02471	-0.00052	0.02471
42	0.06991	-0.04328	0.08223	132	-0.01777	-0.00905	0.01994
43	0.08993	-0.03063	0.09500	133	-0.00657	-0.01517	0.01653
44	0.09057	-0.01178	0.09133	134	0.00595	-0.01749	0.01847
45	0.07191	0.00921	0.07249	135	0.01680	-0.01566	0.02297
46	0.03822	0.02760	0.04715	136	0.02363	-0.01031	0.02578
47	-0.00271	0.03907	0.03916	137	0.02517	-0.00279	0.02532
48	-0.04127	0.04082	0.05805	138	0.02141	0.00516	0.02202
49	-0.06820	0.03236	0.07549	139	0.01343	0.01190	0.01795
50	-0.07693	0.01580	0.07854	140	0.00305	0.01617	0.01645
51	-0.06545	-0.00454	0.06561	141	-0.00763	0.01729	0.01890
52	-0.03705	-0.02318	0.04370	142	-0.01667	0.01526	0.02260
53	0.00036	-0.03494	0.03495	143	-0.02262	0.01062	0.02499
54	0.03623	-0.03648	0.05142	144	-0.02472	0.00432	0.02510
55	0.06034	-0.02733	0.06624	145	-0.02293	-0.00252	0.02306
56	0.06577	-0.01022	0.06656	146	-0.01779	-0.00882	0.01985
57	0.05116	0.00962	0.05206	147	-0.01031	-0.01370	0.01714
58	0.02129	0.02598	0.03359	148	-0.00168	-0.01660	0.01668
59	-0.01423	0.03367	0.03655	149	0.00687	-0.01727	0.01859
60	-0.04391	0.03021	0.05330	150	0.01432	-0.01582	0.02133
61	-0.05812	0.01679	0.06049	151	0.01989	-0.01257	0.02354
62	-0.05234	-0.00201	0.05238	152	0.02317	-0.00806	0.02453
63	-0.02888	-0.01966	0.03493	153	0.02403	-0.00284	0.02420
64	0.00385	-0.03000	0.03024	154	0.02263	0.00249	0.02277
65	0.03412	-0.02938	0.04503	155	0.01933	0.00744	0.02072
66	0.05112	-0.01812	0.05424	156	0.01463	0.01162	0.01868
67	0.04888	-0.00045	0.04888	157	0.00905	0.01477	0.01732
68	0.02856	0.01697	0.03322	158	0.00310	0.01677	0.01705
69	-0.00187	0.02754	0.02761	159	-0.00275	0.01761	0.01782
70	-0.03063	0.02728	0.04101	160	-0.00814	0.01737	0.01918
71	-0.04665	0.01640	0.04945	161	-0.01282	0.01619	0.02066
72	-0.04392	-0.00068	0.04392	162	-0.01663	0.01428	0.02192
73	-0.02387	-0.01709	0.02936	163	-0.01950	0.01182	0.02280
74	0.00514	-0.02623	0.02673	164	-0.02144	0.00901	0.02326
75	0.03121	-0.02448	0.03966	165	-0.02252	0.00603	0.02331
76	0.04376	-0.01270	0.04557	166	-0.02284	0.00302	0.02304
77	0.03791	0.00407	0.03813	167	-0.02254	0.00012	0.02254
78	0.01646	0.01876	0.02495	168	-0.02175	-0.00260	0.02190
79	-0.01128	0.02520	0.02761	169	-0.02059	-0.00508	0.02121
80	-0.03348	0.02080	0.03942	170	-0.01920	-0.00728	0.02054
81	-0.04084	0.00760	0.04155	171	-0.01769	-0.00919	0.01994
82	-0.03055	-0.00856	0.03172	172	-0.01615	-0.01082	0.01944
83	-0.00740	-0.02067	0.02195	173	-0.01465	-0.01217	0.01905
84	0.01832	-0.02353	0.02982	174	-0.01326	-0.01327	0.01876
85	0.03542	-0.01606	0.03889	175	-0.01202	-0.01415	0.01856
86	0.03666	-0.00173	0.03670	176	-0.01097	-0.01482	0.01844
87	0.02187	0.01306	0.02547	177	-0.01013	-0.01532	0.01837
88	-0.00211	0.02178	0.02189	178	-0.00952	-0.01567	0.01833
89	-0.02455	0.02072	0.03213	179	-0.00915	-0.01586	0.01831
				180	-0.00902	-0.01593	0.01831

IEN RADIATION PATTERN FOR THETA=PI/2

H/LMDA=0.5000 B/LMDA= 0.250 OMEGA= 9.92

50 ELEMENT ARRAY

PHI DEG	ER(THETA,PHI) RE	IM	ABSVAL	PHI DEG	ER(THETA,PHI) RE	IM	ABSVAL
0	-0.14202	0.30772	0.33892	90	-0.02279	0.00082	0.02281
1	-0.14230	0.30742	0.33876	91	-0.01576	-0.01192	0.01976
2	-0.14312	0.30649	0.33826	92	-0.00176	-0.01927	0.01935
3	-0.14443	0.30486	0.33734	93	0.01293	-0.01793	0.02210
4	-0.14616	0.30240	0.33588	94	0.02177	-0.00853	0.02338
5	-0.14820	0.29896	0.33368	95	0.02091	0.00471	0.02143
6	-0.15040	0.29432	0.33052	96	0.01081	0.01583	0.01917
7	-0.15258	0.28825	0.32614	97	-0.00399	0.01991	0.02031
8	-0.15451	0.28049	0.32023	98	-0.01695	0.01515	0.02274
9	-0.15594	0.27077	0.31246	99	-0.02243	0.00370	0.02273
10	-0.15658	0.25884	0.30251	100	-0.01813	-0.00939	0.02041
11	-0.15609	0.24448	0.29006	101	-0.00601	-0.01838	0.01934
12	-0.15415	0.22750	0.27481	102	0.00863	-0.01941	0.02124
13	-0.15040	0.20784	0.25655	103	0.01950	-0.01209	0.02295
14	-0.14452	0.18551	0.23516	104	0.02207	0.00040	0.02207
15	-0.13623	0.16069	0.21066	105	0.01536	0.01274	0.01995
16	-0.12532	0.13372	0.18326	106	0.00229	0.01973	0.01987
17	-0.11170	0.10513	0.15339	107	-0.01166	0.01854	0.02190
18	-0.09542	0.07567	0.12178	108	-0.02079	0.00972	0.02295
19	-0.07674	0.04626	0.08960	109	-0.02148	-0.00305	0.02170
20	-0.05612	0.01798	0.05893	110	-0.01359	-0.01461	0.01995
21	-0.03424	-0.00795	0.03515	111	-0.00038	-0.02037	0.02037
22	-0.01205	-0.03029	0.03260	112	0.01291	-0.01815	0.02227
23	0.00936	-0.04788	0.04879	113	0.02119	-0.00892	0.02299
24	0.02871	-0.05974	0.06628	114	0.02141	0.00370	0.02173
25	0.04471	-0.06518	0.07904	115	0.01364	0.01493	0.02022
26	0.05615	-0.06400	0.08514	116	0.00090	0.02064	0.02066
27	0.06204	-0.05651	0.08392	117	-0.01211	0.01885	0.02241
28	0.06179	-0.04361	0.07563	118	-0.02075	0.01034	0.02319
29	0.05533	-0.02680	0.06148	119	-0.02212	-0.00179	0.02219
30	0.04324	-0.00805	0.04398	120	-0.01590	-0.01328	0.02072
31	0.02675	0.01034	0.02868	121	-0.00438	-0.02027	0.02073
32	0.00774	0.02605	0.02718	122	0.00852	-0.02054	0.02224
33	-0.01146	0.03701	0.03874	123	0.01860	-0.01418	0.02339
34	-0.02832	0.04173	0.05043	124	0.02277	-0.00334	0.02301
35	-0.04046	0.03961	0.05662	125	0.01990	0.00848	0.02164
36	-0.04605	0.03108	0.05555	126	0.01107	0.01770	0.02088
37	-0.04414	0.01766	0.04754	127	-0.00095	0.02167	0.02169
38	-0.03496	0.00175	0.03501	128	-0.01261	0.01939	0.02313
39	-0.02002	-0.01371	0.02427	129	-0.02068	0.01171	0.02377
40	-0.00193	-0.02577	0.02584	130	-0.02309	0.00088	0.02311
41	0.01597	-0.03206	0.03582	131	-0.01941	-0.01011	0.02189
42	0.03020	-0.03133	0.04351	132	-0.01080	-0.01843	0.02136
43	0.03784	-0.02376	0.04468	133	0.00040	-0.02210	0.02211
44	0.03723	-0.01109	0.03884	134	0.01141	-0.02042	0.02339
45	0.02844	0.00373	0.02868	135	0.01965	-0.01398	0.02412
46	0.01339	0.01715	0.02176	136	0.02341	-0.00445	0.02383
47	-0.00443	0.02586	0.02623	137	0.02206	0.00595	0.02284
48	-0.02074	0.02765	0.03457	138	0.01612	0.01500	0.02202
49	-0.03146	0.02206	0.03843	139	0.00703	0.02091	0.02206
50	-0.03380	0.01060	0.03542	140	-0.00330	0.02269	0.02293
51	-0.02707	-0.00353	0.02730	141	-0.01287	0.02022	0.02397
52	-0.01308	-0.01631	0.02091	142	-0.02002	0.01416	0.02452
53	0.00430	-0.02399	0.02438	143	-0.02365	0.00576	0.02434
54	0.02009	-0.02425	0.03149	144	-0.02339	-0.00346	0.02364
55	0.02961	-0.01701	0.03415	145	-0.01951	-0.01198	0.02290
56	0.02997	-0.00452	0.03031	146	-0.01283	-0.01858	0.02258
57	0.02102	0.00921	0.02295	147	-0.00449	-0.02244	0.02288
58	0.00556	0.01966	0.02043	148	0.00427	-0.02321	0.02360
59	-0.01143	0.02331	0.02596	149	0.01231	-0.02102	0.02436
60	-0.02434	0.01890	0.03081	150	0.01870	-0.01636	0.02484
61	-0.02881	0.00794	0.02988	151	0.02284	-0.00995	0.02491
62	-0.02329	-0.00570	0.02398	152	0.02448	-0.00264	0.02462
63	-0.00970	-0.01710	0.01966	153	0.02366	0.00475	0.02413
64	0.00714	-0.02205	0.02318	154	0.02070	0.01150	0.02367
65	0.02113	-0.01870	0.02821	155	0.01605	0.01705	0.02342
66	0.02712	-0.00827	0.02835	156	0.01027	0.02107	0.02344
67	0.02287	0.00524	0.02346	157	0.00394	0.02338	0.02371
68	0.00999	0.01656	0.01934	158	-0.00243	0.02400	0.02412
69	-0.00656	0.02123	0.02222	159	-0.00840	0.02308	0.02456
70	-0.02032	0.01734	0.02672	160	-0.01364	0.02086	0.02493
71	-0.02587	0.00645	0.02666	161	-0.01795	0.01764	0.02517
72	-0.02099	-0.00700	0.02213	162	-0.02122	0.01371	0.02526
73	-0.00769	-0.01746	0.01908	163	-0.02343	0.00938	0.02523
74	0.00860	-0.02054	0.02227	164	-0.02464	0.00490	0.02512
75	0.02114	-0.01492	0.02587	165	-0.02495	0.00050	0.02496
76	0.02472	-0.00298	0.02490	166	-0.02451	-0.00367	0.02479
77	0.01787	0.01019	0.02057	167	-0.02347	-0.00747	0.02463
78	0.00352	0.01888	0.01920	168	-0.02199	-0.01085	0.02452
79	-0.01218	0.01930	0.02282	169	-0.02021	-0.01377	0.02445
80	-0.02248	0.01126	0.02514	170	-0.01825	-0.01623	0.02443
81	-0.02295	-0.00172	0.02302	171	-0.01625	-0.01825	0.02443
82	-0.01343	-0.01389	0.01932	172	-0.01428	-0.01987	0.02447
83	0.00187	-0.01984	0.01993	173	-0.01242	-0.02115	0.02452
84	0.01621	-0.01691	0.02343	174	-0.01072	-0.02212	0.02458
85	0.02328	-0.00640	0.02414	175	-0.00924	-0.02284	0.02464
86	0.01999	0.00696	0.02116	176	-0.00800	-0.02337	0.02470
87	0.00786	0.01717	0.01888	177	-0.00702	-0.02373	0.02474
88	-0.00766	0.01961	0.02105	178	-0.00631	-0.02396	0.02478
89	-0.01965	0.01319	0.02367	179	-0.00588	-0.02409	0.02480
				180	-0.00573	-0.02413	0.02480

6. The Two-Element Array

The two-element broadside and unilateral endfire arrays are included in the tables of Section 5. In the former the distance between the elements is given the values $b/\lambda = 0.25$ and 0.5 with $V_2 = V_1 = 1$ assigned or $I_2 = I_1 = 1$ specified; in the latter $b/\lambda = 0.25$ and V_2 and V_1 or I_2 and I_1 are given magnitudes of one but shifted 90° in phase. In this section the formulas of Section 5 are specialized to $N = 2$, but the distance b between the elements is allowed to vary from $b/\lambda = 0.1$ to $b/\lambda = 3.0$ for the broadside array with $V_1 = V_2 = 1$ and $I_1 = I_2 = 1$ and the bilateral endfire array with $V_1 = -V_2 = 1$ and $I_1 = -I_2 = 1$. Since the driving-point admittances of both elements are the same in these cases, only values for element 1 are listed in Table 6.1 for the broadside array and in Table 6.2 for the bilateral endfire array. Self- and mutual admittances are given in Table 6.3; self- and mutual impedances in Table 6.4. Tables of radiation patterns are not required since the distributions of current in the two elements are necessarily identical as a result of symmetry so that the horizontal ($\theta = \pi/2$) field pattern is independent of the distribution of current. It has the form

$$f_B(\pi/2, \Phi) = -2C \cos\left(\frac{k_0 b}{2} \cos \Phi\right) \tag{6.1}$$

for the broadside array, and the form

$$f_E(\pi/2, \Phi) = 2jC \sin\left(\frac{k_0 b}{2} \cos \Phi\right) \tag{6.2}$$

for the bilateral endfire array, where

$$C = 2 - \frac{k_0 h \sin k_0 h}{1 - \cos k_0 h} + \frac{j\zeta_0 \Psi_{dR} Y_1}{2\pi} \times \left[\frac{\sin k_0 h - k_0 h \cos k_0 h}{1 - \cos k_0 h}\right] \tag{6.3}$$

and $Y_1 = I_1(0)/V_1$ is the driving-point admittance of element 1.

TABLE 6.1

DRIVING-POINT ADMITTANCES AND IMPEDANCES OF TWO-ELEMENT BROADSIDE ARRAYS

$h/\lambda = 0.250$; $\Omega = 8.53$

FOR SPECIFIED BASE VOLTAGES							FOR SPECIFIED BASE CURRENTS						
B/LMDA	V01		Y01		Z01		B/LMDA	IZ(0)1		Y01		Z01	
	RE	IM	RE	IM	RE	IM		RE	IM	RE	IM	RE	IM
0.100	1.00	0.0	6.094	-0.218	163.882	5.866	0.100	1.00	0.0	6.094	-0.218	160.293	24.676
0.125	1.00	0.0	6.425	-0.024	155.629	0.578	0.125	1.00	0.0	6.425	-0.024	153.573	17.779
0.150	1.00	0.0	6.769	0.177	147.637	-3.865	0.150	1.00	0.0	6.769	0.177	146.794	11.772
0.175	1.00	0.0	7.129	0.382	139.871	-7.502	0.175	1.00	0.0	7.129	0.382	139.960	6.629
0.200	1.00	0.0	7.511	0.589	132.323	-10.371	0.200	1.00	0.0	7.511	0.589	133.095	2.326
0.225	1.00	0.0	7.921	0.793	124.993	-12.510	0.225	1.00	0.0	7.921	0.793	126.234	-1.160
0.250	1.00	0.0	8.365	0.990	117.892	-13.952	0.250	1.00	0.0	8.365	0.990	119.419	-3.854
0.275	1.00	0.0	8.850	1.174	111.036	-14.730	0.275	1.00	0.0	8.850	1.174	112.694	-5.782
0.300	1.00	0.0	9.384	1.337	104.444	-14.876	0.300	1.00	0.0	9.384	1.337	106.104	-6.972
0.325	1.00	0.0	9.974	1.466	98.137	-14.420	0.325	1.00	0.0	9.974	1.466	99.699	-7.453
0.350	1.00	0.0	10.628	1.545	92.142	-13.392	0.350	1.00	0.0	10.628	1.545	93.525	-7.257
0.375	1.00	0.0	11.350	1.551	86.486	-11.822	0.375	1.00	0.0	11.350	1.551	87.632	-6.419
0.400	1.00	0.0	12.141	1.456	81.199	-9.741	0.400	1.00	0.0	12.141	1.456	82.066	-4.978
0.425	1.00	0.0	12.989	1.222	76.314	-7.182	0.425	1.00	0.0	12.989	1.222	76.875	-2.973
0.450	1.00	0.0	13.869	0.807	71.862	-4.180	0.450	1.00	0.0	13.869	0.807	72.103	-0.451
0.475	1.00	0.0	14.730	0.168	67.880	-0.775	0.475	1.00	0.0	14.730	0.168	67.794	2.540
0.500	1.00	0.0	15.494	-0.719	64.402	2.989	0.500	1.00	0.0	15.494	-0.719	63.988	5.943
0.525	1.00	0.0	16.058	-1.846	61.462	7.064	0.525	1.00	0.0	16.C58	-1.846	60.724	9.702
0.550	1.00	0.0	16.316	-3.146	59.092	11.393	0.550	1.00	0.0	16.316	-3.146	58.031	13.749
0.575	1.00	0.0	16.197	-4.497	57.323	15.914	0.575	1.00	0.0	16.197	-4.497	55.938	18.016
0.600	1.00	0.0	15.699	-5.745	56.177	20.557	0.600	1.00	0.0	15.699	-5.745	54.463	22.428
0.625	1.00	0.0	14.899	-6.757	55.671	25.247	0.625	1.00	0.0	14.899	-6.757	53.618	26.907
0.650	1.00	0.0	13.922	-7.458	55.812	29.899	0.650	1.00	0.0	13.922	-7.458	53.402	31.370
0.675	1.00	0.0	12.897	-7.845	56.595	34.428	0.675	1.00	0.0	12.897	-7.845	53.805	35.734
0.700	1.00	0.0	11.922	-7.963	58.003	38.741	0.700	1.00	0.0	11.922	-7.963	54.807	39.917
0.725	1.00	0.0	11.055	-7.876	60.001	42.751	0.725	1.00	0.0	11.055	-7.876	56.372	43.837
0.750	1.00	0.0	10.318	-7.650	62.541	46.368	0.750	1.00	0.0	10.318	-7.650	58.454	47.417
0.775	1.00	0.0	9.713	-7.336	65.557	49.513	0.775	1.00	0.0	9.713	-7.336	60.995	50.589
C.800	1.00	0.0	9.229	-6.974	68.968	52.118	0.800	1.00	0.0	9.229	-6.974	63.923	53.292
0.825	1.00	0.0	8.850	-6.591	72.682	54.126	0.825	1.00	0.0	8.850	-6.591	67.161	55.479
C.850	1.00	0.0	8.561	-6.203	76.594	55.502	0.850	1.00	0.0	8.561	-6.203	70.622	57.114
0.875	1.00	0.0	8.346	-5.822	80.596	56.226	0.875	1.00	0.0	8.346	-5.822	74.216	58.178
0.900	1.00	0.0	8.193	-5.454	84.579	56.303	0.900	1.00	0.0	8.193	-5.454	77.850	58.665
0.925	1.00	0.0	8.092	-5.101	88.435	55.755	0.925	1.00	0.0	8.092	-5.101	81.435	58.588
0.950	1.00	0.0	8.034	-4.767	92.065	54.626	0.950	1.00	0.0	8.034	-4.767	84.883	57.973
0.975	1.00	0.0	8.012	-4.450	95.383	52.975	0.975	1.00	0.0	8.012	-4.450	88.117	56.858
1.000	1.00	0.0	8.023	-4.152	98.316	50.875	1.000	1.00	0.0	8.023	-4.152	91.066	55.296
1.025	1.00	0.0	8.061	-3.871	100.810	48.406	1.025	1.00	0.0	8.061	-3.871	93.673	53.346
1.050	1.00	0.0	8.124	-3.607	102.825	45.656	1.050	1.00	0.0	8.124	-3.607	95.891	51.076
1.075	1.00	0.0	8.209	-3.360	104.340	42.712	1.075	1.00	0.0	8.209	-3.360	97.688	48.558
1.100	1.00	0.0	8.314	-3.130	105.349	39.662	1.100	1.00	0.0	8.314	-3.130	99.042	45.864
1.125	1.00	0.0	8.438	-2.916	105.862	36.586	1.125	1.00	0.0	8.438	-2.916	99.946	43.070
1.150	1.00	0.0	8.581	-2.720	105.898	33.561	1.150	1.00	0.0	8.581	-2.720	100.404	40.244
1.175	1.00	0.0	8.742	-2.540	105.489	30.652	1.175	1.00	0.0	8.742	-2.540	100.428	37.453
1.200	1.00	0.0	8.919	-2.379	104.673	27.919	1.200	1.00	0.0	8.919	-2.379	100.042	34.760
1.225	1.00	0.0	9.113	-2.237	103.494	25.409	1.225	1.00	0.0	9.113	-2.237	99.277	32.217
1.250	1.00	0.0	9.323	-2.117	101.999	23.164	1.250	1.00	0.0	9.323	-2.117	98.167	29.875
1.275	1.00	0.0	9.549	-2.021	100.238	21.214	1.275	1.00	0.0	9.549	-2.021	96.755	27.773
1.300	1.00	0.0	9.788	-1.951	98.262	19.584	1.300	1.00	0.0	9.788	-1.951	95.086	25.945
1.325	1.00	0.0	10.040	-1.910	96.122	18.289	1.325	1.00	0.0	10.C40	-1.910	93.205	24.418
1.350	1.00	0.0	10.302	-1.903	93.870	17.338	1.350	1.00	0.0	10.302	-1.903	91.163	23.210
[illegible]	[illegible]	[illegible]	[illegible]	[illegible]	[illegible]	[illegible]	[illegible]	[illegible]	[illegible]	[illegible]	[illegible]	[illegible]	[illegible]
1.400	1.00	0.0	10.839	-2.002	89.220	16.476	1.400	1.00	0.0	10.839	-2.002	86.790	21.793
1.425	1.00	0.0	11.103	-2.115	86.916	16.554	1.425	1.00	0.0	11.103	-2.115	84.557	21.589
1.450	1.00	0.0	11.353	-2.273	84.685	16.957	1.450	1.00	0.0	11.353	-2.273	82.355	21.714
1.475	1.00	0.0	11.581	-2.478	82.568	17.667	1.475	1.00	0.0	11.581	-2.478	80.230	22.155
1.500	1.00	0.0	11.775	-2.726	80.603	18.662	1.500	1.00	0.0	11.775	-2.726	78.223	22.894
1.525	1.00	0.0	11.925	-3.013	78.826	19.916	1.525	1.00	0.0	11.925	-3.013	76.374	23.907
1.550	1.00	0.0	12.020	-3.329	77.269	21.399	1.550	1.00	0.0	12.020	-3.329	74.718	25.168
1.575	1.00	0.0	12.052	-3.662	75.961	23.078	1.575	1.00	0.0	12.052	-3.662	73.286	26.643
1.600	1.00	0.0	12.018	-3.996	74.925	24.914	1.600	1.00	0.0	12.018	-3.996	72.105	28.296
1.625	1.00	0.0	11.917	-4.316	74.181	26.868	1.625	1.00	0.0	11.917	-4.316	71.197	30.089
1.650	1.00	0.0	11.756	-4.606	73.742	28.895	1.650	1.00	0.0	11.756	-4.606	70.576	31.977
-1.675	1.00	0.0	11.544	-4.853	73.615	30.951	1.675	1.00	0.0	11.544	-4.853	70.252	33.919
1.700	1.00	0.0	11.294	-5.048	73.802	32.987	1.700	1.00	0.0	11.294	-5.048	70.228	35.867
1.725	1.00	0.0	11.020	-5.185	74.296	34.957	1.725	1.00	0.0	11.020	-5.185	70.501	37.778
1.750	1.00	0.0	10.737	-5.264	75.084	36.814	1.750	1.00	0.0	10.737	-5.264	71.059	39.606
1.775	1.00	0.0	10.458	-5.289	76.144	38.513	1.775	1.00	0.0	10.458	-5.289	71.886	41.309
1.800	1.00	0.0	10.192	-5.266	77.447	40.014	1.800	1.00	0.0	10.192	-5.266	72.956	42.848
1.825	1.00	0.0	9.946	-5.200	78.959	41.280	1.825	1.00	0.0	9.946	-5.200	74.240	44.188
1.850	1.00	0.0	9.727	-5.100	80.638	42.282	1.850	1.00	0.0	9.727	-5.100	75.703	45.299
1.875	1.00	0.0	9.536	-4.974	82.439	42.998	1.875	1.00	0.0	9.536	-4.974	77.306	46.157
1.900	1.00	0.0	9.375	-4.827	84.312	43.412	1.900	1.00	0.0	9.375	-4.827	79.005	46.747
1.925	1.00	0.0	9.244	-4.667	86.206	43.520	1.925	1.00	0.0	9.244	-4.667	80.755	47.058
1.950	1.00	0.0	9.142	-4.497	88.071	43.325	1.950	1.00	0.0	9.142	-4.497	82.512	47.089
1.975	1.00	0.0	9.068	-4.323	89.859	42.841	1.975	1.00	0.0	9.068	-4.323	84.230	46.845
2.000	1.00	0.0	9.019	-4.147	91.523	42.086	2.000	1.00	0.0	9.019	-4.147	85.868	46.340
2.025	1.C0	0.0	8.995	-3.973	93.024	41.089	2.025	1.00	0.0	8.995	-3.973	87.385	45.593
2.050	1.00	0.0	8.994	-3.803	94.327	39.884	2.050	1.00	0.0	8.994	-3.803	88.748	44.629
2.075	1.00	0.0	9.013	-3.638	95.406	38.508	2.075	1.00	0.0	9.013	-3.638	89.926	43.480
2.100	1.00	0.0	9.053	-3.481	96.239	37.003	2.100	1.00	0.0	9.053	-3.481	90.895	42.178
2.125	1.00	0.0	9.110	-3.332	96.816	35.413	2.125	1.00	0.0	9.110	-3.332	91.638	40.761
2.150	1.00	0.0	9.185	-3.194	97.130	33.778	2.150	1.00	0.0	9.185	-3.194	92.143	39.267
2.175	1.00	0.0	9.275	-3.068	97.185	32.143	2.175	1.00	0.0	9.275	-3.068	92.406	37.735
2.200	1.00	0.0	9.380	-2.954	96.989	30.545	2.200	1.00	0.0	9.380	-2.954	92.428	36.203
2.225	1.00	0.0	9.499	-2.855	96.556	29.022	2.225	1.00	0.0	9.499	-2.855	92.216	34.708
2.250	1.00	0.0	9.629	-2.772	95.905	27.607	2.250	1.00	0.0	9.629	-2.772	91.783	33.283
2.275	1.00	0.0	9.770	-2.706	95.059	26.329	2.275	1.C0	0.0	9.770	-2.706	91.144	31.961
2.300	1.00	0.0	9.920	-2.659	94.044	25.211	2.300	1.00	0.0	9.920	-2.659	90.322	30.768
2.325	1.00	0.0	10.077	-2.633	92.889	24.274	2.325	1.00	0.0	10.077	-2.633	89.339	29.729
2.350	1.00	0.0	10.239	-2.630	91.623	23.533	2.350	1.00	0.0	10.239	-2.630	88.224	28.864
2.375	1.00	0.0	10.402	-2.650	90.278	22.997	2.375	1.00	0.0	10.402	-2.650	87.005	28.186
2.400	1.00	0.0	10.563	-2.695	88.885	22.675	2.400	1.00	0.0	10.563	-2.695	85.713	27.708
2.425	1.00	0.0	10.718	-2.765	87.477	22.566	2.425	1.00	0.0	10.718	-2.765	84.378	27.434
2.450	1.00	0.0	10.863	-2.861	86.083	22.668	2.450	1.00	0.0	10.863	-2.861	83.031	27.368
2.475	1.00	0.0	10.993	-2.981	84.733	22.975	2.475	1.00	0.0	10.993	-2.981	81.704	27.505
2.500	1.00	0.0	11.104	-3.124	83.457	23.477	2.500	1.00	0.0	11.104	-3.124	80.425	27.839
2.525	1.00	0.0	11.189	-3.285	82.280	24.160	2.525	1.00	0.0	11.189	-3.285	79.222	28.360
2.550	1.00	0.0	11.245	-3.462	81.228	25.005	2.550	1.00	0.0	11.245	-3.462	78.123	29.051
2.575	1.00	0.0	11.270	-3.647	80.322	25.994	2.575	1.00	0.0	11.270	-3.647	77.149	29.896
2.600	1.00	0.0	11.260	-3.835	79.581	27.101	2.600	1.00	0.0	11.260	-3.835	76.323	30.872
2.625	1.00	0.0	11.216	-4.017	79.020	28.301	2.625	1.00	0.0	11.216	-4.017	75.660	31.955
2.650	1.00	0.0	11.140	-4.188	78.651	29.566	2.650	1.00	0.0	11.140	-4.188	75.176	33.119
2.675	1.00	0.0	11.035	-4.340	78.482	30.867	2.675	1.00	0.0	11.035	-4.340	74.878	34.335
2.700	1.00	0.0	10.905	-4.468	78.516	32.172	2.700	1.00	0.0	10.905	-4.468	74.773	35.575
2.725	1.00	0.0	10.758	-4.569	78.750	33.450	2.725	1.00	0.0	10.758	-4.569	74.860	36.808
2.750	1.00	0.0	10.598	-4.641	79.178	34.671	2.750	1.00	0.0	10.598	-4.641	75.136	38.005
2.775	1.00	0.0	10.432	-4.681	79.789	35.804	2.775	1.00	0.0	10.432	-4.681	75.592	39.138
2.800	1.00	0.0	10.267	-4.693	80.567	36.823	2.800	1.00	0.0	10.267	-4.693	76.216	40.179
2.825	1.00	0.0	10.108	-4.676	81.491	37.701	2.825	1.00	0.0	10.108	-4.676	76.990	41.103
2.850	1.00	0.0	9.958	-4.635	82.537	38.419	2.850	1.00	0.0	9.958	-4.635	77.893	41.889
2.875	1.00	0.0	9.822	-4.573	83.677	38.958	2.875	1.00	0.0	9.822	-4.573	78.902	42.519
2.900	1.00	0.0	9.701	-4.492	84.879	39.306	2.900	1.00	0.0	9.701	-4.492	79.989	42.979
2.925	1.00	0.0	9.597	-4.398	86.114	39.458	2.925	1.00	0.0	9.597	-4.398	81.127	43.259
2.950	1.00	0.0	9.512	-4.292	87.347	39.411	2.950	1.00	0.0	9.512	-4.292	82.286	43.356
2.975	1.00	0.0	9.445	-4.178	88.548	39.171	2.975	1.00	0.0	9.445	-4.178	83.436	43.270
3.000	1.00	0.0	9.397	-4.059	89.684	38.745	3.000	1.00	0.0	9.397	-4.059	84.549	43.005

TABLE 6.1

DRIVING-POINT ADMITTANCES AND IMPEDANCES OF TWO-ELEMENT BROADSIDE ARRAYS

$h/\lambda = 0.500$; $\Omega = 9.92$

	FOR SPECIFIED BASE VOLTAGES						FOR SPECIFIED BASE CURRENTS							
B/LMDA	V01		Y01		Z01		B/LMDA	IZ(0)1		Y01		Z01		
	RE	IM	RE	IM	RE	IM		RE	IM	RE	IM	RE	IM	
0.100	1.00	0.0	1.288	2.262	190.178	-333.825	0.100	1.00	0.0	1.288	2.262	319.192	-381.913	
0.125	1.00	0.0	1.356	2.250	196.523	-326.015	0.125	1.00	0.0	1.356	2.250	324.478	-366.018	
0.150	1.00	0.0	1.422	2.220	204.583	-319.373	0.150	1.00	0.0	1.422	2.220	332.830	-351.077	
0.175	1.00	0.0	1.485	2.175	214.103	-313.573	0.175	1.00	0.0	1.485	2.175	343.559	-336.608	
0.200	1.00	0.0	1.543	2.116	225.002	-308.500	0.200	1.00	0.0	1.543	2.116	356.387	-322.369	
0.225	1.00	0.0	1.594	2.044	237.298	-304.181	0.225	1.00	0.0	1.594	2.044	371.279	-308.258	
0.250	1.00	0.0	1.636	1.959	251.071	-300.758	0.250	1.00	0.0	1.636	1.959	388.366	-294.277	
0.275	1.00	0.0	1.664	1.865	266.433	-298.492	0.275	1.00	0.0	1.664	1.865	407.915	-280.532	
0.300	1.00	0.0	1.677	1.762	283.497	-297.786	0.300	1.00	0.0	1.677	1.762	430.308	-267.252	
0.325	1.00	0.0	1.671	1.654	302.326	-299.207	0.325	1.00	0.0	1.671	1.654	456.042	-254.837	
0.350	1.00	0.0	1.644	1.546	322.853	-303.525	0.350	1.00	0.0	1.644	1.546	485.699	-243.929	
0.375	1.00	0.0	1.596	1.443	344.758	-311.717	0.375	1.00	0.0	1.596	1.443	519.907	-235.522	
0.400	1.00	0.0	1.527	1.351	367.280	-324.921	0.400	1.00	0.0	1.527	1.351	559.219	-231.105	
0.425	1.00	0.0	1.442	1.276	389.000	-344.252	0.425	1.00	0.0	1.442	1.276	603.890	-232.821	
0.450	1.00	0.0	1.344	1.221	407.675	-370.413	0.450	1.00	0.0	1.344	1.221	653.439	-243.571	
0.475	1.00	0.0	1.239	1.189	420.323	-403.080	0.475	1.00	0.0	1.239	1.189	705.977	-266.882	
0.500	1.00	0.0	1.135	1.179	423.842	-440.268	0.500	1.00	0.0	1.135	1.179	757.367	-306.212	
0.525	1.00	0.0	1.036	1.190	416.168	-478.190	0.525	1.00	0.0	1.036	1.190	800.736	-363.361	
0.550	1.00	0.0	0.946	1.219	397.470	-512.110	0.550	1.00	0.0	0.946	1.219	827.350	-436.161	
0.575	1.00	0.0	0.868	1.261	370.431	-537.984	0.575	1.00	0.0	0.868	1.261	829.676	-516.935	
0.600	1.00	0.0	0.804	1.313	339.224	-553.848	0.600	1.00	0.0	0.804	1.313	805.507	-593.967	
0.625	1.00	0.0	0.754	1.371	307.931	-560.047	0.625	1.00	0.0	0.754	1.371	759.787	-656.194	
0.650	1.00	0.0	0.717	1.432	279.438	-558.436	0.650	1.00	0.0	0.717	1.432	702.112	-697.701	
0.675	1.00	0.0	0.691	1.494	255.198	-551.346	0.675	1.00	0.0	0.691	1.494	642.155	-718.627	
0.700	1.00	0.0	0.677	1.554	235.561	-540.883	0.700	1.00	0.0	0.677	1.554	586.624	-722.920	
0.725	1.00	0.0	0.672	1.612	220.237	-528.649	0.725	1.00	0.0	0.672	1.612	538.783	-715.585	
0.750	1.00	0.0	0.674	1.666	208.656	-515.736	0.750	1.00	0.0	0.674	1.666	499.419	-701.014	
0.775	1.00	0.0	0.683	1.717	200.190	-502.827	0.775	1.00	0.0	0.683	1.717	467.979	-682.442	
0.800	1.00	0.0	0.698	1.763	194.260	-490.322	0.800	1.00	0.0	0.698	1.763	443.383	-662.021	
0.825	1.00	0.0	0.718	1.804	190.376	-478.430	0.825	1.00	0.0	0.718	1.804	424.464	-641.080	
0.850	1.00	0.0	0.742	1.842	188.141	-467.245	0.850	1.00	0.0	0.742	1.842	410.168	-620.394	
0.875	1.00	0.0	0.768	1.874	187.242	-456.794	0.875	1.00	0.0	0.768	1.874	399.610	-600.381	
0.900	1.00	0.0	0.798	1.902	187.437	-447.064	0.900	1.00	0.0	0.798	1.902	392.084	-581.245	
0.925	1.00	0.0	0.829	1.926	188.539	-438.024	0.925	1.00	0.0	0.829	1.926	387.036	-563.062	
0.950	1.00	0.0	0.862	1.945	190.410	-429.636	0.950	1.00	0.0	0.862	1.945	384.040	-545.833	
0.975	1.00	0.0	0.897	1.960	192.945	-421.862	0.975	1.00	0.0	0.897	1.960	382.769	-529.527	
1.000	1.00	0.0	0.932	1.971	196.065	-414.667	1.000	1.00	0.0	0.932	1.971	382.979	-514.089	
1.025	1.00	0.0	0.968	1.977	199.715	-408.022	1.025	1.00	0.0	0.968	1.977	384.489	-499.465	
1.050	1.00	0.0	1.004	1.979	203.855	-401.907	1.050	1.00	0.0	1.004	1.979	387.165	-485.600	
1.075	1.00	0.0	1.040	1.976	208.460	-396.312	1.075	1.00	0.0	1.040	1.976	390.916	-472.449	
1.100	1.00	0.0	1.075	1.969	213.512	-391.235	1.100	1.00	0.0	1.075	1.969	395.684	-459.978	
1.125	1.00	0.0	1.109	1.958	219.002	-386.688	1.125	1.00	0.0	1.109	1.958	401.435	-448.164	
1.150	1.00	0.0	1.141	1.942	224.924	-382.693	1.150	1.00	0.0	1.141	1.942	408.159	-437.005	
1.175	1.00	0.0	1.172	1.922	231.274	-379.287	1.175	1.00	0.0	1.172	1.922	415.863	-426.513	
1.200	1.00	0.0	1.200	1.897	238.046	-376.523	1.200	1.00	0.0	1.200	1.897	424.570	-416.728	
1.225	1.00	0.0	1.224	1.869	245.226	-374.468	1.225	1.00	0.0	1.224	1.869	434.311	-407.711	
1.250	1.00	0.0	1.244	1.837	252.792	-373.209	1.250	1.00	0.0	1.244	1.837	445.125	-399.556	
1.275	1.00	0.0	1.260	1.801	260.700	-372.848	1.275	1.00	0.0	1.260	1.801	457.050	-392.396	
1.300	1.00	0.0	1.269	1.763	268.883	-373.504	1.300	1.00	0.0	1.269	1.763	470.114	-386.405	
1.325	1.00	0.0	1.273	1.724	277.239	-375.309	1.325	1.00	0.0	1.273	1.724	484.327	-381.803	
1.350	1.00	0.0	1.271	1.684	285.614	-378.397	1.350	1.00	0.0	1.271	1.684	499.654	-378.869	
1.375	1.00	0.0	1.261	1.644	293.802	-382.896	1.375	1.00	0.0	1.261	1.644	516.003	-377.932	
1.400	1.00	0.0	1.245	1.606	301.530	-388.901	1.400	1.00	0.0	1.245	1.606	533.189	-379.371	
1.425	1.00	0.0	1.222	1.571	308.457	-396.448	1.425	1.00	0.0	1.222	1.571	550.898	-383.593	
1.450	1.00	0.0	1.194	1.541	314.188	-405.485	1.450	1.00	0.0	1.194	1.541	568.655	-390.997	
1.475	1.00	0.0	1.161	1.516	318.298	-415.828	1.475	1.00	0.0	1.161	1.516	585.788	-401.911	
1.500	1.00	0.0	1.124	1.498	320.385	-427.147	1.500	1.00	0.0	1.124	1.498	601.430	-416.504	
1.525	1.00	0.0	1.085	1.487	320.134	-438.951	1.525	1.00	0.0	1.085	1.487	614.550	-434.675	
1.550	1.00	0.0	1.045	1.483	317.385	-450.630	1.550	1.00	0.0	1.045	1.483	624.057	-455.958	
1.575	1.00	0.0	1.006	1.487	312.191	-461.517	1.575	1.00	0.0	1.006	1.487	628.963	-479.460	
1.600	1.00	0.0	0.969	1.496	304.831	-470.981	1.600	1.00	0.0	0.969	1.496	628.582	-503.895	
1.625	1.00	0.0	0.935	1.512	295.780	-478.524	1.625	1.00	0.0	0.935	1.512	622.715	-527.733	
1.650	1.00	0.0	0.905	1.533	285.635	-483.840	1.650	1.00	0.0	0.905	1.533	611.738	-549.435	
1.675	1.00	0.0	0.880	1.557	275.025	-486.842	1.675	1.00	0.0	0.880	1.557	596.547	-567.713	
1.700	1.00	0.0	0.860	1.584	264.524	-487.637	1.700	1.00	0.0	0.860	1.584	578.388	-581.719	
1.725	1.00	0.0	0.845	1.614	254.599	-486.473	1.725	1.00	0.0	0.845	1.614	558.619	-591.116	
1.750	1.00	0.0	0.835	1.644	245.580	-483.677	1.750	1.00	0.0	0.835	1.644	538.499	-596.026	
1.775	1.00	0.0	0.830	1.674	237.670	-479.604	1.775	1.00	0.0	0.830	1.674	519.048	-596.900	
1.800	1.00	0.0	0.829	1.704	230.964	-474.586	1.800	1.00	0.0	0.829	1.704	500.994	-594.371	
1.825	1.00	0.0	0.833	1.732	225.473	-468.923	1.825	1.00	0.0	0.833	1.732	484.787	-589.125	
1.850	1.00	0.0	0.840	1.759	221.156	-462.863	1.850	1.00	0.0	0.840	1.759	470.649	-581.816	
1.875	1.00	0.0	0.851	1.784	217.941	-456.604	1.875	1.00	0.0	0.851	1.784	458.637	-573.017	
1.900	1.00	0.0	0.865	1.806	215.739	-450.302	1.900	1.00	0.0	0.865	1.806	448.699	-563.205	
1.925	1.00	0.0	0.882	1.826	214.458	-444.074	1.925	1.00	0.0	0.882	1.826	440.721	-552.757	
1.950	1.00	0.0	0.901	1.843	214.006	-438.008	1.950	1.00	0.0	0.901	1.843	434.560	-541.966	
1.975	1.00	0.0	0.921	1.857	214.303	-432.169	1.975	1.00	0.0	0.921	1.857	430.063	-531.055	
2.000	1.00	0.0	0.943	1.868	215.274	-426.606	2.000	1.00	0.0	0.943	1.868	427.084	-520.189	
2.025	1.00	0.0	0.966	1.876	216.856	-421.358	2.025	1.00	0.0	0.966	1.876	425.490	-509.491	
2.050	1.00	0.0	0.989	1.881	218.993	-416.454	2.050	1.00	0.0	0.989	1.881	425.166	-499.056	
2.075	1.00	0.0	1.013	1.883	221.637	-411.922	2.075	1.00	0.0	1.013	1.883	426.012	-488.956	
2.100	1.00	0.0	1.037	1.881	224.747	-407.788	2.100	1.00	0.0	1.037	1.881	427.946	-479.250	
2.125	1.00	0.0	1.060	1.876	228.286	-404.080	2.125	1.00	0.0	1.060	1.876	430.902	-469.992	
2.150	1.00	0.0	1.082	1.868	232.221	-400.826	2.150	1.00	0.0	1.082	1.868	434.828	-461.232	
2.175	1.00	0.0	1.103	1.857	236.519	-398.062	2.175	1.00	0.0	1.103	1.857	439.681	-453.026	
2.200	1.00	0.0	1.123	1.843	241.145	-395.824	2.200	1.00	0.0	1.123	1.843	445.426	-445.432	
2.225	1.00	0.0	1.140	1.826	246.060	-394.158	2.225	1.00	0.0	1.140	1.826	452.035	-438.524	
2.250	1.00	0.0	1.154	1.806	251.219	-393.111	2.250	1.00	0.0	1.154	1.806	459.477	-432.382	
2.275	1.00	0.0	1.166	1.785	256.565	-392.735	2.275	1.00	0.0	1.166	1.785	467.719	-427.107	
2.300	1.00	0.0	1.174	1.761	262.028	-393.086	2.300	1.00	0.0	1.174	1.761	476.718	-422.813	
2.325	1.00	0.0	1.179	1.737	267.522	-394.216	2.325	1.00	0.0	1.179	1.737	486.410	-419.633	
2.350	1.00	0.0	1.179	1.712	272.938	-396.174	2.350	1.00	0.0	1.179	1.712	496.708	-417.713	
2.375	1.00	0.0	1.176	1.687	278.146	-398.993	2.375	1.00	0.0	1.176	1.687	507.489	-417.216	
2.400	1.00	0.0	1.168	1.662	282.995	-402.687	2.400	1.00	0.0	1.168	1.662	518.582	-418.303	
2.425	1.00	0.0	1.157	1.640	287.310	-407.240	2.425	1.00	0.0	1.157	1.640	529.759	-421.133	
2.450	1.00	0.0	1.141	1.619	290.906	-412.591	2.450	1.00	0.0	1.141	1.619	540.731	-425.834	
2.475	1.00	0.0	1.123	1.601	293.596	-418.633	2.475	1.00	0.0	1.123	1.601	551.142	-432.486	
2.500	1.00	0.0	1.102	1.587	295.206	-425.201	2.500	1.00	0.0	1.102	1.587	560.576	-441.089	
2.525	1.00	0.0	1.079	1.577	295.597	-432.076	2.525	1.00	0.0	1.079	1.577	568.579	-451.535	
2.550	1.00	0.0	1.054	1.570	294.688	-438.992	2.550	1.00	0.0	1.054	1.570	574.693	-463.582	
2.575	1.00	0.0	1.029	1.568	292.466	-445.655	2.575	1.00	0.0	1.029	1.568	578.502	-476.842	
2.600	1.00	0.0	1.005	1.571	289.006	-451.773	2.600	1.00	0.0	1.005	1.571	579.692	-490.790	
2.625	1.00	0.0	0.981	1.577	284.461	-457.077	2.625	1.00	0.0	0.981	1.577	578.102	-504.804	
2.650	1.00	0.0	0.960	1.587	279.052	-461.355	2.650	1.00	0.0	0.960	1.587	573.759	-518.217	
2.675	1.00	0.0	0.941	1.600	273.043	-464.466	2.675	1.00	0.0	0.941	1.600	566.889	-530.391	
2.700	1.00	0.0	0.924	1.616	266.717	-466.347	2.700	1.00	0.0	0.924	1.616	557.893	-540.793	
2.725	1.00	0.0	0.911	1.634	260.348	-467.014	2.725	1.00	0.0	0.911	1.634	547.292	-549.038	
2.750	1.00	0.0	0.900	1.653	254.176	-466.546	2.750	1.00	0.0	0.900	1.653	535.667	-554.920	
2.775	1.00	0.0	0.894	1.673	248.398	-465.066	2.775	1.00	0.0	0.894	1.673	523.589	-558.404	
2.800	1.00	0.0	0.890	1.693	243.163	-462.725	2.800	1.00	0.0	0.890	1.693	511.572	-559.596	
2.825	1.00	0.0	0.889	1.714	238.571	-459.684	2.825	1.00	0.0	0.889	1.714	500.040	-558.709	
2.850	1.00	0.0	0.892	1.734	234.679	-456.104	2.850	1.00	0.0	0.892	1.734	489.317	-556.017	
2.875	1.00	0.0	0.897	1.752	231.511	-452.130	2.875	1.00	0.0	0.897	1.752	479.626	-551.817	
2.900	1.00	0.0	0.905	1.770	229.065	-447.896	2.900	1.00	0.0	0.905	1.770	471.106	-546.406	
2.925	1.00	0.0	0.915	1.786	227.321	-443.517	2.925	1.00	0.0	0.915	1.786	463.828	-540.061	
2.950	1.00	0.0	0.927	1.800	226.247	-439.091	2.950	1.00	0.0	0.927	1.800	457.810	-533.028	
2.975	1.00	0.0	0.941	1.812	225.804	-434.697	2.975	1.00	0.0	0.941	1.812	453.035	-525.522	
3.000	1.00	0.0	0.956	1.821	225.952	-430.406	3.000	1.00	0.0	0.956	1.821	449.463	-517.724	

TABLE 6.2

DRIVING-POINT ADMITTANCES AND IMPEDANCES OF TWO-ELEMENT BILATERAL ENDFIRE ARRAYS

$h/\lambda = 0.250$; $\Omega = 8.53$

FOR SPECIFIED BASE VOLTAGES							FOR SPECIFIED BASE CURRENTS					
B/LMCA	V01		Y01		Z01		B/LMCA	IZ(0)1		Y01		Z01	
	RE	IM	RE	IM	RE	IM		RE	IM	RE	IM	RE	IM
0.100	1.00	0.0	5.204	-27.360	6.709	35.274	0.100	1.00	0.0	5.204	-27.360	6.381	34.431
0.125	1.00	0.0	5.170	-21.326	10.736	44.288	0.125	1.00	0.0	5.170	-21.327	10.082	42.995
0.150	1.00	0.0	5.188	-17.404	15.731	52.770	0.150	1.00	0.0	5.188	-17.404	14.598	50.998
0.175	1.00	0.0	5.236	-14.643	21.652	60.550	0.175	1.00	0.0	5.236	-14.643	19.877	58.317
0.200	1.00	0.0	5.304	-12.589	28.422	67.458	0.200	1.00	0.0	5.304	-12.589	25.840	64.838
0.225	1.00	0.0	5.386	-10.997	35.920	73.343	0.225	1.00	0.0	5.386	-10.997	32.387	70.460
0.250	1.00	0.0	5.478	-9.721	43.994	78.077	0.250	1.00	0.0	5.478	-9.721	39.400	75.103
0.275	1.00	0.0	5.578	-8.673	52.458	81.564	0.275	1.00	0.0	5.578	-8.673	46.740	78.708
0.300	1.00	0.0	5.685	-7.792	61.108	83.752	0.300	1.00	0.0	5.685	-7.792	54.261	81.240
0.325	1.00	0.0	5.799	-7.038	69.727	84.630	0.325	1.00	0.0	5.799	-7.038	61.809	82.694
0.350	1.00	0.0	5.919	-6.384	78.102	84.238	0.350	1.00	0.0	5.919	-6.384	69.230	83.091
0.375	1.00	0.0	6.044	-5.807	86.033	82.657	0.375	1.00	0.0	6.044	-5.807	76.378	82.480
0.400	1.00	0.0	6.176	-5.293	93.346	80.008	0.400	1.00	0.0	6.176	-5.293	83.118	80.932
0.425	1.00	0.0	6.314	-4.831	99.898	76.440	0.425	1.00	0.0	6.314	-4.831	89.332	78.543
0.450	1.00	0.0	6.458	-4.411	105.583	72.121	0.450	1.00	0.0	6.458	-4.411	94.921	75.421
0.475	1.00	0.0	6.609	-4.027	110.336	67.229	0.475	1.00	0.0	6.609	-4.027	99.810	71.688
0.500	1.00	0.0	6.769	-3.674	114.125	61.939	0.500	1.00	0.0	6.769	-3.673	103.945	67.471
0.525	1.00	0.0	6.936	-3.346	116.953	56.419	0.525	1.00	0.0	6.936	-3.346	107.298	62.899
0.550	1.00	0.0	7.113	-3.042	118.850	50.823	0.550	1.00	0.0	7.113	-3.042	109.858	58.097
0.575	1.00	0.0	7.301	-2.758	119.867	45.286	0.575	1.00	0.0	7.301	-2.758	111.637	53.187
0.600	1.00	0.0	7.499	-2.494	120.072	39.926	0.600	1.00	0.0	7.499	-2.494	112.659	48.278
0.625	1.00	0.0	7.710	-2.247	119.542	34.839	0.625	1.00	0.0	7.710	-2.247	112.966	43.471
0.650	1.00	0.0	7.936	-2.018	118.361	30.101	0.650	1.00	0.0	7.936	-2.018	112.609	38.855
0.675	1.00	0.0	8.176	-1.807	116.615	25.772	0.675	1.00	0.0	8.176	-1.807	111.646	34.506
0.700	1.00	0.0	8.433	-1.614	114.390	21.896	0.700	1.00	0.0	8.433	-1.614	110.143	30.487
0.725	1.00	0.0	8.708	-1.442	111.770	18.504	0.725	1.00	0.0	8.708	-1.442	108.168	26.851
0.750	1.00	0.0	9.003	-1.292	108.833	15.616	0.750	1.00	0.0	9.003	-1.292	105.791	23.640
0.775	1.00	0.0	9.318	-1.168	105.656	13.241	0.775	1.00	0.0	9.318	-1.168	103.085	20.883
0.800	1.00	0.0	9.655	-1.074	102.311	11.380	0.800	1.00	0.0	9.655	-1.074	100.120	18.603
0.825	1.00	0.0	10.012	-1.016	98.862	10.030	0.825	1.00	0.0	10.012	-1.016	96.965	16.811
0.850	1.00	0.0	10.389	-1.000	95.374	9.180	0.850	1.00	0.0	10.389	-1.000	93.690	15.511
0.875	1.00	0.0	10.782	-1.034	91.904	8.816	0.875	1.00	0.0	10.782	-1.034	90.358	14.701
0.900	1.00	0.0	11.185	-1.127	88.508	8.918	0.900	1.00	0.0	11.185	-1.127	87.033	14.372
0.925	1.00	0.0	11.589	-1.287	85.236	9.463	0.925	1.00	0.0	11.589	-1.287	83.775	14.506
0.950	1.00	0.0	11.982	-1.521	82.138	10.426	0.950	1.00	0.0	11.982	-1.521	80.641	15.081
0.975	1.00	0.0	12.345	-1.834	79.258	11.774	0.975	1.00	0.0	12.345	-1.834	77.682	16.071
1.000	1.00	0.0	12.657	-2.225	76.638	13.475	1.000	1.00	0.0	12.657	-2.225	74.949	17.442
1.025	1.00	0.0	12.895	-2.688	74.318	15.490	1.025	1.00	0.0	12.895	-2.688	72.485	19.155
1.050	1.00	0.0	13.038	-3.204	72.333	17.776	1.050	1.00	0.0	13.038	-3.204	70.330	21.168
1.075	1.00	0.0	13.066	-3.749	70.712	20.287	1.075	1.00	0.0	13.066	-3.749	68.519	23.433
1.100	1.00	0.0	12.974	-4.289	69.483	22.971	1.100	1.00	0.0	12.974	-4.289	67.078	25.899
1.125	1.00	0.0	12.765	-4.792	68.665	25.775	1.125	1.00	0.0	12.765	-4.792	66.030	28.510
1.150	1.00	0.0	12.456	-5.225	68.270	28.638	1.150	1.00	0.0	12.456	-5.225	65.388	31.209
1.175	1.00	0.0	12.073	-5.568	68.304	31.501	1.175	1.00	0.0	12.073	-5.568	65.157	33.935
1.200	1.00	0.0	11.645	-5.809	68.763	34.300	1.200	1.00	0.0	11.645	-5.809	65.336	36.630
1.225	1.00	0.0	11.203	-5.948	69.633	36.973	1.225	1.00	0.0	11.203	-5.948	65.911	39.233
1.250	1.00	0.0	10.770	-5.994	70.891	39.457	1.250	1.00	0.0	10.770	-5.994	66.863	41.686
1.275	1.00	0.0	10.365	-5.961	72.502	41.696	1.275	1.00	0.0	10.365	-5.961	68.163	43.935
1.300	1.00	0.0	9.999	-5.863	74.425	43.635	1.300	1.00	0.0	9.999	-5.863	69.771	45.931
1.325	1.00	0.0	9.680	-5.715	76.605	45.230	1.325	1.00	0.0	9.680	-5.715	[illegible]	[illegible]
1.350	1.00	0.0	9.408	-5.532	78.987	[illegible]	[illegible]	[illegible]	[illegible]	9.408	-5.532	73.730	49.000
[illegible]	[illegible]	[illegible]	9.183	-5.325	81.491	47.257	1.375	1.00	0.0	9.183	-5.325	75.971	50.012
1.400	1.00	0.0	9.003	-5.103	84.062	47.650	1.400	1.00	0.0	9.003	-5.103	78.309	50.652
1.425	1.00	0.0	8.864	-4.874	86.624	47.627	1.425	1.00	0.0	8.864	-4.874	80.682	50.913
1.450	1.00	0.0	8.763	-4.642	89.110	47.199	1.450	1.00	0.0	8.763	-4.642	83.029	50.800
1.475	1.00	0.0	8.697	-4.411	91.455	46.392	1.475	1.00	0.0	8.697	-4.411	85.292	50.327
1.500	1.00	0.0	8.660	-4.186	93.601	45.240	1.500	1.00	0.0	8.660	-4.186	87.416	49.519
1.525	1.00	0.0	8.652	-3.967	95.499	43.788	1.525	1.00	0.0	8.652	-3.967	89.354	48.406
1.550	1.00	0.0	8.669	-3.757	97.110	42.086	1.550	1.00	0.0	8.669	-3.757	91.062	47.028
1.575	1.00	0.0	8.709	-3.557	98.404	40.190	1.575	1.00	0.0	8.709	-3.557	92.508	45.429
1.600	1.00	0.0	8.771	-3.368	99.362	38.156	1.600	1.00	0.0	8.771	-3.368	93.666	43.657
1.625	1.00	0.0	8.852	-3.191	99.977	36.042	1.625	1.00	0.0	8.852	-3.191	94.518	41.762
1.650	1.00	0.0	8.951	-3.027	100.249	33.906	1.650	1.00	0.0	8.951	-3.027	95.056	39.795
1.675	1.00	0.0	9.068	-2.878	100.189	31.798	1.675	1.00	0.0	9.068	-2.878	95.280	37.806
1.700	1.00	0.0	9.200	-2.744	99.814	29.769	1.700	1.00	0.0	9.200	-2.744	95.197	35.842
1.725	1.00	0.0	9.348	-2.627	99.147	27.861	1.725	1.00	0.0	9.348	-2.627	94.821	33.949
1.750	1.00	0.0	9.509	-2.528	98.218	26.113	1.750	1.00	0.0	9.509	-2.528	94.173	32.168
1.775	1.00	0.0	9.683	-2.450	97.059	24.556	1.775	1.00	0.0	9.683	-2.450	93.276	30.535
1.800	1.00	0.0	9.868	-2.394	95.706	23.217	1.800	1.00	0.0	9.868	-2.394	92.162	29.081
1.825	1.00	0.0	10.062	-2.362	94.196	22.115	1.825	1.00	0.0	10.062	-2.362	90.862	27.834
1.850	1.00	0.0	10.261	-2.357	92.568	21.265	1.850	1.00	0.0	10.261	-2.357	89.412	26.813
1.875	1.00	0.0	10.464	-2.381	90.859	20.676	1.875	1.00	0.0	10.464	-2.381	87.848	26.035
1.900	1.00	0.0	10.666	-2.436	89.109	20.351	1.900	1.00	0.0	10.666	-2.436	86.210	25.508
1.925	1.00	0.0	10.861	-2.523	87.356	20.290	1.925	1.00	0.0	10.861	-2.523	84.534	25.238
1.950	1.00	0.0	11.045	-2.642	85.635	20.486	1.950	1.00	0.0	11.045	-2.642	82.858	25.222
1.975	1.00	0.0	11.211	-2.794	83.983	20.929	1.975	1.00	0.0	11.211	-2.794	81.220	25.456
2.000	1.00	0.0	11.352	-2.975	82.431	21.604	2.000	1.00	0.0	11.352	-2.975	79.654	25.928
2.025	1.00	0.0	11.460	-3.182	81.012	22.494	2.025	1.00	0.0	11.460	-3.182	78.193	26.623
2.050	1.00	0.0	11.531	-3.409	79.752	23.575	2.050	1.00	0.0	11.531	-3.409	76.867	27.522
2.075	1.00	0.0	11.559	-3.647	78.677	24.823	2.075	1.00	0.0	11.559	-3.647	75.704	28.600
2.100	1.00	0.0	11.543	-3.888	77.809	26.207	2.100	1.00	0.0	11.543	-3.888	74.728	29.831
2.125	1.00	0.0	11.480	-4.121	77.164	27.696	2.125	1.00	0.0	11.480	-4.121	73.957	31.183
2.150	1.00	0.0	11.376	-4.336	76.755	29.257	2.150	1.00	0.0	11.376	-4.336	73.406	32.626
2.175	1.00	0.0	11.234	-4.525	76.591	30.852	2.175	1.00	0.0	11.234	-4.525	73.085	34.124
2.200	1.00	0.0	11.062	-4.681	76.672	32.444	2.200	1.00	0.0	11.062	-4.681	73.000	35.641
2.225	1.00	0.0	10.869	-4.799	76.998	33.996	2.225	1.00	0.0	10.869	-4.799	73.148	37.142
2.250	1.00	0.0	10.663	-4.877	77.557	35.471	2.250	1.00	0.0	10.663	-4.877	73.524	38.591
2.275	1.00	0.0	10.454	-4.916	78.335	36.833	2.275	1.00	0.0	10.454	-4.916	74.116	39.953
2.300	1.00	0.0	10.250	-4.917	79.312	38.048	2.300	1.00	0.0	10.250	-4.917	74.907	41.197
2.325	1.00	0.0	10.055	-4.885	80.461	39.087	2.325	1.00	0.0	10.055	-4.885	75.876	42.293
2.350	1.00	0.0	9.876	-4.823	81.752	39.926	2.350	1.00	0.0	9.876	-4.823	76.995	43.216
2.375	1.00	0.0	9.716	-4.738	83.149	40.545	2.375	1.00	0.0	9.716	-4.738	78.235	43.945
2.400	1.00	0.0	9.577	-4.633	84.616	40.929	2.400	1.00	0.0	9.577	-4.633	79.562	44.466
2.425	1.00	0.0	9.460	-4.512	86.112	41.074	2.425	1.00	0.0	9.460	-4.512	80.943	44.769
2.450	1.00	0.0	9.366	-4.381	87.599	40.978	2.450	1.00	0.0	9.366	-4.381	82.341	44.849
2.475	1.00	0.0	9.294	-4.243	89.037	40.650	2.475	1.00	0.0	9.294	-4.243	83.720	44.708
2.500	1.00	0.0	9.244	-4.101	90.389	40.101	2.500	1.00	0.0	9.244	-4.101	85.047	44.355
2.525	1.00	0.0	9.215	-3.958	91.622	39.353	2.525	1.00	0.0	9.215	-3.958	86.288	43.804
2.550	1.00	0.0	9.205	-3.816	92.708	38.428	2.550	1.00	0.0	9.205	-3.816	87.414	43.071
2.575	1.00	0.0	9.214	-3.677	93.621	37.356	2.575	1.00	0.0	9.214	-3.677	88.400	42.180
2.600	1.00	0.0	9.241	-3.543	94.344	36.169	2.600	1.00	0.0	9.241	-3.543	89.225	41.158
2.625	1.00	0.0	9.285	-3.416	94.863	34.899	2.625	1.00	0.0	9.285	-3.416	89.871	40.031
2.650	1.00	0.0	9.344	-3.297	95.172	33.581	2.650	1.00	0.0	9.344	-3.297	90.327	38.832
2.675	1.00	0.0	9.417	-3.188	95.270	32.249	2.675	1.00	0.0	9.417	-3.188	90.588	37.590
2.700	1.00	0.0	9.504	-3.090	95.160	30.937	2.700	1.00	0.0	9.504	-3.090	90.651	36.339
2.725	1.00	0.0	9.603	-3.004	94.853	29.675	2.725	1.00	0.0	9.603	-3.004	90.521	35.107
2.750	1.00	0.0	9.712	-2.932	94.361	28.491	2.750	1.00	0.0	9.712	-2.932	90.206	33.924
2.775	1.00	0.0	9.831	-2.876	93.701	27.412	2.775	1.00	0.0	9.831	-2.876	89.716	32.817
2.800	1.00	0.0	9.957	-2.836	92.894	26.458	2.800	1.00	0.0	9.957	-2.836	89.070	31.810
2.825	1.00	0.0	10.089	-2.814	91.962	25.649	2.825	1.00	0.0	10.089	-2.814	88.284	30.924
2.850	1.00	0.0	10.225	-2.811	90.931	24.999	2.850	1.00	0.0	10.225	-2.811	87.382	30.177
2.875	1.00	0.0	10.361	-2.828	89.824	24.519	2.875	1.00	0.0	10.361	-2.828	86.386	29.584
2.900	1.00	0.0	10.495	-2.866	88.670	24.215	2.900	1.00	0.0	10.495	-2.866	85.322	29.154
2.925	1.00	0.0	10.624	-2.925	87.495	24.091	2.925	1.00	0.0	10.624	-2.925	84.215	28.895
2.950	1.00	0.0	10.744	-3.005	86.326	24.145	2.950	1.00	0.0	10.744	-3.005	83.091	28.809
2.975	1.00	0.0	10.851	-3.105	85.188	24.374	2.975	1.00	0.0	10.851	-3.105	81.977	28.895
3.000	1.00	0.0	10.941	-3.222	84.106	24.769	3.000	1.00	0.0	10.941	-3.222	80.898	29.148

TABLE 6.2

DRIVING-POINT ADMITTANCES AND IMPEDANCES OF TWO-ELEMENT BILATERAL ENDFIRE ARRAYS

$h/\lambda = 0.500$; $\Omega = 9.92$

	FOR SPECIFIED BASE VOLTAGES							FOR SPECIFIED BASE CURRENTS					
B/LMDA	VO1		YO1		ZO1		B/LMDA	IZ(O)1		YO1		ZO1	
	RE	IM	RE	IM	RE	IM		RE	IM	RE	IM	RE	IM
0.100	1.00	0.0	0.130	1.198	89.351	-824.989	0.100	1.00	0.0	0.130	1.198	528.766	-1948.170
0.125	1.00	0.0	0.174	1.341	95.349	-733.351	0.125	1.00	0.0	0.174	1.341	419.142	-1492.784
0.150	1.00	0.0	0.221	1.460	101.327	-669.458	0.150	1.00	0.0	0.221	1.460	370.296	-1240.260
0.175	1.00	0.0	0.269	1.562	107.104	-621.727	0.175	1.00	0.0	0.269	1.562	344.337	-1077.514
0.200	1.00	0.0	0.318	1.650	112.669	-584.396	0.200	1.00	0.0	0.318	1.650	329.351	-962.782
0.225	1.00	0.0	0.368	1.726	118.053	-554.216	0.225	1.00	0.0	0.368	1.726	320.463	-876.885
0.250	1.00	0.0	0.418	1.792	123.299	-529.190	0.250	1.00	0.0	0.418	1.792	315.326	-809.709
0.275	1.00	0.0	0.468	1.850	128.450	-508.013	0.275	1.00	0.0	0.468	1.850	312.680	-755.395
0.300	1.00	0.0	0.518	1.900	133.544	-489.791	0.300	1.00	0.0	0.518	1.900	311.793	-710.296
0.325	1.00	0.0	0.569	1.944	138.618	-473.889	0.325	1.00	0.0	0.569	1.944	312.216	-672.021
0.350	1.00	0.0	0.619	1.981	143.704	-459.844	0.350	1.00	0.0	0.619	1.981	313.662	-638.933
0.375	1.00	0.0	0.670	2.013	148.836	-447.312	0.375	1.00	0.0	0.670	2.013	315.945	-609.873
0.400	1.00	0.0	0.720	2.039	154.042	-436.030	0.400	1.00	0.0	0.720	2.039	318.941	-583.996
0.425	1.00	0.0	0.771	2.060	159.352	-425.796	0.425	1.00	0.0	0.771	2.060	322.572	-560.674
0.450	1.00	0.0	0.822	2.076	164.798	-416.455	0.450	1.00	0.0	0.822	2.076	326.790	-539.426
0.475	1.00	0.0	0.872	2.087	170.409	-407.887	0.475	1.00	0.0	0.872	2.087	331.571	-519.883
0.500	1.00	0.0	0.922	2.094	176.218	-399.998	0.500	1.00	0.0	0.922	2.094	336.909	-501.757
0.525	1.00	0.0	0.972	2.095	182.260	-392.720	0.525	1.00	0.0	0.972	2.095	342.815	-484.821
0.550	1.00	0.0	1.022	2.092	188.570	-386.005	0.550	1.00	0.0	1.022	2.092	349.314	-468.896
0.575	1.00	0.0	1.070	2.083	195.187	-379.822	0.575	1.00	0.0	1.070	2.083	356.445	-453.842
0.600	1.00	0.0	1.118	2.069	202.152	-374.159	0.600	1.00	0.0	1.118	2.069	364.258	-439.555
0.625	1.00	0.0	1.163	2.049	209.508	-369.022	0.625	1.00	0.0	1.163	2.049	372.818	-425.958
0.650	1.00	0.0	1.207	2.024	217.300	-364.435	0.650	1.00	0.0	1.207	2.024	382.206	-413.006
0.675	1.00	0.0	1.248	1.994	225.572	-360.446	0.675	1.00	0.0	1.248	1.994	392.513	-400.682
0.700	1.00	0.0	1.284	1.957	234.366	-357.125	0.700	1.00	0.0	1.284	1.957	403.849	-389.002
0.725	1.00	0.0	1.317	1.915	243.717	-354.572	0.725	1.00	0.0	1.317	1.915	416.337	-378.020
0.750	1.00	0.0	1.343	1.868	253.646	-352.918	0.750	1.00	0.0	1.343	1.868	430.114	-367.836
0.775	1.00	0.0	1.362	1.817	264.152	-352.332	0.775	1.00	0.0	1.362	1.817	445.324	-358.607
0.800	1.00	0.0	1.374	1.762	275.196	-353.021	0.800	1.00	0.0	1.374	1.762	462.115	-350.562
0.825	1.00	0.0	1.376	1.705	286.685	-355.233	0.825	1.00	0.0	1.376	1.705	480.621	-344.015
0.850	1.00	0.0	1.368	1.647	298.440	-359.245	0.850	1.00	0.0	1.368	1.647	500.938	-339.390
0.875	1.00	0.0	1.350	1.591	310.175	-365.346	0.875	1.00	0.0	1.350	1.591	523.086	-337.237
0.900	1.00	0.0	1.323	1.538	321.458	-373.798	0.900	1.00	0.0	1.323	1.538	546.946	-338.240
0.925	1.00	0.0	1.285	1.491	331.700	-384.776	0.925	1.00	0.0	1.285	1.491	572.183	-343.207
0.950	1.00	0.0	1.240	1.452	340.156	-398.278	0.950	1.00	0.0	1.240	1.452	598.136	-353.018
0.975	1.00	0.0	1.188	1.422	345.986	-414.027	0.975	1.00	0.0	1.188	1.422	623.713	-368.506
1.000	1.00	0.0	1.133	1.403	348.375	-431.397	1.000	1.00	0.0	1.133	1.403	647.320	-390.239
1.025	1.00	0.0	1.076	1.395	346.718	-449.393	1.025	1.00	0.0	1.076	1.395	666.911	-418.232
1.050	1.00	0.0	1.020	1.397	340.802	-466.762	1.050	1.00	0.0	1.020	1.397	680.223	-451.628
1.075	1.00	0.0	0.968	1.410	330.937	-482.195	1.075	1.00	0.0	0.968	1.410	685.243	-488.525
1.100	1.00	0.0	0.920	1.431	317.930	-494.597	1.100	1.00	0.0	0.920	1.431	680.801	-526.106
1.125	1.00	0.0	0.878	1.459	302.924	-503.294	1.125	1.00	0.0	0.878	1.459	667.028	-561.160
1.150	1.00	0.0	0.843	1.492	287.146	-508.114	1.150	1.00	0.0	0.843	1.492	645.418	-590.825
1.175	1.00	0.0	0.815	1.528	271.687	-509.321	1.175	1.00	0.0	0.815	1.528	618.426	-613.224
1.200	1.00	0.0	0.795	1.567	257.359	-507.465	1.200	1.00	0.0	0.795	1.567	588.808	-627.709
1.225	1.00	0.0	0.781	1.607	244.668	-503.217	1.225	1.00	0.0	0.781	1.607	559.020	-634.702
1.250	1.00	0.0	0.775	1.647	233.851	-497.239	1.250	1.00	0.0	0.775	1.647	530.878	-635.298
1.275	1.00	0.0	0.774	1.685	224.945	-490.113	1.275	1.00	0.0	0.774	1.685	505.501	-630.862
1.300	1.00	0.0	0.778	1.722	217.861	-482.309	1.300	1.00	0.0	0.778	1.722	483.415	-622.731
1.325	1.00	0.0	0.787	1.756	212.446	-474.181	1.325	1.00	0.0	0.787	1.756	464.730	-612.058
1.350	1.00	0.0	0.800	1.788	208.517	-465.990	1.350	1.00	0.0	0.800	1.788	449.309	-599.760
1.375	1.00	0.0	0.817	1.817	205.894	-457.914	1.375	1.00	0.0	0.817	1.817	436.881	-586.524
1.400	1.00	0.0	0.837	1.842	204.408	-450.075	1.400	1.00	0.0	0.837	1.842	427.129	-572.845
1.425	1.00	0.0	0.859	1.864	203.912	-442.552	1.425	1.00	0.0	0.859	1.864	419.736	-559.067
1.450	1.00	0.0	0.883	1.882	204.279	-435.394	1.450	1.00	0.0	0.883	1.882	414.412	-545.422
1.475	1.00	0.0	0.909	1.897	205.405	-428.631	1.475	1.00	0.0	0.909	1.897	410.903	-532.065
1.500	1.00	0.0	0.936	1.909	207.202	-422.282	1.500	1.00	0.0	0.936	1.909	408.995	-519.092
1.525	1.00	0.0	0.965	1.916	209.602	-416.359	1.525	1.00	0.0	0.965	1.916	408.513	-506.567
1.550	1.00	0.0	0.993	1.920	212.549	-410.873	1.550	1.00	0.0	0.993	1.920	409.316	-494.528
1.575	1.00	0.0	1.022	1.920	215.998	-405.833	1.575	1.00	0.0	1.022	1.920	411.294	-483.003
1.600	1.00	0.0	1.050	1.916	219.915	-401.257	1.600	1.00	0.0	1.050	1.916	414.365	-472.013
1.625	1.00	0.0	1.078	1.909	224.270	-397.161	1.625	1.00	0.0	1.078	1.909	418.467	-461.581
1.650	1.00	0.0	1.105	1.898	229.038	-393.574	1.650	1.00	0.0	1.105	1.898	423.557	-451.736
1.675	1.00	0.0	1.129	1.883	234.195	-390.529	1.675	1.00	0.0	1.129	1.883	429.609	-442.513
1.700	1.00	0.0	1.152	1.865	239.715	-388.068	1.700	1.00	0.0	1.152	1.865	436.607	-433.965
1.725	1.00	0.0	1.172	1.844	245.568	-386.244	1.725	1.00	0.0	1.172	1.844	444.541	-426.159
1.750	1.00	0.0	1.189	1.819	251.713	-385.120	1.750	1.00	0.0	1.189	1.819	453.406	-419.181
1.775	1.00	0.0	1.202	1.792	258.098	-384.766	1.775	1.00	0.0	1.202	1.792	463.191	-413.144
1.800	1.00	0.0	1.211	1.763	264.650	-385.258	1.800	1.00	0.0	1.211	1.763	473.875	-408.186
1.825	1.00	0.0	1.216	1.733	271.275	-386.676	1.825	1.00	0.0	1.216	1.733	485.420	-404.472
1.850	1.00	0.0	1.215	1.702	277.846	-389.098	1.850	1.00	0.0	1.215	1.702	497.751	-402.199
1.875	1.00	0.0	1.210	1.671	284.204	-392.587	1.875	1.00	0.0	1.210	1.671	510.749	-401.588
1.900	1.00	0.0	1.199	1.642	290.151	-397.182	1.900	1.00	0.0	1.199	1.642	524.229	-402.877
1.925	1.00	0.0	1.184	1.614	295.457	-402.880	1.925	1.00	0.0	1.184	1.614	537.929	-406.306
1.950	1.00	0.0	1.164	1.589	299.865	-409.622	1.950	1.00	0.0	1.164	1.589	551.485	-412.092
1.975	1.00	0.0	1.140	1.569	303.110	-417.271	1.975	1.00	0.0	1.140	1.569	564.429	-420.391
2.000	1.00	0.0	1.112	1.553	304.941	-425.607	2.000	1.00	0.0	1.112	1.553	576.192	-431.250
2.025	1.00	0.0	1.083	1.541	305.163	-434.323	2.025	1.00	0.0	1.083	1.541	586.130	-444.559
2.050	1.00	0.0	1.053	1.536	303.660	-443.041	2.050	1.00	0.0	1.053	1.536	593.577	-459.998
2.075	1.00	0.0	1.022	1.535	300.436	-451.345	2.075	1.00	0.0	1.022	1.535	597.932	-477.021
2.100	1.00	0.0	0.992	1.540	295.620	-458.824	2.100	1.00	0.0	0.992	1.540	598.748	-494.863
2.125	1.00	0.0	0.964	1.550	289.460	-465.122	2.125	1.00	0.0	0.964	1.550	595.831	-512.610
2.150	1.00	0.0	0.939	1.564	282.300	-469.974	2.150	1.00	0.0	0.939	1.564	589.288	-529.304
2.175	1.00	0.0	0.917	1.581	274.529	-473.237	2.175	1.00	0.0	0.917	1.581	579.534	-544.065
2.200	1.00	0.0	0.899	1.601	266.542	-474.887	2.200	1.00	0.0	0.899	1.601	567.230	-556.208
2.225	1.00	0.0	0.884	1.624	258.695	-475.008	2.225	1.00	0.0	0.884	1.624	553.181	-565.311
2.250	1.00	0.0	0.874	1.647	251.282	-473.758	2.250	1.00	0.0	0.874	1.647	538.221	-571.231
2.275	1.00	0.0	0.867	1.672	244.521	-471.344	2.275	1.00	0.0	0.867	1.672	523.117	-574.066
2.300	1.00	0.0	0.865	1.696	238.556	-467.991	2.300	1.00	0.0	0.865	1.696	508.504	-574.094
2.325	1.00	0.0	0.866	1.720	233.465	-463.920	2.325	1.00	0.0	0.866	1.720	494.857	-571.695
2.350	1.00	0.0	0.870	1.743	229.278	-459.332	2.350	1.00	0.0	0.870	1.743	482.497	-567.296
2.375	1.00	0.0	0.877	1.764	225.983	-454.404	2.375	1.00	0.0	0.877	1.764	471.610	-561.315
2.400	1.00	0.0	0.888	1.784	223.547	-449.286	2.400	1.00	0.0	0.888	1.784	462.277	-554.135
2.425	1.00	0.0	0.900	1.802	221.921	-444.100	2.425	1.00	0.0	0.900	1.802	454.506	-546.090
2.450	1.00	0.0	0.915	1.817	221.048	-438.945	2.450	1.00	0.0	0.915	1.817	448.253	-537.462
2.475	1.00	0.0	0.932	1.830	220.871	-433.900	2.475	1.00	0.0	0.932	1.830	443.451	-528.481
2.500	1.00	0.0	0.950	1.841	221.332	-429.029	2.500	1.00	0.0	0.950	1.841	440.012	-519.331
2.525	1.00	0.0	0.969	1.849	222.379	-424.383	2.525	1.00	0.0	0.969	1.849	437.850	-510.160
2.550	1.00	0.0	0.989	1.854	223.963	-420.005	2.550	1.00	0.0	0.989	1.854	436.879	-501.088
2.575	1.00	0.0	1.009	1.856	226.039	-415.933	2.575	1.00	0.0	1.009	1.856	437.021	-492.211
2.600	1.00	0.0	1.029	1.855	228.566	-412.199	2.600	1.00	0.0	1.029	1.855	438.205	-483.612
2.625	1.00	0.0	1.049	1.852	231.506	-408.836	2.625	1.00	0.0	1.049	1.852	440.370	-475.362
2.650	1.00	0.0	1.068	1.846	234.823	-405.878	2.650	1.00	0.0	1.068	1.846	443.463	-467.527
2.675	1.00	0.0	1.086	1.837	238.480	-403.358	2.675	1.00	0.0	1.086	1.837	447.438	-460.172
2.700	1.00	0.0	1.103	1.826	242.439	-401.312	2.700	1.00	0.0	1.103	1.826	452.253	-453.365
2.725	1.00	0.0	1.118	1.812	246.659	-399.778	2.725	1.00	0.0	1.118	1.812	457.868	-447.179
2.750	1.00	0.0	1.131	1.796	251.091	-398.798	2.750	1.00	0.0	1.131	1.796	464.243	-441.695
2.775	1.00	0.0	1.141	1.778	255.680	-398.412	2.775	1.00	0.0	1.141	1.778	471.332	-437.001
2.800	1.00	0.0	1.148	1.758	260.359	-398.661	2.800	1.00	0.0	1.148	1.758	479.078	-433.198
2.825	1.00	0.0	1.153	1.738	265.048	-399.585	2.825	1.00	0.0	1.153	1.738	487.412	-430.395
2.850	1.00	0.0	1.154	1.717	269.655	-401.213	2.850	1.00	0.0	1.154	1.717	496.240	-428.711
2.875	1.00	0.0	1.152	1.696	274.068	-403.565	2.875	1.00	0.0	1.152	1.696	505.441	-428.268
2.900	1.00	0.0	1.146	1.675	278.162	-406.643	2.900	1.00	0.0	1.146	1.675	514.860	-429.186
2.925	1.00	0.0	1.137	1.656	281.802	-410.421	2.925	1.00	0.0	1.137	1.656	524.298	-431.574
2.950	1.00	0.0	1.125	1.638	284.841	-414.847	2.950	1.00	0.0	1.125	1.638	533.514	-435.515
2.975	1.00	0.0	1.110	1.623	287.138	-419.829	2.975	1.00	0.0	1.110	1.623	542.221	-441.049
3.000	1.00	0.0	1.093	1.610	288.563	-425.234	3.000	1.00	0.0	1.093	1.610	550.096	-448.154

TABLE 6.3

TWO-ELEMENT ARRAYS

VEN

H/LMDA= 0.250 OMEGA= 8.53

SELF AND MUTUAL ADMITTANCES

B/LMDA	I, K	YIK IN MMHO RE	IM	I, K	YIK IN MMHO RE	IM
0.100	1, 1	5.6489	-13.7888	1, 2	0.4452	13.5707
0.125	1, 1	5.7976	-10.6752	1, 2	0.6279	10.6513
0.150	1, 1	5.9783	-8.6132	1, 2	0.7904	8.7904
0.175	1, 1	6.1826	-7.1303	1, 2	0.9463	7.5126
0.200	1, 1	6.4076	-6.0003	1, 2	1.1035	6.5890
0.225	1, 1	6.6534	-5.1020	1, 2	1.2677	5.8948
0.250	1, 1	6.9214	-4.3657	1, 2	1.4437	5.3557
0.275	1, 1	7.2141	-3.7494	1, 2	1.6362	4.9235
0.300	1, 1	7.5347	-3.2277	1, 2	1.8495	4.5643
0.325	1, 1	7.8867	-2.7864	1, 2	2.0878	4.2520
0.350	1, 1	8.2735	-2.4195	1, 2	2.3548	3.9642
0.375	1, 1	8.6974	-2.1278	1, 2	2.6531	3.6793
0.400	1, 1	9.1582	-1.9185	1, 2	2.9824	3.3749
0.425	1, 1	9.6512	-1.8044	1, 2	3.3376	3.0267
0.450	1, 1	10.1633	-1.8023	1, 2	3.7053	2.6090
0.475	1, 1	10.6697	-1.9295	1, 2	4.0603	2.0977
0.500	1, 1	11.1314	-2.1963	1, 2	4.3628	1.4772
0.525	1, 1	11.4972	-2.5958	1, 2	4.5610	0.7503
0.550	1, 1	11.7147	-3.0937	1, 2	4.6015	-0.0520
0.575	1, 1	11.7486	-3.6273	1, 2	4.4481	-0.8692
0.600	1, 1	11.5989	-4.1192	1, 2	4.0998	-1.6256
0.625	1, 1	11.3045	-4.5018	1, 2	3.5941	-2.2547
0.650	1, 1	10.9287	-4.7381	1, 2	2.9932	-2.7200
0.675	1, 1	10.5364	-4.8261	1, 2	2.3605	-3.0192
0.700	1, 1	10.1775	-4.7886	1, 2	1.7444	-3.1743
0.725	1, 1	9.8814	-4.6550	1, 2	1.1731	-3.2173
0.750	1, 1	9.6605	-4.4708	1, 2	0.6575	-3.1790
0.775	1, 1	9.5157	-4.2519	1, 2	0.1974	-3.0842
0.800	1, 1	9.4419	-4.0241	1, 2	-0.2128	-2.9502
0.825	1, 1	9.4312	-3.8033	1, 2	-0.5808	-2.7876
0.850	1, 1	9.4748	-3.6016	1, 2	-0.9140	-2.6017
0.875	1, 1	9.5637	-3.4282	1, 2	-1.2179	-2.3940
0.900	1, 1	9.6888	-3.2904	1, 2	-1.4960	-2.1634
0.925	1, 1	9.8404	-3.1941	1, 2	-1.7489	-1.9074
0.950	1, 1	10.0076	-3.1438	1, 2	-1.9740	-1.6230
0.975	1, 1	10.1786	-3.1420	1, 2	-2.1661	-1.3081
1.000	1, 1	10.3400	-3.1885	1, 2	-2.3170	-0.9630
1.025	1, 1	10.4783	-3.2792	1, 2	-2.4172	-0.5914
1.050	1, 1	10.5806	-3.4056	1, 2	-2.4569	-0.2015
1.075	1, 1	10.6374	-3.5544	1, 2	-2.4289	0.1942
1.100	1, 1	10.6439	-3.7096	1, 2	-2.3300	0.5796
1.125	1, 1	10.6016	-3.8539	1, 2	-2.1633	0.9376
1.150	1, 1	10.5185	-3.9723	1, 2	-1.9374	1.2527
1.175	1, 1	10.4071	-4.0539	1, 2	-1.6655	1.5138
1.200	1, 1	10.2821	-4.0939	1, 2	-1.3631	1.7150
1.225	1, 1	10.1579	-4.0928	1, 2	-1.0448	1.8554
1.250	1, 1	10.0465	-4.0559	1, 2	-0.7233	1.9385
1.275	1, 1	9.9566	-3.9908	1, 2	-0.4080	1.9699
1.300	1, 1	9.8936	-3.9067	1, 2	-0.1056	1.9559
1.325	1, 1	9.8597	-3.8127	1, 2	0.1802	1.9025
1.350	1, 1	9.8547	-3.7175	1, 2	0.4469	1.8148
1.375	1, 1	9.8763	-3.6286	1, 2	0.6932	1.6967
1.400	1, 1	9.9209	-3.5525	1, 2	0.9177	1.5509
1.425	1, 1	9.9835	-3.4942	1, 2	1.1191	1.3796
1.450	1, 1	10.0584	-3.4576	1, 2	1.2949	1.1842
1.475	1, 1	10.1388	-3.4447	1, 2	1.4423	0.9668
1.500	1, 1	10.2179	-3.4561	1, 2	1.5574	0.7298
1.525	1, 1	10.2886	-3.4900	1, 2	1.6363	0.4772
1.550	1, 1	10.3446	-3.5430	1, 2	1.6753	0.2142
1.575	1, 1	10.3808	-3.6093	1, 2	1.6714	-0.0522
1.600	1, 1	10.3943	-3.6821	1, 2	1.6235	-0.3141
1.625	1, 1	10.3845	-3.7538	1, 2	1.5327	-0.5626
1.650	1, 1	10.3535	-3.8170	1, 2	1.4023	-0.7895
1.675	1, 1	10.3057	-3.8657	1, 2	1.2379	-0.9877
1.700	1, 1	10.2469	-3.8959	1, 2	1.0466	-1.1519
1.725	1, 1	10.1839	-3.9055	1, 2	0.8361	-1.2791
1.750	1, 1	10.1232	-3.8963	1, 2	0.6140	-1.3682
1.775	1, 1	10.0704	-3.8696	1, 2	0.3873	-1.4198
1.800	1, 1	10.0297	-3.8297	1, 2	0.1618	-1.4359
1.825	1, 1	10.0039	-3.7811	1, 2	-0.0577	-1.4189
1.850	1, 1	9.9941	-3.7287	1, 2	-0.2673	-1.3715
1.875	1, 1	10.0001	-3.6775	1, 2	-0.4641	-1.2963
1.900	1, 1	10.0205	-3.6316	1, 2	-0.6453	-1.1957
1.925	1, 1	10.0528	-3.5948	1, 2	-0.8087	-1.0720
1.950	1, 1	10.0937	-3.5698	1, 2	-0.9517	-0.9275
1.975	1, 1	10.1393	-3.5584	1, 2	-1.0717	-0.7646
2.000	1, 1	10.1853	-3.5612	1, 2	-1.1662	-0.5861
2.025	1, 1	10.2277	-3.5776	1, 2	-1.2327	-0.3955
2.050	1, 1	10.2624	-3.6057	1, 2	-1.2688	-0.1970
2.075	1, 1	10.2863	-3.6425	1, 2	-1.2731	0.0045
2.100	1, 1	10.2975	-3.6842	1, 2	-1.2450	0.2035
2.125	1, 1	10.2952	-3.7264	1, 2	-1.1852	0.3942
2.150	1, 1	10.2802	-3.7651	1, 2	-1.0955	0.5710
2.175	1, 1	10.2544	-3.7963	1, 2	-0.9793	0.7287
2.200	1, 1	10.2208	-3.8176	1, 2	[illegible]	[illegible]
2.225	1, 1	10.1836	-3.8269	1, 2	-0.6851	0.9719
2.250	1, 1	10.1462	-3.8244	1, 2	-0.5171	1.0526
2.275	1, 1	10.1123	-3.8108	1, 2	-0.3421	1.1047
2.300	1, 1	10.0850	-3.7882	1, 2	-0.1646	1.1288
2.325	1, 1	10.0664	-3.7591	1, 2	0.0110	1.1257
2.350	1, 1	10.0576	-3.7266	1, 2	0.1812	1.0969
2.375	1, 1	10.0591	-3.6938	1, 2	0.3428	1.0440
2.400	1, 1	10.0701	-3.6636	1, 2	0.4929	0.9690
2.425	1, 1	10.0894	-3.6387	1, 2	0.6290	0.8738
2.450	1, 1	10.1148	-3.6210	1, 2	0.7487	0.7604
2.475	1, 1	10.1438	-3.6121	1, 2	0.8497	0.6312
2.500	1, 1	10.1737	-3.6123	1, 2	0.9298	0.4888
2.525	1, 1	10.2017	-3.6215	1, 2	0.9872	0.3362
2.550	1, 1	10.2252	-3.6387	1, 2	1.0202	0.1769
2.575	1, 1	10.2420	-3.6618	1, 2	1.0277	0.0148
2.600	1, 1	10.2507	-3.6887	1, 2	1.0093	-0.1458
2.625	1, 1	10.2506	-3.7164	1, 2	0.9657	-0.3007
2.650	1, 1	10.2420	-3.7424	1, 2	0.8981	-0.4454
2.675	1, 1	10.2261	-3.7639	1, 2	0.8087	-0.5761
2.700	1, 1	10.2047	-3.7791	1, 2	0.7006	-0.6893
2.725	1, 1	10.1801	-3.7868	1, 2	0.5774	-0.7826
2.750	1, 1	10.1550	-3.7865	1, 2	0.4428	-0.8540
2.775	1, 1	10.1316	-3.7787	1, 2	0.3007	-0.9027
2.800	1, 1	10.1122	-3.7643	1, 2	0.1550	-0.9283
2.825	1, 1	10.0985	-3.7451	1, 2	0.0093	-0.9311
2.850	1, 1	10.0914	-3.7231	1, 2	-0.1332	-0.9121
2.875	1, 1	10.0913	-3.7005	1, 2	-0.2695	-0.8723
2.900	1, 1	10.0980	-3.6792	1, 2	-0.3970	-0.8132
2.925	1, 1	10.1106	-3.6614	1, 2	-0.5131	-0.7363
2.950	1, 1	10.1278	-3.6484	1, 2	-0.6157	-0.6435
2.975	1, 1	10.1478	-3.6414	1, 2	-0.7027	-0.5368
3.000	1, 1	10.1687	-3.6408	1, 2	-0.7722	-0.4187

TABLE 6.3

TWO-ELEMENT ARRAYS

VEN

H/LMDA= 0.500 OMEGA= 9.92

SELF AND MUTUAL ADMITTANCES

B/LMDA	I, K	YIK IN MMHO RE	YIK IN MMHO IM	I, K	YIK IN MMHO RE	YIK IN MMHO IM
0.100	1, 1	0.7091	1.7298	1, 2	0.5793	0.5317
0.125	1, 1	0.7653	1.7954	1, 2	0.5909	0.4544
0.150	1, 1	0.8216	1.8402	1, 2	0.6006	0.3799
0.175	1, 1	0.8771	1.8686	1, 2	0.6080	0.3065
0.200	1, 1	0.9307	1.8829	1, 2	0.6126	0.2330
0.225	1, 1	0.9810	1.8849	1, 2	0.6133	0.1588
0.250	1, 1	1.0267	1.8759	1, 2	0.6091	0.0835
0.275	1, 1	1.0661	1.8574	1, 2	0.5983	0.0072
0.300	1, 1	1.0976	1.8310	1, 2	0.5794	-0.0694
0.325	1, 1	1.1198	1.7988	1, 2	0.5512	-0.1451
0.350	1, 1	1.1317	1.7635	1, 2	0.5125	-0.2177
0.375	1, 1	1.1328	1.7279	1, 2	0.4631	-0.2849
0.400	1, 1	1.1238	1.6951	1, 2	0.4035	-0.3439
0.425	1, 1	1.1063	1.6679	1, 2	0.3353	-0.3921
0.450	1, 1	1.0826	1.6485	1, 2	0.2611	-0.4276
0.475	1, 1	1.0557	1.6379	1, 2	0.1837	-0.4494
0.500	1, 1	1.0286	1.6363	1, 2	0.1062	-0.4574
0.525	1, 1	1.0040	1.6425	1, 2	0.0316	-0.4526
0.550	1, 1	0.9838	1.6551	1, 2	-0.0380	-0.4364
0.575	1, 1	0.9693	1.6719	1, 2	-0.1010	-0.4109
0.600	1, 1	0.9610	1.6909	1, 2	-0.1568	-0.3779
0.625	1, 1	0.9587	1.7102	1, 2	-0.2048	-0.3391
0.650	1, 1	0.9618	1.7282	1, 2	-0.2452	-0.2961
0.675	1, 1	0.9695	1.7436	1, 2	-0.2781	-0.2499
0.700	1, 1	0.9806	1.7556	1, 2	-0.3038	-0.2016
0.725	1, 1	0.9940	1.7636	1, 2	-0.3225	-0.1518
0.750	1, 1	1.0085	1.7673	1, 2	-0.3344	-0.1011
0.775	1, 1	1.0228	1.7668	1, 2	-0.3394	-0.0501
0.800	1, 1	1.0360	1.7624	1, 2	-0.3376	0.0004
0.825	1, 1	1.0469	1.7546	1, 2	-0.3289	0.0499
0.850	1, 1	1.0549	1.7443	1, 2	-0.3133	0.0973
0.875	1, 1	1.0593	1.7324	1, 2	-0.2911	0.1418
0.900	1, 1	1.0601	1.7201	1, 2	-0.2625	0.1823
0.925	1, 1	1.0572	1.7085	1, 2	-0.2281	0.2176
0.950	1, 1	1.0511	1.6986	1, 2	-0.1889	0.2468
0.975	1, 1	1.0425	1.6913	1, 2	-0.1459	0.2691
1.000	1, 1	1.0325	1.6870	1, 2	-0.1006	0.2839
1.025	1, 1	1.0220	1.6860	1, 2	-0.0542	0.2911
1.050	1, 1	1.0121	1.6882	1, 2	-0.0083	0.2908
1.075	1, 1	1.0036	1.6931	1, 2	0.0360	0.2833
1.100	1, 1	0.9972	1.7001	1, 2	0.0776	0.2694
1.125	1, 1	0.9934	1.7083	1, 2	0.1155	0.2497
1.150	1, 1	0.9922	1.7169	1, 2	0.1493	0.2252
1.175	1, 1	0.9936	1.7252	1, 2	0.1783	0.1967
1.200	1, 1	0.9973	1.7324	1, 2	0.2023	0.1650
1.225	1, 1	1.0027	1.7381	1, 2	0.2212	0.1308
1.250	1, 1	1.0093	1.7418	1, 2	0.2348	0.0950
1.275	1, 1	1.0165	1.7434	1, 2	0.2430	0.0580
1.300	1, 1	1.0237	1.7427	1, 2	0.2458	0.0207
1.325	1, 1	1.0301	1.7401	1, 2	0.2432	-0.0163
1.350	1, 1	1.0354	1.7358	1, 2	0.2353	-0.0522
1.375	1, 1	1.0391	1.7302	1, 2	0.2223	-0.0864
1.400	1, 1	1.0408	1.7239	1, 2	0.2043	-0.1180
1.425	1, 1	1.0407	1.7176	1, 2	0.1818	-0.1463
1.450	1, 1	1.0386	1.7117	1, 2	0.1554	-0.1707
1.475	1, 1	1.0350	1.7068	1, 2	0.1258	-0.1905
1.500	1, 1	1.0301	1.7034	1, 2	0.0936	-0.2052
1.525	1, 1	1.0246	1.7017	1, 2	0.0600	-0.2145
1.550	1, 1	1.0190	1.7017	1, 2	0.0257	-0.2184
1.575	1, 1	1.0138	1.7033	1, 2	-0.0082	-0.2168
1.600	1, 1	1.0094	1.7064	1, 2	-0.0409	-0.2101
1.625	1, 1	1.0063	1.7106	1, 2	-0.0717	-0.1985
1.650	1, 1	1.0047	1.7153	1, 2	-0.0999	-0.1827
1.675	1, 1	1.0045	1.7202	1, 2	-0.1249	-0.1631
1.700	1, 1	1.0058	1.7248	1, 2	-0.1463	-0.1404
1.725	1, 1	1.0084	1.7287	1, 2	-0.1639	-0.1151
1.750	1, 1	1.0119	1.7316	1, 2	-0.1773	-0.0878
1.775	1, 1	1.0160	1.7332	1, 2	-0.1864	-0.0592
1.800	1, 1	1.0203	1.7336	1, 2	-0.1912	-0.0299
1.825	1, 1	1.0244	1.7326	1, 2	-0.1915	-0.0005
1.850	1, 1	1.0279	1.7305	1, 2	-0.1875	0.0284
1.875	1, 1	1.0306	1.7275	1, 2	-0.1793	0.0562
1.900	1, 1	1.0323	1.7239	1, 2	-0.1670	0.0823
1.925	1, 1	1.0328	1.7200	1, 2	-0.1509	0.1060
1.950	1, 1	1.0320	1.7163	1, 2	-0.1315	0.1268
1.975	1, 1	1.0303	1.7130	1, 2	-0.1093	0.1442
2.000	1, 1	1.0276	1.7104	1, 2	-0.0848	0.1579
2.025	1, 1	1.0244	1.7089	1, 2	-0.0587	0.1674
2.050	1, 1	1.0209	1.7084	1, 2	-0.0317	0.1727
2.075	1, 1	1.0175	1.7090	1, 2	-0.0045	0.1736
2.100	1, 1	1.0145	1.7105	1, 2	0.0222	0.1704
2.125	1, 1	1.0122	1.7129	1, 2	0.0477	0.1631
2.150	1, 1	1.0107	1.7158	1, 2	0.0715	0.1521
2.175	1, 1	1.0102	1.7189	1, 2	0.0930	0.1378
2.200	1, 1	1.0106	1.7219	1, 2	0.1119	0.1206
2.225	1, 1	1.0120	1.7246	1, 2	0.1277	0.1010
2.250	1, 1	1.0140	1.7268	1, 2	0.1402	0.0794
2.275	1, 1	1.0165	1.7282	1, 2	0.1493	0.0565
2.300	1, 1	1.0193	1.7287	1, 2	0.1548	0.0326
2.325	1, 1	1.0221	1.7284	1, 2	0.1565	0.0084
2.350	1, 1	1.0246	1.7273	1, 2	0.1547	-0.0156
2.375	1, 1	1.0266	1.7255	1, 2	0.1492	-0.0388
2.400	1, 1	1.0280	1.7232	1, 2	0.1403	-0.0609
2.425	1, 1	1.0285	1.7207	1, 2	0.1281	-0.0812
2.450	1, 1	1.0283	1.7181	1, 2	0.1131	-0.0992
2.475	1, 1	1.0273	1.7158	1, 2	0.0956	-0.1146
2.500	1, 1	1.0257	1.7139	1, 2	0.0760	-0.1270
2.525	1, 1	1.0237	1.7126	1, 2	0.0549	-0.1361
2.550	1, 1	1.0213	1.7121	1, 2	0.0328	-0.1417
2.575	1, 1	1.0190	1.7122	1, 2	0.0103	-0.1438
2.600	1, 1	1.0168	1.7131	1, 2	-0.0120	-0.1424
2.625	1, 1	1.0151	1.7146	1, 2	-0.0337	-0.1375
2.650	1, 1	1.0139	1.7164	1, 2	-0.0540	-0.1295
2.675	1, 1	1.0134	1.7185	1, 2	-0.0728	-0.1185
2.700	1, 1	1.0135	1.7207	1, 2	-0.0894	-0.1049
2.725	1, 1	1.0142	1.7226	1, 2	-0.1036	-0.0891
2.750	1, 1	1.0155	1.7243	1, 2	-0.1151	-0.0714
2.775	1, 1	1.0172	1.7254	1, 2	-0.1237	-0.0524
2.800	1, 1	1.0192	1.7259	1, 2	-0.1292	-0.0325
2.825	1, 1	1.0211	1.7259	1, 2	-0.1317	-0.0121
2.850	1, 1	1.0229	1.7252	1, 2	-0.1310	0.0083
2.875	1, 1	1.0245	1.7241	1, 2	-0.1272	0.0283
2.900	1, 1	1.0255	1.7225	1, 2	-0.1204	0.0472
2.925	1, 1	1.0261	1.7207	1, 2	-0.1109	0.0649
2.950	1, 1	1.0261	1.7189	1, 2	-0.0988	0.0807
2.975	1, 1	1.0255	1.7172	1, 2	-0.0844	0.0944
3.000	1, 1	1.0244	1.7158	1, 2	-0.0682	0.1056

TABLE 6.4

TWO-ELEMENT ARRAYS

IEN

H/LMDA= 0.250 OMEGA= 8.53

SELF AND MUTUAL IMPEDANCES

B/LMDA	I, K	ZIK IN OHMS RE	IM	I, K	ZIK IN OHMS RE	IM
0.100	1, 1	83.337	29.553	1, 2	76.956	-4.878
0.125	1, 1	81.827	30.387	1, 2	71.745	-12.608
0.150	1, 1	80.696	31.385	1, 2	66.098	-19.613
0.175	1, 1	79.918	32.473	1, 2	60.041	-25.844
0.200	1, 1	79.467	33.582	1, 2	53.628	-31.256
0.225	1, 1	79.311	34.650	1, 2	46.923	-35.810
0.250	1, 1	79.409	35.625	1, 2	40.010	-39.478
0.275	1, 1	79.717	36.463	1, 2	32.977	-42.245
0.300	1, 1	80.183	37.134	1, 2	25.921	-44.106
0.325	1, 1	80.754	37.621	1, 2	18.945	-45.073
0.350	1, 1	81.378	37.917	1, 2	12.147	-45.174
0.375	1, 1	82.005	38.030	1, 2	5.627	-44.450
0.400	1, 1	82.592	37.977	1, 2	-0.526	-42.955
0.425	1, 1	83.103	37.785	1, 2	-6.229	-40.758
0.450	1, 1	83.512	37.485	1, 2	-11.409	-37.936
0.475	1, 1	83.802	37.114	1, 2	-16.008	-34.574
0.500	1, 1	83.967	36.707	1, 2	-19.978	-30.764
0.525	1, 1	84.011	36.300	1, 2	-23.287	-26.599
0.550	1, 1	83.945	35.923	1, 2	-25.913	-22.174
0.575	1, 1	83.787	35.602	1, 2	-27.849	-17.585
0.600	1, 1	83.561	35.353	1, 2	-29.098	-12.925
0.625	1, 1	83.292	35.189	1, 2	-29.674	-8.282
0.650	1, 1	83.005	35.113	1, 2	-29.604	-3.743
0.675	1, 1	82.726	35.120	1, 2	-28.920	0.614
0.700	1, 1	[illegible]	[illegible]	[illegible]	[illegible]	[illegible]
0.725	1, 1	82.270	35.344	1, 2	-25.898	8.493
0.750	1, 1	82.123	35.525	1, 2	-23.668	11.889
0.775	1, 1	82.040	35.736	1, 2	-21.045	14.853
0.800	1, 1	82.022	35.948	1, 2	-18.098	17.345
0.825	1, 1	82.063	36.145	1, 2	-14.902	19.334
0.850	1, 1	82.156	36.313	1, 2	-11.534	20.801
0.875	1, 1	82.287	36.440	1, 2	-8.071	21.738
0.900	1, 1	82.442	36.518	1, 2	-4.591	22.147
0.925	1, 1	82.605	36.547	1, 2	-1.170	22.041
0.950	1, 1	82.762	36.527	1, 2	2.121	21.446
0.975	1, 1	82.900	36.465	1, 2	5.217	20.393
1.000	1, 1	83.008	36.369	1, 2	8.059	18.927
1.025	1, 1	83.079	36.250	1, 2	10.594	17.095
1.050	1, 1	83.111	36.122	1, 2	12.780	14.954
1.075	1, 1	83.103	35.995	1, 2	14.584	12.562
1.100	1, 1	83.060	35.882	1, 2	15.982	9.983
1.125	1, 1	82.988	35.790	1, 2	16.958	7.280
1.150	1, 1	82.896	35.726	1, 2	17.508	4.518
1.175	1, 1	82.793	35.694	1, 2	17.635	1.759
1.200	1, 1	82.689	35.695	1, 2	17.353	-0.935
1.225	1, 1	82.594	35.725	1, 2	16.683	-3.508
1.250	1, 1	82.515	35.780	1, 2	15.652	-5.906
1.275	1, 1	82.459	35.854	1, 2	14.296	-8.081
1.300	1, 1	82.429	35.938	1, 2	12.657	-9.993
1.325	1, 1	82.425	36.024	1, 2	10.780	-11.607
1.350	1, 1	82.447	36.105	1, 2	8.717	-12.895
1.375	1, 1	82.490	36.173	1, 2	6.519	-13.840
1.400	1, 1	82.550	36.222	1, 2	4.241	-14.430
1.425	1, 1	82.620	36.251	1, 2	1.938	-14.662
1.450	1, 1	82.692	36.257	1, 2	-0.337	-14.543
1.475	1, 1	82.761	36.241	1, 2	-2.531	-14.086
1.500	1, 1	82.820	36.206	1, 2	-4.597	-13.313
1.525	1, 1	82.864	36.157	1, 2	-6.490	-12.249
1.550	1, 1	82.890	36.098	1, 2	-8.172	-10.930
1.575	1, 1	82.897	36.036	1, 2	-9.611	-9.393
1.600	1, 1	82.886	35.977	1, 2	-10.780	-7.680
1.625	1, 1	82.857	35.926	1, 2	-11.661	-5.837
1.650	1, 1	82.816	35.886	1, 2	-12.240	-3.909
1.675	1, 1	82.766	35.862	1, 2	-12.514	-1.944
1.700	1, 1	82.713	35.855	1, 2	-12.485	0.013
1.725	1, 1	82.661	35.863	1, 2	-12.160	1.914
1.750	1, 1	82.616	35.887	1, 2	-11.557	3.719
1.775	1, 1	82.581	35.922	1, 2	-10.695	5.387
1.800	1, 1	82.559	35.964	1, 2	-9.603	6.883
1.825	1, 1	82.551	36.011	1, 2	-8.311	8.177
1.850	1, 1	82.558	36.056	1, 2	-6.854	9.243
1.875	1, 1	82.577	36.096	1, 2	-5.271	10.061
1.900	1, 1	82.607	36.128	1, 2	-3.603	10.619
1.925	1, 1	82.644	36.148	1, 2	-1.889	10.910
1.950	1, 1	82.685	36.156	1, 2	-0.173	10.933
1.975	1, 1	82.725	36.151	1, 2	1.505	10.695
2.000	1, 1	82.761	36.134	1, 2	3.107	10.206
2.025	1, 1	82.789	36.108	1, 2	4.596	9.485
2.050	1, 1	82.807	36.076	1, 2	5.940	8.554
2.075	1, 1	82.815	36.040	1, 2	7.111	7.440
2.100	1, 1	82.811	36.004	1, 2	8.083	6.174
2.125	1, 1	82.797	35.972	1, 2	8.840	4.789
2.150	1, 1	82.775	35.947	1, 2	9.369	3.321
2.175	1, 1	82.746	35.930	1, 2	9.661	1.806
2.200	1, 1	82.714	35.922	1, 2	9.714	0.281
2.225	1, 1	82.682	35.925	1, 2	9.534	-1.217
2.250	1, 1	82.653	35.937	1, 2	9.129	-2.654
2.275	1, 1	82.630	35.957	1, 2	8.514	-3.996
2.300	1, 1	82.614	35.983	1, 2	7.707	-5.214
2.325	1, 1	82.608	36.011	1, 2	6.732	-6.282
2.350	1, 1	82.610	36.040	1, 2	5.615	-7.176
2.375	1, 1	82.620	36.066	1, 2	4.385	-7.880
2.400	1, 1	82.638	36.087	1, 2	3.075	-8.379
2.425	1, 1	82.660	36.101	1, 2	1.718	-8.667
2.450	1, 1	82.686	36.108	1, 2	0.345	-8.740
2.475	1, 1	82.712	36.107	1, 2	-1.008	-8.602
2.500	1, 1	82.736	36.097	1, 2	-2.311	-8.258
2.525	1, 1	82.755	36.082	1, 2	-3.533	-7.722
2.550	1, 1	82.768	36.061	1, 2	-4.646	-7.010
2.575	1, 1	82.775	36.038	1, 2	-5.626	-6.142
2.600	1, 1	82.774	36.015	1, 2	-6.451	-5.143
2.625	1, 1	82.766	35.993	1, 2	-7.105	-4.038
2.650	1, 1	82.752	35.975	1, 2	-7.576	-2.857
2.675	1, 1	82.733	35.963	1, 2	-7.855	-1.628
2.700	1, 1	82.712	35.957	1, 2	-7.939	-0.382
2.725	1, 1	82.691	35.958	1, 2	-7.831	0.851
2.750	1, 1	82.671	35.965	1, 2	-7.535	2.041
2.775	1, 1	82.654	35.977	1, 2	-7.062	3.161
2.800	1, 1	82.643	35.994	1, 2	-6.427	4.185
2.825	1, 1	82.637	36.013	1, 2	-5.647	5.090
2.850	1, 1	82.638	36.033	1, 2	-4.744	5.856
2.875	1, 1	82.644	36.051	1, 2	-3.742	6.467
2.900	1, 1	82.655	36.066	1, 2	-2.666	6.912
2.925	1, 1	82.671	36.077	1, 2	-1.544	7.182
2.950	1, 1	82.688	36.082	1, 2	-0.403	7.274
2.975	1, 1	82.706	36.082	1, 2	0.730	7.187
3.000	1, 1	82.723	36.076	1, 2	1.825	6.928

TABLE 6.4

TWO-ELEMENT ARRAYS

IEN

H/LMDA= 0.500 OMEGA= 9.92

SELF AND MUTUAL IMPEDANCES

B/LMDA	I, K	ZIK IN OHMS RE	IM	I, K	ZIK IN OHMS RE	IM
0.100	1, 1	423.979	-1165.042	1, 2	-104.787	783.129
0.125	1, 1	371.810	-929.401	1, 2	-47.332	563.383
0.150	1, 1	351.563	-795.669	1, 2	-18.733	444.592
0.175	1, 1	343.948	-707.061	1, 2	-0.389	370.453
0.200	1, 1	342.869	-642.576	1, 2	13.518	320.207
0.225	1, 1	345.871	-592.571	1, 2	25.408	284.314
0.250	1, 1	351.847	-551.993	1, 2	36.520	257.716
0.275	1, 1	360.298	-517.964	1, 2	47.617	237.431
0.300	1, 1	371.051	-488.774	1, 2	59.258	221.522
0.325	1, 1	384.129	-463.429	1, 2	71.913	208.592
0.350	1, 1	399.681	-441.431	1, 2	86.019	197.502
0.375	1, 1	417.926	-422.698	1, 2	101.981	187.175
0.400	1, 1	439.081	-407.551	1, 2	120.139	176.446
0.425	1, 1	463.231	-396.747	1, 2	140.659	163.926
0.450	1, 1	490.115	-391.499	1, 2	163.325	147.927
0.475	1, 1	518.774	-393.382	1, 2	187.203	126.501
0.500	1, 1	547.138	-403.985	1, 2	210.229	97.773
0.525	1, 1	571.776	-424.091	1, 2	228.961	60.730
0.550	1, 1	588.332	-452.529	1, 2	239.018	16.367
0.575	1, 1	593.060	-485.389	1, 2	236.616	-31.546
0.600	1, 1	584.882	-516.761	1, 2	220.625	-77.206
0.625	1, 1	566.303	-541.076	1, 2	193.484	-115.118
0.650	1, 1	542.159	-555.354	1, 2	159.953	-142.347
0.675	1, 1	517.334	-559.655	1, 2	124.821	-158.972
0.700	1, 1	495.237	-555.961	1, 2	91.387	-166.959
0.725	1, 1	477.561	-546.802	1, 2	61.223	-168.783
0.750	1, 1	464.767	-534.425	1, 2	34.653	-166.589
0.775	1, 1	456.652	-520.525	1, 2	11.328	-161.918
0.800	1, 1	452.749	-506.291	1, 2	-9.366	-155.730
0.825	1, 1	452.543	-492.547	1, 2	-28.078	-148.533
0.850	1, 1	455.553	-479.892	1, 2	-45.385	-140.502
0.875	1, 1	461.348	-468.809	1, 2	-61.738	-131.572
0.900	1, 1	469.515	-459.742	1, 2	-77.431	-121.503
0.925	1, 1	479.610	-453.134	1, 2	-92.573	-109.927
0.950	1, 1	491.088	-449.426	1, 2	-107.048	-96.408
0.975	1, 1	503.241	-449.016	1, 2	-120.472	-80.511
1.000	1, 1	515.150	-452.164	1, 2	-132.170	-61.925
1.025	1, 1	525.700	-458.849	1, 2	-141.211	-40.617
1.050	1, 1	533.694	-468.615	1, 2	-146.529	-16.986
1.075	1, 1	538.079	-480.487	1, 2	-147.163	8.038
1.100	1, 1	538.242	-493.042	1, 2	-142.558	33.064
1.125	1, 1	534.231	-504.662	1, 2	-132.796	56.498
1.150	1, 1	526.788	-513.915	1, 2	-118.630	76.910
1.175	1, 1	517.145	-519.869	1, 2	-101.282	93.355
1.200	1, 1	506.689	-522.219	1, 2	-82.119	105.491
1.225	1, 1	496.666	-521.206	1, 2	-62.355	113.496
1.250	1, 1	488.002	-517.427	1, 2	-42.877	117.871
1.275	1, 1	481.276	-511.629	1, 2	-24.226	119.233
1.300	1, 1	476.765	-504.568	1, 2	-6.650	118.163
1.325	1, 1	474.529	-496.931	1, 2	9.798	115.127
1.350	1, 1	474.482	-489.315	1, 2	25.172	110.446
1.375	1, 1	476.442	-482.228	1, 2	39.561	104.296
1.400	1, 1	480.159	-476.108	1, 2	53.030	96.737
1.425	1, 1	485.317	-471.330	1, 2	65.581	87.737
1.450	1, 1	491.534	-468.210	1, 2	77.122	77.212
1.475	1, 1	498.346	-466.988	1, 2	87.443	65.077
1.500	1, 1	505.213	-467.798	1, 2	96.218	51.294
1.525	1, 1	511.532	-470.621	1, 2	103.019	35.946
1.550	1, 1	516.687	-475.243	1, 2	107.371	19.285
1.575	1, 1	520.129	-481.232	1, 2	108.834	1.771
1.600	1, 1	521.473	-487.954	1, 2	107.108	-15.941
1.625	1, 1	520.591	-494.657	1, 2	102.124	-33.076
1.650	1, 1	517.648	-500.586	1, 2	94.090	-48.850
1.675	1, 1	513.078	-505.113	1, 2	83.469	-62.600
1.700	1, 1	507.498	-507.842	1, 2	70.890	-73.877
1.725	1, 1	501.580	-508.637	1, 2	57.039	-82.479
1.750	1, 1	495.952	-507.604	1, 2	42.546	-88.423
1.775	1, 1	491.119	-505.022	1, 2	27.929	-91.878
1.800	1, 1	487.435	-501.279	1, 2	13.559	-93.093
1.825	1, 1	485.104	-496.799	1, 2	-0.317	-92.326
1.850	1, 1	484.200	-492.008	1, 2	-13.551	-89.808
1.875	1, 1	484.693	-487.303	1, 2	-26.056	-85.715
1.900	1, 1	486.464	-483.041	1, 2	-37.765	-80.164
1.925	1, 1	489.325	-479.532	1, 2	-48.604	-73.226
1.950	1, 1	493.023	-477.029	1, 2	-58.463	-64.937
1.975	1, 1	497.246	-475.723	1, 2	-67.183	-55.332
2.000	1, 1	501.638	-475.720	1, 2	-74.554	-44.469
2.025	1, 1	505.810	-477.025	1, 2	-80.320	-32.466
2.050	1, 1	509.372	-479.527	1, 2	-84.206	-19.529
2.075	1, 1	511.972	-482.989	1, 2	-85.960	-5.967
2.100	1, 1	513.347	-487.057	1, 2	-85.401	7.806
2.125	1, 1	513.367	-491.301	1, 2	-82.464	21.309
2.150	1, 1	512.058	-495.268	1, 2	-77.230	34.036
2.175	1, 1	509.607	-498.545	1, 2	-69.927	45.519
2.200	1, 1	506.328	-500.820	1, 2	-60.902	55.388
2.225	1, 1	502.608	-501.917	1, 2	-50.573	63.394
2.250	1, 1	498.849	-501.807	1, 2	-39.372	69.424
2.275	1, 1	495.419	-500.587	1, 2	-27.699	73.479
2.300	1, 1	492.611	-498.453	1, 2	-15.893	75.640
2.325	1, 1	490.634	-495.664	1, 2	-4.224	76.031
2.350	1, 1	489.603	-492.505	1, 2	7.106	74.791
2.375	1, 1	489.550	-489.265	1, 2	17.939	72.050
2.400	1, 1	490.429	-486.219	1, 2	28.152	67.916
2.425	1, 1	492.132	-483.612	1, 2	37.627	62.479
2.450	1, 1	494.492	-481.648	1, 2	46.239	55.814
2.475	1, 1	497.296	-480.484	1, 2	53.846	47.997
2.500	1, 1	500.294	-480.210	1, 2	60.282	39.121
2.525	1, 1	503.215	-480.848	1, 2	65.365	29.312
2.550	1, 1	505.786	-482.335	1, 2	68.907	18.753
2.575	1, 1	507.762	-484.527	1, 2	70.741	7.685
2.600	1, 1	508.949	-487.201	1, 2	70.744	-3.589
2.625	1, 1	509.236	-490.083	1, 2	68.866	-14.721
2.650	1, 1	508.611	-492.872	1, 2	65.148	-25.345
2.675	1, 1	507.164	-495.282	1, 2	59.725	-35.110
2.700	1, 1	505.073	-497.079	1, 2	52.820	-43.714
2.725	1, 1	502.580	-498.109	1, 2	44.712	-50.929
2.750	1, 1	499.955	-498.308	1, 2	35.712	-56.613
2.775	1, 1	497.460	-497.702	1, 2	26.129	-60.701
2.800	1, 1	495.325	-496.397	1, 2	16.247	-63.199
2.825	1, 1	493.726	-494.552	1, 2	6.314	-64.157
2.850	1, 1	492.778	-492.364	1, 2	-3.461	-63.653
2.875	1, 1	492.534	-490.042	1, 2	-12.907	-61.775
2.900	1, 1	492.983	-487.796	1, 2	-21.877	-58.610
2.925	1, 1	494.063	-485.818	1, 2	-30.235	-54.243
2.950	1, 1	495.662	-484.272	1, 2	-37.852	-48.757
2.975	1, 1	497.628	-483.285	1, 2	-44.593	-42.237
3.000	1, 1	499.780	-482.939	1, 2	-50.316	-34.785